CALCULUS

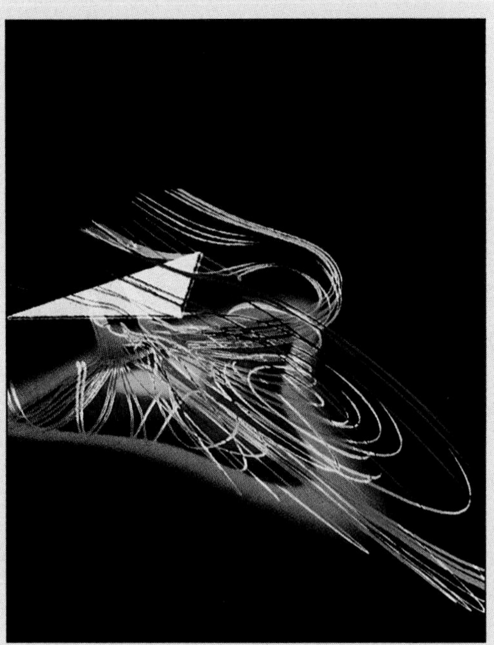

About the cover: This computer-generated figure shows a delta airplane wing with two propulsive jets close to the ground. Also shown are path lines of particles starting from the jet exit. These lines are colored by height of particles above the ground. (Courtesy of NASA, Ames Research Center)

CALCULUS

Ross L. Finney
Massachusetts Institute of Technology

George B. Thomas, Jr.
Massachusetts Institute of Technology

ADDISON–WESLEY PUBLISHING COMPANY

Reading, Massachusetts ▪ Menlo Park, California
New York ▪ Don Mills, Ontario ▪ Wokingham, England ▪ Amsterdam
Bonn ▪ Sydney ▪ Singapore ▪ Tokyo ▪ Madrid ▪ San Juan

Editor in Chief: Robert H. Price, Jr.
Design and Production Director: Susan Trowbridge
Sponsoring Editor: David F. Pallai
Developmental Editor: David M. Chelton
Managing Editor: Barbara Pendergast
Production Supervisor: Marion E. Howe
Copy Editors: Fannie Toldi, Jacqueline Dormitzer
Proofreaders: Marie Hergt, Phyllis Coyne
Text Designer: Catherine L. Johnson
Cover Designer: Marshall Henrichs
Art Consultant: Richard Morton
Art Coordinators: Jennifer DeCristoforo, Janice Mello
Illustrators: Richard Morton, Mike Prendergast, Lisa Sparks, Boston Graphics, Inc.
Electronic Illustrators: Wolfram Research Inc.
Electronic Production Consultant: Mona Zeftel
Photo Researcher: Susan Van Etten
Permissions Editor: Mary Dyer
Manufacturing Supervisor: Roy Logan
Advertising Manager: Eileen Spingler
Marketing Manager: Susan Howell
Production Services Manager: Herbert Nolan
Compositors: Typo·Graphics, Black Dot Graphics
Color Separators: Black Dot Graphics, Color Response
Printer: R. R. Donnelley & Sons

Library of Congress Cataloging-in-Publication Data

Finney, Ross L.
Calculus / Ross L. Finney and George B. Thomas Jr.
p. cm.
Includes index.
ISBN 0-201-19343-4
1. Calculus. I. Thomas, George Brinton, 1914– . II. Title.
QA303.F44 1989
515—dc20 89-6851
CIP

Copyright © 1990 by Addison–Wesley Publishing Company, Inc. All rights reserved. No part of this publication may be reproduced, stored in a retrieval system, or transmitted, in any form or by any means, electronic, mechanical, photocopying, recording, or otherwise, without the prior written permission of the publisher. Printed in the United States of America.

ABCDEFGHIJ–DO–99876543210

Preface

Audience and Prerequisites

This book provides everything necessary for the standard three-semester or four-quarter calculus sequence in the first two years of college. Its purpose, besides making it possible to learn the mathematics of calculus, is to teach students how to use it effectively and to show how knowing calculus can pay off no matter what professions they decide to enter later on. The applications described here are real and their presentations are self-contained. Students will not need any previous experience with the fields from which the applications are drawn. The prerequisites are the usual exposure to algebra and trigonometry, and Chapter 1 reviews what is needed for getting started.

Mathematical Level

Although the writing style in this book is livelier and less formal than that of traditional calculus texts, the appropriate level of rigor has been maintained. We try to explain things carefully but without belaboring the obvious and without answering questions students aren't yet ready to ask. For example, we state the max-min theorem for continuous functions and use it to prove the Mean Value Theorem, but we do not prove the max-min theorem itself or explore the properties of the real number system on which it depends. We use absolute values to control function outputs in Chapter 1 and show how limit theorems are proved with ϵ's and δ's at the end of Chapter 2, but save the more complicated ϵ–δ arguments for the appendices.

Content

Although the topic coverage is traditional in outline, a great deal is new about this book. The spectrum of applications is unusually broad and the book offers more opportunity for calculator and computer work than any previous standard text. For instance, Section 1.7 introduces some of the important functions on a scientific calculator, including e^x and $\ln x$, and discusses what calculus and computation have to do with one another (a great deal, as it turns out). The book continues this theme in appropriate places as the mathematics develops. It takes advantage of the four-color graphing capabilities of *Mathematica** to show surfaces in their entirety and in various stages of construction. It spends less than the usual amount of time on analytic geometry and uses the time saved to introduce the notion of mathematical modeling, first in connection with initial value problems in Section 4.6 and again in connection with exponential and logarithmic functions in Section 5.8. Modeling is a recurrent theme of the book, as you will see, for example, if you look at Sections 6.9 and 12.2.

Exercises

Each exercise set is graded to run from routine at the beginning to more challenging toward the end. Within this framework, the exercises generally follow the order of presentation in the text. Most sets contain applications and many contain **calculator exercises,** including occasional exercises for those who may have access to graphing calculators or computer graphers. The sections on Newton's method, numerical integration, and numerical methods for solving differential equations contain **optional computer programs** in BASIC. Each chapter concludes with **practice exercises** that cover the chapter's main points.

Applications

© Timothy Eagan/Woodfin Camp and Assoc.

See page 426.

The applications come from all over. From the life and social sciences, there are applications to determining lung volume (p. 446), restoring normal blood flow in arteries (p. 221), spreading innovation (p. 554), and predicting the consequences of human evolution (p. 479). From business and economics, there are applications to inventory control (p. 315), marginal cost and revenue (pp. 158 and 255), and the determination of production levels that are likely to maximize profit or loss (p. 255). From science and engineering, we discuss solar-powered cars (p. 345), colliding galaxies (p. 272), sequential vs. binary search (p. 491), coasting ships (p. 480), and warping railroad tracks (p. 76). From everyday life, we apply calculus to everything from the shroud of Turin and art forgery (p. 481), to baseball (p. 789), the recent Armenian earthquake (p. 471), human-powered flight (p. 160), the Concorde's sonic booms (p. 826), the cost of home electricity (p. 346), and the purchasing power of the dollar (p. 361).

Enhancements to Learning

In addition to the attention we have paid to writing style, content, and computer graphics, we have included a number of other features to make calculus more accessible and easier to use. The book is printed in full color throughout and contains more than 1500 figures. There are **drawing lessons,** step-by-step descriptions of **solution procedures,** and **flowcharts** for problem-solving. The steps in many of the worked examples are accom-

---

**Mathematica* is a trademark of Wolfram Research Inc.

DRAWING LESSON

How to Draw Quadric Surfaces

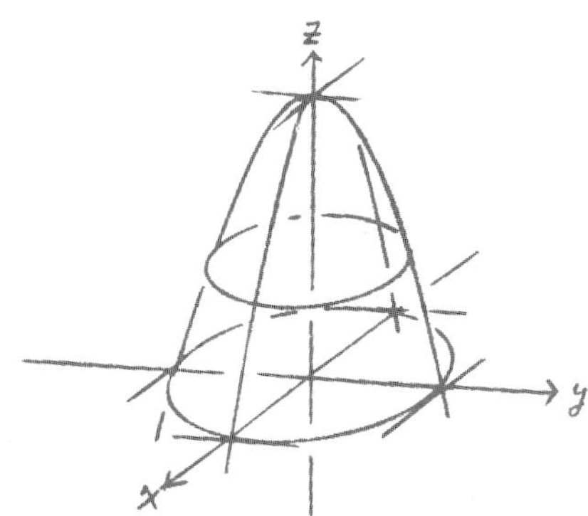

See page 763.

panied by **explanatory notes,** and the burden of exposition is carried by art in the body of the text when we feel that pictures and text together will convey ideas better than words alone (see p. 383 for example). There are entertaining **historical notes** and, from time to time, helpful **marginal lists** and comments.

Supplements for the Instructor

Complete Solutions Manual This supplement contains the worked-out solutions for *all* the exercises in the text.

Complete Answer Book Contains the answers to all exercises in the text.

Computerized Test Generator (AWTest) Based on the learning objectives of the text, this easy to use, algorithm-based system allows the instructor to generate tests or quizzes. Questions are available in open-ended, multiple choice, and true/false formats. This system also generates two- and three-dimensional figures. AWTest is available for the IBM PC* and is free to adopters.

Printed Test Bank Three versions of tests for each chapter in the text are included in multiple choice, open-ended, and true/false formats. The Test Bank also contains printed answer keys and student worksheets for each test.

Transparency Masters Includes a selection of key definitions, theorems, proofs, formulas, tables, and figures for creating overhead projections.

---

*IBM is a registered trademark. PC is a registered trademark.

Software-Related Supplements for the Instructor and the Student

Calculus Toolkit 2.0 Consisting of thirty-three programs ranging from functions to vector fields, this software enables the instructor and students to use the microcomputer as an "electronic chalkboard." Three-dimensional graphics are incorporated where appropriate. The Calculus Toolkit is available for both the Apple* and the IBM PC.

Computer Explorations in Calculus Activities and worksheets (using the Calculus Toolkit) explore concepts of calculus on the computer.

Student Edition of MathCAD This software package is a very powerful free-form scratchpad. When you input equations, MathCAD automatically calculates and displays your results as numbers or graphs. It also allows you to plot the results, annotate your work with text, and print your entire document (available for the IBM PC).

Mathematical Modeling with MathCAD Contains activities using MathCAD.

Exploring Calculus This software contains thirty-four programs that illustrate calculus principles by using twenty graphical displays, quizzes, demonstrations, tutorials and games. Prepared by John Fraleigh, University of Rhode Island (available for IBM PC).

Master Grapher and 3D Grapher A powerful interactive graphing utility for functions, polar equations, parametric equations, and other functions in two and three variables. Prepared by Franklin Demana and Bert Waits, both of Ohio State University. Available for MacIntosh, Apple, and the IBM PC.

Math PRO This interactive problem generator provides drill and practice exercises and includes "help" screens. Prepared by Joseph Mazur of Marlboro College, it is available for the IBM PC.

Supplements for the Student

Study Guide By Maurice Weir, Naval Postgraduate School. Organized to correspond with the text, this workbook in a semi-programmed format increases student proficiency.

Student Solutions Manual By Michael Schneider and Thomas Cochran, both of Belleville Area College. This manual is designed for the student and contains carefully worked-out solutions to all of the odd-numbered exercises in the text.

---

*Apple is a registered trademark.

Acknowledgments

We would like to express our thanks and appreciation for the many valuable contributions of the people who reviewed this book as it developed through its various stages:

Donna Beers, Simmons College
Donald Cohen, SUNY Agricultural and Technical College, Cobleskill
Joseph Egar, Cleveland State University
Theresa Farnum, Willamette University
Arnold Insel, Illinois State University
David Johnson, Lehigh University
Howard Jones, Lansing Community College
Cecilia Knoll, Florida Institute of Technology
H. Elton Lacey, Texas A&M University
Marcella Laddon, Monterey Peninsula College
James Lang, Valencia Community College
Stanley Lukawecki, Clemson University
Edward T. Migliore, Monterey Peninsula College
Lois Miller, Golden West College
Arthur Moore, Orange Coast College
James Osterburg, University of Cincinnati
Judith Palagallo, University of Akron
Michael Schneider, Belleville Area College
Charles Slavin, University of Maine at Orono
Sally Thomas, Orange Coast College
Stan VerNooy, University of Oregon
Loyd Wilcox, Golden West College
Lee Zia, University of New Hampshire

We would particularly like to express our gratitude to Gerald Egerer, Department of Economics, Sonoma State University, who taught us about applications of calculus to business and economics, and to Peter Lindstrom, Math-Tech Division, North Lake College, whose criticism and suggestions for applications contributed to the book's early chapters.

We want to express our special appreciation for the generous advice and help given to us by Thomas Cochran and Michael Schneider, Department of Mathematics, Belleville Area College, as they developed the text's answer and solution manuals.

We owe special thanks to Gary Rockswold, Department of Mathematics, Mankato State University, for proofreading the book in galley pages.

We would also like to thank Maureen Emberley and Laura Finney, who keyboarded the entire manuscript, and Romy Bakhru, for help in paginating the index.

We also wish to express our thanks to the many other contributors whose names we have not been able to mention.

Any errors that appear are the responsibility of the authors. We will appreciate having these brought to our attention.

Lincoln, Mass. R.L.F.
Magnolia, Mass. G.B.T., Jr.

Contents

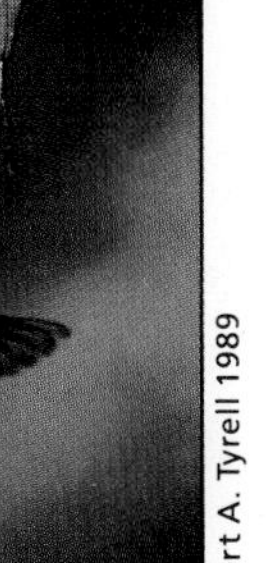

© Robert A. Tyrell 1989

See page 102.

Associated Press

See page 127.

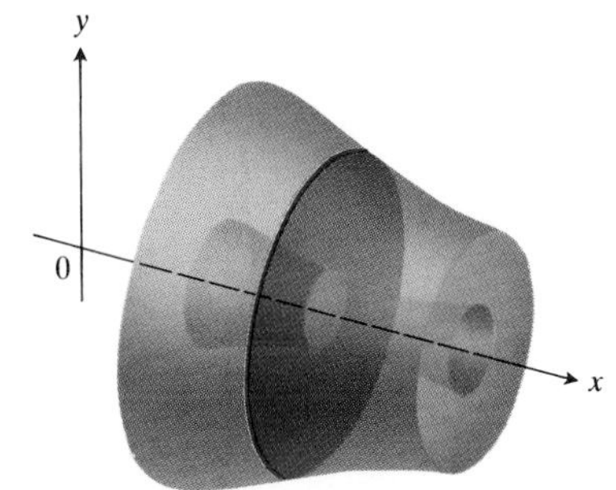

See page 383.

Wide World Photos

See page 514.

See page 573.

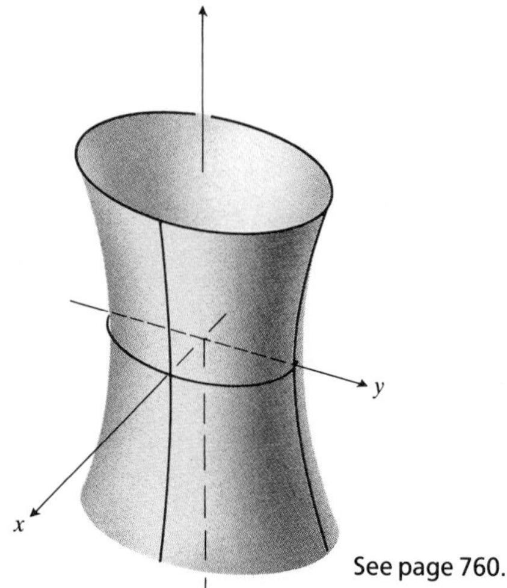

See page 760.

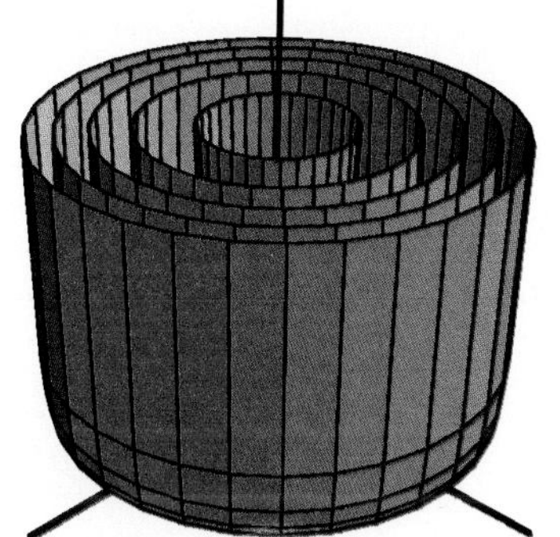

Generated by Mathematica

See page 823.

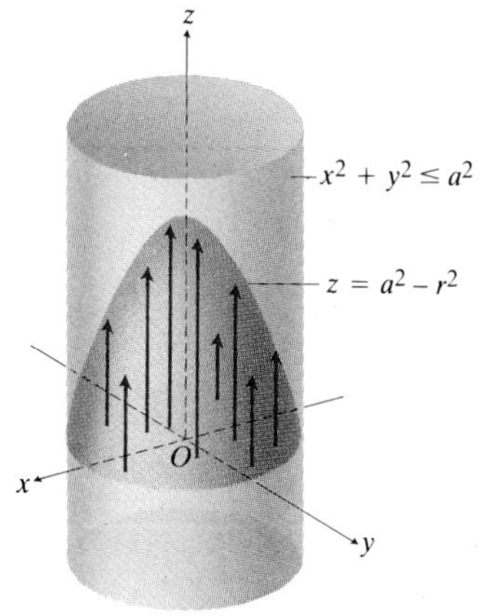

See page 949.

15 Vector Analysis 941

16 Preview of Differential Equations 1005

Appendices A–1

Answers A–29

Index I–1

A Brief Table of Integrals T–1

Prologue: What Is Calculus?

Calculus is the mathematics of motion and change. Where there is motion or growth, where variable forces are at work producing acceleration, calculus is the right mathematics to apply. This was true in the beginnings of the subject, and it is true today.

Calculus was first created to meet the mathematical needs of the scientists of the seventeenth century. Differential calculus dealt with the problem of calculating rates of change. It enabled people to define slopes of curves, to calculate the velocities and accelerations of moving bodies, to find the firing angle that gave a cannon its greatest range, and to predict the times when planets would be closest together or farthest apart. Integral calculus dealt with the problem of determining a function from information about its rate of change. It enabled people to calculate the future location of a body from its present position and a knowledge of the forces acting on it, to find the areas of irregular regions in the plane, to measure the lengths of curves, and to locate the centers of mass of arbitrary solids.

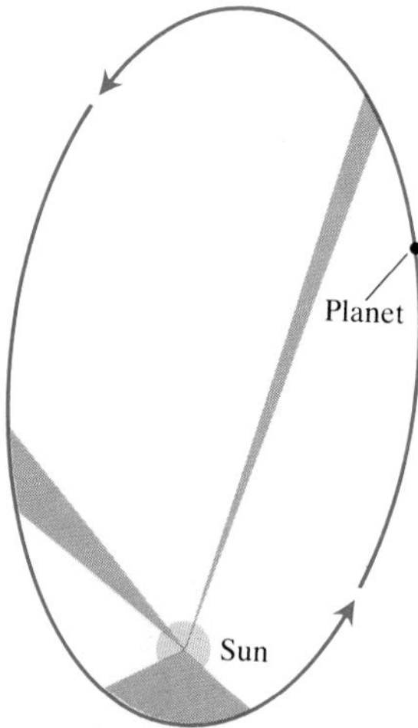

P.1 A planet moving about its sun. The shaded regions have equal areas. According to Kepler's second law, the planet takes the same amount of time to traverse the curved outer boundary of each region. The planet therefore moves faster near the sun than it does farther away.

Before the mathematical developments that culminated in the great discoveries of Sir Isaac Newton (1642–1727) and Baron Gottfried Wilhelm Leibniz (1646–1716), it took the astronomer Johannes Kepler (1571–1630) twenty years of concentration, record-keeping, and arithmetic to discover the three laws of planetary motion that now bear his name:

1. Each planet travels in an ellipse that has one focus at the sun (Fig. P.1).
2. The radius vector from the sun to a planet sweeps out equal areas in equal intervals of time.
3. The squares of the periods of revolution of the planets about the sun are proportional to the cubes of their orbits' semimajor axes. If T is the length of a planet's year and a is the semimajor axis of its orbit, then the ratio T^2/a^3 has the same constant value for all planets in the solar system.

With calculus, deriving Kepler's laws from Newton's laws of motion is but an afternoon's work. Kepler described how the solar system worked—Newton, with his calculus, explained why.

Today, calculus and its extensions in mathematical analysis are far reaching indeed, and the physicists, mathematicians, and astronomers who first invented the subject would surely be amazed and delighted, as we hope you will be, to see what a profusion of problems it solves and what a wide

P.2 Calculus helped us predict that Saturn's rings would be ellipses; it also enabled us to send a camera out there to check. The photographs shown here were taken on October 18, 1980 by NASA's Voyager 1 (left) and on July 12, 1981 by Voyager 2 (right). Notice how the features in Saturn's northern hemisphere and the brightness of its rings changed in the nine months between pictures. (Courtesy of NASA.)

range of fields now use it in the mathematical models that bring understanding about the universe and the world around us.

Economists use calculus to forecast global trends. Oceanographers use calculus to formulate theories about ocean currents and meteorologists use it to describe the flow of air in the upper atmosphere. Biologists use calculus to forecast population size and to describe the way predators like foxes interact with their prey. Medical researchers use calculus to design ultrasound and x-ray equipment for scanning the internal organs of the body. Space scientists use calculus to design rockets and explore distant planets. Psychologists use calculus to understand optical illusions in visual perception. Physicists use calculus to design inertial navigation systems and to study the nature of time and the universe. Hydraulic engineers use calculus to find safe closure patterns for valves in pipelines. Electrical engineers use it to design stroboscopic flash equipment and to solve the differential equations that describe current flow in computers. Sports equipment manufacturers use calculus to design tennis rackets and baseball bats. Stock market analysts use calculus to predict prices and assess interest rate risk. Physiologists use calculus to describe electrical impulses in neurons in the human nervous system. Drug companies use calculus to determine profitable inventory levels and timber companies use it to decide the most profitable time to harvest trees. The list is practically endless, for almost every professional field today uses calculus in some way.

"The calculus was the first achievement of modern mathematics," wrote John von Neumann (1903–1957), one of the great mathematicians of the present century, "and it is difficult to overestimate its importance. I think it defines more unequivocally than anything else the inception of modern mathematics; and the system of mathematical analysis, which is its logical development, still constitutes the greatest technical advance in exact thinking."*

**World of Mathematics*, Vol. 4 (New York: Simon and Schuster, 1960), "The Mathematician," by John von Neumann, pp. 2053–2063.

CALCULUS

1 Prerequisites for Calculus

© Dexter Randall Richards/Photography International

Slope

OVERVIEW This chapter reviews the most important things you need to know to start learning calculus. The emphasis is on functions and graphs, the main building blocks of calculus.

In calculus, functions are the major tools for describing the real world in mathematical terms, from temperature variations to planetary motions, from business cycles to brain waves, and from population growth to heartbeat patterns. Many functions have particular importance because of the kind of behavior they describe. For example, trigonometric functions describe cyclic, repetitive activity; exponential and logarithmic functions describe growth and decay; and polynomial functions can approximate these and most other functions. We shall examine some of these functions in this chapter and meet others in later chapters.

We shall learn that calculus can help us determine the graph of a function more easily than straightforward point-plotting, and that it can tell us many things about a graph and its equation that plotting cannot reveal. First, however, in this chapter, we shall review the things we can learn from a graph just by using algebra. For those interested in calculators and computers, we have included a section describing some of the ways calculators can extend our knowledge of functions, from comparing different functions to revealing hidden patterns in their behavior.

1.1 Coordinates and Graphs in the Plane

To assign coordinates to points in a plane, we start with two number lines that cross at their zero points at right angles. Each line represents the real numbers, which are the numbers that can be represented by decimals. Figure 1.1 shows the usual way of drawing the lines, with one line horizontal and the other vertical. The horizontal line is called the **x-axis** and the vertical line the **y-axis.** The point at which the lines cross is the **origin.**

On the x-axis, the positive number a lies a units to the right of the origin, and the negative number $-a$ lies a units to the left of the origin. On the y-axis, the positive number b lies b units above the origin while the negative number $-b$ lies b units below the origin.

With the axes in place, we assign a pair (a, b) of real numbers to each point P in the plane. The number a is the number at the foot of the perpendicular from P to the x-axis. The number b is the number at the foot of the perpendicular from P to the y-axis. Figure 1.2 shows the construction. The notation (a, b) is read "a b."

The coordinates defined here are often called **Cartesian** coordinates, after their chief inventor, René Descartes (1596–1650).

The number a from the x-axis is the **x-coordinate** of P. The number b from the y-axis is the **y-coordinate** of P. The pair (a, b) is the **coordinate pair** of the point P. It is an **ordered pair,** with the x-coordinate first and y-coordinate second. To show that P has the coordinate pair (a, b), we sometimes write the P and (a, b) together: $P(a, b)$.

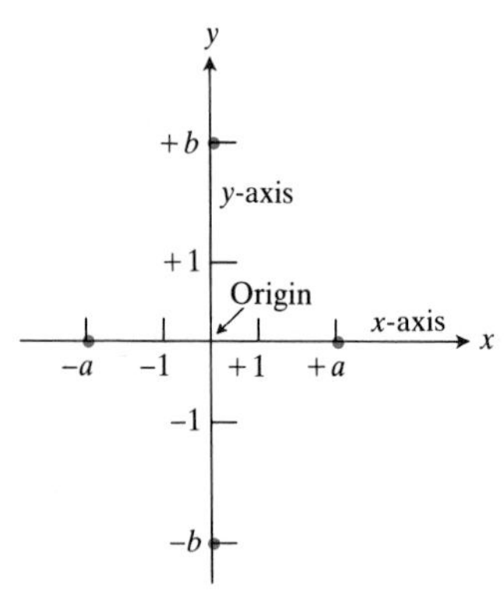

1.1 In Cartesian coordinates, the scaling on each axis is symmetric about the origin.

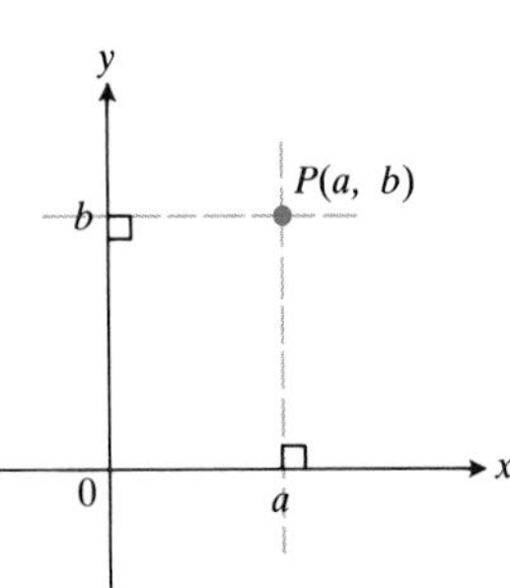

1.2 The pair (a, b) corresponds to the point where the perpendicular to the x-axis at a crosses the perpendicular to the y-axis at b.

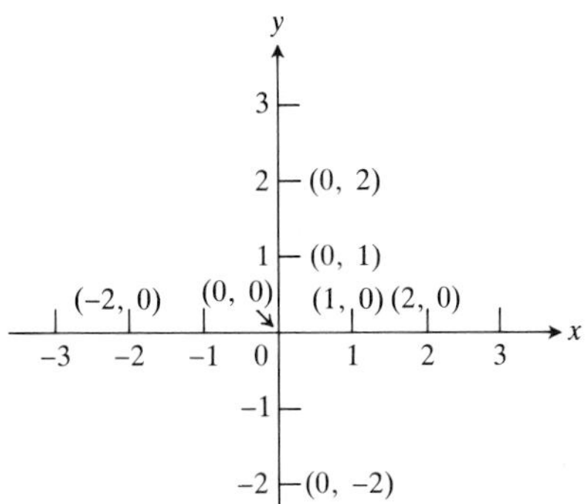

1.3 Points on the axes can be labeled in two ways.

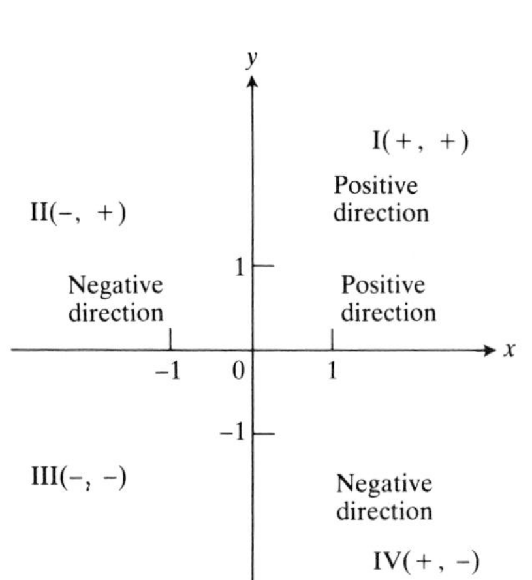

1.4 Directions along the axes: The values of x and y increase in the positive direction and decrease in the negative direction. Roman numerals label the quadrants.

The construction that assigns an ordered pair of real numbers to each point in the plane can be reversed to assign a point in the plane to each ordered pair of real numbers. The point assigned to the pair (a, b) is the point where the perpendicular to the x-axis at a crosses the perpendicular to the y-axis at b. Thus, the assignment of coordinates is a one-to-one correspondence between the points of the plane and the set of all ordered pairs of real numbers. Every point has a pair and every pair has a point, so to speak.

The points on the coordinate axes now have two kinds of numerical labels: single numbers from the axes and paired numbers from the plane. How do the numbers match up? See Fig. 1.3. As you can see, every point on the x-axis has y-coordinate zero and every point on the y-axis has x-coordinate zero. The origin is the point (0, 0).

Directions and Quadrants

Motion from left to right along the x-axis is said to be motion in the **positive x-direction.** Motion from right to left is in the **negative x-direction.** Along the y-axis, the positive direction is up, and the negative direction is down.

The origin divides the x-axis into the **positive x-axis** to the right of the origin and the **negative x-axis** to the left of the origin. Similarly, the origin divides the y-axis into the **positive y-axis** and the **negative y-axis.** The axes divide the plane into four regions called **quadrants,** numbered I, II, III, and IV (Fig1.4).

A Word about Scales

When we plot data in the coordinate plane or graph formulas whose variables have different units of measure, we do not need to use the same scale on the two axes. There is no reason to place the two 1's on the axes the same number of millimeters or whatever from the origin.

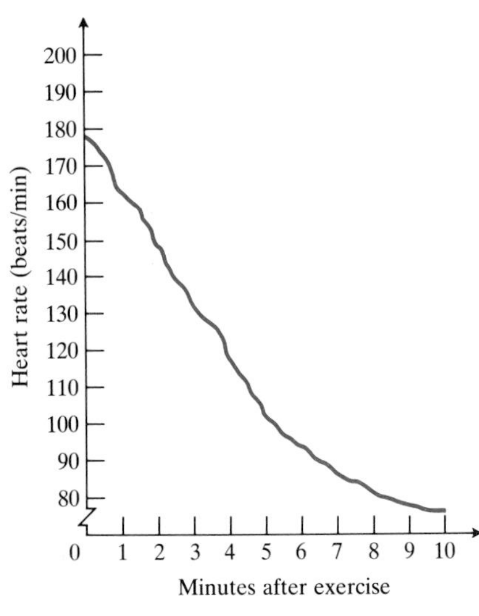

1.5 How the heartbeat returns to a normal rate after running. (Adapted from James F. Fixx's *The Complete Book of Running* (1977). New York: Random House.)

EXAMPLE 1 The graph in Fig. 1.5 shows how long it takes the heart to return to normal after running. The length that shows one minute on the horizontal axis is the same length that shows ten beats on the vertical axis. ■

When we graph functions whose variables do not represent physical measurements and when we draw figures in the coordinate plane to study their geometry or trigonometry, we shall assume that the scales on the axes we draw are the same. One unit of distance up and down in the plane will then look the same as one unit of distance right and left. As on a surveyor's map or a scale drawing, line segments that are supposed to have the same length will look as if they do.

Distance between Points

The coordinates of two points can tell us how far apart the two points are.

If the points lie on a horizontal line, the distance between them is the usual number-line distance between their x-coordinates (Fig. 1.6a).

If the points lie on a vertical line, the distance between them is the number-line distance between their y-coordinates (Fig. 1.6b).

If the line joining the points is not parallel to either coordinate axis, we calculate the distance between the points with the Pythagorean theorem (Fig. 1.6c). The resulting formula works for the other cases as well, so there is only one formula to remember:

Distance Formula for Points in the Plane

The distance between $P(x_1, y_1)$ and $Q(x_2, y_2)$ is

$$d = \sqrt{(x_2 - x_1)^2 + (y_2 - y_1)^2}. \tag{1}$$

EXAMPLE 2 The distance between $P(-1, 2)$ and $Q(3, 4)$ is

$$\sqrt{(3 - (-1))^2 + (4 - 2)^2} = \sqrt{(4)^2 + (2)^2} = \sqrt{20} = \sqrt{4 \cdot 5} = 2\sqrt{5}.$$

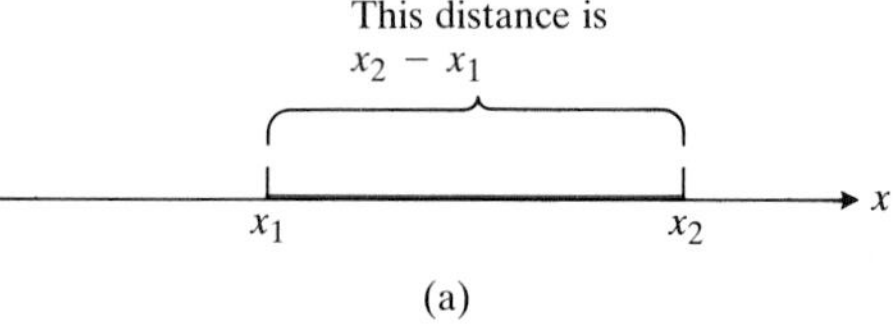

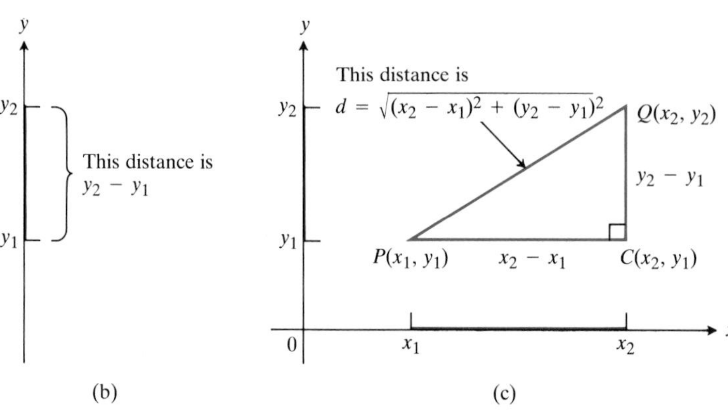

1.6 To calculate the distance between $P(x_1, y_1)$ and $Q(x_2, y_2)$, apply the Pythagorean theorem to triangle PCQ.

Graphs of Equations

The points (x, y) whose coordinates satisfy an equation like $y = x^2$ make up the *graph* of the equation in the xy-plane. Graphs give us a practical way to picture equations as lines or curves.

EXAMPLE 3 Graph the equation $y = x^2$ for values of x from $x = -2$ to $x = 2$.

These are the steps we take to graph an equation.

Solution STEP 1: Make a table of xy-pairs that satisfy the equation $y = x^2$.

x	$y = x^2$
-2	4
-1	1
0	0
1	1
2	4

STEP 2: Plot the points (x, y) whose coordinates appear in the table.

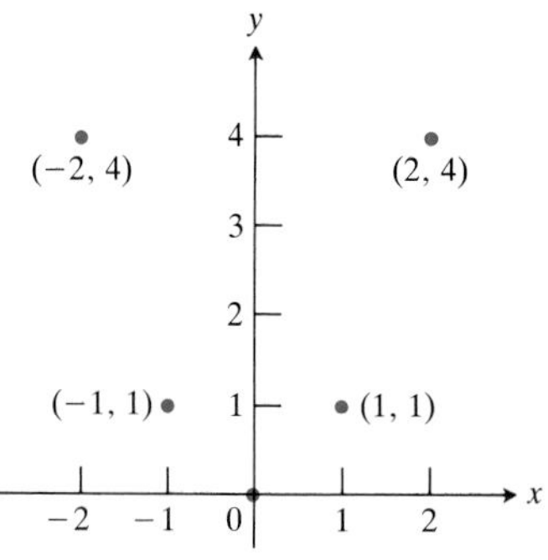

STEP 3: Draw a smooth curve through the plotted points. Label the curve with its equation.

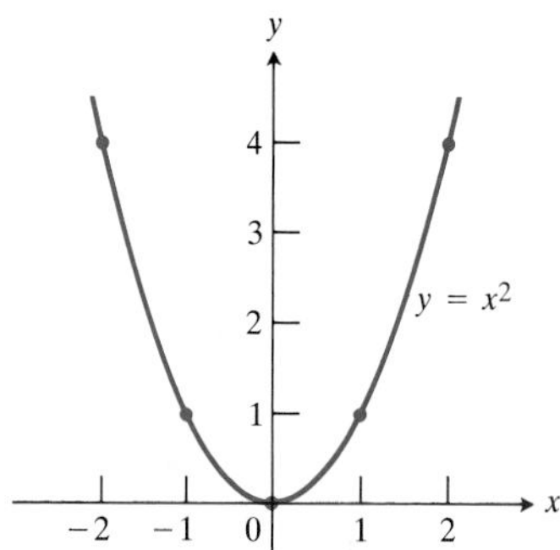

How do we know that the graph of $y = x^2$ doesn't look like one of *these* curves:

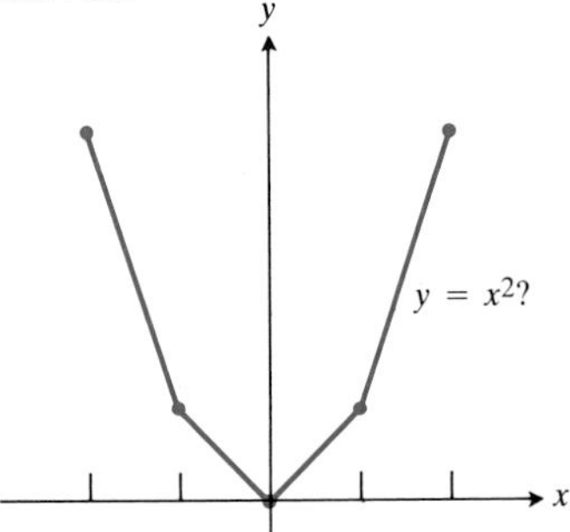

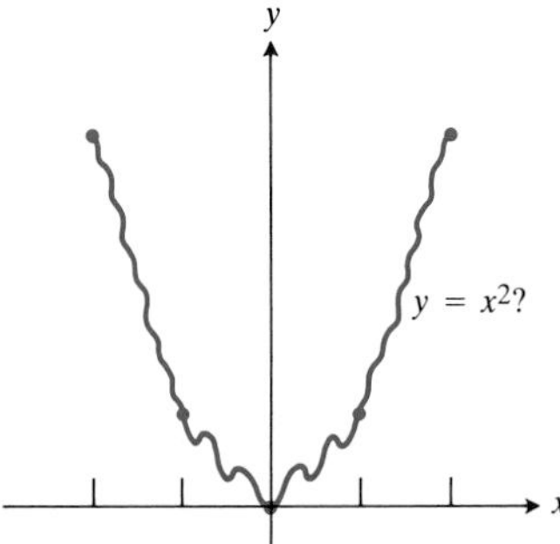

To find out, we might plot more points. But how would we then connect *them?* The basic question still remains: How do we know for sure what the graph does between the points we plot?

The answer lies in calculus, as we shall see in Chapter 4. There we shall learn to use a marvelous mathematical tool called a *derivative* to find a curve's exact shape between plotted points. Meanwhile, we shall have to settle for plotting individual points and connecting them as best we can.

Symmetry

We can use coordinate formulas to describe important symmetries in the coordinate plane. Figure 1.7 shows how this is done.

EXAMPLE 4 Symmetric points:

$P(5, 2)$ and $Q(5, -2)$	symmetric about the x-axis
$P(5, 2)$ and $R(-5, 2)$	symmetric about the y-axis
$P(5, 2)$ and $S(-5, -2)$	symmetric about the origin

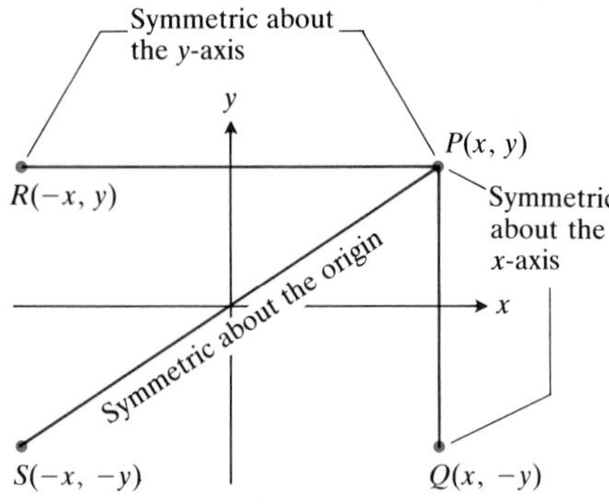

1.7 The coordinate formulas for symmetry with respect to the origin and axes in the coordinate plane.

The coordinate relations in Fig. 1.7 provide the following symmetry tests for graphs.

Symmetry Tests for Graphs

1. *Symmetry about the x-axis:*
   If the point (x, y) lies on the graph, then the point $(x, -y)$ lies on the graph (Fig. 1.8a).
2. *Symmetry about the y-axis:*
   If the point (x, y) lies on the graph, the point $(-x, y)$ lies on the graph (Fig. 1.8b).
3. *Symmetry about the origin:*
   If the point (x, y) lies on the graph, the point $(-x, -y)$ lies on the graph (Fig. 1.8c).

1.8 Symmetry tests for graphs.

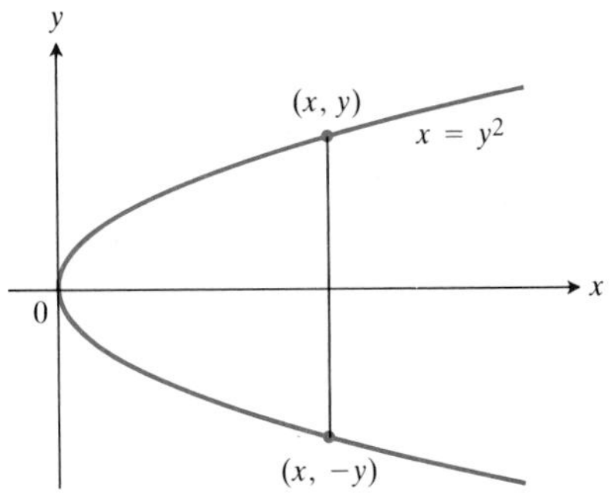

(a) Symmetry about the x-axis.

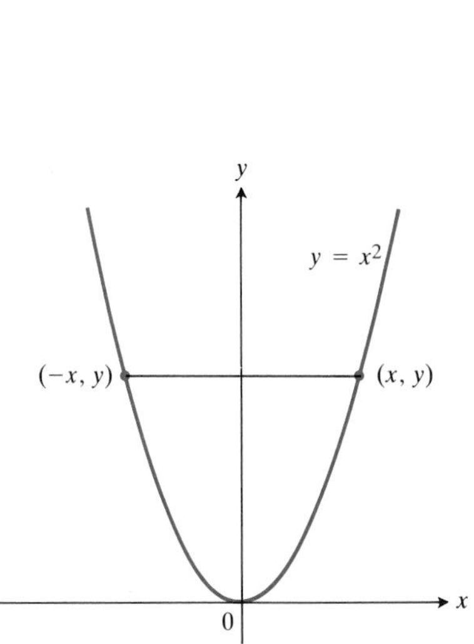

(b) Symmetry about the y-axis.

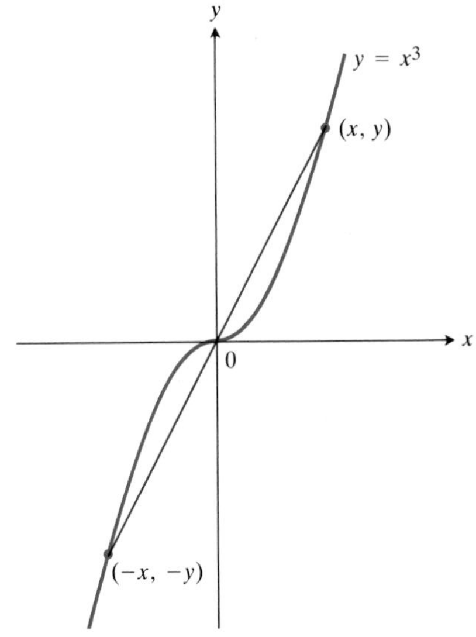

(c) Symmetry about the origin.

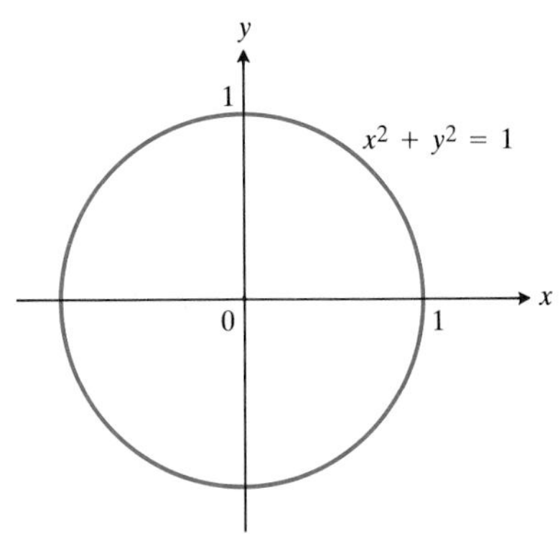

1.9 The graph of the equation $x^2 + y^2 = 1$ is the circle of radius 1, centered at the origin. It is symmetric about both axes and about the origin. Example 5 shows how to predict these symmetries before you graph the equation.

EXAMPLE 5 The graph of $x^2 + y^2 = 1$ has all three of the symmetries listed above (Fig. 1.9).

1. Symmetry about the x-axis:
$$\begin{aligned}(x, y) \text{ on the graph} &\Rightarrow x^2 + y^2 = 1 && (\Rightarrow \text{ means "implies"})\\ &\Rightarrow x^2 + (-y)^2 = 1 && (y^2 = (-y)^2)\\ &\Rightarrow (x, -y) \text{ on the graph.}\end{aligned}$$

2. Symmetry about the y-axis:
$$\begin{aligned}(x, y) \text{ on the graph} &\Rightarrow x^2 + y^2 = 1\\ &\Rightarrow (-x)^2 + y^2 = 1 && (x^2 = (-x)^2)\\ &\Rightarrow (-x, y) \text{ on the graph.}\end{aligned}$$

3. Symmetry about the origin:
$$\begin{aligned}(x, y) \text{ on the graph} &\Rightarrow x^2 + y^2 = 1\\ &\Rightarrow (-x)^2 + (-y)^2 = 1\\ &\Rightarrow (-x, -y) \text{ on the graph.}\end{aligned}$$

■

Intercepts and More about Graphing

The easiest points to find on an equation's graph are the points where the graph touches or crosses the axes. These points (when they exist) are the graph's **intercepts.**

We find the ***x*-intercepts** by setting y equal to 0 in the equation and solving for x. We find the ***y*-intercepts** by setting x equal to 0 and solving for y.

EXAMPLE 6 Find the intercepts of the graph of the equation $y = x^2 - 1$.

Solution

The x-intercepts:
$$\begin{aligned} y &= x^2 - 1 && \text{(Write down the given equation.)}\\ 0 &= x^2 - 1 && \text{(Let } y = 0.)\\ x^2 &= 1 && \text{(Solve for } x.)\\ x &= 1, -1. \end{aligned}$$

The y-intercepts:
$$\begin{aligned} y &= x^2 - 1 && \text{(Write down the given equation again.)}\\ y &= (0)^2 - 1 && \text{(Let } x = 0.)\\ y &= -1. && \text{(Solve for } y\text{—not much to do in this case.)} \end{aligned}$$

There are two x-intercepts ($x = 1$ and $x = -1$) and one y-intercept ($y = -1$). ■

Symmetry and intercepts can provide additional guidance in graphing.

EXAMPLE 7 Graph $y = x^2 - 1$.

Solution STEP 1: Check for symmetry. The graph is symmetric about the y-axis ($y = (-x)^2 - 1$ whenever $y = x^2 - 1$).

STEP 2: Find the intercepts (if any). From Example 6, the intercepts are $x = 1, x = -1$, and $y = -1$.

STEP 3: Make a short table of xy-pairs that satisfy the equation $y = x^2 - 1$. Include the intercepts.

STEP 4: Plot the points (x, y) from the table. Connect them with a smooth curve that is symmetric about the y-axis.

x	$y = x^2 - 1$
−3	8
−2	3
−1	0
0	−1
1	0
2	3
3	8

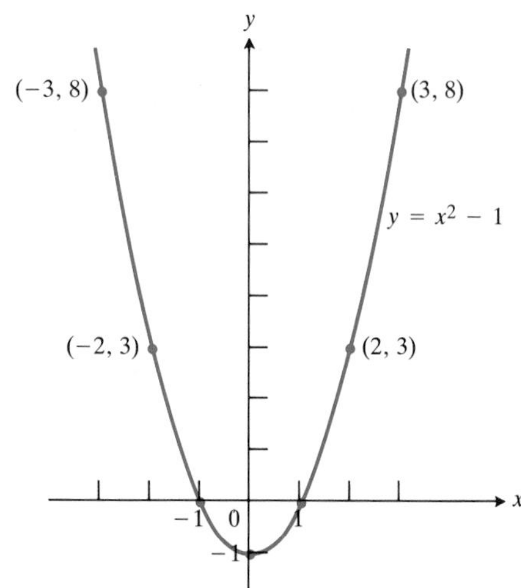

Exercises 1.1

For each point $P(x, y)$ in Exercises 1–12, use symmetry tests to find the point Q symmetric to P across the x-axis, the point R symmetric to P across the y-axis, and the point S symmetric to P through the origin.

1. $(3, 1)$ **2.** $(2, 7/2)$

3. $(-2, 2)$ **4.** $(-2, 1)$

5. $(-1, -1)$ **6.** $(2, -1)$

7. $(1, 0)$ **8.** $(-1, 0)$

9. $(\sqrt{3}, \sqrt{3})$ **10.** $(1, -\sqrt{2})$

11. $(0, \pi)$ **12.** $(-3, -1.5)$

In Exercises 13–18, find the distances between the given points.

13. $(1, 0)$ and $(0, 1)$ **14.** $(2, 4)$ and $(-1, 0)$

15. $(2\sqrt{3}, 4)$ and $(-\sqrt{3}, 1)$ **16.** $(2, 1)$ and $(1, -1/3)$

17. (a, b) and $(0, 0)$ **18.** $(0, y)$ and $(x, 0)$

Test the equations in Exercises 19–26 to find what symmetries their graphs have.

19. $y = -x^2$ **20.** $x = 4 - y^2$

21. $y = 1/x^2$ **22.** $y = 1/(x^2 + 1)$

23. $xy = 1$ **24.** $xy^2 = 1$

25. $x^2y^2 = 1$ **26.** $x^2 + 4y^2 = 1$

Find the intercepts and graph the equations in Exercises 27–36.

27. $y = x + 1$ **28.** $y = -x + 1$

29. $y = -x^2$ **30.** $y = 4 - x^2$

31. $x = -y^2$ **32.** $x = 1 - y^2$

33. $x^2 + y^2 = 4$ **34.** $4x^2 + 4y^2 = 1$

35. $y = \sqrt{x}$ **36.** $y = \sqrt{x + 4}$

37. What acute angle does the line through the points $(0, 0)$ and $(1, 1)$ make with the positive x-axis?

38. A rectangle with sides parallel to the axes has vertices at $(3, -2)$ and $(-4, -7)$.
a) Find the coordinates of the other two vertices.
b) Find the area of the rectangle.

39. The rectangle in Fig. 1.10 has sides parallel to the axes. It is three times as long as it is wide. Its perimeter is 56 units. Find the coordinates of the vertices A, B, and C.

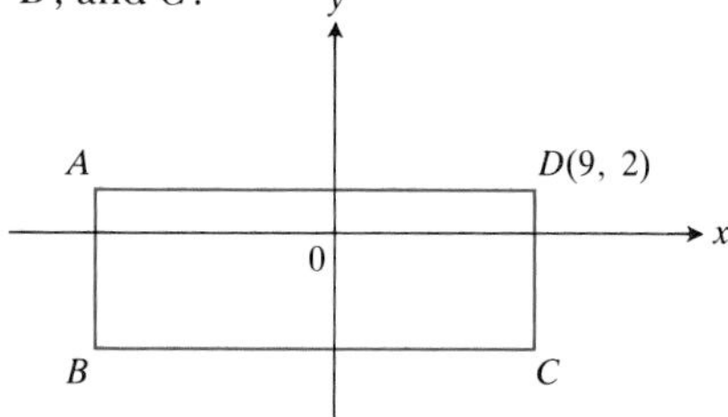

1.10 The rectangle in Exercise 39.

40. A circle in quadrant II is tangent to both axes. It touches the y-axis at $(0, 3)$.
a) At what point does it meet the x-axis?
b) Find the coordinates of the center of the circle.

41. The line through the points $(1, 1)$ and $(2, 0)$ cuts the y-axis at the point $(0, b)$. Find b by using similar triangles.

42. A 90° rotation counterclockwise about the origin takes $(2, 0)$ to $(0, 2)$ and $(0, 3)$ to $(-3, 0)$, as shown in Fig. 1.11. Where does the rotation take each of the following points?
a) $(4, 1)$ b) $(-2, -3)$ c) $(2, -5)$
d) $(x, 0)$ e) $(0, y)$ f) (x, y)
g) What point is taken to $(10, 3)$?

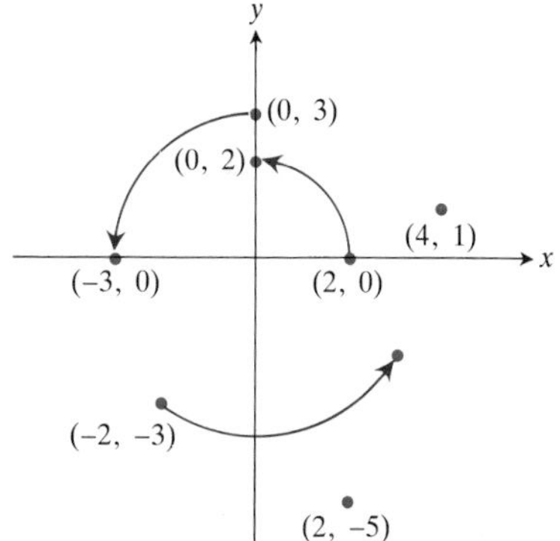

1.11 The points moved by the 90° rotation in Exercise 42.

Decompression Stops

Scuba divers often have to make decompression stops on their way to the surface after deep or long dives. A stop, which may vary from a few minutes to more than an hour,

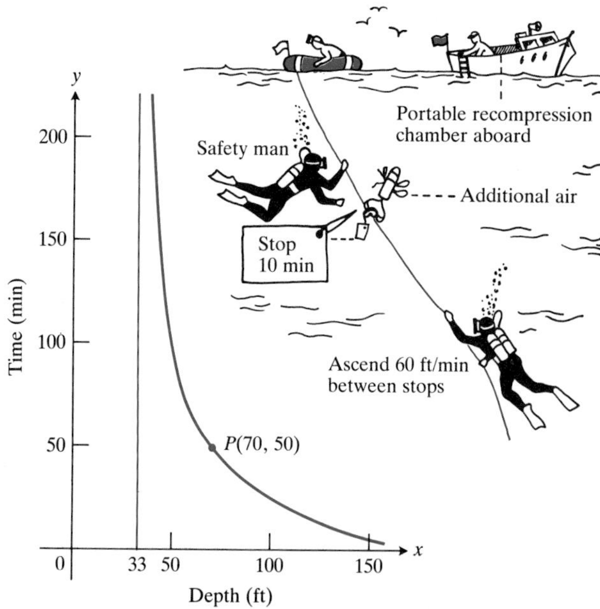

1.12 The coordinates of the points on the curve show how long scuba divers breathing compressed air can safely stay below the surface without a decompression stop. (Data from *U.S. Navy Diving Manual,* NAVSHIPS 250-538.)

provides time for the safe release of nitrogen and other gases absorbed by the tissues and blood while the body was under pressure. Dives to 33 ft or less do not require decompression stops. Dives to greater depths can be made without return stops if the diver does not stay down too long. The graph in Fig. 1.12 shows the longest times that a diver breathing compressed air may spend at various depths and still surface directly (at 60 ft per min). The times shown are total lengths of the dives, not just the times spent at maximum depth.

The coordinates of the point P in Fig. 1.12 are $x = 70$ and $y = 50$, so a diver going to 70 ft (but no deeper) can return safely without stopping if the total dive time is 50 min or less. Decompression stops are needed for dives plotted above or to the right of the curve, because their times exceed the corresponding limits set for their depths. Decompression stops are not needed for dives plotted on the curve or for dives plotted below or to the left of it, because the lengths of these dives do not exceed the limits for their depths. Exercises 43–44 are about these time limits.

43. You have been working at 100 ft below the surface for 1 hr. Do you need a decompression stop on the way up?

44. Which of the following dives need decompression stops and which do not?
a) $(40, 100)$ b) $(100, 40)$ c) $(70, 100)$ d) $(50, 50)$

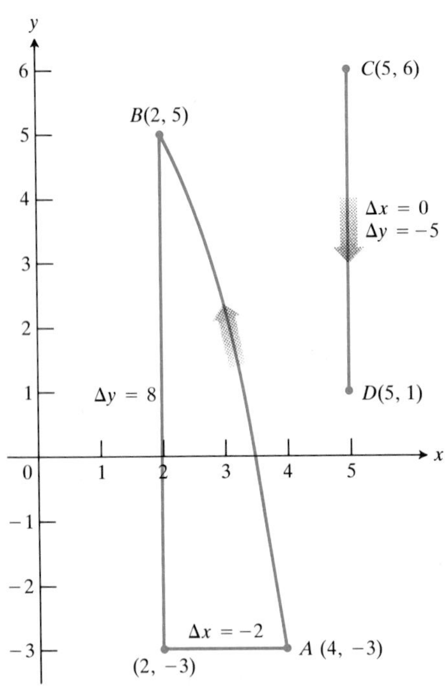

1.13 From A to B, $\Delta x = -2$ and $\Delta y = 8$. From C to D, $\Delta x = 0$ and $\Delta y = -5$.

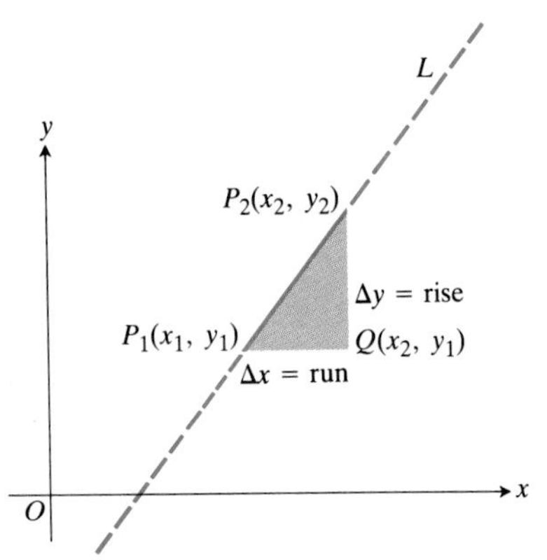

1.14 The slope of the line is

$$m = \frac{\Delta y}{\Delta x} = \frac{\text{rise}}{\text{run}}.$$

1.2 Slope, and Equations for Lines

One of the many reasons calculus has proved so useful over the years is that it is the right mathematics for relating a quantity's rate of change to its graph. Explaining this relationship is one of the goals of this book. Our basic plan is first to define what we mean by the slope of a line. Then we define the slope of a curve at each point on the curve. Later we shall relate the slope of a curve to a rate of change. Just how this is done will become clear as the book goes on. Our first step is to find a practical way to calculate the slopes of lines.

Increments

When a particle moves from one point to another in the plane, the net changes in its coordinates are calculated by subtracting the coordinates of the point where it starts from the coordinates of the point where it stops.

EXAMPLE 1 From $A(4, -3)$ to $B(2, 5)$ the net changes in coordinates (Fig. 1.13) are

$$\Delta x = 2 - 4 = -2, \quad \Delta y = 5 - (-3) = 8.$$ ■

The symbols Δx and Δy in Example 1 are read "delta x" and "delta y." They denote net changes or **increments** in the variables x and y. The letter Δ is a capital Greek "dee," for "difference." Neither Δx nor Δy denotes multiplication; Δx is not "delta times x" nor is Δy "delta times y."

DEFINITION

When a particle moves from (x_1, y_1) to (x_2, y_2), the **increments** are

$$\Delta x = x_2 - x_1 \quad \text{and} \quad \Delta y = y_2 - y_1. \tag{1}$$

EXAMPLE 2 From $C(5, 6)$ to $D(5, 1)$, the increments are (Fig. 1.13)

$$\Delta x = 5 - 5 = 0, \quad \Delta y = 1 - 6 = -5.$$ ■

Slopes of Nonvertical Lines

All lines except vertical lines have slopes. We calculate slopes from changes in coordinates. Once we see how this is done, we shall also see why vertical lines are an exception.

To begin, let L be a nonvertical line in the plane. Let $P_1(x_1, y_1)$ and $P_2(x_2, y_2)$ be two points on L (Fig. 1.14). We call $\Delta y = y_2 - y_1$ the **rise** from P_1 to P_2 and $\Delta x = x_2 - x_1$ the **run** from P_1 to P_2. Since L is not vertical, $\Delta x \neq 0$

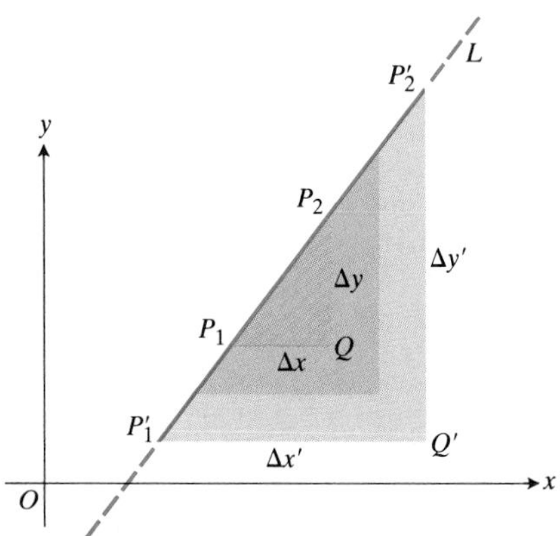

1.15 Because triangles P_1QP_2 and $P_1'QP_2'$ are similar, $\Delta y/\Delta x = \Delta y'/\Delta x'$.

and we may define the **slope** of L to be $\Delta y/\Delta x$, the amount of rise per unit of run. It is conventional to denote the slope by the letter m.

DEFINITION

The **slope** of a nonvertical line is
$$m = \frac{\text{rise}}{\text{run}} = \frac{\Delta y}{\Delta x} = \frac{y_2 - y_1}{x_2 - x_1}. \tag{2}$$

Suppose that instead of choosing the points P_1 and P_2 to calculate the slope in Eq. (2), we choose a different pair of points P_1' and P_2' on L and calculate
$$m' = \frac{y_2' - y_1'}{x_2' - x_1'} = \frac{\Delta y'}{\Delta x'}. \tag{3}$$

Will we get the same value for the slope? In other words, will $m' = m$? The answer is yes. The numbers m and m' are equal because they are the ratios of corresponding sides of similar triangles (Fig. 1.15):
$$m' = \frac{\Delta y'}{\Delta x'} = \frac{\Delta y}{\Delta x} = m. \tag{4}$$

The slope of a line depends only on how steeply the line rises or falls and not on the points we use to calculate it.

A line that goes uphill as x increases (Fig.1.16) has a positive slope. A line that goes downhill as x increases, like the line in Fig. 1.17, has a negative slope. A horizontal line has slope zero. The points on it all have the same y-coordinate, so $\Delta y = 0$.

The formula $m = \Delta y/\Delta x$ does not apply to vertical lines because Δx is zero along a vertical line. We express this by saying that vertical lines have no slope or that the slope of a vertical line is undefined.

Railroads and Highways

Civil engineers calculate the slope of a roadbed, by calculating the ratio of the distance it rises or falls to the distance it runs horizontally. They call this ratio the **grade** of the roadbed, usually written as a percentage. Along the coast, railroad grades are usually less than 2%. In the mountains, they may go as high as 4%. Highway grades are usually less than 5%.

In analytic geometry we calculate slopes the same way, but we usually do not express them as percentages.

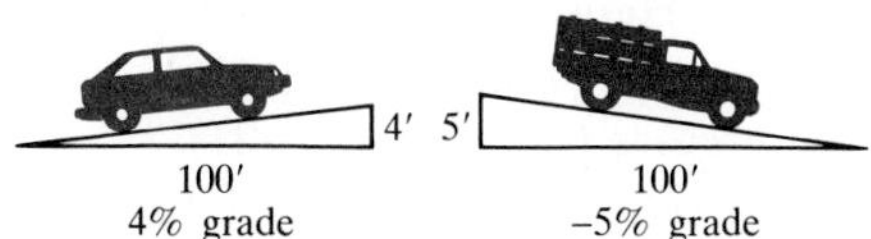

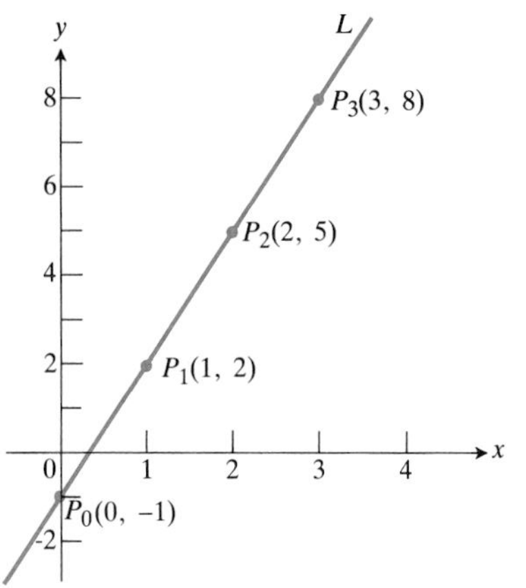

1.16 The slope of this line is
$$m = \frac{\Delta y}{\Delta x} = \frac{(5-2)}{(2-1)} = 3.$$
This means that $\Delta y = 3\,\Delta x$ for every change of position on the line. (Compare the coordinates of the marked points.)

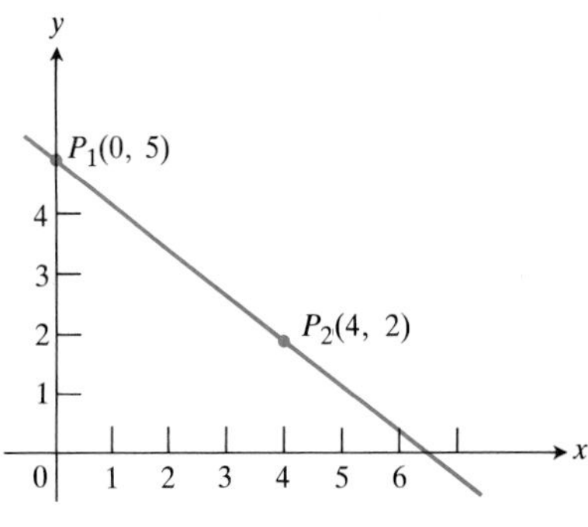

1.17 The slope of this line is
$$m = \frac{\Delta y}{\Delta x} = \frac{(2-5)}{(4-0)} = \frac{-3}{4}.$$
This means that y decreases 3 units every time x increases 4 units.

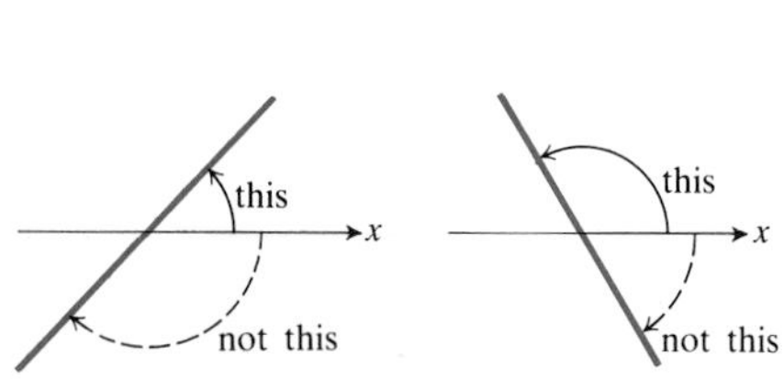

Angles of inclination are measured counterclockwise from the x-axis.

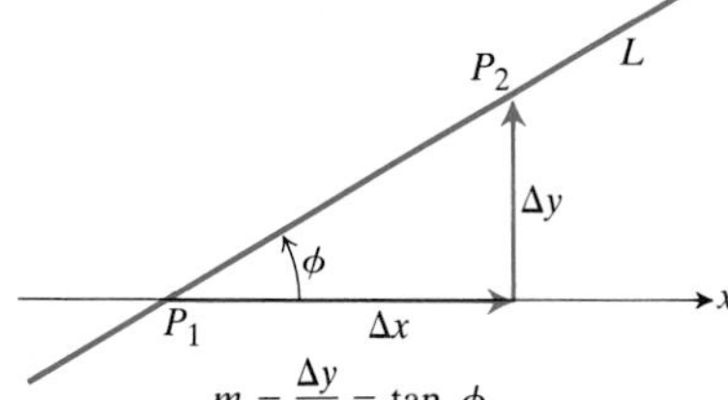

1.19 The slope of a nonvertical line is the tangent of its angle of inclination.

Angles of Inclination

The **angle of inclination** of a line that crosses the x-axis is the smallest angle we get when we measure counterclockwise from the x-axis around the point of intersection (Fig. 1.18). The angle of inclination of a horizontal line is taken to be 0°. Thus, angles of inclination may have any measure from 0° up to but not including 180°.

The slope of a line is the tangent of the line's angle of inclination. Figure 1.19 shows why this is true. If m denotes the slope and ϕ the angle, then

$$m = \tan\phi. \tag{5}$$

The World's Steepest Streets

The world's steepest street is Baldwin Street in Dunedin, near the southeast corner of New Zealand's South Island; it has a slope of 0.790 and an angle of inclination of 38.3° (as steep as a flight of stairs). The runners-up are both in San Francisco, Filbert Street and 22nd Street, with slopes of 0.613 and angles of 31.5°.

EXAMPLE 3 The slopes of lines become increasingly large as their angles of inclination approach 90° from either side. A few keystrokes on a calculator will show you that, after rounding,

ϕ approaching 90° from below	ϕ approaching 90° from above
tan 89.9° = 573	tan 90.1° = −573
tan 89.99 = 5730	tan 90.01 = −5730
tan 89.999 = 57300	tan 90.001 = −57300
tan 89.9999 = 573000	tan 90.0001 = −573000

We say that the slope of a line "becomes infinite" as its angle of inclination approaches 90°. However, vertical lines themselves have no slope (Fig. 1.20). ■

Lines That Are Parallel or Perpendicular

Parallel lines have equal angles of inclination. Hence, if they are not vertical, parallel lines have the same slope. Conversely, lines with equal slopes have equal angles of inclination and are therefore parallel (Fig. 1.21).

If neither of two perpendicular lines L_1 and L_2 is vertical, their slopes m_1 and m_2 are related by the equation $m_1 m_2 = -1$. Figure 1.22 shows why:

$$m_1 = \tan\phi_1 = \frac{a}{h}, \quad \text{while} \quad m_2 = -\frac{h}{a}. \tag{6}$$

Hence,

$$m_2 = -\frac{1}{m_1} \quad \text{and} \quad m_1 m_2 = -1. \tag{7}$$

1.20 When $\phi = 90°$, $\tan\phi$ is undefined. Vertical lines have equal angles of inclination but no slope.

EXAMPLE 4 The slope of a line perpendicular to a line of slope 3/4 is −4/3. ■

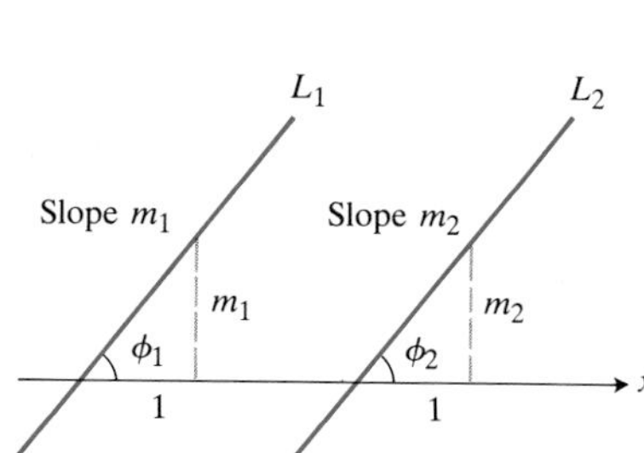

1.21 If $m_1 = m_2$, then $\phi_1 = \phi_2$ and the lines are parallel.

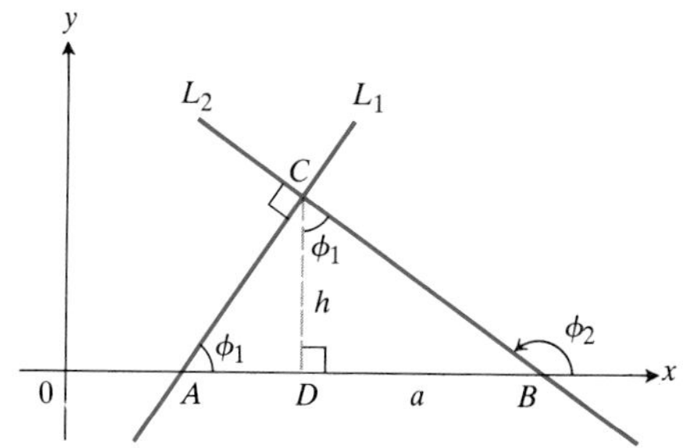

1.22 $\triangle ADC$ is similar to $\triangle CDB$. Hence ϕ_1 is also the upper angle in $\triangle CDB$. From the sides of $\triangle CDB$, we read $\tan \phi_1 = a/h$.

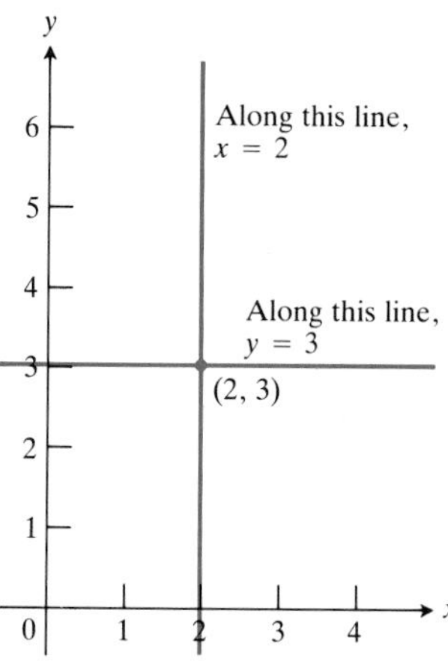

1.23 The standard equations for the horizontal and vertical lines through the point (2, 3) are $x = 2$ and $y = 3$.

Equations for Lines

DEFINITION

An **equation for a line** is an equation that is satisfied by the coordinates of every point on the line but is not satisfied by the coordinates of points that lie elsewhere.

Horizontal and Vertical Lines

The standard equations for the horizontal and vertical lines through a point (a, b) are simply $y = b$ and $x = a$ (Fig. 1.23). A point (x, y) lies on the horizontal line through (a, b) if and only if $y = b$. It lies on the vertical line through (a, b) if and only if $x = a$.

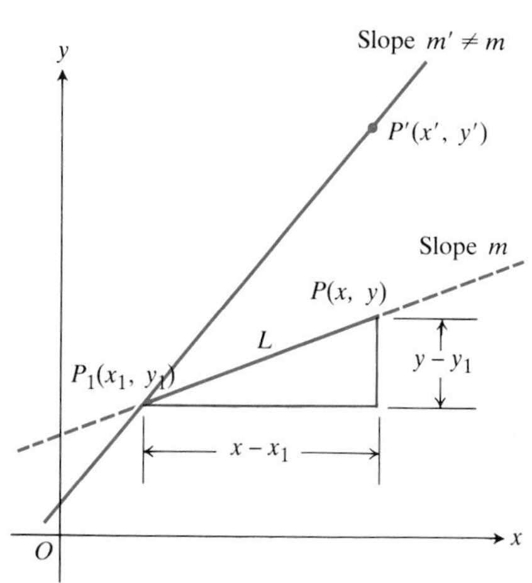

1.24 If L is the line through $P_1(x_1, y_1)$ whose slope is m, then other points $P(x, y)$ lie on this line if and only if slope $PP_1 = m$. This fact gives us the point–slope equation for L.

Point–Slope Equations

To write an equation for a line L that is not vertical, it is enough to know its slope m and the coordinates of a point $P_1(x_1, y_1)$ on it. If $P(x, y)$ is any other point on L (Fig. 1.24), then $x \neq x_1$ and we can write the slope of L as

$$\frac{y - y_1}{x - x_1}. \tag{8}$$

We can then set this expression equal to m to get

$$\frac{y - y_1}{x - x_1} = m. \tag{9}$$

Multiplying both sides of Eq. (9) by $x - x_1$ gives us the more useful equation

$$y - y_1 = m(x - x_1). \tag{10}$$

Equation (10) is an equation for L, as we can check right away. Every point (x, y) on L satisfies the equation—even the point (x_1, y_1). What about the points not on L? If $P'(x', y')$ is a point not on L (Fig. 1.24), then the slope m' of $P'P_1$ is different from m, and the coordinates x' and y' of P' do not satisfy Eqs. (9) and (10).

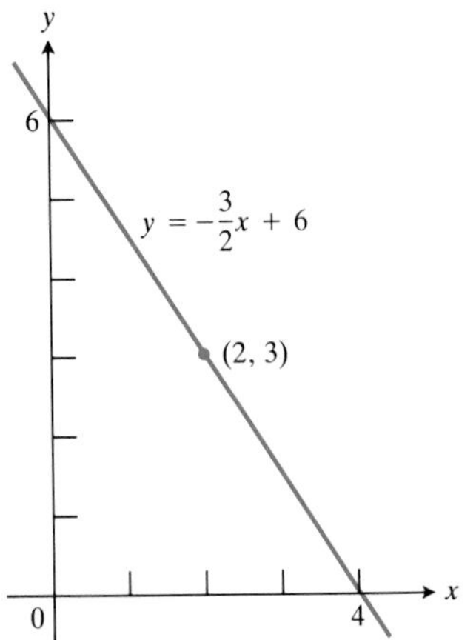

1.25 The line in Example 5.

DEFINITION

> The equation $$y - y_1 = m(x - x_1) \tag{11}$$
> is the **point–slope equation** of the line that passes through the point (x_1, y_1) with slope m.

EXAMPLE 5 Write an equation for the line that passes through the point (2, 3) with slope $-3/2$.

Solution

$$y - y_1 = m(x - x_1) \qquad \left(\text{Start with the general point–slope equation, Eq. (11)}\right)$$

$$y - 3 = -\frac{3}{2}(x - 2) \qquad (\text{Take } m = -3/2 \text{ and } (x_1, y_1) = (2, 3).)$$

$$y = -\frac{3}{2}x + 3 + 3 \qquad (\text{Solve for } y.)$$

$$y = -\frac{3}{2}x + 6.$$

This is an equation for the line (Fig. 1.25). ■

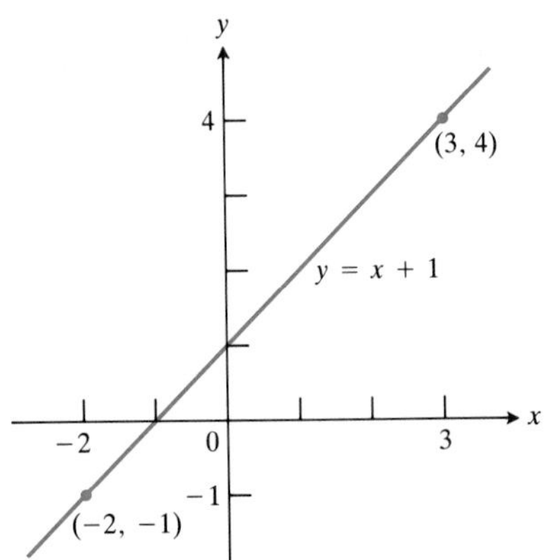

1.26 The line in Example 6.

EXAMPLE 6 Write an equation for the line through $(-2, -1)$ and $(3, 4)$.

Solution We first calculate the slope and then use Eq. (11).

$$m = \frac{-1 - 4}{-2 - 3} = \frac{-5}{-5} = 1.$$

The (x_1, y_1) in Eq. (11) can be either $(-2, -1)$ or $(3, 4)$:

With $(x_1, y_1) = (-2, -1)$	With $(x_1, y_1) = (3, 4)$
$y - (-1) = 1 \cdot (x - (-2))$	$y - 4 = 1 \cdot (x - 3)$
$y + 1 = x + 2$	$y - 4 = x - 3$
$y = x + 1.$	$y = x + 1.$

Same result

Either way, $y = x + 1$ is an equation for the line (Fig. 1.26). ■

Slope–Intercept Equations

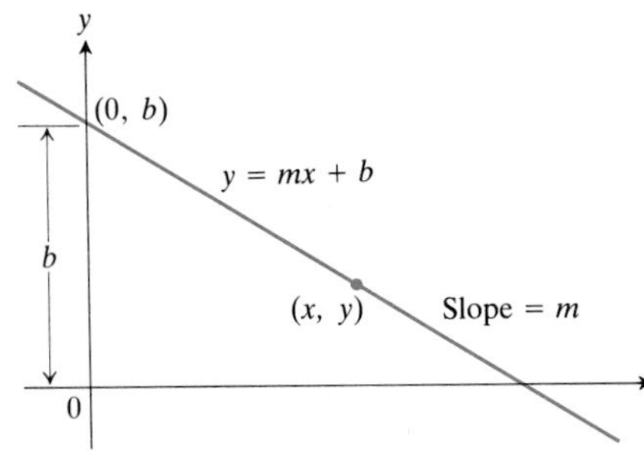

1.27 The line with slope m and y-intercept b. Its standard equation is $y = mx + b$.

Figure 1.27 shows a line with slope m and y-intercept b. If we take $(x_1, y_1) = (0, b)$ in the point–slope equation for the line, we find that

$$y - b = m(x - 0). \tag{12}$$

When rearranged, this becomes

$$y = mx + b. \tag{13}$$

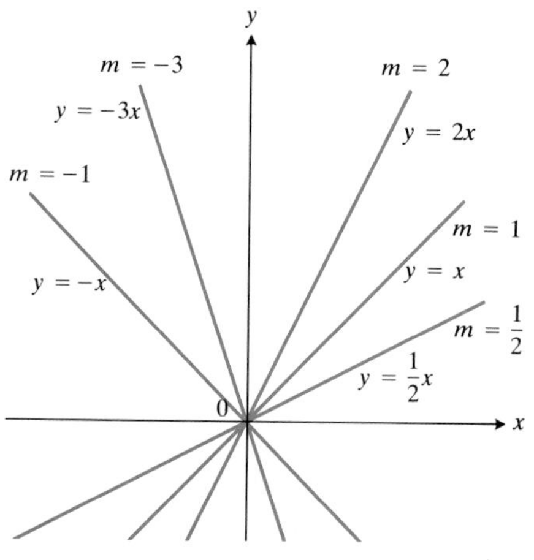

1.28 The line $y = mx$ passes through the origin with slope m.

DEFINITION

> The equation $$y = mx + b \tag{14}$$ is the **slope–intercept equation** of the line with slope m and y-intercept b.

EXAMPLE 7 The slope–intercept equation of the line with slope 2 and y-intercept 5 is

$$y = 2x + 5.$$

■

EXAMPLE 8 *Lines through the origin.* When a line passes through the origin, its y-intercept is $b = 0$ and the equation $y = mx + b$ simplifies to $y = mx$. The line $y = (1/2)x$ passes through the origin with slope 1/2. The line $y = x$ passes through the origin with slope 1. The 1 doesn't show in the equation because $1 \cdot x$ is just x and that's what we normally write. Similarly, the line $y = -x$ is the line through the origin with slope -1 (Fig. 1.28).

■

EXAMPLE 9 *Celsius vs. Fahrenheit.* The standard equation for converting Celsius temperature to Fahrenheit temperature is a slope–intercept equation. If we plot Fahrenheit temperature against Celsius temperature in a coordinate plane, the points we plot always lie along a straight line (Fig. 1.29). The line passes through the point (0, 32) because $F = 32$ when $C = 0$. It also passes through the point (100, 212) because $F = 212$ when $C = 100$.

This is enough information to make a formula for F in terms of C. The line's slope is

$$m = \frac{212 - 32}{100 - 0} = \frac{180}{100} = \frac{9}{5}.$$

The F-intercept of the line is

$$b = 32.$$

The resulting slope–intercept equation for the line,

$$F = \frac{9}{5}C + 32, \tag{15}$$

is the formula we seek.

■

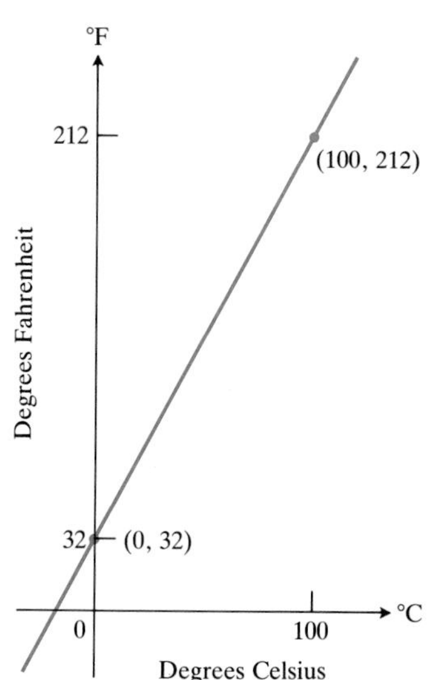

1.29 Fahrenheit versus Celsius temperature.

EXAMPLE 10 Find the slope and y-intercept of the line $8x + 5y = 20$.

Solution Solve the equation for y to put the equation in slope–intercept form. Then read the slope and y-intercept from the equation:

$$\begin{aligned} 8x + 5y &= 20 \\ 5y &= -8x + 20 \\ y &= -\frac{8}{5}x + 4. \end{aligned}$$

The slope is $m = -8/5$. The y-intercept is $b = 4$.

■

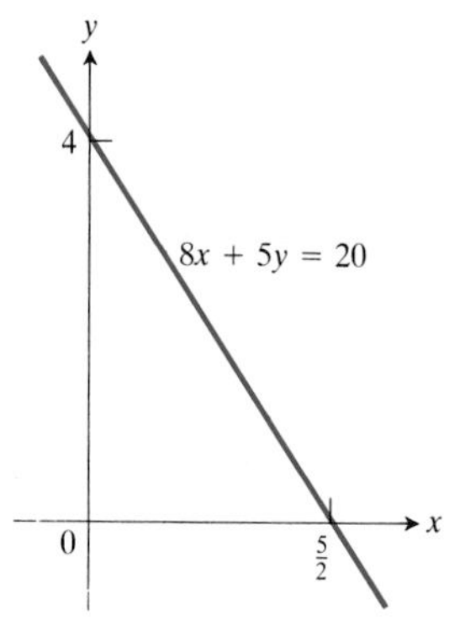

1.30 To graph $8x + 5y = 20$, mark the intercepts and draw a line through the marked points.

Quick Graphing

A quick way to graph a line that crosses both axes is to mark the intercepts and draw a line through the marked points. The method fails only if the line passes through the origin, giving one intercept instead of two, or if the intercepts are hard to plot.

EXAMPLE 11 Graph the line $8x + 5y = 20$.

Solution STEP 1: Find the x-intercept by setting $y = 0$ to obtain $8x = 20$, or $x = 5/2$.

STEP 2: Find the y-intercept by setting $x = 0$ to obtain $5y = 20$, or $y = 4$.

STEP 3: Plot the intercepts and draw the line (Fig. 1.30). ■

The Distance from a Point to a Line

To calculate the distance from a point $P(x_1, y_1)$ to a line L, we find the point $Q(x_2, y_2)$ at the foot of the perpendicular from P to L and calculate the distance from P to Q. The next example shows how this is done.

EXAMPLE 12 Find the distance from the point $P(2, 1)$ to the line L: $y = x + 2$.

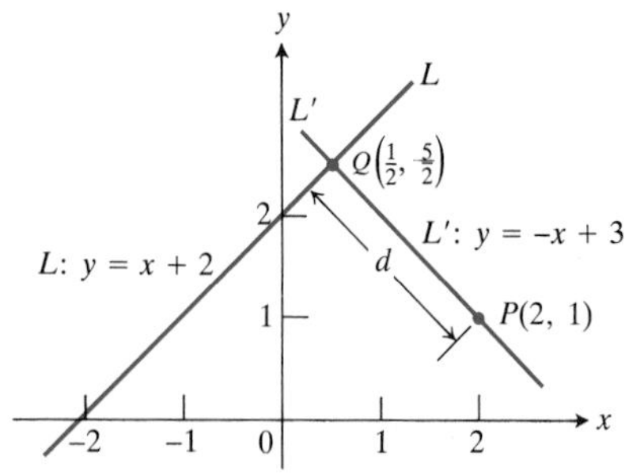

1.31 The distance d from P to L is measured along the line L' perpendicular to L. We can calculate it from the coordinates of P and Q.

Solution We solve the problem in three steps (Fig. 1.31): (1) Find an equation for the line L' through P perpendicular to L; (2) find the point Q where L' meets L; and (3) calculate the distance between P and Q.

STEP 1: We find an equation for the line L' through $P(2, 1)$ perpendicular to L. The slope of L: $y = x + 2$ is $m = 1$. The slope of L' is therefore $m' = -1/1 = -1$. We set $(x_1, y_1) = (2, 1)$ and $m = -1$ in Eq. (11) to find L':

$$y - 1 = -1(x - 2)$$
$$y = -x + 2 + 1$$
$$y = -x + 3.$$

STEP 2: Find the point Q by solving the equations for L and L' simultaneously. To find the x-coordinate of Q, we equate the two expressions for y:

$$x + 2 = -x + 3$$
$$2x = 1$$
$$x = \frac{1}{2}.$$

We can now obtain the y-coordinate by substituting $x = 1/2$ in the equation for either line. We choose $y = x + 2$ arbitrarily and find

$$y = \frac{1}{2} + 2 = \frac{5}{2}.$$

The coordinates of Q are $(1/2, 5/2)$.

STEP 3: We calculate the distance between $P(2, 1)$ and $Q(1/2, 5/2)$:

$$d = \sqrt{\left(2 - \frac{1}{2}\right)^2 + \left(1 - \frac{5}{2}\right)^2} = \sqrt{\left(\frac{3}{2}\right)^2 + \left(-\frac{3}{2}\right)^2} = \sqrt{\frac{18}{4}} = \frac{3}{2}\sqrt{2}.$$

The distance from P to L is $(3/2)\sqrt{2}$. ■

The General Linear Equation

The equation

$$Ax + By = C \quad (A \text{ and } B \text{ not both zero}) \tag{16}$$

is called the **general linear equation** because its graph is always a line and because every line has an equation in this form. We shall not take the time to prove this, but notice that all the equations in this section can be arranged in this form. A few of them are in this form already.

Slope

1. The slope of the line through $P_1(x_1, y_1)$ and $P_2(x_2, y_2)$, $x_1 \neq x_2$, is

$$m = \frac{\text{rise}}{\text{run}} = \frac{y_2 - y_1}{x_2 - x_1} = \frac{\Delta y}{\Delta x}.$$

2. $m = \tan \phi$ (ϕ is the angle of inclination).
3. Vertical lines have no slope.
4. Horizontal lines have slope 0.
5. For lines that are neither horizontal nor vertical it is handy to remember:

   a) they are parallel $\Leftrightarrow m_2 = m_1$;

   b) they are perpendicular $\Leftrightarrow m_2 = -1/m_1$.

   (The symbol $\Leftrightarrow$ is read "if and only if.")

Equations for Lines

$x = a$	Vertical line through (a, b)
$y = b$	Horizontal line through (a, b)
$y = mx + b$	Slope–intercept equation
$y - y_1 = m(x - x_1)$	Point–slope equation
$Ax + By = C$	General linear equation (A and B not both zero)

Exercises 1.2

In Exercises 1–6, a particle moves from A to B. Find the net changes Δx and Δy in the particle's coordinates.

1. $A(-1, 1), B(1, 2)$
2. $A(1, 2), B(-1, -1)$
3. $A(-3, 2), B(-1, -2)$
4. $A(-1, -2), B(-3, 2)$
5. $A(-3, 1), B(-8, 1)$
6. $A(0, 4), B(0, -2)$

Plot the points A and B in Exercises 7–12. Then find the slope (if any) of the line they determine. Also find the slope (if any) of the lines perpendicular to line AB.

7. $A(1, -2), B(2, 1)$
8. $A(-1, 2), B(-2, -1)$
9. $A(-2, -1), B(1, -2)$
10. $A(2, -1), B(-2, 1)$
11. $A(2, 3), B(-1, 3)$
12. $A(1, 2), B(1, -3)$

In Exercises 13–18, find an equation for (a) the vertical line and (b) the horizontal line through the given point.

13. $(2, 3)$
14. $(-1, 4/3)$
15. $(-7, -2)$
16. $(5, -1)$
17. $(0, -\sqrt{2})$
18. $(-\pi, 0)$

In Exercises 19–24, write an equation for the line that passes through the point P and has slope m.

19. $P(1, 1), \quad m = 1$
20. $P(1, -1), \quad m = -1$
21. $P(-1, 1), \quad m = 1$
22. $P(-1, 1), \quad m = -1$
23. $P(0, b), \quad m = 2$
24. $P(a, 0), \quad m = -2$

In Exercises 25–30, find an equation for the line through the two points.

25. $(0, 0), \quad (2, 3)$
26. $(1, 1), \quad (2, 1)$
27. $(1, 1), \quad (1, 2)$
28. $(-2, 1), \quad (2, -2)$
29. $(-2, 0), \quad (-2, -2)$
30. $(1, 3), \quad (3, 1)$

In Exercises 31–36, write an equation for the line with the given slope m and y-intercept b.

31. $m = 3, \quad b = -2$
32. $m = -1, \quad b = 2$
33. $m = 1, \quad b = \sqrt{2}$
34. $m = -1/2, \quad b = -3$
35. $m = -5, \quad b = 2.5$
36. $m = 1/3, \quad b = -1$

In Exercises 37–42, find the x- and y-intercepts of the line. Then use the intercepts to graph the line.

37. $3x + 4y = 12$
38. $x + y = 2$
39. $4x - 3y = 12$
40. $2x - y = 4$
41. $y = 2x + 4$
42. $x + 2y = -4$

In Exercises 43–48, find an equation for the line through P perpendicular to L. Then find the distance from P to L.

43. $P(0, 0), \quad L: y = -x + 2$
44. $P(0, 0), \quad L: x + \sqrt{3}y = 3$
45. $P(1, 2), \quad L: x + 2y = 3$
46. $P(-2, 2), \quad L: 2x + y = 4$
47. $P(3, 6), \quad L: x + y = 3$
48. $P(-2, 4), \quad L: x = 5$

In Exercises 49–52, find an equation for the line through P parallel to L.

49. $P(2, 1), \quad L: y = x + 2$
50. $P(0, 0), \quad L: y = 3x - 5$
51. $P(1, 0), \quad L: 2x + y = -2$
52. $P(1, 1), \quad L: x + y = 1$

Increments and Motion

53. A particle starts at $A(-2, 3)$ and its coordinates change by increments $\Delta x = 5$, $\Delta y = -6$. Find its new position.

54. A particle starts at $A(6, 0)$ and its coordinates change by increments $\Delta x = -6$, $\Delta y = 0$. Find its new position.

55. The coordinates of a particle change by $\Delta x = 5$ and $\Delta y = 6$ as it moves from $A(x, y)$ to $B(3, -3)$. Find x and y.

56. A particle started at $A(1, 0)$, circled the origin once counterclockwise, and returned to $A(1, 0)$. What were the net changes in its coordinates?

Applications

57. *Insulation.* By measuring slopes in Fig. 1.32, find the temperature change in degrees per inch for
a) gypsum wall board
b) fiber glass insulation
c) wood sheathing

58. *Insulation.* Which of the materials listed in Exercise 57 is the best insulator? The poorest? Explain.

59. *Pressure under water.* The pressure p experienced by a diver under water is related to the diver's depth d by an equation of the form $p = kd + 1$ (k a constant). When $d = 0$ meters, the pressure is 1 atmosphere. The pressure at 100 meters is about 10.94 atmospheres. Find the pressure at 50 meters.

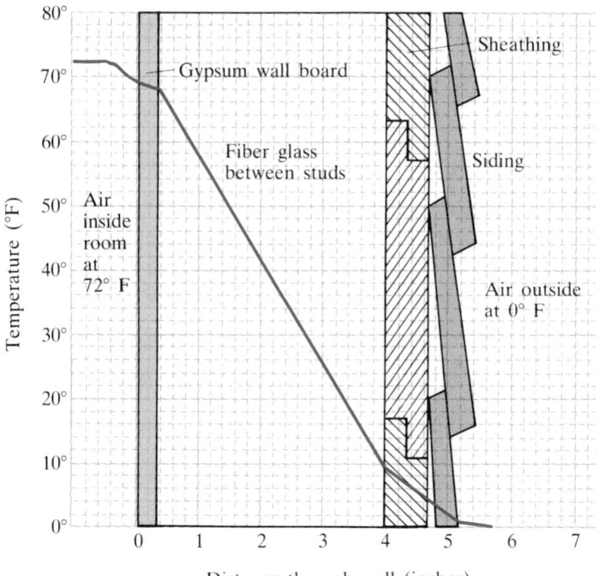

1.32 The temperature changes in the wall in Exercise 57. (Source: *Differentiation*, by W. U. Walton et al., Project CALC, Education Development Center, Inc. Newton, Mass. (1975), p. 25.)

The Mount Washington Cog Railway locomotive *Ammonoosuc* pushing a single car up the steepest part of the road. The locomotive is about 120 years old. (Source: *American Heritage of Invention and Technology*, Spring 1986, p. 6.)

60. *Reflected light.* A ray of light comes in along the line $x + y = 1$ above the x-axis and reflects off the x-axis. The angle of departure is equal to the angle of arrival (Fig. 1.33). Write an equation for the line along which the departing light travels.

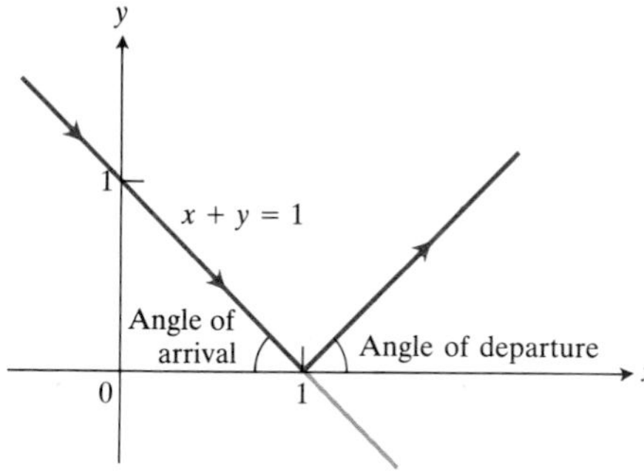

1.33 The path of the light ray in Exercise 60.

61. *Fahrenheit vs. Celsius.* Is there a temperature at which a Fahrenheit thermometer gives the same reading as a Celsius thermometer? If so, what is it? (See Example 9.)

62. CALCULATOR *The Mt. Washington Cog Railway.* The steepest part of the Mt. Washington Cog Railway in New Hampshire has a phenomenal 37.1% grade. At this point, the passengers in the front of the car are 14 ft above those in the rear. About how far apart are the front and rear rows of seats?

Geometry

63. Three different parallelograms have vertices at $(-1, 1)$, $(2, 0)$, and $(2, 3)$. Sketch them and give the coordinates of the missing vertices.

64. CALCULATOR How large a slope can you calculate with your calculator? To find out, continue the list of tangent values in Example 3. The best we could do was $\tan(89.999\,99999\,99) = 57295\,77951\,31$ and $\tan(-89.999\,99999\,99) = -57295\,77951\,31$.

65. For what value of k is the line $2x + ky = 3$ perpendicular to the line $x + y = 1$? For what values of k are the lines parallel?

66. Find the line that passes through the point $(1, 2)$ and the point of intersection of the lines $x + 2y = 3$ and $2x - 3y = -1$.

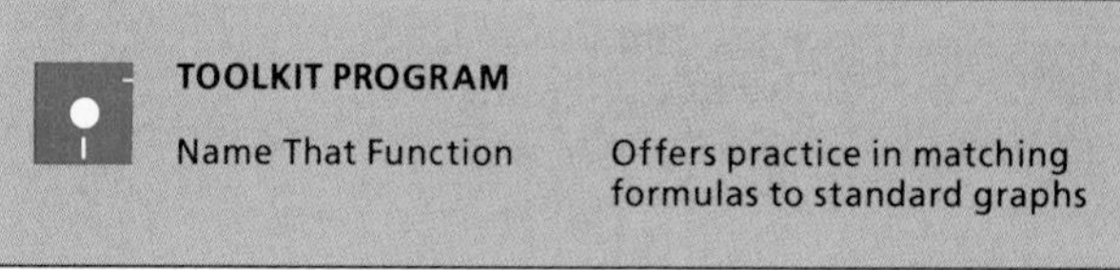

TOOLKIT PROGRAM

Name That Function	Offers practice in matching formulas to standard graphs

1.3 Functions and Their Graphs

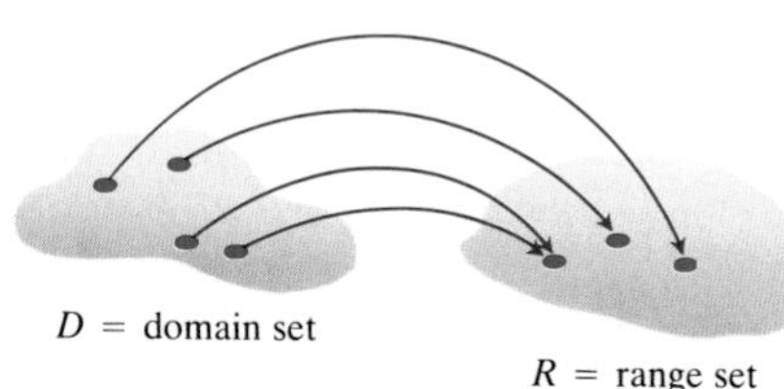

1.34 A function from set D to set R assigns an element of R to each element in D.

The values of one variable quantity often depend on the values of another. For example:

The pressure in the boiler of a power plant depends on the steam temperature.

The rate at which the water drains from your bathtub when you pull the plug depends on how deep the water is.

The area of a circle depends on its radius.

In each of these examples, the value of one variable quantity, which we might call y, depends on the value of another variable quantity, which we might call x. Since the value of y in each case is completely determined by the value of x, we say that y is a function of x.

In mathematics, any rule that assigns to each element in one set some element from another set is called a **function.** The sets may be sets of numbers, sets of number pairs, sets of points, or sets of objects of any kind. The sets do not have to be the same. All the function has to do is assign some element from the second set to each element in the first set (Fig. 1.34). Thus a function is like a machine that assigns an output to every allowable input. The inputs make up the **domain** of the function. The outputs make up the function's **range** (Fig. 1.35).

1.35 A flow diagram for a function f.

DEFINITION

A **function** from a set D to a set R is a rule that assigns a single element of R to each element in D.

The word *single* in the definition of function does not mean that there must be only one element in the function's range, although this can happen for some functions. It means that each input from the domain is assigned exactly one output from the range, no more and no less. In other words, each input appears just once in the list of input–output pairs defined by the function.

Euler invented a symbolic way to say "y is a function of x" by writing

$$y = f(x) \tag{1}$$

which we read "y equals f of x." This notation is shorter than the verbal statements that say the same thing. It also lets us give different functions different names by changing the letters we use. To say that boiler pressure is a function of steam temperature, we can write $p = f(t)$. To say that the area of a circle is a function of its radius, we can write $A = g(r)$. (Here we use a g because we just used f for something else.) We have to know what the variables p, t, A, and r mean, of course, for these equations to make sense.

Euler

Leonhard Euler (1707–1783), the dominant mathematical figure of his century, was also an astronomer, physicist, botanist, chemist, and expert in Oriental languages. He was the first scientist to give the function concept the prominence in his work that it has today. The most prolific mathematician who ever lived, his collected books and papers now fill 70 volumes. His introductory algebra text is still read in English translation.

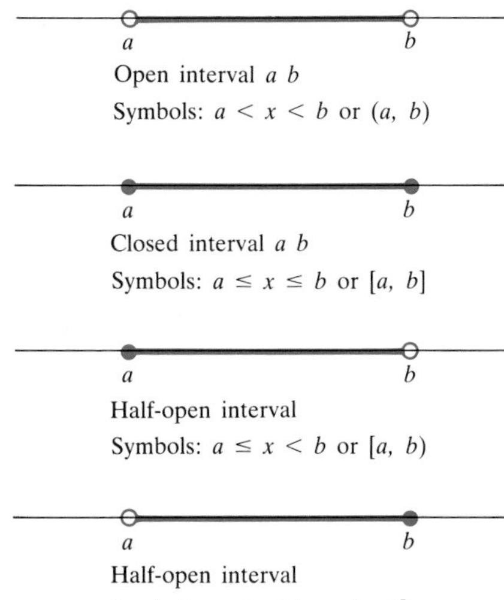

1.36 The domains and ranges of many functions are finite intervals like these.

Real-Valued Functions of a Real Variable

In most of our work, functions have domains and ranges that are sets of real numbers. Such functions are called **real-valued functions of a real variable,** and are usually defined by formulas or equations.

In addition to giving a function a useful name, the notation $y = f(x)$ gives a way to denote specific values of a function. For instance, the value of f at $x = 5$ is $f(5)$ ("f of 5").

EXAMPLE 1 The formula $A = \pi r^2$ gives the area of a circle as a function of its radius, where we would say that $A(r) = \pi r^2$. The area of a circle of radius $r = 2$ is $A(2) = \pi(2)^2 = 4\pi$.

In the context of geometry, the domain of the function $A = \pi r^2$ is the set of all possible radii—in this case the set of all positive real numbers. The range is also the set of positive real numbers. ■

EXAMPLE 2 *The function $y = x^2$.* The formula $y = x^2$ defines the number y to be the square of the number x. If $x = 5$, then $y = (5)^2 = 25$.

The domain is the set of allowable x-values—in this case the set of all real numbers. The range, which consists of the resulting y-values, is the set of nonnegative real numbers. ■

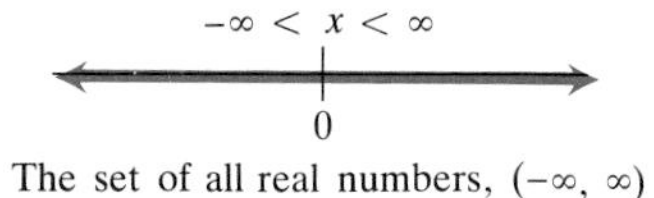

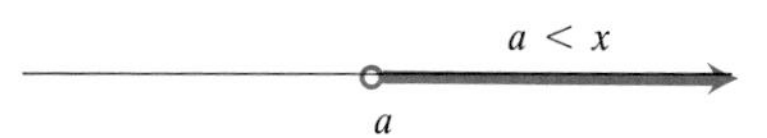

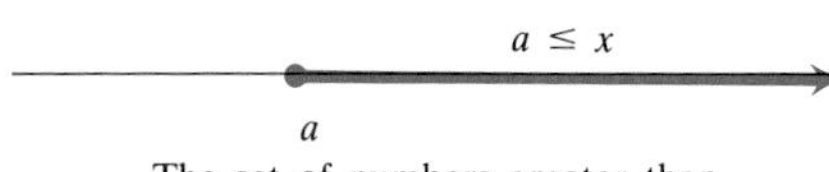

The set of numbers greater than or equal to a, $[a, \infty)$

$x < b$

b

The set of numbers less than b, $(-\infty, b)$

$x \le b$

b

The set of numbers less than or equal to b, $(-\infty, b]$

1.37 Rays on the number line and the line itself are called *infinite intervals.* The symbol ∞ (infinity) in the notation is used merely for convenience; it is not to be taken as a suggestion that there is a number ∞.

Domains and Ranges Are Often Intervals

The domains and ranges of many functions are intervals of real numbers (Fig. 1.36). The set of all real numbers that lie *strictly between* two fixed numbers a and b is an **open interval.** The interval is "open" at each end because it contains neither of its endpoints. Intervals that contain both endpoints are **closed.** Intervals that contain one endpoint but not both are **half-open.** Half-open intervals could just as well be called half-closed, but no one seems to call them that.

The domains and ranges of functions can also be **infinite intervals** (Fig. 1.37).

EXAMPLE 3 In each of the following functions, the domain is taken to be the largest set of real x-values for which the formula gives real y-values. We shall say more about this convention in a moment.

Function	Domain	Range
$y = x^2$	$-\infty < x < \infty$	$0 \le y$
$y = \sqrt{1 - x^2}$	$-1 \le x \le 1$	$0 \le y \le 1$
$y = \dfrac{1}{x}$	$x \ne 0$	$y \ne 0$
$y = \sqrt{x}$	$0 \le x$	$0 \le y$
$y = \sqrt{4 - x}$	$x \le 4$	$0 \le y$

The formula $y = x^2$ gives a real y-value for any real number x.

The formula $y = \sqrt{1 - x^2}$ gives a real y-value for every value of x in the closed interval from -1 to 1. Beyond this domain, the quantity $1 - x^2$ is negative and its square root is not a real number. (Complex numbers of the form $a + bi$, where $i = \sqrt{-1}$, are excluded from our consideration until Chapter 9.)

The formula $y = 1/x$ gives a real y-value for every x except $x = 0$. We cannot divide 1 (or any other number, for that matter) by 0.

The formula $y = \sqrt{x}$ gives a real y-value only when x is positive or zero. The number $y = \sqrt{x}$ is not a real number when x is negative. The domain of $y = \sqrt{x}$ is therefore the interval $x \geq 0$.

In $y = \sqrt{4 - x}$, the quantity $4 - x$ cannot be negative. That is, $4 - x$ must be greater than or equal to 0. In symbols,

$$0 \leq 4 - x \qquad \text{or} \qquad x \leq 4. \tag{2}$$

The formula $y = \sqrt{4 - x}$ gives a real y-value for any x less than or equal to 4. ■

Independent and Dependent Variables, a Warning about Division by 0, and a Convention about Domains

The variable x in a function $y = f(x)$ is called the **independent variable,** or **argument,** of the function. The variable y, whose value depends on x, is called the **dependent variable.**

We must keep two restrictions in mind when we define functions. First, we *never divide by 0*. When we see $y = 1/x$, we must think $x \neq 0$. Zero is not in the domain of the function. When we see $y = 1/(x - 2)$, we must think "$x \neq 2$."

The second restriction is that we shall deal exclusively with real-valued functions (except for a very short while later in the book). We may therefore have to restrict our domains when we have square roots (or fourth roots, or other even roots). If $y = \sqrt{1 - x^2}$, we should think "x^2 must not be greater than 1. The domain must not extend beyond the interval $-1 \leq x \leq 1$."

We observe a convention about the domains of functions defined by formulas. If the domain is not stated explicitly, then the domain is automatically the largest set of x-values for which the formula gives real y-values. If we wish to exclude values from this domain, we must say so. The formula $y = x^2$ gives real y-values for every real x. Therefore, writing

$$y = x^2 \qquad (x \text{ not restricted})$$

tells everyone that the intended domain is $-\infty < x < \infty$. To exclude negative values from the domain, we would limit the equation $y = x^2$ by writing the inequality $x \geq 0$ beside it:

$$y = x^2, \quad x \geq 0. \qquad (x \text{ restricted})$$

Graphs and Graphing

The points in the plane whose coordinates (x, y) are the input–output pairs of a function make up the **graph** of the function. Thus, the graph of a function defined by an equation in x and y is the graph of the equation itself. We draw these graphs the usual way: We make a table of matching xy-pairs and connect the points (x, y) with a smooth curve.

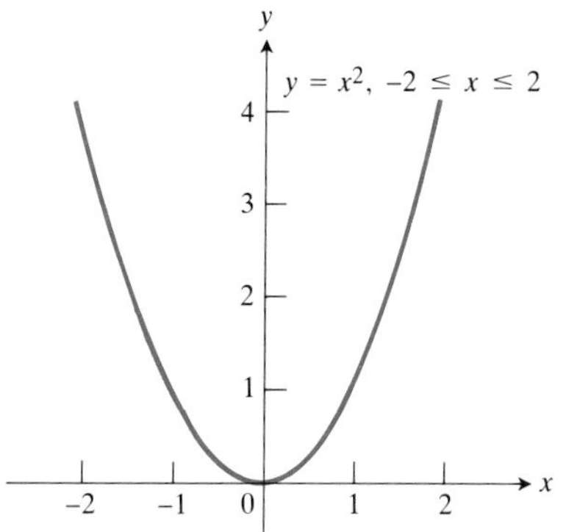

1.38 The graph of the function $y = x^2$ over the interval $-2 \leq x \leq 2$.

EXAMPLE 4 The graph of the function $y = x + 2$ is the line $y = x + 2$. It is the set of points (x, y) in which y equals $x + 2$. ■

EXAMPLE 5 The graph of the function $y = x^2$ is the graph of the equation $y = x^2$.

In Example 3 in Section 1.1, we graphed this function over the interval $-2 \leq x \leq 2$ (Fig. 1.38). We cannot graph the function over its entire domain because the domain is infinite. However, we can imagine what the graph looks like by extending the picture we already have. As x moves outside the interval $-2 \leq x \leq 2$, the values of $y = x^2$ grow faster than x, so we expect the graph to rise sharply on either side. Indeed it does, as we can see in Fig. 1.39. ■

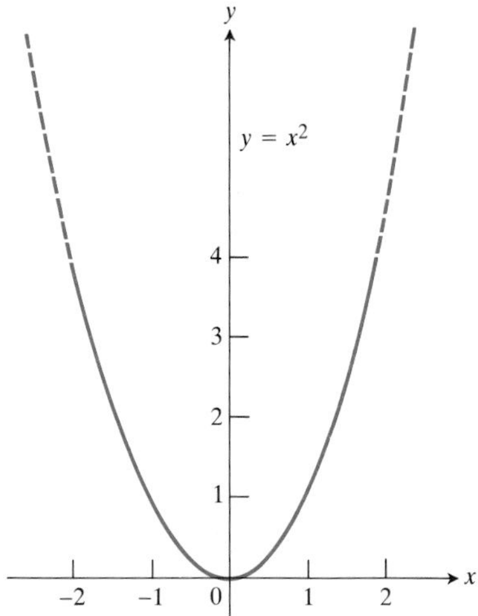

1.39 The graph in Fig. 1.38 extended.

EXAMPLE 6 Graph the function $y = \sqrt{4 - x}$.

Solution The domain of $y = \sqrt{4 - x}$ is the interval $4 - x \geq 0$ or $x \leq 4$. The graph starts at $x = 4$ and runs to the left.

The value $x = 4$ is also the x-intercept. The y-intercept is

$$y = \sqrt{4 - 0} = \sqrt{4} = 2.$$

The graph has none of the symmetries we usually look for, so we are ready to make a table of values, plot points, and draw the curve.

STEP 1: Make a table of values that includes intercepts and domain endpoints.

x	$4 - x$	$\sqrt{4 - x}$
4.0	0	0
3.75	0.25	0.5
2.0	2.0	$\sqrt{2} \approx 1.4$
0	4	2
-2	6	$\sqrt{6} \approx 2.4$

STEP 2: Plot the points from the table. Connect them with a smooth curve. Plot a few more points if you're not sure how the curve goes.

The steps for graphing functions are the same as those for graphing equations, except that in graphing functions we also include domain endpoints in the table of values.

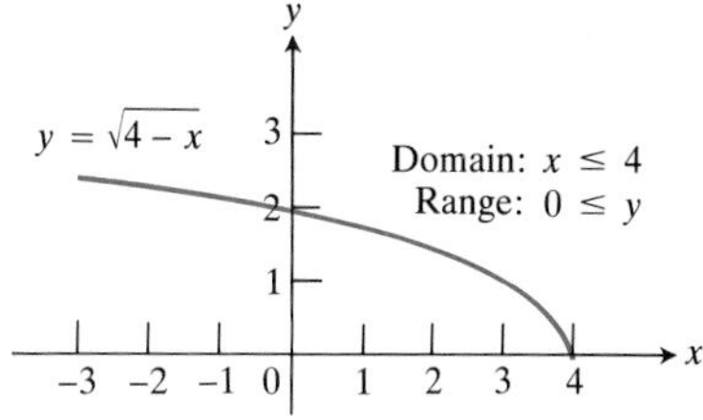

■

Figure 1.40 shows the graphs of functions frequently used in calculus.

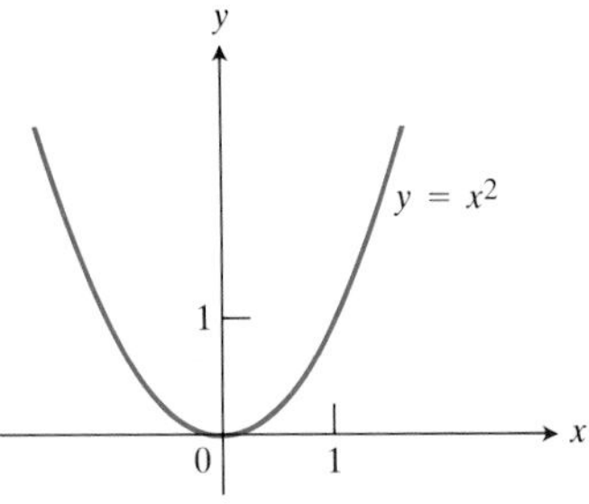

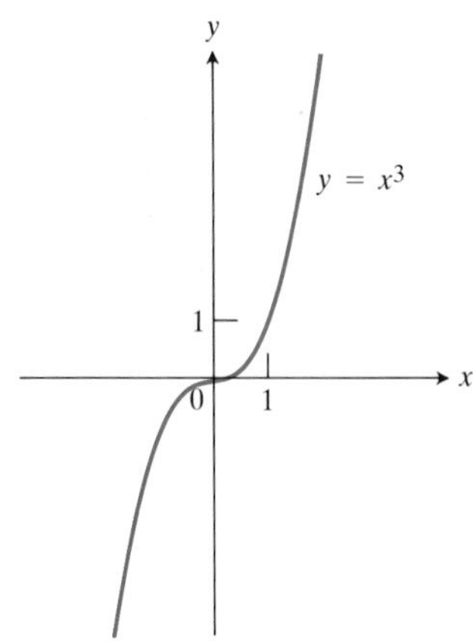

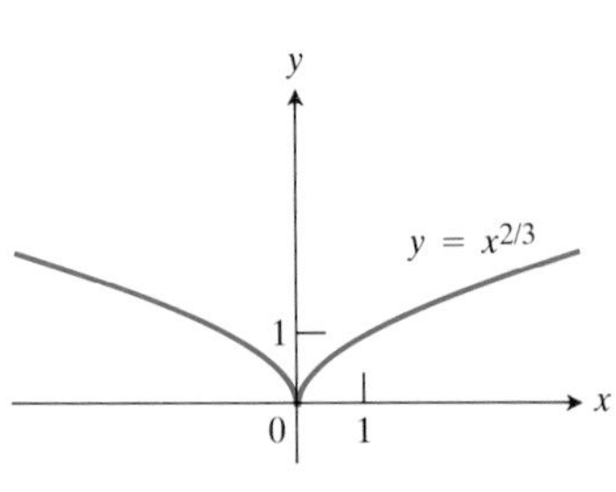

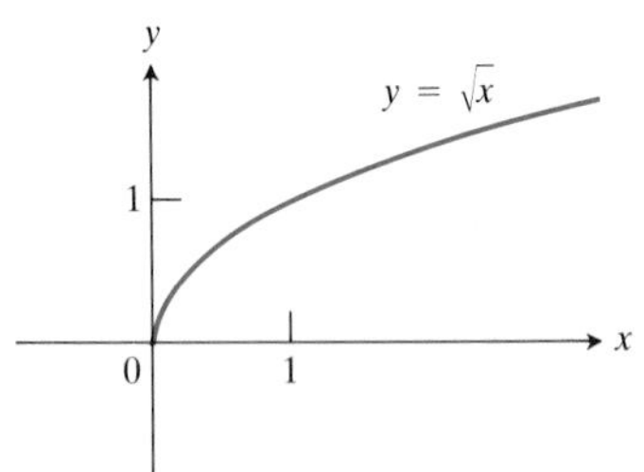

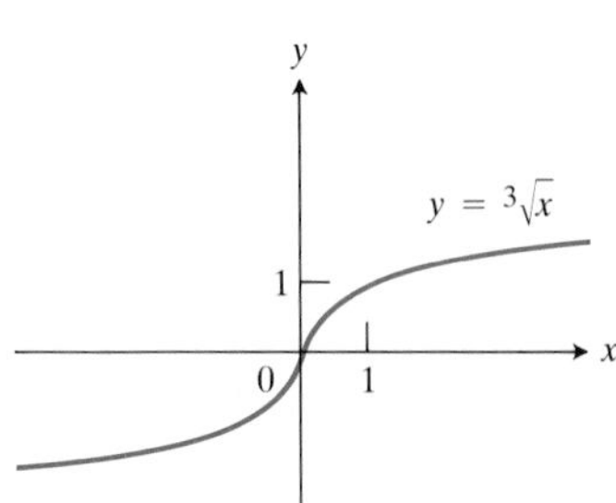

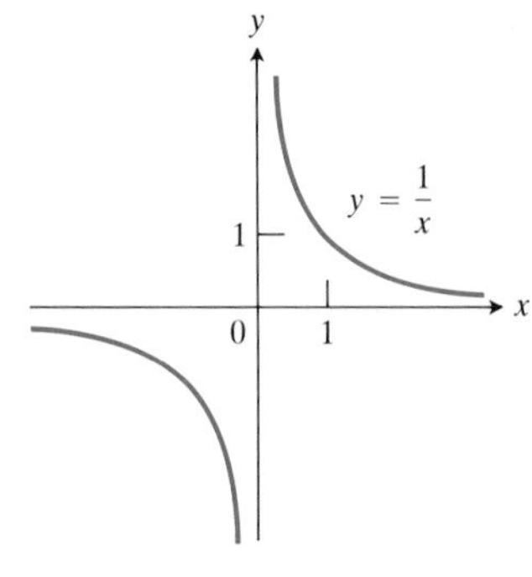

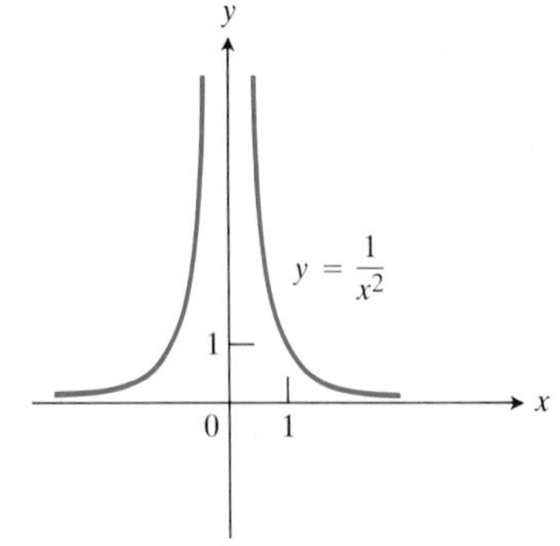

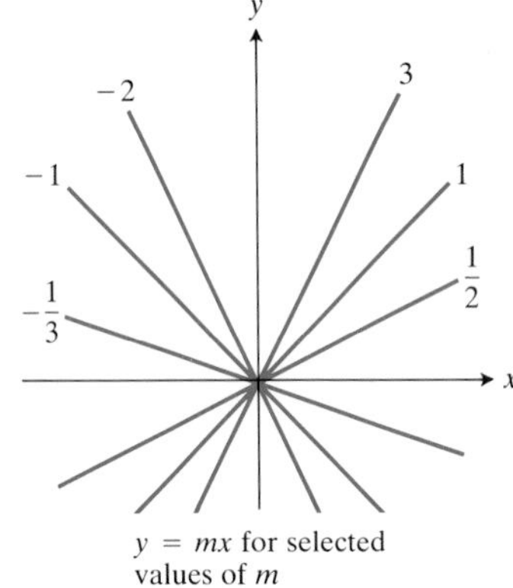

$y = mx$ for selected values of m

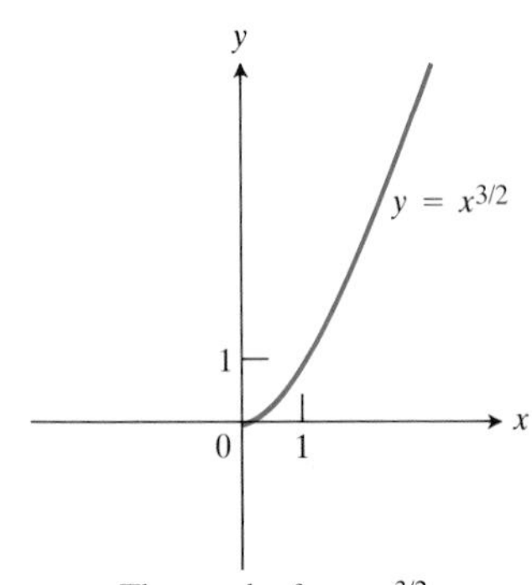

The graph of $y = x^{3/2}$ stops at the origin

1.40 Reference table of useful graphs.

Even Functions and Odd Functions

DEFINITION

A function $y = f(x)$ is an **even** function of x if $f(-x) = f(x)$ for every x in the function's domain. It is an **odd** function of x if $f(-x) = -f(x)$ for every x in the function's domain.

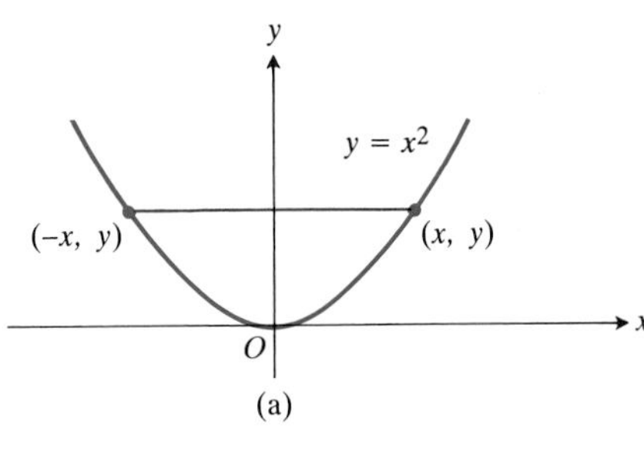

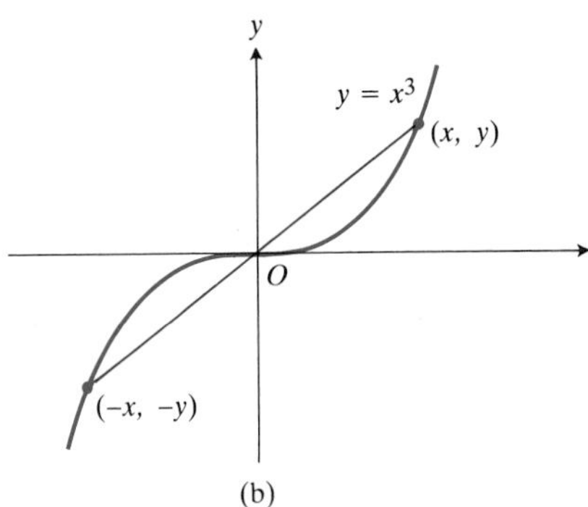

1.41 (a) The graph of an even function is symmetric about the y-axis.
(b) The graph of an odd function is symmetric about the origin.

The names even and odd come from powers of x. If y equals an even power of x, as in $y = x^2$ or $y = x^4$, it is an even function of x (because $(-x)^2 = x^2$ and $(-x)^4 = x^4$). If y equals an odd power of x, as in $y = x$ or $y = x^3$, it is an odd function of x (because $(-x)^1 = -x$ and $(-x)^3 = -x^3$).

Saying that a function $y = f(x)$ is even is equivalent to saying that its graph is symmetric about the y-axis. Since $f(-x) = f(x)$, the point (x, y) lies on the curve if and only if the point $(-x, y)$ lies on the curve (Fig. 1.41a).

Saying that a function $y = f(x)$ is odd is equivalent to saying that its graph is symmetric with respect to the origin. Since $f(-x) = -f(x)$, the point (x, y) lies on the curve if and only if the point $(-x, -y)$ lies on the curve (Fig. 1.41b).

The graphs of equations that contain only even powers of x will be symmetric about the y-axis. However, there is no corresponding rule for odd powers and symmetry about the origin.

EXAMPLE 7 *Even, odd, and neither*

$f(x) = x^2$	Even function: $(-x)^2 = x^2$ for all x Symmetry about y-axis
$f(x) = x^2 + 1$	Even function: $(-x)^2 + 1 = x^2 + 1$ for all x Symmetry about the y-axis (Fig. 1.42a)
$f(x) = x$	Odd function: $(-x) = -(x)$ for all x Symmetry about the origin
$f(x) = x + 1$	Not odd: $f(-x) = -x + 1$, but $-f(x) = -x - 1$. The two are not equal Not even, either: $(-x) + 1 \neq x + 1$. See Fig. 1.42(b)

1.42 (a) When we add 1 to $y = x^2$, the resulting function is still even and its graph is still symmetric about the y-axis. (b) When we add 1 to $y = x$, the resulting function is no longer odd. The symmetry about the origin is lost.

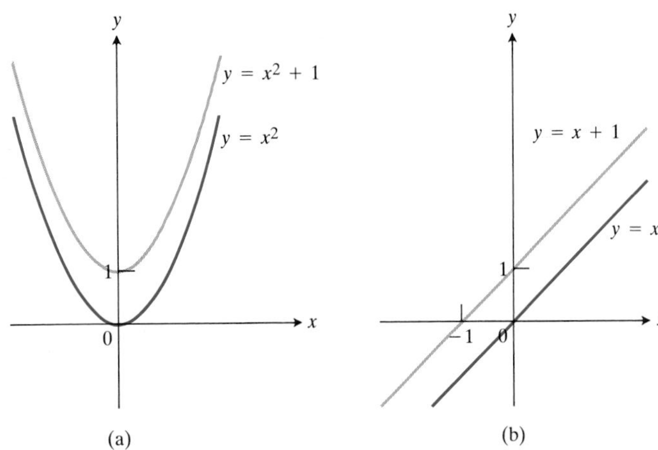

Integer-Valued Functions

The greatest integer less than or equal to a number x is called the *greatest integer in x*. Because each real number x corresponds to only one greatest integer, the greatest integer in x is a function of x. The symbol for it is $\lfloor x \rfloor$, which is read "the greatest integer in x." Figure 1.43 shows the graph of $y = \lfloor x \rfloor$.

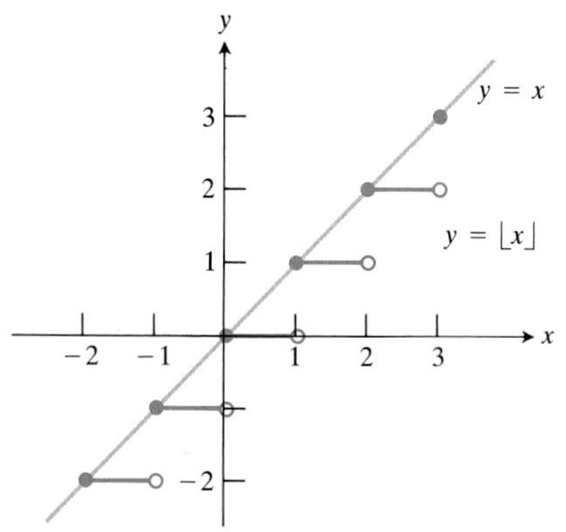

1.43 The graph of $y = \lfloor x \rfloor$ and its relation to the line $y = x$. As the figure shows, $\lfloor x \rfloor$ is less than or equal to x, so it provides an integer floor for x.

EXAMPLE 8 *Values of* $y = \lfloor x \rfloor$

Positive	$\lfloor 1.9 \rfloor = 1$, $\lfloor 2.0 \rfloor = 2$, $\lfloor 2.4 \rfloor = 2$
Zero	$\lfloor 0.5 \rfloor = 0$, $\lfloor 0 \rfloor = 0$
Negative	$\lfloor -1.2 \rfloor = -2$, $\lfloor -0.5 \rfloor = -1$

Notice that if x is negative, $\lfloor x \rfloor$ may have a larger absolute value than x does. ■

Another common notation for the greatest integer function is $y = [x]$. The notation $y = \lfloor x \rfloor$ comes from computer science, where it is used to denote the result of rounding x down to the nearest integer. You can think of it as the integer "floor" for x and the notation is chosen to suggest just that:

$$\lfloor x \rfloor \qquad \text{integer floor for } x.$$

The companion notation

$$\lceil x \rceil \qquad \text{integer ceiling for } x$$

is used to denote the result of rounding x up to the nearest integer. It denotes the least integer greater than or equal to x (Fig. 1.44).

Notice, for later reference, that the integer floor and ceiling functions exhibit points called *discontinuities,* where the functions jump from one value to the next without taking on the intermediate values. They jump like this at every integer value of x.

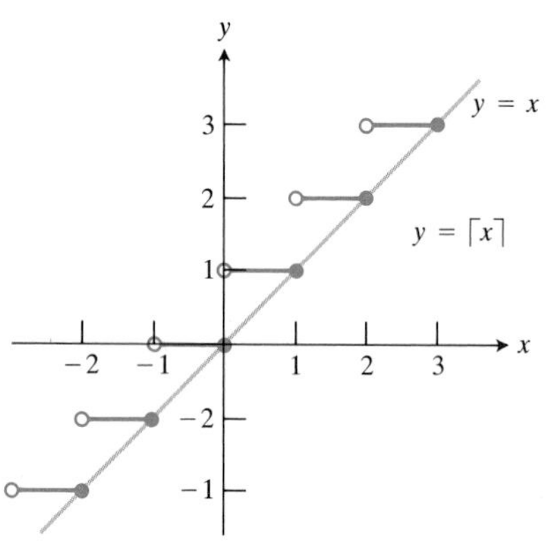

1.44 The graph of $y = \lceil x \rceil$ and its relation to the line $y = x$. As the figure shows, $\lceil x \rceil$ is greater than or equal to x, so it provides an integer ceiling for x.

Functions Defined in Pieces

While some functions are defined by single formulas, others are defined by applying different formulas to different parts of their domains.

EXAMPLE 9 The values of the function

$$y = f(x) = \begin{cases} -x \text{ if } x < 0, \\ x^2 \text{ if } 0 \le x \le 1, \\ 1 \text{ if } x > 1, \end{cases}$$

are given by the formula $y = -x$ when $x < 0$, by the formula $y = x^2$ when $0 \le x \le 1$, and by the formula $y = 1$ when $x > 1$. The function is *just one function,* however, whose domain is the entire real line (Fig. 1.45). ■

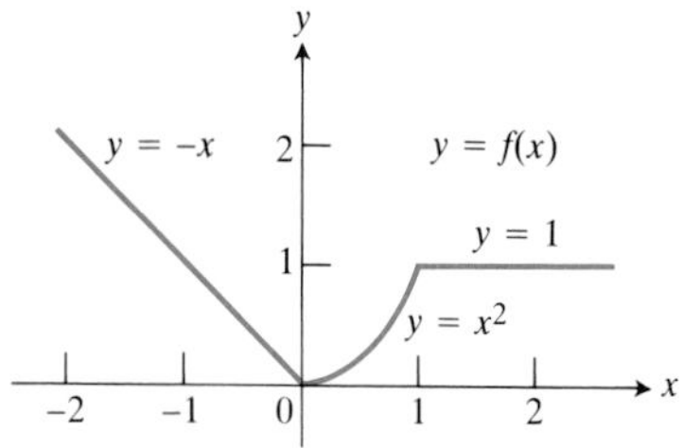

1.45 To graph the function $y = f(x)$ shown here, we apply different formulas to different parts of its domain.

EXAMPLE 10 Suppose that the graph of a function $y = f(x)$ consists of the line segments shown in Fig. 1.46. Write a formula for f.

Solution We find formulas for the segments from (0, 0) to (1, 1) and from (1, 0) to (2, 1) and piece them together in the manner of Example 9.

1.46 The graph of the function $y = f(x)$ (Example 10) shown here consists of two line segments. The segment on the left contains the endpoint at the origin (shown by a heavy dot) but does not contain the endpoint (1, 1). The segment on the right contains both endpoints.

Segment from (0, 0) to (1, 1). The line through (0, 0) and (1, 1) has slope $m = (1 - 0)/(1 - 0) = 1$ and y-intercept $b = 0$. Its slope–intercept equation is therefore $y = x$. The segment from (0, 0) to (1, 1) that includes the point (0, 0) but not the point (1, 1) is the graph of the function $y = x$ restricted to the half-open interval $0 \le x < 1$, namely,

$$y = x, \qquad 0 \le x < 1.$$

Segment from (1, 0) to (2, 1). The line through (1, 0) and (2, 1) has slope $m = (1 - 0)/(2 - 1) = 1$ and passes through the point (1, 0). The corresponding point–slope equation for the line is therefore

$$y - 0 = 1(x - 1), \qquad \text{or} \qquad y = x - 1.$$

The segment from (1, 0) to (2, 1) that includes both endpoints is the graph of $y = x - 1$ restricted to the closed interval $1 \le x \le 2$, namely,

$$y = x - 1, \qquad 1 \le x \le 2.$$

Formula for the function $y = f(x)$ shown in Fig. 1.46. We obtain a formula for f on the interval $0 \le x \le 2$ by combining the formulas we obtained for the two segments of the graph:

$$f(x) = \begin{cases} x & \text{for } 0 \le x < 1, \\ x - 1 & \text{for } 1 \le x \le 2. \end{cases}$$

EXAMPLE 11 The domain of the "step" function $y = g(x)$ graphed in Fig. 1.47 is the closed interval $0 \le x \le 3$. Find a formula for $g(x)$.

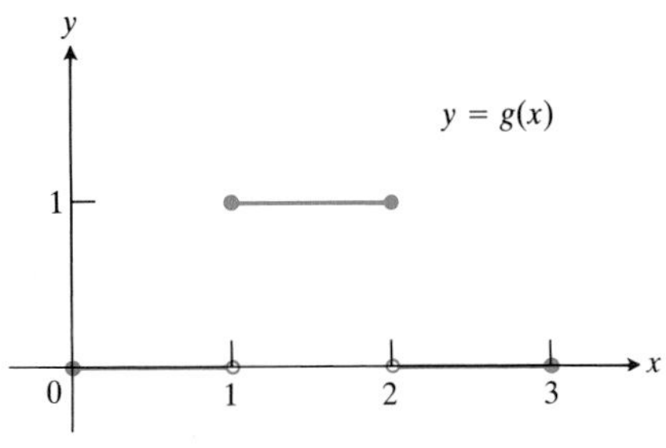

1.47 Functions like the one graphed here are called *step functions.* Example 11 shows how to write a formula for g.

Solution The graph consists of three horizontal line segments. The left segment is the half-open interval $0 \le x < 1$ on the x-axis, which we may think of as a portion of the line $y = 0$:

$$y = 0, \qquad 0 \le x < 1.$$

The second segment is the portion of the line $y = 1$ that lies over the closed interval $1 \le x \le 2$:

$$y = 1, \qquad 1 \le x \le 2.$$

The third segment is the half-open interval $2 < x \le 3$ on the line $y = 0$:

$$y = 0, \qquad 2 < x \le 3.$$

The values of g are therefore given by the three-piece formula

$$g(x) = \begin{cases} 0 & \text{for } 0 \le x < 1, \\ 1 & \text{for } 1 \le x \le 2, \\ 0 & \text{for } 2 < x \le 3. \end{cases}$$

Sums, Differences, Products, and Quotients

The sum $f + g$ of two functions of x is itself a function of x, defined at any point x that lies in both domains. The same holds for the differences $f - g$ and $g - f$, the product $f \cdot g$, and the quotients f/g and g/f, as long as we exclude any points that require division by zero.

EXAMPLE 12 *The sum, differences, products, and quotients of the functions f and g defined by the formulas*

$$f(x) = \sqrt{x} \quad \text{and} \quad g(x) = \sqrt{1-x}.$$

Function	Formula	Domain
f	$f(x) = \sqrt{x}$	$0 \le x$
g	$g(x) = \sqrt{1-x}$	$x \le 1$
$f + g$	$(f+g)(x) = f(x) + g(x) = \sqrt{x} + \sqrt{1-x}$	$0 \le x \le 1$ (The intersection of the domains of f and g)
$f - g$	$(f-g)(x) = f(x) - g(x) = \sqrt{x} - \sqrt{1-x}$	$0 \le x \le 1$
$g - f$	$(g-f)(x) = g(x) - f(x) = \sqrt{1-x} - \sqrt{x}$	$0 \le x \le 1$
$f \cdot g$	$(f \cdot g)(x) = f(x)\,g(x) = \sqrt{x(1-x)}$	$0 \le x \le 1$
f/g	$\dfrac{f}{g}(x) = \dfrac{f(x)}{g(x)} = \sqrt{\dfrac{x}{x-1}}$	$0 \le x < 1$ ($x = 1$ excluded)
g/f	$\dfrac{g}{f}(x) = \dfrac{g(x)}{f(x)} = \sqrt{\dfrac{1-x}{x}}$	$0 < x \le 1$ ($x = 0$ excluded)

See Fig. 1.48.

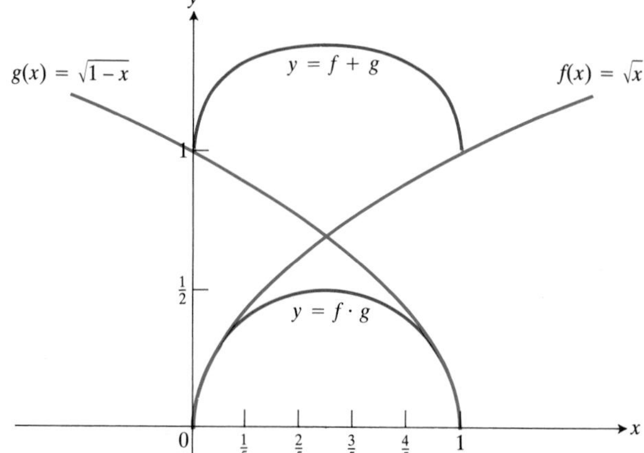

1.48 The domain of the function $f + g$ is the intersection of the domains of f and g, the interval [0, 1] on the x-axis where these domains overlap. This interval is also the domain of the function $f \cdot g$. See Example 12.

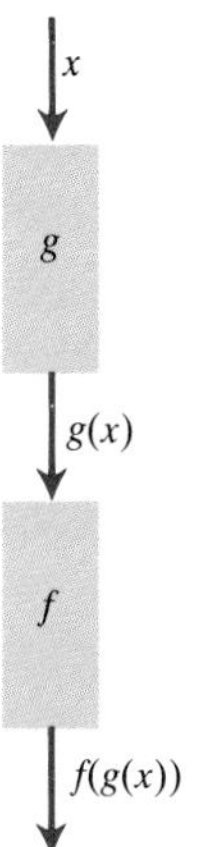

1.49 Two functions can be composed when the range of the first lies in the domain of the second.

Composition of Functions

Suppose that the outputs of a function g can be used as inputs of a function f. We can then hook g and f together to form a new function whose inputs are the inputs of g and whose outputs are the numbers $f(g(x))$, as in Fig. 1.49. We say that the function $f(g(x))$ (pronounced "f of g of x") is the **composite** of g and f. It is made by *composing* g and f in the order first g, then f. The usual "stand-alone" notation for this composite is $f \circ g$, which is read as "f of g." Thus, the value of $f \circ g$ at x is $(f \circ g)(x) = f(g(x))$.

EXAMPLE 13 Find a formula for $f(g(x))$ if $g(x) = x^2$ and $f(x) = x - 7$. Then find the value of $f(g(2))$.

Solution To find $f(g(x))$, we replace x in the formula for $f(x)$ by the expression given for $g(x)$:

$$f(x) = x - 7$$
$$f(g(x)) = g(x) - 7 = x^2 - 7.$$

We then find the value of $f(g(2))$ by substituting 2 for x:

$$f(g(2)) = (2)^2 - 7 = 4 - 7 = -3.$$ ■

Changing the order in which functions are composed usually changes the result. In Example 13, we composed $g(x) = x^2$ and $f(x) = x - 7$ in the order first g, then f, obtaining the function $f \circ g$ whose value at x was $f(g(x)) = x^2 - 7$. In the next example, we see what happens when we reverse the order to obtain the function $g \circ f$.

EXAMPLE 14 Find a formula for $g(f(x))$ if $g(x) = x^2$ and $f(x) = x - 7$. Then find $g(f(2))$.

Solution To find $g(f(x))$, we replace x in the formula for $g(x)$ by the expression for $f(x)$:

$$g(x) = x^2$$
$$g(f(x)) = (f(x))^2 = (x - 7)^2.$$

To find $g(f(2))$, we then substitute 2 for x:

$$g(f(2)) = (2 - 7)^2 = (-5)^2 = 25.$$ ■

In the notation for composite functions, the parentheses tell which function comes first:

The notation $f(g(x))$ says "first g, then f." To calculate $f(g(2))$, calculate $g(2)$ and then apply f.

The notation $g(f(x))$ says "first f, then g." To calculate $g(f(2))$, calculate $f(2)$ and then apply g.

Exercises 1.3

In Exercises 1–8, find the domain and range of each function.

1. $y = 2\sqrt{x}$ **2.** $y = 1 + \sqrt{x}$

3. $y = -\sqrt{x}$ **4.** $y = \sqrt{-x}$

5. $y = \sqrt{x + 4}$ **6.** $y = \sqrt{x - 2}$

7. $y = \dfrac{1}{x - 2}$ **8.** $y = \dfrac{1}{x + 2}$

Say whether the functions in Exercises 9–20 are even, odd, or neither. Try to answer without writing anything down (except the answer).

9. $y = x$ **10.** $y = x^2$

11. $y = x^3$ **12.** $y = x^4$

13. $y = x + 1$ **14.** $y = x + x^2$

15. $y = x^2 + 1$ **16.** $y = x + x^3$

17. $y = \dfrac{1}{x^2 - 1}$ **18.** $y = \dfrac{1}{x - 1}$

19. $y = \dfrac{x}{x^2 - 1}$ **20.** $y = \dfrac{x^2}{x^2 - 1}$

Find the domains and ranges of the functions in Exercises 21–32. Then graph the functions. What symmetries, if any, do the graphs have?

21. $y = 2x^2$ **22.** $y = -x^2$

23. $y = x^2 - 9$ **24.** $y = 4 - x^2$

25. $y = x^3$ **26.** $y = -\sqrt[3]{x}$

27. $y = \sqrt{x + 1}$ **28.** $y = \sqrt{9 - x}$

29. $y = -\dfrac{1}{x}$ **30.** $y = -\dfrac{1}{x^2}$

31. $y = 1 + \dfrac{1}{x}$ **32.** $y = 1 + \dfrac{1}{x^2}$

33. Consider the function $y = 1/\sqrt{x}$.
a) Can x be negative?
b) Can $x = 0$?
c) What is the domain of the function?

34. Consider the function $y = \sqrt{(1/x) - 1}$.
a) Can x be negative?
b) Can $x = 0$?
c) Can x be greater than 1?
d) What is the domain of the function?

35. Which of the graphs in Fig. 1.50 could be the graph of
a) $y = x^2 - 1$? Why?
b) $y = (x - 1)^2$? Why?

36. Which of the graphs in Fig. 1.50 could *not* be the graph of $y = 4x^2$? Why?

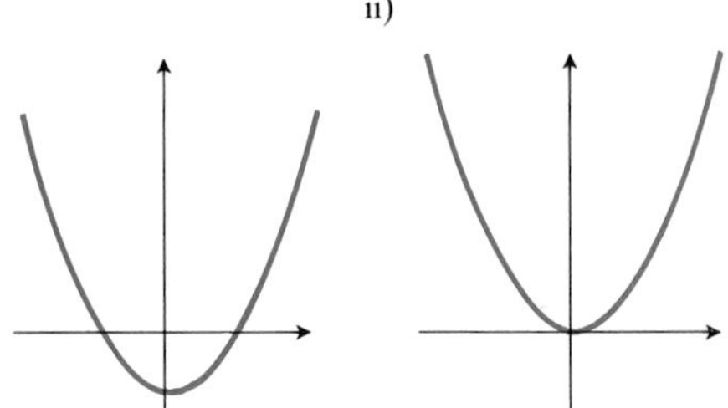

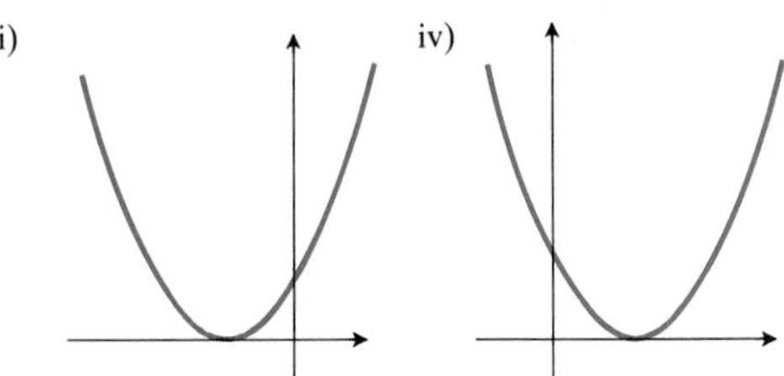

1.50 The graphs for Exercises 35 and 36.

37. For what values of x does (a) $\lfloor x \rfloor = 0$? (b) $\lceil x \rceil = 0$?

38. Does $\lfloor x \rfloor$ ever equal $\lceil x \rceil$? Explain.

39. Graph each function over the given interval.
a) $y = x - \lfloor x \rfloor$, $-3 \le x \le 3$
b) $y = \lfloor x \rfloor - \lceil x \rceil$, $-3 \le x \le 3$.

40. *Integer parts of decimals.* When x is positive or zero, $\lfloor x \rfloor$ is the integer part of the decimal representation of x. What is the corresponding description of $\lceil x \rceil$ when x is negative or zero?

41. Make a table of values with $x = 0$, 1, and 2, and graph the function
$$y = \begin{cases} x, & 0 \le x \le 1 \\ 2 - x, & 1 \le x \le 2 \end{cases}$$

42. Make a table of values with $x = 0$, 1, and 2 and graph the function
$$y = \begin{cases} 1 - x, & 0 \le x \le 1 \\ 2 - x, & 1 < x \le 2 \end{cases}$$

Graph the functions in Exercises 43–46.

43. $y = \begin{cases} 3 - x, & x \le 1, \\ 2x, & 1 < x \end{cases}$ **44.** $y = \begin{cases} 1/x, & x < 0, \\ x, & 0 \le x \end{cases}$

45. $y = \begin{cases} 1, & x < 5, \\ 0, & 5 \le x \end{cases}$ **46.** $y = \begin{cases} 1, & x < 0, \\ \sqrt{x}, & x \ge 0 \end{cases}$

47. Find formulas for the functions graphed in the following figures.

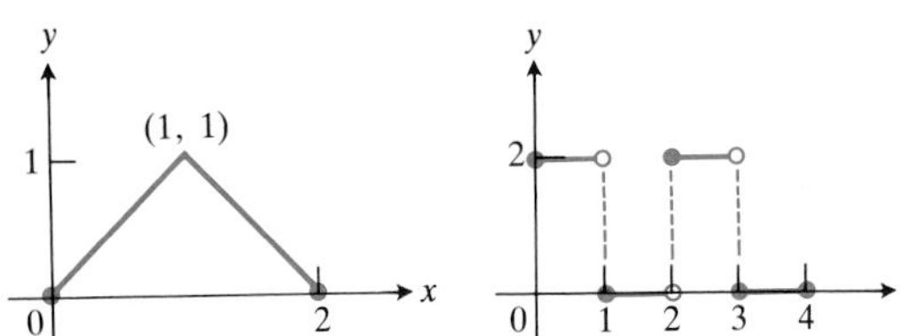

48. Find formulas for the functions graphed in the following figures.

a) 

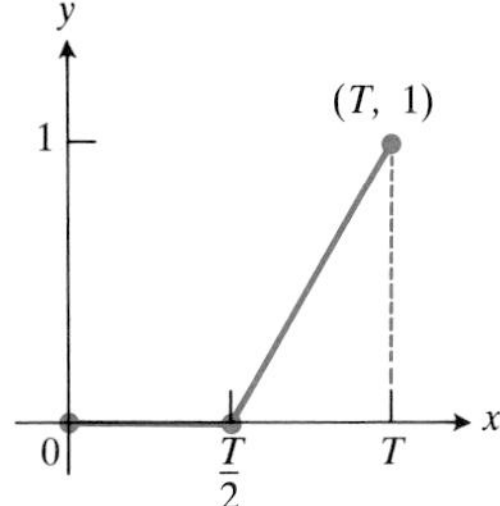

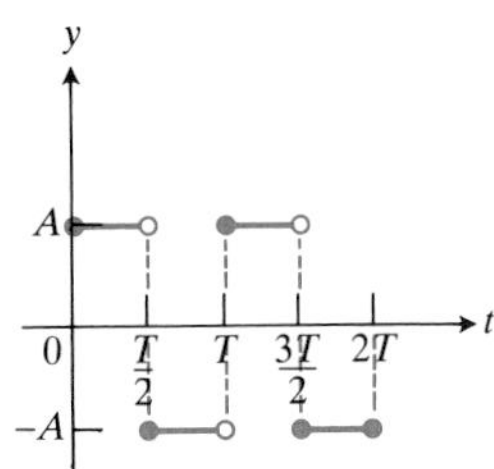

In Exercises 49 and 50 find the domains of f and g and the corresponding domains of $f+g, f-g, f\cdot g, f/g$, and g/f.

49. $f(x)=x, \quad g(x)=\sqrt{x-1}$

50. $f(x)=\sqrt{x+1}, \quad g(x)=\sqrt{x-1}$

51. If $f(x)=x+5$ and $g(x)=x^2-3$, find

a) $f(g(0))$, b) $g(f(0))$,
c) $f(g(x))$, d) $g(f(x))$,
e) $f(f(-5))$, f) $g(g(2))$,
g) $f(f(x))$, h) $g(g(x))$.

52. If $f(x)=x+1$ and $g(x)=x-1$, find

a) $f(g(0))$ b) $g(f(0))$
c) $f(g(1))$ d) $g(f(1))$
e) $f(g(x))$ f) $g(f(x))$

53. Copy and complete the following table.

	$g(x)$	$f(x)$	$f\circ g(x)$
a)	$x-7$	$\sqrt{x}$	
b)	$x+2$	$3x$	
c)		$\sqrt{x-5}$	$\sqrt{x^2-5}$
d)	$\dfrac{x}{x-1}$	$\dfrac{x}{x-1}$	
e)		$1+\dfrac{1}{x}$	x
f)	$\dfrac{1}{x}$		x

54. a) If $f(x)=1/x$, find $f(x+2)-f(2)$.
b) If $F(t)=4t-3$, find $F(t+1)-F(1)$.

55. *A magic trick.* You may have heard of a magic trick that goes like this: Take any number. Add 5. Double the result. Subtract 6. Divide by 2. Subtract 2. Now, tell me your answer, and I'll tell you what you started with.

Pick a number and try it.

You can see what is going on if you let x be your original number and follow the steps to make a formula $f(x)$ for the number you end up with.

TOOLKIT PROGRAMS

Function Evaluator	Makes tables of values of functions
Super*Grapher	Graphs functions

1.4 Shifts, Circles, and Parabolas

In this section, we show how to change an equation to move its graph up or down or to the right or left in the coordinate plane. We practice mostly with circles and parabolas (because they make useful examples in calculus), but the methods we use apply to other curves as well. We shall have more to say about parabolas and circles when we discuss conic sections in Chapter 10.

How to Shift a Graph

To shift the graph of a function $y=f(x)$ straight up, we add a positive constant to the right-hand side of the formula $y=f(x)$.

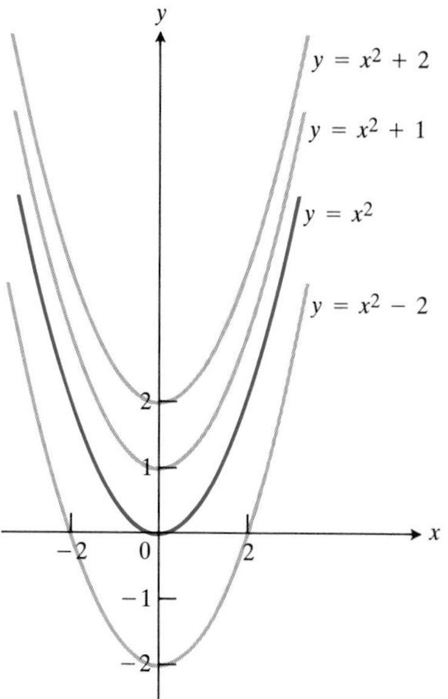

1.51 To shift the graph of $f(x) = x^2$ up (or down), we add positive constants to (or subtract them from) the formula for f.

EXAMPLE 1 Adding 1 to the right-hand side of the formula $y = x^2$ to get $y = x^2 + 1$ shifts the graph up 1 unit (Fig. 1.51). ■

To shift the graph of a function $y = f(x)$ straight down, we subtract a positive constant from the right-hand side of the formula $y = f(x)$.

EXAMPLE 2 Subtracting 2 from the right-hand side of the formula $y = x^2$ to get $y = x^2 - 2$ shifts the graph down 2 units (Fig. 1.51). ■

To shift the graph of $y = f(x)$ to the left, we add a positive constant to x.

EXAMPLE 3 Adding 3 to x in $y = x^2$ to get $y = (x + 3)^2$ shifts the graph 3 units to the left (Fig. 1.52). ■

To shift the graph of $y = f(x)$ to the right, we subtract a positive constant from x.

EXAMPLE 4 Subtracting 2 from x in $y = x^2$ to get $y = (x - 2)^2$ shifts the graph 2 units to the right (Fig. 1.52). ■

Shift Formulas ($c > 0$)

Vertical shifts

$y = f(x) + c$ or $y - c = f(x)$	Shifts the graph of f up c units
$y = f(x) - c$ or $y + c = f(x)$	Shifts the graph of f down c units

Horizontal shifts

$y = f(x + c)$	Shifts the graph of f left c units
$y = f(x - c)$	Shifts the graph of f right c units

The formula for a vertical shift upward is often written as $y - c = f(x)$ instead of $y = f(x) + c$. You can get the new form by subtracting c from both sides of the equation $y = f(x) + c$. But there is also another way to

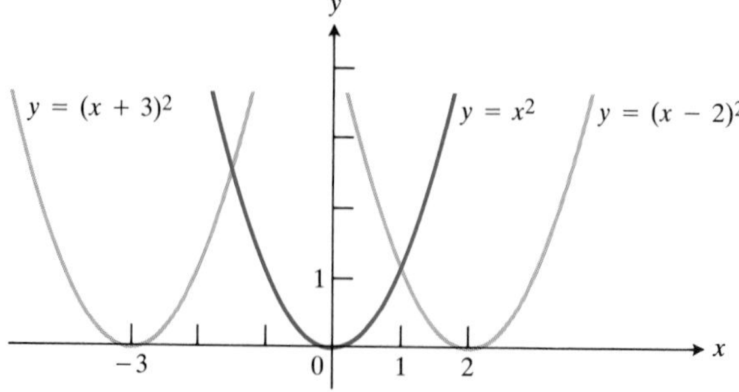

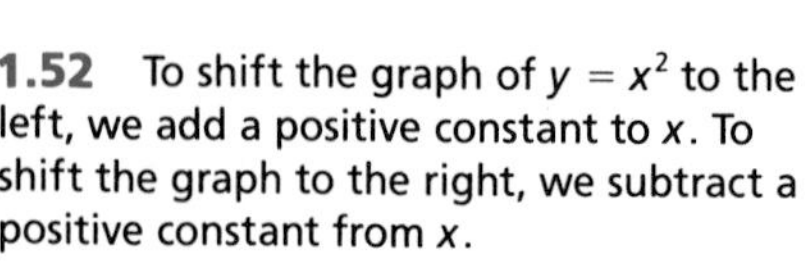
1.52 To shift the graph of $y = x^2$ to the left, we add a positive constant to x. To shift the graph to the right, we subtract a positive constant from x.

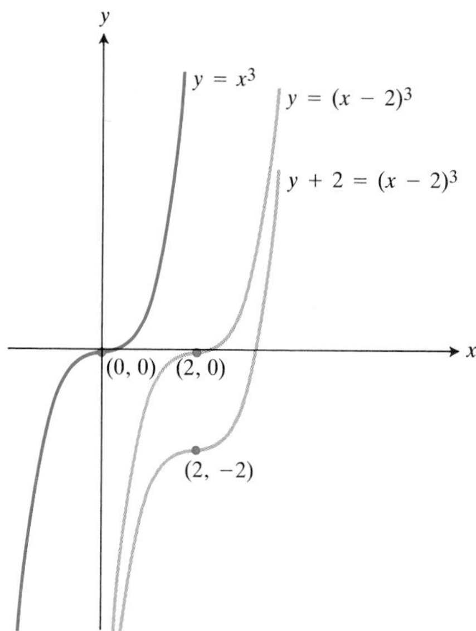

1.53 The graph of $y = x^3$ shifted 2 units right and then 2 units down.

think of it. For $y - c$ to match a given x, the value of y has to be c units larger than before. The graph of y vs. x therefore must be c units higher.

Similarly, the formula for a vertical downward shift of c units is often written as $y + c = f(x)$.

EXAMPLE 5 At each value of x, the graph of $y - 1 = x^2$ is 1 unit higher than the graph of $y = x^2$. See Fig. 1.51, where the graph is labeled $y = x^2 + 1$. ■

EXAMPLE 6 The graph of $y + 2 = (x - 2)^3$ lies 2 units below the graph of $y = x^3$ and 2 units to the right (Fig. 1.53). ■

Equations for Circles in the Plane

DEFINITION

A **circle** is the set of points in a plane whose distance from a given fixed point in the plane is a constant. The fixed point is the **center** of the circle. The constant distance is the **radius** of the circle.

To write an equation for the circle of radius a centered at the point $C(h, k)$, we let $P(x, y)$ denote a typical point on the circle (Fig. 1.54). The statement that CP equals a then becomes

$$\sqrt{(x-h)^2 + (y-k)^2} = a \qquad \begin{pmatrix}\text{The distance from } (x, y)\\ \text{to } (h, k) \text{ equals } a.\end{pmatrix}$$

or

$$(x-h)^2 + (y-k)^2 = a^2. \qquad \text{(Both sides squared.)} \qquad (1)$$

If $CP = a$, then Eq. (1) holds. If Eq. (1) holds, then $CP = a$. Equation (1) is therefore an equation for the circle.

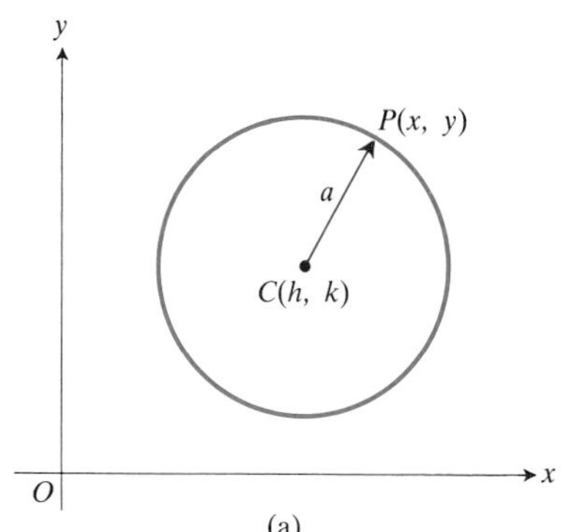

(a)

1.54 The standard equation for the circle shown here is $(x - h)^2 + (y - k)^2 = a^2$.

The Standard Equation for the Circle of Radius *a* Centered at the Point (*h*, *k*)

$$(x-h)^2 + (y-k)^2 = a^2 \qquad (2)$$

EXAMPLE 7 The standard equation for the circle of radius 2 centered at the point (3, 4) is

$$(x-3)^2 + (y-4)^2 = (2)^2$$

or

$$(x-3)^2 + (y-4)^2 = 4.$$

There is no need to square out the x and y terms in this equation. In fact, it is better not to do so. The present form reveals the circle's center and radius. ■

EXAMPLE 8 Find the center and radius of the circle

$$(x - 1)^2 + (y + 5)^2 = 3.$$

Solution Comparing

$$(x - h)^2 + (y - k)^2 = a^2 \quad \text{with} \quad (x - 1)^2 + (y + 5)^2 = 3$$

shows that

$$-h = -1 \quad \text{or} \quad h = 1$$

$$-k = 5 \quad \text{or} \quad k = -5$$

$$a^2 = 3 \quad \text{or} \quad a = \sqrt{3}.$$

The center is the point $(h, k) = (1, -5)$. The radius is $a = \sqrt{3}$. ■

For circles centered at the origin, h and k are 0 and Eq. (2) simplifies to $x^2 + y^2 = a^2$.

The Standard Equation for the Circle of Radius *a* Centered at the Origin

$$x^2 + y^2 = a^2 \tag{3}$$

The circle of radius 1 unit centered at the origin has a special name, *the unit circle.*

Notice that the circle $(x - h)^2 + (y - k)^2 = a^2$ is the same as the circle $x^2 + y^2 = a^2$ with its center shifted from the origin to the point (h, k). The shift formulas we have been using for graphs of functions apply to equations of any kind. Shifts to the right and up are accomplished by subtracting positive values of h and k. Shifts to the left and down are accomplished by subtracting negative values of h and k.

EXAMPLE 9 If the circle $x^2 + y^2 = 25$ is shifted two units to the left and three units up, its new equation is

$$(x - (-2))^2 + (y - 3)^2 = 25$$

or

$$(x + 2)^2 + (y - 3)^2 = 25.$$

As Eq. (2) says it should be, this is the equation of the circle of radius 5 centered at $(h, k) = (-2, 3)$. ■

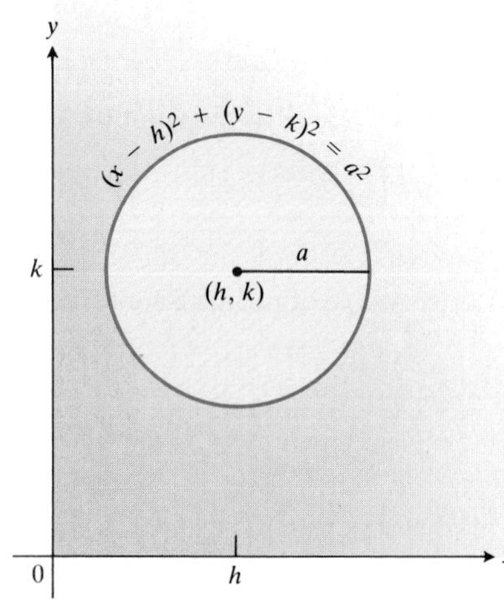

1.55 The interior and exterior of the circle $(x - h)^2 + (y - k)^2 = a^2$.

The points that lie inside the circle $(x - h)^2 + (y - k)^2 = a^2$ are the points less than a units from (h, k). They satisfy the inequality

$$(x - h)^2 + (y - k)^2 < a^2. \tag{4}$$

They make up the region we call the **interior** of the circle (Fig. 1.55).

The circle's **exterior** consists of the points that lie more than a units from (h, k). These points satisfy the inequality

$$(x - h)^2 + (y - k)^2 > a^2. \tag{5}$$

Tip for Drawing Circles: Circle first, axes later

1.

The basic shape:

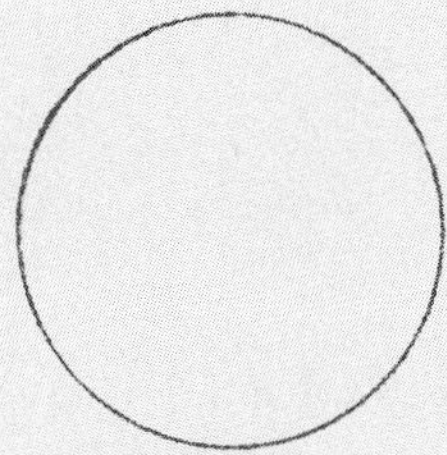

2.

The basic shape with axes in different positions:

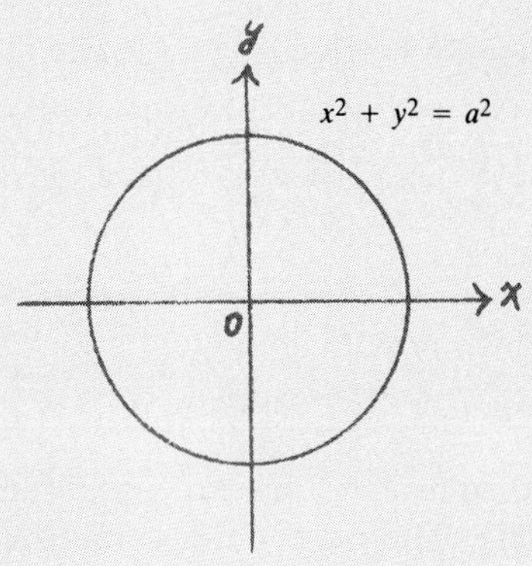

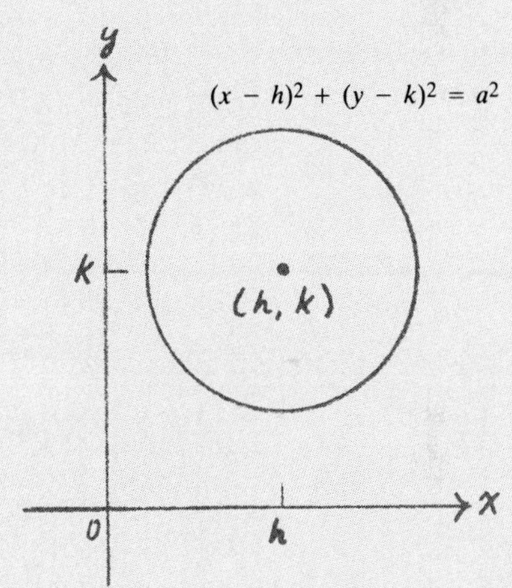

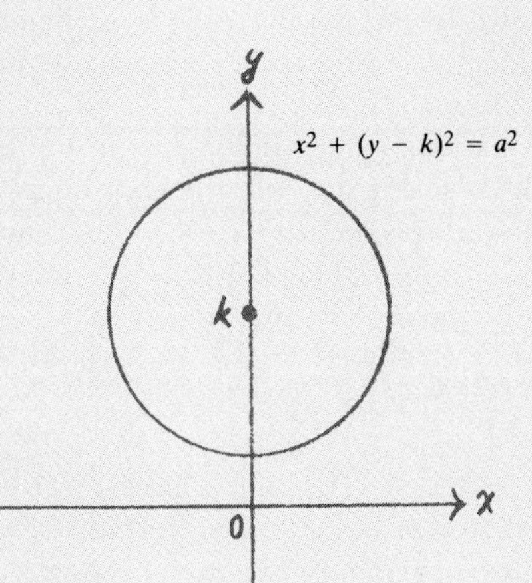

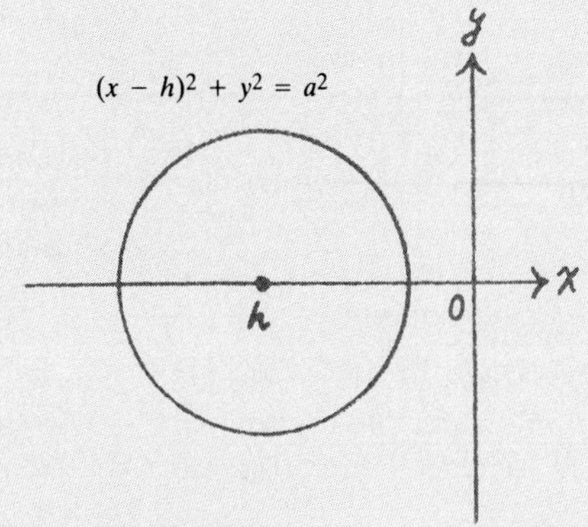

EXAMPLE 10

Inequality	Region
$x^2 + y^2 < 1$	Interior of the unit circle
$x^2 + y^2 \le 1$	Unit circle plus its interior
$x^2 + y^2 > 1$	Exterior of the unit circle
$x^2 + y^2 \ge 1$	Unit circle plus its exterior

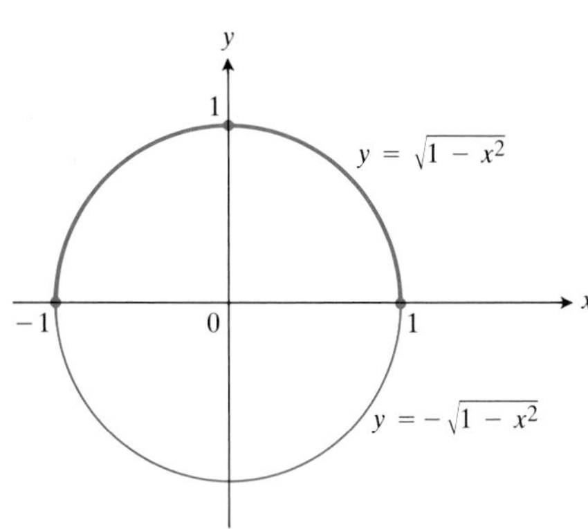

1.56 The unit circle $x^2 + y^2 = 1$ is made from the graphs of two functions, not one, over the interval $-1 \le x \le 1$. The circle contains too much to be the graph of one function all by itself.

EXAMPLE 11 The circle $x^2 + y^2 = 1$ is not the graph of any function of x. If we attempt to solve the equation for y we find that

$$y^2 = 1 - x^2 \quad \text{or} \quad y = \pm\sqrt{1 - x^2}.$$

For each value of x between $x = -1$ and $x = 1$, the formula gives *two* values of y instead of the required single value.

When taken separately, however, the formulas

$$y = \sqrt{1 - x^2} \quad \text{and} \quad y = -\sqrt{1 - x^2}$$

do define functions of x. Each formula gives exactly one value of y for each possible value of x. The graph of $y = \sqrt{1 - x^2}$ is the upper half of the circle $x^2 + y^2 = 1$. The graph of $y = -\sqrt{1 - x^2}$ is the lower half of the circle. The graphs overlap at $x = \pm 1$ (Fig. 1.56).

Equations for Parabolas

DEFINITION

A **parabola** is the set of points in a plane that are equidistant from a given fixed point and fixed line. The fixed point is the **focus** of the parabola. The fixed line is the **directrix** of the parabola.

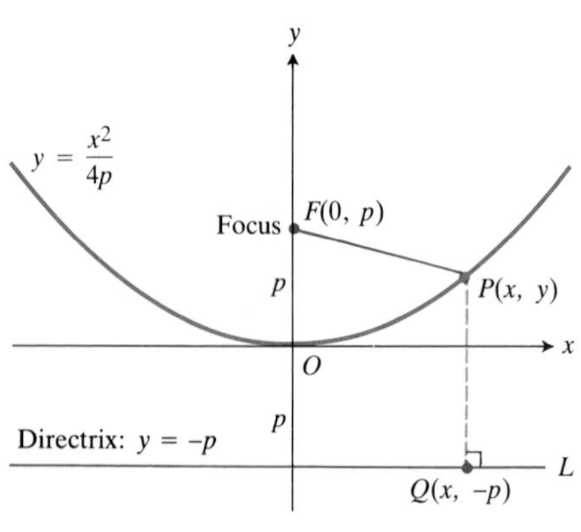

1.57 The parabola $y = x^2/4p$. Notice that the parabola passes halfway between the directrix and focus.

If the focus F lies on the directrix L, the parabola is the line through F perpendicular to L. We consider this to be a degenerate case and assume henceforth that F does not lie on L.

Parabolas have simple equations when their foci (pronounced "foe-sigh") and directrices ("di*REC*tri-seas") straddle the coordinate axes. For example, suppose the focus lies at the point $F(0, p)$ on the positive y-axis and that the directrix is the line $y = -p$ (Fig. 1.57). In the notation of the figure, a point $P(x, y)$ lies on the parabola if and only if

$$PF = PQ. \tag{6}$$

From the distance formula,

$$PF = \sqrt{(x - 0)^2 + (y - p)^2} = \sqrt{x^2 + (y - p)^2} \tag{7}$$

and

$$PQ = \sqrt{(x - x)^2 + (y - (-p))^2} = \sqrt{(y + p)^2}. \tag{8}$$

How the Astronomer Kepler Used String to Draw Parabolas

Kepler's method for drawing a parabola (with more modern tools) requires a string the length of a T square and a table whose edge can serve as the parabola's directrix. Pin one end of the string to the point where you want the focus to be and the other end to the upper end of the T square. Then, holding the string taut against the T square with a pencil, slide the T square along the table's edge. As the T square moves, the pencil will trace a parabola: PF will always equal PB because the string has the constant length AB. (More about Kepler in Section 12.5.)

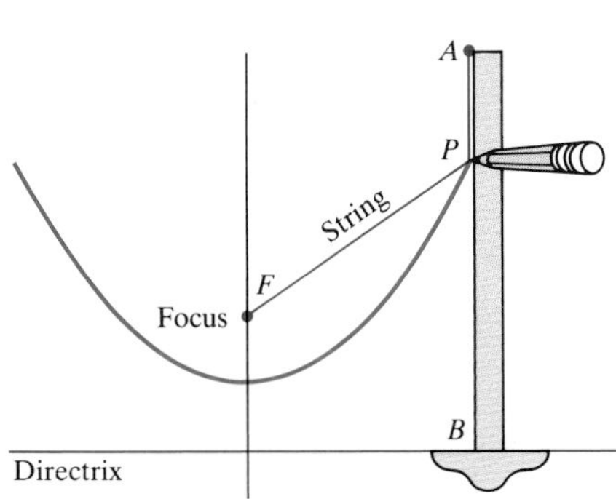

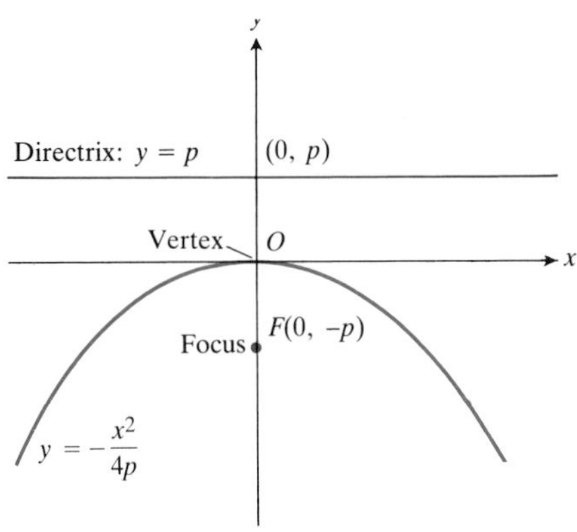

1.58 The parabola $y = -x^2/4p$.

When we equate these expressions, square, and simplify, we get

$$y = \frac{x^2}{4p}. \qquad \text{(Standard equation)} \qquad (9)$$

If we did not already know from the geometry that the parabola opened upward, we could tell from the equation. The curve rises as x moves away from the origin. The equation also reveals the parabola's symmetry about the y-axis. We call the y-axis the **axis** of the parabola (short for "axis of symmetry").

The point where a parabola crosses its axis, midway between the focus and directrix, is called the **vertex** of the parabola. The vertex of the parabola $y = x^2/4p$ lies at the origin.

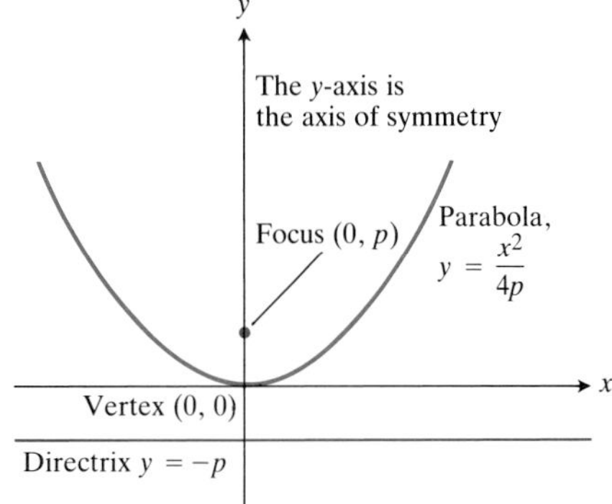

EXAMPLE 12 The standard equation of the parabola with vertex at the origin and focus at the point $(0, p) = (0, 1/4)$ is

$$y = \frac{x^2}{4(1/4)} \qquad \text{or} \qquad y = x^2.$$

■

EXAMPLE 13 Find the focus and directrix of the parabola

$$y = \frac{x^2}{8}.$$

Solution Find the value of p in the standard equation:

$$y = \frac{x^2}{8} \quad \text{is} \quad y = \frac{x^2}{4p} \quad \text{with} \quad 4p = 8 \quad \text{or} \quad p = 2.$$

Then find the focus and directrix for this value of p:

Focus: $(0, p)$ or $(0, 2)$

Directrix: $(y = -p)$ or $y = -2$.

■

If the parabola opens downward (Fig. 1.58), with its focus at $F(0, -p)$ and its directrix the line $y = p$, then Eq. (9) becomes

$$y = -\frac{x^2}{4p}. \qquad \text{(A parabola that opens downward.)} \qquad (10)$$

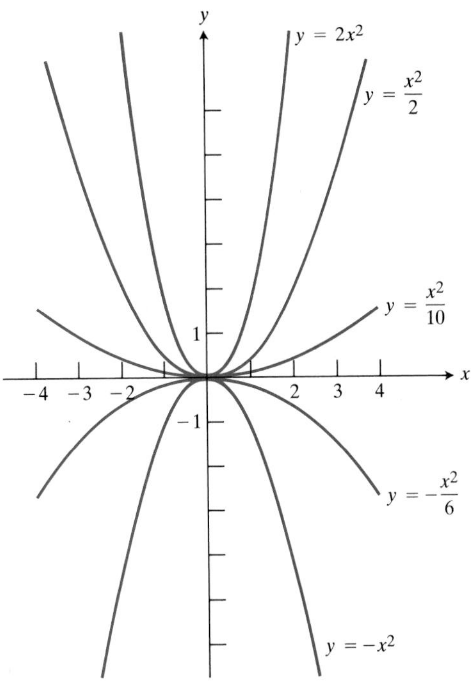

1.59 In addition to determining the direction in which the parabola $y = ax^2$ opens, the number a is a scaling factor. The parabola widens as a approaches 0 and narrows as a becomes numerically large.

From Eqs. (9) and (10), we can see that any equation like $y = 3x^2$ or $y = -5x^2$ that has the form

$$y = ax^2 \tag{11}$$

is the equation of a parabola that is symmetric about the y-axis and whose vertex lies at the origin. The parabola opens upward if a is positive and downward if a is negative (Fig. 1.59).

The Parabola $y = ax^2 + bx + c$

To shift the graph of the parabola $y = ax^2$ horizontally, we rewrite its equation as

$$y = a(x - h)^2. \tag{12}$$

This shifts the parabola to the right if h is positive and to the left if h is negative.

To shift the parabola vertically as well, we change the equation to

$$y - k = a(x - h)^2. \tag{13}$$

This shifts the parabola up if k is positive and down if k is negative.

Normally, there would be no point in squaring out the right-hand side of Eq. (13). In this case, however, we can learn something from doing so, because the resulting equation, when rearranged, takes the form

$$y = ax^2 + bx + c. \tag{14}$$

This tells us that the graph of every equation of the form $y = ax^2 + bx + c$ is a shifted version of the parabola $y = ax^2$. Why? Because the steps that

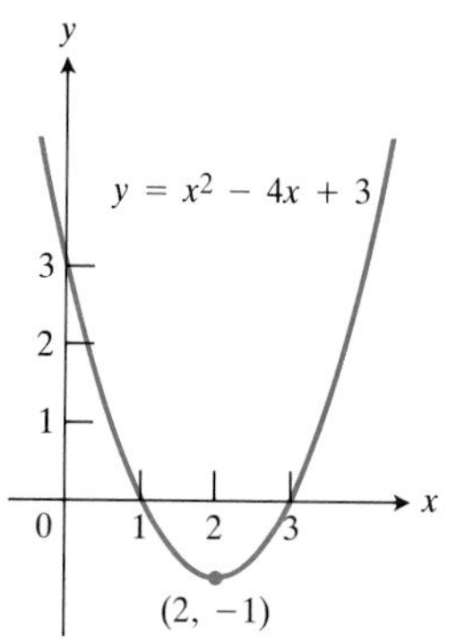

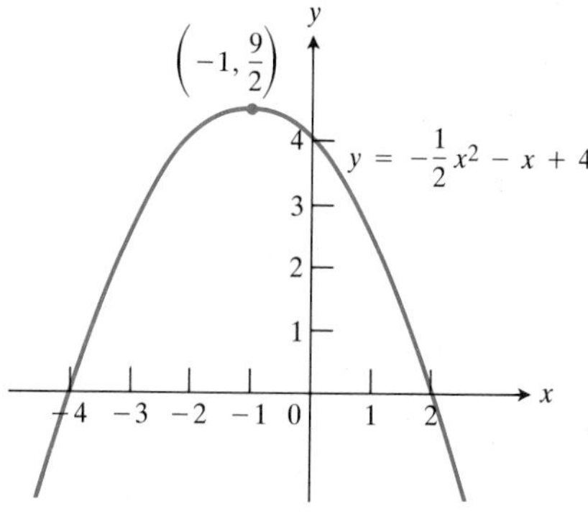

1.60 Typical examples of the parabola $y = ax^2 + bx + c$. In Chapter 4, we shall learn a quick way to locate the vertices of parabolas like these, using calculus.

take us from Equation (13) to (14) can be reversed to take us from (14) back to (13).

> **The graph of $y = ax^2 + bx + c$**
>
> The graph of $y = ax^2 + bx + c$ is a parabola whose axis is parallel to the y-axis. The parabola opens upward if $a > 0$ and downward if $a < 0$. (See Fig. 1.60 for typical examples.)

Parabolas That Open to the Right or Left

If we interchange x and y in the formula $y = ax^2$, we obtain the equation

$$x = ay^2. \tag{15}$$

With the roles of x and y now reversed, the graph is a parabola whose axis is the x-axis. The vertex still lies at the origin. The parabola opens to the right if $a > 0$ and to the left if $a < 0$ (Fig. 1.61).

EXAMPLE 14 The formula $x = y^2$ gives x as a function of y but does *not* give y as a function of x. If we solve for y, we find that

$$y = \pm\sqrt{x}.$$

For each positive value of x we get *two* values of y instead of the required single value.

When taken separately, however, the formulas

$$y = \sqrt{x} \qquad \text{and} \qquad y = -\sqrt{x}$$

do define functions of x. Each formula gives exactly one value of y for each possible value of x. The graph of $y = \sqrt{x}$ is the upper half of the parabola

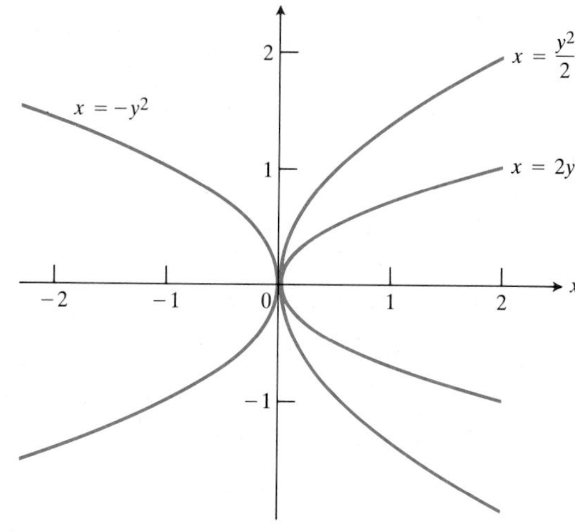

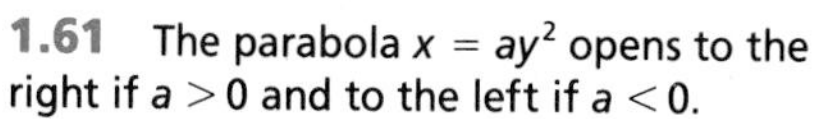

1.61 The parabola $x = ay^2$ opens to the right if $a > 0$ and to the left if $a < 0$.

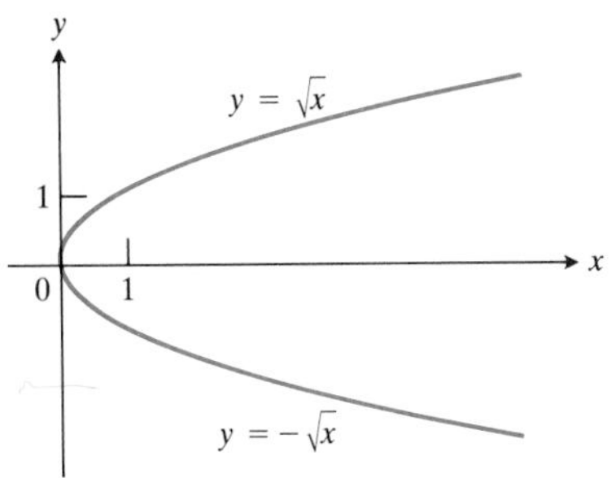

1.62 The graphs of the functions $y = \sqrt{x}$ and $y = -\sqrt{x}$ join at the origin to make the graph of the equation $x = y^2$.

$x = y^2$. The graph of $y = -\sqrt{x}$ is the lower half. The two graphs meet at the origin (Fig. 1.62). ■

The Reflective Properties of Parabolas

The chief application of parabolas involves their use as reflectors of light and radio waves. Rays originating at a parabola's focus are reflected out of the parabola parallel to the parabola's axis (in Fig. 1.61, the x-axis). Similarly, rays coming in parallel to the axis are reflected toward the focus. We shall see why this is so when we get into the calculus.

This property is used in parabolic mirrors and telescopes, in automobile headlamps, in spotlights of all kinds, in parabolic radar and microwave antennas, and in television dish receivers. Parabolas are also used in bridge construction, wind tunnel photography, and submarine tracking.

Exercises 1.4

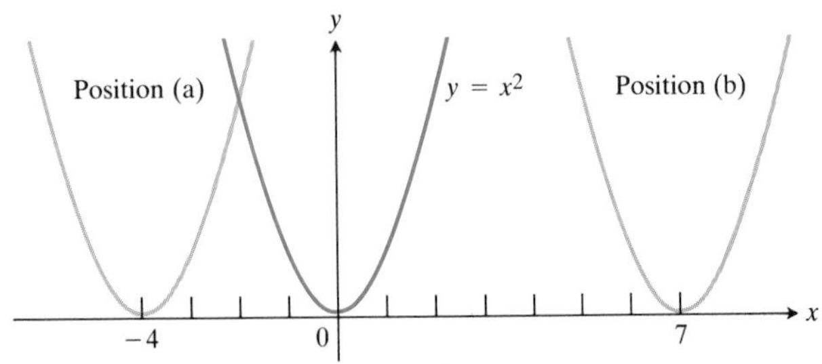

1.63 The parabolas in Exercise 1.

1. Figure 1.63 shows the graph of $y = x^2$ shifted to two
2. Figure 1.64 shows the graph of $y = x^2$ shifted to two new positions. Write equations for the new graphs.
3. Match the equations listed below to the graphs in Fig. 1.65.

a) $y + 4 = (x - 1)^2$ b) $y - 2 = (x - 2)^2$
c) $y - 2 = (x + 2)^2$ d) $y + 2 = (x + 3)^2$

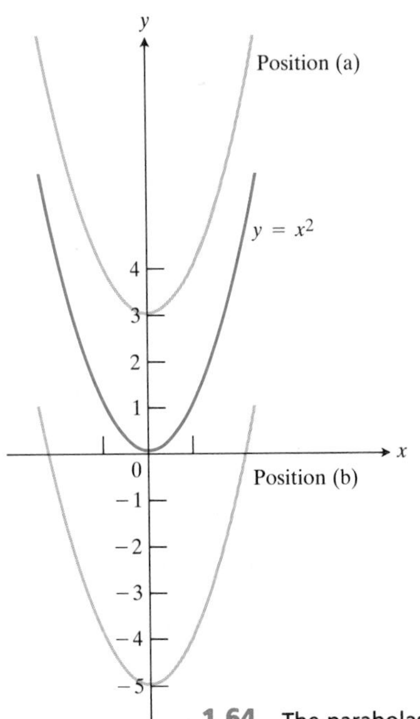

1.64 The parabolas in Exercise 2.

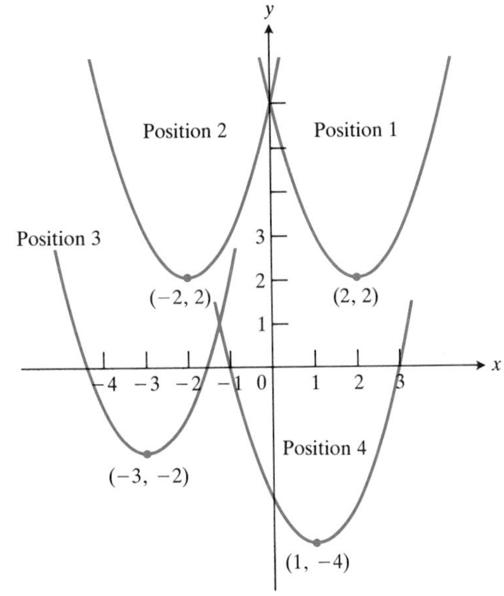

1.65 The parabolas in Exercise 3.

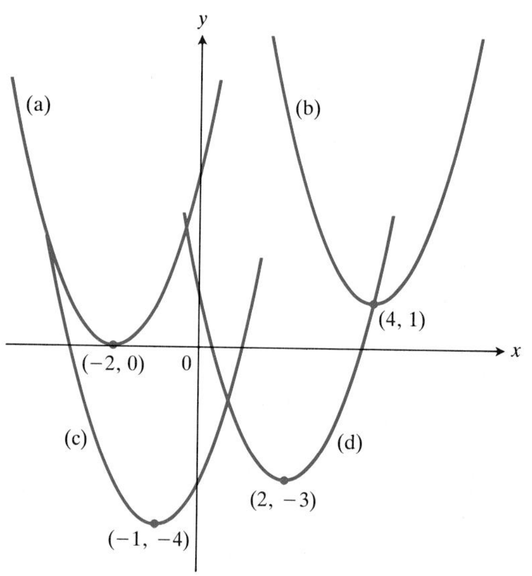

1.66 The parabolas in Exercise 4.

4. Figure 1.66 shows the graph of $y = x^2$ shifted to four new positions. Write an equation for each new graph.

In Exercises 5–8, find an equation for the circle with the given center $C(h, k)$ and radius a. Then draw the circle in the coordinate plane.

5. $C(0, 2)$, $a = 2$ **6.** $C(-2, 0)$, $a = 3$

7. $C(3, -4)$, $a = 5$ **8.** $C(1, 1)$, $a = \sqrt{2}$

Write equations for the circles in Exercises 9–12.

9. **10.**

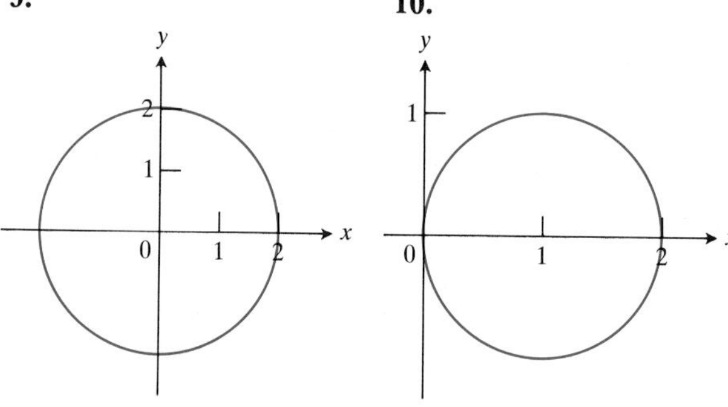

11. **12.**

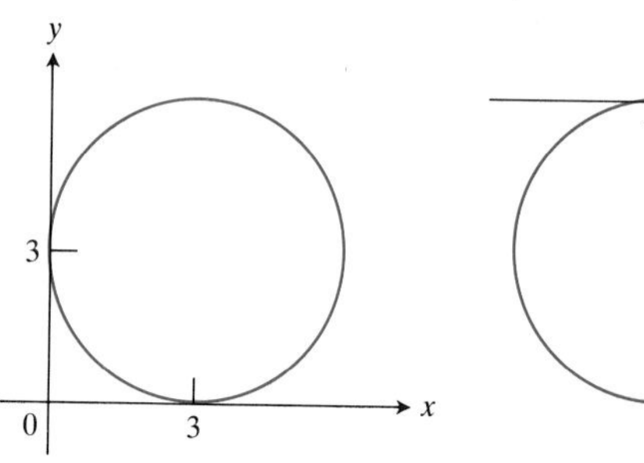

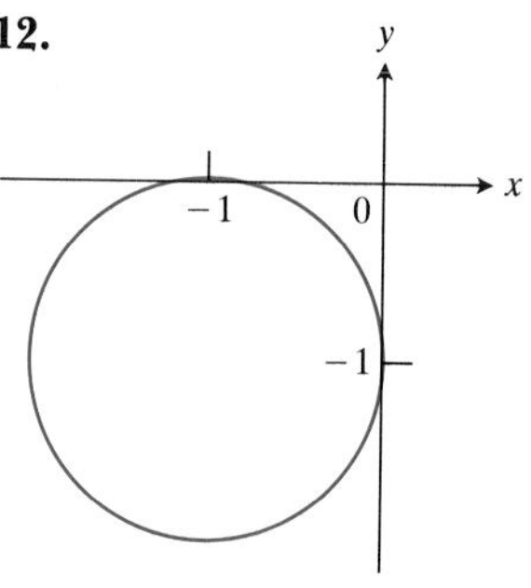

Describe the regions defined by the inequalities and pairs of inequalities in Exercises 13 and 14.

13. a) $x^2 + y^2 > 1$
b) $x^2 + y^2 < 4$
c) the inequalities in (a) and (b) together.

14. a) $x^2 + y^2 \geq 1$
b) $x^2 + y^2 \leq 4$
c) the inequalities in (a) and (b) together.

15. Write an inequality that describes the points that lie inside the circle with center $C(-2, -1)$ and radius $a = \sqrt{6}$.

16. Write an inequality that describes the points that lie outside the circle with center $C(-4, 2)$ and radius $a = 4$.

In Exercises 17–20, write an equation for the parabola with the given focus and directrix.

17. Focus: $(0, 4)$ Directrix: $y = -4$

18. Focus: $\left(0, \frac{1}{4}\right)$ Directrix: $y = -\frac{1}{4}$

19. Focus: $(0, -3)$ Directrix: $y = 3$

20. Focus: $\left(0, -\frac{1}{2}\right)$ Directrix: $y = \frac{1}{2}$

Find the foci and directrices of the parabolas in Exercises 21–24.

21. $y = 4x^2$
22. $y = x^2/3$
23. $y = -3x^2$
24. $y = -x^2/4$

Exercises 25–34 tell how many units and in what directions the graphs of the given equations are to be shifted. Give an equation for the shifted graph in each case.

25. $y = x^2$ Down 3, left 2
26. $x^2 + y^2 = 5$ Up 3, left 4
27. $y + 1 = (x + 1)^2$ Up 1, right 1
28. $y = x^3$ Right 1
29. $y = \sqrt{x}$ Left 4
30. $y = 2x - 7$ Up 7
31. $y - 5 = \frac{1}{2}(x + 1)$ Down 5, right 1
32. $x = y^2$ Left 1
33. $y = \sqrt{-x}$ Right 9
34. $y = 1/x$ Up 1, right 1

35. The line $y = mx$ is shifted horizontally and vertically to make it pass through the point (x_0, y_0). What is the line's new point–slope equation?

36. The line $y = mx$ is shifted vertically to make it pass through the point $(0, b)$. What is the line's new slope–intercept equation?

Find the coordinates of the points at which the lines and parabolas in Exercises 37–40 intersect.

37. **38.**

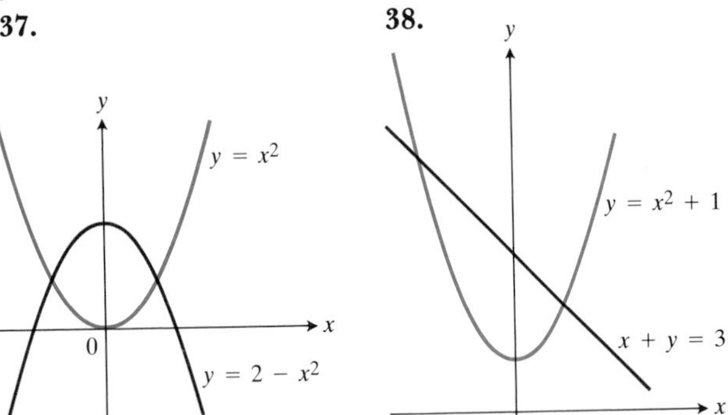

39. **40.**

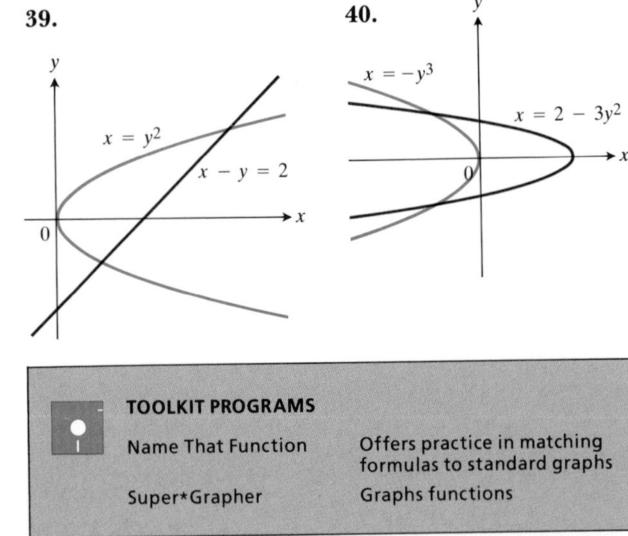

1.5 A Review of Trigonometric Functions

In surveying, navigation, and astronomy, we measure angles in degrees, but in calculus it is usually best to use radians. We shall see why when we study the derivatives of trigonometric functions in Section 3.4. In the present section, we use radians and degrees together so that you can practice relating the two. We also review the trigonometry you will need for calculus and its applications.

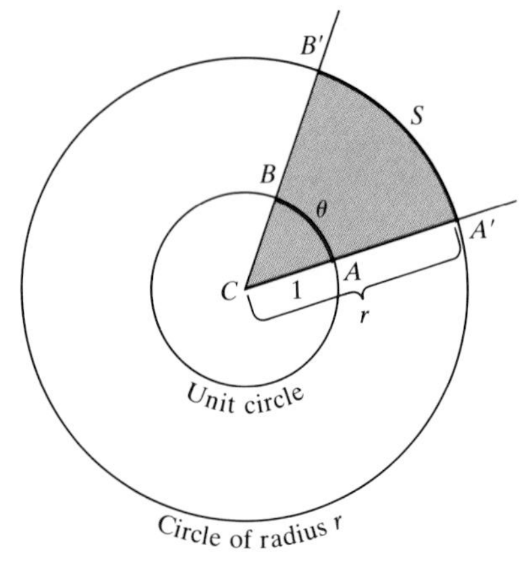

1.67 The radian measure of angle *ACB* is the length θ of arc *AB* on the unit circle centered at *C*. The value of θ can be found from any other circle, however, as the ratio of *s* to *r*.

Radian Measure

The **radian measure** of the angle ACB at the center of the unit circle (Fig. 1.67) equals the length of the arc that the angle cuts from the unit circle.

If angle ACB cuts an arc $A'B'$ from a second circle centered at C, then circular sector $A'CB'$ will be similar to circular sector ACB. In particular,

$$\frac{\text{Length of arc } A'B'}{\text{Radius of second circle}} = \frac{\text{Length of arc } AB}{\text{Radius of first circle}} \tag{1}$$

In the notation of Fig. 1.67, Eq. (1) says that

$$\frac{s}{r} = \frac{\theta}{1} = \theta \qquad \text{or} \qquad \theta = \frac{s}{r}. \tag{2}$$

When you know r and s, you can calculate the angle's radian measure θ from this equation. Notice that the units of length for r and s cancel out and that radian measure, like degree measure, is a dimensionless number.

We find the relation between degree measure and radian measure by observing that a semicircle of radius r, which we know has length $s = \pi r$, subtends a central angle of 180°. Therefore,

$$180° = \pi \text{ radians.} \tag{3}$$

We can restate this relation in several useful ways.

> Degrees to radians:
>
> 1 degree makes $\dfrac{\pi}{180}$ radians (about 0.02 rad)
>
> To change degrees to radians, multiply degrees by $\dfrac{\pi}{180}$.
>
> Radians to degrees:
>
> 1 radian makes $\dfrac{180}{\pi}$ degrees (about 57°)
>
> To change radians to degrees, multiply radians by $\dfrac{180}{\pi}$.

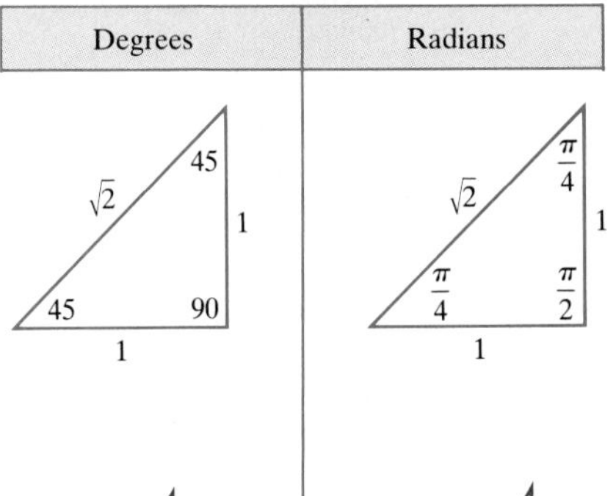

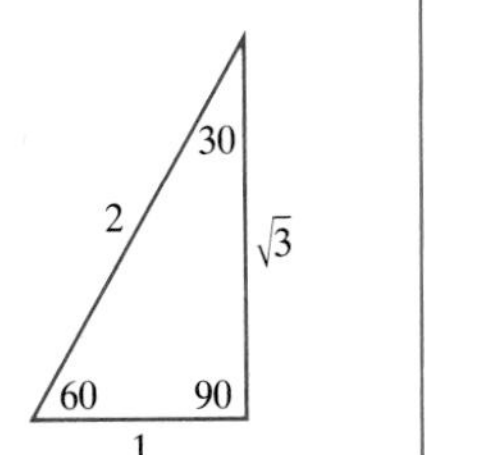

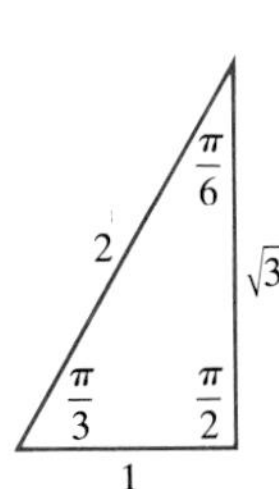

1.68 The angles of two common triangles, in degrees and radians.

EXAMPLE 1 *Conversions*

Change 45° to radians:	$45 \cdot \dfrac{\pi}{180} = \dfrac{\pi}{4}$ rad
Change 90° to radians:	$90 \cdot \dfrac{\pi}{180} = \dfrac{\pi}{2}$ rad
Change $\dfrac{\pi}{6}$ radians to degrees:	$\dfrac{\pi}{6} \cdot \dfrac{180}{\pi} = 30°$
Change $\dfrac{\pi}{3}$ radians to degrees:	$\dfrac{\pi}{3} \cdot \dfrac{180}{\pi} = 60°$

See Fig. 1.68. ■

EXAMPLE 2 An acute angle whose vertex lies at the center of a circle of radius 6 subtends an arc of length 2π (Fig. 1.69). Find the angle's radian measure.

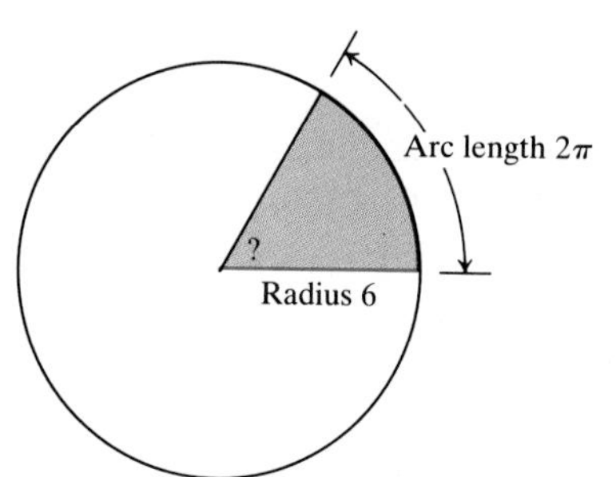

1.69 What is the radian measure of this angle? See Example 2.

Solution

$$\theta = \frac{s}{r} = \frac{2\pi}{6} = \frac{\pi}{3}. \qquad \text{(Eq. (2) with } s = 2\pi,\ r = 6.)$$

The angle's measure is $\pi/3$ radians. ■

The equation $\theta = s/r$ is sometimes written

$$s = r\theta. \tag{4}$$

This equation gives a handy way to find s when you know r and θ.

EXAMPLE 3 An angle of $3\pi/4$ radians lies at the center of a circle of radius 8. How large an arc does it subtend?

Solution

$$s = r\theta = 8 \cdot \frac{3\pi}{4} = 6\pi. \qquad \begin{pmatrix}\text{Eq. (4) with } r = 8 \\ \text{and } \theta = 3\pi/4.\end{pmatrix}$$

The arc is 6π units long. ■

EXAMPLE 4 How long is the arc subtended by a central angle of 120° in a circle of radius 4?

Solution The equation $s = r\theta$ holds only when the angle is measured in radians, so we must find the angle's radian measure before finding s:

$$\theta = 120 \cdot \frac{\pi}{180} = \frac{2\pi}{3} \text{ rad} \qquad \text{(Convert to radians.)}$$

$$s = r\theta = 4 \cdot \frac{2\pi}{3} = \frac{8\pi}{3} \qquad \text{(Then find } s = r\theta.\text{)}$$

The arc is $8\pi/3$ units long. ■

When angles are used to describe counterclockwise rotations, our measurements can go arbitrarily far beyond 2π radians or 360°. Similarly, angles that describe clockwise rotations can have negative measures of all sizes (Fig. 1.70).

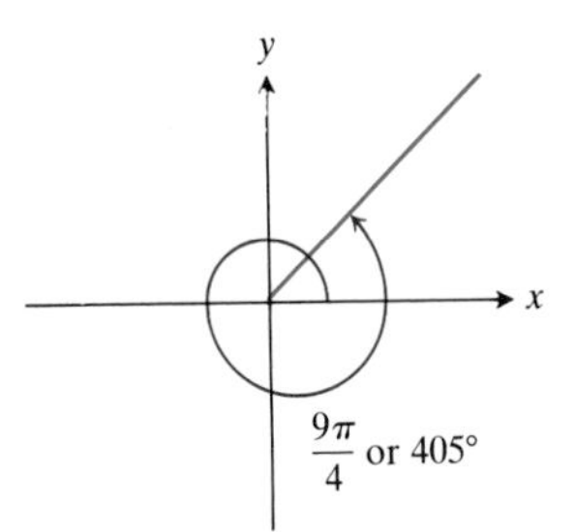

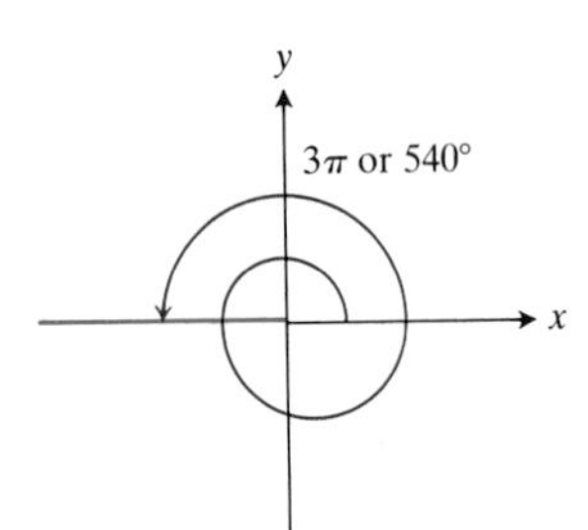

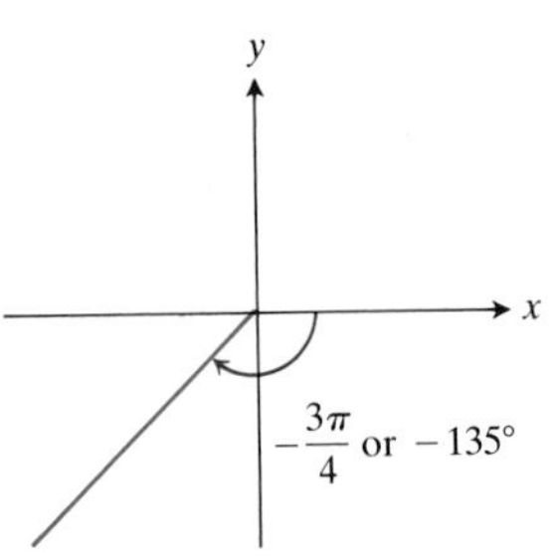

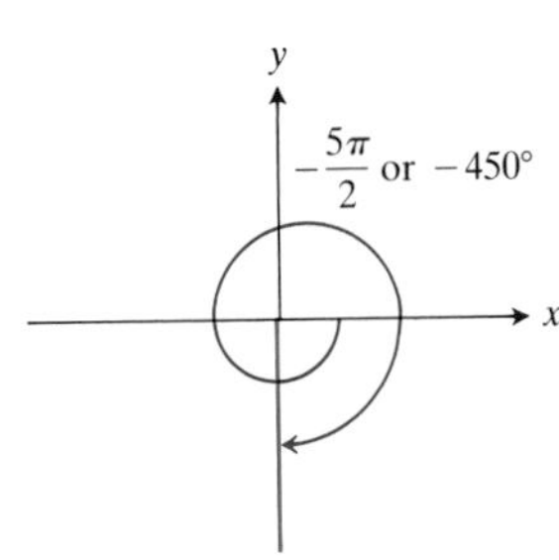

1.70 Angles can have any measure.

The Six Basic Trigonometric Functions

When an angle of measure θ is placed in standard position at the center of a circle of radius r (Fig. 1.71), the six basic trigonometric functions of θ are defined in the following way:

$$\begin{array}{llll} \text{Sine:} & \sin\theta = \dfrac{y}{r} & \text{Cosecant:} & \csc\theta = \dfrac{r}{y} \\ \text{Cosine:} & \cos\theta = \dfrac{x}{r} & \text{Secant:} & \sec\theta = \dfrac{r}{x} \\ \text{Tangent:} & \tan\theta = \dfrac{y}{x} & \text{Cotangent:} & \cot\theta = \dfrac{x}{y} \end{array} \qquad (5)$$

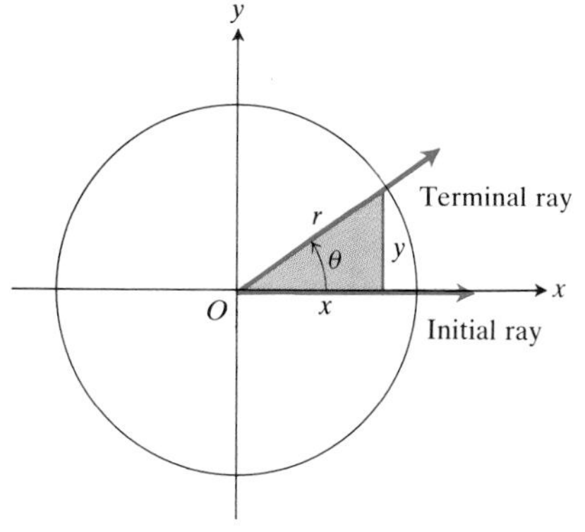

1.71 An angle θ in standard position.

As you can see, $\tan\theta$ and $\sec\theta$ are not defined if $x = 0$. In terms of radian measure, this means they are not defined when θ is $\pm\pi/2, \pm3\pi/2, \ldots$. Similarly, $\cot\theta$ and $\csc\theta$ are not defined for values of θ for which $y = 0$, namely $\theta = 0, \pm\pi, \pm2\pi, \ldots$. Notice also that

$$\tan\theta = \frac{\sin\theta}{\cos\theta}, \qquad \csc\theta = \frac{1}{\sin\theta},$$
$$\sec\theta = \frac{1}{\cos\theta}, \qquad \cot\theta = \frac{1}{\tan\theta} \tag{6}$$

whenever the quotients on the right-hand sides are defined.

Because $x^2 + y^2 = r^2$ (Pythagorean theorem),

$$\cos^2\theta + \sin^2\theta = \frac{x^2}{r^2} + \frac{y^2}{r^2} = \frac{x^2 + y^2}{r^2} = 1. \tag{7}$$

The equation $\cos^2\theta + \sin^2\theta = 1$, true for all values of θ, is probably the most frequently used identity in trigonometry.

The coordinates of the point $P(x, y)$ in Fig. 1.71 can be expressed in terms of r and θ as

$$x = r\cos\theta \qquad (\text{because } x/r = \cos\theta)$$
$$y = r\sin\theta \qquad (\text{because } y/r = \sin\theta) \tag{8}$$

We shall use these equations when we study circular motion and when we work with polar coordinates. Notice that if $\theta = 0$ in Fig. 1.71 then $x = r$ and $y = 0$, so

$$\cos 0 = 1 \quad \text{and} \quad \sin 0 = 0.$$

If $\theta = \pi/2$, we have $x = 0$ and $y = r$. Hence,

$$\cos\frac{\pi}{2} = 0 \quad \text{and} \quad \sin\frac{\pi}{2} = 1.$$

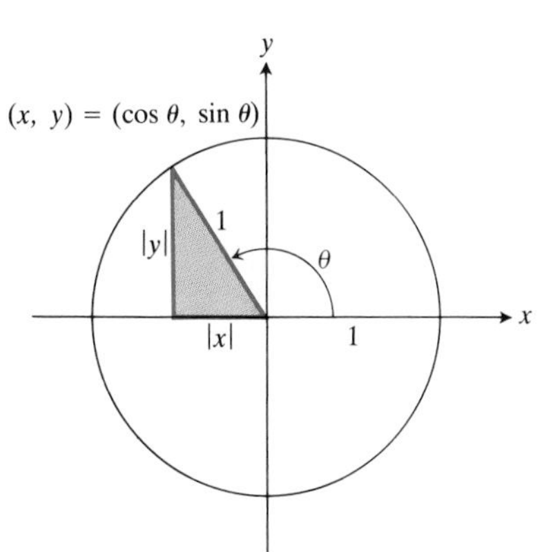

1.72 The acute reference triangle for an angle θ.

Calculating Sines and Cosines

If the circle in Fig. 1.71 has radius $r = 1$ unit, Eqs. (8) simplify to

$$x = \cos\theta, \qquad y = \sin\theta.$$

We can therefore calculate the values of the cosine and sine from the acute reference triangle made by dropping a perpendicular from the point $P(x, y)$ to the x-axis (Fig. 1.72). The numerical values of x and y are read from the triangle's sides. The signs of x and y are determined by the quadrant in which the triangle lies.

EXAMPLE 5 Find the sine and cosine of $2\pi/3$ radians.

Solution

STEP 1: Draw the angle in standard position and write in the lengths of the sides of the reference triangle.

STEP 2: Find the coordinates of the point P where the angle's terminal ray cuts the circle:

$$\cos\frac{2\pi}{3} = x\text{-coordinate of } P$$
$$= -\frac{1}{2}$$

$$\sin\frac{2\pi}{3} = y\text{-coordinate of } P$$
$$= \frac{\sqrt{3}}{2}$$

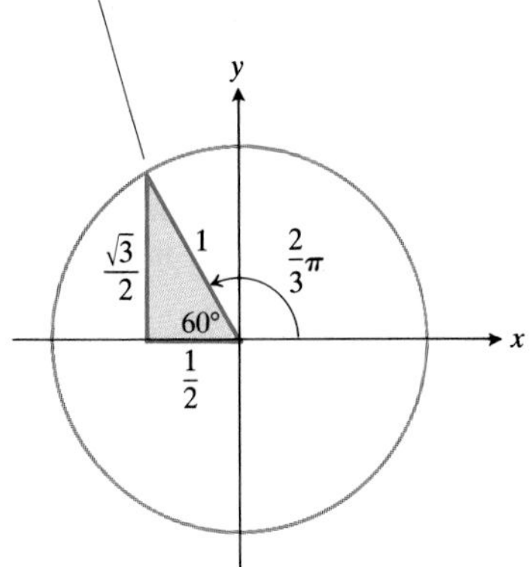

EXAMPLE 6 Find the sine and cosine of $-\pi/4$ radians.

Solution

STEP 1: Draw the angle in standard position and write in the lengths of the sides of the reference triangle.

STEP 2: Find the coordinates of the point P where the angle's terminal ray cuts the circle:

$$\cos -\frac{\pi}{4} = x\text{-coordinate of } P$$
$$= \frac{\sqrt{2}}{2}.$$

$$\sin -\frac{\pi}{4} = y\text{-coordinate of } P$$
$$= -\frac{\sqrt{2}}{2}.$$

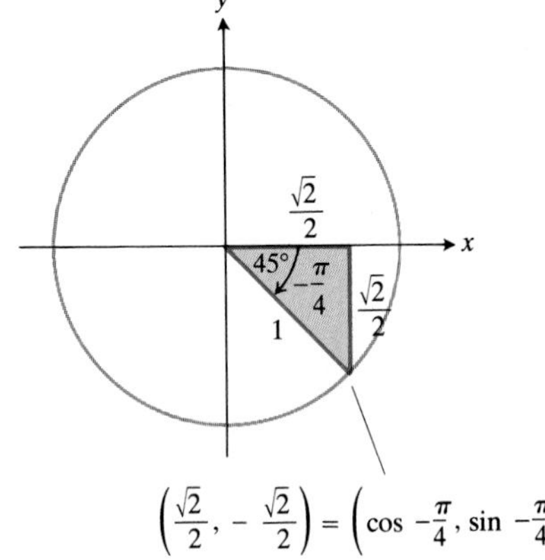

Table 1.1 gives the values of the sine, cosine, and tangent for selected values of θ.

TABLE 1.1
Values of sin θ, cos θ, and tan θ for selected values of θ

Degrees	-180	-135	-90	-45	0	45	90	135	180
θ (radians)	$-\pi$	$-3\pi/4$	$-\pi/2$	$-\pi/4$	0	$\pi/4$	$\pi/2$	$3\pi/4$	π
sin θ	0	$-\sqrt{2}/2$	-1	$-\sqrt{2}/2$	0	$\sqrt{2}/2$	1	$\sqrt{2}/2$	0
cos θ	-1	$-\sqrt{2}/2$	0	$\sqrt{2}/2$	1	$\sqrt{2}/2$	0	$-\sqrt{2}/2$	-1
tan θ	0	1		-1	0	1		-1	0

Periodicity

When an angle of measure θ and an angle of measure $\theta + 2\pi$ are in standard position, their terminal rays coincide. The two angles therefore have the same trigonometric-function values:

$$\begin{aligned}
\cos(\theta + 2\pi) &= \cos\theta \\
\sin(\theta + 2\pi) &= \sin\theta \\
\tan(\theta + 2\pi) &= \tan\theta \\
\cot(\theta + 2\pi) &= \cot\theta \\
\sec(\theta + 2\pi) &= \sec\theta \\
\csc(\theta + 2\pi) &= \csc\theta.
\end{aligned} \tag{9}$$

Similarly, $\cos(\theta - 2\pi) = \cos\theta$, $\sin(\theta - 2\pi) = \sin\theta$, and so on.

From another point of view, Eqs. (9) tell us that if we start at any particular value $\theta = \theta_0$ and let θ increase or decrease steadily, we see the values of the trigonometric functions start to repeat after any interval of length 2π. We describe this behavior by saying that the six basic trigonometric functions are **periodic** and that they repeat after a fixed **period** of θ-values (in this case 2π).

DEFINITION

A function $f(x)$ is **periodic** with **period** $p > 0$ if $f(x + p) = f(x)$ for every value of x.

EXAMPLE 7 Equations (9) tell us that the six basic trigonometric functions are periodic with a period of 2π. Other periods include 4π, 6π, and so on (positive integer multiples of 2π). ■

The importance of periodic functions stems from the fact that much of the behavior we study in science is periodic. Brain waves and heartbeats are periodic, as are household voltage and electric current. The electromagnetic field that heats food in a microwave oven is periodic, as are cash flows in seasonal businesses and the behavior of rotational machinery. The seasons are periodic—so is the weather. The phases of the moon are periodic, as are the motions of the planets. There is strong evidence that the ice ages are periodic, with a period of 90,000–100,000 years.

If so many things are periodic, why limit our discussion to trigonometric functions? The answer lies in a surprising and beautiful theorem from advanced calculus that says that every periodic function we want to use in mathematical modeling can be written as an algebraic combination of sines and cosines. Thus, once we learn the calculus of sines and cosines, we will know everything we need to know to model the mathematical behavior of periodic phenomena.

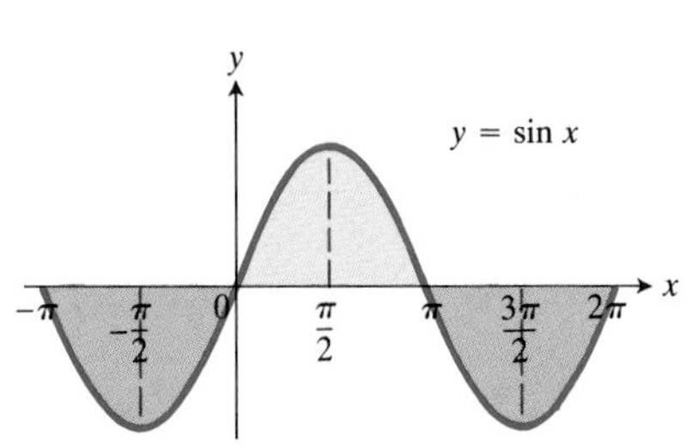

Function: $y = \sin x$
Domain: $-\infty < x < \infty$
Range: $-1 \le y \le 1$

(a)

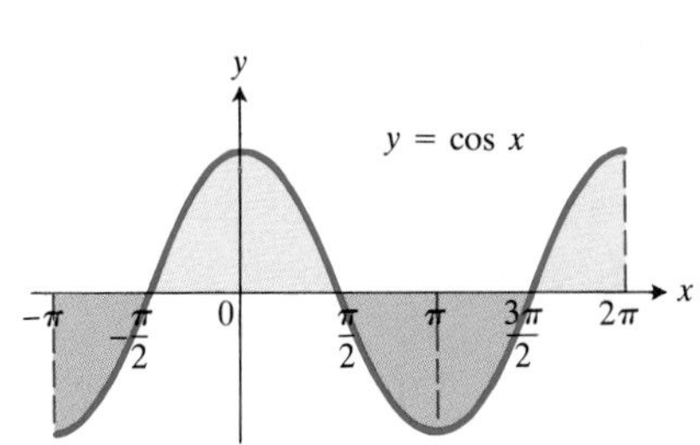

Function: $y = \cos x$
Domain: $-\infty < x < \infty$
Range: $-1 \le y \le 1$

(b)

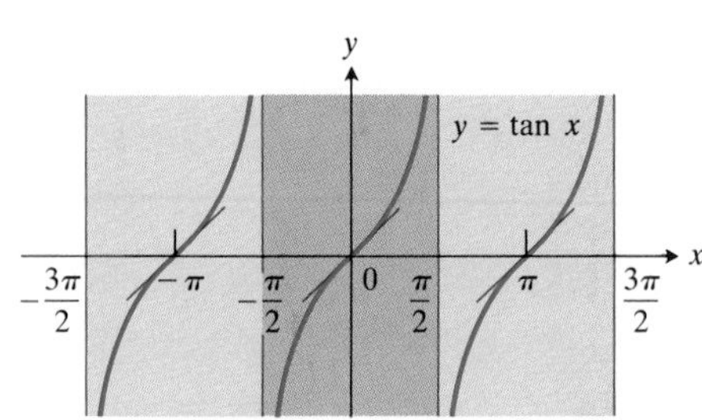

Function: $y = \tan x$
Domain: All real numbers except odd integer multiples of $\pi/2$.
Range: $-\infty < y < \infty$

(c)

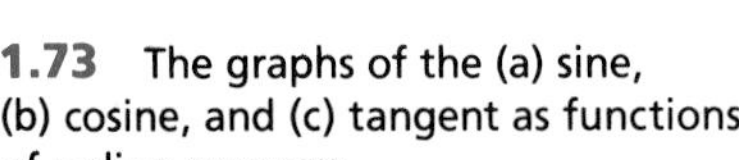

1.73 The graphs of the (a) sine, (b) cosine, and (c) tangent as functions of radian measure.

Graphs of Trigonometric Functions

When we graph trigonometric functions in the coordinate plane, we usually denote the independent variable (radians) by x instead of θ. Figure 1.73 shows the graphs of $y = \sin x$, $y = \cos x$, and $y = \tan x$.

Notice that the graph of

$$\tan x = \frac{\sin x}{\cos x}$$

"blows up" whenever x nears an odd-integer multiple of $\pi/2$. These are the points for which $\sin x = 1$ and $\cos x = 0$. Notice, too, how the periodicity of the sine, cosine, and tangent appears in the graphs. Choose any starting point, and each graph repeats after an interval of length 2π.

1.74 The graphs of the (a) secant, (b) cosecant, and (c) cotangent as functions of radian measure.

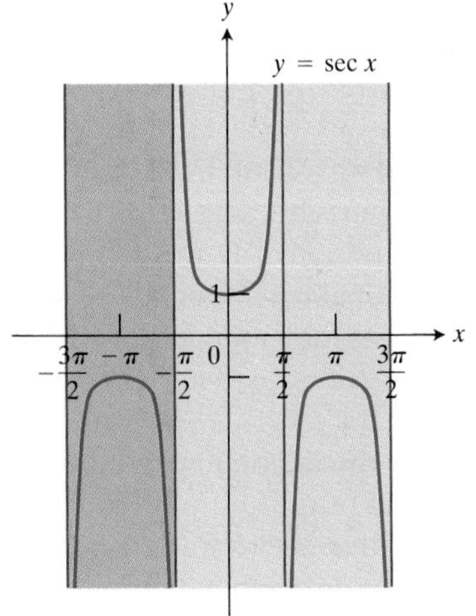

Domain: $x \ne \pm\frac{\pi}{2}, \pm\frac{3\pi}{2}, \ldots$
Range: $y \le -1$ and $y \ge 1$

(a)

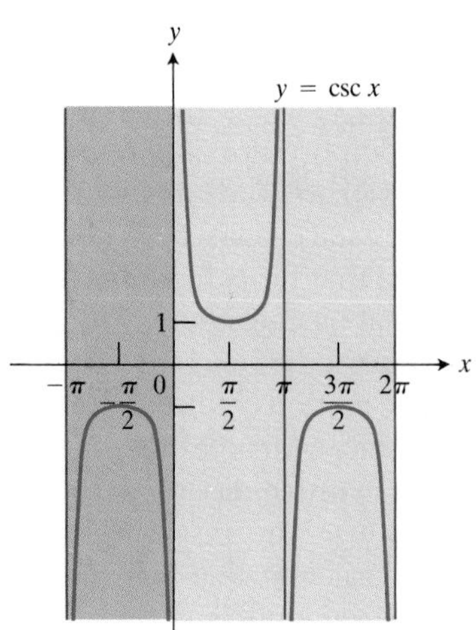

Domain: $x \ne 0, \pm\pi, \pm 2\pi, \ldots$
Range: $y \le -1$ and $y \ge 1$

(b)

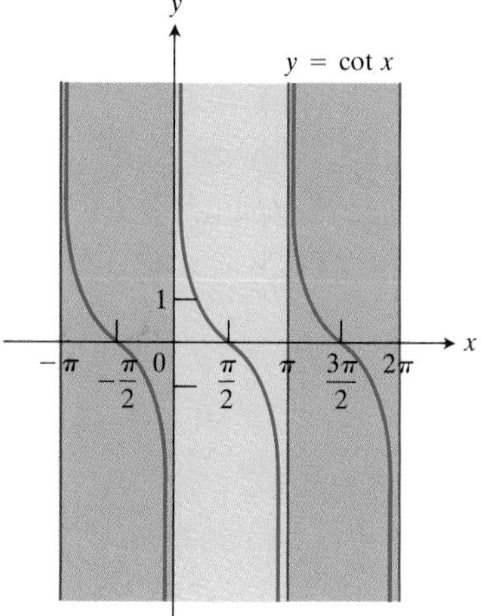

Domain: $x \ne 0, \pm\pi, \pm 2\pi, \ldots$
Range: $-\infty < y < \infty$

(c)

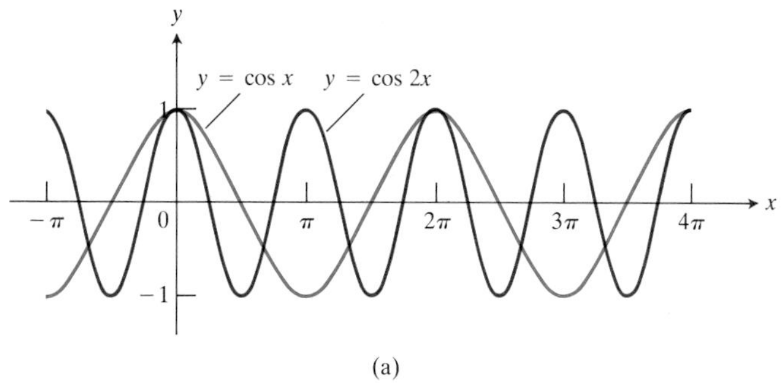

(a)

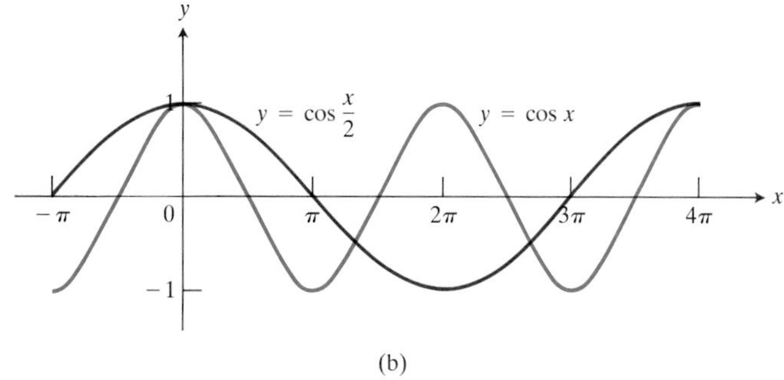

(b)

1.75 (a) Multiplying x by a number greater than 1 speeds the cosine up. (b) Multiplying x by a number less than 1 slows the cosine down.

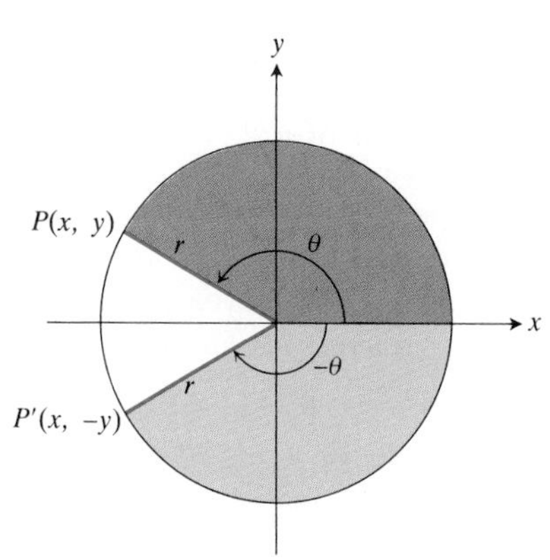

1.76 Angles of opposite sign.

Figure 1.74 shows the graphs of the secant, cosecant, and cotangent functions. Notice again how the periodicity appears in the graphs.

Figure 1.75 shows the graphs of $y = \cos 2x$ and $y = \cos(x/2)$ plotted against the graph of $y = \cos x$. Multiplying x by 2 speeds the cosine up and shortens the minimum period from 2π to π. Multiplying x by 1/2 slows the cosine down and lengthens its minimum period from 2π to 4π.

Odd vs. Even

Figure 1.76 shows two angles that have the same magnitude but opposite signs. By symmetry, the points where the terminal rays cross the circle have the same x-coordinate and their y-coordinates differ only in sign. Hence,

$$\cos(-\theta) = \frac{x}{r} = \cos\theta, \qquad \text{(The cosine is an even function.)} \qquad (10a)$$

$$\sin(-\theta) = \frac{-y}{r} = -\sin\theta. \qquad \text{(The sine is an odd function.)} \qquad (10b)$$

EXAMPLE 8

$$\cos\left(-\frac{\pi}{3}\right) = \cos\frac{\pi}{3} = \frac{1}{2},$$

$$\sin\left(-\frac{\pi}{3}\right) = -\sin\frac{\pi}{3} = -\frac{\sqrt{3}}{2}.$$

■

As for the other basic trigonometric functions, the secant is even and the cosecant, tangent, and cotangent are odd. For the secant and tangent,

$$\sec(-\theta) = \frac{1}{\cos(-\theta)} = \frac{1}{\cos\theta} = \sec\theta. \tag{11}$$

$$\tan(-\theta) = \frac{\sin(-\theta)}{\cos(-\theta)} = \frac{-\sin\theta}{\cos\theta} = -\tan\theta. \tag{12}$$

Similar calculations show that the contangent and cosecant are odd, as we ask you to verify in Exercises 65 and 66.

Shift Formulas

If you look once again at Fig. 1.73, you will see that the cosine curve is the same as the sine curve shifted $\pi/2$ units to the left. Also, the sine curve is the same as the cosine curve shifted $\pi/2$ units to the right. In symbols,

$$\sin\left(x + \frac{\pi}{2}\right) = \cos x, \qquad \cos\left(x - \frac{\pi}{2}\right) = \sin x. \tag{13}$$

Figure 1.77(a) shows the cosine shifted to the left $\pi/2$ units to become the reflection of the sine curve across the x-axis. Next to it, Fig. 1.77(b) shows the sine curve shifted $\pi/2$ units to the right to become the reflection of the cosine curve across the x-axis. In symbols,

$$\cos\left(x + \frac{\pi}{2}\right) = -\sin x, \qquad \sin\left(x - \frac{\pi}{2}\right) = -\cos x. \tag{14}$$

EXAMPLE 9 The builders of the Trans-Alaska Pipeline used insulated pads to keep the heat from the hot oil in the pipeline from melting the permanently frozen soil beneath. To design the pads, it was necessary to take into account the variation in air temperature throughout the year. The variation was represented in the calculations by a *general sine function* of the form

$$f(x) = A\sin\left[\frac{2\pi}{B}(x - C)\right] + D,$$

1.77 (a) The reflection of the sine as a shifted cosine. (b) The reflection of the cosine as a shifted sine.

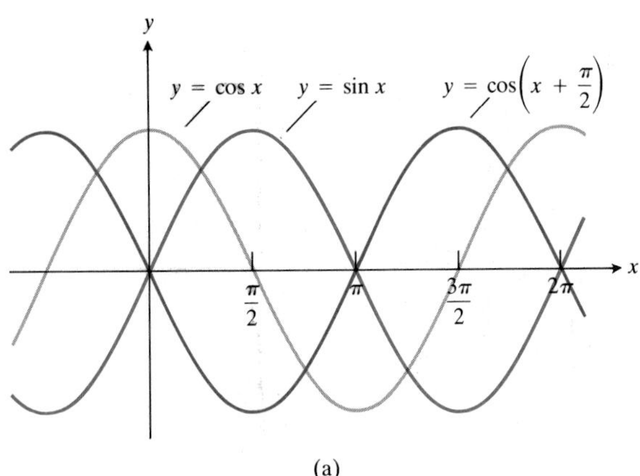

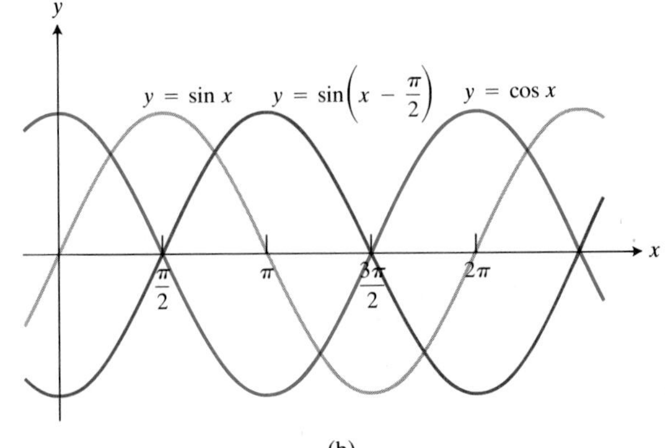

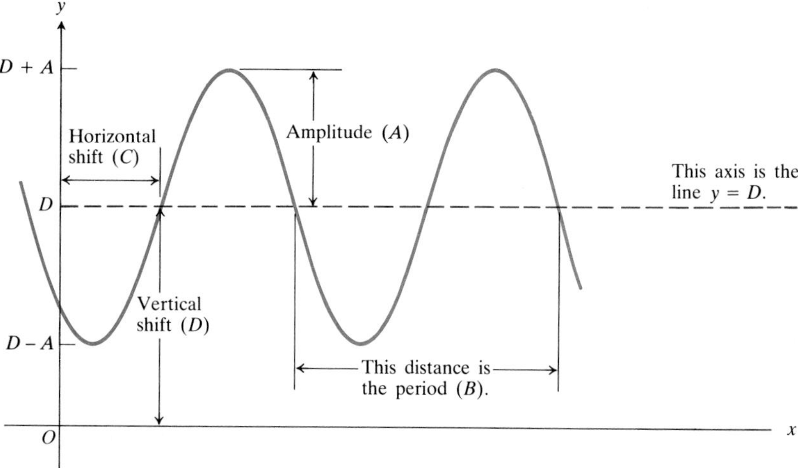

1.78 The general sine curve $y = A \sin[(2\pi/B)(x - C)] + D$, shown for A, B, C, and D positive.

where $|A|$ is the *amplitude*, $|B|$ is the *period*, C is the *horizontal shift*, and D is the *vertical shift* (Fig. 1.78).

Figure 1.79 shows how we can use such a function to represent temperature data. The data points in the figure are plots of the mean air temperature for Fairbanks, Alaska, based on records of the National Weather Service from 1941 to 1970. The sine function used to fit the data is

$$f(x) = 37 \sin\left[\frac{2\pi}{365}(x - 101)\right] + 25,$$

where f is temperature in degrees Fahrenheit, and x is the number of the day counting from the beginning of the year. The fit is remarkably good.

Angle Sum and Difference Formulas

As you may recall from an earlier course,

$$\cos(A + B) = \cos A \cos B - \sin A \sin B, \tag{15}$$

$$\sin(A + B) = \sin A \cos B + \cos A \sin B. \tag{16}$$

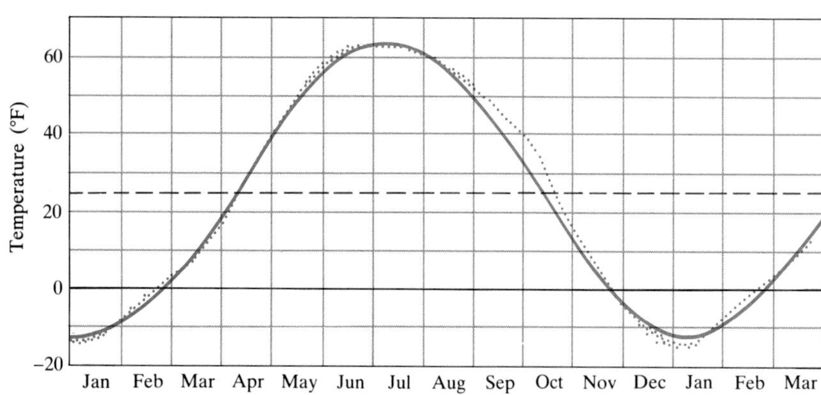

1.79 Normal mean air temperatures at Fairbanks, Alaska, plotted as data points. The approximating sine function is

$$f(x) = 37 \sin\left[\frac{2\pi}{365}(x - 101)\right] + 25.$$

(*Source*: "Is the Curve of Temperature Variation a Sine Curve?" by B. M. Lando and C. A. Lando. *The Mathematics Teacher,* 7:6, Fig. 2, p. 535 (September 1977).)

These formulas hold for all angles A and B.

If we replace B by $-B$ in Eqs. (15) and (16), we get

$$\begin{aligned}\cos(A - B) &= \cos A \cos(-B) - \sin A \sin(-B)\\ &= \cos A \cos B - \sin A(-\sin B)\\ &= \cos A \cos B + \sin A \sin B, \end{aligned} \tag{17}$$

$$\begin{aligned}\sin(A - B) &= \sin A \cos(-B) + \cos A \sin(-B)\\ &= \sin A \cos B + \cos A(-\sin B)\\ &= \sin A \cos B - \cos A \sin B \end{aligned} \tag{18}$$

Double-angle (Half-angle) Formulas

As you will see later on, it is sometimes possible to simplify a calculation by changing trigonometric functions of θ into trigonometric functions of 2θ. There are four basic formulas for doing this, called **double-angle formulas.** The first two come from setting A and B equal to θ in Eqs. (15) and (16):

$$\cos 2\theta = \cos^2\theta - \sin^2\theta, \qquad \text{(Eq. (15) with } A = B = \theta) \tag{19}$$

$$\sin 2\theta = 2 \sin \theta \cos \theta. \qquad \text{(Eq. (16) with } A = B = \theta) \tag{20}$$

The other two double-angle formulas come from the equations

$$\cos^2\theta + \sin^2\theta = 1, \qquad \cos^2\theta - \sin^2\theta = \cos 2\theta.$$

We add to get

$$2 \cos^2\theta = 1 + \cos 2\theta,$$

subtract to get

$$2 \sin^2\theta = 1 - \cos 2\theta,$$

and divide by 2 to get

$$\cos^2\theta = \frac{1 + \cos 2\theta}{2} \tag{21}$$

$$\sin^2\theta = \frac{1 - \cos 2\theta}{2} \tag{22}$$

When θ is replaced by $\theta/2$ in Eqs. (21) and (22), the resulting formulas are called **half-angle formulas.** Some books refer to Eqs. (21) and (22) by this name as well.

Where to Look for Other Formulas

Additional information is available in

1. Appendix 1 of the present book.
2. *CRC Standard Mathematical Tables* (any recent edition). Boca Raton, Florida: CRC Press, Inc.

Tips for Graphing sines and cosines: Curve first, scaled axes later

1.

The one basic sine and cosine curve:

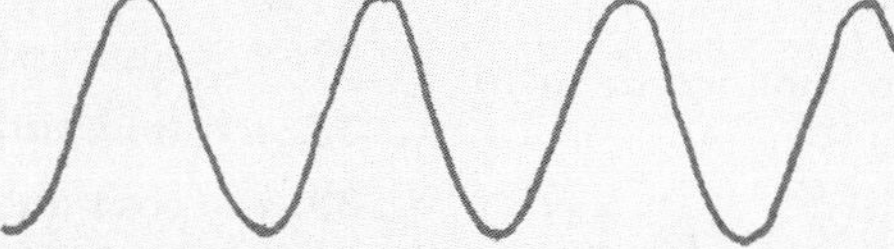

2.

The basic curve with approximately scaled axes in different positions:

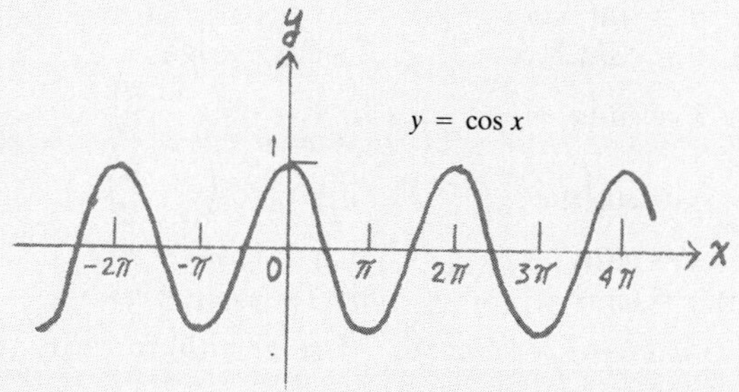

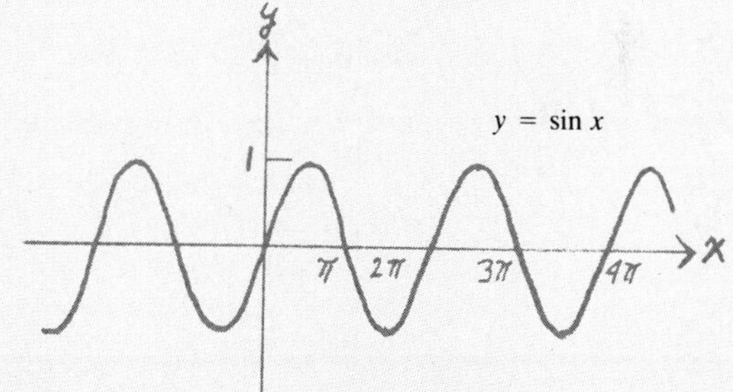

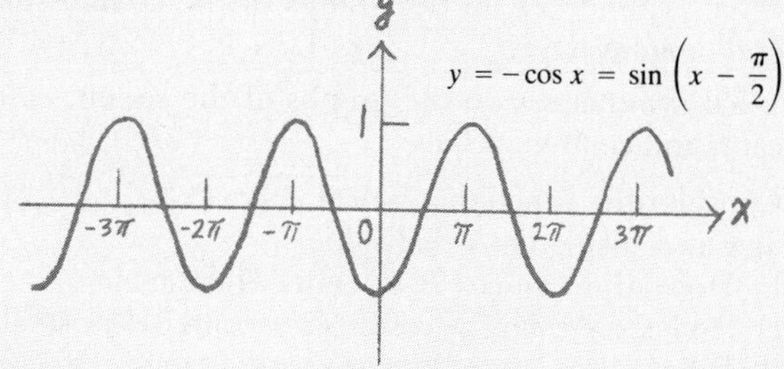

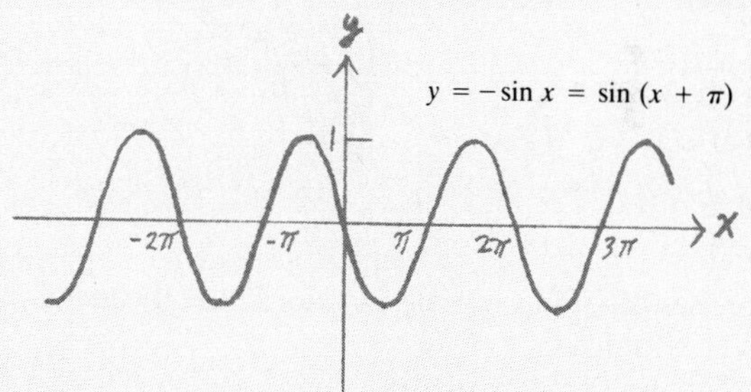

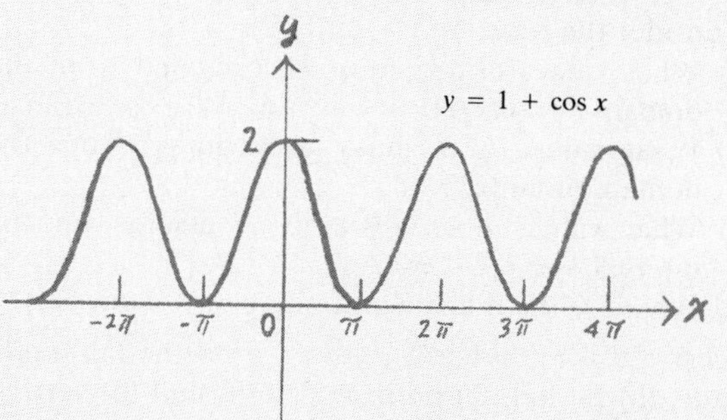

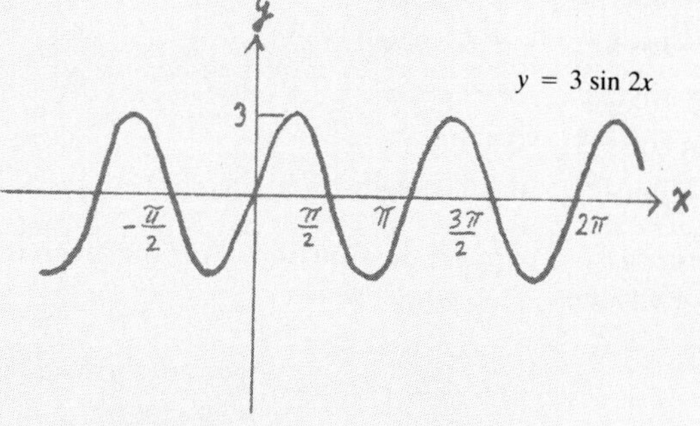

Exercises 1.5

Exercises 1–8 give angles in degrees. Change them to radians.

1. $120°$ 2. $150°$ 3. $270°$ 4. $405°$
5. $-45°$ 6. $-60°$ 7. $-180°$ 8. $-540°$

Exercises 9–16 give angles in radians. Change them to degrees.

9. $\frac{\pi}{6}$ 10. $\frac{3\pi}{4}$ 11. $\frac{5\pi}{4}$ 12. $\frac{9\pi}{4}$
13. $-\frac{3\pi}{2}$ 14. $-\frac{7\pi}{4}$ 15. $-\frac{7\pi}{2}$ 16. -4π

Exercises 17–24 give angles in radians. Find the sine, cosine, tangent, cotangent, secant, and cosecant of each angle (when defined).

17. a) $\frac{\pi}{6}$ b) $-\frac{\pi}{6}$
18. a) $\frac{\pi}{4}$ b) $-\frac{\pi}{4}$
19. a) $\frac{\pi}{3}$ b) $-\frac{\pi}{3}$
20. a) $\frac{2\pi}{3}$ b) $-\frac{2\pi}{3}$
21. a) $\frac{3\pi}{4}$ b) $-\frac{3\pi}{4}$
22. a) $\frac{5\pi}{6}$ b) $-\frac{5\pi}{6}$
23. a) 0 b) π
24. a) $\frac{\pi}{2}$ b) $\frac{3\pi}{2}$

In Exercises 25–30, graph the functions over the given intervals of x-values.

25. $y = \sin 2x, \quad -2\pi \le x \le 2\pi$
26. $y = \sin(x/2), \quad 0 \le x \le 4\pi$
27. $y = \cos 3x, \quad -\pi \le x \le \pi$
28. $y = \sin 3x, \quad -\pi \le x \le \pi$
29. $y = \cos(x/3), \quad 0 \le x \le 6\pi$
30. $y = \sin(x/3), \quad 0 \le x \le 6\pi$

In Exercises 31–34, graph the functions over the given interval of x-values.

31. $y = 1 + \sin x, \quad -2\pi \le x \le 2\pi$
32. $y = 1 + \cos x, \quad -2\pi \le x \le 2\pi$
33. $y = 1 + \cos 2x, \quad -\pi \le x \le \pi$
34. $y = (1 + \cos 2x)/2, \quad -\pi \le x \le \pi$

In Exercises 35–42, sketch the graphs of the functions together over the given intervals of x-values.

35. $y = \sin x$ and $y = \sin 2x, \quad -2\pi \le x \le 2\pi$
36. $y = \sin x$ and $y = \sin(x/2), \quad 0 \le x \le 4\pi$
37. $y = \cos x$ and $y = \cos 3x, \quad 0 \le x \le 2\pi$
38. $y = \sin x$ and $y = \sin 3x, \quad 0 \le x \le 2\pi$
39. $y = \cos x$ and $y = \cos(x/3), \quad 0 \le x \le 6\pi$
40. $y = \sin x$ and $y = \sin(x/3), \quad 0 \le x \le 6\pi$
41. $y = \cos x$ and $y = \sec x, \quad -3\pi/2 \le x \le 3\pi/2$
42. $y = \sin x$ and $y = \csc x, \quad -\pi \le x \le \pi$

43. Which equations have the same graph?
a) $y = \sin x$ b) $y = \cos x$
c) $y = \sin(-x)$ d) $y = \cos(-x)$
e) $y = -\sin x$ f) $y = -\cos x$
g) $y = \sin\left(x + \frac{\pi}{2}\right)$ h) $y = \sin\left(x - \frac{\pi}{2}\right)$
i) $y = \cos\left(x + \frac{\pi}{2}\right)$ j) $y = \cos\left(x - \frac{\pi}{2}\right)$
k) $y = \cos(x + \pi)$ l) $y = \cos(x - \pi)$
m) $y = \sin(x + \pi)$ n) $y = \sin(x - \pi)$

44. *Two more useful identities.* Use Eqs. (5) to verify the following identities.
a) $\sec^2\theta = 1 + \tan^2\theta$ b) $\csc^2\theta = 1 + \cot^2\theta$

45. What symmetries do the graphs of the cosine, sine, and tangent have?

46. What symmetries do the graphs of the secant, cosecant, and cotangent have?

47. Consider the function $y = \sqrt{(1 + \cos 2x)/2}$.
a) Can x take on any real value?
b) How large can $\cos 2x$ become? How small?
c) How large can $(1 + \cos 2x)/2$ become? How small?
d) What are the domain and range of $y = \sqrt{(1 + \cos 2x)/2}$?

48. Consider the function $y = \tan(x/2)$.
a) What values of $x/2$ must be excluded from the domain of $\tan(x/2)$?
b) What values of x must be excluded from the domain of $\tan(x/2)$?
c) What values does $y = \tan(x/2)$ assume on the interval $-\pi < x < \pi$?
d) What are the domain and range of $y = \tan(x/2)$?

49. *Temperature in Fairbanks, Alaska.* Find the (a) amplitude, (b) period, (c) horizontal shift, and (d) vertical shift of the general sine function

$$f(x) = 37 \sin\left[\frac{2\pi}{365}(x - 101)\right] + 25.$$

50. *Temperature in Fairbanks, Alaska.* Use the equation in Exercise 49 to approximate the answers to the following questions about the temperature in Fairbanks, Alaska, shown in Fig. 1.79. Assume that the year has 365 days.
   a) What are the highest and lowest mean daily temperatures shown?
   b) What is the average of the highest and lowest mean daily temperatures shown? Why is this average the vertical shift of the function?

51. What happens if you take $A = B$ in Eq. (17)? Does the result agree with something you already know?

52. What happens if you take $B = \pi/2$ in Eqs. (17) and (18)? Do these results agree with something you already know?

53. What happens if you take $B = \pi/2$ in Eqs. (15) and (16)? Do these results agree with something you already know?

54. What happens if you take $B = \pi$ in Eqs. (15) and (16)? in Eqs. (17) and (18)?

55. Evaluate $\cos 15°$ as $\cos(45° - 30°)$.

56. Evaluate $\sin 75°$ as $\sin(45° + 30°)$.

57. Evaluate $\sin \frac{7\pi}{12}$ (radians) as $\sin\left(\frac{\pi}{4} + \frac{\pi}{3}\right)$.

58. Evaluate $\cos \frac{10\pi}{24}$ (radians) as $\cos\left(\frac{\pi}{4} + \frac{\pi}{6}\right)$.

Use double-angle formulas to find the function values in Exercises 59–62 (angles in radians).

59. $\cos^2 \frac{\pi}{8}$

60. $\cos^2 \frac{\pi}{12}$

61. $\sin^2 \frac{\pi}{12}$

62. $\sin^2 \frac{\pi}{8}$

63. *The tangent sum formula.* The standard formula for the tangent of the sum of two angles is

$$\tan(A + B) = \frac{\tan A + \tan B}{1 - \tan A \ \tan B}.$$

Derive the formula by writing $\tan(A + B)$ as $\sin(A + B)/\cos(A + B)$ and applying Eqs. (15) and (16).

64. Derive a formula for $\tan(A - B)$ by replacing B by $-B$ in the formula for $\tan(A + B)$ in Exercise 63.

Even vs. Odd

65. a) Show that $\cot x$ is an odd function of x.
   b) Show that the quotient of an even function and an odd function is always odd (on their common domain).

66. a) Show that $\csc x$ is an odd function of x.
   b) Show that the reciprocal of an odd function (when defined) is odd.

67. a) Show that the product $y = \sin x \ \cos x$ is an odd function of x.
   b) Show that the product of an even function and an odd function is always odd (on their common domain).

68. a) Show that the function $y = \sin^2 x$ is an even function of x (even though the sine itself is odd).
   b) Show that the square of an odd function is always even.
   c) Show that the product of any two odd functions is even (on their common domain).

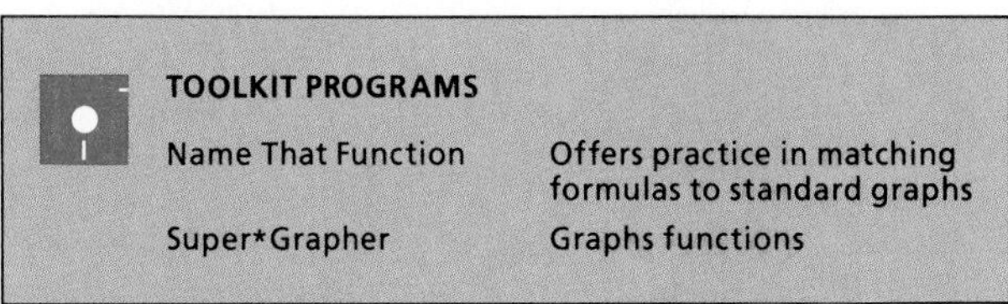

TOOLKIT PROGRAMS	
Name That Function	Offers practice in matching formulas to standard graphs
Super*Grapher	Graphs functions

1.6 Absolute Value or Magnitude, and Target Values

As you may already know, the absolute value of a number is its numerical size or magnitude, and the function that sends each number to its magnitude is called the absolute-value function. In calculus, we use absolute values to measure differences between numbers, to create formulas that define intervals, and to control function values. This section gives the details.

The Absolute Value Function

The **absolute value** or **magnitude** of a number x, denoted by $|x|$ (read "the absolute value of x"), is defined by the formula

$$|x| = \begin{cases} x \text{ if } x \geq 0 \\ -x \text{ if } x < 0. \end{cases} \tag{1}$$

The vertical lines in the symbol $|x|$ are called *absolute value bars.*

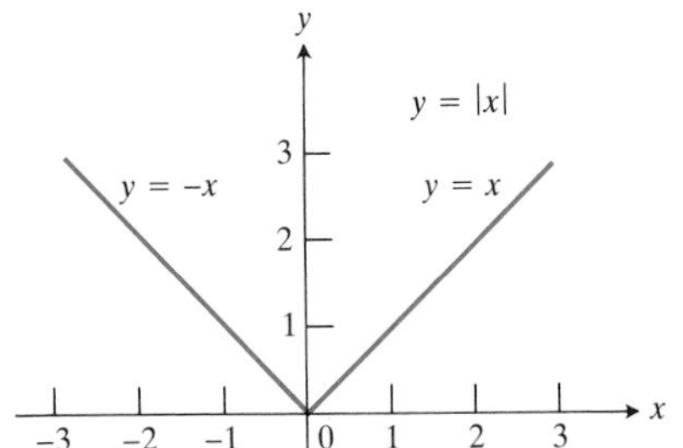

1.80 The absolute value function.

EXAMPLE 1

$$|3| = 3$$
$$|0| = 0$$
$$|-5| = 5$$

■

The function $y = |x|$ is called *the absolute value function.* Its graph lies along the line $y = x$ when $x \geq 0$ and along the line $y = -x$ when $x < 0$ (Fig. 1.80).

Another way to find the absolute value of a number is to square the number and then take the positive square root:

$$|x| = \sqrt{x^2}. \tag{2}$$

EXAMPLE 2

$$|2| = \sqrt{2^2} = \sqrt{4} = 2$$
$$|-2| = \sqrt{(-2)^2} = \sqrt{4} = 2$$

■

To solve an equation that contains absolute values, we write equivalent equations without absolute values, then solve as usual.

EXAMPLE 3 Solve the equation $|2x - 3| = 7$.

Solution The equation says that $2x - 3 = \pm 7$, so there are two possibilities:

$2x - 3 = 7$	$2x - 3 = -7$	(Equivalent equations without absolute values.)
$2x = 10$	$2x = -4$	(Solve as usual.)
$x = 5$	$x = -2.$	

The equation $|2x - 3| = 7$ has two solutions: $x = 5$ and $x = -2$. ■

To graph a formula $y = f(x)$ that contains absolute values, divide the x-axis into intervals on which the absolute values can be removed. Then graph as usual.

EXAMPLE 4 Graph the function

$$f(x) = |x + 1| + |x - 3|.$$

Solution The points where the expressions inside the bars change sign are

$x = -1$ (for $x + 1$)

$x = 3$ (for $x - 3$).

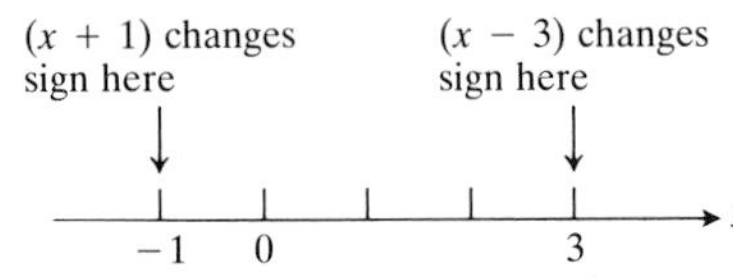

These points divide the x-axis into intervals on which we can write absolute-value-free formulas for $f(x)$:

For $x < -1$: Here, $x + 1 < 0$ and $x - 3 < 0$, so

$$\begin{aligned} f(x) &= |x+1| + |x-3| \\ &= -(x+1) - (x-3) \\ &= -x - 1 - x + 3 = 2 - 2x. \end{aligned}$$

For $-1 \le x \le 3$: Here, $x + 1 \ge 0$ and $x - 3 \le 0$, so

$$\begin{aligned} f(x) &= |x+1| + |x-3| \\ &= x + 1 - (x-3) \\ &= x + 1 - x + 3 = 4 \end{aligned}$$

For $x > 3$: Here, $x + 1 > 0$ and $x - 3 > 0$, so

$$\begin{aligned} f(x) &= |x+1| + |x+3| \\ &= x + 1 + x - 3 = 2x - 2. \end{aligned}$$

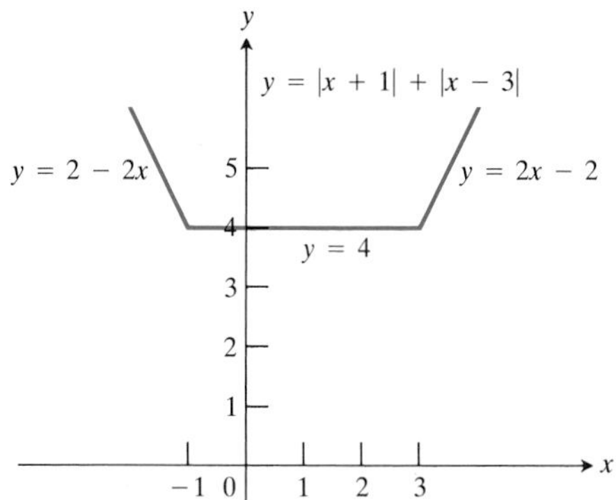

1.81 To graph $f(x) = |x + 1| + |x - 3|$, we divide the x-axis into intervals on which we can replace the formula for f with expressions that do not contain absolute values. Example 4 shows how this is done.

Thus,

$$f(x) = \begin{cases} 2 - 2x, & x < -1 \\ 4, & -1 \le x \le 3 \\ 2x - 2, & x > 3. \end{cases}$$

With the absolute value signs now removed, we can graph f in the usual way (Fig. 1.81). ■

When we do arithmetic with absolute values, we can always use the following rules:

Arithmetic with Absolute Values

1. $|-a| = |a|$ — A number and its negative have the same absolute value.
2. $|ab| = |a|\,|b|$ — The absolute value of a product is the product of the absolute values.
3. $\left|\dfrac{a}{b}\right| = \dfrac{|a|}{|b|}$ — The absolute value of a quotient is the quotient of the absolute values.

EXAMPLE 5

$$|-\sin x| = |\sin x|$$

$$|-2(x+5)| = |-2|\,|x+5| = 2|x+5|$$

$$\left|\frac{3}{x}\right| = \frac{|3|}{|x|} = \frac{3}{|x|}$$

■

The absolute value of a sum of two numbers is never larger than the sum of their absolute values. When we put this in symbols, we get the important triangle inequality.

> **The Triangle Inequality**
>
> $$|a + b| \leq |a| + |b| \qquad \text{for all numbers } a \text{ and } b \tag{3}$$

EXAMPLE 6 The number $|a + b|$ is less than $|a| + |b|$ if a and b have different signs. In all other cases, $|a + b|$ equals $|a| + |b|$.

$$|-3 + 5| = |2| = 2 < |-3| + |5| = 8$$

$$|3 + 5| = |8| = 8 = |3| + |5|$$

$$|-3 + 0| = |-3| = 3 = |-3| + |0|$$

$$|-3 - 5| = |-8| = 8 = |-3| + |-5|.$$

Notice that absolute value bars in expressions such as $|-3 + 5|$ also work like parentheses: We do the arithmetic inside *before* we take the absolute value. ■

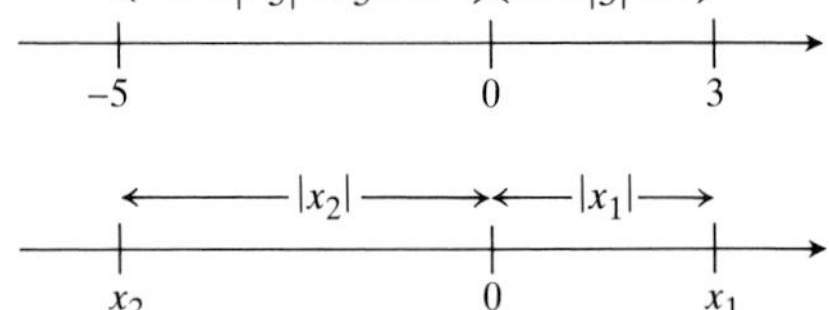

1.82 Absolute values give distances between points on the number line.

Absolute Values and Distance

The numbers $|a - b|$ and $|b - a|$ are always equal because

$$|a - b| = |(-1)(b - a)| = |-1|\,|b - a| = |b - a|. \tag{4}$$

They give the distance between the points a and b on the number line (Fig. 1.82).

> **Number-line Distance**
>
> $|a - b| = |b - a|$ for all numbers a and b. This number is the distance between a and b on the number line.

Absolute Values and Intervals

The connection between absolute values and distance gives us a new way to write formulas for intervals.

The inequality $|a| < 5$ says that the distance from a to the origin is less than 5. This is the same as saying that a lies between -5 and 5 on the number line. In symbols,

$$|a| < 5 \quad \Leftrightarrow \quad -5 < a < 5. \tag{5}$$

The set of numbers a with $|a| < 5$ is the open interval from -5 to 5 (Fig. 1.83).

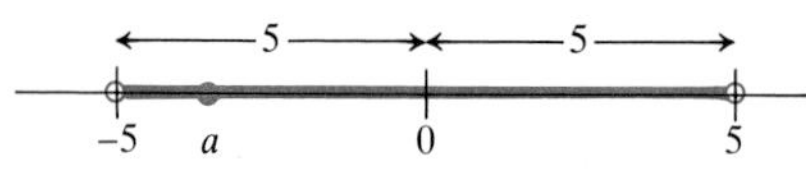

1.83 $|a| < 5$ means $-5 < a < 5$.

The general rule is this:

Relation between Intervals and Absolute Values

If D is any positive number, then

$$|a| < D \quad \Leftrightarrow \quad -D < a < D, \tag{6}$$

$$|a| \leq D \quad \Leftrightarrow \quad -D \leq a \leq D. \tag{7}$$

EXAMPLE 7 What values of x satisfy the inequality $|x - 5| < 9$?

Solution Change

$$|x - 5| < 9$$

to $\quad -9 < x - 5 < 9 \qquad$ (Eq. (6) with $a = x - 5$ and $D = 9$)

to $\quad -9 + 5 < x < 9 + 5 \qquad$ (Adding a positive number to both sides of an inequality gives an equivalent inequality. Adding 5 here isolated the x.)

or $\quad -4 < x < 14.$

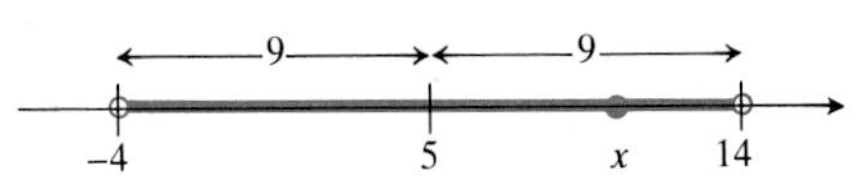

1.84 $|x - 5| < 9$ means $-4 < x < 14$.

The steps we just took are reversible, so the values of x that satisfy the inequality $|x - 5| < 9$ are the numbers in the interval $-4 < x < 14$ (Fig. 1.84). ■

EXAMPLE 8 Describe the interval $-3 < x < 5$ with an absolute value inequality of the form $|x - x_0| < D$.

Solution We average the endpoint values to find the interval's midpoint:

$$\text{midpoint } x_0 = \frac{-3 + 5}{2} = \frac{2}{2} = 1.$$

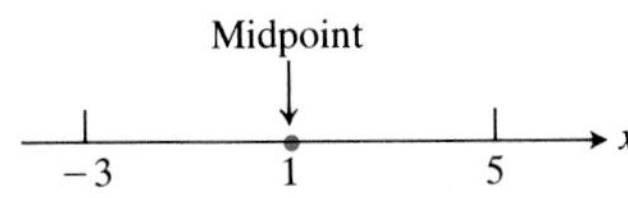

The midpoint lies 4 units away from each endpoint. The interval therefore consists of the points that lie within 4 units of the midpoint, or the points x with

$$|x - 1| < 4.$$ ■

The midpoint of the interval (a, b) is

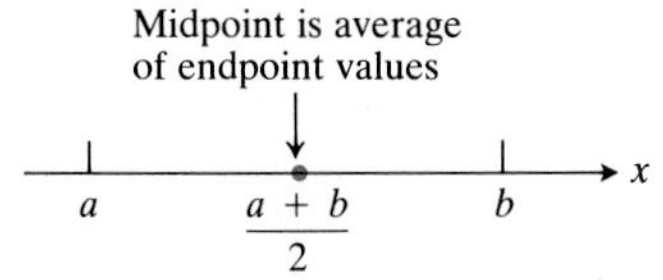

EXAMPLE 9 What values of x satisfy the inequality $\left|\frac{2x}{3}\right| \leq 1$?

Solution Change

$$\left|\frac{2x}{3}\right| \leq 1$$

to $\quad -1 \leq \frac{2x}{3} \leq 1 \qquad$ (Eq. (7) with $a = 2x/3$ and $D = 1$)

to $\quad -3 \leq 2x \leq 3 \qquad$ (Muliplying both sides of an inequality by a positive number gives an equivalent inequality.)

to $-\frac{3}{2} \le x \le \frac{3}{2}$. (Dividing both sides of an inequality by a positive number gives an equivalent inequality.)

The original inequality holds for x in the closed interval from $-3/2$ to $3/2$ (Fig. 1.85). ■

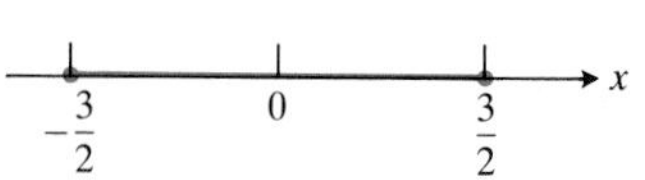

1.85 The inequality $|2x/3| \le 1$ holds on the interval $-3/2 \le x \le 3/2$.

EXAMPLE 10 What values of x satisfy the inequality

$$\left|5 - \frac{2}{x}\right| < 1?$$

Solution Change

$$\left|5 - \frac{2}{x}\right| < 1$$

to $-1 < 5 - \frac{2}{x} < 1$ (Eq. (6) with $a = (5 - 2/x)$ and $D = 1$)

to $-6 < -\frac{2}{x} < -4$ (Subtracting a positive number, in this case 5, from both sides of an inequality gives an equivalent inequality.)

to $4 < \frac{2}{x} < 6$ (Multiplying both sides of an inequality by -1 reverses the inequality.)

to $2 < \frac{1}{x} < 3$ (Divide by 2.)

to $\frac{1}{3} < x < \frac{1}{2}$. (Take reciprocals. When the numbers involved have the same sign, taking reciprocals reverses an inequality)

The original inequality holds if and only if x lies between $\frac{1}{3}$ and $\frac{1}{2}$. ■

Keeping Function Outputs near Target Values

We sometimes want the outputs of a function $y = f(x)$ to lie near a particular target value y_0. This need can come about in different ways. A gas-station attendant, asked for \$5.00 worth of gas, will try to pump the gas to the nearest cent. A mechanic grinding a 3.385-inch cylinder bore will not let the bore exceed this value by more than .002 in. A pharmacist making ointments will measure the ingredients to the nearest milligram. A photographer making a long exposure will set a timer to get the right amount of light to the film—one second either way might not matter, but ten seconds might.

So, the question becomes: How accurate do our machines and instruments have to be to keep the outputs within useful bounds? When we express this question with mathematical symbols, we ask: How closely must we control x to keep $y = f(x)$ within an acceptable interval about some particular target value y_0? The following examples show how to answer this question.

EXAMPLE 11 *Controlling a linear function.* How close to $x_0 = 4$ must we hold x to be sure that $y = 2x - 1$ lies within 2 units of $y_0 = 7$?

Solution We are asked: For what values of x is $|y - 7| < 2$? To find the answer, we first express $|y - 7|$ in terms of x:

$$|y - 7| = |(2x - 1) - 7| = |2x - 8|.$$

The question then becomes: What values of x satisfy the inequality $|2x - 8| < 2$? To find out, we change

$$|2x - 8| < 2$$

to $\quad -2 < 2x - 8 < 2$

to $\quad 6 < 2x < 10$

to $\quad 3 < x < 5.$

To keep y within 2 units of $y_0 = 7$, we must keep x within 1 unit of $x_0 = 4$ (Fig. 1.86). ■

1.86 Keeping x between 3 and 5 will keep $y = 2x - 1$ between $y = 5$ and $y = 9$.

EXAMPLE 12 *Controlling the area of a circle.* In what interval about $r_0 = 10$ must we hold r to be sure that $A = \pi r^2$ lies within π square units of $A_0 = 100\pi$?

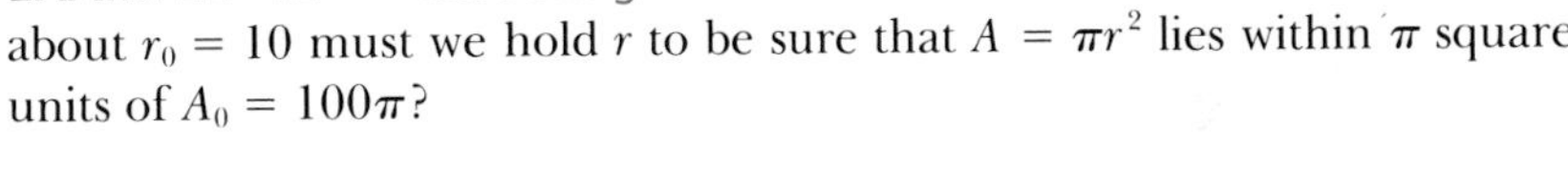

Solution We want to know the values of r for which $|A - A_0| < \pi$. To find them, change the inequality from

$$|A - A_0| < \pi$$

to $\quad |\pi r^2 - 100\pi| < \pi$

to $\quad -\pi < \pi r^2 - 100\pi < \pi$

to $\quad -1 < r^2 - 100 < 1$

to $\quad 99 < r^2 < 101$

to $\quad \sqrt{99} < r < \sqrt{101}. \qquad \left(\begin{array}{l}\text{For } a, b, \text{ and } c \text{ nonnegative,} \\ a < b < c \Rightarrow \sqrt{a} < \sqrt{b} < \sqrt{c}.\end{array}\right)$

The interval of possible radii is the open interval from $r = \sqrt{99}$ to $r = \sqrt{101}$ (Fig. 1.87). ■

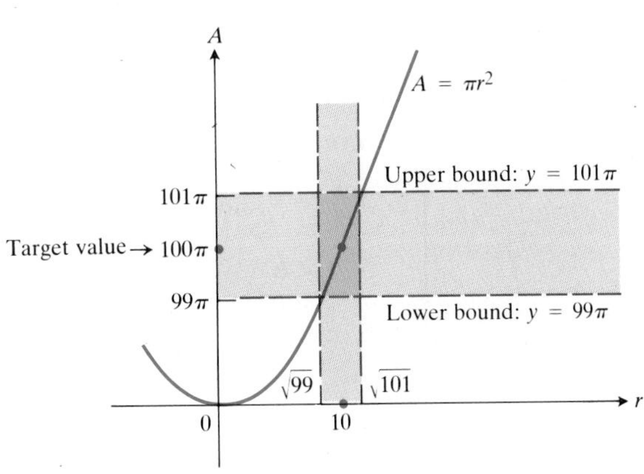

1.87 Keeping r between $\sqrt{99}$ and $\sqrt{101}$ will keep πr^2 between 99π and 101π.

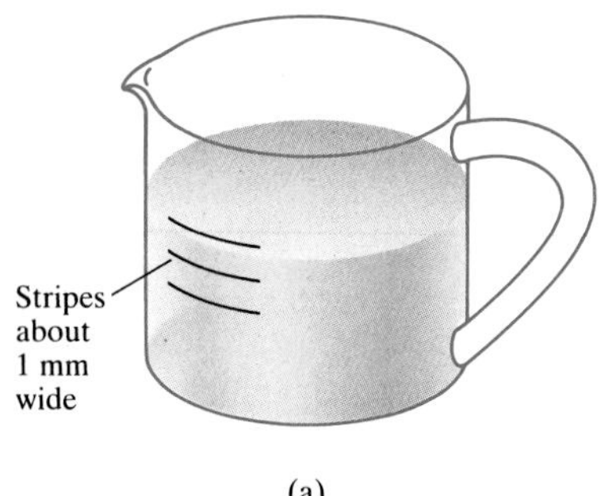

(a)

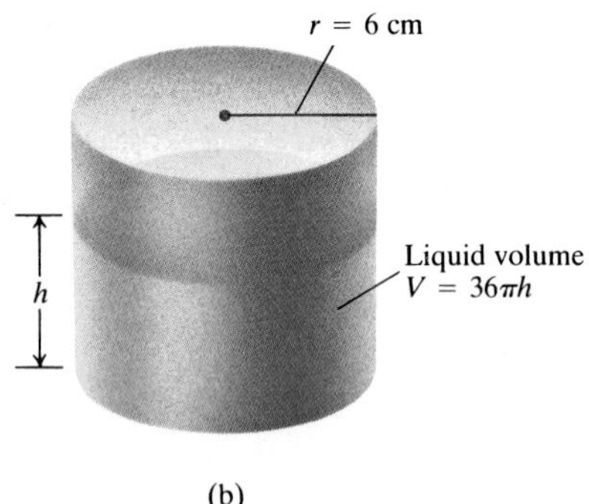

(b)

1.88 A typical 1-liter measuring cup (a) modeled in (b) as a right circular cylinder of radius $r = 6$ cm. To get a liter of water to the nearest 1%, how accurately must we measure h? See Example 13.

EXAMPLE 13 *Why the stripes on a 1-liter kitchen measuring cup are about a millimeter wide.* The interior of a typical 1-L measuring cup is a right circular cylinder of radius 6 cm (Fig. 1.88). The volume of water we put in the cup is therefore a function of the level h to which the cup is filled, the formula being

$$V = \pi(6)^2 h = 36\pi h.$$

How closely do we have to measure h to measure out one liter of water (1000 cm^3) with an error of no more than 1% (10 cm^3)?

Solution In terms of V and h, we want to know in what interval of values to hold h to make V satisfy the inequality

$$|V - 1000| = |36\pi h - 1000| \leq 10.$$

To find out, we change

$$|36\pi h - 1000| \leq 10$$

to $\quad -10 \leq 36\pi h - 1000 \leq 10$

to $\quad 990 \leq 36\pi h \leq 1010$

to $\quad \dfrac{990}{36\pi} \leq h \leq \dfrac{1010}{36\pi}$

to $\quad \underset{\begin{pmatrix}\text{rounded up,}\\\text{to be safe}\end{pmatrix}}{8.8} \leq h \leq \underset{\begin{pmatrix}\text{rounded down,}\\\text{to be safe}\end{pmatrix}}{8.9}.$ (Values found with a calculator.)

The interval in which we should hold h is about one tenth of a centimeter wide (one millimeter). With stripes one millimeter wide, we can therefore expect to measure a liter of water with an accuracy of one percent, which is more than enough accuracy for cooking. ■

Properties of Inequalities Used in This Section

1. $a < b \quad \Rightarrow \quad a + c < b + c$
2. $a < b \quad \Rightarrow \quad a - c < b - c$
3. $a < b$ and $c > 0 \quad \Rightarrow \quad ac < bc$
4. $a < b$ and $c > 0 \quad \Rightarrow \quad \dfrac{a}{c} < \dfrac{b}{c}$
5. $a < b \quad \Rightarrow \quad -b < -a$
6. $a < b \quad \Rightarrow \quad \dfrac{1}{b} < \dfrac{1}{a}$ (requires a and b both positive or both negative)

These hold for $\leq$ as well for $<$.

Exercises 1.6

1. If $2 < x < 6$, which of the following statements about x are true and which are false?
   a) $0 < x < 4$ b) $0 < x - 2 < 4$
   c) $1 < \frac{x}{2} < 3$ d) $\frac{1}{6} < \frac{1}{x} < \frac{1}{2}$
   e) $1 < \frac{6}{x} < 3$ f) $|x - 4| < 2$
   g) $-6 < -x < 2$ h) $-6 < -x < -2$

2. If $-1 < y - 5 < 1$, which of the following statements about y are true and which are false?
   a) $4 < y < 6$ b) $|y - 5| < 1$
   c) $y > 4$ d) $y < 6$
   e) $0 < y - 4 < 2$ f) $2 < \frac{y}{2} < 3$
   g) $\frac{1}{6} < \frac{1}{y} < \frac{1}{4}$ h) $-6 < y < -4$

Find the absolute values in Exercises 3–6.

3. $|-3|$
4. $|2 - 7|$
5. $|-2 + 7|$
6. $|1.1 - 5.2|$

Solve the equations in Exercises 7–12.

7. $|x| = 2$
8. $|x - 3| = 7$
9. $|2x + 5| = 4$
10. $|1 - x| = 1$
11. $|8 - 3x| = 9$
12. $\left|\frac{x}{2} - 1\right| = 1$

In Exercises 13–20, match each absolute value inequality with the interval it determines.

13. $|x + 3| < 1$ a) $-2 < x < 1$
14. $|x - 5| < 2$ b) $-1 < x < 3$
15. $\left|\frac{x}{2}\right| < 1$ c) $3 < x < 7$
16. $|1 - x| < 2$ d) $-\frac{5}{2} < x < -\frac{3}{2}$
17. $|2x - 5| \le 1$ e) $-2 < x < 2$
18. $|2x + 4| < 1$ f) $-4 < x < 4$
19. $\left|\frac{x - 1}{2}\right| < 1$ g) $-4 < x < -2$
20. $\left|\frac{2x + 1}{3}\right| < 1$ h) $2 \le x \le 3$
    i) $-2 \le x \le 2$

The inequalities in Exercises 21–32 define intervals. Describe each interval with inequalities that do not involve absolute values.

21. $|y| < 2$
22. $|y| \le 2$
23. $|y - 1| \le 2$
24. $|y + 2| < 1$
25. $|3y - 7| < 2$
26. $|2y + 5| < 1$
27. $\left|\frac{y}{3}\right| \le 10$
28. $\left|\frac{2}{3}y\right| \le \frac{2}{3}$
29. $\left|\frac{y}{2} - 1\right| \le 1$
30. $\left|2 - \frac{y}{2}\right| < \frac{1}{2}$
31. $|1 - y| < \frac{1}{10}$
32. $\left|\frac{7 - 3y}{2}\right| < 1$

Describe the intervals in Exercises 33–36 with absolute value inequalities of the form $|x - x_0| < D$. It may help to draw a picture of the interval first.

33. $3 < x < 9$
34. $-3 < x < 9$
35. $-5 < x < 3$
36. $-7 < x < -1$

Each of Exercises 37–42 gives a function $y = f(x)$, a number E, and a target value y_0. In what interval must we hold x in each case to be sure that $y = f(x)$ lies within E units of y_0?

37. $y = x^2$, $E = 0.1$, $y_0 = 100$
38. $y = x^2 - 5$, $E = 1$, $y_0 = 11$
39. $y = \sqrt{x - 7}$, $E = 0.1$, $y_0 = 4$
40. $y = \sqrt{19 - x}$, $E = 1$, $y_0 = 3$
41. $y = 120/x$, $E = 1$, $y_0 = 5$
42. $y = 1/4x$, $E = 1/2$, $y_0 = 1$

Each of Exercises 43–46 gives a function $y = f(x)$, a number E, a point x_0, and a target value y_0. In what interval about x_0 must we hold x in each case to be sure that $y = f(x)$ lies within E units of y_0? Describe the interval with an absolute value inequality of the form $|x - x_0| < D$.

43. $y = x + 1$, $E = 0.5$, $x_0 = 3$, $y_0 = 4$
44. $y = 2x - 1$, $E = 1$, $x_0 = -2$, $y_0 = -5$
45. $y = -(x/2) + 1$, $E = 1/2$, $x_0 = 6$, $y_0 = -2$
46. $y = -2x - 2$, $E = 0.2$, $x_0 = -3$, $y_0 = 4$
47. **CALCULATOR** *Grinding engine cylinders.* Before contracting to grind engine cylinders to a cross-section area of 9 in^2, you want to know how much deviation from the ideal cylinder diameter of $x_0 = 3.385$ in. you can allow and still have the area come within 0.01 in^2 of the required 9 in^2. To find out, you let $A = \pi(x/2)^2$ and look for the interval in which you must hold x to make $|A - 9| \le 0.01$. What interval do you find?
48. *Manufacturing electrical resistors.* Ohm's law for electrical circuits, like the one shown in Fig. 1.89, states that $V = RI$. In this equation, V is a constant voltage, I is the current in amperes, and R is the resistance in

1.89 The current in Exercise 48.

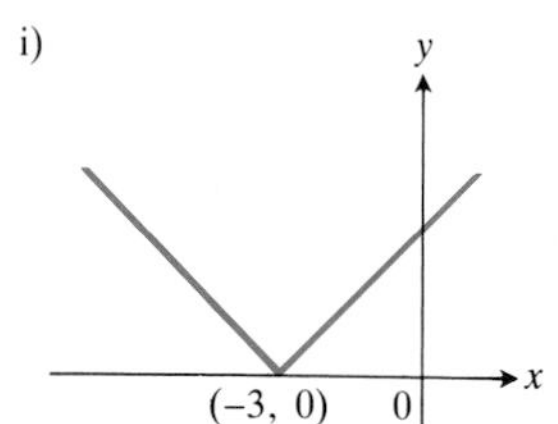

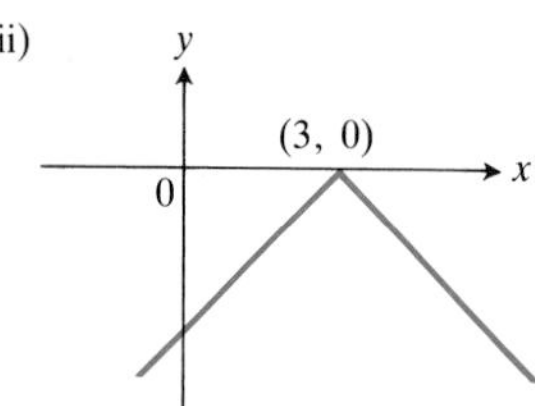

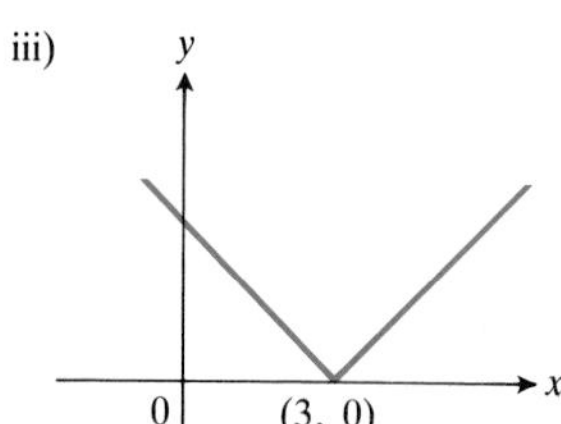

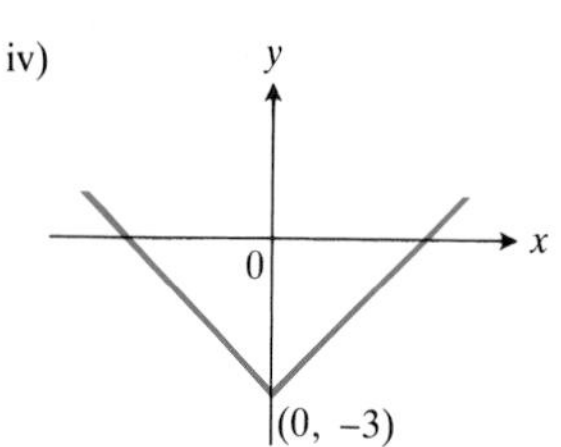

1.90 The graphs for Exercise 49.

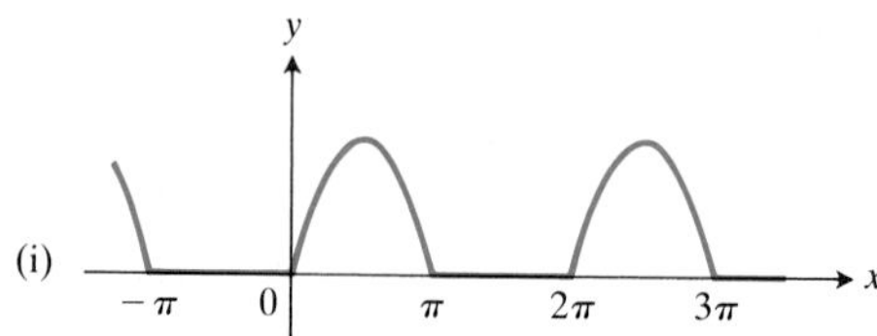

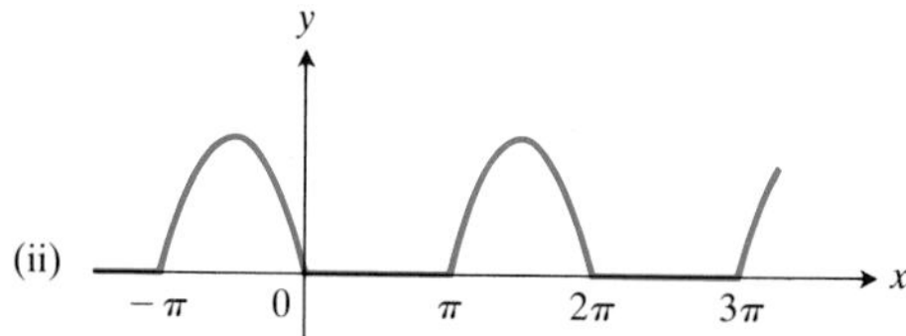

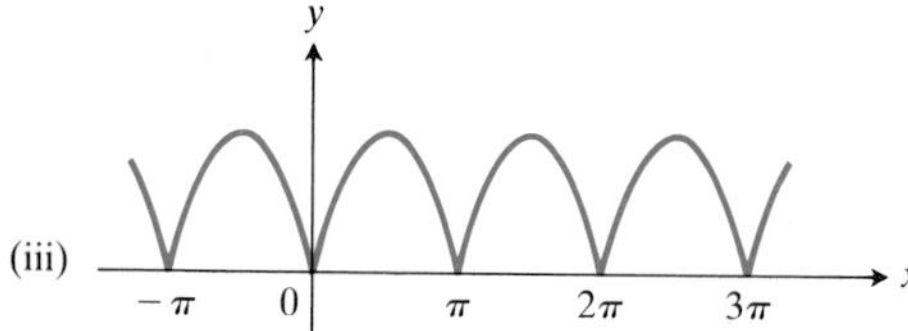

1.91 The graphs for Exercise 50.

ohms. Your firm has been asked to supply the resistors for a circuit in which V will be 120 volts and I is to be 5 ± 0.1 amperes. In what interval does R have to lie for I to be within 0.1 amperes of the target value $I_0 = 5$?

49. Match the following functions with the graphs in Fig. 1.90.

a) $y = |x - 3|$ b) $y = |x| - 3$

c) $y = |x + 3|$ d) $y = -|x - 3|$

50. Match the following functions with the graphs in Fig. 1.91

a) $y = |\sin x|$ b) $y = (\sin x + |\sin x|)/2$

c) $y = (|\sin x| - \sin x)/2$

Graph the functions in Exercises 51–54.

51. $y = \dfrac{|x|}{x}$ **52.** $y = \dfrac{|x - 1|}{x - 1}$

53. $y = \dfrac{x - |x|}{2}$ **54.** $y = \dfrac{x + |x|}{2}$

55. Do not fall into the trap $|-a| = a$. The equation does not hold for all values of a.

a) Find a value of a for which $|-a| \neq a$.

b) For what values of a does the equation $|-a| = a$ hold?

56. For what values of x does $|1 - x|$ equal $1 - x$? For what values of x does it equal $x - 1$?

57. Compare the domains and ranges of the functions $y = \sqrt{x^2}$ and $y = (\sqrt{x})^2$.

58. Find $f(x)$ if $g(x) = \sqrt{x}$ and $(g \circ f)(x) = |x|$.

59. Find $g(x)$ if $f(x) = x^2 + 2x + 1$ and $(g \circ f)(x) = |x + 1|$.

60. Find functions $f(x)$ and $g(x)$ whose composites satisfy the two equations

$$(g \circ f)(x) = |\sin x| \text{ and } (f \circ g)(x) = (\sin \sqrt{x})^2.$$

Graph the functions in Exercises 61–64.

61. $f(x) = |x + 1| + 2|x - 3|$ (*Hint:* Use three intervals.)

62. $f(x) = |x + 2| + |x - 1|$

63. $f(x) = |x| + |x - 1| + |x - 3|$ (*Hint:* Use four intervals.)

64. $f(x) = |x + 2| + |x| + |x + 1|$

65. *The Best Location for a Factory Assembly Table* (adapted from *Fantastiks of Mathematiks,* Cliff Sloyer, Janson Publications, Inc., Providence, R.I., 1986). Because of a design change, the parts produced by three machines along a factory aisle (shown here as the x-axis)

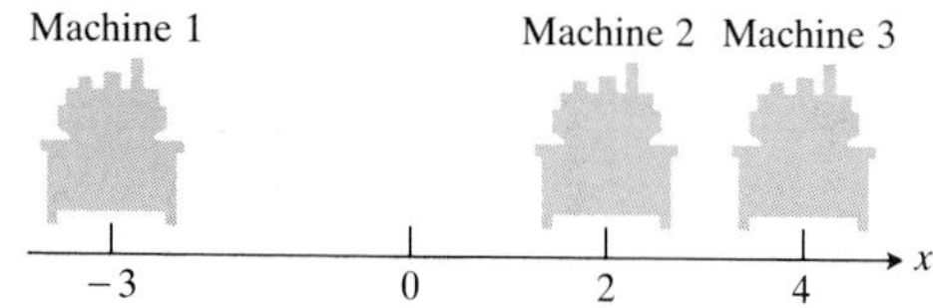

are to go to a nearby table for assembly before they undergo further processing. Each assembly takes one part from each machine and there is a fixed cost per foot for moving each part. As the plant's production engineer, you have been asked to find a location for the assembly table that will keep the total cost of moving the parts at a minimum.

To solve the problem, you let x represent the table's location and look for the value of x that minimizes the sum

$$d(x) = |x + 3| + |x - 2| + |x - 4|$$

of the distances from the table to the three machines. Since the cost of moving the parts to the assembly table is proportional to the total distance the parts travel, any value of x that minimizes d will minimize the cost.

Complete the job now by graphing $d(x)$ to find its smallest value. Then say where you would put the table.

66. *Best location (Continuation of Exercise 65).* You solved the table-location problem in Exercise 65 so well that your manager has asked you to solve a similar problem at a neighboring plant. This time there are four machines instead of three

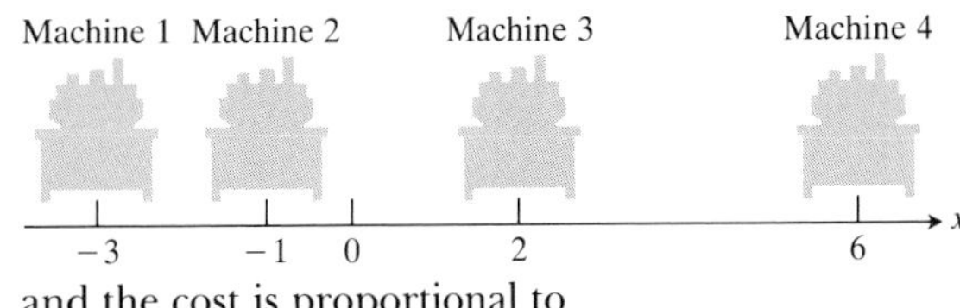

and the cost is proportional to

$$d(x) = |x + 3| + |x + 1| + |x - 2| + |x - 6|.$$

Where should the assembly table go now?

67. *Best location (Continuation of Exercise 66).* As the result of another design change, the assembly in the plant in Exercise 66 is to use twice as many parts from Machine 1 as before, and three times as many parts from Machine 3 as before. The total cost of moving parts from the four machines to the assembly table is now proportional to the "weighted" distance

$$d(x) = 2|x + 3| + |x + 1| + 3|x - 2| + |x - 6|.$$

What is the minimum value of this new function? Where should the table go?

1.7 Calculators and Calculus

In this section, we look at some of the important functions on a scientific calculator, including e^x, $\ln x$, and some of the trigonometric functions. We also discuss how to use a calculator to explore various properties of functions, such as the possibility that two functions are equal or differ by some constant value. We then encounter a mysterious pattern (that will later be explained by calculus) and explore the way Fermat wanted to use slopes of lines to define slopes of curves.

Evaluating Functions

Calculators enable us to evaluate most of the functions in calculus to six or more decimal places with only a few key presses. Calculators work either with algebraic notation or with reverse polish notation. Here are two typical calculations.

EXAMPLE 1 *Algebraic notation.* To find the sine of 2 radians on a Casio fx-7000G, set the calculator in radian mode and press [SIN] [2] [EXE].

Key	Operation	Display
[SIN]	Choose the function.	SIN
[2]	Enter the input value.	SIN 2
[EXE]	Execute the calculation.	0.9092974268

On another algebraic machine, the appropriate key sequence might be [SIN] [2] [=] or simply [2] [SIN]. There can be quite a bit of variation. ■

EXAMPLE 2 *Reverse polish notation.* To find the sine of 2 radians on an HP 28S, set the calculator in radian mode and press [2] [ENTER] [SIN].

Key	Operation	Display
[2]	Put 2 in the command line.	2
[ENTER]	Enter it in stack level 1.	2
[SIN]	Find the sine of the number in stack level 1.	0.909297426826

■

Questions of Accuracy

Although machines can calculate only the first few digits of an infinite decimal like

$$\sin 2 = 0.\ 90929\ 74268\ 26\ldots,$$

they give estimates to more decimal places than you usually find in printed tables. They are also easier to carry and faster to use than tables.

However, when your inputs require the calculator to do arithmetic with numbers that are very small or very large, the errors associated with rounding and truncation (clipping without rounding) may produce a sizable error in the final result. Sometimes we can avoid round-off and truncation errors by restructuring the calculation. For example, suppose we want to evaluate the quotient

$$Q = \frac{\dfrac{1}{x + \Delta x} - \dfrac{1}{x}}{\Delta x} \tag{1}$$

when $x = 0.7$ and $\Delta x = 10^{-8}$ (we shall see why we might want to do this when we get to Chapter 3). A direct evaluation gives

$$\begin{aligned} Q &= \frac{1.4285\ 714088 - 1.4285\ 71429}{10^{-8}} \\ &= -\frac{0.0000\ 00021}{10^{-8}} = -2.1 \end{aligned} \tag{2}$$

The numbers originally in the numerator of Q are precise to ten digits but their difference is precise to only two digits. We get a bad cancellation and a relatively imprecise answer.

We get a better result if we combine the fractions in the numerator of Q before we calculate, to get

$$Q = \frac{1}{\Delta x}\left(\frac{x - (x + \Delta x)}{(x + \Delta x)x}\right) = \frac{1}{\Delta x}\left(\frac{-\Delta x}{(x + \Delta x)x}\right) = -\frac{1}{(x + \Delta x)x}. \tag{3}$$

Now, evaluation gives

$$Q = -\frac{1}{(0.7 + 10^{-8})(0.7)} = -2.0408\ 16298, \tag{4}$$

which is correct to 10 digits. If we just subtract, as we did in Eq. (2), we lose 8 digits and cannot hope to be anywhere near right.

At other times, the errors are an unavoidable consequence of how the calculator works, although some calculators work better than others. When we calculated the tangent of 89.99999° on three popular models, we got three different answers:

Calculator A: $\tan 89.99999° = 5729545$

Calculator B: $\tan 89.99999° = 5729577$

Calculator C: $\tan 89.99999° = 5729580.$

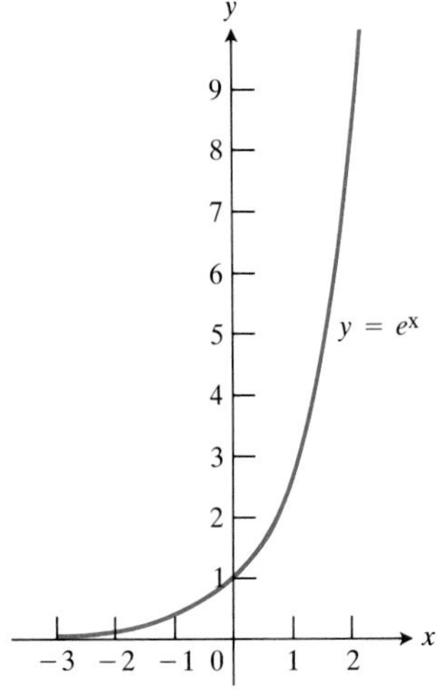

1.92 The graph of $y = e^x$.
Domain: $-\infty < x < \infty$
Range: $y > 0$

Calculator B was the only one to return an answer that was correct in all seven places.

In this book, we shall not ask for calculations that push machines over the edge this way, unless the purpose of the exercise is to explore a calculator's limitations or to reveal the risk of attempting a particular kind of computation.

Exponential and Logarithmic Functions In addition to having keys for trigonometric functions and functions like x^2, $\sqrt{x}$, and $1/x$, scientific calculators have keys for

e^x (the exponential function e to the x)

$\ln x$ (the natural logarithm of x)

$\log x$ (the base-10 logarithm of x).

The number e is about 2.7 1828 1828 and the function e^x plays a key role in describing growth, as does the natural logarithm function. The graphs of $y = e^x$ and $y = \ln x$ are shown in Figs. 1.92 and 1.93. Since $\log x = (\ln x)/\ln 10$, as we shall explain later in the book, the graph of $\log x$ is just a scaled-down version of the graph of $\ln x$. We shall study these and a number of other so-called transcendental functions in Chapter 7.

As you look at Figs. 1.92 and 1.93, notice the difference in the rates at which e^x and $\ln x$ grow as x increases. The function e^x grows rapidly—exponentially, in fact (which is where the adverb comes from). As we shall see, e^x grows faster as x increases than any power of x, even $x^{1,000,000}$. In contrast, the natural logarithm $\ln x$ grows more slowly as x increases than any fractional power of x, even $x^{1/1,000,000}$. As we shall see in Chapter 7, the functions e^x and $\ln x$ provide the standards by which we measure the growth rates of other functions.

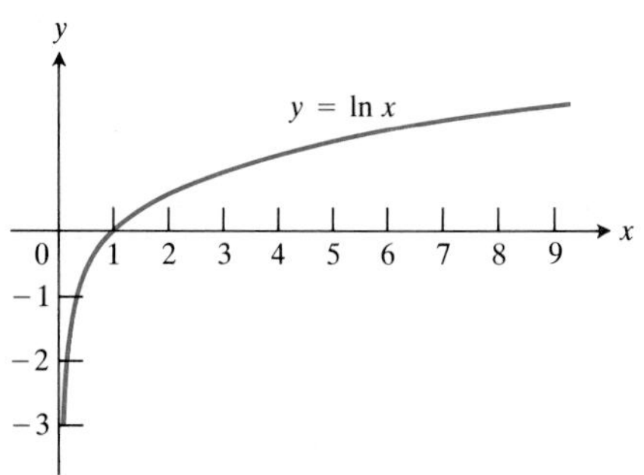

1.93 The graph of $y = \ln x$.
Domain: $x > 0$
Range: $-\infty < y < \infty$

Solving Equations

Calculus provides the mathematical justification for many of the things we do with a calculator, including the ways we solve equations.

The solutions of an equation $f(x) = 0$ are the x-coordinates of the points where the graph of the equation $y = f(x)$ touches or crosses the x-axis. On a graphing calculator we can try to find these points by displaying the graph of f and reading the coordinates of the points of contact.

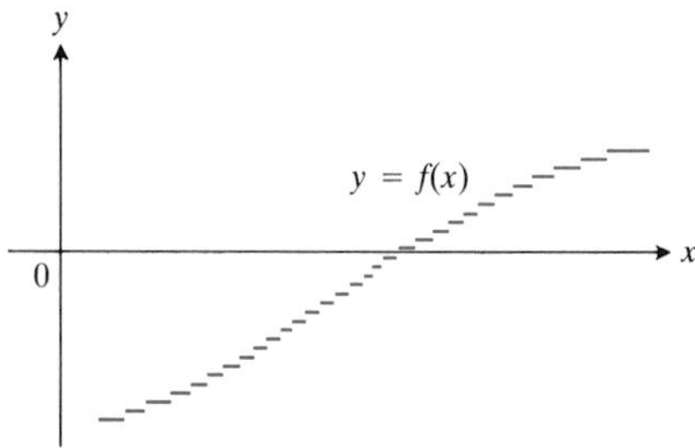

1.94 The graph of f steps across the axis without touching it. Hence, f changes from negative to positive without becoming zero in between. The equation $f(x) = 0$ has no solution.

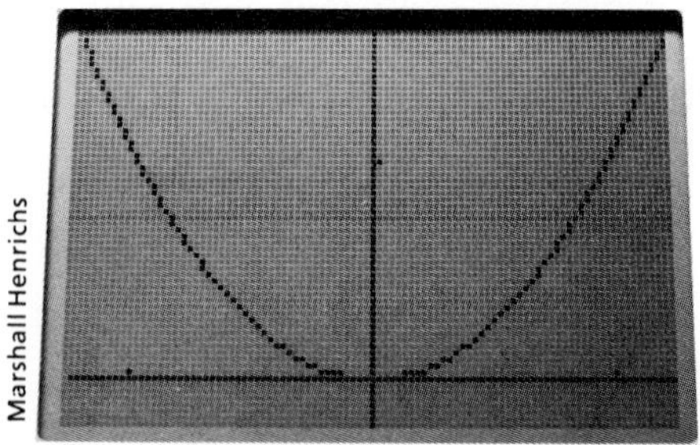

1.95 On a graphing calculator, the graph of $y = x^2 + 0.0001$ appears to pass through the origin even though y is never zero.

At least, that's the idea. But is there really a solution there? Does a curve that crosses the axis really touch the axis as it crosses? Perhaps it steps across the axis instead, without touching it, like the miniature step function in Fig. 1.94.

Also, does a curve that appears to be tangent to the axis really touch it? The function $y = x^2 + 0.0001$ is never zero, but this fact is certainly not revealed by the calculator screen in Fig. 1.95.

We answer these and many other questions about graphs by applying the theorems of calculus, as you will see in the next few chapters. Calculus will tell us when there is a solution (and when there isn't) and how many solutions to look for. It will also furnish us with a first-rate method for calculating solutions to as many decimal places as we please.

Comparing Function Values

In the 1640's, Henry Bond, a British teacher of surveying and navigation, discovered a close agreement between the tables of values of two functions used in navigation. In 1645, he published his conjecture that the two functions were equal, in Norwood's *Epitome of Navigation*. The conjecture became widely known, and in 1666 the mathematician Nicolas Mercator (no relation to the map maker) offered a prize to anyone who could settle it.

It was finally settled by James Gregory in 1668, but with a geometric proof so long that even the genius Edmund Halley (of comet fame) found it nearly too tiresome to read. The first intelligible proof involved calculus and was published by Newton's university mathematics teacher Isaac Barrow in 1669. One of the functions in question was

$$f(x) = \ln\left|\tan\left(\frac{x}{2} + \frac{\pi}{4}\right)\right|.$$

Barrow accomplished his proof by showing that the other function was

$$g(x) = \ln|\sec x + \tan x|,$$

which can be transformed into f with trigonometric identities.

The point here is that Bond was led to conjecture the equality of the two functions by comparing tables of their values. Today, we can make such comparisons at will with a calculator. If we suspect that two functions f and g are equal, or that they differ, say, by some constant value, we can test the possibility by calculating $f(x) - g(x)$ for a number of values of x. If the difference varies from one value of x to another, our suspicion is unfounded. But if the calculated differences are equal, or very nearly so, we may be on to something. Of course, knowing that $f(x) - g(x) = C$ for a few values of x does not prove equality for all values of x. But it makes the possibility worth investigating and gives a target value for C. You will find some nice examples in Exercises 7–12.

Successive Square Roots

Calculators can also reveal interesting behavior on the part of an individual function. If we start with a positive number like $x = 2$, and press the square root key repeatedly, here is what we find:

$$
\begin{aligned}
x_0 &= 2 & x_7 &= \sqrt{x_6} = 1.005429901 \\
x_1 &= \sqrt{x_0} = 1.414213562 & x_8 &= \sqrt{x_7} = 1.002711275 \\
x_2 &= \sqrt{x_1} = 1.189207115 & x_9 &= \sqrt{x_8} = 1.00135472 \\
x_3 &= \sqrt{x_2} = 1.090507733 & x_{10} &= \sqrt{x_9} = 1.000677131 \\
x_4 &= \sqrt{x_3} = 1.044273782 & x_{11} &= \sqrt{x_{10}} = 1.000338508 \\
x_5 &= \sqrt{x_4} = 1.021897149 & x_{12} &= \sqrt{x_{11}} = 1.00016924 \\
x_6 &= \sqrt{x_5} = 1.010889286 & x_{13} &= \sqrt{x_{12}} = 1.000084616.
\end{aligned}
$$

We could continue this further, but two patterns are already emerging. One is that the numbers seem to be approaching the value 1. Indeed they must, as we shall see in Chapter 9. The other pattern, less obvious perhaps, but more intriguing, is that each keypress divides the decimal part of the number nearly in half. This behavior, too, is explained by calculus, as we shall see in Chapter 3.

These two phenomena have nothing particular to do with the starting number 2. Taking successive square roots of any number greater than 1 brings you ever closer to 1 by dividing the decimal part roughly in half with each keypress.

Is This Idea Any Good?

One of the ways Fermat tried to define the slope of a curve at a point P was to run a secant line through P and a nearby point Q and then watch the slope of the secant line as Q approached P (see Fig. 1.96). Every time he did this, the secant slopes seemed to approach a particular value that depended only on the location of P, so he wanted to call this value the slope of the curve at P. Was this a good idea?

To find out, we might try it at a particular point on a particular curve, say at the point $P(0, 0)$ on the curve $y = \sin x$. In the notation of Fig. 1.97, the point Q is the point $(\Delta x, \sin \Delta x)$. The slope of the secant line is therefore

$$m_{\text{sec}} = \frac{\sin \Delta x - 0}{\Delta x - 0} = \frac{\sin \Delta x}{\Delta x}. \tag{5}$$

What happens to the value of this ratio as Q approaches P along the curve?

1.96 Fermat wanted to find out whether the slope of the secant shown here would approach a fixed value as Q approached P. If it did, he would call this value the slope of the curve at P.

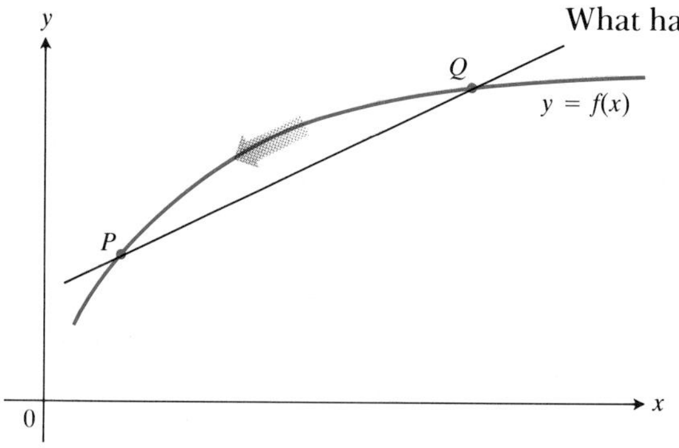

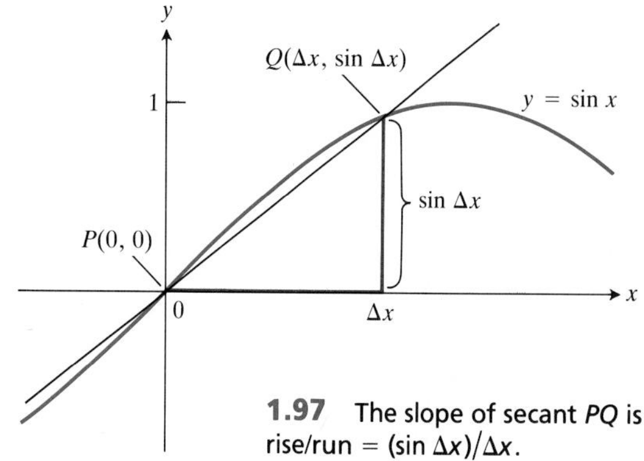

1.97 The slope of secant PQ is rise/run $= (\sin \Delta x)/\Delta x$.

To see what to expect, we might try it with a calculator, starting with $\Delta x = 1$ and dividing by 10 each time to get $\Delta x = 0.1, 0.01$, and so on. If we do this with the calculator in radian mode, we get the following values.

Δx (radians)	$(\sin \Delta x)/\Delta x$	
1.0	.8414709848	
.1	.998334166	
.01	.999998333	
.001	.9999998	
.0001	.999999	
.00001	1.0	(The calculator thinks that sin(0.00001) = 0.00001.)
.00000 1	1	
.00000 01	1	
.00000 001	1	
.00000 0001	1	
.00000 00001	1	
.00000 00000 1	0	(Δx is now so small that the calculator thinks $\sin(\Delta x) = 0$.)

What are we to make of these numbers? It looks as if the values of $(\sin \Delta x)/\Delta x$ are steadily approaching 1. At least they do so until we reach the limit of our calculator's ability to distinguish $\sin \Delta x$ from Δx. Then the numbers remain at 1 until we reach the limit of the calculator's ability to distinguish $\sin \Delta x$ from zero.

If the calculator could work with arbitrarily many decimal places, would the ratios approach 1 forever? Would it be a good idea to define the slope of the sine curve at $x = 0$ to be 1? The answers, obtained from calculus, as you will soon see, are both "yes."

Exercises 1.7

1. a) For how large a value of x can your calculator evaluate e^x?
   b) Calculate e^x for $x = -1, -10, -100$, and -1000. How far to the left of zero can your calculator evaluate e^x without getting 0 or an error message?

2. a) For how large a value of x can your calculator evaluate $\ln x$? How far above the x-axis is the graph of $y = \ln x$ at this point?
   b) The function $y = \ln x$ is defined for all positive values of x. What is the smallest value of x for which your calculator can find $\ln x$? How far below the x-axis is the graph of $y = \ln x$ at this point? How close is the graph to the y-axis there?

3. **GRAPHING CALCULATOR** Graph $y = \sin x$ and estimate the value x^* where the graph first crosses the positive x-axis. To 9 decimal places, $\pi = 3.141592654$.

4. **GRAPHING CALCULATOR** The equation $x^3 - x - 1 = 0$ has one solution between $x = 1$ and $x = 2$. Graph $y = x^3 - x - 1$ over the interval $1 \le x \le 2$ and estimate the value x^* where the graph crosses the x-axis. To 9 decimal places, $x^* = 1.324717957$.

5. Evaluate the composite function
$$y = e^{(\ln x)}$$
for various values of x. What do you find? (This will be explained in Chapter 5.)

6. Evaluate the composite
$$y = \ln(e^x)$$
for various values of x. What do you find? (This will be explained in Chapter 5.)

In Exercises 7–12, test the functions to indicate whether they could differ by a constant value on some domain of x-values. If they appear to do so, see if you can prove that they really do and find an appropriate domain for x.

7. $\dfrac{x}{x+1}$ and $\dfrac{-1}{x+1}$

8. $\dfrac{x^2+3}{x^2+1}$ and $\dfrac{2}{x^2+1}$

9. $\tan x \sin 2x$ and $-2\cos^2 x$

10. $\cos^2 x$ and $\cos 2x$ **11.** $\ln 2x$ and $\ln x$

12. x and $\sqrt{x^2}$

13. *Successive square roots*

a) Enter the number 3 into your calculator. Then press the square-root key repeatedly, pausing between keypresses to read the display. With each keypress you will find the decimal part of the display approximately halved.

b) Repeat Part (a) with other numbers greater than 1.

14. *Continuation of Exercise 13.* One way to describe the halving of the decimal parts of the square roots in Exercise 13 is to say that the number-line distance between 1 and the square root is approximately halved each time. What happens if you start with a positive number that is less than 1 instead of greater than 1? Do successive square roots approach 1 the same way? Try it with $x = 0.5$.

15. *Successive tenth roots.* If you have a $\boxed{\sqrt[x]{\ }}$ key or some other key that will enable you to calculate tenth roots, calculate successive tenth roots of 2, pausing to view each new display. What pattern do you see? (This, too, will be explained in Chapter 3.)

16. *Continuation of Exercise 15.* Repeat Exercise 15, starting with 0.5 instead of 2. What pattern do you see now?

17. *A lightning-fast estimate of $\pi/2$.* Set your calculator in radian mode and enter the number $x_0 = 1$. Then calculate the succesive numbers $x_1, x_2, \ldots$ in the following list.

$$\begin{aligned} x_0 &= 1 \\ x_1 &= 1 + \cos 1 \\ x_2 &= x_1 + \cos x_1 \\ x_3 &= x_2 + \cos x_2 \\ x_4 &= x_3 + \cos x_3 \\ &\vdots \\ x_n &= x_{n-1} + \cos x_{n-1} \quad \begin{pmatrix}\text{Formula for} \\ \text{generating the} \\ \text{sequence.}\end{pmatrix} \end{aligned}$$

The numbers you calculate will soon begin to repeat at the value $x^* = 1.5707\ 96327$, which is $\pi/2$ to nine decimal places. Figure 1.98 explains why.

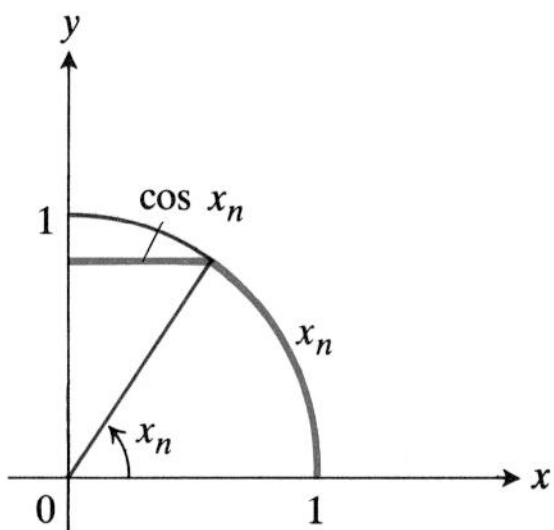

1.98 The length $\pi/2$ of the circular arc is approximated by $x_n + \cos x_n$ (Exercise 17).

18. *The function x^x (for calculators with an $\boxed{x^y}$ key).* The rules of exponents tell us that $a^0 = 1$ if a is any number different from zero. They also tell us that $0^n = 0$ if n is any positive number.

If we tried to extend these rules to include the case 0^0, we would get conflicting results. The first rule would say $0^0 = 1$ while the second would say $0^0 = 0$.

We are not dealing with a question of right or wrong here. Neither rule applies as it stands, so there is no contradiction to deal with. We could, in fact, define 0^0 to have any value we wanted as long as we could sell other people on our idea.

What value would you like 0^0 to have? Here are some calculations that might help you decide: Calculate x^x for $x = 0.1, 0.01, 0.001$, and so on, as far as your calculator can go. Write down the value you get each time. What pattern do you see?

We shall say more about the function $y = x^x$ in Chapter 7.

19. *Radians vs. Degrees.* What happens if you use degrees instead of radians to calculate the slope of the curve $y = \sin x$ at $x = 0$? To find out, set your calculator in degree mode and make a table of values for

$$m_{\text{sec}} = \frac{\sin \Delta x}{\Delta x}.$$

Use $\Delta x = 0.1, 0.01, 0.001$, and so on, as far as your calculator can go. What do you find? Now multiply each value of m_{sec} in your table by the number $180/\pi$. What do you find?

20. *The slope of $y = \ln x$ at $x = 1$.* In this exercise, we use Fermat's idea to find a numerical candidate for the slope of the curve $y = \ln x$ at the point $P(1, 0)$ where the graph crosses the x-axis (Fig. 1.99).

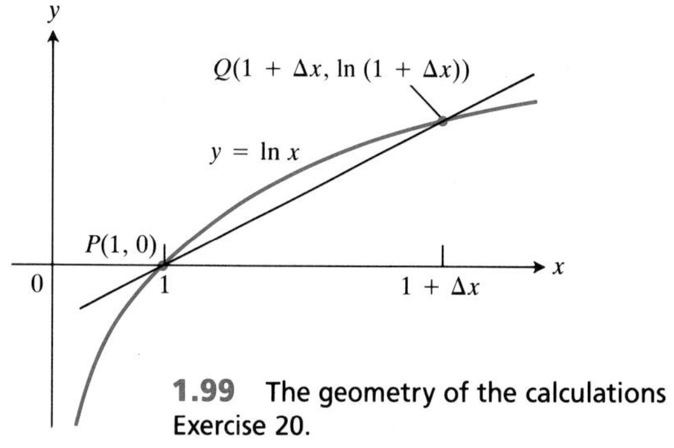

1.99 The geometry of the calculations in Exercise 20.

Make a table of values of

$$m_{\text{sec}} = \frac{\ln(1 + \Delta x) - 0}{(1 + \Delta x) - 1} = \frac{\ln(1 + \Delta x)}{\Delta x}$$

for $\Delta x = 0.1, 0.01, 0.001$, and so on, as far as your calculator can go. As far as you can tell, what would be a good value for the curve's slope at $x = 1$?

Stairways

Exercises 21–23 are about the stairway shown in Fig. 1.100. The slope of the stairway can be calculated, from the riser height R and the tread width T, as R/T. The manual from which the drawing was adapted defines a stairway as a stepped footway having a slope not less than 5:16 or 31 1/4%, and not greater than 9:8 or 112 1/2%. (The man ual goes on to say that below these limits footways become ramps. Above these limits, footways become stepladders.)

21. A common angle for household stairs is 40°. Find the slope of a 40° stairway.

22. If the treads on a 40° stairway are 9 in. wide, about how high are the risers?

23. (An inv or arc key is helpful but not necessary.) What are the maximum and minimum angles in Fig. 1.100 allowed by the definition of *stairway*?

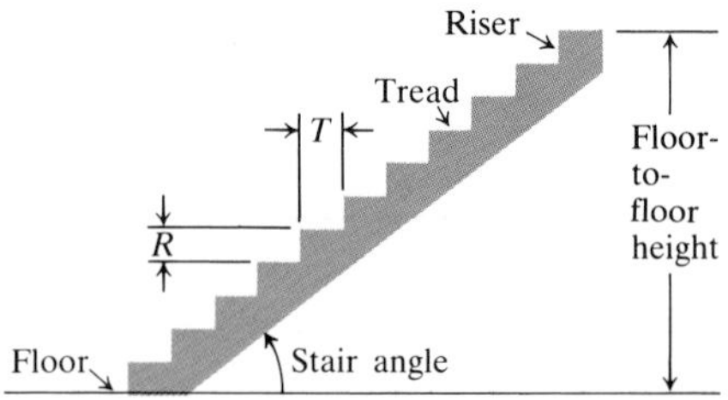

1.100 The stairway in Exercises 21–23.

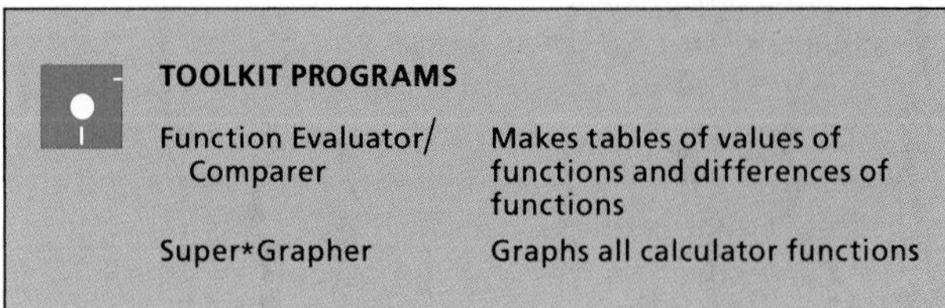

Review Questions

1. How do you find the distance between two points in the xy-plane? between a point and a line in the plane? Give examples.

2. What are the basic steps in graphing an equation in x and y? Illustrate them.

3. What coordinate tests determine whether a graph in the xy-plane is symmetric with respect to the coordinate axes or the origin? Give examples.

4. How can you write the equation for a line if you know the coordinates of two points on the line? The line's slope and the coordinates of one point on the line? The line's slope and y-intercept? Give examples.

5. What are the standard equations for lines perpendicular to the coordinate axes?

6. How are the slopes of mutually perpendicular lines related? Give examples.

7. When a line is not vertical, what is the relation between its slope and its angle of inclination?

8. What is a function? Give examples. How do you graph a real-valued function of a real variable?

9. Name some typical functions and draw their graphs.

10. What is an even function? An odd function? What symmetries do the graphs of such functions have? Give examples. Give an example of a function that is neither odd nor even.

11. When is it possible to compose one function with another? Give examples of composites and their values at various points. Does the order in which functions are composed ever matter?

12. How can you write an equation for a circle in the xy-plane if you know its radius a and the coordinates (h, k) of its center? Give examples.

13. What inequality is satisfied by the coordinates of the points that lie inside the circle of radius a centered at (h, k)? What inequality is satisfied by the coordinates of the points that lie outside this circle?

14. The graph of a function $y = f(x)$ in the xy-plane is shifted 5 units to the left and then 3 units straight up. Write an equation for the new graph.

15. What is a parabola? What are typical equations for parabolas? What reflective property do parabolas have?

16. How do you convert between degree measure and radian measure? Give examples.

17. Graph the six basic trigonometric functions as functions of radian measure. What symmetries do the graphs have?

18. What does it mean for a function $y = f(x)$ to be periodic? Give examples of functions with various periods. Name some real-world phenomena that we model with periodic functions.

19. List the angle sum and difference formulas for the sine and cosine functions.

20. List the four basic double-angle formulas for sines and cosines.

21. Define the function $y = |x|$. Give examples of numbers and their absolute values. How are $|-a|$, $|ab|$, $|a/b|$, and $|a+b|$ related to $|a|$ and $|b|$?

22. How are absolute values used to describe intervals of real numbers?

23. Show by example how absolute values are used to control function values.

Practice Exercises

In Exercises 1–4, find the points that are symmetric to the given point (a) across the x-axis, (b) across the y-axis, and (c) across the origin.

1. $(1, 4)$ **2.** $(2, -3)$ **3.** $(-4, 2)$ **4.** $(-2, -2)$

Test the equations in Exercises 5–8 to find out whether their graphs are symmetric with respect to the axes or the origin.

5. a) $y = x$ b) $y = x^2$

6. a) $y = x^3$ b) $y = x^4$

7. a) $x^2 - y^2 = 4$ b) $x - y = 4$

8. a) $y = x^{1/3}$ b) $y = x^{2/3}$

Find equations for the vertical and horizontal lines through the points in Exercises 9–12.

9. $(1, 3)$ **10.** $(2, 0)$ **11.** $(0, -3)$ **12.** (x_0, y_0)

In Exercises 13–20, write an equation for the line that passes through point P with slope m. Then use the equation to find the line's intercepts and graph the line.

13. $P(2, 3)$, $m = 2$ **14.** $P(2, 3)$, $m = 0$

15. $P(1, 0)$, $m = -1$ **16.** $P(0, 1)$, $m = -1$

17. $P(1, -6)$, $m = 3$ **18.** $P(-2, 0)$, $m = 1$

19. $P(-1, 2)$, $m = -\frac{1}{2}$ **20.** $P(3, 1)$, $m = \frac{1}{3}$

In Exercises 21–24, find an equation for the line through the two points.

21. $(-2, -2)$, $(1, 3)$ **22.** $(-3, 6)$, $(1, -2)$

23. $(2, -1)$, $(4, 4)$ **24.** $(3, 3)$, $(-2, 5)$

In Exercises 25–28, find an equation for the line with the given slope m and y-intercept b.

25. $m = \frac{1}{2}$, $b = 2$ **26.** $m = -3$, $b = 3$

27. $m = -2$, $b = -1$ **28.** $m = 2$, $b = 0$

In Exercises 29–32: (a) Find an equation for the line through P parallel to L. (b) Then find an equation for the line through P perpendicular to L and the distance from P to L.

29. $P(6, 0)$, $L: 2x - y = -2$

30. $P(3, 1)$, $L: y = x + 2$

31. $P(4, -12)$, $L: 4x + 3y = 12$

32. $P(0, 1)$, $L: y = -\sqrt{3}x - 3$

To be sure you are familiar with the functions in Exercises 33–48, see if you can sketch their graphs from memory. State the function's domain and range with each graph.

33. $y = x^2$ **34.** $y = x^3$ **35.** $y = x^{2/3}$

36. $y = \sqrt{x}$ **37.** $y = \sqrt[3]{x}$ **38.** $y = \frac{1}{x}$

39. $y = \frac{1}{x^2}$ **40.** $y = x^{3/2}$ **41.** $y = \lfloor x \rfloor$

42. $y = \lceil x \rceil$ **43.** $y = \cos x$ **44.** $y = \sin x$

45. $y = \tan x$ **46.** $y = \cot x$ **47.** $y = \sec x$

48. $y = \csc x$

In Exercises 49–54, say whether the functions are even, odd, or neither.

49. a) $y = \cos x$ b) $y = -\cos x$ c) $y = 1 - \cos x$

50. a) $y = \sin x$ b) $y = -\sin x$ c) $y = 1 - \sin x$

51. a) $y = x^2 + 1$ b) $y = x$ c) $y = x(x^2 + 1)$

52. a) $y = x^3$ b) $y = -x$ c) $y = -x^4$

53. a) $y = \sec x$ b) $y = \tan x$ c) $y = \sec x \tan x$

54. a) $y = \csc x$ b) $y = \cot x$ c) $y = \csc x \cot x$

55. Graph the function $y = x - \lfloor x \rfloor$. Is the function periodic? If so, what is its smallest period?

56. Graph the function $y = \lceil x \rceil - \lfloor x \rfloor$. Is the function periodic? If so, what is its smallest period?

Graph the functions in Exercises 57–60.

57. $y = \begin{cases} \sqrt{-x}, & -4 \leq x \leq 0 \\ \sqrt{x}, & 0 \leq x \leq 4 \end{cases}$

58. $y = \begin{cases} -x - 2, & -2 \leq x \leq -1 \\ x, & -1 \leq x \leq 1 \\ -x + 2, & 1 \leq x \leq 2 \end{cases}$

59. $y = \begin{cases} \sin x, & 0 \leq x \leq 2\pi \\ 0, & 2\pi < x \end{cases}$

60. $y = \begin{cases} \cos x, & 0 \leq x \leq 2\pi \\ 0, & 2\pi < x \end{cases}$

Write "two-part" formulas for the functions graphed in Exercises 61 and 62.

61.

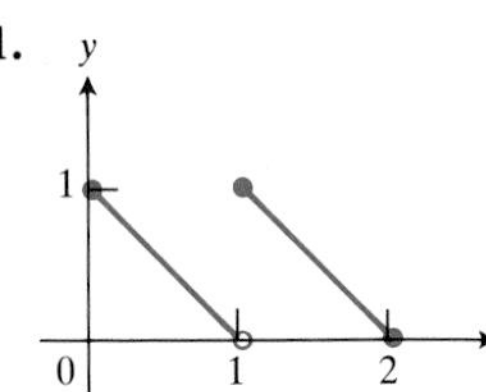

62.

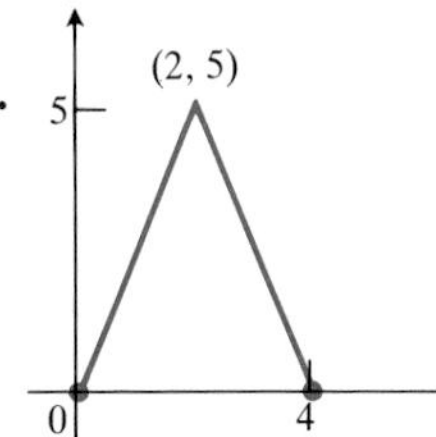

In Exercises 63 and 64, find the domains and ranges of f, g, $f + g$, $f \cdot g$, f/g, and g/f. Also, find the domains and ranges of the composites $f \circ g$ and $g \circ f$.

63. $f(x) = \dfrac{1}{x}$, $g(x) = \dfrac{1}{\sqrt{x}}$

64. $f(x) = \sqrt{x}$, $g(x) = \sqrt{1 - x}$

Exercises 65–70 tell how many units and in what directions the graphs of the given equations are to be shifted. Give an equation for the shifted graph in each case.

65. $y = x^2$ a) Down 1, right 2 b) Left 2

66. $x = y^2$ a) Left 2 b) Up 2, right 2

67. $y = \sin x$ a) Right $\dfrac{\pi}{2}$ b) Up 1, right π

68. $y = \cos x$ a) Left π b) Down 1, left $\dfrac{\pi}{2}$

69. $y = \sqrt{x}$ a) Right 2 b) Left 2

70. $y = x^3$ a) Up 1, right 1 b) Left 1

In Exercises 71–74, write an equation for the circle with the given center (h, k) and radius a.

71. $(h, k) = (1, 1)$, $a = 1$ **72.** $(h, k) = (2, 0)$, $a = 5$

73. $(h, k) = (2, -3)$, $a = \dfrac{1}{2}$ **74.** $(h, k) = (-3, 0)$, $a = 3$

Identify the centers and radii of the circles in Exercises 75–78.

75. $(x - 3)^2 + (y + 5)^2 = 16$

76. $x^2 + (y - 5)^2 = 2$

77. $(x + 1)^2 + (y - 7)^2 = 121$

78. $(x + 4)^2 + (y + 1)^2 = 81$

Write inequalities to describe the regions in Exercises 79 and 80.

79. a) The interior of the circle of radius 1 centered at the origin.
b) The region consisting of the circle plus its interior.

80. a) The exterior of the circle of radius 2 centered at the point (1, 1).
b) The region consisting of the circle plus its exterior.

In Exercises 81–84, write an equation for the parabola with the given focus and directrix.

81. Focus: (0, 2) Directrix: $y = -2$

82. Focus: $\left(0, \dfrac{1}{16}\right)$ Directrix: $y = -\dfrac{1}{16}$

83. Focus: $\left(0, -\dfrac{1}{4}\right)$ Directrix: $y = \dfrac{1}{4}$

84. Focus: (0, −7) Directrix: $y = 7$

Find the foci and directrices of the parabolas in Exercises 85–88.

85. $y = x^2/2$ **86.** $y = x^2/12$

87. $y = -x^2/4$ **88.** $y = -x^2$

89. Change from degrees to radians:
a) 30° b) 45° c) −120° d) −150°

90. Change from radians to degrees:
a) $\dfrac{3\pi}{2}$ b) $-\dfrac{\pi}{4}$ c) $\dfrac{3\pi}{4}$ d) $-\dfrac{5\pi}{4}$

Find the sine, cosine, tangent, cotangent, secant, and cosecant of the angles in Exercises 91 and 92. The angles are given in radian measure.

91. a) $\dfrac{\pi}{3}$ b) $-\dfrac{\pi}{3}$ c) $\dfrac{2\pi}{3}$ d) $-\dfrac{2\pi}{3}$

92. a) $\dfrac{\pi}{4}$ b) $-\dfrac{\pi}{4}$ c) $\dfrac{3\pi}{4}$ d) $-\dfrac{3\pi}{4}$

93. Graph the following functions side by side over the interval $0 \leq x \leq 2\pi$.
a) $y = \cos 2x$ b) $y = 1 + \cos 2x$ c) $y = \cos^2 x$

94. Graph the following functions side by side over the interval $0 \leq x \leq 2\pi$.
a) $y = \cos 2x$ b) $y = -\cos 2x$
c) $y = 1 - \cos 2x$ d) $y = \sin^2 x$

95. Evaluate cos 15° as cos (60° − 45°).

96. Evaluate sin 75° as sin (45° + 30°).

97. Evaluate $\sin \dfrac{\pi}{12}$ as $\sin\left(\dfrac{\pi}{3} - \dfrac{\pi}{4}\right)$.

98. Evaluate $\cos \frac{\pi}{12}$ as $\cos\left(\frac{\pi}{4} - \frac{\pi}{6}\right)$.

99. Find $\cos^2 \frac{\pi}{6}$

a) by finding $\cos \frac{\pi}{6}$ and squaring.

b) by using a double-angle formula.

100. Find $\sin^2 \frac{\pi}{4}$

a) by finding $\sin \frac{\pi}{4}$ and squaring.

b) by using a double-angle formula.

Solve the equations in Exercises 101–104.

101. $|x - 1| = \frac{1}{2}$ **102.** $|2 - 3x| = 1$

103. $\left|\frac{2x}{5} + 1\right| = 7$ **104.** $\left|\frac{5 - x}{2}\right| = 7$

Describe the intervals in Exercises 105–108 with inequalities that do not involve absolute values.

105. $|x + 2| \le \frac{1}{2}$ **106.** $|2x - 7| \le 3$

107. $\left|y - \frac{2}{5}\right| < \frac{3}{5}$ **108.** $\left|8 - \frac{y}{2}\right| < 1$

Describe the intervals in Exercises 109–112 with absolute value inequalities of the form $|x - x_0| < D$ or $|y - y_0| < E$.

109. $3 < x < 11$ **110.** $-4 < x < 0$

111. $-1 < y < 7$ **112.** $-1 < y < 6$

Each of Exercises 113–116 gives a function $y = f(x)$, a number E, and a target value y_0. In each case, find the interval of x-values for which $y = f(x)$ lies within E units of y_0. Then describe the interval with an absolute value inequality of the form $|x - x_0| < D$.

	$f(x)$	E	y_0
113.	$2x - 3$	2	1
114.	$4 - x$	2	0
115.	$\sqrt{x + 2}$	1	4
116.	$\sqrt{\frac{x + 1}{2}}$	$\frac{1}{2}$	1

Graph the equations in Exercises 117–122.

117. $y = -|x|$ **118.** $y = |x - 2|$

119. $y = |x| - 2$ **120.** $y = |x - 2|/(x - 2)$

121. $y = |x + 2| + x$ **122.** $y = |x + 2| + |x|$

123. Match the following equations with the graphs in Fig. 1.101.

a) $y = |\cos x|$

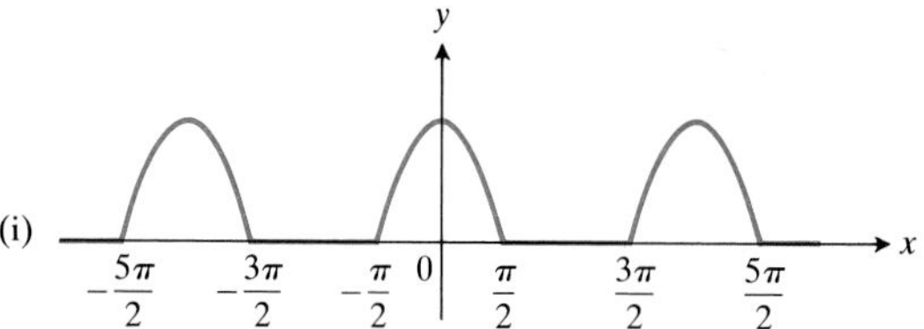

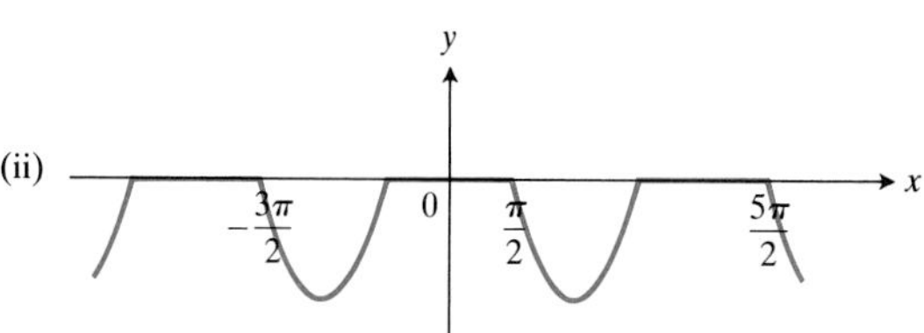

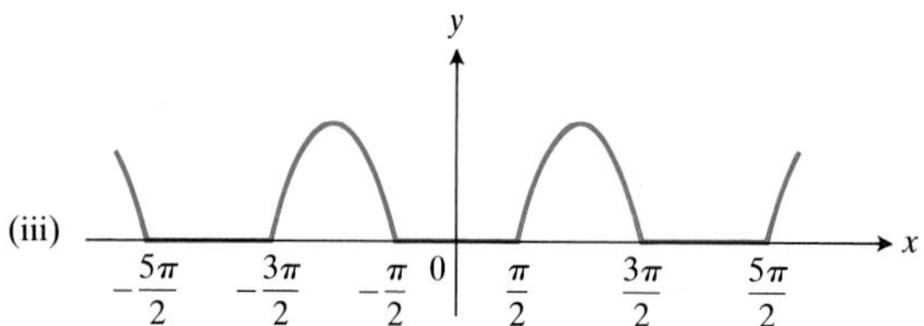

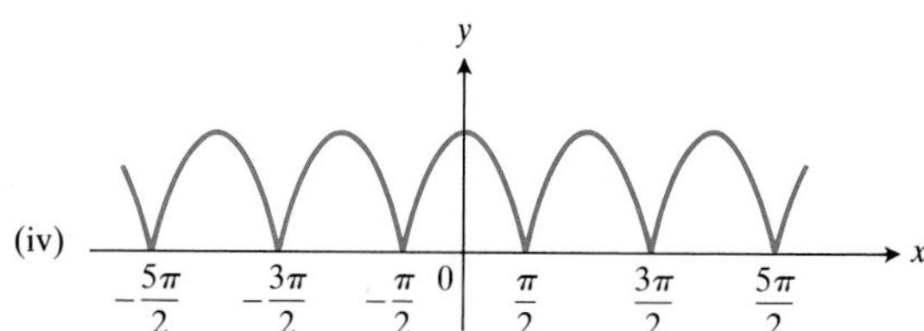

1.101 The graphs for Exercise 123.

b) $y = \frac{\cos x + |\cos x|}{2}$

c) $y = \frac{|\cos x| - \cos x}{2}$

d) $y = \frac{\cos x - |\cos x|}{2}$

124. Graph the equation $|x| + |y| = 1$. (*Hint:* Work quadrant by quadrant, each time replacing the equation by an equivalent equation without absolute values. In the first quadrant, the equation becomes $x + y = 1$. In the second, it becomes $-x + y = 1$, and so on.)

125. *Controlling the flow from a draining tank.* Torricelli's law says that if you drain a tank like the one in Fig. 1.102, the rate y at which the water runs out is a constant times the square root of the water's depth. As

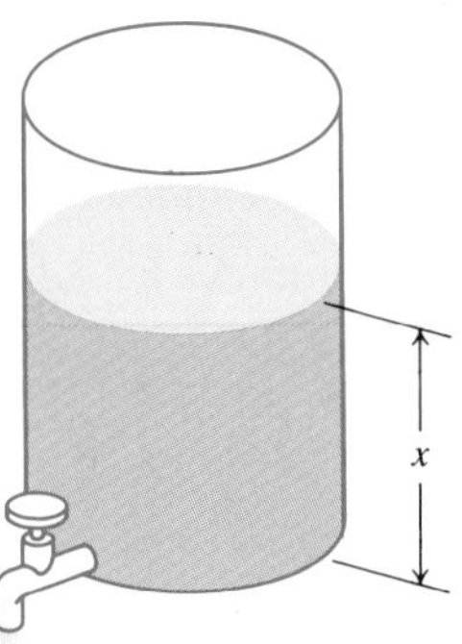

1.102 The tank in Exercise 125.

the tank drains, x decreases, and so does y, but y decreases less rapidly than x. The value of the constant depends on the size of the exit valve.

Suppose that for the tank in question, $y = \sqrt{x}/2$. You are trying to maintain a constant exit rate of $y_0 = 1$ ft^3/min by refilling the tank with a hose from time to time. How deep must you keep the water to hold the rate to within 0.2 ft^3/min of $y_0 = 1$? Within 0.1 ft^3/min of $y_0 = 1$? In other words, in what interval must you keep x to hold y within 0.2 (or 0.1) units of $y_0 = 1$?

Remark: What if we want to know how long it will take the tank to drain if we do not refill it? We cannot answer such a question with the usual equation time = amount/rate, because the rate changes as the tank drains. We could always open the valve, sit down with a watch, and wait; but with a large tank or a reservoir that might take hours or even days. With calculus, we will be able to find the answer in just a minute or two, as you will see if you do Exercise 45 in Section 4.6.

126. *Dimension changes in equipment.* As you probably know, most metals expand when heated and contract when cooled, and people sometimes have to take this into account in their work. Boston and Maine Railroad crews try to lay track at temperatures as close to 65°F as they can, so the track won't expand too much in the summer or shrink too much in the winter. Surveyors have to correct their measurements for temperature when they use steel measuring tapes.

The dimensions of a piece of laboratory equipment are often so critical that the machine shop in which it is made has to be held at the same temperature as the laboratory where the part is to be installed. And, once the piece is installed, the laboratory must continue to be held at that temperature.

A typical aluminum bar that is 10 cm wide at 70°F will be

$$y = 10 + (t - 70) \times 10^{-4}$$

centimeters wide at a nearby temperature t. As t rises above 70, the bar's width increases; as t falls below 70, the bar's width decreases.

Suppose you had a bar like this made for a gravity-wave detector you were building. You need the width of the bar to stay within 0.0005 cm of the ideal 10 cm. How close to 70°F must you maintain the temperature of your laboratory to achieve this? In other words, how close to $t_0 = 70$ must you keep t to be sure that y lies within 0.0005 of $y_0 = 10$?

Wide World Photos

If you don't think expansion can make a difference, look at these railroad tracks warped by 113° heat in Perugia, Italy. (Source: *The Boston Globe,* July 26, 1988, page 1.)

2 Limits and Continuity

NASA

No matter how good a lens you use, there will be a limit to what you can see.

OVERVIEW This chapter shows how limits of function values are defined and calculated.

Calculus is built on the notion of limit. The rules for calculating limits are straightforward and most of the limits we need can be found with a combination of direct substitution and algebra. Proving that the calculation rules always work, however, is a more subtle affair that requires a formal definition of limit. We present this definition in Section 2.5 and show there how it is used to justify the rules.

One of the most important uses of limits in calculus is to test functions for continuity. Continuous functions are widely used in science because they serve to model an enormous range of natural behavior. In Section 2.4, we will see what makes continuous functions special. We shall work mainly with continuous functions in this book.

2.1 Limits

If you invest \$100 at a fixed annual rate of 6% and interest is added to your account k times a year, the amount of money you will have when the year is up is

$$A = 100\left(1 + \frac{0.06}{k}\right)^k.$$

If the interest is added quarterly ("compounded quarterly," bankers say), then $k = 4$ and A will be about 106.14. By the end of the year, you will have earned \$6.14. If the interest is added monthly instead, then $k = 12$ and A will be about \$106.17. You will have earned \$6.17. The interest might be added daily ($k = 365$) or even more frequently, say by the hour, by the minute, or even by the second. But there is still a limit to how much you will earn, and the limit to the nearest tenth of a cent is \$6.184, as you will see in Chapter 7.

When you cough, your windpipe contracts to increase the speed of the air going out. Naturally, there is a limit to how fast the air can go—contract the windpipe a bit and the air speeds up, but contract it too much and the air will be choked off.

When a new species is introduced into a region, its population grows rapidly at first but then begins to level off as the members encounter predators and restrictions on food and space. As the days pass, the population, growing ever more slowly, approaches a limiting size, called the *carrying capacity* of the environment.

These and many other examples convince us that we live in a world of limits. There is a limit to how far a cannon can fire a shell. There is a limit to how fast we can brake for a red light. There is a limit to how much weight we can lift. There is a limit to how long a battery will last. And this is where the calculus comes in. Most of the limits that interest us can be viewed as numerical limits to values of functions. And, as you will see in this and subsequent chapters, calculus is the right mathematics for finding the limiting values of functions.

Examples of Limits

To start off on a slightly different tack, one of the important things to know about a function f is how its outputs will change when the inputs change. If the inputs get closer and closer to some specific value c, for example, will the outputs get closer to some specific value L? If they do, we want to know that, because it means we can control the outputs by controlling the inputs, the way we did in Section 1.6.

To talk sensibly about this, we need the language of limits. We develop this in two stages. First, we define limit informally, look at examples, and learn the calculation rules that have been discovered over the years to be the most useful ones to know. Then, in Section 2.5, when we have worked with enough examples for the formal definition to make sense, we define limit more precisely and examine the mathematics behind our calculations.

DEFINITION

Informal Definition of Limit

If the values of a function f of x approach the value L as x approaches c, we say f has **limit** L as x approaches c and we write

$\lim_{x \to c} f(x) = L$ (Read "The limit of f of x as x approaches c equals L".)

EXAMPLE 1 As Table 2.1 and Fig. 2.1 suggest,

$$\lim_{x \to 2} (2x + 1) = 5$$ ■

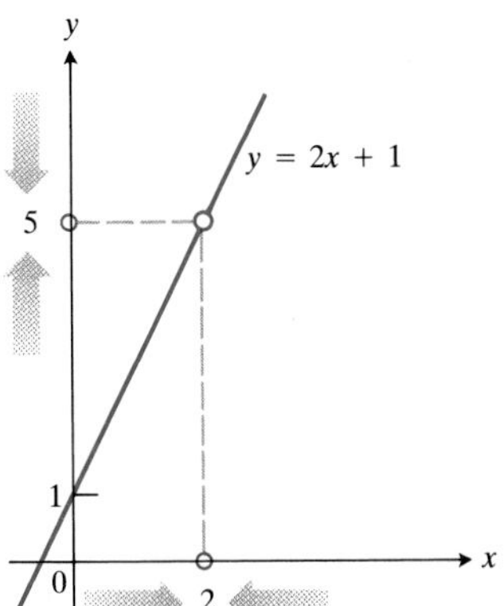

2.1 As the values of x approach 2, the values of $y = 2x + 1$ approach 5.

TABLE 2.1
Values of $f(x) = 2x + 1$ as $x \to 2$

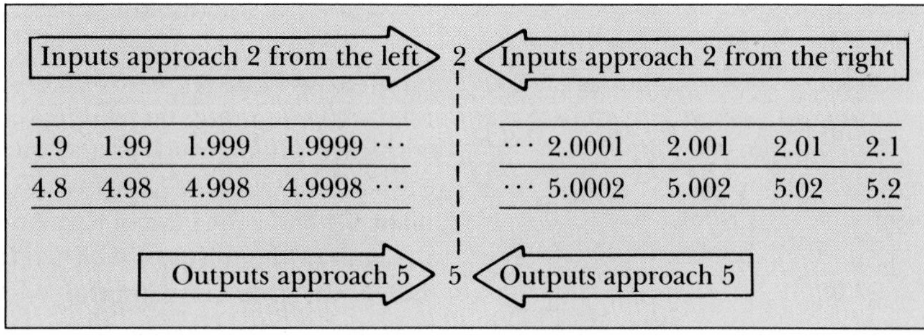

Inputs approach 2 from the left				2	Inputs approach 2 from the right			
1.9	1.99	1.999	1.9999 ···		··· 2.0001	2.001	2.01	2.1
4.8	4.98	4.998	4.9998 ···		··· 5.0002	5.002	5.02	5.2
Outputs approach 5				5	Outputs approach 5			

EXAMPLE 2 If f is the **identity function** $f(x) = x$, then for any value of c

$$\lim_{x \to c} f(x) = \lim_{x \to c} (x) = c$$

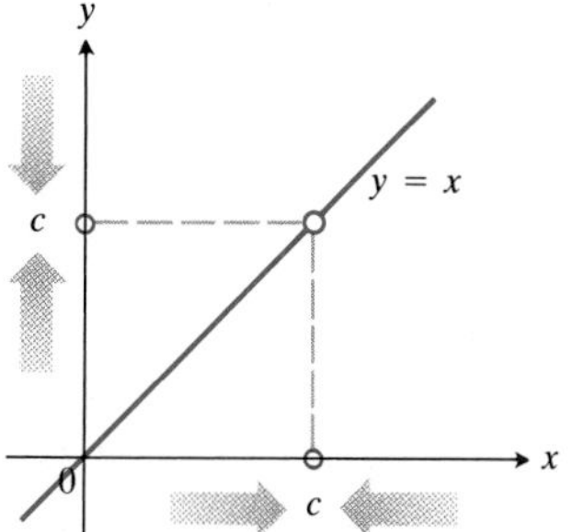

■

EXAMPLE 3 If f is the **constant function** $f(x) = k$ (the function whose outputs have the constant value k), then for any value of c

$$\lim_{x \to c} f(x) = \lim_{x \to c} (k) = k.$$

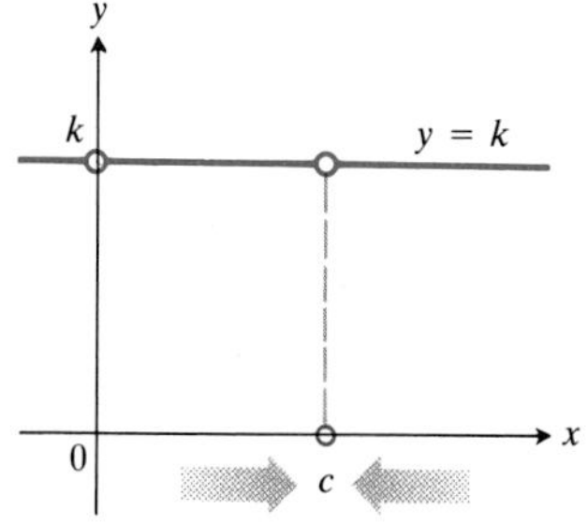

■

EXAMPLE 4 If c is any number,

$$\lim_{x\to c} 2x^2 = 2c^2.$$

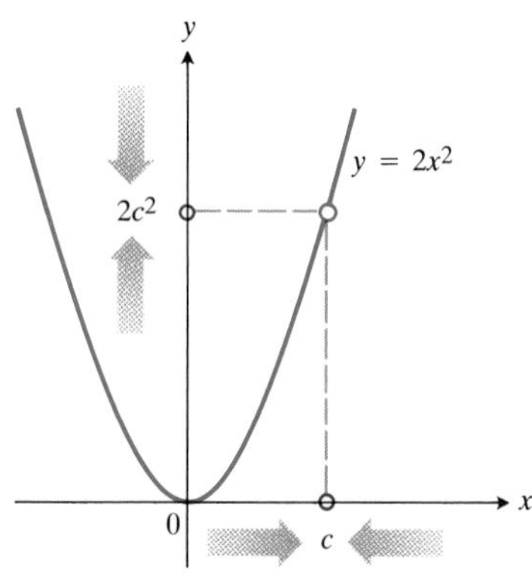

EXAMPLE 5 $\lim_{x\to -2} \frac{4}{x^2} = \frac{4}{(-2)^2} = 1$

x	$\frac{4}{x^2}$ (rounded)
−1.9	1.1080
−1.99	1.0100
−1.999	1.0010
−1.9999	1.0001
⋮	⋮
−2	1
⋮	⋮
−2.0001	0.99990
−2.001	0.99900
−2.01	0.99007
−2.1	0.90702

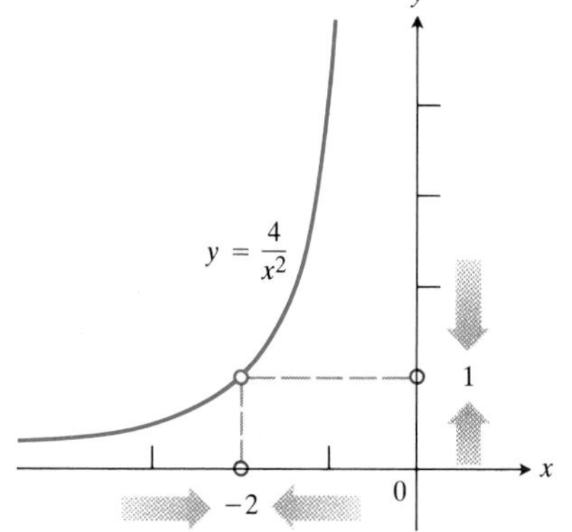

We could go on like this to make a huge list of functions and their limits, but we would be missing the boat if we did, because there is a more constructive (and much easier) way to proceed. For instance, once we know the limits of two particular functions as x approaches some value c, we automatically know the limit of their sum—it is the sum of their limits. Similarly, the limit of the difference of two functions is the difference of their limits, and so on, as described in the following theorem.

Properties of Limits

THEOREM 1

Properties of Limits

If $\lim_{x\to c} f_1(x) = L_1$ and $\lim_{x\to c} f_2(x) = L_2$, then

1. *Sum Rule:* $\lim\,[f_1(x) + f_2(x)] = L_1 + L_2$
2. *Difference Rule:* $\lim\,[f_1(x) - f_2(x)] = L_1 - L_2$
3. *Product Rule:* $\lim f_1(x) \cdot f_2(x) = L_1 \cdot L_2$
4. *Constant Multiple Rule:* $\lim k \cdot f_2(x) = k \cdot L_2$ (any number k)
5. *Quotient Rule:* $\lim \frac{f_1(x)}{f_2(x)} = \frac{L_1}{L_2}$ if $L_2 \neq 0$.

The limits are all taken as $x \to c$, and L_1 and L_2 are real numbers.

In words, the formulas in Theorem 1 say:

1. The limit of the sum of two functions is the sum of their limits.
2. The limit of the difference of two functions is the difference of their limits.
3. The limit of a product of two functions is the product of their limits.
4. The limit of a constant times a function is the constant times the limit of the function.
5. The limit of a quotient of two functions is the quotient of their limits, provided the denominator does not tend to zero.

We have included a formal proof of Theorem 1 in Appendix 3. Informally, we can paraphrase the theorem in terms that make it highly reasonable: When x is close to c, $f_1(x)$ is close to L_1 and $f_2(x)$ is close to L_2. Then we naturally think that $f_1(x) + f_2(x)$ is close to $L_1 + L_2$; $f_1(x) - f_2(x)$ is close to $L_1 - L_2$; $f_1(x)f_2(x)$ is close to L_1L_2; $kf_2(x)$ is close to kL_2; and $f_1(x)/f_2(x)$ is close to L_1/L_2 if L_2 is not zero.

What keeps this discussion from being a proof is that the word *close* is vague. Phrases like *arbitrarily close to* and *sufficiently close to* might seem at first to improve the argument, but what are really needed are the formal definitions and arguments developed for the purpose by the great European mathematicians of the nineteenth century. You will see what we mean if you read Section 2.5.

In the meantime, here are some examples of what Theorem 1 can do for us.

EXAMPLE 6 We know from Examples 2 and 3 that $\lim_{x\to c} x = c$ and $\lim_{x\to c} k = k$. The various parts of Theorem 1 now let us combine these results to calculate other limits:

a) $\lim_{x\to c} x^2 = \lim_{x\to c} x \cdot x = c \cdot c = c^2$ (Product)

b) $\lim_{x\to c} (x^2 + 5) = \lim_{x\to c} x^2 + \lim_{x\to c} 5$ (Sum)

$= c^2 + 5$ (from (a))

c) $\lim_{x\to c} 4x^2 = 4 \lim_{x\to c} x^2$ (Constant Multiple)

$= 4c^2$ (from (a))

d) $\lim_{x\to c} (4x^2 - 3) = \lim_{x\to c} 4x^2 - \lim_{x\to c} 3$ (Difference)

$= 4c^2 - 3$ (from (c))

e) $\lim_{x\to c} x^3 = \lim_{x\to c} x^2 \cdot x = c^2 \cdot c = c^3$ (Product and (a))

f) $\lim_{x\to c} (x^3 + 4x^2 - 3) = \lim_{x\to c} x^3 + \lim_{x\to c} (4x^2 - 3)$ (Sum)

$= c^3 + 4c^2 - 3$ (e and d)

g) $\lim_{x\to c} \dfrac{x^3 + 4x^2 - 3}{x^2 + 5} = \dfrac{\lim_{x\to c} (x^3 + 4x^2 - 3)}{\lim_{x\to c} (x^2 + 5)}$ (Quotient)

$= \dfrac{c^3 + 4c^2 - 3}{c^2 + 5}.$ (f and b) ■

Example 6 shows the remarkable strength of Theorem 1. From the two simple observations that $\lim_{x\to c} x = c$ and $\lim_{x\to c} k = k$ we can immediately work our way to limits of all polynomials and most **rational functions** (ratios of polynomials). As in Part (f) of Example 6, the limit of any polynomial $f(x)$ as x approaches c is $f(c)$, the number we get when we substitute $x = c$. As in Part (g) of Example 6, the limit of the ratio $f(x)/g(x)$ of two polynomials is $f(c)/g(c)$, provided $g(c)$ is different from 0.

THEOREM 2

Limits of Polynomials Can Be Found by Substitution

If $f(x) = a_n x^n + a_{n-1}x^{n-1} + \cdots + a_0$ is any polynomial function, then

$$\lim_{x\to c} f(x) = f(c) = a_n c^n + a_{n-1}c^{n-1} + \cdots + a_0. \tag{1}$$

THEOREM 3

Limits of (Many But Not All) Rational Functions Can Be Found by Substitution

If $f(x)$ and $g(x)$ are polynomials, then

$$\lim_{x\to c} \frac{f(x)}{g(x)} = \frac{f(c)}{g(c)} \quad \text{(provided } g(c) \neq 0\text{)}. \tag{2}$$

EXAMPLE 7

a) $\lim_{x\to 3} x^2(2 - x) = \lim_{x\to 3} (2x^2 - x^3) = 2(3)^2 - (3)^3 = 18 - 27 = -9.$

b) Same limit, found another way:

$$\lim_{x\to 3} x^2(2 - x) = \lim_{x\to 3} x^2 \cdot \lim_{x\to 3} (2 - x) = (3)^2 \cdot (2 - 3) = 9 \cdot (-1) = -9. \quad \blacksquare$$

EXAMPLE 8

$$\lim_{x\to 2} \frac{x^2 + 2x + 4}{x + 2} = \frac{(2)^2 + 2(2) + 4}{2 + 2} = \frac{12}{4} = 3. \quad \blacksquare$$

In Example 8, we can use Eq. (2) to find the limit of $f(x)/g(x)$ because the value of the denominator, $g(x) = x + 2$, is different from zero when $x = 2$. In the next example, the denominator is zero when $x = 2$, so we cannot apply Eq. (2) directly. We have to rewrite the fraction $f(x)/g(x)$ first.

EXAMPLE 9

Find

$$\lim_{x\to 2} \frac{x^3 - 8}{x^2 - 4}.$$

Solution The denominator is 0 when $x = 2$, so we cannot calculate the limit by substitution. However, if we factor the numerator and denominator we find that

$$\frac{x^3 - 8}{x^2 - 4} = \frac{(x - 2)(x^2 + 2x + 4)}{(x - 2)(x + 2)} = \frac{x^2 + 2x + 4}{x + 2}.$$

Is it really all right to cancel terms like this? Yes, because 2 is not in the domain of the original function $(x^3 - 8)/(x^2 - 4)$, so we're not dividing by zero in disguise when we cancel the two $(x - 2)$s.

With the $(x - 2)$ out of the way, we can now find the limit by substitution:

$$\lim_{x \to 2} \frac{x^3 - 8}{x^2 - 4} = \lim_{x \to 2} \frac{x^2 + 2x + 4}{x + 2} \qquad \text{(The algebra we just did)}$$

$$= \frac{(2)^2 + 2(2) + 4}{2 + 2} \qquad \text{(Eq. (2) now applies.)}$$

$$= \frac{12}{4} = 3.$$

■

Example 9 illustrates an important point about limits: The limit of a function $f(x)$ as x approaches c *never* depends on what happens when $x = c$. The limit, if it exists at all, is *entirely* determined by the values that f has when $x \neq c$. In Example 9, the quotient $f(x) = (x^3 - 8)/(x^2 - 4)$ is not even defined at $x = 2$. Yet it has a limit as x approaches 2 and this limit is 3.

Here is another example.

EXAMPLE 10

$$\lim_{x \to 5} \frac{x^2 - 25}{3(x - 5)} \qquad \left(\begin{array}{l}\text{Substitution will not give the limit} \\ \text{because } x - 5 = 0 \text{ when } x = 5.\end{array}\right)$$

$$= \lim_{x \to 5} \frac{(x + 5)(x - 5)}{3(x - 5)} \qquad \left(\begin{array}{l}\text{We factor the numerator to see} \\ \text{if } (x - 5) \text{ is a factor. It is. We} \\ \text{cancel the } (x - 5)\text{'s, leaving}\ldots\end{array}\right)$$

$$= \lim_{x \to 5} \frac{x + 5}{3} \qquad \left(\begin{array}{l}\ldots\text{an equivalent form whose limit} \\ \text{we can now find by substitution.}\end{array}\right)$$

$$= \frac{5 + 5}{3} = \frac{10}{3}.$$

■

In Example 10, as in Example 9, we found the limit of a function as x approached a value at which the function was undefined. In the next example, the function is defined at every value of x, but the function's limit as x approaches 2 is not the same as the function's value at $x = 2$.

EXAMPLE 11

If

$$f(x) = \begin{cases} x, & x \neq 2 \\ 3, & x = 2, \end{cases}$$

then

$$\lim_{x \to 2} f(x) = \lim_{x \to 2} (x) = 2 \quad \text{while} \quad f(2) = 3.$$

As always, the limit is determined by the function's approach behavior, not by what happens at $x = 2$ (Fig. 2.2). ■

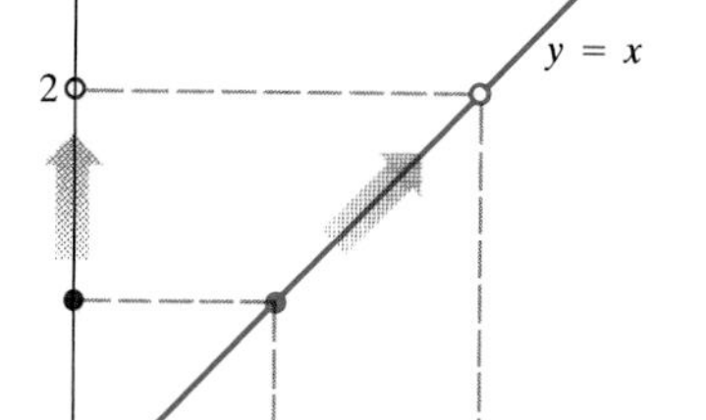

2.2 The graph of

$$f(x) = \begin{cases} x, & x \neq 2 \\ 3, & x = 2. \end{cases}$$

Notice that $f(x) \to 2$ as $x \to 2$ even though $f(2)$ itself is 3.

Examples 9–11 show how special polynomial functions really are. If $f(x)$ is a polynomial function, then $\lim_{x\to c} f(x)$ is always $f(c)$ itself. Functions with this remarkable property are extremely useful in mathematics, as we shall see in Section 2.4.

Right-hand Limits and Left-hand Limits

Sometimes the values of a function $f(x)$ tend to different limits as x approaches a number c from different sides. When this happens, we call the limit of f as x approaches c from the right the **right-hand limit** of f at c, and the limit as x approaches c from the left the **left-hand limit** of f at c.

The notation for the right-hand limit is

$$\lim_{x\to c^+} f(x). \qquad \text{("the limit of } f \text{ as } x \text{ approaches } c \text{ from the right")}$$

The $(+)$ is there to say that x approaches c through values above c on the number line.

The notation for the left-hand limit is

$$\lim_{x\to c^-} f(x). \qquad \text{("the limit of } f \text{ as } x \text{ approaches } c \text{ from the left")}$$

The $(-)$ is there to say that x approaches c through values below c on the number line.

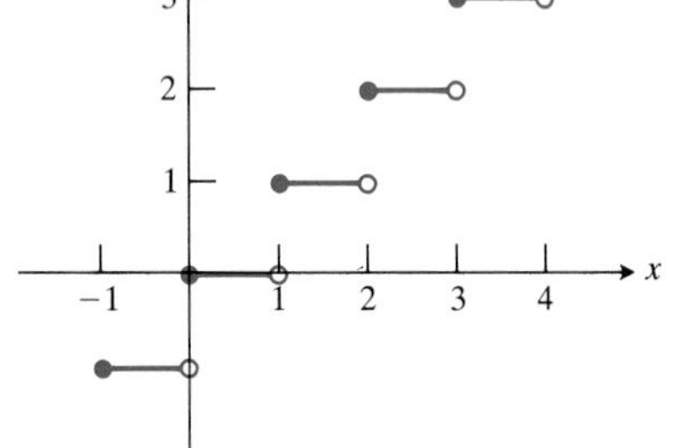

2.3 At each integer, the greatest integer function $y = \lfloor x \rfloor$ has different right-hand and left-hand limits.

EXAMPLE 12 The greatest-integer function $f(x) = \lfloor x \rfloor$ has different right-hand and left-hand limits at each integer. As we can see in Fig. 2.3,

$$\lim_{x\to 3^+} \lfloor x \rfloor = 3 \quad \text{and} \quad \lim_{x\to 3^-} \lfloor x \rfloor = 2.$$

The limit of $\lfloor x \rfloor$ as x approaches an integer n from above is n, while the limit as x approaches n from below is $n - 1$. ■

One-sided vs. Two-sided Limits

We sometimes call $\lim_{x\to c} f(x)$ the **two-sided** limit of f at c to distinguish it from the **one-sided** right-hand and left-hand limits of f at c. If the two one-sided limits of $f(x)$ exist at c and are equal, their common value is the two-sided limit of f at c. Conversely, if the two-sided limit of f at c exists, the two one-sided limits exist and have the same value as the two-sided limit.

Relationship between One-sided and Two-sided Limits

A function $f(x)$ has a limit as x approaches c if and only if the right-hand and left-hand limits at c exist and are equal. In symbols,

$$\lim_{x\to c} f(x) = L \quad \Leftrightarrow \quad \lim_{x\to c^+} f(x) = L \quad \text{and} \quad \lim_{x\to c^-} f(x) = L. \tag{3}$$

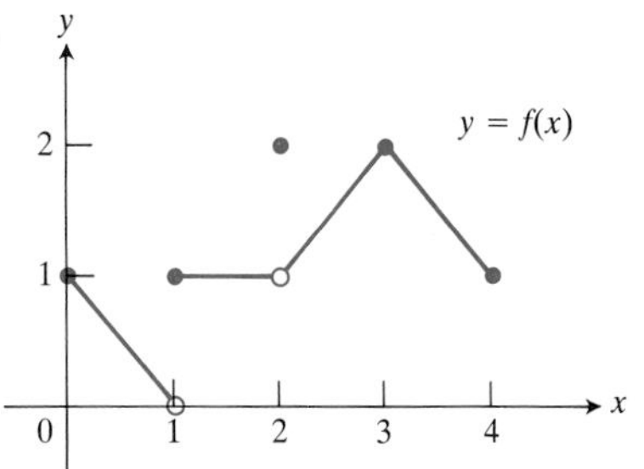

2.4 Example 13 discusses the limit properties of the function $y = f(x)$ graphed here.

The implications in Eq. (3) will be proved formally in Section 2.5, but you can see what is going on if you look at the next examples.

EXAMPLE 13 All the following statements about the function $y = f(x)$ graphed in Fig. 2.4 are true.

At $x = 0$: $\lim_{x\to 0^+} f(x) = 1$.

At $x = 1$: $\lim_{x\to 1^-} f(x) = 0$ even though $f(1) = 1$,

$\lim_{x\to 1^+} f(x) = 1$,

$f(x)$ has no limit as $x \to 1$. (The right- and left-hand limits at 1 are not equal.)

At $x = 2$: $\lim_{x\to 2^-} f(x) = 1$,

$\lim_{x\to 2^+} f(x) = 1$,

$\lim_{x\to 2} f(x) = 1$ even though $f(2) = 2$.

At $x = 3$: $\lim_{x\to 3^-} f(x) = \lim_{x\to 3^+} f(x) = \lim_{x\to 3} f(x) = f(3) = 2$.

At $x = 4$: $\lim_{x\to 4^-} f(x) = 1$.

At every other point c between 0 and 4, $f(x)$ has a limit as $x \to c$. ■

EXAMPLE 14 The greatest-integer function $f(x) = \lfloor x \rfloor$ has no limit as x approaches 3. As we saw in Example 12, $\lim_{x\to 3^+} \lfloor x \rfloor = 3$ while $\lim_{x\to 3^-} \lfloor x \rfloor = 2$. Since the right-hand and left-hand limits of f at 3 are not equal, the function has no single limiting value as x approaches 3. ■

EXAMPLE 15 Show that $\lim_{x\to 0} |x| = 0$.

Solution We prove that $\lim_{x\to 0} |x| = 0$ by showing that the right-hand and left-hand limits are both 0:

$$\lim_{x\to 0^+} |x| = \lim_{x\to 0^+} x = 0. \quad (|x| = x \quad \text{if} \quad x > 0)$$

$$\begin{aligned} \lim_{x\to 0^-} |x| &= \lim_{x\to 0^-} (-x) && (|x| = -x \quad \text{if} \quad x < 0) \\ &= -\lim_{x\to 0^-} x && \text{(A limit property)} \\ &= -0 \\ &= 0. \end{aligned}$$

■

In the examples we have seen so far, the functions that failed to have limits at various points did so because the right-hand and left-hand limits at the points were not equal. The function in the next example fails to have a limit because neither the right-hand limit nor the left-hand limit exists at all.

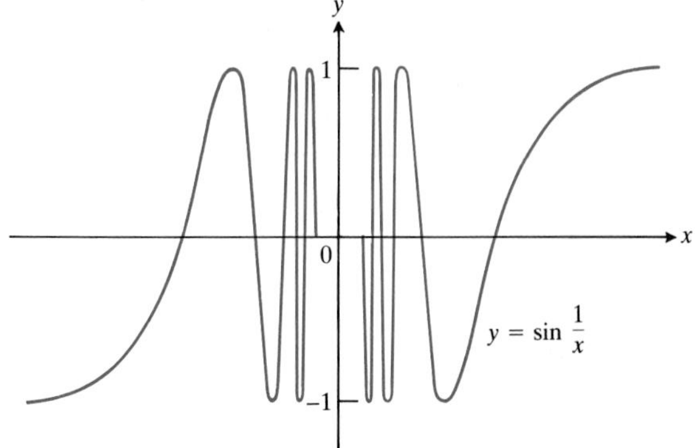

2.5 The function $y = \sin(1/x)$ has neither a right-hand nor a left-hand limit as x approaches 0.

EXAMPLE 16 Show that the function $y = \sin(1/x)$ has no limit as x approaches 0 from either side (Fig. 2.5).

Solution As x approaches 0, its reciprocal, $1/x$, becomes infinite and the values of $\sin(1/x)$ cycle repeatedly from -1 to 1. Thus there is no single number L that the function's values all get close to as x approaches zero. This is true even if we restrict x to positive values or to negative values. The function has neither a right-hand limit nor a left-hand limit as x approaches 0. ■

Exercises 2.1

Find the limits in Exercises 1–20.

1. $\lim_{x\to2} 2x$

2. $\lim_{x\to0} 2x$

3. $\lim_{x\to4} 4$

4. $\lim_{x\to-2} 4$

5. $\lim_{x\to1} (3x - 1)$

6. $\lim_{x\to1/3} (3x - 1)$

7. $\lim_{x\to5} x^2$

8. $\lim_{x\to2} x(2 - x)$

9. $\lim_{x\to0} 5(2x - 1)$

10. $\lim_{x\to1} x(2x - 1)$

11. $\lim_{x\to-1} 3x(2x - 1)$

12. $\lim_{x\to-1} 3x^2(2x - 1)$

13. $\lim_{x\to-2} (x + 3)^{171}$

14. $\lim_{x\to-4} (x + 3)^{1984}$

15. $\lim_{x\to2} 5(2x - 1)(x + 1)$

16. $\lim_{x\to6} 8(x - 5)(x - 7)$

17. $\lim_{x\to-1} (x^2 - 2x + 1)$

18. $\lim_{x\to0} (x^2 - 3x - 18)$

19. $\lim_{x\to1} (x^3 + 3x^2 - 2x - 17)$

20. $\lim_{x\to-2} (x^3 - 2x^2 + 4x + 8)$

Find the limits in Exercises 21–30.

21. $\lim_{x\to2} \frac{x + 3}{x + 6}$

22. $\lim_{x\to5} \frac{4}{x - 7}$

23. $\lim_{x\to-1} \frac{x + 3}{x^2 + 3x + 1}$

24. $\lim_{x\to-5} \frac{x^2}{5 - x}$

25. $\lim_{y\to2} \frac{y^2 + 5y + 6}{y + 2}$

26. $\lim_{y\to0} \frac{4 - y}{3 + y^3}$

27. $\lim_{y\to-3} \frac{y^2 + 4y + 3}{y^2 - 3}$

28. $\lim_{y\to3} \frac{1}{y^4 - 9y^2 + y}$

29. $\lim_{x\to-1} \frac{x^3 - 5x + 7}{-x^3 + x^2 - x + 1}$

30. $\lim_{t\to2} \frac{3t^3 - 10t - 3}{t^4 - 14}$

Find the limits in Exercises 31–44 by rewriting the fractions first.

31. $\lim_{x\to1} \frac{x - 1}{x^2 - 1}$

32. $\lim_{x\to1} \frac{x^2 + x - 2}{x^2 - 1}$

33. $\lim_{x\to-5} \frac{x^2 + 3x - 10}{x + 5}$

34. $\lim_{x\to-3} \frac{x + 3}{x^2 + 4x + 3}$

35. $\lim_{x\to-2} \frac{x^2 + x - 2}{x^2 - 4}$

36. $\lim_{x\to2} \frac{x^2 - 7x + 10}{x - 2}$

37. $\lim_{t\to1} \frac{t^2 - 3t + 2}{t^2 - 1}$

38. $\lim_{t\to2} \frac{t^2 - 3t + 2}{t^2 - 4}$

39. $\lim_{x\to5} \frac{x - 5}{x^2 - 25}$

40. $\lim_{x\to-5} \frac{x + 5}{x^2 - 25}$

41. $\lim_{x\to2} \frac{2x - 4}{x^3 - 2x^2}$

42. $\lim_{x\to0} \frac{5x^3 + 8x^2}{3x^4 - 16x^2}$

43. $\lim_{t\to 1} \frac{t^3 - 1}{t - 1}$

44. $\lim_{a\to 2} \frac{a^3 - 8}{a^4 - 16}$

45. Let $f(x) = \begin{cases} 3 - x, & x < 2 \\ \frac{x}{2} + 1, & x > 2 \end{cases}$

a) Find $\lim_{x\to 2^+} f(x)$ and $\lim_{x\to 2^-} f(x)$.

b) Does $\lim_{x\to 2} f(x)$ exist? If so, what is it? If not, why not?

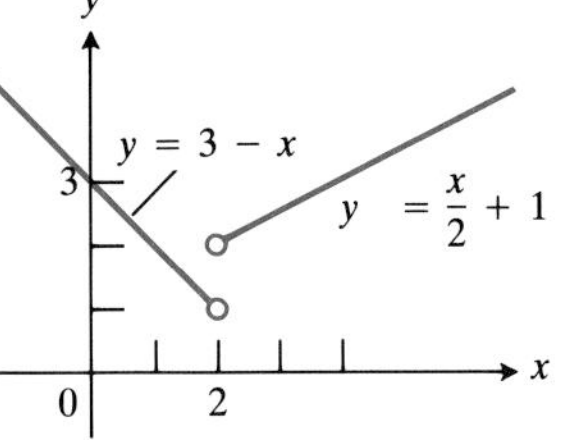

46. Let $f(x) = \begin{cases} 3 - x, & x < 2 \\ 2, & x = 2 \\ \frac{x}{2}, & x > 2 \end{cases}$

a) Find $\lim_{x\to 2^+} f(x)$ and $\lim_{x\to 2^-} f(x)$.

b) Does $\lim_{x\to 2} f(x)$ exist? If so, what is it? If not, why not?

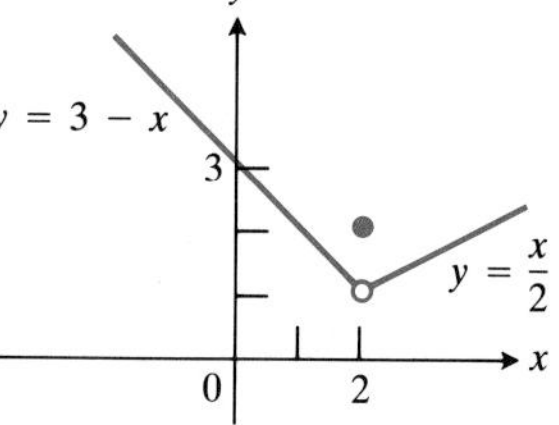

47. Which of the following statements are true of the function $y = f(x)$ graphed here?

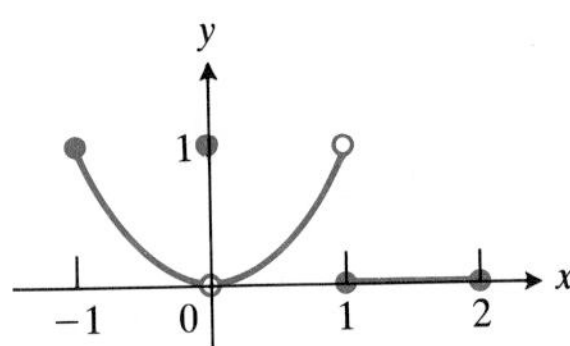

a) $\lim_{x\to -1^+} f(x) = 1$

b) $\lim_{x\to 0^-} f(x) = 0$

c) $\lim_{x\to 0^-} f(x) = 1$

d) $\lim_{x\to 0^-} f(x) = \lim_{x\to 0^+} f(x)$

e) $\lim_{x\to 0} f(x)$ exists

f) $\lim_{x\to 0} f(x) = 0$

g) $\lim_{x\to 0} f(x) = 1$

h) $\lim_{x\to 1} f(x) = 1$

i) $\lim_{x\to 1} f(x) = 0$

j) $\lim_{x\to 2^-} f(x) = 2$

48. Which of the following statements are true of the function graphed here?

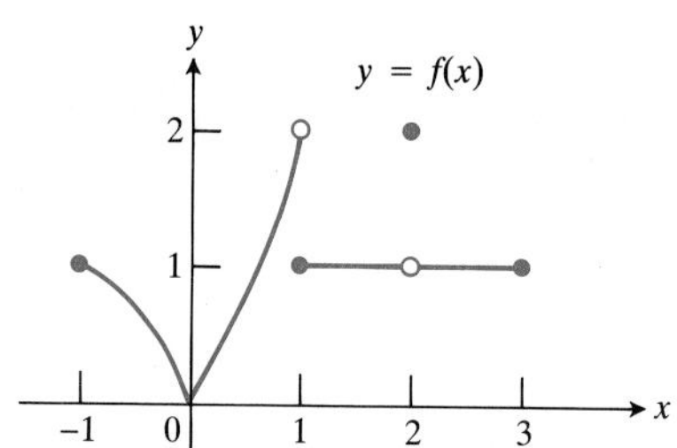

a) $\lim_{x\to -1^+} f(x) = 1$

b) $\lim_{x\to 2} f(x)$ does not exist.

c) $\lim_{x\to 2} f(x) = 2$

d) $\lim_{x\to 1^-} f(x) = 2$

e) $\lim_{x\to 1^+} f(x) = 1$

f) $\lim_{x\to 1} f(x)$ does not exist.

g) $\lim_{x\to 0^+} f(x) = \lim_{x\to 0^-} f(x)$

h) $\lim_{x\to c} f(x)$ exists at every c in $(-1, 1)$.

i) $\lim_{x\to c} f(x)$ exists at every c in $(1, 3)$.

49. a) Graph $f(x) = \begin{cases} x^3, & x \neq 1 \\ 0, & x = 1 \end{cases}$

b) Find $\lim_{x\to 1^-} f(x)$ and $\lim_{x\to 1^+} f(x)$.

c) Does $\lim_{x\to 1} f(x)$ exist? If so, what is it? If not, why not?

50. a) Graph $f(x) = \begin{cases} 1 - x^2, & x \neq 1 \\ 2, & x = 1 \end{cases}$

b) Find $\lim_{x\to 1^+} f(x)$ and $\lim_{x\to 1^-} f(x)$.

c) Does $\lim_{x\to 1} f(x)$ exist? If so, what is it? If not, why not?

Graph the two functions in Exercises 51 and 52. Then answer these questions:

a) At what points c in the domain of f does $\lim_{x\to c} f(x)$ exist?

b) At what points does only the left-hand limit exist?

c) At what points does only the right-hand limit exist?

51. $f(x) = \begin{cases} \sqrt{1 - x^2} & \text{if } 0 \le x < 1, \\ 1 & \text{if } 1 \le x < 2, \\ 2 & \text{if } x = 2. \end{cases}$

52. $f(x) = \begin{cases} x & \text{if } -1 \le x < 0, \text{ or } 0 < x \le 1, \\ 1 & \text{if } x = 0, \\ 0 & \text{if } x < -1, \text{ or } x > 1. \end{cases}$

53. Let $f(x) = \begin{cases} 0, & x \le 0 \\ \sin \frac{1}{x}, & x > 0. \end{cases}$

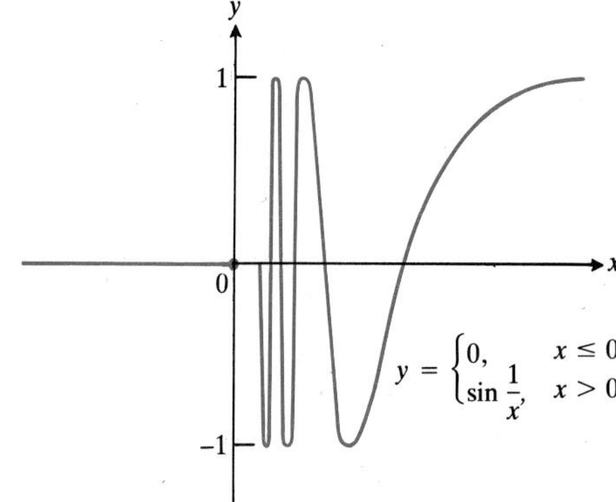

a) Does $\lim_{x\to 0^+} f(x)$ exist? If so, what is it?

b) Does $\lim_{x\to 0^-} f(x)$ exist? If so, what is it?

c) Does $\lim_{x\to 0} f(x)$ exist? If so, what is it? If not, why not?

54. Let $f(x) = \begin{cases} 0, & x = 0 \\ x \sin \dfrac{1}{x}, & x \neq 0. \end{cases}$

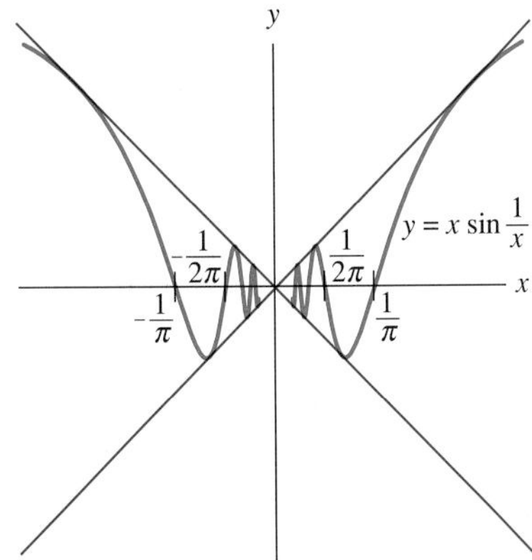

(Generated by Mathematica)

a) Does $\lim_{x\to 0^+} f(x)$ exist? If so, what is it?
b) Does $\lim_{x\to 0^-} f(x)$ exist? If so, what is it?
c) Does $\lim_{x\to 0} f(x)$ exist? If so, what is it? If not, why not?

Find the limits of the greatest-integer function in Exercises 55–58.

55. $\lim_{x\to 0^+} \lfloor x \rfloor$ **56.** $\lim_{x\to 0^-} \lfloor x \rfloor$

57. $\lim_{x\to 0.5} \lfloor x \rfloor$ **58.** $\lim_{x\to 2^-} \lfloor x \rfloor$

Find the limits in Exercises 59–60.

59. $\lim_{x\to 0^+} \dfrac{x}{|x|}$

60. $\lim_{x\to 0^-} \dfrac{x}{|x|}$

61. Suppose $\lim_{x\to c} f(x) = 5$ and $\lim_{x\to c} g(x) = -2$. Find

a) $\lim_{x\to c} f(x)g(x)$ b) $\lim_{x\to c} 2f(x)g(x)$

62. Suppose $\lim_{x\to 4} f(x) = 0$ and $\lim_{x\to 4} g(x) = -3$. Find

a) $\lim_{x\to 4} (g(x) + 3)$ b) $\lim_{x\to 4} x f(x)$
c) $\lim_{x\to 4} g^2(x)$ d) $\lim_{x\to 4} \dfrac{g(x)}{f(x) - 1}$

63. Suppose $\lim_{x\to b} f(x) = 7$ and $\lim_{x\to b} g(x) = -3$. Find

a) $\lim_{x\to b} (f(x) + g(x))$ b) $\lim_{x\to b} f(x) \cdot g(x)$
c) $\lim_{x\to b} 4g(x)$ d) $\lim_{x\to b} f(x)/g(x)$

64. Suppose $\lim_{x\to -2} p(x) = 4$, $\lim_{x\to -2} r(x) = 0$, and $\lim_{x\to -2} s(x) = -3$. Find

a) $\lim_{x\to -2} (p(x) + r(x) + s(x))$
b) $\lim_{x\to -2} p(x) \cdot r(x) \cdot s(x)$

65. a) CALCULATOR Estimate the value of

$$\lim_{x\to 1} \frac{x^2 - 1}{x - 1}$$

by taking $x = 1.1, 1.01, 1.001$, and so on as far as your calculator can go.

b) Find the limit's exact value.

66. a) CALCULATOR Estimate the value of

$$\lim_{x\to 3} \frac{x^2 - 2x - 3}{x^2 - 4x + 3}$$

by taking $x = 3.1, 3.01, 3.001$, and so on as far as your calculator can go.

b) Find the limit's exact value.

67. CALCULATOR By taking $x = 1.1, 1.01, 1.001$, and so on as far as your calculator can go, estimate the value of

$$\lim_{x\to 1} \frac{\ln(x^2)}{\ln x}.$$

(Chapter 7 will explain what is going on here.)

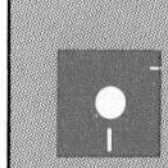

TOOLKIT PROGRAM

Limit Problems Offers practice in finding limits

2.2 The Sandwich Theorem and (sin θ)/θ

One of the most useful facts in calculus is that $\lim_{\theta\to 0} (\sin \theta)/\theta = 1$ when θ is measured in radians. This beautiful and simple result turns out to be the key to measuring the rates at which all trigonometric functions of θ change their values as θ changes, as we shall see in Chapter 3.

This limit is not the kind we can evaluate by substituting $\theta = 0$, so we have to find it in a more subtle way. What we do is sandwich the fraction (sin θ)/θ between the number 1 and a fraction that is known to approach 1 as θ approaches 0. This tells us that (sin θ)/θ approaches 1 as well.

The theorem we rely on for this argument is the Sandwich Theorem.

THEOREM 4

The Sandwich Theorem

Suppose that

$$g(x) \le f(x) \le h(x)$$

for all $x \neq c$ in some interval about c and that

$$\lim_{x\to c} g(x) = \lim_{x\to c} h(x) = L.$$

Then

$$\lim_{x\to c} f(x) = L.$$

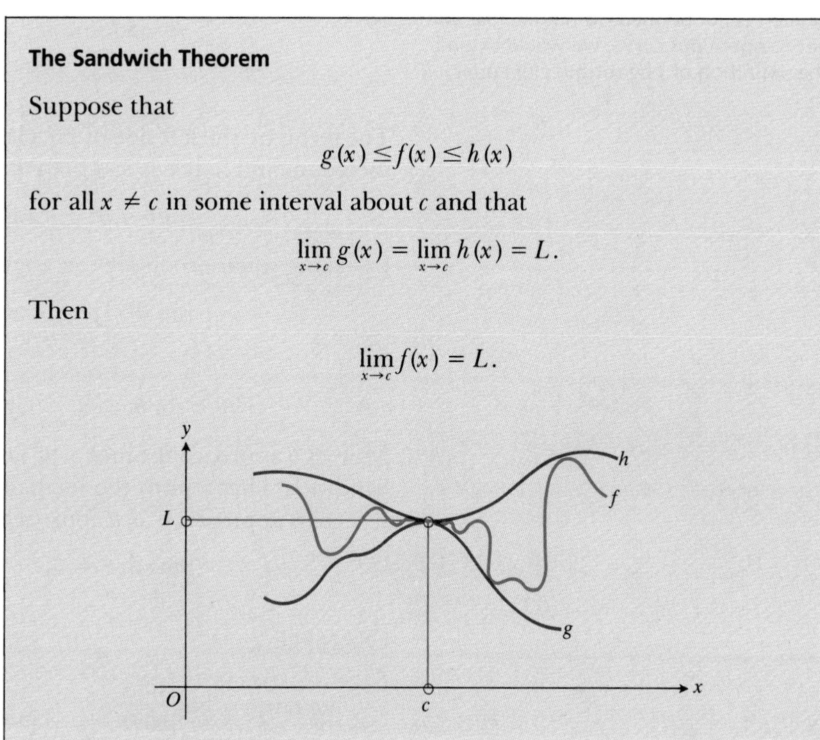

The idea is that if the values of f are sandwiched between the values of two functions that approach L, then the values of f approach L, too. We have included a proof of the theorem in Appendix 2.

The following example shows how the Sandwich Theorem is typically used to calculate limits.

EXAMPLE 1

$$\lim_{\theta\to 0} \sin\theta = 0 \qquad \text{and} \qquad \lim_{\theta\to 0} \cos\theta = 1.$$

To calculate these limits, we picture θ as the radian measure of an angle in standard position (Fig. 2.6). The circle in the figure is a unit circle, so $|\theta|$ also equals the length of the circular arc AP. The length of the line segment AP is therefore less than $|\theta|$.

Triangle APQ is a right triangle with legs of length

$$QP = |\sin\theta|, \qquad AQ = 1 - \cos\theta.$$

From the Pythagorean theorem and the fact that $AP < |\theta|$, we get

$$\sin^2\theta + (1 - \cos\theta)^2 = (AP)^2 < \theta^2. \tag{1}$$

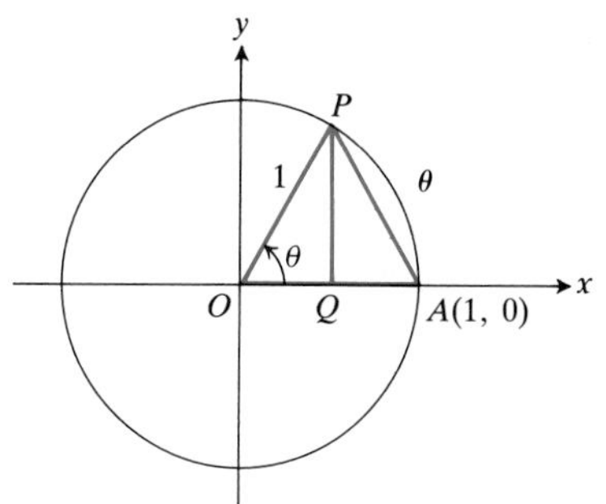

2.6 From the geometry of this figure, drawn for $\theta > 0$, we get the inequality

$$\sin^2\theta + (1 - \cos\theta)^2 < \theta^2.$$

This inequality is the chief step in showing that $\sin\theta \to 0$ and $\cos\theta \to 1$ as $\theta \to 0$ (Example 1).

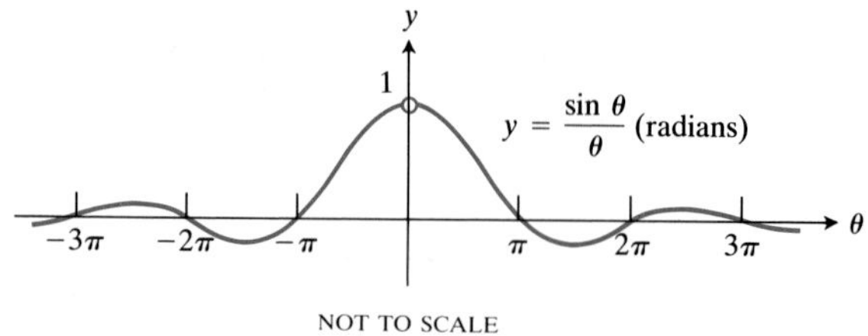

2.7 The graph of $f(\theta) = (\sin\theta)/\theta$ in the Cartesian θy-plane has a "hole" at $\theta = 0$ where f is not defined. To fill the hole and get a continuous curve, we would extend the definition of f by setting $f(0)$ equal to 1.

The terms on the left side of Eq. (1) are both positive, so each is smaller than their sum and hence is less than θ^2:

$$\sin^2\theta < \theta^2 \quad \text{and} \quad (1 - \cos\theta)^2 < \theta^2. \tag{2}$$

By taking square roots we can see this is equivalent to saying that

$$|\sin\theta| < |\theta| \quad \text{and} \quad |1 - \cos\theta| < |\theta| \tag{3}$$

or

$$-|\theta| < \sin\theta < |\theta| \quad \text{and} \quad -|\theta| < 1 - \cos\theta < |\theta|. \tag{4}$$

Now let θ approach 0. Since $-|\theta|$ and $|\theta|$ both approach 0, we may apply the Sandwich Theorem to the inequalities in (4) and conclude that $\sin\theta$ and $1 - \cos\theta$ approach 0 as θ approaches 0. Hence,

$$\lim_{\theta\to 0} \sin\theta = 0 \quad \text{and} \quad \lim_{\theta\to 0} \cos\theta = 1.$$

EXAMPLE 2

$$\begin{aligned}
\lim_{x\to 0} \tan x &= \lim_{x\to 0} \frac{\sin x}{\cos x} && \left(\tan x = \frac{\sin x}{\cos x}\right) \\
&= \frac{\lim_{x\to 0} \sin x}{\lim_{x\to 0} \cos x} && \text{(a property of limits)} \\
&= \frac{0}{1} = 0. && \text{(values from Example 1)}
\end{aligned}$$

We now extend the results in Example 1 to show that $\lim_{\theta\to 0} (\sin\theta)/\theta = 1$ when θ is measured in radians. Figure 2.7 shows the graph of $(\sin\theta)/\theta$ and Table 2.2 shows values of the function for θ near zero. During the proof we draw on a formula from geometry that says that the area cut from a unit circle by a central angle of θ radians is $\theta/2$. Figure 2.8 shows where this formula comes from.

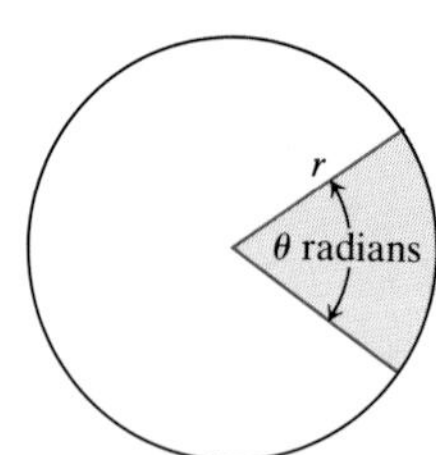

$$\begin{aligned}
\text{Sector area} &= \frac{\theta}{2\pi} \cdot \text{Circle area} \\
&= \frac{\theta}{2\pi} \cdot \pi r^2 \\
&= \frac{1}{2} r^2 \theta \quad \text{(Usual form)} \\
&= \frac{\theta}{2} \quad (\text{If } r = 1)
\end{aligned}$$

2.8 The formula for the area of a sector of a unit circle is $A = \theta/2$.

TABLE 2.2

θ (radians)	$\sin\theta$	$(\sin\theta)/\theta$
0.1	0.09983 34166 5	0.99833 41665
0.01	.00999 98333 34	.99998 33334
0.001	.00099 99998 333	.99999 98333
0.0001	.00009 99999 9983	.99999 99983

EXAMPLE 3 If θ is measured in radians, then

$$\lim_{\theta \to 0} \frac{\sin \theta}{\theta} = 1. \tag{5}$$

Our plan is to show that the right-hand and left-hand limits are both 1. We will then know that the two-sided limit is 1 as well.

To show that the right-hand limit is 1, we begin with values of θ that are positive and less than $\pi/2$ (Fig. 2.9). We compare the areas of $\triangle OAP$, sector OAP, and $\triangle OAT$ and note that

$$\text{Area } \triangle OAP < \text{Area sector } OAP < \text{Area } \triangle OAT. \tag{6}$$

We can express these areas in terms of θ as follows:

$$\text{Area } \triangle OAP = \frac{1}{2} \text{base} \times \text{height} = \frac{1}{2}(1)(\sin \theta) = \frac{1}{2} \sin \theta, \tag{7}$$

$$\text{Area sector } OAP = \frac{1}{2} r^2 \theta = \frac{1}{2}(1)^2 \theta = \frac{\theta}{2}, \tag{8}$$

$$\text{Area } \triangle OAT = \frac{1}{2} \text{base} \times \text{height} = \frac{1}{2}(1)(\tan \theta) = \frac{1}{2} \tan \theta, \tag{9}$$

Equation (8) is where the radian measurement comes in: The area of sector OAP is $\theta/2$ only if θ is measured in radians.

so that

$$\frac{1}{2} \sin \theta < \frac{1}{2} \theta < \frac{1}{2} \tan \theta. \tag{10}$$

The inequality in (10) will go the same way if we divide all three terms by the positive number $(1/2)\sin \theta$:

$$1 < \frac{\theta}{\sin \theta} < \frac{1}{\cos \theta}. \tag{11}$$

We next take reciprocals in (11), which reverses the inequalities:

$$\cos \theta < \frac{\sin \theta}{\theta} < 1. \tag{12}$$

Because $\cos \theta$ approaches 1 as θ approaches 0, the Sandwich Theorem tells us that

$$\lim_{\theta \to 0^+} \frac{\sin \theta}{\theta} = 1. \tag{13}$$

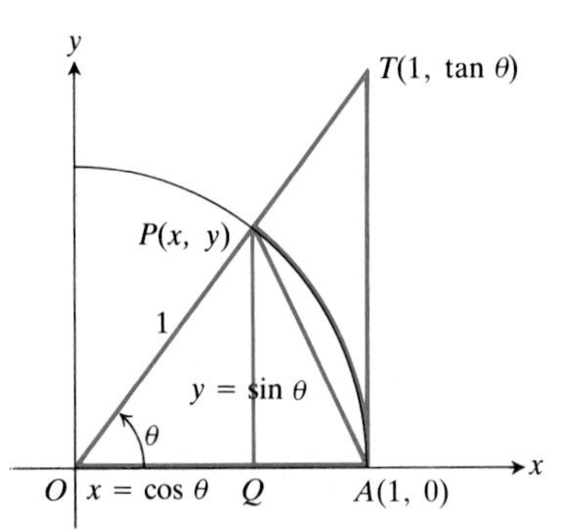

2.9 Area $\triangle OAP <$ area sector $OAP <$ area $\triangle OAT$.

The limit in Eq. (13) is a right-hand limit because we have been dealing with values of θ between 0 and $\pi/2$, but we obtain the same limit for $(\sin \theta)/\theta$ as θ approaches 0 from the left. For if $\theta = -\alpha$ and α is positive, then

$$\frac{\sin \theta}{\theta} = \frac{\sin(-\alpha)}{-\alpha} = \frac{-\sin(\alpha)}{-\alpha} = \frac{\sin \alpha}{\alpha}. \tag{14}$$

Therefore,

$$\lim_{\theta \to 0^-} \frac{\sin \theta}{\theta} = \lim_{\alpha \to 0^+} \frac{\sin \alpha}{\alpha} = 1. \tag{15}$$

Together, Eqs. (13) and (15) imply that $\lim_{\theta \to 0} (\sin \theta)/\theta = 1$. ■

Knowing the limit of $(\sin\theta)/\theta$ helps us in calculating a number of related limits.

EXAMPLE 4

$$\lim_{x\to 0}\frac{\sin 3x}{x} \qquad \text{(Eq. (5) does not apply. Substitute } \theta = 3x\text{.)}$$

$$= \lim_{\theta\to 0}\frac{\sin\theta}{\theta/3} \qquad (x = \theta/3 \text{ and } \theta\to 0 \text{ as } x\to 0.)$$

$$= \lim_{\theta\to 0} 3\,\frac{\sin\theta}{\theta} \qquad \text{(Eq. (5) applies now.)}$$

$$= 3\lim_{\theta\to 0}\frac{\sin\theta}{\theta} = 3\cdot 1 = 3.$$

EXAMPLE 5

$$\lim_{x\to 0}\frac{\tan x}{x} = \lim_{x\to 0}\frac{\sin x}{x}\,\frac{1}{\cos x} \qquad \left(\tan x = \frac{\sin x}{\cos x}\right)$$

$$= \lim_{x\to 0}\frac{\sin x}{x}\cdot\lim_{x\to 0}\frac{1}{\cos x} \qquad \text{(A property of limits)}$$

$$= 1\cdot 1 \qquad \text{(Known values)}$$

$$= 1.$$

Exercises 2.2

Find the limits in Exercises 1–12.

1. $\lim_{x\to 0}\dfrac{1}{\cos x}$

2. $\lim_{x\to 0}(2\sin x + 3\cos x)$

3. $\lim_{x\to 0}\dfrac{1+\sin x}{1+\cos x}$

4. $\lim_{x\to 0}\dfrac{x^2+1}{1-\sin x}$

5. $\lim_{x\to 0^+}\dfrac{x}{\sin x}$

6. $\lim_{x\to 0^-}\dfrac{x}{\tan x}$

7. $\lim_{x\to 0}\dfrac{\sin 2x}{x}$

8. $\lim_{x\to 0}\dfrac{x}{\sin 3x}$

9. $\lim_{x\to 0}\dfrac{\tan 2x}{2x}$

10. $\lim_{x\to 0}\dfrac{\tan 2x}{x}$

11. $\lim_{x\to 0}\dfrac{\sin x}{2x^2-x}$

12. $\lim_{x\to 0}\dfrac{x+\sin x}{x}$

13. *Sandwich Theorem.* The inequality

$$1-\frac{x^2}{6}<\frac{\sin x}{x}<1$$

holds when x is measured in radians and $-1<x<1$. Use this inequality to calculate $\lim_{x\to 0}(\sin x)/x$. In Chapter 9, we shall see where this inequality comes from.

14. *Sandwich Theorem.* As we saw in Example 3 and again in Exercise 13, we can sometimes use the Sandwich

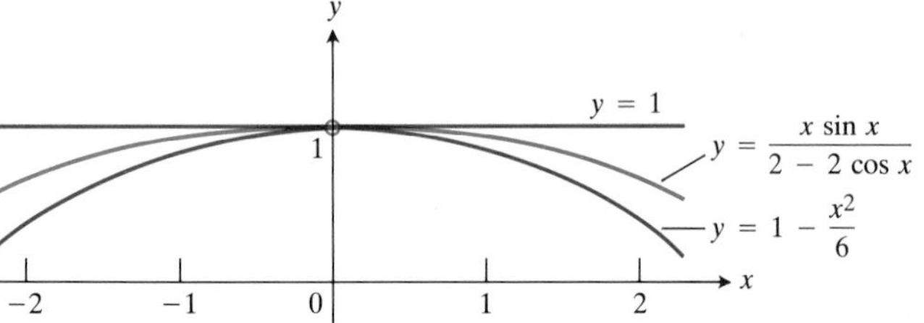

2.10 For all $x \neq 0$, the graph of $f(x) = (x\sin x)/(2-2\cos x)$ lies between $y = 1$ and $y = 1 - x^2/6$ (Exercise 14).

Theorem to calculate the limit of a fraction whose numerator and denominator both approach zero. Another example is the fraction

$$f(x) = (x\sin x)/(2-2\cos x),$$

which satisfies the inequality

$$1-\frac{x^2}{6}<\frac{x\sin x}{2-2\cos x}<1$$

when x is an angle in radians close to zero (Fig. 2.10). Use this inequality to find $\lim_{x\to 0} f(x)$. Inequalities like this come from infinite series (Chapter 9).

15. a) **CALCULATOR** Estimate the value of

$$\lim_{x\to 0} \frac{1 - \cos x}{x^2}$$

by taking $x = 0.1, 0.01, 0.001$, and so on as far as your calculator can go.

b) *Sandwich Theorem.* Use the inequality

$$\frac{1}{2} - \frac{x^2}{24} < \frac{1 - \cos x}{x^2} < \frac{1}{2}$$

(Fig. 2.11) to find the exact value of the limit in (a). Inequalities like this come from infinite series (Chapter 9).

16. The area formula $A = (1/2)r^2\theta$ derived in Fig. 2.8 for radian measure has to be changed if the angle is measured in degrees. What should the new formula be?

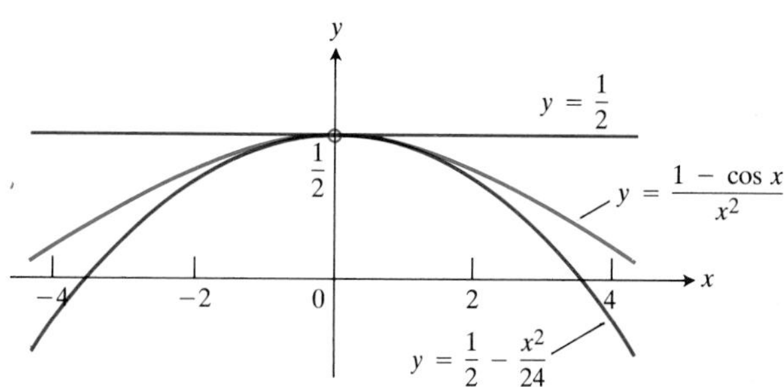

2.11 For all $x \neq 0$, the graph of $f(x) = (1 - \cos x)/x^2$ lies between the line $y = 1/2$ and the parabola $y = (1/2) - x^2/24$ (Exercise 15).

2.3 Limits Involving Infinity

In this section, we describe what it means for the values of a function to approach infinity and what it means for a function $f(x)$ to have a limit as x approaches infinity. Although there is no real number *infinity*, the word *infinity* is useful for describing how some functions behave when their domains or ranges exceed all bounds.

x	$\frac{1}{x}$
⋮	⋮
1000	0.001
10	0.1
0.1	10
0.001	1000
⋮	⋮
0	Undefined
⋮	⋮
−0.001	−1000
−0.1	−10
−10	−0.1
−1000	−0.001
⋮	⋮

2.12 The graph of $y = 1/x$.

Limits as $x \to \infty$ or $x \to -\infty$

The function

$$f(x) = \frac{1}{x} \tag{1}$$

is defined for all real numbers except $x = 0$.

As Fig. 2.12 suggests,

a) $1/x$ is small and positive when x is large and positive.
b) $1/x$ is large and positive when x is small and positive.
c) $1/x$ is large and negative when x is small and negative.
d) $1/x$ is small and negative when x is large and negative.

We summarize these facts by saying:

a) As x tends to ∞, $1/x$ approaches 0.
b) As x approaches 0 from the right, $1/x$ tends to ∞.
c) As x approaches 0 from the left, $1/x$ tends to $-\infty$.
d) As x tends to $-\infty$, $1/x$ approaches 0.

The symbol ∞, **infinity,** does not represent any real number. We cannot use ∞ in arithmetic in the usual way, but it is convenient to be able to say things like "the limit of $1/x$ as x approaches infinity is 0."

Calculation Rules for Functions with Finite Limits as $x \to \pm\infty$

Our strategy is again the one that worked so well in Section 2.1. We find the limits of two "basic" functions as $x \to \infty$ and $x \to -\infty$ and then, to find everything else, we use a theorem about limits of algebraic combinations. In Section 2.1, the basic functions were the constant function $y = k$ and the identity function $y = x$. Here, the basic functions are $y = k$ and the reciprocal $y = 1/x$. Our presentation continues to be informal.

EXAMPLE 1 If $f(x) = 3$ is the constant function whose outputs have the constant value 3, then

$$\lim_{x \to \infty} f(x) = \lim_{x \to \infty} (3) = 3$$

$$\lim_{x \to -\infty} f(x) = \lim_{x \to -\infty} (3) = 3.$$

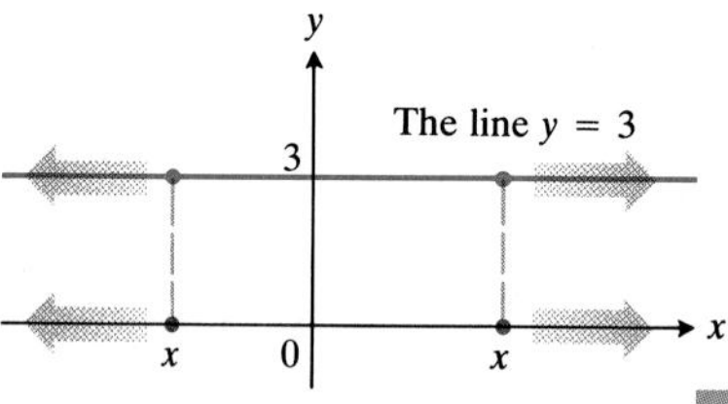

EXAMPLE 2 What we saw in Example 1 holds for any number k: If f is the constant function with $f(x) = k$, then

$$\lim_{x \to \infty} f(x) = \lim_{x \to \infty} (k) = k$$

$$\lim_{x \to -\infty} f(x) = \lim_{x \to -\infty} (k) = k.$$

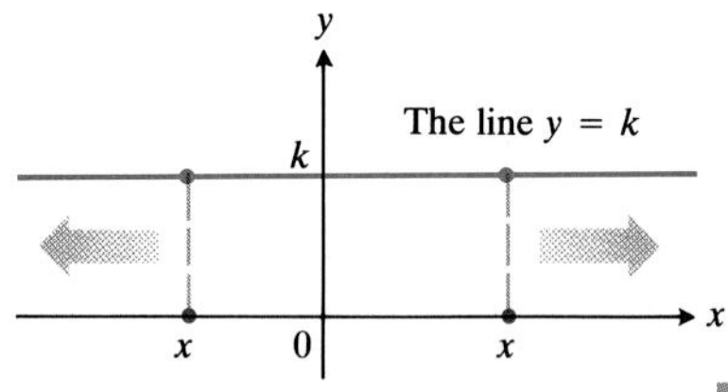

EXAMPLE 3

$$\lim_{x \to \infty} \frac{1}{x} = 0$$

As $x \to \infty$, the graph of $1/x$ gets arbitrarily close to the x-axis, in the following sense: No matter how small a positive number you name, the value of $1/x$ eventually gets smaller.

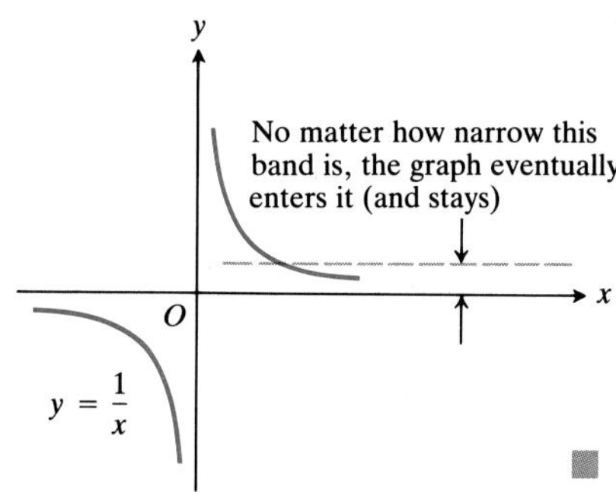

EXAMPLE 4

$$\lim_{x \to -\infty} \frac{1}{x} = 0$$

Once again, the graph gets arbitrarily close to the x-axis, this time rising toward it from below.

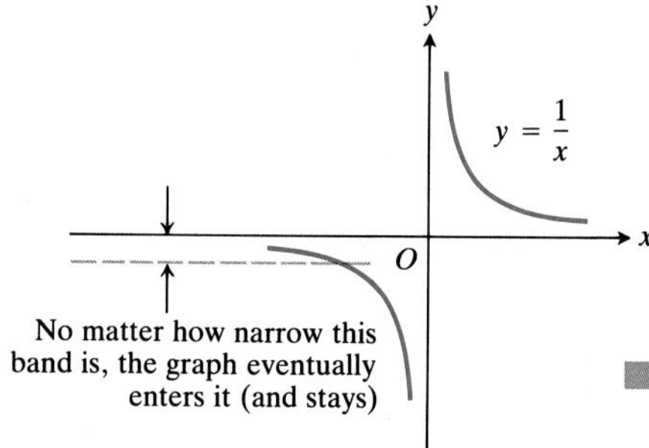

Believe it or not, we now have enough specific information to calculate the limits of a wide variety of rational functions as $x \to \pm\infty$. To do so, we simply use the limit properties listed in the following theorem.

THEOREM 5

Properties of Finite Limits as $x \to \pm\infty$

If

$$\lim_{x\to\infty} f_1(x) = L_1 \quad \text{and} \quad \lim_{x\to\infty} f_2(x) = L_2$$

and L_1 and L_2 are (finite) real numbers, then

1. *Sum Rule:* $\lim_{x\to\infty} [f_1(x) + f_2(x)] = L_1 + L_2$
2. *Difference Rule:* $\lim_{x\to\infty} [f_1(x) - f_2(x)] = L_1 - L_2$
3. *Product Rule:* $\lim_{x\to\infty} f_1(x) \cdot f_2(x) = L_1 \cdot L_2$
4. *Constant Multiple Rule:* $\lim_{x\to\infty} k \cdot f_1(x) = k \cdot L_1$ (any number k)
5. *Quotient Rule:* $\lim_{x\to\infty} \dfrac{f_1(x)}{f_2(x)} = \dfrac{L_1}{L_2}$ if $L_2 \neq 0$.

These properties hold for $x \to -\infty$ as well as $x \to \infty$.

These properties are just like the properties we stated in Section 2.1 for limits as $x \to c$, and we use them the same way.

EXAMPLE 5

$$\lim_{x\to\infty}\left(5 + \frac{1}{x}\right) = \lim_{x\to\infty} 5 + \lim_{x\to\infty}\frac{1}{x} \qquad \text{(Sum Rule)}$$

$$= 5 + 0 = 5. \qquad \text{(Known values)}$$

EXAMPLE 6

$$\lim_{x\to-\infty}\frac{4}{x^2} = \lim_{x\to-\infty} 4 \cdot \lim_{x\to-\infty}\frac{1}{x} \cdot \lim_{x\to-\infty}\frac{1}{x} \qquad \text{(Product Rule)}$$

$$= 4 \cdot 0 \cdot 0 = 0. \qquad \text{(Known values)}$$

EXAMPLE 7 $\lim_{x\to\infty}\dfrac{\sin x}{x} = 0$

The Sandwich Theorem, stated in Section 2.2 for limits as $x \to c$, also holds for limits as $x \to \pm\infty$.

Notice first that $\sin x$ lies between -1 and 1. Therefore, for all positive values of x,

$$-\frac{1}{x} \leq \frac{\sin x}{x} \leq \frac{1}{x}.$$

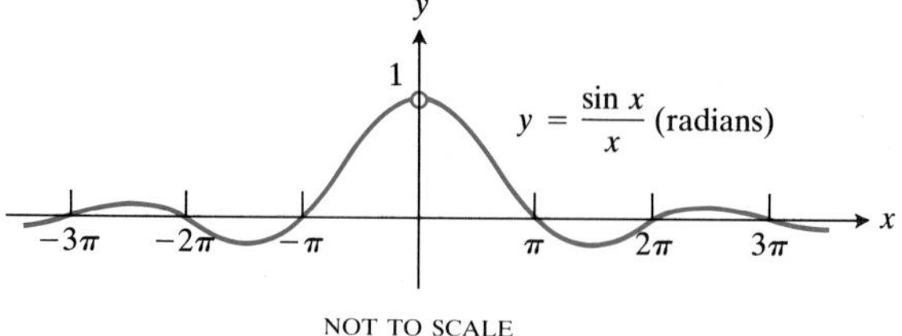

2.13 The graph of $y = (\sin x)/x$ oscillates about the x-axis. The amplitude of the oscillation decreases toward zero as $x \to \pm\infty$.

Because $-1/x$ and $1/x$ both approach 0 as $x \to \infty$, a modified version of the Sandwich Theorem tells us that $(\sin x)/x$ approaches 0 as well (Fig. 2.13). ■

Limits of Rational Functions as $x \to \pm\infty$

To find the limit of a rational function as $x \to \pm\infty$ (when the limit exists), we divide the numerator and denominator by the highest power of x in the denominator.

EXAMPLE 8 *A rational function in which the numerator and denominator have the same degree.*

$$\lim_{x\to\infty} \frac{-x}{7x+4} = \lim_{x\to\infty} \frac{-1}{7+(4/x)} \qquad \left(\begin{array}{l}\text{Divide numerator and denominator}\\ \text{by highest power of } x \text{ in denominator,}\\ \text{in this case } x.\end{array}\right)$$

$$= \frac{-1}{7+0} \qquad \text{(Known values)}$$

$$= -\frac{1}{7}.$$ ■

EXAMPLE 9 *Another rational function in which the numerator and denominator have the same degree.*

$$\lim_{x\to-\infty} \frac{2x^2-x+3}{3x^2+5} = \lim_{x\to-\infty} \frac{2-(1/x)+(3/x^2)}{3+(5/x^2)} \qquad \left(\begin{array}{l}\text{Divide numerator and}\\ \text{denominator by } x^2.\end{array}\right)$$

$$= \frac{2-0+0}{3+0} \qquad \text{(Known values)}$$

$$= \frac{2}{3}.$$ ■

EXAMPLE 10 *A rational function in which the degree of the numerator is less than the degree of the denominator.*

$$\lim_{x\to\infty} \frac{5x+2}{2x^3-1} = \lim_{x\to\infty} \frac{(5/x^2)+(2/x^3)}{2-(1/x^3)} \qquad \left(\begin{array}{l}\text{Divide numerator and}\\ \text{denominator by } x^3\end{array}\right)$$

$$= \frac{0+0}{2-0} \qquad \text{(Known values)}$$

$$= 0.$$ ■

EXAMPLE 11 *Sums and Products.*

a) $$\lim_{x\to\infty}\left(\frac{-x}{7x+4}+\frac{5x+2}{2x^3-1}\right)=\lim_{x\to\infty}\frac{-x}{7x+4}+\lim_{x\to\infty}\frac{5x+2}{2x^3-1} \qquad \text{(Sum Rule)}$$

$$= -\frac{1}{7}+0=-\frac{1}{7}. \qquad \text{(Known values)}$$

b) $$\lim_{x\to\infty}\left(\frac{-x}{7x+4}\cdot\frac{5x+2}{2x^3-1}\right)=\lim_{x\to\infty}\frac{-x}{7x+4}\cdot\lim_{x\to\infty}\frac{5x+2}{2x^3-1} \qquad \text{(Product Rule)}$$

$$= -\frac{1}{7}\cdot 0=0. \qquad \text{(Known values)}$$

We could have multiplied the original fractions together first, to express the product as a single rational function, but it would have taken longer to get the answer that way. ■

The *degree* of $a_nx^n + a_{n-1}x^{n-1} + \cdots + a_1x + a_0$ is n, the largest exponent. The *leading coefficient* is a_n.

The rule for finding limits of rational functions as $x \to \pm\infty$ is this: If the numerator and denominator have the same degree, the limit is the ratio of the leading coefficients. If the degree of the numerator is less than the degree of the denominator, the limit is zero. If the degree of the numerator is greater than the degree of the denominator, the limit is infinite, as we shall see in a moment.

Lim $f(x) = \infty$ or lim $f(x) = -\infty$

As suggested by the behavior of $1/x$ as $x \to 0$, we sometimes want to say such things as

$$\lim_{x\to c} f(x)=\infty, \qquad (1)$$

$$\lim_{x\to c^+} f(x)=\infty, \qquad (2)$$

$$\lim_{x\to c^-} f(x)=\infty, \qquad (3)$$

$$\lim_{x\to \infty} f(x)=\infty, \qquad (4)$$

$$\lim_{x\to -\infty} f(x)=\infty. \qquad (5)$$

In every instance, we mean that the value of $f(x)$ eventually exceeds any positive real number B. That is, for any real number B no matter how large, the values of f eventually satisfy the condition

$$f(x) > B. \qquad (6)$$

Similarly, we write

$$\lim_{x\to c} f(x)=-\infty, \quad \lim_{x\to c^+} f(x)=-\infty, \quad \lim_{x\to c^-} f(x)=-\infty \qquad (7)$$

to say that no matter how large the negative number $-B$ may be numerically, the values of f eventually satisfy the condition

$$f(x) < -B. \qquad (8)$$

EXAMPLE 12

$$\lim_{x\to 0^+} \frac{1}{x} = \infty$$

$$\lim_{x\to 0^-} \frac{1}{x} = -\infty$$

EXAMPLE 13

$$\lim_{x\to 0^+} \frac{1}{x^2} = \infty$$

$$\lim_{x\to 0^-} \frac{1}{x^2} = \infty$$

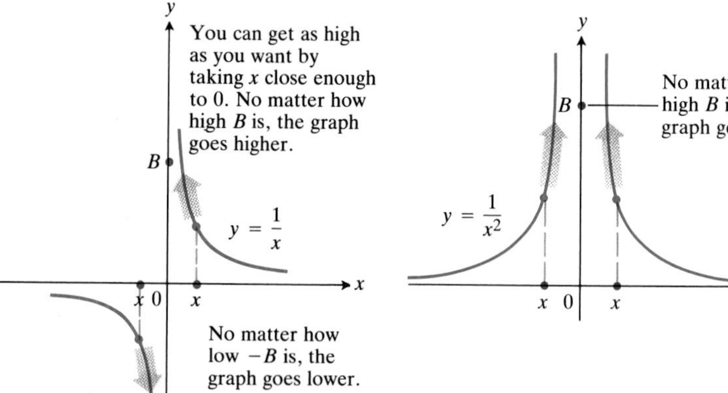

EXAMPLE 14

$$\lim_{x\to 1^+} \frac{1}{x-1} = \infty$$

$$\lim_{x\to 1^-} \frac{1}{x-1} = -\infty$$

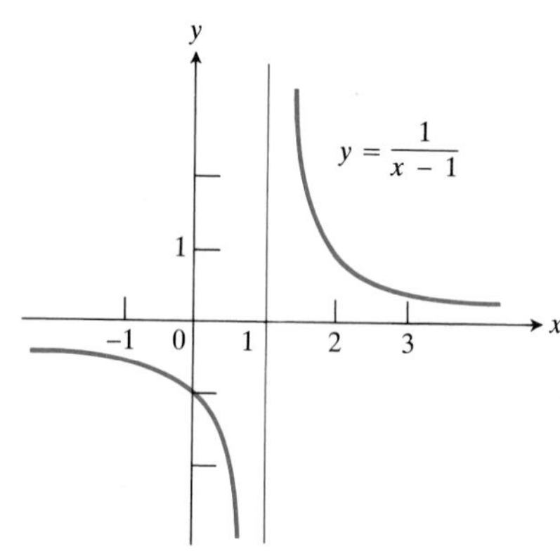

The idea here is to think about the number $x - 1$. As x approaches 1 from above, $x - 1$ approaches 0 from above. The reciprocal $1/(x - 1)$ stays positive and increases beyond all bounds.

On the other hand, if x approaches 1 from below, $x - 1$ remains negative and approaches 0 from below. Its reciprocal $1/(x - 1)$ is negative as well, and approaches $-\infty$. ■

EXAMPLE 15 *A rational function in which the degree of the numerator is greater than the degree of the denominator.*

a) $$\lim_{x\to\infty} \frac{2x^2 - 3}{7x + 4} = \lim_{x\to\infty} \frac{2x - (3/x)}{7 + (4/x)}$$ (Divide by the highest power of x in the denominator.)

$$= \infty$$ (The numerator approaches ∞ and the denominator approaches 7, so their ratio approaches ∞.)

b) $\lim_{x\to-\infty} \frac{2x^2 - 3}{7x + 4} = \lim_{x\to-\infty} \frac{2x - (3/x)}{7 + (4/x)}$ (Divide by the highest power of x in the denominator.)

$= -\infty$ (The numerator approaches $-\infty$ and the denominator approaches 7, so their ratio approaches $-\infty$.) ■

The procedure in Example 15 is typical. To find the limit as $x \to \pm\infty$ of a rational function in which the degree of the numerator exceeds the degree of the denominator, divide by the highest power of x in the denominator. The limit of the new denominator will then be finite, and the limit of the new numerator will be infinite. The limit of the ratio can be either $+\infty$ or $-\infty$, depending on the signs assumed by the numerator and denominator as x becomes numerically large.

Here is another example.

EXAMPLE 16 *A rational function in which the degree of the numerator is greater than the degree of the denominator.*

a) $\lim_{x\to\infty} \frac{-4x^3 + 7x}{2x^2 - 3x - 10} = \lim_{x\to\infty} \frac{-4x + (7/x)}{2 - (3/x) - (10/x^2)}$ (Divide by the highest power of x in the denominator.)

$= -\infty$ (Numerator $\to -\infty$. Denominator $\to 2$. Ratio $\to -\infty$.)

b) $\lim_{x\to-\infty} \frac{-4x^3 + 7x}{2x^2 - 3x - 10} = \lim_{x\to-\infty} \frac{-4x + (7/x)}{2 - (3/x) - (10/x^2)}$ (Divide by the highest power of x in the denominator.)

$= \infty$ (Numerator $\to \infty$. Denominator $\to 2$. Ratio $\to \infty$.) ■

Summary for Rational Functions

a) $\lim_{x\to\pm\infty} \frac{f(x)}{g(x)}$ is 0 if $\deg(f) < \deg(g)$

b) $\lim_{x\to\pm\infty} \frac{f(x)}{g(x)}$ is finite if $\deg(f) = \deg(g)$

c) $\lim_{x\to\pm\infty} \frac{f(x)}{g(x)}$ is infinite if $\deg(f) > \deg(g)$

Changing Variables with Substitutions

Sometimes a change of variable can turn an unfamiliar expression into one whose limit we know how to find. Here are two examples.

EXAMPLE 17

$\lim_{x\to\infty} \sin\frac{1}{x} = \lim_{\theta\to 0^+} \sin\theta$ (Substitute $\theta = 1/x$. Then $\theta \to 0^+$ as $x \to \infty$.)

$= 0.$ (Known value) ■

EXAMPLE 18

$$\lim_{x\to-\infty}\left(1+\frac{2}{x}\right)\left(\cos\frac{1}{x}\right) = \lim_{\theta\to0^-}(1+2\theta)(\cos\theta) \qquad \left(\text{Substitute } \theta=\frac{1}{x}.\ \text{Then } \theta\to0^- \text{ as } x\to-\infty.\right)$$

$$= \lim_{\theta\to0^-}(1+2\theta)\lim_{\theta\to0^-}\cos\theta \qquad (\text{Product Rule})$$

$$= 1\cdot1 = 1. \qquad (\text{Known values})$$

Exercises 2.3

Find the limits of the functions defined by the expressions in Exercises 1–18 (a) as $x\to\infty$ and (b) as $x\to-\infty$.

1. $\dfrac{2x+3}{5x+7}$
2. $\dfrac{2x^3+7}{x^3-x^2+x+7}$
3. $\dfrac{x+1}{x^2+3}$
4. $\dfrac{3x+7}{x^2-2}$
5. $\dfrac{1-12x^2}{4x^2+12}$
6. $\dfrac{2x^2+3}{-x^2+x}$
7. $\dfrac{3x^2-6x}{4x-8}$
8. $\dfrac{x^4}{x^3+1}$
9. $\dfrac{1}{x^3-4x+1}$
10. $\dfrac{10x^5+x^4+31}{x^6}$
11. $\dfrac{7x^3}{x^3-3x^2+6x}$
12. $\dfrac{9x^4+x}{2x^4+5x^2-x+6}$
13. $\dfrac{-2x^3-2x+3}{3x^3+3x^2-5x}$
14. $\dfrac{-x^4}{x^4-7x^3+7x^2+9}$
15. $\left(\dfrac{-x}{x+1}\right)\left(\dfrac{x^2}{5+x^2}\right)$
16. $\left(\dfrac{2}{x}+1\right)\left(\dfrac{5x^2-1}{x^2}\right)$
17. $\left(\dfrac{1-x^2}{1+2x^2}\right)\left(\dfrac{8x^2+7x}{4x^2}\right)$
18. $\left(\dfrac{x-3}{x^2-5x+4}\right)\left(\dfrac{x}{x-1}\right)$

Find the limits in Exercises 19–28.

19. $\displaystyle\lim_{x\to0^+}\frac{1}{3x}$
20. $\displaystyle\lim_{x\to0^+}\frac{5}{2x}$
21. $\displaystyle\lim_{x\to2^+}\frac{1}{x-2}$
22. $\displaystyle\lim_{x\to2^-}\frac{1}{x-2}$
23. $\displaystyle\lim_{x\to2^+}\frac{x}{x-2}$
24. $\displaystyle\lim_{x\to2^-}\frac{x}{x-2}$
25. $\displaystyle\lim_{x\to-3^+}\frac{1}{x+3}$
26. $\displaystyle\lim_{x\to-3^-}\frac{1}{x+3}$
27. $\displaystyle\lim_{x\to-3^+}\frac{x}{x+3}$
28. $\displaystyle\lim_{x\to-3^-}\frac{x}{x+3}$

Find the limits in Exercises 29–36.

29. $\displaystyle\lim_{t\to\infty}\frac{t^2-2t+3}{2t^2+5t-3}$
30. $\displaystyle\lim_{r\to-\infty}\frac{3r}{1-2r^2}$
31. $\displaystyle\lim_{s\to\infty}\left(\frac{1}{s^4}+\frac{1}{s^2}+1\right)$
32. $\displaystyle\lim_{y\to-\infty}\frac{9y^3-7y}{y^3+1}$
33. $\displaystyle\lim_{z\to\infty}\frac{z^3-7z+10}{2z^2-7}$
34. $\displaystyle\lim_{x\to\infty}\frac{8x^{23}-7x^2+5}{2x^{23}+x^{22}}$
35. $\displaystyle\lim_{t\to\infty}\left(\frac{t^2+5}{3t^2-2}-\frac{5t-7}{6t+8}\right)$
36. $\displaystyle\lim_{y\to-\infty}\left(\frac{y+2}{y+3}-\frac{y^3+2}{y^3+3}\right)$

In Exercises 37–40, use the graphs to find the values of the limits.

37. $\displaystyle\lim_{x\to2^+}\frac{1}{x^2-4}$

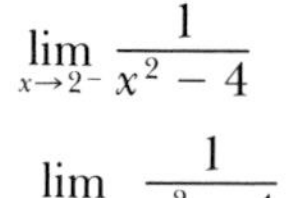

$\displaystyle\lim_{x\to2^-}\frac{1}{x^2-4}$

$\displaystyle\lim_{x\to-2^+}\frac{1}{x^2-4}$

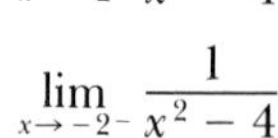

$\displaystyle\lim_{x\to-2^-}\frac{1}{x^2-4}$

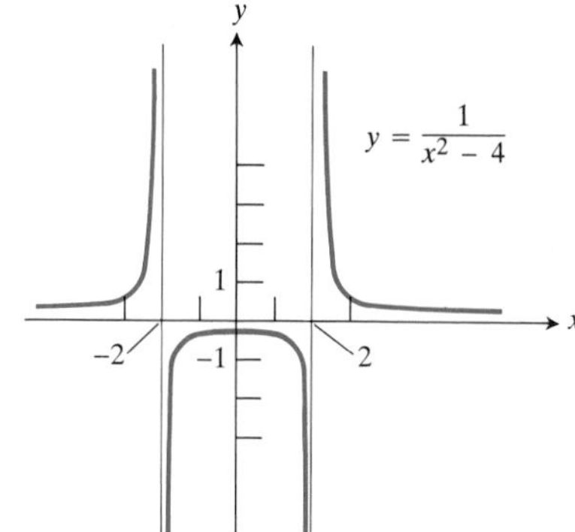

38. $\displaystyle\lim_{x\to1^+}\frac{x}{x^2-1}$

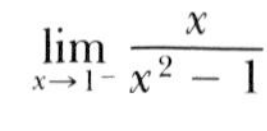

$\displaystyle\lim_{x\to1^-}\frac{x}{x^2-1}$

$\displaystyle\lim_{x\to-1^+}\frac{x}{x^2-1}$

$\displaystyle\lim_{x\to-1^-}\frac{x}{x^2-1}$

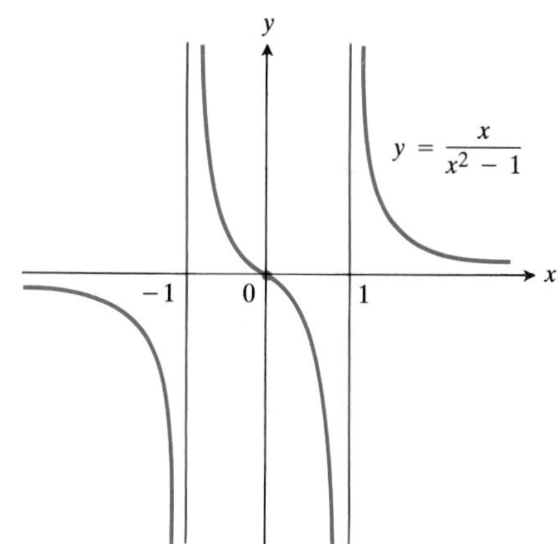

39. $\lim_{x\to -2^+} \dfrac{x^2 - 1}{2x + 4}$

$\lim_{x\to -2^-} \dfrac{x^2 - 1}{2x + 4}$

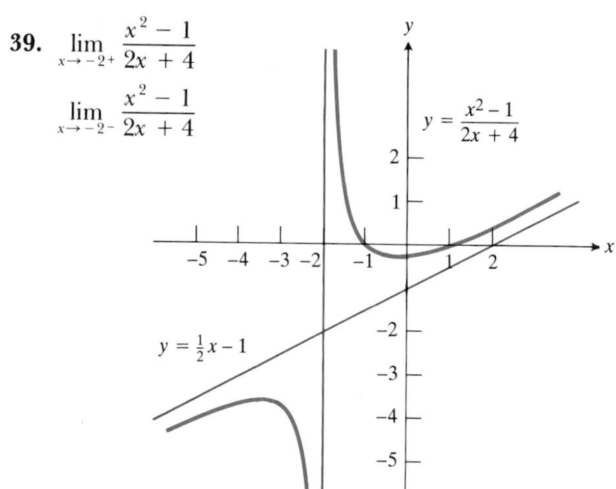

40. $\lim_{x\to 0^+} \left(x^2 + \dfrac{4}{x}\right)$

$\lim_{x\to 0^-} \left(x^2 + \dfrac{4}{x}\right)$

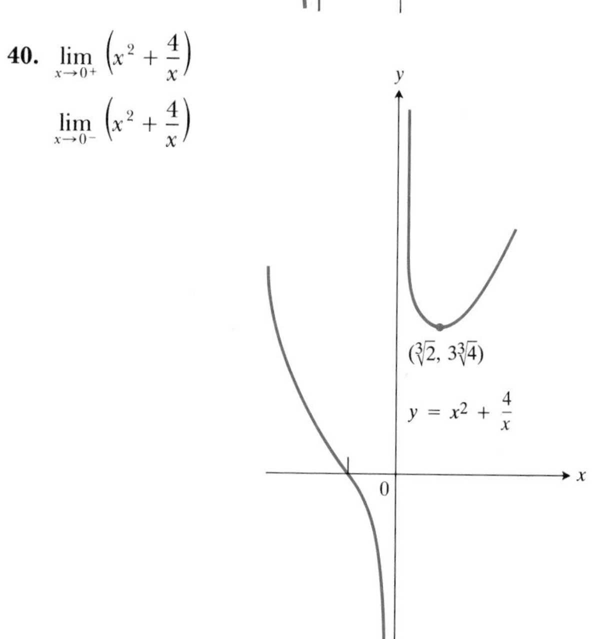

Find the limits in Exercises 41–48.

41. $\lim_{x\to 0^+} \dfrac{\lfloor x \rfloor}{x}$

42. $\lim_{x\to 0^-} \dfrac{\lfloor x \rfloor}{x}$

43. $\lim_{x\to \infty} \dfrac{|x|}{|x| + 1}$

44. $\lim_{x\to -\infty} \dfrac{x}{|x|}$

45. $\lim_{x\to 0^+} \dfrac{1}{\sin x}$

46. $\lim_{x\to 0^-} \dfrac{1}{\sin x}$

47. $\lim_{x\to (\pi/2)^+} \dfrac{1}{\cos x}$

48. $\lim_{x\to (\pi/2)^-} \dfrac{1}{\cos x}$

49. Let $f(x) = \begin{cases} \dfrac{1}{x}, & x < 0 \\ -1, & x \geq 0. \end{cases}$

Find $\lim f(x)$ as $x \to -\infty, 0^-, 0^+$, and ∞.

50. Let $f(x) = \begin{cases} \dfrac{x-2}{x-1}, & x \leq 0 \\ \dfrac{1}{x^2}, & x > 0. \end{cases}$

Find $\lim f(x)$ as $x \to -\infty, 0^-, 0^+$, and ∞.

Find the limits in Exercises 51–56.

51. $\lim_{x\to\infty} \left(2 + \dfrac{\sin x}{x}\right)$

52. $\lim_{x\to -\infty} \dfrac{\sin x}{x}$

53. $\lim_{x\to\infty} \left(1 + \cos \dfrac{1}{x}\right)$

54. $\lim_{x\to\infty} x \sin \dfrac{1}{x}$

55. $\lim_{x\to\infty} \dfrac{\sin 2x}{x}$

56. $\lim_{x\to\infty} \dfrac{\cos(1/x)}{1 + (1/x)}$

As we mentioned, the Sandwich Theorem in Section 2.2 also holds for limits as $x \to \pm\infty$. Use the theorem to find the limits in Exercises 57 and 58.

57. Find $\lim_{x\to\infty} f(x)$ and $\lim_{x\to -\infty} f(x)$ if

$$\frac{2x - 3}{x} < f(x) < \frac{2x^2 + 5x}{x^2}.$$

58. *The greatest integer function.* Find $\lim_{x\to\infty} \lfloor x \rfloor / x$ and $\lim_{x\to -\infty} \lfloor x \rfloor / x$ given that

$$\frac{x - 1}{x} < \frac{\lfloor x \rfloor}{x} \leq 1 \quad (x \neq 0).$$

59. **CALCULATOR** *Limits of ratios of logarithms.*

a) By taking $x = 10, 100, 1000$, and so on as far as your calculator can go, estimate the value of

$$\lim_{x\to\infty} \frac{\ln(x + 1)}{\ln x}.$$

b) Does the 1 in $\ln(x + 1)$ really matter? Suppose you have 999 there instead. What do you get for the value of

$$\lim_{x\to\infty} \frac{\ln(x + 999)}{\ln x}?$$

c) Estimate the value of

$$\lim_{x\to\infty} \frac{\ln x^2}{\ln x}.$$

d) Estimate the value of

$$\lim_{x\to\infty} \frac{\ln x}{\log x}.$$

The behavior you see here will all be explained in Chapter 7.

TOOLKIT PROGRAMS

Limit Problems	Provides practice in determining when limits exist and in finding them when they do
Super*Grapher	Graphs all the functions in this section

2.4 Continuous Functions

© Robert A. Tyrrell 1989

The motion of a hummingbird's wing is continuous but so rapid that we can see the wing's shape only in a discrete photograph.

When Jim Fixx drew the heart-rate curve in Fig. 1.5, he plotted data that came from measurements in the field and fitted the plotted points with an unbroken curve. In doing so, he assumed that the values he was plotting were the values of a continuous function, a function whose outputs varied continuously with the inputs and did not jump from one value to another without taking on all the values in between.

Continuous functions are the functions we normally use in the equations that describe numerical relations in the world around us. They are the functions we use to find a cannon's maximum range or a planet's closest approach to the sun. They are also the functions we use to describe how a body moves through space or how the speed of a chemical reaction changes with time. In fact, so many observable physical processes proceed continuously that throughout the eighteenth and nineteenth centuries it rarely occurred to anyone to look for any other kind of behavior. It came as quite a surprise when the physicists of the 1920s discovered that the vibrating atoms in a hydrogen molecule can oscillate only at discrete energy levels, that light comes in particles, and that, when heated, atoms emit light in discrete frequencies and not in continuous spectra.

As a result of these and other discoveries, and because of the heavy use of discrete functions in computer science and statistics, the issue of continuity has become one of practical as well as theoretical importance. As scientists, we need to know when continuity is called for, what it is, and how to test for it.

The Definition of Continuity

A function $y = f(x)$ that can be graphed over each interval of its domain with one continuous motion of the pen is an example of a **continuous function.** The height of the graph over the interval varies continuously with x. At each interior point of the function's domain, like the point c in Fig. 2.14, the function value $f(c)$ is the limit of the function values on either side; that is,

$$f(c) = \lim_{x \to c} f(x).$$

The function value at each endpoint is also the limit of the nearby function values. At the left endpoint a in Fig. 2.14,

$$f(a) = \lim_{x \to a^+} f(x).$$

At the right endpoint b,

$$f(b) = \lim_{x \to b^-} f(x).$$

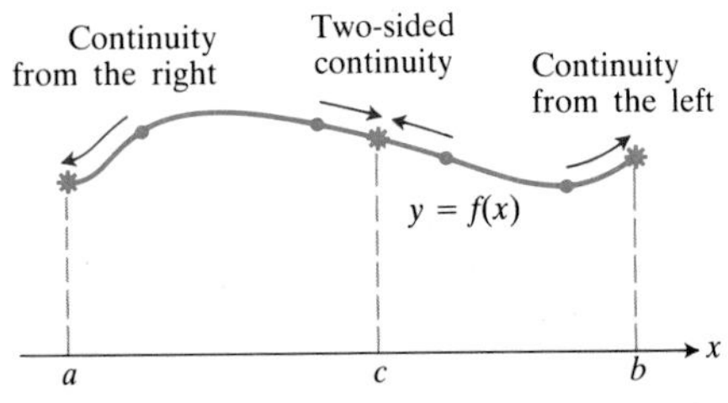

2.14 Continuity at points a, b, and c.

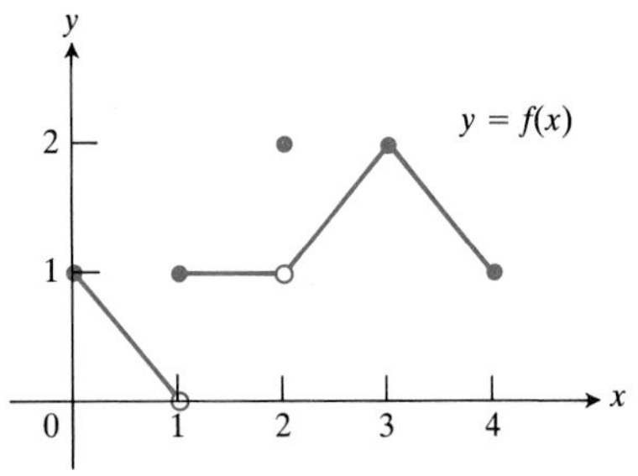

2.15 Discontinuity at $x = 1$ and $x = 2$.

To be specific, let us look at the function in Fig. 2.15, whose limits we investigated in Example 13 in Section 2.1.

EXAMPLE 1 The function in Fig. 2.15 is continuous at every point in its domain except $x = 1$ and $x = 2$. At these points there are breaks in the graph. Note the relation between the limit of f and the value of f at each point of the function's domain.

Points of discontinuity:

$$\text{At } x = 1: \quad \lim_{x\to 1} f(x) \text{ does not exist.}$$

$$\text{At } x = 2: \quad \lim_{x\to 2} f(x) = 1, \text{ but } 1 \neq f(2).$$

Points at which f is continuous:

$$\text{At } x = 0: \quad \lim_{x\to 0^+} f(x) = f(0).$$

$$\text{At } x = 4: \quad \lim_{x\to 4^-} f(x) = f(4).$$

$$\text{At every point } 0 < c < 4 \text{ except } x = 1, 2: \quad \lim_{x\to c} f(x) = f(c).$$

We now come to the formal definition of continuity at a point in a function's domain. In the definition we distinguish between continuity at an endpoint (which involves a one-sided limit) and continuity at an interior point (which involves a two-sided limit).

DEFINITIONS

Continuity at an Interior Point

A function $y = f(x)$ is continuous at an interior point c of its domain if

$$\lim_{x\to c} f(x) = f(c). \tag{1}$$

Continuity at an Endpoint

A function $y = f(x)$ is continuous at a left endpoint a of its domain if

$$\lim_{x\to a^+} f(x) = f(a). \tag{2}$$

A function $y = f(x)$ is continuous at a right endpoint b of its domain if

$$\lim_{x\to b^-} f(x) = f(b). \tag{3}$$

Continuous Function

A function is continuous if it is continuous at each point of its domain.

Discontinuity at a Point

If a function f is not continuous at a point c, we say that f is discontinuous at c and call c a point of discontinuity of f.

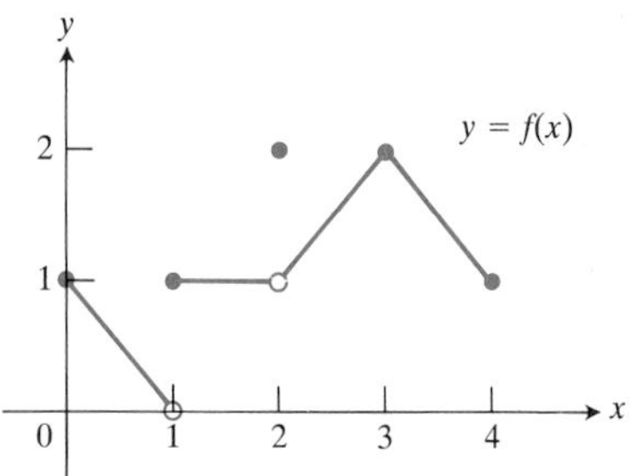

2.16 This function is continuous at $x = 0$, 3, and 4, and discontinuous at $x = 1$ and 2.

How to Test for Continuity at a Point

To test for continuity at a point, we apply the following test.

> **The Continuity Test**
>
> The function $y = f(x)$ is continuous at $x = c$ if and only if *all three* of the following statements are true:
>
> 1. $f(c)$ exists (c lies in the domain of f).
> 2. $\lim_{x\to c} f(x)$ exists (f has a limit as $x \to c$).
> 3. $\lim_{x\to c} f(x) = f(c)$ (the limit equals the function value).

(The limit in the continuity test is to be two-sided if c is an interior point of the domain of f; it is to be the appropriate one-sided limit if c is an endpoint of the domain.)

EXAMPLE 2 When applied to the function $y = f(x)$ of Example 1 at the points $x = 0$, 1, 2, 3, and 4, the continuity test gives the following results. (The graph of f is reproduced here as Fig. 2.16.)

a) f is continuous at $x = 0$ because
   i) $f(0)$ exists (it equals 1),
   ii) $\lim_{x\to 0^+} f(x) = 1$ (f has a limit as $x \to 0^+$),
   iii) $\lim_{x\to 0^+} = f(0)$ (the limit equals the function value).

b) f is discontinuous at $x = 1$ because $\lim_{x\to 1} f(x)$ does not exist. The function fails part (2) of the test. (The right-hand and left-hand limits exist at $x = 1$, but they are not equal.)

c) f is discontinuous at $x = 2$ because $\lim_{x\to 2} f(x) \neq f(2)$. The function fails part (3) of the test.

d) f is continuous at $x = 3$ because
   i) $f(3)$ exists (it equals 2),
   ii) $\lim_{x\to 3} f(x) = 2$ (f has a limit as $x \to 3$),
   iii) $\lim_{x\to 3} f(x) = f(3)$ (the limit equals the function value).

e) f is continuous at $x = 4$ because
   i) $f(4)$ exists (it equals 1),
   ii) $\lim_{x\to 4^-} f(x) = 1$ (f has a limit as $x \to 4^-$),
   iii) $\lim_{x\to 4^-} f(x) = f(4)$ (the limit equals the function value). ■

EXAMPLE 3 The function $y = 1/x$ is continuous at every value of x except $x = 0$. The function is not defined at $x = 0$ and therefore fails part (1) of the continuity test at $x = 0$. See Fig. 2.17. ■

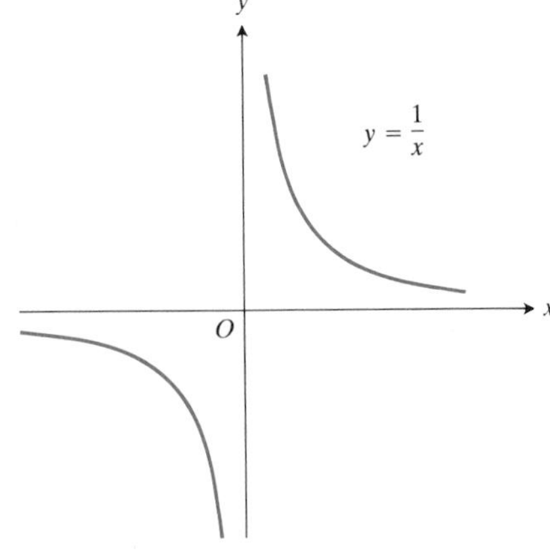

2.17 The function $y = 1/x$ is continuous at every point except $x = 0$.

2.19 The graph of $f(x) = \dfrac{x^2 + x - 6}{x^2 - 4}$.

The extended function

$$f(x) = \begin{cases} \dfrac{x^2 + x - 6}{x^2 - 4} & \text{if } x \neq 2, \\ \dfrac{5}{4} & \text{if } x = 2, \end{cases} \tag{4}$$

is continuous at $x = 2$ because $\lim_{x\to 2} f(x)$ exists and equals $f(2)$ (Fig. 2.19).

The function in Eq. (4) is called the **continuous extension** of the original function to the point $x = 2$. ■

Composites of Continuous Functions Are Continuous

All composites of continuous functions are continuous. This means that composites like

$$y = \sin\sqrt{x} \qquad \text{and} \qquad y = |\cos x|$$

are continuous at every point at which they are defined. The idea is that if $f(x)$ is continuous at $x = c$ and $g(x)$ is continuous at $x = f(c)$, then $g \circ f$ is continuous at $x = c$ (Fig. 2.20).

THEOREM 7

> If f is continuous at c and g is continuous at $f(c)$, then the composite $g \circ f$ is continuous at c.

For an outline of the proof of Theorem 7 see Problem 6 in Appendix 3.

EXAMPLE 11 Show that the function

$$y = \left|\frac{x \sin x}{x^2 + 2}\right|$$

is continuous at every value of x.

Solution The function y is the composite of the continuous functions

$$f(x) = \frac{x \sin x}{x^2 + 2} \qquad \text{and} \qquad g(x) = |x|.$$

The function f is continuous by Theorem 6, the function g by Example 8, and their composite $g \circ f$ by Theorem 7. ■

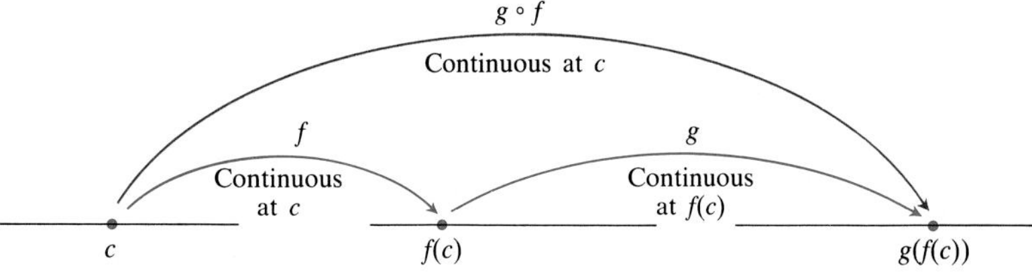

2.20 Composites of continuous functions are continuous.

If a composite function $g \circ f$ is continuous at a point $x = c$, its limit as $x \to c$ is $g(f(c))$.

EXAMPLE 12

a) $\lim_{x \to 1} \sin \sqrt{x - 1} = \sin \sqrt{1 - 1} = \sin 0 = 0$

b) $\lim_{x \to 0} |1 + \cos x| = |1 + \cos 0| = |1 + 1| = 2$ ■

Continuous Functions Have Important Properties

We study continuous functions because they are useful in mathematics and its applications. It turns out that every continuous function is some other function's derivative, as we shall see in Chapter 5. The ability to recover a function from information about its derivative is one of the great powers given to us by calculus. Thus, given a formula $v(t)$ for the velocity of a moving body as a continuous function of time, we shall be able, with the calculus of Chapters 3, 4, and 5, to produce a formula $s(t)$ that tells how far the body has traveled from its starting point at any instant.

In addition, a function that is continuous at every point of a closed interval $[a, b]$ has an absolute maximum value and an absolute minimum value on this interval. We always look for these values when we graph a function, and we shall see the role they play in problem solving (Chapter 4) and in the development of the integral calculus (Chapters 5 and 6).

Finally, a function f that is continuous at every point of a closed interval $[a, b]$ assumes every value between $f(a)$ and $f(b)$. We shall see some consequences of this in a moment.

The proofs of these properties require a detailed knowledge of the real number system and we shall not give them here.

THEOREM 8

The Max-Min Theorem for Continuous Functions

If f is continuous at every point of a closed interval $[a, b]$, then f takes on both an absolute maximum value M and an absolute minimum value m somewhere in that interval. That is, for some numbers x_1 and x_2 in $[a, b]$ we have $f(x_1) = m$, $f(x_2) = M$, and $m \le f(x) \le M$ at every other point x of the interval.

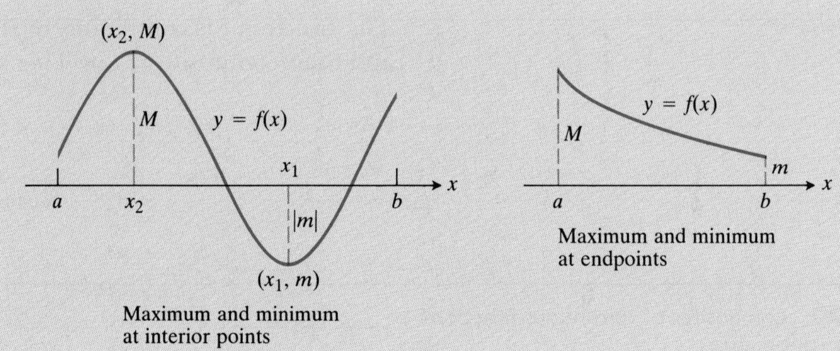

Maximum and minimum at interior points

Maximum and minimum at endpoints

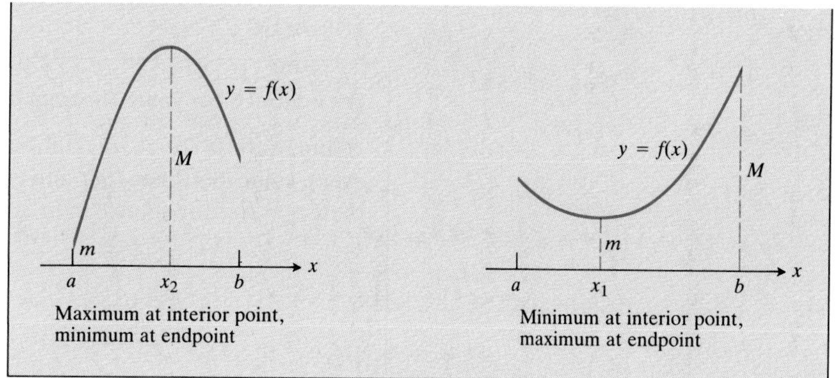

Maximum at interior point, minimum at endpoint

Minimum at interior point, maximum at endpoint

EXAMPLE 13 On the interval $-\pi/2 \leq x \leq \pi/2$, the cosine takes on a maximum value of 1 (once) and a minimum value of 0 (twice). The sine takes on a maximum value of 1 and a minimum value of -1.

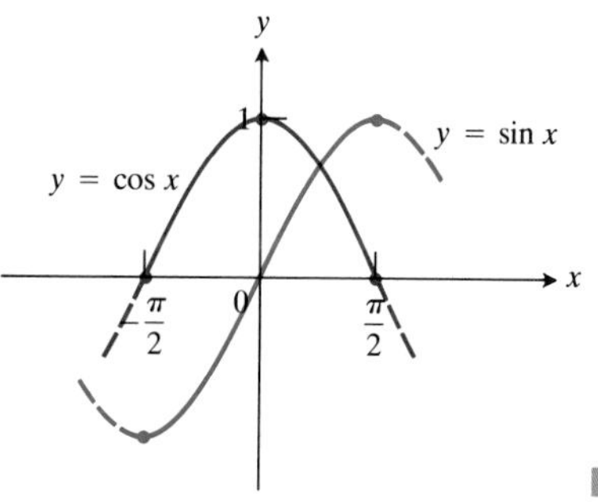

EXAMPLE 14 On an open interval, a continuous function need not have either a maximum or a minimum value. The function $f(x) = x$ has neither a largest nor a smallest value on the interval $0 < x < 1$.

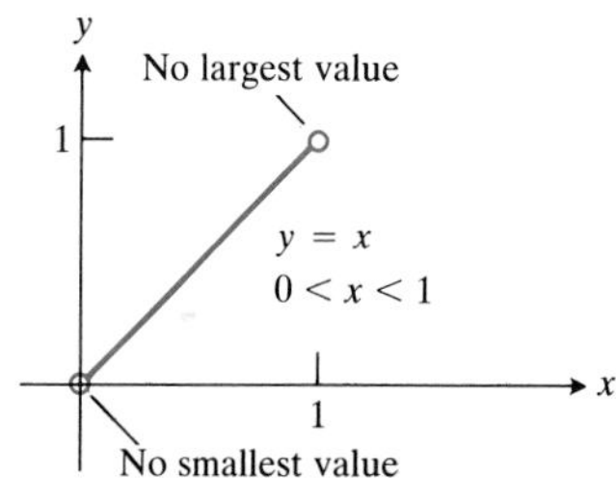

EXAMPLE 15 Even a single point of discontinuity can keep a function from having either a maximum or a minimum value on a closed interval. The function

$$y = \begin{cases} x + 1, & -1 \leq x < 0 \\ 0 & x = 0 \\ x - 1, & 0 < x \leq 1 \end{cases}$$

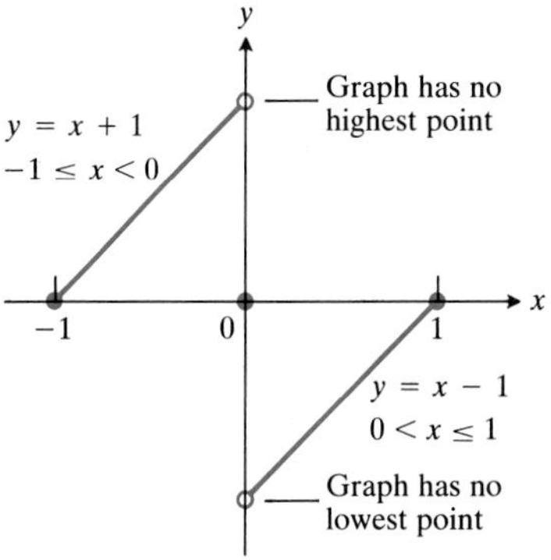

is continuous at every point of the interval $-1 \leq x \leq 1$ except $x = 0$, yet its graph over the interval has neither a highest nor a lowest point.

THEOREM 9

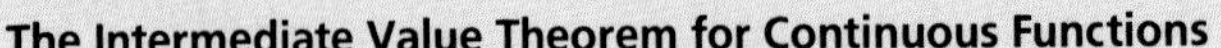

The Intermediate Value Theorem for Continuous Functions

A function $y = f(x)$ that is continuous on a closed interval $[a, b]$ takes on every value between $f(a)$ and $f(b)$. In other words, if $f(a) \leq y_0 \leq f(b)$ then $y_0 = f(c)$ for some c in $[a, b]$.

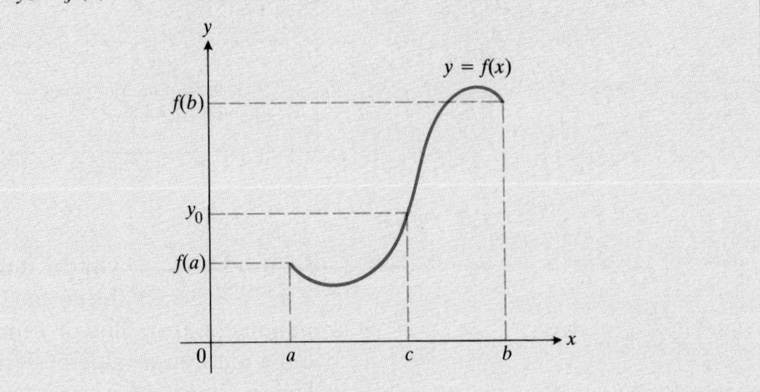

A Consequence for Graphing: Connectivity Suppose we want to graph a function $y = f(x)$ that is continuous throughout some interval I on the x-axis. Theorem 9 tells us that the graph of f over I will never move from one y-value to another without taking on the y-values in between. The graph of f over I will be **connected:** it will consist of a single, unbroken curve, like the graph of $y = \sin x$. The graph of f will not have jumps like the graph of the greatest-integer function or separate branches like the graph of $y = \tan x$.

A Consequence for Root Finding Suppose that $f(x)$ is continuous at every point of a closed interval $[a, b]$ and that $f(a)$ and $f(b)$ differ in sign. Then zero lies between $f(a)$ and $f(b)$, so there is at least one number c between a and b where $f(c) = 0$. In other words, if f is continuous and $f(a)$ and $f(b)$ differ in sign, then the equation $f(x) = 0$ has at least one solution in the open interval (a, b).

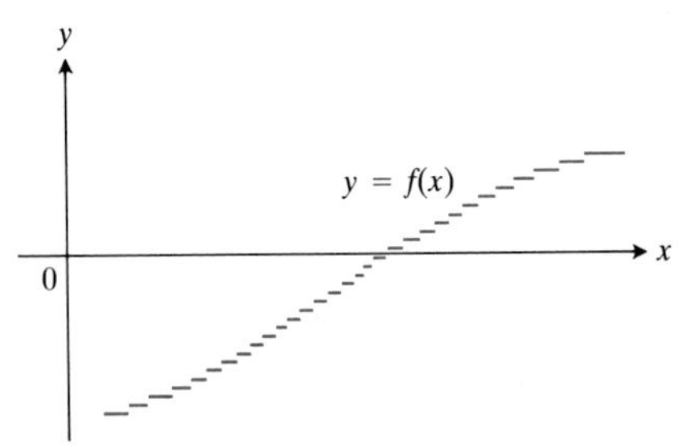

2.21 The graph of a continuous function never has steps like these.

This property of continuous functions enables us to locate roots by graphing. Between any point where the graph lies above the x-axis and any point where it lies below, there must be at least one point where it actually intersects the axis. The graph of a continuous function of x never steps across the x-axis the way the step function in Fig. 2.21 does. When we see the graph of a continuous function cross the axis on a computer screen, there really is a root there. In Chapter 3, we shall see how to find the value of such a root.

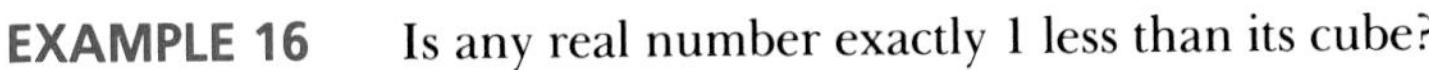

EXAMPLE 16 Is any real number exactly 1 less than its cube?

Solution Any such number must satisfy the equation $x = x^3 - 1$ or $x^3 - x - 1 = 0$. Hence, we are looking for a zero value of the function

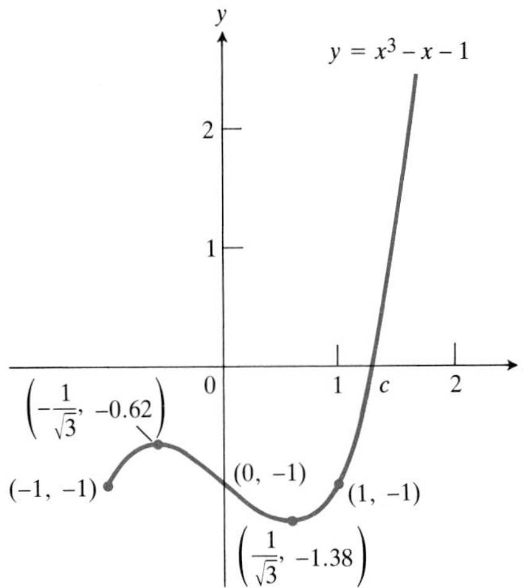

2.22 The graph of $f(x) = x^3 - x - 1$ crosses the x-axis between $x = 1$ and $x = 2$.

$f(x) = x^3 - x - 1$. By trial we find that $f(1) = -1$ and $f(2) = 5$. We conclude that the equation $f(x) = 0$ has at least one solution $x = c$ between 1 and 2. At this point, $c^3 - c - 1 = 0$, or $c = c^3 - 1$. So, yes, there is a number that is one less than its cube. In Section 3.8, where we study root finding, we shall find that c is about 1.32 (Fig. 2.22). ■

Concluding Remarks

For any function $y = f(x)$ it is important to distinguish between continuity at $x = c$ and having a limit as $x \to c$. The limit, $\lim_{x\to c} f(x)$, is where the function is headed as $x \to c$. Continuity is the property of arriving at the point where $f(x)$ has been heading when x actually gets to c. (Someone is home when you get there, so to speak.) If the limit is what you expect as $x \to c$, and the number $f(c)$ is what you get when $x = c$, then the function is continuous at c if you get what you expect.

Finally, remember the test for continuity at a point:

1. Does $f(c)$ exist?
2. Does $\lim_{x\to c} f(x)$ exist?
3. Does $\lim_{x\to c} f(x) = f(c)$?

For f to be continuous at $x = c$, all three answers must be *yes*.

Exercises 2.4

Exercises 1–6 are about the function graphed in Fig. 2.23, whose formula is

$$f(x) = \begin{cases} x^2 - 1 & -1 \le x < 0, \\ 2x, & 0 \le x < 1, \\ 1, & x = 1, \\ -2x + 4, & 1 < x < 2, \\ 0, & 2 < x \le 3. \end{cases}$$

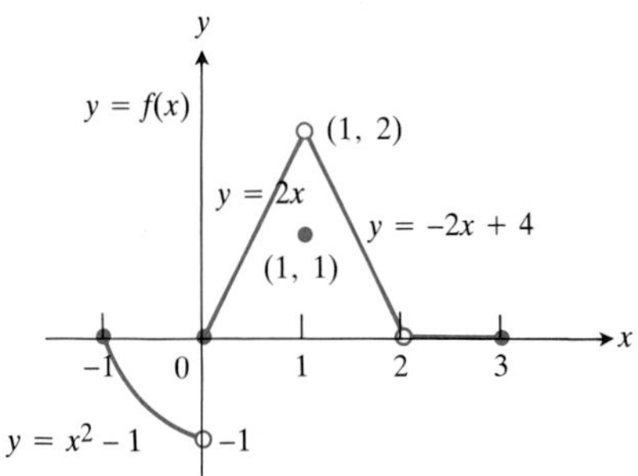

2.23 The function $y = f(x)$ for Exercises 1–6.

1. a) Does $f(-1)$ exist?
b) Does $\lim_{x\to -1^+} f(x)$ exist?
c) Does $\lim_{x\to -1^+} f(x) = f(-1)$?
d) Is f continuous at $x = -1$?

2. a) Does $f(1)$ exist?
b) Does $\lim_{x\to 1} f(x)$ exist?
c) Does $\lim_{x\to 1} f(x) = f(1)$?
d) Is f continuous at $x = 1$?

3. a) Is f defined at $x = 2$? (Look at the definition of f.)
b) Is f continuous at $x = 2$?

4. At what values of x is f continuous?

5. a) What is the value of $\lim_{x\to 2} f(x)$?
b) What value should be assigned to $f(2)$ to make f continuous at $x = 2$?

6. To what new value should $f(1)$ be changed to make f continuous at $x = 1$?

At which points are the functions in the following exercises in Section 2.1 continuous?

7. Exercise 45

8. Exercise 46

9. Exercise 47

10. Exercise 48

11. Exercise 49

12. Exercise 50

13. At what points is the function

$$f(x) = \begin{cases} 0, & x < 0, \\ 1, & 0 \leq x \leq 1, \\ 0, & 1 < x, \end{cases}$$

continuous? (*Hint:* Graph the function.)

14. Let $f(x)$ be defined by

$$f(x) = \begin{cases} 1 & \text{for } x < 0, \\ \sqrt{1 - x^2} & \text{for } 0 \leq x \leq 1, \\ x - 1 & \text{for } x > 1. \end{cases}$$

Is f continuous? (*Hint:* Graph the function.)

Find the points, if any, at which the functions in Exercises 15–24 are *not* continuous.

15. $y = \dfrac{1}{x - 2}$

16. $y = \dfrac{1}{(x + 2)^2}$

17. $y = \dfrac{x}{x + 1}$

18. $y = \dfrac{x + 1}{x^2 - 4x + 3}$

19. $y = |x - 1|$

20. $y = \dfrac{x + 3}{x^2 - 3x - 10}$

21. $y = \dfrac{x^3 - 1}{x^2 - 1}$

22. $y = \dfrac{1}{x^2 + 1}$

23. $y = \dfrac{\cos x}{x}$

24. $y = \dfrac{|x|}{x}$

25. The function $f(x)$ is defined by $f(x) = (x^2 - 1)/(x - 1)$ when $x \neq 1$ and by $f(1) = 2$. Is f continuous at $x = 1$? Explain.

26. Define $g(3)$ in a way that extends $g(x) = (x^2 - 9)/(x - 3)$ to be continuous at $x = 3$.

27. Define $h(2)$ in a way that extends $h(x) = (x^2 + 3x - 10)/(x - 2)$ to be continuous at $x = 2$.

28. Define $f(1)$ in a way that extends $f(x) = (x^3 - 1)/(x^2 - 1)$ to be continuous at $x = 1$.

29. Define $g(4)$ in a way that extends $g(x) = (x^2 - 16)/(x^2 - 3x - 4)$ to be continuous at $x = 4$.

30. How should $f(2)$ be redefined in Fig. 2.16 to make the function continuous at $x = 2$?

31. What value should be assigned to a to make the function

$$f(x) = \begin{cases} x^2 - 1, & x < 3, \\ 2ax, & x \geq 3, \end{cases}$$

continuous at $x = 3$?

32. What value should be assigned to b to make the function

$$g(x) = \begin{cases} x^3, & x < 1/2, \\ bx^2, & x \geq 1/2, \end{cases}$$

continuous at $x = 1/2$?

Find the limits in Exercises 33–36.

33. $\lim\limits_{x \to 0} \dfrac{1 + \cos x}{2}$

34. $\lim\limits_{x \to 0} \cos\left(1 - \dfrac{\sin x}{x}\right)$

35. $\lim\limits_{x \to 0} \tan x$

36. $\lim\limits_{x \to 0} \sin\left(\dfrac{\pi}{2} \cos(\tan x)\right)$

37. At what values of x does the function in Fig. 2.16 take on its maximum value? Does the function take on a minimum value? Explain.

38. At what values (if any) does the function in Fig. 2.23 take on a maximum value? A minimum value?

39. Does the function $y = x^2$ have a maximum value on the open interval $-1 < x < 1$? A minimum value? Explain.

40. On the closed interval $0 \leq x \leq 1$ the greatest-integer function $y = \lfloor x \rfloor$ takes on a minimum value $m = 0$ and a maximum value $M = 1$. It does so even though it is discontinuous at $x = 1$. Does this violate the Max-Min Theorem? Why?

41. A continuous function $y = f(x)$ is known to be negative at $x = 0$ and positive at $x = 1$. Why does the equation $f(x) = 0$ have at least one solution between $x = 0$ and $x = 1$? Illustrate with a sketch.

42. Assuming $y = \cos x$ to be continuous, show that the equation $\cos x = x$ has at least one solution. (*Hint:* Show that the equation $\cos x - x = 0$ has at least one solution.)

TOOLKIT PROGRAM

Continuity at a Point	Provides practice with the three-step definition of continuity at a point, with multi-line function definitions, and with one-sided limits

2.5 Defining Limits Formally with Epsilons and Deltas

We have just spent four sections calculating limits. Our basic tools were the Sandwich Theorem and the theorems that gave the calculation rules for sums, differences, products, and quotients. With these theorems in hand, we started with sensible assumptions about the limits of constant functions and the identity function and worked our way to limits of rational functions. In every case, the calculations were straightforward and the results made sense.

The only problem is that we do not yet know why the theorems we used to get our results are true. The entire calculus depends on these theorems and we haven't a clue about why they hold. If we were to try to prove them now, however, we would quickly realize that we never said what a limit really is (except by example) and that we do not have a definition good enough to establish even the simple facts that

$$\lim_{x\to x_0} (k) = k \qquad \text{and} \qquad \lim_{x\to x_0} (x) = x_0. \tag{1}$$

To establish these facts and to understand why the limit theorems hold, we must take one final step, and define *limit* formally. This section is where we take that step.

As it turns out, the mathematics we need to make the notion of limit precise enough to be useful is the same mathematics we used in Section 1.6 to study target values of functions. There, we had a function $y = f(x)$, a target value y_0, and an upper bound E on the amount of error we could allow the output value y to have. We wanted to know how close we had to keep x to a particular value x_0 so that y would lie within E units of y_0. In symbols, we were asking for a value of D that would make the inequality $|x - x_0| < D$ imply the inequality $|y - y_0| < E$. The number D described the amount by which x could differ from x_0 and still give y-values that approximated y_0 with an error less than E.

In the limit discussions that follow, we shall use the traditional Greek letters δ (delta) and ϵ (epsilon) in place of the English letters D and E. These are the letters that Cauchy and Weierstrass used in their pioneering work on continuity in the nineteenth century. In their arguments, δ meant "différence" (French for *difference*) and ϵ meant "erreur" (French for *error*).

As you read along, please keep in mind that the purpose of this section is not to calculate limits of particular functions. We already know how to do that. The purpose here is to develop a technical definition of limit that is good enough to establish the limit theorems on which our calculations depend.

The Definition of Limit

Suppose we are watching the values of a function $f(x)$ as x approaches x_0 (without taking on the value x_0 itself). What do we have to know about the values of f to say that they have a particular number L as their limit? What observable pattern in their behavior would guarantee their eventual approach to L?

Certainly we want to be able to say that $f(x)$ stays within one tenth of a unit of L as soon as x stays within a certain radius r_1 of x_0, as shown here:

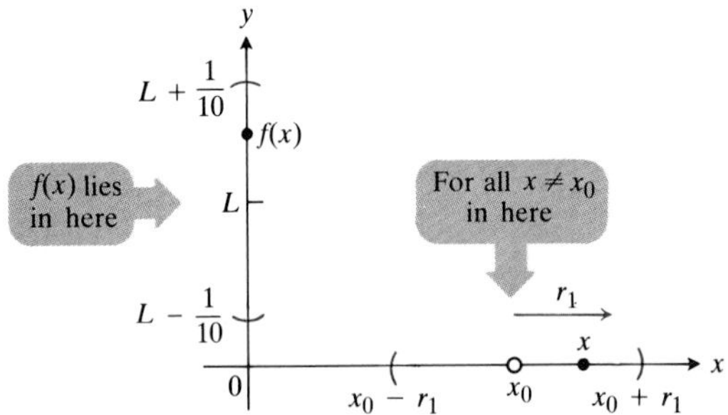

But that in itself is not enough, because as x continues on its course toward x_0, what is to prevent $f(x)$ from jittering about within the interval from $L - 1/10$ to $L + 1/10$ without tending toward L?

We need to say also that as x continues toward x_0, the number $f(x)$ has to get still closer to L. We might say this by requiring $f(x)$ to lie within 1/100 of a unit of L for all values of x within some smaller radius r_2 of x_0:

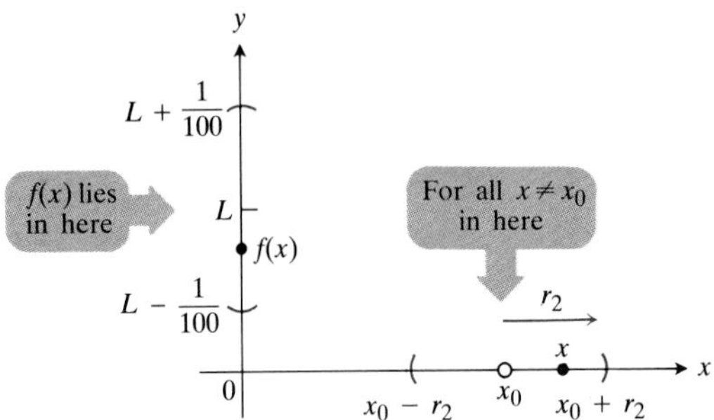

But this is not enough either. What if $f(x)$ skips about within the interval from $L - 1/100$ to $L + 1/100$ from then on, without heading toward L?

We had better require that $f(x)$ lie within 1/1000 of a unit of L after a while. That is, for all values of x within some still smaller radius r_3 of x_0, all the values of $y = f(x)$ should lie in the interval

$$L - \frac{1}{1000} < y < L + \frac{1}{1000},$$

as shown here:

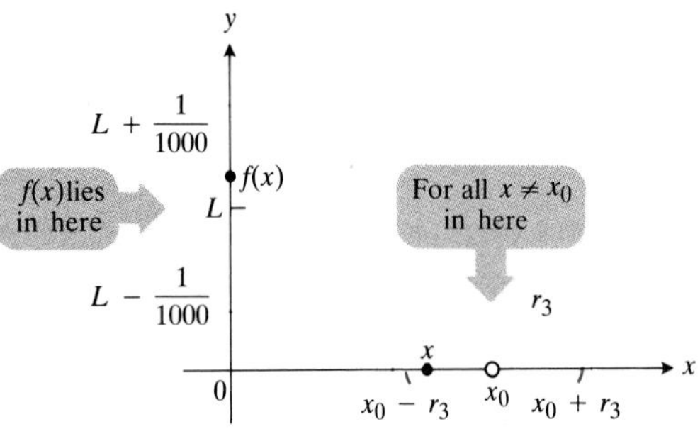

This still does not guarantee that $f(x)$ will now move toward L as x approaches x_0. Even if $f(x)$ has not skipped about before, it might start now. We need more.

We need to require that for *every* interval about L, no matter how small, we can find an interval of numbers about x_0 whose f-values all lie within that interval about L. In other words, given any positive radius ϵ about L, there should exist some positive radius δ about x_0 such that for all x within δ units of x_0 (except $x_0 = c$ itself) the values $y = f(x)$ lie within ϵ units of L:

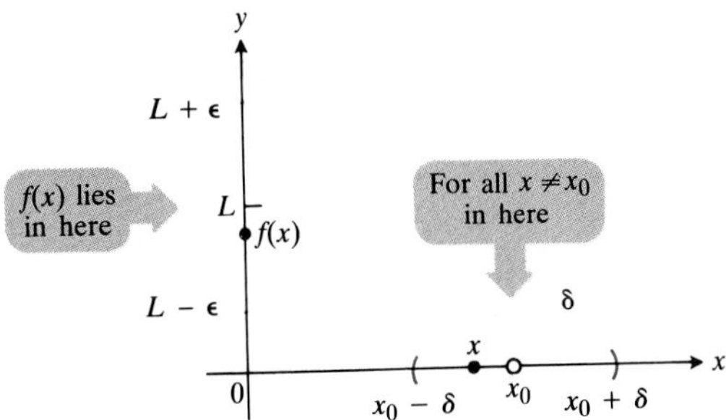

If f satisfies these requirements we will say that

$$\lim_{x \to x_0} f(x) = L.$$

Here, at last, is a mathematical way to say "the closer x gets to x_0, the closer $y = f(x)$ must get to L."

DEFINITION

The **limit** of $f(x)$ as x approaches x_0 is the number L if the following criterion holds:

Given any radius $\epsilon > 0$ about L there exists a radius $\delta > 0$ about x_0 such that for all x

$$0 < |x - x_0| < \delta \quad \text{implies} \quad |f(x) - L| < \epsilon. \tag{2}$$

To return to the notions of error and difference, we might think of machining something like a generator shaft to a close tolerance. We try for diameter L, but since nothing is perfect, we must be satisfied to get the diameter $f(x)$ somewhere between $L - \epsilon$ and $L + \epsilon$. The δ is the measure of how accurate our control setting for x must be to guarantee this degree of accuracy in the diameter of the shaft.

Examples—Testing the Definition

Whenever someone proposes a new definition, it is a good idea to see if it gives results that are consistent with past experience. For instance, our experience tells us that as x approaches 1, the number $5x - 3$ approaches $5 - 3 = 2$. If our new definition were to lead to some other result, we would want to throw the definition out and look for a new one. The following three examples are included in part to show that the definition in Eq. (2) gives the kinds of results we want.

EXAMPLE 1 *Testing the definition.* Show that

$$\lim_{x\to 1} (5x - 3) = 2.$$

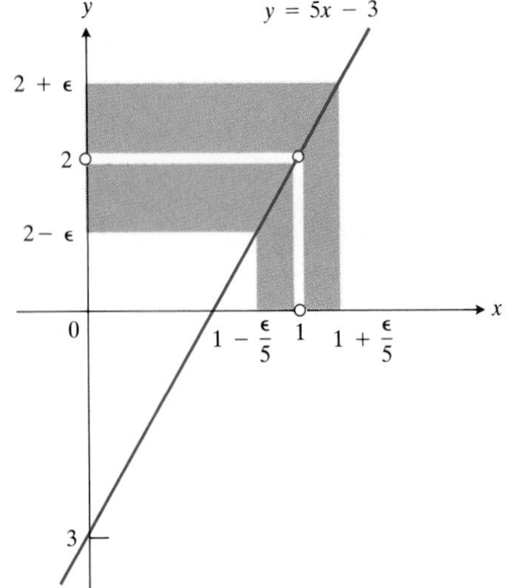

2.24 For the function $f(x) = 5x - 3$, we find that $|x - 1| < \epsilon/5$ will guarantee $|f(x) - 2| < \epsilon$.

Solution In the definition of limit, we set $x_0 = 1, f(x) = 5x - 3$, and $L = 2$. To show that $\lim_{x\to 1} (5x - 3) = 2$, we need to show that for any number $\epsilon > 0$ there exists a number $\delta > 0$ such that for all x,

$$0 < |x - 1| < \delta \quad \Rightarrow \quad |(5x - 3) - 2| < \epsilon. \tag{3}$$

(The symbol $\Rightarrow$ is read "implies.") To find a suitable value for δ, we change the ϵ-inequality from

$$|(5x - 3) - 2| < \epsilon$$

to

$$|5x - 5| < \epsilon$$

to

$$5|x - 1| < \epsilon$$

to

$$|x - 1| < \epsilon/5.$$

The last line here tells us that the original ϵ-inequality, and hence the implication in (3), will hold if we choose $\delta = \epsilon/5$ (Fig. 2.24).

The value $\delta = \epsilon/5$ is not the only value that will make the implication in (3) hold. Any smaller positive δ will do as well. The definition does not ask for a "best" δ, just one that will work. ■

EXAMPLE 2 *Establishing a basic fact.* Show that for any number x_0

$$\lim_{x\to x_0} (x) = x_0.$$

Solution In the definition of limit, we set $f(x) = x$ and $L = x_0$. To show that $\lim_{x\to x_0} (x) = x_0$ we must show that for any $\epsilon > 0$ there exists a $\delta > 0$ such that for all x,

$$0 < |x - x_0| < \delta \quad \Rightarrow \quad |x - x_0| < \epsilon.$$

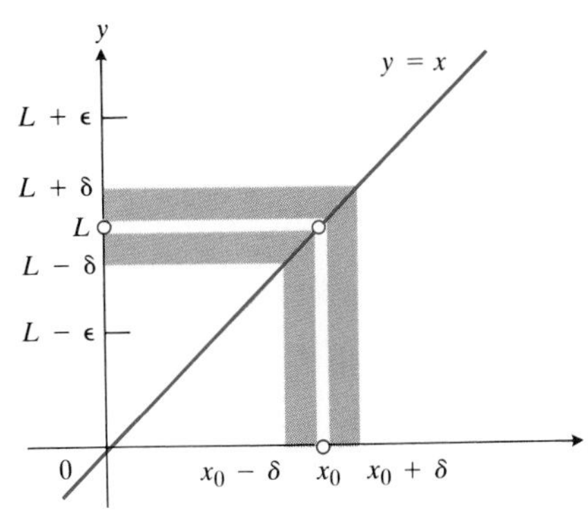

2.25 For the function $f(x) = x$, we find that $|x - x_0| < \delta$ will guarantee $|f(x) - x_0| < \epsilon$ whenever $\delta \leq \epsilon$.

The implication will hold if δ is ϵ itself or any smaller positive number (Fig. 2.25). ■

EXAMPLE 3 *Establishing another basic fact.* Let $f(x) = k$ be the function whose outputs have the constant value k. Show that for any number x_0

$$\lim_{x \to x_0} f(x) = k.$$

Solution In the definition of limit, we set $f(x) = k$ and $L = k$. To show that $\lim_{x \to x_0} f(x) = k$, we must show that for any $\epsilon > 0$ there exists a $\delta > 0$ such that for all x,

$$0 < |x - x_0| < \delta \quad \Rightarrow \quad |k - k| < \epsilon.$$

This implication will hold for any positive δ, because $|k - k| = 0$ is less than every positive ϵ for all x (Fig. 2.26). ■

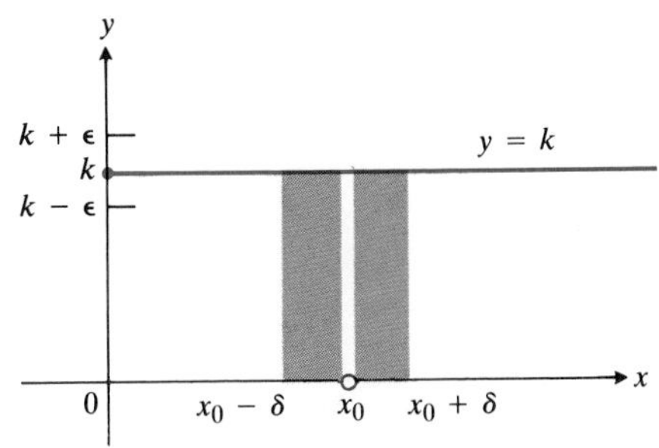

2.26 For the function $f(x) = k$, we find $|f(x) - k| < \epsilon$ for any positive δ.

Finding Deltas for Given Epsilons

Here are four numerical examples of finding δ's for given ϵ's.

EXAMPLE 4 Find the largest $\delta > 0$ such that if $0 < |x + 2| < \delta$ then $|3x + 6| < 0.03$.

Solution We rewrite the inequality $|3x + 6| < 0.03$ to see what it says about $|x + 2|$:

$$|3x + 6| < 0.03$$

$$3|x + 2| < 0.03 \qquad \text{(Factor out a 3.)}$$

$$|x + 2| < 0.01 \qquad \text{(Divide by 3.)}$$

We now see that $|3x + 6| < 0.03$ if and only if $|x + 2| < 0.01$. The largest δ is 0.01. If $|x + 2| < 0.01$ then $|3x + 6| < 0.03$. If $|x + 2| \geq 0.01$, then $|3x + 6| \geq 0.03$. ■

The interval of x-values for which $|f(x) - L| < \epsilon$ is not always symmetric about x_0. When this happens, we take δ to be the distance from x_0 to the nearer endpoint.

EXAMPLE 5 Find the largest interval symmetric about $x_0 = 5$ lying in the interval $2 < x < 10$.

Solution We draw a picture to show $x_0 = 5$ in the interval:

The endpoint nearer $x_0 = 5$ is the point $x = 2$, which is 3 units away. We can take δ as large as 3 (but no larger) and have the symmetric interval

$$5 - \delta < x < 5 + \delta$$

lie in the interval $2 < x < 10$. The interval we seek is the interval $2 < x < 8$. ■

EXAMPLE 6 Find the largest interval about $x_0 = 5$ for which

$$|\sqrt{x-1} - 2| < 1.$$

Then find a value of δ for which

$$|x - 5| < \delta \quad \Rightarrow \quad |\sqrt{x-1} - 2| < 1.$$

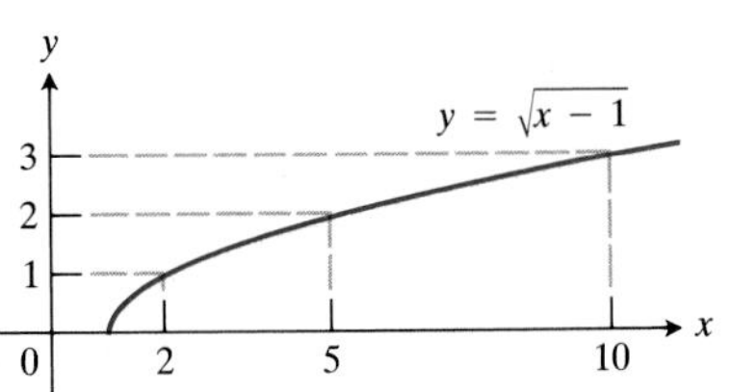

Solution We rewrite the inequality $|\sqrt{x-1} - 2| < 1$ in the form $a < x < b$:

$$|\sqrt{x-1} - 2| < 1$$

$$-1 < \sqrt{x-1} - 2 < 1 \qquad \text{(Remove absolute value bars.)}$$

$$1 < \sqrt{x-1} < 3 \qquad \text{(Add 2.)}$$

$$1 < x - 1 < 9 \qquad \text{(Square.)}$$

$$2 < x < 10. \qquad \text{(Add 1.)}$$

The original inequality holds for all x in the interval $2 < x < 10$.

As in Example 5, we may take δ to be 3, the distance from $x_0 = 5$ to the nearer endpoint $a = 2$. If δ is 3 or any smaller positive number, the inequality $|x - 5| < \delta$ will automatically place x between 2 and 10 to make $|\sqrt{x-1} - 2| < 1$. ■

EXAMPLE 7 Find the largest interval about $x_0 = \frac{1}{2}$ for which

$$\left|\frac{1}{x} - 2\right| < 0.01.$$

Then find a value of δ for which

$$\left|x - \frac{1}{2}\right| < \delta \quad \Rightarrow \quad \left|\frac{1}{x} - 2\right| < 0.01.$$

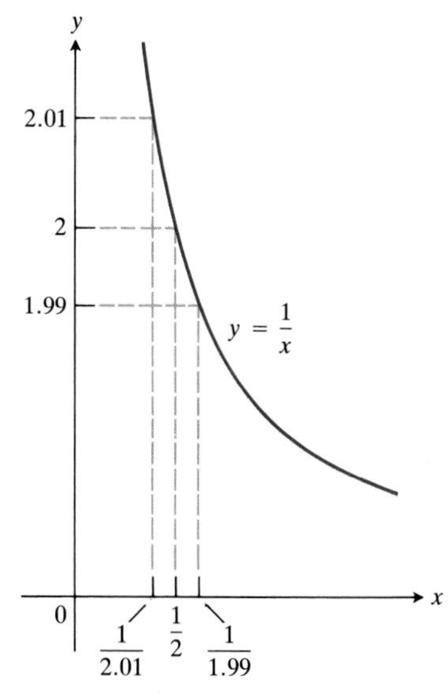

Solution We rewrite the inequality $\left|\frac{1}{x} - 2\right| < 0.01$ in the form $a < x < b$:

$$\left|\frac{1}{x} - 2\right| < 0.01$$

$$-0.01 < \frac{1}{x} - 2 < 0.01 \qquad \text{(Remove absolute value bars.)}$$

$$1.99 < \frac{1}{x} < 2.01 \qquad \text{(Add 2.)}$$

$$\frac{1}{2.01} < x < \frac{1}{1.99}. \qquad \text{(Take reciprocals.)}$$

The original inequality holds for all x in the interval from 1/2.01 up to 1/1.99.

Again, we can take δ to be the distance from x_0 to the nearer endpoint, in this case the distance from $x_0 = 1/2$ to the nearer of the points 1/2.01 and 1/1.99. To find out which is nearer, we calculate the distances and compare:

$$\text{Distance from } \frac{1}{2} \text{ to } \frac{1}{2.01}: \quad \frac{1}{2} - \frac{1}{2.01} = \frac{2.01 - 2}{4.02} = \frac{0.01}{4.02} = \frac{1}{402}.$$

$$\text{Distance from } \frac{1}{2} \text{ to } \frac{1}{1.99}: \quad \frac{1}{1.99} - \frac{1}{2} = \frac{2 - 1.99}{3.98} = \frac{0.01}{3.98} = \frac{1}{398}.$$

The smaller of these is 1/402. If δ has this value (or any smaller positive value), then the inequality $|x - (1/2)| < \delta$ will automatically place x between 1/2.01 and 1/1.99 to make $|(1/x) - 2| < 0.01$. ■

How Limit Theorems Are Proved

Although we shall not ask you to prove limit theorems yourself, we want to show how a typical proof goes, if only to support our claim that having a precise definition of limit now makes it possible for us to prove the limit theorems on which calculus depends. Our example will be the proof of the sum rule for limits (the first part of Theorem 1 from Section 2.1). You can find a proof of the rest of Theorem 1 in Appendix 2.

Theorem 1, Part 1: If $\lim_{x\to x_0} f_1(x) = L_1$ and $\lim_{x\to x_0} f_2(x) = L_2$, then $\lim_{x\to x_0} (f_1(x) + f_2(x)) = L_1 + L_2$.

PROOF To show that $\lim_{x\to x_0} (f_1(x) + f_2(x)) = L_1 + L_2$, we must show that for any $\epsilon > 0$ there exists a $\delta > 0$ such that for all x

$$0 < |x - x_0| < \delta \quad \Rightarrow \quad |f_1(x) + f_2(x) - (L_1 + L_2)| < \epsilon. \tag{4}$$

Suppose, then, that ϵ is a positive number. The number $\epsilon/2$ is positive, too, and because $\lim_{x\to x_0} f_1(x) = L_1$ we know that there is a $\delta_1 > 0$ such that for all x

$$0 < |x - x_0| < \delta_1 \quad \Rightarrow \quad |f_1(x) - L_1| < \frac{\epsilon}{2}. \tag{5}$$

Because $\lim_{x\to x_0} f_2(x) = L_2$, there is also a $\delta_2 > 0$ such that for all x

$$0 < |x - x_0| < \delta_2 \quad \Rightarrow \quad |f_2(x) - L_2| < \frac{\epsilon}{2}. \tag{6}$$

Now, either δ_1 equals δ_2 or it doesn't. If δ_1 equals δ_2, the implications in (5) and (6) both hold true for their common value δ. Taken together, (5) and (6) then say that, for all x, $0 < |x - x_0| < \delta$ implies

$$
\begin{aligned}
&|f_1(x) + f_2(x) - (L_1 + L_2)| \\
&= |(f_1(x) - L_1) + (f_2(x) - L_2)| \\
&\le |(f_1(x) - L_1)| + |f_2(x) - L_2| && \text{(Triangle inequality)} \\
&< \frac{\epsilon}{2} + \frac{\epsilon}{2} && \left(\begin{array}{l}\text{The implications in (5) and (6) both} \\ \text{hold for } \delta \text{ because } \delta = \delta_1 = \delta_2.\end{array}\right) \\
&< \epsilon.
\end{aligned}
$$

If $\delta_1 \neq \delta_2$, let δ be the smaller of δ_1 and δ_2. The implications in (5) and (6) then both hold for all x such that $0 < |x - x_0| < \delta$. As before,

$$|f_1(x) + f_2(x) - (L_1 + L_2)| < \epsilon.$$

Either way, we know that given any $\epsilon > 0$ there exists a $\delta > 0$ such that for all x

$$0 < |x - x_0| < \delta \quad \Rightarrow \quad |f_1(x) + f_2(x) - (L_1 + L_2)| < \epsilon.$$

According to the $\epsilon - \delta$ definition of limit, then,

$$\lim_{x \to x_0} (f(x_1) + f(x_2)) = L_1 + L_2.$$ ■

The Relation between One-sided and Two-sided Limits

The formal definitions of right-hand and left-hand limits go like this:

DEFINITION

Right-hand Limit: $\lim_{x \to x_0^+} f(x) = L$

The limit of $f(x)$ as x approaches x_0 from the right is the number L if the following criterion holds:

Given any radius $\epsilon > 0$ about L there exists a radius $\delta > 0$ to the right of x_0 such that for all x

$$x_0 < x < x_0 + \delta \text{ implies } |f(x) - L| < \epsilon. \tag{7}$$

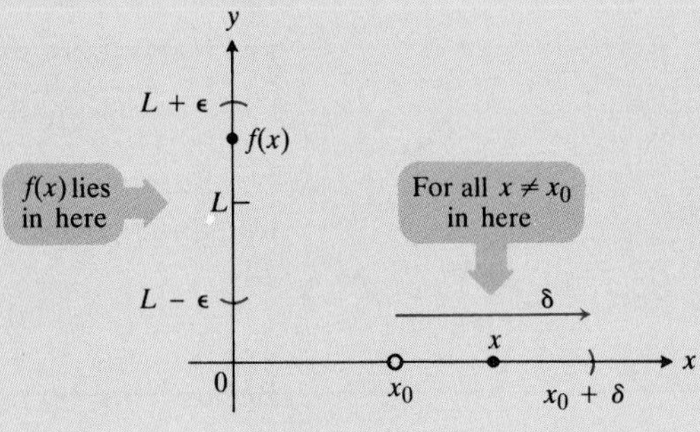

DEFINITION

Left-hand Limit: $\lim_{x \to x_0^-} f(x) = L$

The limit of $f(x)$ as x approaches x_0 from the left is the number L if the following criterion holds:

Given any radius $\epsilon > 0$ about L there exists a radius $\delta > 0$ to the left of x_0 such that for all x

$$x_0 - \delta < x < x_0 \text{ implies } |f(x) - L| < \epsilon. \tag{8}$$

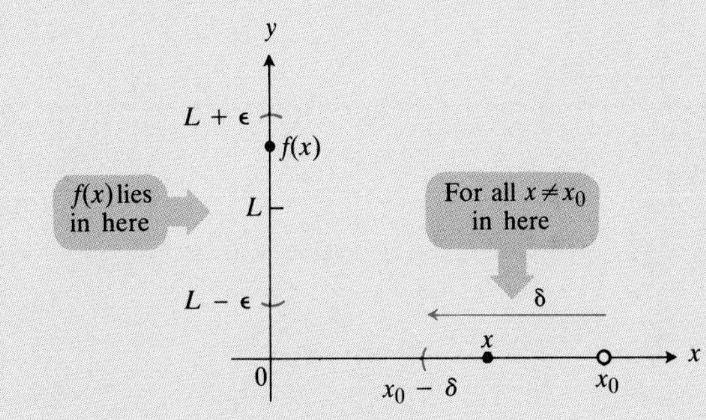

By comparing Eqs. (7) and (8) with Eq. (2), we can see the relation between the one-sided limits just defined and the two-sided limit defined earlier. If we subtract x_0 from the δ-inequalities in Eqs. (7) and (8), they become

$$0 < x - x_0 < \delta \quad \Rightarrow \quad |f(x) - L| < \epsilon \tag{9}$$

and

$$-\delta < x - x_0 < 0 \quad \Rightarrow \quad |f(x) - L| < \epsilon. \tag{10}$$

Together, Eqs. (9) and (10) say the same thing as

$$0 < |x - x_0| < \delta \quad \Rightarrow \quad |f(x) - L| < \epsilon, \tag{11}$$

which is Eq. (2) in the definition of limit. In other words, $f(x)$ has limit L at x_0 if and only if the right-hand and left-hand limits of f at x_0 exist and equal L.

Exercises 2.5

In Exercises 1–8, sketch the interval (a, b) on the x-axis with the point x_0 inside. Then find the largest value of $\delta > 0$ such that $|x - x_0| < \delta$ implies $a < x < b$.

1. $a = 1,\ b = 7,\ x_0 = 5$

2. $a = 1,\ b = 7,\ x_0 = 2$

3. $a = -7/2,\ b = -1/2,\ x_0 = -3$

4. $a = -7/2,\ b = -1/2,\ x_0 = -3/2$

5. $a = -5,\ b = 3,\ x_0 = 1$

6. $a = -5,\ b = 3,\ x_0 = -2$

7. $a = 4/9,\ b = 4/7,\ x_0 = 1/2$

8. $a = 2.7591,\ b = 3.2391,\ x_0 = 3$

Use the graphs in Exercises 9–14 to find a $\delta > 0$ such that for all x,

$$0 < |x - x_0| < \delta \quad \Rightarrow \quad |f(x) - L| < \epsilon.$$

9.

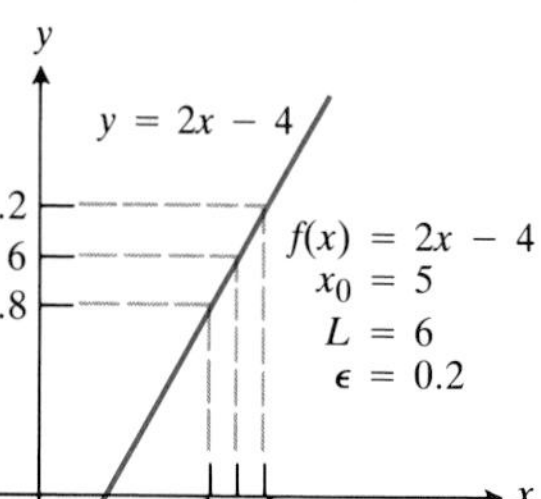

10.

$f(x) = -\frac{3}{2}x + 3$
$x_0 = -3$
$L = 7.5$
$\epsilon = 0.15$

$y = -\frac{3}{2}x + 3$

7.65
7.5
7.35
−3.1
−3
−2.9
0

NOT TO SCALE

11.

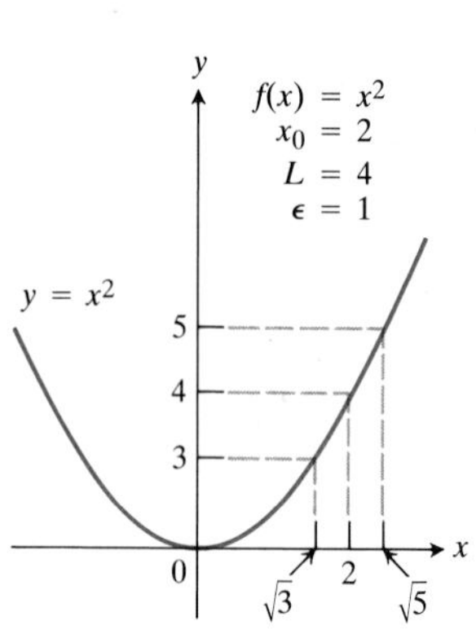

12.

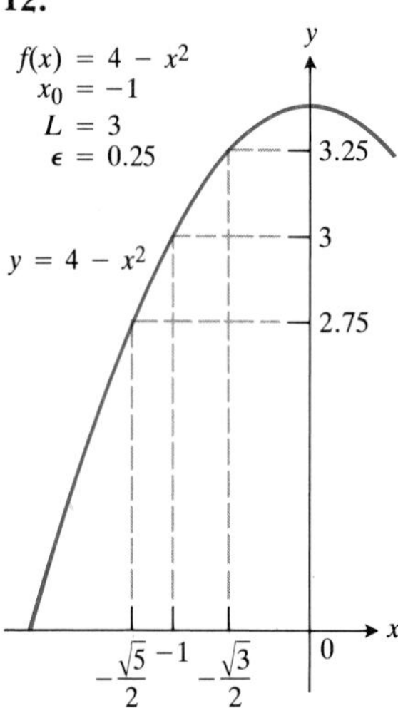

13.

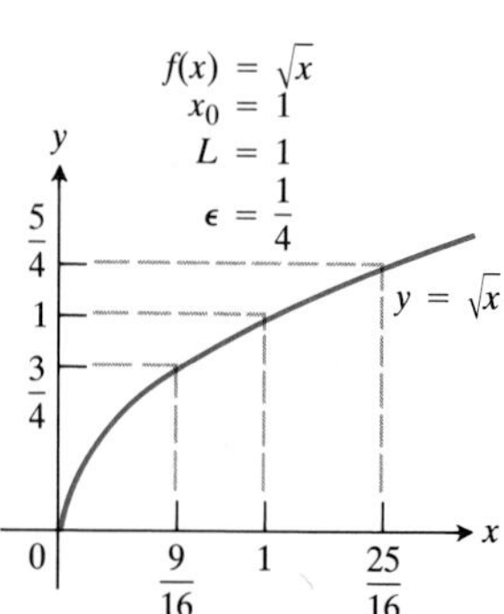

14.

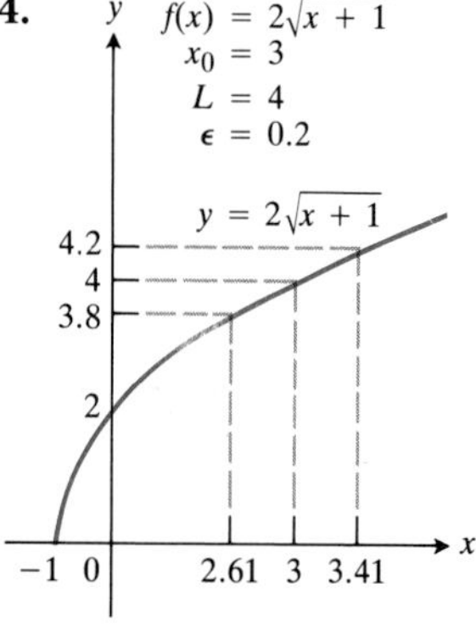

For each function $y = f(x)$ and number $\epsilon > 0$ in Exercises 15–22, find the set of x-values for which $|f(x) - 4| < \epsilon$.

15. $f(x) = x + 1,\ \epsilon = 0.01$

16. $f(x) = 2x - 2,\ \epsilon = 0.02$

17. $f(x) = x^2 - 5,\ \epsilon = 0.05$

18. $f(x) = 4 - x^2,\ \epsilon = 0.04$

19. $f(x) = \sqrt{19 - x},\ \epsilon = 0.03$

20. $f(x) = \sqrt{x + 1},\ \epsilon = 0.1$

21. $f(x) = 1/x,\ \epsilon = 0.1$

22. $f(x) = \dfrac{1}{x - 2},\ \epsilon = 0.5$

Each of Exercises 23–32 gives a function $f(x)$, a point x_0, and a positive number ϵ. Find $L = \lim_{x \to x_0} f(x)$. Then find a number $\delta > 0$ such that for all x

$$0 < |x - x_0| < \delta \quad \Rightarrow \quad |f(x) - L| < \epsilon.$$

23. $f(x) = 2x + 3,\ x_0 = 1,\ \epsilon = 0.01$

24. $f(x) = 3 - 2x,\ x_0 = 3,\ \epsilon = 0.02$

25. $f(x) = 4x - 2,\ x_0 = 1/2,\ \epsilon = 0.02$

26. $f(x) = -3x - 2,\ x_0 = -1,\ \epsilon = 0.03$

27. $f(x) = \dfrac{x^2 - 4}{x - 2},\ x_0 = 2,\ \epsilon = 0.05$

28. $f(x) = \dfrac{x^2 + 6x + 5}{x + 5},\ x_0 = -5,\ \epsilon = 0.05$

29. $f(x) = \sqrt{x - 7},\ x_0 = 11,\ \epsilon = 0.01$

30. $f(x) = \sqrt{1 - 5x},\ x_0 = -3,\ \epsilon = 0.5$

31. $f(x) = 4/x,\ x_0 = 2,\ \epsilon = 0.4$

32. $f(x) = 4/x,\ x_0 = 1/2,\ \epsilon = 0.04$

In Exercises 33 and 34, find the largest $\delta > 0$ such that for all x

$$0 < |x - 4| < \delta \quad \Rightarrow \quad |f(x) - 5| < \epsilon.$$

33. $f(x) = 9 - x$; $\epsilon = 0.01, 0.001, 0.0001$, arbitrary $\epsilon > 0$

34. $f(x) = |3x - 7|$; $\epsilon = 0.003, 0.0003$, arbitrary $\epsilon > 0$

35. Given $\epsilon > 0$, find an interval $I = (5, 5 + \delta),\ \delta > 0$, such that if x lies in I then $\sqrt{x - 5} < \epsilon$. What limit is being verified?

36. Given $\epsilon > 0$, find an interval $I = (4 - \delta, 4),\ \delta > 0$, such that if x lies in I then $\sqrt{4 - x} < \epsilon$. What limit is being verified?

37. Graph the function

$$f(x) = \begin{cases} 4 - 2x, & x < 1, \\ 6x - 4, & x \geq 1. \end{cases}$$

Then, given $\epsilon > 0$, find the largest δ for which $f(x)$ lies between $y = 2 - \epsilon$ and $y = 2 + \epsilon$ for x in the interval $I = (1 - \delta, 1 + \delta)$.

38. Let $f(x) = |x - 5|/(x - 5)$. Find the set of x-values for which

$$1 - \epsilon < f(x) < 1 + \epsilon, \quad \text{for } \epsilon = 4, 2, 1, \text{ and } 1/2.$$

(*Hint:* First graph f for $x > 5$ and $x < 5$.)

39. Define what it means to say that $\lim_{x\to 2} f(x) = 5$.

40. Define what it means to say that $\lim_{x\to 0} g(x) = k$.

41. Suppose $0 < \epsilon < 4$. Find the largest $\delta > 0$ with the property that $f(x) = x^2$ lies between $y = 4 - \epsilon$ and $y = 4 + \epsilon$ for all x in the interval $I = (2 - \delta, 2 + \delta)$. What happens to δ as ϵ decreases toward 0? Graph δ as a function of ϵ.

TOOLKIT PROGRAM

Limit Definition	Enables you to work with the definition graphically, to choose appropriate δ's for given ϵ's

Review Questions

1. You have been asked to calculate the limit of a function $f(x)$ as x approaches a finite number c. What theorems are available for calculating the limit? Give examples to show how the theorems are used.
2. What is the relation between one-sided and two-sided limits? How is this relation sometimes used to calculate a limit or to prove that a limit does not exist? Give examples.
3. **CALCULATOR** You used a calculator to estimate $\lim_{\theta\to 0} (\sin \theta)/\theta$ by evaluating the quotient at $\theta = 0.1, 0.01, 0.001$, and so on. You found, to your surprise, that instead of approaching 1 the quotients approached $0.0174532925 \approx \pi/180$. What did you do wrong?
4. What is the procedure for finding the limit of a rational function of x as x approaches $\pm\infty$? When is the limit zero? Finite and different from zero? Infinite?
5. What test can you apply to find out whether a function $y = f(x)$ is continuous at point $x = c$? Give examples of functions that are continuous at $x = 0$. Give examples of functions that fail to be continuous at $x = 0$ for various reasons. (They don't have to be examples from the book. You may make up your own.)
6. What can be said about the continuity of polynomial functions and rational functions?
7. What can be said about the continuity of composites of continuous functions?
8. What are the important theorems about continuous functions? Can functions that are not continuous be expected to have the properties guaranteed by these theorems? Give examples.
9. What are the formal definitions of (two-sided) limit, right-hand limit, and left-hand limit?

Practice Exercises

Find the limits in Exercises 1–10.

1. $\lim_{x\to 1} 5x - 4$
2. $\lim_{x\to -3} |7x + 10|$
3. $\lim_{x\to -2} x^2(x + 1)$
4. $\lim_{x\to 3} (x + 2)(x - 5)$
5. $\lim_{x\to 1} \dfrac{x + 1}{x + 2}$
6. $\lim_{x\to 1} \dfrac{x - 1}{x + 2}$
7. $\lim_{x\to 3} \dfrac{x - 3}{x^2}$
8. $\lim_{x\to -1} \dfrac{x^2 + 1}{3x^2 - 2x + 5}$

9. $\lim_{x \to -2} \left(\frac{x}{x+1}\right)\left(\frac{3x+5}{x^2+x}\right)$

10. $\lim_{x \to 1} \left(\frac{1}{x+1}\right)\left(\frac{x+6}{x}\right)\left(\frac{3-x}{7}\right)$

Find the limits in Exercises 11–16.

11. $\lim_{x \to -1} \frac{x^2 - x - 2}{x + 1}$

12. $\lim_{x \to 2} \frac{x^2 - x - 2}{x - 2}$

13. $\lim_{x \to 1} \frac{x^2 - 1}{x - 1}$

14. $\lim_{x \to -5} \frac{x^2 + 3x - 10}{x + 5}$

15. $\lim_{x \to 2} \frac{x - 2}{x^2 + x - 6}$

16. $\lim_{x \to 1} \frac{x^2 - 2x + 1}{x^3 - 2x^2 + x}$

Find the limits in Exercises 17–22.

17. $\lim_{x \to \infty} \frac{2x + 3}{5x + 7}$

18. $\lim_{x \to -\infty} \frac{2x^2 + 3}{5x^2 + 7}$

19. $\lim_{x \to -\infty} \frac{x^2 - 4x + 8}{3x^3}$

20. $\lim_{x \to \infty} \frac{1}{x^2 - 7x + 1}$

21. $\lim_{x \to -\infty} \frac{x^2 - 7x}{x + 1}$

22. $\lim_{x \to \infty} \frac{x^4 + x^3}{12x^3 + 128}$

Find the limits in Exercises 23–26.

23. $\lim_{x \to 3^+} \frac{1}{x - 3}$

24. $\lim_{x \to 3^-} \frac{1}{x - 3}$

25. $\lim_{x \to 0^+} \frac{1}{x^2}$

26. $\lim_{x \to 0^-} \frac{1}{|x|}$

Find the limits in Exercises 27–30.

27. $\lim_{x \to 0} \frac{\sin 2x}{4x}$

28. $\lim_{x \to 0} \frac{x + \sin x}{x}$

29. $\lim_{x \to \pi} \sin\left(\frac{x + \pi}{4}\right) \cos\left(\frac{x - \pi}{4}\right)$

30. $\lim_{x \to \pi} \frac{x + \sin x}{x + \cos x}$

In Exercises 31 and 32, use the graphs to help find the limits.

31. a) $\lim_{x \to -2^+} \frac{x + 3}{x + 2}$ b) $\lim_{x \to -2^-} \frac{x + 3}{x + 2}$

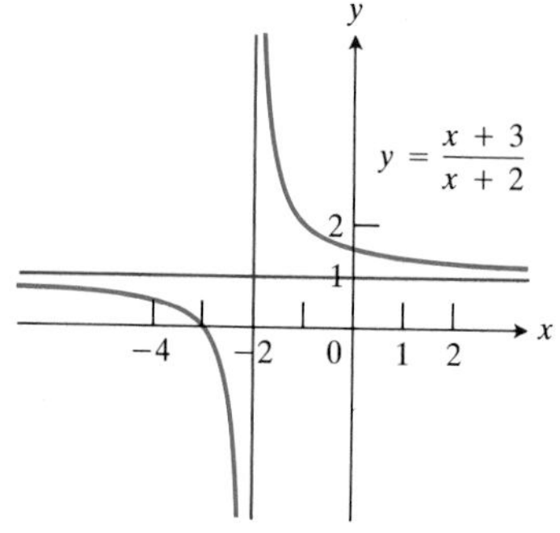

32. a) $\lim_{x \to 2^+} \frac{x - 1}{x^2(x - 2)}$ b) $\lim_{x \to 2^-} \frac{x - 1}{x^2(x - 2)}$

c) $\lim_{x \to 0^+} \frac{x - 1}{x^2(x - 2)}$ d) $\lim_{x \to 0^-} \frac{x - 1}{x^2(x - 2)}$

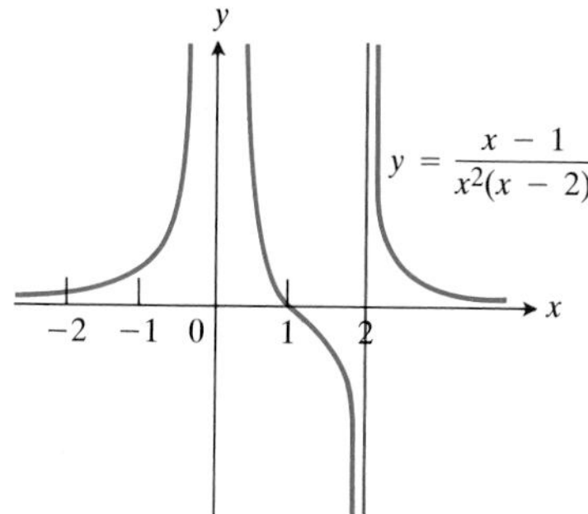

33. Answer the following questions about the function

$$f(x) = \begin{cases} 1, & x \leq -1 \\ -x, & -1 < x < 0 \\ 1, & x = 0 \\ -x, & 0 < x < 1 \\ 1, & x \geq 1 \end{cases}$$

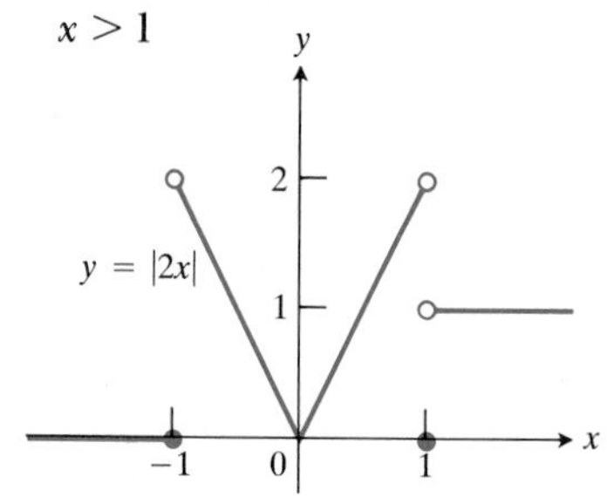

a) Find the right-hand and left-hand limits of f at -1, 0, and 1.

b) Does f have a limit as x approaches -1? 0? 1? If so, what is it? If not, why not?

c) At which of the points $x = -1$, 0, 1, if any, is f continuous?

34. Repeat the questions in Exercise 33 for the function

$$f(x) = \begin{cases} 0, & x \leq -1 \\ |2x|, & -1 < x < 1 \\ 0, & x = 1 \\ 1, & x > 1 \end{cases}$$

35. Let $f(x) = \begin{cases} -x, & x < 1 \\ x - 1, & x > 1. \end{cases}$

a) Graph f.
b) Find the right-hand and left-hand limits of f at $x = 1$.
c) What value, if any, should be assigned to $f(1)$ to make f continuous at $x = 1$?

36. Repeat Exercise 35 for the function

$$f(x) = \begin{cases} 3x^2, & x < 1 \\ 4 - x^2, & x > 1. \end{cases}$$

37. Suppose that $f(x)$ and $g(x)$ are defined for all x and that $\lim_{x\to c} f(x) = -7$ and $\lim_{x\to c} g(x) = 0$. Find the limit as $x \to c$ of the following functions.

a) $3f(x)$ b) $(f(x))^2$ c) $f(x) \cdot g(x)$
d) $\dfrac{f(x)}{g(x) - 7}$ e) $\cos(g(x))$ f) $|f(x)|$

38. Suppose that $f(x)$ and $g(x)$ are defined for all x and that $\lim_{x\to 0} f(x) = \frac{1}{2}$ and $\lim_{x\to 0} g(x) = \sqrt{2}$. Find the limits as $x \to 0$ of the following functions.

a) $-g(x)$ b) $g(x) \cdot f(x)$ c) $f(x) + g(x)$
d) $1/f(x)$ e) $x + f(x)$ f) $\dfrac{f(x) \cdot \sin x}{x}$

39. Use the inequality

$$0 \le \left|\sqrt{x}\sin\frac{1}{x}\right| \le \sqrt{x}$$

to find $\lim_{x\to 0^+} \sqrt{x}\sin(1/x)$. See Fig. 2.27.

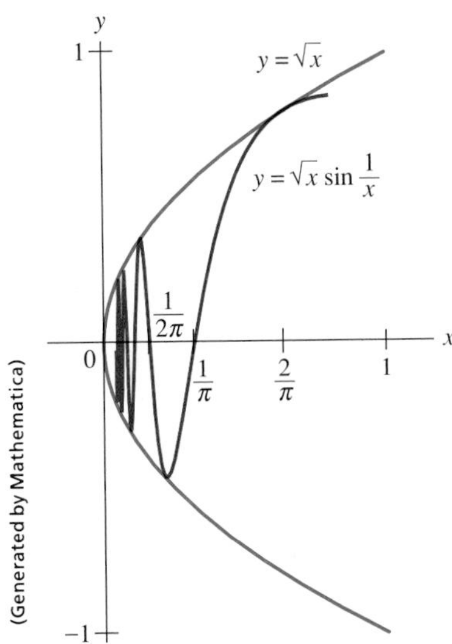

2.27 The graph of $y = \sqrt{x}\sin(1/x)$ (Exercise 39).

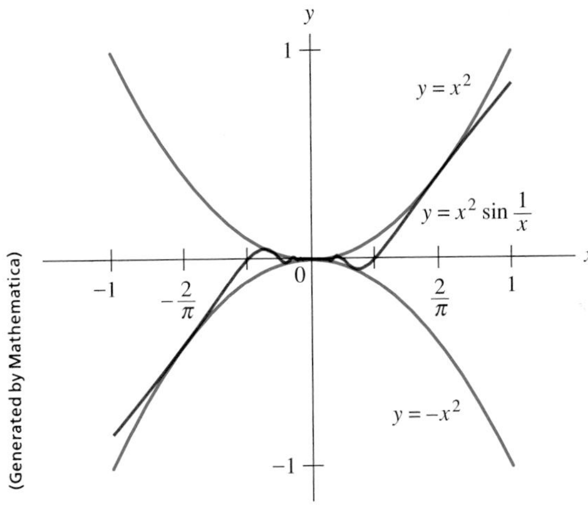

2.28 The graph of $y = x^2\sin(1/x)$ (Exercise 40).

40. Use the inequality

$$0 \le \left|x^2 \sin\frac{1}{x}\right| \le x^2$$

to find $\lim_{x\to 0} x^2 \sin(1/x)$. See Fig. 2.28.

Use the Sandwich Theorem to find the limits in Exercises 41 and 42.

41. $\displaystyle\lim_{x\to\infty} \frac{\sin x}{\sqrt{x}}$ **42.** $\displaystyle\lim_{x\to\infty} \frac{\cos x}{\sqrt{x}}$

Given that $\lim_{x\to\infty} (\sin x)/x = \lim_{x\to\infty} (\cos x)/x = 0$, find the limits in Exercises 43 and 44.

43. $\displaystyle\lim_{x\to\infty} \frac{x + \sin x}{x}$ **44.** $\displaystyle\lim_{x\to\infty} \frac{x + \sin x}{x + \cos x}$

45. Let $f(x) = \begin{cases} \dfrac{x^2 + 2x - 15}{x - 3}, & x \ne 3 \\ k, & x = 3. \end{cases}$

What value, if any, should be assigned to k to make f continuous at $x = 3$?

46. Let $f(x) = \begin{cases} \dfrac{\sin x}{2x}, & x \ne 0 \\ k, & x = 0. \end{cases}$

What value, if any, should be assigned to k to make f continuous at $x = 0$?

47. The function $y = 1/x$ does not take on either a maximum or a minimum on the interval $0 < x < 1$ even though the function is continuous on this interval. Does this contradict the Max-Min Theorem for continuous functions? Why?

48. What are the maximum and minimum values of the function $y = |x|$ on the interval $-1 \le x < 1$? Note that the interval is not closed. Is this consistent with the Max-Min Theorem for continuous functions? Why?

49. True, or false? If $y = f(x)$ is continuous, with $f(1) = 0$ and $f(2) = 3$, then f takes on the value 2.5 at some point between $x = 1$ and $x = 2$. Explain.

50. Show that there is at least one value of x for which $x + \cos x = 0$.

The Definition of Limit

51. Define what it means to say that

$$\lim_{x \to 1} f(x) = 3.$$

52. Define what it means to say that

$$\lim_{x \to 0} \frac{\sin x}{x} = 1.$$

Wrong descriptions of limit. Show by example that the statements in Exercises 53 and 54 are wrong.

53. The number L is the limit of $f(x)$ as x approaches x_0 if $f(x)$ gets closer to L as x approaches x_0.

54. The number L is the limit of $f(x)$ as x approaches x_0 if, given any $\epsilon > 0$, there is a value of x for which $|f(x) - L| < \epsilon$.

55. The function $f(x) = 2x - 3$ is continuous at $x = 2$. Given a positive number ϵ, how small must δ be for $|x - 2| < \delta$ to imply $|f(x) - 1| < \epsilon$?

56. The function $f(x) = |x|$ is continuous at $x = 0$. Given a positive number ϵ, how small must δ be for $|x - 0| < \delta$ to imply $|f(x) - 0| < \epsilon$?

Each of Exercises 57–64 gives a function $f(x)$, a point x_0, and a positive number ϵ. Find $L = \lim_{x \to x_0} f(x)$. Then find a number $\delta > 0$ such that for all x

$$0 < |x - x_0| < \delta \quad \Rightarrow \quad |f(x) - L| < \epsilon.$$

57. $f(x) = 5x - 10, \quad x_0 = 3, \quad \epsilon = 0.05$

58. $f(x) = 5x - 10, \quad x_0 = 2, \quad \epsilon = 0.05$

59. $f(x) = 5x - 10, \quad x_0 = 1, \quad \epsilon = 0.05$

60. $f(x) = 5x - 10, \quad x_0 = 0, \quad \epsilon = 0.05$

61. $f(x) = \sqrt{x - 5}, \quad x_0 = 9, \quad \epsilon = 1$

62. $f(x) = \sqrt{2x - 3}, \quad x_0 = 2, \quad \epsilon = 1/2$

63. $f(x) = 2/x, \quad x_0 = 2, \quad \epsilon = 0.1$

64. $f(x) = 1/(4x), \quad x_0 = 1/4, \quad \epsilon = 1/20$

3 Derivatives

Wide World Photos

Acceleration and velocity are derivatives.

OVERVIEW

Derivatives are the functions we use to measure the rates at which things change. We define derivatives as limiting values of average changes, just as we define slopes of curves as limiting values of slopes of secants. Now that we can calculate limits, we can calculate derivatives.

The notion of derivative is one of the most important ideas in calculus. Any subject area that uses calculus, and most subjects do, has applications of derivatives. This chapter shows how to calculate derivatives quickly, without having to do tedious limit calculations. The key result is the Chain Rule, which shows how to calculate derivatives of composite functions.

Applications of derivatives will be discussed mainly in the next chapter. However, to give some idea of the usefulness and importance of derivatives, we have included applications to estimation and root finding, as well as applications to the life sciences, to economics, and to the study of motion along a straight line.

3.1 Slopes, Tangent Lines, and Derivatives

In this section, we get our first view of the role calculus plays in describing how rapidly things change. Our point of departure is the coordinate plane of Descartes and Fermat. The plane is the natural place to draw curves and calculate the slopes of lines and it is from the slopes of lines that we find the slopes of curves. Once we can do that, we can do two really important things: we can find tangent lines for curves and we can find formulas for rates of change.

These formulas for the rates at which functions change define new functions called *derivatives*. Derivatives are calculated with limits. In this section we develop the ideas of slope and derivative, and see how some typical calculations go. In the next section, we show how to calculate derivatives rapidly. With this facility at our command, we can calculate the velocities, accelerations, and other rates of change in Section 3.3.

Average Rates of Change

We encounter average rates of change in such forms as average speeds (distance traveled divided by elapsed time, say, in miles per hour), growth rates of populations (in percent per year), and average monthly rainfall (in inches per month). The **average rate of change** in a quantity over a period of time is the amount of change divided by the time it takes.

Experimental biologists often want to know the rates at which populations grow under controlled laboratory conditions. Figure 3.1 shows data from a fruit fly–growing experiment, the setting for our first example.

EXAMPLE 1 *The average growth rate of a laboratory population.* The graph in Fig. 3.1 shows how the number of fruit flies *(Drosophila)* grew in a controlled 50-day experiment. The graph was made by counting flies at regular intervals, plotting a point for each count, and drawing a smooth curve through the plotted points.

There were 150 flies on day 23 and 340 flies on day 45. This gave an increase of $340 - 150 = 190$ flies in $45 - 23 = 22$ days. The average rate of change in the population from day 23 to day 45 was therefore

$$\textit{Average rate of change:}\quad \frac{\Delta p}{\Delta t} = \frac{340 - 150}{45 - 23} = \frac{190}{22} \approx 9 \text{ flies/day.} \tag{1}$$

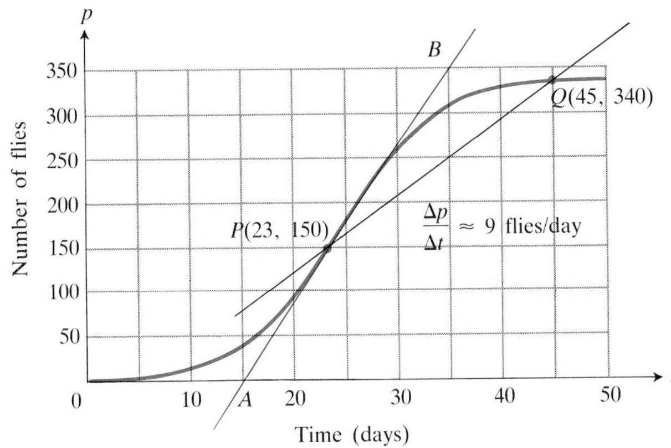

3.1 Growth of a fruit fly population in a controlled experiment. (*Source*: A. J. Lotka, *Elements of Mathematical Biology.* Dover, New York (1956), p. 69.)

The average rate of change in Eq. (1) is also the slope of the secant line through the two points

$$P(23, 150) \quad \text{and} \quad Q(45, 340)$$

on the population curve. (A line through two points on a curve is called a **secant** to the curve.) We can calculate the slope of the secant PQ from the coordinates of P and Q:

$$\textit{Secant slope:} \qquad \frac{\Delta p}{\Delta t} = \frac{340 - 150}{45 - 23} = \frac{190}{22} \approx 9 \text{ flies/day.} \tag{2}$$

By comparing Eqs. (1) and (2) we can see that the average rate of change in (1) is the same number as the slope in (2), units and all. We can always think of an average rate of change as the slope of a secant line.

In addition to knowing the average rate at which the population grew from day 23 to day 45, we may also want to know how fast the population was growing on day 23 itself. To find out we can watch the slope of the secant PQ change as we back Q along the curve toward P. The results for four positions of Q are shown in Fig. 3.2.

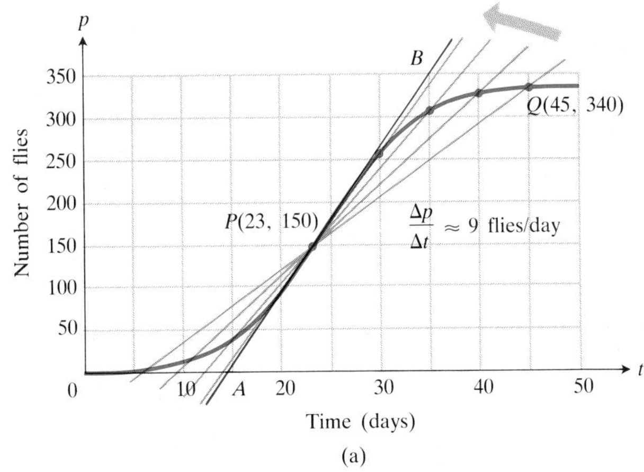

Q	Slope of $PQ = \Delta p/\Delta t$ (flies/day)
(45, 340)	$(340 - 150)/(45 - 23) \approx 9$
(40, 330)	$(330 - 150)/(40 - 23) \approx 13$
(35, 310)	$(310 - 150)/(35 - 23) \approx 15$
(30, 265)	$(265 - 150)/(30 - 23) \approx 16.4$

(b)

3.2 (a) Four secants to the fruit fly graph of Figure 3.1, through the point $P(23, 150)$. (b) The slopes of the four secants.

In terms of geometry, what we see as Q approaches P along the curve is this: The secant PQ approaches the tangent line AB that we drew by eye at P. This means that within the limitations of our drawing, the slopes of the secants approach the slope of the tangent, which we calculate from the coordinates of A and B to be

$$\frac{350 - 0}{35 - 15} = 17 \text{ flies/day.}$$

In terms of population change, what we see as Q approaches P is this: The average growth rates for increasingly smaller time intervals approach the slope of the tangent to the curve at P (17 flies per day). The slope of the tangent line is therefore the number we take as the rate at which the fly population was changing on day $t = 23$.

Defining Slopes and Tangent Lines

The moral of the fruit fly story would seem to be that we should define the rate at which the value of the function $y = f(x)$ is changing with respect to x at any particular value $x = x_1$ to be the slope of the tangent to the curve $y = f(x)$ at $x = x_1$. But how are we to define the tangent line at an arbitrary point P on the curve and deduce its slope from the formula $y = f(x)$?

The answer that Fermat finally found in 1629 proved to be one of that century's major contributions to calculus. We still use his method of defining tangents to produce formulas for slopes of curves and rates of change. It goes like this:

1. We start with what we *can* calculate, namely the slope of a secant through P and a point Q nearby on the curve.
2. We find the limiting value of the secant slope (if it exists) as Q approaches P along the curve.
3. We take this number to be the slope of the curve at P and define the tangent to the curve at P to be the line through P with this slope.

How do you find a tangent to a curve?

It is hard to overestimate how important the answer to this question was to the scientists of the early seventeenth century. In optics, the angle at which a ray of light strikes the surface of a lens is defined in terms of the tangent to the surface. In physics, the direction of a body's motion at any point of its path is along the tangent to the path. In geometry, the angle between two intersecting curves is the angle between their tangents at the point of intersection.

EXAMPLE 2 Find the slope of the parabola $y = x^2$ at the point $P(2, 4)$. Write an equation for the tangent to the parabola at this point.

Solution We begin with a secant line that passes through $P(2, 4)$ and a neighboring point $Q(2 + h, (2 + h)^2)$ on the curve (Fig. 3.3). We then write an expression for the slope of the secant line and find the limiting value of this slope as Q approaches P along the curve.

The slope of the secant PQ is

$$\begin{aligned}\text{Secant slope} = \frac{\Delta y}{\Delta x} &= \frac{(2 + h)^2 - 2^2}{h} \\ &= \frac{h^2 + 4h + 4 - 4}{h} \\ &= \frac{h^2 + 4h}{h} = h + 4.\end{aligned}$$

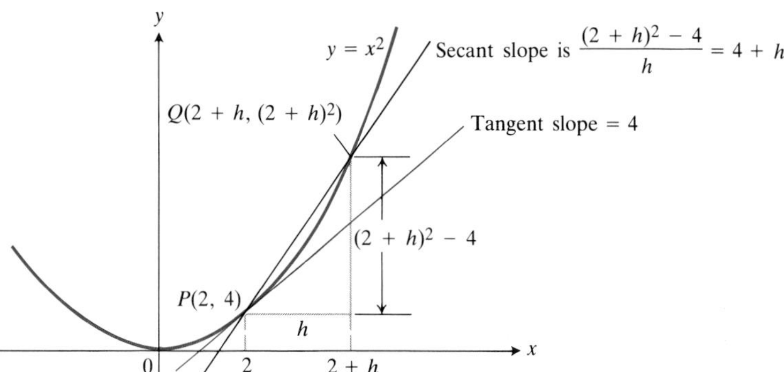

3.3 The slope of the secant *PQ* approaches 4 as *Q* approaches *P* along the curve.

The limit of the secant slope as Q approaches P along the curve is

$$\lim_{Q \to P} (\text{Secant slope}) = \lim_{h \to 0} (h + 4) = 4.$$

The tangent to the parabola at P is the line through P with slope 4:

Point: $(2, 4)$

Slope: $m = 4$

Equation: $y - 4 = 4(x - 2)$

$$y = 4x - 8 + 4$$

$$y = 4x - 4.$$

The mathematics we just used to find the slope of the parabola $y = x^2$ at the point $P(2, 4)$ will find the slope of the parabola at any other point, too. Here's how it works.

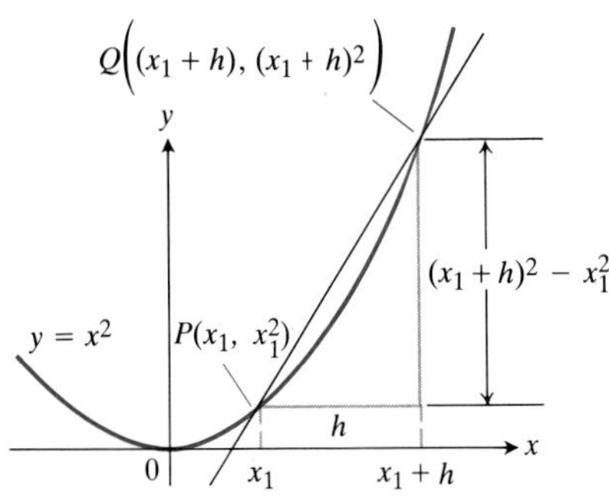

3.4 The slope of the secant *PQ* is $[(x_1 + h)^2 - x_1^2]/(h) = 2x_1 + h$.

EXAMPLE 3 Find the slope of the parabola $y = x^2$ at any point on the curve.

Solution Let $P(x_1, x_1^2)$ be the point. In the notation of Fig. 3.4, the slope of the secant line through P and any nearby point $Q(x_1 + h, (x_1 + h)^2)$ is

$$\text{Secant slope} = \frac{\Delta y}{\Delta x} = \frac{(x_1 + h)^2 - x_1^2}{h^2}$$

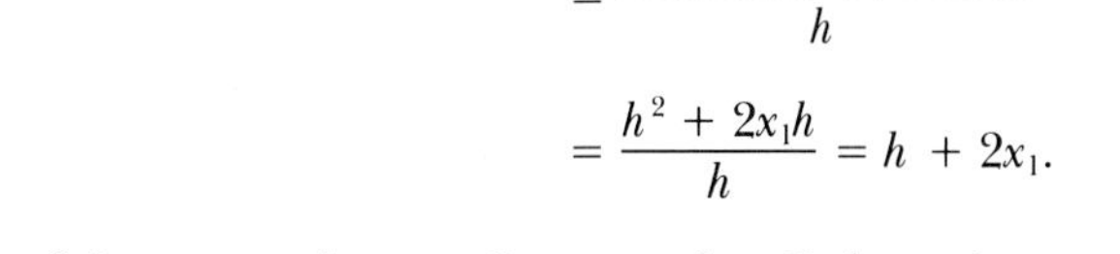

$$= \frac{h^2 + 2x_1 h + x_1^2 - x_1^2}{h}$$

$$= \frac{h^2 + 2x_1 h}{h} = h + 2x_1.$$

The limit of the secant slope as Q approaches P along the curve is

$$\lim_{Q \to P} (\text{Secant slope}) = \lim_{h \to 0} (h + 2x_1) = 2x_1.$$

Since x_1 can be any value of x, we can omit the subscript 1. At any point (x, y) on the parabola, the slope is

$$m = 2x.$$

When $x = 2$, for example, the slope is 4, as in Example 2.

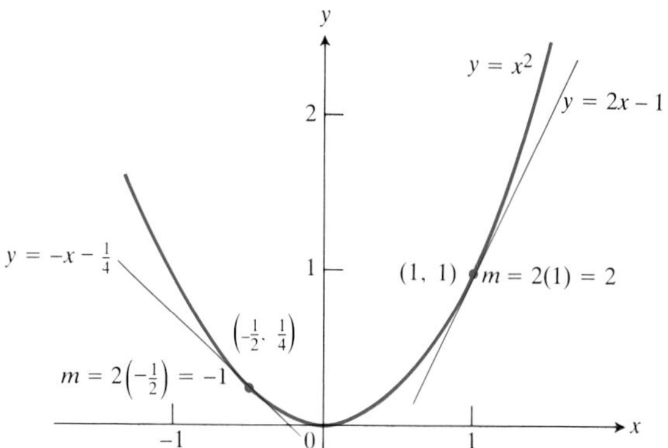

3.5 The slope of the tangent at a point (x, y) on the parabola $y = x^2$ is $m = 2x$.

The next example shows how to use the slope formula $m = 2x$ from Example 3 to find equations for tangent lines.

EXAMPLE 4 Find equations for the tangents to the curve $y = x^2$ at the points $(-1/2, 1/4)$ and $(1, 1)$.

Solution We use the slope formula $m = 2x$ from Example 3 to find the point–slope equation for each line.

Tangent at $(-1/2, 1/4)$ Point: $(-1/2, 1/4)$

Slope: $m = 2x = 2(-1/2) = -1$

Equation: $y - 1/4 = -1(x - (-1/2))$

$y - 1/4 = -x - 1/2$

$y = -x - 1/4$

Tangent at $(1, 1)$ Point: $(1, 1)$

Slope: $m = 2x = 2(1) = 2$

Equation: $y - 1 = 2(x - 1)$

$y - 1 = 2x - 2$

$y = 2x - 1$

See Fig. 3.5. ■

The Derivative of a Function

The function $m = 2x$ that gives the slope of the parabola $y = x^2$ at x is the derivative of the function $y = x^2$.

To find the derivative of an arbitrary function $y = f(x)$ (when the function has one—we'll come back to that), we simply repeat for f the steps we took in Examples 2 and 3 for x^2. We start with an arbitrary point $P(x, f(x))$ on the graph of f, as in Fig. 3.6. The slope of the secant line through P and a nearby point $Q(x + h, f(x + h))$ is then

$$\text{Secant slope} = \frac{\Delta y}{\Delta x} = \frac{f(x + h) - f(x)}{h}. \tag{3}$$

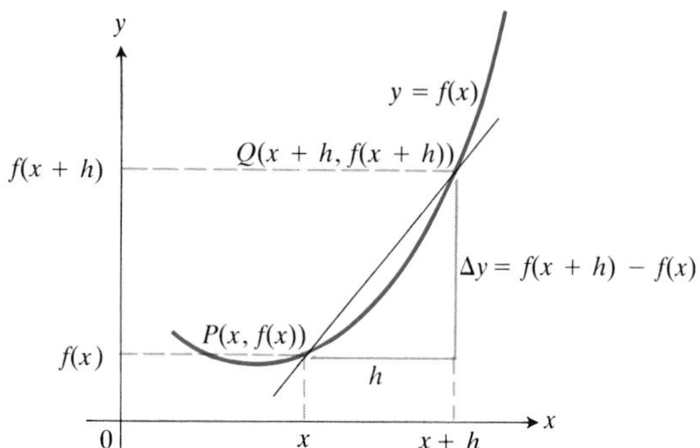

3.6 The slope of the line PQ is $\dfrac{f(x+h)-f(x)}{h}$.

The slope of the graph at P is the limit of the secant slope as Q approaches P along the graph. We find this limit from the formula for f by calculating the limit

$$\lim_{h\to 0}\frac{f(x+h)-f(x)}{h}. \tag{4}$$

The limit in Eq. (4) is itself a function of x. We denote it by f' ("f prime") and call it the derivative of f. Its domain is a subset of the domain of f. For most of the functions in this book, f' will be defined at all or all but a few of the points where f is defined.

DEFINITIONS

The Derivative of a Function

The **derivative** of a function f is the function f' whose value at x is defined by the equation

$$f'(x)=\lim_{h\to 0}\frac{f(x+h)-f(x)}{h}. \tag{5}$$

The fraction $(f(x+h)-f(x))/h$ is the **difference quotient** for f at x.

Differentiable at a Point

A function that has a derivative at a point x is said to be **differentiable at x.**

Differentiable Function

A function that is differentiable at every point of its domain is called **differentiable.**

The Slope and Tangent

When the number $f'(x)$ exists it is called the **slope** of the curve $y=f(x)$ at x. The line through the point $(x, f(x))$ with slope $f'(x)$ is the **tangent** to the curve at x.

The most common notations for the derivative of a function $y = f(x)$, besides $f'(x)$, are

y'	("y prime")	(Nice and brief)
$\dfrac{dy}{dx}$	("$d\,y\,d\,x$")	(Names the variables and has a "d" for derivative)
$\dfrac{df}{dx}$	("$d\,f\,d\,x$")	(Emphasizes the function's name)
$D_x(f)$	("$D\,x$ of f")	(Emphasizes the idea that taking the derivative is an operation performed on f)
$\dfrac{d}{dx}(f)$	("$d\,dx$ of f")	(Ditto)

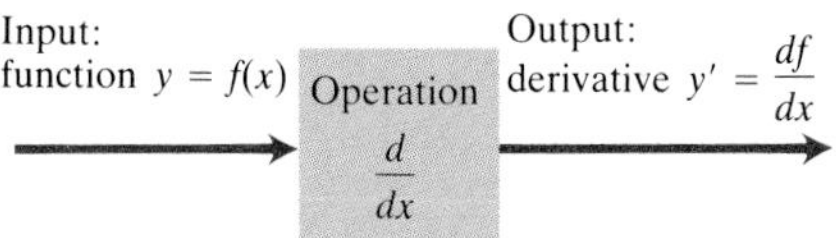

3.7 Flow diagram for the operation of taking a derivative with respect to x.

We also read dy/dx as "the derivative of y with respect to x" and df/dx as "the derivative of f with respect to x." See Fig. 3.7.

The Slopes of Lines

We now have two definitions for the slope of a line $y = mx + b$: the number m and, at each point, the derivative of the function $f(x) = mx + b$. Whenever we bring in a new definition, it is a good idea to be sure that the new and old definitions agree on objects to which they both apply. We do this in the next example.

EXAMPLE 5 The derivative of the function $f(x) = mx + b$ is the slope of the line $y = mx + b$.

The idea is to show that the derivative of $f(x) = mx + b$ has the constant value m. To see that this is so, we calculate the limit in Eq. (5) with $f(x) = mx + b$. The calculation takes four steps:

STEP 1: Write out $f(x)$ and $f(x + h)$:

$$f(x) = mx + b$$

$$f(x + h) = m(x + h) + b = mx + mh + b.$$

STEP 2: Subtract $f(x)$ from $f(x + h)$:

$$f(x + h) - f(x) = mh.$$

STEP 3: Divide by h:

$$\frac{f(x + h) - f(x)}{h} = \frac{mh}{h} = m.$$

STEP 4: Take the limit as $h \to 0$:

$$\lim_{h\to 0} \frac{f(x + h) - f(x)}{h} = \lim_{h\to 0} (m) = m.$$

The derivative of f does indeed have the constant value m. ■

Typical Derivative Calculations

EXAMPLE 6 Find dy/dx if $y = 1/x$.

Solution We take $f(x) = 1/x$, $f(x+h) = 1/(x+h)$, and form the quotient

$$\frac{f(x+h)-f(x)}{h} = \frac{\dfrac{1}{x+h}-\dfrac{1}{x}}{h}$$

$$= \frac{1}{h}\cdot\frac{x-(x+h)}{x(x+h)} \tag{6}$$

$$= \frac{1}{h}\cdot\frac{-h}{x(x+h)} = \frac{-1}{x(x+h)}.$$

We then take the limit as $h \to 0$:

$$\frac{dy}{dx} = \lim_{h\to 0}\frac{f(x+h)-f(x)}{h} = \lim_{h\to 0}\frac{-1}{x(x+h)} = \frac{-1}{x(x+0)} = -\frac{1}{x^2}.$$

■

EXAMPLE 7 Find $f'(x)$ if $f(x) = x^2 - 2x$.

The algebraic steps we use to calculate $f'(x)$ directly from the definition are always the same:

1. Write out $f(x)$ and $f(x+h)$.
2. Subtract $f(x)$ from $f(x+h)$.
3. Divide by h.
4. Take the limit as $h \to 0$.

Solution We subtract $f(x) = x^2 - 2x$ from

$$f(x+h) = (x+h)^2 - 2(x+h) = x^2 + 2hx + h^2 - 2x - 2h$$

and form the difference quotient

$$\frac{f(x+h)-f(x)}{h} = \frac{(x^2+2hx+h^2-2x-2h)-(x^2-2x)}{h}$$

$$= \frac{h^2+2hx-2h}{h}$$

$$= h + 2x - 2.$$

We then take the limit as $h \to 0$:

$$f'(x) = \lim_{h\to 0}\frac{f(x+h)-f(x)}{h}$$

$$= \lim_{h\to 0}(h+2x-2) = 2x-2.$$

The derivative of $x^2 - 2x$ is $2x - 2$. ■

EXAMPLE 8 Show that the derivative of $y = \sqrt{x}$ is $\dfrac{dy}{dx} = \dfrac{1}{2\sqrt{x}}$.

Solution We use Eq. (5) with $f(x+h) = \sqrt{x+h}$ and $f(x) = \sqrt{x}$ to form the quotient

$$\frac{f(x+h)-f(x)}{h} = \frac{\sqrt{x+h}-\sqrt{x}}{h}. \tag{7}$$

Unfortunately, this will involve division by 0 if we try to calculate the limit by replacing h with 0. We therefore look for an equivalent expression in which this difficulty does not arise. If we rationalize the numerator in Eq. (7), we find

$$\begin{aligned}\frac{\sqrt{x+h}-\sqrt{x}}{h} &= \frac{\sqrt{x+h}-\sqrt{x}}{h}\cdot\frac{\sqrt{x+h}+\sqrt{x}}{\sqrt{x+h}+\sqrt{x}} \\ &= \frac{(x+h)-x}{h(\sqrt{x+h}+\sqrt{x})} = \frac{1}{\sqrt{x+h}+\sqrt{x}}.\end{aligned} \tag{8}$$

Now as h approaches 0, the denominator in the final form approaches $\sqrt{x}+\sqrt{x} = 2\sqrt{x}$, which is positive because $x > 0$. Therefore,

$$\frac{dy}{dx} = \lim_{h\to 0}\frac{\sqrt{x+h}-\sqrt{x}}{h} = \lim_{h\to 0}\frac{1}{\sqrt{x+h}+\sqrt{x}} = \frac{1}{2\sqrt{x}}. \tag{9}$$

■

EXAMPLE 9 Find an equation for the tangent to the curve $y = \sqrt{x}$ at $x = 4$.

Solution The slope at $x = 4$ is the value of the function's derivative there. Equation (9) gives $dy/dx = 1/(2\sqrt{x})$, so the slope is

$$\left.\frac{1}{2\sqrt{x}}\right|_{x=4} = \frac{1}{2\sqrt{4}} = \frac{1}{4}.$$

The tangent is the line through the point $P(4, \sqrt{4}) = (4, 2)$ with slope 1/4:

Point: $(4, 2)$

Slope: $\frac{1}{4}$

Equation:
$$\begin{aligned} y - 2 &= \frac{1}{4}(x-4) \\ y &= \frac{1}{4}x - 1 + 2 \\ &= \frac{1}{4}x + 1.\end{aligned}$$

The tangent is the line $y = (1/4)x + 1$ (Fig. 3.8). ■

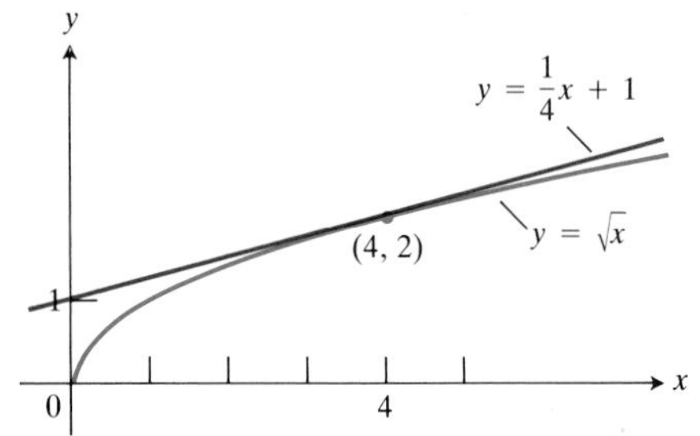

3.8 The curve $y = \sqrt{x}$ and its tangent at (4, 2). The slope of the tangent is found by evaluating dy/dx at $x = 4$.

Derivatives Have the Intermediate Value Property

It comes in handy now and then to know that derivatives have the intermediate value property: If f has a derivative at every point of a closed interval $[a, b]$, then f' assumes every value between $f'(a)$ and $f'(b)$. We shall refer to the property briefly in Chapter 4 but will make no attempt to prove it. There are proofs in more advanced texts.

This property of derivatives allows a partial answer to the question, When is a function defined on an interval the derivative of some other function throughout that interval? The (partial) answer is, Only when it has the intermediate value property. No step functions, for example.

The question of when a function is a derivative is one of the central questions in all calculus, and Newton and Leibniz's answer to this question revolutionized the world of mathematics. We shall see what their answer was when we get to Chapter 5.

Differentiable Functions Are Continuous

A function is continuous at every point at which it has a derivative.

THEOREM 1

If f has a derivative at $x = c$, then f is continuous at $x = c$.

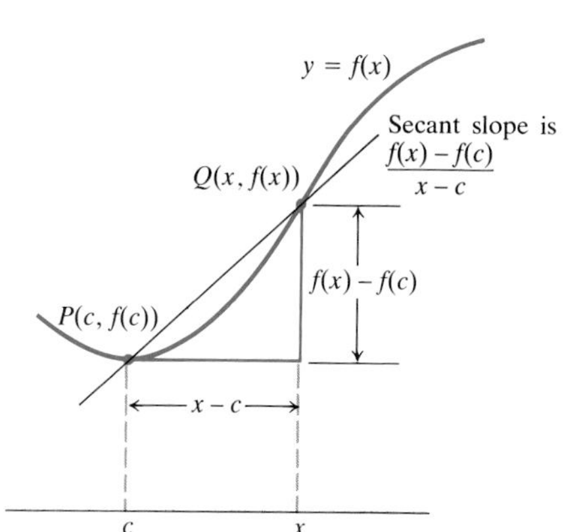

3.9 Figure for the proof that a function is continuous at every point at which it has a derivative.

PROOF Our task is to show that $\lim_{x\to c} f(x) = f(c)$ or, equivalently, that

$$\lim_{x\to c} [f(x) - f(c)] = 0. \tag{10}$$

To this end, we let $P(c, f(c))$ be a point on the graph of f and let $Q(x, f(x))$ be a point nearby (Fig. 3.9). The slope of the secant PQ is

$$\text{Secant slope} = \frac{f(x) - f(c)}{x - c}.$$

By definition, the derivative of f at c is the limiting value of this slope as Q approaches P along the curve, which means in this case, the limit as $x \to c$:

$$f'(c) = \lim_{x\to c} \frac{f(x) - f(c)}{x - c}. \tag{11}$$

Why should the mere existence of this limit imply that $[f(x) - f(c)] \to 0$ as $x \to c$? Because, with the denominator $x - c$ going to zero, the quotient can have a finite limit only if the numerator goes to zero at the same time. Indeed, this is exactly what we find if we apply the Limit Product Rule from Chapter 2:

$$\begin{aligned}
\lim_{x\to c} [f(x) - f(c)] &= \lim_{x\to c} \left[(x - c)\frac{f(x) - f(c)}{x - c}\right] && \text{(Algebra)} \\
&= \lim_{x\to c} (x - c) \cdot \lim_{x\to c} \frac{f(x) - f(c)}{x - c} && \left(\begin{matrix}\text{Limit Product} \\ \text{Rule}\end{matrix}\right) \\
&= 0 \cdot f'(c) = 0. && \text{(Known values)}
\end{aligned}$$

■

Differentiable on a Closed Interval— One-sided Derivatives

A function $y = f(x)$ is *differentiable on a closed interval* $[a, b]$ if it has a derivative at every interior point and if the limits

$$\lim_{h\to 0^+} \frac{f(a + h) - f(a)}{h} \qquad \textit{(Right-hand derivative at } a\textit{)}$$

$$\lim_{h\to 0^-} \frac{f(b + h) - f(b)}{h} \qquad \textit{(Left-hand derivative at } b\textit{)}$$

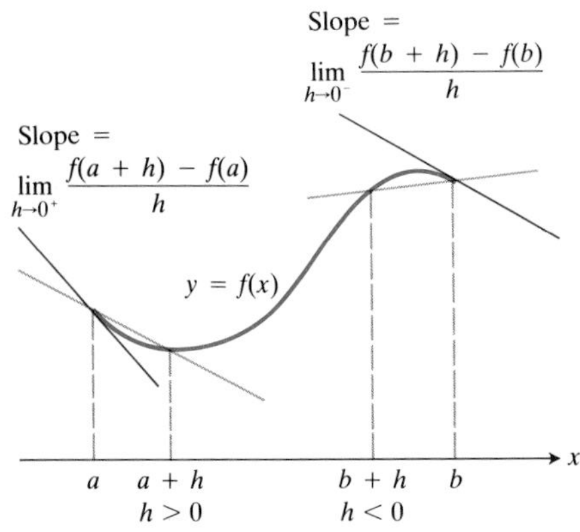

3.10 Derivatives at endpoints are one-sided limits.

exist at the endpoints. In the right-hand derivative, h is positive and $a + h$ approaches a from the right. In the left-hand derivative, h is negative and $b + h$ approaches b from the left (Fig. 3.10).

Right-hand and left-hand derivatives may be defined at any point of a function's domain.

The usual relation between one-sided and two-sided limits holds for derivatives. A function has a (two-sided) derivative at a point if and only if the function's right-hand and left-hand derivatives are defined and are equal at that point.

EXAMPLE 10 The function $y = |x|$ has no derivative with respect to x at $x = 0$ even though it has a derivative with respect to x at every other point. The reason is that the right-hand and left-hand derivatives of $y = |x|$ are not equal at $x = 0$:

$$\begin{aligned}\lim_{h\to 0^+} \frac{|0+h|-|0|}{h} &= \lim_{h\to 0^+} \frac{|h|}{h}\\ &= \lim_{h\to 0^+} \frac{h}{h} \qquad \text{(Because } h > 0\text{)}\\ &= \lim_{h\to 0^+} 1 = 1,\end{aligned}$$

$$\begin{aligned}\lim_{h\to 0^-} \frac{|0+h|-|0|}{h} &= \lim_{h\to 0^-} \frac{|h|}{h}\\ &= \lim_{h\to 0^-} \frac{-h}{h} \qquad (|h| = -h \text{ because } h < 0)\\ &= \lim_{h\to 0^-} (-1) = -1.\end{aligned}$$

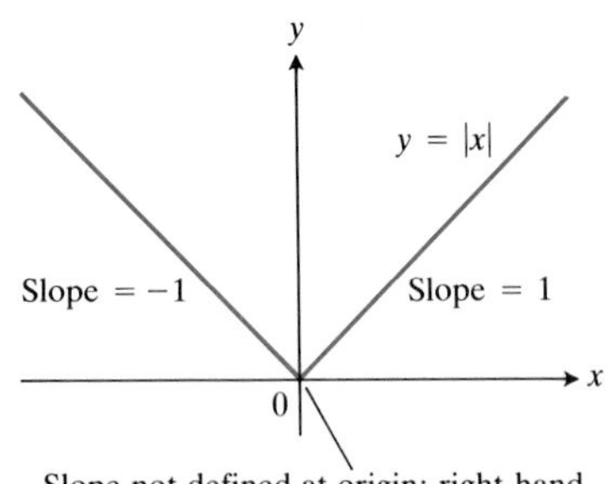

3.11 The absolute value function has a derivative at every point except the origin.

To the right of the origin, $|x| = x$, and the slope of the curve is 1 at every point. To the left of the origin, $|x| = -x$, and the slope is -1 at every point (Fig. 3.11). ■

EXAMPLE 11 Show that the following function has no derivative at $x = 0$.

$$y = \begin{cases} x^2, & x \le 0 \\ 2x, & x > 0 \end{cases}$$

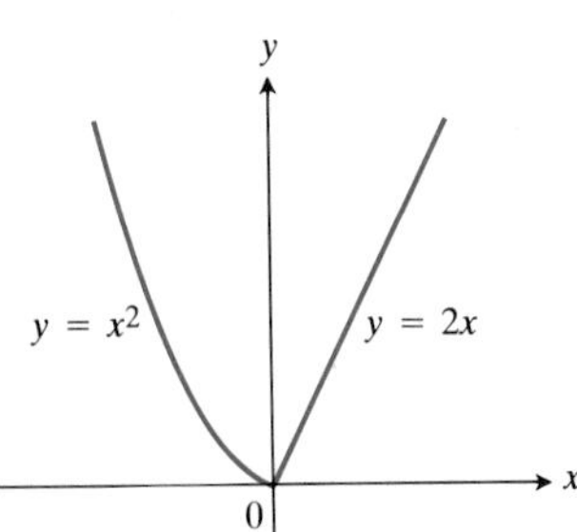

Solution There is no derivative at the origin because the right-hand and left-hand derivatives are different there. The slope of the parabola on the left is $2(0) = 0$ (from Example 3). The slope of the line on the right is 2. ■

When Does a Function *Not* Have a Derivative at a Point?

As we know, a function has a derivative at a point x_0 if the slopes of the secant lines through $P(x_0, f(x_0))$ and a nearby point Q on the graph approach a limit as Q approaches P. Whenever the secants fail to take up a limiting position as Q approaches P, the derivative does not exist. Typically, a function whose graph is otherwise smooth will fail to have a derivative at a point where the graph has

1. *A corner* where the one-sided derivatives differ.

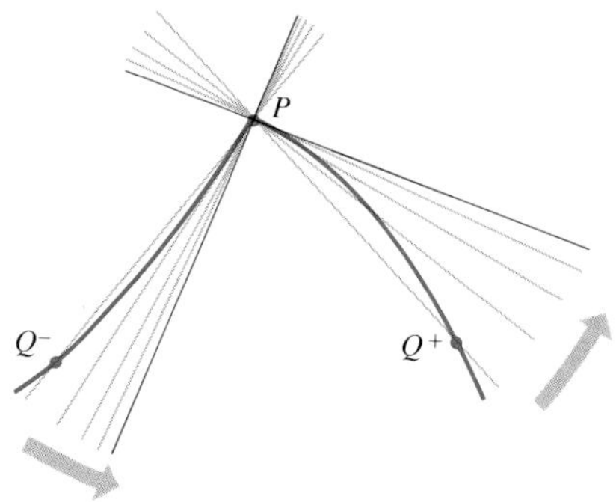

2. *A jump* that makes one of the one-sided derivatives infinite (here the left derivative).

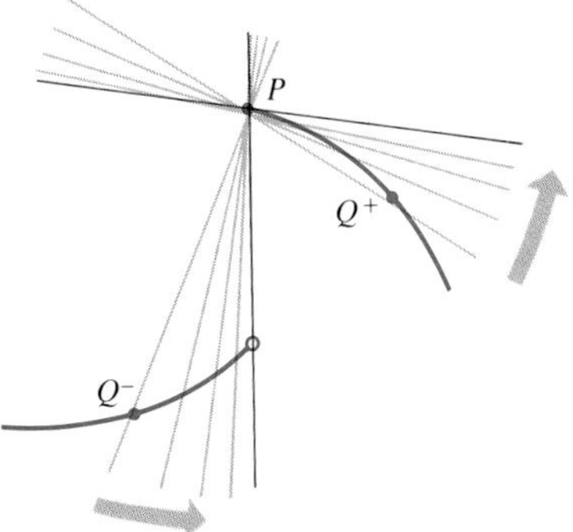

3. A vertical tangent.

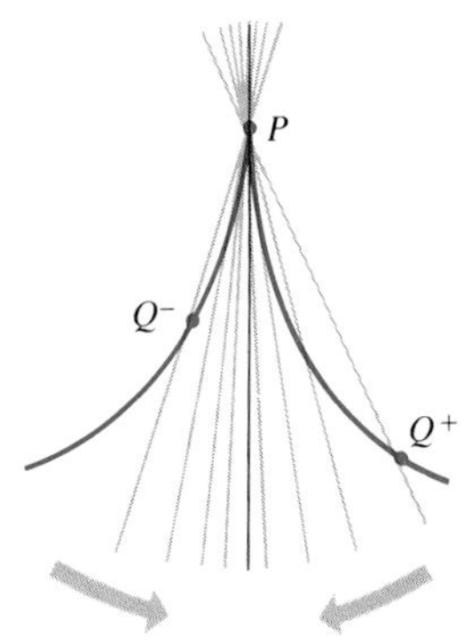

A function will also fail to have a derivative at any point of discontinuity (this from Theorem 1). Thus, the greatest-integer function $y = \lfloor x \rfloor$ is not differentiable at any integer value of x.

On the other hand, a function like $y = |x|$ can be continuous at a point and still not have a derivative there, as we saw in Example 10.

How rough can the graph of a function be?

The absolute value function fails to be differentiable at a single point. Using a similar idea, we can use a saw-tooth graph to define a continuous function that fails to have a derivative at infinitely many points.

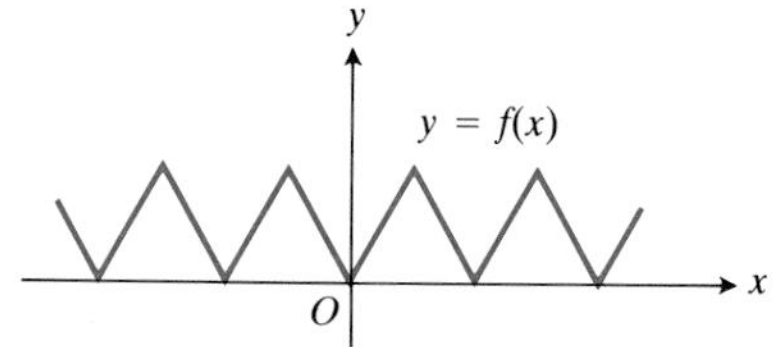

But can a continuous function fail to have a derivative at *every* point?

The answer, surprisingly enough, is yes, as Karl Weierstrass (1815–1897) found in 1872. One of his formulas (there are many like it) was

$$f(x) = \sum_{n=0}^{\infty} \left(\frac{2}{3}\right)^n \cos(9^n \pi x),$$

a formula that expresses f as an infinite sum of cosines with increasingly shorter periods. By adding wiggles to wiggles infinitely many times, so to speak, the formula produces a graph that is too bumpy to have a tangent anywhere.

Continuous curves that fail to have a tangent anywhere play an important role in chaos theory, in part because there is no way to measure the length of such a curve. We'll see what length has to do with derivatives when we get to Section 6.4.

What Functions Are Differentiable?

Most of the functions we have worked with so far are differentiable. Polynomials are differentiable, as are rational functions and trigonometric functions. Composites of differentiable functions are differentiable, and so are sums, differences, products, powers, and quotients of differentiable functions. We shall explain all this as the chapter continues.

Exercises 3.1

In Exercises 1–20, use Eq. (5) to find the derivative dy/dx of the given function $y = f(x)$. Then find the slope of the curve $y = f(x)$ at $x = 3$ and write an equation for the tangent there.

1. $y = 3x^2$

2. $y = \dfrac{x^2}{2}$

3. $y = 2x^2 - 5$

4. $y = x^2 - x$

5. $y = x^2 - 6x$

6. $y = x^2 - 6$

7. $y = \dfrac{2}{x}$

8. $y = \dfrac{1}{x + 1}$

9. $y = \dfrac{x}{x + 1}$

10. $y = \dfrac{1}{2x + 1}$

11. $y = x + \dfrac{9}{x}$

12. $y = x - \dfrac{1}{x}$

13. $y = 1 + \sqrt{x}$

14. $y = \sqrt{x + 1}$

15. $y = \sqrt{2x}$

16. $y = \sqrt{2x + 3}$

17. $y = \dfrac{1}{\sqrt{x}}$

18. $y = \dfrac{1}{\sqrt{2x + 3}}$

19. $y = x^3$

20. $y = x^4$

In Exercises 21–26, find an equation for the tangent to the curve at the given point. Then sketch the curve and tangent together.

21. $y = x^2 + 1, \quad (2, 5)$

22. $y = -x^2, \quad (1, -1)$

23. $y = 4 - x^2, \quad (-1, 3)$

24. $y = (x - 1)^2 + 1, \quad (1, 1)$

25. $y = \sqrt{x}, \quad (1, 1)$

26. $y = \dfrac{1}{x^2}, \quad (-1, 1)$

Which of the graphs in Exercises 27–30 suggest a function $y = f(x)$ that is

a) continuous at every point of its domain?
b) differentiable at every point of its domain?
c) both (a) and (b)?
d) neither (a) nor (b)?

Explain in each case.

27.

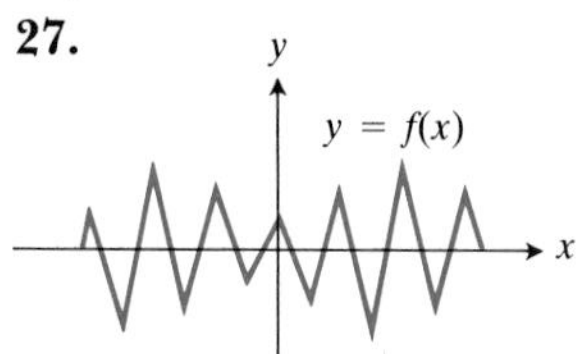

28.

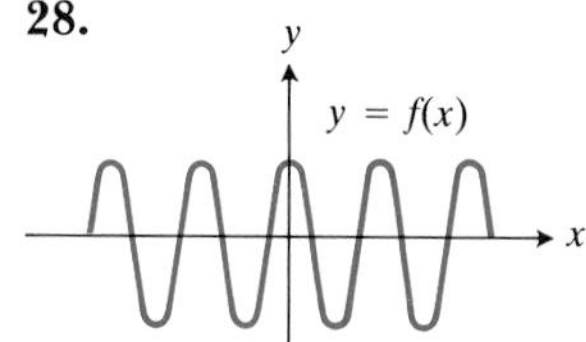

29. **30.**

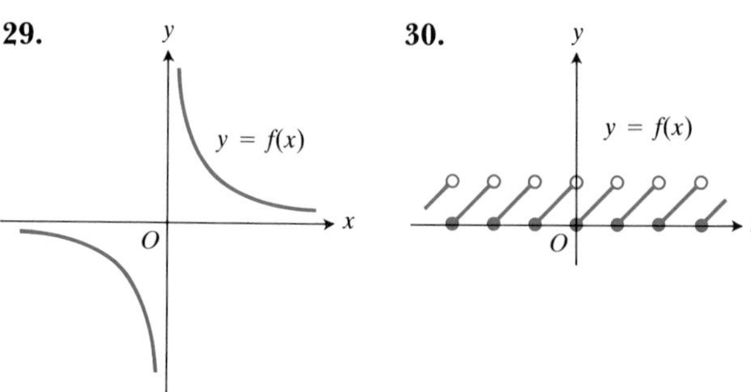

Use right-hand and left-hand derivatives to show that the functions graphed in Exercises 31–34 are not differentiable at the indicated point P.

31.

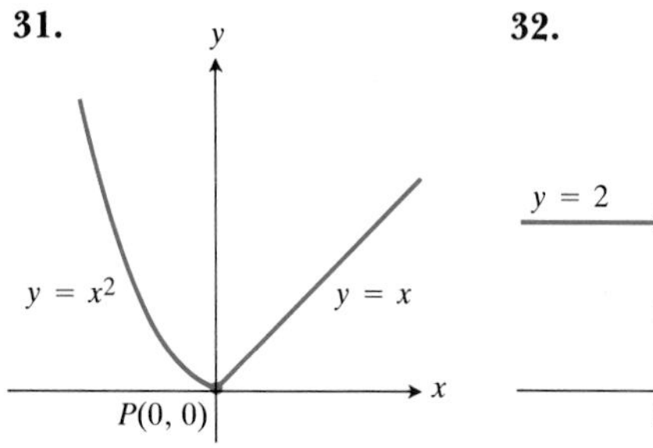

32.

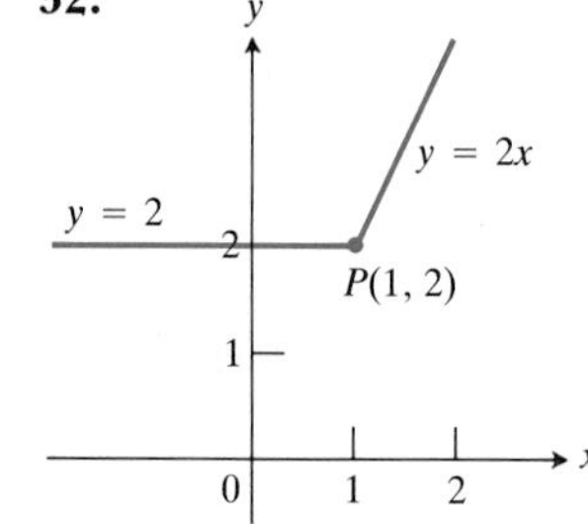

33.

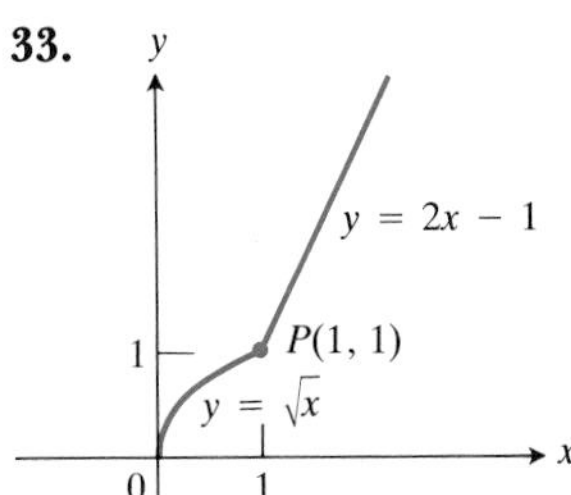

34.

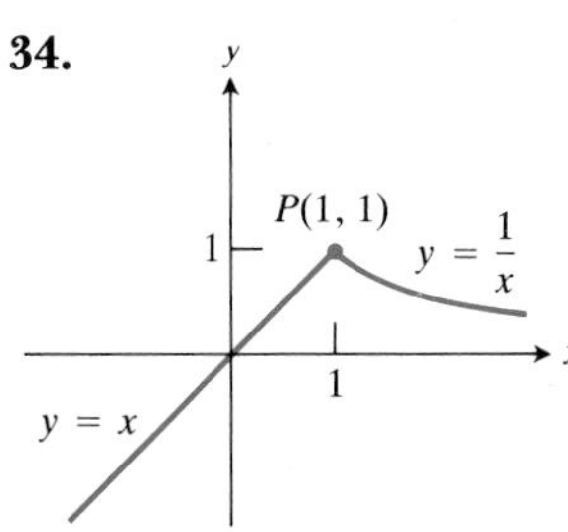

35. Show that the function

$$f(x) = \begin{cases} 0, & -1 \leq x < 0 \\ 1, & 0 \leq x \leq 1 \end{cases}$$

is not the derivative of any function on the interval $-1 \leq x \leq 1$. (*Hint:* Does f have the intermediate value property on the interval? What does the graph look like?)

36. Show that the greatest integer function $y = \lfloor x \rfloor$ is not the derivative of any function throughout the interval $-\infty < x < \infty$.

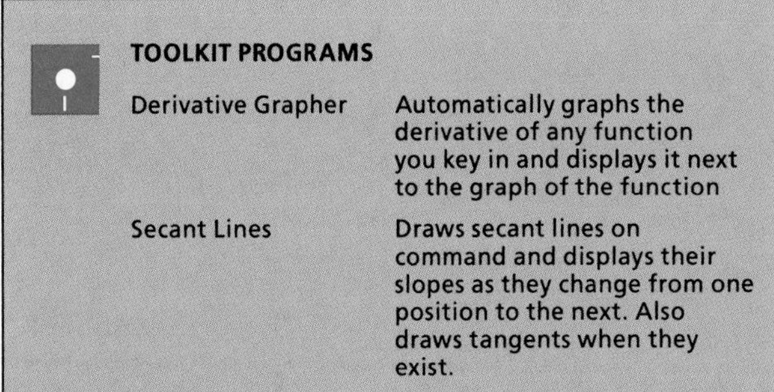

3.2 Differentiation Rules

The process of calculating a derivative is called differentiation. The goal of this section is to show how to differentiate functions rapidly—without having to apply the definition each time. It will then be an easy matter to calculate the velocities, accelerations, and other important rates of change we will encounter in Section 3.3.

Integer Powers, Multiples, Sums, and Differences

The first rule of differentiation is that the derivative of every constant function is zero. In short,

RULE 1

Derivative of a Constant

If c is a constant, then

$$\frac{d}{dx}(c) = 0.$$

The reason for this rule is the following calculation. If $f(x) = c$ is a function with a constant value c, then

$$\lim_{h \to 0} \frac{f(x+h) - f(x)}{h} = \lim_{h \to 0} \frac{c - c}{h} = \lim_{h \to 0} 0 = 0.$$

This is another way to say that the values of constant functions never change and that the slope of a horizontal line is zero at every point. See Fig. 3.12.

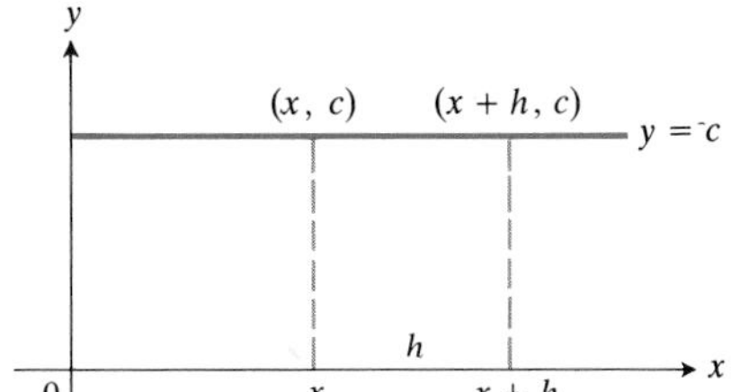

3.12 The slope of the graph of y = constant is zero at every point.

The next rule is about derivatives of positive integer powers of x.

RULE 2

Power Rule for Positive Integer Powers of x

If n is a positive integer, then

$$\frac{d}{dx}(x^n) = nx^{n-1}. \tag{1}$$

To apply the power rule, we subtract 1 from the original exponent (n) and multiply the result by n.

EXAMPLE 1

$$\frac{d}{dx}(x) = \frac{d}{dx}(x^1) = 1 \cdot x^0 = 1$$

$$\frac{d}{dx}(x^2) = 2x^1 = 2x$$

$$\frac{d}{dx}(x^3) = 3x^2$$

$$\frac{d}{dx}(x^4) = 4x^3$$

$$\frac{d}{dx}(x^5) = 5x^4$$

■

PROOF OF RULE 2 We set $f(x) = x^n$ and find the limit as h approaches zero of

$$\frac{f(x+h)-f(x)}{h} = \frac{(x+h)^n - x^n}{h}. \tag{2}$$

Since n is a positive integer, we can apply the algebra formula

$$a^n - b^n = (a-b)(a^{n-1} + a^{n-2}b + \cdots + ab^{n-2} + b^{n-1}),$$

with $(a-b) = h$, $a = x+h$, and $b = x$, to replace the expression $(x+h)^n - x^n$ in Eq. (2) by a form that is divisible by h. The resulting division tells us that

$$\begin{aligned}\frac{f(x+h)-f(x)}{h} &= \frac{(x+h)^n - x^n}{h} \\ &= \frac{(h)[(x+h)^{n-1} + (x+h)^{n-2}x + \cdots + (x+h)x^{n-2} + x^{n-1}]}{h} \\ &= \underbrace{[(x+h)^{n-1} + (x+h)^{n-2}x + \cdots + (x+h)x^{n-2} + x^{n-1}]}_{n \text{ terms, each with limit } x^{n-1} \text{ as } h \to 0}.\end{aligned} \tag{3}$$

Hence

$$\frac{d}{dx}(x^n) = \lim_{h \to 0} \frac{f(x+h) - f(x)}{h} = nx^{n-1}.$$ ■

The next rule says that when a differentiable function is multiplied by a constant its derivative is multiplied by the same constant.

RULE 3

The (Constant) Multiple Rule

If u is a differentiable function of x and c is a constant, then

$$\frac{d}{dx}(cu) = c\frac{du}{dx}. \tag{4}$$

In particular, if n is a positive integer then

$$\frac{d}{dx}(cx^n) = cnx^{n-1}. \tag{5}$$

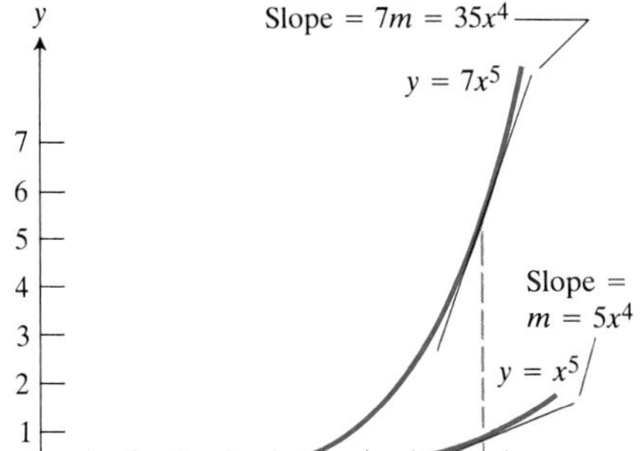

3.13 The graphs of $y = x^5$ and the stretched curve $y = 7x^5$. Multiplying the y-coordinates by 7 multiplies the slopes by 7.

EXAMPLE 2 The derivative

$$\frac{d}{dx}(7x^5) = 7 \cdot 5x^4 = 35x^4$$

says that if we stretch the graph of $y = x^5$ by multiplying each y-coordinate by 7, then we multiply the slope at each point by 7. See Fig. 3.13. ■

EXAMPLE 3 *(A useful special case)* The derivative of the negative of any function is the negative of the function's derivative:

$$\begin{aligned} \frac{d}{dx}(-u) &= \frac{d}{dx}(-1 \cdot u) \\ &= -1 \cdot \frac{d}{dx}(u) \qquad \text{(Eq. (4) with } c = -1) \\ &= -\frac{du}{dx}. \end{aligned}$$ ■

PROOF OF RULE 3

$$\begin{aligned} \frac{d}{dx}cu &= \lim_{h \to 0} \frac{cu(x+h) - cu(x)}{h} \qquad \begin{pmatrix}\text{Derivative definition} \\ \text{applied to } f(x) = cu(x)\end{pmatrix} \\ &= c \lim_{h \to 0} \frac{u(x+h) - u(x)}{h} \qquad \text{(Limit property)} \\ &= c\frac{du}{dx}. \qquad (u \text{ is differentiable.}) \end{aligned}$$ ■

The next rule says that the derivative of the sum or difference of two differentiable functions is the sum or difference of their derivatives.

RULE 4

The Sum and Difference Rule

If u and v are differentiable functions of x then their sum and difference are differentiable at every point where u and v are both differentiable. At such points,

1. $\dfrac{d}{dx}(u + v) = \dfrac{du}{dx} + \dfrac{dv}{dx}$, 2. $\dfrac{d}{dx}(u - v) = \dfrac{du}{dx} - \dfrac{dv}{dx}$.

Similar equations hold for more than two functions, as long as the number of functions involved is finite.

EXAMPLE 4

1. $y = x^4 + 12x$

$$\frac{dy}{dx} = \frac{d}{dx}(x^4) + \frac{d}{dx}(12x)$$
$$= 4x^3 + 12$$

2. $y = \dfrac{7x^2}{3} - 5$

$$\frac{dy}{dx} = \frac{d}{dx}\frac{(7x^2)}{3} - \frac{d}{dx}(5)$$
$$= \frac{7}{3} \cdot 2x - 0 = \frac{14}{3}x$$

3. $y = x^3 + 3x^2 - 5x + 1$

$$\frac{dy}{dx} = \frac{d}{dx}(x^3) + \frac{d}{dx}(3x^2) - \frac{d}{dx}(5x) + \frac{d}{dx}(1)$$
$$= 3x^2 + 3 \cdot 2x - 5 + 0$$
$$= 3x^2 + 6x - 5$$

■

Notice that we can differentiate any polynomial term by term, the way we differentiated the polynomials in Example 4.

PROOF OF RULE 4 To prove Part 1, we apply the derivative definition with $f(x) = u(x) + v(x)$:

$$\frac{d}{dx}[u(x) + v(x)] = \lim_{h\to 0} \frac{[u(x+h) + v(x+h)] - [u(x) + v(x)]}{h}$$
$$= \lim_{h\to 0} \left[\frac{u(x+h) - u(x)}{h} + \frac{v(x+h) - v(x)}{h}\right]$$
$$= \lim_{h\to 0} \frac{u(x+h) - u(x)}{h} + \lim_{h\to 0} \frac{v(x+h) - v(x)}{h}$$
$$= \frac{du}{dx} + \frac{dv}{dx}.$$

The proof of Part 2 is similar. ■

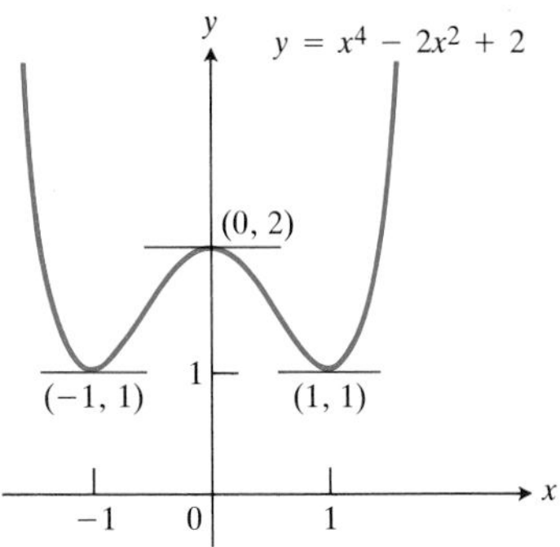

3.14 The curve $y = x^4 - 2x^2 + 2$ and its horizontal tangents. The tangents were located by setting dy/dx equal to zero and solving for x, as in Example 5.

EXAMPLE 5 *Finding horizontal tangents.* Does the curve $y = x^4 - 2x^2 + 2$ have any horizontal tangents? If so, where?

Solution The horizontal tangents, if any, occur where the slope dy/dx is zero. To find these points, we

1. Calculate dy/dx: $\quad \dfrac{dy}{dx} = \dfrac{d}{dx}(x^4 - 2x^2 + 2) = 4x^3 - 4x,$

2. Solve the equation $\dfrac{dy}{dx} = 0$ for x:

$$4x^3 - 4x = 0$$
$$4x(x^2 - 1) = 0$$
$$x = 0, 1, -1.$$

The curve in question has horizontal tangents at $x = 0$, 1, and -1. The corresponding points on the curve (found from the equation $y = x^4 - 2^2 + 2$) are (0, 2), (1, 1), and (−1, 1). See Fig. 3.14. ■

Second and Higher Order Derivatives

The derivative

$$y' = \frac{dy}{dx}$$

is the *first derivative* of y with respect to x. The first derivative may also be a differentiable function of x. If so, its derivative,

$$y'' = \frac{dy'}{dx} = \frac{d}{dx}\left(\frac{dy}{dx}\right) = \frac{d^2y}{dx^2},$$

is called the *second derivative* of y with respect to x. If y'' ("y double prime") is differentiable, its derivative,

$$y''' = \frac{dy''}{dx} \qquad (\text{"}y\text{ triple prime"}),$$

is the *third derivative* of y with respect to x. The names continue as you imagine they would, with

$$y^{(n)} = \frac{d}{dx}y^{(n-1)} \qquad (\text{"}y\text{ super } n\text{"})$$

denoting the n*th derivative* of y with respect to x.

EXAMPLE 6 The first four derivatives of $y = x^3 - 3x^2 + 2$ are

First derivative:	$y' = 3x^2 - 6x$
Second derivative:	$y'' = 6x - 6$
Third derivative:	$y''' = 6$
Fourth derivative:	$y^{(4)} = 0.$

The function has derivatives of all orders, but the fifth and subsequent order derivatives are all zero. ■

Products

While the derivative of the sum of two functions is the sum of their derivatives and the derivative of the difference of two functions is the difference of their derivatives, the derivative of the product of two functions is *not* the product of their derivatives. The derivative of a product is a sum of *two* products, as we now explain.

RULE 5

The Product Rule

The product of two differentiable functions u and v is differentiable and

$$\frac{d}{dx}(uv) = u\frac{dv}{dx} + v\frac{du}{dx}. \tag{6}$$

As with the Sum and Difference Rule, the Product Rule is understood to hold only at values of x where u and v both have derivatives. At such a value of x, the derivative of the product uv is u times the derivative of v plus v times the derivative of u.

PROOF OF RULE 5

$$\frac{d}{dx}(uv) = \lim_{h\to 0}\frac{u(x+h)v(x+h) - u(x)v(x)}{h}.$$

To change this fraction into an equivalent one that contains difference quotients for the derivatives of u and v, we subtract and add $u(x+h)v(x)$ in the numerator. Then,

$$\frac{d}{dx}(uv) = \lim_{h\to 0}\frac{u(x+h)v(x+h) - u(x+h)v(x) + u(x+h)v(x) - u(x)v(x)}{h}$$

$$= \lim_{h\to 0}\left[u(x+h)\frac{v(x+h)-v(x)}{h} + v(x)\frac{u(x+h)-u(x)}{h}\right]$$

$$= \lim_{h\to 0} u(x+h)\cdot\lim_{h\to 0}\frac{v(x+h)-v(x)}{h} + v(x)\cdot\lim_{h\to 0}\frac{u(x+h)-u(x)}{h}.$$

As h approaches zero, $u(x+h)$ approaches $u(x)$ because u, being differentiable at x, is continuous at x. The two fractions approach the values of du/dx at x and dv/dx at x. In short,

$$\frac{d}{dx}(uv) = u\frac{dv}{dx} + v\frac{du}{dx}.$$

■

EXAMPLE 7 Find the derivative of $y = (x^2+1)(x^3+3)$.

Solution From the Product Rule with

$$u = x^2 + 1, \qquad v = x^3 + 3,$$

we find

$$\frac{d}{dx}[(x^2 + 1)(x^3 + 3)] = (x^2 + 1)(3x^2) + (x^3 + 3)(2x)$$
$$= 3x^4 + 3x^2 + 2x^4 + 6x$$
$$= 5x^4 + 3x^2 + 6x. \qquad \blacksquare$$

This particular example can be done as well (perhaps better) by multiplying out the original expression for y and differentiating the resulting polynomial. We do that now as a check. From

$$y = (x^2 + 1)(x^3 + 3) = x^5 + x^3 + 3x^2 + 3,$$

we obtain

$$\frac{dy}{dx} = 5x^4 + 3x^2 + 6x,$$

in agreement with our first calculation.

There are times, however, when the Product Rule *must* be used, as the next example shows.

EXAMPLE 8 Let $y = uv$ be the product of the functions u and v, and suppose that

$$u(2) = 3, \qquad u'(2) = -4, \qquad v(2) = 1, \qquad \text{and} \qquad v'(2) = 2.$$

Find $y'(2)$.

Solution From the Product Rule, in the form

$$y' = (uv)' = uv' + vu',$$

we have

$$y'(2) = u(2)v'(2) + v(2)u'(2) = (3)(2) + (1)(-4) = 6 - 4 = 2. \qquad \blacksquare$$

Quotients

Just as the derivative of the product of two differentiable functions is not the product of their derivatives, the derivative of the quotient of two functions is not the quotient of their derivatives. What happens instead is this:

RULE 6

The Quotient Rule

At a point where $v \neq 0$, the quotient $y = u/v$ of two differentiable functions is differentiable and

$$\frac{d}{dx}\left(\frac{u}{v}\right) = \frac{v\dfrac{du}{dx} - u\dfrac{dv}{dx}}{v^2}. \qquad (7)$$

As with the earlier combination rules, the Quotient Rule holds only at values of x at which u and v both have derivatives.

PROOF OF RULE 6

$$\frac{d}{dx}\left(\frac{u}{v}\right) = \lim_{h\to 0} \frac{\dfrac{u(x+h)}{v(x+h)} - \dfrac{u(x)}{v(x)}}{h}$$

$$= \lim_{h\to 0} \frac{v(x)u(x+h) - u(x)v(x+h)}{hv(x+h)v(x)}.$$

To change the last fraction into an equivalent one that contains the difference quotients for the derivatives of u and v, we subtract and add $v(x)u(x)$ in the numerator. This allows us to continue with

$$\frac{d}{dx}\left(\frac{u}{v}\right) = \lim_{h\to 0} \frac{v(x)u(x+h) - v(x)u(x) + v(x)u(x) - u(x)v(x+h)}{hv(x+h)v(x)}$$

$$= \lim_{h\to 0} \frac{v(x)\dfrac{u(x+h) - u(x)}{h} - u(x)\dfrac{v(x+h) - v(x)}{h}}{v(x+h)v(x)}.$$

Taking the limit in the numerator and denominator now gives the Quotient Rule. ■

EXAMPLE 9 Find the derivative of $y = \dfrac{x^2 - 1}{x^2 + 1}$.

Solution We apply the Quotient Rule with $u = x^2 - 1$ and $v = x^2 + 1$:

$$\frac{dy}{dx} = \frac{(x^2+1)\cdot 2x - (x^2-1)\cdot 2x}{(x^2+1)^2}$$

$$= \frac{2x^3 + 2x - 2x^3 + 2x}{(x^2+1)^2}$$

$$= \frac{4x}{(x^2+1)^2}.$$ ■

Negative Integer Powers of x

The rule for differentiating negative powers of x is the same as the rule for differentiating positive powers of x.

RULE 7

Power Rule for Negative Integer Powers of x

If n is a negative integer and $x \neq 0$, then

$$\frac{d}{dx}(x^n) = nx^{n-1}. \tag{8}$$

PROOF OF RULE 7 The proof uses the Quotient Rule in a clever way. If n is a negative integer, then $n = -m$ where m is a positive integer. Hence, $x^n = x^{-m} = 1/x^m$ and

$$\begin{aligned}
\frac{d}{dx}(x^n) &= \frac{d}{dx}\left(\frac{1}{x^m}\right) \\
&= \frac{x^m \cdot \frac{d}{dx}(1) - 1 \cdot \frac{d}{dx}(x^m)}{(x^m)^2} && \left(\begin{array}{l}\text{Quotient Rule} \\ \text{with } u = 1 \text{ and } v = x^m\end{array}\right) \\
&= \frac{0 - mx^{m-1}}{x^{2m}} && \left(\begin{array}{l}\text{Since } m > 0, \\ \frac{d}{dx}(x^m) = mx^{m-1}\end{array}\right) \\
&= -mx^{-m-1} \\
&= nx^{n-1}. && (\text{Changing back to } n)
\end{aligned}$$

■

EXAMPLE 10

$$\frac{d}{dx}\left(\frac{1}{x}\right) = \frac{d}{dx}(x^{-1}) = (-1)x^{-2} = -\frac{1}{x^2}$$

$$\frac{d}{dx}\left(\frac{4}{x^3}\right) = 4\frac{d}{dx}(x^{-3}) = 4(-3)x^{-4} = -\frac{12}{x^4}$$

■

EXAMPLE 11 Find an equation for the tangent to the curve

$$y = x + \frac{2}{x}$$

at the point (1, 3).

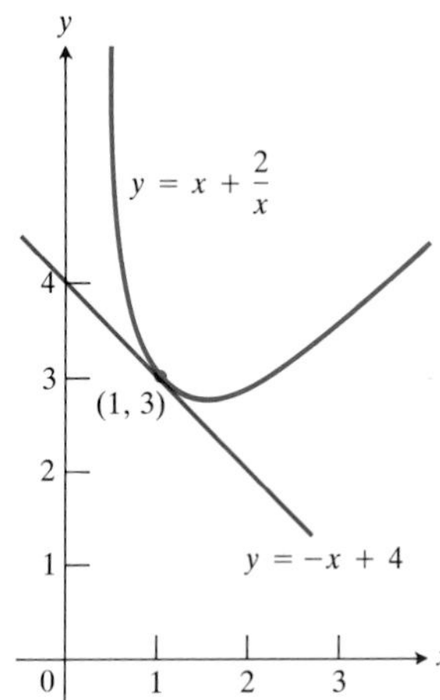

3.15 The slope of the tangent to the curve $y = x + (2/x)$ at the point (1, 3) is found by evaluating y' at $x = 1$. (The curve also has a portion in the third quadrant, not shown here. Don't worry about how to graph functions like this right now; we'll come to that in Chapter 4.)

Solution The slope of the curve is

$$\begin{aligned}
\frac{dy}{dx} &= \frac{d}{dx}(x) + 2\frac{d}{dx}\left(\frac{1}{x}\right) \\
&= 1 + 2\left(-\frac{1}{x^2}\right) && (\text{From Example 10}) \\
&= 1 - \frac{2}{x^2}.
\end{aligned}$$

The slope at $x = 1$ is

$$\left.\frac{dy}{dx}\right|_{x=1} = \left[1 - \frac{2}{x^2}\right]_{x=1} = 1 - 2 = -1.$$

The line through (1, 3) with slope $m = -1$ is

$$\begin{aligned}
y - 3 &= (-1)(x - 1) && (\text{Point–slope equation}) \\
y &= -x + 1 + 3 \\
y &= -x + 4.
\end{aligned}$$

See Fig. 3.15.

■

Choosing Which Rules to Use

The choice of which rules to use in solving a differentiation problem can make a difference in how much work you have to do. Here is an example.

EXAMPLE 12 Do not use the Quotient Rule to find the derivative of

$$y = \frac{(x-1)(x^2-2x)}{x^4}.$$

Instead, expand the numerator and divide by x^4:

$$y = \frac{(x-1)(x^2-2x)}{x^4} = \frac{x^3-3x^2+2x}{x^4} = x^{-1} - 3x^{-2} + 2x^{-3}.$$

Then use the Sum and Power Rules:

$$\frac{dy}{dx} = -x^{-2} - 3(-2)x^{-3} + 2(-3)x^{-4} = -\frac{1}{x^2} + \frac{6}{x^3} - \frac{6}{x^4}.$$ ■

Exercises 3.2

In Exercises 1–10, find dy/dx and d^2y/dx^2 when y equals the given expression. Try to answer without writing anything down (except the answer).

1. x **2.** $-x$

3. x^2 **4.** $-10x^2$

5. $-x^2+3$ **6.** $\frac{x^3}{3} - x$

7. $2x+1$ **8.** x^2+x+1

9. $\frac{x^3}{3} + \frac{x^2}{2} + x$ **10.** $1 - x + x^2 - x^3$

In Exercises 11–16, find $y' = dy/dx$ and $y'' = d^2y/dx^2$ when y equals the given expression.

11. $x^4 - 7x^3 + 2x^2 + 15$ **12.** $5x^3 - 3x^5$

13. $4x^2 - 8x + 1$ **14.** $\frac{x^4}{4} - \frac{x^3}{3} + \frac{x^2}{2} - x + 3$

15. $2x^4 - 4x^2 - 8$ **16.** $3x^7 - 7x^3 + 21x^2$

Find all the nonzero derivatives of the functions in Exercises 17–20.

17. $y = x^2 - x$ **18.** $y = \frac{x^3}{3} + \frac{x^2}{2} - 5$

19. $y = \frac{x^4}{2} - \frac{3}{2}x^2 - x$ **20.** $y = \frac{x^5}{120}$

In Exercises 21–28, find dy/dx when y equals the given expression. Find each derivative two ways: (a) by applying the Product Rule and (b) by multiplying the factors to produce a sum of simpler terms to differentiate. In these examples, (b) tends to be faster, but it is not always possible to multiply first this way. For functions like $y = x \sin x$, encountered later, the Product Rule offers the only way.

21. $(3x-1)(2x+5)$ **22.** $(5-x)(4-2x)$

23. $(x+1)(x^2+1)$ **24.** $(x+1)(3-x^2)$

25. $x^2(x^3-1)$ **26.** $x^2\left(x+5+\frac{1}{x}\right)$

27. $(x-1)(x^2+x+1)$ **28.** $\left(x+\frac{1}{x}\right)\left(x-\frac{1}{x}\right)$

In Exercises 29–36, find dy/dx when y equals the given expression.

29. $\frac{x-1}{x+7}$ **30.** $\frac{2x+5}{3x-2}$

31. $\frac{2x+1}{x^2-1}$ **32.** $\frac{x^2-3}{x+1/2}$

33. $(1-x)(1+x^2)^{-1}$ **34.** $(2x-7)^{-1}(x+5)$

35. $\frac{x^2}{1-x^3}$ **36.** $\frac{x^2-1}{x^2+x-2}$

In Exercises 37–44, find $y' = dy/dx$ and $y'' = d^2y/dx^2$ when y equals the given expression.

37. $\frac{3}{x^2}$ **38.** $-\frac{1}{x}$ **39.** $\frac{1}{2x^2}$

40. $\frac{5}{3x}$ **41.** $\frac{5}{x^4}$ **42.** $-\frac{3}{x^7}$

43. $x + 1 + \frac{1}{x}$ **44.** $\frac{12}{x} - \frac{4}{x^3} + \frac{1}{x^4}$

In Exercises 45–48, find dy/dx when y equals the given expression.

45. $\frac{x^3+7}{x}$ **46.** $\frac{x^2+5x-1}{x^2}$

47. $\frac{(x-1)(x^2+x+1)}{x^3}$ **48.** $\frac{(x^2+x)(x^2-x+1)}{x^4}$

49. Suppose u and v are functions of x that are differentiable at $x = 0$ and that

$$u(0) = 5, \quad u'(0) = -3, \quad v(0) = -1, \quad v'(0) = 2.$$

Find the values of the following derivatives at $x = 0$.

a) $\frac{d}{dx}(uv)$ b) $\frac{d}{dx}\left(\frac{u}{v}\right)$ c) $\frac{d}{dx}\left(\frac{v}{u}\right)$ d) $\frac{d}{dx}(7v - 2u)$

50. Suppose u and v are differentiable functions of x and that

$$u(1) = 2, \quad u'(1) = 0, \quad v(1) = 5, \quad v'(1) = 0.$$

Find the values of the following derivatives at $x = 1$.

a) $\frac{d}{dx}(uv)$ b) $\frac{d}{dx}\left(\frac{u}{v}\right)$ c) $\frac{d}{dx}\left(\frac{v}{u}\right)$ d) $\frac{d}{dx}(7v - 2u)$

51. Which of the following numbers is the slope of the line tangent to the curve $y = x^2 + 5x$ at $x = 3$?

a) 24 b) $-5/2$ c) 11 d) 8

52. Which of the following numbers is the slope of the line $3x - 2y + 12 = 0$?

a) 6 b) 3 c) 3/2 d) 2/3

53. Find the equation of the line perpendicular to the tangent to the curve $y = x^3 - 3x + 1$ at the point (2, 3).

54. Find the tangents to the curve $y = x^3 + x$ at the points where the slope is 4. What is the smallest slope on the curve? At what value of x does the curve have this slope?

55. Find the points on the curve $y = 2x^3 - 3x^2 - 12x + 20$ where the tangent is parallel to the x-axis.

56. Find the x- and y-intercepts of the line that is tangent to the curve $y = x^3$ at the point $(-2, -8)$.

57. Find the tangents to *Newton's Serpentine*,

$$y = \frac{4x}{x^2 + 1}$$

at the origin and the point (1, 2).

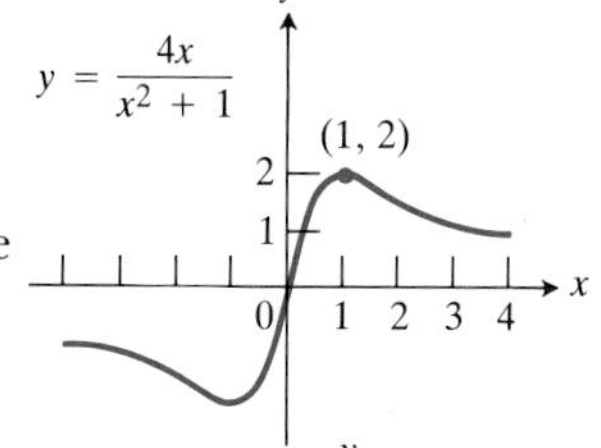

58. Find the tangent to the *Witch of Agnesi*

$$y = \frac{8}{4 + x^2}$$

at the point (2, 1).

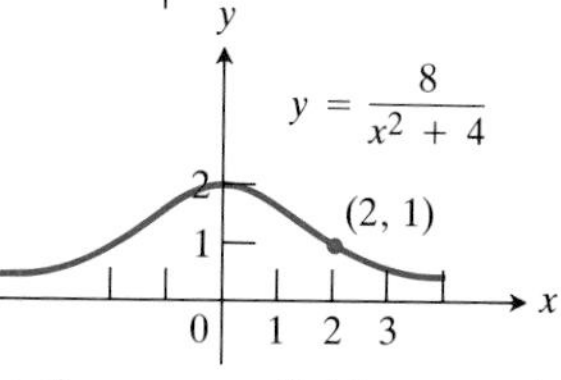

There is a nice story about the name of this curve in the historical note on Agnesi in Chapter 10.

59. *Cylinder pressure.* If the gas in a cylinder is maintained at a constant temperature T, the pressure P is related to the volume V by a formula of the form

$$P = \frac{nRT}{V - nb} - \frac{an^2}{V^2}$$

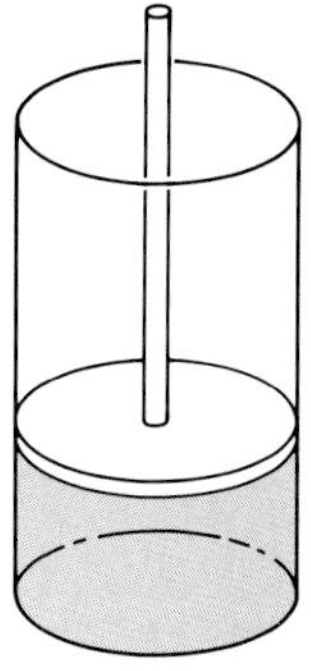

in which a, b, n, and R are constants. Find dP/dV.

60. *Free fall.* When a rock falls from rest near the surface of the earth, the distance it covers during the first few seconds is given by the equation

$$s = 4.9t^2.$$

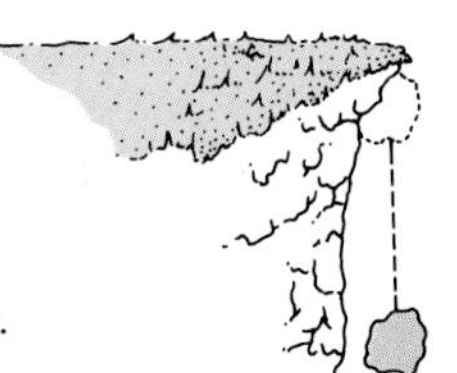

In this equation, s is the distance in meters and t is the elapsed time in seconds. Find ds/dt and d^2s/dt^2.

61. *The body's reaction to medicine.* The reaction of the body to a dose of medicine can often be represented by an equation of the form

$$R = M^2\left(\frac{C}{2} - \frac{M}{3}\right),$$

where C is a positive constant and M is the amount of medicine absorbed in the blood. If the reaction is a change in blood pressure, R is measured in millimeters of mercury. If the reaction is a change in temperature, R is measured in degrees, and so on.

Find dR/dM. This derivative, as a function of M, is called the sensitivity of the body to the medicine. In Chapter 4, we shall see how to find the amount of medicine to which the body is most sensitive. (*Source: Some Mathematical Models in Biology*, Revised Edition, R. M. Thrall, J. A. Mortimer, K. R. Rebman, R. F. Baum, eds., December 1967, PB-202 364, p. 221; distributed by N.T.I.S., U.S. Department of Commerce.)

When we work with functions of a single variable in mathematics, we normally call the independent variable x and the dependent variable y. Applied fields use many different letters, however. Here are some examples.

TOOLKIT PROGRAM

Derivative Grapher — Key in any function $y = f(x)$. The program will automatically graph f and f' together.

3.3 Velocity, Speed, and Other Rates of Change

In this section, we see how derivatives provide the mathematics we need to understand the way things change in the world around us. With derivatives, we can describe the rates at which water reservoirs empty, populations change, rocks fall, the economy changes, and an athlete's blood sugar varies with exercise. We begin with free fall, the kind of fall that takes place in a vacuum near the surface of the earth.

Free Fall

Near the surface of the earth, all bodies fall with the same constant acceleration. The distance a body falls after it is released from rest is a constant multiple of the square of the time elapsed. At least, that is what happens when the body falls in a vacuum, where there is no air to slow it down. The square-of-time rule also holds for dense, heavy objects like rocks, ball bearings, and steel tools during the first few seconds of their fall through air, before their velocities build up to where air resistance begins to matter. When air resistance is absent, or insignificant, and the only force acting on a falling body is the force of gravity, we call the way the body falls *free fall.*

The equation we write to say that the distance an object falls from rest is proportional to the square of the time elapsed is

$$s = \frac{1}{2}gt^2. \tag{1}$$

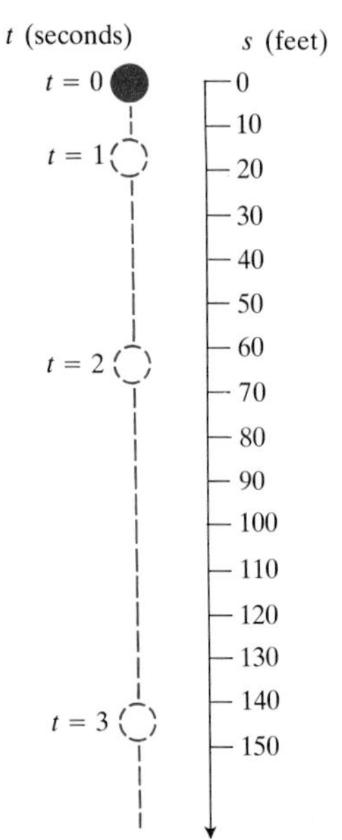

3.16 Distance fallen by a ball bearing released from rest at $t = 0$ sec.

In this equation, s is distance, t is time, and g, as we shall see in a moment, is the constant acceleration given to an object by the force of gravity.

EXAMPLE 1 The value of g in the equation $s = (1/2)gt^2$ depends on the units used to measure t and s. With t in seconds (the usual unit),

$$g = 32 \text{ ft/sec}^2 \qquad s = \frac{1}{2}(32)t^2 = 16t^2 \qquad (s \text{ in feet}),$$

$$g = 9.80 \text{ m/sec}^2 \qquad s = \frac{1}{2}(9.80)t^2 = 4.9t^2 \qquad (s \text{ in meters}),$$

$$g = 980 \text{ cm/sec}^2 \qquad s = \frac{1}{2}(980)t^2 = 490t^2 \qquad (s \text{ in centimeters}).$$

The abbreviation ft/sec^2 is read "feet per second squared" or "feet per second per second." The other units for g are "meters per second squared" and "centimeters per second squared." ■

EXAMPLE 2 Figure 3.16 shows the free fall of a heavy ball bearing released from rest at time $t = 0$. During the first 2 sec, the ball falls

$$s(2) = 16(2)^2 = 16 \cdot 4 = 64 \text{ ft}.$$ ■

EXAMPLE 3 How long did it take the ball bearing in Fig. 3.16 to fall the first 14.7 m?

Solution The free-fall equation for s in meters and t in seconds is

$$s = 4.9t^2$$

(from Example 1). To find the time it took the ball bearing to cover the first 14.7 m, we substitute $s = 14.7$ and solve for t.

$$14.7 = 4.9t^2$$

$$t^2 = \frac{14.7}{4.9} = 3$$

$$t = \sqrt{3}. \qquad \begin{pmatrix}\text{Time increases from } t = 0 \text{ so}\\ \text{we ignore the negative root.}\end{pmatrix}$$

It took the ball $t = \sqrt{3}$ or about 1.732 seconds to fall the first 14.7 m. ■

Velocity

Suppose we have a body moving along a coordinate line and we know that its position at time t is $s = f(t)$.

Position at time t . . . and at time $t + \Delta t$

$s = f(t)$ $s + \Delta s = f(t + \Delta t)$ s

As the body moves along, it has a velocity at each particular instant and we want to find out what that velocity is. The information we seek must somehow be contained in the formula $s = f(t)$, but how do we find it?

We reason like this: In the interval from any time t to the slightly later time $t + \Delta t$, the body moves from position $s = f(t)$ to position

$$s + \Delta s = f(t + \Delta t). \tag{2}$$

The body's net change in position, or *displacement,* for this short time interval is

$$\Delta s = f(t + \Delta t) - f(t). \tag{3}$$

The body's average velocity for the time interval is Δs divided by Δt.

DEFINITION

The **average velocity** of a body moving along a line from position $s = f(t)$ to position $s = f(t + \Delta t)$ is

$$v_{av} = \frac{\text{displacement}}{\text{travel time}} = \frac{\Delta s}{\Delta t} = \frac{f(t + \Delta t) - f(t)}{\Delta t}. \tag{4}$$

To find the body's velocity at the exact instant t, we take the limit of the average velocity over the interval from t to $t + \Delta t$ as the interval gets shorter and shorter and Δt shrinks to zero. Here is where the derivative comes in. For, as we now know, this limit is the derivative of f with respect to t.

DEFINITION

> **Instantaneous velocity** is the derivative of position. If the position function of a body moving along a line is $s = f(t)$, the body's velocity at time t is
>
> $$v(t) = \frac{ds}{dt} = \lim_{\Delta t \to 0} \frac{f(t + \Delta t) - f(t)}{\Delta t}. \tag{5}$$
>
> When people use the word **velocity** alone, they usually mean instantaneous velocity.

EXAMPLE 4

Fig. 3.17 is a time-to-distance graph of a 1989 Ford Thunderbird SC. The slope of the secant PQ is the average speed for the ten-second interval from $t = 5$ to $t = 15$ sec, in this case 40 m/sec or 144 km/h. The slope of the tangent at P is the speedometer reading at $t = 5$ sec, about 20 m/sec or 72 km/h. The car's top speed is 235 km/h (146 mph). (*Source: Car and Driver*, March 1989.) ■

3.17

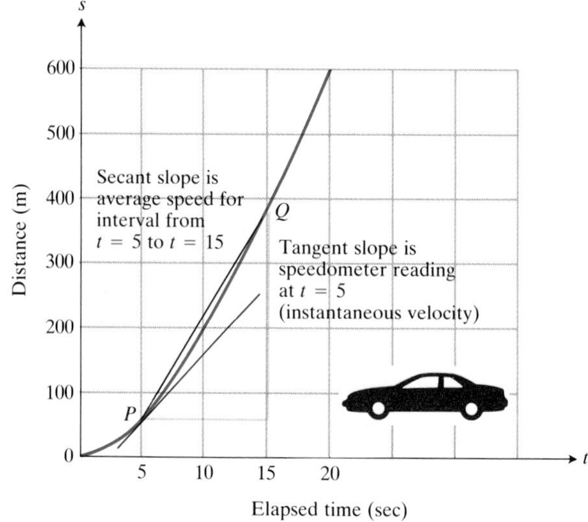

EXAMPLE 5 *Velocity during free fall.*

Free fall equation from Example 1 $s = \frac{1}{2}gt^2$	Corresponding velocity equation $v = \frac{ds}{dt} = gt$
$s = 16t^2$	$v = 32t$
$s = 4.9t^2$	$v = 9.8t$
$s = 490t^2$	$v = 980t$

■

EXAMPLE 6 From Example 5, the velocity of the falling ball bearing t seconds after release is $v = 32$ ft/sec (Fig. 3.16).

$$\text{At } t = 2: \quad v = 32(2) = 64 \text{ ft/sec},$$

$$\text{At } t = 3: \quad v = 32(3) = 96 \text{ ft/sec}.$$

■

EXAMPLE 7 A dynamite blast blows a heavy rock straight up with a launch velocity of 160 ft/sec (about 109 mph)(Fig. 3.18a). It reaches a height of $s = 160t - 16t^2$ ft after t sec.

a) How high does the rock go?

b) How fast is the rock traveling when it is 256 ft above the ground on the way up? On the way down?

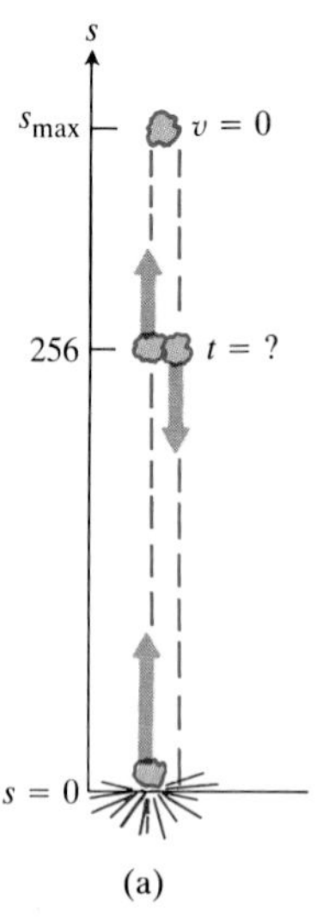

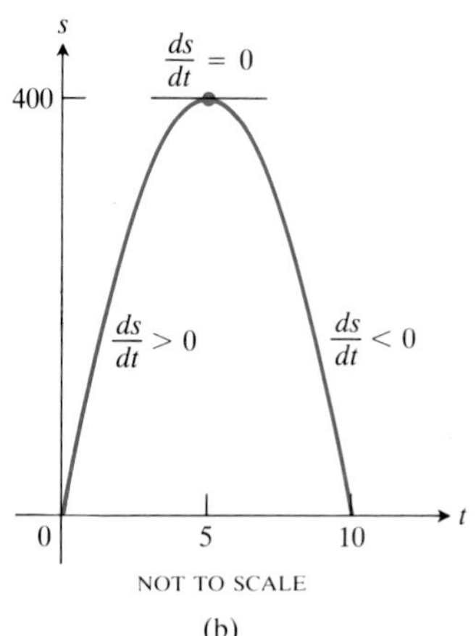

3.18 The graph of the rock's height s as a function of time. (This curve is not the path of the rock itself.)

Solution To find how high the rock goes, we find the value of s when the rock's velocity is zero. The velocity is

$$v = \frac{ds}{dt} = \frac{d}{dt}(160t - 16t^2)$$
$$= 160 - 32t \text{ ft/sec}.$$

The velocity is zero when

$$160 - 32t = 0 \quad \text{or} \quad t = 5 \text{ sec}.$$

The rock's height at $t = 5$ sec is

$$s_{max} = s(5) = 160(5) - 16(5)^2 = 800 - 400 = 400 \text{ ft}.$$

(b) To find the rock's velocity at 256 ft on the way up and again on the way down, we find the two values of t for which

$$s(t) = 160t - 16t^2 = 256. \tag{6}$$

To solve Eq. (6) we write

$$16t^2 - 160t + 256 = 0$$
$$16(t^2 - 10t + 16) = 0$$
$$16(t - 2)(t - 8) = 0$$
$$t = 2 \text{ sec}, \; t = 8 \text{ sec}.$$

The rock is 256 ft above the ground 2 sec after the explosion and again 8 sec after the explosion. The rock's velocities at these times are

$$v(2) = 160 - 32(2) = 160 - 64 = 96 \text{ ft/sec},$$

$$v(8) = 160 - 32(8) = 160 - 256 = -96 \text{ ft/sec}.$$

Why is the downward velocity negative? It has to do with how we set up the coordinate system. Since s measures height from the ground up, changes in s are positive as the rock rises and negative as the rock falls (Fig. 3.18b). ■

Speed

If we drive over to a friend's house and back at 30 mph, say, the speedometer will show 30 on the way over but it will not show -30 on the way back. The speedometer always shows speed, and speed is the absolute value of velocity. Speed measures the rate of forward progress regardless of direction.

DEFINITION

Speed is the absolute value of velocity.

EXAMPLE 8 When the rock in Example 7 passed the 256-ft mark, its forward speed was 96 ft/sec on the way up and $|-96| = 96$ ft/sec again on the way down. ■

Acceleration

In studies of motion along a coordinate line, we usually assume that the body's position function $s = f(t)$ has a second derivative as well as a first. The first derivative gives the body's velocity as a function of time; the second derivative gives the body's acceleration. Thus, the velocity is how fast the position is changing and the acceleration is how fast the velocity is changing. The acceleration tells how quickly the body picks up or loses speed.

DEFINITION

Acceleration is the derivative of velocity. If a body's position at time t is $s = f(t)$, then the body's acceleration at time t is

$$a = \frac{dv}{dt} = \frac{d^2s}{dt^2}. \tag{7}$$

EXAMPLE 9 The acceleration of the rock in Example 7 is

$$a = \frac{dv}{dt} = \frac{d}{dt}(160 - 32t) = 0 - 32 = -32 \text{ ft/sec}^2.$$

The minus sign confirms that the acceleration is downward, in the negative s direction. Whether the rock is going up or down, it is subject to the same constant downward pull of gravity. ■

Other Rates of Change

Average and instantaneous rates give us the right language for many other applications.

EXAMPLE 10 The number of gallons $g(t)$ of water in a reservoir at time t(min) can be regarded as a differentiable function of t. If the volume of water changes by the amount Δg in the interval from time t to time $t + \Delta t$, then

$$\frac{\Delta g}{\Delta t}(\text{gal/min}) = \begin{array}{l}\text{average rate of change}\\ \text{for the time interval,}\end{array}$$

$$\frac{dg}{dt} = \lim_{\Delta t \to 0} \frac{\Delta g}{\Delta t} = \begin{array}{l}\text{instantaneous rate}\\ \text{of change at time } t.\end{array}$$ ■

Although it is natural to think of rates of change in terms of motion and time, there is no need to be so restrictive. We can define the average rate of change of any function over any interval of its domain as the change in the function divided by the length of the interval. We can then go on to define the instantaneous rate of change as the limit of average change as the length of the interval goes to zero.

DEFINITION

Rates of Change

The **average rate of change** of a function $f(x)$ over the interval from x to $x + h$ is

$$\text{Average rate of change} = \frac{f(x+h) - f(x)}{h}.$$

The **(instantaneous) rate of change** of f at x is the derivative

$$f'(x) = \lim_{h \to 0} \frac{f(x+h) - f(x)}{h},$$

provided the limit exists.

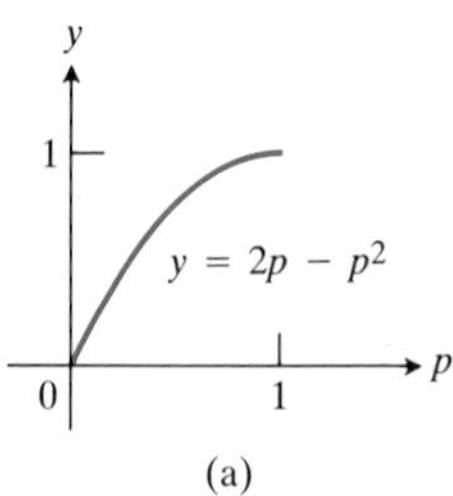

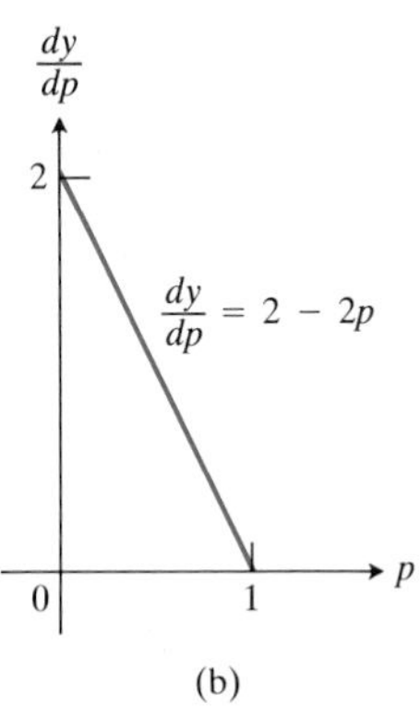

3.19 (a) The frequency $y = 2p - p^2$ of smooth skin in garden peas as a function of the prevalence p of the dominant smooth-skin gene. (b) The graph of dy/dp shows, among other things, how much y will change in response to a small change in p—a lot, if p is near 0, but not much if p is near 1.

Notice the use of the word *instantaneous* even when x does not represent time. It is conventional to do that.

EXAMPLE 11 The Austrian monk Gregor Johan Mendel (1822–1888), working with garden peas and other plants, provided the first scientific explanation of hybridization. His careful records showed that if p (a number between zero and one) is the frequency of the gene for smooth skin in peas (dominant) and $(1 - p)$ is the frequency of the gene for wrinkled skin in peas, then the proportion of smooth-skinned peas in the population at large is

$$y = 2p(1 - p) + p^2 = 2p - p^2.$$

The graph of y versus p in Fig. 3.19(a) suggests that the value of y is more sensitive to a change in p when p is small than it is to a change in p when p is large. Indeed, this is born out by the derivative graph in Fig. 3.19(b), which shows that dy/dp is close to 2 when p is near zero and close to 0 when p is near 1.

We will be able to say more about how sensitive functions are to changes in their variables when we get to Section 3.7. ■

Derivatives in Economics

Economists often call the derivative of a function the **marginal value** of the function.

EXAMPLE 12 *Marginal cost.* Suppose it costs a company $c(x)$ dollars to produce x tons of steel in a week. It costs more to produce $x + h$ tons per week, and the cost difference, divided by h, is the average increase in cost per ton per week:

$$\frac{c(x + h) - c(x)}{h} = \text{average increase in cost.} \tag{8}$$

The limit of the ratio as $h \to 0$ is the marginal cost when x tons of steel are produced:

$$c'(x) = \lim_{h \to 0} \frac{c(x + h) - c(x)}{h} = \text{marginal cost.} \tag{9}$$

How are we to interpret this derivative? First of all, it is the slope of the graph of c at the point marked P in Fig. 3.20. But there is more.

Figure 3.21 shows an enlarged view of the curve and its tangent at P. We can see that if the company, currently producing x tons, increases production by one ton, then the incremental cost $\Delta c = c(x + 1) - c(x)$ of producing that one ton is approximately $c'(x)$. That is,

$$\Delta c \approx c'(x) \quad \text{when} \quad \Delta x = 1. \tag{10}$$

Herein lies the economic importance of marginal cost. It estimates the cost of producing one unit beyond the present production level; that is, it approximates the cost of producing one more car, one more radio, one more washing machine, whatever. ■

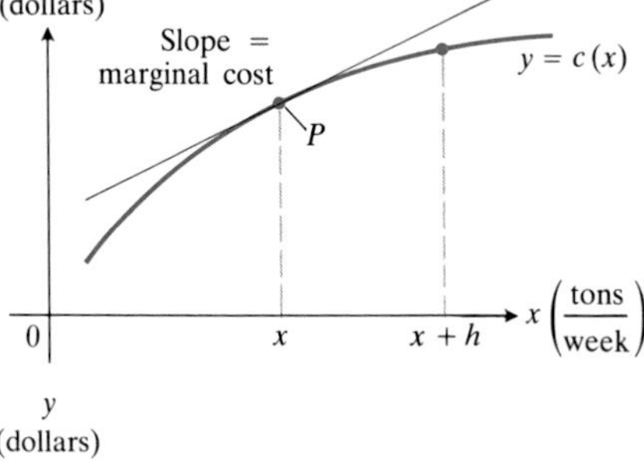

3.20 Weekly steel production: $c(x)$ is the cost of producing x tons per week. The cost of producing an additional h tons is $c(x + h) - c(x)$.

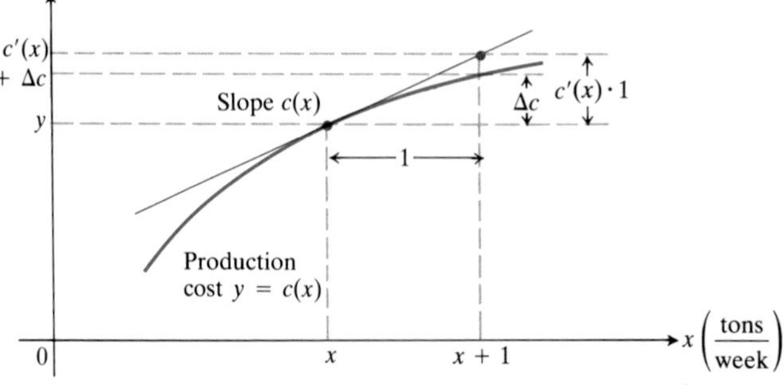

3.21 As weekly steel production increases from x to $x + 1$ tons, the cost curve rises by the amount Δc. The tangent line rises by the amount slope · run $= c'(x) \cdot 1 = c'(x)$. Since $\Delta c/\Delta x \approx c'(x)$, we have $\Delta c \approx c'(x)$ when $\Delta x = 1$.

EXAMPLE 13 *Marginal cost (continued).* Suppose it costs

$$c(x) = x^3 - 6x^2 + 15x$$

dollars to produce x stoves and your shop is currently producing 10 stoves a day. About how much extra will it cost to produce one more stove a day?

Solution The cost of producing one more stove a day when 10 are produced is about $c'(10)$. Since

$$\begin{aligned} c'(x) &= \frac{d}{dx}(x^3 - 6x^2 + 15x) \\ &= 3x^2 - 12x + 15, \\ c'(10) &= 3(100) - 12(10) + 15 \\ &= 195. \end{aligned}$$

The additional cost will be about \$195. ■

EXAMPLE 14 *Marginal revenue.* If

$$r(x) = x^3 - 3x^2 + 12x$$

gives the dollar revenue from selling x thousand candy bars, the marginal revenue when x thousand are sold is

$$\begin{aligned} r'(x) &= \frac{d}{dx}(x^3 - 3x^2 + 12x) \\ &= 3x^2 - 6x + 12. \end{aligned}$$

As with marginal cost, the marginal-revenue function estimates the increase in revenue that will result from selling one additional unit. If you currently sell 10 thousand candy bars a week, you can expect your revenue to increase by about

$$\begin{aligned} r'(10) &= 3(100) - 6(10) + 12 \\ &= \$252 \end{aligned}$$

if you increase sales to 11 thousand bars a week. ■

☐ Estimating f' from a Graph of f

When we record data in the laboratory or in the field, we are often recording the values of a function $y = f(x)$. We might be recording the pressure in a gas as a function of volume, or the size of a population as a function of time. To see what the function looks like, we usually plot the data points and fit them with a curve. Even if we have no formula for the function $y = f(x)$ from which to calculate the derivative $y' = f'(x)$, we can still graph f': We estimate the slopes on the graph of f and plot these slopes. The following examples show how this is done and what we can learn from the graph of f'.

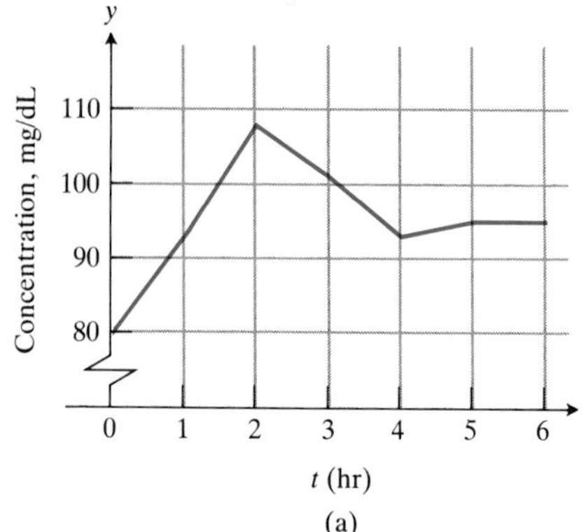

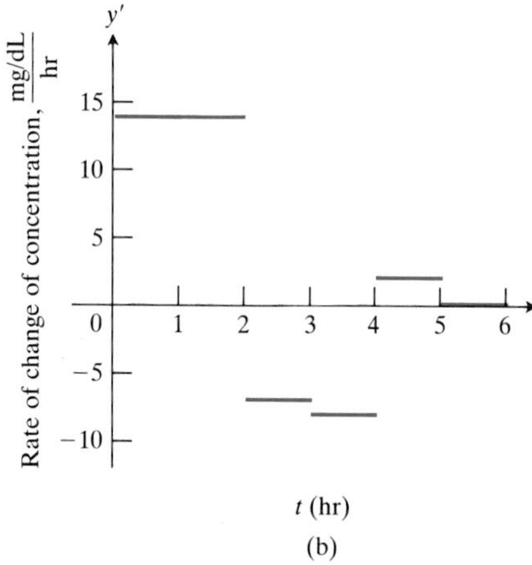

3.22 (a) The sugar concentration in the blood of a *Daedalus* pilot during a 6-hr pre-flight endurance test. (b) The derivative of the pilot's blood sugar concentration shows how rapidly the concentration rose and fell during various portions of the test. (*Source: The Daedalus Project: Physiological Problems and Solutions,* Ethan R. Nadel and Steven R. Bussolari, *American Scientist,* Vol. 76, No. 4, July–August 1988, p. 358.)

EXAMPLE 15 On April 23, 1988, the human-powered airplane *Daedalus* flew a record-breaking 119 km from Crete to the island of Santorini in the Aegean Sea, southeast of mainland Greece. During the 6-hr endurance tests before the flight, researchers monitored the prospective pilots' blood–sugar concentrations. The concentration graph for one of the athlete-pilots is shown in Fig. 3.22(a), where the concentration in milligrams/deciliter is plotted against time in hours.

The graph is made of line segments connecting data points. The constant slope of each segment gives an estimate of the derivative of the concentration between measurements. We calculated the slopes from the coordinate grid and plotted the derivative as a step function in Fig. 3.22(b). To make the plot for the first hour, for instance, we observed that the concentration increased from about 79 mg/dl to 93 mg/dl. The net increase was $\Delta y = 93 - 79 = 14$ mg/dl. Dividing this by $\Delta t = 1$ hr gave the rate of change:

$$\frac{\Delta y}{\Delta t} = \frac{14}{1} = 14 \text{ mg/dl per hour.}$$

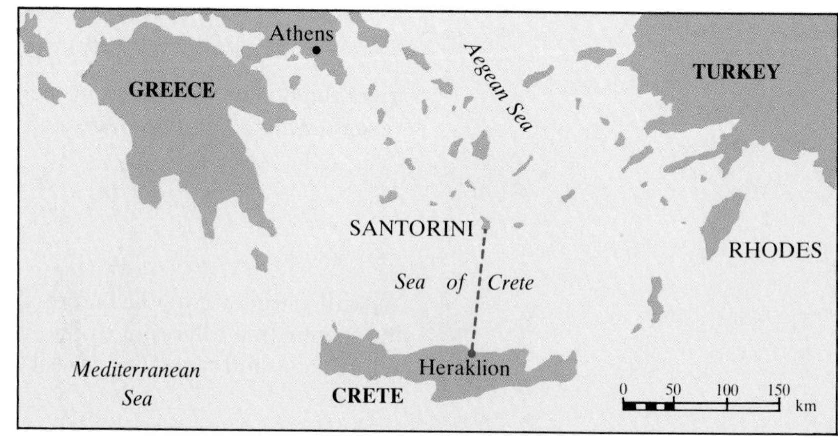

The *Daedalus* flying over the Aegean Sea on April 23, 1988.

Wide World Photos

When we have so many data that the graph we get by connecting the data points looks like a smooth curve, we may also wish to plot the derivative as a smooth curve. The next example shows how this is done.

EXAMPLE 16 Graph the derivative of the function f in Fig. 3.23(a).

Solution First, we draw a pair of coordinate axes, marking the horizontal axis in x-units and the vertical axis in slope units (Fig. 3.23b). Next, we estimate the slope of the graph of f in y-units per x-unit at frequent intervals, plotting the corresponding points against the new axes. We then connect the plotted points with a smooth curve.

From the graph of $y' = f'(x)$ we can see at a glance

1. Where f's rate of change is positive, negative, or zero.
2. The rough size of the growth rate at any x and its size in relation to the size of $f(x)$.
3. Where the rate of change itself is increasing or decreasing. ■

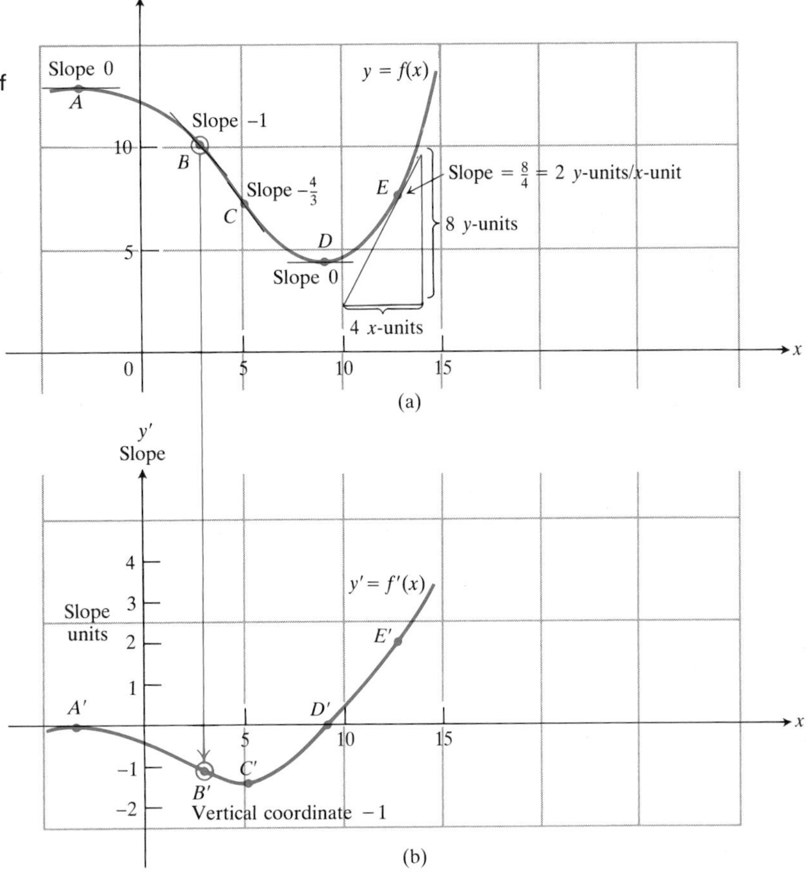

3.23 We made the graph of $y' = f'(x)$ in (b) by plotting slopes from the graph of $y = f(x)$ in (a). The vertical coordinate of B' is the slope at B, and so on. The graph of $y' = f'(x)$ is a visual record of how the slope of f changes with x.

Exercises 3.3

The equations in Exercises 1–8 give the position $s = f(t)$ of a body moving along a coordinate line; s is in meters and t in seconds.

a) Find the body's displacement and average velocity for the time interval from $t = 0$ to $t = 2$.

b) Find the body's velocity and acceleration at time $t = 2$.

1. $s = 0.8t^2$ (free fall on the moon)

2. $s = 1.86t^2$ (free fall on Mars)

3. $s = 2t^2 + 5t - 3$

4. $s = t^2 - 3t + 2$

5. $s = 4 - 2t - t^2$

6. $s = 3 - 2t^2$

7. $s = 4t + 3$

8. $s = (1/2)gt^2 + v_0t + s_0$ (g, v_0, s_0 constants)

9. The equations for free fall at the surfaces of Mars and Jupiter (s in meters, t in seconds) are Mars, $s = 1.86t^2$; Jupiter, $s = 11.44t^2$. How long would it take a rock falling from rest to reach a velocity of 16.6 m/sec (about 100 km/h) on each planet?

10. A rock thrown vertically upward from the surface of the moon at a velocity of 24 m/sec (about 86 km/h) reaches a height of $s = 24t - 0.8t^2$ meters in t seconds.

a) Find the rock's velocity and acceleration. (The acceleration in this case is the acceleration of gravity on the moon.)

b) How long did it take the rock to reach its highest point?

c) How high did the rock go?

d) How long did it take the rock to reach half its maximum height?

e) How long was the rock aloft?

11. On the earth, in the absence of air, the rock in Exercise 10 would reach a height of $s = 24t - 4.9t^2$ meters in t seconds. How high would the rock go?

12. A 45-caliber bullet fired straight up from the surface of the moon would reach a height of $s = 832t - 2.6t^2$ feet after t seconds. On the earth, in the absence of air, its height would be $s = 832t - 16t^2$ feet after t seconds. How long would it take the bullet to get back down in each case?

13. When a bactericide was added to a nutrient broth in which bacteria were growing, the bacterium population continued to grow for a while, but then stopped growing and began to decline. The size of the population at time t (hours) was $b(t) = 10^6 + 10^4t - 10^3t^2$. Find the growth rates at (a) $t = 0$; (b) $t = 5$; and (c) $t = 10$ hours.

14. The number of gallons of water in a tank t minutes after the tank has started to drain is $Q(t) = 200(30 - t)^2$. How fast is the water running out at the end of 10 min? What is the average rate at which the water flows out during the first 10 min?

15. *Marginal cost.* Suppose that the dollar cost of producing x washing machines is $c(x) = 2000 + 100x - 0.1x^2$.

a) Find the average cost of producing 100 washing machines.

b) Find the marginal cost when 100 washing machines are produced.

c) Show that the marginal cost when 100 washing machines are produced is approximately the cost of producing one more washing machine after the first 100 have been made, by calculating the latter cost directly.

16. *Marginal revenue.* Suppose the revenue from selling x custom-made office desks is

$$r(x) = 2000\left(1 - \frac{1}{x+1}\right)$$

dollars.

a) Find the marginal revenue when x desks are produced.

b) Use the function $r'(x)$ to estimate the increase in revenue that will result from increasing production from 5 desks a week to 6 desks a week.

c) Find the limit of $r'(x)$ as $x \to \infty$. How would you interpret this number?

17. The position of a body at time t sec is $s = t^3 - 6t^2 + 9t$ m. Find the body's acceleration each time the velocity is zero.

18. The velocity of a body at time t sec is $v = 2t^3 - 9t^2 + 12t - 5$ m/sec. Find the body's speed each time the acceleration is zero.

19. When a model rocket is launched, the propellant burns for a few seconds, accelerating the rocket upward. After burnout, the rocket coasts upward for a while and then begins to fall. A small explosive charge pops out a parachute shortly after the rocket starts down. The parachute slows the rocket to keep it from breaking when it lands.

Figure 3.24 shows velocity data from the flight of a model rocket. Use the data to answer the following.

a) How fast was the rocket climbing when the engine stopped?

b) For how many seconds did the engine burn?

c) When did the rocket reach its highest point? What was its velocity then?

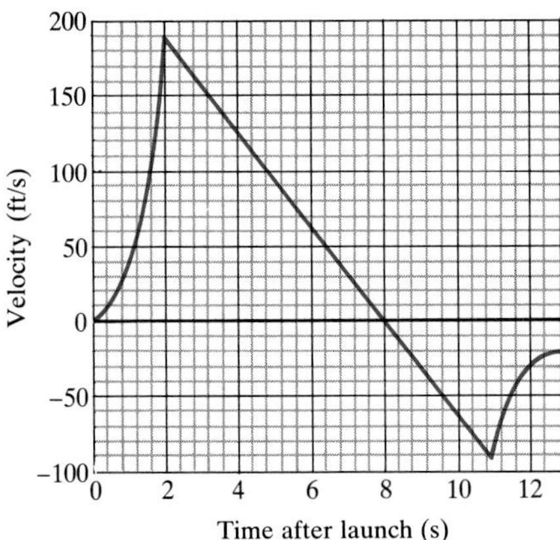

3.24 Velocity of the model rocket in Exercise 19.

d) When did the parachute pop out? How fast was the rocket falling then?
e) How long did the rocket fall before the parachute opened?
f) When was the rocket's acceleration greatest? When was the acceleration constant?

20. *Pisa by parachute (continuation of Exercise 19).* The accompanying photograph shows Mike McCarthy parachuting from the top of the Tower of Pisa on August 5, 1988. Make a rough sketch to show the shape of the graph of his downward velocity during the jump.

Mike McCarthy of London jumped from the Tower of Pisa, then opened his parachute in what he said was a world record low-level parachute jump of 179 feet. (*Source: Boston Globe*, Saturday, Aug. 6, 1988.)

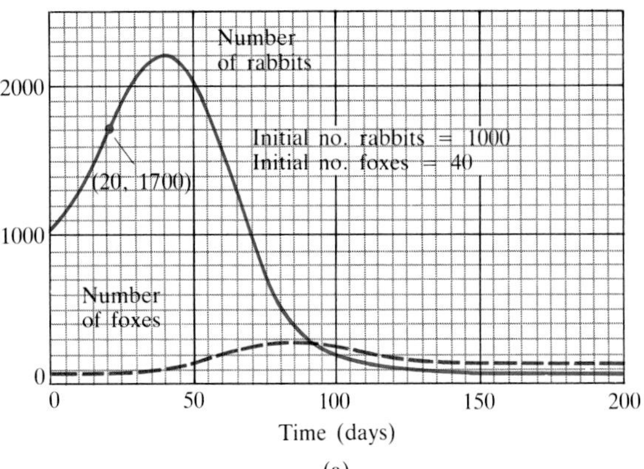

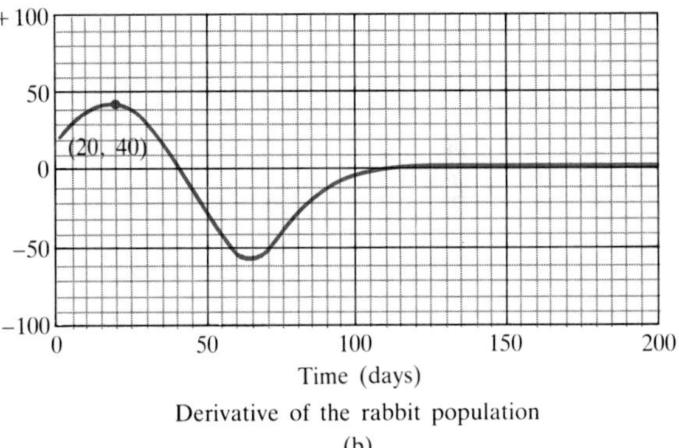

3.25 Rabbits and foxes in an arctic predator–prey food chain. (*Source: Differentiation,* by W. U. Walton et al., Project CALC, Education Development Center, Inc., Newton, Mass. (1975), p. 86.)

Exercises 21 and 22 are about the graphs in Fig. 3.25. The graphs in part (a) show the numbers of rabbits and foxes in a small arctic population. They are plotted as functions of time for 200 days. The number of rabbits increases at first, as the rabbits reproduce. But the foxes prey on the rabbits and, as the number of foxes increases, the rabbit population levels off and then drops. Figure 3.25(b) shows the graph of the derivative of the rabbit population. We made it by plotting slopes, as in Example 16.

21. a) What is the value of the derivative of the rabbit population in Fig. 3.25 when the number of rabbits is largest? Smallest?
b) What is the size of the rabbit population in Fig. 3.25 when its derivative is largest? Smallest?

22. In what units should the slopes of the rabbit and fox population curves be measured?

Match the graphs of the functions in Exercises 23–26 with the graphs of the derivatives from the following list:

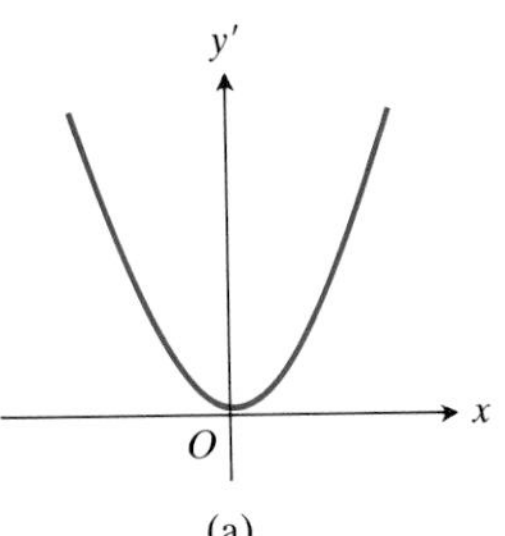
(a)

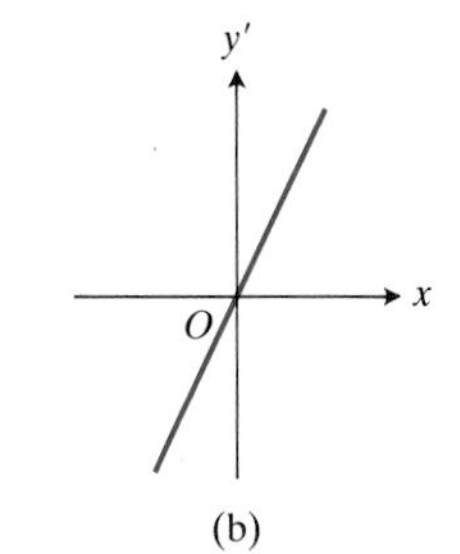
(b)

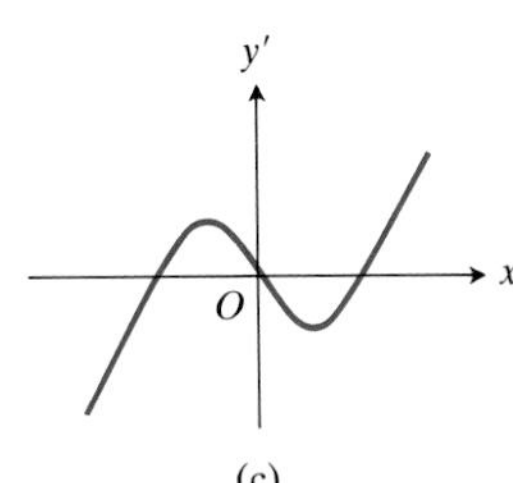
(c)

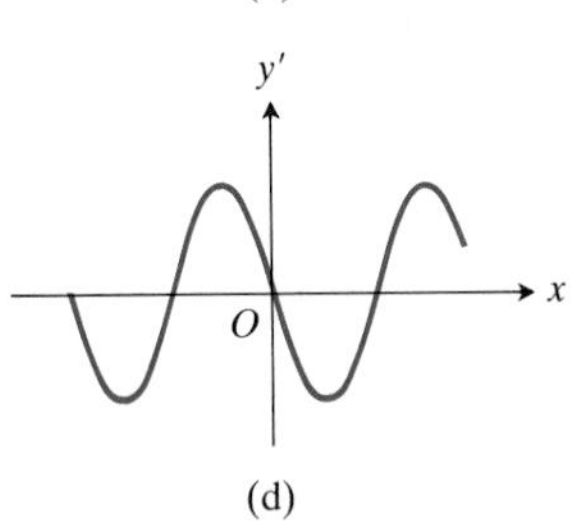
(d)

23.

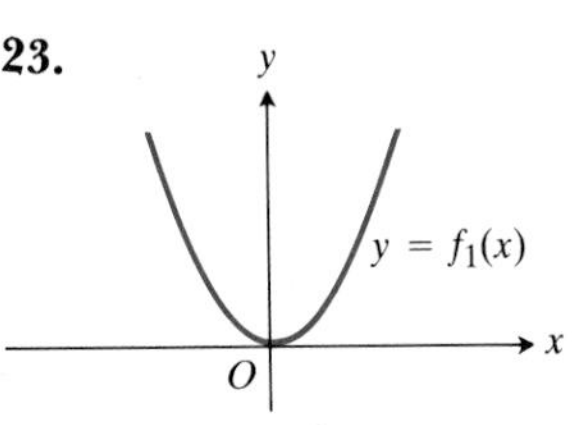

24.

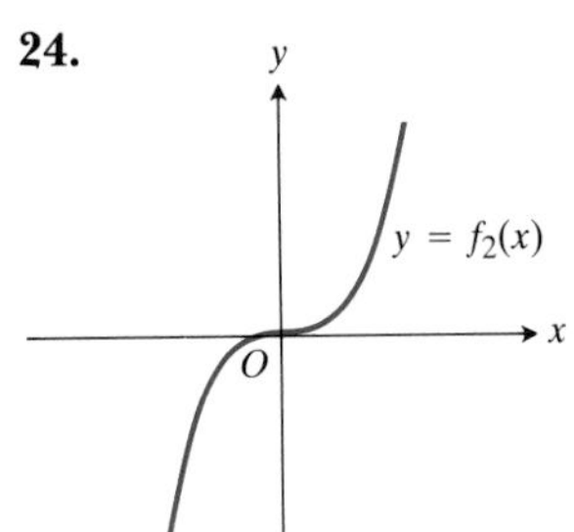

25.

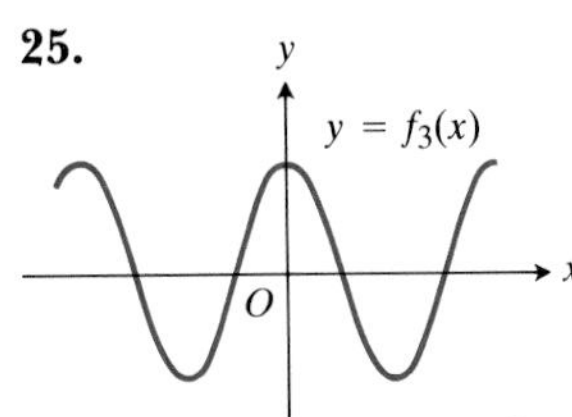

26.

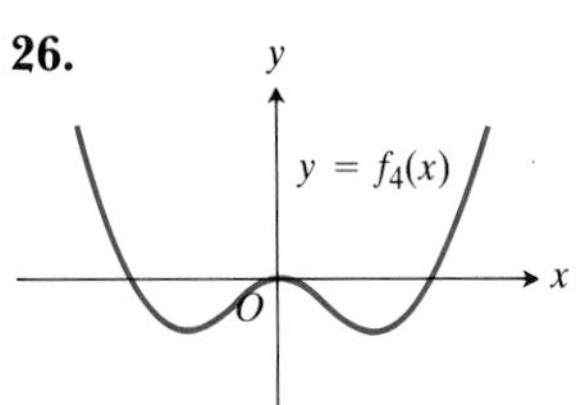

27. The graph of the function $y = f(x)$ in Fig. 3.26 is made of line segments joined end to end.

a) Graph the function's derivative.

b) At what values of x between $x = -3$ and $x = 7$ is the derivative not defined?

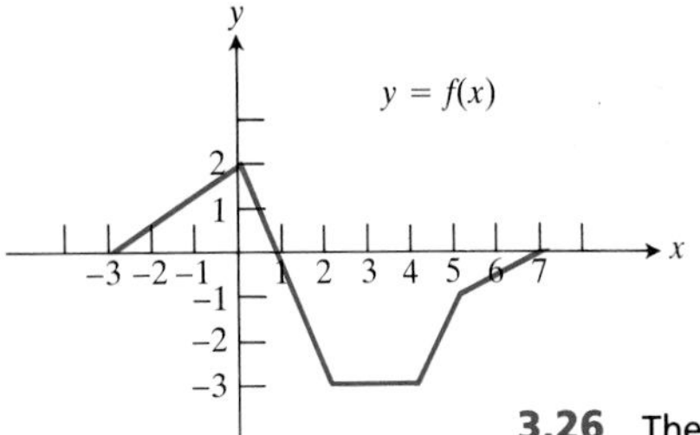

3.26 The graph for Exercise 27.

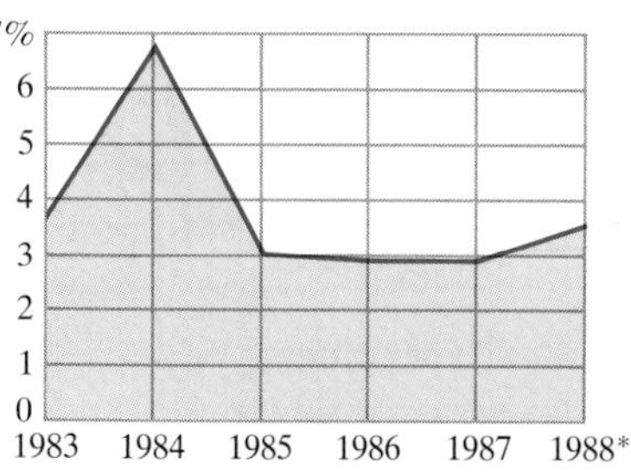

3.27 The graph for Exercise 28.

28. *Growth in the economy.* The graph in Fig. 3.27 shows the values of the U.S. Gross National Product (GNP) $y = f(t)$ for the years 1983–1988 (*Source: The Wall Street Journal*, July 11, 1988, page 1.) Graph the function's derivative (where defined).

29. *Fruit flies (Example 1, Section 3.1 continued).* Populations starting out in closed environments grow slowly at first, when there are relatively few members, then more rapidly as the number of reproducing individuals increases and resources are still abundant, then slowly again as the population reaches the carrying capacity of the environment.

a) Use the graphical technique of Example 16 to graph the derivative of the fruit fly population introduced in Section 3.1. The graph of the population is reproduced here as Fig. 3.28. What units should be used on the horizontal and vertical axes for the derivative's graph?

b) During what days does the population seem to be increasing fastest? Slowest?

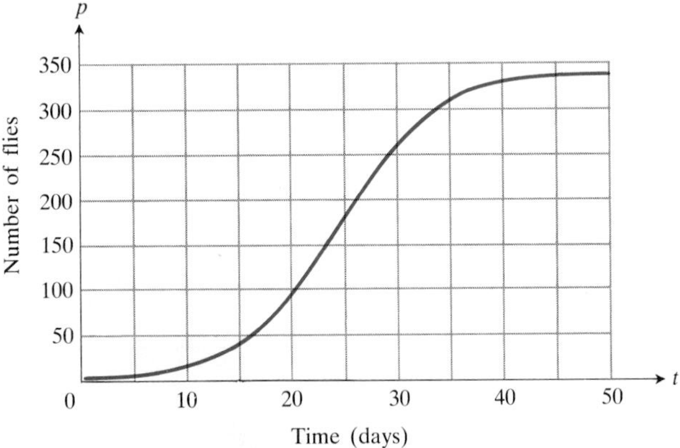

3.28 The graph for Exercise 29.

TOOLKIT PROGRAM

Derivative Grapher — Key in any function $y = f(x)$. The program will automatically graph f and f' together.

3.4 Derivatives of Trigonometric Functions

As we mentioned in Section 1.5, trigonometric functions are important because so many of the phenomena we want information about are periodic (heart rhythms, earthquakes, tides, weather). Continuous periodic functions can always be expressed in terms of sines and cosines, so the derivatives of sines and cosines play a key role in describing and predicting important changes. This section shows how to differentiate the six basic trigonometric functions.

A New Limit

We begin by showing that

$$\lim_{h \to 0} \frac{\cos h - 1}{h} = 0. \tag{1}$$

We could have done this earlier, but we didn't need it then.

To prove Eq. (1), we divide the identity

$$\frac{1 - \cos 2\theta}{2} = \sin^2\theta \quad \text{(Section 1.5, Eq. 22)}$$

by θ so that we get

$$\frac{1 - \cos 2\theta}{2\theta} = \frac{\sin \theta}{\theta} \cdot \sin \theta.$$

Letting $\theta \to 0$ then gives

$$\begin{aligned} \lim_{\theta \to 0} \frac{1 - \cos 2\theta}{2\theta} &= \lim_{\theta \to 0} \frac{\sin \theta}{\theta} \cdot \lim_{\theta \to 0} \sin \theta && \text{(Limit Product Rule)} \\ &= 1 \cdot 0 = 0. && \text{(Known values)} \end{aligned}$$

Replacing 2θ by h gives Eq. (1).

The Derivative of the Sine

The derivative of $y = \sin x$ is the limit

$$\frac{dy}{dx} = \lim_{h \to 0} \frac{\sin(x + h) - \sin x}{h}. \tag{2}$$

To calculate this limit, we combine Eq. (1) with two results from our earlier work:

1. $\sin(x + h) = \sin x \cos h + \cos x \sin h$ (Section 1.5, Eq. 16),
2. $\lim_{h \to 0} \frac{\sin h}{h} = 1$ (Section 2.2).

Then, taking all limits as $h \to 0$, we have

$$\begin{aligned}
\frac{dy}{dx} &= \lim \frac{\sin(x+h) - \sin x}{h} \\
&= \lim \frac{\sin x \cos h + \cos x \sin h - \sin x}{h} \\
&= \lim \frac{\sin x (\cos h - 1) + \cos x \sin h}{h} \qquad (3) \\
&= \lim \sin x \cdot \lim \frac{\cos h - 1}{h} + \lim \cos x \cdot \lim \frac{\sin h}{h} \\
&= \sin x \cdot 0 + \cos x \cdot 1 \\
&= \cos x .
\end{aligned}$$

In short,

$$\frac{d}{dx} \sin x = \cos x . \qquad (4)$$

The sine and its derivative obey all the usual differentiation rules.

EXAMPLE 1

a) $y = x^2 - \sin x$ $\quad \dfrac{dy}{dx} = 2x - \dfrac{d}{dx}(\sin x)$ (Difference Rule)

$$= 2x - \cos x$$

b) $y = x^2 \sin x$ $\quad \dfrac{dy}{dx} = x^2 \dfrac{d}{dx}(\sin x) + 2x \sin x$ (Product Rule)

$$= x^2 \cos x + 2x \sin x$$

c) $y = \dfrac{\sin x}{x}$ $\quad \dfrac{dy}{dx} = \dfrac{x \cdot \dfrac{d}{dx}(\sin x) - \sin x \cdot 1}{x^2}$ (Quotient Rule)

$$= \frac{x \cos x - \sin x}{x^2}$$

■

In case you are still wondering why calculus uses radian measure when the rest of the world seems to use degrees, the answer is provided by the argument that the derivative of the sine is the cosine. The derivative of $\sin x$ is $\cos x$ *only* if x is measured in radians. The argument requires that when h is a small increment in x,

$$\lim_{h \to 0} (\sin h)/h = 1.$$

This is true only for radian measure, as we discussed in Section 2.2.

Now that we know that the sine is differentiable, we also know that it is continuous. The same holds for the other trigonometric functions in this section. Each one is differentiable at every point in its domain and is therefore continuous at every point in its domain.

The Derivative of the Cosine

To calculate the derivative of the cosine, we use the equation

$$\cos(x + h) = \cos x \cos h - \sin x \sin h. \quad \text{(Section 1.5, Eq. 15)}$$

Then, with all limits taken as $h \to 0$, we have

$$\begin{aligned}
\frac{dy}{dx} &= \lim \frac{\cos(x + h) - \cos x}{h} && \text{(Derivative definition)} \\
&= \lim \frac{\cos x \cos h - \sin x \sin h - \cos x}{h} \\
&= \lim \frac{\cos x (\cos h - 1) - \sin x \sin h}{h} \\
&= \cos x \cdot \lim \frac{\cos h - 1}{h} - \sin x \cdot \lim \frac{\sin h}{h} \\
&= \cos x \cdot 0 - \sin x \cdot 1 && \text{(Eq. 1 again)} \\
&= -\sin x.
\end{aligned}$$

In short,

$$\frac{d}{dx} \cos x = -\sin x. \quad (5)$$

Notice the minus sign. The derivative of the sine is the cosine, but the derivative of the cosine is *minus* the sine (Fig. 3.29).

EXAMPLE 2

a) $y = 5x + \cos x$

$$\begin{aligned}
\frac{dy}{dx} &= \frac{d}{dx}(5x) + \frac{d}{dx}(\cos x) && \text{(Sum Rule)} \\
&= 5 - \sin x
\end{aligned}$$

b) $y = \sin x \cos x$

$$\begin{aligned}
\frac{dy}{dx} &= \sin x \frac{d}{dx}(\cos x) + \cos x \frac{d}{dx}(\sin x) && \text{(Product Rule)} \\
&= \sin x (-\sin x) + \cos x (\cos x) \\
&= \cos^2 x - \sin^2 x
\end{aligned}$$

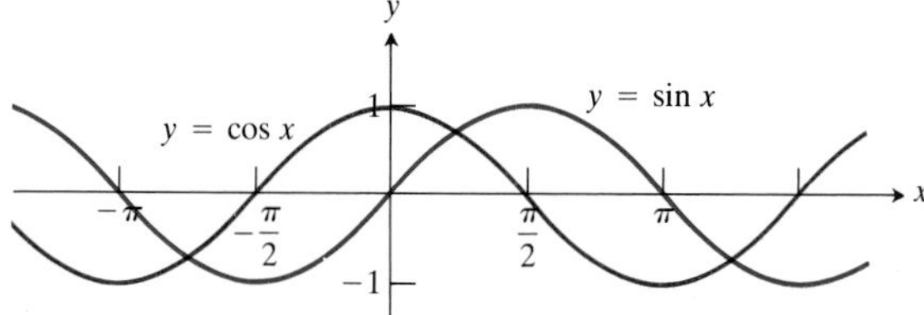

3.29 The sine and cosine plotted together. The sine gives the cosine's slope; the cosine gives the negative of the sine's slope.

c) $y = \dfrac{\cos x}{1 - \sin x}$

$$\begin{aligned}\frac{dy}{dx} &= \frac{(1 - \sin x)\frac{d}{dx}(\cos x) - \cos x \frac{d}{dx}(1 - \sin x)}{(1 - \sin x)^2} && \text{(Quotient Rule)} \\ &= \frac{(1 - \sin x)(-\sin x) - \cos x(0 - \cos x)}{(1 - \sin x)^2} \\ &= \frac{1 - \sin x}{(1 - \sin x)^2} && (\sin^2 x + \cos^2 x = 1) \\ &= \frac{1}{1 - \sin x}\end{aligned}$$

■

Simple Harmonic Motion

The motion of a weight bobbing up and down on the end of a spring is a *simple harmonic motion.* The next example describes a case in which there are no opposing forces like air friction to slow the motion down.

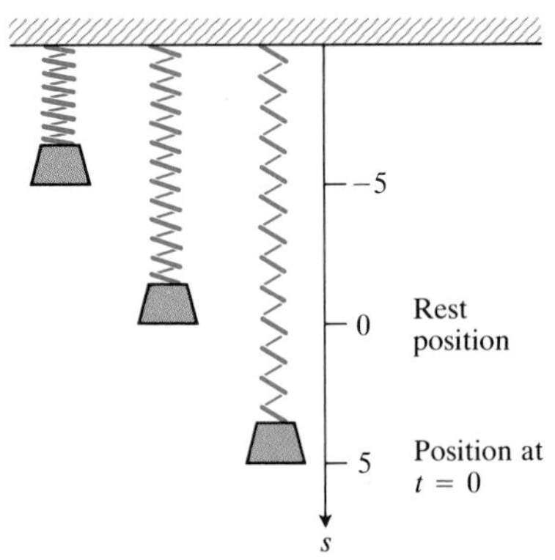

3.30 The weighted spring in Example 3.

EXAMPLE 3 A weight hanging from a spring (Fig. 3.30) is stretched 5 units beyond its rest position and released at time $t = 0$ to bob up and down. Its position at any later time t is

$$s = 5 \cos t.$$

What are its velocity and acceleration at time t?

Solution We have

Position: $s = 5 \cos t$

Velocity: $\dfrac{ds}{dt} = \dfrac{d}{dt}(5 \cos t) = 5\dfrac{d}{dt}(\cos t) = -5 \sin t$

Acceleration: $\dfrac{dv}{dt} = \dfrac{d}{dt}(-5 \sin t) = -5\dfrac{d}{dt}(\sin t) = -5 \cos t.$

Here is what we can learn from these equations:

1. As time passes, the weight moves up and down between $s = 5$ and $s = -5$ on the s-axis. The amplitude of the motion is 5. The period of the motion is 2π.
2. The function $\sin t$ attains its greatest magnitude (1) when $\cos t = 0$, as the graphs of the sine and cosine show (Fig. 3.29). Hence, the weight's speed, $|v| = 5|\sin t|$, is greatest when $\cos t = 0$, that is, when $s = 0$.

   The weight's speed is zero when $\sin t = 0$. This occurs when $\cos t = \pm 1$, at the end points of the interval of motion.
3. The acceleration, $a = -5 \cos t$, is zero only at the origin, where $\cos t = 0$. When the weight is anywhere else, the spring is either pulling on it or pushing on it. The acceleration is greatest in magnitude at the points farthest from the origin, where $\cos t = \pm 1$. ■

The Derivatives of the Other Basic Functions

Because sin x and cos x are differentiable functions of x, the related functions

$$\tan x = \frac{\sin x}{\cos x} \qquad \sec x = \frac{1}{\cos x}$$

$$\cot x = \frac{\cos x}{\sin x} \qquad \csc x = \frac{1}{\sin x}$$

are differentiable at every value of x at which they are defined. Their derivatives, calculated from the quotient rule, are given by the following formulas.

$$\frac{d}{dx} \tan x = \sec^2 x \quad (6) \qquad \frac{d}{dx} \sec x = \sec x \tan x \quad (7)$$

$$\frac{d}{dx} \cot x = -\csc^2 x \quad (8) \qquad \frac{d}{dx} \csc x = -\csc x \cot x \quad (9)$$

Notice the minus signs in the equations for the cotangent and cosecant.

To show how a typical calculation goes, we derive Eq. (6). The other derivations are left as exercises.

EXAMPLE 4 Find dy/dx if $y = \tan x$.

Solution

$$\frac{d}{dx} \tan x = \frac{d}{dx}\left(\frac{\sin x}{\cos x}\right) = \frac{\cos x \frac{d}{dx}(\sin x) - \sin x \frac{d}{dx}(\cos x)}{\cos^2 x}$$

$$= \frac{\cos x \cos x - \sin x(-\sin x)}{\cos^2 x} = \frac{\cos^2 x + \sin^2 x}{\cos^2 x}$$

$$= \frac{1}{\cos^2 x} = \sec^2 x$$

■

EXAMPLE 5 Find y'' if $y = \sec x$.

Solution

$$y = \sec x$$

$$y' = \sec x \tan x \qquad \text{(Eq. 8)}$$

$$y'' = \frac{d}{dx}(\sec x \tan x)$$

$$= \sec x \frac{d}{dx}(\tan x) + \tan x \frac{d}{dx}(\sec x)$$

$$= \sec x(\sec^2 x) + \tan x(\sec x \tan x)$$

$$= \sec^3 x + \sec x \tan^2 x$$

■

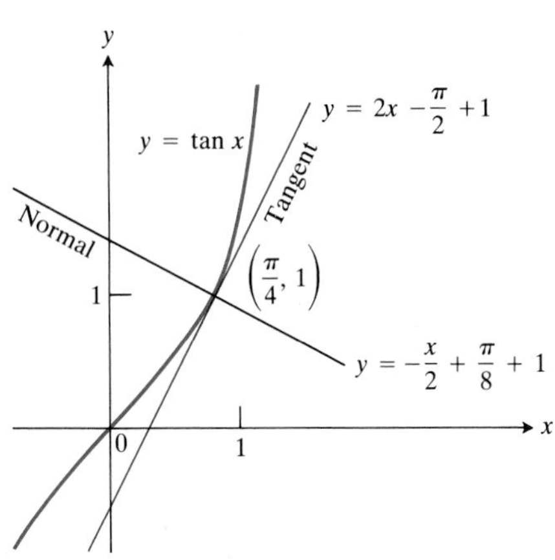

3.31 Example 7 shows how to find equations for the tangent and normal lines shown here.

The Word *Normal*

When analytic geometry was developed in the seventeenth century, European scientists still wrote about their work and ideas in Latin, the one language that all educated Europeans could read and understand. The word *normalis,* which scholars used for perpendicular in Latin, became *normal* when they discussed geometry in English.

EXAMPLE 6

a) $\frac{d}{dx}(3x + \cot x) = 3 + \frac{d}{dx}(\cot x) = 3 - \csc^2 x$

b) $\frac{d}{dx}\left(\frac{2}{\sin x}\right) = \frac{d}{dx}(2 \csc x) = 2\frac{d}{dx}(\csc x)$

$$= 2(-\csc x \cot x) = -2 \csc x \cot x$$

EXAMPLE 7 Find the lines that are tangent and normal (perpendicular) to the curve $y = \tan x$ at the point $(\pi/4, 1)$ (Fig. 3.31).

Solution The slope of the curve at $(\pi/4, 1)$ is the value of $dy/dx = \sec^2 x$ at $x = \pi/4$:

$$\left.\frac{dy}{dx}\right|_{x=\pi/4} = \sec^2\left(\frac{\pi}{4}\right) = (\sqrt{2})^2 = 2.$$

The tangent is the line

$$y - 1 = 2\left(x - \frac{\pi}{4}\right) \qquad \left(\begin{array}{l} y - y_0 = m(x - x_0) \\ \text{with } y_0 = 1,\ m = 2,\ x_0 = \frac{\pi}{4} \end{array}\right)$$

$$y = 2x - \frac{\pi}{2} + 1.$$

The normal has slope $m = -1/2$, so its point–slope equation is

$$y - 1 = -\frac{1}{2}\left(x - \frac{\pi}{4}\right)$$

$$y = -\frac{x}{2} + \frac{\pi}{8} + 1.$$

Exercises 3.4

In Exercises 1–24, find dy/dx when y equals the given expression.

1. $1 + x - \cos x$

2. $2 \sin x - \tan x$

3. $\frac{1}{x} + 5 \sin x$

4. $x^2 - \sec x$

5. $\csc x - 5x + 7$

6. $2x + \cot x$

7. $x \sec x$

8. $x \csc x$

9. $x^2 \cot x$

10. $4 - x^2 \sin x$

11. $3x + x \tan x$

12. $x \sin x + \cos x$

13. $\sin x \sec x$

14. $\sec x \csc x$

15. $\tan x \cot x$

16. $\cos x(1 + \sec x)$

17. $\frac{4}{\cos x}$

18. $5 + \frac{1}{\tan x}$

19. $\frac{\cos x}{x}$

20. $\frac{2}{\csc x} - \frac{1}{\sec x}$

21. $\frac{x}{1 + \cos x}$

22. $\frac{\sin x + \cos x}{\cos x}$

23. $\frac{\cot x}{1 + \cot x}$

24. $\frac{\cos x}{1 + \sin x}$

25. Find y'' if $y = \csc x$.

26. Find $y^{(4)} = d^4y/dx^4$ if
a) $y = \sin x$,
b) $y = \cos x$.

In Exercises 27–30, find equations for the lines that are tangent and normal to the curve $y = f(x)$ at the given point.

27. $y = \sin x$, $(0, 0)$

28. $y = \tan x$, $(0, 0)$

29. $y = \cos x$, $(\pi, -1)$

30. $y = 1 + \cos x$, $(\pi/2, 1)$

31. Show that the graphs of $y = \sec x$ and $y = \cos x$ have horizontal tangents at $x = 0$.

32. Show that the graphs of $y = \tan x$ and $y = \cot x$ never have horizontal tangents.

Do the graphs of the functions in Exercises 33–36 have any horizontal tangents in the interval $0 \leq x \leq 2\pi$? If so, where? If not, why not?

33. $y = x + \sin x$

34. $y = 2x + \sin x$

35. $y = x + \cos x$

36. $y = x + 2 \cos x$

37. *Simple harmonic motion.* The equations in (a) and (b) give the position $s = f(t)$ of a body moving along a coordinate line. Find each body's velocity, speed, and acceleration at time $t = \pi/4$.
a) $s = 2 - 2 \sin t$
b) $s = \sin t + \cos t$

38. *Running machinery too fast.* Suppose that a mass like a piston is moving straight up and down and that its position at time t seconds is

$$s = A \cos(2\pi bt).$$

The value of A in this equation is the amplitude of the motion, and b is the frequency (number of times the piston moves up and down each second). What effect does doubling the frequency have on the piston's velocity and acceleration? (Once you find out, you will know why machinery breaks when you run it too fast.)

39. Find equations for the lines that are tangent and normal to the curve $y = \sqrt{2} \cos x$ at the point $(\pi/4, 1)$.

40. Find the points on the curve $y = \tan x$, $-\pi/2 < x < \pi/2$, where the tangent is parallel to the line $y = 2x$.

Find equations for the horizontal tangents to the graphs in Exercises 41 and 42.

41. $y = \cot x - \sqrt{2} \csc x$, $0 < x < \pi$

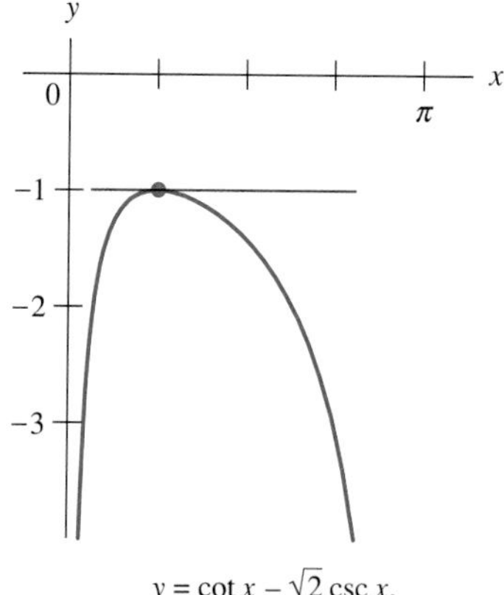

$y = \cot x - \sqrt{2} \csc x$, $0 < x < \pi$

(Generated by Mathematica)

42. $y = \tan x + 3 \cot x - 3$, $0 < x < \pi/2$

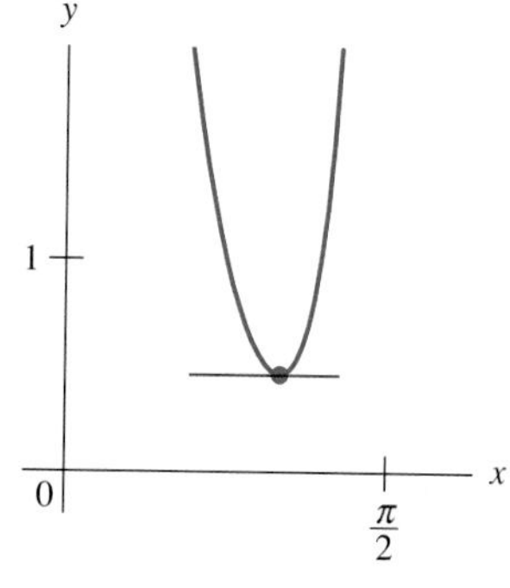

$y = \tan x + 3 \cot x - 3$, $0 < x < \pi/2$

(Generated by Mathematica)

43. **GRAPHING CALCULATOR** Graph $y = \tan x$ and its derivative together over the interval $-\pi/2 < x < \pi/2$.

44. **GRAPHING CALCULATOR** Graph $y = \cot x$ and its derivative together for $0 < x < \pi$.

45. **CALCULATOR** Although $\lim_{h\to 0} (1 - \cos h)/h = 0$, it turns out that

$$\lim_{h\to 0} \frac{1 - \cos h}{h^2} \neq 0.$$

What do you think the limit is? Try finding out by taking $h = 1, 0.1, 0.01$, and so on, as far as your calculator can go. (L'Hôpital's rule in Section 7.5 will provide a quick way to confirm your answer.)

46. **CALCULATOR** *Radians vs. Degrees.* What happens if you use degrees instead of radians to calculate

$$\lim_{h\to 0} \frac{\sin h}{h}?$$

To find out, set your calculator in degree mode and make a table of values of $(\sin h)/h$ for $h = 0.1, 0.01, 0.001$, and so on, as far as your calculator can go. Now multiply each entry in the table by $180/\pi$. What do you find? What would the derivative of $\sin x$ be if x were measured in degrees instead of radians? (Look at Eq. (3) to find out.)

47. Derive Eq. (7) by writing $\sec x = 1/\cos x$ and differentiating with respect to x.

48. Derive Eq. (8) by writing $\cot x = (\cos x)/(\sin x)$ and differentiating with respect to x.

49. Derive Eq. (9) by writing $\csc x = 1/\sin x$ and differentiating with respect to x.

3.5 The Chain Rule

We now know how to differentiate $\sin x$ and $x^2 - 4$, but how do we differentiate a composite like $\sin(x^2 - 4)$? The answer is, with the Chain Rule, which says that the derivative of the composite of two differentiable functions is the product of their derivatives. The Chain Rule is probably the most widely used differentiation rule in mathematics. This section describes the rule and how to use it.

Introductory Examples

A few examples will show what is going on.

EXAMPLE 1 The function $y = 6x - 10 = 2(3x - 5)$ is the composite of the functions $y = 2u$ and $u = 3x - 5$. How are the derivatives of these three functions related?

Solution We have $\dfrac{dy}{dx} = 6$, $\dfrac{dy}{du} = 2$, $\dfrac{du}{dx} = 3$. Since $6 = 2 \cdot 3$,

$$\frac{dy}{dx} = \frac{dy}{du}\frac{du}{dx}.$$

■

EXAMPLE 2 A particle moves along the line $y = 5u - 2$ in the uy-plane in such a way that its u-coordinate at time t is $u = 3t$. Find dy/dt (Fig. 3.32).

Solution As a function of t,

$$y = 5u - 2 = 5(3t) - 2 = 15t - 2.$$

Therefore,

$$\frac{dy}{dt} = \frac{d}{dt}(15t - 2) = 15.$$

Notice that $dy/du = 5$, $du/dt = 3$, and

$$\frac{dy}{dt} = 5 \cdot 3 = \frac{dy}{du}\frac{du}{dt}.$$

■

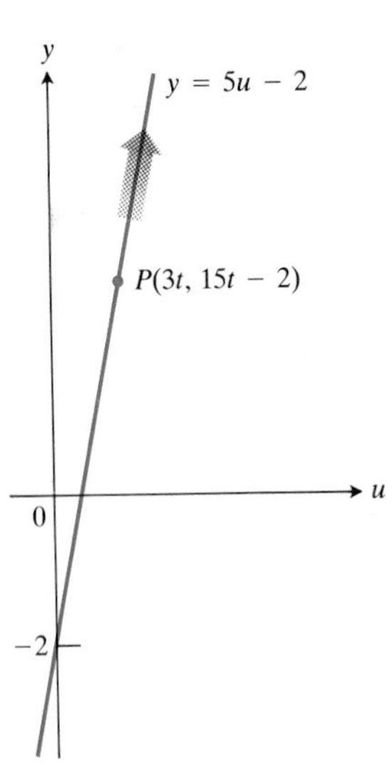

3.32 A particle moving along the line $y = 5u - 2$ in such a way that $u = 3t$.

EXAMPLE 3 The function $y = 9x^2 + 6x + 1 = (3x + 1)^2$ is the composite of $y = u^2$ and $u = 3x + 1$. The derivatives involved are

$$\frac{dy}{dx} = \frac{d}{dx}(9x^2 + 6x + 1) = 18x + 6 = 6(3x + 1) = 6u$$

$$\frac{dy}{du} = \frac{d}{du}(u^2) = 2u$$

$$\frac{du}{dx} = \frac{d}{dx}(3x + 1) = 3.$$

3.33 When wheel A takes x turns, wheel B takes u turns and wheel C takes y turns. By comparing circumferences we see that $dy/du = 1/2$ and $du/dx = 3$. What is dy/dx?

Once again,

$$\frac{dy}{dx} = \frac{dy}{du}\frac{du}{dx}.$$

EXAMPLE 4 In the gear train in Fig. 3.33, the ratios of the radii of gears A, B, and C are 3:1:2. If gear A turns x times, then gear B turns $u = 3x$ times and gear C turns $y = u/2 = (3/2)x$ times. In terms of derivatives,

$$\frac{du}{dx} = 3 \qquad (B \text{ turns at 3 times } A\text{'s rate.})$$

$$\frac{dy}{du} = \frac{1}{2}. \qquad \begin{pmatrix} C \text{ turns at three-halves } A\text{'s rate,} \\ \text{three-halves of a turn for } A\text{'s one.} \end{pmatrix}$$

In this example, too, we can calculate dy/dx by multiplying dy/du by du/dx:

$$\frac{dy}{dx} = \frac{3}{2} = \frac{1}{2}\cdot 3 = \frac{dy}{du}\frac{du}{dx}.$$

The Chain Rule

The preceding examples all work because the derivative of a composite $f \circ g$ of two differentiable functions is the product of their derivatives. This is the observation we state formally as the Chain Rule. As in Section 1.3, the notation $f \circ g$ ("f of g") denotes the composite of the functions f and g, with f following g. The value of $f \circ g$ at a point x is $(f \circ g)(x) = f(g(x))$.

The Chain Rule (First Form)

Suppose that $f \circ g$ is the composite of the differentiable functions $y = f(u)$ and $u = g(x)$. Then $f \circ g$ is a differentiable function of x whose derivative at each value of x is

$$(f \circ g)'_{\text{at } x} = f'_{\text{at } u = g(x)} \cdot g'_{\text{at } x}. \tag{1}$$

In short,

$$(f \circ g)'(x) = f'(g(x)) \cdot g'(x). \tag{2}$$

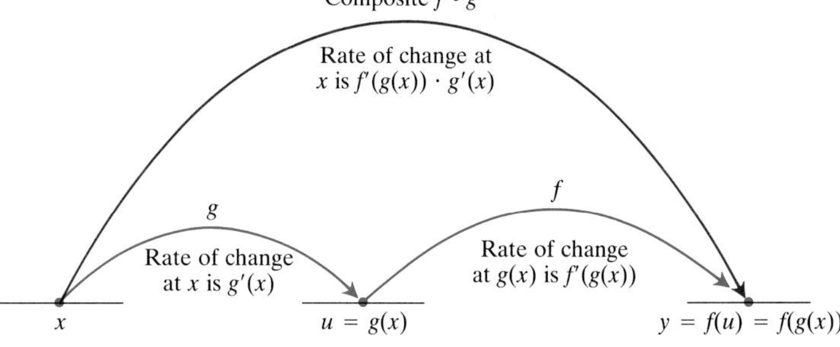

Rates of change multiply: The derivative of f of g at x is the derivative of f at $g(x)$ times the derivative of g at x

Equations (1) and (2) name the function involved as well as the dependent and independent variables. Once we know what the functions are, as we usually do in any particular example, we can get by with writing the Chain Rule a shorter way.

Chain Rule (Shorter Form)

If y is a differentiable function of u and u is a differentiable function of x, then y is a differentiable function of x and

$$\left.\frac{dy}{dx}\right|_{\text{at } x} = \left.\frac{dy}{du}\right|_{\text{at } u(x)} \cdot \left.\frac{du}{dx}\right|_{\text{at } x} \tag{3}$$

Equation (3) still tells how each derivative is to be evaluated. When we don't need to be told that, we can get along with an even shorter form.

Chain Rule (Shortest Form)

If y is a differentiable function of u and u is a differentiable function of x, then

$$\frac{dy}{dx} = \frac{dy}{du}\frac{du}{dx}. \tag{4}$$

You might think it would be a relatively easy matter to prove the Chain Rule by starting with the derivative definition the way we started the proofs of the Product and Quotient Rules. Unfortunately, this is the way the *hard* proof starts. The (relatively) easy proof starts with an equation we won't get to until Section 3.7. We have therefore placed the proof of the Chain Rule in Appendix 3, to be looked at later. We'll direct your attention to it when the time comes.

Like different instruments in a doctor's bag, each form of the Chain Rule makes some task a little easier. We shall use them all in the examples that follow. But remember—they all express the same one rule: The derivative of a composite of differentiable functions is the product of their derivatives.

EXAMPLE 5 If $f(u) = \sin u$ and $u = g(x) = x^2 - 4$, find $(f \circ g)'$ at $x = 2$.

Solution Equation (1) for the Chain Rule gives

$$\begin{aligned}
(f \circ g)'_{\text{at } x=2} &= f'_{\text{at } u=g(2)} \cdot g'_{\text{at } x=2} \\
&= \frac{d}{dx}(\sin u)_{\text{at } u=0} \cdot \frac{d}{dx}(x^2 - 4)_{\text{at } x=2} \\
&= \cos u \Big|_{u=0} \cdot 2x \Big|_{x=2} \\
&= 1 \cdot 4 = 4.
\end{aligned}$$

EXAMPLE 6 Find dy/dx at $x = 0$ if $y = \cos u$ and $u = (\pi/2) - 3x$.

Solution With Eq. (3) this time,

$$\left.\frac{dy}{dx}\right|_{x=0} = \left.\frac{dy}{du}\right|_{u=\pi/2} \cdot \left.\frac{du}{dx}\right|_{x=0}$$

$$= -\sin u\Big|_{u=\pi/2} \cdot (-3) = 3\sin\frac{\pi}{2} = 3 \cdot 1 = 3.$$

EXAMPLE 7 Express dy/dx in terms of x if $y = u^3$ and $u = x^2 - 1$.

Solution

$$\frac{dy}{dx} = \frac{dy}{du}\frac{du}{dx} \qquad \text{(Eq. (4))}$$

$$= 3u^2 \cdot 2x$$

$$= 3(x^2 - 1)^2 \cdot 2x \qquad (u = x^2 - 1)$$

$$= 6x(x^2 - 1)^2$$

Integer Powers of Differentiable Functions

The Chain Rule enables us to differentiate powers like $y = \sin^5 x$ and $y = (2x + 1)^{-3}$ because these powers are composites:

$$y = \sin^5 x \text{ is } u^5 \quad \text{with} \quad u = \sin x.$$

$$y = (2x + 1)^{-3} \text{ is } u^{-3} \quad \text{with} \quad u = 2x + 1.$$

If $y = u^n$ and u is any differentiable function of x and $y = u^n$, then the Chain Rule in the form

$$\frac{dy}{dx} = \frac{dy}{du} \cdot \frac{du}{dx}$$

gives

$$\frac{dy}{dx} = \frac{d}{du}(u^n) \cdot \frac{du}{dx}$$

$$= nu^{n-1}\frac{du}{dx}. \qquad \left(\begin{array}{l}\text{Differentiating } u^n \text{ with respect to } u \text{ itself} \\ \text{gives } nu^{n-1} \text{ by Rule 2 in Section 3.2.}\end{array}\right)$$

Integer Powers of a Differentiable Function

If u^n is an integer power of a differentiable function $u(x)$, then u^n is differentiable and

$$\frac{d}{dx}u^n = nu^{n-1}\frac{du}{dx}. \tag{5}$$

EXAMPLE 8

a) $\dfrac{d}{dx}\sin^5 x = 5\sin^4 x\dfrac{d}{dx}(\sin x)$ $\quad\begin{pmatrix}\text{Eq. (5) with}\\ u = \sin x,\ n = 5\end{pmatrix}$

$= 5\sin^4 x\cos x$

b) $\dfrac{d}{dx}(2x+1)^{-3} = -3(2x+1)^{-4}\dfrac{d}{dx}(2x+1)$ $\quad\begin{pmatrix}\text{Eq. (5) with}\\ u = 2x+1,\ n = -3\end{pmatrix}$

$= -3(2x+1)^{-4}(2)$

$= -6(2x+1)^{-4}$ ■

The "Inside-Outside" Rule

It sometimes helps to think about the Chain Rule the following way. If $y = f(g(x))$, Eq. (2) tells us that

$$\frac{dy}{dx} = f'(g(x)) \cdot g'(x). \tag{6}$$

In words, Eq. (6) says: To find dy/dx, differentiate the "outside" function f and leave the "inside" $g(x)$ alone; then multiply by the derivative of the inside.

EXAMPLE 9

a) $\dfrac{d}{dx}\overbrace{\sin}^{\text{outside}}\underbrace{(x^2+x)}_{\text{inside}} = \overbrace{\cos}^{\text{derivative of the outside}}\underbrace{(x^2+x)}_{\text{inside left alone}} \cdot \underbrace{(2x+1)}_{\text{derivative of the inside}}$ ■

b) $\dfrac{d}{dx}\overbrace{\tan}^{\text{outside}}\underbrace{\sqrt{x}}_{\text{inside}} = \overbrace{\sec^2}^{\text{derivative of outside}}\underbrace{\sqrt{x}}_{\text{inside left alone}}\ \underbrace{\dfrac{1}{2\sqrt{x}}}_{\text{derivative of inside}} = \dfrac{\sec^2\sqrt{x}}{2\sqrt{x}}.$ ■

Repeated Use

We sometimes have to use the Chain Rule two or more times to get the job done. Here is an example.

EXAMPLE 10

a) $\dfrac{d}{dx}\cos^2 3x = 2\cos 3x \cdot \dfrac{d}{dx}(\cos 3x)$ (Power (Chain) Rule)

$= 2\cos 3x(-\sin 3x)\dfrac{d}{dx}(3x)$ (Chain Rule again)

$= 2\cos 3x(-\sin 3x)(3)$

$= -6\cos 3x\sin 3x.$

$$\text{b) } \frac{d}{dx}\sin(1 + \tan 2x) = \cos(1 + \tan 2x) \cdot \frac{d}{dx}(1 + \tan 2x) \quad \text{(Chain Rule)}$$
$$= \cos(1 + \tan 2x) \cdot \sec^2 2x \cdot \frac{d}{dx}(2x) \quad \text{(Chain Rule)}$$
$$= 2\cos(1 + \tan 2x)\sec^2 2x$$

Derivative Formulas that Include the Chain Rule

Many of the derivative formulas you will encounter in your scientific work already include the Chain Rule.

If f is a differentiable function of u and u is a differentiable function of x, then substituting $y = f(u)$ in the Chain Rule formula

$$\frac{dy}{dx} = \frac{dy}{du}\frac{du}{dx}$$

leads to

$$\frac{d}{dx}f(u) = f'(u)\frac{du}{dx}. \tag{7}$$

When we spell this out for the functions whose derivatives we have studied so far, we get the formulas in Table 3.1.

EXAMPLE 11

$$\text{a) } \frac{d}{dx}\sin(-x) = \cos(-x)\frac{d}{dx}(-x) \qquad \left(\frac{d}{dx}\sin u = \cos u\frac{du}{dx} \text{ with } u = -x\right)$$
$$= -\cos(-x)$$

$$\text{b) } \frac{d}{dx}\tan\left(\frac{1}{x}\right) = \sec^2\frac{1}{x} \cdot \frac{d}{dx}\left(\frac{1}{x}\right) \qquad \left(\frac{d}{dx}\tan u = \sec^2 u\frac{du}{dx} \text{ with } u = 1/x\right)$$
$$= \sec^2\frac{1}{x} \cdot \left(-\frac{1}{x^2}\right)$$
$$= -\frac{1}{x^2}\sec^2\frac{1}{x}$$

TABLE 3.1
Derivative Formulas that Include the Chain Rule

$$\frac{d}{dx}u^n = nu^{n-1}\frac{du}{dx} \quad (n \text{ an integer})$$

$\frac{d}{dx}\sin u = \cos u\frac{du}{dx}$	$\frac{d}{dx}\sec u = \sec u \tan u\frac{du}{dx}$
$\frac{d}{dx}\cos u = -\sin u\frac{du}{dx}$	$\frac{d}{dx}\cot u = -\csc^2 u\frac{du}{dx}$
$\frac{d}{dx}\tan u = \sec^2 u\frac{du}{dx}$	$\frac{d}{dx}\csc u = -\csc u \cot u\frac{du}{dx}$

$$\text{c) } \frac{d}{dx}\cos(\cos x) = -\sin(\cos x)\cdot\frac{d}{dx}(\cos x) \qquad \left(\begin{array}{l}\dfrac{d}{dx}\cos u = -\sin u\dfrac{du}{dx} \\ \text{with } u = \cos x\end{array}\right)$$

$$= -\sin(\cos x)\cdot(-\sin x)$$

$$= \sin x\,\sin(\cos x)$$

Melting Ice Cubes

In mathematics, we tend to use letters like f, g, x, y, and u for functions and variables. However, applied fields use letters like V, for volume, and s, for side, that come from the names of the things being modeled. The letters in the Chain Rule then change, too, as in the next example.

Grant Heilman

3.34 This ice cube lost a fourth of its volume during the first hour. How much longer will it last? Example 12 shows how to use the Chain Rule to find out.

EXAMPLE 12 *The melting ice cube.* How long will it take an ice cube like the one in Fig. 3.34 to melt?

Solution As with all applications to science, we start with a mathematical model. Let us assume that the cube has side length s, so that its volume is $V = s^3$, and that V and s are differentiable functions of time t. Suppose also that the cube's volume decreases at a rate that is proportional to its surface area. This assumption seems reasonable enough when we think that the melting takes place at the surface: Changing the surface changes the amount of ice exposed to melt. In mathematical terms,

$$\frac{dV}{dt} = -k(6s^2).$$

The minus sign is there because the volume is decreasing. We assume that the proportionality factor k is constant. (It probably depends on many things, like the relative humidity of the surrounding air, the air temperature, and the incidence or absence of sunlight, to name only a few.)

Finally, we need at least one more piece of information: How long will it take the ice cube to melt some specific percent? We have nothing to guide us unless we make one or more observations, but now let us assume a particular set of conditions in which the cube lost 1/4 of its volume during the first hour. (You could use letters instead of precise numbers: say $n\%$ in h hr. Then your answer would be in terms of n and h.) Now to work.

Mathematically, we now have the following problem.

Given: $V = s^3$ and $\dfrac{dV}{dt} = a - k(6s^2)$

$V = V_0$ when $t = 0$

$V = (3/4)V_0$ when $t = 1$ hr

Find: The value of t when $V = 0$

We apply the Chain Rule to differentiate $V = s^3$ with respect to t:

$$\frac{dV}{dt} = 3s^2\frac{ds}{dt}.$$

We set this equal to the given rate, $-k(6s^2)$, to get

$$3s^2\frac{ds}{dt} = -6ks^2,$$

$$\frac{ds}{dt} = -2k.$$

The side length is *decreasing* at the constant rate of $2k$ units per hour (centimeters, inches, whatever). Thus, if the initial length of the cube's side is s_0, the length of its side one hour later is $s_1 = s_0 - 2k$. This equation tells us that

$$2k = s_0 - s_1.$$

The melting time is the value of t that makes $2kt = s_0$. Hence,

$$t_{\text{melt}} = \frac{s_0}{2k} = \frac{s_0}{s_0 - s_1} = \frac{1}{1 - (s_1/s_0)}.$$

But

$$\frac{s_1}{s_0} = \frac{\left(\frac{3}{4}V_0\right)^{1/3}}{(V_0)^{1/3}} = \left(\frac{3}{4}\right)^{1/3} \approx 0.91.$$

Therefore,

$$t_{\text{melt}} \approx \frac{1}{1 - 0.91} \approx 11 \text{ hr}.$$

If 1/4 of the cube melts in 1 hr, it will take nearly 10 hr more for the rest of it to melt.

REMARK If we were natural scientists who were really interested in testing our model, we could collect some data and compare them with the results of the mathematics. One practical application might lie in analyzing the proposal to tow large icebergs from polar waters to offshore locations near southern California where the melting ice would provide fresh water. As a first approximation, we might assume that the iceberg is a large cube, or a pyramid, or a sphere. ■

Exercises 3.5

In Exercises 1–18, find dy/dx when y equals the given expression.

1. $\sin(x + 1)$

2. $\sin(x/2)$

3. $\cos 5x$

4. $\cos(-2x)$

5. $\sin(2\pi x/5)$

6. $\cos(3\pi x/2)$

7. $\tan(2 - x)$

8. $\tan 5(x - 1)$

9. $\sec(2x - 1)$

10. $\sec(x^2 + 1)$

11. $\csc(x^2 + 7x)$

12. $\csc(1 - 2x)$

13. $\cot(3x + \pi)$

14. $\cot(1/x)$

15. $\sec^2 x - \tan^2 x$

16. $\csc^2 x - \cot^2 x$

17. $\sin^3 x$

18. $\cos^3 x$

In Exercises 19–28, find dy/dx when y equals the given expression.

19. $(2x + 1)^5$

20. $(4 - 3x)^9$

21. $(x + 1)^{-3}$

22. $(2x + 1)^{-3}$

23. $\left(1 - \frac{x}{7}\right)^{-7}$

24. $\left(\frac{x}{2} - 1\right)^{-10}$

25. $\left(1 + x - \frac{1}{x}\right)^{3}$

26. $\left(\frac{x}{5} + \frac{1}{5x}\right)^{5}$

27. $(x^2 + 2x + 3)^3$

28. $(x^2 + 2x + 3)^{-3}$

In Exercises 29–36, find dy/dx when y equals the given expression.

29. $\cos(\sin x)$

30. $\sin(\sin x)$

31. $1 + x + \sec^2 x$

32. $x^2 \cos\left(\frac{1}{x}\right)$

33. $(\csc x + \cot x)^{-1}$

34. $-(\sec x + \tan x)^{-1}$

35. $\sin\left(\frac{x-2}{x+3}\right)$

36. $\left(\frac{\sin x}{1 + \cos x}\right)^2$

In Exercises 37–42, find dy/dx when y equals the given expression. You will need to use the Chain Rule two or three times in each case.

37. $\sin^2(3x - 2)$

38. $\sec^2 5x$

39. $(1 + \cos 2x)^2$

40. $(1 - \tan(x/2))^{-2}$

41. $\sin(\cos(2x - 5))$

42. $(1 + \cos^2 7x)^3$

Find y'' in Exercises 43–46.

43. $y = \tan x$

44. $y = 9\tan(x/3)$

45. $y = \cot x$

46. $y = \cot(3x - 1)$

In Exercises 47–52, find the value of $(f \circ g)'$ at the given value of x.

47. $f(u) = u^5 + 1,\ u = g(x) = \sqrt{x},\ x = 1$

48. $f(u) = 1 - \frac{1}{u},\ u = g(x) = \frac{1}{1-x},\ x = -1$

49. $f(u) = \cot\frac{\pi u}{10},\ u = g(x) = 5\sqrt{x},\ x = 1$

50. $f(u) = u + \frac{1}{\cos^2 u},\ u = g(x) = \pi x,\ x = 1/4$

51. $f(u) = \frac{2u}{u^2 + 1},\ u = g(x) = 10x^2 + x + 1,\ x = 0$

52. $f(u) = \left(\frac{u-1}{u+1}\right)^2,\ u = g(x) = \frac{1}{x^2} - 1,\ x = -1$

What happens if you can write a function as a composite in different ways? Do you get the same derivative each time? The Chain Rule says you should. Try it with the functions in Exercises 53–56.

53. Find dy/dx if $y = \cos(6x + 2)$ by writing y as a composite with
a) $y = \cos u$ and $u = 6x + 2$
b) $y = \cos 2u$ and $u = 3x + 1$

54. Find dy/dx if $y = \sin(x^2 + 1)$ by writing y as a composite with
a) $y = \sin(u + 1)$ and $u = x^2$
b) $y = \sin u$ and $u = x^2 + 1$

55. Find dy/dx if $y = x$ by writing y as the composite of
a) $y = (u/5) + 7$ and $u = 5x - 35$
b) $y = 1 + (1/u)$ and $u = 1/(x - 1)$

56. Find dy/dx if $y = \sin(\sin(2x))$ by writing y as the composite of
a) $y = \sin u$ and $u = \sin 2x$
b) $y = \sin(\sin u)$ and $u = 2x$

57. Find ds/dt if $s = \cos\theta$ and $d\theta/dt = 5$ when $\theta = 3\pi/2$.

58. Find dy/dt if $y = x^2 + 7x - 5$ and $dx/dt = 1/3$ when $x = 1$.

59. What is the largest value the slope of the curve $y = \sin(x/2)$ can ever have?

60. Write an equation for the tangent to the curve $y = \sin mx$ at the origin.

61. Find the lines that are tangent and normal to the curve $y = 2\tan(\pi x/4)$ at $x = 1$.

62. *Orthogonal curves.* Two curves are said to cross at right angles if their tangents are perpendicular at the crossing point. The technical word for "crossing at right angles" is *orthogonal.* Show that the curves $y = \sin 2x$ and $y = -\sin(x/2)$ are orthogonal at the origin.

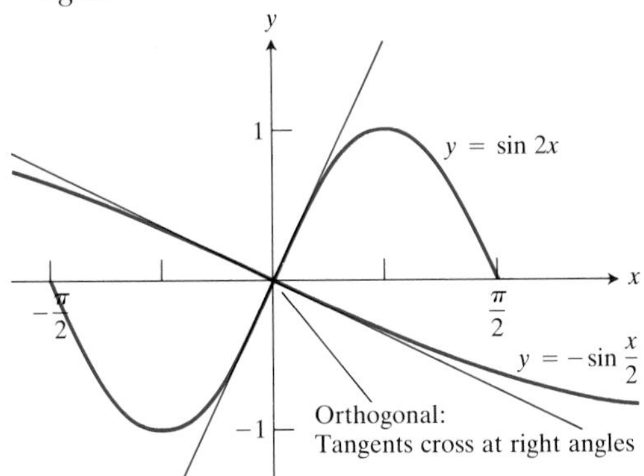

63. Suppose that functions f and g and their derivatives have the following values at $x = 2$ and $x = 3$:

x	$f(x)$	$g(x)$	$f'(x)$	$g'(x)$
2	8	2	$\frac{1}{3}$	-3
3	3	-4	2π	5

Find the values of the following derivatives.

a) $\frac{d}{dx}\{2f(x)\}$ at $x = 2$

b) $\frac{d}{dx}\{f(x) + g(x)\}$ at $x = 3$

c) $\frac{d}{dx}\{f(x) \cdot g(x)\}$ at $x = 3$

d) $\frac{d}{dx}\left\{\frac{f(x)}{g(x)}\right\}$ at $x = 2$

e) $\frac{d}{dx}\{f(g(x))\}$ at $x = 2$

f) $\frac{d}{dx}\left\{\sqrt{f(x)}\right\}$ at $x = 2$

g) $\frac{d}{dx}\left\{\frac{1}{g^2(x)}\right\}$ at $x = 3$

h) $\frac{d}{dx}\left\{\sqrt{f^2(x) + g^2(x)}\right\}$ at $x = 2$

64. Show that the function $y = \cos(2x + B)$ satisfies the equation

$$y'' = -4y$$

whatever the value of B. Equations like $y'' = -4y$ come up whenever we design loudspeakers and audio amplifiers, and their solutions are always cosines. We shall say more about equations like this in Chapter 16.

65. When this picture (above right) was taken, the radius of the soap bubble was 10 cm and increasing at the rate of 1/2 cm/sec. How fast was the volume changing then?

To find out, start with the equation $V = (4/3)\pi r^3$ for the volume of a sphere. Then assume that r is a differentiable function of time t and use the Chain Rule to differentiate both sides of the equation with respect to t. Then substitute $r = 10$ and $dr/dt = 1/2$ to find the value of dV/dt when the picture was taken.

If we know how fast a bubble's radius is changing, we can use the Chain Rule to find out how fast its volume is changing.

66. *Temperatures in Fairbanks, Alaska.* The graph in Fig. 3.35, taken from Example 9 in Section 1.5, shows the average temperature in Fairbanks, Alaska, during a typical 365-day year. The equation that gives the temperature on day x is

$$y = 37 \sin\left[\frac{2\pi}{365}(x - 101)\right] + 25.$$

a) On what day is the temperature increasing the fastest?

b) **CALCULATOR** About how many degrees per day is the temperature increasing when it is increasing at its fastest?

3.35 Normal mean air temperatures at Fairbanks, Alaska, plotted as data points. The approximating sine function is

$$f(x) = 37 \sin\left[\frac{2\pi}{365}(x - 101)\right] + 25.$$

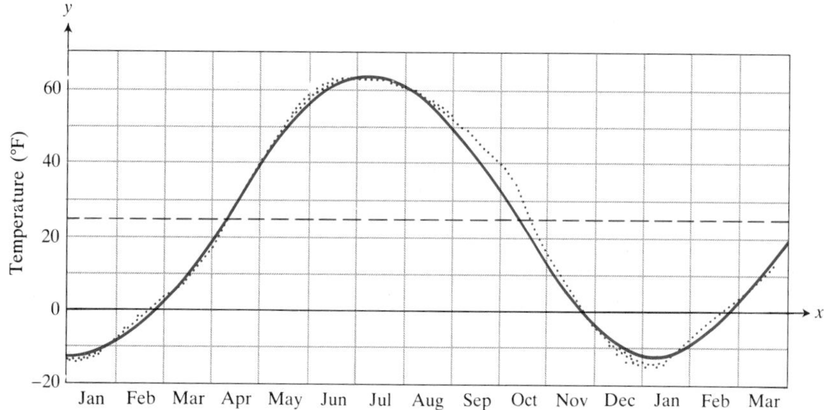

3.6 Implicit Differentiation and Fractional Powers

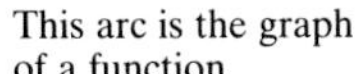

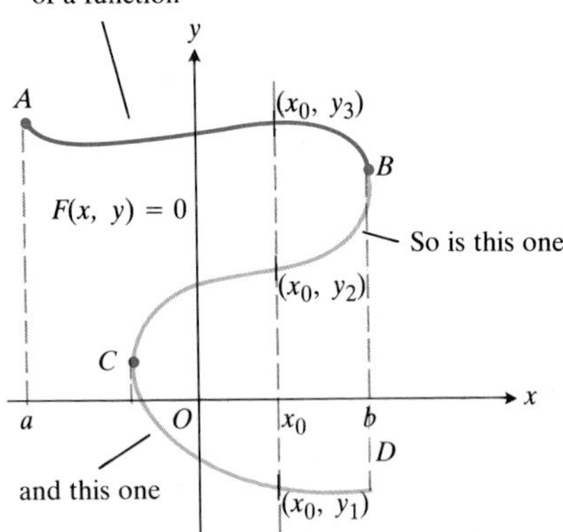

3.36 As a whole, the curve $F(x, y) = 0$ is not the graph of a function of x. Some of the vertical lines that cross it intersect it more than once. However, the curve can be divided into separate arcs that *are* the graphs of functions of x.

When an equation like $y^5 + \sin xy = 0$ defines y as a differentiable function of x but does not let us solve for y in terms of x, we can still find dy/dx with a technique called implicit differentiation. This section describes the technique and uses it to extend the Power Rule for differentiation to include fractional exponents.

Implicit Differentiation

The graph of an equation $F(x, y) = 0$ is not the graph of a function of x if some of the vertical lines that cross it intersect it more than once. For example, the numbers y_0, y_1, and y_2 in Fig. 3.36 all correspond to the same x-value, $x = x_0$. However, various parts of the curve $F(x, y) = 0$ may well be the graphs of functions of x. The arc AB in Fig. 3.36 is the graph of a function of x, and so are arcs BC and CD.

Sometimes we can find explicit formulas for the functions defined by an equation $F(x, y) = 0$ but usually we cannot. The equation $F(x, y) = 0$ has defined the functions *implicitly* but not *explicitly*.

When may we expect the functions defined by an equation $F(x, y) = 0$ to be differentiable? The answer is, when their graphs are smooth enough to have a tangent at every point, as they will, for instance, if the formula for F is an algebraic combination of powers of x and y (a theorem from advanced mathematics).

EXAMPLE 1 Find dy/dx if $y^2 = x$.

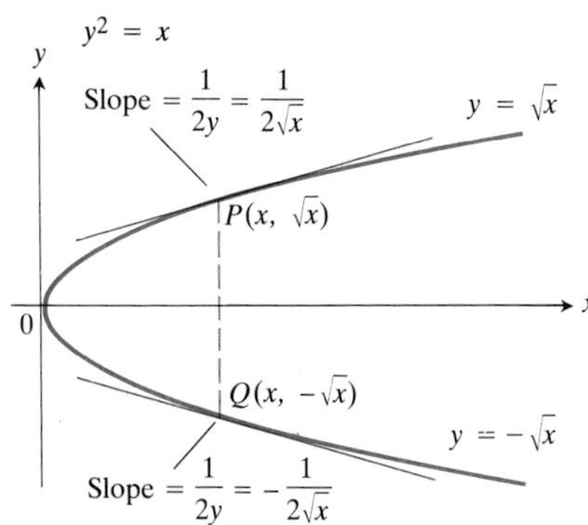

3.37 The equation $y^2 - x = 0$, or $y^2 = x$ as it is usually written, defines two differentiable functions of x on the interval $x \geq 0$. Example 1 shows how you can calculate the derivatives of these functions without solving the equation $y^2 = x$ for y.

Solution The equation $y^2 = x$ defines two differentiable functions of x, namely $y = \sqrt{x}$ and $y = -\sqrt{x}$ (Fig. 3.37). We know how to find the derivative of each of these, from Example 8 in Section 3.1. But suppose we knew only that the equation $y^2 = x$ defined y as one or more differentiable functions of x without knowing exactly what these functions were. Could we still find dy/dx?

The answer in this case is yes. To find dy/dx we simply differentiate both sides of the equation $y^2 = x$ with respect to x, treating y as a differentiable but otherwise unknown function of x. When we do this, we get

$$y^2 = x$$

$$2y\frac{dy}{dx} = 1 \qquad \left(\begin{array}{l}\text{The Chain Rule gives}\\ (d/dx)y^2 = 2y\,(dy/dx).\end{array}\right)$$

$$\frac{dy}{dx} = \frac{1}{2y}.$$

How does this compare with what happens when we solve $y^2 = x$ for y first and then differentiate?

With $y = \sqrt{x}$	With $y = -\sqrt{x}$
$\dfrac{dy}{dx} = \dfrac{1}{2\sqrt{x}}$	$\dfrac{dy}{dx} = -\dfrac{1}{2\sqrt{x}}$
$= \dfrac{1}{2y}$	$= \dfrac{1}{2(-\sqrt{x})}$
	$= \dfrac{1}{2y}$

In both cases, the derivative is given by the formula we obtained without solving for y, so the two methods agree. ■

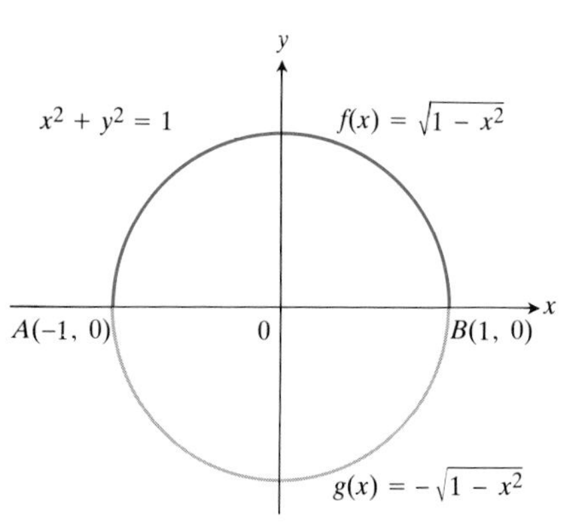

3.38 The graph of the equation $F(x, y) = x^2 + y^2 - 1 = 0$ is the complete circle $x^2 + y^2 = 1$. The upper semicircle AB is the graph of the function $f(x) = \sqrt{1 - x^2}$. The lower semicircle AB is the graph of $g(x) = -\sqrt{1 - x^2}$.

EXAMPLE 2 The graph of $F(x, y) = x^2 + y^2 - 1 = 0$ is the circle $x^2 + y^2 = 1$. Taken as a whole, the circle is not the graph of any single function of x (Fig. 3.38). Each x in the interval $-1 < x < 1$ gives two values of y, namely $y = \sqrt{1 - x^2}$ and $y = -\sqrt{1 - x^2}$.

The upper and lower semicircles are the graphs of the functions $f(x) = \sqrt{1 - x^2}$ and $g(x) = -\sqrt{1 - x^2}$. These functions are differentiable because they are composites of differentiable functions. The quickest way to find their derivatives, however, is not to differentiate the square-root formulas but to differentiate both sides of the original equation, treating y as a differentiable but otherwise unknown function of x:

$$x^2 + y^2 = 1$$

$$\frac{d}{dx}(x^2) + \frac{d}{dx}(y^2) = \frac{d}{dx}(1)$$

$$2x + 2y\frac{dy}{dx} = 0$$

$$\frac{dy}{dx} = -\frac{x}{y}.$$

This formula for dy/dx is simpler than either of the formulas we would get by differentiating f and g, and holds for all points on the curve above or below the x-axis. It is also easy to evaluate at any such point. At $(\sqrt{2}/2, \sqrt{2}/2)$, for instance,

$$\frac{dy}{dx} = -\frac{\sqrt{2}/2}{\sqrt{2}/2} = -1.$$

■

To calculate the derivatives of other implicitly defined functions we simply proceed as in Examples 1 and 2: We treat y as a differentiable (but otherwise unknown) function of x and apply the already familiar rules of differentiation to differentiate both sides of the defining equation. This procedure is called *implicit differentiation.*

EXAMPLE 3 Find dy/dx if $2y = x^2 + \sin y$.

Solution

$$2y = x^2 + \sin y$$

$$\frac{d}{dx}(2y) = \frac{d}{dx}(x^2) + \frac{d}{dx}(\sin y) \quad \begin{pmatrix}\text{Differentiate both}\\ \text{sides with respect to } x.\end{pmatrix}$$

$$2\frac{dy}{dx} = 2x + \cos y \frac{dy}{dx}$$

$$2\frac{dy}{dx} - \cos y \frac{dy}{dx} = 2x \qquad \text{(Collect terms with } dy/dx.$$

$$(2 - \cos y)\frac{dy}{dx} = 2x \qquad \text{Factor out } dy/dx \ldots$$

$$\frac{dy}{dx} = \frac{2x}{2 - \cos y}. \qquad \ldots \text{ and divide.)}$$

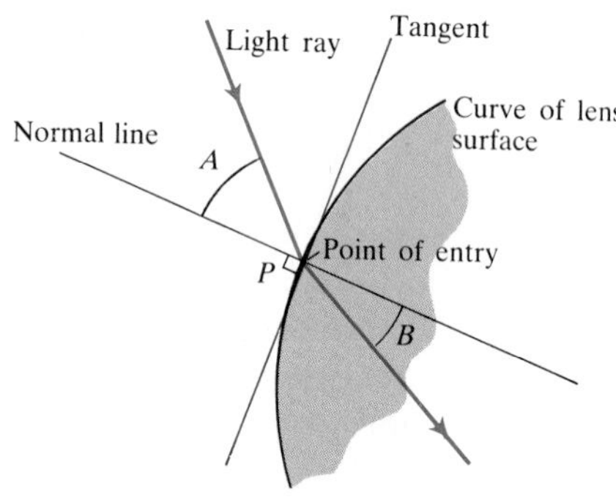

3.39 The profile or cutaway view of a lens, showing the bending (refraction) of a light ray as it passes through the lens surface.

Implicit Differentiation Takes Four Steps

1. Differentiate both sides of the equation with respect to x.
2. Collect the terms with dy/dx on one side of the equation.
3. Factor out dy/dx.
4. Solve for dy/dx by dividing.

Lenses, Tangents, and Normal Lines

In the law that describes how light changes direction as it enters a lens, the important angles are the angles the light makes with the line perpendicular to the surface of the lens at the point of entry (angles A and B in Fig. 3.39). This line is called the *normal to the surface* at the point of entry. In a profile view of a lens like the one in Fig. 3.39, the normal is the line perpendicular to the tangent to the profile curve at the point of entry.

The profiles of lenses are often described by quadratic curves like the one in Fig. 3.40. When they are, we can use implicit differentiation to find the tangents and normals.

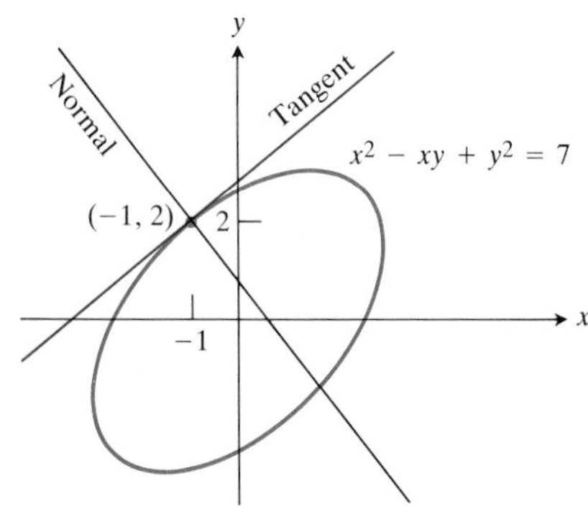

3.40 The graph of $x^2 - xy + y^2 = 7$ is an ellipse. The slope of the curve at the point $(-1, 2)$ is $(dy/dx)_{(-1,2)} = 4/5$.

EXAMPLE 4 Find the tangent and normal to the curve $x^2 - xy + y^2 = 7$ at the point $(-1, 2)$. (See Fig. 3.40.)

Solution We first use implicit differentiation to find dy/dx:

$$x^2 - xy + y^2 = 7$$

$$\frac{d}{dx}(x^2) - \frac{d}{dx}(xy) + \frac{d}{dx}(y^2) = \frac{d}{dx}(7) \quad \begin{pmatrix}\text{Differentiate both sides}\\ \text{with respect to } x \ldots\end{pmatrix}$$

$$2x - \left(x\frac{dy}{dx} + y\frac{dx}{dx}\right) + 2y\frac{dy}{dx} = 0 \quad \begin{pmatrix}\ldots \text{ treating } xy \text{ as a product}\\ \text{and } y^2 \text{ as a power.}\end{pmatrix}$$

$$(2y - x)\frac{dy}{dx} = y - 2x \quad \text{(Collect terms.)}$$

$$\frac{dy}{dx} = \frac{y - 2x}{2y - x} \quad \text{(Solve as usual.)}$$

We then evaluate the derivative at $x = -1, y = 2$, to obtain

$$\left.\frac{dy}{dx}\right|_{(-1,2)} = \left.\frac{y - 2x}{2y - x}\right|_{(-1,2)} = \frac{2 - 2(-1)}{2(2) - (-1)} = \frac{4}{5}.$$

The tangent to the curve at $(-1, 2)$ is

$$y - 2 = \frac{4}{5}(x - (-1))$$

$$y = \frac{4}{5}x + \frac{14}{5}.$$

The normal to the curve at $(-1, 2)$ is

$$y - 2 = -\frac{5}{4}(x + 1)$$

$$y = -\frac{5}{4}x + \frac{3}{4}.$$

■

Using Implicit Differentiation to Find Derivatives of Higher Order

Implicit differentiation can also produce derivatives of higher order. Here is an example.

EXAMPLE 5 Find d^2y/dx^2 if $2x^3 - 3y^2 = 7$.

Solution To start, we differentiate both sides of the equation with respect to x to find $y' = dy/dx$:

$$2x^3 - 3y^2 = 7$$

$$\frac{d}{dx}(2x^3) - \frac{d}{dx}(3y^2) = \frac{d}{dx}(7)$$

$$6x^2 - 6yy' = 0 \qquad (1)$$

$$x^2 - yy' = 0$$

$$y' = \frac{x^2}{y} \qquad \text{(when } y \neq 0\text{).}$$

We now apply the Quotient Rule to find y'':

$$y'' = \frac{d}{dx}\left(\frac{x^2}{y}\right) = \frac{2xy - x^2y'}{y^2} = \frac{2x}{y} - \frac{x^2}{y^2}y'. \qquad (2)$$

Finally, we substitute $y' = x^2/y$ to express y'' in terms of x and y:

$$y'' = \frac{2x}{y} - \frac{x^2}{y^2}\left(\frac{x^2}{y}\right) = \frac{2x}{y} - \frac{x^4}{y^3}. \qquad (3)$$

The second derivative is not defined at $y = 0$ but is given by Eq. (3) when $y \neq 0$. ■

Fractional Powers of Differentiable Functions

We know that the Power Rule

$$\frac{d}{dx}u^n = nu^{n-1}\frac{du}{dx} \tag{4}$$

holds when n is an integer. Our goal now is to show that it holds when n is a fraction. We will then be able to differentiate functions like

$$y = x^{4/3} \qquad \text{and} \qquad y = (\cos x)^{-1/5}$$

that were beyond our reach before.

Power Rule for Fractional Exponents

If n is any rational number, then

$$\frac{d}{dx}x^n = nx^{n-1}, \tag{5}$$

provided $x \neq 0$ if $n - 1 < 0$ (i.e., $n < 1$).

If n is a rational number and u is a differentiable function of x, then u^n is a differentiable function of x and

$$\frac{d}{dx}u^n = nu^{n-1}\frac{du}{dx}, \tag{6}$$

provided $u \neq 0$ if $n < 1$.

The restrictions $x \neq 0$ if $n < 1$ and $u \neq 0$ if $n < 1$ are there to protect against inadvertent attempts to divide by zero. There is nothing mysterious about these restrictions. They come up quite naturally in practice, as the next example shows.

EXAMPLE 6

a) $\frac{d}{dx}(x^{1/2}) = \frac{1}{2}x^{-1/2} = \frac{1}{2\sqrt{x}}$ $\left(\text{Eq. 5 with } n = \frac{1}{2}\right)$

function defined for $x \geq 0$ — derivative defined only for $x > 0$

b) $\frac{d}{dx}(x^{1/5}) = \frac{1}{5}x^{-4/5}$ $\left(\text{Eq. 5 with } n = \frac{1}{5}\right)$

function defined for all x — derivative not defined at $x = 0$

c) $\frac{d}{dx}(1 - x^2)^{1/2} = \frac{1}{2}(1 - x^2)^{-1/2}(-2x)$ $\left(\text{Eq. 6 with } u = 1 - x^2 \text{ and } n = \frac{1}{2}\right)$

function defined for $-1 \leq x \leq 1$

$$= \frac{-x}{(1 - x^2)^{1/2}}$$

derivative defined only for $-1 < x < 1$ ■

Helga von Koch's Snowflake Curve (1904)

Start with an equilateral triangle, calling it Curve 1. On the middle third of each side, build an equilateral triangle pointing outward. Then erase the old middle thirds. Call the expanded curve Curve 2. Now put equilateral triangles, again pointing them outward, on the middle thirds of the sides of Curve 2. Erase the old middle thirds to make Curve 3. Repeat the process, as shown, to define an infinite sequence of plane curves. The limit curve of the sequence is Koch's snowflake curve.

The snowflake curve is too rough to have a tangent at any point. In other words, the equation $F(x, y) = 0$ defining the curve does not define y as a differentiable function of x or x as a differentiable function of y at any point.

The snowflake curve plays an important role in chaos theory. We shall encounter it again when we study length in Section 6.4.

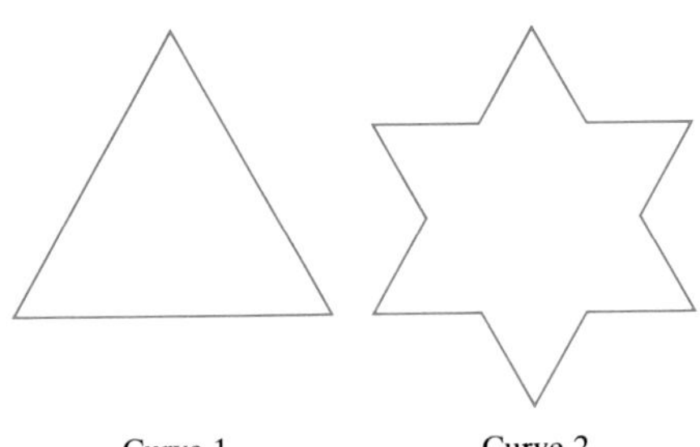

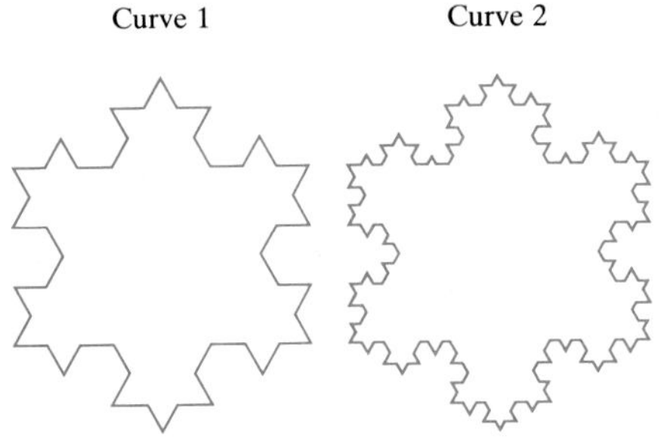

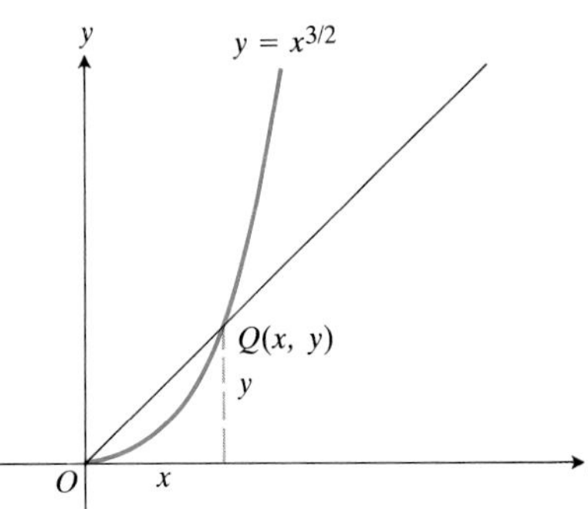

3.41 The graph of $y = x^{3/2}$. The slope of the curve at $x = 0$ is $\lim_{Q \to O} m_{OQ} = 0$.

The derivatives of the functions $x^{4/3}$ and $(\cos x)^{-1/5}$ are defined wherever the functions themselves are defined, as we see in the next example.

EXAMPLE 7

a) $\dfrac{d}{dx} x^{4/3} = \dfrac{4}{3} x^{1/3}$

b) $$\begin{aligned}\frac{d}{dx}(\cos x)^{-1/5} &= -\frac{1}{5} \cos x^{-6/5} \frac{d}{dx}(\cos x) \\ &= -\frac{1}{5}(\cos x)^{-6/5}(-\sin x) \\ &= \frac{1}{5} \sin x \,(\cos x)^{-6/5}\end{aligned}$$

EXAMPLE 8 Find the derivative of $y = x^{3/2}$ at $x = 0$.

Solution When the graph of a function stops abruptly at a point, as the graph of $y = x^{3/2}$ does at $x = 0$ (Fig. 3.41), we calculate its derivative as a one-sided limit. The Power Rule still applies, giving in this case

$$\left.\frac{dy}{dx}\right|_{x=0} = \left.\frac{3}{2} x^{1/2}\right|_{x=0} = 0.$$

We can see why this equation holds by looking at the geometry of the curve. The slope of a typical secant line through the origin and a point $Q(x, y)$ on the curve is

$$m_{OQ} = \frac{y - 0}{x - 0} = \frac{x^{3/2}}{x} = x^{1/2}.$$

As Q approaches the origin from the right, m_{OQ} approaches zero, in agreement with the result from the Power Rule.

PROOF OF THE POWER RULE FOR FRACTIONAL EXPONENTS We prove Eq. (5) first and then apply the Chain Rule to get Eq. (6).

To prove Eq. (5), let p and q be integers with $q > 0$ and suppose that $y = x^{p/q}$. Then

$$y^q = x^p.$$

This equation is an algebraic combination of powers of x and y, so the advanced theorem we mentioned earlier assures us that y is a differentiable function of x. Since p and q are integers (for which we already have the Power Rule), we can differentiate both sides of the equation with respect to x and obtain

$$qy^{q-1}\frac{dy}{dx} = px^{p-1}.$$

Hence, if $y \neq 0$,

$$\frac{dy}{dx} = \frac{p\,x^{p-1}}{q\,y^{q-1}} = \frac{p}{q}\frac{x^{p-1}}{(x^{p/q})^{q-1}} = \frac{p}{q}\frac{x^{p-1}}{x^{p-p/q}} = \frac{p}{q}x^{(p/q)-1}.$$

This proves Eq. (5).

To prove Eq. (6) we let $y = u^{p/q}$ and apply the Chain Rule in the form

$$\frac{dy}{dx} = \frac{dy}{du}\frac{du}{dx}.$$

From Eq. (5), $(d/du)u^{p/q} = (p/q)u^{(p/q)-1}$. Hence

$$\frac{dy}{dx} = \frac{p}{q}u^{(p/q)-1}\frac{du}{dx}$$

and we're done. ■

Exercises 3.6

Find dy/dx in Exercises 1–18.

1. $y = x^{9/4}$

2. $y = x^{-3/5}$

3. $y = \sqrt[3]{x}$

4. $y = \sqrt[4]{x}$

5. $y = (2x + 5)^{-1/2}$

6. $y = (1 - 6x)^{2/3}$

7. $y = x\sqrt{x^2 + 1}$

8. $y = \dfrac{x}{\sqrt{x^2 + 1}}$

9. $x^2y + xy^2 = 6$

10. $x^3 + y^3 = 18xy$

11. $2xy + y^2 = x + y$

12. $x^3 - xy + y^3 = 1$

13. $x^2y^2 = x^2 + y^2$

14. $(3x + 7)^2 = 2y^3$

15. $y^2 = \dfrac{x - 1}{x + 1}$

16. $x^2 = \dfrac{x - y}{x + y}$

17. $y = \sqrt{1 - \sqrt{x}}$

18. $y = 3(2x^{-1/2} + 1)^{-1/3}$

Find dy/dx in Exercises 19–26.

19. $y = \sqrt{1 + \cos 2x}$

20. $y = \sqrt{\sec 2x}$

21. $y = 3(\csc x)^{3/2}$

22. $y = [\sin(x + 5)]^{5/4}$

23. $x = \tan y$

24. $x = \sin y$

25. $x + \tan(xy) = 0$

26. $x + \sin y = xy$

In Exercises 27–32, use implicit differentiation to find dy/dx and then d^2y/dx^2.

27. $x^2 + y^2 = 1$

28. $x^{2/3} + y^{2/3} = 1$

29. $y^2 = x^2 + 2x$

30. $y^2 + 2y = 2x + 1$

31. $y + 2\sqrt{y} = x$

32. $xy + y^2 = 1$

In Exercises 33–36, find the lines that are (a) tangent and (b) normal to the curve at the given point.

33. $x^2 + xy - y^2 = 1, \ (2, 3)$

34. $x^2 + y^2 = 25, \ (3, -4)$

35. $x^2y^2 = 9, \ (-1, 3)$

36. $y^2 - 2x - 4y - 1 = 0, \ (-2, 1)$

37. Find the two points where the curve $x^2 + xy + y^2 = 7$ crosses the x-axis, and show that the tangents to the curve at these points are parallel. What is the common slope of these tangents?

38. Find points on the curve $x^2 + xy + y^2 = 7$ (a) where the tangent is parallel to the x-axis and (b) where the tangent is parallel to the y-axis. (In the latter case, dy/dx is not defined, but dx/dy is. What value does dx/dy have at these points?)

39. Assume that the equation $2xy + \pi \sin y = 2\pi$ defines y as a differentiable function of x. Find dy/dx when $x = 1$ and $y = \pi/2$.

40. Find an equation for the tangent to the curve $x \sin 2y = y \cos 2x$ at the point $(\pi/4, \pi/2)$.

41. *The Eight Curve.* Find the slopes of the figure-eight–shaped curve

$$y^4 = y^2 - x^2$$

at the two points shown on the graph.

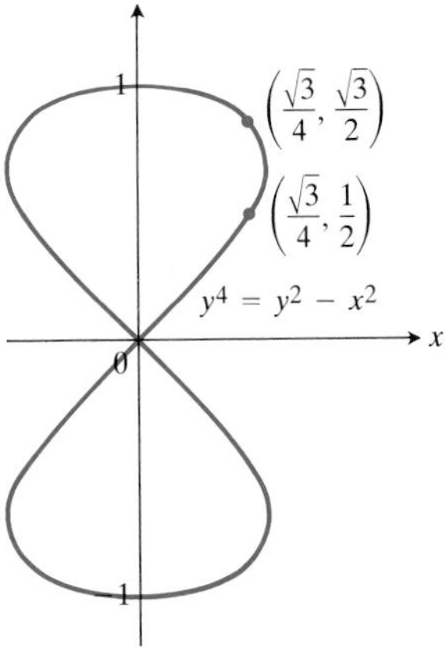

42. *The Cissoid of Diocles (dates from about 200 BC).* Find equations for the tangent and normal to the Cissoid of Diocles,

$$y^2(2 - x) = x^3,$$

at the point (1, 1).

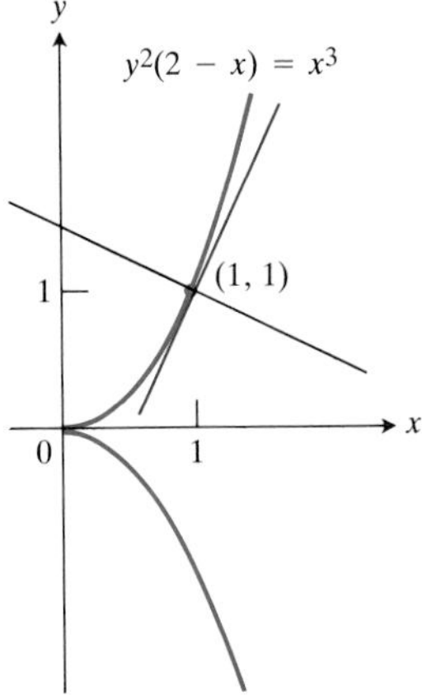

43. Which of the following could be true if $f''(x) = x^{-1/3}$?

a) $f(x) = \frac{3}{2}x^{2/3} - 3$ b) $f(x) = \frac{9}{10}x^{5/3} - 7$

c) $f'''(x) = -\frac{1}{3}x^{-4/3}$ d) $f'(x) = \frac{3}{2}x^{2/3} + 6$

44. *Orthogonal Curves.* Two curves are *orthogonal* at a point of intersection if their tangents there cross at right angles. Show that the curves $2x^2 + 3y^2 = 5$ and $y^2 = x^3$ are orthogonal at (1, 1) and (1, −1).

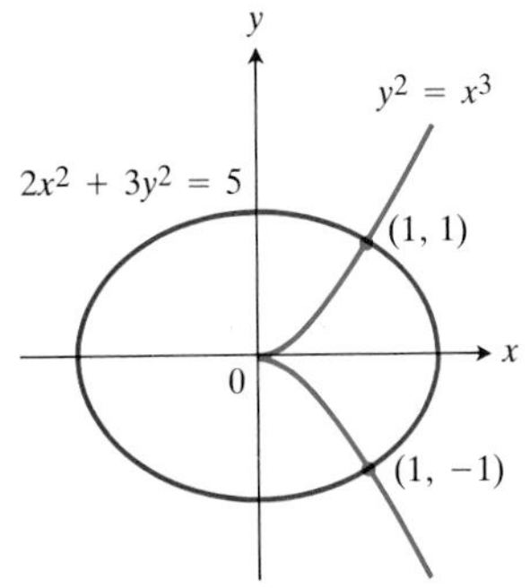

45. The position of a body moving along a coordinate line at time t is $s = \sqrt{1 + 4t}$, with s in meters and t in seconds. Find the body's velocity and acceleration when $t = 6$ sec.

46. The velocity of a falling body is $v = k\sqrt{s}$ meters per second (k a constant) at the instant the body has fallen s meters from its starting point. Show that the body's acceleration is constant.

3.7 Linear Approximations and Differentials

Sometimes we can approximate complicated functions with simpler ones that give the accuracy we want for specific applications without being so hard to work with. It is important to know how to do this, and in this section we study the simplest of the useful approximations. For reasons that will be clear in a moment, the approximation is called a linearization.

We also introduce a new symbol, dx, for an increment in a variable x. This symbol is called the differential of x. In the physical sciences, it is used more frequently than Δx. In mathematics, differentials are used to estimate changes in function values, as we shall see toward the end of this section.

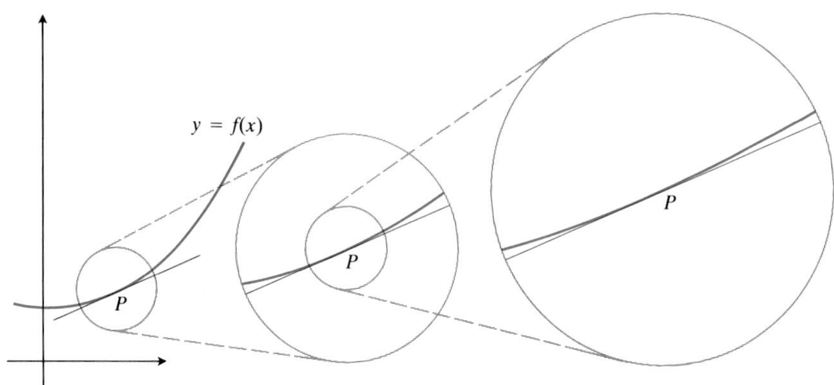

3.42 Successive enlargements show a close fit between a curve and its tangent line.

Linearizations Are Linear Replacement Formulas

As you can see in Fig. 3.42, the tangent to a curve $y = f(x)$ lies close to the curve near the point of tangency. For a brief interval to either side, the y-values along the tangent line give good approximations to the y-values on the curve. Therefore, to simplify the expression for the function near this point, we propose to replace the formula for f over this interval by the formula for its tangent line.

In the notation of Fig. 3.43, the tangent passes through the point $P(a, f(a))$ with slope $f'(a)$, so its point–slope equation is

$$y - f(a) = f'(a)(x - a)$$

or

$$y = f(a) + f'(a)(x - a). \tag{1}$$

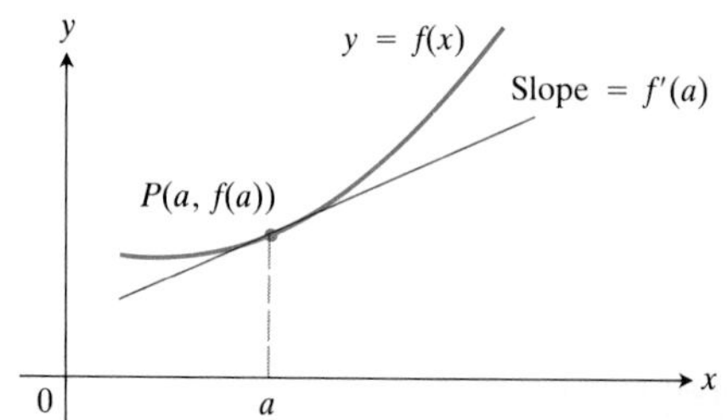

3.43 The equation of the tangent line is $y = f(a) + f'(a)(x - a)$.

Thus, the tangent line is the graph of the function

$$L(x) = f(a) + f'(a)(x - a). \tag{2}$$

For as long as the line remains close to the graph of f, $L(x)$ will give a good approximation to $f(x)$.

DEFINITIONS

Linearization and Standard Linear Approximation

If $y = f(x)$ is differentiable at $x = a$, then

$$L(x) = f(a) + f'(a)(x - a) \tag{3}$$

is the **linearization** of f at a. The approximation

$$f(x) \approx L(x)$$

is the **standard linear approximation** of f at a.

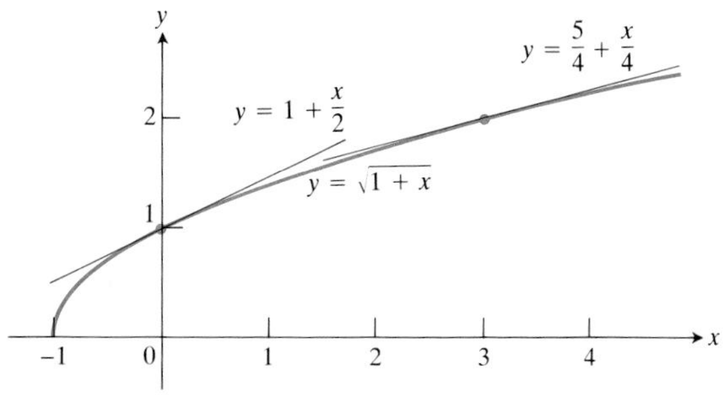

3.44 The graph of $y = \sqrt{1 + x}$ and its linearizations at $x = 0$ and $x = 3$.

EXAMPLE 1 Find the linearization of $f(x) = \sqrt{1 + x}$ at $x = 0$.

Solution We evaluate Eq. (3) for $f(x) = \sqrt{1 + x}$ and $a = 0$.

The derivative of f is

$$f'(x) = \frac{1}{2}(1 + x)^{-1/2} = \frac{1}{2\sqrt{1 + x}}.$$

Its value at $x = 0$ is $1/2$. We substitute this along with $a = 0$ and $f(0) = 1$ into Eq. (3):

$$L(x) = f(a) + f'(a)(x - a) = 1 + \frac{1}{2}(x - 0) = 1 + \frac{x}{2}.$$

The linearization of $\sqrt{1 + x}$ at $x = 0$ is $L(x) = 1 + \frac{x}{2}$. See Fig. 3.44. ■

In case you are wondering how close the approximation

$$\sqrt{1 + x} \approx 1 + \frac{x}{2}$$

really is, we can try a few values with a calculator:

$$\sqrt{1.2} \approx 1 + \frac{0.2}{2} = 1.10 \qquad \text{(accurate to 2 decimals)},$$

$$\sqrt{1.05} \approx 1 + \frac{0.05}{2} = 1.025 \qquad \text{(accurate to 3 decimals)},$$

$$\sqrt{1.005} \approx 1 + \frac{0.005}{2} = 1.00250 \quad \text{(accurate to 5 decimals)}.$$

The approximation becomes more accurate as we move toward the center, $x = 0$, and less accurate as we move away. As Fig. 3.44 suggests, the approximation will probably be too crude to be useful if we move out as far, say, as $x = 3$. To approximate $\sqrt{1 + x}$ near $x = 3$, we had best find its linearization at $x = 3$.

EXAMPLE 2 Find the linearization of $f(x) = \sqrt{1 + x}$ at $x = 3$.

Solution We evaluate Eq. (3) for $f(x) = \sqrt{1 + x}$, $f'(x) = 1/(2\sqrt{1 + x})$, and $a = 3$. With

$$f(3) = 2, \; f'(3) = \frac{1}{2\sqrt{1 + 3}} = \frac{1}{4},$$

Eq. (3) gives

$$L(x) = 2 + \frac{1}{4}(x - 3) = 2 + \frac{x}{4} - \frac{3}{4} = \frac{5}{4} + \frac{x}{4}.$$

Thus, near $x = 3$,

$$\sqrt{1 + x} \approx \frac{5}{4} + \frac{x}{4}.$$

■

At $x = 3.2$, the linearization we just obtained gives

$$\sqrt{1 + x} = \sqrt{1 + 3.2} \approx \frac{5}{4} + \frac{3.2}{4} = 1.250 + 0.800 = 2.050,$$

which differs from $\sqrt{4.2} = 2.04939$ by less than one thousandth. The linearization from Example 1 gives

$$\sqrt{1 + x} = \sqrt{1 + 3.2} \approx 1 + \frac{3.2}{2} = 1 + 1.6 = 2.6,$$

a result that is off by more than 25%. The linearization at $x = 3$ is obviously the one to use for values of $\sqrt{1 + x}$ near 3.

Do not be misled by our calculations here into thinking that whatever we do with a linearization is better done with a calculator. In practice, we would never use a linearization to find the value of a particular square root. That is not what linearizations are for. The utility of the linearizations in Examples 1 and 2 lies in their ability to replace the complicated formula $\sqrt{1 + x}$ by a simpler formula. If we have to work with $\sqrt{1 + x}$ for values of x close to 0, and can tolerate the small amount of error involved, we can safely work with $1 + (x/2)$ instead. Of course, we then need to know just how much error there really is. We shall look at that in a moment but the full answer won't come until Chapter 9.

EXAMPLE 3 The most important linearization for replacing roots and powers is

$$(1 + x)^k \approx 1 + kx \quad \text{(any number } k\text{).} \tag{4}$$

(See Exercise 22.) The approximation is good for values of x near zero. For instance, when x is numerically small,

$$\begin{aligned}
\sqrt{1 + x} &\approx 1 + \frac{x}{2} \\
\frac{1}{1 - x} &= (1 - x)^{-1} \approx 1 + (-1)(-x) = 1 + x \\
\sqrt[3]{1 + 5x^4} &= (1 + 5x^4)^{1/3} \approx 1 + \frac{1}{3}(5x^4) = 1 + \frac{5}{3}x^4 \\
\frac{1}{\sqrt{1 - x^2}} &= (1 - x^2)^{-1/2} \approx 1 + \left(-\frac{1}{2}\right)(-x^2) = 1 + \frac{1}{2}x^2.
\end{aligned} \tag{5}$$

■

Trigonometric functions have delightfully simple linearizations at the origin.

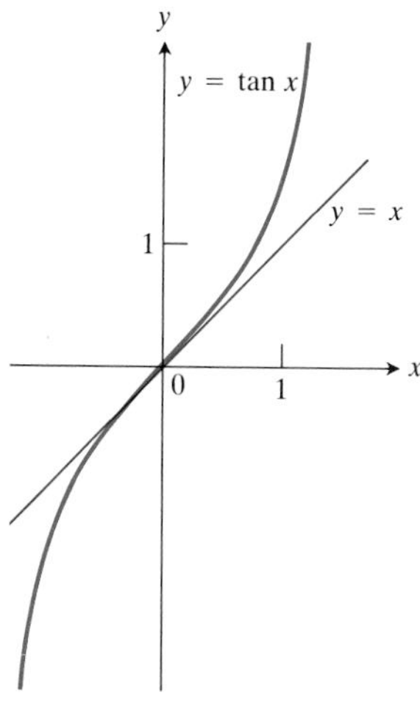

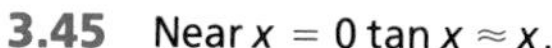

3.45 Near $x = 0 \tan x \approx x$.

EXAMPLE 4 Find the linearization of $f(x) = \tan x$ at $x = 0$.

Solution We use the equation

$$L(x) = f(a) + f'(a)(x - a)$$

with $f(x) = \tan x$ and $a = 0$. Since

$$f(0) = \tan(0) = 0, \qquad f'(0) = \sec^2(0) = 1,$$

we have $L(x) = 0 + 1(x - 0) = x$. Near $x = 0$ (Fig. 3.45),

$$\tan x \approx x.$$

■

In Exercises 13 and 14, you will be asked to derive the linearizations of $\sin x$ and $\cos x$ at $x = 0$.

Linearizations at $x = 0$

Function $f(x)$	Linearization $L(x)$
$\sin x$	x
$\cos x$	1
$\tan x$	x

Of course, trigonometric functions have linearizations at other points as well, any point where they are differentiable.

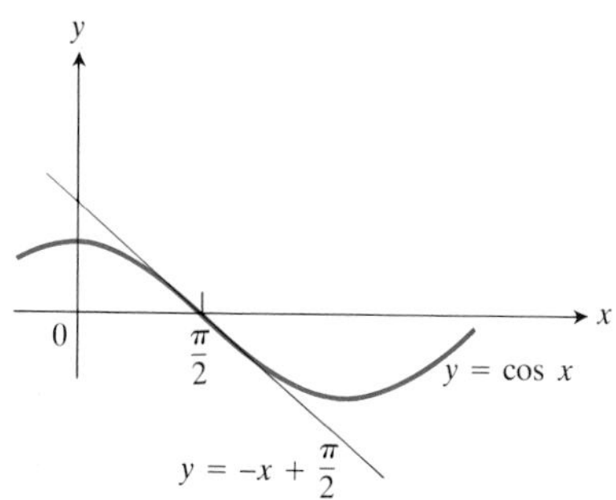

3.46 The graph of $y = \cos x$ and its linearization at $x = \pi/2$. Near $x = \pi/2$, $\cos x \approx x + (\pi/2)$.

EXAMPLE 5 Find the linearization of $f(x) = \cos x$ at $x = \pi/2$.

Solution We use the equation

$$L(x) = f(a) + f'(a)(x - a)$$

with $f(x) = \cos x$ and $a = \pi/2$. Since

$$f(\pi/2) = \cos(\pi/2) = 0, \qquad f'(\pi/2) = -\sin(\pi/2) = -1,$$

the linearization is

$$L(x) = 0 - 1\left(x - \frac{\pi}{2}\right) = -x + \frac{\pi}{2}.$$

See Fig. 3.46. ■

Estimating Change with Differentials

Suppose we know the value of a differentiable function $f(x)$ at a particular point x_0 and want to predict how much this value will change if we move nearby to the point $x_0 + h$. If h is small, f and its linearization L at x_0 will change by nearly the same amount. Since the values of L are always simple to calculate, calculating the change in L gives a practical way to estimate the change in f.

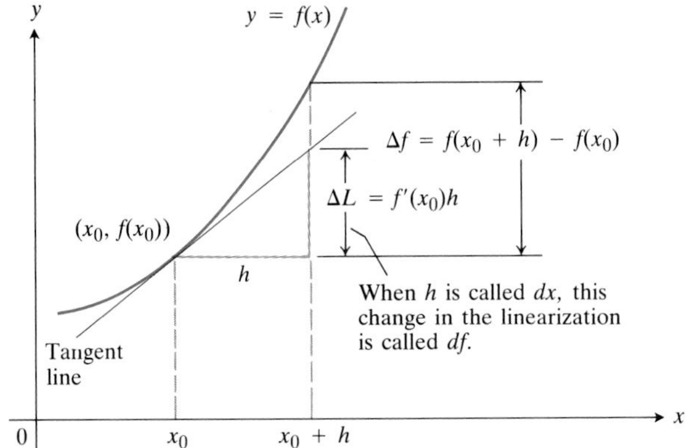

3.47 If h is small, the change in the linearization of f is nearly the same as the change in f.

In the notation of Fig. 3.47, the change in f is

$$\Delta f = f(x_0 + h) - f(x_0).$$

The corresponding change in L is

$$\begin{aligned} \Delta L &= L(x_0 + h) - L(x_0) \\ &= f(x_0) + f'(x_0)[(x_0 + h) - x_0] - f(x_0) \\ &= f'(x_0)h. \end{aligned} \tag{6}$$

The formula for Δf is usually as hard to work with as the formula for f. The formula for ΔL, however, is always simple to work with. As you can see, the change in L is just a constant times h.

The change $\Delta L = f'(x_0)h$ is usually described with the more suggestive notation

$$df = f'(x_0)\,dx, \tag{7}$$

in which df denotes the change in the linearization of f that results from the change dx in x. We call dx the **differential** of x and df the corresponding **differential** of f.

If $y = f(x)$ and we divide both sides of the equation $dy = f'(x)\,dx$ by dx, we obtain the familiar equation

$$\frac{df}{dx} = f'(x).$$

This equation now says that we may regard the derivative df/dx as a quotient of differentials. In many calculations, it is convenient to be able to think this way. For example, in writing the Chain Rule as

$$\frac{dy}{dx} = \frac{dy}{du}\frac{du}{dx},$$

we can think of the derivatives on the right as quotients in which the du's cancel to produce the fraction on the left. This gives a quick check on whether we remembered the rule correctly.

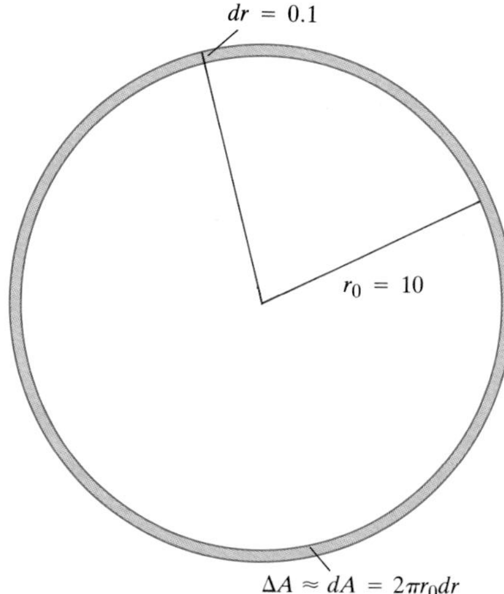

3.48 When dr is small compared to r_0, as it is when $dr = 0.1$ and $r_0 = 10$, the differential $dA = 2\pi r_0\, dr$ gives a good estimate of ΔA. See Example 6.

EXAMPLE 6 The radius of a circle increases from an initial value of $r_0 = 10$ by an amount $dr = 0.1$ (Fig. 3.48). Estimate the corresponding increase in the circle's area $A = \pi r^2$ by calculating dA. Compare dA with the true change ΔA.

Solution To calculate dA, we apply Eq. (7) to the function $A = \pi r^2$:

$$dA = A'(r_0)\, dr = 2\pi r_0 dr.$$

We then substitute the values $r_0 = 10$ and $dr = 0.1$:

$$dA = 2\pi(10)(0.1) = 2\pi.$$

The estimated change is 2π square units.

A direct calculation of ΔA gives

$$\Delta A = \pi(10.1)^2 - \pi(10)^2 = (102.01 - 100)\pi = \underbrace{2\pi}_{dA} + \underbrace{0.01\pi}_{\text{error}}.$$

The error in the estimate dA is 0.01π square units. As a percentage of the circle's original area, the error is quite small, as we can see from the following calculation:

$$\frac{\text{Error}}{\text{Original area}} = \frac{0.01\pi}{100\pi} = 0.01\%.$$ ■

Absolute, Relative, and Percentage Change

What is the difference again between Δf and df? The increment Δf is the change in f; the differential df is the change in the linearization of f. Unlike Δf, the differential df is always simple to calculate, and it gives a good estimate of Δf when the change in x is small.

As we move from x_0 to a nearby point, we can describe the corresponding change in the value of f in three ways:

	True	Estimate
Absolute change	Δf	df
Relative change	$\dfrac{\Delta f}{f(x_0)}$	$\dfrac{df}{f(x_0)}$
Percentage change	$\dfrac{\Delta f}{f(x_0)} \times 100$	$\dfrac{df}{f(x_0)} \times 100$

EXAMPLE 7 Estimate the percentage change that will occur in the are of a circle if its radius increases from $r_0 = 10$ units to 10.1 units.

Solution From the preceding table we have

$$\text{Estimated percentage change} = \frac{dA}{A(r_0)} \times 100.$$

With $dA = 2\pi$ (from Example 6) and $A(r_0) = 100\pi$, the formula gives

$$\frac{dA}{A(r_0)} \times 100 = \frac{2\pi}{100\pi} \times 100 = 2\%.$$

NASA

3.49 If we underestimated the radius of the earth by 528 ft during a calculation of the earth's surface area, we would leave out an area the size of the state of Maryland. (See Example 9.)

EXAMPLE 8 Suppose the earth were a perfect sphere and we determined its radius to be 3959 ± 0.1 miles. What effect would the tolerance of ± 0.1 have on our estimate of the earth's surface area?

Solution The surface area of a sphere of radius r is $S = 4\pi r^2$. The uncertainty in the calculation of S that arises from measuring r with a tolerance of dr miles is about

$$dS = \left(\frac{dS}{dr}\right) dr = 8\pi r\, dr.$$

With $r = 3959$ and $dr = 0.1$,

$$dS = 8\pi(3959)(0.1) = 9950 \text{ square miles}$$

to the nearest square mile, which is about the area of the state of Maryland (Fig. 3.49). In absolute terms this might seem like a large error. However, 9950 mi^2 is a relatively small error when compared to the calculated surface area of the earth:

$$\frac{dS}{\text{Calculated } S} = \frac{9950}{4\pi(3959)^2} \approx \frac{9950}{196{,}961{,}284} \approx .005\%.$$

EXAMPLE 9 About how accurately should we measure the radius r of a sphere to calculate the surface area $S = 4\pi r^2$ within 1% of its true value?

Solution We want any inaccuracy in our measurement to be small enough to make the corresponding increment ΔS in the surface area satisfy the inequality

$$|\Delta S| \le \frac{1}{100} S = \frac{4\pi r^2}{100}. \tag{8}$$

We replace ΔS in this inequality with

$$dS = \left(\frac{dS}{dr}\right) dr = 8\pi r\, dr.$$

This gives

$$|8\pi r\, dr| \le \frac{4\pi r^2}{100} \quad \text{and} \quad |dr| \le \frac{1}{8\pi r} \cdot \frac{4\pi r^2}{100} = \frac{1}{2}\frac{r}{100}.$$

We should measure the radius with an error dr that is no more than 0.5% of the true value.

EXAMPLE 10 *Unclogging arteries.* In the late 1830s, the French physiologist Jean Poiseuille ("pwa·*zoy*") discovered the formula we use today to predict how much the radius of a partially clogged artery has to be expanded to restore normal flow. His formula,

$$V = kr^4, \tag{9}$$

says that the volume V of fluid flowing through a small pipe or tube in a

Angiography—An opaque dye is injected into a partially blocked artery to make the inside visible under x-rays. This reveals the location and severity of the blockage.

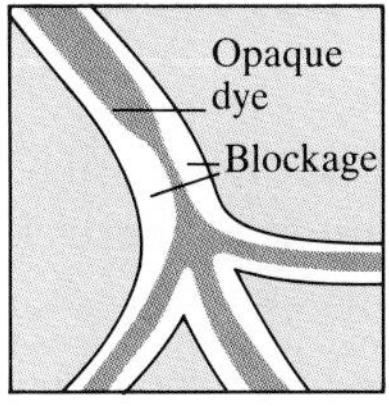

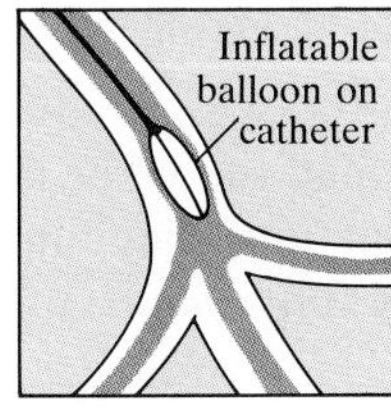

Angioplasty—A ballon-tipped catheter is inflated inside the artery to widen it at the blockage site.

unit of time at a fixed pressure is a constant times the fourth power of the tube's radius r. How will a 10% increase in r affect V?

Solution The differentials of r and V are related by the equation

$$dV = \frac{dV}{dr}\,dr = 4kr^3dr.$$

Hence,

$$\frac{dV}{V} = \frac{4kr^3dr}{kr^4} = 4\frac{dr}{r}. \quad \text{(Dividing by } V = kr^4\text{)}$$

The relative change in V is four times the relative change in r, so a 10% increase in r will produce a 40% increase in the flow. ■

Sensitivity

The equation $df = f'(x)\,dx$ tells how sensitive the output of f is to a change in input at different values of x. The larger the value of f' at x, the greater is the effect of a given change dx.

EXAMPLE 11 You want to calculate the height of a bridge from the equation $s = 16t^2$ by timing how long it takes a heavy stone you drop to splash into the water below. How sensitive will your calculation be to a 0.1 sec error in measuring the time?

Solution The size of ds in the equation

$$ds = 32t\,dt$$

depends on how big t is. If $t = 2$ sec, the error caused by $dt = 0.1$ is only

$$ds = 32(2)(0.1) = 6.4 \text{ ft.}$$

Three seconds later, at $t = 5$ sec, the error caused by the same dt is

$$ds = 32(5)(0.1) = 16 \text{ ft.}$$ ■

The Error in the Approximation $\Delta f \approx f'(a)\Delta x$

How well does the quantity $f'(a)\Delta x$ estimate the true increment $\Delta f = f(a + \Delta x) - f(a)$? We measure the error by subtracting one from the other:

$$\begin{aligned}\text{Approximation error} &= \Delta f - f'(a)\Delta x \\ &= f(a + \Delta x) - f(a) - f'(a)\Delta x \\ &= \underbrace{\left(\frac{f(a + \Delta x) - f(a)}{\Delta x} - f'(a)\right)}_{\text{call this part } \epsilon} \Delta x, \qquad (10) \\ &= \epsilon \cdot \Delta x.\end{aligned}$$

As $\Delta x \to 0$, the difference quotient

$$\frac{f(a + \Delta x) - f(a)}{\Delta x}$$

approaches $f'(a)$ (remember the definition of $f'(a)$), so the quantity in parentheses becomes a very small number (which is why we called it ϵ). In fact,

$$\epsilon \to 0 \quad \text{as} \quad \Delta x \to 0.$$

Thus, when Δx is small, the approximation error $\epsilon \Delta x$ is smaller still.

$$\underbrace{\Delta f}_{\substack{\text{true}\\ \text{change}}} = \underbrace{f'(a)\Delta x}_{\substack{\text{estimated}\\ \text{change}}} + \underbrace{\epsilon \Delta x}_{\text{error}} \tag{11}$$

While we do not know exactly how small the error is and will not be able to make much progress on this front until later in Chapter 9, there is something worth noting here, namely the *form* taken by the equation.

> If $y = f(x)$ is differentiable at $x = a$, and x changes from a to $a + \Delta x$, the change Δy in f is given by an equation of the form
>
> $$\Delta y = f'(a)\Delta x + \epsilon \Delta x \tag{12}$$
>
> in which $\epsilon \to 0$ as $\Delta x \to 0$.

Surprising as it may seem, just knowing the form of Eq. (12) enables us to bring the proof of the Chain Rule to a successful conclusion. You can find out what we mean by turning to Appendix 3.

Derivatives in Differential Notation

Every formula like

$$\frac{d(u + v)}{dx} = \frac{du}{dx} + \frac{dv}{dx}$$

has a corresponding differential formula like

$$d(u + v) = du + dv$$

that comes from multiplying both sides by dx.

TABLE 3.2
Formulas for Differentials

$dc = 0$	$d(\sin u) = \cos u \, du$
$d(cu) = c\,du$	$d(\cos u) = -\sin u \, du$
$d(u + v) = du + dv$	$d(\tan u) = \sec^2 u \, du$
$d(uv) = u\,dv + v\,du$	$d(\cot u) = \csc^2 u \, du$
$d\left(\frac{u}{v}\right) = \frac{v\,du - u\,dv}{v^2}$	$d(\sec u) = \sec u \tan u \, du$
$d(u^n) = nu^{n-1}du$	$d(\csc u) = -\csc u \cot u \, du$

To find dy when y is a differentiable function of x, we may either find dy/dx and multiply by dx or use one of the formulas in Table 3.2.

EXAMPLE 12

a) $d(3x^2 - 6) = 6x\,dx$

b) $d(\cos 3x) = -\sin 3x\,d(3x) = -3\sin 3x\,dx$

c)
$$d\frac{x}{x+1} = \frac{(x+1)\,dx - x\,d(x+1)}{(x+1)^2} = \frac{x\,dx + dx - x\,dx}{(x+1)^2} = \frac{dx}{(x+1)^2}$$

Notice that a differential on one side of an equation always calls for a differential on the other side of the equation. Thus, we never have $dy = 3x^2$ but, instead, $dy = 3x^2 dx$.

Exercises 3.7

In Exercises 1–6, find the linearization $L(x)$ of $f(x)$ at $x = a$.

1. $f(x) = x^4$ at $x = 1$
2. $f(x) = x^{-1}$ at $x = 2$
3. $f(x) = x^3 - x$ at $x = 1$
4. $f(x) = x^3 - 2x + 3$ at $x = 2$
5. $f(x) = \sqrt{x}$ at $x = 4$
6. $f(x) = \sqrt{x^2 + 9}$ at $x = -4$

You want linearizations that will replace the functions in Exercises 7–12 over intervals that include the given points x_0. To make your subsequent work as simple as possible, you want to center each linearization not at x_0 but at a nearby integer $x = a$ at which the given function and its derivative are easy to evaluate. What linearization do you use in each case?

7. $f(x) = x^2 + 2x,\ x_0 = 0.1$
8. $f(x) = x^{-1},\ x_0 = 0.6$
9. $f(x) = 2x^2 + 4x - 3,\ x_0 = -0.9$
10. $f(x) = 1 + x,\ x_0 = 8.1$
11. $f(x) = \sqrt[3]{x},\ x_0 = 8.5$
12. $f(x) = \dfrac{x}{x+1},\ x_0 = 1.3$

In Exercises 13–18, find the linearization $L(x)$ of the given function at $x = a$. Then graph f and L together near $x = a$.

13. $f(x) = \sin x$ at $x = 0$
14. $f(x) = \cos x$ at $x = 0$
15. $f(x) = \sin x$ at $x = \pi$
16. $f(x) = \cos x$ at $x = -\pi/2$
17. $f(x) = \tan x$ at $x = \pi/4$
18. $f(x) = \sec x$ at $x = \pi/4$
19. Use the formula $(1 + x)^k \approx 1 + kx$ to find linear approximations of the following functions for values of x near zero.

    a) $(1 + x)^2$
    b) $\dfrac{1}{(1 + x)^5}$
    c) $\dfrac{2}{1 - x}$
    d) $(1 - x)^6$
    e) $3(1 + x)^{1/3}$
    f) $\dfrac{1}{\sqrt{1 + x}}$

20. *Faster than a calculator.* Use the approximation $(1 + x)^k \approx 1 + kx$ to estimate

    a) $(1.002)^{100}$
    b) $\sqrt[3]{1.009}$

21. Find the linearization of $f(x) = \sqrt{x + 1} + \sin x$ at $x = 0$. How is it related to the individual linearizations for $\sqrt{x + 1}$ and $\sin x$?
22. We know from the Power Rule that the equation
$$\frac{d}{dx}(1 + x)^k = k(1 + x)^{k-1}$$
holds for every rational number k. In Chapter 7, we shall show that it holds for every irrational number as well. Assuming this result for now, verify Eq. (4) by showing that the linearization of $f(x) = (1 + x)^k$ at $x = 0$ is $L(x) = 1 + kx$ for any number k.

23. CALCULATOR In Section 1.7, we promised to explain what happened when you took successive square roots of 2 by entering 2 in your calculator and pressing the square-root key repeatedly. As you may recall, or as you will find if you try it now, each keypress divides the decimal part of the display approximately in half. The explanation comes from the fact that the linearization of $\sqrt{1+x}$ is $1 + (x/2)$. The x is the decimal part of the display $(1.x)$ and each new square root is about $1 + (x/2)$.

If you have not done so already, enter 2 into your calculator and press $\boxed{\sqrt{\ }}$ repeatedly to see what happens.

24. CALCULATOR If you have not already done so, turn to Section 1.7 and do Exercise 14 there. If you have a key that will enable you to calculate tenth roots, do Exercise 15 as well.

In Exercises 25–30, each function $f(x)$ changes value when x changes from x_0 to $x_0 + dx$. Find

a) the change $\Delta f = f(x_0 + dx) - f(x_0)$;

b) the value of the estimate $df = f'(x_0)dx$; and

c) the error $|\Delta f - df|$.

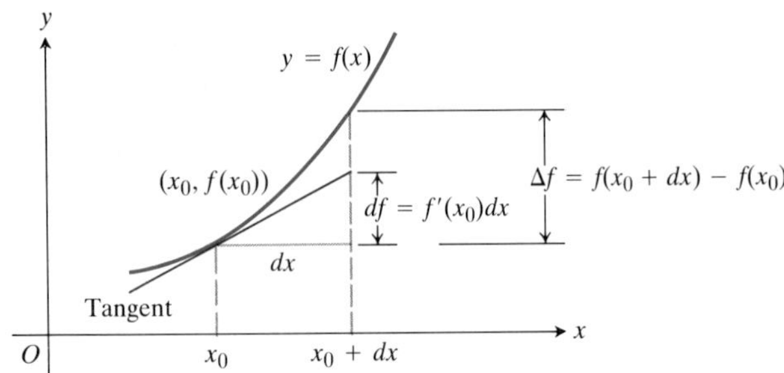

25. $f(x) = x^2 + 2x,\ \ x_0 = 0,\ dx = 0.1$

26. $f(x) = 2x^2 + 4x - 3,\ x_0 = -1,\ dx = 0.1$

27. $f(x) = x^3 - x,\ x_0 = 1,\ dx = 0.1$

28. $f(x) = x^4,\ x_0 = 1,\ dx = 0.1$

29. $f(x) = x^{-1},\ x_0 = 0.5,\ dx = 0.1$

30. $f(x) = x^3 - 2x + 3,\ x_0 = 2,\ dx = 0.1$

In Exercises 31–36, write a differential formula that estimates the given change in volume or surface area.

31. The change in the volume $V = (4/3)\pi r^3$ of a sphere when the radius changes from r_0 to $r_0 + dr$.

32. The change in the surface area $S = 4\pi r^2$ of a sphere when the radius changes from r_0 to $r_0 + dr$.

33. The change in the volume $V = x^3$ of a cube when the edge lengths change from x_0 to $x_0 + dx$.

34. The change in the surface area $S = 6x^2$ of a cube when the edge lengths change from x_0 to $x_0 + dx$.

35. The change in the volume $V = \pi r^2 h$ of a right circular cylinder when the radius changes from r_0 to $r_0 + dr$ and the height does not change.

36. The change in the lateral surface area $S = 2\pi rh$ of a right circular cylinder when the height changes from h_0 to $h_0 + dh$ and the radius does not change.

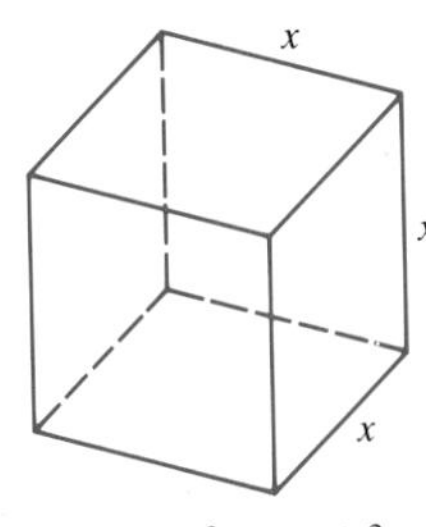

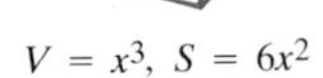
$V = x^3,\ S = 6x^2$

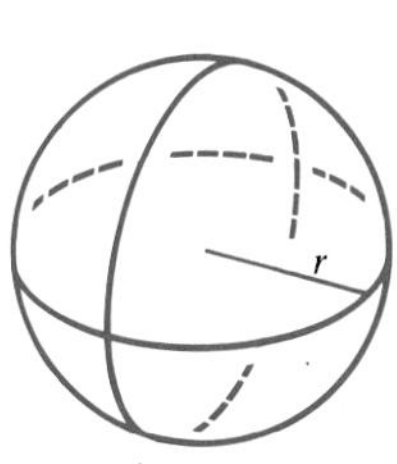

$V = \frac{4}{3}\pi r^3,\ S = 4\pi r^2$

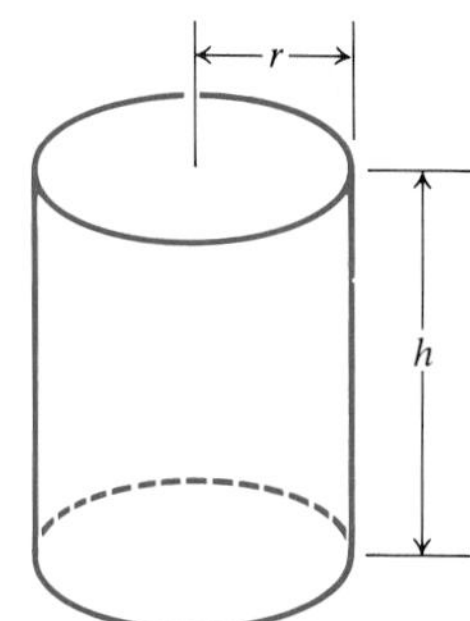

$V = \pi r^2 h,\ S = 2\pi rh$

37. The radius of a circle is increased from 2.00 to 2.02 m.

a) Estimate the resulting change in area.

b) Express the estimate in (a) as a percentage of the circle's original area.

38. The diameter of a tree was 10 in. During the following year, the circumference grew 2 in. About how much did the tree's diameter grow? the tree's cross section area?

39. The edge of a cube is measured as 10 cm with an error of 1%. The cube's volume is to be calculated from this measurement. Estimate the percentage error in the volume calculation.

40. About how accurately should you measure the side of a square to be sure of calculating the area within 2% of its true value?

41. The diameter of a sphere is measured as 100 ± 1 cm and the volume is calculated from this measurement. Estimate the percentage error in the volume calculation.

42. Estimate the allowable percentage error in measuring the diameter d of a sphere if the volume is to be calculated correctly to within 3%.

43. The height and radius of a right circular cylinder are equal, so the cylinder's volume is $V = \pi h^3$. The volume is to be calculated from a measurement of h and must be calculated with an error of no more than 1% of the true value. Find approximately the greatest error that can be tolerated in the measurement of h, expressed as a percentage of h.

44. a) About how accurately must the interior diameter of a 10-m-high cylindrical storage tank be measured to calculate the tank's volume to within 1% of its true value?

b) About how accurately must the tank's exterior diameter be measured to calculate the amount of paint it will take to paint the side of the tank within 5% of the true amount?

45. A manufacturer contracts to mint coins for the federal government. How much variation dr in the radius of the coins can be tolerated if the coins are to weigh within 1/1000 of their ideal weight? Assume that the thickness does not vary.

46. *Continuation of Example* By what percentage should r be increased to increase V by 50%?

47. *Continuation of Example 11.* Show that a 5% error in measuring t will cause about a 10% error in calculating $s(t)$ from the equation $s = 16t^2$.

48. *The effect of flight maneuvers on the heart.* The amount of work done by the heart's main pumping chamber, the left ventricle, is given by the equation

$$W = PV + \frac{Vpv^2}{2g},$$

where W is the work per unit time, P is the average blood pressure, V is the volume of blood pumped out during the unit of time, p is the density of the blood, v is the average velocity of the exiting blood, and g is the acceleration of gravity.

When P, V, p and v remain constant, W becomes a function of g and the equation takes the simplified form

$$W = a + \frac{b}{g} \quad (a, b \text{ constant}) \qquad (13)$$

As a member of NASA's medical team, you want to know how sensitive W is to apparent changes in g caused by flight maneuvers, and this depends on the initial value of g. As part of your investigation, you decide to compare the effect on W of a given change dg on the moon, where $g = 5.2$ ft/sec^2, with the effect the same change dg would have on Earth, where $g = 32$ ft/sec^2. You use Eq. (13) to find the ratio of dW_{moon} to dW_{Earth}. What do you find?

Using Linearizations to Solve Equations

49. Let $g(x) = \sqrt{x} + \sqrt{1+x} - 4$.

a) Find $g(3) < 0$ and $g(4) > 0$ to show (by the Intermediate Value Theorem, Section 2.4) that the equation $g(x) = 0$ has a solution between $x = 3$ and $x = 4$.

b) To estimate the solution of $g(x) = 0$, replace the square roots by their linearizations at $x = 3$ and solve the resulting linear equation.

c) **CALCULATOR** Check your estimate in the original equation.

50. Carry out the following steps to estimate the solution of $2 \cos x = \sqrt{1+x}$.

a) Let $f(x) = 2\cos x - \sqrt{1+x}$. Find $f(0) > 0$ and $f(\pi/2) < 0$ to show that $f(x)$ has a zero between 0 and $\pi/2$.

b) Find the linearizations of $\cos x$ at $x = \pi/4$ and $\sqrt{1+x}$ at $x = 0.69$.

c) **CALCULATOR** To estimate the solution of the original equation, replace $\cos x$ and $\sqrt{1+x}$ by their linearizations from (b) and solve the resulting linear equation for x. Check your estimate in the original equation.

Derivatives in Differential Form

In Exercises 51–62, find dy.

51. $y = x^3 - 3x$

52. $y = x\sqrt{1-x^2}$

53. $y = 2x/(1+x^2)$

54. $y = (3x^2 - 1)^{3/2}$

55. $y + xy - x = 0$

56. $xy^2 + x^2y - 4 = 0$

57. $y = \sin(5x)$

58. $y = \cos(x^2)$

59. $y = 4\tan(x/2)$

60. $y = \sec(x^2 - 1)$

61. $y = 3\csc(1 - (x/3))$

62. $y = 2\cot\sqrt{x}$

3.8 Newton's Method for Approximating Solutions of Equations

When exact formulas are not available for solving an equation $f(x) = 0$, we can turn to numerical techniques from calculus for approximating the solution we seek. One of these techniques is Newton's method or, as it is more accurately called, the Newton–Raphson method, discussed in this section. The method is based on the idea of using a tangent line to replace the graph of $y = f(x)$ near the points where f is zero. Once again, we see that linearization is the key to solving a practical problem.

If you have access to a computer, you can use the program at the end of this section to do the arithmetic. If not, you can still see how the technique works and the beginning exercises at the end of the section can still be worked by hand.

Algorithm, Recursion, and Iteration

It is customary to call a specified sequence of computational steps like the one in Newton's method an *algorithm.* When an algorithm proceeds by repeating a given set of steps over and over, using the answer from the previous step as the input for the next, the algorithm is called *recursive* and each repetition is called an *iteration.* Newton's method is one of the really fast recursive techniques for finding roots.

The Procedure for Newton's Method

1. Guess a first approximation to a root of the equation $f(x) = 0$. A graph of $y = f(x)$ will help.
2. Use the first approximation to get a second, the second to get a third, and so on. To go from the nth approximation x_n to the next approximation x_{n+1}, use the formula

$$x_{n+1} = x_n - \frac{f(x_n)}{f'(x_n)}, \qquad (1)$$

where $f'(x_n)$ is the derivative of f at x_n.

We first show how the method works and then go to the theory behind it.

In our first example we find decimal approximations to $\sqrt{2}$ by estimating the positive root of the equation $f(x) = x^2 - 2 = 0$.

EXAMPLE 1 Find the positive root of the equation $f(x) = x^2 - 2 = 0$.

Solution With $f(x) = x^2 - 2$ and $f'(x) = 2x$, Eq. (1) becomes

$$x_{n+1} = x_n - \frac{x_n^2 - 2}{2x_n}. \qquad (2)$$

To use our calculator efficiently, we rewrite Eq. (2) in a form that uses fewer arithmetic operations:

$$x_{n+1} = x_n - \frac{x_n^2 - 2}{2x_n} = \frac{x_n^2 + 2}{2x_n} = \frac{x_n}{2} + \frac{1}{x_n}. \qquad (3)$$

The equation

$$x_{n+1} = \frac{x_n}{2} + \frac{1}{x_n} \tag{4}$$

enables us to go from each approximation to the next by the following steps:

	Operation	Result	Example
$\boxed{x}$	Enter x	x	1.5
$\boxed{\frac{1}{x}}$ $\boxed{\text{STO}}$	Store the reciprocal	$\frac{1}{x}$	0.66667
$\boxed{\frac{1}{x}}$	Take the reciprocal of the display again	x	1.5
$\boxed{\div}$ $\boxed{2}$	Divide by 2	$\frac{x}{2}$	0.75
$\boxed{+}$ $\boxed{\text{RCL}}$ $\boxed{=}$	Add memory to display	$\frac{x}{2} + \frac{1}{x}$	1.41667

With the starting value $x_0 = 1$, we get the results in the first column of the following table. (To five decimal places, $\sqrt{2} = 1.41421$.)

	Error	Number of correct figures
$x_0 = 1$	-0.41421	1
$x_1 = 1.5$	$+0.08579$	1
$x_2 = 1.41667$	0.00245	3
$x_3 = 1.41422$	0.00001	5

Newton's method is the method used by most calculators to calculate roots because it converges so fast (more about this later). If the arithmetic in the table in Example 1 had been carried to 13 decimal places rather than 5, then going one step further would have given $\sqrt{2}$ correctly to more than 10 decimal places.

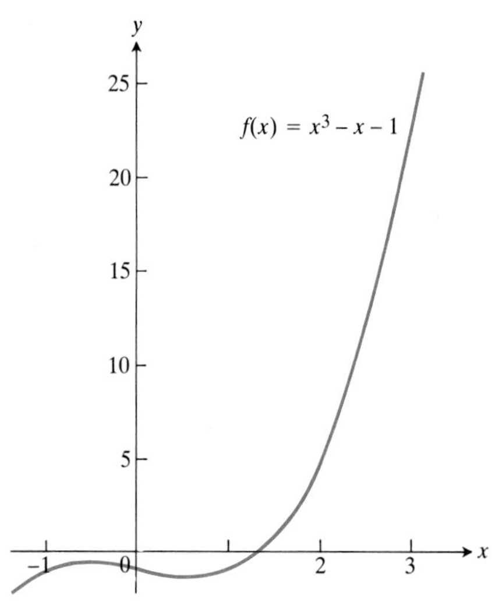

3.50 The graph of $f(x) = x^3 - x - 1$ crosses the x-axis just once, at a point between $x = 1$ and $x = 2$.

EXAMPLE 2 Find the x-coordinate of the point where the curve $y = x^3 - x$ crosses the horizontal line $y = 1$.

Solution The curve crosses the line when $x^3 - x = 1$ or $x^3 - x - 1 = 0$. When does $f(x) = x^3 - x - 1$ equal zero? The graph of f (Fig. 3.50) shows a single root, located between $x = 1$ and $x = 2$. We apply Newton's method to f with the starting value $x_0 = 1$. The results are displayed in Table 3.3 and Fig. 3.51.

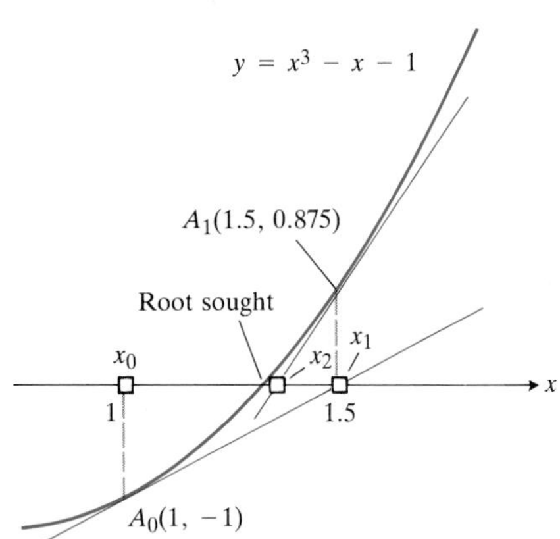

3.51 The first three x-values in Table 3.3.

TABLE 3.3
The result of applying Newton's method to $f(x) = x^3 - x - 1$ with $x_0 = 1$

n	x_n	$f(x_n)$	$f'(x_n)$	$x_{n+1} = x_n - \dfrac{f(x_n)}{f'(x_n)}$
0	1	-1	2	1.5
1	1.5	0.875	5.75	1.347826087
2	1.347826087	0.100682174	4.449905482	1.325200399
3	1.325200399	0.002058363	4.268468293	1.324718174
4	1.324718174	0.000000925	4.264634722	1.324717957
5	1.324717957	-5×10^{-10}	4.264632997	1.324717957

At $n = 5$ we come to the result $x_5 = x_4 = 1.324717957$. When $x_{n+1} = x_n$, Eq. (1) shows that $f(x_n) = 0$. Hence we have found a solution of $f(x) = 0$ to nine decimals, or so it appears. (Our calculator shows only ten figures, and we cannot guarantee the accuracy of the ninth decimal, though we believe it to be correct.) ■

In Fig. 3.52, we have indicated that the process in Example 2 might have started at the point $B(3, 23)$ on the curve, with $x_0 = 3$. Point B is quite far from the x-axis, but the tangent at B crosses the x-axis at about $C(2.11, 0)$, so x_1 is still an improvement over x_0. If we use Eq. (1) repeatedly as before, with $f(x) = x^3 - x - 1$ and $f'(x) = 3x^2 - 1$, we confirm the nine-place solution $x_6 = x_5 = 1.324717957$ in six steps.

The curve in Fig. 3.52 has a high turning point at $x = -1/\sqrt{3}$ and a low turning point at $x = +1/\sqrt{3}$. We would not expect good results from Newton's method if we were to start with x_0 between these points, but we can start any place to the right of $x = 1/\sqrt{3}$ and get the answer. It would not be very clever to do so, but we could even begin far to the right of B, for example with $x_0 = 10$. It takes a bit longer, but the process still converges to the same answer as before.

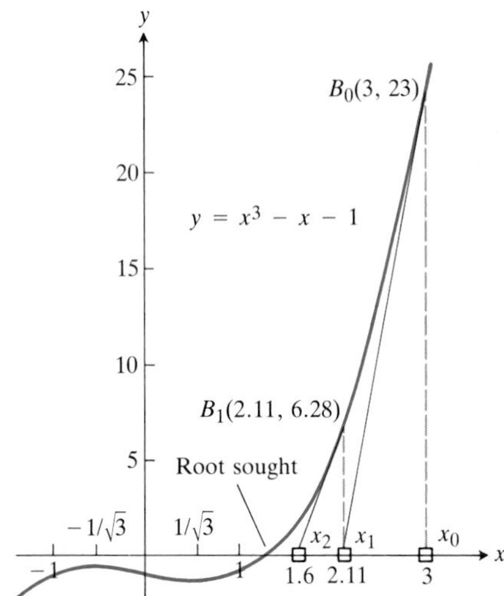

3.52 Any starting value x_0 to the right of $x = 1/\sqrt{3}$ will lead to the root.

What Is the Theory Behind the Method?

It is this: We use the tangent to approximate the graph of $y = f(x)$ near the point $P(x_n, y_n)$, where $y_n = f(x_n)$ is small, and we let x_{n+1} be the value of x where that tangent line crosses the x-axis. (We assume that the slope $f'(x_n)$ of the tangent is not zero.) The equation of the tangent is

$$y - y_n = f'(x_n)(x - x_n). \tag{5}$$

We put $y_n = f(x_n)$ and $y = 0$ into Eq. (5) and solve for x to get

$$x = x_n - \frac{f(x_n)}{f'(x_n)}.$$

See Fig. 3.53.

From another point of view, every time we replace the graph of f by one of its tangent lines we are replacing the function by one of its linearizations L. We then solve $L(x) = 0$ to estimate the solution of $f(x) = 0$.

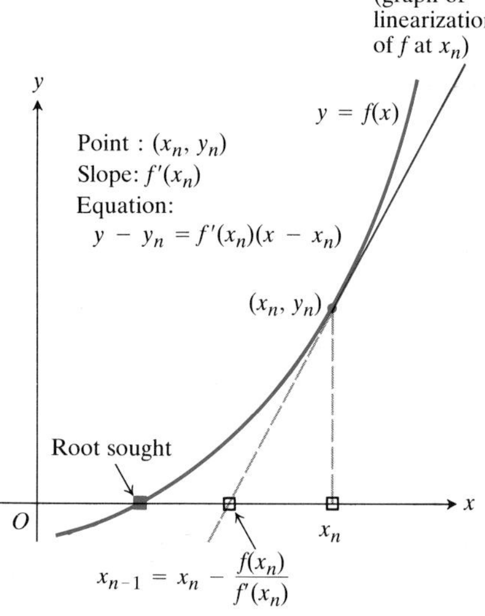

3.53 The geometry of the successive steps of Newton's method. From x_n we go up to the curve and follow the tangent line down to find x_{n+1}.

Strengths and Limitations of the Method

Newton's method does not work if $f'(x_n) = 0$. In that case, choose a new starting point. Of course, it may happen that $f(x) = 0$ and $f'(x) = 0$ have a common root. To detect whether this is so, we could first find the solutions of $f'(x) = 0$ and check the value of $f(x)$ at such places. Or we could graph f and f' together on a computer to look for places where the graphs might cross the x-axis together.

Newton's method does not always converge. For instance, if

$$f(x) = \begin{cases} \sqrt{x - r} & \text{for } x \geq r, \\ -\sqrt{r - x} & \text{for } x < r, \end{cases} \tag{6}$$

the graph will be like that shown in Fig. 3.54. If we begin with $x_0 = r - h$, we get $x_1 = r + h$, and successitve approximations go back and forth between these two values. No amount of iteration will bring us any closer to the root r than our first guess.

If Newton's method does converge, it converges to a root of $f(x)$. However, the method may converge to a root different from the expected one if the starting value is not close enough to the root sought. Figure 3.55 shows two ways this might happen.

When will Newton's method converge? A result from advanced calculus says that if the inequality

$$\left| \frac{f(x)f''(x)}{[f'(x)]^2} \right| < 1 \tag{7}$$

holds for all values of x in an interval about a root r of f, then the method will converge to r for any starting value x_0 in that interval. This is a *sufficient*, but not a necessary, condition. The method can (and does) converge in some cases when there is no interval about r in which the inequality (7) holds. Newton's method will always work if the curve $y = f(x)$ is convex

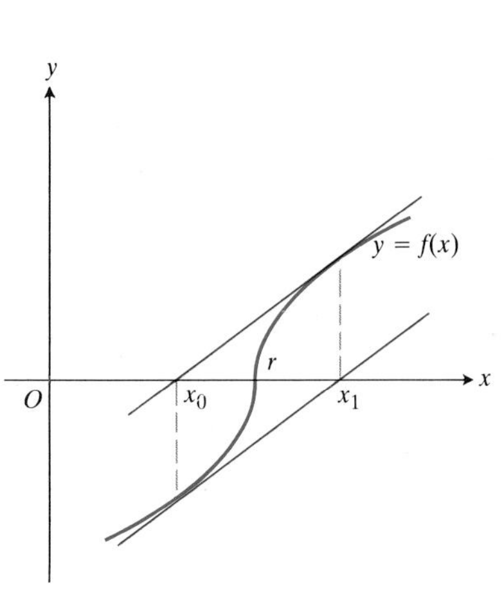

3.54 The graph of a function for which Newton's method fails to converge.

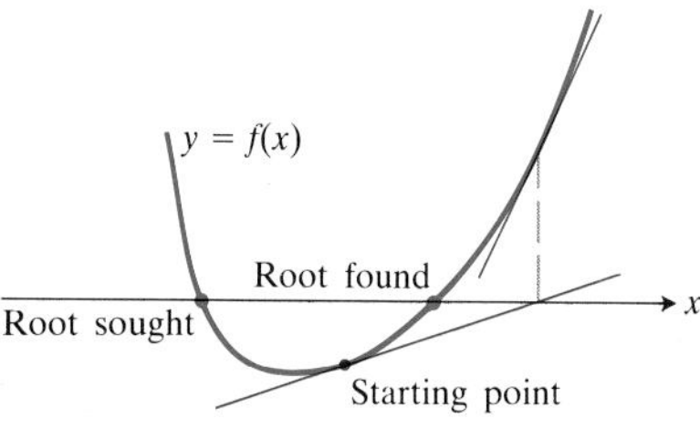

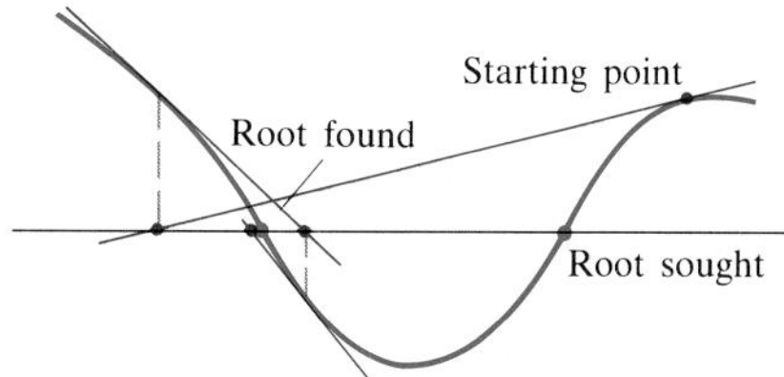

3.55 Newton's method may miss the root you want if you start too far away.

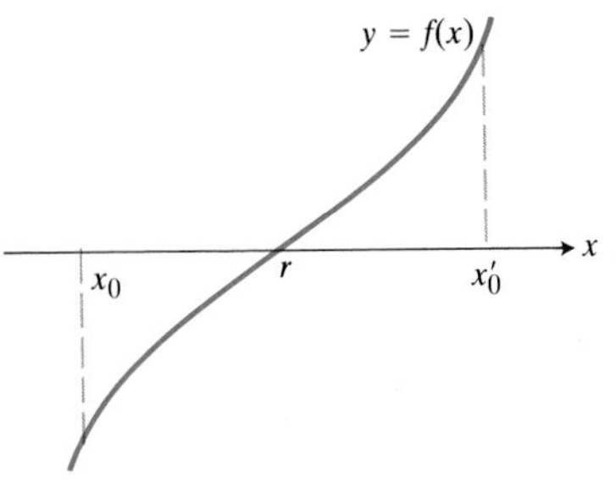

3.56 The curve $y = f(x)$ is convex toward the axis between x_0 and r and between x_0' and r. Newton's method will converge to r from either starting point.

("bulges") toward the axis in the interval between x_0 and the root sought. See Fig. 3.56.

The speed with which Newton's method converges to a root r is expressed by the advanced-calculus formula

$$|r - x_{n+1}| \le \frac{1}{2}\frac{|\max f''|}{|\min f'|}(r - x_n)^2, \tag{8}$$

where "max" and "min" refer to the maximum and minimum values f'' and f' have in an interval surrounding r. The formula says that the error at step $n + 1$ is no greater than a constant times the square of the error at step n. That might not seem like much to write home about, but think of what it says. If the constant $(1/2)|\max f''|/|\min f'|$ is less than or equal to 1, and $|r - x_n| \le 10^{-3}$, then $|r - x_{n+1}| \le 10^{-6}$. *In a single step,* the method takes us to within less than one millionth of the root. Little wonder that Newton's method, when it applies, is the method of choice for modern computers.

A Computer Program for Newton's Method

You may already be familiar with BASIC as a computer language, but if you are not, you will still be able to follow the next examples. For convenience, the steps in a program are usually labeled 10, 20, 30, and so on, leaving the numbers in between for corrections if we discover a flaw in our first attempt to write the program. The next example shows a program that uses Newton's method to estimate where the graphs of $y = \sin x$ and $y = x^2$ intersect. A quick sketch (Fig. 3.57) shows them to intersect at zero and again at some point near $x = 1$. That is the point we need Newton's method for.

EXAMPLE 3 Write and run a BASIC program based on Newton's method for estimating a solution of the equation $\sin x = x^2$.

Solution The calculations to be made are these:

1. Start with $x_0 = 1$.
2. We want to make the function $f(x) = \sin x - x^2$ small, zero if possible, and its derivative is $f'(x) = \cos x - 2x$.
3. Given any x_n, the next value of x is

$$x_{n+1} = x_n - \frac{f(x_n)}{f'(x_n)}.$$

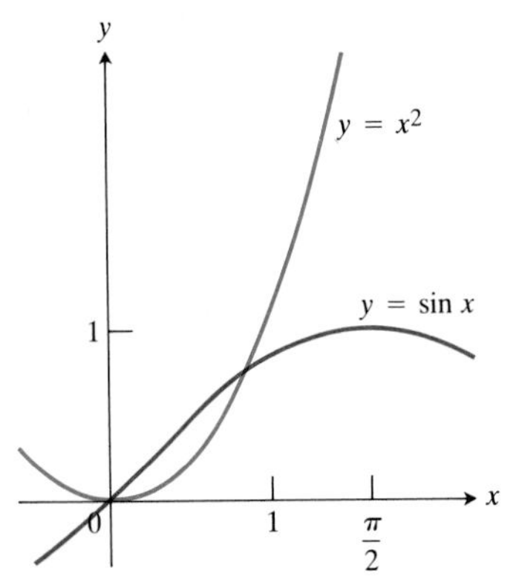

3.57 The curves $y = x^2$ and $y = \sin x$ cross at $x = 0$ and again near $x = 1$. Example 3 uses Newton's method to find this second solution of the equation $\sin x = x^2$.

The computer does not recognize x^2 unless we write it as x ^ 2, as shown in line 30 of the program that follows. Likewise, 10 ^ (−6) in line 80 is how we write 10^{-6} in the program. The asterisk (*) in 2*x indicates multiplication—the machine does not recognize $2x$ as 2 times x.

Now, let's look again at line 80. We don't know how many iterations may be needed to find a "sufficiently accurate" estimate of the root of $f(x) = 0$, so we arbitrarily tell the computer to stop when it reaches a value of x for which the absolute value of f is less than 10^{-6}. The entire command in line 80 is known as an "IF...THEN" conditional. If the condition is satisfied, the machine goes to line 100; but if the condition is not satisfied, the

TABLE 3.4
Estimates of the positive solution of $\sin x - x^2 = 0$.

n	x_n	$f(x_n)$	$f(x_n)/f'(x_n)$
0	1	−0.158529	0.108604
1	0.891396	−0.0166374	0.0144114
2	0.876984	−2.88069E − 04	2.58475E − 04
3	0.876726	−1.19209E − 07	1.07031E − 07

machine proceeds to line 90, which sends it back to line 30 for another round of calculation. Here is the program.

PROGRAM	COMMENTS
10 LET X = 1	The starting value of x is 1.
20 LET N = 0	We start counting the steps.
30 LET F = SIN(X) − X^2	The function is $f(x) = \sin x - x^2$.
40 LET G = COS(X) − 2*X	We let G stand for f'.
50 PRINT N; X; F; F/G	So we can see the values at each step.
60 X = X − F/G	The next x will be $x_n - f(x_n)/f'(x_n)$.
70 N = N + 1	Numbers the new x.
80 IF ABS(F) < 10^(−6) THEN 100	Conditional command discussed in text.
90 GOTO 30	If the condition ABS(F) < 10^(−6) is not yet satisfied, do it again.
100 END	The program stops only after ABS(F) < 10^(−6).

Table 3.4 shows what happens when you run the program. ■

Each time you want to change the starting value in the program in Example 3, you have to rewrite line 10. To avoid this, and make the program easier to use, you can replace line 10 with

```
10 INPUT X.
```

Then when you run the program, the computer will first return a question mark, your cue to enter the starting value you want. Once you have done so (and pressed ENTER) the program will complete the run.

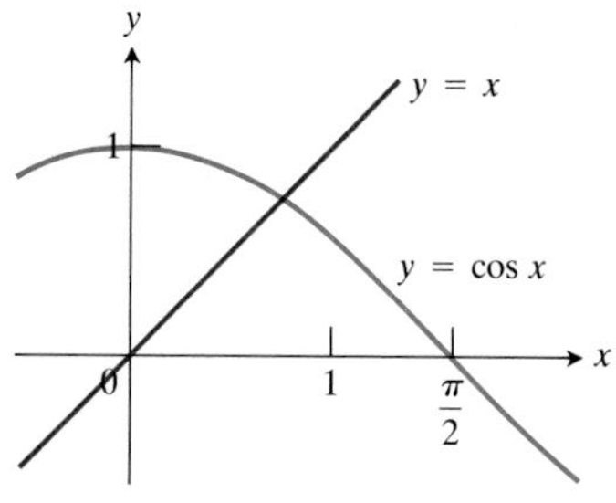

3.58 Example 4 shows how to modify the program in Example 3 to find where $\cos x = x$.

EXAMPLE 4 To run the program in Example 3 for other functions, you have only to change lines 30 and 40. To estimate where $x = \cos x$ (Fig. 3.58), you would use

```
30 LET F = X − COS(X)
40 LET G = 1 + SIN(X).
```

TABLE 3.5
Estimates of the solution of $\cos x = x$.

n	x_n	$f(x_n)$	$f(x_n)/f'(x_n)$
0	1	0.459698	0.249636
1	0.750364	0.0189231	0.011251
2	0.739113	4.65512E − 05	2.78145E − 05
3	0.739085	−5.96046E − 08	−3.56144E − 08

Running the program with a starting value of $x = 1$ would then give the results in Table 3.5. ■

Exercises 3.8

1. Find the two displayed solutions of the equation

$$x^2 + x - 1 = 0.$$

Start with $x_0 = 1$ and $x_0 = -1$ and find x_2 in each case.

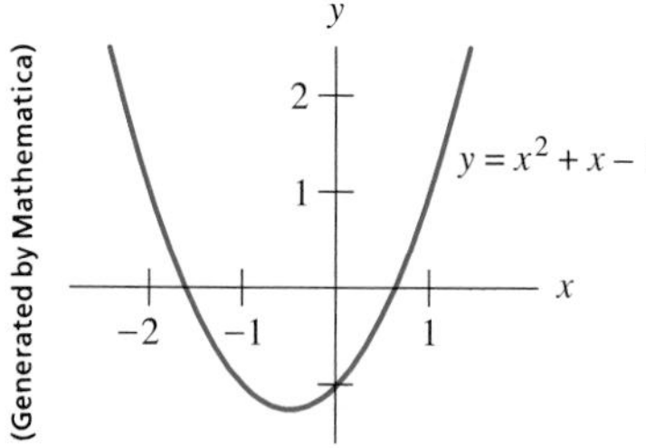

2. Find the displayed solution of the equation

$$x^3 + x - 1 = 0.$$

Start with $x_0 = 1$ and find x_2.

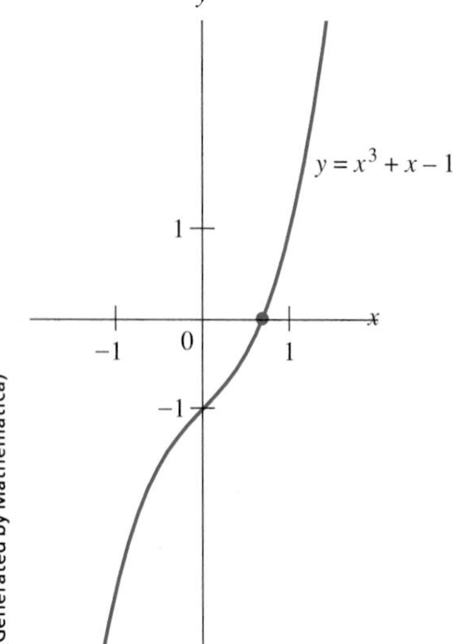

3. Find the two displayed roots of the function

$$f(x) = x^4 + x - 3.$$

Start with $x_0 = 1$ and $x_0 = -1$ and find x_2 in each case.

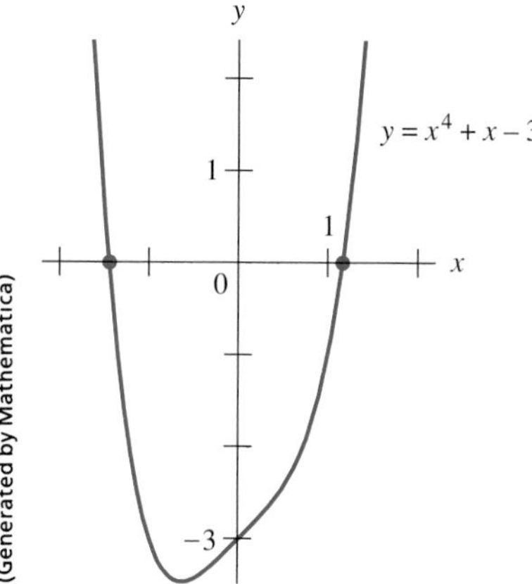

4. Find the displayed roots of the function

$$f(x) = 2x - x^2 + 1.$$

Start with $x_0 = 0$ and $x_0 = 2$ and find x_2 in each case.

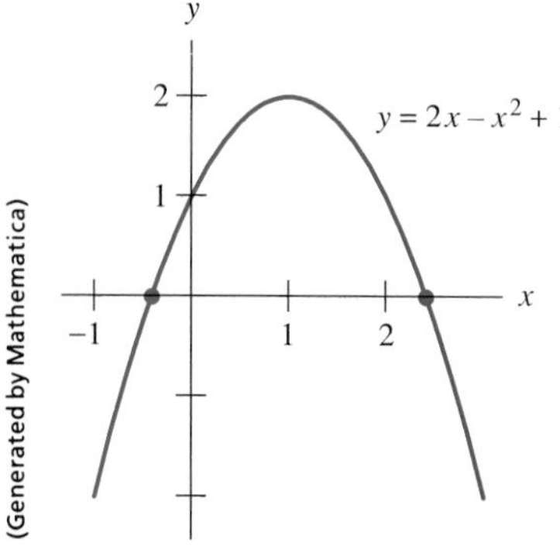

5. Find the positive fourth root of 2 by solving the equation $x^4 - 2 = 0$. Start with $x_0 = 1$ and find x_2.

6. Find the negative fourth root of 2 by solving the equation $x^4 - 2 = 0$. Start with $x_0 = -1$ and find x_2.

7. Suppose your first guess is lucky, in the sense that x_0 is a root of $f(x) = 0$. What happens to x_1 and later approximations?

8. You plan to estimate $\pi/2$ to five decimal places by solving the equation $\cos x = 0$ by Newton's method. Does it matter what your starting value is? Explain.

9. *Oscillation.* Show that if $h > 0$, applying Newton's method to

$$f(x) = \begin{cases} \sqrt{x}, & x \geq 0 \\ \sqrt{-x}, & x < 0 \end{cases}$$

leads to $x_1 = -h$ if $x_0 = h$ and to $x_1 = h$ if $x_0 = -h$. Draw a picture that shows what is going on.

10. *Approximations that get worse and worse.* Apply Newton's method to $f(x) = x^{1/3}$ with $x_0 = 1$, and calculate x_1, x_2, x_3, and x_4. Find a formula for $|x_n|$. What happens to $|x_n|$ as $n \to \infty$? Draw a picture that shows what is going on.

11. **CALCULATOR** Use the Intermediate Value Theorem from Section 2.4 to show that $f(x) = x^3 + 2x - 4$ has a root between $x = 1$ and $x = 2$. Then find the root to five decimal places.

12. a) Explain why the following four statements ask for the same information:
    i) Find the roots of $f(x) = x^3 - 3x - 1$.
    ii) Find the x-coordinates of the intersections of the curve $y = x^3$ with the line $y = 3x + 1$.
    iii) Find the x-coordinates of the points where the curve $y = x^3 - 3x$ crosses the horizontal line $y = 1$.
    iv) Find the values of x where the derivative of $g(x) = (1/4)x^4 - (3/2)x^2 - x + 5$ equals zero.

    b) Sketch the graph of $f(x) = x^3 - 3x - 1$ over the interval $-2 \leq x \leq 2$.

    c) **CALCULATOR** Find the positive root of $f(x) = x^3 - 3x - 1$ to five decimal places.

    d) **CALCULATOR** Find the two negative roots of $f(x) = x^3 - 3x - 1$ to five decimal places.

13. **CALCULATOR** Estimate π to five decimal places by applying Newton's method to the equation $\tan x = 0$ with $x_0 = 3$. Remember to use radians.

14. **CALCULATOR** *Locating a planet.* To calculate a planet's space coordinates, we have to solve equations like $x = 1 + 0.5 \sin x$. Graphing the function $f(x) = x - 1 - 0.5 \sin x$ suggests that the function has a root near $x = 1.5$. Use one application of Newton's method to improve this estimate. That is, start with $x_0 = 1.5$ and find x_1. (The value of the root is 1.49870 to five decimal places.) Remember to use radians.

15. **CALCULATOR** (Programmable feature helpful but not necessary.) *The sonobuoy problem.* From C. O. Wilde's *The Contraction Mapping Principle*, UMAP Unit 326 (Arlington, Mass.: COMAP, Inc.). In submarine location problems it is often necessary to find the submarine's closest point of approach (CPA) to a sonobuoy (sound detector) in the water. Suppose that the submarine travels on a parabolic path $y = x^2$ and that the buoy is located at the point $(2, -1/2)$. As we shall see in Section 4.5, Exercise 20, the value of x that minimizes the distance between the point (x, x^2) and the point $(2, -1/2)$ is a solution of the equation

$$\frac{1}{x^2 + 1} = x.$$

Solve this equation by Newton's method to find the CPA to five decimal places.

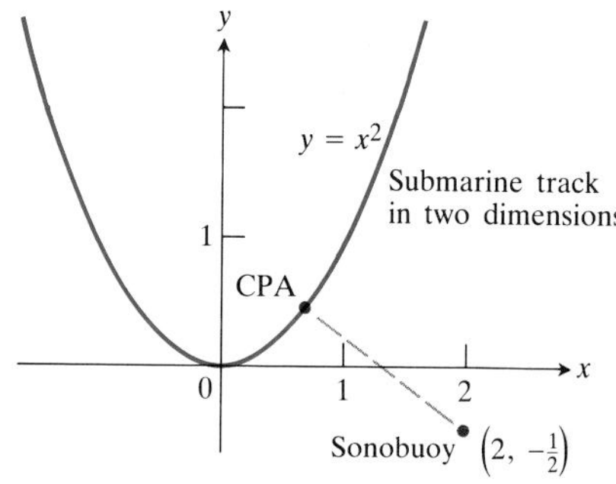

16. **CALCULATOR** *Finding an ion concentration.* While trying to find the acidity of a saturated solution of magnesium hydroxide in hydrochloric acid, you derive the equation

$$\frac{3.64 \times 10^{-11}}{[OH_3^-]^2} = [OH_3^-] + 3.6 \times 10^{-4}$$

for the hydronium ion concentration $[OH_3^-]$. To find the value of $[OH_3^-]$, you set $x = 10^4[OH_3^-]$ and convert the equation to

$$x^3 + 3.6x^2 - 36.4 = 0.$$

You then solve this by Newton's method. What do you get for x? (Make it good to two decimal places.) For $[OH_3^-]$?

COMPUTER or PROGRAMMABLE CALCULATOR

17. The curve $y = \tan x$ crosses the line $y = 2x$ somewhere between $x = 0$ and $x = \pi/2$. Where?

18. Find the two real roots of the equation

$$x^4 - 2x^3 - x^2 - 2x + 2 = 0.$$

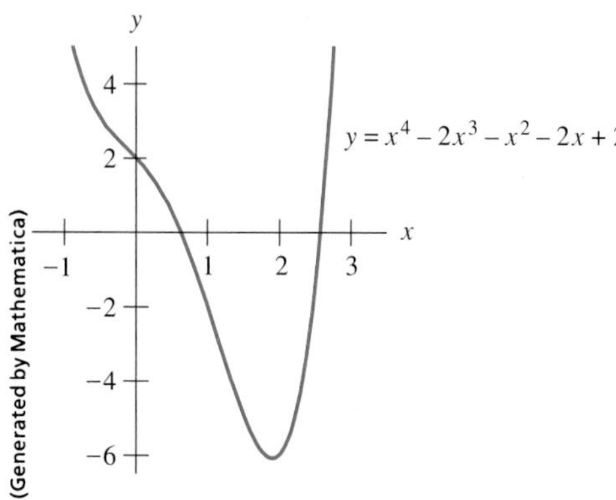

19. Find the four roots of $2x^4 - 4x^2 + 1 = 0$.

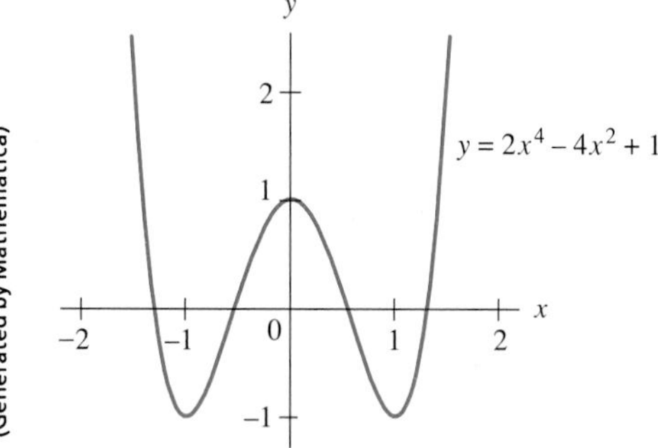

20. *Curves that are almost flat near the root.* Some curves are so flat that Newton's method stops some distance from the root. Try Newton's method on $f(x) = (x - 1)^{40}$ with a starting value of $x_0 = 2$. How close does the computer come to the root $x = 1$?

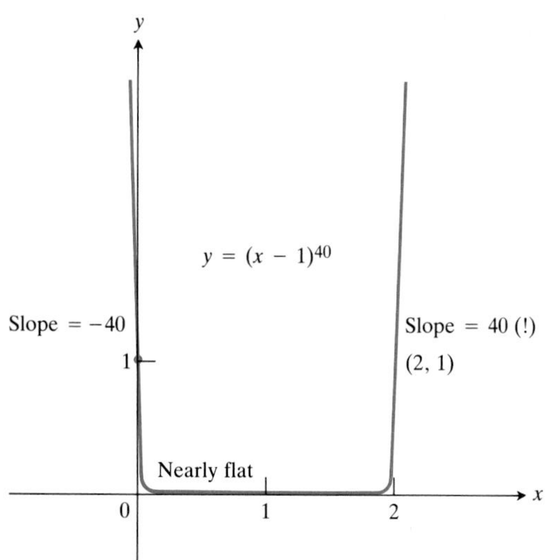

21. *Finding a root different from the one sought.* All three roots of $f(x) = x^4 - x^2$ can be found by starting Newton's method near $x = -\sqrt{2}/2$.

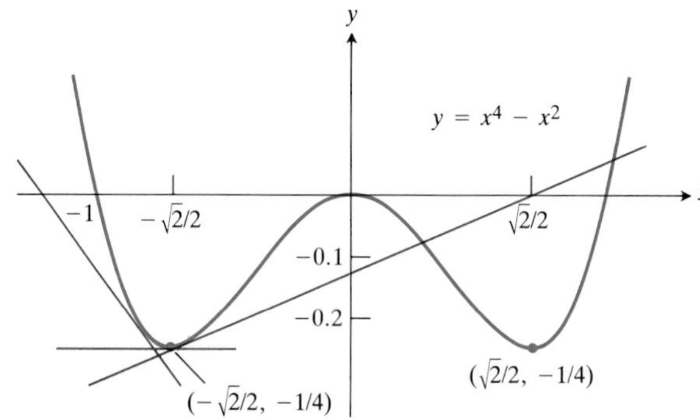

The method will find the root $x = -1$ if x_0 is far enough to the left of $-\sqrt{2}/2$. If x_0 is too close to $-\sqrt{2}/2$, the computer will encounter a zero slope or a value of x_1 too large to handle. There is a zone just to the right of $-\sqrt{2}/2$ where values of x_0 lead to $x = 1$ instead of $x = 0$.

Likewise, selected values of x_0 near $\sqrt{2}/2$ will lead to $x = 1$, $x = 0$, $x = -1$, or no root at all.

Try it.

TOOLKIT PROGRAMS

Root Finder — Graphs any function you key in and approximates its roots by any of four methods you select

Review Questions

1. What is a derivative? a right-hand derivative? a left-hand derivative? How are they related? Give examples.
2. What geometric significance do derivatives have?
3. How is the differentiability of a function at a point related to its continuity (if any) at a point?
4. Give some examples of differentiable functions.
5. When does a function typically *not* have a derivative at a point?
6. What rules do you know for calculating derivatives? Give examples.
7. Explain how the three formulas

   a) $\frac{d(x^n)}{dx} = nx^{n-1}$,

   b) $\frac{d(cu)}{dx} = c\frac{du}{dx}$,

   c) $\frac{d(u+v)}{dx} = \frac{du}{dx} + \frac{dv}{dx}$

   let us differentiate any polynomial.
8. What formula do we need, in addition to the three listed in Question 7, to differentiate rational functions?
9. What is a second derivative? a third derivative? How many derivatives do the functions you know have? Give examples.
10. When a body moves along a coordinate line and its position $s(t)$ is a differentiable function of t, how do you define the body's velocity, speed, and acceleration? Give an example.
11. Besides velocity, speed, and acceleration, what other rates of change are found with derivatives?
12. What are the derivatives of the six basic trigonometric functions? How does their calculation depend on radian measure?
13. When is the composite of two functions differentiable at a point? What do you need to know to calculate its derivative there? Give examples.
14. What is implicit differentiation and what is it good for? Give examples.
15. What is the linearization $L(x)$ of a function $f(x)$ at a point $x = a$? What is required of f at a for the linearization to exist? How are linearizations used? Give examples.
16. If x moves from x_0 to a nearby value $x_0 + dx$, how do we estimate the corresponding change in the value of a differentiable function $f(x)$? How do we estimate the relative change? the percentage change? Give an example.
17. How are derivatives expressed in differential notation? Give examples.
18. Describe Newton's method for solving equations. Give an example. What is the theory behind the method? What are some of the things to watch out for when you use the method?

Practice Exercises

In Exercises 1–34, find dy/dx.

1. $y = x^5 - \frac{1}{8}x^2 + \frac{1}{4}x$
2. $y = 3 - 7x^3 + 3x^7$
3. $y = (x+1)^2(x^2+2x)$
4. $y = (2x-5)(4-x)^{-1}$
5. $y = 2\sin x\cos x$
6. $y = \sin x - x\cos x$
7. $y = \frac{x}{x+1}$
8. $y = \frac{2x+1}{2x-1}$
9. $y = (x^3+1)^{-4/3}$
10. $y = (x^2-8x)^{-1/2}$
11. $y = \cos(1-2x)$
12. $y = \cot\frac{2}{x}$
13. $y = (x^2+x+1)^3$
14. $y = \left(-1 - \frac{x}{2} - \frac{x^2}{4}\right)^2$
15. $y = \sqrt{2u+u^2},\ u = 2x+3$

16. $y = \dfrac{-u}{1+u}, \quad u = \dfrac{1}{x}$

17. $xy + y^2 = 1$

18. $xy + 2x + 3y = 1$

19. $x^2 + xy + y^2 - 5x = 2$

20. $x^3 + 4xy - 3y^2 = 2$

21. $5x^{4/5} + 10y^{6/5} = 15$

22. $\sqrt{xy} = 1$

23. $y^2 = \dfrac{x}{x+1}$

24. $y^2 = \sqrt{\dfrac{1+x}{1-x}}$

25. $y^2 = \dfrac{(5x^2+2x)^{3/2}}{3}$

26. $y = \dfrac{3}{(5x^2+2x)^{3/2}}$

27. $y = \sqrt{x} + 1 + \dfrac{1}{\sqrt{x}}$

28. $y = x\sqrt{2x+1}$

29. $y = \sec(1 + 3x)$

30. $y = \sec^2(1 + 3x)$

31. $y = \cot x^2$

32. $y = x^2 \cos 5x$

33. $y = \sqrt{\dfrac{1-x}{1+x^2}}$

34. $y^2 = \dfrac{x^2-1}{x^2+1}$

35. a) Graph the function

$$f(x) = \begin{cases} x, & 0 \le x \le 1 \\ 2 - x, & 1 < x \le 2. \end{cases}$$

b) Is f continuous at $x = 1$?
c) Is f differentiable at $x = 1$?
Explain.

36. a) Find the values of the left-hand and right-hand derivatives of

$$f(x) = \begin{cases} \sin 2x, & x \le 0 \\ mx, & x > 0 \quad (m \text{ constant}) \end{cases}$$

at $x = 0$.
b) For what value of m, if any, is f differentiable at $x = 0$?

37. Find the points on the curve $y = 2x^3 - 3x^2 - 12x + 20$ where the tangent is parallel to the x-axis.

38. The line normal to the curve $y = x^2 + 2x - 3$ at $(1, 0)$ intersects the curve at what other point?

39. The position at time $t \ge 0$ of a particle moving along a coordinate line is

$$s(t) = 10 \cos(t + \pi/4).$$

a) What is the particle's starting position ($t = 0$)?
b) What are the points farthest to the left and right of the origin reached by the particle?
c) Find the particle's velocity and acceleration at the points in (b).
d) When does the particle first reach the origin? What are its velocity, speed, and acceleration then?

40. On Earth, you can easily shoot a paper clip 64 ft into the air with a rubber band. In t seconds after firing, the paper clip is $s = 64t - 16t^2$ ft above your hand.
a) How long does it take the paper clip to reach its maximum height? With what velocity does it leave your hand?
b) On the moon, the same force will send the paper clip to a height of $s(t) = 64t - 2.6t^2$ ft in t seconds. About how long will it take the paper clip to reach its maximum height and how high will it go?

41. Figure 3.59 shows a multiflash photograph of two balls falling from rest. The rulers in the figure are marked in centimeters. Use the equation $s = 490t^2$ to answer the following questions.
a) How long did it take the balls to fall the first 160 cm? What was their average velocity for the period?
b) How fast were the balls falling when they reached the 160 cm mark? What was their acceleration then?
c) About how fast was the light flashing (flashes per second)?

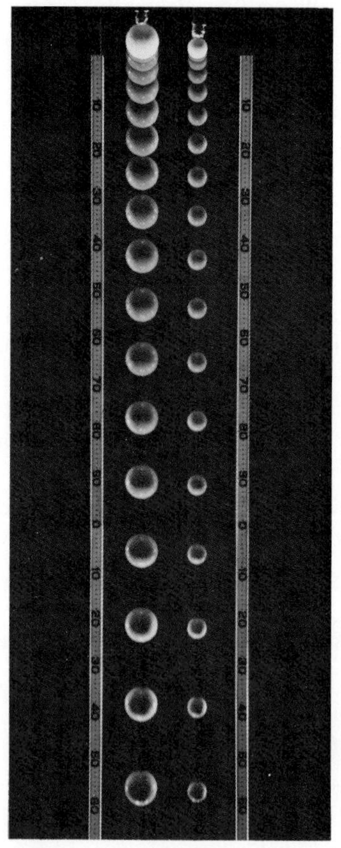

Education Development Center, Inc., Newton, Mass.

3.59 Two balls falling from rest.

42. The following data give the coordinates s of a moving body for various values of t. Plot s versus t on coordinate paper and sketch a smooth curve through the given points. Assuming that this smooth curve represents the motion of the body, estimate the velocity at (a) $t = 1.0$; (b) $t = 2.5$; (c) $t = 2.0$.

s (in ft)	10	38	58	70	74	70	58	38	10
t (in sec)	0	0.5	1.0	1.5	2.0	2.5	3.0	3.5	4.0

43. The graphs in Fig. 3.60 show the distance traveled (miles), velocity (mph), and acceleration (mph/sec) for each second of a 2-minute automobile trip. Which graph shows

a) distance? b) velocity? c) acceleration?

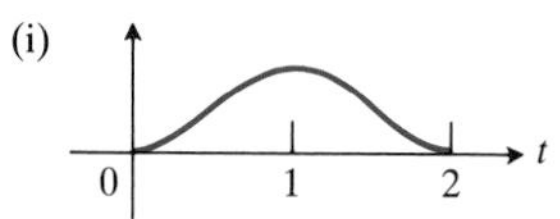

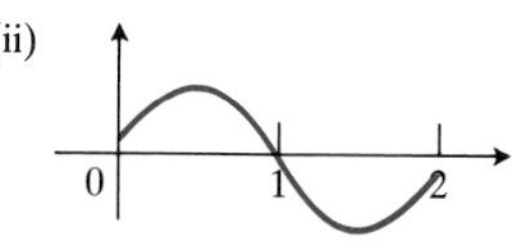

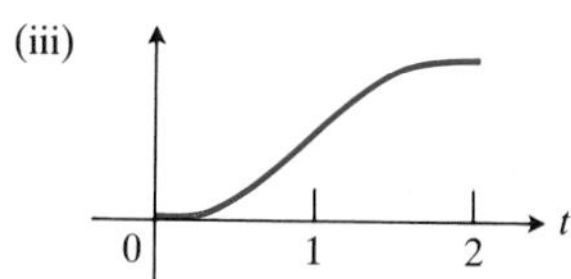

3.60 The graphs for Exercise 43.

44. The graph in Fig. 3.61 shows the position $s(t)$ of a truck traveling on a highway. The truck starts at $t = 0$ and returns 15 hours later at $t = 15$.

a) Use the technique described at the end of Section 3.3 to graph the truck's velocity $v = ds/dt$ for $0 \le t \le 15$. Then repeat the process, with the velocity curve, to graph the truck's acceleration dv/dt.

b) Suppose $s(t) = 15t^2 - t^3$. Graph ds/dt and d^2s/dt^2 and compare your graphs with those in (a).

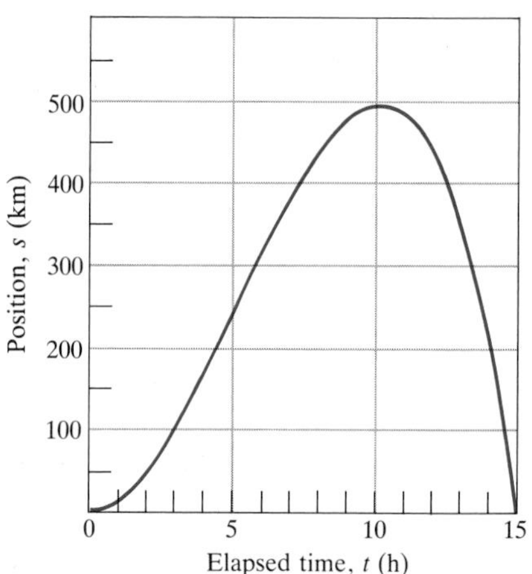

3.61 The time vs. position graph for the truck in Exercise 44.

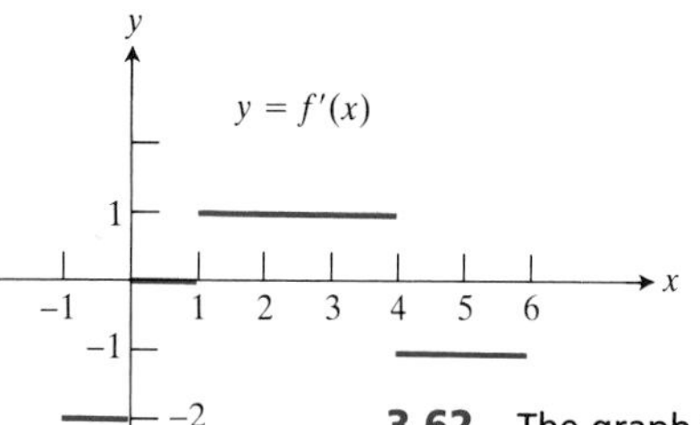

3.62 The graph for Exercise 45.

45. Use the following information to graph the function $y = f(x)$ for $-1 \le x \le 6$.

i) The graph of f is made of line segments joined end to end.

ii) The graph starts at the point $(-1, 2)$

iii) The derivative of f, where defined, is the step function shown in Fig. 3.62.

46. Repeat Exercise 45, supposing that the graph starts at $(-1, 0)$ instead of $(-1, 2)$.

47. If a hemispherical bowl of radius 10 in. is filled with water to a depth of x in., the volume of water is given by $V = \pi[10 - (x/3)]x^2$. Find the rate of increase of the volume per inch increase of depth.

48. A bus will hold 60 people. The number x of people per trip who use the bus is related to the fare charged (p dollars) by the law $p = [3 - (x/40)]^2$. Write an expression for the total revenue $r(x)$ per trip received by the bus company. What number of people per trip will make the marginal revenue dr/dx equal to zero? What is the corresponding fare? (This is the fare that maximizes the revenue, so the bus company should probably rethink its fare policy.)

In Exercises 49 and 50, find an equation for

a) the horizontal tangent and

b) the tangent to the curve at the indicated point P.

49.

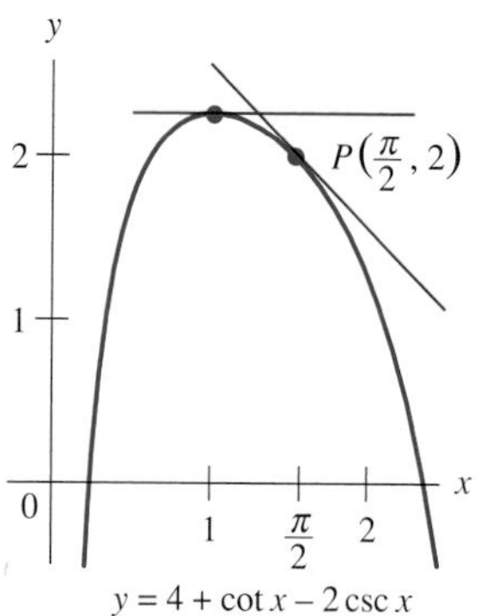

50.

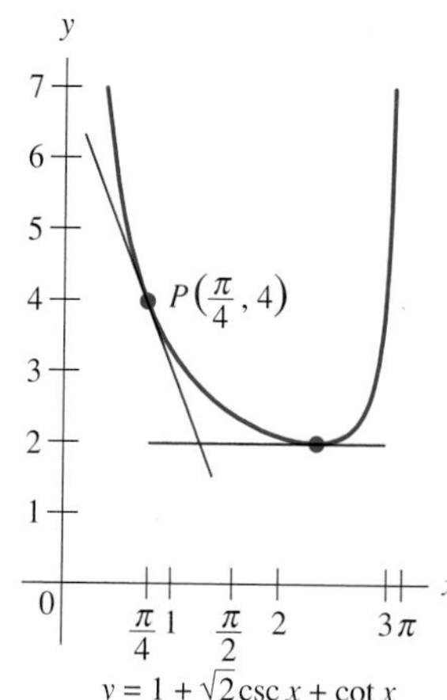

(Generated by Mathematica)

51. The accompanying graph of $y = \sin(x - \sin x)$ suggests that the curve might have horizontal tangents at the x-axis. Does it?

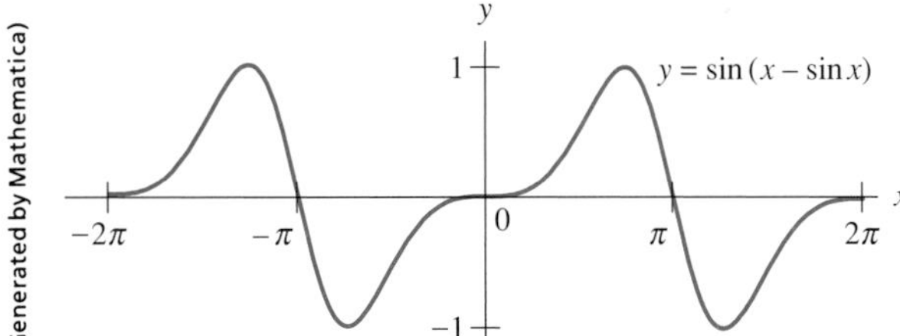

52. The figure shows a boat 1 km offshore, sweeping the shore with a search light. The light turns at the constant rate $d\theta/dt = -3/5$ radians per second. (This rate is called the light's *angular velocity*.)

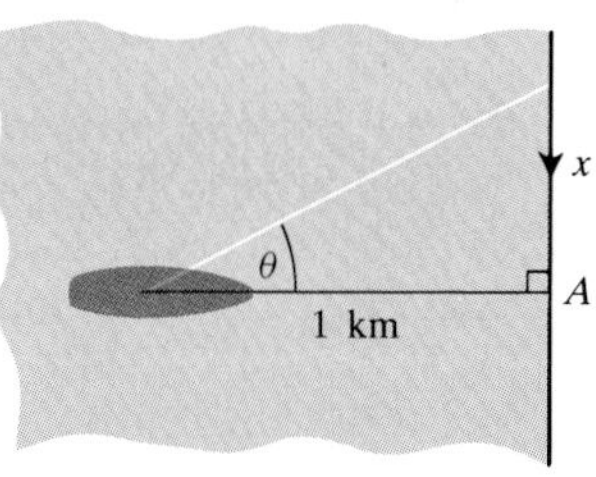

a) Express x (see the figure) in terms of θ.

b) Differentiate both sides of the equation you obtained in (a) with respect to t. Then substitute $d\theta/dt = -3/5$. This will express dx/dt (the rate at which the light moves along the shore) as a function of θ.

c) How fast (m/sec) is the light moving along the shore when it reaches point A?

d) How many revolutions per minute is 0.6 radian per second?

53. Suppose that functions f and g and their derivatives have the following values at $x = 0$ and $x = 1$.

x	$f(x)$	$g(x)$	$f'(x)$	$g'(x)$
0	1	1	5	1/3
1	3	-4	$-1/3$	$-8/3$

Find the derivatives of the following combinations at the given value of x.

a) $5f(x) - g(x),\ x = 1$ b) $f(x)g^3(x),\ x = 0$

c) $\dfrac{f(x)}{g(x) + 1},\ x = 1$ d) $f(g(x)),\ x = 0$

e) $g(f(x)),\ x = 0$ f) $(x + f(x))^{3/2},\ x = 1$

g) $f(x + g(x)),\ x = 0$

54. Suppose that $f(x) = x^2$ and $g(x) = |x|$. Then the composites

$$(f \circ g)(x) = |x|^2 = x^2 \quad \text{and} \quad (g \circ f)(x) = |x^2| = x^2$$

are both differentiable at $x = 0$ even though g is not differentiable at $x = 0$. Does this contradict the Chain Rule? Explain.

55. If the identity $\sin(x + a) = \sin x \cos a + \cos x \sin a$ is differentiated with respect to x, is the resulting equation also an identity? Does this principle apply to the equation $x^2 - 2x - 8 = 0$? Explain.

56. Find dy/dt at $t = 0$ if $y = 3 \sin 2x$ and $x = t^2 + \pi$.

57. Find ds/du at $u = 2$ if $s = t^2 + 5t$ and $t = (u^2 + 2u)^{1/3}$.

58. Find dw/ds at $s = 0$ if $w = \sin(\sqrt{r} - 2)$ and $r = 8 \sin(s + \pi/6)$.

59. Find the points where the tangent to the curve $y = \sqrt{x}$ at $x = 4$ crosses the coordinate axes.

60. What horizontal line crosses the curve $y = \sqrt{x}$ at a 45° angle?

61. Find the lines that are tangent and normal to the curve at the given point.

a) $x^2 + 2y^2 = 9$ at $(1, 2)$

b) $x^3 + y^2 = 2$ at $(1, 1)$

c) $xy + 2x - 5y = 2$ at $(3, 2)$

62. Which of the following statements could be true if $f''(x) = x^{1/3}$?

I. $f(x) = \dfrac{9}{28}x^{7/3} + 9$ II. $f'(x) = \dfrac{9}{28}x^{7/3} - 2$

III. $f'(x) = \dfrac{3}{4}x^{4/3} + 6$ IV. $f(x) = \dfrac{3}{4}x^{4/3} - 4$

a) I only b) III only

c) II and IV only c) I and III only

63. The designer of a 30-ft-diameter spherical hot-air balloon wishes to suspend the gondola 8 ft below the bottom of the balloon with suspension cables tangent to the surface of the balloon. Two of the cables are shown running from the top edges of the gondola to their points of tangency, $(-12, -9)$ and $(12, -9)$. How wide must the gondola be?

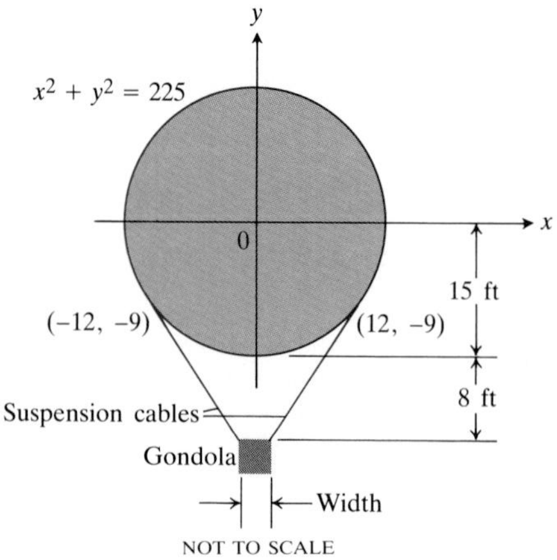

64. *What determines the fundamental frequency of a vibrating piano string?* We measure the frequencies at which wires vibrate in cycles (trips back and forth) per second. The unit of measure is a *hertz*: 1 cycle per second. Middle A on a piano has a frequency of 440 hertz. For any given wire, the fundamental frequency y is a function of four variables:

r: the radius of the wire
l: the length
d: the density of the wire
T: the tension (force) holding the wire taut.

With r and l in centimeters, d in grams per cubic centimeter, and T in dynes (it takes about 100,000 dynes to lift an apple), the fundamental frequency of the wire is

$$y = \frac{1}{2rl}\sqrt{\frac{T}{\pi d}}.$$

If we keep all the variables fixed except one, then y can be alternately thought of as four different functions of one variable, $y(r)$, $y(l)$, $y(d)$, and $y(T)$. How would changing each variable then affect the string's fundamental frequency? To find out, calculate $y'(r)$, $y'(l)$, $y'(d)$, and $y'(T)$.

65. Find d^2y/dx^2 by implicit differentiation:

a) $x^3 + y^3 = 1$ b) $y^2 = 1 - \dfrac{2}{x}$

66. a) By differentiating $x^2 - y^2 = 1$ implicitly, show that $dy/dx = x/y$.

b) Then show that $d^2y/dx^2 = -1/y^3$.

67. Find d^2y/dx^2 if

a) $y = \sqrt{2x + 7}$ b) $x^2 + y^2 = 1$

68. If $y^3 + y = 8x - 6$, find d^2y/dx^2 at the point (1, 1).

69. Find the linearizations of

a) $\tan x$ at $x = -\pi/4$

b) $\sec x$ at $x = -\pi/4$.

Graph the curves and linearizations together.

70. A useful linear approximation to

$$\frac{1}{1 + \tan x}$$

can be obtained by combining the approximations

$$\frac{1}{1+x} \approx 1 - x \quad \text{and} \quad \tan x \approx x$$

to get

$$\frac{1}{1 + \tan x} \approx 1 - x.$$

Show that this is the standard linear approximation of $1/(1 + \tan x)$.

71. Let $f(x) = \sqrt{1 + x} + \sin x - 0.5$.

a) Find $f(-\pi/4) < 0$ and $f(0) > 0$, to show that the equation $f(x) = 0$ has a solution between $-\pi/4$ and 0.

b) To estimate the solution of $f(x) = 0$, replace $\sqrt{1 + x}$ and $\sin x$ by their linearizations at $x = 0$ and solve the resulting linear equation.

c) **CALCULATOR** Check your estimate in the original equation.

72. Let

$$f(x) = \frac{2}{1 - x} + \sqrt{1 + x} - 3.1.$$

a) Find $f(0) < 0$ and $f(0.5) > 0$ to show that the equation $f(x) = 0$ has a solution between $x = 0$ and $x = 0.5$.

b) To estimate the solution of the equation $f(x) = 0$, replace $2/(1 - x)$ and $\sqrt{1 + x}$ by their linearizations at $x = 0$ and solve the resulting linear equation.

c) **CALCULATOR** Check your estimate in the original equation.

73. Write a formula that estimates the change that occurs in the volume of a right circular cone when the radius changes from r_0 to $r_0 + dr$ and the height does not change.

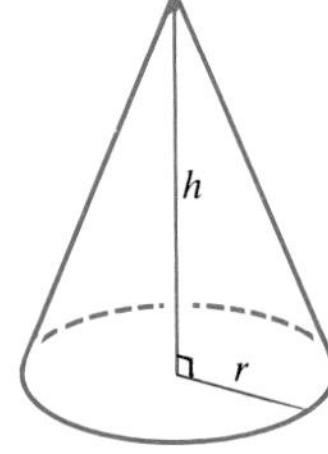

$V = \frac{1}{3}\pi r^2 h$

$S = \pi r\sqrt{r^2 + h^2}$

(Lateral surface area)

74. Write a formula that estimates the change that occurs in the lateral surface area of a cone when the height changes from h_0 to $h_0 + dh$ and the radius does not change.

75. a) How accurately should you measure the edge of a cube to be reasonably sure of calculating the cube's surface area with an error of no more than 2%?

b) Suppose the edge is measured with the accuracy required in (a). About how accurately can the cube's volume be calculated from the edge measurement? To find out, estimate the percentage error in the volume calculation that would result from using the edge measurement.

76. The circumference of the equator of a sphere is measured as 10 cm with a possible error of 0.4 cm. The measurement is then used to calculate the radius. The radius is then used to calculate the surface area and volume of the sphere. Estimate the percentage errors in the calculated values of (a) the radius, (b) the surface area, and (c) the volume.

77. To find the height of a tree, you measure the angle from the ground to the treetop from a point 100 ft

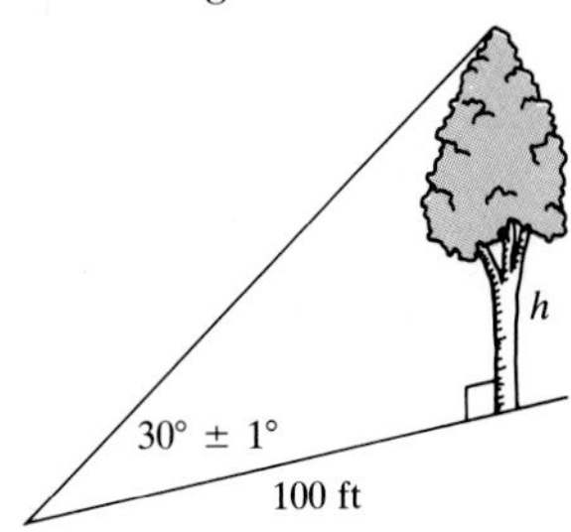

away from the base. The best figure you can get with the equipment at hand is $30° \pm 1°$. About how much error could the tolerance of $\pm 1°$ create in the calculated height? Remember to work in radians.

78. To find the height of a lamppost, you stand a 6-ft pole 20 ft from the lamp and measure the length a of its shadow. The figure you get for a is 15 ft, give or take an inch. Calculate the height of the lamppost from the value $a = 15$ and estimate the possible error in the result.

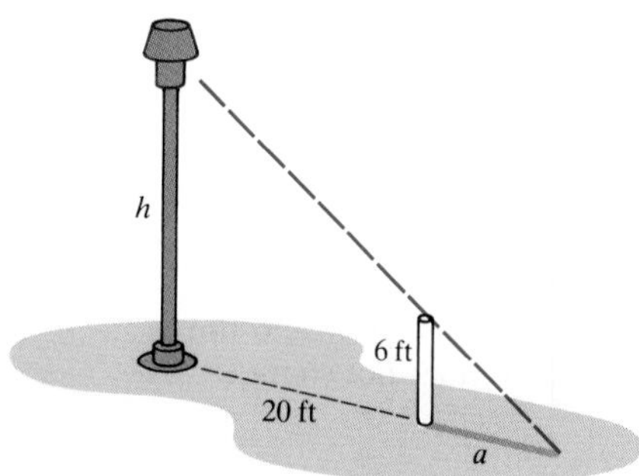

79. *The linearization is the best linear approximation.* Suppose that $y = f(x)$ is differentiable at $x = a$ and that $g(x) = m(x - a) + c$ (m and c constants). If the error $E(x) = f(x) - g(x)$ were small enough near $x = a$, we might think of using g as a linear approximation of f instead of the linearization $L(x) = f(a) + f'(a)(x - a)$. Show that if we impose on g the conditions

1. $E(a) = 0$ (The approximation error is zero at $x = a$.)
2. $\lim_{x \to a} \frac{E(x)}{x - a} = 0$ (The error is negligible when compared with $(x - a)$.),

then $g(x) = f(a) + f'(a)(x - a)$. Thus, the linearization gives the only linear approximation whose error is both zero at $x = a$ and negligible in comparison with $(x - a)$.

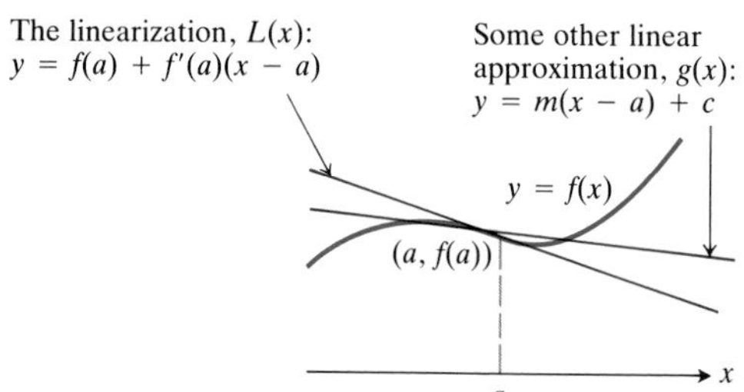

80. The volume $y = x^3$ of a cube with edges of length x increases an amount Δy when x increases by an amount Δx. Show with a sketch how to represent Δy geometrically as the sum of the volumes of

a) three slabs of dimensions x by x by Δx
b) three bars of dimensions x by Δx by Δx
c) one cube of dimensions Δx by Δx by Δx.

The differential formula $dy = 3x^2 dx$ estimates the change in y with the three slabs.

CALCULATOR

81. Use Newton's method to solve the equation $x^3 = 4$.

82. Use Newton's method to find where the curve $y = -x^3 + 3x + 4$ crosses the x-axis.

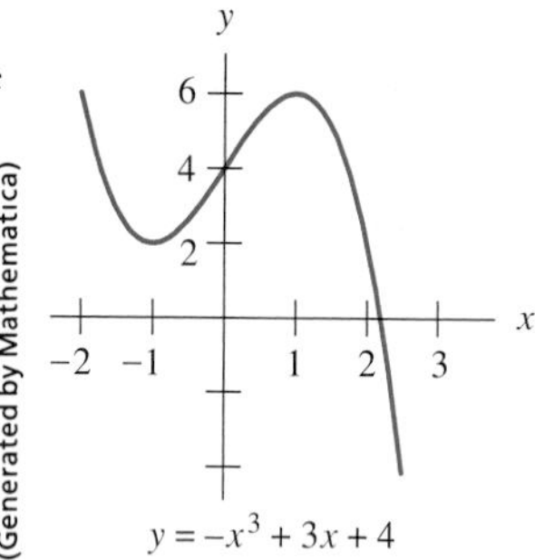

83. Solve the equation $\sec x = 4$ on the interval $0 \le x \le \pi/2$.

84. Use the Intermediate Value Theorem from Section 2.4 to show that $f(x) = x^4 - x^3 - 75$ has a root between $x = 3$ and $x = 4$. Then find the root.

COMPUTER or PROGRAMMABLE CALCULATOR

85. Solve the equation $2 \cos x - \sqrt{1 + x} = 0$.

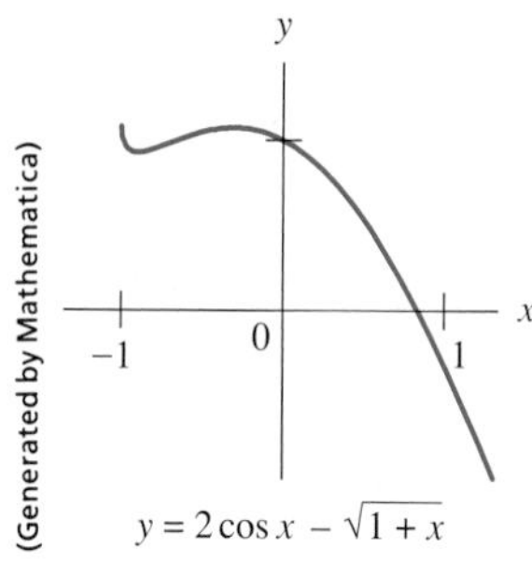

86. Find the approximate values of r_1 through r_4 in the factorization

$$8x^4 - 14x^3 - 9x^2 + 11x - 1 = 8(x - r_1)(x - r_2)(x - r_3)(x - r_4).$$

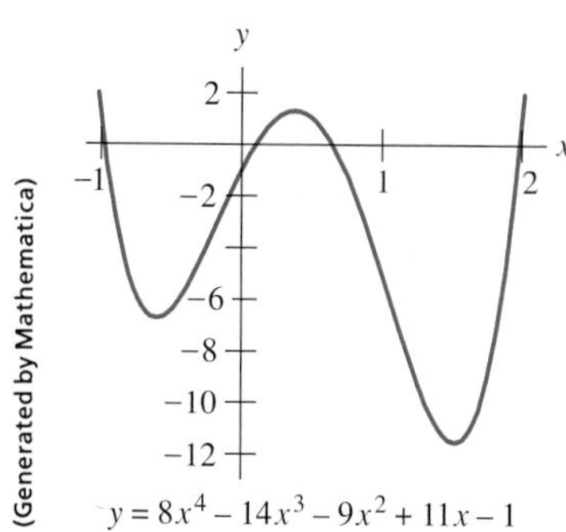

4 Applications of Derivatives

V. Clevenger/H. Armstrong Roberts
© John Blaustein/Woodfin Camp and Assoc.

Mt. Everest, the highest point on Earth, and Hell's Canyon, Snake River Idaho, the deepest gorge on Earth.

OVERVIEW

If you know the rate at which a function is changing, you can often calculate the rates at which functions closely related to it are changing at the same time. This chapter shows how to do that. It also shows how to use derivatives to make reliable graphs of polynomials and rational functions, how to find the largest and smallest values that a differentiable function assumes on a closed interval, and how to find a function $y = f(x)$ when all you know about it initially is its derivative and its value at a particular point. The key to recovering functions from their derivatives is The Mean Value Theorem, a theorem whose corollaries provide the gateway to the so-called integral calculus we shall begin studying in Chapter 5.

Many of the applications in this chapter involve writing down equations that model something in the real world we want to find out about, and the steps we go through each time to get the answers we want are almost always the same. We close the chapter by showing the role that these steps play in a general process called modeling and how we, as scientists, can use these steps to formulate ideas and test them against reality.

4.1 Related Rates of Change

How fast does the radius change when you blow air into a spherical soap bubble at the rate of 10 cm^3/sec? How fast does the water level drop when a cylindrical tank is drained at the rate of 3 liters/sec?

Questions like these ask us to calculate the rate at which one variable changes from the rate at which another variable is known to change. To calculate that rate, we write an equation that relates the two variables and differentiate it to get an equation that relates the rate we seek to the rate we know.

EXAMPLE 1 *The soap bubble.* How fast does the radius of a spherical soap bubble change when you blow air into it at the rate of 10 cm^3/sec?

Solution We are given the rate at which the volume is changing and are asked for the rate at which the radius is changing.

We think abstractly at first, picturing the bubble as a sphere whose volume V and radius r are differentiable functions of time t. The equation that relates V and r is

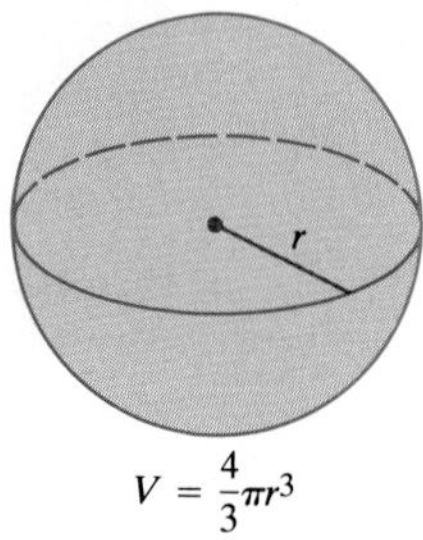

$$V = \frac{4}{3}\pi r^3. \tag{1}$$

To find an equation that relates the rate we seek (dr/dt) to the rate we know (dV/dt), we use the Chain Rule to differentiate both sides of Eq. (1) with respect to t:

$$\frac{dV}{dt} = \frac{d}{dt}\left(\frac{4}{3}\pi r^3\right) = \frac{4}{3}\pi\frac{d}{dt}(r^3) = \frac{4}{3}\pi \cdot 3r^2\frac{dr}{dt} = 4\pi r^2\frac{dr}{dt}. \tag{2}$$

We are told that

$$\frac{dV}{dt} = 10. \quad \text{(Air is blown in at the rate of 10 cm}^3\text{/sec.)}$$

We substitute this value in Eq. (2) and solve for dr/dt:

$$10 = 4\pi r^2 \frac{dr}{dt} \qquad \frac{dr}{dt} = \frac{10}{4\pi r^2}. \tag{3}$$

We see from Eq. (3) that the rate at which r changes at any particular time depends on how big r is at the time. When r is small, dr/dt will be large; when r is large, dr/dt will be small:

$$\text{At } r = 1 \text{ cm:} \qquad \frac{dr}{dt} = \frac{10}{4\pi} \approx 0.8 \text{ cm/sec,}$$

$$\text{At } r = 10 \text{ cm:} \qquad \frac{dr}{dt} = \frac{10}{400\pi} \approx 0.008 \text{ cm/sec.}$$

■

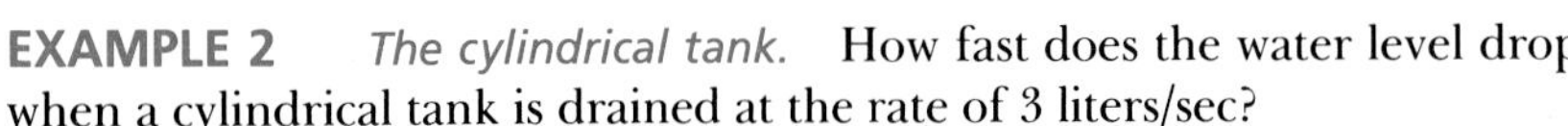

EXAMPLE 2 *The cylindrical tank.* How fast does the water level drop when a cylindrical tank is drained at the rate of 3 liters/sec?

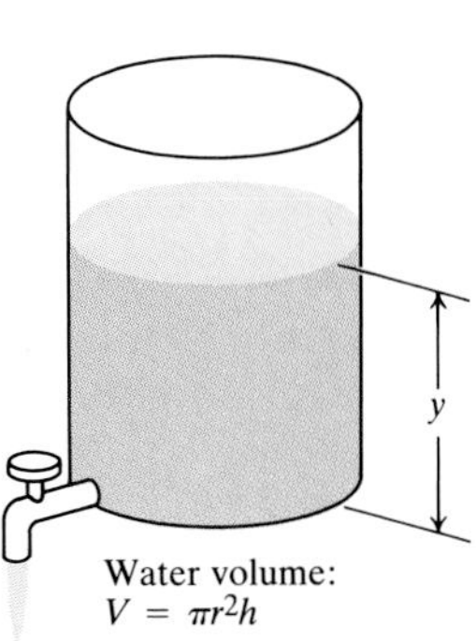

4.1 The cylindrical tank in Example 2.

Solution We draw a picture of a partially filled cylindrical tank, calling its radius r and the height of the water h (Fig. 4.1). We call the volume of water in the tank V.

The radius r is a constant, but V and h change with time. We think of V and h as differentiable functions of time and use t to represent time. The derivatives dV/dt and dh/dt give the rates at which V and h change. We are told that

$$\frac{dV}{dt} = -3, \qquad \text{(The tank is drained at the rate of 3 liters/sec.)}$$

and we are asked for

$$\frac{dh}{dt}. \qquad \text{(How fast does the water level drop?)}$$

To answer the question, we first write an equation that relates V and h:

$$V = \pi r^2 h. \qquad \text{(The tank is cylindrical.)}$$

We then differentiate both sides with respect to t to get an equation that relates dh/dt to dV/dt:

$$\frac{dV}{dt} = \pi r^2 \frac{dh}{dt}.$$

We substitute the known value $dV/dt = -3$ and solve for dh/dt:

$$\frac{dh}{dt} = -\frac{3}{\pi r^2}. \tag{4}$$

The water level is dropping at the constant rate of $3/\pi r^2$ liters/sec. ■

EXAMPLE 3 *A rising balloon.* A hot-air balloon, rising straight up from a level field, is tracked by a range finder 500 ft from the point of lift-off. At the moment the range finder's elevation angle is $\pi/4$, the angle is increasing at the rate of 0.14 radians/min. How fast is the balloon rising?

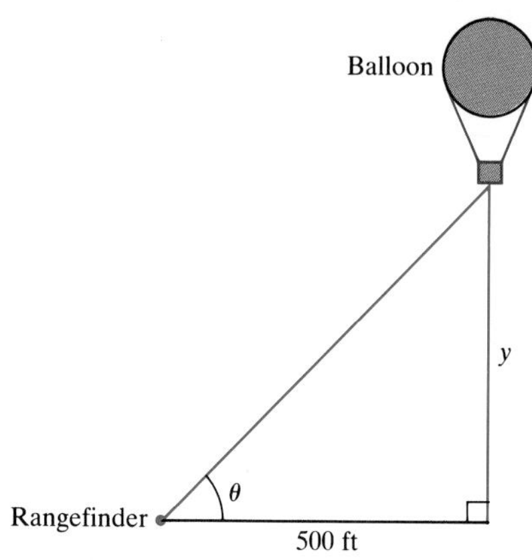

4.2 If $d\theta/dt = 0.14$ when $\theta = \pi/4$, what is the value of dy/dt when $\theta = \pi/4$? See Example 3.

© Susan Van Etten

Balloons rising above a field in Kenya on a still day.

Solution We answer the question in six steps.

STEP 1: *We draw a picture and name the variables and constants* (Fig. 4.2). The variables in the picture are

θ = the angle the range finder makes with the ground (radians)

y = the height of the balloon (feet).

We let t represent time and assume θ and y to be differentiable functions of t.

The one constant in the picture is the distance from the range finder to the point of liftoff (500 ft). There is no need to give it a special symbol.

STEP 2: *We write down the additional numerical information:*

$$\frac{d\theta}{dt} = 0.14 \text{ rad/min} \qquad \text{when} \quad \theta = \frac{\pi}{4}.$$

STEP 3: *We write down what we are asked to find:* We are asked to find dy/dt when $\theta = \pi/4$.

STEP 4: *We write an equation that relates the variables:* The equation that relates y to θ is

$$\frac{y}{500} = \tan\theta \qquad \text{or} \qquad y = 500 \tan\theta.$$

STEP 5: *We differentiate with respect to t to find how dy/dt* (which we want) *is related to dθ/dt* (which we know):

$$\frac{dy}{dt} = 500 \sec^2\theta \, \frac{d\theta}{dt}.$$

STEP 6: *We evaluate with* $\theta = \pi/4$ and $d\theta/dt = 0.14$ *to find dy/dt*:

$$\frac{dy}{dt} = 500(\sqrt{2})^2(0.14) = (1000)(0.14) = 140. \qquad \left(\sec\frac{\pi}{4} = \sqrt{2}\right)$$

At the moment in question, the balloon is rising at the rate of 140 ft/min. ■

Strategy for Related Rate Problems

1. *Draw a picture and name the variables and constants.* Use t for time. Assume all variables are differentiable functions of t.
2. *Write down the numerical information* (in terms of the symbols you have chosen).
3. *Write down what you are asked to find* (usually a rate, expressed as a derivative).
4. *Write an equation that relates the variables.* You may have to combine two or more equations to get a single equation that relates the variable whose rate you want to the variable whose rate you know.
5. *Differentiate* with respect to t to express the rate you want in terms of the rate and variables whose values you know.
6. *Evaluate.*

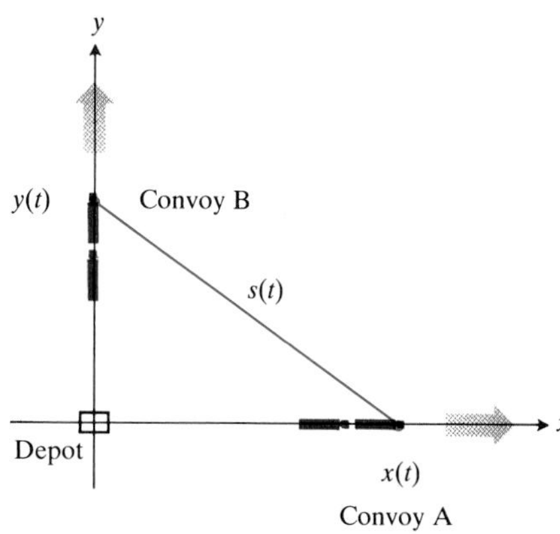

4.3 If you know where the convoys are and how fast they are moving, you can calculate how fast the distance between them is growing (Example 4).

EXAMPLE 4 *Truck convoys.* Two truck convoys leave a depot, convoy A traveling east at 40 mph and convoy B traveling north at 30 mph. How fast is the distance between the convoys changing 6 min later, when convoy A is 4 mi from base and convoy B is 3 mi from base?

Solution We carry out the steps of the basic strategy.

STEP 1: *Picture and variables*: We picture the convoys in the coordinate plane, using the positive x-axis as the eastbound highway and the positive y-axis as the northbound highway (Fig. 4.3). We let t represent time and set

$$x(t) = \text{position of convoy A},$$

$$y(t) = \text{position of convoy B},$$

$$s(t) = \text{distance between convoys}.$$

We assume x, y, and s to be differentiable functions of t.

STEP 2: *Numerical information:* At the time in question,

$$x = 4 \text{ mi}, \qquad y = 3 \text{ mi}, \qquad \frac{dx}{dt} = 40 \text{ mph}, \qquad \frac{dy}{dt} = 30 \text{ mph}.$$

STEP 3: *To find:* $\dfrac{ds}{dt}$.

STEP 4: *How the variables are related:* $s^2 = x^2 + y^2$. (Pythagorean theorem)

(The equation $s = \sqrt{x^2 + y^2}$ would also work.)

STEP 5: *Differentiate with respect to t:*

$$2s\frac{ds}{dt} = 2x\frac{dx}{dt} + 2y\frac{dy}{dt}$$

$$\frac{ds}{dt} = \frac{1}{s}\left(x\frac{dx}{dt} + y\frac{dy}{dt}\right).$$

STEP 6: *Evaluate* with $x = 4$, $y = 3$, $(dx/dt) = 40$, $(dy/dt) = 30$:

$$\frac{ds}{dt} = \frac{1}{\sqrt{4^2 + 3^2}}(4(40) + 3(30)) = \frac{1}{5}(160 + 90) = \frac{250}{5} = 50.$$

At the moment in question, the distance between the convoys is growing at the rate of 50 mph. ■

EXAMPLE 5 *Relief from a heart attack.* A heart attack victim has been given a blood vessel dilator to lower the pressure against which the heart has to pump. For a short while after the drug is administered, the radii of the affected blood vessels will increase at about 1% per minute. According to Poiseuille's law, $V = kr^4$ (Section 3.7, Example 10), what percentage rate of increase can we expect in the blood flow over the next few minutes (all other things being equal)?

Solution STEP 1: *Picture and variables:* We really don't need a picture, and the variables r and V are already named. It remains only to assume that r and V are differentiable functions of time t.

STEP 2: *Numerical information:*

$$\frac{dr/dt}{r} = \frac{1}{100}. \qquad (r \text{ increases at } 1\%/\text{min.})$$

STEP 3: *To find:*

$$\frac{dV/dt}{V}. \qquad \text{(What percentage increase can we expect in the blood flow?)}$$

STEP 4: *How the variables are related:* $V = kr^4$.

STEP 5: *Differentiate* (and in this case divide as well) to find how $(dV/dt)/V$ is related to $(dr/dt)/r$:

$$\frac{dV}{dt} = 4kr^3\frac{dr}{dt}$$

$$\frac{dV/dt}{V} = \frac{4kr^3}{kr^4}\frac{dr}{dt} = 4\frac{dr/dt}{r}. \qquad (\text{Divide by } V = kr^4.)$$

STEP 6: *Evaluate* by substituting $(dr/dt)/r = 1/100$ to find $(dV/dt)/V$:

$$\frac{dV/dt}{V} = 4\left(\frac{1}{100}\right) = \frac{4}{100}.$$

The blood flow will increase 4%/min. ■

Exercises 4.1

1. The radius r and area $A = \pi r^2$ of a circle are differentiable functions of t. Write an equation that relates dA/dt to dr/dt.

2. The radius r and surface area $S = 4\pi r^2$ of a sphere are differentiable functions of t. Write an equation that relates dS/dt to dr/dt.

3. The side length x and volume $V = x^3$ of a cube are differentiable functions of t. Write an equation that relates dV/dt to dx/dt.

4. The radius r and volume $V = (1/3)\pi r^2 h$ of a right circular cone are differentiable functions of t. How is dV/dt related to dr/dt if h is constant?

5. The height h and volume $V = (1/3)\pi r^2 h$ of a right circular cone are differentiable functions of t. How is dV/dt related to dh/dt if r is constant?

6. Let $x(t)$ and $y(t)$ be differentiable functions of t and let $s = \sqrt{x^2 + y^2}$ be the distance between the points $(x, 0)$ and $(0, y)$ in the xy-plane. How is ds/dt related to dx/dt and dy/dt?

7. *Heating a plate.* When a circular plate of metal is heated in an oven, its radius increases at the rate of 0.01 cm/min. At what rate is the plate's area increasing when the radius is 50 cm?

8. *Changing voltage.* Ohm's law for electrical circuits like the one here states that $V = IR$, where V is the voltage, I is the current in amperes, and R is the resistance in ohms. Suppose that V is increasing at the rate of 1 volt/sec while I is decreasing at the rate of 1/3 amp/sec. Let t denote time in seconds.
   a) What is the value of dV/dt?
   b) What is the value of dI/dt?
   c) What equation relates dR/dt to dV/dt and dI/dt?
   d) Find the rate at which R is changing when $V = 12$ volts and $I = 2$ amp. Is R increasing or decreasing?

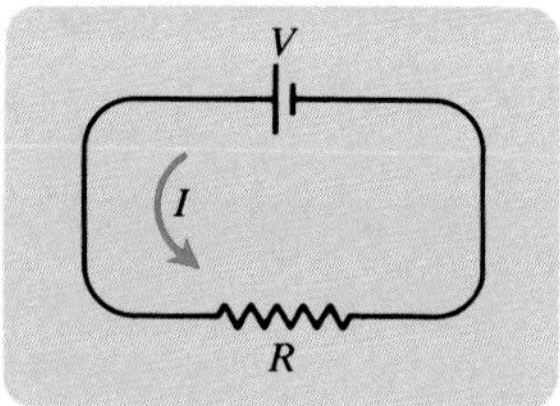

9. *Changing dimensions in a rectangle.* The length l of a rectangle is decreasing at the rate of 2 cm/sec while the width w is increasing at the rate of 2 cm/sec. When $l = 12$ cm and $w = 5$ cm, find the rates of change of (a) the area, (b) the perimeter, and (c) the lengths of the diagonals of the rectangle. Which of these quantities are decreasing and which are increasing?

10. *Commercial air traffic.* Two commercial jets at 40,000 ft are flying at 520 mph along straight-line courses that cross at right angles. How fast is the distance between the planes closing when plane A is 5 mi from the intersection point and plane B is 12 mi from the intersection point?

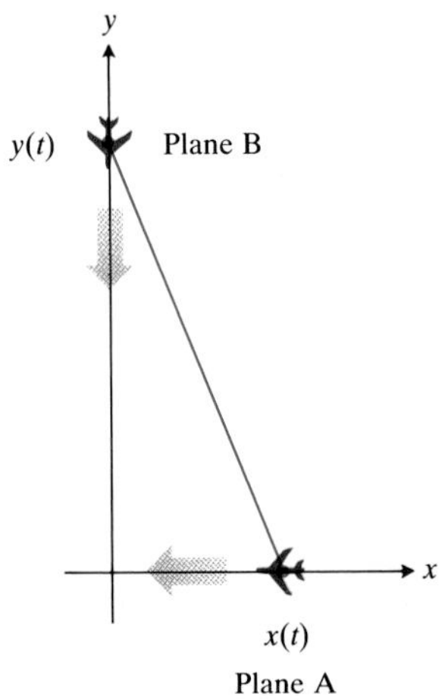

11. *A sliding ladder.* A 13-ft ladder is leaning against a house when its base starts to slide away. By the time the base is 12 ft from the house, the base is moving at the rate of 5 ft/sec. How fast is the top of the ladder sliding down the wall then? How fast is the area of the triangle formed by the ladder, wall, and ground changing?

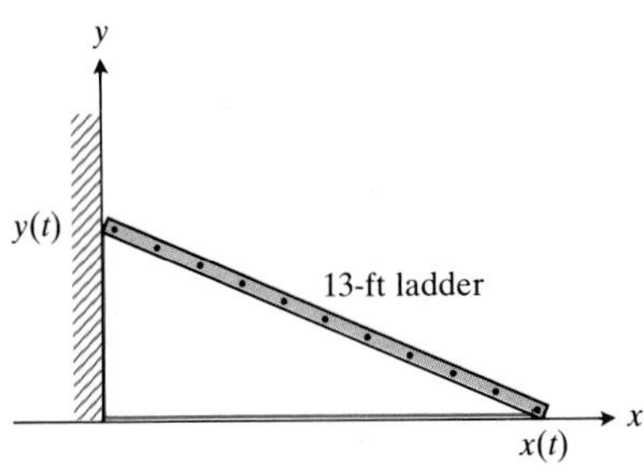

12. *A shrinking lollipop.* A spherical Tootsie Roll Pop you are sucking on is giving up mass at a steady rate of 800 mg/min. How fast will the radius be decreasing when the Tootsie Roll Pop is 20 mm across?

13. *Boring a cylinder.* The mechanics at Lincoln Automotive are reboring a 6-in.-deep cylinder to fit a new piston. The machine they are using increases the cylinder's radius one-thousandth of an inch every 3 min. How rapidly is the cylinder volume increasing when the bore (diameter) is 3.80 in.?

14. *A growing sand pile.* Sand falls from a conveyor belt at the rate of 10 ft^3/min onto a conical pile. The radius of the base of the pile is always equal to half the pile's height. How fast is the height growing when the pile is 5 ft high?

15. *A growing raindrop.* Suppose that a drop of mist is a perfect sphere and that, through condensation, the drop picks up moisture at a rate proportional to its surface area. Show that under these circumstances the drop's radius increases at a constant rate.

16. *The radius of an inflating balloon.* A spherical balloon is inflated with helium at the rate of 100π ft^3/min. How fast is the balloon's radius increasing at the instant the radius is 5 ft? How fast is the surface area increasing?

17. *Hauling in a dinghy.* A dinghy is pulled toward a dock by a rope from the bow through a ring on the dock 6 ft above the bow. If the rope is hauled in at the rate of 2 ft/sec, how fast is the boat approaching the dock when 10 ft of rope are out?

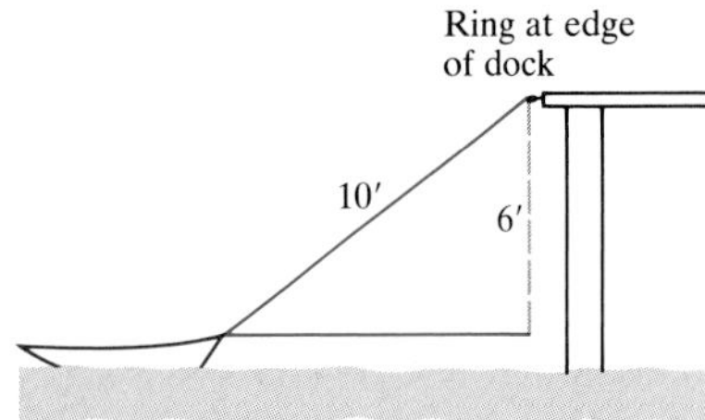

18. *A balloon and a bicycle.* A balloon is rising vertically above a level, straight road at a constant rate of 1 ft/sec. Just when the balloon is 65 ft above the ground, a bicycle passes under it, going 17 ft/sec. How fast is the distance between the bicycle and balloon increasing 3 sec later?

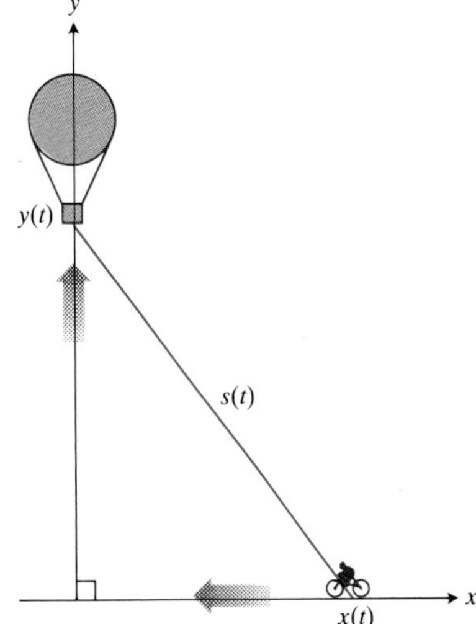

19. *Making coffee.* Coffee is draining from a full conical filter basket into a cylindrical coffee pot at the rate of 10 in^3/min. How fast is the level in the pot rising when the height of coffee in the filter is 5 in.? How fast is the level in the filter cone falling?

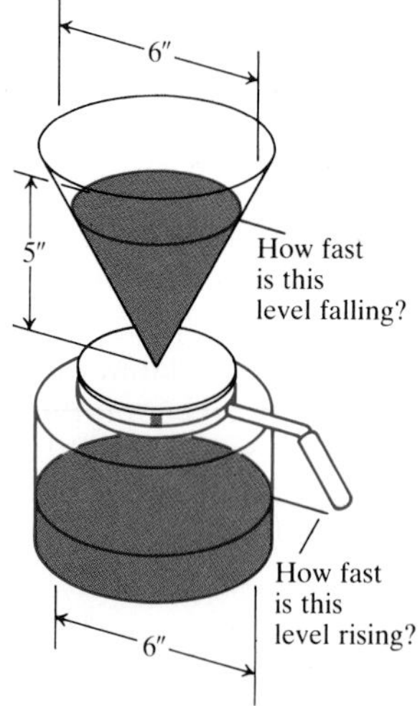

20. *Blood flow.* Cold water has the effect of contracting the blood vessels in the hands, and the radius of a typical vein might decrease at the rate of 20%/min. According to Poiseuille's law, $V = kr^4$ (see Example 5), at what percentage rate can we expect the volume of blood flowing through that vein to decrease?

21. *Cardiac output.* In the late 1860s, Adolfe Fick, a professor of physiology in the Faculty of Medicine in Würtzberg, Germany, developed the method we use today for measuring how much blood your heart pumps in a minute. Your cardiac output as you read this sentence is probably about 7 liters a minute. At rest it is likely to be a bit under 6 L/min. If you are a trained marathon runner running a marathon, your cardiac output can be as high as 30 L/min.

Your cardiac output can be calculated with the formula

$$y = \frac{Q}{D},$$

where Q is the number of milliliters of CO_2 you exhale in a minute and D is the difference between the CO_2 concentration (ml/L) in the blood pumped to the lungs and the CO_2 concentration in the blood returning from the lungs. With $Q = 233$ ml/min and $D = 97 - 56 = 41$ ml/min,

$$y = \frac{233 \text{ ml/min}}{41 \text{ ml/L}} \approx 5.95 \text{ L/min},$$

close to the 6 L/min that most people have at basal (resting) conditions. (Data courtesy of J. Kenneth Herd, M.D., Quillen Dishner College of Medicine, East Tennessee State University.)

Suppose that when $Q = 233$ and $D = 41$, we also know that D is decreasing at the rate of 2 units a minute but that Q remains unchanged. What is happening to the cardiac output?

22. *Filling a conical tank.* Water runs into the conical tank shown here at the rate of 9 ft^3/min. How fast is the water level rising when the water is 6 ft deep?

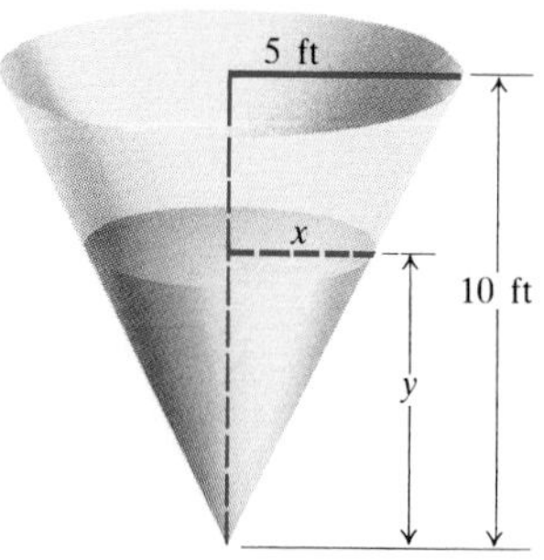

23. *Moving along a parabola.* A particle moves along the parabola $y = x^2$ in the first quadrant in such a way that its x-coordinate increases at a steady 10 m/sec. How fast is the angle of inclination θ of the line joining the particle to the origin changing when $x = 3$ m? What is the limiting value of $d\theta/dt$ as $x \to \infty$?

24. *Cost, revenue, and profit.* A company can manufacture x items at a cost of $c(x)$ dollars, a sales revenue of $r(x)$ dollars, and a profit of $p(x) = r(x) - c(x)$ dollars (everything in thousands). Find the rates of change of cost, revenue, and profit for the following values of x and dx/dt.

a) $r(x) = 9x$, $c(x) = x^3 - 6x + 15x$, and $dx/dt = 0.1$ when $x = 2$

b) $r(x) = 70x$, $c(x) = x^3 - 6x^2 + 45/x$, and $dx/dt = 0.05$ when $x = 1.5$

25. *A moving shadow.* A man 6 ft tall walks at the rate of 5 ft/sec toward a streetlight that is 16 ft above the

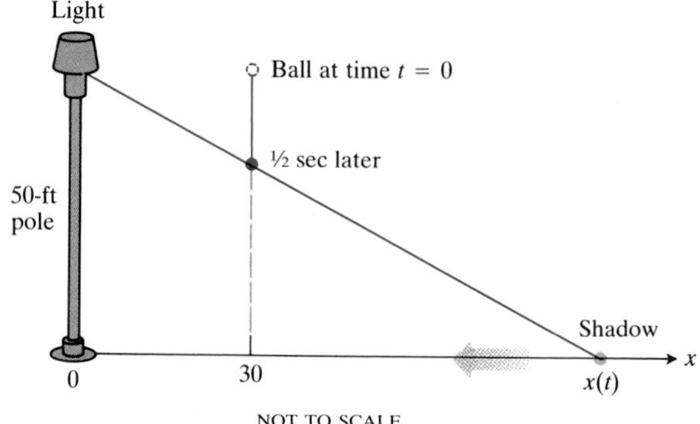

ground. At what rate is the tip of his shadow moving? At what rate is the length of his shadow changing when he is 10 ft from the base of the light?

26. *Another moving shadow.* A light shines from the top of a pole 50 ft high. A ball is dropped from the same height from a point 30 ft away from the light. How fast is the shadow of the ball moving along the ground 1/2 sec later? (Assume the ball falls a distance $s = 16t^2$ ft in t sec.)

27. *Flying a kite.* A girl flies a kite at a height of 300 ft, the wind carrying the kite horizontally away from her at a rate of 25 ft/sec. How fast must she let out the string when the kite is 500 ft away from her?

28. *A melting ice layer.* A spherical iron ball 8 in. in diameter is coated with a layer of ice of uniform thickness. If the ice melts at the rate of 10 in^3/min, how fast is the thickness of the ice decreasing when it is 2 in. thick? How fast is the outer surface area of ice decreasing?

29. *Highway patrol.* A highway patrol plane flies 3 mi above a level, straight road at a steady ground speed of 120 mph. The pilot sees an oncoming car and determines with radar that the line-of-sight distance from the plane to the car is 5 mi and decreasing at the rate of 160 mph. Find the car's speed along the highway.

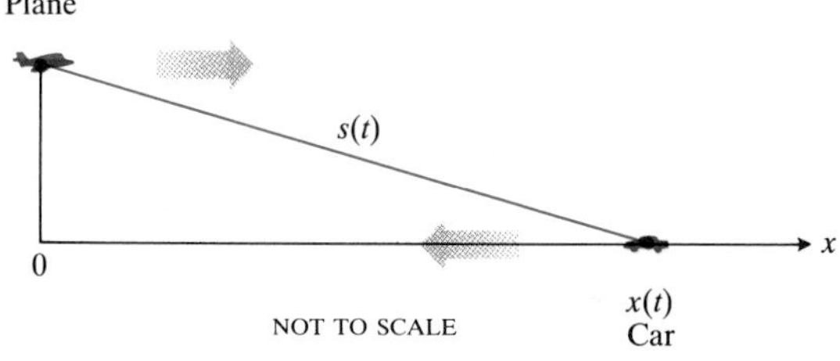

30. *The sun's shadow.* On a morning when the sun will pass directly overhead, the shadow of an 80-ft building on level ground is 60 ft long. At the moment in question, the angle θ the sun makes with the ground is increasing at the rate of 0.27°/min. At what rate is the shadow decreasing? (Remember to use radians. Express your answer in inches per minute, to the nearest tenth.)

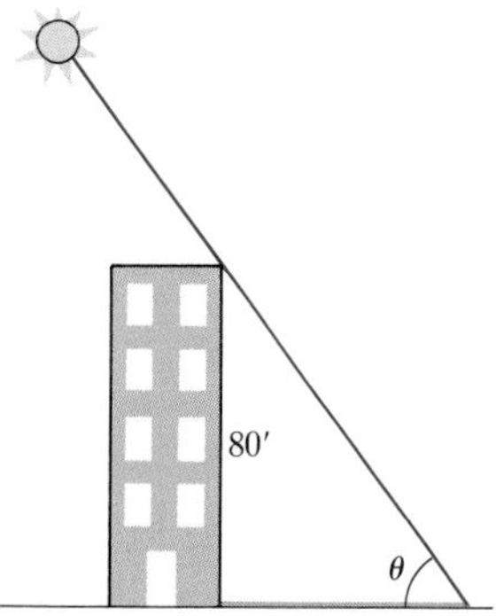

31. *Ships.* Two ships are steaming straight away from a point O along routes that make a 120° angle. Ship A moves at 14 knots (nautical miles per hour—a nautical mile is 2000 yd). Ship B moves at 21 knots. How fast are the ships moving apart when $OA = 5$ and $OB = 3$?

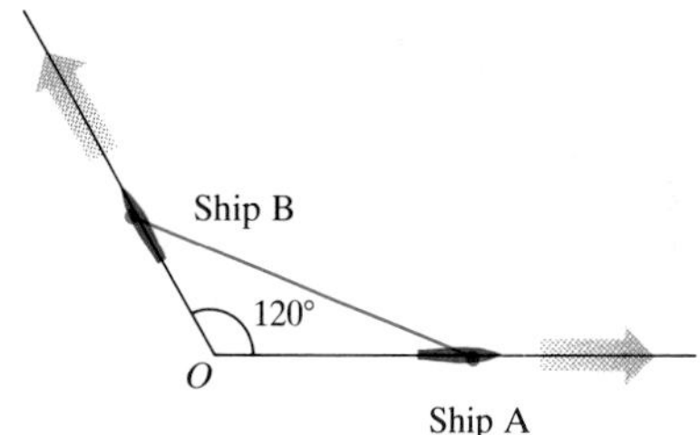

4.2 Maxima, Minima, and the Mean Value Theorem

Differential calculus is the mathematics of working with derivatives. One of the things we can do with derivatives is find where functions take on their maximum and minimum values. In this section we lay the theoretical ground for finding these extreme values, as they are called. This will enable us to graph polynomials and rational functions effectively (in Sections 4.3 and 4.4) and to solve important optimization problems that come up in other fields (Section 4.5).

We also introduce the Mean Value Theorem, one of the most exciting and influential theorems in calculus.

Maxima and Minima—Relative vs. Absolute

Figure 4.4 shows a point c where a function $y = f(x)$ has a maximum value. If we move to either side of c, the function values get smaller and the curve falls away. When we take in more of the curve, however, we find that f assumes an even larger value at d. Thus, $f(c)$ is not the absolute maximum value of f on the interval $[a, b]$ but only a relative or local maximum value.

DEFINITIONS

A function f has a **local maximum** value at an interior point c of its domain if $f(x) \leq f(c)$ for all x in some open interval I about c. The function has an **absolute maximum** value at c if $f(x) \leq f(c)$ for all x in the domain.

Similarly, f has a **local minimum** value at an interior point c of its domain if $f(x) \geq f(c)$ for all x in an open interval I about c. The function has an **absolute minimum** value at c if $f(x) \geq f(c)$ for all x in the domain.

The definitions of local maximum and local minimum are extended to endpoints of the function's domain by requiring the intervals I to be appropriate half-open intervals containing the endpoints.

Notice that an absolute maximum is also a local maximum because it is the largest value in its immediate neighborhood as well as overall. Hence a list of all local maxima will include the absolute maximum if there is one. Similarly, an absolute minimum, when it exists, is also a local minimum. A list of all local minima will include the absolute minimum if there is one.

The First Derivative Theorem

In Fig. 4.4 two extreme values of f occur at endpoints of the function's domain, one occurs at a point where f' fails to exist, and two occur at interior

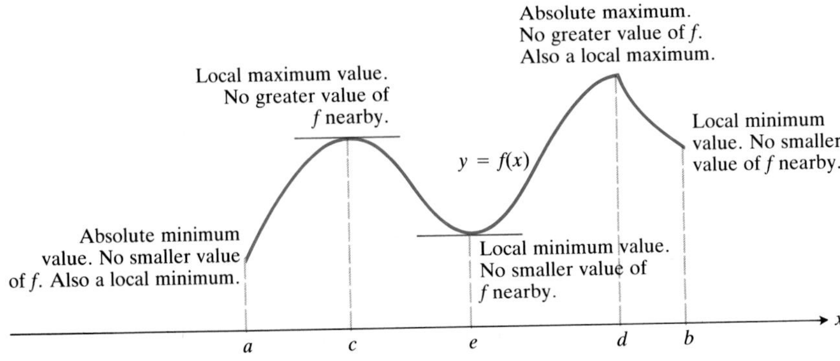

4.4 How to classify maxima and minima.

points where $f' = 0$. This is typical for a function defined on a closed interval. As the following theorem says, a function's first derivative is always zero at an interior point where the function has a local extreme value. Hence the only places where a function f can ever have an extreme value are

1. Interior points where f' is zero,
2. Interior points where f' does not exist,
3. Endpoints of the function's domain.

We shall see the importance of this observation as the chapter continues.

THEOREM 1

The First Derivative Theorem for Local Extreme Values

If a function f has a local maximum or a local minimum value at an interior point c of an interval where it is defined, and if f' is defined at c, then

$$f'(c) = 0.$$

PROOF You may not have seen an argument like the one we are about to use, so we shall explain its form first. We want to show that $f'(c) = 0$, and our plan is to do that indirectly by showing first that $f'(c)$ cannot be positive and second that $f'(c)$ cannot be negative either. Why does that show $f'(c) = 0$? Because, in the entire real number system, only one number is neither positive nor negative, and that number is zero.

To be specific, suppose f has a local maximum value at $x = c$, so that $f(x) \leq f(c)$ for all values of x near c (Fig. 4.5). Since c is an interior point of f's domain, the limit

$$\lim_{x \to c} \frac{f(x) - f(c)}{x - c} \qquad (1)$$

defining $f'(c)$ is two-sided. This means that the right-hand and left-hand limits both exist at $x = c$, and both equal $f'(c)$.

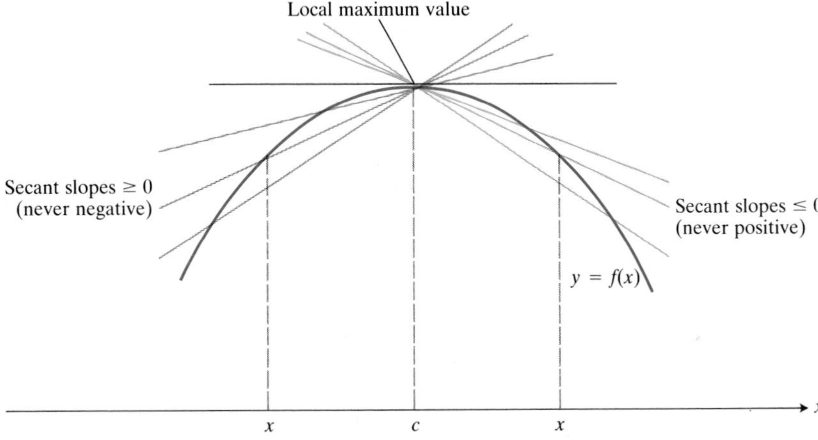

4.5 A curve with a local maximum value. The slope at c, simultaneously the limit of nonpositive numbers and nonnegative numbers, is zero.

When we examine these limits separately, we find that

$$\lim_{x \to c^+} \frac{f(x) - f(c)}{x - c} \leq 0 \tag{2}$$

because, immediately to the right of c, $f(x) \leq f(c)$ and $x - c > 0$. Similarly,

$$\lim_{x \to c^-} \frac{f(x) - f(c)}{x - c} \geq 0 \tag{3}$$

because, immediately to the left of c, $f(x) \leq f(c)$ and $x - c < 0$.

The inequality in (2) says that $f'(c)$ cannot be greater than zero, whereas (3) says that $f'(c)$ cannot be less than zero. So $f'(c) = 0$.

This proves the theorem for local maximum values. To prove it for local minimum values, simply replace f by $-f$ and run through the argument again. ■

Rolle's Theorem

There is strong geometric evidence that between any two points where a smooth curve crosses the x-axis there is a point on the curve where the tangent is horizontal. A 300-year-old theorem of Michel Rolle (1652–1719) assures us that this is indeed the case.

When the French mathematician Michel Rolle published his theorem in 1691, his goal was to show that between every two zeros of a polynomial function there always lies a zero of the polynomial we now know to be the function's derivative. (The version of the theorem we have proved here is not restricted to polynomials.)

Rolle distrusted the new methods of calculus, however, and spent a great deal of time and energy denouncing its use and attacking l'Hôpital's all too popular (he felt) calculus book. It is ironic that Rolle is known today only for his inadvertent contribution to a field he tried to suppress.

THEOREM 2

Rolle's Theorem

Suppose that $y = f(x)$ is continuous at every point of the closed interval $[a, b]$ and differentiable at every point of its interior (a, b). If

$$f(a) = f(b) = 0,$$

then there is at least one number c between a and b at which

$$f'(c) = 0.$$

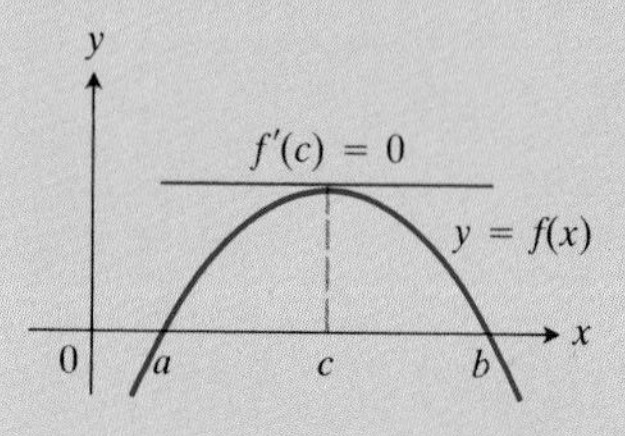

PROOF OF ROLLE'S THEOREM We know from Section 2.4 that a continuous function defined on a closed interval assumes absolute maximum and minimum values on the interval. The question is, where? Theorem 1 tells us there are only three kinds of places to look:

1. At interior points where f' is zero,
2. At interior points where f' does not exist,
3. At the endpoints of the function's domain, in this case a and b.

By hypothesis, f has a derivative at every interior point. That rules out (2), leaving us with interior points where $f' = 0$ and with the two endpoints a and b.

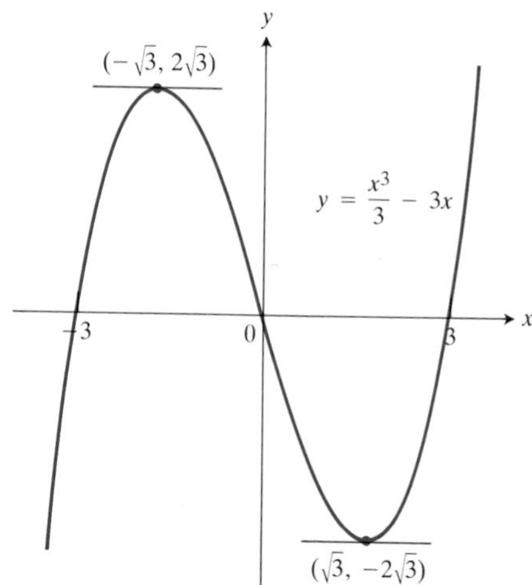

4.6 As predicted by Rolle's theorem, this smooth curve has horizontal tangents between the points where it crosses the x-axis.

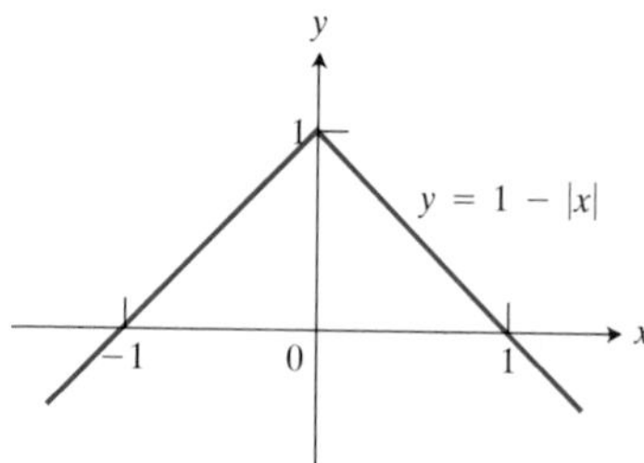

4.7 This curve has no horizontal tangent between the points where it crosses the x-axis.

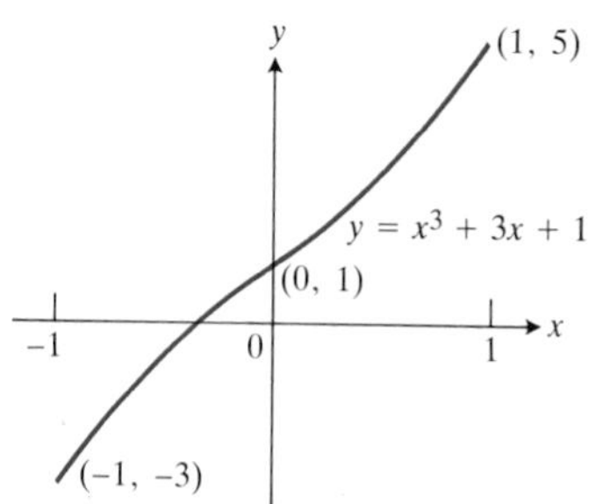

4.8 The only real zero of the polynomial $y = x^3 - 3x + 1$ is the one shown here between -1 and 0.

If either the maximum or the minimum occurs at a point c inside the interval, then $f'(c) = 0$ by Theorem 1, and we have found a point for Rolle's theorem.

If both the maximum and the minimum occur at the endpoints a and b where f is zero, then zero is the maximum value of f as well as the minimum value of f. In other words, for every value of x,

$$0 = \min(f) \leq f(x) \leq \max(f) = 0.$$

So f is zero throughout the interval, and, because f has a constant value, its derivative is zero throughout the interval. In other words, $f'(c) = 0$ at every interior point. Either way, we find a point c in (a, b) where $f'(c)$ is zero. This concludes the proof. ■

EXAMPLE 1 The polynomial function

$$f(x) = \frac{x^3}{3} - 3x$$

graphed in Fig. 4.6 is continuous at every point of the interval $-3 \leq x \leq 3$ and differentiable at every point of the interval $-3 < x < 3$. Since $f(-3) = f(3) = 0$, Rolle's theorem says that f' must be zero at least once in the open interval between $a = -3$ and $b = 3$. In fact, $f'(x) = x^2 - 3$ is zero twice in this interval, once at $x = -\sqrt{3}$ and again at $x = \sqrt{3}$. ■

EXAMPLE 2 As the function $f(x) = 1 - |x|$ shows (Fig. 4.7), the differentiability of f is essential to Rolle's theorem. If we allow even one interior point in (a, b) where f is not differentiable, there may be no horizontal tangent to the curve. ■

Finding Solutions of Equations

When we solve equations by Newton's method on a calculator, we may also want to know how many solutions there are in a given interval $[a, b]$. With Rolle's theorem we can sometimes find out.

Suppose, for example, that

1. f is continuous on $[a, b]$ and differentiable on (a, b),
2. $f(a)$ and $f(b)$ have opposite signs,
3. $f' \neq 0$ between a and b.

Then f has exactly one zero between a and b: It cannot have more than one because f' would then have a zero too. Yet it has at least one, by the Intermediate Value Theorem of Section 2.4.

EXAMPLE 3 The function $f(x) = x^3 + 3x + 1$ is continuous and differentiable on the interval $-1 \leq x \leq 1$, $f(-1) = -3$ and $f(1) = 5$ have opposite signs, and $f'(x) = 3x^2 + 3$ is never zero. Therefore the equation $x^3 + 3x + 1 = 0$ has only one real solution in the interval $-1 \leq x \leq 1$ (Fig. 4.8).

In this case we can also tell that f has only one zero, period. If f had more than one, f' would have a zero, and it doesn't. ■

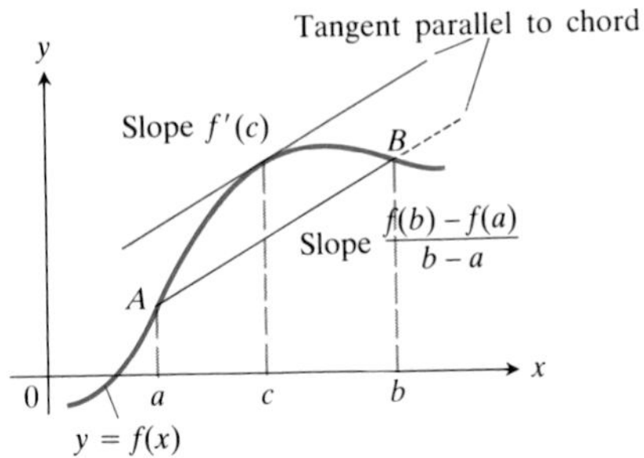

4.9 Geometrically, the Mean Value Theorem says that somewhere between A and B the curve has at least one tangent parallel to chord AB.

The Mean Value Theorem

The Mean Value Theorem is Rolle's theorem for a chord instead of an interval. You will see what we mean if you look at Fig. 4.9. The figure shows the graph of a differentiable function f defined on an interval $a \le x \le b$. There is a point on the curve where the tangent is parallel to the chord AB. In Rolle's theorem, the line AB is the x-axis and $f'(c) = 0$. Here the line AB is a chord joining the endpoints of the curve above a and b, and $f'(c)$ is the slope of the chord.

THEOREM 3

The Mean Value Theorem

If $y = f(x)$ is continuous at every point of the closed interval $[a, b]$ and differentiable at every point of its interior (a, b), then there is at least one number c between a and b at which

$$\frac{f(b) - f(a)}{b - a} = f'(c). \tag{4}$$

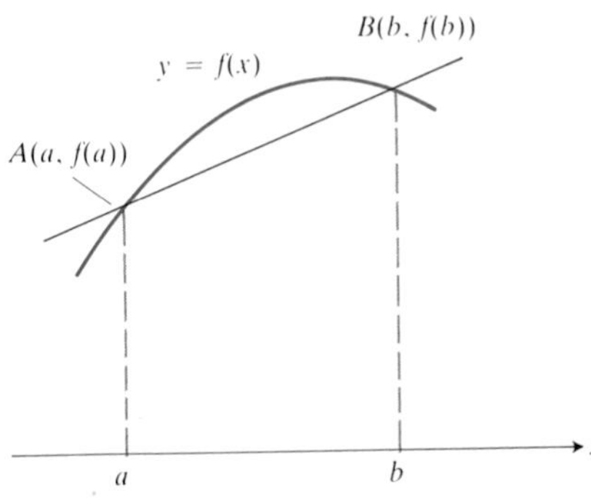

4.10 The graph of f, and the chord AB over the interval $a \le x \le b$.

PROOF If we graph f over $[a, b]$ and draw the line through the endpoints $A[a, f(a)]$ and $B[b, f(b)]$, the figure we get, Fig. 4.10, resembles the one we drew for Rolle's theorem. The difference is that the line AB need not be the x-axis because $f(a)$ and $f(b)$ may not be zero. We cannot apply Rolle's theorem directly to f, but we can apply it to the function that measures the vertical distance between the graph of f and the line AB. This, it turns out, will tell us what we want to know about the derivative of f.

The line AB is the graph of the function

$$g(x) = f(a) + \frac{f(b) - f(a)}{b - a}(x - a) \tag{5}$$

(point–slope equation), and the formula for the vertical distance between the graphs of f and g at x is

$$d(x) = f(x) - g(x) = f(x) - f(a) - \frac{f(b) - f(a)}{b - a}(x - a). \tag{6}$$

Figure 4.11 shows the graphs of f, g, and d together.

The function d satisfies the hypotheses of Rolle's theorem on the interval $[a, b]$. It is continuous on $[a, b]$ because f and g are. Both $d(a)$ and $d(b)$ are zero because the graphs of f and g pass through A and B.

Therefore $d' = 0$ at some point c between a and b. To see what this says about f', we differentiate both sides of Eq. (6) with respect to x and set $x = c$. This gives

$$d'(x) = f'(x) - \frac{f(b) - f(a)}{b - a}, \qquad \text{(Derivative of Eq. (6)...)}$$

$$d'(c) = f'(c) - \frac{f(b) - f(a)}{b - a}, \qquad \text{(... with } x = c\text{)}$$

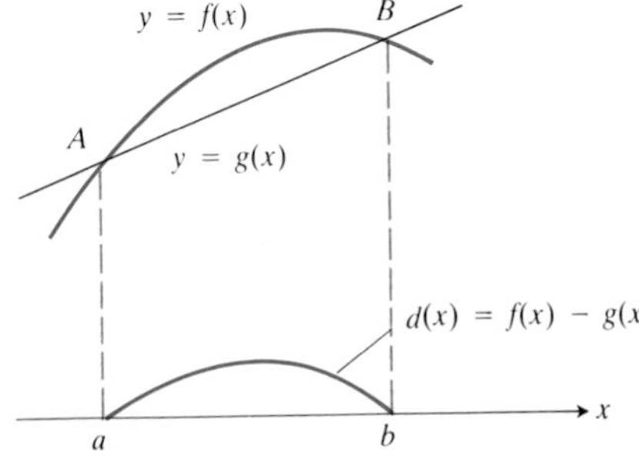

4.11 The chord AB in Fig. 4.10 is the graph of the function $g(x)$. The function $d(x) = f(x) - g(x)$ gives the vertical distance between the graphs of f and g at x.

$$0 = f'(c) - \frac{f(b) - f(a)}{b - a}, \qquad (d'(c) = 0)$$

$$f'(c) = \frac{f(b) - f(a)}{b - a}, \qquad \text{(Rearranged)}$$

which is what we set out to prove. ■

If $f'(x)$ is continuous on $[a, b]$, then the Max–Min Theorem for continuous functions in Section 2.4 tells us that f' has an absolute maximum value $\max f'$ and an absolute minimum value $\min f'$ on the interval. Since the number $f'(c)$ can neither exceed $\max f'$ nor be less than $\min f'$, the equation

$$\frac{f(b) - f(a)}{b - a} = f'(c) \tag{7}$$

gives us the inequality

$$\min f' \le \frac{f(b) - f(a)}{b - a} \le \max f'. \tag{8}$$

The importance of the Mean Value Theorem lies in the estimates that sometimes come from Eq. (8) and in the mathematical conclusions that come from Eq. (7), one of which we shall see in a moment.

We usually do not know any more about the number c than the theorem tells us, which is that c exists. In a few cases we can satisfy our curiosity about the identity of c, as in the next example. Keep in mind, however, that our ability to identify c is the exception rather than the rule, and the importance of the Mean Value Theorem lies elsewhere.

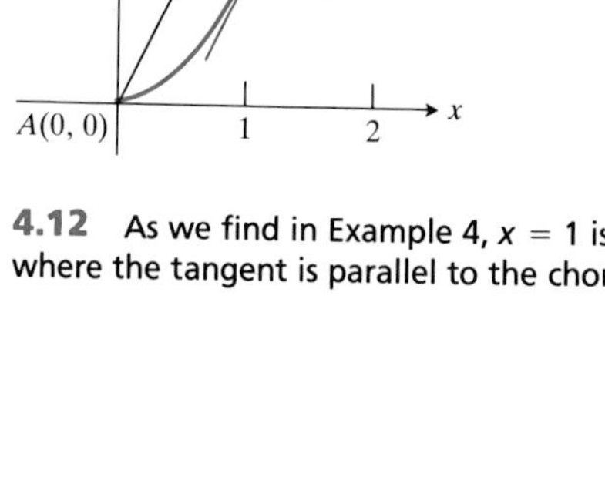

4.12 As we find in Example 4, $x = 1$ is where the tangent is parallel to the chord.

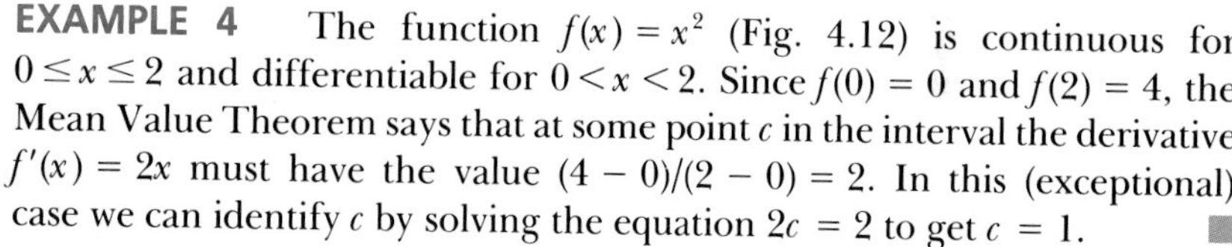

EXAMPLE 4 The function $f(x) = x^2$ (Fig. 4.12) is continuous for $0 \le x \le 2$ and differentiable for $0 < x < 2$. Since $f(0) = 0$ and $f(2) = 4$, the Mean Value Theorem says that at some point c in the interval the derivative $f'(x) = 2x$ must have the value $(4 - 0)/(2 - 0) = 2$. In this (exceptional) case we can identify c by solving the equation $2c = 2$ to get $c = 1$. ■

EXAMPLE 5 The function $y = \sqrt{1 - x^2}$ (Fig. 4.13) satisfies the hypotheses (and conclusion) of the Mean Value Theorem on the interval $-1 \le x \le 1$. It is continuous on the closed interval, and its derivative

$$y' = \frac{-x}{\sqrt{1 - x^2}}$$

is defined at every interior point. The graph has a horizontal tangent at $x = 0$.

Notice that the function is not differentiable at $x = -1$ and $x = 1$. It does not need to be for the theorem to apply. ■

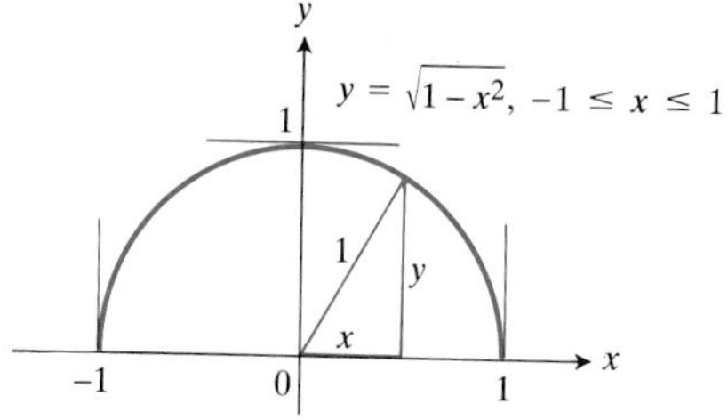

4.13 The function $y = \sqrt{1 - x^2}$ satisfies the hypotheses (and conclusion) of the Mean Value Theorem on the interval $[-1, 1]$ despite the presence of vertical tangents at $x = 1$ and $x = -1$.

Physical Interpretations

If we think of $(f(b) - f(a))/(b - a)$ as the average change in f over $[a, b]$ and $f'(c)$ as an instantaneous change, then the Mean Value Theorem says that the instantaneous change at some interior point must equal the average change over the entire interval.

EXAMPLE 6 If a car takes 8 sec to drive 352 ft, its average velocity for the 8-sec interval is $352/8 = 44$ ft/sec, or 30 mph. At some point during the acceleration, the theorem says, the speedometer must read exactly 30. ■

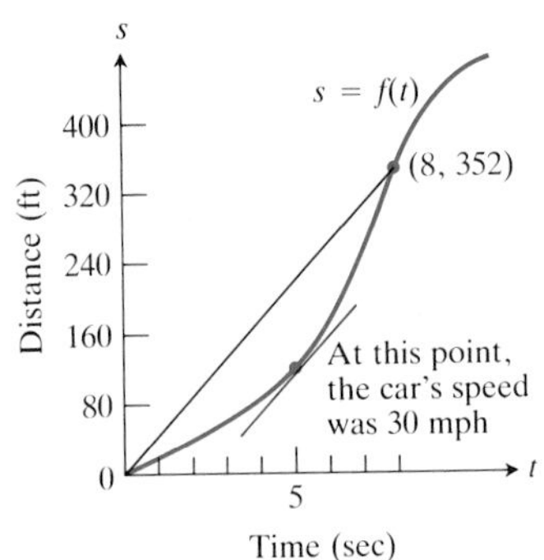

The First Corollary

The Mean Value Theorem is famous for, among other things, three important corollaries. The first, which we shall get to in a moment, says exactly when graphs rise and fall. The second, which we shall come to in Section 4.6, says that only constant functions can have zero derivatives. The third, also in Section 4.6, says that functions with identical derivatives must differ at most by a constant value.

You may have noticed that differentiable functions all seem to increase when their derivatives are positive and to decrease when their derivatives are negative. The corollary we are about to prove says that this is always true. To prove the corollary, we need precise definitions of *increasing* and *decreasing*.

DEFINITIONS

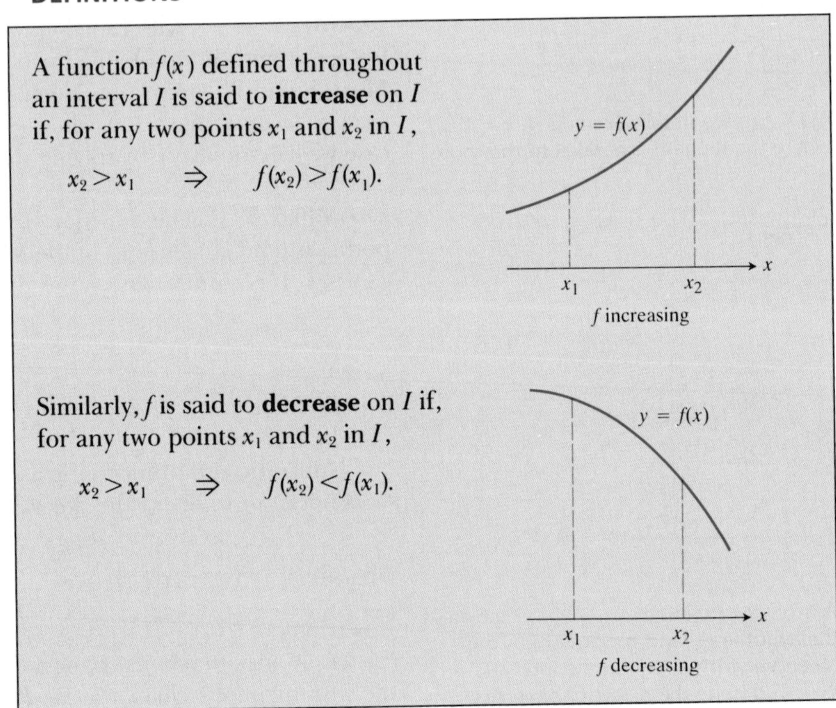

A function $f(x)$ defined throughout an interval I is said to **increase** on I if, for any two points x_1 and x_2 in I,

$$x_2 > x_1 \quad \Rightarrow \quad f(x_2) > f(x_1).$$

Similarly, f is said to **decrease** on I if, for any two points x_1 and x_2 in I,

$$x_2 > x_1 \quad \Rightarrow \quad f(x_2) < f(x_1).$$

COROLLARY 1

The First Derivative Test for Increasing and Decreasing: f increases when $f' > 0$ and decreases when $f' < 0$.

Suppose that f is continuous at each point of the closed interval $[a, b]$ and differentiable at each point of its interior (a, b). If $f' > 0$ at each point of (a, b), then f increases throughout $[a, b]$. If $f' < 0$ at each point of (a, b), then f decreases throughout $[a, b]$. In either case, f is one-to-one.

PROOF Let x_1 and x_2 be any two numbers in $[a, b]$ with $x_1 < x_2$. Apply the Mean Value Theorem to f on $[x_1, x_2]$:

$$f(x_2) - f(x_1) = f'(c)(x_2 - x_1) \tag{9}$$

for some c between x_1 and x_2. The sign of the right-hand side of Eq. (9) is the same as the sign of $f'(c)$ because $x_2 - x_1$ is positive. Therefore

$$f(x_2) > f(x_1) \qquad \text{if } f'(x) \text{ is positive on } (a, b)$$

(f is increasing) and

$$f(x_2) < f(x_1) \qquad \text{if } f'(x) \text{ is negative on } (a, b)$$

(f is decreasing). In either case, $x_1 \neq x_2$ implies that $f(x_1) \neq f(x_2)$, so f is one-to-one. ■

Notice that we were also able to show that a function has to be one-to-one on any interval where its derivative is positive or its derivative is negative. Knowing this will pay off later on, in Chapter 7.

EXAMPLE 7 The function $y = x^2$ decreases on $(-\infty, 0)$, where the derivative $y' = 2x$ is negative, and increases on $(0, \infty)$, where the derivative is positive. In between, $y' = 0$ and the tangent to the curve is horizontal.

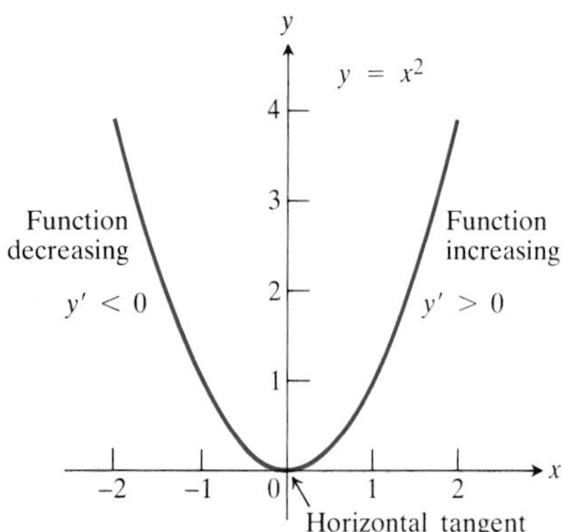

■

EXAMPLE 8 The graph of $y = \tan x$ has infinitely many separate pieces called branches. Two of them are shown here. On each branch $y' = \sec^2 x$ is positive and y is an increasing function of x.

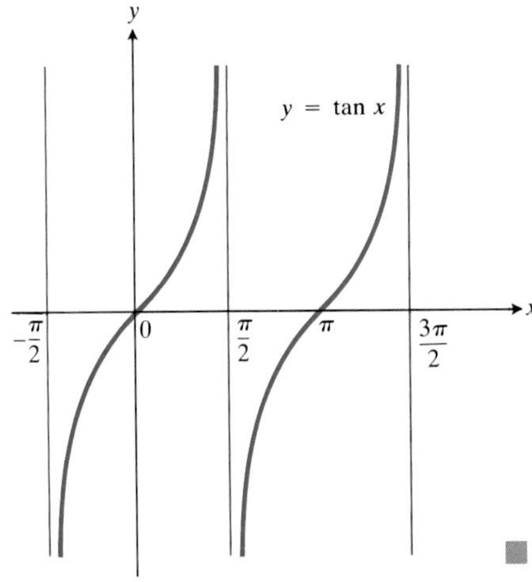

Corollary 1 provides one of our most useful tools for graphing, as we shall see in Section 4.3.

Exercises 4.2

Rolle's Theorem

Show that the equations in Exercises 1–4 have exactly one solution in the given interval.

1. $x^4 + 3x + 1 = 0, \quad -2 \le x \le -1$
2. $-x^3 - 3x + 1 = 0, \quad 0 \le x \le 1$
3. $x - \dfrac{2}{x} = 0, \quad 1 \le x \le 3$
4. $2x - \cos x = 0, \quad -\pi \le x \le \pi$
5. a) Plot the zeros of each polynomial on a line together with the zeros of its first derivative:
   i) $y = x^2 - 4$
   ii) $y = x^2 + 8x + 15$
   iii) $y = x^3 - 3x^2 + 4 = (x + 1)(x - 2)^2$
   iv) $y = x^3 - 33x^2 + 216x = x(x - 9)(x - 24)$
   What pattern do you see?
   b) Use Rolle's theorem to prove that between every two zeros of the polynomial $x^n + a_{n-1}x^{n-1} + \cdots + a_1x + a_0$ there lies a zero of the polynomial
   $$nx^{n-1} + (n - 1)a_{n-1}x^{n-2} + \cdots + a_1.$$
6. The function
   $$y = f(x) = \begin{cases} x & \text{if } 0 \le x < 1, \\ 0 & \text{if } x = 1, \end{cases}$$
   is zero at $x = 0$ and at $x = 1$. Its derivative, $y' = 1$, is different from zero at every point between 0 and 1. Why doesn't that contradict Rolle's theorem?

The Mean Value Theorem

Find the value or values of c that satisfy the equation
$$\frac{f(b) - f(a)}{b - a} = f'(c)$$
in the conclusion of the Mean Value Theorem for the functions and intervals in Exercises 7–10.

7. $f(x) = x^2 + 2x - 1, \quad 0 \le x \le 1$
8. $f(x) = x^{2/3}, \quad 0 \le x \le 1$
9. $f(x) = x + \dfrac{1}{x}, \quad \dfrac{1}{2} \le x \le 2$
10. $f(x) = \sqrt{x - 1}, \quad 1 \le x \le 3$
11. *Speeding.* A trucker handed in a ticket at a toll booth, showing that in 2 hr the truck had covered 159 mi on a toll road on which the speed limit was 65 mph. The trucker was cited for speeding. Why?
12. *Temperature change.* It took 20 sec for a thermometer to rise from 10°F to 212°F when it was taken from a freezer and placed in boiling water. Show that somewhere along the way the mercury was rising at exactly 10.1°F/sec.

© Paul Lipke/Trireme Trust

13. *Triremes.* Classical accounts tell us that a 170-oar trireme like the one shown here once covered 184 sea miles in 24 hr. Show that at some point during this feat the trireme's speed exceeded 7.5 knots.

14. Suppose that the derivative of a differentiable function $f(x)$ is never zero on the interval $0 \le x \le 1$. Show that $f(0) \ne f(1)$.

15. Show that for any numbers a and b
$$|\sin b - \sin a| \le |b - a|.$$

16. Suppose that f is differentiable for $a \le x \le b$ and that $f(b) < f(a)$. Show that f' is negative at some point between a and b.

17. Show that $y = 1/x$ decreases on any interval on which it is defined.

18. Show that $y = 1/x^2$ increases on any interval to the left of the origin and decreases on any interval to the right of the origin.

Make the estimates in Exercises 19 and 20 by applying the inequality
$$\min f' \le \frac{f(b) - f(a)}{b - a} \le \max f'.$$

19. **CALCULATOR** Suppose that $f'(x) = 1/(1 + x^4 \cos x)$ for $0 \le x \le 0.1$ and that $f(0) = 1$. Estimate $f(0.1)$.

20. **CALCULATOR** Suppose that $f'(x) = 1/(1 - x^4)$ for $0 \le x \le 0.1$ and that $f(0) = 2$. Estimate $f(0.1)$.

21. Suppose that $f(0) = 3$ and that $f'(x) = 0$ for all x. Use the Mean Value Theorem to show that $f(x)$ must be 3 for all x.

22. Suppose that $f'(x) = 2$ and that $f(0) = 5$. Use the Mean Value Theorem to show that $f(x) = 2x + 5$ at every value of x.

4.3 Curve Sketching with y' and y''

The problem with graphing a curve $y = f(x)$ without calculus is that we usually have no way to know what the curve does between the points we plot. Even if we have a computer plot hundreds of points and connect them with short line segments (which is all a computer does), the verification that the graph really looks like what's on the screen must still come from calculus. The computer can only suggest what *might* be true.

In this section we see how to use derivatives to tell what graphs do between the points we plot.

The First Derivative

When we know that a function has a derivative at every point of an interval, we also know that it is continuous throughout the interval (Section 3.1) and that its graph over the interval is connected (Section 2.4). Thus the graphs of $y = \sin x$ and $y = \cos x$ remain unbroken however far extended, as do the graphs of polynomials. The graphs of $y = \tan x$ and $y = 1/x^2$ break only at points where the functions are undefined. On every interval that avoids

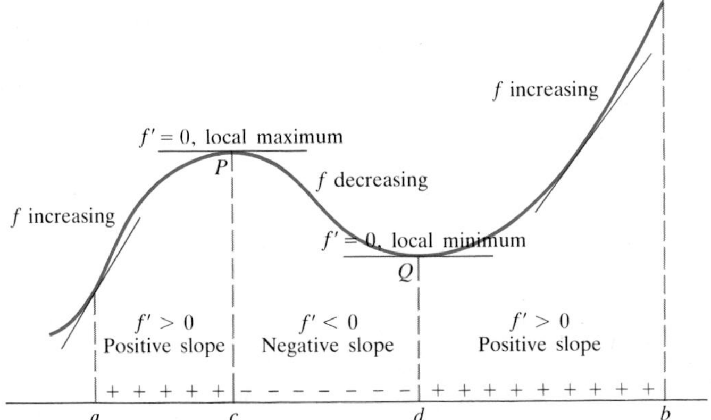

4.14 The function $y = f(x)$ increases on (a, c) where $f' > 0$, decreases on (c, d) where $f' < 0$, and increases again on (d, b). The transitions are marked by horizontal tangents.

these points, the functions are differentiable, so they are continuous and have connected graphs.

We gain additional information about the shape of a function's graph when we know where the function's first derivative is positive, negative, or zero. For, as we saw in Section 4.2, this tells us where the graph is rising or falling or has a horizontal tangent (Fig. 4.14).

There are two things to watch out for here, however. A curve may have a horizontal tangent without having a local maximum or minimum, and a curve may have a local maximum or minimum without having a horizontal tangent.

EXAMPLE 1 The curves $y = x^3$ and $y = -x^3$ have horizontal tangents at the origin without having maxima or minima there.

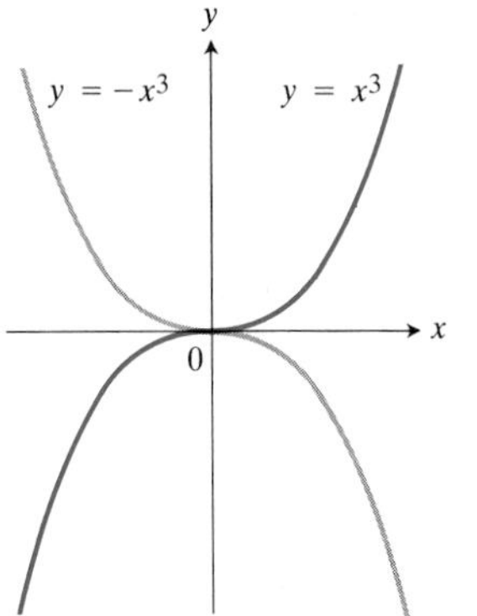

■

EXAMPLE 2 The function $y = |x|$ takes on a minimum value at $x = 0$ without having a horizontal tangent there.

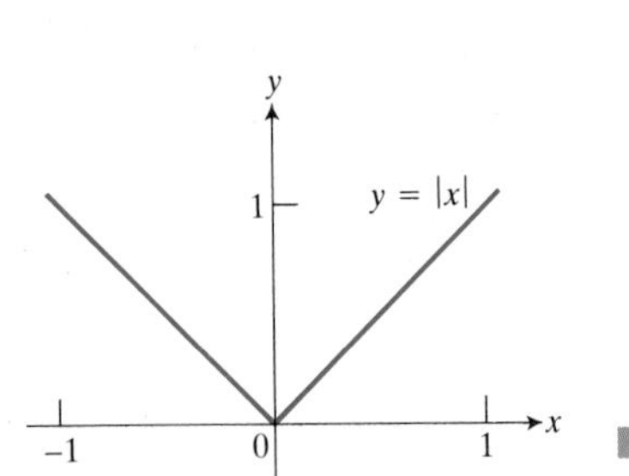

■

Concavity

As you can see in Fig. 4.15, the function $y = x^3$ increases as x increases, but the portions of the curve defined on the intervals $(-\infty, 0)$ and $(0, \infty)$ turn in different ways. If we come in from the left toward the origin along the

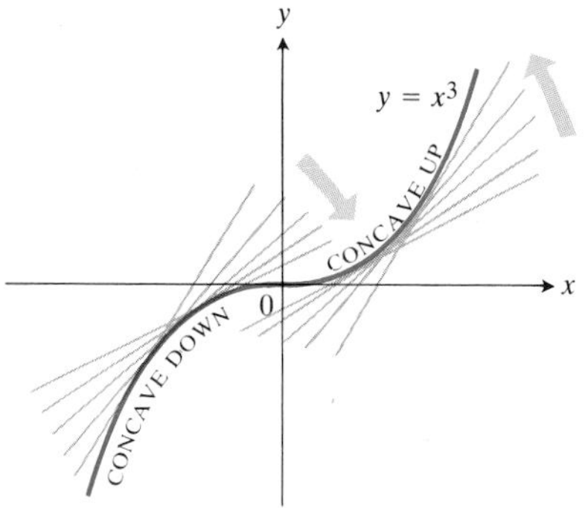

4.15 The graph of $y = x^3$ is concave down on the left, concave up on the right.

curve, the curve turns to our right and falls below its tangents. As we leave the origin, the curve turns to our left and rises above its tangents.

To put it another way, the slopes of the tangents to the curve $y = x^3$ decrease as we approach the origin from the left and increase as we move from the origin into the first quadrant.

We say that the curve $y = x^3$ is concave down on the interval $(-\infty, 0)$, where y' decreases, and concave up on $(0, \infty)$, where y' increases.

DEFINITION

> The graph of a differentiable function $y = f(x)$ is **concave up** on an interval where y' is increasing and **concave down** on an interval where y' is decreasing.

If a function $y = f(x)$ has a second derivative as well as a first (as do most of the functions we deal with in this text), we can apply Corollary 1 of the Mean Value Theorem (Section 4.2) to the function $f' = y'$ to conclude that y' decreases if $y'' < 0$ and increases if $y'' > 0$. We therefore have a test that we can apply to the formula $y = f(x)$ to determine the concavity of its graph. It is called the second derivative test for concavity.

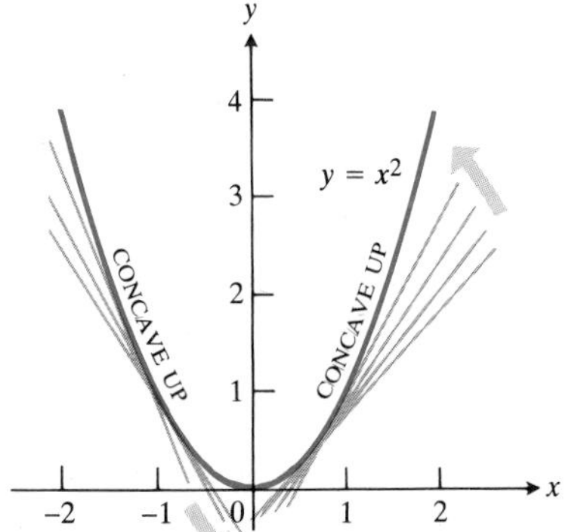

4.16 The graph of $y = x^2$ is concave up. The tangent turns counterclockwise as x increases; y' is increasing.

The Second Derivative Test for Concavity

The graph of $y = f(x)$ is

concave down on any interval where $y'' < 0$,

concave up on any interval where $y'' > 0$.

The idea is that if $y'' < 0$, then y' decreases as x increases and the tangent turns clockwise. Conversely, if $y'' > 0$, then y' increases as x increases and the tangent turns counterclockwise.

EXAMPLE 3 The curve $y = x^2$ (Fig. 4.16) is concave up on the entire x-axis because its second derivative $y'' = 2$ is always positive. ■

EXAMPLE 4 The curve $y = x^3$ in Fig. 4.15 is concave down to the left of the origin, where its second derivative $y'' = 6x$ is negative, and concave up to the right of the origin, where its second derivative $y'' = 6x$ is positive. ■

Points of Inflection

When we graph the position of a moving body as a function of time, we want to be able to see where the acceleration changes sign. These points are easy to find because they are the points where the curve changes concavity. In mathematics such points are called points of inflection.

DEFINITION

> A point on the graph of a differentiable function where the concavity changes is called a **point of inflection.**

Thus a point of inflection on a twice-differentiable curve is a point where y'' is positive on one side and negative on the other. At such a point y'' is zero because derivatives have the intermediate value property.

> At a point of inflection on the graph of a twice-differentiable function, $y'' = 0$.

EXAMPLE 5 The graph of the simple harmonic motion $y = \sin x$ shown here changes concavity at $x = 0$ and $x = \pi$, where the acceleration $y'' = -\sin x$ is zero.

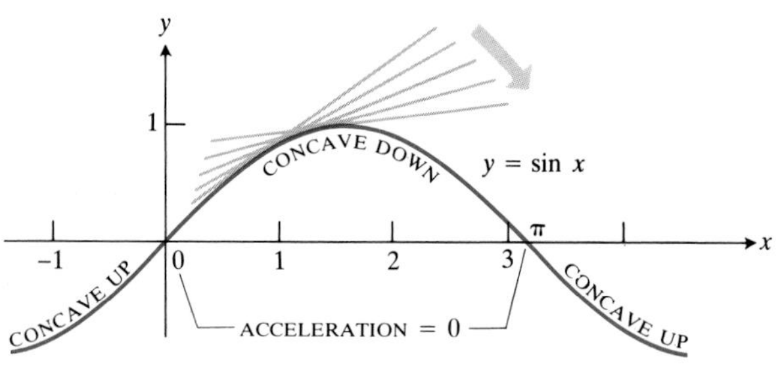

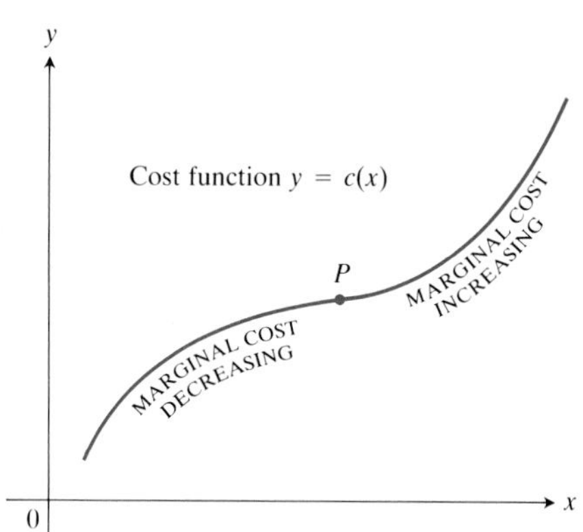

4.17 The point of inflection on a typical cost curve separates the interval of decreasing marginal cost from the interval of increasing marginal cost. This is the point where the marginal cost is smallest.

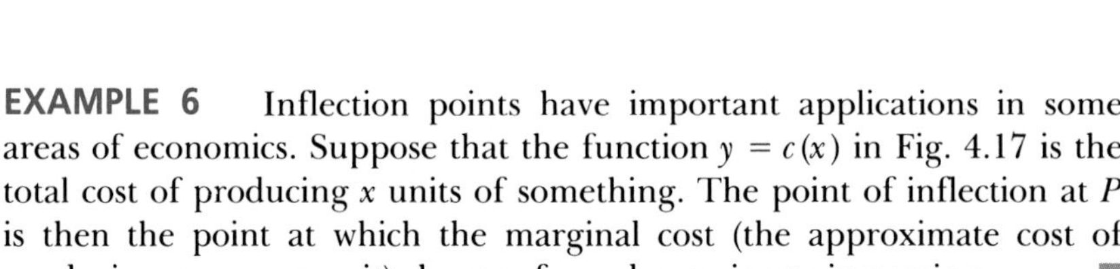

EXAMPLE 6 Inflection points have important applications in some areas of economics. Suppose that the function $y = c(x)$ in Fig. 4.17 is the total cost of producing x units of something. The point of inflection at P is then the point at which the marginal cost (the approximate cost of producing one more unit) changes from decreasing to increasing.

It is possible for y'' to be zero at a point that is *not* a point of inflection, and you have to watch out for this. Also, a point of inflection may occur where y'' fails to exist.

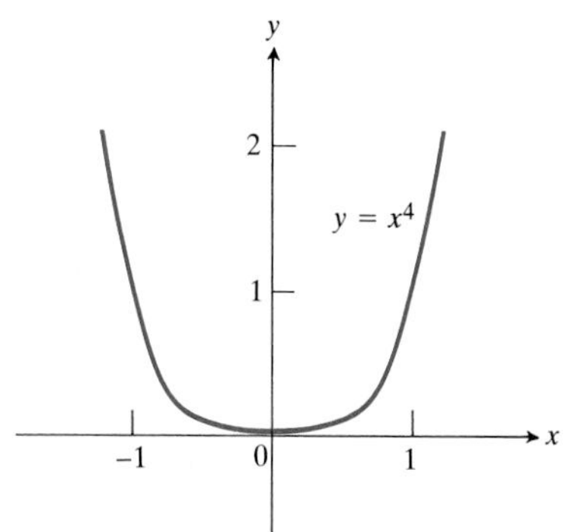

4.18 The graph of $y = x^4$ has no inflection point at the origin, even though $y''(0) = 0$.

EXAMPLE 7 *No inflection where $y'' = 0$.* The curve $y = x^4$ (Fig. 4.18) has no inflection point at $x = 0$. Even though $y'' = 12x^2$ is zero there, it does not change sign.

EXAMPLE 8 *An inflection point where y'' does not exist.* The curve $y = x^{1/3}$ (Fig. 4.19) has a point of inflection at $x = 0$, but y'' does not exist there. The formulas for y' and y'' are

$$y' = \frac{1}{3}x^{-2/3}, \qquad y'' = -\frac{2}{9}x^{-5/3}.$$

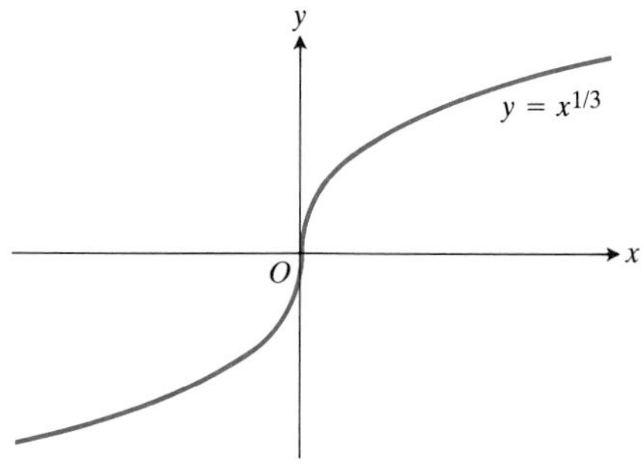

4.19 The graph of $y = x^{1/3}$ shows that a point where y'' fails to exist can be a point of inflection.

The curve is concave up for $x < 0$, where $y'' > 0$ and y' is increasing, and concave down for $x > 0$, where $y'' < 0$ and y' is decreasing. ■

Graphing with y' and y''

We now apply what we have learned to graph a cubic equation. The steps we take are those of a general procedure we recommend for graphing.

EXAMPLE 9 Graph the function

$$y = x^3 - 3x^2 + 4 = (x + 1)(x - 2)^2.$$

Solution STEP 1: *We find y' and y'':*

$$y = x^3 - 3x^2 + 4,$$
$$y' = 3x^2 - 6x,$$
$$y'' = 6x - 6.$$

STEP 2: *We find where y' is positive, negative, and zero.* This will show where the curve may have local maxima and minima, and where the curve is rising and falling.

When factored, $y' = 3x(x - 2)$, so its zeros occur at $x = 0$ and $x = 2$. To find the sign of y' elsewhere, we draw a picture like this:

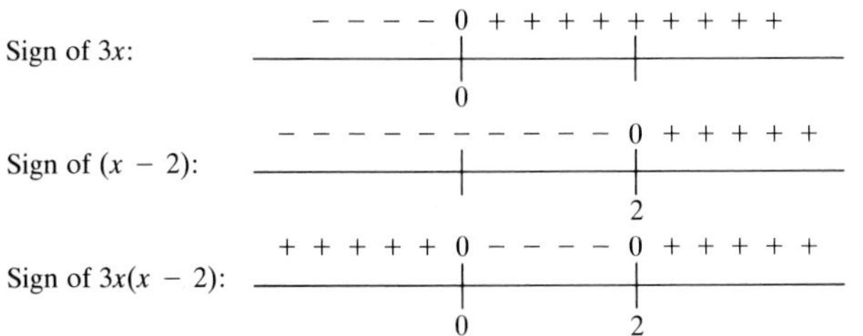

The sign pattern for y' tells us that the curve rises as it comes in from the left toward $x = 0$, falls from $x = 0$ to $x = 2$, and rises again to the right of $x = 2$. The curve has a local maximum at $x = 0$, where y' changes from $+$ to $-$, and a local minimum at $x = 2$, where y' changes from $-$ to $+$. There are no other extreme values: y is differentiable at all values of x, and its domain has no endpoints.

STEP 3: *We find where y'' is positive, negative, and zero.* This will tell us about concavity and possible inflection points.

When factored,

$$y'' = 6(x - 1),$$

which is positive if $x > 1$, negative if $x < 1$, and zero if $x = 1$. The curve is concave down to the left of 1 and concave up to the right of 1. It has an inflection point at $x = 1$, where y'' changes sign.

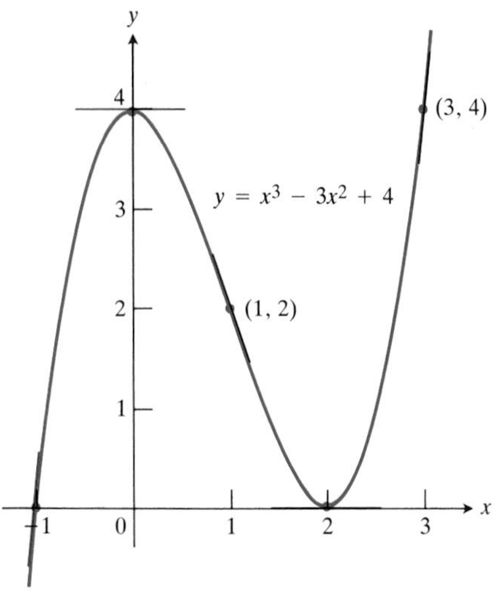

4.20 The graph of $y = x^3 - 3x^2 + 4$, based on the values of y' and y''.

STEP 4: *We make a summary table.* We include the values of y, y', and y'' at the intercepts and at the other important points. We summarize what we have learned about the curve's behavior.

x	y	y'	y''	Behavior
-1	0	9	-12	Rising, concave down
0	4	0	-6	Local maximum
1	2	-3	0	Falling, inflection point
2	0	0	6	Local minimum
3	4	9	12	Rising, concave up

STEP 5: *We draw the graph.* To do this, we plot the points from the table and sketch the tangents at these points. We then use the information about rise, fall, and concavity to draw the rest of the curve (Fig. 4.20). ■

Steps in Graphing y = f(x)

1. Find y' and y''.
2. Find where y' is positive, negative, and zero.
3. Find where y'' is positive, negative, and zero.
4. Make a summary table.
5. Draw the graph.

A Useful Shortcut—The Second Derivative Test for Local Maxima and Minima

Instead of looking at how the sign of y' changes at a point where $y' = 0$, we can often use the following test to determine whether there is a local maximum or minimum at the point.

The Second Derivative Test for Local Maxima and Minima

If $f'(c) = 0$ and $f''(c) < 0$, then f has a local maximum at $x = c$.

If $f'(c) = 0$ and $f''(c) > 0$, then f has a local minimum at $x = c$.

Notice that the test requires us to know y'' only at c itself, and not in an interval about c. This makes the test easy to apply. That's the good news. The bad news is that the test fails if $y'' = 0$ or if y'' fails to exist or is hard to find.

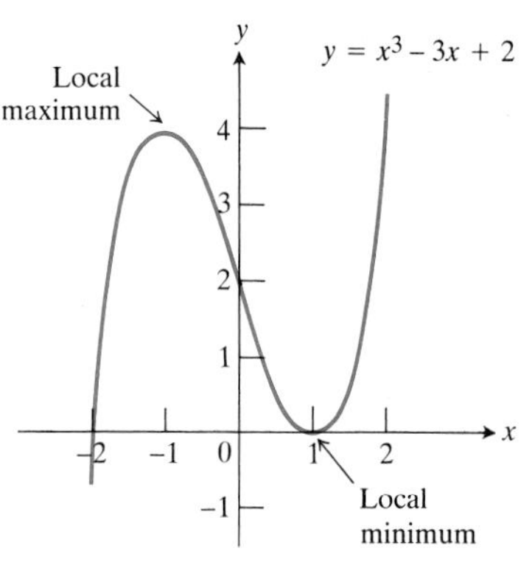

4.21 The graph of $y = x^3 - 3x + 2$.

EXAMPLE 10 Find all maxima and minima of the function

$$y = x^3 - 3x + 2$$

on the interval $-\infty < x < \infty$.

Solution The domain has no endpoints and the function is differentiable at every point. Therefore extreme values can occur only where the first derivative,

$$y' = 3x^2 - 3 = 3(x - 1)(x + 1),$$

equals zero, which means at $x = 1$ and $x = -1$. The second derivative,

$$y'' = 6x,$$

is positive at $x = 1$ and negative at $x = -1$. Hence $y(1) = 0$ is a local minimum value and $y(-1) = 4$ is a local maximum value (Fig. 4.21). ■

EXAMPLE 11 The second derivative test does not identify the local minimum of the function $y = x^4$ at $x = 0$. The second derivative, $y'' = 12x^2$, is zero when $x = 0$, so the test does not apply. ■

Summary

What Derivatives Tell Us about Graphs		
$y = f(x)$ Differentiable ⇒ smooth, connected; may rise and fall	$y = f(x)$ $y' > 0$ ⇒ rises from left to right; may be wavy	$y = f(x)$ $y' < 0$ ⇒ falls from left to right; may be wavy
or $y'' > 0$ ⇒ concave up throughout; no waves	or $y'' < 0$ ⇒ concave down throughout; no waves	y'' changes sign Inflection point
+ − or − + y' changes sign ⇒ local maximum or local minimum	$y' = 0$ and $y'' < 0$ Local maximum	$y' = 0$ and $y'' > 0$ Local minimum

Exercises 4.3

Find the inflection points and local maximum and minimum values of the functions graphed in Exercises 1–4. Identify the intervals on which the graphs are rising, falling, concave up, and concave down.

1. $y = \frac{x^3}{3} - \frac{x^2}{2} - 2x + \frac{1}{3}$

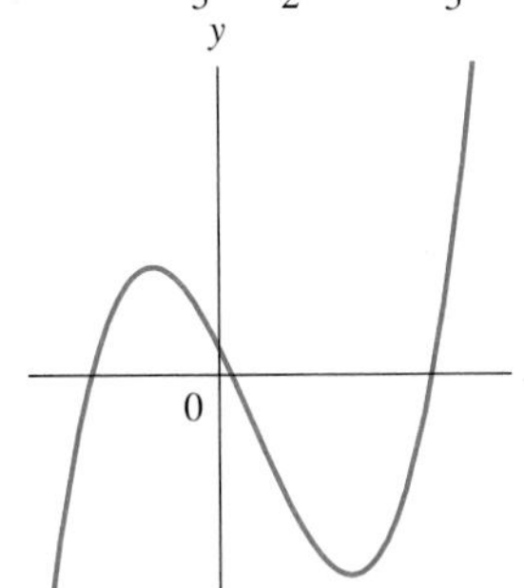

(Generated by Mathematica)

2. $y = \frac{x^4}{4} - 2x + 4$

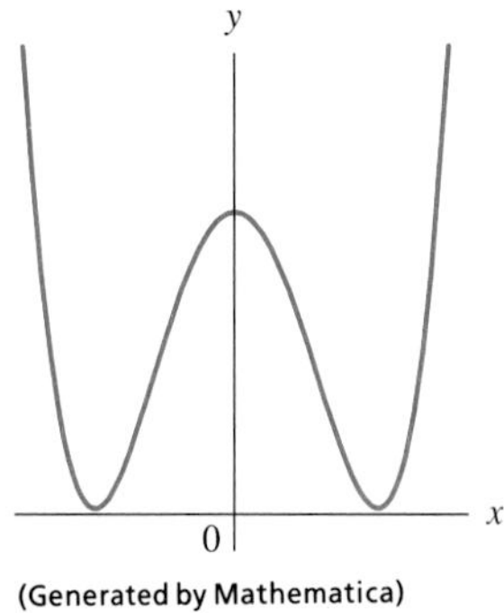

(Generated by Mathematica)

3. $y = \sin|x|, \ -2\pi \le x \le 2\pi$

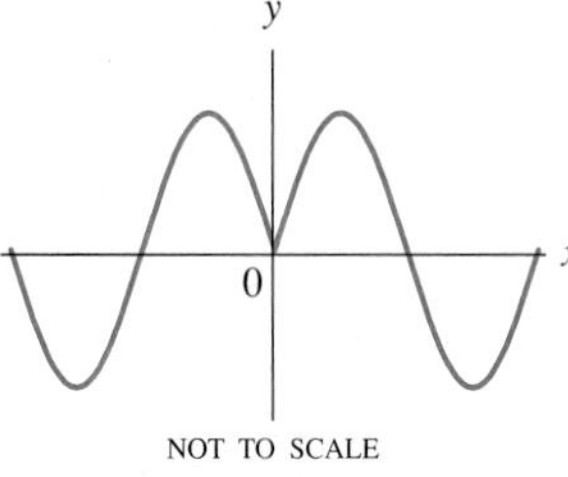

(Generated by Mathematica)

4. $y = x + \sin 2x, \ -\frac{2\pi}{3} \le x \le \frac{2\pi}{3}$

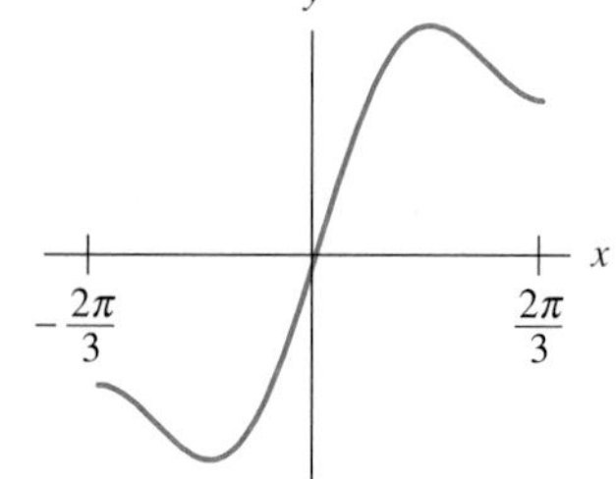

(Generated by Mathematica)

Use the steps listed in the text to graph the functions in Exercises 5–20. Indicate inflection points and local maxima and minima.

5. $y = x^2 - 4x + 3$
6. $y = 6 - 2x - x^2$
7. $y = 2x - x^2$
8. $y = x^2 - x + 1$
9. $y = x^3 - 3x + 3$
10. $y = 3x^2 - 2x^3$
11. $y = 4 + 3x - x^3$
12. $y = x^3 - x$
13. $y = x^3 - 6x^2 + 9x + 1$
14. $y = -2x^3 + 6x^2 - 3$
15. $y = \frac{x^3}{9} - 3x + 4$
16. $y = \frac{x^3}{3} - \frac{x^2}{2} - 6x$
17. $y = 2x^4 - 4x^2 + 1$
18. $y = x^4 - 2x^2$
19. $y = x + \sin x, \quad 0 \le x \le 2\pi$
20. $y = x - \sin x, \quad 0 \le x \le 2\pi$

Velocity and acceleration. Each of the graphs in Exercises 21 and 22 is the graph of the position function $y = s(t)$ of a body moving back and forth on a coordinate line. At approximately what times is each body's (a) velocity equal to zero? (b) acceleration equal to zero?

21.

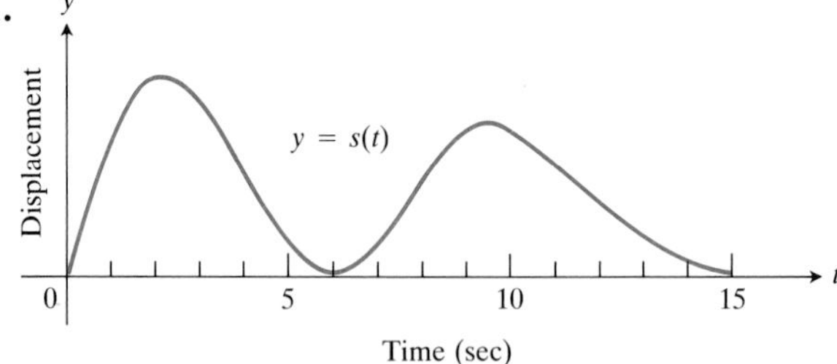

22.

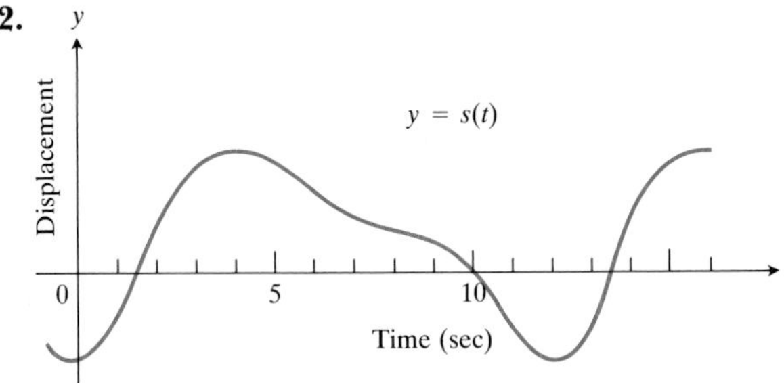

23. Find the coordinates of the inflection points on the graph of the function $y = 3/(x^2 + 3)$.

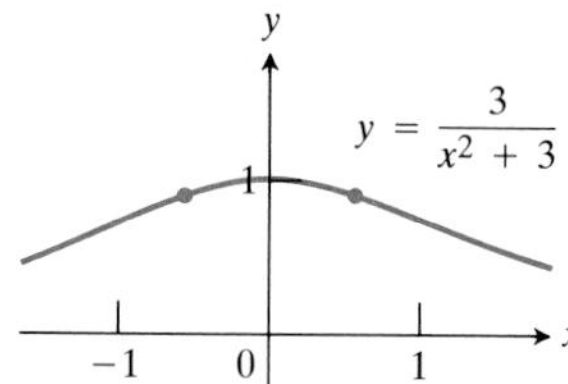

24. *Marginal cost.* The accompanying graph shows the hypothetical cost $y = c(x)$ of manufacturing x items. At approximately what production level x_0 does the marginal cost change from decreasing to increasing?

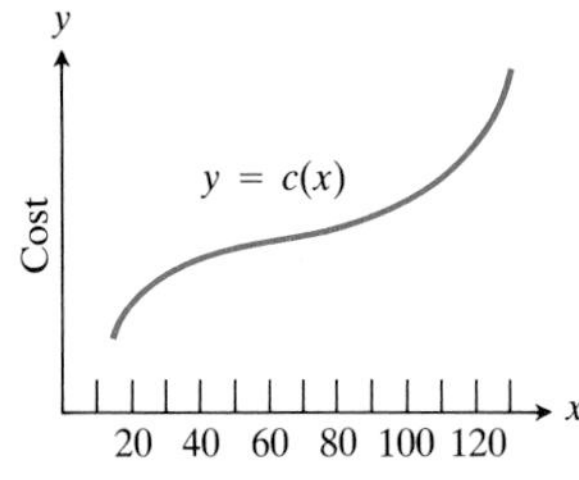

25. Sketch a smooth curve $y = f(x)$ through the origin with the properties that $f'(x) < 0$ for $x < 0$, and $f'(x) > 0$ for $x > 0$.

26. Sketch a smooth curve $y = f(x)$ through the origin with the properties that $f''(x) < 0$ for $x < 0$, and $f''(x) > 0$ for $x > 0$.

27. Sketch a continuous curve $y = f(x)$ having the following characteristics:

$f(-2) = 8,$	$f'(2) = f'(-2) = 0,$
$f(0) = 4,$	$f'(x) < 0 \quad \text{for } \lvert x \rvert < 2,$
$f(2) = 0,$	$f''(x) < 0 \quad \text{for } x < 0,$
$f'(x) > 0 \quad \text{for } \lvert x \rvert > 2,$	$f''(x) > 0 \quad \text{for } x > 0.$

28. Sketch a continuous curve $y = f(x)$ with the following properties. Label coordinates where possible.

x	y	Curve
$x < 2$		Falling, concave up
2	1	Horizontal tangent
$2 < x < 4$		Rising, concave up
4	4	Inflection point
$4 < x < 6$		Rising, concave down
6	7	Horizontal tangent
$x > 6$		Falling, concave down

29. Suppose that the derivative of the function $y = f(x)$ is

$$y' = (x - 1)^2(x - 2).$$

At what points, if any, does the graph of f have a local minimum, local maximum, or point of inflection? (*Hint:* Draw the sign pattern for y'.)

30. Suppose that the derivative of the function $y = f(x)$ is

$$y' = (x - 1)^2(x - 2)(x - 4).$$

At what points, if any, does the graph of f have a local minimum, local maximum, or point of inflection?

Find the local maximum and minimum values of the functions in Exercises 31 and 32. (In Section 4.4 we shall see how to graph functions like these.)

31. $y = x + \dfrac{1}{x}$

32. $y = \dfrac{x}{2} + \dfrac{1}{2x - 1}$

33. If $f(x)$ is a differentiable function and $f'(c) = 0$ at an interior point c of f's domain, must f have a local maximum or minimum at $x = c$? Explain.

34. If $f(x)$ is a twice-differentiable function and $f'(c) = 0$ at an interior point c of f's domain, must the graph of f have an inflection point at $x = c$? Explain.

35. *Quadratic curves.* True, or false? A quadratic curve $y = ax^2 + bx + c$ never has an inflection point. (*Hint:* What is the corresponding formula for y''?)

36. *Cubic curves.* True, or false? A cubic curve $y = ax^3 + bx^2 + cx + d$, $a \neq 0$, always has one inflection point. (*Hint:* What is the corresponding formula for y''?)

Computer Grapher or Graphing Calculator

37. a) Graph the function

$$f(x) = \frac{x^4}{3} - 2x^2 - 1$$

and find the coordinates of its inflection points and local extremes.

b) Use Newton's method to find the roots of f to as many decimal places as your machine will allow.

38. *Linearizations at inflection points.* Linearizations fit particularly well at points of inflection. You will see what we mean if you graph the following examples.

a) $f(x) = \sin x$ and its linearization $L(x) = x$ at $x = 0$

b) *Newton's serpentine.* $f(x) = 4x/(x^2 + 1)$ and its linearization $L(x) = 4x$ at $x = 0$

c) *Newton's serpentine.* $f(x) = 4x/(x^2 + 1)$ and its linearization $L(x) = -(x/2) + 3\sqrt{3}/2$ at the point $(\sqrt{3}, \sqrt{3})$

TOOLKIT PROGRAMS

Super*Grapher	Graphs any function
Root Finder	Includes Newton's method
Secant Lines	Draws tangents to graphs

4.4 Graphing Rational Functions—Asymptotes and Dominant Terms

In Section 4.3 we graphed polynomials after looking at rise, fall, extreme values, concavity, and points of inflection. But it takes more than that to graph most rational functions effectively; we have to look at asymptotes, symmetry, and dominant terms as well. This section, in which we graph some of the more common rational functions, gives the details.

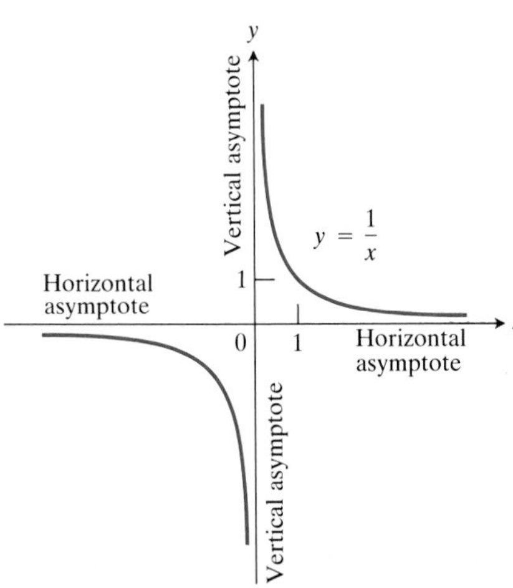

4.22 The coordinate axes are asymptotes of both branches of the hyperbola $y = 1/x$.

Horizontal and Vertical Asymptotes

If the distance between the graph of a function and some fixed line approaches zero as the graph moves farther and farther from the origin, we say that the graph approaches the line asymptotically and that the line is an asymptote of the graph.

EXAMPLE 1 The coordinate axes are asymptotes of the curve $y = 1/x$ (Fig. 4.22). The x-axis is an asymptote of the curve on the right because

$$\lim_{x\to\infty} \frac{1}{x} = 0$$

and on the left because

$$\lim_{x\to-\infty} \frac{1}{x} = 0.$$

The y-axis is an asymptote of the curve both above and below because

$$\lim_{x\to0^+} \frac{1}{x} = \infty \qquad \text{and} \qquad \lim_{x\to0^-} \frac{1}{x} = -\infty.$$

DEFINITIONS

A line $y = b$ is a **horizontal asymptote** of the graph of a function $y = f(x)$ if either

$$\lim_{x\to\infty} f(x) = b \qquad \text{or} \qquad \lim_{x\to-\infty} f(x) = b.$$

A line $x = a$ is a **vertical asymptote** of the graph if either

$$\lim_{x\to a^+} f(x) = \pm\infty \qquad \text{or} \qquad \lim_{x\to a^-} f(x) = \pm\infty.$$

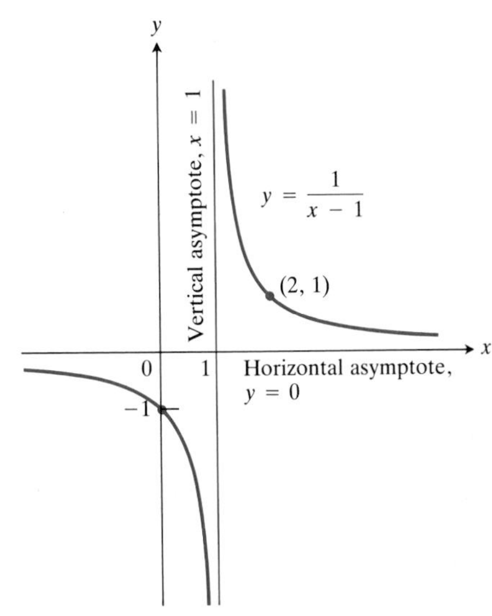

4.23 The lines $y = 0$ and $x = 1$ are asymptotes of the curve $y = 1/(x - 1)$.

EXAMPLE 2 Find the asymptotes of the curve

$$y = \frac{1}{x-1}.$$

Solution This is the curve $y = 1/x$ shifted 1 unit to the right, so its asymptotes are the x-axis and the line $x = 1$ (Fig. 4.23). Indeed,

$$\lim_{x\to\infty} \frac{1}{x-1} = 0, \quad \lim_{x\to-\infty} \frac{1}{x-1} = 0, \quad \lim_{x\to1^+} \frac{1}{x-1} = \infty, \quad \lim_{x\to1^-} \frac{1}{x-1} = -\infty.$$

EXAMPLE 3 Find the asymptotes of the curve

$$y = \frac{x}{x-1}.$$

Solution The asymptotes are quickly revealed if we divide $(x - 1)$ into x:

$$\begin{array}{r} 1 \\ x-1\,\overline{)\,x} \\ \underline{x-1} \\ 1 \end{array}$$

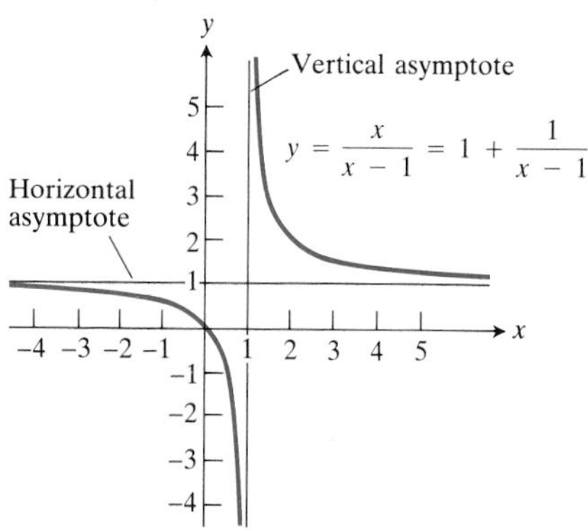

4.24 The lines $y = 1$ and $x = 1$ are asymptotes of the curve $y = x/(x - 1)$.

This enables us to rewrite the formula for y as

$$y = 1 + \frac{1}{x - 1}.$$

From this we see that the curve in question is the graph of $y = 1/x$ shifted 1 unit up and 1 unit to the right. The asymptotes are therefore the lines $x = 1$ and $y = 1$ (Fig. 4.24). Indeed, if we apply the asymptote definitions, we find that

$$\lim_{x\to\infty}\left(1 + \frac{1}{x - 1}\right) = 1 \qquad \lim_{x\to-\infty}\left(1 + \frac{1}{x - 1}\right) = 1$$

$$\lim_{x\to1^+}\left(1 + \frac{1}{x - 1}\right) = \infty \qquad \lim_{x\to1^-}\left(1 + \frac{1}{x - 1}\right) = -\infty.$$

EXAMPLE 4 Find the asymptotes of the curve

$$y = -\frac{8}{x^2 - 4}.$$

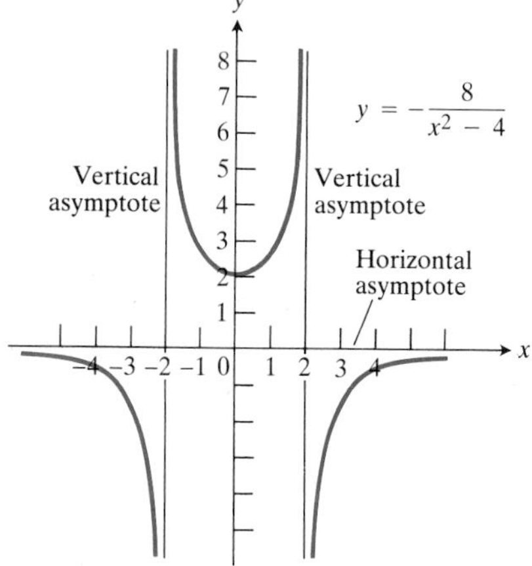

4.25 The graph of $y = -8/(x^2 - 4)$. Notice that the curve approaches the x-axis from only one side. Asymptotes do not have to be two-sided.

Solution The line $y = 0$ is an asymptote on the right because $y\to0$ as $x\to+\infty$, and again on the left because $y\to0$ as $x\to-\infty$ (Fig. 4.25).

The line $x = 2$ is an asymptote because

$$\lim_{x\to2^-}\frac{-8}{x^2 - 4} = \infty$$

and again because

$$\lim_{x\to2^+}\frac{-8}{x^2 - 4} = -\infty.$$

Similarly, the line $x = -2$ is an asymptote because $y\to\infty$ as $x\to-2^+$, and again because $y\to-\infty$ as $x\to-2^-$. There are no other asymptotes because y has a finite limit at every other point.

Oblique Asymptotes

If the degree of the numerator of a rational function is one greater than the degree of the denominator, the graph has an **oblique asymptote,** that is, an asymptote that is neither vertical nor horizontal.

EXAMPLE 5 Find the asymptotes of the curve

$$y = \frac{x^2 - 3}{2x - 4}.$$

Solution To find the asymptotes, oblique and otherwise, we divide $(2x - 4)$ into $(x^2 - 3)$:

$$\begin{array}{r} \frac{x}{2} + 1 \\ 2x - 4 \overline{) x^2 - 3} \\ \underline{x^2 - 2x} \\ 2x - 3 \\ \underline{2x - 4} \\ 1 \end{array}$$

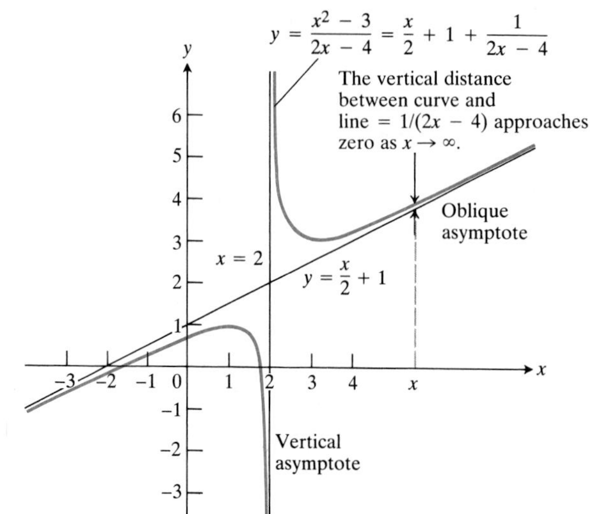

4.26 The graph of $y = (x^2 - 3)/(2x - 4)$.

This tells us that

$$y = \frac{x^2 - 3}{2x - 4} = \underbrace{\frac{x}{2} + 1}_{\text{Linear}} + \underbrace{\frac{1}{2x - 4}}_{\substack{\text{Remainder} \\ \text{goes to 0} \\ \text{as } x \to \pm\infty}}. \tag{1}$$

From this representation we see that the line

$$y = \frac{x}{2} + 1$$

is an asymptote of the curve:

$$\lim_{x \to \infty} \frac{1}{2x - 4} = 0, \quad \text{so} \quad y = \frac{x^2 - 3}{2x - 4} \to \frac{x}{2} + 1.$$

Equation (1) also reveals the presence of a vertical asymptote at $x = 2$. As $x \to 2^+$, $1/(2x - 4) \to \infty$, and as $x \to 2^-$, $1/(2x - 4) \to -\infty$. See Fig. 4.26. ■

Dominant Terms

Of all the observations we can make quickly about the function

$$y = \frac{x^2 - 3}{2x - 4},$$

probably the most useful is that

$$\frac{x^2 - 3}{2x - 4} = \frac{x}{2} + 1 + \frac{1}{2x - 4}. \quad \text{(Obtained by dividing as in Eq. (1).)}$$

This tells us immediately that

$$y \approx \frac{x}{2} + 1, \quad \text{for } x \text{ numerically large,}$$

$$y \approx \frac{1}{2x - 4}, \quad \text{for } x \text{ close to 2.}$$

If we want to know quickly how the function "goes," this is the way to find out. It goes like $y = (x/2) + 1$ when x is large, and like $y = 1/(2x - 4)$ when x is close to 2.

We say that $(x/2) + 1$ **dominates** when x is large and that $1/(2x - 4)$ dominates when x is close to 2.

Dominant terms like these are the key to predicting the function's behavior over different portions of the x-axis.

EXAMPLE 6 For

$$y = \frac{x + 3}{x + 2} = 1 + \frac{1}{x + 2}, \tag{2}$$

we have

$$y \approx 1, \quad |x| \text{ large,}$$

$$y \approx \frac{1}{x + 2}, \quad x \text{ near } -2.$$

Graphing with Asymptotes and Dominant Terms

When combined with information about symmetry and about a function's derivatives, asymptotes and dominant terms tell us what we need to know to graph rational functions effectively.

EXAMPLE 7 Graph the function

$$y = \frac{x^2 + 1}{x}.$$

Solution We find out all we can about symmetry, dominant terms and asymptotes, rise and fall, and concavity. Then we draw the graph.

1. *Symmetry:* The function is odd, so its graph is symmetric with respect to the origin.
2. *Dominant terms:* We divide x into $x^2 + 1$ to express the function as

$$y = x + \frac{1}{x}. \tag{3}$$

This tells us that

$$y \approx x, \quad x \text{ numerically large,}$$

$$y \approx \frac{1}{x}, \quad x \text{ close to 0.}$$

Checklist for Graphing a Rational Function

1. Symmetry
2. Dominant terms
3. Asymptotes
4. Rise, fall, extreme values
5. Concavity
6. Sketch dominant terms, asymptotes
7. Graph the function

3. *Asymptotes:* Equation (3) also reveals the curve's asymptotes:

$y = x,$ (Oblique asymptote, because $\frac{1}{x} \to 0$ as $x \to \infty$ or $x \to -\infty$)

$x = 0.$ (Vertical asymptote, because $\frac{1}{x} \to \infty$ as $x \to 0^+$ and $\frac{1}{x} \to -\infty$ as $x \to 0^-$)

4. *Rise and fall:* The first derivative (from Eq. 3) is

$$y' = 1 - \frac{1}{x^2}.$$

The derivative is zero when

$$y' = 1 - \frac{1}{x^2} = 0 \quad \text{or} \quad x = \pm 1.$$

The sign pattern is

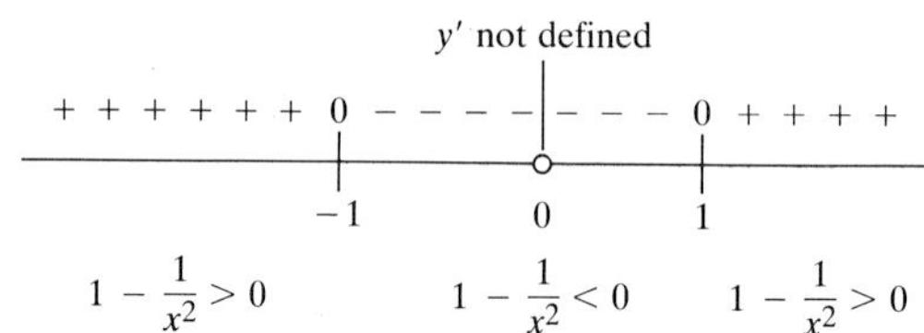

The curve rises on $(-\infty, -1)$, falls on $(-1, 0)$ and $(0, 1)$, and rises again on $(1, \infty)$. There is a local maximum at $x = -1$ and a local minimum at $x = 1$.

5. *Concavity:* The second derivative,

$$y'' = \frac{2}{x^3},$$

is negative if $x < 0$ and positive if $x > 0$. The curve is therefore concave down on $(-\infty, 0)$ and concave up on $(0, \infty)$.

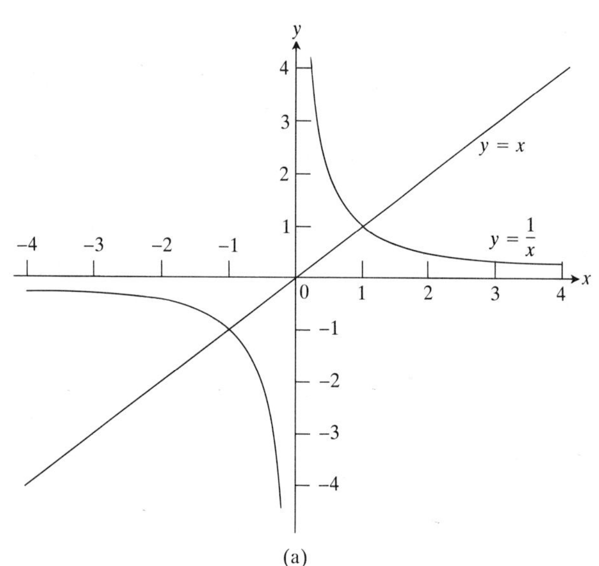

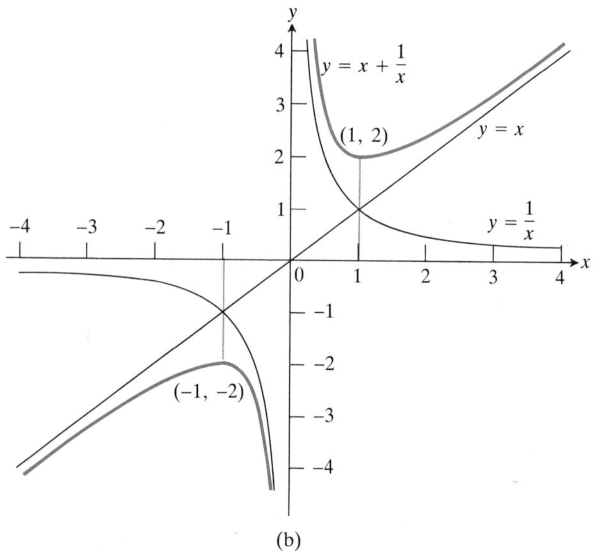

4.27 Stages in graphing $y = x + 1/x$: (a) asymptotes and dominant terms; (b) the completed graph.

6. *Sketch the asymptotes and dominant terms:* We do so in Fig. 4.27(a).
7. *Graph the function:* We plot the points where the local extreme values occur and draw short horizontal tangents. We then sketch in a curve that fits these, the asymptotes, and the graphs of the dominant terms, and has the other properties we discovered (Fig. 4.27b). ■

Exercises 4.4

Use the steps in the checklist above, when they apply, to graph the rational functions in Exercises 1–28.

1. $y = \dfrac{1}{x-2}$
2. $y = \dfrac{1}{x+1}$
3. $y = \dfrac{1}{2x+4}$
4. $y = \dfrac{3}{x-3}$
5. $y = \dfrac{x+3}{x+2}$
6. $y = \dfrac{x}{x+1}$
7. $y = \dfrac{x+1}{x-1}$
8. $y = \dfrac{x-4}{x-5}$
9. $y = \dfrac{x^2-1}{x}$
10. $y = \dfrac{x^2+4}{2x}$
11. $y = \dfrac{x^4+1}{x^2}$
12. $y = \dfrac{x^3+1}{x^2}$
13. $y = \dfrac{1}{x^2-1}$
14. $y = \dfrac{x^2}{x^2-1}$
15. $y = -\dfrac{x^2-2}{x^2-1}$
16. $y = \dfrac{x^2-4}{x^2-2}$
17. $y = \dfrac{x^2}{x-1}$
18. $y = -\dfrac{x^2}{x+1}$
19. $y = \dfrac{x^2-4}{x-1}$
20. $y = -\dfrac{x^2-4}{x+1}$
21. $y = \dfrac{x^2-x+1}{x-1}$
22. $y = -\dfrac{x^2-x+1}{x-1}$
23. $y = \dfrac{-x^2+2x-2}{x-1}$
24. $y = \dfrac{x^2+x-6}{2x-2}$
25. $y = \dfrac{x}{x^2-1}$
26. $y = \dfrac{x-1}{x^2(x-2)}$
27. $y = \dfrac{8}{x^2+4}$ (Agnesi's witch)
28. $y = \dfrac{4x}{x^2+4}$ (Newton's serpentine)
29. The word *asymptote* derives from an old Greek word for "never touching." In practice, however, a curve may cross one of its asymptotes a finite number of times or even infinitely often, as does the curve $y = 2 + (\sin x)/x$, $x > 0$, graphed here.

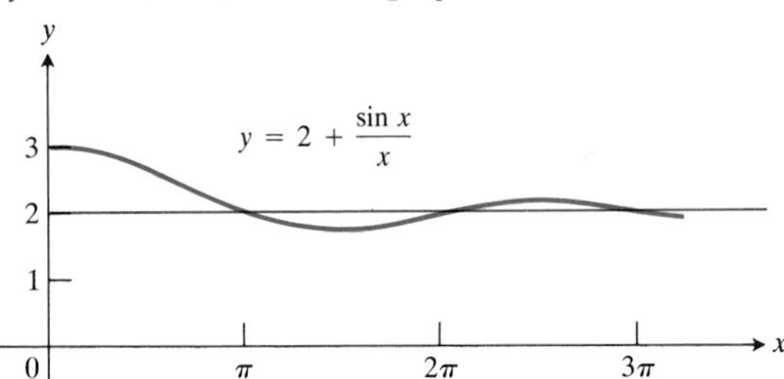

Show that the slope of the curve nevertheless approaches the slope of the asymptote as $x \to \infty$.

30. *Symmetry*
    a) Suppose that an odd function is known to be increasing for $x > 0$. What can be said of its behavior for $x < 0$?
    b) uppose that an even function is known to be increasing on the interval $-\infty < x < 0$. What can be said about its behavior for $x > 0$?

Computer Grapher or Graphing Calculator

If you have a computer grapher or graphing calculator, try graphing the following functions.

31. $y = x + 4/x^2$
32. $y = (x+1)/(x^2+1)$
33. $y = (x^2+1)/(x^3-4x)$
34. $y = (x-1)/(x^3-2x^2)$
35. $y = x + \sin 2x$
36. $y = x^2 + 3\sin 2x$

TOOLKIT PROGRAM

Super*Grapher	Graphs functions together, so you can graph a function along with its dominant terms in different colors

4.5 Optimization

To optimize something means to make it as useful or effective as possible. In the mathematical models in which we use differentiable functions to describe the things that interest us, this usually means finding where some function has its greatest or smallest value. What is the size of the most profitable production run? What is the best shape for an oil can? What is the stiffest beam we can cut from a 12-in. log?

In this section we show where such functions come from and how to find their extreme values.

Critical Points and Endpoints

Our basic tool is the observation we made in Section 4.2 about local maxima and minima. There we discovered that the extreme values of any function f whatever can occur only at

1. Interior points where $f' = 0$,
2. Interior points where f' does not exist,
3. Endpoints of the function's domain.

None of these points is necessarily the location of an extreme value, but these are the only candidates.

The points where $f' = 0$ or fails to exist are called the **critical points** of f. Thus the only points worth considering in the search for a function's extreme values are critical points and endpoints.

Continuous Functions on Closed Intervals

To find the absolute maximum and minimum values of a continuous function on a closed interval, evaluate the function at the critical points and endpoints and take the largest and smallest of these values.

Most applications call for finding the absolute maximum value or absolute minimum value of a continuous function on a closed interval. The Max–Min Theorem in Section 2.4 assures us they exist. The number of points where these values might occur is usually so small that we can simply list the points and compute the corresponding function values to see what the maximum and minimum are and where they are taken on.

EXAMPLE 1 Find the absolute maximum and minimum values of $y = x^{2/3}$ on the interval $-2 \leq x \leq 3$.

Solution We evaluate the function at the critical points and endpoints and take the largest and smallest of these values.

The first derivative,

$$y' = \frac{2}{3}x^{-1/3} = \frac{2}{3\sqrt[3]{x}},$$

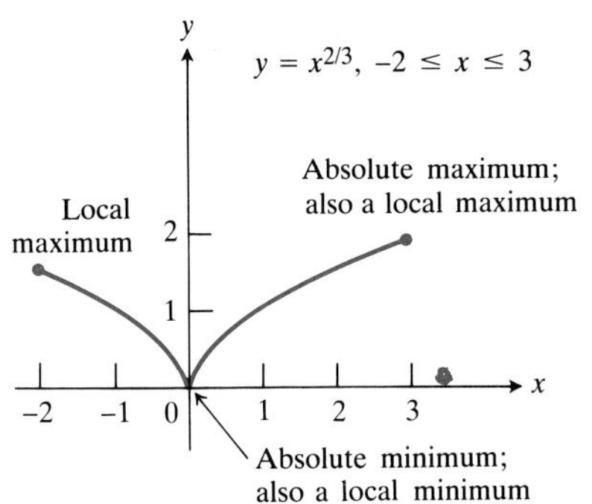

4.28 The extreme values of $y = x^{2/3}$ on the interval $-2 \le x \le 3$.

has no zeros but is undefined at $x = 0$. The values of the function at this one critical point and at the endpoints are

$$\text{Critical point value:} \quad f(0) = 0$$
$$\text{Endpoint values:} \quad f(-2) = (-2)^{2/3} = 4^{1/3}, \qquad f(3) = (3)^{2/3} = 9^{1/3}.$$

We conclude that the function's maximum value is $9^{1/3}$, taken on at $x = 3$. The minimum value is 0, taken on at $x = 0$ (Fig. 4.28). ■

Applied Examples from Mathematics

EXAMPLE 2 *Products of numbers.* Find two positive numbers whose sum is 20 and whose product is as large as possible.

Solution If one number is x, the other is $(20 - x)$. Their product is

$$f(x) = x(20 - x) = 20x - x^2.$$

We want the value or values of x that make $f(x)$ as large as possible. The domain of f is the closed interval $0 \le x \le 20$.

We evaluate f at the critical points and endpoints. The first derivative,

$$f'(x) = 20 - 2x,$$

is defined at every point of the interval $0 \le x \le 20$ and is zero only at $x = 10$. The values of f at this one critical point and at the endpoints are

$$\text{Critical point value:} \quad f(10) = 20(10) - (10)^2 = 100$$
$$\text{Endpoint values:} \quad f(0) = 0, \qquad f(20) = 0.$$

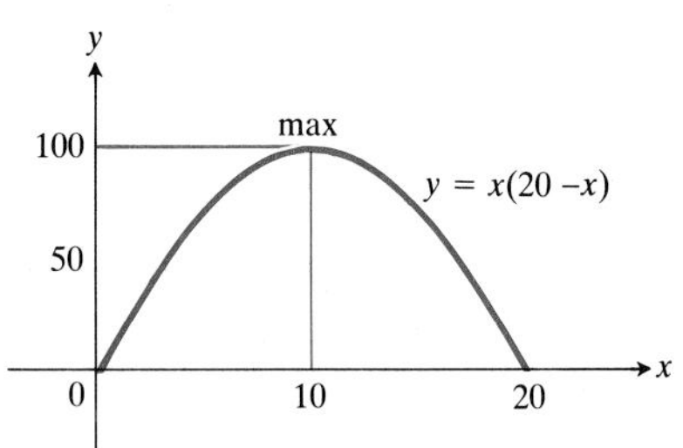

4.29 The product of x and $(20 - x)$ reaches a maximum value of 100 when $x = 10$.

We conclude that the maximum value is $f(10) = 100$. The corresponding numbers are $x = 10$ and $(20 - 10) = 10$ (Fig. 4.29). ■

EXAMPLE 3 *Geometry.* A rectangle is to be inscribed in a semicircle of radius 2. What is the largest area the rectangle can have and what are its dimensions?

Solution To describe the dimensions of the rectangle, we place the circle and rectangle in the coordinate plane (Fig. 4.30). The length, height, and area of the rectangle can then be expressed in terms of the position x of the lower right-hand corner:

Length: $2x$

Height: $\sqrt{4 - x^2}$

Area: $2x \cdot \sqrt{4 - x^2}$.

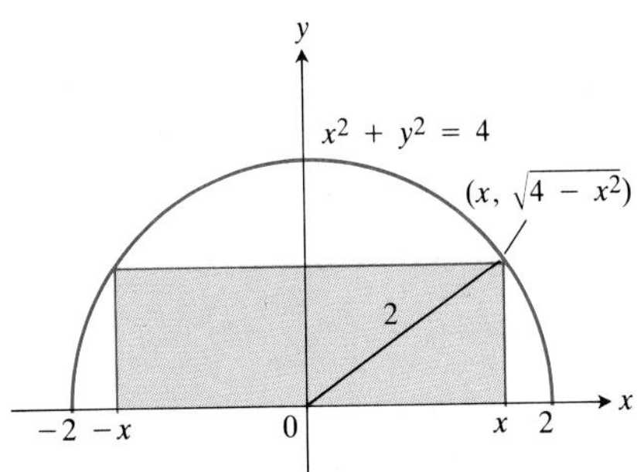

4.30 The rectangle and semicircle in Example 3.

Our mathematical goal is now to find the absolute maximum value of the continuous function

$$A(x) = 2x\sqrt{4 - x^2}$$

on the interval $0 \le x \le 2$. We do this by examining the values of A at the critical points and endpoints.

The Least and the Greatest

Many problems of the seventeenth century that motivated the development of the calculus were maxima and minima problems. Often these problems came from research in physics, such as finding the maximum range of a cannon. Galileo showed that the maximum range of a cannon is obtained with a firing angle of 45 degrees above the horizontal. He also found formulas for predicting maximum heights reached by projectiles fired at various angles to the ground. Pierre de Fermat worked on other problems of maxima and minima, culminating in his principle of least time (see Exercise 37). This was generalized by Sir William Hamilton in his principle of least action—one of the most powerful underlying ideas in physics.

The derivative

$$\frac{dA}{dx} = \frac{-2x^2}{\sqrt{4-x^2}} + 2\sqrt{4-x^2}$$

is not defined when $x = 2$ and is equal to zero when

$$\begin{aligned} \frac{-2x^2}{\sqrt{4-x^2}} + 2\sqrt{4-x^2} &= 0 \\ -2x^2 + 2(4-x^2) &= 0 \\ 8 - 4x^2 &= 0 \\ x^2 &= 2 \\ x &= \pm\sqrt{2}. \end{aligned}$$

For $0 \le x \le 2$ we therefore have

Critical point value: $A(\sqrt{2}) = 2\sqrt{2}\sqrt{4-2} = 4,$

Endpoint values: $A(0) = 0, \quad A(2) = 0.$

The area has a maximum value of 4 when the rectangle is $2x = 2\sqrt{2}$ units long by $\sqrt{4-x^2} = \sqrt{2}$ units high. ■

Applied Examples from Industry and Business

EXAMPLE 4 *Metal fabrication.* An open-top box is to be made by cutting small congruent squares from the corners of a 12-by-12-in. sheet of tin and bending up the sides. How large should the squares cut from the corners be to make the box hold as much as possible?

Solution We start with a picture, the way we do when we solve a problem in related rates (Fig. 4.31). In the figure the corner squares are x in. on a side. The volume of the box is a function of this variable:

$$V(x) = x(12-2x)^2 = 144x - 48x^2 + 4x^3.$$

The domain of V is the interval $0 \le x \le 6$.

4.31 To make an open box, squares are cut from the corners of a square sheet of tin (a) and the sides are bent up (b). What value of x gives the largest volume?

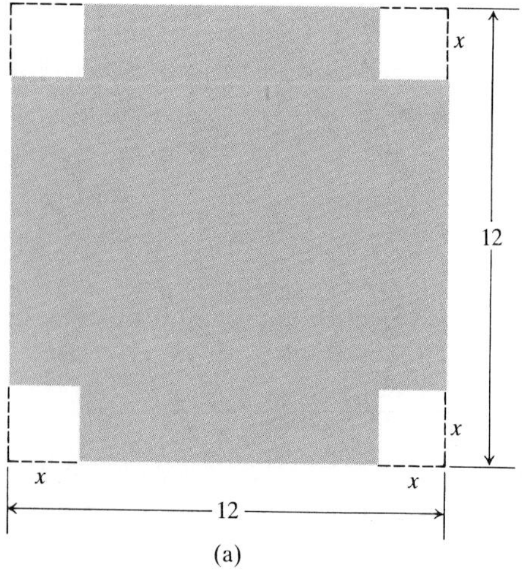

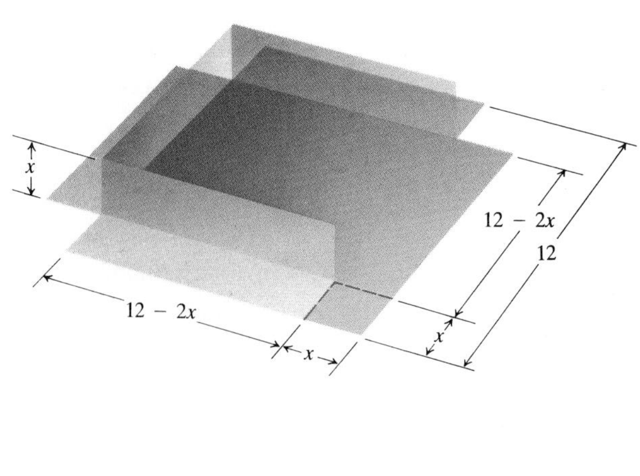

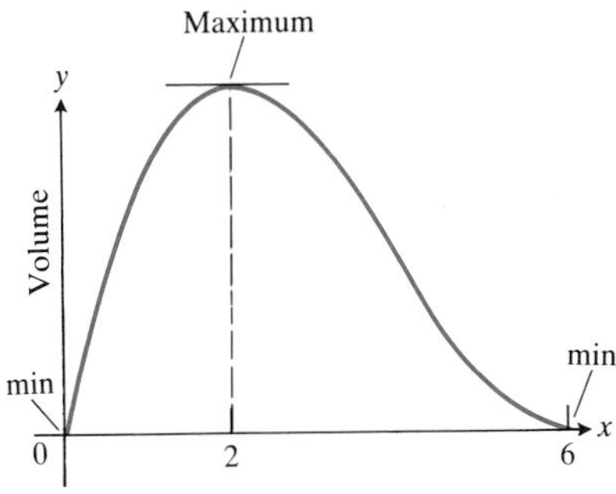

4.32 The volume of the box in Fig. 4.31, graphed as a function of x.

We evaluate V at the critical points and endpoints. The first derivative is

$$\frac{dV}{dx} = 144 - 96x + 12x^2 = 12(12 - 8x + x^2) = 12(2 - x)(6 - x).$$

Of the two zeros, $x = 2$ and $x = 6$, only $x = 2$ lies in the interior of the function's domain and makes the critical-point list. The values of V at this one critical point and two endpoints are

Critical point value: $V(2) = 128$

Endpoint values: $V(0) = 0, \quad V(6) = 0.$

The maximum volume is 128 in^3. The cut-out squares should be 2 in. on a side. See Fig. 4.32 for a graph of V. ■

EXAMPLE 5 *Product design.* You have been asked to design a one-liter oil can shaped like a right circular cylinder. What dimensions will use the least material?

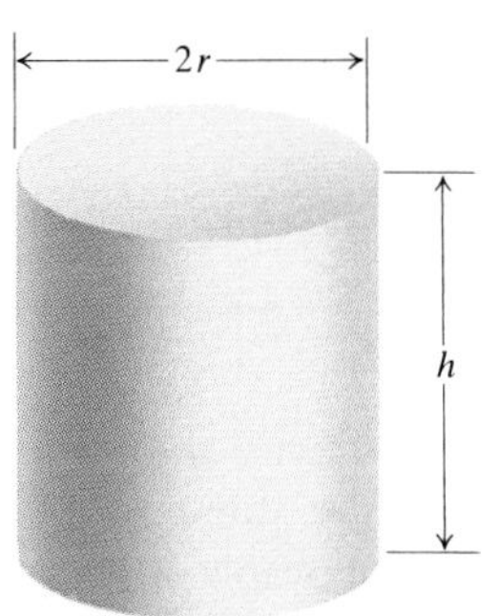

4.33 This one-liter can uses the least material when $h = 2r$ (Example 5).

Solution We picture the can as a right circular cylinder with height h and diameter $2r$ (Fig. 4.33). If r and h are measured in centimeters and the volume is expressed as 1000 cm^3, then r and h are related by the equation

$$\pi r^2 h = 1000. \tag{1}$$

How shall we interpret the phrase "least material"? One possibility is to ignore the thickness of the material and the waste in manufacturing. Then we ask for dimensions r and h that make the total surface area

$$A = \underbrace{2\pi r^2}_{\text{Cylinder ends}} + \underbrace{2\pi rh}_{\text{Cylinder wall}} \tag{2}$$

as small as possible while satisfying the constraint $\pi r^2 h = 1000$. (Exercise 18 describes one way we might take waste into account.)

What kind of oil can do we expect? Not a tall, thin one like a 6-ft. pipe, nor a short, wide one like a covered pizza pan. We expect something in between.

We are not quite ready to follow the procedure of the earlier examples because Eq. (2) gives A as a function of two variables and our procedure calls for A to be a function of a single variable. However, Eq. (1) can be solved to express either r or h in terms of the other.

Solving for h is easier, so we take

$$h = \frac{1000}{\pi r^2}. \tag{3}$$

This changes the formula for A to

$$A = 2\pi r^2 + 2\pi rh = 2\pi r^2 + 2\pi r\frac{1000}{\pi r^2} = 2\pi r^2 + \frac{2000}{r}. \tag{4}$$

Our mathematical goal is to find the minimum value of A on the open interval $r > 0$. To do so, we shall have to look at the second derivative as well as the first, just as we did when we graphed rational functions in Section

4.4. Here the rational function is a function of r instead of a function of x, but we analyze it the same way.

Since A is differentiable throughout its domain, and its domain has no endpoints, it can have a minimum value only where its first derivative is zero.

$$A = 2\pi r^2 + \frac{2000}{r}$$

$$\frac{dA}{dr} = 4\pi r - \frac{2000}{r^2} \qquad \text{(Differentiate.)}$$

$$4\pi r - \frac{2000}{r^2} = 0 \qquad \text{(Set equal to zero.)}$$

$$4\pi r^3 = 2000 \qquad \text{(Rearrange.)}$$

$$r = \sqrt[3]{\frac{500}{\pi}}. \qquad \text{(Solve for } r\text{.)}$$

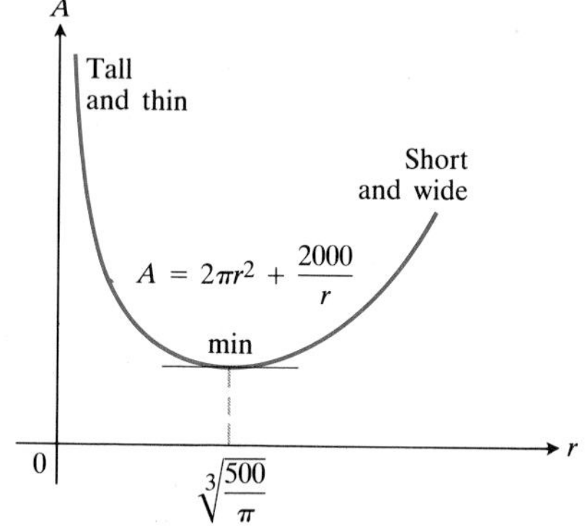

4.34 The graph of $A = 2\pi r^2 + 2000/r$ is concave up.

So something happens at $r = \sqrt[3]{500/\pi}$, but what?

To find out, we calculate d^2A/dr^2:

$$\frac{dA}{dr} = 4\pi r - \frac{2000}{r^2},$$

$$\frac{d^2A}{dr^2} = 4\pi + \frac{4000}{r^3}.$$

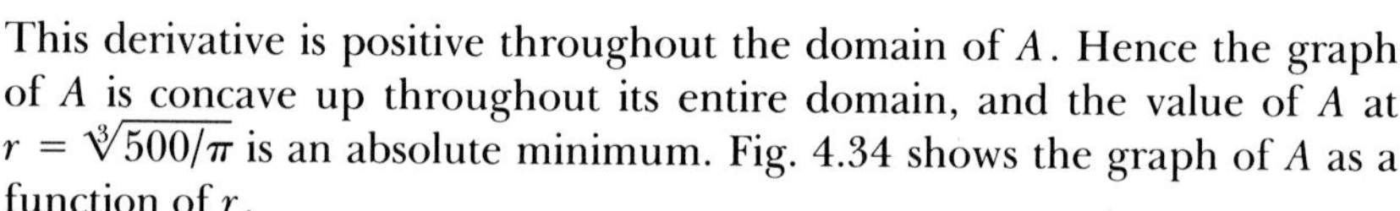

This derivative is positive throughout the domain of A. Hence the graph of A is concave up throughout its entire domain, and the value of A at $r = \sqrt[3]{500/\pi}$ is an absolute minimum. Fig. 4.34 shows the graph of A as a function of r.

When $r = \sqrt[3]{500/\pi}$,

$$h = \frac{1000}{\pi r^2} = 2\sqrt[3]{500/\pi} = 2r. \qquad \text{(After some arithmetic)}$$

Thus the most efficient can is one in which the height equals the diameter. With a calculator we find

$$r = 5.42 \text{ cm}, \quad h = 10.84 \text{ cm},$$

to the nearest hundredth. ■

EXAMPLE 6 *Maximizing profit.* Suppose a manufacturer can sell x items a week for a revenue of $r(x) = 200x - 0.01x^2$ cents, and it costs $c(x) = 50x + 20{,}000$ cents to make x items. Is there a most profitable number of items to make each week? If so, what is it?

Solution Profit is revenue minus cost, so the weekly profit on x items is

$$p(x) = r(x) - c(x) = 150x - 0.01x^2 - 20{,}000.$$

We want to find the maximum value (if any) of p on the open interval $x > 0$.

The function is differentiable at every point, and the domain in question has no endpoints. If there is a maximum value, it will therefore occur where

$$\frac{dp}{dx} = 150 - 0.02x = 0,$$

which means at

$$x = \frac{150}{0.02} = 7500.$$

The second derivative of p is

$$\frac{d^2p}{dx^2} = \frac{d}{dx}(150 - 0.02x) = -0.02,$$

which is negative for all x. The graph of p is concave down, so the critical point $x = 7500$ is the location of an absolute maximum.

To answer the questions, then, there *is* a production level for maximum profit, and that level is $x = 7500$ units per week. ■

Cost and Revenue in Economics

Here we want to point out two of the many places where calculus makes an important contribution to economic theory. The first has to do with the relationship between profit, revenue (money received), and cost.

Suppose that

$r(x) =$ the revenue from selling x items,

$c(x) =$ the cost of producing the x items,

$p(x) = r(x) - c(x) =$ the profit from selling x items.

The marginal revenue and cost at this production level (x items) are

$\frac{dr}{dx} =$ marginal revenue,

$\frac{dc}{dx} =$ marginal cost.

The first theorem is about the relationship of p to these derivatives.

THEOREM 4

Maximum profit (if any) occurs at a production level at which marginal revenue equals marginal cost.

PROOF OF THEOREM 4 We assume that $r(x)$ and $c(x)$ are differentiable for all $x > 0$, so if $p(x) = r(x) - c(x)$ has a maximum value, it occurs at a pro-

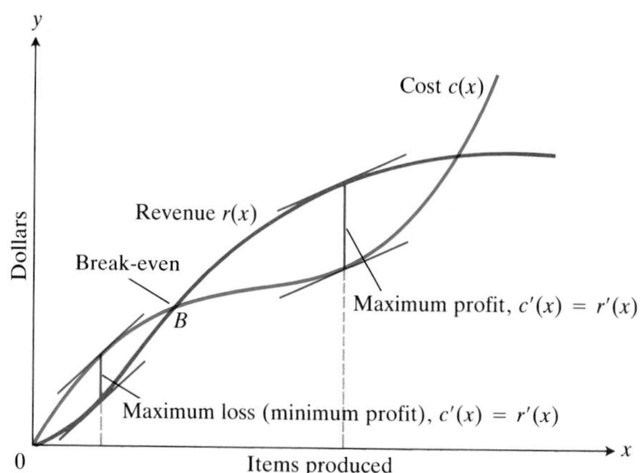

4.35 The graph of a typical cost function starts concave down and later turns concave up. It crosses the revenue curve at the break-even point B. To the left of B, the company operates at a loss. To the right, the company operates at a profit, with the maximum profit occurring where $c'(x) = r'(x)$. Farther to the right, cost exceeds revenue (perhaps because of a combination of market saturation and rising labor and material costs) and production levels become unprofitable again.

duction level at which $p'(x) = 0$. Since $p'(x) = r'(x) - c'(x)$, $p'(x) = 0$ implies

$$r'(x) - c'(x) = 0 \quad \text{or} \quad r'(x) = c'(x).$$

This concludes the proof (Fig. 4.35). ■

What guidance do we get from this theorem? We know that a production level at which $p'(x) = 0$ need not be a level of maximum profit. It might be a level of minimum profit, for example. But if we are making financial projections for our company, we should look for production levels at which marginal cost seems to equal marginal revenue. If there is a most profitable production level, it will be one of these.

EXAMPLE 7 Suppose that

$$r(x) = 9x \quad \text{and} \quad c(x) = x^3 - 6x^2 + 15x,$$

where x represents thousands of units. Is there a production level that maximizes profit? If so, what is it?

Solution

$$r(x) = 9x, \quad c(x) = x^3 - 6x^2 + 15x \qquad \text{(Find } r'(x) \text{ and } c'(x).)$$

$$r'(x) = 9, \quad c'(x) = 3x^2 - 12x + 15$$

$$3x^2 - 12x + 15 = 9 \qquad \text{(Set them equal.)}$$

$$3x^2 - 12x + 6 = 0 \qquad \text{(Rearrange.)}$$

$$x^2 - 4x + 2 = 0$$

$$x = \frac{4 \pm \sqrt{16 - 4 \cdot 2}}{2} \qquad \left(\begin{matrix}\text{Solve for } x \text{ with the} \\ \text{quadratic formula.}\end{matrix}\right)$$

$$= \frac{4 \pm 2\sqrt{2}}{2}$$

$$= 2 \pm \sqrt{2}$$

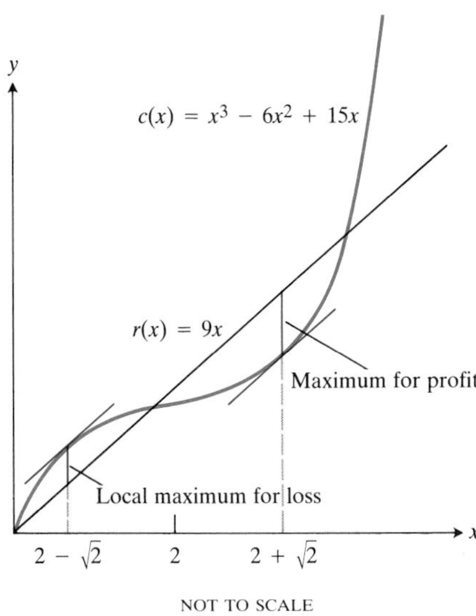

4.36 The cost and revenue curves for Example 7.

The possible production levels for maximum profit are $x = 2 + \sqrt{2}$ thousand units and $x = 2 - \sqrt{2}$ thousand units. A quick glance at the graphs in Fig. 4.36 or at the corresponding values of r and c shows $x = 2 + \sqrt{2}$ to be a point of maximum profit and $x = 2 - \sqrt{2}$ to be a local maximum for loss. ■

Another way to look for optimal production levels is to look for levels that minimize the average cost of the units produced. The next theorem gives a way to find them.

THEOREM 5

> The production level (if any) at which average cost is smallest is a level at which the average cost equals the marginal cost.

PROOF OF THEOREM 5 We start with

$$c(x) = \text{cost of producing } x \text{ items, } x > 0,$$

$$\frac{c(x)}{x} = \text{average cost of producing } x \text{ items,}$$

assumed differentiable.

If the average cost can be minimized, it will be at a production level at which

$$\frac{d}{dx}\left(\frac{c(x)}{x}\right) = 0$$

$$\frac{x\,c'(x) - c(x)}{x^2} = 0 \qquad \text{(Quotient Rule)}$$

$$x\,c'(x) - c(x) = 0 \qquad \text{(Multiplied by } x^2\text{)}$$

$$\underbrace{c'(x)}_{\text{Marginal cost}} = \underbrace{\frac{c(x)}{x}}_{\text{Average cost}}.$$

This completes the proof. ■

Again we have to be careful about what the theorem does and does not say. It does not say that there is a production level of minimum average cost, but it does say where to look to see if there is one. Look for production levels at which average cost and marginal cost are equal. Then check to see if any of them gives a minimum average cost.

EXAMPLE 8 Suppose $c(x) = x^3 - 6x^2 + 15x$ (x in thousands of units). Is there a production level that minimizes average cost? If so, what is it?

Solution Theorem 5 tells us to look for production levels at which average cost equals marginal cost:

$$\begin{aligned} &\text{Cost:} && c(x) = x^3 - 6x^2 + 15x, \\ &\text{Marginal cost:} && c'(x) = 3x^2 - 12x + 15, \\ &\text{Average cost:} && \frac{c(x)}{x} = x^2 - 6x + 15, \end{aligned}$$

$$\begin{aligned} 3x^2 - 12x + 15 &= x^2 - 6x + 15 \quad (\text{MC} = \text{AC}) \\ 2x^2 - 6x &= 0 \\ 2x(x - 3) &= 0 \\ x = 0 \quad &\text{or} \quad x = 3. \end{aligned}$$

Since $x > 0$, the only production level that might minimize average cost is $x = 3$ thousand units.

We check the derivatives:

$$\frac{c(x)}{x} = x^2 - 6x + 15, \qquad (\text{Average cost})$$

$$\frac{d}{dx}\left(\frac{c(x)}{x}\right) = 2x - 6,$$

$$\frac{d^2}{dx^2}\left(\frac{c(x)}{x}\right) = 2 > 0.$$

The second derivative is positive, so the graph of the average cost is concave up. Therefore the value $x = 3$, where the first derivative is zero, gives an absolute minimum. ■

Modeling Discrete Phenomena with Differentiable Functions

In case you are wondering how we can use differentiable functions $c(x)$ and $r(x)$ to describe the cost and revenue that come from producing a number of items x that can only be an integer, here is the rationale.

When x is large, we can reasonably fit the cost and revenue data with smooth curves $c(x)$ and $r(x)$ that are defined not only at integer values of x but at the values in between. Once we have these differentiable functions, which are supposed to behave like the real cost and revenue when x is an integer, we can apply calculus to come to conclusions about their values. We then translate these mathematical conclusions into inferences about the real world that we hope will have predictive value. When they do, as is the case with the economic theory here, we say that the functions give a good model of reality.

What do we do when our calculus tells us that the best production level is a value of x that isn't an integer, as it did in Example 7 when it said that $x = 2 + \sqrt{2}$ thousand units would be the production level for maximum profit? The answer is to use the nearest convenient integer. For $x = 2 + \sqrt{2}$ thousand, we might use 3414, or perhaps 3410 or 3420 if we ship in boxes of 10.

Strategy for Solving Max-Min Problems

1. *Draw a picture.* Label the parts that are important in the problem.
2. *Write an equation.* Write an equation for the quantity whose maximum or minimum value you want. If you can, express the quantity as a function of a single variable, say $y = f(x)$. This may require some algebra and the use of information from the statement of the problem. Note the domain in which the values of x are to be found.
3. *Test the critical points and endpoints.* The extreme value of f will be found among the values f takes on at the endpoints of the domain and at the points where f' is zero or fails to exist. List the values of f at these points. If f has an absolute maximum or minimum on its domain, it will appear on the list. You may have to examine the sign pattern of f' or the sign of f'' to decide whether a given value represents a maximum, a minimum, or neither.

Exercises 4.5

1. The sum of two nonnegative numbers is 20. Find the numbers (a) if the sum of their squares is to be as large as possible; (b) if one number plus the square root of the other is to be as large as possible.

2. What is the largest possible area for a right triangle whose hypotenuse is 5 cm long?

3. What is the smallest perimeter possible for a rectangle whose area is 16 in^2?

4. Show that among all rectangles with an 8-ft perimeter, the one with the largest area is a square.

5. Figure 4.37 shows a rectangle inscribed in an isosceles right triangle whose hypotenuse is 2 units long.

a) Express the y-coordinate of P in terms of x. (You might start by writing an equation for the line AB.)

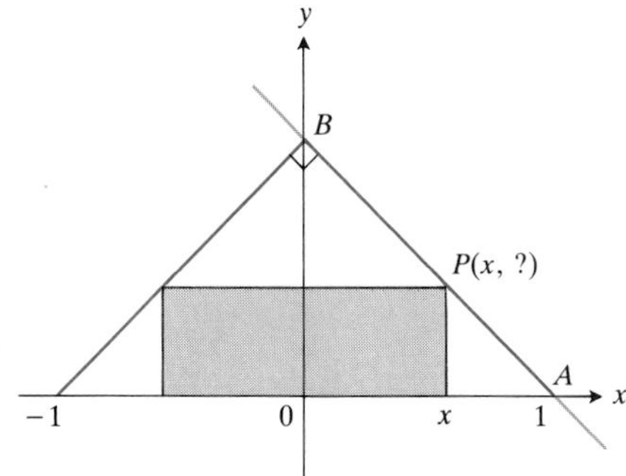

4.37 The rectangle in Exercise 5.

b) Express the area of the rectangle in terms of x.

c) What is the largest area the rectangle can have?

6. A rectangle has its base on the x-axis and its upper two vertices on the parabola $y = 12 - x^2$. What is the largest area the rectangle can have?

7. You are planning to make an open rectangular box from an 8-in. × 15-in. piece of cardboard by cutting squares from the corners and folding up the sides. What are the dimensions of the box of largest volume you can make this way?

8. You are planning to close off a corner of the first quadrant with a line segment 20 units long running from $(a, 0)$ to $(0, b)$. Show that the area of the triangle enclosed by the segment is largest when $a = b$.

9. A rectangular plot of farmland will be bounded on one side by a river and on the other three sides by a single-strand electric fence. With 800 m of wire at your disposal, what is the largest area you can enclose?

10. A 216-m^2 rectangular pea patch is to be enclosed by a fence and divided into two equal parts by another fence parallel to one of the sides. What dimensions for the outer rectangle will require the smallest total length of fence? How much fence will be needed?

11. *The lightest steel holding tank.* Your iron works has contracted to design and build a 500-ft^3, square-based, open-top, rectangular steel holding tank for a paper company. The tank is to be made by welding half-inch-thick stainless steel plates together along their edges. As the production engineer, your job is to find dimensions for the base and height that will make the tank weigh as little as possible. What dimensions do you tell the shop to use?

12. *Catching rain water.* An 1125-ft^3 open-top rectangular tank with a square base x ft on a side and y ft deep is to be built with its top flush with the ground to catch runoff water. The costs associated with the tank involve not only the material from which the tank is made but also an excavation charge proportional to the product xy. If the total cost is

$$c = 5(x^2 + 4xy) + 10xy,$$

what values of x and y will minimize it?

13. You are designing a poster to contain 50 in^2 of printing with margins of 4 in. each at top and bottom and 2 in. at each side. What overall dimensions will minimize the amount of paper used?

14. The height of an object moving vertically is given by

$$s = -16t^2 + 96t + 112,$$

with s in feet and t in seconds. Find (a) the object's velocity when $t = 0$, (b) its maximum height, and (c) its velocity when $s = 0$.

15. Two sides of a triangle have lengths a and b, and the angle between them is θ. What value of θ will maximize the triangle's area? (*Hint:* $A = (1/2)ab \sin \theta$.)

16. Find the largest possible value of $s = 2x + y$ if x and y are side lengths in a right triangle whose hypotenuse is $\sqrt{5}$ units long.

17. What are the dimensions of the lightest open-top right circular cylindrical can that will hold a volume of 1000 cm^3? Compare the result here with the result in Example 5.

18. You are designing 1000-cm^3 right circular cylindrical cans whose manufacture will take waste into account. There is no waste in cutting the aluminum for the sides, but the tops and bottoms of radius r will be cut from squares that measure $2r$ units on a side. The total amount of aluminum used up by each can will therefore be

$$A = 8r^2 + 2\pi rh$$

rather than the $A = 2\pi r^2 + 2\pi rh$ in Example 5. In Example 5 the ratio of h to r for the most economical cans was 2 to 1. What is the ratio now?

19. The U. S. Postal service will accept a box for domestic shipment only if the sum of the length and girth (distance around) does not exceed 108 in. Find the dimensions of the largest acceptable box with a square end.

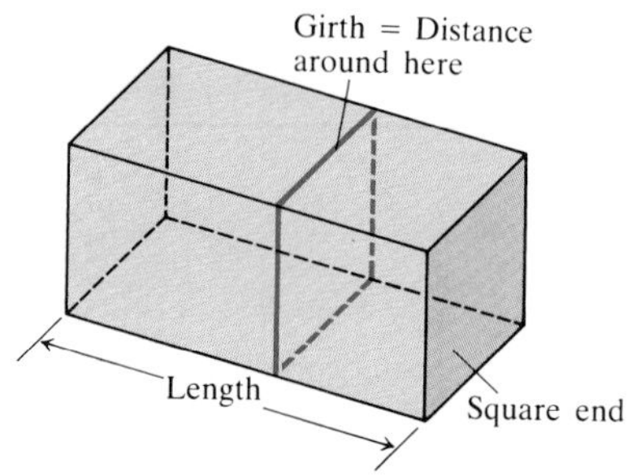

20. *Conclusion of the sonobuoy problem (Exercise 15, Section 3.8).* Show that the value of x that minimizes the square of the distance, and hence the distance, between the points (x, x^2) and $(2, -1/2)$ in Fig. 4.38 is a solution of the equation $x = 1/(x^2 + 1)$.

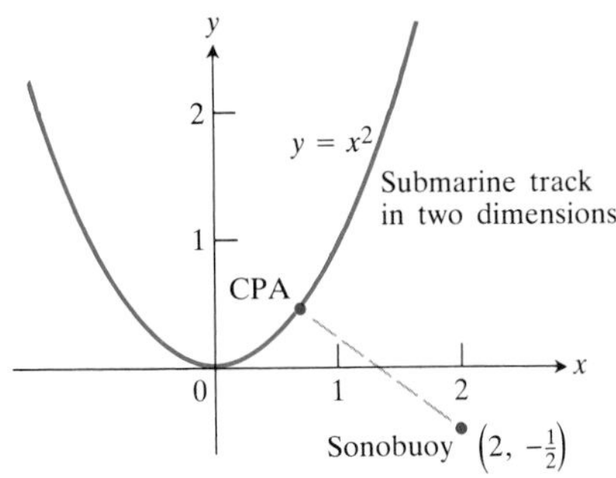

4.38 The submarine track and sonobuoy in Exercise 20. CPA = closest point of approach.

21. Compare the answers to the following two construction problems.

a) A rectangular sheet of perimeter 36 cm and dimensions x cm by y cm is to be rolled into the cylinder in Fig. 4.39(a). What values of x and y give the largest volume?

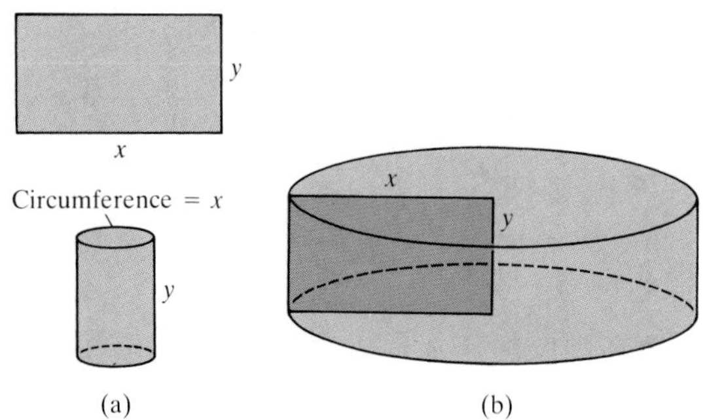

4.39 The rectangular sheet and cylinders in Exercise 21.

b) The rectangular sheet of perimeter 36 cm and dimensions x by y is to be revolved about one of the sides of length y to sweep out the cylinder in Fig. 4.39(b). What values of x and y give the largest volume?

22. A right triangle whose hypotenuse is $\sqrt{3}$ meters long is revolved about one of its legs to generate a right circular cone. Find the radius, height, and volume of the cone of greatest volume that can be made this way.

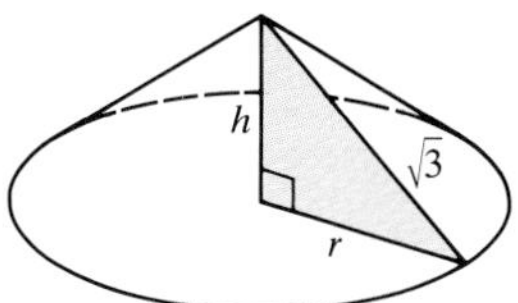

23. What value of a makes

$$f(x) = x^2 + \frac{a}{x}$$

have
a) a local minimum at $x = 2$?
b) a point of inflection at $x = 1$?

24. Show that

$$f(x) = x^2 + \frac{a}{x}$$

cannot have a local maximum for any value of a.

25. What values of a and b make

$$f(x) = x^3 + ax^2 + bx$$

have
a) a local maximum at $x = -1$ and a local minimum at $x = 3$?
b) a local minimum at $x = 4$ and a point of inflection at $x = 1$?

26. Find the volume of the largest right circular cone that can be inscribed in a sphere of radius 3.

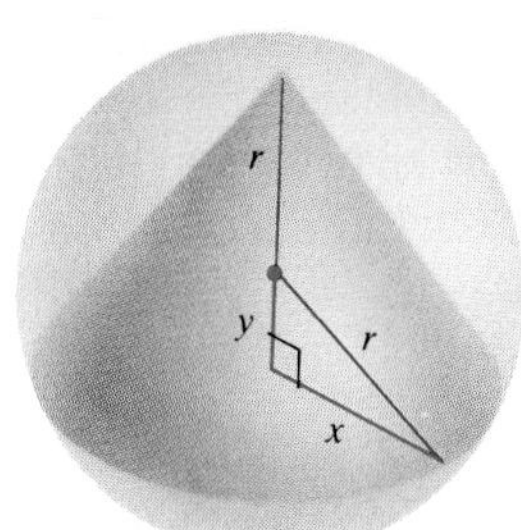

27. *The strength of a beam.* The strength of a rectangular beam is proportional to its width times the square of its depth. Find the dimensions of the strongest beam that can be cut from a 12-in.-diameter log.

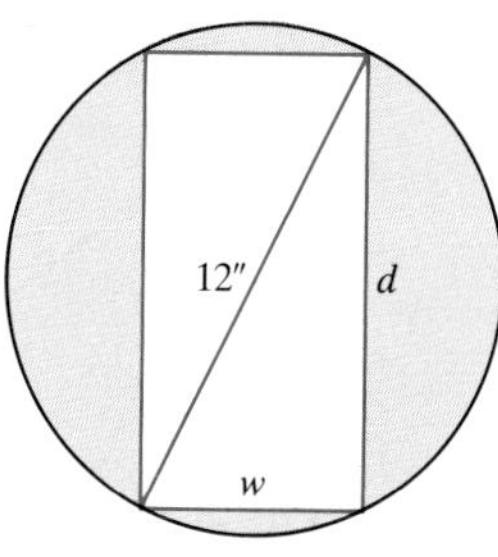

28. *The stiffness of a beam.* The stiffness of a rectangular beam is proportional to its width times the cube of its depth. Find the dimensions of the stiffest beam that can be cut from a 12-in.-diameter log. Compare your answers with the dimensions of the beam in Exercise 27.

29. The function $y = \cot x - \sqrt{2}\csc x$ has an absolute maximum value on the interval $0 < x < \pi$. Find it. (The graph may be found in Exercise 41, Section 3.4.)

30. The function $y = \tan x + \sqrt{3}\cot x$ has an absolute minimum value on the interval $0 < x < \pi/2$. Find it. (The graph may be found in Exercise 42, Section 3.4.)

31. How close does the curve $y = \sqrt{x}$ come to the point $(3/2, 0)$? (*Hint:* If you minimize the *square* of the distance, you can avoid square roots.)

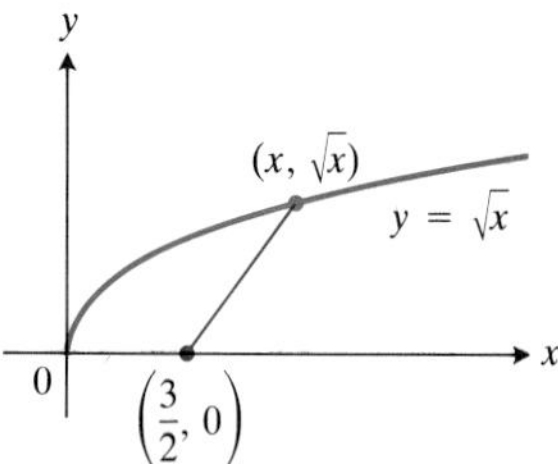

32. How close does the semicircle $y = \sqrt{16 - x^2}$ come to the point $(1, \sqrt{3})$?

33. Is the function $f(x) = x^2 - x + 1$ ever negative? How do you know?

34. Is the function $f(x) = 3 + 4\cos x + \cos 2x$ ever negative? How do you know?

35. The trough in Fig. 4.40 is to be made to the dimensions shown. Only the angle θ can be varied. What value of θ will give the trough its maximum volume?

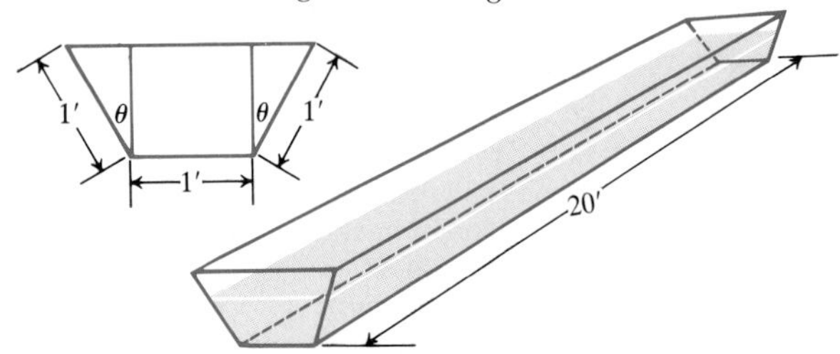

4.40 The trough in Exercise 35.

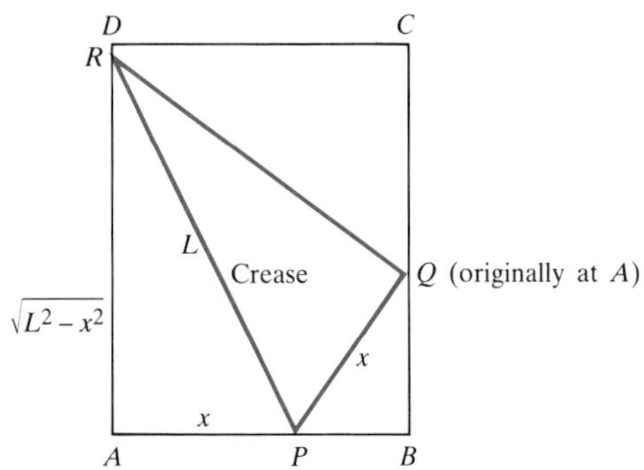

4.41 The paper in Exercise 36.

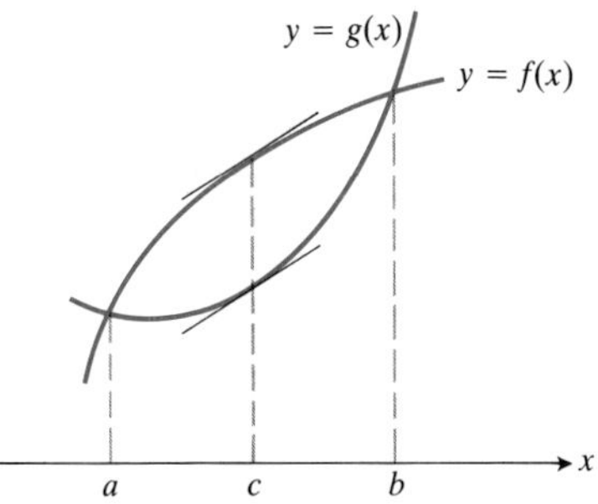

4.43 The graphs for Exercise 38.

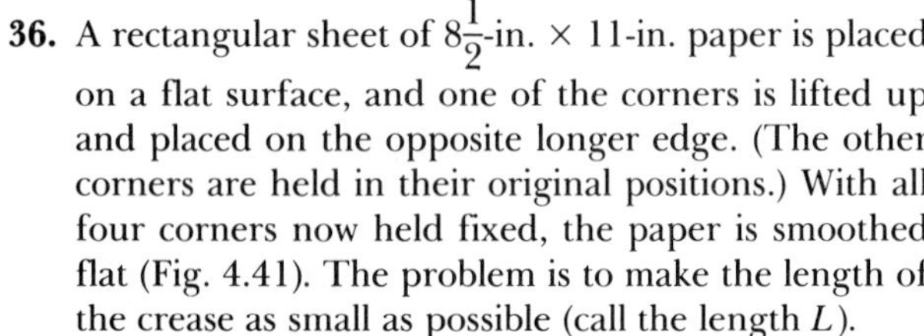

36. A rectangular sheet of $8\frac{1}{2}$-in. × 11-in. paper is placed on a flat surface, and one of the corners is lifted up and placed on the opposite longer edge. (The other corners are held in their original positions.) With all four corners now held fixed, the paper is smoothed flat (Fig. 4.41). The problem is to make the length of the crease as small as possible (call the length L).

a) Try it with paper.

b) Show that $L^2 = 2x^3/(2x - 8.5)$.

c) Minimize L^2.

d) **CALCULATOR** Find the minimum value of L to the nearest tenth of an inch.

37. Fermat's principle in optics states that light always travels from one point to another along a path that minimizes the travel time. Figure 4.42 shows light from a source A reflected by a plane mirror to a receiver at point B. Show that for the light to obey Fermat's principle, the angle of incidence must equal the angle of reflection.

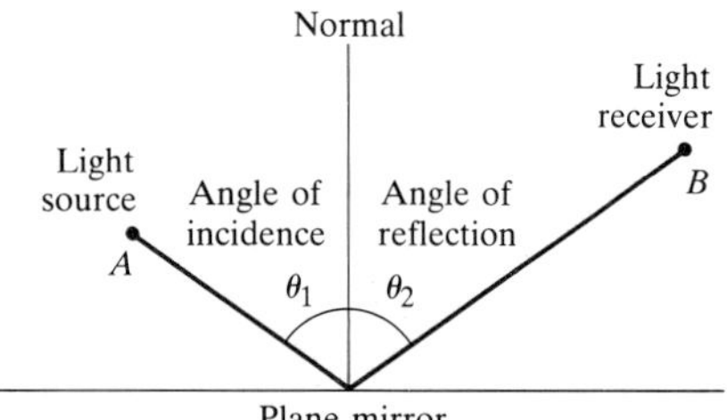

4.42 In studies of light reflection, the angles of incidence and reflection are measured from the line normal to the reflecting surface. Exercise 37 asks you to show that if light obeys Fermat's "least-time" principle, then $\theta_1 = \theta_2$.

38. Let $f(x)$ and $g(x)$ be the differentiable functions graphed in Fig. 4.43. Point c is the point where the vertical distance between the curves is the greatest. Show that the tangents to the curves at $x = c$ have to be parallel.

39. *Tin pest.* Metallic tin, when kept below 13°C for a while, becomes brittle and crumbles to a gray powder. Tin objects eventually crumble to this gray powder spontaneously if kept in a cold climate for years. The Europeans who saw the tin organ pipes in their churches crumble away years ago called the change *tin pest* because it seemed to be contagious. And indeed it was, for the gray powder is a catalyst for its own formation.

A *catalyst* for a chemical reaction is a substance that controls the rate of the reaction without undergoing any permanent change in itself. An *autocatalytic reaction* is one whose product is a catalyst for its own formation. Such a reaction may proceed slowly at first if the amount of catalyst present is small, and slowly again at the end, when most of the original substance is used up. But in between, when both the substance and its product are abundant, the reaction proceeds at a faster pace.

In some cases it is reasonable to assume that the rate $v = dx/dt$ of the reaction is proportional both to the amount of the original substance present and to the amount of product. That is, v may be considered to be a function of x alone, and

$$v = kx(a - x) = kax - kx^2,$$

where

$x =$ the amount of product,

$a =$ the amount of substance at the beginning,

$k =$ a positive constant.

At what value of x does the rate v have a maximum? What is the maximum value of v?

40. Suppose that at time t the position of a particle moving on the x-axis is $x = (t - 1)(t - 4)^4$.
a) When is the particle at rest?
b) During what time interval does the particle move to the left?
c) What is the fastest the particle goes while moving to the left?

Medicine

41. *How we cough.* When we cough, the trachea (TRAY-*kee-uh*, windpipe) contracts to increase the velocity of the air going out. This raises the questions of how much it should contract to maximize the velocity and whether it really contracts that much when we cough.

Under reasonable assumptions about the elasticity of the tracheal wall and about how the air near the wall is slowed by friction, the average flow velocity v can be modeled by the equation

$$v = c(r_0 - r)r^2 \frac{\text{cm}}{\text{sec}}, \qquad \frac{r_0}{2} \le r \le r_0,$$

where r_0 is the rest radius of the trachea in centimeters and c is a positive constant whose value depends in part on the length of the trachea.

Show that v has its maximum value when $r = (2/3)r_0$, that is, when the trachea is about 33% contracted. The remarkable fact is that x-ray photographs confirm that the trachea contracts about this much during a cough.

42. *Sensitivity to medicine (continuation of Exercise 61, Section 3.2).* Find the amount of medicine to which the body is most sensitive by finding the value of M that maximizes the derivative dR/dM.

Business

43. It costs a manufacturer c dollars each to manufacture and distribute a certain item. If the items sell at x dollars each, the number sold is given by $n = a/(x - c) + b(100 - x)$, where a and b are certain positive constants. What selling price will bring a maximum profit?

44. You operate a tour service that offers the following rates:
a) $200 per person if 50 people (the minimum number to book the tour) go on the tour.
b) For each additional person, up to a maximum of 80 people total, everyone's charge is reduced by $2.

It costs you $6000 (a fixed cost) plus $32 per person to conduct the tour. How many people does it take to maximize your profit?

45. *The best quantity to order.* One of the formulas for inventory management says that the average weekly cost of ordering, paying for, and holding merchandise is

$$A(q) = \frac{km}{q} + cm + \frac{hq}{2},$$

where q is the quantity you order when things run low (shoes, radios, brooms, or whatever the item might be), k is the cost of placing an order (the same, no matter how often you order), c is the cost of one item (a constant), m is the number of items sold each week (a constant), and h is the weekly holding cost per item (a constant that takes into account things such as space, utilities, insurance, and security). Your job, as the inventory manager for your store, is to find the quantity that will minimize $A(q)$. What is it? (The formula you get for the answer is called the *Wilson lot size formula.*)

46. *Continuation of Exercise 45.* Shipping costs sometimes depend on order size. When they do, it is more realistic to replace k by $k + bq$, the sum of k and a constant multiple of q. What is the most economical quantity to order now?

Economics

47. Use Theorem 4 to show that if $r(x) = 6x$ and $c(x) = x^3 - 6x^2 + 15x$ are your revenue and cost functions, then your operation will never be profitable and the best you can do is break even (have revenue equal cost).

48. Suppose $c(x) = x^3 - 20x^2 + 20{,}000x$ is the cost of manufacturing x items. Use Theorem 5 to find a production level that will minimize the average cost of making x items.

TOOLKIT PROGRAMS

Function Evaluator	Evaluates functions at up to 500 points in an interval, enabling you to make informed guesses about zeros and extremes
Super*Grapher	Graphs functions singly or together

4.6 Antiderivatives, Initial Value Problems, and Mathematical Modeling

One of the early accomplishments of calculus was predicting the future position of a moving body from one of its known locations and a formula for its velocity function. Today we view this as one of a number of occasions on which we recover a function from one of its known values and a formula for its rate of change. It is a routine process today, thanks to calculus, to calculate a factory's future output from its present output and its production rate or to predict a population's future size from its present size and its growth rate.

The process of finding a function from one of its known values and its derivative $f(x)$ has two steps. The first is to find a formula that gives all the functions that could possibly have f as a derivative. These functions are the so-called antiderivatives of f, and the formula that gives them all is the general antiderivative of f. The second step is then to use the known function value to select the particular antiderivative we want from the general antiderivative.

Finding a formula that gives all of a function's antiderivatives might seem like an impossible task, or at least to require a little magic, but this is not the case at all. It turns out that if we can find even one of a function's antiderivatives, then we can find them all, thanks to the remaining two corollaries of the Mean Value Theorem of Section 4.2.

The present section begins with these two corollaries. It then shows how to "reverse" known differentiation formulas to find general antiderivatives and uses this technique to solve a number of differential equations. It closes with a brief discussion of mathematical modeling, the process by which we, as scientists, use mathematics to learn about reality.

The Mean Value Theorem's Second and Third Corollaries

The first corollary of the Mean Value Theorem in Section 4.2 gave the first derivative test for rise and fall. The second corollary says that only constant functions can have zero derivatives. The third says that functions with identical derivatives can only differ by a constant value.

COROLLARY 2

If $F' = 0$, then F is a constant.

If $F'(x) = 0$ for all x in (a, b), then F has a constant value throughout (a, b). In other words, there is a constant C such that $F(x) = C$ for all x in (a, b).

Corollary 2 is the converse of the rule that says the derivative of a constant is zero. While the derivatives of nonconstant functions may be zero from time to time, the only functions whose derivatives are zero throughout an entire interval are the functions that are constant on the interval.

PROOF OF COROLLARY 2 We want to show that F has a constant value throughout the interval (a, b). The way we do so is to show that if x_1 and x_2 are any two points in (a, b), then $F(x_2) = F(x_1)$.

Suppose, then, that x_1 and x_2 are two points in (a, b), numbered from left to right, say, so that $x_1 < x_2$. Then F satisfies the hypotheses of the Mean Value Theorem on the closed interval $[x_1, x_2]$. It is differentiable at every point of the interval and hence continuous at every point of the interval. Therefore

$$\frac{F(x_2) - F(x_1)}{x_2 - x_1} = F'(c)$$

at some point c between x_1 and x_2. Since $F' = 0$ throughout (a, b), this translates into

$$\frac{F(x_2) - F(x_1)}{x_2 - x_1} = 0, \quad F(x_2) - F(x_1) = 0, \quad \text{and} \quad F(x_2) = F(x_1). \qquad \blacksquare$$

COROLLARY 3

> **Functions with identical derivatives can differ only by a constant.**
>
> If $F_1'(x) = F_2'(x)$ at each point x of an open interval (a, b), then there is a constant C such that
>
> $$F_1(x) = F_2(x) + C$$
>
> for all x in (a, b).

Corollary 3 says that the only way two functions can have identical rates of change throughout an interval is for their values on the interval to differ by a constant. For example, we know that the derivative of the function x^2 is $2x$. Therefore every other function whose derivative is $2x$ is given by the formula $F(x) = x^2 + C$ for some value of C. No other functions have $2x$ as their derivative.

PROOF OF COROLLARY 3 Since $F_1'(x) = F_2'(x)$ at each point of (a, b), the derivative of the function $F = F_1 - F_2$ at each point is

$$F'(x) = F_1'(x) - F_2'(x) = 0.$$

Therefore F has a constant value C throughout (a, b) (from Corollary 2), so $F_1(x) - F_2(x) = C$. That is, $F_1(x) = F_2(x) + C$ at each point (a, b). $\blacksquare$

Finding Antiderivatives

As we mentioned in the introduction, a function $F(x)$ is an **antiderivative** of a function $f(x)$ over an interval I if $F'(x) = f(x)$ at every point of I. Once we have found one antiderivative F of f, Corollary 3 of the Mean Value Theorem tells us that all others are given by the formula

$$y = F(x) + C.$$

We call $F(x) + C$ the **general antiderivative** of f over the interval I. Each particular antiderivative of f is given by this formula for some value of C. The constant C is called the **arbitrary constant** in the formula. Thus once we have a particular antiderivative F of f, the general antiderivative of f is F plus an arbitrary constant.

In the examples that follow, the interval I will be the natural domain of f unless we say otherwise.

The use of the letters F and f is conventional in this context, even though F and f are pronounced the same way in normal speech. To distinguish between the two, we recommend saying "cap eff" for F and "little eff" for f.

We can find many of the antiderivatives we need in scientific work by reversing derivative formulas we already know. The next example shows you what we mean.

EXAMPLE 1

Function $f(x)$	General antiderivative $F(x) + C$	Reversed derivative formula
$\cos x$	$\sin x + C$	$\frac{d}{dx}\sin x = \cos x$
$\cos 2x$	$\frac{\sin 2x}{2} + C$	$\frac{d}{dx}\frac{\sin 2x}{2} = \cos 2x$
$3x^2$	$x^3 + C$	$\frac{d}{dx}x^3 = 3x^2$
$\frac{1}{2\sqrt{x}}$	$\sqrt{x} + C$	$\frac{d}{dx}\sqrt{x} = \frac{1}{2\sqrt{x}}$
$\frac{1}{x^2}$	$-\frac{1}{x} + C$	$\frac{d}{dx}\left(-\frac{1}{x}\right) = \frac{1}{x^2}$ ■

There are also some useful general rules.

EXAMPLE 2 *General Rules*

Function	General antiderivative	Source
1. $k\frac{du}{dx}$ (k constant)	$ku + C$	Constant Multiple Rule
2. $\frac{du}{dx} + \frac{dv}{dx}$	$u + v + C$	Sum Rule
3. $\frac{du}{dx} - \frac{dv}{dx}$	$u - v + C$	Difference Rule
4. x^n $(n \neq -1)$	$\frac{x^{n+1}}{n+1} + C$	Power Rule
5. $\sin kx$	$-\frac{\cos kx}{k} + C$	Chain Rule
6. $\cos kx$	$\frac{\sin kx}{k} + C$	Chain Rule ■

The next examples show how to apply the rules in Example 2.

EXAMPLE 3 *We can find antiderivatives term by term.*

Function	General antiderivative	Source
$10x$	$5x^2 + C$	(Constant Multiple and Power Rules . . .)
$10x - x^2$	$5x^2 - \frac{x^3}{3} + C$	(. . . along with the Difference Rule . . .)
$10x - x^2 + 2$	$5x^2 - \frac{x^3}{3} + 2x + C$	(. . . and Sum Rule)

EXAMPLE 4 *Fractional powers are handled the same way as integer powers.*

Function	General antiderivative	Source
$\sqrt{x} = x^{1/2}$	$\frac{x^{3/2}}{3/2} + C = \frac{2}{3}x^{3/2} + C$	(Power Rule with $n = 1/2$)
$\frac{1}{\sqrt{x}} = x^{-1/2}$	$\frac{x^{1/2}}{1/2} + C = 2x^{1/2} + C$	(Power Rule with $n = -1/2$)

EXAMPLE 5 *The k in Rules 5 and 6 can be any real number different from zero.*

Function	General antiderivative	Source
$6 \sin 3x$	$6 \cdot \frac{-\cos 3x}{3} + C = -2 \cos 3x + C$	(Rule 5 with $k = 3$)
$5 \cos \frac{x}{2}$	$5 \cdot \frac{\sin (x/2)}{1/2} + C = 10 \sin \frac{x}{2} + C$	(Rule 6 with $k = \frac{1}{2}$)
$\cos 2\pi x$	$\frac{\sin 2\pi x}{2\pi} + C$	(Rule 6 with $k = 2\pi$)

Differential Equations and Initial Value Problems

The problem of finding a function y of x when we know its derivative

$$\frac{dy}{dx} = f(x)$$

and its value y_0 at a particular point x_0 is called an **initial value problem.** We solve the problem in two steps. First we find the general antiderivative of f,

$$y = F(x) + C. \tag{1}$$

Then we use the fact that $y = y_0$ when $x = x_0$ to find the right value of C. In this case $y_0 = F(x_0) + C$, so $C = y_0 - F(x_0)$, and the solution of the initial value problem is the function

$$y = F(x) + (y_0 - F(x_0)). \tag{2}$$

This function has the right derivative, because

$$\frac{dy}{dx} = \frac{d}{dx}F(x) + \frac{d}{dx}(y_0 - F(x_0)) = f(x) + 0 = f(x). \tag{3}$$

It also has the right value when $x = x_0$, because

$$y\Big|_{x=x_0} = F(x_0) + (y_0 - F(x_0)) = y_0. \tag{4}$$

An equation like

$$\frac{dy}{dx} = f(x) \tag{5}$$

that has a derivative in it is called a **differential equation.** Equation (5) gives dy/dx as a function of x. A more complicated differential equation might involve y as well as x:

$$\frac{dy}{dx} = 2xy^2. \tag{6}$$

It might also involve higher-order derivatives:

$$\frac{d^2y}{dx^2} - \frac{dy}{dx} + 5y = 3. \tag{7}$$

We shall see how to solve equations like (6) and (7) when we get to Chapter 16. For the time being, we shall steer away from such complications.

In the language of differential equations, the general antiderivative $y = F(x) + C$ of the function $f(x)$ is called the **general solution** of the equation $dy/dx = f(x)$. We solve the initial value problem by using the **initial condition** that $y = y_0$ when $x = x_0$ to find the **particular solution** $y = F(x) + (y_0 - F(x_0))$.

EXAMPLE 6 *Finding velocity from acceleration.* The acceleration of gravity near the surface of the earth is 9.8 m/sec². This means that the velocity v of a body falling freely in a vacuum changes at the rate of

$$\frac{dv}{dt} = 9.8 \text{ m/sec}^2.$$

If the body is dropped from rest, what will its velocity be t seconds after it is released?

Solution In mathematical terms, we want to solve the initial value problem that consists of

The differential equation: $\dfrac{dv}{dt} = 9.8$

The initial condition: $v = 0$ when $t = 0$.

To solve it, we first use what we know about antiderivatives to find the general solution of the differential equation $dv/dt = 9.8$:

$$v = 9.8t + C.$$

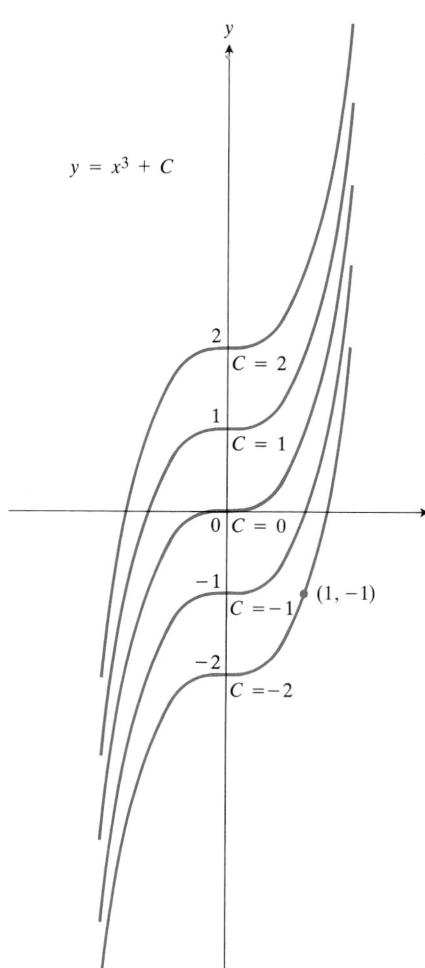

4.44 The curves $y = x^3 + C$ fill the coordinate plane without overlapping. In Example 7 we identify the curve $y = x^3 - 2$ as the one that passes through the point $(1, -1)$.

Then we use the initial condition to find the right value of C for this particular problem:

$$v = 9.8t + C,$$

$$0 = 9.8(0) + C, \qquad (v = 0 \text{ when } t = 0)$$

$$C = 0.$$

The velocity of the falling body t seconds after release is

$$v(t) = 9.8t \text{ m/sec.}$$

EXAMPLE 7 Find the curve whose slope at the point (x, y) is $3x^2$ if the curve is required to pass through the point $(1, -1)$.

Solution In mathematical language, we are asked to solve the initial value problem that consists of

The differential equation: $\dfrac{dy}{dx} = 3x^2$

The initial condition: $y = -1$ when $x = 1$.

To solve it, we first use what we know about antiderivatives to find the general solution of the differential equation:

$$y = x^3 + C.$$

Then we substitute $x = 1$ and $y = -1$ to find C:

$$y = x^3 + C,$$

$$-1 = (1)^3 + C,$$

$$C = -2.$$

The curve we want is $y = x^3 - 2$ (Fig. 4.44).

EXAMPLE 8 Suppose that the marginal cost of manufacturing an item when x thousand items are produced is

$$\frac{dc}{dx} = 10 + 3\sqrt{x}$$

thousand dollars and that the cost of producing 4 thousand items is 60 thousand dollars. What will it cost to produce 9 thousand items?

Solution We have

The differential equation: $\dfrac{dc}{dx} = 10 + 3\sqrt{x}$

The initial condition: $c = 60$ when $x = 4$.

The general solution of the differential equation is

$$c(x) = 10x + 3\frac{x^{3/2}}{3/2} + C, \qquad \left(\text{Power Rule with } n = \frac{1}{2}\right)$$

$$c(x) = 10x + 2x^{3/2} + C.$$

We substitute $x = 4$, $c(4) = 60$ to find the right value for C:

$$60 = 10(4) + 2(4)^{3/2} + C,$$
$$60 = 40 + 2(8) + C,$$
$$C = 60 - 40 - 16 = 4.$$

The cost of producing x thousand items is therefore

$$c(x) = 10x + 2x^{3/2} + 4.$$

The cost of producing 9 thousand items is

$$c(9) = 90 + 2(9)^{3/2} + 4 = 94 + 2(27) = 148$$

thousand dollars. ■

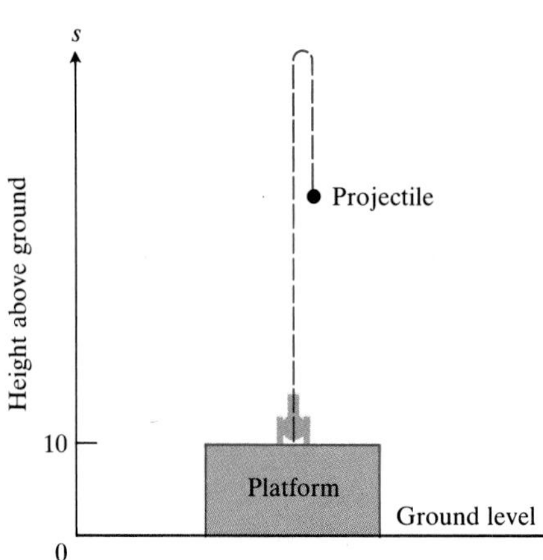

4.45 The sketch for modeling the projectile motion in Example 9.

Some problems require us to solve two or more differential equations in a row. Here is an example.

EXAMPLE 9 A heavy projectile is fired straight up from a platform 10 ft above the ground, with an initial velocity of 160 ft/sec. Assume that the only force affecting the projectile during its flight is from gravity, which produces a downward acceleration of 32 ft/sec^2. Find an equation for the projectile's height above the ground as a function of time t if $t = 0$ when the projectile is fired.

Solution To model the problem, we draw a figure (Fig. 4.45) and let $s(t)$ denote the projectile's height above the ground at time t. We assume s to be a twice-differentiable function of t, so that

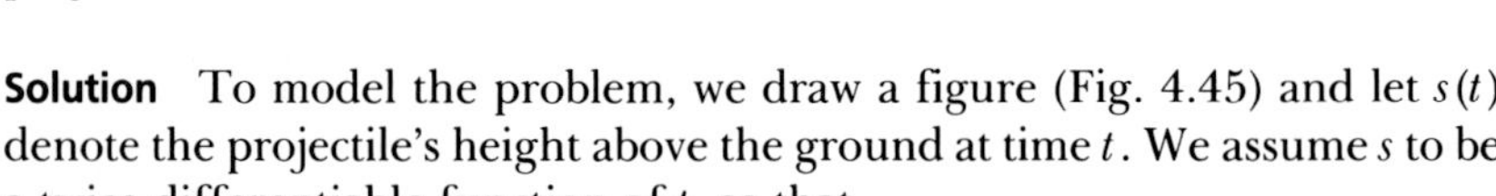

$$v = \frac{ds}{dt} \quad \text{and} \quad a = \frac{dv}{dt} = \frac{d^2s}{dt^2}.$$

Since gravity acts in the negative s direction, the direction of decreasing s in our model, the initial value problem to solve is

The differential equation: $\dfrac{dv}{dt} = -32$ ft/sec^2

The initial conditions: $v(0) = 160$ ft/sec, $s(0) = 10$ ft.

We find v from the equation

$$v = \frac{ds}{dt} = -32t + C_1 \qquad \text{(General antiderivative of } -32\text{)}$$

and s from the equation

$$s = -16t^2 + C_1t + C_2. \qquad \text{(General antiderivative of } -32t + C_1\text{)}$$

The appropriate values of C_1 and C_2 are determined by the initial conditions:

$$C_1 = v(0) = 160, \qquad C_2 = s(0) = 10.$$

The projectile's height above the ground at time t is

$$s(t) = -16t^2 + 160t + 10.$$ ■

Mathematical Modeling

The development of a mathematical model usually takes four steps: First we observe something in the real world (a ball bearing falling from rest or the trachea contracting during a cough, for example) and construct a system of mathematical variables and relationships that imitate some of its important features. We build a mathematical metaphor for what we see. Next we apply (usually) existing mathematics to the variables and relationships in the model to draw conclusions about them. After that we translate the mathematical conclusions into "news" about the system under study. Finally we check the news against observation to see if the model has predictive value. We also investigate the possibility that the model applies to other systems. The really good models are the ones that lead to conclusions that are consistent with observation, that have predictive value and broad application, and that are not too hard to use.

The natural cycle of mathematical imitation, deduction, interpretation, and comparison is shown in the following diagrams.

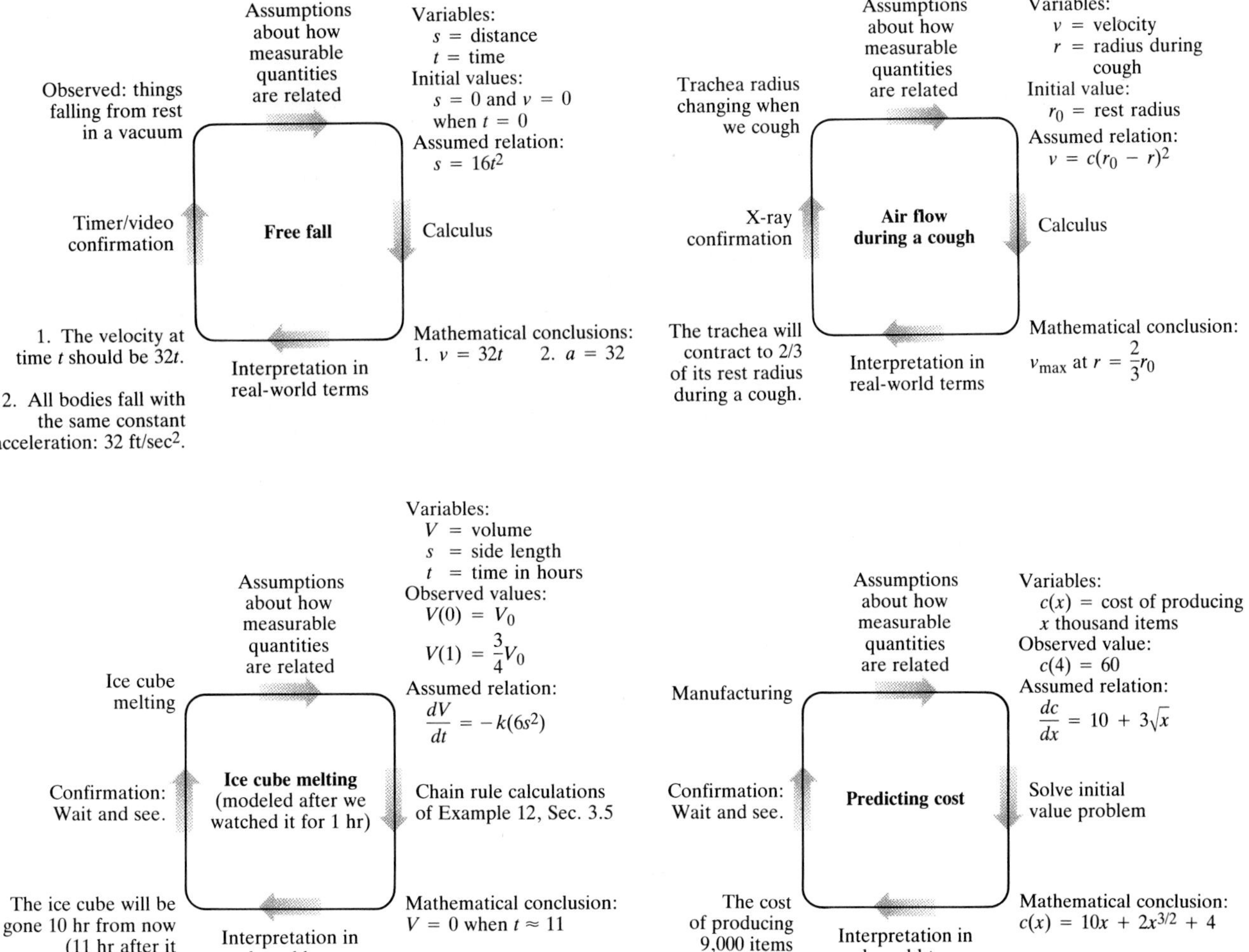

Computer Simulation

When a system we want to study is complicated, we can sometimes experiment first to see how the system behaves under different circumstances. But if this is not possible (the experiments might be expensive, time-consuming, or dangerous), we might run a series of simulated experiments on a computer—experiments that behave like the real thing, without the disadvantages. Thus we might model the effects of atomic war, the effect of waiting a year longer to harvest trees, the effect of crossing particular breeds of cattle, or the effect of reducing atmospheric ozone by 1%, all without having to pay the consequences or wait to see how things work out.

We also bring computers in when the model we want to use has too many calculations to be practical any other way. NASA's space-flight models are run on computers—they have to be to generate course corrections on time. If you want to model the behavior of galaxies that contain billions and billions of stars, a computer offers the only possible way. One of the most spectacular computer simulations in recent years, carried out by Alar Toomre at MIT, explained a peculiar galactic shape that was not consistent with our previous ideas about how galaxies are formed.

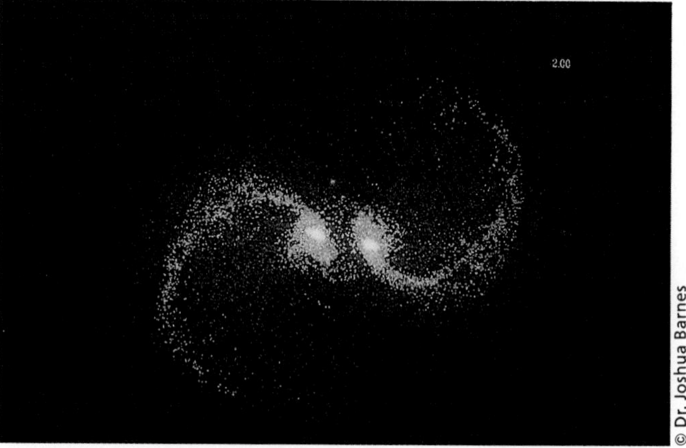

© Dr. Joshua Barnes

The galaxies had acquired their odd shapes, Toomre deduced, by passing through one another.

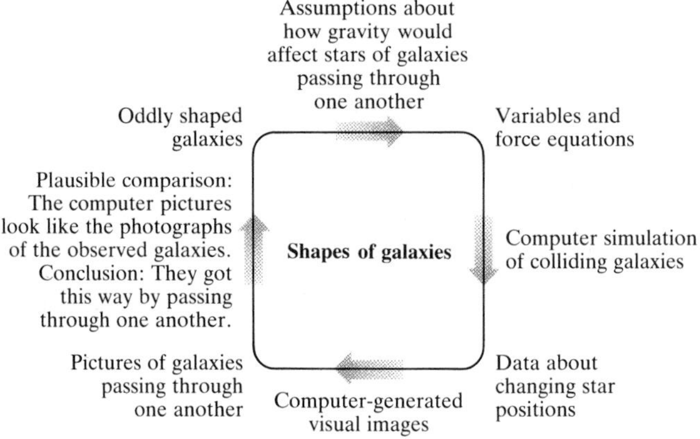

Models in Biology

You may have noticed that we haven't mentioned models in biology yet. The reason is that most mathematical models of life processes use either exponential functions or logarithms, functions that will not make their first appearance until Section 5.8. Typical of the models we shall study there is the model for unchecked bacterial growth. The basic assumption is that at any time t the rate dy/dt at which the population is changing is proportional to the number $y(t)$ of bacteria present. If the population's original size is y_0, this leads to the initial value problem

Differential equation: $\dfrac{dy}{dt} = ky$

Initial condition: $y = y_0$ when $t = 0$.

As you will see, the solution turns out to be $y = y_0 e^{kt}$, so the modeling cycle looks like this:

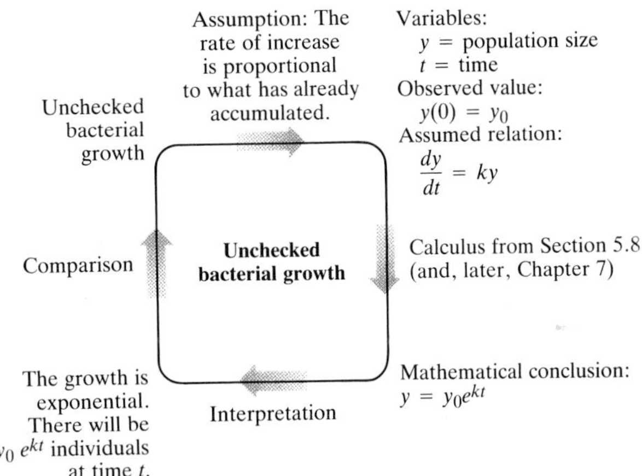

This model is one of the really good models we talked about earlier, because it applies to so many of the phenomena we want to forecast and understand: cell growth, heat transfer, radioactive decay, the flow of electrical current, and the accumulation of capital by compound interest, to mention only a few. We shall see how all of this works by the time we are through with Chapter 7.

Exercises 4.6

Find the general antiderivatives of the functions in Exercises 1–18. Do as many as you can without writing anything down (except the answer).

1. a) $2x$ b) 3 c) $2x + 3$

2. a) $6x$ b) -2 c) $6x - 2$

3. a) $3x^2$ b) x^2 c) $x^2 + 2x + 1$

4. a) $8x^7$ b) x^7 c) $x^7 - 6x + 8$

5. a) $-3x^{-4}$ b) x^{-4} c) $x^{-4} + 2x + 3$

6. a) $\dfrac{1}{x^2}$ b) $\dfrac{5}{x^2}$ c) $2 - \dfrac{5}{x^2}$

7. a) $\dfrac{3}{2}\sqrt{x}$ b) $4\sqrt{x}$ c) $x^2 - 4\sqrt{x}$

8. a) $\frac{1}{2}x^{-1/3}$ b) $-\frac{1}{2}x^{-3/2}$ c) $-\frac{3}{2}x^{-5/2}$

9. a) $\frac{2}{3}x^{-1/3}$ b) $\frac{1}{3}x^{-2/3}$ c) $-\frac{1}{3}x^{-4/3}$

10. a) $x^{-1/3}$ b) $x^{-2/3}$ c) $x^{-4/3}$

11. a) $-\sin 3x$ b) $3 \sin x$ c) $3 \sin x - \sin 3x$

12. a) $\pi \cos \pi x$ b) $\frac{\pi}{2} \cos \frac{\pi x}{2}$ c) $\cos \frac{\pi x}{2}$

13. a) $\sec^2 x$ b) $5 \sec^2 5x$ c) $\sec^2 5x$

14. a) $\csc^2 x$ b) $7 \csc^2 7x$ c) $\csc^2 7x$

15. a) $\sec x \tan x$ b) $2 \sec 2x \tan 2x$ c) $4 \sec 2x \tan 2x$

16. a) $\csc x \cot x$ b) $8 \csc 4x \cot 4x$ c) $\csc 4x \cot 4x$

17. $(\sin x - \cos x)^2$ (*Hint:* $2 \sin x \cos x = \sin 2x$)

18. $(1 + 2 \cos x)^2$ (*Hint:* $2 \cos^2 x = 1 + \cos 2x$)

19. Suppose that $1 - \sqrt{x}$ is an antiderivative of $f(x)$ and that $x + 2$ is an antiderivative of $g(x)$. Find the *general* antiderivatives of the following functions.
a) $f(x)$
b) $g(x)$
c) $-f(x)$
d) $-g(x)$
e) $f(x) + g(x)$
f) $3f(x) - 2g(x)$
g) $x + f(x)$
h) $g(x) - 4$

20. Repeat Exercise 19, assuming that e^x is an antiderivative of $f(x)$ and that $x \sin x$ is an antiderivative of $g(x)$.

Solve the initial value problems in Exercises 21–32 for y as a function of x.

21. $\frac{dy}{dx} = 2x - 7, \quad y = 0$ when $x = 2$

22. $\frac{dy}{dx} = 10 - x, \quad y = -1$ when $x = 0$

23. $\frac{dy}{dx} = x^2 + 1, \quad y = 1$ when $x = 0$

24. $\frac{dy}{dx} = x^2 + \sqrt{x}, \quad y = 1$ when $x = 1$

25. $\frac{dy}{dx} = -5/x^2, \quad x > 0; \quad y = 3$ when $x = 5$

26. $\frac{dy}{dx} = \frac{1}{x^2} + x, \quad x > 0; \quad y = 1$ when $x = 2$

27. $\frac{dy}{dx} = 3x^2 + 2x + 1, \quad y = 0$ when $x = 1$

28. $\frac{dy}{dx} = 9x^2 - 4x + 5, \quad y = 0$ when $x = -1$

29. $\frac{dy}{dx} = 1 + \cos x, \quad y = 4$ when $x = 0$

30. $\frac{dy}{dx} = \cos x + \sin x, \quad y = 1$ when $x = \pi$

31. $\frac{d^2y}{dx^2} = 2 - 6x, \quad y = 1$ and $\frac{dy}{dx} = 4$ when $x = 0$

32. $\frac{d^2y}{dx^2} = 6; \quad y = 5, \quad \frac{dy}{dx} = 0,$ and $\frac{d^2y}{dx^2} = -8$ when $x = 0$

Exercises 33 and 34 give the velocity and initial position of a body moving along a coordinate line. Find the body's position at time t.

33. $v = 9.8t, \quad s = 10, \quad$ when $t = 0$

34. $v = \sin t, \quad s = 0, \quad$ when $t = 0$

Exercises 35 and 36 give the acceleration, initial velocity, and initial position of a body moving along a coordinate line. Find the body's position at time t.

35. $a = 32, \quad v = 20,$ and $s = 0$ when $t = 0$

36. $a = \sin t, \quad v = -1,$ and $s = 1$ when $t = 0$

37. Find the curve in the xy-plane that passes through the point (9, 4) and whose slope at each point is $3\sqrt{x}$.

38. a) Find a function $y = f(x)$ with the following properties:
i) $\frac{d^2y}{dx^2} = 6x$;
ii) Its graph in the xy-plane passes through the point (0, 1) and has a horizontal tangent there.

b) How many functions like this are there? How do you know?

39. *Revenue from marginal revenue.* Suppose that the marginal revenue when x thousand units are sold is

$$\frac{dr}{dx} = 3x^2 - 6x + 12$$

dollars per unit. Find the revenue function $r(x)$ given that $r(0) = 0$.

40. *Cost from marginal cost.* Suppose that the marginal cost of manufacturing an item when x thousand items are produced is

$$\frac{dc}{dx} = 3x^2 - 12x + 15$$

dollars per item. Find the cost function $c(x)$ if $c(0) = 400$.

41. On the moon the acceleration of gravity is 1.6 m/sec^2. If a rock is dropped into a crevasse, how fast will it be going just before it hits bottom 30 sec later?

42. A rocket lifts off the surface of Earth with a constant acceleration of 20 m/sec^2. How fast will the rocket be going 1 min later?

43. With approximately what velocity do you enter the water if you dive from a 10-m platform? (Use $g = 9.8$ m/sec^2.)

44. CALCULATOR The acceleration of gravity near the surface of Mars is 3.72 m/sec^2. If a rock is blasted straight up from the surface with an initial velocity of 93 m/sec (about 208 mph), how high does it go? (*Hint:* When is the velocity zero?)

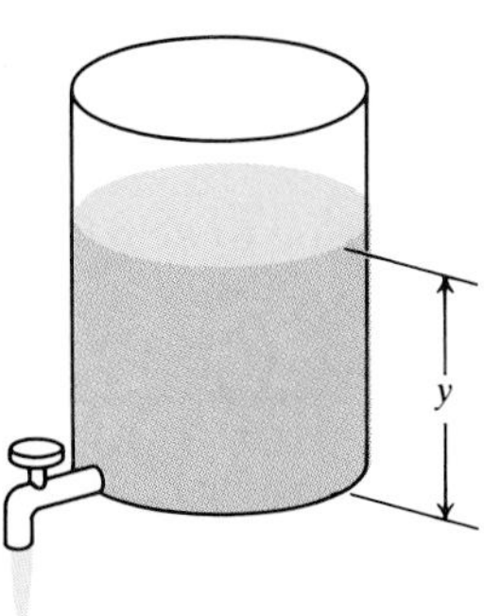

4.46 The tank in Exercises 45 and 46.

45. *How long will it take a tank to drain?* If we open a valve to drain the water from a cylindrical tank (Fig. 4.46), the water will flow fast when the tank is full but slow down as the tank drains. It turns out that the rate at which the water level drops is proportional to the square root of the water's depth. In the notation of Fig. 4.46, this means that

$$\frac{dy}{dt} = -k\sqrt{y}. \tag{8}$$

The value of k depends on the acceleration of gravity and the cross-section areas of the tank and drain hole. Equation (8) has a minus sign because y decreases with time. To solve Eq. (8), rewrite it as

$$\frac{1}{\sqrt{y}}\frac{dy}{dt} = -k \tag{9}$$

and carry out the following steps.

a) Find the general antiderivative of each side of Eq. (9).

b) Set the antiderivatives in (a) equal and combine their arbitrary constants into a single arbitrary constant. (Nothing is achieved by having two when one will do.) This will give an equation that relates y directly to t.

46. *Continuation of Exercise 45.* a) Suppose t is measured in minutes and $k = 1/10$. Find y as a function of t if $y = 9$ ft when $t = 0$. (b) How long does it take the tank to drain if the water is 9 ft deep to start with?

47. Figure 4.47 shows the graph of a function $y = f(x)$ that solves one of the following initial value problems. Which one? How do you know?

a) $dy/dx = 2x, \quad y(1) = 0$

b) $dy/dx = x^2, \quad y(1) = 1$

c) $dy/dx = 2x + 2, \quad y(1) = 1$

d) $dy/dx = 2x, \quad y(1) = 1$.

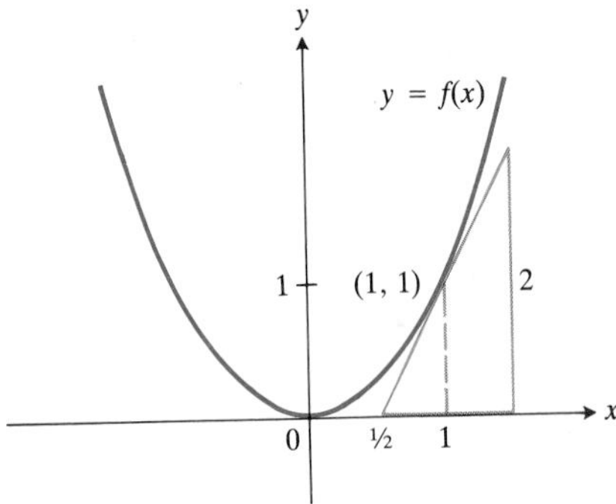

4.47 The function in Exercise 47.

TOOLKIT PROGRAM

Antiderivatives and Direction Fields

Graphs solutions of the initial value problem

$$y' = f(x, y), \qquad y(x_0) = y_0.$$

You can see the effect of changing x_0 and y_0 by graphing different solutions together.

Review Questions

1. Outline a general method for solving related rate problems. Illustrate with an example.
2. State the first derivative test for local extreme values.
3. What does it mean for a function $y = f(x)$ to have an absolute or local maximum or minimum value?
4. How do you find the local and absolute maximum and minimum values of a function $y = f(x)$?
5. What are the hypotheses and conclusion of Rolle's theorem? How does the theorem sometimes help you to tell how many solutions an equation has in a given interval?
6. What are the hypotheses and conclusion of the Mean Value Theorem? What physical interpretation does the theorem sometimes have? Give an example.
7. This chapter gives three important corollaries of the Mean Value Theorem. State each one and describe how it is used.
8. How do you test a function to find out where its graph is concave up or concave down? What is an inflection point? What physical significance do inflection points sometimes have?
9. List the steps you would take to graph a polynomial $y = f(x)$. How does calculus tell you the shape of the graph between the points you plot? Give an example.
10. List the steps you would take to graph a rational function $y = f(x)$. Illustrate with an example.
11. Outline a general method for solving max–min problems. Illustrate with an example.
12. What guidance do you get from calculus about finding production levels that maximize profit? That minimize average manufacturing cost?
13. What is an antiderivative of a function $y = f(x)$? When a function has an antiderivative, how do we find its general antiderivative? Illustrate with an example.
14. What general rules can you call on to help find antiderivatives? Show, by example, how they are used.
15. What is an initial value problem? How do you solve one? Illustrate with an example.
16. How can you predict cost from marginal cost and revenue from marginal revenue?

Practice Exercises

1. The radius of a circle is changing at the rate of $-2/\pi$ m/sec. At what rate is the circle's area changing when $r = 10$ m?
2. The coordinates of a particle moving in the metric xy-plane are differentiable functions of time t with $dx/dt = -1$ m/sec and $dy/dt = -5$ m/sec. How fast is the particle approaching the origin as it passes through the point (5, 12)?
3. The volume of a cube is increasing at the rate of 1200 cm^3/min at the instant its edges are 20 cm long. At what rate are the edges changing at that instant?
4. A point moves smoothly along the curve $y = x^{3/2}$ in the first quadrant in such a way that its distance from the origin increases at the constant rate of 11 units per second. Find dx/dt when $x = 3$.
5. Water drains from the conical tank in Fig. 4.48 at the rate of 5 ft^3/min. (a) What is the relation between the variables h and r in the figure? (b) How fast is the water level dropping when $h = 6$ ft?
6. Two cars are approaching an intersection along straight highways that cross at right angles, car A moving at 36 mph and car B at 50 mph. At what rate

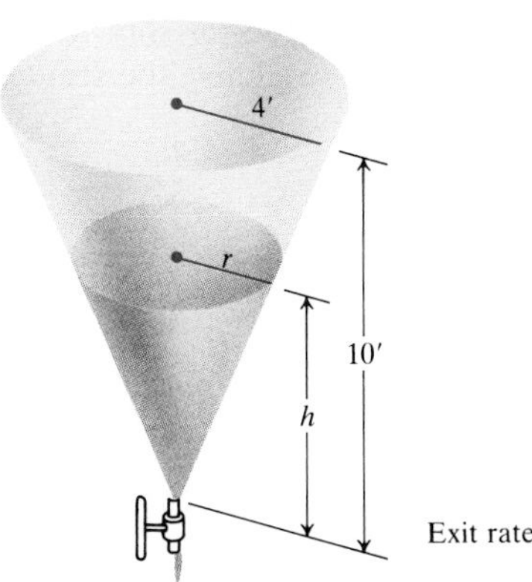

4.48 Exercise 5 asks how fast the water level in this tank is dropping when the water is 6 ft deep.

is the straight-line distance between the cars changing when car A is 5 mi and car B is 12 mi from the intersection?

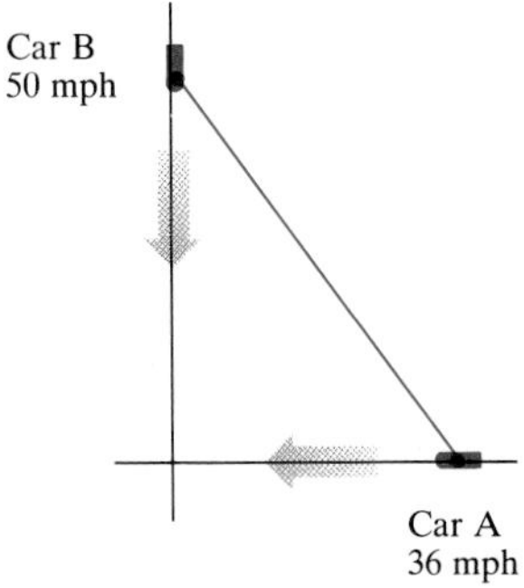

7. You are videotaping a race from a stand 132 ft from the track, following a car that is traveling at 180 mph (264 ft/sec). How fast will your camera angle θ be changing when the car is right in front of you? A half-second later?

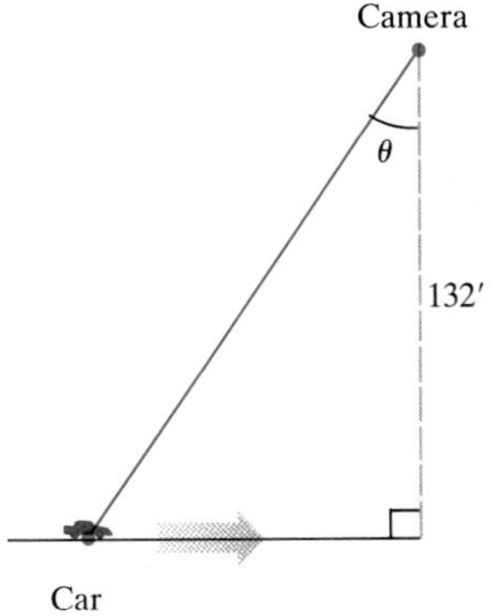

8. As television cable is pulled from a large spool to be strung from the telephone poles along a street, it unwinds from the spool in layers of constant radius. If the truck pulling the cable moves at a steady 6 ft/sec (a touch over 4 mph), how fast is the spool turning when the layer of radius 1.2 ft is being unwound?

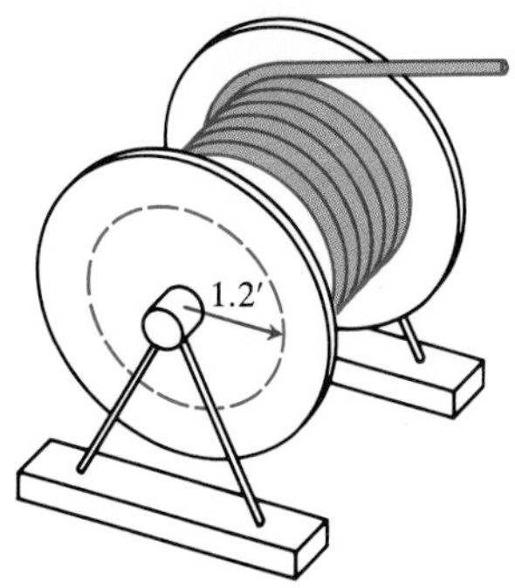

9. Show that the equation $x^4 + 2x^2 - 2 = 0$ has exactly one solution on the interval $0 \leq x \leq 1$.

10. If $f'(x) \leq 2$ for all x, what is the most f can increase on the interval $0 \leq x \leq 6$?

11. Show that $y = x/(x + 1)$ increases on every interval in its domain.

12. Show that $y = x^3 + 2x$ has no maximum or minimum value.

Graph the polynomials in Exercises 13–20.

13. $y = x^2 - 2x - 3$

14. $y = 9x - x^2$

15. $y = 4x^3 - x^4$

16. $y = x^2 - x^3/6$

17. $y = x^3 - 3x^2 + 3$

18. $y = 2x^3 - 12x^2 + 18x$

19. $y = -x^3 + 6x^2 - 9x + 3$

20. $y = (1/8)(x - 3)(x + 3)^2 = (1/8)(x^3 + 3x^2 - 9x - 27)$

21. Suppose that the first derivative of $y = f(x)$ is

$$y' = 6(x + 1)(x - 2)^2.$$

At what points, if any, does the graph of f have a local maximum, local minimum, or point of inflection?

22. Suppose that the first derivative of $y = f(x)$ is

$$y' = 6x(x + 1)(x - 2).$$

At what points, if any, does the graph of f have a local maximum, local minimum, or point of inflection?

23. What value of b will make the graph of $y = x^3 + bx^2 - 5x + 7$ have a point of inflection at $x = 1$?

24. Use the following information to find the values of a, b, and c in the formula

$$f(x) = \frac{x + a}{bx^2 + cx + 2}.$$

i) The values of a, b, and c are either 0 or 1.
ii) The graph of f passes through the point $(-1, 0)$.
iii) The line $y = 1$ is an asymptote of the graph of f.

25. At which of the five points on the graph of $y = f(x)$ shown here (a) are y' and y'' both negative? (b) is y' negative and y'' positive?

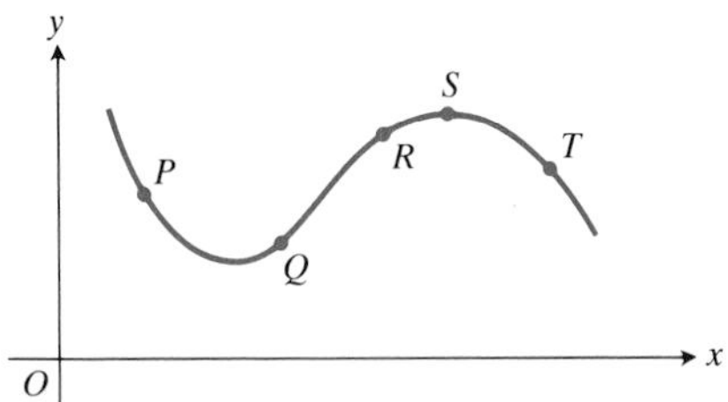

26. Here is the graph of the fruit fly population again. On approximately what day did the population's growth rate change from increasing to decreasing?

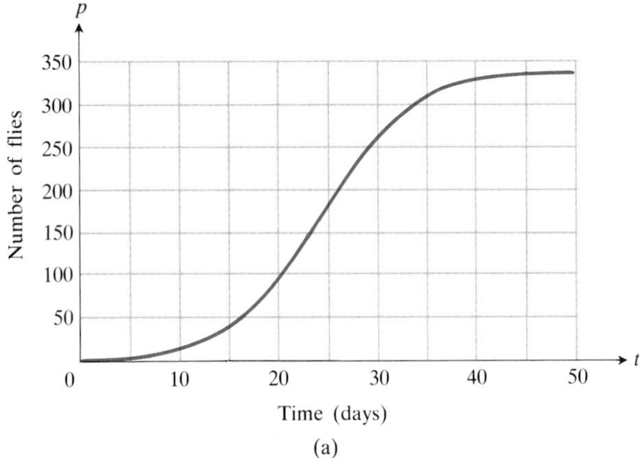

(a)

Q	Slope of $PQ = \Delta p/\Delta t$ (flies/day)
(45, 340)	$(340 - 150)/(45 - 23) \approx 9$
(40, 330)	$(330 - 150)/(40 - 23) \approx 13$
(35, 310)	$(310 - 150)/(35 - 23) \approx 15$
(30, 265)	$(265 - 150)/(30 - 23) \approx 16.4$

(b)

27. Find the maximum and minimum values of $f(x) = 4x^3 - 12x^2 + 9x$ on the interval $0 \le x \le 2$ and say where they are taken on.

28. Find the largest product you can get from two positive numbers whose sum is 36. What are the numbers?

29. If the perimeter of the circular sector shown here is 100 ft, what values of r and s will give the sector the greatest area?

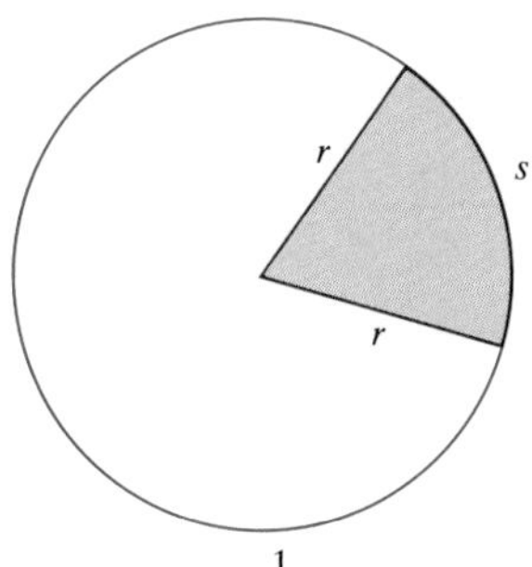

Area: $A = \frac{1}{2}rs$

Perimeter: $p = s + 2r$

30. An isosceles triangle has its vertex at the origin and its base parallel to the x-axis with the vertices above the axis on the curve $y = 27 - x^2$. Find the largest area the triangle can have.

31. Find the dimensions of the largest open storage bin with a square base and vertical sides that can be made from 108 ft^2 of sheet steel. (Neglect the thickness of the steel and assume there is no waste.)

32. A customer has asked you to design an open-top rectangular stainless steel vat. It is to have a square base and a volume of 32 ft^3, to be welded from quarter-inch plate, and to weigh no more than necessary. What dimensions do you recommend?

33. Find the height and radius of the largest right circular cylinder that can be put in a sphere of radius $r = \sqrt{3}$

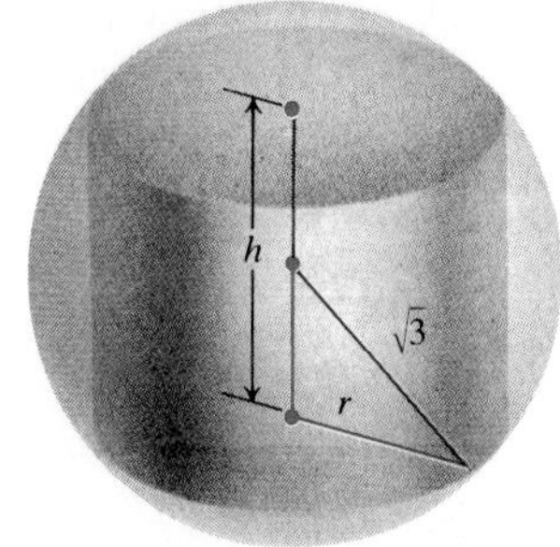

34. Figure 4.49 shows two right circular cones, one upside down inside the other. The two bases are parallel, and the vertex of the smaller cone lies at the center of the larger cone's base. What values of r and h will give the smaller cone the largest possible volume?

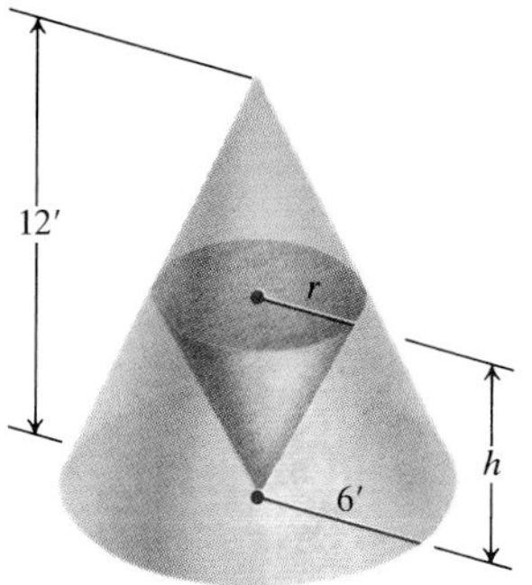

4.49 The cone in Exercise 34.

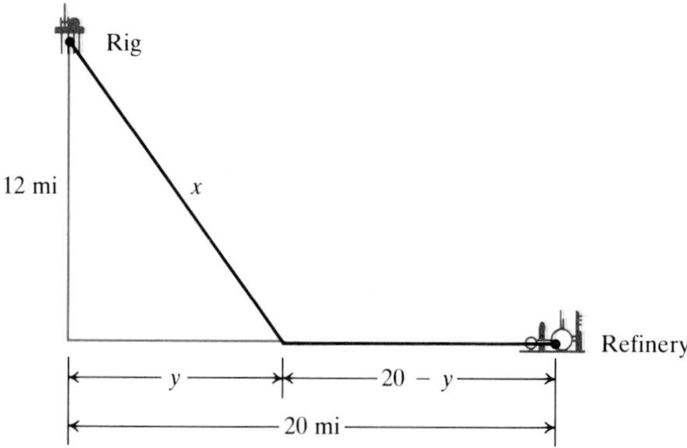

4.50 Diagram for the pipe in Exercise 35.

35. A drilling rig 12 mi off shore is to be connected by a pipe to a refinery on shore, 20 mi down the coast from the rig (Fig. 4.50). If underwater pipe costs \$50,000 per mile and land-based pipe costs \$30,000 per mile, what values of x and y give the least expensive connection?

36. An athletic field is to be built in the shape of a rectangle x units long capped by semicircular regions of radius r at the two ends. The field is to be bounded by a 400-m racetrack. What values of x and r will give the rectangle the largest possible area?

37. Your company can manufacture x hundred grade A tires and y hundred grade B tires a day, where $0 \le x \le 4$ and

$$y = \frac{40 - 10x}{5 - x}.$$

Your profit on Grade A tires is twice your profit on grade B tires. What is the most profitable number of each kind of tire to make?

38. The positions of two particles on the s-axis are $s_1 = \sin t$ and $s_2 = \sin(t + \pi/3)$. What is the farthest apart the particles ever get?

39. The formula $F(x) = 3x + C$ gives a different function for each value of C. All of these functions, however, have the same derivative with respect to x, namely $F'(x) = 3$. Are these the only differentiable functions whose derivative is 3? Could there be any others? Explain.

40. Show that

$$\frac{d}{dx}\left(\frac{x}{x+1}\right) = \frac{d}{dx}\left(-\frac{1}{x+1}\right)$$

even though

$$\frac{x}{x+1} \ne -\frac{1}{x+1}.$$

Doesn't this contradict Corollary 3 of the Mean Value Theorem? Explain.

41. Find the general antiderivatives of the following functions.

a) 0 b) 1 c) x d) x^2
e) x^{10} f) x^{-2} g) x^{-5} h) $x^{5/2}$
i) $x^{4/3}$ j) $x^{3/4}$ k) $x^{1/2}$ l) $x^{-1/2}$
m) $x^{-3/7}$ n) $x^{-7/3}$

42. Find the general antiderivatives of the following functions.

a) $\sin x$ b) $\cos x$ c) $\sec x \tan x$
d) $-\csc^2 x$ e) $\sec^2 x$ f) $-\csc x \cot x$

Find the general antiderivatives of the functions in Exercises 43–58.

43. $3x^2 + 5x - 7$

44. $\dfrac{1}{x^2} + x + 1$

45. $\sqrt{x} + \dfrac{1}{\sqrt{x}}$

46. $\sqrt[3]{x} + \sqrt[4]{x}$

47. $3\cos 5x$

48. $8\sin(x/2)$

49. $3\sec^2 3x$

50. $4\csc^2 2x$

51. $\dfrac{1}{2} - \cos x$

52. $3x^5 + 16\cos 8x$

53. $\sec\dfrac{x}{3}\tan\dfrac{x}{3} + 5$

54. $1 - \csc x \cot x$

55. $\tan^2 x$ (*Hint:* $\tan^2 x = \sec^2 x - 1$)

56. $\cot^2 x$ (*Hint:* $\cot^2 x = \csc^2 x - 1$)

57. $2\sin^2 x$ (*Hint:* $2\sin^2 x = 1 - \cos 2x$)

58. $\cos^2 x - \sin^2 x$ (*Hint:* $\cos^2 x - \sin^2 x = \cos 2x$)

Solve the initial value problems in Exercises 59–64.

59. $\dfrac{dy}{dx} = 1 + x + \dfrac{x^2}{2}$, $y = 1$ when $x = 0$

60. $\dfrac{dy}{dx} = 4x^3 - 21x^2 + 14x - 7, \quad y = 1$ when $x = 1$

61. $\dfrac{dy}{dx} = \dfrac{x^2+1}{x^2}, \quad y = -1$ when $x = 1$

62. $\dfrac{dy}{dx} = \left(x + \dfrac{1}{x}\right)^2, \quad y = 1$ when $x = 1$

63. $\dfrac{d^2y}{dx^2} = -\sin x, \quad y = 0$ and $\dfrac{dy}{dx} = 1$ when $x = 0$

64. $\dfrac{d^2y}{dx^2} = \cos x, \quad y = -1$ and $\dfrac{dy}{dx} = 0$ when $x = 0$

65. Does any function $y = f(x)$ satisfy all of the following conditions? If so, what is it? If not, why not?
i) $d^2y/dx^2 = 0$ for all x
ii) $dy/dx = 1$ when $x = 0$
iii) $y = 0$ when $x = 0$.

66. Find an equation for the curve in the xy-plane that passes through the point $(1, -1)$ if its slope at x is always $3x^2 + 2$.

67. You sling a shovelful of dirt up from the bottom of a 17-ft hole with an initial velocity of 32 ft/sec. Is that enough speed to get the dirt out of the hole, or had you better duck?

68. The acceleration of a particle moving along a coordinate line is $d^2s/dt^2 = 2 + 6t$ m/sec^2. At $t = 0$, the velocity is 4 m/sec. Find the velocity as a function of t. Then find how far the particle moves during the first second of its trip, from $t = 0$ to $t = 1$.

69. **CALCULATOR** Show that the equation $x^3 + x - 1 = 0$ has exactly one solution and use Newton's method to find it to three decimal places.

70. *Estimating reciprocals without division.* Newton's method in Section 3.8 can be used to estimate the reciprocal of a positive number a without ever dividing by a, by taking $f(x) = (1/x) - a$. For example, if $a = 3$ the function involved is $f(x) = (1/x) - 3$.
a) Graph $y = (1/x) - 3$. Where does the graph cross the x-axis?
b) Show that the recursion formula in Newton's method in this case is
$$x_{n+1} = x_n(2 - 3x_n),$$
so indeed there is no division.

71. **COMPUTER GRAPHER** If you have access to a computer or a graphing calculator, try these graphs to see how they go:
a) $y = 3x(x-1)(x+1)^2$
b) $y = 2\cos x + \cos^2 x$

TOOLKIT PROGRAMS

Derivative Grapher	Graphs functions and their derivatives side by side
Root Finder	Graphs functions and estimates roots by Newton's method (and others)
Super*Grapher	Graphs functions in color

5 Integration

H. Armstrong Roberts

We calculate the areas of shapes like these with calculus.

OVERVIEW This chapter introduces the second main branch of calculus, the branch called integral calculus. Integral calculus is the mathematics we use to find lengths, areas, and volumes of irregular shapes; to calculate the average values of functions; and to predict future population sizes and future costs of living. In this chapter, we set the stage for these and other applications.

The development of integral calculus starts from the calculation of areas. We show how to use antiderivatives in these calculations, and how this technique leads to a natural definition of area as a limit of finite sums. The limits used to define areas are special cases of a kind of limit called a *definite integral.* Presenting the properties of definite integrals is a central goal of this chapter.

The single most important concept in this chapter is the connection between definite integrals and derivatives. The discovery of this connection (called the fundamental theorem of integral calculus) by Leibniz and Newton turned calculus into the most important applied mathematics in the world. The rest of the chapter describes the basic methods for finding antiderivatives and using them to evaluate integrals.

In the final section of this chapter we introduce the natural logarithm function and its inverse, the exponential function. We examine these functions in more detail in Chapter 7, but we discuss them here to complete our basic picture of how to use definite integrals to define new functions and to illustrate important applications in business, finance, and the life sciences.

5.1 Calculus and Area

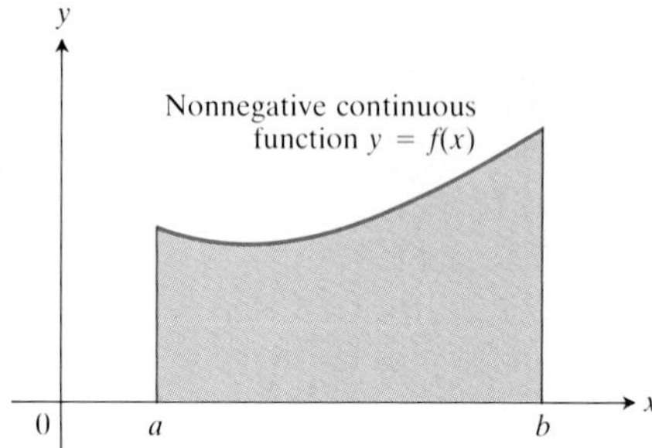

5.1 We can use integral calculus to find the areas of regions like the shaded one here.

Integral calculus is the mathematics we use to define and calculate the areas of irregularly shaped regions like the cross sections of machine parts and airplane wings. This section explains what calculus and area have to do with one another and shows how our ability to find antiderivatives enables us to calculate the areas of regions like the one in Fig. 5.1.

Regions Bounded by Curves

To find the area of a triangle, we use the formula $A = (1/2)bh$, area equals one-half base times height. To find the areas of more general polygonal regions, we can divide them into triangles, then add the areas of the triangles (Fig. 5.2). But we get stuck if we try to calculate the area of a circle this way. No matter how many triangles we draw inside the circle, their straight edges never quite match the curve of the circle, and some of the circle's interior remains uncovered.

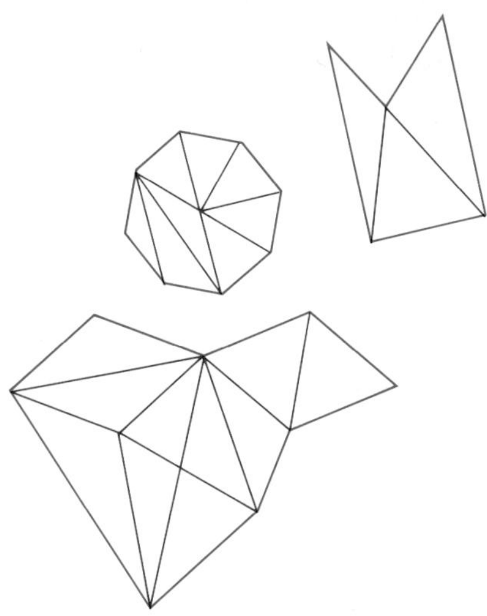

5.2 We find the areas of plane regions with polygonal boundaries by dividing the regions into triangles. The answer is the same for every triangulation.

The Greeks of the fifth century B.C. overcame this problem by filling the circle with an infinite sequence of increasingly fine regular polygons, exhausting the circle's area, so to speak, step by step (Fig. 5.3). They then took the circle's area to be the limit of the areas of these polygons. A decreasing sequence of circumscribed polygons would have worked as well (Fig. 5.4).

The difficulty with applying this approach to more general curves is not the involvement of limits (at least, that's not a difficulty for us). Rather, it is the complication associated with finding workable formulas for the areas of the inscribed polygons, which in an arbitrary curve can assume irregular shapes. We can avoid this difficulty if, instead of working in the abstract plane of Euclidean geometry, we work in the coordinate plane of Descartes and Fermat. For then we can approximate the region under a curve with rectangles whose numerical dimensions, and hence areas, are given by the curve itself.

You will see what we mean if you look at the curve and rectangles in Fig. 5.5. The rectangle areas, added up, approximate the area between the

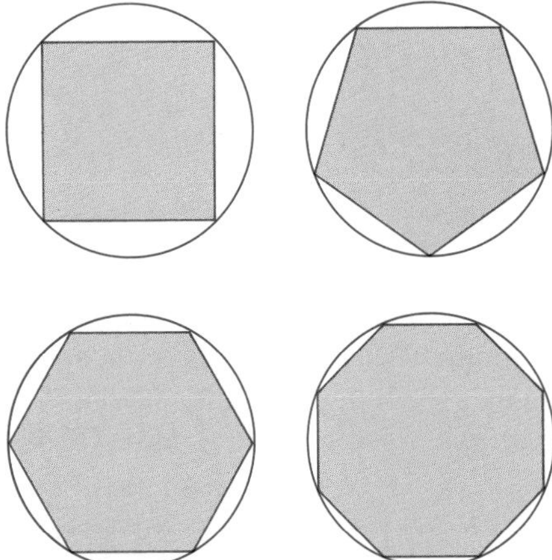

5.3 An increasing sequence of regular polygons inscribed in a circle. Approximations like these were the basis of the method used in classical Greek times to determine the area of a circle.

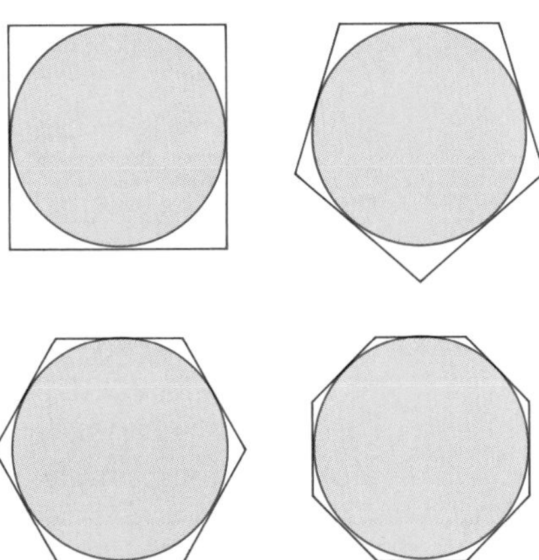

5.4 Circumscribed polygons work too. In this case the polygons' areas approach the circle's area from above instead of below.

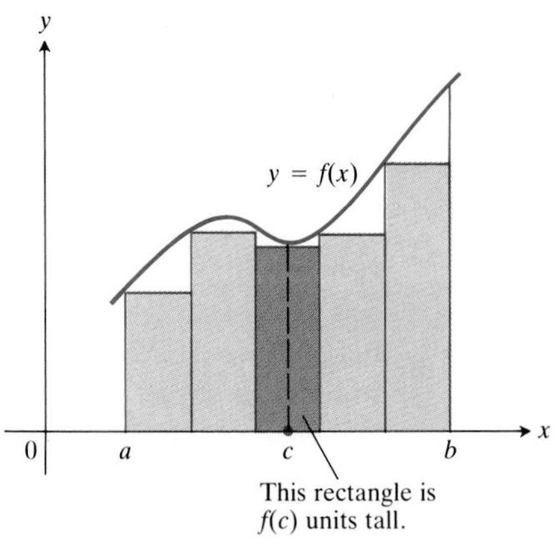

5.5 If we approximate the region under the curve $y = f(x)$ from $x = a$ to $x = b$ with inscribed rectangles that reach from the x-axis up to the curve, then the height of each rectangle is the value of f at some point along the rectangle's base.

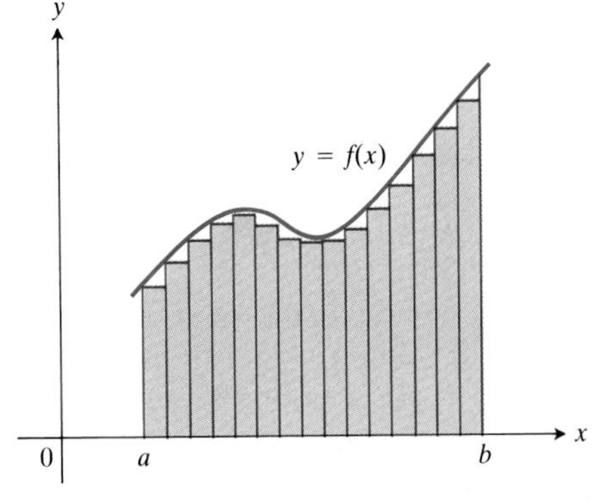

5.6 The more rectangles we use, the better the approximation becomes (provided all the rectangles become narrower as we go along).

curve $y = f(x)$ and the x-axis over the interval from $x = a$ to $x = b$. The area of each rectangle, base times height, is the base length times some particular function value, a value we can find from the formula $y = f(x)$.

Notice how the approximations improve as the rectangles become thinner and more numerous (Fig. 5.6). With each refinement, we get closer to filling up the region whose area we want to find. To finish the job, all we need is

1. A way to write formulas for sums of large numbers of terms, and
2. A way to find the numerical limits of such sums as the number of terms tends to infinity (when the limits exist, that is).

As soon as we know how to take these two steps, we shall be able to define and calculate all the areas we want.

What we find, when we take these steps, will also be surprising. We shall be able to do much more than just calculate areas. And, thanks to a great breakthrough in calculation discovered by Leibniz and Newton, the calculations involved will all be straightforward.

The Area under the Graph of a Nonnegative Continuous Function

Here is a preview of how we shall be able to calculate area once we have made the necessary mathematical arrangements. We all know what we want area to be like, so let us suppose for the moment that the forthcoming mathematical definition in Section 5.3 gives us everything we want. (It will.) Let $y = f(x)$ be a nonnegative continuous function, like the one graphed here, for example:

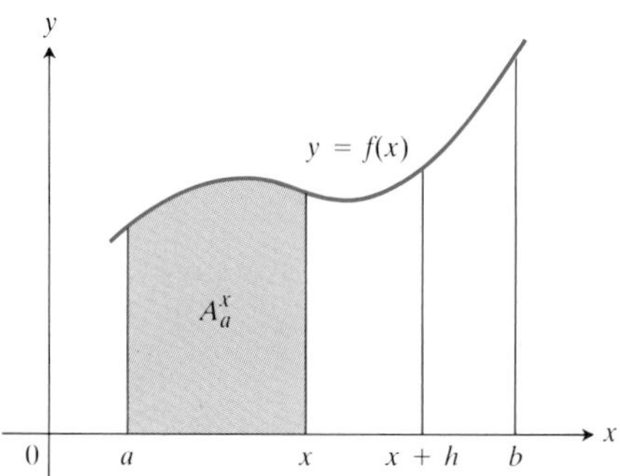

Let A_a^x denote the area of the region that lies between the curve and the x-axis from a up to x. We'll call this the area under the curve from a to x:

$$A_a^x = \text{the area under the curve from } a \text{ to } x.$$

Here a is a constant and x a variable. The area A_a^x, whose value changes with each new value of x, is a function of x.

If we move from x to a nearby point $x + h$, the amount of area we add is A_x^{x+h}, the area under the curve from x to $x + h$. These two areas combine to give the area from a to $x + h$, so

$$A_a^x + A_x^{x+h} = A_a^{x+h} \tag{1}$$

and

$$A_x^{x+h} = A_a^{x+h} - A_a^x. \tag{2}$$

Now, the area under the curve from x to $x + h$ can be trapped between the areas of two rectangles of base length h, as shown here:

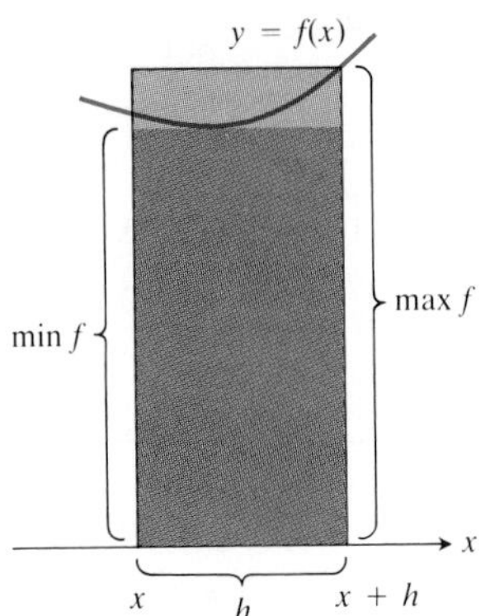

The height of the shorter rectangle is min f, the minimum value of f on the interval from x to $x+h$. The area of this rectangle is therefore $h \cdot \min f$ (base times height). The height of the taller rectangle is max f, the maximum value of f on the interval from x to $x+h$. The area of this rectangle is therefore $h \cdot \max f$.

We record the observation that the area under the curve from x to $x+h$ lies between the areas of these rectangles by writing the following inequality:

$$h \cdot \min f \leq A_a^{x+h} - A_a^x \leq h \cdot \max f. \tag{3}$$

Dividing by h gives

$$\min f \leq \frac{A_a^{x+h} - A_a^x}{h} \leq \max f. \tag{4}$$

The fraction in the middle of the inequality in Eq. (4) is Fermat's difference quotient for the derivative of the area function A_a^x. It is the value of the function at $x+h$ minus the value of the function at x, all divided by h. We can therefore calculate the derivative of the area function at x by finding the limit of this quotient as h goes to zero.

As h goes to zero, the interval from x to $x+h$ gets shorter and shorter. As it does so, the values of max f and min f both approach the value of f at x (remember f is continuous). Hence, by the Sandwich Theorem of Section 2.2, the difference quotient approaches $f(x)$ as well. In symbols,

$$\lim_{h \to 0} \frac{A_a^{x+h} - A_a^x}{h} = f(x). \tag{5}$$

We are thus led to the astonishing conclusion that, when f is a nonnegative continuous function of x, the area under its graph from a to x is a differentiable function of x whose derivative at x is $f(x)$:

$$\frac{d}{dx} A_a^x = f(x). \tag{6}$$

Among other things, this means that we can find an explicit formula for A_a^x whenever we can solve the following initial value problem:

Differential equation: $\frac{d}{dx} A_a^x = f(x)$

Initial condition: $A_a^x = 0$ when $x = a$.

The initial condition comes from the observation that the area under the graph from a point a to the same point a is zero. There is no area under a graph of zero length.

We solve the initial value problem in the usual way. We find an antiderivative $F(x)$ of $f(x)$ to get

$$A_a^x = F(x) + C. \tag{7}$$

We then find the right value of C from the initial condition, by setting x equal to a:

$$A_a^a = F(a) + C \qquad (x = a \text{ in Eq. 7})$$

$$0 = F(a) + C \qquad (A_a^x = 0 \text{ when } x = a)$$

$$C = -F(a). \qquad (\text{Solved for } C)$$

The area under the curve $y = f(x)$ from a to x is therefore

$$A_a^x = F(x) - F(a). \tag{8}$$

Here, then, is the relation between area and calculus. We calculate area with antiderivatives.

How to Find the Area under the Graph of a Nonnegative Continuous Function $y = f(x)$

STEP 1: Find an antiderivative $F(x)$ of $f(x)$. (Any antiderivative will do.)

STEP 2: Calculate the number $F(b) - F(a)$. This number will be the area under the curve from a to b.

Besides the fact that we have yet to define area (we shall do so in Section 5.3), there are two practical questions to face here: How do we know f *has* an antiderivative, and how do we find one when f does? We shall take care of the existence when we get to Section 5.4. As for finding antiderivatives, there is no need to worry. We shall get better at that as we learn more calculus. Here, now, are some examples.

EXAMPLE 1 Find the area under the curve $y = x^2$ from $x = 0$ to $x = 1$.

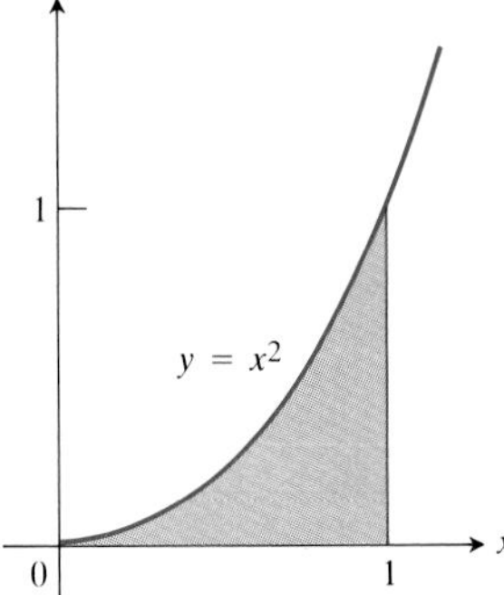

Solution In this example $f(x) = x^2$, $a = 0$, and $b = 1$. We find the area in two steps.

STEP 1: Find an antiderivative $F(x)$ of $f(x) = x^2$. Any antiderivative will do, so we can choose the simplest one:

$$F(x) = \frac{x^3}{3}.$$

STEP 2: Calculate $F(1) - F(0)$:

$$F(1) - F(0) = \frac{(1)^3}{3} - \frac{(0)^3}{3} = \frac{1}{3} - 0 = \frac{1}{3}.$$

The area is 1/3. ■

If we use a different antiderivative in Example 1, say

$$F(x) = \frac{x^3}{3} + 4,$$

we still get 1/3 for the answer:

$$F(1) - F(0) = \left(\frac{(1)^3}{3} + 4\right) - \left(\frac{(0)^3}{3} + 4\right) = \frac{1}{3} + 4 - 0 - 4 = \frac{1}{3}.$$

Notice how the 4 cancels out. Any constant we add to the $x^3/3$ will cancel out the same way when we subtract.

EXAMPLE 2 *(Continuation of Example 1)* The area under the parabola $y = x^2$ from $x = 1$ to $x = 2$ is

$$F(2) - F(1) = \frac{(2)^3}{3} - \frac{(1)^3}{3}$$
$$= \frac{8}{3} - \frac{1}{3} = \frac{7}{3}.$$

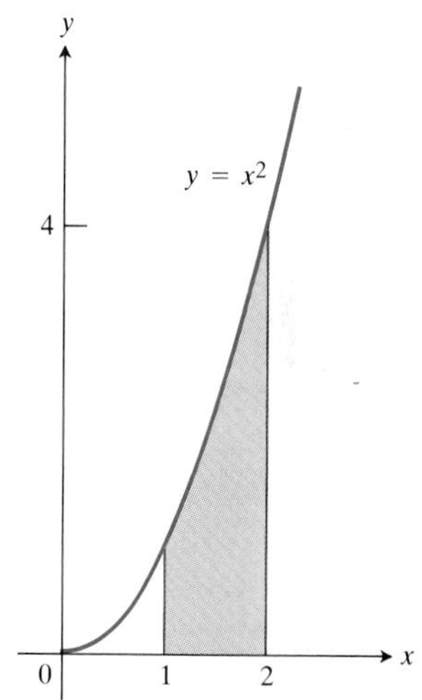

■

EXAMPLE 3 Find the area under one arch of the curve $y = \cos x$.

Solution STEP 1: Find an antiderivative of $y = \cos x$:

$$F(x) = \sin x. \qquad \text{(Simplest one)}$$

STEP 2: Calculate $F\left(\frac{\pi}{2}\right) - F\left(-\frac{\pi}{2}\right)$:

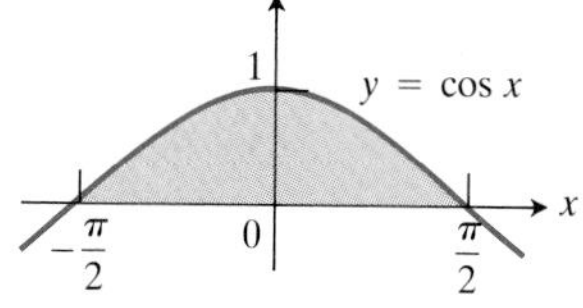

$$F\left(\frac{\pi}{2}\right) - F\left(-\frac{\pi}{2}\right) = \sin\left(\frac{\pi}{2}\right) - \sin\left(-\frac{\pi}{2}\right) = 1 - (-1) = 2.$$

The area is 2 square units. ■

What Lies Ahead

If all we wanted to do was find areas, we would be nearly done now, but the idea of approximating things with small manageable pieces the way we approximated regions with rectangles extends to thousands of other situations. We calculate the volumes of large objects by slicing them like loaves of bread and adding up the volumes of the slices. We find the lengths of curves by approximating small pieces of the curve with line segments and adding the lengths of the line segments. The same idea extends to finding the areas of surfaces and the forces against dams, and to calculating how much work it takes to serve a tennis ball or lift a satellite into orbit. In each case, as we shall see in the chapters to come, we find the answer by finding the endpoint values of some antiderivative $F(x)$ and calculating their difference, $F(b) - F(a)$. The antiderivatives vary from problem to problem, but the general procedure remains the same.

Exercises 5.1

Use antiderivatives to find the areas of the shaded regions in Exercises 1–16.

1.

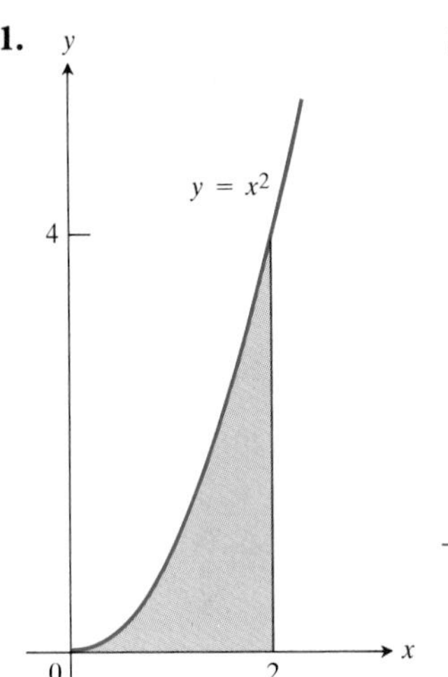

2.

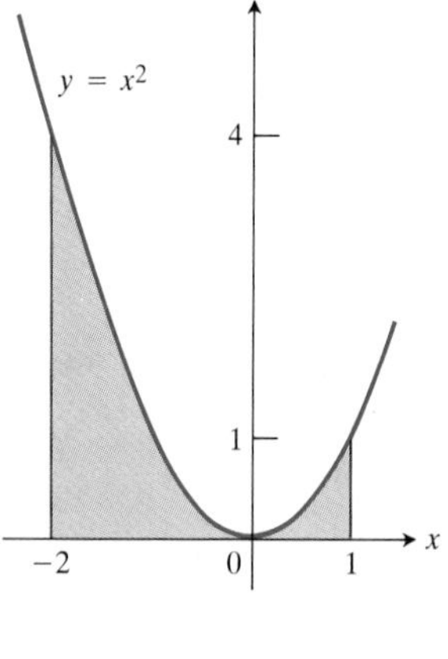

3.

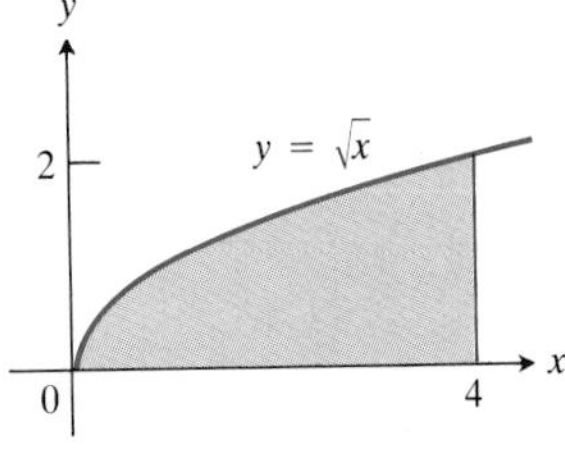

4.

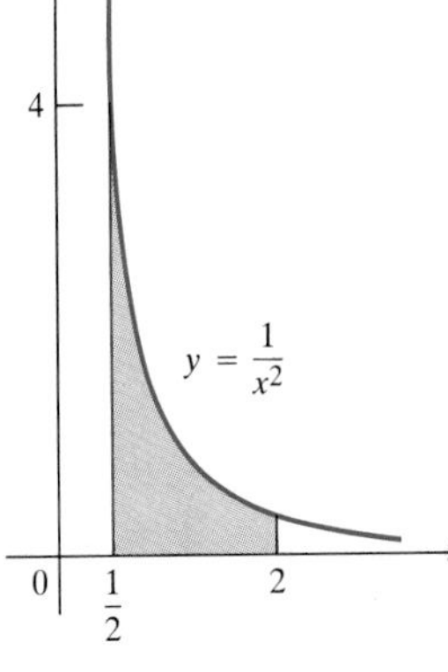

5.

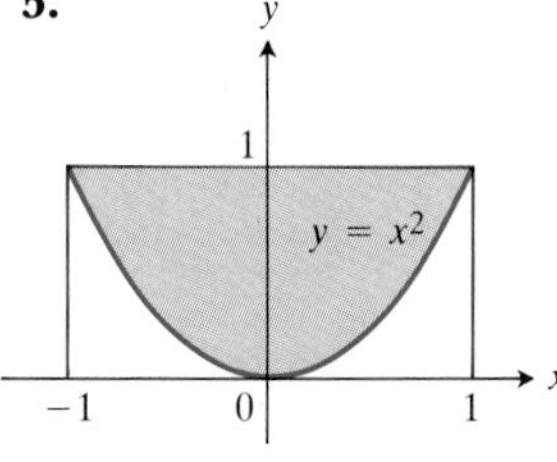

6.

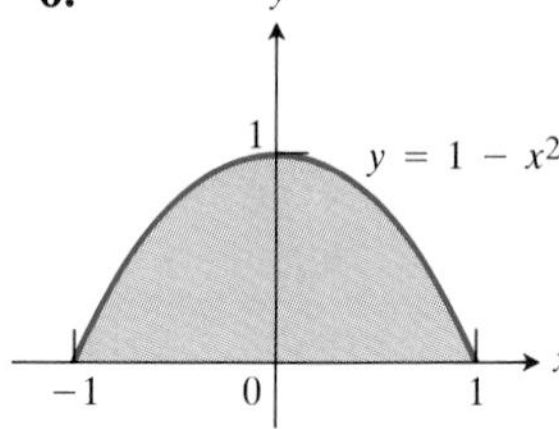

7.

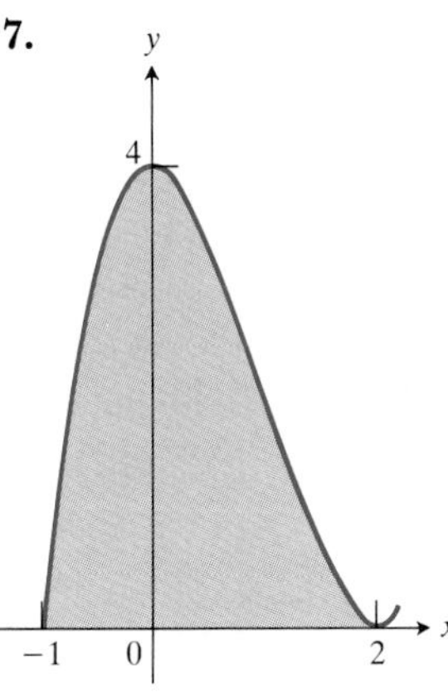

8.

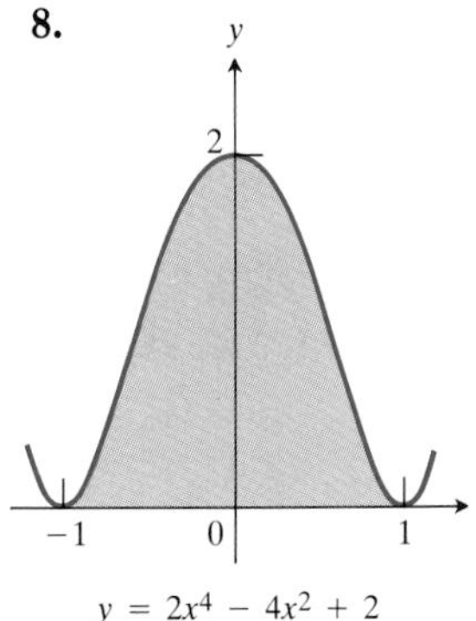

9.

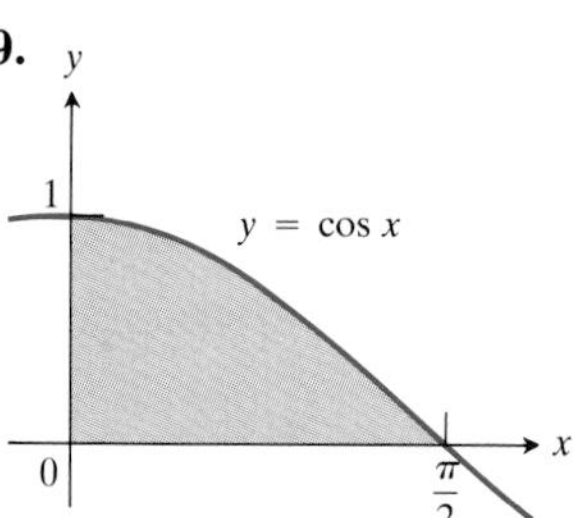

10.

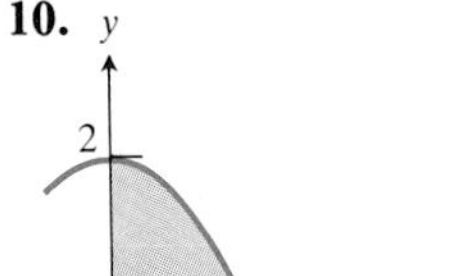

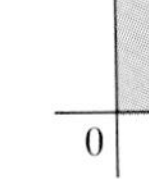

11.

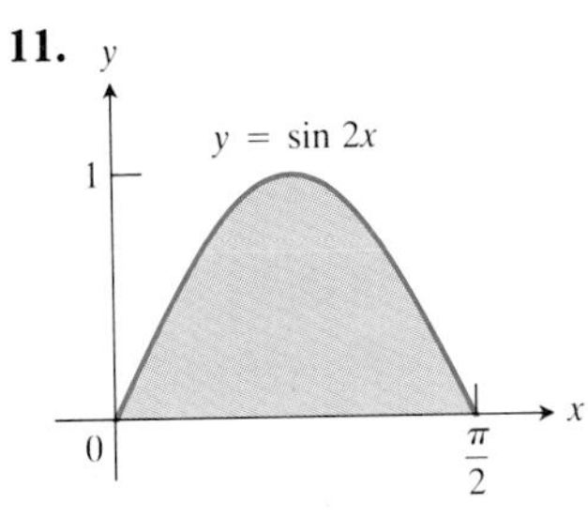

12.

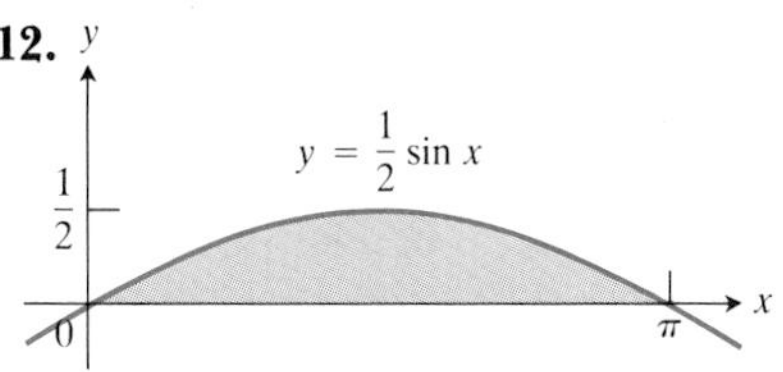

13.

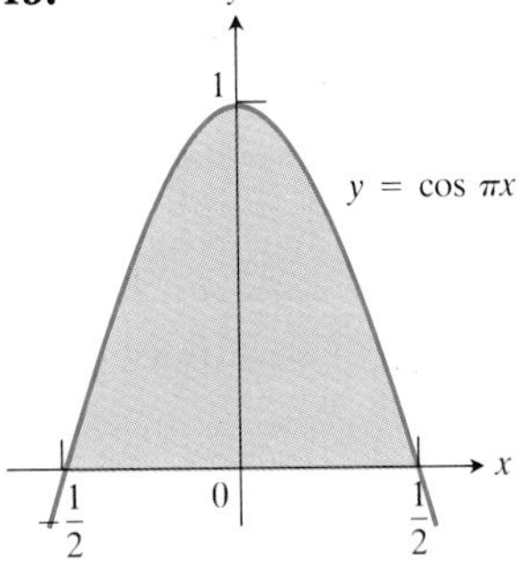

14.

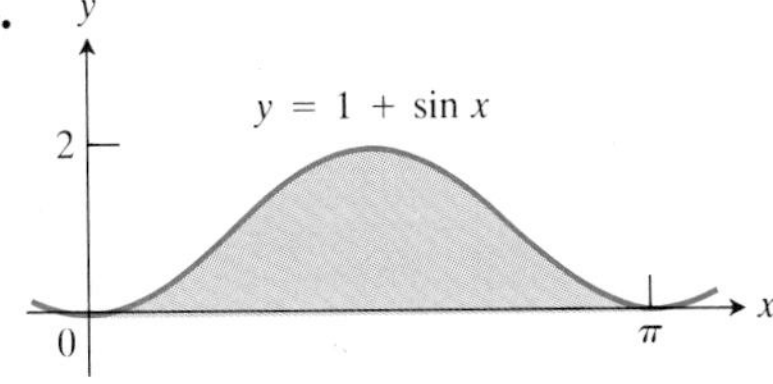

15.

16.

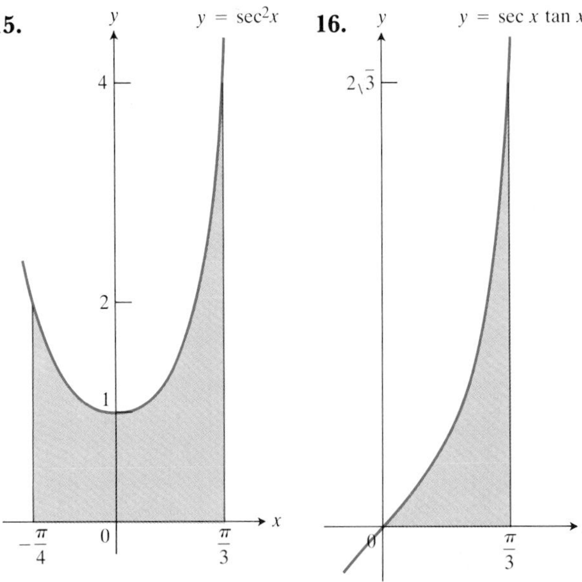

17. Let b be any positive number and n any rational number other than -1. Find a formula for the area under the curve $y = x^n$ from $x = 0$ to $x = b$.

18. Whenever we find a new way to calculate something, it is a good idea to be sure that the new and old ways agree on the objects to which they both apply. If you use an antiderivative to find the area of the triangle in Fig. 5.7, will you still get $A = (1/2)bh$? Try it and find out.

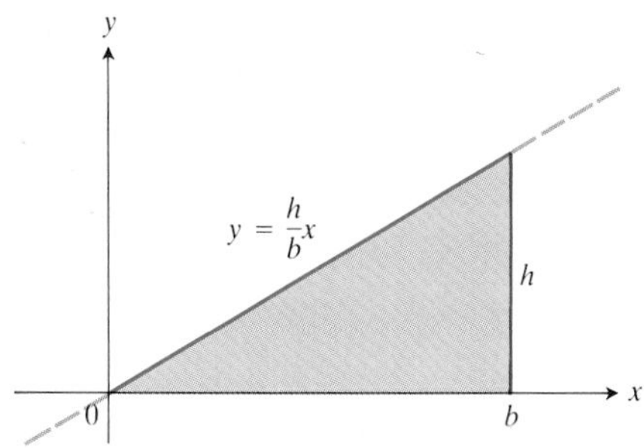

5.7 Is the area of this triangle still $\frac{1}{2}bh$? See Exercise 18.

5.2 Formulas for Finite Sums

As we saw in Section 5.1, we need a notation for sums of large numbers of terms. We develop that notation here.

Sigma Notation for Finite Sums

The letter we use to indicate sums is Σ (pronounced "sigma"), the Greek letter for S.

EXAMPLE 1

The sum	In sigma notation	One way to read the notation
$a_1 + a_2$	$\sum_{k=1}^{2} a_k$	The sum of a sub k from k equals 1 to k equals 2.
$a_1 + a_2 + a_3$	$\sum_{k=1}^{3} a_k$	The sum of a sub k from k equals 1 to k equals 3.
$a_1 + a_2 + a_3 + a_4$	$\sum_{k=1}^{4} a_k$	The sum of a sub k from k equals 1 to k equals 4.
$a_1 + a_2 + \cdots + a_n$	$\sum_{k=1}^{n} a_k$	The sum of a sub k from k equals 1 to k equals n. ■

As you can see, the notation remains compact no matter how many terms are being added. It is just what we need for writing sums that involve millions and millions of terms.

There are many different ways to read the notation $\sum_{k=1}^{n} a_k$, all equally good. Some people say, "Summation from k equals 1 to n of a sub k." Others say, "Summation a k from k equals 1 to n." Still others, "The sum of the a sub k's as k goes from 1 to n," and so on. Take your pick.

DEFINITIONS

Sigma Notation for Finite Sums

The symbol

$$\sum_{k=1}^{n} a_k$$

denotes the sum of the n terms

$$a_1 + a_2 + \cdots + a_{n-1} + a_n .$$

The variable k is the **index of summation.** The values of k run through the integers from 1 to n. The a's are the **terms** of the sum: a_1 is the first term, a_2 is the second term, a_k is the ***k*th term,** and a_n is the nth and last term. The number n is the **upper limit of summation;** the number 1 is the **lower limit of summation.**

Here are some numerical examples.

EXAMPLE 2

The sum in sigma notation	The sum written out—one term for each value of k	The value of the sum
$\sum_{k=1}^{5} k$	$1 + 2 + 3 + 4 + 5$	15
$\sum_{k=1}^{3} (-1)^k k$	$(-1)^1(1) + (-1)^2(2) + (-1)^3(3)$	$-1 + 2 - 3 = -2$
$\sum_{k=1}^{2} \frac{k}{k+1}$	$\frac{1}{1+1} + \frac{2}{2+1}$	$\frac{1}{2} + \frac{2}{3} = \frac{7}{6}$

EXAMPLE 3 Find the value of $\sum_{k=1}^{3} \sin\left(\frac{k\pi}{2}\right)$.

Solution

$$\sum_{k=1}^{3} \sin\left(\frac{k\pi}{2}\right) = \sin\left(\frac{1 \cdot \pi}{2}\right) + \sin\left(\frac{2 \cdot \pi}{2}\right) + \sin\left(\frac{3 \cdot \pi}{2}\right) \quad \begin{pmatrix}\text{A term for each} \\ \text{value of } k\end{pmatrix}$$

$$= 1 + 0 + (-1) = 0 \quad \begin{pmatrix}\text{The terms evaluated} \\ \text{and added}\end{pmatrix}$$

The lower limit of summation does not have to be 1; it could be some other integer.

EXAMPLE 4

a) $$\sum_{k=0}^{2} \frac{1}{2^k} = \frac{1}{2^0} + \frac{1}{2^1} + \frac{1}{2^2}$$
$$= \frac{1}{1} + \frac{1}{2} + \frac{1}{4}$$
$$= \frac{7}{4}$$

b) $$\sum_{k=-3}^{-1} (k + 1) = (-3 + 1) + (-2 + 1) + (-1 + 1)$$
$$= -2 + -1 + 0$$
$$= -3$$

Algebra Rules for Finite Sums

When you work with finite sums, you can always use the following rules.

Algebra Rules

1. Sum Rule: $\sum_{k=1}^{n} (a_k + b_k) = \sum_{k=1}^{n} a_k + \sum_{k=1}^{n} b_k$

2. Difference Rule: $\sum_{k=1}^{n} (a_k - b_k) = \sum_{k=1}^{n} a_k - \sum_{k=1}^{n} b_k$

3. Constant Multiple Rule: $\sum_{k=1}^{n} ca_k = c \cdot \sum_{k=1}^{n} a_k$ (Any number c)

4. Constant Value Rule: $\sum_{k=1}^{n} a_k = n \cdot c$ if a_k has the constant value c.

There are no surprises in this list of rules, but the formal proofs require a technique called mathematical induction (Appendix 4).

EXAMPLE 5

a) $\sum_{k=1}^{n} (k - k^2) = \sum_{k=1}^{n} k - \sum_{k=1}^{n} k^2$ (Difference Rule)

b) $\sum_{k=1}^{n} -a_k = \sum_{k=1}^{n} -1 \cdot a_k = -1 \cdot \sum_{k=1}^{n} a_k = -\sum_{k=1}^{n} a_k$ (Constant Multiple Rule)

c) $\sum_{k=1}^{3} (k + 4) = \sum_{k=1}^{3} k + \sum_{k=1}^{3} 4$ (Sum Rule)

$= (1 + 2 + 3) + (3 \cdot 4) = 6 + 12 = 18$ ■

Standard Formulas for Sums

Over the years people have discovered a variety of formulas for the values of finite sums. The most famous of these are the formula for the sum of the first n integers (which Gauss discovered at age 5) and the formulas for the sums of the squares and cubes of the first n integers.

The first n integers: $$\sum_{k=1}^{n} k = \frac{n(n+1)}{2} \tag{1}$$

The first n squares: $$\sum_{k=1}^{n} k^2 = \frac{n(n+1)(2n+1)}{6} \tag{2}$$

The first n cubes: $$\sum_{k=1}^{n} k^3 = \left(\frac{n(n+1)}{2}\right)^2 \tag{3}$$

Notice the relationship between the first sum and the third.

EXAMPLE 6

a) $\displaystyle\sum_{k=1}^{5} k = 1 + 2 + \cdots + 5 = \frac{5(5+1)}{2} = \frac{5 \cdot 6}{2} = 15$

b) $\displaystyle\sum_{k=1}^{5} k^2 = 1 + 4 + \cdots + 25 = \frac{5(6)(2 \cdot 5 + 1)}{6} = 5 \cdot 11 = 55$

c) $\displaystyle\sum_{k=1}^{5} k^3 = 1 + 8 + \cdots + 125 = \left(\frac{5(5+1)}{2}\right)^2 = 15^2 = 225$ ■

EXAMPLE 7 Using values from Example 6, we have

a) $\displaystyle\sum_{k=1}^{5} \frac{k}{3} = \frac{1}{3}\sum_{k=1}^{5} k^3 = \frac{225}{3} = 75,$

b) $\displaystyle\sum_{k=1}^{5} (k - k^2) = \sum_{k=1}^{5} k - \sum_{k=1}^{5} k^2 = 15 - 55 = -40.$ ■

Exercises 5.2

Write the sums in Exercises 1–16 without sigma notation. Then evaluate them.

1. $\displaystyle\sum_{k=1}^{4} \frac{1}{k}$

2. $\displaystyle\sum_{k=1}^{4} \frac{12}{k}$

3. $\displaystyle\sum_{k=1}^{3} (k+2)$

4. $\displaystyle\sum_{k=1}^{5} (2k-1)$

5. $\displaystyle\sum_{k=0}^{4} \frac{k}{4}$

6. $\displaystyle\sum_{k=-2}^{2} 3k$

7. $\displaystyle\sum_{k=1}^{5} (2k-2)$

8. $\displaystyle\sum_{k=1}^{4} (3k-2)$

9. $\displaystyle\sum_{k=1}^{2} \frac{6k}{k+1}$

10. $\displaystyle\sum_{k=1}^{3} \frac{k-1}{k}$

11. $\displaystyle\sum_{k=1}^{5} k(k-1)(k-2)$

12. $\displaystyle\sum_{k=0}^{3} (1-k)(2-k)$

13. $\displaystyle\sum_{k=1}^{4} \cos k\pi$

14. $\displaystyle\sum_{k=1}^{3} \sin\frac{\pi}{k}$

15. $\displaystyle\sum_{k=1}^{4} (-1)^k$

16. $\displaystyle\sum_{k=1}^{4} (-1)^{k+1}$

Express the sums in Exercises 17–24 in sigma notation.

17. $1 + 2 + 3 + 4 + 5 + 6$

18. $1 + 4 + 9 + 16$

19. $\frac{1}{2} + \frac{1}{4} + \frac{1}{8} + \frac{1}{16}$

20. $1 + \frac{1}{2} + \frac{1}{3} + \frac{1}{4} + \frac{1}{5}$

21. $1 - \frac{1}{2} + \frac{1}{3} - \frac{1}{4} + \frac{1}{5}$

22. $2 + 4 + 6 + 8 + 10 + 12$

23. $\frac{1}{5} - \frac{2}{5} + \frac{3}{5} - \frac{4}{5} + \frac{5}{5}$

24. $-\frac{1}{5} + \frac{2}{5} - \frac{3}{5} + \frac{4}{5} - \frac{5}{5}$

Use algebra and the formulas in Eqs. (1)–(3) to evaluate the sums in Exercises 25–32.

25. $\displaystyle\sum_{k=1}^{10} k$

26. $\displaystyle\sum_{k=1}^{7} 2k$

27. $\displaystyle\sum_{k=1}^{6} -k^2$

28. $\displaystyle\sum_{k=1}^{6} (k^2+5)$

29. $\sum_{k=1}^{5} k(k-5)$

30. $\sum_{k=1}^{7} (2k-8)$

31. $\sum_{k=1}^{100} k^3 - \sum_{k=1}^{99} k^3$

32. $\left(\sum_{k=1}^{7} k\right)^2 - \sum_{k=1}^{7} k^3$

33. Which of the following express $1 + 2 + 4 + 8 + 16 + 32$ in sigma notation?

a) $\sum_{k=1}^{6} 2^{k-1}$ b) $\sum_{k=0}^{5} 2^k$ c) $\sum_{k=-1}^{4} 2^{k+1}$

34. Which formula is not equivalent to the others?

a) $\sum_{k=-1}^{1} \frac{(-1)^k}{k+2}$ b) $\sum_{k=0}^{2} \frac{(-1)^k}{k+1}$

c) $\sum_{k=1}^{3} \frac{(-1)^k}{k}$ d) $\sum_{k=2}^{4} \frac{(-1)^{k-1}}{k-1}$

35. Suppose that $\sum_{k=1}^{n} a_k = -5$ and $\sum_{k=1}^{n} b_k = 6$. Find the values of

a) $\sum_{k=1}^{n} 3a_k$ b) $\sum_{k=1}^{n} \frac{b_k}{6}$

c) $\sum_{k=1}^{n} (a_k + b_k)$ d) $\sum_{k=1}^{n} (a_k - b_k)$

e) $\sum_{k=1}^{n} (b_k - 2a_k)$

36. Suppose that $\sum_{k=1}^{n} a_k = 0$ and $\sum_{k=1}^{n} b_k = 1$. Find the values of

a) $\sum_{k=1}^{n} 8a_k$ b) $\sum_{k=1}^{n} 250b_k$

c) $\sum_{k=1}^{n} (a_k + 1)$ d) $\sum_{k=1}^{n} (b_k - 1)$

37. Use a summation formula to find the number of cans in this supermarket display.

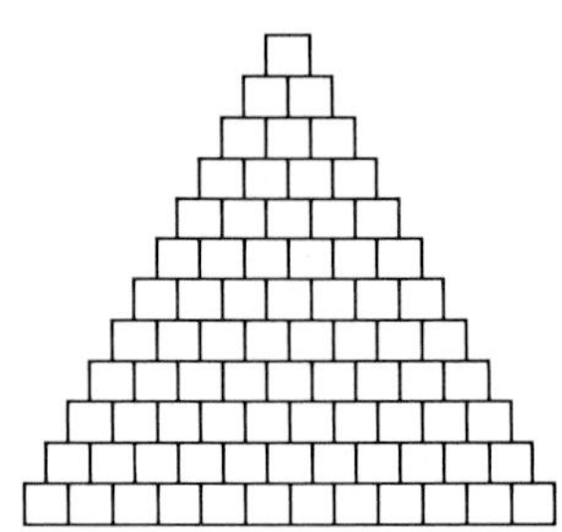

38. Use a summation formula to find the number of oranges in this fruit-market display.

© Susan Van Etten

Use arithmetic rules and the formula for the value of $\Sigma_{k=1}^{n} k$ to establish the formulas in Exercises 39 and 40.

39. $\sum_{k=1}^{n} (2k-1) = n^2$

40. $\sum_{k=1}^{n} k + \sum_{k=1}^{n-1} k = n^2$

Proofs without Words

Here are three informal pictorial proofs of summation formulas. See if you can tell what is going on in each case.

41.

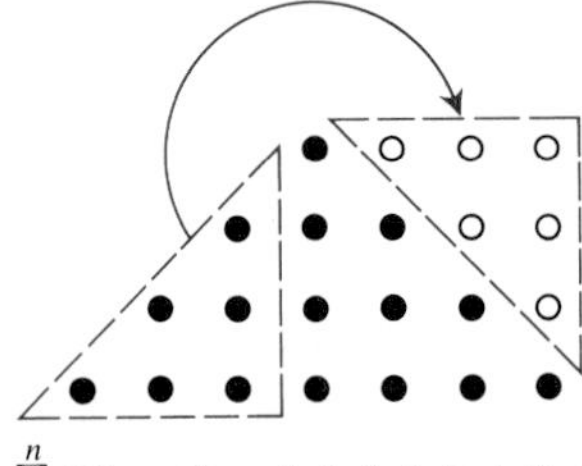

$\sum_{k=1}^{n} (2k-1) = 1 + 3 + 5 + 7 + \cdots + (2n-1) = n^2$

42.

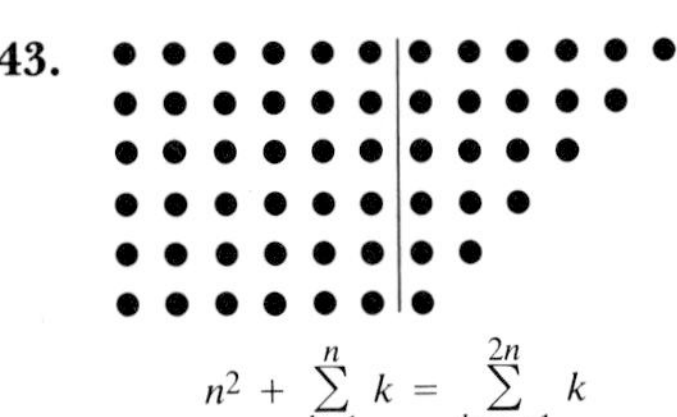

$\sum_{k=1}^{n} k + \sum_{k=1}^{n-1} k = n^2$

43.

$n^2 + \sum_{k=1}^{n} k = \sum_{k=n+1}^{2n} k$

5.3 Definite Integrals

In this section we develop the mathematics that supports the area calculations in Section 5.1. We do this by defining a limit of sums called the definite integral of a function $y = f(x)$ over an interval $[a, b]$, a limit that exists whenever f is continuous, regardless of the numerical signs of the values of f. In the special case where f is nonnegative, the definite integral of f from a to b is also the number we call the area under the curve $y = f(x)$ from a to b.

Definite integrals define and calculate many things besides areas. We use them to find the volumes of solids, the lengths of curves, the forces against dams, and the average values of periodic functions. The list is nearly endless, and it grows with every passing year.

You may recall that in Section 5.1 we said much the same thing about antiderivatives. The connection is that those quantities—length, volume, force, and so on—once defined as definite integrals are then calculated with antiderivatives, just the way area is. The connection between antiderivatives and definite integrals is beautiful, deep, and surprising, and its discovery by Newton and Leibniz in the mid-seventeenth century changed the mathematical world forever. We shall see for ourselves what that connection is when we get to the next section.

Riemann Sums

We begin with an arbitrary continuous function $y = f(x)$ defined over a closed interval $a \le x \le b$. Like the function graphed in Fig. 5.8, it may have negative values as well as positive values.

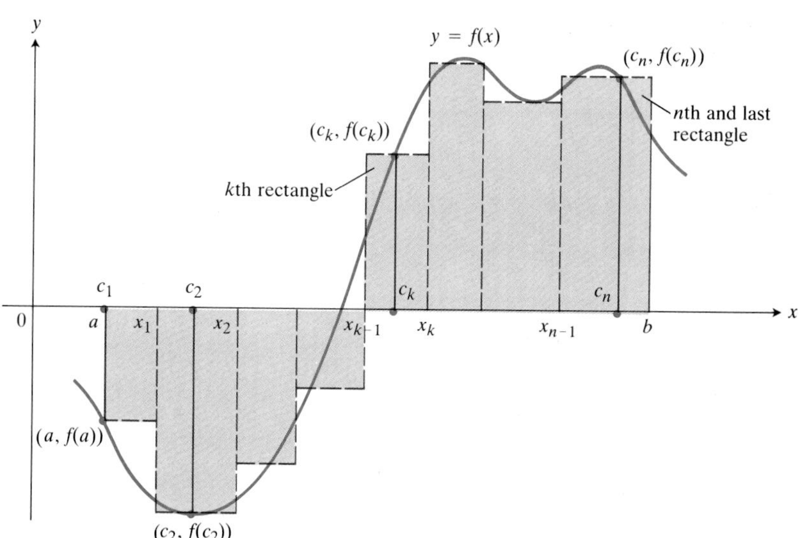

5.8 The graph of a typical function $y = f(x)$ over a closed interval $[a, b]$. The rectangles approximate the region between the graph of the function and the x-axis.

We partition the interval $[a, b]$ into n subintervals by choosing $n - 1$ points, say $x_1, x_2, \ldots, x_{n-1}$, between a and b subject only to the condition that

$$a < x_1 < x_2 < \cdots < x_{n-1} < b. \tag{1}$$

To make the notation consistent, we usually denote a by x_0 and b by x_n. The set

$$P = \{x_0, x_1, \ldots, x_n\} \tag{2}$$

is then called a **partition** of $[a, b]$.

The partition P defines n closed **subintervals**

$$[x_0, x_1], [x_1, x_2], \ldots, [x_{n-1}, x_n]. \tag{3}$$

The typical closed subinterval $[x_{k-1}, x_k]$ is called the ***k*th subinterval** of P.

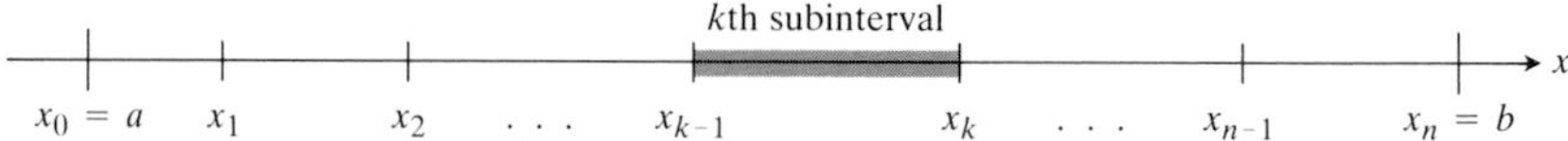

The length of the kth subinterval is $\Delta x_k = x_k - x_{k-1}$.

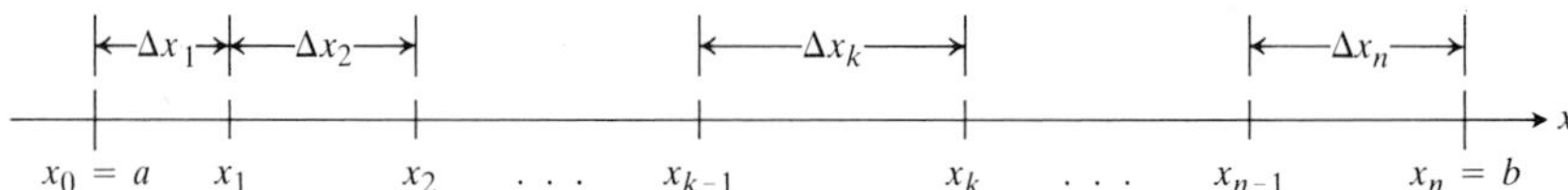

On each subinterval we stand a vertical rectangle that reaches from the x-axis to the curve $y = f(x)$. The exact height of the rectangle does not matter as long as its top or base touches the curve at some point $(c_k, f(c_k))$. See Fig. 5.8 again.

If $f(c_k)$ is positive, the number $f(c_k)\Delta x_k$ = height × base is the area of the rectangle. If $f(c_k)$ is negative, then $f(c_k)\Delta x_k$ is the negative of the area. In any case, we add the n products $f(c_k)\Delta x_k$ to form the sum

$$S_P = \sum_{k=1}^{n} f(c_k)\Delta x_k. \tag{4}$$

This sum, which depends on P and the choice of the numbers c_k, is called a **Riemann sum for f on the interval $[a, b]$,** in honor of the German mathematician Georg Friedrich Bernhard Riemann (1826–1866), who studied the limits of such sums.

EXAMPLE 1 Riemann sums for $f(x) = \sin \pi x$ on the interval $[0, 3/2]$.

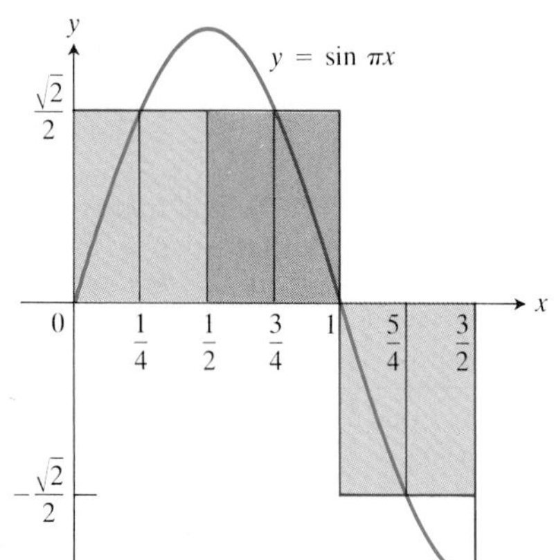

Partition:

$$P = \left\{0, \frac{1}{2}, 1, \frac{3}{2}\right\}$$

Subinterval lengths:

$$\Delta x_k = \frac{1}{2}, \qquad k = 1, 2, 3$$

Choice of c_k: the interval midpoints

$$c_1 = \frac{1}{4}, \qquad c_2 = \frac{3}{4}, \qquad c_3 = \frac{5}{4}$$

The Riemann sum:

$$\sum_{k=1}^{3} f(c_k)\Delta x_k = \sum_{k=1}^{3} \sin(\pi c_k)\cdot\frac{1}{2} = \frac{1}{2}\sum_{k=1}^{3} \sin(\pi c_k) = \frac{1}{2}\left(\sin\frac{\pi}{4} + \sin\frac{3\pi}{4} + \sin\frac{5\pi}{4}\right)$$

$$= \frac{1}{2}\left(\frac{\sqrt{2}}{2} + \frac{\sqrt{2}}{2} - \frac{\sqrt{2}}{2}\right) = \frac{\sqrt{2}}{4}$$

If instead of choosing the c_k's to be midpoints we choose each c_k to be the left-hand endpoint of its interval, then

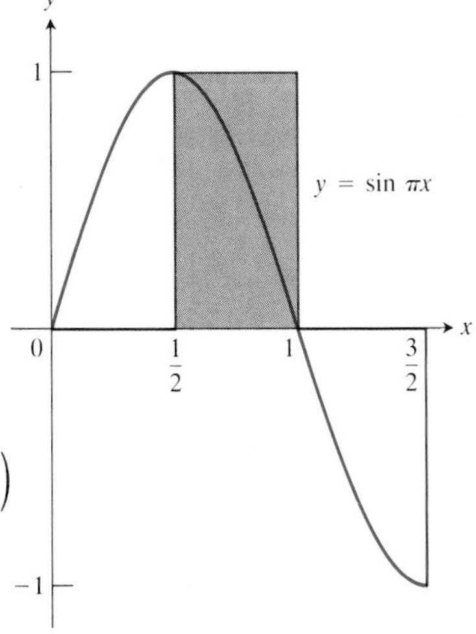

$$c_1 = 0,\ c_2 = \frac{1}{2},\ c_3 = 1,$$

and the Riemann sum is

$$\sum_{k=1}^{3} \sin(\pi c_k)\Delta x_k = \frac{1}{2}\sum_{k=1}^{3} \sin(\pi c_k)$$

$$= \frac{1}{2}\left(\sin 0 + \sin\frac{\pi}{2} + \sin\pi\right)$$

$$= \frac{1}{2}(0 + 1 + 0) = \frac{1}{2}.$$

If we choose each c_k to be the right-hand endpoint of its subinterval, then

$$c_1 = \frac{1}{2},\ c_2 = 1,\ c_3 = \frac{3}{2},$$

and the Riemann sum is

$$\sum_{k=1}^{3} \sin(\pi c_k)\Delta x_k = \frac{1}{2}\left(\sin\frac{\pi}{2} + \sin\pi + \sin\frac{3\pi}{2}\right) = \frac{1}{2}(1 + 0 - 1) = 0.$$ ■

What happens to the Riemann sums as the number of points in the partition increases and the partition becomes finer? As Fig. 5.9 suggests, the rectangles involved overlap the region between the curve and the x-axis with increasing accuracy, and we should find the sums approaching a limiting value of some kind. To make this idea precise, we need to define what it

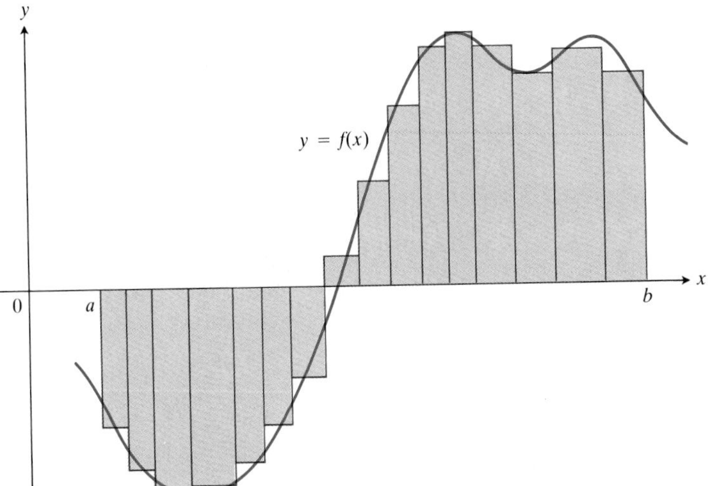

5.9 The curve of Fig. 5.8 with rectangles from a finer partition.

means for partitions to become finer and for Riemann sums to have a limit. We accomplish this with the following definitions.

DEFINITION

The **norm** of a partition is the length of the partition's longest subinterval. If the partition is denoted by P, its norm is denoted by putting double bars around the P:

$$\|P\|. \quad \text{(The norm of } P\text{)}$$

EXAMPLE 2 The set $P = \left\{0, \frac{1}{4}, \frac{2}{3}, 1, \frac{3}{2}, 2\right\}$ is a partition of the interval $[0, 2]$.

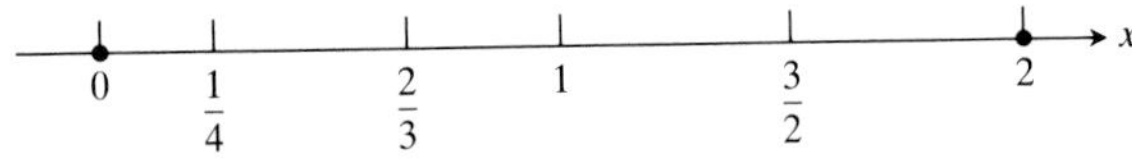

The subintervals in the partition are

$$\left[0, \frac{1}{4}\right], \quad \left[\frac{1}{4}, \frac{2}{3}\right], \quad \left[\frac{2}{3}, 1\right], \quad \left[1, \frac{3}{2}\right], \quad \left[\frac{3}{2}, 2\right].$$

The lengths of the subintervals are

$$\Delta x_1 = \frac{1}{4}, \quad \Delta x_2 = \frac{2}{3} - \frac{1}{4} = \frac{5}{12}, \quad \Delta x_3 = \frac{1}{3}, \quad \Delta x_4 = \frac{1}{2}, \quad \Delta x_5 = \frac{1}{2}.$$

The norm of the subdivision is 1/2, the longest of these lengths. As you can see, there are two subintervals of length 1/2. There can be more than one longest subinterval. ■

The way we say that successive partitions of an interval become finer is to say that their norms approach zero.

DEFINITION

The Definite Integral as a Limit of Riemann Sums

Let $f(x)$ be a function defined on a closed interval $[a, b]$. We say that the **limit** of the Riemann sums $\sum_{k=1}^{n} f(c_k)\Delta x_k$ on $[a, b]$ as $\|P\| \to 0$ is the number I if the following condition is satisfied:

Given any positive number ϵ, there exists a positive number δ such that for every partition P of $[a, b]$

$$\|P\| < \delta \quad \text{implies} \quad \left| \sum_{k=1}^{n} f(c_k)\Delta x_k - I \right| < \epsilon \tag{5}$$

for any choice of the numbers c_k in the subintervals $[x_{k-1}, x_k]$.

If the limit exists, we write

$$\lim_{\|P\| \to 0} \sum_{k=1}^{n} f(c_k)\Delta x_k = I. \tag{6}$$

We call I the **definite integral** of f over $[a, b]$, we say that f is **integrable** over $[a, b]$, and we say that the Riemann sums of f on $[a, b]$ **converge** to I.

The amazing fact is that despite the potential for variety in the Riemann sums $\Sigma f(c_k)\Delta x_k$ as the partitions change and the c_k's are chosen at random in the intervals of each new partition, the sums always have a limit as $\|P\| \to 0$ when f is continuous on $[a, b]$. The existence of this limit, of the definite integral of a continuous function on a closed interval, blithely assumed by the mathematicians of the seventeenth and eighteenth centuries, was finally established, once and for all, by Georg Riemann in 1854.

THEOREM 1

The Existence of Definite Integrals

All continuous functions are integrable. That is, if a function $y = f(x)$ is continuous on an interval $[a, b]$, then its definite integral over $[a, b]$ exists.

You can find a current version of Riemann's proof of this theorem in most advanced calculus books.

Theorem 1 says nothing about how to calculate definite integrals. Except for a few special cases, that takes another theorem, and we shall get to it in the next section.

Finally, Theorem 1 speaks only about continuous functions. Many discontinuous functions are integrable as well, but we shall not deal with this here.

Terminology

There is a fair amount of terminology to learn in connection with definite integrals.

The definite integral of a function $f(x)$ over an interval $[a, b]$ is usually denoted by the symbol

$$\int_a^b f(x)\,dx, \tag{7}$$

which is read as "the integral of f of x d x from a to b." The symbol $\int$ is an **integral sign.** Leibniz chose it because it resembled the S in the German word for *summation.*

When we find the value of $\int_a^b f(x)\,dx$, we say that we have **evaluated the integral** and that we have **integrated** f from a to b. We call $[a, b]$ the **interval of integration.** The numbers a and b are the **limits of integration,** a being the **lower limit of integration** and b the **upper limit of integration.** The function f is the **integrand** of the integral. The variable x is the **variable of integration.**

The value of the definite integral of a function over any particular interval depends on the function and not on the letter we choose to represent its independent variable. If we decide to use t or u instead of x, we simply write the integral as

$$\int_a^b f(t)\,dt \quad \text{or} \quad \int_a^b f(u)\,du \quad \text{instead of} \quad \int_a^b f(x)\,dx.$$

No matter how we write the integral, it is still the same number, defined as a limit of Riemann sums. Since it does not matter what letter we use, the variable of integration is called a **dummy variable.**

EXAMPLE 3 Express the limit of Riemann sums

$$\lim_{\|P\|\to 0} \sum_{k=1}^{n} (3c_k^2 - 2c_k + 5)\,\Delta x_k$$

as an integral if P denotes a partition of the interval $[-1, 3]$.

Solution The function being evaluated at c_k in each term of the sum is $f(x) = 3x^2 - 2x + 5$. The interval being partitioned is $[-1, 3]$. The limit is therefore the integral of f from -1 to 3:

$$\lim_{\|P\|\to 0} \sum_{k=1}^{n} (3c_k^2 - 2c_k + 5)\,\Delta x_k = \int_{-1}^{3} (3x^2 - 2x + 5)\,dx. \qquad \blacksquare$$

Constant Functions

Integrals of constant functions are always easy to evaluate.

THEOREM 2

If $f(x) = c$ has the constant value c on the interval $[a, b]$, then

$$\int_a^b f(x)\,dx = \int_a^b c\,dx = c(b - a). \tag{8}$$

Theorem 2 says that the integral of a constant function over a closed interval is the constant times the length of the interval.

PROOF OF THEOREM 2 The Riemann sums of f on $[a, b]$ all have the constant value

$$\begin{aligned} \sum f(c_k)\,\Delta x_k &= \sum c \cdot \Delta x_k \\ &= c \cdot \sum \Delta x_k \qquad \text{(Constant multiple rule for sums)} \\ &= c(b - a). \qquad \text{(The } \Delta x_k\text{'s add up to the length of } [a, b].) \end{aligned}$$

The limit of these sums, the integral they converge to, therefore has this value too. ■

EXAMPLE 4

$$\int_1^4 5\,dx = 5(4 - 1) = 5(3) = 15$$ ■

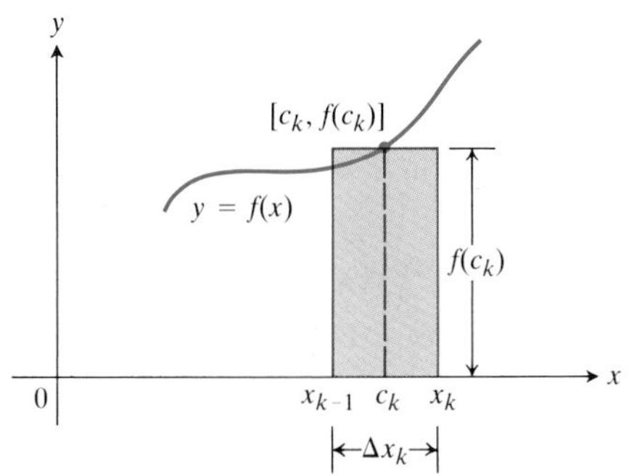

5.10 If $f(x) \geq 0$, then each term in the Riemann sum $\Sigma\, f(c_k)\,\Delta x_k$ is a rectangle area.

Area Is Strictly a Special Case

If an integrable function $y = f(x)$ is nonnegative throughout an interval $[a, b]$, as in Fig. 5.10, each term $f(c_k)\Delta x_k$ is the area of a rectangle reaching from the x-axis up to the curve $y = f(x)$. The Riemann sum

$$\sum_{k=1}^{n} f(c_k)\,\Delta x_k,$$

which is the sum of the areas of these rectangles, gives an estimate of the area of the region between the curve and the x-axis from a to b. Since the rectangles give an increasingly good approximation of the region as we use subdivisions with smaller and smaller norms, we call the limiting value

$$\lim_{\|P\| \to 0} \sum f(c_k)\,\Delta x_k = \int_a^b f(x)\,dx$$

the area under the curve.

DEFINITION

> If $y = f(x)$ is nonnegative and integrable over a closed interval $[a, b]$, then the integral of f from a to b is the **area** of the region between the graph of f and the x-axis from a to b. We sometimes call this number the **area under the curve $y = f(x)$ from a to b.**

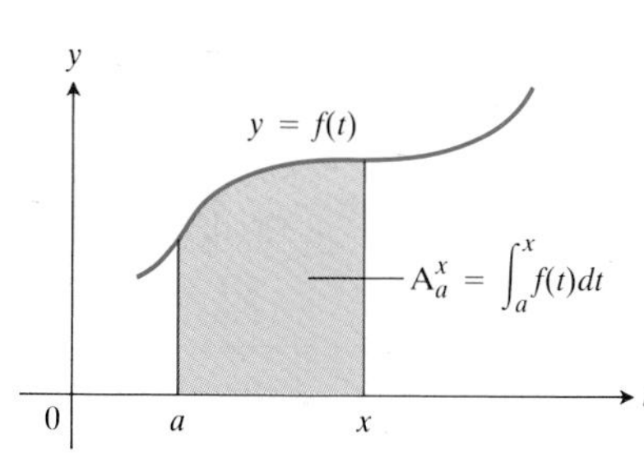

5.11 The area under the graph of f from a to x is defined and calculated as an integral.

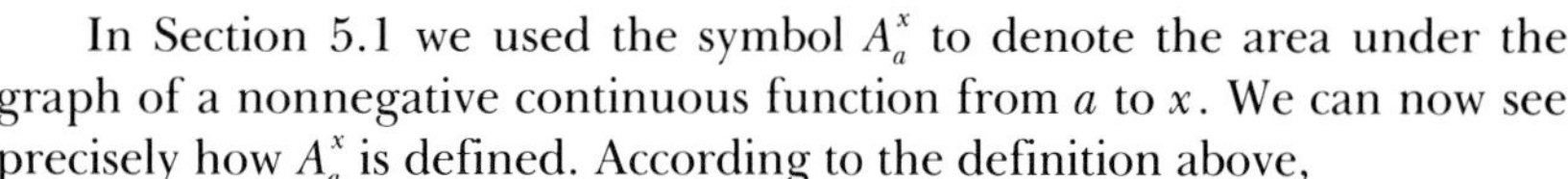

In Section 5.1 we used the symbol A_a^x to denote the area under the graph of a nonnegative continuous function from a to x. We can now see precisely how A_a^x is defined. According to the definition above,

$$A_a^x = \int_a^x f(t)\,dt. \tag{9}$$

(See Fig. 5.11.) In case you are wondering why we assumed f to be continuous back in Section 5.1, when integrability is all we need to define the integral in Eq. (9), the answer lies in the argument that A_a^x was a differentiable function of x whose derivative at x was $f(x)$. We used the continuity of f to prove that.

EXAMPLE 5 Find the value of the integral

$$\int_{-2}^{2} \sqrt{4 - x^2}\, dx$$

by regarding it as the area under the graph of an appropriately chosen function.

Solution We graph the integrand $f(x) = \sqrt{4 - x^2}$ over the interval of integration $[-2, 2]$ and see that the graph is a semicircle of radius 2. The area between the semicircle and the x-axis is

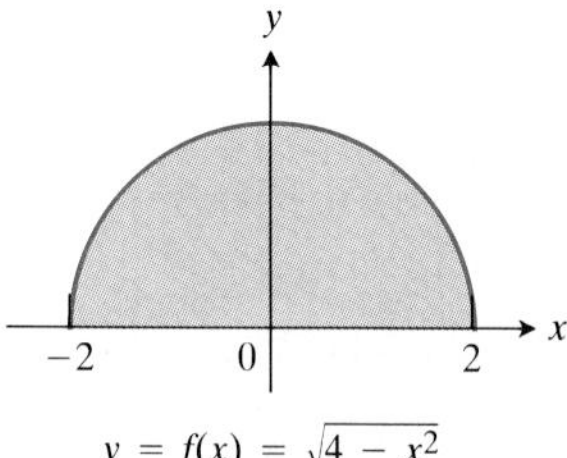

$$\text{Area} = \frac{1}{2} \cdot \pi r^2 = \frac{1}{2}\pi (2)^2 = 2\pi.$$

Because the area is also the value of the integral of f from -2 to 2,

$$\int_{-2}^{2} \sqrt{4 - x^2}\, dx = 2\pi.$$

If an integrable function $y = f(x)$ is nonpositive, the terms $f(c_k)\,\Delta x_k$ in the Riemann sums for f over an interval $[a, b]$ are all negatives of rectangle areas. The limit of the Riemann sums, the integral of f from a to b, is therefore the negative of the area between the graph of f and the x-axis.

$$\int_a^b f(x)\, dx = -(\text{the area}) \text{ if } f(x) \le 0. \tag{10}$$

Or, turning this around,

$$\text{Area} = -\int_a^b f(x)\, dx \text{ when } f(x) \le 0. \tag{11}$$

EXAMPLE 6

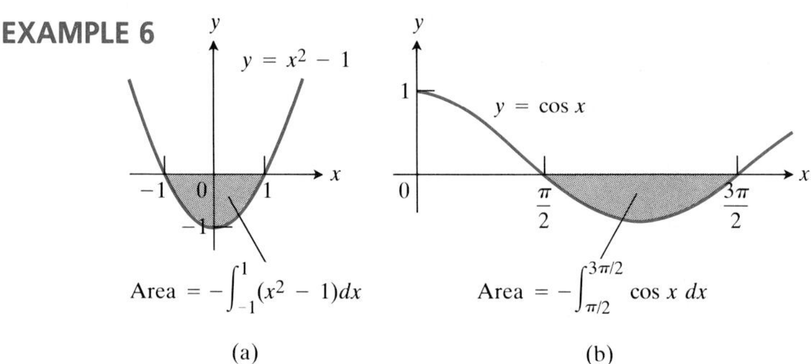

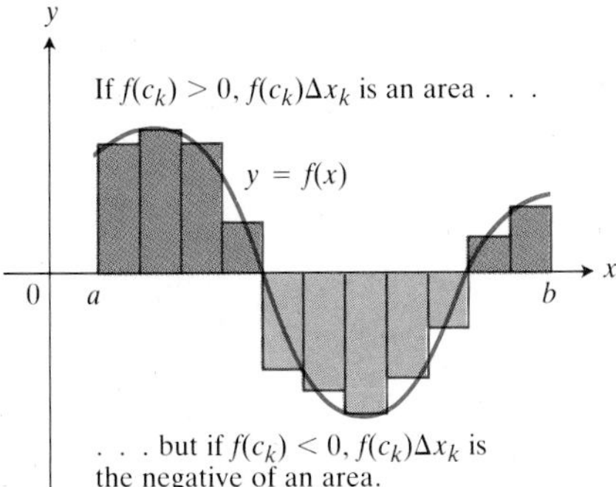

5.12 If f has negative as well as positive values, the Riemann sums for f add the areas of the rectangles that lie above the axis to the negatives of the areas of the rectangles that lie below the axis. The resulting cancellation reduces the value of the sums' limit to something less than the total area.

If an integrable function $y = f(x)$ has both positive and negative values on an interval $[a, b]$, then the Riemann sums for f on $[a, b]$ add the areas of the rectangles that lie above the x-axis to the negatives of the areas of the rectangles that lie below the x-axis, as in Fig. 5.12. The resulting cancellation reduces the sums, so their limiting value is a number whose magnitude is less than the total area between the curve and the x-axis. The value of the integral is the area above the axis minus the area below the axis.

For any integrable function,

$$\int_a^b f(x)\,dx = (\text{area above } x\text{-axis}) - (\text{area below } x\text{-axis}). \tag{12}$$

EXAMPLE 7

$$\int_0^{2\pi} \sin x\,dx = 0$$

The integral is the area of region A above the x-axis minus the area of region B below the x-axis. The areas are the same, so the integral is zero.

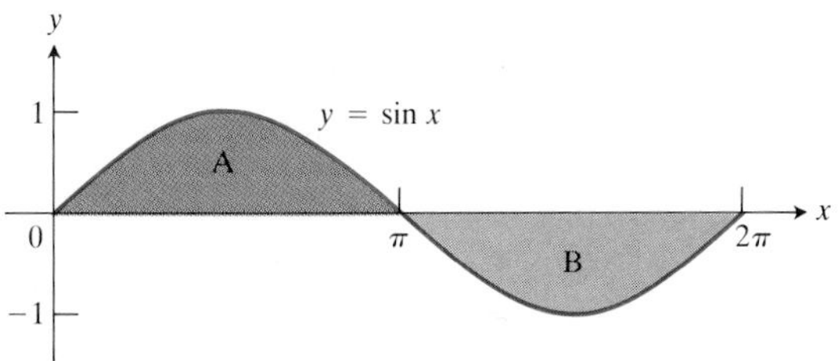

Useful Rules for Working with Integrals

We often want to add and subtract definite integrals, multiply them by constants, and compare them with other definite integrals. We do this with the rules in Table 5.1. All the rules except the first two follow from the way integrals are defined as limits of Riemann sums. The sums have these properties, so their limits do too. For example, Rule 3 says that the integral of k times a function is k times the integral of the function. This is true because

$$\begin{aligned}\int_a^b kf(x)\,dx &= \lim_{\|P\|\to 0} \sum_{i=1}^{n} kf(c_i)\,\Delta x_i \\ &= \lim_{\|P\|\to 0} k \sum_{i=1}^{n} f(c_i)\,\Delta x_i \qquad (13) \\ &= k \lim_{\|P\|\to 0} \sum_{i=1}^{n} f(c_i)\,\Delta x_i = k \int_a^b f(x)\,dx.\end{aligned}$$

Rule 1 is a definition. We want the integral of a function over an interval of zero length to be zero.

TABLE 5.1
Rules for definite integrals

1. The zero rule: $\int_a^a f(x)\,dx = 0.$ (A definition)

2. Reversing the order of integration changes an integral's sign: $\int_b^a f(x)\,dx = -\int_a^b f(x)\,dx.$ (Also a definition)

3. The constant multiple rule: $\int_a^b kf(x)\,dx = k\int_a^b f(x)\,dx.$ (Any number k)

4. Special case of Rule 3: $\int_a^b -f(x)\,dx = -\int_a^b f(x)\,dx.$ (Take $k = -1$ in Rule 3.)

5. The sum rule: If f and g are integrable on $[a, b]$, then $f + g$ is integrable on $[a, b]$ and
$$\int_a^b [f(x) + g(x)]\,dx = \int_a^b f(x)\,dx + \int_a^b g(x)\,dx.$$

6. The difference rule: If f and g are integrable on $[a, b]$, then $f - g$ is integrable on $[a, b]$ and
$$\int_a^b [f(x) - g(x)]\,dx = \int_a^b f(x)\,dx - \int_a^b g(x)\,dx.$$

7. Integrals of nonnegative functions are nonnegative: $f(x) \geq 0$ on $[a, b]$ $\Rightarrow$ $\int_a^b f(x)\,dx \geq 0.$

8. The domination rule: $g(x) \geq f(x)$ on $[a, b]$ $\Rightarrow$ $\int_a^b g(x)\,dx \geq \int_a^b f(x)\,dx.$

9. The max-min rule: If $b > a$, then
$$\underbrace{\min f \cdot (b - a)}_{\text{Lower bound}} \leq \int_a^b f(x)\,dx \leq \underbrace{\max f \cdot (b - a)}_{\text{Upper bound}},$$
where max and min refer to the values of f on $[a, b]$.

10. The interval addition rule: $\int_a^b f(x)\,dx + \int_b^c f(x)\,dx = \int_a^c f(x)\,dx.$

The only requirement is that f be integrable on the intervals joining a, b, and c (Fig. 5.13).

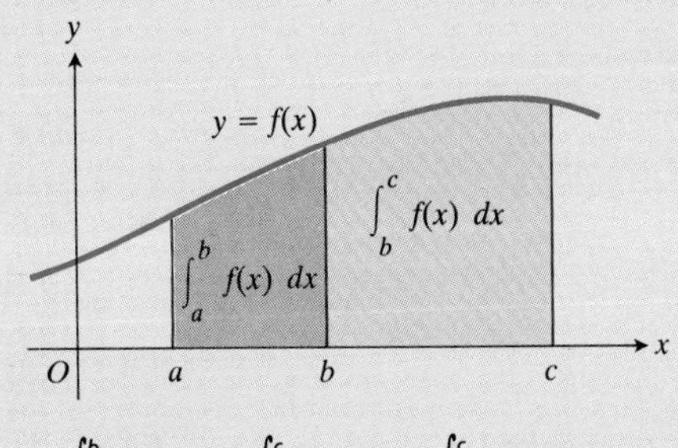

5.13

$$\int_a^b f(x)\,dx - \int_d^c f(x)\,dx = \int_a^c f(x)\,dx.$$

Similarly,

$$\int_b^c f(x)\,dx = \int_a^c f(x)\,dx - \int_a^b f(x)\,dx.$$

11. The interval subtraction rule (Rule 10 in another form): $\int_b^c f(x)\,dx = \int_a^c f(x)\,dx - \int_a^b f(x)\,dx.$

EXAMPLE 8 Suppose that f, g, and h are integrable, that

$$\int_{-1}^{1} f(x)\,dx = 5, \quad \int_{1}^{4} f(x)\,dx = -2, \quad \int_{-1}^{1} h(x)\,dx = 7,$$

and that $g(x) \geq f(x)$ on $[-1, 1]$. Then

1. $\displaystyle\int_{4}^{1} f(x)\,dx = -\int_{1}^{4} f(x)\,dx = -(-2) = 2,$ (Rule 2)

2. $\displaystyle\int_{-1}^{1} [2f(x) + 3h(x)]\,dx = 2\int_{-1}^{1} f(x)\,dx + 3\int_{-1}^{1} h(x)\,dx$

   $\displaystyle= 2(5) + 3(7) = 31,$ (Rules 3 and 5)

3. $\displaystyle\int_{-1}^{1} [f(x) - h(x)]\,dx = 5 - 7 = -2,$ (Rule 6)

4. $\displaystyle\int_{-1}^{1} g(x)\,dx \geq 5,$ (Because $g(x) \geq f(x)$ on $[-1, 1]$—Rule 8)

5. $\displaystyle\int_{-1}^{4} f(x)\,dx = \int_{-1}^{1} f(x)\,dx + \int_{1}^{4} f(x)\,dx = 5 + (-2) = 3.$ (Rule 10) ■

The Average Value of a Function

If we divide the inequality in the max-min rule for definite integrals by $(b - a)$, we get the inequality

$$\min f \leq \frac{1}{b-a}\int_{a}^{b} f(x)\,dx \leq \max f. \tag{14}$$

If f is continuous, the Intermediate Value Theorem in Section 2.4 says that f must assume every value between $\min f$ and $\max f$. In particular, f must assume the value

$$\frac{1}{b-a}\int_{a}^{b} f(x)\,dx.$$

THEOREM 3

The Mean Value Theorem for Definite Integrals

If f is continuous on the closed interval $[a, b]$, then, at some point c in the interval $[a, b]$,

$$f(c) = \frac{1}{b-a}\int_{a}^{b} f(x)\,dx. \tag{15}$$

The number on the right-hand side of Eq. (15) is called the mean value or average value of f on the interval $[a, b]$. We shall see some of the applications of average values in Section 5.4.

DEFINITION

The **average value** of f on $[a, b]$ is $\dfrac{1}{b-a}\displaystyle\int_a^b f(x)\,dx.$ (16)

Notice that the average value of f on $[a, b]$ is the integral of f divided by the length of the interval.

EXAMPLE 9 If f is continuous and nonnegative on $[a, b]$, its average value is the height of a rectangle whose area,

$$f(c)(b-a) = \int_a^b f(x)\,dx,$$

is the area under the graph of f from a to b.

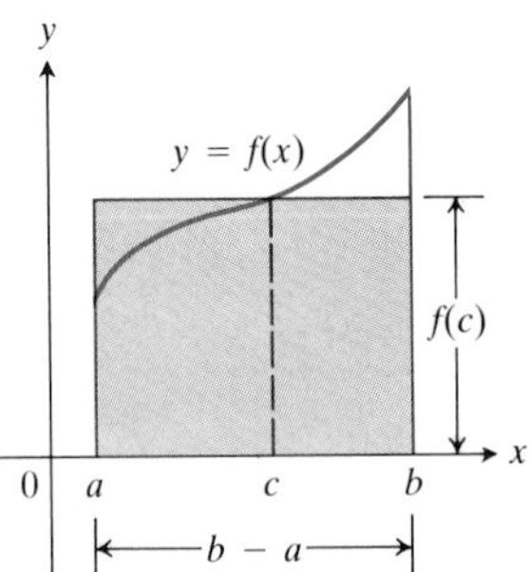

EXAMPLE 10 Find the average value of $f(x) = \sqrt{4-x^2}\,dx$ on the interval $[-2, 2]$.

Solution

$$\text{Av. val. of } f \text{ on } [-2, 2] = \frac{1}{2-(-2)}\int_{-2}^{2}\sqrt{4-x^2}\,dx \qquad \text{(Eq. 16)}$$

$$= \frac{1}{4}(2\pi) = \frac{\pi}{2} \qquad \begin{pmatrix}\text{Value from}\\ \text{Example 5}\end{pmatrix}$$

The average value of $f(x) = \sqrt{4-x^2}$ on $[-2, 2]$ is $\pi/2$.

Exercises 5.3

Express the limits in Exercises 1–12 as definite integrals.

1. $\displaystyle\lim_{\|P\|\to 0}\sum_{k=1}^{n} c_k^2\,\Delta x_k$, where P is a partition of $[0, 2]$

2. $\displaystyle\lim_{\|P\|\to 0}\sum_{k=1}^{n} 2c_k^3\,\Delta x_k$, where P is a partition of $[-1, 0]$

3. $\displaystyle\lim_{\|P\|\to 0}\sum_{k=1}^{n} (c_k^2 - 3c_k)\,\Delta x_k$, where P is a partition of $[-7, 5]$

4. $\displaystyle\lim_{\|P\|\to 0}\sum_{k=1}^{n} \frac{1}{c_k}\,\Delta x_k$, where P is a partition of $[1, 4]$

5. $\displaystyle\lim_{\|P\|\to 0}\sum_{k=1}^{n} \frac{1}{1-c_k}\,\Delta x_k$, where P is a partition of $[2, 3]$

6. $\displaystyle\lim_{\|P\|\to 0}\sum_{k=1}^{n} \sqrt{4-c_k^2}\,\Delta x_k$, where P is a partition of $[0, 1]$

7. $\displaystyle\lim_{\|P\|\to 0}\sum_{k=1}^{n} \cos c_k\,\Delta x_k$, where P is a partition of $[0, 4]$

8. $\displaystyle\lim_{\|P\|\to 0}\sum_{k=1}^{n} \tan c_k\,\Delta x_k$, where P is a partition of $[0, \pi/4]$

9. $\displaystyle\lim_{\|P\|\to 0}\sum_{k=1}^{n} \sec c_k\,\Delta x_k$, where P is a partition of $[-\pi/4, 0]$

10. $\displaystyle\lim_{\|P\|\to 0}\sum_{k=1}^{n} \sin^3 c_k\,\Delta x_k$, where P is a partition of $[-\pi, \pi]$

11. $\displaystyle\lim_{\|P\|\to 0}\sum_{k=1}^{n} \sin(3c_k)\,\Delta x_k$, where P is a partition of $[-\pi, 0]$

12. $\lim_{\|P\|\to 0} \sum_{k=1}^{n} \cot(\pi c_k)\,\Delta x_k$, where P is a partition of $[\pi/4, 3\pi/4]$

In Exercises 13–18, express the area of the shaded region as an integral (without bothering to evaluate the integral).

13. 14.

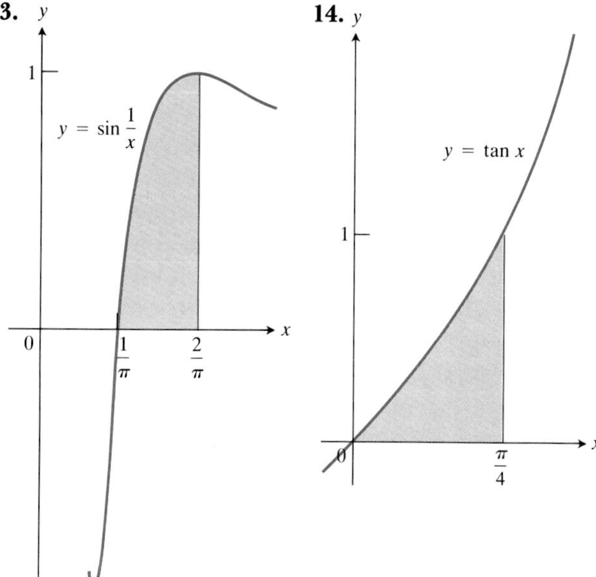

15.

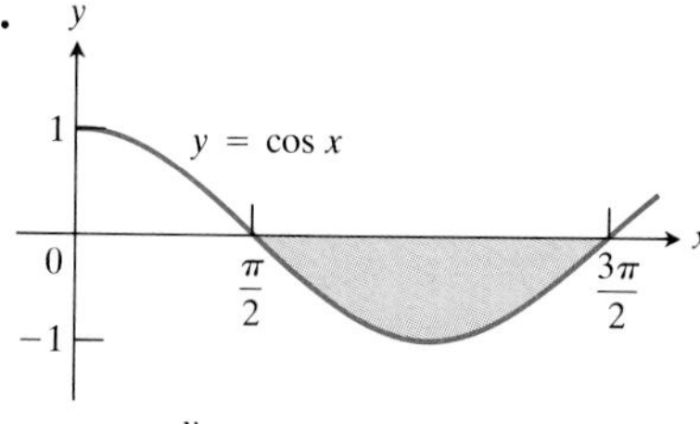

16.

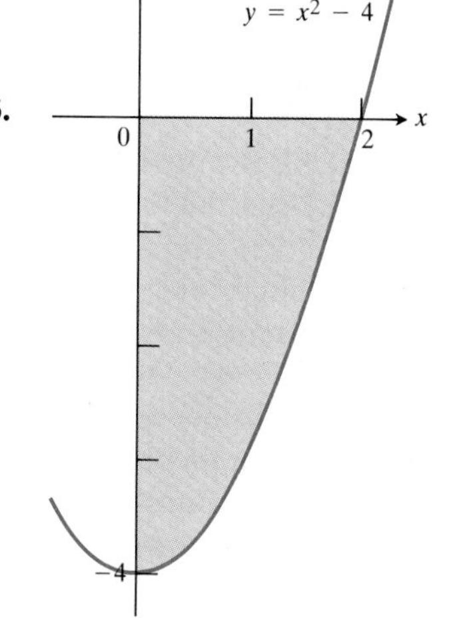

17.

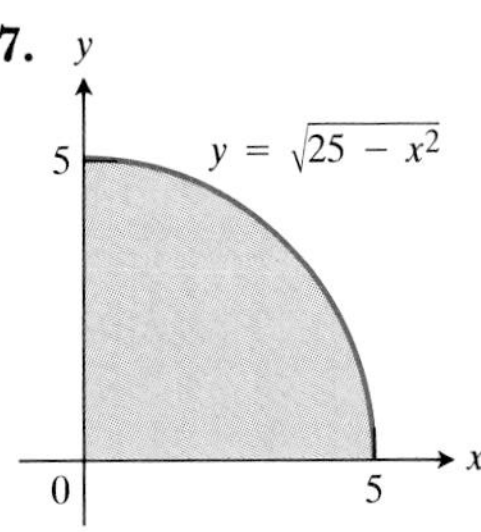

18.

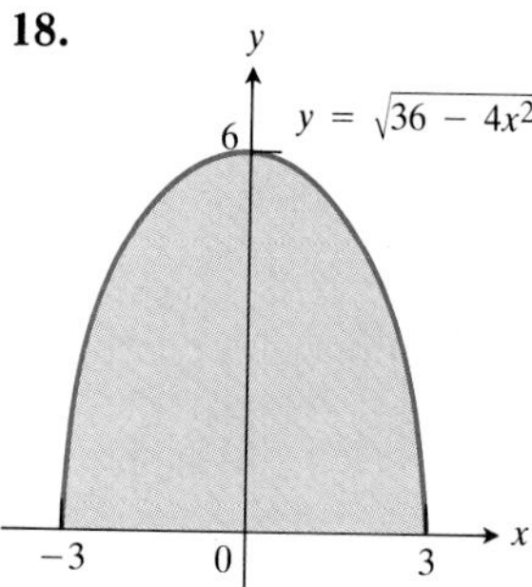

In Exercises 19–22, find the value of each integral by regarding it as the area under the graph of an appropriately chosen function and using an area formula from plane geometry.

19. $\int_{-1}^{1} \sqrt{1 - x^2}\,dx$

20. $\int_{0}^{2} \sqrt{4 - x^2}\,dx$

21. $\int_{-1}^{1} (1 - |x|)\,dx$

22. $\int_{-1}^{1} (1 + \sqrt{1 - x^2})\,dx$

23. Suppose f and g are continuous and that

$$\int_1^2 f(x)\,dx = -4, \quad \int_1^5 f(x)\,dx = 6, \quad \int_1^5 g(x)\,dx = 8.$$

Use the rules in Table 5.1 to find

a) $\int_2^2 g(x)\,dx$

b) $\int_5^1 g(x)\,dx$

c) $\int_1^2 3f(x)\,dx$

d) $\int_2^5 f(x)\,dx$

e) $\int_1^5 [f(x) - g(x)]\,dx$

f) $\int_1^5 [4f(x) - g(x)]\,dx$.

24. Suppose f and h are continuous and that

$$\int_1^9 f(x)\,dx = -1, \quad \int_7^9 f(x)\,dx = 5, \quad \int_7^9 h(x)\,dx = 4.$$

Use the rules in Table 5.1 to find

a) $\int_1^9 -2f(x)\,dx$

b) $\int_7^9 [f(x) + h(x)]\,dx$

c) $\int_7^9 [2f(x) - 3h(x)]\,dx$

d) $\int_9^1 f(x)\,dx$

e) $\int_1^7 f(x)\,dx$

f) $\int_9^7 [h(x) - f(x)]\,dx$.

25. Suppose $\int_1^2 f(x)\,dx = 5$. Find

a) $\int_1^2 f(u)\,du$, b) $\int_1^2 f(z)\,dz$, c) $\int_2^1 f(t)\,dt$.

26. Suppose f is continuous and that

$$\int_0^3 f(x)\,dx = 3, \quad \int_0^4 f(z)\,dz = 7.$$

Find

$$\int_3^4 f(y)\,dy.$$

27. Use Rule 9 in Table 5.1 to find upper and lower bounds for the value of

$$\int_0^1 \frac{1}{1+x^2}\,dx.$$

28. *Continuation of Exercise 27.* Use Rule 9 in Table 5.1 to find upper and lower bounds for the values of

$$\int_0^{1/2} \frac{1}{1+x^2}\,dx \quad \text{and} \quad \int_{1/2}^1 \frac{1}{1+x^2}\,dx.$$

Then add these to arrive at an improved estimate of

$$\int_0^1 \frac{1}{1+x^2}\,dx.$$

29. Suppose f is continuous and that

$$\int_1^2 f(x)\,dx = 4.$$

Show that $f(x) = 4$ at least once on the interval $[1, 2]$.

30. Show that the value of

$$\int_0^1 \sin^2 x\,dx$$

cannot possibly be 2.

COMPUTER GRAPHER If you have access to a Riemann sum grapher like the one in *The Calculus Toolkit,* graph the upper-endpoint-value and lower-endpoint-value Riemann sums for the integrals in Exercises 31–36. In each case try it with $n = 4, 10, 20$, and 50 subintervals of equal length.

31. $\int_0^1 (1-x)\,dx = \frac{1}{2}$

32. $\int_0^1 (x^2+1)\,dx = \frac{4}{3}$

33. $\int_{-\pi}^{\pi} \cos x\,dx = 0$

34. $\int_0^{\pi/4} \sec^2 x\,dx = 1$

35. $\int_{-1}^1 |x|\,dx = 1$

36. $\int_1^2 \frac{1}{x}\,dx = \ln 2 \approx 0.6931471806$

TOOLKIT PROGRAM

Integration — Draws curve and rectangles for upper-endpoint and lower-endpoint Riemann sums. Also gives the trapezoidal sum for comparison (see Section 5.7).

5.4 The Fundamental Theorems of Integral Calculus

This section presents the two fundamental theorems of integral calculus. The first theorem says that the definite integral of a continuous function is a differentiable function of its upper limit of integration, and tells us what the value of that derivative is. The second theorem tells us that the definite integral of a continuous function from a to b can be found from any one of the function's antiderivatives F as the number $F(b) - F(a)$. The discovery by Newton and Leibniz of these astonishing connections between integration and differentiation started the mathematical development that fueled the scientific revolution for the next 200 years and constitutes what is still regarded as the most important computational discovery in the history of the Western world.

The First Fundamental Theorem

If $f(t)$ is an integrable function, its integral from any fixed number a to another number x defines a function F whose value at x is

$$F(x) = \int_a^x f(t)\,dt. \tag{1}$$

For example, if f is nonnegative and x lies to the right of a, the area under the graph of f from a to x is

$$A_a^x = \int_a^x f(t)\,dt. \tag{2}$$

The variable x in the function A_a^x is the upper limit of integration of an integral, but the function is just like any other function. For each value of the input x there is an output A_a^x, in this case the value of the integral of f from a to x.

The formula

$$F(x) = \int_a^x f(t)\,dt$$

gives an important way to define new functions in science and provides an especially useful way to describe solutions of differential equations (more about this later). The reason for our mentioning the formula now is that this formula makes the connection between integrals and derivatives. For, if f is any continuous function whatever, F is a differentiable function of x and, even more important, its derivative, dF/dx, is f itself. At every value of x,

$$\frac{d}{dx}\int_a^x f(t)\,dt = f(x). \tag{3}$$

If you were being sent to a desert island and could take only one equation with you, Eq. (3) might well be your choice. It says that the differential equation $dF/dx = f$ has a solution for any continuous function f. It says that every continuous function f is the derivative of some other function, namely, $\int_a^x f(t)\,dt$. It says that every continuous function has an antiderivative. Equation (3) is so important that we call it the First Fundamental Theorem of Calculus.

THEOREM 4

The First Fundamental Theorem of Calculus

If f is continuous on $[a, b]$, then the function

$$F(x) = \int_a^x f(t)\,dt \tag{4}$$

has a derivative at every point in $[a, b]$ and

$$\frac{dF}{dx} = \frac{d}{dx}\int_a^x f(t)\,dt = f(x). \tag{5}$$

COROLLARY

The Existence of Antiderivatives of Continuous Functions

If f is continuous on $[a, b]$, then there exists a function F whose derivative on $[a, b]$ equals f.

PROOF OF THE COROLLARY Take $F(x) = \int_a^x f(t)\,dt$. The integral exists by Theorem 1, Section 5.3, and $dF/dx = f$ by Theorem 4 above. ■

PROOF OF THE FIRST FUNDAMENTAL THEOREM We prove Theorem 4 by applying the definition of derivative directly to the function $F(x)$. This means writing out Fermat's difference quotient,

$$\frac{F(x+h) - F(x)}{h}, \tag{6}$$

and showing that its limit as $h \to 0$ is the number $f(x)$.

When we replace $F(x+h)$ and $F(x)$ by their defining integrals, the numerator in Eq. (6) becomes

$$F(x+h) - F(x) = \int_a^{x+h} f(t)\,dt - \int_a^x f(t)\,dt. \tag{7}$$

The interval subtraction rule for integrals (Table 5.1, preceding section) simplifies this to

$$\int_x^{x+h} f(t)\,dt, \tag{8}$$

so that Eq. (6) becomes

$$\frac{F(x+h) - F(x)}{h} = \frac{1}{h}[F(x+h) - F(x)] = \frac{1}{h}\int_x^{x+h} f(t)\,dt. \tag{9}$$

According to the Mean Value Theorem for definite integrals, Theorem 3 in the preceding section, the value of the entire expression on the right-hand side of Eq. (9) is one of the values taken on by f in the interval joining x and $x+h$. That is, for some number c in this interval,

$$\frac{1}{h}\int_x^{x+h} f(t)\,dt = f(c). \tag{10}$$

We can therefore find out what happens to $(1/h)$ times the integral as $h \to 0$ by watching what happens to $f(c)$ as $h \to 0$.

What does happen to $f(c)$ as $h \to 0$? As $h \to 0$, the endpoint $x+h$ approaches x, taking c along with it like a bead on a wire:

x c $x+h$ or $x+h$ c x

So c approaches x, and, since f is continuous at x, $f(c)$ approaches $f(x)$:

$$\lim_{h \to 0} f(c) = f(x). \tag{11}$$

Going back to the beginning, then, we have

$$\begin{aligned}
\frac{dF}{dx} &= \lim_{h\to 0} \frac{F(x+h) - F(x)}{h} && \text{(The definition of derivative)}\\
&= \lim \frac{1}{h}\int_x^{x+h} f(t)\,dt && \text{(Eq. 9)}\\
&= \lim_{h\to 0} f(c) && \text{(Eq. 10)}\\
&= f(x). && \text{(Eq. 11)}
\end{aligned}$$

This concludes the proof. ■

EXAMPLE 1

$$\frac{d}{dx}\int_{-\pi}^{x} \cos t\,dt = \cos x \qquad \text{(Eq. (5) with } f(t) = \cos t)$$

$$\frac{d}{dx}\int_{0}^{x} \frac{1}{1+t^2}\,dt = \frac{1}{1+x^2} \qquad \left(\text{Eq. (5) with } f(t) = \frac{1}{1+t^2}\right)$$

■

EXAMPLE 2 Find dy/dx if

$$y = \int_1^{x^2} \cos t\,dt.$$

Solution Notice that the upper limit of integration is not x but x^2. To find dy/dx we must therefore treat y as the composite of

$$y = \int_1^u \cos t\,dt \quad \text{and} \quad u = x^2$$

and apply the Chain Rule:

$$\begin{aligned}
\frac{dy}{dx} &= \frac{dy}{du}\frac{du}{dx}. && \text{(Chain Rule)}\\
&= \frac{d}{du}\int_1^u \cos t\,dt \cdot \frac{du}{dx} && \text{(Substitute the formula for } y.)\\
&= \cos u \cdot \frac{du}{dx} && \text{(Eq. (5) with } f(t) = \cos t)\\
&= \cos x^2 \cdot 2x && (u = x^2)\\
&= 2x \cos x^2. && \text{(Usual form)}
\end{aligned}$$

■

EXAMPLE 3 Show that the function

$$y = \int_0^x \tan t\,dt + 5$$

solves the initial value problem

Differential equation: $\dfrac{dy}{dx} = \tan x$

Initial condition: $y = 5$ when $x = 0$.

Solution The function satisfies the differential equation because

$$\frac{d}{dx}\left(\int_0^x \tan t\, dt + 5\right) = \tan x + 0 = \tan x. \qquad \text{(Eq. (5) with } f(t) = \tan t\text{)}$$

It fulfills the initial condition because

$$y(0) = \int_0^0 \tan t\, dt + 5 = 0 + 5. \qquad \text{(Table 5.1, Rule 1)}$$

The Integral Evaluation Theorem

We now come to the theorem that tells how to evaluate definite integrals with antiderivatives, a theorem so useful it is called the Second Fundamental Theorem of Calculus.

To evaluate $\int_a^b f(x)\, dx$:

1. Find an antiderivative F of f.
2. Calculate the number $F(b) - F(a)$.

This number will be $\int_a^b f(x)\, dx$. Any antiderivative will do.

THEOREM 5

The Integral Evaluation Theorem (Second Fundamental Theorem of Calculus)

If f is continuous at every point of $[a, b]$ and F is any antiderivative of f on $[a, b]$, then

$$\int_a^b f(x)\, dx = F(b) - F(a). \tag{12}$$

Theorem 5 says that to evaluate the definite integral of a continuous function f from a to b, all we need do is find an antiderivative F of f and calculate the number $F(b) - F(a)$. The existence of the antiderivative is assured by the First Fundamental Theorem.

PROOF OF THE INTEGRAL EVALUATION THEOREM To prove the theorem, we use Corollary 3 of the Mean Value Theorem for derivatives (Section 4.6), which says that functions with identical derivatives can differ only by a constant. We already know one function whose derivative equals f, namely,

$$G(x) = \int_a^x f(t)\, dt.$$

Therefore, if F is any other such function, then

$$F(x) = G(x) + C \tag{13}$$

throughout $[a, b]$ for some constant C. When we use Eq. (13) to calculate $F(b) - F(a)$, we find that

$$F(b) - F(a) = [G(b) + C] - [G(a) + C] = G(b) - G(a)$$

$$= \int_a^b f(t)\, dt - \int_a^a f(t)\, dt = \int_a^b f(t)\, dt - 0 = \int_a^b f(t)\, dt.$$

This establishes Eq. (12) and concludes the proof.

Notation

The usual notation for the number $F(b) - F(a)$ is $F(x)]_a^b$ or $[F(x)]_a^b$, depending on whether F has one or more terms. As you will see, this notation provides a compact "recipe" for the evaluation.

Write $F(x)]_a^b$ for $F(b) - F(a)$ when $F(x)$ has a single term.

Write $[F(x)]_a^b$ for $F(b) - F(a)$ when $F(x)$ has more than one term.

EXAMPLE 4

$$\int_0^{\pi} \cos x \, dx = \sin x \Big]_0^{\pi} \qquad \text{(Simplest antiderivative of } \cos x\text{)}$$

$$= \sin \pi - \sin 0 = 0 - 0 = 0$$

EXAMPLE 5

$$\int_0^2 x^2 \, dx = \frac{x^3}{3}\Big]_0^2 \qquad \text{(Simplest antiderivative of } x^2\text{)}$$

$$= \frac{2^3}{3} - \frac{0^3}{3} = \frac{8}{3}$$

EXAMPLE 6

$$\int_{-2}^2 (4 - x^2) \, dx = \left[4x - \frac{x^3}{3}\right]_{-2}^2 \qquad \text{(Simplest antiderivative of } 4x - x^3/3\text{)}$$

$$= \left[4(2) - \frac{(2)^3}{3}\right] - \left[4(-2) - \frac{(-2)^3}{3}\right] = \left[8 - \frac{8}{3}\right] - \left[-8 + \frac{8}{3}\right]$$

$$= 8 - \frac{8}{3} + 8 - \frac{8}{3} = 16 - \frac{16}{3} = \frac{32}{3}$$

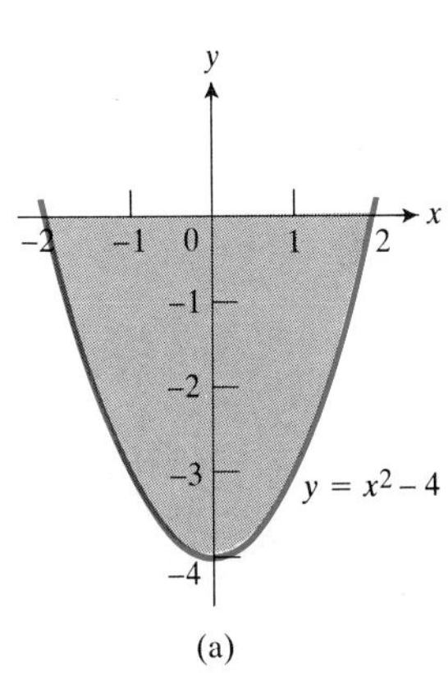

(a)

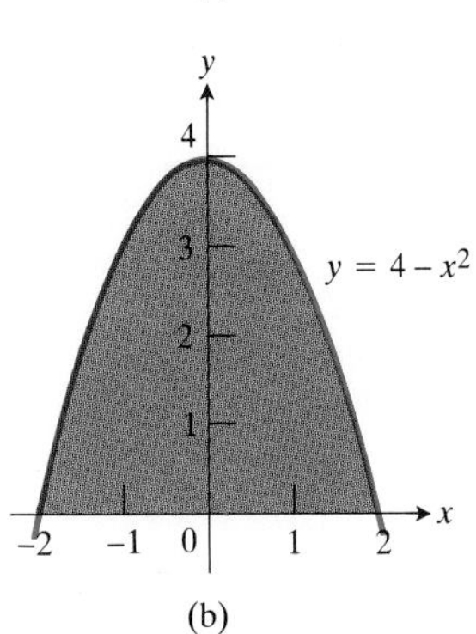

(b)

5.14 The graphs in (a) and (b) enclose the same amount of area with the x-axis, but the definite integrals of the functions from -2 to 2 differ in sign.

Theorem 5 justifies the way we calculated areas with antiderivatives in Section 5.1. The area under the graph of a nonnegative function $y = f(x)$ from $x = a$ to $x = b$ is $\int_a^b f(x)\,dx$ (from the definition of area in Section 5.3), and the value of this integral is $F(b) - F(a)$.

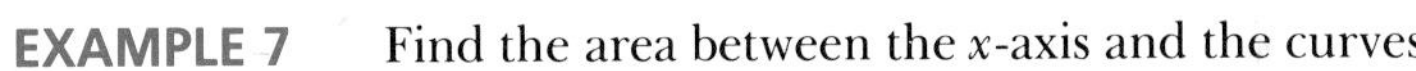

EXAMPLE 7 Find the area between the x-axis and the curves

a) $y = 4 - x^2$ b) $y = x^2 - 4$

for $-2 \le x \le 2$.

Solution We graph the curves over $[-2, 2]$ to see where the function values are positive and negative (Fig. 5.14).

a) Since $y = 4 - x^2 \ge 0$ on $[-2, 2]$, the area between the curve and the x-axis from -2 to 2 is

$$\text{Area} = \int_{-2}^2 (4 - x^2)\,dx = \left[4x - \frac{x^3}{3}\right]_{-2}^2 = \frac{32}{3}. \qquad \begin{pmatrix}\text{Value from} \\ \text{Example 6}\end{pmatrix}$$

b) Since $y = x^2 - 4 \le 0$ on $[-2, 2]$, the area between the curve and the x-axis from -2 to 2 is the negative of the integral of $x^2 - 4$ from -2 to 2:

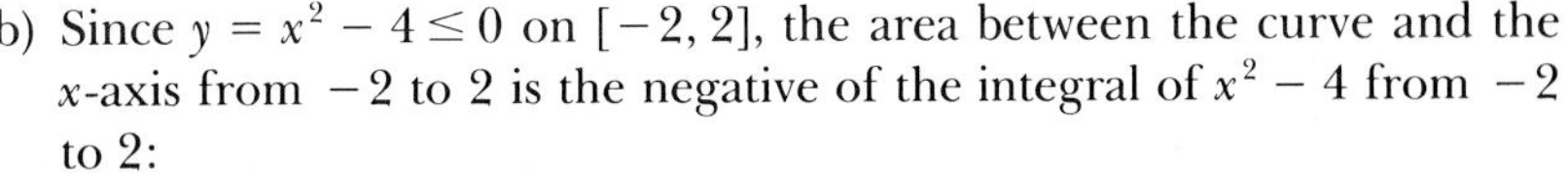

$$\text{Area} = -\int_{-2}^2 (x^2 - 4)\,dx = \int_{-2}^2 -(x^2 - 4)\,dx \qquad \text{(Rule 4, Table 5.1)}$$

$$= \int_{-2}^2 (4 - x^2)\,dx = \frac{32}{3}. \qquad \text{(Value from above)}$$

When the graph of $y = f(x)$ crosses the x-axis between $x = a$ and $x = b$, we find the area between the graph and the axis from a to b by taking the following steps.

Steps for Finding Area When f Has Both Positive and Negative Values on $[a, b]$

STEP 1: Find the points where $f = 0$.

STEP 2: Use the zeros of f to partition $[a, b]$ into subintervals.

STEP 3: Integrate f over each subinterval.

STEP 4: Add the absolute values of the results.

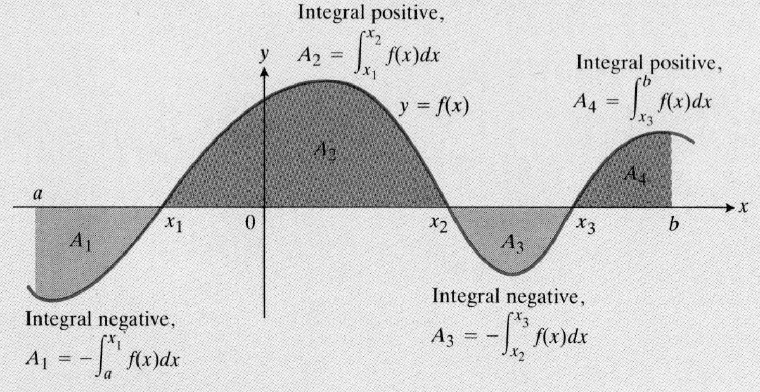

Here the area between the graph of f and the x-axis from a to b is found with four integrations. This is less work than it seems because we have to find the antiderivatives of f only once.

We illustrate these steps in the next example.

EXAMPLE 8 Find the area of the region between the x-axis and the curve

$$y = x^3 - 4x, \quad -2 \le x \le 2.$$

Solution

STEP 1: *The zeros of y.* We factor the formula for y to find where y is zero:

$$y = x^3 - 4x = x(x^2 - 4) = x(x - 2)(x + 2).$$

The zeros occur at $x = -2$, 0, and 2.

STEP 2: *The intervals of integration.* The points $x = -2$, 0, and 2 partition $[-2, 2]$ into two subintervals, $[-2, 0]$ and $[0, 2]$.

STEP 3: *The integrations.*

$$\int_{-2}^{0} (x^3 - 4x)\,dx = \left[\frac{x^4}{4} - 2x^2\right]_{-2}^{0} = [0] - [4 - 8] = 4$$

$$\int_{0}^{2} (x^3 - 4x)\,dx = \left[\frac{x^4}{4} - 2x^2\right]_{0}^{2} = [4 - 8] - [0] = -4$$

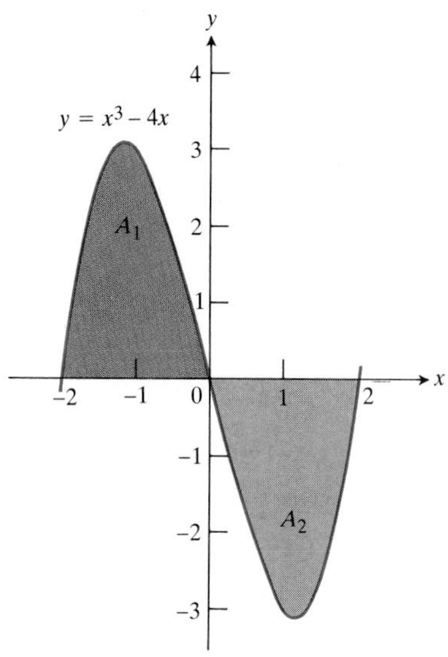

5.15 The graph of $y = x^3 - 4x$ from $x = -2$ to $x = 2$.

STEP 4: *The absolute values added.*

$$\text{Area of region} = |4| + |-4| = 4 + 4 = 8$$

Figure 5.15 shows the graph of $y = x^3 - 4x$ over $[-2, 2]$. The first integral in Step 3 gives the area A_1. The second integral gives the negative of the area A_2. The sum of the integrals' absolute values gives $A_1 + A_2$. ■

EXAMPLE 9 *Cost from marginal cost.* The fixed cost of starting a manufacturing run and producing the first 10 units is \$200. After that the marginal cost at x units output is

$$\frac{dc}{dx} = \frac{1000}{x^2}.$$

Find the total cost of producing the first 100 units.

Solution If $c(x)$ is the cost of x units, then

$$\underbrace{c(100)}_{\substack{\text{cost of}\\ \text{100 units}}} = \underbrace{200}_{\substack{\text{startup}\\ \text{first 10}}} + \underbrace{c(100) - c(10)}_{\substack{\text{cost of units}\\ \text{11–100}}}$$

$$= 200 + \int_{10}^{100} \frac{dc}{dx}\,dx$$

$$= 200 + \int_{10}^{100} \frac{1000}{x^2}\,dx \qquad \left(\frac{dc}{dx} = \frac{1000}{x^2}\right)$$

$$= 200 + 1000\int_{10}^{100} \frac{1}{x^2}\,dx \qquad \text{(Rule 3, Table 5.1)}$$

$$= 200 + 1000\left[-\frac{1}{x}\right]_{10}^{100} = 200 + 1000\left[-\frac{1}{100} + \frac{1}{10}\right]$$

$$= 200 - 10 + 100 = 290.$$

The total cost of producing the first 100 units is \$290. ■

☐ Average Daily Inventory

The notion of a function's average value is used in economics to study things like average daily inventory. If $I(x)$ is the number of radios, tires, shoes, or whatever product a firm has on hand on day x (we call $I(x)$ an **inventory function**), the average value of I over a time period $a \leq x \leq b$ is the firm's average daily inventory for the period.

DEFINITION

If $I(x)$ is the number of items on hand on day x, the **average daily inventory** of these items for the period $a \leq x \leq b$ is

$$I_{\text{av}} = \frac{1}{b-a}\int_a^b I(x)\,dx. \qquad (14)$$

If h is the dollar cost of holding one item per day, the **average daily holding cost** for the period $a \leq x \leq b$ is $I_{\text{av}} \cdot h$.

EXAMPLE 10 Suppose a wholesaler receives a shipment of 1200 cases of chocolate bars every 30 days. The chocolate is sold to retailers at a steady rate, and x days after the shipment arrives, the inventory of cases still on hand is $I(x) = 1200 - 40x$. Find the average daily inventory. Also find the average daily holding cost for the chocolate if the cost of holding one case is 3¢ a day.

Solution The average daily inventory is

$$I_{av} = \frac{1}{30-0}\int_0^{30}(1200 - 40x)\,dx = \frac{1}{30}\Big[1200x - 20x^2\Big]_0^{30} = 600.$$

The average daily holding cost for the chocolate is the dollar cost of holding one case times the average daily inventory. This works out to \$18 a day:

$$\text{Average daily holding cost} = (0.03)(600) = 18.$$

Exercises 5.4

Use the Second Fundamental Theorem to evaluate the integrals in Exercises 1–16.

1. $\displaystyle\int_1^2 (2x + 5)\,dx$
2. $\displaystyle\int_0^1 8x^3\,dx$
3. $\displaystyle\int_0^3 (4 - x^2)\,dx$
4. $\displaystyle\int_0^1 (x^2 - 2x + 3)\,dx$
5. $\displaystyle\int_0^1 (x^2 + \sqrt{x})\,dx$
6. $\displaystyle\int_0^5 x^{3/2}\,dx$
7. $\displaystyle\int_1^{32} x^{-6/5}\,dx$
8. $\displaystyle\int_{-2}^{-1} \frac{2}{x^2}\,dx$
9. $\displaystyle\int_0^{\pi} \sin x\,dx$
10. $\displaystyle\int_0^{\pi} (1 + \cos x)\,dx$
11. $\displaystyle\int_0^{\pi/3} 2\sec^2 x\,dx$
12. $\displaystyle\int_{\pi/6}^{5\pi/6} \csc^2 x\,dx$
13. $\displaystyle\int_{\pi/4}^{3\pi/4} \csc x\cot x\,dx$
14. $\displaystyle\int_0^{\pi/3} 4\sec x\tan x\,dx$
15. $\displaystyle\int_{-1}^{1} (r + 1)^2\,dr$ (*Hint:* Square first.)
16. $\displaystyle\int_9^4 \frac{1-\sqrt{u}}{\sqrt{u}}\,du$ (*Hint:* Divide first.)

In Exercises 17–20, find the total area of the region between the curve and the x-axis.

17. $y = 2 - x,\ 0 \le x \le 3$
18. $y = 3x^2 - 3,\ -2 \le x \le 2$
19. $y = x^3 - 3x^2 + 2x,\ 0 \le x \le 2$
20. $y = x^3 - 4x,\ -2 \le x \le 2$

Find the areas of the shaded regions in Exercises 21–24.

21.

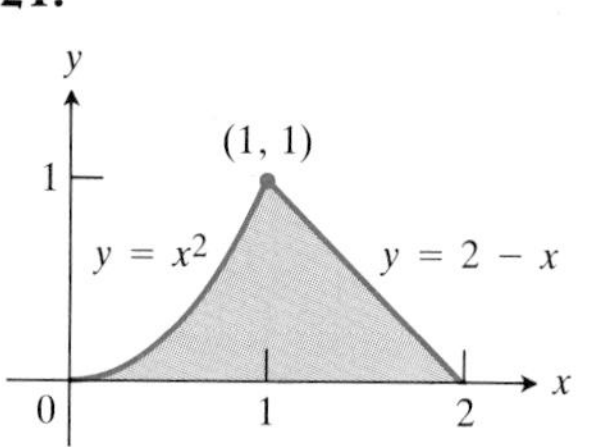

22.

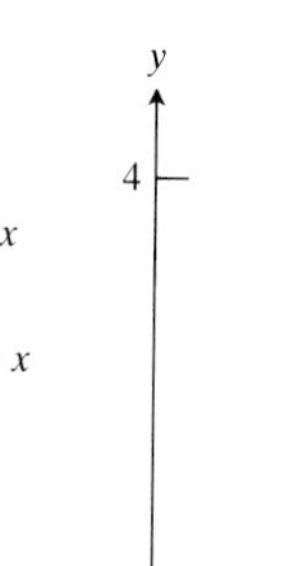

23.

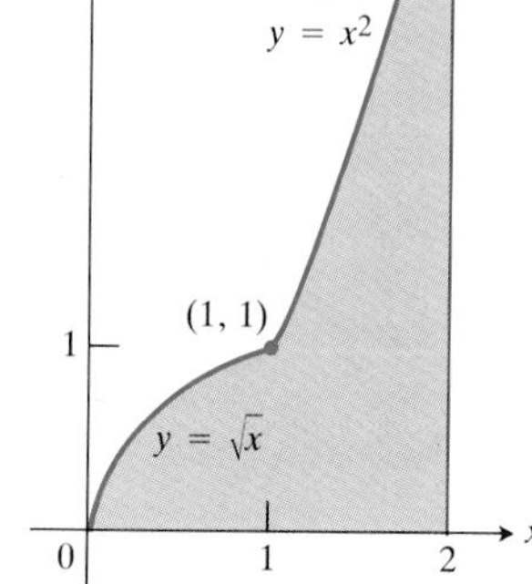

24.

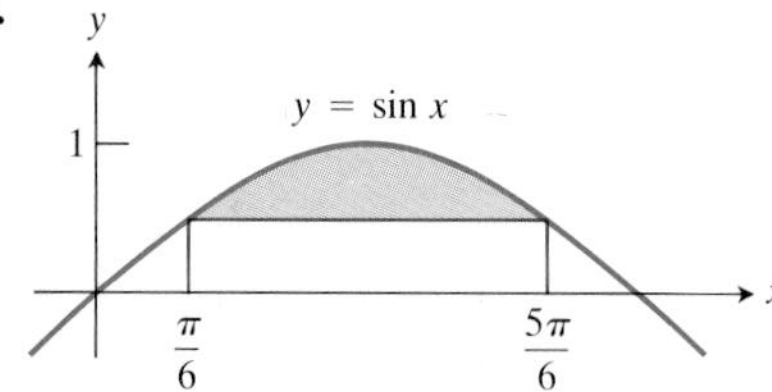

Find the derivatives in Exercises 25–26 two ways: (a) by evaluating the integral and differentiating the result and (b) by applying the First Fundamental Theorem.

25. $\displaystyle\frac{d}{dx}\int_0^x \cos t\,dt$
26. $\displaystyle\frac{d}{dx}\int_1^{\sin x} 3t^2\,dt$

Find dy/dx in Exercises 27–30.

27. $y = \int_0^x \sqrt{1 + t^2}\, dt$

28. $y = \int_1^x \frac{1}{t}\, dt,\ x > 0$

29. $y = \int_0^{\sqrt{x}} \sin(t^2)\, dt$

30. $y = \int_0^{2x} \cos t\, dt$

Each of the following functions solves one of the initial value problems in Exercises 31–34. Which function solves which problem?

a) $y = \int_1^x \frac{1}{t}\, dt - 3$

b) $y = \int_0^x \sec t + 4$

c) $y = \int_{-1}^x \sec t + 4$

d) $y = \int_\pi^x \frac{1}{t}\, dt - 3$

31. $\frac{dy}{dx} = \frac{1}{x},\ y(\pi) = -3$

32. $y' = \sec x,\ y(-1) = 4$

33. $y' = \sec x,\ y(0) = 4$

34. $y' = \frac{1}{x},\ y(1) = -3$

35. Show that if k is a positive constant, then the area between the x-axis and one arch of the curve $y = \sin kx$ is always $2/k$.

36. *Archimedes' area formula for parabolas.* Archimedes (287–212 B.C.), inventor, military engineer, physicist, and the greatest mathematician of classical times, discovered that the area under a parabolic arch like the one shown here is always two-thirds the base times the height.

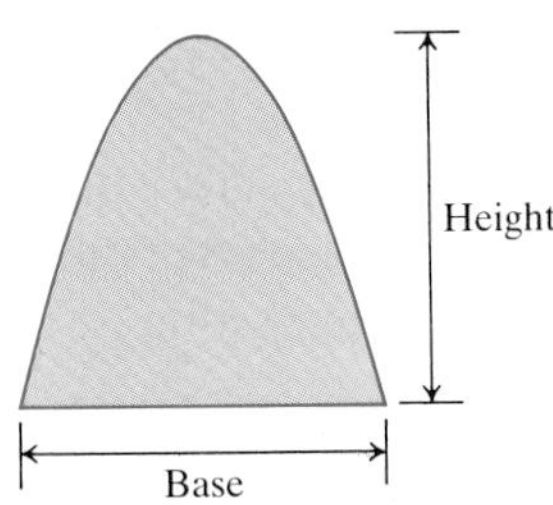

a) Find the area under the parabolic arch

$$y = 6 - x - x^2,\ -3 \le x \le 2.$$

b) Find the height of the arch. (Where does y have its maximum value?)

c) Show that the area is two-thirds the base times the height.

37. *Cost from marginal cost.* The marginal cost of printing a poster when x posters have been printed is

$$\frac{dc}{dx} = \frac{1}{2\sqrt{x}}$$

dollars. Find (a) $c(100) - c(1)$, the cost of printing posters 2–100; (b) $c(400) - c(100)$, the cost of printing posters 101–400.

38. *Revenue from marginal revenue.* Suppose that a company's marginal revenue from the manufacture and sale of egg beaters is

$$\frac{dr}{dx} = 2 - 2/(x + 1)^2,$$

where r is measured in thousands of dollars and x in thousands of units. How much money should the company expect from a production run of $x = 3$ thousand egg beaters? To find out, integrate the marginal revenue from $x = 0$ to $x = 3$.

39. Solon Container receives 450 drums of plastic pellets every 30 days. The inventory function (drums on hand as a function of days) is $I(x) = 450 - x^2/2$. Find the average daily inventory. If the holding cost for one drum is 2¢ per day, find the average daily holding cost.

40. Mitchell Mailorder receives a shipment of 600 cases of athletic socks every 60 days. The number of cases on hand x days after the shipment arrives is $I(x) = 600 - 20\sqrt{15x}$. Find the average daily inventory. If the holding cost for one case is 1/2¢ per day, find the average daily holding cost.

41. For what value of x is

$$\int_a^x f(t)\, dt$$

sure to be zero?

42. Suppose $\int_1^x f(t)\, dt = x^2 - 2x + 1$. Find $f(x)$. (*Hint*: Differentiate both sides of the equation with respect to x.)

43. Find $f(4)$ if $\int_0^x f(t)\, dt = x \cos \pi x$.

44. Find the linearization of

$$f(x) = 2 + \int_0^x \frac{10}{1 + t}\, dt$$

at $x = 0$.

TOOLKIT PROGRAM

Derivative Grapher — Graphs $f(x)$ and $\int_a^x f(t)\, dt$ together (among other things)

5.5 Indefinite Integrals

Because antiderivatives make it possible to evaluate definite integrals with arithmetic, we shall be working with antiderivatives a lot. We therefore need a notation that will make antiderivatives easier to describe and work with. This section introduces the notation and shows how to use it.

The Indefinite Integral of a Function

We call the set of all antiderivatives of a function the indefinite integral of the function, according to the following definition.

DEFINITION

> If the function $f(x)$ is a derivative, then the set of all antiderivatives of f is called the **indefinite integral** of f, denoted by the symbols
>
> $$\int f(x)\,dx.$$
>
> As in definite integrals, the symbol $\int$ is called an **integral sign.** The function f is the **integrand** of the integral and x is the **variable of integration.**

Since every continuous function has an antiderivative (Corollary of the First Fundamental Theorem), every continuous function has an indefinite integral.

Once we have found an antiderivative $F(x)$ of a function $f(x)$, the other antiderivatives of f differ from F only by a constant (Corollary 3 of the Mean Value Theorem). We indicate this in the new notation by writing

$$\int f(x)\,dx = F(x) + C. \tag{1}$$

The constant C is called the **constant of integration** or the **arbitrary constant,** and Eq. (1) is read, "The indefinite integral of f with respect to x is $F(x) + C$." When we find $F(x) + C$, we say that we have **evaluated** the indefinite integral.

> To evaluate $\int f(x)\,dx$:
>
> STEP 1: Find an antiderivative $F(x)$ of $f(x)$.
>
> STEP 2: Add C (the constant of integration). Then
>
> $$\int f(x)\,dx = F(x) + C.$$

EXAMPLE 1 Evaluate $\int (x^2 - 2x + 5)\,dx$.

Solution

An antiderivative of $f(x) = x^2 - 2x + 5$; The constant of integration

$$\int (x^2 - 2x + 5)\,dx = \overbrace{\frac{x^3}{3} - x^2 + 5x}^{\text{An antiderivative of } f(x)=x^2-2x+5} + \underbrace{C}_{\text{The constant of integration}}$$

■

TABLE 5.2
Integration Formulas

1. $\int x^n\,dx = \dfrac{x^{n+1}}{n+1} + C \qquad (n \neq -1)$	
2. $\int \sin kx\,dx = -\dfrac{\cos kx}{k} + C$	3. $\int \cos kx\,dx = \dfrac{\sin kx}{k} + C$
4. $\int \sec^2 x\,dx = \tan x + C$	5. $\int \csc^2 x\,dx = -\cot x + C$
6. $\int \sec x \tan x\,dx = \sec x + C$	7. $\int \csc x \cot x\,dx = -\csc x + C$

The formulas for general antiderivatives in Section 4.6 translate into the formulas for evaluating indefinite integrals listed in Table 5.2.

In case you are wondering why the integrals of the tangent, cotangent, secant, and cosecant are not listed here, the answer is that the usual formulas for them require logarithms. We know that these functions do have indefinite integrals on intervals where they are continuous, but we shall have to wait until Chapters 7 and 8 to see what the integrals are.

EXAMPLE 2 *Selected Integrals from Table 5.2*

a) $\int x^5\,dx = \dfrac{x^6}{6} + C$ (Formula 1)

b) $\int \sin 2x\,dx = -\dfrac{\cos 2x}{2} + C$ (Formula 2 with $k = 2$)

c) $\int \cos \dfrac{x}{2}\,dx = \int \cos \dfrac{1}{2}x\,dx = \dfrac{\sin (1/2)x}{1/2} + C$ (Formula 3 with $k = 1/2$)

$$= 2 \sin \frac{x}{2} + C$$

■

The formulas in Table 5.2 hold because, in each case, the derivative of the function $F(x) + C$ on the right is the integrand $f(x)$ on the left. Finding an integral formula can sometimes be a difficult task, but *checking* an integral formula, once found, is relatively easy: Differentiate the right-hand side. If the derivative is the integrand, the formula is correct; otherwise it is wrong.

EXAMPLE 3

RIGHT: $\int x \cos x\,dx = x \sin x + \cos x + C$

Reason: The derivative of the right-hand side is the integrand:

$$\frac{d}{dx}(x \sin x + \cos x + C) = x \cos x + \sin x - \sin x + 0 = x \cos x.$$

WRONG: $$\int x \cos x \, dx = x \sin x + C$$

Reason: The derivative of the right-hand side is not the integrand:

$$\frac{d}{dx}(x \sin x + C) = x \cos x + \sin x + 0 \neq x \cos x.$$

Do not worry about where the (correct) integral formula in Example 3 comes from right now. There is a nice technique in Chapter 8 for evaluating integrals like this.

Rules of Algebra

Among the things we know about antiderivatives are these:

1. The general antiderivative of a derivative dF/dx is $F(x) + C$.
2. When we differentiate an antiderivative of a function $f(x)$ with respect to x, we get f back again.
3. A function is an antiderivative of a constant multiple kf of a function f if and only if it is k times an antiderivative of f.
4. In particular, a function is an antiderivative of $-f$ if and only if it is the negative of an antiderivative of f.
5. A function is an antiderivative of a sum $f(x) + g(x)$ if and only if it is the sum of an antiderivative of f and an antiderivative of g.
6. A function is an antiderivative of a difference $f(x) - g(x)$ if and only if it is an antiderivative of f minus an antiderivative of g.

These observations can be expressed very nicely in the notation of indefinite integrals (Table 5.3).

TABLE 5.3
Rules for Indefinite Integrals

1.	$\int \frac{dF}{dx} \, dx = F(x) + C$	
2.	$\frac{d}{dx} \int f(x) \, dx = f(x)$	
3.	$\int k f(x) \, dx = k \int f(x) \, dx$	(Provided k is a constant. Does not work if k varies with x.)
4.	$\int -f(x) \, dx = -\int f(x) \, dx$	(Special case of Rule 3)
5.	$\int [f(x) + g(x)] \, dx = \int f(x) \, dx + \int g(x) \, dx$	
6.	$\int [f(x) - g(x)] \, dx = \int f(x) \, dx - \int g(x) \, dx$	

EXAMPLE 4

$$\int \cos x \, dx = \int \frac{d}{dx}(\sin x)\, dx = \sin x + C \qquad \text{(Rule 1 with } F(x) = \sin x\text{)}$$ ■

EXAMPLE 5

$$\frac{d}{dx}\int \tan x \, dx = \tan x \qquad \text{(Rule 2 with } f(x) = \tan x\text{)}$$

We know this even without knowing how to evaluate the integral. ■

EXAMPLE 6 *Rewriting the constant of integration.*

$$\begin{aligned} \int 5 \sec x \tan x \, dx &= 5 \int \sec x \tan x \, dx && \left(\begin{array}{l}\text{Table 5.3, Rule 3, with} \\ k = 5, f(x) = \sec x \tan x\end{array}\right) \\ &= 5(\sec x + C) && \text{(Table 5.2, Rule 6)} \\ &= 5 \sec x + 5C && \text{(First form)} \\ &= 5 \sec x + C' && \text{(Shorter form)} \\ &= 5 \sec x + C. && \text{(Usual form—no prime)} \end{aligned}$$ ■

What about all the different forms in Example 6? Each one of them gives all the antiderivatives of $f(x) = 5 \sec x \tan x$, so each answer is correct. But the least complicated of the three, and the usual choice, is

$$\int 5 \sec x \tan x \, dx = 5 \sec x + C.$$

The general rule, in practice, is this:

> If $F'(x) = f(x)$ and k is a constant, then
>
> $$k \int f(x)\, dx = \int k f(x)\, dx = k F(x) + C. \qquad (2)$$

EXAMPLE 7

$$\int 8 \cos x \, dx = 8 \sin x + C \qquad \left(\begin{array}{l}\text{Eq. (2) with } k = 8, f(x) = \cos x, \\ F(x) = \sin x\end{array}\right)$$ ■

EXAMPLE 8 *When we integrate term by term, we combine the constants of integration into a single arbitrary constant.* Evaluate $\int (x^2 - 2x + 5)\, dx$.

Solution This is the integral of Example 1, which we evaluated by finding an antiderivative of $(x^2 - 2x + 5)$ and adding an arbitrary constant:

$$\int (x^2 - 2x + 5)\, dx = \underbrace{\frac{x^3}{3} - x^2 + 5x}_{\text{Antiderivative}} + \underbrace{C}_{\text{Arbitrary constant}}.$$

Suppose, however, we evaluate the integral term by term by applying the sum and difference rules (Rules 5 and 6 in Table 5.3). Then, instead of getting one constant of integration, we get three:

$$\int (x^2 - 2x + 5)\,dx = \int x^2\,dx - \int 2x\,dx + \int 5\,dx$$

$$= \frac{x^3}{3} + C_1 - x^2 + C_2 + 5x + C_3.$$

This formula certainly does give all the antiderivatives of $x^2 - 2x + 5$. But it is more complicated than it needs to be. If we were to combine C_1, C_2, and C_3 into a single constant $C = C_1 + C_2 + C_3$, the formula would simplify to

$$\frac{x^3}{3} - x^2 + 5x + C$$

and *still* give all the antiderivatives there are. For this reason we recommend that you go right to the final form even if you elect to integrate term by term. Write

$$\int (x^2 - 2x + 5)\,dx = \int x^2\,dx - \int 2x\,dx + \int 5\,dx = \frac{x^3}{3} - x^2 + 5x + C.$$

Find the simplest antiderivative you can for each part, then add the constant at the end. ■

The Integrals of $\sin^2 x$ and $\cos^2 x$

We can sometimes use trigonometric identities to transform indefinite integrals we do not know how to evaluate into indefinite integrals we do know how to evaluate. Among the examples you should know about, important because of how frequently they arise in applications, are the integral formulas for $\sin^2 x$ and $\cos^2 x$. You need not remember the formulas themselves, but try to remember how they can be derived when you need them.

EXAMPLE 9

$$\int \sin^2 x\,dx = \int \frac{1 - \cos 2x}{2}\,dx \qquad \left(\text{Because } \sin^2 x = \frac{1 - \cos 2x}{2}\right)$$

$$= \frac{1}{2}\int (1 - \cos 2x)\,dx = \frac{1}{2}\int dx - \frac{1}{2}\int \cos 2x\,dx$$

$$= \frac{1}{2}x - \frac{1}{2}\frac{\sin 2x}{2} + C = \frac{x}{2} - \frac{\sin 2x}{4} + C$$ ■

EXAMPLE 10

$$\int \cos^2 x\,dx = \int \frac{1 + \cos 2x}{2}\,dx \qquad \left(\cos^2 x = \frac{1 + \cos 2x}{2}\right)$$

$$= \frac{x}{2} + \frac{\sin 2x}{4} + C \qquad \left(\begin{array}{l}\text{As in Example 9, but}\\ \text{with a sign change}\end{array}\right)$$ ■

Solving Initial Value Problems with Indefinite Integrals

As you know, we solve initial value problems with antiderivatives. We now look at the solutions in the language of indefinite integration. The first example is based on Example 6 in Section 4.6. You need not look back, however, unless you are interested in the modeling that gave rise to the problem. The solution we give here is mathematically self-contained.

EXAMPLE 11 *Finding velocity from acceleration.* As a function of elapsed time t, the velocity $v(t)$ of a body falling from rest in a vacuum near the surface of the earth satisfies

Differential equation: $$\frac{dv}{dt} = 9.8$$ (The acceleration is 9.8 m/sec^2.)

Initial condition: $v = 0$ when $t = 0$. (The velocity is 0 at the start.)

Find v as a function of t.

Solution We find the general solution of the differential equation by integrating both sides of it with respect to t:

$$\frac{dv}{dt} = 9.8 \qquad \text{(Differential equation)}$$

$$\int \frac{dv}{dt}\,dt = \int 9.8\,dt \qquad \text{(Integral equation)}$$

$$v + C_1 = 9.8t + C_2 \qquad \text{(Integrals evaluated)}$$

$$v = 9.8t + C_2 - C_1 \qquad \text{(Solved for } v\text{)}$$

$$v = 9.8t + C. \qquad \text{(Constants combined as one)}$$

This last equation tells us that the body's velocity t seconds into the fall is $9.8t + C$ m/sec for some value of C. What value? We find out from the initial condition:

$$v = 9.8t + C$$

$$0 = 9.8(0) + C \qquad (v = 0 \text{ when } t = 0)$$

$$C = 0.$$

Conclusion: The body's velocity t seconds into the fall is

$$v = 9.8t + 0 = 9.8t \text{ m/sec.}$$

■

In the next example we have to integrate a second derivative twice to find the function we are after. The first integration,

$$\int \frac{d^2y}{dx^2}\,dx = \frac{dy}{dx} + C, \tag{3}$$

gives the function's first derivative. The second integration gives the function.

EXAMPLE 12 Solve the following initial value problem for y as a function of x:

Differential equation: $\dfrac{d^2y}{dx^2} = 6x - 2$

Initial conditions: $\dfrac{dy}{dx} = 0$ and $y = 10$ when $x = 1$.

Solution We integrate the differential equation with respect to x to find dy/dx:

$$\int \frac{d^2y}{dx^2}\,dx = \int (6x - 2)\,dx$$

$$\frac{dy}{dx} = 3x^2 - 2x + C_1. \qquad \text{(Constants of integration combined as } C_1\text{)}$$

We apply the first initial condition to find C_1:

$$0 = 3(1)^2 - 2(1) + C_1 \qquad \left(\frac{dy}{dx} = 0 \text{ when } x = 1\right)$$

$$C_1 = -3 + 2 = -1.$$

This completes the formula for dy/dx:

$$\frac{dy}{dx} = 3x^2 - 2x - 1.$$

We integrate dy/dx with respect to x to find y:

$$\int \frac{dy}{dx}\,dx = \int (3x^2 - 2x - 1)\,dx$$

$$y = x^3 - x^2 - x + C_2. \qquad \text{(Constants of integration combined as } C_2\text{)}$$

We apply the second initial condition to find C_2:

$$10 = (1)^3 - (1)^2 - 1 + C_2 \qquad (y = 10 \text{ when } x = 1)$$

$$10 = -1 + C_2$$

$$C_2 = 11.$$

This completes the formula for y as a function of x:

$$y = x^3 - x^2 - x + 11.$$

When we find a function by integrating its first derivative, we have one constant of integration, as in Example 11. When we find a function from its second derivative, we have to deal with two constants of integration, one from each integration, as in Example 12. If we were to find a function from its third derivative, we would have to find the values of three constants of integration, and so on. In each case the values of the constants are determined by the problem's initial conditions. Each time we integrate, we need an initial condition to tell us the value of C.

Exercises 5.5

Evaluate the integrals in Exercises 1–30.

1. $\int x^3\,dx$

2. $\int 7\,dx$

3. $\int (x+1)\,dx$

4. $\int (6-6x)\,dx$

5. $\int 3\sqrt{x}\,dx$

6. $\int \frac{4}{x^2}\,dx$

7. $\int x^{-1/3}\,dx$

8. $\int (1-4x^{-3})\,dx$

9. $\int (5x^2+2x)\,dx$

10. $\int \left(\frac{x^2}{2}+\frac{x^3}{3}\right)dx$

11. $\int (2x^3-5x+7)\,dx$

12. $\int (1-x^2-3x^5)\,dx$

13. $\int 2\cos x\,dx$

14. $\int 5\sin\theta\,d\theta$

15. $\int \sin\frac{x}{3}\,dx$

16. $\int 3\cos 5x\,dx$

17. $\int 3\csc^2 x\,dx$

18. $\int \frac{\sec^2 x}{3}\,dx$

19. $\int \frac{\csc x\cot x}{2}\,dx$

20. $\int \frac{2}{5}\sec x\tan x\,dx$

21. $\int (4\sec x\tan x-2\sec^2 x)\,dx$

22. $\int \frac{1}{2}(\csc^2 x-\csc x\cot x)\,dx$

23. $\int (\sin 2x-\csc^2 x)\,dx$

24. $\int (2\cos 2x-3\sin 3x)\,dx$

25. $\int 4\sin^2 y\,dy$

26. $\int \frac{\cos^2 x}{7}\,dx$

27. $\int \sin x\cos x\,dx$

(Hint: $2\sin x\cos x=\sin 2x$)

28. $\int (1-\cos^2 t)\,dt$

29. $\int (1+\tan^2\theta)\,d\theta$

(Hint: $1+\tan^2\theta=\sec^2\theta$)

30. $\int \frac{1+\cot^2 x}{2}\,dx$

Show that the integral formulas in Exercises 31–34 are correct by showing that the derivatives of the right-hand sides are the integrands in the integrals on the left-hand sides. (In Section 5.6 we shall see where formulas like these come from.)

31. $\int (7x-2)^3\,dx=\frac{(7x-2)^4}{28}+C$

32. $\int \sec^2 5x\,dx=\frac{\tan 5x}{5}+C$

33. $\int \frac{1}{(x+1)^2}\,dx=-\frac{1}{x+1}+C$

34. $\int \frac{1}{(x+1)^2}\,dx=\frac{x}{x+1}+C'$

35. Right, or wrong? Say which for each formula.

a) $\int x\sin x\,dx=\frac{x^2}{2}\sin x+C$

b) $\int x\sin x\,dx=-x\cos x+C$

c) $\int x\sin x\,dx=-x\cos x+\sin x+C$

36. Right, or wrong? Say which for each formula.

a) $\int (2x+1)^2\,dx=\frac{(2x+1)^3}{3}+C$

b) $\int 3(2x+1)^2\,dx=(2x+1)^3+C$

c) $\int 6(2x+1)^2\,dx=(2x+1)^3+C$

Solve the initial value problems in Exercises 37–44 for y as a function of x.

37. Differential equation: $\frac{dy}{dx}=3\sqrt{x}$
Initial condition: $y=4$ when $x=9$

38. Differential equation: $\frac{dy}{dx}=\frac{1}{2\sqrt{x}}$
Initial condition: $y=0$ when $x=4$

39. Differential equation: $\frac{dy}{dx}=-\pi\sin\pi x$
Initial condition: $y=0$ when $x=0$

40. Differential equation: $\frac{dy}{dx}=\frac{1}{2}\sec x\tan x$
Initial condition: $y=1$ when $x=0$

41. Differential equation: $\frac{d^2y}{dx^2} = 0$

Initial conditions: $\frac{dy}{dx} = 2$ and $y = 0$ when $x = 0$

42. Differential equation: $\frac{d^2y}{dx^2} = \frac{2}{x^3}$

Initial conditions: $\frac{dy}{dx} = 1$ and $y = 1$ when $x = 1$

43. Differential equation: $\frac{d^2y}{dx^2} = \frac{3x}{8}$

Initial conditions: $\frac{dy}{dx} = 3$ and $y = 4$ when $x = 4$

44. Differential equation: $\frac{d^3y}{dx^3} = 6$

Initial conditions: $\frac{d^2y}{dx^2} = -8$, $\frac{dy}{dx} = 0$, and $y = 5$ when $x = 0$

45. *Stopping a car in time.* You are driving along a highway at a steady 60 mph (88 ft/sec) when you see an accident ahead and slam on the brakes. What constant deceleration is required to stop your car in 242 ft? To find out, carry out the following steps:

STEP 1: Solve the initial value problem

Differential equation: $\frac{d^2s}{dt^2} = -k$ (k constant)

Initial conditions: $\frac{ds}{dt} = 88$ and $s = 0$ when $t = 0$.

(Measuring time and distance from when the brakes are applied)

STEP 2: Find the value of t that makes $ds/dt = 0$. (The answer will involve k.)

STEP 3: Find the value of k that makes $s = 242$ for the value of t you found in Step 2.

46. *Motion along a coordinate line.* A particle moves along a coordinate line with acceleration $a = d^2s/dt^2 = 15\sqrt{t} - (3/\sqrt{t})$, subject to the conditions that $ds/dt = 4$ and $s = 0$ when $t = 1$. Find

a) the velocity $v = ds/dt$ in terms of t,

b) the position s in terms of t.

47. CALCULATOR *The hammer and the feather.* When Apollo 15 astronaut David Scott dropped a hammer and a feather on the moon to demonstrate that in a vacuum all bodies fall with the same (constant) acceleration, he dropped them from about 4 ft above the ground. The television footage of the event shows the hammer and feather falling more slowly than on earth, where, in a vacuum, they would have taken only half a second to fall the four feet. How long did it take the hammer and feather to fall the four feet on the moon? To find out, solve the following initial value problem for s as a function of t. Then find the value of t that makes s equal 4.

Differential equation: $\frac{d^2s}{dt^2} = 5.2 \text{ ft/sec}^2$

Initial conditions: $\frac{ds}{dt} = 0$ and $s = 0$ when $t = 0$

48. *The standard equation for free fall.* The standard equation for free fall near the surface of every planet is

$$s(t) = \frac{1}{2}gt^2 + v_0t + s_0,$$

where $s(t)$ is the body's position on the line of fall, g is the planet's (constant) acceleration of gravity, v_0 is the body's initial velocity, and s_0 is the body's initial position. Derive this equation by solving the following initial value problem:

Differential equation: $\frac{d^2s}{dt^2} = g$

Initial conditions: $\frac{ds}{dt} = v_0$ and $s = s_0$ when $t = 0$.

TOOLKIT PROGRAM

Antiderivatives and Direction Fields	Graphs solutions of the initial value problem $y' = f(x, y)$, $y(x_0) = y_0$. You can see the effect of changing x_0 and y_0 by graphing different solutions together.

5.6 Integration by Substitution—Running the Chain Rule Backward

A change of variable can often turn an unfamiliar integral into one we can evaluate. The method for doing this is called the substitution method of integration. It is the principal method by which integrals are evaluated. This section shows how and why the method works.

The Generalized Power Rule in Integral Form

When u is a differentiable function of x and n is a rational number different from -1, the Chain Rule tells us that

$$\frac{d}{dx}\left(\frac{u^{n+1}}{n+1}\right) = u^n \frac{du}{dx}. \tag{1}$$

This same equation, from another point of view, says that $u^{n+1}/(n+1)$ is one of the antiderivatives of $u^n(du/dx)$. The set of all antiderivatives of $u^n(du/dx)$ is therefore

$$\int \left(u^n \frac{du}{dx}\right) dx = \frac{u^{n+1}}{n+1} + C. \tag{2}$$

The integral on the left-hand side of this equation is usually written in the simpler "differential" form,

$$\int u^n \, du, \tag{3}$$

obtained by treating the dx's as differentials that cancel. Combining Eqs. (2) and (3) then gives the following rule.

> If u is any differentiable function of x,
>
> $$\int u^n \, du = \frac{u^{n+1}}{n+1} + C. \tag{4}$$

Whenever we can cast an integral in the form

$$\int u^n \, du$$

with u a differentiable function of x and du the differential of u, we can integrate with respect to u in the usual way to evaluate the integral as $[u^{n+1}/(n+1)] + C$.

EXAMPLE 1 Evaluate $\int (x+2)^5 \, dx$.

Solution We can put the integral in the form

$$\int u^5 \, du$$

by substituting

$$u = x + 2, \quad du = d(x+2) = dx.$$

Then

$$\int (x+2)^5 \, dx = \int u^5 \, du \qquad \text{(Substitute } u = x+2,\ du = dx.\text{)}$$

$$= \frac{u^6}{6} + C \qquad \text{(Integrate, using Eq. (4) with } n = 5.\text{)}$$

$$= \frac{(x+2)^6}{6} + C. \qquad \text{(Replace } u \text{ by } x+2.\text{)}$$

■

EXAMPLE 2

$$\int \sqrt{1 + x^2} \cdot 2x\, dx = \int u^{1/2}\, du \qquad \left(\begin{array}{l}\text{Substitute } u = 1 + x^2,\\ du = 2x\, dx.\end{array}\right)$$

$$= \frac{u^{(1/2)+1}}{(1/2) + 1} + C \qquad \left(\begin{array}{l}\text{Integrate, using Eq. (4)}\\ \text{with } n = \frac{1}{2}.\end{array}\right)$$

$$= \frac{2}{3} u^{3/2} + C \qquad \text{(Simpler form)}$$

$$= \frac{2}{3}(1 + x^2)^{3/2} + C \qquad \text{(Replace } u \text{ by } 1 + x^2.)$$

■

EXAMPLE 3 *Adjusting the integrand by a constant to put it in standard form.*

$$\int \sqrt{4x - 1}\, dx = \int u^{1/2} \cdot \frac{1}{4}\, du \qquad \left(\begin{array}{l}\text{Substitute } u = 4x - 1,\\ du = 4\, dx,\ \frac{1}{4}\, du = dx.\end{array}\right)$$

$$= \frac{1}{4} \int u^{1/2}\, du \qquad \left(\begin{array}{l}\text{With the } \frac{1}{4} \text{ out front, the integral}\\ \text{is now in standard form.}\end{array}\right)$$

$$= \frac{1}{4} \frac{u^{3/2}}{3/2} + C \qquad \left(\text{Integrate, using Eq. (4) with } n = \frac{1}{2}.\right)$$

$$= \frac{1}{6} u^{3/2} + C \qquad \text{(Simpler form)}$$

$$= \frac{1}{6}(4x - 1)^{3/2} + C \qquad \text{(Replace } u \text{ by } 4x - 1.)$$

■

Sines and Cosines

If u is a differentiable function of x, then $\sin u$ is a differentiable function of x. The Chain Rule gives the derivative of $\sin u$ as

$$\frac{d}{dx} \sin u = \cos u \frac{du}{dx}. \tag{5}$$

From another point of view, however, this same equation says that $\sin u$ is one of the antiderivatives of the product $\cos u \cdot (du/dx)$. The set of all antiderivatives of the product is therefore

$$\int \left(\cos u \frac{du}{dx}\right) dx = \sin u + C. \tag{6}$$

A formal cancellation of the dx's in the integral on the left leads to the following rule.

> If u is a differentiable function of x, then
>
> $$\int \cos u\, du = \sin u + C. \tag{7}$$

Equation (7) says that whenever we can cast an integral in the form

$$\int \cos u \, du,$$

we can integrate with respect to u in the usual way to evaluate the integral as $\sin u + C$.

EXAMPLE 4

$$\int \cos(7x + 5)\, dx = \int \cos u \cdot \frac{1}{7}\, du \qquad \left(\begin{array}{l}\text{Substitute } u = 7x + 5,\\ du = 7\, dx,\ \frac{1}{7}\, du = dx.\end{array}\right)$$

$$= \frac{1}{7}\int \cos u \, du \qquad \left(\begin{array}{l}\text{With the } \frac{1}{7} \text{ out front, the inte-}\\ \text{gral is now in standard form.}\end{array}\right)$$

$$= \frac{1}{7} \sin u + C \qquad \text{(Integrate with respect to } u.)$$

$$= \frac{1}{7} \sin(7x + 5) + C \qquad \text{(Replace } u \text{ by } 7x + 5.)$$

■

The companion formula for the integral of $\sin u$ when u is a differentiable function of x is

$$\int \sin u \, du = -\cos u + C. \tag{8}$$

EXAMPLE 5

$$\int x^2 \sin(x^3)\, dx = \int \sin u \cdot \frac{1}{3}\, du \qquad \left(\begin{array}{l}\text{Substitute } u = x^3,\\ du = 3x^2\, dx,\ \frac{1}{3}\, du = x^2\, dx.\end{array}\right)$$

$$= \frac{1}{3}\int \sin u \, du$$

$$= \frac{1}{3}(-\cos u) + C \qquad \text{(Integrate with respect to } u.)$$

$$= -\frac{1}{3}\cos(x^3) + C \qquad \text{(Replace } u \text{ by } x^3.)$$

■

The Chain-Rule formulas for the derivatives of the tangent, cotangent, secant, and cosecant of a differentiable function u of x lead to the following integrals.

$$\int \sec^2 u \, du = \tan u + C \tag{9}$$

$$\int \csc^2 u \, du = -\cot u + C \tag{10}$$

$$\int \sec u \tan u \, du = \sec u + C \tag{11}$$

$$\int \csc u \cot u \, du = -\csc u + C \tag{12}$$

In each formula we assume u to be a differentiable function of x. Each formula can be checked by differentiating the right-hand side with respect to x. In each case the Chain Rule applies to produce the integrand on the left.

EXAMPLE 6

$$\int \frac{1}{\cos^2 2x}\,dx = \int \sec^2 2x\,dx \qquad \left(\sec 2x = \frac{1}{\cos 2x}\right)$$

$$= \int \sec^2 u \cdot \frac{1}{2}\,du \qquad \left(\text{Substitute } u = 2x,\ du = 2\,dx,\ dx = \frac{1}{2}\,du.\right)$$

$$= \frac{1}{2}\int \sec^2 u\,du$$

$$= \frac{1}{2}\tan u + C \qquad \text{(Integrate, using Eq. (9).)}$$

$$= \frac{1}{2}\tan 2x + C \qquad \text{(Replace } u \text{ by } 2x.\text{)}$$

■

The Substitution Method of Integration

The substitutions we have been using in the examples are all instances of a general rule:

$$\int f(g(x)) \cdot g'(x)\,dx = \int f(u)\,du \qquad \text{1. Substitute } u = g(x),\ du = g'(x)\,dx.$$

$$= F(u) + C \qquad \text{2. Evaluate by finding an antiderivative of } f(u). \text{ (Any one will do.)}$$

$$= F(g(x)) + C \qquad \text{3. Substitute back.}$$

These three steps are the steps of the substitution method of integration.

The Substitution Method of Integration

To evaluate the integral

$$\int f(g(x))g'(x)\,dx$$

when f and g' are continuous functions, carry out the following steps:

STEP 1: Substitute $u = g(x)$ and $du = g'(x)\,dx$ to obtain the integral

$$\int f(u)\,du.$$

STEP 2: Integrate with respect to u.

STEP 3: Replace u by $g(x)$ in the result.

EXAMPLE 7

$$\int (x^2 + 2x - 3)^2(x + 1)\,dx = \int u^2 \cdot \frac{1}{2}\,du \qquad \left(\begin{aligned}&\text{Substitute } u = x^2 + 2x - 3,\\ &du = 2x\,dx + 2\,dx\\ &\quad = 2(x+1)\,dx,\\ &\frac{1}{2}\,du = (x+1)\,dx.\end{aligned}\right)$$

$$= \frac{1}{2}\int u^2\,du$$

$$= \frac{1}{2}\cdot\frac{u^3}{3} + C = \frac{1}{6}u^3 + C \qquad \left(\begin{aligned}&\text{Integrate with}\\ &\text{respect to } u.\end{aligned}\right)$$

$$= \frac{1}{6}(x^2 + 2x - 3)^3 + C \qquad \left(\begin{aligned}&\text{Replace } u \text{ by}\\ &x^2 + 2x - 3.\end{aligned}\right)$$

■

EXAMPLE 8

$$\int \sin^4 x \cos x\,dx = \int u^4\,du \qquad (\text{Substitute } u = \sin x,\ du = \cos x\,dx.)$$

$$= \frac{u^5}{5} + C \qquad (\text{Integrate with respect to } u.)$$

$$= \frac{\sin^5 x}{5} + C \qquad (\text{Replace } u \text{ by } \sin x.)$$

■

There is often more than one way to make a successful substitution, as the next example shows.

EXAMPLE 9 Evaluate $\displaystyle\int \frac{2z\,dz}{\sqrt[3]{z^2+1}}$.

Solution We can use the substitution method of integration as an exploratory tool: Substitute for the most troublesome part of the integrand and see how things work out. For the integral here, we might try $u = z^2 + 1$ or we might even press our luck and take u to be the entire cube root. Here is what happens in each case.

SOLUTION 1: Substitute $u = z^2 + 1$.

$$\int \frac{2z\,dz}{\sqrt[3]{z^2+1}} = \int \frac{du}{u^{1/3}} \qquad (\text{Substitute } u = z^2 + 1,\ du = 2z\,dz.)$$

$$= \int u^{-1/3}\,du \qquad (\text{In the form } \textstyle\int u^n\,du.)$$

$$= \frac{u^{2/3}}{2/3} + C \qquad (\text{Integrate with respect to } u.)$$

$$= \frac{3}{2}u^{2/3} + C$$

$$= \frac{3}{2}(z^2 + 1)^{2/3} + C \qquad (\text{Replace } u \text{ by } z^2 + 1.)$$

SOLUTION 2: Substitute $u = \sqrt[3]{z^2+1}$ instead.

$$\int \frac{2z\,dz}{\sqrt[3]{z^2+1}} = \int \frac{3u^2\,du}{u} \qquad \left(\begin{array}{l}\text{Substitute } u = \sqrt[3]{z^2+1},\\ u^3 = z^2+1,\ 3u^2\,du = 2z\,dz.\end{array}\right)$$

$$= 3\int u\,du$$

$$= 3\cdot\frac{u^2}{2} + C \qquad \text{(Integrate with respect to } u.\text{)}$$

$$= \frac{3}{2}(z^2+1)^{2/3} + C \qquad \text{(Replace } u \text{ by } (z^2+1)^{1/3}.\text{)}$$

■

Substitution in Definite Integrals

The formula for evaluating definite integrals by substitution first appeared in a book by Isaac Barrow (1630–1677), Newton's mathematics teacher at Cambridge University. It looks like this:

Substitution In Definite Integrals

THE FORMULA	HOW TO USE IT	
$\int_a^b f(g(x))\cdot g'(x)\,dx = \int_{g(a)}^{g(b)} f(u)\,du$	Substitute $u = g(x)$, $du = g'(x)\,dx$, and integrate from $g(a)$ to $g(b)$.	(13)

To use the formula, make the same u-substitution you would use to evaluate the corresponding indefinite integral. Then integrate with respect to u from the value u has at $x = a$ to the value u has at $x = b$.

EXAMPLE 10

$$\int_0^{\pi/4} \tan x \sec^2 x\,dx = \int_0^1 u\,du \qquad \left(\begin{array}{l}\text{Substitute } u = \tan x,\ du = \sec^2 x\,dx,\\ \text{and integrate from } \tan 0 = 0\\ \text{to } \tan\frac{\pi}{4} = 1.\end{array}\right)$$

$$= \left.\frac{u^2}{2}\right]_0^1 \qquad \text{(Evaluate the definite integal.)}$$

$$= \frac{(1)^2}{2} - \frac{(0)^2}{2}$$

$$= \frac{1}{2}$$

■

We do not have to use Eq. (13) if we do not want to. We can always transform the integral as an indefinite integral, integrate, change back to x, and use the original x limits. In the next example we evaluate a definite integral both ways—with Eq. (13) and without.

EXAMPLE 11 Evaluate $\int_{-1}^{1} 3x^2\sqrt{x^3+1}\,dx$.

Solution We have two choices here:

METHOD 1: Transform the integral and evaluate it with transformed limits.

$$\int_{-1}^{1} 3x^2\sqrt{x^3+1}\,dx = \int_0^2 \sqrt{u}\,du \qquad \begin{pmatrix}\text{Substitute } u = x^3+1,\\ du = 3x^2\,dx \text{ and integrate}\\ \text{from } u(-1)=0 \text{ to } u(1)=2.\end{pmatrix}$$

$$= \frac{2}{3}u^{3/2}\Big]_0^2 \qquad \text{(Integrate.)}$$

$$= \frac{2}{3}\left[2^{3/2} - 0^{3/2}\right] = \frac{2}{3}[2\sqrt{2}] = \frac{4\sqrt{2}}{3}$$

METHOD 2: Transform the integral as an indefinite integral, integrate, change back to x, and use the original x-limits.

$$\int 3x^2\sqrt{x^3+1}\,dx = \int \sqrt{u}\,du \qquad \begin{pmatrix}\text{Substitute } u = x^3+1,\\ du = 3x^2\,dx.\end{pmatrix}$$

$$= \frac{2}{3}u^{3/2} + C \qquad \text{(Integrate with respect to } u\text{.)}$$

$$= \frac{2}{3}(x^3+1)^{3/2} + C \qquad \text{(Replace } u \text{ by } x^3+1\text{.)}$$

$$\int_{-1}^{1} 3x^2\sqrt{x^3+1}\,dx = \frac{2}{3}(x^3+1)^{3/2}\Big]_{-1}^{1} \qquad \begin{pmatrix}\text{Using the indefinite}\\ \text{integral just found}\end{pmatrix}$$

$$= \frac{2}{3}\left[((1)^3+1)^{3/2} - ((-1)^3+1)^{3/2}\right]$$

$$= \frac{2}{3}\left[2^{3/2} - 0^{3/2}\right] = \frac{2}{3}[2\sqrt{2}] = \frac{4\sqrt{2}}{3}$$

■

Which method is better—evaluating the transformed integral with transformed limits or transforming back to use the original limits of integration? In Example 11 the first method seems easier, but that is not always the case. As a rule, it is best to know both methods and use whichever one seems better at the time.

Exercises 5.6

Evaluate the indefinite integrals in Exercises 1–10 by using the given substitutions to reduce the integrals to standard form.

1. $\displaystyle\int \sin 3x\,dx,\; u = 3x$

2. $\displaystyle\int x\sin(2x^2)\,dx,\; u = 2x^2$

3. $\displaystyle\int \sec 2x \tan 2x\,dx,\; u = 2x$

4. $\displaystyle\int \left(1-\cos\frac{t}{2}\right)^2 \sin\frac{t}{2}\,dt,\; u = 1-\cos\frac{t}{2}$

5. $\displaystyle\int 28(7x-2)^3\,dx,\; u = 7x-2$

6. $\displaystyle\int 4x^3(x^4-1)^2\,dx,\; u = x^4-1$

7. $\displaystyle\int \frac{9r^2\,dr}{\sqrt{1-r^3}},\; u = 1-r^3$

8. $\int 12(y^4 + 4y^2 + 1)^2(y^3 + 2y)\,dy,\; u = y^4 + 4y^2 + 1$

9. $\int \csc^2 2\theta \cot 2\theta\, d\theta$,
   a) Using $u = \cot 2\theta$ b) Using $u = \csc 2\theta$

10. $\int \frac{dx}{\sqrt{5x}}$,
   a) Using $u = 5x$ b) Using $u = \sqrt{5x}$

Evaluate the definite integrals in Exercises 11–18 by using the given substitutions.

11. $\int_0^{1/2} \frac{dx}{(2x+1)^3},\; u = 2x + 1$

12. $\int_0^1 \sqrt{5x+4}\,dx,\; u = 5x + 4$

13. $\int_0^{\pi/6} \frac{\sin 2x}{\cos^2 2x}\,dx,\; u = \cos 2x$

14. $\int_{\pi/6}^{\pi/2} \sin^2\theta \cos\theta\, d\theta,\; u = \sin\theta$

15. $\int_{-1}^{1} x\sqrt{1-x^2}\,dx,\; u = 1 - x^2$

16. $\int_1^4 \frac{dy}{2\sqrt{y}(1+\sqrt{y})},\; u = 1 + \sqrt{y}$

17. $\int_{-\pi/2}^{\pi/2} \frac{\cos x}{(2+\sin x)^2}\,dx,\; u = 2 + \sin x$

18. $\int_{\pi^2/4}^{\pi^2} \frac{\sin\sqrt{x}}{\sqrt{x}}\,dx,\; u = \sqrt{x}$

Evaluate the indefinite integrals in Exercises 19–28.

19. $\int \frac{dx}{(1-x)^2}$ 20. $\int \frac{4y}{\sqrt{2y^2+1}}\,dy$

21. $\int \sec^2(x+2)\,dx$ 22. $\int \sec^2\left(\frac{x}{4}\right)dx$

23. $\int 8r(r^2-1)^{1/3}\,dr$ 24. $\int x^4(7-x^5)^3\,dx$

25. $\int \sec\left(\theta + \frac{\pi}{2}\right)\tan\left(\theta + \frac{\pi}{2}\right)d\theta$

26. $\int \sqrt{\tan x}\,\sec^2 x\,dx$ 27. $\int \frac{6x^3}{\sqrt[4]{1+x^4}}\,dx$

28. $\int (s^3 + 2s^2 - 5s + 6)^2(3s^2 + 4s - 5)\,ds$

Evaluate the definite integrals in Exercises 29–48.

29. a) $\int_0^3 \sqrt{y+1}\,dy$ b) $\int_{-1}^0 \sqrt{y+1}\,dy$

30. a) $\int_0^1 r\sqrt{1-r^2}\,dr$ b) $\int_{-1}^1 r\sqrt{1-r^2}\,dr$

31. a) $\int_0^{\pi/4} \tan x \sec^2 x\,dx$ b) $\int_{-\pi/4}^0 \tan x \sec^2 x\,dx$

32. a) $\int_0^1 x^3(1+x^4)^3\,dx$ b) $\int_{-1}^1 x^3(1+x^4)^3\,dx$

33. a) $\int_0^1 \frac{x^3}{\sqrt{x^4+9}}\,dx$ b) $\int_{-1}^0 \frac{x^3}{\sqrt{x^4+9}}\,dx$

34. a) $\int_{-1}^1 \frac{x}{(1+x^2)^2}\,dx$ b) $\int_0^1 \frac{x}{(1+x^2)^2}\,dx$

35. a) $\int_0^{\sqrt{7}} x(x^2+1)^{1/3}\,dx$ b) $\int_{-\sqrt{7}}^0 x(x^2+1)^{1/3}\,dx$

36. a) $\int_0^{\pi} 3\cos^2 x \sin x\,dx$ b) $\int_{2\pi}^{3\pi} 3\cos^2 x \sin x\,dx$

37. a) $\int_0^{\pi/6} (1 - \cos 3x)\sin 3x\,dx$

   b) $\int_{\pi/6}^{\pi/3} (1 - \cos 3x)\sin 3x\,dx$

38. a) $\int_0^{\sqrt{3}} \frac{4x}{\sqrt{x^2+1}}\,dx$ b) $\int_{-\sqrt{3}}^{\sqrt{3}} \frac{4x}{\sqrt{x^2+1}}\,dx$

39. a) $\int_0^{2\pi} \frac{\cos x}{\sqrt{2+\sin x}}\,dx$ b) $\int_{-\pi}^{\pi} \frac{\cos x}{\sqrt{2+\sin x}}\,dx$

40. a) $\int_{-\pi/2}^0 \frac{\sin x}{(3+\cos x)^2}\,dx$ b) $\int_0^{\pi/2} \frac{\sin x}{(3+\cos x)^2}\,dx$

41. $\int_0^1 \sqrt{t^5+2t}\,(5t^4+2)\,dt$ 42. $\int_1^4 \frac{dy}{2\sqrt{y}(1+\sqrt{y})^2}$

43. $\int_0^{\pi/2} \cos^3 2x \sin 2x\,dx$ 44. $\int_{-\pi/4}^{\pi/4} \tan^2 x \sec^2 x\,dx$

45. $\int_0^{\pi} \frac{8\sin t}{\sqrt{5-4\cos t}}\,dt$

46. $\int_0^{\pi/4} (1 - \sin 2t)^{3/2}\cos 2t\,dt$

47. $\int_0^1 15x^2\sqrt{5x^3+4}\,dx$

48. $\int_0^1 (y^3 + 6y^2 - 12y + 5)(y^2 + 4y - 4)\, dy$

Find the total areas of the shaded regions in Exercises 49 and 50.

49. 50.

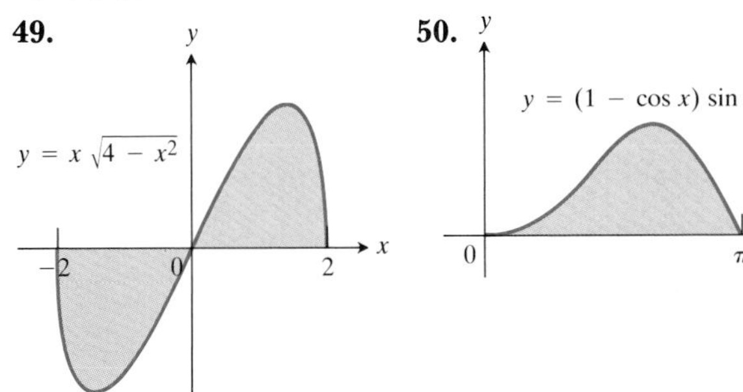

Solve the initial value problems in Exercises 51–54.

51. $\dfrac{ds}{dt} = 24t(3t^2 - 1)^3,\ s = 0$ when $t = 0$

52. $\dfrac{dy}{dx} = 4x(x^2 + 8)^{-1/3},\ y = 0$ when $x = 0$

53. $\dfrac{ds}{dt} = 6\sin(t + \pi),\ s = 0$ when $t = 0$

54. $\dfrac{d^2s}{dt^2} = -4\sin\left(2t - \dfrac{\pi}{2}\right),\ \dfrac{ds}{dt} = 100$ and $s = 0$ when $t = 0$

Sequences of Substitutions

If you do not know what substitution to make, try reducing the integral step by step, using a trial substitution to simplify the integral a bit, then another to simplify it some more. You will see what we mean if you try the sequences of substitutions in Exercises 55 and 56.

55. $\displaystyle\int_0^{\pi/4} \frac{18\tan^2 x\sec^2 x}{(2 + \tan^3 x)^2}\, dx$

a) $u = \tan x$, followed by $v = u^3$, then by $w = 2 + v$
b) $u = \tan^3 x$ followed by $v = 2 + u$
c) $u = 2 + \tan^3 x$

56. $\displaystyle\int \sqrt{1 + \sin^2(x - 1)}\,\sin(x - 1)\cos(x - 1)\, dx$

a) $u = x - 1$, followed by $v = \sin u$, then by $w = 1 + v^2$
b) $u = \sin(x - 1)$ followed by $v = 1 + u^2$
c) $u = 1 + \sin^2(x - 1)$

57. It looks as if we can integrate $2\sin x\cos x$ with respect to x in three different ways:

a) $$\int 2\sin x\cos x\, dx = \int 2u\, du \qquad \begin{pmatrix} u = \sin x, \\ du = \cos x\, dx \end{pmatrix}$$
$$= u^2 + C_1 = \sin^2 x + C_1;$$

b) $$\int 2\sin x\cos x\, dx = \int -2u\, du \qquad \begin{pmatrix} u = \cos x, \\ du = -\sin x\, dx, \\ -du = \sin x\, dx \end{pmatrix}$$
$$= -u^2 + C_2 = -\cos^2 x + C_2;$$

c) $$\int 2\sin x\cos x\, dx = \int \sin 2x\, dx \qquad \begin{pmatrix} 2\sin x\cos x \\ = \sin 2x \end{pmatrix}$$
$$= -\frac{\cos 2x}{2} + C_3.$$

Can all three integrations be correct? Explain.

5.7 Numerical Integration

As we have seen, the ideal way to evaluate a definite integral

$$\int_a^b f(x)\, dx$$

is to find a formula $F(x)$ for one of the antiderivatives of $f(x)$ and calculate the number $F(b) - F(a)$. But some antiderivatives are hard to find and still others, like the antiderivatives of $(\sin x)/x$ and $\sqrt{1 + x^4}$, have no elementary formulas. We do not mean merely that no one has yet succeeded in finding simple expressions for evaluating the antiderivatives of $(\sin x)/x$ and $\sqrt{1 + x^4}$. We mean it has been proved that no such expressions exist.

Whatever the reason, when we cannot evaluate a definite integral with an antiderivative, we turn to numerical methods like the trapezoidal rule and Simpson's rule described in this section. These rules enable us to estimate an integral's value to as many decimal places as we please whenever we want. They also enable us to calculate integrals with reasonable accuracy from numerical tables of function values. This comes in handy when the only information we have about a function is a set of specific values measured in the laboratory or in the field.

The Trapezoidal Rule

The trapezoidal rule for estimating the value of

$$\int_a^b f(x)\,dx$$

is based on approximating the region between the graph of f and the x-axis with n trapezoids of equal width (Fig. 5.16.) The trapezoids have the common base length $h = (b - a)/n$, and the side of each trapezoid runs from the x-axis up (or down) to the curve. The areas of the trapezoids add up to

$$\begin{aligned} T &= \frac{1}{2}(y_0 + y_1)h + \frac{1}{2}(y_1 + y_2)h + \cdots \\ &\quad + \frac{1}{2}(y_{n-2} + y_{n-1})h + \frac{1}{2}(y_{n-1} + y_n)h \\ &= h\left(\frac{1}{2}y_0 + y_1 + y_2 + \cdots + y_{n-1} + \frac{1}{2}y_n\right) \qquad (1) \\ &= \frac{h}{2}(y_0 + 2y_1 + 2y_2 + \cdots + 2y_{n-1} + y_n), \end{aligned}$$

where

$$y_0 = f(a),\quad y_1 = f(x_1),\quad \ldots,\quad y_{n-1} = f(x_{n-1}),\quad y_n = f(b).$$

The trapezoidal rule says: Use T to estimate the integral of f from a to b.

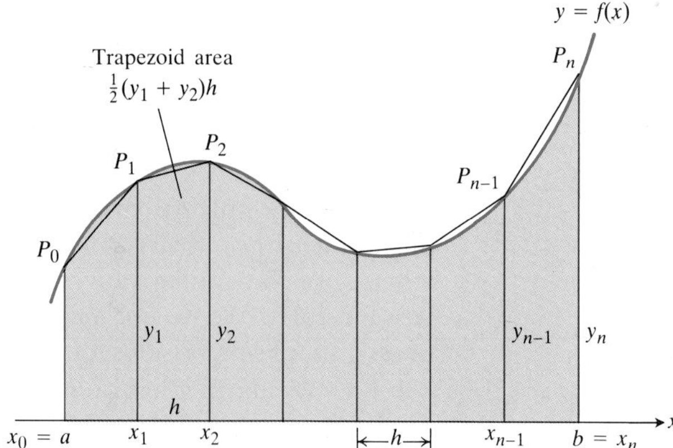

5.16 The trapezoidal rule approximates short stretches of curve with line segments. To estimate the shaded area, we add the areas of the trapezoids made by joining the ends of these segments to the x-axis.

The Trapezoidal Rule

To approximate
$$\int_a^b f(x)\,dx,$$
use
$$T = \frac{h}{2}(y_0 + 2y_1 + 2y_2 + \cdots + 2y_{n-1} + y_n) \tag{2}$$
(n subintervals of length $h = (b - a)/n$).

TABLE 5.4

x	$y = x^2$
1	1
$\frac{5}{4}$	$\frac{25}{16}$
$\frac{6}{4}$	$\frac{36}{16}$
$\frac{7}{4}$	$\frac{49}{16}$
2	4

EXAMPLE 1 Use the trapezoidal rule with $n = 4$ to estimate
$$\int_1^2 x^2\,dx.$$
Compare the estimate with the exact value of the integral.

Solution The exact value of the integral is
$$\int_1^2 x^2\,dx = \frac{x^3}{3}\bigg]_1^2 = \frac{8}{3} - \frac{1}{3} = \frac{7}{3}.$$

To find the trapezoidal approximation, we divide the interval of integration into four subintervals of equal length and list the values of $y = x^2$ at the endpoints and subdivision points (see Table 5.4). We then evaluate Eq. (2) with $n = 4$ and $h = 1/4$:
$$\begin{aligned} T &= \frac{h}{2}(y_0 + 2y_1 + 2y_2 + 2y_3 + y_4) \\ &= \frac{1}{8}\left(1 + 2\left(\frac{25}{16}\right) + 2\left(\frac{36}{16}\right) + 2\left(\frac{49}{16}\right) + 4\right) = \frac{75}{32} = 2.34375. \end{aligned}$$

The approximation overestimates the area by about half a percent of its true value. Each trapezoid contains slightly more than the corresponding strip under the curve (Fig. 5.17). ■

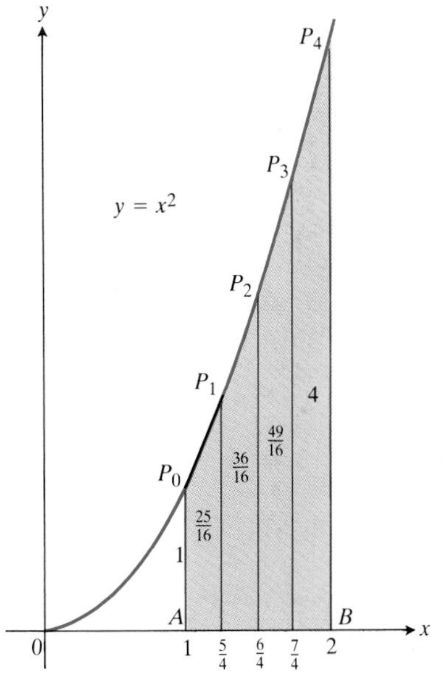

5.17 The trapezoidal approximation of the area under the graph of $y = x^2$ from $x = 1$ to $x = 2$ is a slight overestimate.

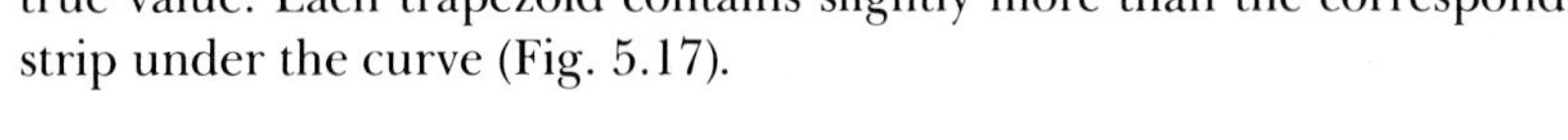

Controlling the Error in the Trapezoidal Approximation

Pictures suggest that the error
$$E_T = T - \int_a^b f(x)\,dx \tag{3}$$
in the trapezoidal approximation will go down as the **step size** h decreases, because the trapezoids fit the curve better as their number increases. A theorem from advanced calculus assures us that this will always be the case if f has a continuous second derivative.

The Error Estimate for the Trapezoidal Rule

If f'' is continuous and M is any upper bound for the values of $|f''|$ on $[a, b]$, then
$$|E_T| \le \frac{b-a}{12}h^2 M. \tag{4}$$

Although theory tells us there will always be a smallest safe value of M, in practice we can hardly ever find it. Instead, we find the best value we can and go on to estimate $|E_T|$ from there. This may seem sloppy, but it works. To make $|E_T|$ small for a given M, we just make h small.

EXAMPLE 2 Find an upper bound for the error in the approximation found in Example 1 for the value of

$$\int_1^2 x^2\,dx.$$

Solution We first find an upper bound M for the magnitude of the second derivative of $f(x) = x^2$ on the interval $1 \le x \le 2$. Since $f''(x) = 2$ for all x, we may safely take $M = 2$. With $b - a = 1$ and $h = 1/4$, Eq. (4) gives

$$|E_T| \le \frac{b-a}{12}h^2M = \frac{1}{12}\left(\frac{1}{4}\right)^2(2) = \frac{1}{96}.$$

This is precisely what we find when we subtract $T = 75/32$ from $\int_1^2 x^2\,dx = 7/3$, since $7/3 - 75/32 = -1/96$. Here we are able to give the error *exactly*, but this is exceptional. ■

EXAMPLE 3 The trapezoidal rule is used to estimate the value of

$$\int_0^1 x \sin x\,dx$$

when $n = 10$ steps. Find an upper bound for the error in the estimate.

Solution We use the formula

$$|E_T| \le \frac{b-a}{12}h^2M$$

with $b = 1$, $a = 0$, and $h = 1/n = 1/10$. This gives

$$|E_T| \le \frac{1}{12}\left(\frac{1}{10}\right)^2 M = \frac{1}{1200}M.$$

The number M can be any upper bound for the values of $|f''|$ on $[0, 1]$. To choose a value for M, we calculate f'' to see how big it might be. A straightforward differentiation gives

$$f'' = \cos x + (1 - x)\sin x.$$

Hence, by the triangle inequality,

$$|f''| \le |\cos x| + |1 - x||\sin x| \le 1 + (1)(1) = 2$$

because $0 \le x \le 1$ and $|\cos x|$ and $|\sin x|$ never exceed 1. We can safely take $M = 2$. Therefore

$$|E_T| \le \frac{1}{1200}(2) = \frac{1}{600} < 0.00167.$$

The error is no greater than 1.67×10^{-3}.

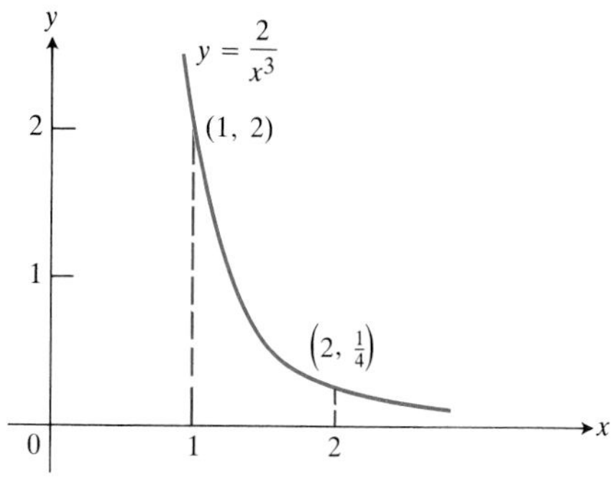

5.18 The continuous function $y = 2/x^3$ has its maximum value on [1, 2] at $x = 1$.

For greater accuracy we would not try to improve M but would take more steps. With $n = 100$ steps, for example, $h = 1/100$ and

$$|E_T| \leq \frac{1}{12}\left(\frac{1}{100}\right)^2 (2) < 1.67 \times 10^{-5}.$$

EXAMPLE 4 As we shall see in Section 5.8, the value of ln 2 ("log two"), the natural logarithm of 2, is given by the integral

$$\ln 2 = \int_1^2 \frac{1}{x}\,dx.$$

How many subintervals (steps) should be used in the trapezoidal rule to approximate the integral, and hence the value of ln 2, with an error of absolute value less than 10^{-4}?

Solution To determine n, the number of subintervals, we use Eq. (4) with

$$b - a = 2 - 1 = 1, \qquad h = \frac{b-a}{n} = \frac{1}{n},$$

$$f''(x) = \frac{d^2}{dx^2}(x^{-1}) = 2x^{-3} = \frac{2}{x^3}.$$

Then

$$|E_T| \leq \frac{b-a}{12} h^2 \max|f''(x)| = \frac{1}{12}\left(\frac{1}{n}\right)^2 \max\left|\frac{2}{x^3}\right|,$$

where max refers to [1, 2].

This is one of the rare cases where we can find the exact value of $\max|f''|$. On [1, 2], $y = 2/x^3$ decreases steadily from a maximum of $y = 2$ to a minimum of $y = 1/4$ (Fig. 5.18). Therefore

$$|E_T| \leq \frac{1}{12}\left(\frac{1}{n}\right)^2 \cdot 2 = \frac{1}{6n^2}.$$

The error's absolute value will therefore be less than 10^{-4} if

$$\frac{1}{6n^2} < 10^{-4}, \qquad \frac{10^4}{6} < n^2, \qquad \frac{100}{\sqrt{6}} < n, \qquad \text{or} \qquad 40.83 < n.$$

The first integer beyond 40.83 is $n = 41$. With $n = 41$ subdivisions we can guarantee calculating ln 2 with an error of magnitude less than 10^{-4}. Any larger n will work, too.

Simpson's Rule

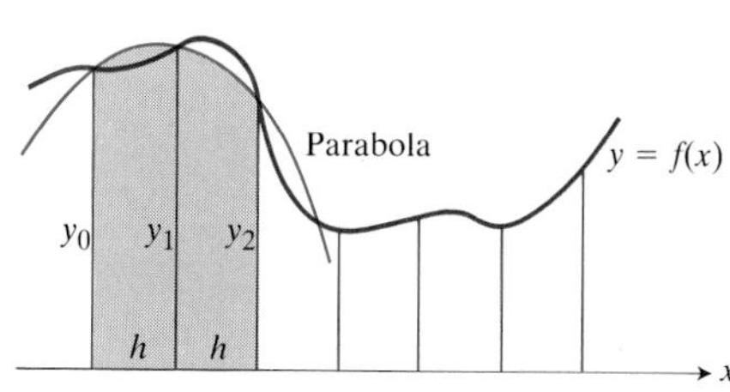

5.19 Simpson's rule approximates short stretches of curve with parabolas.

Simpson's rule is based on approximating curves with parabolas instead of line segments. The shaded area under the parabola in Fig. 5.19 is

$$A_p = \frac{h}{3}(y_0 + 4y_1 + y_2). \tag{5}$$

Applying this formula successively along a continuous curve $y = f(x)$ from $x = a$ to $x = b$ leads to an estimate of $\int_a^b f(x)\,dx$ that is generally more accurate than T for a given step size h.

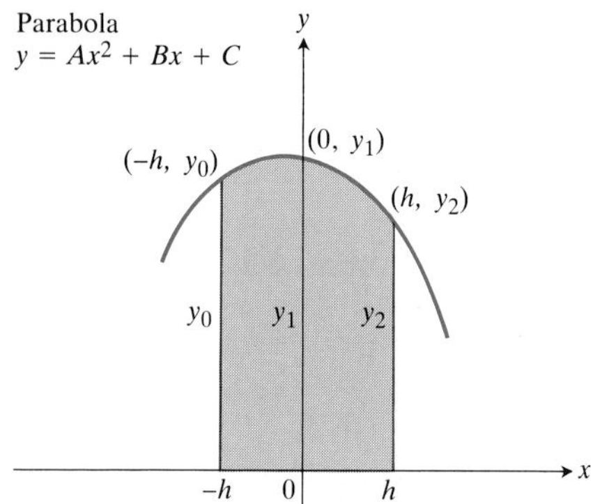

5.20 By integrating from $-h$ to h, the shaded area is found to be

$$A_p = \frac{h}{3}(y_0 + 4y_1 + y_2).$$

We can derive the formula for A_p in the following way. To simplify the algebra, we use the coordinate system shown in Fig. 5.20. The area under the parabola is the same no matter where the y-axis is, as long as we preserve the vertical scale. The parabola has an equation of the form

$$y = Ax^2 + Bx + C,$$

so the area under it from $x = -h$ to $x = h$ is

$$\begin{aligned} A_p &= \int_{-h}^{h} (Ax^2 + Bx + C)\,dx = \frac{Ax^3}{3} + \frac{Bx^2}{2} + Cx\bigg]_{-h}^{h} \\ &= \frac{2Ah^3}{3} + 2Ch = \frac{h}{3}(2Ah^2 + 6C). \end{aligned} \tag{6}$$

Since the curve passes through the three points $(-h, y_0)$, $(0, y_1)$, and (h, y_2), we also have

$$y_0 = Ah^2 - Bh + C, \qquad y_1 = C, \qquad y_2 = Ah^2 + Bh + C,$$

from which we obtain

$$\begin{aligned} C &= y_1, \\ Ah^2 - Bh &= y_0 - y_1, \\ Ah^2 + Bh &= y_2 - y_1, \\ 2Ah^2 &= y_0 + y_2 - 2y_1. \end{aligned} \tag{7}$$

Hence expressing the area A_p in terms of y_0, y_1, and y_2, we have

$$A_p = \frac{h}{3}(2Ah^2 + 6C) = \frac{h}{3}((y_0 + y_2 - 2y_1) + 6y_1) = \frac{h}{3}(y_0 + 4y_1 + y_2). \tag{8}$$

Simpson's rule follows from applying the formula for A_p to successive pieces of the curve $y = f(x)$ between $x = a$ and $x = b$. Each separate piece of the curve, covering an x-subinterval of width $2h$, is approximated by an arc of a parabola through its ends and midpoint. The areas under the parabolic arcs are then added to give Simpson's rule.

Simpson's Rule

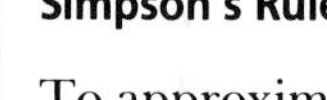

To approximate $$\int_a^b f(x)\,dx,$$

use $$S = \frac{h}{3}(y_0 + 4y_1 + 2y_2 + 4y_3 + \cdots + 2y_{n-2} + 4y_{n-1} + y_n) \tag{9}$$

(n even, $h = (b - a)/n$).

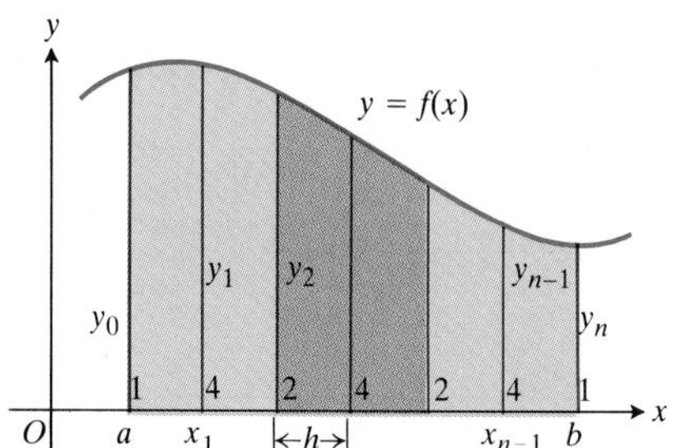

5.21 The y's in Eq. (9) are the values of f at the points of subdivision.

The y's in Eq. (9) are the values of $y = f(x)$ at the points

$$a, \quad x_1 = a + h, \quad x_2 = a + 2h, \quad \ldots, \quad x_{n-1} = a + (n-1)h, \quad b$$

that subdivide $[a, b]$ into n equal subintervals of length $h = (b - a)/n$ (Fig. 5.21). The number n must be even to apply the rule because each parabolic arc uses two subintervals.

Simpson's One-Third Rule

The rule

$$\text{Area} = \frac{h}{3}(y_0 + 4y_1 + y_2)$$

for calculating area by replacing curves by parabolas, Eq. (5) in the text, was discovered long before Thomas Simpson (1720–1761) was born. It is another of history's beautiful quirks that one of the ablest mathematicians of eighteenth-century England is remembered not for his successful texts and his contributions to mathematical analysis but for a rule that was never his, that he never laid claim to, and that bears his name only because he happened to mention it in one of his books.

TABLE 5.5

x	$y = 5x^4$
0	0
$\frac{1}{4}$	$\frac{5}{256}$
$\frac{2}{4}$	$\frac{80}{256}$
$\frac{3}{4}$	$\frac{405}{256}$
1	5

Calculus and Computers

Here is another example of calculus having something important to say about computation. It is easy to implement the trapezoidal rule and Simpson's rule on a computer. But that in itself is not enough. We need to know how many steps to take to achieve the accuracy we want, and the guidance for *that* comes from calculus.

Error Control for Simpson's Rule

The Simpson's rule error,

$$E_S = S - \int_a^b f(x)\,dx, \qquad (10)$$

decreases with the step size too. The inequality for controlling it, however, assumes f to have a continuous fourth derivative instead of merely a continuous second derivative. The formula, once again from advanced calculus, is this:

The Error Estimate for Simpson's Rule

If $f^{(4)}$ is continuous and M is any upper bound for the values of $|f^{(4)}|$ on $[a, b]$, then

$$|E_S| \le \frac{b-a}{180} h^4 M. \qquad (11)$$

As with the trapezoidal rule, we can almost never find the smallest possible value of M. We just find the best value we can and go on from there to estimate $|E_S|$.

EXAMPLE 5 Use Simpson's rule with $n = 4$ to approximate

$$\int_0^1 5x^4\,dx.$$

What estimate does Eq. (11) give for the error in the approximation?

Solution Again we have chosen an integral whose exact value we can calculate directly:

$$\int_0^1 5x^4\,dx = x^5\Big]_0^1 = 1.$$

To find the Simpson approximation, we partition the interval of integration into four subintervals and evaluate $f(x) = 5x^4$ at the partition points (Table 5.5). We then evaluate Eq. (9) with $n = 4$ and $h = 1/4$:

$$S = \frac{h}{3}(y_0 + 4y_1 + 2y_2 + 4y_3 + y_4)$$

$$= \frac{1}{12}\left(0 + 4\left(\frac{5}{256}\right) + 2\left(\frac{80}{256}\right) + 4\left(\frac{405}{256}\right) + 5\right) = 1.00260. \qquad \text{(Rounded)}$$

To estimate the error, we first find an upper bound M for the magnitude of the fourth derivative of $f(x) = 5x^4$ on the interval $0 \le x \le 1$. Since the fourth derivative has the constant value $f^{(4)}(x) = 120$, we may safely take $M = 120$. With $(b - a) = 1$ and $h = 1/4$, Eq. (11) then gives

$$|E_S| \le \frac{b-a}{180} h^4 M = \frac{1}{180}\left(\frac{1}{4}\right)^4 (120) = \frac{1}{384} < 0.00261.$$ ■

Which Rule Gives Better Results?

The answer lies in the error-control formulas for the two rules:

$$|E_T| \le \frac{b-a}{12} h^2 M, \qquad |E_S| \le \frac{b-a}{180} h^4 M. \tag{12}$$

The M's of course mean different things, the first being an upper bound on $|f''|$ and the second an upper bound on $|f^{(4)}|$. But there is more than that going on here. The factor $(b-a)/180$ in the Simpson formula is one-fifteenth of the factor $(b-a)/12$ in the trapezoidal formula. More important still, the Simpson formula has an h^4 while the trapezoidal formula has only an h^2. If h is one-tenth, then h^2 is a hundredth but h^4 is only a ten-thousandth. If both M's are 1 for example, and $b - a = 1$, then, with $h = 1/10$,

$$|E_T| \le \frac{1}{12}\left(\frac{1}{10}\right)^2 \cdot 1 \le \frac{1}{1200}, \tag{13}$$

while

$$|E_S| \le \frac{1}{180}\left(\frac{1}{10}\right)^4 \cdot 1 \le \frac{1}{1{,}800{,}000} = \frac{1}{1500} \cdot \frac{1}{1200}. \tag{14}$$

For roughly the same amount of computational effort, we get better accuracy with Simpson's rule—at least in this case.

The h^2 versus h^4 is the key. If h is less than 1, then h^4 can be significantly smaller than h^2. On the other hand, if h equals 1, there is no difference between h^2 and h^4. If h is greater than 1, the value of h^4 may be significantly larger than the value of h^2. In the latter two cases, the error-control formulas offer little help. We have to go back to the geometry of the curve $y = f(x)$ to see whether trapezoids or parabolas, if either, are going to give a better fit.

Working with Numerical Data

The next example shows how to use Simpson's rule to estimate the integral of a function from values measured in the laboratory or in the field even when we have no formula for the function. The trapezoidal rule can be used the same way.

EXAMPLE 6 A town wants to drain and fill the small swamp shown in Fig. 5.22. The swamp averages 5 ft deep. About how many cubic yards of dirt will it take to fill the area after the swamp is drained?

Solution To calculate the volume of the swamp, we estimate the surface area and multiply by 5. To estimate the area, we use Simpson's rule with $h = 20$ ft and the y's equal to the distances measured across the swamp, as shown in Fig. 5.22:

$$\begin{aligned} S &= \frac{h}{3}(y_0 + 4y_1 + 2y_2 + 4y_3 + 2y_4 + 4y_5 + y_6) \\ &= \frac{20}{3}(146 + 488 + 152 + 216 + 80 + 120 + 13) = 8100. \end{aligned}$$

The volume is about $(8100)(5) = 40{,}500$ ft^3 or 1500 yd^3. ■

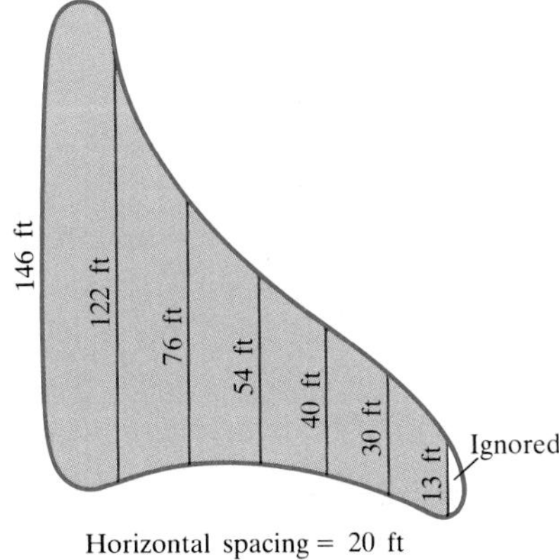

5.22 The swamp in Example 6.

Round-off Errors

Although decreasing the step size h reduces the error in the Simpson and trapezoidal approximations in theory, it may fail to do so in practice. When h is very small, say $h = 10^{-5}$, the round-off errors in the arithmetic required to evaluate S and T may accumulate to such an extent that the error formulas no longer describe what is going on. Shrinking h below a certain size can actually make things worse rather than better. While this will not be an issue in the present book, you should consult a text on numerical analysis for alternative methods if you are having problems with round-off.

Computer Programs

Here are computer programs for approximating the value of

$$\int_a^b f(x)\,dx.$$

We begin with a BASIC program for the trapezoidal approximation, then modify it for Simpson's rule.

PROGRAM	COMMENT
10 DEF FNF (X) = F(X)	Key in the formula for $f(x)$. Use lines 10 to 39 as needed.
40 INPUT A	Lower limit of integration
50 INPUT B	Upper limit of integration
60 INPUT C	Number of data we want printed
70 FOR M = 1 TO C	Each value of M gives one trapezoidal approximation.
80 LET N = 10*M	Number of subintervals in Mth approximation
90 LET S = 0	Begin at 0 for each new M.
100 LET H = (B − A)/N	Width of each subinterval
110 FOR X=A TO B−H STEP H	The left-hand endpoints of the n subintervals (CHANGE for Simpson's rule)
120 LET S=S+(FNF(X) + FNF(X+H))/2	Cumulative trapezoidal sum (CHANGE for Simpson's rule)
130 NEXT X	Returns to line 110 for next step
140 PRINT N; H*S	
150 NEXT M	Returns to line 70 with new M.
160 END	

Table 5.6 shows the results for $f(x) = 1/x$, $a = 1$, $b = 2$, and $c = 5$. The listed numbers are the trapezoidal approximations to

$$\ln 2 = \int_1^2 \frac{1}{x}\,dx$$

TABLE 5.6
Trapezoidal approximations of $\ln 2 = \int_1^2 (1/x)\,dx$

n	$T(n)$	|Error| less than. . .
10	.6937713	.0006242
20	.6933035	.0001564
30	.6932168	.0000697
40	.6931864	.0000393
50	.6931723	.0000252

For comparison: $\ln 2 = 0.6931471805\ldots$

TABLE 5.7
Simpson's rule approximations of ln 2 = $\int_1^2 (1/x)\,dx$

n	$S(n)$	|Error| less than . . .
10	.6931502	.0000031
20	.6931476	.0000005
30	.6931475	.0000004
40	.6931474	.0000003
50	.6931473	.0000002

For comparison: ln 2 = 0.6931471805 . . .

for $n = 10, 20, 30, 40$, and 50. Notice that with $n = 30$ we already have the accuracy required in Example 4. The error-control formula in Eq. (4) always gives a safe "worst-case" value; the results in practice are often better.

To change the program for trapezoidal approximations to Simpson's rule, do the following:

REVISED COMMANDS:	COMMENT
110 FOR X=A TO B−2*H STEP 2*H	We now group the subintervals in pairs.
120 LET S=S+(FNF(X)+ 4*FNF(X + H) + FNF(X + 2*H))/3	Cumulative Simpson sum

Table 5.7 shows the result of using the Simpson program with $f(x) = 1/x$, $a = 1$, $b = 2$ and $c = 5$ to approximate $\ln 2 = \int_1^2 (1/x)dx$ with $n = 10$, 20, 30, 40, and 50 subintervals. Notice the improvement over the trapezoidal rule: The approximation for $n = 50$ rounds accurately to six places instead of three.

Exercises 5.7

Use (a) the trapezoidal rule and (b) Simpson's rule to approximate the integrals in Exercises 1–6 with $n = 4$. Then (c) find the integral's exact value for comparison.

1. $\int_0^2 x\,dx$ **2.** $\int_0^2 x^2\,dx$ **3.** $\int_0^2 x^3\,dx$

4. $\int_1^2 \frac{1}{x^2}\,dx$ **5.** $\int_0^4 \sqrt{x}\,dx$ **6.** $\int_0^{\pi} \sin x\,dx$

7. Use Eq. (4) to estimate the error in using the trapezoidal rule with $n = 10$ to estimate the value of

$$\ln 2 = \int_1^2 \frac{1}{x}\,dx.$$

Compare your answer with the result in Table 5.6.

8. Use Eq. (11) to estimate the error in using Simpson's rule with $n = 10$ to estimate the value of

$$\ln 2 = \int_1^2 \frac{1}{x}\,dx.$$

Compare your answer with the result in Table 5.7.

In Exercises 9–14, estimate the minimum number of subdivisions needed to approximate the integrals with an error of absolute value less than 10^{-4} by (a) the trapezoidal rule and (b) Simpson's rule.

9. $\int_0^2 x\,dx$ **10.** $\int_0^2 x^2\,dx$ **11.** $\int_0^2 x^3\,dx$

12. $\int_1^2 \frac{1}{x^2}\,dx$ **13.** $\int_1^4 \sqrt{x}\,dx$ **14.** $\int_0^{\pi} \sin x\,dx$

15. As the fish-and-game warden of your township, you are responsible for stocking the town pond with fish before fishing season. The average depth of the pond is 20 ft. You plan to start the season with one fish per 1000 ft^3. You intend to have at least 25% of the opening day's fish population left at the end of the season. What is the maximum number of licenses the town can sell if the average seasonal catch is 20 fish per license?

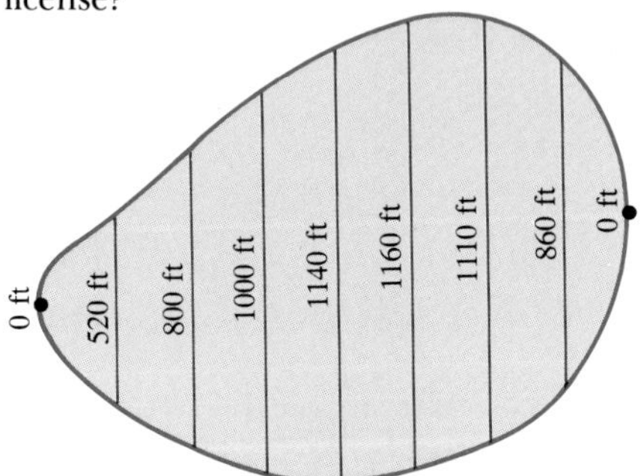

Horizontal spacing = 200 ft

16. CALCULATOR The design of a new airplane requires a gasoline tank of constant cross-section area in each wing. A scale drawing of a cross section is shown here. The tank must hold 5000 lb of gasoline that has a density of 42 lb/ft^3. Estimate the length of the tank.

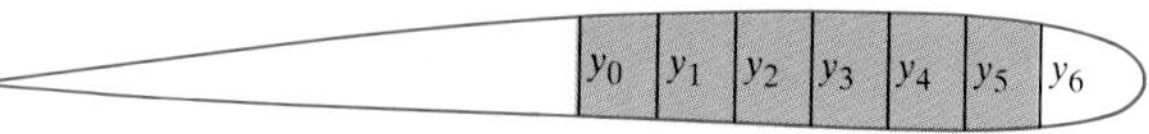

$y_0 = 1.5$ ft, $y_1 = 1.6$ ft, $y_2 = 1.8$ ft, $y_3 = 1.9$ ft,
$y_4 = 2.0$ ft, $y_5 = y_6 = 2.1$ ft Horizontal spacing = 1 ft

17. **CALCULATOR** A vehicle's aerodynamic drag is determined in part by its cross-section area and, all other things being equal, engineers try to make this area as small as possible. Use Simpson's rule to estimate the cross-section area of James Worden's solar-powered Solectria car at M.I.T. from the diagram below.

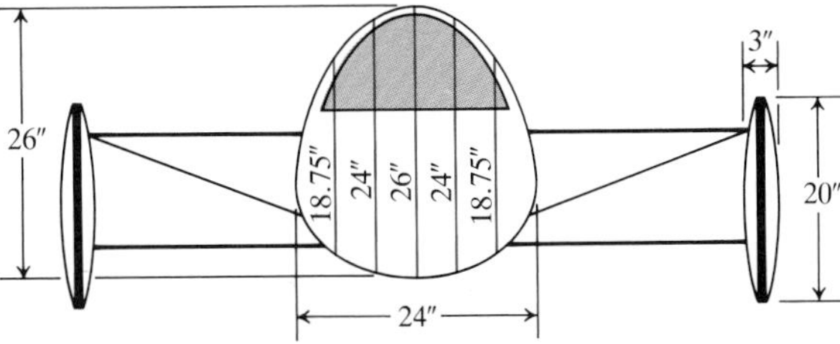

18. **CALCULATOR** *The dye-dilution technique for measuring cardiac output.* Instead of measuring a patient's cardiac output with exhaled carbon dioxide, as in Exercise 21, Section 4.1, a doctor may prefer to use the dye-dilution technique described here. You start by injecting 5–10 mg of dye in a main vein near the heart. The dye is drawn into the right side of the heart and then pumped through the lungs and out the left side of the heart into the aorta, where its concentration is measured each second as the blood flows past. The data in Table 5.8 and the plot in Fig. 5.23 show the response of a healthy, resting patient to an injection of 5.6 mg of dye.

The patient's cardiac output is calculated by dividing the area under the concentration curve into the number of milligrams of dye and multiplying the result by 60:

$$\text{Cardiac output} = \frac{\text{milligrams of dye}}{\text{area under curve}} \times 60. \quad (15)$$

You can see why if you check the units in which these quantities are measured. The dye is in milligrams, the area is in (milligrams/liter) × seconds, and

$$\frac{\text{mg}}{\frac{\text{mg}}{\text{L}} \cdot \text{sec}} \cdot 60 = \text{mg} \cdot \frac{\text{L}}{\text{mg} \cdot \text{sec}} \cdot 60 = \frac{\text{L}}{\text{sec}} \cdot 60 = \frac{\text{L}}{\text{min}}.$$

a) Use the trapezoidal rule and the data in Table 5.8 to calculate the area under the concentration curve in Fig. 5.23.

b) Then use Eq. (15) to calculate the patient's cardiac output.

TABLE 5.8
Dye-dilution data for Exercise 18

Seconds after injection t	Dye concentration (adjusted for recirculation) $c(t)$	Seconds after injection t	Dye concentration (adjusted for recirculation) $c(t)$
1	0	16	18.5
2	0	17	14.5
3	0	18	11.5
4	0	19	9.1
5	0	20	7.3
6	1.5	21	5.7
7	38.0	22	4.5
8	67.0	23	3.6
9	80.0	24	2.8
10	73.0	25	2.3
11	61.0	26	1.8
12	48.0	27	1.4
13	36.0	28	1.1
14	29.0	29	0.9
15	23.0	30	0

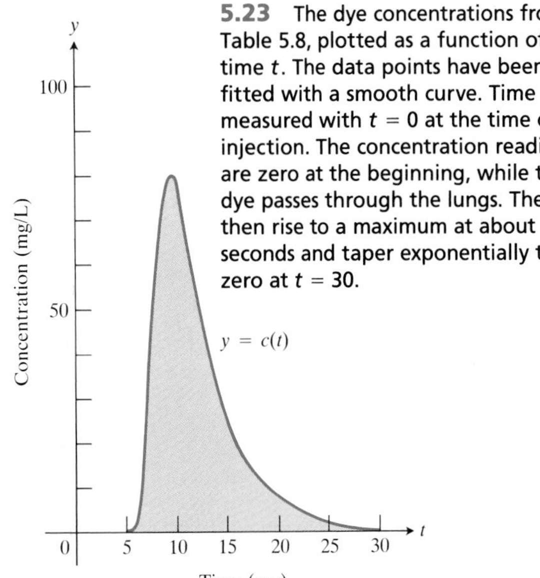

5.23 The dye concentrations from Table 5.8, plotted as a function of time t. The data points have been fitted with a smooth curve. Time is measured with $t = 0$ at the time of injection. The concentration readings are zero at the beginning, while the dye passes through the lungs. They then rise to a maximum at about $t = 9$ seconds and taper exponentially to zero at $t = 30$.

Electric Energy Consumption

The Louisiana Power and Light Company tries to forecast the demand for electricity throughout the day so that it can have enough generators on line at any given time to carry the load. Boilers take a while to fire up, so the company has to know in advance what the load is going to be if service is not to be interrupted. Like all power companies, Louisiana Power measures electrical demand in kilowatt-hours (KWH on your electric bill). A 1500-watt space heater, running for 10 hours, for instance, uses 15,000 watt-hours or 15 kilowatt-hours of electricity. A kilowatt, like a horsepower, is a unit of power. A kilowatt-hour is a unit of energy, and energy is what Louisiana Power, despite its name, sells.

Table 5.9 and the graph in Fig. 5.24 show the results of a 1984 residential-load study of Louisiana Power residential customers with electric heating, for typical weekdays and weekend days in January. A typical customer used 56.60 kwh on a weekday that January and 53.46 kwh on a weekend day. At the time of the study, 53.60 kwh cost a customer \$3.40.

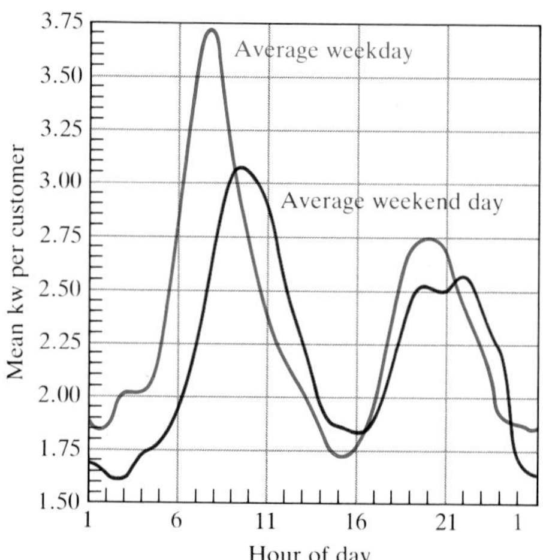

5.24 Louisiana Power and Light Company 1984 residental-load profile, residential customers with electric heating, averaged days for January. The curves plot kilowatts against hours. The areas under the curves give kilowatt-hours. See Table 5.9 for data.

19. **CALCULATOR** Find the weekday energy consumption of a typical residential customer by using the trapezoidal rule to estimate the area under the weekday power curve. To save time, you might try using only the data for the even-numbered hours or the data for the odd-numbered hours. How close do you come to Louisiana Power's own estimate if you do that? (If you use the even-numbered hours, be sure to count hour 2 twice so that you cover a complete 24-hr period. If you use odd-numbered hours, use hour 1 twice.)

TABLE 5.9
Residential electric loads (Exercises 19 and 20)

Hour of day	Weekday kw	Weekend kw
1	1.88	1.69
2	1.88	1.64
3	2.02	1.63
4	2.02	1.73
5	2.25	1.80
6	2.76	1.97
7	3.60	2.25
8	3.66	2.68
9	3.05	3.05
10	2.70	3.05
11	2.38	2.88
12	2.17	2.55
13	2.02	2.25
14	1.82	1.95
15	1.72	1.87
16	1.77	1.83
17	1.97	1.90
18	2.43	2.17
19	2.68	2.46
20	2.75	2.52
21	2.65	2.50
22	2.40	2.57
23	2.21	2.40
24	1.90	2.22

20. **CALCULATOR** *Continuation of Exercise 19.* Use the trapezoidal rule to find the daily weekend energy consumption.

21. **CALCULATOR** *Usable values of the sine-integral function.* The sine-integral function,

$$\mathrm{Si}(x) = \int_0^x \frac{\sin t}{t}\,dt, \quad (\text{"Sine integral of } x\text{"})$$

is one of the many functions in engineering whose formulas cannot be simplified. There simply is no elementary formula for the antiderivative of $(\sin t)/t$. The values of $\mathrm{Si}(x)$, however, are readily produced by numerical integration.

Although the notation does not show it explicitly, the function being integrated is

$$f(t) = \begin{cases} \dfrac{\sin t}{t}, & t \neq 0, \\ 1, & t = 0, \end{cases}$$

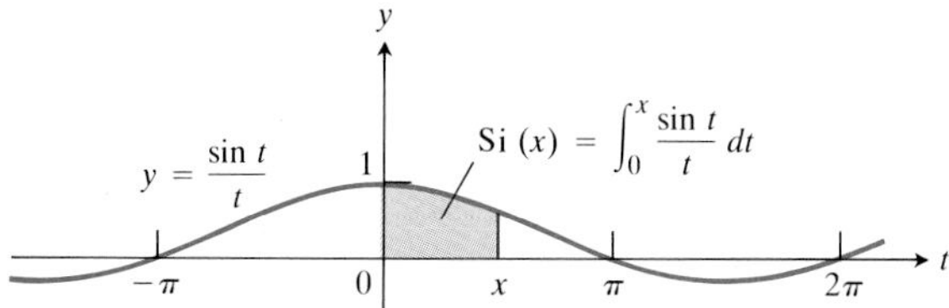

5.25 The continuous extension of $y = (\sin t)/t$. The sine integral function $\mathrm{Si}(x)$ is the subject of Exercises 21 and 24.

the continuous extension of $(\sin t)/t$ to the interval $[0, x]$. The function has derivatives of all orders at every point of its domain. Its graph is smooth (Fig. 5.25) and you can expect good results from Simpson's rule.

a) Use the fact that $|f^{(4)}| \leq 1$ on $[0, \pi/2]$ to give an upper bound for the error that will occur if

$$\mathrm{Si}\left(\frac{\pi}{2}\right) = \int_0^{\pi/2} \frac{\sin t}{t}\,dt$$

is estimated by Simpson's rule with $n = 4$.

b) Estimate $\mathrm{Si}(\pi/2)$ by Simpson's rule with $n = 4$.

c) Express the error bound you found in (a) as a percentage of the value you found in (b).

Computer or Programmable Calculator

Use the text's computer program for Simpson's rule to estimate the values of the integrals in Exercises 22–25.

22. $\displaystyle\int_{-1}^{1} 2\sqrt{1 - x^2}\,dx$ (The exact value is π.)

23. $\displaystyle\int_{0}^{1} \sqrt{1 + x^4}\,dx$ (A nonelementary integral that came up in Newton's research)

24. $\displaystyle\int_{0}^{\pi/2} \frac{\sin x}{x}\,dx$ (The integral from Exercise 21. To avoid division by zero, you may have to start the integration at a small positive number like 10^{-6} instead of 0.)

25. $\displaystyle\int_{0}^{\pi/2} \sin(x^2)\,dx$ (An integral associated with the diffraction of light)

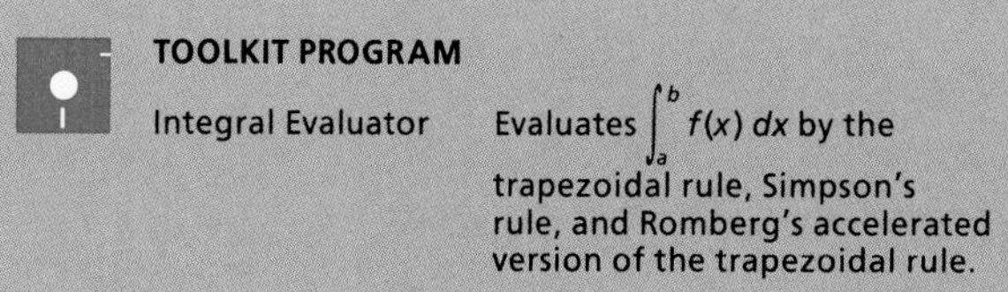
TOOLKIT PROGRAM

Integral Evaluator — Evaluates $\int_a^b f(x)\,dx$ by the trapezoidal rule, Simpson's rule, and Romberg's accelerated version of the trapezoidal rule.

5.8 A Brief Introduction to Logarithms and Exponentials

You may recall our saying that the exponential model for bacterial growth, reproduced here from Section 4.6, is one of the really good models in

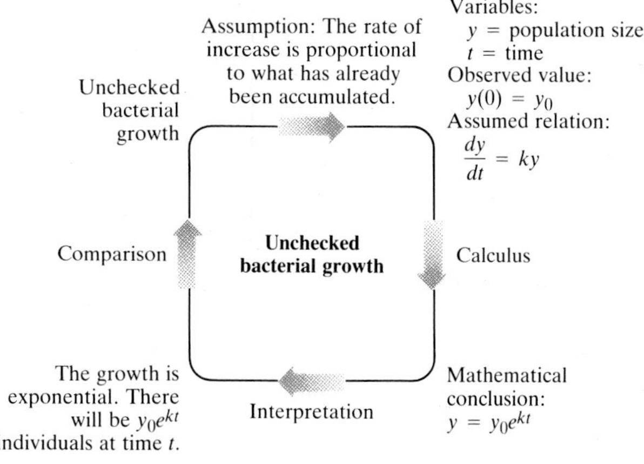

science because it applies to so many phenomena people want to forecast and understand. Whenever you have a quantity y whose rate of change over time is proportional to the amount of y present, you have a function that satisfies the differential equation

$$\frac{dy}{dt} = ky.$$

If, in addition, $y = y_0$ when $t = 0$, the function is none other than the exponential function

$$y = y_0 e^{kt}.$$

This section shows where the exponential function comes from and how it can be used to solve problems in the life sciences and in business and economics.

The key to understanding the exponential function is its inverse, the natural logarithm function, so we introduce that first. You will find detailed treatments of exponential and logarithmic functions, along with additional applications, in Chapter 7.

The Natural Logarithm Function

The natural logarithm of a positive number x, written as $\ln x$, is the value of an integral.

DEFINITION

The Natural Logarithm Function

$$\ln x = \int_1^x \frac{1}{t}\,dt, \qquad x > 0 \tag{1}$$

If $x > 1$, then $\ln x$ is the area under the curve $y = 1/t$ from $t = 1$ to $t = x$ (Fig. 5.26). If x is less than 1 (but still positive), $\ln x$ gives the negative of the area under the curve from $t = x$ to $t = 1$. The function is not defined for $x \le 0$. The natural logarithm of 1 itself is zero because

$$\ln 1 = \int_1^1 \frac{1}{t}\,dt = 0. \qquad \text{(Upper and lower limits equal)} \tag{2}$$

Notice that we show the graph of $y = 1/x$ in Fig. 5.26 but use the function $y = 1/t$ in the integral. Using x for everything would have us writing

$$\ln x = \int_1^x \frac{1}{x}\,dx. \tag{3}$$

This has too many x's to make sense, so we changed the dummy variable to t.

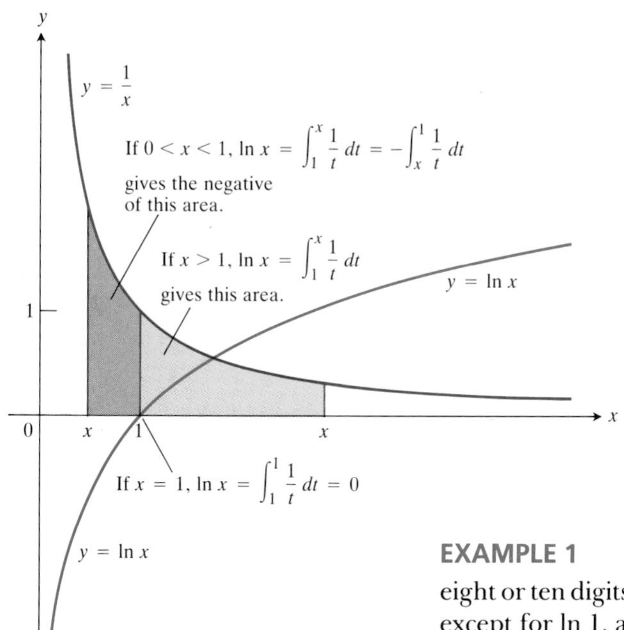

5.26 The graph of $y = \ln x$ and its relation to the function $y = 1/x$, $x > 0$. The graph rises above the x-axis as x moves from 1 to the right and falls below the axis as x moves from 1 to the left.

EXAMPLE 1 You can find values of $\ln x$ on most scientific calculators to eight or ten digits by entering x and pressing $\boxed{\ln x}$. Typical values, rounded except for ln 1, are

$$\begin{aligned} \ln 0.5 &= -0.693, \\ \ln 1 &= 0, \qquad \text{(Exact)} \\ \ln 2 &= 0.693, \\ \ln 10 &= 2.303, \\ \ln 1000 &= 6.908. \end{aligned} \tag{4}$$

The logarithm of x grows very slowly as $x \to \infty$, a point we shall return to in Section 7.6. ■

Historically the importance of logarithms came from the improvement they brought to arithmetic. The revolutionary properties of logarithms made possible the calculations behind the great seventeenth-century advances in offshore navigation and celestial mechanics. Logarithms are still important today, but mainly for other reasons.

Rules for Logarithms

The following rules hold for any positive numbers a and x and for any exponent n.

1. $\ln ax = \ln a + \ln x$
2. $\ln \dfrac{a}{x} = \ln a - \ln x$
3. $\ln \dfrac{1}{x} = -\ln x$ (Rule 2 with $a = 1$)
4. $\ln x^n = n \ln x$

EXAMPLE 2

$$\ln 6 = \ln(2 \cdot 3) = \ln 2 + \ln 3 \qquad \text{(Rule 1)}$$

$$\ln 4 - \ln 5 = \ln\left(\frac{4}{5}\right) = \ln 0.8 \qquad \text{(Rule 2)}$$

$$\ln \frac{1}{8} = -\ln 8 \qquad \text{(Rule 3)}$$

$$= -\ln 2^3 = -3 \ln 2 \qquad \text{(Rule 4)}$$

■

The Derivative of $y = \ln x$

By the First Fundamental Theorem of Calculus, Section 5.4,

$$\frac{d}{dx} \ln x = \frac{d}{dx} \int_1^x \frac{1}{t}\, dt = \frac{1}{x}.$$

For every positive value of x, therefore,

$$\frac{d}{dx} \ln x = \frac{1}{x}. \qquad (5)$$

If u is a differentiable function of x whose values are positive, so that $\ln u$ is defined, then applying the Chain Rule

$$\frac{dy}{dx} = \frac{dy}{du} \cdot \frac{du}{dx}$$

to the function $y = \ln u$ gives

$$\frac{d}{dx} \ln u = \frac{d}{du} \ln u \cdot \frac{du}{dx} = \frac{1}{u} \frac{du}{dx}. \qquad (6)$$

$$\frac{d}{dx} \ln u = \frac{1}{u} \frac{du}{dx}, \qquad u > 0 \qquad (7)$$

EXAMPLE 3 $\frac{d}{dx} \ln 2x = \frac{1}{2x} \frac{d}{dx}(2x) = \frac{1}{2x}(2) = \frac{1}{x}$ ■

Notice the remarkable occurrence in Example 3. The function $y = \ln 2x$ has the same derivative as the function $y = \ln x$. This is true of $y = \ln kx$ for any constant k:

$$\frac{d}{dx} \ln kx = \frac{1}{kx} \cdot \frac{d}{dx}(kx) = \frac{1}{kx}(k) = \frac{1}{x}. \qquad (8)$$

EXAMPLE 4 Equation (7) with $u = x^2 + 3$ gives

$$\frac{d}{dx} \ln(x^2 + 3) = \frac{1}{x^2 + 3} \cdot \frac{d}{dx}(x^2 + 3) = \frac{1}{x^2 + 3} \cdot 2x = \frac{2x}{x^2 + 3}.$$

■

The Integral $\int \frac{1}{u}\, du,\ u > 0$

Equation (7) leads to the integral formula

$$\int \frac{1}{u}\, du = \ln u + C, \qquad u > 0. \tag{9}$$

We shall see what to do when $u < 0$ when we get to Chapter 7. It is not really a problem—we just want to spend our time on something else right now.

EXAMPLE 5

$$\begin{aligned} \int \frac{2x}{x^2+3}\, dx &= \int \frac{du}{u} && \text{(Substitute } u = x^2 + 3,\ du = 2x\, dx.) \\ &= \ln u + C && \text{(Eq. 9)} \\ &= \ln(x^2+3) + C && \text{(Replace } u \text{ by } x^2 + 3.) \end{aligned}$$

■

EXAMPLE 6

$$\begin{aligned} \int_0^3 \frac{2x}{x^2+1}\, dx &= \ln(x^2+3)\Big]_0^3 && \begin{pmatrix}\text{Antiderivative}\\ \text{from Example 5}\end{pmatrix} \\ &= \ln(3^2+3) - \ln(0+3) && \\ &= \ln 12 - \ln 3 && \text{(OK answer)} \\ &= \ln \frac{12}{3} && \text{(Rule 2)} \\ &= \ln 4 && \text{(More useful answer)} \end{aligned}$$

■

The Exponential Function $\exp(x)$

As you may recall, a function $y = f(x)$ is one-to-one if different input values for x always give different output values for y. The function $y = x$ is one-to-one because different x-values always give different y-values. The function $y = \sin x$ is not one-to-one because different x-values sometimes give the same y-value. The sines of 0 and π both have the value 0, even though 0 and π are not the same number.

The function $y = \ln x$ is one-to-one because its derivative, $y' = 1/x$, is positive at every point in the function's domain (Corollary 1 of the Mean Value Theorem, Section 4.2). The natural logarithm function therefore has an inverse. We call the inverse the exponential function of x, abbreviated $\exp(x)$.

DEFINITION

The function $y = \exp(x)$, **the exponential function of** x, is the inverse of the natural logarithm function, $y = \ln x$.

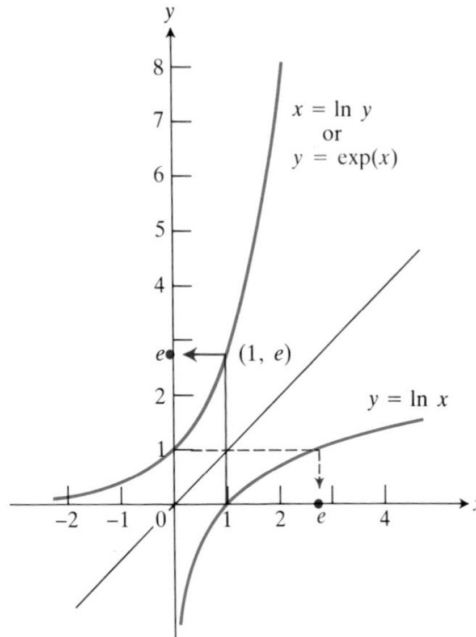

5.27 The graphs of $y = \ln x$ and its inverse $y = \exp(x)$. The number e is the one number whose natural logarithm is 1.

When functions are inverses of one another, their composites in either order give the identity function. For the exponential and natural logarithm, this means that

$$\exp(\ln x) = x, \quad x > 0, \tag{10}$$

$$\ln(\exp(x)) = x, \quad \text{all } x. \tag{11}$$

We shall have more to say about these equations in a while.

The Graph of $y = \exp(x)$

In a sense, the graph of $y = \ln x$ is already a graph of $y = \exp(x)$, with the xy-pairs plotted in reverse. The usual graph of the exponential function, however, is obtained by reflecting the graph of $y = \ln x$ across the line $y = x$ to turn the pairs around the right way. The reflection interchanges the x and y to place the independent variable along the horizontal axis in the normal way (Fig. 5.27).

The Number e

The number e, which you may have worked with before, is the one number whose natural logarithm is 1. (See Fig. 5.27 again.) Although e is not a rational number, it is possible to find its value with a computer to as many decimal places as we please by using the formula

$$e = \lim_{n \to \infty} \left(2 + \frac{1}{2} + \frac{1}{6} + \cdots + \frac{1}{n!}\right). \tag{12}$$

To fifteen places,

$$e = 2.7\ 1828\ 1828\ 45\ 90\ 45.$$

(This layout makes the number easy to remember.)

The Function $y = e^x$

We raise the number e to a rational power x in the usual way:

$$e^2 = e \cdot e, \quad e^{-2} = \frac{1}{e^2}, \quad e^{1/2} = \sqrt{e},$$

and so on. Since e is positive, the number e^x ("e to the x") is always positive. This means that e^x has a logarithm. When we take its logarithm, here is what we find:

$$\begin{aligned} \ln e^x &= x \ln e && \text{(Rule 4)} \\ &= x \cdot 1 = x. && (\ln e = 1) \end{aligned}$$

But we also have

$$\ln(\exp(x)) = x$$

from Eq. (11). Combining these results gives

$$e^x = \exp(x) \text{ when } x \text{ is rational.} \tag{13}$$

Equation (13) provides a way to complete the definition of e^x to include irrational values of x. The function $\exp(x)$ is defined for all x, so we can use it to give a value to e^x at points where e^x had no previous value.

DEFINITION

For every real number x, $\quad e^x = \exp(x)$. (14)

EXAMPLE 7 You can find values of e^x on most scientific calculators to eight or ten digits by entering x and pressing $\boxed{e^x}$ or $\boxed{\text{INV}}\boxed{\ln x}$. Typical values (rounded) are

$$e^{-1} = 0.3679 \qquad e^2 = 7.3891$$

$$e^{10} = 22026 \qquad e^{100} = 2.6881 \times 10^{43}.$$

The values of e^x grow very rapidly as $x \to \infty$, another point we shall return to in Section 7.6. ■

Equations Involving ln x and e^x

When we replace $\exp(x)$ by e^x in Eqs. (10) and (11), we get the two most important rules for combining $\ln x$ and e^x.

For all positive x: $\quad e^{\ln x} = x \quad$ (15)

For all x: $\quad \ln(e^x) = x \quad$ (16)

If you did Exercises 6 and 7 in Section 1.7 on your calculator, these equations explain what you found. If you did not do them, you might want to do them now, along with the next example.

EXAMPLE 8

a) $\ln e^2 = 2$ b) $\ln e^{-1} = -1$

c) $\ln\sqrt{e} = \frac{1}{2}$ d) $\ln e^{\sin x} = \sin x$

e) $e^{\ln 2} = 2$ f) $e^{\ln(x^2+1)} = x^2 + 1$

g) $e^{3\ln 2} = e^{\ln 2^3} = e^{\ln 8} = 8$ (One way)

h) $e^{3\ln 2} = \left(e^{\ln 2}\right)^3 = 2^3 = 8$ (Another way) ■

EXAMPLE 9 Find y if $\ln y = 3t + 5$.

Solution Exponentiate:

$$e^{\ln y} = e^{3t+5} \qquad \text{(Eq. 15)}$$

$$y = e^{3t+5}$$ ■

EXAMPLE 10 Find k if $e^{2k} = 10$.

Solution Take the logarithm of both sides:

$$e^{2k} = 10$$

$$\ln e^{2k} = \ln 10 \qquad \text{(Eq. 16)}$$

$$2k = \ln 10$$

$$k = \frac{1}{2} \ln 10$$

■

Two Useful Operating Rules

1. To remove logarithms from an equation, exponentiate both sides.
2. To remove exponentials, take the logarithm of both sides.

As we shall see in Chapter 7, the function e^x obeys the familiar rules of exponents from algebra.

Rules of Exponents

1. For any number x: $$e^{-x} = \frac{1}{e^x} \qquad (17)$$
2. For any x_1 and x_2: $$e^{x_1 + x_2} = e^{x_1} \cdot e^{x_2} \qquad (18)$$

EXAMPLE 11

$$e^{-\ln 2} = \frac{1}{e^{\ln 2}} = \frac{1}{2} \qquad \text{(Eq. 17)}$$

$$e^{4t + \ln 5} = e^{4t} \cdot e^{\ln 5} \qquad \text{(Eqs. 18 and 15)}$$

$$= e^{4t} \cdot 5 = 5e^{4t}$$

■

The Derivative of $y = e^x$

As it turns out, the function $y = e^x$ is not only continuous at every value of x but also differentiable at every value of x. Although we shall have to wait until Chapter 7 to see exactly why this is so, we can calculate the derivative in a straightforward way right now.

Starting with $y = e^x$, we have, in order,

$$y = e^x$$

$$\ln y = \ln e^x \qquad \text{(Take logarithms of both sides)}$$

$$\ln y = x \qquad \text{(Because } \ln e^x = x\text{)}$$

$$\frac{d}{dx}\ln y = \frac{d}{dx}x \qquad \begin{pmatrix}\text{Implicit differentiation—assumes} \\ e^x \text{ to be differentiable}\end{pmatrix}$$

$$\frac{1}{y}\frac{dy}{dx} = 1 \qquad \text{(Eq. (7) with } u = y\text{)}$$

$$\frac{dy}{dx} = y$$

$$\frac{dy}{dx} = e^x \qquad \text{(Replace } y \text{ by } e^x.\text{)}$$

The startling conclusion we draw from this sequence of equations is that the function $y = e^x$ is its own derivative. Never before have we encountered such a function. No matter how many times we differentiate it, we always get the function back. Constants times e^x are the only other functions to behave this way.

$$\frac{d}{dx}e^x = e^x \tag{19}$$

EXAMPLE 12

$$\frac{d}{dx}(5e^x) = 5\frac{d}{dx}e^x = 5e^x$$

The Chain Rule extends Eq. (19) in the usual way to a more general form.

If u is any differentiable function of x, then

$$\frac{d}{dx}e^u = e^u\frac{du}{dx}. \tag{20}$$

EXAMPLE 13 Equation (20) with $u = -x$:

$$\frac{d}{dx}e^{-x} = e^{-x}\frac{d}{dx}(-x) = e^{-x}(-1) = -e^{-x}.$$

EXAMPLE 14 Equation (20) with $u = \sin x$:

$$\frac{d}{dx}e^{\sin x} = e^{\sin x}\frac{d}{dx}(\sin x) = e^{\sin x} \cdot \cos x.$$

The integral-formula equivalent of Eq. (20) is

$$\int e^u \, du = e^u + C. \tag{21}$$

EXAMPLE 15

$$\int_0^{\ln 2} e^{3x}\,dx = \int_0^{3\ln 2} e^u \cdot \frac{1}{3}\,du \qquad \left(\begin{matrix} u = 3x,\, du = 3\,dx,\, \frac{1}{3}du = dx, \\ u(0) = 0,\, u(\ln 2) = 3\ln 2 \end{matrix}\right)$$

$$= \frac{1}{3}\int_0^{3\ln 2} e^u\,du = \frac{1}{3}e^u\Big]_0^{3\ln 2} = \frac{1}{3}\left[e^{3\ln 2} - e^0\right] = \frac{1}{3}[8-1] = \frac{7}{3}$$

EXAMPLE 16

$$\int_0^{\pi/2} e^{\sin x}\cos x\,dx = e^{\sin x}\Big]_0^{\pi/2} \qquad \text{(Antiderivative from Example 14)}$$

$$= e^{\sin(\pi/2)} - e^{\sin 0} = e^1 - e^0 = e - 1$$

The Law of Exponential Change

In many instances in biology and economics, some positive quantity y grows or decreases at a rate that at any given time t is proportional to the amount that is present. If we also know the initial amount y_0 at time $t = 0$, we can find y by solving the initial value problem

Differential equation: $\dfrac{dy}{dt} = ky$ (22)

Initial condition: $y = y_0$ when $t = 0$.

The constant k is positive if y is increasing and negative if y is decreasing.

To solve Eq. (22), we divide through by y to get

$$\frac{1}{y}\frac{dy}{dt} = k$$

and integrate both sides with respect to t to get

$$\ln y = kt + C. \tag{23}$$

We then solve this for y by exponentiating:

$$e^{\ln y} = e^{kt + C}$$

$$y = e^{kt} \cdot e^C$$

$$y = Ae^{kt} \qquad \text{(Write } A \text{ for } e^C.)$$

We find the value of A from the initial condition:

$$y_0 = Ae^{k(0)} = A \cdot 1 = A.$$

The solution of the initial value problem is $y = y_0 e^{kt}$. We call this equation the law of exponential change.

The Law of Exponential Change

$$y = y_0 e^{kt} \tag{24}$$

EXAMPLE 17 *The growth of a cell.* In an ideal environment, the mass m of a cell will grow exponentially, at least early on. Nutrients pass quickly through the cell wall, and growth is limited only by the metabolism within the cell, which in turn depends on the mass of participating molecules. If we make the reasonable assumption that, at each instant of time, the cell's growth rate dm/dt is proportional to the mass that has already been accumulated, then

$$\frac{dm}{dt} = km \qquad \text{and} \qquad m = m_0 e^{kt}.$$

There are limitations, of course, and in any particular case we would expect this equation to provide reliable information only for values of m below a certain size. ■

EXAMPLE 18 *Birth rates and population growth.* Strictly speaking, the number of individuals in a population (of people, plants, foxes, or whatever) is a discontinuous function of time because it takes on discrete values. However, as soon as the number of individuals becomes large enough, it may safely be described with a continuous and even differentiable function. If we assume that the proportion of reproducing individuals remains constant and assume a constant fertility, then at any instant t the birth rate is proportional to the number $y(t)$ of individuals present. If, further, we neglect departures, arrivals, and deaths, the growth rate dy/dt will be the same as the birth rate ky. In other words,

$$\frac{dy}{dt} = ky.$$

Once again, we find that $y = y_0 e^{kt}$. ■

The Incidence of Disease

One model for the way diseases spread assumes that the rate dy/dt at which the number of infected people changes is proportional to the number y itself. The more infected people there are, the faster the disease will spread. The fewer there are, the slower it will spread. Once again,

$$y = y_0 e^{kt}.$$

EXAMPLE 19 In the course of any given year, the number y of cases of a disease is reduced by 20%. If there are 10,000 cases today, how many years will it take to reduce the number of cases to 1000?

Solution The equation we use is $y = y_0 e^{kt}$, and there are three things to find:

1. The value of y_0,
2. The value of k,
3. The value of t that makes $y = 1000$.

STEP 1: *The value of* y_0. We are free to count time from anywhere we want. If we start counting from today, then $y = 10{,}000$ when $t = 0$, so $y_0 = 10{,}000$. Our equation becomes

$$y = 10{,}000e^{kt}.$$

STEP 2: *The value of k.* When $t = 1$, the number of cases will be 80% of its present value, or 8000. Hence

$$\begin{aligned} 10{,}000e^{k(1)} &= 8000 \\ e^k &= 0.8 \\ \ln e^k &= \ln 0.8 \\ k &= \ln 0.8. \end{aligned}$$

At any given time t, therefore,

$$y = 10{,}000e^{(\ln 0.8)t}.$$

Two down and one to go.

STEP 3: *The value of t that makes* $y = 1000$. Set y equal to 1000 and solve for t:

$$\begin{aligned} 10{,}000e^{(\ln 0.8)t} &= 1000 \\ e^{(\ln 0.8)t} &= 0.1 && \text{(Divide by 10,000.)} \\ (\ln 0.8)t &= \ln 0.1 && \text{(Take logs of both sides.)} \\ t &= \frac{\ln 0.1}{\ln 0.8} = 10.32. && \text{(With a calculator, rounded)} \end{aligned}$$

It will take a little more than 10 years to reduce the number of cases to 1000. ■

Continuously Compounded Interest

If you invest an amount A_0 of money at a fixed annual interest rate r and interest is added to your account k times a year, it turns out that the amount of money you will have at the end of t years is

$$A_t = A_0\left(1 + \frac{r}{k}\right)^{kt}. \tag{25}$$

The money might be added ("compounded," bankers say) monthly ($k = 12$), weekly ($k = 52$), daily ($k = 365$), or even more frequently, say by the hour or by the minute. But there is still a limit to how much you will earn that way, and the limit is

$$\lim_{k\to\infty} A_t = \lim_{k\to\infty} A_0\left(1 + \frac{r}{k}\right)^{kt} = A_0 e^{rt}. \tag{26}$$

(See Section 7.5.)

The resulting formula for the amount of money in your account after t years is $A(t) = A_0 e^{rt}$.

The Continuous Compound Interest Formula

$$A(t) = A_0 e^{rt} \tag{27}$$

Interest paid according to this formula is said to be **compounded continuously.** The number r is called the **continuous interest rate.**

EXAMPLE 20 Suppose you deposit \$621 in a bank account that pays 6% compounded continuously. How much money will you have in the account 8 years later?

Solution We use Eq. (27) with $A_0 = 621$, $r = 0.06$, and $t = 8$:

$$A(8) = 621\, e^{(0.06)(8)} = 621\, e^{0.48} = 1003.58 \qquad \text{(calculator, nearest cent)}$$

Had the bank paid interest quarterly ($k = 4$ in Eq. (25)), the amount in your account would have been an even \$1000. Thus the effect of continuous compounding, as compared with quarterly compounding, has been an addition of \$3.58. A bank might decide it would be worth this additional amount to be able to advertise, "We compound your money every second, night and day—better than that, we compound the interest continuously." ■

The Rules of 70 and 72

How long does it take to double money at a continuous interest rate of 7%? To find out, we set $A_0 e^{0.07t}$ equal to $2A_0$ and solve for t:

$$\begin{aligned} A_0 e^{0.07t} &= 2A_0 \\ e^{0.07t} &= 2 && \text{(Divide out } A_0.) \\ \ln e^{0.07t} &= \ln 2 && \text{(Take logs of both sides.)} \\ 0.07t &= \ln 2 \\ t = \frac{\ln 2}{0.07} &\approx \frac{0.693}{0.07} = \frac{69.3}{7} \approx 10. && \text{(Calculator)} \end{aligned}$$

It takes about 10 years for money to double at 7%.

A similar calculation using i in place of 7 tells us that it takes about

$$t = \frac{69.3}{i}$$

years for money to double at i%.

The number 69.3 is not convenient for mental arithmetic, so most people use 70 instead. This gives rise to the so-called Rule of 70.

The Rule of 70

It takes about $y = 70/i$ years for money to double at i percent.

The Rule of 70 gives quick results that are usually accurate enough for the work at hand. Some people prefer to use a "Rule of 72" because 72 has more divisors than 70.

> **The Rule of 72**
>
> It takes about $y = 72/i$ years for money to double at i percent.

The Rules of 70 and 72 can be used "backward" to estimate the interest rate needed to double money in a given period of time.

EXAMPLE 21 What interest rate do you need to double your money in 6 years?

Solution

$$\text{Rule of 72:} \qquad i = \frac{72}{y} = \frac{72}{6} = 12\%$$

$$\text{Rule of 70:} \qquad i = \frac{70}{y} = \frac{70}{6} = 11\frac{2}{3}\%$$

Either way, it takes a rate of about 12% to double money in 6 years. ■

Strictly speaking, the Rules of 70 and 72 refer to continuous compounding and not, say, to annual or quarterly compounding. As Example 21 shows, however, the results of the different frequencies of compounding are much the same, and the Rules of 70 and 72 have good predictive value for any of the frequencies currently in use.

The Consumer Price Index

As you know, prices often change (they usually go up) and it is important to know what things will cost in the years ahead. The economists in the U.S. Department of Labor measure the cost of living with a number called the **consumer price index (CPI).** This index is a weighted average of the costs of food, clothing, housing, transportation, medical care, personal care, and entertainment.

The current index, set at 100 (an arbitrary choice for convenient arithmetic) in 1984, was 112.1 in March of 1987 and 116.5 in March of 1988. What you could buy for \$1.00 in 1984 cost roughly \$1.12 in 1987 and \$1.16 in 1988.

One of the predictors of price change assumes that the rate dp/dt at which the CPI changes is proportional to p, so that

$$\frac{dp}{dt} = kp. \tag{28}$$

The constant k is called the **continuous rate of inflation.** In newspapers k is usually expressed as a percent. If $k = 0.04$, for example, the U.S. Bureau of

Labor Statistics reports an inflation rate of 4%. The solution of Eq. (28) is

$$p(t) = p_0 e^{kt}. \tag{29}$$

If the CPI is p_0 at time $t = 0$, then t years later it will be $p(t) = p_0 e^{kt}$.

EXAMPLE 22 The CPI in March 1988 was 116.5. What will its value be 10 years later, in March 1998, if the inflation rate is a constant 4%?

Solution We take $p_0 = 116.5$ and $k = 0.04$ in Eq. (29) and find $p(10)$:

$$\begin{aligned} p(10) &= 116.5e^{(0.04)(10)} = 116.5e^{0.4} \\ &= (116.5)(1.49) = 173.6. \qquad \text{(Calculator, rounded)} \end{aligned}$$

What conclusions can we draw from this? One conclusion is that from 1988 to 1998 prices will rise

$$\frac{173.6 - 116.5}{116.5} \times 100 = 49\%.$$

Another is that by 1998, relative to March 1984, when the index was 100, prices will have risen

$$\frac{173.6 - 100}{100} \times 100 = 73.6\%.$$

(Now you see why the index is set at 100 in the base year. Subtracting 100 from any later value automatically gives the percentage change.)

Still another conclusion is that by 1998 it will cost nearly $1.74 to buy what $1.00 bought in 1984. ■

EXAMPLE 23 About how many years will it take the CPI to increase 50% if the inflation rate is a steady 4%?

Solution We want to know when $p_0 e^{0.04t}$ will equal $1.5p_0$, so we set these equal and solve for t:

$$\begin{aligned} p_0 e^{0.04t} &= 1.5p_0 \\ e^{0.04t} &= 1.5 \\ \ln e^{0.04t} &= \ln 1.5 \\ 0.04t &= \ln 1.5 \\ t &= \frac{\ln 1.5}{0.04} \approx \frac{0.405}{0.04} \approx 10.1. \end{aligned}$$

With 4% inflation, it will take a little more than 10 years for the cost of living to increase 50%. ■

The Purchasing Power of the Dollar

The higher the consumer price index, the less a dollar will buy, and economists call the number $100/p$ the **purchasing power of the dollar.**

In 1984, when the CPI was 100, the purchasing power of the dollar was $100/100 = 1.00$. Two years later, in 1986, it was $100/112.1 = 0.89$, and a year after that it was $100/116.5 = 0.86$.

EXAMPLE 24 Assuming the 4% inflation rate of Example 22, what will the purchasing power of the dollar be in 1998?

Solution We divide the CPI for 1998 into 100:

$$\frac{100}{1998 \text{ CPI}} = \frac{100}{173.6} = 0.57. \qquad \text{(CPI value from Example 22, rounded)}$$

This represents a change of -0.43, or a loss in purchasing power of 43% since 1984. ■

Exercises 5.8

The logarithms in Exercises 1–10 can all be expressed in terms of ln 2 and ln 3. For example, $\ln 1.5 = \ln(3/2) = \ln 3 - \ln 2$. See if you can do the others.

1. $\ln 1.5$
2. $\ln 16$
3. $\ln 4/9$
4. $\ln 12$
5. $\ln(1/2)$
6. $\ln(1/3)$
7. $\ln 4.5$
8. $\ln \sqrt[3]{9}$
9. $\ln 3\sqrt{2}$
10. $\ln \sqrt{13.5}$

Use the fact that the functions $y = \ln x$ and $y = e^x$ are inverses of each other to simplify the expressions in Exercises 11–16.

11. $e^{\ln 7}$
12. $e^{-\ln 7}$
13. $\ln e^2$
14. $e^{3 \ln 2}$
15. $e^{2+\ln 3}$
16. $e^{-2 \ln 3}$

Find dy/dx in Exercises 17–28.

17. $y = \ln 3x$
18. $y = \ln(x/2)$
19. $y = \ln(x^2)$
20. $y = (\ln x)^2$
21. $y = \ln(1/x)$
22. $y = \ln(10/x)$
23. $y = \ln(x + 2)$
24. $y = \ln(2x + 2)$
25. $y = \ln(2 - \cos x)$
26. $y = \ln(x^2 + 1)$
27. $y = \ln(\ln x)$
28. $y = x \ln x - x$

Find dy/dx in Exercises 29–38.

29. $y = 2e^x$
30. $y = e^{2x}$
31. $y = e^{-x}$
32. $y = e^{-5x}$
33. $y = e^{2x/3}$
34. $y = e^{-x/4}$
35. $y = xe^x - e^x$
36. $y = x^2e^x - xe^x$
37. $y = e^{\sqrt{x}}$
38. $y = e^{(x^2)}$

Evaluate the integrals in Exercises 39–46 by using the given substitutions.

39. $\displaystyle\int_0^3 \frac{1}{x+1}\,dx, \quad u = x + 1$

40. $\displaystyle\int_0^4 \frac{2x\,dx}{x^2+9}, \quad u = x^2 + 9$

41. $\displaystyle\int_{\ln 3}^{\ln 5} e^{2x}\,dx, \quad u = 2x$

42. $\displaystyle\int_0^{\ln 2} e^{-x}\,dx, \quad u = -x$

43. $\displaystyle\int_0^1 (1 + e^x)\,e^x\,dx, \quad u = 1 + e^x$

44. $\displaystyle\int_e^{e^2} \frac{dx}{x \ln x}, \quad u = \ln x$

45. $\displaystyle\int_1^4 \frac{e^{\sqrt{x}}\,dx}{2\sqrt{x}}, \quad u = \sqrt{x}$

46. $\displaystyle\int_0^{\pi} \frac{\sin x}{2 - \cos x}\,dx, \quad u = 2 - \cos x$

Evaluate the integrals in Exercises 47–54.

47. $\displaystyle\int_1^{e^2} \frac{1}{x}\,dx$
48. $\displaystyle\int_1^{e} \frac{2}{x}\,dx$
49. $\displaystyle\int_{\ln 2}^{\ln 3} e^x\,dx$
50. $\displaystyle\int_{-1}^{1} e^{(x+1)}\,dx$
51. $\displaystyle\int_2^4 \frac{dx}{x+2}$
52. $\displaystyle\int_{-1}^{0} \frac{8\,dx}{2x+3}$
53. $\displaystyle\int_{-1}^{1} 2xe^{-x^2}\,dx$
54. $\displaystyle\int_0^1 \frac{x\,dx}{4x^2+1}$

In Exercises 55–58, solve for k.

55. $e^{2k} = 4$
56. $e^{5k} = \dfrac{1}{4}$
57. $100e^{10k} = 200$
58. $100e^k = 1$

In Exercises 59–64, solve for t.

59. $e^t = 1$

60. $e^{kt} = \frac{1}{2}$

61. $e^{-0.3t} = 27$

62. $e^{-0.01t} = 1000$

63. $4e^{-0.1t} = 20$

64. $e^{(\ln 0.2)t} = 0.4$

In Exercises 65–68, solve for y.

65. $\ln y = 2t + 4$

66. $\ln y = -t + 5$

67. $\ln(y - 40) = 5t$

68. $\ln(1 - 2y) = t$

CALCULATOR HELPFUL

The answers to many of the following exercises come out in terms of logarithms and exponentials. A calculator will enable you to express the answers in decimal form.

69. *Cholera bacteria.* Suppose that the bacteria in a colony can grow unchecked, by the law of exponential change. The colony starts with 1 bacterium and doubles every half hour. How many bacteria will the colony contain at the end of 24 hr? (Under favorable laboratory conditions, the number of cholera bacteria can double every 30 min. In an infected person, many bacteria are destroyed, but this example helps explain why a person who feels well in the morning may be dangerously ill by evening.)

70. *Growth of bacteria.* A colony of bacteria is grown under ideal conditions in a laboratory so that the population increases exponentially with time. At the end of 3 hr there are 10,000 bacteria. At the end of 5 hr there are 40,000. How many bacteria were present initially?

71. *The incidence of a disease (continuation of Example 18).* Suppose that in any given year the number of cases can be reduced by 25% instead of 20%.

a) How long will it take to reduce the number of cases to 1000?

b) How long will it take to eradicate the disease, that is, reduce the number of cases to less than 1?

72. *John Napier's question.* John Napier, who invented natural logarithms, was the first person to answer the question, what happens if you invest an amount of money at 100% interest, compounded continuously?

a) What does happen?

b) How long does it take to triple your money?

c) How much can you earn in a year?

73. *Benjamin Franklin's will.* The Franklin Technical Institute of Boston owes its existence to a provision in a codicil to the will of Benjamin Franklin. In part it reads:

> I was born in Boston, New England and owe my first instruction in Literature to the free Grammar Schools established there: I have therefore already considered those schools in my Will. . . . I have considered that among Artisans good Apprentices are most likely to make good citizens . . . I wish to be useful even after my Death, if possible, in forming and advancing other young men that may be serviceable to their Country in both Boston and Philadelphia. To this end I devote Two thousand Pounds Sterling, which I give, one thousand thereof to the Inhabitants of the Town of Boston in Massachusetts, and the other thousand to the inhabitants of the City of Philadelphia, in Trust and for the Uses, Interests and Purposes hereinafter mentioned and declared.

Franklin's plan was to lend money to young apprentices at 5% interest with the provision that each borrower should pay each year

> . . . with the yearly Interest, one tenth part of the Principal, which sums of Principal and Interest shall be again let to fresh Borrowers. . . . If this plan is executed and succeeds as projected without interruption for one hundred Years, the Sum will then be one hundred and thirty-one thousand Pounds of which I would have the Managers of the Donation to the Inhabitants of the Town of Boston, then lay out at their discretion one hundred thousand Pounds in Public Works. . . . The remaining thirty-one thousand Pounds, I would have continued to be let out on Interest in the manner above directed for another hundred Years. . . . At the end of this second term if no unfortunate accident has prevented the operation the sum will be Four Millions and Sixty-one Thousand Pounds.

It was not always possible to find as many borrowers as Franklin had planned, but the managers of the trust did the best they could; they lent money to medical students as well as to others. At the end of 100 years from the reception of the Franklin gift, in January 1894, the fund had grown from 1000 pounds to almost exactly 90,000 pounds. In 100 years the original capital had multiplied about 90 times instead of the 131 times Franklin had imagined.

What rate of interest, compounded continuously for 100 years, would have multiplied Benjamin Franklin's original capital by 90?

74. In Benjamin Franklin's estimate that the original 1000 pounds would grow to 131,000 in 100 years, he was using an annual rate of 5% and compounding once each year. What rate of interest per year when compounded continuously for 100 years would multiply the original amount by 131?

75. *The U.S. population.* The Museum of Science in Boston, Massachusetts, displays the running total of the U.S. population. On December 27, 1988, it was increasing the total by 1 every 21 seconds, which works out to an instantaneous rate of 1,502,743 people a year. Working from 246,605,103, the display's population figure for 2:40 P.M. that day, this gives $k = 0.00609$, or 0.609%. Assuming this value of k for the next 10 years, what will the U.S. population be at 2:40 P.M. Boston time on December 27, 1998?

76. *The Rule of 70.* Use the Rule of 70 to answer these questions:

a) How long does it take to double money at 5% interest? At 7% interest?

b) What interest rate do you need to double money in 5 years? In 20 years?

77. The Consumer Price Index in March 1988 was 116.5. Assuming a constant inflation rate of 2.8% from then on, about when would you expect the index to double? (*Hint*: Use the Rule of 70.)

78. Repeat Exercise 77, but now use Eq. (28) directly instead of the Rule of 70.

79. *The purchasing power of the dollar.* Assuming a constant inflation rate of 4.4% starting in March 1988, when the Consumer Price Index was 116.5, what will the purchasing power of the dollar be in March 1990? March 1992? March 1994?

80. *Inflation and the CPI.* You have just seen a newspaper headline saying that the consumer price index rose 4% last year. Assuming this was caused by a constant inflation rate, what was that rate?

81. *Runaway inflation.* At the end of 1988 the consumer price index in Brazil was increasing at the continuous annual rate of 800% (yes, it was). How many days does it take prices to double at this rate?

To find out, solve the equation

$$p_0 e^{8t} = 2p_0$$

for t and convert your answer to days, rounding your answer to the nearest day.

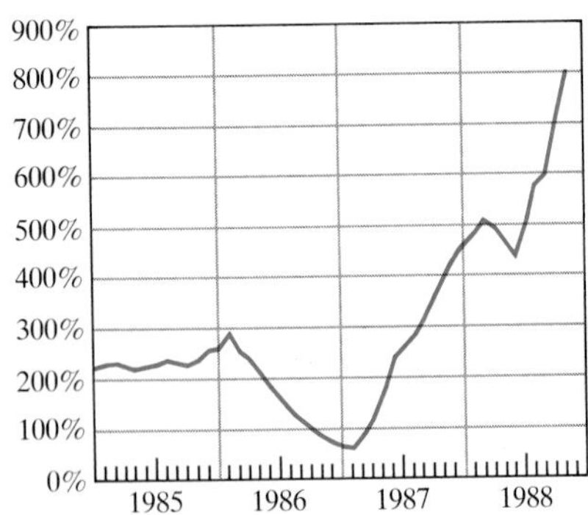

(Source: *Wall Street Journal*, Thursday, December 8, 1988, p. 1.)

82. *The purchasing power of the dollar.* How long will it take the purchasing power of the dollar to fall to half its present value if the inflation rate holds at a steady 4%?

83. *Oil depletion.* Suppose the amount of oil pumped from one of the canyon wells in Whittier, California, decreases at the continuous rate of 10% a year. When will the well's output fall to a fifth of its present value?

84. *Continuous price discounting.* To encourage buyers to place 100-unit orders, your firm's sales department applies a continuous discount that makes the unit price a function $p(x)$ of the number of units x ordered. The discount decreases the price at the rate of \$0.01 per unit ordered. The price per unit for a 100-unit order is $p(100) = \$20.09$.

a) Find $p(x)$ by solving the following initial value problem:

Differential equation: $\dfrac{dp}{dx} = -\dfrac{1}{100}x,$

Initial condition: $p(100) = 20.09.$

b) Find the unit price $p(10)$ for a 10-unit order and the unit price $p(90)$ for a 90-unit order.

c) The sales department has asked you to find out if it is discounting so much that the firm's revenue, $r(x) = x \cdot p(x)$, will actually be less for a 100-unit order than, say, for a 90-unit order. Reassure them by showing that r has its maximum value at $x = 100$.

d) **COMPUTER GRAPHER or GRAPHING CALCULATOR.** Graph the revenue function $r(x) = xp(x)$ for $0 \le x \le 200$.

85. *Subindices of the Consumer Price Index.* There are many "subindices" of the Consumer Price Index—separate indices for food, rent, medical care, and so on. Each index has its own inflation rate. From March 1987 to March 1988 the costs of food, rent, and medical care rose at 3%, 5%, and 6.4%, respectively. At these rates

a) how long will it take costs to increase 50%?

b) how long will it take the costs to double? (Use the Rule of 70.)

Variable Inflation Rates

When the inflation rate k varies with time instead of being constant, the formula $p = p_0 e^{kt}$ in Eq. (29) no longer gives the solution of the equation $dp/dt = kp$. The corrected formula, as you will see in Chapter 16, is

$$p(t) = p_0 e^{\int_0^t k(\tau)\,d\tau}. \tag{30}$$

Use this formula to answer the questions in Exercises 86 and 87.

86. CALCULATOR Suppose that $p_0 = 100$ and that $k(t) = 0.04/(1 + t)$, a rate that starts at 4% when $t = 0$ but decreases steadily as the years pass.

a) Find $\int_0^9 k(\tau)\,d\tau$.

b) Use Eq. (30) above to find the value of p when $t = 9$ yr.

87. CALCULATOR Suppose that $p_0 = 100$ and that the inflation rate is $k(t) = 1 + 1.3t$ (as it was in Brazil during the first few months of 1987).

a) What does the formula in Eq. (30) look like in this case?

b) Find $p(1)$ and $p(2)$, the 1-yr and 2-yr percentage increases in the associated consumer price index.

Review Questions

1. How can you use antiderivatives to find areas? Give an example.
2. How are finite sums written in sigma notation? Give examples.
3. What is a partition of an interval? What is the norm of a partition? Give examples.
4. What is a Riemann sum? Give an example.
5. What does $\int_a^b f(x)\,dx$ mean? Is it a number? How is it defined? What is it called? When does it exist?
6. What is the relation between definite integrals and area? Does a definite integral have to represent an area?
7. State eleven rules for working with definite integrals (Table 5.1). Give a specific example of each rule.
8. What is the average value of a function $f(x)$ over an interval $[a, b]$? Give an example.
9. State the Mean Value Theorem for Definite Integrals.
10. State the two fundamental theorems of integral calculus. What are they good for? Illustrate each theorem with an example.
11. What is an indefinite integral? How are indefinite integrals evaluated? What corollary of the Mean Value Theorem for derivatives makes the evaluation possible?
12. What specific formulas do you know for evaluating indefinite integrals (standard forms, so to speak)?
13. How does integration by substitution work? Does it apply to definite integrals as well as indefinite integrals? Give specific examples.
14. What numerical methods are available for estimating the values of definite integrals that cannot be evaluated directly with antiderivatives? What formulas sometimes help to determine the accuracy of these methods? How do you know if you are using a step size small enough to get the accuracy you want? Give an example.
15. How is the function $y = \ln x$ defined? What are the rules for doing arithmetic with natural logarithms?
16. What is the derivative of $y = \ln x$? What integrals involve logarithms? Give examples.
17. What is the function $y = \exp(x)$? What is number e? How is the function $y = e^x$ defined?
18. What derivatives and integrals are associated with the function $y = e^x$?
19. What initial value problem is solved by the function $y = y_0 e^{kt}$?
20. How does the exponential function arise in the life sciences and in business and economics? Give examples.

Practice Exercises

Use standard formulas to evaluate the sums in Exercises 1–4.

1. a) $\sum_{k=1}^{100} k$ b) $\sum_{k=1}^{100} k^2$ c) $\sum_{k=1}^{100} k^3$

2. a) $\sum_{k=1}^{10} (k+2)$ b) $\sum_{k=1}^{10} (2k-12)$

3. a) $\sum_{k=1}^{6} \left(k^2 - \frac{1}{6}\right)$ b) $\sum_{k=1}^{6} k(k+1)$

4. a) $\sum_{k=1}^{5} (k^3 - 45)$ b) $\sum_{k=1}^{6} \left(\frac{k^3}{7} - \frac{k}{7}\right)$

5. Evaluate the following sums.

a) $\sum_{k=1}^{3} 2^{k-1}$ b) $\sum_{k=0}^{4} (-1)^k \cos k\pi$

c) $\sum_{k=-1}^{2} k(k+1)$ d) $\sum_{k=1}^{4} \frac{(-1)^{k+1}}{k(k+1)}$

6. Express the following sums in sigma notation.

a) $1 + 2 + 4 + 8$ b) $1 + \frac{1}{3} + \frac{1}{9} + \frac{1}{27} + \frac{1}{81}$

c) $1 - 2 + 3 - 4 + 5$ d) $\frac{5}{2} + \frac{5}{4} + \frac{5}{6}$

7. How many cubes were stacked to make this solid, square-based pyramid?

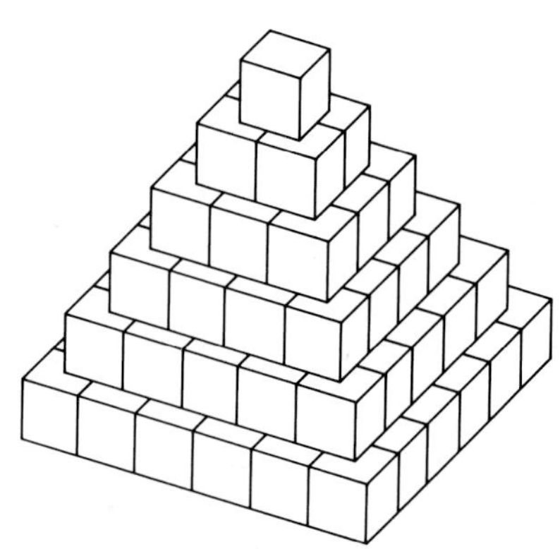

8. Here is a pictorial "proof" that $\sum_{k=1}^{n} k = \frac{n(n+1)}{2}$. See if you can tell what is going on.

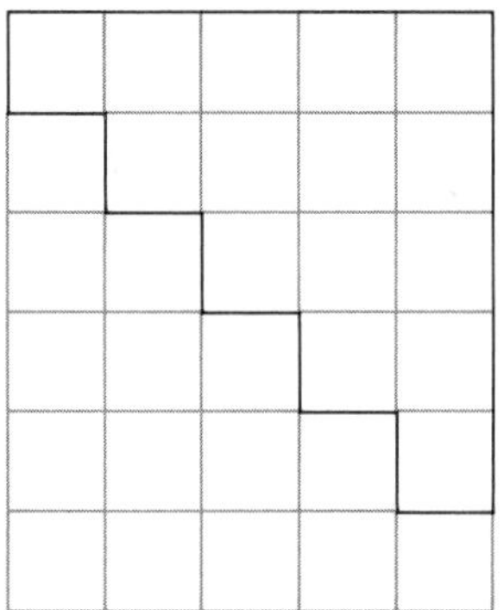

Express the limits in Exercises 9 and 10 as definite integrals.

9. $\lim_{\|P\|\to 0} \sum_{k=1}^{n} \frac{1}{c_k} \Delta x_k$, where P is a partition of $[1, 2]$.

10. $\lim_{\|P\|\to 0} \sum_{k=1}^{n} e^{c_k} \Delta x_k$, where P is a partition of $[0, 1]$.

11. Suppose $\int_{-2}^{2} f(x)\,dx = 4, \int_{2}^{5} f(x)\,dx = 3, \int_{-2}^{5} g(x)\,dx = 2.$ Which, if any, of the following statements are true, and which, if any, are false?

a) $\int_{5}^{2} f(x)\,dx = -3$

b) $\int_{-2}^{5} (f(x) + g(x)) = 9$

c) $f(x) \le g(x)$ on the interval $-2 \le x \le 5$

12. Suppose $\int_{0}^{1} f(x)\,dx = \pi$. Find

a) $\int_{0}^{1} f(t)\,dt$ b) $\int_{1}^{0} f(y)\,dy$ c) $\int_{0}^{1} -3f(z)\,dz$.

Find the areas of the shaded regions in Exercises 13–16.

13.

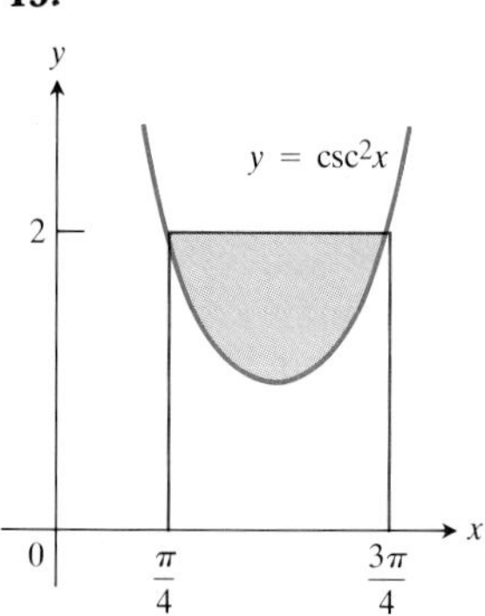

14.

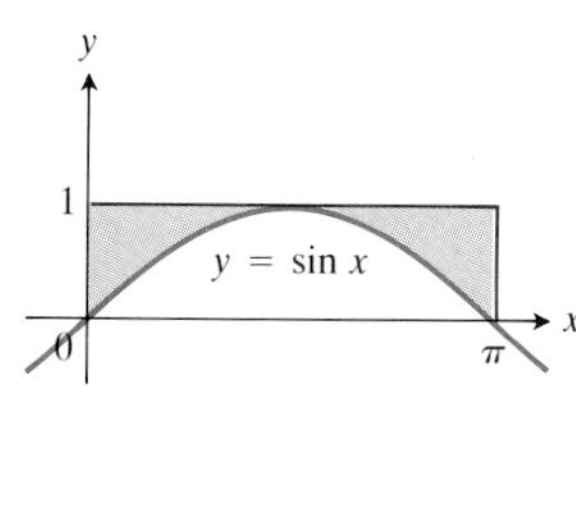

15.

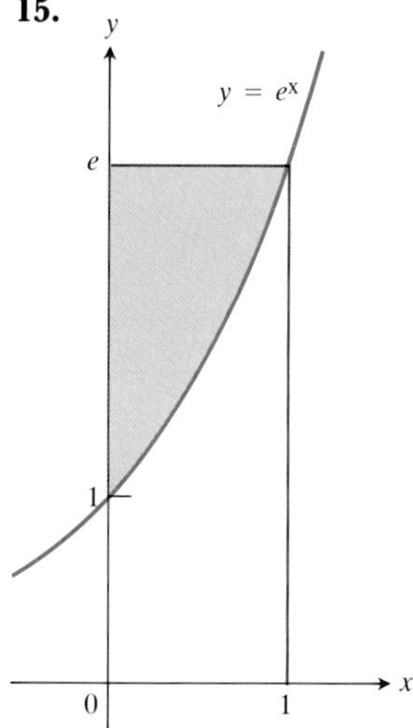

16.

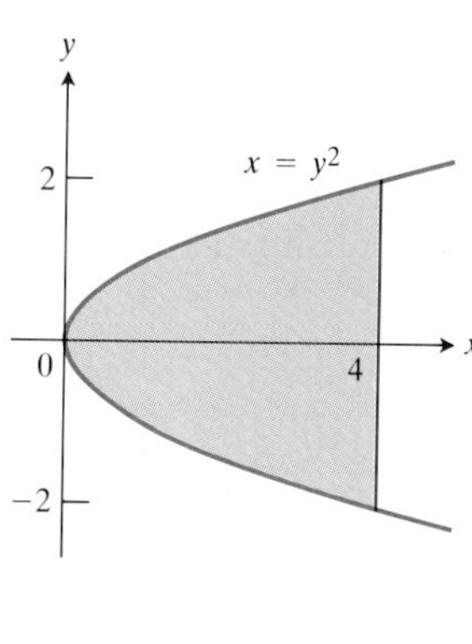

In Exercises 17 and 18, find the total area between the curve and the x-axis.

17. $y = 4 - x,\ 0 \le x \le 6$

18. $y = \cos x,\ -\pi \le x \le \pi$

Evaluate the integrals in Exercises 19–44.

19. $\int_{-1}^{1} (3x^2 - 4x + 7)\, dx$

20. $\int_{0}^{1} (8s^3 - 12s^2 + 5)\, ds$

21. $\int_{1}^{2} \frac{4}{x^2}\, dx$

22. $\int_{1}^{27} x^{-4/3}\, dx$

23. $\int_{1}^{4} \frac{dt}{t\sqrt{t}}$

24. $\int_{0}^{2} 3\sqrt{4x + 1}\, dx$

25. $\int_{0}^{1} \frac{36\, dx}{(2x + 1)^3}$

26. $\int_{1}^{2} \left(x + \frac{1}{x^2}\right) dx$

27. $\int_{0}^{\pi} \sin 5\theta\, d\theta$

28. $\int_{0}^{\pi} \cos 5t\, dt$

29. $\int_{0}^{\pi/3} \sec^2\theta\, d\theta$

30. $\int_{\pi/4}^{3\pi/4} \csc^2 x\, dx$

31. $\int_{\pi}^{3\pi} \cot^2 \frac{x}{6}\, dx$

32. $\int_{0}^{\pi} \tan^2 \frac{\theta}{3}\, d\theta$

33. $\int_{-\pi/3}^{0} \sec x \tan x\, dx$

34. $\int_{\pi/4}^{3\pi/4} \csc x \cot x\, dx$

35. $\int_{0}^{\pi/2} 5(\sin x)^{3/2} \cos x\, dx$

36. $\int_{-1}^{1} 2x \sin(1 - x^2)\, dx$

37. $\int_{4}^{8} \frac{1}{t}\, dt$

38. $\int_{0}^{2} \frac{2}{x + 1}\, dx$

39. $\int_{0}^{2} \frac{x\, dx}{x^2 + 5}$

40. $\int_{0}^{\pi} \frac{\cos x}{3 - \sin x}\, dx$

41. $\int_{\ln 3}^{\ln 4} e^x\, dx$

42. $\int_{0}^{\ln 3} e^{2x}\, dx$

43. $\int_{0}^{\pi/4} e^{\tan x} \sec^2 x\, dx$

44. $\int_{0}^{\pi/3} e^{\sec x} \sec x \tan x\, dx$

Evaluate the integrals in Exercises 45–48.

45. $\int_{2}^{3} \left(t - \frac{2}{t}\right)\left(t + \frac{2}{t}\right) dt$ (*Hint*: Multiply first.)

46. $\int_{-1}^{0} (1 - 3w)^2 dw$ (*Hint*: Square first.)

47. $\int_{-4}^{0} |x|\, dx$ (*Hint*: Write it without absolute values first.)

48. $\int_{1}^{e} \frac{x + 1}{x}\, dx$ (*Hint:* Divide first.)

Use the given substitutions to evaluate the integrals in Exercises 49–52.

49. $\int_{-\pi/2}^{\pi/2} 15 \sin^4 3x \cos 3x\, dx,\quad u = \sin 3x$

50. $\int_{0}^{1} \frac{e^x}{1 + e^x}\, dx,\quad u = 1 + e^x$

51. $\int_{0}^{\pi/2} \frac{3 \sin x \cos x}{\sqrt{1 + 3 \sin^2 x}}\, dx,\quad u = 1 + 3 \sin^2 x$

52. $\displaystyle\int_0^1 (2x-1)e^{(x^2-x)}\,dx, \quad u = x^2 - x$

In Exercises 53–60, find dy/dx.

53. $y = \ln\sqrt{x}$

54. $y = \ln\left(\dfrac{e^x}{2}\right)$

55. $y = \ln(3x^2+6)$

56. $y = \ln(1+e^x)$

57. $y = \dfrac{1}{e^x}$

58. $y = xe^{-x}$

59. $y = e^{(1+\ln x)}$

60. $y = \displaystyle\int_0^x \ln t\,dt$

Verify the integral formulas in Exercises 61–64 by differentiating their right-hand sides.

61. $\displaystyle\int \frac{\ln 5x}{x}\,dx = \frac{1}{2}(\ln 5x)^2 + C$

62. $\displaystyle\int x^2\ln x\,dx = \frac{x^3}{3}\ln x - \frac{x^3}{9} + C$

63. $\displaystyle\int e^x\sin x\,dx = \frac{e^x}{2}(\sin x - \cos x) + C$

64. $\displaystyle\int xe^x\,dx = xe^x - e^x + C$

65. Simplify:

a) $\ln e^{2x}$ b) $\ln 2e$ c) $\ln\dfrac{1}{e}$

66. Simplify:

a) $e^{2\ln 2}$ b) $e^{-\ln 4}$ c) $e^{\ln(\ln x)}$

Solve for y in Exercises 67–70.

67. $\ln(y^2+y) - \ln y = x, \quad y > 0$

68. $\ln(y-4) = -4t$

69. $e^{2y} = 4x^2, \quad x > 0$

70. $e^{-0.1y} = \dfrac{1}{2}$

71. CALCULATOR *The equations* $e^{\ln x} = x$ *and* $\ln e^x = x$. Evaluate the composites $e^{\ln x}$ and $\ln e^x$ for $x = 0.01$, 0.1, 1, 10, and 100, watching the calculator's display as you work through each evaluation.

72. CALCULATOR There are other logarithms besides the natural logarithm, among them the base-10 logarithm whose values can be found to 8 or 10 digits on most calculators by pressing [log] instead of [ln x]. Try calculating the ratio $(\ln x)/(\log x)$ for various values of x, noting the results as you go along. What happens? When you think you know, try it for a few more values of x. Then calculate ln 10. Section 7.3 will explain the connection.

73. Which of the following methods could be used successfully to prepare the integral

$$\int 3x^2(x^3-1)^5\,dx$$

for evaluation?

a) Expand $(x^3-1)^5$ and multiply the result by $3x^2$ to get a polynomial to integrate term by term.

b) Factor $3x^2$ out front to get an integral of the form $3x^2\int u^5\,du$.

c) Substitute $u = x^3 - 1$ to get an integral of the form $\int u^5\,du$.

74. The substitution $u = \tan x$ gives

$$\int \sec^2 x\tan x\,dx = \int \tan x\cdot\sec^2 x\,dx = \int u\,du = \frac{u^2}{2} + C = \frac{\tan^2 x}{2} + C.$$

The substitution $u = \sec x$ gives

$$\int \sec^2 x\tan x\,dx = \int \sec x\cdot\sec x\tan x\,dx = \int u\,du = \frac{u^2}{2} + C = \frac{\sec^2 x}{2} + C.$$

Can both integrations be correct? Explain.

75. Suppose that $y = f(x)$ is continuous and positive throughout the interval $[0, 1]$. Suppose also that for every value of x in this interval the area between the graph of f and the subinterval $[0, x]$ is $\sin x$. Find $f(x)$.

76. Suppose that f has a positive derivative for all values of x and that $f(1) = 0$. Which of the following statements must be true of the function

$$g(x) = \int_0^x f(t)\,dt?$$

a) g is a differentiable function of x.

b) g is a continuous function of x.

c) The graph of g has a horizontal tangent at $x = 1$.

d) g has a local maximum at $x = 1$.

e) g has a local minimum at $x = 1$.

f) The graph of g has an inflection point at $x = 1$.

g) The graph of dg/dx crosses the x-axis at $x = 1$.

77. Suppose $F(x)$ is an antiderivative of $f(x) = \sqrt{1+x^4}$. Express $\displaystyle\int_0^1 \sqrt{1+x^4}\,dx$ in terms of F.

78. Show that $y = x^2 + \displaystyle\int_1^x \frac{1}{t}\,dt + 1$ solves the initial value problem

Differential equation: $y'' = 2 - \dfrac{1}{x^2}$

Initial conditions: $y = 2$ and $y' = 3$ when $x = 1$.

79. The acceleration of a particle moving back and forth along a line is $d^2s/dt^2 = \pi^2 \cos \pi t$ m/sec². If $s = 0$ m and $v = 8$ m/sec when $t = 0$, find the value of s when $t = 1$.

80. Solve the following initial value problems.

a) Differential equation: $y^{(4)} = \cos x$
   Initial conditions: $y = 3, y' = 2, y'' = 1$, and $y''' = 0$ when $x = 0$

b) Differential equation: $\dfrac{dy}{dx} = \dfrac{4x}{(1 + x^2)^2}$
   Initial condition: $y = 0$ when $x = 0$

c) Differential equation: $\dfrac{dy}{dt} = -ky$
   Initial condition: $y = y_0$ when $t = 0$

81. *Stopping a motorcycle.* The State of Illinois Cycle Rider Safety Program requires riders to be able to brake from 30 mph (44 ft/sec) to 0 in 45 ft. What constant deceleration does it take to do that? To find out, carry out these steps:

STEP 1: Solve the following initial value problem. The answer will involve k.

Differential equation: $\dfrac{d^2s}{dt^2} = -k$

Initial conditions: $ds/dt = 44$ and $s = 0$ when $t = 0$

STEP 2: Find the time t^* when $ds/dt = 0$. The answer will still involve k.

STEP 3: Solve the equation $s(t^*) = 45$ for k.

82. Which of the following graphs shows the solution of the initial value problem

$$\frac{dy}{dx} = 2x, \quad y = 4 \text{ when } x = 1?$$

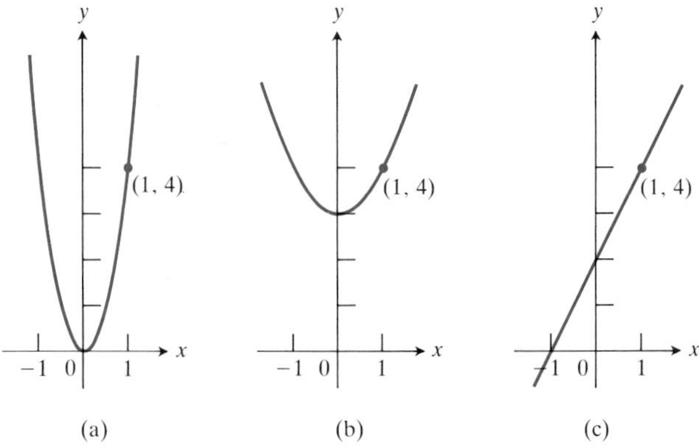

83. *Household electricity.* We model the voltage V in our homes with the sine function

$$V = V_{max} \sin 120\pi t,$$

which expresses V in volts as a function of time t in seconds. The function runs through 60 cycles each second. The number V_{max} is the **peak voltage.**

To measure the voltage effectively, we use an instrument that measures the average value of the square of the voltage over a 1-sec interval:

$$V_{rms} = \sqrt{(V^2)_{av}}.$$

The subscript "rms" (read the letters separately) stands for "root mean square."

It turns out that

$$V_{rms} = V_{max}/\sqrt{2}. \quad (1)$$

The familiar phrase "115 volts ac" means that the rms voltage is 115. The peak value, obtained from Eq. (1) as $V_{max} = 115\sqrt{2}$, is about 163 volts.

a) Find the average value of V^2 over a 1-sec interval. Then find V_{rms} and verify Eq. (1).

b) The circuit that runs your electric stove is rated 240 volts rms. What is the peak value of the allowable voltage?

84. CALCULATOR Compute the average value of the temperature function

$$f(x) = 37 \sin\left[\frac{2\pi}{365}(x - 101)\right] + 25$$

for a 365-day year. This is one way to estimate the annual mean air temperature in Fairbanks, Alaska. The National Weather Service's official figure, a numerical average of the daily normal mean air temperatures for the year, is 25.7°F, which is slightly higher than the average value of $f(x)$. Figure 1.79 shows why.

85. Let f be a function that is differentiable on $[a, b]$. In Chapter 1 we defined the average rate of change of f on $[a, b]$ to be

$$\frac{f(b) - f(a)}{b - a}$$

and the instantaneous rate of change of f at x to be $f'(x)$. In this chapter we defined the average value of a function. For the new definition of average to be consistent with the old one, we should have

$$\frac{f(b) - f(a)}{b - a} = \text{average value of } f' \text{ on } [a, b].$$

Show that this is the case.

86. Find the average value of

a) $y = \sqrt{3x}$ over the interval $0 \le x \le 3$,

b) $y = \sqrt{ax}$ over the interval $0 \le x \le a$.

Calculator Exercises (Mostly)

The following exercises involve numerical integration and applications of logarithms and exponentials. You had best use a calculator. The main exceptions are the exercises based on the Rule of 70.

87. What step size h would you use to be sure of estimating the value of

$$\ln 3 = \int_1^3 \frac{1}{x}\,dx$$

by Simpson's rule with an error of no more than 10^{-4} in absolute value?

88. A brief calculation shows that if $0 \le x \le 1$, then the second derivative of $f(x) = \sqrt{1+x^4}$ lies between 0 and 8. Based on this, about how many subdivisions would you need to estimate the integral of f from 0 to 1 with an error no greater than 10^{-3} in absolute value?

89. A direct calculation shows that

$$\int_0^{\pi} 2\sin^2 x\,dx = \pi.$$

How close do you come to this value by using the trapezoidal rule with $n = 6$? Simpson's rule with $n = 6$? Try them and find out.

90. You are planning to use Simpson's rule to estimate the value of the integral

$$\int_1^2 f(x)\,dx$$

with an error magnitude less than 10^{-5}. You have determined that $|f^{(4)}(x)| \le 3$ throughout the interval of integration. How many subdivisions should you use to assure the required accuracy? (Remember that for Simpson's rule the number has to be even.)

91. *A new parking lot.* To meet the demand for parking, your town has allocated the area shown in Fig. 5.28. As the town engineer, you have been asked by the town council to find out if the lot can be built for \$11,000. The cost to clear the land will be \$0.10 a square foot, and the lot will cost \$2.00 a square foot to pave. Can the job be done for \$11,000?

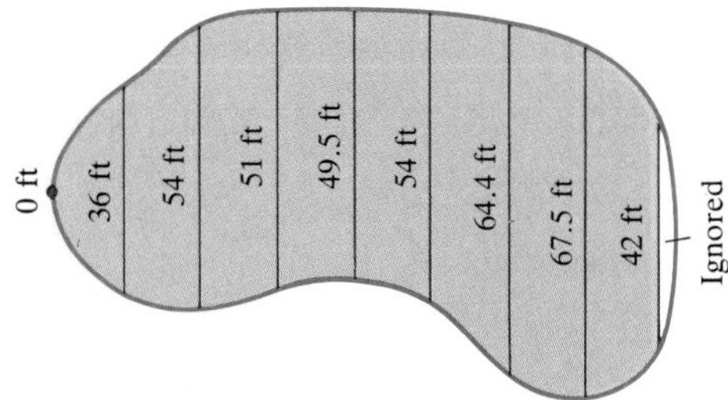

5.28

92. *Appreciation.* A violin made in 1785 by John Betts, one of England's finest violin makers, cost \$250 in 1924 and sold for \$7500 in 1988. Assuming a constant rate of appreciation, what was that rate?

93. *Working under water.* The intensity $L(x)$ of light x feet beneath the surface of the ocean satisfies the differential equation

$$\frac{dL}{dx} = -kx.$$

As a diver, you know from experience that diving to 18 ft in the Caribbean Sea cuts the intensity in half. You cannot work without artificial light when the intensity falls below a tenth of the surface value. About how deep can you expect to work without artificial light?

94. *The purchasing power of the dollar.* How long will it take the purchasing power of the dollar to decrease to three fourths of its present value if the annual inflation rate holds steady at 4%? (*Hint:* How long will it take the Consumer Price Index to increase to four thirds its present value at this inflation rate?)

95. *Inflation in West Germany.* The consumer price index (1967 = 100) in West Germany was 175.8 in 1980 and 211.9 in 1986.

a) Assuming a constant rate of inflation, what was it?
b) What would you expect the index to be in 1992 if this rate continued?

96. *Inflation in Italy.* In contrast to West Germany (Exercise 95), the index in Italy was 295.5 in 1980 and 480.1 in 1986.

a) Assuming a constant rate of inflation, what was it?
b) What would you expect the index to be in 1992 if this rate were to continue?

97. *The Rule of 70 for depletion.* The Rule of 70 applies to deflation and depletion, too, to estimate how many years it will take something that decreases at a constant rate to reach one half its present value. To see why, solve the equation

$$A_0 e^{-kt} = \frac{1}{2}A_0$$

to show that $t = (\ln 2)/k$. Thus if $k = i/100$ is given as $i\%$,

$$t = \frac{\ln 2}{(i/100)} = \frac{100 \ln 2}{i} \approx \frac{69.3}{i} \approx \frac{70}{i}.$$

As a rule, it will take about $70/i$ years for the amount in question to decline from A_0 to $A_0/2$.

98. *Continuation of Exercise 97.* Use the Rule of 70 to tell about how long it will take

a) for prices to decline to half their present level at a constant deflation rate of 5%;
b) for oil reserves to deplete to half their present volume at a constant depletion rate of 7%.

6 Applications of Definite Integrals

© Susan Van Etten

We find the volume of a dome by modeling it as a solid of revolution.

OVERVIEW The importance of integral calculus stems from the fact that thousands of things we want to know can be calculated with integrals. This includes areas between curves, volumes and surface areas of solids, lengths of curves, the amount of work it takes to pump liquids up from below ground, the forces against flood gates, and the coordinates of the points where solid objects will balance. We can define all these things in natural ways as limits of Riemann sums of continuous functions on closed intervals. Then we can evaluate these limits by applying techniques of integral calculus.

There is a clear pattern to how we go about defining and calculating the integrals in our applications, a pattern that, once learned, enables us to define new integrals whenever we need them. In this chapter we look at specific applications first; then, in the concluding section, we examine the pattern and show how it leads to integrals in new situations.

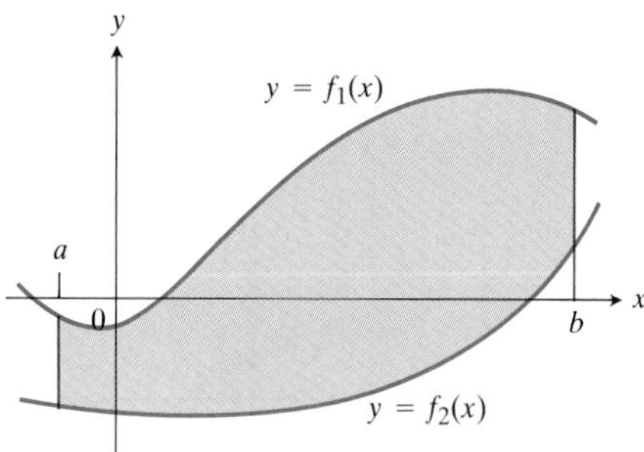

6.1 A typical region between two curves.

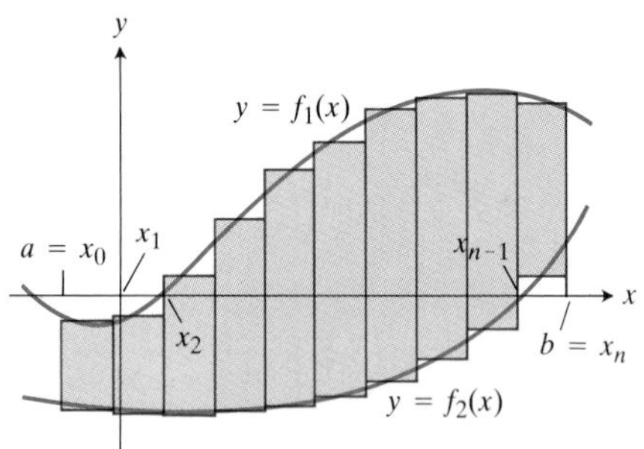

6.2 To derive a formula for the region's area, we think of approximating the region with rectangles perpendicular to the x-axis.

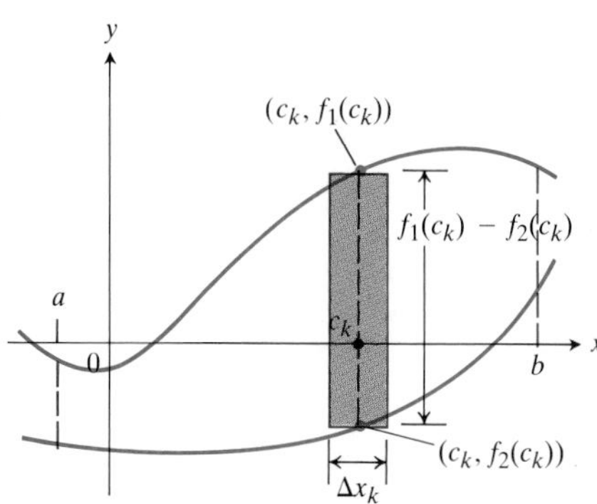

6.3 A typical rectangle is $f_1(c_k) - f_2(c_k)$ units high by Δx_k units wide.

6.1 Areas between Curves

This section shows how to find the area of a region in the coordinate plane by integrating the functions that define the region's boundaries.

The Basic Formula, Derived from Riemann Sums

Suppose that functions $f_1(x)$ and $f_2(x)$ are continuous and that $f_1(x) \geq f_2(x)$ throughout an interval $a \leq x \leq b$ (Fig. 6.1). The region between the curves $y = f_1(x)$ and $y = f_2(x)$ from a to b might accidentally have a shape that would let us find its area with a formula from geometry, but we usually have to find the area with an integral instead.

To see what that integral should be, we start by approximating the region with vertical rectangles. Figure 6.2 shows a typical approximation, based on a partition $P = \{x_0, x_1, \ldots, x_n\}$ of the interval $[a, b]$. The kth rectangle, shown in detail in Fig. 6.3, is Δx_k units wide and runs from the point $(c_k, f_2(c_k))$ on the lower curve to the point $(c_k, f_1(c_k))$ on the upper curve. Its area, height $\times$ width, is $(f_1(c_k) - f_2(c_k))\, \Delta x_k$.

We approximate the area of the region by adding the areas of the rectangles from a to b:

$$\text{Rectangle area sum} = \sum_{k=1}^{n} (f_1(c_k) - f_2(c_k))\, \Delta x_k. \tag{1}$$

This sum is a Riemann sum for the difference function $(f_1 - f_2)$ over the closed interval $[a, b]$.

Since f_1 and f_2 are continuous, two things will happen as we subdivide $[a, b]$ more finely and let the norm of the partition go to zero. The rectangles will approximate the region with increasing geometric accuracy and the Riemann sums in (1) will approach a limit. We therefore define this limit, the definite integral of $(f_1 - f_2)$ from a to b, to be the area of the region.

DEFINITION

If functions f_1 and f_2 are continuous and if $f_1(x) \geq f_2(x)$ throughout the interval $a \leq x \leq b$, then the **area of the region between the curves** $y = f_1(x)$ and $y = f_2(x)$ from a to b is the integral of $(f_1 - f_2)$ from a to b:

$$\text{Area} = \int_a^b (f_1(x) - f_2(x))\, dx. \tag{2}$$

To apply Eq. (2), we take the following steps.

How to Find the Area between Two Curves

1. *Graph the curves together.* This tells you which is f_1 (upper curve) and which is f_2 (lower curve). It also helps find the limits of integration if you do not already know them.
2. *Find the limits of integration.*
3. *Write a formula for $f_1(x) - f_2(x)$.* Simplify it if you can.
4. *Integrate $f_1(x) - f_2(x)$ from a to b.* The number you get is the area.

EXAMPLE 1 Find the area between the curves $y = \cos x$ and $y = -\sin x$ from 0 to $\pi/2$.

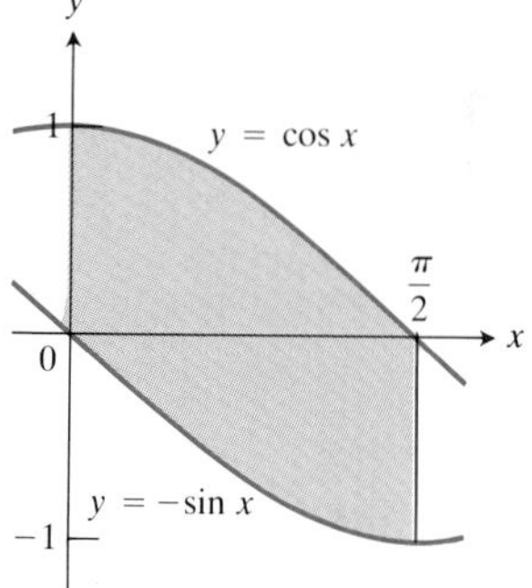

Solution STEP 1: *The graphs.* We graph the curves together. The upper curve is $y = \cos x$, so we take $f_1(x) = \cos x$ in the area formula. The lower curve is $y = -\sin x$, so $f_2(x) = -\sin x$.

STEP 2: *The limits of integration.* They are already given: $a = 0$ and $b = \pi/2$.

STEP 3: *The formula for $f_1(x) - f_2(x)$.* From Step 1,

$$\begin{aligned} f_1(x) - f_2(x) &= \cos x - (-\sin x) \\ &= \cos x + \sin x. \end{aligned}$$

STEP 4: *Integrate $f_1(x) - f_2(x)$ from $a = 0$ to $b = \pi/2$:*

$$\begin{aligned} \int_0^{\pi/2} (\cos x + \sin x)\, dx &= \Big[\sin x - \cos x\Big]_0^{\pi/2} \\ &= [1 - 0] - [0 - 1] = 2. \end{aligned}$$

The area between the curves is 2. ■

Curves That Cross

When a region is determined by curves that cross, the crossing points give the limits of integration.

EXAMPLE 2 Find the area of the region enclosed by the parabola $y = 2 - x^2$ and the line $y = -x$.

Solution STEP 1: *The graphs.*
We graph the curves together. The upper curve is $y = 2 - x^2$, so $f_1(x) = 2 - x^2$. The lower curve is $y = -x$, so $f_2(x) = -x$. The x-coordinates of the points where the parabola and line cross are the limits of integration. We find them in Step 2.

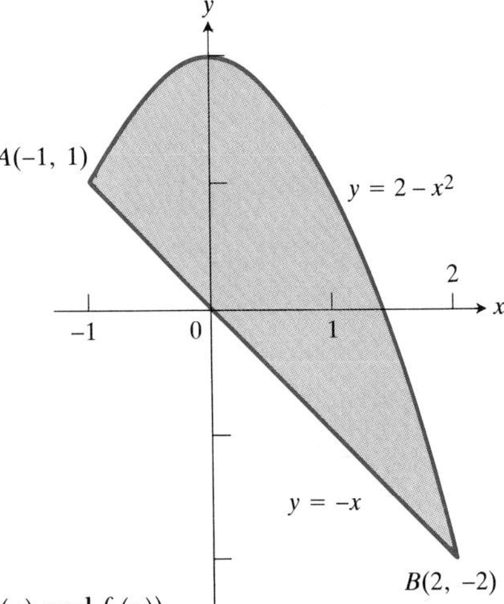

STEP 2: *The limits of integration.*
We find the limits of integration by solving the equations $y = 2 - x^2$ and $y = -x$ simultaneously for x:

$$
\begin{aligned}
2 - x^2 &= -x && \text{(Equate } f_1(x) \text{ and } f_2(x)\text{)} \\
x^2 - x - 2 &= 0 && \text{(Transpose)} \\
(x + 1)(x - 2) &= 0 && \text{(Factor)} \\
x = -1, \quad x &= 2. && \text{(Solve)}
\end{aligned}
$$

The region runs from $x = -1$ on the left to $x = 2$ on the right. The limits of integration are $a = -1$ and $b = 2$.

STEP 3: *The formula for $f_1(x) - f_2(x)$.*

$$
\begin{aligned}
f_1(x) - f_2(x) &= (2 - x^2) - (-x) = 2 - x^2 + x \\
&= 2 + x - x^2. \qquad \begin{pmatrix}\text{Rearranged—a} \\ \text{matter of taste}\end{pmatrix}
\end{aligned}
$$

STEP 4. *Integrate.*

$$
\begin{aligned}
\int_a^b [f_1(x) - f_2(x)]\,dx &= \int_{-1}^{2} (2 + x - x^2)\,dx = \left[2x + \frac{x^2}{2} - \frac{x^3}{3}\right]_{-1}^{2} \\
&= \left(4 + \frac{4}{2} - \frac{8}{3}\right) - \left(-2 + \frac{1}{2} + \frac{1}{3}\right) \\
&= 6 + \frac{3}{2} - \frac{9}{3} = \frac{9}{2}.
\end{aligned}
$$

The area of the region is 9/2. ■

Boundaries with Changing Formulas

If the formula for one of the bounding curves changes at some point across the region, you may have to add two or more integrals to find the area.

EXAMPLE 3 Find the area of the region in the first quadrant bounded above by the curve $y = \sqrt{x}$ and below by the x-axis and the line $y = x - 2$.

Solution STEP 1: *The graphs.* We graph the curves together. The entire upper boundary of the region consists of the curve $y = \sqrt{x}$, so $f_1(x) = \sqrt{x}$. The lower boundary consists of two curves, first $y = 0$ for $0 \le x \le 2$ and then $y = x - 2$ for $2 \le x \le 4$. Hence the formula for $f_2(x)$ changes from $f_2(x) = 0$ for $0 \le x \le 2$ to $f_2(x) = x - 2$ for $2 \le x \le 4$.

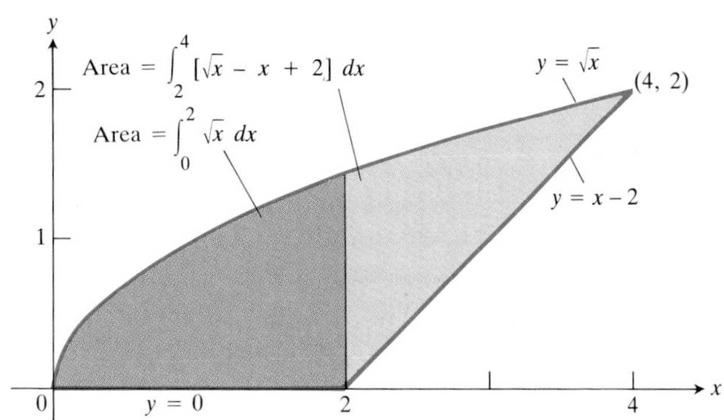

STEP 2: *The limits of integration.* The limits of integration for the pair $f_1(x) = \sqrt{x}$ and $f_2(x) = 0$ are $a = 0$ and $b = 2$.

For the pair $f_1(x) = \sqrt{x}$ and $f_2(x) = x - 2$, the left-hand limit is $a = 2$ and the right-hand limit is the x-coordinate of the upper point where the line crosses the parabola. To find it, we solve the equations $y = \sqrt{x}$ and $y = x - 2$ simultaneously for x:

$$\sqrt{x} = x - 2 \qquad \left(\begin{array}{l}\text{Equate } f_1(x) = \sqrt{x} \text{ and}\\ f_2(x) = x - 2\end{array}\right)$$

$$x = (x - 2)^2 = x^2 - 4x + 4 \qquad \text{(Square)}$$

$$x^2 - 5x + 4 = 0 \qquad \text{(Rearrange)}$$

$$(x - 1)(x - 4) = 0 \qquad \text{(Factor)}$$

$$x = 1, \quad x = 4. \qquad \text{(Solve)}$$

The value $x = 1$ does not satisfy the equation $\sqrt{x} = x - 2$. It is an extraneous root introduced by squaring. The value $x = 4$ gives our upper limit of integration.

STEP 3: *The formulas for $f_1(x) - f_2(x)$.*

For $0 \le x \le 2$: $\quad f_1(x) - f_2(x) = \sqrt{x} - 0 = \sqrt{x}$,

For $2 \le x \le 4$: $\quad f_1(x) - f_2(x) = \sqrt{x} - (x - 2) = \sqrt{x} - x + 2$.

STEP 4: *Integrate.* We have two integrals to evaluate. Their sum is the area.

$$\text{Area} = \int_0^4 [f_1(x) - f_2(x)]\, dx = \int_0^2 \sqrt{x}\, dx + \int_2^4 (\sqrt{x} - x + 2)\, dx$$

$$= \left[\frac{2}{3}x^{3/2}\right]_0^2 + \left[\frac{2}{3}x^{3/2} - \frac{x^2}{2} + 2x\right]_2^4$$

$$= \frac{2}{3}(2)^{3/2} + \left(\frac{2}{3}(4)^{3/2} - \frac{16}{2} + 8\right) - \left(\frac{2}{3}(2)^{3/2} - \frac{4}{2} + 4\right) = \frac{2}{3}(8) - 2 = \frac{10}{3}.$$

The area of the region is 10/3. ■

Integrating with Respect to *y*

When a region's bounding curves are described by giving x as a function of y, the basic formula changes.

For regions like these,

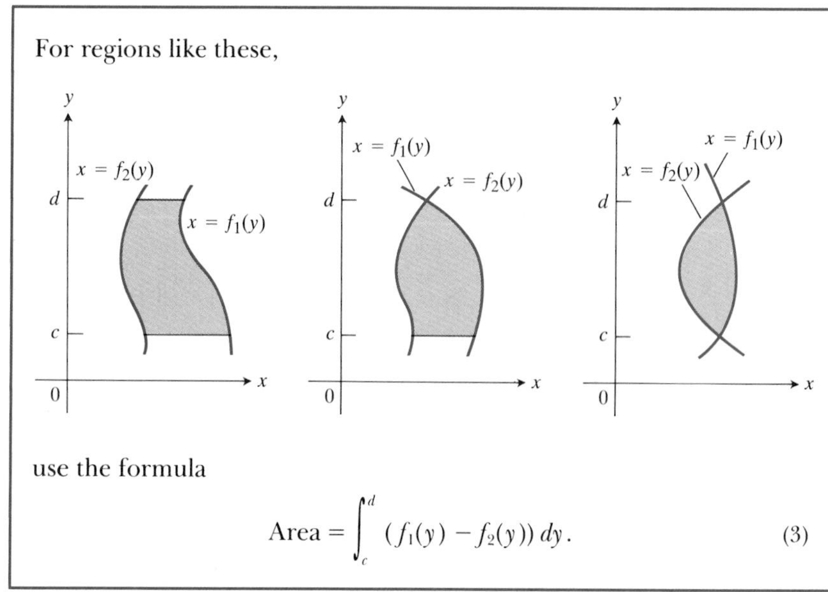

use the formula

$$\text{Area} = \int_c^d (f_1(y) - f_2(y))\, dy. \qquad (3)$$

The only difference between this formula and the one in Eq. (2) is that we are now integrating with respect to y instead of x. We can sometimes save time by doing so. The basic steps are the same as before.

EXAMPLE 4 *The area in Example 3, found by a single integration with respect to y.* Find the area of the region between the curves $x = y^2$ and $x = y + 2$ in the first quadrant.

Solution STEP 1: *The graphs.* We graph the curves together. The right-hand curve is $x = y + 2$, so $f_1(y) = y + 2$. The left-hand curve is $x = y^2$, so $f_2(y) = y^2$.

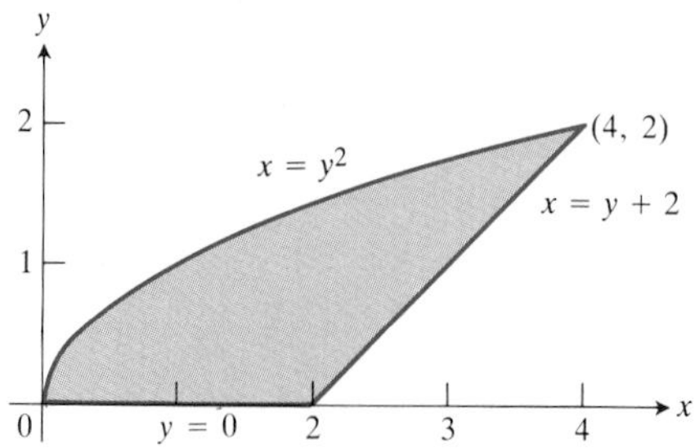

STEP 2: *The limits of integration.* The lower limit of integration is $y = 0$. The upper limit is the y-coordinate of the upper point where the line crosses the

parabola. We find it by solving the equations $x = y + 2$ and $x = y^2$ simultaneously for y:

$$
\begin{aligned}
y + 2 &= y^2 && (\text{Equate } f_1(y) = y + 2 \text{ and } f_2(y) = y^2) \\
y^2 - y - 2 &= 0 && (\text{Rearrange}) \\
(y + 1)(y - 2) &= 0 && (\text{Factor}) \\
y = -1, \quad y &= 2. && (\text{Solve})
\end{aligned}
$$

The upper limit of integration is 2. (The value $y = -1$ gives the point of intersection *below* the x-axis.)

STEP 3: *The formula for* $f_1(y) - f_2(y)$.

$$f_1(y) - f_2(y) = y + 2 - y^2 = 2 + y - y^2. \qquad \begin{pmatrix}\text{Rearrangement} \\ \text{a matter of taste}\end{pmatrix}$$

STEP 4: *Integrate.*

$$
\begin{aligned}
\text{Area} &= \int_a^b (f_1(y) - f_2(y))\, dy = \int_0^2 (2 + y - y^2)\, dy \\
&= \left[2y + \frac{y^2}{2} - \frac{y^3}{3}\right]_0^2 = 4 + \frac{4}{2} - \frac{8}{3} = \frac{10}{3}.
\end{aligned}
$$

The area of the region is 10/3. This is the result of Example 3, found with less work. ■

Combining Integrals with Formulas from Geometry

Sometimes the fastest way to find the area of a region is to combine calculus and geometry.

EXAMPLE 5 *The area of the region in Examples 3 and 4 found the fastest way.* Find the area of the region in Examples 3 and 4.

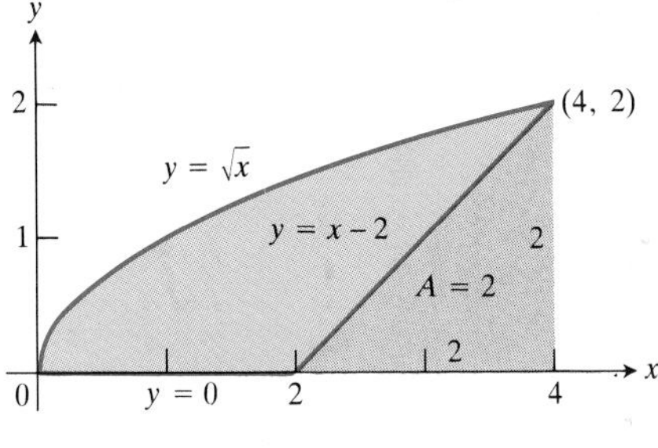

Solution The area we want to find is the area between the curve $y = \sqrt{x}$, $0 \le x \le 4$, and the x-axis, *minus* the area of a triangle with base 2 and height 2:

$$
\begin{aligned}
\text{Area} &= \int_0^4 \sqrt{x}\, dx - \frac{1}{2}(2)(2) = \frac{2}{3}x^{3/2}\Big]_0^4 - 2 \\
&= \frac{2}{3}(8) - 0 - 2 = \frac{10}{3}.
\end{aligned}
$$

■

Moral of Examples 3–5 It is sometimes easier to find the area between two curves by integrating with respect to y instead of x. Examine each region beforehand to determine which method, if either, is easier. Sketching the region may also reveal how to use geometry to simplify your work.

Exercises 6.1

Find the areas of the shaded regions in Exercises 1–6.

1.

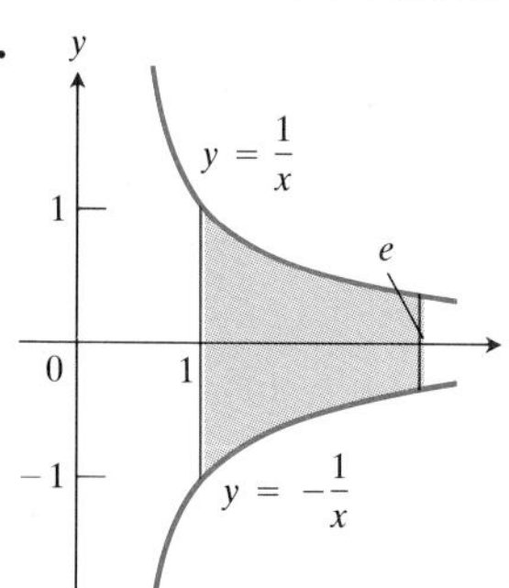

2.

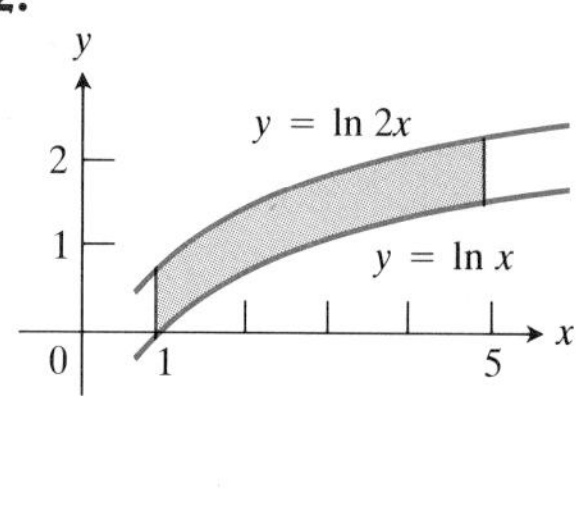

3.

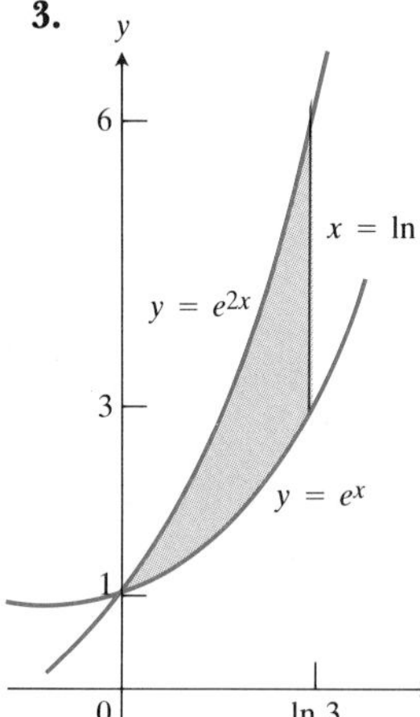

4.

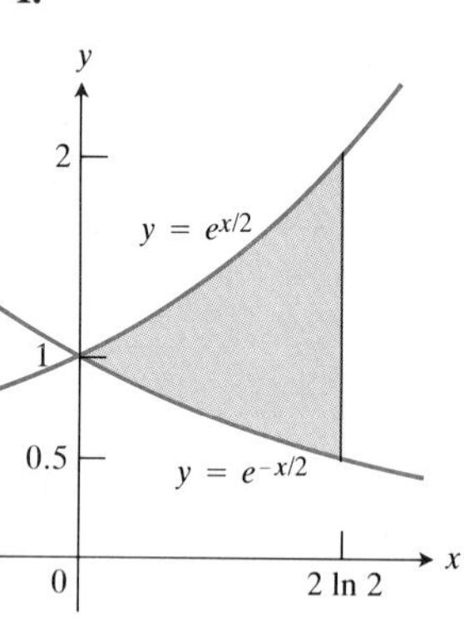

5.

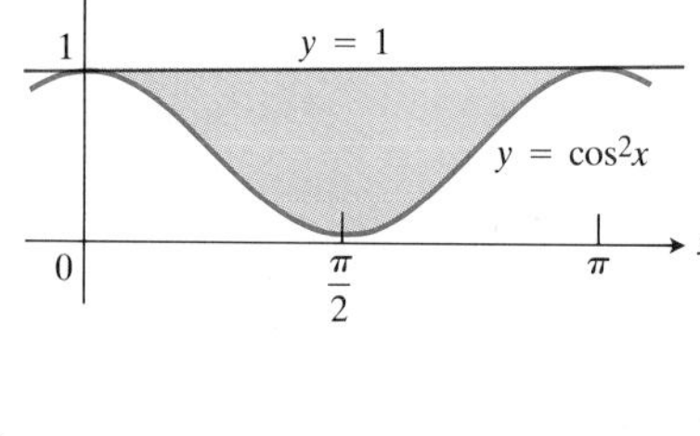

6.

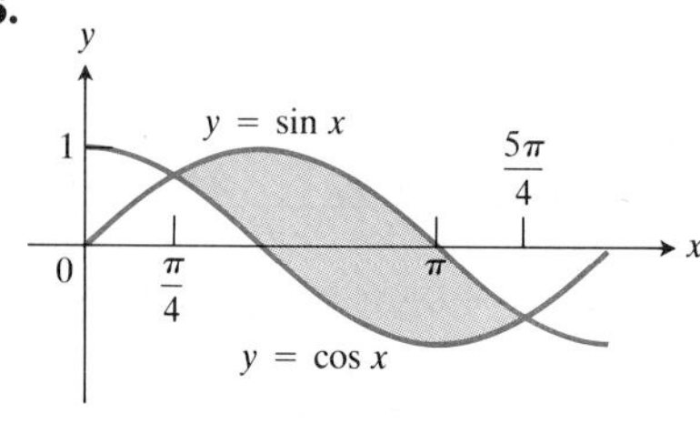

Find the areas of the regions enclosed by the lines and curves in Exercises 7–22.

7. The curve $y = x^2 - 2$ and the line $y = 2$

8. The x-axis and the curve $y = 2x - x^2$

9. The curve $y^2 = x$ and the line $x = 4$

10. The curve $y = 2x - x^2$ and the line $y = -3$

11. The curve $y = x^2$ and the line $y = x$

12. The curve $x = 3y - y^2$ and the line $x + y = 3$

13. The curve $x = y^2$ and the line $x = y + 2$

14. The curve $y = x^4$ and the line $y = 8x$

15. The curve $y = x^2 - 2x$ and the line $y = x$

16. The curve $x = 10 - y^2$ and the line $x = 1$

17. The curves $x = y^2, \quad x = -2y^2 + 3$

18. The curves $y = x^2, \quad y = -x^2 + 4x$

19. The line $y = x$ and the curve $y = 2 - (x - 2)^2$

20. The curves $y = 7 - 2x^2$ and $y = x^2 + 4$

21. The curves $y = \cos(\pi x/2), \quad y = 1 - x^2$

22. The curve $y = \sin(\pi x/2)$ and the line $y = x$

Find the areas of the shaded regions in Exercises 23–26.

23.

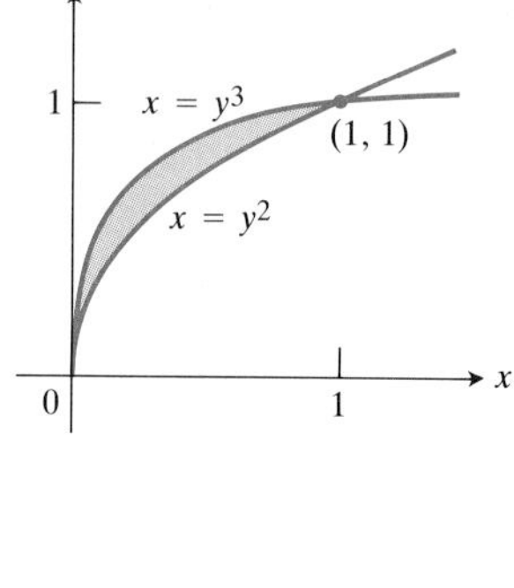

24.

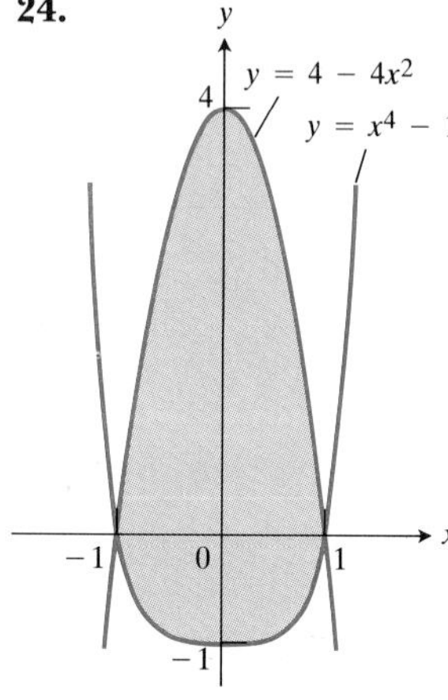

25.

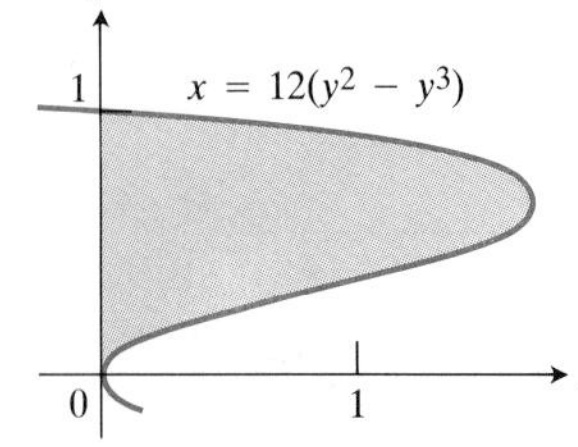

26.

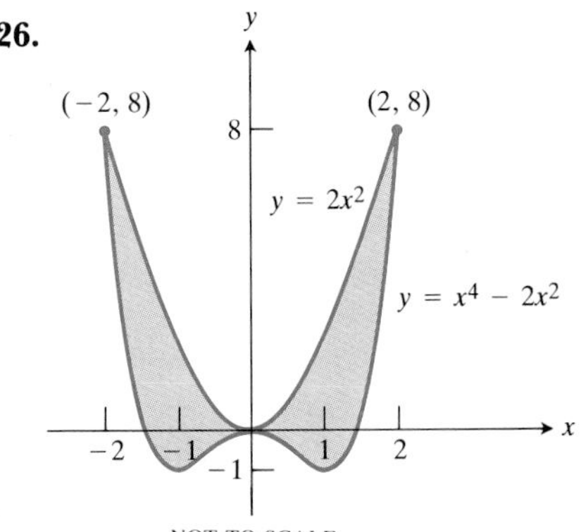

27. Find the area of the "triangular" region bounded by the y-axis and the curves $y = \sin x$ and $y = \cos x$ in the first quadrant.

28. Find the area of the region between the curve $y = 3 - x^2$ and the line $y = -1$ by integrating with respect to (a) x; (b) y.

29. The area of the region between the curve $y = x^2$ and the line $y = 4$ is divided into two equal portions by the line $y = c$.

a) Find c by integrating with respect to y. (This puts c into the limits of integration.)

b) Find c by integrating with respect to x. (This puts c into the integrand as well.)

30. Figure 6.4 shows triangle AOC inscribed in the region cut from the parabola $y = x^2$ by the line $y = a^2$. Find the limit of the ratio of the area of the triangle to the area of the parabolic region as a approaches zero.

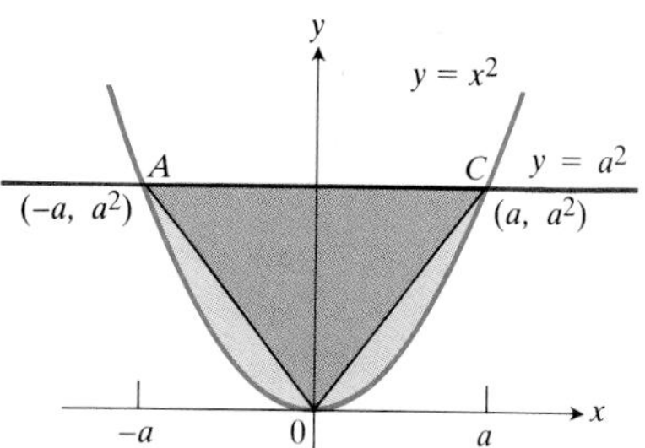

6.4 The Figure for Exercise 30

31. Suppose that the area between the continuous curve $y = f(x)$ shown here and the x-axis from $x = a$ to $x = b$ is 4 square units. Find the area between the curves $y = f(x)$ and $y = 2f(x)$ from $x = a$ to $x = b$.

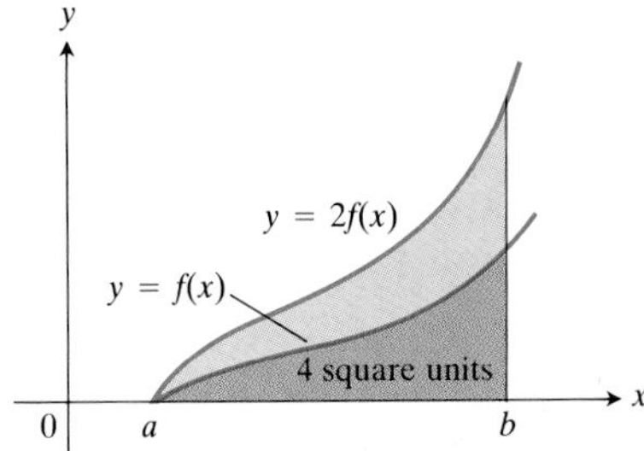

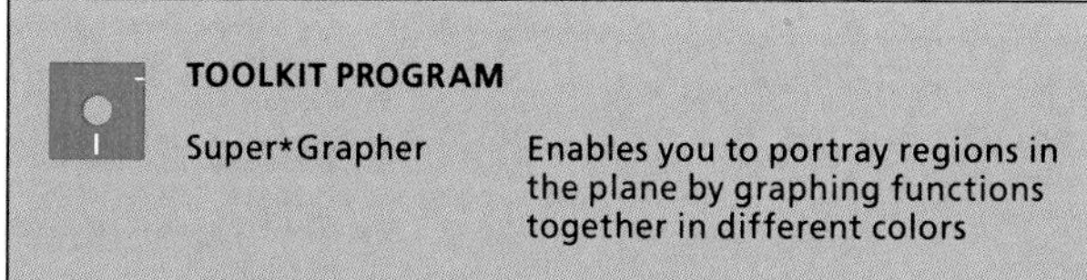

TOOLKIT PROGRAM

Super*Grapher — Enables you to portray regions in the plane by graphing functions together in different colors

6.2 Volumes of Solids of Revolution—Disks and Washers

Solids of revolution are solids whose shapes can be generated by revolving plane regions about axes. Thread spools are solids of revolution; so are hand weights and billiard balls. Solids of revolution sometimes have

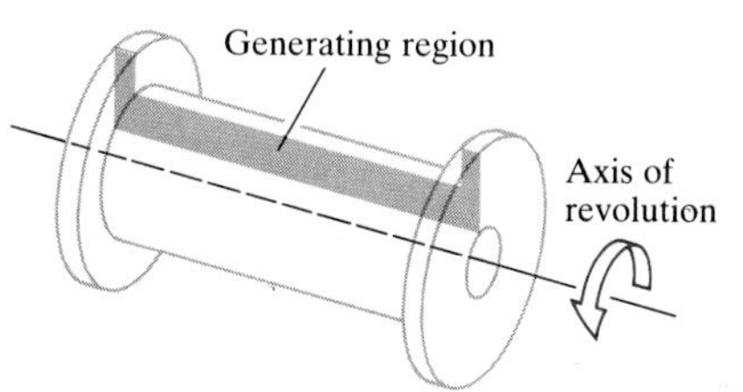

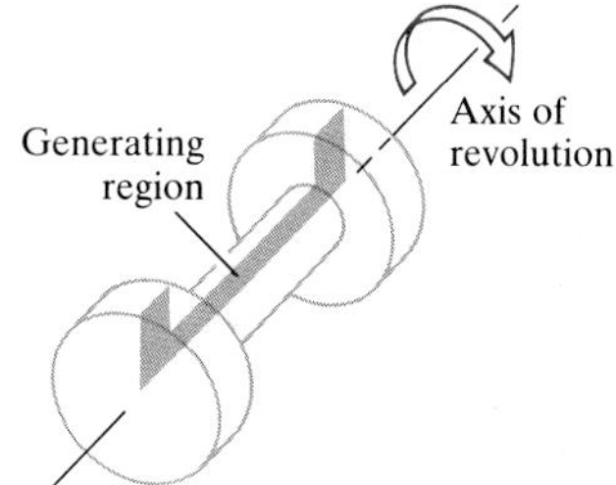

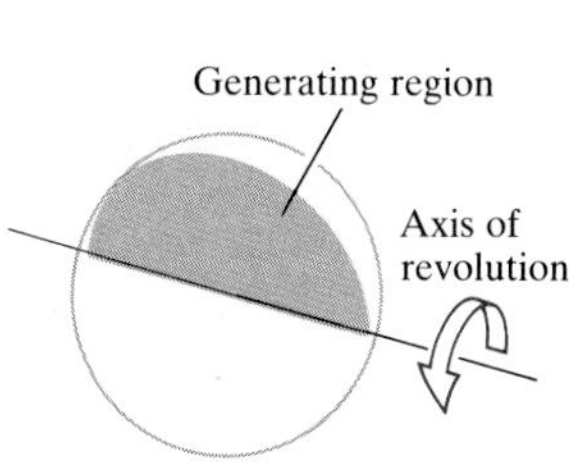

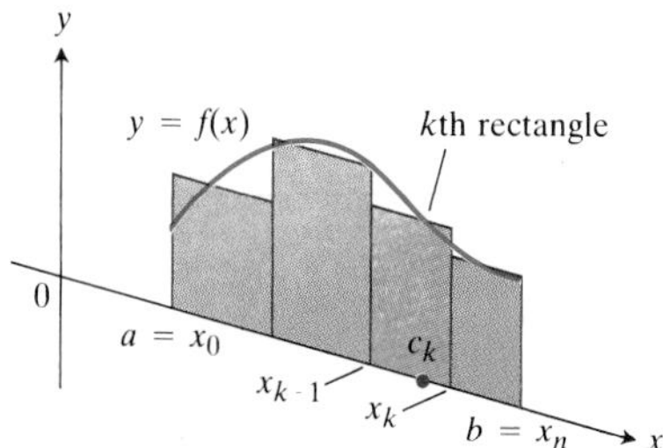

6.5 The first step in developing an integral to calculate the volume of a solid of revolution is to approximate the generating region with rectangles.

volumes we can find with formulas from geometry, as we can the volume of the billiard ball. But, likely as not, we want to find the volume of a blimp instead, or to predict the weight of a part we are going to have turned on a lathe. In cases like these, formulas from geometry are of little help and we must turn to calculus for the answers. In this section and the next we show how to find these answers.

The Disk Method

As we said, a solid of revolution is generated by revolving a region about an axis. We usually assume the axis lies in the same plane as the region. If we can set things up so that the region is the region between the graph of a continuous function $y = f(x)$, $a \leq x \leq b$, and the x-axis, and the axis of revolution is the x-axis, then we can define and calculate the resulting solid's volume in the following way.

We begin by approximating the region itself with vertical rectangles based on a partition of the closed interval $[a, b]$ (Fig. 6.5). We then imagine the rectangles to be revolved about the x-axis along with the region. Each rectangle generates a solid disk (Fig. 6.6). The disks, taken together, approximate the solid of revolution. The approximation is similar to what we would get if we sliced up the original solid like a loaf of bread and reshaped each slice into a disk.

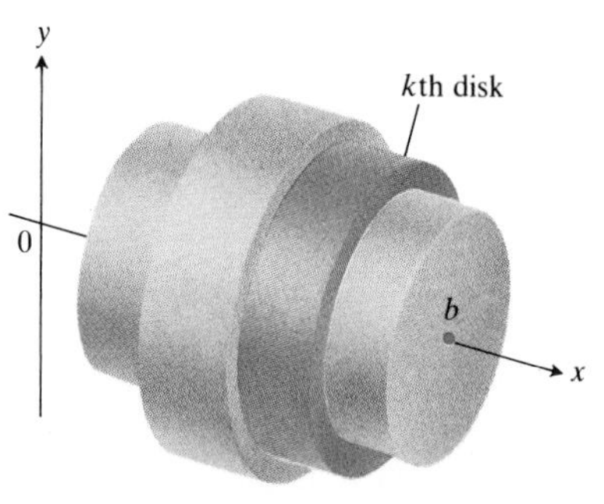

6.6 The rectangles revolve around the x-axis to generate solid disks whose volumes, when added, approximate the volume of the solid.

If we focus on a typical rectangle and the disk it generates (Fig. 6.7), we see that the disk is a right circular cylinder with height Δx_k and radius $f(c_k)$. We can therefore calculate the volume of the disk with the geometry formula $V = \pi r^2 h$:

$$\text{Disk volume} = \pi \times (\text{radius})^2 \times \text{height} = \pi(f(c_k))^2 \Delta x_k. \tag{1}$$

The volumes of all the disks together add up to

$$\text{Disk volume sum} = \sum_{k=1}^{n} \pi(f(c_k))^2 \Delta x_k. \tag{2}$$

The expression on the right-hand side of Eq. (2) is a Riemann sum for the function πf^2 on the closed interval $[a, b]$. As the norm of the partition of $[a, b]$ approaches zero, two things happen simultaneously. First, the rectangles approximating the revolved region fit the region with increasing accuracy and the disks they generate fit the solid of revolution with increasing accuracy. Second, the Riemann sums in Eq. (2) approach the integral of πf^2 from a to b. We therefore define the volume of the solid of revolution to be the value of this integral.

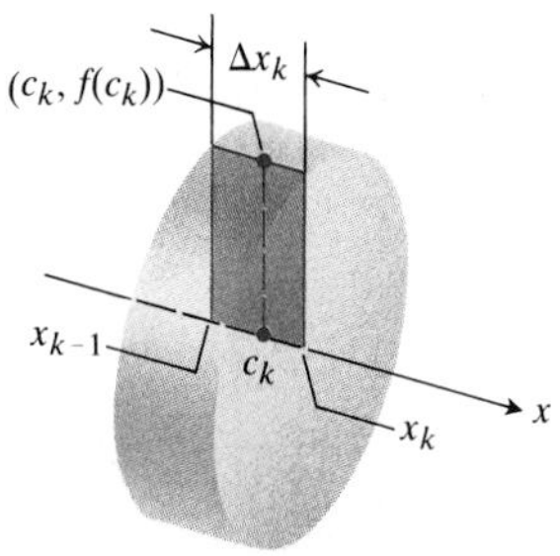

6.7 Enlarged view of the kth disk from Fig. 6.6. Its volume is base area × height = $\pi[f(c_k)]^2 \Delta x_k$.

DEFINITION

Volume of a Solid of Revolution (Rotation about the *x*-Axis)

The volume of the solid generated by revolving the region between the graph of a continuous function $y = f(x)$ and the x-axis from $x = a$ to $x = b$ about the x-axis is

$$\text{Volume} = \int_a^b \pi(\text{radius})^2\, dx = \int_a^b \pi(f(x))^2\, dx. \tag{3}$$

To Apply Eq. (3)

1. Square the expression for the radius function $f(x)$.
2. Multiply by π.
3. Integrate from a to b.

EXAMPLE 1 The region between the curve $y = \sqrt{x}$, $0 \le x \le 4$, and the x-axis is revolved about the x-axis to generate the shape in Fig. 6.8. Find its volume.

Solution

$$\text{Volume} = \int_a^b \pi(\text{radius})^2\,dx \qquad \text{(Eq. 3)}$$

$$= \int_0^4 \pi(\sqrt{x})^2\,dx \qquad \left(\begin{array}{l}\text{The radius function is}\\ f(x) = \sqrt{x},\ 0 \le x \le 4.\end{array}\right)$$

$$= \pi\int_0^4 x\,dx = \pi\frac{x^2}{2}\bigg]_0^4 = \pi\frac{(4)^2}{2} = 8\pi$$

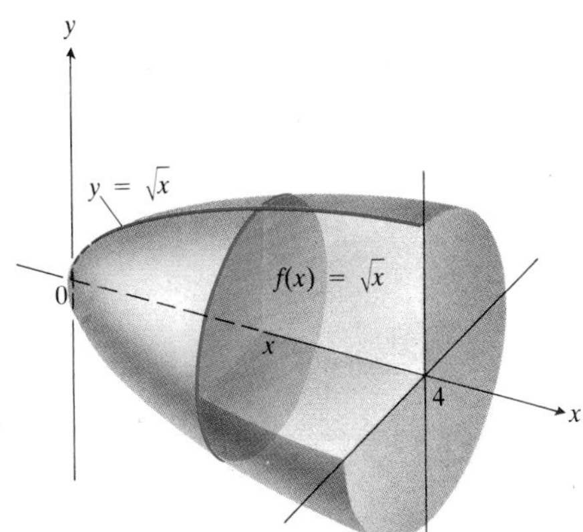

6.8 A slice perpendicular to the axis of the solid in Example 1. The radius at x is (x) = √x.

The volume formula in Eq. (3) is consistent with all the standard formulas from geometry. If we use Eq. (3) to calculate the volume of a sphere of radius a, for instance, we get $(4/3)\pi a^3$, just as we should. The next example shows how the calculation goes.

EXAMPLE 2 The region enclosed by the semicircle $y = \sqrt{a^2 - x^2}$ and the x-axis is revolved about the x-axis to generate a sphere (Fig. 6.9). Find the volume of the sphere.

Solution

$$\text{Volume} = \int_{-a}^{a} \pi(\text{radius})^2\,dx \qquad \left(\begin{array}{l}\text{Eq. (3) with limits of}\\ \text{integration } -a \text{ and } a\end{array}\right)$$

$$= \int_{-a}^{a} \pi\left(\sqrt{a^2 - x^2}\right)^2 dx \qquad \left(\begin{array}{l}\text{The radius at } x \text{ is}\\ f(x) = \sqrt{a^2 - x^2}.\end{array}\right)$$

$$= \pi\int_{-a}^{a} (a^2 - x^2)\,dx = \pi\left[a^2x - \frac{x^3}{3}\right]_{-a}^{a}$$

$$= \pi\left(a^3 - \frac{a^3}{3}\right) - \pi\left(-a^3 + \frac{a^3}{3}\right)$$

$$= \pi\left(\frac{2a^3}{3}\right) - \pi\left(-\frac{2a^3}{3}\right) = \frac{4}{3}\pi a^3$$

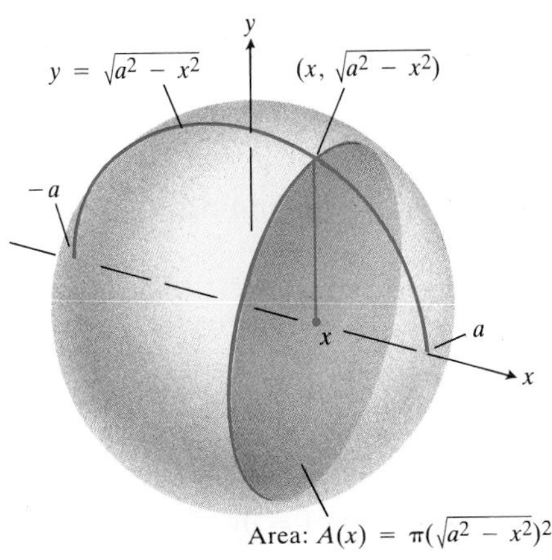

6.9 The sphere generated by revolving the semicircle y = √(a² − x²) about the -axis. The radius at x is f(x) = √(a² − x²).

The axis of revolution in the next example is not the x-axis, but the rule for calculating the volume is the same: Integrate $\pi(\text{radius})^2$ between appropriate limits.

EXAMPLE 3 Find the volume generated by revolving the region bounded by $y = \sqrt{x}$ and the lines $y = 1$ and $x = 4$ about the line $y = 1$.

Solution We draw a figure that shows the region and the radius at a typical point on the axis of revolution (Fig. 6.10). The region runs from $x = 1$ to $x = 4$. At each x in the interval $1 \le x \le 4$, the cross-section radius is

$$\text{Radius at } x = \sqrt{x} - 1.$$

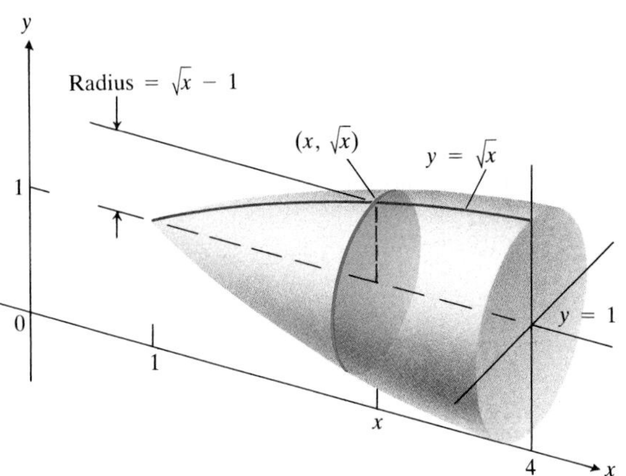

6.10 The solid swept out by revolving the region bounded by $y = \sqrt{x}$, $x = 4$, and $y = 1$ about the line $y = 1$.

Therefore

$$
\begin{aligned}
\text{Volume} &= \int_1^4 \pi(\text{radius})^2\,dx = \int_1^4 \pi(\sqrt{x} - 1)^2\,dx \\
&= \pi \int_1^4 (x - 2\sqrt{x} + 1)\,dx \\
&= \pi \left[\frac{x^2}{2} - 2 \cdot \frac{2}{3} x^{3/2} + x\right]_1^4 \\
&= \frac{7\pi}{6}. \qquad \begin{pmatrix}\text{Arithmetic} \\ \text{omitted}\end{pmatrix}
\end{aligned}
$$

To find the volume of the solid generated by revolving the region between a curve $x = f(y)$ and the y-axis from $y = c$ to $y = d$ about the y-axis, use Eq. (3) with x replaced by y.

Volume of a Solid of Revolution (Rotation about the *y*-Axis)

$$\text{Volume} = \int_c^d \pi(\text{radius})^2\,dy = \int_c^d \pi(f(y))^2\,dy \tag{4}$$

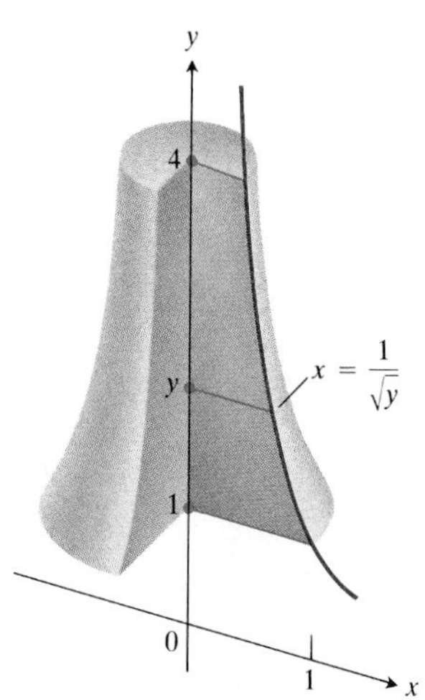

6.11 The solid in Example 4.

EXAMPLE 4 The region between the curve $x = 1/\sqrt{y}$, $1 \le y \le 4$, is revolved about the y-axis to generate a solid (Fig. 6.11). Find the volume of the solid.

Solution

$$
\begin{aligned}
\text{Volume} &= \int_1^4 \pi(\text{radius})^2\,dy = \int_1^4 \pi\left(\frac{1}{\sqrt{y}}\right)^2 dy \qquad \begin{pmatrix}\text{The radius function} \\ \text{is } f(y) = 1/\sqrt{y}.\end{pmatrix} \\
&= \pi \int_1^4 \frac{1}{y}\,dy = \pi \ln y \Big]_1^4 = \pi \ln 4 - 0 \qquad (\ln 1 = 0) \\
&= 2\pi \ln 2 \qquad (\ln 4 = \ln 2^2 = 2 \ln 2)
\end{aligned}
$$

The Washer Method

If the region we revolve to generate a solid does not border on the axis of revolution,

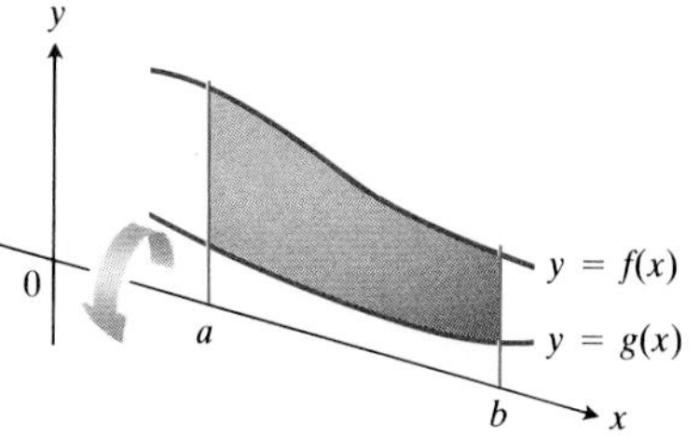

. . . the solid has a hole in it.

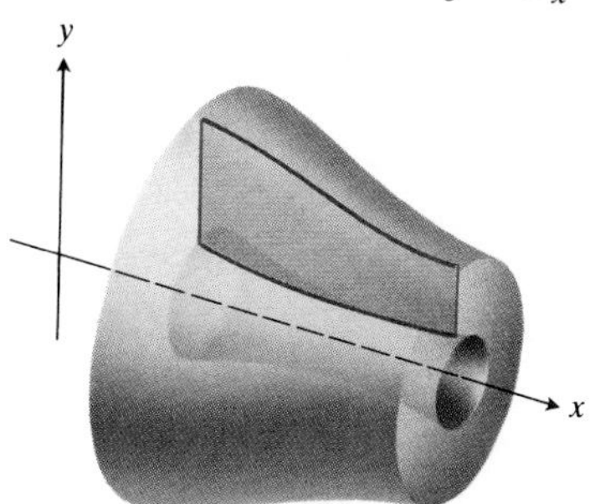

The cross sections perpendicular to the axis of revolution are washers instead of disks.

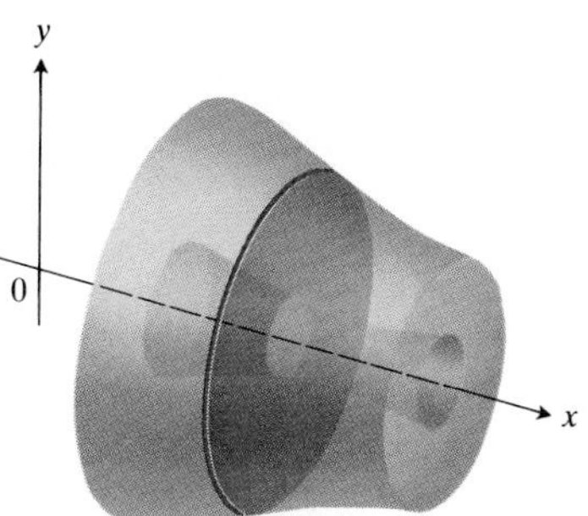

If we fill in the hole, the volume is

$$\int_a^b \pi(f(x))^2\,dx.$$

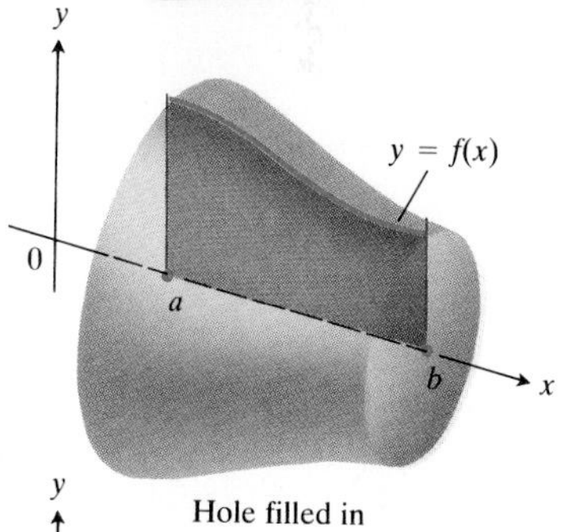

Hole filled in

The volume of the hole itself is

$$\int_a^b \pi(g(x))^2\,dx.$$

Volume of the hole

The volume of the original solid is therefore

$$\begin{aligned}\text{Volume} &= (\text{volume with hole filled in}) - (\text{volume of hole})\\ &= \int_a^b \pi(f(x))^2\,dx - \int_a^b \pi(g(x))^2\,dx = \int_a^b \pi(f^2(x) - g^2(x))\,dx. \quad (5)\end{aligned}$$

If we return to the cross sections to look at the typical washer at x, we see that it has two radii, an outer one and an inner one:

Outer washer radius: $R(x) = f(x)$

Inner washer radius: $r(x) = g(x)$.

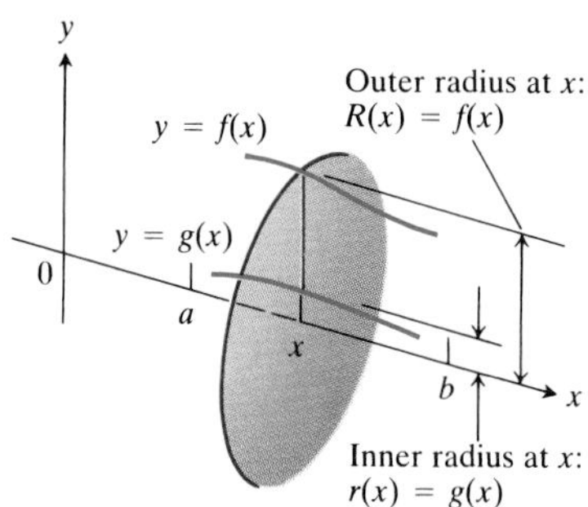

Equation (5) can be described in terms of these radii as

$$\text{Volume} = \int_a^b \pi((\text{outer radius})^2 - (\text{inner radius})^2)\, dx$$
$$= \int_a^b \pi(R^2(x) - r^2(x))\, dx. \tag{6}$$

Washer Method for Calculating Volumes

$$\text{Volume} = \int_a^b \pi(R^2(x) - r^2(x))\, dx \tag{7}$$

$R(x) =$ outer radius, $r(x) =$ inner radius

Notice that the function being integrated in Eq. (7) is $\pi(R^2 - r^2)$, not $\pi(R - r)^2$. Also notice that Eq. (7) turns into the disk-method formula if the inner radius $r(x)$ is zero throughout the interval $a \le x \le b$. The disk-method formula is a special case of what we have here.

We apply Eq. (7) with the steps shown in the following example. (We shall list the steps separately again after the example.)

EXAMPLE 5 The region bounded by the curve $y = x^2 + 1$ and the line $y = -x + 3$ is revolved about the x-axis to generate a solid. Find the volume of the solid.

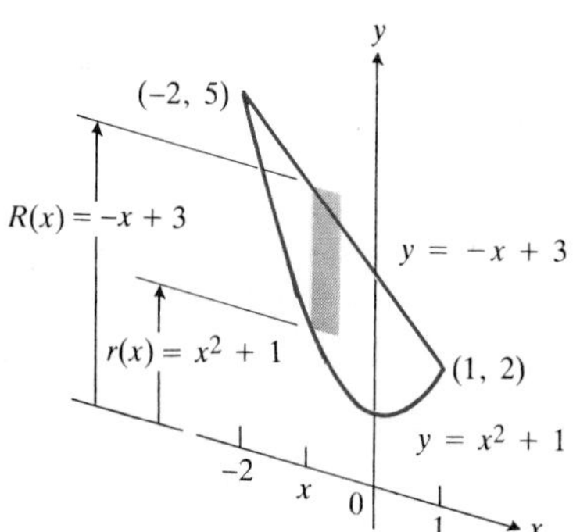

6.12 The region bounded by the curve $y = x^2 + 1$ and the line $y = -x + 3$ with a thin rectangle perpendicular to the axis of revolution.

Solution STEP 1: Draw the region and find the limits of integration (Fig. 6.12). The limits of integration are the x-coordinates of the points where the parabola and line cross. We find them by solving the equations $y = x^2 + 1$ and $y = -x + 3$ simultaneously for x:

$$x^2 + 1 = -x + 3$$
$$x^2 + x - 2 = 0$$
$$(x + 2)(x - 1) = 0$$
$$x = -2, \quad x = 1.$$

STEP 2: Draw a thin rectangle across the region perpendicular to the axis of revolution (the orange rectangle in Fig. 6.12).

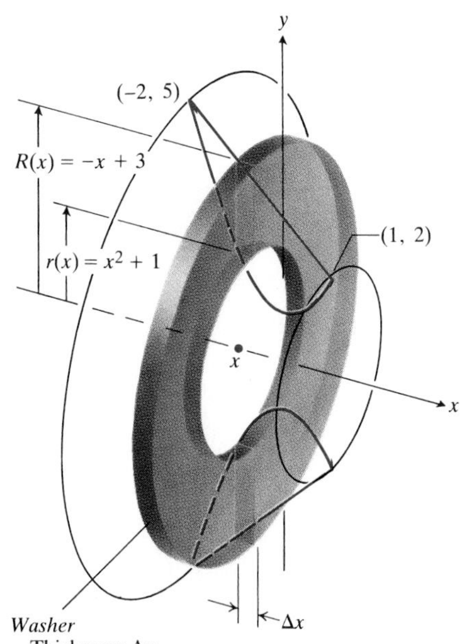

6.13 The inner and outer radii of the washer swept out by revolving the rectangle in Fig. 6.12 about the x-axis are the distances of the rectangle's two ends from the axis of revolution.

STEP 3: Find the outer and inner radii of the washer that would be swept out by the rectangle if it were revolved about the x-axis along with the region. (We drew the washer in Fig. 6.13, but in your own work you need not do that.) These radii are the distances of the two ends of the rectangle from the axis of revolution. You can read these distances from the formulas for the bounding curves:

Outer radius: $R(x) = -x + 3$

Inner radius: $r(x) = x^2 + 1$.

STEP 4: Evaluate the volume integral.

$$\text{Volume} = \int_a^b \pi(R^2(x) - r^2(x))\,dx \qquad \text{(Eq. 7)}$$

$$= \int_{-2}^{1} \pi((-x + 3)^2 - (x^2 + 1)^2)\,dx \qquad \left(\begin{array}{l}\text{Values from}\\ \text{Steps 1 and 3}\end{array}\right)$$

$$= \int_{-2}^{1} \pi(8 - 6x - x^2 - x^4)\,dx \qquad \left(\begin{array}{l}\text{Expressions squared}\\ \text{and combined}\end{array}\right)$$

$$= \pi\left[8x - 3x^2 - \frac{x^3}{3} - \frac{x^5}{5}\right]_{-2}^{1} = \frac{117\pi}{5} \qquad \text{(Arithmetic omitted)}$$

The volume of the solid is $117\pi/5$. ■

As illustrated in Example 5, the steps we take to implement the washer method are these:

How to Find Volumes by the Washer Method

STEP 1: Draw the region and find the limits of integration.

STEP 2: Draw a thin rectangle across the region perpendicular to the axis of revolution.

STEP 3: Find the distances of the ends of the rectangle from the axis of revolution. These give the radius functions for the volume integral.

STEP 4: Integrate to find the volume.

To find the volume of a solid generated by revolving a region about the y-axis, we use the steps listed above and integrate with respect to y instead of x. The next example shows how to do this.

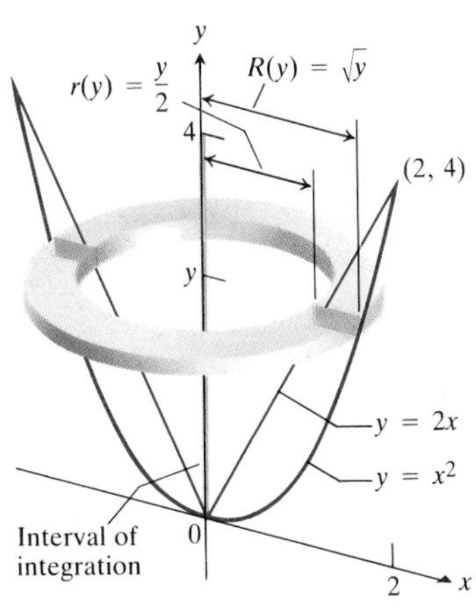

6.14 The region bounded by the parabola $y = x^2$ and the line $y = 2x$. The thin rectangle perpendicular to the axis of revolution sweeps out a washer whose radii are given by the formulas for the curves.

EXAMPLE 6 The region bounded by the parabola $y = x^2$ and the line $y = 2x$ in the first quadrant is revolved about the y-axis to generate a solid. Find the volume of the solid.

Solution STEP 1: Draw the region (Fig. 6.14) and find the limits of integration. The curves $y = x^2$ and $y = 2x$ cross at (0, 0) and (2, 4), so we integrate from $y = 0$ to $y = 4$.

Solids of Revolution

We can generate a solid of revolution by revolving a plane region about an axis. To calculate the volume of the solid, we integrate the square of the radius of the solid along an interval on the axis and multiply by pi.

If the axis of revolution is the x-axis, we integrate along an interval of the x-axis (Example 1).

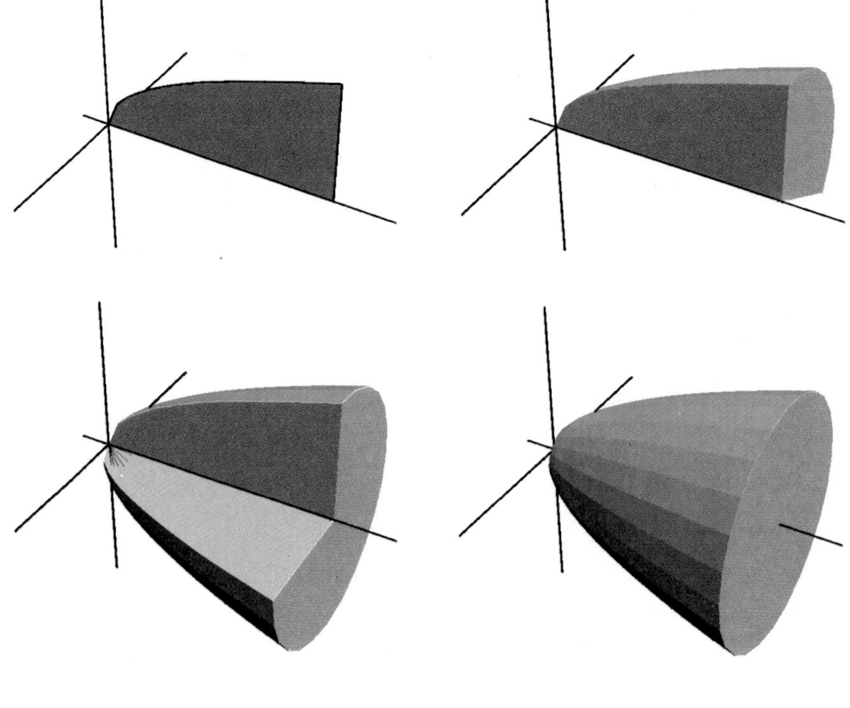

If the axis of revolution is parallel to the x-axis, we also integrate along an interval of the x-axis (Example 3).

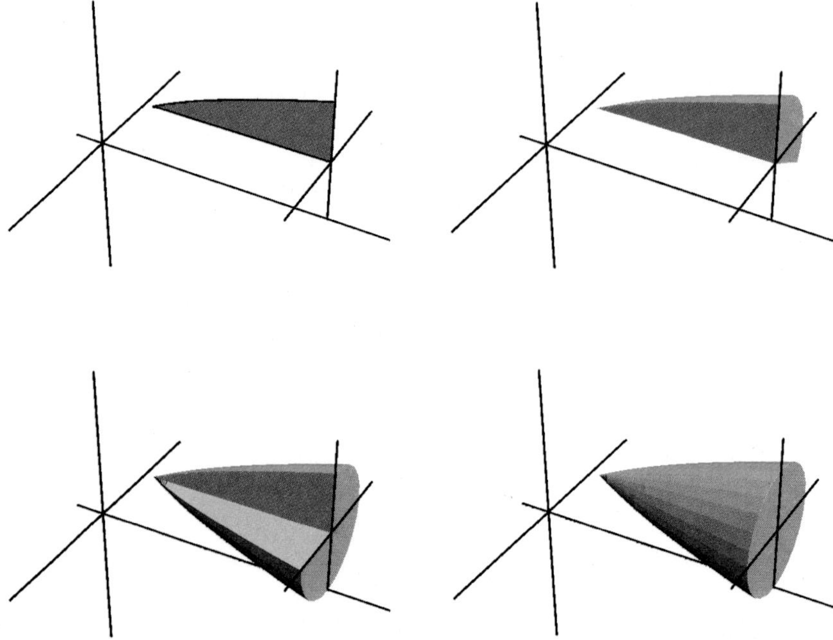

(Generated by Mathematica)

If the axis of revolution is the y-axis, we integrate along an interval of the y-axis (Example 4).

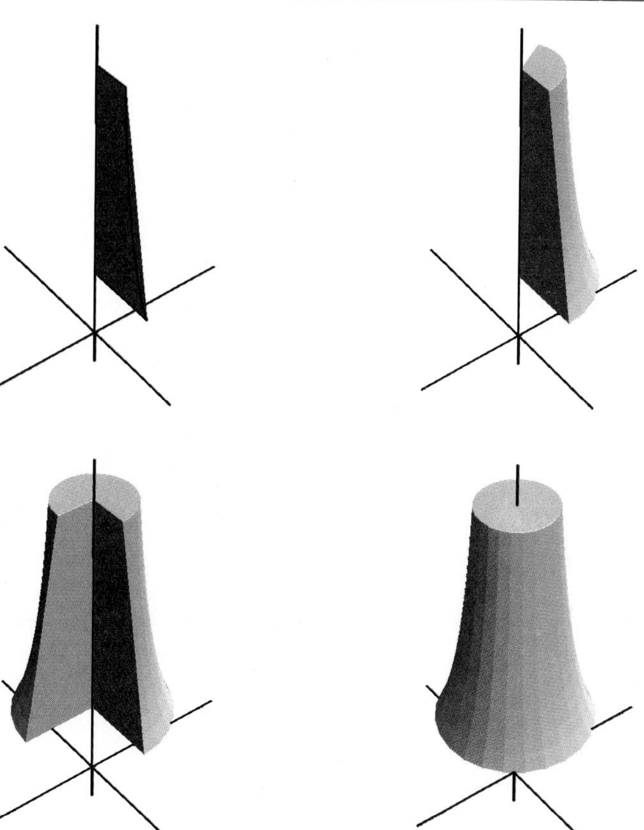

If the axis of revolution is not part of the region being revolved, the solid has a hole in it. We integrate along the coordinate axis parallel to the axis of revolution (Example 5).

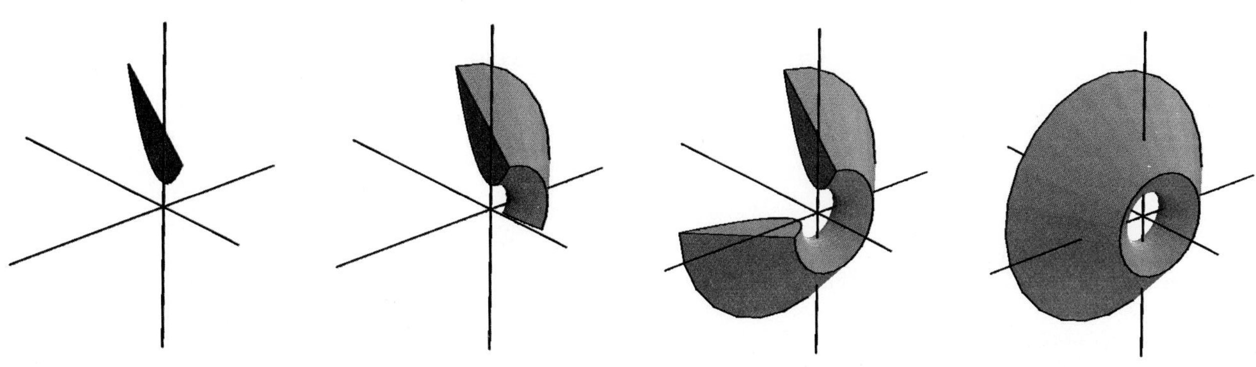

STEP 2: Draw a thin rectangle across the region perpendicular to the axis of revolution (the rectangle in Fig. 6.14).

STEP 3: Find the distances of the ends of the rectangle from the axis of revolution. From Fig. 6.14,

$$R(y) = \sqrt{y}, \qquad r(y) = \frac{y}{2}.$$

STEP 4: Integrate to find the volume.

$$\text{Volume} = \int_c^d \pi(R^2(y) - r^2(y))\,dy \qquad \text{(Eq. (7) with } y \text{ in place of } x)$$

$$= \int_0^4 \pi\left((\sqrt{y})^2 - \left(\frac{y}{2}\right)^2\right) dy \qquad \text{(Values from Steps 1 and 3)}$$

$$= \pi\int_0^4 \left(y - \frac{y^2}{4}\right) dy = \pi\left[\frac{y^2}{2} - \frac{y^3}{12}\right]_0^4 = \frac{8}{3}\pi$$

If the axis of revolution is not one of the coordinate axes, the rule for finding the volume is still the same: Find formulas for the solid's outer and inner radii and integrate $\pi(R^2 - r^2)$ between appropriate limits. In the next example we revolve a region about a line parallel to the y-axis. We would handle a region revolved about a line parallel to the x-axis in a similar way.

EXAMPLE 7 The region between the parabola $y = x^2$ and the line $y = 2x$ is revolved about the line $x = 2$ parallel to the y-axis. Find the volume swept out.

Solution STEP 1: Draw the region (Fig. 6.15) and find the limits of integration. The line crosses the parabola at (0, 0) and (2, 4), so we integrate from $y = 0$ to $y = 4$.

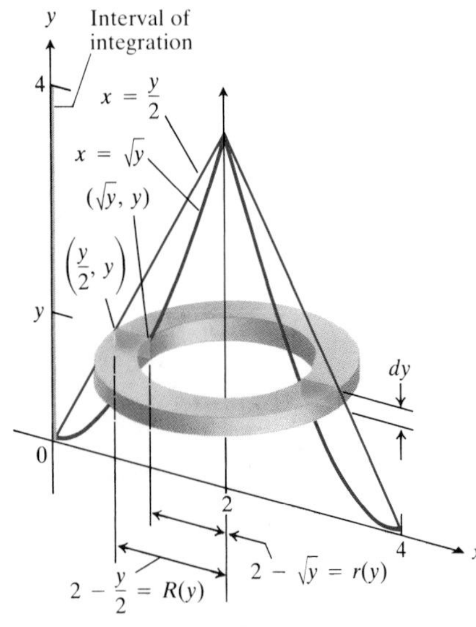

6.15 The inner and outer radii of the washer are measured, as always, as distances from the axis of revolution.

STEP 2: Draw a thin rectangle across the region perpendicular to the axis of revolution (the rectangle in Fig. 6.15).

STEP 3: Find the distances of the ends of the rectangle from the axis of revolution. From Fig. 6.15,

$$R(y) = 2 - \frac{y}{2}, \qquad r(y) = 2 - \sqrt{y}.$$

STEP 4: Integrate with respect to y to find the volume.

$$\text{Volume} = \int_c^d \pi(R^2(y) - r^2(y))\,dy \qquad \left(\begin{array}{l}\text{Eq. (7) with } y\\ \text{in place of } x\end{array}\right)$$

$$= \int_0^4 \pi\left(\left(2 - \frac{y}{2}\right)^2 - (2 - \sqrt{y})^2\right) dy \qquad \left(\begin{array}{l}\text{Values from}\\ \text{Steps 1 and 3}\end{array}\right)$$

$$= \pi\int_0^4 \left(\frac{y^2}{4} - 3y + 4\sqrt{y}\right) dy \qquad \left(\begin{array}{l}\text{Expressions squared}\\ \text{and combined}\end{array}\right)$$

$$= \pi\left[\frac{y^3}{12} - \frac{3y^2}{2} + \frac{8}{3}y^{3/2}\right]_0^4 = \frac{8}{3}\pi \qquad \text{(Arithmetic omitted)}$$

Exercises 6.2

Use the disk method to find the volumes of the solids generated by revolving the regions bounded by the lines and curves in Exercises 1–8 about the x-axis.

1. $x + y = 2, \quad y = 0, \quad x = 0$

2. $y = x^2, \quad y = 0, \quad x = 2$

3. $y = \sqrt{9 - x^2}, \quad y = 0$

4. $y = x - x^2, \quad y = 0$

5. $y = x^3, \quad y = 0, \quad x = 2$

6. $y = e^x, \quad y = 0, \quad x = 0, \quad x = \ln 2$

7. $y = \sqrt{\cos x}, \quad 0 \le x \le \pi/2, \quad y = 0, \quad x = 0$

8. $y = \sec x, \quad y = 0, \quad x = -\pi/4, \quad x = \pi/4$

Use the disk method to find the volumes of the solids generated by revolving about the y-axis the regions bounded by the lines and curves in Exercises 9–16.

9. $y = x/2, \quad y = 2, \quad x = 0$

10. $x = \sqrt{4 - y}, \quad x = 0, \quad y = 0$

11. $x = \sqrt{5}\,y^2, \quad x = 0, \quad y = -1, \quad y = 1$

12. $x = 1 - y^2, \quad x = 0$

13. $x = y^{3/2}, \quad x = 0, \quad y = 2$

14. $x = \sqrt{2 \sin 2y}, \quad 0 \le x \le \pi/2, \quad x = 0$

15. $x = 2/\sqrt{y + 1}, \quad x = 0, \quad y = 0, \quad y = 3$

16. $x = 2/(y + 1), \quad x = 0, \quad y = 0, \quad y = 1$

Use the washer method to find the volumes of the solids generated by revolving about the x-axis the regions bounded by the lines and curves in Exercises 17–24.

17. $y = x, \quad y = 1, \quad x = 0$

18. $y = 2x, \quad y = x, \quad x = 1$

19. $y = x^2, \quad y = 4, \quad x = 0$

20. $y = x^2 + 3, \quad y = 4$

21. $y = x^2 + 1, \quad y = x + 3$

22. $y = 4 - x^2, \quad y = 2 - x$

23. $y = \sec x, \quad y = \sqrt{2}, \quad -\pi/4 \le x \le \pi/4$

24. $y = 2/\sqrt{x}, \quad y = 2, \quad x = 4$

Use the washer method to find the volumes of the solids generated by revolving the regions bounded by the lines and curves in Exercises 25–30 about the y-axis.

25. $y = x - 1, \quad y = 1, \quad x = 1$

26. $y = x - 1, \quad y = 0, \quad x = 1$

27. $y = x^2, \quad y = 0, \quad x = 2$

28. $y = x, \quad y = \sqrt{x}$

29. The semicircle $x = \sqrt{25 - y^2}$ and the y-axis

30. The semicircle $x = \sqrt{25 - y^2}$ and the line $x = 4$

Find the volume of the solid generated by revolving the shaded region about the indicated axis in Exercises 31–34.

31. The x-axis

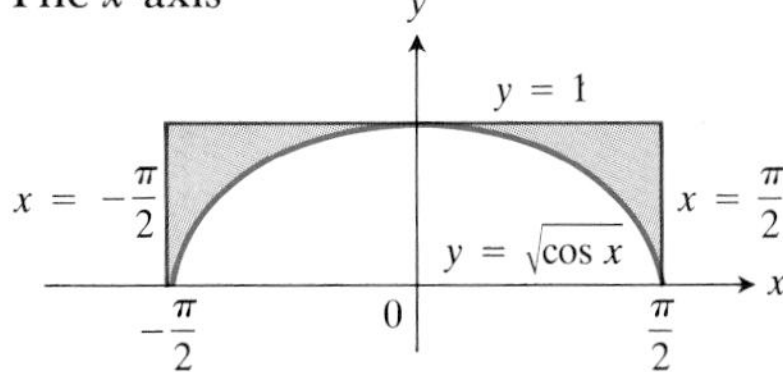

32. The x-axis

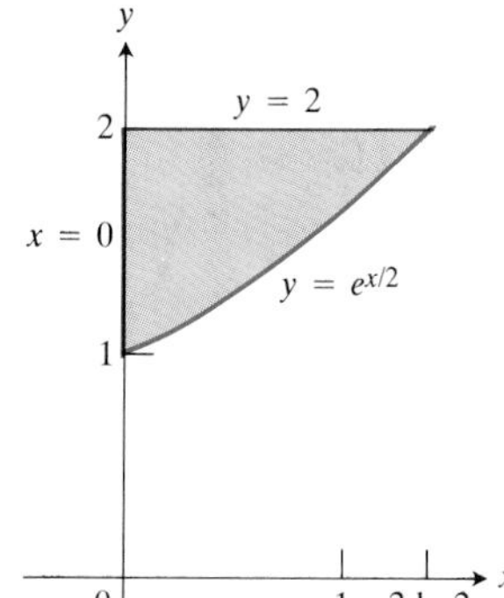

33. The y-axis

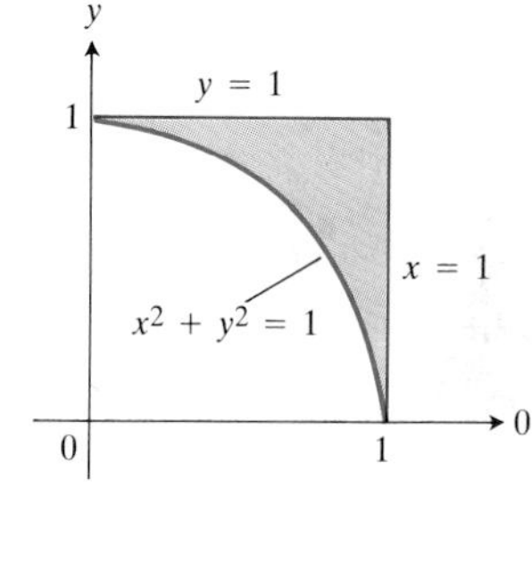

34. a) The x-axis
b) The y-axis

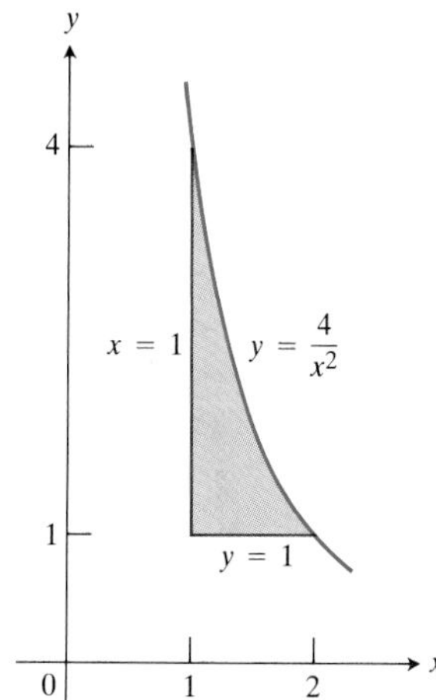

35. Find the volume of the solid generated by revolving the region bounded by $y = \sqrt{x}$ and the lines $y = 2$ and $x = 0$
a) about the x-axis
b) about the y-axis
c) about the line $y = 2$
d) about the line $x = 4$.

36. Find the volume of the solid generated by revolving the triangular region bounded by the lines $y = 2x$, $y = 0$, and $x = 0$
a) about the line $x = 1$
b) about the line $x = 2$.

37. Find the volume of the solid generated by revolving the region bounded by the parabola $y = x^2$ and the line $y = 1$
a) about the line $y = 1$
b) about the line $y = 2$
c) about the line $y = -1$.

38. By integration, find the volume of the solid generated by revolving the triangular region with vertices $(0, 0)$, $(b, 0)$, $(0, h)$
a) about the x-axis
b) about the y-axis.

39. *Minimizing a volume.* The arch $y = \sin x$, $0 \leq x \leq \pi$, is revolved about the line $y = c$ to generate the solid shown in Fig. 6.16. Find the value of c that minimizes this volume.

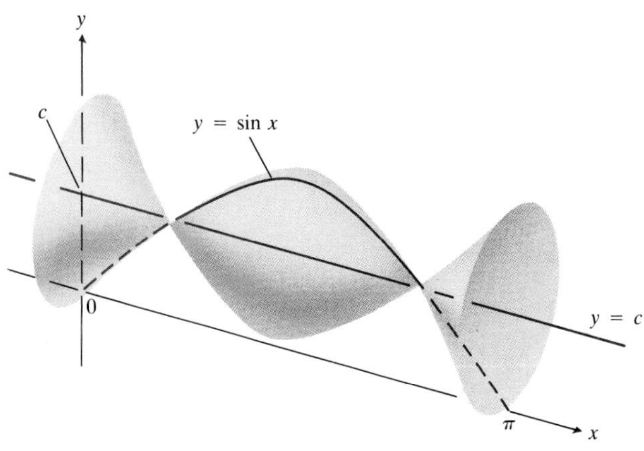

6.16 Exercise 39 asks for the value of c that minimizes this volume.

40. *Designing a plumb bob.* Having been asked to design a brass plumb bob that will weigh in the neighborhood of 190 gm, you decide to shape it like the solid of revolution shown here. What is the volume of the solid? If you specify a brass that weighs 8.5 gm/cm^3, about how much will the bob weigh?

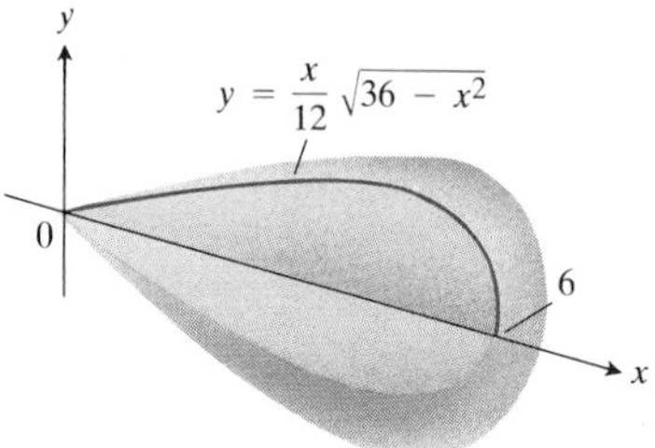

41. *Designing a wok.* You are designing a wok frying pan that will be shaped like a spherical bowl with handles.

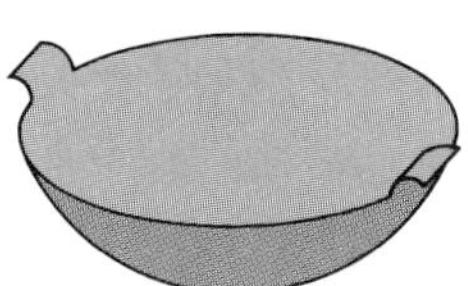

A bit of experimentation at home persuades you that you can get one that holds about 3 L if you make it 9 cm deep and give the sphere a radius of 16 cm. To be sure, you picture the wok as a solid of revolution and calculate the volume with an integral. Your picture looks like this:

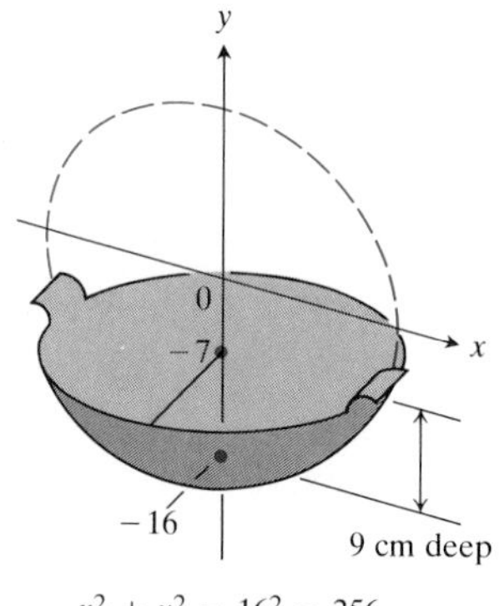

What volume do you really get?

TOOLKIT PROGRAM

Super*Grapher	Enables you to portray regions in the plane by graphing functions in different colors

6.3 Cylindrical Shells—An Alternative to Washers

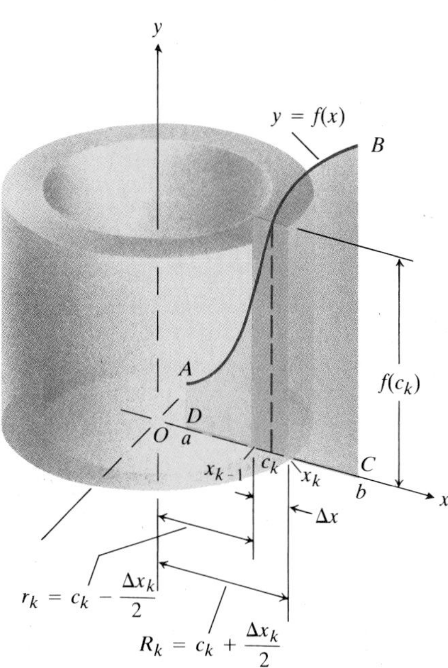

6.17 The solid swept out by revolving region *ABCD* about the *y*-axis can be approximated with cylindrical shells like the one shown here.

If the rectangular strips that approximate a region being revolved about an axis lie parallel to the axis instead of perpendicular to it, they sweep out cylindrical shells. Cylindrical shells are sometimes easier to work with than washers because the formula they lead to does not require squaring.

The Basic Shell Formula

We arrive at the cylindrical shell volume formula in the following way.

Suppose we revolve the tinted region *ABCD* in Fig. 6.17 about the y-axis to generate a solid. To find the volume of the solid, we first approximate the region with vertical rectangles based on a partition of the closed interval $[a, b]$ on which the region stands. The rectangles run parallel to the y-axis, the axis of revolution.

Figure 6.17 also shows a typical approximating rectangle. Its dimensions are $f(c_k)$ by Δx_k. The point c_k is chosen to be the midpoint of the interval from x_{k-1} to x_k. Since it does not matter where the c_k's are chosen in their intervals when we find limits of Riemann sums, we are free to choose each c_k as we please. The resulting formula in this case will be less cumbersome if we stay with midpoints. You will see why in just a moment.

Again with reference to Fig. 6.17, the cylindrical shell swept out by revolving the rectangle about the y-axis is a solid cylinder with these dimensions:

Shell height: $f(c_k)$

Inner radius: $r_k = c_k - \dfrac{\Delta x_k}{2}$

Outer radius: $R_k = c_k + \dfrac{\Delta x_k}{2}$

Base ring area:
$$\begin{aligned} A_k &= \pi R_k^2 - \pi r_k^2 = \pi(R_k + r_k)(R_k - r_k) \\ &= \pi(2c_k)\Delta x_k \end{aligned} \tag{1}$$

Shell volume:
$$\begin{aligned} V_k &= \text{base ring area} \times \text{shell height} \\ &= 2\pi c_k f(c_k)\Delta x_k . \end{aligned}$$

The advantage of choosing c_k to be the midpoint of its interval becomes clear in the formula for A_k. With this choice, $R_k + r_k$ equals $2c_k$; without this choice, it doesn't.

The volumes of all the shells generated by the partition of $[a, b]$ add up to

$$\text{Shell volume sum} = \sum_{k=1}^{n} 2\pi c_k f(c_k)\Delta x_k . \tag{2}$$

The volume of the solid swept out by revolving region $ABCD$ about the y-axis is taken to be the limit of the shell volume sums as the norm of the partition of $[a, b]$ goes to zero. If f is continuous, the limit exists and can be found by integrating the product $2\pi x f(x)$ from $x = a$ to $x = b$.

The Shell Method (Axis the y-Axis)

Suppose $y = f(x)$ is continuous throughout an interval $a \le x \le b$ that does not cross the y-axis. Then, the volume of the solid generated by revolving the region between the graph of f and the interval $a \le x \le b$ about the y-axis is found by integrating $2\pi x f(x)$ with respect to x from a to b.

$$\text{Volume} = \int_a^b 2\pi \begin{pmatrix}\text{shell}\\ \text{radius}\end{pmatrix}\begin{pmatrix}\text{shell}\\ \text{height}\end{pmatrix} dx = \int_a^b 2\pi x f(x)\, dx. \qquad (3)$$

One way to remember Eq. (3) is to imagine that a cylindrical shell of average circumference $2\pi x$, height $f(x)$, and thickness dx has been cut along a generating rectangle and rolled flat like a sheet of tin (Fig. 6.18). The sheet is almost a rectangular solid of dimensions $2\pi x$ by $f(x)$ by dx. Hence the shell's volume is about $2\pi x f(x)\, dx$. Equation (3) says that the volume of the complete solid is the integral of $2\pi x f(x)\, dx$ from a to b.

How to Find a Volume by the Shell Method

To apply Eq. (3), take these steps:

STEP 1: Sketch the region.

STEP 2: Draw a thin rectangle across the region parallel to the y-axis (the axis of revolution). The radius of the cylindrical shell swept out by the rectangle is x. The height of the shell is $f(x)$, the height of the rectangle. The width of the rectangle is dx. Add this information to the picture. (We also sketched the shell, but you need not do that in your own work.)

STEP 3: Find limits of integration a and b that include all possible rectangles like this from one end of the region to the other.

STEP 4: Integrate the product $2\pi x f(x)$ with respect to x from a to b to find the volume.

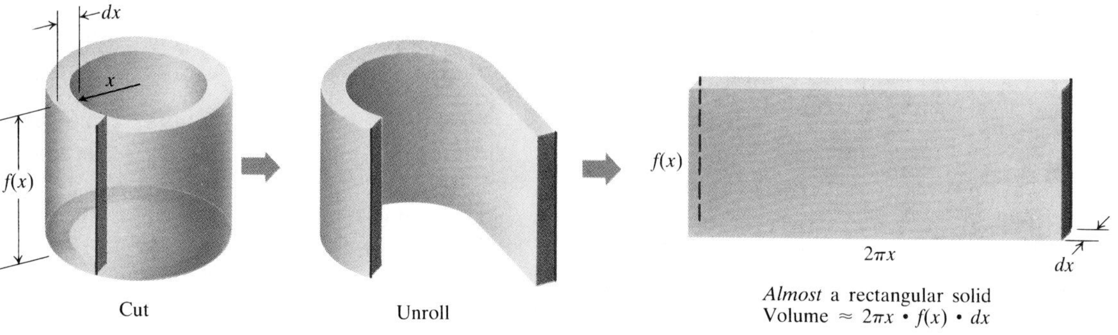

6.18 How to remember the integral formula for cylindrical shells.

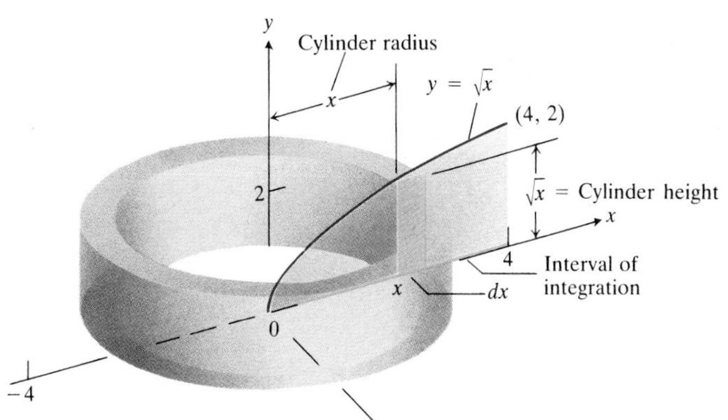

6.19 The region in Example 1, together with a thin rectangle parallel to the axis of revolution (the y-axis, in this case). The rectangle sweeps out a cylindrical shell.

EXAMPLE 1 The region bounded by the curve $y = \sqrt{x}$, the x-axis, and the line $x = 4$ is revolved about the y-axis to generate a solid. Find the volume of the solid.

Solution STEP 1: Sketch the region (Fig. 6.19).

STEP 2: Draw a thin rectangle across the region parallel to the y-axis. Label the rectangle's height $\sqrt{x}$, its width dx, and its distance from the y-axis x. (We added all this to Fig. 6.19.)

STEP 3: Find limits of integration that include all rectangles like the one in Step 2. The limits in this case are $x = 0$ and $x = 4$.

STEP 4: Integrate to find the volume.

$$\text{Volume} = \int_a^b 2\pi x f(x)\,dx = \int_0^4 2\pi x\sqrt{x}\,dx \qquad \begin{pmatrix}\text{Eq. (3) with values}\\ \text{from Steps 2 and 3}\end{pmatrix}$$

$$= 2\pi\int_0^4 x^{3/2}\,dx = 2\pi\left[\frac{2}{5}x^{5/2}\right]_0^4 = \frac{128\pi}{5} \qquad \begin{pmatrix}\text{Arithmetic}\\ \text{omitted}\end{pmatrix}$$

The volume is $128\pi/5$. ■

To use shells to find the volume of a solid generated by revolving a region about the x-axis instead of the y-axis, use Eq. (3) with y in place of x. Except for changes in notation, the steps we take to implement the new formula are the same as before.

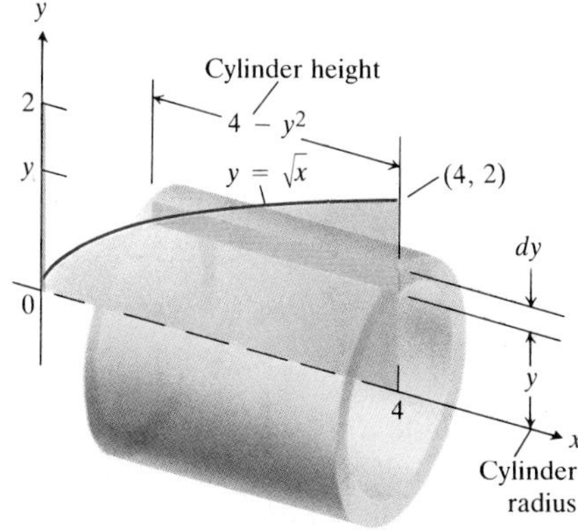

6.20 The region in Example 2, together with a thin rectangle parallel to the axis of revolution.

EXAMPLE 2 The region bounded by the curve $y = \sqrt{x}$, the x-axis, and the line $x = 4$ is revolved about the x-axis to generate a solid. Find the volume of the solid.

Solution STEP 1: Sketch the region (Fig. 6.20).

STEP 2: Draw a thin rectangle across the region parallel to the x-axis (the axis of revolution). Describe the rectangle's height as a function of y. Label the rectangle's width dy and its distance from the x-axis y. (We added all this to Fig. 6.20.)

STEP 3: Find limits of integration that include all the rectangles like the one in Step 2. The rectangles run from $y = 0$, the smallest value of y in the region, to $y = 2$, the largest value of y in the region.

STEP 4: Integrate to find the volume.

$$\text{Volume} = \int_c^d 2\pi\binom{\text{shell}}{\text{radius}}\binom{\text{shell}}{\text{height}}\,dy = \int_0^2 2\pi y(4-y^2)\,dy$$

$$= 2\pi\int_0^2 (4y - y^3)\,dy = 2\pi\left[2y^2 - \frac{y^4}{4}\right]_0^2 = 8\pi, \quad \binom{\text{Arithmetic}}{\text{omitted}}$$

in agreement with the disk-method calculation in Example 1, Section 6.2. ■

If the axis of revolution is a line parallel to one of the coordinate axes, we use the same steps as before. The only added complication is that the expression for the radius of the typical cylinder is no longer simply x or y.

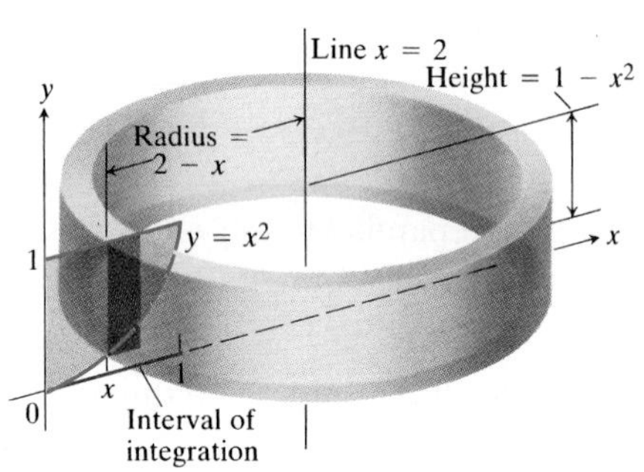

6.21 The region in Example 3. As always, the radius of the cylindrical shell is measured as a distance from the axis of revolution.

EXAMPLE 3 The region bounded by the parabola $y = x^2$, the y-axis, and the line $y = 1$ is revolved about the line $x = 2$ to generate a solid. Find the volume of the solid.

Solution STEP 1: Sketch the region (Fig. 6.21).

STEP 2: Draw a thin rectangle across the region parallel to the line $x = 2$ (the axis of revolution). Describe the rectangle's height as a function of x. Label the rectangle's width dy. Describe the rectangle's distance from the line $x = 2$ as a function of x. This is the radius of the typical cylindrical shell the rectangle sweeps out. (We added all this to Fig. 6.21.)

STEP 3: Find limits of integration that take in all rectangles like the one in Step 2. The limits are $x = 0$ and $x = 1$, the extreme values of x in the region.

STEP 4: Integrate to find the volume.

$$\text{Volume} = \int_a^b 2\pi\binom{\text{shell}}{\text{radius}}\binom{\text{shell}}{\text{height}}\,dx = \int_0^1 2\pi(2-x)(1-x^2)\,dx$$

$$= 2\pi\int_0^1 (2 - x - 2x^2 + x^3)dx = 2\pi\left[2x - \frac{x^2}{2} - \frac{2}{3}x^3 + \frac{x^4}{4}\right]_0^1 = \frac{13\pi}{6}$$ ■

The disk, washer, and shell methods for calculating the volumes of solids of revolution always agree. We illustrate this agreement.

EXAMPLE 4 The disk enclosed by the circle $x^2 + y^2 = 4$ is revolved about the y-axis to generate a solid sphere. A hole of diameter 2 is then bored through the sphere along the y-axis. Find the volume of the "cored" sphere.

Solution The cored sphere could have been generated by revolving the shaded region in Fig. 6.22(a) about the y-axis. Thus there are three methods we might use to find the volume: disks, washers, and shells.

METHOD 1: *Disks and subtraction.* Figure 6.22(b) shows the solid sphere with the core pulled out. The core is a circular cylinder with spherical end caps. Our plan is to subtract the volume of the core from the volume of the sphere.

6.22 (a) The tinted region generates the volume of revolution. (b) An exploded view showing the sphere with the core removed. (c) A phantom view showing a cross-section slice of the sphere with the core removed. (d) Filling the volume with cylindrical shells.

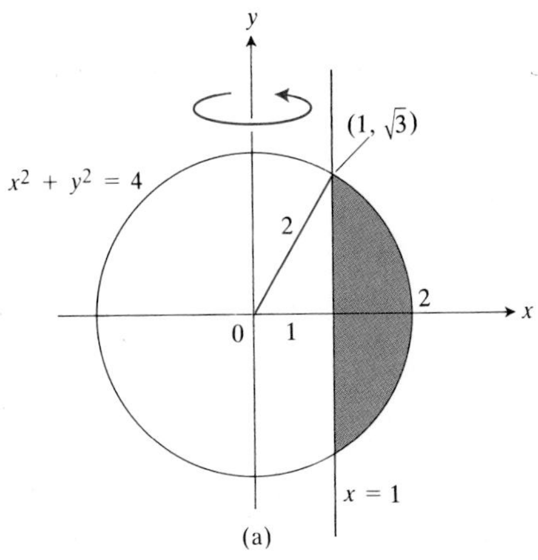

The volume of the cored sphere is the volume swept out by the shaded region as it revolves about the y-axis.

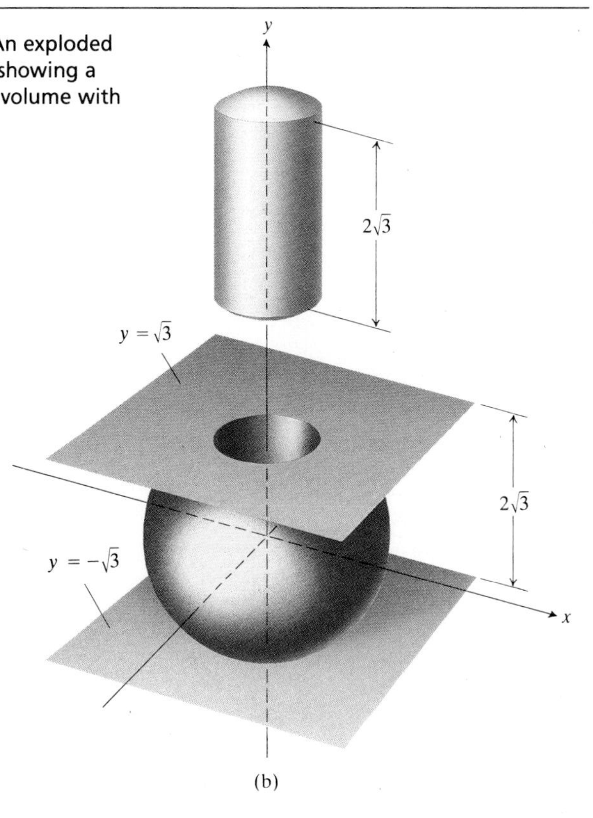

With the method of disks, we can calculate the volume of the hole left by the core and subtract it from the volume of the truncated sphere.

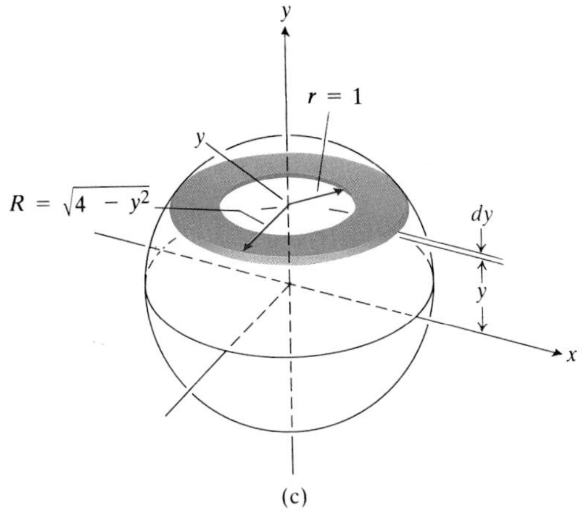

With the method of washers, we can calculate the volume of the cored sphere directly by modeling it as a stack of washers perpendicular to the y-axis.

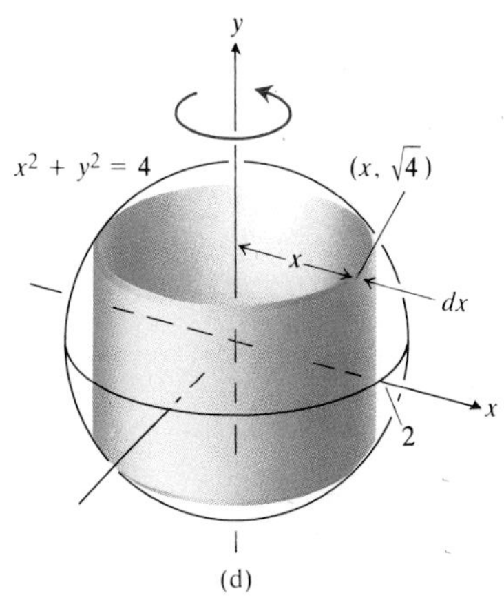

With the method of cylinders, we can calculate the volume of the cored sphere directly by modeling it as a union of cylindrical shells parallel to the y-axis.

TABLE 6.1
Washers versus Shells

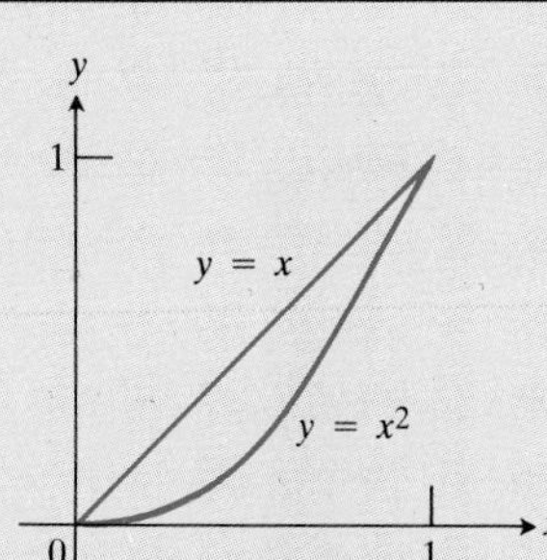

The region bounded by
$y = x, \quad y = x^2 \quad (x = \sqrt{y})$

REVOLVED ABOUT X-AXIS	REVOLVED ABOUT Y-AXIS	
Washer: $\pi(R^2 - r^2)\,dx$	Washer: $\pi(R^2 - r^2)\,dy$	

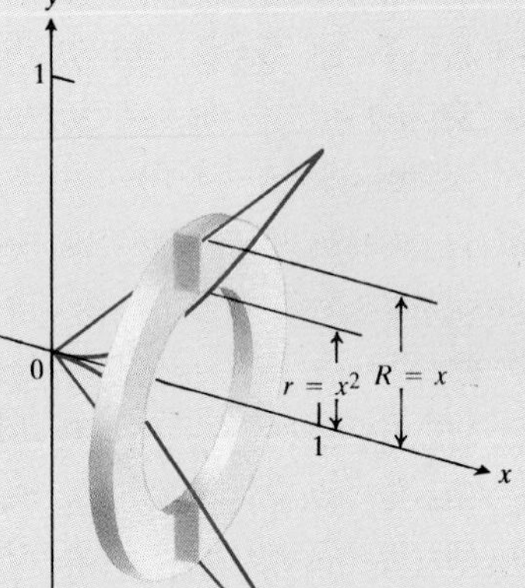

$$V = \int_{x=0}^{x=1} \pi((x)^2 - (x^2)^2)dx = \frac{2\pi}{15}$$

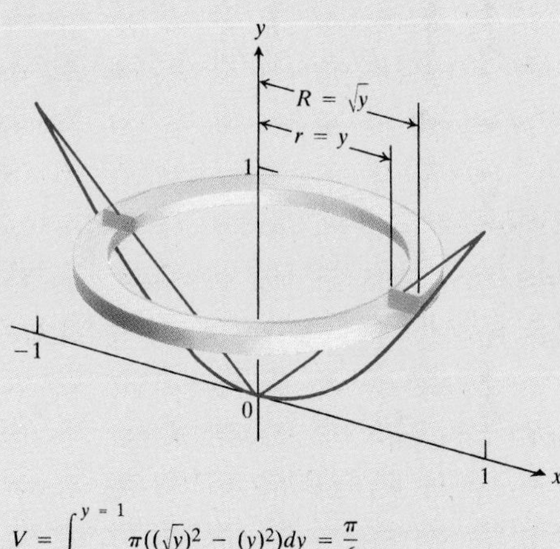

$$V = \int_{y=0}^{y=1} \pi((\sqrt{y})^2 - (y)^2)dy = \frac{\pi}{6}$$

RECTANGLE PERPENDICULAR TO AXIS: USE WASHERS

Shell: 2π(radius)(height) dy

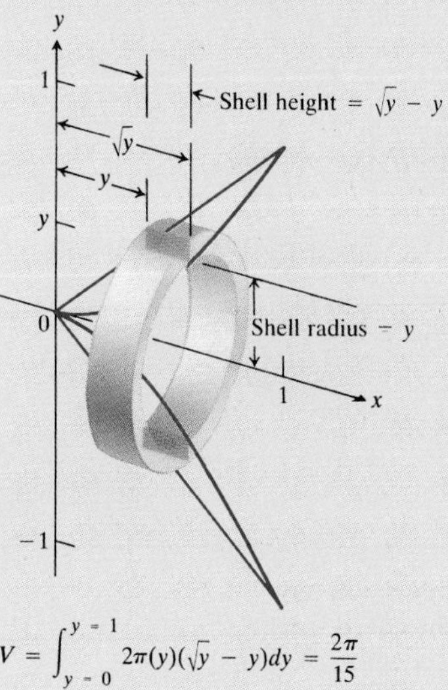

$$V = \int_{y=0}^{y=1} 2\pi(y)(\sqrt{y} - y)dy = \frac{2\pi}{15}$$

Shell: 2π(radius)(height) dx

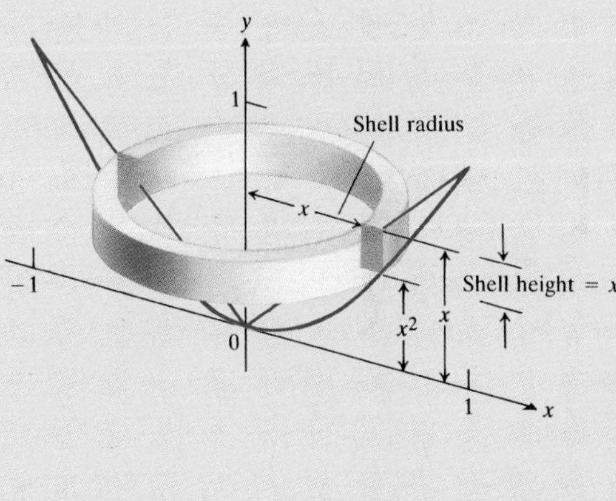

RECTANGLE PARALLEL TO AXIS: USE SHELLS.

We can simplify matters by imagining that the two caps have already been sliced off the sphere by planes perpendicular to the y-axis at $y = \sqrt{3}$ and $y = -\sqrt{3}$. With the caps removed, the truncated sphere has volume T, say. From this we subtract the volume of the hole, a right circular cylinder of radius 1 and height $2\sqrt{3}$.

The volume of the hole is

$$H = \pi(1)^2(2\sqrt{3}) = 2\pi\sqrt{3}.$$

The truncated sphere (before drilling) is a solid of revolution whose cross sections perpendicular to the y-axis are disks. The radius of a typical disk is $\sqrt{4 - y^2}$. Therefore,

$$T = \int_{-\sqrt{3}}^{\sqrt{3}} \pi(\text{radius})^2\,dy = \int_{-\sqrt{3}}^{\sqrt{3}} \pi(4 - y^2)\,dy = \pi\left[4y - \frac{y^3}{3}\right]_{-\sqrt{3}}^{\sqrt{3}} = 6\pi\sqrt{3}.$$

The volume of the cored sphere is

$$T - H = 6\pi\sqrt{3} - 2\pi\sqrt{3} = 4\pi\sqrt{3}.$$

METHOD 2: *Washers.* The cored sphere is a solid of revolution whose cross sections perpendicular to the y-axis are washers (Fig. 6.22c). The radii of a typical washer are

Outer radius: $R = \sqrt{4 - y^2}$

Inner radius: $r = 1.$

The volume of the cored sphere is therefore

$$V = \int_{-\sqrt{3}}^{\sqrt{3}} \pi(R^2 - r^2)\,dy = \int_{-\sqrt{3}}^{\sqrt{3}} \pi(4 - y^2 - 1)\,dy = \pi\left[3y - \frac{y^3}{3}\right]_{-\sqrt{3}}^{\sqrt{3}} = 4\pi\sqrt{3}.$$

METHOD 3: *Cylindrical shells.* We model the volume of the cored sphere with cylindrical shells like the one in Fig. 6.22(d). The typical shell has radius x, height $2\sqrt{4 - x^2}$, and thickness dx. The volume of the cored sphere is

$$V = \int_1^2 2\pi\begin{pmatrix}\text{shell}\\ \text{radius}\end{pmatrix}\begin{pmatrix}\text{shell}\\ \text{height}\end{pmatrix}dx = \int_1^2 4\pi x\sqrt{4 - x^2}\,dx$$

$$= 4\pi\left[-\frac{1}{3}(4 - x^2)^{3/2}\right]_1^2 \qquad \begin{pmatrix}\text{After substituting } u = 4 - x^2,\\ \text{integrating, and substituting back}\end{pmatrix}$$

$$= 0 - 4\pi\left[-\frac{1}{3}(4 - 1)^{3/2}\right] = 4\pi\sqrt{3}.$$ ■

Table 6.1 summarizes the methods of finding volumes with washers and shells.

Exercises 6.3

Use the shell method to find the volumes of the solids generated by revolving the regions bounded by the curves and lines in Exercises 1–6 about the y-axis.

1. $y = x, \quad y = -x/2, \quad \text{and} \quad x = 2$
2. $y = \sqrt{x}, \quad y = 0, \quad \text{and} \quad x = 4$
3. $y = x^2 + 1, \quad y = 0, \quad x = 0, \quad \text{and} \quad x = 1$
4. $y = 2x - 1, \quad y = \sqrt{x}, \quad \text{and} \quad x = 0$
5. $y = 1/x, \quad y = 0, \quad x = 1/2, \quad x = 2$
6. $y = 1/x^2, \quad y = 0, \quad x = 1/2, \quad x = 2$

Use the shell method to find the volumes of the solids generated by revolving the regions bounded by the curves and lines in Exercises 7–12 about the x-axis.

7. $y = |x|$ and $y = 1$
8. $y = x$, $y = 1$, and $x = 2$
9. $y = \sqrt{x}$, $y = 0$, and $y = x - 2$
10. $y = 2$, $x = -y$, and $x = \sqrt{y}$
11. The parabola $x = 2y - y^2$ and the y-axis
12. The parabola $x = 2y - y^2$ and the line $y = x$

Find the volumes of the solids generated by revolving the shaded regions in Exercises 13–16 about the indicated axes.

13. The y-axis

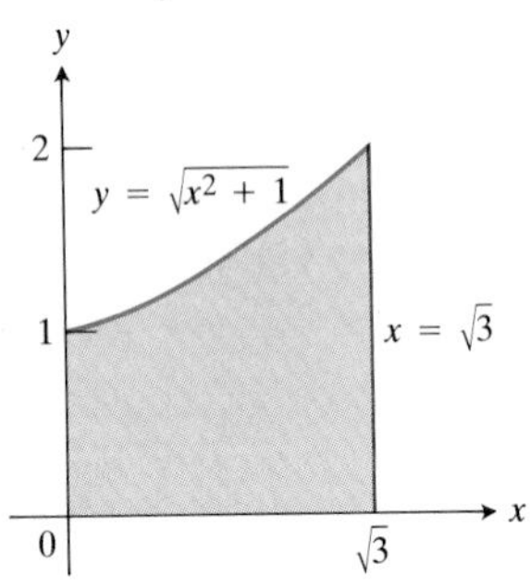

14. The y-axis

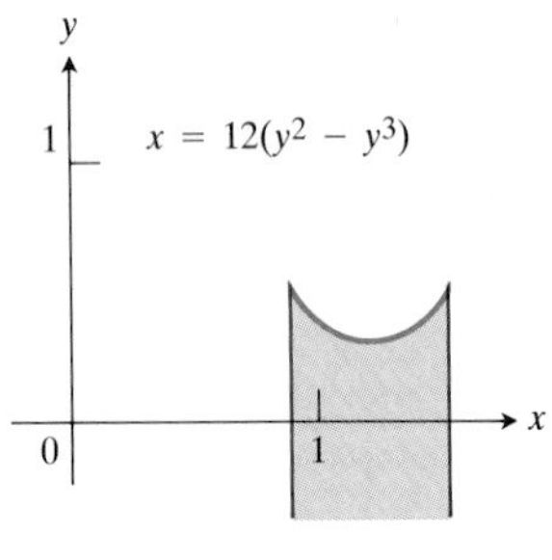

15. The x-axis

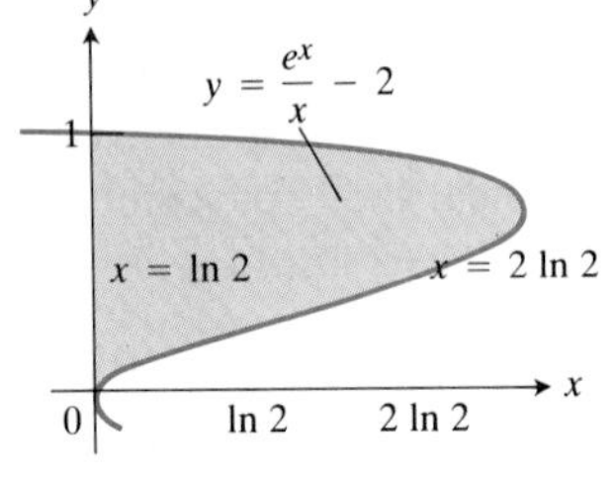

16. The x-axis

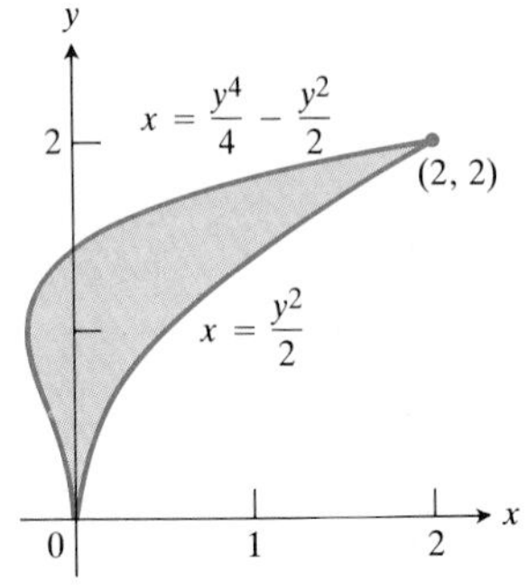

In Exercises 17–24, find the volumes of the solids generated by revolving the regions about the given axes.

17. The triangle with vertices $(1, 1)$, $(1, 2)$, and $(2, 2)$ about (a) the x-axis; (b) the y-axis
18. The region in the first quadrant bounded by the curve $x = y - y^3$ and the y-axis about (a) the x-axis; (b) the y-axis
19. The region in the first quadrant bounded by $x = y - y^3$, $x = 1$, and $y = 1$ about (a) the x-axis; (b) the y-axis; (c) the line $x = 1$; (d) the line $y = 1$
20. The triangular region bounded by the lines $2y = x + 4$, $y = x$, and $x = 0$ about (a) the x-axis; (b) the y-axis; (c) the line $x = 4$; (d) the line $y = 8$
21. The region in the first quadrant bounded by $y = x^3$ and $y = 4x$ about (a) the x-axis; (b) the line $y = 8$
22. The region bounded by $y = \sqrt{x}$ and $y = x^2/8$ about (a) the x-axis; (b) the y-axis
23. The region bounded by $y = 2x - x^2$ and $y = x$ about (a) the y-axis; (b) the line $x = 1$
24. The region bounded by $y = \sqrt{x}$, $y = 2$, $x = 0$ about (a) the x-axis; (b) the y-axis; (c) the line $x = 4$; (d) the line $y = 2$

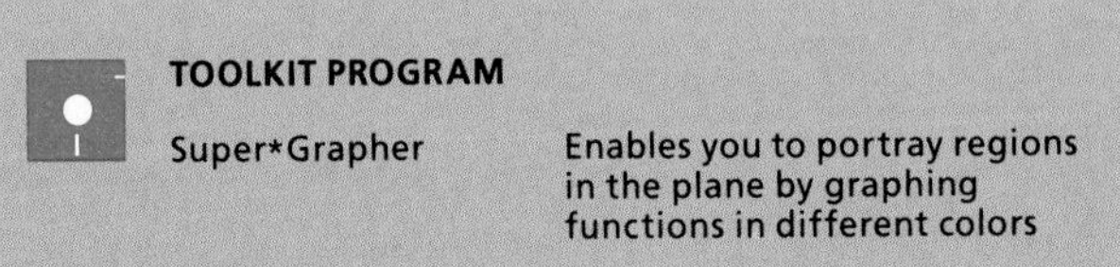

6.4 Lengths of Curves in the Plane

We approximate the length of a curved path in the plane the way we use a ruler to estimate the length of a curved road on a map, by measuring from point to point with straight line segments and adding the results. There is a limit to the accuracy of such an estimate, however, imposed in part by how accurately we measure and in part by how many line segments we are willing to use.

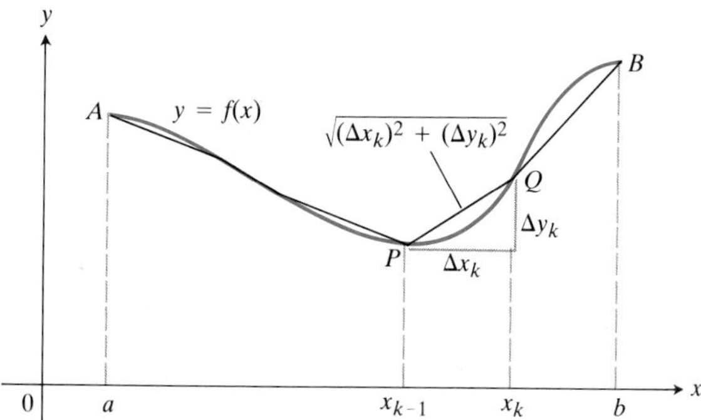

6.23 A typical segment PQ of a polygonal path approximating the curve AB.

With calculus we can usually do a better job because we can imagine using straight line segments as short as we please, each set of segments making a polygonal path that fits the curve more tightly than before. When we proceed this way, with a smooth enough curve, the lengths of the polygonal paths approach a limit we can calculate with an integral. In this section we shall see what that integral is. We shall also see what happens if, instead of being smooth, the curve is Helga von Koch's snowflake curve.

The Basic Formula

Suppose the curve whose length we want to find is the graph of the function $y = f(x)$ from $x = a$ to $x = b$. We partition the closed interval $[a, b]$ in the usual way and connect the corresponding points on the curve with line segments (Fig. 6.23). The line segments, taken together, form a polygonal path that approximates the curve.

The length of a typical line segment PQ (shown in the figure) is

$$\sqrt{(\Delta x_k)^2 + (\Delta y_k)^2}. \tag{1}$$

The length of the curve is therefore approximated by the sum

$$\sum_{k=1}^{n} \sqrt{(\Delta x_k)^2 + (\Delta y_k)^2}. \tag{2}$$

We expect the approximation to improve as the partition $[a, b]$ becomes finer, and we would like to show that the sums in (2) approach a calculable limit as the norm of the partition goes to zero. To show this, we rewrite the sum in (2) in a form to which we can apply the Integral Existence Theorem from Chapter 5. Our starting point, oddly enough, is the Mean Value Theorem for derivatives.

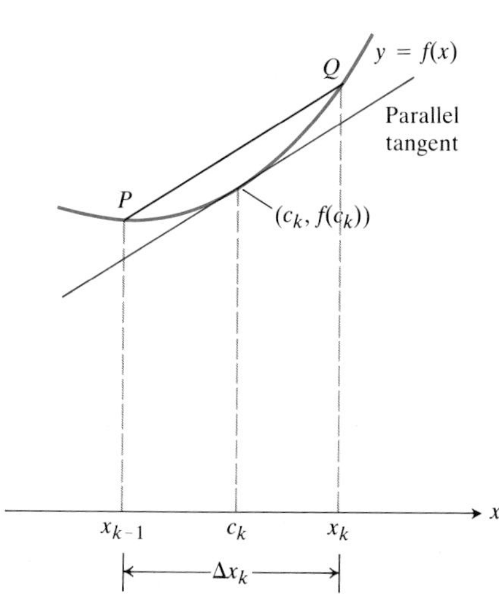

6.24 Enlargement of the arc PQ in Fig. 6.23.

Suppose that f has a derivative that is continuous at every point of $[a, b]$. Then, by the Mean Value Theorem, there is a point $(c_k, f(c_k))$ on the curve between P and Q where the tangent is parallel to the segment PQ (Fig. 6.24). At this point

$$f'(c_k) = \frac{\Delta y_k}{\Delta x_k}, \tag{3}$$

so that

$$\Delta y_k = f'(c_k)\Delta x_k. \tag{4}$$

With this substitution for Δy_k, the sums in (2) take the form

$$\sum_{k=1}^{n} \sqrt{(\Delta x_k)^2 + (f'(c_k)\Delta x_k)^2} = \sum_{k=1}^{n} \sqrt{1 + (f'(c_k))^2}\, \Delta x_k. \tag{5}$$

The sums on the right are Riemann sums for the continuous function $\sqrt{1 + (f'(x))^2}$ on the interval $[a, b]$. They therefore converge to the integral of this function as the norm of the partition of the interval goes to zero. We define this integral to be the length of the curve from a to b.

DEFINITION

If the function f has a continuous first derivative throughout the interval $a \leq x \leq b$, the **length of the curve $y = f(x)$ from a to b** is the number

$$L = \int_a^b \sqrt{1 + \left(\frac{dy}{dx}\right)^2}\, dx. \tag{6}$$

EXAMPLE 1 Find the length of the curve

$$y = \frac{4\sqrt{2}}{3}x^{3/2} - 1, \qquad 0 \leq x \leq 1.$$

Solution We use Eq. (6) with $a = 0$, $b = 1$, and

$$y = \frac{4\sqrt{2}}{3}x^{3/2} - 1$$

$$\frac{dy}{dx} = \frac{4\sqrt{2}}{3} \cdot \frac{3}{2}x^{1/2} = 2\sqrt{2}x^{1/2}$$

$$1 + \left(\frac{dy}{dx}\right)^2 = 1 + (2\sqrt{2}x^{1/2})^2 = 1 + 8x.$$

The length of the curve from $x = 0$ to $x = 1$ is

$$L = \int_0^1 \sqrt{1 + \left(\frac{dy}{dx}\right)^2}\, dx = \int_0^1 \sqrt{1 + 8x}\, dx \qquad \left(\text{Eq. (6) with } a = 0, b = 1\right)$$

$$= \frac{2}{3} \cdot \frac{1}{8}(1 + 8x)^{3/2}\Big]_0^1 = \frac{13}{6}. \qquad \left(\begin{array}{l}\text{Substitute } u = 1 + 8x, \text{ integrate,} \\ \text{and replace } u \text{ by } 1 + 8x.\end{array}\right)$$

■

Dealing with Discontinuities in *dy/dx*

At a point on a curve where dy/dx fails to exist, dx/dy may exist and we may be able to find the curve's length by interchanging x and y in Eq. (6). The revised formula looks like this:

$$L = \int_c^d \sqrt{1 + \left(\frac{dx}{dy}\right)^2}\, dy. \tag{7}$$

To use Eq. (7), we express x as a function of y, calculate dx/dy, and proceed as before to square, add 1, take the square root, and integrate.

EXAMPLE 2 Find the length of the curve $y = (x/2)^{2/3}$ from $x = 0$ to $x = 2$.

Solution The derivative

$$\frac{dy}{dx} = \frac{2}{3}\left(\frac{x}{2}\right)^{-1/3} = \frac{2}{3}\left(\frac{2}{x}\right)^{1/3}$$

is not defined at $x = 0$, so we cannot find the curve's length with Eq. (6).

We therefore rewrite the equation to express x in terms of y:

$$y = \left(\frac{x}{2}\right)^{2/3}$$

$$y^{3/2} = \frac{x}{2} \qquad \text{(Raise both sides to the power 3/2.)}$$

$$x = 2y^{3/2}. \qquad \text{(Solve for } x\text{.)}$$

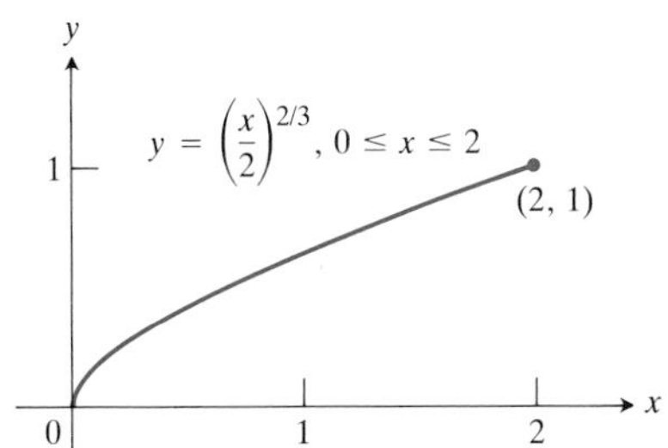

6.25 The graph of $y = (x/2)^{2/3}$ from $x = 0$ to $x = 2$ is also the graph of $x = 2y^{3/2}$ from $y = 0$ to $y = 1$, a function that has a continuous first derivative. We may therefore use Eq. (7) with $x = 2y^{3/2}$ to find the curve's length.

From this we see that the curve whose length we want is also the graph of $x = 2y^{3/2}$ from $y = 0$ to $y = 1$ (Fig. 6.25).

The derivative

$$\frac{dx}{dy} = 2 \cdot \frac{3}{2} y^{1/2} = 3y^{1/2}$$

is continuous throughout the interval $0 \le y \le 1$. We may therefore find the curve's length by setting

$$\left(\frac{dx}{dy}\right)^2 = (3y^{1/2})^2 = 9y$$

in Eq. (7) and integrating from $y = 0$ to $y = 1$:

$$L = \int_c^d \sqrt{1 + \left(\frac{dx}{dy}\right)^2}\, dy = \int_0^1 \sqrt{1 + 9y}\, dy \qquad \text{(Eq. 7)}$$

$$= \frac{1}{9} \cdot \frac{2}{3}(1 + 9y)^{3/2}\Big]_0^1 \qquad \left(\begin{array}{l}\text{Substitute } u = 1 + 9y,\ du/9 = dy,\\ \text{integrate, and substitute back.}\end{array}\right)$$

$$= \frac{2}{27}(10\sqrt{10} - 1) = 2.27. \qquad \text{(To two places with a calculator)}$$

■

The Short Differential Formula

The equations

$$L = \int_a^b \sqrt{1 + \left(\frac{dy}{dx}\right)^2}\, dx \quad \text{and} \quad L = \int_c^d \sqrt{1 + \left(\frac{dx}{dy}\right)^2}\, dy \tag{8}$$

are often written with differentials instead of derivatives. This is done formally by thinking of the derivatives as quotients of differentials and bringing the dx and dy inside the radicals to cancel the denominators. In the first integral we have

$$\sqrt{1 + \left(\frac{dy}{dx}\right)^2}\, dx = \sqrt{1 + \frac{dy^2}{dx^2}}\, dx = \sqrt{dx^2 + \frac{dy^2}{dx^2}dx^2} = \sqrt{dx^2 + dy^2}.$$

In the second integral we have

$$\sqrt{1+\left(\frac{dx}{dy}\right)^2}\,dy = \sqrt{1+\frac{dx^2}{dy^2}}\,dy = \sqrt{dy^2+\frac{dx^2}{dy^2}dy^2} = \sqrt{dx^2+dy^2}. \qquad (10)$$

Thus the integrals in (8) reduce to a single differential formula:

$$L = \int_a^b \sqrt{dx^2+dy^2}. \qquad (11)$$

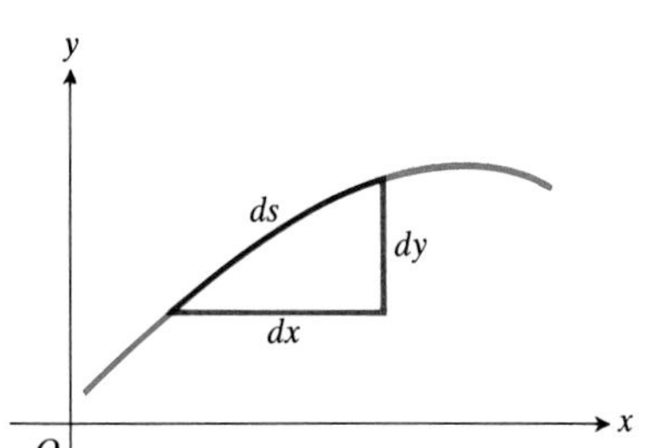

6.26 Diagram for remembering the equation $ds = \sqrt{dx^2 + dy^2}$.

Of course, dx and dy must be expressed in terms of a common variable, and appropriate limits of integration must be found before the integration in Eq. (11) is performed.

We can shorten Eq. (11) still further. Think of dx and dy as two sides of a small triangle whose hypotenuse is

$$ds = \sqrt{dx^2+dy^2} \qquad (12)$$

(Fig. 6.26). The differential ds is then regarded as a differential of arc length that can be integrated between appropriate limits to give the length of the curve. With $\sqrt{dx^2 + dy^2}$ set equal to ds, the integral in Eq. (11) simply becomes the integral of ds.

DEFINITION

The Arc Length Differential and the Differential Formula for Arc Length

$$ds = \sqrt{dx^2+dy^2} \qquad\qquad L = \int ds$$

Arc length differential — Differential formula for arc length

Curves with Infinite Length

As you may recall from Section 3.6, Helga von Koch's snowflake curve K is the limit curve of an infinite sequence $C_1, C_2, \ldots, C_n, \ldots$ of "triangular" polygonal curves. The first four curves in the sequence look like this:

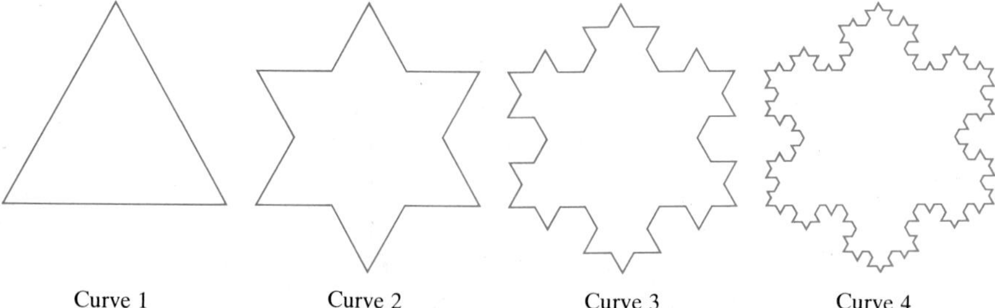

Each time we introduce a new vertex in the construction process, it remains as a vertex in all subsequent curves and becomes a point on the limit curve K. This means that each of the C's is itself a polygonal approximation of K—the endpoints of its sides all belonging to K. The length of K should therefore be the limit of the lengths of the curves C_n. At least, that is what it should be if we stick to the definition of length we developed for smooth curves.

What, then, is the limit of the lengths of the curves C_n? If the original equilateral triangle C_1 has sides of length 1, the total length of C_1 is 3. To make C_2 from C_1, we replace each side of C_1 by four segments, each of which is one-third as long as the original side.

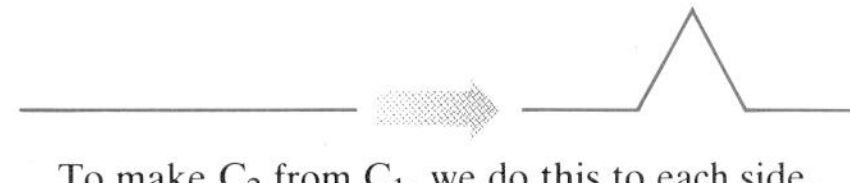

To make C_2 from C_1, we do this to each side.

The total length of C_2 is therefore 3(4/3). To get the length of C_3, we multiply by 4/3 again. We do so again to get the length of C_4. By the time we get out to C_n, we have a curve of length $3(4/3)^{n-1}$.

Curve Number	1	2	3	...	n	...
Length	3	$3\left(\frac{4}{3}\right)$	$3\left(\frac{4}{3}\right)^2$	...	$3\left(\frac{4}{3}\right)^{n-1}$	...

The length of C_{10} is nearly 40 and the length of C_{100} is greater than 7,000,000,000,000. The lengths grow too rapidly to have any finite limit. Therefore the snowflake curve has no length, or, if you prefer, infinite length.

What went wrong? Nothing. The formulas we derived for length are for the graphs of functions with continuous first derivatives, curves that are smooth enough to have a continuously turning tangent at every point. Helga von Koch's snowflake curve is too rough for that, and our derivative-based formulas do not apply.

Benoit Mandelbrot's theory of fractals has proved to be a rich source of curves with infinite length, curves that when magnified prove to be as rough and varied as they looked before magnification. Like coastlines on an ocean, no amount of magnification smooths them out.

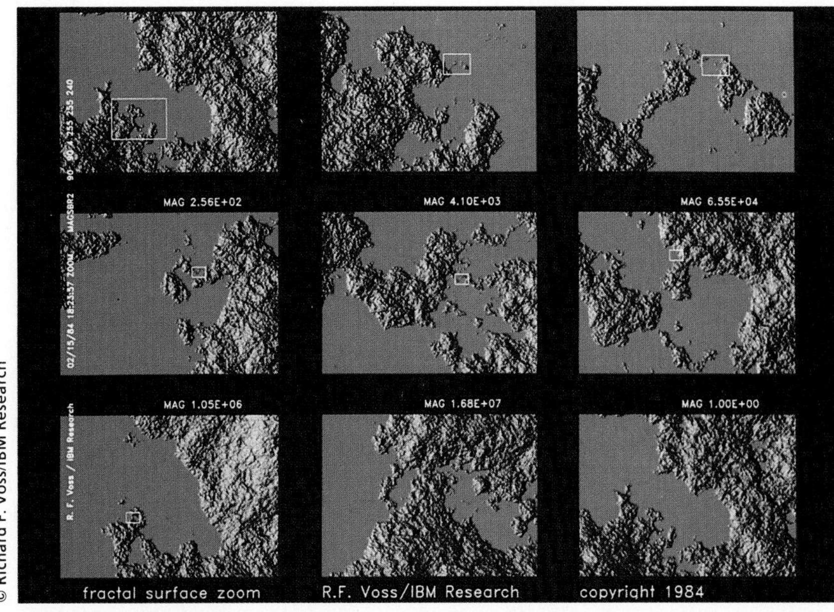

© Richard F. Voss/IBM Research

Exercises 6.4

Find the lengths of the curves in Exercises 1–10.

1. $y = (1/3)(x^2 + 2)^{3/2}$ from $x = 0$ to $x = 3$
2. $y = x^{3/2}$ from $x = 0$ to $x = 4$
3. $9x^2 = 4y^3$ from $(0, 0)$ to $(2\sqrt{3}, 3)$
4. $y = x^{2/3}$ from $x = 0$ to $x = 4$
5. $y = (x^3/3) + 1/(4x)$ from $x = 1$ to $x = 3$ (*Hint:* $1 + (dy/dx)^2$ is a perfect square.)
6. $y = (x^{3/2}/3) - x^{1/2}$ from $x = 0$ to $x = 9$ (*Hint:* $1 + (dy/dx)^2$ is a perfect square.)
7. $x = (y^4/4) + 1/(8y^2)$ from $y = 1$ to $y = 2$ (*Hint:* $1 + (dx/dy)^2$ is a perfect square.)
8. $x = (y^3/6) + 1/(2y)$ from $y = 2$ to $y = 3$ (*Hint:* $1 + (dx/dy)^2$ is a perfect square.)
9. $y = (e^x + e^{-x})/2$ from $x = -\ln 2$ to $x = \ln 2$ (*Hint:* $1 + (dy/dx)^2$ is a perfect square.)
10. $x = y^2 - (1/8)\ln y$ from $y = 1$ to $y = 3$ (*Hint:* $1 + (dx/dy)^2$ is a perfect square.)
11. *The length of an astroid.* The graph of the equation $x^{2/3} + y^{2/3} = 1$ is one of a family of curves called *astroids* (not "asteroids") because of their starlike shapes. Find the length of this particular astroid by finding the length of the first quadrant portion, $y = (1 - x^{2/3})^{3/2}$, $0 \le x \le 1$, and multiplying by 4.

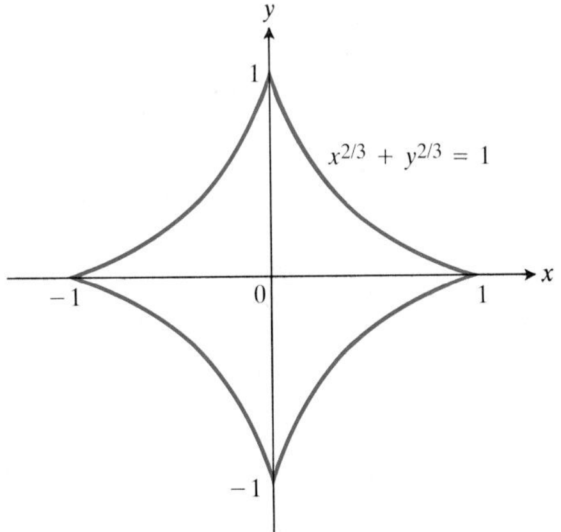

12. Find the length of the curve
$$y = \int_0^x \sqrt{\cos 2t}\, dt$$
from $x = 0$ to $x = \pi/4$. (*Hint:* For $0 \le x \le \pi/4$, $1 + \cos 2x = 2\cos^2 x$.)
13. Find a curve through the origin whose length is
$$L = \int_0^4 \sqrt{1 + \frac{1}{4x}}\, dx.$$
(*Hint:* Solve the initial value problem $dy/dx = \sqrt{1/4x}$, $y(0) = 0$.)
14. Find a curve through the point $(0, 1)$ whose length from $x = 0$ to $x = 1$ is
$$L = \int_0^1 \sqrt{1 + e^{2x}}\, dx.$$
15. Find a curve through the point $(1, 0)$ whose length from $x = 1$ to $x = 2$ is
$$L = \int_1^2 \sqrt{1 + \frac{1}{x^2}}\, dx.$$
16. Without evaluating either integral, show why
$$2\int_{-1}^{1} \sqrt{1 - x^2}\, dx = \int_{-1}^{1} \frac{1}{\sqrt{1 - x^2}}\, dx.$$
(*Hint:* Interpret one integral as an area and the other as a length.) (*Source:* Peter A. Lindstrom, *Mathematics Magazine,* Volume 45, Number 1, January 1972, page 47.)

Write integrals for the lengths of the curves in Exercises 17–18. Do not be concerned about how to evaluate the integrals right now. We shall see how to do that when we get to Chapter 8.

17. $y = \ln(1 - x^2)$, $0 \le x \le 1/2$
18. $y = \ln(\cos x)$, $0 \le x \le \pi/3$

You may have wondered why the curves we have been working with have such unusual formulas. The reason is that the square root $\sqrt{1 + (dy/dx)^2}$ in the arc-length integral almost never produces a function whose anti-derivative we can find. In fact, this square root is a famous source of nonelementary integrals. Most arc-length integrals have to be evaluated numerically, as in the exercises that follow.

19. **CALCULATOR** Your metal fabrication company is bidding for a contract to make sheets of corrugated iron roofing like the one shown in Fig. 6.27. The cross sections of the corrugated sheets are to conform to the curve
$$y = \sin \frac{3\pi}{20}x, \quad 0 \le x \le 20 \text{ in.}$$
If the roofing is to be stamped from flat sheets by a process that does not stretch the material, how wide should the original material be? To find out, use Simpson's rule with $n = 10$ to find the approximate length of the sine curve.

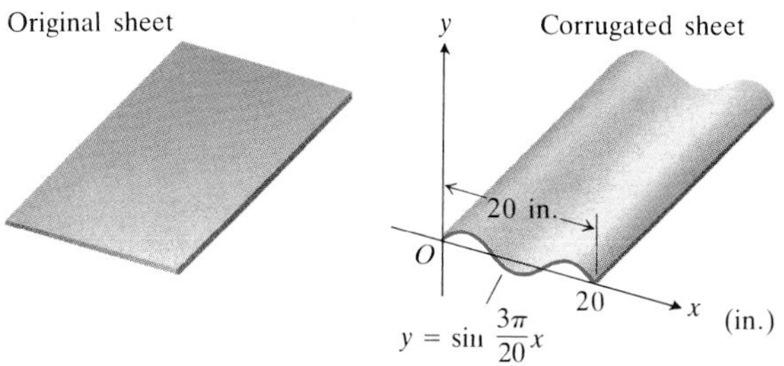

6.27 How wide does the original sheet have to be? See Exercise 19.

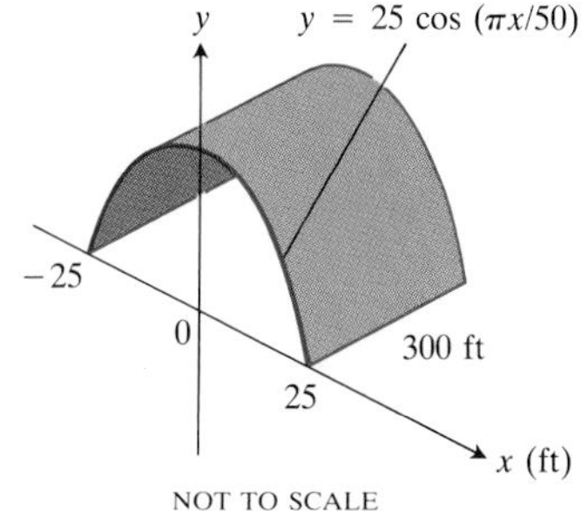

6.28 The tunnel in Exercise 20.

20. **CALCULATOR** Your engineering firm is bidding for the contract to construct the tunnel shown in Fig. 6.28. The tunnel is 300 ft long and 50 ft wide at the base. The cross section is shaped like one arch of the curve $y = 25 \cos(\pi x/50)$. Upon completion, the tunnel's inside surface (excluding the roadway) will be treated with a waterproof sealer that costs \$1.75 per ft^2 to apply. How much will it cost to apply the sealer? (*Hint:* Use numerical integration to find the length of the cosine curve.)

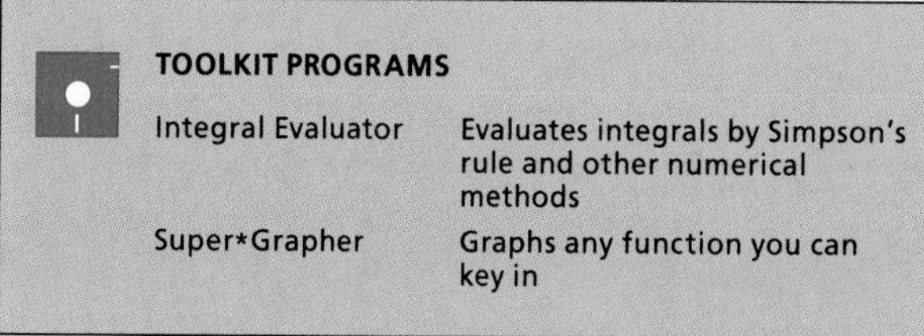

TOOLKIT PROGRAMS	
Integral Evaluator	Evaluates integrals by Simpson's rule and other numerical methods
Super*Grapher	Graphs any function you can key in

6.5 Areas of Surfaces of Revolution

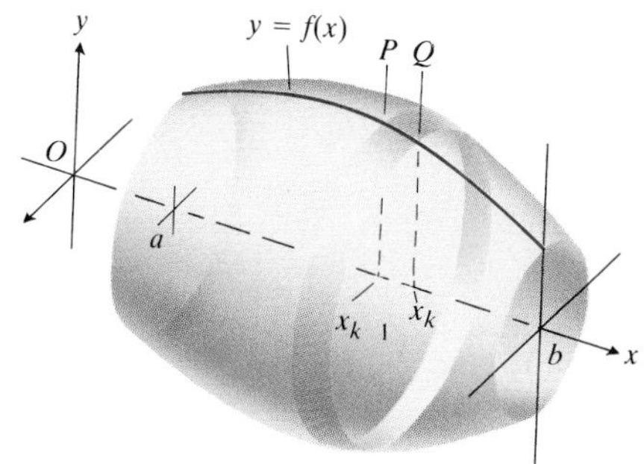

6.29 The surface swept out by revolving the curve $y = f(x)$, $a \leq x \leq b$, about the x-axis is a union of bands like the one swept out by the arc PQ.

When you jump rope, the rope sweeps out a surface in the space around you, a surface called a surface of revolution. As you can imagine, the area of this surface depends on the rope's length and on how far away each segment of the rope swings. This section explores the relation between the area of a surface of revolution and the length and reach of the curve that generates it.

The Basic Formula

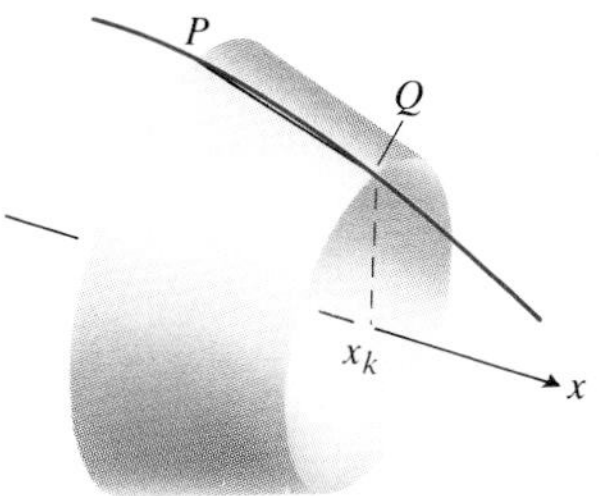

6.30 The straight line segment joining P and Q sweeps out a frustum of a cone.

Suppose we want to find the area of the surface swept out by revolving the graph of the function $y = f(x)$, $a \leq x \leq b$, about the x-axis. We partition the closed interval $[a, b]$ in the usual way and use the points in the partition to divide the graph into short arcs. Figure 6.29 shows a typical arc PQ and the band it sweeps out as part of the graph of f.

As the arc PQ revolves about the x-axis, the line segment joining P and Q sweeps out part of a cone whose axis lies along the x-axis (magnified view in Fig. 6.30). A piece of a cone like this is called a frustum of the cone, *frustum* being Latin for "piece." The surface area of the frustum approximates the surface area of the band swept out by the arc PQ.

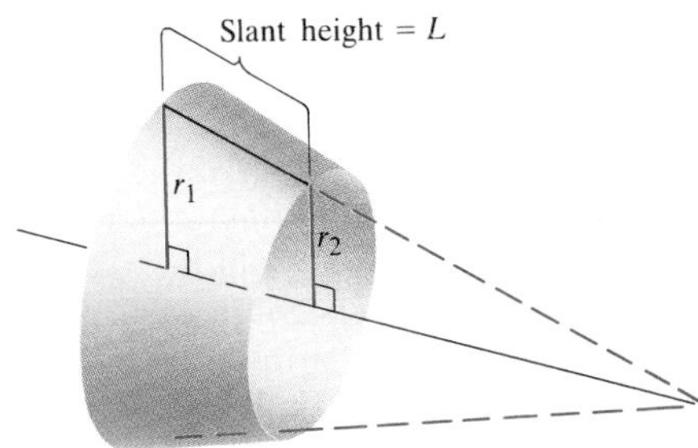

6.31 The important dimensions of the frustum in Fig. 6.30.

The surface area of the frustum of a cone (see Fig. 6.31) is 2π times the average of the base radii times the slant height:

$$\text{Frustum surface area} = 2\pi \cdot \frac{r_1 + r_2}{2} \cdot L = \pi(r_1 + r_2)L. \tag{1}$$

For the frustum swept out by the segment PQ (Fig. 6.32), this works out to be

$$\text{Frustum surface area} = \pi(f(x_{k-1}) + f(x_k))\sqrt{(\Delta x_k)^2 + (\Delta y_k)^2}. \tag{2}$$

The area of our original surface, being the sum of the areas of the bands swept out by arcs like arc PQ, is approximated by the frustum area sum

$$\sum_{k=1}^{n} \pi(f(x_{k-1}) + f(x_k))\sqrt{(\Delta x_k)^2 + (\Delta y_k)^2}. \tag{3}$$

We expect the approximation to improve as the partition of $[a, b]$ becomes finer, and we would like to show that the sums in (3) approach a calculable limit as the norm of the partition goes to zero.

To show this, we try to rewrite the sum in (3) as the Riemann sum of some function over the interval from a to b. As in the calculation of arc length, we begin by appealing to the Mean Value Theorem for derivatives.

Suppose as before that f has a derivative that is continuous at every point of $[a, b]$. Then, by the Mean Value Theorem, there is a point $(c_k, f(c_k))$ on the curve between P and Q where the tangent is parallel to the segment PQ (Fig. 6.33). At this point,

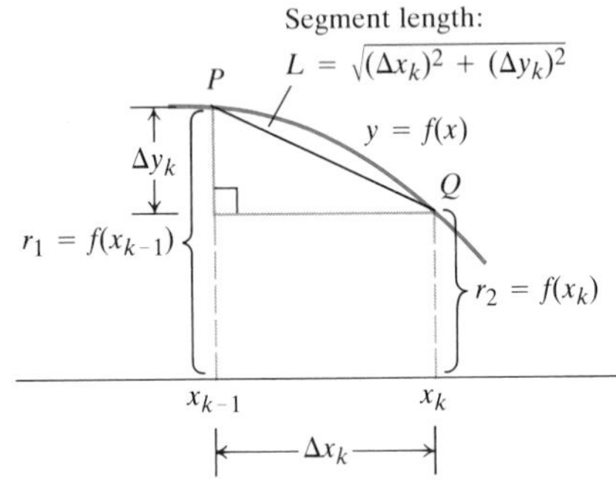

6.32 The important dimensions associated with the arc and segment PQ.

$$f'(c_k) = \frac{\Delta y_k}{\Delta x_k}, \tag{4}$$

so

$$\Delta y_k = f'(c_k)\Delta x_k. \tag{5}$$

With this substitution for Δy_k, the sums in (3) take the form

$$\sum_{k=1}^{n} \pi(f(x_{k-1}) + f(x_k))\sqrt{(\Delta x_k)^2 + (f'(c_k)\Delta x_k)^2}$$

$$= \sum_{k=1}^{n} \pi(f(x_{k-1}) + f(x_k))\sqrt{1 + (f'(c_k))^2}\,\Delta x_k. \tag{6}$$

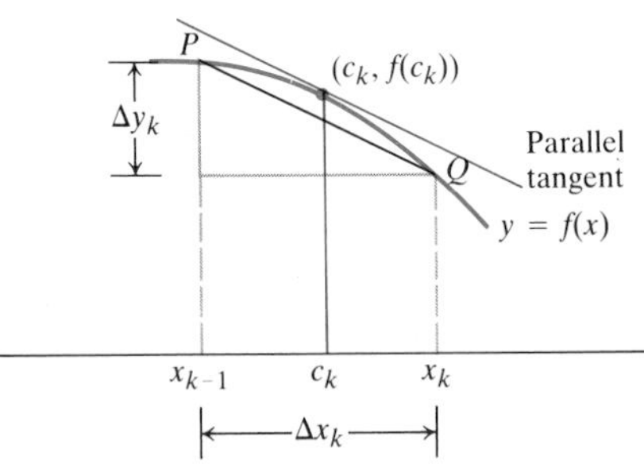

6.33 If f' is continuous, the Mean Value Theorem guarantees the existence of a point on arc PQ where the tangent is parallel to segment PQ.

At this point there is both good news and bad.

The bad news is that the sums in (6) are not the Riemann sums of any function because the points x_{k-1}, x_k, and c_k are not the same and there is no way to make them the same. The good news is that this does not matter. A theorem from advanced calculus assures us that as the norm of the subdivision of $[a, b]$ goes to zero the sums in (3) converge to

$$\int_a^b 2\pi f(x)\sqrt{1 + (f'(x))^2}\,dx \tag{7}$$

just the way we want them to. We therefore define this integral to be the area of the surface swept out by the graph of f from a to b.

6.34 Example 1 calculates the area of this surface.

DEFINITION

If the function f has a continuous first derivative throughout the interval $a \le x \le b$, the **area of the surface** generated by revolving the curve $y = f(x)$ about the x-axis is the number

$$S = \int_a^b 2\pi y \sqrt{1 + \left(\frac{dy}{dx}\right)^2}\, dx. \tag{8}$$

Notice that the square root in Eq. (8) is the same one that appears in the formula for arc length. More about that later.

EXAMPLE 1 Find the area of the surface generated by revolving the curve $y = 2\sqrt{x}$, $1 \le x \le 2$, about the x-axis (Fig. 6.34).

Solution We evaluate the formula

$$\text{Surface area} = \int_a^b 2\pi y \sqrt{1 + \left(\frac{dy}{dx}\right)^2}\, dx \quad \text{(Eq. 8)}$$

with

$$a = 1, \qquad b = 2, \qquad y = 2\sqrt{x}, \qquad \frac{dy}{dx} = \frac{1}{\sqrt{x}},$$

$$\sqrt{1 + \left(\frac{dy}{dx}\right)^2} = \sqrt{1 + \left(\frac{1}{\sqrt{x}}\right)^2} = \sqrt{1 + \frac{1}{x}} = \sqrt{\frac{x+1}{x}} = \frac{\sqrt{x+1}}{\sqrt{x}}.$$

With these substitutions,

$$\text{Surface area} = \int_1^2 2\pi \cdot 2\sqrt{x}\frac{\sqrt{x+1}}{\sqrt{x}}\, dx = 4\pi \int_1^2 \sqrt{x+1}\, dx$$

$$= 4\pi \cdot \frac{2}{3}(x+1)^{3/2}\Big]_1^2 = \frac{8\pi}{3}(3\sqrt{3} - 2\sqrt{2}).$$

Revolution about the y-Axis

If the axis of revolution is the y-axis, we use the formula we get from interchanging x and y in Eq. (8):

Revolution about the y-Axis

If $x = f(y)$ has a continuous first derivative throughout the interval $c \le y \le d$, the area of the surface S generated by revolving the curve $x = f(y)$, $c \le y \le d$, about the y-axis is

$$S = \int_c^d 2\pi x \sqrt{1 + \left(\frac{dx}{dy}\right)^2}\, dy. \tag{9}$$

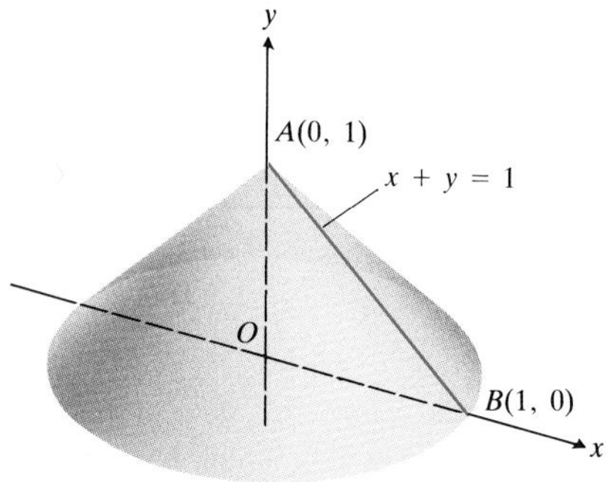

6.35 Revolving line segment AB about the y-axis generates a cone whose lateral surface area we can now calculate two different ways. See Example 2.

EXAMPLE 2 The line segment $x = 1 - y$, $0 \le y \le 1$, is revolved about the y-axis to generate the cone in Fig. 6.35. Find its lateral surface area.

Solution Here we have a calculation we can check with a formula from geometry:

$$\text{Lateral surface area} = \frac{\text{base circumference}}{2} \times \text{slant height} = \pi\sqrt{2}.$$

To see how Eq. (9) gives the same result, we take

$$c = 0, \qquad d = 1, \qquad x = 1 - y, \qquad \frac{dx}{dy} = -1,$$

$$\sqrt{1 + \left(\frac{dx}{dy}\right)^2} = \sqrt{1 + (-1)^2} = \sqrt{2}$$

and calculate

$$\text{Surface area} = \int_c^d 2\pi x\sqrt{1 + \left(\frac{dx}{dy}\right)^2}\,dy = \int_0^1 2\pi(1 - y)\sqrt{2}\,dy \qquad \text{(Eq. 9)}$$

$$= 2\pi\sqrt{2}\left[y - \frac{y^2}{2}\right]_0^1 = 2\pi\sqrt{2}\left(1 - \frac{1}{2}\right) = \pi\sqrt{2}.$$

The results agree, as they should. ■

The Short Differential Form

The equations

$$S = \int_a^b 2\pi y\sqrt{1 + \left(\frac{dy}{dx}\right)^2}\,dx \quad \text{and} \quad S = \int_c^d 2\pi x\sqrt{1 + \left(\frac{dx}{dy}\right)^2}\,dy \qquad (10)$$

are often written in terms of the arc length differential $ds = \sqrt{dx^2 + dy^2}$ as

$$S = \int_a^b 2\pi y\,ds \quad \text{and} \quad S = \int_c^d 2\pi x\,ds. \qquad (11)$$

In the first of these, y is the distance from the x-axis to an element of arc length ds. In the second, x is the distance from the y-axis to an element of arc length ds. In both cases the integrals have the form

$$S = \int 2\pi(\text{radius})(\text{band width}) = \int 2\pi\rho\,ds, \qquad (12)$$

where ρ is the radius from the axis of revolution to an element of arc length ds (Fig. 6.36).

If you wish to remember only one formula for surface area, you might make it the short differential form.

Short Differential Form

$$S = \int 2\pi\rho\,ds \qquad (13)$$

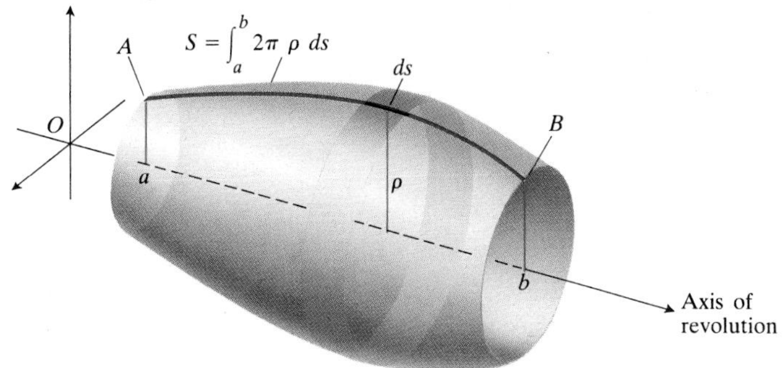

6.36 The area of the surface swept out by revolving arc AB about the axis shown here is $\int_a^b 2\pi\rho\,ds$. The exact expression depends on the formulas for ρ and ds.

In any particular problem you would then express the radius function ρ and the arc length differential ds in terms of a common variable and supply limits of integration for that variable.

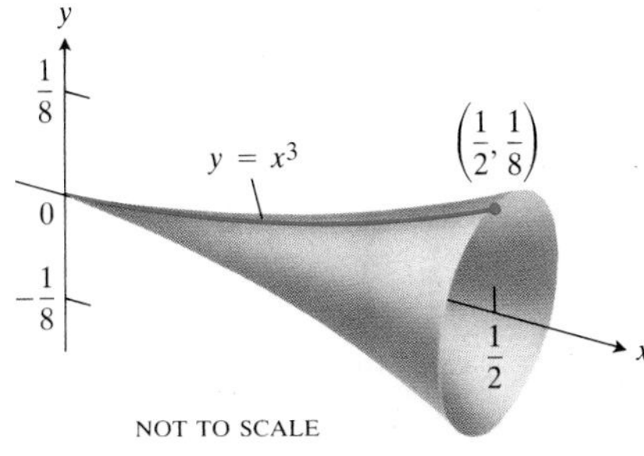

6.37 The surface generated by revolving the curve $y = x^3$, $0 \le x \le 1/2$, about the x-axis could be the design for a champagne glass. The surface area is calculated in Example 3.

EXAMPLE 3 Find the area of the surface generated by revolving the curve $y = x^3$, $0 \le x \le 1/2$, about the x-axis (Fig. 6.37).

Solution We start with the short differential form:

$$
\begin{aligned}
S &= \int 2\pi\,\rho\,ds \\
&= \int 2\pi y\,ds \qquad \left(\begin{array}{l}\text{For revolution about the } x\text{-axis,}\\ \text{the radius function is } \rho = y.\end{array}\right) \\
&= \int 2\pi y\sqrt{dx^2 + dy^2}. \qquad \left(ds = \sqrt{dx^2 + dy^2}\right)
\end{aligned}
$$

We then decide whether to express dy in terms of dx or dx in terms of dy. The original form of the equation, $y = x^3$, makes its easier to express dy in terms of dx, so we continue the calculation with

$$
y = x^3, \qquad dy = 3x^2\,dx, \qquad \text{and} \qquad \sqrt{dx^2 + dy^2} = \sqrt{dx^2 + (3x^2\,dx)^2} = \sqrt{1 + 9x^4}\,dx.
$$

With these substitutions, x becomes the variable of integration, and we continue with

$$
\begin{aligned}
S &= \int_{x=0}^{x=1/2} 2\pi y\sqrt{dx^2 + dy^2} = \int_0^{1/2} 2\pi x^3\sqrt{1 + 9x^4}\,dx \\
&= 2\pi \cdot \frac{1}{36} \cdot \frac{2}{3}(1 + 9x^4)^{3/2}\Big]_0^{1/2} \qquad \left(\begin{array}{l}\text{Substitute } u = 1 + 9x^4,\ du/36 = x^3\,dx,\\ \text{integrate, and substitute back.}\end{array}\right) \\
&= \frac{\pi}{27}\left[\left(1 + \frac{9}{16}\right)^{3/2} - 1\right] \\
&= \frac{\pi}{27}\left[\left(\frac{25}{16}\right)^{3/2} - 1\right] = \frac{\pi}{27}\left(\frac{125}{64} - 1\right) = \frac{61\pi}{1728}.
\end{aligned}
$$

As with arc length calculations, even the simplest curves can give you a workout. ■

Exercises 6.5

Find the areas of the surfaces generated by revolving the curves in Exercises 1–14 about the axes indicated.

1. $y = x/2$, $0 \le x \le 4$, about the x-axis. Check your result with a formula from geometry, as in Example 2.
2. $y = x/2$, $0 \le x \le 4$, about the y-axis. Check your result with a formula from geometry, as in Example 2.
3. $y = (x/2) + (1/2)$, $1 \le x \le 3$, about the x-axis. Check your result with the geometry formula in Eq. (1).
4. $y = (x/2) + (1/2)$, $1 \le x \le 3$, about the y-axis. Check your result with the geometry formula in Eq. (1).
5. $y = x^3/9$, $0 \le x \le 2$, about the x-axis
6. $y = \sqrt{x}$, $3/4 \le x \le 15/4$, about the x-axis
7. $y = \sqrt{2x - x^2}$, $0 \le x \le 2$, about the x-axis
8. $y = \sqrt{x + 1}$, $1 \le x \le 5$, about the x-axis
9. $x = y^3/3$, $0 \le y \le 1$, about the y-axis
10. $y = 2\sqrt{4 - x}$, $1 \le x \le 4$, about the y-axis

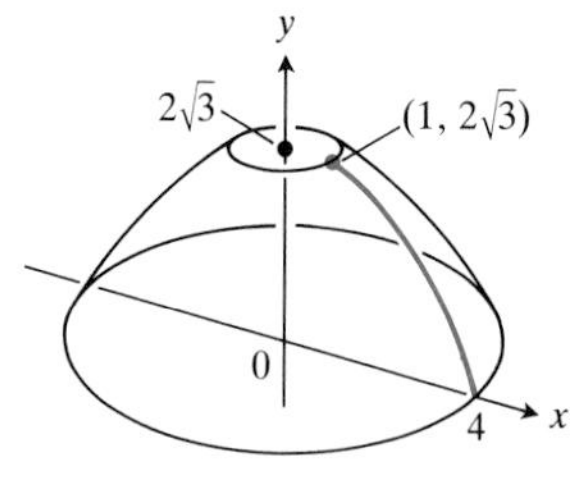

11. $x = \sqrt{2y - 1}$, $1/2 \le y \le 1$, about the y-axis

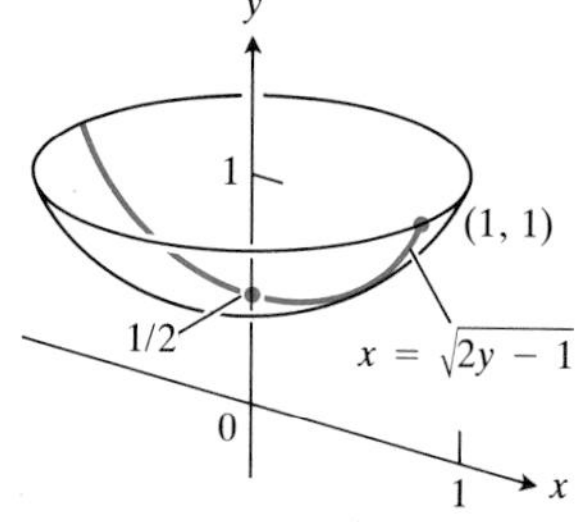

12. $x = (e^y + e^{-y})/2$, $0 \le y \le \ln 2$, about the y-axis

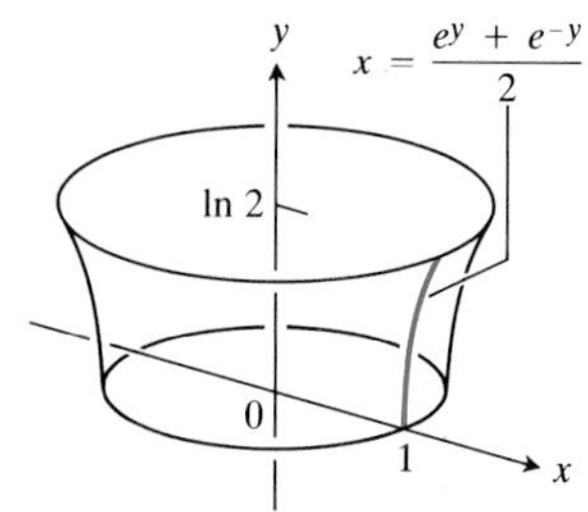

13. $x = (y^4/4) + 1/(8y^2)$, $1 \le y \le 2$, about the x-axis
(*Hint:* Express $ds = \sqrt{dx^2 + dy^2}$ in terms of dy and evaluate the integral $S = \int 2\pi y\, ds$ with appropriate limits.)
14. $y = (1/3)(x^2 + 2)^{3/2}$, $0 \le x \le 3$, about the y-axis
(*Hint:* Express $ds = \sqrt{dx^2 + dy^2}$ in terms of dx and evaluate the integral $S = \int 2\pi x\, ds$ with appropriate limits.)
15. Use an integral to find the surface area of the sphere generated by revolving the semicircle $y = \sqrt{1 - x^2}$, $-1 \le x \le 1$, about the x-axis. Check your result with a formula from geometry.
16. Write an integral for the area of the surface generated by revolving the curve $y = \cos x$, $-\pi/2 \le x \le \pi/2$, about the x-axis. We shall see how to evaluate the integral when we get to Chapter 8.
17. *The surface of an astroid.* Find the area of the surface generated by revolving the portion of the astroid $x^{2/3} + y^{2/3} = 1$ shown below about the x-axis. (*Hint:* Revolve the first quadrant portion $y = (1 - x^{2/3})^{3/2}$, $0 \le x \le 1$, about the x-axis and double your result.)

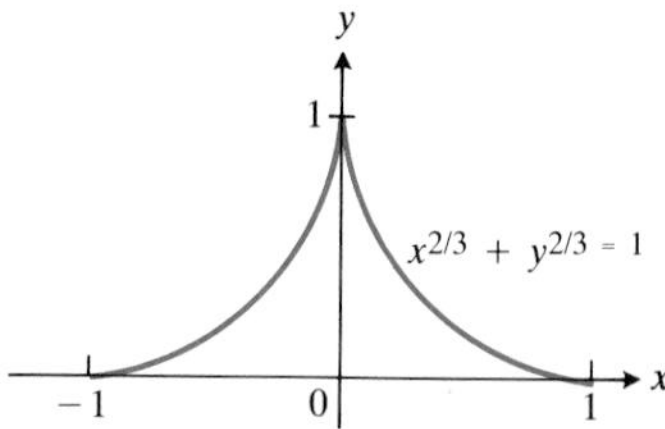

18. *Enameling woks.* Your company has decided to put out a deluxe version of the successful wok you designed in Exercise 41 of Section 6.2. The plan is to coat it inside with white enamel and outside with blue

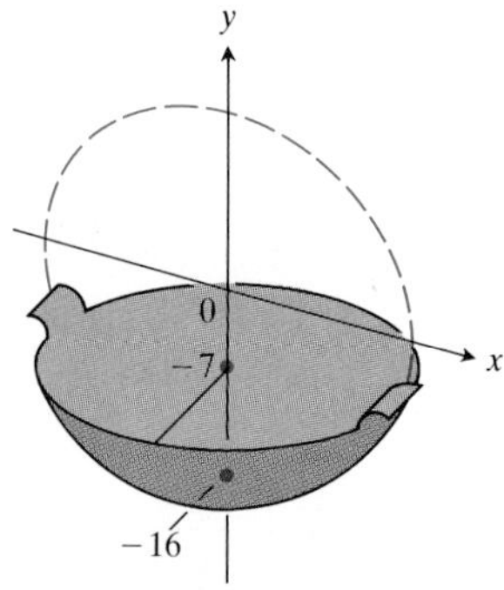

enamel. Each enamel will be sprayed on a millimeter thick before baking. Manufacturing wants to know how much enamel it will take for a production run of 5,000 woks. What do you tell them? (Neglect waste and unused material. Answer in liters.)

19. *Slicing bread.* Did you know that if you cut a spherical loaf of bread into slices of equal width, each slice will have the same amount of crust? To see why, suppose the semicircle $y = \sqrt{r^2 - x^2}$ in Fig. 6.38 is revolved about the x-axis to generate a sphere. Let AB be an arc of the semicircle that lies above an interval of length h on the x-axis. Show that the area swept out by AB does not depend on the location of the interval. (It does depend on the length of the interval.)

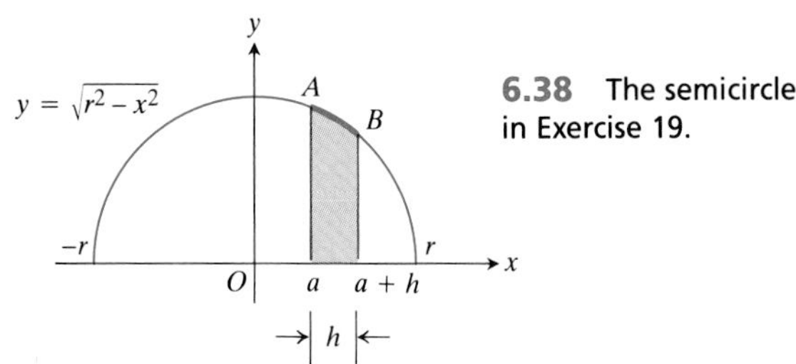

6.38 The semicircle in Exercise 19.

6.6 Work

In everyday life, *work* describes any activity that takes muscular or mental effort. In science, however, the term is used in a narrower sense that involves the application of a force to a body and the body's subsequent displacement. This section shows how to calculate work. The applications run from stretching springs and pumping liquids from subterranean tanks to lifting satellites into orbit.

The Constant-Force Formula for Work

We begin with a definition.

DEFINITION

When a body moves a distance d along a straight line as the result of being acted on by a force that has a constant magnitude F in the direction of the motion, the **work** W done by the force in moving the body is F times d.

$$W = Fd \tag{1}$$

We can see right away that there is a considerable difference between what we are used to calling work and what this formula says work is. If you push a car down a street, you are doing work, both by our own reckoning and according to Eq. (1). But if you push against the car and the car does not move, Eq. (1) says you are doing no work, no matter how hard or how long you push.

Work is measured in foot-pounds, newton-meters, or whatever force-distance unit is appropriate to the occasion.

EXAMPLE 1 It takes a force of about 1 newton (1N) to lift an apple from a table. If you lift it 1 meter, you have done about 1 newton-meter (N · m) of work on the apple. ■

The Variable-Force Integral Formula for Work

If the force you apply varies along the way, as it will if you are lifting a leaking bucket or compressing a spring, the formula $W = Fd$ has to be replaced by an integral formula that takes the variation in F into account. It takes calculus to measure the work done by a variable force.

Suppose that the force performing the work varies continuously along a line that we can take to be the x-axis and that the force is represented by the function $F(x)$. We are interested in the work done along an interval from $x = a$ to $x = b$. We partition the closed interval $[a, b]$ in the usual way and choose an arbitrary point c_k in each subinterval $[x_{k-1}, x_k]$.

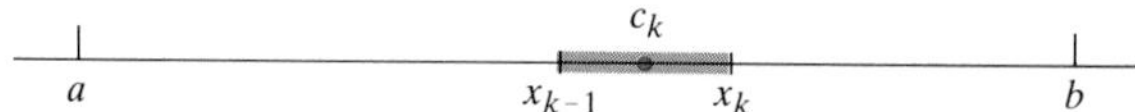

If the subinterval is short enough, F, being continuous, will not vary much from x_{k-1} to x_k. The amount of work done by the force from x_{k-1} to x_k will be nearly equal to $F(c_k)$ times distance Δx_k, as it would be if we could apply Eq. (1). The total work done from a to b is thus approximated by the Riemann sum

$$\sum_{k=1}^{n} F(c_k)\Delta x_k. \tag{2}$$

We expect the approximations to improve as the norm of the partition goes to zero, so we define the work done by the force from a to b to be the integral of F from a to b.

DEFINITION

The **work** done by a continuous force $F(x)$ directed along the x-axis from $x = a$ to $x = b$ is

$$W = \int_a^b F(x)\,dx. \tag{3}$$

EXAMPLE 2 A leaky 5-lb bucket is lifted from the ground into the air by pulling in 20 ft of rope at a constant speed (Fig. 6.39). The rope weighs 0.08 lb/ft. The bucket starts with 2 gal of water (16 lb) and leaks at a constant rate. It finishes draining just as it reaches the top. How much work was spent

a) lifting the water alone?

b) lifting the water and bucket together?

c) lifting the water, bucket, and rope?

6.39 The leaky bucket in Example 2.

Solution

a) ***The water alone.*** The force required to lift the water is the water's weight, which varies steadily from 16 to 0 lb over the 20-ft lift. When the bucket is x ft off the ground, the water weighs

$$F(x) = \underbrace{16}_{\text{Original weight of water}}\underbrace{\left(\frac{20-x}{20}\right)}_{\text{Proportion left at elevation } x} = 16\left(1-\frac{x}{20}\right) = 16 - \frac{4x}{5} \text{ lb.}$$

The work done is

$$W = \int_a^b F(x)\,dx \qquad \text{(Use Eq. (3) for variable forces.)}$$

$$= \int_0^{20}\left(16 - \frac{4x}{5}\right)dx = \left[16x - \frac{2x^2}{5}\right]_0^{20} = 320 - 160 = 160 \text{ ft} \cdot \text{lb.}$$

b) ***The water and bucket together.*** According to Eq. (1), it takes $5 \times 20 = 100$ ft · lb to lift a 5-lb weight 20 ft. Therefore

$$160 + 100 = 260 \text{ ft} \cdot \text{lb}$$

of work were spent lifting the water and bucket together.

c) ***The water, bucket, and rope.*** Now the total weight at level x is

$$F(x) = \underbrace{\left(16 - \frac{4x}{5}\right)}_{\text{Variable weight of water}} + \underbrace{5}_{\text{Constant weight of bucket}} + \underbrace{\overset{\text{lb/ft}}{(0.08)}\overset{\text{ft}}{(20 - x)}}_{\text{Weight of rope paid out at elevation } x}.$$

The work lifting the rope is

$$\text{Work on rope} = \int_0^{20}(0.08)(20-x)\,dx = \int_0^{20}(1.6 - 0.08x)\,dx$$

$$= \left[1.6x - 0.04x^2\right]_0^{20} = 32 - 16 = 16 \text{ ft} \cdot \text{lb.}$$

The total work for the water, bucket, and rope combined is

$$160 + 100 + 16 = 276 \text{ ft} \cdot \text{lb.}$$

■

Hooke's Law for Springs: $F = kx$

Hooke's law says that the amount of force F it takes to stretch or compress a spring x length units from its natural length is proportional to x. In symbols,

$$F = kx. \tag{4}$$

The number k, measured in force units per unit length, is a constant characteristic of the spring, called the **spring constant.** Hooke's law (Eq. 4) holds

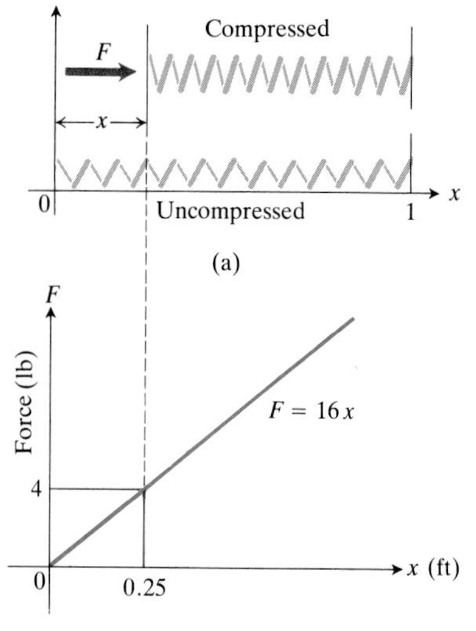

6.40 The force F required to hold a spring under compression increases linearly as the spring is compressed.

as long as the force doesn't distort the metal in the spring. We shall assume that the forces in this section are too small to do that.

EXAMPLE 3 Find the work required to compress a spring from its natural length of 1 ft to a length of 0.75 ft if the spring constant is $k = 16$ lb/ft.

Solution We picture the uncompressed spring laid out along the x-axis with its movable end at the origin and its fixed end at $x = 1$ ft (Fig. 6.40). This enables us to describe the force required to compress the spring from 0 to x with the formula $F = 16x$. As the spring is compressed from 0 to 0.25 ft, the force varies from

$$F(0) = 16 \cdot 0 = 0 \text{ lb} \quad \text{to} \quad F(0.25) = 16 \cdot 0.25 = 4 \text{ lb}.$$

The work done by F over this interval is

$$W = \int_0^{0.25} 16x\,dx = 8x^2\Big]_0^{0.25} = 0.5 \text{ ft} \cdot \text{lb}. \qquad \begin{pmatrix}\text{Eq. (3) with } a = 0,\\ b = 0.25, F(x) = 16x\end{pmatrix}$$

EXAMPLE 4 A spring has a natural length of 1 m. A force of 24 N stretches the spring to a length of 1.8 m.

a) Find the spring constant k.
b) How much work will it take to stretch the spring 2 m beyond its natural length?
c) How far will a 45-N force stretch the spring?

Solution

a) *The spring constant.* We find it from Eq. (4). A force of 24 N stretches the spring 0.8 m, so

$$24 = k(0.8) \qquad (\text{Eq. (4) with } F = 24, x = 0.8)$$

$$k = 24/0.8 = 30 \text{ N/m}.$$

b) *The work to stretch the spring 2 m.* We imagine the unstressed spring hanging along the x-axis with its free end at $x = 0$ (Fig. 6.41). Then the force required to stretch the spring x m beyond its natural length is the force required to pull the free end of the spring x units from the origin. Hooke's law with $k = 30$ tells us this force is

$$F(x) = 30x.$$

The work required to apply this force from $x = 0$ m to $x = 2$ m is

$$W = \int_0^2 30x\,dx = 15x^2\Big]_0^2 = 60 \text{ N} \cdot \text{m}.$$

c) *How far will a 45-N force stretch the spring?* We substitute $F = 45$ in the equation $F = 30x$ to find

$$45 = 30x \quad \text{and} \quad x = 1.5 \text{ m}.$$

A 45-N force will stretch the spring 1.5 m. No calculus is required to find this.

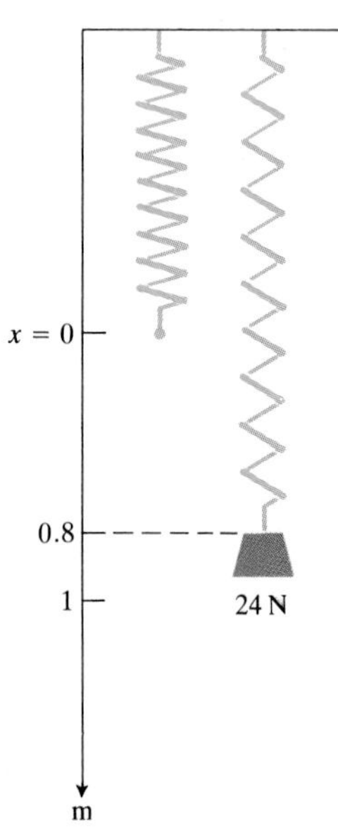

6.41 A 24-newton weight stretches this spring 0.8 m beyond its unstressed length.

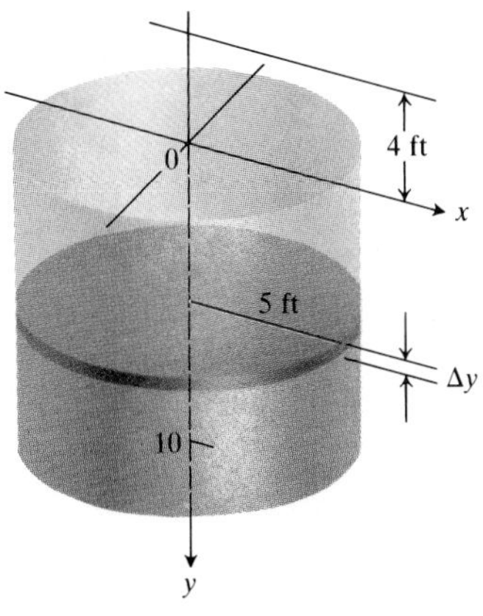

6.42 To find the work it takes to pump the water from a tank, think of lifting the water out one slab at a time. Pointing the y-axis down makes y measure the distance from the slab to the top of the tank.

Pumping Liquids from Containers—Do-it-yourself Integrals

To find how much work it takes to pump all or part of the liquid from a container, imagine lifting the liquid out one horizontal slab at a time and applying the equation $W = Fd$ to each slab. The integral we get each time depends on the weight of the liquid and the cross-section dimensions of the container, but the way we find the integral is the same for all containers. The next two examples show what to do.

EXAMPLE 5 How much work does it take to pump the water from a full upright right circular cylindrical tank of radius 5 ft and height 10 ft to a level 4 ft above the top of the tank?

Solution We draw coordinate axes (Fig. 6.42) and imagine the water divided into thin slabs by planes perpendicular to the y-axis at the points of a partition of the interval $[0, 10]$.

The typical slab between the planes at y and $y + \Delta y$ has a volume of approximately

$$\Delta V = \pi(\text{radius})^2(\text{thickness}) = \pi(5)^2\Delta y = 25\pi\,\Delta y \text{ ft}^3.$$

The force $F(y)$ required to lift this slab is its weight,

$$F(y) = w\,\Delta V = 25\pi w\,\Delta y \text{ lb},$$

where w is the weight of a cubic foot of water (we can substitute for w later).

The distance through which $F(y)$ must act is about $(y + 4)$ ft, so the work done in lifting this slice 4 ft above the top of the bowl is about

$$\Delta W = 25\pi w(y + 4)\Delta y \text{ ft} \cdot \text{lb}.$$

The work of lifting all the slabs of water is about

$$\sum_0^{10} \Delta W = \sum_0^{10} 25\pi w(y + 4)\Delta y \text{ ft} \cdot \text{lb}.$$

This is a Riemann sum for the function $25\pi w(y + 4)$ on the interval from $y = 0$ to $y = 10$. The work of pumping the tank dry is the limit of these sums as the norm of the partition goes to zero:

$$\begin{aligned} \text{Work} &= \int_0^{10} 25\pi w(y + 4)\,dy = 25\pi w \int_0^{10} (y + 4)\,dy \\ &= 25\pi w\left[\frac{y^2}{2} + 4y\right]_0^{10} = 25\pi w(50 + 40) = 2250\pi w \\ &= 2250\pi(62.5) \qquad \text{(Water weighs 62.5 lb/ft}^3\text{.)} \\ &= 441{,}786 \text{ ft} \cdot \text{lb}. \qquad \text{(Nearest ft} \cdot \text{lb, with a calculator)} \end{aligned}$$

A 1-horsepower pump, rated at 550 ft · lb per second, could empty the tank in a little less than 14 min. ■

Weight-Density

Weight per unit volume is called **weight-density.** Typical values (lb/ft^3) are

gasoline	42
mercury	849
milk	64.5
olive oil	57
seawater	64
water	62.5.

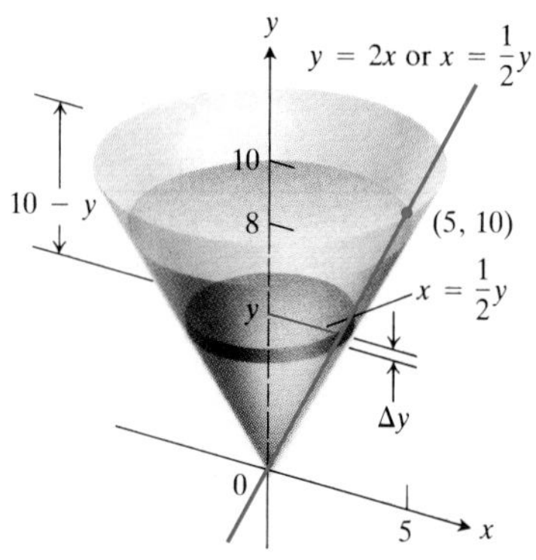

6.43 The salad oil in Example 6.

EXAMPLE 6 The inverted conical tank in Fig. 6.43 is filled to within 2 ft of the top with salad oil weighing 57 lb/ft^3. How much work does it take to pump the oil to the rim of the tank?

Solution We imagine the oil divided into thin slabs by planes perpendicular to the y-axis at the points of a partition of the interval $[0, 8]$.

The typical slab between the planes at y and $y + \Delta y$ has a volume of about

$$\Delta V = \pi(\text{radius})^2(\text{thickness}) = \pi\left(\frac{1}{2}y\right)^2 \Delta y = \frac{\pi}{4}y^2 \Delta y \text{ ft}^3.$$

The force $F(y)$ required to lift this slab is its weight,

$$F(y) = 57\Delta V = \frac{57\pi}{4}y^2\Delta y \text{ lb.} \qquad \begin{pmatrix}\text{Weight = weight per} \\ \text{unit volume} \times \text{volume}\end{pmatrix}$$

The distance through which $F(y)$ must act to lift this slab to the level of the rim of the cone is about $(10 - y)$ft, so the work done lifting the slab is about

$$\Delta W = \frac{57\pi}{4}(10 - y)y^2\Delta y \text{ ft} \cdot \text{lb}.$$

The work done lifting all the slabs from $y = 0$ to $y = 8$ to the rim is about

$$\sum_0^8 \frac{57\pi}{4}(10 - y)y^2\Delta y \text{ ft} \cdot \text{lb}.$$

This is a Riemann sum for the function $(57\pi/4)(10 - y)y^2$ on the interval from $y = 0$ to $y = 8$. The work of pumping the oil to the rim is the limit of these sums as the norm of the partition goes to zero.

$$\text{Work} = \int_0^8 \frac{57\pi}{4}(10 - y)y^2\,dy = \frac{57\pi}{4}\int_0^8 (10y^2 - y^3)\,dy$$

$$= \frac{57\pi}{4}\left[\frac{10y^3}{3} - \frac{y^4}{4}\right]_0^8 = 30{,}561 \text{ ft} \cdot \text{lb} \qquad \begin{pmatrix}\text{With a calculator,} \\ \text{rounded}\end{pmatrix}$$

How to Find Work Done during Pumping

1. Draw a figure with a coordinate system.
2. Find the weight of a thin horizontal slab of liquid.
3. Find the work to lift the slab to its destination.
4. Integrate the work expression from the top to the bottom of the liquid.

Exercises 6.6

1. The workers in Example 2 changed to a larger bucket that held 5 gal (40 lb) of water, but the new bucket had an even larger leak so that it, too, was empty by the time it reached the top. Assuming that the water leaked out at a steady rate, how much work was done lifting the water? (Do not include the rope and bucket.)

2. The bucket in Example 2 is hauled up twice as fast so that there is still 1 gal (8 lb) of water left when the bucket reaches the top. How much work is done lifting the water this time? (Do not include the rope and bucket.)

3. A mountain climber is about to haul up a 50-m length of hanging rope. How much work will it take if the rope weighs 0.74 newtons per meter?

4. A model rocket engine burned up its 2-oz fuel cartridge lifting the rocket to 170 ft. Assuming the fuel burned at a steady rate, how much work was spent just lifting fuel?

5. An electric elevator with a motor at the top has a multistrand cable weighing 4 lb/ft. One hundred eighty feet of cable are paid out when the car is at the first floor and effectively zero ft are out when the car is at the top floor. How much work does the motor do just lifting the cable when it takes the car from the first floor to the top?

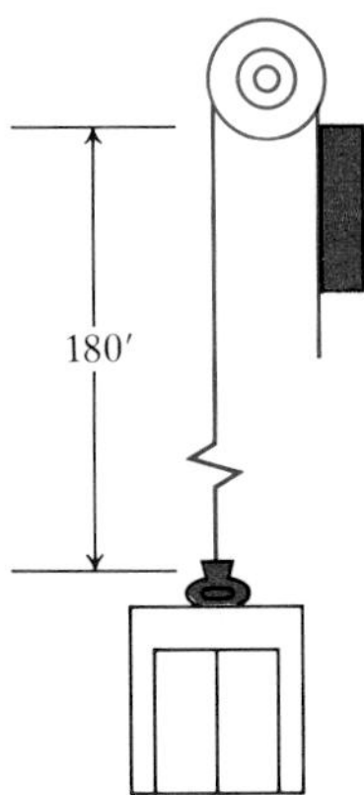

6. A bag of sand originally weighing 144 lb was lifted at a constant rate. The sand leaked out at a steady rate, and the sand was half gone by the time the bag had been lifted 18 ft. How much work was done lifting the sand this far? (Neglect the weight of the bag and lifting equipment.)

7. If a force of 6 N stretches a spring 0.4 m beyond its natural length, how much work is done?

8. If a force of 90 N stretches a spring 1 m beyond its natural length, how much work does it take to stretch the spring 5 m beyond its natural length?

9. A 10,000-lb force compressed a spring from its natural length of 12 in. to a length of 11 in. How much work did it do in compressing the spring
a) the first half-inch?
b) the second half-inch?

10. A bathroom scale is compressed 1/16 in. when a 150-lb person stands on it. Assuming the scale behaves like a spring, how much does someone who compresses the scale 1/8 in. weigh? How much work does it take to compress the scale 1/8 in?

11. The rectangular tank shown here with its top at ground level is used to catch run-off water.

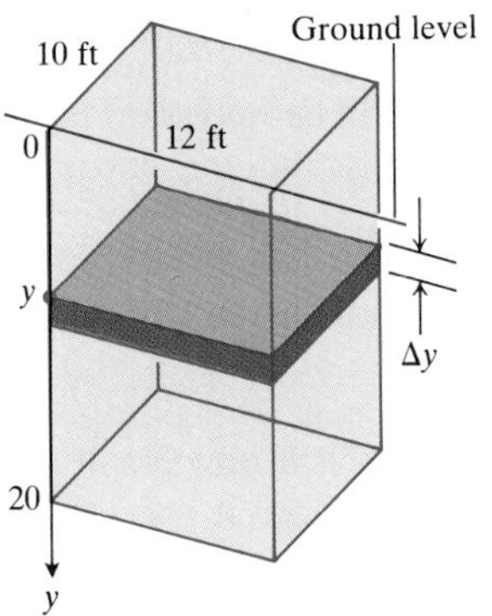

a) How much work does it take to empty the tank by pumping the water back to ground level once the tank is full?
b) If the water is pumped to ground level with a (5/11)-hp motor (work output 250 ft · lb/sec), how long will it take to empty the full tank?
c) Show that the pump in part (b) will lower the water level 10 ft (half way) during the first 25 min of pumping.

12. The full rectangular cistern (rain water storage tank) shown here with its top 10 ft below ground level is to be emptied for inspection by pumping its contents to ground level.

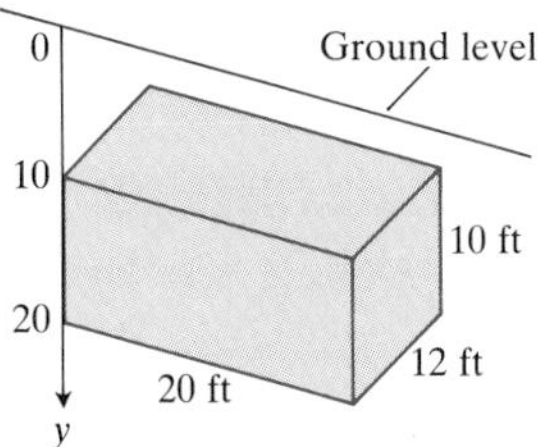

a) How much work will it take to empty the cistern?
b) How long will it take a (1/2)-hp pump, rated at 225 ft · lb/sec, to pump the tank dry?
c) How long will it take the pump in part (b) to empty the tank half way? (It will be less than half the time required to empty the tank completely.)

13. How much work would it take to pump the water from the tank in Example 5 to the level of the top of the tank (instead of 4 ft higher)?

14. Suppose that instead of being completely full, the tank in Example 5 is only half full. How much work does it take to pump the water that's left to a level 4 ft above the top of the tank?

15. A vertical right circular cylindrical tank measures 30 ft high and 20 ft in diameter. It is full of kerosene weighing 51.2 lb/ft^3. How much work does it take to pump the kerosene to the level of the top of the tank?

16. The cylindrical tank shown here can be filled by

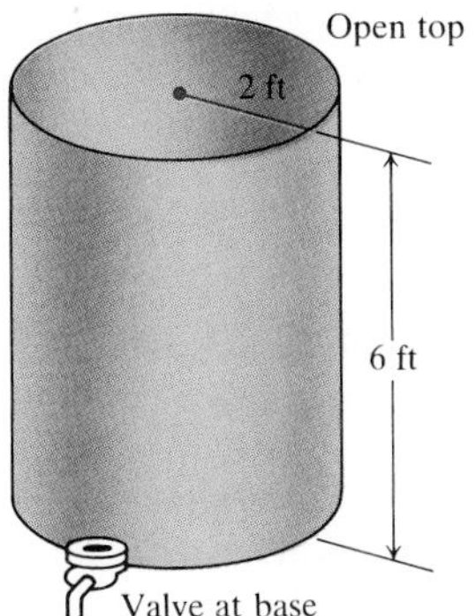

pumping water from a lake 15 ft below the bottom of the tank. There are two ways to go about it. One is to pump the water through a hose to a valve in the bottom of the tank. The other is to attach the hose to the rim of the tank and let the water pour in. Which way will be faster?

17. a) Suppose the cone in Example 6 contained milk (weight density 64.5 lbs/ft^3) instead of salad oil. How much work would it have taken to pump the contents to the rim?

b) How much work would it have taken to pump the oil in Example 6 to a level 3 ft above the cone's rim?

18. **CALCULATOR** The truncated conical container shown here is full of strawberry milkshake that weighs (4/9) oz/in^3.

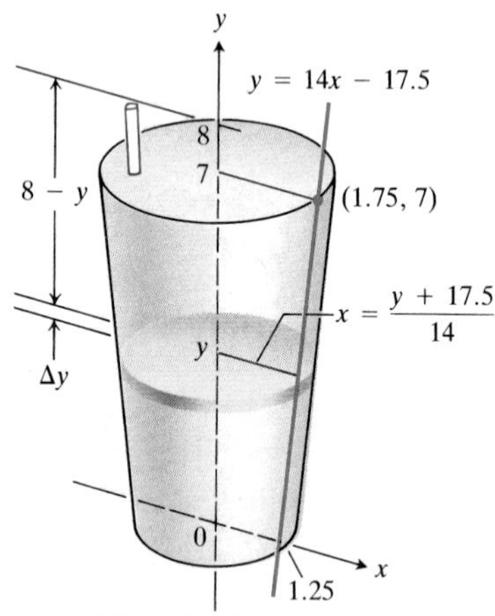

As you can see, the container is 7 in. deep, 2.5 in. across at the base, and 3.5 in. across at the top (a standard size at *Brigham's* in Boston). The straw sticks up an inch above the top. About how much work does it take to suck up the milkshake through the straw (neglecting friction)? Answer in inch-ounces.

19. To design the interior surface of a huge stainless steel tank, you revolve the curve $y = x^2$, $0 \leq x \leq 4$, about the y-axis. The container, with dimensions in meters, is to be filled with seawater, which weighs 10,000 newtons per cubic meter. How much work will it take to empty the tank by pumping the water to the tank's top?

20. We model pumping from spherical containers the way we do from others, with the axis of integration along the vertical axis of the sphere. We drew Fig. 6.44 to help find out how much work it takes to pump the water from a full hemispherical bowl of radius 5 ft

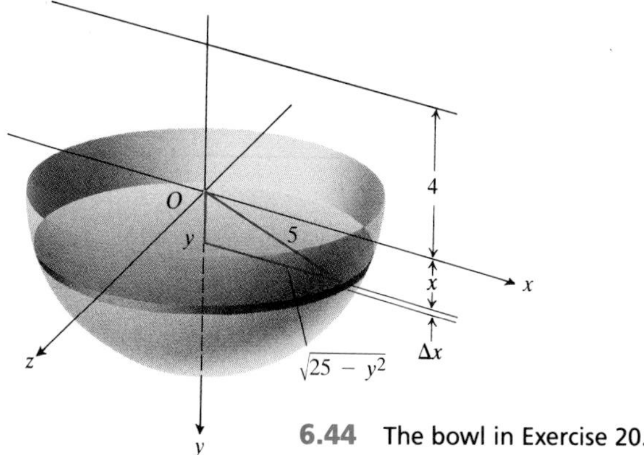

6.44 The bowl in Exercise 20.

to a height 4 ft above the top of the bowl. How much work does it take?

21. You are in charge of the evacuation and repair of the storage tank shown in Fig. 6.45. The tank is a hemisphere of radius 10 ft and is full of benzene weighing 56 lb/ft^3. A firm you contacted says it can empty the tank for 1/2¢ per foot-pound of work. Find the work required to empty the tank by pumping the benzene to an outlet 2 ft above the top of the tank. If you have $5000 budgeted for the job, can you afford to hire the firm?

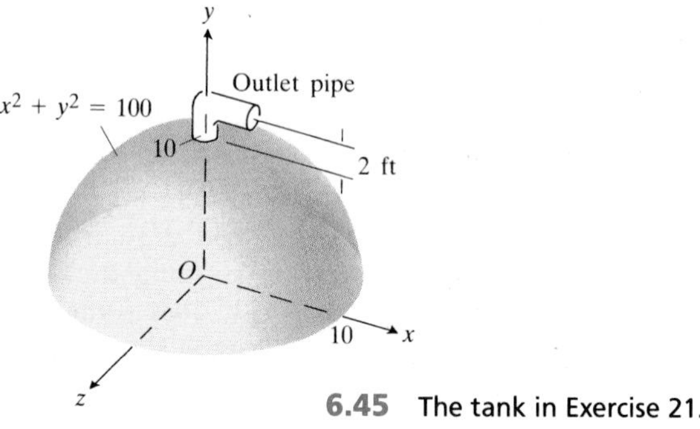

6.45 The tank in Exercise 21.

22. Your town has decided to drill a well to increase its water supply. As the town engineer, you have determined that a water tower will be necessary to provide the pressure needed for distribution, and you have designed the system in Fig. 6.46. The water is to be pumped from a 300-ft well through a vertical 4-in. pipe into the base of a cylindrical tank 20 ft in diameter and 25 ft high. The base of the tank will be 60 ft above ground. The pump is a 3-hp pump, rated at 1650 ft · lb/sec. How long will it take to fill the tank the first time? (Include the time it takes to fill the pipe.)

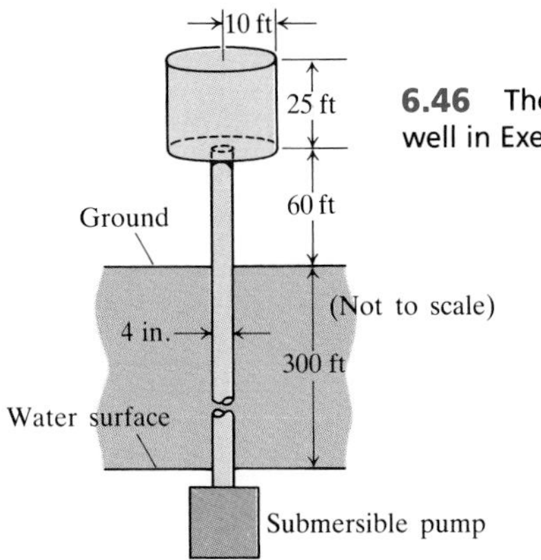

6.46 The water tower and well in Exercise 22.

23. CALCULATOR *Putting a satellite in orbit.* The strength of the earth's gravitational field varies with the distance r from the earth's center, and the magnitude of the gravitational force experienced by a satellite of mass m during and after launch is

$$F(r) = \frac{mMG}{r^2}.$$

Here, $M = 5.975 \times 10^{24}$ kg is the earth's mass, $G = 6.6720 \times 10^{-11}$ Nm^{2}kg^{-2} is the universal gravitational constant, and r is measured in meters. The number of newton-meters of work it takes to lift a 1000-kg satellite from the earth's surface to a circular orbit 35,780 km above the earth's center is therefore given by the integral

$$\text{Work} = \int_{6,370,000}^{35,780,000} \frac{mMG}{r^2}\,dr.$$

Evaluate the integral. The lower limit of integration is the earth's radius in meters at the launch site. This calculation does not take into account energy spent by the launch vehicle or energy spent bringing the satellite to orbit velocity.

24. CALCULATOR *Forcing electrons together.* Two electrons r meters apart repel each other with a force of

$$F = \frac{23 \times 10^{-29}}{r^2}$$

newtons.

a) Suppose one electron is held fixed at the point $(1, 0)$ on the x-axis (units in meters). How much work does it take to move a second electron along the x-axis from the point $(-1, 0)$ to the origin?

b) Suppose an electron is held fixed at each of the points $(-1, 0)$ and $(1, 0)$. How much work does it take to move a third electron along the x-axis from $(5, 0)$ to $(3, 0)$?

6.7 Fluid Pressures and Fluid Forces

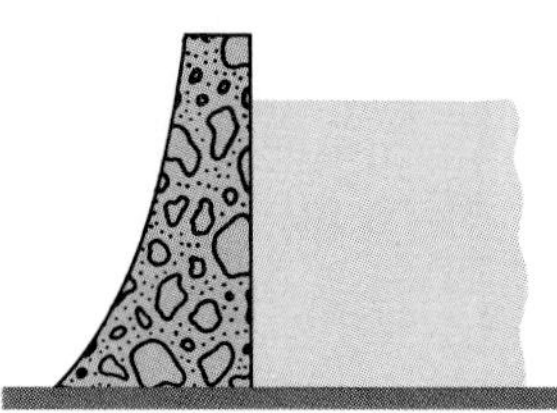

6.47 To withstand the increasing pressure, dams are built thicker as they go down.

We make dams thicker at the bottom than at the top (Fig. 6.47) because the pressure against them increases with depth. The deeper the water, the thicker the dam has to be.

It is a remarkable fact that the pressure at any point on the dam depends only on how far below the surface the point is and not on how much the surface happens to be tilted at that point. The pressure, in pounds per square foot at a point h feet below the surface, is always $62.5\,h$. The number 62.5 is the weight-density of water in pounds per cubic foot.

The formula, pressure $= 62.5\,h$, makes sense when you think of the units involved: Pounds per square foot equals pounds per cubic foot times feet:

$$\frac{\text{lb}}{\text{ft}^2} = \frac{\text{lb}}{\text{ft}^3} \times \text{ft}. \tag{1}$$

As you can see, this equation depends only on units and not on what fluid is involved. The pressure h feet below the surface of any fluid is the fluid's weight-density times the depth.

Weight-Density

A fluid's weight-density is its weight per unit volume. Typical values (lb/ft^3) are

gasoline	42
mercury	849
milk	64.5
molasses	100
olive oil	57
seawater	64
water	62.5.

The Pressure-Depth Equation

In a fluid that is standing still, the pressure at depth h is the fluid's weight-density times h:

$$p = wh. \tag{2}$$

In this section we use the equation $p = wh$ to derive a formula for the total force exerted by a fluid against all or part of a vertical or horizontal containing wall.

The Constant-Depth Formula for Force

In a container of fluid with a flat horizontal base, the total force exerted by the fluid against the base can be calculated by multiplying the area of the base by the pressure at that level. We can do this because total force equals force per unit area (pressure) times area. If F, p, and A are the total force, pressure, and area, then

$$\begin{aligned} F &= \text{total force} = \text{force per unit area} \times \text{area} \\ &= \text{pressure} \times \text{area} = pA \\ &= whA. \qquad (p = wh \text{ from Eq. (2)}) \end{aligned}$$

Total Force on a Constant-Depth Surface

$$F = pA = whA \tag{3}$$

EXAMPLE 1 These two containers are filled with water to the same depth and have the same base area. The total force is therefore the same on the bottom of each container. The containers' shapes do not matter here. The only things that count are depth and base area. ■

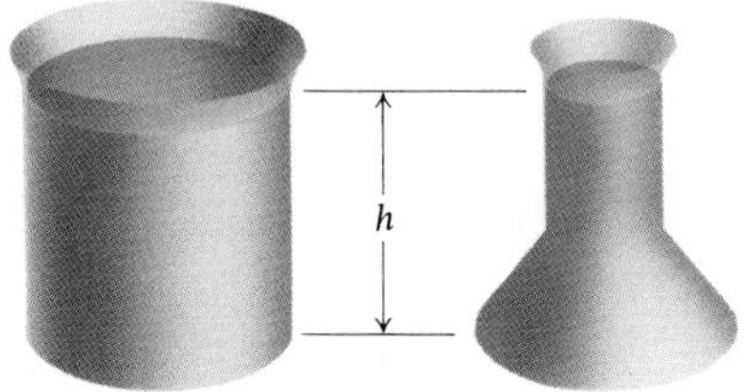

EXAMPLE 2 *The Great Molasses Flood.* At 1:00 P.M. on January 15, 1919, an unusually warm day, a 90-ft-high, 90-ft-diameter, cylindrical metal tank in which the Puritan Distilling Company was storing molasses at the corner of Foster and Commercial streets in Boston's North End exploded. The molasses flooded into the streets, 30 ft deep, trapping pedestrians and horses, knocking down buildings, and oozing into homes. It was eventually tracked all over town and even into the suburbs on people's shoes. The cleanup went on for weeks.

Given that the molasses weighed 100 lb/ft^3, what was the total force exerted by the molasses against the bottom of the tank at the time it blew? As

suming the tank was full, we can find out from Eq. (3):

$$\text{Total force} = wh\,A = (100)(90)(\pi(45)^2) = 57{,}255{,}526 \text{ lb.} \quad \text{(Rounded)}$$

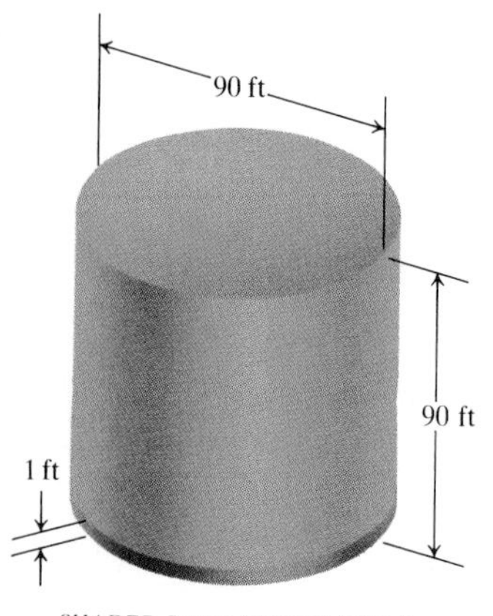

6.48 Schematic drawing of the molasses tank in Example 2. How much force did the bottom foot of wall have to withstand when the tank was full? It takes an integral to find out.

How about the force against the walls of the tank? For example, what was the total force against the bottom foot-wide band of tank wall (Fig. 6.48)?

The area of the band was

$$A = \pi\,dh = \pi(90)(1) = 90\pi \text{ ft}^2.$$

The tank was 90 ft deep, so the pressure near the bottom was about

$$p = wh = (100)(90) = 9000 \text{ lb/ft}^2.$$

Therefore the total force against the band was about

$$F = wh\,A = (9000)(90\pi) = 2{,}544{,}690 \text{ lb.} \qquad \text{(Rounded)}$$

But this is not exactly right. The top of the band was 89 ft below the surface, not 90, and the pressure there was less. To find out exactly what the force on the band was, we need to take the variation of the pressure across the band into account, and this means using calculus.

The Variable-Depth Integral for the Force against a Submerged Vertical Wall

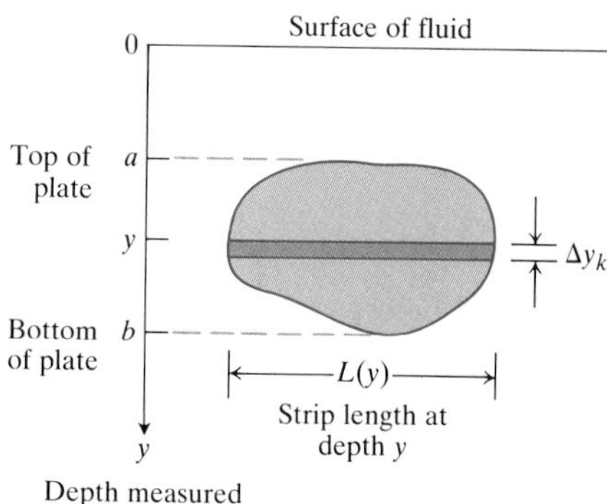

6.49 The force exerted by the fluid against one side of an approximating horizontal strip y units beneath the surface is about

$F = (\text{pressure})(\text{area}) = (wy)(L(y)\Delta y)$.

Suppose we want to find the force against one side of a submerged vertical plate whose surface looks like the shaded region in Fig. 6.49. The region runs from a units below the surface to b units below the surface. We have chosen to measure depth with the y-axis, and the region's width at depth y is $L(y)$.

We partition the closed interval $[a, b]$ in the usual way and imagine the region to be cut into thin horizontal strips by planes perpendicular to the y-axis at the points of the partition. The typical strip from y to $y + \Delta y$ is Δy units wide by $L(y)$ units long. We assume $L(y)$ to be continuous throughout the closed interval $[a, b]$.

The pressure varies across the strip from top to bottom, just as it did in the molasses tank. But, if the strip is narrow enough, the pressure will remain close to its top-edge value wy (w being the weight-density of the liquid we are working with, and y the depth of the top edge). The total force against one side of the strip will therefore be about

$$\begin{aligned}\Delta F &= (\text{pressure along top edge})(\text{area}) \\ &= (wy)(L(y)\Delta y) = wy\,L(y)\Delta y.\end{aligned} \tag{4}$$

The force against the entire wall will be about

$$\sum_a^b \Delta F = \sum_a^b wy\,L(y)\Delta y. \tag{5}$$

The sum on the right-hand side of Eq. (5) is a Riemann sum for the continuous function $wy\,L(y)$ on the closed interval $[a, b]$. We expect the approxi-

mations to improve as the norm of the partition of $[a, b]$ goes to zero, so we define the total force against the wall to be the limit of these sums as the norm goes to zero.

DEFINITION

The Integral for Fluid Force

Suppose a submerged vertical plate running from depth $y = a$ to depth $y = b$ in a fluid of weight-density w is $L(y)$ units across at depth y, as measured along the plate. Then the total force of the fluid against one side of the plate is

$$F = \int_a^b wy\, L(y)\, dy. \tag{6}$$

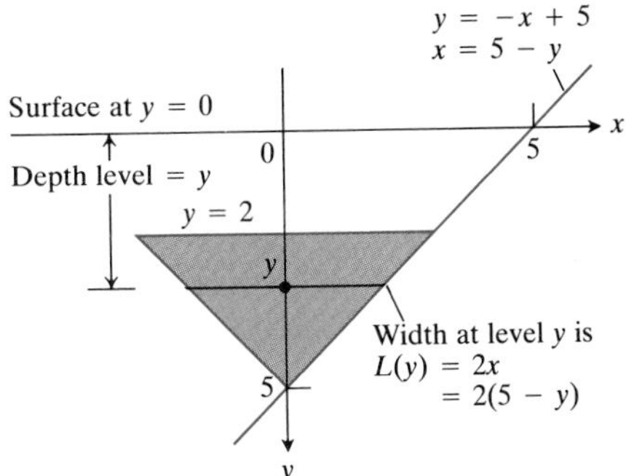

6.50 The important dimensions of the plate in Example 3.

EXAMPLE 3 A flat triangular plate is submerged vertically, base up, 2 ft below the surface of a swimming pool (Fig. 6.50). Find the fluid force against one side of the plate.

Solution We see from Fig. 6.50 that the plate runs from $y = 2$ to $y = 5$ and that its width at depth y is $L(y) = 2(5 - y)$. Therefore Eq. (6) gives

$$\text{Force} = \int_a^b wy\, L(y)\, dy = \int_2^5 (62.5)(y)(2)(5 - y)\, dy \qquad \begin{pmatrix}\text{For water,}\\ w = 62.5\end{pmatrix}$$

$$= 125 \int_2^5 (5y - y^2)\, dy = 125\left[\frac{5}{2}y^2 - \frac{1}{3}y^3\right]_2^5 = 1687.5 \text{ lb.}$$

A coordinate system with the positive y-axis pointing down is not always the most convenient one to use. It is sometimes better to put the origin at the bottom of the plate instead of at the fluid's surface and have the y-axis point up. This changes the factor y in the integrand to some other expression for the depth, but the rest of the integral remains the same. Here is an example.

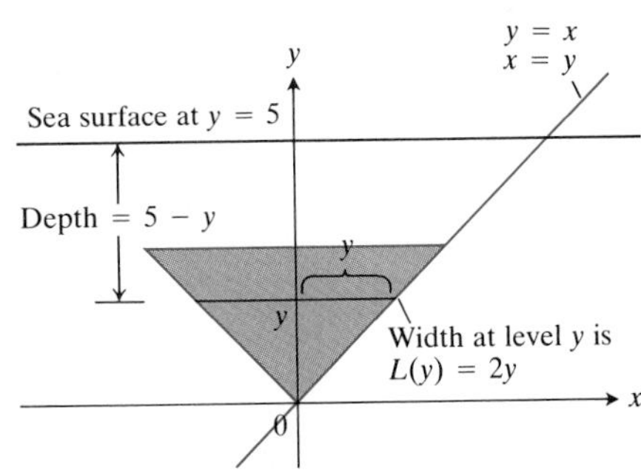

6.51 Example 4 calculates the force on the plate in Example 3 by using this more convenient coordinate system.

EXAMPLE 4 Find the force against one side of the plate in Example 3 by using the coordinate system in Fig. 6.51.

Solution The plate now runs from $y = 0$ to $y = 3$, and the length of a thin horizontal strip at level y is $L(y) = 2y$. The depth of this strip beneath the surface is $(5 - y)$. The force against the plate is therefore

$$\text{Force} = \int_a^b w \cdot \text{depth} \cdot L(y)\, dy \qquad \text{(Modified Eq. 6)}$$

$$= \int_0^3 (62.5)(5 - y)(2y)\, dy = 125 \int_0^3 (5y - y^2)\, dy$$

$$= 125\left[\frac{5}{2}y^2 - \frac{1}{3}y^3\right]_0^3 = 1687.5 \text{ lb.} \qquad \text{(As before)}$$

> **Strategy for Finding Fluid Force**
>
> Whatever coordinate system you use, you can always find the fluid force against one side of a submerged vertical plate or wall by taking these steps:
>
> 1. Find expressions for the length and depth of a typical thin horizontal strip.
> 2. Multiply their product by the fluid's weight-density w and integrate over the interval of depths occupied by the plate or wall.

EXAMPLE 5 We can now calculate exactly the force exerted by the molasses against the bottom 1-ft band of the Puritan Distilling Company's storage tank when the tank was full.

The tank was a right circular cylindrical tank, 90 ft high and 90 ft in diameter. Using a coordinate system with the origin at the bottom of the tank and the y-axis pointing up, we find that the typical horizontal strip at level y has

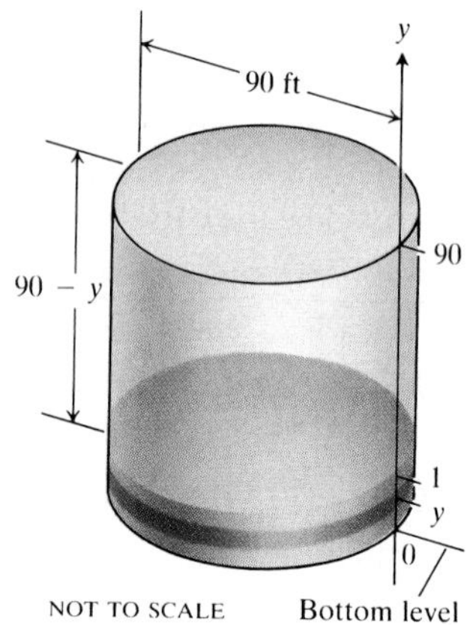

Strip depth: $(90 - y)$

Strip length: $\pi \times$ tank diameter $= 90\pi$.

The force against the band is therefore

$$\text{Force} = \int_0^1 w\,(\text{depth})(\text{length})\,dy$$

$$= \int_0^1 100(90 - y)(90\pi)\,dy \qquad \begin{pmatrix}\text{For molasses,}\\ w = 100\end{pmatrix}$$

$$= 9000\pi \int_0^1 (90 - y)\,dy = 9000\pi\left[90y - \frac{1}{2}y^2\right]_0^1$$

$$= 9000\pi(89.5) = 2{,}530{,}553 \text{ lb.} \qquad \text{(Rounded)}$$

As expected, the force is slightly less than the constant-depth estimate following Example 2.

Exercises 6.7

1. Suppose the triangular plate in Fig. 6.51 is 4 ft beneath the surface instead of 2 ft. What is the force on one side of the plate now?

2. What was the total force against the side wall of the Puritan Distilling Company's molasses tank when the tank was full? Half full?

3. The vertical ends of a watering trough are inverted isosceles triangles like the one shown here. What is the force on each end of the trough when the trough is full? Does it matter how long the trough is?

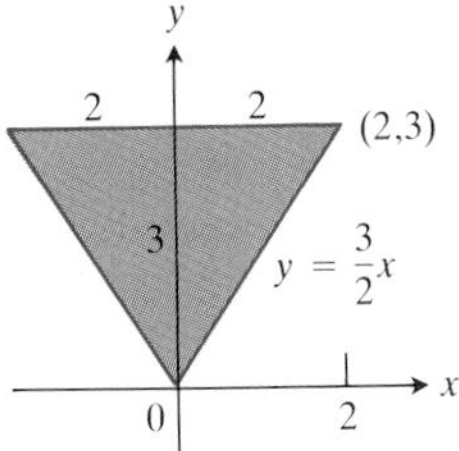

4. What is the force on each end of the trough in Exercise 3 if the water level is lowered 1 ft?

5. The triangular plate shown here is submerged vertically, 1 ft below the surface of the water.

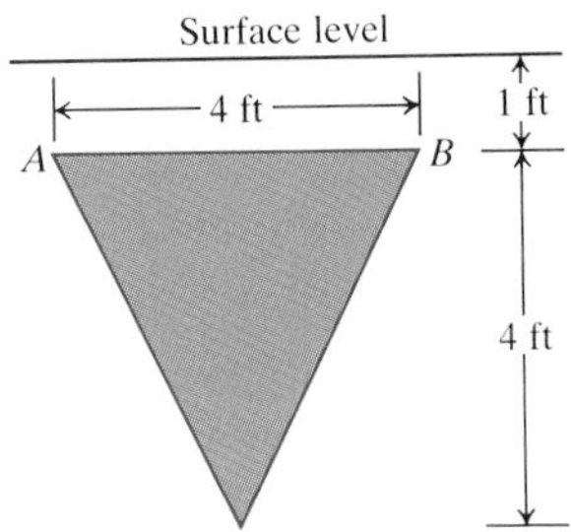

Find the force on one side of the plate.

6. The triangular plate in Exercise 5 is revolved 180° about the line AB so that part of it sticks up above the surface. What force does the water exert on one face of the plate now?

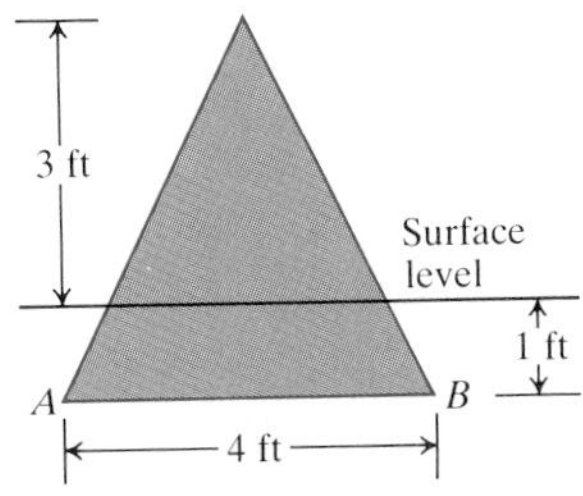

7. A semicircular plate is submerged straight down in the water with its diameter at the surface. Find the force exerted by the water on one side of the plate.

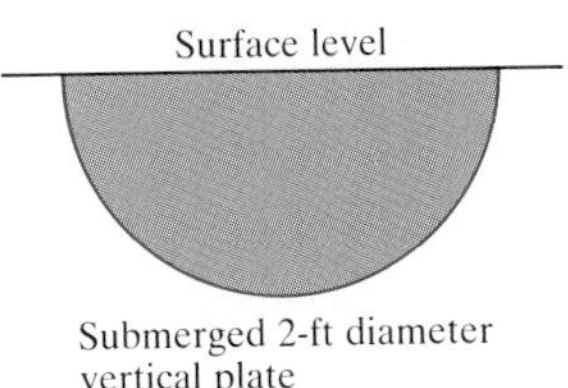

8. A rectangular fish tank of interior dimensions $2 \times 2 \times 4$ ft is filled to within 2 in. of the top with water. Find the force against the sides and ends of the tank.

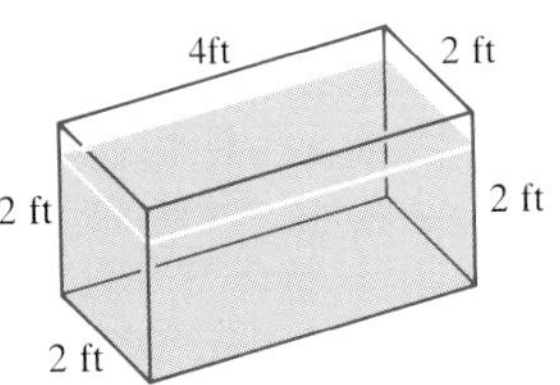

9. The viewing portion of the rectangular glass window in a typical fish tank at the New England Aquarium in Boston, Massachusetts, is 63 in. wide and runs from 0.5 in. below the water's surface to 33.5 in. below the surface. Find the force against this portion of the window. The weight-density of sea water is 64 lb/ft^3. (In case you were wondering, the glass is 3/4 in. thick and the tank walls extend 4 in. above the water to keep the fish from jumping out.)

10. **CALCULATOR** A rectangular milk carton measures $3.75 \times 3.75 \times 7.75$ in. Find the force of the milk on one side when the carton is full.

11. A tank truck hauls milk in a 6-ft diameter horizontal right circular cylindrical tank. What is the force on the end of the tank when the tank is half full?

12. The cubical metal tank shown here is used to store liquids. It has a parabolic gate, held in place by bolts and designed to withstand a force of 160 lb without rupturing. The liquid you plan to store has a weight-density of 50 lb/ft^3.
    a) What will be the force on the gate when the liquid is 2 ft deep?

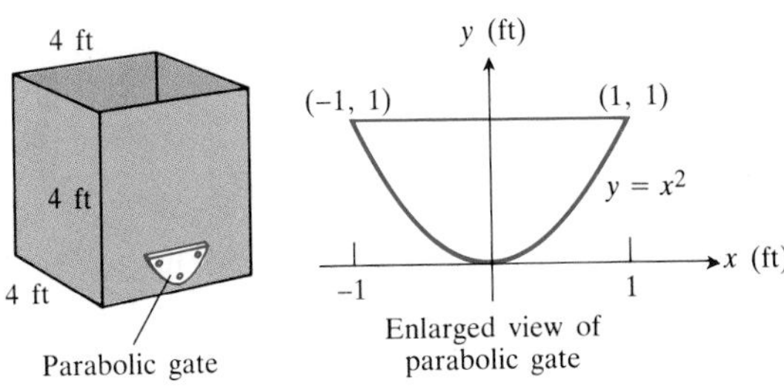

b) What is the maximum height to which the container can be filled without exceeding the design limitation on the gate?

13. **CALCULATOR** The end plates in the trough shown here were designed to withstand a force of 6667 lb. How many cubic feet of water can the tank hold without exceeding design limitations?

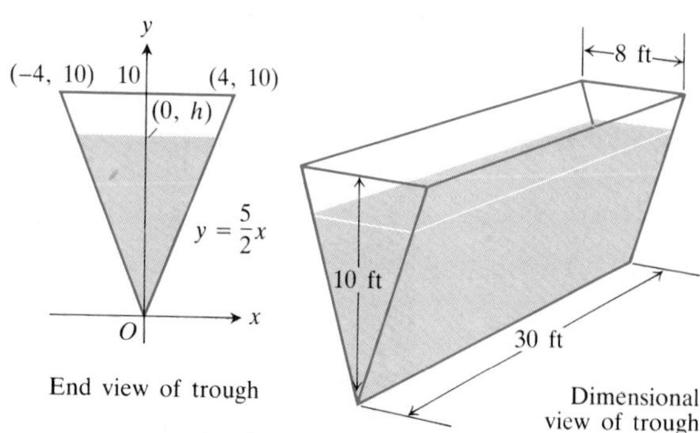

End view of trough

Dimensional view of trough

14. Water is running into the rectangular swimming pool shown here at the rate of 1000 ft^3/hr.

a) Find the force against the triangular drain plate after 9 hr of filling.

b) The plate is designed to withstand a force of 520 lb. How high can the pool be filled without exceeding this design limitation?

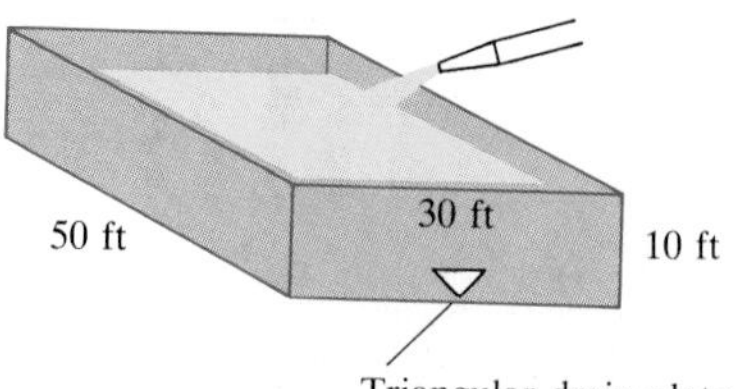

Triangular drain plate

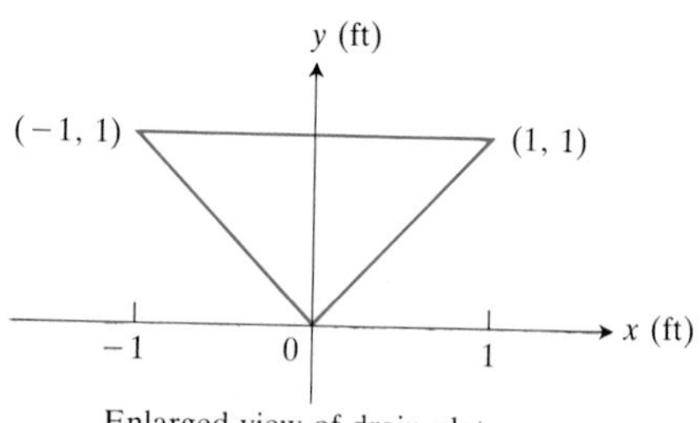

Enlarged view of drain plate

15. A vertical rectangular plate is submerged in a fluid with its top edge parallel to the fluid's surface. Show that the force on one side of the plate is the average value of the pressure up and down the plate times the area of the plate.

6.8 Centers of Mass

Many structures and mechanical systems behave as if their masses were concentrated at a single point called the center of mass. It is important to know how to locate this point, and it turns out that doing so is basically a mathematical undertaking. We do it with calculus, and this section shows how. For the moment we shall deal only with one- and two-dimensional shapes. Three-dimensional shapes are best done with the so-called multiple integrals of Chapter 14.

Masses along a Line

If we imagine masses m_1, m_2, and m_3 placed on a rigid x-axis, and the axis supported by a fulcrum at the origin, the resulting system might or might not balance.

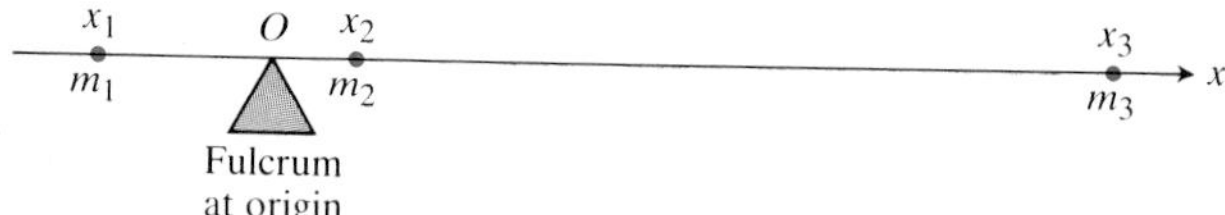

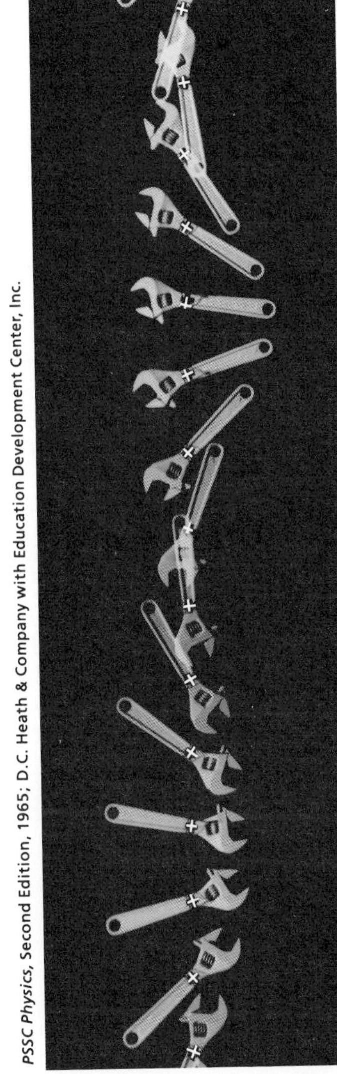
PSSC Physics, Second Edition, 1965; D.C. Heath & Company with Education Development Center, Inc.

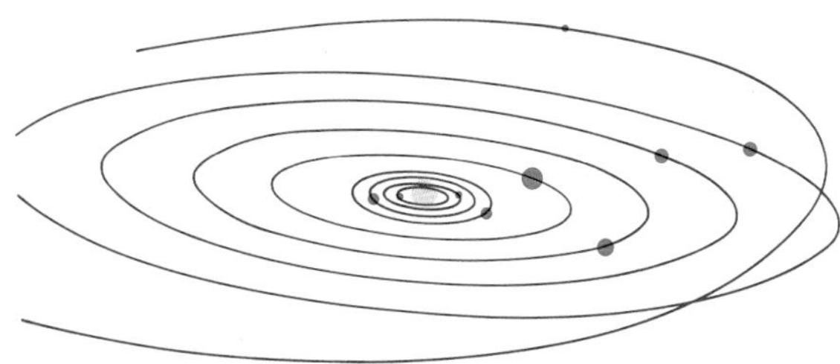

© Timothy Eagan/Woodfin Camp and Assoc.

Mark S. Gerson/F.P.G.

(Left) The motion of a falling wrench seems haphazard until we notice that the wrench is simply turning about its center of mass as the center falls in a straight line. (Top) The planets, asteroids, and comets of our solar system revolve about their collective center of mass. (It lies inside the sun.) (Center) This rock balances because its center of mass lies directly above the point where the rock touches the ground. (Right) These acrobats stay in place because their collective center of mass lies directly above the wire.

Mass versus Weight

Weight is the force that results from gravity pulling on a mass. If an object of mass m is placed in a location where the acceleration of gravity is g, the object's weight there is

$$F = mg$$

(a version of Newton's second law).

Each mass m_k exerts a downward force $m_k g$ equal to the magnitude of the mass times the acceleration of gravity. Each of these forces has a tendency to turn the axis about the origin, the way somebody's weight might turn a seesaw. This turning effect, called a **torque,** is measured by multiplying the force $m_k g$ by the signed distance x_k from the mass to the origin. Masses to the left of the origin exert a negative (counterclockwise) torque. Masses to the right of the origin exert a positive (clockwise) torque.

We use the sum of the torques to measure the tendency of a system to rotate about the origin. This sum is called the **system torque.**

$$\text{System torque} = m_1 g x_1 + m_2 g x_2 + m_3 g x_3 \tag{1}$$

The system will balance if and only if its net torque is zero.

If we factor out the g in Eq. (1), we see that the system torque is

$$g(m_1x_1 + m_2x_2 + m_3x_3). \tag{2}$$

A feature of the environment (the factor g); A feature of the system (the sum $m_1x_1 + m_2x_2 + m_3x_3$)

Thus the torque is the product of the gravitational acceleration g, which is a feature of the environment in which the system happens to reside, and the number $(m_1x_1 + m_2x_2 + m_3x_3)$, which is a feature of the system itself, a constant that stays the same no matter where the system is placed.

The number $(m_1x_1 + m_2x_2 + m_3x_3)$ is called the **moment of the system about the origin.**

$$M_O = \text{Moment of system about origin} = \sum m_k x_k \tag{3}$$

(We shift to sigma notation here to allow for sums with more terms. If you want a quick way to read $\Sigma m_k x_x$, try "summation m k x k".)

We usually want to know where to place the fulcrum to make the system balance, that is, at what point $\bar{x}$ to place it to make the torque zero.

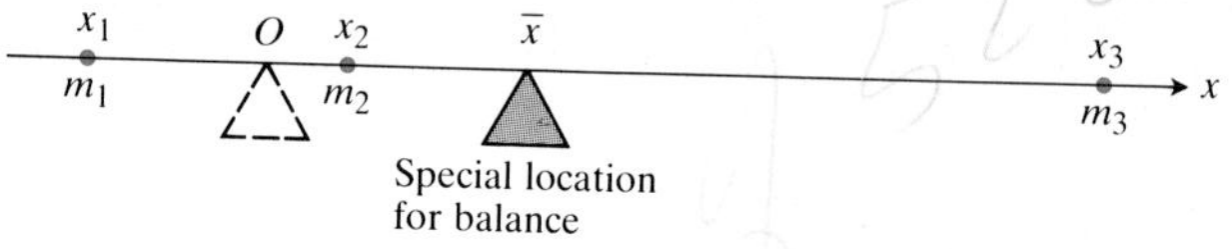

The torque of each mass about the fulcrum in this special location is

$$\text{Torque of } m_k \text{ about } \bar{x} = \begin{pmatrix}\text{signed distance}\\ \text{of } m_k \text{ from } \bar{x}\end{pmatrix} \cdot \begin{pmatrix}\text{downward}\\ \text{force}\end{pmatrix}$$

$$= (x_k - \bar{x}) \cdot m_k g. \tag{4}$$

When we write down the equation that says that the sum of these torques is zero, we get an equation we can solve for $\bar{x}$:

$$\sum (x_k - \bar{x})m_k g = 0 \qquad \text{(Sum of the torques equals zero)}$$

$$g\sum (x_k - \bar{x})m_k = 0 \qquad \text{(Constant multiple rule for sums)}$$

$$\sum (m_k x_k - \bar{x}\, m_k) = 0 \qquad \text{(} g \text{ divided out, } m_k \text{ distributed)}$$

$$\sum m_k x_k - \sum \bar{x}\, m_k = 0 \qquad \text{(Difference rule for sums)}$$

$$\sum m_k x_k = \bar{x} \sum m_k \qquad \text{(Rearranged, constant multiple rule again)}$$

$$\bar{x} = \frac{\sum m_k x_k}{\sum m_k}. \qquad \text{(Solved for } \bar{x}\text{)}$$

This last equation tells us to find $\bar{x}$ by dividing the system's moment about the origin by the system's total mass:

$$\bar{x} = \frac{\sum x_k m_k}{\sum m_k} = \frac{\text{System moment about origin}}{\text{System mass}}. \tag{5}$$

The point $\bar{x}$ is called the system's **center of mass.**

Wires and Thin Rods

In many applications, we want to know the center of mass of a rod or a thin strip of metal. In cases like these, where we can assume the distribution of mass is continuous, the summation signs in our formulas become integrals in a manner we shall now describe.

Imagine a long, thin strip lying along the x-axis from $x = a$ to $x = b$ and cut into small pieces of mass Δm_k by a partition of the interval $[a, b]$.

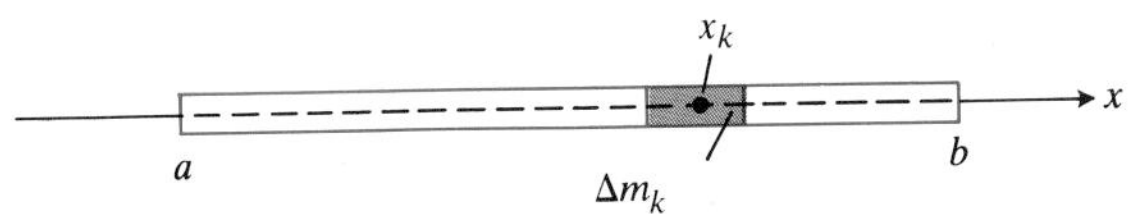

Each piece is Δx units long and lies approximately x_k units from the origin. Now observe three things.

First, the strip's center of mass $\bar{x}$ is nearly the same as the center of mass of the system of point masses we would get by attaching each mass Δm_k to the point x_k:

$$\bar{x} \approx \frac{\text{System moment}}{\text{System mass}}. \tag{6}$$

Second, the moment of each piece of the strip about the origin is approximately $x_k \Delta m_k$, so the system moment is approximately the sum of the $x_k \Delta m_k$:

$$\text{System moment} \approx \sum x_k \Delta m_k. \tag{7}$$

Third, if the density of the strip at x_k is $\delta(x_k)$, expressed in terms of mass per unit length, and δ is continuous, then Δm_k is approximately equal to $\delta(x_k)\Delta x$ (mass per unit length times length):

$$\Delta m_k \approx \delta(x_k)\Delta x. \tag{8}$$

Combining these three observations gives

$$\bar{x} \approx \frac{\text{System moment}}{\text{System mass}} \approx \frac{\sum x_k \Delta m_k}{\sum \Delta m_k} \approx \frac{\sum x_k\, \delta(x_k)\Delta x}{\sum \delta(x_k)\Delta x}. \tag{9}$$

The sum in the numerator of the last quotient in (9) is a Riemann sum for the continuous function $x\,\delta(x)$ over the closed interval $[a, b]$. The sum in the denominator is a Riemann sum for the function $\delta(x)$ over this interval. We expect the approximations in (9) to improve as the strip is partitioned ever more finely and are led to the equation

$$\bar{x} = \frac{\displaystyle\int_a^b x\,\delta(x)\,dx}{\displaystyle\int_a^b \delta(x)\,dx}. \tag{10}$$

This is the formula we use to calculate $\bar{x}$.

Density

A material's density is defined to be its mass per unit volume. In practice, however, we tend to use units we can conveniently measure. For wires, rods, and narrow strips we use mass per unit length. For flat sheets and plates we use mass per unit area.

Moment, Mass, and Center of Mass of a Thin Rod or Strip along the *x*-Axis

Moment about the origin: $$M_O = \int_a^b x\,\delta(x)\,dx \qquad \text{(11a)}$$

Mass: $$M = \int_a^b \delta(x)\,dx \qquad \text{(11b)}$$

Center of mass: $$\bar{x} = \frac{M_O}{M} \qquad \text{(11c)}$$

Equation (11c) says that to find the center of mass of a rod or a thin strip, we divide its moment about the origin by its mass.

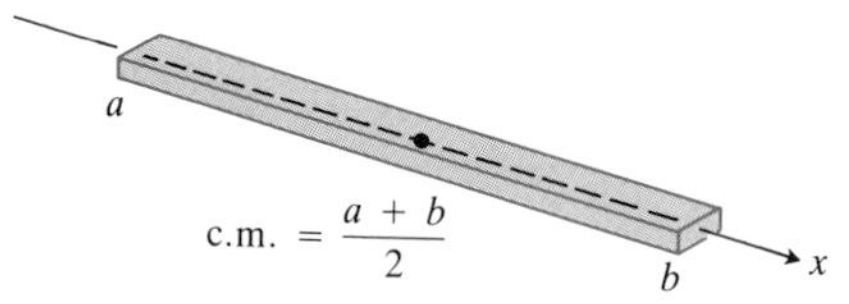

6.52 The center of mass of a straight, thin rod or strip of constant density lies half way between its ends.

EXAMPLE 1 *A useful result.* Show that the center of mass of a straight, thin strip or rod of constant density is always located half way between its two ends.

Solution We model the strip as a portion of the x-axis from $x = a$ to $x = b$ (Fig. 6.52). Our goal is to show that $\bar{x} = (a + b)/2$, the point half way between a and b.

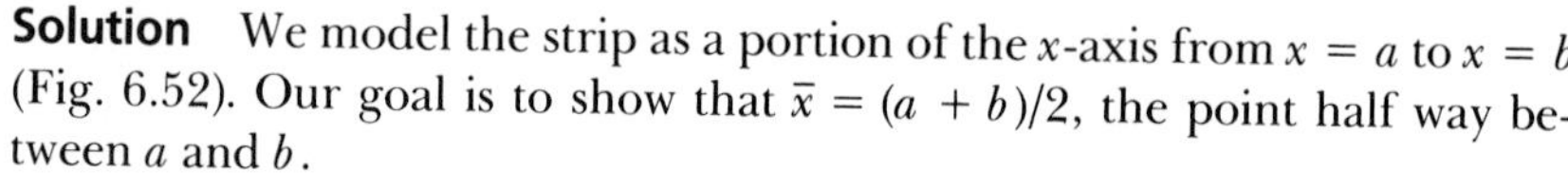

The key is the density's having a constant value. This enables us to regard the function $\delta(x)$ in the integrals in Eqs. (11) as a constant (call it δ), with the result that

$$M_O = \int_a^b \delta x\,dx = \delta \int_a^b x\,dx = \delta\left[\frac{1}{2}x^2\right]_a^b = \frac{\delta}{2}(b^2 - a^2)$$

$$M = \int_a^b \delta\,dx = \delta \int_a^b dx = \delta\Big[x\Big]_a^b = \delta(b - a)$$

$$\bar{x} = \frac{M_O}{M} = \frac{\frac{\delta}{2}(b^2 - a^2)}{\delta(b - a)} = \frac{a + b}{2}. \qquad \text{(The } \delta\text{'s cancel.)}$$

EXAMPLE 2 *A variable density.* The 10-meter-long rod in Fig. 6.53 thickens from left to right so that its density, instead of being constant, is $\delta(x) = 1 + (x/10)$ kg/m. Find the rod's center of mass.

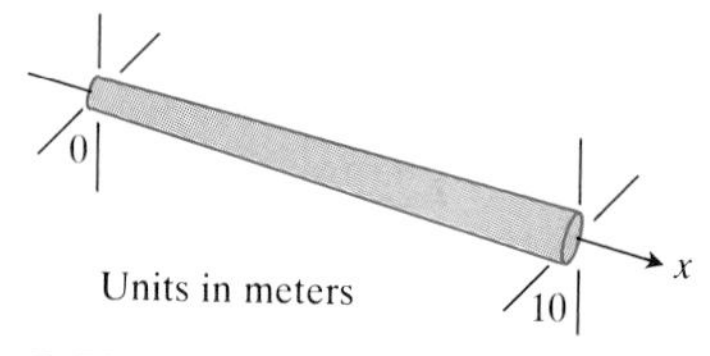

6.53 We can treat a rod of variable thickness as a rod of variable density. See Example 2.

Solution The rod's moment about the origin (Eq. 11a) is

$$M_O = \int_0^{10} x\,\delta(x)\,dx = \int_0^{10} x\left(1 + \frac{x}{10}\right)dx = \int_0^{10}\left(x + \frac{x^2}{10}\right)dx$$

$$= \left[\frac{x^2}{2} + \frac{x^3}{30}\right]_0^{10} = 50 + \frac{100}{3} = \frac{250}{3}\ \text{kg}\cdot\text{m}. \qquad \left(\begin{array}{l}\text{The units of a moment}\\ \text{are mass} \times \text{length.}\end{array}\right)$$

The rod's mass (Eq. 11b) is

$$M = \int_0^{10} \delta(x)\,dx = \int_0^{10}\left(1 + \frac{x}{10}\right)dx = \left[x + \frac{x^2}{20}\right]_0^{10} = 10 + 5 = 15\ \text{kg}.$$

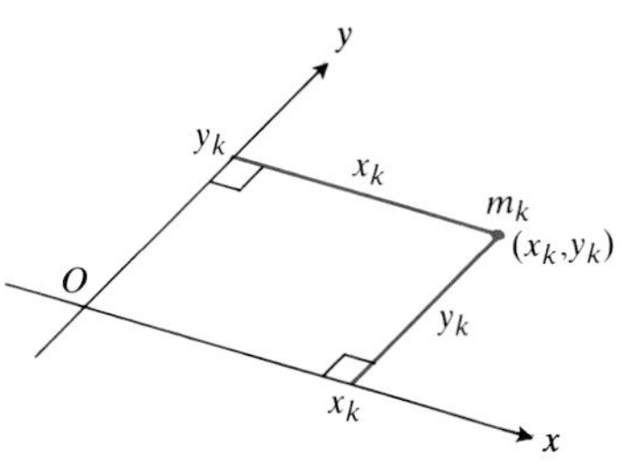

6.54 The mass m_k has a moment about each axis.

The center of mass (Eq. 11c) is located at the point

$$\bar{x} = \frac{M_O}{M} = \frac{250}{3} \cdot \frac{1}{15} = \frac{50}{9} \approx 5.56 \text{ m}.$$

Masses Distributed over a Plane Region

Suppose we have a finite collection of masses located in the coordinate plane, the mass m_k being located at the point (x_k, y_k) (see Fig. 6.54). The total mass of the system is

$$\text{System mass:} \qquad M = \sum m_k .$$

Each mass m_k has a moment about each axis. Its moment about the x-axis is $m_k y_k$, and its moment about the y-axis is $m_k x_k$. The moments of the entire system about the two axes are

$$\text{Moment about } x\text{-axis:} \qquad M_x = \sum m_k y_k ,$$

$$\text{Moment about } y\text{-axis:} \qquad M_y = \sum m_k x_k .$$

The x-coordinate of the system's center of mass is defined to be

$$\bar{x} = \frac{M_y}{M} = \sum m_k x_k \Big/ \sum m_k . \tag{12}$$

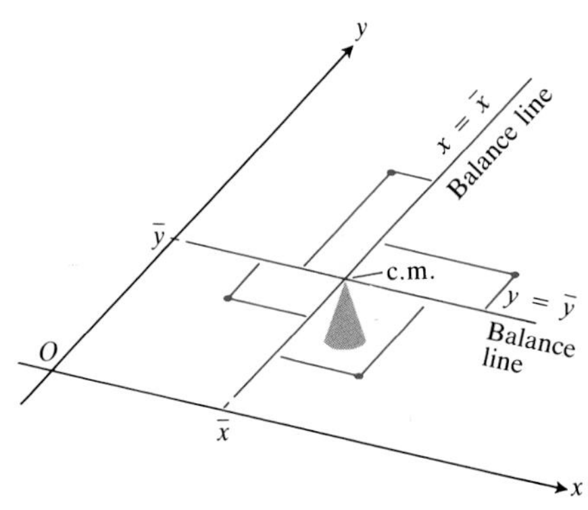

6.55 A two-dimensional array of masses balances on its center of mass.

With this choice of $\bar{x}$, as in the one-dimensional case, the system balances about the line $x = \bar{x}$ (Fig. 6.55).

The y-coordinate of the system's center of mass is defined to be

$$\bar{y} = \frac{M_x}{M} = \sum m_k y_k \Big/ \sum m_k . \tag{13}$$

With this choice of $\bar{y}$, the system balances about the line $y = \bar{y}$ as well. The torques exerted by the masses about the line $y = \bar{y}$ cancel out. Thus, as far as balance is concerned, the system behaves as if all its mass were at the single point $(\bar{x}, \bar{y})$. We call this point the system's center of mass.

Thin, Flat Plates

In many applications, we need to find the center of mass of a thin, flat plate: a disk of aluminum, say, or a triangular sheet of steel. In such cases we assume the distribution of mass to be continuous, and the formulas we use to calculate $\bar{x}$ and $\bar{y}$ contain integrals instead of finite sums. The integrals arise in the following way.

Imagine the plate occupying a region in the xy-plane, cut into thin strips parallel to one of the axes (in Fig. 6.56, the y-axis). The center of mass of a typical strip is $(\tilde{x}, \tilde{y})$. (The symbol ~ over the x and y is a *tilde,* pronounced to rhyme with "Hilda." Thus $\tilde{x}$ is read "x tilde.") We treat the strip's mass Δm as if it were concentrated at $(\tilde{x}, \tilde{y})$. The moment of the strip about the y-axis is then $\tilde{x}\Delta m$, and the moment of the strip about the x-axis is $\tilde{y}\Delta m$. Equations (12) and (13) then become

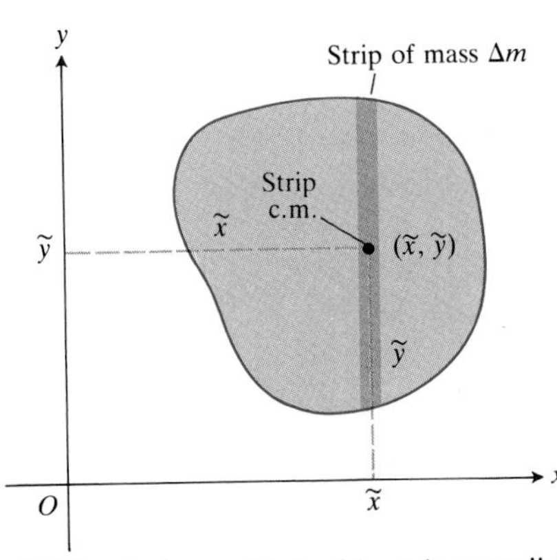

6.56 A plate cut into thin strips parallel to the y-axis. The moment exerted by a typical strip about each axis is the moment its mass Δm would exert if concentrated at the strip's center of mass $(\tilde{x}, \tilde{y})$

$$\bar{x} = \frac{M_y}{M} = \frac{\sum \tilde{x}\Delta m}{\sum \Delta m}, \qquad \bar{y} = \frac{M_x}{M} = \frac{\sum \tilde{y}\Delta m}{\sum \Delta m}. \tag{14}$$

As in the one-dimensional case, the sums in the numerator and denominator are Riemann sums for integrals and approach these integrals as limiting values as the strips into which the plate is cut become narrower and narrower. We write these integrals symbolically as

$$\bar{x} = \frac{\int \tilde{x}\, dm}{\int dm} \quad \text{and} \quad \bar{y} = \frac{\int \tilde{y}\, dm}{\int dm} \tag{15}$$

Moments, Mass, and Center of Mass of a Thin Plate Covering a Region in the *xy*-Plane

Moment about the x-axis: $M_x = \int \tilde{y}\, dm$

Moment about the y-axis: $M_y = \int \tilde{x}\, dm$ (16)

Mass: $M = \int dm$

Center of mass: $\bar{x} = \dfrac{M_y}{M}, \quad \bar{y} = \dfrac{M_x}{M}$

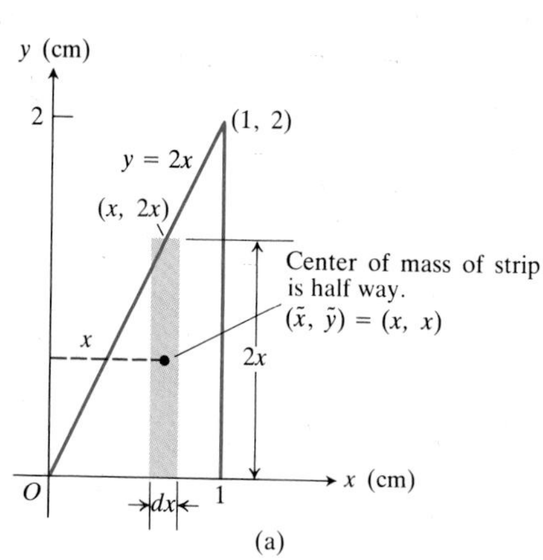

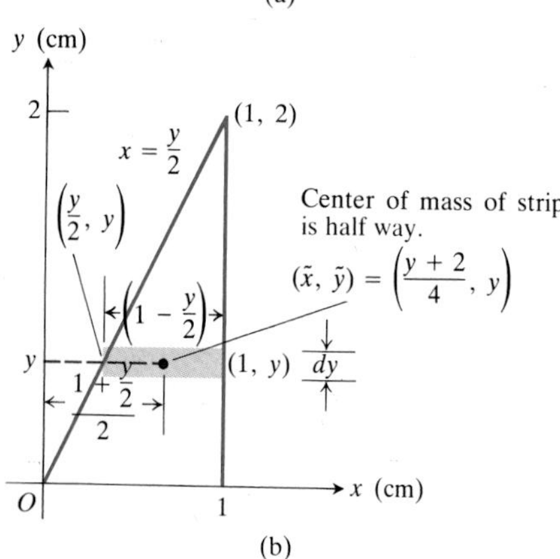

6.57 Two ways to model the calculation of the moment M_y of the triangular plate in Example 3.

To evaluate these integrals, we picture the plate in the coordinate plane and stretch a strip of mass parallel to one of the coordinate axes. We then express the strip's mass dm and the coordinates $(\tilde{x}, \tilde{y})$ of the strip's center of mass in terms of x or y. Finally, we integrate $\tilde{y}\, dm$, $\tilde{x}\, dm$, and dm between limits of integration determined by the plate's location in the plane.

EXAMPLE 3 The triangular plate shown in Fig. 6.57, bounded by the lines $y = 0$, $y = 2x$, and $x = 1$, has a constant density of $\delta = 3$ gm/cm^2. Find (a) the plate's moment M_y about the y-axis, (b) the plate's mass M, and (c) the x-coordinate of the plate's center of mass.

Solution METHOD 1: *Vertical strips* (Fig. 6.57a).

a) The moment M_y: The typical vertical strip has

center of mass (c.m.): $(\tilde{x}, \tilde{y}) = (x, x)$,

length: $2x$,

width: dx,

area: $dA = 2x\, dx$,

mass: $dm = \delta\, dA = 3 \cdot 2x\, dx = 6x\, dx$,

distance of c.m. from y-axis: $\tilde{x} = x$.

The moment of the strip about the y-axis is

$$\tilde{x}\, dm = x \cdot 6x\, dx = 6x^2\, dx.$$

The moment of the plate about the y-axis is therefore

$$M_y = \int \tilde{x}\, dm = \int_0^1 6x^2\, dx = 2x^3\Big]_0^1 = 2 \text{ gm} \cdot \text{cm}.$$

b) The plate's mass:

$$M = \int dm = \int_0^1 6x\, dx = 3x^2\Big]_0^1 = 3 \text{ gm}.$$

c) The x-coordinate of the plate's center of mass:

$$\bar{x} = \frac{M_y}{M} = \frac{2 \text{ gm} \cdot \text{cm}}{3 \text{ gm}} = \frac{2}{3} \text{ cm}.$$

By a similar computation we could find M_x and $\bar{y} = M_x/M$.

METHOD 2: *Horizontal strips* (Fig. 6.57b).

a) The moment M_y: The typical horizontal strip has

center of mass (c.m.): $(\tilde{x}, \tilde{y}) = \left(\frac{1}{2}\left(1 + \frac{y}{2}\right), y\right) = \left(\frac{y+2}{4}, y\right)$,

length: $1 - \frac{y}{2} = \frac{2-y}{2}$,

width: dy,

area: $dA = \frac{2-y}{2}\, dy$,

mass: $dm = \delta\, dA = 3 \cdot \frac{2-y}{2}\, dy$,

distance of c.m. to y-axis: $\tilde{x} = \frac{y+2}{4}$.

The moment of the strip about the y-axis is

$$\tilde{x}\, dm = \frac{y+2}{4} \cdot 3 \cdot \frac{2-y}{2}\, dy = \frac{3}{8}(4 - y^2)\, dy.$$

The moment of the plate about the y-axis is

$$M_y = \int \tilde{x}\, dm = \int_0^2 \frac{3}{8}(4 - y^2)\, dy = \frac{3}{8}\left[4y - \frac{y^3}{3}\right]_0^2 = \frac{3}{8}\left(\frac{16}{3}\right) = 2 \text{ gm} \cdot \text{cm}.$$

b) The plate's mass:

$$M = \int dm = \int_0^2 \frac{3}{2}(2 - y)\, dy = \frac{3}{2}\left[2y - \frac{y^2}{2}\right]_0^2 = \frac{3}{2}(4 - 2) = 3 \text{ gm}.$$

c) The x-coordinate of the plate's center of mass:

$$\bar{x} = \frac{M_y}{M} = \frac{2 \text{ gm} \cdot \text{cm}}{3 \text{ gm}} = \frac{2}{3} \text{ cm}.$$

By a similar computation, we could find M_x and $\bar{y}$. ■

How to Find a Plate's Center of Mass

1. Picture the plate in the xy-plane.
2. Sketch a strip of mass parallel to one of the coordinate axes and find its dimensions.
3. Find the strip's mass dm and center of mass $(\tilde{x}, \tilde{y})$.
4. Integrate $\tilde{y}\, dm$, $\tilde{x}\, dm$, and dm to find M_x, M_y, and M.
5. Divide the moments by the mass to calculate $\bar{x}$ and $\bar{y}$.

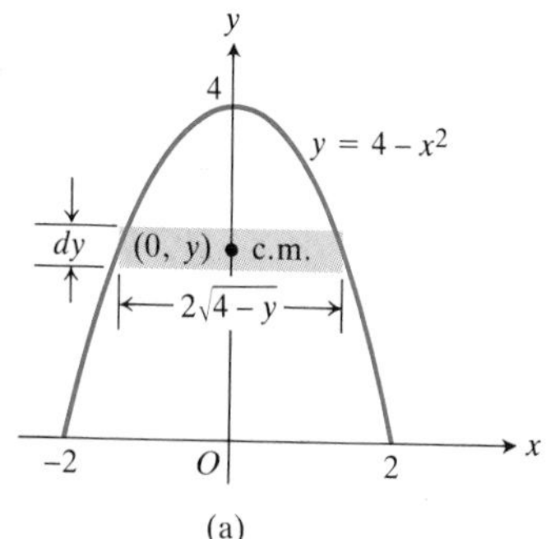

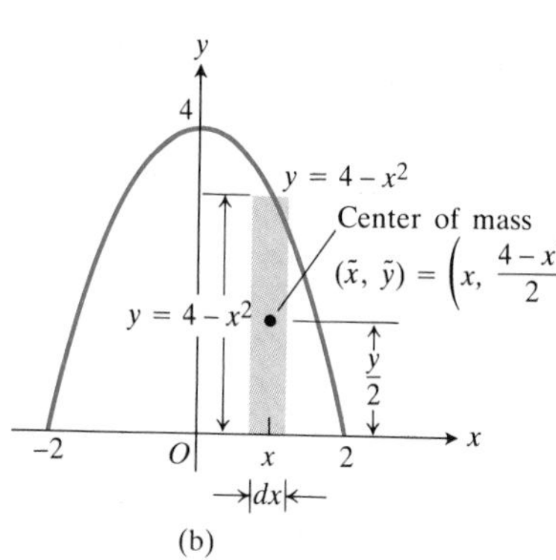

6.58 Modeling the problem in Example 4 with horizontal strips (a) leads to a difficult integration, so we model with vertical

EXAMPLE 4 Find the center of mass of a thin plate of constant density δ covering the region bounded above by the parabola $y = 4 - x^2$ and below by the x-axis (Fig. 6.58).

Solution Since the plate is symmetric about the y-axis and its density is constant, the center of mass lies on the y-axis. This means that $\bar{x} = 0$. It remains to find $\bar{y} = M_x/M$.

A trial calculation with horizontal strips (Fig. 6.58a) leads to a difficult integration:

$$M_x = \int_0^4 2\delta y\sqrt{4 - y}\, dy.$$

We therefore model the distribution of mass with vertical strips instead (6.58b). The typical vertical strip has

center of mass (c.m.): $(\tilde{x}, \tilde{y}) = \left(x, \dfrac{4 - x^2}{2}\right)$,

length: $4 - x^2$,

width: dx,

area: $dA = (4 - x^2)\, dx$,

mass: $dm = \delta\, dA = \delta(4 - x^2)\, dx$,

distance from c.m. to x-axis: $\tilde{y} = \dfrac{4 - x^2}{2}$.

The moment of the strip about the x-axis is

$$\tilde{y}\, dm = \frac{4 - x^2}{2} \cdot \delta(4 - x^2)\, dx = \frac{\delta}{2}(4 - x^2)^2\, dx.$$

The moment of the plate about the x-axis is

$$M_x = \int \tilde{y}\, dm = \int_{-2}^{2} \frac{\delta}{2}(4 - x^2)^2\, dx = \frac{256}{15}\delta. \tag{17}$$

The mass of the plate is

$$M = \int dm = \int_{-2}^{2} \delta(4 - x^2)\, dx = \frac{32}{3}\delta.$$

Therefore

$$\bar{y} = \frac{M_x}{M} = \frac{\frac{256}{15}\delta}{\frac{32}{3}\delta} = \frac{8}{5}.$$

The plate's center of mass is the point

$$(\bar{x}, \bar{y}) = \left(0, \frac{8}{5}\right).$$

EXAMPLE 5 *Variable density.* Find the center of mass of the plate in Example 4 if the density at any point (x, y) is $\delta = 2x^2$, twice the square of the distance from the point to the y-axis.

Solution The mass distribution is still symmetric about the y-axis, so

$$\bar{x} = 0.$$

With $\delta = 2x^2$, Eqs. (17) and (18) become

$$M_x = \int \tilde{y}\,dm = \int_{-2}^{2} \frac{\delta}{2}(4 - x^2)^2\,dx = \int_{-2}^{2} x^2(4 - x^2)^2\,dx = \frac{2048}{105} \qquad (17')$$

$$M = \int dm = \int_{-2}^{2} \delta(4 - x^2)\,dx = \int_{-2}^{2} 2x^2(4 - x^2)\,dx = \frac{256}{15}. \qquad (18')$$

Therefore

$$\bar{y} = \frac{M_x}{M} = \frac{2048}{105} \cdot \frac{15}{256} = \frac{8}{7}.$$

The plate's new center of mass is

$$(\bar{x}, \bar{y}) = \left(0, \frac{8}{7}\right).$$

Centers of Gravity, Homogeneity, Uniformity, and Centroids

As you read elsewhere, you will find some variety in the vocabulary used in connection with centers of mass.

When physicists discuss the effects of a constant gravitational force on a system of masses, they may call the center of mass the **center of gravity.**

Material that has a constant density δ is also said to be **homogeneous,** or to be **uniform,** or to have **uniform density.**

When the density function is constant, it cancels out of the numerator and denominator of the formulas for $\bar{x}$ and $\bar{y}$. This happened in nearly every example in this section. As far as $\bar{x}$ and $\bar{y}$ were concerned, δ might as well have been 1. Thus, when the density is constant, the location of the center of mass is a feature of the geometry of the object and not of the material from which it is made. In such cases engineers may call the center of mass the **centroid** of the shape, as in "Find the centroid of a triangle or a solid cone." To do so, we just set δ equal to 1 and proceed to find $\bar{x}$ and $\bar{y}$ as before, by dividing moments by masses.

Exercises 6.8

1. Two children are balancing on a seesaw. The 80-lb child is 5 ft from the fulcrum. How far from the fulcrum is the 100-lb child?

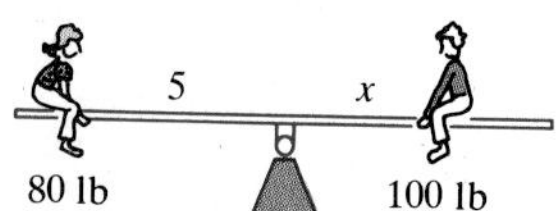

2. The ends of two thin steel rods of equal length are welded together to make a right-angled frame. Locate the frame's center of mass. (*Hint:* Where is the center of mass of each rod?)

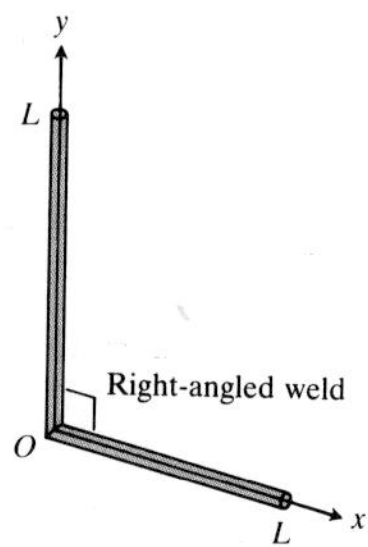

Exercises 3–6 give the density functions of thin rods lying along various intervals of the x-axis. Find each rod's moment about the origin and center of mass.

3. $\delta(x) = 4,\ 0 \le x \le 2$
4. $\delta(x) = 1 + (x/3),\ 0 \le x \le 3$
5. $\delta(x) = \left(1 + \dfrac{x}{4}\right)^2,\ 0 \le x \le 4$
6. $\delta(x) = \begin{cases} 2, & 0 \le x \le 3 \\ 1, & 3 \le x \le 6 \end{cases}$

In Exercises 7–18, find the center of mass of a thin plate of constant density δ covering the given region.

7. The triangular region bounded below by the x-axis and on the sides by the lines $y = 2x + 2$ and $y = -2x + 2$
8. The region bounded by the parabola $y = x^2$ and the line $y = 4$
9. The region bounded by the y-axis and the curve $x = y - y^3,\ 0 \le y \le 1$
10. The region bounded by the parabola $y = x - x^2$ and the line $y = -x$
11. The region bounded by the parabola $x = y^2 - y$ and the line $y = x$
12. The region bounded by the parabola $y = 25 - x^2$ and the x-axis
13. The region bounded by the x-axis and the curve $y = \cos x,\ -\pi/2 \le x \le \pi/2$
14. The region between the x-axis and the curve $y = \sec x,\ -\pi/4 \le x \le \pi/4$
15. The region bounded by the parabolas $y = 2x^2 - 4x$ and $y = 2x - x^2$

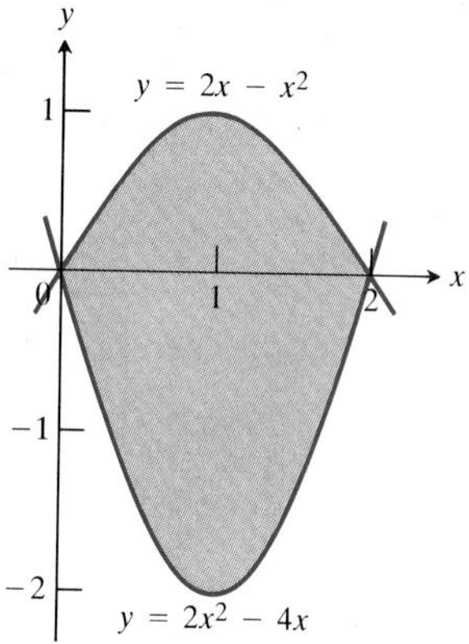

16. a) The region cut from first quadrant by the circle $x^2 + y^2 = 9$
    b) The region bounded by the x-axis and the semicircle $y = \sqrt{9 - x^2}$

    Compare your answer with the answer in (a).
17. The "triangular" region in the first quadrant between the circle $x^2 + y^2 = 9$ and the lines $x = 3$ and $y = 3$. (*Hint:* Use geometry to find the area.)
18. The region bounded above by the curve $y = 1/x^2$, below by the curve $y = -1/x^2$, and on the left and right by the lines $x = 1$ and $x = 2$

It can be shown that the centroid of a triangle always lies at the intersection of the medians, one third of the way from the midpoint of each side toward the opposite vertex. Use this to find the centroids of the triangles whose vertices are given in Exercises 19–22. (*Hint:* Draw each triangle first.)

19. $(-1, 0), (1, 0), (0, 3)$
20. $(0, 0), (1, 0), (0, 1)$
21. $(0, 0), (a, 0), (0, a)$
22. $(0, 0), (a, 0), (0, b)$

Centroid

23. Find the center of mass of a thin plate covering the region between the x-axis and the curve $y = 2/x^2$, $1 \le x \le 2$, if the density is $\delta(x) = x^3$.

24. Find the center of mass of a thin plate covering the region bounded below by the parabola $y = x^2$ and above by the line $y = x$ if the density is $\delta(x) = 12x$.

The Theorems of Pappus

In the third century an Alexandrian Greek named Pappus discovered two formulas that relate centers of mass to surfaces and volumes of revolution. These formulas, easy to remember, provide useful shortcuts to a number of otherwise lengthy calculations.

THEOREM 1

If a plane region is revolved once about an axis in the plane that does not pass through the region's interior, then the volume of the solid swept out by the region is equal to the region's area times the distance traveled by the region's center of mass. In symbols,

$$V = 2\pi \bar{y} A.$$

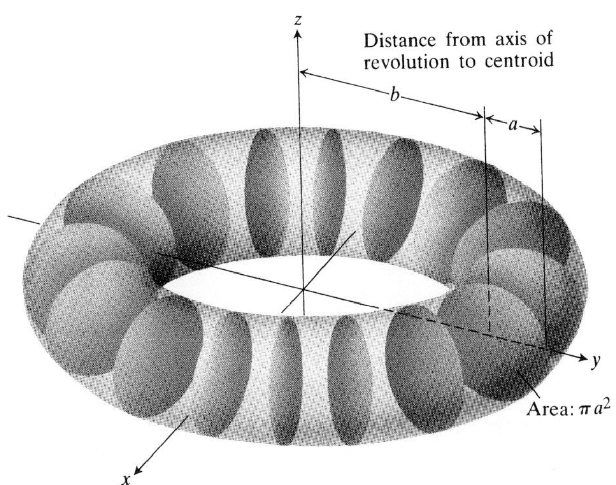

6.59 The volume swept out by the revolving disk is $(2\pi b)(\pi a^2)$.

THEOREM 2

If an arc of a plane curve is revolved once about a line in the plane that does not cut through the interior of the arc, then the area of the surface swept out by the arc is equal to the length of the arc times the distance traveled by the arc's center of mass. In symbols,

$$S = 2\pi \bar{y} L.$$

EXAMPLE 1 The volume of the torus (doughnut) generated by revolving a circle of radius a about an axis in its plane at a distance $b \geq a$ from its center (see Fig. 6.59) is

$$V = (2\pi b)(\pi a^2) = 2\pi^2 b a^2.$$ ■

EXAMPLE 2 The surface area of the torus in Example 1 is

$$S = (2\pi b)(2\pi a) = 4\pi^2 ba.$$ ■

25. The square region with vertices (0, 2), (2, 0), (4, 2) and (2, 4) is revolved about the x-axis to generate a solid. Find the volume and surface area of the solid.

26. Use a theorem of Pappus to find the volume generated by revolving about the line $x = 5$ the triangular region bounded by the coordinate axes and the line $2x + y = 6$. (The centroid of a triangle lies at the intersection of the medians, one third of the way from the midpoint of each side toward the opposite vertex.)

27. Find the volume of the torus generated by revolving the circle $(x - 2)^2 + y^2 = 1$ about the y-axis.

28. Use the theorems of Pappus to find the lateral surface area and the volume of a right circular cone.

29. Use the second theorem of Pappus and the fact that the surface area of a sphere of radius a is $4\pi a^2$ to find the center of mass of the semicircle $y = \sqrt{a^2 - x^2}$.

30. As found in Exercise 29, the center of mass of the semicircle $y = \sqrt{a^2 - x^2}$ lies at the point $(0, 2a/\pi)$. Find the area of the surface swept out by revolving the semicircle about the line $y = a$.

31. Use the first theorem of Pappus and the fact that the volume of a sphere of radius a is $V = (4/3)\pi a^3$ to find the center of mass of the region enclosed by the x-axis and the semicircle $y = \sqrt{a^2 - x^2}$.

32. As found in Exercise 31, the center of mass of the region enclosed by the x-axis and the semicircle $y = \sqrt{a^2 - x^2}$ lies at the point $(0, 4a/3\pi)$. Find the volume of the solid generated by revolving this region about the line $y = -a$.

33. The region of Exercise 32 is revolved about the line $y = x - a$ to generate a solid. Find the volume of the solid.

34. As found in Exercise 29, the center of mass of the semicircle $y = \sqrt{a^2 - x^2}$ lies at the point $(0, 2a/\pi)$.the semicircle about the line $y = x - a$.

6.9 The Basic Idea. Other Modeling Applications

There is a pattern to what we have been doing in the preceding sections. In each section we wanted to measure something that was modeled or described by one or more continuous functions. In Section 6.1 it was the area between the graphs of two continuous functions. In Section 6.2 it was the volume of the solid defined by revolving the graph of a continuous function about an axis. In Section 6.6 it was the work done by a force directed along the x-axis whose magnitude was given by a continuous function, and so on. In each case we responded by partitioning the interval on which the function or functions were defined and approximating what we wanted to measure with Riemann sums over the interval. We then used the integral defined by the limit of the Riemann sums to define and calculate what we wanted to measure. You will see what we mean if you look at Table 6.2.

TABLE 6.2
The phases of developing an integral to calculate something

Phase 1	Phase 2	Phase 3
We describe or model something we want to measure in terms of one or more continuous functions defined on a closed interval $[a, b]$.	We partition $[a, b]$ into subintervals of length Δx_k and choose a point c_k in each subinterval. We approximate what we want to measure with a finite sum. We identify the sum as a Riemann sum of a continuous function over $[a, b]$.	The approximations improve as the norm of the partition goes to zero. The Riemann sums approach a limiting integral. We use the integral to define and calculate what we originally wanted to measure.
The area between the curves $y = f_1(x)$, $y = f_2(x)$, on $[a, b]$ when $f_2(x) \leq f_1(x)$	$\sum (f_1(c_k) - f_2(c_k))\Delta x_k$	$\text{Area} = \int_a^b (f_1(x) - f_2(x))\,dx$
The volume of the solid defined by revolving the curve $y = f(x)$, $a \leq x \leq b$, about the x-axis	$\sum \pi f^2(c_k)\Delta x_k$	$\text{Volume} = \int_a^b \pi f^2(x)\,dx$
The length of a continuously differentiable curve $y = f(x)$, $a \leq x \leq b$	$\sum \sqrt{1 + (f'(c_k))^2}\Delta x_k$	$\text{Length} = \int_a^b \sqrt{1 + (f'(x))^2}\,dx$
The work done by a variable force $F(x)$ directed along the x-axis from a to b	$\sum F(c_k)\Delta x_k$	$\text{Work} = \int_a^b F(x)\,dx$

Literally thousands of things in biology, chemistry, economics, engineering, finance, geology, medicine, and other fields (the list would fill pages) are modeled and calculated by exactly this process.

In this section we review the process and look at more of the important integrals it leads to.

Volumes of Arbitrary Solids—Slicing

Now that we can find the areas of regions bounded by smooth curves, we can find the volumes of a wide variety of cylinders. As we shall now see, this enables us to define the volumes of many new solids.

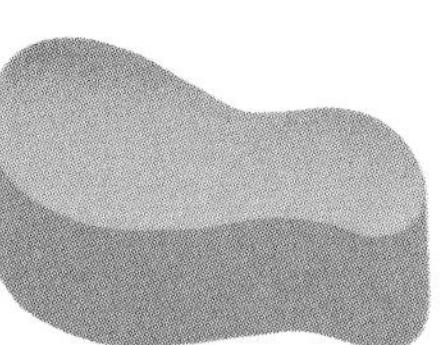

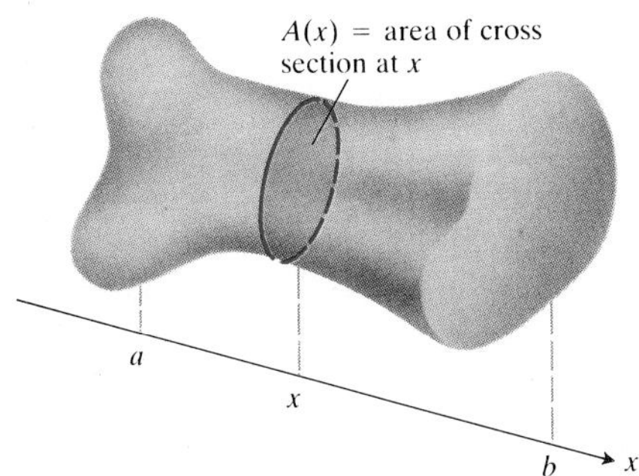

6.60 If the area of the cross section is a continuous function of x, we can find the volume of the solid in the way explained in the text.

Suppose we want to find the volume of a solid like the one shown in Fig. 6.60. The solid lies between planes perpendicular to the x-axis at $x = a$ and $x = b$. Each cross section of the solid by a plane perpendicular to the x-axis is a region whose area we know how to find. Specifically, at each point x in the closed interval $[a, b]$ the cross section of the solid is a region $R(x)$ whose area is $A(x)$. This makes A a real-valued function of x. If it is also a continuous function of x, we can use it to define and calculate the volume of the solid as an integral in the following way.

We partition the interval $[a, b]$ in the usual manner and slice the solid, as we would a loaf of bread, by planes perpendicular to the x-axis at the partition points. The kth slice, the one between the planes at x_{k-1} and x_k, has approximately the same volume as the cylinder between these two planes based on the region $R(x_k)$ (Fig. 6.61). The volume of this cylinder is

$$\begin{aligned} V_k &= \text{base area} \times \text{height} \\ &= A(x_k) \times (\text{distance between the planes at } x_{k-1} \text{ and } x_k) \\ &= A(x_k)\Delta x_k. \end{aligned} \tag{1}$$

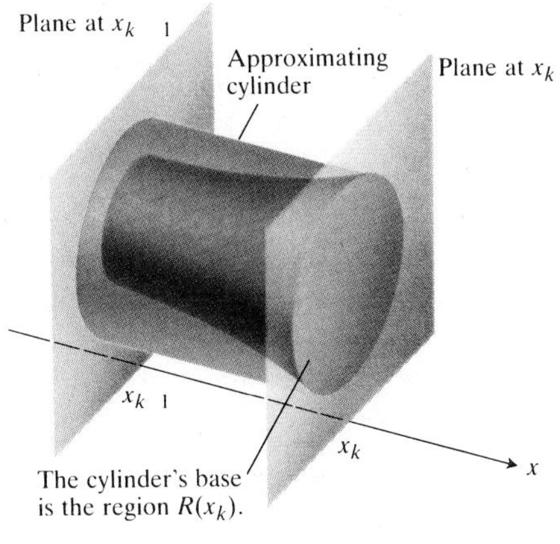

6.61 Enlarged view of the slice of the solid between the planes at x_{k-1} and x_k and its approximating cylinder.

The volume of the solid is therefore approximated by the cylinder volume sum

$$\sum_{k=1}^{n} A(x_k)\Delta x_k. \tag{2}$$

This is a Riemann sum for the function $A(x)$ on the closed interval $[a, b]$. We expect the approximations we get from sums like these to improve as the norm of the partition of $[a, b]$ goes to zero, so we define their limiting integral to be the volume of the solid.

DEFINITION

The **volume** of a solid of known cross-section area $A(x)$ from $x = a$ to $x = b$ is the integral of A from a to b,

$$\text{Volume} = \int_a^b A(x)\,dx. \tag{3}$$

Notice that the new formula is consistent with the disk and washer formulas for solids of revolution (a good sign). In the disk formula,

$$V = \int_a^b \pi(f(x))^2\,dx,$$

the cross section at x is a disk of radius $f(x)$ whose area is

$$A(x) = \pi(f(x))^2.$$

In the washer formula,

$$V = \int_a^b \pi(R^2(x) - r^2(x))\,dx,$$

the cross section at x is a washer of inner radius $r(x)$ and outer radius $R(x)$ whose area is

$$A(x) = \pi R^2(x) - \pi r^2(x) = \pi(R^2(x) - r^2(x)).$$

To apply Eq. (3), we take the following steps.

The Method of Slicing

STEP 1: Sketch the solid and a typical cross section.

STEP 2: Find a formula for $A(x)$.

STEP 3: Find the limits of integration.

STEP 4: Integrate $A(x)$ to find the volume.

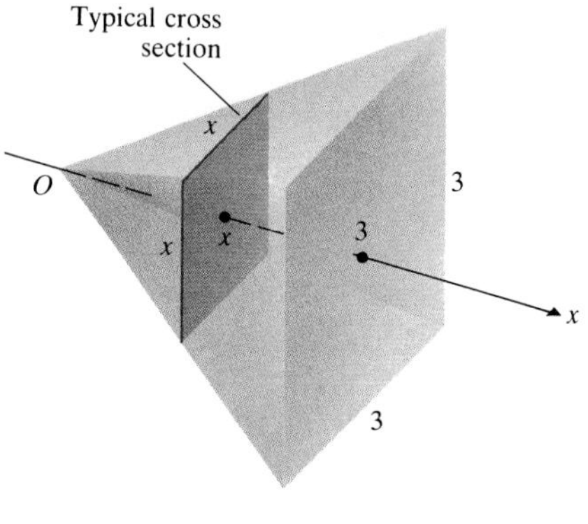

6.62 The cross sections of the pyramid in Example 1 are squares.

EXAMPLE 1 A pyramid 3 m high has a square base that is 3 m on a side. The cross section of the pyramid perpendicular to the altitude x units down from the vertex is a square x units on a side. Find the volume of the pyramid.

Solution STEP 1: *A sketch.* We draw the pyramid with its altitude along the x-axis and its vertex at the origin and include a typical cross section (Fig. 6.62).

STEP 2: *A formula for A(x).* The cross section at x is a square x meters on a side, so its area is

$$A(x) = x^2.$$

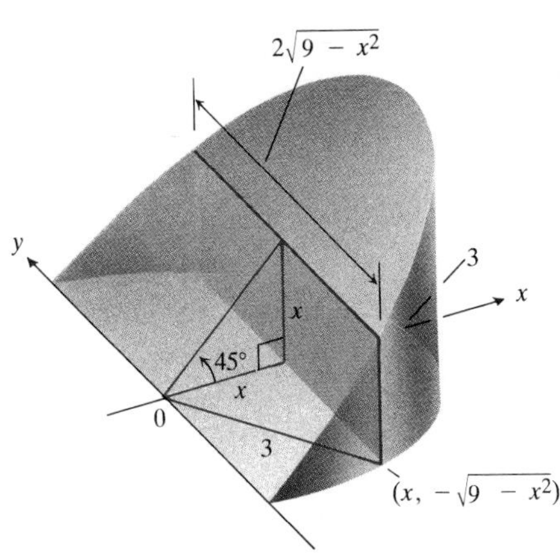

6.63 The wedge of Example 2, sliced perpendicular to the x-axis. The cross sections are rectangles.

STEP 3: *The limits of integration.* The squares go from $x = 0$ to $x = 3$.

STEP 4: *The volume:*

$$\begin{aligned}\text{Volume} &= \int_a^b A(x)\,dx \qquad \text{(Eq. 3)}\\ &= \int_0^3 x^2\,dx\\ &= \frac{x^3}{3}\bigg]_0^3\\ &= 9.\end{aligned}$$

The volume is 9 m^3.

EXAMPLE 2 A curved wedge is cut from a cylinder of radius 3 by two planes. One plane is perpendicular to the axis of the cylinder. The second plane crosses the first plane at a 45° angle at the center of the cylinder. Find the volume of the wedge.

Solution STEP 1: *A sketch.* We draw the wedge and sketch a typical cross section perpendicular to the x-axis (Fig. 6.63).

STEP 2: *The formula for A(x).* The cross section at x is a rectangle of area

$$A(x) = (\text{height})(\text{width}) = (x)(2\sqrt{9 - x^2}) = 2x\sqrt{9 - x^2}.$$

STEP 3: *The limits of integration.* The rectangles run from $x = 0$ to $x = 3$.

STEP 4: *The volume:*

$$\begin{aligned}\text{Volume} &= \int_a^b A(x)\,dx = \int_0^3 2x\sqrt{9 - x^2}\,dx \qquad \text{(Eq. 3)}\\ &= -\frac{2}{3}(9 - x^2)^{3/2}\bigg]_0^3 = 0 + \frac{2}{3}(9)^{3/2} \qquad \begin{pmatrix}\text{Substitute } u = 9 - x^2,\\ du = -2x\,dx, \text{ integrate,}\\ \text{and substitute back.}\end{pmatrix}\\ &= 18.\end{aligned}$$

EXAMPLE 3 *Cavalieri's theorem.* Cavalieri's theorem says that two solids with equal altitudes and identical parallel cross sections have the same volume (Fig. 6.64). We can see this immediately from Eq. (3) because the crosssection area function $A(x)$ is the same in each case.

Bonaventura Cavalieri (1598–1647)

Cavalieri, a student of Galileo's whom Galileo told to study calculus, discovered that if two plane regions can be arranged to lie over the same interval of the x-axis in such a way that they have identical cross sections at every point, then the regions have the same area. The theorem was good enough to win Cavalieri a chair at the University of Bologna in 1629. The solid-geometry version in Example 3, which Cavalieri never proved, was given his name by later geometers.

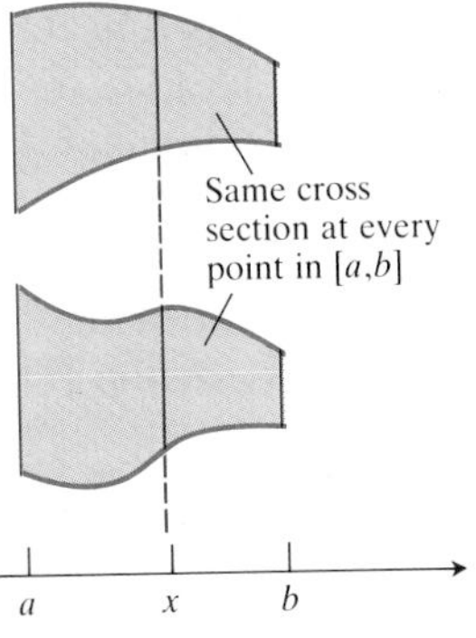

Position Shift and Distance Traveled

The total distance traveled by a body moving up and down a coordinate line is found by integrating the absolute value of the body's velocity (that is, the body's speed) over the time interval of the motion.

To see why, partition the time interval $a \le t \le b$ into subintervals in the usual way and let Δt_k denote the length of the kth interval. If Δt_k is small enough, the body's velocity $v(t)$ will not change much from t_{k-1} to t_k and the

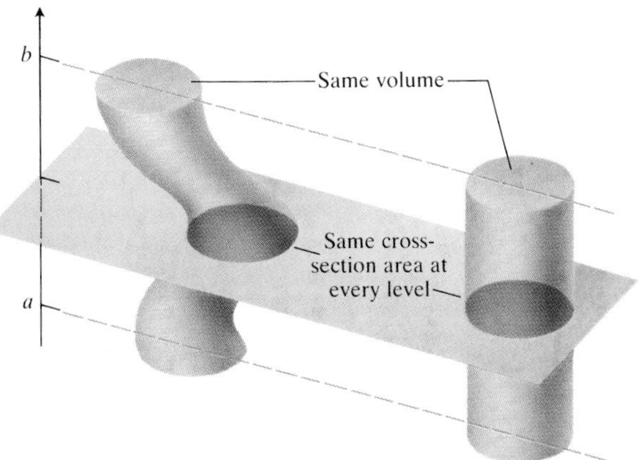

6.64 These two solids have the same volume even though the one on the left looks bigger. You can illustrate this yourself with two stacks of coins.

If $s(t)$ is a body's position on a coordinate line at time t, then

$$\frac{ds}{dt} = v = \text{velocity}$$

and

$$\left|\frac{ds}{dt}\right| = |v| = \text{speed.}$$

right-hand endpoint value $v(t_k)$ will give a good approximation of the velocity throughout the interval. Accordingly, the change in the body's position coordinate during the kth time interval will be about

$$v(t_k)\Delta t_k. \tag{4}$$

The change will be positive if $v(t_k)$ is positive and negative if $v(t_k)$ is negative.

In either case, the amount of distance traveled by the body during the kth interval will be about

$$|v(t_k)|\Delta t_k. \tag{5}$$

The total trip distance will be about

$$\sum_{k=1}^{n} |v(t_k)|\Delta t_k. \tag{6}$$

The sum in (6) is a Riemann sum for the speed $|v(t)|$ on the interval $[a, b]$. We expect the approximations we get from sums like these to improve as the norm of the partition of $[a, b]$ goes to zero. It therefore looks as if we should be able to calculate the total distance traveled by the body by integrating the body's speed from a to b. In practice, this turns out to be just the right thing to do. The mathematical model predicts the distance every time.

$$\text{Distance traveled} = \int_a^b |v(t)|\, dt \tag{7}$$

If we wish instead to predict how far up or down the line from its initial position a body will end up when a trip is over, we integrate v instead of its absolute value.

To understand why, let $s(t)$ be the body's position at time t and let F be an antiderivative of v. Then

$$s(t) = F(t) + C$$

Conclusions

1. To find distance traveled, integrate speed
2. To find position shift, integrate velocity.

for some constant C. The shift in the body's position caused by the trip from $t = a$ to $t = b$ is

$$s(b) - s(a) = (F(b) + C) - (F(a) + C) = F(b) - F(a) = \int_a^b v(t)\,dt. \tag{8}$$

$$\text{Position shift} = \int_a^b v(t)\,dt \tag{9}$$

EXAMPLE 4 The velocity of a body moving along a line from $t = 0$ to $t = 3\pi/2$ seconds was

$$v(t) = 5\cos t \text{ m/sec}.$$

Find the total distance traveled and the shift in the body's position.

Solution

$$\text{Distance traveled} = \int_0^{3\pi/2} |5\cos t|\,dt \qquad \text{(Distance is the integral of speed.)}$$

$$= \int_0^{\pi/2} 5\cos t\,dt + \int_{\pi/2}^{3\pi/2} (-5\cos t)\,dt$$

$$= 5\sin t\Big]_0^{\pi/2} - 5\sin t\Big]_{\pi/2}^{3\pi/2} = 5(1-0) - 5(-1-1)$$

$$= 5 + 10 = 15 \text{ m}.$$

$$\text{Position shift} = \int_0^{3\pi/2} 5\cos t\,dt \qquad \text{(Shift is the integral of velocity.)}$$

$$= 5\sin t\Big]_0^{3\pi/2} = 5(-1) - 5(0) = -5 \text{ m}.$$

During the trip, the body traveled 5 m forward and 10 m backward for a total distance of 15 m. This shifted the body 5 m to the left (Fig. 6.65). ■

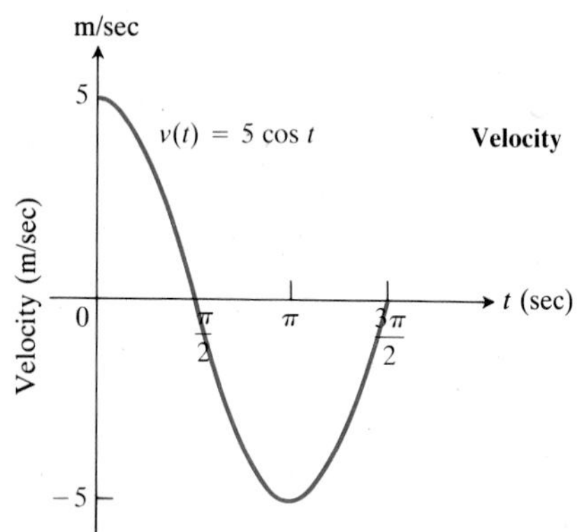

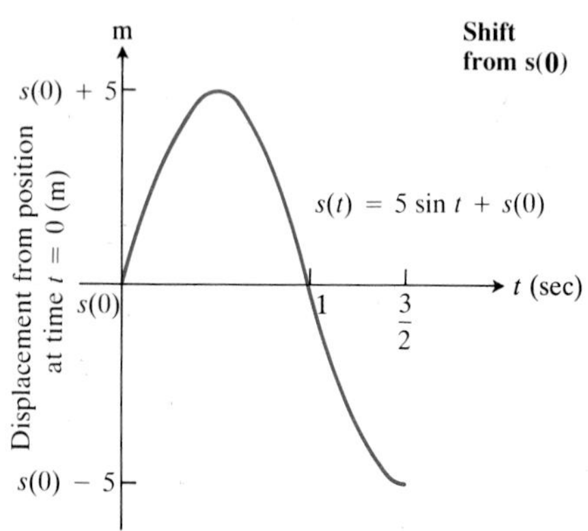

6.65 The position and velocity of the body in Example 4. The velocity is positive at first, and the corresponding displacement is positive. But the body stops at $t = \pi/2$ and reverses direction. By time $t = 3\pi/2$, the body lies 5 m to the left of where it started.

Delesse's Rule

As you may know, the sugar in an apple starts turning into starch as soon as the apple is picked, and the longer the apple sits around, the starchier it becomes. You can tell fresh apples from stale by both flavor and consistency.

To find out how much starch is in a given apple, we can look at a thin slice under a microscope. The cross sections of the starch granules will show up clearly, and it is easy to estimate the proportion of the viewing area they occupy. This two-dimensional proportion will be the same as the three-dimensional proportion of uncut starch granules in the apple itself. The apparently magical equality of these proportions was first discovered by a French geologist, Achille Ernest Delesse, in the 1840s. Its explanation lies in the notion of average value.

Achille Ernest Delesse was a mining engineer interested in determining the composition of rocks. To find out how much of a particular mineral a rock contained, he cut it through, polished an exposed face, and covered the face with transparent waxed paper, trimmed to size. He then traced on the paper the exposed portions of the mineral that interested him. After weighing the paper, he cut out the mineral traces and weighed them. The ratio of the weights gave not only the proportion of the surface occupied by the mineral but, more important, the proportion of the rock occupied by the mineral. Delesse described his method in an article entitled "A mechanical procedure for determining the composition of rocks," in the *Annales des Mines,* 13, 1848, pp. 379–388. His method is still used by petroleum geologists today. A two-dimensional analogue of it is used to determine the porosities of the ceramic filters that extract organic molecules in chemistry laboratories and screen out microbes in water purifiers.

Suppose we want to find the proportion of some granular material in a solid and that the sample we have chosen to analyze is a cube whose edges have length L. We picture the cube with an x-axis along one edge and imagine slicing the cube with planes perpendicular to points of the interval $[0, L]$. Call the proportion of the area of the slice at x occupied by the granular material of interest (starch, in our apple example) $r(x)$ and assume r is a continuous function of x.

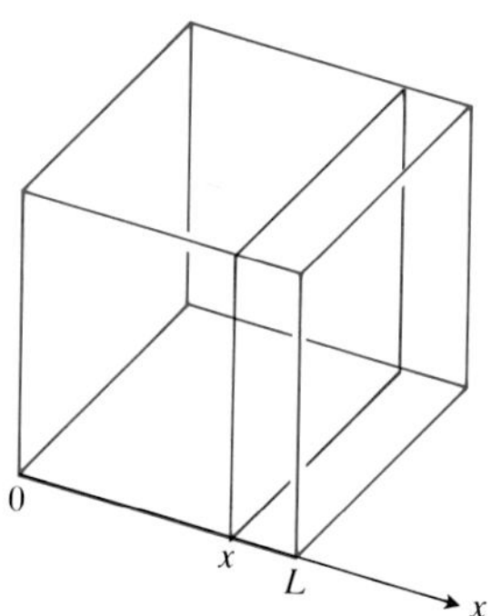

Now partition the interval $[0, L]$ into subintervals in the usual way. Imagine the cube sliced into thin slices by planes at the subdivision points. The length Δx_k of the kth subinterval is the distance between the planes at x_{k-1} and x_k. If the planes are close enough together, the sections cut from the grains by the planes will resemble cylinders with bases in the plane at x_k. The proportion of granular material between the planes will therefore be about the same as the proportion of cylinder base area in the plane at x_k, which in turn will be about $r(x_k)$. Thus the amount of granular material in the slab between the two planes will be about

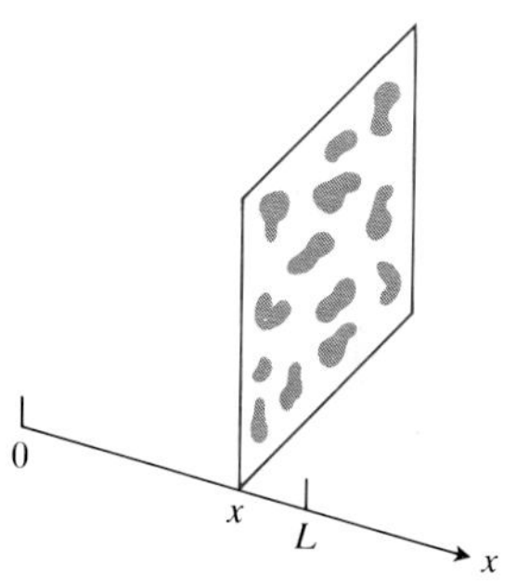

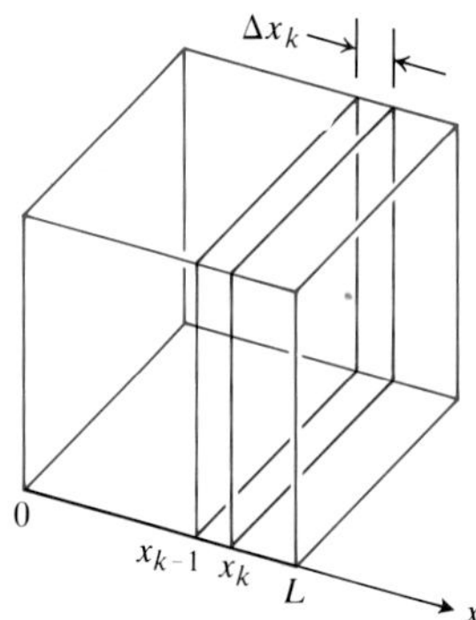

$$\text{(Proportion)} \times \text{(slab volume)} = r(x_k)L^2\Delta x_k. \tag{10}$$

The amount of granular material in the entire sample cube will be about

$$\sum_{k=1}^{n} r(x_k)L^2\,\Delta x_k. \tag{11}$$

This sum is a Riemann sum for the function $r(x)L^2$ over the interval $[0, L]$. We expect the approximations by sums like these to improve as the norm of the subdivision of $[0, L]$ goes to zero and therefore expect the integral

$$\int_a^b r(x)L^2\,dx \tag{12}$$

to give the amount of granular material in the sample cube.

We can then obtain the proportion of granular material in the sample by dividing this amount by the cube's volume L^3. If we have chosen our sample well, this will also be the proportion of granular material in the solid from which the sample was taken. Putting it all together, we get

$$\begin{aligned}\text{Proportion of granular material in solid} &= \text{Proportion of granular material in the sample cube} \\ &= \frac{\int_0^L r(x)L^2\,dx}{L^3} = \frac{L^2\int_0^L r(x)\,dx}{L^3} = \frac{1}{L}\int_0^L r(x)\,dx \qquad (13) \\ &= \text{average value of } r(x) \text{ over } [0, L] \\ &= \text{proportion of area occupied by granular material in a typical cross section.}\end{aligned}$$

This is Delesse's rule. Once we have found $\bar{r}$, the average of $r(x)$ over $[0, L]$, we have found the proportion of granular material in the solid.

In practice, $\bar{r}$ is found by averaging over a number of cross sections. There are several things to watch out for in the process. In addition to the possibility that the granules cluster in ways that make representative samples difficult to find, there is the possibility we might not recognize a granule's trace for what it is. Some cross sections of normal red blood cells look like disks and ovals, but others look surprisingly like outlines of dumbbells. We do not want to dismiss the dumbbells as experimental error the way one research group we know of did a few years ago.

Useless Integrals

Some of the integrals we get from forming Riemann sums do what we want, but others do not. It all depends on how we choose to model the problems we want to solve. Some choices are good—others are not. Here is an example.

We use the surface area formula

$$\text{Surface area} = \int_a^b 2\pi f(x)\sqrt{1 + \left(\frac{df}{dx}\right)^2}\,dx \qquad (14)$$

because it has predictive value and always gives results consistent with information from other sources. In other words, the model we used to derive the formula was a good one.

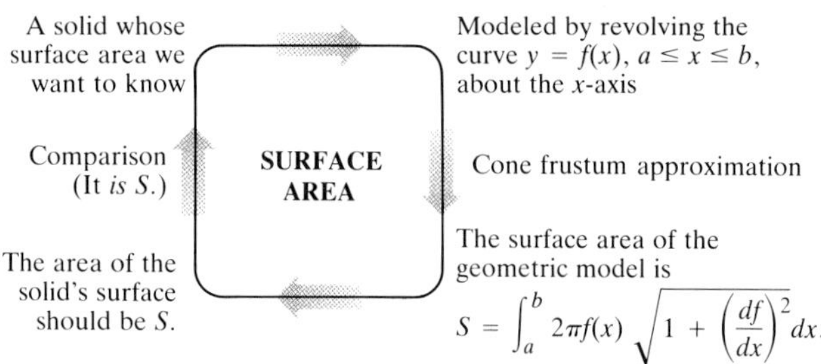

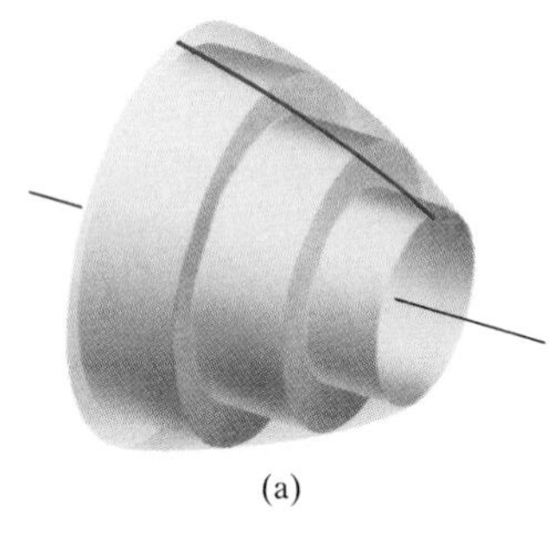

(a)

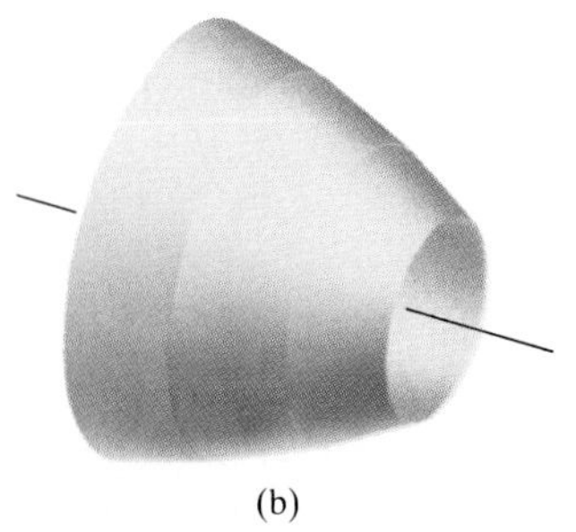

(b)

6.66 Why not use cylindrical bands (a) instead of conical bands (b) to approximate surface area?

Why not find the surface area by approximating with cylindrical bands instead of conical bands, as suggested in Fig. 6.66? The Riemann sums we get this way converge just as nicely as the ones based on conical bands, and the resulting integral is simpler. Instead of Eq. (14), we get

$$\text{Surface area candidate} = \int_a^b 2\pi f(x)\,dx. \tag{15}$$

After all, we might argue, we used cylinders to derive good volume formulas, so why not use them again to derive surface-area formulas?

The answer is that the formula in Eq. (15) has no predictive value and almost never gives results consistent with experience. The comparison step in the model fails for this formula.

There is a moral here: Just because we end up with a nice-looking integral does not mean it will do what we want. Constructing an integral is not enough—we have to test it too.

Exercises 6.9

Find the volumes of the solids in Exercises 1–8.

1. The solid lies between planes perpendicular to the x-axis at $x = 0$ and $x = 4$. The cross sections perpendicular to the axis on the interval $0 \le x \le 4$ are squares whose diagonals run from the parabola $y = -\sqrt{x}$ to the parabola $y = \sqrt{x}$.
2. The solid lies between planes perpendicular to the x-axis at $x = -1$ and $x = 1$. The cross sections perpendicular to the x-axis between these planes are squares whose diagonals run from the semicircle $y = -\sqrt{1 - x^2}$ to the semicircle $y = \sqrt{1 - x^2}$.
3. The solid lies between planes perpendicular to the x-axis at $x = -1$ and $x = 1$. The cross sections perpendicular to the axis between these planes are squares with edges running from the semicircle $y = -\sqrt{1 - x^2}$ to the semicircle $y = \sqrt{1 - x^2}$.

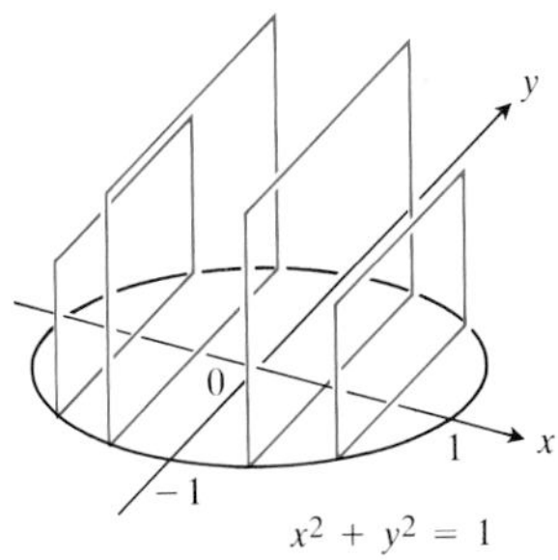

4. The solid lies between the planes perpendicular to the x-axis at $x = -1$ and $x = 1$. The cross sections perpendicular to the x-axis are circular disks whose diameters run from the parabola $y = x^2$ to the parabola $y = 2 - x^2$.
5. The solid lies between planes perpendicular to the x-axis at $x = 1$ and $x = 2$. The cross sections perpendicular to the x-axis are circular disks with diameters running from the x-axis up to the curve $y = 2/\sqrt{x}$.
6. The solid lies between planes perpendicular to the x-axis at $x = 0$ and $x = 2$. The cross sections perpendicular to the x-axis are circular disks with diagonals running from the x-axis up to the parabola $y = \sqrt{5}\,x^2$.
7. The base of the solid is the disk $x^2 + y^2 \le 1$. The cross sections by planes perpendicular to the y-axis between $y = -1$ and $y = 1$ are isosceles right triangles with one leg in the disk.

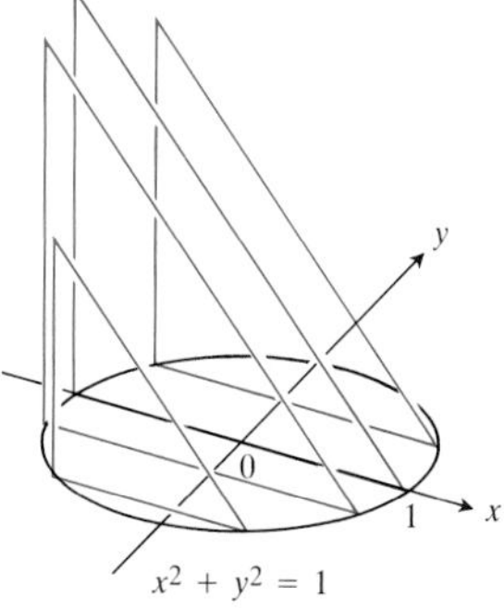

8. The base of the solid is the region between the curve $y = 2\sqrt{\sin x}$ and the interval $0 \le x \le \pi$ of the x-axis. The cross sections perpendicular to the x-axis are equilateral triangles.

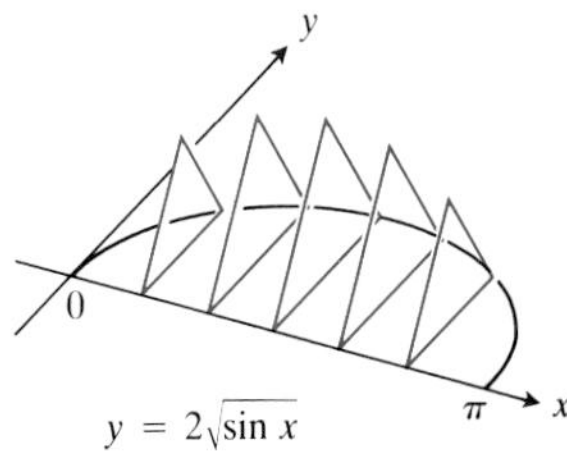

9. A twisted solid is generated as follows: We are given a line L in space and a square of side s in a plane perpendicular to L. One vertex of the square is on L. As this vertex moves a distance h along L, the square turns through a full revolution about L. Find the volume of the solid generated by this motion. What would the volume be if the square had turned two revolutions in moving the same distance along L?

10. Prove Cavalieri's original theorem (historical note, this section), assuming that each region is the region between the graphs of two continuous functions over the interval $a \leq x \leq b$.

In Exercises 11–18, the function $v(t)$ is the velocity in meters per second of a body moving along a coordinate line. (a) Graph v as a function of t to see where it is positive and negative. Then find (*b*) the total distance traveled by the body during the given time interval and (c) the shift in the body's position.

11. $v(t) = 5 \cos t,\ 0 \leq t \leq 2\pi$

12. $v(t) = \sin \pi t,\ 0 \leq t \leq 2$

13. $v(t) = 6 \sin 3t,\ 0 \leq t \leq \pi/2$

14. $v(t) = 4 \cos 2t,\ 0 \leq t \leq \pi$

15. $v(t) = 49 - 9.8t,\ 0 \leq t \leq 10$

16. $v(t) = 8 - 1.6t,\ 0 \leq t \leq 10$

17. $v(t) = 6t^2 - 18t + 12 = 6(t - 1)(t - 2),\ 0 \leq t \leq 2$

18. $v(t) = 6t^2 - 18t + 12 = 6(t - 1)(t - 2),\ 0 \leq t \leq 3$

19. Figure 6.67 shows the velocity graphs of four bodies moving on coordinate lines. Find the distance traveled by each body and the position shift for the given time interval.

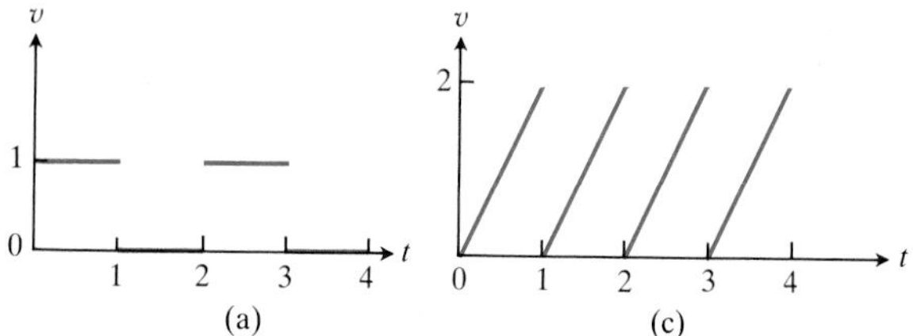

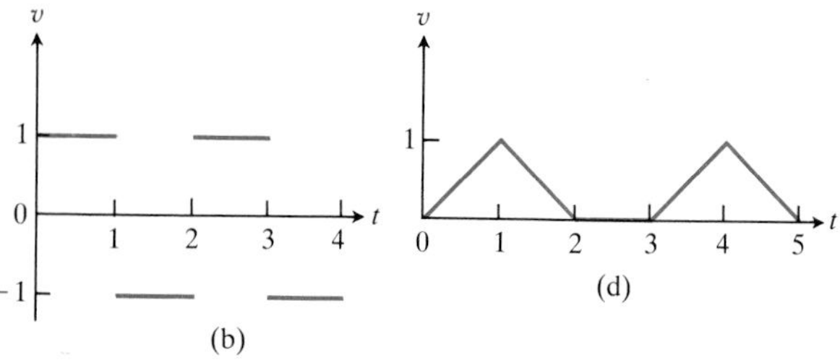

6.67 The graphs for Exercise 19.

20. CALCULATOR Table 6.3 shows the velocity of a model train engine moving back and forth on a track for 10 sec. Use Simpson's rule to find the resulting position shift and total distance traveled.

TABLE 6.3
Selected velocities of the model train engine in Exercise 20

Time (sec)	Velocity (in. per sec)	Time (sec)	Velocity (in. per sec)
0	0	6	−11
1	12	7	−6
2	22	8	2
3	10	9	6
4	−5	10	0
5	−13		

21. The photograph below shows a grid superimposed on the polished face of a piece of granite. Use the grid and Delesse's rule to estimate the proportion of shrimp-colored granular material in the rock.

© Susan Van Etten

22. The photograph below shows a grid superimposed on a microscopic view of a stained section of human lung tissue. The clear spaces between the cells are cross sections of the lung's air sacks (called *aveolae*—accent on the second syllable). Use the grid and Delesse's rule to estimate the proportion of air space in the lung.

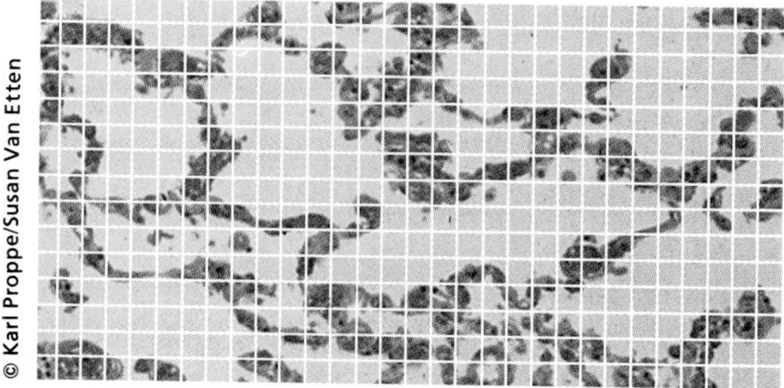

© Karl Proppe/Susan Van Etten

23. *Modeling surface area.* The lateral surface area of the cone swept out by revolving the line segment $y = x/\sqrt{3}$, $0 \le x \le \sqrt{3}$, about the x-axis should be (1/2)(base circumference)(slant height) = $(1/2)(2\pi)(2) = 2\pi$. What do you get if you use Eq. (15) with $f(x) = x/\sqrt{3}$?

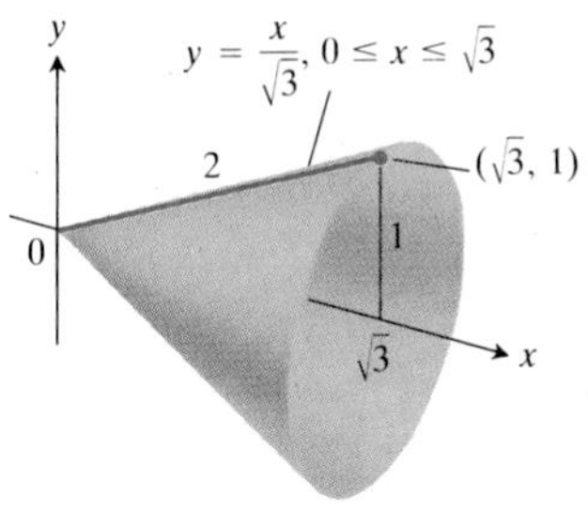

24. *Modeling surface area.* The only surface for which Eq. (15) gives the area we want is a cylinder. Show that Eq. (15) gives $S = 2\pi rh$ for the cylinder swept out by revolving the line segment $y = r$, $0 \le x \le h$, about the x-axis.

TOOLKIT PROGRAM

Derivative Grapher — Graphs functions and their integrals and derivatives together

Review Questions

1. How do you define and calculate the area between the graphs of two continuous functions? Give an example.
2. How do you define and calculate the volume of a solid of revolution by
   a) the disk method?
   b) the washer method?
   c) the method of cylindrical shells?
   Give examples.
3. How do you define and calculate the length of the graph of a continuously differentiable function over a closed interval? Give an example. What about functions that aren't continuously differentiable?
4. How do you define and calculate the area of the surface swept out by revolving the graph of a continuously differentiable function $y = f(x)$, $a \le x \le b$, about the x-axis? Give an example.
5. How do you define and calculate the work done by a force directed along a portion of the x-axis? How do you calculate the work it takes to pump liquid from a tank? Give examples.
6. How do you calculate the force exerted by a liquid against a portion of vertical wall? Give an example.
7. How do you locate the center of mass of a straight, narrow rod or strip of material? Give an example. If the density of the material is constant, you can tell right away where the center of mass is. Where is it?
8. How do you locate the center of mass of a thin flat plate of material? Give an example.
9. Suppose that you know the velocity of a body that will move back and forth along a coordinate line tomorrow from time $t = a$ to time $t = b$. How can you calculate in advance how much the motion will shift the body's position? How do you predict the total distance the body will travel?
10. How do you define and calculate volumes of solids by the method of slicing? Give an example. How is the method of slicing related to the disk and washer methods?
11. What does Delesse's rule say? Give an example.
12. There is a basic pattern to the way we constructed integrals in this chapter. What is it?

Practice Exercises

Find the areas of the regions enclosed by the curves and lines in Exercises 1–14.

1. $y = x,\ y = 1/x^2,\ x = 2$
2. $y = x,\ y = 1/\sqrt{x},\ x = 2$
3. $y = x + 1,\ y = 3 - x^2$
4. $\sqrt{x} + \sqrt{y} = 1,\ x = 0,\ y = 0$
5. $x = 2y^2,\ x = 0,\ y = 3$
6. $4x = y^2 - 4,\ 4x = y + 16$
7. $y^2 = 4x,\ y = 4x - 2$
8. $x = 4 - y^2,\ x = 0,\ y = 0$
9. $y = \sin x,\ y = x,\ 0 \le x \le \pi/4$
10. $y = |\cos x|,\ y = 1,\ -\pi/2 \le x \le \pi/2$
11. $y = 2 \sin x,\ y = \sin 2x,\ 0 \le x \le \pi$
12. $y = 8 \cos x,\ y = \sec^2 x,\ -\pi/3 \le x \le \pi/3$
13. The "triangular" region bounded on the left by $x + y = 2$, on the right by $y = x^2$, and above by $y = 2$
14. The "triangular" region bounded on the left by $y = \sqrt{x}$, on the right by $y = 6 - x$, and below by $y = 1$

Find the areas of the shaded regions in Exercises 15 and 16.

15.

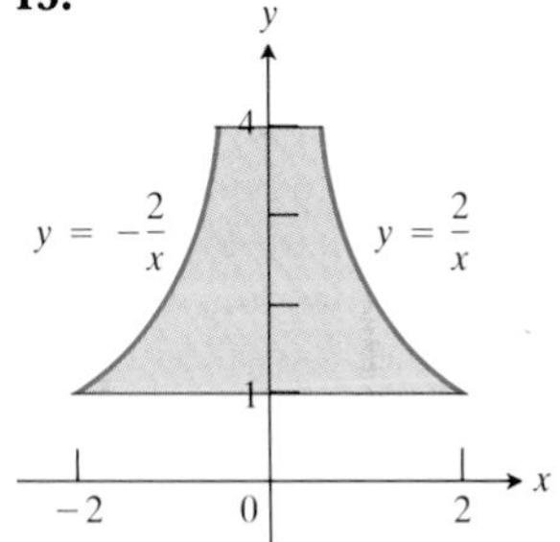

16.

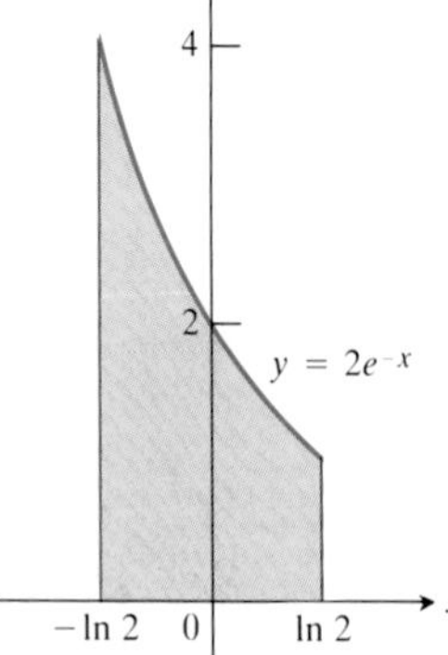

17. Find the volume of the solid generated by revolving the region bounded by the x-axis, the curve $y = 3x^4$, and the lines $x = 1$ and $x = -1$ about (a) the x-axis; (b) the y-axis.

18. Find the volume of the solid generated by revolving the region in the first quadrant bounded by the x-axis, the parabola $y = x^2$, and the line $x = 3$ about (a) the x-axis; (b) the y-axis.

19. Find the volume of the solid generated by revolving the region bounded on the left by the parabola $x = y^2 + 1$ and on the right by the line $x = 5$ about (a) the x-axis; (b) the y-axis; (c) the line $x = 5$.

20. Find the volume of the solid generated by revolving the region bounded by the parabola $y^2 = 4x$ and the line $y = x$ about (a) the x-axis; (b) the y-axis; (c) the line $x = 2$; (d) the line $y = 2$.

21. Find the volume of the solid generated by revolving the "triangular" region bounded by the x-axis, the line $x = \pi/3$, and the curve $y = \tan x$ in the first quadrant about the x-axis.

22. Find the volume of the solid generated by revolving the region bounded by the curve $y = \sin x$ and the lines $x = 0$, $x = \pi$, and $y = 2$ about the line $y = 2$.

23. Find the volume of the solid spindle generated by revolving the region bounded by the x-axis, the curve $y = 1/\sqrt{x}$, and the lines $x = 1$ and $x = 16$ about the x-axis.

24. Find the volume of the solid generated by revolving the region bounded by the curve $y = e^{x/2}$ and the lines $x = \ln 3$ and $y = 1$ about the x-axis.

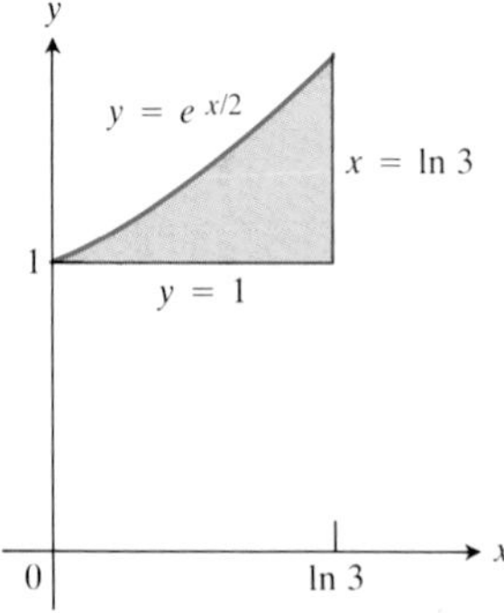

25. A round hole of radius $\sqrt{3}$ ft is bored through the center of a sphere of radius 2 ft. Find the volume cut out.

26. **CALCULATOR** The profile of a football resembles the ellipse shown here. Find the volume of the football to the nearest cubic inch.

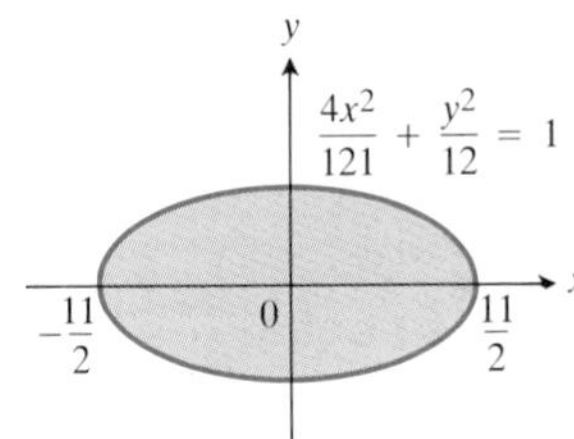

Find the lengths of the curves in Exercises 27–30.

27. $y = x^{1/2} - (1/3)x^{3/2}, \quad 0 \le x \le 3$
(*Hint:* $1 + (dy/dx)^2$ is a perfect square.)

28. $x = y^{2/3}, \quad 1 \le y \le 8$

29. $y = (x^2/8) - \ln x, \quad 4 \le x \le 8$

30. $y = e^x + (1/4)\,e^{-x}, \quad 0 \le x \le \ln 3$

In Exercises 31–36, find the areas of the surfaces generated by revolving the curves about the given axes.

31. $y = \sqrt{2x + 1}, \quad 0 \le x \le 12, \quad x\text{-axis}$

32. $y = x^3/9, \quad -1 \le x \le 1, \quad x\text{-axis}$

33. $y = (1/3)x^{3/2} - x^{1/2}, \quad 0 \le x \le 3, \quad x\text{-axis}$

34. $y = e^x + (1/4)e^{-x}, \quad -\ln 4 \le x \le \ln 2, \quad x\text{-axis}$

35. $y = (1/3)(x^2 + 2)^{3/2}, \quad 0 \le x \le 1, \quad y\text{-axis}$
(*Hint:* Express $ds = \sqrt{dx^2 + dy^2}$ in terms of dx and evaluate $\int 2\pi x\, ds$.)

36. $y = x^2, \quad 0 \le y \le 2, \quad y\text{-axis}$

37. A rock climber is about to haul up 10 kg of equipment that has been hanging beneath her on 40 m of rope that weighs 0.8 newtons per meter. How much work will it take? (*Hint:* Solve for the rope and equipment separately; then add.)

38. You drove an 800-gal tank truck from the base to the summit of Mt. Washington and discovered on arrival that the tank was only half full. You started out with a full tank, climbed at a steady rate, and took 50 min to accomplish the 4750-ft elevation change. Assuming that the water leaked out at a steady rate, how much work was spent in carrying water to the top? Do not count the work done in getting yourself and the truck there. Water weighs 8 lb/U.S. gal.

39. If a force of 20 lb is required to hold a spring 1 ft beyond its unstressed length, how much work does it take to stretch the spring this far? How much work does it take to stretch the spring an additional foot?

40. A force of 2 N will stretch a rubber band 2 cm. Assuming Hooke's law applies, how far will a 4-N force stretch the rubber band? How much work does it take to stretch the rubber band this far?

41. A reservoir, shaped like an inverted right circular cone 20 ft across the top and 8 ft deep, is full of water. How much work does it take to pump the water to a level 6 ft above the top?

42. *Continuation of Exercise 41.* The reservoir is filled to a depth of 5 ft, and the water is to be pumped to the same level as the top. How much work does it take?

43. The vertical triangular plate shown here is the end plate of a triangular watering trough full of water. What is the force against the end of the plate?

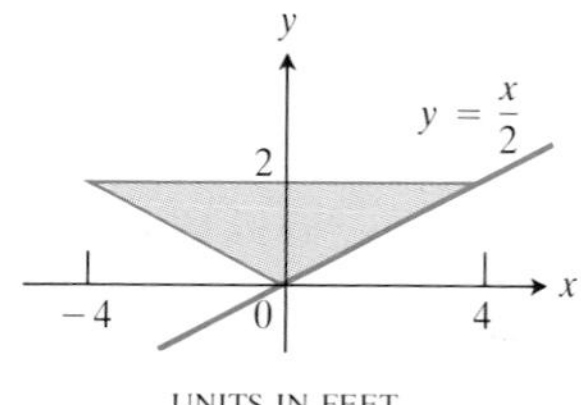

44. The vertical trapezoidal plate shown here is the end plate of a trough of maple syrup weighing 75 lb/ft^3. What is the force against the end of the trough when there are 10 in. of syrup in the trough?

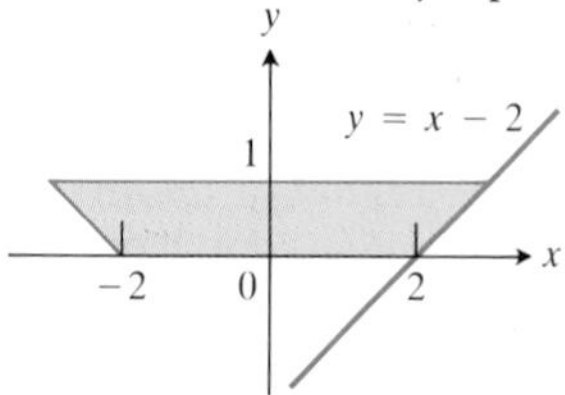

45. A flat vertical gate in the face of a dam is shaped like the parabolic region between the curve $y = 4x^2$ and the line $y = 4$, with measurements in feet. The top of the gate lies 5 ft below the surface of the water. Find the force against the gate.

46. **CALCULATOR** A standard olive oil can measures 5.75 by 3.5 by 10 in. Find the fluid force against the base and each side of the can when the can is full.

47. Find the center of mass of a thin, flat plate of constant density covering the region enclosed by the parabolas $y = 2x^2$ and $y = 3 - x^2$.

48. Find the center of mass of a thin, flat plate of constant density covering the region enclosed by the x-axis, the lines $x = 2$ and $x = -2$, and the parabola $y = x^2$.

49. Find the center of mass of a thin, flat plate of constant density covering the "triangular" region in the first quadrant bounded by the y-axis, the parabola $y = x^2/4$, and the line $y = 4$.

50. Find the center of mass of a thin, flat plate of density $\delta = 3$ covering the region enclosed by the parabola $y^2 = x$ and the line $x = 2y$.

51. Find the centroid of the trapezoid in Exercise 44.

52. Find the centroid of the triangle in Exercise 43.

53. Find the center of mass of a thin plate of constant density covering the region between the curve $y = 1/\sqrt{x}$ and the x-axis from $x = 1$ to $x = 16$.

54. *Continuation of Exercise 53.* Now find the center of mass, assuming that the density is $\delta(x) = 4/\sqrt{x}$.

Find the volumes of the solids in Exercises 55–58.

55. The solid lies between planes perpendicular to the x-axis at $x = 0$ and $x = 1$. The cross sections perpendicular to the x-axis between these planes are circular disks whose diameters run from the parabola $y = x^2$ to the parabola $y = \sqrt{x}$.

56. The base of the solid is the region in the first quadrant between the line $y = x$ and the parabola $y = 2\sqrt{x}$. The cross sections of the solid perpendicular to the x-axis are equilateral triangles whose bases stretch from the line to the curve.

57. The solid lies between planes perpendicular to the x-axis at $x = \pi/4$ and $x = 5\pi/4$. The cross sections between these planes are circular disks whose diameters run from the curve $y = 2\cos x$ to the curve $y = 2\sin x$.

58. The solid lies between planes perpendicular to the x-axis at $x = 0$ and $x = 6$. The cross sections between these planes are squares whose bases run from the x-axis up to the curve $x^{1/2} + y^{1/2} = \sqrt{6}$.

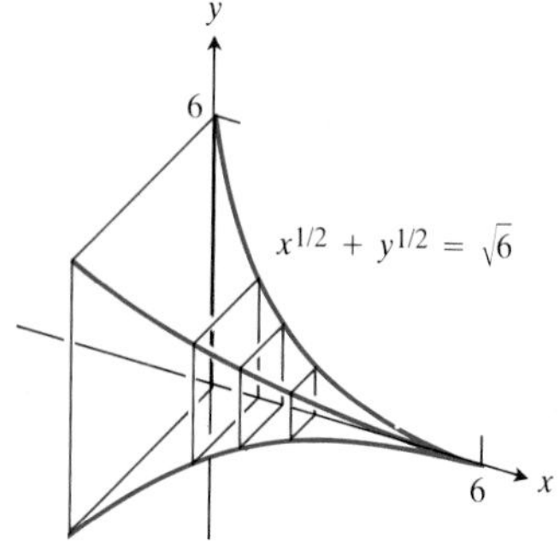

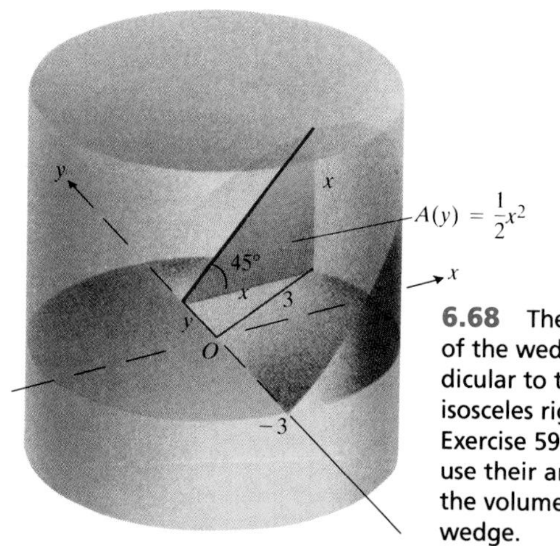

6.68 The cross sections of the wedge perpendicular to the y-axis are isosceles right triangles. Exercise 59 asks you to use their areas to find the volume of the wedge.

59. The volume of the wedge in Section 6.9, Example 2, could have been found by drawing a picture like the one in Fig. 6.68, taking cross sections perpendicular to the y-axis, and integrating with respect to y. Find the volume this way.

60. A solid lies between planes perpendicular to the x-axis at $x = 0$ and $x = 12$. The cross sections by planes perpendicular to the x-axis for $0 \le x \le 12$ are circular disks whose diameters run from the line $y = x/2$ to the line $y = x$. Use Cavalieri's theorem (Section 6.9, Example 3) to explain why the solid has the same volume as a right circular cone with base radius 3 and altitude 12.

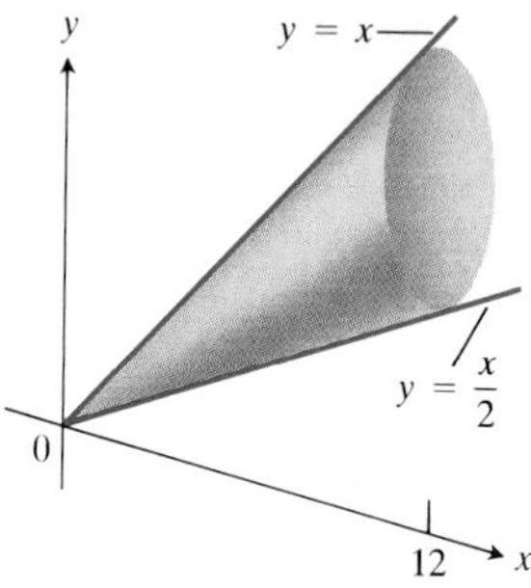

In Exercises 61–64, the function $v(t)$ is the velocity in feet per second of a body moving on a coordinate line. (a) Graph v as a function of t to see where it is positive and negative. Then find (b) the total distance traveled by the body during the given time interval and (c) how much the trip shifted the body's position.

61. $v(t) = (t/2) - 1,\ 0 \le t \le 6$

62. $v(t) = t^2 - 8t + 12 = (t - 2)(t - 6),\ 0 \le t \le 6$

63. $v(t) = 5\cos t,\ 0 \le t \le 3\pi/2$

64. $v(t) = \pi \sin \pi t,\ 0 \le t \le 3/2$

7 The Calculus of Transcendental Functions

© Laurie Platt Winfrey, Inc.

The proportion of carbon-14 in a painting's organic material is a decreasing exponential function of the painting's age. We can date the painting by working backward from how much carbon-14 is left.

OVERVIEW Many of the functions we work with in mathematics and science are inverses of one another. The functions $\ln x$ and e^x are probably the most famous function–inverse pair but there are others nearly as important. The trigonometric functions, when suitably restricted, have important inverses and there are other useful pairs of logarithmic and exponential functions. Less widely known are the hyperbolic functions and their inverses, functions that arise when we study hanging cables, heat flow, and the friction encountered by objects falling through the air. We shall describe all of these functions in this chapter and look at the kinds of problems they were designed to solve.

7.1 Inverse Functions and Their Derivatives

In Section 5.8, where we defined the exponential function e^x as the inverse of the natural logarithm function $\ln x$, we saw that the inverse of a function can easily be as important as the function itself. This is not an isolated case and we have reached the point in our development of calculus where we need the inverses of some of the other functions we have been working with. In this section, we review what it means for one function to be the inverse of another and look at what this says about the formulas, graphs, and derivatives of function–inverse pairs. In the process, we shall explain why defining e^x as the inverse of $\ln x$ automatically made it differentiable.

One-to-one Functions Have Inverses

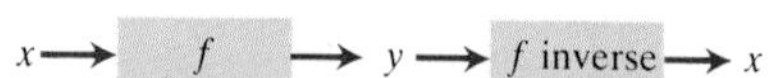

7.1 The inverse of a function f sends every output of f back to the input from which it came.

As you know, a function is a rule that assigns a number from its range to each number in its domain. The numbers in the domain are called inputs and the numbers assigned to them outputs. Functions like $y = \sin x$ and $y = x^2$ can assign the same output to different inputs. The sines of 0 and π are both 0 and the squares of -1 and 1 are both 1. Other functions, however, like $y = \sqrt{x}$ and $y = (x/4) + 3$, never assign an output number more than once. Functions like these are called **one-to-one.**

Since each output of a one-to-one function comes from just one input, a one-to-one function can be reversed to send the outputs back to the inputs from which they came (Fig. 7.1).

The function defined by reversing a one-to-one function f is called the **inverse** of f. The symbol for the inverse is f^{-1}, read "f inverse." The -1 in f^{-1} is *not* an exponent: $f^{-1}(x)$ does not mean $1/f(x)$.

$x \rightarrow f \rightarrow f(x) \rightarrow f^{-1} \rightarrow x = f^{-1}(f(x))$

$y \rightarrow f^{-1} \rightarrow f^{-1}(y) \rightarrow f \rightarrow y = f(f^{-1}(y))$

7.2 If $y = f(x)$ is a one-to-one function, then $f^{-1}(f(x)) = x$ and $f(f^{-1}(y)) = y$. Each of the composites $f^{-1} \circ f$ and $f \circ f^{-1}$ is the identity function on its domain.

As you can see in Fig. 7.2, the result of composing f and f^{-1} in either order is the **identity function,** the function that assigns each number to itself. This gives us a way to test whether two functions f and g are inverses of one another. Compute $f \circ g$ and $g \circ f$. If both composites are identity functions, then f and g are inverses of one another; otherwise they are not. If f squares every number in its domain, g had better take square roots or it isn't the inverse of f.

What Functions Have Inverses?

Increasing functions have inverses and decreasing functions have inverses, as you will be asked to show in Exercise 20. In fact, these are the only continuous functions to have inverses, although we shall not prove this.

The Graphs of Inverses

How is the graph of the inverse of a function related to the graph of the function? If the function is increasing, say, its graph rises from left to right like the graph in Fig. 7.3a. To read the graph, we start at the point x on the

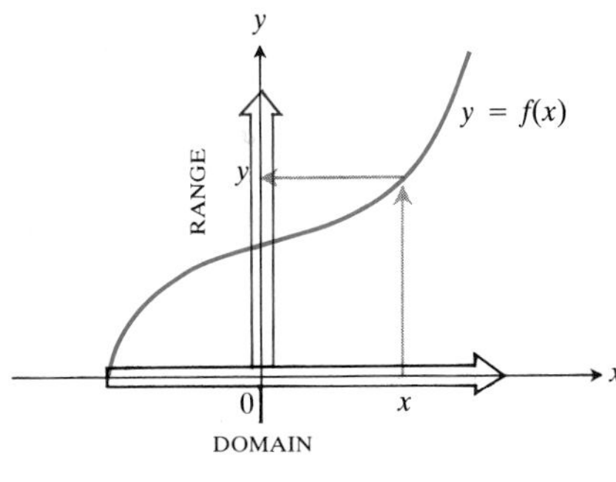

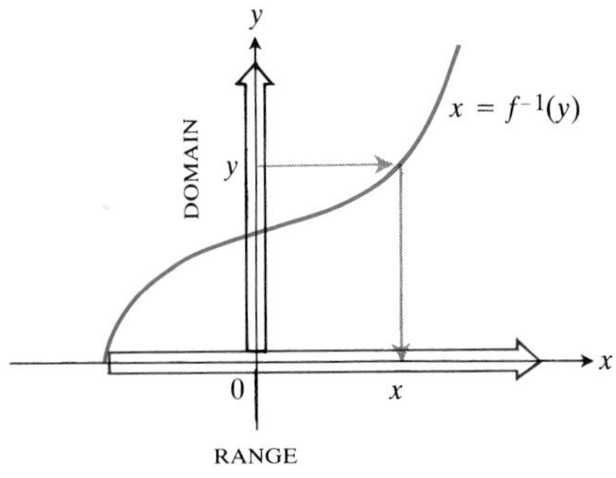

7.3 (a) To find the value of f at x, we go up to the curve and over to the y-axis. (b) The graph of f is also the graph of f^{-1}. To find the y that gives x, we go over to the curve and down to the x-axis. Notice that the domain of f^{-1} is the range of f and the range of f^{-1} is the domain of f.

x-axis, go up to the graph and then over to the y-axis to read the value of y. If we start with y and want to find the x from which it came, we reverse the process (Fig. 7.3b).

The graph of f is already the graph of f^{-1}, although it is not drawn in the normal way with the domain axis horizontal and the range axis vertical. To draw the graph of f^{-1} the way we are used to seeing graphs, we have to reconstruct it from the graph of f in the following way. We rotate the graph of f (Fig. 7.4a) counterclockwise to make the y-axis horizontal and the x-axis vertical (Fig. 7.4b). Then we reflect the graph across the vertical axis as if the axis were a mirror (Fig. 7.4c) to make the y-axis point to the right. Finally, we write x for y and y for x (Fig. 7.4d). We now have a normal-looking graph of f^{-1} as a function of x.

A Quicker Way We can now see a quicker way to graph f^{-1}: What we did in going from (a) to (d) in Fig. 7.4 amounted to solving for x in terms of y and then switching the letters x and y. This has the same effect as reflecting the graph of $y = f(x)$ across the line $y = x$. It also tells us exactly how to express f^{-1} as a function of x.

> **To Express f^{-1} as a Function of x**
>
> 1. Solve the equation $y = f(x)$ for x in terms of y.
> 2. Switch x and y. The resulting formula will be $y = f^{-1}(x)$.

EXAMPLE 1 Find the inverse of $y = \frac{1}{2}x + 1$, expressed as a function of x.

Solution

STEP 1: Solve for x in terms of y:

$$y = \frac{1}{2}x + 1$$

$$2y = x + 2$$

$$x = 2y - 2$$

STEP 2: Switch x and y: $y = 2x - 2$

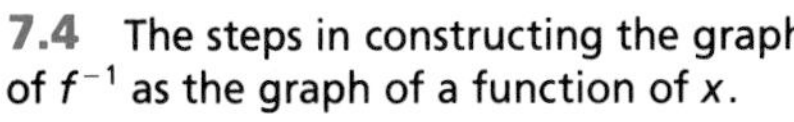
7.4 The steps in constructing the graph of f^{-1} as the graph of a function of x.

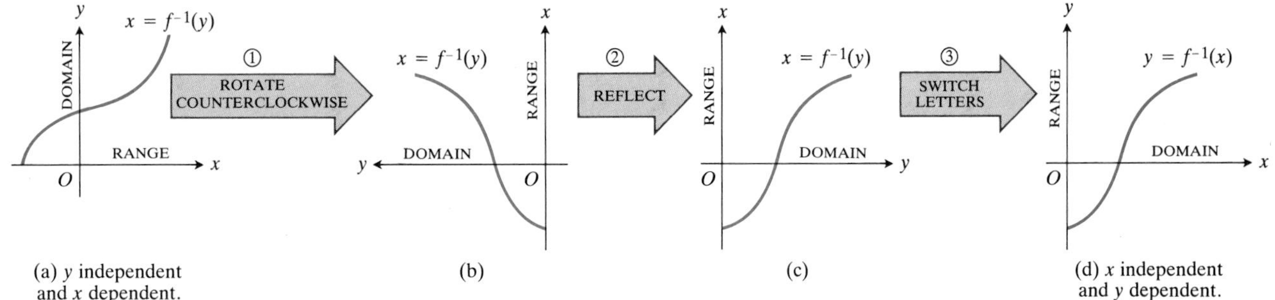

(a) y independent and x dependent.

(b)

(c)

(d) x independent and y dependent.

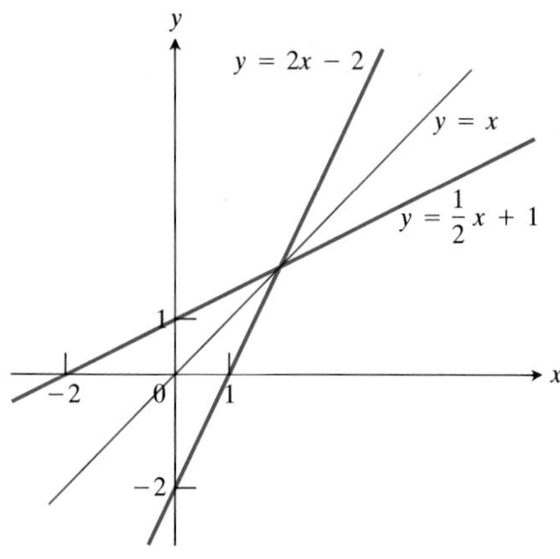

7.5 Graphing the functions $y = (1/2)x + 1$ and $y = 2x - 2$ together shows the graphs' symmetry with respect to the line $y = x$. The graph of a function and its inverse always have this symmetry.

The inverse of the function $f(x) = (1/2)x + 1$ is the function $f^{-1}(x) = 2x - 2$. To check, we verify that both composites give the identity function:

$$f^{-1}(f(x)) = 2\left(\frac{1}{2}x + 1\right) - 2 = x + 2 - 2 = x,$$

$$f(f^{-1}(x)) = \frac{1}{2}(2x - 2) + 1 = x - 1 + 1 = x.$$

See Fig. 7.5. ■

EXAMPLE 2 Find the inverse of the function $y = x^2, x \geq 0$, expressed as a function of x.

Solution

STEP 1: Solve for x in terms of y: $y = x^2$

$$\sqrt{y} = \sqrt{x^2} = x \quad (\sqrt{x^2} = x \text{ because } x \geq 0)$$

STEP 2: Switch x and y: $y = \sqrt{x}$

The inverse of the function $y = x^2$, $x \geq 0$ is the function $y = \sqrt{x}$. See Fig. 7.6.

Notice that, unlike the restricted function $y = x^2, x \geq 0$, the unrestricted function $y = x^2$ is not one-to-one and therefore has no inverse. ■

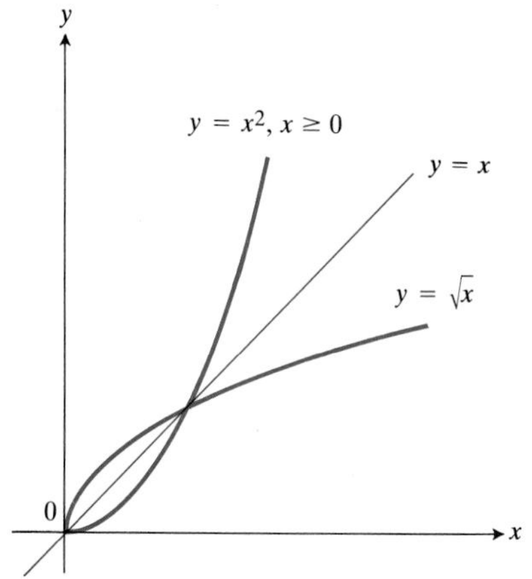

7.6 The functions $y = \sqrt{x}$ and $y = x^2$, $x \geq 0$, are inverses of one another. Again, we see the symmetry with respect to the line $y = x$. The graphs are mirror images of one another across the line.

Derivatives of Inverses of Differentiable Functions

The relation between the graphs of f and f^{-1} suggests an important relationship between their derivatives:

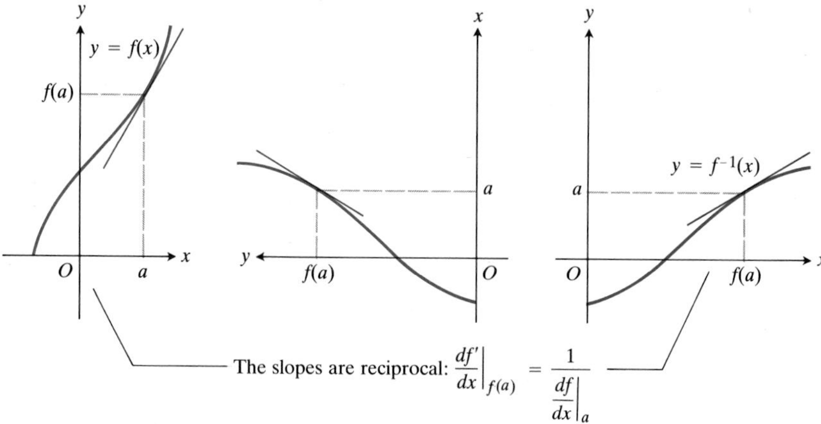

If $f'(a) \neq 0$, then f^{-1} has a derivative at the point $f(a)$ and its value there is $1/f'(a)$. Naturally, we have to impose some mathematical conditions on f to make sure this conclusion holds. The usual conditions, from advanced calculus, are these.

THEOREM 1

The Derivative Rule for Inverses

If f is differentiable at every point of an interval I, and df/dx is never zero on I, then f^{-1} is differentiable at every interior point of the interval $f(I)$. The value of df^{-1}/dx at any particular image point $f(a)$ is the reciprocal of the value of df/dx at a.

$$\left.\frac{df^{-1}}{dx}\right|_{x=f(a)} = 1\Big/\left(\left.\frac{df}{dx}\right|_{x=a}\right) \tag{1}$$

EXAMPLE 3 Verify Eq. (1) for $f(x) = \frac{1}{2}x + 1$ and its inverse $f^{-1}(x) = 2x - 2$.

Solution We have

$$\frac{d}{dx}\left(f^{-1}(x)\right) = \frac{d}{dx}(2x - 2) = 2, \qquad \frac{d}{dx}(f(x)) = \frac{d}{dx}\left(\frac{1}{2}x + 1\right) = \frac{1}{2}.$$

Therefore, regardless of the values of a and $f(a)$, we have

$$\frac{df^{-1}}{dx} = \frac{1}{\dfrac{df}{dx}} = \frac{1}{1/2} = 2.$$

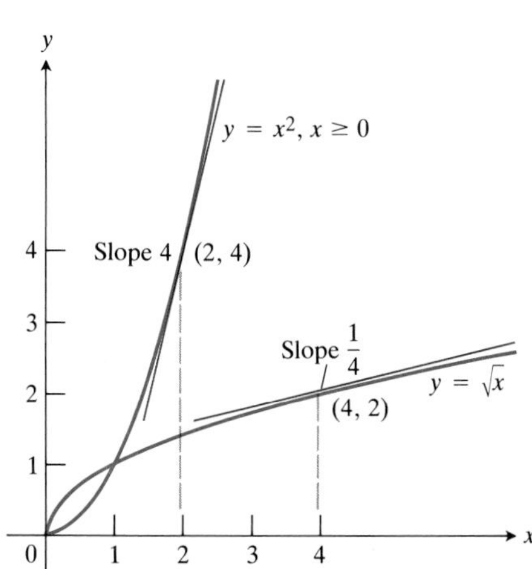

7.7 The graphs of inverse functions have reciprocal slopes at corresponding points.

EXAMPLE 4 Verify Eq. (1) for $f(x) = x^2$, $x \geq 0$, and its inverse $f^{-1}(x) = \sqrt{x}$ at the point $f(2) = 2^2 = 4$ in the domain of f^{-1} (Fig. 7.7).

Solution

$$\left.\frac{df^{-1}}{dx}\right|_{x=f(2)} = \left.\frac{d}{dx}\left(\sqrt{x}\right)\right|_{x=4} = \left.\left(\frac{1}{2\sqrt{x}}\right)\right|_{x=4} = \frac{1}{4}$$

$$\left.\frac{df}{dx}\right|_{x=2} = \left.\frac{d}{dx}(x^2)\right|_{x=2} = \left.(2x)\right|_{x=2} = 4.$$

Thus,

$$\left.\frac{df^{-1}}{dx}\right|_{x=f(2)} = \frac{1}{\left.\dfrac{df}{dx}\right|_{x=2}}.$$

The Function e^x Is Differentiable

The derivation of the formula

$$\frac{d}{dx}e^x = e^x \tag{2}$$

in Section 5.8 assumed e^x to be differentiable and did not prove the differentiability itself. You can find where the assumption was made if you look at the argument again. In the light of Theorem 1, we can now see why e^x is differentiable. It is the inverse of the function $y = \ln x$, a function whose derivative $d(\ln x)/dx = 1/x$ is never zero. It therefore has a derivative at every interior point of its own domain, the real line.

Exercises 7.1

In Exercises 1–4:

a) Find the inverse f^{-1} of the function f, expressed as a function of x.

b) Graph f and f^{-1} together.

c) Verify Eq. (1) by evaluating df/dx at $x = a$ and df^{-1}/dx at $x = f(a)$.

1. $f(x) = 2x + 3, \quad a = -1$

2. $f(x) = 5 - 4x, \quad a = 1/2$

3. $f(x) = (1/5)x + 7, \quad a = -1$

4. $f(x) = 2x^2, \ x \geq 0, \quad a = 5$

In Exercises 5–8, find $f^{-1}(x)$.

5. $f(x) = x^2 + 1, \quad x \geq 0$

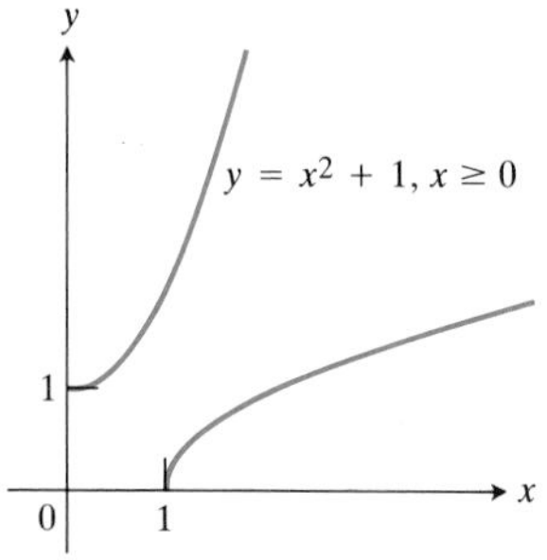

6. $f(x) = x^2, \quad x \leq 0$

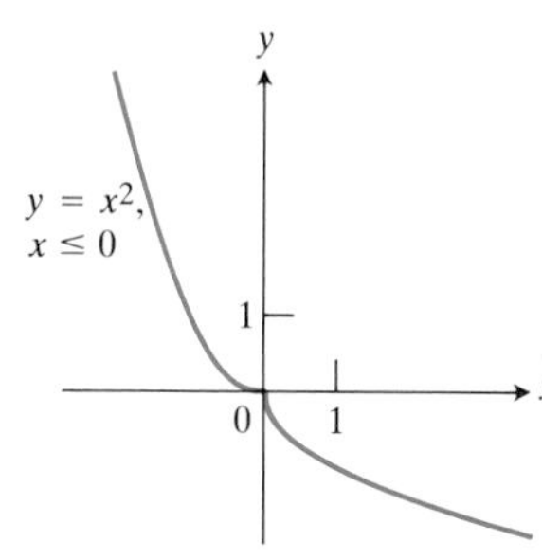

7. $f(x) = x^3 - 1$

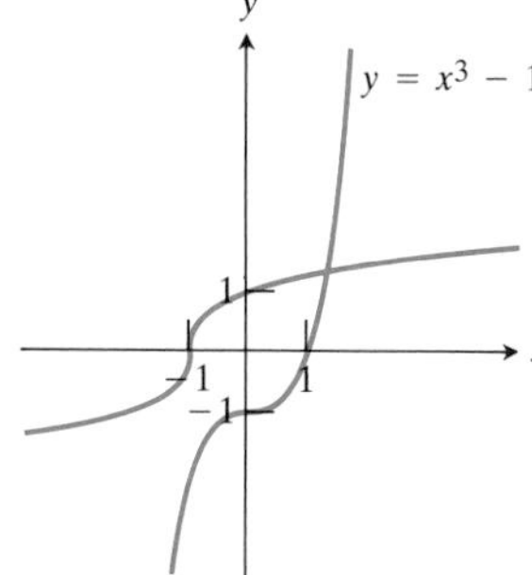

8. $f(x) = x^2 - 2x + 1, \quad x \geq 1$

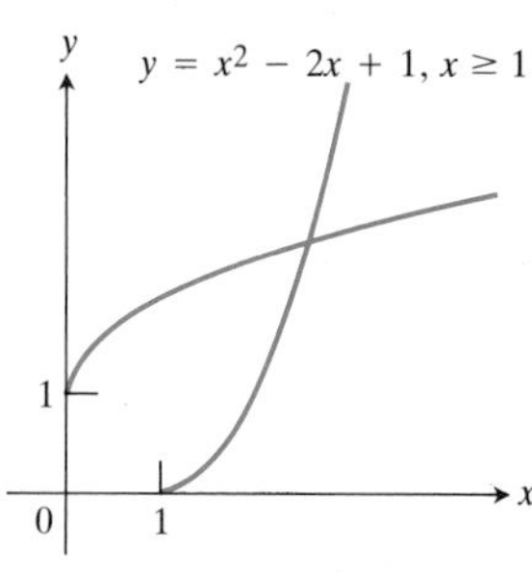

In Exercises 9–16, find $f^{-1}(x)$ and show that $f(f^{-1}(x)) = f^{-1}(f(x)) = x$.

9. $f(x) = x^5$

10. $f(x) = x^4, \quad x \geq 0$

11. $f(x) = x^3 + 1$

12. $f(x) = (1/2)x - 7/2$

13. $f(x) = 1/x^2, \quad x > 0$

14. $f(x) = 1/x^3, \quad x \neq 0$

15. $f(x) = (x + 1)^2, \quad x \geq 1$

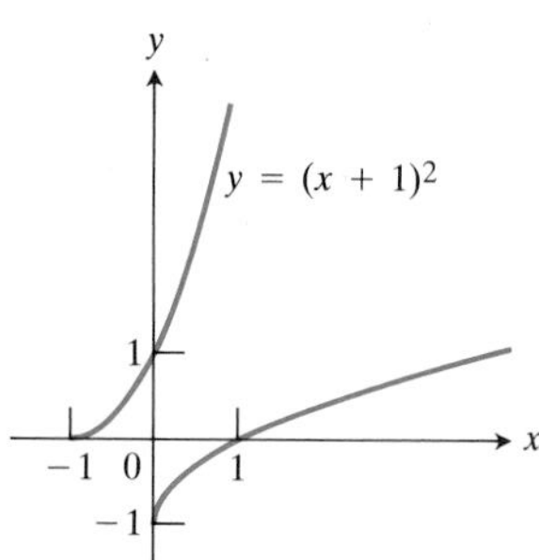

16. $f(x) = x^{2/3}, \quad x \geq 0$

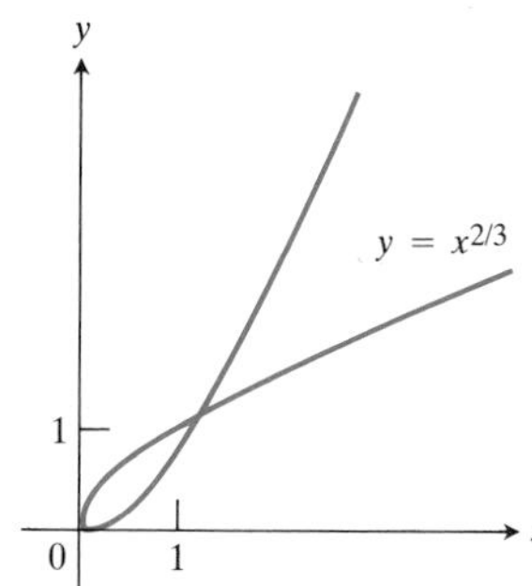

17. Graph $y = x^3$ and $y = x^{1/3}$ together over the interval $-2 \leq x \leq 2$ and sketch the tangents at $(1, 1)$ and $(-1, -1)$. What lines are tangent to the curves at $x = 0$?

18. Graph the curve $y = 1/x, x > 0$ and notice its symmetry about the line $y = x$. Find the inverse of the function $f(x) = 1/x$.

19. One of the virtues of Eq. (1) is that it enables us to find values of df^{-1}/dx even when we do not have an explicit formula for the derivative. As a case in point, let $f(x) = x^2 - 4x - 3, \ x > 2$ and find the value of df^{-1}/dx at the point $x = -3 = f(4)$.

20. *Increasing functions and decreasing functions.* As we saw in Section 4.2, a function $f(x)$ increases on its domain if for any two points x_1 and x_2 in the domain,

$$x_2 > x_1 \quad \Rightarrow \quad f(x_2) > f(x_1).$$

Similarly, a function decreases on its domain if for any two points x_1 and x_2 in the domain,

$$x_2 > x_1 \quad \Rightarrow \quad f(x_2) < f(x_1).$$

Show that increasing functions and decreasing functions are one-to-one. That is, show that $x_2 \neq x_1$ always implies $f(x_2) \neq f(x_1)$.

TOOLKIT PROGRAM

Picard's Fixed Point Method

In addition to being a powerful equation solver, this program enables you to toggle back and forth between the graphs of a function and its inverse.

7.2 ln x, e^x, and Logarithmic Differentiation

In this section, we reexamine our first function–inverse pair, $\ln x$ and e^x, and substantiate the claims we made in Section 5.8. In the process, we also show how to evaluate $\int(1/u)\,du$ when u is negative and describe a new technique, called logarithmic differentiation, for simplifying the work of differentiating products, quotients, and powers.

In the late 1500s, a Scottish baron, John Napier, invented a device called the **logarithm** that simplified arithmetic by replacing multiplication by addition. The equation that accomplished this was

$$\ln ax = \ln a + \ln x.$$

To multiply two positive numbers a and x, you looked up their logarithms in a table, added the logarithms, found the sum in the body of the table, and read the table backward to find the product ax.

Having the table was the key, of course, and Napier spent the last 20 years of his life working on a table he never finished (while the astronomer Tycho Brahe waited in vain for the information he needed to speed his calculations). The table was completed after Napier's death (and Brahe's) by Napier's friend Henry Briggs in London. Logarithms subsequently became widely known as Briggs's logarithms (what else?) and some books on navigation still refer to them by this name.

Napier also invented an artillery piece that could hit a cow a mile away. Horrified by the weapon's accuracy, he stopped production and suppressed the cannon's design.

The Natural Logarithm Function

The natural logarithm of a positive number x is the number

$$\ln x = \int_1^x \frac{1}{t}\,dt, \quad x > 0. \tag{1}$$

The logarithm of x is positive if $x > 1$, zero when $x = 1$, and negative when $0 < x < 1$.

The function $\ln x$ is differentiable and

$$\frac{d}{dx}\ln x = \frac{1}{x}. \tag{2}$$

Being differentiable, the function is also continuous.

The Chain Rule extends Eq. (2) to give

$$\frac{d}{dx}\ln u = \frac{1}{u}\frac{du}{dx} \tag{3}$$

for any positive differentiable function $u(x)$.

EXAMPLE 1

$$\frac{d}{dx}\ln(3x^2+4) = \frac{1}{3x^2+4}\frac{d}{dx}(3x^2+4) = \frac{6x}{3x^2+4} \qquad ■$$

Logarithms became famous because of the improvements they brought to arithmetic, improvements made possible by the rules presented in Table 7.1. Nowadays we do our complicated arithmetic with calculators, but the

TABLE 7.1
Rules of arithmetic for logarithms

For any positive numbers a and x and for any exponent n,

1. $\ln ax = \ln a + \ln x$	2. $\ln \frac{a}{x} = \ln a - \ln x$
3. $\ln \frac{1}{x} = -\ln x$ $\begin{pmatrix}\text{Rule 2 with}\\ a = 1\end{pmatrix}$	4. $\ln x^n = n \ln x$ (The *Exponent Rule*)

rules themselves are as important as ever, albeit for other reasons. For the moment, there is the added restriction that the exponent n in Rule 4 be a rational number. You will see why when we derive the rule.

You saw how the rules worked in Section 5.8. One useful consequence of Rule 4 is that $\ln \sqrt[m]{x} = \ln x^{1/m} = (1/m) \ln x$:

$$\ln \sqrt[m]{x} = \frac{1}{m} \ln x \tag{4}$$

EXAMPLE 2

$$\ln \sqrt{\cos x} = \frac{1}{2} \ln \cos x \qquad \ln \sqrt[3]{x+1} = \frac{1}{3} \ln(x+1)$$ ■

The derivations of the rules in Table 7.1 go like this:

PROOF that ln *ax* = ln *a* + ln *x* The argument is unusual—and elegant. It starts by showing that $\ln ax$ and $\ln x$ have the same derivative. According to Corollary 3 of the Mean Value Theorem, then, the functions must differ by a constant, which means that

$$\ln ax = \ln x + C$$

for some C. With this much accomplished, it remains only to show that C equals $\ln a$.

The equality of the derivatives of $\ln ax$ and $\ln x$ comes from the fact that

$$\frac{d}{dx} \ln ax = \frac{1}{ax} \frac{d}{dx}(ax) = \frac{1}{ax} \cdot a = \frac{1}{x}, \qquad \text{(Eq. (3) with } u = ax) \tag{5}$$

which is the derivative of $\ln x$. Therefore,

$$\ln ax = \ln x + C \tag{6}$$

for some constant C. Equation (6) holds for all positive values of x, so it must hold for $x = 1$. Hence,

$$\begin{aligned} \ln(a \cdot 1) &= \ln 1 + C \\ \ln a &= 0 + C && (\ln 1 = 0) \\ C &= \ln a. && \text{(Rearranged)} \end{aligned}$$

Substituting $C = \ln a$ in Eq. (6) gives the equation we wanted to prove:

$$\ln ax = \ln a + \ln x \tag{7}$$ ■

PROOF that $\ln \frac{a}{x} = \ln a - \ln x$ We get this from Eq. (7) in two stages. Equation (7) with x replaced by $1/a$ gives

$$\ln a + \ln \frac{1}{a} = \ln a \cdot \frac{1}{a} = \ln 1 = 0,$$

so that

$$\ln \frac{1}{a} = -\ln a. \tag{8}$$

Equation (7) with a replaced by $1/a$ then gives

$$\ln\frac{x}{a} = \ln\left(x\cdot\frac{1}{a}\right) = \ln x + \ln\frac{1}{a} = \ln x - \ln a. \qquad \blacksquare$$

PROOF that $\ln x^n = n \ln x$ (assuming n rational) We use the same-derivative argument again. For all positive values of x,

$$\begin{aligned} \frac{d}{dx}\ln x^n &= \frac{1}{x^n}\frac{d}{dx}(x^n) && \text{(Eq. (3) with } u = x^n) \\ &= \frac{1}{x^n}nx^{n-1} && \left(\begin{array}{l}\text{Here is where we need } n \text{ to be rational, at} \\ \text{least for now. We have proved the power} \\ \text{rule only for rational exponents.}\end{array}\right) \\ &= n\cdot\frac{1}{x} = \frac{d}{dx}(n\ln x). \end{aligned}$$

Since $\ln x^n$ and $n \ln x$ have the same derivative,

$$\ln x^n = n \ln x + C$$

for some constant C. Taking x to be 1 identifies C as zero, and we're done. $\blacksquare$

As for using the rule $\ln x^n = n \ln x$ for irrational values of n, go right ahead and do so. It does hold for all n, and there is no need to pretend otherwise. From the point of view of mathematical development, however, we want you to be aware that the rule is far from proved. The situation is much worse than not knowing how to take the derivative of x^n when n is irrational. We do not even know how to raise a number like 2 to an irrational power like $\sqrt{3}$. Even your calculator cannot do so. Calculators work only with rational numbers and only with a finite assortment of them at that. There is no room in a calculator for an infinite decimal.

There is hope, however, for besides 0 and 1 there is one other number we *can* raise to irrational powers, and that is the number e. The function e^x is defined for all values of x, irrational as well as rational. This, as we shall see in Section 7.3, will save the day.

The Graph of ln x

The derivative $d(\ln x)/dx = 1/x$ is positive and continuous at every point in the function's domain, so $\ln x$ is an increasing function of x. The graph is connected and rises steadily from left to right with a continuously turning tangent. The second derivative

$$\frac{d^2(\ln x)}{dx^2} = \frac{d}{dx}\left(\frac{1}{x}\right) = -\frac{1}{x^2}$$

is always negative, so the graph is concave down throughout.

We know that ln 1 is zero and we can estimate ln 2 numerically to show that

$$\ln 2 = \int_1^2 \frac{1}{t}\,dt \approx 0.69.$$

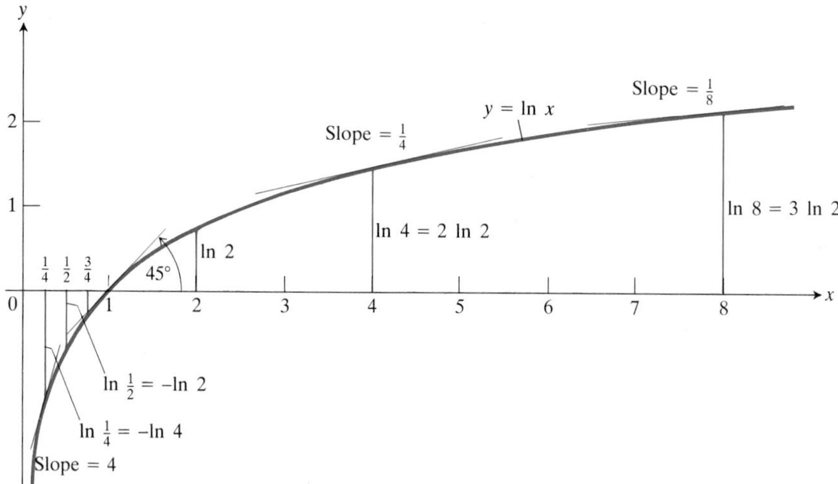

7.8 The graph of $y = \ln x$.

The arithmetic rules in Table 7.1 then give

$$\ln 4 = \ln 2^2 = 2 \ln 2 \approx 1.38 \qquad \ln 8 = \ln 2^3 = 3 \ln 2 \approx 2.07$$

$$\ln \frac{1}{2} = \ln 2^{-1} = -\ln 2 \approx -0.69 \qquad \ln \frac{1}{4} = \ln 4^{-1} = -\ln 4 \approx -1.38$$

and so on. We plot the points and connect them with a smooth curve with the guidance of tangent lines. The results are shown in Fig. 7.8.

How high does the graph go and how low? Since $\ln 2 > (1/2)$,

$$\ln 2^n = n \ln 2 > n\left(\frac{1}{2}\right) = \frac{n}{2} \qquad \ln 2^{-n} = -n \ln 2 < -n\left(\frac{1}{2}\right) = -\frac{n}{2}.$$

The logarithm of x tends to ∞ as x tends to ∞ and tends to $-\infty$ as x approaches zero from above. In short,

$$\lim_{x\to\infty} \ln x = \infty \quad \text{and} \quad \lim_{x\to 0^+} \ln x = -\infty. \tag{9}$$

The domain of $\ln x$ is the set of positive real numbers. The range of $\ln x$ is the entire real line.

Logarithmic Differentiation

The derivatives of functions given by formulas that involve products, quotients, and powers can often be found quickly if we take the natural logarithm of both sides before differentiating. This enables us to use the rules in Table 7.1 to simplify the formulas before we go to work. The process, called **logarithmic differentiation,** is illustrated in the next example.

EXAMPLE 3 Find dy/dx if

$$y = \frac{(x^2 + 1)(x + 3)^{1/2}}{x - 1}, \quad x > 1.$$

Logarithmic differentiation has four steps:

1. Take logs of both sides and simplify.
2. Differentiate implicitly.
3. Solve for dy/dx.
4. Substitute for y.

Solution We take the natural logarithm of both sides and simplify the result with the rules in Table 7.1:

$$\begin{aligned} \ln y &= \ln\left(\frac{(x^2+1)(x+3)^{1/2}}{x-1}\right) \\ &= \ln(x^2+1)(x+3)^{1/2} - \ln(x-1) && \text{(Rule 2)} \\ &= \ln(x^2+1) + \ln(x+3)^{1/2} - \ln(x-1) && \text{(Rule 1)} \\ &= \ln(x^2+1) + \frac{1}{2}\ln(x+3) - \ln(x-1). && \text{(Rule 4)} \end{aligned}$$

We then take derivatives of both sides with respect to x, using implicit differentiation on the left:

$$\frac{1}{y}\frac{dy}{dx} = \frac{1}{x^2+1}\cdot 2x + \frac{1}{2}\cdot\frac{1}{x+3} - \frac{1}{x-1}.$$

Next, we solve for dy/dx:

$$\frac{dy}{dx} = y\left(\frac{2x}{x^2+1} + \frac{1}{2x+6} - \frac{1}{x-1}\right).$$

Finally, we substitute for y:

$$\frac{dy}{dx} = \frac{(x^2+1)(x+3)^{1/2}}{x-1}\left(\frac{2x}{x^2+1} + \frac{1}{2x+6} - \frac{1}{x-1}\right).$$ ■

The Integral $\int \frac{1}{u}\,du$

If u is a positive differentiable function of x, we know from Section 5.8 that

$$\int \frac{1}{u}\,du = \ln u + C. \tag{10}$$

The expression on the right could also be written as $\ln|u| + C$ because $u = |u|$ if u is positive.

If u is a negative differentiable function of x, then $-u$ is a positive differentiable function of x and Eq. (10) applies to give

$$\int \frac{1}{u}\,du = \int \frac{1}{-u}\,d(-u) = \ln(-u) + C. \tag{11}$$

The expression on the right could also be written as $\ln|u| + C$ because $-u = |u|$ if u is negative. Whether u is positive or u is negative, the integral of $(1/u)\,du$ is $\ln|u| + C$. This removes the restriction we had in Section 5.8, that u be positive.

$$\int \frac{1}{u}\,du = \ln|u| + C \tag{12}$$

EXAMPLE 4

$$\int_0^2 \frac{2x}{x^2-5}\,dx = \int_{-5}^{-1} \frac{du}{u} = \ln|u|\Big]_{-5}^{-1} \qquad \begin{pmatrix}\text{Substitute } u = x^2-5,\ du = 2x\,dx, \\ u(0) = -5,\ u(2) = -1 \text{ in Eq. (12)}\end{pmatrix}$$

$$= \ln|-1| - \ln|-5| = \ln 1 - \ln 5 = -\ln 5$$

■

The Integrals of tan *x* and cot *x*

Equation (12) tells us at last how to integrate the tangent and cotangent functions. For the tangent,

$$\int \tan x\,dx = \int \frac{\sin x}{\cos x}\,dx = \int \frac{-du}{u} \qquad \begin{pmatrix}\text{Substitute } u = \cos x, \\ du = -\sin x\,dx\end{pmatrix}$$

$$= -\int \frac{du}{u} = -\ln|u| + C \qquad \text{(Eq. (12))}$$

$$= -\ln|\cos x| + C = \ln\frac{1}{|\cos x|} + C \qquad \text{(Table 7.1, Rule 3)}$$

$$= \ln|\sec x| + C.$$

For the cotangent,

$$\int \cot x\,dx = \int \frac{\cos x\,dx}{\sin x} = \int \frac{du}{u} \qquad \begin{pmatrix}\text{Substitute } u = \sin x, \\ du = \cos x\,dx\end{pmatrix}$$

$$= \ln|u| + C = \ln|\sin x| + C. \qquad \text{(Eq. (12))}$$

The general formulas are

$$\int \tan u\,du = -\ln|\cos u| + C = \ln|\sec u| + C \qquad (13)$$

$$\int \cot u\,du = \ln|\sin u| + C \qquad (14)$$

EXAMPLE 5

$$\int_0^{\pi/6} \tan 2x\,dx = \int_0^{\pi/3} \tan u \cdot \frac{du}{2} = \frac{1}{2}\int_0^{\pi/3} \tan u\,du \qquad \begin{pmatrix}\text{Substitute } u = 2x, \\ dx = du/2,\ u(0) = 0, \\ u(\pi/6) = \pi/3\end{pmatrix}$$

$$= \frac{1}{2}\ln|\sec u|\Big]_0^{\pi/3} = \frac{1}{2}(\ln 2 - \ln 1) = \frac{1}{2}\ln 2 \qquad \text{(Eq. (13))}$$

■

The Exponential Function exp(*x*) = e^x

The natural logarithm function, being an increasing function of x with domain the set of positive real numbers and range the entire real line, has an inverse whose domain is the entire real line and whose range is the set of positive real numbers. We call the inverse **exp(*x*)**, the exponential function of x.

$$\exp(x) = \ln^{-1} x. \qquad (15)$$

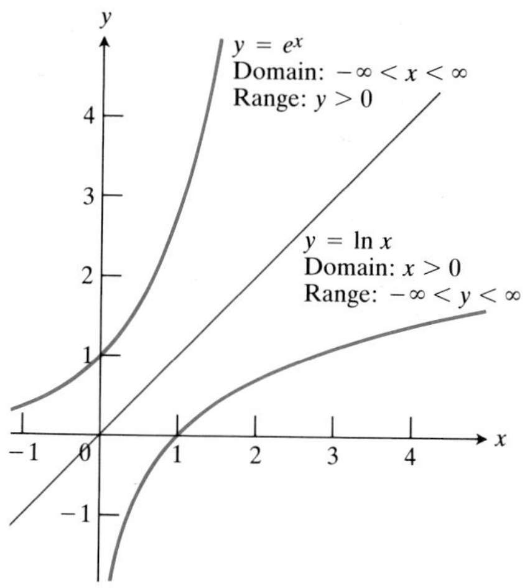

7.9 The graphs of $y = e^x$ and $y = \ln x$. Notice the symmetry.

Since ln 2 is less than 1 and ln 4 is greater than 1, the Intermediate Value Theorem for Continuous Functions assures us of at least one number between 2 and 4 whose logarithm is exactly 1. In fact, there is only one such number because $\ln x$ is one-to-one. We call that number e. To fifteen decimal places,

$$e = 2.7\ 1828\ 1828\ 45\ 90\ 45.$$

As we found in Section 5.8, $e^x = \exp(x)$ when x is rational, so we defined e^x to be $\exp(x)$ for irrational values of x as well.

$$\text{For all real values of } x, \quad e^x = \exp(x) = \ln^{-1}(x). \tag{16}$$

Because e^x is the inverse of $\ln x$, its graph can be obtained by reflecting the graph of $\ln x$ across the line $y = x$ (Fig. 7.9). As we can see,

$$\lim_{x\to\infty} e^x = \infty \quad \text{and} \quad \lim_{x\to-\infty} e^x = 0. \tag{17}$$

Laws of Exponents

Even though e^x is defined in a seemingly roundabout way as $\ln^{-1}x$, it obeys the familiar laws of exponents from algebra.

Laws of Exponents

For all real numbers x, x_1, and x_2,

1. $e^{x_1} \cdot e^{x_2} = e^{x_1 + x_2}$
2. $e^{-x} = \dfrac{1}{e^x}$
3. $\dfrac{e^{x_1}}{e^{x_2}} = e^{x_1 - x_2}$

PROOF of Law 1 Let

$$y_1 = e^{x_1} \quad \text{and} \quad y_2 = e^{x_2}. \tag{18}$$

Then,

$$\begin{aligned}
x_1 = \ln y_1 \quad \text{and} \quad x_2 &= \ln y_2 && \text{(Take logs of both sides of Eqs. 18)}\\
x_1 + x_2 &= \ln y_1 + \ln y_2 = \ln y_1 y_2 && \text{(Table 7.1, Rule 1)}\\
e^{x_1+x_2} &= e^{\ln y_1 y_2} && \text{(Exponentiate)}\\
&= y_1 y_2 && (e^{\ln u} = u)\\
&= e^{x_1} e^{x_2}.
\end{aligned}$$

■

Laws 2 and 3 follow from Law 1, as you will be asked to show in Exercises 51 and 52.

EXAMPLE 6

a) $e^{x+\ln 2} = e^x \cdot e^{\ln 2} = 2e^x$ (Law 1)

b) $e^{-\ln x} = \dfrac{1}{e^{\ln x}} = \dfrac{1}{x}$ (Law 2)

c) $\dfrac{e^{2x}}{e} = e^{2x-1}$ (Law 3)

■

Exercises 7.2

Use logarithmic differentiation to find dy/dx in Exercises 1–12. In each case, assume y to be positive.

1. $y = \sqrt{x(x+1)}$
2. $y = \sqrt{\dfrac{x}{x+1}}$
3. $y = \sqrt{x+3}\,\sin x$
4. $y = \dfrac{\tan x}{\sqrt{2x+1}}$
5. $y = x(x+1)(x+2)$
6. $y = \dfrac{1}{x(x+1)(x+2)}$
7. $y = \dfrac{x+5}{x\cos x}$
8. $y = \dfrac{x\sin x}{\sqrt{\sec x}}$
9. $y = \dfrac{x\sqrt{x^2+1}}{(x+1)^{2/3}}$
10. $y = \sqrt{\dfrac{(x+1)^{10}}{(2x+1)^5}}$
11. $y = \sqrt[3]{\dfrac{x(x-2)}{x^2+1}}$
12. $y = \sqrt[3]{\dfrac{x(x+1)(x-2)}{(x^2+1)(2x+3)}}$

Evaluate the integrals in Exercises 13–30.

13. $\displaystyle\int_{-3}^{-2} \frac{dx}{x}$
14. $\displaystyle\int_{-9}^{-4} \frac{dx}{2x}$
15. $\displaystyle\int_{-1}^{0} \frac{3\,dx}{3x-2}$
16. $\displaystyle\int_{-1}^{0} \frac{dx}{2x+3}$
17. $\displaystyle\int_{3}^{4} \frac{dx}{x-5}$
18. $\displaystyle\int_{2}^{5} \frac{dx}{1-x}$
19. $\displaystyle\int_{0}^{3} \frac{2x\,dx}{x^2-25}$
20. $\displaystyle\int_{0}^{1} \frac{8x\,dx}{4x^2-5}$
21. $\displaystyle\int_{0}^{\pi} \frac{\sin x}{2-\cos x}\,dx$ (Let $u = 2 - \cos x$.)
22. $\displaystyle\int_{0}^{\pi/3} \frac{4\sin x}{1-4\cos x}\,dx$
23. $\displaystyle\int_{1}^{2} \frac{2\ln x}{x}\,dx$ (Let $u = \ln x$.)
24. $\displaystyle\int_{e}^{e^2} \frac{1}{x\ln x}\,dx$
25. $\displaystyle\int_{-1}^{1} \frac{e^x}{1+e^x}\,dx$ (Let $u = e^x + 1$.)
26. $\displaystyle\int_{\ln(\pi/6)}^{\ln(\pi/2)} 2e^x\cos(e^x)\,dx$ (Let $u = e^x$.)
27. $\displaystyle\int_{0}^{\pi/2} \tan\frac{x}{2}\,dx$
28. $\displaystyle\int_{-\pi/2}^{-\pi/4} \cot x\,dx$
29. $\displaystyle\int_{\pi/2}^{\pi} 2\cot\frac{x}{3}\,dx$
30. $\displaystyle\int_{0}^{\pi/12} 6\tan 3x\,dx$

Find the limits in Exercises 31–36.

31. $\displaystyle\lim_{x\to\infty} \ln\frac{1}{x}$
32. $\displaystyle\lim_{x\to 0^+} \ln\frac{1}{x}$
33. $\displaystyle\lim_{x\to\infty} e^{-x}$
34. $\displaystyle\lim_{x\to -\infty} e^{-x}$
35. $\displaystyle\lim_{x\to -\infty} \ln(2+e^x)$
36. $\displaystyle\lim_{x\to\infty} \int_{x}^{2x} \frac{1}{t}\,dt$

37. Find the area between the curve $y = \tan x$ and the x-axis from $x = 0$ to $x = \pi/3$.

38. The region between the curve $y = \sqrt{\cot x}$ and the x-axis from $x = \pi/6$ to $x = \pi/2$ is revolved about the x-axis to generate a solid. Find the volume of the solid.

39. Find the linearization of $f(x) = e^x$ at $x = 0$.

40. Find the linearization of $f(x) = x + e^{4x}$ at $x = 0$.

41. *The linearization of* $\ln(1+x)$. Instead of linearizing $\ln x$ at $x = 1$ we linearize $\ln(1+x)$ at $x = 0$. We get a simpler expression that way, namely $L(x) = x$. Derive this formula.

42. *Calculating values of* $\ln x$ *with Simpson's rule*. Although linearizations are good for replacing the logarithm function over short intervals, when it comes to estimating *particular* values of $\ln x$, Simpson's rule is better.

    As a case in point, the values of $\ln(1.2)$ and $\ln(0.8)$ to five places are
    $$\ln(1.2) = 0.18232, \quad \ln(0.8) = -0.22314.$$
    Estimate $\ln(1.2)$ and $\ln(0.8)$ first with the formula
    $$\ln(1+x) \approx x$$
    and then with Simpson's rule using $n = 2$. Impressive, no?

43. Find the maximum value of $f(x) = x^2\ln(1/x)$.

44. Find the maximum and minimum values of the periodic function $f(x) = e^{\sin x}$.

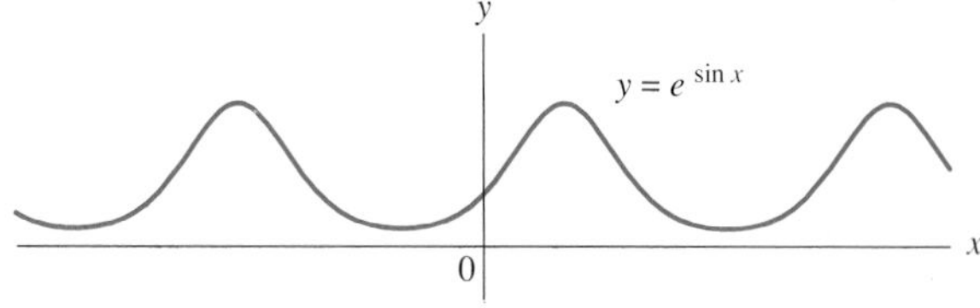

Solve the initial value problems in Exercises 45–46.

45. Differential equation: $\frac{dy}{dx} = \cos x\, e^{\sin x}$

Initial condition: $y = 0$ when $x = 0$

46. Differential equation: $\frac{dy}{dx} = 1 + \frac{1}{x}$

Initial condition: $y = 3$ when $x = 1$

47. A body moves along a coordinate line with acceleration $d^2s/dt^2 = 4/(4-t)^2$. When $t = 0$, the body's velocity is 2 m/sec. Find the total distance traveled by the body from time $t = 1$ sec to time $t = 2$ sec.

48. Show that, for any number $a > 1$,

$$\int_1^a \ln x\, dx + \int_0^{\ln a} e^y\, dy = a \ln a.$$

(*Hint:* Look at Fig. 7.10.)

49. CALCULATOR *The value of e.* Find e as accurately as your calculator allows by taking $x_1 = 1$ and using Newton's method to solve the equation $\ln x - 1 = 0$.

50. CALCULATOR *The inverse relation between* e^x *and* $\ln x$. If you did not already do so in Section 1.7, find out how good your calculator is at evaluating the composites

$$e^{\ln x} \quad \text{and} \quad \ln(e^x).$$

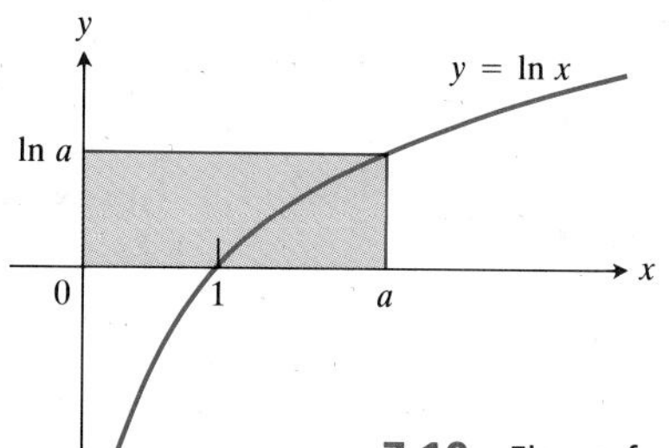

7.10 Figure for Exercise 48.

51. *Laws of exponents.* Set $x_1 = x$ and $x_2 = -x$ in the equation $e^{x_1}e^{x_2} = e^{x_1+x_2}$ to show that $e^{-x} = 1/e^x$.

52. *Laws of exponents.* Use the first and second laws of exponents together to show that $e^{x_1}/e^{x_2} = e^{x_1 - x_2}$.

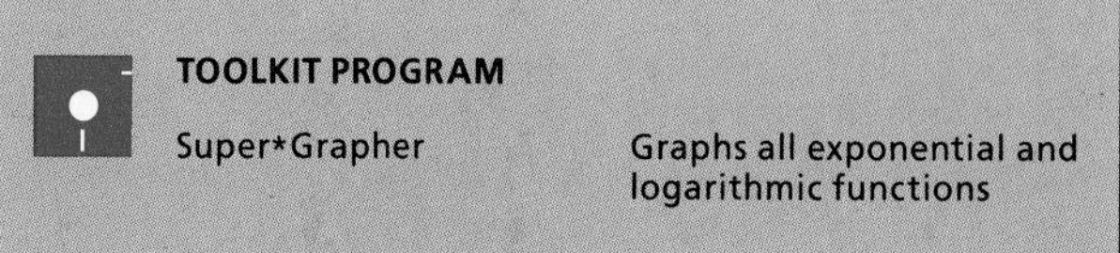

TOOLKIT PROGRAM	
Super*Grapher	Graphs all exponential and logarithmic functions

7.3 Other Exponential and Logarithmic Functions

While we have not yet devised a way to raise positive numbers to any but rational powers, we have an exception in the number e. The definition $e^x = \ln^{-1} x$ defines e^x for every real value of x, irrational as well as rational. In this section, we show how this good fortune enables us to raise any other positive number to an arbitrary power and thus to define an exponential function $y = a^x$ for any positive number a. We also prove the power rule for differentiation in its final form (good for all exponents) and define functions like x^x and $(\sin x)^{\tan x}$ that involve raising the values of one function to powers given by another.

Just as we can now see that e^x is but one of many exponential functions, $\ln x$ is one of many logarithmic functions, the others being the inverses of the functions a^x. These new logarithmic functions have important applications in science and engineering.

The Function a^x

If a is a positive number and x is any number whatever, we define the function a^x ("a to the x") by the equation

$$a^x = e^{x \ln a}.$$

Transcendental Numbers and Transcendental Functions

The numbers e and π are different from numbers like -2 and $\sqrt{3}$ in that they are not solutions of polynomial equations. Euler called e and π transcendental because, as he said, "they transcend the power of algebraic methods." Today, we call a function $y = f(x)$ *transcendental* if it satisfies no equation of the form

$$P_n(x)y^n + \cdots + P_1(x)y + P_0(x) = 0$$

in which the coefficients $P_0(x)$, $P_1(x)$, $\ldots$, $P_n(x)$ are polynomials in x.

Functions that do satisfy such an equation are called *algebraic.* For instance, $y = 1/\sqrt{(x+1)}$ is algebraic because it satisfies the equation $(x+1)y^2 - 1 = 0$. Here the polynomial coefficients are $P_2(x) = x + 1$, $P_1(x) = 0$, and $P_0(x) = -1$. Polynomials and rational functions are algebraic, and all sums, products, quotients, powers, and roots of algebraic functions are algebraic.

The six basic trigonometric functions are transcendental, as are the inverse trigonometric functions and the exponential and logarithmic functions that are the main subject of the present chapter.

It is natural to think of $e^{x \ln a}$ as being the same as $(e^{\ln a})^x$ and this, in turn, is $(a)^x$ because $e^{\ln a} = a$.

DEFINITION

The Function $y = a^x$

If a is a positive number, then $a^x = e^{x \ln a}$. (1)

EXAMPLE 1 $2^{\sqrt{3}} = e^{\sqrt{3} \ln 2}$ $\quad 2^{\pi} = e^{\pi \ln 2}$ ■

EXAMPLE 2 The number x^n can now be defined for any positive number x and any real number n:

$$x^n = e^{n \ln x}. \qquad \text{(Eq. (1) with } a = x \text{ and } x = n)$$ ■

EXAMPLE 3 The n in the equation $\ln x^n = n \ln x$ no longer has to be a rational number. It can be any real number, as long as $x > 0$:

$$\begin{aligned} \ln x^n &= \ln(e^{n \ln x}) && \text{(Example 2)} \\ &= n \ln x. && \begin{pmatrix} \ln e^u = u \text{ for any } u\text{, in} \\ \text{particular for } u = n \ln x \end{pmatrix} \end{aligned}$$ ■

The exponential function a^x obeys all the standard laws of exponents (Table 7.2). (We omit the proofs.)

Law 3, together with the definition of x^n as $e^{n \ln x}$, enables us to prove the Power Rule for differentiation in its final form. Differentiating x^n with respect to x gives

$$\begin{aligned} \frac{d}{dx} x^n &= \frac{d}{dx} e^{n \ln x} && \text{(Definition of } x^n) \\ &= e^{n \ln x} \cdot \frac{d}{dx}(n \ln x) && \text{(Chain Rule for } e^u) \\ &= x^n \cdot \frac{n}{x} && \text{(The definition again)} \\ &= n x^{n-1}. && \text{(Table 7.2, No. 3)} \end{aligned}$$

In short, as long as $x > 0$,

$$\frac{d}{dx} x^n = n x^{n-1}. \qquad (2)$$

The Chain Rule extends Eq. (3) to the Power Rule's final form.

Power Rule (Final Form)

If u is a positive differentiable function of x and n is any real number, then u^n is a differentiable function of x and

$$\frac{d}{dx} u^n = n u^{n-1} \frac{du}{dx}. \qquad (3)$$

TABLE 7.2
Laws of exponents ($a > 0$, any x and y)

1. $a^x \cdot a^y = a^{x+y}$
2. $a^{-x} = \dfrac{1}{a^x}$
3. $\dfrac{a^x}{a^y} = a^{x-y}$
4. $(a^x)^y = a^{(xy)} = (a^y)^x$

EXAMPLE 4

$$\frac{d}{dx}x^{\sqrt{2}} = \sqrt{2}\,x^{\sqrt{2}-1} \qquad (x > 0)$$

$$\frac{d}{dx}(\sin x)^{\pi} = \pi(\sin x)^{\pi-1}\cos x \qquad (\sin x > 0)$$

■

The Derivative of a^x

We find the derivative of a^x the way we found the derivative of x^n, starting with the definition $a^x = e^{x \ln a}$:

$$\begin{aligned}\frac{d}{dx}a^x &= \frac{d}{dx}e^{x\ln a}\\ &= e^{x\ln a}\cdot\frac{d}{dx}(x\ln a) \qquad \text{(Chain Rule)}\\ &= a^x \ln a.\end{aligned}$$

> If $a > 0$, then $\frac{d}{dx}a^x = a^x \ln a$. (4)

EXAMPLE 5 $\frac{d}{dx}3^x = 3^x \ln 3$ ■

Equation (4) shows why the function e^x is the exponential function preferred in calculus. If $a = e$, then $\ln a = 1$ and Eq. (4) simplifies to

$$\frac{d}{dx}e^x = e^x.$$

Equation (4) also comes in a more general form, based on the Chain Rule.

> If $a > 0$ and u is a differentiable function of x, then a^u is a differentiable function of x and
>
> $$\frac{d}{dx}a^u = a^u \ln a \frac{du}{dx} \qquad (5)$$

EXAMPLE 6

a) $\frac{d}{dx}3^{-x} = 3^{-x}\ln 3\frac{d}{dx}(-x) = -3^{-x}\ln 3$

b) $\frac{d}{dx}3^{\sin x} = 3^{\sin x}\ln 3\frac{d}{dx}(\sin x) = 3^{\sin x}(\ln 3)\cos x$ ■

In practice, there is little reason to memorize Eq. (5) because the derivative of a^u can always be found by logarithmic differentiation. The main reason for producing the formula here is to derive from it the companion integration formula. We shall do that in a moment.

EXAMPLE 7 Use logarithmic differentiation to find the derivative of $y = 3^{\sin x}$.

Solution

$$y = 3^{\sin x}$$

$$\ln y = \ln 3^{\sin x} = \sin x \ln 3$$

$$\frac{1}{y}\frac{dy}{dx} = \cos x \ln 3$$

$$\frac{dy}{dx} = y \cos x \ln 3 = 3^{\sin x} \cos x \ln 3$$

As you can see, this agrees with the result in Example 6. ■

Other Power Functions

The ability to raise positive numbers to arbitrary real powers makes it possible to define functions like x^x and $x^{\ln x}$ for $x > 0$. We find the derivatives of functions like these by logarithmic differentiation.

EXAMPLE 8 Find dy/dx if $y = x^x,\ x > 0$.

Solution

$$y = x^x$$

$$\ln y = \ln x^x = x \ln x$$

$$\frac{1}{y}\frac{dy}{dx} = x \cdot \frac{1}{x} + 1 \cdot \ln x = 1 + \ln x$$

$$\frac{dy}{dx} = y(1 + \ln x) = (1 + \ln x)x^x$$

■

The Graph of a^x

From the formula

$$\frac{d}{dx}a^x = a^x \ln a,$$

we see that the derivative of a^x is positive if $a > 1$ and negative if $0 < a < 1$. Thus, a^x is an increasing function of x if $a > 1$ and a decreasing function of x if $0 < a < 1$. In either case, a^x is one-to-one. The second derivative,

$$\frac{d^2}{dx^2}a^x = \frac{d}{dx}(a^x \ln a) = \ln a \frac{d}{dx}(a^x)$$

$$= \ln a\ (a^x \ln a) = (\ln a)^2 a^x,$$

is always positive, so the graph is concave up. Figure 7.11 shows the graphs of $y = 2^x$ (increasing and one-to-one) and $y = (1/2)^x$ (decreasing and one-to-one).

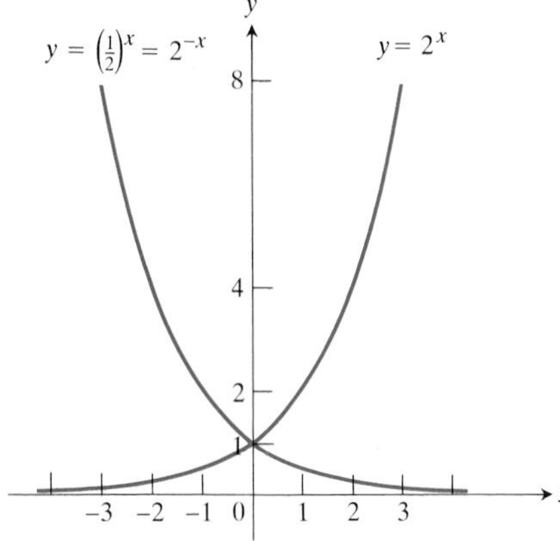

7.11 The graphs of $y = 2^x$ and $y = (1/2)^x$. Both functions are one-to-one. As $x \to \infty$, $2^x \to \infty$ and $(1/2)^x \to 0$. Similarly, as $x \to -\infty$, $2^x \to 0$ and $(1/2)^x \to \infty$.

The Integral of a^u

If $a \neq 1$ so that $\ln a \neq 0$, we can divide both sides of Eq. (5) by $\ln a$ to obtain

$$a^u \frac{du}{dx} = \frac{1}{\ln a} \frac{d}{dx}(a^u). \tag{6}$$

Integrating with respect to x then gives

$$\int a^u \frac{du}{dx}\,dx = \int \frac{1}{\ln a} \frac{d}{dx}(a^u)\,dx = \frac{1}{\ln a} \int \frac{d}{dx}(a^u)\,dx = \frac{1}{\ln a} a^u + C. \tag{7}$$

Writing the first integral in differential form then gives

$$\int a^u\,du = \frac{a^u}{\ln a} + C \tag{8}$$

EXAMPLE 9

$$\int 2^x\,dx = \frac{2^x}{\ln 2} + C \qquad \text{(Eq. (8) with } a = 2, u = x) \qquad \blacksquare$$

EXAMPLE 10

$$\int 2^{\sin x} \cos x\,dx = \int 2^u\,du = \frac{2^u}{\ln 2} + C = \frac{2^{\sin x}}{\ln 2} + C \qquad (u = \sin x \text{ in Eq. (8)}) \qquad \blacksquare$$

Base *a* Logarithms

As we saw earlier, if a is any positive number other than 1, the function a^x is one-to-one and has a nonzero derivative at every point. It therefore has a differentiable inverse. We call the inverse the **base *a* logarithm of *x*** and denote it by $\mathbf{\log_a x}$.

DEFINITION

$$\log_a x = \text{inverse of } a^x \qquad (a > 0,\ a \neq 1)$$

Because the logarithm and exponential are inverses of one another, their composites in either order give the identity function.

$$\log_a(a^x) = x \quad \text{(for all } x) \tag{9}$$

$$a^{(\log_a x)} = x \quad \text{(for each positive } x) \tag{10}$$

EXAMPLE 11 $\log_2(2^5) = 5 \qquad \log_{10}(10^{-7}) = -7$ ■

EXAMPLE 12 $2^{(\log_2 3)} = 3 \qquad 10^{(\log_{10} 4)} = 4$ ■

The Evaluation of $\log_a x$

The values of $\log_a x$ can be calculated from the natural logarithms of a and x with the following formula.

$$\log_a x = \frac{\ln x}{\ln a} \tag{11}$$

This formula can be derived from Eq. (10):

$$a^{(\log_a x)} = x \qquad \text{(Eq. (10))}$$

$$\ln a^{(\log_a x)} = \ln x \qquad \text{(Take the natural logarithm of both sides.)}$$

$$\log_a x \cdot \ln a = \ln x \qquad \text{(The Exponent Rule)}$$

$$\log_a x = \frac{\ln x}{\ln a}. \qquad \text{(Solved for } \log_a x\text{)}$$

EXAMPLE 13

$$\log_{10} 2 = \frac{\ln 2}{\ln 10} \approx \frac{0.69}{2.30} = 0.3$$ ■

TABLE 7.3
Rules of arithmetic for base *a* logarithms

1. $\log_a uv = \log_a u + \log_a v$
2. $\log_a \frac{u}{v} = \log_a u - \log_a v$
3. $\log_a u^n = n \log_a u$

EXAMPLE 14 The ratio of $\ln x$ to $\log_{10} x$ has a constant value:

$$\frac{\ln x}{\log_{10} x} = \ln 10 = 2.302585093. \qquad \text{(To 10 digits)}$$

This explains what you discovered in Chapter 5, Practice Exercise 72, and Section 2.3, Exercise 59(d). ■

EXAMPLE 15 Most calculators have keys for $\log_{10} x$ and $\ln x$. To find logarithms to other bases, we use Eq. (11).

To find $\log_2 x$, find $\ln x$ and divide by $\ln 2$: $\quad \log_2 5 = \frac{\ln 5}{\ln 2}$

To find $\ln x$ given $\log_2 x$, multiply by $\ln 2$: $\quad \ln 5 = \log_2 5 \cdot \ln 2$ ■

The rules of arithmetic are the same for $\log_a x$, as for $\ln x$ (Table 7.3). These rules come from dividing the corresponding rules for the natural logarithm by $\ln a$. For example,

$$\ln uv = \ln u + \ln v \qquad \text{(Rule 1 for natural logarithms . . .)}$$

$$\frac{\ln uv}{\ln a} = \frac{\ln u}{\ln a} + \frac{\ln v}{\ln a} \qquad \text{(. . . divided by } \ln a \text{ . . .)}$$

$$\log_a uv = \log_a u + \log_a v. \qquad \text{(. . . gives Rule 1 for base } a \text{ logarithms.)}$$

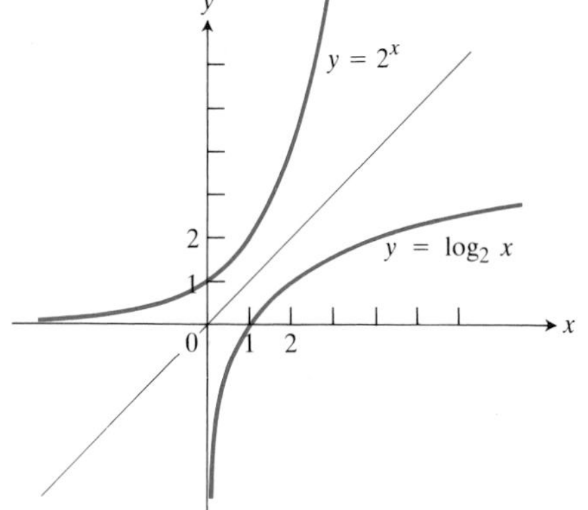

7.12 The graph of 2^x and its inverse, $\log_2 x$. As $x \to \infty$, $\log_2 x \to \infty$. As $x \to 0^+$, $\log_2 x \to -\infty$.

The Graph of $\log_a x$

The graph of $\log_a x$ is obtained by reflecting the graph of a^x across the line $y = x$. It looks very much like a scaled version of the graph of $\ln x$ and indeed it is, because of Eq. (11). See Fig. 7.12.

The Derivative of $\log_a u$

To find the derivative of a base a logarithm, we first convert it to a natural logarithm. If u is a positive differentiable function of x, then

$$\frac{d}{dx}(\log_a u) = \frac{d}{dx}\left(\frac{\ln u}{\ln a}\right) = \frac{1}{\ln a}\frac{d}{dx}(\ln u) = \frac{1}{\ln a}\cdot\frac{1}{u}\frac{du}{dx}.$$

$$\frac{d}{dx}\log_a u = \frac{1}{\ln a}\cdot\frac{1}{u}\frac{du}{dx} \qquad (12)$$

EXAMPLE 16

$$\frac{d}{dx}\log_{10}(3x+1) = \frac{1}{\ln 10}\cdot\frac{1}{3x+1}\frac{d}{dx}(3x+1) = \frac{3}{(\ln 10)(3x+1)}$$ ■

Integrals Involving $\log_a x$

To evaluate integrals involving base a logarithms, we convert them to natural logarithms.

EXAMPLE 17

$$\int \frac{\log_2 x}{x}\,dx = \frac{1}{\ln 2}\int \frac{\ln x}{x}\,dx \qquad \left(\log_2 x = \frac{\ln x}{\ln 2}\right)$$

$$= \frac{1}{\ln 2}\int u\,du \qquad \left(u = \ln x,\ du = \frac{1}{x}dx\right)$$

$$= \frac{1}{\ln 2}\frac{u^2}{2} + C = \frac{1}{\ln 2}\frac{(\ln x)^2}{2} + C = \frac{(\ln x)^2}{2\ln 2} + C$$ ■

Wide World Photos

The town of Spitak in Soviet Armenia stood less than 5 km from the epicenter of the 1988 earthquake. Just 4 min after the magnitude 6.8 main shock struck, an aftershock of magnitude 5.8 collapsed many of the weakened buildings. (*Source: Science* Vol. 243, 13 Jan. 1989, p. 170)

Base 10 Logarithms

Base 10 logarithms, often called **common logarithms,** appear in many scientific formulas. For example, earthquake intensity is often reported on the **Richter scale.** Here the formula is

$$\text{Magnitude } R = \log_{10}\left(\frac{a}{T}\right) + B, \qquad (13)$$

where a is the amplitude of the ground motion in microns at the receiving station, T is the period of the seismic wave in seconds, and B is an empirical factor that allows for the weakening of the seismic wave with increasing distance from the epicenter of the earthquake.

EXAMPLE 18 For an earthquake 10,000 km from the receiving station, $B = 6.8$. If the recorded vertical ground motion is $a = 10$ microns and the period is $T = 1$ second, the earthquake's magnitude is

$$R = \log_{10}\left(\frac{10}{1}\right) + 6.8 = 1 + 6.8 = 7.8.$$

An earthquake of this magnitude does great damage near its epicenter. ■

The pH scale for measuring the acidity of a solution is a logarithmic scale. The pH value (hydrogen potential) of the solution is the common logarithm of the reciprocal of the solution's hydronium ion concentration, $[H_3O^+]$:

$$pH = \log_{10}\frac{1}{[H_3O^+]} = -\log_{10}[H_3O^+]. \tag{14}$$

The hydronium ion concentration is measured in moles per liter. Vinegar has a pH of 3, distilled water a pH of 7, seawater a pH of 8.15 and household ammonia a pH of 12. The total scale ranges from about 0.1 for normal hydrochloric acid to 14 for a normal solution of sodium hydroxide.

Another example of the use of common logarithms is the db ("dee bee") scale for measuring loudness in decibels. If I is the intensity of sound in watts per square meter, the decibel level of the sound is

$$\text{Sound level} = 10\log_{10}(I \times 10^{12})\text{ db}. \tag{15}$$

If you ever wondered why doubling the power of your audio amplifier increased the sound level by only a few decibels, Eq. (15) provides the answer. As the following calculation shows, doubling I adds only about 3 db.

EXAMPLE 19 Doubling I in Eq. (15) adds about 3 db: Writing log for $\log_{10}$ (a common practice) we have

$$\begin{aligned}
\text{Sound level with } I \text{ doubled} &= 10\log(2I \times 10^{12}) && \begin{pmatrix}\text{Eq. (15) with}\\ 2I \text{ for } I\end{pmatrix}\\
&= 10\log(2 \cdot I \times 10^{12})\\
&= 10\log 2 + 10\log(I \times 10^{12})\\
&= \text{original sound level} + 10\log 2\\
&\approx \text{original sound level} + 3 && (\log_{10}2 \approx 0.30)
\end{aligned}$$

■

REMARK ON NOTATION Many advanced texts and research publications in mathematics use $\log x$, with no base specified, to represent the natural logarithm $\ln x$. Most texts in the physical sciences use $\log x$ to represent $\log_{10}x$. Most calculators use $\ln x$ for the natural logarithm and $\log x$ for base 10 logarithms. Computers, however, may use LOG(X) for the natural logarithm. One then evaluates $\log_{10}x$ as (LOG(X))/(LOG(10)).

Most Foods Are Acidic (ph < 7).

Food	pH value
Bananas	4.5–4.7
Grapefruit	3.0–3.3
Oranges	3.0–4.0
Limes	1.8–2.0
Milk	6.3–6.6
Soft drinks	2.0–4.0
Spinach	5.1–5.7

Typical Sound Levels

Threshold of hearing	0 db
Rustle of leaves	10 db
Average whisper	20 db
Quiet automobile	50 db
Ordinary conversation	65 db
Pneumatic drill 10 feet away	90 db
Threshold of pain	120 db

Exercises 7.3

In Exercises 1–14, find dy/dx.

1. $y = x^{\pi}$
2. $y = x^{1+\sqrt{2}}$
3. $y = x^{-\sqrt{2}}$
4. $y = x^{1-e}$
5. $y = 2^x$
6. $y = 8^x$
7. $y = 2^{-x}$
8. $y = 9^{-x}$
9. $y = 2^{\sec x}$
10. $y = 2^{\tan x}\ln 2$
11. $y = x^{\ln x},\ x > 0$
12. $y = x^{(x+1)}$
13. $y = (\sin x)^{\tan x}$
14. $y = (\sqrt{x})^x$

Evaluate the integrals in Exercises 15–26.

15. $\int_0^1 3x^{\sqrt{3}}\,dx$
16. $\int_0^1 x^{\sqrt{2}}\,dx$

17. $\int_0^3 (\sqrt{2}+1)x^{\sqrt{2}}\,dx$ **18.** $\int_1^e x^{\ln 2 - 1}\,dx$

19. $\int_0^1 5^x\,dx$ **20.** $\int_{-1}^0 2^x\,dx$

21. $\int_0^1 \frac{1}{2^x}\,dx$ **22.** $\int_{-1}^1 2^{(x+1)}\,dx$

23. $\int_{-1}^0 4^{-x}\ln 2\,dx$ **24.** $\int_{-2}^0 5^{-x}\,dx$

25. $\int_1^{\sqrt{2}} x2^{x^2}\,dx$ **26.** $\int_0^{\pi/2} 2^{\cos x}\sin x\,dx$

Express the numbers in Exercises 27–32 in terms of natural logarithms and simplify.

27. $\log_4 16$ **28.** $\log_8 32$

29. $\log_5 0.04$ **30.** $\log_{0.5} 4$

31. $\log_2 4$ **32.** $\log_4 2$

Express the ratios in Exercises 33–36 as ratios of natural logarithms.

33. $\dfrac{\log_2 x}{\log_3 x}$ **34.** $\dfrac{\log_2 x}{\log_8 x}$

35. $\dfrac{\log_9 x}{\log_3 x}$ **36.** $\dfrac{\log_{\sqrt{10}} x}{\log_{\sqrt{2}} x}$

Solve the equations in Exercises 37–38 for x.

37. $3^{\log_3 7} + 2^{\log_2 5} = 5^{\log_5 x}$ **38.** $8^{\log_8 3} - e^{\ln 5} = x^2 - 7^{\log_7 3x}$

In Exercises 39–50, find dy/dx.

39. $y = \log_4 x$ **40.** $y = \log_4 x^2$

41. $y = \log_{10} e^x$ **42.** $y = \log_5 \sqrt{x}$

43. $y = \ln 2 \cdot \log_2 x$ **44.** $y = \log_2(1/x)$

45. $y = \log_{10}\sqrt{x+1}$ **46.** $y = \log_2(3x+1)$

47. $y = 1/\log_2 x$ **48.** $y = \ln 10^x$

49. $y = \log_5(x+1)^2$ **50.** $y = \log_7 \sin x$

Evaluate the integrals in Exercises 51–58.

51. $\int_1^{10} \frac{\log_{10} x}{x}\,dx$ **52.** $\int_1^4 \frac{\log_2 x}{x}\,dx$

53. $\int_1^4 \frac{\ln 2 \log_2 x}{x}\,dx$ **54.** $\int_1^e \frac{2\ln 10 \log_{10} x}{x}\,dx$

55. $\int_0^2 \frac{\log_2(x+2)}{x+2}\,dx$ **56.** $\int_{1/10}^{10} \frac{\log_{10}(10x)}{x}\,dx$

57. $\int_0^9 \frac{2\log_{10}(x+1)}{x+1}\,dx$ **58.** $\int_2^3 \frac{2\log_2(x-1)}{x-1}\,dx$

Find the limits in Exercises 59–62.

59. a) $\lim_{x\to\infty} \log_2 x$ b) $\lim_{x\to\infty} \log_2(1/x)$

60. a) $\lim_{x\to 0^+} \log_{10} x$ b) $\lim_{x\to 0^+} \log_{10}(1/x)$

61. a) $\lim_{x\to\infty} 3^x$ b) $\lim_{x\to\infty} 3^{-x}$

62. a) $\lim_{x\to -\infty} 3^x$ b) $\lim_{x\to -\infty} 3^{-x}$

63. Find dy/dx if
a) $y = 2^{\ln x}$ b) $y = \ln 2^x$
c) $y = \ln x^2$ d) $y = (\ln x)^2$

64. **CALCULATOR** The curves $y = x^2$ and $y = 2^x$ intersect at $x = 2$ and $x = 4$. There is also a third intersection between $x = -1$ and $x = 0$ (Fig. 7.13). Find its coordinates, as accurately as your calculator will allow, by applying Newton's method with $f(x) = 2^x - x^2$.

65. *Blood pH.* The pH of human blood normally falls between 7.37 and 7.44. Find the corresponding bounds for $[H_3O^+]$.

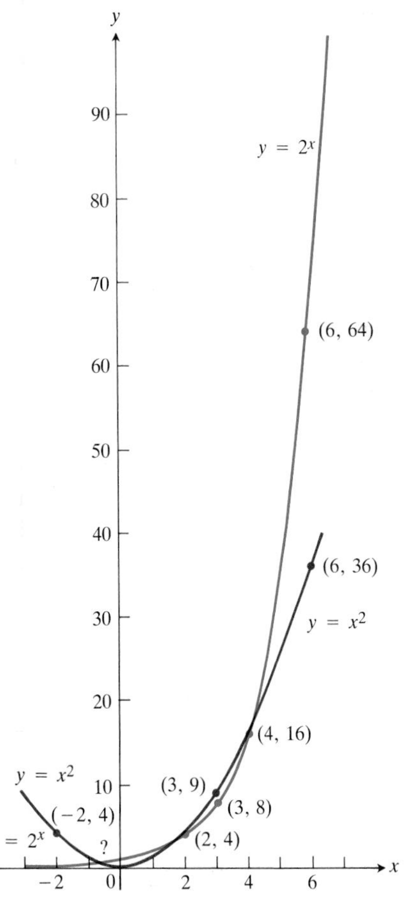

7.13 The curves $y = x^2$ and $y = 2^x$ intersect at $x = 2$, $x = 4$, and $x = ?$ See Exercise 64.

66. *Brain fluid pH.* The cerebrospinal fluid in the brain has a hydronium ion concentration of about $[H_3O^+] = 4.8 \times 10^{-8}$ moles per liter. What is the pH?

67. *Audio amplifiers.* By what factor k do you have to multiply the intensity I of the sound from your audio amplifier to add 10 db to the sound level?

68. *Conversion factors.* Show that the equation for converting base 10 logarithms to base 2 logarithms is

$$\log_2 x = \frac{\ln 10}{\ln 2} \log_{10} x.$$

69. *Continuation of Exercise 68.* Show that the equation for converting base a logarithms to base b logarithms is

$$\log_b x = \frac{\ln a}{\ln b} \log_a x.$$

TOOLKIT PROGRAM

Super*Grapher	Graphs all exponential and logarithmic functions

7.4 Growth and Decay

In this section we derive the law of exponential change without requiring the quantities involved to be positive and look at some of the important applications we did not have time to present in Section 5.8.

The Law of Exponential Change: $y = y_0 e^{kt}$

Now that we can evaluate $\int(1/u)\,du$ for any continuous function u, we can derive the law of exponential change without requiring the quantities involved to be positive. The derivation is much the same as before.

Suppose the quantity y we are interested in—velocity, temperature, electrical current, whatever—grows or decreases at a rate that at any given time t is proportional to the amount present. If we also know the amount present at time $t = 0$, call it y_0, we can find y as a function of t by solving the following initial value problem:

Differential equation: $\quad \dfrac{dy}{dt} = ky \qquad (1)$

Initial condition: $\quad y = y_0$ when $t = 0$.

If y is increasing, then k is positive and Eq. (1) is how we say that the rate of growth is proportional to what has already been accumulated. If y is decreasing, then k is negative and Eq. (1) is how we say that the rate of decay is proportional to the amount still left.

We see right away that the constant function $y = 0$ is a solution of Eq. (1). To find what the nonzero solutions are, divide by y to get, in order,

$$\frac{1}{y}\frac{dy}{dt} = k$$

$$\ln|y| = kt + C \qquad \left(\begin{array}{l}\text{Integrate with respect to } t \text{, using the}\\ \text{equation } \int (1/u)\,du = \ln|u| + C\end{array}\right)$$

$$|y| = e^{kt+C} \qquad \text{(Exponentiate)}$$

$$|y| = e^C \cdot e^{kt} \qquad \text{(A law of exponents)}$$

$$y = \pm e^C e^{kt} \qquad \text{(Eliminate absolute values)}$$

$$y = Ae^{kt}. \qquad \text{(A more convenient name for } \pm e^c\text{)}$$

By allowing A to take on the value 0 as well as the arbitrary positive and negative values given by the expression $\pm e^c$, we can include the additional solution $y = 0$ in the formula.

We find the right value of A for the initial value problem by solving for A when $y = y_0$ and $t = 0$:

$$y_0 = A\,e^{k \cdot 0} = A.$$

The solution of the initial value problem is

$$y = y_0\,e^{kt}.$$

This is the same equation we derived in Section 5.8, only now we do not have to confine our attention to quantities that never become negative.

The Law of Exponential Change (Unrestricted) $\quad y = y_0\,e^{kt} \quad$ (2)

Radioactivity

For radon gas, t is measured in days and $k = 0.18$. For radium-226, which used to be painted on watch dials to make them glow at night (a dangerous practice), t is measured in years and $k = 4.3 \times 10^{-2}$. (Incidentally, the decay of radium in the earth's crust is the source of the radon we find in our basements.)

When a radioactive atom emits some of its mass as radiation, the remainder of the atom re-forms to make an atom of some new substance. This process of radiation and change is called **radioactive decay,** and an element whose atoms go through this process spontaneously is called **radioactive.** Thus, radioactive carbon-14 decays into nitrogen; radium, through a number of intervening radioactive steps, decays into lead.

Experiments have shown that at any given time the rate at which a radioactive element decays (as measured by the number of nuclei that change per unit time) is approximately proportional to the number of radioactive nuclei present. Thus, the decay of a radioactive element is described by the equation $dy/dt = -ky$. If y_0 is the number of radioactive nuclei present at time zero, the number still present at any later time t will be $y = y_0\,e^{-kt}$.

Radioactive Decay Equation $\quad y = y_0\,e^{-kt} \quad$ (3)

EXAMPLE 1 *The half-life of a radioactive element.* The **half-life** of a radioactive element is the time required for half of the radioactive nuclei present in a sample to decay. It is a remarkable fact that the half-life is a constant that does not depend on the number of radioactive nuclei initially present in the sample.

To see why, let y_0 be the number of radioactive nuclei initially present in the sample. Then the number y present at any later time t will be

$$y = y_0\,e^{-kt}.$$

We seek the value of t at which

$$y_0\,e^{-kt} = \frac{1}{2}y_0,$$

for this will be the time when the number of radioactive nuclei present equals half the original number. The y_0's cancel in this equation to give

$$e^{-kt} = \frac{1}{2}, \qquad -kt = \ln\frac{1}{2} = -\ln 2, \qquad \text{and} \qquad t = \frac{\ln 2}{k}.$$

This value of t is the half-life of the element. It depends only on the value of k; the number y_0 does not enter in. ■

$$\text{Half-life} = \frac{\ln 2}{k} \tag{4}$$

EXAMPLE 2 *Polonium-210.* The effective radioactive lifetime of polonium-210 is so short we measure it in days rather than years. The number of radioactive atoms remaining after t days in a sample that starts with y_0 radioactive atoms is

$$y = y_0 e^{-4.95 \times 10^{-3} t}.$$

Find the element's half-life.

Solution

$$\begin{aligned} \text{Half-life} &= \frac{\ln 2}{k} && \text{(Eq. 4)} \\ &= \frac{\ln 2}{4.95 \times 10^{-3}} && \left(\begin{array}{l}\text{The } k \text{ from polonium's} \\ \text{decay equation}\end{array}\right) \\ &= 140 \text{ days} && \text{(With a calculator, rounded)} \end{aligned}$$

■

EXAMPLE 3 *Carbon-14.* People who do carbon-14 dating use a figure of 5700 years for its half-life (more about carbon-14 dating in the exercises). Find the age of a sample in which 10% of the radioactive nuclei originally present have decayed.

Solution We use the decay equation $y = y_0 e^{-kt}$. There are two things to find:

1. The value of k
2. The value of t that makes $y_0 e^{-kt} = (9/10)y_0$ or $e^{-kt} = 9/10$.

STEP 1: *The value of k*: We find it from the half-life equation:

$$\begin{aligned} k &= \frac{\ln 2}{\text{Half-life}} && \text{(Eq. (4) turned around)} \\ &= \frac{\ln 2}{5700} = 1.2 \times 10^{-4} \end{aligned}$$

STEP 2: The value of t that makes $e^{-kt} = 9/10$:

$$\begin{aligned} e^{-1.2 \times 10^{-4} t} &= 0.9 \\ -1.2 \times 10^{-4} t &= \ln(0.9) && \text{(Logs of both sides)} \\ t &= \frac{\ln(0.9)}{-1.2 \times 10^{-4}} = 878 \text{ years} && \text{(Calculator, rounded)} \end{aligned}$$

The sample is about 878 years old. ■

Resistance Proportional to Velocity

In some cases it makes sense to assume that the resistance encountered by a moving object, like a car coasting to a stop, is proportional to the velocity of the object's motion. The slower the object moves, the less its forward progress is resisted by the air through which it passes. We can describe this in mathematical terms if we picture the object as a mass m moving along a coordinate line with position $s(t)$ and velocity $v(t)$ at time t. The magnitude of the resisting force opposing the motion is then $m(dv/dt)$ and we can write

$$m\frac{dv}{dt} = -kv \quad (k > 0) \tag{5}$$

to say that the force decreases in proportion to velocity. If we rewrite (5) as

$$\frac{dv}{dt} = -\frac{k}{m}v,$$

we can see that the solution is $v = v_0 e^{-(k/m)t}$.

Resistance Proportional to Velocity	$v = v_0 e^{-(k/m)t}$	(6)

What can we learn from this equation? For one thing, we can see that if m is something large, like the mass of a 20,000-ton ore boat in Lake Erie, it will take a long time for the velocity to get near zero. For another, we can integrate the equation to find s.

Suppose a body is coasting to a stop and the only force acting on it is a resistance proportional to its speed. How far will it coast? To find out, we start with Eq. (6) and solve the initial value problem

Differential equation: $\dfrac{ds}{dt} = v_0 e^{-(k/m)t}$

Initial condition: $s = 0$ when $t = 0$.

Integrating with respect to t gives

$$s = -\frac{v_0 m}{k} e^{-(k/m)t} + C \tag{7}$$

Substituting $s = 0$ when $t = 0$ gives

$$0 = -\frac{v_0 m}{k} + C \quad \text{and} \quad C = \frac{v_0 m}{k}.$$

The body's position at time t is therefore

$$s(t) = -\frac{v_0 m}{k} e^{-(k/m)t} + \frac{v_0 m}{k} = \frac{v_0 m}{k}(1 - e^{-(k/m)t}). \tag{8}$$

To find how far the body will coast, we find the limit of $s(t)$ as $t \to \infty$. Since $\lim_{t\to\infty} e^{-(k/m)t} = 0$, the limit of $s(t)$ is $v_0 m/k$.

$$\text{Distance coasted} = \frac{v_0 m}{k} \tag{9}$$

This is an ideal figure, of course. Only in mathematics can time stretch to infinity. The number $v_0 m/k$ is only an upper bound (albeit a useful one). It is true to life in one respect, at least—if m is large the body will coast a long way before it stops. That is why ocean liners have to be docked by tugboats. Any liner of conventional design entering a slip with enough speed to steer would smash into the pier before it could stop.

EXAMPLE 4 For a 192-lb ice skater, the k in Eq. (6) is about 1/3 kg/sec and $m = 192/32 = 6$ slugs. How long will it take the skater to coast from 11 ft/sec (7.5 mph) to 1 ft/sec? How far will the skater coast before coming to a complete stop?

Solution We answer the first question by solving Eq. (6) for t:

$$
\begin{aligned}
11e^{-t/18} &= 1 && \left(\text{Eq. (6) with } k = 1/3, m = 6, \; v_0 = 11, v = 1\right) \\
e^{-t/18} &= 1/11 \\
-t/18 &= \ln(1/11) = -\ln 11 \\
t &= 18 \ln 11 = 43 \text{ sec.} && \text{(Calculator, rounded)}
\end{aligned}
$$

We answer the second question with Eq. (9):

$$\text{Distance coasted} = \frac{v_0 m}{k} = \frac{11 \cdot 6}{(1/3)} = 198 \text{ ft.}$$

■

Heat Transfer: Newton's Law of Cooling

Hot chocolate left in a tin cup cools to the temperature of the surrounding air. A hot silver ingot immersed in water cools to the temperature of the surrounding water. In situations like these, the rate at which an object's temperature is changing at any given time is roughly proportional to the difference between its temperature and the temperature of the surrounding medium. This observation is called **Newton's law of cooling,** although it applies to warming as well. It can be written as an equation in the following way.

If $T(t)$ is the temperature of the object at time t, and T_s is the surrounding temperature, then

$$\frac{dT}{dt} = -k(T - T_s). \tag{10}$$

Since T_s is constant, this is the same as $dy/dt = -ky$ with $y = (T - T_s)$. Hence, the solution of Eq. (10) is

$$y = y_0 e^{-kt}$$

or

$$T - T_s = (T_0 - T_s)e^{-kt}, \tag{11}$$

where T_0 is the value of T at time zero.

Newton's Law of Cooling $\quad T - T_s = (T_0 - T_s)e^{-kt} \quad$ (12)

EXAMPLE 5 A hard-boiled egg at 98°C is put in a sink of 18°C water to cool. After 5 minutes, the egg's temperature is found to be 38°C. Assuming that the water has not warmed appreciably, how much longer will it take the egg to reach 20°C?

Solution We find how long it would take the egg to cool from 98°C to 20°C and subtract the 5 minutes that have already elapsed.

According to Eq. (12), the egg's temperature t minutes after it is put in the sink is

$$T - 18 = (98 - 18)e^{-kt}, \quad \text{or} \quad T = 18 + 80e^{-kt}.$$

To find k, we use the information that $T = 38$ when $t = 5$. This gives

$$38 = 18 + 80e^{-5k}$$

$$e^{-5k} = \frac{1}{4}$$

$$-5k = \ln\frac{1}{4} = -\ln 4$$

$$k = \frac{1}{5}\ln 4 = 0.28$$

(to two decimal places). The egg's temperature at time t is

$$T = 18 + 80e^{-0.28t}.$$

When will $T = 20$? When

$$20 = 18 + 80e^{-0.28t}$$

$$80e^{-0.28t} = 2$$

$$e^{-0.28t} = \frac{1}{40}$$

$$-0.28t = \ln\frac{1}{40} = -\ln 40$$

$$t = \frac{\ln 40}{0.28} = 13 \text{ min.} \qquad \text{(To two digits)}$$

The egg's temperature will reach 20°C 13 minutes after it is put in to cool. Since it took 5 minutes to reach 38°C, it will take 8 more to reach 20°C. ■

Exercises 7.4

CALCULATOR Most of these exercises should be done with a calculator.

1. *Human evolution continues.* The analysis of tooth shrinkage by C. Loring Brace and colleagues at the University of Michigan's Museum of Anthropology indicates that human tooth size is continuing to decrease and that the evolutionary process did not come to a halt some 30,000 years ago as many scientists contend. In northern Europeans, for example, tooth size reduction now has a rate of 1% per 1000 years.

a) If t represents time in years and y represents tooth size, use the condition that $y = 0.99y_0$ when

$t = 1000$ to find the value of k in the equation $y = y_0 e^{kt}$. Then use this value of k to answer the following questions.

b) In about how many years will human teeth be 90% of their present size?

c) What will be our descendants' tooth size 20,000 years from now (as a percentage of our present tooth size)?

(Source: *LSA Magazine,* Spring 1989, Volume 12, Number 2, p. 19, Ann Arbor, MI)

2. *Atmospheric pressure.* The earth's atmospheric pressure p is often modeled by assuming that the rate dp/dh at which p changes with the altitude h above sea level is proportional to p. Suppose that the pressure at sea level is 1013 millibars (about 14.7 pounds per square inch) and that the pressure at an altitude of 20 km is 90 millibars.

a) Solve the equation $dp/dh = kp$ (k a constant) to express p in terms of h. Determine the values of k and the constant of integration from the given initial conditions.

b) What is the atmospheric pressure at $h = 50$ km?

c) At what altitude is the pressure equal to 900 millibars?

3. *First order chemical reactions.* In some chemical reactions, the rate at which the amount of a substance changes with time is proportional to the amount present. For the change of δ-glucono lactone into gluconic acid, for example,

$$\frac{dy}{dt} = -0.6y$$

when t is measured in hours. If there are 100 grams of δ-glucono lactone present when $t = 0$, how many grams will be left after the first hour?

4. *The inversion of sugar.* The processing of raw sugar has a step called "inversion" that changes the sugar's molecular structure. Once the process has begun, the rate of change of the amount of raw sugar is proportional to the amount of raw sugar remaining. If 1000 kg of raw sugar reduces to 800 kg of raw sugar during the first 10 hours, how much raw sugar will remain after another 14 hours?

5. *Radon gas.* The decay equation for (radioactive) radon gas is $y = y_0 e^{-0.18t}$, with t in days. About how long will it take the radon in a sealed sample of air to fall to 90% of its original value?

6. *Polonium-210.* The half-life of polonium is 140 days but your sample will not be useful to you after 95% of the radioactive nuclei present on the day the sample arrives has disintegrated. About how many days after the sample arrives will you be able to use the polonium?

7. For a 145-lb cyclist on a 15-lb bicycle on level ground, the k in Eq. (6) is about 1/5 and $m = 160/32 = 5$ slugs. The cyclist starts coasting at 22 ft/sec (15 mph).

a) About how far will the cyclist coast before reaching a complete stop?

b) How long will it take the cyclist's speed to drop to 1 ft/sec?

8. For a 56,000-ton Iowa class battleship, $m = 1{,}750{,}000$ slugs and the k in Eq. (6) might be 3000. Suppose the battleship loses power when it is moving at a speed of 22 ft/sec (13.2 knots).

a) About how far will the ship coast before it is dead in the water?

b) About how long will it take the ship's speed to drop to 1 ft/sec?

9. *Cooling cocoa.* Suppose that a cup of cocoa cooled from 90°C to 60°C after 10 minutes in a room whose temperature was 20°C. Use Newton's law of cooling to answer the following questions.

a) How much longer would it take the cocoa to cool to 35°C?

b) Instead of being left to stand in the room, the cup of 90°C cocoa is put in a freezer whose temperature is −15°C. How long will it take the cocoa to cool from 90°C to 35°C?

10. *Body of unknown temperature.* A body of unknown temperature was placed in a room that was held at 30°F. After 10 minutes, the body's temperature was 0°F, and 20 minutes after the body was placed in the room the body's temperature was 15°F. Use Newton's law of cooling to estimate the body's initial temperature.

11. *Surrounding medium of unknown temperature.* A pan of warm water (46°C) was put in a refrigerator. Ten minutes later, the water's temperature was 39°C; 10 minutes after that, it was 33°C. Use Newton's law of cooling to estimate how cold the refrigerator was.

12. *Silver cooling in air.* The temperature of an ingot of silver is 60°C above room temperature right now. Twenty minutes ago, it was 70°C above room temperature. How far above room temperature will the silver be 15 minutes from now? Two hours from now? When will the silver be 10°C above room temperature?

13. *Voltage in a discharging capacitor.* Suppose that electricity is draining from a capacitor at a rate that is proportional to the voltage $V(t)$ across its terminals and that, if t is measured in seconds,

$$\frac{dV}{dt} = -\frac{1}{40}V.$$

Solve this equation for V, using V_0 to denote the value of V when $t = 0$. How long will it take the voltage to drop to 10% of its original value?

14. *The mean life of a radioactive nucleus.* Physicists using the radioactivity equation $y = y_0 e^{-kt}$ call the number $1/k$ the *mean life* of a radioactive nucleus. The mean life of a radon nucleus is about $1/0.18 = 5.6$ days. The mean life of a carbon-14 nucleus is more than 8000 years. Show that 95% of the radioactive nuclei originally present in a sample will disintegrate within three mean lifetimes, i.e., by time $t = 3/k$. Thus, the mean life of a nucleus gives a quick way to estimate how long the radioactivity of a sample will last.

Carbon-14 Dating

The half-lives of radioactive elements can sometimes be used to date events from the Earth's past. The ages of rocks more than 2 billion years old have been measured by the extent of the radioactive decay of uranium (half-life 4.5 billion years!). In a living organism, the ratio of radioactive carbon, carbon-14, to ordinary carbon stays fairly constant during the lifetime of the organism, being approximately equal to the ratio in the organism's surroundings at the time. After the organism's death, however, no new carbon is ingested, and the proportion of carbon-14 in the organism's remains decreases as the carbon-14 decays. Since the half-life of carbon-14 is known to be about 5700 years, it is possible to estimate the age of organic remains by comparing the proportion of carbon-14 they contain with the proportion assumed to have been in the organism's environment at the time it lived. Archeologists have dated shells (which contain $CaCO_3$), seeds, and wooden artifacts this way. The estimate of 15,500 years for the age of the cave paintings at Lascaux, France, is based on carbon-14 dating. After generations of controversy, the shroud of Turin was shown by carbon-14 dating in 1988 to have been made later than 1200 A.D.

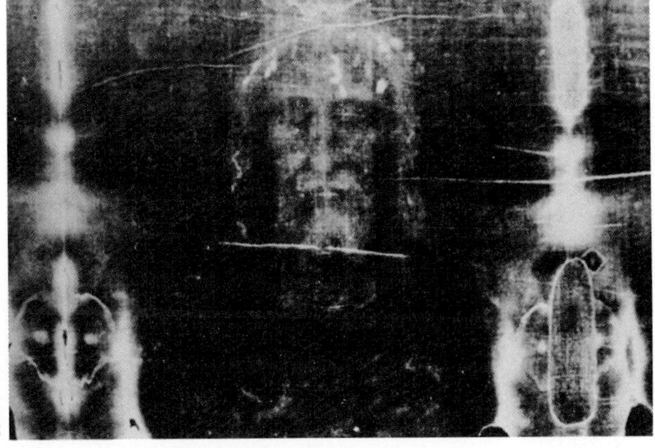

Wide World Photos

The Shroud of Turin, long believed to be the burial cloth of Christ, is now known to be medieval. It was probably made between 1260 and 1390 A.D.—certainly not before 1200 A.D. The dating procedures, carried out independently by laboratories in England, Switzerland, and the United States, used up only a few square centimeters of the cloth. (*Source: Science* 21, October 1988, Vol. 242, p. 378)

15. The charcoal from a tree killed in the volcanic eruption that formed Crater Lake in Oregon contained 44.5% of the carbon-14 found in living matter. About how old is Crater Lake?

16. To see the effect of a relatively small error in the estimate of the amount of carbon-14 in a sample being dated, consider this hypothetical situation:

a) A fossilized bone found in central Illinois in the year 2000 A.D. is found to contain 17% of its original carbon-14 content. Estimate the year the animal died.

b) Repeat (a) assuming 18% instead of 17%.

c) Repeat (a) assuming 16% instead of 17%.

17. *Art forgery.* A painting attributed to Vermeer (1632–1675) should have contained no more than 96.2% of its original carbon-14, but it contained 99.5% instead. What is the earliest date at which the painting could have been made?

TOOLKIT PROGRAM

Super*Grapher	Graphs exponential functions of all kinds

7.5 Indeterminate Forms and l'Hôpital's Rule

In the late seventeenth century, John Bernoulli discovered a rule for calculating limits of fractions whose numerators and denominators both approach zero. The rule is known today as **l'Hôpital's rule,** named after Guillaume François Antoine de l'Hôpital (1661–1704), Marquis de St.

Mesme, a French nobleman who wrote the first introductory differential calculus text, in which the rule first appeared.

L'Hôpital's rule gives fast results even when other methods are slow or unavailable. This section introduces the rule, uses it to show that $\lim_{n\to\infty}(1 + (1/n))^n = e$, and sets the stage for the growth-rate comparisons in Section 7.6.

The Indeterminate Form 0/0

If functions f and g are continuous at $x = a$ but $f(a) = g(a) = 0$, the limit

$$\lim_{x\to a} \frac{f(x)}{g(x)} \tag{1}$$

cannot be evaluated by substituting $x = a$, since this produces 0/0, a meaningless expression known as an **indeterminate form.**

As we have seen, the value of the limit in Eq. (1) is hard to predict:

$$\lim_{x\to 2} \frac{x^2 - 4}{x - 2} = \lim_{x\to 2} (x + 2) = 4$$

$$\lim_{x\to 0} \frac{\sin x}{x} = 1$$

$$\lim_{x\to 0} \frac{1 - \cos x}{x} = 0.$$

The limit

$$f'(a) = \lim_{x\to a} \frac{f(x) - f(a)}{x - a}$$

from which we calculate derivatives always produces the indeterminate form 0/0. Our success in calculating derivatives suggests that we might turn things around and use derivatives to calculate limits that lead to indeterminate forms. For example, knowing the derivative of $\sin x$ would let us find

$$\lim_{x\to 0} \frac{\sin x}{x} = \lim_{x\to 0} \frac{\sin x - \sin 0}{x - 0} = \frac{d}{dx}(\sin x)\bigg|_{x=0} = \cos 0 = 1.$$

L'Hôpital's rule gives an explicit connection between derivatives and limits that lead to the indeterminate form 0/0.

THEOREM 1

L'Hôpital's Rule (First Form)

Suppose that $f(a) = g(a) = 0$, that $f'(a)$ and $g'(a)$ exist, and that $g'(a) \neq 0$. Then

$$\lim_{x\to a} \frac{f(x)}{g(x)} = \frac{f'(a)}{g'(a)}.$$

PROOF Working backward from $f'(a)$ and $g'(a)$, which are themselves limits, we have

$$\frac{f'(a)}{g'(a)} = \frac{\lim_{x\to a} \dfrac{f(x)-f(a)}{x-a}}{\lim_{x\to a} \dfrac{g(x)-g(a)}{x-a}} = \lim_{x\to a} \frac{\dfrac{f(x)-f(a)}{x-a}}{\dfrac{g(x)-g(a)}{x-a}}$$

$$= \lim_{x\to a} \frac{f(x)-f(a)}{g(x)-g(a)} = \lim_{x\to a} \frac{f(x)-0}{g(x)-0} = \lim_{x\to a} \frac{f(x)}{g(x)}.$$

■

EXAMPLE 1

a) $\displaystyle \lim_{x\to 0} \frac{3x - \sin x}{x} = \left.\frac{3 - \cos x}{1}\right|_{x=0} = 2$

b) $\displaystyle \lim_{x\to 0} \frac{\sqrt{1+x} - 1}{x} = \left.\frac{\dfrac{1}{2\sqrt{1+x}}}{1}\right|_{x=0} = \frac{1}{2}$

c) $\displaystyle \lim_{x\to 0} \frac{x - \sin x}{x^3} = \left.\frac{1 - \cos x}{3x^2}\right|_{x=0} = ?$ ■

Notice that to apply l'Hôpital's rule to f/g we divide the derivative of f by the derivative of g. Do not fall into the trap of taking the derivative of f/g. The quotient to use is f'/g', not $(f/g)'$.

What can we do about the limit in Example 1(c)? The first form of l'Hôpital's rule does not tell us what the limit is because the derivative of $g(x) = x^3$ is zero at $x = 0$. However, a stronger form of l'Hôpital's rule from advanced calculus says that whenever the rule gives 0/0 we can apply it again, repeating the process until we get a different result. With this stronger rule we can finish the work begun in Example 1(c):

$$\lim_{x\to 0} \frac{x - \sin x}{x^3} = \lim_{x\to 0} \frac{1 - \cos x}{3x^2} \qquad \left(\text{Still } \frac{0}{0}\text{; apply the rule again.}\right)$$

$$= \lim_{x\to 0} \frac{\sin x}{6x} \qquad \left(\text{Still } \frac{0}{0}\text{; apply the rule again.}\right)$$

$$= \lim_{x\to 0} \frac{\cos x}{6} = \frac{1}{6}. \qquad \text{(A different result. Stop.)}$$

THEOREM 2

L'Hôpital's Rule (Stronger Form)

Suppose that $f(x_0) = g(x_0) = 0$ and that the functions f and g are both differentiable on an open interval (a, b) that contains the point x_0. Suppose also that $g' \neq 0$ at every point in (a, b) except possibly x_0. Then

$$\lim_{x\to x_0} \frac{f(x)}{g(x)} = \lim_{x\to x_0} \frac{f'(x)}{g'(x)}, \tag{2}$$

provided the limit on the right exists.

EXAMPLE 2

$$\lim_{x\to 0}\frac{\sqrt{1+x}-1-(x/2)}{x^2} \qquad \left(\frac{0}{0}\right)$$

$$= \lim_{x\to 0}\frac{(1/2)(1+x)^{-1/2}-(1/2)}{2x} \qquad \left(\text{Still } \frac{0}{0}\right)$$

$$= \lim_{x\to 0}\frac{-(1/4)(1+x)^{-3/2}}{2} = -\frac{1}{8}$$

When you apply l'Hôpital's rule, look for a change from 0/0 to something else. This is where the limit's value is revealed.

EXAMPLE 3

$$\lim_{x\to 0}\frac{1-\cos x}{x+x^2} \qquad \left(\frac{0}{0}\right)$$

$$= \lim_{x\to 0}\frac{\sin x}{1+2x} = \frac{0}{1} = 0$$

If we continue to differentiate in an attempt to apply l'Hôpital's rule once more, we get

$$\lim_{x\to 0}\frac{1-\cos x}{x+x^2} = \lim_{x\to 0}\frac{\sin x}{1+2x} = \lim_{x\to 0}\frac{\cos x}{2} = \frac{1}{2},$$

which is wrong.

If we reach a point where one of the derivatives is zero and the other is not, then the limit in question is either zero, as in Example 3, or infinity, as in the next example.

EXAMPLE 4

$$\lim_{x\to 0^+}\frac{\sin x}{x^2} \qquad \left(\frac{0}{0}\right)$$

$$= \lim_{x\to 0^+}\frac{\cos x}{2x} = \infty$$

A Misnamed Rule and the First Differential Calculus Text

In 1694 John Bernoulli agreed to accept a retainer of 300 pounds per year from his former pupil l'Hôpital to solve problems for him and keep him up to date on calculus. One of the problems was the so-called 0/0 problem, which Bernoulli solved as agreed. When l'Hôpital published his notes on calculus in book form in 1696, the 0/0 rule appeared as a theorem. L'Hôpital acknowledged his debt to Bernoulli and, to avoid claiming authorship of the book's entire contents, had the book appear without bearing his own name. Bernoulli nevertheless accused l'Hôpital of plagiarism, an accusation inadvertently supported after l'Hôpital's death by the publisher's promotion of the book as l'Hôpital's. By 1721, Bernoulli, a man so jealous he once threw his son Daniel out of the house for accepting a mathematics prize from the French Academy of Sciences, claimed to have been the author of the entire work. As puzzling and fickle as ever, history accepted Bernoulli's claim (until recently), but named the rule after l'Hôpital.

The Forms ∞/∞ and $\infty \cdot 0$

In more advanced books it is proved that l'Hôpital's rule applies to the indeterminate form ∞/∞ as well as 0/0. If $f(x)$ and $g(x)$ both approach infinity as x approaches a, then

$$\lim_{x\to a}\frac{f(x)}{g(x)} = \lim_{x\to a}\frac{f'(x)}{g'(x)},$$

provided the limit on the right exists. In the notation $x \to a$, a may be either finite or infinite.

EXAMPLE 5

$$\lim_{x\to(\pi/2)^-} \frac{\tan x}{1+\tan x} \qquad \left(\frac{\infty}{\infty}\right)$$

$$= \lim_{x\to(\pi/2)^-} \frac{\sec^2 x}{\sec^2 x} = 1$$

The form $\infty \cdot 0$ can sometimes be handled by using algebra to get $0/0$ or ∞/∞ instead. Here again, we do not mean to suggest that there is a number $\infty \cdot 0$ any more than we mean to suggest that there is a number $0/0$ or ∞/∞. These forms are not numbers but descriptions of limits.

EXAMPLE 6

$$\lim_{x\to\infty} x \sin\frac{1}{x} \qquad (\infty \cdot 0)$$

$$= \lim_{h\to 0^+} \frac{1}{h}\sin h \qquad (\text{Set } h = 1/x)$$

$$= 1. \qquad (\text{A known result})$$

The Forms 1^∞ and 0^0

The forms 1^∞ and 0^0 can sometimes be handled by taking logarithms first. The idea is to calculate the limit of the logarithm and exponentiate that limit.

If: $\lim_{x\to a} \ln f(x) = L,$

Then: $\lim_{x\to a} f(x) = \lim_{x\to a} e^{\ln f(x)} = e^L.$ (3)

EXAMPLE 7 Show that

$$\lim_{x\to 0^+} (1+x)^{1/x} = e. \tag{4}$$

Solution We let $f(x) = (1+x)^{1/x}$ and find $\lim_{x\to 0^+} \ln f(x)$. Since

$$\ln f(x) = \ln(1+x)^{1/x} = \frac{1}{x}\ln(1+x),$$

l'Hopital's rule gives

$$\lim_{x\to 0^+} \ln f(x) = \lim_{x\to 0^+} \frac{\ln(1+x)}{x} \qquad \left(\frac{0}{0}\right)$$

$$= \lim_{x\to 0^+} \frac{\frac{1}{1+x}}{1} = 1.$$

Therefore,

$$\lim_{x\to 0^+} f(x) = e^1 = e. \qquad (\text{Eq. (3) with } L = 1).$$

An Alternative Definition of e

The limit

$$\lim_{n\to\infty}\left(1+\frac{1}{n}\right)^n = e$$

is the special case of Eq. (4) obtained by restricting x to be the reciprocal of a positive integer. It gives a way to define e without logarithms. However, the proof that the limit exists would then have to be different from the logarithm-based proof in Example 7.

FLOWCHART 7.1 l'Hôpital's Rule

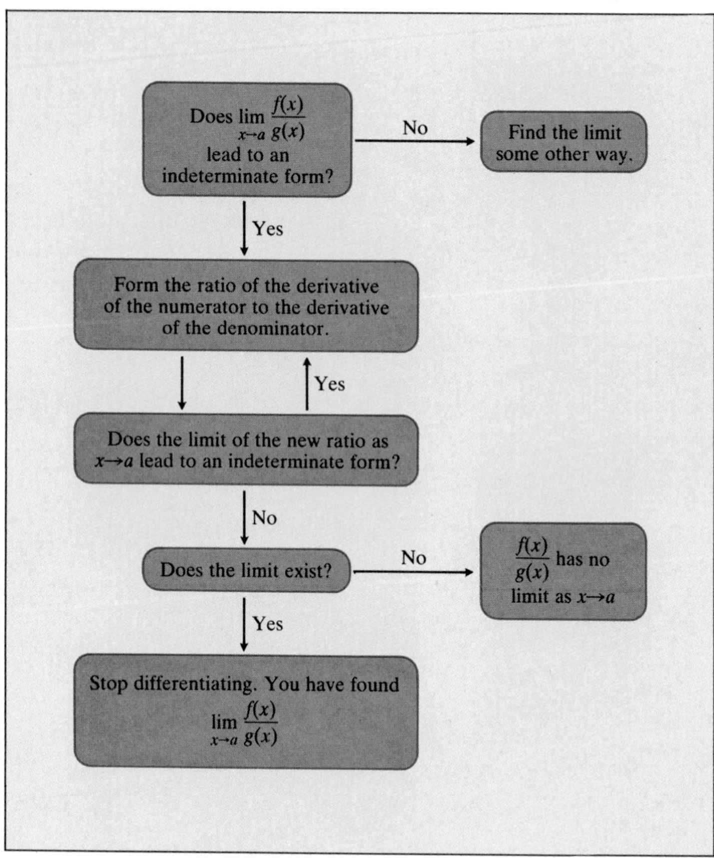

Exercises 7.5

Use l'Hôpital's rule to find the limits in Exercises 1–18.

1. $\lim_{x\to 2} \frac{x-2}{x^2-4}$
2. $\lim_{t\to\infty} \frac{6t+5}{3t-8}$
3. $\lim_{x\to 1} \frac{x^3-1}{4x^3-x-3}$
4. $\lim_{x\to\pi/2} \frac{2x-\pi}{\cos x}$
5. $\lim_{t\to 0} \frac{\sin t^2}{t}$
6. $\lim_{x\to 0} \frac{\sin 5x}{x}$
7. $\lim_{x\to 0} \frac{3^{\sin x}-1}{x}$
8. $\lim_{x\to 0} \frac{(1/2)^x-1}{x}$
9. $\lim_{x\to\infty} \frac{5x^2-3x}{7x^2+1}$
10. $\lim_{t\to 0} \frac{\cos t-1}{t^2}$
11. $\lim_{x\to\pi/2} \frac{1-\sin x}{1+\cos 2x}$
12. $\lim_{x\to\pi/2} \left(\frac{\pi}{2}-x\right)\tan x$
13. $\lim_{x\to 0} \frac{2x}{x+7\sqrt{x}}$
14. $\lim_{x\to\infty} \frac{x-2x^2}{3x^2+5x}$
15. $\lim_{t\to 0} \frac{10(\sin t-t)}{t^3}$
16. $\lim_{x\to 0} \frac{x(1-\cos x)}{x-\sin x}$
17. $\lim_{x\to 0} \left(\frac{1}{\sin x}-\frac{1}{x}\right)$ (*Hint:* Add the fractions first.)
18. $\lim_{x\to 0} \left(\frac{1}{x}-\frac{1}{\sqrt{x}}\right)$ (*Hint:* Add the fractions first.)

Use the technique of Example 7 to find the limits in Exercises 19–22.

19. $\lim_{x\to 0^+} x^{(1/\ln x)}$
20. $\lim_{x\to 0^+} x^x$
21. $\lim_{x\to 1^+} x^{1/(x-1)}$
22. $\lim_{x\to 0} (e^x+x)^{1/x}$

23. Which is correct, (a) or (b)? Explain.

a) $\lim_{x\to 3}\frac{x-3}{x^2-3}=\lim_{x\to 3}\frac{1}{2x}=\frac{1}{6}$

b) $\lim_{x\to 3}\frac{x-3}{x^2-3}=\frac{0}{6}=0$

24. L'Hôpital's rule does not help with

$$\lim_{x\to(\pi/2)^-}\frac{\sec x}{\tan x}.$$

Try it—you just keep on going. Find the limit some other way.

25. Let $A(t)$ be the area of the region in the first quadrant enclosed by the coordinate axes, the curve $y=e^{-x}$, and the line $x=t>0$ (Fig. 7.14). Let $V(t)$ be the volume of the solid generated by revolving the region about the x-axis. Find the following limits.

a) $\lim_{t\to\infty}A(t)$

b) $\lim_{t\to\infty}V(t)/A(t)$

c) $\lim_{t\to 0^+}V(t)/A(t)$

26. Let
$$f(x)=\begin{cases}x+2, & x\neq 0\\ 0, & x=0\end{cases}$$
$$g(x)=\begin{cases}x+1, & x\neq 0\\ 0, & x=0.\end{cases}$$

Show that

$$\lim_{x\to 0}\frac{f'(x)}{g'(x)}=1 \quad\text{but}\quad \lim_{x\to 0}\frac{f(x)}{g(x)}=2.$$

Doesn't this contradict l'Hôpital's rule?

27. **CALCULATOR** Find out how close you can come to

$$e=\lim_{n\to\infty}\left(1+\frac{1}{n}\right)^n=2.718281828459045\ldots$$

with your calculator by taking $n=10, 10^2, 10^3, \ldots$. You can expect the approximations to approach e at first but on some calculators they will move away again as round-off errors take their toll.

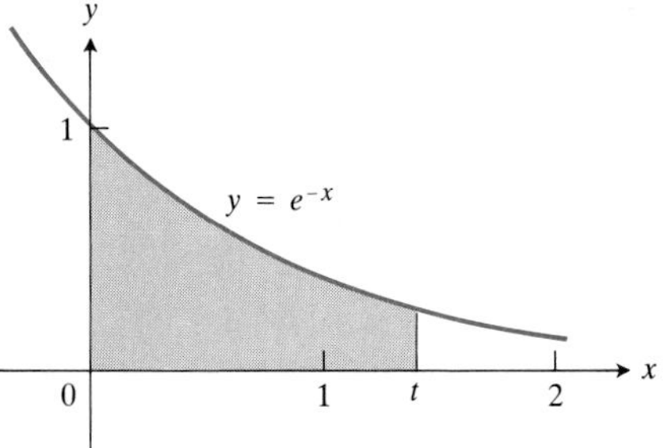

7.14 The region in Exercise 25.

28. **CALCULATOR** If you did not do so in Section 3.4, try to guess the value of

$$\lim_{h\to 0}\frac{1-\cos h}{h^2}$$

by taking $h=1, 0.1, 0.01$, and so on, as far as your calculator can go. Then confirm your guess by finding the limit with l'Hôpital's rule.

29. *Continuously compounded interest continued.* The introduction to Section 2.1 says that the limit to how much you can earn in a year by investing $100 at 6% compound interest is

$$\lim_{k\to\infty}100\left(1+\frac{0.06}{k}\right)^k=106.184. \qquad \begin{pmatrix}\text{Nearest tenth}\\ \text{of a cent}\end{pmatrix}$$

a) Use Eq. (4) of the present section with $x=1/k$ to confirm the statement in Section 5.8, Eq. (26), that

$$\lim_{k\to\infty}A_0\left(1+\frac{r}{k}\right)^{kt}=A_0e^{rt}.$$

b) **CALCULATOR** Find $100e^{0.06}$ to three decimal places.

7.6 The Rates at Which Functions Grow

This section shows how to compare the rates at which functions of x grow as x becomes large and introduces the so-called little-oh and big-oh notation sometimes used to describe the results of these comparisons. The exponential function e^x grows so rapidly as $x\to\infty$, and the logarithmic function $\ln x$ grows so slowly, that they set the standards by which all other functions are judged.

We shall restrict our attention to functions whose values eventually become and remain positive as $x\to\infty$. (Functions with negative values are compared by taking absolute values first.)

Relative Rates of Growth

You may have noticed that exponential functions like 2^x and e^x seem to grow more rapidly as x gets large than the polynomials and rational functions we graphed in Chapter 4. These exponentials certainly grow more rapidly than x itself, as Figs. 7.9 and 7.11 show, and you can see 2^x outgrowing x^2 as x increases in Fig. 7.13. In fact, as $x \to \infty$, the functions 2^x and e^x grow faster than any power of x, even $x^{1,000,000}$ (Exercise 9).

To get a feeling for how rapidly the values of $y = e^x$ grow with increasing x, think of graphing the function on a large blackboard, with the axes scaled in centimeters. At $x = 1$ cm, the graph is $e^1 \approx 3$ cm above the x-axis. At $x = 6$ cm, the graph is $e^6 \approx 403$ cm ≈ 4 m high (it is about to go through the ceiling if it hasn't done so already). At $x = 10$ cm, the graph is $e^{10} \approx 22{,}026$ cm ≈ 220 m high, higher than most buildings. At $x = 24$ cm, the graph is more than halfway to the moon, and at $x = 43$ cm from the origin, the graph is high enough to reach past the nearest neighboring star, Proxima Centauri:

$$\begin{aligned} e^{43} &\approx 4.7 \times 10^{18} \text{ cm} \\ &= 4.7 \times 10^{13} \text{ km} \\ &\approx 1.57 \times 10^{8} \text{ light seconds} \qquad \begin{pmatrix}\text{Light travels at 300,000} \\ \text{km/sec in a vacuum}\end{pmatrix} \qquad (1) \\ &\approx 5.0 \text{ light years} \end{aligned}$$

The distance to Proxima Centauri is about 4.3 light years. Yet, with $x = 43$ cm from the origin, the graph is still less than 2 feet to the right of the y-axis.

In contrast, logarithmic functions like $y = \log_2 x$ and $y = \ln x$ grow more slowly as $x \to \infty$ than any positive power of x (Exercise11). With axes scaled in centimeters, you have to go more than 4 light-years out on the x-axis to find a point where the graph of $y = \ln x$ is even $y = 43$ cm high.

These important comparisons of exponential, polynomial, and logarithmic functions can be made precise by defining what it means for a function $f(x)$ to grow faster than another function $g(x)$ as $x \to \infty$.

DEFINITION

Rates of Growth as $x \to \infty$

1. f **grows faster than** g as $x \to \infty$ if $\lim_{x\to\infty} \dfrac{f(x)}{g(x)} = \infty$ (2)

   or, equivalently, if $\lim_{x\to\infty} \dfrac{g(x)}{f(x)} = 0.$ (3)

2. If f grows faster than g as $x \to \infty$, we also say that g **grows slower than** f as $x \to \infty$.
3. f and g **grow at the same rate** as $x \to \infty$ if

$$\lim_{x\to\infty} \frac{f(x)}{g(x)} = L \neq 0. \qquad (L \text{ finite and not zero}) \qquad (4)$$

According to these definitions, $y = 2x$ does not grow faster than $y = x$. The two functions grow at the same rate because

$$\lim_{x\to\infty} \frac{2x}{x} = \lim_{x\to\infty} 2 = 2,$$

which is a finite nonzero limit. The reason for this apparent disregard of common sense is that we want "f grows faster than g" to mean that for large x-values, g is negligible when compared to f.

EXAMPLE 1 e^x grows faster than x^2 as $x \to \infty$ because

$$\lim_{x\to\infty} \frac{e^x}{x^2} = \infty,$$

as we can see by two applications of l'Hôpital's rule:

$$\lim_{x\to\infty} \frac{e^x}{x^2} = \lim_{x\to\infty} \frac{e^x}{2x} = \lim_{x\to\infty} \frac{e^x}{2} = \infty.$$ ■

EXAMPLE 2 3^x grows faster than 2^x as $x \to \infty$ because

$$\lim_{x\to\infty} \frac{3^x}{2^x} = \lim_{x\to\infty} \left(\frac{3}{2}\right)^x = \infty.$$ ■

EXAMPLE 3 As Example 2 may have suggested, exponential functions never grow at the same rate as $x \to \infty$. If $a > b > 0$, then a^x grows faster than b^x. Since $(a/b) > 1$,

$$\lim_{x\to\infty} \frac{a^x}{b^x} = \lim_{x\to\infty} \left(\frac{a}{b}\right)^x = \infty.$$ ■

EXAMPLE 4 x^2 grows faster than $\ln x$ as $x \to \infty$ because

$$\begin{aligned} \lim_{x\to\infty} \frac{x^2}{\ln x} &= \lim_{x\to\infty} \frac{2x}{1/x} \qquad \text{(l'Hôpital's rule)} \\ &= \lim_{x\to\infty} 2x^2 = \infty. \end{aligned}$$ ■

EXAMPLE 5 $\ln x$ grows slower than x as $x \to \infty$ because

$$\begin{aligned} \lim_{x\to\infty} \frac{\ln x}{x} &= \lim_{x\to\infty} \frac{1/x}{1} \qquad \text{(l'Hôpital's rule)} \\ &= \lim_{x\to\infty} \frac{1}{x} = 0. \end{aligned}$$ ■

EXAMPLE 6 In contrast to exponential functions, logarithmic functions always grow at the same rate as $x \to \infty$. For any bases a and b,

$$\lim_{x\to\infty} \frac{\log_a x}{\log_b x} = \lim_{x\to\infty} \frac{\ln x/\ln a}{\ln x/\ln b} = \frac{\ln b}{\ln a}.$$

The limiting ratio is always finite and never zero. ■

Growing at the Same Rate Is a Transitive Relation

If f grows at the same rate as g as $x \to \infty$, and g grows at the same rate as h as $x \to \infty$, then f grows at the same rate as h as $x \to \infty$. The reason is that

$$\lim_{x\to\infty} \frac{f}{g} = L_1 \quad \text{and} \quad \lim_{x\to\infty} \frac{g}{h} = L_2$$

together imply

$$\lim_{x\to\infty} \frac{f}{h} = \lim_{x\to\infty} \frac{f}{g} \cdot \frac{g}{h} = L_1 L_2.$$

If L_1 and L_2 are finite and nonzero, then so is $L_1 L_2$.

EXAMPLE 7 Show that $\sqrt{x^2 + 5}$ and $(2\sqrt{x} - 1)^2$ grow at the same rate as $x \to \infty$.

Solution We show that the functions grow at the same rate by showing that they both grow at the same rate as x:

$$\lim_{x\to\infty} \frac{\sqrt{x^2+5}}{x} = \lim_{x\to\infty} \sqrt{1 + \frac{5}{x^2}} = 1$$

$$\lim_{x\to\infty} \frac{(2\sqrt{x} - 1)^2}{x} = \lim_{x\to\infty} \left(\frac{2\sqrt{x} - 1}{\sqrt{x}}\right)^2 = \lim_{x\to\infty} \left(2 - \frac{1}{\sqrt{x}}\right)^2 = 4.$$ ■

Order of Magnitude as $x \to \infty$

Here we introduce the "little-oh" and "big-oh" notation, invented by number theorists a hundred years ago and now commonplace in mathematical analysis and computer science.

DEFINITION

> A function f is **of smaller order than** g as $x \to \infty$ if $\lim_{x\to\infty} \frac{f(x)}{g(x)} = 0$.
>
> We indicate this by writing $\boldsymbol{f = o(g)}$ ("f is little-oh of g")

Notice that saying $f = o(g)$ as $x \to \infty$ is another way to say that f grows slower than g as $x \to \infty$.

EXAMPLE 8

$$\ln x = o(x) \text{ as } x \to \infty \quad \text{because} \quad \lim_{x\to\infty} \frac{\ln x}{x} = 0$$

$$x^2 = o(x^3 + 1) \text{ as } x \to \infty \quad \text{because} \quad \lim_{x\to\infty} \frac{x^2}{x^3+1} = 0$$ ■

If $f(x)/g(x)$ fails to approach zero as $x \to \infty$ but the ratio remains bounded, we say that f is at most the order of g as $x \to \infty$. What we mean, exactly, is that $f(x)/g(x)$ stays less than or equal to some integer M for x sufficiently large.

DEFINITION

A function f is **of at most the order of** g as $x \to \infty$ if there is a positive integer M for which

$$\frac{f(x)}{g(x)} \leq M,$$

for x sufficiently large. We indicate this by writing $\boldsymbol{f = O(g)}$ ("f is big-oh of g")

EXAMPLE 9

$$x + \sin x = O(x) \text{ as } x \to \infty \quad \text{because} \quad \frac{x + \sin x}{x} \leq 2 \text{ for } x \text{ sufficiently large.}$$

■

EXAMPLE 10

$$e^x + x^2 = O(e^x) \text{ as } x \to \infty \quad \text{because} \quad \frac{e^x + x^2}{e^x} \to 1 \text{ as } x \to \infty.$$

■

EXAMPLE 11

$$x = O(e^x) \text{ as } x \to \infty \quad \text{because} \quad \frac{x}{e^x} \to 0 \text{ as } x \to \infty.$$

■

If you look at the definitions again, you will see that $f = o(g)$ implies $f = O(g)$. Also, if f and g grow at the same rate, then $f = O(g)$ and $g = O(f)$, as you will be asked to confirm in Exercise 15.

Sequential versus Binary Search

Computer scientists sometimes measure the efficiency of an algorithm by counting the number of steps a computer has to take to use the algorithm to do something. There can be significant differences in how efficiently algorithms perform, even if they are designed to accomplish the same task. These differences are often described in big-oh notation. Here is an example.

Webster's *Third New International Dictionary* lists about 26,000 words that begin with the letter *a*. One way to look up a word, or find out it is not there, is to read through the list one word at a time from the beginning until you either find your word or determine that it is not there. This method, called sequential search, makes no particular use of the words' alphabetical arrangement. You can be sure of getting an answer this way, but it might take 26,000 steps.

Another way to find the word or determine it is not there is to go straight to the middle of the list (give or take a few words). If you do not find the word, then go to the middle of the half that contains it and forget about the half that does not. (You know which half contains it because you know the list is ordered alphabetically.) This method eliminates roughly 13,000 words in a single step. If you do not find your word on the second try, then jump to the middle of the half that contains it. Continue this way until you have either found your word or divided the list in half so many times there are no words left. How many times do you have to divide the list to find the word or determine that it is not there? At most 15, because

$$2^{14} < 26{,}000 < 2^{15}.$$

That certainly beats a possible 26,000.

For a list of length n, a sequential search algorithm takes on the order of n steps to find a word or determine that it is not in the list. A binary search, as the second algorithm is called, takes on the order of $\log_2 n$ steps. The reason is that if $2^{m-1} < n \le 2^m$, then $m - 1 < \log_2 n \le m$ and the number of bisections required to narrow the list down to one word will be no more than $m = \lceil \log_2 n \rceil$, the smallest integer greater than or equal to $\log_2 n$.

Big-oh notation provides a compact way to say all this. The number of steps in a sequential search of an ordered list is $O(n)$ while the number of steps in a binary search is $O(\log_2 n)$. In our example, there is a big difference between the two and the difference can only increase with n because n grows faster than $\log_2 n$ as $n \to \infty$.

> A sequential search takes $O(n)$ steps.
>
> A binary search takes $O(\log_2 n)$ steps.

Exercises 7.6

1. Which of the following functions grow slower than e^x as $x \to \infty$?

a) $x + 3$ b) $x^3 - 3x + 1$
c) $\sqrt{x}$ d) 4^x
e) $(5/2)^x$ f) $\ln x$
g) $\log_{10} x$ h) e^{-x}
i) e^{x+1} j) $(1/2)e^x$

2. Which of the following functions grow faster than x^2 as $x \to \infty$? Which grow at the same rate as x^2? Which grow slower?

a) $x^2 + 4x$ b) $x^3 + 3$
c) x^5 d) $15x + 3$
e) $\sqrt{x^4 + 5x}$ f) $(x + 1)^2$
g) $\ln x$ h) $\ln(x^2)$
i) $\ln(10^x)$ j) 2^x

3. Which of the following functions grow at the same rate as $\ln x$ as $x \to \infty$?

a) $\log_3 x$ b) $\log_2 x^2$
c) $\log_{10}\sqrt{x}$ d) $1/x$
e) $1/\sqrt{x}$ f) e^{-x}
g) x h) $5 \ln x$
i) 2 j) $\sin x$

4. Order the following functions from fastest growing to slowest growing as $x \to \infty$.

a) e^x b) x^x
c) $(\ln x)^x$ d) $e^{x/2}$

5. Show that $\sqrt{10x + 1}$ and $\sqrt{x + 1}$ grow at the same rate as $x \to \infty$ by showing that they both grow at the same rate as $\sqrt{x}$ as $x \to \infty$.

6. Show that $\sqrt{x^4 + x}$ and $\sqrt{x^4 - x^3}$ grow at the same rate as $x \to \infty$ by showing that they both grow at the same rate as x^2 as $x \to \infty$.

7. True, or false? As $x \to \infty$,

a) $x = o(x)$ b) $x = o(x + 5)$
c) $x = O(x + 5)$ d) $x = O(2x)$
e) $e^x = o(e^{2x})$ f) $x + \ln x = O(x)$
g) $\ln x = o(\ln 2x)$ h) $\sqrt{x^2 + 5} = O(x)$

8. True, or false? As $x \to \infty$,

a) $\dfrac{1}{x + 3} = O\left(\dfrac{1}{x}\right)$ b) $\dfrac{1}{x} + \dfrac{1}{x^2} = O\left(\dfrac{1}{x}\right)$
c) $\dfrac{1}{x} - \dfrac{1}{x^2} = o\left(\dfrac{1}{x}\right)$ d) $2 + \cos x = O(2)$
e) $e^x + x = O(e^x)$ f) $x \ln x = o(x^2)$
g) $\ln(\ln x) = O(\ln x)$ h) $\ln(x) = o(\ln(x^2 + 1))$

9. Show that e^x grows faster as $x \to \infty$ than x^n for any positive integer n, even $x^{1{,}000{,}000}$. (*Hint:* What is the nth derivative of x^n?)

10. *The function e^x outgrows any polynomial.* Show that e^x grows faster as $x \to \infty$ than any polynomial

$$a_n x^n + a_{n-1} x^{n-1} + \cdots + a_1 x + a_0.$$

11. a) Show that $\ln x$ grows slower as $x \to \infty$ than $x^{1/n}$ for any positive integer n, even $x^{1/1{,}000{,}000}$.
    b) **CALCULATOR** Although the values of $x^{1/1{,}000{,}000}$ eventually overtake the values of $\ln x$, you have to go way out on the x-axis before this happens. Find a value of x greater than 1 for which $x^{1/1{,}000{,}000} > \ln x$. You might start by observing that when $x > 1$ the equality $\ln x = x^{1/1{,}000{,}000}$ is equivalent to $\ln \ln x = (\ln x)/1{,}000{,}000$.

12. *The function* $\ln x$ *grows slower than any polynomial.* Show that $\ln x$ grows slower as $x \to \infty$ than any nonconstant polynomial.

13. Suppose you have four different algorithms for solving the same problem and each algorithm takes a number of steps equal to one of the functions listed here:

$$n, \quad n \log_2 n, \quad n^2, \quad n(\log_2 n)^2.$$

Which of the algorithms, if any, is the most efficient in the long run?

14. **CALCULATOR** Suppose you are looking for an item in an ordered list one million items long. How many steps might it take to find that item with a sequential search? A binary search?

15. Show that if functions f and g grow at the same rate as $x \to \infty$ then $f = O(g)$ and $g = O(f)$.

16. *Simpson's rule and the trapezoidal rule.* The definitions in the present section can be made more general by lifting the restriction that $x \to \infty$ and considering limits as $x \to a$ for any real number a. Show that the error E_S in the Simpson's-rule approximation of a definite integral is $O(h^4)$ as $h \to 0$ while the error E_T in the trapezoidal-rule approximation is $O(h^2)$. This gives another way to explain the relative accuracies of the two approximation methods.

7.7 The Inverse Trigonometric Functions

The inverse trigonometric functions arise in problems that require finding angles from side measurements in triangles. They also provide antiderivatives for a wide variety of functions and hence appear in solutions to a number of differential equations that arise in mathematics, engineering, and physics. In this section we show how the functions are defined, graphed, and evaluated. In the next section, we look at their derivatives and integrals.

The Arc Sine

The function $y = \sin x$ is not one-to-one; it runs through its full range of values from -1 to 1 twice on every interval of length 2π. However, if we restrict the domain of the sine to the interval from $-\pi/2$ to $\pi/2$, we find that the restricted function

$$y = \sin x, \quad -\pi/2 \le x \le \pi/2, \tag{1}$$

is one-to-one. It therefore has an inverse, which we denote by

$$y = \sin^{-1} x. \tag{2}$$

This equation is read "y equals the arc sine of x" or "y equals arc sine x" and is often written as

$$y = \arcsin x. \tag{3}$$

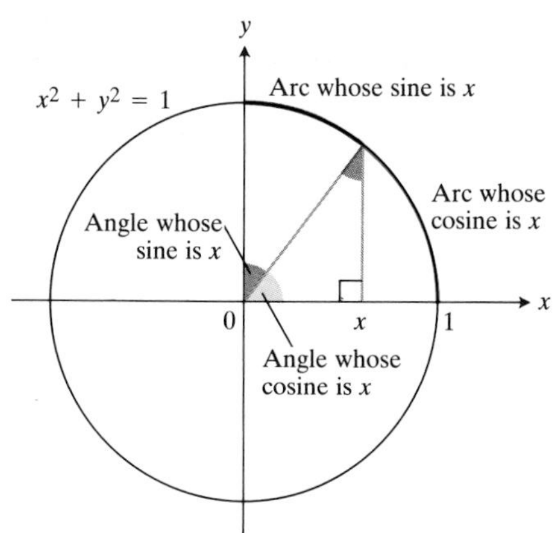

7.15 Arcs on the unit circle whose lengths represent $\sin^{-1}x$ and $\cos^{-1}x$.

In case you are wondering what the "arc" is doing there, look at Fig. 7.15, which gives the geometric interpretation of $y = \sin^{-1}x$ when y is positive. If $x = \sin y$, then y is the *arc* on the unit circle whose sine is x. For every

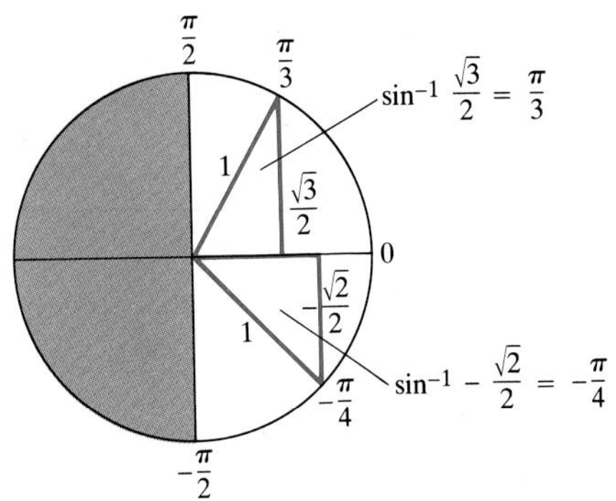

7.16 The angle whose measure is $y = \sin^{-1}x$ ranges from $-\pi/2$ to $\pi/2$.

value of x in the interval $[-1, 1]$, $y = \sin^{-1}x$ is the number in the interval $[-\pi/2, \pi/2]$ whose sine is x. For instance (Fig. 7.16),

$$\sin^{-1}0 = 0 \qquad \text{because} \quad \sin 0 = 0,$$

$$\sin^{-1}\sqrt{3}/2 = \pi/3 \qquad \text{because} \quad \sin \pi/3 = \sqrt{3}/2,$$

$$\sin^{-1}1 = \pi/2 \qquad \text{because} \quad \sin \pi/2 = 1,$$

$$\sin^{-1}(-\sqrt{2}/2) = -\pi/4 \qquad \text{because} \quad \sin(-\pi/4) = -\sqrt{2}/2.$$

The graph of $y = \sin^{-1}x$ is shown in Fig. 7.17. The gray curve in the figure is the reflection of the graph of $y = \sin x$ across the line $y = x$, and so it is the graph of $x = \sin y$. The graph of $y = \sin^{-1}x$ is the portion of this curve between $y = -\pi/2$ and $y = \pi/2$.

The -1 in $y = \sin^{-1}x$ is not an exponent; it means "inverse," not "reciprocal." The *reciprocal* of $\sin x$ is

$$(\sin x)^{-1} = \frac{1}{\sin x} = \csc x.$$

The graph of the arc sine in Fig. 7.17 is symmetric about the origin because the graph of $x = \sin y$ is symmetric about the origin. Algebraically, this means that

$$\sin^{-1}(-x) = -\sin^{-1}x \tag{4}$$

for every x in the domain of the arc sine, which is another way to say that the function $y = \sin^{-1}x$ is odd.

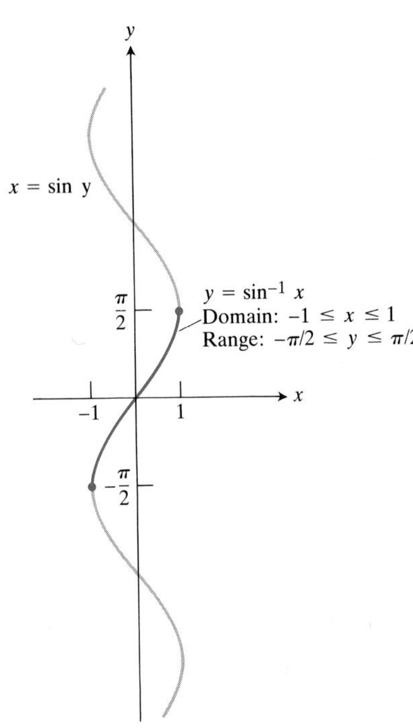

7.17 The graph of $y = \sin^{-1}x$.

The Arc Cosine

Like the sine function, the cosine function $y = \cos x$ is not one-to-one, but its restriction to the interval $[0, \pi]$,

$$y = \cos x, \quad 0 \le x \le \pi, \tag{5}$$

is one-to-one. The restricted function therefore has an inverse,

$$y = \cos^{-1}x, \tag{6}$$

which we call the **arc cosine** of x. For each value of x in the interval $[-1, 1]$, $y = \cos^{-1}x$ is the number in the interval $[0, \pi]$ whose cosine is x. The graph of $y = \cos^{-1}x$ is shown in Fig. 7.18.

As we can see from Fig. 7.19, the arc cosine of x satisfies the identity

$$\cos^{-1}x + \cos^{-1}(-x) = \pi, \tag{7}$$

or

$$\cos^{-1}(-x) = \pi - \cos^{-1}x. \tag{8}$$

We can also see from the triangle in Fig.7.20 that for $x > 0$,

$$\sin^{-1}x + \cos^{-1}x = \pi/2 \tag{9}$$

because $\sin^{-1}x$ and $\cos^{-1}x$ are then complementary angles in a right triangle whose hypotenuse is 1 unit long and one of whose legs is x units long.

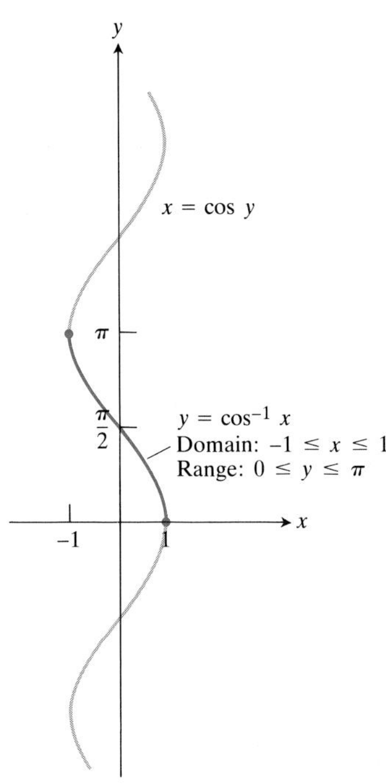

7.18 The graph of

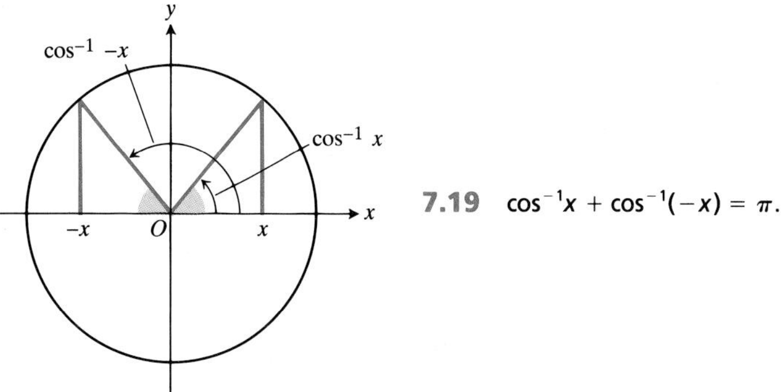

7.19 $\cos^{-1}x + \cos^{-1}(-x) = \pi$.

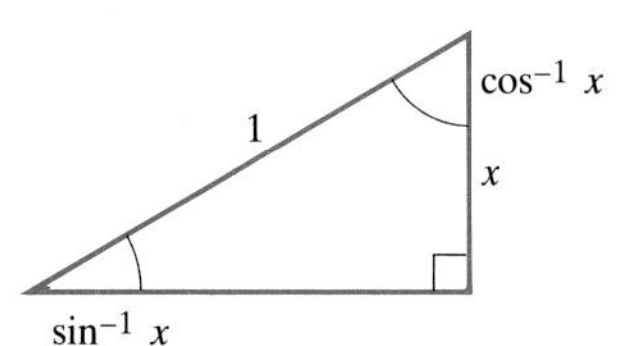

7.20 In this figure,

$\sin^{-1}x + \cos^{-1}x = \pi/2$.

Equation (9) holds for the other values of x in $[-1, 1]$ as well, but we cannot draw this conclusion from the geometry of the triangle in Fig. 7.20. It is, however, a consequence of Eqs. (4) and (8).

The Inverses of tan *x*, sec *x*, csc *x*, cot *x*

The other four basic trigonometric functions, $y = \tan x$, $y = \sec x$, $y = \csc x$, and $y = \cot x$, also have inverses when suitably restricted. The inverse of

$$y = \tan x, \qquad -\pi/2 < x < \pi/2, \tag{10}$$

is denoted by

$$y = \tan^{-1}x. \tag{11}$$

The domain of the arc tangent is the entire real line, and the range is the open interval $(-\pi/2, \pi/2)$. For every value of x, $y = \tan^{-1}x$ is the angle between $-\pi/2$ and $\pi/2$ whose tangent is x. The graph of $y = \tan^{-1}x$ is shown in Fig. 7.21.

The graph of $y = \tan^{-1}x$ is symmetric about the origin because it is a branch of the graph of $x = \tan y$ that is symmetric about the origin. Algebraically, this means that

$$\tan^{-1}(-x) = -\tan^{-1}x. \tag{12}$$

Like the arc sine, the arc tangent is an odd function of x.

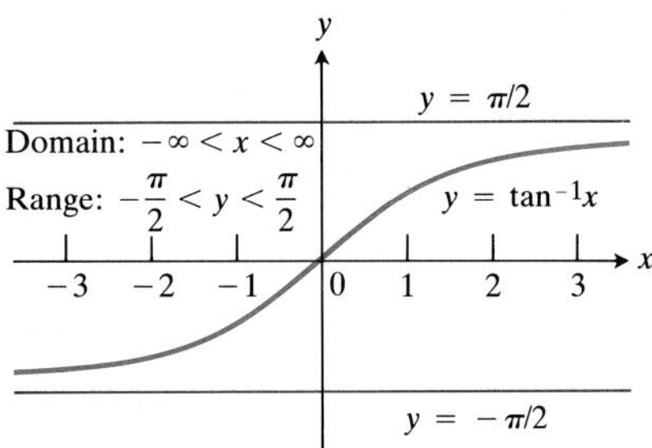

7.21 The branch chosen for $y = \tan^{-1}x$ is the one through the origin.

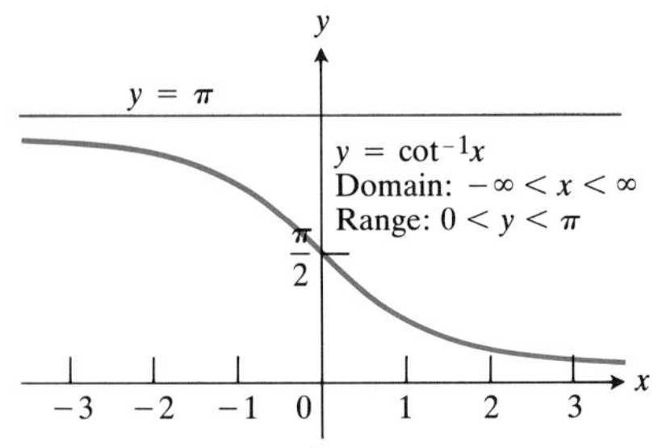

7.22 The graph of

$y = \cot^{-1}x = \pi/2 - \tan^{-1}x$.

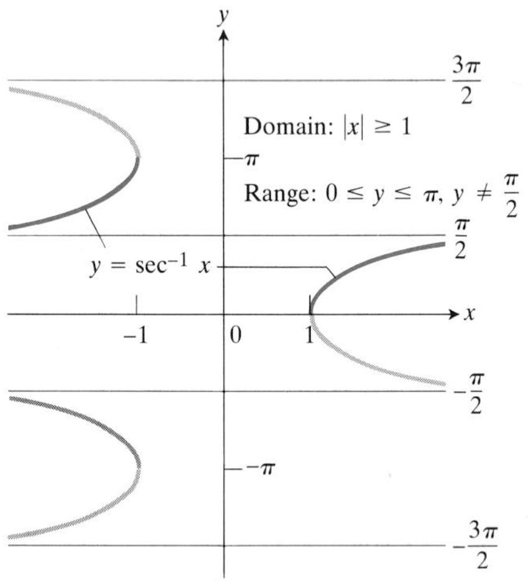

7.23 $y = \sec^{-1}x = \cos^{-1}(1/x)$ is defined for $|x| \geq 1$.

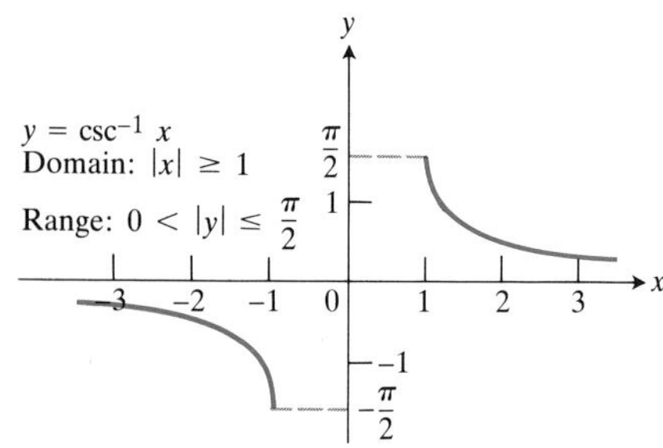

7.24 The graph of

$y = \csc^{-1}x = \sin^{-1}(1/x)$.

The inverses of the (restricted) functions

$$y = \cot x, \qquad 0 < x < \pi, \tag{13}$$

$$y = \sec x, \qquad 0 \leq x \leq \pi, \quad x \neq \pi/2, \tag{14}$$

$$y = \csc x, \qquad -\pi/2 \leq x \leq \pi/2, \quad x \neq 0, \tag{15}$$

are chosen to be the functions graphed in Figs. 7.22, 7.23, and 7.24. They are chosen this way to satisfy the relationships

$$\cot^{-1}x = \pi/2 - \tan^{-1}x, \tag{16}$$

$$\sec^{-1}x = \cos^{-1}(1/x), \tag{17}$$

$$\csc^{-1}x = \sin^{-1}(1/x). \tag{18}$$

We shall not dwell on these relationships, but they are handy for finding the values of $\cot^{-1}x$, $\sec^{-1}x$, and $\csc^{-1}x$ on a calculator that gives only $\tan^{-1}x$, $\cos^{-1}x$, and $\sin^{-1}x$.

WARNING ABOUT THE ARC SECANT Some writers choose $\sec^{-1}x$ to lie between 0 and $\pi/2$ when x is positive and between $-\pi$ and $-\pi/2$ when x is negative (hence as a negative angle in the third quadrant, as shown by the gray curve in Fig. 7.23). This has the advantage of simplifying the formula for the derivative of $\sec^{-1}x$ but the disadvantage of failing to satisfy Eq. (17) when x is negative. Also, some mathematical tables give third-quadrant values for $\sec^{-1}x$ instead of the second-quadrant values used in this book. Watch out for this when you use tables.

Right-Triangle Interpretations

The right-triangle interpretations of the inverse trigonometric functions in Fig. 7.25 can be useful in integration problems that require substitutions. We shall use some of them in Chapter 8.

EXAMPLE 1 Find $\csc\alpha$, $\cos\alpha$, $\sec\alpha$, $\tan\alpha$, and $\cot\alpha$ if

$$\alpha = \sin^{-1}\frac{\sqrt{3}}{2}. \tag{19}$$

Solution Equation (19) tells us that

$$\sin\alpha = \frac{\sqrt{3}}{2}.$$

We draw a reference triangle with vertical $\sqrt{3}$ and hypotenuse 2 (Fig. 7.26). The length of the remaining side is

$$\sqrt{(2)^2 - (\sqrt{3})^2} = \sqrt{4 - 3} = 1.$$

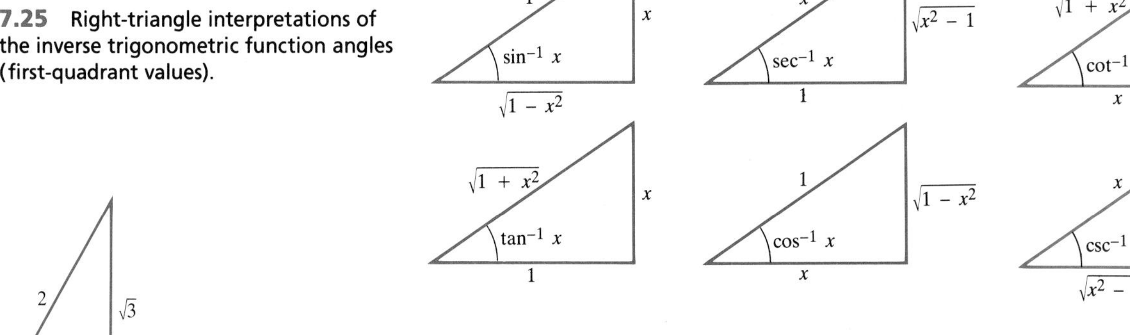

7.25 Right-triangle interpretations of the inverse trigonometric function angles (first-quadrant values).

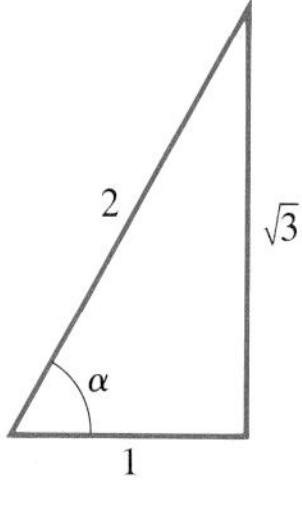

7.26 If $\alpha = \sin^{-1}(\sqrt{3}/2)$, then the values of the trigonometric functions of α can be read from this triangle.

The values we want can be read as ratios of side lengths from the completed triangle:

$$\csc\alpha = \frac{2}{\sqrt{3}} = \frac{2\sqrt{3}}{3}, \qquad \cos\alpha = \frac{1}{2}, \qquad \sec\alpha = 2,$$

$$\tan\alpha = \frac{\sqrt{3}}{1} = \sqrt{3}, \qquad \cot\alpha = \frac{1}{\sqrt{3}} = \frac{\sqrt{3}}{3}.$$

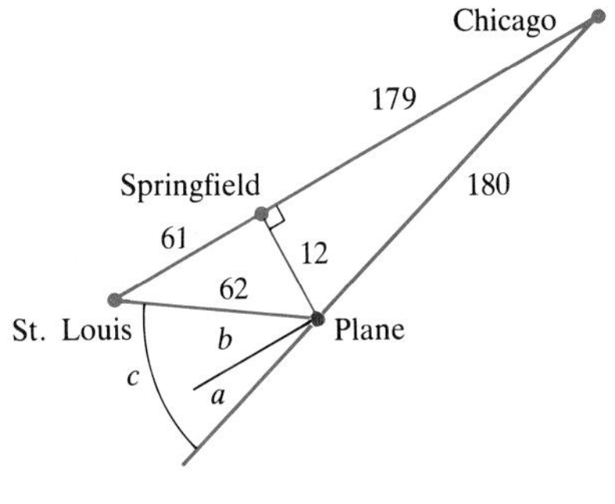

7.27 Diagram for drift correction (Example 2), with distances rounded to the nearest mile (drawing not to scale).

EXAMPLE 2 *Drift correction.* During an airplane flight from Chicago to St. Louis the navigator determines that the plane is 12 miles off course, as shown in Fig. 7.27. Find the angle a for a course parallel to the original, correct course, the angle b, and the correction angle $c = a + b$.

Solution

$$a = \sin^{-1}\frac{12}{180} \approx 0.067 \text{ radians} \approx 3.8°,$$

$$b = \sin^{-1}\frac{12}{62} \approx 0.195 \text{ radians} \approx 11.2°,$$

$$c = a + b \approx 15°.$$

Exercises 7.7

Use reference triangles like those in Figs. 7.25 and 7.26 to find the angles in Exercises 1–12.

1. a) $\tan^{-1}1$ b) $\tan^{-1}\sqrt{3}$ c) $\tan^{-1}\left(\frac{1}{\sqrt{3}}\right)$

2. a) $\tan^{-1}(-1)$ b) $\tan^{-1}(-\sqrt{3})$ c) $\tan^{-1}\left(\frac{-1}{\sqrt{3}}\right)$

3. a) $\sin^{-1}\left(\frac{1}{2}\right)$ b) $\sin^{-1}\left(\frac{1}{\sqrt{2}}\right)$ c) $\sin^{-1}\left(\frac{\sqrt{3}}{2}\right)$

4. a) $\sin^{-1}\left(\frac{-1}{2}\right)$ b) $\sin^{-1}\left(\frac{-1}{\sqrt{2}}\right)$ c) $\sin^{-1}\left(\frac{-\sqrt{3}}{2}\right)$

5. a) $\cos^{-1}\left(\frac{1}{2}\right)$ b) $\cos^{-1}\left(\frac{1}{\sqrt{2}}\right)$ c) $\cos^{-1}\left(\frac{\sqrt{3}}{2}\right)$

6. a) $\cos^{-1}\left(\frac{-1}{2}\right)$ b) $\cos^{-1}\left(\frac{-1}{\sqrt{2}}\right)$ c) $\cos^{-1}\left(\frac{-\sqrt{3}}{2}\right)$

7. a) $\sec^{-1}\sqrt{2}$ b) $\sec^{-1}\left(\frac{2}{\sqrt{3}}\right)$ c) $\sec^{-1}2$

8. a) $\sec^{-1}(-\sqrt{2})$ b) $\sec^{-1}\left(\frac{-2}{\sqrt{3}}\right)$ c) $\sec^{-1}(-2)$

9. a) $\csc^{-1}\sqrt{2}$ b) $\csc^{-1}\left(\frac{2}{\sqrt{3}}\right)$ c) $\csc^{-1}2$

10. a) $\csc^{-1}(-\sqrt{2})$ b) $\csc^{-1}\left(\frac{-2}{\sqrt{3}}\right)$ c) $\csc^{-1}(-2)$

11. a) $\cot^{-1}1$ b) $\cot^{-1}\sqrt{3}$ c) $\cot^{-1}\left(\frac{1}{\sqrt{3}}\right)$

12. a) $\cot^{-1}(-1)$ b) $\cot^{-1}(-\sqrt{3})$ c) $\cot^{-1}\left(\frac{-1}{\sqrt{3}}\right)$

13. Given that $\alpha = \sin^{-1}(1/2)$, find $\cos\alpha$, $\tan\alpha$, $\sec\alpha$, $\csc\alpha$.

14. Given that $\alpha = \cos^{-1}(-1/2)$, find $\sin\alpha$, $\tan\alpha$, $\sec\alpha$, $\csc\alpha$.

Evaluate the expressions in Exercises 15–34.

15. $\sin\left(\cos^{-1}\frac{\sqrt{2}}{2}\right)$

16. $\tan\left(\sin^{-1}\left(-\frac{1}{2}\right)\right)$

17. $\sec\left(\cos^{-1}\frac{1}{2}\right)$

18. $\cot\left(\sin^{-1}\left(-\frac{1}{2}\right)\right)$

19. $\csc(\sec^{-1}2)$

20. $\cos(\tan^{-1}(-\sqrt{3}))$

21. $\cos(\cot^{-1}1)$

22. $\csc\left(\sin^{-1}\left(-\frac{\sqrt{2}}{2}\right)\right)$

23. $\cot(\cos^{-1}0)$

24. $\sec\left(\tan^{-1}\left(-\frac{1}{2}\right)\right)$

25. $\tan(\sec^{-1}1)$

26. $\sin(\csc^{-1}(-1))$

27. $\sin^{-1}(1) - \sin^{-1}(-1)$

28. $\tan^{-1}(1) - \tan^{-1}(-1)$

29. $\sec^{-1}(2) - \sec^{-1}(-2)$

30. $\sin(\sin^{-1}0.735)$

31. $\cos(\sin^{-1}0.8)$

32. $\tan^{-1}(\tan\pi/3)$

33. $\cos^{-1}(-\sin\pi/6)$

34. $\sec^{-1}(\sec(-30°))$ (The answer is *not* $-30°$.)

Find the limits in Exercises 35–42. (If in doubt, look at the function's graph.)

35. $\lim_{x\to 1^-}\sin^{-1}x$

36. $\lim_{x\to -1^+}\cos^{-1}x$

37. $\lim_{x\to\infty}\tan^{-1}x$

38. $\lim_{x\to -\infty}\tan^{-1}x$

39. $\lim_{x\to\infty}\sec^{-1}x$

40. $\lim_{x\to -\infty}\sec^{-1}x$

41. $\lim_{x\to\infty}\csc^{-1}x$

42. $\lim_{x\to -\infty}\csc^{-1}x$

43. You are sitting in a classroom next to the wall looking at the blackboard at the front of the room. The blackboard is 12 ft long and starts 3 ft from the wall you are sitting next to. Show that your viewing angle is

$$\alpha = \cot^{-1}\frac{x}{15} - \cot^{-1}\frac{x}{3}$$

if you are x ft from the front wall.

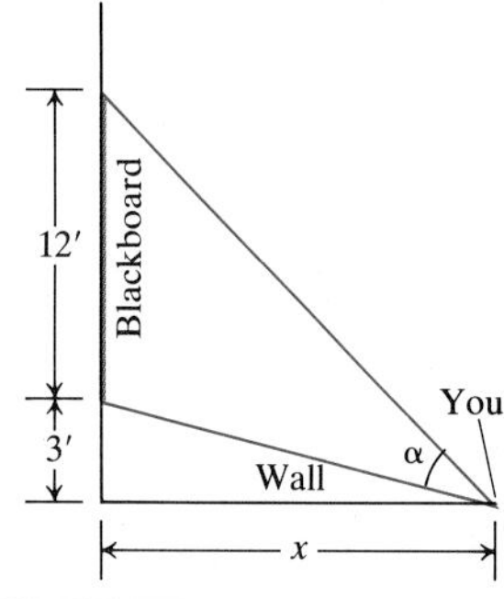

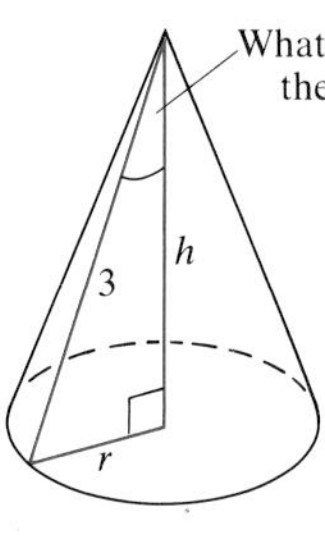

7.28 The cone in Exercise 45.

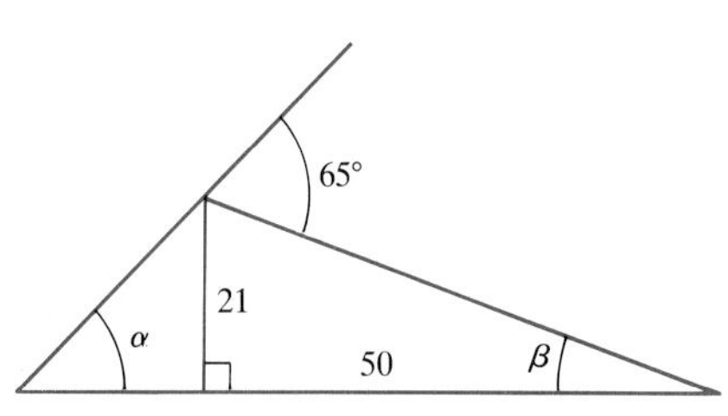

7.29 Find α. (*Hint:* $\alpha + \beta = 65°$.)

44. The region between the curve $y = \sec^{-1}x$ and the x-axis between $x = 1$ and $x = 2$ is revolved about the y-axis to generate a solid. Find the volume of the solid. (*Hint:* Use washers.)

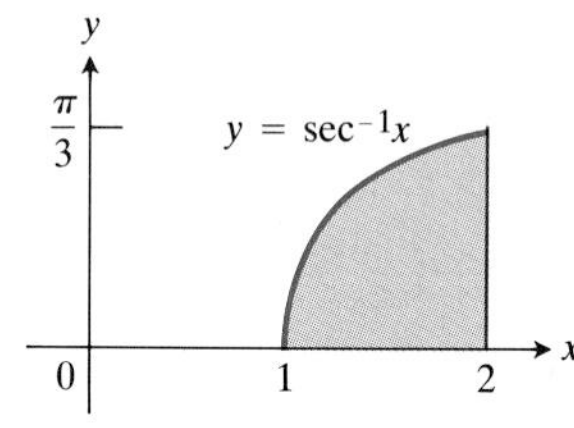

45. CALCULATOR Figure 7.28 shows a cone whose slant height is 3 m. How large should the indicated angle be to maximize the cone's volume?

46. CALCULATOR Find the angle α in Fig. 7.29.

47. CALCULATOR Find the values of $\cot^{-1}2$, $\sec^{-1}(1.5)$, and $\csc^{-1}(1.5)$.

48. Here is an informal proof that

$$\tan^{-1}1 + \tan^{-1}2 + \tan^{-1}3 = \pi.$$

See if you can tell what is going on. (*Source:* Edward M. Harris, *Behold! Sums of Arctan,* The College Mathematics Journal, Vol. 18, No. 2, Page 141.)

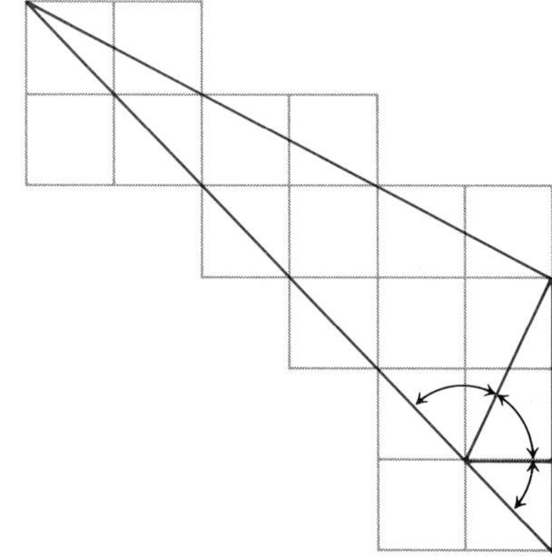

7.8 Derivatives of Inverse Trigonometric Functions; Related Integrals

In this section, we list the standard formulas for the derivatives of the inverse trigonometric functions (Table 7.4), show how they are derived, and discuss their companion integral formulas. As we shall see, the restrictions on the domains of the inverse trigonometric functions show up in natural ways as restrictions on the domains of the derivatives.

EXAMPLE 1

a) $$\frac{d}{dx}\sin^{-1}x^2 = \frac{1}{\sqrt{1-(x^2)^2}}\cdot\frac{d}{dx}(x^2) = \frac{2x}{\sqrt{1-x^4}}$$

b) $$\frac{d}{dx}\tan^{-1}\sqrt{x+1} = \frac{1}{1+(\sqrt{x+1})^2}\cdot\frac{d}{dx}(\sqrt{x+1})$$
$$= \frac{1}{x+2}\cdot\frac{1}{2\sqrt{x+1}} = \frac{1}{2\sqrt{x+1}(x+2)}$$

c) $$\frac{d}{dx}\sec^{-1}(3x) = \frac{1}{|3x|\sqrt{(3x)^2-1}}\cdot\frac{d}{dx}(3x)$$
$$= \frac{3}{|3x|\sqrt{9x^2-1}} = \frac{1}{|x|\sqrt{9x^2-1}}.$$ ■

To show how the derivative formulas in Table 7.4 may be derived, we shall prove Formulas 1 and 5.

The Derivative of $y = \sin^{-1}u$

We know that the function $x = \sin y$ is differentiable in the open interval $-\pi/2 < y < \pi/2$ and that its derivative, the cosine, is positive there. The derivative rule for inverses in Section 7.1 therefore assures us that the inverse function $y = \sin^{-1}x$ is differentiable throughout the interval $-1 < x < 1$. We cannot expect it to be differentiable at $x = 1$ or $x = -1$,

TABLE 7.4
Derivatives

1. $\frac{d(\sin^{-1}u)}{dx} = \frac{du/dx}{\sqrt{1-u^2}}, \quad -1 < u < 1$
2. $\frac{d(\cos^{-1}u)}{dx} = -\frac{du/dx}{\sqrt{1-u^2}}, \quad -1 < u < 1$
3. $\frac{d(\tan^{-1}u)}{dx} = \frac{du/dx}{1+u^2}$
4. $\frac{d(\cot^{-1}u)}{dx} = -\frac{du/dx}{1+u^2}$
5. $\frac{d(\sec^{-1}u)}{dx} = \frac{du/dx}{|u|\sqrt{u^2-1}}, \quad |u| > 1$
6. $\frac{d(\csc^{-1}u)}{dx} = \frac{-du/dx}{|u|\sqrt{u^2-1}}, \quad |u| > 1$

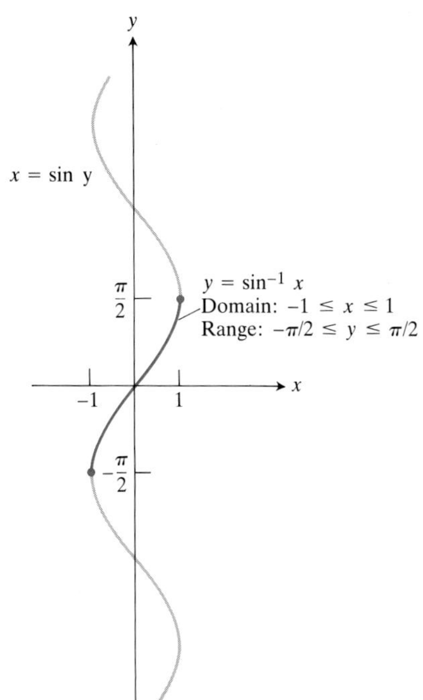

7.30 The graph of $y = \sin^{-1}x$.

however, because the tangents to the graph are vertical at these points (see Fig. 7.30).

To calculate the derivative of $y = \sin^{-1}x$, we differentiate both sides of the equation $\sin y = x$ with respect to x:

$$\begin{aligned} \sin y &= x, \\ \frac{d}{dx}\sin y &= 1, \\ \cos y\frac{dy}{dx} &= 1. \end{aligned} \tag{1}$$

We then divide through by $\cos y$ (>0 for $-\pi/2 < y < \pi/2$) to get

$$\frac{dy}{dx} = \frac{1}{\cos y} = \frac{1}{\sqrt{1-\sin^2 y}} = \frac{1}{\sqrt{1-x^2}}. \tag{2}$$

The derivative of $y = \sin^{-1}x$ with respect to x is

$$\frac{d}{dx}\sin^{-1}x = \frac{1}{\sqrt{1-x^2}}. \tag{3}$$

If u is a differentiable function of x, we apply the Chain Rule in the form

$$\frac{dy}{dx} = \frac{dy}{du}\frac{du}{dx}$$

to $y = \sin^{-1}u$ to obtain

$$\frac{d}{dx}\sin^{-1}u = \frac{1}{\sqrt{1-u^2}}\frac{du}{dx}. \tag{4}$$

The Derivative of $y = \sec^{-1}u$

We begin by differentiating both sides of the equation $\sec y = x$ with respect to x:

$$\begin{aligned} \sec y &= x, \\ \frac{d}{dx}\sec y &= 1, \\ \sec y\tan y\frac{dy}{dx} &= 1, \\ \frac{dy}{dx} &= \frac{1}{\sec y\tan y}. \end{aligned} \tag{5}$$

To express the result in terms of x, we use the relations

$$\sec y = x \qquad \text{and} \qquad \tan y = \pm\sqrt{\sec^2 y - 1} = \pm\sqrt{x^2-1}.$$

Hence,

$$\frac{dy}{dx} = \pm\frac{1}{x\sqrt{x^2-1}}.$$

7.31 The graph of $y = \sec^{-1}x$.

What do we do about the sign? A glance at Fig. 7.31 shows that the slope of the graph of $y = \sec^{-1}x$ is always positive. Therefore,

$$\frac{d}{dx}\sec^{-1}x = \begin{cases} \dfrac{1}{x\sqrt{x^2-1}} & \text{if } x > 1 \\ -\dfrac{1}{x\sqrt{x^2-1}} & \text{if } x < -1. \end{cases} \tag{6}$$

With absolute values, we can write Eq. (6) as a single formula:

$$\frac{d}{dx}\sec^{-1}x = \frac{1}{|x|\sqrt{x^2-1}}, \quad |x| > 1. \tag{7}$$

We can then apply the Chain Rule to obtain

$$\frac{d}{dx}\sec^{-1}u = \frac{1}{|u|\sqrt{u^2-1}}\frac{du}{dx}, \quad |u| > 1, \tag{8}$$

where u is a differentiable function of x.

Integration Formulas

We might expect the six derivative formulas in Table 7.4 to lead to six new integration formulas, but the only three that matter are shown in Table 7.5.

TABLE 7.5
Integrals Leading to Inverse Trigonometric Functions

1. $\displaystyle\int \frac{du}{\sqrt{1-u^2}} = \sin^{-1}u + C$ (Valid for $u^2 < 1$)

3. $\displaystyle\int \frac{du}{1+u^2} = \tan^{-1}u + C$ (Valid for all u)

5. $\displaystyle\int \frac{du}{u\sqrt{u^2-1}} = \int \frac{d(-u)}{(-u)\sqrt{u^2-1}} = \sec^{-1}|u| + C = \cos^{-1}\left|\frac{1}{u}\right| + C$ (Valid for $u^2 > 1$)

EXAMPLE 2 Evaluate

a) $\displaystyle\int_0^1 \frac{dx}{1+x^2}$ b) $\displaystyle\int_{2/\sqrt{3}}^{\sqrt{2}} \frac{dx}{x\sqrt{x^2-1}}$.

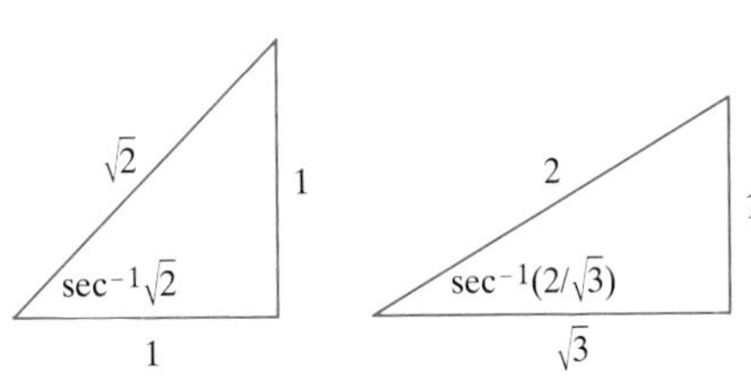

An isosceles right triangle: $\sec^{-1}\sqrt{2} = \frac{\pi}{4}$

A 30-60-90 triangle: $\sec^{-1}\frac{2}{\sqrt{3}} = \frac{\pi}{6}$

7.32 Triangles you can use for finding $\sec^{-1}\sqrt{2}$ and $\sec^{-1}(2/\sqrt{3})$.

Solution

a) $\displaystyle\int_0^1 \frac{dx}{1+x^2} = \tan^{-1}x\Big]_0^1 = \tan^{-1}1 - \tan^{-1}0 = \frac{\pi}{4} - 0 = \frac{\pi}{4}$

b) $\displaystyle\int_{2/\sqrt{3}}^{\sqrt{2}} \frac{dx}{x\sqrt{x^2-1}} = \sec^{-1}x\Big]_{2/\sqrt{3}}^{\sqrt{2}} = \frac{\pi}{4} - \frac{\pi}{6} = \frac{\pi}{12}$

If you do not see right away how to evaluate $\sec^{-1}x$ at the limits of integration, draw triangles like the ones in Fig. 7.32. ■

EXAMPLE 3 Evaluate

$$\int \frac{x^2\,dx}{\sqrt{1-x^6}}.$$

Solution The resemblance between the given integral and the standard form

$$\int \frac{du}{\sqrt{1-u^2}} = \sin^{-1}u + C$$

suggests the substitution

$$u = x^3, \quad du = 3x^2\,dx.$$

Indeed,

$$\int \frac{x^2\,dx}{\sqrt{1-x^6}} = \frac{1}{3}\int \frac{3x^2\,dx}{\sqrt{1-(x^3)^2}} = \frac{1}{3}\int \frac{du}{\sqrt{1-u^2}}$$

$$= \frac{1}{3}\sin^{-1}u + C = \frac{1}{3}\sin^{-1}(x^3) + C.$$

■

EXAMPLE 4 Evaluate

$$\int \frac{dx}{\sqrt{9-x^2}}.$$

Solution This is the integral for the arc sine of x with a 9 in place of the 1. To get the 1 back, we factor a 9 from $9 - x^2$ and bring it outside the radical as a 3:

$$\sqrt{9-x^2} = \sqrt{9\left(1-\frac{x^2}{9}\right)} = 3\sqrt{1-\left(\frac{x}{3}\right)^2}.$$

Notice that we also write $x^2/9$ as $(x/3)^2$. Then

$$\int \frac{dx}{\sqrt{9-x^2}} = \int \frac{dx}{3\sqrt{1-(x/3)^2}}.$$

We now substitute

$$u = \frac{x}{3} \quad \text{and} \quad du = \frac{dx}{3}, \quad \text{or} \quad dx = 3\,du.$$

This gives

$$\int \frac{dx}{\sqrt{9-x^2}} = \int \frac{dx}{3\sqrt{1-(x/3)^2}} = \int \frac{3\,du}{3\sqrt{1-u^2}}$$

$$= \int \frac{du}{\sqrt{1-u^2}} = \sin^{-1}u + C$$

$$= \sin^{-1}\left(\frac{x}{3}\right) + C.$$

■

Exercises 7.8

Find dy/dx in Exercises 1–18.

1. $y = \cos^{-1} x^2$
2. $y = \cos^{-1}(1/x)$
3. $y = 5\tan^{-1} 3x$
4. $y = \cot^{-1}\sqrt{x}$
5. $y = \sin^{-1}(x/2)$
6. $y = \sin^{-1}(1 - x)$
7. $y = \sec^{-1} 5x$
8. $y = (1/3)\tan^{-1}(x/3)$
9. $y = \csc^{-1}(x^2 + 1)$
10. $y = \cos^{-1} 2x$
11. $y = \csc^{-1}\sqrt{x} + \sec^{-1}\sqrt{x}$
12. $y = \csc^{-1}\dfrac{1}{x},\ x > 0$
13. $y = \cot^{-1}\sqrt{x - 1}$
14. $y = x\sqrt{1 - x^2} - \cos^{-1} x$
15. $y = \sqrt{x^2 - 1} - \sec^{-1} x$
16. $y = \cot^{-1}\dfrac{1}{x} - \tan^{-1} x$
17. $y = 2x\tan^{-1} x - \ln(x^2 + 1)$
18. $y = \ln(x^2 + 1) - 2x + 2\tan^{-1} x$

Evaluate the integrals in Exercises 19–30.

19. $\displaystyle\int_0^{1/2} \frac{dx}{\sqrt{1 - x^2}}$
20. $\displaystyle\int_{-1}^{1} \frac{dx}{1 + x^2}$
21. $\displaystyle\int_{\sqrt{2}}^{2} \frac{dx}{x\sqrt{x^2 - 1}}$
22. $\displaystyle\int_{-2}^{-\sqrt{2}} \frac{dx}{x\sqrt{x^2 - 1}}$
23. $\displaystyle\int_{-1}^{0} \frac{4\,dx}{1 + x^2}$
24. $\displaystyle\int_{\sqrt{3}/3}^{\sqrt{3}} \frac{6\,dx}{1 + x^2}$
25. $\displaystyle\int_0^{\sqrt{2}/2} \frac{x\,dx}{\sqrt{1 - x^4}}$
26. $\displaystyle\int_0^{1/4} \frac{dx}{\sqrt{1 - 4x^2}}$
27. $\displaystyle\int_{1/\sqrt{3}}^{1} \frac{dx}{x\sqrt{4x^2 - 1}}$
28. $\displaystyle\int_0^{1} \frac{x}{1 + x^4}\,dx$
29. $\displaystyle\int_0^{1} \frac{4x\,dx}{\sqrt{4 - x^4}}$
30. $\displaystyle\int_0^{1} \frac{dx}{\sqrt{4 - x^2}}$

Use the indicated substitutions to evaluate the integrals in Exercises 31–36.

31. $\displaystyle\int_{\sqrt[4]{2}}^{\sqrt{2}} \frac{2x\,dx}{x^2\sqrt{x^4 - 1}},\quad u = x^2$
32. $\displaystyle\int_0^{2} \frac{dx}{1 + (x - 1)^2},\quad u = x - 1$
33. $\displaystyle\int_1^{\sqrt{3}} \frac{2\,dx}{(1 + x^2)\tan^{-1} x},\quad u = \tan^{-1} x$
34. $\displaystyle\int_0^{\ln\sqrt{3}} \frac{e^x\,dx}{1 + e^{2x}},\quad u = e^x$
35. $\displaystyle\int_2^{4} \frac{dx}{2x\sqrt{x - 1}},\quad u^2 = x$
36. $\displaystyle\int_{-\pi/2}^{\pi/2} \frac{2\cos x\,dx}{1 + \sin^2 x},\quad u = \sin x$

Use l'Hôpital's rule to find the limits in Exercises 37–40.

37. $\displaystyle\lim_{x\to 0} \frac{\sin^{-1} x}{x}$
38. $\displaystyle\lim_{x\to 0} \frac{\sin^{-1} x}{x^3}$
39. $\displaystyle\lim_{x\to 0} \frac{\tan^{-1} x}{x}$
40. $\displaystyle\lim_{x\to 0} \frac{\tan^{-1} x}{x^3}$

41. Find the volume of this solid of revolution.

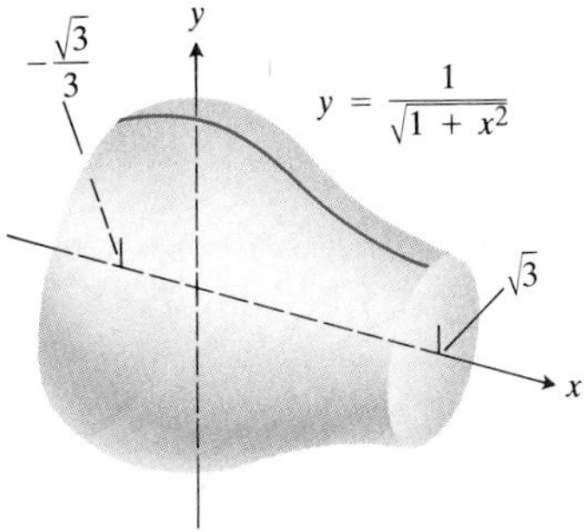

42. Find the length of the curve $y = 1/\sqrt{1 - x^2}$, $-\pi/3 \le x \le \pi/3$.
43. *Continuation of Exercise 43, Section 7.7.* You want to move your chair to a position along the wall that will maximize your viewing angle α. About how far from the front wall should you sit?
44. Find the center of mass of a thin plate of constant density $\delta = 1$ bounded by the curves $y = 1/(1 + x^2)$ and $y = -1/(1 + x^2)$ and by the lines $x = 0$ and $x = 1$.

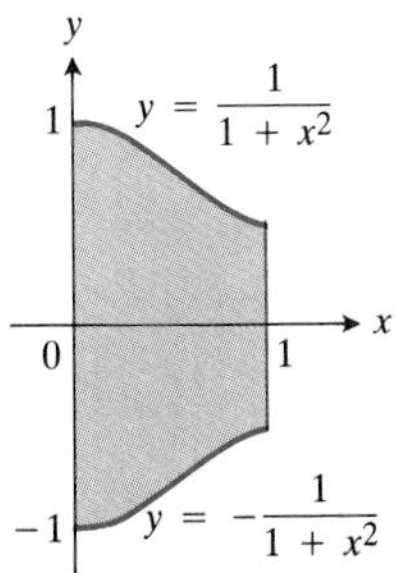

45. Find the value of x that maximizes θ in the figure at the right. How large is θ at that point? Begin by showing that

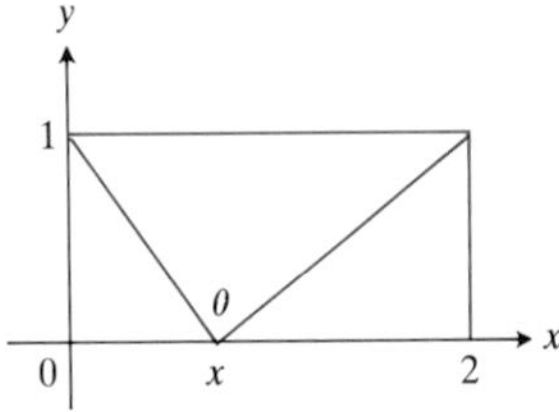

$$\theta = \pi - \cot^{-1}x - \cot^{-1}(2 - x).$$

46. Find the linearizations of $\sin^{-1}x$, $\cos^{-1}x$, $\tan^{-1}x$, and $\cot^{-1}x$ at $x = 0$.

Solve the initial value problems in Exercises 47–50.

47. Differential equation: $\dfrac{dy}{dx} = \dfrac{1}{x\sqrt{x^2 - 1}}$

Initial condition: $y = \pi$ when $x = 2$

48. Differential equation: $\dfrac{dy}{dx} = \dfrac{1}{\sqrt{1 - x^2}}$

Initial condition: $y = 1$ when $x = 0$

49. Differential equation: $\dfrac{dy}{dx} = -\dfrac{1}{\sqrt{1 - x^2}}$

Initial condition: $y = \pi/2$ when $x = -\sqrt{2}/2$

50. Differential equation: $\dfrac{dy}{dx} = -\dfrac{1}{1 + x^2}$

Initial condition: $y = \pi/2$ when $x = 0$

51. **CALCULATOR** Use Simpson's rule with $n = 6$ to estimate the value of

$$\sin^{-1}0.6 = \int_0^{0.6} \frac{dx}{\sqrt{1 - x^2}}.$$

For reference, $\sin^{-1}0.6 = 0.64350$ to five places.

52. **CALCULATOR** Use Simpson's rule with $n = 8$ to estimate the value of

$$\pi = 4\int_0^1 \frac{1}{1 + x^2}\,dx.$$

53. Can the integrations in (a) and (b) both be correct? Explain.

a) $\displaystyle\int \frac{dx}{\sqrt{1 - x^2}} = \sin^{-1}x + C$

b) $\displaystyle\int \frac{dx}{\sqrt{1 - x^2}} = -\int -\frac{dx}{\sqrt{1 - x^2}} = -\cos^{-1}x + C$

54. a) Show that the functions

$$f(x) = \sin^{-1}\frac{x - 1}{x + 1} \quad \text{and} \quad g(x) = 2\tan^{-1}\sqrt{x},$$

both defined for $x \geq 0$, have the same derivative and therefore that

$$f(x) = g(x) + C. \qquad (9)$$

b) Find C. (*Hint:* Evaluate both sides of Eq. 9 for a particular value of x.)

Verify the derivative formulas in Exercises 55–58.

55. $\dfrac{d(\cos^{-1}u)}{dx} = -\dfrac{du/dx}{\sqrt{1 - u^2}}, \quad -1 < u < 1$

56. $\dfrac{d(\tan^{-1}u)}{dx} = \dfrac{du/dx}{1 + u^2}$

57. $\dfrac{d(\cot^{-1}u)}{dx} = -\dfrac{du/dx}{1 + u^2}$

58. $\dfrac{d(\csc^{-1}u)}{dx} = \dfrac{-du/dx}{|u|\sqrt{u^2 - 1}}, \quad |u| > 1$

TOOLKIT PROGRAMS

Derivative Grapher	Graphs functions and their derivatives side by side
Super*Grapher	Graphs all the functions in this section

7.9 Hyperbolic Functions

It can be shown that every function defined on an interval centered at the origin can be written in a unique way as the sum of one even function and one odd function. For an arbitrary function f, the decomposition is

$$f(x) = \underbrace{\frac{f(x) + f(-x)}{2}}_{f_{\text{even}}} + \underbrace{\frac{f(x) - f(-x)}{2}}_{f_{\text{odd}}}.$$

If we write e^x this way, we get

$$e^x = \underbrace{\frac{e^x + e^{-x}}{2}}_{\text{Even part}} + \underbrace{\frac{e^x - e^{-x}}{2}}_{\text{Odd part}}.$$

The even and odd parts of e^x, called the hyperbolic cosine and hyperbolic sine of x, respectively, turn out to be important functions in their own right. They describe the motions of waves in elastic solids, the shapes of hanging electric power lines, and the temperature distributions in metal cooling fins. They even come up in the general theory of relativity. The designers of the Gateway Arch to the West in St. Louis used a hyperbolic cosine function to predict the arch's internal forces and then shaped the arch like an upside-down hyperbolic cosine curve.

This section introduces the calculus of hyperbolic functions.

Definitions and Identities

The hyperbolic cosine and hyperbolic sine functions are defined by the equations in Table 7.6.

TABLE 7.6

Hyperbolic cosine of x:	$\cosh x = \dfrac{e^x + e^{-x}}{2}$
Hyperbolic sine of x:	$\sinh x = \dfrac{e^x - e^{-x}}{2}$

The notation $\cosh x$ is often read "kosh x," which rhymes with "gosh x," and $\sinh x$ is pronounced as if spelled "cinch x" or "shine x." They all sound a bit strange.

Four additional hyperbolic functions are defined in terms of $\cosh x$ and $\sinh x$, as shown in Table 7.7. See Figs. 7.33 and 7.34 for graphs.

TABLE 7.7

Hyperbolic tangent:	$\tanh x = \dfrac{\sinh x}{\cosh x} = \dfrac{e^x - e^{-x}}{e^x + e^{-x}}$
Hyperbolic cotangent:	$\coth x = \dfrac{\cosh x}{\sinh x} = \dfrac{e^x + e^{-x}}{e^x - e^{-x}}$
Hyperbolic secant:	$\operatorname{sech} x = \dfrac{1}{\cosh x} = \dfrac{2}{e^x + e^{-x}}$
Hyperbolic cosecant:	$\operatorname{csch} x = \dfrac{1}{\sinh x} = \dfrac{2}{e^x - e^{-x}}$

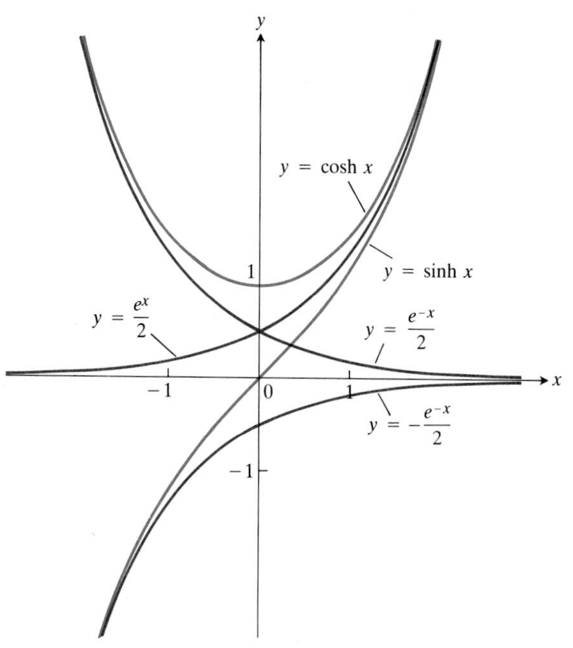

7.33 The graphs of the hyperbolic sine and cosine.

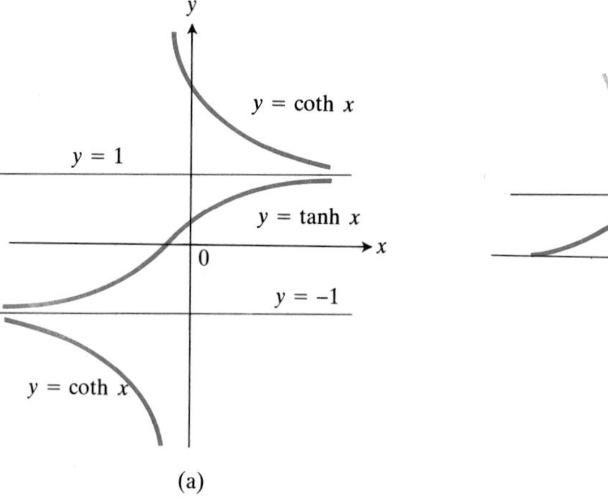

7.34 The graphs of the hyperbolic tangent, cotangent, secant, and cosecant, and their relations to the graphs of the hyperbolic sine and cosine.

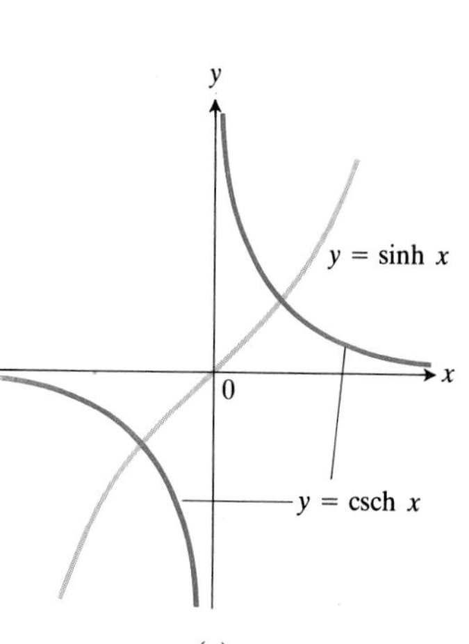

Identities

Hyperbolic functions satisfy the identities shown in Table 7.8. Except for differences in sign, these are the identities we already know for trigonometric functions. We shall say more about the analogy between hyperbolic and trigonometric functions in the exercises.

TABLE 7.8

$\sinh 2x = 2\sinh x \cosh x$	$\cosh^2 x - \sinh^2 x = 1$
$\cosh 2x = \cosh^2 x + \sinh^2 x$	$\tanh^2 x = 1 - \operatorname{sech}^2 x$
$\cosh^2 x = \dfrac{\cosh 2x + 1}{2}$	$\coth^2 x = 1 + \operatorname{csch}^2 x$
$\sinh^2 x = \dfrac{\cosh 2x - 1}{2}$	

Derivatives and Integrals

The six hyperbolic functions, being rational combinations of the differentiable functions e^x and e^{-x}, have derivatives at every point at which they are defined. Finding formulas for the derivatives is a straightforward exercise in the sum, difference, and quotient rules of differentiation. Again, there are similarities with trigonometric functions. These derivatives are shown in Table 7.9.

TABLE 7.9

$\dfrac{d}{dx}(\sinh u) = \cosh u \dfrac{du}{dx}$	$\dfrac{d}{dx}(\coth u) = -\operatorname{csch}^2 u \dfrac{du}{dx}$
$\dfrac{d}{dx}(\cosh u) = \sinh u \dfrac{du}{dx}$	$\dfrac{d}{dx}(\operatorname{sech} u) = -\operatorname{sech} u \tanh u \dfrac{du}{dx}$
$\dfrac{d}{dx}(\tanh u) = \operatorname{sech}^2 u \dfrac{du}{dx}$	$\dfrac{d}{dx}(\operatorname{csch} u) = -\operatorname{csch} u \coth u \dfrac{du}{dx}$

These derivative formulas produce the integral formulas shown in Table 7.10.

TABLE 7.10

$\int \sinh u \, du = \cosh u + C$	$\int \operatorname{csch}^2 u \, du = -\coth u + C$
$\int \cosh u \, du = \sinh u + C$	$\int \operatorname{sech} u \tanh u \, du = -\operatorname{sech} u + C$
$\int \operatorname{sech}^2 u \, du = \tanh u + C$	$\int \operatorname{csch} u \coth u \, du = -\operatorname{csch} u + C$

EXAMPLE 1

$$\int \coth 5x\,dx = \int \frac{\cosh 5x}{\sinh 5x}\,dx = \frac{1}{5}\int \frac{du}{u} \qquad (u = \sinh 5x)$$
$$= \frac{1}{5}\ln|u| + C = \frac{1}{5}\ln|\sinh 5x| + C.$$

■

EXAMPLE 2

$$\int_0^1 \sinh^2 x\,dx = \int_0^1 \frac{\cosh 2x - 1}{2}\,dx \qquad \text{(Table 7.8)}$$
$$= \frac{1}{2}\int_0^1 (\cosh 2x - 1)\,dx = \frac{1}{2}\left[\frac{\sinh 2x}{2} - x\right]_0^1$$
$$= \frac{\sinh 2}{4} - \frac{1}{2} = 0.40672 \qquad \left(\begin{matrix}\text{Calculator,}\\ \text{rounded}\end{matrix}\right)$$

■

As in Example 2, hyperbolic functions are usually evaluated with a calculator. Use the [hyp], [cos], [sin], and [tan] keys, or [e^x].

EXAMPLE 3

$$\int_0^{\ln 2} 4e^x \sinh x\,dx = \int_0^{\ln 2} 4\,e^x \frac{e^x - e^{-x}}{2}\,dx = \int_0^{\ln 2} (2e^{2x} - 2)\,dx$$
$$= \left[e^{2x} - 2x\right]_0^{\ln 2} = (e^{2\ln 2} - 2\ln 2) - (1 - 0)$$
$$= 4 - 2\ln 2 - 1 = 3 - 2\ln 2$$

■

The Inverses of the Hyperbolic Functions

All hyperbolic functions have inverses that are useful in integration. Since

$$\frac{d}{dx}\sinh x = \cosh x \tag{1}$$

is positive for every value of x, the hyperbolic sine is an increasing function of x and therefore has an inverse, which we denote by

$$y = \sinh^{-1} x. \tag{2}$$

This equation is read "y equals the arc hyperbolic sine of x" or "y equals the inverse hyperbolic sine of x." For every value of x in the interval $-\infty < x < \infty$, the value of $y = \sinh^{-1} x$ is the number whose hyperbolic sine is x. The graphs of $y = \sinh x$ and $y = \sinh^{-1} x$ are shown in Fig. 7.35(a).

The function $y = \cosh x$ is not one-to-one, as we can see quickly enough from its graph in Fig. 7.33. But the restricted function

$$y = \cosh x, \qquad x \geq 0, \tag{3}$$

is one-to-one and therefore has an inverse, denoted by

$$y = \cosh^{-1} x. \tag{4}$$

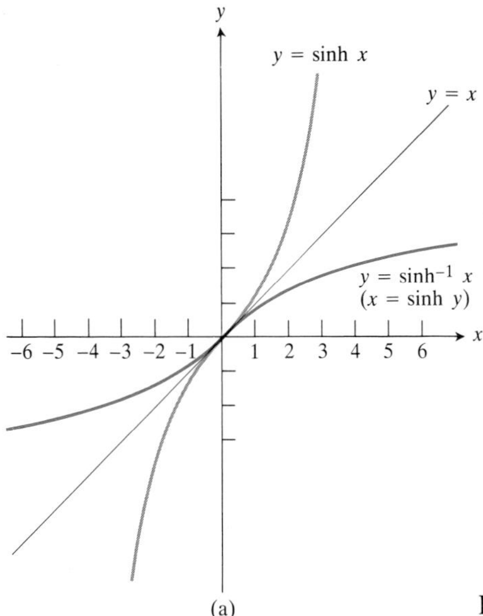

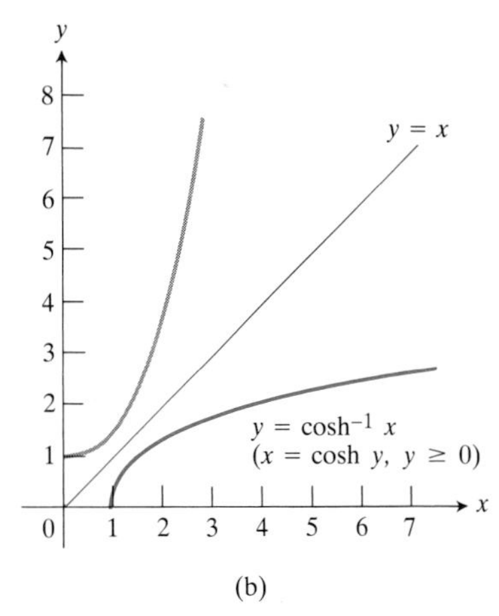

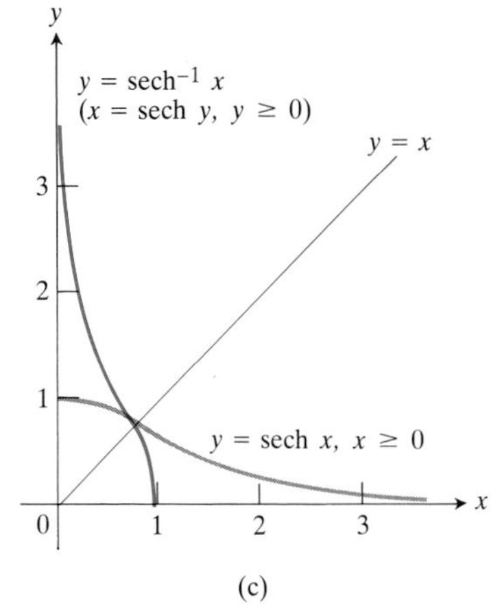

7.35 The graphs of $y = \sinh^{-1}x$, $y = \cosh^{-1}x$, and $y = \operatorname{sech}^{-1}x$. Note the symmetries about the line $y = x$.

For every value of $x \geq 1$, $y = \cosh^{-1}x$ is the number in the interval $0 \leq y < \infty$ whose hyperbolic cosine is x. The graphs of $y = \cosh x$, $x \geq 0$, and $y = \cosh^{-1}x$ are shown in Fig. 7.35(b).

Like $y = \cosh x$, the function $y = \operatorname{sech} x = 1/\cosh x$ fails to be one-to-one, but its restriction to nonnegative values of x does have an inverse, denoted by

$$y = \operatorname{sech}^{-1}x. \tag{5}$$

For every value of x in the interval $(0, 1]$, $y = \operatorname{sech}^{-1}x$ is the nonnegative number whose hyperbolic secant is x. The graphs of $y = \operatorname{sech} x$, $x \geq 0$, and $y = \operatorname{sech}^{-1}x$ are shown in Fig. 7.35(c).

The hyperbolic tangent, cotangent, and cosecant are one-to-one on their domains and therefore have inverses, denoted by

$$y = \tanh^{-1}x, \quad y = \coth^{-1}x, \quad y = \operatorname{csch}^{-1}x. \tag{6}$$

These are graphed in Fig. 7.36.

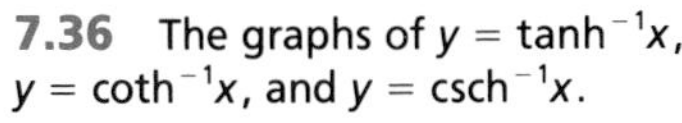

7.36 The graphs of $y = \tanh^{-1}x$, $y = \coth^{-1}x$, and $y = \operatorname{csch}^{-1}x$.

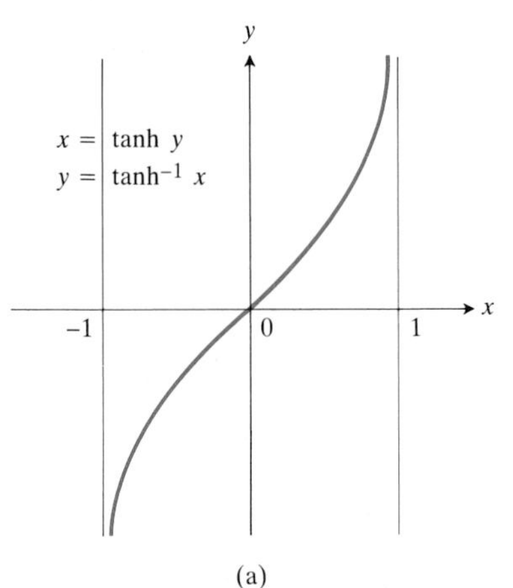

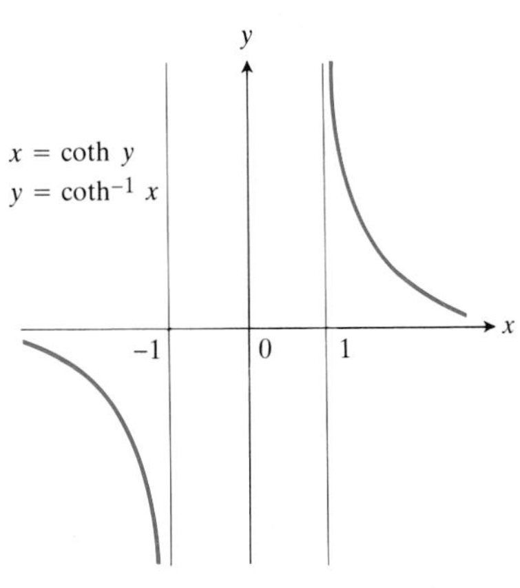

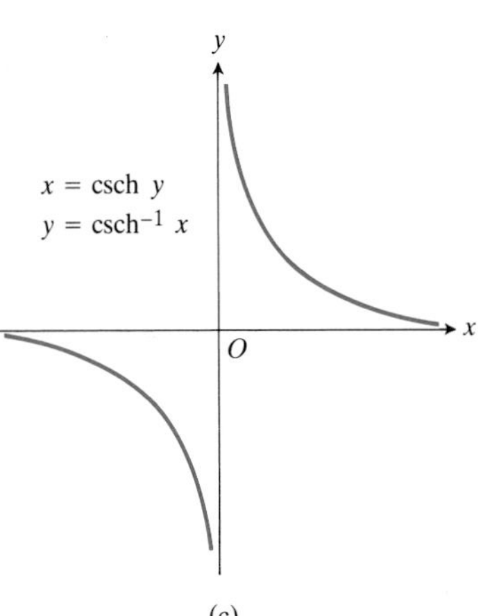

TABLE 7.11

$\operatorname{sech}^{-1}x = \cosh^{-1}\dfrac{1}{x}$	$\operatorname{csch}^{-1}x = \sinh^{-1}\dfrac{1}{x}$	$\coth^{-1}x = \tanh^{-1}\dfrac{1}{x}.$

Useful Identities

The inverse hyperbolic secant, cosecant, and cotangent satisfy the identities shown in Table 7.11. These identities come in handy when we have to calculate the values of $\operatorname{sech}^{-1}x$, $\operatorname{csch}^{-1}x$, and $\coth^{-1}x$ on a calculator that gives only $\cosh^{-1}x$, $\sinh^{-1}x$, and $\tanh^{-1}x$.

Derivatives and Integrals

The chief merit of the inverse hyperbolic functions lies in their usefulness in integration. You will see why after we have derived the formulas for their derivatives, in Table 7.12.

The restrictions $|u| < 1$ and $|u| > 1$ on the derivative formulas for $\tanh^{-1}u$ and $\coth^{-1}u$ come from the natural restrictions on the values of these functions. (See Figs. 7.36a and b.) The distinction between $|u| < 1$ and $|u| > 1$ becomes important when we convert the derivative formulas into integral formulas. If $|u| < 1$, the integral of $1/(1-u^2)$ is $\tanh u + C$. If $|u| > 1$, the integral is $\coth u + C$.

EXAMPLE 4 Show that if u is a differentiable function of x whose values are greater than 1, then

$$\frac{d}{dx}\cosh^{-1}u = \frac{1}{\sqrt{u^2-1}}\frac{du}{dx}.$$

TABLE 7.12

$$\frac{d(\sinh^{-1}u)}{dx} = \frac{1}{\sqrt{1+u^2}}\frac{du}{dx},$$

$$\frac{d(\cosh^{-1}u)}{dx} = \frac{1}{\sqrt{u^2-1}}\frac{du}{dx}, \quad u > 1$$

$$\frac{d(\tanh^{-1}u)}{dx} = \frac{1}{1-u^2}\frac{du}{dx}, \quad |u| < 1,$$

$$\frac{d(\coth^{-1}u)}{dx} = \frac{1}{1-u^2}\frac{du}{dx}, \quad |u| > 1,$$

$$\frac{d(\operatorname{sech}^{-1}u)}{dx} = \frac{-du/dx}{u\sqrt{1-u^2}}, \quad 0 < u < 1,$$

$$\frac{d(\operatorname{csch}^{-1}u)}{dx} = \frac{-du/dx}{|u|\sqrt{1+u^2}}, \quad u \neq 0.$$

TABLE 7.13

$\displaystyle\int \frac{du}{\sqrt{1+u^2}} = \sinh^{-1} u + C$	$\displaystyle\int \frac{du}{u\sqrt{1-u^2}} = -\operatorname{sech}^{-1}\lvert u\rvert + C = -\cosh^{-1}\left(\frac{1}{\lvert u\rvert}\right) + C$
$\displaystyle\int \frac{du}{\sqrt{u^2-1}} = \cosh^{-1} u + C, \quad u > 1$	$\displaystyle\int \frac{du}{u\sqrt{1+u^2}} = -\operatorname{csch}^{-1}\lvert u\rvert + C = -\sinh^{-1}\left(\frac{1}{\lvert u\rvert}\right) + C$
$\displaystyle\int \frac{du}{1-u^2} = \begin{cases} \tanh^{-1} u + C & \text{if } \lvert u\rvert < 1 \\ \coth^{-1} u + C & \text{if } \lvert u\rvert > 1 \end{cases}$	

Solution First we find the derivative of $y = \cosh^{-1} x$ for $x > 1$:

$$y = \cosh^{-1} x$$

$$x = \cosh y \qquad \text{(Invert)}$$

$$1 = \sinh y \frac{dy}{dx} \qquad \text{(Differentiation with respect to } x\text{)}$$

$$\frac{dy}{dx} = \frac{1}{\sinh y} = \frac{1}{\sqrt{\cosh^2 y - 1}} \qquad \text{(Since } x > 1,\ y > 0 \text{ and } \sinh y > 0\text{)}$$

$$= \frac{1}{\sqrt{x^2-1}}. \qquad (\cosh y = x)$$

In short,

$$\frac{d}{dx}\cosh^{-1} x = \frac{1}{\sqrt{x^2-1}}.$$

The Chain Rule gives the final result:

$$\frac{d}{dx}\cosh^{-1} u = \frac{1}{\sqrt{u^2-1}}\frac{du}{dx}.$$ ■

The derivative formulas in Table 7.12 lead to the integration formulas in Table 7. 13.

EXAMPLE 5

$$\int_0^1 \frac{2\,dx}{\sqrt{1+4x^2}} = \int_0^2 \frac{du}{\sqrt{1+u^2}} = \sinh^{-1} u\Big]_0^2 \qquad (u = 2x,\ du = 2\,dx)$$

$$= \sinh^{-1} 2 - \sinh^{-1} 0$$

$$= \sinh^{-1} 2 - 0 = 1.4436 \qquad \text{(Calculator, rounded)}$$ ■

As in Example 5, inverse hyperbolic functions are usually evaluated with a calculator. On calculators without [inv] and [hyp] keys, we use formulas that express the inverse hyperbolic functions in terms of natural logarithms. We have included these in Table 7.14 in the exercises.

Exercises 7.9

Each of Exercises 1–4 gives a value of $\sinh x$ or $\cosh x$. Use the definitions and the identity $\cosh^2 x - \sinh^2 x = 1$ to find the values of the remaining five hyperbolic functions.

1. $\sinh x = -\dfrac{3}{4}$

2. $\sinh x = \dfrac{4}{3}$

3. $\cosh x = \dfrac{17}{15}, \quad x > 0$

4. $\cosh x = \dfrac{13}{5}, \quad x > 0$

Rewrite the expressions in Exercises 5–10 in terms of exponentials and simplify the results as much as you can.

5. $2\cosh(\ln x)$

6. $\sinh(2\ln x)$

7. $\cosh 5x + \sinh 5x$

8. $(\sinh x + \cosh x)^4$

9. $\cosh 3x - \sinh 3x$

10. $\ln(\cosh x + \sinh x) + \ln(\cosh x - \sinh x)$

11. Starting with the identities

$$\sinh(x + y) = \sinh x \cosh y + \cosh x \sinh y$$
$$\cosh(x + y) = \cosh x \cosh y + \sinh x \sinh y,$$

which you may assume, show that

a) $\sinh 2x = 2\sinh x \cosh x$

b) $\cosh 2x = \cosh^2 x + \sinh^2 x$.

12. Starting with the definitions of $\cosh x$ and $\sinh x$, show that

$$\cosh^2 x - \sinh^2 x = 1.$$

Find dy/dx in Exercises 13–24.

13. $y = \sinh 3x$

14. $y = \dfrac{1}{2}\sinh(2x + 1)$

15. $y = 2\tanh\dfrac{x}{2}$

16. $y = x - \tanh x$

17. $y = \ln(\operatorname{sech} x)$

18. $y = \ln(\operatorname{csch} x)$

19. $y = \ln(\operatorname{csch} x + \coth x)$

20. $y = x\cosh x - \sinh x$

21. $y = \dfrac{1}{2}\ln|\tanh x|$

22. $y = \tan^{-1}(\sinh x)$

23. a) $y = \cosh^2 x$ b) $y = \sinh^2 x$ c) $y = \dfrac{1}{2}\cosh 2x$

24. $y = (x^2 + 1)\operatorname{sech}(\ln x)$ (*Hint:* Express in terms of exponentials and simplify before differentiating.)

Find dy/dx in Exercises 25–36.

25. $y = \sinh^{-1} 2x$

26. $y = 2\cosh^{-1}\sqrt{x}$

27. $y = (1 - x)\tanh^{-1} x$

28. $y = (1 - x^2)\coth^{-1} x$

29. $y = x\operatorname{sech}^{-1} x$

30. $y = x^2\operatorname{csch}^{-1} x^2$

31. $y = \sinh^{-1}(\tan x)$

32. $y = \cosh^{-1}(\sec x), \quad -\pi/2 < x < \pi/2$

33. $y = \tanh^{-1}(\sin x), \quad -\pi/2 < x < \pi/2$

34. $y = \coth^{-1}(\sec x), \quad -\pi/2 < x < \pi/2$

35. $y = \operatorname{sech}^{-1}(\sin x), \quad 0 < x < \pi/2$

36. $y = \operatorname{csch}^{-1}(\tan x), \quad 0 < x < \pi/2$

Evaluate the integrals in Exercises 37–48.

37. $\displaystyle\int_{-1}^{1} \cosh 5x\, dx$

38. $\displaystyle\int_{-1}^{0} \cosh(2x + 1)\, dx$

39. $\displaystyle\int_{-3}^{3} \sinh x\, dx$

40. $\displaystyle\int_{-\pi}^{\pi} \tanh 2x\, dx$

41. $\displaystyle\int_{0}^{1/2} 4e^x \cosh x\, dx$

42. $\displaystyle\int_{0}^{1/2} 4e^{-x}\sinh x\, dx$

43. $\displaystyle\int_{1}^{2} \frac{\cosh(\ln x)}{x}\, dx$

44. $\displaystyle\int_{0}^{\ln 2} \frac{\sinh x}{\cosh x}\, dx$

45. $\displaystyle\int_{0}^{\ln 3} \operatorname{sech}^2 x\, dx$

46. $\displaystyle\int_{0}^{\ln 2} \tanh^2 x\, dx$

47. $\displaystyle\int_{0}^{4} \frac{\cosh\sqrt{x}}{\sqrt{x}}\, dx$

48. $\displaystyle\int_{\ln 2}^{\ln 3} \operatorname{csch}^2 x\, dx$

Evaluating Inverse Hyperbolic Functions and Related Integrals

When hyperbolic function keys are not available on a calculator, it is still possible to evaluate the inverse hyperbolic functions by expressing them first in terms of logarithms. The conversion formulas are listed in Table 7.14.

TABLE 7.14
Logarithm Formulas for Evaluating Inverse Hyperbolic Functions

$\sinh^{-1} x = \ln(x + \sqrt{x^2 + 1})$,	$-\infty < x < \infty$
$\cosh^{-1} x = \ln(x + \sqrt{x^2 - 1})$,	$x \geq 1$
$\tanh^{-1} x = \dfrac{1}{2}\ln\dfrac{1 + x}{1 - x}$,	$\lvert x\rvert < 1$
$\operatorname{sech}^{-1} x = \ln\left(\dfrac{1 + \sqrt{1 - x^2}}{x}\right)$,	$0 < x \leq 1$
$\operatorname{csch}^{-1} x = \ln\left(\dfrac{1}{x} + \dfrac{\sqrt{1 + x^2}}{\lvert x\rvert}\right)$,	$x \neq 0$
$\coth^{-1} x = \dfrac{1}{2}\ln\dfrac{x + 1}{x - 1}$,	$\lvert x\rvert > 1$

Use the formulas in Table 7.14 to express the numbers in Exercises 49–60 in terms of natural logarithms.

49. $\sinh^{-1} 0$

50. $\sinh^{-1}(-4/3)$

51. $\cosh^{-1}(5/3)$

52. $\cosh^{-1}(2/\sqrt{3})$

53. $\tanh^{-1}(1/2)$

54. $\tanh^{-1}(-3/5)$

55. $\coth^{-1}(5/4)$

56. $\coth^{-1}(-2)$

57. $\operatorname{sech}^{-1}(3/5)$

58. $\operatorname{sech}^{-1}(4/5)$

59. $\operatorname{csch}^{-1}(5/12)$

60. $\operatorname{csch}^{-1}(-1/\sqrt{3})$

Evaluate the integrals in Exercises 61–68 in terms of

a) inverse hyperbolic functions

b) natural logarithms using the formulas in Table 7.14.

61. $\displaystyle\int_0^1 \frac{dx}{\sqrt{1+x^2}}$

62. $\displaystyle\int_{4/5}^1 \frac{dx}{x\sqrt{1-x^2}}$

63. $\displaystyle\int_{5/4}^{5/3} \frac{dx}{\sqrt{x^2-1}}$

64. $\displaystyle\int_0^{1/2} \frac{dx}{1-x^2}$

65. $\displaystyle\int_{5/4}^{2} \frac{dx}{1-x^2}$

66. $\displaystyle\int_0^{2\sqrt{3}} \frac{dx}{\sqrt{4+x^2}}$ (Set $x = 2u$)

67. $\displaystyle\int_1^{2} \frac{dx}{x\sqrt{4+x^2}}$ (Set $x = 2u$)

68. $\displaystyle\int_0^{\pi} \frac{\cos x\, dx}{\sqrt{1+\sin^2 x}}$ (Set $u = \sin x$)

In Exercises 69 and 70, find the volumes of the solids generated by revolving the shaded regions about the x-axis.

69. **70.**

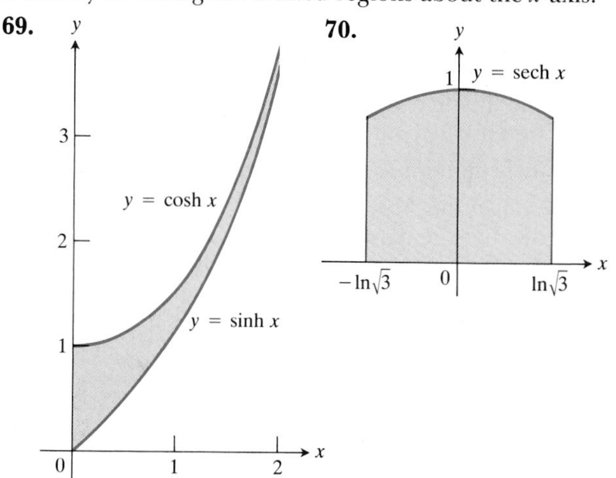

71. Use the formula

$$\int \operatorname{sech} x\, dx = \sin^{-1}(\tanh x) + C$$

to find the centroid of the shaded region in Exercise 70.

72. Find the volume of the solid generated by revolving the shaded region about the line $y = 1$.

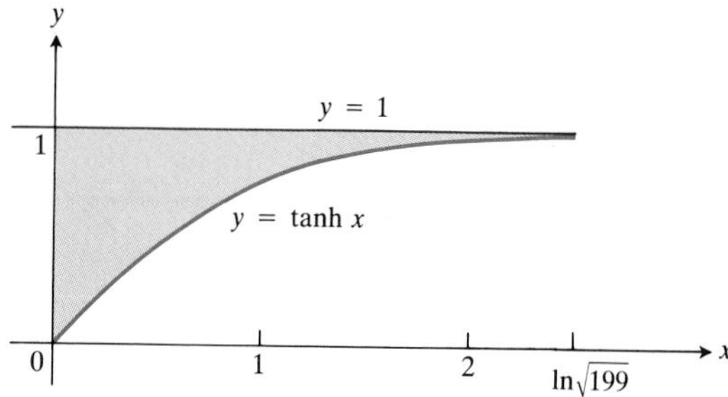

73. Show that if a function f is defined on an interval centered at the origin (so that $f(-x)$ is defined whenever $f(x)$ is defined), then

$$f(x) = \frac{f(x)+f(-x)}{2} + \frac{f(x)-f(-x)}{2}.$$

Then show that the functions defined by the formulas $(f(x)+f(-x))/2$ and $(f(x)-f(-x))/2$ are, respectively, even and odd.

74. What does the equation

$$f(x) = \frac{f(x)+f(-x)}{2} + \frac{f(x)-f(-x)}{2}$$

look like when f itself is (a) an even function, (b) an odd function?

75. *Skydiving.* If a body of mass m falling from rest under the action of gravity encounters an air resistance proportional to the square of the velocity, then the body's velocity t seconds into the fall satisfies the differential equation

$$m\frac{dv}{dt} = mg - kv^2,$$

where k is a constant that depends on the body's aerodynamic properties and the density of the air. (We assume that the fall is short enough so that the variation in the air's density will not affect the outcome.)

a) Show that

$$v = \sqrt{\frac{mg}{k}}\tanh\left(\sqrt{\frac{gk}{m}}\,t\right)$$

satisfies the differential equation and the initial condition that $v = 0$ when $t = 0$.

b) Find the body's *limiting velocity*, $\lim_{t\to\infty} v$.

Wide World Photos

Skydivers can vary their limiting velocities by changing the amount of body area opposing the fall. Their velocities can vary from 95 to 180 miles per hour.

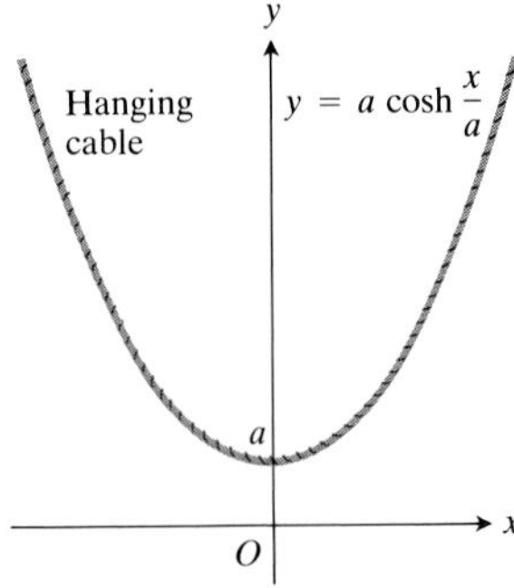

7.37 A hanging cable typically has this shape.

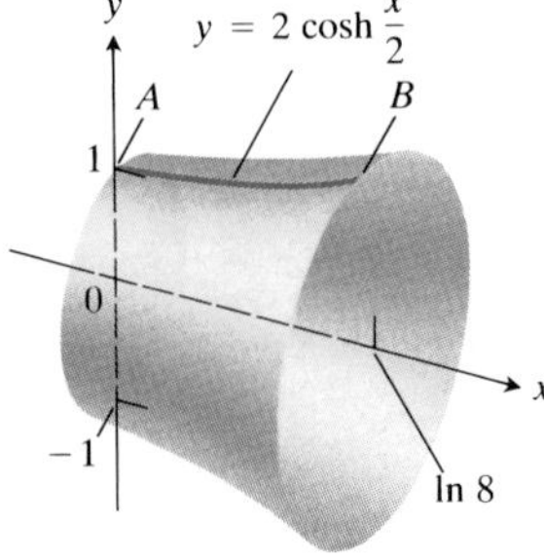

7.38 The minimal surface in Exercise 79.

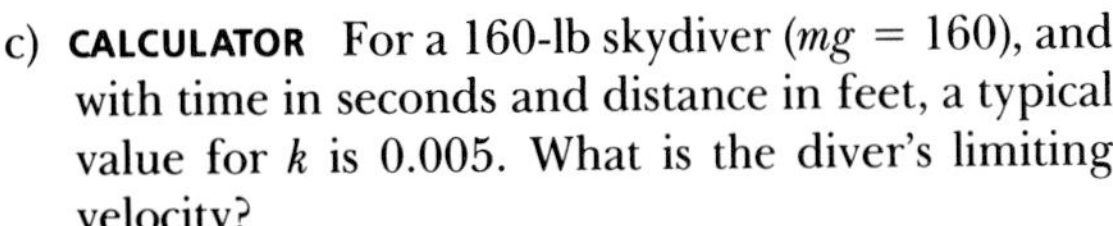

c) **CALCULATOR** For a 160-lb skydiver ($mg = 160$), and with time in seconds and distance in feet, a typical value for k is 0.005. What is the diver's limiting velocity?

76. *Accelerations whose magnitudes are proportional to displacement.* Suppose that the position of a body moving along a coordinate line at time t is
a) $s(t) = a \cos kt + b \sin kt$
b) $s(t) = a \cosh kt + b \sinh kt$.
Show that in both cases the acceleration d^2s/dt^2 is proportional to s, but that in the first case it is always directed toward the origin while in the second case it is directed away from the origin.

Hanging Cables and Hyperbolic Cosines

It can be shown that clotheslines, chains, telephone lines, and electric power cables that are strung from one support to another always hang in the shape of a hyperbolic cosine curve (Fig. 7.37). In contrast, the cables of suspension bridges, which do not hang freely but support a uniform load per horizontal foot, hang in parabolas. Hyperbolic cosine curves are sometimes called **chain curves** or **catenaries,** the latter deriving from the Latin *catena,* for "chain."

77. Find the length of the curve

$$y = 10 \cosh(x/10), \quad -10 \ln 10 \le x \le 10 \ln 10$$

78. Find the area between the curve in Exercise 77 and the x-axis.

79. *Minimal surfaces.* Find the area of the surface swept out by revolving the curve $y = 2\cosh(x/2)$, $0 \le x \le \ln 8$, about the x-axis (Fig. 7.38).

It can be shown that, of all continuously differentiable curves joining the points A and B in Fig. 7.38, the hyperbolic cosine curve $y = 2\cosh(x/2)$ generates the surface of least area. If you made a rigid wire frame of the circles through A and B and dipped it in a soap solution, the surface spanning the circles would be the one generated by the hyperbolic cosine.

80. Show that the function $y = a \cosh(x/a)$ solves the initial value problem

Differential equation: $y'' = \sqrt{1 + (y')^2}$

Initial conditions: $y'(0) = 0$ and $y(0) = a$.

(By analyzing the forces in hanging cables, we can show that the curves they hang in always satisfy the differential equation and initial conditions given here. That is how we know that cables hang in hyperbolic cosines.)

The Hyperbolic in Hyperbolic Functions

In case you are wondering where the name *hyperbolic* comes from, here is the answer: Just as $x = \cos u$ and

$y = \sin u$ are identified with points (x, y) on the unit circle, the functions $x = \cosh u$ and $y = \sinh u$ are identified with points (x, y) on the right-hand branch of the unit hyperbola, $x^2 - y^2 = 1$.

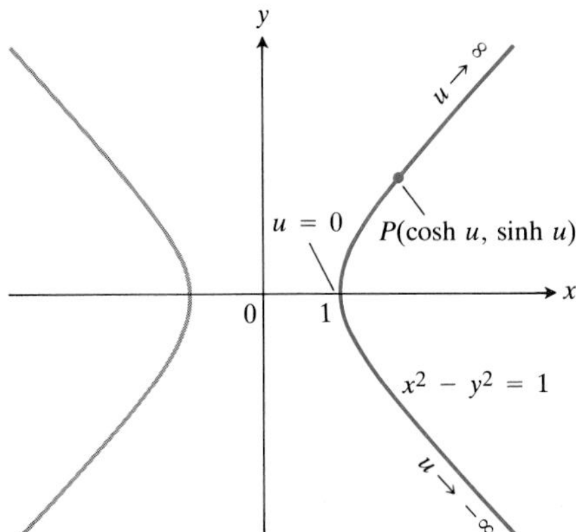

81. Another analogy between hyperbolic and circular functions is that the variable u in the coordinates $(\cosh u, \sinh u)$ for the points of the right-hand branch of the hyperbola $x^2 - y^2 = 1$ is twice the area of the sector AOP pictured in Fig. 7.39. To see why this is so, carry out the following steps.
   a) Show that the area $A(u)$ of sector AOP is given by the formula
   $$A(u) = \frac{1}{2}\cosh u \sinh u - \int_1^{\cosh u} \sqrt{x^2 - 1}\, dx.$$
   b) Differentiate both sides of the equation in (a) to show that
   $$A'(u) = \frac{1}{2}.$$
   c) Solve this last equation for $A(u)$. What is the value of $A(0)$?

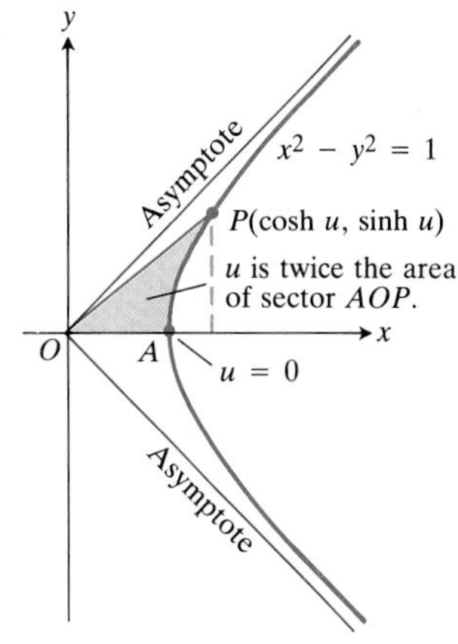

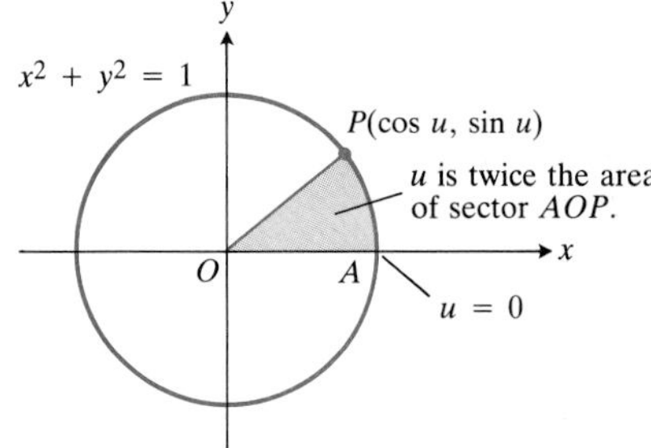

7.39 One of the analogies between hyperbolic and circular functions is revealed by these two diagrams. See Exercise 81.

Review Questions

1. What is the inverse of a one-to-one function? How do you tell when functions are inverses of one another? Give examples of functions that are inverses of one another; that are not inverses of one another.
2. What continuous functions have inverses?
3. How are the graphs, domains, and ranges of functions and their inverses related? Give an example.
4. How are the derivatives of functions and their inverses related? Give an example.

5. What are the rules for doing arithmetic with logarithms?
6. What is logarithmic differentiation? Give an example.
7. What integrals lead naturally to logarithms? Give an example.
8. Graph the functions $\ln x$ and e^x together. What are the domains and ranges of these functions? Can e^x ever equal zero?
9. What laws of exponents do exponential functions obey?
10. How are the functions a^x and $\log_a x$ defined? What are their derivatives? What integrals are associated with these functions? Give examples.
11. How is $\log_a x$ evaluated in terms of $\ln x$? How is the graph of $\log_a x$ related to the graph of $\ln x$?
12. What does the graph of the function a^x look like if $a > 1$? If $0 < a < 1$?
13. Describe some of the applications of base-10 logarithms.
14. Where does the equation $y = y_0 e^{kt}$ come from? Describe some of the applications of this equation.
15. How are limits found with l'Hôpital's rule? Give examples. How do you know when to stop using the rule?
16. How do you compare the growth rates of functions of x as $x \to \infty$? Give examples.
17. Explain little-oh and big-oh notation. Give examples.
18. Graph the six basic inverse trigonometric functions. Indicate their domains and ranges.
19. What are the derivatives of the inverse trigonometric functions? What integrals are associated with these functions? Give examples.
20. How are the six basic hyperbolic functions defined? Graph them and indicate their domains and ranges.
21. What identities are associated with hyperbolic functions?
22. What are the derivatives of the six basic hyperbolic functions? What are the corresponding integrals? Give examples.
23. What is special about the hyperbolic cosine curves?
24. Graph the six inverse hyperbolic functions and give their domains and ranges.
25. What are the derivatives of the inverse hyperbolic functions? Give examples of integrals that lead naturally to inverse hyperbolic functions?

Practice Exercises

Find the derivatives with respect to x of the functions whose formulas appear in Exercises 1–18.

1. $\ln(\cos x)$
2. $\ln(\sin x)$
3. $\ln(\cos^{-1} x)$
4. $\ln(\sin^{-1} x)$
5. $\log_2(x^2)$
6. $\log_5(x-7)$
7. $e^{\tan^{-1} x}$
8. $e^{\cot^{-1} x}$
9. 8^{-x}
10. 9^x
11. $\sin^{-1}(\sqrt{1-x})$
12. $\cos^{-1}(1/x) - \csc^{-1} x,\ x > 0$
13. $\tan^{-1}(\tan 2x)$
14. $(1+x^2)\cot^{-1} 2x$
15. $x \tan^{-1} x - \dfrac{1}{2}\ln x$
16. $x \cos^{-1} x - \sqrt{1-x^2}$
17. $2\sqrt{x-1}\,\sec^{-1}\sqrt{x}$
18. $\csc^{-1}(\sec x),\ 0 < x < \pi/2$

Use logarithmic differentiation to find dy/dx in Exercises 19–24.

19. $y = \dfrac{2(x^2+1)}{\sqrt{\cos 2x}}$
20. $y = \sqrt[10]{\dfrac{3x+4}{2x-4}}$
21. $y = \left(\dfrac{(x+5)(x-1)}{(x-2)(x+3)}\right)^5$
22. $y = x^{\ln x},\ x > 0$
23. $y = (1+x^2)\,e^{\tan^{-1} x}$
24. $y = \dfrac{2x\,2^x}{\sqrt{x^2+1}}$

Find the derivatives with respect to x of the functions whose formulas appear in Exercises 25–30.

25. $x - \coth x$
26. $x \sinh x - \cosh x$
27. $\ln(\operatorname{csch} x) + x \coth x$
28. $\ln(\operatorname{sech} x) + x \tanh x$
29. $\sin^{-1}(\tanh x)$
30. $\tan^{-1}(\sinh x)$

Find the derivatives with respect to x of the functions whose formulas appear in Exercises 31–36.

31. $\sqrt{1+x^2}\sinh^{-1}x$ **32.** $\sqrt{x^2-1}\cosh^{-1}x$

33. $1-\tanh^{-1}(1/x),\ |x|>1$

34. $\coth^{-1}(\csc x),\ 0<x<\pi/2$

35. $\operatorname{sech}^{-1}(\cos 2x),\ 0<x<\pi/4$

36. $\operatorname{csch}^{-1}(\cot x),\ 0<x<\pi/2$

Evaluate the integrals in Exercises 37–54.

37. $\int_{-1}^{1}\frac{dx}{3x-4}$ **38.** $\int_{1}^{e}\frac{\sqrt{\ln x}}{x}\,dx$

39. $\int_{0}^{\pi}\tan\frac{x}{3}\,dx$ **40.** $\int_{1/6}^{1/4}2\cot \pi x\,dx$

41. $\int_{0}^{4}\frac{2x\,dx}{x^2-25}$ **42.** $\int_{-\pi/3}^{\pi/3}\frac{\sec x+\tan x}{\sec x}\,dx$

43. $\int_{0}^{\pi/4}\frac{\sec x\tan x+\sec^2 x}{\sec x+\tan x}\,dx\quad (u=\sec x+\tan x)$

44. $\int_{-\pi/2}^{\pi/2}\frac{\cos x}{2-\sin x}\,dx$ **45.** $\int_{1}^{8}\frac{\log_4 x}{x}\,dx$

46. $\int_{1}^{e}\frac{8\ln 3\log_3 x}{x}\,dx$ **47.** $\int_{0}^{1}x\,3^{(x^2)}\,dx$

48. $\int_{0}^{\pi/4}2^{\tan x}\sec^2 x\,dx$ **49.** $\int_{-1/2}^{1/2}\frac{3\,dx}{\sqrt{1-x^2}}$

50. $\int_{1}^{1+(\sqrt{2}/2)}\frac{dx}{\sqrt{1-(x-1)^2}}\quad (u=x-1)$

51. $\int_{-1}^{1}\frac{1}{1+x^2}\,dx$

52. $\int_{1}^{3}\frac{2\,dx}{\sqrt{x}(1+x)}\quad (u=\sqrt{x})$

53. $\int_{1/2}^{3/4}\frac{dx}{\sqrt{x}\sqrt{1-x}}\quad (u=\sqrt{x})$

54. $\int_{\sqrt{2}/3}^{2/3}\frac{dx}{x\sqrt{9x^2-1}}\quad (u=3x)$

Evaluate the integrals in Exercises 55–60.

55. $\int_{0}^{\ln 2}4e^x\cosh x\,dx$ **56.** $\int_{0}^{\ln 2}\frac{\sinh x}{1+\cosh x}\,dx$

57. $\int_{-\ln 3}^{\ln 3}3\sqrt{\cosh 2x+1}\,dx$ **58.** $\int_{1}^{2}\frac{5\operatorname{sech}^2(\ln x)}{x}\,dx$

59. $\int_{2}^{4}10\operatorname{csch}^2 x\coth x\,dx$ **60.** $\int_{0}^{\ln\sqrt{2}}4\operatorname{sech}^4 x\tanh x\,dx$

Evaluate the integrals in Exercises 61–66 in terms of

a) inverse hyperbolic functions
b) natural logarithms using the formulas in Table 7.14 in Exercises 7.9.

61. $\int_{0}^{\pi/2}\frac{\sin x\,dx}{\sqrt{1+\cos^2 x}}$ **62.** $\int_{\sqrt{2}}^{\sqrt{17}}\frac{2x\,dx}{\sqrt{x^4-1}}$

63. $\int_{1/5}^{1/2}\frac{4\tanh^{-1}x}{1-x^2}\,dx$ **64.** $\int_{\pi/6}^{\pi/4}\frac{\cos x\,dx}{\sin x\sqrt{1+\sin^2 x}}$

65. $\int_{3/5}^{4/5}\frac{2\operatorname{sech}^{-1}x}{x\sqrt{1-x^2}}\,dx$ **66.** $\int_{\sqrt{8}}^{\sqrt{3}}\frac{e^{\coth^{-1}x}}{1-x^2}\,dx$

67. The function $f(x)=e^x+x$, being differentiable and one-to-one, has a differentiable inverse, $f^{-1}(x)$. Find the value of df^{-1}/dx at the point $f(\ln 2)$.

68. Find the area between the curve $y=2(\ln x)/x$ and the x-axis from $x=1$ to $x=e$.

69. The functions $\ln 5x$ and $\ln 3x$ differ by a constant. What constant?

70. Show that the area between the curve $y=1/x$ from $x=1$ to $x=2$ is the same as the area under the curve from $x=10$ to $x=20$.

71. Find the area between each of the curves $y=(2\log_2 x)/x$ and $y=(2\log_4 x)/x$ and the interval $1\le x\le e$ of the x-axis. What is the ratio of these areas?

72. **CALCULATOR** Find the following numbers as accurately as your calculator will allow using only [ln x], [÷], and [×] (and the number keys, of course).
a) $\log_{10}5$ b) $\log_2 3$ c) $\log_7 2$

73. **CALCULATOR** What is the age of a sample of charcoal in which 90% of the carbon-14 originally present has decayed?

74. **CALCULATOR** *Californium-252.* What costs \$27 million per gram and can be used to treat brain cancer, analyze coal for its sulfur content, and detect explosives in luggage? The answer is Californium-252, a radioactive isotope so rare that only 8 g of it have been made in the western world since its discovery by Glenn Seaborg in 1950. The half-life of the isotope is 2.645 years—long enough for a useful service life and short enough to have a high radioactivity per unit mass. One microgram of the isotope releases 170 million neutrons per second.

a) What is the value of k in the decay equation for this isotope?

b) What is the isotope's mean life? (See Section 7.4, Exercise 14.)

c) How long will it take 95% of a sample's radioactive nuclei to disintegrate?

75. **CALCULATOR** *Cooling a pie.* A deep-dish apple pie, whose internal temperature was 220°F when removed from the oven, was set out on a 40° breezy porch to cool. Fifteen minutes later, the pie's internal temperature was 180°F. How long did it take the pie to cool from there to 70°F?

76. *Transport through a cell membrane.** Under some conditions the result of the movement of a dissolved substance across a cell's membrane is described by the equation

$$\frac{dy}{dt} = k\frac{A}{V}(c - y).$$

In this equation, y is the concentration of the substance inside the cell and dy/dt is the rate with which y changes over time. The letters k, A, V, and c stand for constants, k being the *permeability coefficient* (a property of the membrane), A the surface area of the membrane, V the cell's volume, and c the concentration of the substance outside the cell. The equation says that the rate at which the concentration changes within the cell is proportional to the difference between it and the outside concentration.

a) Solve the equation for $y(t)$, using y_0 to denote $y(0)$.

b) Find the steady-state concentration, $\lim_{t\to\infty} y(t)$.

Find the limits in Exercises 77–82.

77. $\lim\limits_{t\to 0}\dfrac{t - \ln(1 + 2t)}{t^2}$ **78.** $\lim\limits_{x\to 0}\dfrac{\sin x}{e^x - x - 1}$

79. $\lim\limits_{x\to 0}\dfrac{x \sin x}{1 - \cos x}$ **80.** $\lim\limits_{x\to 1}\dfrac{\log_4 x}{\log_2 x}$

81. $\lim\limits_{x\to 0}\dfrac{2^{\sin x} - 1}{e^x - 1}$

82. $\lim\limits_{x\to 0} x \ln x$ (*Hint:* $x \ln x = (\ln x)/(1/x)$)

Use logarithms and l'Hôpital's rule to find the limits in Exercises 83–86.

83. $\lim\limits_{x\to\infty} x^{1/x}$ **84.** $\lim\limits_{x\to\infty} x^{1/x^2}$

85. $\lim\limits_{x\to\infty}\left(1 + \dfrac{3}{x}\right)^x$ **86.** $\lim\limits_{x\to 0}\left(1 + \dfrac{3}{x}\right)^x$

87. Compare the rates of growth as $x \to \infty$ of

a) x and $5x$

b) $x + \dfrac{1}{x}$ and x

c) $x^2 + x$ and $x^2 - x$

88. Compare the rates of growth as $x \to \infty$ of

a) $\ln 2x$, $\ln x^2$, and $\ln(x + 2)$

b) $x^{\ln x}$ and $x^{\log_2 x}$

c) $\left(\dfrac{1}{2}\right)^x$ and $\left(\dfrac{1}{3}\right)^x$

89. True, or false?

a) $\dfrac{1}{x^2} + \dfrac{1}{x^4} = O\left(\dfrac{1}{x^2}\right)$

b) $\dfrac{1}{x^2} + \dfrac{1}{x^4} = O\left(\dfrac{1}{x^4}\right)$

c) $\sqrt{x^2 + 1} = O(x)$

90. True, or false?

a) $\ln x = o(x)$

b) $\ln \ln x = o(\ln x)$

c) $x = o(x + \ln x)$

91. Show that

$$\tan^{-1} x + \tan^{-1}\frac{1}{x} = \text{constant}.$$

Find the constant.

92. Use Fig. 7.40 to show that

$$\int_0^{\pi/2} \sin x\, dx = \frac{\pi}{2} - \int_0^1 \sin^{-1} x\, dx.$$

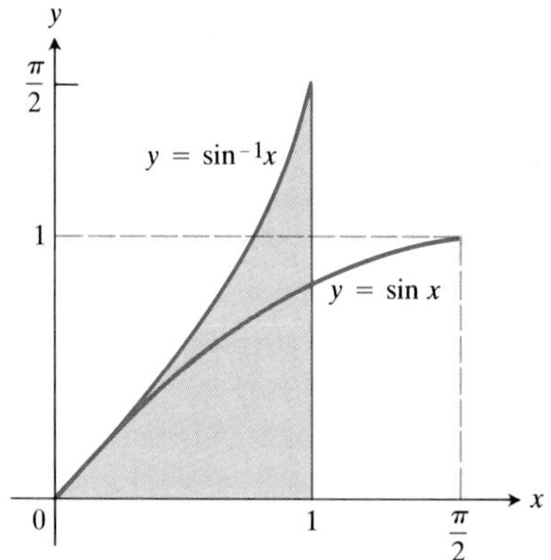

7.40 Figure for Exercise 92.

93. **CALCULATOR** *Locating a solar station.* You are under contract to build a solar station at ground level on the east-west line between the two buildings shown in

*Based on *Some Mathematical Models in Biology*, Revised Edition, R. M. Thrall, J. A. Mortimer, K. R. Rebman, R. F. Baum, eds., December 1967, PB–202 364, pp. 101–103; distributed by N.T.I.S., U.S. Department of Commerce.

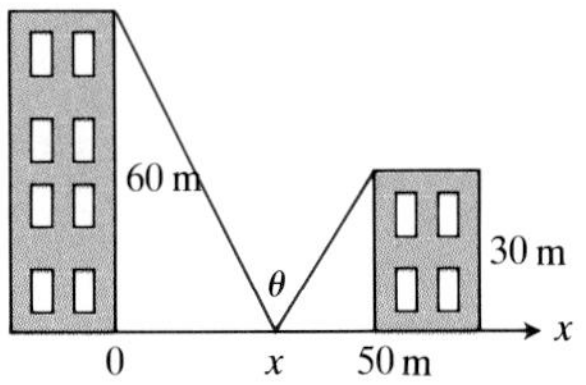

7.41 Figure for Exercise 93.

Fig. 7.41. How far from the taller building should you place the station to maximize the number of hours it will be in the sun on a day when the sun passes directly overhead? Begin by observing that

$$\theta = \pi - \cot^{-1}\frac{x}{60} - \cot^{-1}\frac{50 - x}{30}.$$

Then find the value of x that maximizes θ.

94. *The best branching angles for blood vessels and pipes.* When a smaller pipe branches off from a larger one in a flow system, we may want it to run off at an angle that is best from some energy-saving point of view. We might require, for instance, that energy loss due to friction be minimized along the section AOB shown in Fig. 7.42. In this diagram, B is a given point to be reached by the smaller pipe, A is a point in the larger pipe upstream from B, and O is the point where the branching occurs. A law due to Poiseuille states that the loss of energy due to friction in nonturbulent flow is proportional to the length of the path and inversely proportional to the fourth power of the radius. Thus, the loss along AO is $(kd_1)/R^4$ and along OB is $(kd_2)/r^4$, where k is a constant, d_1 is the length of AO, d_2 is the length of OB, R is the radius of the larger pipe, and r is the radius of the smaller pipe. The angle θ is to be chosen to minimize the sum of these two losses:

$$L = k\frac{d_1}{R^4} + k\frac{d_2}{r^4}.$$

In our model, we assume that $AC = a$ and $BC = b$ are fixed. Thus we have the relations

$$d_1 + d_2 \cos\theta = a, \quad d_2 \sin\theta = b,$$

so that

$$d_2 = b \csc\theta$$

and

$$d_1 = a - d_2 \cos\theta = a - b \cot\theta.$$

We can express the total loss L as a function of θ:

$$L = k\left(\frac{a - b\cot\theta}{R^4} + \frac{b\csc\theta}{r^4}\right).$$

a) Show that the critical value of θ for which $dL/d\theta$ equals zero is

$$\theta_c = \cos^{-1}\frac{r^4}{R^4}.$$

b) **CALCULATOR** If the ratio of the pipe radii is $r/R = 5/6$, estimate to the nearest degree the optimal branching angle given in part (a).

The mathematical analysis described here is also used to explain the angles at which arteries branch in an animal's body. See Edward Batschelet's *Introduction to Mathematics for Life Scientists,* 2nd ed. (New York: Springer-Verlag, 1976).

95. *Tractor trailers and the tractrix.* When a tractor trailer turns into a cross street, its rear wheels follow a curve like the one in Fig. 7.43. (This is why the rear wheels

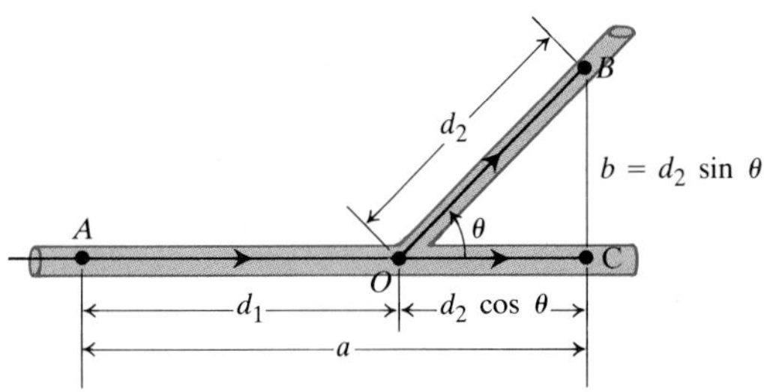

7.42 The smaller pipe OB branches away from the larger AOC at an angle θ that minimizes the friction loss along AO and OB. The optimum angle is found to be $\theta_c = \cos^{-1}(r^4/R^4)$, where r is the radius of the smaller pipe and R is the radius of the larger (Exercise 94).

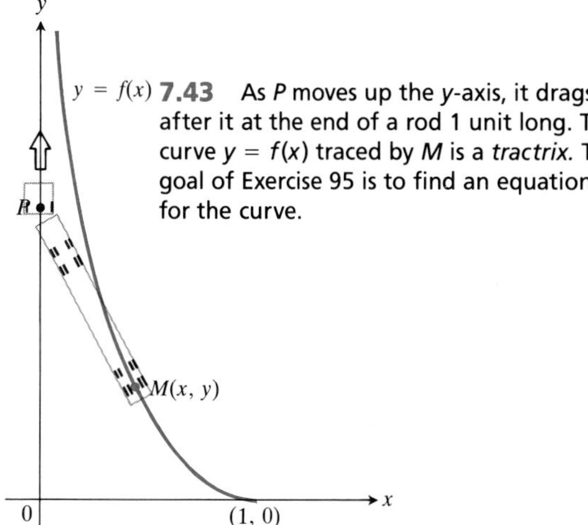

7.43 As P moves up the y-axis, it drags M after it at the end of a rod 1 unit long. The curve $y = f(x)$ traced by M is a *tractrix.* The goal of Exercise 95 is to find an equation for the curve.

sometimes ride up on the sidewalk.) We can find an equation for the curve if we picture the rear wheels as a mass M at the point $(1, 0)$ on the x-axis attached by a rod of unit length to a point P representing the cab at the origin. As the point P moves up the y-axis, it drags M along behind it. The curve traced by M, called a *tractrix* from the Latin word *tractum* for drag, can be shown to be the graph of the function $y = f(x)$ that solves the initial value problem

Differential equation: $\dfrac{dy}{dx} = -\dfrac{1}{x\sqrt{1-x^2}} + \dfrac{x}{\sqrt{1-x^2}}$

Initial condition: $y = 0$ when $x = 1$.

Solve the initial value problem to find an equation for the curve. (You need an inverse hyperbolic function.)

8 Techniques of Integration

Scala/Art Resource

The mathematical shape called "Gabriel's Horn" has a finite volume but an infinite area.

OVERVIEW We have seen how integrals arise in modeling real phenomena and in measuring objects in the world around us, and we know in theory how integrals are evaluated with antiderivatives. The more sophisticated our models become, however, the more involved our integrals become. We need to know how to rearrange these integrals into forms we can work with. The goal of this chapter is to show how to change unfamiliar integrals into integrals we can recognize, find in a table, or evaluate with a computer.

8.1 Basic Integration Formulas

As we saw in Section 5.5, indefinite integration is the reverse of differentiation. So, to evaluate $\int f(x)\,dx$, we need to find an antiderivative function $F(x)$ whose derivative is $f(x)$ or whose differential is $f(x)\,dx$. The list of integrals in Table 8.1 includes the integrals we have worked with so far. You will find more extensive integral tables at the back of this book. We recommend that you do not consult these more extensive tables until Section 8.6 so that you can learn through practice the algebraic and trigonometric procedures needed for understanding how best to use these tables.

TABLE 8.1
A brief table of standard integral forms

Definition	$\int du = u + C$	
Constant multiple	$\int a\,du = au + C$	
Sum	$\int (du + dv) = \int du + \int dv$	
Powers	$\int u^n\,du = \dfrac{u^{n+1}}{n+1} + C, n \neq -1$	$\int \dfrac{du}{u} = \ln\|u\| + C$
Trigonometric	$\int \cos u\,du = \sin u + C$	$\int \sin u\,du = -\cos u + C$
	$\int \tan u\,du = \ln\|\sec u\| + C$	$\int \cot u\,du = \ln\|\sin u\| + C$
	$\int \sec u\,du = \ln\|\sec u + \tan u\| + C$	$\int \csc u\,du = -\ln\|\csc u + \cot u\| + C$
Exponential	$\int e^u\,du = e^u + C$	$\int a^u\,du = \dfrac{1}{\ln a}a^u + C; a > 0, a \neq 1$
Special algebraic forms	$\int \dfrac{du}{1+u^2} = \tan^{-1}u + C$	$\int \dfrac{du}{\sqrt{1-u^2}} = \sin^{-1}u + C$
	$\int \dfrac{du}{u\sqrt{u^2-1}} = \sec^{-1}\|u\| + C$	
Integration by parts (from the product rule for derivatives)	$\int u\,dv = uv - \int v\,du$	

Three of the formulas in Table 8.1 are new:

$$\int u\,dv = uv - \int v\,du \qquad \begin{pmatrix}\text{Integration by parts,}\\ \text{Section 8.2}\end{pmatrix}$$

$$\int \sec u\,du = \ln|\sec u + \tan u| + C \qquad \text{(Example 6, Eq. (1))}$$

$$\int \csc u\,du = -\ln|\csc u + \cot u| + C \qquad \text{(Eq. 2)}$$

The latter two are included to complete the list of integrals of the six trigonometric functions: sine, cosine, tangent, cotangent, secant, and cosecant. The formula for integration by parts comes from integrating both sides of the formula $d(uv) = u\,dv + v\,du$. It is so important we study it in detail in Section 8.2.

Algebraic Procedures and Trigonometric Identities

The most frequently used integration technique is *substitution.* We look at the function to be integrated and try to match it with a standard form.

EXAMPLE 1 *Substitution.* Evaluate $\displaystyle\int \frac{dx}{1 + 4x^2}$.

Solution The nearest standard form is $\displaystyle\int \frac{du}{1 + u^2} = \tan^{-1}u + C$. Since $4x^2 = u^2$ if $u = 2x$, we substitute

$$u = 2x, \qquad du = 2\,dx, \qquad \text{and} \qquad dx = \frac{1}{2}\,du.$$

Then,

$$\int \frac{dx}{1 + 4x^2} = \int \frac{(1/2)\,du}{1 + u^2} = \frac{1}{2}\int \frac{du}{1 + u^2}$$

$$= \frac{1}{2}\tan^{-1}u + C = \frac{1}{2}\tan^{-1}2x + C.$$

■

EXAMPLE 2 *Completing the square.* Evaluate $\displaystyle\int \frac{dx}{x^2 + 2x + 2}$.

Solution We complete the square (Appendix 1) to write the denominator in the form

$$x^2 + 2x + 2 = (x^2 + 2x + 1) + (2 - 1) = (x + 1)^2 + 1.$$

The substitution $u = x + 1$ then reduces our integral to

$$\int \frac{dx}{x^2 + 2x + 2} = \int \frac{dx}{(x + 1)^2 + 1} = \int \frac{du}{u^2 + 1}$$

$$= \tan^{-1}u + C = \tan^{-1}(x + 1) + C.$$

■

EXAMPLE 3 *Expanding a power and using a trigonometric identity.*

Evaluate $\int (\sec x + \tan x)^2\, dx$.

Solution We expand the integrand and get

$$(\sec x + \tan x)^2 = \sec^2 x + 2 \sec x \tan x + \tan^2 x.$$

The first two terms on the right-hand side of this equation are old friends; we can integrate them at once. How about $\tan^2 x$? There is an identity that connects it with $\sec^2 x$:

$$\tan^2 x + 1 = \sec^2 x, \qquad \tan^2 x = \sec^2 x - 1.$$

We replace $\tan^2 x$ by $\sec^2 x - 1$ and get

$$\begin{aligned}\int (\sec x + \tan x)^2\, dx &= \int (\sec^2 x + 2 \sec x \tan x + \sec^2 x - 1)\, dx \\ &= 2\int \sec^2 x\, dx + 2\int \sec x \tan x\, dx - \int 1\, dx \\ &= 2 \tan x + 2 \sec x - x + C.\end{aligned}$$

We'll do more with trigonometric techniques in Sections 8.3 and 8.4.

EXAMPLE 4 *Reducing an improper fraction.* Evaluate $\int \frac{(x-2)^3}{x^2-4}\, dx$.

Solution The numerator and denominator have a common factor $x - 2$. The integrand is not defined for $x = 2$, so we can divide both numerator and denominator by $x - 2$ because it is not zero in the domain of the quotient:

$$\frac{(x-2)^3}{x^2-4} = \frac{(x-2)^3}{(x-2)(x+2)} = \frac{(x-2)^2}{x+2} = \frac{x^2-4x+4}{x+2}.$$

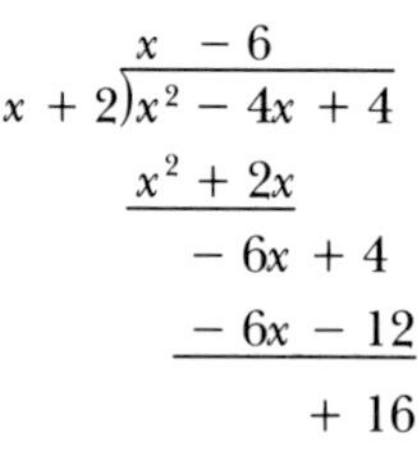

$$\begin{array}{r} x - 6 \\ x+2\overline{)x^2 - 4x + 4} \\ \underline{x^2 + 2x} \\ -6x + 4 \\ \underline{-6x - 12} \\ +16 \end{array}$$

The result is an improper fraction, in which the degree of the numerator is equal to or greater than the degree of the denominator. To integrate such fractions, we first divide them, getting a quotient plus a remainder that is a proper fraction:

$$\frac{x^2-4x+4}{x+2} = x - 6 + \frac{16}{x+2}.$$

Therefore,

$$\int \frac{(x-2)^3}{x^2-4}\, dx = \int \left(x - 6 + \frac{16}{x+2}\right) dx = \frac{x^2}{2} - 6x + 16 \ln|x+2| + C.$$

EXAMPLE 5 *Separating a fraction.* Evaluate $\int \frac{3x+2}{\sqrt{1-x^2}}\, dx$.

Solution We first separate the integrand to get

$$\int \frac{3x+2}{\sqrt{1-x^2}}\, dx = 3\int \frac{x\, dx}{\sqrt{1-x^2}} + 2\int \frac{dx}{\sqrt{1-x^2}}.$$

In the first of these new integrals we substitute

$$u = 1 - x^2, \qquad du = -2x\,dx, \qquad \text{and} \qquad x\,dx = -\frac{1}{2}du.$$

$$\int 3\frac{x\,dx}{\sqrt{1-x^2}} = 3\int \frac{(-1/2)\,du}{\sqrt{u}} = -\frac{3}{2}\int u^{-1/2}\,du$$

$$= -\frac{3}{2}\frac{u^{1/2}}{1/2} + C_1 = -3\sqrt{1-x^2} + C_1.$$

The second of the new integrals is a standard form,

$$\int \frac{dx}{\sqrt{1-x^2}} = \sin^{-1}x + C_2.$$

Combining these results gives

$$\int \frac{3x+2}{\sqrt{1-x^2}}dx = -3\sqrt{1-x^2} + 2\sin^{-1}x + C.$$

■

EXAMPLE 6 *The secant integral.* Show that

$$\int \sec x\,dx = \ln|\sec x + \tan x| + C. \tag{1}$$

Solution

$$\int \sec x\,dx = \int (\sec x)(1)\,dx = \int \sec x \frac{\sec x + \tan x}{\sec x + \tan x}dx$$

$$= \int \frac{\sec^2 x + \sec x \tan x}{\sec x + \tan x}dx$$

$$= \int \frac{du}{u} \qquad \begin{pmatrix} u = \sec x + \tan x \\ du = \sec^2 x + \sec x \tan x \end{pmatrix}$$

$$= \ln|u| + C = \ln|\sec x + \tan x| + C.$$

■

With cosecants and cotangents in place of secants and tangents, the method of Example 6 leads to the companion formula

$$\int \csc x\,dx = -\ln|\csc x + \cot x| + C. \tag{2}$$

Mercator's World Map

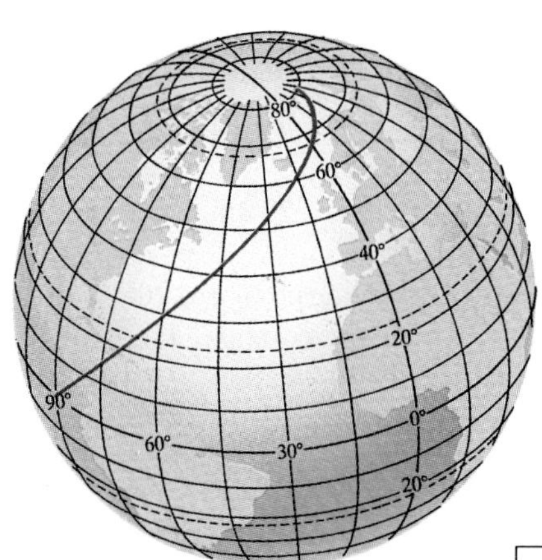

8.1 A flight with a constant bearing of 45° E of N from the Galápagos Islands in the Pacific to Franz Josef Land in the Arctic Ocean as it appears on a globe.

The integral of the secant plays an important role in making maps for compass navigation. The easiest course for a sailor or pilot to steer is a course whose compass heading is constant. This might be a course of 45° (northeast), for example, or a course of 225° (southwest), or whatever. Such a course will be along a spiral that winds around the globe toward one of the poles (Fig. 8.1) unless the course runs due north or south or lies parallel to the equator.

In 1569, Gerhard Krämer, a Flemish surveyor and geographer known to us by his Latinized last name, Mercator, made a world map on which all

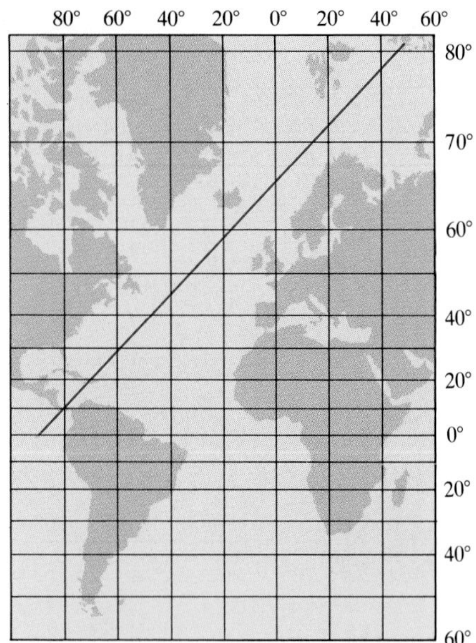

8.2 The flight of Fig. 8.1 traced on a Mercator map.

spirals of constant compass heading appeared as straight lines (Fig. 8.2). A sailor could then read the compass heading for a voyage between any two points from the direction of a straight line connecting them on Mercator's map.

If you look closely at Fig. 8.2, you will see that the vertical lines of longitude that meet at the poles on a globe have been spread apart to lie parallel on the map. The horizontal lines of latitude that are shown every 10° are also parallel, as they are on the globe, but they are not evenly spaced. The spacing between them increases toward the poles.

The scaling factor by which horizontal distances are increased at a fixed latitude $\theta°$ is precisely $\sec\theta$, and this scaling factor increases with the latitude θ. If R is the radius of the globe being modeled (Fig. 8.3), then the distance on the map between the lines representing the equator and the latitude $\alpha°$ is R times the integral of the secant from zero to α:

$$D = R\int_0^\alpha \sec x\,dx. \tag{3}$$

Therefore, the map distance between the two latitude lines on the same side of the equator, say at $\alpha°$ and $\beta°$ ($\alpha < \beta$) is

$$R\int_0^\beta \sec x\,dx - R\int_0^\alpha \sec x\,dx = R\int_\alpha^\beta \sec x\,dx = R\ln|\sec x + \tan x|\Big]_\alpha^\beta. \tag{4}$$

EXAMPLE 7 Suppose that the equatorial length of a Mercator map just matches the equator of a globe of radius 25 cm. Then Eq. (4) gives the spacing on the map between the equator and latitude 20° north as

$$25\int_0^{20°} \sec x\,dx = 25\ln|\sec x + \tan x|\Big]_0^{20°} \approx 9\text{ cm}.$$

The spacing between 60° and 80° north is given by Eq. (4) as

$$25\int_{60°}^{80°} \sec x\,dx = 25\ln|\sec x + \tan x|\Big]_{60°}^{80°} \approx 28\text{ cm}.$$

As you can see, the map distance between latitude lines that are 20° apart is considerably greater near the pole than it is near the equator. The navigational properties of a Mercator map are achieved at the expense of a considerable distortion of distance. ■

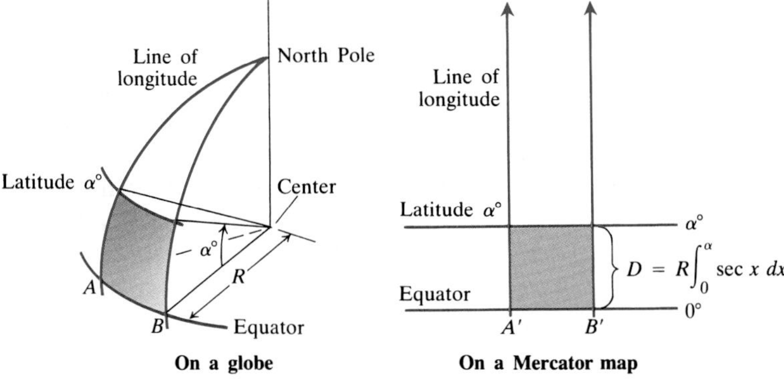

8.3 Lines of latitude and longitude.

Exercises 8.1

Evaluate the integrals in Exercises 1–20 by using the given substitutions to reduce the integral to a standard form.

1. $\displaystyle\int_0^1 \frac{16x\,dx}{\sqrt{8x^2+1}}, \quad u = 8x^2 + 1$

2. $\displaystyle\int_0^{\pi/2} \frac{3\cos x\,dx}{\sqrt{1+3\sin x}}, \quad u = 1 + 3\sin x$

3. $\displaystyle\int_0^1 \frac{16x\,dx}{8x^2+2}, \quad u = 8x^2 + 2$

4. $\displaystyle\int_4^9 \frac{dx}{x-\sqrt{x}}, \quad u = \sqrt{x}$

5. $\displaystyle\int 4x\tan x^2\,dx, \quad u = x^2$

6. $\displaystyle\int \cot(3-7x)\,dx, \quad u = 3 - 7x$

7. $\displaystyle\int_{-\pi}^{\pi} \sec\frac{x}{3}\,dx, \quad u = \frac{x}{3}$

8. $\displaystyle\int x\sec(x^2-5)\,dx, \quad u = x^2 - 5$

9. $\displaystyle\int_{3\pi/2}^{7\pi/4} \csc(x-\pi)\,dx, \quad u = x - \pi$

10. $\displaystyle\int \frac{1}{x^2}\csc\frac{1}{x}\,dx, \quad u = \frac{1}{x}$

11. $\displaystyle\int_0^{\sqrt{\ln 2}} 2x\,e^{x^2}\,dx, \quad u = x^2$

12. $\displaystyle\int_{\pi/2}^{\pi} \sin x\,e^{\cos x}\,dx, \quad u = \cos x$

13. $\displaystyle\int_{-1}^{0} 3^{x+1}\,dx, \quad u = x + 1$

14. $\displaystyle\int_1^2 \frac{2^{\ln x}}{x}\,dx, \quad u = \ln x$

15. $\displaystyle\int_1^3 \frac{6\,dy}{\sqrt{y}\,(1+y)}, \quad u = \sqrt{y}$

16. $\displaystyle\int_{-1}^{0} \frac{4\,dx}{1+(2x+1)^2}, \quad u = 2x + 1$

17. $\displaystyle\int_0^{1/6} \frac{dx}{\sqrt{1-9x^2}}, \quad u = 3x$

18. $\displaystyle\int_0^1 \frac{dx}{\sqrt{4-x^2}}, \quad u = x/2$

19. $\displaystyle\int_{2/5\sqrt{3}}^{2/5} \frac{6\,dx}{x\sqrt{25x^2-1}}, \quad u = 5x$

20. $\displaystyle\int_{-6}^{-3\sqrt{2}} \frac{dx}{x\sqrt{x^2-9}}, \quad 3u = x$

Evaluate the integrals in Exercises 21–26 by completing the square and using substitutions to reduce them to standard forms.

21. $\displaystyle\int \frac{dx}{\sqrt{-x^2+4x-3}}$

22. $\displaystyle\int \frac{dx}{\sqrt{2x-x^2}}$

23. $\displaystyle\int_1^2 \frac{8\,dx}{x^2-2x+2}$

24. $\displaystyle\int_2^4 \frac{2\,dx}{x^2-6x+10}$

25. $\displaystyle\int \frac{dx}{(x+1)\sqrt{x^2+2x}}$

26. $\displaystyle\int \frac{dx}{(x-2)\sqrt{x^2-4x+3}}$

Evaluate the integrals in Exercises 27–30.

27. $\displaystyle\int_{\pi/4}^{3\pi/4} (\csc x - \cot x)^2\,dx$

28. $\displaystyle\int_0^{\pi/4} (\sec x + 4\cos x)^2\,dx$

29. $\displaystyle\int_{\pi/6}^{\pi/3} (\csc x - \sec x)(\sin x + \cos x)\,dx$

30. $\displaystyle\int_0^{\pi/2} (\sin 3x\cos 2x - \cos 3x\sin 2x)\,dx$

Evaluate the integrals in Exercises 31–34.

31. $\displaystyle\int_0^1 \frac{x}{x+1}\,dx$

32. $\displaystyle\int_0^1 \frac{x^2}{x^2+1}\,dx$

33. $\displaystyle\int_{\sqrt{2}}^{3} \frac{2x^3}{x^2-1}\,dx$

34. $\displaystyle\int_{-1}^{3} \frac{4x^2-7}{2x+3}\,dx$

Evaluate the integrals in Exercises 35–38 by separating the fractions.

35. $\displaystyle\int_0^{\sqrt{3}/2} \frac{1-x}{\sqrt{1-x^2}}\,dx$

36. $\displaystyle\int_2^5 \frac{x+2\sqrt{x-1}}{2x\sqrt{x-1}}\,dx$

37. $\displaystyle\int_0^{\pi/4} \frac{1+\sin x}{\cos^2 x}\,dx$

38. $\displaystyle\int_0^{1/2} \frac{2-8x}{1+4x^2}\,dx$

39. Find the length of the curve $y = \ln(\cos x)$, $0 \le x \le \pi/3$.

40. Find the length of the curve $y = \ln(\sec x)$, $0 \le x \le \pi/4$.

41. Find the centroid of the region bounded by the x-axis, the curve $y = \sec x$, and the lines $x = -\pi/4$, $x = \pi/4$.

42. Find the area of the region bounded above by $y = 2\cos x$ and below by $y = \sec x$, $-\pi/4 \le x \le \pi/4$.

CALCULATOR or TABLES How far apart should the lines of latitude in Exercises 43 and 44 be on the Mercator map in Example 7?

43. Latitudes 30° and 45° north (New Orleans, Louisianna, and Minneapolis, Minnesota).

44. Latitudes 45° and 60° north (Salem, Oregon, and Seward, Alaska).

Evaluate the integrals in Exercises 45–60 by finding substitutions that reduce them to standard forms.

45. $\int_0^{\pi/2} 3\sqrt{\sin x}\cos x\,dx$

46. $\int_{\pi/6}^{\pi/2} \cot^3 x\csc^2 x\,dx$

47. $\int_{-\pi}^{0} \frac{\sin x\,dx}{2+\cos x}$

48. $\int_0^2 \frac{x\,dx}{4x^2+1}$

49. $\int_0^{1/4} \sec \pi x\,dx$

50. $\int_{1/4}^{3/4} \csc \pi x\,dx$

51. $\int_0^{\pi/3} e^{\tan x}\sec^2 x\,dx$

52. $\int_{\ln^2 2}^{\ln^2 3} \frac{e^{\sqrt{x}}\,dx}{\sqrt{x}}$

53. $\int_1^4 \frac{2^{\sqrt{x}}\,dx}{2\sqrt{x}}$

54. $\int_0^1 10^{2x}\,dx$

55. $\int_0^{\sqrt{3}/3} \frac{9\,dx}{1+9x^2}$

56. $\int_0^{\ln\sqrt{3}} \frac{e^x\,dx}{1+e^{2x}}$

57. $\int_0^{1/4} \frac{2\,dx}{\sqrt{1-4x^2}}$

58. $\int_0^{1/\sqrt{2}} \frac{2x\,dx}{\sqrt{1-x^4}}$

59. $\int_{1/\sqrt{2}}^{1} \frac{dx}{x\sqrt{4x^2-1}}$

60. $\int_{\ln(2/\sqrt{3})}^{\ln 2} \frac{dx}{\sqrt{e^{2x}-1}}$

8.2 Integration by Parts

Integration by parts is a technique used mainly for simplifying integrals of the form

$$\int f(x)\,g(x)\,dx \tag{1}$$

in which f can be differentiated repeatedly to become zero and g can be integrated repeatedly without difficulty. The integral

$$\int xe^x\,dx$$

is such an integral because $f(x) = x$ can be differentiated twice to become zero and $g(x) = e^x$ can be integrated repeatedly without difficulty. Integration by parts also applies to integrals like

$$\int e^x \sin x\,dx,$$

in which each part of the integrand appears again after repeated differentiation or integration.

In this section, we describe integration by parts and show how to apply it.

The Formula

The formula for integration by parts comes from the Product Rule,

$$\frac{d}{dx}(uv) = u\frac{dv}{dx} + v\frac{du}{dx}.$$

In its differential form, the rule becomes

$$d(uv) = u\,dv + v\,du,$$

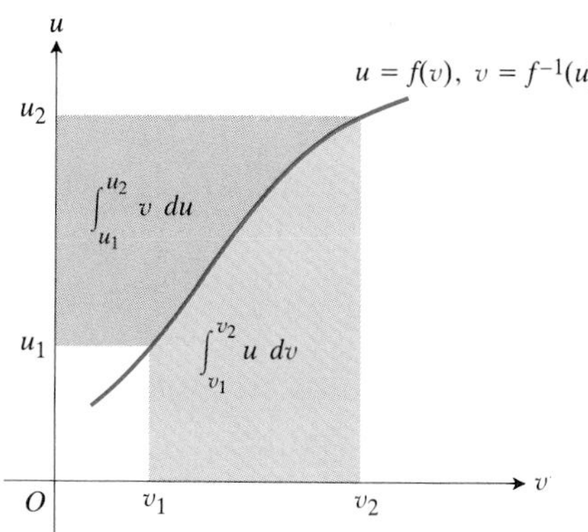

8.4 The area of the blue region, $\int_{v_1}^{v_2} u\,dv$, is equal to the area of the large rectangle, u_2v_2, minus the areas of the small rectangle, u_1v_1, and the gray region, $\int_{u_1}^{u_2} v\,du$. In symbols,

$$\int_{v_1}^{v_2} u\,dv = (u_2v_2 - u_1v_1) - \int_{u_1}^{u_2} v\,du.$$

When and How to use Integration by Parts

When: If substitution doesn't work, try integration by parts.

How: Start with an integral of the form

$$\int f(x)\,g(x)\,dx.$$

Match this with an integral of the form

$$\int u\,dv$$

by choosing dv to be part of the integrand including dx and possibly $f(x)$ or $g(x)$.

Guideline for choosing u and dv: The formula

$$\int u\,dv = uv - \int v\,du$$

gives a new integral on the right side of the equation. If the new integral is more complex than the original one, try a different choice.

which is then written as

$$u\,dv = d(uv) - v\,du$$

and integrated to give the following formula.

The Integration-by-Parts Formula

$$\int u\,dv = uv - \int v\,du. \tag{2}$$

The integration-by-parts formula expresses one integral, $\int u\,dv$, in terms of a second integral, $\int v\,du$. With a proper choice of u and v, the second integral may be easier to evaluate than the first. This is the reason for the importance of the formula. When faced with an integral we cannot handle, we can replace it by one with which we might have more success.

The equivalent formula for definite integrals is

$$\int_{v_1}^{v_2} u\,dv = (u_2\,v_2 - u_1\,v_1) - \int_{u_1}^{u_2} v\,du. \tag{3}$$

Figure 8.4 shows how the different parts of the formula may be interpreted as areas.

EXAMPLE 1 Evaluate $\displaystyle\int x\cos x\,dx.$

Solution We use the formula $\displaystyle\int u\,dv = uv - \int v\,du$ with

$$u = x, \qquad dv = \cos x\,dx,$$

$$du = dx, \qquad v = \sin x \qquad \text{(Simplest function with } dv = \cos x\,dx\text{)}.$$

Then

$$\int x\cos x\,dx = x\sin x - \int \sin x\,dx = x\sin x + \cos x + C.$$

■

Before we go on to new examples, let's look at the choices available in Example 1.

EXAMPLE 2 *Example 1 revisited.* To apply integration by parts to

$$\int x\cos x\,dx = \int f(x)\,g(x)\,dx = \int u\,dv$$

we have four possible choices:

1. Let $u = 1$ and $dv = f(x)\,g(x)\,dx = x\cos x\,dx$.
2. Let $u = f(x) = x$ and $dv = g(x)\,dx = \cos x\,dx$.
3. Let $u = f(x)\,g(x) = x\cos x$ and $dv = dx$.
4. Let $u = g(x) = \cos x$ and $dv = x\,dx$.

Let's examine these one at a time.

Choice 1 won't do because we don't know how to integrate $dv = x \cos x\, dx$ to get v.

Choice 2 worked well. We did that in Example 1.

Choice 3 leads to $du = \dfrac{du}{dx}\, dx = (-x \sin x + \cos x)\, dx$, and $v = x$ with a new integral

$$\int v\, du = \int (-x^2 \sin x + x \cos x)\, dx.$$

This is worse than the integral we started with.

Choice 4 leads to $du = -\sin x\, dx$ and $v = \int x\, dx = x^2/2$, so the new integral is

$$\int v\, du = -\int \frac{x^2}{2} \sin x\, dx.$$

This is also harder than the original because we have gone from $x \cos x$ to $x^2 \sin x$ in the integrand.

Summary. Keep in mind that the object is to go from the given integral ($\int u\, dv$) to a new integral ($\int v\, du$) that is simpler. (Integration by parts does not always work, so we can't always achieve that goal.) ■

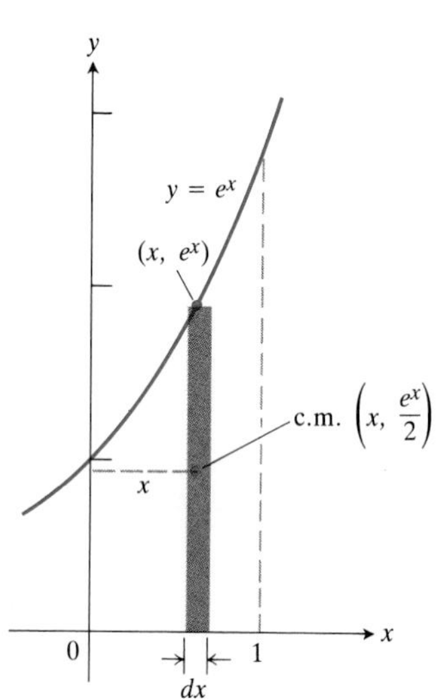

8.5 The moment of the strip about the y-axis is $x\, \delta\, dA = \delta\, xe^x\, dx$.

EXAMPLE 3 Find the moment about the y-axis of a thin plate of constant density δ covering the region in the first quadrant bounded by the curve $y = e^x$ and the line $x = 1$ (Fig. 8.5).

Solution A typical vertical strip has

center of mass (c.m.):	$(\tilde{x}, \tilde{y}) = \left(x, \dfrac{e^x}{2}\right)$,
length:	e^x,
width:	dx,
area:	$da = e^x\, dx$,
mass:	$dm = \delta\, dA = \delta\, e^x\, dx$.

The moment of the strip about the y-axis is therefore

$$\tilde{x}\, dm = x \cdot \delta\, e^x\, dx = \delta\, xe^x\, dx.$$

The moment of the plate about the y-axis is $\quad M_y = \int \tilde{x}\, dm = \delta \int_0^1 xe^x\, dx.$

To evaluate this integral we use the formula $\int u\, dv = uv - \int v\, du$ with

$$u = x, \qquad dv = e^x\, dx,$$

$$du = dx, \qquad v = e^x \qquad \text{(Simplest function with } dv = e^x\, dx\text{)}.$$

Then

$$\int xe^x\,dx = xe^x - \int e^x\,dx,$$

so

$$\int_0^1 xe^x\,dx = xe^x\Big]_0^1 - \int_0^1 e^x\,dx = e - \Big[e^x\Big]_0^1 = e - [e - 1] = 1.$$

The moment of the plate about the y-axis is

$$M_y = \delta\int_0^1 xe^x\,dx = \delta\cdot 1 = \delta.$$

EXAMPLE 4 Evaluate $\displaystyle\int \ln x\,dx$.

Solution We use the formula $\displaystyle\int u\,dv = uv - \int v\,du$ with

$$
\begin{aligned}
u &= \ln x, && \text{(Simplifies when differentiated)}\\
du &= \frac{1}{x}\,dx, &&\\
dv &= dx, && \text{(Easy to integrate)}\\
v &= x && \text{(Simplest function with } dv = dx\text{)}.
\end{aligned}
$$

Then

$$\int \ln x\,dx = x\ln x - \int x\cdot\frac{1}{x}\,dx = x\ln x - \int dx = x\ln x - x + C.$$

Repeated Use

Sometimes we have to use integration by parts more than once to obtain an answer.

EXAMPLE 5 Evaluate $\displaystyle\int x^2e^x\,dx$.

Solution We use the formula $\displaystyle\int u\,dv = uv - \int v\,du$ with

$$u = x^2, \qquad dv = e^x\,dx, \qquad v = e^x, \qquad du = 2x\,dx.$$

This gives

$$\int x^2e^x\,dx = x^2e^x - 2\int xe^x\,dx.$$

It takes a second integration by parts to find the integral on the right. As in Example 3, its value is $xe^x - e^x + C'$. Hence

$$\int x^2e^x\,dx = x^2e^x - 2xe^x + 2e^x + C.$$

Solving for the Unknown Integral

Integrals like the one in the next example occur in electrical engineering problems. Their evaluation requires two integrations by parts, followed by solving for the unknown integral.

EXAMPLE 6 Evaluate $\int e^x \cos x \, dx$.

Solution We first use the formula $\int u \, dv = uv - \int v \, du$ with

$$u = e^x, \qquad dv = \cos x \, dx, \qquad v = \sin x, \qquad du = e^x \, dx.$$

Then

$$\int e^x \cos x \, dx = e^x \sin x - \int e^x \sin x \, dx. \tag{4}$$

The second integral is like the first, except it has $\sin x$ in place of $\cos x$. To evaluate it, we use integration by parts with

$$u = e^x, \qquad dv = \sin x \, dx, \qquad v = -\cos x, \qquad du = e^x \, dx.$$

Then

$$\int e^x \cos x \, dx = e^x \sin x - \left(-e^x \cos x - \int (-\cos x)(e^x \, dx)\right)$$

$$= e^x \sin x + e^x \cos x - \int e^x \cos x \, dx.$$

The unknown integral now appears on both sides of the equation. Combining the two expressions gives

$$2 \int e^x \cos x \, dx = e^x \sin x + e^x \cos x.$$

Dividing by 2 and adding a constant of integration gives

$$\int e^x \cos x \, dx = \frac{e^x \sin x + e^x \cos x}{2} + C.$$

Our choice of $u = e^x$ and $dv = \sin x \, dx$ in the second integration may have seemed arbitrary but it wasn't. In theory, we could have chosen $u = \sin x$ and $dv = e^x \, dx$. Doing so, however, would have turned Eq. (4) into

$$\int e^x \cos x \, dx = e^x \sin x - \left(e^x \sin x - \int e^x \cos x \, dx\right)$$

$$= \int e^x \cos x \, dx.$$

The resulting identity is correct, but useless. *Moral:* Once you have decided on what to differentiate and integrate in circumstances like these, stick with them. General formulas for the integrals of $e^{ax} \cos bx$ and the closely related $e^{ax} \sin bx$ can be found in the integral table at the end of this book. ■

Tabular Integration

We have seen that integrals of the form $\int f(x)\,g(x)\,dx$, in which f can be differentiated repeatedly to become zero and g can be integrated repeatedly without difficulty, are natural candidates for integration by parts. However, if many repetitions are required, the calculations can be cumbersome. In situations like this, there is a way to organize the calculations that saves a great deal of work. It is called **tabular integration** and is illustrated in the following examples.

EXAMPLE 7 Evaluate $\int x^2 e^x\,dx$ by tabular integration.

Solution With $f(x) = x^2$ and $g(x) = e^x$, we list

$f(x)$ and its derivatives		$g(x)$ and its integrals
x^2	$(+)$	e^x
$2x$	$(-)$	e^x
2	$(+)$	e^x
0		e^x.

We add the products of the functions connected by the arrows, with the middle sign changed, to obtain

$$\int x^2 e^x\,dx = x^2 e^x - 2xe^x + 2e^x + C.$$

EXAMPLE 8 Evaluate $\int x^3 \sin x\,dx$ by tabular integration.

Solution With $f(x) = x^3$ and $g(x) = \sin x$, we list

$f(x)$ and its derivatives		$g(x)$ and its integrals
x^3	$(+)$	$\sin x$
$3x^2$	$(-)$	$-\cos x$
$6x$	$(+)$	$-\sin x$
6	$(-)$	$\cos x$
0		$\sin x$.

Again we add the products of the functions connected by the arrows, with every other sign changed, to obtain

$$\int x^3 \sin x\,dx = -x^3\cos x + 3x^2\sin x + 6x\cos x - 6\sin x + C.$$

Exercises 8.2

Evaluate the integrals in Exercises 1–24.

1. $\int x \sin x\, dx$
2. $\int x \cos 2x\, dx$
3. $\int x^2 \sin x\, dx$
4. $\int x^2 \cos x\, dx$
5. $\int_1^2 x \ln x\, dx$
6. $\int x^3 \ln x\, dx$
7. $\int \tan^{-1} x\, dx$
8. $\int \sin^{-1} x\, dx$
9. $\int x \sec^2 x\, dx$
10. $\int 4x \sec^2 2x\, dx$
11. $\int x^3 e^x\, dx$
12. $\int x^4 e^{-x}\, dx$
13. $\int (x^2 - 5x) e^x\, dx$
14. $\int (x^2 + x + 1) e^x\, dx$
15. $\int x^5 e^x\, dx$
16. $\int x^2 e^{4x}\, dx$
17. $\int_0^{\pi/2} x^2 \sin 2x\, dx$
18. $\int_0^{\pi/2} x^3 \cos 2x\, dx$
19. $\int_1^2 x \sec^{-1} x\, dx$
20. $\int_1^4 \sec^{-1}\sqrt{x}\, dx$
21. $\int e^x \sin x\, dx$
22. $\int e^{-x} \cos x\, dx$
23. $\int e^{2x} \cos 3x\, dx$
24. $\int e^{-2x} \sin 2x\, dx$

25. Find the area of the region enclosed by the x-axis and the curve $y = x \sin x$ for (a) $0 \le x \le \pi$, (b) $\pi \le x \le 2\pi$.

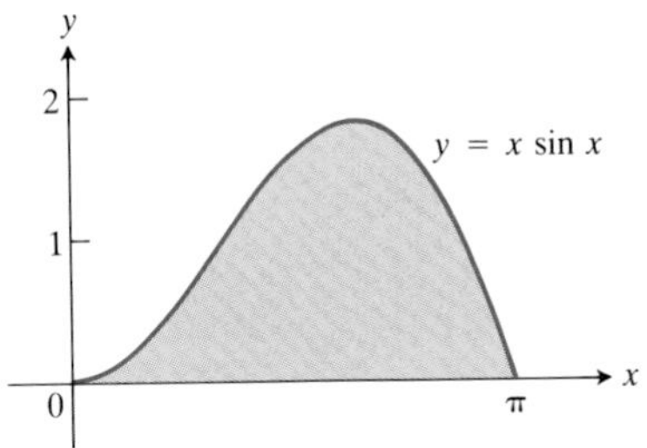

8.6 The region in Exercise 28.

26. Use cylindrical shells to find the volume swept out by revolving the region bounded by $x = 0$, $y = 0$, and $y = \cos x$, $0 \le x \le \pi/2$, about the y-axis.

27. Find the volume swept out by revolving about the y-axis the region in the first quadrant bounded by the coordinate axes, the curve $y = e^{-x}$, and the line $x = 1$.

28. Find the volume swept out when the region in the first quadrant bounded by the x-axis and the curve $y = x \sin x$, $0 \le x \le \pi$ (Fig. 8.6), is revolved (a) about the x-axis, (b) about the line $x = \pi$.

29. Find the moment about the y-axis of a thin plate of density $\delta = (1 + x)$ covering the region bounded by the x-axis and the curve $y = \sin x$, $0 \le x \le \pi$.

30. Although we usually drop the constant of integration in determining v as $\int dv$ in integration by parts, choosing the constant to be different from zero can occasionally be helpful. As a case in point, evaluate

$$\int x \tan^{-1} x\, dx$$

with $u = \tan^{-1} x$ and $v = (x^2/2) + 1/2$.

8.3 Trigonometric Integrals

Trigonometric integrals involve algebraic combinations of the six basic trigonometric functions. In principle, we can always express such integrals in terms of sines and cosines, but it is often simpler to work with other functions, as in the integral

$$\int \sec^2 x\, dx = \tan x + C.$$

The general idea is to use identities to transform the integrals we have to find into integrals that are easier to work with.

Products of Sines and Cosines

We begin with integrals of the form

$$\int \sin^m x \cos^n x \, dx,$$

where m and n are nonnegative integers (positive or zero). We can divide the work into three cases.

Case 1: m is odd

Case 2: m is even and n is odd

Case 3: m and n are both even

In each case we can use a trigonometric identity to transform the integral into a more convenient form.

CASE 1 If m is odd, we write m as $2k + 1$ and use the identity $\sin^2 x = 1 - \cos^2 x$ to obtain

$$\sin^m x = \sin^{2k+1} x = (\sin^2 x)^k \sin x = (1 - \cos^2 x)^k \sin x. \tag{1}$$

Then we combine the single $\sin x$ with dx in the integral and set $\sin x \, dx$ equal to $-d(\cos x)$.

EXAMPLE 1 Evaluate $\int \sin^3 x \cos^2 x \, dx$.

Solution

$$\begin{aligned}
\int \sin^3 x \cos^2 x \, dx &= \int \sin^2 x \cos^2 x \sin x \, dx \\
&= \int (1 - \cos^2 x) \cos^2 x (-d(\cos x)) \\
&= \int (1 - u^2)(u^2)(-du) \qquad (u = \cos x) \\
&= \int (u^4 - u^2) \, du \\
&= \frac{u^5}{5} - \frac{u^3}{3} + C \\
&= \frac{\cos^5 x}{5} - \frac{\cos^3 x}{3} + C.
\end{aligned}$$

■

CASE 2 If m is even and n is odd in $\int \sin^m x \cos^n x \, dx$, we write n as $2k + 1$ and use the identity $\cos^2 x = 1 - \sin^2 x$ to obtain

$$\cos^n x = \cos^{2k+1} x = (\cos^2 x)^k \cos x = (1 - \sin^2 x)^k \cos x.$$

We then combine the single $\cos x$ with dx and set $\cos x \, dx$ equal to $d(\sin x)$.

EXAMPLE 2 Evaluate $\int \cos^5 x\,dx$.

Solution

$$\begin{aligned}\int \cos^5 x\,dx &= \int \cos^4 x \cos x\,dx = \int (1-\sin^2 x)^2\,d(\sin x)\\ &= \int (1-u^2)^2\,du \qquad (u=\sin x)\\ &= \int (1-2u^2+u^4)\,du\\ &= u - \frac{2}{3}u^3 + \frac{1}{5}u^5 + C = \sin x - \frac{2}{3}\sin^3 x + \frac{1}{5}\sin^5 x + C.\end{aligned}$$

■

CASE 3 If both m and n are even in $\int \sin^m x \cos^n x\,dx$, we substitute

$$\sin^2 x = \frac{1-\cos 2x}{2}, \qquad \cos^2 x = \frac{1+\cos 2x}{2} \tag{2}$$

to reduce the integrand to one in lower powers of $\cos 2x$.

EXAMPLE 3 Evaluate $\int \sin^2 x \cos^4 x\,dx$.

Solution

$$\begin{aligned}\int \sin^2 x \cos^4 x\,dx &= \int \left(\frac{1-\cos 2x}{2}\right)\left(\frac{1+\cos 2x}{2}\right)^2 dx\\ &= \frac{1}{8}\int (1-\cos 2x)(1+2\cos 2x+\cos^2 2x)\,dx\\ &= \frac{1}{8}\int (1+\cos 2x-\cos^2 2x-\cos^3 2x)\,dx. \qquad \left(\begin{array}{l}\text{After some}\\ \text{algebra}\end{array}\right)\end{aligned}$$

For the term involving $\cos^2 2x$ we use

$$\begin{aligned}\int \cos^2 2x\,dx &= \frac{1}{2}\int (1+\cos 4x)\,dx\\ &= \frac{1}{2}\left(x+\frac{1}{4}\sin 4x\right). \qquad \left(\begin{array}{l}\text{Omitting the constant of}\\ \text{integration until the final result}\end{array}\right)\end{aligned}$$

For the $\cos^3 2x$ term we have

$$\begin{aligned}\int \cos^3 2x\,dx &= \int (1-\sin^2 2x)\cos 2x\,dx \qquad \left(\begin{array}{l}u=\sin 2x,\\ du = 2\cos 2x\,dx\end{array}\right)\\ &= \frac{1}{2}\int (1-u^2)\,du = \frac{1}{2}\left(\sin 2x - \frac{1}{3}\sin^3 2x\right). \qquad \left(\begin{array}{l}\text{Again}\\ \text{omitting } C\end{array}\right)\end{aligned}$$

Combining everything and simplifying we get

$$\int \sin^2 x \cos^4 x\,dx = \frac{1}{16}\left(x - \frac{1}{4}\sin 4x + \frac{1}{3}\sin^3 2x\right) + C.$$

■

FLOW CHART 8.1. $\int \sin^m x \cos^n x\, dx$ **(m, n nonnegative integers)**

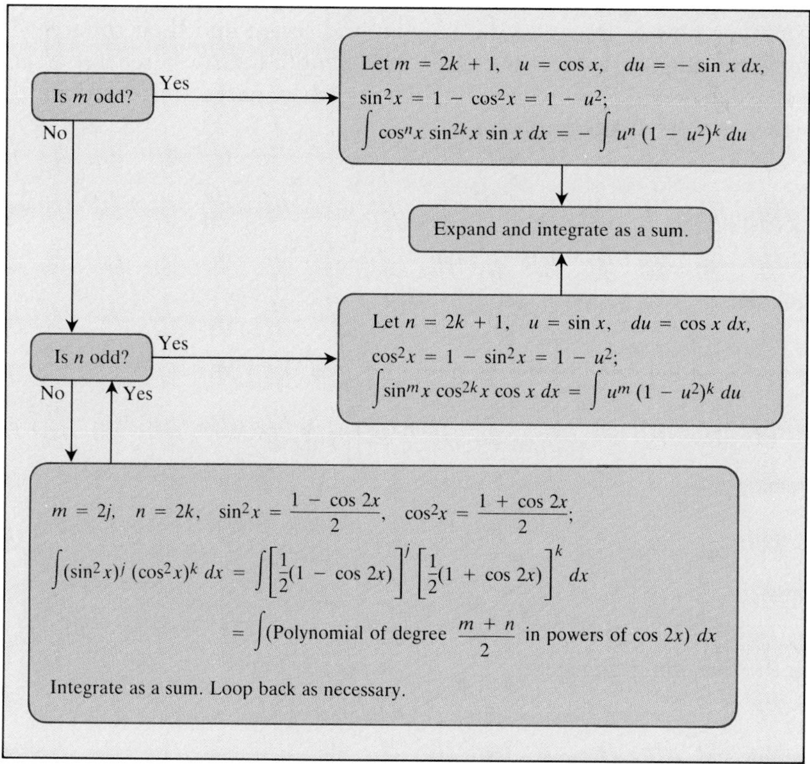

Eliminating Square Roots

In the next example, we use the identity $\cos^2\theta = (1 + \cos 2\theta)/2$ to eliminate a square root.

EXAMPLE 4 Evaluate $\displaystyle\int_0^{\pi/4} \sqrt{1 + \cos 4x}\, dx$.

Solution To eliminate the square root we use the identity

$$\cos^2\theta = \frac{1 + \cos 2\theta}{2}, \qquad \text{or} \qquad 1 + \cos 2\theta = 2\cos^2\theta.$$

With $\theta = 2x$, this becomes

$$1 + \cos 4x = 2\cos^2 2x.$$

Therefore,

$$\begin{aligned}
\int_0^{\pi/4} \sqrt{1 + \cos 4x}\, dx &= \int_0^{\pi/4} \sqrt{2\cos^2 2x}\, dx = \int_0^{\pi/4} \sqrt{2}\sqrt{\cos^2 2x}\, dx \\
&= \sqrt{2}\int_0^{\pi/4} |\cos 2x|\, dx = \sqrt{2}\int_0^{\pi/4} \cos 2x\, dx \qquad \left(\begin{array}{l}\text{Because} \\ \cos 2x \ge 0 \\ \text{on } [0, \pi/4]\end{array}\right) \\
&= \sqrt{2}\left[\frac{\sin 2x}{2}\right]_0^{\pi/4} = \frac{\sqrt{2}}{2}[1 - 0] = \frac{\sqrt{2}}{2}.
\end{aligned}$$

■

Integrals of Powers of tan x and sec x

We know how to integrate the tangent and secant and their squares. To integrate higher powers we use the identities $\tan^2 x = \sec^2 x - 1$ and $\sec^2 x = \tan^2 x + 1$, and integrate by parts when necessary to reduce the higher powers to lower powers.

EXAMPLE 5 Evaluate $\displaystyle\int \tan^4 x\, dx$.

Solution

$$\begin{aligned}\int \tan^4 x\, dx &= \int \tan^2 x \cdot \tan^2 x\, dx = \int \tan^2 x \cdot (\sec^2 x - 1)\, dx\\ &= \int \tan^2 x \sec^2 x\, dx - \int \tan^2 x\, dx\\ &= \int \tan^2 x \sec^2 x\, dx - \int (\sec^2 x - 1)\, dx\\ &= \int \tan^2 x \sec^2 x\, dx - \int \sec^2 x\, dx + \int dx.\end{aligned}$$

In the first integral, we let

$$u = \tan x, \qquad du = \sec^2 x\, dx$$

and have

$$\int u^2\, du = \frac{1}{3}u^3 + C'.$$

The remaining integrals are standard forms, so

$$\int \tan^4 x\, dx = \frac{1}{3}\tan^3 x - \tan x + x + C.$$

■

EXAMPLE 6 Evaluate $\displaystyle\int \sec^3 x\, dx$.

Solution We integrate by parts, using

$$u = \sec x, \qquad dv = \sec^2 x\, dx, \qquad v = \tan x, \qquad du = \sec x \tan x\, dx.$$

Then

$$\begin{aligned}\int \sec^3 x\, dx &= \sec x \tan x - \int (\tan x)(\sec x \tan x\, dx)\\ &= \sec x \tan x - \int (\sec^2 x - 1)\sec x\, dx \qquad (\tan^2 x = \sec^2 x - 1)\\ &= \sec x \tan x + \int \sec x\, dx - \int \sec^3 x\, dx.\end{aligned}$$

Combining the two secant-cubed integrals gives

$$2\int \sec^3 x\, dx = \sec x \tan x + \int \sec x\, dx$$

and

$$\int \sec^3 x\, dx = \frac{1}{2}\sec x \tan x + \frac{1}{2}\ln|\sec x \tan x| + C.$$

Products of Sines and Cosines

The integrals

$$\int \sin mx \sin nx\, dx, \qquad \int \sin mx \cos nx\, dx, \qquad \text{and} \qquad \int \cos mx \cos nx\, dx$$

arise in many places where trigonometric functions are applied to problems in mathematics and science. We can evaluate these integrals through integration by parts, but two such integrations are required in each case. It is simpler to use the identities

$$\sin mx \sin nx = \frac{1}{2}[\cos(m-n)x - \cos(m+n)x] \tag{3}$$

$$\sin mx \cos nx = \frac{1}{2}[\sin(m-n)x + \sin(m+n)x] \tag{4}$$

$$\cos mx \cos nx = \frac{1}{2}[\cos(m-n)x + \cos(m+n)x]. \tag{5}$$

These come from combining the identities

$$\cos(A+B) = \cos A \cos B - \sin A \sin B \tag{6}$$

$$\cos(A-B) = \cos A \cos B + \sin A \sin B \tag{7}$$

$$\sin(A+B) = \sin A \cos B + \cos A \sin B \tag{8}$$

$$\sin(A-B) = \sin A \cos B - \cos A \sin B \tag{9}$$

For example, if we take $A = mx$ and $B = nx$ in Eqs. (6) and (7), add, and divide by 2, we get Eq. (5). We get Eq. (3) by subtracting (6) from (7) and dividing by 2. To get Eq. (4), add Eqs. (8) and (9) and divide by 2.

EXAMPLE 7 Evaluate $\int \sin 3x \cos 5x\, dx$.

Solution From Eq. (4) with $m = 3$ and $n = 5$ we get

$$\begin{aligned}\int \sin 3x \cos 5x\, dx &= \frac{1}{2}\int [\sin(-2x) + \sin 8x]\, dx \\ &= \frac{1}{2}\int (\sin 8x - \sin 2x)\, dx \\ &= -\frac{\cos 8x}{16} + \frac{\cos 2x}{4} + C.\end{aligned}$$

Integrals of Odd Functions

In many applications, we integrate functions over intervals that are symmetric about the origin. In an amazing number of cases, these integrals are zero. For example,

$$\int_{-1}^{1} 2x\,dx = x^2\Big]_{-1}^{1} = 1 - 1 = 0,$$

$$\int_{-\pi}^{\pi} \sin x\,dx = -\cos x\Big]_{-\pi}^{\pi} = -\cos\pi + \cos(-\pi) = -(-1) + (-1) = 0,$$

$$\int_{-a}^{a} \frac{\sin x}{\cos^2 x}\,dx = \frac{1}{\cos x}\Big]_{-a}^{a} = \frac{1}{\cos a} - \frac{1}{\cos(-a)} = 0.$$

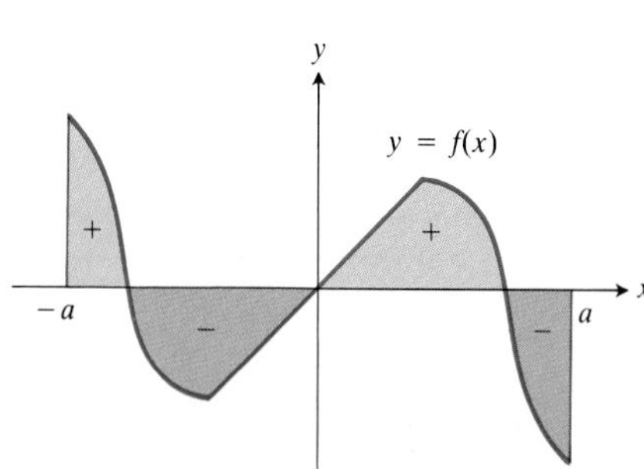

8.7 The graph of an odd function over an interval that is symmetric about the origin encloses as much area above the axis as below. The integral of the function from $-a$ to a is therefore zero.

These three integrals are zero because they are integrals of odd functions over intervals that are symmetric about the origin. For each function, $f(-x) = -f(x)$. Hence, the amount of area enclosed above the x-axis to the left of the origin equals the amount of area enclosed below the x-axis to the right of the origin and cancels it out in integration. Similarly, the area enclosed beneath the x-axis to the left of the origin cancels the area enclosed above the x-axis to the right of the origin (Fig. 8.7).

What functions are odd?

Odd powers of x: $\quad x,\ x^{-1},\ x^3,\ x^{-3}$

Sines: $\quad \sin ax,\ a \neq 0$

Odd integer powers of odd functions: $\quad \sin^3 x,\ \dfrac{1}{\sin^5 x}$

The product of an odd function with an even function: $\quad \cos x \sin 3x,\ x^2 \sin x$

The quotient of an odd function and an even function: $\quad \tan x = \dfrac{\sin x}{\cos x}$

Definite Integrals of Even Functions

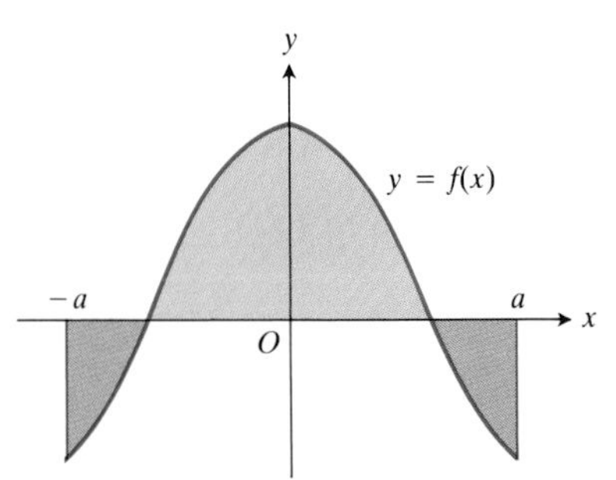

8.8 The graph of an even function is symmetric about the y-axis, so

$$\int_{-a}^{0} f(x)\,dx = \int_{0}^{a} f(x)\,dx$$

and

$$\int_{-a}^{a} f(x)\,dx = 2\int_{0}^{a} f(x)\,dx.$$

The definite integral of an even function $f(x)$ over $[-a, a]$ is twice the value of the integral of f over $[0, a]$ (Fig. 8.8). This is because the integral of f from $-a$ to 0 has the same value as the integral of f from 0 to a:

$$\int_{-a}^{0} f(x)\,dx = \int_{a}^{0} f(-u)(-du) \qquad (x = -u,\quad dx = -du)$$

$$= -\int_{a}^{0} f(u)\,du \qquad (f \text{ even} \iff f(-u) = f(u))$$

$$= \int_{0}^{a} f(u)\,du.$$

This observation saves time when the antiderivative of f is more easily evaluated at 0 than at $-a$.

What functions are even?

Constants

Cosines: $\cos ax$ (any number a)

Products and quotients of even functions: $x^2\cos x$, $\dfrac{\sin^2 x}{\cos x}$

A product of two odd functions: $x \sin x$

Roots of even functions: $\sqrt{x^2+1}$, $\sqrt[3]{1-x^2}$

EXAMPLE 8

$$\int_{-\pi/4}^{\pi/4} \cos x\,dx = 2\int_0^{\pi/4} \cos x\,dx = 2\Big[\sin x\Big]_0^{\pi/4} = 2\left[\frac{\sqrt{2}}{2} - 0\right] = \sqrt{2}.$$

Exercises 8.3

Evaluate the integrals in Exercises 1–14.

1. $\displaystyle\int_0^{\pi/2} \sin^5 x\,dx$

2. $\displaystyle\int_0^{\pi} \sin^5 \frac{x}{2}\,dx$

3. $\displaystyle\int_{-\pi/2}^{\pi/2} \cos^3 x\,dx$

4. $\displaystyle\int_0^{\pi/6} 3\cos^5 3x\,dx$

5. $\displaystyle\int_0^{\pi/2} \sin^7 y\,dy$

6. $\displaystyle\int_0^{\pi/2} 7\cos^7 t\,dt$

7. $\displaystyle\int_0^{\pi} 8\sin^4 x\,dx$

8. $\displaystyle\int_0^{1} 8\cos^4 2\pi x\,dx$

9. $\displaystyle\int_{-\pi/4}^{\pi/4} 16\sin^2 x\cos^2 x\,dx$

10. $\displaystyle\int_0^{\pi} 8\sin^4 y\cos^2 y\,dy$

11. $\displaystyle\int_0^{\pi/2} 35\sin^4 x\cos^3 x\,dx$

12. $\displaystyle\int_0^{\pi} \sin 2x\cos^2 2x\,dx$

13. $\displaystyle\int_0^{\pi/4} 8\cos^3 2\theta\sin 2\theta\,d\theta$

14. $\displaystyle\int_0^{\pi/2} \sin^2 2\theta\cos^3 2\theta\,d\theta$

Evaluate the integrals in Exercises 15–22.

15. $\displaystyle\int_0^{2\pi} \sqrt{\frac{1-\cos x}{2}}\,dx$

16. $\displaystyle\int_0^{\pi} \sqrt{1-\cos 2x}\,dx$

17. $\displaystyle\int_0^{\pi} \sqrt{1-\sin^2 t}\,dt$

18. $\displaystyle\int_0^{\pi} \sqrt{1-\cos^2\theta}\,d\theta$

19. $\displaystyle\int_{-\pi/4}^{\pi/4} \sqrt{1+\tan^2 x}\,dx$

20. $\displaystyle\int_{-\pi/4}^{\pi/4} \sqrt{\sec^2 x - 1}\,dx$

21. $\displaystyle\int_0^{\pi/2} \theta\sqrt{1-\cos 2\theta}\,d\theta$

22. $\displaystyle\int_{-\pi}^{\pi} (1-\cos^2 t)^{3/2}\,dt$

Evaluate the integrals in Exercises 23–32.

23. $\displaystyle\int_{-\pi/3}^{0} 2\sec^3 x\,dx$

24. $\displaystyle\int e^x\sec^3 e^x\,dx$

25. $\displaystyle\int_0^{\pi/4} \sec^4\theta\,d\theta$

26. $\displaystyle\int_0^{\pi/12} 3\sec^4 3x\,dx$

27. $\displaystyle\int_{\pi/4}^{\pi/2} \csc^4\theta\,d\theta$

28. $\displaystyle\int_{\pi/2}^{\pi} 3\csc^4\frac{\theta}{2}\,d\theta$

29. $\displaystyle\int_0^{\pi/4} 4\tan^3 x\,dx$

30. $\displaystyle\int_{-\pi/4}^{\pi/4} 6\tan^4 x\,dx$

31. $\displaystyle\int_{\pi/6}^{\pi/3} \cot^3 x\,dx$

32. $\displaystyle\int_{\pi/4}^{\pi/2} 8\cot^4 t\,dt$

Evaluate the integrals in Exercises 33–38.

33. $\displaystyle\int_{-\pi}^{0} \sin 3x\cos 2x\,dx$

34. $\displaystyle\int_0^{\pi/2} \sin 2x\cos 3x\,dx$

35. $\displaystyle\int_{-\pi}^{\pi} \sin 3x\sin 3x\,dx$

36. $\displaystyle\int_0^{\pi/2} \sin x\cos x\,dx$

37. $\displaystyle\int_0^{\pi} \cos 3x\cos 4x\,dx$

38. $\displaystyle\int_{-\pi/2}^{\pi/2} \cos x\cos 7x\,dx$

39. Which integrals are zero and which are not? (You can do most of these without writing anything down.)

a) $\int_{-\pi}^{\pi} \sin x \cos^2 x \, dx$ b) $\int_{-L}^{L} \sqrt[3]{\sin x} \, dx$

c) $\int_{-\pi/4}^{\pi/4} x \sec x \, dx$ d) $\int_{-\pi/2}^{\pi/2} x \sin x \, dx$

e) $\int_{-a}^{a} \sin mx \cos mx \, dx, \quad m \neq 0$

f) $\int_{-\pi/2}^{\pi/2} \cos^3 x \, dx$ g) $\int_{-\ln 2}^{\ln 2} x(e^x + e^{-x}) \, dx$

h) $\int_{-\pi/2}^{\pi/2} \sin x \sin 2x \, dx$ i) $\int_{-a}^{a} (e^x \sin x + e^{-x} \sin x) dx$

40. Which integrals are zero and which are not? (You can do most of these without writing anything down.)

a) $\int_{-1}^{1} \sin 3x \cos 5x \, dx$ b) $\int_{-a}^{a} x\sqrt{a^2 - x^2} \, dx$

c) $\int_{-\pi/4}^{\pi/4} \tan^3 x \, dx$ d) $\int_{-\pi/2}^{\pi/2} x \cos x \, dx$

e) $\int_{-\pi}^{\pi} \sin^5 x \, dx$ f) $\int_{-\pi}^{\pi} \cos^5 x \, dx$

g) $\int_{-\pi/2}^{\pi/2} \sin^2 x \cos x \, dx$ h) $\int_{-\pi/4}^{\pi/4} \sec x \tan x \, dx$

i) $\int_{-1}^{1} \frac{\sin x \, dx}{e^x + e^{-x}}$

41. Which integrals in Exercise 39 have even integrands? Evaluate these integrals.

42. Which integrals in Exercise 40 have even integrands? Evaluate these integrals.

43. Show that

$$\int \csc x \, dx = -\ln|\csc x + \cot x| + C.$$

(*Hint:* Repeat the derivation in Section 8.1, Example 6, with cofunctions.)

44. Use the result in Exercise 43 to show that

$$\int \csc^3 x \, dx = -\frac{1}{2} \csc x \cot x - \frac{1}{2} \ln|\csc x + \cot x| + C.$$

8.4 Trigonometric Substitutions

We now embark on a three-step program that will enable us (in theory, at least) to integrate all rational functions of x. The first step is to study substitutions that change binomials like $a^2 + x^2$, $a^2 - x^2$, and $x^2 - a^2$ into single squared terms. The second step will be to simplify integrals involving $ax^2 + bx + c$ by completing the square and then replacing the resulting sums and differences of squares by single squared terms. The third and last step, taken in Section 8.5, will be to express rational functions of x as sums of polynomials (which we already know how to integrate), fractions with linear-factored denominators (which become logarithms or fractions when integrated), and fractions with quadratic denominators (which we will be able to integrate by the techniques of the present section).

Trigonometric Substitutions for Combining Squares

Trigonometric substitutions enable us to replace the binomials

$$a^2 + x^2, \qquad a^2 - x^2, \qquad \text{and} \qquad x^2 - a^2$$

by single squared terms and thereby transforming a number of important integrals into integrals we can recognize or can find in a table. The most commonly used substitutions, $x = a \tan\theta$, $x = a \sin\theta$, and $x = a \sec\theta$, come from the reference triangles in Fig. 8.9.

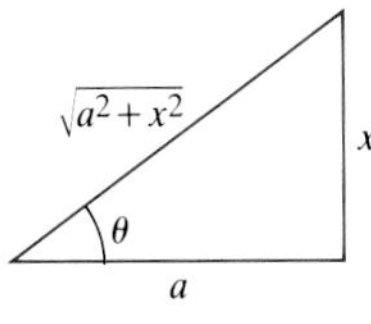

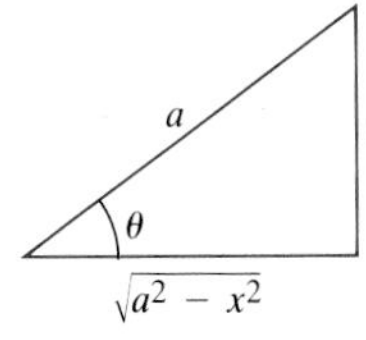

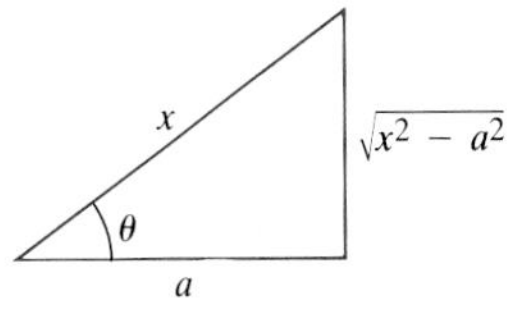

$x = a \tan \theta$ $\quad$ $x = a \sin \theta$ $\quad$ $x = a \sec \theta$

$\sqrt{a^2+x^2} = a|\sec \theta|$ $\quad$ $\sqrt{a^2-x^2} = a|\cos \theta|$ $\quad$ $\sqrt{x^2-a^2} = a|\tan \theta|$

8.9 Reference triangles for trigonometric substitutions that change binomials into single squared terms.

With $x = a \tan \theta$,

$$a^2 + x^2 = a^2 + a^2\tan^2\theta = a^2(1 + \tan^2\theta) = a^2\sec^2\theta. \tag{1}$$

With $x = a \sin \theta$,

$$a^2 - x^2 = a^2 - a^2\sin^2\theta = a^2(1 - \sin^2\theta) = a^2\cos^2\theta. \tag{2}$$

With $x = a \sec \theta$,

$$x^2 - a^2 = a^2\sec^2\theta - a^2 = a^2(\sec^2\theta - 1) = a^2\tan^2\theta. \tag{3}$$

$x = a \tan \theta$	replaces $a^2 + x^2$	by $a^2\sec^2\theta$,	(4)
$x = a \sin \theta$	replaces $a^2 - x^2$	by $a^2\cos^2\theta$,	(5)
$x = a \sec \theta$	replaces $x^2 - a^2$	by $a^2\tan^2\theta$.	(6)

When we make substitutions, we always want them to be reversible so that we can change back to the original variables when we're done. For example, if $x = a \tan \theta$, we want to be able to set

$$\theta = \tan^{-1}\frac{x}{a}$$

after the integration takes place. If $x = \sin \theta$, we want to be able to set

$$\theta = \sin^{-1}\frac{x}{a}$$

when we're done, and similarly for $x = a \sec \theta$.

As we know from Section 7.7, the functions in these substitutions have inverses only for selected values of θ (Fig. 8.10). For reversibility,

$$x = a \tan \theta \text{ requires } \theta = \tan^{-1}\frac{x}{a} \text{ with } -\frac{\pi}{2} < \theta < \frac{\pi}{2},$$

$$x = a \sin \theta \text{ requires } \theta = \sin^{-1}\frac{x}{a} \text{ with } -\frac{\pi}{2} \le \theta \le \frac{\pi}{2},$$

$$x = a \sec \theta \text{ requires } \theta = \sec^{-1}\frac{x}{a} \text{ with } \begin{cases} 0 \le \theta < \dfrac{\pi}{2} & \text{if } \dfrac{x}{a} \ge 1, \\ \dfrac{\pi}{2} < \theta \le \pi & \text{if } \dfrac{x}{a} \le -1. \end{cases}$$

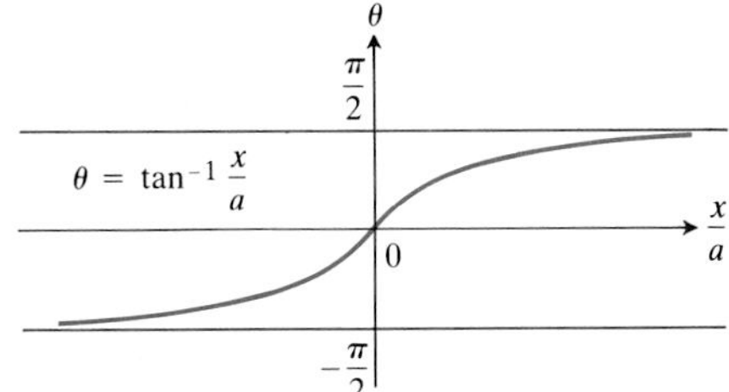

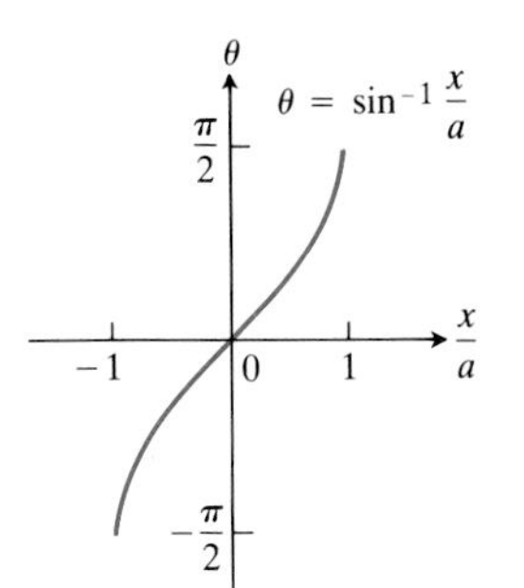

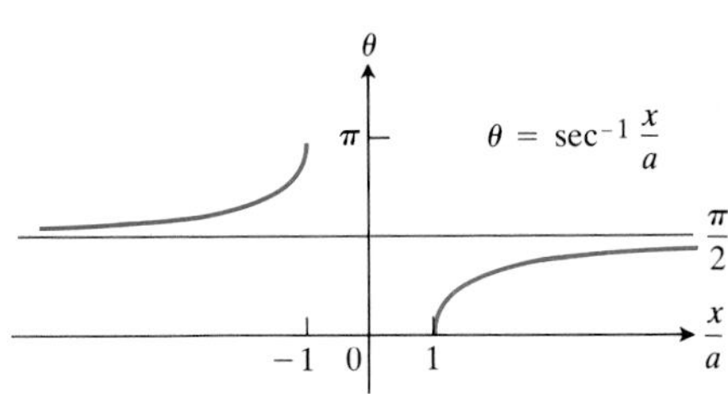

8.10 The arc tangent, arc sine, and arc secant of x/a, graphed as functions of x/a.

EXAMPLE 1 Evaluate $\displaystyle\int \frac{dx}{\sqrt{4+x^2}}$.

Solution We set

$$x = 2\tan\theta, \qquad dx = 2\sec^2\theta\, d\theta, \qquad -\frac{\pi}{2} < \theta < \frac{\pi}{2}$$

$$4 + x^2 = 4 + 4\tan^2\theta = 4(1+\tan^2\theta) = 4\sec^2\theta.$$

Then

$$\begin{aligned}
\int \frac{dx}{\sqrt{4+x^2}} &= \int \frac{2\sec^2\theta\, d\theta}{\sqrt{4\sec^2\theta}} = \int \frac{\sec^2\theta\, d\theta}{|\sec\theta|} && (\sqrt{\sec^2\theta} = |\sec\theta|)\\
&= \int \sec\theta\, d\theta && \left(\sec\theta > 0 \quad \text{for} \quad -\frac{\pi}{2} < \theta < \frac{\pi}{2}\right)\\
&= \ln|\sec\theta + \tan\theta| + C\\
&= \ln\left|\frac{\sqrt{4+x^2}}{2} + \frac{x}{2}\right| + C && \text{(From Fig. 8.11)}\\
&= \ln\left|\sqrt{4+x^2} + x\right| + C'. && \text{(Taking } C' = C - \ln 2)
\end{aligned}$$

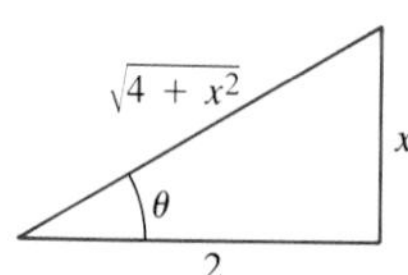

8.11 Reference triangle for $x = 2\tan\theta$ (Example 1).

Notice how we expressed $\ln|\sec\theta + \tan\theta|$ in terms of x: We drew a reference triangle for the original substitution $x = 2\tan\theta$ and read the ratios from the triangle. ■

EXAMPLE 2 Evaluate $\displaystyle\int \frac{x^2\, dx}{\sqrt{9-x^2}}$.

Solution To replace $9 - x^2$ by a single squared term, we set

$$x = 3\sin\theta, \qquad dx = 3\cos\theta\, d\theta, \qquad -\frac{\pi}{2} < \theta < \frac{\pi}{2}$$

$$9 - x^2 = 9(1 - \sin^2\theta) = 9\cos^2\theta.$$

Then

$$\begin{aligned}
\int \frac{x^2\, dx}{\sqrt{9-x^2}} &= \int \frac{9\sin^2\theta \cdot 3\cos\theta\, d\theta}{|3\cos\theta|}\\
&= 9\int \sin^2\theta\, d\theta && \left(\cos\theta > 0 \text{ for } -\frac{\pi}{2} < \theta < \frac{\pi}{2}\right)\\
&= 9\int \frac{1-\cos 2\theta}{2}\, d\theta = \frac{9}{2}\left(\theta - \frac{\sin 2\theta}{2}\right) + C\\
&= \frac{9}{2}(\theta - \sin\theta\cos\theta) + C\\
&= \frac{9}{2}\left(\sin^{-1}\frac{x}{3} - \frac{x}{3}\cdot\frac{\sqrt{9-x^2}}{3}\right) + C && \text{(Fig. 8.12)}\\
&= \frac{9}{2}\sin^{-1}\frac{x}{3} - \frac{x}{2}\sqrt{9-x^2} + C.
\end{aligned}$$

■

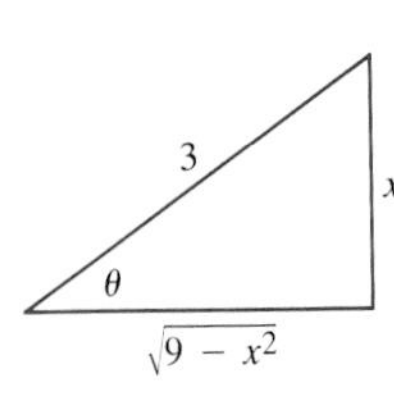

8.12 Reference triangle for $x = 3\sin\theta$ (Example 2).

EXAMPLE 3 Evaluate $\displaystyle\int \frac{dx}{\sqrt{x^2 - 25}}$.

Solution To replace $x^2 - 25$ by a single squared term, we set

$$x = 5 \sec \theta, \quad dx = 5 \sec \theta \tan \theta \, d\theta, \quad \theta = \sec^{-1}\frac{x}{5}$$

$$x^2 - 25 = 25 \sec^2\theta - 25 = 25(\sec^2\theta - 1) = 25 \tan^2\theta.$$

With this substitution, we have

$$0 < \theta < \frac{\pi}{2} \text{ for } \frac{x}{5} > 1 \qquad \text{and} \qquad \frac{\pi}{2} < \theta < \pi \text{ for } \frac{x}{5} < -1$$

(Fig. 8.10). Then,

$$\begin{aligned}
\int \frac{dx}{\sqrt{x^2 - 25}} &= \int \frac{5 \sec \theta \tan \theta \, d\theta}{\sqrt{25 \tan^2\theta}} \\
&= \int \frac{\sec \theta \tan \theta \, d\theta}{|\tan \theta|} \qquad (\sqrt{\tan^2\theta} = |\tan \theta|) \\
&= \pm \int \sec \theta \, d\theta = \pm \ln |\sec \theta + \tan \theta| + C \\
&= \pm \ln \left| \frac{x}{5} \pm \frac{\sqrt{x^2 - 25}}{5} \right| + C. \qquad \text{(Fig. 8.13)} \\
&= \pm \ln \left| x \pm \sqrt{x^2 - 25} \right| + C'. \qquad (C' = C \pm \ln 5)
\end{aligned}$$

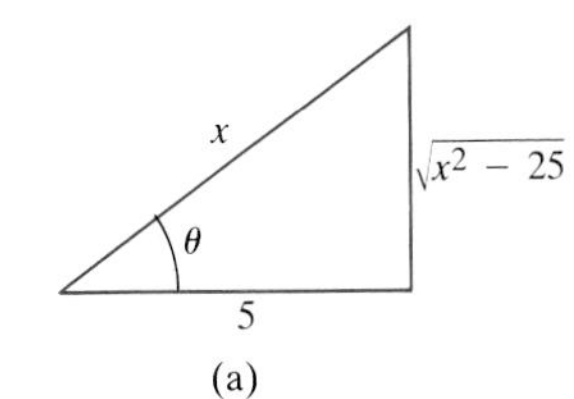

(a)

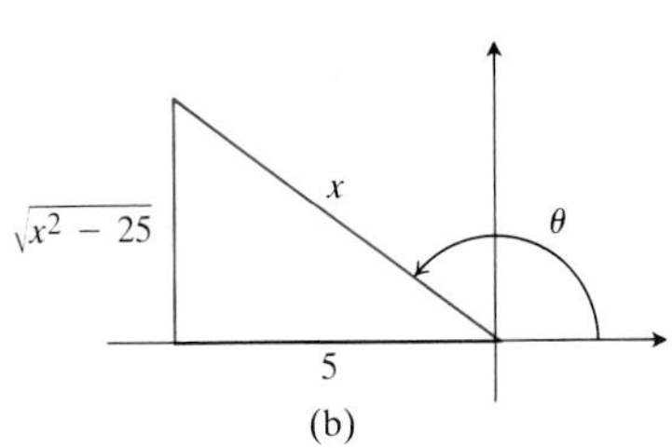

(b)

8.13 (a) Reference triangle for $x = 5 \sec \theta$, $0 < \theta < \pi/2$, $x > 5$; (b) reference triangle for $x = 5 \sec \theta$, $\pi/2 < \theta < \pi$, $x < -5$.

What do we do about the signs? When $0 < \theta < \pi/2$, the tangent and secant are both positive and both signs are $(+)$. When $\pi/2 < \theta < \pi$, the tangent and secant are both negative and both signs are $(-)$. Therefore,

$$\int \frac{dx}{\sqrt{x^2 - 25}} = \begin{cases} \ln \left| x + \sqrt{x^2 - 25} \right| + C' \\ \text{or} \\ -\ln \left| x - \sqrt{x^2 - 25} \right| + C'. \end{cases} \tag{7}$$

Fortunately, we do not have to live with this two-line formula because the two logarithmic expressions on the right-hand side differ only by a constant:

$$\begin{aligned}
-\ln \left| x - \sqrt{x^2 - 25} \right| &= \ln \left| \frac{1}{x - \sqrt{x^2 - 25}} \right| \\
&= \ln \left| \frac{1}{x - \sqrt{x^2 - 25}} \cdot \frac{x + \sqrt{x^2 - 25}}{x + \sqrt{x^2 - 25}} \right| \\
&= \ln \left| \frac{x + \sqrt{x^2 - 25}}{25} \right| \qquad \left(\begin{aligned} &x^2 - (\sqrt{x^2 - 25})^2 \\ &= x^2 - x^2 + 25 = 25. \end{aligned} \right) \\
&= \ln \left| x + \sqrt{x^2 - 25} \right| - \ln 25.
\end{aligned}$$

Therefore,

$$\int \frac{dx}{\sqrt{x^2 - 25}} = \ln \left| x + \sqrt{x^2 - 25} \right| + C.$$

■

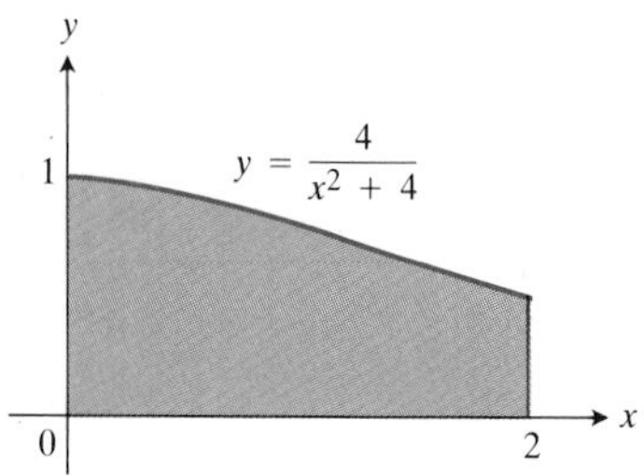

8.14 The region in Example 4.

EXAMPLE 4 Find the volume of the solid generated by revolving the region bounded by the curve $y = 4/(x^2 + 4)$, the x-axis, and the lines $x = 0$ and $x = 2$, about the x-axis.

Solution We sketch the region (Fig. 8.14) and use the disk method from Section 6.2:

$$\text{Volume} = \int_0^2 \pi(f(x))^2\, dx = 16\pi \int_0^2 \frac{dx}{(x^2+4)^2}.$$

To evaluate the integral, we set

$$x = 2\tan\theta, \qquad dx = 2\sec^2\theta\, d\theta, \qquad \theta = \tan^{-1}\frac{x}{2}$$

$$x^2 + 4 = 4\tan^2\theta + 4 = 4(\tan^2\theta + 1) = 4\sec^2\theta$$

(Fig. 8.15). With these substitutions, we have $\theta = 0$ when $x = 0$ and $\theta = \pi/4$ when $x = 2$.

$$\text{Volume} = 16\pi \int_{x=0}^{x=2} \frac{dx}{(x^2+4)^2} = 16\pi \int_{\theta=0}^{\theta=\pi/4} \frac{2\sec^2\theta\, d\theta}{(4\sec^2\theta)^2}$$

$$= 16\pi \int_0^{\pi/4} \frac{2\sec^2\theta\, d\theta}{16\sec^4\theta} = \pi \int_0^{\pi/4} 2\cos^2\theta\, d\theta$$

$$= \pi \int_0^{\pi/4} (1 + \cos 2\theta)\, d\theta = \pi\left[\theta + \frac{\sin 2\theta}{2}\right]_0^{\pi/4} \qquad \begin{pmatrix} 2\cos^2\theta = \\ 1 + \cos 2\theta \end{pmatrix}$$

$$= \pi\left[\frac{\pi}{4} + \frac{1}{2}\right] = 4.04. \qquad \text{(Calculator, rounded)}$$

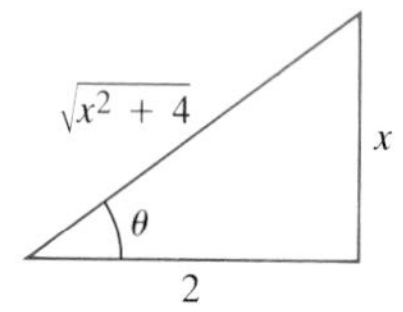

8.15 Reference triangle for $x = 2\tan\theta$ (Example 4). We have $\theta = 0$ when $x = 0$ and $\theta = \pi/4$ when $x = 2$.

Integrals involving $ax^2 + bx + c$, $a \neq 0$

We handle these by first completing the square:

$$ax^2 + bx + c = a\left(x^2 + \frac{b}{a}x + \left(\frac{b}{2a}\right)^2\right) + \left(c - \frac{b^2}{4a}\right) = a\left(x + \frac{b}{2a}\right)^2 + \left(c - \frac{b^2}{4a}\right).$$

We then substitute

$$u = x + \frac{b}{2a}, \qquad x = u - \frac{b}{2a}, \qquad dx = du.$$

Examples 5 and 6 illustrate the method.

EXAMPLE 5 Evaluate $\displaystyle\int \frac{dx}{\sqrt{2x - x^2}}.$

Solution First we do the necessary algebra:

$$2x - x^2 = -(x^2 - 2x + 1) + 1 = 1 - (x-1)^2.$$

Then we substitute $u = x - 1$ and $du = dx$ to get

$$\int \frac{dx}{\sqrt{2x - x^2}} = \int \frac{du}{\sqrt{1 - u^2}} = \sin^{-1} u + C = \sin^{-1}(x - 1) + C.$$

EXAMPLE 6 Evaluate $\displaystyle\int \frac{dx}{4x^2 + 4x + 2}$.

Solution

$$4x^2 + 4x + 2 = 4\left(x^2 + x + \frac{1}{4}\right) + 2 - \frac{4}{4} = 4\left(x + \frac{1}{2}\right)^2 + 1 = (2x + 1)^2 + 1.$$

Hence if we let $u = 2x + 1$ we get

$$\int \frac{dx}{4x^2 + 4x + 2} = \int \frac{dx}{(2x + 1)^2 + 1} = \frac{1}{2} \int \frac{du}{u^2 + 1}$$

$$= \frac{1}{2} \tan^{-1} u + C = \frac{1}{2} \tan^{-1}(2x + 1) + C.$$

Two Useful Formulas

Integrals of the form $\int du/(u^2 + a^2)$ and $\int du/\sqrt{a^2 - u^2}$ arise so often in applications of integration that many people find it saves time to memorize formulas for evaluating them.

$$\int \frac{du}{u^2 + a^2} = \frac{1}{a} \tan^{-1} \frac{u}{a} + C \qquad (8)$$

$$\int \frac{du}{\sqrt{a^2 - u^2}} = \sin^{-1} \frac{u}{a} + C \qquad (9)$$

We can derive Eq. (8) by substituting $u = a \tan \theta$ and Eq. (9) by substituting $u = a \sin \theta$.

EXAMPLE 7

a) $$\int \frac{dx}{(x + 1)^2 + 4} = \frac{1}{2} \tan^{-1} \frac{x + 1}{2} + C \qquad \begin{pmatrix}\text{Eq. (8) with } u = x + 1, \\ du = dx,\ a = 2\end{pmatrix}$$

b) $$\int \frac{dx}{\sqrt{3 - 4x^2}} = \frac{1}{2} \int \frac{du}{\sqrt{a^2 - u^2}} \qquad \left(u = 2x,\ \frac{du}{2} = dx,\ a = \sqrt{3}\right)$$

$$= \frac{1}{2} \sin^{-1} \frac{u}{a} + C \qquad \text{(Eq. 9)}$$

$$= \frac{1}{2} \sin^{-1} \frac{2x}{\sqrt{3}} + C$$

Exercises 8.4

Evaluate the integrals in Exercises 1–16.

1. $\displaystyle\int_{-2}^{2} \frac{dx}{4 + x^2}$

2. $\displaystyle\int_{0}^{2} \frac{dx}{8 + 2x^2}$

3. $\displaystyle\int_{0}^{3/2} \frac{dx}{\sqrt{9 - x^2}}$

4. $\displaystyle\int_{0}^{1/2\sqrt{2}} \frac{2\,dx}{\sqrt{1 - 4x^2}}$

5. $\displaystyle\int \frac{dx}{\sqrt{x^2-4}}$

6. $\displaystyle\int \frac{3\,dx}{\sqrt{9x^2-1}}$

7. $\displaystyle\int_0^{\sqrt{3}/2} \frac{2\,dy}{1+4y^2}$

8. $\displaystyle\int_0^{1/3} \frac{3\,dy}{\sqrt{1+9y^2}}$

9. $\displaystyle\int_0^{3\sqrt{2}/4} \frac{dx}{\sqrt{9-4x^2}}$

10. $\displaystyle\int_0^5 \sqrt{25-x^2}\,dx$

11. $\displaystyle\int_{1/\sqrt{3}}^1 \frac{2\,dz}{z\sqrt{4z^2-1}}$

12. $\displaystyle\int_{8/\sqrt{3}}^8 \frac{24\,dx}{x\sqrt{x^2-16}}$

13. $\displaystyle\int_0^2 \frac{dx}{\sqrt{4+x^2}}$

14. $\displaystyle\int_0^1 \frac{x^3\,dx}{\sqrt{x^2+1}}$

15. $\displaystyle\int_1^2 \frac{6\,dx}{\sqrt{4-(x-1)^2}}$

16. $\displaystyle\int \frac{\sqrt{1-x^2}}{x^2}\,dx$

Evaluate the integrals in Exercises 17–24.

17. $\displaystyle\int_1^3 \frac{dy}{y^2-2y+5}$

18. $\displaystyle\int_1^4 \frac{dy}{y^2-2y+10}$

19. $\displaystyle\int_1^{3/2} \frac{(x-1)\,dx}{\sqrt{2x-x^2}}$

20. $\displaystyle\int \frac{(x-2)\,dx}{\sqrt{5+4x-x^2}}$

21. $\displaystyle\int \frac{dx}{\sqrt{x^2-2x}}$

22. $\displaystyle\int \frac{dx}{\sqrt{x^2+2x}}$

23. $\displaystyle\int_{-2}^2 \frac{(x+2)\,dx}{\sqrt{x^2+4x+13}}$

24. $\displaystyle\int_0^1 \frac{(1-x)\,dx}{\sqrt{8+2x-x^2}}$

Evaluate the integrals in Exercises 25 and 26.

25. $\displaystyle\int \frac{4x^2\,dx}{(1-x^2)^{3/2}}$

26. $\displaystyle\int_0^1 \frac{4\,dx}{(4-x^2)^{3/2}}$

27. Find the area of the region cut from the first quadrant by the elliptical curve $3y=\sqrt{9-x^2}$.

28. Find the average value of the function $f(x) = 4/(x^2-4x+8)$ over the interval $[2, 4]$.

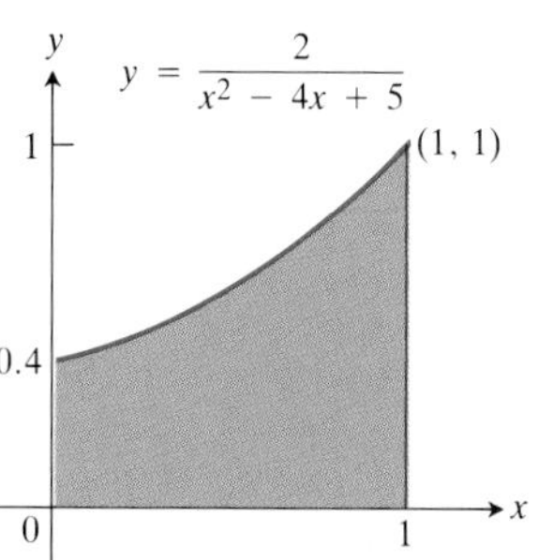

8.16 The region in Exercise 29.

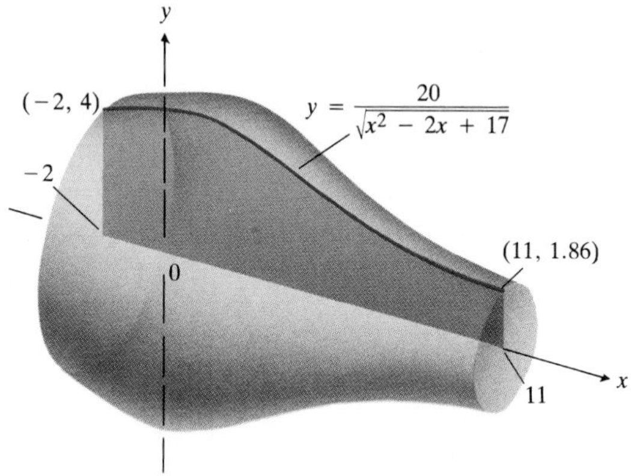

8.17 The solid in Exercise 30.

29. **CALCULATOR** Find the area of the region in the first quadrant bounded by the coordinate axes, the line $x = 1$, and the curve $y = 2/(x^2 - 4x + 5)$ (Fig. 8.16).

30. **CALCULATOR** Find the volume of the solid generated by revolving the region bounded by the x-axis, the curve $y = 20/\sqrt{x^2-2x+17}$, and the lines $x = -2$ and $x = 11$, about the x-axis (Fig. 8.17).

8.5 Rational Functions and Partial Fractions

A theorem from advanced algebra (mentioned later in more detail) says that every rational function, no matter how complicated, comes from adding smaller fractions that we can integrate with techniques we already know. The first step toward integrating a rational function is therefore to reverse the addition to get the smaller fractions back. The technique for doing so is the **method of partial fractions,** discussed in this section.

EXAMPLE 1 *An example of the kind of addition we want to reverse.*

$$\frac{2}{x+1}+\frac{3}{x-3}=\frac{2(x-3)+3(x+1)}{(x+1)(x-3)}=\frac{5x-3}{x^2-2x-3} \tag{1}$$

Adding the fractions on the left produces the fraction on the right. The reverse process consists of finding constants A and B such that

$$\frac{5x-3}{x^2-2x-3}=\frac{A}{x+1}+\frac{B}{x-3}. \tag{2}$$

(Pretend for a moment, we don't know that $A = 2$ and $B = 3$ will work.) We call the fractions $A/(x+1)$ and $B/(x-3)$ **partial fractions** because their denominators are only part of the original denominator $x^2 - 2x - 3$. We call A and B **undetermined coefficients** until proper values for them have been found.

To evaluate undetermined coefficients,

1. Clear the equation of fractions.
2. Equate the coefficients of like powers of x.
3. Solve for the coefficients.

To find A and B, we first clear Eq. (2) of fractions, obtaining

$$5x-3=A(x-3)+B(x+1)=(A+B)x-3A+B.$$

This will be an identity in x if and only if the coefficients of like powers of x on the two sides are equal:

$$A+B=5, \qquad -3A+B=-3.$$

These equations determine the values of A and B to be $A = 2$, $B = 3$. ■

EXAMPLE 2 *Two linear factors in the denominator.* Evaluate

$$\int \frac{5x-3}{(x+1)(x-3)}\,dx.$$

Solution From Example 1,

$$\int \frac{5x-3}{(x+1)(x-3)}\,dx=\int \frac{2}{x+1}\,dx+\int \frac{3}{x-3}\,dx$$
$$=2\ln|x+1|+3\ln|x-3|+C.$$ ■

EXAMPLE 3 *A repeated linear factor in the denominator.* Express

$$\frac{6x+7}{(x+2)^2}$$

as a sum of partial fractions.

Solution Since the denominator has a repeated linear factor, $(x+2)^2$, we must express the fraction in the form

$$\frac{6x+7}{(x+2)^2}=\frac{A}{x+2}+\frac{B}{(x+2)^2}. \tag{3}$$

Clearing Eq. (3) of fractions gives

$$6x+7=A(x+2)+B=Ax+(2A+B).$$

Matching coefficients of like terms gives $A = 6$ and

$$7 = 2A + B = 12 + B \qquad \text{or} \qquad B = -5.$$

Hence,

$$\frac{6x + 7}{(x + 2)^2} = \frac{6}{x + 2} - \frac{5}{(x + 2)^2}.$$

■

EXAMPLE 4 *An improper fraction (degree of the numerator greater than or equal to the degree of the denominator).* Express

$$\frac{2x^3 - 4x^2 - x - 3}{x^2 - 2x - 3}$$

as a sum of partial fractions.

Solution First, we divide the denominator into the numerator to get a polynomial plus a proper fraction. Then we write the proper fraction as a sum of partial fractions. Long division gives

$$\begin{array}{r} 2x \\ x^2 - 2x - 3 \overline{\smash{)}\, 2x^3 - 4x^2 - x - 3} \\ \underline{2x^3 - 4x^2 - 6x } \\ 5x - 3 \end{array}$$

Hence

$$\begin{aligned} \frac{2x^3 - 4x^2 - x - 3}{x^2 - 2x - 3} &= 2x + \frac{5x - 3}{x^2 - 2x - 3} && \text{(Result of the division)} \\ &= 2x + \frac{2}{x + 1} + \frac{3}{x - 3}. && \begin{pmatrix}\text{Proper fraction expanded} \\ \text{as in Example 1}\end{pmatrix} \end{aligned}$$

■

EXAMPLE 5 *An irreducible quadratic factor in the denominator.* Express

$$\frac{-2x + 4}{(x^2 + 1)(x - 1)^2}$$

as a sum of partial fractions.

Solution The denominator has an irreducible quadratic factor as well as a repeated linear factor, so we write

$$\frac{-2x + 4}{(x^2 + 1)(x - 1)^2} = \frac{Ax + B}{x^2 + 1} + \frac{C}{x - 1} + \frac{D}{(x - 1)^2}. \tag{4}$$

Notice the numerator over $x^2 + 1$. For quadratic factors, we use first-degree numerators, not constant numerators. Clearing the equation of fractions gives

$$\begin{aligned} -2x + 4 &= (Ax + B)(x - 1)^2 + C(x - 1)(x^2 + 1) + D(x^2 + 1) \\ &= (A + C)x^3 + (-2A + B - C + D)x^2 \\ &\quad + (A - 2B + C)x + (B - C + D). \end{aligned}$$

Equating coefficients of like terms gives

$$\begin{aligned}
&\text{Coefficients of } x^3: && 0 = A + C \\
&\text{Coefficients of } x^2: && 0 = -2A + B - C + D \\
&\text{Coefficients of } x^1: && -2 = A - 2B + C \\
&\text{Coefficients of } x^0: && 4 = B - C + D.
\end{aligned}$$

We solve these equations simultaneously to find the values of A, B, C, and D:

$$\begin{aligned}
&-4 = -2A, \quad A = 2 && \text{(Subtract fourth equation from second)} \\
&C = -A = -2 && \text{(From the first equation)} \\
&B = 1 && (A = 2 \text{ and } C = -2 \text{ in third equation}) \\
&D = 4 - B + C = 1. && \text{(From the fourth equation)}
\end{aligned}$$

We substitute these values into Eq. (4), obtaining

$$\frac{-2x + 4}{(x^2 + 1)(x - 1)^2} = \frac{2x + 1}{x^2 + 1} - \frac{2}{x - 1} + \frac{1}{(x - 1)^2}.$$

■

EXAMPLE 6 Evaluate $\displaystyle\int \frac{-2x + 4}{(x^2 + 1)(x - 1)^2}\, dx.$

Solution We expand the integrand by partial fractions, as in Example 5, and integrate the terms of the expansion:

$$\begin{aligned}
\int \frac{-2x + 4}{(x^2 + 1)(x - 1)^2}\, dx &= \int \left(\frac{2x + 1}{x^2 + 1} - \frac{2}{x - 1} + \frac{1}{(x - 1)^2}\right) dx \qquad \text{(Example 5)} \\
&= \int \left(\frac{2x}{x^2 + 1} + \frac{1}{x^2 + 1} - \frac{2}{x - 1} + \frac{1}{(x - 1)^2}\right) dx \\
&= \ln(x^2 + 1) + \tan^{-1} x - 2 \ln |x - 1| - \frac{1}{x - 1} + C.
\end{aligned}$$

■

General Description of the Method

Success in writing a rational function $f(x)/g(x)$ as a sum of partial fractions depends on two things:

1. *The degree of $f(x)$ must be less than the degree of $g(x)$.* (If it isn't, divide and work with the remainder term.)
2. *We must know the factors of $g(x)$.* (In theory, any polynomial with real coefficients can be written as a product of real linear factors and real quadratic factors. In practice, the factors may be hard to find.)

If these two conditions are met we can take the following steps.

STEP 1: Let $x - r$ be a linear factor of $g(x)$. Suppose $(x - r)^m$ is the highest power of $x - r$ that divides $g(x)$. Then assign the sum of m partial fractions to this factor, as follows:

$$\frac{A_1}{x - r} + \frac{A_2}{(x - r)^2} + \cdots + \frac{A_m}{(x - r)^m}.$$

Do this for each distinct linear factor of $g(x)$.

STEP 2: Let $x^2 + px + q$ be a quadratic factor of $g(x)$. Suppose

$$(x^2 + px + q)^n$$

is the highest power of this factor that divides $g(x)$. Then, to this factor, assign the sum of the n partial fractions:

$$\frac{B_1x + C_1}{x^2 + px + q} + \frac{B_2x + C_2}{(x^2 + px + q)^2} + \cdots + \frac{B_nx + C_n}{(x^2 + px + q)^n}.$$

Do this for each distinct quadratic factor of $g(x)$ that cannot be factored into linear factors with real coefficients.

STEP 3: Set the original fraction $f(x)/g(x)$ equal to the sum of all these partial fractions. Clear the resulting equation of fractions and arrange the terms in decreasing powers of x.

STEP 4: Equate the coefficients of corresponding powers of x and solve the resulting equations for the undetermined coefficients.

Proofs that $f(x)/g(x)$ can be written as a sum of partial fractions as described here are given in advanced algebra texts.

The Substitution $z = \tan(x/2)$

The substitution

$$z = \tan\frac{x}{2} \tag{5}$$

reduces the problem of integrating any rational function of $\sin x$ and $\cos x$ to a problem involving a rational function of z. This in turn can be integrated by partial fractions. Thus the substitution (5) is a very powerful tool. It is cumbersome, however, and used only when simpler methods fail.

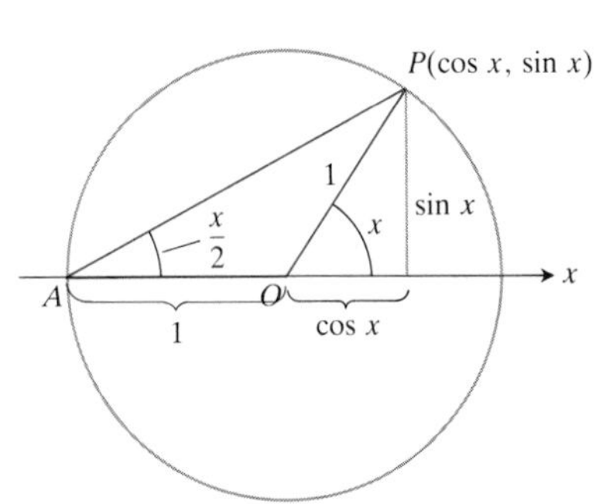

8.18 From this figure, we can read the relation

$$\tan(x/2) = \frac{(\sin x)}{(1 + \cos x)}.$$

Figure 8.18 shows how $\tan(x/2)$ expresses a rational function of $\sin x$ and $\cos x$. To see the effect of the substitution, we calculate

$$\cos x = 2\cos^2\frac{x}{2} - 1 = \frac{2}{\sec^2(x/2)} - 1 = \frac{2}{1 + \tan^2(x/2)} - 1 = \frac{2}{1 + z^2} - 1$$

or

$$\cos x = \frac{1 - z^2}{1 + z^2}, \tag{6}$$

and

$$\sin x = 2\sin\frac{x}{2}\cos\frac{x}{2} = 2\,\frac{\sin(x/2)}{\cos(x/2)}\cdot\cos^2\frac{x}{2}$$

$$= 2\tan\frac{x}{2}\cdot\frac{1}{\sec^2(x/2)} = \frac{2\tan(x/2)}{1 + \tan^2(x/2)}$$

or

$$\sin x = \frac{2z}{1 + z^2}. \tag{7}$$

Finally, $x = 2\tan^{-1}z$, so that

$$dx = \frac{2\,dz}{1 + z^2}. \tag{8}$$

EXAMPLE 7

$$\int \frac{1}{1+\cos x}\,dx = \int \frac{1+z^2}{2}\frac{2\,dz}{1+z^2} \qquad \begin{pmatrix}\text{Eqs. (6) and (8)}\\ \text{plus some algebra}\end{pmatrix}$$

$$= \int dz = z + C = \tan\frac{x}{2} + C \qquad \text{(Eq. 5)}$$

EXAMPLE 8

$$\int \frac{1}{2+\sin x}\,dx = \int \frac{1+z^2}{2+2z+2z^2}\frac{2\,dz}{1+z^2} \qquad \begin{pmatrix}\text{Eqs. (7) and (8)}\\ \text{plus some algebra}\end{pmatrix}$$

$$= \int \frac{dz}{z^2+z+1} = \int \frac{dz}{(z+1/2)^2+3/4} \qquad \begin{pmatrix}\text{Complete the}\\ \text{square}\end{pmatrix}$$

$$= \int \frac{du}{u^2+a^2} \qquad \begin{pmatrix}u = z + 1/2,\\ a = \sqrt{3}/2\end{pmatrix}$$

$$= \frac{1}{a}\tan^{-1}\frac{u}{a} + C \qquad \text{(Section 8.4, Eq. (8))}$$

$$= \frac{2}{\sqrt{3}}\tan^{-1}\frac{2z+1}{\sqrt{3}} + C \qquad \begin{pmatrix}u = z + 1/2,\\ a = \sqrt{3}/2\end{pmatrix}$$

$$= \frac{2}{\sqrt{3}}\tan^{-1}\frac{1+2\tan(x/2)}{\sqrt{3}} + C \qquad \left(z = \tan\frac{x}{2}\right)$$

Exercises 8.5

Expand the quotients in Exercises 1–8 by partial fractions.

1. $\dfrac{5x-13}{(x-3)(x-2)}$

2. $\dfrac{5x-7}{x^2-3x+2}$

3. $\dfrac{x+4}{(x+1)^2}$

4. $\dfrac{2x+2}{x^2-2x+1}$

5. $\dfrac{x+1}{x^2(x-1)}$

6. $\dfrac{z}{z^3-z^2-6z}$

7. $\dfrac{x^2+8}{x^2-5x+6}$ (Remember to divide first.)

8. $\dfrac{3}{x^2(x^2+9)}$

Evaluate the integrals in Exercises 9–20.

9. $\displaystyle\int_0^{1/2} \frac{dx}{1-x^2}$

10. $\displaystyle\int_1^2 \frac{dx}{x^2+2x}$

11. $\displaystyle\int \frac{x+4}{x^2+5x-6}\,dx$

12. $\displaystyle\int \frac{2x+1}{x^2-7x+12}\,dx$

13. $\displaystyle\int_4^8 \frac{y\,dy}{y^2-2y-3}$

14. $\displaystyle\int \frac{y^3+4y^2}{y^3+y}\,dy$

15. $\displaystyle\int \frac{dt}{t^3+t^2-2t}$

16. $\displaystyle\int \frac{x+3}{2x^3-8x}\,dx$

17. $\displaystyle\int \frac{x^3\,dx}{x^2+2x+1}$

18. $\displaystyle\int_2^6 \frac{x^3\,dx}{x^2-2x+1}$

19. $\displaystyle\int \frac{dx}{(x^2-1)^2}$

20. $\displaystyle\int \frac{x^2\,dx}{(x-1)(x^2+2x+1)}$

Evaluate the integrals in Exercises 21–30.

21. $\displaystyle\int_0^{2\sqrt{2}} \frac{x^3\,dx}{x^2+1}$

22. $\displaystyle\int_0^1 \frac{x^4+2x}{x^2+1}\,dx$

23. $\displaystyle\int_1^2 \frac{dy}{y^3+y}$

24. $\displaystyle\int_0^1 \frac{dx}{(x+1)(x^2+1)}$

25. $\displaystyle\int_0^{\sqrt{3}} \frac{5x^2\,dx}{x^2+1}$

26. $\displaystyle\int \frac{3t^2+t+4}{t^3+t}\,dt$

27. $\displaystyle\int \frac{4x+4}{x^2(x^2+1)}\,dx$

28. $\displaystyle\int_{-1}^0 \frac{x^3-x}{(x^2+1)(x-1)^2}\,dx$

29. $\displaystyle\int_{-1}^0 \frac{2x\,dx}{(x^2+1)(x-1)^2}$

30. $\displaystyle\int_1^2 \frac{x^3+x^2+2x-1}{x^2(x^2+x-1)}\,dx$

Evaluate the integrals in Exercises 31–36.

31. $\displaystyle\int_0^1 \frac{x^3+1}{x^2+1}\,dx$

32. $\displaystyle\int_0^1 \frac{x^4+2x}{x^2+1}\,dx$

33. $\displaystyle\int_0^1 \frac{y^2+2y+1}{(y^2+1)^2}\,dy$

34. $\displaystyle\int_{-1/2}^{1/2} \frac{8x^2+8x+2}{(4x^2+1)^2}\,dx$

35. $\displaystyle\int_0^1 \frac{2r^3+3r^2+5r+2}{r^2+r+1}\,dr$

36. $\displaystyle\int_0^1 \frac{dx}{(x^2+1)^2}$

37. Find the volume of the solid generated by revolving about the x-axis the region bounded by the x-axis, the lines $x = 1/2$ and $x = 5/2$, and the curve $y = 3/\sqrt{3x - x^2}$ (Fig. 8.19).

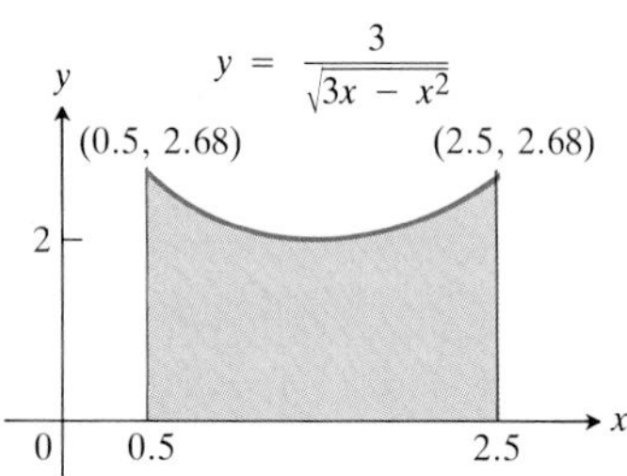

8.19 The region in Exercise 37.

38. Find the length of the curve $y = \ln(1 - x^2)$, $0 \le x \le 1/2$.

39. CALCULATOR *Social diffusion.* Sociologists sometimes use the phrase "social diffusion" to describe the way information spreads through a population. The information might be a rumor, a cultural fad, or news about a technical innovation. In a sufficiently large population, the number of people x who have the information is treated as a differentiable function of time t and the rate of diffusion, dx/dt, is assumed to be proportional to the number of people who have the information times the number of people who do not. This leads to the equation

$$\frac{dx}{dt} = kx(N - x),$$

where N is the number of people in the population.

Suppose t is measured in days, $k = 1/250$, and two people start a rumor at time $t = 0$ in a population of $N = 1000$ people.

a) Find x as a function of t.

b) When will half the population have heard the rumor? (This is when the rumor will be spreading the fastest.)

40. *Second-order chemical reactions.* Many chemical reactions are the result of the interaction of two molecules that undergo a change to produce a new product. The rate of the reaction typically depends on the concentrations of the two kinds of molecules. If a is the amount of substance A, and b is the amount of substance B at time $t = 0$, and if x is the amount of product at time t, then the rate of formation of x may be given by the differential equation

$$\frac{dx}{dt} = k(a - x)(b - x),$$

or

$$\frac{1}{(a - x)(b - x)}\frac{dx}{dt} = k$$

(k is a constant for the reaction). Integrate both sides of this equation to obtain a relation between x and t (a) if $a = b$, and (b) if $a \ne b$. Assume in each case that $x = 0$ when $t = 0$.

The Substitution $z = \tan(x/2)$

Use the substitution $z = \tan(x/2)$ to evaluate the integrals in Exercises 41–48. Integrals like these arise when we calculate the average angular velocity of the output shaft of a universal joint when the input and output shafts are not aligned.

41. $\displaystyle\int_0^{\pi/2} \frac{dx}{1 + \sin x}$

42. $\displaystyle\int_{\pi/3}^{\pi/2} \pi/2\,\frac{dx}{1 - \cos x}$

43. $\displaystyle\int \frac{dx}{1 - \sin x}$

44. $\displaystyle\int_0^{\pi/2} \frac{dx}{2 + \cos x}$

45. $\displaystyle\int \frac{\cos x\,dx}{1 - \cos x}$

46. $\displaystyle\int \frac{dx}{1 + \sin x + \cos x}$

47. $\displaystyle\int \frac{dx}{\sin x - \cos x}$

48. $\displaystyle\int_{\pi/2}^{2\pi/3} \frac{dx}{\sin x + \tan x}$

TOOLKIT PROGRAM

Partial Fractions	Generates partial fraction problems one at a time. You can choose whether the problems are to contain linear factors only or quadratic factors as well, and whether to allow repeated roots. The program provides help on request.

8.6 Using Integral Tables

The numbered integration formulas at the back of this book are stated in terms of constants a, b, c, m, n, and so on. These constants can usually assume any real value and need not be integers. Occasional limitations on their values are stated along with the formulas. Formula 5 requires $n \neq -1$, for example, and Formula 11 requires $n \neq -2$.

The formulas also assume that the constants do not take on values that require dividing by zero or taking even roots of negative numbers. For example, Formula 8 assumes $a \neq 0$, and Formula 13(a) cannot be used unless b is negative.

The following examples show how the formulas are commonly used.

EXAMPLE 1 Evaluate $\displaystyle\int x(2x+5)^{-1}\,dx.$

Solution We use Formula 8 (not 7, which requires $n \neq -1$),

$$\int x(ax+b)^{-1}\,dx = \frac{x}{a} - \frac{b}{a^2}\ln|ax+b| + C,$$

with $a = 2$ and $b = 5$:

$$\int x(2x+5)^{-1}\,dx = \frac{x}{2} - \frac{5}{4}\ln|2x+5| + C.$$

■

EXAMPLE 2 Evaluate $\displaystyle\int \frac{dx}{x\sqrt{2x+4}}.$

Solution We use Formula 13(b),

$$\int \frac{dx}{x\sqrt{ax+b}} = \frac{1}{\sqrt{b}}\ln\left|\frac{\sqrt{ax+b}-\sqrt{b}}{\sqrt{ax+b}+\sqrt{b}}\right| + C, \qquad \text{if } b > 0,$$

with $a = 2$ and $b = 4$:

$$\int \frac{dx}{x\sqrt{2x+4}} = \frac{1}{\sqrt{4}}\ln\left|\frac{\sqrt{2x+4}-\sqrt{4}}{\sqrt{2x+4}+\sqrt{4}}\right| + C$$

$$= \frac{1}{2}\ln\left|\frac{\sqrt{2x+4}-2}{\sqrt{2x+4}+2}\right| + C.$$

Formula 13(a), which requires $b < 0$, would not have been appropriate here. It *is* appropriate, however, in the next example. ■

EXAMPLE 3 Evaluate $\displaystyle\int \frac{dx}{x\sqrt{2x-4}}.$

Solution We use Formula 13(a),

$$\int \frac{dx}{x\sqrt{ax+b}} = \frac{2}{\sqrt{-b}}\tan^{-1}\sqrt{\frac{ax+b}{-b}} + C, \qquad \text{if } b < 0,$$

with $a = 2$ and $b = -4$:

$$\int \frac{dx}{x\sqrt{2x-4}} = \frac{2}{\sqrt{-(-4)}}\tan^{-1}\sqrt{\frac{2x-4}{-(-4)}} + C = \tan^{-1}\sqrt{\frac{x-2}{2}} + C. \quad \blacksquare$$

EXAMPLE 4 Evaluate $\displaystyle\int \frac{dx}{x^2\sqrt{2x-4}}$.

Solution We begin with Formula 15,

$$\int \frac{dx}{x^2\sqrt{ax+b}} = -\frac{\sqrt{ax+b}}{bx} - \frac{a}{2b}\int \frac{dx}{x\sqrt{ax+b}} + C,$$

with $a = 2$ and $b = -4$:

$$\int \frac{dx}{x^2\sqrt{2x-4}} = -\frac{\sqrt{2x-4}}{-4x} + \frac{2}{2\cdot 4}\int \frac{dx}{x\sqrt{2x-4}} + C.$$

We then use Formula 13(a) to evaluate the integral on the right (Example 3) to obtain

$$\int \frac{dx}{x^2\sqrt{2x-4}} = \frac{\sqrt{2x-4}}{4x} + \frac{1}{4}\tan^{-1}\sqrt{\frac{x-2}{2}} + C. \quad \blacksquare$$

EXAMPLE 5 Evaluate $\displaystyle\int x \sin^{-1}x \, dx$.

Solution We use Formula 99,

$$\int x^n \sin^{-1}ax\, dx = \frac{x^{n+1}}{n+1}\sin^{-1}ax - \frac{a}{n+1}\int \frac{x^{n+1}\,dx}{\sqrt{1-a^2x^2}}, \qquad n \neq -1,$$

with $n = 1$ and $a = 1$:

$$\int x \sin^{-1}x\, dx = \frac{x^2}{2}\sin^{-1}x - \frac{1}{2}\int \frac{x^2\,dx}{\sqrt{1-x^2}}.$$

The integral on the right is found in the table as Formula 33,

$$\int \frac{x^2}{\sqrt{a^2-x^2}}\,dx = \frac{a^2}{2}\sin^{-1}\frac{x}{a} - \frac{1}{2}x\sqrt{a^2-x^2} + C,$$

with $a = 1$:

$$\int \frac{x^2\,dx}{\sqrt{1-x^2}} = \frac{1}{2}\sin^{-1}x - \frac{1}{2}x\sqrt{1-x^2} + C.$$

The combined result is

$$\begin{aligned}\int x \sin^{-1}x\, dx &= \frac{x^2}{2}\sin^{-1}x - \frac{1}{2}\left(\frac{1}{2}\sin^{-1}x - \frac{1}{2}x\sqrt{1-x^2}\right) + C \\ &= \left(\frac{x^2}{2} - \frac{1}{4}\right)\sin^{-1}x + \frac{1}{4}x\sqrt{1-x^2} + C. \quad \blacksquare\end{aligned}$$

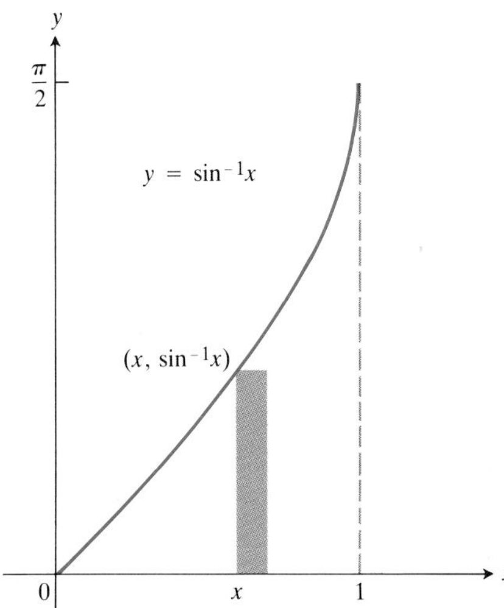

8.20 The region in Example 6.

EXAMPLE 6 The region in the first quadrant enclosed by the x-axis, the line $x = 1$, and the curve $y = \sin^{-1}x$ is revolved about the y-axis to generate a solid. Find the volume of the solid.

Solution We sketch the region (Fig. 8.20) and decide on the method of cylindrical shells. The volume is

$$V = \int_0^1 2\pi(\text{radius})(\text{height})\,dx$$

$$= \int_0^1 2\pi x \sin^{-1}x\,dx$$

$$= 2\pi\left[\left(\frac{x^2}{2} - \frac{1}{4}\right)\sin^{-1}x + \frac{1}{4}x\sqrt{1 - x^2}\right]_0^1 \qquad \text{(Example 5)}$$

$$= 2\pi\left[\left(\frac{1}{4}\right)\left(\frac{\pi}{2}\right) + 0\right] - 2\pi[0 + 0] = \frac{\pi^2}{4}.$$

Reduction Formulas

Our integral tables, and those that are more extensive, include a collection of reduction formulas, such as

$$\int (\ln x)^n\,dx = x(\ln x)^n - n\int (\ln x)^{n-1}\,dx \tag{1}$$

$$\int \sec^n x\,dx = \frac{1}{n-1}\sec^{n-2}x\,\tan x - \frac{n-2}{n-1}\int \sec^{n-2}x\,dx \tag{2}$$

$$\int \tan^n x\,dx = \frac{1}{n-1}\tan^{n-1}x - \int \tan^{n-2}x\,dx. \tag{3}$$

A **reduction formula** expresses an integral containing a power of some function in terms of an integral in which that power is replaced by a lower power, thus reducing the exponent. By applying such a formula repeatedly, we can eventually express the original integral in terms of a power low enough to be evaluated directly.

EXAMPLE 7 Evaluate $\displaystyle\int \tan^5 x\,dx.$

Solution We apply Eq. (3) with $n = 5$ to get

$$\int \tan^5 x\,dx = \frac{1}{4}\tan^4 x - \int \tan^3 x\,dx.$$

We then apply Eq. (3) again, with $n = 3$, to evaluate the remaining integral:

$$\int \tan^3 x\,dx = \frac{1}{2}\tan^2 x - \int \tan x\,dx = \frac{1}{2}\tan^2 x + \ln|\cos x| + C.$$

The combined result is

$$\int \tan^5 x\,dx = \frac{1}{4}\tan^4 x - \frac{1}{2}\tan^2 x - \ln|\cos x| + C.$$

Reduction formulas are derived by integration by parts.

EXAMPLE 8 Show that for any positive integer n

$$\int (\ln x)^n \, dx = x(\ln x)^n - n \int (\ln x)^{n-1} \, dx .$$

Solution We use the integration-by-parts formula

$$\int u \, dv = uv - \int v \, du$$

with

$$u = (\ln x)^n, \qquad du = n(\ln x)^{n-1} \frac{dx}{x}, \qquad dv = dx, \qquad v = x,$$

to obtain

$$\int (\ln x)^n \, dx = x(\ln x)^n - n \int (\ln x)^{n-1} \, dx .$$

A General Procedure

There is no surefire way to evaluate every indefinite integral, but the procedure in Flow Chart 8.2 may help.

FLOW CHART 8.2. Procedure for Indefinite Integration

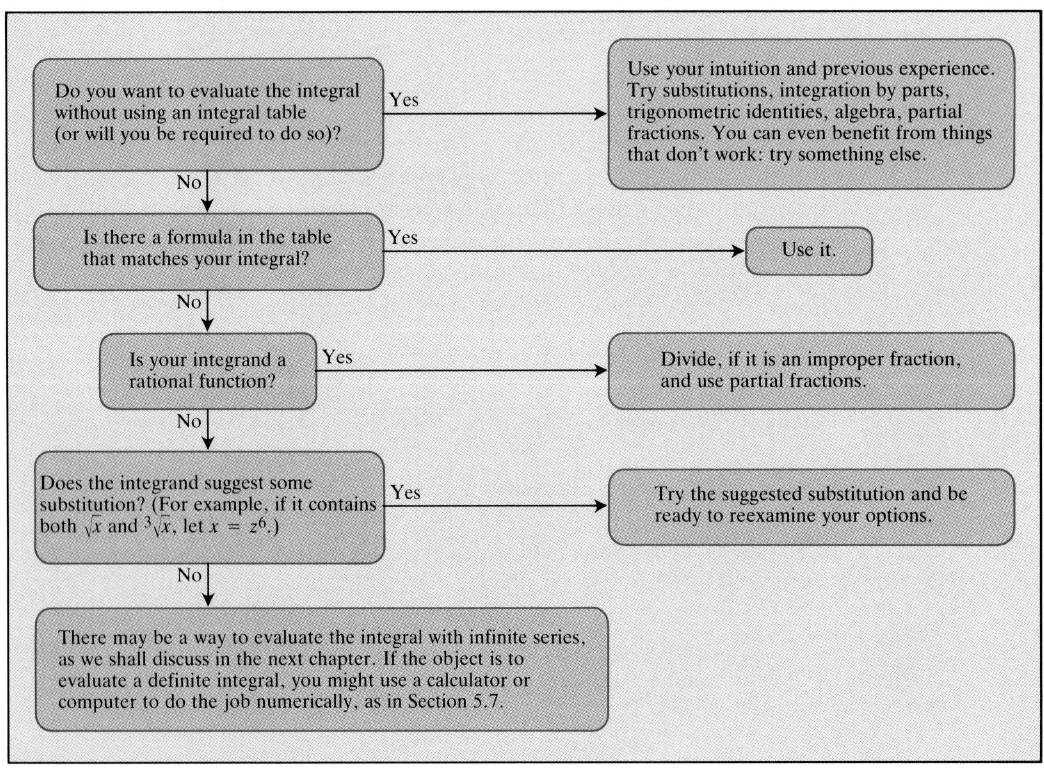

Nonelementary Integrals

The development of computer programs that evaluate indefinite integrals by symbolic manipulation has led to a renewed interest in determining which integrals can be expressed as finite algebraic combinations of elementary functions (the functions we have been studying) and which require infinite series (Chapter 9) or numerical methods for their evaluation. Examples of the latter include the error function

$$\operatorname{erf}(x) = \frac{2}{\sqrt{\pi}} \int_0^x e^{-t^2}\, dt$$

and integrals like

$$\int \sin x^2\, dx \qquad \text{and} \qquad \int \sqrt{1 + x^4}\, dx$$

that arise in engineering and physics. These, and a number of others, like

$$\int \frac{e^x}{x}\, dx, \qquad \int e^{(e^x)}\, dx, \qquad \int \frac{1}{\ln x}\, dx, \qquad \int \ln(\ln x)\, dx, \qquad \int \frac{\sin x}{x}\, dx,$$

$$\int \sqrt{1 - k^2 \sin^2 x}\, dx, \qquad 0 < k < 1,$$

look so easy they tempt us to try them just to see how they turn out. It can be proved, however, that there is no way to express these integrals as finite combinations of elementary functions. The same applies to integrals that can be changed into these by substitutions. None of the integrals you are asked to evaluate in the present chapter falls into this category, but you may encounter nonelementary integrals from time to time in your other work.

Exercises 8.6

Use the integral formulas at the back of this book to evaluate the integrals in Exercises 1–16.

1. $\displaystyle\int_0^{\infty} e^{-x^2} dx$

2. $\displaystyle\int x \cos^{-1} x\, dx$

3. $\displaystyle\int_6^9 \frac{dx}{x\sqrt{x - 3}}$

4. $\displaystyle\int_0^{1/2} x \tan^{-1} 2x\, dx$

5. $\displaystyle\int \frac{dx}{(9 - x^2)^2}$

6. $\displaystyle\int_4^{10} \frac{\sqrt{4x + 9}}{x^2}\, dx$

7. $\displaystyle\int_3^{11} \frac{dx}{x^2\sqrt{7 + x^2}}$

8. $\displaystyle\int \frac{dx}{x^2\sqrt{7 - x^2}}$

9. $\displaystyle\int_{-2}^{-\sqrt{2}} \frac{\sqrt{x^2 - 2}}{x}\, dx$

10. $\displaystyle\int_{-\pi/12}^{\pi/4} \frac{dx}{5 + 4 \sin 2x}$

11. $\displaystyle\int \frac{dx}{4 + 5 \sin 2x}$

12. $\displaystyle\int_3^6 \frac{x}{\sqrt{x - 2}}\, dx$ (*Hint:* The n in Formula 7 need not be an integer.)

13. $\displaystyle\int x\sqrt{2x - 3}\, dx$

14. $\displaystyle\int \frac{\sqrt{3x - 4}}{x}\, dx$

15. $\displaystyle\int_0^{\infty} x^{10} e^{-x}\, dx$

16. $\displaystyle\int_0^1 x^2 \tan^{-1} x\, dx$

In Exercises 17–20, use the substitution to change the integral into one you can find in the tables. Then evaluate the integral.

17. $\displaystyle\int_0^1 \sin^{-1}\sqrt{x}\,dx, \quad u = \sqrt{x}$

18. $\displaystyle\int_{3/4}^1 \frac{\cos^{-1}\sqrt{x}}{\sqrt{x}}\,dx, \quad u = \sqrt{x}$

19. $\displaystyle\int_0^{1/2} \frac{\sqrt{x}}{\sqrt{1-x}}\,dx, \quad u = \sqrt{x}$

20. $\displaystyle\int_{\pi/4}^{\pi/2} \cot x\sqrt{1-\sin^2 x}\,dx, \quad u = \sin x$

21. Find the centroid of the region cut from the first quadrant by the curve $y = 1/\sqrt{x+1}$ and the line $x = 3$.

22. A thin plate of constant density $\delta = 1$ occupies the region enclosed by the curve $y = 36/(2x+3)$ and the line $x = 3$ in the first quadrant. Find the moment of the plate about the y-axis.

Exercises 23–26 refer to the formulas at the back of this book by number.

23. Verify Formula 55 for $x > 0$ by differentiating the right side.

24. Verify Formula 76 by differentiating the right side.

25. Verify Formula 9 by integrating

$$\int \frac{x}{(ax+b)^2}\,dx$$

with the substitution $u = ax + b$.

26. Verify Formula 46 by integrating

$$\int \frac{dx}{x^2\sqrt{x^2-a^2}}$$

with the substitution $x = a \sec u$.

Use reduction formulas to evaluate the integrals in Exercises 27–40.

27. $\displaystyle\int_{-\pi}^{\pi} \cos^4 x\,dx$

28. $\displaystyle\int_{\pi/6}^{\pi/3} \cos^6 3x\,dx$

29. $\displaystyle\int_0^{\pi} \sin^4 x\,dx$

30. $\displaystyle\int_0^{\pi/4} \sin^5 2x\,dx$

31. $\displaystyle\int_0^{\pi/8} \tan^3 2x\,dx$

32. $\displaystyle\int_0^{\pi/4} \tan^4\left(\frac{x}{2}\right)dx$

33. $\displaystyle\int_{\pi/4}^{3\pi/4} \cot^4 x\,dx$

34. $\displaystyle\int_{\pi/8}^{3\pi/8} \cot^4 2x\,dx$

35. $\displaystyle\int_{-\pi/3}^{\pi/3} \sec^4 x\,dx$

36. $\displaystyle\int_0^{\pi/4} \sec^5 x\,dx$

37. $\displaystyle\int_{\pi/4}^{\pi/2} \csc^4 x\,dx$

38. $\displaystyle\int_{\pi/4}^{\pi/2} \csc^5 x\,dx$

39. $\displaystyle\int_1^3 16x^3(\ln x)^2\,dx$

40. $\displaystyle\int_1^e (\ln x)^3\,dx$

Evaluate the integrals in Exercises 41–44 by making a trigonometric substitution and then applying a reduction formula.

41. $\displaystyle\int_0^1 (x^2+1)^{-3/2}\,dx$

42. $\displaystyle\int_0^1 (x^2+1)^{3/2}\,dx$

43. $\displaystyle\int_0^{3/5} \frac{dx}{(1-x^2)^3}$

44. $\displaystyle\int_1^2 \frac{(x^2-1)^{3/2}}{x}\,dx$

45. Verify the formula

$$\int \sin^n x\,dx = -\frac{\sin^{n-1}x\cos x}{n} + \frac{n-1}{n}\int \sin^{n-2}x\,dx$$

by differentiating the right-hand side and combining the results to get $\sin^n x$.

46. Verify the formula

$$\int \csc^n x\,dx = -\frac{\csc^{n-2}x\cot x}{n-1} + \frac{n-2}{n-1}\int \csc^{n-2}x\,dx$$

by differentiating the right-hand side and combining the results to get $\csc^n x$.

47. a) Derive the formula

$$\int x^n \sin x\,dx = -x^n\cos x + n\int x^{n-1}\cos x\,dx.$$

b) What is the corresponding formula for $x^n \sin ax$?

48. a) Derive the formula

$$\int x^n \cos x\,dx = x^n\sin x - n\int x^{n-1}\sin x\,dx.$$

b) What is the corresponding formula for $x^n \cos ax$?

8.7 Improper Integrals

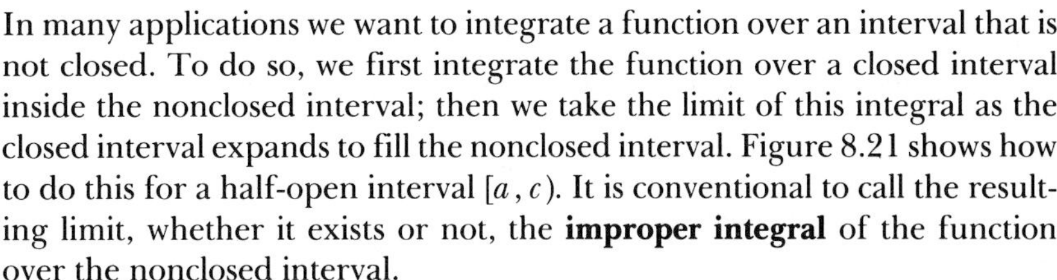

In many applications we want to integrate a function over an interval that is not closed. To do so, we first integrate the function over a closed interval inside the nonclosed interval; then we take the limit of this integral as the closed interval expands to fill the nonclosed interval. Figure 8.21 shows how to do this for a half-open interval $[a, c)$. It is conventional to call the resulting limit, whether it exists or not, the **improper integral** of the function over the nonclosed interval.

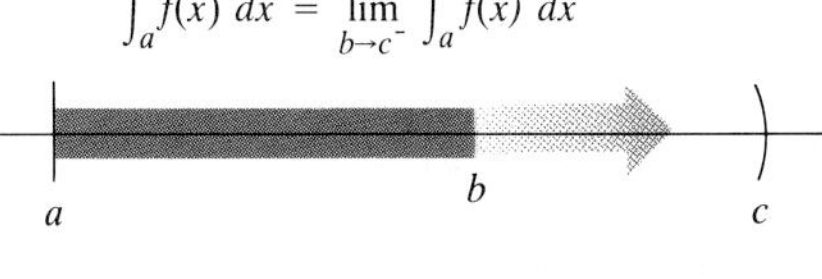

8.21 To integrate over a half-open interval $[a, c)$, we (1) integrate over a closed interval $[a, b]$ inside $[a, c)$ and (2) take the limit as $[a, b]$ expands to fill $[a, c)$.

How can we tell if the improper integral exists? That is the basic question and the one we address in this section. Once the integral is known to exist, its value, if not immediately apparent, can be found by numerical methods.

Convergence and Divergence

In lifting-line theory in aerodynamics, the function

$$f(x) = \sqrt{\frac{1 + x}{1 - x}} \tag{1}$$

needs to be integrated over the interval from $x = 0$ to $x = 1$. The function is not defined at $x = 1$, although it is defined and continuous everywhere else in $[0, 1]$. To integrate f from zero to one, we integrate f from zero to a positive number b less than 1 and take the limit of the resulting definite integral as b approaches 1. If the limit exists, we define the integral of f from zero to 1 to be this value and write

$$\int_0^1 \sqrt{\frac{1 + x}{1 - x}}\,dx = \lim_{b\to 1^-} \int_0^b \sqrt{\frac{1 + x}{1 - x}}\,dx. \tag{2}$$

In this case, we also say that the integral

$$\int_0^1 \sqrt{\frac{1 + x}{1 - x}}\,dx$$

converges and say that the area under the curve $y = \sqrt{(1 + x)/(1 - x)}$ from 0 to 1 is the value of the integral. If the limit in (2) does not exist, we say that the integral **diverges.**

EXAMPLE 1 Determine whether

$$\int_0^1 \sqrt{\frac{1 + x}{1 - x}}\,dx$$

converges or diverges.

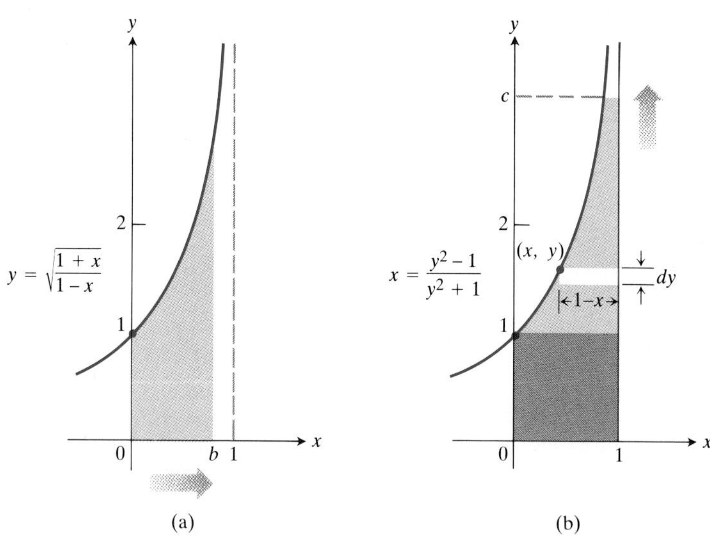

8.22 (a) To evaluate

$$\int_0^1 \sqrt{(1+x)/(1-x)}\,dx,$$

we evaluate

$$\int_0^b \sqrt{(1+x)/(1-x)}\,dx$$

and let b approach 1 from below. (b) If we treat the curve as the graph of the function $x = (y^2 - 1)/(y^2 + 1)$, we can make an equivalent calculation by integrating from 1 to c with respect to y and letting c approach infinity. We then add 1 to the result to account for the area of the gray square.

Solution See Fig. 8.22(a). Multiplying numerator and denominator by $\sqrt{1+x}$ gives

$$\int \sqrt{\frac{1+x}{1-x}}\,dx = \int \frac{1+x}{\sqrt{1-x^2}}\,dx = \int \frac{1}{\sqrt{1-x^2}}\,dx + \int \frac{x}{\sqrt{1-x^2}}\,dx$$
$$= \sin^{-1}x - \sqrt{1-x^2} + C.$$

Therefore,

$$\lim_{b\to 1^-}\int_0^b \sqrt{\frac{1+x}{1-x}}\,dx = \lim_{b\to 1^-}\left[\sin^{-1}x - \sqrt{1-x^2}\right]_0^b$$
$$= \lim_{b\to 1^-}[\sin^{-1}b - \sqrt{1-b^2} + 1]$$
$$= \sin^{-1}1 - 0 + 1 = (\pi/2) + 1.$$

The integral converges to $(\pi/2) + 1$.

The computation above integrates vertical strips with respect to x. We can get the same result by integrating horizontal strips with respect to y (Fig. 8.22b). In this case,

$$dA = (1-x)\,dy = \frac{2}{y^2+1}\,dy \qquad \left(\text{using } x = \frac{y^2-1}{y^2+1}\right)$$

and the portion of the area above the shaded square is

$$\int_1^\infty \frac{2}{y^2+1}\,dy = \lim_{c\to\infty}\int_1^c \frac{2}{y^2+1}\,dy$$
$$= \lim_{c\to\infty}\left[2\tan^{-1}y\right]_1^c$$
$$= \lim_{c\to\infty} 2\tan^{-1}c - 2\cdot\frac{\pi}{4} = 2\cdot\frac{\pi}{2} - \frac{\pi}{2} = \frac{\pi}{2}.$$

Including the shaded square, the area is $(\pi/2) + 1$, in agreement with our first calculation. ■

The notation

$$\int_a^b f(x)\,dx$$

for improper integrals is the same as the notation for definite integrals. In any given case it is usually a simple matter to tell whether a particular integral is to be calculated as an ordinary definite integral or as a limit. If a and b are finite and f is continuous at every point of $[a, b]$, the integral is an ordinary definite integral. If f becomes infinite at one or more points in the interval of integration, or one or both of the limits of integration is infinite, then the designated integral is improper and is to be calculated as a limit.

DEFINITION

$\int_a^b f(x)\,dx$ denotes an **improper integral** if

1. f becomes infinite at one or more points of the interval of integration, or
2. one or both of the limits of integration is infinite, or
3. both (1) and (2) hold.

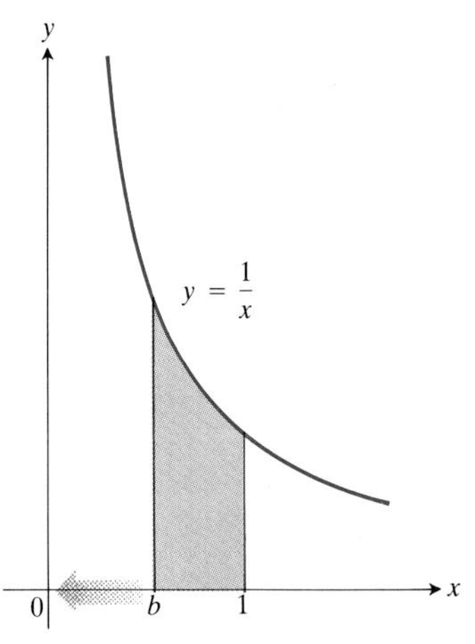

8.23 $\int_0^1 \frac{dx}{x} = \lim_{b\to 0^+} \int_b^1 \frac{dx}{x}.$

EXAMPLE 2 In the integral $\displaystyle\int_0^1 \frac{dx}{x}$, the function

$$f(x) = \frac{1}{x}$$

becomes infinite at $x = 0$. We cut off the point $x = 0$ and start our integration at some positive number $b < 1$ (Fig. 8.23).

$$\int_b^1 \frac{dx}{x} = \ln x\Big]_b^1 = \ln 1 - \ln b = \ln \frac{1}{b}$$

We then investigate the behavior of the integral as b approaches zero from the right. Since

$$\lim_{b\to 0^+} \int_b^1 \frac{dx}{x} = \lim_{b\to 0^+} \left(\ln \frac{1}{b}\right) = +\infty,$$

we say that the integral from $x = 0$ to $x = 1$ *diverges.* ■

The method to be used when the function f becomes infinite at an interior point of the interval of integration is illustrated in the following example.

EXAMPLE 3 In the integral $\displaystyle\int_0^3 \frac{dx}{(x-1)^{2/3}}$, the function

$$f(x) = \frac{1}{(x-1)^{2/3}}$$

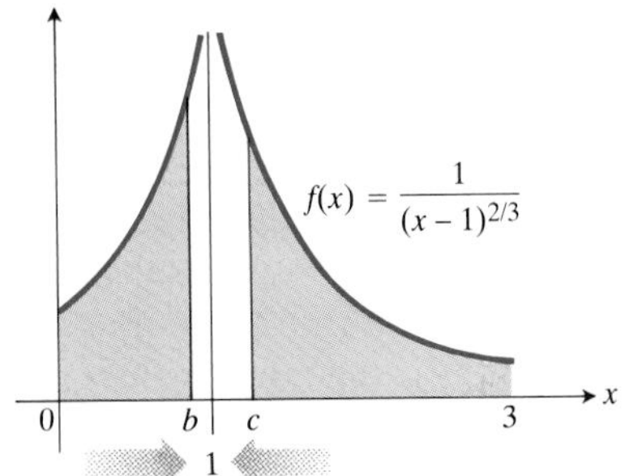

8.24 $\int_0^3 f(x)\,dx$
$= \lim_{b\to1^-}\int_0^b f(x)\,dx + \lim_{c\to1^+}\int_c^3 f(x)\,dx.$

becomes infinite at $x = 1$, which lies between the limits of integration 0 and 3. In such a case, we again cut out the point where $f(x)$ becomes infinite. This time we integrate from 0 to b, where b is slightly less than 1, and start again on the other side of 1 at c and integrate from c to 3 (Fig. 8.24). This gives two integrals to investigate:

$$\int_0^b \frac{dx}{(x-1)^{2/3}} \quad \text{and} \quad \int_c^3 \frac{dx}{(x-1)^{2/3}}.$$

If the first of these has a definite limit as $b \to 1^-$ and if the second also has a definite limit as $c \to 1^+$, then we say that the integral of f from 0 to 3 *converges* and that its value is

$$\int_0^3 \frac{dx}{(x-1)^{2/3}} = \lim_{b\to1^-}\int_0^b \frac{dx}{(x-1)^{2/3}} + \lim_{c\to1^+}\int_c^3 \frac{dx}{(x-1)^{2/3}}.$$

If either limit fails to exist, we say that the integral of f from 0 to 3 *diverges*. For this example,

$$\lim_{b\to1^-}\int_0^b (x-1)^{-2/3}\,dx = \lim_{b\to1^-}[3(b-1)^{1/3} - 3(0-1)^{1/3}] = +3$$

and

$$\lim_{c\to1^+}\int_c^3 (x-1)^{-2/3}\,dx = \lim_{c\to1^+}[3(3-1)^{1/3} - 3(c-1)^{1/3}] = 3\sqrt[3]{2}.$$

Since both limits exist and are finite, the integral of f converges and its value is $3 + 3\sqrt[3]{2}$. ■

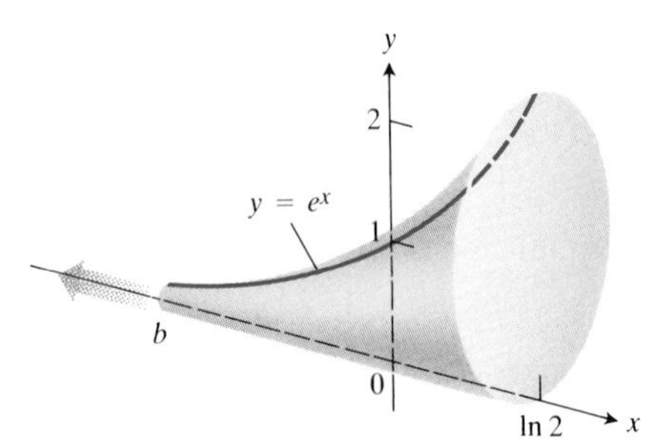

8.25 The calculation in Example 4 shows that this infinite solid horn has a finite volume.

EXAMPLE 4 Each cross section of the solid infinite horn in Fig. 8.25, cut by a plane perpendicular to the x-axis for $-\infty < x \le \ln 2$, is a circular disk with one diameter reaching from the x-axis to the curve $y = e^x$. Find the volume of the horn.

Solution The area of a typical cross section is

$$A(x) = \pi(\text{radius})^2 = \pi\left(\frac{1}{2}y\right)^2 = \frac{\pi}{4}e^{2x}.$$

We define the volume of the horn to be the limit as $b \to -\infty$ of the volume of the portion from b to ln 2. As in Section 6.9 (the method of slicing), the volume of this portion is

$$V = \int_b^{\ln 2} A(x)\,dx = \int_b^{\ln 2} \frac{\pi}{4}e^{2x}\,dx = \frac{\pi}{8}e^{2x}\Big]_b^{\ln 2} = \frac{\pi}{8}\left(e^{\ln 4} - e^{2b}\right) = \frac{\pi}{8}\left(4 - e^{2b}\right).$$

As $b \to -\infty$, $e^{2b} \to 0$ and $V \to (\pi/8)(4-0) = \pi/2$. The volume of the horn is $\pi/2$. ■

The Integral $\int_1^\infty dx/x^p$

The convergence of the integral $\int_1^\infty \frac{dx}{x^p}$ depends on the value of the exponent p. The next example illustrates this with $p = 1$ and $p = 2$.

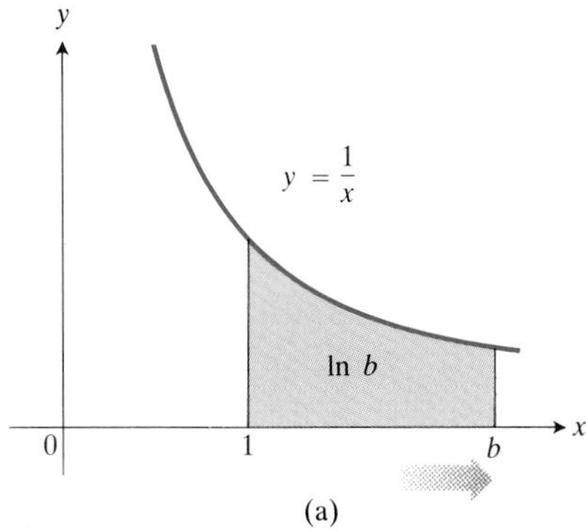

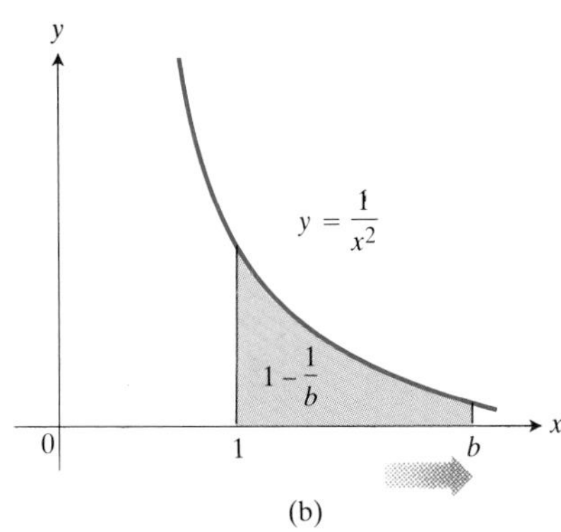

8.26 (a) The shaded area, ln b, does not approach a finite limit as $b \to \infty$. (b) The shaded area, $1 - (1/b)$, approaches 1 as $b \to \infty$.

EXAMPLE 5 Do the integrals

$$\int_1^\infty \frac{dx}{x} \quad \text{and} \quad \int_1^\infty \frac{dx}{x^2}$$

converge, or diverge?

Solution The curves

$$y = \frac{1}{x} \quad \text{and} \quad y = \frac{1}{x^2}$$

both approach the x-axis as $x \to \infty$ (Fig. 8.26). In the first case,

$$\int_1^b \frac{dx}{x} = \ln x \Big]_1^b = \ln b, \qquad \text{so that} \qquad \lim_{b\to\infty} \int_1^b \frac{dx}{x} = \lim_{b\to\infty} \ln b = \infty.$$

Therefore,

$$\int_1^\infty \frac{dx}{x} = \infty$$

and the integral *diverges*.

In the second case,

$$\int_1^b \frac{dx}{x^2} = -\frac{1}{x}\Big]_1^b = 1 - \frac{1}{b}, \qquad \text{so that} \qquad \int_1^\infty \frac{dx}{x^2} = \lim_{b\to\infty} \int_1^b \frac{dx}{x^2} = \lim_{b\to\infty}\left(1 - \frac{1}{b}\right) = 1.$$

The integral converges and its value is 1. ■

Generally, the integral

$$\int_1^\infty \frac{dx}{x^p}$$

converges when $p > 1$ but diverges when $p \le 1$ (Exercise 43).

The Domination Test for Convergence and Divergence

Sometimes we can determine whether an improper integral converges without having to evaluate it. Instead, we compare it to an integral whose convergence or divergence we already know. This is the case with the next example, an integral important in probability theory.

EXAMPLE 6 Determine whether the improper integral $\displaystyle\int_1^\infty e^{-x^2}\,dx$ converges or diverges.

Solution Even though we cannot find any simpler expression for

$$I(b) = \int_1^b e^{-x^2}\,dx$$

(the integral is nonelementary), we can show that $I(b)$ has a finite limit as $b \to \infty$.

The function $I(b)$ represents the area between the x-axis and the curve $y = e^{-x^2}$ from $x = 1$ to $x = b$. It is an increasing function of b. Therefore, it

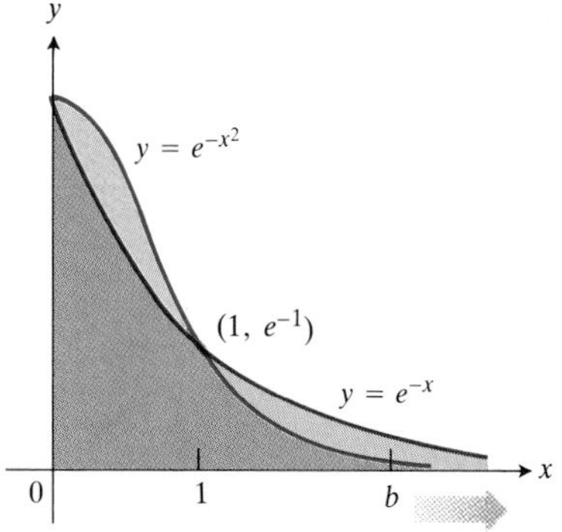

8.27 The graphs of $y = e^{-x^2}$ and $y = e^{-x}$.

either becomes infinite as $b \to \infty$ or has a finite limit as $b \to \infty$. It does not become infinite, because for every value of $b \geq 1$,

$$\begin{aligned} I(b) &= \int_1^b e^{-x^2}\,dx \leq \int_1^b e^{-x}\,dx \qquad \text{(Fig. 8.27)} \\ &= e^{-1} - e^{-b} < e^{-1} < 0.37. \qquad \text{(Calculator, rounded)} \end{aligned}$$

Hence

$$\int_1^{\infty} e^{-x^2}\,dx = \lim_{b\to\infty} \int_1^{b} e^{-x^2}\,dx$$

converges to a definite finite value. We have not calculated what the value is, but we know that it is less than 0.37. ■

A positive function f **dominates** a positive function g as $x \to \infty$ if $g(x) \leq f(x)$ for all values of x beyond some point a. For instance, $f(x) = e^{-x}$ dominates $g(x) = e^{-x^2}$ as $x \to \infty$ because $e^{-x^2} \leq e^{-x}$ for all $x > a = 1$.

If f dominates g as $x \to \infty$, then

$$\int_a^b g(x)\,dx \leq \int_a^b f(x)\,dx, \qquad b > a,$$

and from this it can be argued as in Example 6 that

$$\int_a^{\infty} g(x)\,dx \text{ converges if } \int_a^{\infty} f(x) \text{ converges.}$$

Turning this around says that

$$\int_a^{\infty} f(x)\,dx \text{ diverges if } \int_a^{\infty} g(x) \text{ diverges.}$$

We state these results as a theorem and then give examples.

THEOREM 1

Domination Test for Convergence and Divergence of Improper Integrals

If $0 \leq g(x) \leq f(x)$ for all $x > a$, then

1. $\displaystyle\int_a^{\infty} g(x)\,dx$ converges if $\displaystyle\int_a^{\infty} f(x)\,dx$ converges,

2. $\displaystyle\int_a^{\infty} f(x)\,dx$ diverges if $\displaystyle\int_a^{\infty} g(x)\,dx$ diverges.

Theorem 1 assumes that f and g are integrable over every finite interval $[a, b]$, which will be true, for instance, if the functions are continuous.

EXAMPLE 7

a) $\displaystyle\int_1^{\infty} \frac{1}{e^{2x}}\,dx$ converges because $\dfrac{1}{e^{2x}} < \dfrac{1}{e^{x}}$ and $\displaystyle\int_1^{\infty} \frac{1}{e^{x}}\,dx$ converges.

b) $\displaystyle\int_1^\infty \frac{1}{\sqrt{x}}\,dx$ diverges because $\dfrac{1}{\sqrt{x}} \geq \dfrac{1}{x}$ for $x \geq 1$ and $\displaystyle\int_1^\infty \frac{1}{x}\,dx$ diverges.

c) $\displaystyle\int_1^\infty \left(\frac{1}{x} + \frac{1}{x^2}\right) dx$ diverges because $\dfrac{1}{x} + \dfrac{1}{x^2} > \dfrac{1}{x}$ and $\displaystyle\int_1^\infty \frac{1}{x}\,dx$ diverges. ■

The Limit Comparison Test

Another useful result, which we shall not prove, is the limit comparison test for the convergence and divergence of improper integrals. (You will find a similar result for infinite series in Chapter 9.) It goes like this:

THEOREM 2

Limit Comparison Test for Convergence and Divergence of Improper Integrals

Suppose that $f(x)$ and $g(x)$ are positive functions and that

$$\lim_{x\to\infty} \frac{f(x)}{g(x)} = L, \qquad 0 < L < \infty.$$

Then, $\displaystyle\int_a^\infty f(x)\,dx$ and $\displaystyle\int_a^\infty g(x)\,dx$ both converge or both diverge.

Like Theorem 1, Theorem 2 assumes that f and g are integrable over every finite interval $[a, b]$.

In the language of Section 7.6, Theorem 2 says that *if two functions grow at the same rate as $x \to \infty$, then their integrals from a to ∞ behave alike: they both converge or both diverge.* This does not mean that their integrals must have the same value, as the next example shows.

EXAMPLE 8 Compare

$$\int_1^\infty \frac{dx}{x^2} \quad \text{and} \quad \int_1^\infty \frac{dx}{1+x^2}$$

with the limit comparison test.

Solution With

$$f(x) = \frac{1}{x^2}, \qquad g(x) = \frac{1}{1+x^2},$$

we have

$$\lim_{x\to\infty} \frac{1}{x^2}\,\frac{1+x^2}{1} = \lim_{x\to\infty} \frac{1+x^2}{x^2} = \lim_{x\to\infty} \left(\frac{1}{x^2} + 1\right) = 0 + 1 = 1,$$

as a positive finite limit. Therefore,

$$\int_1^\infty \frac{dx}{1+x^2} \text{ converges because } \int_1^\infty \frac{dx}{x^2} \text{ converges.}$$

The integrals converge to different values, however:

$$\int_1^\infty \frac{dx}{x^2} = 1,$$

from Example 5, and

$$\int_1^\infty \frac{dx}{1+x^2} = \lim_{b\to\infty} \int_1^b \frac{dx}{1+x^2} \qquad \text{(the definition)}$$

$$= \lim_{b\to\infty} \left[\tan^{-1} b - \tan^{-1} 1\right] = \frac{\pi}{2} - \frac{\pi}{4} = \frac{\pi}{4}.$$

EXAMPLE 9

$$\int_1^\infty \frac{3}{e^x+5}\,dx \text{ converges because } \int_1^\infty \frac{1}{e^x}\,dx \text{ converges}$$

and

$$\lim_{x\to\infty} \frac{1}{e^x} \cdot \frac{e^x+5}{3} = \lim_{x\to\infty} \frac{e^x+5}{3e^x} = \lim_{x\to\infty} \left(\frac{1}{3} + \frac{5}{3e^x}\right) = \frac{1}{3} + 0 = \frac{1}{3},$$

a positive finite limit. As far as the convergence of the improper integral is concerned, $3/(e^x+5)$ behaves like $1/e^x$.

EXAMPLE 10

$$\int_3^\infty \frac{1}{e^{2x} - 10e^x}\,dx \text{ converges because } \int_3^\infty \frac{1}{e^{2x}}\,dx \text{ converges}$$

and

$$\lim_{x\to\infty} \frac{1}{e^{2x}} \cdot \frac{e^{2x}-10e^x}{1} = \lim_{x\to\infty} \left(1 - \frac{10}{e^x}\right) = 1 - 0 = 1,$$

a positive finite limit. As far as the convergence of the improper integral is concerned, $1/(e^{2x} - 10e^x)$ behaves like $1/e^{2x}$.

Concluding Remarks

We know that

$$\int_1^\infty \frac{1}{x^2}\,dx$$

converges, but what about integrals like

$$\int_2^\infty \frac{1}{x^2}\,dx \qquad \text{and} \qquad \int_{100}^\infty \frac{1}{x^2}\,dx?$$

The answer is that they converge, too. The existence of the limits

$$\lim_{b\to\infty} \int_2^b \frac{1}{x^2}\,dx \qquad \text{and} \qquad \lim_{b\to\infty} \int_{100}^b \frac{1}{x^2}\,dx$$

does not depend on the starting points $a = 2$ and $a = 100$, but only on the values of $1/x^2$ as x approaches infinity. Indeed, we find for any positive a that

$$\lim_{b\to\infty} \int_a^b \frac{1}{x^2}\,dx = \lim_{b\to\infty} \left(-\frac{1}{b} + \frac{1}{a}\right) = \frac{1}{a}.$$

The *value* of the limit depends on the value of a, but the *existence* of the limit does not. The limit exists for any positive number a.

Similarly, the integrals

$$\int_2^\infty \frac{1}{x}\,dx \quad \text{and} \quad \int_{100}^\infty \frac{1}{x}\,dx$$

both diverge. We find for any positive number a that

$$\lim_{b\to\infty} \int_a^b \frac{1}{x}\,dx = \lim_{b\to\infty}(\ln b - \ln a) = \infty.$$

The convergence and divergence of integrals that are improper only because the upper limit is infinite never depend on the lower limit of integration.

In our final example, we show that an improper integral may diverge without becoming infinite.

EXAMPLE 11 The integral

$$\int_0^b \cos x\,dx = \sin b$$

takes all values between -1 and $+1$ as b varies between $2n\pi - \pi/2$ and $2n\pi + \pi/2$, where n is any integer. Hence,

$$\lim_{b\to\infty} \int_0^b \cos x\,dx$$

does not exist. We might say that this integral "diverges by oscillation." ■

Exercises 8.7

Evaluate the integrals in Exercises 1–10.

1. $\displaystyle\int_0^\infty \frac{dx}{x^2+1}$

2. $\displaystyle\int_0^1 \frac{dx}{\sqrt{x}}$

3. $\displaystyle\int_{-1}^1 \frac{dx}{x^{2/3}}$

4. $\displaystyle\int_1^\infty \frac{dx}{x^{1.001}}$

5. $\displaystyle\int_0^4 \frac{dx}{\sqrt{4-x}}$

6. $\displaystyle\int_0^1 \frac{dx}{\sqrt{1-x^2}}$

7. $\displaystyle\int_0^1 \frac{dx}{x^{0.999}}$

8. $\displaystyle\int_{-\infty}^2 \frac{dx}{4-x}$

9. $\displaystyle\int_2^\infty \frac{2}{x^2-x}\,dx$

10. $\displaystyle\int_0^\infty \frac{dx}{(1+x)\sqrt{x}}$

In Exercises 11–40, determine whether the integrals converge or diverge. (In some cases, you may not need to evaluate the integral to decide. Name any tests you use.)

11. $\displaystyle\int_1^\infty \frac{dx}{\sqrt{x}}$

12. $\displaystyle\int_1^\infty \frac{dx}{x^3}$

13. $\displaystyle\int_1^\infty \frac{dx}{x^3+1}$

14. $\displaystyle\int_0^\infty \frac{dx}{x^3}$

15. $\displaystyle\int_0^\infty \frac{dx}{x^{3/2}+1}$

16. $\displaystyle\int_0^\infty \frac{dx}{1+e^x}$

17. $\displaystyle\int_0^{\pi/2} \tan x\,dx$

18. $\displaystyle\int_{-1}^1 \frac{dx}{x^2}$

19. $\displaystyle\int_{-1}^1 \frac{dx}{x^{2/5}}$

20. $\displaystyle\int_0^\infty \frac{dx}{\sqrt{x}}$

21. $\displaystyle\int_2^{\infty} \frac{dx}{\sqrt{x-1}}$

22. $\displaystyle\int_1^{\infty} \frac{5}{x}\,dx$

23. $\displaystyle\int_0^{2} \frac{dx}{1-x^2}$

24. $\displaystyle\int_2^{\infty} \frac{dx}{(x+1)^2}$

25. $\displaystyle\int_0^{\infty} \frac{dx}{\sqrt{x^6+1}}$

26. $\displaystyle\int_{-1}^{1} \frac{dx}{\sqrt[3]{x}}$

27. $\displaystyle\int_0^{\infty} x^2 e^{-x}\,dx$

28. $\displaystyle\int_1^{\infty} \frac{\sqrt{x+1}}{x^2}\,dx$

29. $\displaystyle\int_{\pi}^{\infty} \frac{2+\cos x}{x}\,dx$

30. $\displaystyle\int_1^{\infty} \frac{\ln x}{x}\,dx$

31. $\displaystyle\int_6^{\infty} \frac{1}{\sqrt{x+5}}\,dx$

32. $\displaystyle\int_1^{\infty} \frac{dx}{\sqrt{2x+10}}$

33. $\displaystyle\int_2^{\infty} \frac{2}{x^2-1}\,dx$

34. $\displaystyle\int_1^{\infty} \frac{1}{e^{\ln x}}\,dx$

35. $\displaystyle\int_2^{\infty} \frac{1}{\ln x}\,dx$

36. $\displaystyle\int_1^{\infty} \frac{1}{\sqrt{e^x-x}}\,dx$

37. $\displaystyle\int_1^{\infty} \frac{1}{e^x-2^x}\,dx$

38. $\displaystyle\int_2^{\infty} \frac{1}{x^3-5}\,dx$

39. $\displaystyle\int_0^{\infty} \frac{dx}{\sqrt{x+x^4}}$

(*Hint:* Compare the integral with $\int dx/\sqrt{x}$ for x near zero and with $\int dx/x^2$ for large x.)

40. $\displaystyle\int_0^{\infty} e^{-x}\cos x\,dx$

41. *Estimating the value of a convergent improper integral whose domain is infinite.* Show that

$$\int_3^{\infty} e^{-3x}\,dx = \frac{1}{3}e^{-9} < 0.000042,$$

and hence that $\int_3^{\infty} e^{-x^2}\,dx < 0.000042$. Therefore, $\int_0^{\infty} e^{-x^2}\,dx$ can be replaced by $\int_0^3 e^{-x^2}\,dx$ without introducing more than this much error. Evaluate this last integral by Simpson's Rule with $n = 6$. (This illustrates one method by which a convergent improper integral may be approximated numerically.)

42. *The infinite paint can or Gabriel's horn.* As Example 5 shows, the integral $\int_1^{\infty} (dx/x)$ diverges. This means that the integral

$$\int_1^{\infty} 2\pi\frac{1}{x}\sqrt{1+\frac{1}{x^4}}\,dx,$$

which measures the *surface area* of the solid of revolution traced out by revolving the curve $y = 1/x$, $1 \le x$, about the x-axis, diverges also. For, by comparing the two integrals, we see that, for every finite value $b > 1$,

$$\int_1^{b} 2\pi\frac{1}{x}\sqrt{1+\frac{1}{x^4}}\,dx > \int_1^{b} \frac{1}{x}\,dx.$$

However, the integral

$$\int_1^{\infty} \pi\left(\frac{1}{x}\right)^2 dx$$

for the *volume* of the solid converges. Calculate it. This solid of revolution is sometimes described as a can that does not hold enough paint to cover its outside surface (Fig. 8.28).

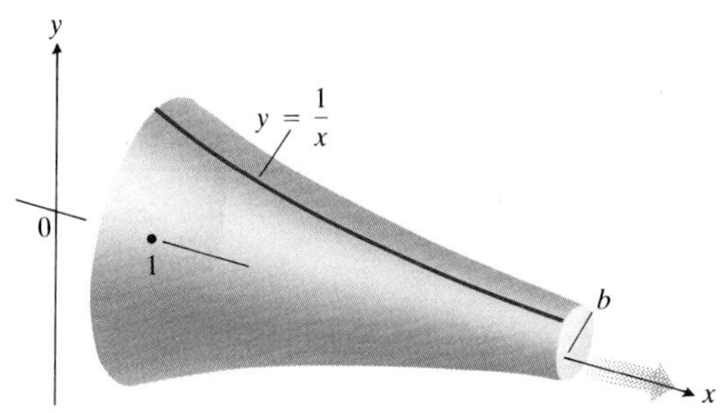

8.28 Extending this solid to the right indefinitely produces an "infinite" solid with a finite volume but an infinite surface area. To learn why, read Exercise 42.

43. Show that

$$\int_1^{\infty} \frac{dx}{x^p} = \frac{1}{p-1} \quad \text{when} \quad p > 1,$$

but that the integral is infinite when $p < 1$. Example 5 shows what happens when $p = 1$.

44. Find the values of p for which each integral converges:

a) $\displaystyle\int_1^{2} \frac{dx}{x(\ln x)^p}$, b) $\displaystyle\int_2^{\infty} \frac{dx}{x(\ln x)^p}$.

Exercises 45–48 are about the region in the first quadrant between the curve $y = e^{-x}$ and the x-axis.

45. Find the area of the region.

46. Find the centroid of the region.

47. Find the volume swept out by revolving the region about the y-axis.

48. Find the volume swept out by revolving the region about the x-axis.

49. Find the area of the region that lies between the curves $y = \sec x$ and $y = \tan x$ for $0 \le x \le \pi/2$.

50. Show that the area of the region between the curve $y = 1/(1+x^2)$ and the entire x-axis is the same as the area of the unit disk, $x^2 + y^2 \le 1$.

Review Questions

1. What are the general methods for finding indefinite integrals?
2. What substitution(s) would you consider if the integrand contained the following terms?

   a) $\sqrt{x^2+9}$ b) $\sqrt{x^2-9}$

   c) $\sqrt{9-x^2}$ d) $\sin^3 x \cos^2 x$

   e) $\sin^2 x \cos^2 x$
3. What method(s) would you try if the integrand contained the following terms?

   a) $\sin^{-1} x$ b) $\ln x$

   c) $\sqrt{1+2x-x^2}$ d) $x \sin x$

   e) $\dfrac{2x+3}{x^2-5x+6}$ f) $\sin 5x \cos 3x$

   g) $\dfrac{1-\sqrt{x}}{1+\sqrt[4]{x}}$ h) $x\sqrt{2x+3}$
4. What is a reduction formula? How are reduction formulas used? Illustrate by using the formula

$$\int \tan^n ax = \frac{\tan^{n-1} ax}{a(n-1)} - \int \tan^{n-2} ax\, dx$$

to evaluate

$$\int_0^{\pi/2} \tan^3\left(\frac{x}{2}\right) dx.$$

5. Discuss two types of improper integral. Define convergence and divergence of each type. How do you test for convergence and divergence of improper integrals? Give examples.

Practice Exercises

Evaluate the integrals in Exercises 1–42.

1. $\displaystyle\int_0^{\pi/2} \frac{\cos x\, dx}{\sqrt{1+\sin x}}$
2. $\displaystyle\int_{\pi^2/16}^{\pi^2/9} 3\ \pi/4 \frac{\tan\sqrt{x}}{2\sqrt{x}}\, dx$
3. $\displaystyle\int_{-1}^{1} \frac{2y}{y^4+1}\, dy$
4. $\displaystyle\int \sec x \tan x\, e^{\sec x}\, dx$
5. $\displaystyle\int_0^{\sqrt{2}/2} \frac{\sin^{-1} x}{\sqrt{1-x^2}}\, dx$
6. $\displaystyle\int_1^{e} \frac{dx}{x(2+\ln x)}$
7. $\displaystyle\int_{\pi/4}^{\pi/3} \frac{dx}{2\sin x \cos x}$
8. $\displaystyle\int_0^{\pi/6} \frac{2\, dt}{\cos^2 t - \sin^2 t}$
9. $\displaystyle\int \frac{x+4}{x^2+1}\, dx$
10. $\displaystyle\int \frac{x+2}{\sqrt{1-x^2}}\, dx$
11. $\displaystyle\int x^2 \ln x\, dx$
12. $\displaystyle\int_0^1 \ln(x+1)\, dx$
13. $\displaystyle\int x^5 \sin x\, dx$
14. $\displaystyle\int \frac{2\tan^{-1} x}{x^2}\, dx$
15. $\displaystyle\int e^x \cos 2x\, dx$
16. $\displaystyle\int e^{-x} \sin x\, dx$
17. $\displaystyle\int \sin^3 y\, dy$
18. $\displaystyle\int \sin^3 y \cos^2 y\, dy$
19. $\displaystyle\int \sin^4 x \cos^2 x\, dx$
20. $\displaystyle\int \sin^3 x \cos^3 x\, dx$
21. $\displaystyle\int_0^{\pi} \sqrt{\frac{1+\cos 2x}{2}}\, dx$
22. $\displaystyle\int_{\pi/4}^{3\pi/4} \sqrt{\cot^2 t + 1}\, dt$
23. $\displaystyle\int_0^{\pi/3} \tan^3 t\, dt$
24. $\displaystyle\int_{-\pi/4}^{\pi/4} 6\sec^4 t\, dt$
25. $\displaystyle\int_0^3 \frac{dz}{(16+z^2)^{3/2}}$
26. $\displaystyle\int_0^{\sqrt{3}} \frac{z^3+z}{\sqrt{1+z^2}}\, dz$
27. $\displaystyle\int \frac{dx}{x^2\sqrt{1-x^2}}$
28. $\displaystyle\int \frac{x^2\, dx}{\sqrt{1-x^2}}$

29. $\displaystyle\int_{5/4}^{5/3} \frac{12\,dx}{(x^2-1)^{3/2}}$

30. $\displaystyle\int_{5}^{6} \frac{dx}{\sqrt{x^2-9}}$

31. $\displaystyle\int_{1/3}^{1} \frac{3\,dx}{9x^2-6x+5}$

32. $\displaystyle\int_{-1}^{-1/2} \frac{dx}{\sqrt{-2x-x^2}}$

33. $\displaystyle\int_{0}^{1} \frac{dx}{(x+1)\sqrt{x^2+2x}}$

34. $\displaystyle\int_{-2}^{-1} \frac{2\,dx}{x^2+4x+5}$

35. $\displaystyle\int_{2}^{6} \frac{x^3+x^2}{x^2+x-2}\,dx$

36. $\displaystyle\int_{2}^{3} \frac{x^3+1}{x^3-x}\,dx$

37. $\displaystyle\int \frac{x\,dx}{(x-1)^2}$

38. $\displaystyle\int \frac{8\,dx}{x^3(x+2)}$

39. $\displaystyle\int \frac{4\,dx}{x^3+4x}$

40. $\displaystyle\int_{0}^{1} \frac{x\,dx}{x^4-16}$

41. $\displaystyle\int_{3}^{\infty} \frac{2\,dx}{x^2-2x}$

42. $\displaystyle\int_{1}^{\infty} \frac{\ln x\,dx}{x^2}$

43. Find the moment of inertia about the x-axis of a thin plate of constant density $\delta = 2$ covering the region enclosed by the lines $x = 1$ and $y = 1$ and the curve $y = \ln x$ in the first quadrant (Fig. 8.29).

44. Find the area of the region in the first quadrant bounded by the x-axis, the line $x = 2$, and the curve $y = \ln x$.

45. Find the volume of the solid generated by revolving about the y-axis the region bounded by the x-axis and the curve $y = 3x\sqrt{1-x}$, $0 \le x \le 1$ (Fig. 8.30).

46. Find the length of the segment of the parabola $y = x^2$ that extends from $x = 0$ to $x = \sqrt{3}/2$.

47. *Autocatalytic reactions.* The equation that describes the autocatalytic reaction in Section 4.5, Exercise 39, can be written as $dx/dt = kx(a - x)$. Read Exercise 39 for background (there is no need to do the exercise). Then solve the differential equation above to find x as a function of t. Assume that $x = x_0$ when $t = 0$.

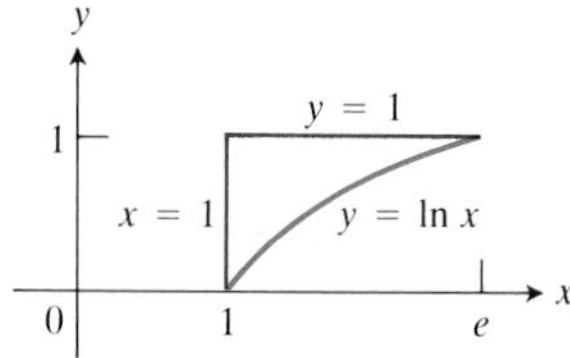

8.29 The region in Exercise 43.

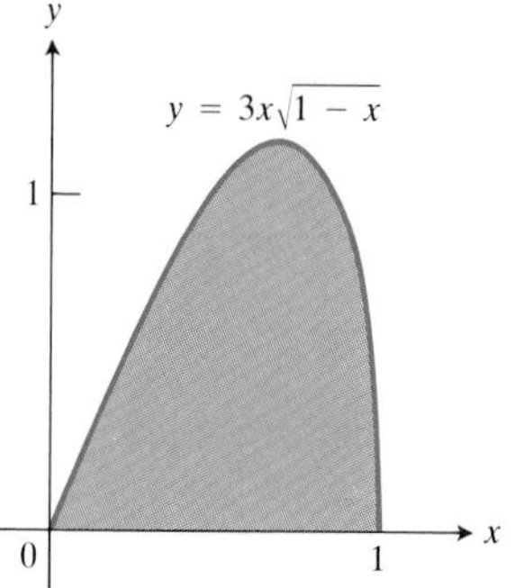

8.30 The region in Exercise 45.

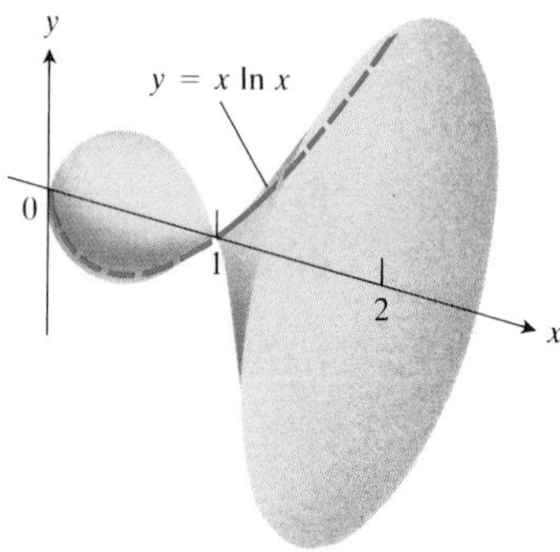

8.31 The solid of revolution in Exercise 50.

48. Find the centroid of the region bounded by the curves $y = \pm(1-x^2)^{-1/2}$ and the lines $x = 0$ and $x = 1$.

49. The infinite region bounded by the coordinate axes and the curve $y = -\ln x$ in the first quadrant is revolved about the x-axis to generate a solid. Find the volume of the solid.

50. The region between the curve $y = x \ln x$ and the x-axis from $x = 0$ to $x = 2$ is revolved about the x-axis to generate a solid (Fig. 8.31). Find the volume of the solid.

9 Infinite Series

Dick Morton

The sequence 1, 1, 2, 3, 5, 8, . . . , in which each term is the sum of the preceding two, occurs frequently in nature. Leonardo Fibonacci (1170–1240) used it to describe the sizes of successive generations in an ideal rabbit population. The numbers of clockwise spirals and counterclockwise spirals in the seeds of a sunflower, though different, are always a pair of adjacent Fibonacci numbers.

OVERVIEW In this chapter, we study infinite polynomials called power series and, as a product of our study, develop one of the most remarkable formulas in all of mathematics. The formula does two things for us. It shows how to calculate the value of an infinitely differentiable function like e^x at any point, just from its value and the values of its derivatives at the origin. And, as if that were not enough, the formula gives us polynomial approximations of differentiable functions of any order we want, along with their error formulas, all in a single equation. Power series have many additional uses. They provide an efficient way to evaluate nonelementary integrals and they solve differential equations that give insight into heat flow, vibration, chemical diffusion, and signal transmission. What you will learn here sets the stage for the roles played by series of functions of all kinds in science and mathematics.

9.1 Limits of Sequences of Numbers

© Chuck Fishman/Woodfin Camp and Assoc.

The ratios of the successive terms of the Fibonacci sequence converge to $(1 + \sqrt{5})/2$, the ratio the ancient Greeks used in proportioning their beautiful temples and public buildings (they called it the "golden" ratio). Here we see the ratio, about 1.6 to 1, as the ratio of width to height in the rectangular front of the Parthenon.

This section describes what it means for an infinite sequence of numbers to have a limit and shows how to find the limits of many of the sequences that arise in mathematics and applied fields.

Informally, a sequence is an ordered collection of things, but in this chapter the things will usually be numbers. We have seen sequences before, such as sequences $x_0, x_1, \cdots, x_n, \cdots$ of numerical approximations generated by Newton's method and the sequence $A_3, A_4, \cdots, A_n, \cdots$ of areas of n-sided regular polygons used to define the area of a circle. These sequences have limits, but many equally important sequences do not. The sequence $1, 2, 3, \cdots, n, \cdots$ of positive integers has no limit, nor does the sequence $2, 3, 5, 7, 11, 13, \cdots$ of prime numbers. We need to know when sequences do and do not have limits, and how to find the limits when they exist. As with functions, we usually find out with theorems based on a formal definition. You will see a close parallel between what we do here and what we did with limits of functions in Chapter 2.

Definitions and Notation

Our starting point is a formal definition of sequence.

DEFINITION

An **infinite sequence** (or **sequence**) of numbers is a function whose domain is the set of integers greater than or equal to some integer n_0.

Usually n_0 is 1 and the domain is the set of all positive integers. But sometimes we want to start our sequences elsewhere. We might take $n_0 = 0$, for instance, when we begin Newton's method, or take $n_0 = 3$ when we work with n-sided polygons.

Sequences are defined the way other functions are, some typical rules being

$$a(n) = n - 1, \quad a(n) = 1 - \frac{1}{n}, \quad a(n) = \frac{\ln n}{n^2}. \tag{1}$$

To indicate that the domains are sets of integers, we use a letter like n from the middle of the alphabet for the independent variable, instead of the x, y, z, and t used so widely in other contexts. The formulas in the defining rules, however, like the ones above, are often valid for domains much larger than the set of positive integers. This can be an advantage, as we shall see.

The number $a(n)$ is the **nth term** of the sequence, or the **term with index n.** For example, if $a(n) = (n - 1)/n$, then the terms are

First term	Second term	Third term		nth term	
$a(1) = 0$	$a(2) = \frac{1}{2}$,	$a(3) = \frac{2}{3}$,	$\dots$,	$a(n) = \frac{n-1}{n}$.	(2)

When we use the subscript notation a_n for $a(n)$, the sequence in (2) becomes

$$a_1 = 0, \quad a_2 = \frac{1}{2}, \quad a_3 = \frac{2}{3}, \quad \dots, \quad a_n = \frac{n-1}{n}. \tag{3}$$

To describe sequences, we often write the first few terms as well as a formula for the nth term.

EXAMPLE 1

We write	For the sequence whose defining rule is
$0, 1, 2, \dots, n - 1, \dots$	$a_n = n - 1$
$1, \frac{1}{2}, \frac{1}{3}, \dots, \frac{1}{n}, \dots$	$a_n = \frac{1}{n}$
$1, -\frac{1}{2}, \frac{1}{3}, -\frac{1}{4}, \dots, (-1)^{n+1}\frac{1}{n}, \dots$	$a_n = (-1)^{n+1}\frac{1}{n}$
$0, \frac{1}{2}, \frac{2}{3}, \frac{3}{4}, \dots, \frac{n-1}{n}, \dots$	$a_n = \frac{n-1}{n}$
$0, -\frac{1}{2}, \frac{2}{3}, -\frac{3}{4}, \dots, (-1)^{n+1}\left(\frac{n-1}{n}\right), \dots$	$a_n = (-1)^{n+1}\left(\frac{n-1}{n}\right)$
$3, 3, 3, \dots, 3, \dots$	$a_n = 3$ ■

Notation We refer to the sequence whose nth term is a_n with the notation $\{a_n\}$ ("the sequence a sub n"). The second sequence in Example 1 is $\{1/n\}$ ("the sequence 1 over n"); the last sequence is $\{3\}$ ("the sequence 3").

Convergence and Divergence

As Fig. 9.1 shows, the sequences of Example 1 exhibit different kinds of behavior. The sequences $\{1/n\}$, $\{(-1)^{n+1}(1/n)\}$, and $\{(n - 1)/n\}$ each seem to approach single limiting values as n increases, and $\{3\}$ is already at a limiting value from the very first. On the other hand, terms of $\{(-1)^{n+1}(n - 1)/n\}$ seem to accumulate near two different values, -1 and 1, while the terms of $\{n - 1\}$ become increasingly large and do not accumulate anywhere.

To distinguish sequences that approach a unique limiting value L, as n increases, from those that do not, we say that those former sequences *converge*, according to the following definition.

9.1 The sequences of Example 1 are graphed here in two different ways: by plotting the numbers a_n on a horizontal axis and by plotting the points (n, a_n) in the coordinate plane.

The terms $a_n = n - 1$ eventually surpass every integer, so the sequence $\{a_n\}$ diverges, . . .

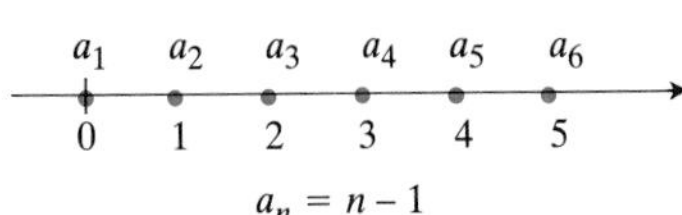

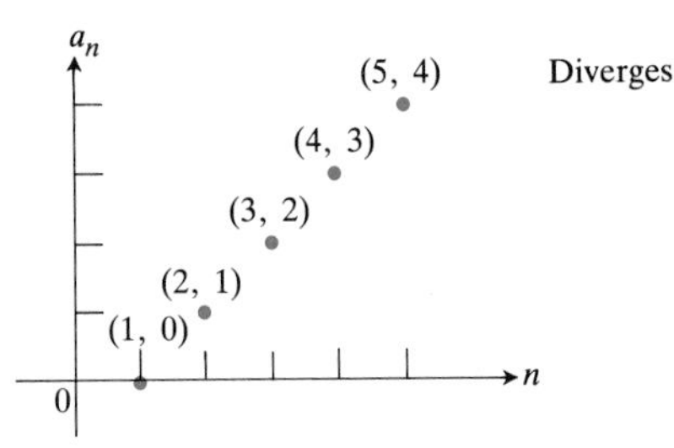

. . . however, the terms $a_n = 1/n$ decrease steadily and get arbitrarily close to 0 as n increases, so the sequence $\{a_n\}$ converges to 0.

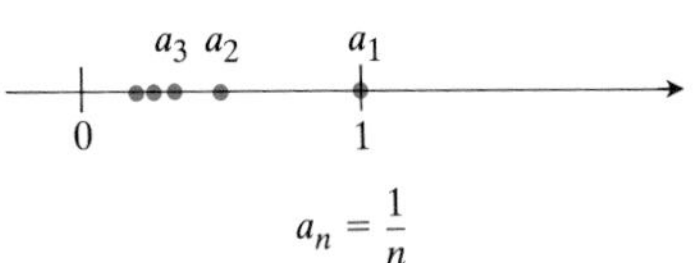

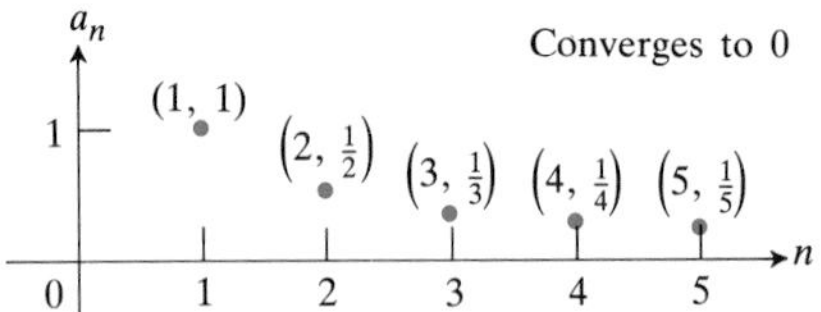

The terms $a_n = (-1)^{n+1}(1/n)$ alternate in sign but still converge to 0.

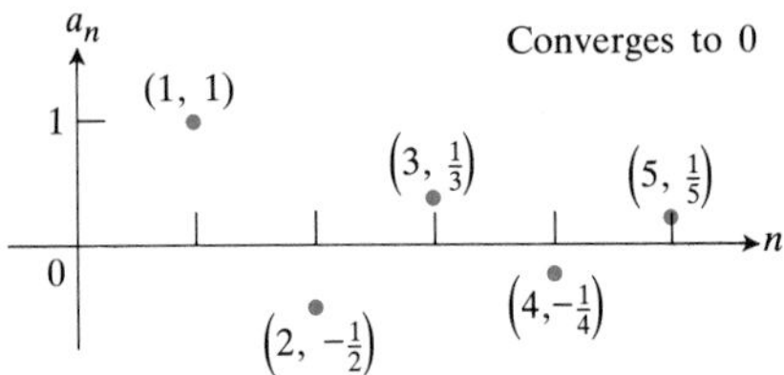

The terms $a_n = (n - 1)/n$ approach 1 steadily and get arbitrarily close as n increases, so the sequence $\{a_n\}$ converges to 1.

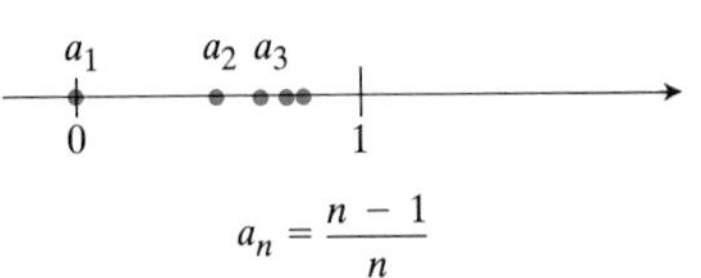

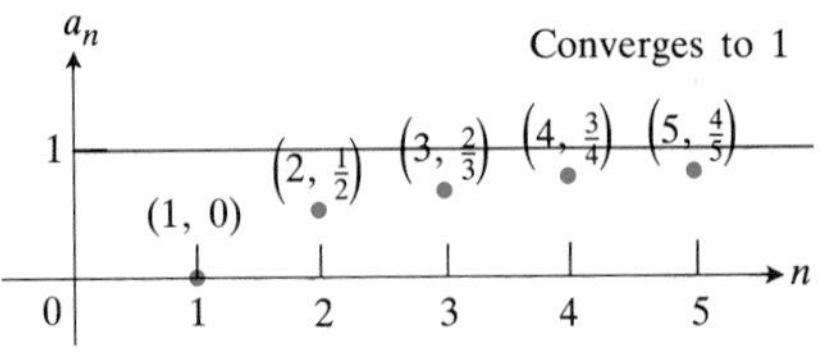

The terms $a_n = (-1)^{n+1}(n - 1)/n$ alternate in sign. The positive terms approach 1. But the negative terms approach -1 as n increases, so the sequence $\{a_n\}$ diverges.

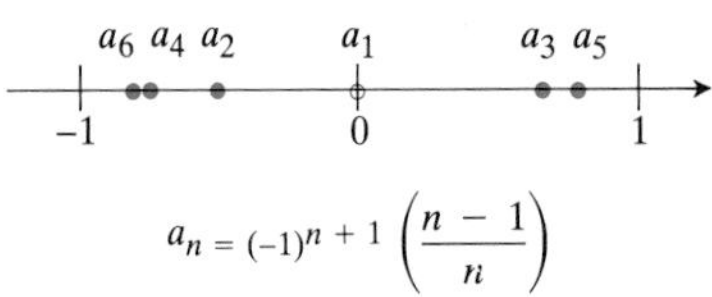

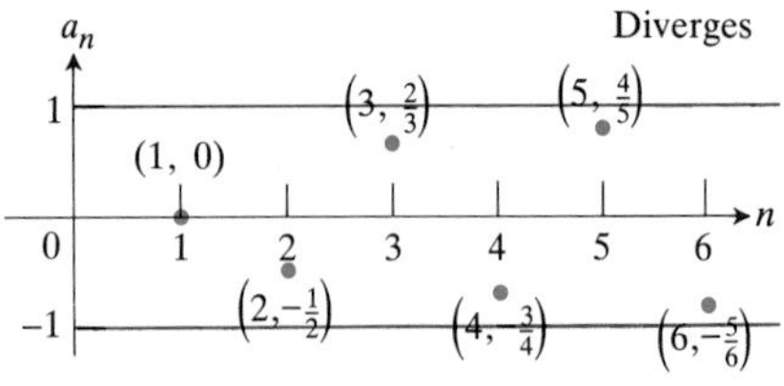

The terms in the sequence of constants $a_n = 3$ have the same value regardless of n; so we say the sequence $\{a_n\}$ converges to 3.

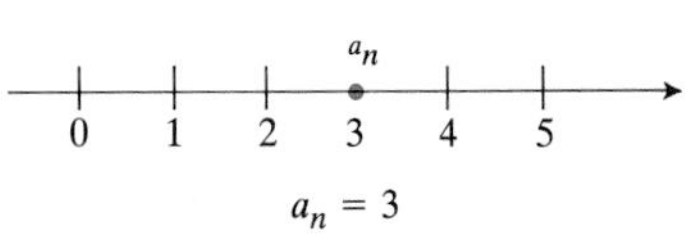

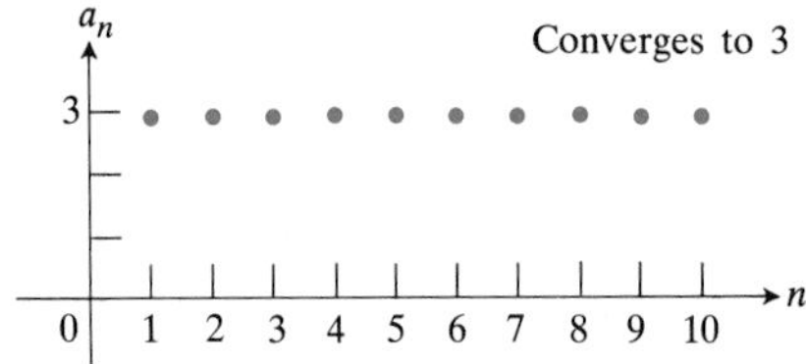

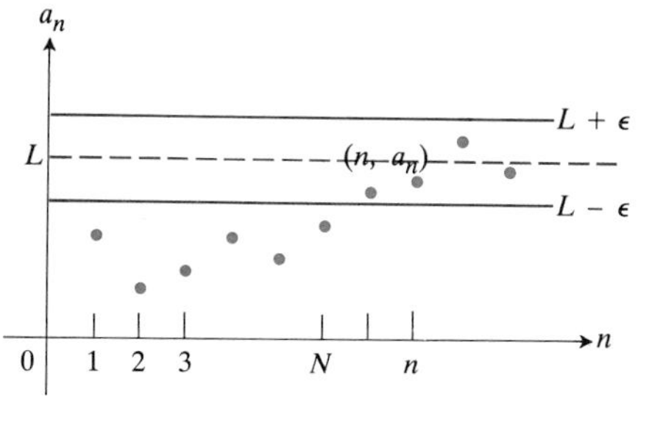

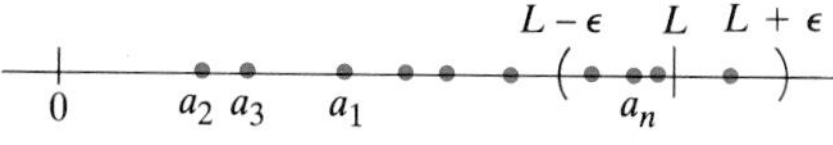

9.2 $a_n \to L$ if L is a horizontal asymptote of the sequence of points $\{(n, a_n)\}$. In this figure, all the a_n's after a_N lie within ϵ of L.

DEFINITIONS

> The sequence $\{a_n\}$ **converges** to the number L if to every positive number ϵ there corresponds an integer N such that for all n
>
> $$n > N \Rightarrow |a_n - L| < \epsilon. \tag{4}$$
>
> If no such limit exists, we say that $\{a_n\}$ **diverges.**
>
> If $\{a_n\}$ converges to L, we write $\lim_{n\to\infty} a_n = L$ or simply $a_n \to L$, and call L the **limit** of the sequence. (See Fig. 9.2.)

EXAMPLE 2 Show that $\{1/n\}$ converges to 0.

Solution We set $a_n = 1/n$ and $L = 0$ in the definition of convergence. To show that $1/n \to 0$, we must show that for any $\epsilon > 0$ there exists an integer N such that for all n

$$n > N \quad \Rightarrow \quad \left|\frac{1}{n} - 0\right| < \epsilon. \tag{5}$$

This implication will hold for all n for which

$$\frac{1}{n} < \epsilon \quad \text{or, equivalently,} \quad n > \frac{1}{\epsilon}.$$

Pick an integer N greater than $1/\epsilon$. Then any n greater than N will automatically be greater than $1/\epsilon$ and the implication in (5) will hold. ■

EXAMPLE 3 Show that if k is any number then $\{k\}$ converges to k.

Solution We set $a_n = k$ and $L = k$ in the definition of convergence. To show that $a_n \to k$, we must show that for any $\epsilon > 0$ there exists an integer N such that for all n

$$n > N \quad \Rightarrow \quad |k - k| < \epsilon.$$

This implication holds for any integer N because $|k - k| = 0$ is less than every positive ϵ for all n. ■

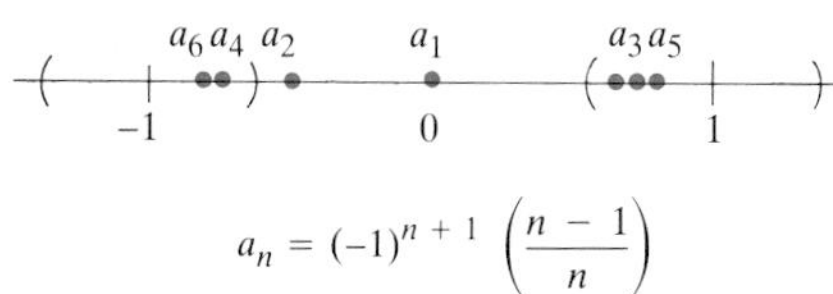

Neither the ϵ-interval about 1 nor the ϵ-interval about -1 contains a complete tail of the sequence.

EXAMPLE 4 Show that $\{(-1)^{n+1}(n - 1)/n\}$ diverges.

Solution Take a positive ϵ smaller than 1 so that the bands shown in Fig. 9.3 about the lines $y = 1$ and $y = -1$ do not overlap. Any $\epsilon < 1$ will do. Convergence to 1 would require every point of the graph beyond a certain index N to lie inside the upper band, but this will never happen. As soon as a point (n, a_n) lies within the upper band, every alternate point starting with $(n + 1, a_{n+1})$ will lie within the lower band. Hence the sequence cannot converge to 1. Likewise, it cannot converge to -1. On the other hand, because the terms of the sequence get alternately closer to 1 and -1, they never accumulate near any other value. Therefore, the sequence diverges. ■

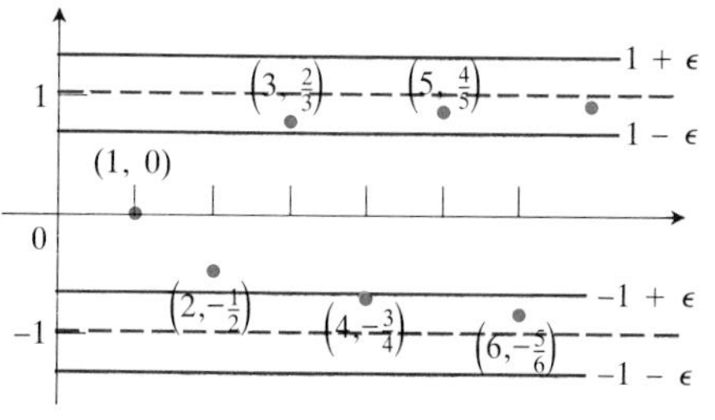

9.3 The sequence $\{(-1)^{n+1}(n - 1)/n\}$ diverges.

A **tail** of a sequence $\{a_n\}$ is the collection of all terms whose indices are greater than some integer N. In other words, a tail is one of the sets $\{a_n \mid n > N\}$. Another way to say that $a_n \to L$ is to say that every ϵ-interval about L contains a tail of the sequence. The convergence or divergence of a sequence has nothing to do with how a sequence starts out. It depends only on how the tails behave.

The behavior of $\{(-1)^{n+1}(n - 1)/n\}$ is qualitatively different from that of $\{n - 1\}$, which diverges because it outgrows every real number L. To describe the behavior of $\{n - 1\}$ we write

$$\lim_{n \to \infty} (n - 1) = \infty.$$

In speaking of infinity as a limit of a sequence $\{a_n\}$, we do not mean that the difference between a_n and infinity becomes small as n increases. We mean that a_n becomes numerically large as n increases.

Limits are Unique

A sequence cannot converge to two different limits.

> If a sequence $\{a_n\}$ converges, then its limit is unique.

The argument goes like this: If $\{a_n\}$ converged to two different limits L_1 and L_2 we could take ϵ to be the positive number $|L_1 - L_2|/2$. Because $a_n \to L_1$, there would exist an integer N_1 such that

$$n > N_1 \quad \Rightarrow \quad |a_n - L_1| < \epsilon.$$

There would also exist an integer N_2 such that

$$n > N_2 \quad \Rightarrow \quad |a_n - L_2| < \epsilon.$$

For n greater than both N_1 and N_2, we would then have

$$\begin{aligned} |L_1 - L_2| &= |L_1 - a_n + a_n - L_2| \leq |L_1 - a_n| + |a_n - L_2| \qquad \begin{pmatrix}\text{Triangle} \\ \text{inequality}\end{pmatrix} \\ &< \epsilon + \epsilon = 2(|L_1 - L_2|/2) = |L_1 - L_2|. \end{aligned}$$

But this is absurd; no number is less than itself. Hence, if a sequence converges its limit is unique.

Recursive Definitions

So far, we have calculated each a_n directly from the value of n. But sequences are often defined **recursively** by giving

1. The value of the first term (or first few terms).
2. A rule, called a **recursion formula,** for calculating any later term from terms that precede it.

The initial value $a_1 = 1$ and the recursion formula $a_{n+1} = a_n + 1$ defines the sequence of positive integers recursively. The terms of the factorial sequence $1, 2, 6, 24, \cdots, n!, \cdots$ are usually calculated recursively with the

Factorial notation

The notation $n!$ ("n factorial") means the product $1 \cdot 2 \cdot 3 \cdot \cdots \cdot n$ of the integers from 1 to n. Notice that $(n + 1)! = (n + 1) \cdot n!$. Thus, $4! = 1 \cdot 2 \cdot 3 \cdot 4 = 24$ and $5! = 1 \cdot 2 \cdot 3 \cdot 4 \cdot 5 = 5 \cdot 4! = 120$. We also define $0!$ to be 1. Factorials grow even faster than exponentials, as the following table suggests.

n	e^n (rounded)	$n!$
1	3	1
5	148	120
10	22,026	3,628,800
20	4.8×10^8	2.4×10^{18}

formula $(n + 1)! = (n + 1)n!$. The initial value $x_0 = 1$ and the recursion formula $x_{n+1} = x_n - (\sin x_n - x_n^2)/(\cos x_n - 2x_n)$ define a sequence that converges to the solution of the equation $\sin x = x^2$ (Section 3.8, Example 3). Recursion formulas arise regularly in computer programming and we shall encounter them again when we preview numerical methods for solving differential equations (Section 16.7).

Subsequences

If the terms of one sequence occur in their given order among the terms of a second sequence, we call the first sequence a **subsequence** of the second.

EXAMPLE 5 *Subsequences of the sequence of positive integers.*

a) The sequence $2, 4, 6, \ldots, 2n, \ldots$ of even integers.

b) The sequence $1, 3, 5, \ldots, 2n + 1, \ldots$ of odd integers.

c) The sequence $2, 3, 5, 7, 11, \ldots$ of primes. ■

Subsequences are important for two reasons. First, if a sequence $\{a_n\}$ converges to a limit L, then all subsequences also converge to L. If we know a sequence converges, it may be quicker for us to find or estimate its limit by choosing a rapidly convergent subsequence. The second reason is related to the first: If any subsequence of the original sequence diverges, or if two subsequences have different limits, then the original sequence diverges. The sequence $\{(-1)^{n-1}\}$ diverges because the subsequence $1, 1, 1, \ldots$ of odd numbered terms converges to 1 while the subsequence $-1, -1, -1, \ldots$ of even numbered terms converges to a different limit.

Useful Theorems

The study of limits would be a cumbersome business if we had to answer every question about convergence by applying the definition directly. Fortunately, three theorems will make this process largely unnecessary from now on. The first two are practically the same as Theorems 1 and 2 in Chapter 2.

THEOREM 1

If $\lim_{n\to\infty} a_n = A$ and $\lim_{n\to\infty} b_n = B$, then

1. *Sum Rule:* $\lim \{a_n + b_n\} = A + B$
2. *Difference Rule:* $\lim \{a_n - b_n\} = A - B$
3. *Product Rule:* $\lim \{a_n \cdot b_n\} = A \cdot B$
4. *Constant Multiple Rule:* $\lim \{k \cdot b_n\} = k \cdot B$ (Any number k)
5. *Quotient Rule:* $\lim \dfrac{a_n}{b_n} = \dfrac{A}{B}$ if $B \neq 0$.

The limits are all taken as $n\to\infty$, and A and B are real numbers.

By combining Theorem 1 with Examples 2 and 3, we can proceed immediately to

$$\lim_{n\to\infty}\left(-\frac{1}{n}\right) = -1\cdot\lim_{n\to\infty}\frac{1}{n} = -1\cdot 0 = 0,$$

$$\lim_{n\to\infty}\left(\frac{n-1}{n}\right) = \lim_{n\to\infty}\left(1-\frac{1}{n}\right) = \lim_{n\to\infty} 1 - \lim_{n\to\infty}\frac{1}{n} = 1-0 = 1,$$

$$\lim_{n\to\infty}\frac{5}{n^2} = 5\cdot\lim_{n\to\infty}\frac{1}{n}\cdot\lim_{n\to\infty}\frac{1}{n} = 5\cdot 0\cdot 0 = 0,$$

$$\lim_{n\to\infty}\frac{4-7n^6}{n^6+3} = \lim_{n\to\infty}\frac{(4/n^6)-7}{1+(3/n^6)} = \frac{0-7}{1+0} = -7.$$

One general consequence of Theorem 1 is that every nonzero multiple of a divergent sequence $\{a_n\}$ diverges. For, suppose $\{ca_n\}$ were to converge for some number $c \neq 0$. Then $(1/c)\{ca_n\} = \{a_n\}$ would converge by the Constant Multiple Rule—but it doesn't.

The next theorem is the sequence version of the Sandwich Theorem of Section 2.2.

THEOREM 2

The Sandwich Theorem for Sequences

If $a_n \le b_n \le c_n$ for all n beyond some index N, and if $\lim a_n = \lim c_n = L$, then $\lim b_n = L$ also.

An immediate consequence of Theorem 2 is that, if $|b_n| \le c_n$ and $c_n \to 0$, then $b_n \to 0$ because $-c_n \le b_n \le c_n$. We use this fact in the next example.

EXAMPLE 6 Since $1/n \to 0$, we know that

a) $\frac{\cos n}{n} \to 0$ because $0 \le \left|\frac{\cos n}{n}\right| = \frac{|\cos n|}{n} \le \frac{1}{n}$;

b) $\frac{1}{2^n} \to 0$ because $0 \le \frac{1}{2^n} \le \frac{1}{n}$;

c) $(-1)^n\frac{1}{n} \to 0$ because $0 \le \left|(-1)^n\frac{1}{n}\right| \le \frac{1}{n}$. ■

The application of Theorems 1 and 2 is broadened by a theorem stating that applying a continuous function to a convergent sequence produces a convergent sequence. We state the theorem without proof.

THEOREM 3

If $a_n \to L$ and if f is a function that is continuous at L and defined at all the a_n's, then $f(a_n) \to f(L)$.

n	$2^{1/n}$
1	2
2	1.414213562
4	1.189207115
10	1.071773463
100	1.006955555
1000	1.000693387
10000	1.000069317

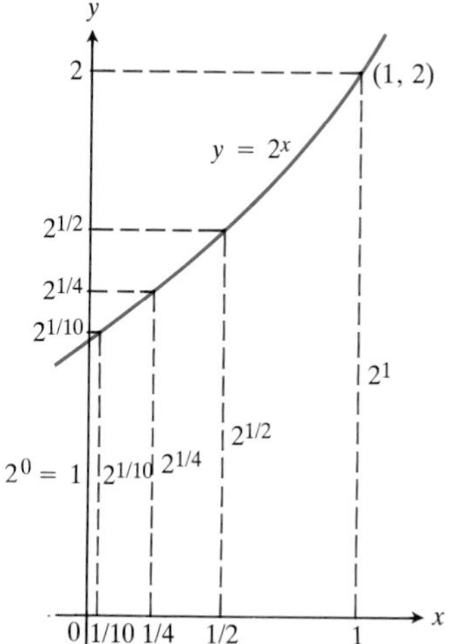

9.4 Since 2^x is continuous at $x = 0$ as $n \to \infty$, $x = 1/n \to 0$ and $y = 2^{1/n} \to 2^0 = 1$.

EXAMPLE 7 Show that $\sqrt{(n+1)/n} \to 1$.

Solution We know that $(n+1)/n \to 1$. Taking $f(x) = \sqrt{x}$ and $L = 1$ in Theorem 3 therefore gives $\sqrt{(n+1)/n} \to \sqrt{1} = 1$. ■

EXAMPLE 8 Show that $2^{1/n} \to 1$ (Fig. 9.4).

Solution We know that $1/n \to 0$. Taking $f(x) = 2^x$ and $L = 0$ in Theorem 3 therefore gives $2^{1/n} \to 2^0 = 1$. ■

In Section 1.7, we saw that entering 2 in a calculator and pressing the square root key repeatedly produces a succession of numbers that approach 1. The result of Example 8 now tells us why. The successive square roots form a subsequence $2^{1/2}, 2^{1/4}, 2^{1/8}, 2^{1/16}, \cdots$ of the sequence $\{2^{1/n}\}$. The sequence converges to 1, so the subsequence does too.

The next theorem enables us to use l'Hôpital's rule to find the limits of some sequences. We state and prove the theorem first, then show how to apply it.

THEOREM 4

Suppose that $f(x)$ is a function defined for all $x \ge n_0$ and $\{a_n\}$ is a sequence such that $a_n = f(n)$ when $n \ge n_0$. If

$$\lim_{x\to\infty} f(x) = L, \qquad \text{then} \qquad \lim_{n\to\infty} a_n = L.$$

PROOF Suppose that $\lim_{x\to\infty} f(x) = L$. Then for each positive number ϵ there is a number M such that

$$x > M \quad \Rightarrow \quad |f(x) - L| < \epsilon.$$

Let N be an integer greater than M and greater than or equal to n_0. Then

$$n > N \;\Rightarrow\; a_n = f(n) \;\text{ and }\; |a_n - L| = |f(n) - L| < \epsilon. \qquad ■$$

EXAMPLE 9 Show that $\lim_{n\to\infty} (\ln n)/n = 0$.

Solution The function $(\ln x)/x$ is defined for all $x \ge 1$ and agrees with the given sequence at positive integers. Therefore $\lim_{n\to\infty} (\ln n)/n$ will equal $\lim_{x\to\infty} (\ln x)/x$ if the latter exists. A single application of l'Hôpital's rule shows that

$$\lim_{x\to\infty} \frac{\ln x}{x} = \lim_{x\to\infty} \frac{1/x}{1} = \frac{0}{1} = 0.$$

We conclude that $\lim_{n\to\infty} (\ln n)/n = 0$. ■

When we use l'Hôpital's rule to find the limit of a sequence, we often treat n as a continuous real variable and differentiate directly with respect to n. This saves us from having to rewrite the formula for a_n as we did in Example 9.

EXAMPLE 10 Find $\lim_{n\to\infty}(2^n/5n)$.

Solution By l'Hôpital's rule,

$$\lim_{n\to\infty}\frac{2^n}{5n}=\lim_{n\to\infty}\frac{2^n\cdot\ln 2}{5}=\infty.$$

Limits That Arise Frequently

The limits in Table 9.1 are useful and arise frequently. The first limit is from Example 9. The others are derived in Appendix 5.

TABLE 9.1

1. $\lim_{n\to\infty}\frac{\ln n}{n}=0$		2. $\lim_{n\to\infty}\sqrt[n]{n}=1$	
3. $\lim_{n\to\infty}x^{1/n}=1$	$(x>0)$	4. $\lim_{n\to\infty}x^n=0$	$(\lvert x\rvert<1)$
5. $\lim_{n\to\infty}\left(1+\frac{x}{n}\right)^n=e^x$	(Any x)	6. $\lim_{n\to\infty}\frac{x^n}{n!}=0$	(Any x)

In Eqs. (3)–(6), x remains fixed while $n\to\infty$.

EXAMPLE 11 *Limits from Table 9.1.*

1. $\frac{\ln(n^2)}{n}=\frac{2\ln n}{n}\to 2\cdot 0=0$ (Eq. 1)
2. $\sqrt[n]{n^2}=n^{2/n}=(n^{1/n})^2\to(1)^2=1$ (Eq. 2)
3. $\sqrt[n]{3n}=3^{1/n}(n^{1/n})\to 1\cdot 1=1.$ (Eq. (3) with $x=3$, and Eq. (2))
4. $\left(-\frac{1}{2}\right)^n\to 0$ $\left(\text{Eq. (4) with } x=-\frac{1}{2}\right)$
5. $\left(\frac{n-2}{n}\right)^n=\left(1-\frac{2}{n}\right)^n\to e^{-2}$ (Eq. (5) with $x=-2$)
6. $\frac{100^n}{n!}\to 0$ (Eq. (6) with $x=100$)

Exercises 9.1

Each of Exercises 1–4 gives a formula for the nth term a_n of a sequence $\{a_n\}$. Find the values of a_1, a_2, a_3, and a_4.

1. $a_n=\frac{1-n}{n^2}$

2. $a_n=\frac{1}{n!}$

3. $a_n=\frac{(-1)^{n+1}}{2n-1}$

4. $a_n=2+(-1)^n$

Each of Exercises 5–10 gives the first term or two of a sequence and a recursion formula for the remaining terms. Write out the first six terms of each sequence.

5. $x_1=1,\quad x_{n+1}=x_n+(1/2^n)$

6. $x_1=1,\quad x_{n+1}=x_n/(n+1)$

7. $x_1=2,\quad x_{n+1}=x_n/2$

8. $x_1 = -2, \quad x_{n+1} = nx_n/(n+1)$

9. $x_1 = x_2 = 1, \quad x_{n+2} = x_n + x_{n+1}$

10. $x_1 = 1, \quad x_n = x_1 + x_2 + \cdots + x_{n-1}$

Which of the sequences $\{a_n\}$ in Exercises 11–26 converge and which diverge? Find the limit of each convergent sequence.

11. $a_n = 2 + (0.1)^n$

12. $a_n = \dfrac{n + (-1)^n}{n}$

13. $a_n = \dfrac{1 - 2n}{1 + 2n}$

14. $a_n = \dfrac{2n + 1}{1 - 3n}$

15. $a_n = \dfrac{n}{10}$

16. $a_n = 1 + (-1)^n$

17. $a_n = \dfrac{1 - 5n^4}{n^4 + 8n^3}$

18. $a_n = \dfrac{n + 3}{n^2 + 5n + 6}$

19. $a_n = \dfrac{n^2 - 2n + 1}{n - 1}$

20. $a_n = \dfrac{1 - n^3}{70 - 4n^2}$

21. $a_n = 5$

22. $a_n = \dfrac{10^{n+1}}{10^n}$

23. $a_n = 5^n$

24. $a_n = (-1)^n\left(1 - \dfrac{1}{n}\right)$

25. $a_n = \left(\dfrac{n+1}{2n}\right)\left(1 - \dfrac{1}{n}\right)$

26. $a_n = \left(2 - \dfrac{1}{2^n}\right)\left(3 + \dfrac{1}{2^n}\right)$

Find the limits of the sequences $\{a_n\}$ whose nth terms appear in Exercises 27–30.

27. $a_n = \dfrac{(-1)^{n+1}}{2n - 1}$

28. $a_n = \left(-\dfrac{1}{2}\right)^n$

29. $a_n = \dfrac{\sin n}{n}$

30. $a_n = \dfrac{\sin^2 n}{2^n}$

Find the limits of the sequences whose nth terms appear in Exercises 31–34.

31. $a_n = \sqrt{\dfrac{2n}{n+1}}$

32. $a_n = \sin\left(\dfrac{\pi}{2} + \dfrac{1}{n}\right)$

33. $a_n = \tan^{-1} n$

34. $a_n = \ln n - \ln(n + 1)$

Find the limits, if they exist, of the sequences whose nth terms appear in Exercises 35–38.

35. $a_n = \dfrac{n}{2^n}$

36. $a_n = \dfrac{3^n}{n^3}$

37. $a_n = \dfrac{\ln(n+1)}{n}$

38. $a_n = \dfrac{\ln n}{\ln 2n}$

Find the limits, if they exist, of the sequences whose nth terms appear in Exercises 39–58.

39. $a_n = 8^{1/n}$

40. $a_n = (0.03)^{1/n}$

41. $a_n = \left(1 + \dfrac{7}{n}\right)^n$

42. $a_n = \left(1 - \dfrac{1}{n}\right)^n$

43. $a_n = \dfrac{1}{(0.9)^n}$

44. $a_n = \dfrac{(n+1)!}{n!}$

45. $a_n = \sqrt[n]{10n}$

46. $a_n = \sqrt[n]{n^2}$

47. $a_n = \left(\dfrac{3}{n}\right)^{1/n}$

48. $a_n = (n + 4)^{1/(n+4)}$

49. $a_n = \dfrac{\ln n}{n^{1/n}}$

50. $a_n = \sqrt[n]{4^n n}$

51. $a_n = \left(\dfrac{1}{3}\right)^n$

52. $a_n = \dfrac{1}{\sqrt{2^n}}$

53. $a_n = \dfrac{1}{n!}$

54. $a_n = \dfrac{(-4)^n}{n!}$

55. $a_n = \left(\dfrac{1}{n}\right)^{1/\ln n}$

56. $a_n = \dfrac{n!}{2^n \cdot 3^n}$

57. $a_n = \dfrac{n!}{10^{6n}}$

58. $a_n = \dfrac{3^n \cdot 6^n}{2^{-n} \cdot n!}$

CALCULATOR In Exercises 59–62, experiment with a calculator to identify a value of N that will make the inequality hold for $n \geq N$.

59. $|\sqrt[n]{0.5} - 1| < 10^{-3}$

60. $|\sqrt[n]{n} - 1| < 10^{-3}$

61. $(0.9)^n < 10^{-3}$

62. $2^n/n! < 10^{-7}$

63. **CALCULATOR** *A recursive definition of $\pi/2$.* If you start with $x_1 = 1$ and define the subsequent terms of $\{x_n\}$ by the rule $x_n = x_{n-1} + \cos x_{n-1}$, you generate a sequence that converges rapidly to $\pi/2$. Try it. Figure 9.5 explains what is going on.

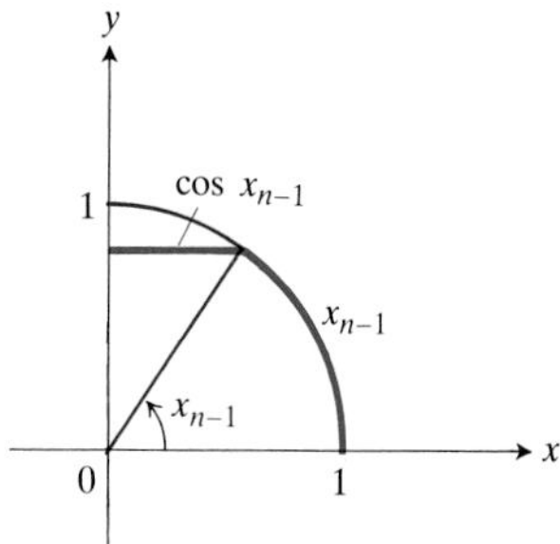

9.5 The length $\pi/2$ of the circular arc is approximated by $x_{n-1} + \cos x_{n-1}$ (Exercise 63).

64. **CALCULATOR** *A recursive solution of the equation $\cos x = x$.* If you start with $x_1 = 1$ and define the subsequent terms of $\{x_n\}$ by the rule $x_{n+1} = \cos x_n$, your

sequence will converge to a solution of the equation $\cos x = x$. Try it. Figure 9.6 shows what is going on.

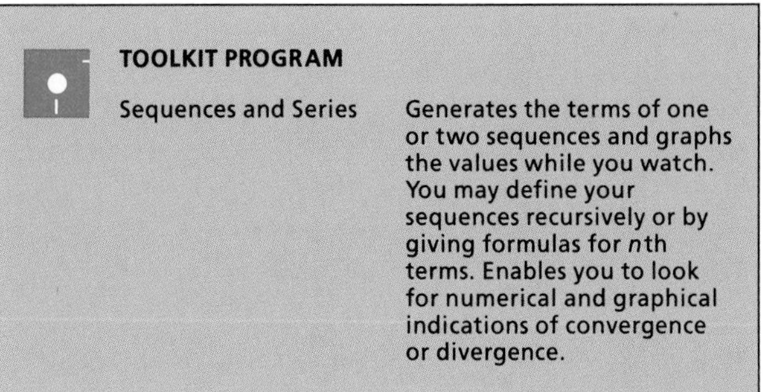

TOOLKIT PROGRAM

Sequences and Series Generates the terms of one or two sequences and graphs the values while you watch. You may define your sequences recursively or by giving formulas for nth terms. Enables you to look for numerical and graphical indications of convergence or divergence.

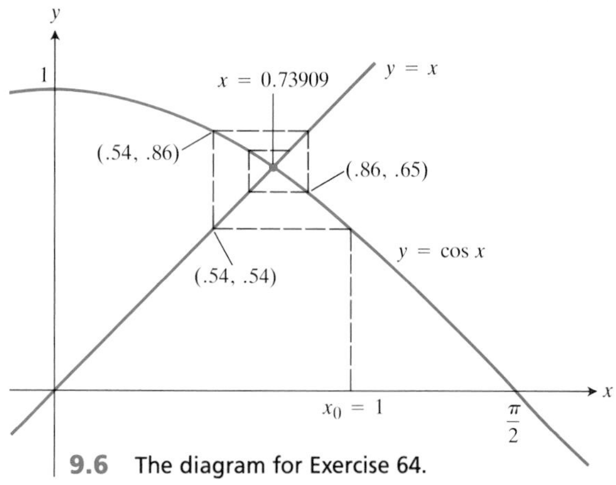

9.6 The diagram for Exercise 64.

9.2 Infinite Series

In mathematics and science we often use infinite polynomials like

$$1 + x + x^2 + x^3 + \cdots + x^n + \cdots$$

to represent functions (the series above represents $1/(1-x)$ for $|x| < 1$), evaluate nonelementary integrals, and solve differential equations. For any particular value of x such a polynomial is calculated as an infinite sum of constants, a sum we call an infinite series. The goal of this and the next three sections is to learn to work with infinite series. Then, in the three sections after that, we shall build on what we have learned to study infinite polynomials as infinite series of powers of x.

Let us begin by asking how to assign meaning to an expression like

$$1 + \frac{1}{2} + \frac{1}{4} + \frac{1}{8} + \frac{1}{16} + \cdots.$$

The way to do so is not to try to add all the terms at once (we can't) but rather to add the terms one at a time from the beginning and look for a pattern in how these partial sums grow. When we do this, we find:

Partial sum		Value	
first:	$s_1 = 1$	1	
second:	$s_2 = 1 + \frac{1}{2}$	$2 - \frac{1}{2}$	
third:	$s_3 = 1 + \frac{1}{2} + \frac{1}{4}$	$2 - \frac{1}{4}$	
⋮			
nth:	$s_n = 1 + \frac{1}{2} + \frac{1}{4} + \cdots + \frac{1}{2^{n-1}}$	$2 - \frac{1}{2^{n-1}}$	(After some algebra)

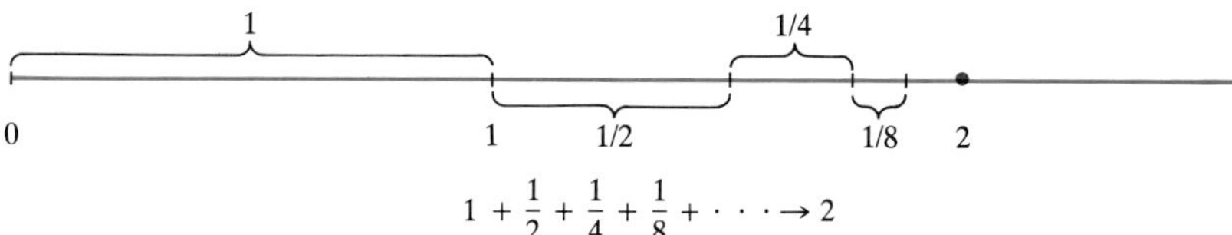

$$1 + \frac{1}{2} + \frac{1}{4} + \frac{1}{8} + \cdots \to 2$$

9.7 As the lengths 1, 1/2, 1/4, 1/8, . . . , are added one by one, the sum approaches 2.

Indeed there *is* a pattern. The partial sums form a sequence whose nth term is

$$s_n = 2 - \frac{1}{2^{n-1}}$$

and this sequence converges to 2. We therefore say

$$\text{“the sum of the series } 1 + \frac{1}{2} + \frac{1}{4} + \cdots + \frac{1}{2^{n-1}} + \cdots \text{ is 2.”}$$

Is the sum of any finite number of terms in the series 2? No. Can we actually add an infinite number of terms one by one? No. But we can still define their sum by defining it to be the limit of the sequence of partial sums as $n \to \infty$, in this case 2 (Fig. 9.7). Our knowledge of sequences and limits enables us to break away from the confines of finite sums.

DEFINITIONS

Given a sequence of numbers $\{a_n\}$, an expression of the form

$$a_1 + a_2 + a_3 + \cdots + a_n + \cdots \tag{1}$$

is called an **infinite series.** The number a_n is called the ***n*th term** of the series. The sequence $\{s_n\}$ defined by

$$\begin{aligned} s_1 &= a_1 \\ s_2 &= a_1 + a_2 \\ &\vdots \\ s_n &= a_1 + a_2 + \cdots + a_n = \sum_{k=1}^{n} a_k \end{aligned} \tag{2}$$

is the **sequence of partial sums** of the series, the number s_n being the ***n*th partial sum.** If the sequence of partial sums converges to a limit L, we say that the series **converges** and that its **sum** is L. In this case, we also write

$$a_1 + a_2 + \cdots + a_n + \cdots = \sum_{n=1}^{\infty} a_n = L. \tag{3}$$

If the sequence of partial sums of the series does not converge, we say that the series **diverges.**

When we first begin to study a given series $a_1 + a_2 + \cdots + a_n + \cdots$, we might not know whether it converges, or diverges. In either case, it is convenient to use sigma notation to write the series as

$$\sum_{n=1}^{\infty} a_n, \quad \sum_{k=1}^{\infty} a_k, \quad \text{or} \quad \sum a_n .$$

The first of these is read "summation from n equals 1 to infinity of a_n"; the second as "summation from k equals 1 to infinity of a_k"; and the third as "summation a_n."

Geometric Series

Geometric series are series of the form

$$a + ar + ar^2 + \cdots + ar^{n-1} + \cdots = \sum_{n=1}^{\infty} ar^{n-1} \tag{4}$$

in which a and r are fixed real numbers and $a \neq 0$. The **ratio** r can be positive, as in

$$1 + \frac{1}{2} + \frac{1}{4} + \cdots + \frac{1}{2^{n-1}} + \cdots, \tag{5}$$

or negative, as in

$$1 - \frac{1}{3} + \frac{1}{9} - \cdots + (-1)^{n-1}\frac{1}{3^{n-1}} + \cdots. \tag{6}$$

If $r = 1$, the nth partial sum of the geometric series in (4) is

$$s_n = a + a(1) + a(1)^2 + \cdots + a(1)^{n-1} = na$$

and the series diverges because $\lim_{n\to\infty} s_n = \pm\infty$. If $r \neq 1$, we can determine the convergence or nonconvergence of the series in the following way. We multiply the nth partial sum

$$s_n = a + ar + ar^2 + \cdots + ar^{n-1}$$

by r, obtaining

$$rs_n = ar + ar^2 + \cdots + ar^{n-1} + ar^n.$$

We then subtract s_n from rs_n. Most of the terms on the right cancel when we do this, leaving only

$$s_n - rs_n = a - ar^n \quad \text{or} \quad s_n(1 - r) = a(1 - r^n). \tag{7}$$

We solve for s_n, obtaining

$$s_n = \frac{a(1 - r^n)}{(1 - r)}, \quad (r \neq 1). \tag{8}$$

If $|r| < 1$, then $r^n \to 0$ as $n \to \infty$ (as we saw in Section 9.1), and $s_n \to a/(1 - r)$. In other words, the series converges to $a/(1 - r)$. If $|r| > 1$, then $|r^n| \to \infty$ and the series diverges.

> If $|r| < 1$, the geometric series converges and
>
> $$\sum_{n=1}^{\infty} ar^{n-1} = \frac{a}{1 - r}. \tag{9}$$
>
> If $|r| \geq 1$, the series diverges.

EXAMPLE 1 The geometric series with $a = 1/9$ and $r = 1/3$:

$$\frac{1}{9} + \frac{1}{27} + \frac{1}{81} + \cdots = \frac{1}{9}\left(1 + \frac{1}{3} + \frac{1}{3^2} + \cdots\right) = \frac{1/9}{1 - 1/3} = \frac{1}{6}.$$ ■

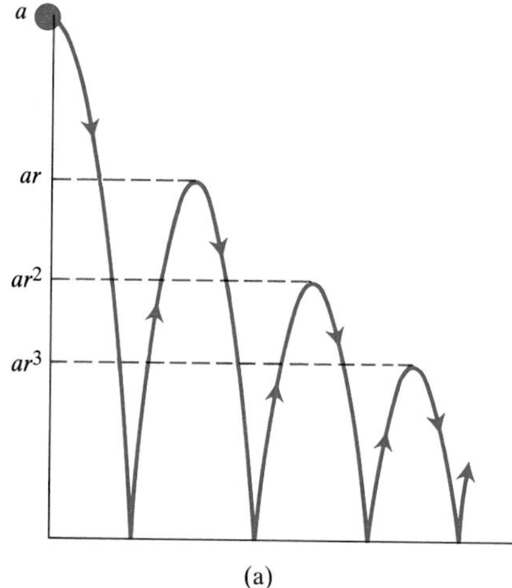

(a)

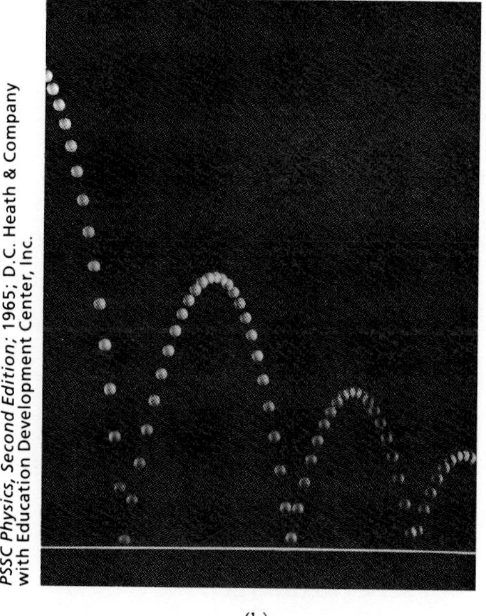

PSSC Physics, Second Edition; 1965; D.C. Heath & Company with Education Development Center, Inc.

(b)

9.8 (a) The height of each rebound is reduced by the factor r. (b) A stroboscopic photo of a bouncing ball.

EXAMPLE 2 The geometric series with $a = 4$ and $r = -1/2$:

$$4 - 2 + 1 - \frac{1}{2} + \frac{1}{4} - \cdots = 4\left(1 - \frac{1}{2} + \frac{1}{4} - \frac{1}{8} + \frac{1}{16} - \cdots\right)$$

$$= \frac{4}{1 + (1/2)} = \frac{8}{3}.$$

■

EXAMPLE 3 You drop a ball from a meters above a flat surface. Each time the ball hits after falling a distance h, it rebounds a distance rh, where r is positive but less than 1. Find the total distance the ball travels up and down (Fig. 9.8).

Solution The total distance is

$$s = \underbrace{a + 2ar + 2ar^2 + 2ar^3 + \cdots}_{\text{This sum is } 2ar/(1-r)} = a + \frac{2ar}{1 - r} = a\frac{1 + r}{1 - r}.$$

If $a = 6$ m and $r = 2/3$, for instance, the distance is

$$s = 6\,\frac{1 + (2/3)}{1 - (2/3)} = 6\,\frac{5/3}{1/3} = 30 \text{ m}.$$

■

Repeating Decimals

We use geometric series to explain why repeating decimals represent rational numbers.

EXAMPLE 4 Express the repeating decimal $5.23\,23\,23 \cdots$ as the ratio of two integers.

Solution

$$5.23\,23\,23 \cdots = 5 + \frac{23}{100} + \frac{23}{(100)^2} + \frac{23}{(100)^3} + \cdots$$

$$= 5 + \frac{23}{100}\underbrace{\left(1 + \frac{1}{100} + \left(\frac{1}{100}\right)^2 + \cdots\right)}_{1/(1-0.01)}$$

$$= 5 + \frac{23}{100}\left(\frac{1}{0.99}\right) = 5 + \frac{23}{99} = \frac{518}{99}.$$

■

Sums of Other Convergent Series

For geometric series we get a closed form for the nth partial sum and, from that, the formula $s = a/(1 - r)$. Unfortunately, such formulas are rare. What we usually have to do is test the series for convergence and then find some way to estimate the sums of the series that pass the test. Sometimes a convergent series represents a known function like $\sin x$ or $\ln x$. In other cases we might need a computer or calculator to estimate the sum.

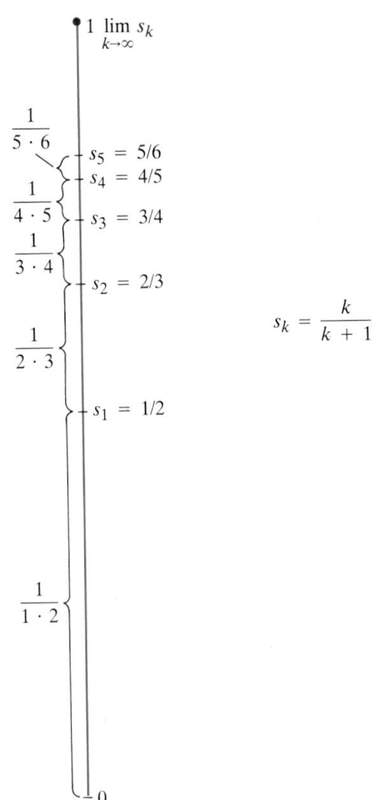

9.9 The sum of the first k terms of the series

$$\sum_{n=1}^{\infty} \frac{1}{n(n+1)}$$

is $k/(k+1)$ and the sum of the series is

$$\lim_{k\to\infty} \frac{k}{k+1} = 1.$$

The next example, however, is still one of those rare cases in which we can find the series' sum from a formula for s_n.

EXAMPLE 5 Determine whether $\Sigma_{n=1}^{\infty} [1/n(n+1)]$ converges. If it does, find the sum.

Solution We look for a pattern in the sequence of partial sums that might lead us to a closed expression for s_k. The key to success here, as in the integration

$$\int \frac{dx}{x(x+1)} = \int \frac{dx}{x} - \int \frac{dx}{x+1},$$

is partial fractions. The observation that

$$\frac{1}{k(k+1)} = \frac{1}{k} - \frac{1}{k+1} \tag{10}$$

permits us to write the partial sum

$$\sum_{n=1}^{k} \frac{1}{n(n+1)} = \frac{1}{1\cdot 2} + \frac{1}{2\cdot 3} + \cdots + \frac{1}{k\cdot(k+1)}$$

as

$$s_k = \left(\frac{1}{1} - \frac{1}{2}\right) + \left(\frac{1}{2} - \frac{1}{3}\right) + \cdots + \left(\frac{1}{k} - \frac{1}{k+1}\right). \tag{11}$$

Removing parentheses and cancelling the terms of opposite sign collapses the sum to

$$s_k = 1 - \frac{1}{k+1}. \tag{12}$$

We then see that $s_k \to 1$ as $k \to \infty$. The series converges, and its sum is 1 (Fig. 9.9):

$$\sum_{n=1}^{\infty} \frac{1}{n(n+1)} = 1.$$

■

Divergent Series

Geometric series with $|r| \geq 1$ aren't the only series to diverge.

EXAMPLE 6 The series

$$\sum_{n=1}^{\infty} n^2 = 1 + 4 + 9 + \cdots + n^2 + \cdots$$

diverges because the partial sums grow beyond every number L. After $n = 1$, the number $s_n = 1 + 4 + 9 + \cdots + n^2$ is greater than n^2. ■

EXAMPLE 7 The series

$$\sum_{n=1}^{\infty} \frac{n+1}{n} = \frac{2}{1} + \frac{3}{2} + \frac{4}{3} + \cdots + \frac{n+1}{n} + \cdots$$

diverges because the partial sums eventually outgrow every preassigned number. Each term is greater than 1, and so the sum of n terms is greater than n. ■

A series can diverge without its partial sums becoming large. The partial sums can oscillate between two extremes, for instance, as they do in the next example.

EXAMPLE 8 The series $\Sigma_{n=1}^{\infty}(-1)^{n+1}$ diverges because its partial sums alternate between 1 and 0:

$$s_1 = (-1)^2 = 1,$$

$$s_2 = (-1)^2 + (-1)^3 = 1 - 1 = 0,$$

$$s_3 = (-1)^2 + (-1)^3 + (-1)^4 = 1 - 1 + 1 = 1,$$

and so on. ■

The *n*th-Term Test for Divergence

There is a test for detecting the kind of divergence that occurs in Examples 6, 7, and 8.

The *n*th-Term Test for Divergence

If $\lim_{n\to\infty} a_n \neq 0$, or if $\lim_{n\to\infty} a_n$ fails to exist, then $\Sigma_{n=1}^{\infty} a_n$ diverges.

When we apply the *n*th-Term Test to the series in Examples 6, 7, and 8, we find that

$$\sum_{n=1}^{\infty} n^2 \quad \text{diverges because } n^2 \to \infty,$$

$$\sum_{n=1}^{\infty} \frac{n+1}{n} \quad \text{diverges because } \frac{n+1}{n} \to 1,$$

$$\sum_{n=1}^{\infty} (-1)^{n+1} \quad \text{diverges because } \lim_{n\to\infty} (-1)^{n+1} \text{ does not exist.}$$

The reason the *n*th-Term Test works is that $\lim_{n\to\infty} a_n$ must equal zero if $\Sigma\, a_n$ converges. To see why, let

$$s_n = a_1 + a_2 + \cdots + a_n$$

and suppose that $\Sigma\, a_n$ converges to S; that is, $s_n \to S$. When n is large, so is $n - 1$, and both s_n and s_{n-1} are close to S. Their difference, a_n, must then be close to zero. More formally, $a_n = s_n - s_{n-1} \to S - S = 0$.

EXAMPLE 9 Determine whether each series converges, or diverges. If it converges, find its sum.

a) $\displaystyle\sum_{n=1}^{\infty} \frac{n}{2n+5}$ b) $\displaystyle\sum_{n=1}^{\infty} \frac{5(-1)^n}{4^n}$

Solution

a) $\lim_{n\to\infty} \frac{n}{2n+5} = \frac{1}{2} \neq 0$. The series diverges by the nth-Term Test.

b) This is a geometric series with $a = -5/4$ and $r = -1/4$. It converges to

$$\frac{a}{1-r} = \frac{-5/4}{1+(1/4)} = -1.$$

■

A Necessary Condition for Convergence

We often state the nth-Term Test for divergence a shorter way.

> If $\Sigma_{n=1}^{\infty} a_n$ converges, then $a_n \to 0$.

CAUTION: This does *not* mean that Σa_n converges if $a_n \to 0$. A series Σa_n may diverge even though $a_n \to 0$. Thus, $\lim a_n = 0$ is a *necessary* but *not a sufficient* condition for the series Σa_n to converge.

EXAMPLE 10 The series

$$1 + \underbrace{\frac{1}{2} + \frac{1}{2}}_{2 \text{ terms}} + \underbrace{\frac{1}{4} + \frac{1}{4} + \frac{1}{4} + \frac{1}{4}}_{4 \text{ terms}} + \cdots + \underbrace{\frac{1}{2^n} + \frac{1}{2^n} + \cdots + \frac{1}{2^n}}_{2^n \text{ terms}} + \cdots$$

diverges even though its terms form a sequence that converges to 0. ■

Whenever we have two convergent series we can add them, subtract them, and multiply them by constants to make other convergent series. The next theorem gives the details.

THEOREM 5

> If $\Sigma a_n = A$ and $\Sigma b_n = B$, then
>
> 1. *Sum Rule:* $\sum (a_n + b_n) = A + B$
> 2. *Difference Rule:* $\sum (a_n - b_n) = A - B$
> 3. *Constant Multiple Rule:* $\sum ka_n = k\sum a_n = kA$ (Any number k)

As corollaries of Theorem 5 we have

1. Every nonzero constant multiple of a divergent series diverges.
2. If Σa_n converges and Σb_n diverges then $\Sigma (a_n + b_n)$ and $\Sigma (a_n - b_n)$ both diverge.

The proofs resemble the proofs of similar theorems discussed earlier, and we omit them.

EXAMPLE 11

a) $$\sum_{n=1}^{\infty} \frac{4}{2^{n-1}} = 4\sum_{n=1}^{\infty} \frac{1}{2^{n-1}} = 4\,\frac{1}{1-\frac{1}{2}} = 8$$

b) $$\sum_{n=1}^{\infty} \frac{3^{n-1}-1}{6^{n-1}} = \sum_{n=1}^{\infty} \frac{1}{2^{n-1}} - \sum_{n=1}^{\infty} \frac{1}{6^{n-1}} = 2 - \frac{1}{1-\frac{1}{6}} = 2 - \frac{6}{5} = \frac{4}{5}$$ ■

We can always add a finite number of terms to a series or delete a finite number of terms from a series without altering its convergence or divergence. If $\Sigma_{n=1}^{\infty} a_n$ converges, then $\Sigma_{n=k}^{\infty} a_n$ converges for any $k > 1$ and

$$\sum_{n=1}^{\infty} a_n = a_1 + a_2 + \cdots + a_{k-1} + \sum_{n=k}^{\infty} a_n. \tag{13}$$

Conversely, if $\Sigma_{n=k}^{\infty} a_n$ converges for any $k > 1$, then $\Sigma_{n=1}^{\infty} a_n$ converges. Thus,

$$\sum_{n=1}^{\infty} \frac{1}{5^n} = \frac{1}{5} + \frac{1}{25} + \frac{1}{125} + \sum_{n=4}^{\infty} \frac{1}{5^n} \tag{14}$$

and

$$\sum_{n=4}^{\infty} \frac{1}{5^n} = \sum_{n=1}^{\infty} \frac{1}{5^n} - \frac{1}{5} - \frac{1}{25} - \frac{1}{125}. \tag{15}$$

We can always shift the indexing of a series without altering its convergence. For example, we can write the geometric series $1 + (1/2) + (1/4) + \cdots = \Sigma_{n=1}^{\infty} 1/2^{n-1}$ as

$$\sum_{n=0}^{\infty} \frac{1}{2^n}, \quad \sum_{n=5}^{\infty} \frac{1}{2^{n-5}}, \quad \text{or even} \quad \sum_{n=-4}^{\infty} \frac{1}{2^{n+4}}.$$

Exercises 9.2

In Exercises 1–6, find a formula for the nth partial sum of each series and use it to find the series' sum if the series converges.

1. $2 + \frac{2}{3} + \frac{2}{9} + \frac{2}{27} + \cdots + \frac{2}{3^{n-1}} + \cdots$

2. $\frac{9}{100} + \frac{9}{100^2} + \frac{9}{100^3} + \cdots + \frac{9}{100^n} + \cdots$

3. $1 - \frac{1}{2} + \frac{1}{4} - \frac{1}{8} + \cdots + (-1)^{n-1}\frac{1}{2^{n-1}} + \cdots$

4. $1 - 2 + 4 - 8 + \cdots + (-1)^{n-1}2^{n-1} + \cdots$

5. $\frac{1}{2\cdot 3} + \frac{1}{3\cdot 4} + \frac{1}{4\cdot 5} + \cdots + \frac{1}{(n+1)(n+2)} + \cdots$

6. $\frac{5}{1\cdot 2} + \frac{5}{2\cdot 3} + \frac{5}{3\cdot 4} + \cdots + \frac{5}{n(n+1)} + \cdots$

In Exercises 7–14, write out the first few terms of the series to show how the series starts. Then find the sum of the series.

7. $\sum_{n=0}^{\infty} \frac{1}{4^n}$

8. $\sum_{n=2}^{\infty} \frac{1}{4^n}$

9. $\sum_{n=1}^{\infty} \frac{7}{4^n}$

10. $\sum_{n=0}^{\infty} (-1)^n \frac{5}{4^n}$

11. $\sum_{n=0}^{\infty} \left(\frac{5}{2^n} + \frac{1}{3^n}\right)$

12. $\sum_{n=0}^{\infty} \left(\frac{5}{2^n} - \frac{1}{3^n}\right)$

13. $\sum_{n=0}^{\infty} \left(\frac{1}{2^n} + \frac{(-1)^n}{5^n}\right)$

14. $\sum_{n=0}^{\infty} \left(\frac{2^{n+1}}{5^n}\right)$

Use partial fractions to find the sums of the series in Exercises 15–18.

15. $\displaystyle\sum_{n=1}^{\infty} \frac{4}{(4n-3)(4n+1)}$

16. $\displaystyle\sum_{n=1}^{\infty} \frac{1}{(4n-3)(4n+1)}$

17. $\displaystyle\sum_{n=3}^{\infty} \frac{4}{(4n-3)(4n+1)}$

18. $\displaystyle\sum_{n=1}^{\infty} \frac{2n+1}{n^2(n+1)^2}$

Which series in Exercises 19–32 converge and which diverge? If the series converges, find its sum.

19. $\displaystyle\sum_{n=0}^{\infty} \left(\frac{1}{\sqrt{2}}\right)^n$

20. $\displaystyle\sum_{n=1}^{\infty} \ln \frac{1}{n}$

21. $\displaystyle\sum_{n=1}^{\infty} (-1)^{n+1} \frac{3}{2^n}$

22. $\displaystyle\sum_{n=1}^{\infty} (\sqrt{2})^n$

23. $\displaystyle\sum_{n=0}^{\infty} \cos n\pi$

24. $\displaystyle\sum_{n=0}^{\infty} \frac{\cos n\pi}{5^n}$

25. $\displaystyle\sum_{n=0}^{\infty} e^{-2n}$

26. $\displaystyle\sum_{n=1}^{\infty} \frac{n^2+1}{n}$

27. $\displaystyle\sum_{n=1}^{\infty} (-1)^{n+1} n$

28. $\displaystyle\sum_{n=1}^{\infty} \frac{2}{10^n}$

29. $\displaystyle\sum_{n=0}^{\infty} \frac{2^n - 1}{3^n}$

30. $\displaystyle\sum_{n=1}^{\infty} \left(1 - \frac{1}{n}\right)^n$

31. $\displaystyle\sum_{n=0}^{\infty} \frac{n!}{1000^n}$

32. $\displaystyle\sum_{n=0}^{\infty} \frac{1}{x^n}, \quad |x| > 1$

The series in Exercises 33 and 34 are geometric series. Find a and r in each case.

33. $\displaystyle\frac{1}{1+x} = \sum_{n=0}^{\infty} (-1)^n x^n, \quad |x| < 1$

34. $\displaystyle\frac{1}{1+x^2} = \sum_{n=0}^{\infty} (-1)^n x^{2n}, \quad |x| < 1$

35. A ball is dropped from a height of 4 m. Each time it strikes the pavement after falling from a height of h m it rebounds to a height of $0.75h$ m. Find the total distance the ball travels up and down.

36. *Continuation of Exercise 35.* Find the total number of seconds the ball in Exercise 35 is traveling. (*Hint:* The formula $s = 4.9t^2$ gives $t = \sqrt{s/4.9}$.)

37. Write the repeating decimal $0.234\,234\,234\cdots$ as a geometric series and express the rational number it represents as the ratio of two integers.

38. Express the rational number represented by the number $1.24\,123\,123\,123\cdots$, in which the 123 repeats, as the ratio of two integers.

39. The series in Exercise 5 can be written as

$$\sum_{n=1}^{\infty} \frac{1}{(n+1)(n+2)} \quad \text{and} \quad \sum_{n=-1}^{\infty} \frac{1}{(n+3)(n+4)}.$$

Write it as a sum beginning with (a) $n = -2$, (b) $n = 0$, (c) $n = 5$.

40. The series in Exercise 6 can be written as

$$\sum_{n=1}^{\infty} \frac{5}{n(n+1)} \quad \text{and} \quad \sum_{n=0}^{\infty} \frac{5}{(n+1)(n+2)}.$$

Write it as a sum beginning with (a) $n = -1$, (b) $n = 3$, (c) $n = 20$.

41. Figure 9.10 shows the first five of an infinite sequence of squares. The outermost square has an area of 4 m². Each of the other squares is obtained by joining the midpoints of the sides of the square before it. Find the sum of the areas of all the squares.

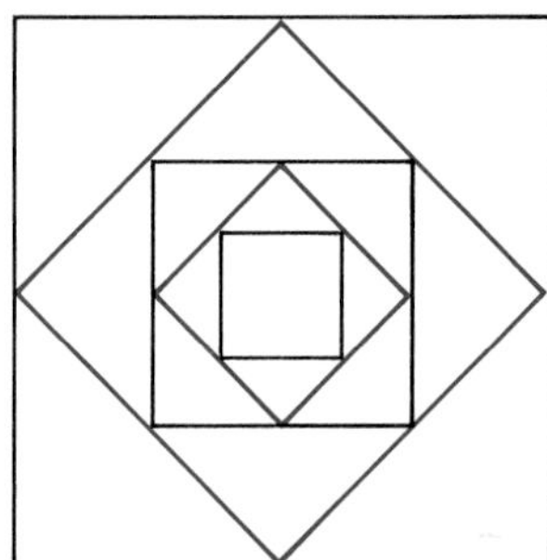

9.10 The first five squares in Exercise 41.

42. Figure 9.11 shows the first three rows of a sequence of rows of semicircles. There are 2^n semicircles in the nth

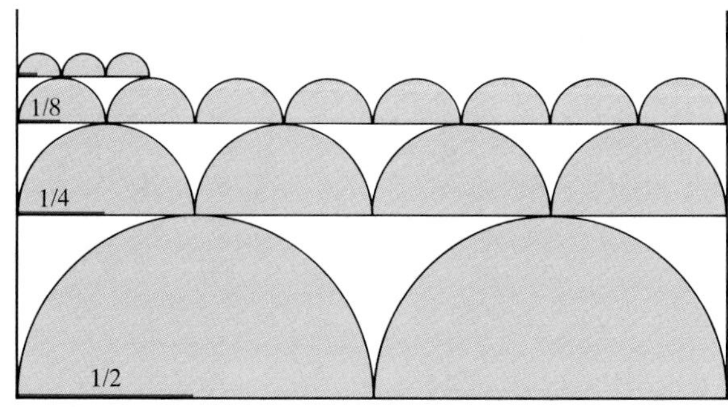

9.11 The semicircles in Exercise 42.

row, each of radius $1/2^n$. Find the sum of the areas of all the semicircles.

43. Make up an example of two divergent series whose term-by-term sum converges.

44. Show by example that $\Sigma (a_n/b_n)$ may diverge even though Σa_n and Σb_n converge and no b_n equals 0.

45. Show by example that $\Sigma (a_n/b_n)$ may converge to something other than A/B even when $A = \Sigma a_n$, $B = \Sigma b_n \neq 0$, and no b_n equals 0.

46. Show that if Σa_n converges, and $a_n \neq 0$ for all n, then $\Sigma (1/a_n)$ diverges.

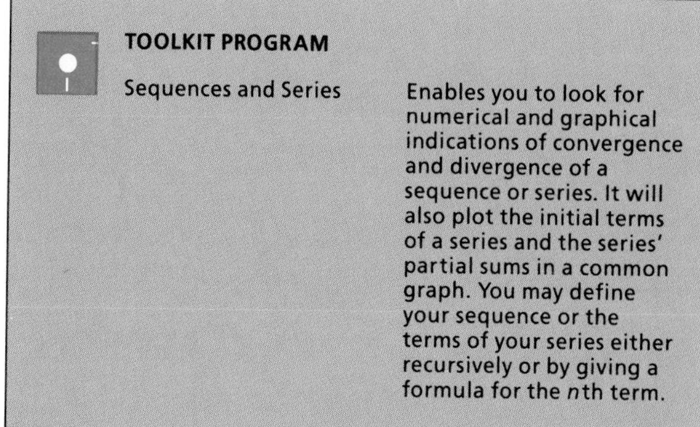

9.3 Series without Negative Terms: Comparison and Integral Tests

Given a series Σa_n, we have two questions:

1. Does the series converge?
2. If it converges, what is its sum?

Much of the rest of this chapter is devoted to answering the first question. But as a practical matter, the second question is just as important, and we come back to that question later.

In this section and the next we study series that do not have negative terms. The reason for this restriction is that the partial sums of these series form nondecreasing sequences, and nondecreasing sequences *that are bounded from above* always converge. To show that a series of nonnegative terms converges, we need only show that there is some number beyond which the partial sums never go.

It may first seem to be a drawback that this approach establishes the fact of convergence without producing the sum of the series in question. Surely it would be better to compute sums of series directly from nice formulas for their partial sums. But in most cases such formulas are not available, and in their absence we have to turn instead to the two-step procedure of first establishing convergence and then approximating the sum.

Nondecreasing Sequences

Suppose that Σa_n is an infinite series and that $a_n \geq 0$ for every n. Then, when we calculate the partial sums s_1, s_2, s_3, and so on, we see that each one is greater than or equal to its predecessor because $s_{n+1} = s_n + a_n$:

$$s_1 \leq s_2 \leq s_3 \leq \cdots \leq s_n \leq s_{n+1} \leq \cdots. \tag{1}$$

A sequence $\{s_n\}$ with the property that $s_n \leq s_{n+1}$ for every n is called a **nondecreasing sequence.**

There are two kinds of nondecreasing sequences—those that increase beyond any finite bound and those that don't. The former diverge to infinity, so we turn our attention to the other kind: those that do not grow beyond all bounds. Such a sequence is said to be **bounded from above,** and any number M such that $s_n \leq M$ for all n is called an **upper bound** of the sequence.

EXAMPLE 1 If $s_n = n/(n+1)$, then 1 is an upper bound and so is any number greater than 1. No number smaller than 1 is an upper bound, so for this sequence 1 is the **least upper bound.** ■

A nondecreasing sequence bounded from above always has a least upper bound, but we shall not prove this fact. We prove that if L is the least upper bound, then the sequence converges to L. The following argument shows why L is the limit.

Suppose we plot the points $(1, s_1), (2, s_2), \ldots, (n, s_n)$ in the xy-plane. If M is an upper bound of the sequence, all these plotted points will lie on or below the line $y = M$ (Fig. 9.12). The line $y = L$ is the lowest such line. None of the points (n, s_n) lies above $y = L$ but some do lie above any lower line $y = L - \epsilon$, if ϵ is a positive number. The sequence converges to L because

a) $s_n \leq L$ for *all* values of n and

b) given any $\epsilon > 0$, there exists at least one integer N for which $s_N > L - \epsilon$.

The fact that $\{s_n\}$ is a nondecreasing sequence tells us further that

$$s_n \geq s_N > L - \epsilon \qquad \text{for all } n \geq N.$$

This means that *all* the numbers s_n beyond the Nth one lie within ϵ of L. This is precisely the condition for L to be the limit of the sequence s_n.

The facts for nondecreasing sequences are summarized in the following theorem.

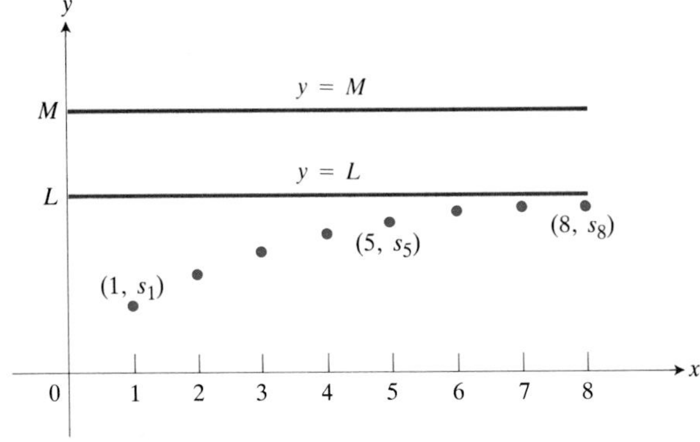

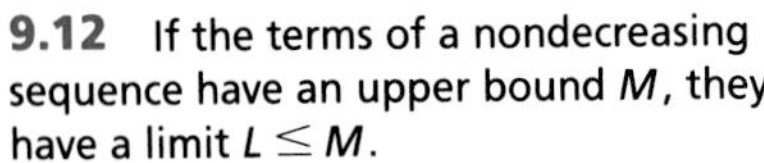
9.12 If the terms of a nondecreasing sequence have an upper bound M, they have a limit $L \leq M$.

THEOREM 6

The Nondecreasing Sequence Theorem

A nondecreasing sequence converges if and only if its terms are bounded from above. If all the terms are less than or equal to M then the limit of the sequence is less than or equal to M as well.

Theorem 6 tells us that we can show that a series $\Sigma\, a_n$ of nonnegative terms converges if we can show that its partial sums are bounded from above. The question, of course, is how to find out in any particular instance whether the s_n's have an upper bound.

Sometimes we can show that the s_n's are bounded above by showing that each one is less than or equal to the corresponding partial sum of a series that is already known to converge. The next example shows how this can happen.

EXAMPLE 2 The series

$$\sum_{n=0}^{\infty} \frac{1}{n!} = 1 + \frac{1}{1!} + \frac{1}{2!} + \frac{1}{3!} + \cdots \tag{2}$$

converges because its terms are all positive and less than or equal to the corresponding terms of

$$1 + \sum_{n=0}^{\infty} \frac{1}{2^n} = 1 + 1 + \frac{1}{2} + \frac{1}{2^2} + \cdots. \tag{3}$$

To see how this relationship leads to an upper bound for the partial sums of $\Sigma_{n=0}^{\infty}\,(1/n!)$, let

$$s_n = 1 + \frac{1}{1!} + \frac{1}{2!} + \cdots + \frac{1}{n!},$$

and observe that, for each n,

$$s_n \le 1 + 1 + \frac{1}{2} + \frac{1}{2^2} + \cdots + \frac{1}{2^n} < 1 + \sum_{n=0}^{\infty} \frac{1}{2^n} = 1 + \frac{1}{1 - (1/2)} = 3.$$

Thus the partial sums of $\Sigma_{n=0}^{\infty}\,(1/n!)$ are all less than 3. Therefore, $\Sigma_{n=0}^{\infty}\,(1/n!)$ converges.

The fact that 3 is an upper bound for the partial sums of $\Sigma_{n=0}^{\infty}\,(1/n!)$ does not necessarily mean that the series converges to 3. As we shall see in Section 9.7, the series converges to e. ■

EXAMPLE 3 *The harmonic series.* The series

$$\sum_{1}^{\infty} \frac{1}{n} = 1 + \frac{1}{2} + \frac{1}{3} + \cdots + \frac{1}{n} + \cdots$$

is called **the harmonic series.** The series diverges because there is no upper bound for its sequence of partial sums. To see why, imagine grouping the

The terms in a harmonic series correspond to the stops on a vibrating string that produce multiples of the fundamental frequency. For example, 1/2 produces the first harmonic at a frequency that is twice the fundamental frequency; 1/3 produces a frequency that is three times the fundamental; and so on. This combination of frequencies produces musical harmony; hence the name of the series.

terms of the series in the following way:

$$1 + \frac{1}{2} + \underbrace{\left(\frac{1}{3} + \frac{1}{4}\right)}_{>\frac{2}{4} = \frac{1}{2}} + \underbrace{\left(\frac{1}{5} + \frac{1}{6} + \frac{1}{7} + \frac{1}{8}\right)}_{>\frac{4}{8} = \frac{1}{2}} + \underbrace{\left(\frac{1}{9} + \frac{1}{10} + \cdots + \frac{1}{16}\right)}_{>\frac{8}{16} = \frac{1}{2}} + \cdots.$$

The sum of the first two terms is 1.5. The sum of the next two terms is $1/3 + 1/4$ which is greater than $1/4 + 1/4 = 1/2$ (because 1/3 is greater than 1/4). The sum of the next four terms is $1/5 + 1/6 + 1/7 + 1/8$ which is greater than $1/8 + 1/8 + 1/8 + 1/8 = 1/2$. The sum of the next eight terms is $1/9 + 1/10 + 1/11 + 1/12 + 1/13 + 1/14 + 1/15 + 1/16$ which is greater than $8/16 = 1/2$. The sum of the next 16 terms is greater than $16/32 = 1/2$, and so on. In general, the sum of the 2^n terms ending with $1/2^{n+1}$ is greater than $2^n/2^{n+1} = 1/2$. The sequence of partial sums is not bounded: If $n = 2^k$ the partial sum s_n is greater the $k/2$. The harmonic series diverges. (We shall see later that the nth partial sum is slightly greater than $\ln(n + 1)$).

Notice that the nth–Term Test for divergence does not detect the divergence of the harmonic series. The nth term, $1/n$, goes to zero but the series still diverges. ■

Comparison Test for Convergence

We established the convergence of the series in Example 2 by comparing it with a series that was already known to converge. This kind of comparison is typical of a procedure called the Comparison Test for convergence of series of nonnegative terms.

Comparison Test for Series of Nonnegative Terms

Let $\Sigma\, a_n$ be a series with no negative terms.

a) **Test for convergence.** The series $\Sigma\, a_n$ converges if there is a convergent series $\Sigma\, c_n$ with $a_n \le c_n$ for all $n > n_0$, for some positive integer n_0.

b) **Test for divergence.** The series $\Sigma\, a_n$ diverges if there is a divergent series of nonnegative terms $\Sigma\, d_n$ with $a_n \ge d_n$ for all $n > n_0$.

In part (a), the partial sums of the series $\Sigma\, a_n$ are bounded above by

$$M = a_1 + a_2 + \cdots + a_{n_0} + \sum_{n=n_0+1}^{\infty} c_n.$$

They therefore form a nondecreasing sequence with a limit L that is less than or equal to M.

In part (b), the partial sums for $\Sigma\, a_n$ are not bounded from above. If they were, the partial sums for $\Sigma\, d_n$ would be bounded by

$$M' = d_1 + d_2 + \cdots + d_{n_0} + \sum_{n=n_0+1}^{\infty} a_n$$

and $\Sigma\, d_n$ would have to converge instead of diverge.

To apply the Comparison Test to a series, we do not have to include the early terms of the series. We can start the test with any index N, provided we include all the terms of the series being tested from there on.

EXAMPLE 4 We can establish the convergence of the series

$$5 + \frac{2}{3} + 1 + \frac{1}{7} + \frac{1}{2} + \frac{1}{3!} + \frac{1}{4!} + \cdots + \frac{1}{k!} + \cdots$$

by ignoring the first four terms and comparing the remainder with the convergent geometric series

$$\sum_{n=1}^{\infty} \frac{1}{2^n} = \frac{1}{2} + \frac{1}{4} + \frac{1}{8} + \cdots .$$

■

To apply the Comparison Test, we need to have on hand a list of series we already know about. Here is what we know so far:

Convergent series	Divergent series
Geometric series with $\|r\| < 1$	Geometric series with $\|r\| \geq 1$
Telescoping series like $\sum \frac{1}{n(n+1)}$	The harmonic series $\sum_{1}^{\infty} \frac{1}{n}$
The series $\sum_{0}^{\infty} \frac{1}{n!}$	Any series $\Sigma\, a_n$ with $\lim a_n \neq 0$

The next test, the Integral Test, will add some series to these lists.

The Integral Test

We introduce the Integral Test with a specific example, a series that is related to the harmonic series, but where the nth term is $1/n^2$ instead of $1/n$.

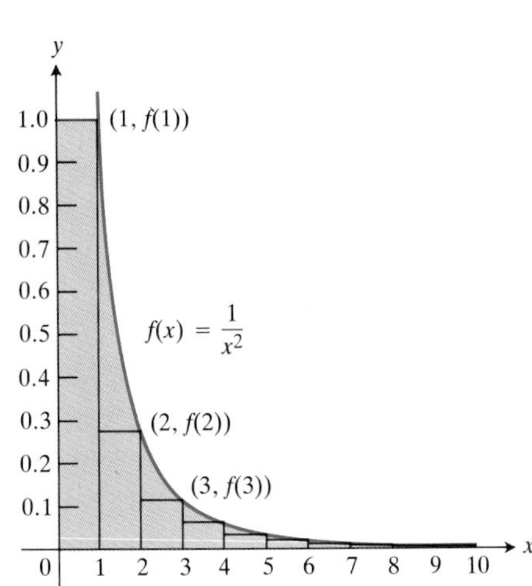

9.13 When $f(x) = 1/x^2$,
$$S_n = f(1) + f(2) + \cdots + f(n) < f(1) + \int_1^n f(x)\,dx = 1 + \left(\frac{1}{1} - \frac{1}{n}\right) < 2.$$

EXAMPLE 5 Does the series

$$\sum_{n=1}^{\infty} \frac{1}{n^2} = 1 + \frac{1}{4} + \frac{1}{9} + \frac{1}{16} + \cdots + \frac{1}{n^2} + \cdots \tag{4}$$

converge, or does it diverge?

Solution When we studied improper integrals in Chapter 8, we learned that

$$\int_1^{\infty} \frac{1}{x}\,dx \text{ diverges} \quad \text{and} \quad \int_1^{\infty} \frac{1}{x^2}\,dx \text{ converges.}$$

If we can show that the sequence of partial sums of our present series is bounded above, we can conclude that the series converges. Figure 9.13 suggests how we can find an upper bound for

$$s_n = \frac{1}{1^2} + \frac{1}{2^2} + \frac{1}{3^2} \cdots + \frac{1}{n^2} = f(1) + f(2) + f(3) + \cdots + f(n),$$

where $f(x) = 1/x^2$.

The first term of the series is $f(1) = 1$, which we can interpret as the area of a rectangle of height $1/1^2$ and base the length of the interval $[0, 1]$ on the x-axis. The next rectangle, over the interval $[1, 2]$, has area $f(2) = 1/2^2$. That rectangle lies below the curve $y = 1/x^2$, so

$$f(2) = \frac{1}{2^2} < \int_1^2 \frac{1}{x^2}\,dx.$$

In the same way,

$$f(3) = \frac{1}{3^2} < \int_2^3 \frac{1}{x^2}\,dx, \quad \cdots, \quad f(n) = \frac{1}{n^2} < \int_{n-1}^n \frac{1}{x^2}\,dx. \tag{5}$$

Adding these inequalities, we get

$$f(2) + f(3) + \cdots + f(n) < \int_1^n (1/x^2)\,dx = -\frac{1}{x}\bigg]_1^n = 1 - \frac{1}{n}.$$

To get s_n we must add $f(1) = 1$. When we do so, we have

$$s_n = f(1) + f(2) + f(3) + \cdots + f(n) < 2 - \frac{1}{n} < 2.$$

The sequence of partial sums $\{s_n\}$ is an increasing sequence that is bounded above, so it has a limit, and the given series converges. Its sum is known to be $\pi^2/6$, which is about 1.64493. ■

We now state and prove the Integral Test in more general terms.

Integral Test

Let $a_n = f(n)$ where $f(x)$ is a continuous, positive, decreasing function of x for all $x \geq 1$. Then the series $\Sigma\, a_n$ and the integral $\int_1^\infty f(x)\,dx$ both converge or both diverge.

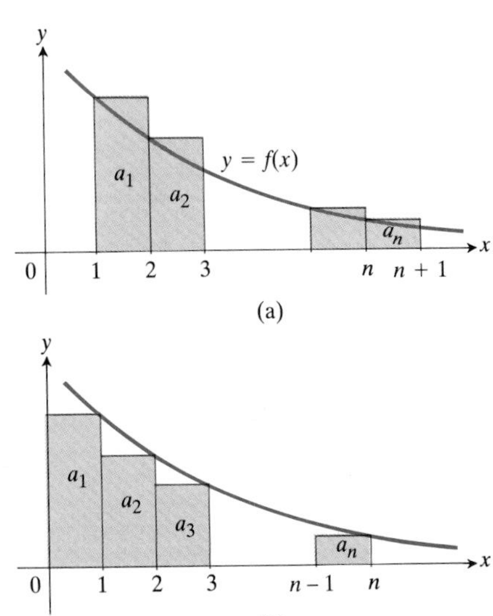

9.14 Subject to the conditions of the Integral Test, the series $\Sigma_{n=1}^\infty a_n$ and the integral $\int_1^\infty f(x)\,dx$ both converge or both diverge.

PROOF of the Integral Test. We start with the assumption that f is a decreasing function with $f(n) = a_n$ for every n. This leads us to observe that the rectangles in Fig. 9.14(a), which have areas $a_1, a_2, \ldots, a_n$, collectively enclose more area than that under the curve $y = f(x)$ from $x = 1$ to $x = n + 1$. That is,

$$\int_1^{n+1} f(x)\,dx \leq a_1 + a_2 + \cdots + a_n.$$

In Fig. 9.14(b) the rectangles have been faced to the left instead of to the right. If we momentarily disregard the first rectangle, of area a_1, we see that

$$a_2 + a_3 + \cdots + a_n \leq \int_1^n f(x)\,dx.$$

If we include a_1, we have

$$a_1 + a_2 + \cdots + a_n \leq a_1 + \int_1^n f(x)\,dx.$$

Combining these results gives

$$\int_1^{n+1} f(x)\,dx \leq a_1 + a_2 + \cdots + a_n \leq a_1 + \int_1^n f(x)\,dx. \tag{6}$$

If the integral $\int_1^\infty f(x)\,dx$ is finite, the right-hand inequality shows that $\Sigma\, a_n$ is also finite. But if $\int_1^\infty f(x)\,dx$ is infinite, then the left-hand inequality shows that the series is also infinite.

Hence the series and the integral are both finite or both infinite. ■

EXAMPLE 6 *The p-series.* If p is a real constant, the series

$$\sum_{n=1}^{\infty} \frac{1}{n^p} = \frac{1}{1^p} + \frac{1}{2^p} + \frac{1}{3^p} + \cdots + \frac{1}{n^p} + \cdots \tag{7}$$

converges if $p > 1$ and diverges if $p \le 1$. To prove this, let $f(x) = 1/x^p$. Then, if $p > 1$, we have $-p + 1 < 0$ and

$$\int_1^\infty x^{-p}\,dx = \lim_{b\to\infty} \frac{x^{-p+1}}{-p+1}\bigg]_1^b = \frac{1}{1-p}\lim_{b\to\infty}(b^{-p+1} - 1) = \frac{1}{p-1},$$

which is finite. Hence the p-series converges if p is greater than 1.

If $p = 1$, we have

$$1 + \frac{1}{2} + \frac{1}{3} + \cdots + \frac{1}{n} + \cdots,$$

which we already know diverges. Or, by the Integral Test,

$$\int_1^\infty x^{-1}\,dx = \lim_{b\to\infty} \ln x\bigg]_1^b = \infty,$$

and, since the integral diverges, the series diverges.

Finally, if $p < 1$, then the terms of the p-series are greater than the corresponding terms of the divergent harmonic series. Hence the p-series diverges, by the Comparison Test.

We have convergence for $p > 1$ but divergence for every other value. ■

The Limit Comparison Test

We now present a more powerful form of the Comparison Test, known as the Limit Comparison Test. It is particularly handy when we deal with series in which a_n is a rational function of n. The next example will show you what we mean.

EXAMPLE 7 Do the following series converge, or diverge?

a) $\displaystyle\sum_{n=2}^{\infty} \frac{2n}{n^2 - n + 1}$ b) $\displaystyle\sum_{n=2}^{\infty} \frac{2n^3 + 100n^2 + 1000}{(1/8)n^6 - n + 2}$

Solution In determining convergence or divergence, only the tails count. And, when n is very large, the highest powers of n in numerator and denominator are what count the most. So, in (a), we reason this way:

$$a_n = \frac{2n}{n^2 - n + 1}$$

behaves about like $2n/n^2 = 2/n$, and, by comparison with $\Sigma\, 1/n$, we guess that $\Sigma\, a_n$ diverges. In (b), we reason that a_n will behave about like

$2n^3/(1/8)n^6 = 16/n^3$ and, by comparison with $\Sigma\, 1/n^3$, a p-series with $p = 3$, we guess that the series converges.

To be more precise, in part (a) we take

$$a_n = \frac{2n}{n^2 - n + 1} \quad \text{and} \quad d_n = \frac{1}{n}$$

and look at the ratio $$\frac{a_n}{d_n} = \frac{2n^2}{n^2 - n + 1} = \frac{2}{1 - \left(\frac{1}{n}\right) + \left(\frac{1}{n^2}\right)}.$$

Clearly, as $n \to \infty$ the limit is 2: $\lim \dfrac{a_n}{d_n} = 2.$

This means that, in particular, if we take $\epsilon = 1$ in the definition of limit, we know there is an integer N such that a_n/d_n is within 1 unit of this limit for all $n \geq N$:

$$2 - 1 \leq a_n/d_n \leq 2 + 1 \qquad \text{for} \qquad n \geq N.$$

Thus $a_n \geq d_n$ for $n \geq N$. Therefore, by the Comparison Test, $\Sigma\, a_n$ diverges because $\Sigma\, d_n$ diverges.

In part (b), if we let $c_n = 1/n^3$, we can show that $\lim \dfrac{a_n}{c_n} = 16.$

Taking $\epsilon = 1$ in the definition of limit, we can conclude that there is an index N' such that a_n/c_n is between 15 and 17 when $n \geq N'$. Since $\Sigma\, c_n$ converges, so also does $\Sigma\, 17c_n$ and thus $\Sigma\, a_n$. ■

Our rather rough guesswork paved the way for successful choices of comparison series. We make all of this more precise in the following Limit Comparison Test.

Limit Comparison Test

a) **Test for convergence.** If $a_n \geq 0$ for $n \geq n_0$ and there is a convergent series $\Sigma\, c_n$ such that $c_n > 0$ and

$$\lim \frac{a_n}{c_n} < \infty, \tag{8}$$

then $\Sigma\, a_n$ converges.

b) **Test for divergence.** If $a_n \geq 0$ for $n \geq n_0$ and there is a divergent series $\Sigma\, d_n$ such that $d_n > 0$ and

$$\lim \frac{a_n}{d_n} > 0, \tag{9}$$

then $\Sigma\, a_n$ diverges.

A simpler version of the Limit Comparison Test combines parts (a) and (b) in the following way.

Simplified Limit Comparison Test

If the terms of the two series $\Sigma\, a_n$ and $\Sigma\, b_n$ are positive for $n \geq n_0$, and the limit of a_n/b_n is finite and positive, then both series converge or both diverge.

The Simplified Limit Comparison Test is the one we use most often.

EXAMPLE 8 Do the following series converge, or diverge?

a) $\frac{3}{4}+\frac{5}{9}+\frac{7}{16}+\frac{9}{25}+\cdots=\sum_{n=1}^{\infty}\frac{2n+1}{(n+1)^2}$

b) $\frac{1}{2}+\frac{2}{3}+\frac{3}{4}+\frac{4}{5}+\frac{5}{6}+\cdots=\sum_{n=1}^{\infty}\frac{n}{n+1}$

c) $\frac{101}{3}+\frac{102}{10}+\frac{103}{29}+\cdots=\sum_{n=1}^{\infty}\frac{100+n}{n^3+2}$

d) $\frac{1}{1}+\frac{1}{3}+\frac{1}{7}+\cdots=\sum_{n=1}^{\infty}\frac{1}{2^n-1}$

Solution

a) Let $a_n=(2n+1)/(n^2+2n+1)$ and $d_n=1/n$. Then

$$\sum d_n \text{ diverges} \qquad \text{and} \qquad \lim\frac{a_n}{d_n}=\lim\frac{2n^2+n}{n^2+2n+1}=2,$$

so $\Sigma\, a_n$ diverges.

b) Let $b_n=n/(n+1)$. Then $\lim b_n=1\neq 0$, so $\Sigma\, b_n$ diverges, by the nth Term Test.

c) Let $a_n=(100+n)/(n^3+2)$. When n is large, this ought to compare with $n/n^3=1/n^2$, so we let $c_n=1/n^2$ and apply the Limit Comparison Test:

$$\sum c_n \text{ converges} \qquad \text{and} \qquad \lim\frac{a_n}{c_n}=\lim\frac{n^3+100n^2}{n^3+2}=1,$$

so $\Sigma\, a_n$ converges.

d) Let $a_n=1/(2^n-1)$ and $c_n=1/2^n$. (We reason that 2^n-1 behaves somewhat like 2^n when n is large.) Then

$$\frac{a_n}{c_n}=\frac{2^n}{2^n-1}=\frac{1}{1-(1/2)^n}\to 1 \qquad \text{as } n\to\infty.$$

Because $\Sigma\, c_n$ converges, we conclude that $\Sigma\, a_n$ does also. ■

Exercises 9.3

Which series in Exercises 1–24 converge and which diverge? Give reasons for your answers.

1. $\sum_{n=1}^{\infty}\frac{1}{10^n}$
2. $\sum_{n=1}^{\infty}\frac{n}{n+2}$
3. $\sum_{n=1}^{\infty}\frac{\sin^2 n}{2^n}$
4. $\sum_{n=1}^{\infty}\frac{5}{n}$
5. $\sum_{n=1}^{\infty}\frac{1+\cos n}{n^2}$
6. $\sum_{n=1}^{\infty}-\frac{1}{8^n}$
7. $\sum_{n=2}^{\infty}\frac{\ln n}{n}$
8. $\sum_{n=1}^{\infty}\frac{1}{n\sqrt{n}}$
9. $\sum_{n=1}^{\infty}\frac{2^n}{3^n}$
10. $\sum_{n=0}^{\infty}\frac{-2}{n+1}$
11. $\sum_{n=1}^{\infty}\frac{1}{1+\ln n}$
12. $\sum_{n=1}^{\infty}\frac{1}{2n-1}$
13. $\sum_{n=1}^{\infty}\frac{2^n}{n+1}$
14. $\sum_{n=1}^{\infty}\left(\frac{n}{3n+1}\right)^n$
15. $\sum_{n=1}^{\infty}\frac{1}{\sqrt{n^3+2}}$

16. $\sum_{n=1}^{\infty} \frac{\sqrt{n}}{\ln n}$ 17. $\sum_{n=1}^{\infty} \frac{n}{n^2+1}$ 18. $\sum_{n=1}^{\infty} \frac{1}{n\sqrt[n]{n}}$

19. $\sum_{n=1}^{\infty} \left(1+\frac{1}{n}\right)^n$ 20. $\sum_{n=1}^{\infty} \frac{\sqrt{n}}{n^2+1}$ 21. $\sum_{n=1}^{\infty} \frac{1-n}{n \cdot 2^n}$

22. $\sum_{n=1}^{\infty} \frac{1}{(\ln 2)^n}$ 23. $\sum_{n=1}^{\infty} \frac{1}{3^{n-1}+1}$

24. $\sum_{n=1}^{\infty} \frac{10n+1}{n(n+1)(n+2)}$

25. **CALCULATOR** There is absolutely no empirical evidence for the divergence of the harmonic series even though we know it diverges. The partial sums, which satisfy the inequality

$$\ln(n+1) = \int_1^{n+1} \frac{1}{x}\,dx \le 1 + \frac{1}{2} + \cdots + \frac{1}{n}$$
$$\le 1 + \int_1^n \frac{1}{x}\,dx = 1 + \ln n$$

(Eq. 6), just grow too slowly. To see what we mean, suppose you had started with $s_1 = 1$ the day the universe was formed, thirteen billion years ago, and added a new term every *second*. About how large would s_n be today?

26. There are no values of x for which $\Sigma_{n=1}^{\infty} (1/nx)$ converges. Why?

27. Show that if $\Sigma_{n=1}^{\infty} a_n$ is a convergent series of nonnegative numbers then $\Sigma_{n=1}^{\infty} (a_n/n)$ converges.

28. Show that if $\Sigma\, a_n$ and $\Sigma\, b_n$ are convergent series with $a_n \ge 0$ and $b_n \ge 0$, then $\Sigma\, a_n b_n$ converges. (*Hint:* From some integer on, $0 \le a_n$ and $b_n < 1$ so $a_n b_n \le a_n + b_n$.)

29. *Nonincreasing sequences.* A sequence of numbers $\{s_n\}$ in which $s_n \ge s_{n+1}$ for every n is called a **nonincreasing sequence.** A sequence $\{s_n\}$ is bounded from below if there is a finite constant M with $M \le s_n$ for every n. Such a number M is called a lower bound for the sequence. Deduce from Theorem 6 that a nonincreasing sequence that is bounded from below converges, and that a nonincreasing sequence that is not bounded from below diverges.

30. *The Cauchy condensation test.* The Cauchy condensation test says: Let $\{a_n\}$ be a nonincreasing sequence ($a_n \ge a_{n+1}$ for all n) of positive terms that converges to 0. Then $\Sigma\, a_n$ converges if and only if $\Sigma\, 2^n a_{2^n}$ converges. For example, $\Sigma\,(1/n)$ diverges because $\Sigma\, 2^n \cdot (1/2^n) = \Sigma\, 1$ diverges.

Use the Cauchy condensation test to show that

a) $\sum_{n=2}^{\infty} \frac{1}{n \ln n}$ diverges.

b) $\sum_{n=1}^{\infty} \frac{1}{n^p}$ converges if $p > 1$ and diverges if $p \le 1$

9.4 Series with Nonnegative Terms: Ratio and Root Tests

Convergence tests that depend on comparing one series with another series or with an integral are called **extrinsic** tests. They are very useful, but there are reasons to look for tests that do not require comparison. As a practical matter, we may not be able to find the series or function we need to make a comparison work. And, in principle, all the information about a given series should be contained in its own terms. We therefore turn our attention to **intrinsic** tests—those that depend only on the series at hand.

The Ratio Test

Our first intrinsic test, the Ratio Test, measures the rate of growth (or decline) of a series by examining the ratio a_{n+1}/a_n. For a geometric series, this rate of growth is a constant, and the series converges if and only if its ratio is less than 1 in absolute value. But even if the ratio is not constant, we may be able to find a geometric series for comparison, as in the next example.

EXAMPLE 1 Let $a_1 = 1$ and define a_{n+1} to be $a_{n+1} = \dfrac{n}{2n+1} a_n$.

Does the series $\Sigma\, a_n$ converge, or diverge?

Solution We begin by writing out a few terms of the series:

$$a_1 = 1, \qquad a_2 = \frac{1}{3}a_1 = \frac{1}{3}, \qquad a_3 = \frac{2}{5}a_2 = \frac{1\cdot 2}{3\cdot 5}, \qquad a_4 = \frac{3}{7}a_3 = \frac{1\cdot 2\cdot 3}{3\cdot 5\cdot 7}.$$

Each term is somewhat less than 1/2 the term before it, because $n/(2n+1)$ is less than 1/2. Therefore the terms of the given series are less than or equal to the terms of the geometric series

$$1 + \left(\frac{1}{2}\right) + \left(\frac{1}{4}\right) + \cdots + \left(\frac{1}{2}\right)^{n-1} + \cdots$$

that converges to 2. So our series also converges, and its sum is less than 2. Computer data show that the sum is approximately 1.57079 6327. ■

The series in Example 1 converges rapidly, as the following computer data suggest.

n	s_n
5	1.56075
10	1.57055
15	1.57078 9894
20	1.57079 6149
25	1.57079 6322
30	1.57079 6327
35	1.57079 6327

In providing the Ratio Test, we shall make a comparison with appropriate geometric series as in the example above, but when we *apply* it we do not actually make a direct comparison.

The Ratio Test

Let $\Sigma\, a_n$ be a series with positive terms, and suppose that

$$\lim_{n\to\infty} \frac{a_{n+1}}{a_n} = \rho \qquad \text{(Greek letter rho).}$$

Then

a) the series *converges* if $\rho < 1$,

b) the series *diverges* if $\rho > 1$,

c) the series *may converge or it may diverge* if $\rho = 1$. (The test provides no information.)

PROOF of the Ratio Test

a) $\rho < 1$. Let r be a number between ρ and 1. Then the number $\epsilon = r - \rho$ is positive. Since

$$\frac{a_{n+1}}{a_n} \to \rho,$$

a_{n+1}/a_n must lie within ϵ of ρ when n is large enough, say for all $n \geq N$. In particular,

$$\frac{a_{n+1}}{a_n} < \rho + \epsilon = r, \qquad \text{when } n > N.$$

That is,

$$\begin{aligned} a_{N+1} &< ra_N, \\ a_{N+2} &< ra_{N+1} < r^2 a_N, \\ a_{N+3} &< ra_{N+2} < r^3 a_N, \\ &\vdots \\ a_{N+m} &< ra_{N+m-1} < r^m a_N. \end{aligned}$$

These inequalities show that the terms of our series, after the Nth term, approach zero more rapidly than the terms in a geometric series with ratio $r < 1$. More precisely, consider the series $\Sigma\, c_n$, where $c_n = a_n$ for $n = 1, 2, \ldots, N$ and $c_{N+1} = ra_N, c_{N+2} = r^2 a_N, \ldots, \quad c_{N+m} = r^m a_N, \ldots$. Now $a_n \le c_n$ for all n, and

$$\begin{aligned}\sum_{n=1}^{\infty} c_n &= a_1 + a_2 + \cdots + a_{N-1} + a_N + ra_N + r^2 a_N + \cdots \\ &= a_1 + a_2 + \cdots + a_{N-1} + a_N(1 + r + r^2 + \cdots).\end{aligned}$$

The geometric series $1 + r + r^2 + \cdots$ converges because $|r| < 1$, so $\Sigma\, c_n$ converges. Since $a_n \le c_n$, $\Sigma\, a_n$ also converges.

b) $\rho > 1$. From some index M on,

$$\frac{a_{n+1}}{a_n} > 1 \qquad \text{and} \qquad a_M < a_{M+1} < a_{M+2} < \cdots.$$

The terms of the series do not approach zero as n becomes infinite, and the series diverges by the nth-Term Test.

c) $\rho = 1$. The two series

$$\sum_{n=1}^{\infty} \frac{1}{n} \qquad \text{and} \qquad \sum_{n=1}^{\infty} \frac{1}{n^2}$$

show that some other test for convergence must be used when $\rho = 1$.

$$\text{For } \sum_{n=1}^{\infty} \frac{1}{n}: \quad \frac{a_{n+1}}{a_n} = \frac{1/(n+1)}{1/n} = \frac{n}{n+1} \to 1.$$

$$\text{For } \sum_{n=1}^{\infty} \frac{1}{n^2}: \quad \frac{a_{n+1}}{a_n} = \frac{1/(n+1)^2}{1/n^2} = \left(\frac{n}{n+1}\right)^2 \to 1^2 = 1.$$

In both cases $\rho = 1$, yet the first series diverges while the second converges. ■

The Ratio Test is often effective when the terms of the series contain factorials of expressions involving n or expressions raised to the nth power or combinations, as in the next example.

EXAMPLE 2 Use the Ratio Test to investigate the convergence or divergence of the following series.

a) $\displaystyle\sum_{n=1}^{\infty} \frac{n!n!}{(2n)!}$ b) $\displaystyle\sum_{n=1}^{\infty} \frac{4^n n!n!}{(2n)!}$ c) $\displaystyle\sum_{n=0}^{\infty} \frac{2^n + 5}{3^n}$

Solution

a) If $a_n = n!n!/(2n)!$, then $a_{n+1} = (n+1)!(n+1)!/(2n+2)!$, and

$$\begin{aligned}\frac{a_{n+1}}{a_n} &= \frac{(n+1)!(n+1)!(2n)!}{n!n!(2n+2)(2n+1)(2n)!} \\ &= \frac{(n+1)(n+1)}{(2n+2)(2n+1)} = \frac{n+1}{4n+2} \to \frac{1}{4}.\end{aligned}$$

The series converges because $\rho = 1/4$ is less than 1.

b) If $a_n = 4^n n! n!/(2n)!$, then

$$\frac{a_{n+1}}{a_n} = \frac{4^{n+1}(n+1)!(n+1)!}{(2n+2)(2n+1)(2n)!} \times \frac{(2n)!}{4^n n! n!}$$
$$= \frac{4(n+1)(n+1)}{(2n+2)(2n+1)} = \frac{2(n+1)}{2n+1} \to 1.$$

Because the limit is $\rho = 1$, we cannot decide on the basis of the Ratio Test alone whether the series converges or diverges. However, when we note that $a_{n+1}/a_n = (2n+2)/(2n+1)$, we conclude that a_{n+1} is always greater than a_n because $(2n+2)/(2n+1)$ is always greater than 1. Therefore, all terms are greater than or equal to $a_1 = 2$, and the nth term does not go to zero as n tends to infinity. Hence, by the nth-Term Test, the series diverges.

c) For the series $\Sigma_{n=0}^{\infty} (2^n + 5)/3^n$,

$$\frac{a_{n+1}}{a_n} = \frac{(2^{n+1}+5)/3^{n+1}}{(2^n+5)/3^n} = \frac{1}{3} \cdot \frac{2^{n+1}+5}{2^n+5} = \frac{1}{3} \cdot \left(\frac{2 + 5 \cdot 2^{-n}}{1 + 5 \cdot 2^{-n}}\right) \to \frac{1}{3} \cdot \frac{2}{1} = \frac{2}{3}.$$

The series converges because $\rho = 2/3$ is less than 1.

This does *not* mean that 2/3 is the sum of the series. In fact,

$$\sum_{n=0}^{\infty} \frac{2^n+5}{3^n} = \sum_{n=0}^{\infty} \left(\frac{2}{3}\right)^n + \sum_{n=0}^{\infty} \frac{5}{3^n} = \frac{1}{1-(2/3)} + \frac{5}{1-(1/3)} = \frac{21}{2}.$$ ■

The *n*th-Root Test

We return to the question "Does $\Sigma\, a_n$ converge?" When there is a simple formula for a_n, we can try one of the tests we already have. But consider the following example.

EXAMPLE 3 Let $a_n = f(n)/2^n$, where

$$f(n) = \begin{cases} n & \text{if } n \text{ is a prime number,} \\ 1 & \text{otherwise.} \end{cases}$$

Does $\Sigma\, a_n$ converge?

Solution We write out several terms of the series:

$$\sum a_n = \frac{1}{2} + \frac{2}{4} + \frac{3}{8} + \frac{1}{16} + \frac{5}{32} + \frac{1}{64} + \frac{7}{128} + \cdots + \frac{f(n)}{2^n} + \cdots.$$

Clearly, this is not a geometric series. The nth term approaches zero as $n \to \infty$, so we don't know that the series diverges. The Integral Test doesn't look promising. The Ratio Test produces

$$\frac{a_{n+1}}{a_n} = \frac{1}{2}\frac{f(n+1)}{f(n)} = \begin{cases} \dfrac{1}{2} & \text{if neither } n \text{ nor } n+1 \text{ is a prime,} \\[2ex] \dfrac{1}{2n} & \text{if } n \text{ is a prime} \geq 3, \\[2ex] \dfrac{n+1}{2} & \text{if } n+1 \text{ is a prime } \geq 5. \end{cases}$$

The ratio is sometimes close to zero, sometimes is very large, and sometimes is 1/2. It has no limit because there are infinitely many primes. A test that will answer the question (affirmatively—yes, the series does converge) is the nth-Root Test. To apply it, we consider the following:

$$\sqrt[n]{a_n} = \frac{\sqrt[n]{f(n)}}{2} = \begin{cases} \dfrac{\sqrt[n]{n}}{2} & \text{if } n \text{ is a prime,} \\ \dfrac{1}{2} & \text{otherwise.} \end{cases}$$

Therefore,

$$\frac{1}{2} \leq \sqrt[n]{a_n} \leq \frac{\sqrt[n]{n}}{2}$$

and $\lim \sqrt[n]{a_n} = 1/2$ by the Sandwich Theorem. Because this limit is less than 1, the nth-Root Test tells us that the given series converges, as we shall now see. ■

The *n*th-Root Test

Let $\Sigma\, a_n$ be a series with $a_n \geq 0$ for $n \geq n_0$, and suppose that $\sqrt[n]{a_n} \to \rho$. Then

a) the series *converges* if $\rho < 1$,

b) the series *diverges* if $\rho > 1$,

c) the test is *not conclusive* if $\rho = 1$.

PROOF of the *n*th-Root Test

a) $\boldsymbol{\rho < 1}$**.** Choose an $\epsilon > 0$ so small that $\rho + \epsilon < 1$ also. Since $\sqrt[n]{a_n} \to \rho$, the terms $\sqrt[n]{a_n}$ eventually get closer than ϵ to ρ. In other words, there exists an index $N \geq n_0$ such that

$$\sqrt[n]{a_n} < \rho + \epsilon \qquad \text{when } n \geq N.$$

Then it is also true that

$$a_n < (\rho + \epsilon)^n \qquad \text{for } n \geq N.$$

Now, $\Sigma_{n=N}^{\infty} (\rho + \epsilon)^n$, a geometric series with ratio $(\rho + \epsilon) < 1$, converges. By comparison, $\Sigma_{n=N}^{\infty} a_n$ converges, from which it follows that

$$\sum_{n=1}^{\infty} a_n = a_1 + \cdots + a_{N-1} + \sum_{n=N}^{\infty} a_n$$

converges.

b) $\boldsymbol{\rho > 1}$**.** For all indices beyond some integer M, we have $\sqrt[n]{a_n} > 1$, so that $a_n > 1$ for $n > M$. The terms of the series do not converge to zero. The series diverges by the nth-Term Test.

c) $\boldsymbol{\rho = 1}$**.** The series $\Sigma_{n=1}^{\infty} (1/n)$ and $\Sigma_{n=1}^{\infty} (1/n^2)$ show that the test is not conclusive when $\rho = 1$. The first series diverges and the second converges, but in both cases $\sqrt[n]{a_n} \to 1$. ■

EXAMPLE 4 One of the following series converges and the other diverges. Which does which?

a) $\sum_{n=1}^{\infty} \frac{n^2}{2^n}$ b) $\sum_{n=1}^{\infty} \frac{2^n}{n^2}$

Solution Series (a) converges because $\sqrt[n]{a_n} = \frac{\sqrt[n]{n^2}}{2} \to \frac{1}{2} < 1$. But series (b) diverges because $\sqrt[n]{b_n} = \frac{2}{\sqrt[n]{n^2}} \to 2 > 1$. ■

Exercises 9.4

Which series in Exercises 1–26 converge and which diverge? Give reasons for your answers.

1. $\sum_{n=1}^{\infty} \frac{n^2}{2^n}$
2. $\sum_{n=1}^{\infty} \frac{n!}{10^n}$
3. $\sum_{n=1}^{\infty} \frac{n^{10}}{10^n}$
4. $\sum_{n=1}^{\infty} n^2 e^{-n}$
5. $\sum_{n=1}^{\infty} \left(\frac{n-2}{n}\right)^n$
6. $\sum_{n=1}^{\infty} \frac{2+(-1)^n}{1.25^n}$
7. $\sum_{n=1}^{\infty} n!e^{-n}$
8. $\sum_{n=1}^{\infty} \frac{(-2)^n}{3^n}$
9. $\sum_{n=1}^{\infty} \left(1-\frac{3}{n}\right)^n$
10. $\sum_{n=1}^{\infty} \left(1-\frac{1}{n^2}\right)^n$
11. $\sum_{n=1}^{\infty} \frac{\ln n}{n^3}$
12. $\sum_{n=1}^{\infty} \left(\frac{1}{n}-\frac{1}{n^2}\right)$
13. $\sum_{n=1}^{\infty} \frac{\ln n}{n}$
14. $\sum_{n=1}^{\infty} \frac{n \ln n}{2^n}$
15. $\sum_{n=1}^{\infty} \frac{(n+1)(n+2)}{n!}$
16. $\sum_{n=1}^{\infty} e^{-n}(n^3)$
17. $\sum_{n=1}^{\infty} \frac{(n+3)!}{3!n!3^n}$
18. $\sum_{n=1}^{\infty} -\frac{n^2}{2^n}$
19. $\sum_{n=1}^{\infty} \frac{1}{(2n+1)!}$
20. $\sum_{n=1}^{\infty} \frac{n!}{n^n}$
21. $\sum_{n=2}^{\infty} \frac{n}{(\ln n)^n}$
22. $\sum_{n=2}^{\infty} \frac{1}{(\ln n)^2}$
23. $\sum_{n=1}^{\infty} \frac{n!}{(n+2)!}$
24. $\sum_{n=1}^{\infty} \frac{3^n}{n^3 2^n}$
25. $\sum_{n=1}^{\infty} \frac{n!}{(2n+1)!}$
26. $\sum_{n=1}^{\infty} \frac{(n!)^n}{n^{(n^2)}}$

Do the series $\sum_{n=1}^{\infty} a_n$ defined by the formulas in Exercises 27–35 converge, or diverge? Give reasons for your answers.

27. $a_1 = 2, \quad a_{n+1} = \frac{1+\sin n}{n} a_n$
28. $a_1 = \frac{1}{3}, \quad a_{n+1} = \frac{3n-1}{2n+5} a_n$
29. $a_1 = 3, \quad a_{n+1} = \frac{n}{n+1} a_n$
30. $a_1 = 2, \quad a_{n+1} = \frac{2}{n} a_n$
31. $a_1 = -1, \quad a_{n+1} = \frac{1+\ln n}{n} a_n$
32. $a_1 = \frac{1}{2}, \quad a_{n+1} = \frac{n+\ln n}{n+10} a_n$
33. $a_n = \frac{2^n n! n!}{(2n)!}$
34. $a_n = \frac{(3n)!}{n!(n+1)!(n+2)!}$
35. $a_1 = 1, \quad a_{n+1} = \frac{n(n+1)}{(n+2)(n+3)} a_n$ (*Hint:* Write out several terms, see what factors cancel, and then generalize.)
36. Neither the Ratio Test nor the nth-Root Test helps with p-series. Try them on $\sum_{n=1}^{\infty} (1/n^p)$ to see what happens.

9.5 Alternating Series and Absolute Convergence

A series in which the terms are alternately positive and negative is called an **alternating series.** Here are three examples:

$$1 - \frac{1}{2} + \frac{1}{3} - \frac{1}{4} + \frac{1}{5} - \cdots + \frac{(-1)^{n+1}}{n} + \cdots \tag{1}$$

$$-2 + 1 - \frac{1}{2} + \frac{1}{4} - \frac{1}{8} + \cdots + \frac{(-1)^n 4}{2^n} + \cdots \tag{2}$$

$$1 - 2 + 3 - 4 + 5 - 6 + \cdots + (-1)^{n-1} n + \cdots \tag{3}$$

Series (1), called the **alternating harmonic series,** converges, as we shall see in a moment. Series (2), a geometric series with ratio $r = -1/2$, converges to $-2/(1 + 1/2) = -4/3$. Series (3) diverges because the nth term does not approach zero.

We prove the convergence of the alternating harmonic series by applying a general result known as the Alternating Series Theorem.

This theorem is also known as Leibniz's theorem.

THEOREM 7

The Alternating Series Theorem

The series

$$\sum_{n=1}^{\infty} (-1)^{n+1} a_n = a_1 - a_2 + a_3 - a_4 + \cdots \tag{4}$$

converges if all three of the following conditions are satisfied:

1. The a_n's are all positive.
2. $a_n \geq a_{n+1}$ for all n.
3. $a_n \to 0$.

PROOF If n is an even integer, say $n = 2m$, then the sum of the first n terms is

$$\begin{aligned} S_{2m} &= (a_1 - a_2) + (a_3 - a_4) + \cdots + (a_{2m-1} - a_{2m}) \\ &= a_1 - (a_2 - a_3) - (a_4 - a_5) - \cdots - (a_{2m-2} - a_{2m-1}) - a_{2m}. \end{aligned} \tag{5}$$

The first equality shows that S_{2m} is the sum of m nonnegative terms, since each term in parentheses is positive or zero. Hence $S_{2m+2} \geq S_{2m}$, and the sequence $\{S_{2m}\}$ is nondecreasing. The second equality shows that $S_{2m} \leq a_1$. Since $\{S_{2m}\}$ is nondecreasing and bounded from above, it has a limit, say

$$\lim_{m \to \infty} S_{2m} = L. \tag{6}$$

If n is an odd integer, say $n = 2m + 1$, then the sum of the first n terms is

$$S_{2m+1} = S_{2m} + a_{2m+1}.$$

Since $a_n \to 0$,

$$\lim_{m \to \infty} a_{2m+1} = 0$$

and, as $m \to \infty$,

$$S_{2m+1} = S_{2m} + a_{2m+1} \to L + 0 = L. \tag{7}$$

When we combine the results of (6) and (7), we get

$$\lim_{n \to \infty} S_n = L.$$

■

EXAMPLE 1 The *alternating harmonic series*

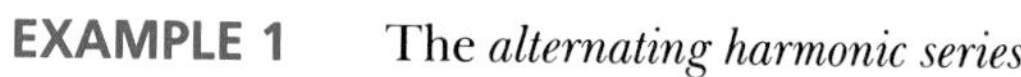

$$\sum_{n=1}^{\infty} (-1)^{n+1} \frac{1}{n} = 1 - \frac{1}{2} + \frac{1}{3} - \frac{1}{4} + \cdots$$

satisfies the three requirements of the theorem; therefore it converges. ■

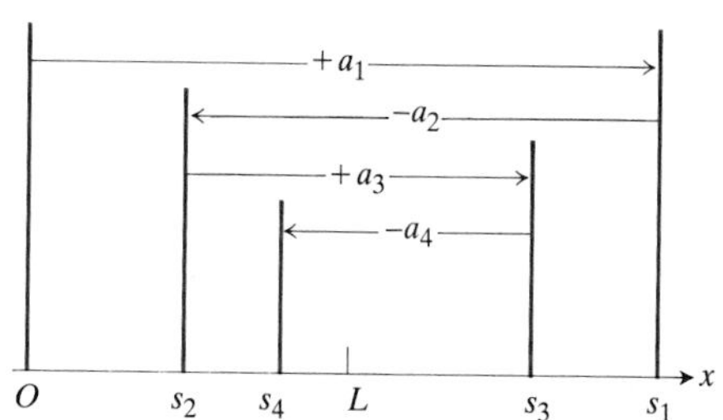

9.15 The partial sums of an alternating series that satisfies the hypotheses of Theorem 7 straddle their limit.

We can use a graphical interpretation of the partial sums to gain added insight into the way in which an alternating series converges to its limit L when the three conditions of the theorem are satisfied. Starting from the origin on a scale of real numbers (Fig. 9.15), we lay off the positive distance $s_1 = a_1$. To find the point corresponding to $s_2 = a_1 - a_2$, we must back up a distance equal to a_2. Since $a_2 \leq a_1$, we do not back up any farther than O at most. Next we go forward a distance a_3 and mark the point corresponding to $s_3 = a_1 - a_2 + a_3$. Since $a_3 \leq a_2$, we go forward by an amount that is no greater than the previous backward step; that is, $s_3 \leq s_1$. We continue in this seesaw fashion, backing up or going forward as the signs in the series demand. But each forward or backward step is shorter than (or at most the same size as) the preceding step, because $a_{n+1} \leq a_n$. And since the nth term approaches zero as n increases, the size of step we take forward or backward gets smaller and smaller. We thus oscillate across the limit L, but the amplitude of oscillation continually decreases and approaches zero as its limit. The even-numbered partial sums $s_2, s_4, s_6, \ldots, s_{2m}$ increase toward L, while the odd-numbered sums $s_1, s_3, s_5, \ldots, s_{2m+1}$ decrease toward L. The limit L is between any two successive sums s_n and s_{n+1} and hence differs from s_n by an amount less than a_{n+1}.

Because

$$|L - s_n| < a_{n+1} \qquad \text{for every } n, \tag{8}$$

we can make useful estimates of the sums of convergent alternating series.

THEOREM 8

The Alternating Series Estimation Theorem

If the alternating series $\sum_{n=1}^{\infty} (-1)^{n+1} a_n$

satisfies the three conditions of Theorem 7, then

$$s_n = a_1 - a_2 + \cdots + (-1)^{n+1} a_n$$

approximates the sum L of the series with an error whose absolute value is less than a_{n+1}, the numerical value of the first unused term. Furthermore, the remainder, $L - s_n$, has the same sign as the first unused term.

We leave the determination of the sign of the remainder for Exercise 45.

EXAMPLE 2 We first try Theorem 8 on an alternating series whose sum we already know, namely the geometric series

$$\sum_{n=0}^{\infty} (-1)^n \frac{1}{2^n} = 1 - \frac{1}{2} + \frac{1}{4} - \frac{1}{8} + \frac{1}{16} - \frac{1}{32} + \frac{1}{64} - \frac{1}{128} \,\vdots + \frac{1}{256} - \cdots.$$

The theorem says that if we truncate the series after the eighth term, we throw away a total that is positive and less than 1/256. The sum of the first eight terms is 0.6640625. The sum of the series is

$$\frac{1}{1-(-1/2)} = \frac{1}{3/2} = \frac{2}{3}.$$

The difference,

$$\frac{2}{3} - 0.6640625 = 0.0026041666\ldots,$$

is positive and less than

$$\frac{1}{256} = 0.00390625.$$

■

Absolute Convergence

DEFINITION

A series $\Sigma\, a_k$ **converges absolutely** (is **absolutely convergent**) if the corresponding series of absolute values, $\Sigma\, |a_k|$, is convergent.

The geometric series

$$1 - \frac{1}{2} + \frac{1}{4} - \frac{1}{8} + \cdots$$

converges absolutely because the corresponding series of absolute values

$$1 + \frac{1}{2} + \frac{1}{4} + \frac{1}{8} + \cdots$$

converges. But the alternating harmonic series, although it converges, does not converge absolutely because the corresponding series of absolute values is the (divergent) harmonic series.

DEFINITION

A series that converges but does not converge absolutely **converges conditionally.**

The alternating harmonic series converges conditionally.

The concept of absolute convergence is important because, first, we have many good tests for convergence of series of positive terms. Second, if a series converges absolutely, then it converges. That is the thrust of the next theorem.

THEOREM 9

The Absolute Convergence Theorem

If $\Sigma_{n=1}^{\infty} |a_n|$ converges, then $\Sigma_{n=1}^{\infty} a_n$ converges.

PROOF of Theorem 9 For each n,

$$-|a_n| \le a_n \le |a_n|,$$

so

$$0 \le a_n + |a_n| \le 2|a_n|.$$

If $\Sigma_{n=1}^{\infty} |a_n|$ converges, then $\Sigma_{n=1}^{\infty} 2|a_n|$ converges and, by the Comparison Test, the nonnegative series $\Sigma_{n=1}^{\infty} (a_n + |a_n|)$ converges. The equality $a_n = (a_n + |a_n|) - |a_n|$ now lets us express $\Sigma_{n=1}^{\infty} a_n$ as the difference of two convergent series:

$$\sum_{n=1}^{\infty} a_n = \sum_{n=1}^{\infty} (a_n + |a_n| - |a_n|) = \sum_{n=1}^{\infty} (a_n + |a_n|) - \sum_{n=1}^{\infty} |a_n|.$$

Therefore, $\Sigma_{n=1}^{\infty} a_n$ converges. ■

We can rephrase Theorem 9 to say that *every absolutely convergent series converges.* However, the converse statement is false. Many convergent series do not converge absolutely. The convergence of many series depends on the series' having infinitely many positive and negative terms arranged in a particular order.

EXAMPLE 3 For $\displaystyle\sum_{n=1}^{\infty} (-1)^{n+1}\frac{1}{n^2} = 1 - \frac{1}{4} + \frac{1}{9} - \frac{1}{16} + \cdots$, the corresponding series of absolute values is

$$\sum_{n=1}^{\infty} \frac{1}{n^2} = 1 + \frac{1}{4} + \frac{1}{9} + \frac{1}{16} + \cdots.$$

The series converges because it is a p-series with $p = 2 > 1$. Therefore

$$\sum_{n=1}^{\infty} (-1)^{n+1} \frac{1}{n^2}$$

converges absolutely. Therefore

$$\sum_{n=1}^{\infty} (-1)^{n+1} \frac{1}{n^2}$$

converges. ■

EXAMPLE 4 For $\sum_{n=1}^{\infty} \frac{\sin n}{n^2} = \frac{\sin 1}{1} + \frac{\sin 2}{4} + \frac{\sin 3}{9} + \cdots$, the corresponding series of absolute values is

$$\sum_{n=1}^{\infty} \left| \frac{\sin n}{n^2} \right| = \frac{|\sin 1|}{1} + \frac{|\sin 2|}{4} + \cdots,$$

which converges by comparison with $\Sigma_{n=1}^{\infty} (1/n^2)$ because $|\sin n| \leq 1$ for every n. The original series converges absolutely; therefore it converges. ■

Alternating p-series

When p is a positive constant, the sequence $\{1/n^p\}$ is a decreasing sequence with limit zero. Therefore the alternating p-series

$$\sum_{n=1}^{\infty} \frac{(-1)^{n-1}}{n^p} = 1 - \frac{1}{2^p} + \frac{1}{3^p} - \frac{1}{4^p} + \cdots, \qquad p > 0$$

converges.

For $p > 1$ the series converges absolutely. For $0 < p \leq 1$ the series converges conditionally. For example,

$$1 - \frac{1}{\sqrt{2}} + \frac{1}{\sqrt{3}} - \frac{1}{\sqrt{4}} + \cdots$$

converges conditionally, while

$$1 - \frac{1}{2^{3/2}} + \frac{1}{3^{3/2}} - \frac{1}{4^{3/2}} + \cdots$$

converges absolutely.

Preview of Power Series

In the next section we shall take up the subject of power series, series in which the nth term is a constant times x^n. We investigate the convergence of such series with the Ratio Test or nth-Root Test, as in the next example.

EXAMPLE 5 Find all values of x for which the series

$$x - \frac{x^2}{2} + \frac{x^3}{3} - \frac{x^4}{4} + \cdots + (-1)^{n-1}\frac{x^n}{n} + \cdots$$

converges.

Solution The series converges absolutely for all values of x for which

$$\lim_{n\to\infty} \sqrt[n]{\left|(-1)^{n-1}\frac{x^n}{n}\right|} = \lim_{n\to\infty} \frac{|x|}{\sqrt[n]{n}} = |x|$$

is less than 1.

At $x = 1$, the series is the alternating harmonic series, which converges. At $x = -1$, the series is the negative of the harmonic series and diverges. For $|x| > 1$, the nth term of the series does not approach zero, and the series diverges. Hence the series converges only for $-1 < x \leq 1$. ■

FLOWCHART 9.1 Procedure for Determining Convergence

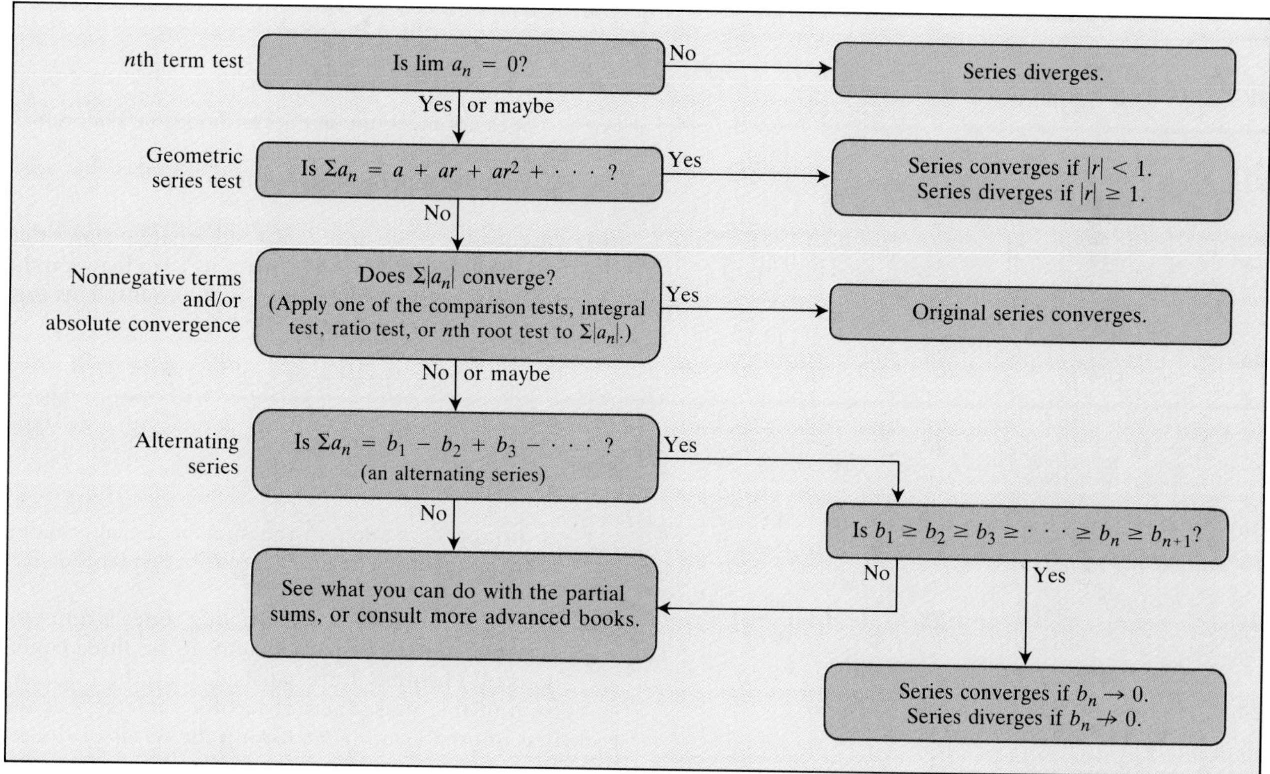

Exercises 9.5

Which of the alternating series in Exercises 1–10 converge and which diverge?

1. $\sum_{n=1}^{\infty} (-1)^{n+1} \frac{1}{n^2}$

2. $\sum_{n=2}^{\infty} (-1)^{n+1} \frac{1}{\ln n}$

3. $\sum_{n=1}^{\infty} (-1)^{n+1}$

4. $\sum_{n=1}^{\infty} (-1)^{n+1} \frac{10^n}{n^{10}}$

5. $\sum_{n=1}^{\infty} (-1)^{n+1} \frac{\sqrt{n}+1}{n+1}$

6. $\sum_{n=1}^{\infty} (-1)^{n+1} \frac{\ln n}{n}$

7. $\sum_{n=1}^{\infty} (-1)^{n+1} \frac{1}{n^{3/2}}$

8. $\sum_{n=1}^{\infty} (-1)^{n+1} \frac{\ln n}{\ln n^2}$

9. $\sum_{n=1}^{\infty} (-1)^{n+1} \frac{3\sqrt{n}+1}{\sqrt{n}+1}$

10. $\sum_{n=1}^{\infty} (-1)^{n} \ln\left(1+\frac{1}{n}\right)$

Which of the series in Exercises 11–36 converge absolutely, which converge conditionally, and which diverge?

11. $\sum_{n=1}^{\infty} (-1)^{n+1}(0.1)^n$

12. $\sum_{n=1}^{\infty} (-1)^{n+1}\frac{1}{\sqrt{n}}$

13. $\sum_{n=1}^{\infty} (-1)^{n+1}\frac{n}{n^3+1}$

14. $\sum_{n=1}^{\infty} \frac{n!}{2^n}$

15. $\sum_{n=1}^{\infty} (-1)^n\frac{1}{n+3}$

16. $\sum_{n=1}^{\infty} (-1)^n\frac{\sin n}{n^2}$

17. $\sum_{n=1}^{\infty} (-1)^{n+1}\frac{3+n}{5+n}$

18. $\sum_{n=2}^{\infty} (-1)^n\frac{1}{\ln n^3}$

19. $\sum_{n=1}^{\infty} (-1)^{n+1}\frac{1+n}{n^2}$

20. $\sum_{n=1}^{\infty} \frac{(-2)^{n+1}}{n+5^n}$

21. $\sum_{n=1}^{\infty} n^2(2/3)^n$

22. $\sum_{n=1}^{\infty} (-1)^{n+1}(\sqrt[n]{10})$

23. $\sum_{n=1}^{\infty} (-1)^n\frac{\tan^{-1} n}{n^2+1}$

24. $\sum_{n=2}^{\infty} (-1)^{n+1}\frac{1}{n \ln n}$

25. $\sum_{n=1}^{\infty} \left(\frac{1}{n}-\frac{1}{2n}\right)$

26. $\sum_{n=1}^{\infty} (-1)^{n+1}\frac{(0.1)^n}{n}$

27. $\sum_{n=1}^{\infty} (-1)^n\frac{n}{n+1}$

28. $\sum_{n=1}^{\infty} \frac{(-1)^n}{1+\sqrt{n}}$

29. $\sum_{n=1}^{\infty} \frac{-1}{n^2+2n+1}$

30. $\sum_{n=1}^{\infty} (5)^{-n}$

31. $\sum_{n=1}^{\infty} \frac{(-100)^n}{n!}$

32. $\sum_{n=1}^{\infty} (-1)^n\left(\frac{\ln n}{\ln n^2}\right)^n$

33. $\sum_{n=1}^{\infty} \frac{\cos n\pi}{n\sqrt{n}}$

34. $\sum_{n=1}^{\infty} \frac{\cos n\pi}{n}$

35. $\sum_{n=1}^{\infty} \frac{(-1)^n}{\sqrt{n}+\sqrt{n+1}}$

36. $\sum_{n=1}^{\infty} \frac{(-1)^{n+1}(n!)^2}{(2n)!}$

In Exercises 37–40, estimate the error involved in using the sum of the first four terms to approximate the sum of the entire series.

37. $\sum_{n=1}^{\infty} (-1)^{n+1}\frac{1}{n}$

38. $\sum_{n=1}^{\infty} (-1)^{n+1}\frac{1}{10^n}$

39. $\ln(1.01) = \sum_{n=1}^{\infty} (-1)^{n+1}\frac{(0.01)^n}{n}$

40. $\frac{1}{1+t} = \sum_{n=0}^{\infty} (-1)^n t^n, \quad 0<t<1$

CALCULATOR Approximate the sums in Exercises 41 and 42 with an error of magnitude less than 5×10^{-6}.

41. $\sum_{n=0}^{\infty} (-1)^n\frac{1}{(2n)!}$ (The sum is cos 1, the cosine of one radian.)

42. $\sum_{n=0}^{\infty} (-1)^n\frac{1}{n!}$ (The sum is $\frac{1}{e}$.)

43. a) The series

$$\frac{1}{3}-\frac{1}{2}+\frac{1}{9}-\frac{1}{4}+\frac{1}{27}-\frac{1}{8}+\cdots+\frac{1}{3^n}-\frac{1}{2^n}+\cdots$$

does not meet one of the conditions of Theorem 7. Which one?

b) Find the sum of the series in (a).

44. CALCULATOR The limit L of an alternating series that satisfies the conditions of Theorem 7 lies between the values of any two consecutive partial sums. This suggests using the average

$$\frac{s_n+s_{n+1}}{2}=s_n+\frac{1}{2}a_{n+1}$$

to estimate L. Compute

$$s_{20}+\frac{1}{2}\cdot\frac{1}{21}$$

as an approximation to the sum of the alternating harmonic series. The exact sum is $\ln 2 = 0.6931\ldots$.

45. Show that whenever an alternating series is approximated by one of its partial sums, if the three conditions of Theorem 7 are satisfied, then the *remainder* (sum of the unused terms) has the same sign as the first unused term. (*Hint:* Group the terms of the remainder in consecutive pairs.)

46. Show that the sum of the first $2n$ terms of the series

$$1-\frac{1}{2}+\frac{1}{2}-\frac{1}{3}+\frac{1}{3}-\frac{1}{4}+\frac{1}{4}-\frac{1}{5}+\frac{1}{5}-\frac{1}{6}+\cdots$$

is the same as the sum of the first n terms of the series

$$\frac{1}{1\cdot 2}+\frac{1}{2\cdot 3}+\frac{1}{3\cdot 4}+\frac{1}{4\cdot 5}+\frac{1}{5\cdot 6}+\cdots.$$

Do these series converge? What is the sum of the first $2n+1$ terms of the first series? If the series converge, what is their sum?

9.6 Power Series

Now that we know how to test infinite series for convergence we can study the infinite polynomials we mentioned at the beginning of Section 9.2. We call these infinite polynomials power series because they are defined as infinite series of powers of some variable, in our case x. Like polynomials, power series can be added, subtracted, multiplied, differentiated, and integrated to give new power series. They can be divided, too, but we shall not go into that. Almost any function with infinitely many derivatives can be represented by a power series, as long as the derivatives don't become too large. The series for e^x, $\sin x$, $\cos x$, $\ln(1 + x)$, $\tan^{-1}x$, and so on, enable us to approximate the values of these functions as accurately as we please. We shall show you what these series are as the chapter continues.

Power Series and Convergence

We begin with the formal definition.

DEFINITIONS

A **power series** is a series of the form

$$\sum_{n=0}^{\infty} c_n x^n = c_0 + c_1 x + c_2 x^2 + \cdots + c_n x^n + \cdots \tag{1}$$

or

$$\sum_{n=0}^{\infty} c_n (x - a)^n = c_0 + c_1(x - a) + c_2(x - a)^2 + \cdots + c_n (x - a)^n + \cdots \tag{2}$$

in which the **center** a and the **coefficients** $c_0, c_1, c_2, \ldots c_n, \ldots$ are constants.

Equation (1) is the special case obtained by taking $a = 0$ in Eq. (2).

EXAMPLE 1 Taking all the coefficients to be 1 in Eq. (1) gives the geometric power series

$$\sum_{n=0}^{\infty} x^n = 1 + x + x^2 + \cdots + x^n + \cdots .$$

The series converges to $1/(1 - x)$ when $|x| < 1$. We express this fact by writing

$$\frac{1}{1 - x} = 1 + x + x^2 + \cdots + x^n + \cdots \quad \text{for} \quad -1 < x < 1. \tag{3}$$

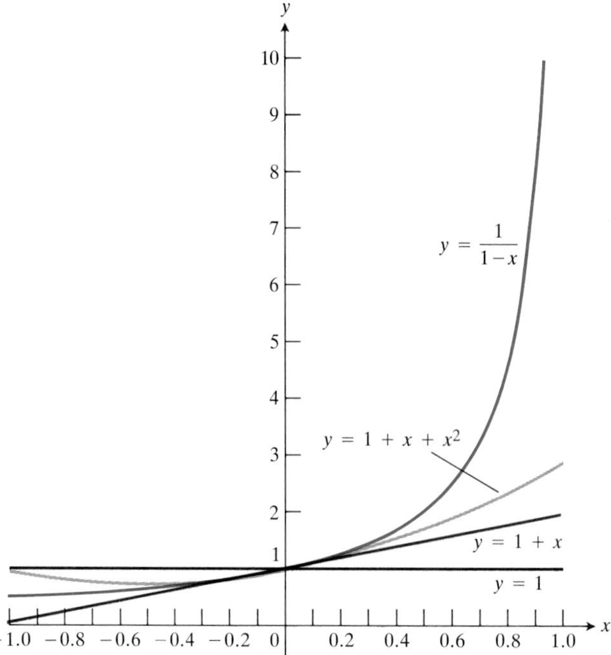

9.16 The graphs of $y = 1/(1 - x)$ and the approximating polynomials $P_0(x) = 1$, $P_1(x) = 1 + x$, and $P_2(x) = 1 + x + x^2$.

Up to now we have used Eq. (3) as a formula to give us the sum of the series on the right. We now change the focus: We think of the partial sums of the series on the right as polynomials $P_n(x)$ that we can use to approximate the function on the left. For values of x near zero, we need take only a few terms of the series to get a good approximation. As we move toward $x = 1$, or -1, we must take more terms. Figure 9.16 shows the graphs of $y = 1/(1 - x)$, and the approximating polynomials $y = P_n(x)$ for $n = 0$, 1, and 2. ■

EXAMPLE 2 The power series

$$1 - \frac{1}{2}(x - 2) + \frac{1}{4}(x - 2)^2 + \cdots + \left(-\frac{1}{2}\right)^n (x - 2)^n + \cdots \tag{4}$$

matches Eq. (2) with $a = 2$, $c_0 = 1$, $c_1 = -1/2$, $c_2 = 1/4, \ldots, c_n = (-1/2)^n$. This is a geometric series with first term 1 and ratio $r = -\dfrac{x - 2}{2}$. The series converges for $\left|\dfrac{x - 2}{2}\right| < 1$ or $0 < x < 4$. The sum is

$$\frac{1}{1 - r} = \frac{1}{1 + \dfrac{x - 2}{2}} = \frac{2}{x},$$

so we write

$$\frac{2}{x} = 1 - \frac{(x - 2)}{2} + \frac{(x - 2)^2}{4} - \cdots + \left(-\frac{1}{2}\right)^n (x - 2)^n + \cdots, \qquad 0 < x < 4.$$

We use series (4) to generate polynomial approximations to $2/x$ (or $1/x$ if we divide by 2) for values of x near 2. We would keep the powers of $(x - 2)$ and write the approximating polynomials as

$$P_0(x) = 1, \quad P_1(x) = 1 - \frac{1}{2}(x - 2), \quad P_2(x) = 1 - \frac{1}{2}(x - 2) + \frac{1}{4}(x - 2)^2$$

and so on. The higher powers of $(x - 2)$ decrease rapidly as n increases when $|x - 2|$ is small. ■

EXAMPLE 3 For what values of x do the following series converge?

a) $\displaystyle\sum_{n=1}^{\infty} (-1)^{n-1} \frac{x^n}{n} = x - \frac{x^2}{2} + \frac{x^3}{3} - \cdots$

b) $\displaystyle\sum_{n=1}^{\infty} (-1)^{n-1} \frac{x^{2n-1}}{2n-1} = x - \frac{x^3}{3} + \frac{x^5}{5} - \cdots$

c) $\displaystyle\sum_{n=0}^{\infty} \frac{x^n}{n!} = 1 + x + \frac{x^2}{2!} + \frac{x^3}{3!} + \cdots$

d) $\displaystyle\sum_{n=0}^{\infty} n!\,x^n = 1 + x + 2!\,x^2 + 3!\,x^3 + \cdots$

Solution Apply the ratio test to the series $\Sigma\,|u_n|$, where u_n is the nth term of the series in question.

a) $\displaystyle\left|\frac{u_{n+1}}{u_n}\right| = \frac{n}{n+1}|x| \to |x|$. The series converges absolutely for $|x| < 1$. It diverges if $|x| > 1$ because the nth term does not converge to zero. At $x = 1$, we get the alternating harmonic series $1 - 1/2 + 1/3 - 1/4 + \cdots$, which converges. At $x = -1$ we get $-1 - 1/2 - 1/3 - 1/4 - \cdots$, which is the negative of the harmonic series; it diverges. Series (a) converges for $-1 < x \leq 1$ and diverges elsewhere.

b) $\displaystyle\left|\frac{u_{n+1}}{u_n}\right| = \frac{2n-1}{2n+1}x^2 \to x^2$. The series converges absolutely for $x^2 < 1$. It diverges for $x^2 > 1$ because the nth term does not converge to zero. At $x = 1$ the series becomes $1 - 1/3 + 1/5 - 1/7 + \cdots$, which converges by the Alternating Series Theorem. It also converges at $x = -1$ because it again is an alternating series that satisfies the conditions for convergence. The value at $x = -1$ is the negative of the value at $x = 1$.

c) $\displaystyle\left|\frac{u_{n+1}}{u_n}\right| = \left|\frac{x^{n+1}}{(n+1)!} \cdot \frac{n!}{x^n}\right| = \frac{|x|}{n+1} \to 0$ for every x. The series converges absolutely for all x.

d) $\displaystyle\left|\frac{u_{n+1}}{u_n}\right| = \left|\frac{(n+1)!\,x^{n+1}}{n!\,x^n}\right| = (n+1)\,|x| \to \infty$ unless $x = 0$. The series diverges for all values of x except $x = 0$. The nth term does not approach zero unless $x = 0$. ■

Example 3 illustrates how we usually test a power series for convergence, and the kinds of results we'll get.

How to Test a Power Series for Convergence

STEP 1: Use the Ratio Test (or nth Root Test) to find the interval where the series converges absolutely. Ordinarily, this is an open interval

$$|x - a| < h \qquad \text{or} \qquad a - h < x < a + h.$$

In some instances, as in Example 3(c), the series converges for all values of x. These instances are not uncommon. In rare cases, the series may converge only at a single point, as in Example 3(d).

STEP 2: If the interval of absolute convergence is finite, test for convergence or divergence at each of the two endpoints as in Example 2 and Examples 3(a) and (b). Neither the Ratio Test nor the nth Root Test is useful at these points. Use a Comparison Test, the Integral Test, or the Alternating Series Theorem.

STEP 3: If the interval of absolute convergence is $a - h < x < a + h$, the series diverges (it does not even converge conditionally) for $|x - a| > h$, because for those values of x the nth term does not approach zero.

In the next section we shall see how series in powers of $(x - a)$ are generated by the values of a function f and its derivatives at $x = a$. Since $1/x$, $\sqrt{x}$, and $\ln x$ exist and have derivatives of all orders at $x = 1$, the power series we use to represent them can be in powers of $x - 1$. But they do not have power series representations in powers of x because they have no derivatives at $x = 0$.

To simplify the notation, the next theorem deals with the convergence of series of the form $\sum a_n x^n$. For series of the form $\sum a_n (x - a)^n$ we can replace $x - a$ by x' and apply the results to the series $\sum a_n (x')^n$.

THEOREM 10

The Convergence Theorem for Power Series

If
$$\sum_{n=0}^{\infty} a_n x^n = a_0 + a_1 x + a_2 x^2 + \cdots \tag{5}$$

converges for $x = c$ $(c \neq 0)$, then it converges absolutely for all $|x| < |c|$. If the series diverges for $x = d$, then it diverges for all $|x| > |d|$.

PROOF Suppose the series

$$\sum_{n=0}^{\infty} a_n c^n \tag{6}$$

converges. Then

$$\lim_{n\to\infty} a_n c^n = 0.$$

Hence, there is an integer N such that $|a_n c^n| < 1$ for all $n \geq N$. That is,

$$|a_n| < \frac{1}{|c|^n} \qquad \text{for } n \geq N. \tag{7}$$

Now take any x such that $|x| < |c|$ and consider

$$|a_0| + |a_1 x| + \cdots + |a_{N-1} x^{N-1}| + |a_N x^N| + |a_{N+1} x^{N+1}| + \cdots.$$

There is only a finite number of terms prior to $|a_N x^N|$, and their sum is finite. Starting with $|a_N x^N|$ and beyond, the terms are less than

$$\left|\frac{x}{c}\right|^N + \left|\frac{x}{c}\right|^{N+1} + \left|\frac{x}{c}\right|^{N+2} + \cdots \tag{8}$$

because of (7). But the series in (8) is a geometric series with ratio $r = |x/c|$, which is less than 1, since $|x| < |c|$. Hence the series (8) converges, so the original series (6) converges absolutely. This proves the first half of the theorem.

The second half of the theorem follows from the first. If the series diverges at $x = d$ and converges at a value x_0 with $|x_0| > |d|$, we may take $c = x_0$ in the first half of the theorem and conclude that the series converges absolutely at d. But the series cannot converge absolutely and diverge at one and the same time. Hence, if it diverges at d, it diverges for all $|x| > |d|$. ■

The Radius and Interval of Convergence

The examples we have looked at, and the theorem we just proved, lead to the conclusion that a power series always behaves in exactly one of the following three ways.

Possible Behavior of $\Sigma\, c_n(x - a)^n$

1. The series converges at $x = a$ and diverges elsewhere.
2. There is a positive number h such that the series diverges for $|x - a| > h$ but converges absolutely for $|x - a| < h$. The series may or may not converge at either of the endpoints $x = a - h$ and $x = a + h$.
3. The series converges absolutely for every x.

In case 2, the set of points at which the series converges is a finite interval, called the **interval of convergence.** We know from past examples that the interval may be open, half-open, or closed, depending on the particular series. But no matter which kind of interval it is, h is called the **radius of convergence** of the series, and $a + h$ is the least upper bound of the set of points at which the series converges. The convergence is absolute at every point in the interior of the interval. If $a = 0$, the interval is centered at the origin. If a power series converges absolutely for all values of x, we say that its radius of convergence is infinite. If it converges only at $x = a$, we say that the radius of convergence is zero.

Term-by-Term Differentiation of Power Series

A theorem from advanced calculus tells us that a power series can be differentiated term by term at each point in the interior of its interval of convergence.

A Word of Caution

Term-by-term differentiation might not work for other kinds of series. For example, the trigonometric series

$$\sum_{n=1}^{\infty} \frac{\sin(n!x)}{n^2}$$

converges for every x. But if we differentiate term by term we get the series

$$\sum_{n=1}^{\infty} \frac{n!\cos(n!x)}{n^2},$$

which diverges for all x.

THEOREM 11

The Term-by-Term Differentiation Theorem

If $\Sigma c_n(x-a)^n$ converges for $a-h<x<a+h$ for some $h>0$, it defines a function f:

$$f(x) = \sum_{n=0}^{\infty} c_n(x-a)^n, \qquad a-h<x<a+h.$$

Such a function f has derivatives of all orders inside the interval of convergence. We can obtain the derivatives by differentiating the original series term by term:

$$f'(x) = \sum_{n=0}^{\infty} n c_n (x-a)^{n-1},$$

$$f''(x) = \sum_{n=0}^{\infty} n(n-1) c_n (x-a)^{n-2},$$

and so on. Each of these derived series converges at every interior point of the interval of convergence of the original series.

Here is an example of how to apply term-by-term differentiation.

EXAMPLE 4 Is there a more familiar name for the function f defined by the power series

$$f(x) = \sum_{n=1}^{\infty} \frac{(-1)^n x^{2n-1}}{2n-1} = x - \frac{x^3}{3} + \frac{x^5}{5} - \cdots, \quad -1 \le x \le 1?$$

Solution We differentiate the original series term by term and get

$$f'(x) = \sum_{n=1}^{\infty} \frac{(2n-1)(-1)^n x^{2n-2}}{2n-1} = \sum_{n=1}^{\infty} (-1)^n x^{2n-2} = 1 - x^2 + x^4 - x^6 + \cdots.$$

This is a geometric series with first term 1 and ratio $-x^2$, so

$$f'(x) = \frac{1}{1-(-x^2)} = \frac{1}{1+x^2}.$$

We can now integrate $f'(x) = 1/(1+x^2)$ to get

$$\int f'(x)\,dx = \int \frac{dx}{1+x^2} = \tan^{-1}x + C.$$

The series for $f(x)$ is zero when $x=0$, so $C=0$. Hence

$$f(x) = x - \frac{x^3}{3} + \frac{x^5}{5} - \frac{x^7}{7} + \cdots = \tan^{-1}x, \quad -1 \le x \le 1. \tag{9}$$

The function is the restriction of $\tan^{-1}x$ to the interval $[-1, 1]$. ■

Notice that the original series in Example 4 converges at both endpoints of the original interval of convergence, but the differentiated series converges only inside the interval. This is all that the theorem guarantees.

Term-by-Term Integration of Power Series

Another theorem from advanced calculus states that a power series can also be integrated term by term throughout its interval of convergence.

THEOREM 12

The Term-by-Term Integration Theorem

If $\Sigma c_n(x-a)^n$ converges when $a-h<x<a+h$ for some $h>0$ and $f(x)=\sum_{n=0}^{\infty} c_n(x-a)^n$ for $a-h<x<a+h$, then the series $\sum c_n \frac{(x-a)^{n+1}}{n+1}$ converges for $a-h<x<a+h$ and

$$\int f(x)\,dx = \sum_{n=0}^{\infty} c_n \frac{(x-a)^{n+1}}{n+1} + C \quad \text{for } a-h<x<a+h.$$

EXAMPLE 5 The series

$$\frac{1}{1+t} = 1 - t + t^2 - t^3 \cdots$$

converges on the open interval $-1<t<1$. Therefore,

$$\begin{aligned}\ln(1+x) &= \int_0^x \frac{1}{1+t}\,dt = t - \frac{t^2}{2} + \frac{t^3}{3} - \frac{t^4}{4} \cdots \Big]_0^x \\ &= x - \frac{x^2}{2} + \frac{x^3}{3} - \frac{x^4}{4} + \cdots, \qquad -1<x<1.\end{aligned}$$

As you know, the latter series also converges at $x=1$, but that was not guaranteed by the theorem.

Multiplication of Power Series

Still another theorem from advanced calculus states that power series can be multiplied term by term.

THEOREM 13

The Series Multiplication Theorem for Power Series

If both $\Sigma a_n x^n$ and $\Sigma b_n x^n$ converge absolutely for $|x|<h$, and

$$c_n = a_0 b_n + a_1 b_{n-1} + a_2 b_{n-2} + \cdots + a_{n-1} b_1 + a_n b_0 = \sum_{k=0}^{n} a_k b_{n-k}, \tag{10}$$

then the series $\Sigma c_n x^n$ also converges absolutely for $|x|<h$, and

$$(\Sigma a_n x^n)\cdot(\Sigma b_n x^n) = \Sigma c_n x^n. \tag{11}$$

EXAMPLE 6 Multiply the geometric series

$$\sum_{n=0}^{\infty} x^n = 1 + x + x^2 + \cdots + x^n + \cdots = \frac{1}{1-x}, \text{ for } |x| < 1,$$

by itself to get the power series for $1/(1-x)^2$, for $|x| < 1$.

Solution Let

$$A(x) = \sum_{n=0}^{\infty} a_n x^n = 1 + x + x^2 + \cdots + x^n + \cdots = 1/(1-x)$$

$$B(x) = \sum_{n=0}^{\infty} b_n x^n = 1 + x + x^2 + \cdots + x^n + \cdots = 1/(1-x)$$

and

$$c_n = \underbrace{a_0 b_n + a_1 b_{n-1} + \cdots + a_k b_{n-k} + \cdots a_n b_0}_{n+1 \text{ terms}} = \underbrace{1 + 1 + \cdots + 1}_{n+1 \text{ ones}} = n + 1.$$

Then, by the series multiplication theorem,

$$A(x) \cdot B(x) = \sum_{n=0}^{\infty} c_n x^n = \sum_{n=0}^{\infty} (n+1) x^n$$
$$= 1 + 2x + 3x^2 + 4x^3 + \cdots + (n+1)x^n + \cdots$$

is the series for $1/(1-x)^2$. The series all converge absolutely for $|x| < 1$.

Exercises 9.6

Each of Exercises 1–24 gives a formula for the nth term of a series. For what values of x does each series (a) converge and (b) converge absolutely?

1. x^n
2. $(x+5)^n$
3. $(-1)^n(x+1)^n$
4. $(x-2)^n/n$
5. $(x-2)^n/10^n$
6. $(2x)^n$
7. $nx^n/(n+2)$
8. $(-1)^n(x+2)^n/n$
9. $x^n/(n\sqrt{n})$
10. $(x-1)^n/\sqrt{n}$
11. $(-1)^n x^n/n!$
12. $3^n x^n/n!$
13. $x^{2n+1}/n!$
14. $(x-3)^{2n+1}/n!$
15. $x^n/\sqrt{n^2+3}$
16. $(-1)^n x^n/\sqrt{n^2+3}$
17. $nx^n/(n^2+1)$
18. $n(x-3)^n$
19. $\sqrt{n}\, x^n/3^n$
20. $\sqrt[n]{n}\,(x-1)^n$
21. $(1+1/n)^n x^n$
22. $(\ln n)\, x^n$
23. $n^n x^n$
24. $n!(x-4)^n$

25. Where does the geometric series

$$1 - \frac{1}{2}(x-3) + \frac{1}{4}(x-3)^2 + \cdots + \left(-\frac{1}{2}\right)^n (x-3)^n + \cdots$$

converge? What is the sum? What series do you get if you differentiate the given series term by term? Where does that series converge? What is its sum?

26. If you integrate the series of Exercise 25 term by term, what new series do you get? Where does the new series converge, and what is another name for its sum?

27. The series for $\tan x$,

$$\tan x = x + \frac{x^3}{3} + \frac{2x^5}{15} + \frac{17x^7}{315} + \cdots,$$

converges for $-\pi/2 < x < \pi/2$.

a) Find the first four terms of the series for $\ln|\sec x|$. For what values of x should the series converge?

b) Find the first three terms of the series for $\sec^2 x$. For what values of x should this series converge?

c) Check your result in (b) by squaring the series given for $\sec x$ in Exercise 28.

28. The series for $\sec x$,

$$\sec x = 1 + \frac{x^2}{2} + \frac{5}{24}x^4 + \frac{61}{720}x^6 + \cdots,$$

converges for $-\pi/2 < x < \pi/2$.

a) Find the first four terms of a series for the function $\ln|\sec x + \tan x|$. For what values of x should the series converge?

b) Find the first three terms of a series for $\sec x \tan x$. For what values of x should the series converge?

c) Check your result in (b) by multiplying the series for $\sec x$ by the series given for $\tan x$ in Exercise 27.

9.7 Taylor Series and Maclaurin Series

This section shows how to find power series for functions that have infinitely many derivatives and shows how to control the errors involved in using the partial sums of these series as polynomial approximations of the functions they represent. As a special case, we shall at last see how to control the errors in linearizations. One of the important features of Taylor series is that they enable us to extend the domains of functions to include the complex numbers. We shall go into this briefly at the end.

DEFINITIONS

Let f be a function with derivatives of all orders throughout some interval containing a as an interior point. Then the **Taylor series generated by f at a** is

$$\sum_{k=0}^{\infty} \frac{f^{(k)}(a)}{k!}(x-a)^k = f(a) + f'(a)(x-a) + \frac{f''(a)}{2!}(x-a)^2 + \cdots + \frac{f^{(n)}(a)}{n!}(x-a)^n + \cdots \qquad (1)$$

and the **Maclaurin series generated by f** is

$$\sum_{k=0}^{\infty} \frac{f^{(k)}(0)}{k!}x^k = f(0) + f'(0)x + \frac{f''(0)}{2!}x^2 + \cdots + \frac{f^{(n)}(0)}{n!}x^n + \cdots, \qquad (2)$$

the Taylor series generated by f at $x = 0$.

Once we have found the Taylor series generated by the function f at a particular a, we can apply our usual tests to find where the series converges; usually in some interval $(a - h, a + h)$ or for all x. When the series does converge, we ask "Does it converge to $f(x)$?"

EXAMPLE 1 Find the Taylor series generated by $f(x) = 1/x$ at $a = 2$. Where, if anywhere, does the series converge to $1/x$?

Solution We need to compute $f(2), f'(2), f''(2)$, and so on. Taking derivatives we get

$$
\begin{aligned}
f(x) &= x^{-1}, & f(2) &= 2^{-1} = \frac{1}{2},\\
f'(x) &= -x^{-2}, & f'(2) &= -\frac{1}{2^2},\\
f''(x) &= 2!\,x^{-3}, & \frac{f''(2)}{2!} &= 2^{-3} = \frac{1}{2^3},\\
f'''(x) &= -3!\,x^{-4}, & \frac{f'''(2)}{3!} &= -\frac{1}{2^4},\\
&\vdots\\
f^{(n)}(x) &= (-1)^n n!\,x^{-(n+1)}, & \frac{f^{(n)}(2)}{n!} &= \frac{(-1)^n}{2^{n+1}}.
\end{aligned}
$$

The Taylor series is

$$
\begin{aligned}
f(2) + f'(2)(x-2) &+ \frac{f''(2)}{2!}(x-2)^2 + \frac{f'''(2)}{3!}(x-2)^3 + \cdots + \frac{f^{(n)}(2)}{n!}(x-2)^n + \cdots \\
&= \frac{1}{2} - \frac{(x-2)}{2^2} + \frac{(x-2)^2}{2^3} - \cdots + (-1)^n \frac{(x-2)^n}{2^{n+1}} + \cdots .
\end{aligned}
$$

This is a geometric series with first term 1/2 and ratio $r = -(x-2)/2$. It converges absolutely for $|x-2| < 2$ and its sum is

$$\frac{1/2}{1 + (x-2)/2} = \frac{1}{2 + (x-2)} = \frac{1}{x}.$$

In this example the Taylor series generated by $f(x) = 1/x$ at $a = 2$ converges to $1/x$ for $|x-2| < 2$ or $0 < x < 4$. ■

Taylor Polynomials

The linearization of a function f at a point a where f is differentiable is the polynomial

$$P_1(x) = f(a) + f'(a)(x-a).$$

If f has derivatives of higher order at a then it has higher order polynomial approximations as well, one for each available derivative. These polynomials are called the Taylor polynomials of f.

DEFINITION

Let f be a function with derivatives of order k for $k = 1, 2, \ldots, N$ in some interval containing a as an interior point. Then for any integer n from 0 through N the **Taylor polynomial** of order n generated by f at a is the polynomial

$$P_n(x) = f(a) + f'(a)(x-a) + \frac{f''(a)}{2!}(x-a)^2 + \cdots + \frac{f^{(k)}(a)}{k!}(x-a)^k + \cdots + \frac{f^{(n)}(a)}{n!}(x-a)^n. \tag{3}$$

We speak of a Taylor polynomial of *order n* rather than *degree n* because $f^{(n)}(a)$ may be zero. The first two Taylor polynomials of $\cos x$ at $x = 0$, for example, are $P_0(x) = 1$ and $P_1(x) = 1$. The first order polynomial has degree zero, not one.

Special Property of Taylor Polynomials

What's so special about Taylor polynomials? We answer by looking for polynomials whose values at $x = a$ are equal to $f(a)$ and whose derivatives at $x = a$ are $f'(a), f''(a)$ and so on. With that object in mind, we start with a polynomial of order n in powers of $x - a$ with undetermined coefficients c_0, $c_1, c_2, \ldots, c_n$:

$$P(x) = c_0 + c_1(x - a) + c_2(x - a)^2 + c_3(x - a)^3 + \cdots + c_n(x - a)^n. \tag{4}$$

Its derivatives of order 1, 2, ..., n are

$$\begin{aligned}
P'(x) &= c_1 + 2c_2(x - a) + 3c_3(x - a)^2 + \cdots + nc_n(x - a)^{n-1}\\
P''(x) &= 2c_2 + (3)(2)c_3(x - a) + \cdots + n(n - 1)c_n(x - a)^{n-2}\\
P'''(x) &= (3!)c_3 + \cdots + n(n - 1)(n - 2)c_n(x - a)^{n-3}\\
&\vdots\\
P^{(n)}(x) &= (n!)\, c_n
\end{aligned}$$

When we substitute $x = a$, the terms with $x - a$ become 0. We also want $P(a) = f(a)$, $P'(a) = f'(a)$, $P''(a) = f''(a), \ldots, P^{(n)}(a) = f^{(n)}(a)$. This leads to the equations:

$$P(a) = f(a) = c_0, \qquad P'(a) = f'(a) = c_1, \qquad P''(a) = f''(a) = (2!)c_2,$$
$$P'''(a) = f'''(a) = (3!)c_3, \quad \ldots, \quad P^{(n)}(a) = f^{(n)}(a) = (n!)c_n.$$

When we determine the coefficients c_k in this way we get

$$c_0 = f(a), \qquad c_1 = f'(a), \qquad c_2 = \frac{f''(a)}{2!}, \qquad c_3 = \frac{f'''(a)}{3!}, \quad \ldots, \quad c_n = \frac{f^{(n)}(a)}{n!}.$$

Substituting these values for $c_0, c_1, c_2, \ldots, c_n$ in Eq. (4) gives the Taylor polynomial $P_n(x)$ in Eq. (3). The special property of the Taylor polynomials is just this:

> The Taylor polynomial of order n, and its first n derivatives, have the same values as f and its corresponding derivatives at $x = a$.

A function that has derivatives of all orders at $x = a$ generates a Taylor polynomial for every $n \geq 0$.

EXAMPLE 2 Find the Taylor polynomials generated by $f(x) = e^x$ at $a = 0$.

Solution Expressed in terms of x, the given function and its derivatives are

$$f(x) = e^x, \qquad f'(x) = e^x, \quad \ldots, \quad f^{(n)}(x) = e^x,$$

so

$$f(0) = e^0 = 1, \qquad f'(0) = 1, \quad \ldots, \quad f^{(n)}(0) = 1,$$

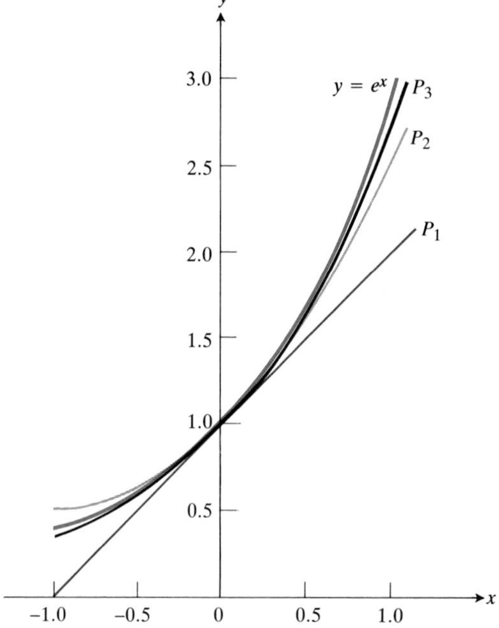

9.17 The graph of $f(x) = e^x$ and its Taylor polynomials $P_1(x) = 1 + x$, $P_2(x) = 1 + x + x^2/2!$, and $P_3(x) = 1 + x + x^2/2! + x^3/3!$. Notice the close agreement near the center $x = 0$.

and

$$P_n(x) = 1 + x + \frac{x^2}{2!} + \frac{x^3}{3!} + \cdots + \frac{x^n}{n!}.$$

See Fig. 9.17. ■

EXAMPLE 3 Find the Taylor polynomials generated by $f(x) = \cos x$ at $a = 0$.

Solution The cosine and its derivatives are

$$\begin{aligned} f(x) &= \cos x, & f'(x) &= -\sin x, \\ f''(x) &= -\cos x, & f^{(3)}(x) &= \sin x, \\ &\vdots & &\vdots \\ f^{(2n)}(x) &= (-1)^n \cos x, & f^{(2n+1)}(x) &= (-1)^{n+1} \sin x. \end{aligned}$$

When $x = 0$, the cosines are 1 and the sines are 0, so

$$f^{(2n)}(0) = (-1)^n, \qquad f^{(2n+1)}(0) = 0.$$

Notice that the Taylor polynomials of order $2n$ and of order $2n + 1$ are identical:

$$P_{2n}(x) = P_{2n+1}(x) = 1 - \frac{x^2}{2!} + \frac{x^4}{4!} - \cdots + (-1)^n \frac{x^{2n}}{(2n)!}.$$

Figure 9.18 shows how well these polynomials approximate $y = \cos x$ near $x = 0$. Only the right-hand portions of the graphs are shown because the graphs are symmetric about the y-axis. ■

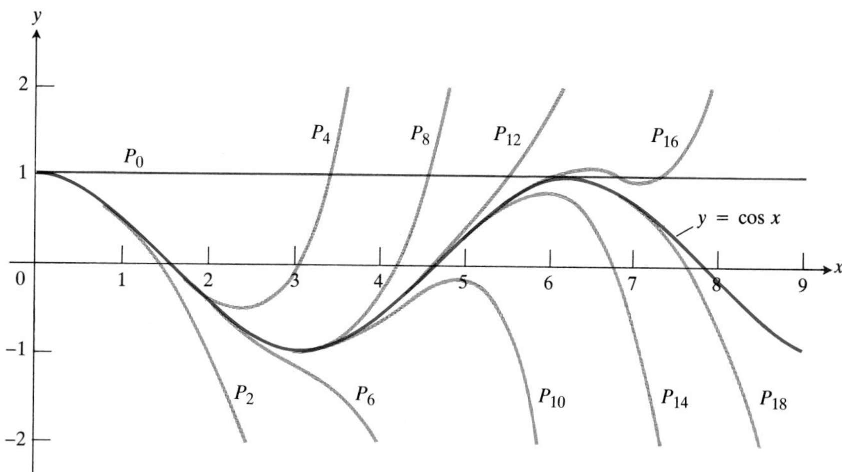

9.18 The polynomials

$$P_{2n}(x) = \sum_{k=0}^{n} ((-1)^k x^{2k}/(2k)!)$$

converge to $\cos x$ as $n \to \infty$. Notice how we can deduce the behavior of $\cos x$ arbitrarily far away solely from the values of the cosine and its derivatives at $a = 0$.

Taylor's Theorem with Remainder

In the preceding examples there was a Taylor polynomial of order n for $n = 0, 1, 2, \ldots$. As $n \to \infty$, the Taylor polynomials become the partial sums of the Taylor series generated by f at a. The next theorem helps us find out whether these sums converge to $f(x)$ in some interval $(a - h, a + h)$ or, perhaps, for all x.

THEOREM 14

Taylor's Theorem

If f and its first n derivatives $f', f'', \ldots, f^{(n)}$ are continuous on $[a, b]$ or on $[b, a]$, and $f^{(n)}$ is differentiable on (a, b) or on (b, a), then there exists a number c between a and b such that

$$f(b) = f(a) + f'(a)(b - a) + \frac{f''(a)}{2!}(b - a)^2 + \cdots$$

$$+ \frac{f^{(n)}(a)}{n!}(b - a)^n + \frac{f^{(n+1)}(c)}{(n + 1)!}(b - a)^{n+1}.$$

We prove the theorem assuming $a < b$. The proof for the case $a > b$ is nearly the same.

PROOF (for $a < b$) The Taylor polynomial

$$P_n(x) = f(a) + f'(a)(x - a) + \frac{f''(a)}{2!}(x - a)^2 + \cdots + \frac{f^{(n)}(a)}{n!}(x - a)^n$$

and its first n derivatives match the function f and its first n derivatives at $x = a$. We do not disturb that matching if we add another term of the form

$K(x-a)^{n+1}$, where K is any constant, because such a term and its first n derivatives are all equal to zero at $x = a$. The new function

$$\phi_n(x) = P_n(x) + K(x-a)^{n+1}$$

and its first n derivatives still agree with f and its first n derivatives at $x = a$.

We now choose the particular value of K that makes the curve $y = \phi_n(x)$ agree with the original curve $y = f(x)$ at $x = b$. This can be done: We need only satisfy

$$f(b) = P_n(b) + K(b-a)^{n+1} \qquad \text{or} \qquad K = \frac{f(b) - P_n(b)}{(b-a)^{n+1}}. \tag{5}$$

With K defined by Eq. (5), the function

$$F(x) = f(x) - \phi_n(x) \tag{6}$$

measures the difference between the original function f and the approximating function ϕ_n for each x in $[a, b]$.

We now use Rolle's theorem. First, because $F(a) = F(b) = 0$ and both F and F' are continuous on $[a, b]$, we know that

$$F'(c_1) = 0 \qquad \text{for some } c_1 \text{ in } (a, b).$$

Next, because $F'(a) = F'(c_1) = 0$ and both F' and F'' are continuous on $[a, c_1]$, we know that

$$F''(c_2) = 0 \qquad \text{for some } c_2 \text{ in } (a, c_1).$$

Rolle's theorem, applied successively to $F'', F''', \ldots, F^{(n-1)}$ implies the existence of

$$\begin{aligned} &c_3 \text{ in } (a, c_2) && \text{such that } F'''(c_3) = 0, \\ &c_4 \text{ in } (a, c_3) && \text{such that } F^{(4)}(c_4) = 0, \\ &\vdots \\ &c_n \text{ in } (a, c_{n-1}) && \text{such that } F^{(n)}(c_n) = 0. \end{aligned}$$

Finally, because $F^{(n)}$ is continuous on $[a, c_n]$ and differentiable on (a, c_n) and $F^{(n)}(a) = F^{(n)}(c_n) = 0$, Rolle's theorem implies that there is a number c_{n+1} in (a, c_n) such that

$$F^{(n+1)}(c_{n+1}) = 0. \tag{7}$$

If we differentiate

$$F(x) = f(x) - P_n(x) - K(x-a)^{n+1}$$

$n + 1$ times, we get

$$F^{(n+1)}(x) = f^{(n+1)}(x) - 0 - (n+1)!\,K. \tag{8}$$

Equations (7) and (8) together give

$$K = \frac{f^{(n+1)}(c)}{(n+1)!} \qquad \text{for some number } c = c_{n+1} \text{ in } (a, b). \tag{9}$$

Equations (5) and (9) give

$$\frac{f(b) - P_n(b)}{(b-a)^{n+1}} = \frac{f^{(n+1)}(c)}{(n+1)!} \quad \text{or} \quad f(b) = P_n(b) + \frac{f^{(n+1)}(c)}{(n+1)!}(b-a)^{n+1}.$$

This concludes the proof. ■

When we apply Taylor's theorem, we usually want to hold a fixed and treat b as an independent variable. Taylor's formula is easier to use in circumstances like these if we change b to x. Here is how the theorem reads with this change.

If f has derivatives of all orders in an open interval I containing a, then for each positive integer n and for each x in I,

$$f(x) = f(a) + f'(a)(x-a) + \frac{f''(a)}{2!}(x-a)^2 + \cdots + \frac{f^{(n)}(a)}{n!}(x-a)^n + R_n(x), \tag{10}$$

where

$$R_n(x) = \frac{f^{(n+1)}(c)}{(n+1)!}(x-a)^{n+1} \quad \text{for some } c \text{ between } a \text{ and } x. \tag{11}$$

When we state Taylor's theorem this way, it says that for each x in I,

$$f(x) = P_n(x) + R_n(x). \tag{12}$$

Pause for a moment to think about how remarkable this equation is. For any value of n we want, the equation gives both a polynomial approximation of f of that order and a formula for the error involved in using that approximation over the interval I.

Equation (10) is called **Taylor's formula.** The function $R_n(x)$ is called the **remainder of order n** or the **error term** for the approximation of f by $P_n(x)$ over I. When $R_n(x) \to 0$ as $n \to \infty$, for all x in I, we say that the Taylor series generated by f at $x = a$ **converges** to f on I and we write

$$f(x) = \sum_{k=0}^{\infty} \frac{f^{(k)}(a)}{k!}(x-a)^k. \tag{13}$$

EXAMPLE 4 *The Maclaurin series for e^x.* Show that the Taylor series generated by $f(x) = e^x$ at $x = 0$ converges to $f(x)$ for every real value of x.

Solution The function has derivatives of all orders throughout the interval $-\infty < x < \infty$. Equations (10) and (11) with $f(x) = e^x$ and $a = 0$ give

$$e^x = 1 + x + \frac{x^2}{2!} + \cdots + \frac{x^n}{n!} + R_n(x), \quad \text{(Polynomial from Example 2)}$$

and

$$R_n(x) = \frac{e^c}{(n+1)!}x^{n+1} \quad \text{for some } c \text{ between 0 and } x.$$

Since e^x is an increasing function of x, e^c lies between $e^0 = 1$ and e^x. When x is negative, so is c, and $e^c < 1$. When x is zero, $e^x = 1$ and $R_n(x) = 0$. When x is positive, so is c, and $e^c < e^x$. Thus,

$$|R_n(x)| \leq \frac{|x|^{n+1}}{(n+1)!} \qquad \text{when } x \leq 0,$$

and

$$|R_n(x)| < e^x \frac{x^{n+1}}{(n+1)!} \qquad \text{when } x > 0.$$

Finally, because

$$\lim_{n\to\infty} \frac{x^{n+1}}{(n+1)!} = 0 \qquad \text{for every } x, \qquad \text{(Section 9.1)}$$

it is also true that $\lim_{n\to\infty} R_n(x) = 0$ and the series converges to e^x for every x.

$$e^x = \sum_{k=0}^{\infty} \frac{x^k}{k!} = 1 + x + \frac{x^2}{2!} + \cdots + \frac{x^k}{k!} + \cdots. \tag{14}$$

■

Estimating the Remainder

It is often possible to estimate $R_n(x)$ as we did in Example 4. This method of estimation is so convenient that we state it as a theorem for future reference.

THEOREM 15

The Remainder Estimation Theorem

If there are positive constants M and r such that $|f^{(n+1)}(t)| \leq Mr^{n+1}$ for all t between a and x, inclusive, then the remainder term $R_n(x)$ in Taylor's theorem satisfies the inequality

$$|R_n(x)| \leq M \frac{r^{n+1}|x-a|^{n+1}}{(n+1)!}.$$

If these conditions hold for every n and all the other conditions of Taylor's theorem are satisfied by $f(x)$, then the series converges to $f(x)$.

In the simplest examples, we can take $r = 1$ provided f and all its derivatives are bounded by some constant M. But if $f(x) = 2\cos(3x)$, each time we differentiate we get a factor of 3, so we could take $r = 3$ and $M = 2$.

We are now ready to look at some examples of how the Remainder Estimation Theorem and Taylor's theorem can be used together to settle questions of convergence. As you will see, they can also be used to determine the accuracy with which a function is approximated by one of its Taylor polynomials.

EXAMPLE 5 *The Maclaurin series for* $\sin x$. Show that the Maclaurin series for $\sin x$ converges to $\sin x$ for all x.

Solution The function and its derivatives are

$$\begin{aligned} f(x) &= \sin x, & f'(x) &= \cos x, \\ f''(x) &= -\sin x, & f'''(x) &= -\cos x, \\ &\vdots \\ f^{(2k)}(x) &= (-1)^k \sin x, & f^{(2k+1)}(x) &= (-1)^k \cos x, \end{aligned}$$

so

$$f^{(2k)}(0) = 0 \quad \text{and} \quad f^{(2k+1)}(0) = (-1)^k.$$

The series has only odd-powered terms and, for $n = 2k + 1$, Taylor's theorem gives

$$\sin x = x - \frac{x^3}{3!} + \frac{x^5}{5!} - \cdots + \frac{(-1)^k x^{2k+1}}{(2k+1)!} + R_{2k+1}(x).$$

All the derivatives of $\sin x$ have absolute values less than or equal to 1 so we can apply the Remainder Estimation Theorem with $M = 1$ and $r = 1$ to obtain

$$|R_{2k+1}(x)| \le 1 \cdot \frac{|x|^{2k+2}}{(2k+2)!}.$$

Since $(|x|^{2k+2}/(2k+2)!) \to 0$ as $k \to \infty$, whatever the value of x, $R_{2k+1}(x) \to 0$, and the Maclaurin series for $\sin x$ converges to $\sin x$ for every x.

$$\sin x = \sum_{k=0}^{\infty} \frac{(-1)^k x^{2k+1}}{(2k+1)!} = x - \frac{x^3}{3!} + \frac{x^5}{5!} - \frac{x^7}{7!} + \cdots. \tag{15}$$

■

EXAMPLE 6 *The Maclaurin series for* $\cos x$. Show that the Maclaurin series for $\cos x$ converges to $\cos x$ for every value of x.

Solution We add the remainder term to the Taylor polynomial for $\cos x$ in Example 3 to obtain Taylor's formula for $\cos x$ with $n = 2k$:

$$\cos x = 1 - \frac{x^2}{2!} + \frac{x^4}{4!} - \cdots + (-1)^k \frac{x^{2k}}{(2k)!} + R_{2k}(x).$$

Because the derivatives of the cosine have absolute value less than or equal to 1, the Remainder Estimation Theorem with $M = 1$ and $r = 1$ gives

$$|R_{2k}(x)| \le 1 \cdot \frac{|x|^{2k+1}}{(2k+1)!}.$$

For every value of x, $R_{2k} \to 0$ as $k \to \infty$. Therefore, the series converges to $\cos x$ for every value of x.

$$\cos x = \sum_{k=0}^{\infty} \frac{(-1)^k x^{2k}}{(2k)!} = 1 - \frac{x^2}{2!} + \frac{x^4}{4!} - \frac{x^6}{6!} + \cdots. \tag{16}$$

■

Brook Taylor (1685–1731) did not invent Taylor series, and Maclaurin series were not developed by Colin Maclaurin (1698–1746). James Gregory was working with Taylor series when Taylor was only a few years old, and he published the Maclaurin series for $\tan x$, $\sec x$, $\tan^{-1}x$, and $\sec^{-1}x$ ten years before Maclaurin was born. At about the same time, Nicolaus Mercator discovered the Maclaurin series for $\ln(1+x)$.

Taylor was not aware of Gregory's work when he published his book *Methodus incrementorum directa et inversa* in 1715, which contained what we now call Taylor series. Maclaurin quoted Taylor's work in a calculus book he wrote in 1742. The book popularized series representations of functions and although Maclaurin never claimed to have discovered them, Taylor series centered at $a = 0$ became known as Maclaurin series. History evened things up in the end. Maclaurin, a brilliant mathematician, was the original discoverer of the rule for solving systems of equations that we now call Cramer's rule.

EXAMPLE 7 Find the Maclaurin series for $\cos 2x$ and show that it converges to $\cos 2x$ for every value of x.

Solution We replace the x in Eq. (16) by $2x$, to obtain

$$\cos 2x = \sum_{k=0}^{\infty} \frac{(-1)^k(2x)^{2k}}{(2k)!} = 1 - \frac{(2x)^2}{2!} + \frac{(2x)^4}{4!} - \frac{(2x)^6}{6!} + \cdots.$$

The Maclaurin series for $\cos x$ converges to $\cos x$ for every value of x and therefore converges for every value of $2x$. ■

Truncation Error

Here are some examples of how to use the Remainder Estimation Theorem to estimate truncation error.

EXAMPLE 8 Calculate e with an error of less than 10^{-6}.

Solution We can use the result of Example 4 with $x = 1$ to write

$$e = 1 + 1 + \frac{1}{2!} + \cdots + \frac{1}{n!} + R_n(1),$$

with

$$R_n(1) = e^c \frac{1}{(n+1)!} \qquad \text{for some } c \text{ between 0 and 1.}$$

For the purposes of this example, we assume that we know that $e < 3$. Hence, we are certain that

$$\frac{1}{(n+1)!} < R_n(1) < \frac{3}{(n+1)!}$$

because $1 < e^c < 3$.

By experiment we find that $1/9! > 10^{-6}$, while $3/10! < 10^{-6}$. Thus we should take $(n+1)$ to be at least 10, or n to be at least 9. With an error of less than 10^{-6},

$$e = 1 + 1 + \frac{1}{2} + \frac{1}{3!} + \cdots + \frac{1}{9!} = 2.718282. \qquad \text{(rounded)}$$ ■

EXAMPLE 9 For what values of x can we replace $\sin x$ by $x - (x^3/3!)$ with an error of magnitude no greater than 3×10^{-4}?

Solution Here we can take advantage of the fact that the Maclaurin series for $\sin x$ is an alternating series for every nonzero value of x. According to the Alternating Series Estimation Theorem (Section 9.5), the error in truncating

$$\sin x = x - \frac{x^3}{3!} \;\vdots\; + \frac{x^5}{5!} - \cdots$$

after $(x^3/3!)$ is no greater than

$$\left|\frac{x^5}{5!}\right| = \frac{|x|^5}{120}.$$

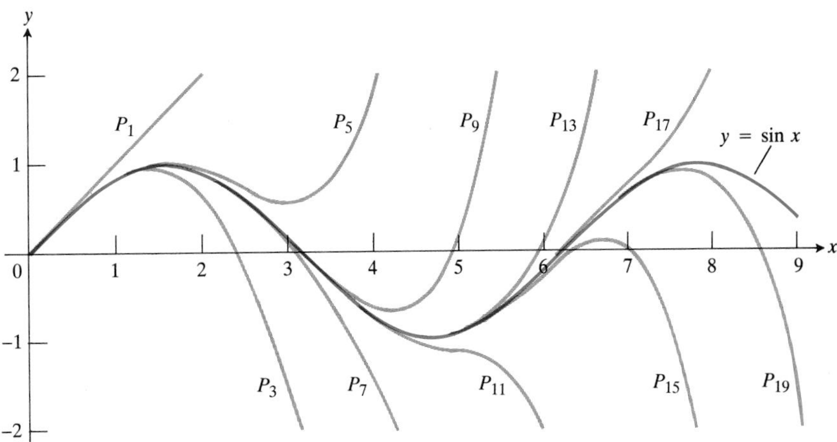

9.19 The polynomials

$$P_{2n+1}(x) = \sum_{k=0}^{n} ((-1)^k x^{2k+1}/(2k+1)!)$$

converge to sin x as $n \to \infty$.

Therefore the error will be less than or equal to 3×10^{-4} if

$$\frac{|x|^5}{120} < 3 \times 10^{-4} \quad \text{or} \quad |x| < \sqrt[5]{360 \times 10^{-4}} \approx 0.514. \quad \left(\begin{array}{l}\text{Rounded}\\\text{down, to be}\\\text{safe}\end{array}\right)$$

The Alternating Series Estimation Theorem tells us something that the Remainder Estimation Theorem does not: namely, that the estimate $x - (x^3/3!)$ for $\sin x$ is an underestimate when x is positive because then $x^5/120$ is positive.

Figure 9.19 shows the graph of $\sin x$, along with the graphs of a number of its approximating Taylor polynomials. Notice that the graph of $P_3(x) = x - (x^3/3!)$ is almost indistinguishable from the sine curve when $-1 \le x \le 1$. However, it crosses the x-axis at $\pm\sqrt{6} \approx \pm 2.45$, whereas the sine curve crosses the axis at $\pm\pi \approx \pm 3.14$.

You might wonder how the estimate given by the Remainder Estimation Theorem compares with the one just obtained from the Alternating Series Estimation Theorem. If we write

$$\sin x = x - \frac{x^3}{3!} + R_3,$$

then the Remainder Estimation Theorem gives

$$|R_3| \le 1 \cdot \frac{|x|^4}{4!} = \frac{|x|^4}{24},$$

which is not very good. But when we recognize that $x - x^3/3! = 0 + x + 0x^2 - x^3/3! + 0x^4$ is the Taylor polynomial of order 4 as well as of order 3, then we have

$$\sin x = x - \frac{x^3}{3!} + 0 + R_4,$$

and the Remainder Estimation Theorem with $M = r = 1$ gives

$$|R_4| \le 1 \cdot \frac{|x|^5}{5!} = \frac{|x|^5}{120}.$$

This is what we had from the Alternating Series Estimation Theorem. ■

Combining Taylor Series

On common intervals of convergence, Taylor series can be added, subtracted, and multiplied by constants just as other series can, and the results are once again Taylor series. The Taylor series for $f(x) + g(x)$ is the sum of the Taylor series for $f(x)$ and $g(x)$ because the nth derivative of $f + g$ is $f^{(n)} + g^{(n)}$, and so on. Thus we obtain the Maclaurin series for $(1 + \cos 2x)/2$ by adding 1 to the Maclaurin series for $\cos 2x$ and dividing the combined results by 2, and the Maclaurin series for $\sin x + \cos x$ is the term by term sum of the Maclaurin series for $\sin x$ and $\cos x$.

$e^{i\theta} = \cos\theta + i\sin\theta$

As you may recall, a complex number is a number of the form $a + bi$, where a and b are real numbers and $i = \sqrt{-1}$. If we substitute $x = i\theta$ (θ real) in the Maclaurin series for e^x and use the relations

$$i^2 = -1, \qquad i^3 = i^2 i = -i, \qquad i^4 = i^2 i^2 = 1, \qquad i^5 = i^4 i = i,$$

and so on, to simplify the result, we obtain

$$\begin{aligned} e^{i\theta} &= 1 + \frac{i\theta}{1!} + \frac{i^2\theta^2}{2!} + \frac{i^3\theta^3}{3!} + \frac{i^4\theta^4}{4!} + \frac{i^5\theta^5}{5!} + \frac{i^6\theta^6}{6!} + \cdots \\ &= \left(1 - \frac{\theta^2}{2!} + \frac{\theta^4}{4!} - \frac{\theta^6}{6!} + \cdots\right) + i\left(\theta - \frac{\theta^3}{3!} + \frac{\theta^5}{5!} - \cdots\right) = \cos\theta + i\sin\theta. \end{aligned}$$

This does not *prove* that $e^{i\theta} = \cos\theta + i\sin\theta$ because we haven't yet defined what it means to raise e to an imaginary power. But it does say how we ought to define $e^{i\theta}$ to be consistent with other things we know.

DEFINITION

For any real number θ, $\quad e^{i\theta} = \cos\theta + i\sin\theta. \qquad (18)$

Equation (18), called **Euler's formula,** enables us to define e^{a+bi} to be $e^a \cdot e^{bi}$ for any complex number $a + bi$.

One of the amazing consequences of Euler's formula is the equation

$$e^{i\pi} = -1.$$

When written in the form $e^{i\pi} + 1 = 0$, this equation combines the five most important constants in mathematics.

Exercises 9.7

In Exercises 1–8, find the Taylor polynomials of orders 0, 1, 2, and 3, generated by f at a.

1. $f(x) = \ln x, \quad a = 1$ **2.** $f(x) = \ln(1 + x), \quad a = 0$

3. $f(x) = 1/x, \quad a = 2$ **4.** $f(x) = 1/(x + 2), \quad a = 0$

5. $f(x) = \sin x, \quad a = \pi/4$ **6.** $f(x) = \cos x, \quad a = \pi/4$

7. $f(x) = \sqrt{x}, \quad a = 4$ **8.** $f(x) = \sqrt{x + 4}, \quad a = 0$

Find the Maclaurin series for the functions in Exercises 9–18.

9. e^{-x} **10.** $e^{x/2}$ **11.** $\sin 3x$

12. $5 \cos \pi x$ **13.** $\cos(-x)$ **14.** $x \sin x$

15. $\cosh x = (e^x + e^{-x})/2$

16. $\sinh x = (e^x - e^{-x})/2$

17. $(x^2/2) - 1 + \cos x$

18. $\cos^2 x$ (*Hint*: $\cos^2 x = (1 + \cos 2x)/2$)

Quadratic Approximations

Write out Taylor's formula (Eq. 10) with $n = 2$ and $a = 0$ for the functions in Exercises 19–24. This will give you the quadratic approximations of these functions at $x = 0$ and the associated error terms.

19. $f(x) = \dfrac{1}{1+x}$ **20.** $f(x) = \sqrt{1+x}$

21. $f(x) = \ln(1+x)$

22. $f(x) = (1+x)^k$ (Any number k)

23. $f(x) = \sin x$ **24.** $f(x) = \cos x$

25. Use the Taylor series generated by e^x at $x = a$ to show that

$$e^x = e^a\left(1 + (x-a) + \frac{(x-a)^2}{2!} + \cdots\right).$$

26. Find the Taylor series generated by e^x at $a = 1$. Compare your answer with the formula in Exercise 25.

27. For approximately what values of x can you replace $\sin x$ by $x - (x^3/6)$ with an error of magnitude no greater than 5×10^{-4}?

28. If $\cos x$ is replaced by $1 - (x^2/2)$ and $|x| < 0.5$, what estimate can be made of the error? Does $1 - (x^2/2)$ tend to be too large, or too small?

29. How close is the approximation $\sin x = x$ when $|x| < 10^{-3}$? For which of these values of x is $x < \sin x$?

30. The estimate $\sqrt{1+x} = 1 + (x/2)$ is used when x is small. Estimate the error when $|x| < 0.01$.

31. The approximation $e^x = 1 + x + (x^2/2)$ is used when x is small. Use the Remainder Estimation Theorem to estimate the error when $|x| < 0.1$.

32. When $x < 0$, the series for e^x is an alternating series. Use the Alternating Series Estimation Theorem to estimate the error that results from replacing e^x by $1 + x + (x^2/2)$ when $-0.1 < x < 0$. Compare with Exercise 31.

33. Estimate the error in the approximation $\sinh x = x + (x^3/3!)$ when $|x| < 0.5$. (*Hint:* Use R_4, not R_3.)

34. When $0 \le h \le 0.01$, show that e^h may be replaced by $1 + h$ with an error of magnitude no greater than 0.6% of h. Use $e^{0.01} = 1.01$.

Each of the series in Exercises 35 and 36 is the value of the Maclaurin series of a function $f(x)$ at some point. What function and what point? What is the sum of the series?

35. $(0.1) - \dfrac{(0.1)^3}{3!} + \dfrac{(0.1)^5}{5!} - \cdots + \dfrac{(-1)^k(0.1)^{2k+1}}{(2k+1)!} + \cdots$

36. $1 - \dfrac{\pi^2}{4^2 \cdot 2!} + \dfrac{\pi^4}{4^4 \cdot 4!} - \cdots + \dfrac{(-1)^k(\pi)^{2k}}{4^{2k} \cdot (2k!)} - \cdots$

37. Differentiate the Maclaurin series for $\sin x$, $\cos x$, and e^x term by term and compare your results with the Maclaurin series for $\cos x$, $\sin x$, and e^x.

38. Integrate the Maclaurin series for $\sin x$, $\cos x$, and e^x term by term and compare your results with the Maclaurin series for $\cos x$, $\sin x$, and e^x.

39. Multiply the Maclaurin series for e^x and $\sin x$ together to find the first five nonzero terms of the Maclaurin series for $e^x \sin x$.

40. Multiply the Maclaurin series for e^x and $\cos x$ together to find the first five nonzero terms of the Maclaurin series for $e^x \cos x$.

41. Use the Maclaurin series for $\sin x$ and the Alternating Series Estimation Theorem to show that

$$1 - \frac{x^2}{6} < \frac{\sin x}{x} < 1 \quad \text{for} \quad 0 < |x| < 1.$$

(This is the inequality in Section 2.2, Exercise 13.)

42. Use the Maclaurin series for $\cos x$ and the Alternating Series Estimation Theorem to show that

$$\frac{1}{2} - \frac{x^2}{24} < \frac{1 - \cos x}{x^2} < \frac{1}{2} \quad \text{for} \quad 0 < |x| < 1.$$

(This is the inequality in Section 2.2, Exercise 15.)

43. Use Eq. (18) to write the following powers of e in the form $a + bi$.

a) $e^{-i\pi}$ b) $e^{i\pi/4}$ c) $e^{-i\pi/2}$

44. *Euler's identities* Use Eq. (18) to show that

$$\cos\theta = \frac{e^{i\theta} + e^{-i\theta}}{2} \quad \text{and} \quad \sin\theta = \frac{e^{i\theta} - e^{-i\theta}}{2i}.$$

45. Establish the equations in Exercise 44 by combining the formal Maclaurin series for $e^{i\theta}$ and $e^{-i\theta}$.

46. When a and b are real, we define $e^{(a+ib)x}$ to be $e^{ax}(\cos bx + i \sin bx)$. From this definition, show that

$$\frac{d}{dx}e^{(a+ib)x} = (a + ib)e^{(a+ib)x}.$$

47. Two complex numbers, $a + ib$ and $c + id$, are equal if and only if $a = c$ and $b = d$. Use this fact to evaluate

$$\int e^{ax} \cos bx\, dx \quad \text{and} \quad \int e^{ax} \sin bx\, dx$$

from

$$\int e^{(a+ib)x}\, dx = \frac{a - ib}{a^2 + b^2} e^{(a+ib)x} + C,$$

where $C = C_1 + iC_2$ is a complex constant of integration.

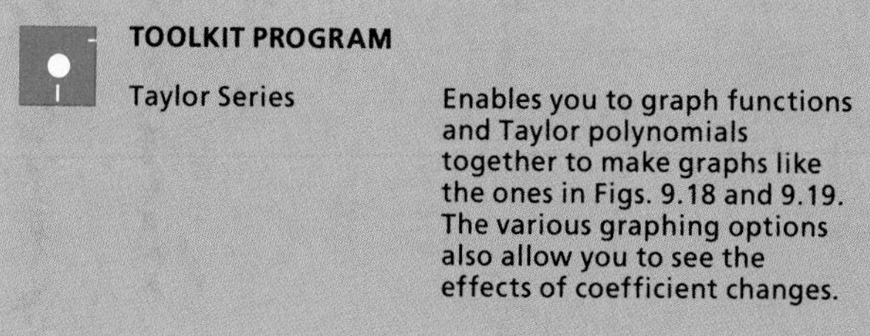

9.8 Further Calculations with Taylor Series

This section introduces the binomial series, discusses how to choose the best center for a Taylor series, and shows how series are sometimes used to evaluate nonelementary integrals. At the end of the section there is a table of frequently used Maclaurin series.

The Binomial Series

The Maclaurin Series generated by $f(x) = (1 + x)^m$, when m is constant, is

$$1 + mx + \frac{m(m-1)}{2!}x^2 + \frac{m(m-1)(m-2)}{3!}x^3 + \cdots + \frac{m(m-1)(m-2)\cdots(m-k+1)}{k!}x^k + \cdots. \tag{1}$$

This series, called the **binomial series,** converges absolutely for $|x| < 1$. To derive the series, we first list the function and its derivatives:

$$\begin{aligned} f(x) &= (1+x)^m, \\ f'(x) &= m(1+x)^{m-1}, \\ f''(x) &= m(m-1)(1+x)^{m-2}, \\ f'''(x) &= m(m-1)(m-2)(1+x)^{m-3}, \\ &\vdots \\ f^{(k)}(x) &= m(m-1)(m-2)\cdots(m-k+1)(1+x)^{m-k}. \end{aligned}$$

We then evaluate these at $x = 0$ and substitute in the Maclaurin series formula to obtain

$$1 + mx + \frac{m(m-1)}{2!}x^2 + \cdots + \frac{m(m-1)(m-2)\cdots(m-k+1)}{k!}x^k + \cdots. \tag{2}$$

If m is an integer greater than or equal to zero, the series stops after $(m + 1)$ terms because the coefficients from $k = m + 1$ on are zero.

If m is not a positive integer or zero, the series is infinite and converges for $|x| < 1$. To see why, let u_k be the term involving x^k. Then apply the ratio test for absolute convergence to see that

$$\left|\frac{u_{k+1}}{u_k}\right| = \left|\frac{m-k}{k+1}x\right| \to |x| \qquad \text{as } k \to \infty.$$

Our derivation of the binomial series shows only that it is generated by $(1 + x)^m$ and converges for $|x| < 1$. The derivation does not show that the series actually converges to $(1 + x)^m$. It does, but we shall assume that part without proof.

For $-1 < x < 1$,

$$(1 + x)^m = 1 + \sum_{k=1}^{\infty} \binom{m}{k} x^k \tag{3}$$

where

$$\binom{m}{1} = m, \quad \binom{m}{2} = \frac{m(m-1)}{2!},$$

and

$$\binom{m}{k} = \frac{m(m-1)(m-2)\cdots(m-k+1)}{k!} \qquad \text{for } k \geq 3.$$

EXAMPLE 1 Show that when $m = -1$, Eq. (3) gives the geometric series

$$\frac{1}{1+x} = 1 - x + x^2 - x^3 + \cdots + (-1)^k x^k + \cdots. \tag{4}$$

Solution When $m = -1$,

$$\binom{-1}{1} = -1, \quad \binom{-1}{2} = \frac{-1(-2)}{2!} = 1,$$

and

$$\binom{-1}{k} = \frac{-1(-2)(-3)\cdots(-1-k+1)}{k!} = (-1)^k \left(\frac{k!}{k!}\right) = (-1)^k.$$

With these coefficient values, Eq. (3) becomes

$$(1 + x)^{-1} = 1 + \sum_{k=1}^{\infty} (-1)^k x^k = 1 - x + x^2 - x^3 + \cdots + (-1)^k x^k + \cdots,$$

which is Eq. (4).

Choosing Centers for Taylor Series

Taylor's formula,

$$f(x) = f(a) + f'(a)(x-a) + \frac{f''(a)}{2!}(x-a)^2 + \cdots$$
$$+ \frac{f^{(n)}(a)}{n!}(x-a)^n + \frac{f^{(n+1)}(c)}{(n+1)!}(x-a)^{n+1}, \tag{5}$$

expresses the value of f at x in terms of f and its derivatives at a. In numerical computations, we therefore need a to be a point where we know the values of f and its derivatives. We also need a to be close enough to the values of x we are interested in to make $(x - a)^{n+1}$ so small we can neglect the remainder.

EXAMPLE 2 What value of a might we choose in Taylor's formula (Eq. 5) to compute sin 35° efficiently?

Solution The radian measure for 35° is $35\pi/180$. We could choose $a = 0$ and use the series

$$\sin x = x - \frac{x^3}{3!} + \frac{x^5}{5!} - \cdots + (-1)^n \frac{x^{2n+1}}{(2n+1)!} + 0 \cdot x^{2n+2} + R_{2n+2}(x). \tag{6}$$

Alternatively, we could choose $a = \pi/6$ (which corresponds to 30°) and use the series

$$\sin x = \sin\frac{\pi}{6} + \cos\frac{\pi}{6}\left(x - \frac{\pi}{6}\right) - \sin\frac{\pi}{6}\frac{(x - \pi/6)^2}{2!} - \cos\frac{\pi}{6}\frac{(x - \pi/6)^3}{3!}$$

$$+ \cdots + \sin\left(\frac{\pi}{6} + n\frac{\pi}{2}\right)\frac{(x - \pi/6)^n}{n!} + R_n(x).$$

The remainder in Eq. (6) satisfies the inequality

$$|R_{2n+2}(x)| \le \frac{|x|^{2n+3}}{(2n+3)!},$$

which tends to zero as $n \to \infty$, no matter how large $|x|$ may be. We could therefore calculate sin 35° by placing

$$x = \frac{35\pi}{180} \approx 0.6108652$$

in the approximation

$$\sin x \approx x - \frac{x^3}{6} + \frac{x^5}{120} - \frac{x^7}{5040}.$$

This gives a truncation error of magnitude no greater than 3.3×10^{-8}, since

$$\left| R_8\left(\frac{35\pi}{180}\right)\right| < \frac{(0.611)^9}{9!} < 3.3 \times 10^{-8}.$$

By using the series with $a = \pi/6$, we could obtain equal accuracy with a smaller exponent n, but at the expense of introducing $\cos \pi/6 = \sqrt{3}/2$ as one of the coefficients. In this series, with $a = \pi/6$, we would take $x = 35\pi/180$, and the quantity $(x - a)$ would be

$$x - \frac{\pi}{6} = \frac{35\pi}{180} - \frac{30\pi}{180} = \frac{5\pi}{180} \approx 0.0872665,$$

which decreases rapidly as it is raised to higher powers. ■

Evaluating Nonelementary Integrals

Maclaurin series are often used to express nonelementary integrals in terms of series.

EXAMPLE 3 Express $\int \sin x^2\,dx$ as a power series.

Solution From the series for $\sin x$ we obtain

$$\sin x^2 = x^2 - \frac{x^6}{3!} + \frac{x^{10}}{5!} - \frac{x^{14}}{7!} + \frac{x^{18}}{9!} - \cdots.$$

Therefore,

$$\int \sin x^2\,dx = C + \frac{x^3}{3} - \frac{x^7}{7\cdot 3!} + \frac{x^{11}}{11\cdot 5!} - \frac{x^{15}}{15\cdot 7!} + \frac{x^{19}}{19\cdot 9!} - \cdots.$$

EXAMPLE 4 Estimate $\int_0^1 \sin x^2\,dx$ with an error of less than 0.001.

Solution From the indefinite integral in Example 3,

$$\int_0^1 \sin x^2\,dx = \frac{1}{3} - \frac{1}{7\cdot 3!} + \frac{1}{11\cdot 5!} - \frac{1}{15\cdot 7!} + \frac{1}{19\cdot 9!} - \cdots.$$

The series alternates, and we find by experiment that

$$\frac{1}{11\cdot 5!} \approx 0.00076$$

is the first term to be numerically less than 0.001. The sum of the preceding two terms gives

$$\int_0^1 \sin x^2\,dx \approx \frac{1}{3} - \frac{1}{42} \approx 0.310.$$

With two more terms we could estimate

$$\int_0^1 \sin x^2\,dx \approx 0.310268$$

with an error of less than 10^{-6}, and with only one term beyond that we have

$$\int_0^1 \sin x^2\,dx \approx \frac{1}{3} - \frac{1}{42} + \frac{1}{1320} - \frac{1}{75600} + \frac{1}{6894720} \approx 0.310268303,$$

with an error of less than 10^{-9}. To guarantee this accuracy with the error formula for the Trapezoidal Rule would require using about 13,000 subintervals.

Arctangents

In Section 9.6, Example 4, we found a series for $\tan^{-1}x$ by differentiating to get

$$\frac{d}{dx}\tan^{-1}x = \frac{1}{1+x^2} = 1 - x^2 + x^4 - x^6 + \cdots$$

and integrating to get

$$\tan^{-1}x = x - \frac{x^3}{3} + \frac{x^5}{5} - \frac{x^7}{7} + \cdots .$$

However, we did not prove the term-by-term integration theorem on which this conclusion depended. We now derive the series again by integrating both sides of the finite formula

$$\frac{1}{1+t^2} = 1 - t^2 + t^4 - t^6 + \cdots + (-1)^n t^{2n} + \frac{(-1)^{n+1}t^{2n+2}}{1+t^2}, \tag{7}$$

in which the last term comes from adding the remaining terms as a geometric series with first term $a = t^{2n+2}$ and ratio $r = -t^2$. Integrating both sides of Eq. (7) from $t = 0$ to $t = x$ gives

$$\tan^{-1}x = x - \frac{x^3}{3} + \frac{x^5}{5} - \frac{x^7}{7} + \cdots + (-1)^n \frac{x^{2n+1}}{2n+1} + R,$$

where

$$R = \int_0^x \frac{(-1)^{n+1}t^{2n+2}}{1+t^2}\,dt .$$

The denominator of the integrand is greater than or equal to 1; hence

$$|R| \le \int_0^{|x|} t^{2n+2}\,dt = \frac{|x|^{2n+3}}{2n+3}.$$

If $|x| \le 1$, the right side of this inequality approaches zero as $n \to \infty$. Therefore R also approaches zero and we have

$$\tan^{-1}x = \sum_{n=0}^{\infty} \frac{(-1)^n x^{2n+1}}{2n+1}.$$

We take this route instead of finding the Maclaurin series directly because the formulas for the higher order derivatives of $\tan^{-1}x$ are unmanageable.

$$\tan^{-1}x = x - \frac{x^3}{3} + \frac{x^5}{5} - \frac{x^7}{7} + \cdots, \quad |x| \le 1. \tag{8}$$

When we put $x = 1$ and $\tan^{-1} 1 = \pi/4$ in Eq. (8) we get **Leibniz's formula:**

$$\frac{\pi}{4} = 1 - \frac{1}{3} + \frac{1}{5} - \frac{1}{7} + \frac{1}{9} - \cdots + \frac{(-1)^n}{2n+1} + \cdots .$$

This series converges too slowly to be a good source of decimal approximations of π. It is better to use a formula like

$$\pi = 48 \tan^{-1}\frac{1}{18} + 32 \tan^{-1}\frac{1}{57} - 20 \tan^{-1}\frac{1}{239}, \tag{9}$$

which uses values of x closer to zero.

Frequently Used Maclaurin Series

$$\frac{1}{1-x} = 1 + x + x^2 + \cdots + x^n + \cdots = \sum_{n=0}^{\infty} x^n, \quad |x| < 1$$

$$\frac{1}{1+x} = 1 - x + x^2 - \cdots + (-x)^n + \cdots = \sum_{n=0}^{\infty} (-1)^n x^n, \quad |x| < 1$$

$$e^x = 1 + x + \frac{x^2}{2!} + \cdots + \frac{x^n}{n!} + \cdots = \sum_{n=0}^{\infty} \frac{x^n}{n!}, \quad |x| < \infty$$

$$\sin x = x - \frac{x^3}{3!} + \frac{x^5}{5!} - \cdots + (-1)^n \frac{x^{2n+1}}{(2n+1)!} + \cdots = \sum_{n=0}^{\infty} \frac{(-1)^n x^{2n+1}}{(2n+1)!}, \quad |x| < \infty$$

$$\cos x = 1 - \frac{x^2}{2!} + \frac{x^4}{4!} - \cdots + (-1)^n \frac{x^{2n}}{(2n)!} + \cdots = \sum_{n=0}^{\infty} \frac{(-1)^n x^{2n}}{(2n)!}, \quad |x| < \infty$$

$$\ln(1+x) = x - \frac{x^2}{2} + \frac{x^3}{3} - \cdots + (-1)^{n-1} \frac{x^n}{n} + \cdots = \sum_{n=1}^{\infty} \frac{(-1)^{n-1} x^n}{n}, \quad -1 < x \le 1$$

$$\ln \frac{1+x}{1-x} = 2 \tanh^{-1} x = 2\left(x + \frac{x^3}{3} + \frac{x^5}{5} + \cdots + \frac{x^{2n+1}}{2n+1} + \cdots\right) = 2 \sum_{n=0}^{\infty} \frac{x^{2n+1}}{2n+1}, \quad |x| < 1$$

$$\tan^{-1} x = x - \frac{x^3}{3} + \frac{x^5}{5} - \cdots + (-1)^n \frac{x^{2n+1}}{2n+1} + \cdots = \sum_{n=0}^{\infty} \frac{(-1)^n x^{2n+1}}{2n+1}, \quad |x| \le 1$$

Binomial Series

$$(1+x)^m = 1 + mx + \frac{m(m-1)x^2}{2!} + \frac{m(m-1)(m-2)x^3}{3!} + \cdots + \frac{m(m-1)(m-2)\cdots(m-k+1)x^k}{k!} + \cdots$$

$$= 1 + \sum_{k=1}^{\infty} \binom{m}{k} x^k, \quad |x| < 1$$

where

$$\binom{m}{1} = m, \quad \binom{m}{2} = \frac{m(m-1)}{2!}, \quad \binom{m}{k} = \frac{m(m-1)\cdots(m-k+1)}{k!} \quad \text{for } k \ge 3.$$

NOTE: It is customary to define $\binom{m}{0}$ to be 1 and to take $x^0 = 1$ (even in the usually excluded case where $x = 0$) in order to write the binomial series compactly as

$$(1+x)^m = \sum_{k=0}^{\infty} \binom{m}{k} x^k, \quad |x| < 1.$$

If m is a *positive integer,* the series terminates at x^m, and the result converges for all x.

Exercises 9.8

What Taylor series would you use to represent the functions in Exercises 1–6 near the given values of x? (There may be more than one good answer.) Write out the first four nonzero terms of the series you choose.

1. $\cos x$ near $x = 1$

2. $\sin x$ near $x = 6.3$

3. e^x near $x = 0.4$

4. $\ln x$ near $x = 1.3$

5. $\cos x$ near $x = 69$

6. $\tan^{-1} x$ near $x = 2$

CALCULATOR Use series to evaluate the integrals in Exercises 7–10 as accurately as your calculator will allow.

7. $\displaystyle\int_0^{0.1} \frac{\sin x}{x}\,dx$

8. $\displaystyle\int_0^{0.1} e^{-x^2}\,dx$

9. $\displaystyle\int_0^{0.1} \sqrt{1 + x^4}\,dx$ (Use the binomial series with x^4 in place of x.)

10. $\displaystyle\int_0^{1} \frac{1 - \cos x}{x^2}\,dx$

11. Replace x by $-x$ in the Maclaurin series for $\ln(1 + x)$ to obtain a series for $\ln(1 - x)$. Then subtract this from the Maclaurin series for $\ln(1 + x)$ to show that for $|x| < 1$

$$\ln\frac{1 + x}{1 - x} = 2\left(x + \frac{x^3}{3} + \frac{x^5}{5} + \cdots\right).$$

12. How many terms of the Maclaurin series for $\ln(1 + x)$ should you add to be sure of calculating $\ln(1.1)$ with an error of magnitude less than 10^{-8}?

13. According to the Alternating Series Estimation Theorem, how many terms of the Maclaurin series for $\tan^{-1} 1$ would you have to add to be sure of finding $\pi/4$ with an error of magnitude less than 10^{-3}?

14. Show that the Maclaurin series for $\tan^{-1} x$ diverges for $|x| > 1$.

15. **CALCULATOR** About how many terms of the Maclaurin series for $\tan^{-1} x$ would you have to use to evaluate each term on the right-hand side of the equation

$$\pi = 48 \tan^{-1}\frac{1}{18} + 32 \tan^{-1}\frac{1}{57} - 20 \tan^{-1}\frac{1}{239}$$

with an error of magnitude less than 10^{-6}? In contrast, the convergence of $\sum_{n=1}^{\infty} (1/n^2)$ to $\pi^2/6$ is so slow that even 50 terms will not yield two-place accuracy.

Review Questions

1. Define infinite sequence (sequence), infinite series (series), and sequence of partial sums of a series.
2. Define convergence for (a) sequences, (b) series.
3. Which of the following statements are true, and which are false?
   a) If a sequence does not converge, then it diverges.
   b) If a sequence $\{a_n\}$ does not converge, then a_n tends to infinity as n does.
   c) If a series does not converge, then its nth term does not approach zero as n tends to infinity.
   d) If the nth term of a series does not approach zero as n tends to infinity, then the series diverges.
   e) If a sequence $\{a_n\}$ converges, then there is a number L such that a_n lies within 1 unit of L (i) for all values of n, (ii) for all but a finite number of values of n.
   f) If all partial sums of a series are less than some constant L, then the series converges.
   g) If a series converges, then its partial sums s_n are bounded (that is, $m \le s_n \le M$ for some constants m and M).
4. What tests do you know for the convergence and divergence of infinite series?
5. Under what circumstances do you know for sure that a bounded sequence converges?
6. Define absolute convergence and conditional convergence for infinite series. Give examples of series that (a) converge absolutely, (b) converge conditionally, (c) diverge.
7. What are geometric series? Under what circumstances do they converge? diverge?

8. What test would you apply to decide whether an alternating series converges? Give examples of convergent and of divergent alternating series. How can you estimate the error involved in using a partial sum to estimate the sum of a convergent alternating series?
9. What is a power series? How do you test a power series for convergence? What kinds of results can you get? Give examples.
10. What are the basic facts about term-by-term differentiation and integration of power series? Give examples.
11. What is the Taylor series generated by a function $f(x)$ at a point $x = a$? What do you need to know about f to construct the series? Give an example.
12. What is the Maclaurin series generated by a function $f(x)$? Give an example.
13. What are Taylor polynomials and what good are they? Give examples.
14. What is Taylor's formula and what does Taylor's theorem say about it? When does the Taylor series generated by f at $x = a$ converge to f?
15. What does Taylor's formula tell us about the error in a linearization?
16. What are the Maclaurin series for e^x, $\sin x$, $\cos x$, $1/(1+x)$, $1/(1-x)$, $1/(1+x^2)$, $\tan^{-1}x$, and $\ln(1+x)$? How do you estimate the errors involved in replacing these series by their partial sums? Give examples.
17. What do you take into account when choosing a center for a Taylor series? Illustrate with examples.
18. How can you sometimes use series to evaluate nonelementary integrals? Give an example.

Practice Exercises

Determine which of the sequences $\{a_n\}$ in Exercises 1–12 converge and which diverge. Find the limit of each convergent sequence.

1. $a_n = 1 + \dfrac{(-1)^n}{n}$
2. $a_n = \dfrac{1-2^n}{2^n}$
3. $a_n = \cos\dfrac{n\pi}{2}$
4. $a_n = \left(\dfrac{3}{n}\right)^{1/n}$
5. $a_n = \dfrac{\ln(n^2)}{n}$
6. $a_n = \left(\dfrac{n+5}{n}\right)^n$
7. $a_n = \sqrt[n]{\dfrac{3^n}{n}}$
8. $a_n = \dfrac{1}{3^{2n-1}}$
9. $a_n = \dfrac{(-4)^n}{n!}$
10. $a_n = \dfrac{\ln(2n+1)}{n}$
11. $a_n = \dfrac{(n+1)!}{n!}$
12. $a_n = \dfrac{n^2-n}{2n^2+n}$

Find the sums of the series in Exercises 13–16.

13. $\displaystyle\sum_{n=1}^{\infty} \ln\left(\frac{n}{n+1}\right)$
14. $\displaystyle\sum_{n=2}^{\infty} \frac{-2}{n(n+1)}$
15. $\displaystyle\sum_{n=0}^{\infty} e^{-n}$
16. $\displaystyle\sum_{n=1}^{\infty} (-1)^n \frac{3}{4^n}$

Which of the series in Exercises 17–28 converge absolutely, which converge conditionally, and which diverge? Give reasons for your answers.

17. $\displaystyle\sum_{n=1}^{\infty} \frac{1}{\sqrt{n}}$
18. $\displaystyle\sum_{n=1}^{\infty} \frac{-5}{n}$
19. $\displaystyle\sum_{n=1}^{\infty} \frac{(-1)^n}{\sqrt{n}}$
20. $\displaystyle\sum_{n=1}^{\infty} \frac{1}{2n^3}$
21. $\displaystyle\sum_{n=1}^{\infty} \frac{(-1)^n}{\ln(n+1)}$
22. $\displaystyle\sum_{n=1}^{\infty} \frac{1}{n(\ln n)^2}$
23. $\displaystyle\sum_{n=1}^{\infty} \frac{(-1)^n}{n\sqrt{n^2+1}}$
24. $\displaystyle\sum_{n=1}^{\infty} \frac{(-1)^n 3n^2}{n^3+1}$
25. $\displaystyle\sum_{n=1}^{\infty} \frac{n+1}{n!}$
26. $\displaystyle\sum_{n=1}^{\infty} \frac{(-1)^n (n^2+1)}{2n^2+n-1}$
27. $\displaystyle\sum_{n=1}^{\infty} \frac{(-3)^n}{n!}$
28. $\displaystyle\sum_{n=1}^{\infty} \frac{2^n 3^n}{n^n}$

Each of Exercises 29–34 gives a formula for the nth term of a series. For what values of x does each series (a) converge and (b) converge absolutely?

29. $\dfrac{(x+2)^n}{3^n \cdot n}$
30. $\dfrac{(x-1)^{2n-2}}{(2n-1)!}$

31. $\dfrac{x^n}{n^n}$

32. $\dfrac{n+1}{2n+1}\dfrac{(x-1)^n}{2^n}$

33. $\dfrac{(-1)^{n-1}(x-1)^n}{n^2}$

34. $\dfrac{x^n}{\sqrt{n}}$

Each of the series in Exercises 35–40 is the value of the Maclaurin series of a function $f(x)$ at a particular point. What function and what point? What is the sum of the series?

35. $1 - \dfrac{1}{4} + \dfrac{1}{16} - \cdots + (-1)^n \dfrac{1}{4^n} + \cdots$

36. $\dfrac{2}{3} - \dfrac{4}{18} + \dfrac{8}{81} - \cdots + (-1)^{n-1}\dfrac{2^n}{n\,3^n} + \cdots$

37. $\pi - \dfrac{\pi^3}{3!} + \dfrac{\pi^5}{5!} - \cdots + (-1)^n \dfrac{\pi^{2n+1}}{(2n+1)!} + \cdots$

38. $1 - \dfrac{\pi^2}{9 \cdot 2!} + \dfrac{\pi^4}{81 \cdot 4!} - \cdots + (-1)^n \dfrac{\pi^{2n}}{4^{2n}(2n!)} + \cdots$

39. $1 + \ln 2 + \dfrac{(\ln 2)^2}{2!} + \cdots + \dfrac{(\ln n)^n}{n!} + \cdots$

40. $\dfrac{1}{\sqrt{3}} - \dfrac{1}{9\sqrt{3}} + \dfrac{1}{45\sqrt{3}} - \cdots + (-1)^{n-1}\dfrac{1}{(2n-1)(\sqrt{3})^{2n-1}} + \cdots$

In Exercises 41–42, find the first four nonzero terms of the Taylor series for the function $f(x)$ at the center $x = a$.

41. $f(x) = \sqrt{3 + x^2}$ at $x = -1$

42. $f(x) = 1/(1 - x)$ at $x = 2$

43. Multiply the Maclaurin series for $\sin x$ and $2 \cos x$ together to get the first four nonzero terms of the Maclaurin series for $\sin 2x$.

44. Use the formula $\sin^2 x = (1 - \cos 2x)/2$ to get the Maclaurin series for $\sin^2 x$ from the Maclaurin series for $\cos 2x$.

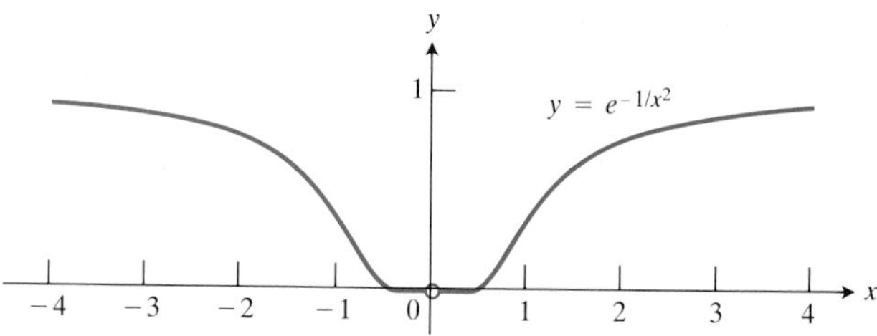

9.20 The graph of the continuous extension of $y = e^{-1/x^2}$ is so flat at the origin that all of its derivatives there are zero (Exercise 47).

CALCULATOR Use series to approximate the values of the integrals in Exercises 45 and 46 with an error of magnitude less than 10^{-8}.

45. $\displaystyle\int_0^{1/2} e^{-x^3}\,dx$

46. $\displaystyle\int_0^{1/2} \frac{\tan^{-1} x}{x}\,dx$

47. *A convergent Taylor series that converges to its generating function only at its center.* It can be shown (though not simply) that the function f defined by the rule

$$f(x) = \begin{cases} 0 & \text{if } x = 0 \\ e^{-1/x^2} & \text{if } x \neq 0 \end{cases}$$

has derivatives of all orders at $x = 0$ (Fig. 9.20) and that $f^{(n)}(0) = 0$ for all n.

a) Find the Maclaurin series generated by f. At what values of x does the series converge? At what values of x does the series converge to $f(x)$?

b) Write out Taylor's formula (Section 9.7, Eq. 10) for f, taking $a = 0$ and assuming $x \neq 0$. What does the formula tell you about the value of $R_n(x)$?

10 Plane Curves and Polar Coordinates

Grant Heilman

Every planet traces an ellipse about its sun.

OVERVIEW The study of motion has been important since ancient times, and, as you know, calculus gives us the mathematics we need to describe it. In this chapter, we extend our ability to analyze motion by showing how to keep track of the position of a moving body as a function of time. We do this with parametric equations. We study them in the coordinate plane here, and then extend our work to three dimensions in Chapters 11 and 12. We begin our study, though, by developing equations for conic sections, since these are the paths traveled by planets, satellites, and other bodies (even electrons) whose motions are driven by inverse square forces. Planetary motion is best described in polar coordinates (another of Newton's inventions, although James Bernoulli usually gets the credit because he published first), so we will spend our remaining time finding out what curves, derivatives, and integrals look like in this new coordinate system.

10.1 Conic Sections and Quadratic Equations

This section shows how the conic sections that originated in Greek geometry are described today as the graphs of quadratic equations in the coordinate plane. The Greeks of Plato's time described these curves as the curves formed by cutting a double cone with a plane (Fig. 10.1); hence the name "conic section." We begin by reviewing briefly the equations for parabolas and circles, then continue on to ellipses and hyperbolas.

The mathematics of conic sections is just what we need to describe the paths of planets, comets, moons, asteroids, satellites, or anything else that is moved through space by gravitational forces. Once we know that the path of a moving body is a conic section, we immediately have information about its velocity and the force that drives it, as we shall see in Chapter 12.

Equations from the Distance Formula—Circles and Parabolas

As we saw in Section 1.4, the distance between two points (x_1, y_1) and (x_2, y_2) in the coordinate plane is

$$d = \sqrt{(x_2 - x_1)^2 + (y_2 - y_1)^2}.$$

We used this formula to derive the standard equation for circles centered at the origin and the standard equations for parabolas with vertices at the origin.

The Standard Equation for a Circle of Radius *a* Centered at the Origin

$$x^2 + y^2 = a^2$$

TABLE 10.1
The standard equations for parabolas with vertices at the origin ($p > 0$)

Equation	Focus	Directrix	Axis	Direction
$y = x^2/4p$	$(0, p)$	$y = -p$	y-axis	opens up
$y = -x^2/4p$	$(0, -p)$	$y = p$	y-axis	opens down
$x = y^2/4p$	$(p, 0)$	$x = -p$	x-axis	opens to right
$x = -y^2/4p$	$(-p, 0)$	$x = p$	x-axis	opens to left

The horizontal and vertical shift formulas then gave us equations for circles and a variety of parabolas in other locations.

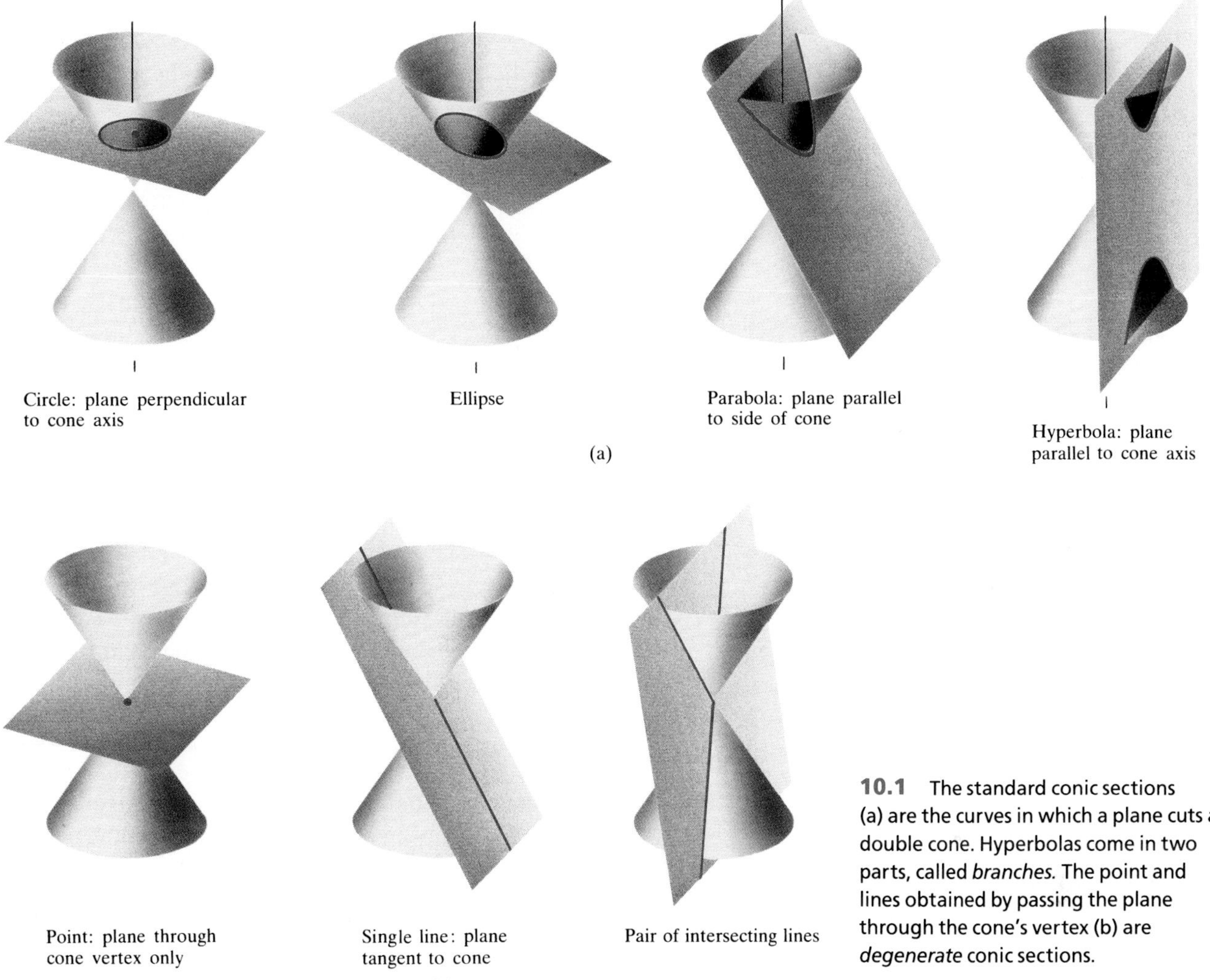

10.1 The standard conic sections (a) are the curves in which a plane cuts a double cone. Hyperbolas come in two parts, called *branches.* The point and lines obtained by passing the plane through the cone's vertex (b) are *degenerate* conic sections.

Ellipses

The equations we use for ellipses come from the distance formula, too.

DEFINITIONS

An **ellipse** is the set of points in a plane whose distances from two fixed points in the plane have a constant sum. The two fixed points are the **foci** of the ellipse.

The quickest way to construct an ellipse uses the definition. Put a loop of string around two tacks F_1 and F_2, pull the string taut with a pencil point

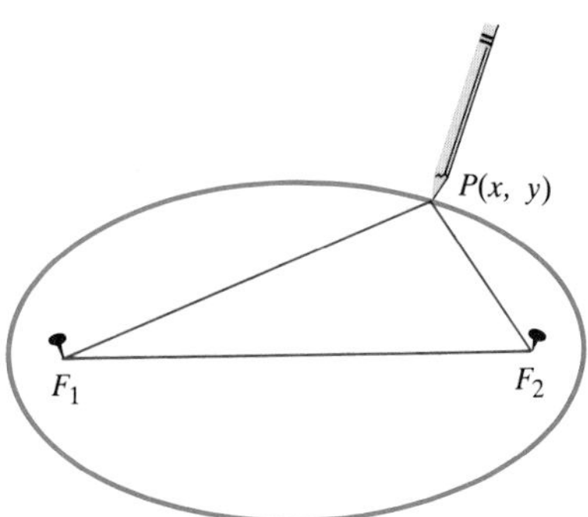

10.2 How to construct an ellipse.

P, and move the pencil around to trace a curve (Fig. 10.2). The curve is an ellipse because the sum $PF_1 + PF_2$, being equal to the length of the loop minus the distance between the tacks, has a constant value. The foci of the ellipse lie at F_1 and F_2.

DEFINITIONS

The line through the foci of an ellipse is the **focal axis.** The point on this line halfway between the foci is the ellipse's **center.** The points where the focal axis and the ellipse cross are the ellipse's **vertices.**

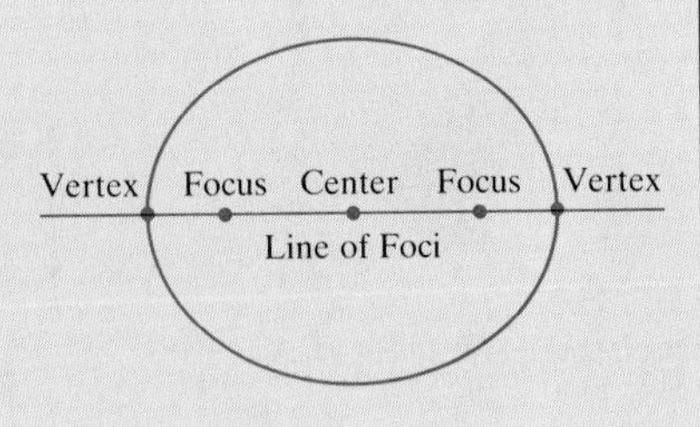

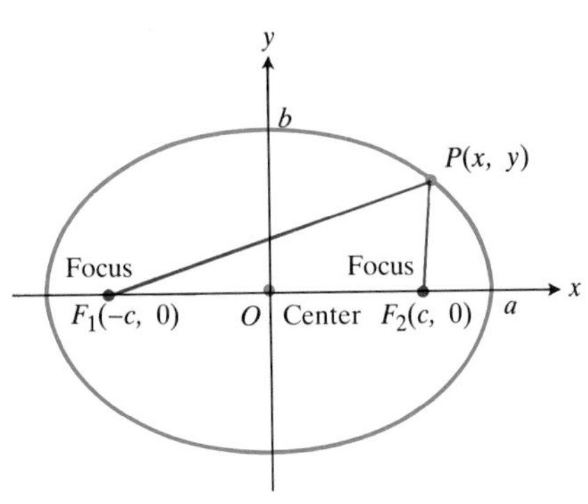

10.3 The ellipse defined by the equation $PF_1 + PF_2 = 2a$ is the graph of the equation $(x^2/a^2) + (y^2/b^2) = 1$.

If the foci are $F_1(-c, 0)$ and $F_2(c, 0)$ (Fig. 10.3), and the sum of the distances $PF_1 + PF_2$ is denoted by $2a$, then the coordinates of a point P on the ellipse satisfy the equation

$$\sqrt{(x+c)^2 + y^2} + \sqrt{(x-c)^2 + y^2} = 2a. \tag{1}$$

To simplify this equation, we move the second radical to the right-hand side, square, isolate the remaining radical, and square again, obtaining

$$\frac{x^2}{a^2} + \frac{y^2}{a^2 - c^2} = 1. \tag{2}$$

Since the sum $PF_1 + PF_2$ is greater than the length F_1F_2 (triangle inequality for triangle PF_1F_2), the number $2a$ is greater than $2c$. Accordingly, a is greater than c and the number $a^2 - c^2$ in Eq. (2) is positive.

The algebraic steps taken to arrive at Eq. (2) can be reversed to show that every point P whose coordinates satisfy an equation of this form with $0 < c < a$ also satisfies the equation $PF_1 + PF_2 = 2a$. Thus, a point lies on the ellipse if and only if its coordinates satisfy Eq. (2).

If we let b denote the positive square root of $(a^2 - c^2)$,

$$b = \sqrt{a^2 - c^2}, \tag{3}$$

then $a^2 - c^2 = b^2$ and Eq. (2) takes the more compact form

$$\frac{x^2}{a^2} + \frac{y^2}{b^2} = 1. \tag{4}$$

Equation (4) reveals that this ellipse is symmetric with respect to the origin and both coordinate axes. It lies inside the rectangle bounded by the lines $x = \pm a$ and $y = \pm b$. It crosses the axes at the points $(\pm a, 0)$ and $(0, \pm b)$. The tangents at these points are perpendicular to the axes, because the slope

$$\frac{dy}{dx} = -\frac{b^2 x}{a^2 y}$$

is zero when $x = 0$ and infinite when $y = 0$. These observations are the basis of the following drawing lesson.

DRAWING LESSON

How to graph the ellipse $\dfrac{x^2}{a^2} + \dfrac{y^2}{b^2} = 1$

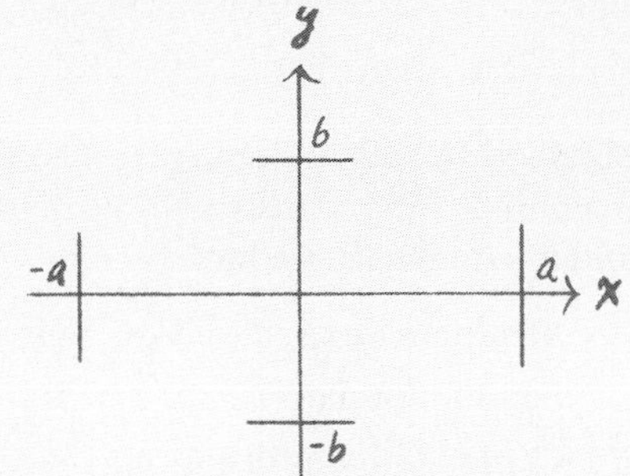

1.
Mark the intercepts $(\pm a, 0)$ and $(0, \pm b)$ with line segments perpendicular to the coordinate axes.

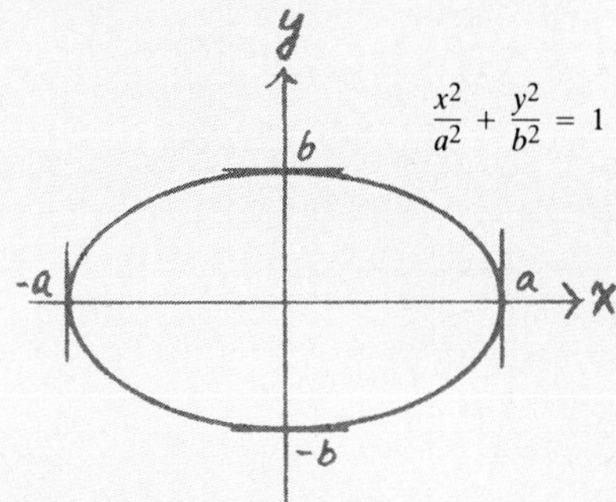

2.
Use the segments as tangent lines to guide your drawing.

Major and Minor Axes The **major axis** of the ellipse described by Eq. (4) is the line segment of length $2a$ joining the intercepts $(\pm a, 0)$. The **minor axis** is the line segment of length $2b$ joining the intercepts $(0, \pm b)$. The number a itself is called the **semimajor axis** and the number b the **semiminor axis.** The number c, which can be found from Eq. (3) as

$$c = \sqrt{a^2 - b^2}, \tag{5}$$

is sometimes called the **center-to-focus distance** of the ellipse (because that's what it is).

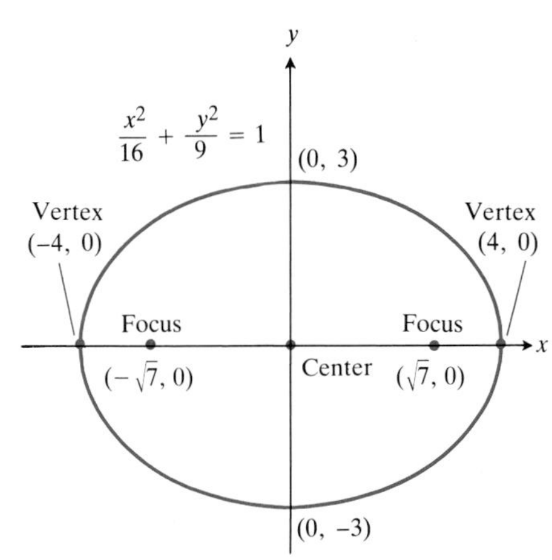

10.4 The major axis of $(x^2/16) + (y^2/9) = 1$ is horizontal.

EXAMPLE 1 *Major axis horizontal.* The ellipse

$$\frac{x^2}{16} + \frac{y^2}{9} = 1 \tag{6}$$

(Fig. 10.4) has

Semimajor axis: $a = \sqrt{16} = 4$

Semiminor axis: $b = \sqrt{9} = 3$

Center-to-focus distance: $c = \sqrt{16 - 9} = \sqrt{7}$

Foci: $(\pm c, 0) = (\pm\sqrt{7}, 0)$

Vertices: $(\pm a, 0) = (\pm 4, 0)$

Center: $(0, 0)$. ■

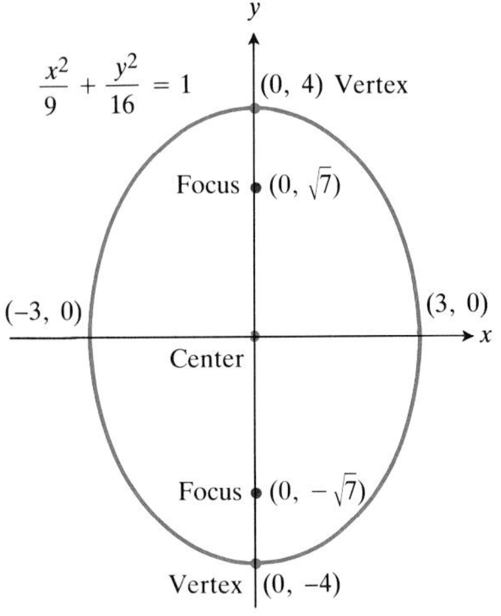

10.5 The major axis of $(x^2/9) + (y^2/16) = 1$ is vertical.

EXAMPLE 2 *Major axis vertical.* The ellipse

$$\frac{x^2}{9} + \frac{y^2}{16} = 1, \tag{7}$$

obtained by interchanging x and y in Eq. (6), represents an ellipse with its major axis vertical instead of horizontal (Fig. 10.5). With a^2 still equal to 16 and b^2 equal to 9, we have

Semimajor axis: $a = \sqrt{16} = 4$

Semiminor axis: $b = \sqrt{9} = 3$

Center-to-focus distance: $c = \sqrt{16 - 9} = \sqrt{7}$

Foci: $(0, \pm c) = (0, \pm\sqrt{7})$

Vertices: $(0, \pm a) = (0, \pm 4)$

Center: $(0, 0)$. ■

There is never any cause for confusion in analyzing equations like (6) and (7). We simply find the intercepts on the coordinate axes; then we know which way the major axis runs because it is the longer of the two axes. The center always lies at the origin and the foci always lie on the major axis.

Standard Equations for Ellipses Centered at the Origin

Foci on the x-axis: $\dfrac{x^2}{a^2} + \dfrac{y^2}{b^2} = 1 \quad (a > b)$

Center-to-focus distance: $c = \sqrt{a^2 - b^2}$

Foci: $(\pm c, 0)$

Vertices: $(\pm a, 0)$

Foci on the y-axis: $\dfrac{x^2}{b^2} + \dfrac{y^2}{a^2} = 1 \quad (a > b)$

Center-to-focus distance: $c = \sqrt{a^2 - b^2}$

Foci: $(0, \pm c)$

Vertices: $(0, \pm a)$

In each case, a is the semimajor axis and b is the semiminor axis.

Hyperbolas

DEFINITIONS

A **hyperbola** is the set of points in a plane whose distances from two fixed points in the plane have a constant difference. The two fixed points are the **foci** of the hyperbola.

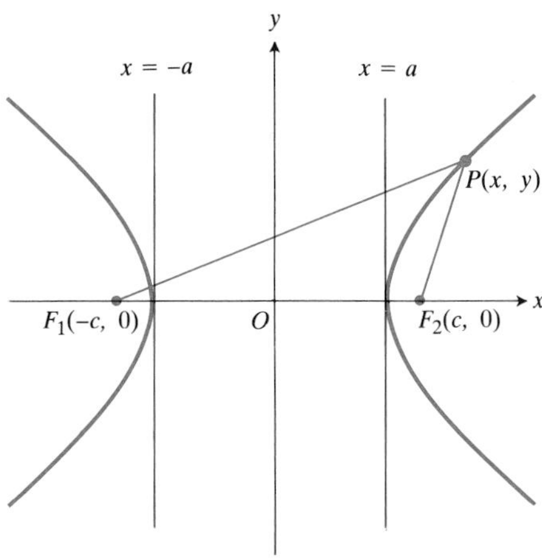

10.6 Hyperbolas have two branches. For points on the right-hand branch of the hyperbola shown here, $PF_1 - PF_2 = 2a$. For points on the left-hand branch, $PF_2 - PF_1 = 2a$.

If the foci are $F_1(-c, 0)$ and $F_2(c, 0)$ (Fig. 10.6) and the constant difference is $2a$, then a point (x, y) lies on the hyperbola if and only if

$$\sqrt{(x+c)^2+y^2} - \sqrt{(x-c)^2+y^2} = \pm 2a. \tag{8}$$

To simplify this equation, we move the second radical to the right-hand side, square, isolate the remaining radical, and square again, obtaining

$$\frac{x^2}{a^2} + \frac{y^2}{a^2-c^2} = 1. \tag{9}$$

So far, this is just like the equation for an ellipse. But now $a^2 - c^2$ is negative because $2a$, being the difference of two sides of triangle PF_1F_2, is less than $2c$, the third side.

The algebraic steps taken to arrive at Eq. (9) can be reversed to show that every point P whose coordinates satisfy an equation of this form with $0 < a < c$ also satisfies Eq. (8). Thus, a point lies on the hyperbola if and only if its coordinates satisfy Eq. (9).

If we let b denote the positive square root of $c^2 - a^2$,

$$b = \sqrt{c^2 - a^2}, \tag{10}$$

then $a^2 - c^2 = -b^2$ and Eq. (9) takes the more compact form

$$\frac{x^2}{a^2} - \frac{y^2}{b^2} = 1. \tag{11}$$

The only difference between Eq. (11) and the equation for an ellipse is the minus sign in the equation and the new relation

$$c^2 = a^2 + b^2 \tag{12}$$

given by Eq. (10).

Like the ellipse, the hyperbola is symmetric with respect to the origin and both coordinate axes. It crosses the x-axis at the points $(\pm a, 0)$. The tangents at these points are vertical because the derivative

$$\frac{dy}{dx} = \frac{b^2x}{a^2y}$$

is infinite when $y = 0$. The hyperbola has no y-intercepts; in fact, no part of the curve lies between the lines $x = -a$ and $x = a$.

DEFINITIONS

The line through the foci of a hyperbola is the **focal axis.** The point on this line halfway between the foci is the hyperbola's **center.** The points where the hyperbola and focal axis cross are the hyperbola's **vertices.**

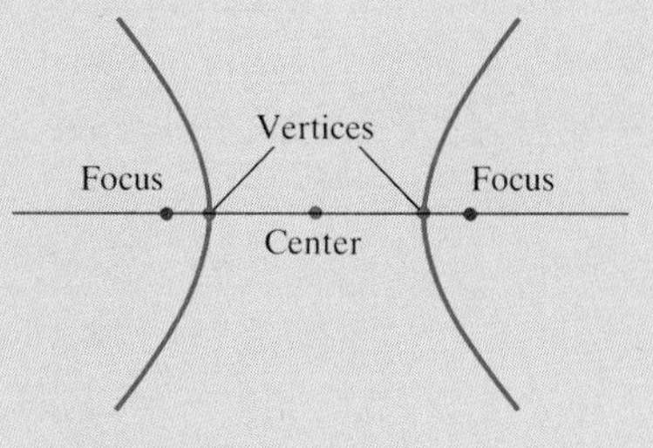

Asymptotes of Hyperbolas—Graphing As we saw in Section 4.4, the distance between a curve and some fixed line may approach zero as the curve moves farther and farther from the origin. If this happens, we call the line an **asymptote** of the curve. It turns out that the hyperbola

$$\frac{x^2}{a^2} - \frac{y^2}{b^2} = 1$$

has two asymptotes, these being the lines

$$y = \pm\frac{b}{a}x. \tag{13}$$

These asymptotes give us the guidance we need to graph hyperbolas quickly. (See facing pages.)

EXAMPLE 3 *Foci on the x-axis.* The equation

$$\frac{x^2}{4} - \frac{y^2}{5} = 1 \tag{14}$$

is Eq. (11) with $a^2 = 4$ and $b^2 = 5$ (Fig. 10.7a). We have

Center-to-focus distance: $c = \sqrt{a^2 + b^2} = \sqrt{4 + 5} = 3$

Foci: $(\pm c, 0) = (\pm 3, 0)$

Vertices: $(\pm a, 0) = (\pm 2, 0)$

Center: $(0, 0)$

Asymptotes: $y = \pm\frac{b}{a}x$ or $y = \pm\frac{\sqrt{5}}{2}x$. ■

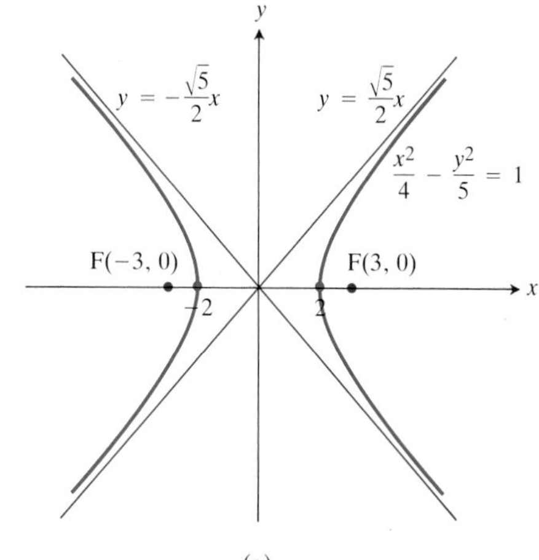

(a)

EXAMPLE 4 *Foci on the y-axis.* The hyperbola

$$\frac{y^2}{4} - \frac{x^2}{5} = 1, \tag{15}$$

obtained by interchanging x and y in Eq. (14), represents a hyperbola with vertices on the y-axis (Fig. 10.7b). With a^2 still equal to 4 and b^2 equal to 5, we have

Center-to-focus distance: $c = \sqrt{a^2 + b^2} = \sqrt{4 + 5} = 3$

Foci: $(0, \pm c) = (0, \pm 3)$

Vertices: $(0, \pm a) = (0, \pm 2)$

Center: $(0, 0)$

Asymptotes: $y = \pm\frac{a}{b}x$ or $y = \pm\frac{2}{\sqrt{5}}x$.

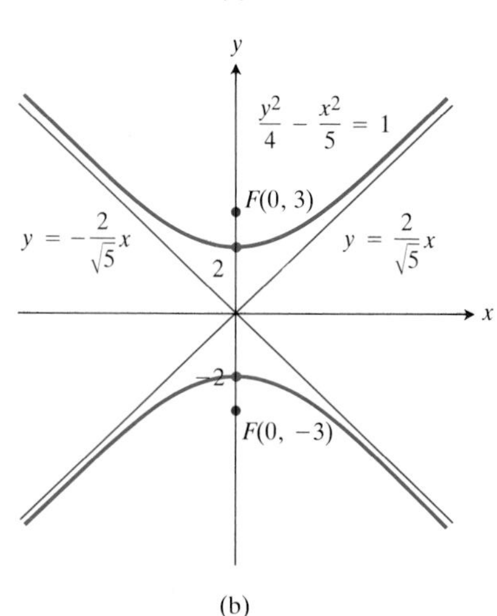

(b)

10.7 (a) The hyperbola in Example 3. (b) The hyperbola in Example 4.

Notice the revised formula for the asymptotes of this hyperbola. When you interchange x and y, the equation $y = \pm(b/a)x$ becomes $x = \pm(b/a)y$ or $y = \pm(a/b)x$. ■

DRAWING LESSON

How to graph the hyperbola $\dfrac{x^2}{a^2} - \dfrac{y^2}{b^2} = 1$

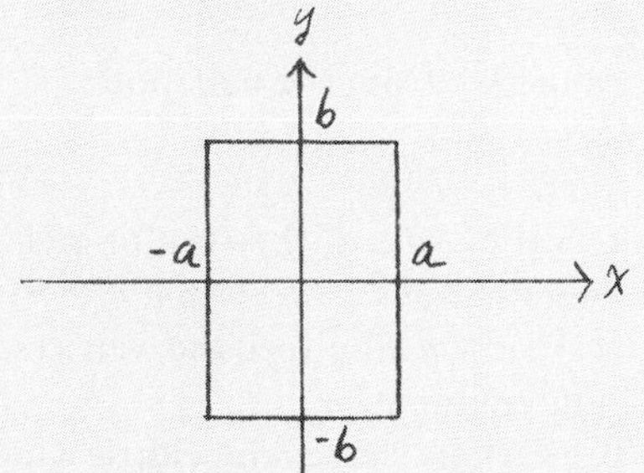

1.

Mark the points $(\pm a, 0)$ and $(0, \pm b)$ with line segments and complete the rectangle they determine.

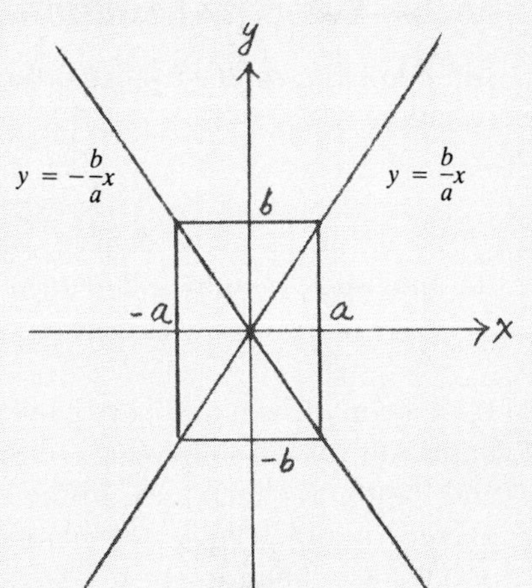

2.

Sketch the asymptotes by extending the rectangle's diagonals.

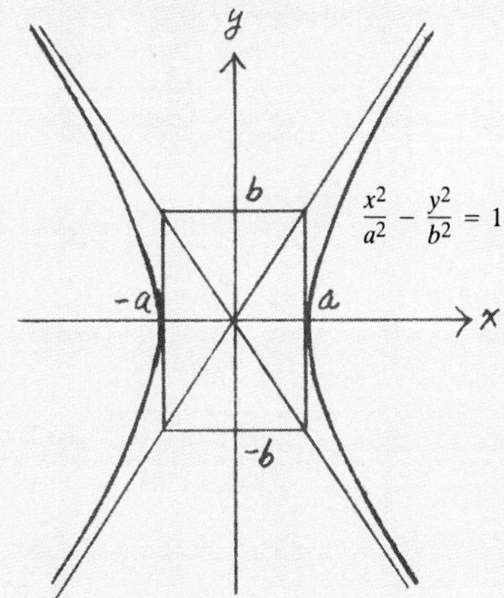

3.

Use the rectangle and asymptotes to guide your drawing.

Standard Equations for Hyperbolas Centered at the Origin

Foci on the x-axis: $\dfrac{x^2}{a^2} - \dfrac{y^2}{b^2} = 1$

Center-to-focus distance: $c = \sqrt{a^2 + b^2}$

Foci: $(\pm c, 0)$

Vertices: $(\pm a, 0)$

Asymptotes: $y = \pm \dfrac{b}{a} x$

Foci on the y-axis: $\dfrac{y^2}{a^2} - \dfrac{x^2}{b^2} = 1$

Center-to-focus distance: $c = \sqrt{a^2 + b^2}$

Foci: $(0, \pm c)$

Vertices: $(0, \pm a)$

Asymptotes: $y = \pm \dfrac{a}{b} x$

Notice the difference in the asymptote equations (b/a in the first, a/b in the second).

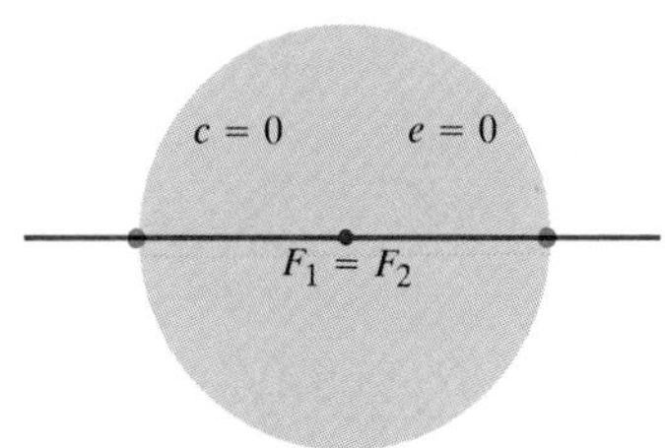

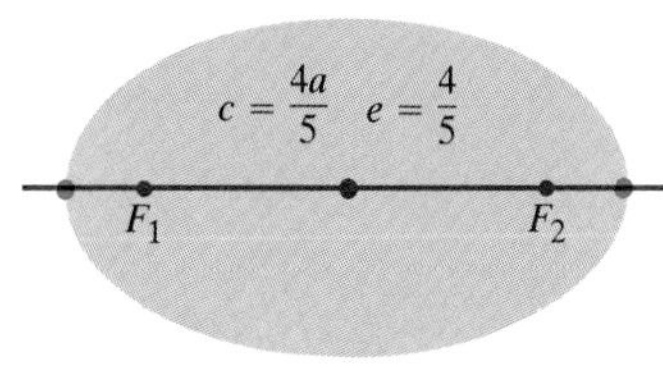

F_1 $c = a$ $e = 1$ F_2

10.8 The ellipse changes from a circle to a line segment as c increases from 0 to a.

Classifying Conic Sections by Eccentricity—The Focus–Directrix Equation

Although the center-to-focus distance c does not appear in the standard equation

$$\frac{x^2}{a^2} + \frac{y^2}{b^2} = 1 \quad (a > b)$$

for an ellipse, we may still determine the value of c from the equation

$$c = \sqrt{a^2 - b^2}.$$

If we keep a fixed and vary c over the interval $0 \le c \le a$, the resulting ellipses will vary in shape (Fig. 10.8). They are circles if $c = 0$ (so that $a = b$) and flatten as c increases. In the extreme case $c = a$, the foci and vertices overlap and the ellipse degenerates into a line segment.

We use the ratio of c to a to describe the various shapes the ellipse can take. We call this ratio the ellipse's eccentricity.

DEFINITION

The **eccentricity** of the ellipse $(x^2/a^2) + (y^2/b^2) = 1$ $(a > b)$ is the number

$$e = \frac{c}{a} = \frac{\sqrt{a^2 - b^2}}{a}. \tag{16}$$

TABLE 10.2
Eccentricities of planetary orbits

Mercury	0.21	Saturn	0.06
Venus	0.01	Uranus	0.05
Earth	0.02	Neptune	0.01
Mars	0.09	Pluto	0.25
Jupiter	0.05		

The planets in the solar system revolve around the sun in elliptical orbits with the sun at one focus. Most of the planets, including the earth, have orbits that are nearly circular, as can be seen from the eccentricities in Table 10.2. Pluto, however, has a fairly eccentric orbit, with $e = 0.25$, as does Mercury, with $e = 0.21$. Other members of the solar system have orbits that are even more eccentric. Icarus, an asteroid about 1 mile wide that revolves around the sun every 409 Earth days, has an orbital eccentricity of 0.83 (Fig. 10.9).

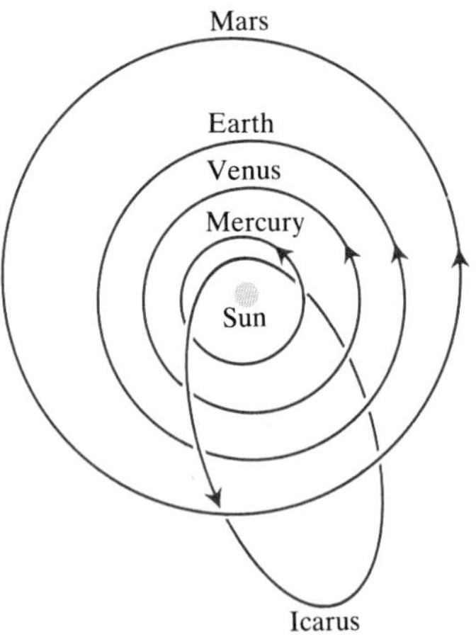

10.9 The orbit of the asteroid Icarus is highly eccentric. The earth's orbit is so nearly circular that both its foci lie inside the sun.

EXAMPLE 5 The orbit of Halley's comet is an ellipse 36.18 astronomical units long by 9.12 astronomical units wide. (One *astronomical unit* (AU) is the semimajor axis of the earth's orbit, about 92,600,000 miles.) Its eccentricity is

$$e = \frac{\sqrt{a^2 - b^2}}{a} = \frac{\sqrt{(36.18/2)^2 - (9.12/2)^2}}{(36.18/2)} = \frac{\sqrt{(18.09)^2 - (4.56)^2}}{18.09}$$

$$= 0.97. \qquad \text{(Rounded, with a calculator)}$$

Whereas a parabola has one focus and one directrix, each ellipse has two foci and two directrices. These are the lines perpendicular to the major axis at distances $\pm a/e$ from the center. The parabola has the property that

$$PF = 1 \cdot PD \tag{17}$$

for any point P on it, where F is the focus and D is the point nearest P on

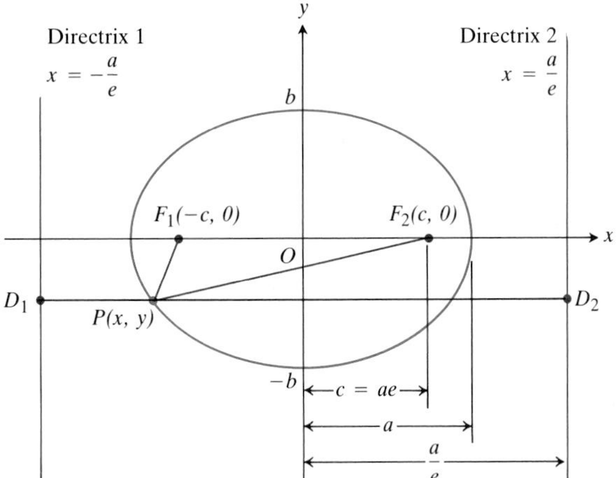

10.10 The foci and directrices of the ellipse $(x^2/a^2) + (y^2/b^2) = 1$. Directrix 1 corresponds to focus F_1, and Directrix 2 to focus F_2.

the directrix. For an ellipse, it can be shown that the equations that replace (17) are

$$PF_1 = e \cdot PD_1, \qquad PF_2 = e \cdot PD_2. \tag{18}$$

Here, e is the eccentricity, P is any point on the ellipse, F_1 and F_2 are the foci, and D_1 and D_2 are the points on the directrices nearest P (Fig. 10.10).

In each equation in (18) the directrix and focus must correspond; that is, if we use the distance from P to F_1, we must also use the distance from P to the directrix at the same end of the ellipse. Thus, the directrix $x = -a/e$ corresponds to $F_1(-c, 0)$ and the directrix $x = a/e$ corresponds to $F_2(c, 0)$.

We define the eccentricity of a hyperbola with the same formula we use for the ellipse, $e = c/a$, only in this case c equals $\sqrt{a^2 + b^2}$ instead of $\sqrt{a^2 - b^2}$. In contrast to the eccentricity of an ellipse, the eccentricity of a hyperbola is always greater than 1.

DEFINITION

The **eccentricity** of the hyperbola $(x^2/a^2) - (y^2/b^2) = 1$ is the number

$$e = \frac{c}{a} = \frac{\sqrt{a^2 + b^2}}{a}. \tag{19}$$

In both ellipse and hyperbola, the eccentricity is the ratio of the distance between the foci to the distance between the vertices (because $c/a = 2c/2a$).

$$\text{Eccentricity} = \frac{\text{Distance between foci}}{\text{Distance between vertices}}$$

In an ellipse, the foci are closer together than the vertices and the ratio is less than 1. In a hyperbola, the foci are farther apart than the vertices and the ratio is greater than 1.

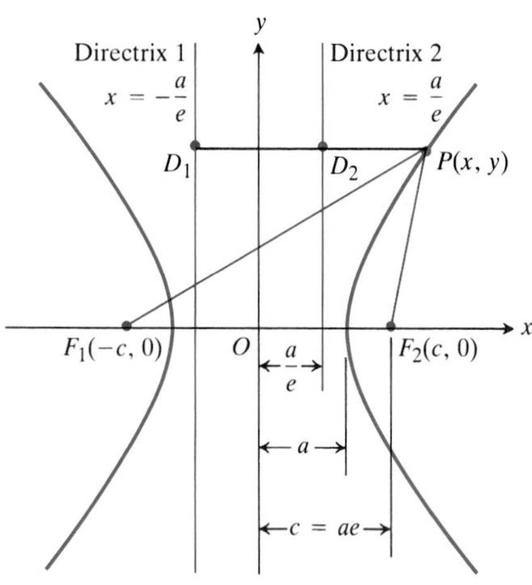

10.11 The foci and directrices of the hyperbola $(x^2/a^2) - (y^2/b^2) = 1$. No matter where P lies on the hyperbola, $PF_1 = e \cdot PD_1$ and $PF_2 = e \cdot PD_2$.

EXAMPLE 6 Find the eccentricity of the hyperbola $9x^2 - 16y^2 = 144$.

Solution We divide both sides of the hyperbola's equation by 144 to put it in standard form, obtaining

$$\frac{9x^2}{144} - \frac{16y^2}{144} = 1 \quad \text{and} \quad \frac{x^2}{16} - \frac{y^2}{9} = 1.$$

With $a^2 = 16$ and $b^2 = 9$, we find that

$$c = \sqrt{a^2 + b^2} = \sqrt{16 + 9} = 5,$$

so

$$e = \frac{c}{a} = \frac{5}{4}.$$

■

As with the ellipse, it can be shown that the lines $x = \pm a/e$ act as directrices for the hyperbola and that

$$PF_1 = e \cdot PD_1 \quad \text{and} \quad PF_2 = e \cdot PD_2. \tag{20}$$

Here P is any point on the hyperbola, F_1 and F_2 are the foci, and D_1 and D_2 are the points nearest P on the directrices (Fig. 10.11).

To complete the picture, we now define the eccentricity of a parabola to be $e = 1$. Equations (17), (18), and (20) then have the common form $PF = e \cdot PD$.

DEFINITION

The **eccentricity** of a parabola is $e = 1$.

The "focus–directrix" equation $PF = e \cdot PD$ unites the parabola, ellipse, and hyperbola in the following way. Suppose that the distance PF of a point P from a fixed point F (the focus) is a constant multiple of its distance from a fixed line (the directrix). That is, suppose

$$PF = e \cdot PD, \tag{21}$$

where e is the constant of proportionality. Then the path traced by P is

a) a *parabola* if $e = 1$,

b) an *ellipse* of eccentricity e if $e < 1$, and

c) a *hyperbola* of eccentricity e if $e > 1$.

Equation (21) may not look like much to get excited about. There are no coordinates in it and when we try to translate it into coordinate form it translates in different ways, depending on the numerical size of e. At least, that is what happens in the Cartesian plane. However, in the polar coordinate plane, as we shall see in Section 10.7, the equation $PF = e \cdot PD$ translates into a single equation regardless of the size of e, an equation so simple that it has been the equation of choice for astronomers and space scientists for nearly 300 years.

Statuary Hall in the U.S. capitol building is elliptical. A person standing at one focus can carry on a whispered conversation with someone at the other focus even when the room is noisy.

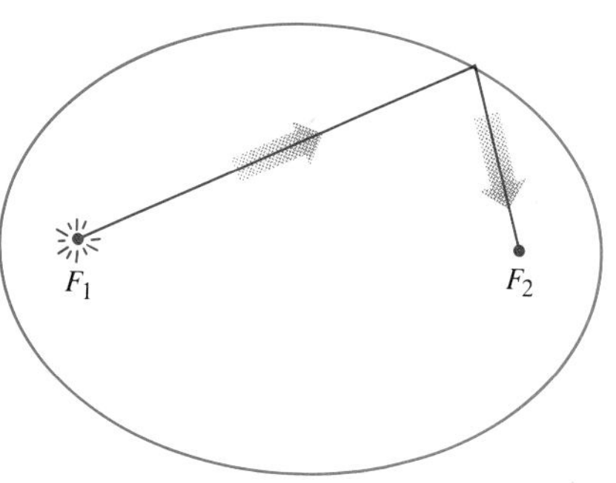

10.12 An elliptical mirror (shown here in profile) reflects light from one focus to the other.

Reflective Properties

Like the parabola, ellipses and hyperbolas have reflective properties that are important in science and engineering. If an ellipse is revolved about its major axis to generate a surface (the surface is called an *ellipsoid*), and the interior is silvered to produce a mirror, light from one focus will be reflected to the other focus (Fig. 10.12). Ellipsoids reflect sound the same way, and this property is used to construct *whispering galleries,* rooms in which a person standing at one focus can hear a whisper from the other focus. Ellipsoids also appear in instruments used to study aircraft noise in wind tunnels (sound at one focus can be received at the other focus with relatively little interference from other sources).

Light directed toward one focus of a hyperbolic mirror is reflected toward the other focus (Fig. 10.13). This property of hyperbolas is combined with the reflective properties of parabolas and ellipses in designing modern telescopes (Fig. 10.14)

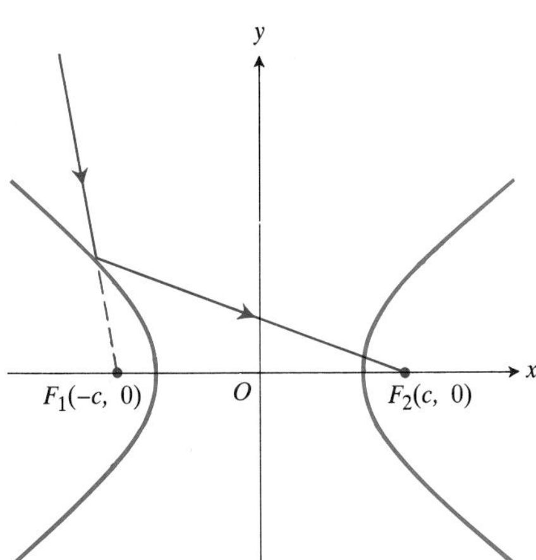

10.13 In this profile of a hyperbolic mirror, light coming toward focus F_1 is reflected toward focus F_2.

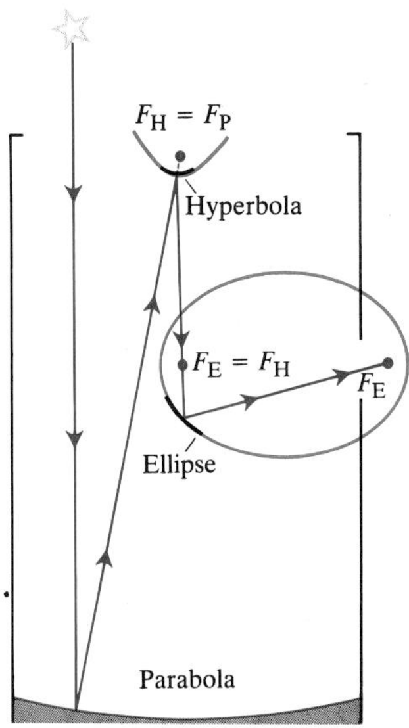

10.14 In this schematic drawing of a reflecting telescope, starlight reflects off a primary parabolic mirror toward the mirror's focus F_P. It is then reflected by a small hyperbolic mirror, whose focus is $F_H = F_P$, toward the second focus of the hyperbola, $F_E = F_H$. Since this focus is shared by an ellipse, the light is reflected by the elliptical mirror to the ellipse's second focus to be seen by an observer.

Other Applications

Ellipses appear in airplane wings (British Spitfire) and sometimes in gears designed for racing bicycles. Stereo systems often have elliptical styli, and water pipes are sometimes designed with elliptical cross sections to allow for expansion when the water freezes. The triggering mechanisms in some lasers are elliptical, and stones on a beach become more and more elliptical as they are ground down by waves. There are also applications of ellipses to fossil formation. The ellipsolith, once thought to be a separate species, is now known to be an elliptically deformed nautilus.

Hyperbolic paths arise in Einstein's theory of relativity and form the basis for the (unrelated) LORAN radio navigation system. (LORAN is short for Long Range Navigation.) Hyperbolas also form the basis for a new system the Burlington Northern Railroad is developing, for using synchronized electronic signals from satellites to track freight trains. In a recent test in Minnesota, computers aboard Burlington Northern locomotives were able to track trains to within one mile per hour of their speed and to within 150 feet of their actual location.

Exercises 10.1

Match each of the parabolas in Exercises 1–4 with one of the following equations:

$$y = x^2/2 \qquad y = -x^2/6 \qquad x = y^2/8 \qquad x = -y^2/4$$

Then find the parabola's focus and directrix.

1.

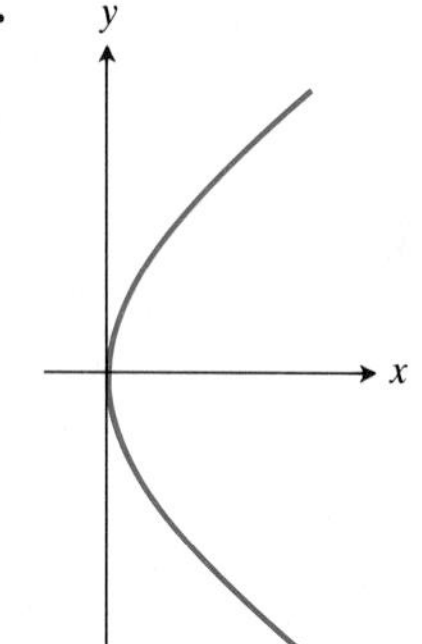

2.

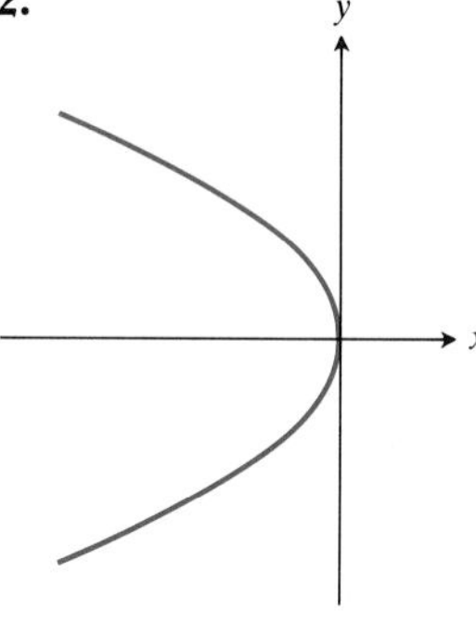

3.

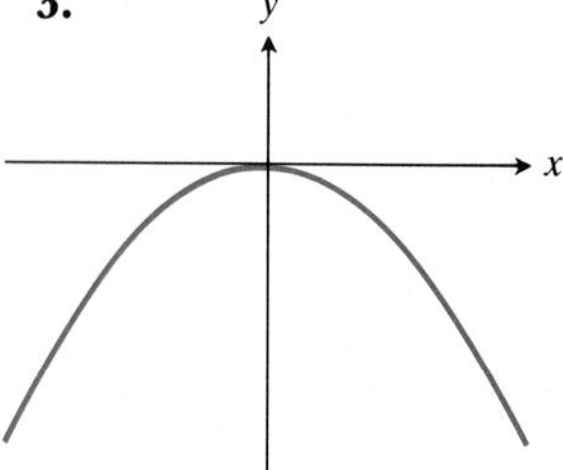

4.

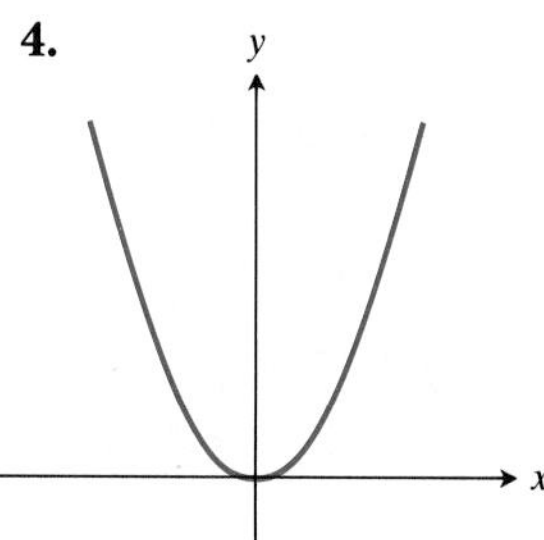

Match each of the conic sections in Exercises 5–8 with one of the following equations:

$$\frac{x^2}{4} + \frac{y^2}{9} = 1 \qquad \frac{x^2}{2} + y^2 = 1 \qquad \frac{y^2}{4} - x^2 = 1 \qquad \frac{x^2}{4} - \frac{y^2}{9} = 1$$

Then find each conic section's foci, eccentricity, and directrices. If the conic section is a hyperbola, find its asymptotes as well.

5.

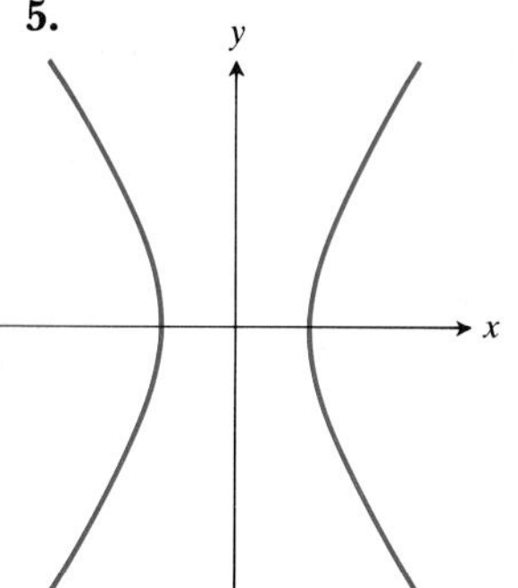

6.

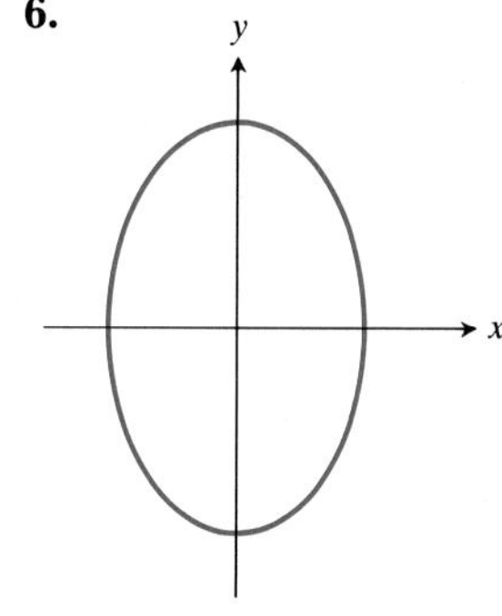

7.

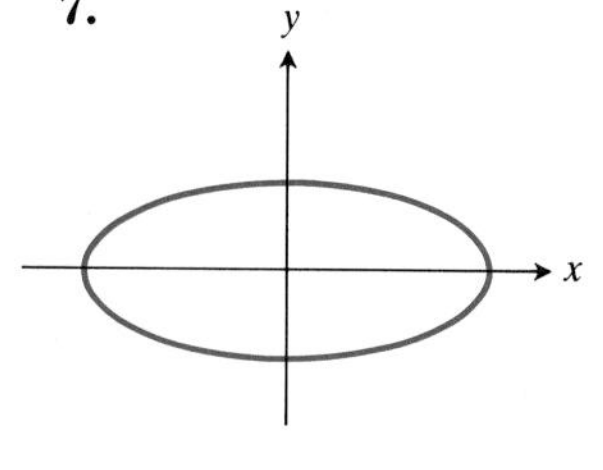

8.

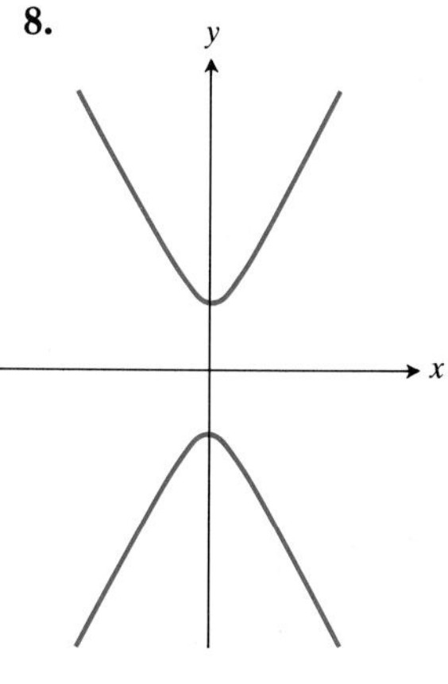

Exercises 9–16 give equations for ellipses. Put each equation in standard form and find the ellipse's eccentricity. Then sketch the ellipse. Include the foci in your sketch.

9. $16x^2 + 25y^2 = 400$
10. $7x^2 + 16y^2 = 112$
11. $2x^2 + y^2 = 2$
12. $2x^2 + y^2 = 4$
13. $3x^2 + 2y^2 = 6$
14. $9x^2 + 10y^2 = 90$
15. $6x^2 + 9y^2 = 54$
16. $169x^2 + 25y^2 = 4225$

Exercises 17–24 give equations for hyperbolas. Put each equation in standard form and find the hyperbola's eccentricity and asymptotes. Then sketch the hyperbola. Include the asymptotes and foci in your sketch.

17. $x^2 - y^2 = 1$
18. $9x^2 - 16y^2 = 144$
19. $y^2 - x^2 = 8$
20. $y^2 - x^2 = 4$
21. $8x^2 - 2y^2 = 16$
22. $y^2 - 3x^2 = 3$
23. $8y^2 - 2x^2 = 16$
24. $64x^2 - 36y^2 = 2304$

25. *Archimedes' formula for the volume of a parabolic solid.* The region enclosed by the parabola $y = (4h/b^2)x^2$ and the line $y = h$ is revolved about the y-axis to generate a solid (Fig. 10.15). Show that the volume of the solid is three-halves the volume of the corresponding cone.

26. *Suspension bridge cables hang in parabolas.* Figure 10.16 shows a cable of a suspension bridge supporting a uniform load of w pounds per horizontal foot. It can be shown that if H is the horizontal tension in the cable at the origin, then the curve of the cable satisfies the differential equation

$$\frac{dy}{dx} = \frac{w}{H}x.$$

Show that the cable hangs in a parabola by solving this equation with the condition that $y = 0$ when $x = 0$.

27. Draw an ellipse of eccentricity 4/5.

28. Draw the orbit of Pluto to scale.

29. Find the dimensions of the rectangle of largest area that can be inscribed in the ellipse $x^2 + 4y^2 = 4$ with its sides parallel to the coordinate axes. What is the area of the rectangle?

30. Find the center of mass of a thin homogeneous plate that is bounded below by the x-axis and above by the ellipse $(x^2/9) + (y^2/16) = 1$.

31. The region bounded on the left by the y-axis, on the right by the hyperbola $x^2 - y^2 = 1$, and above and below by the lines $y = \pm 3$, is revolved about the y-axis to generate a solid. Find the volume of the solid.

32. The curve $y = \sqrt{x^2 + 1}$, $0 \le x \le \sqrt{2}$, which is part of the upper branch of the hyperbola $y^2 - x^2 = 1$, is revolved about the x-axis to generate a surface. Find the area of the surface.

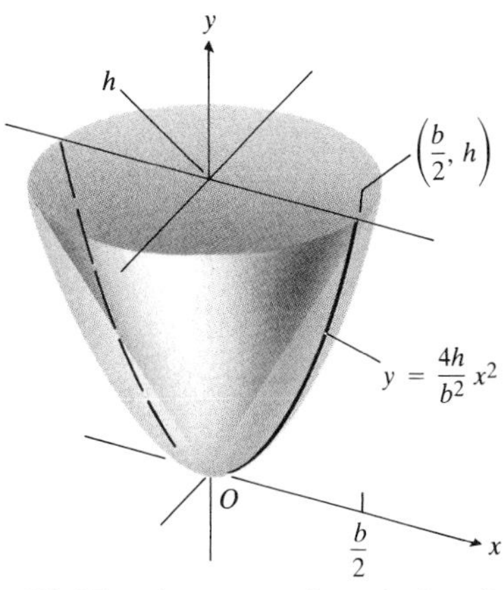

10.15 The cone and parabolic solid in Exercise 25.

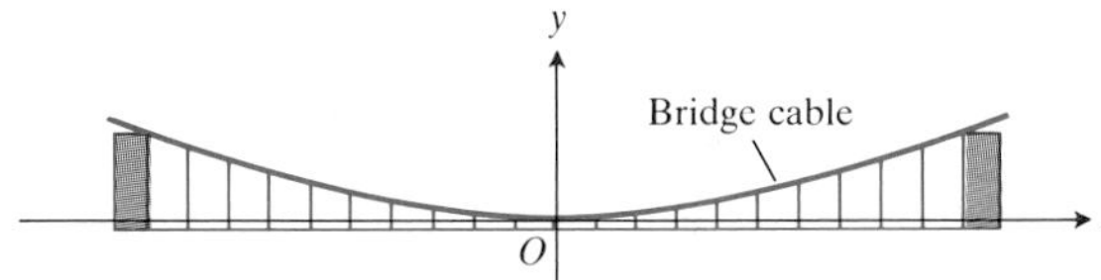

10.16 The suspension-bridge cable in Exercise 26.

33. The circular waves in the photograph below were made by touching the surface of a ripple tank, first at A and then at B. As the waves expanded, their point of intersection seemed to trace a hyperbola. Did it really do that?

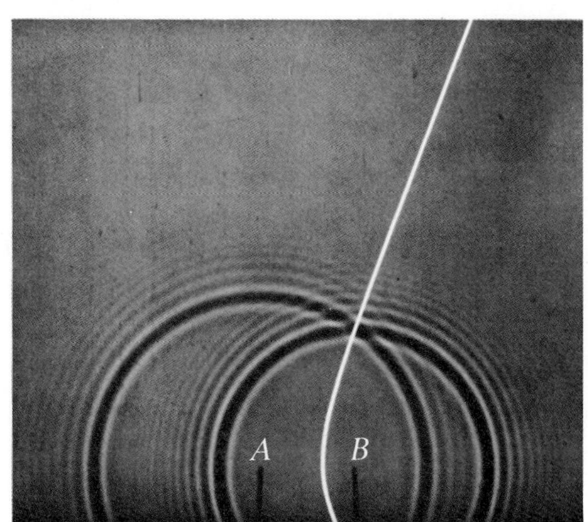

(Photograph from *PSSC Physics, Second Edition;* 1965; D.C. Heath & Company with Education Development Center, Inc. *NCFMF Book of Film Notes,* 1974; The MIT Press with Education Development Center, Inc., Newton, Massachusetts.)

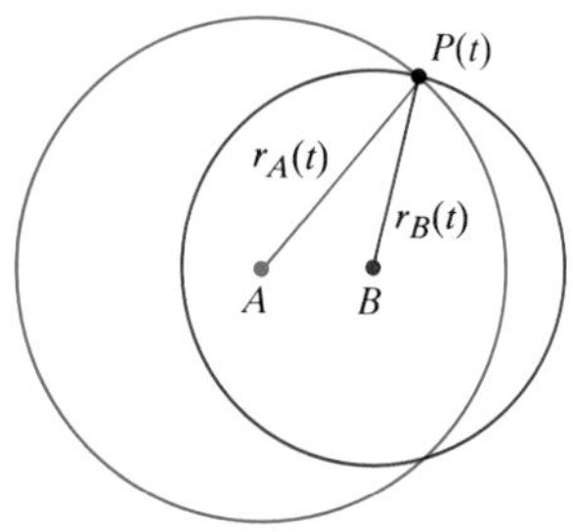

10.17 The wave model for Exercise 33.

To find out, we can model the waves with circles centered at A and B (Fig. 10.17). At time t, the point P is $r_A(t)$ units from A and $r_B(t)$ units from B. Since the radii of the circles increase at a constant rate, the rate at which the waves are traveling is

$$\frac{dr_A}{dt} = \frac{dr_B}{dt}.$$

Conclude from this equation that $r_A - r_B$ has a constant value, so that P must lie on a hyperbola with foci at A and B.

34. *The reflective property of parabolas.* Figure 10.18 shows a typical point $P(x_0, y_0)$ on the parabola $y^2 = 4px$. The line L is tangent to the parabola at P. The parabola's focus lies at $F(p, 0)$. The ray L' extending from P to the right is parallel to the x-axis. We show that light from F to P will be reflected out along L' by showing that β equals α. Establish this equality by taking the following steps.

1. Show that $\tan\beta = 2p/y_0$.
2. Show that $\tan\phi = y_0/(x_0 - p)$.
3. Use the identity

$$\tan\alpha = \frac{\tan\phi - \tan\beta}{1 + \tan\phi\tan\beta}$$

to show that $\tan\alpha = 2p/y_0$.

Since the angles involved are both acute, $\tan\beta = \tan\alpha$ implies $\beta = \alpha$.

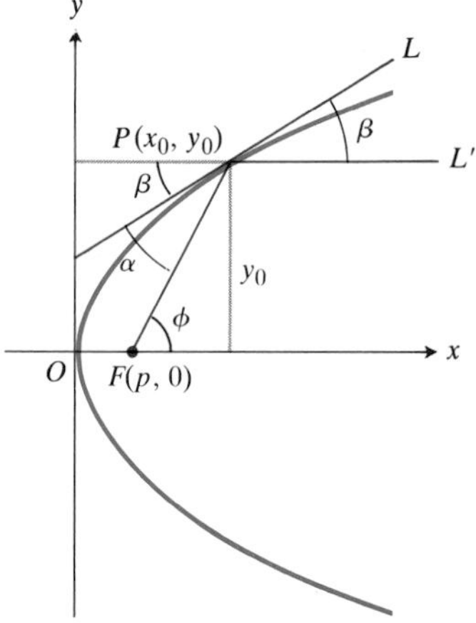

10.18 A parabolic reflector sends all light from the focus out parallel to the parabola's axis (Exercise 34).

10.2 The Graphs of Quadratic Equations in *x* and *y*

In this section, we establish one of the most amazing results in analytic geometry, which is that the Cartesian graph of any equation of the form

$$Ax^2 + Bxy + Cy^2 + Dx + Ey + F = 0, \tag{1}$$

in which A, B, and C are not all zero is nearly always a conic section. The only exceptions are the case in which the graph consists of two parallel lines and the case in which there is no graph at all. It is conventional to call all graphs of Eq. (1), curved or not, **quadratic curves.**

The Cross-Product Term

You may have noticed that the term Bxy did not appear in the equations for the conic sections in Section 10.1. This happened because the axes of the conic sections ran parallel to (in fact, coincided with) the coordinate axes.

TABLE 10.3
Examples of quadratic curves

$Ax^2 + Bxy + Cy^2 + Dx + Ey + F = 0$

	A	B	C	D	E	F	Equation	Remarks
Circle	1		1			−4	$x^2 + y^2 = 4$	$A = C$
Parabola			1	−9			$y^2 = 9x$	Quadratic in y, linear in x
Ellipse	4		9			−36	$4x^2 + 9y^2 = 36$	A, C have same sign, $A \neq C$
Hyperbola	1		−1			−1	$x^2 - y^2 = 1$	A, C have opposite signs
One line (still a conic section)	1						$x^2 = 0$	y-axis
Intersecting lines (still a conic section)		1		1	−1	−1	$xy + x - y - 1 = 0$	Factors to $(x - 1)(y + 1) = 0$, so $x = 1$, $y = -1$.
Parallel lines (not a conic section)	1			−3		2	$x^2 - 3x + 2 = 0$	Factors to $(x - 1)(x - 2) = 0$, so $x = 1$, $x = 2$.
Point	1		1				$x^2 + y^2 = 0$	The origin
No graph	1					1	$x^2 = -1$	No graph

To see what happens when the parallelism is absent, let us write an equation for a hyperbola with $a = 3$ and foci at $F_1(-3, -3)$ and $F_2(3, 3)$ (Fig. 10. 19). The equation $|PF_1 - PF_2| = 2a$ then becomes $|PF_1 - PF_2| = 2(3) = 6$ and

$$\sqrt{(x + 3)^2 + (y + 3)^2} - \sqrt{(x - 3)^2 + (y - 3)^2} = \pm 6.$$

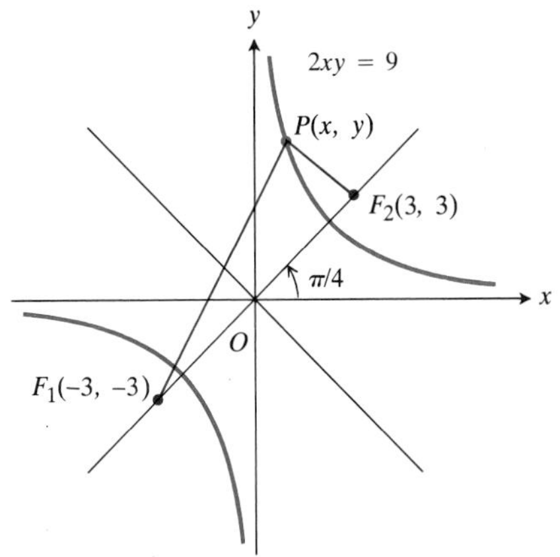

10.19 The focal axis of the hyperbola $2xy = 9$ makes an angle of $\pi/4$ radians with the positive x-axis.

When we transpose one radical, square, solve for the radical that still appears, and square again, this reduces to

$$2xy = 9, \tag{2}$$

which is a special case of Eq. (1) in which the cross-product term is present. The asymptotes of the hyperbola in Eq. (2) are the x- and y-axes, and the focal axis makes an angle of $\pi/4$ radians with the positive x-axis. As in this example, the cross-product term is present in Eq. (1) only when the axes of the conic are tilted.

Rotating the Coordinate Axes to Eliminate the Cross-Product Term

To eliminate the xy term from the equation of a conic, we rotate the coordinate axes to eliminate the "tilt" in the axes of the conic. The equations for the rotations we use are derived in the following way. In the notation of

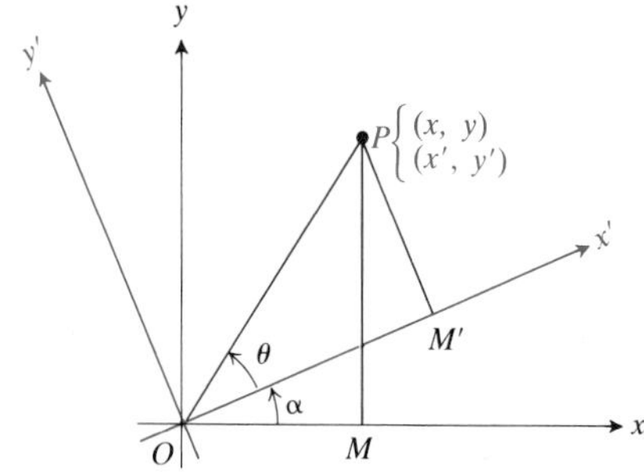

10.20 A counterclockwise rotation through angle α about the origin.

Fig. 10.20, which shows a counterclockwise rotation about the origin through an angle α,

$$
\begin{aligned}
x &= OM = OP\cos(\theta+\alpha) = OP\cos\theta\cos\alpha - OP\sin\theta\sin\alpha,\\
y &= MP = OP\sin(\theta+\alpha) = OP\cos\theta\sin\alpha + OP\sin\theta\cos\alpha.
\end{aligned}
\tag{3}
$$

Since

$$OP\cos\theta = OM' = x' \quad \text{and} \quad OP\sin\theta = M'P = y', \tag{4}$$

the equations in (3) reduce to

Equations for Rotating the Coordinate Axes

$$
\begin{aligned}
x &= x'\cos\alpha - y'\sin\alpha,\\
y &= x'\sin\alpha + y'\cos\alpha.
\end{aligned}
\tag{5}
$$

EXAMPLE 1 The x- and y-axes are rotated through an angle of $\pi/4$ radians about the origin. Find an equation for the hyperbola $2xy = 9$ in the new coordinates.

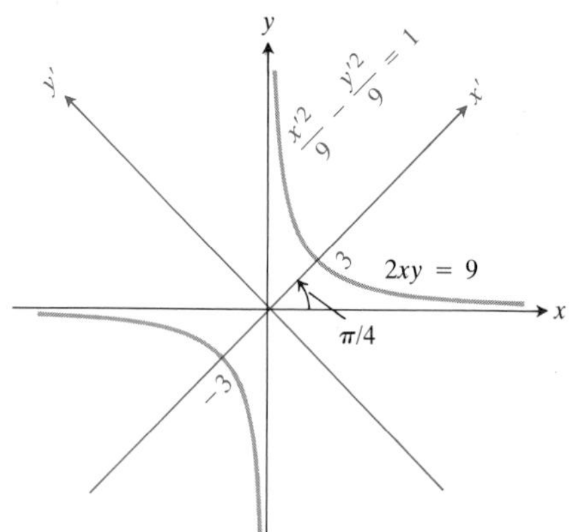

10.21 The hyperbola in Example 1.

Solution Since $\cos \pi/4 = \sin \pi/4 = 1/\sqrt{2}$, we substitute

$$x = \frac{x' - y'}{\sqrt{2}}, \qquad y = \frac{x' + y'}{\sqrt{2}}$$

from Eqs. (5) into the equation $2xy = 9$ and obtain

$$
\begin{aligned}
2\left(\frac{x'-y'}{\sqrt{2}}\right)\left(\frac{x'+y'}{\sqrt{2}}\right) &= 9\\
x'^2 - y'^2 &= 9\\
\frac{x'^2}{9} - \frac{y'^2}{9} &= 1.
\end{aligned}
$$

See Fig. 10.21. ■

If we apply the rotation equations in (5) to the general quadratic equation (1), we obtain a new quadratic equation

$$A'x'^2 + B'x'y' + C'y'^2 + D'x' + E'y' + F' = 0. \tag{6}$$

The new coefficients are related to the old ones by the equations

$$
\begin{aligned}
A' &= A\cos^2\alpha + B\cos\alpha\sin\alpha + C\sin^2\alpha,\\
B' &= B\cos 2\alpha + (C - A)\sin 2\alpha,\\
C' &= A\sin^2\alpha - B\sin\alpha\cos\alpha + C\cos^2\alpha,\\
D' &= D\cos\alpha + E\sin\alpha,\\
E' &= -D\sin\alpha + E\cos\alpha,\\
F' &= F.
\end{aligned}
\tag{7}
$$

These equations show, among other things, that if we start with an equation for a curve in which the cross-product term is present ($B \neq 0$), we can find a rotation angle α that produces an equation in which no cross-product term appears ($B' = 0$). To find α, we put $B' = 0$ in the second equation in (7) and solve the resulting equation,

$$B \cos 2\alpha + (C - A)\sin 2\alpha = 0,$$

for α. In practice, this means finding α from one of the two equations

$$\cot 2\alpha = \frac{A - C}{B} \quad \text{or} \quad \tan 2\alpha = \frac{B}{A - C}. \tag{8}$$

EXAMPLE 2 The coordinate axes are to be rotated through an angle α to produce an equation for the curve

$$x^2 + xy + y^2 - 6 = 0 \tag{9}$$

that has no cross-product term. Find α and the new equation.

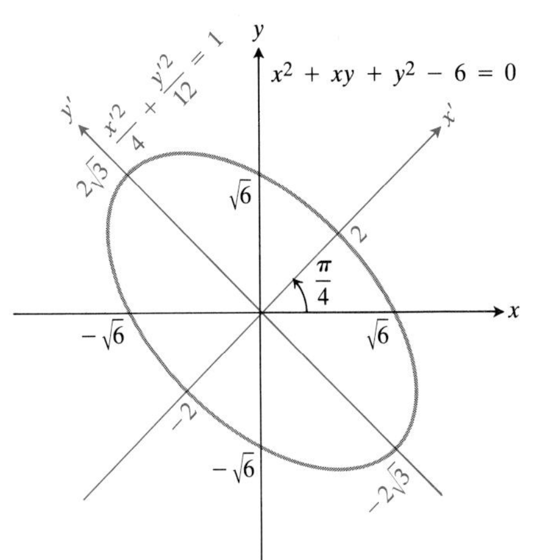

10.22 The ellipse in Example 2.

Solution 1 *Using Eqs. (8), (7), and (6).* Equation (9) has $A = B = C = 1$. We substitute these values into Eq. (8) to find α:

$$\cot 2\alpha = \frac{A - C}{B} = \frac{1 - 1}{1} = 0, \qquad 2\alpha = \frac{\pi}{2}, \qquad \alpha = \frac{\pi}{4}.$$

Substituting $\alpha = \pi/4$, $A = B = C = 1$, $D = E = 0$, and $F = -6$ into Eqs. (7) gives

$$A' = \frac{3}{2}, \qquad B' = 0, \qquad C' = \frac{1}{2}, \qquad D' = E' = 0, \qquad F' = -6.$$

Equation (6) then gives

$$\frac{3}{2}x'^2 + \frac{1}{2}y'^2 - 6 = 0, \qquad \text{or} \qquad \frac{x'^2}{4} + \frac{y'^2}{12} = 1.$$

This is the equation of an ellipse with foci on the new y'-axis (Fig. 10.22).

Solution 2 *Using the rotation equations directly.* This method brings in fewer formulas but requires more arithmetic. We begin by finding α as in Solution 1 and substitute the value found, $\alpha = \pi/4$, into the rotation equations to get

$$x = x' \cos \alpha - y' \sin \alpha = \frac{\sqrt{2}}{2}x' - \frac{\sqrt{2}}{2}y',$$

$$y = x' \sin \alpha + y' \cos \alpha = \frac{\sqrt{2}}{2}x' + \frac{\sqrt{2}}{2}y'.$$

We then substitute the primed expressions for x and y in the original equation, $x^2 + xy + y^2 - 6 = 0$. This gives

$$\left(\frac{\sqrt{2}}{2}x' - \frac{\sqrt{2}}{2}y'\right)^2 + \left(\frac{\sqrt{2}}{2}x' - \frac{\sqrt{2}}{2}y'\right)\left(\frac{\sqrt{2}}{2}x' + \frac{\sqrt{2}}{2}y'\right) + \left(\frac{\sqrt{2}}{2}x' + \frac{\sqrt{2}}{2}y'\right)^2 - 6 = 0,$$

or, skipping over the arithmetic,

$$\frac{x'^2}{4} + \frac{y'^2}{12} = 1.$$

■

The Graphs of Quadratic Equations

We now return to our analysis of the graph of the general quadratic equation.

Since axes may always be rotated to eliminate the cross-product term, there is no loss of generality in assuming that this has been done, and our equation has the form

$$Ax^2 + Cy^2 + Dx + Ey + F = 0. \tag{10}$$

Equation 10 represents

a) a *circle* if $A = C \neq 0$ (special cases: the graph is a point or there is no graph at all);

b) a *parabola* if Eq. (10) is quadratic in one variable and linear in the other;

c) an *ellipse* if A and C are both positive or both negative (special cases: a single point or no graph at all);

d) a *hyperbola* if A and C have opposite signs (special case: a pair of intersecting lines);

e) a *straight line* if A and C are zero and at least one of D and E is different from zero;

f) *one or two straight lines* if the left-hand side of Eq. (10) can be factored into the product of two linear factors.

The Discriminant

There is a quick way to tell whether the graph of the equation

$$Ax^2 + Bxy + Cy^2 + Dx + Ey + F = 0 \tag{11}$$

is a parabola, ellipse, or hyperbola. The test does not require us to eliminate the xy term first.

As we have seen, if B is not zero, then rotating the axes through the angle α determined by the equation

$$\cot 2\alpha = \frac{A - C}{B} \tag{12}$$

will change Eq. (11) into the equivalent form

$$A'x'^2 + C'y'^2 + D'x' + E'y' + F' = 0 \tag{13}$$

without a cross-product term.

Now, the graph of Eq. (13) is a (real or degenerate)

a) *parabola* if A' or $C' = 0$; that is, if $A'C' = 0$;

b) *ellipse* if A' and C' have the same sign; that is, if $A'C' > 0$;

c) *hyperbola* if A' and C' have opposite signs; that is, if $A'C' < 0$.

It can also be verified, by using Eqs. (7), that for any rotation of axes,

$$B^2 - 4AC = B'^2 - 4A'C'. \tag{14}$$

This means that the quantity $B^2 - 4AC$ is not changed by a rotation. But when we rotate through the angle α given by Eq. (12), B' becomes zero, so that

$$B^2 - 4AC = -4A'C'.$$

Since the curve is a parabola if $A'C' = 0$, an ellipse if $A'C' > 0$, and a hyperbola if $A'C' < 0$, the curve must be

a) a *parabola* if $B^2 - 4AC = 0$,

b) an *ellipse* if $B^2 - 4AC < 0$, (15)

c) a *hyperbola* if $B^2 - 4AC > 0$.

The number $B^2 - 4AC$ is called the **discriminant** of Eq. (11). What we have just seen is that the graph of Eq. (11) is a parabola if the discriminant is zero, an ellipse if the discriminant is negative, and a hyperbola if the discriminant is positive (with the understanding that occasional degenerate cases may arise).

EXAMPLE 3

a) $3x^2 - 6xy + 3y^2 + 2x - 7 = 0$ represents a parabola because

$$B^2 - 4AC = (-6)^2 - 4 \cdot 3 \cdot 3 = 36 - 36 = 0.$$

b) $x^2 + xy + y^2 - 1 = 0$ represents an ellipse because

$$B^2 - 4AC = (1)^2 - 4 \cdot 1 \cdot 1 = -3 < 0.$$

c) $xy - y^2 - 5y + 1 = 0$ represents a hyperbola because

$$B^2 - 4AC = (1)^2 - 4(0)(-1) = 1 > 0.$$ ■

Exercises 10.2

Use the discriminant $B^2 - 4AC$ to decide whether the equations in Exercises 1–16 represent parabolas, ellipses, or hyperbolas.

1. $x^2 - y^2 - 1 = 0$

2. $25x^2 + 9y^2 - 225 = 0$

3. $y^2 - 4x - 4 = 0$

4. $x^2 + y^2 - 10 = 0$

5. $x^2 + 4y^2 - 4x - 8y + 4 = 0$

6. $2x^2 - y^2 + 4xy - 2x + 3y = 6$

7. $x^2 + 4xy + 4y^2 - 3x = 6$

8. $x^2 + y^2 + 3x - 2y = 10$

9. $xy + y^2 - 3x = 5$

10. $3x^2 + 6xy + 3y^2 - 4x + 5y = 12$

11. $x^2 - y^2 = 1$

12. $2x^2 + 3y^2 - 4x = 7$

13. $x^2 - 3xy + 3y^2 + 6y = 7$

14. $25x^2 - 4y^2 - 350x = 0$

15. $6x^2 + 3xy + 2y^2 + 17y + 2 = 0$

16. $3x^2 + 12xy + 12y^2 + 435x - 9y + 72 = 0$

In Exercises 17–26, rotate the coordinate axes to change the given equation into an equation that has no cross-product (xy) term. Then identify the graph of the equation. (The new equations will vary with the size and direction of the rotation you choose.)

17. $xy = 2$

18. $x^2 + xy + y^2 = 1$

19. $3x^2 + 2\sqrt{3}xy + y^2 - 8x + 8\sqrt{3}y = 0$

20. $x^2 - \sqrt{3}xy + 2y^2 = 1$

21. $x^2 - 2xy + y^2 = 2$

22. $x^2 - 3xy + y^2 = 5$

23. $\sqrt{2}x^2 + 2\sqrt{2}xy + \sqrt{2}y^2 - 8x + 8y = 0$

24. $xy - y - x + 1 = 0$

25. $3x^2 + 2xy + 3y^2 = 19$

26. $3x^2 + 4\sqrt{3}xy - y^2 = 7$

27. *The hyperbola $xy = a$.* The hyperbola $xy = 1$ is one of many hyperbolas of the form $xy = a$ that appear in science and mathematics.

a) Rotate the coordinate axes through an angle of 45° to change the equation $xy = 1$ into an equation with no xy term. What is the new equation?

b) Do the same for the equation $xy = a$.

TOOLKIT PROGRAM

Conic Sections — Shows how the coefficients in the equation $Ax^2 + Bxy + Cy^2 + Dx + Ey + F = 0$ change as the conic section is rotated in the xy-plane

10.3 Parametric Equations for Plane Curves

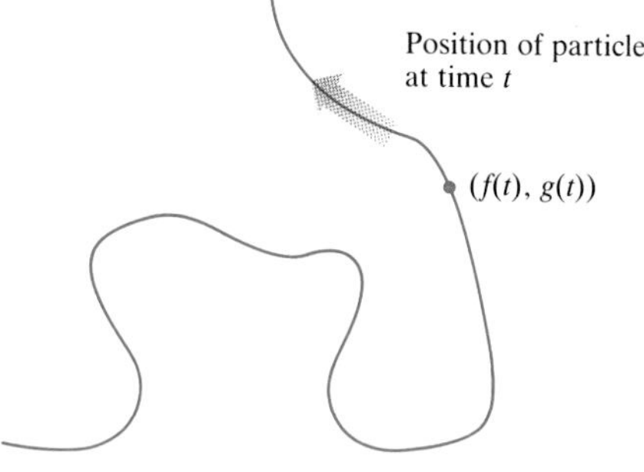

10.23 The path traced by a particle moving in the xy-plane is not always the graph of a function of x.

When the path of a particle moving in the plane is not the graph of a function (Fig. 10.23), we cannot hope to describe it with a Cartesian formula that expresses y directly in terms of x. Instead, we express each of the particle's coordinates as a function of time t and describe the path with a pair of equations, $x = f(t)$ and $y = g(t)$. Indeed, for studying motion, equations like these are preferable to a Cartesian formula for the path because they immediately tell us the particle's position at any time t. They become equations for the motion as well as equations for the path along which the motion takes place. They also enable us to calculate the particle's velocity and acceleration at any time t, as we shall see in Chapter 12. In the present section, we focus on the geometry of curves defined by parametric equations.

DEFINITIONS

If x and y are given as functions

$$x = f(t), \qquad y = g(t) \tag{1}$$

over an interval of t-values, then the set of points $(x, y) = (f(t), g(t))$ defined by these equations is called a **curve** in the coordinate plane. The equations are **parametric equations** for the curve. The variable t is the **parameter** of the curve and its domain I is called the **parameter interval.** If I is a closed interval, $a \leq t \leq b$, the point $(f(a), g(a))$ is the **initial point** of the curve and $(f(b), g(b))$ is the **terminal point** of the curve. When we write parametric equations for a curve in the plane, we say that we have **parametrized** the curve.

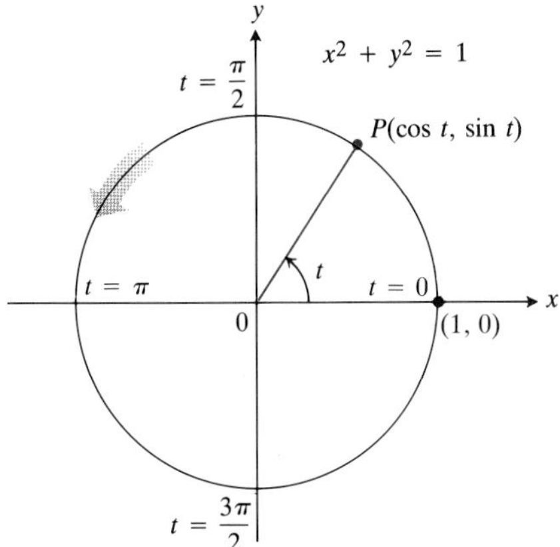

10.24 The equations $x = \cos t$ and $y = \sin t$ describe motion on the unit circle $x^2 + y^2 = 1$. The arrow shows the direction of increasing t.

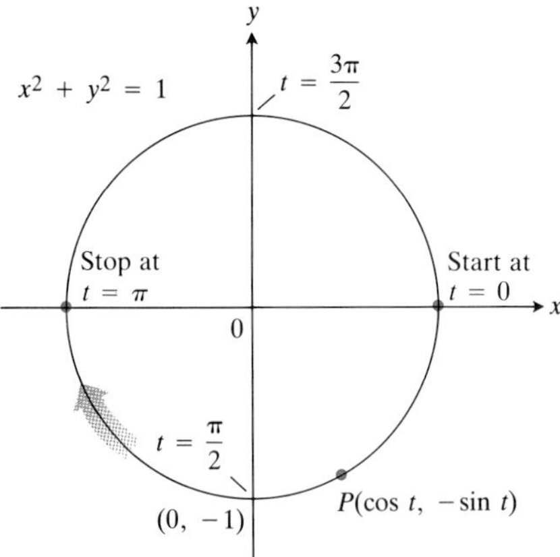

10.25 The point $P(\cos t, -\sin t)$ moves clockwise as t increases from 0 to π.

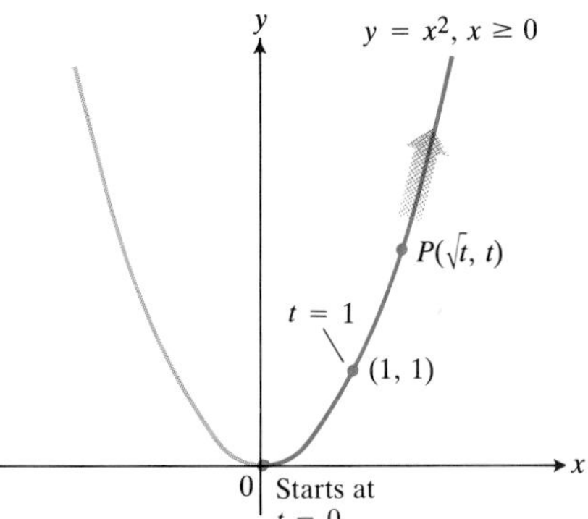

10.26 The equations $x = \sqrt{t}, y = t$ describe the motion of a particle that traces the right-hand half of the parabola $y = x^2$.

In many applications, t denotes time, but it might instead denote an angle (as in some of the following examples) or the distance a particle has traveled from its starting point (as it sometimes will when we study motion later on).

EXAMPLE 1 *The circle $x^2 + y^2 = 1$.* The equations

$$x = \cos t, \qquad y = \sin t, \qquad 0 \leq t \leq 2\pi,$$

describe the position $P(x, y)$ of a particle that moves counterclockwise around the circle $x^2 + y^2 = 1$ as t increases (Fig. 10.24).

We know that the point lies on this circle for every value of t because

$$x^2 + y^2 = \cos^2 t + \sin^2 t = 1.$$

But how much of the circle does the point $P(x, y)$ actually traverse? To find out, we track the motion as t runs from 0 to 2π. The parameter t is the radian measure of the angle that radius OP makes with the positive x-axis. The particle starts at $(1, 0)$, moves up and to the left as t approaches $\pi/2$, and continues around the circle to stop again at $(1, 0)$ when $t = 2\pi$. The particle traces the circle exactly once. ■

EXAMPLE 2 *A semicircle.* The equations

$$x = \cos t, \qquad y = -\sin t, \qquad 0 \leq t \leq \pi,$$

describe the position $P(x, y)$ of a particle that moves clockwise around the circle $x^2 + y^2 = 1$ as t increases from 0 to π.

We know that the point P lies on this circle for all t because its coordinates satisfy the circle's equation. How much of the circle does the particle traverse? To find out, we track the motion as t runs from 0 to π. As in Example 1, the particle starts at $(1, 0)$. But now as t increases y becomes negative, decreasing to -1 when $t = \pi/2$ and then increasing back to 0 as t approaches π. The motion stops at $t = \pi$ with only the lower half of the circle covered (Fig. 10.25). ■

EXAMPLE 3 *Half a parabola.* The position $P(x, y)$ of a particle moving in the xy-plane is given by the equations

$$x = \sqrt{t}, \qquad y = t, \qquad t \geq 0.$$

Identify the path traced by the particle and describe the motion.

Solution We try to identify the path by eliminating t between the equations $x = \sqrt{t}$ and $y = t$. With any luck, this will produce a recognizable algebraic relation between x and y. We find that

$$y = t = (\sqrt{t})^2 = x^2.$$

This means the particle's position coordinates satisfy the equation $y = x^2$, so the particle moves along the parabola $y = x^2$.

It would be a mistake, however, to conclude that the particle's path is the entire parabola $y = x^2$—it is only half the parabola. The particle's x-coordinate is never negative. The particle starts at $(0, 0)$ when $t = 0$ and rises into the first quadrant as t increases (Fig. 10.26). ■

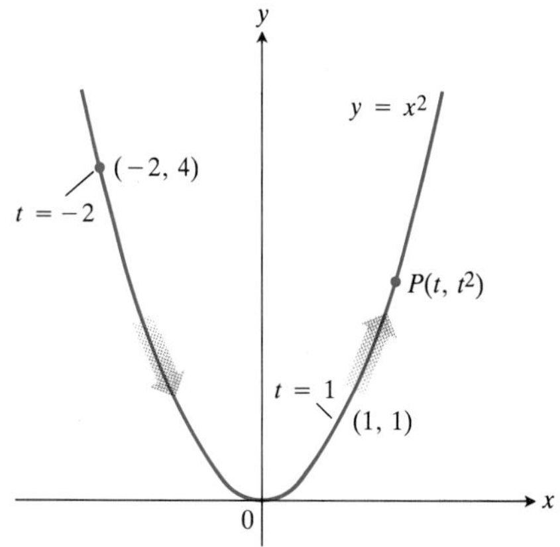

10.27 The path defined by the parametric equations $x = t, y = t^2, -\infty < t < \infty$.

EXAMPLE 4 *A complete parabola.* The position $P(x, y)$ of a particle moving in the xy-plane is given by the equations

$$x = t, \qquad y = t^2, \qquad -\infty < t < \infty.$$

Identify the particle's path and describe the motion.

Solution We identify the path by eliminating t between the equations $x = t$ and $y = t^2$, obtaining

$$y = (t)^2 = x^2.$$

The particle's position coordinates satisfy the equation $y = x^2$, so the particle moves along this curve.

In contrast to Example 3, the particle now traverses the entire parabola. As t increases from $-\infty$ to ∞, the particle comes down the left-hand side, passes through the origin, and moves up the right-hand side (Fig. 10.27). ■

As Example 4 illustrates, the graph of a function $y = f(x)$ has the automatic parametrization $x = x, y = f(x)$. This is so simple we usually don't use it, but the point of view is occasionally helpful.

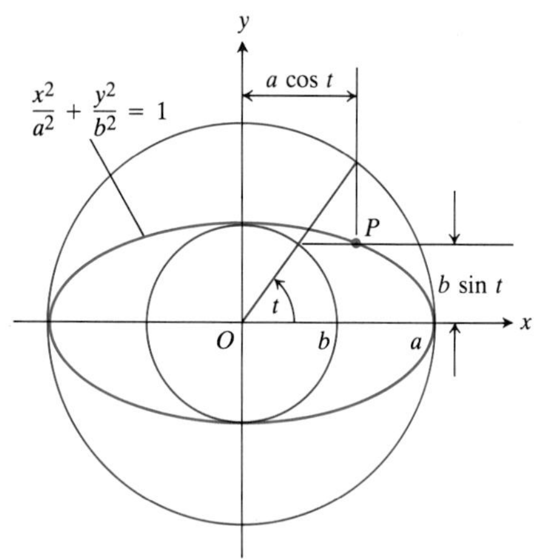

10.28 The coordinates of P are $x = a \cos t, y = b \sin t$.

EXAMPLE 5 *Parametric equations for the ellipse $x^2/a^2 + y^2/b^2 = 1$.* Describe the motion of a particle whose position $P(x, y)$ at time t is given by the equations

$$x = a \cos t, \qquad y = b \sin t, \qquad 0 \le t \le 2\pi.$$

Solution We find a Cartesian equation for the coordinates of the particle by eliminating t between the equations for x and y. Since

$$\frac{x^2}{a^2} + \frac{y^2}{b^2} = \frac{a^2\cos^2 t}{a^2} + \frac{b^2\sin^2 t}{b^2} = \cos^2 t + \sin^2 t = 1,$$

the motion takes place on the ellipse $x^2/a^2 + y^2/b^2 = 1$. The particle begins at $(a, 0)$ when t equals zero and moves counterclockwise around the ellipse, traversing it exactly once as t moves from 0 to 2π (Fig. 10.28). ■

EXAMPLE 6 *Parametric equations for the circle $x^2 + y^2 = a^2$.* The equations

$$x = a \cos t, \qquad y = a \sin t, \qquad 0 \le t \le 2\pi,$$

obtained by taking $b = a$ in Example 5, are parametric equations for the circle $x^2 + y^2 = a^2$. ■

EXAMPLE 7 *Parametric equations for the right-hand branch of the hyperbola $x^2 - y^2 = 1$.* Describe the motion of the particle whose position $P(x, y)$ at time t is given by the equations

$$x = \sec t, \qquad y = \tan t, \qquad -\frac{\pi}{2} < t < \frac{\pi}{2}.$$

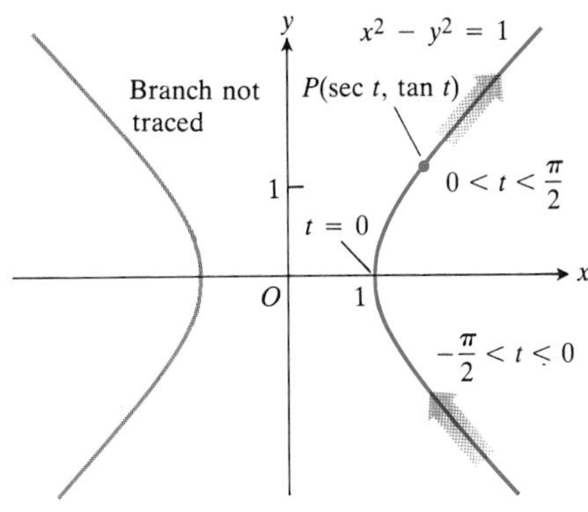

10.29 The equations $x = \sec t, y = \tan t$, $-\pi/2 < t < \pi/2$, are parametric equations for the right-hand branch of the hyperbola $x^2 - y^2 = 1$.

Solution We find a Cartesian equation for the coordinates of P by eliminating t between the equation for x and y. Since

$$x^2 - y^2 = \sec^2 t - \tan^2 t = 1,$$

we see that the motion takes place somewhere on the hyperbola $x^2 - y^2 = 1$. Since $x = \sec t$ is always positive for the parameter values $-\pi/2 < t < \pi/2$, the motion takes place on the hyperbola's right-hand branch. As t moves from $-\pi/2$ to $\pi/2$, the particle comes in along the lower half of the right-hand branch, reaching the vertex $(1, 0)$ at $t = 0$. It then moves into the first quadrant to complete the coverage of the right-hand branch as t approaches $\pi/2$ (Fig. 10.29).

EXAMPLE 8 *Cycloids.* A wheel of radius a rolls along a horizontal straight line without slipping. Find parametric equations for the path (called a **cycloid**) traced by a point P on the wheel's circumference.

Solution We take the line to be the x-axis and suppose that the wheel rolls to the right, with P being at the origin at time $t = 0$. Figure 10.30 shows the wheel a short time later, when the base of the wheel lies at units from the origin. At this time, the wheel's center lies at (at, a) and the coordinates of P are

$$x = at + a\cos\theta, \qquad y = a + a\sin\theta. \tag{2}$$

To express θ in terms of t, we observe that $t + \theta = 3\pi/2$, so that

$$\theta = \frac{3\pi}{2} - t. \tag{3}$$

This makes

$$\begin{aligned} \cos\theta &= \cos\left(\frac{3\pi}{2} - t\right) = -\sin t, \\ \sin\theta &= \sin\left(\frac{3\pi}{2} - t\right) = -\cos t. \end{aligned} \tag{4}$$

The equations we seek are

$$x = at - a\sin t, \qquad y = a - a\cos t. \tag{5}$$

10.30 The position of $P(x, y)$ on the rolling wheel at time t (Example 8).

Pendulum Clocks

One problem with a pendulum clock whose bob swings in a circular arc is that the frequency with which the pendulum swings changes with the amplitude of the swing. The wider the swing, the longer it takes the bob to return to center.

This does not happen if the bob can be made to swing in a cycloid, and, in 1673, Christiaan Huygens (1629–1695), the Dutch mathematician, physicist, and astronomer who discovered the rings of Saturn, designed a pendulum clock whose bob would swing in a cycloid. He hung the bob from a fine wire constrained by "guards" that caused it to draw up as it swung. How were the guards shaped? They were cycloids, too.

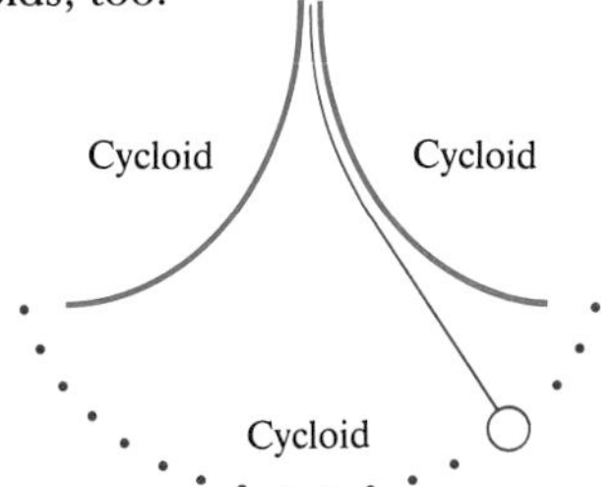

The Witch of Agnesi

Although l'Hôpital wrote the first text on differential calculus, the first text to include differential and integral calculus along with analytic geometry, infinite series, and differential equations was written in the 1740s by the Italian mathematician Maria Gaetana Agnesi (1718–1799). Agnesi, a gifted scholar and linguist whose Latin essay defending higher education for women was published when she was only nine years old, was a well-published scientist by age 20 and an honorary faculty member of the University of Bologna by age 30.

Today, Agnesi is remembered chiefly for a bell-shaped curve called *the witch of Agnesi.* This name, found only in English texts, is the result of a mistranslation. Agnesi's own name for the curve was *versiera* or "turning curve." John Colson, a noted Cambridge mathematician who felt the book so important he learned Italian to translate it "for the benefit of British youth" (he particularly had in mind young women, for whom he hoped Agnesi would be a role model), probably confused versiera with *avversiera,* which means "wife of the devil" and translates into "witch." You can find out more about the witch by doing Exercise 28.

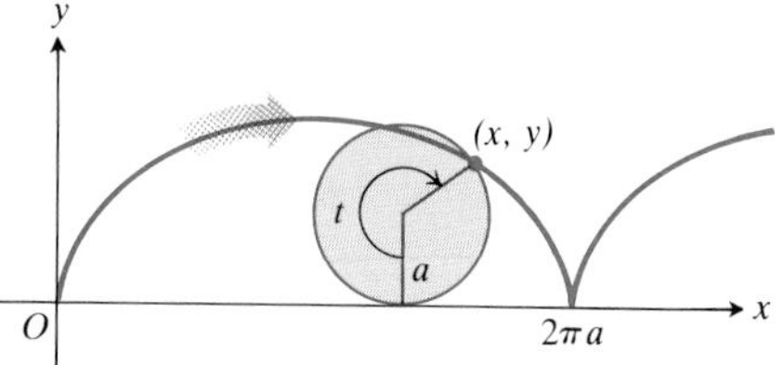

10.31 The cycloid $x = a(t - \sin t)$, $y = a(1 - \cos t)$, shown for $t \geq 0$.

These are usually written with the a factored out:

$$x = a(t - \sin t), \qquad y = a(1 - \cos t). \tag{6}$$

Figure 10.31 shows the first arch of the cycloid and part of the next. ■

Standard Parametric Equations

Circle $x^2 + y^2 = a^2$: $\quad x = a \cos t$, $\; y = a \sin t$, $\; 0 \leq t \leq 2\pi$

Ellipse $\dfrac{x^2}{a^2} + \dfrac{y^2}{b^2} = 1$: $\quad x = a \cos t$, $\; y = b \sin t$, $\; 0 \leq t \leq 2\pi$

Cycloid generated by a circle of radius 1:

$$x = t - \sin t, \qquad y = 1 - \cos t$$

Cycloid generated by a circle of radius a:

$$x = a(t - \sin t), \qquad y = a(1 - \cos t)$$

Exercises 10.3

Exercises 1–24 give parametric equations for the motion of a particle in the xy-plane. Identify the particle's path by finding a Cartesian equation for it. Graph the Cartesian equation. Indicate the portion traced by the particle and the direction of its motion.

1. $x = \cos t, \quad y = \sin t, \quad 0 \leq t \leq \pi$

2. $x = \cos 2t, \quad y = \sin 2t, \quad 0 \leq t \leq \pi$

3. $x = \sin 2\pi t, \quad y = \cos 2\pi t, \quad 0 \leq t \leq 1$

4. $x = \cos(\pi - t), \quad y = \sin(\pi - t), \quad 0 \leq t \leq \pi$

5. $x = 4 \cos t, \quad y = 2 \sin t, \quad 0 \leq t \leq 2\pi$

6. $x = 4 \sin t, \quad y = 2 \cos t, \quad 0 \leq t \leq \pi$

7. $x = 4 \cos t, \quad y = 5 \sin t, \quad 0 \leq t \leq \pi$

8. $x = 4 \sin t, \quad y = 5 \cos t, \quad 0 \leq t \leq 2\pi$

9. $x = 3t, \quad y = 9t^2, \quad -\infty < t < \infty$

10. $x = -\sqrt{t}, \quad y = t, \quad t \geq 0$

11. $x = t, \quad y = \sqrt{t}, \quad t \geq 0$
12. $x = \sec^2 t - 1, \quad y = \tan t, \quad -\pi/2 < t < \pi/2$
13. $x = -\sec t, \quad y = \tan t, \quad -\pi/2 < t < \pi/2$
14. $x = \csc t, \quad y = \cot t, \quad 0 < t < \pi$
15. $x = 2t - 5, \quad y = 4t - 7, \quad -\infty < t < \infty$
16. $x = 1 - t, \quad y = 1 + t, \quad -\infty < t < \infty$
17. $x = t, \quad y = 1 - t, \quad 0 \leq t \leq 1$
18. $x = 3t, \quad y = 2 - 2t, \quad 0 \leq t \leq 1$
19. $x = t, \quad y = \sqrt{1 - t^2}, \quad -1 \leq t \leq 1$
20. $x = t, \quad y = \sqrt{4 - t^2}, \quad 0 \leq t \leq 2$
21. $x = t^2, \quad y = \sqrt{t^4 + 1}, \quad t \geq 0$
22. $x = \sqrt{t + 1}, \quad y = \sqrt{t}, \quad t \geq 0$
23. $x = \cosh t, \quad y = \sinh t, \quad -\infty < t < \infty$
24. $x = 2 \sinh t, \quad y = 2 \cosh t, \quad -\infty < t < \infty$
25. Find parametric equations for the motion of a particle that starts at $(a, 0)$ and traces the circle $x^2 + y^2 = a^2$
    a) once clockwise
    b) once counterclockwise
    c) twice clockwise
    d) twice counterclockwise
    (There are lots of ways to do these, so your answers may not be the same as the ones in the back of the book.)

26. Find parametric equations for the motion of a particle that starts at $(a, 0)$ and traces the ellipse $(x^2/a^2) + (y^2/b^2) = 1$.
    a) once clockwise
    b) once counterclockwise
    c) twice clockwise
    d) twice counterclockwise
    (As in Exercise 25, there are many right answers.)

27. *The involute of a circle.* If a string wound around a fixed circle is unwound while held taut in the plane of the circle, its end P traces an *involute* of the circle. In Fig. 10.32, the circle in question is the unit circle in the xy-plane and the initial position of the tracing point is the point $(1, 0)$ on the x-axis. The unwound portion of the string is tangent to the circle at Q, and t is the radian measure of the angle from the positive x-axis to segment OQ. Derive parametric equations for the involute by expressing the coordinates x and y of P in terms of t.

28. *The witch of Maria Agnesi.* The bell-shaped witch of Maria Agnesi can be constructed in the following way. Start with a circle of radius 1, centered at the point $(0, 1)$ on the y-axis (Fig. 10.33). Choose a point A on the line $y = 2$ and connect it to the origin with a line segment. Call the point where the segment crosses the circle B. Let P be the point where the vertical line through A crosses the horizontal line through B. The witch is the curve traced by P as A moves along the line $y = 2$. Find parametric equations for the witch by expressing the coordinates of P in terms of t, the radian measure of the angle that segment OA makes with the positive x-axis. The following equalities (which you may assume) will help:
    1. $x = AQ$
    2. $y = 2 - AB \sin t$
    3. $AB \cdot OA = (AQ)^2$

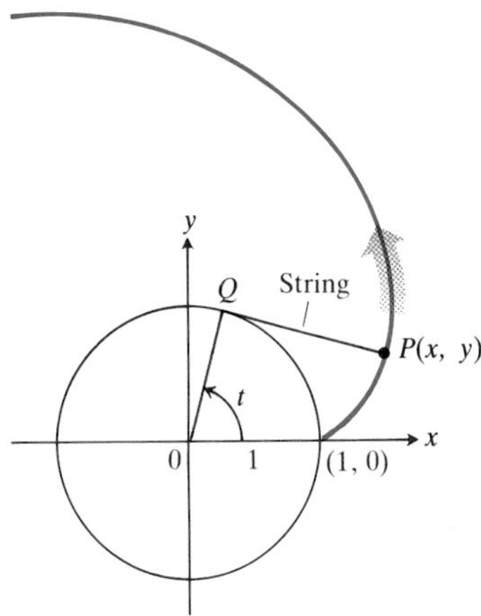

10.32 The involute of a circle of radius 1 (Exercise 27).

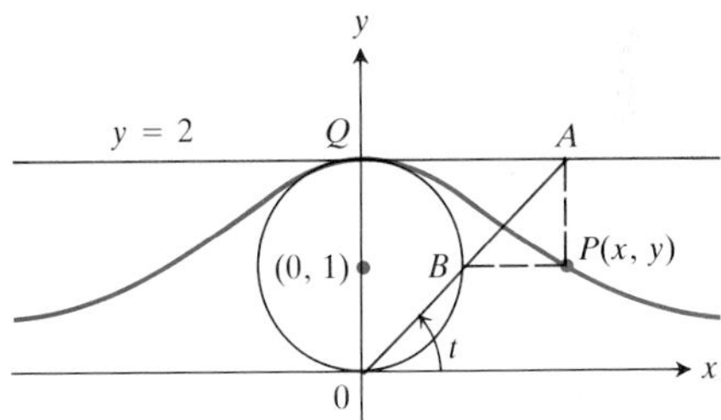

10.33 The witch of Maria Agnesi (Exercise 28).

29. Find the point on the parabola $x = t, y = t^2$ closest to the point $(2, 1/2)$. (*Hint:* Minimize the square of the distance as a function of t.)
30. Find the point on the ellipse $x = 2 \cos t, \ y = \sin t$, $0 \leq t \leq 2\pi$ closest to the point $(3/4, 0)$. (*Hint:* Minimize the square of the distance as a function of t.)

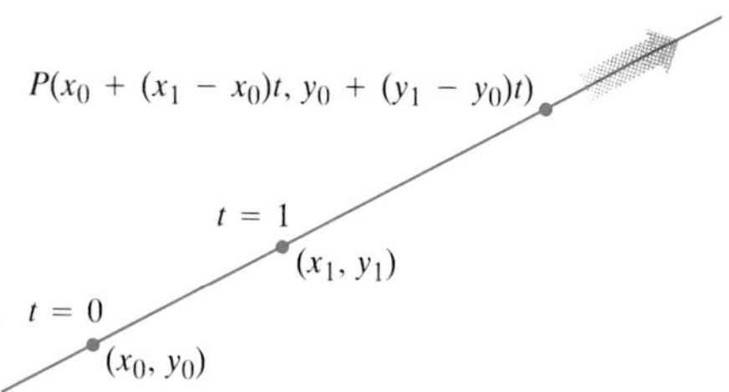

10.34 The parametrized line in Exercise 31. The arrow shows the direction of increasing t.

31. *Parametric equations for lines in the plane (Fig. 10.34).*
a) Show that the equations

$$x = x_0 + (x_1 - x_0)t, \qquad y = y_0 + (y_1 - y_0)t,$$

$-\infty < t < \infty$ are parametric equations for the line through the points (x_0, y_0) and (x_1, y_1).
b) Write parametric equations for the line through a point (x_1, y_1) and the origin.
c) Write parametric equations for the line through $(-1, 0)$ and $(0, 1)$.

Computer Graphing Exercises

If you have access to a parametric equation grapher, graph the following equations over the given parameter intervals. (If you have the *Toolkit* program PARAMETRIC EQUATIONS, plot the equations first on the *X vs. Y* screen and then on the *ALL* screen.)

32. *Ellipse.* $x = 4 \cos t$, $y = 2 \sin t$, over
a) $0 \le t \le 2\pi$
b) $0 \le t \le \pi$
c) $-\pi/2 \le t \le \pi/2$

33. *Hyperbola branch.* $x = \sec t$ (enter as $1/\cos(t)$), $y = \tan t$ (enter as $(\sin(t)/\cos(t))$), over
a) $-1.5 \le t \le 1.5$
b) $-0.5 \le t \le 0.5$
c) $-0.1 \le t \le 0.1$

34. *Parabola.* $x = 2t + 3$, $y = t^2 - 1$, $-2 \le t \le 2$

35. *Cycloid.* $x = t - \sin t$, $y = 1 - \cos t$, over
a) $0 \le t \le 2\pi$
b) $0 \le t \le 4\pi$
c) $\pi \le t \le 3\pi$

36. *Hypocycloid.* $x = \cos^3 t$, $y = \sin^3 t$, over
a) $0 \le t \le 2\pi$
b) $-\pi/2 \le t \le \pi/2$

37. *A nice curve (a deltoid).*

$$x = 2\cos t + \cos 2t,$$
$$y = 2\sin t - \sin 2t,$$
$$0 \le t \le 2\pi.$$

What happens if you replace 2 with -2 in the equations for x and y? Graph the new equations and find out.

38. *An even nicer curve.*

$$x = 3\cos t + \cos 3t,$$
$$y = 3\sin t - \sin 3t,$$
$$0 \le t \le 2\pi.$$

What happens if you replace 3 with -3 in the equations for x and y? Graph the new equations and find out.

39. *Projectile motion.* Graph

$$x = (64\cos\alpha)t,$$
$$y = -16t^2 + (64\sin\alpha)t,$$
$$0 \le t \le 4\sin\alpha,$$

for the following firing angles:
a) $\alpha = \pi/4$ b) $\alpha = \pi/6$ c) $\alpha = \pi/3$
d) $\alpha = \pi/2$ (watch out—here it comes!)

40. *Three beautiful curves.*
a) *Epicycloid:*

$$x = 9\cos t - \cos 9t$$
$$y = 9\sin t - \sin 9t$$
$$0 \le t \le 2\pi$$

b) *Hypocycloid:*

$$x = 8\cos t + 2\cos 4t$$
$$y = 8\sin t - 2\sin 4t$$
$$0 \le t \le 2\pi$$

c) *Hypotrochoid:*

$$x = \cos t + 5\cos 3t$$
$$y = 6\cos t - 5\sin 3t$$
$$0 \le t \le 2\pi$$

TOOLKIT PROGRAMS

Parametric Equations	Traces the curves for $x(t)$, $y(t)$, and $P(x, y)$ in side-by-side displays as t increases through the parameter interval. Also graphs $P(x, y)$ in a separate full-screen display
Super*Grapher	Graphs different sets of parametric equations in a common display in different colors

10.4 The Calculus of Parametric Equations

This section shows how to calculate slopes, lengths, and surface areas associated with parametric curves.

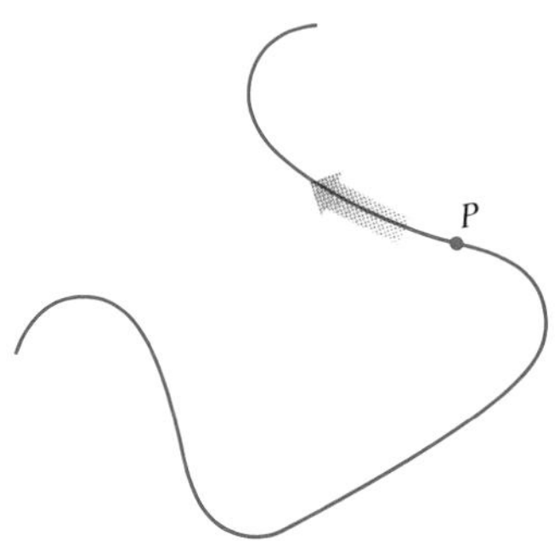

10.35 When the three first derivatives dx/dt, dy/dt, and dy/dx exist at a point P on a parametrized curve, they are related by the Chain Rule:

$$\frac{dy}{dt} = \frac{dy}{dx}\frac{dx}{dt}.$$

Slopes of Parametrized Curves

DEFINITIONS

A parametrized curve $x = f(t)$, $y = g(t)$ is said to be **differentiable at $t = t_0$** if f and g are differentiable at $t = t_0$. The curve is **differentiable** if it is differentiable at every parameter value.

At a point on a differentiable parametrized curve where y is also a differentiable function of x, the derivatives dx/dt, dy/dt, and dy/dx are related by the Chain Rule:

$$\frac{dy}{dt} = \frac{dy}{dx}\frac{dx}{dt} \tag{1}$$

(Fig. 10.35). If $dx/dt \neq 0$, we may divide both sides of this equation by dx/dt to solve for dy/dx.

Formula for Finding dy/dx from dy/dt and dx/dt, ($dx/dt \neq 0$)

$$\frac{dy}{dx} = \frac{dy/dt}{dx/dt} \tag{2}$$

If you want to think in terms of differentials, think of the two dt's canceling out to produce dy/dx. Equation (2) is the equation we use to calculate the slopes of parametrized curves.

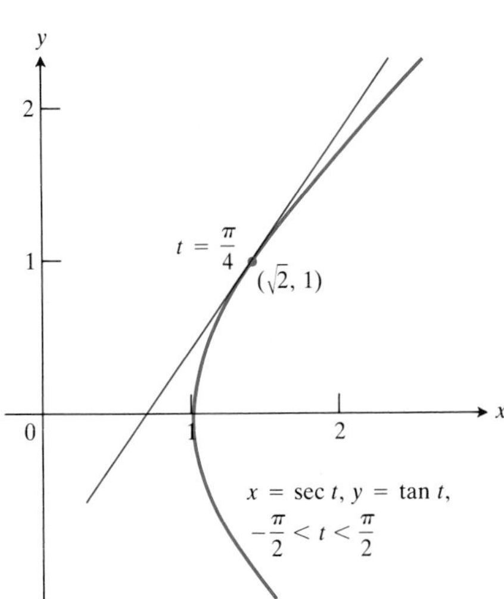

10.36 The hyperbola branch in Example 1.

EXAMPLE 1 Find the tangent to the right-hand hyperbola branch

$$x = \sec t, \qquad y = \tan t, \qquad -\frac{\pi}{2} < t < \frac{\pi}{2}$$

at the point $(\sqrt{2}, 1)$, where $t = \pi/4$ (Fig. 10.36).

Solution The slope of the curve at t is

$$\frac{dy}{dx} = \frac{dy/dt}{dx/dt} \qquad \text{(Eq. 2)}$$

$$= \frac{\sec^2 t}{\sec t \tan t} = \frac{\sec t}{\tan t}.$$

Setting t equal to $\pi/4$ gives

$$\left.\frac{dy}{dx}\right|_{t=\pi/4} = \frac{\sec(\pi/4)}{\tan(\pi/4)} = \frac{\sqrt{2}}{1} = \sqrt{2}.$$

The point–slope equation of the tangent is

$$y - y_0 = m(x - x_0)$$
$$y - 1 = \sqrt{2}(x - \sqrt{2})$$
$$y = \sqrt{2}x - 2 + 1$$
$$y = \sqrt{2}x - 1.$$

The Parametric Formula for d^2y/dx^2

If the parametric equations for a curve define y as a twice-differentiable function of x, we may calculate d^2y/dx^2 as a function of t in the following way:

$$\frac{d^2y}{dx^2} = \frac{d}{dx}(y')$$

$$= \frac{dy'/dt}{dx/dt}. \qquad \text{(Eq. (2) with } y \text{ replaced by } y')$$

Formula for Finding d^2y/dx^2 from dx/dt and $y' = dy/dx$, $(dx/dt \neq 0)$

$$\frac{d^2y}{dx^2} = \frac{dy'/dt}{dx/dt} \qquad (3)$$

The following example illustrates the steps we take to evaluate d^2y/dx^2.

EXAMPLE 2 Find d^2y/dx^2 if $x = t - t^2$ and $y = t - t^3$.

To find d^2y/dx^2 in terms of t,
1. Express $y' = dy/dx$ in terms of t
2. Find dy'/dt
3. Divide dy'/dt by dx/dt.

Solution STEP 1: Express y' in terms of t:

$$y' = \frac{dy}{dx} = \frac{dy/dt}{dx/dt} = \frac{1 - 3t^2}{1 - 2t} \qquad \text{(Eq. (2) with } x = t - t^2, y = t - t^3)$$

STEP 2: Differentiate y' with respect to t:

$$\frac{dy'}{dt} = \frac{d}{dt}\left(\frac{1 - 3t^2}{1 - 2t}\right) = \frac{2 - 6t + 6t^2}{(1 - 2t)^2} \qquad \text{(Quotient Rule)}$$

STEP 3: Divide dy'/dt by dx/dt:

$$\frac{dx}{dt} = \frac{d}{dt}(t - t^2) = 1 - 2t$$

$$\frac{dy'/dt}{dx/dt} = \frac{2 - 6t + 6t^2}{(1 - 2t)^2} \cdot \frac{1}{1 - 2t}$$

$$= \frac{2 - 6t + 6t^2}{(1 - 2t)^3}.$$

■

The Length of a Parametric Curve

We can find an integral for the length of a continuously differentiable parametric curve

$$x = f(t), \qquad y = g(t), \qquad a \le t \le b$$

by rewriting the integral $L = \int ds$ from Section 6.4 in the following way:

$$\text{Length} = \int_{t=a}^{t=b} ds$$

$$= \int_a^b \sqrt{dx^2 + dy^2} \qquad \text{(because } ds = \sqrt{dx^2 + dy^2}\text{)}$$

$$= \int_a^b \sqrt{\left(\frac{dx^2}{dt^2} + \frac{dy^2}{dt^2}\right) dt^2} \tag{4}$$

$$= \int_a^b \sqrt{\left(\frac{dx}{dt}\right)^2 + \left(\frac{dy}{dt}\right)^2}\, dt.$$

The only requirement besides the continuity of the integrand is that the point $P(x, y)$ not trace out any portion of the curve more than once as t moves from a to b.

Parametric Formula for Arc Length

If the functions $x = f(t)$ and $y = g(t)$ have continuous first derivatives with respect to t for $a \le t \le b$, and if the point $P(x, y)$ traces the curve defined by these equations exactly once as t moves from $t = a$ to $t = b$, then the length of the curve is given by the formula

$$\text{Length} = \int_a^b \sqrt{\left(\frac{dx}{dt}\right)^2 + \left(\frac{dy}{dt}\right)^2}\, dt. \tag{5}$$

What if we have two different parametrizations for a curve whose length we want to find—does it matter which one we use? The answer, from advanced calculus, is no. As long as the parametrization we choose meets the conditions preceding Eq. (5), the formula gives the correct length.

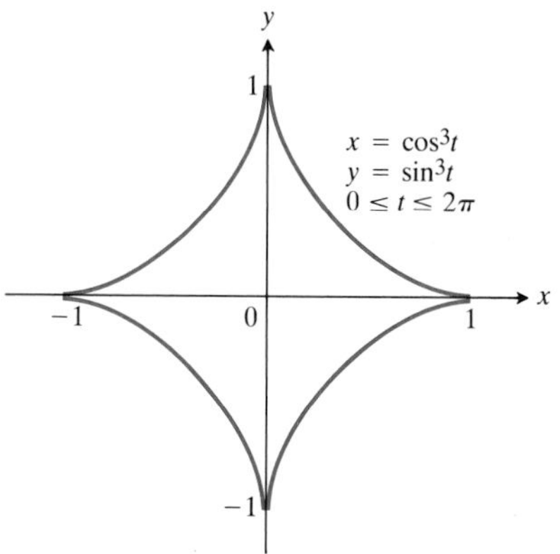

10.37 The astroid in Example 3.

EXAMPLE 3 Find the length of the astroid (Fig. 10.37)

$$x = \cos^3 t, \qquad y = \sin^3 t, \qquad 0 \le t \le 2\pi.$$

Solution We find the length of the first-quadrant portion, $0 \le t \le \pi/2$, and multiply by 4. We have

$$x = \cos^3 t, \qquad y = \sin^3 t$$

$$\left(\frac{dx}{dt}\right)^2 = [3\cos^2 t(-\sin t)]^2 = 9\cos^4 t\,\sin^2 t$$

$$\left(\frac{dy}{dt}\right)^2 = [3\sin^2 t(\cos t)]^2 = 9\sin^4 t\,\cos^2 t$$

$$\begin{aligned}\sqrt{\left(\frac{dx}{dt}\right)^2 + \left(\frac{dy}{dt}\right)^2} &= \sqrt{9\cos^2 t\,\sin^2 t\,\underbrace{(\cos^2 t + \sin^2 t)}_{1}} \\ &= \sqrt{9\cos^2 t\,\sin^2 t} \\ &= 3|\cos t\,\sin t| \\ &= 3\cos t\,\sin t. \qquad (\cos t\,\sin t \ge 0 \text{ for } 0 \le t \le \pi/2)\end{aligned}$$

Therefore,

Length of first-quadrant portion

$$\begin{aligned} &= \int_0^{\pi/2} 3\cos t\,\sin t\,dt \\ &= 3\int_{u=0}^{u=1} u\,du \qquad \left(\begin{matrix} u = \sin t,\ du = \cos t\,dt, \\ u(0) = 0,\ u(\pi/2) = 1 \end{matrix}\right) \\ &= 3\left[\frac{u^2}{2}\right]_0^1 \\ &= \frac{3}{2}.\end{aligned}$$

The length of the complete curve is four times this:

$$\text{Length of complete curve} = 4\left(\frac{3}{2}\right) = 6.$$

■

The Area of a Surface of Revolution

The formula $S = \int 2\pi\rho\,ds$ developed in Section 6.5 for the area of the surface swept out by revolving a continuously differentiable curve about an axis translates into $S = \int 2\pi y\,ds$ if the axis is the x-axis and into $S = \int 2\pi x\,ds$ if the axis is the y-axis. With $ds = \sqrt{(dx/dt)^2 + (dy/dt)^2}\,dt$, these lead to the following formulas.

Parametric Formulas for the Area of a Surface of Revolution

If the functions $x = f(t)$ and $y = g(t)$ have continuous first derivatives with respect to t for $a \le t \le b$, and if the point $P(x, y)$ traces the curve defined by these equations exactly once as t moves from $t = a$ to $t = b$, then the areas of the surfaces generated by revolving the curve about the coordinate axes are

1. Revolution about the x-axis ($y \ge 0$)

$$\text{Area} = \int_a^b 2\pi y \sqrt{\left(\frac{dx}{dt}\right)^2 + \left(\frac{dy}{dt}\right)^2}\, dt \qquad (6)$$

2. Revolution about the y-axis ($x \ge 0$)

$$\text{Area} = \int_a^b 2\pi x \sqrt{\left(\frac{dx}{dt}\right)^2 + \left(\frac{dy}{dt}\right)^2}\, dt \qquad (7)$$

As with length, we can calculate surface area from any convenient parametrization that meets the criteria stated above.

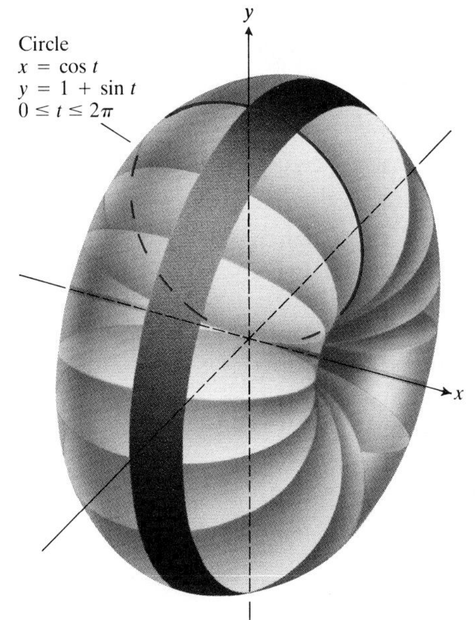

10.38 The surface in Example 4.

EXAMPLE 4 The standard parametric equations for the circle of radius 1 centered at the point (0, 1) in the xy-plane are

$$x = \cos t, \qquad y = 1 + \sin t, \qquad 0 \le t \le 2\pi.$$

Use these equations to find the area of the surface swept out by revolving this circle about the x-axis (Fig. 10.38).

Solution We use Eq. (6) with

$$y = 1 + \sin t, \qquad \frac{dx}{dt} = -\sin t, \qquad \frac{dy}{dt} = \cos t$$

to obtain

$$\text{Area} = \int_a^b 2\pi y \sqrt{\left(\frac{dx}{dt}\right)^2 + \left(\frac{dy}{dt}\right)^2}\, dt \qquad \text{(Eq. (6))}$$

$$= \int_0^{2\pi} 2\pi(1 + \sin t)\underbrace{\sqrt{(-\sin t)^2 + (\cos t)^2}}_{1}\, dt$$

$$= 2\pi \int_0^{2\pi} (1 + \sin t)\, dt$$

$$= 2\pi \Big[t - \cos t\Big]_0^{2\pi}$$

$$= 2\pi\, [(2\pi - 1) - (0 - 1)]$$

$$= 4\pi^2.$$

Exercises 10.4

In Exercises 1–12, find an equation for the line tangent to the curve given at the point defined by the given value of t. Also, find the value of d^2y/dx^2 at this point.

1. $x = 2\cos t,\quad y = 2\sin t;\quad t = \pi/4$

2. $x = \sin 2\pi t,\quad y = \cos 2\pi t;\quad t = -1/6$

3. $x = 4\sin t,\quad y = 2\cos t;\quad t = \pi/4$

4. $x = \cos t,\quad y = \sqrt{3}\cos t;\quad t = 2\pi/3$

5. $x = t,\quad y = \sqrt{t};\quad t = 1/4$

6. $x = \sec^2 t - 1,\quad y = \tan t;\quad t = -\pi/4$

7. $x = \sec t,\quad y = \tan t;\quad t = \pi/6$

8. $x = -\sqrt{t+1},\quad y = \sqrt{3t};\quad t = 3$

9. $x = 2t^2 + 3,\quad y = t^4;\quad t = -1$

10. $x = 1/t,\quad y = -2 + \ln t;\quad t = 1$

11. $x = t - \sin t,\quad y = 1 - \cos t;\quad t = \pi/3$

12. $x = \cos t,\quad y = 1 + \sin t;\quad t = \pi/2$

Find the lengths of the curves in Exercises 13–18.

13. $x = \cos t,\quad y = t + \sin t,\quad 0 \le t \le \pi$

14. $x = t^3,\quad y = 3t^2/2,\quad 0 \le t \le \sqrt{3}$

15. $x = t^2/2,\quad y = (2t+1)^{3/2}/3,\quad 0 \le t \le 4$

16. $x = (2t+3)^{3/2}/3,\quad y = t + t^2/2,\quad 0 \le t \le 3$

17. $x = 8\cos t + 8t\sin t,$
$y = 8\sin t - 8t\cos t,$
$0 \le t \le \pi/2$

18. $x = \ln(\sec t + \tan t) - \sin t,$
$y = \cos t,$
$0 \le t \le \pi/3$ (See Fig. 10.39)

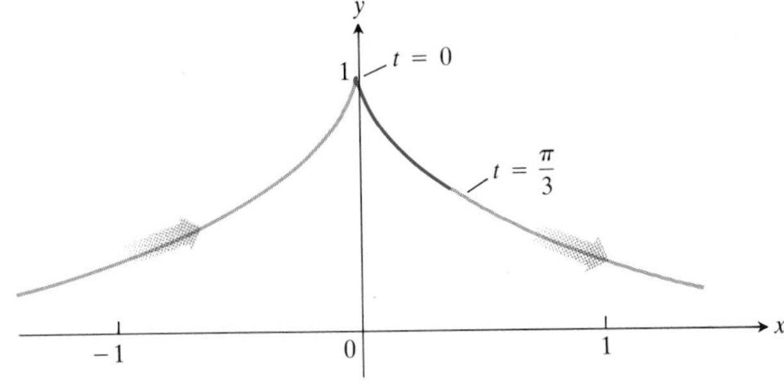

10.39 The curve in Exercise 18 is a portion of a curve that covers the entire x-axis as t runs between $-\pi/2$ and $\pi/2$. The arrows show the direction of increasing t.

Find the areas of the surfaces generated by revolving the curves in Exercises 19–22 about the indicated axes.

19. $x = \cos t,\quad y = 2 + \sin t,\quad 0 \le t \le 2\pi,$
about the x-axis

20. $x = (2/3)t^{3/2},\quad y = 2\sqrt{t},\quad 0 \le t \le \sqrt{3},$
about the y-axis

21. $x = t + \sqrt{2},\quad y = (t^2/2) + \sqrt{2}t,\quad -\sqrt{2} \le t \le \sqrt{2},$
about the y-axis

22. $x = \ln(\sec t + \tan t) - \sin t,$
$y = \cos t,\quad 0 \le t \le \pi/3,$
about the x-axis

23. *A cone frustum.* The line segment joining the points (0, 1) and (2, 2) is revolved about the x-axis to generate a frustum of a cone. Find the surface area of the frustum with the parametric equations

$$x = 2t,\quad y = t + 1,\quad 0 \le t \le 1.$$

Check your result with the geometry formula

$$\text{Area} = \pi(r_1 + r_2)(\text{slant height}).$$

24. *A cone.* The line segment joining the origin to the point (h, r) is revolved about the x-axis to generate a cone of height h and base radius r. Find the cone's surface area with the parametric equations

$$x = ht,\quad y = rt,\quad 0 \le t \le 1.$$

Check your result with the geometry formula

$$\text{Area} = \pi(r)(\text{slant height}).$$

25. *Length is independent of parametrization.* To illustrate the fact that the numbers we get for length do not depend on the way we parametrize our curves (except for the mild restrictions mentioned earlier), calculate the length of the semicircle $y = \sqrt{1 - x^2}$ with these two different parametrizations:

a) $x = \cos 2t,\quad y = \sin 2t,\quad 0 \le t \le \pi/2$
b) $x = \sin \pi t,\quad y = \cos \pi t,\quad -1/2 \le t \le 1/2$

26. *Elliptic integrals.* The length of the ellipse

$$x = a\cos t,\quad y = b\sin t,\quad 0 \le t \le 2\pi$$

turns out to be

$$\text{Length} = 4a\int_0^{\pi/2} \sqrt{1 - e^2\cos^2 t}\, dt,$$

where e is the ellipse's eccentricity. The integral in this formula, called an *elliptic integral*, is nonelementary except when $e = 0$ or 1.

a) **CALCULATOR** Use the Trapezoidal Rule with $n = 10$ to estimate the length of the ellipse when $a = 1$ and $e = 1/2$.

b) Use the fact that the absolute value of the second derivative of $f(t) = \sqrt{1 - e^2\cos^2 t}$ is less than 1, to find an upper bound for the error in the estimate you obtained in (a).

The curves drawn by computer in Exercises 27 and 28 are called *Bowditch curves* or *Lissajous figures*. In each case, find the point in the first quadrant where the tangent to the curve is horizontal and find the equations of the two tangents at the origin.

27.

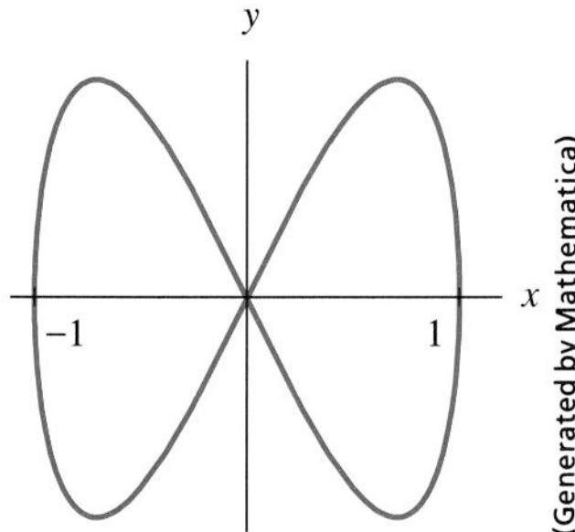

28.

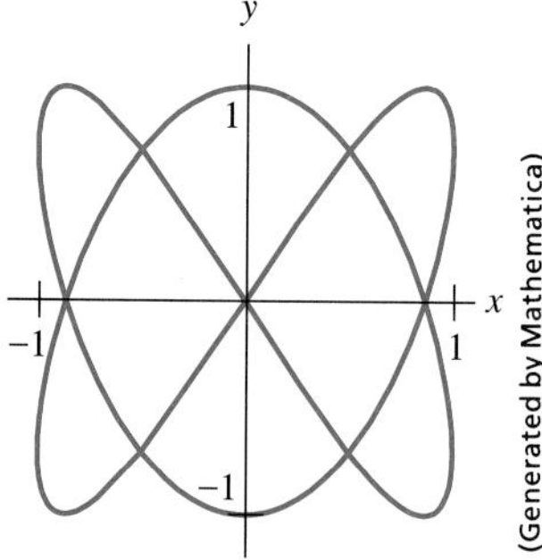

Computer Grapher

Graph the parametric curves in Exercises 29–35 over parameter intervals of your choice. The curves are all Bowditch curves (Lissajous figures), their general formula being

$$x = a \sin(mt + d), \qquad y = b \sin nt,$$

with m and n integers.

29. $x = \sin 2t, \quad y = \sin t$

30. $x = \sin 3t, \quad y = \sin 4t$

31. $x = \sin t, \quad y = \sin 4t$

32. $x = \sin t, \quad y = \sin 5t$

33. $x = \sin 3t, \quad y = \sin 5t$

34. $x = \sin(3t + \pi/2), \quad y = \sin 5t$

35. $x = \sin(3t + \pi/4), \quad y = \sin 5t$

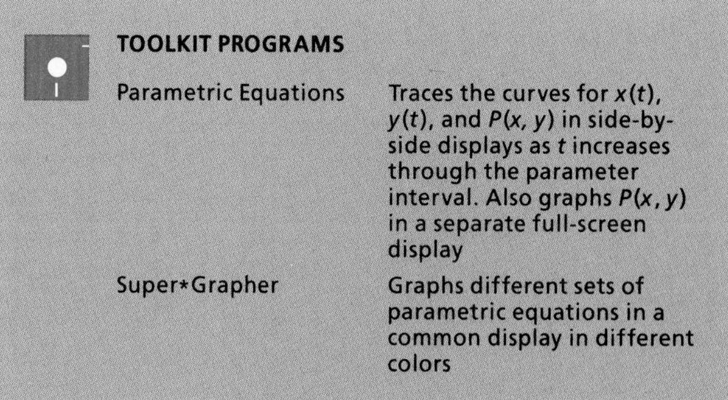

TOOLKIT PROGRAMS

Parametric Equations	Traces the curves for $x(t)$, $y(t)$, and $P(x, y)$ in side-by-side displays as t increases through the parameter interval. Also graphs $P(x, y)$ in a separate full-screen display
Super*Grapher	Graphs different sets of parametric equations in a common display in different colors

10.5 Polar Coordinates

In this section, we define polar coordinates and study their relation to Cartesian coordinates. One of the distinctions between polar and Cartesian coordinates is that while a point in the plane has just one pair of Cartesian coordinates, it has infinitely many pairs of polar coordinates. This has interesting consequences for graphing, as we shall see in the next section. Polar coordinates enable us to describe all conic sections with a single equation, as we shall see in Section 10.7, and the calculus we have done in rectangular coordinates carries over to this new system as well.

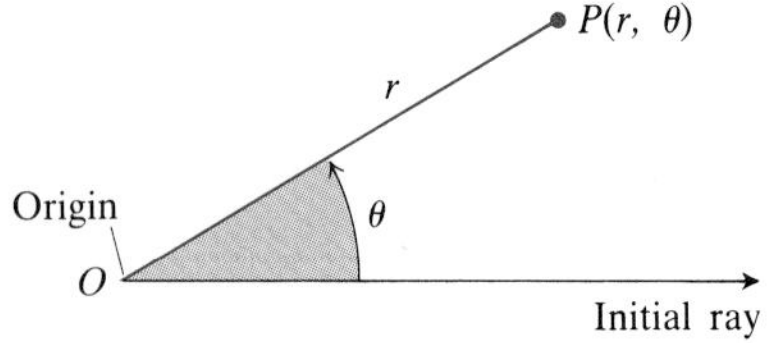

10.40 To define polar coordinates for the plane, we start with an origin and an initial ray.

Definition of Polar Coordinates

To define polar coordinates, we first fix an **origin** O and an **initial ray** from O (Fig. 10.40). Then each point P can be located by assigning to it a **polar coordinate pair** (r, θ), in which the first number, r, gives the directed distance from O to P and the second number, θ, gives the directed angle from the initial ray to the segment OP:

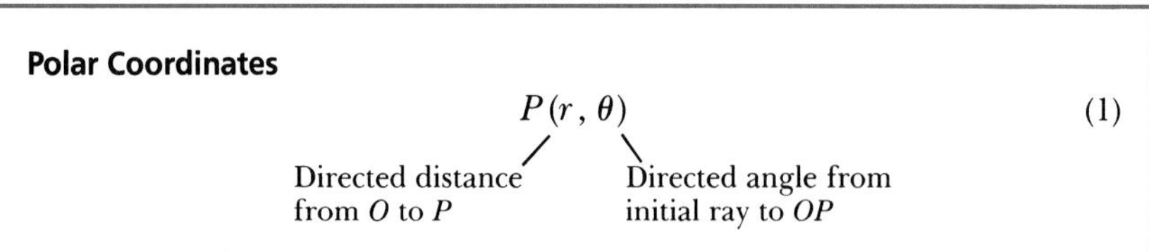

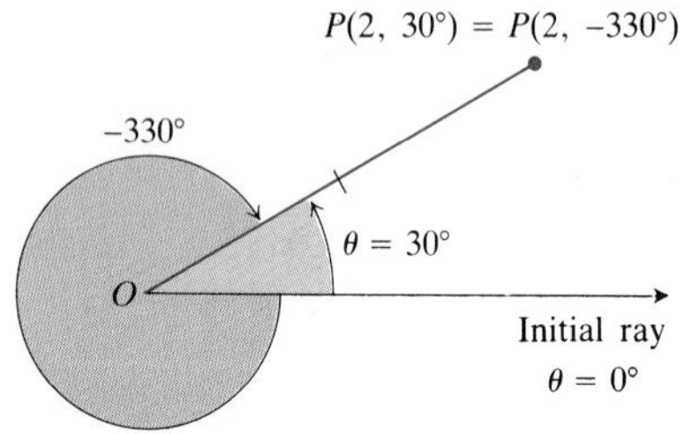

10.41 The ray $\theta = 30°$ is the same as the ray $\theta = -330°$.

As in trigonometry, the angle θ is positive when measured counterclockwise and negative when measured clockwise. But the angle associated with a given point is not unique. For instance, the point 2 units from the origin along the ray $\theta = 30°$ has polar coordinates $r = 2$, $\theta = 30°$. It also has coordinates $r = 2$, $\theta = -330°$, and $r = 2$, $\theta = 390°$ (Fig. 10.41).

Negative Values of r; Changing to Radian Measure

There are occasions when we wish to allow r to be negative. That is why we say "directed distance" in (1). The ray $\theta = 30°$ and the ray $\theta = 210°$ together make a complete line through O (Fig. 10.42). The point $P(2, 210°)$ 2 units from O on the ray $\theta = 210°$ has polar coordinates $r = 2$, $\theta = 210°$. It can be reached by turning 210° counterclockwise from the initial ray and going forward 2 units. It can also be reached by turning 30° counterclockwise from the initial ray and going *backward* two units. So we say that the point also has polar coordinates $r = -2$, $\theta = 30°$.

Whenever the angle between two rays is 180°, the rays make a straight line. We then say that each ray is the **opposite** of the other. Points on the ray $\theta = \alpha$ have polar coordinates (r, α) with $r \geq 0$. Points on the opposite ray, the ray $\theta = \alpha + 180°$, have coordinates (r, α) with $r \leq 0$ (Fig. 10.43).

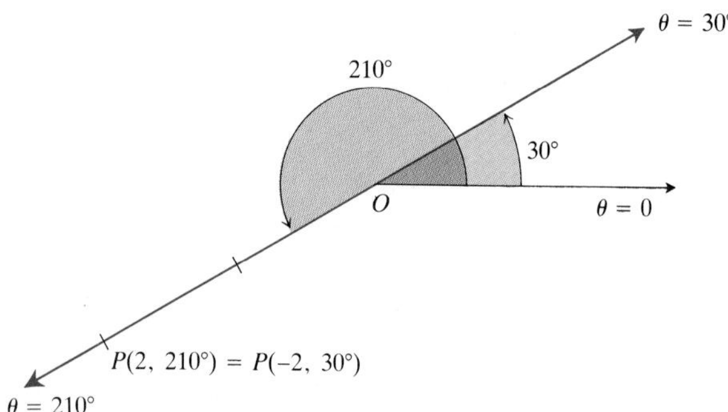

10.42 Points can have polar coordinates with negative r-values.

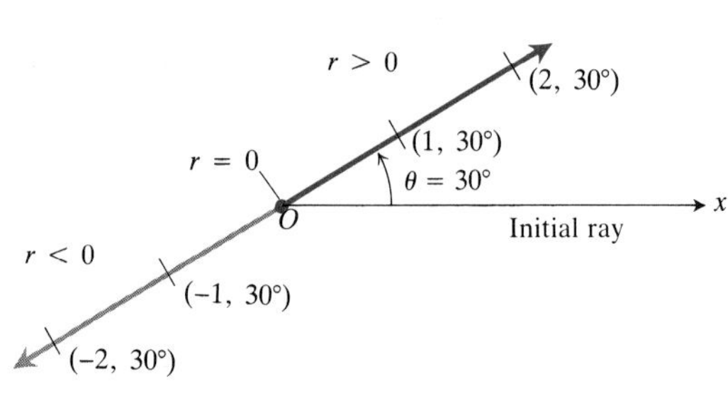

10.43 The ray $\theta = 30°$ and its opposite, the ray $\theta = 210°$, make a straight line.

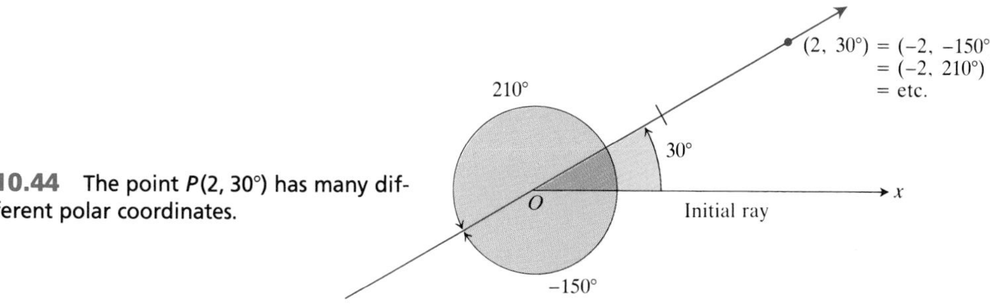

10.44 The point $P(2, 30°)$ has many different polar coordinates.

EXAMPLE 1 Find all the polar coordinates of the point (2, 30°). Express the angles in radians as well as in degrees.

Solution We sketch the initial ray of the coordinate system, draw the ray through the origin that makes a 30° angle with the initial ray, and mark the point (2, 30°) (Fig. 10.44). We then find formulas for the coordinate pairs in which $r = 2$ and $r = -2$ and convert the formulas to radian measure.

For $r = 2$*:* The angles

$$\begin{array}{ll} 30° + 1 \cdot 360° = 390° & 30° - 1 \cdot 360° = -330° \\ 30° + 2 \cdot 360° = 750° & 30° - 2 \cdot 360° = -690° \\ 30° + 3 \cdot 360° = 1110° & 30° - 3 \cdot 360° = -1050° \\ \vdots & \vdots \end{array} \tag{2}$$

all end in the same ray as the angle 30°. Thus, the polar coordinates

$$(2, 30° + n \cdot 360°), \qquad n = 0, \pm 1, \pm 2, \ldots \tag{3}$$

all identify the point (2, 30°).

For $r = -2$*:* Numerous as they are, the coordinates in (3) are not the only polar coordinates of the point (2, 30°). The angles

$$\begin{array}{ll} -150° & -150° \\ -150° + 360° = 210° & -150° - 360° = -510° \\ -150° + 720° = 570° & -150° - 720° = -870° \\ \vdots & \vdots \end{array} \tag{4}$$

all define the ray opposite the ray $\theta = 30°$. Hence, the polar coordinates

$$(-2, -150° + n \cdot 360°), \qquad n = 0, \pm 1, \pm 2, \ldots \tag{5}$$

represent the point (2, 30°) as well.

Radian measure: If we measure angles in radians, the formulas that correspond to (3) and (5) are

$$\left(2, \frac{\pi}{6} + 2n\pi\right), \qquad n = 0, \pm 1, \pm 2, \ldots \tag{6}$$

and

$$\left(-2, -\frac{5\pi}{6} + 2n\pi\right), \qquad n = 0, \pm 1, \pm 2, \ldots. \tag{7}$$

When $n = 0$, these formulas give

$$(2, \pi/6) \quad \text{and} \quad (-2, -5\pi/6).$$

When $n = 1$, they give

$$(2, 13\pi/6) \quad \text{and} \quad (-2, 7\pi/6),$$

and so on. ■

The Use of Radian Measure

Although nothing in the definition of polar coordinates requires the use of radian measure, we shall need to have all angles in radians when we differentiate and integrate trigonometric functions of θ. We shall therefore use radian measure almost exclusively from now on.

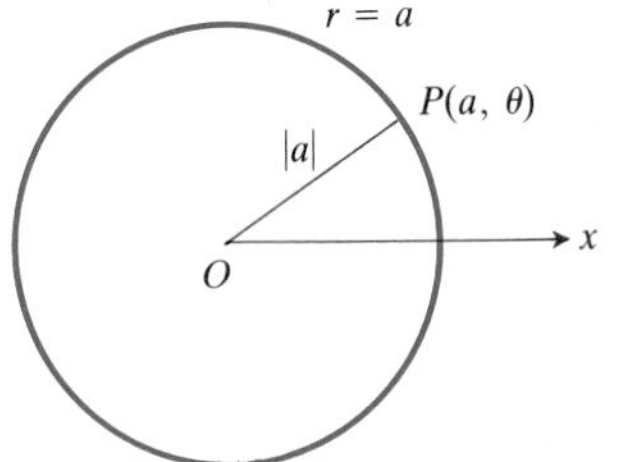

10.45 The polar equation for this circle is $r = a$.

Elementary Coordinate Equations and Inequalities

If we hold r fixed at a constant nonzero value $r = a$, then the point $P(r, \theta)$ lies $|a|$ units from the origin. As θ varies over any interval of length 2π radians, P traces a circle of radius $|a|$ centered at the origin (Fig. 10.45).

Circle of Radius $|a|$ Centered at the Origin

$$r = a \tag{8}$$

EXAMPLE 2 The equation $r = 1$ is an equation for the circle of radius 1 centered at the origin. So is the equation $r = -1$. ■

If we hold θ fixed at a constant value $\theta = \theta_0$ and let r run between $-\infty$ and ∞, the point $P(r, \theta)$ traces a line through the origin that makes an angle θ_0 with the initial ray. The line therefore consists of all the points in the plane that have coordinates of the form (r, θ_0).

Equation for Lines Through the Origin

$$\theta = \theta_0 \tag{9}$$

EXAMPLE 3 One equation for the line in Figs. 10.43 and 10.44 is $\theta = \pi/6$. The equations $\theta = 7\pi/6$ and $\theta = -5\pi/6$ are also equations for this line. ■

Equations of the form $r = a$ and $\theta = \theta_0$ can be combined to define regions, segments, and rays.

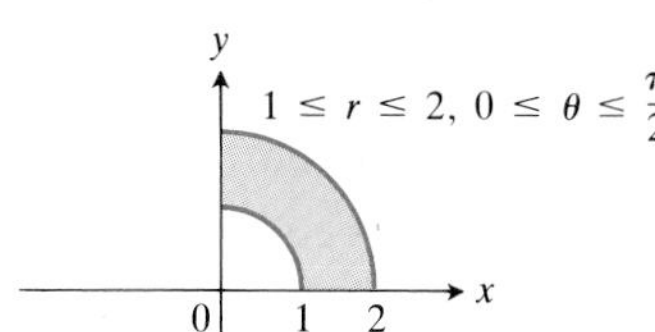

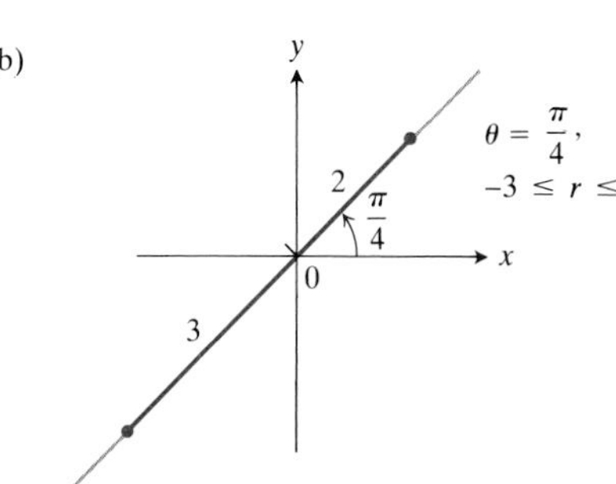

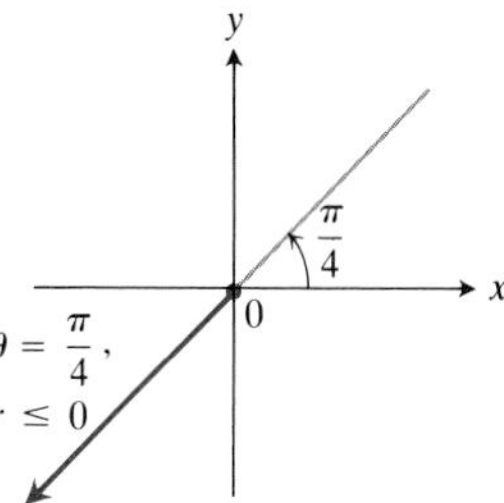

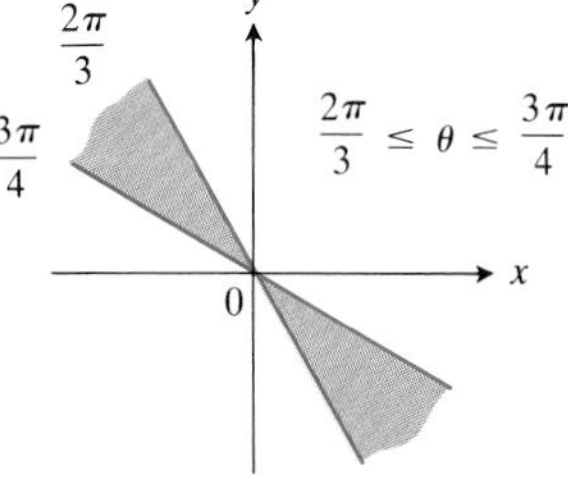

10.46 The graphs of typical inequalities in r and θ.

EXAMPLE 4 Graph the sets of points whose polar coordinates satisfy the following conditions.

a) $1 \le r \le 2$ and $0 \le \theta \le \frac{\pi}{2}$

b) $-3 \le r \le 2$ and $\theta = \frac{\pi}{4}$

c) $r \le 0$ and $\theta = \frac{\pi}{4}$

d) $\frac{2\pi}{3} \le \theta \le \frac{3\pi}{4}$ (no restriction on r)

Solution The graphs are shown in Fig. 10.46. ■

Cartesian versus Polar Coordinates

When we use both polar and Cartesian coordinates in a plane, we place the two origins together and take the initial polar ray to be the positive x-axis. The ray $\theta = \pi/2$, $r > 0$, becomes the positive y-axis (Fig. 10.47). The two sets of coordinates are then related by the following equations.

Equations Relating Polar and Cartesian Coordinates

$$x = r\cos\theta, \qquad y = r\sin\theta, \qquad x^2 + y^2 = r^2, \qquad \frac{y}{x} = \tan\theta \tag{10}$$

Equations (10) are the equations we use to rewrite polar equations in Cartesian form and vice versa.

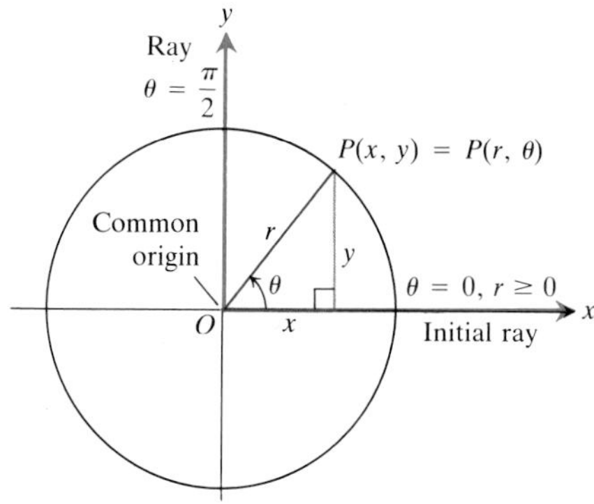

10.47 The usual way to relate polar and Cartesian coordinates.

EXAMPLE 5

Polar equation	Cartesian equivalent
$r\cos\theta = 2$	$x = 2$
$r^2\cos\theta\sin\theta = 4$	$xy = 4$
$r^2\cos^2\theta - r^2\sin^2\theta = 1$	$x^2 - y^2 = 1$
$r = 1 + 2r\cos\theta$	$y^2 - 3x^2 - 4x - 1 = 0$
$r = 1 - \cos\theta$	$x^4 + y^4 + 2x^2y^2 + 2x^3 + 2xy^2 - y^2 = 0$

With some curves, we are better off with polar coordinates; with others, we aren't. ■

EXAMPLE 6 Find a Cartesian equation for the curve

$$r\cos\left(\theta - \frac{\pi}{3}\right) = 3.$$

Solution We use the identity

$$\cos(A - B) = \cos A \cos B + \sin A \sin B$$

with $A = \theta$ and $B = \pi/3$:

$$r \cos\left(\theta - \frac{\pi}{3}\right) = 3,$$

$$r\left(\cos\theta \cos\frac{\pi}{3} + \sin\theta \sin\frac{\pi}{3}\right) = 3,$$

$$r\cos\theta \cdot \frac{1}{2} + r\sin\theta \cdot \frac{\sqrt{3}}{2} = 3,$$

$$\frac{1}{2}x + \frac{\sqrt{3}}{2}y = 3,$$

$$x + \sqrt{3}y = 6.$$

■

EXAMPLE 7 Replace the following polar equations by equivalent Cartesian equations, and identify their graphs.

a) $r \cos\theta = -4$

b) $r^2 = 4r\cos\theta$

c) $r = \dfrac{4}{2\cos\theta - \sin\theta}$

Solution We use the substitutions $r\cos\theta = x, r\sin\theta = y, r^2 = x^2 + y^2$.

a) $r\cos\theta = -4$

The Cartesian equation: $r\cos\theta = -4$

$x = -4$

The graph: Vertical line through $x = -4$ on the x-axis

b) $r^2 = 4r\cos\theta$

The Cartesian equation: $r^2 = 4r\cos\theta$

$x^2 + y^2 = 4x$

$x^2 - 4x + y^2 = 0$

$x^2 - 4x + 4 + y^2 = 4$ (completing the square)

$(x - 2)^2 + y^2 = 4$

The graph: Circle, radius 2, center $(h, k) = (2, 0)$

c) $r = \dfrac{4}{2\cos\theta - \sin\theta}$

The Cartesian equation: $r(2\cos\theta - \sin\theta) = 4$

$2r\cos\theta - r\sin\theta = 4$

$2x - y = 4$

$y = 2x - 4$

The graph: Line, slope $m = 2$, y-intercept $b = -4$

■

Exercises 10.5

NOTE: *All angles are in radians.*

1. Pick out the polar coordinate pairs that label the same point.

a) $(3, 0)$ b) $(-3, 0)$
c) $(-3, \pi)$ d) $(-3, 2\pi)$
e) $(2, 2\pi/3)$ f) $(2, -2\pi/3)$
g) $(2, 7\pi/3)$ h) $(-2, \pi/3)$
i) $(2, -\pi/3)$ j) $(2, \pi/3)$
k) $(-2, -\pi/3)$ l) $(-2, 2\pi/3)$
m) (r, θ) n) $(r, \theta + \pi)$
o) $(-r, \theta + \pi)$ p) $(-r, \theta)$

2. Find the Cartesian coordinates of the points whose polar coordinates are given in parts (a)–(l) of Exercise 1.

3. Plot the following points (given in polar coordinates). Then find all the polar coordinates of each point.

a) $(2, \pi/2)$ b) $(2, 0)$
c) $(-2, \pi/2)$ d) $(-2, 0)$

4. Plot the following points (given in polar coordinates). Then find all the polar coordinates of each point.

a) $(3, \pi/4)$ b) $(-3, \pi/4)$
c) $(3, -\pi/4)$ d) $(-3, -\pi/4)$

5. Find the Cartesian coordinates of the following points (given in polar coordinates).

a) $(\sqrt{2}, \pi/4)$ b) $(1, 0)$
c) $(0, \pi/2)$ d) $(-\sqrt{2}, \pi/4)$
e) $(-3, 5\pi/6)$ f) $(5, \tan^{-1}(4/3))$
g) $(-1, 7\pi)$ h) $(2\sqrt{3}, 2\pi/3)$

6. Find all polar coordinates of the origin.

Graph the sets of points whose polar coordinates satisfy the equations and inequalities in Exercises 7–22.

7. $r = 2$ **8.** $0 \le r \le 2$

9. $r \ge 1$ **10.** $1 \le r \le 2$

11. $0 \le \theta \le \pi/6, \quad r \ge 0$ **12.** $\theta = 2\pi/3, \quad r \le -2$

13. $\theta = \pi/3, \quad -1 \le r \le 3$ **14.** $\theta = 11\pi/4, \quad r \ge -1$

15. $\theta = \pi/2, \quad r \ge 0$ **16.** $\theta = \pi/2, \quad r \le 0$

17. $0 \le \theta \le \pi, \quad r = 1$

18. $0 \le \theta \le \pi, \quad r = -1$

19. $\pi/4 \le \theta \le 3\pi/4, \quad 0 \le r \le 1$

20. $-\pi/4 \le \theta \le \pi/4, \quad -1 \le r \le 1$

21. $-\pi/2 \le \theta \le \pi/2, \quad 1 \le r \le 2$

22. $0 \le \theta \le \pi/2, \quad 1 \le |r| \le 2$

Replace the polar equations in Exercises 23–34 by equivalent Cartesian equations. Then identify the graph.

23. $r \cos\theta = 2$ **24.** $r \sin\theta = -1$

25. $r \sin\theta = 4$ **26.** $r \cos\theta = 0$

27. $r \sin\theta = 0$ **28.** $r \cos\theta = -3$

29. $r \cos\theta + r \sin\theta = 1$ **30.** $r \sin\theta = r \cos\theta$

31. $r^2 = 1$ **32.** $r^2 = 4r \sin\theta$

33. $r = \dfrac{5}{\sin\theta - 2\cos\theta}$ **34.** $r = 4\tan\theta \sec\theta$

Replace the Cartesian equations in Exercises 35–44 by equivalent polar equations.

35. $x = 7$ **36.** $y = 1$

37. $x = y$ **38.** $x - y = 3$

39. $x^2 + y^2 = 4$ **40.** $x^2 - y^2 = 1$

41. $\dfrac{x^2}{9} + \dfrac{y^2}{4} = 1$ **42.** $xy = 2$

43. $y^2 = 4x$ **44.** $x^2 - y^2 = 25\sqrt{x^2 + y^2}$

10.6 Graphing in Polar Coordinates

The graph of the equation

$$F(r, \theta) = 0$$

consists of the points whose polar coordinates in some form satisfy the equation. We say "in some form" because it is a sad fact, but true, that some coordinate pairs of a point on the graph may not satisfy the equation even when others do.

To speed our work, we look for symmetries, for values of θ at which the curve passes through the origin, and for points at which r takes on extreme values. When the curve passes through the origin, we also try to calculate the curve's slope there.

Symmetry and Slope

The three parts of Fig. 10.48 illustrate the standard polar coordinate tests for symmetry.

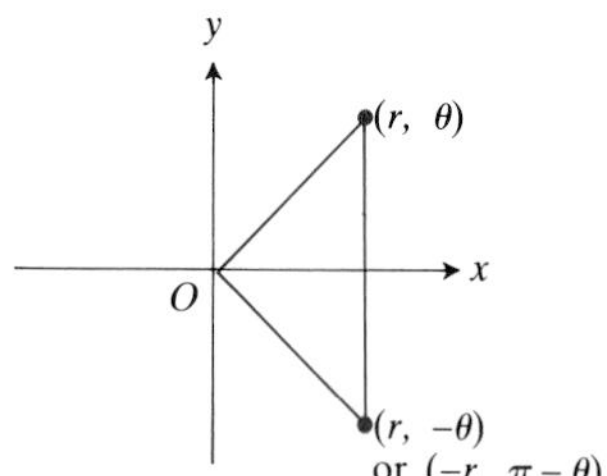

About the x-axis
(a)

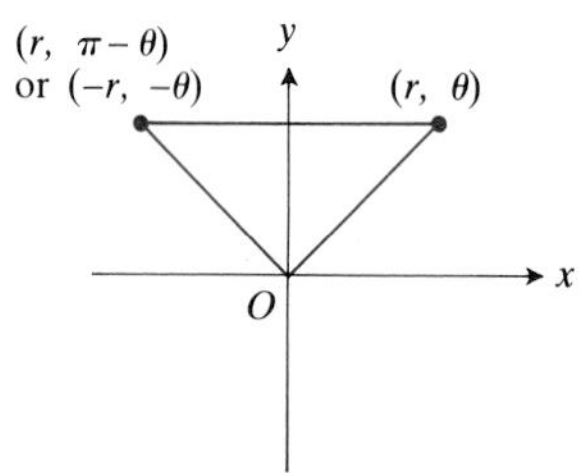

About the y-axis
(b)

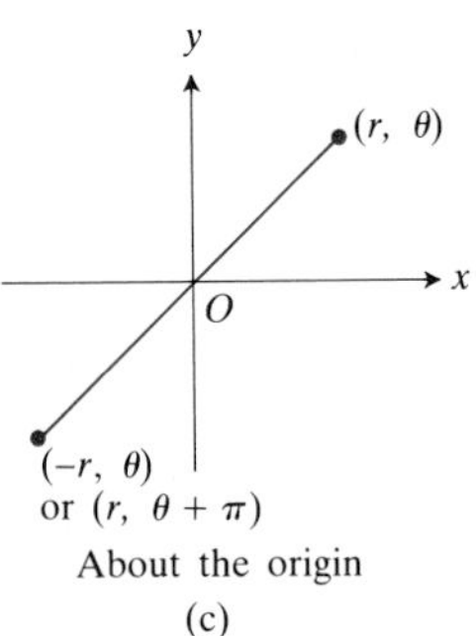

About the origin
(c)

10.48 Three tests for symmetry.

Symmetry Tests for Graphs

1. *Symmetry about the x-axis:* If the point (r, θ) lies on the graph, the point $(r, -\theta)$ lies on the graph (Fig. 10.48a).
2. *Symmetry about the y-axis:* If the point (r, θ) lies on the graph, the point $(r, \pi - \theta)$ or $(-r, -\theta)$ lies on the graph (Fig. 10.48b).
3. *Symmetry about the origin:* If the point (r, θ) lies on the graph, the point $(-r, \theta)$ or $(r, \theta + \pi)$ lies on the graph (Fig. 10.48c).

If a curve has any two of the symmetries listed here, it also has the third (as you will be invited to show in Exercise 30). Thus, if two of the tests are positive there is no need to apply the third test.

The slope of a polar curve $r = f(\theta)$ is not given by the derivative $r' = df/d\theta$, but by a different formula. To see why, and what the formula is, think of the graph of f as the graph of the parametric equations

$$x = r\cos\theta = f(\theta)\cos\theta,$$
$$y = r\sin\theta = f(\theta)\sin\theta. \tag{1}$$

If f is a differentiable function of θ, then so are x and y and, when $dx/d\theta \neq 0$, we may calculate dy/dx from the parametric formula

$$\frac{dy}{dx} = \frac{dy/d\theta}{dx/d\theta} \qquad \text{(Section 10.4, Eq. (2) with } t = \theta)$$

$$= \frac{\dfrac{d}{d\theta}(f(\theta)\cdot\sin\theta)}{\dfrac{d}{d\theta}(f(\theta)\cdot\cos\theta)}$$

$$= \frac{\dfrac{df}{d\theta}\sin\theta + f(\theta)\cos\theta}{\dfrac{df}{d\theta}\cos\theta - f(\theta)\sin\theta} \qquad \text{(Product Rule for Derivatives)}$$

$$= \frac{r'\sin\theta + r\cos\theta}{r'\cos\theta - r\sin\theta}. \qquad (r = f,\ r' = df/d\theta)$$

> **Slope of a Polar Curve**
>
> If $r = f(\theta)$ is differentiable and $dx/d\theta \neq 0$, then the slope dy/dx at the point (r, θ) on the graph of f is given by the formula
>
> $$\text{Slope at } (r, \theta) = \frac{r'\sin\theta + r\cos\theta}{r'\cos\theta - r\sin\theta}. \tag{2}$$

If $r = 0$ when $\theta = \theta_0$, then Eq. (2) reduces to

$$\text{Slope at } (0, \theta_0) = \frac{r'\sin\theta_0}{r'\cos\theta_0} = \frac{\sin\theta_0}{\cos\theta_0} = \tan\theta_0. \tag{3}$$

> **Slopes at the Origin**
>
> If the graph of $r = f(\theta)$ passes through the origin at the value $\theta = \theta_0$, then the slope of the curve there is
>
> $$\text{Slope at } (0, \theta_0) = \tan\theta_0. \tag{4}$$

The reason we say "slope at $(0, \theta_0)$" and not just "slope at the origin" is that a polar curve may pass through the origin more than once, with different slopes at different θ-values. This will not be the case in our first example, however.

EXAMPLE 1 *A cardioid.* Graph the curve

$$r = 1 - \cos\theta.$$

Solution The curve is symmetric about the x-axis because

$$\begin{aligned} (r, \theta) \text{ on the graph} \quad &\Rightarrow \quad r = 1 - \cos\theta \\ &\Rightarrow \quad r = 1 - \cos(-\theta) \qquad \begin{pmatrix}\cos\theta = \\ \cos(-\theta)\end{pmatrix} \\ &\Rightarrow \quad (r, -\theta) \text{ on the graph.} \end{aligned}$$

As θ increases from 0 to π, $\cos\theta$ decreases from 1 to -1, and $r = 1 - \cos\theta$ increases from a minimum value of 0 to a maximum value of 2. As θ continues on from π to 2π, $\cos\theta$ increases from -1 back to 1 and r decreases from 2 back to 0. The curve starts to repeat when $\theta = 2\pi$ because the cosine has period 2π.

The curve leaves the origin with slope $\tan(0) = 0$ and returns to the origin with slope $\tan(2\pi) = 0$.

We make a table of values from $\theta = 0$ to $\theta = \pi$, plot the points, draw a smooth curve through them with a horizontal tangent at the origin, and reflect the curve across the x-axis to complete the graph (Fig. 10.49). The curve is called a *cardioid* because of its heart shape. ■

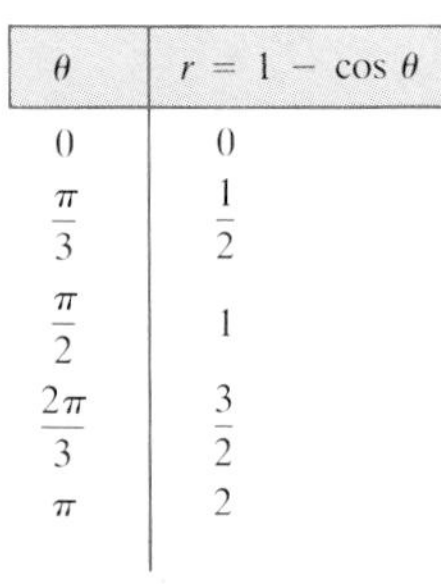

θ	$r = 1 - \cos\theta$
0	0
$\frac{\pi}{3}$	$\frac{1}{2}$
$\frac{\pi}{2}$	1
$\frac{2\pi}{3}$	$\frac{3}{2}$
π	2

(a)

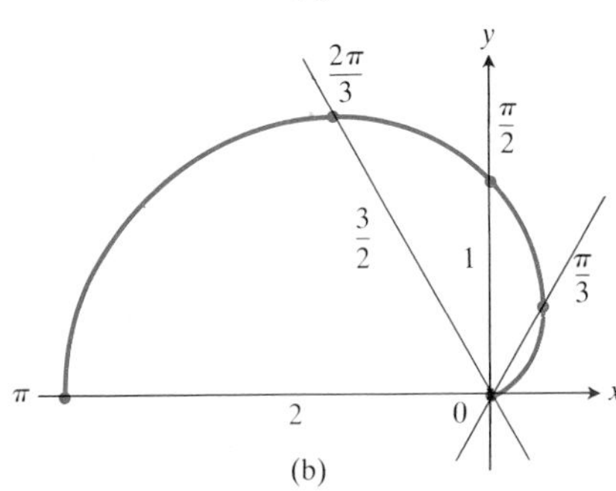

(b)

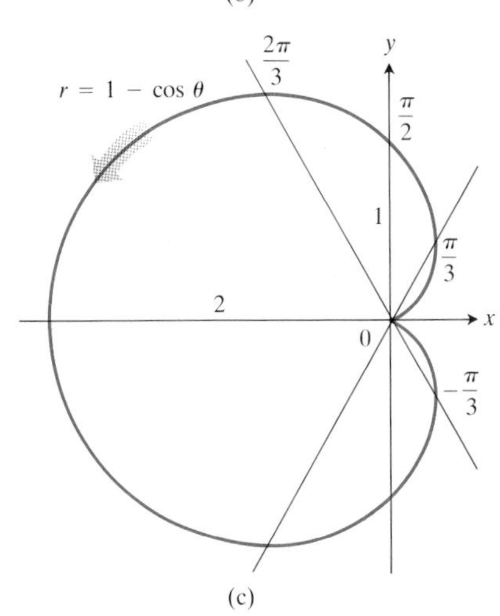

(c)

10.49 The steps in graphing the cardioid $r = 1 - \cos\theta$. The arrow shows the direction of increasing θ.

EXAMPLE 2 Graph the curve $r^2 = 4\cos\theta$.

Solution Although $\cos\theta$ has period 2π, the equation $r^2 = 4\cos\theta$ requires $\cos\theta \geq 0$, so we get the entire graph by running θ from $-\pi/2$ to $\pi/2$. The curve is symmetric about the x-axis because

$$\begin{aligned} (r, \theta) \text{ on the graph} \quad &\Rightarrow \quad r^2 = 4\cos\theta \\ &\Rightarrow \quad r^2 = 4\cos(-\theta) \qquad \begin{pmatrix}\cos\theta = \\ \cos(-\theta)\end{pmatrix} \\ &\Rightarrow \quad (r, -\theta) \text{ on the graph.} \end{aligned}$$

The curve is also symmetric about the origin because

$$\begin{aligned} (r, \theta) \text{ on the graph} \quad &\Rightarrow \quad r^2 = 4\cos\theta \\ &\Rightarrow \quad (-r)^2 = 4\cos\theta \\ &\Rightarrow \quad (-r, \theta) \text{ on the graph.} \end{aligned}$$

Together, these two symmetries imply symmetry about the y-axis.

The curve passes through the origin when $\theta = -\pi/2$ and $\theta = \pi/2$. It has a vertical tangent both times because $\tan\theta$ is infinite.

For each value of θ in the interval between $-\pi/2$ and $\pi/2$, the formula $r^2 = 4\cos\theta$ gives two values of r:

$$r = \pm 2\sqrt{\cos\theta}.$$

We make a short table of values, plot the corresponding points, and use information about symmetry and tangents as a guide in connecting the points with a smooth curve (Fig. 10.50). ■

Faster Graphing

One way to graph a polar equation $r = f(\theta)$ is to make a table of (r, θ) values, plot the corresponding points, and connect them in order of increasing θ. This can work well if enough points have been plotted to reveal all the loops and dimples in the graph. In this section we describe another method

θ	$\cos\theta$	$r = \pm 2\sqrt{\cos\theta}$
0	1	± 2
$\pm\frac{\pi}{6}$	$\frac{\sqrt{3}}{2}$	± 1.9
$\pm\frac{\pi}{4}$	$\frac{1}{\sqrt{2}}$	± 1.7
$\pm\frac{\pi}{3}$	$\frac{1}{2}$	± 1.4
$\pm\frac{\pi}{2}$	0	0

(a)

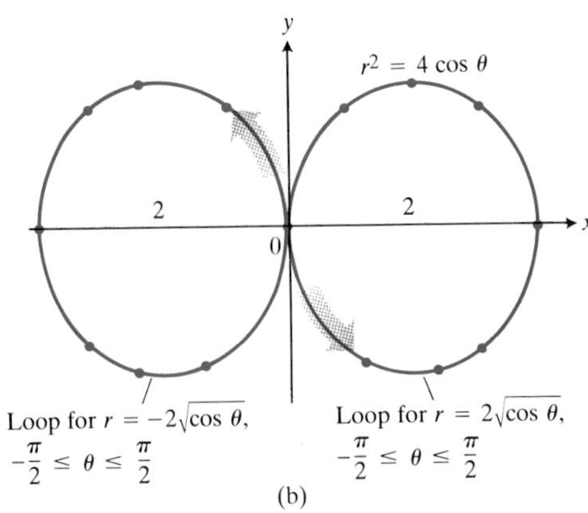

10.50 The graph of $r^2 = 4\cos\theta$. The arrows show the direction of increasing θ. The values of r in the table were found with a calculator and rounded.

of graphing that is usually quicker and more reliable. The steps in the new method are described in the box.

Steps for Quick Graphing

STEP 1: First graph $r = f(\theta)$ in the *Cartesian* $r\theta$-plane (that is, plot the values of θ on a horizontal axis and the corresponding values of r along a vertical axis).

STEP 2: Then use the Cartesian graph as a "table" and guide to sketch the *polar* coordinate graph.

This method is better than simple point plotting because the Cartesian graph, even when hastily drawn, shows at a glance where r is positive, negative, and nonexistent, as well as where r is increasing and decreasing. As examples, we graph $r = 1 + \cos(\theta/2)$ and $r^2 = \sin 2\theta$.

EXAMPLE 3 Graph the curve

$$r = 1 + \cos\frac{\theta}{2}.$$

Solution We first graph r as a function of θ in the Cartesian $r\theta$-plane. Since the cosine has period 2π, we must let θ run from 0 to 4π to produce the entire graph (Fig. 10.51a). The arrows from the θ-axis to the curve give the radius vectors for graphing $r = 1 + \cos(\theta/2)$ in the polar plane (Fig. 10.51b). ■

EXAMPLE 4 *A lemniscate.* Graph the curve $r^2 = \sin 2\theta$.

Solution Here we begin by plotting r^2 (not r) as a function of θ in the Cartesian $r^2\theta$-plane, treating r^2 as a variable that may have negative as well as positive values (Fig. 10.52a). We pass from there to the graph of $r = \pm\sqrt{\sin 2\theta}$ in the $r\theta$-plane (Fig. 10.52b), and then draw the polar graph (Fig. 10.52c). The graph in Fig. 10.52b "covers" the final polar graph in Fig. 10.52c twice. We could have managed with either loop alone, with the two upper halves, or with the two lower halves. The double covering does no harm, however, and we actually learn a little more about the behavior of the function this way. ■

Finding the Points Where Curves Intersect

The fact that we can represent a point in different ways in polar coordinates makes extra care necessary in deciding when a point lies on the graph of a polar equation and in determining the points at which the graphs of polar equations intersect. The problem is that a point of intersection may appear in the equation of one curve with different polar coordinates than it has in the equation of another curve. Thus, solving the equations of two curves simultaneously may not identify all their points of intersection.

How to use Cartesian Graphs to Draw Polar Graphs

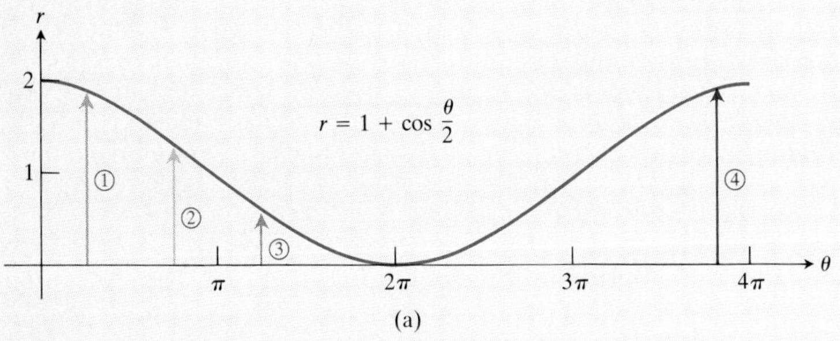

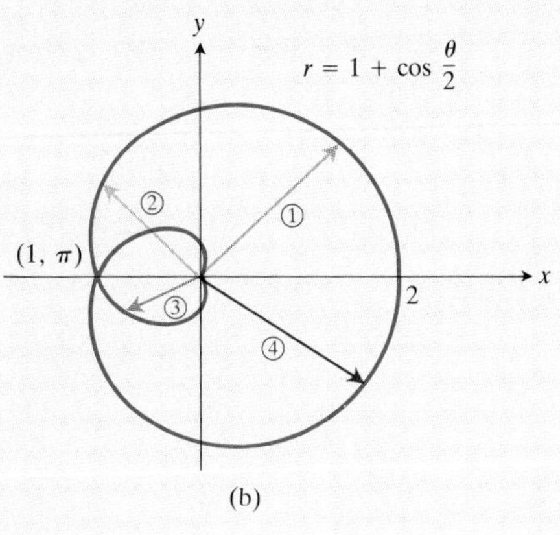

10.51 (a) The graph of $r = 1 + \cos(\theta/2)$ in the Cartesian $r\theta$-plane gives us any number of radius vectors from the θ-axis to the curve. (b) The radius vectors from the previous sketch help us draw the graph in the polar $r\theta$-plane.

10.52 (a) The graph of $r^2 = \sin 2\theta$ in the Cartesian $r^2\theta$-plane includes negative values of the dependent variable r^2 as well as positive values. (b) When we graph r vs θ in the Cartesian $r\theta$-plane we ignore the points where r^2 is negative but plot + and − parts from the points where r^2 is positive. (c) In the polar $r\theta$-plane, the radius vectors from the previous sketch cover the final graph twice.

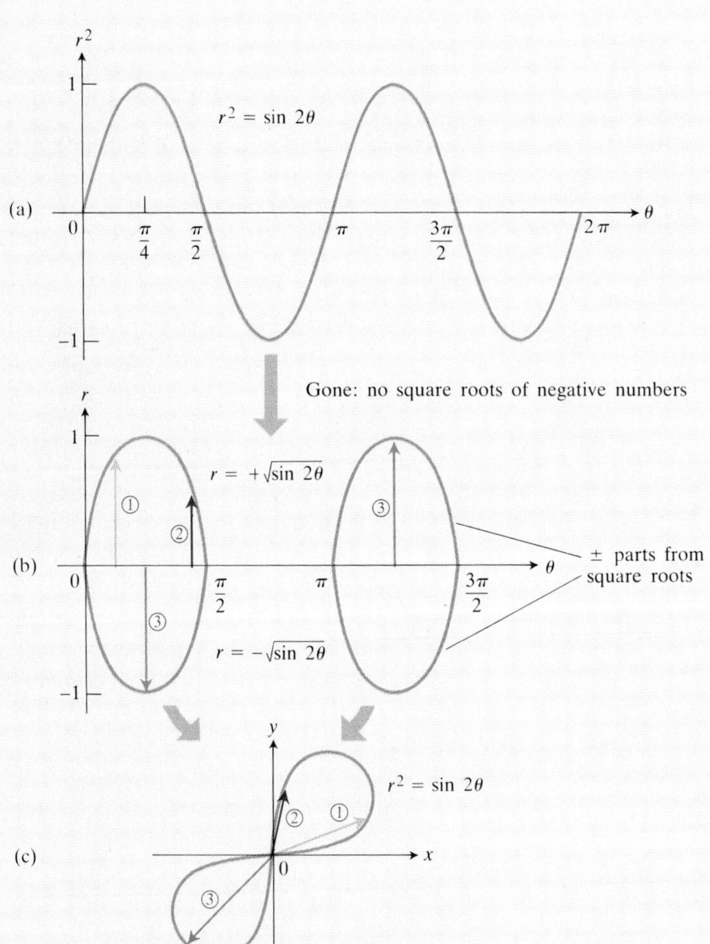

The only sure way to identify all the points of intersection is to graph the equations.

EXAMPLE 5 Show that the point $(2, \pi/2)$ lies on the curve $r = 2\cos 2\theta$.

Solution It may seem at first that the point $(2, \pi/2)$ does not lie on the curve because substituting the given coordinates into the equation gives

$$2 = 2\cos 2\left(\frac{\pi}{2}\right) = 2\cos \pi = -2,$$

which is not a true equality. The magnitude is right, but the sign is wrong. This suggests looking for a pair of coordinates for the given point in which r is negative, for example,

$$\left(-2, -\frac{\pi}{2}\right).$$

When we try these in the equation $r = 2\cos 2\theta$, we find

$$-2 = 2\cos 2\left(-\frac{\pi}{2}\right) = 2(-1) = -2,$$

and the equation is satisfied. The point $(2, \pi/2)$ does lie on the curve after all. ■

EXAMPLE 6 Find the points of intersection of the curves

$$r^2 = 4\cos\theta \qquad \text{and} \qquad r = 1 - \cos\theta.$$

Solution In Cartesian coordinates, we can always find the points where two curves cross by solving their equations simultaneously. In polar coordinates, the story is different. Simultaneous solution may reveal some intersection points without revealing others. In this example, simultaneous solution reveals only two of the four intersection points. The others must be found by graphing.

If we substitute $\cos\theta = r^2/4$ in the equation $r = 1 - \cos\theta$, we get

$$\begin{aligned} r &= 1 - \cos\theta \\ r &= 1 - \frac{r^2}{4} \\ 4r &= 4 - r^2 \\ r^2 + 4r - 4 &= 0 \\ r &= -2 \pm 2\sqrt{2}. \quad \text{(Quadratic Formula)} \end{aligned}$$

The value $r = -2 - 2\sqrt{2}$ has too large an absolute value to give a point on either curve. The values of θ corresponding to $r = -2 + 2\sqrt{2}$ are

$$\begin{aligned} \theta &= \cos^{-1}(1 - r) && \text{(From } r = 1 - \cos\theta\text{)} \\ &= \cos^{-1}(1 - (2\sqrt{2} - 2)) && \text{(Set } r = 2\sqrt{2} - 2\text{)} \\ &= \cos^{-1}(3 - 2\sqrt{2}) \\ &= \pm 80°. && \text{(With a calculator, rounded).} \end{aligned}$$

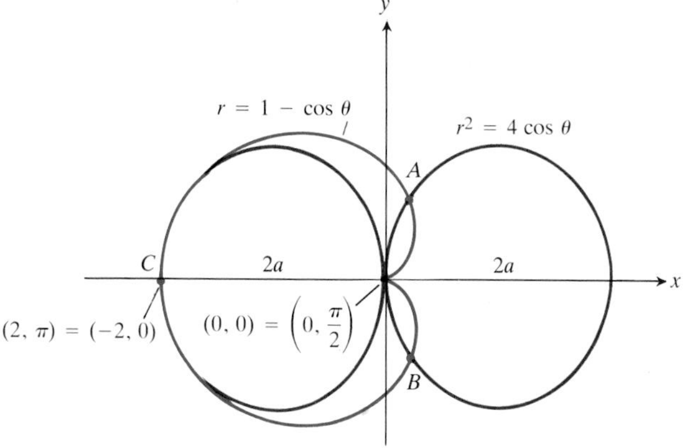

10.53 The four points of intersection of the curves $r = 1 - \cos\theta$ and $r^2 = 4\cos\theta$. Only A and B were found by simultaneous solution. The other two were disclosed by graphing.

We have thus identified two intersection points:

$$(r, \theta) = (2\sqrt{2} - 2, \pm 80°).$$

If we graph the equations $r^2 = 4\cos\theta$ and $r = 1 - \cos\theta$ together (Fig. 10.53), as we can easily do by combining the graphs in Figs. 10.49 and 10.50, we see that the curves also intersect at the origin and at the point $(2, \pi)$. Why weren't the r-values of these points revealed by the simultaneous solution? The answer is that the points $(0, 0)$ and $(2, \pi)$ are not on the curves "simultaneously." They are not reached at the same value of θ on each curve. On the curve $r = 1 - \cos\theta$, the point $(2, \pi)$ is reached when $\theta = \pi$. On the curve $r^2 = 4\cos\theta$, it is reached when $\theta = 0$, where it is identified not by the coordinates $(2, \pi)$, which do not satisfy the equation, but by the coordinates $(-2, 0)$, which do. Similarly, the cardioid reaches the origin when $\theta = 0$, but the curve $r^2 = 4\cos\theta$ reaches the origin when $\theta = \pi/2$. ■

Exercises 10.6

Sketch the curves in Exercises 1–10.

1. $r = 1 + \cos\theta$ **2.** $r = 2 - 2\cos\theta$

3. $r = 1 - \sin\theta$ **4.** $r = 1 + \sin\theta$

5. $r = 2 + \sin\theta$ **6.** $r = 1 + 2\sin\theta$

7. $r^2 = 4\cos 2\theta$ **8.** $r^2 = 4\sin\theta$

9. $r = \theta$ **10.** $r = \sin(\theta/2)$

Use Eq. (2) to find the slopes of the curves in Exercises 11–14 at the given points. Sketch the curves along with their tangents at these points.

11. *Cardioid.* $r = -1 + \cos\theta$; $\theta = \pm\pi/2$

12. *Cardioid.* $r = -1 + \sin\theta$; $\theta = 0, \pi/2, \pi$

13. *Four-leaved rose.* $r = \sin 2\theta$; $\theta = \pm\pi/4$, $\pm 3\pi/4$, and the values of θ at which the curve passes through the origin

14. *Four-leaved rose.* $r = \cos 2\theta$; $\theta = 0$, $\pm\pi/2$, π, and the values of θ at which the curve passes through the origin

Graph the lemniscates in Exercises 15–16.

15. $r^2 = 4\cos 2\theta$ **16.** $r^2 = 4\sin 2\theta$

Graph the limaçons in Exercises 17–20. Limaçon ("leemasahn") is Old French for "snail." You will see why the name is appropriate when you graph the limaçons in Exercise 17. Equations for limaçons have the form $r = a \pm b\cos\theta$ or $r = a \pm b\sin\theta$. There are four basic shapes.

17. *Limaçons with an inner loop*

a) $r = \frac{1}{2} + \cos\theta$

b) $r = \frac{1}{2} + \sin\theta$

18. *Cardioids*

a) $r = 1 - \cos\theta$

b) $r = -1 + \sin\theta$

19. *Dimpled limaçons*

a) $r = \frac{3}{2} + \cos\theta$

b) $r = \frac{3}{2} - \sin\theta$

20. *Convex limaçons*

a) $r = 2 + \cos\theta$

b) $r = -2 + \sin\theta$

21. Sketch the region defined by the inequality $0 \leq r \leq 2 - 2\cos\theta$.

22. Sketch the region defined by the inequality $0 \leq r^2 \leq \cos\theta$.

23. Show that the point $(2, 3\pi/4)$ lies on the curve $r = 2\sin 2\theta$.

24. Show that the point $(1/2, 3\pi/2)$ lies on the curve $r = -\sin(\theta/3)$.

Find the points of intersection of the pairs of curves in Exercises 25–28.

25. $r = 1 + \cos\theta, \quad r = 1 - \cos\theta$

26. $r = 1 + \sin\theta, \quad r = 1 - \sin\theta$

27. $r = 1 - \sin\theta, \quad r^2 = 4\sin\theta$

28. $r^2 = \sqrt{2}\sin\theta, \quad r^2 = \sqrt{2}\cos\theta$

29. *Continuation of Example 6.* The simultaneous solution of the equations

$$r^2 = 4\cos\theta, \tag{5}$$

$$r = 1 - \cos\theta, \tag{6}$$

in the text did not reveal the points $(0, 0)$ and $(2, \pi)$ in which their graphs intersected.

a) We could have found the point $(2, \pi)$, however, by replacing the (r, θ) in Eq. (5) by the equivalent $(-r, \theta + \pi)$, to obtain

$$\begin{aligned} r^2 &= 4\cos\theta \\ (-r)^2 &= 4\cos(\theta + \pi) \\ r^2 &= -4\cos\theta. \end{aligned} \tag{7}$$

Solve Eqs. (6) and (7) simultaneously to show that $(2, \pi)$ is a common solution. (This will still not reveal that the graphs intersect at $(0, 0)$.)

b) The origin is still a special case. (It often is.) One way to handle it is the following: Set $r = 0$ in Eqs. (5) and (6) and solve each equation for a corresponding value of θ. Since $(0, \theta)$ is the origin for *any* θ, this will show that both curves pass through the origin even if they do so for different θ-values.

30. Show that a curve with any two of the symmetries listed at the beginning of the section automatically has the third.

Computer Grapher

Find the points of intersection of the pairs of curves in Exercises 31–34.

31. $r^2 = \sin 2\theta, \quad r^2 = \cos 2\theta$

32. $r = 1 + \cos\frac{\theta}{2}, \quad r = 1 - \sin\frac{\theta}{2}$

33. $r = 1, \quad r = 2\sin 2\theta$

34. $r = 1, \quad r^2 = 2\sin 2\theta$

35. *A rose within a rose.* Graph the equation $r = 1 - 2\sin 3\theta$.

36. Graph *the nephroid of Freeth:*

$$r = 1 + 2\sin\frac{\theta}{2}$$

37. Graph the *roses* $r = \cos m\theta$ for $m = 1/3, 2, 3$, and 7.

38. *Spirals.* Polar coordinates are just the thing for defining spirals. Graph the following spirals:

a) *A logarithmic spiral:* $r = e^{\theta/10}$

b) *A hyperbolic spiral:* $r = 8/\theta$

c) *Equilateral hyperbola:* $r = \pm 10/\sqrt{\theta}$

(Try using different colors for the two branches.)

TOOLKIT PROGRAM

Super*Grapher	Graphs equations of the form $r = f(\theta)$ singly or together in a common display in different colors

10.7 Polar Equations of Conic Sections

Polar coordinates are important in astronomy and astronautical engineering because the ellipses, parabolas, and hyperbolas along which satellites, moons, planets, and comets move can all be described with one general polar equation. In Cartesian coordinates, the equations of conics have different forms, but not so here. This section develops the general equation, along with special equations for lines and circles.

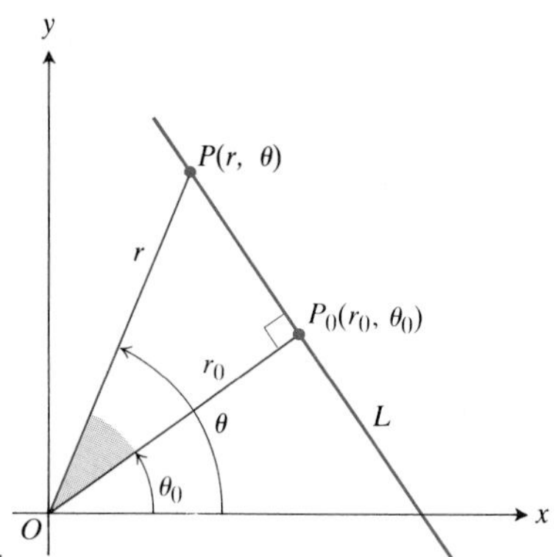

10.54 We can obtain a polar equation for line L by reading the relation $r_0/r = \cos(\theta - \theta_0)$ from triangle OP_0P.

Lines

Suppose the perpendicular from the origin to line L meets L at the point $P_0(r_0, \theta_0)$, with $r_0 \geq 0$ (Fig. 10.54). Then, if $P(r, \theta)$ is any other point on L, the points P, P_0, and O are the vertices of a right triangle, from which we can read the relation

$$\frac{r_0}{r} = \cos(\theta - \theta_0) \tag{1}$$

or

$$r \cos(\theta - \theta_0) = r_0. \tag{2}$$

The coordinates of P_0 satisfy this equation as well.

The Standard Polar Equation for Lines

If the point $P_0(r_0, \theta_0)$ is the foot of the perpendicular from the origin to the line L, and $r_0 \geq 0$, then an equation for L is

$$r \cos(\theta - \theta_0) = r_0. \tag{3}$$

EXAMPLE 1 The standard polar equation for the line shown here is

$$r \cos\left(\theta - \frac{\pi}{3}\right) = 2.$$

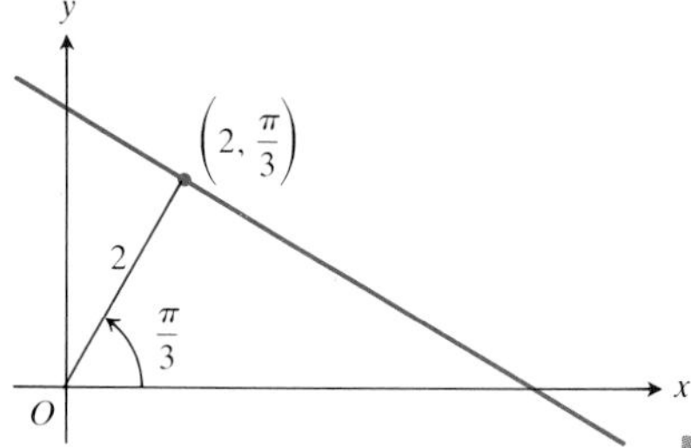

EXAMPLE 2 Use the identity $\cos(A - B) = \cos A \cos B + \sin A \sin B$ to find a Cartesian equation for the line in Example 1.

Solution

$$r \cos\left(\theta - \frac{\pi}{3}\right) = 2$$

$$r\left(\cos\theta \cos\frac{\pi}{3} + \sin\theta \sin\frac{\pi}{3}\right) = 2$$

$$\frac{1}{2} r \cos\theta + \frac{\sqrt{3}}{2} r \sin\theta = 2$$

$$\frac{1}{2} x + \frac{\sqrt{3}}{2} y = 2$$

$$x + \sqrt{3}\, y = 4$$

Circles

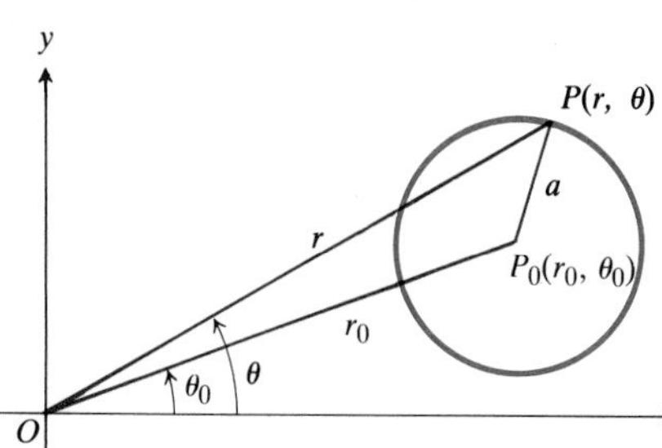

10.55 We can get an equation for this circle by applying the Law of Cosines to triangle OP_0P.

To find a polar equation for the circle of radius a centered at $P_0(r_0, \theta_0)$, we let $P(r, \theta)$ be a point on the circle and apply the Law of Cosines to triangle OP_0P (Fig. 10.55). This gives

$$a^2 = r_0^2 + r^2 - 2r_0 r \cos(\theta - \theta_0). \tag{4}$$

If the circle passes through the origin, then $r_0 = a$ and Eq. (4) simplifies somewhat:

$$a^2 = a^2 + r^2 - 2ar\cos(\theta - \theta_0) \qquad \text{(Eq. (4) with } r_0 = a\text{)}$$

$$r^2 = 2ar\cos(\theta - \theta_0)$$

$$r = 2a\cos(\theta - \theta_0). \tag{5}$$

If the circle's center lies on the positive x-axis, so that $\theta_0 = 0$, Eq. (5) becomes

$$r = 2a\cos\theta. \tag{6}$$

If, instead, the circle's center lies on the positive y-axis, $\theta = \pi/2$, $\cos(\theta - \pi/2) = \sin\theta$, and Eq. (5) becomes

$$r = 2a\sin\theta. \tag{7}$$

Equations for circles through the origin centered on the negative x- and y-axes can be obtained from Eqs. (6) and (7) by replacing r with $-r$.

Polar Equations for Circles Through the Origin Centered on the *x*- and *y*-Axes, Radius *a*

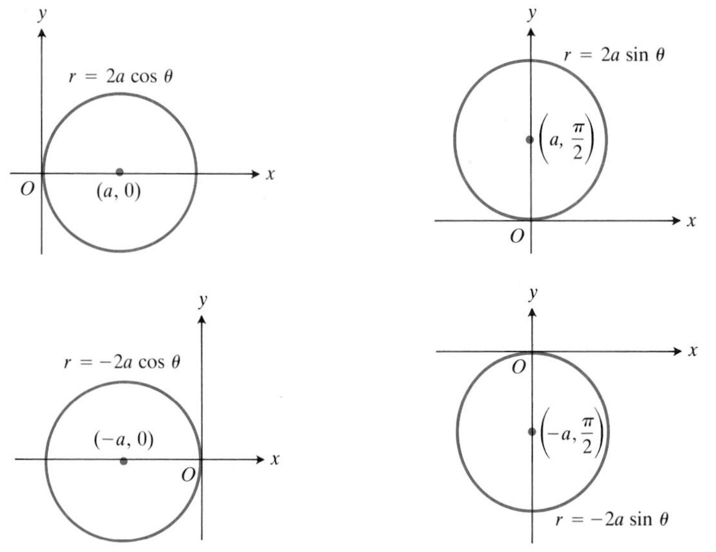

EXAMPLE 3 *Circles through the origin*

Radius	Center (polar coordinates)	Equation
3	$(3, 0)$	$r = 6\cos\theta$
2	$(2, \pi/2)$	$r = 4\sin\theta$
1/2	$(-1/2, 0)$	$r = -\cos\theta$
1	$(-1, \pi/2)$	$r = -2\sin\theta$

■

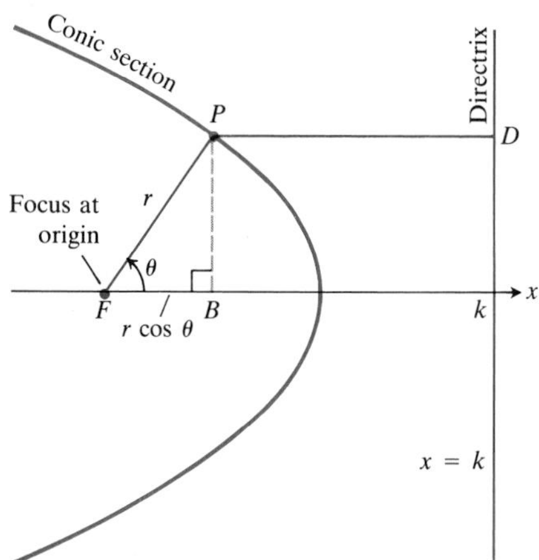

10.56 If a conic section is put in this position, then $PF = r$ and $PD = k - r\cos\theta$.

Ellipses, Parabolas, and Hyperbolas

To find polar equations for ellipses, parabolas, and hyperbolas, we first assume that the conic has one focus at the origin (for the parabola, its only focus) and that the corresponding directrix is the vertical line $x = k$ lying to the right of the origin (Fig. 10.56). This makes

$$PF = r \tag{8}$$

and

$$PD = k - FB = k - r\cos\theta. \tag{9}$$

The conic's focus–directrix equation $PF = e \cdot PD$ then becomes

$$r = e(k - r\cos\theta), \tag{10}$$

which can be solved for r to obtain

$$r = \frac{ke}{1 + e\cos\theta}. \tag{11}$$

This equation represents an ellipse if $0 < e < 1$, a parabola if $e = 1$, and a hyperbola if $e > 1$. And there we have it—ellipses, parabolas, and hyperbolas all with the same basic equation.

EXAMPLE 4 *Typical conics from Eq. (11).*

$e = \frac{1}{2}$: ellipse $\quad r = \dfrac{k}{2 + \cos\theta}$

$e = 1$: parabola $\quad r = \dfrac{k}{1 + \cos\theta}$

$e = 2$: hyperbola $\quad r = \dfrac{2k}{1 + 2\cos\theta}$ ■

You may see some variations of Eq. (11) from time to time, depending on the location of the directrix. If the directrix is the line $x = -k$ to the left of the origin (the origin is still the focus), the equation we get in place of Eq. (11) is

$$r = \frac{ke}{1 - e\cos\theta}. \tag{12}$$

The denominator now has a $(-)$ instead of a $(+)$. If the directrix is either of the lines $y = k$ or $y = -k$, the equations we get have sines in them instead of cosines, as shown in Table 10.4.

EXAMPLE 5 Find an equation for the hyperbola with eccentricity 3/2 and directrix $x = 2$.

Solution We use Eq. (1) in Table 10.4 with $k = 2$ and $e = 3/2$ to get

$$r = \frac{2(3/2)}{1 + (3/2)\cos\theta} \quad \text{or} \quad r = \frac{6}{2 + 3\cos\theta}.$$ ■

TABLE 10.4
Equations for conic sections ($e > 0$)

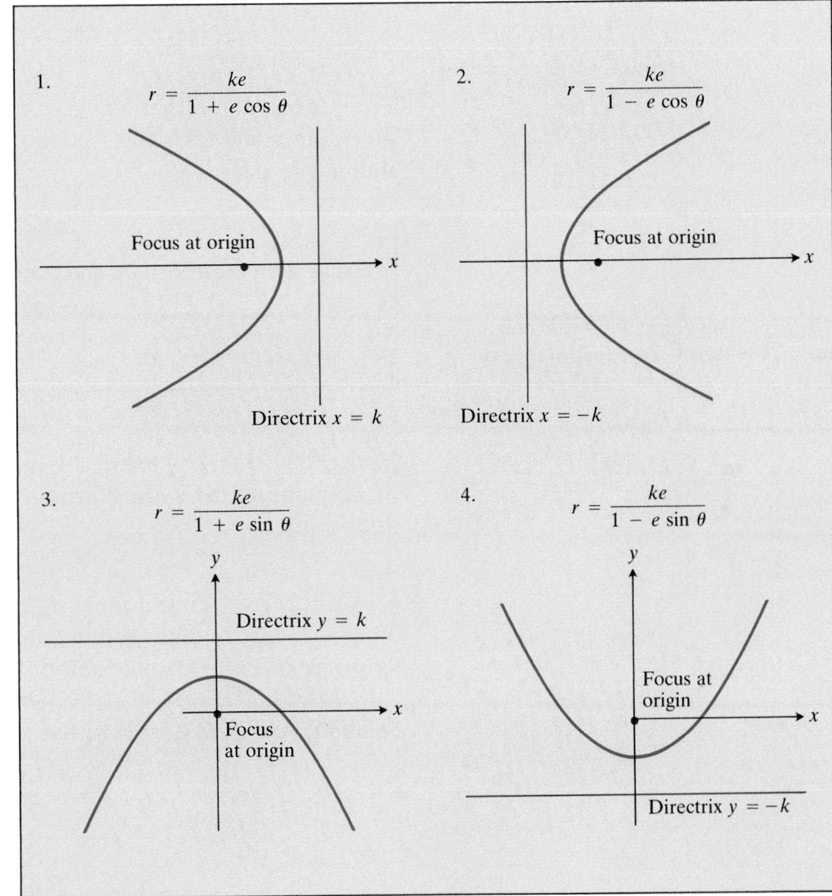

EXAMPLE 6 Find the directrix of the parabola

$$r = \frac{25}{10 + 10 \cos \theta}.$$

Solution We divide the numerator and denominator by 10 to put the equation in standard form:

$$r = \frac{5/2}{1 + \cos \theta}.$$

This is the equation

$$r = \frac{ke}{1 + e \cos \theta}$$

with $k = 5/2$ and $e = 1$. The equation of the directrix is $x = 5/2$. ■

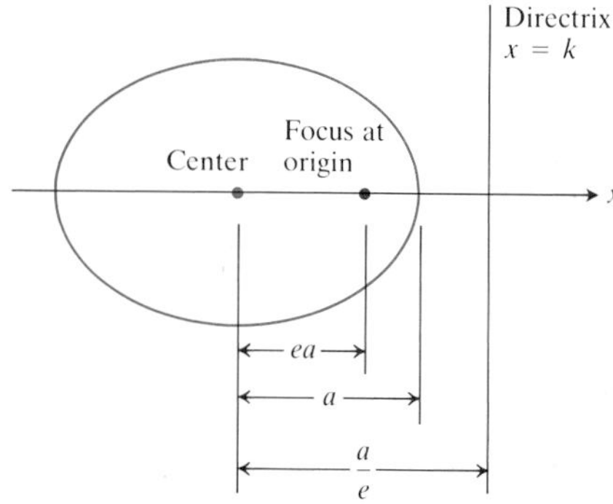

10.57 In an ellipse with semimajor axis a, the focus–directrix distance is $k = (a/e) - ea$, so $ke = a(1 - e^2)$.

From the ellipse diagram in Fig. 10.57, we see that k is related to the eccentricity e and the semimajor axis a by the equation

$$k = \frac{a}{e} - ea. \tag{13}$$

From this, we find that $ke = a(1 - e^2)$. Replacing ke by $a(1 - e^2)$ in Eq. (11) gives the standard polar equation for an ellipse with eccentricity e and semimajor axis a.

Ellipse with Eccentricity e and Semimajor Axis a

$$r = \frac{a(1 - e^2)}{1 + e \cos \theta} \tag{14}$$

Notice that when $e = 0$, Eq. (14) becomes $r = a$, which represents a circle.

Equation (14) is the starting point for calculating planetary orbits in astronomy.

EXAMPLE 7 Find a polar equation for an ellipse with semimajor axis 39.44 AU (astronomical units) and eccentricity 0.25. This is the approximate size of Pluto's orbit around the sun.

Solution We use Eq. (14) with $a = 39.44$ and $e = 0.25$ to find

$$r = \frac{39.44(1 - (0.25)^2)}{1 + 0.25 \cos \theta}$$

$$= \frac{147.9}{4 + \cos \theta}.$$

At its point of closest approach (perihelion), Pluto is

$$r = \frac{147.9}{4 + 1} = 29.58 \text{ AU}$$

from the sun. At its most distant point (aphelion), Pluto is

$$r = \frac{147.9}{4 - 1} = 49.3 \text{ AU}$$

from the sun. ■

EXAMPLE 8 Find the distance from one focus of the ellipse in Example 7 to the associated directrix.

Solution We use Eq. (13) with $a = 39.44$ and $e = 0.25$ to find

$$k = 39.44\left(\frac{1}{0.25} - 0.25\right) = 147.9 \text{ AU}.$$ ■

Exercises 10.7

Find polar and Cartesian equations for the lines in Exercises 1–4.

1.

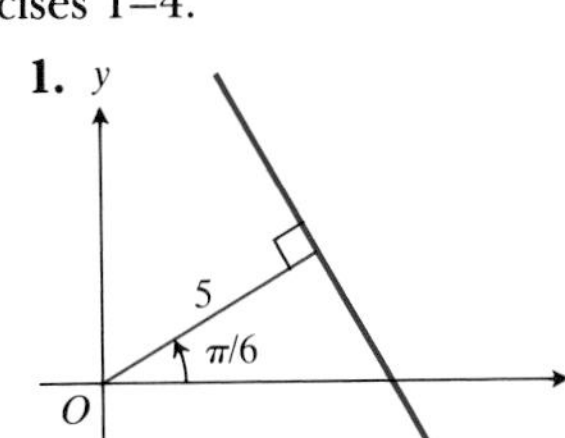

2.

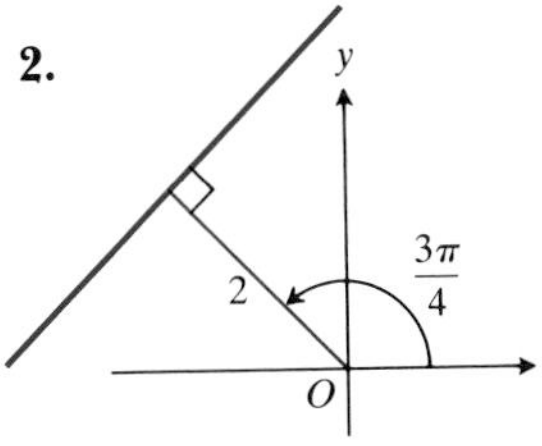

3.

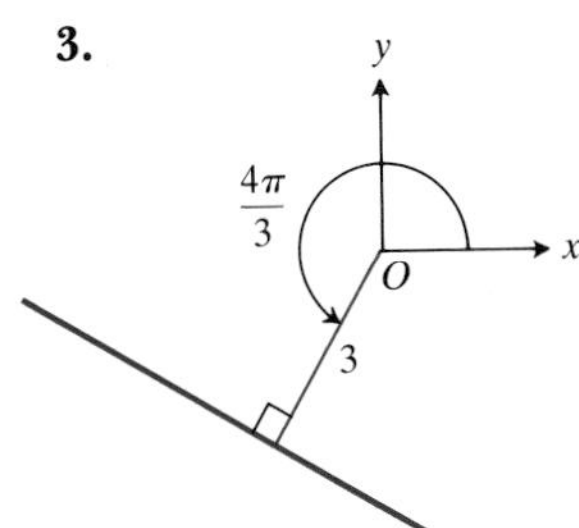

4.

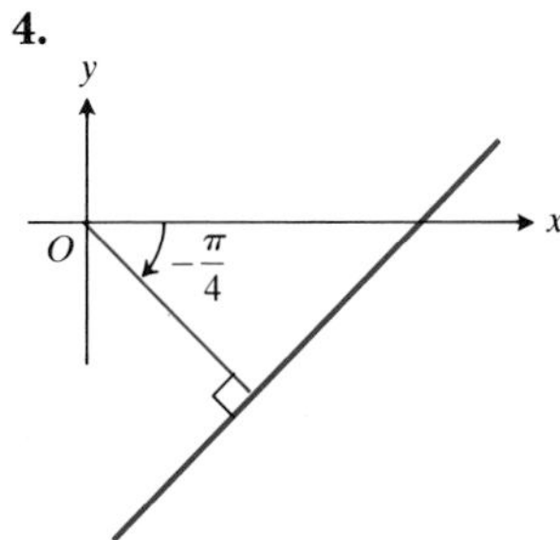

Sketch the lines in Exercises 5–8 and find Cartesian equations for them.

5. $r\cos\left(\theta - \frac{\pi}{4}\right) = \sqrt{2}$

6. $r\cos\left(\theta - \frac{2\pi}{3}\right) = 3$

7. $r\cos\left(\theta - \frac{3\pi}{2}\right) = 1$

8. $r\cos\left(\theta + \frac{\pi}{3}\right) = 2$

Find polar equations for the circles in Exercises 9–12.

9.

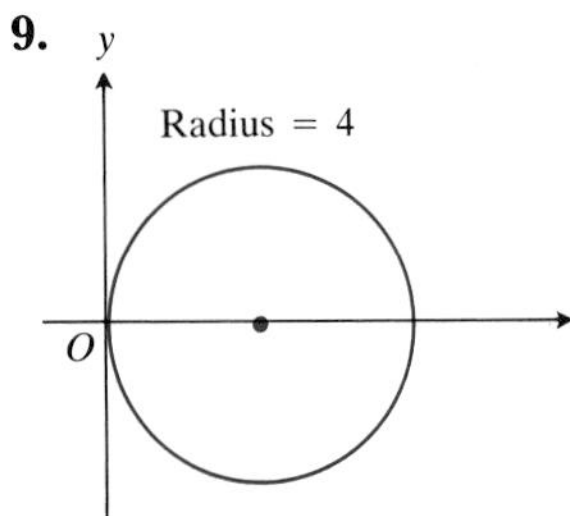

10.

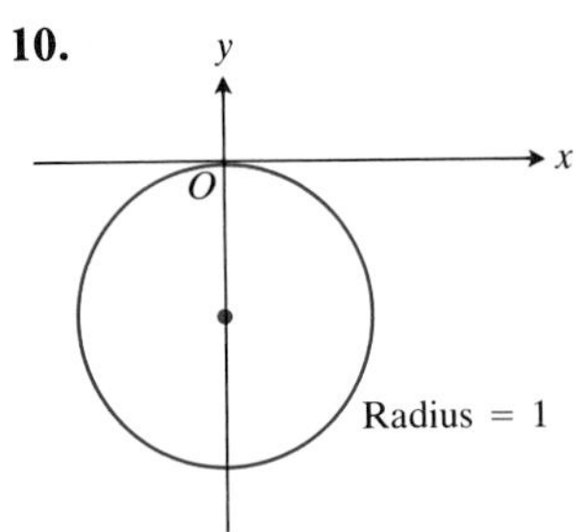

11.

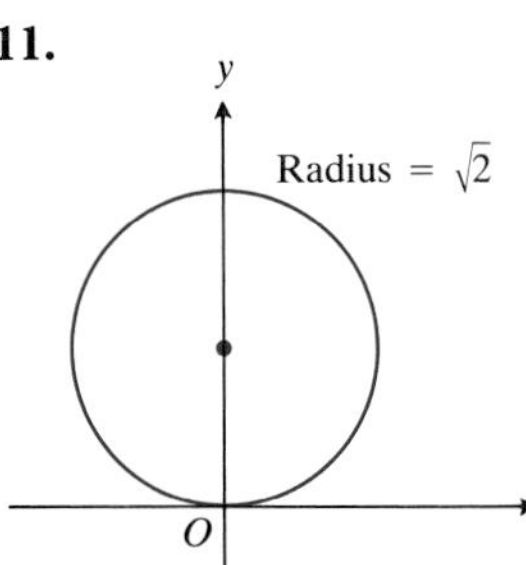

12.

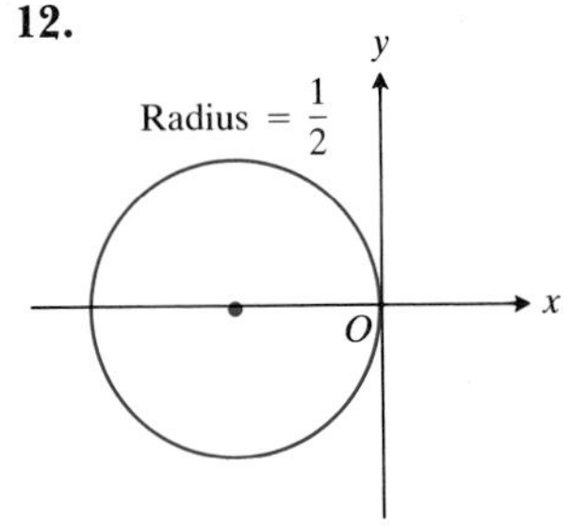

Sketch the circles in Exercises 13–16. Give polar coordinates for their centers and identify their radii.

13. $r = 4\cos\theta$

14. $r = 6\sin\theta$

15. $r = -2\cos\theta$

16. $r = -8\sin\theta$

Exercises 17–24 give the eccentricities of conic sections with one focus at the origin, along with the directrix corresponding to that focus. Find a polar equation for each conic section.

17. $e = 1, \quad x = 2$

18. $e = 1, \quad y = 2$

19. $e = 2, \quad x = 4$

20. $e = 5, \quad y = -6$

21. $e = 1/2, \quad x = 1$

22. $e = 1/4, \quad x = -2$

23. $e = 1/5, \quad y = -10$

24. $e = 1/3, \quad y = 6$

Sketch the parabolas and ellipses in Exercises 25–32. Include the directrix that corresponds to the focus at the origin. Label the vertices with appropriate polar coordinates. Label the centers of the ellipses as well.

25. $r = \dfrac{1}{1 + \cos\theta}$

26. $r = \dfrac{6}{2 + \cos\theta}$

27. $r = \dfrac{25}{10 - 5\cos\theta}$

28. $r = \dfrac{4}{2 - 2\cos\theta}$

29. $r = \dfrac{400}{16 + 8\sin\theta}$

30. $r = \dfrac{12}{3 + 3\sin\theta}$

31. $r = \dfrac{8}{2 - 2\sin\theta}$

32. $r = \dfrac{4}{2 - \sin\theta}$

Sketch the regions defined by the inequalities in Exercises 33–34.

33. $0 \le r \le 2\cos\theta$

34. $-3\cos\theta \le r \le 0$

35. *Perihelion and aphelion.* A planet travels about its sun in an ellipse whose semimajor axis has length a (Fig. 10.58).

a) Show that $r = a(1 - e)$ when the planet is closest to the sun and that $r = a(1 + e)$ when the planet is farthest from the sun.

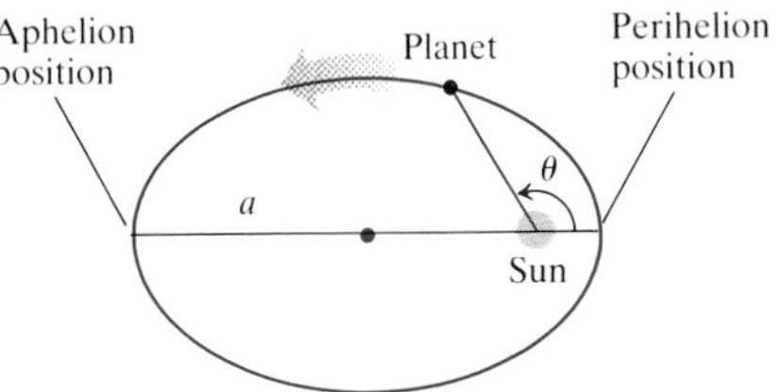

10.58 The positions of a planet closest to and farthest from its sun are called *perihelion* and *aphelion*.

b) Use the data in Table 10.5 to find how close each planet in our solar system comes to the sun and how far away each planet gets from the sun.

36. *Planetary orbits.* In Example 7, we found a polar equation for the orbit of Pluto. Use the data in Table 10.5 to find polar equations for the orbits of the other planets.

TABLE 10.5
Semimajor Axes and Eccentricities of the Planets in Our Solar System

Planet	Semimajor axis (astronomical units)	Eccentricity
Mercury	0.3871	0.2056
Venus	0.7233	0.0068
Earth	1.000	0.0167
Mars	1.524	0.0934
Jupiter	5.203	0.0484
Saturn	9.539	0.0543
Uranus	19.18	0.0460
Neptune	30.06	0.0082
Pluto	39.44	0.2481

37. a) Find Cartesian equations for the curves $r = 2 \sin \theta$ and $r = \csc \theta$.

b) Sketch the curves together and label their points of intersection in both Cartesian and polar coordinates.

38. Repeat Exercise 37 for $r = 2 \cos \theta$ and $r = \sec \theta$.

39. Find a polar equation for the parabola whose focus lies at the origin and whose directrix is the line $r \cos \theta = 4$.

40. Find a polar equation for the parabola whose focus lies at the origin and whose directrix is the line $r \cos(\theta - \pi/2) = 2$.

Computer Grapher

Graph the lines and conic sections in Exercises 41–50.

41. $r = 3 \sec(\theta - \pi/3)$

42. $r = 4 \sec(\theta + \pi/6)$

43. $r = 4 \sin \theta$

44. $r = -2 \cos \theta$

45. $r = 8/(4 + \cos \theta)$

46. $r = 8/(4 + \sin \theta)$

47. $r = 1/(1 - \sin \theta)$

48. $r = 1/(1 + \cos \theta)$

49. $r = 1/(1 + 2 \sin \theta)$

50. $r = 1/(1 + 2 \cos \theta)$

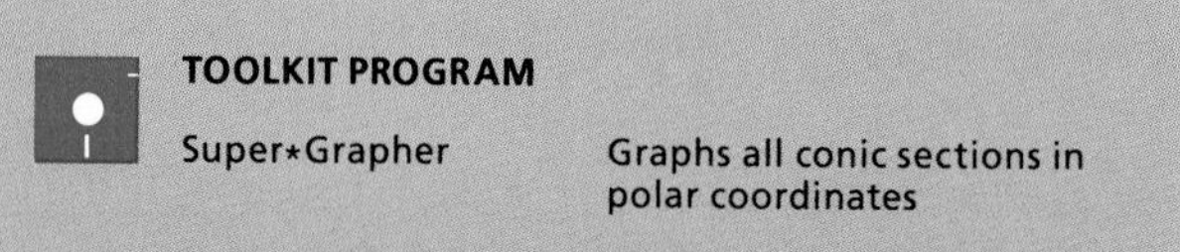

10.8 Integration in Polar Coordinates

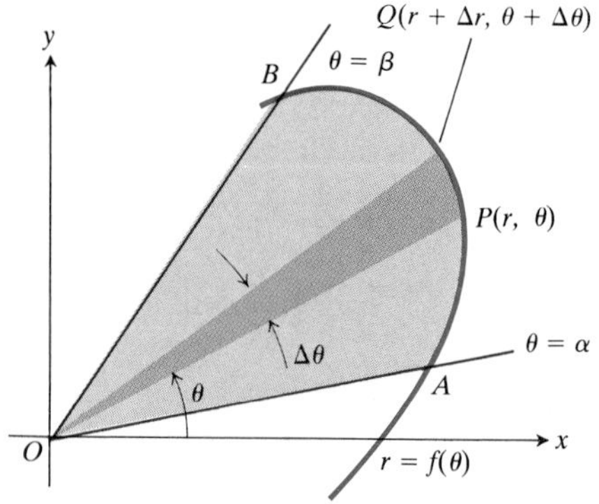

10.59 To derive a formula for the area swept out by the radius *OP* as *P* moves from *A* to *B* along the curve, we divide the area into sectors.

This section shows how to calculate areas of plane regions, lengths of curves, and areas of surfaces of revolution in polar coordinates. The general methods for setting up the integrals are the same as for Cartesian coordinates, although the resulting formulas are somewhat different. For anyone interested in finding centroids, we have included the formulas for doing so at the end of the exercise set.

Area in the Plane

The region *AOB* in Fig. 10.59 is bounded by the rays $\theta = \alpha$, $\theta = \beta$, and the curve $r = f(\theta)$. We divide angle *AOB* into n parts and approximate a typical sector *POQ* by a *circular* sector of radius r and central angle $\Delta\theta$ (Fig. 10.60). Then

$$\text{Area of } POQ \approx \frac{1}{2} r^2 \, \Delta\theta,$$

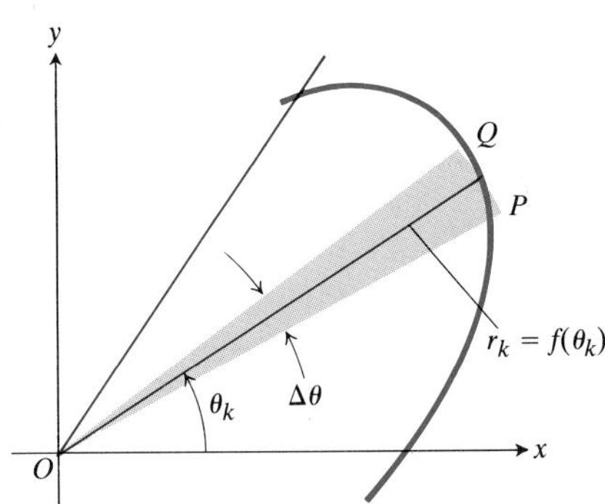

10.60 For some θ_k between θ and $\theta + \Delta\theta$, the area of the shaded circular sector just equals the area of the sector POQ bounded by the curve shown in Fig. 10.59.

and

$$A = \text{Area } AOB \approx \sum_{\theta=\alpha}^{\beta} \frac{1}{2} r^2 \, \Delta\theta.$$

If the function $r = f(\theta)$ is continuous for $\alpha \le \theta \le \beta$, then there is a θ_k between θ and $\theta + \Delta\theta$ such that the circular sector of radius

$$r_k = f(\theta_k)$$

and central angle $\Delta\theta$ gives the *exact* area of POQ (Fig. 10.60). Then the entire area is given exactly by

$$A = \sum \frac{1}{2} r_k^2 \, \Delta\theta = \sum \frac{1}{2} (f(\theta_k))^2 \, \Delta\theta.$$

If we let $\Delta\theta \to 0$, we see that

$$A = \lim_{\Delta\theta \to 0} \sum \frac{1}{2} (f(\theta_k))^2 \, \Delta\theta = \int_\alpha^\beta \frac{1}{2} f^2(\theta) \, d\theta = \int_\alpha^\beta \frac{1}{2} r^2 \, d\theta.$$

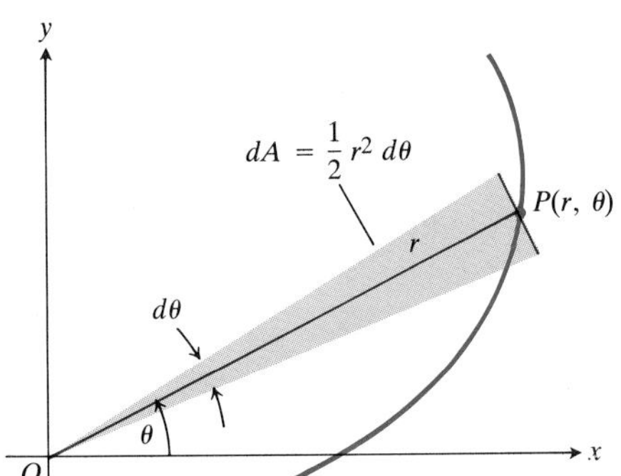

10.61 The area differential dA.

Area Between the Origin and $r = f(\theta)$, $\alpha \le \theta \le \beta$

$$A = \int_\alpha^\beta \frac{1}{2} r^2 \, d\theta. \tag{1}$$

This is the integral of the **area differential** (Fig. 10.61)

$$dA = \frac{1}{2} r^2 \, d\theta. \tag{2}$$

EXAMPLE 1 Find the area of the region enclosed by the cardioid $r = 2(1 + \cos\theta)$.

Solution We graph the cardioid (Fig. 10.62) and determine that the radius OP sweeps out the region exactly once as θ runs from 0 to 2π. The area is therefore

$$\begin{aligned}
\int_{\theta=0}^{\theta=2\pi} \frac{1}{2} r^2 \, d\theta &= \int_0^{2\pi} \frac{1}{2} \cdot 4(1 + \cos\theta)^2 \, d\theta \\
&= \int_0^{2\pi} 2(1 + 2\cos\theta + \cos^2\theta) \, d\theta \\
&= \int_0^{2\pi} \left(2 + 4\cos\theta + 2\frac{1 + \cos 2\theta}{2}\right) d\theta \\
&= \int_0^{2\pi} (3 + 4\cos\theta + \cos 2\theta) \, d\theta \\
&= 3\theta + 4\sin\theta + \frac{\sin 2\theta}{2}\bigg]_0^{2\pi} \\
&= 6\pi - 0 = 6\pi.
\end{aligned}$$

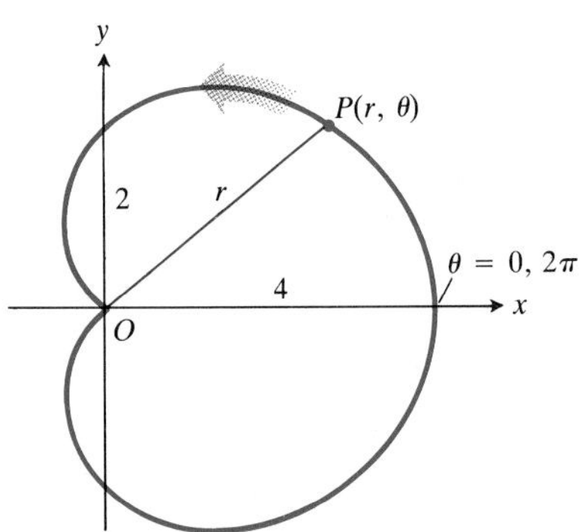

10.62 The cardioid $r = 2(1 + \cos\theta)$.

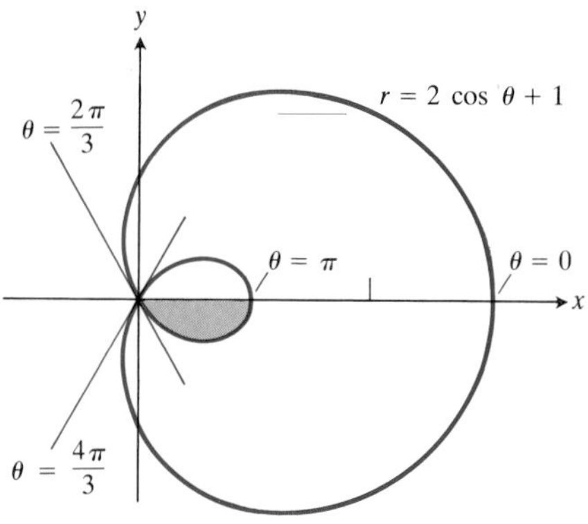

10.63 The limaçon in Example 2.

EXAMPLE 2 Find the area inside the smaller loop of the limaçon

$$r = 2\cos\theta + 1.$$

Solution After sketching the curve (Fig. 10.63), we see that the smaller loop is traced out by the point (r, θ) as θ increases from $\theta = 2\pi/3$ to $\theta = 4\pi/3$. Since the curve is symmetric about the x-axis (the equation is unaltered when we replace θ by $-\theta$), we may calculate the area of the shaded half of the inner loop by integrating from $\theta = 2\pi/3$ to $\theta = \pi$. The area A we seek will be twice the value of the resulting integral:

$$A = 2\int_{2\pi/3}^{\pi} \frac{1}{2}r^2\,d\theta = \int_{2\pi/3}^{\pi} r^2\,d\theta.$$

Since

$$\begin{aligned} r^2 &= (2\cos\theta + 1)^2 \\ &= 4\cos^2\theta + 4\cos\theta + 1 \\ &= 4\cdot\frac{1+\cos 2\theta}{2} + 4\cos\theta + 1 \\ &= 2 + 2\cos 2\theta + 4\cos\theta + 1 \\ &= 3 + 2\cos 2\theta + 4\cos\theta, \end{aligned}$$

we have

$$\begin{aligned} A &= \int_{2\pi/3}^{\pi} (3 + 2\cos 2\theta + 4\cos\theta)\,d\theta \\ &= \Big[3\theta + \sin 2\theta + 4\sin\theta\Big]_{2\pi/3}^{\pi} \\ &= (3\pi) - \left(2\pi - \frac{\sqrt{3}}{2} + 4\cdot\frac{\sqrt{3}}{2}\right) \\ &= \pi - \frac{3\sqrt{3}}{2}. \end{aligned}$$

■

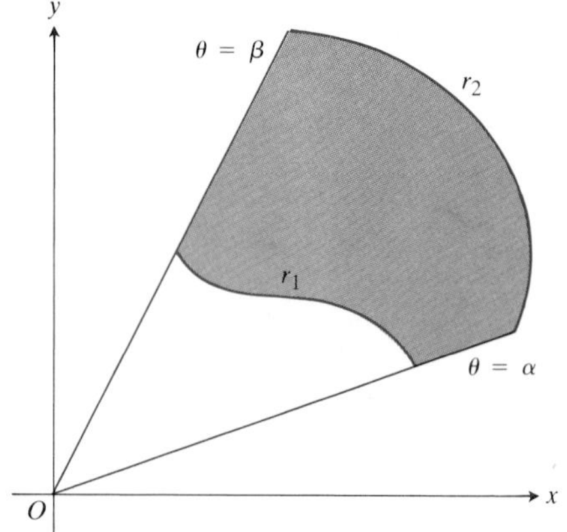

10.64 The area of the shaded region is calculated by subtracting the area of the region between r_1 and the origin from the area of the region between r_2 and the origin.

To find the area of a region like the one in Fig. 10.64, which lies between two polar curves from $\theta = \alpha$ to $\theta = \beta$, we subtract the integral of $(1/2)r_1^2\,d\theta$ from the integral of $(1/2)r_2^2\,d\theta$. This leads to the following formula.

Area of the Region $r_1(\theta) \le r \le r_2(\theta)$, $\alpha \le \theta \le \beta$

$$A = \int_{\alpha}^{\beta} \frac{1}{2}r_2^2\,d\theta - \int_{\alpha}^{\beta} \frac{1}{2}r_1^2\,d\theta = \int_{\alpha}^{\beta} \frac{1}{2}(r_2^2 - r_1^2)\,d\theta \qquad (3)$$

EXAMPLE 3 Find the area of the region that lies inside the circle $r = 1$ and outside the cardioid $r = 1 - \cos\theta$.

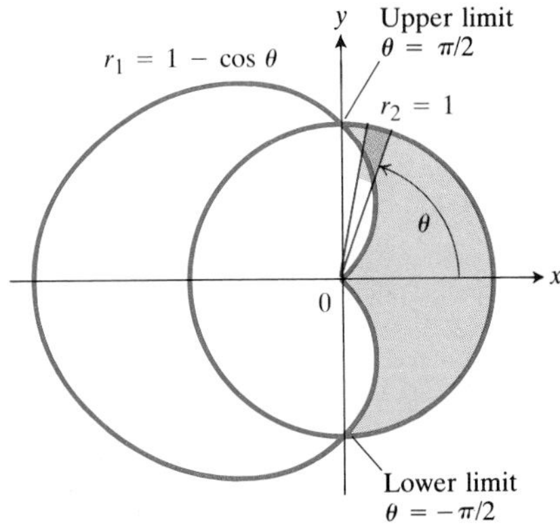

10.65 The region and limits of integration in Example 3.

Solution We sketch the region to determine its boundaries and find the limits of integration (Fig. 10.65). The outer curve is $r_2 = 1$, the inner curve is $r_1 = 1 - \cos\theta$, and θ runs from $-\pi/2$ to $\pi/2$. The area, from Eq. (3), is

$$\begin{aligned}
A &= \int_{-\pi/2}^{\pi/2} \frac{1}{2}(r_2^2 - r_1^2)\, d\theta \\
&= 2\int_0^{\pi/2} \frac{1}{2}(r_2^2 - r_1^2)\, d\theta \qquad \text{(Symmetry)} \\
&= \int_0^{\pi/2} (1 - (1 - 2\cos\theta + \cos^2\theta))\, d\theta \\
&= \int_0^{\pi/2} (2\cos\theta - \cos^2\theta)\, d\theta = \int_0^{\pi/2} \left(2\cos\theta - \frac{\cos 2\theta + 1}{2}\right) d\theta \\
&= \left[2\sin\theta - \frac{\sin 2\theta}{4} - \frac{\theta}{2}\right]_0^{\pi/2} = 2 - \frac{\pi}{4}.
\end{aligned}$$

■

The Length of a Curve

We calculate the length of a curve $r = f(\theta)$, $\alpha \le \theta \le \beta$ by expressing the differential $ds = \sqrt{dx^2 + dy^2}$ in terms of θ and integrating from α to β. To express ds in terms of θ, we first write dx and dy as

$$\begin{aligned}
dx &= d(r\cos\theta) = -r\sin\theta\, d\theta + \cos\theta\, dr, \\
dy &= d(r\sin\theta) = r\cos\theta\, d\theta + \sin\theta\, dr.
\end{aligned} \tag{4}$$

We then square and add (arithmetic omitted) to obtain

$$ds = \sqrt{dx^2 + dy^2} = \sqrt{r^2\, d\theta^2 + dr^2}. \tag{5}$$

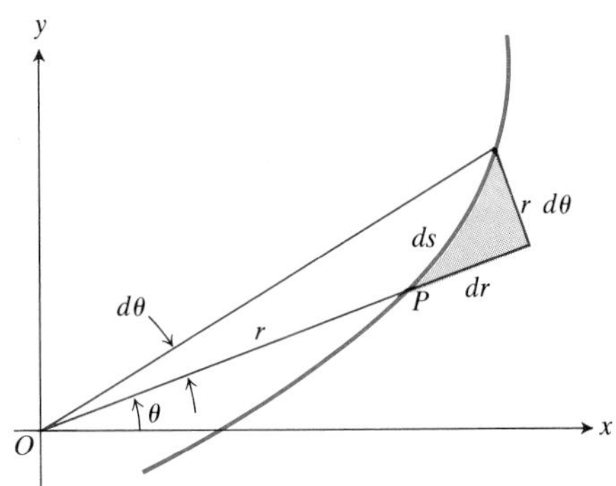

10.66 For arc length, $ds^2 = r^2\, d\theta^2 + dr^2$.

Think of ds as the hypotenuse of a right triangle whose sides are $r\, d\theta$ and dr (Fig. 10.66). For the purpose of evaluation, Eq. (5) is usually written with $d\theta$ factored out:

$$ds = \sqrt{r^2 + \left(\frac{dr}{d\theta}\right)^2}\, d\theta. \tag{6}$$

Length of a Curve

If $r = f(\theta)$ has a continuous first derivative for $\alpha \le \theta \le \beta$ and if the point $P(r, \theta)$ traces the curve $r = f(\theta)$ exactly once as θ runs from α to β, then the length of the curve is given by the formula

$$\text{Length} = \int_\alpha^\beta \sqrt{r^2 + \left(\frac{dr}{d\theta}\right)^2}\, d\theta. \tag{7}$$

EXAMPLE 4 Find the length of the cardioid $r = 1 - \cos\theta$.

Solution We sketch the cardioid to determine the limits of integration (Fig. 10.67). The point $P(r, \theta)$ starts at the origin and traces the curve once,

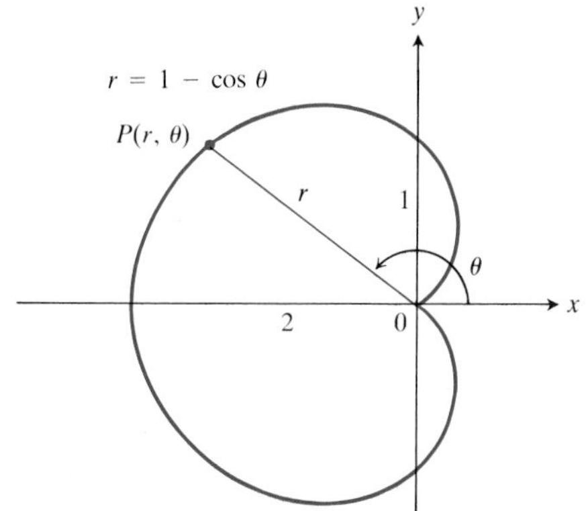

10.67 Example 4 calculates the length of this cardioid.

counterclockwise, as θ runs from 0 to 2π, so these are the values we take for α and β.

With

$$r = 1 - \cos\theta, \qquad \frac{dr}{d\theta} = \sin\theta,$$

we have

$$\begin{aligned} r^2 + \left(\frac{dr}{d\theta}\right)^2 &= (1 - \cos\theta)^2 + (\sin\theta)^2 \\ &= 1 - 2\cos\theta + \underbrace{\cos^2\theta + \sin^2\theta}_{1} \\ &= 2 - 2\cos\theta \end{aligned}$$

and

$$\begin{aligned} \text{Length} &= \int_\alpha^\beta \sqrt{r^2 + \left(\frac{dr}{d\theta}\right)^2}\, d\theta = \int_0^{2\pi} \sqrt{2 - 2\cos\theta}\, d\theta \\ &= \int_0^{2\pi} \sqrt{4\sin^2\frac{\theta}{2}}\, d\theta \qquad \left(1 - \cos\theta = 2\sin^2\frac{\theta}{2}\right) \\ &= \int_0^{2\pi} 2\left|\sin\frac{\theta}{2}\right| d\theta \\ &= \int_0^{2\pi} 2\sin\frac{\theta}{2}\, d\theta \qquad \left(\sin\frac{\theta}{2} \ge 0 \text{ for } 0 \le \theta \le 2\pi\right) \\ &= \left[-4\cos\frac{\theta}{2}\right]_0^{2\pi} = [4 + 4] = 8. \end{aligned}$$

■

The Area of a Surface of Revolution

The formula for the area of a surface of revolution is $S = \int 2\pi\rho\, ds$, just as in rectangular coordinates, but now we express the radius function ρ and the arc length differential ds in terms of r and θ.

Area of a Surface of Revolution

If $r = f(\theta)$ has a continuous first derivative for $\alpha \le \theta \le \beta$ and if the point $P(r, \theta)$ traces the curve $r = f(\theta)$ exactly once as θ runs from α to β, then the area of the surfaces generated by revolving the curve about the x- and y-axes are given by the following formulas:

1. Revolution about the x-axis:

$$\text{Area} = \int_\alpha^\beta 2\pi y\, ds = \int_\alpha^\beta 2\pi r\sin\theta \sqrt{r^2 + \left(\frac{dr}{d\theta}\right)^2}\, d\theta \qquad (8)$$

2. Revolution about the y-axis:

$$\text{Area} = \int_\alpha^\beta 2\pi x\, ds = \int_\alpha^\beta 2\pi r\cos\theta \sqrt{r^2 + \left(\frac{dr}{d\theta}\right)^2}\, d\theta \qquad (9)$$

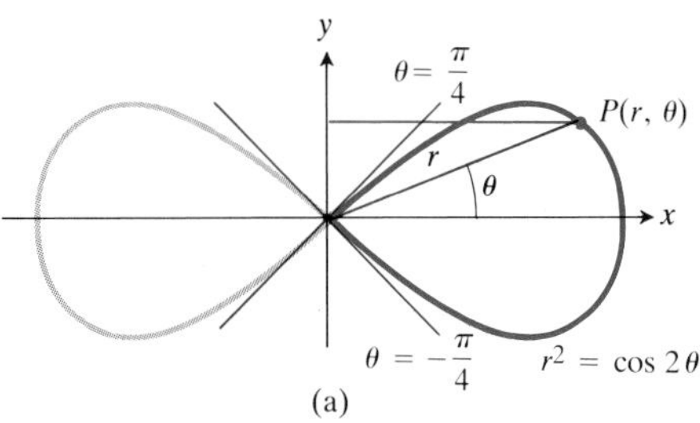

(b)

10.68 The right-hand half of a lemniscate (a) is revolved about the y-axis to generate a surface (b), whose area is calculated in Example 5.

EXAMPLE 5 Find the area of the surface generated by revolving the right-hand loop of the lemniscate $r^2 = \cos 2\theta$ about the y-axis.

Solution We sketch the loop to determine the limits of integration (Fig. 10.68). The point $P(r, \theta)$ traces the curve once, counterclockwise, as θ runs from $-\pi/4$ to $\pi/4$, so these are the values we take for α and β.

We evaluate the surface area integrand

$$2\pi r \cos\theta \sqrt{r^2 + \left(\frac{dr}{d\theta}\right)^2} = 2\pi \cos\theta \sqrt{r^4 + \left(r\frac{dr}{d\theta}\right)^2} \tag{10}$$

in stages. First of all, $r^2 = \cos 2\theta$, so

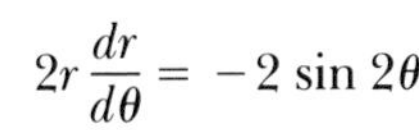

$$2r\frac{dr}{d\theta} = -2\sin 2\theta$$

$$r\frac{dr}{d\theta} = -\sin 2\theta$$

$$\left(r\frac{dr}{d\theta}\right)^2 = \sin^2 2\theta.$$

Since

$$r^4 = (r^2)^2 = \cos^2 2\theta,$$

the square root on the right-hand side of (10) simplifies to

$$\sqrt{r^4 + \left(r\frac{dr}{d\theta}\right)^2} = \sqrt{\cos^2 2\theta + \sin^2 2\theta} = 1.$$

Hence,

$$\begin{aligned} \text{Surface area} &= \int_{\alpha}^{\beta} 2\pi r \cos\theta \sqrt{r^2 + \left(\frac{dr}{d\theta}\right)^2}\, d\theta \\ &= \int_{-\pi/4}^{\pi/4} 2\pi \cos\theta \cdot (1)d\theta \\ &= 2\pi\Big[\sin\theta\Big]_{-\pi/4}^{\pi/4} = 2\pi\left[\frac{\sqrt{2}}{2} + \frac{\sqrt{2}}{2}\right] = 2\pi\sqrt{2}. \end{aligned}$$

■

Exercises 10.8

Find the areas of the regions in Exercises 1–18.

1. Inside the circle $r = \cos\theta$ between the rays $\theta = 0$ and $\theta = \pi/4$
2. Enclosed by the rays $\theta = 0$ and $\theta = \ln 25$ and the curve $r = e^{\theta}$, $0 \le \theta \le \ln 25$
3. Inside the convex limaçon $r = 4 + 2\cos\theta$
4. Inside the cardioid $r = a(1 + \cos\theta)$
5. Inside one leaf of the four-leaved rose $r = \cos 2\theta$
6. Inside the circle $r = 2a \sin\theta$
7. Inside the lemniscate $r^2 = 2a^2 \cos 2\theta$
8. Inside one loop of the lemniscate $r^2 = 4 \sin 2\theta$
9. Inside the six-leaved rose $r^2 = 2 \sin 3\theta$ (Fig. 10.69)

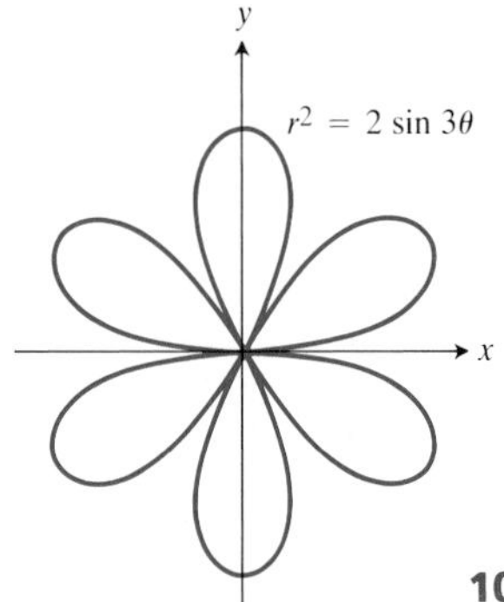

10.69 The rose in Exercise 9.

10. Shared by the circles $r = 2 \cos \theta$ and $r = 2 \sin \theta$

11. Shared by the circles $r = 1$ and $r = 2 \sin \theta$

12. Shared by the circle $r = 2$ and the cardioid $r = 2(1 - \cos \theta)$

13. Shared by the cardioids $r = 2(1 + \cos \theta)$ and $r = 2(1 - \cos \theta)$

14. Inside the lemniscate $r^2 = 6 \cos 2\theta$ and outside the circle $r = \sqrt{3}$

15. Inside the circle $r = 3a \cos \theta$ and outside the cardioid $r = a(1 + \cos \theta)$

16. Inside the circle $r = -2 \cos \theta$ and outside the circle $r = 1$

17. a) Inside the outer loop of the limaçon $r = 2 \cos \theta + 1$, in Example 2
b) Inside the outer loop and outside the inner loop

18. Inside the circle $r = 6$ above the line $r \sin \theta = 3$

Find the lengths of the curves in Exercises 19–23.

19. The spiral $r = \theta^2$, $0 \le \theta \le \sqrt{5}$

20. The curve $r = e^{\theta}/\sqrt{2}$, $0 \le \theta \le \pi$

21. The curve $r = \sec \theta$, $0 \le \theta \le \pi/4$

22. The curve $r = \csc \theta$, $\pi/6 \le \theta \le \pi/2$

23. The cardioid $r = 1 + \cos \theta$

24. As usual, when faced with a new formula, it is a good idea to try it out on familiar objects to be sure it gives results consistent with past experience. Use the length formula in Eq. (7) to calculate the circumferences of the following circles:
a) $r = a$
b) $r = a \cos \theta$
c) $r = a \sin \theta$

Find the areas of the surfaces generated by revolving the curves in Exercises 25–28 about the indicated axes.

25. $r = \sqrt{\cos 2\theta}$, $0 \le \theta \le \pi/4$, y-axis

26. $r = \sqrt{2}\, e^{\theta/2}$, $0 \le \theta \le \pi/2$, x-axis

27. $r^2 = \cos 2\theta$, x-axis

28. $r = 2a \cos \theta$, y-axis

Centroids

Since the centroid of a triangle is located on each median, two-thirds of the way from the vertex to the opposite base, the lever arm for the moment about the x-axis of the thin triangular region in Fig. 10.70 is about $(2/3)r \sin \theta$. Similarly, the lever arm for the moment of the triangular region about the y-axis is about $(2/3)r \cos \theta$.

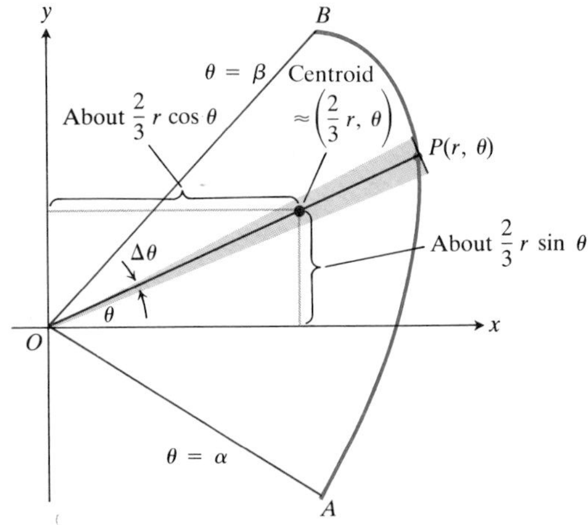

10.70 The moment of the thin triangular sector about the x-axis is approximately

$$\frac{2}{3}r \sin \theta \, dA = \frac{2}{3}r \sin \theta \cdot \frac{1}{2}r^2 d\theta = \frac{1}{3}r^3 \sin \theta \, d\theta.$$

These approximations improve as $\Delta\theta \to 0$ and lead to the following formulas for the coordinates of the centroid of region AOB:

$$\bar{x} = \frac{\int \frac{2}{3}r \cos \theta \cdot \frac{1}{2}r^2 d\theta}{\int \frac{1}{2}r^2 d\theta} = \frac{\frac{2}{3}\int r^3 \cos \theta \, d\theta}{\int r^2 d\theta},$$

$$\bar{y} = \frac{\int \frac{2}{3}r \sin \theta \cdot \frac{1}{2}r^2 d\theta}{\int \frac{1}{2}r^2 d\theta} = \frac{\frac{2}{3}\int r^3 \sin \theta \, d\theta}{\int r^2 d\theta},$$

with limits $\theta = \alpha$ to $\theta = \beta$ on all integrals.

29. Find the centroid of the region enclosed by the cardioid $r = a(1 + \cos \theta)$.

30. Find the centroid of the semicircular region $0 \le r \le a$, $0 \le \theta \le \pi$.

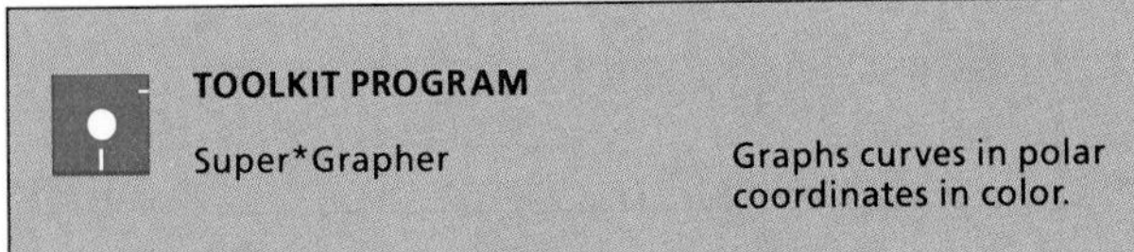

Review Questions

1. Name the conic sections. Where does *conic* come from?
2. What reflective properties do parabolas, ellipses, and hyperbolas have? What applications use these properties?
3. How are ellipses defined in terms of distance? Give typical equations for ellipses. Graph one of the equations and include the ellipse's vertices and foci. How is the eccentricity of an ellipse defined? Sketch an ellipse whose eccentricity is close to 1 and another whose eccentricity is close to 0.
4. How are hyperbolas defined in terms of distance? Give typical equations for hyperbolas. Graph one of the equations and include the hyperbola's vertices, foci, axes, and asymptotes. What values can the eccentricities of hyperbolas have?
5. Explain the equation $PF = e \cdot PD$.
6. What can be said about the graph of the equation
$$Ax^2 + Bxy + Cy^2 + Dx + Ey + F = 0.$$
if A, B, and C are not all zero? What can you tell from the number $B^2 - 4AC$?
7. How do you find a coordinate system in which the new equation for a conic section has no xy-term? Give an example.
8. Give parametric equations for a circle, parabola, and ellipse. In each case, give a parameter domain that covers the conic exactly once.
9. Give parametric equations for one branch of a hyperbola. What is the appropriate domain for the parameter if the curve is to be traced out exactly once by your equations?
10. What is a cycloid? What are typical parametric equations for a cycloid?
11. How do you find dy/dx and d^2y/dx^2 when a curve in the xy-plane is given by parametric equations. Give examples.
12. How do you find the length of a parametric curve? Give an example.
13. How do you find the area of a surface of revolution generated by a parametric curve? Give an example.
14. Make a diagram to show the standard relations between Cartesian coordinates (x, y) and polar coordinates (r, θ). Express each set of coordinates in terms of the other kind.
15. If a point has polar coordinates (r_0, θ_0), what other polar coordinates does the point have?
16. How do you test the graph of the equation $F(r, \theta) = 0$ for symmetry about the origin? About the x-axis? About the y-axis? Give examples.
17. Describe a technique for graphing the equation $r = f(\theta)$ that involves the Cartesian $r\theta$-plane. Give an example.
18. What are the standard equations for lines, circles, ellipses, parabolas, and hyperbolas in polar coordinates? Give examples.
19. How do you find the areas of plane regions in polar coordinates? Give examples.
20. How do you find the length of a curve in polar coordinates? Give an example.
21. How do you find the area of a surface of revolution in polar coordinates? Give an example.

Practice Exercises

Sketch the parabolas in Exercises 1–2. Include the parabola's focus and directrix in each sketch.

1. $x = y^2/8$

2. $y = -x^2/4$

Sketch the ellipses and hyperbolas in Exercises 3–6. Include the vertices, foci, and directrices in your sketches.

3. $16x^2 + 7y^2 = 112$

4. $x^2 + 2y^2 = 4$

5. $3x^2 - y^2 = 3$ **6.** $2y^2 - 8x^2 = 16$

Use the discriminant to decide whether the equations in Exercises 7–10 represent parabolas, ellipses, or hyperbolas.

7. $x^2 + xy + y^2 + x + y + 1 = 0$

8. $x^2 + 3xy + 2y^2 + x + y + 1 = 0$

9. $x^2 + 4xy + 4y^2 + x + y + 1 = 0$

10. $x^2 + 2xy - 2y^2 + x + y + 1 = 0$

Identify the conic sections in Exercises 11–12. Then rotate the coordinate axes to find a new equation for the conic section that has no cross-product term. (The new equations will vary with the size and direction of the rotations you choose.)

11. $2x^2 + xy + 2y^2 - 15 = 0$

12. $x^2 + 2\sqrt{3}xy - y^2 + 4 = 0$

13. Find the eccentricity of the hyperbola $xy = 2$. (*Hint:* Rotate the coordinate axes to eliminate the xy-term. Then get the information you need from the new equation.)

14. Find an equation for the hyperbola with eccentricity 2 and vertices $(2, 0)$ and $(-2, 0)$.

15. Find the volume of the solid generated by revolving the region enclosed by the ellipse $9x^2 + 4y^2 = 36$ about the

a) x-axis b) y-axis.

16. The "triangular" region in the first quadrant bounded by the x-axis, the line $x = 4$, and the hyperbola $9x^2 - 4y^2 = 36$ is revolved about the x-axis to generate a solid. Find the volume of the solid.

17. Find the points on the parabola $x = 2t, y = t^2$ closest to the point $(0, 3)$.

18. *Archimedes' area formula for parabolic arches.* Show that the area of the shaded region in Fig. 10.71 is $A = (2/3)bh$. Thus, the area under a parabolic arch is two-thirds the base times the height.

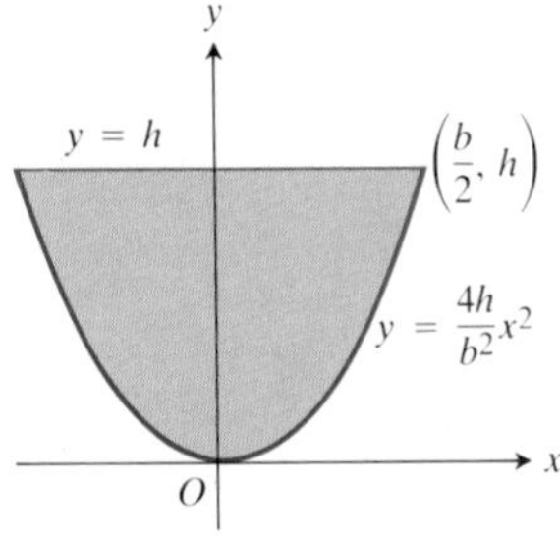

10.71 The parabolic region in Exercise 18.

Exercises 19–24 give parametric equations for the motion of a particle in the xy-plane. Identify the particle's path by finding a Cartesian equation for it. Graph the Cartesian equation and indicate the direction of motion and the portion traced by the particle.

19. $x = t/2,\ y = t + 1;\quad -\infty < t < \infty$

20. $x = \sqrt{t},\ y = 1 - \sqrt{t};\quad t \geq 0$

21. $x = (1/2)\tan t,\ y = (1/2)\sec t;\quad -\pi/2 < t < \pi/2$

22. $x = -2\cos t,\ y = 2\sin t;\quad 0 \leq t \leq \pi$

23. $x = -\cos t,\ y = \cos^2 t;\quad 0 \leq t \leq \pi$

24. $x = 4\cos t,\ y = 9\sin t;\quad 0 \leq t \leq 2\pi$

25. Find parametric equations for the motion of a particle in the xy-plane that traces the ellipse $16x^2 + 9y^2 = 144$ once counterclockwise. (There are many ways to do this, so your answer may not be the same as the one in the back of the book.)

26. Find parametric equations for the motion of a particle that starts at the point $(-2, 0)$ in the xy-plane and traces the circle $x^2 + y^2 = 4$ three times clockwise. (There are many ways to do this.)

In Exercises 27–28, find an equation for the line in the xy-plane that is tangent to the curve at the point corresponding to the given value of t. Also, find the value of d^2y/dx^2 at this point.

27. $x = (1/2)\tan t,\ y = (1/2)\sec t;\quad t = \pi/3$

28. $x = 1 + 1/t^2,\ y = 1 - 3/t;\quad t = 2$

Find the lengths of the curves in Exercises 29–30.

29. $x = e^{2t} - \frac{t}{8},\ y = e^t;\quad 0 \leq t \leq \ln 2$

30. The closed loop in Fig. 10.72.

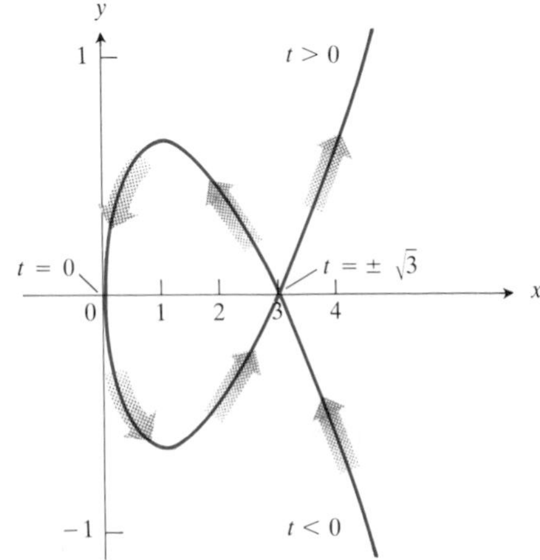

10.72 Exercise 30 refers to the curve $x = t^2, y = (t^3/3) - t$ shown here. The loop starts at $t = -\sqrt{3}$ and ends at $t = \sqrt{3}$.

Find the areas of the surfaces generated by revolving the curves in Exercises 31–32 about the indicated axes.

31. $x = t^2/2,\ y = 2t,\ 0 \le t \le \sqrt{5};\quad x\text{-axis}$

32. $x = t^2 + 1/(2t),\ y = 4\sqrt{t},\ 1/\sqrt{2} \le t \le 1;\quad y\text{-axis}$
(*Hint:* $(dx/dt)^2 + (dy/dt)^2$ is a perfect square.)

Each of the graphs in Exercises 33–40 is the graph of one of the equations (a)–(l) listed below. Find the equation for each graph.

a) $r = \cos 2\theta$ b) $r \cos \theta = 1$

c) $r = \dfrac{6}{1 - 2\cos\theta}$ d) $r = \sin 2\theta$

e) $r = \theta$ f) $r^2 = \cos 2\theta$

g) $r = 1 + \cos\theta$ h) $r = 1 - \sin\theta$

i) $r = \dfrac{2}{1 - \cos\theta}$ j) $r^2 = \sin 2\theta$

k) $r = -\sin\theta$ l) $r = 2\cos\theta + 1$

33. Four-leaved rose

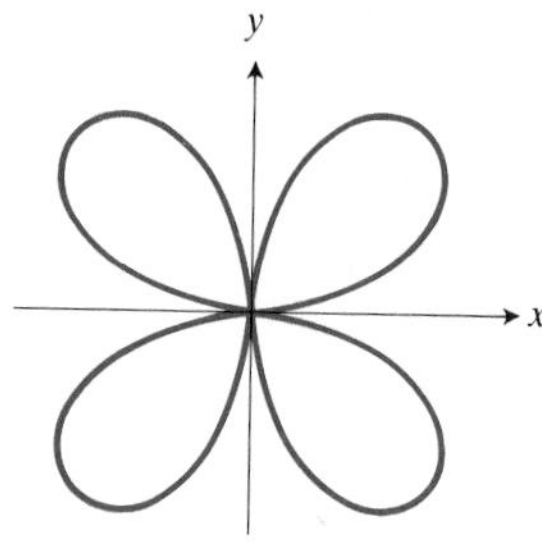

34. Spiral

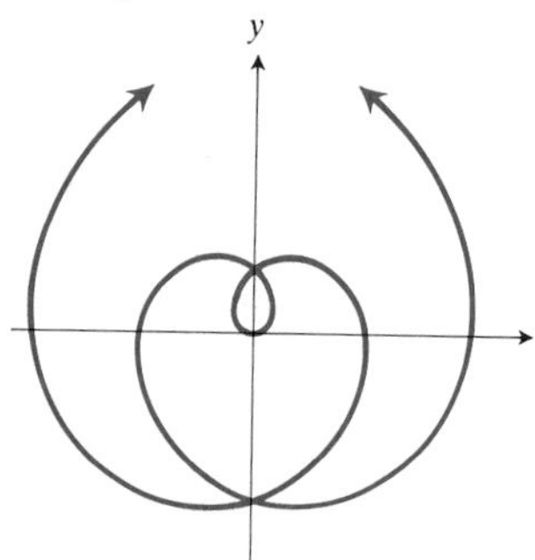

35. Limaçon

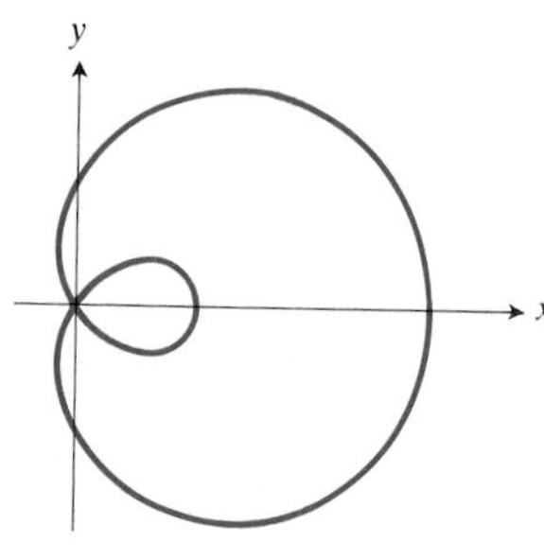

36. Lemniscate

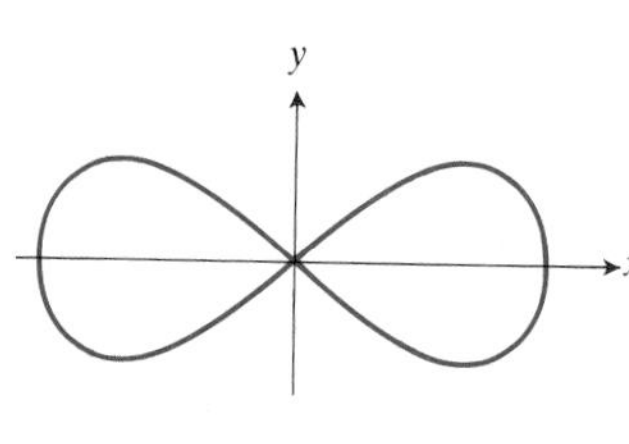

37. Circle

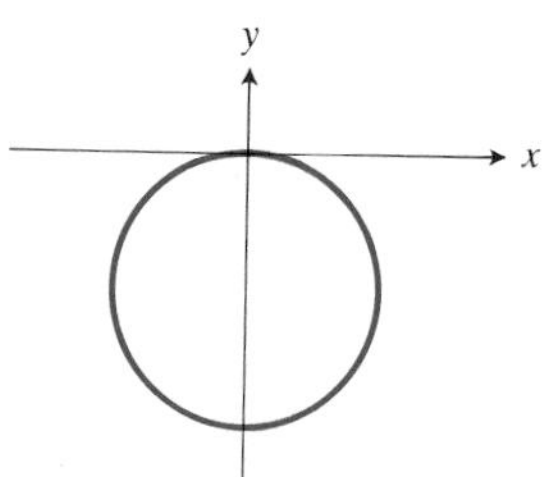

38. Cardioid

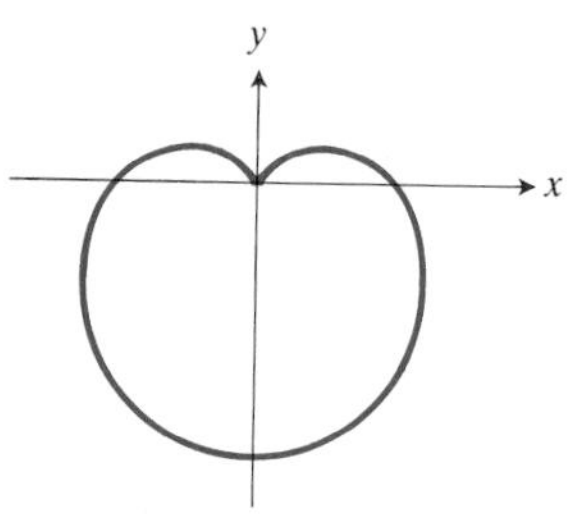

39. Parabola

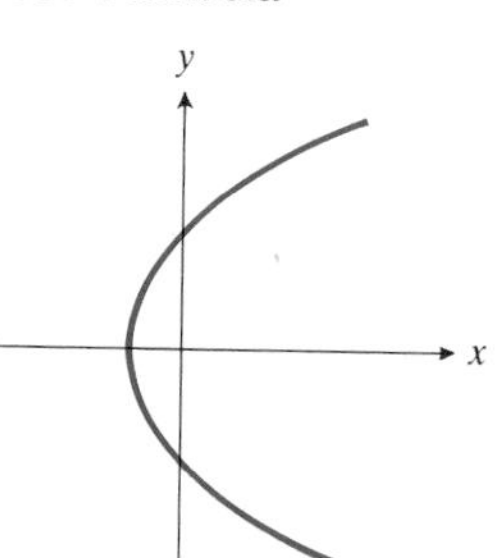

40. Lemniscate

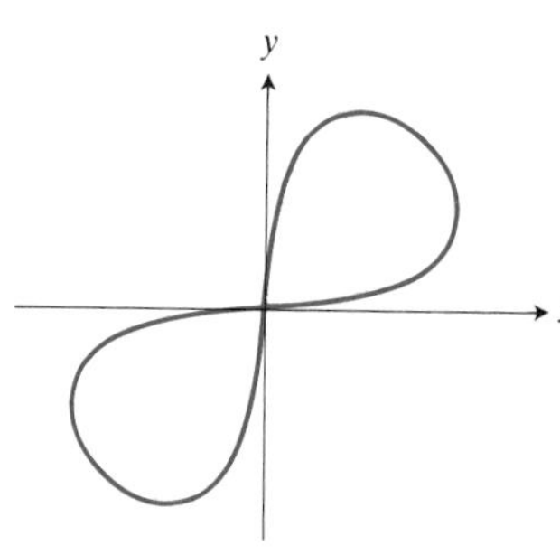

Sketch the lines given by the polar coordinate equations in Exercises 41–42 and find Cartesian equations for them.

41. $r \cos\left(\theta - \dfrac{\pi}{3}\right) = 2\sqrt{3}$

42. $r \cos\left(\theta - \dfrac{3\pi}{4}\right) = \sqrt{2}/2$

Find the centers and radii of the circles given by the polar coordinate equations in Exercises 43–44.

43. $r = 2\sin\theta$

44. $r = -4\cos\theta$

Sketch the regions defined by the polar coordinate inequalities in Exercises 45–46.

45. $0 \le r \le 6\cos\theta$

46. $-4\sin\theta \le r \le 0$

Find the points of intersection of the curves given by the polar coordinate equations in Exercises 47–50.

47. $r = \sin\theta,\ r = 1 + \sin\theta$

48. $r = \cos\theta,\ r = 1 - \cos\theta$

49. $r = 1 + \sin\theta,\ r = -1 + \sin\theta$

50. $r = 1 + \cos\theta,\ r = -1 - \cos\theta$

Sketch the conic sections whose polar coordinate equations are given in Exercises 51–54. Give polar coordinates for the vertices and, in the case of ellipses, for the centers as well.

51. $r = \dfrac{2}{1 + \cos\theta}$

52. $r = \dfrac{8}{2 + \cos\theta}$

53. $r = \dfrac{6}{1 - 2\cos\theta}$

54. $r = \dfrac{12}{3 + \sin\theta}$

Exercises 55–58 give the eccentricities of conic sections with one focus at the origin of the polar coordinate plane, along with the directrix for that focus. Find a polar equation for each conic section.

55. $e = 2,\ r\cos\theta = 2$

56. $e = 1,\ r\cos\theta = -4$

57. $e = 1/2,\ r\sin\theta = 2$

58. $e = 1/3,\ r\sin\theta = -6$

Find equations for the lines that are tangent to the polar coordinate curves in Exercises 59–60 at the origin.

59. The lemniscate $r^2 = \cos 2\theta$

60. The limaçon $r = 2\cos\theta + 1$

61. Find polar coordinate equations for the lines that are tangent to the tips of the petals of the four-leaved rose $r = \sin 2\theta$.

62. Find polar coordinate equations for the lines that are tangent to the cardioid $r = 1 + \sin\theta$ at the points where it crosses the x-axis.

Find the areas of the regions in the polar coordinate plane described in Exercises 63–66.

63. Enclosed by the limaçon $r = 2 - \cos\theta$

64. Enclosed by one leaf of the three-leaved rose $r = \sin 3\theta$

65. Inside the two-leaved rose $r = 1 + \cos 2\theta$ and outside the circle $r = 1$

66. Inside the cardioid $r = 2(1 + \sin\theta)$ and outside the circle $r = 2\sin\theta$

Find the areas of the surfaces generated by revolving the polar coordinate curves in Exercises 67–68 about the indicated axes.

67. $r = \sqrt{\cos 2\theta}$, $0 \le \theta \le \pi/4$, about the x-axis

68. $r^2 = \sin 2\theta$, about the y-axis

Find the lengths of the curves given by the polar coordinate equations in Exercises 69–72.

69. $r = -1 + \cos\theta$

70. $r = \sin\theta$, $0 \le \theta \le \pi$

71. $r = \cos^3(\theta/3)$, $0 \le \theta \le \pi/4$

72. $r = \sqrt{1 + \sin 2\theta}$

73. *Average value.* The average value of the polar coordinate r over the curve $r = f(\theta)$, $\alpha \le \theta \le \beta$, with respect to θ is given by the formula

$$r_{av} = \frac{1}{\beta - \alpha}\int_{\alpha}^{\beta} f(\theta)\, d\theta.$$

Use this formula to find the average value of r with respect to θ over the following curves.

a) The cardioid $r = a(1 - \cos\theta)$

b) The circle $r = a$

c) The circle $r = a\cos\theta$, $-\pi/2 \le \theta \le \pi/2$

74. *Archimedes' spiral.* The graph of an equation of the form $r = a\theta$, where a is a nonzero constant, is called an Archimedes spiral. Show that such a spiral cuts any ray from the origin into congruent segments. In other words, show that the width between successive turns of the spiral remains the same.

75. *The space engineer's formula for eccentricity.* The space engineer's formula for the eccentricity of an elliptical orbit is

$$e = \frac{r_{max} - r_{min}}{r_{max} + r_{min}},$$

where r is the distance from the space vehicle to the focus of the ellipse around which it revolves. Why does the formula work? (*Hint:* You do not need to use calculus. Just think about the definition of eccentricity.)

76. *A satellite orbit.* A satellite is in an orbit that passes over the North and South Poles of the earth. When it is over the South Pole it is at the highest point of its orbit, 1000 miles above the earth's surface. Above the North Pole it is at the lowest point of its orbit, 300 miles above the earth's surface.

a) Assuming that the orbit (with reference to the earth) is an ellipse with one focus at the center of the earth, find its eccentricity. (Take the diameter of the earth to be 8000 miles.)

b) Using the north–south axis of the earth as the x-axis and the center of the earth as origin, find a polar equation for the orbit.

77. Show that the length of the polar coordinate curve $r = 2f(\theta)$, $\alpha \le \theta \le \beta$, is twice the length of the curve $r = f(\theta)$, $\alpha \le \theta \le \beta$.

78. Show that the area of the surface generated by revolving the polar coordinate curve $r = 2f(\theta)$, $\alpha \le \theta \le \beta$, about the x-axis is four times the area of the surface generated by revolving the curve $r = f(\theta)$, $\alpha \le \theta \le \beta$, about the x-axis.

11 Vectors and Analytic Geometry in Space

© Joe Sohm/Uniphoto

The reflector of this radio telescope is the base of a circular paraboloid.

OVERVIEW This chapter introduces vectors and three-dimensional coordinate systems for space. Just as the coordinate plane is the natural place to study functions of a single variable, coordinate space is the place to study functions of two variables (or even more). We establish coordinates in space by adding an axis that measures distance above and below the *xy*-plane. This builds on what we already know without forcing us to start over again.

Equations in three variables define surfaces in space the way equations in two variables define curves in the plane. We use these surfaces to graph functions (not in this chapter, but later), define regions, bound solids, describe walls of containers, and so on. In short, we use them to do all the things we do in the plane, but stepped up one dimension.

Once in space, we can model motion in three dimensions and track the positions of moving bodies with vectors. We can also calculate the directions and magnitudes of their velocities and accelerations and predict the effects of the forces we see working on them. As we shall see in the next chapter, coordinates and vectors are a powerful combination. Coordinates tell us where a moving body is, and vectors tell us what is happening to it.

11.1 Vectors in the Plane

Some of the things we measure are completely determined by their magnitudes. To record mass, length, or time, for example, we need only write down a number and name an appropriate unit of measure. But we need more than that to describe a force, displacement, or velocity, for these quantities have direction as well as magnitude. To describe a force, we need to record the direction in which it acts as well as how large it is. To describe a body's displacement, we have to say in what direction it moved as well as how far it moved. To describe a body's velocity at any given time, we have to know where the body is headed as well as how fast it is going.

Quantities that have direction as well as magnitude are usually represented by arrows that point in the direction of the action and whose lengths give the magnitude of the action in terms of a suitably chosen unit. When we describe these arrows abstractly, as directed line segments in the plane or in space,we call them *vectors*.

DEFINITIONS

A **vector** in the plane is a directed line segment. Two vectors are **equal** or **the same** if they have the same length and direction.

The arrows we use when we draw vectors are understood to represent the same vector if they have the same length, are parallel, and point the same way.

In print, vectors are usually represented by single boldface roman letters, as in **v** ("vector **v**"). The vector defined by the directed line segment from point A to point B, however, is written as $\overrightarrow{AB}$ ("vector AB"). In handwritten work it is customary to draw small arrows above all letters representing vectors. Thus the equation appearing as $\mathbf{v} = \overrightarrow{AB}$ in print would appear as $\vec{v} = \overrightarrow{AB}$ when written by hand.

EXAMPLE 1 The four arrows shown here have the same length and direction. They therefore represent the same vector, and we write

$$\overrightarrow{AB} = \overrightarrow{CD} = \overrightarrow{OP} = \overrightarrow{EF}.$$

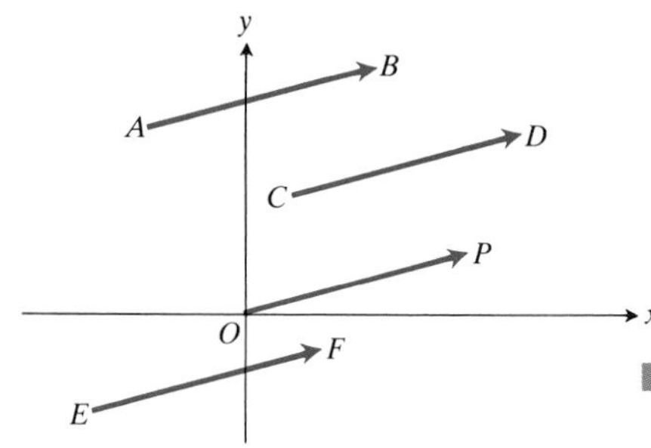

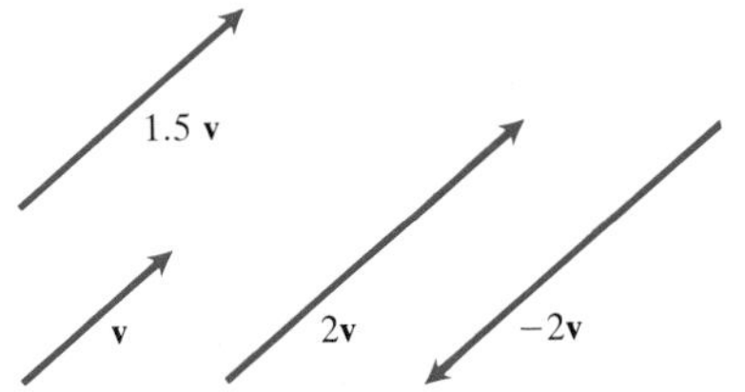

11.1 Scalar multiples of **v**.

Scalars and Scalar Multiples

We scale vectors by multiplying them by real numbers. To double a vector's length, we multiply the vector by 2. To increase a vector's length 50%, we multiply by 1.5. To reverse a vector's direction and double its length at the same time, we multiply by -2.

If c is a number and $\mathbf{v}$ a vector, the direction of $c\mathbf{v}$ agrees with $\mathbf{v}$ if c is positive and is opposite to that of $\mathbf{v}$ if c is negative (Fig. 11.1). Since real numbers work like scaling factors in this context, we tend to call them **scalars** and to call multiples like $c\mathbf{v}$ **scalar multiples.**

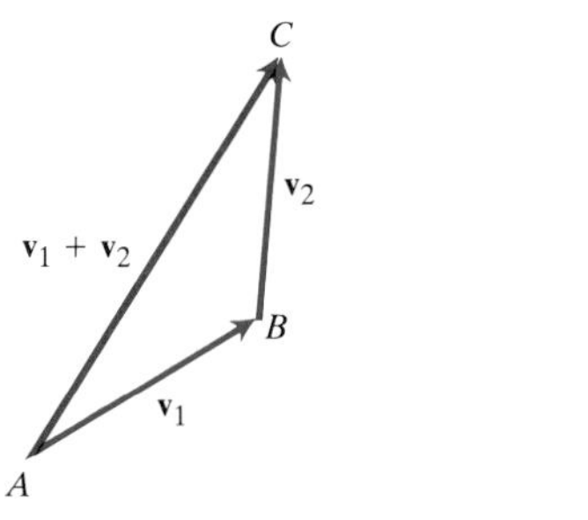

11.2 The sum of $\mathbf{v}_1$ and $\mathbf{v}_2$.

Geometric Addition: The Parallelogram Law

Two vectors $\mathbf{v}_1$ and $\mathbf{v}_2$ may be added geometrically by drawing a representative of $\mathbf{v}_1$, say from A to as B in Fig. 11.2, and then a representative of $\mathbf{v}_2$ starting from the terminal point of $\mathbf{v}_1$. In Fig. 11.2, $\mathbf{v}_2 = \overrightarrow{BC}$. The sum $\mathbf{v}_1 + \mathbf{v}_2$ is then the vector from the initial point A of $\mathbf{v}_1$ to the terminal point C of $\mathbf{v}_2$. That is, if

$$\mathbf{v}_1 = \overrightarrow{AB} \qquad \text{and} \qquad \mathbf{v}_2 = \overrightarrow{BC},$$

then

$$\mathbf{v}_1 + \mathbf{v}_2 = \overrightarrow{AB} + \overrightarrow{BC} = \overrightarrow{AC}.$$

This description of addition is sometimes called the **parallelogram law** of addition because $\mathbf{v}_1 + \mathbf{v}_2$ is given by the diagonal of the parallelogram determined by $\mathbf{v}_1$ and $\mathbf{v}_2$ (Fig. 11.3).

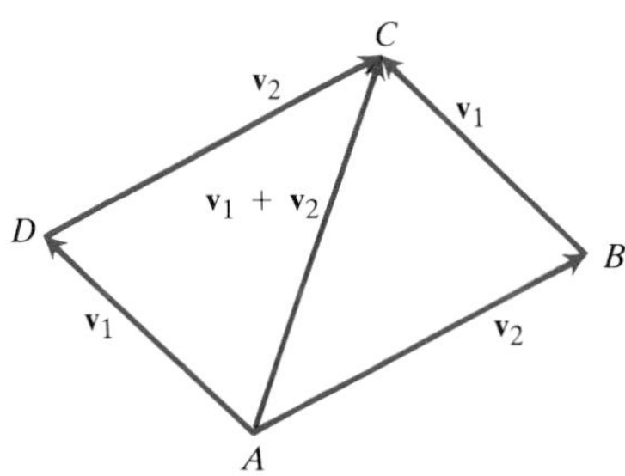

11.3 The parallelogram law of addition. Quadrilateral *ABCD* is a parallelogram because its opposite sides have equal lengths. The law was used by Aristotle to describe the combined action of two forces.

Components

Whenever a vector $\mathbf{v}$ can be written as a sum

$$\mathbf{v} = \mathbf{v}_1 + \mathbf{v}_2,$$

the vectors $\mathbf{v}_1$ and $\mathbf{v}_2$ are said to be **components** of $\mathbf{v}$. We also say that we have **represented** or **resolved v** in terms of $\mathbf{v}_1$ and $\mathbf{v}_2$.

The most common algebra of vectors is based on representing each vector in terms of components parallel to the Cartesian coordinate axes and writing each component as an appropriate multiple of a **basic** vector of unit length. The basic vector in the positive x direction is the vector $\mathbf{i}$ that runs from (0, 0) to (1, 0). The basic vector in the positive y direction is the vector $\mathbf{j}$ from (0, 0) to (0, 1). Then $a\mathbf{i}$, a being a scalar, represents a vector of length $|a|$ parallel to the x-axis, pointing to the right if a is positive and to the left if a is negative. Similarly, $b\mathbf{j}$ is a vector of length $|b|$ parallel to the y-axis, pointing up if b is positive and down if b is negative. Figure 11.4 shows a vector $\mathbf{v} = \overrightarrow{AC}$ resolved into its $\mathbf{i}$- and $\mathbf{j}$-components as the sum

$$\mathbf{v} = a\mathbf{i} + b\mathbf{j}.$$

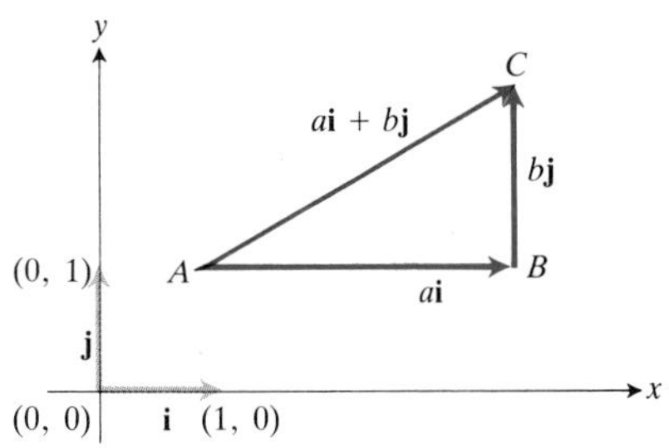

11.4 The basic vectors **i** and **j**. Any vector $\overrightarrow{AC}$ can be expressed as a multiple of **i** plus a multiple of **j**.

DEFINITION

If $\mathbf{v} = a\mathbf{i} + b\mathbf{j}$, the vectors $a\mathbf{i}$ and $b\mathbf{j}$ are the **vector components of v in the directions of i and j.** The numbers a and b are the **scalar components of v in the directions of i and j.**

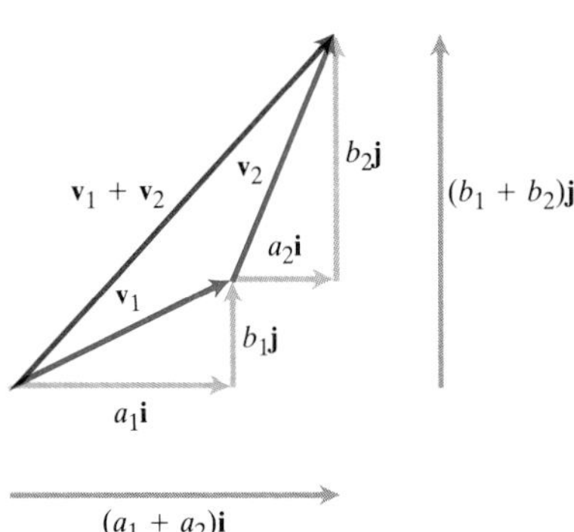

11.5 If $\mathbf{v}_1 = a_1\mathbf{i} + b_1\mathbf{j}$ and $\mathbf{v}_2 = a_2\mathbf{i} + b_2\mathbf{j}$, then $\mathbf{v}_1 + \mathbf{v}_2 = (a_1 + a_2)\mathbf{i} + (b_1 + b_2)\mathbf{j}$.

Components give us a way to define the equality of vectors algebraically.

DEFINITION

Equality of Vectors (Algebraic Definition)

$$a\mathbf{i} + b\mathbf{j} = a'\mathbf{i} + b'\mathbf{j} \quad \Leftrightarrow \quad a = a' \text{ and } b = b' \tag{1}$$

That is, two vectors are equal if and only if their scalar components in the directions of **i** and **j** are equal.

Algebraic Addition

Two vectors may be added algebraically by adding their corresponding scalar components, as shown in Fig. 11.5.

$$\text{If } \mathbf{v}_1 = a_1\mathbf{i} + b_1\mathbf{j}, \text{ and } \mathbf{v}_2 = a_2\mathbf{i} + b_2\mathbf{j}, \text{ then}$$
$$\mathbf{v}_1 + \mathbf{v}_2 = (a_1 + a_2)\mathbf{i} + (b_1 + b_2)\mathbf{j}. \tag{2}$$

EXAMPLE 2

$$(2\mathbf{i} - 4\mathbf{j}) + (5\mathbf{i} + 3\mathbf{j}) = (2 + 5)\mathbf{i} + (-4 + 3)\mathbf{j} = 7\mathbf{i} - \mathbf{j}.$$

Subtraction

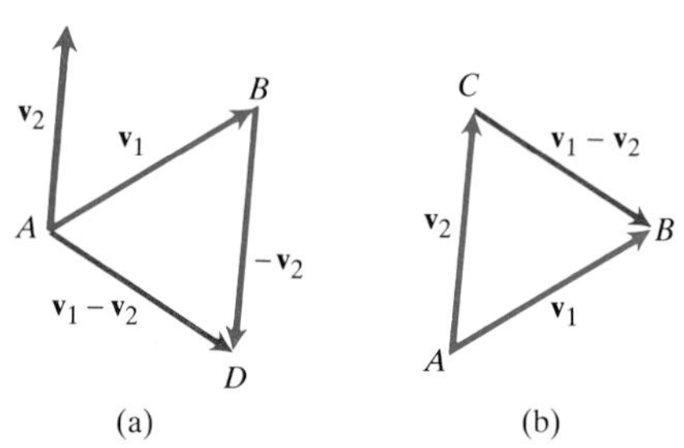

11.6 Two ways to draw $\mathbf{v}_1 - \mathbf{v}_2$: (a) as $\mathbf{v}_1 + (-\mathbf{v}_2)$ and (b) as the vector from the tip of $\mathbf{v}_2$ to the tip of $\mathbf{v}_1$.

The negative of a vector **v** is the vector $-\mathbf{v}$ that has the same length as **v** but points in the opposite direction. To subtract a vector $\mathbf{v}_2$ from a vector $\mathbf{v}_1$, we add $-\mathbf{v}_2$ to $\mathbf{v}_1$. This may be done geometrically by drawing $-\mathbf{v}_2$ from the tip of $\mathbf{v}_1$ and then drawing the vector from the initial point of $\mathbf{v}_1$ to the tip of $-\mathbf{v}_2$, as shown in Fig. 11.6(a), where

$$\overrightarrow{AD} = \overrightarrow{AB} + \overrightarrow{BD} = \mathbf{v}_1 + (-\mathbf{v}_2) = \mathbf{v}_1 - \mathbf{v}_2.$$

Another way to draw $\mathbf{v}_1 - \mathbf{v}_2$ is to draw $\mathbf{v}_1$ and $\mathbf{v}_2$ with a common initial point and then draw $\mathbf{v}_1 - \mathbf{v}_2$ as the vector from the tip of $\mathbf{v}_2$ to the tip of $\mathbf{v}_1$. This is illustrated in Fig. 11.6(b), where

$$\overrightarrow{CB} = \overrightarrow{CA} + \overrightarrow{AB} = -\mathbf{v}_2 + \mathbf{v}_1 = \mathbf{v}_1 - \mathbf{v}_2.$$

Thus, $\overrightarrow{CB}$ is the vector that when added to $\mathbf{v}_2$ gives $\mathbf{v}_1$:

$$\overrightarrow{CB} + \mathbf{v}_2 = (\mathbf{v}_1 - \mathbf{v}_2) + \mathbf{v}_2 = \mathbf{v}_1.$$

In terms of components, vector subtraction follows the algebraic law

$$\mathbf{v}_1 - \mathbf{v}_2 = (a_1 - a_2)\mathbf{i} + (b_1 - b_2)\mathbf{j}, \tag{3}$$

which says that corresponding scalar components are subtracted.

EXAMPLE 3

$$(6\mathbf{i} + 2\mathbf{j}) - (3\mathbf{i} - 5\mathbf{j}) = (6 - 3)\mathbf{i} + (2 - (-5))\mathbf{j} = 3\mathbf{i} + 7\mathbf{j}.$$ ■

We find the components of the vector from a point $P_1(x_1, y_1)$ to a point $P_2(x_2, y_2)$ by subtracting the components of $\overrightarrow{OP_1} = x_1\mathbf{i} + y_1\mathbf{j}$ from the components of $\overrightarrow{OP_2} = x_2\mathbf{i} + y_2\mathbf{j}$.

> The vector from $P_1(x_1, y_1)$ to $P_2(x_2, y_2)$ is
> $$\overrightarrow{P_1P_2} = (x_2 - x_1)\mathbf{i} + (y_2 - y_1)\mathbf{j}. \tag{4}$$

EXAMPLE 4 The vector from P_1 (3, 4) to P_2 (5, 1) is

$$\overrightarrow{P_1P_2} = (5 - 3)\mathbf{i} + (1 - 4)\mathbf{j} = 2\mathbf{i} - 3\mathbf{j}.$$ ■

The Length of a Vector

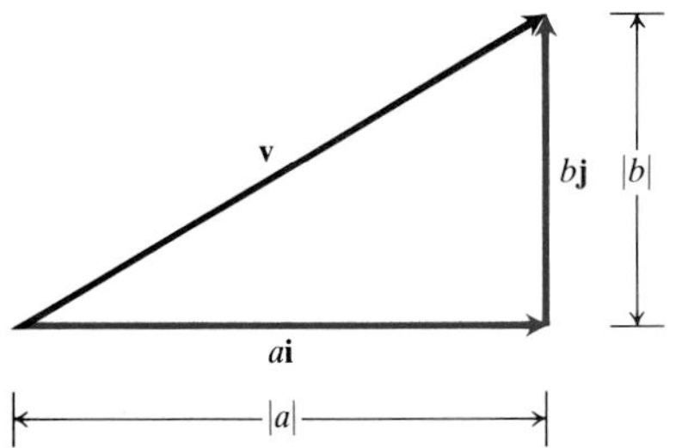

11.7 The length of **v** is $\sqrt{|a|^2 + |b|^2} = \sqrt{a^2 + b^2}$.

We calculate the length of $\mathbf{v} = a\mathbf{i} + b\mathbf{j}$ by representing **v** as the hypotenuse of a right triangle with sides $|a|$ and $|b|$ (Fig. 11.7) and applying the Pythagorean Theorem to get

$$\text{Length of } \mathbf{v} = \sqrt{|a|^2 + |b|^2} = \sqrt{a^2 + b^2}.$$

The usual symbol for the length is $|\mathbf{v}|$, which is read "the length of **v**" or "the magnitude of **v**," the latter being more common in applied fields. The bars are the same as the ones we use for absolute values.

> The **length** or **magnitude** of $\mathbf{v} = a\mathbf{i} + b\mathbf{j}$ is $|\mathbf{v}| = \sqrt{a^2 + b^2}$. (5)

Scalar Multiplication

Scalar multiplication can be accomplished component by component.

> If c is a scalar and $\mathbf{v} = a\mathbf{i} + b\mathbf{j}$ is a vector, then
> $$c\mathbf{v} = c(a\mathbf{i} + b\mathbf{j}) = (ca)\mathbf{i} + (cb)\mathbf{j} \tag{6}$$

To check that the length of $c\mathbf{v}$ is still $|c|$ times the length of **v** when we do scalar multiplication this way, we can calculate the length with Eq. (5):

$$\begin{aligned}
|c\mathbf{v}| &= |(ca)\mathbf{i} + (cb)\mathbf{j}| && \text{(Eq. 6)} \\
&= \sqrt{(ca)^2 + (cb)^2} && \text{(Eq. 5 with } ca \text{ and } cb \text{ in place of } a \text{ and } b) \\
&= \sqrt{c^2(a^2 + b^2)} \\
&= \sqrt{c^2}\sqrt{a^2 + b^2} \\
&= |c|\,|\mathbf{v}|.
\end{aligned}$$

> If c is a scalar and $\mathbf{v}$ is a vector, then
> $$|c\mathbf{v}| = |c|\,|\mathbf{v}|. \tag{7}$$

EXAMPLE 5 If $c = -2$ and $\mathbf{v} = -3\mathbf{i} + 4\mathbf{j}$, then

$$|\mathbf{v}| = |-3\mathbf{i} + 4\mathbf{j}| = \sqrt{(-3)^2 + (4)^2} = \sqrt{9 + 16} = \sqrt{25} = 5$$

$$|-2\,\mathbf{v}| = |(-2)(-3\mathbf{i} + 4\mathbf{j})| = |6\mathbf{i} - 8\mathbf{j}| = \sqrt{(6)^2 + (-8)^2} = \sqrt{36 + 64}$$

$$= \sqrt{100} = 10 = |-2|5 = |c|\,|\mathbf{v}|.$$

The Zero Vector

The vector

$$\mathbf{0} = 0\mathbf{i} + 0\mathbf{j} \tag{8}$$

is called the **zero vector.** It is the only vector whose length is zero, as we can see from the fact that

$$|a\mathbf{i} + b\mathbf{j}| = \sqrt{a^2 + b^2} = 0 \quad \Leftrightarrow \quad a = b = 0.$$

Unit Vectors

Any vector $\mathbf{u}$ whose length is equal to the unit of length used along the coordinate axes is called a **unit vector.** The vectors $\mathbf{i}$ and $\mathbf{j}$ are unit vectors:

$$|\mathbf{i}| = |1\mathbf{i} + 0\mathbf{j}| = \sqrt{1^2 + 0^2} = 1, \qquad |\mathbf{j}| = |0\mathbf{i} + 1\mathbf{j}| = \sqrt{0^2 + 1^2} = 1.$$

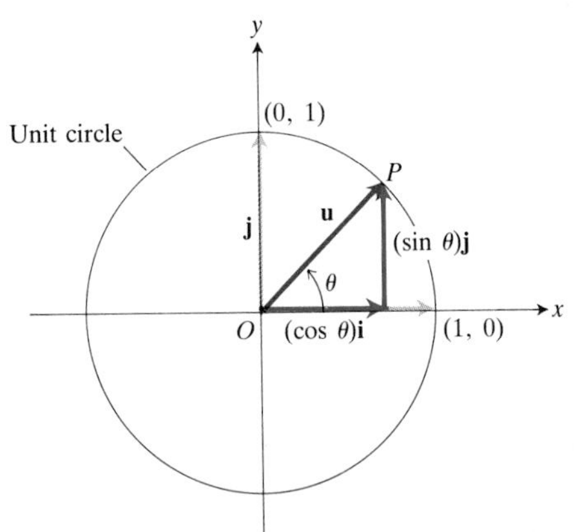

11.8 The unit vector that makes an angle of measure θ with the positive x-axis. Every unit vector has the form

$$\mathbf{u} = (\cos\theta)\mathbf{i} + (\sin\theta)\mathbf{j}$$

for some θ.

If $\mathbf{u}$ is the unit vector obtained by rotating $\mathbf{i}$ through an angle θ in the positive direction, then (Fig. 11.8) $\mathbf{u}$ has a horizontal component $\cos\theta$ and a vertical component $\sin\theta$, so that

$$\mathbf{u} = (\cos\theta)\mathbf{i} + (\sin\theta)\mathbf{j}. \tag{9}$$

If we allow the angle θ in Eq. (9) to vary from 0 to 2π, the point P in Fig. 11.8 traces the unit circle $x^2 + y^2 = 1$ once counterclockwise. Since this takes in all possible directions, every unit vector in the plane is given by Eq. (9) for some value of θ.

In handwritten work it is common to denote unit vectors with small "hats," as in $\hat{\mathrm{u}}$ (pronounced "u hat"). In hat notation, $\mathbf{i}$ and $\mathbf{j}$ become $\hat{\imath}$ and $\hat{\jmath}$.

Direction as a Vector

It is common in subjects like classical electricity and magnetism, which use vectors a great deal, to define the direction of a nonzero vector $\mathbf{v}$ to be the unit vector obtained by dividing $\mathbf{v}$ by its own length.

DEFINITION

> If $\mathbf{v} \neq \mathbf{0}$, the **direction** of $\mathbf{v}$ is the vector $\dfrac{\mathbf{v}}{|\mathbf{v}|}$. $\quad$ (10)

Notice that instead of just saying that $\mathbf{v}/|\mathbf{v}|$ *represents* the direction of $\mathbf{v}$, we say that it *is* the direction of $\mathbf{v}$.

To see that $\mathbf{v}/|\mathbf{v}|$ really is a unit vector, we can calculate its length directly:

$$\begin{aligned}\text{Length of } \frac{\mathbf{v}}{|\mathbf{v}|} &= \left|\frac{\mathbf{v}}{|\mathbf{v}|}\right| \\ &= \frac{1}{|\mathbf{v}|}|\mathbf{v}| \qquad \left(\text{Eq. 7 with } c = \frac{1}{|\mathbf{v}|}\right) \\ &= 1.\end{aligned}$$

The zero vector, having no length, so to speak, has no direction either.

Any nonzero vector can be expressed in terms of its length and direction by using the equation

$$\mathbf{v} = |\mathbf{v}| \cdot \frac{\mathbf{v}}{|\mathbf{v}|} = (\text{length of } \mathbf{v}) \cdot (\text{direction of } \mathbf{v}). \tag{11}$$

EXAMPLE 6 Express $\mathbf{v} = 3\mathbf{i} - 4\mathbf{j}$ in terms of its length and direction.

Solution Length of $\mathbf{v}$: $|\mathbf{v}| = \sqrt{(3)^2 + (-4)^2} = \sqrt{9 + 16} = 5$

Direction of $\mathbf{v}$: $\dfrac{\mathbf{v}}{|\mathbf{v}|} = \dfrac{3\mathbf{i} - 4\mathbf{j}}{5} = \dfrac{3}{5}\mathbf{i} - \dfrac{4}{5}\mathbf{j}$

$$\mathbf{v} = 3\mathbf{i} - 4\mathbf{j} = \underbrace{5}_{\text{Length of } \mathbf{v}}\underbrace{\left(\frac{3}{5}\mathbf{i} - \frac{4}{5}\mathbf{j}\right)}_{\text{Direction of } \mathbf{v}}$$

■

It follows from the definition of direction as a vector that vectors **A** and **B** have the same direction if and only if

$$\frac{\mathbf{A}}{|\mathbf{A}|} = \frac{\mathbf{B}}{|\mathbf{B}|} \quad \text{or} \quad \mathbf{A} = \frac{|\mathbf{A}|}{|\mathbf{B}|}\mathbf{B}. \tag{12}$$

Thus, if **A** and **B** have the same direction, then **A** is a positive scalar multiple of **B**. Conversely, if $\mathbf{A} = k\mathbf{B}$, $k > 0$, then

$$\frac{\mathbf{A}}{|\mathbf{A}|} = \frac{k\mathbf{B}}{|k\mathbf{B}|} = \frac{k}{|k|}\frac{\mathbf{B}}{|\mathbf{B}|} = \frac{k}{k}\frac{\mathbf{B}}{|\mathbf{B}|} = \frac{\mathbf{B}}{|\mathbf{B}|}. \tag{13}$$

Therefore, two nonzero vectors **A** and **B** point in the same direction if and only if **A** is a positive scalar multiple of **B**.

We say that two nonzero vectors **A** and **B** point in *opposite* directions if their directions are opposite in sign:

$$\frac{\mathbf{A}}{|\mathbf{A}|} = -\frac{\mathbf{B}}{|\mathbf{B}|}. \tag{14}$$

From this it follows that **A** and **B** have opposite directions if and only if **A** is a negative scalar multiple of **B**.

EXAMPLE 7

a) Same direction: $\mathbf{A} = 3\mathbf{i} - 4\mathbf{j}$ and $\mathbf{B} = \frac{3}{2}\mathbf{i} - 2\mathbf{j} = \frac{1}{2}\mathbf{A}$

(**B** is a positive scalar multiple of **A**.)

b) Opposite directions: $\mathbf{A} = 3\mathbf{i} - 4\mathbf{j}$ and $\mathbf{B} = -9\mathbf{i} + 12\mathbf{j} = -3\mathbf{A}$

(**B** is a negative scalar mulitple of **A**.) ■

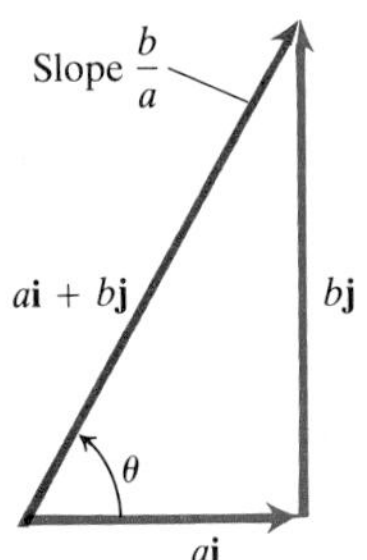

11.9 If $a \neq 0$, the vector $a\mathbf{i} + b\mathbf{j}$ has slope $b/a = \tan\theta$.

Slopes, Tangents, and Normals

Two vectors are said to be **parallel** if they are either positive or negative scalar multiples of one another or, equivalently, if the line segments representing them are parallel. Similarly, a vector is parallel to a line if the segments that represent the vector are parallel to the line. The **slope** of a vector that is not parallel to the y-axis is the slope shared by the lines parallel to the vector. Thus, when $a \neq 0$, the vector $\mathbf{v} = a\mathbf{i} + b\mathbf{j}$ has a well-defined slope, which can be calculated from the components of **v** as the number b/a (Fig. 11.9).

When we say that a vector is **tangent** or **normal** to a curve at a point, we mean that the vector is parallel or normal to the line that is tangent to the curve at the point. The next example shows how to find such vectors.

EXAMPLE 8 Find unit vectors tangent and normal to the curve

$$y = \frac{x^3}{2} + \frac{1}{2}$$

at the point (1, 1).

Solution We find the unit vectors that are parallel and normal to the curve's tangent line at the point (1, 1), shown in Fig. 11.10.

The slope of the line tangent to the curve at (1, 1) is

$$y' = \left.\frac{3x^2}{2}\right|_{x=1} = \frac{3}{2}.$$

We find a unit vector with this slope. The vector $\mathbf{v} = 2\mathbf{i} + 3\mathbf{j}$ has slope 3/2, as does every nonzero multiple of **v**. To find a multiple of **v** that is a unit vector, we divide **v** by its length,

$$|\mathbf{v}| = \sqrt{2^2 + 3^2} = \sqrt{13}.$$

This produces the unit vector

$$\mathbf{u} = \frac{\mathbf{v}}{|\mathbf{v}|} = \frac{2}{\sqrt{13}}\mathbf{i} + \frac{3}{\sqrt{13}}\mathbf{j}.$$

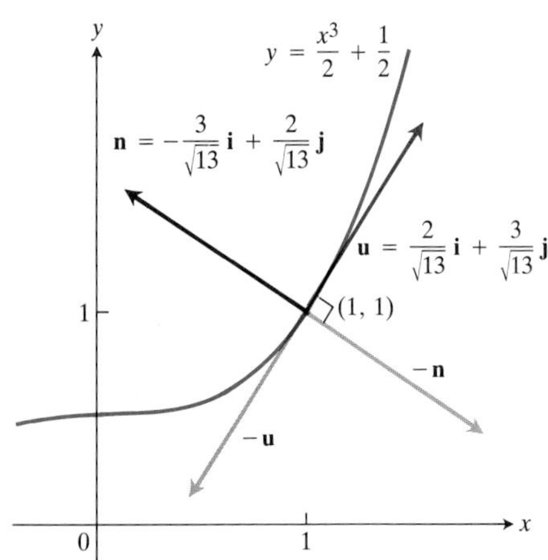

11.10 The unit tangent and normal vectors at the point (1, 1) on the curve $y = (x^3/2) + 1/2$.

The vector **u** is tangent to the curve at (1, 1) because it has the same direction as **v**. Of course, the vector

$$-\mathbf{u} = -\frac{2}{\sqrt{13}}\mathbf{i} - \frac{3}{\sqrt{13}}\mathbf{j},$$

which points in the opposite direction, is also tangent to the curve at (1, 1).

If $\mathbf{v} = a\mathbf{i} + b\mathbf{j}$, then $\mathbf{p} = -b\mathbf{i} + a\mathbf{j}$ and $\mathbf{q} = b\mathbf{i} - a\mathbf{j}$ are perpendicular to $\mathbf{v}$ because their slopes are both $-a/b$, the negative reciprocal of $\mathbf{v}$'s slope.

Without some additional requirement, there is no reason to prefer one of these vectors to the other.

To find unit vectors normal to the curve at (1, 1), we look for unit vectors whose slopes are the negative reciprocal of the slope of $\mathbf{u}$. This is quickly done by interchanging the scalar components of $\mathbf{u}$ and changing the sign of one of them. We obtain

$$\mathbf{n} = -\frac{3}{\sqrt{13}}\mathbf{i} + \frac{2}{\sqrt{13}}\mathbf{j} \quad \text{and} \quad -\mathbf{n} = \frac{3}{\sqrt{13}}\mathbf{i} - \frac{2}{\sqrt{13}}\mathbf{j}.$$

Again, either one will do. The vectors have opposite directions but both are normal to the curve at the point (1, 1). ■

Exercises 11.1

1. The vectors **A**, **B**, and **C** shown below lie in a plane. Copy them on a sheet of paper.

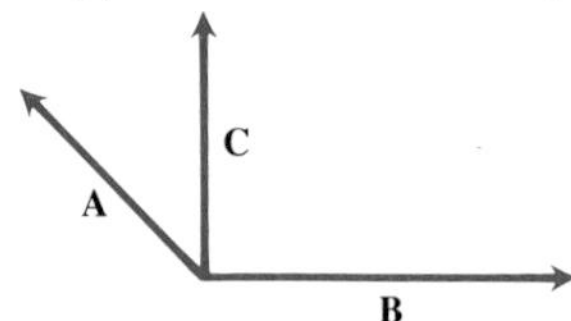

Then, by arranging the vectors head to tail, as in Figs. 11.3, 11.5, and 11.6, sketch

a) $\mathbf{A} + \mathbf{B}$ b) $\mathbf{A} + \mathbf{B} + \mathbf{C}$
c) $\mathbf{A} - \mathbf{B}$ d) $\mathbf{A} - \mathbf{C}$

2. The vectors **A**, **B**, and **C** shown below lie in a plane. Copy them on a sheet of paper.

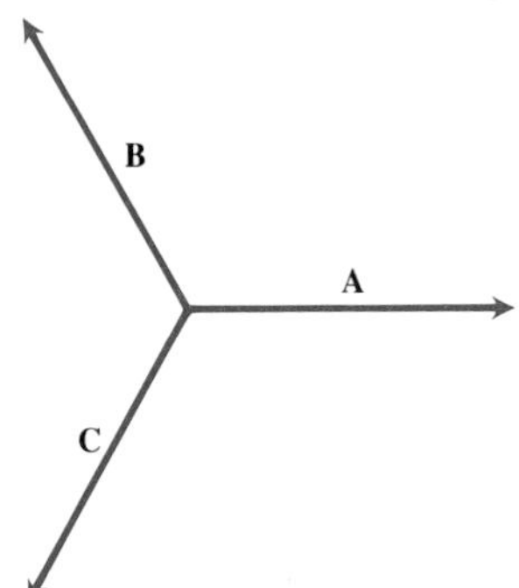

Then, by arranging the vectors head to tail, as in Figs. 11.3, 11.5, and 11.6, sketch

a) $\mathbf{A} - \mathbf{B}$ b) $\mathbf{A} - \mathbf{B} + \mathbf{C}$
c) $2\mathbf{A} - \mathbf{B}$ d) $\mathbf{A} + \mathbf{B} + \mathbf{C}$

Express the vectors in Exercises 3–10 in the form $a\mathbf{i} + b\mathbf{j}$ and sketch them as arrows in the coordinate plane.

3. $\overrightarrow{P_1P_2}$ if P_1 is the point (1, 3) and P_2 is the point (2, −1)

4. $\overrightarrow{OP_3}$ if O is the origin and P_3 is the midpoint of the vector $\overrightarrow{P_1P_2}$ joining $P_1(2, -1)$ and $P_2(-4, 3)$

5. The vector from the point $A(2, 3)$ to the origin

6. The sum of the vectors $\overrightarrow{AB}$ and $\overrightarrow{CD}$, given the four points $A(1, -1)$, $B(2, 0)$, $C(-1, 3)$, and $D(-2, 2)$

7. The unit vectors $\mathbf{u} = (\cos\theta)\mathbf{i} + (\sin\theta)\mathbf{j}$ for $\theta = \pi/6$ and $\theta = 2\pi/3$. Include the circle $x^2 + y^2 = 1$ in your sketch.

8. The unit vectors $\mathbf{u} = (\cos\theta)\mathbf{i} + (\sin\theta)\mathbf{j}$ for $\theta = -\pi/4$ and $\theta = -3\pi/4$. Include the circle $x^2 + y^2 = 1$ in your sketch.

9. The unit vector obtained by rotating $\mathbf{j}$ 120° clockwise about the origin.

10. The unit vector obtained by rotating $\mathbf{i}$ 135° counterclockwise about the origin.

11. Find the unit vectors tangent and normal to the parabola $y = x^2$ at the point (2, 4) (four vectors in all). Sketch them with the parabola.

12. Find the unit vectors tangent and normal to the ellipse $x^2 + 2y^2 = 6$ at the point (2, 1) (four vectors in all). Sketch them with the ellipse.

Use Eq. (11) to express the vectors in Exercises 13–18 in terms of their lengths and directions.

13. $\mathbf{i} + \mathbf{j}$ **14.** $2\mathbf{i} - 3\mathbf{j}$

15. $\sqrt{3}\mathbf{i} + \mathbf{j}$ **16.** $-2\mathbf{i} + 3\mathbf{j}$

17. $5\mathbf{i} + 12\mathbf{j}$ **18.** $-5\mathbf{i} - 12\mathbf{j}$

19. Show that $\mathbf{A} = 3\mathbf{i} + 6\mathbf{j}$ and $\mathbf{B} = -\mathbf{i} - 2\mathbf{j}$ have opposite directions. Sketch **A** and **B** together.

20. Show that $\mathbf{A} = 3\mathbf{i} + 6\mathbf{j}$ and $\mathbf{B} = \frac{1}{2}\mathbf{i} + \mathbf{j}$ have the same direction.

11.2 Cartesian (Rectangular) Coordinates and Vectors in Space

Our goal now is to describe the three-dimensional Cartesian coordinate system and learn our way around in space. This means defining distance, practicing with the arithmetic of vectors (same rules as in the plane but with an extra term), and making connections between sets of points in space and equations and inequalities. Everyone we know finds it harder to draw in three dimensions than in two, so we have included some drawing tips as well. You will find them on pages 726–727. The Cartesian coordinates for space are often called *rectangular coordinates* because the axes that define them meet at right angles.

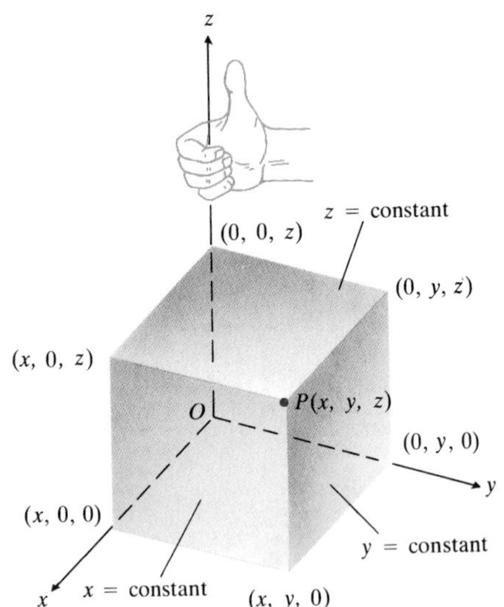

11.11 The Cartesian coordinate system is right-handed.

Cartesian Coordinates

To locate points in space, we use three mutually perpendicular coordinate axes, arranged as in Fig. 11.11. The axes Ox, Oy, and Oz shown there make what is known as a *right-handed* coordinate frame. When you hold your right hand so that the fingers curl from the positive x-axis toward the positive y-axis, your thumb points along the positive z-axis.

The Cartesian coordinates (x, y, z) of a point P in space are the numbers where the planes through P perpendicular to the three axes cut the axes.

Points that lie on the x-axis have their y- and z-coordinates equal to zero. That is, they have coordinates of the form $(x, 0, 0)$. Similarly, points on the y-axis have coordinates of the form $(0, y, 0)$, and points on the z-axis have coordinates of the form $(0, 0, z)$.

The points in a plane perpendicular to the x-axis all have the same x-coordinate, the number at which that plane cuts the x-axis. Similarly, the points in a plane perpendicular to the y-axis have a common y-coordinate, and the points in a plane perpendicular to the z-axis have a common z-coordinate. It is therefore easy to write equations for these planes—we just name the common coordinate's value. The equation $x = 2$ is an equation for the plane perpendicular to the x-axis at $x = 2$. The equation $y = 3$ is an equation for the plane perpendicular to the y-axis at $y = 3$. The equation $z = 5$ is an equation for the plane perpendicular to the z-axis at $z = 5$. Figure 11.12 shows the planes $x = 2$, $y = 3$, and $z = 5$, together with their intersection point $(2, 3, 5)$.

The planes $x = 2$ and $y = 3$ in Fig. 11.12 intersect in a line that runs parallel to the z-axis. This line is described by the *pair* of equations $x = 2$, $y = 3$. A point (x, y, z) lies on the line if and only if x equals 2 and y equals 3. Similarly, the line of intersection of the planes $y = 3$ and $z = 5$ is described by the equation pair $y = 3$, $z = 5$. This line runs parallel to the x-axis. The line of intersection of the planes $x = 2$ and $z = 5$, parallel to the y-axis, is described by the equation pair $x = 2$, $z = 5$.

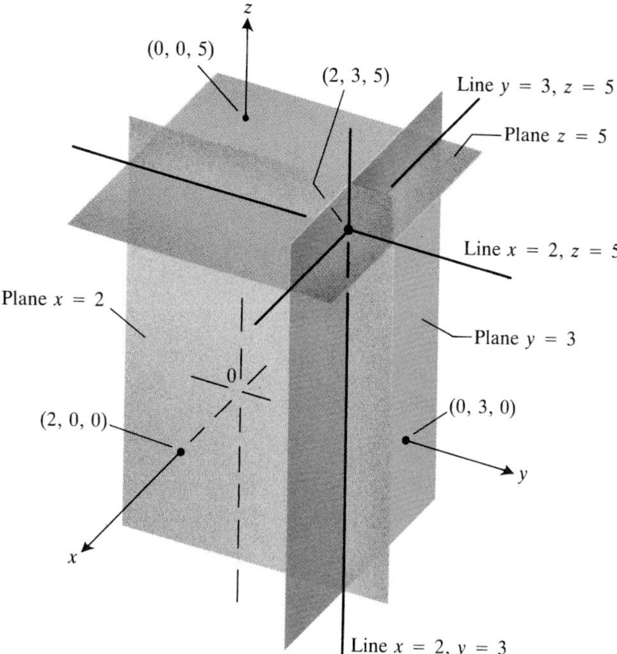

11.12 The planes $x = 2$, $y = 3$, and $z = 5$ determine three lines through the point (2, 3, 5).

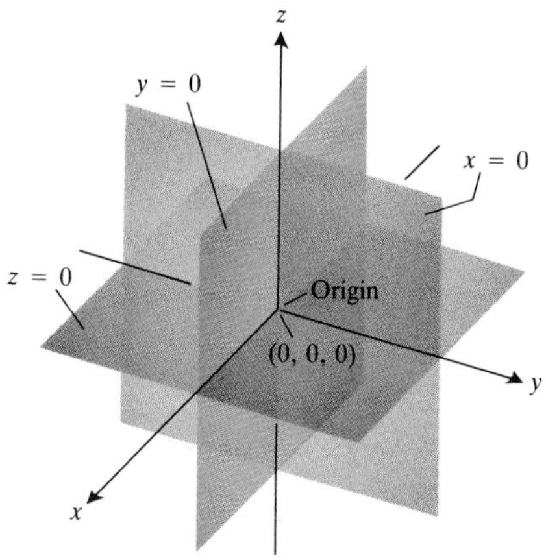

11.13 The planes $x = 0$, $y = 0$, and $z = 0$ are the planes determined by the coordinate axes. They divide space into eight cells called octants.

The planes determined by the three coordinate axes are the **xy-plane,** whose standard equation is $z = 0$; the **yz-plane,** whose standard equation is $x = 0$; and the **xz-plane,** whose standard equation is $y = 0$. They meet in the point (0, 0, 0), which is called the **origin** of the coordinate system (Fig. 11.13).

The three **coordinate planes** $x = 0, y = 0$, and $z = 0$ divide space into eight cells called **octants.** The octant in which the point coordinates are all positive is called the **first octant,** but there is no conventional numbering for the remaining seven octants.

In the following examples, we match a number of coordinate equations and inequalities with the sets of points they define in space.

EXAMPLE 1

Defining equations and inequalities	Verbal description
$z \geq 0$	The half-space consisting of the points on and above the xy-plane.
$x = -3$	The plane perpendicular to the x-axis at $x = -3$. This plane lies parallel to the yz-plane and 3 units behind it.
$z = 0, x \leq 0, y \geq 0$	The second quadrant of the xy-plane.
$x \geq 0, y \geq 0, z \geq 0$	The first octant.
$-1 \leq y \leq 1$	The slab between the planes $y = -1$ and $y = 1$ (planes included).
$y = -2, z = 2$	The line in which the planes $y = -2$ and $z = 2$ intersect. Alternatively, the line through the point $(0, -2, 2)$ parallel to the x-axis.

■

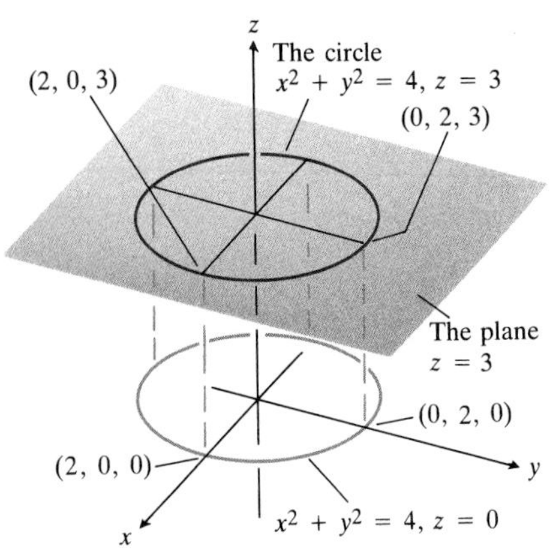

11.14 The circle $x^2 + y^2 = 4$, $z = 3$.

EXAMPLE 2 Identify the set of points $P(x, y, z)$ whose coordinates satisfy the two equations

$$x^2 + y^2 = 4 \quad \text{and} \quad z = 3.$$

Solution The points lie in the horizontal plane $z = 3$ and, in this plane, make up the circle $x^2 + y^2 = 4$. We call this set of points "the circle $x^2 + y^2 = 4$ in the plane $z = 3$" or, more simply, "the circle $x^2 + y^2 = 4$, $z = 3$" (Fig. 11.14). ■

Vectors in Space

The directed line segments we use to represent forces, displacements, and velocities in space are called vectors, just as they are in the plane. The same rules of addition, subtraction, and scalar multiplication apply.

The vectors from the origin to the points (1, 0, 0), (0, 1, 0), and (0, 0, 1) are the basic vectors. We denote them by **i**, **j**, and **k**. The position vector **r** from the origin O to the typical point $P(x, y, z)$ is

$$\mathbf{r} = x\mathbf{i} + y\mathbf{j} + z\mathbf{k}. \tag{1}$$

Addition, Subtraction, and Scalar Multiplication for Vectors in Space

For any vectors $\mathbf{A} = a_1\mathbf{i} + a_2\mathbf{j} + a_3\mathbf{k}$ and $\mathbf{B} = b_1\mathbf{i} + b_2\mathbf{j} + b_3\mathbf{k}$, and for any scalar c,

$$\mathbf{A} + \mathbf{B} = (a_1 + b_1)\mathbf{i} + (a_2 + b_2)\mathbf{j} + (a_3 + b_3)\mathbf{k}$$

$$\mathbf{A} - \mathbf{B} = (a_1 - b_1)\mathbf{i} + (a_2 - b_2)\mathbf{j} + (a_3 - b_3)\mathbf{k}$$

$$c\mathbf{A} = (ca_1)\mathbf{i} + (ca_2)\mathbf{j} + (ca_3)\mathbf{k}.$$

The Vector between Two Points

Often we want to express the vector $\overrightarrow{P_1P_2}$ from the point $P_1(x_1, y_1, z_1)$ to the point $P_2(x_2, y_2, z_2)$ in terms of the coordinates of P_1 and P_2. To do so, we first observe that $\overrightarrow{P_1P_2} = \overrightarrow{P_1O} + \overrightarrow{OP_2}$, as in Fig. 11.15. We then write $\overrightarrow{P_1O}$ as $-\overrightarrow{OP_1}$ and express the results in terms of **i**, **j**, and **k**:

$$\begin{aligned}\overrightarrow{P_1P_2} &= \overrightarrow{P_1O} + \overrightarrow{OP_2}\\ &= \overrightarrow{OP_2} - \overrightarrow{OP_1}\\ &= (x_2\mathbf{i} + y_2\mathbf{j} + z_2\mathbf{k}) - (x_1\mathbf{i} + y_1\mathbf{j} + z_1\mathbf{k})\\ &= (x_2 - x_1)\mathbf{i} + (y_2 - y_1)\mathbf{j} + (z_2 - z_1)\mathbf{k}.\end{aligned} \tag{2}$$

The vector from $P_1(x_1, y_1, z_1)$ to $P_2(x_2, y_2, z_2)$ is

$$\overrightarrow{P_1P_2} = (x_2 - x_1)\mathbf{i} + (y_2 - y_1)\mathbf{j} + (z_2 - z_1)\mathbf{k}. \tag{3}$$

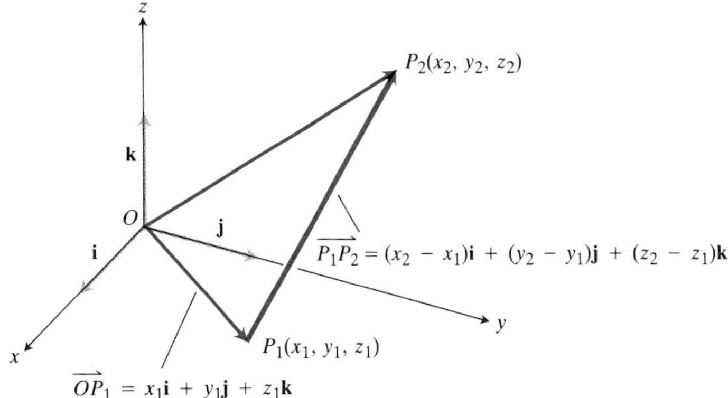

11.15 The vector from P_1 to P_2 is $\overrightarrow{P_1P_2} = (x_2 - x_1)\mathbf{i} + (y_2 - y_1)\mathbf{j} + (z_2 - z_1)\mathbf{k}$.

Length and Direction

As in the plane, the important features of a vector are its length and direction. The length of a vector $a\mathbf{i} + b\mathbf{j} + c\mathbf{k}$ is calculated by applying the Pythagorean theorem twice. In the notation of Fig. 11.16, we have

$$|\overrightarrow{AC}| = |a\mathbf{i} + b\mathbf{j}| = \sqrt{a^2 + b^2},$$

from triangle ABC, and then

$$|a\mathbf{i} + b\mathbf{j} + c\mathbf{k}| = |\overrightarrow{AD}| = \sqrt{|\overrightarrow{AC}|^2 + |\overrightarrow{CD}|^2} = \sqrt{a^2 + b^2 + c^2},$$

from triangle ACD.

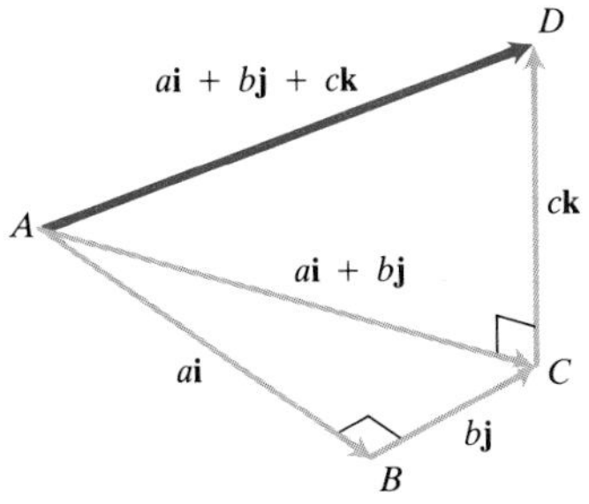

11.16 The length of the vector $\overrightarrow{AD}$ can be determined by applying the Pythagorean theorem to the right triangles ABC and ACD.

> The **length** of the vector $\mathbf{A} = a\mathbf{i} + b\mathbf{j} + c\mathbf{k}$ is
> $$|\mathbf{A}| = |a\mathbf{i} + b\mathbf{j} + c\mathbf{k}| = \sqrt{a^2 + b^2 + c^2}. \qquad (4)$$

EXAMPLE 3 The length of $\mathbf{A} = \mathbf{i} - 2\mathbf{j} + 3\mathbf{k}$ is

$$|\mathbf{A}| = \sqrt{(1)^2 + (-2)^2 + (3)^2} = \sqrt{1 + 4 + 9} = \sqrt{14}.$$ ■

If we multiply a vector $\mathbf{A} = a_1\mathbf{i} + a_2\mathbf{j} + a_3\mathbf{k}$ by a scalar c, the length of $c\mathbf{A}$ is $|c|$ times the length of $\mathbf{A}$, just as in the plane. The reason is the same, as well:

$$c\mathbf{A} = ca_1\mathbf{i} + ca_2\mathbf{j} + ca_3\mathbf{k},$$

so

$$|c\mathbf{A}| = \sqrt{(ca_1)^2 + (ca_2)^2 + (ca_3)^2} = \sqrt{c^2a_1^2 + c^2a_2^2 + c^2a_3^2}$$
$$= |c|\sqrt{a_1^2 + a_2^2 + a_3^2} = |c||\mathbf{A}|. \qquad (5)$$

EXAMPLE 4 If $\mathbf{A}$ is the vector of Example 3, then the length of

$$2\mathbf{A} = 2(\mathbf{i} - 2\mathbf{j} + 3\mathbf{k}) = 2\mathbf{i} - 4\mathbf{j} + 6\mathbf{k}$$

is

$$\sqrt{(2)^2 + (-4)^2 + (6)^2} = \sqrt{4 + 16 + 36} = \sqrt{56}$$
$$= \sqrt{4 \cdot 14} = 2\sqrt{14} = 2|\mathbf{A}|.$$ ■

Again as with vectors in the plane, vectors of unit length are called **unit vectors.** The vectors **i**, **j**, and **k** are unit vectors because

$$|\mathbf{i}| = 1\mathbf{i} + 0\mathbf{j} + 0\mathbf{k} = \sqrt{1^2 + 0^2 + 0^2} = 1,$$

$$|\mathbf{j}| = 0\mathbf{i} + 1\mathbf{j} + 0\mathbf{k} = \sqrt{0^2 + 1^2 + 0^2} = 1,$$

$$|\mathbf{k}| = 0\mathbf{i} + 0\mathbf{j} + 1\mathbf{k} = \sqrt{0^2 + 0^2 + 1^2} = 1.$$

The direction of a nonzero vector **A** is the unit vector obtained by dividing **A** by its length $|\mathbf{A}|$.

$$\text{The \textbf{direction} of } \mathbf{A} \text{ is } \frac{\mathbf{A}}{|\mathbf{A}|}. \tag{6}$$

As in the plane, we can use the equation

$$\mathbf{A} = |\mathbf{A}| \cdot \frac{\mathbf{A}}{|\mathbf{A}|} \tag{7}$$

to express any nonzero vector as a product of its length and direction.

EXAMPLE 5 Express $\mathbf{A} = \mathbf{i} - 2\mathbf{j} + 3\mathbf{k}$ as a product of its length and direction.

Solution

$$\begin{aligned} \mathbf{A} &= |\mathbf{A}| \cdot \frac{\mathbf{A}}{|\mathbf{A}|} && \text{(Eq. 7)} \\ &= \sqrt{14} \cdot \frac{\mathbf{i} - 2\mathbf{j} + 3\mathbf{k}}{\sqrt{14}} && \text{(Values from Example 3)} \\ &= \sqrt{14}\left(\frac{1}{\sqrt{14}}\mathbf{i} - \frac{2}{\sqrt{14}}\mathbf{j} + \frac{3}{\sqrt{14}}\mathbf{k}\right) = (\text{length of } \mathbf{A}) \cdot (\text{direction of } \mathbf{A}) \end{aligned}$$

■

EXAMPLE 6 Find a unit vector **u** in the direction of the vector from $P_1(1, 0, 1)$ to $P_2(3, 2, 0)$.

Solution The vector we want is the direction of $\overrightarrow{P_1P_2}$. To find it, we divide $\overrightarrow{P_1P_2}$ by its own length:

$$\overrightarrow{P_1P_2} = (3 - 1)\mathbf{i} + (2 - 0)\mathbf{j} + (0 - 1)\mathbf{k} = 2\mathbf{i} + 2\mathbf{j} - \mathbf{k},$$

$$|\overrightarrow{P_1P_2}| = \sqrt{(2)^2 + (2)^2 + (-1)^2} = \sqrt{4 + 4 + 1} = \sqrt{9} = 3,$$

$$\mathbf{u} = \frac{\overrightarrow{P_1P_2}}{|\overrightarrow{P_1P_2}|} = \frac{2\mathbf{i} + 2\mathbf{j} - \mathbf{k}}{3} = \frac{2}{3}\mathbf{i} + \frac{2}{3}\mathbf{j} - \frac{1}{3}\mathbf{k}.$$

■

EXAMPLE 7 Find a vector 6 units long in the direction of $\mathbf{A} = 2\mathbf{i} + 2\mathbf{j} - \mathbf{k}$.

Solution The vector we want is

$$6\frac{\mathbf{A}}{|\mathbf{A}|} = 6\frac{2\mathbf{i} + 2\mathbf{j} - \mathbf{k}}{\sqrt{2^2 + 2^2 + (-1)^2}} = 6\frac{2\mathbf{i} + 2\mathbf{j} - \mathbf{k}}{3} = 4\mathbf{i} + 4\mathbf{j} - 2\mathbf{k}.$$ ■

Distance in Space

To find the distance between two points P_1 and P_2 in space, we find the length of $\overrightarrow{P_1P_2}$. Equation (8) is the resulting formula:

The Distance between $P_1(x_1, y_1, z_1)$ and $P_2(x_2, y_2, z_2)$

$$|\overrightarrow{P_1P_2}| = \sqrt{(x_2 - x_1)^2 + (y_2 - y_1)^2 + (z_2 - z_1)^2} \tag{8}$$

EXAMPLE 8 The distance between

$$P_1(2, 1, 5) \quad \text{and} \quad P_2(-2, 3, 0)$$

is

$$\begin{aligned}|\overrightarrow{P_1P_2}| &= \sqrt{(-2-2)^2 + (3-1)^2 + (0-5)^2} = \sqrt{16 + 4 + 25} \\ &= \sqrt{45} = 3\sqrt{5}.\end{aligned}$$ ■

Spheres

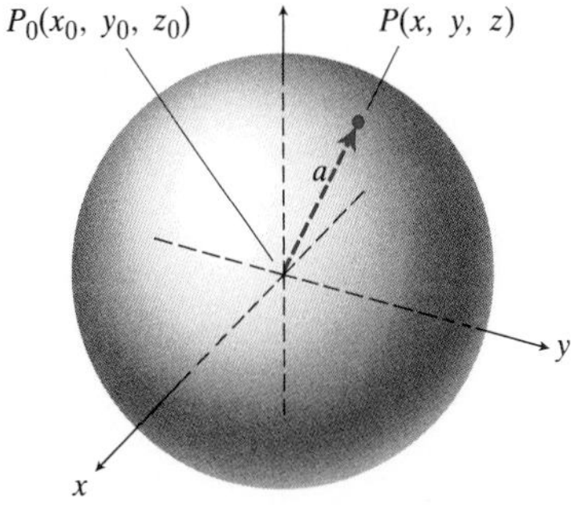

11.17 The standard equation of the sphere of radius a centered at the point (x_0, y_0, z_0) is

$$(x - x_0)^2 + (y - y_0)^2 + (z - z_0)^2 = a^2.$$

We can use Eq. (8) to write equations for spheres. Since a point $P(x, y, z)$ lies on the sphere of radius a centered at $P_0(x_0, y_0, z_0)$ if and only if it lies a units from P_0, it lies on the sphere if and only if

$$|\overrightarrow{P_0P}| = a$$

or

$$(x - x_0)^2 + (y - y_0)^2 + (z - z_0)^2 = a^2.$$

This equation is the *standard equation* for the sphere (Fig. 11.17).

The Standard Equation for the Sphere of Radius a and Center (x_0, y_0, z_0)

$$(x - x_0)^2 + (y - y_0)^2 + (z - z_0)^2 = a^2 \tag{9}$$

EXAMPLE 9 Find the center and radius of the sphere

$$x^2 + y^2 + z^2 + 2x + 4y = 0.$$

Solution Complete the squares in the given equation to obtain

$$x^2 + 2x + 1 + y^2 - 4y + 4 + z^2 = 0 + 1 + 4$$

$$(x + 1)^2 + (y - 2)^2 + z^2 = 5.$$

This is Eq. (9) with $x_0 = -1$, $y_0 = 2$, $z_0 = 0$, and $a = \sqrt{5}$. The center is $(-1, 2, 0)$ and the radius is $\sqrt{5}$. ■

How to Draw Three-dimensional Objects to Look Three-dimensional

1.
Break lines. When one line passes behind another, break it to show that it doesn't touch and that part of it is hidden.

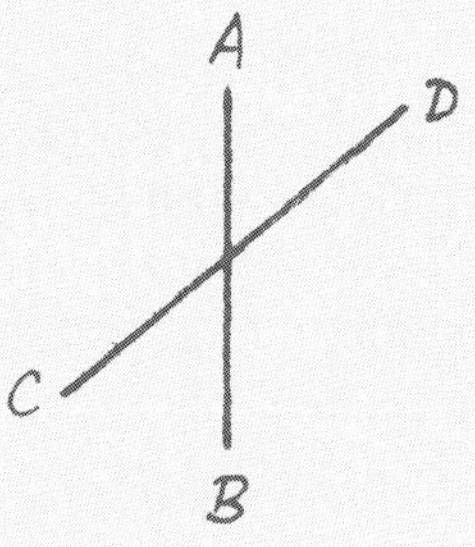

Crossing

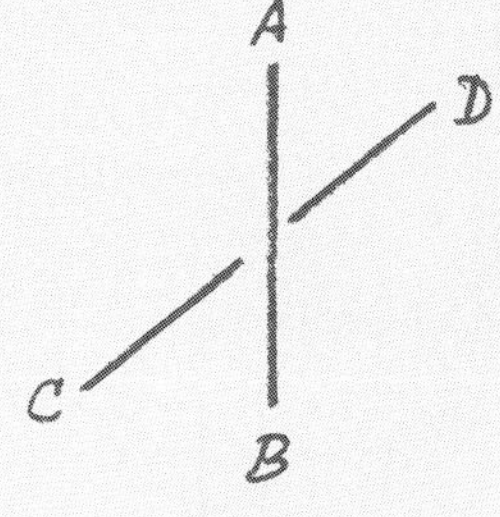

CD behind *AB*

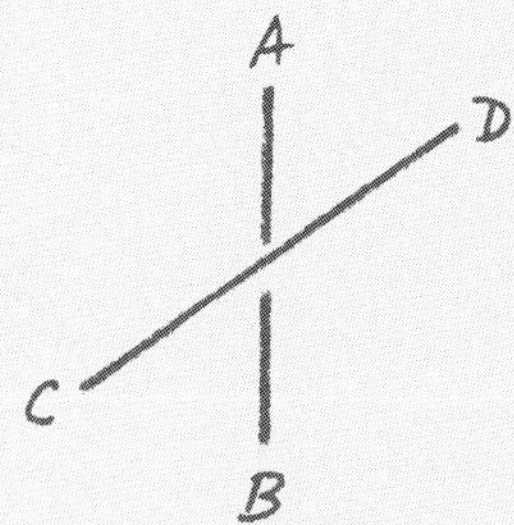

AB behind *CD*

2.
Make the angle between the positive *x*-axis and the positive *y*-axis large enough.

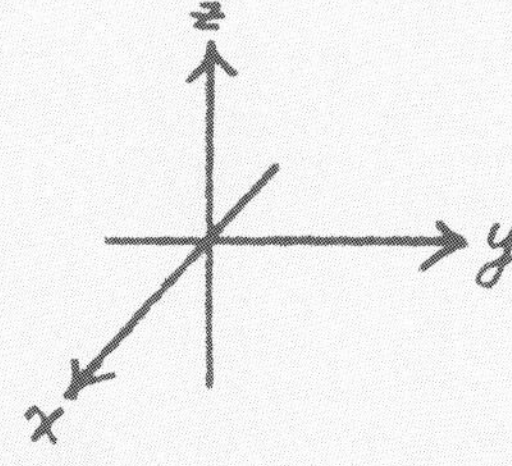

This

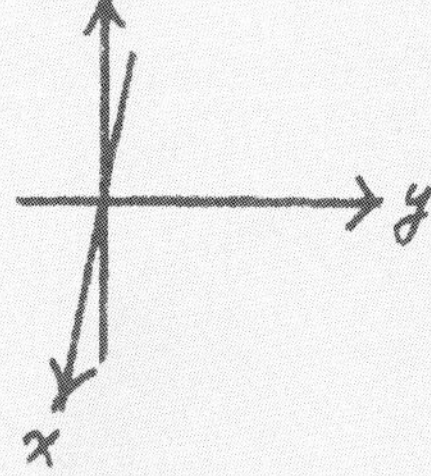

Not this

3.
Draw planes parallel to the coordinate planes as if they were rectangles with sides parallel to the coordinate axes.

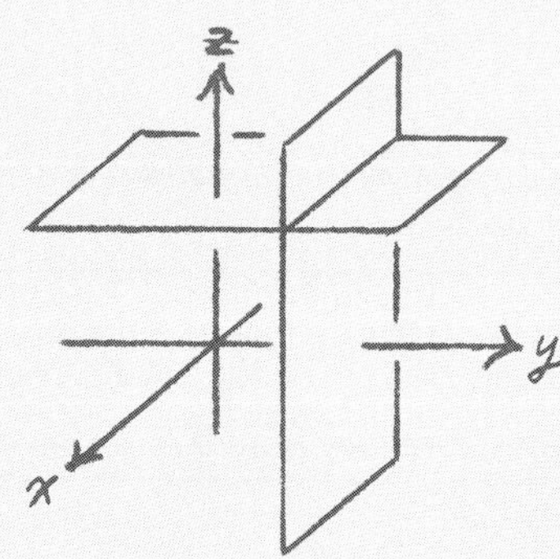

4.
Dot or omit hidden portions of lines. Don't let the line touch the boundary of the parallelogram that represents the plane, unless the line lies in the plane.

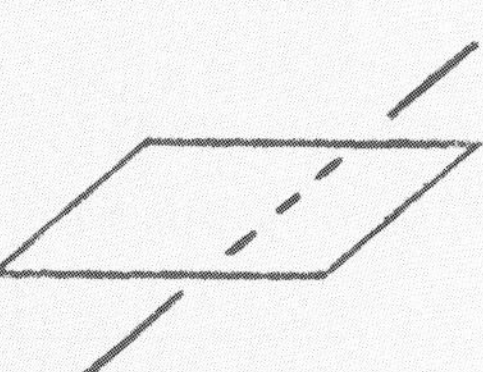
Line below plane

Line above plane

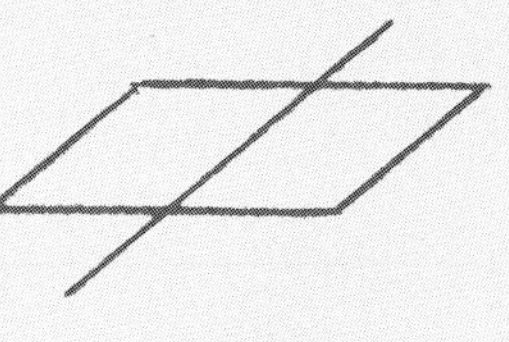
Line *in* plane

5.

Spheres: Draw the sphere first (outline and equator); draw axes, if any, later. Use line breaks and dotted lines.

Hidden part dotted

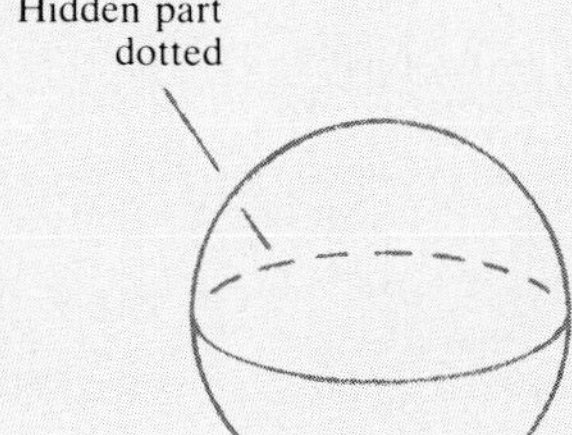

Sphere first

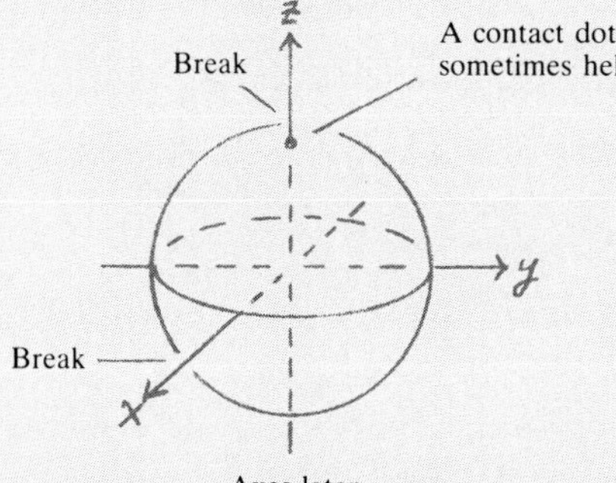

Axes later

6.

A general rule for perspective: Draw the object as if it lies some distance away, below, and to the left.

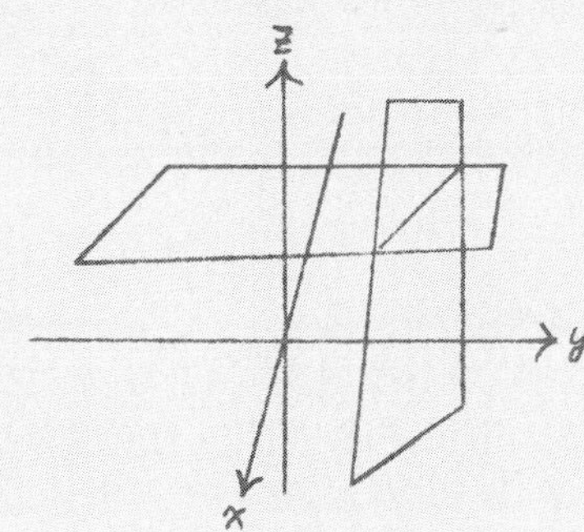

Advice ignored

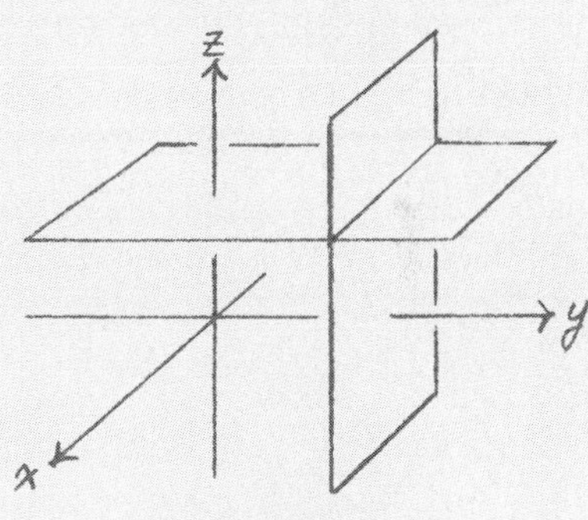

Advice followed

7.

To draw a plane that crosses all three coordinate axes, follow the steps shown here: (a) Sketch the axes and mark the intercepts. (b) Connect the intercepts to form two sides of a parallelogram. (c) Complete the parallelogram and enlarge it by drawing lines parallel to its sides. (d) Darken the exposed parts, break hidden lines, and, if desired, dash hidden portions of the axes. You may wish to erase the smaller parallelogram at this point.

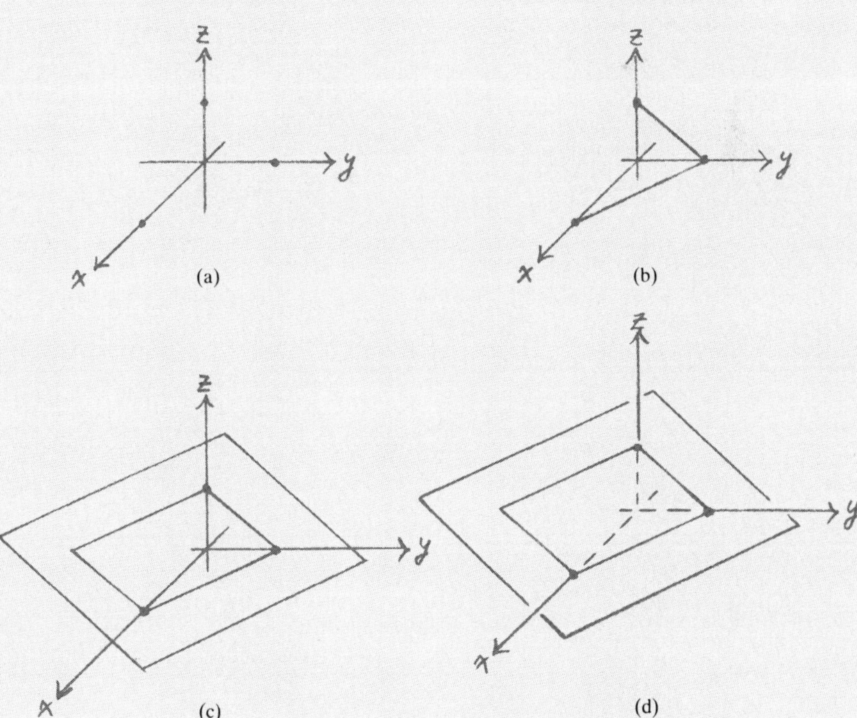

EXAMPLE 10 *Sets bounded by spheres or portions of spheres.*

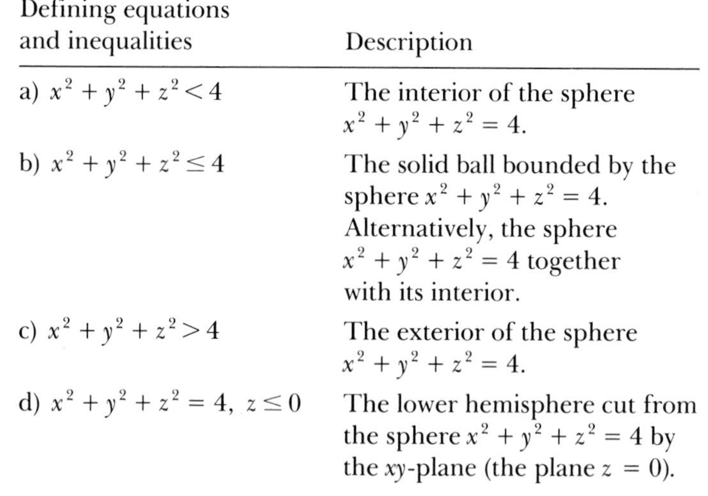

Defining equations and inequalities	Description
a) $x^2 + y^2 + z^2 < 4$	The interior of the sphere $x^2 + y^2 + z^2 = 4$.
b) $x^2 + y^2 + z^2 \leq 4$	The solid ball bounded by the sphere $x^2 + y^2 + z^2 = 4$. Alternatively, the sphere $x^2 + y^2 + z^2 = 4$ together with its interior.
c) $x^2 + y^2 + z^2 > 4$	The exterior of the sphere $x^2 + y^2 + z^2 = 4$.
d) $x^2 + y^2 + z^2 = 4,\ z \leq 0$	The lower hemisphere cut from the sphere $x^2 + y^2 + z^2 = 4$ by the xy-plane (the plane $z = 0$).

Midpoints of Line Segments

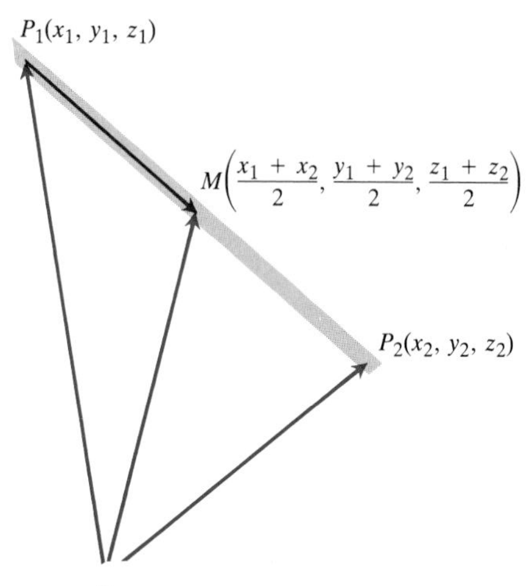

11.18 The coordinates of the point halfway between P_1 and P_2 are found by averaging the coordinates of P_1 and P_2.

The coordinates of the midpoint M of the line segment joining two points $P_1(x_1, y_1, z_1)$ and $P_2(x_2, y_2, z_2)$ are found by averaging the coordinates of P_1 and P_2. That is,

$$M = \left(\frac{x_1 + x_2}{2}, \frac{y_1 + y_2}{2}, \frac{z_1 + z_2}{2}\right). \tag{10}$$

To see why, we have only to observe that these coordinates are the scalar components of the position vector $\overrightarrow{OM}$ (see Fig. 11.18) and that

$$\begin{aligned}\overrightarrow{OM} &= \overrightarrow{OP_1} + \frac{1}{2}(\overrightarrow{P_1P_2}) = \overrightarrow{OP_1} + \frac{1}{2}(\overrightarrow{OP_2} - \overrightarrow{OP_1}) \\ &= \frac{1}{2}(\overrightarrow{OP_1} + \overrightarrow{OP_2}) \\ &= \frac{x_1 + x_2}{2}\mathbf{i} + \frac{y_1 + y_2}{2}\mathbf{j} + \frac{z_1 + z_2}{2}\mathbf{k}.\end{aligned}$$

EXAMPLE 11 The midpoint of the segment joining $P_1(3, -2, 0)$ and $P_2(7, 4, 4)$ is

$$\left(\frac{3 + 7}{2}, \frac{-2 + 4}{2}, \frac{0 + 4}{2}\right) = (5, 1, 2).$$

Tips for Drawing

Some tips for making pictures of three-dimensional objects look three-dimensional are presented on the preceding pages. Sketching with a pencil is safer than sketching with a pen because you can erase, but the advice applies no matter what medium you choose.

Exercises 11.2

In Exercises 1–12, give a geometric description of the set of points in space whose coordinates satisfy the given pairs of equations.

1. $x = 2,\ y = 3$

2. $x = -1,\ z = 0$

3. $y = 0,\ z = 0$

4. $x = 1,\ y = 0$

5. $x^2 + y^2 = 4,\ z = 0$

6. $x^2 + y^2 = 4,\ z = -2$

7. $x^2 + z^2 = 4,\ y = 0$

8. $y^2 + z^2 = 1,\ x = 0$

9. $x^2 + y^2 + z^2 = 1,\ x = 0$

10. $x^2 + y^2 + z^2 = 25,\ y = -4$

11. $x^2 + y^2 + (z + 3)^2 = 25,\ z = 0$

12. $x^2 + (y - 1)^2 + z^2 = 4,\ y = 0$

In Exercises 13–18, describe the sets of points in space whose coordinates satisfy the given inequalities or combinations of equations and inequalities.

13. a) $x \geq 0,\ y \geq 0,\ z = 0$
b) $x \geq 0,\ y \leq 0,\ z = 0$

14. a) $0 \leq x \leq 1$
b) $0 \leq x \leq 1,\ 0 \leq y \leq 1$
c) $0 \leq x \leq 1,\ 0 \leq y \leq 1,\ 0 \leq z \leq 1$

15. a) $x^2 + y^2 + z^2 \leq 1$
b) $x^2 + y^2 + z^2 > 1$

16. a) $x^2 + y^2 \leq 1,\ z = 0$
b) $x^2 + y^2 \leq 1,\ z = 3$
c) $x^2 + y^2 \leq 1$, no restriction on z

17. a) $x^2 + y^2 + z^2 = 1,\ z \geq 0$
b) $x^2 + y^2 + z^2 \leq 1,\ z \geq 0$

18. a) $x = y,\ z = 0$
b) $x = y$, no restriction on z

In Exercises 19–28, describe the given set with a single equation or with a pair of equations.

19. The plane perpendicular to the
a) x-axis at (3, 0, 0),
b) y-axis at (0, −1, 0),
c) z-axis at (0, 0, −2)

20. The plane through the point (3, −1, 2) perpendicular to the
a) x-axis, b) y-axis, c) z-axis

21. The plane through the point (3, −1, 1) parallel to the
a) xy-plane, b) yz-plane, c) xz-plane

22. The circle of radius 2 centered at (0, 0, 0) and lying in the
a) xy-plane, b) yz-plane, c) xz-plane

23. The circle of radius 2 centered at (0, 2, 0) and lying in the
a) xy-plane, b) yz-plane, c) plane $y = 2$

24. The circle of radius 1 centered at (−3, 4, 1) and lying in a plane parallel to the
a) xy-plane, b) yz-plane, c) xz-plane

25. The line through the point (1, 3, −1) parallel to the
a) x-axis, b) y-axis, c) z-axis

26. The set of points in space equidistant from the origin and the point (0, 2, 0)

27. The circle in which the plane through the point (1, 1, 3) perpendicular to the z-axis meets the sphere of radius 5 centered at the origin

28. The set of points in space that lie 2 units from the point (0, 0, 1) and, at the same time, 2 units from the point (0, 0, −1)

Write inequalities to describe the sets in Exercises 29–34.

29. The slab bounded by the planes $z = 0$ and $z = 1$ (planes included)

30. The solid cube in the first octant bounded by the planes $x = 2$, $y = 2$, and $z = 2$

31. The half-space consisting of the points on and below the xy-plane

32. The upper hemisphere of the sphere of radius 1 centered at the origin

33. The (a) interior and (b) exterior of the sphere of radius 1 centered at the point (1, 1, 1)

34. The closed region bounded by the spheres of radius 1 and radius 2 centered at the origin. (*Closed* means the spheres are to be included. Had we wanted the spheres left out, we would have asked for the *open* region bounded by the spheres. This is analogous to the way we use "closed" and "open" to describe intervals: "closed" means endpoints included, "open" means endpoints left out. Closed sets include boundaries; open sets leave them out.)

Find the lengths and directions of the vectors in Exercises 35–46.

35. $2\mathbf{i} + \mathbf{j} - 2\mathbf{k}$

36. $3\mathbf{i} - 6\mathbf{j} + 2\mathbf{k}$

37. $\mathbf{i} + 4\mathbf{j} - 8\mathbf{k}$

38. $9\mathbf{i} - 2\mathbf{j} + 6\mathbf{k}$

39. $5\mathbf{k}$

40. $6\mathbf{i}$

41. $-4\mathbf{j}$

42. $\frac{3}{5}\mathbf{i} + \frac{4}{5}\mathbf{k}$

43. $-\frac{1}{3}\mathbf{j}+\frac{1}{4}\mathbf{k}$

44. $\frac{1}{\sqrt{2}}\mathbf{i}-\frac{1}{\sqrt{2}}\mathbf{k}$

45. $\frac{1}{\sqrt{6}}\mathbf{i}-\frac{1}{\sqrt{6}}\mathbf{j}-\frac{1}{\sqrt{6}}\mathbf{k}$

46. $\frac{\mathbf{i}}{\sqrt{3}}+\frac{\mathbf{j}}{\sqrt{3}}+\frac{\mathbf{k}}{\sqrt{3}}$

In Exercises 47–52, find the length and direction of the vector from point P_1 to point P_2.

47. $P_1(1, 1, 1)$, $P_2(3, 3, 0)$

48. $P_1(-1, 1, 5)$, $P_2(2, 5, 0)$

49. $P_1(1, 4, 5)$, $P_2(4, -2, 7)$

50. $P_1(3, 4, 5)$, $P_2(2, 3, 4)$

51. $P_1(0, 0, 0)$, $P_2(2, -2, -2)$

52. $P_1(5, 3, -2)$, $P_2(0, 0, 0)$

53. Find the vectors whose lengths and directions are given. Try to answer without writing anything down.

Length	Direction
a) 2	$\mathbf{i}$
b) -4	$\mathbf{j}$
c) $\sqrt{3}$	$\mathbf{k}$
d) $\frac{1}{2}$	$\frac{3}{5}\mathbf{j}+\frac{4}{5}\mathbf{k}$
e) 7	$\frac{6}{7}\mathbf{i}-\frac{2}{7}\mathbf{j}+\frac{3}{7}\mathbf{k}$
f) a	$u_1\mathbf{i}+u_2\mathbf{j}+u_3\mathbf{k}$

54. Find a unit vector in the direction of $\mathbf{A} = 12\mathbf{i} - 5\mathbf{k}$.

55. Find a vector 5 units long in the direction of $\mathbf{A} = \mathbf{i} + \mathbf{j} + \mathbf{k}$.

56. Find a vector 1/2 unit long in the direction of $\mathbf{A} = \mathbf{i} - \mathbf{j}$.

Spheres and Distance

57. Find the centers and radii of the following spheres.

a) $(x+2)^2+y^2+(z-2)^2=8$

b) $\left(x+\frac{1}{2}\right)^2+\left(y+\frac{1}{2}\right)^2+\left(z+\frac{1}{2}\right)^2=\frac{21}{4}$

c) $(x-\sqrt{2})^2+(y-\sqrt{2})^2+(z+\sqrt{2})^2=2$

d) $x^2+\left(y+\frac{1}{3}\right)^2+\left(z-\frac{1}{3}\right)^2=\frac{29}{9}$

58. Find equations for the spheres whose centers and radii are given here.

Center	Radius
a) $(1, 2, 3)$	$\sqrt{14}$
b) $(0, -1, 5)$	2
c) $(-2, 0, 0)$	$\sqrt{3}$
d) $(0, -7, 0)$	7

59. Find a formula for the distance from the point $P(x, y, z)$ to the

a) x-axis b) y-axis c) z-axis.

60. Find a formula for the distance from the point $P(x, y, z)$ to the

a) xy-plane b) yz-plane c) xz-plane.

11.3 Dot Products

We now introduce the dot product of two vectors, the first of the two methods we shall learn for multiplying vectors together. Our motivation is the need to calculate the work done by a constant force in displacing a mass. When the force and displacement are represented as vectors, the dot product of the two vectors gives the work done by the force during the displacement.

Dot products, also called *scalar products* because the resulting products are numbers and not vectors, have applications in mathematics as well as in engineering and physics. In this section we present the algebraic and geometric properties on which many of these applications depend. The second kind of vector product, the cross product, will be described in the next section.

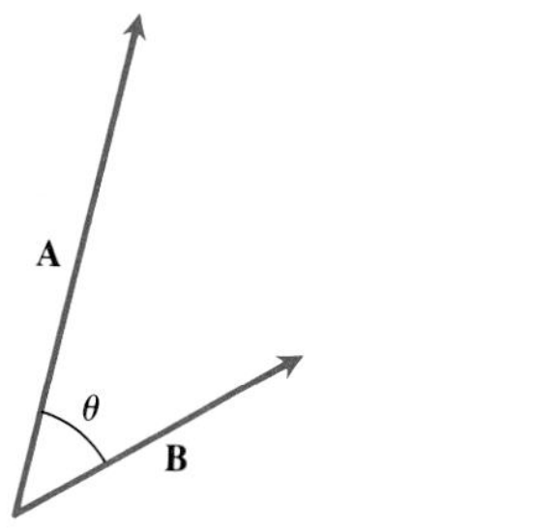

11.19 $\mathbf{A} \cdot \mathbf{B}$ is the number $|\mathbf{A}||\mathbf{B}| \cos \theta$.

DEFINITION

The **scalar product A · B** ("A dot B") or **dot product** of two vectors **A** and **B** is the number

$$\mathbf{A} \cdot \mathbf{B} = |\mathbf{A}||\mathbf{B}| \cos \theta, \tag{1}$$

where θ measures the smaller angle made by **A** and **B** when their initial points coincide (as in Fig. 11.19).

In words, the scalar product of **A** and **B** is the length of **A** times the length of **B** times the cosine of the angle between **A** and **B**. The product is a scalar, not a vector. It is often called the dot product because of the dot in the notation **A · B**.

From Eq. (1) we see that the scalar product of two vectors is positive when the angle between them is acute, negative when the angle is obtuse.

Since the angle a vector **A** makes with itself is zero, and the cosine of the angle is 1,

$$\mathbf{A} \cdot \mathbf{A} = |\mathbf{A}||\mathbf{A}| \cos 0 = |\mathbf{A}||\mathbf{A}|(1) = |\mathbf{A}|^2 \quad \text{or} \quad |\mathbf{A}| = \sqrt{\mathbf{A} \cdot \mathbf{A}}. \tag{2}$$

This gives us a handy way to calculate a vector's length, as we shall see.

Calculation

To calculate **A · B** from the components of **A** and **B**, we let

$$\mathbf{A} = a_1\mathbf{i} + a_2\mathbf{j} + a_3\mathbf{k},$$

$$\mathbf{B} = b_1\mathbf{i} + b_2\mathbf{j} + b_3\mathbf{k},$$

and

$$\mathbf{C} = \mathbf{B} - \mathbf{A} = (b_1 - a_1)\mathbf{i} + (b_2 - a_2)\mathbf{j} + (b_3 - a_3)\mathbf{k}.$$

Then we apply the law of cosines to a triangle whose sides represent the vectors **A**, **B**, and **C** (Fig. 11.20) and obtain

$$|\mathbf{C}|^2 = |\mathbf{A}|^2 + |\mathbf{B}|^2 - 2|\mathbf{A}||\mathbf{B}| \cos \theta,$$

$$|\mathbf{A}||\mathbf{B}| \cos \theta = \frac{|\mathbf{A}|^2 + |\mathbf{B}|^2 - |\mathbf{C}|^2}{2}.$$

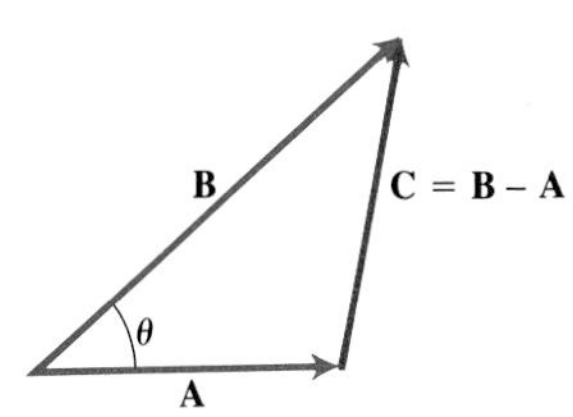

11.20 Equation (3) is obtained by applying the law of cosines to a triangle whose sides represent **A**, **B**, and **C** = **B** − **A**.

The left side of this equation is **A · B**, and we may evaluate the right side by applying Eq. (4) of Section 11.2 to find the lengths of **A**, **B**, and **C**. The result of this algebra is the formula

$$\mathbf{A} \cdot \mathbf{B} = a_1b_1 + a_2b_2 + a_3b_3. \tag{3}$$

Thus, to find the scalar product of two given vectors we multiply their corresponding **i**, **j**, and **k** components and add the results. In particular, from

Eq. (2) we have

$$|\mathbf{A}| = \sqrt{\mathbf{A}\cdot\mathbf{A}} = \sqrt{a_1^2 + a_2^2 + a_3^2}. \tag{4}$$

When we solve Eq. (1) for θ, we get a formula for finding the angle between two nonzero vectors.

The Angle between Two Vectors

The angle between two nonzero vectors **A** and **B** is

$$\theta = \cos^{-1}\left(\frac{\mathbf{A}\cdot\mathbf{B}}{|\mathbf{A}||\mathbf{B}|}\right). \tag{5}$$

Since the values of the arc cosine lie between 0 and π, Eq. (5) automatically gives the smaller of the two angles made by **A** and **B**, when there is one.

EXAMPLE 1 Find the angle between $\mathbf{A} = \mathbf{i} - 2\mathbf{j} - 2\mathbf{k}$ and $\mathbf{B} = 6\mathbf{i} + 3\mathbf{j} + 2\mathbf{k}$.

Solution We use Eq. (5):

$$\mathbf{A}\cdot\mathbf{B} = (1)(6) + (-2)(3) + (-2)(2) = 6 - 6 - 4 = -4$$

$$|\mathbf{A}| = \sqrt{\mathbf{A}\cdot\mathbf{A}} = \sqrt{(1)^2 + (-2)^2 + (-2)^2} = \sqrt{9} = 3$$

$$|\mathbf{B}| = \sqrt{\mathbf{B}\cdot\mathbf{B}} = \sqrt{(6)^2 + (3)^2 + (2)^2} = \sqrt{49} = 7$$

$$\begin{aligned} \theta &= \cos^{-1}\left(\frac{\mathbf{A}\cdot\mathbf{B}}{|\mathbf{A}||\mathbf{B}|}\right) && \text{(Eq. 5)} \\ &= \cos^{-1}\left(\frac{-4}{(3)(7)}\right) \\ &= \cos^{-1}\left(-\frac{4}{21}\right) \\ &= 101^\circ. && \text{(Calculator, rounded)} \end{aligned}$$

■

Laws of Multiplication

From the equation $\mathbf{A}\cdot\mathbf{B} = a_1b_1 + a_2b_2 + a_3b_3$, we can see right away that

$$\mathbf{A}\cdot\mathbf{B} = \mathbf{B}\cdot\mathbf{A}. \tag{6}$$

In other words, scalar multiplication is commutative. We can also see from Eq. (3) that if c is any number, then

$$(c\mathbf{A})\cdot\mathbf{B} = \mathbf{A}\cdot(c\mathbf{B}) = c(\mathbf{A}\cdot\mathbf{B}). \tag{7}$$

If $\mathbf{C} = c_1\mathbf{i} + c_2\mathbf{j} + c_3\mathbf{k}$ is any third vector, then

$$\begin{aligned}\mathbf{A}\cdot(\mathbf{B}+\mathbf{C}) &= a_1(b_1+c_1) + a_2(b_2+c_2) + a_3(b_3+c_3)\\ &= (a_1b_1 + a_2b_2 + a_3b_3) + (a_1c_1 + a_2c_2 + a_3c_3)\\ &= \mathbf{A}\cdot\mathbf{B} + \mathbf{A}\cdot\mathbf{C}.\end{aligned}$$

Hence scalar products obey the distributive law:

$$\mathbf{A}\cdot(\mathbf{B}+\mathbf{C}) = \mathbf{A}\cdot\mathbf{B} + \mathbf{A}\cdot\mathbf{C}. \tag{8}$$

If we combine this with the commutative law, Eq. (6), it is also evident that

$$(\mathbf{A}+\mathbf{B})\cdot\mathbf{C} = \mathbf{A}\cdot\mathbf{C} + \mathbf{B}\cdot\mathbf{C}. \tag{9}$$

Equations (8) and (9) together permit us to multiply sums of vectors by the familiar laws of algebra. For example,

$$(\mathbf{A}+\mathbf{B})\cdot(\mathbf{C}+\mathbf{D}) = \mathbf{A}\cdot\mathbf{C} + \mathbf{A}\cdot\mathbf{D} + \mathbf{B}\cdot\mathbf{C} + \mathbf{B}\cdot\mathbf{D}. \tag{10}$$

Orthogonal Vectors

Two vectors whose scalar product is zero are said to be orthogonal. From the equation $\mathbf{A}\cdot\mathbf{B} = |\mathbf{A}||\mathbf{B}|\cos\theta$, we can see that when neither $|\mathbf{A}|$ nor $|\mathbf{B}|$ is zero, $\mathbf{A}\cdot\mathbf{B}$ is zero if and only if $\cos\theta$ is zero, that is, when θ equals 90°. Hence, for the vectors we are dealing with, "orthogonal" means the same as "perpendicular." (In other scientific contexts in which the word is used, there may be no such geometric interpretation.)

The zero vector $\mathbf{0} = 0\mathbf{i} + 0\mathbf{j} + 0\mathbf{k}$ is orthogonal to every vector because its dot product with every vector is zero.

DEFINITION

Vectors **A** and **B** are **orthogonal** if $\mathbf{A}\cdot\mathbf{B} = 0$. (11)

EXAMPLE 2 The vectors $\mathbf{A} = 3\mathbf{i} - 2\mathbf{j} + \mathbf{k}$ and $\mathbf{B} = 2\mathbf{j} + 4\mathbf{k}$ are orthogonal because

$$\mathbf{A}\cdot\mathbf{B} = (3)(0) - (2)(2) + (1)(4) = 0.$$

Vector Projections and Scalar Components

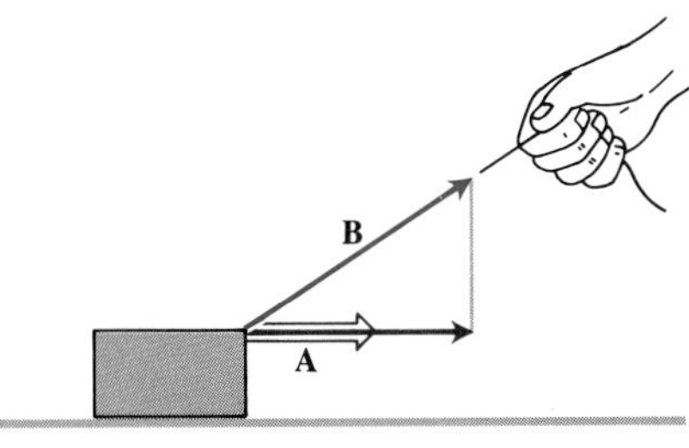

11.21 Pulling on the box has the effect of moving the box in the direction of **A**. The effective force in this direction is represented by the vector projection of **B** onto **A**.

The vector we get by projecting a vector **B** onto the line through a vector **A** is called the **vector projection of B onto A,** sometimes denoted

$$\text{proj}_{\mathbf{A}}\mathbf{B} \qquad \text{("the vector projection of } \mathbf{B} \text{ onto } \mathbf{A}\text{")}.$$

If **B** represents a force, then the vector projection of **B** onto **A** represents the effective force in the direction of **A** (Fig. 11.21).

If the angle between **B** and **A** is acute, the length of the vector projection of **B** onto **A** is $|\mathbf{B}|\cos\theta$. If the angle is obtuse, its cosine is negative and

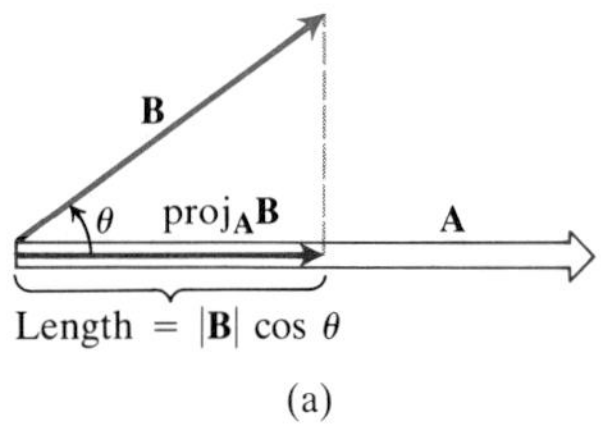

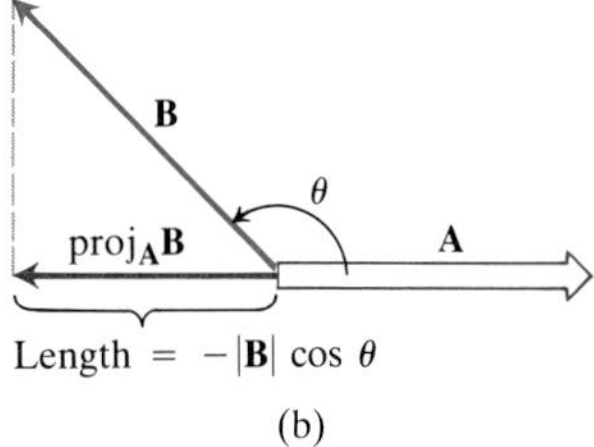

11.22 (a) When θ is acute, the length of the vector projection of **B** onto **A** is $|\mathbf{B}|\cos\theta$, the scalar component of **B** in the direction of **A**. (b) When θ is obtuse, the scalar component of **B** in the direction of **A** is negative and the length of the vector projection of **B** onto **A** is $-|\mathbf{B}|\cos\theta$.

the length of the vector projection of **B** onto **A** is $-|\mathbf{B}|\cos\theta$. In either case, the number $|\mathbf{B}|\cos\theta$ is called the **scalar component of B in the direction of A** (Fig. 11.22).

The scalar component of **B** in the direction of **A** can be found by dividing both sides of the equation $\mathbf{A}\cdot\mathbf{B} = |\mathbf{A}||\mathbf{B}|\cos\theta$ by $|\mathbf{A}|$. This gives

$$|\mathbf{B}|\cos\theta = \frac{\mathbf{A}\cdot\mathbf{B}}{|\mathbf{A}|} = \mathbf{B}\cdot\frac{\mathbf{A}}{|\mathbf{A}|}. \tag{12}$$

Equation (12) says that the scalar component of **B** in the direction of **A** can be obtained by "dotting" **B** with the direction of **A**.

The vector projection of **B** onto **A** is the scalar component of **B** in the direction of **A** times the direction of **A**. If the angle between **A** and **B** is acute, the vector projection has length $|\mathbf{B}|\cos\theta$ and direction $\mathbf{A}/|\mathbf{A}|$. If the angle is obtuse, the vector projection has length $-|\mathbf{B}|\cos\theta$ and direction $-\mathbf{A}/|\mathbf{A}|$. In either case,

$$\text{proj}_{\mathbf{A}}\mathbf{B} = (|\mathbf{B}|\cos\theta)\frac{\mathbf{A}}{|\mathbf{A}|}. \tag{13}$$

Equations (12) and (13) together give a useful way to find $\text{proj}_{\mathbf{A}}\mathbf{B}$:

$$\begin{aligned}
\text{proj}_{\mathbf{A}}\mathbf{B} &= (|\mathbf{B}|\cos\theta)\frac{\mathbf{A}}{|\mathbf{A}|} && \text{(Eq. 13)}\\
&= \left(\frac{\mathbf{A}\cdot\mathbf{B}}{|\mathbf{A}|}\right)\frac{\mathbf{A}}{|\mathbf{A}|} && \text{(Eq. 12)}\\
&= \frac{\mathbf{A}\cdot\mathbf{B}}{\mathbf{A}\cdot\mathbf{A}}\mathbf{A}. && (|\mathbf{A}|^2 = \mathbf{A}\cdot\mathbf{A})
\end{aligned} \tag{14}$$

EXAMPLE 3 Find the vector projection of $\mathbf{B} = 6\mathbf{i} + 3\mathbf{j} + 2\mathbf{k}$ onto $\mathbf{A} = \mathbf{i} - 2\mathbf{j} - 2\mathbf{k}$ and the scalar component of **B** in the direction of **A**.

Solution We find $\text{proj}_{\mathbf{A}}\mathbf{B}$ from Eq. (14):

$$\begin{aligned}
\text{proj}_{\mathbf{A}}\mathbf{B} &= \frac{\mathbf{A}\cdot\mathbf{B}}{\mathbf{A}\cdot\mathbf{A}}\mathbf{A} = \frac{6-6-4}{1+4+4}(\mathbf{i} - 2\mathbf{j} - 2\mathbf{k})\\
&= -\frac{4}{9}(\mathbf{i} - 2\mathbf{j} - 2\mathbf{k}) = -\frac{4}{9}\mathbf{i} + \frac{8}{9}\mathbf{j} + \frac{8}{9}\mathbf{k}.
\end{aligned}$$

We find the scalar component of **B** in the direction of **A** from Eq. (12):

$$\begin{aligned}
|\mathbf{B}|\cos\theta &= \mathbf{B}\cdot\frac{\mathbf{A}}{|\mathbf{A}|} = (6\mathbf{i} + 3\mathbf{j} + 2\mathbf{k})\cdot\left(\frac{1}{3}\mathbf{i} - \frac{2}{3}\mathbf{j} - \frac{2}{3}\mathbf{k}\right)\\
&= 2 - 2 - \frac{4}{3} = -\frac{4}{3}.
\end{aligned}$$

Writing a Vector as a Sum of Orthogonal Vectors

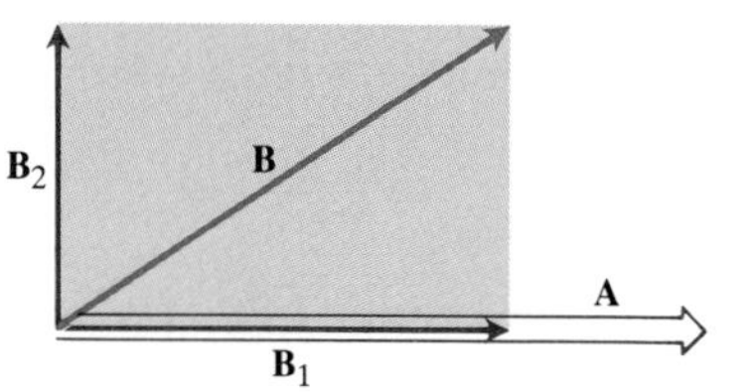

11.23 The vector **B** as the sum of vectors parallel and orthogonal to **A**.

In mechanics, we often want to express a vector **B** as a sum of a vector $\mathbf{B}_1$ parallel to a vector **A** and a vector $\mathbf{B}_2$ orthogonal to **A** (Fig. 11.23). We can do this by writing **B** as a sum of its vector projection on **A** plus whatever is left over, because the leftover part will automatically be orthogonal to **A**.

Where Vectors Came From

Although Aristotle used vectors to describe the effects of forces, the idea of resolving vectors into geometric components parallel to the coordinate axes comes from Descartes. The algebra of vectors we use today was developed simultaneously and independently in the 1870s by Josiah Willard Gibbs (1839–1903), a mathematical physicist at Yale University, and by the English mathematical physicist Oliver Heaviside (1850–1925), the Heaviside of Heaviside layer fame. The works of Gibbs and Heaviside grew out of more complicated mathematical theories developed some years earlier by the Irish mathematician William Hamilton (1805–1865) and the German scientist and geometer Hermann Grassman (1809–1877). Hamilton's quaternions and Grassman's algebraic forms are still in use, but tend to appear in more theoretical work.

Formula for Writing B as a Vector Parallel to A Plus a Vector Orthogonal to A

$$\mathbf{B} = \underbrace{\frac{\mathbf{A}\cdot\mathbf{B}}{\mathbf{A}\cdot\mathbf{A}}\mathbf{A}}_{\text{Parallel to }\mathbf{A}} + \underbrace{\left(\mathbf{B} - \frac{\mathbf{A}\cdot\mathbf{B}}{\mathbf{A}\cdot\mathbf{A}}\mathbf{A}\right)}_{\text{Orthogonal to }\mathbf{A}}. \tag{15}$$

The vector $\mathbf{B}_1$, being the vector projection of $\mathbf{B}$ onto $\mathbf{A}$, is parallel to $\mathbf{A}$, while $\mathbf{B}_2$ can be seen to be orthogonal to $\mathbf{A}$ because $\mathbf{A}\cdot\mathbf{B}_2$ is zero:

$$\mathbf{A}\cdot\mathbf{B}_2 = \mathbf{A}\cdot\left(\mathbf{B} - \frac{\mathbf{A}\cdot\mathbf{B}}{\mathbf{A}\cdot\mathbf{A}}\mathbf{A}\right) = \mathbf{A}\cdot\mathbf{B} - \frac{\mathbf{A}\cdot\mathbf{B}}{\mathbf{A}\cdot\mathbf{A}}\mathbf{A}\cdot\mathbf{A} = \mathbf{A}\cdot\mathbf{B} - \mathbf{A}\cdot\mathbf{B} = 0. \tag{16}$$

EXAMPLE 4 Express $\mathbf{B} = 2\mathbf{i} + \mathbf{j} - 3\mathbf{k}$ as the sum of a vector parallel to $\mathbf{A} = 3\mathbf{i} - \mathbf{j}$ and a vector orthogonal to $\mathbf{A}$.

Solution We use Eq. (15). With

$$\mathbf{A}\cdot\mathbf{B} = 6 - 1 = 5 \quad\text{and}\quad \mathbf{A}\cdot\mathbf{A} = 9 + 1 = 10,$$

Eq. (15) gives

$$\begin{aligned}\mathbf{B} = \frac{\mathbf{A}\cdot\mathbf{B}}{\mathbf{A}\cdot\mathbf{A}}\mathbf{A} + \left(\mathbf{B} - \frac{\mathbf{A}\cdot\mathbf{B}}{\mathbf{A}\cdot\mathbf{A}}\mathbf{A}\right) &= \frac{5}{10}(3\mathbf{i} - \mathbf{j}) + \left(2\mathbf{i} + \mathbf{j} - 3\mathbf{k} - \frac{5}{10}(3\mathbf{i} - \mathbf{j})\right)\\ &= \left(\frac{3}{2}\mathbf{i} - \frac{1}{2}\mathbf{j}\right) + \left(\frac{1}{2}\mathbf{i} + \frac{3}{2}\mathbf{j} - 3\mathbf{k}\right).\end{aligned}$$

Check: The first vector in the sum is parallel to $\mathbf{A}$ because it is $(1/2)\mathbf{A}$. The second vector in the sum is orthogonal to $\mathbf{A}$ because

$$\left(\frac{1}{2}\mathbf{i} + \frac{3}{2}\mathbf{j} - 3\mathbf{k}\right)\cdot(3\mathbf{i} - \mathbf{j}) = \frac{3}{2} - \frac{3}{2} = 0.$$

■

Lines in the Plane and Distances from Points to Lines

Dot products give a new understanding of the equations we write for lines in the plane and a quick way to calculate distances from points to lines.

Suppose L is a line through the point $P_0(x_0, y_0)$ perpendicular to a vector $\mathbf{N} = A\mathbf{i} + B\mathbf{j}$ (Fig. 11.24). Then a point $P(x, y)$ lies on L if and only if

$$\mathbf{N}\cdot\overrightarrow{P_0P} = 0 \qquad\text{or}\qquad A(x - x_0) + B(y - y_0) = 0. \tag{17}$$

When rearranged, the second equation becomes

$$Ax + By = Ax_0 + By_0. \tag{18}$$

Line through $P(x_0, y_0)$ Perpendicular to $\mathbf{N} = A\mathbf{i} + B\mathbf{j}$

$$Ax + By = C, \qquad C = Ax_0 + By_0 \tag{19}$$

Notice how the components of $\mathbf{N}$ become coefficients in the equation $Ax + By = C$.

y
L
$P_0(x_0, y_0)$
$P(x, y)$
$\mathbf{N} = A\mathbf{i} + B\mathbf{j}$
O
x

11.24 A point $P(x, y)$ lies on the line through P_0 perpendicular to $\mathbf{N}$ if and only if $\mathbf{N}\cdot\overrightarrow{P_0P} = 0$.

EXAMPLE 5 Find an equation for the line through $P_0(3, 5)$ perpendicular to $\mathbf{N} = \mathbf{i} + 2\mathbf{j}$.

Solution We use Eq. (19) with $A = 1$, $B = 2$, and $C = (1)(3) + (2)(5) = 13$ to get

$$x + 2y = 13.$$

EXAMPLE 6 Find the distance from the point $S(4, 4)$ to the line L: $x + 3y = 6$.

To find the distance from a point S to the line L: $Ax + By = C$, find

1. a point P on L,
2. $\overrightarrow{PS}$,
3. the direction of $\mathbf{N} = A\mathbf{i} + B\mathbf{j}$.

Then, calculate the distance as

$$\left| \overrightarrow{PS} \cdot \frac{\mathbf{N}}{|\mathbf{N}|} \right|.$$

Solution We find a point P on the line and calculate the distance as the length of the vector projection of $\overrightarrow{PS}$ onto a vector $\mathbf{N}$ perpendicular to the line (Fig. 11.25). Any point on the line will do for P and we can find $\mathbf{N}$ from the coefficients of $x + 3y = 6$ as $\mathbf{N} = \mathbf{i} + 3\mathbf{j}$. With $P = (0, 2)$, say, we then have

$$\overrightarrow{PS} = (4 - 0)\mathbf{i} + (4 - 2)\mathbf{j} = 4\mathbf{i} + 2\mathbf{j}$$

$$\text{Distance from } S \text{ to } L = \left|\text{proj}_{\mathbf{N}}\, \overrightarrow{PS}\right| = \left| \overrightarrow{PS} \cdot \frac{\mathbf{N}}{|\mathbf{N}|} \right|$$

$$= \left| (4\mathbf{i} + 2\mathbf{j}) \cdot \frac{\mathbf{i} + 3\mathbf{j}}{\sqrt{(1)^2 + (3)^2}} \right| = \frac{4 + 6}{\sqrt{10}} = \sqrt{10}.$$

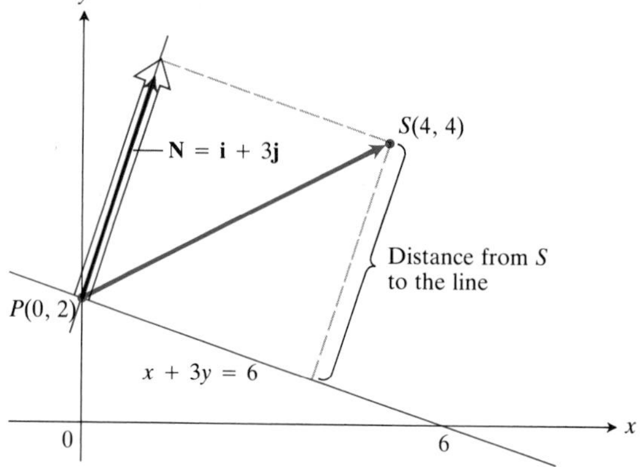

11.25 The distance from point S to the line is the length of the vector projection of $\overrightarrow{PS}$ onto **N**.

Work

To return to our example at the beginning of this section, in mechanics, the work done by a constant force **F** when the point of application undergoes a displacement $\overrightarrow{PQ}$ (Fig. 11.26) is defined to be the dot product of **F** with $\overrightarrow{PQ}$.

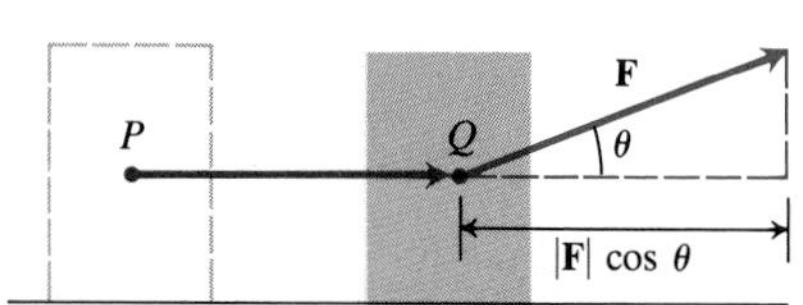

11.26 The work done by a constant force **F** during a displacement $\overrightarrow{PQ}$ is $(|\mathbf{F}| \cos\theta)|\overrightarrow{PQ}|$.

DEFINITION

The **work** done by a constant force **F** acting through a displacement $\overrightarrow{PQ}$ is

$$\text{Work} = \mathbf{F} \cdot \overrightarrow{PQ} = |\mathbf{F}||\overrightarrow{PQ}| \cos\theta. \tag{20}$$

EXAMPLE 7 If $|\mathbf{F}| = 40$ newtons (about 9 pounds), $|\overrightarrow{PQ}| = 3$ m, and $\theta = 60°$, the work done by **F** in acting from P to Q is

$$\begin{aligned} \text{Work} &= |\mathbf{F}||\overrightarrow{PQ}| \cos \theta && \text{(Eq. 20)} \\ &= (40)(3) \cos 60° && \text{(Given values)} \\ &= (120)(1/2) \\ &= 60 \text{ newton-meters.} \end{aligned}$$

■

Exercises 11.3

In Exercises 1–12, find $\mathbf{A} \cdot \mathbf{B}$, $|\mathbf{A}|$, $|\mathbf{B}|$, the cosine of the angle between **A** and **B**, the scalar component of **B** in the direction of **A**, and the vector projection of **B** onto **A**.

1. $\mathbf{A} = 3\mathbf{i} + 2\mathbf{j}$, $\mathbf{B} = 5\mathbf{j} + \mathbf{k}$

2. $\mathbf{A} = \mathbf{i}$, $\mathbf{B} = 5\mathbf{j} - 3\mathbf{k}$

3. $\mathbf{A} = 3\mathbf{i} - 2\mathbf{j} - \mathbf{k}$, $\mathbf{B} = -2\mathbf{j}$

4. $\mathbf{A} = -2\mathbf{i} + 7\mathbf{j}$, $\mathbf{B} = \mathbf{k}$

5. $\mathbf{A} = 5\mathbf{j} - 3\mathbf{k}$, $\mathbf{B} = \mathbf{i} + \mathbf{j} + \mathbf{k}$

6. $\mathbf{A} = \frac{1}{\sqrt{2}}\mathbf{i} + \frac{1}{\sqrt{3}}\mathbf{j} + \frac{1}{\sqrt{6}}\mathbf{k}$, $\mathbf{B} = \frac{1}{\sqrt{2}}\mathbf{j} - \mathbf{k}$

7. $\mathbf{A} = -\mathbf{i} + \mathbf{j}$, $\mathbf{B} = \sqrt{2}\mathbf{i} + \sqrt{3}\mathbf{j} + 2\mathbf{k}$

8. $\mathbf{A} = \mathbf{i} + \mathbf{k}$, $\mathbf{B} = \mathbf{i} + \mathbf{j} + \mathbf{k}$

9. $\mathbf{A} = 2\mathbf{i} - 4\mathbf{j} + \sqrt{5}\mathbf{k}$, $\mathbf{B} = -2\mathbf{i} + 4\mathbf{j} - \sqrt{5}\mathbf{k}$

10. $\mathbf{A} = -5\mathbf{i} + \mathbf{j}$, $\mathbf{B} = 2\mathbf{i} + \sqrt{17}\mathbf{j} + 10\mathbf{k}$

11. $\mathbf{A} = 10\mathbf{i} + 11\mathbf{j} - 2\mathbf{k}$, $\mathbf{B} = 3\mathbf{j} + 4\mathbf{k}$

12. $\mathbf{A} = 2\mathbf{i} + 10\mathbf{j} - 11\mathbf{k}$, $\mathbf{B} = 2\mathbf{i} + 2\mathbf{j} + \mathbf{k}$

13. Write $\mathbf{B} = 3\mathbf{j} + 4\mathbf{k}$ as the sum of a vector parallel to $\mathbf{A} = \mathbf{i} + \mathbf{j}$ and a vector orthogonal to **A**.

14. Write $\mathbf{B} = \mathbf{j} + \mathbf{k}$ as the sum of a vector parallel to $\mathbf{A} = \mathbf{i} + \mathbf{j}$ and a vector orthogonal to **A**.

15. Write $\mathbf{B} = 8\mathbf{i} + 4\mathbf{j} - 12\mathbf{k}$ as the sum of a vector parallel to $\mathbf{A} = \mathbf{i} + 2\mathbf{j} - \mathbf{k}$ and a vector orthogonal to **A**.

16. $\mathbf{B} = \mathbf{i} + (\mathbf{j} + \mathbf{k})$ is already the sum of a vector parallel to **i** and a vector orthogonal to **i**. If you use Eq. (15) with $\mathbf{A} = \mathbf{i}$, do you get $\mathbf{B}_1 = \mathbf{i}$ and $\mathbf{B}_2 = \mathbf{j} + \mathbf{k}$? Try it and find out.

In Exercises 17–20, find an equation for the line in the xy-plane that passes through the given point perpendicular to the given vector. Then sketch the line. Include the vector in your sketch as a vector starting at the origin (as in Fig. 11.24).

17. $P(2, 1)$, $\mathbf{i} + 2\mathbf{j}$

18. $P(-2, 1)$, $\mathbf{i} - \mathbf{j}$

19. $P(-1, 2)$, $-2\mathbf{i} - \mathbf{j}$

20. $P(-1, 2)$, $2\mathbf{i} - 3\mathbf{j}$

In Exercises 21–24, find the distance in the xy-plane from the point to the line.

21. $S(2, 8)$, $x + 3y = 6$

22. $S(0, 0)$, $x + 3y = 6$

23. $S(2, 1)$, $x + y = 1$

24. $S(1, 3)$, $y = -2x$

25. Show that the vectors

$$\mathbf{A} = \frac{1}{\sqrt{3}}(\mathbf{i} - \mathbf{j} + \mathbf{k}), \quad \mathbf{B} = \frac{1}{\sqrt{2}}(\mathbf{j} + \mathbf{k}),$$

$$\mathbf{C} = \frac{1}{\sqrt{6}}(-2\mathbf{i} - \mathbf{j} + \mathbf{k})$$

are orthogonal to one another.

26. Find the vector projections of $\mathbf{D} = \mathbf{i} + \mathbf{j} + \mathbf{k}$ on the vectors **A**, **B**, and **C** of Exercise 25. Then show that **D** is the sum of these vector projections.

27. *Cancellation in dot products is risky.* In real-number multiplication, if $ab_1 = ab_2$ and a is not zero, we can safely cancel the a and conclude that $b_1 = b_2$. Not so for vector multiplication—if $\mathbf{A} \cdot \mathbf{B}_1 = \mathbf{A} \cdot \mathbf{B}_2$ and $\mathbf{A} \neq \mathbf{0}$, it is not safe to conclude that $\mathbf{B}_1 = \mathbf{B}_2$. See if you can come up with an example. Keep it simple: Experiment with **i**, **j**, and **k**.

28. *Sums and differences.* In Fig. 11.27, it looks as if $\mathbf{v}_1 + \mathbf{v}_2$ and $\mathbf{v}_1 - \mathbf{v}_2$ are orthogonal. Is this mere coincidence, or are there circumstances under which we may expect the sum of two vectors to be orthogonal to their difference? Find out by expanding the left-hand side of the equation

$$(\mathbf{v}_1 + \mathbf{v}_2) \cdot (\mathbf{v}_1 - \mathbf{v}_2) = 0.$$

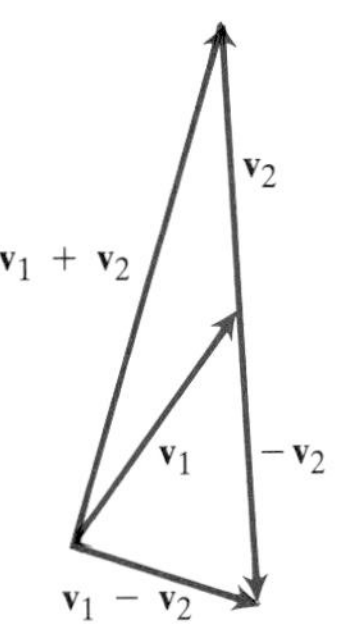

11.27 Exercise 28 is about the vectors in this figure.

CALCULATOR Use a calculator to find the angles in Exercises 29–32 to the nearest tenth of a degree.

29. Find the interior angles of the triangle ABC whose vertices are $A(-1, 0, 2)$, $B(2, 1, -1)$, and $C(1, -2, 2)$.

30. Find the angle between $\mathbf{A} = 2\mathbf{i} + 2\mathbf{j} + \mathbf{k}$ and $\mathbf{B} = 2\mathbf{i} + 10\mathbf{j} - 11\mathbf{k}$.

31. Find the angle between the diagonal of a cube and the diagonal of one of its faces. (*Hint:* Use a cube whose edges represent **i**, **j**, and **k**.)

32. Find the angle between the diagonal of a cube and one of the edges it meets at a vertex.

Work

33. Find the work done by a force $\mathbf{F} = -5\mathbf{k}$ (magnitude 5 newtons) in moving an object along the line from the origin to the point $(1, 1, 1)$ (distance in meters).

34. A locomotive exerted a constant force of 60,000 newtons on a freight train while pulling it 1 km along a straight track. How much work did the locomotive do?

35. How much work does it take to slide a crate 20 m along a loading dock by pulling on it with a 200-newton force at an angle of 30° from the horizontal?

36. The wind passing over a boat's sail exerted a 1000-lb-magnitude force **F** as shown in Fig. 11.28. If the force vector made a 60° angle with the line of the boat's forward motion, how much work did the wind perform in moving the boat forward 1 mi? Answer in foot-pounds.

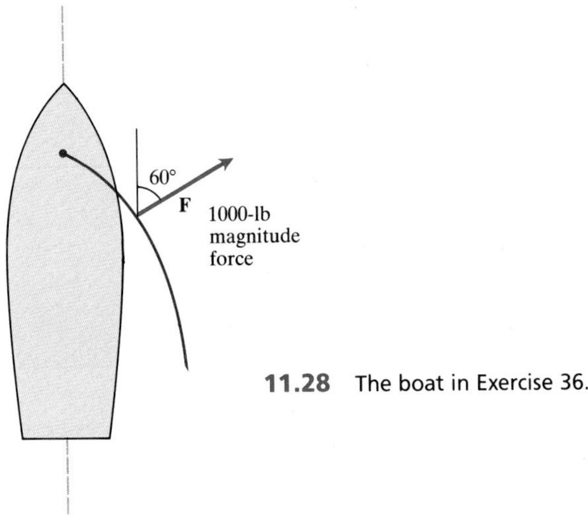

11.28 The boat in Exercise 36.

Angles between Curves

The angles between two differentiable curves at a point of intersection are the angles between the curves' tangents at these points. Find the angles between the curves in Exercises 37–41 at their points of intersection. (You will not need a calculator.)

37. $3x + y = 5, \quad 2x - y = 4$

38. $y = \sqrt{3}x - 1, \quad y = -\sqrt{3}x + 2$

39. $y = \sqrt{(3/4) + x}, \quad y = \sqrt{(3/4) - x}$

40. $y = x^3, \quad y = \sqrt{x}$ (two points of intersection)

41. $y = x^2, \quad y = \sqrt[3]{x}$ (two points of intersection)

11.4 Cross Products

When we turn a bolt by applying a force to a wrench (Fig. 11.29), the torque we produce acts along the axis of the bolt to drive the bolt forward. The magnitude of the torque depends on how far along the wrench the force is applied and on how much of the force is actually perpendicular to the wrench at that point. The number we use to measure the torque's magnitude is a product made up of the length of the vector arm **A** and the scalar component of **B** perpendicular to **A**. In the notation of Fig. 11.29,

$$\text{Magnitude of the torque vector} = |\mathbf{A}||\mathbf{B}|\sin\theta.$$

If we let **n** be a unit vector along the axis of the bolt in the direction of the torque, then the complete description of the torque vector is

$$\text{Torque vector} = \mathbf{n}|\mathbf{A}||\mathbf{B}|\sin\theta.$$

We call the torque the **vector product** of **A** and **B**.

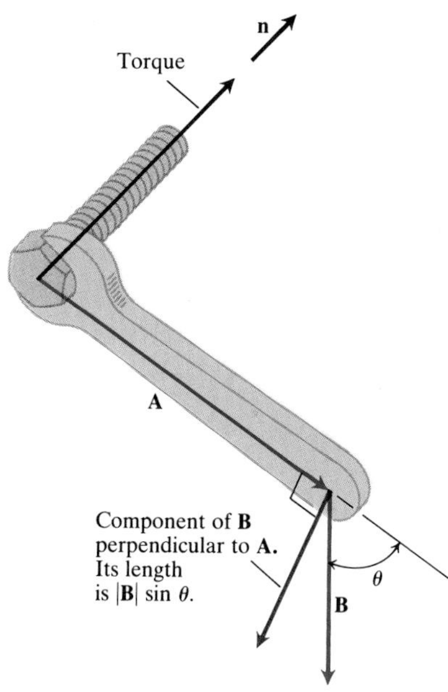

11.29 The torque vector describes the tendency of the force **B** to drive the bolt forward.

Vector products are widely used to describe the effects of forces in studies of electricity, magnetism, fluid mechanics, and planetary motion. The goal of this section is to acquaint you with the mathematical properties of vector products that account for their use in these fields. We shall see a number of these applications ourselves in Chapters 12 and 15, where we study motion in space and the integrals associated with fluid flow. In the next section we shall also see how vector products are combined with scalar products to produce equations for lines and planes in space.

The Vector Product of Two Vectors in Space

When we define vector products in mathematics, we start with two nonzero vectors **A** and **B** in space without requiring them to have any particular physical interpretation. If **A** and **B** are not parallel, they determine a plane. We select a unit vector **n** perpendicular to the plane by the **right-hand rule.** This means we choose **n** to be the unit (normal) vector that points the way your right thumb points when your fingers curl through the angle θ from **A** to **B** (Fig. 11.30). We then define the vector product $\mathbf{A} \times \mathbf{B}$ (pronounced "**A** cross **B**") of **A** and **B** to be the vector

$$\mathbf{A} \times \mathbf{B} = \mathbf{n}|\mathbf{A}||\mathbf{B}|\sin\theta. \tag{1}$$

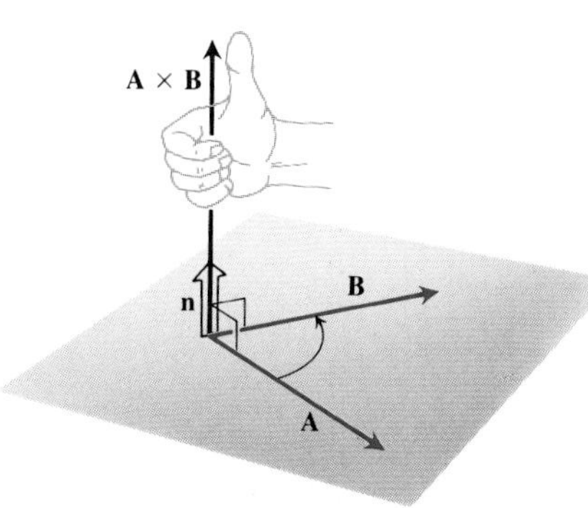

11.30 The construction of $\mathbf{A} \times \mathbf{B}$.

Since $\mathbf{A} \times \mathbf{B}$ is a scalar multiple of **n**, it is perpendicular to both **A** and **B**.

If θ approaches 0° or 180° in Eq. (1), the length of $\mathbf{A} \times \mathbf{B}$ approaches zero. We therefore define $\mathbf{A} \times \mathbf{B}$ to be **0** if **A** and **B** are parallel (and fail to determine a plane). This is consistent with our torque interpretation as well. If the force **B** in Fig. 11.29 is parallel to the wrench, meaning that we are trying to turn the bolt by pushing or pulling straight along the handle of the wrench, the torque produced is **0**.

If one or both of **A** and **B** is zero, we define $\mathbf{A} \times \mathbf{B}$ to be zero as well. Thus, the cross product of two vectors **A** and **B** is zero if and only if **A** and **B** are parallel or one or both of them is the zero vector.

The vector product of **A** and **B** is often called the **cross product** of **A** and **B** because of the cross in the notation $\mathbf{A} \times \mathbf{B}$. In contrast to the dot product $\mathbf{A} \cdot \mathbf{B}$, which is a scalar, the cross product of **A** and **B** is a vector.

A × B vs. B × A

Reversing the order of the factors in a nonzero vector product reverses the direction of the resulting vector. When the fingers of our right hand curl through the angle θ from **B** to **A**, our thumb points the opposite way and the unit vector we choose in forming $\mathbf{B} \times \mathbf{A}$ is the negative of the one we choose in forming $\mathbf{A} \times \mathbf{B}$ (Fig. 11.31). Thus,

$$\mathbf{B} \times \mathbf{A} = -(\mathbf{A} \times \mathbf{B}) \tag{2}$$

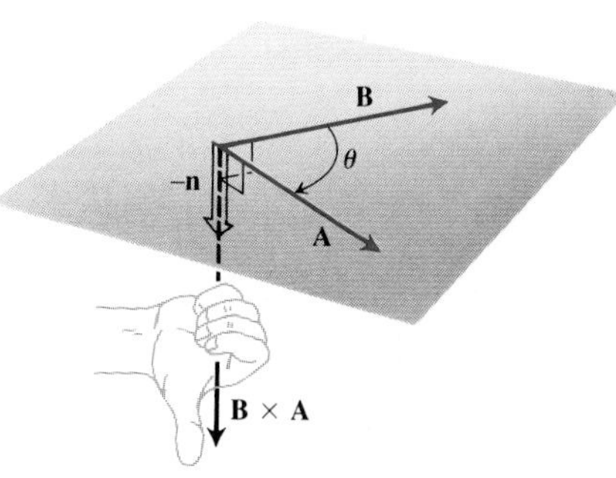

11.31 The construction of $\mathbf{B} \times \mathbf{A}$.

for all vectors **A** and **B**. Unlike the dot product, the cross product is not commutative.

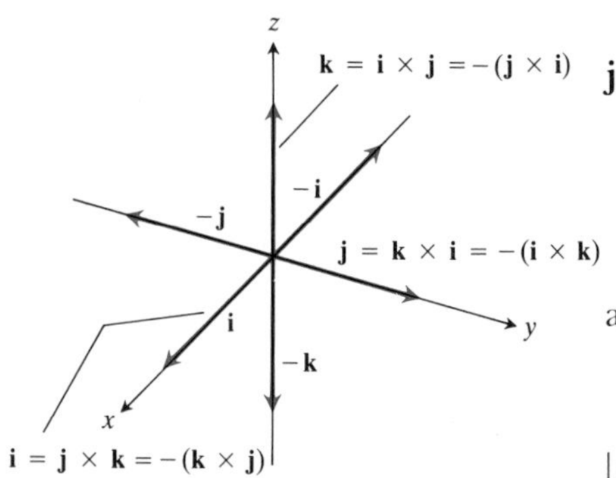

11.32 The pairwise cross products of **i**, **j**, and **k** can be read from this diagram.

When we apply the definition of the cross product to the unit vectors, **i**, **j**, and **k**, we find (Fig. 11.32)

$$\begin{aligned} \mathbf{i} \times \mathbf{j} &= -(\mathbf{j} \times \mathbf{i}) = \mathbf{k}, \\ \mathbf{j} \times \mathbf{k} &= -(\mathbf{k} \times \mathbf{j}) = \mathbf{i}, \\ \mathbf{k} \times \mathbf{i} &= -(\mathbf{i} \times \mathbf{k}) = \mathbf{j}, \end{aligned} \qquad (3)$$

and

$$\mathbf{i} \times \mathbf{i} = \mathbf{j} \times \mathbf{j} = \mathbf{k} \times \mathbf{k} = \mathbf{0}.$$

$|\mathbf{A} \times \mathbf{B}|$ Is the Area of a Parallelogram

Because **n** is a unit vector, the magnitude of $\mathbf{A} \times \mathbf{B}$ is

$$|\mathbf{A} \times \mathbf{B}| = |\mathbf{n}||\mathbf{A}||\mathbf{B}||\sin\theta| = |\mathbf{A}||\mathbf{B}||\sin\theta|. \qquad (4)$$

This is the area of the parallelogram determined by **A** and **B** (Fig. 11.33), $|\mathbf{A}|$ being the base of the parallelogram and $|\mathbf{B}||\sin\theta|$ the height.

The Magnitude of a Torque

Equation (4) is the equation we use to calculate magnitudes of torques.

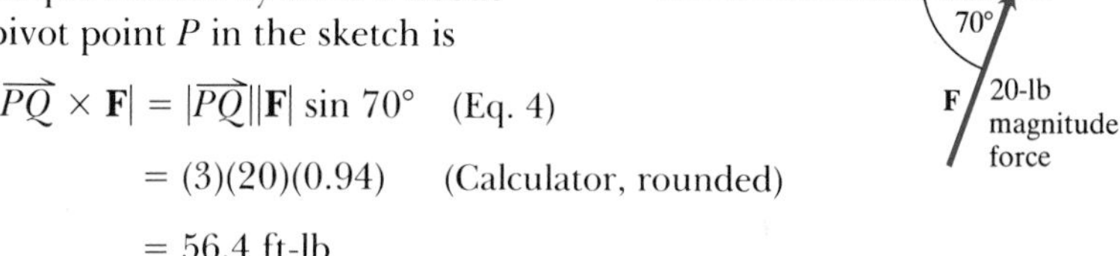

EXAMPLE 1 The magnitude of the torque exerted by force **F** about the pivot point P in the sketch is

$$\begin{aligned} |\overrightarrow{PQ} \times \mathbf{F}| &= |\overrightarrow{PQ}||\mathbf{F}| \sin 70^\circ && \text{(Eq. 4)} \\ &= (3)(20)(0.94) && \text{(Calculator, rounded)} \\ &= 56.4 \text{ ft-lb} \end{aligned}$$

■

The Associative and Distributive Laws

As a rule, cross-product multiplication is not associative because $(\mathbf{A} \times \mathbf{B}) \times \mathbf{C}$ lies in the plane of **A** and **B** whereas $\mathbf{A} \times (\mathbf{B} \times \mathbf{C})$ lies in the plane of **B** and **C**. However, the **scalar distributive law**

$$(r\mathbf{A}) \times (s\mathbf{B}) = (rs)\mathbf{A} \times \mathbf{B} \qquad (5)$$

does hold, as do the **vector distributive laws**

$$\mathbf{A} \times (\mathbf{B} + \mathbf{C}) = \mathbf{A} \times \mathbf{B} + \mathbf{A} \times \mathbf{C} \qquad (6)$$

and

$$(\mathbf{B} + \mathbf{C}) \times \mathbf{A} = \mathbf{B} \times \mathbf{A} + \mathbf{C} \times \mathbf{A}. \qquad (7)$$

As a special case of (5) we have

$$(-\mathbf{A}) \times \mathbf{B} = \mathbf{A} \times (-\mathbf{B}) = -(\mathbf{A} \times \mathbf{B}).$$

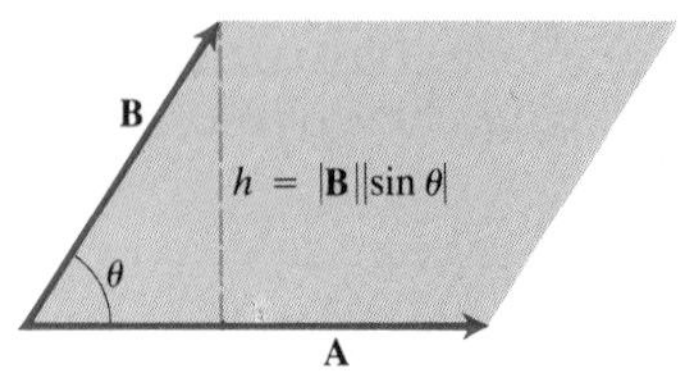

11.33 The parallelogram determined by **A** and **B**. Its area is $|\mathbf{A} \times \mathbf{B}|$.

The scalar distributive law can be verified by applying Eq. (1) to the products on both sides of Eq. (5) and comparing the results. The vector distributive law in Eq. (6) is not easy to prove, however, and we shall assume it here. Equation (7) follows from Eq. (6): just multiply both sides of Eq. (6) by -1 and reverse the orders of the products.

The Determinant Formula for A × B

Our next objective is to show how to calculate the components of $\mathbf{A} \times \mathbf{B}$ from the components of $\mathbf{A}$ and $\mathbf{B}$.

Suppose

$$\mathbf{A} = a_1\mathbf{i} + a_2\mathbf{j} + a_3\mathbf{k}, \qquad \mathbf{B} = b_1\mathbf{i} + b_2\mathbf{j} + b_3\mathbf{k}.$$

Then the distributive laws and the rules for multiplying $\mathbf{i}$, $\mathbf{j}$, and $\mathbf{k}$ tell us that

$$\begin{aligned}
\mathbf{A} \times \mathbf{B} &= (a_1\mathbf{i} + a_2\mathbf{j} + a_3\mathbf{k}) \times (b_1\mathbf{i} + b_2\mathbf{j} + b_3\mathbf{k}) \\
&= a_1b_1\mathbf{i} \times \mathbf{i} + a_1b_2\mathbf{i} \times \mathbf{j} + a_1b_3\mathbf{i} \times \mathbf{k} \\
&\quad + a_2b_1\mathbf{j} \times \mathbf{i} + a_2b_2\mathbf{j} \times \mathbf{j} + a_2b_3\mathbf{j} \times \mathbf{k} \qquad (8) \\
&\quad + a_3b_1\mathbf{k} \times \mathbf{i} + a_3b_2\mathbf{k} \times \mathbf{j} + a_3b_3\mathbf{k} \times \mathbf{k} \\
&= (a_2b_3 - a_3b_2)\mathbf{i} + (a_3b_1 - a_1b_3)\mathbf{j} + (a_1b_2 - a_2b_1)\mathbf{k}.
\end{aligned}$$

The terms at the end of Eq. (8) are the same as the terms in the expansion of the symbolic determinant

$$\begin{vmatrix} \mathbf{i} & \mathbf{j} & \mathbf{k} \\ a_1 & a_2 & a_3 \\ b_1 & b_2 & b_3 \end{vmatrix}$$

We may therefore use the following rule to calculate $\mathbf{A} \times \mathbf{B}$.

> If $\mathbf{A} = a_1\mathbf{i} + a_2\mathbf{j} + a_3\mathbf{k}$ and $\mathbf{B} = b_1\mathbf{i} + b_2\mathbf{j} + b_3\mathbf{k}$, then
>
> $$\mathbf{A} \times \mathbf{B} = \begin{vmatrix} \mathbf{i} & \mathbf{j} & \mathbf{k} \\ a_1 & a_2 & a_3 \\ b_1 & b_2 & b_3 \end{vmatrix} \qquad (9)$$

Equation (9) is remarkable, given that neither the associative nor commutative law holds for cross-product multiplication.

EXAMPLE 2 Find $\mathbf{A} \times \mathbf{B}$ and $\mathbf{B} \times \mathbf{A}$ if

$$\mathbf{A} = 2\mathbf{i} + \mathbf{j} + \mathbf{k}, \qquad \mathbf{B} = -4\mathbf{i} + 3\mathbf{j} + \mathbf{k}.$$

Solution We use Eq. (9) to find $\mathbf{A} \times \mathbf{B}$:

$$\mathbf{A} \times \mathbf{B} = \begin{vmatrix} \mathbf{i} & \mathbf{j} & \mathbf{k} \\ 2 & 1 & 1 \\ -4 & 3 & 1 \end{vmatrix} = \begin{vmatrix} 1 & 1 \\ 3 & 1 \end{vmatrix}\mathbf{i} - \begin{vmatrix} 2 & 1 \\ -4 & 1 \end{vmatrix}\mathbf{j} + \begin{vmatrix} 2 & 1 \\ -4 & 3 \end{vmatrix}\mathbf{k}$$

$$= -2\mathbf{i} - 6\mathbf{j} + 10\mathbf{k}.$$

Determinant Formulas

(For more information, see Appendix 7.)

$$\begin{vmatrix} a & b \\ c & d \end{vmatrix} = ad - bc$$

Example

$$\begin{vmatrix} 2 & 1 \\ -4 & 3 \end{vmatrix} = (2)(3) - (-4)(1) = 6 + 4 = 10$$

$$\begin{vmatrix} a_{11} & a_{12} & a_{13} \\ a_{21} & a_{22} & a_{23} \\ a_{31} & a_{32} & a_{33} \end{vmatrix} = a_{11}\begin{vmatrix} a_{22} & a_{23} \\ a_{32} & a_{33} \end{vmatrix} - a_{12}\begin{vmatrix} a_{21} & a_{23} \\ a_{31} & a_{33} \end{vmatrix} + a_{13}\begin{vmatrix} a_{21} & a_{22} \\ a_{31} & a_{32} \end{vmatrix}$$

Example

$$\begin{vmatrix} -5 & 3 & 1 \\ 2 & 1 & 1 \\ -4 & 3 & 1 \end{vmatrix} = (-5)\begin{vmatrix} 1 & 1 \\ 3 & 1 \end{vmatrix} - (3)\begin{vmatrix} 2 & 1 \\ -4 & 1 \end{vmatrix} + (1)\begin{vmatrix} 2 & 1 \\ -4 & 3 \end{vmatrix}$$

$$= -5(1 - 3) - 3(2 + 4) + 1(6 + 4)$$
$$= 10 - 18 + 10 = 2$$

Equation (2) then gives $\mathbf{B} \times \mathbf{A}$:

$$\mathbf{B} \times \mathbf{A} = -\mathbf{A} \times \mathbf{B} = 2\mathbf{i} + 6\mathbf{j} - 10\mathbf{k}. \qquad \blacksquare$$

EXAMPLE 3 Find a vector perpendicular to the plane of $P(1, -1, 0)$, $Q(2, 1, -1)$, and $R(-1, 1, 2)$.

Solution The vector $\overrightarrow{PQ} \times \overrightarrow{PR}$ is perpendicular to the plane because it is perpendicular to both vectors. In terms of components,

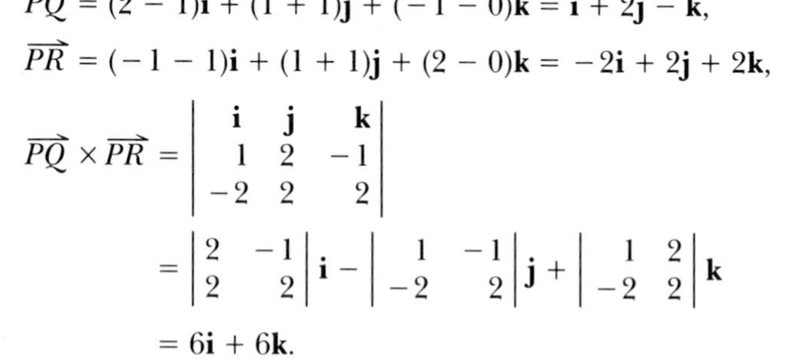

$$\overrightarrow{PQ} = (2-1)\mathbf{i} + (1+1)\mathbf{j} + (-1-0)\mathbf{k} = \mathbf{i} + 2\mathbf{j} - \mathbf{k},$$

$$\overrightarrow{PR} = (-1-1)\mathbf{i} + (1+1)\mathbf{j} + (2-0)\mathbf{k} = -2\mathbf{i} + 2\mathbf{j} + 2\mathbf{k},$$

$$\overrightarrow{PQ} \times \overrightarrow{PR} = \begin{vmatrix} \mathbf{i} & \mathbf{j} & \mathbf{k} \\ 1 & 2 & -1 \\ -2 & 2 & 2 \end{vmatrix}$$

$$= \begin{vmatrix} 2 & -1 \\ 2 & 2 \end{vmatrix}\mathbf{i} - \begin{vmatrix} 1 & -1 \\ -2 & 2 \end{vmatrix}\mathbf{j} + \begin{vmatrix} 1 & 2 \\ -2 & 2 \end{vmatrix}\mathbf{k}$$

$$= 6\mathbf{i} + 6\mathbf{k}. \qquad \blacksquare$$

11.34 The area of triangle *PQR* is half of $|\overrightarrow{PQ} \times \overrightarrow{PR}|$.

EXAMPLE 4 Find the area of the triangle with vertices $P(1, -1, 0)$, $Q(2, 1, -1)$, and $R(-1, 1, 2)$ (Fig. 11.34).

Solution The area of the parallelogram determined by P, Q, and R is

$$|\overrightarrow{PQ} \times \overrightarrow{PR}| = |6\mathbf{i} + 6\mathbf{k}| \qquad \text{(Values from Example 3)}$$

$$= \sqrt{(6)^2 + (6)^2} = \sqrt{2 \cdot 36} = 6\sqrt{2}.$$

The triangle's area is half of this, or $3\sqrt{2}$. $\blacksquare$

EXAMPLE 5 Find a unit vector perpendicular to the plane of $P(1, -1, 0)$, $Q(2, 1, -1)$, and $R(-1, 1, 2)$.

Solution Since $\overrightarrow{PQ} \times \overrightarrow{PR}$ is perpendicular to the plane, its direction $\mathbf{n}$ is a unit vector perpendicular to the plane. In component form,

$$\mathbf{n} = \frac{\overrightarrow{PQ} \times \overrightarrow{PR}}{|\overrightarrow{PQ} \times \overrightarrow{PR}|}$$

$$= \frac{6\mathbf{i} + 6\mathbf{k}}{6\sqrt{2}} \qquad \text{(Values from Examples 3 and 4)}$$

$$= \frac{1}{\sqrt{2}}\mathbf{i} + \frac{1}{\sqrt{2}}\mathbf{k}. \qquad \blacksquare$$

The Test for Parallelism

Since the sines of 0° and 180° are both zero, the cross product $\mathbf{A} \times \mathbf{B} = \mathbf{n}|\mathbf{A}||\mathbf{B}|\sin\theta$ of two nonzero vectors will be zero if and only if the vectors are parallel.

> Nonzero vectors **A** and **B** are parallel if and only if
>
> $$\mathbf{A} \times \mathbf{B} = \mathbf{0}. \tag{10}$$

EXAMPLE 6

Parallel: $\mathbf{A} = \mathbf{i} + 2\mathbf{j} - \mathbf{k}$ and $\mathbf{B} = -2\mathbf{i} - 4\mathbf{j} + 2\mathbf{k}$.

The test:

$$\mathbf{A} \times \mathbf{B} = \begin{vmatrix} \mathbf{i} & \mathbf{j} & \mathbf{k} \\ 1 & 2 & -1 \\ -2 & -4 & 2 \end{vmatrix}$$

$$= \begin{vmatrix} 2 & -1 \\ -4 & 2 \end{vmatrix} \mathbf{i} - \begin{vmatrix} 1 & -1 \\ -2 & 2 \end{vmatrix} \mathbf{j} + \begin{vmatrix} 1 & 2 \\ -2 & -4 \end{vmatrix} \mathbf{k}$$

$$= (4 - 4)\mathbf{i} - (2 - 2)\mathbf{j} + (-4 + 4)\mathbf{k}$$

$$= 0\mathbf{i} - 0\mathbf{j} + 0\mathbf{k} = \mathbf{0}$$

Not parallel: $\mathbf{A} = \mathbf{i} + 2\mathbf{j} - \mathbf{k}$ and $\mathbf{B} = -2\mathbf{i} + 2\mathbf{j} + 2\mathbf{k}$

The test: $\mathbf{A} \times \mathbf{B} = 6\mathbf{i} + 6\mathbf{k} \neq \mathbf{0}$ (Example 3) ■

The Triple Scalar or Box Product

The product $(\mathbf{A} \times \mathbf{B}) \cdot \mathbf{C}$ is called the **triple scalar product** of **A**, **B**, and **C** (in that order). As you can see from the formula

$$(\mathbf{A} \times \mathbf{B}) \cdot \mathbf{C} = |\mathbf{A} \times \mathbf{B}|\,|\mathbf{C}| \cos \theta,$$

the product is the volume of the parallelogram-sided box (parallelepiped) determined by **A**, **B**, and **C** (Fig. 11.35). The number $|\mathbf{A} \times \mathbf{B}|$ is the area of the base parallelogram, and $|\mathbf{C}| \cos \theta$ is the height of the tip of **C** above the plane of **A** and **B**. If θ is greater than 90°, then $\cos \theta$ is negative and we must take the absolute value of $(\mathbf{A} \times \mathbf{B}) \cdot \mathbf{C}$ to get the volume. Because of the geometry here, $(\mathbf{A} \times \mathbf{B}) \cdot \mathbf{C}$ is often called the **box product** of **A**, **B**, and **C**.

By treating the planes of **B** and **C** and of **C** and **A** as the base planes of the parallelepiped determined by **A**, **B**, and **C**, we can see that

$$(\mathbf{A} \times \mathbf{B}) \cdot \mathbf{C} = (\mathbf{B} \times \mathbf{C}) \cdot \mathbf{A} = (\mathbf{C} \times \mathbf{A}) \cdot \mathbf{B}. \tag{11}$$

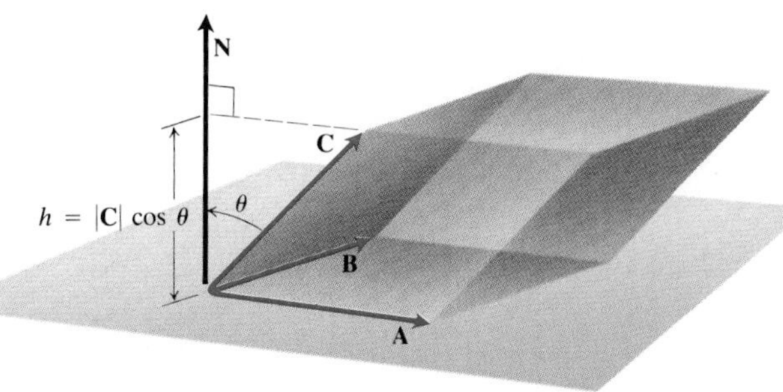

11.35 Except perhaps for sign, the number $(\mathbf{A} \times \mathbf{B}) \cdot \mathbf{C}$ is the volume of the parallelepiped (parallelogram-sided box) shown here.

Since the dot product is commutative, Eq. (11) also gives

$$(\mathbf{A} \times \mathbf{B}) \cdot \mathbf{C} = \mathbf{A} \cdot (\mathbf{B} \times \mathbf{C}). \tag{12}$$

In other words, the dot and cross may be interchanged in a triple scalar product without altering its value.

The triple scalar product can be evaluated as a determinant in the following way:

$$\begin{aligned}
\mathbf{A} \cdot (\mathbf{B} \times \mathbf{C}) &= \mathbf{A} \cdot \left[\begin{vmatrix} b_2 & b_3 \\ c_2 & c_3 \end{vmatrix} \mathbf{i} - \begin{vmatrix} b_1 & b_3 \\ c_1 & c_3 \end{vmatrix} \mathbf{j} + \begin{vmatrix} b_1 & b_2 \\ c_1 & c_2 \end{vmatrix} \mathbf{k} \right] \\
&= a_1 \begin{vmatrix} b_2 & b_3 \\ c_2 & c_3 \end{vmatrix} - a_2 \begin{vmatrix} b_1 & b_3 \\ c_1 & c_3 \end{vmatrix} + a_3 \begin{vmatrix} b_1 & b_2 \\ c_1 & c_2 \end{vmatrix} \\
&= \begin{vmatrix} a_1 & a_2 & a_3 \\ b_1 & b_2 & b_3 \\ c_1 & c_2 & c_3 \end{vmatrix}.
\end{aligned}$$

The number $\mathbf{A} \cdot (\mathbf{B} \times \mathbf{C})$ can be calculated with the formula

$$\mathbf{A} \cdot (\mathbf{B} \times \mathbf{C}) = \begin{vmatrix} a_1 & a_2 & a_3 \\ b_1 & b_2 & b_3 \\ c_1 & c_2 & c_3 \end{vmatrix}. \tag{13}$$

EXAMPLE 7 Find the volume of the box (parallelepiped) determined by $\mathbf{A} = \mathbf{i} + 2\mathbf{j} - \mathbf{k}$, $\mathbf{B} = -2\mathbf{i} + 3\mathbf{k}$, and $\mathbf{C} = 7\mathbf{j} - 4\mathbf{k}$.

Solution The volume is the absolute value of

$$\begin{aligned}
\mathbf{A} \cdot (\mathbf{B} \times \mathbf{C}) &= \begin{vmatrix} 1 & 2 & -1 \\ -2 & 0 & 3 \\ 0 & 7 & -4 \end{vmatrix} \\
&= \begin{vmatrix} 0 & 3 \\ 7 & -4 \end{vmatrix} - 2 \begin{vmatrix} -2 & 3 \\ 0 & -4 \end{vmatrix} - \begin{vmatrix} -2 & 0 \\ 0 & 7 \end{vmatrix} \\
&= -21 - 16 + 14 = -23,
\end{aligned}$$

or 23. ■

Exercises 11.4

In Exercises 1–8, find the length and direction (when defined) of $\mathbf{A} \times \mathbf{B}$ and $\mathbf{B} \times \mathbf{A}$.

1. $\mathbf{A} = 2\mathbf{i} - 2\mathbf{j} - \mathbf{k}, \quad \mathbf{B} = \mathbf{i} - \mathbf{k}$
2. $\mathbf{A} = 2\mathbf{i} + 3\mathbf{j}, \quad \mathbf{B} = -\mathbf{i} + \mathbf{j}$
3. $\mathbf{A} = 2\mathbf{i} - 2\mathbf{j} + 4\mathbf{k}, \quad \mathbf{B} = -\mathbf{i} + \mathbf{j} - 2\mathbf{k}$
4. $\mathbf{A} = \mathbf{i} + \mathbf{j} - \mathbf{k}, \quad \mathbf{B} = \mathbf{0}$
5. $\mathbf{A} = 2\mathbf{i}, \quad \mathbf{B} = -3\mathbf{j}$
6. $\mathbf{A} = \mathbf{i} \times \mathbf{j}, \quad \mathbf{B} = \mathbf{j} \times \mathbf{k}$
7. $\mathbf{A} = -8\mathbf{i} - 2\mathbf{j} - 4\mathbf{k}, \quad \mathbf{B} = 2\mathbf{i} + 2\mathbf{j} + \mathbf{k}$
8. $\mathbf{A} = \frac{3}{2}\mathbf{i} - \frac{1}{2}\mathbf{j} + \mathbf{k}, \quad \mathbf{B} = \mathbf{i} + \mathbf{j} + 2\mathbf{k}$

In Exercises 9–14, sketch the coordinate axes and then include the vectors **A**, **B**, and **A** × **B** as vectors coming out from the origin.

9. $\mathbf{A} = \mathbf{i}, \quad \mathbf{B} = \mathbf{j}$

10. $\mathbf{A} = \mathbf{i} + \mathbf{k}, \quad \mathbf{B} = \mathbf{j}$

11. $\mathbf{A} = \mathbf{i} - \mathbf{k}, \quad \mathbf{B} = \mathbf{j} + \mathbf{k}$

12. $\mathbf{A} = 2\mathbf{i} - \mathbf{j}, \quad \mathbf{B} = \mathbf{i} + 2\mathbf{j}$

13. $\mathbf{A} = \mathbf{i} + 3\mathbf{j} + 2\mathbf{k}, \quad \mathbf{B} = \mathbf{k}$

14. $\mathbf{A} = \mathbf{i} + 2\mathbf{j}, \quad \mathbf{B} = 2\mathbf{j} + \mathbf{k}$

In Exercises 15–18:

a) Find a vector **N** perpendicular to the plane of the points P, Q, and R.

b) Find the area of triangle PQR.

c) Find a unit vector perpendicular to plane PQR.

15. $P(1, -1, 2), \quad Q(2, 0, -1), \quad R(0, 2, 1)$

16. $P(1, 1, 1), \quad Q(2, 1, 3), \quad R(3, -1, 1)$

17. $P(2, -2, 1), \quad Q(3, -1, 2), \quad R(3, -1, 1)$

18. $P(-2, 2, 0), \quad Q(0, 1, -1), \quad R(-1, 2, -2)$

19. Let $\mathbf{A} = 5\mathbf{i} - \mathbf{j} + \mathbf{k}$, $\mathbf{B} = \mathbf{j} - 5\mathbf{k}$, $\mathbf{C} = -15\mathbf{i} + 3\mathbf{j} - 3\mathbf{k}$. Which vectors, if any, are (a) perpendicular, (b) parallel?

20. Let $\mathbf{A} = \mathbf{i} + 2\mathbf{j} - \mathbf{k}$, $\mathbf{B} = -\mathbf{i} + \mathbf{j} + \mathbf{k}$, $\mathbf{C} = \mathbf{i} + \mathbf{k}$, $\mathbf{D} = \mathbf{i} + 2\mathbf{j} - \mathbf{k}$. Which vectors, if any, are (a) perpendicular, (b) parallel?

21. If $\mathbf{A} = 2\mathbf{i} - \mathbf{j}$ and $\mathbf{B} = \mathbf{i} + 3\mathbf{j} - 2\mathbf{k}$, find $\mathbf{A} \times \mathbf{B}$. Then calculate $(\mathbf{A} \times \mathbf{B}) \cdot \mathbf{A}$ and $(\mathbf{A} \times \mathbf{B}) \cdot \mathbf{B}$.

22. Is $(\mathbf{A} \times \mathbf{B}) \cdot \mathbf{A}$ always zero? Explain. What about $(\mathbf{A} \times \mathbf{B}) \cdot \mathbf{B}$?

23. Given vectors **A**, **B**, and **C**, use dot-product and cross-product notation to describe the following vectors:

a) The vector projection of **A** onto **B**

b) A vector orthogonal to **A** and **B**

c) A vector with the length of **A** and the direction of **B**

d) A vector orthogonal to **A** × **B** and **C**

e) A vector in the plane of **B** and **C** perpendicular to **A**

24. *Cancellation is risky in cross products, too.* Find an example to show that $\mathbf{A} \times \mathbf{B} = \mathbf{A} \times \mathbf{C}$ need not imply that **B** equals **C** even if $\mathbf{A} \neq \mathbf{0}$.

In Exercises 25–26, find the magnitude of the torque exerted by **F** on the bolt at P if $|\overrightarrow{PQ}| = 8$ in. and $|\mathbf{F}| = 30$ lb. Answer in foot-pounds.

25. **26.**

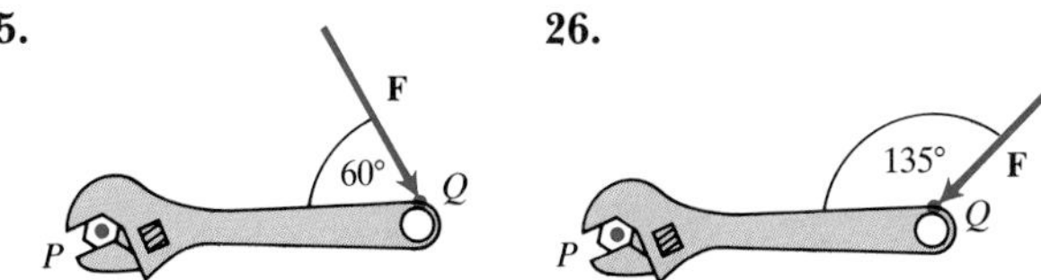

Triple Scalar Products

In Exercises 27–30, verify that $(\mathbf{A} \times \mathbf{B}) \cdot \mathbf{C} = (\mathbf{B} \times \mathbf{C}) \cdot \mathbf{A} = (\mathbf{C} \times \mathbf{A}) \cdot \mathbf{B}$ and find the volume of the box (parallelepiped) determined by **A**, **B**, and **C**.

	A	**B**	**C**
27.	$2\mathbf{i}$	$2\mathbf{j}$	$2\mathbf{k}$
28.	$\mathbf{i} - \mathbf{j} + \mathbf{k}$	$2\mathbf{i} + \mathbf{j} - 2\mathbf{k}$	$-\mathbf{i} + 2\mathbf{j} - \mathbf{k}$
29.	$2\mathbf{i} + \mathbf{j}$	$2\mathbf{i} - \mathbf{j} + \mathbf{k}$	$\mathbf{i} + 2\mathbf{k}$
30.	$\mathbf{i} + \mathbf{j} - 2\mathbf{k}$	$-\mathbf{i} - \mathbf{k}$	$2\mathbf{i} + 4\mathbf{j} - 2\mathbf{k}$

11.5 Lines and Planes in Space

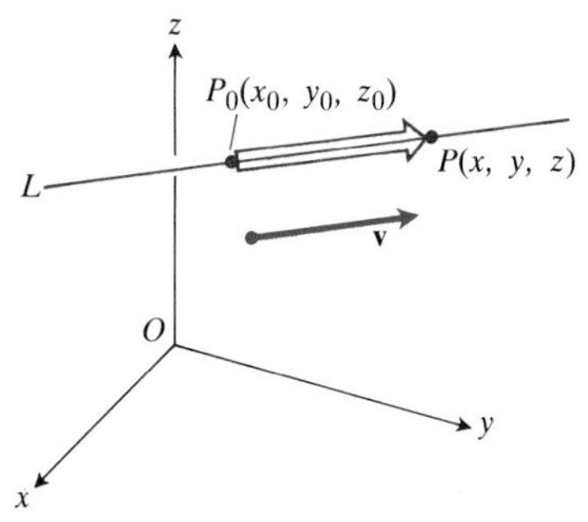

11.36 A point P lies on the line through P_0 parallel to **v** if and only if $\overrightarrow{P_0P}$ is a scalar multiple of **v**.

This section shows how to use scalar and vector products to write equations for lines, line segments, and planes in space.

Equations for Lines and Line Segments

Suppose L is a line in space that passes through a point $P_0(x_0, y_0, z_0)$ and lies parallel to a vector $\mathbf{v} = A\mathbf{i} + B\mathbf{j} + C\mathbf{k}$. Then L is the set of all points $P(x, y, z)$ for which the vector $\overrightarrow{P_0P}$ is parallel to **v** (Fig. 11.36). That is, P lies on L if and only if

$$\overrightarrow{P_0P} = t\mathbf{v} \tag{1}$$

for some number t.

When we write Eq. (1) in terms of components,

$$(x - x_0)\mathbf{i} + (y - y_0)\mathbf{j} + (z - z_0)\mathbf{k} = t(A\mathbf{i} + B\mathbf{j} + C\mathbf{k}), \tag{2}$$

and equate the corresponding components on the two sides, we get three equations in the parameter t:

$$x - x_0 = tA, \qquad y - y_0 = tB, \qquad z - z_0 = tC. \tag{3}$$

When rearranged, these become the standard parametric equations of line L:

Parametric Equations for the Line through $P_0(x_0, y_0, z_0)$ Parallel to $\mathbf{v} = A\mathbf{i} + B\mathbf{j} + C\mathbf{k}$

$$x = x_0 + tA, \qquad y = y_0 + tB, \qquad z = z_0 + tC \tag{4}$$

EXAMPLE 1 Find parametric equations for the line through the point $(-2, 0, 4)$ parallel to the vector $\mathbf{v} = 2\mathbf{i} + 4\mathbf{j} - 2\mathbf{k}$.

Solution With $P_0(x_0, y_0, z_0) = (-2, 0, 4)$ and

$$A\mathbf{i} + B\mathbf{j} + C\mathbf{k} = 2\mathbf{i} + 4\mathbf{j} - 2\mathbf{k},$$

Eqs. (4) become

$$x = -2 + 2t, \qquad y = 4t, \qquad z = 4 - 2t.$$

See Fig. 11.37. ■

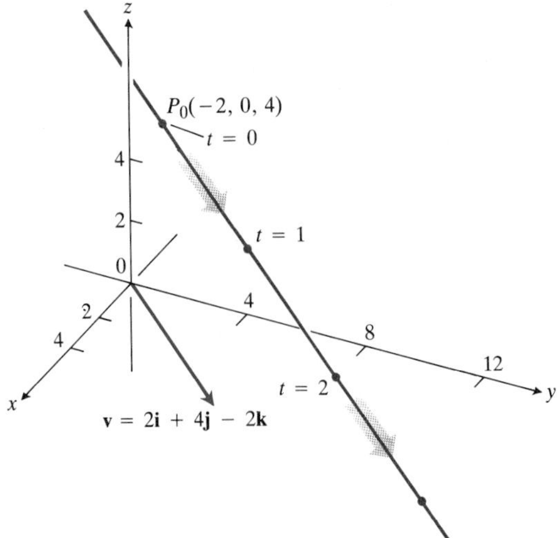

11.37 Selected points and parameter values on the line $x = -2 + 2t$, $y = 4t$, $z = 4 - 2t$. The arrows show the direction of increasing t.

EXAMPLE 2 Find parametric equations for the line through the points $P(-3, 2, -3)$ and $Q(1, -1, 4)$.

Solution The vector

$$\overrightarrow{PQ} = 4\mathbf{i} - 3\mathbf{j} + 7\mathbf{k}$$

is parallel to the line, and Eqs. (4) with $(x_0, y_0, z_0) = (-3, 2, -3)$ give

$$x = -3 + 4t, \qquad y = 2 - 3t, \qquad z = -3 + 7t. \tag{5}$$

We could equally well have chosen $Q(1, -1, 4)$ as the "base point" and written

$$x = 1 + 4t, \qquad y = -1 - 3t, \qquad z = 4 + 7t \tag{6}$$

as equations for the line. The parametrizations in (5) and (6) place you at different points for a given value of t, but each set of equations covers the line completely as t runs from $-\infty$ to ∞. ■

To find equations for the line segment joining two points, we first find equations for the line through the points. We then find the t-values for which the line passes through the points and restrict t to lie within the closed interval bounded by these values. The line equations, together with this added restriction, are called equations for the line segment.

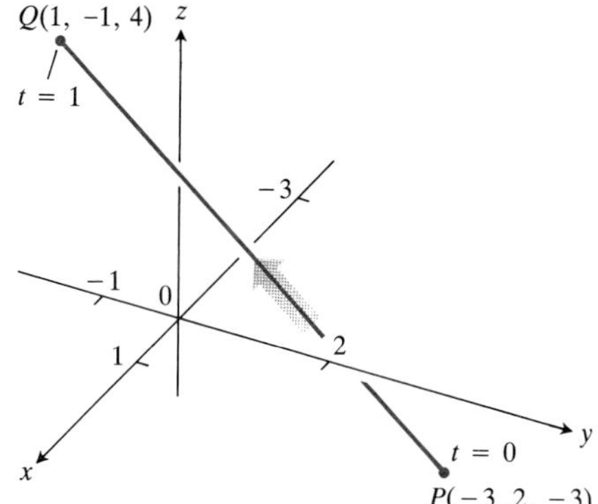

11.38 Example 3 derives parametric equations for the line segment joining P and Q. The arrow shows the direction of increasing t for the parametrization found there.

EXAMPLE 3 Find parametric equations for the line segment joining the points $P(-3, 2, -3)$ and $Q(1, -1, 4)$.

Solution We begin with equations for the line through P and Q, taking them, in this case, from Example 2:

$$x = -3 + 4t, \qquad y = 2 - 3t, \qquad z = -3 + 7t. \tag{7}$$

We notice that the point

$$(x, y, z) = (-3 + 4t, 2 - 3t, -3 + 7t)$$

passes through $P(-3, 2, -3)$ at $t = 0$ and $Q(1, -1, 4)$ at $t = 1$. We add the restriction $0 \le t \le 1$ to Eqs. (7) to get equations for the segment:

$$x = -3 + 4t, \qquad y = 2 - 3t, \qquad z = -3 + 7t, \qquad 0 \le t \le 1.$$

See Fig. 11.38. ■

The Distance from a Point to a Line

To find the distance from a point P to a line L, find the point Q on L closest to P and calculate the distance from P to Q. The next example shows how this is done.

EXAMPLE 4 Find the distance from the point $P(1, 1, 5)$ to the line

$$x = 1 + t, \qquad y = 3 - t, \qquad z = 2t.$$

How to find the distance from a point P to a line L

1. Find the point Q on L closest to P.
2. Calculate the distance from P to Q.

Solution STEP 1: We find the point on the line closest to P. The coordinates of a typical point Q on the line are $Q(1+t, 3-t, 2t)$.

We want the value of t that minimizes the distance from P to Q. To avoid working with square roots, we find this value of t by minimizing the *square* of the distance instead. The formula for the square of the distance from P to Q is

$$\begin{aligned} f(t) &= (1+t-1)^2 + (3-t-1)^2 + (2t-5)^2 \\ &= (t)^2 + (2-t)^2 + (2t-5)^2 \\ &= t^2 + 4 - 4t + t^2 + 4t^2 - 20t + 25 \\ &= 6t^2 - 24t + 29. \end{aligned}$$

The derivative,

$$\frac{df}{dt} = 12t - 24,$$

equals 0 when $t = 2$. This value of t gives a minimum value for f because $d^2f/dt^2 = 12$ is positive. The point Q closest to P is $Q(1+t, 3-t, 2t)_{t=2} = Q(3, 1, 4)$.

STEP 2: We find the distance from $P(1, 1, 5)$ to $Q(3, 1, 4)$:

$$\begin{aligned} d(P, Q) &= \sqrt{(1-3)^2 + (1-1)^2 + (5-4)^2} \\ &= \sqrt{(-2)^2 + (0)^2 + (1)^2} = \sqrt{4+1} = \sqrt{5}. \end{aligned}$$

The distance from P to the line is $\sqrt{5}$. ■

Equations for Planes

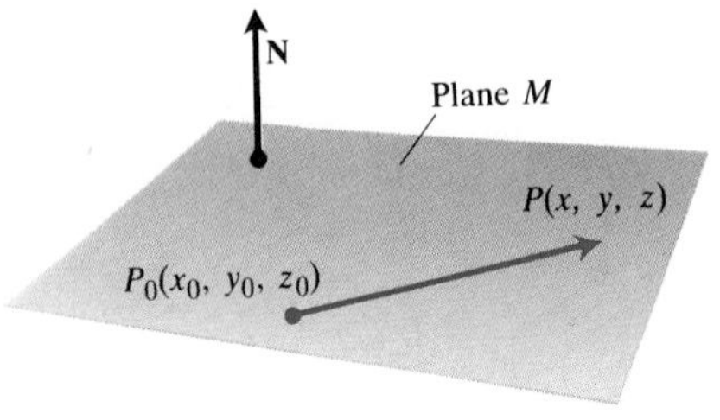

11.39 The standard equation for a plane in space is defined in terms of a vector normal to the plane: a point P lies in the plane through P_0 normal to $\mathbf{N}$ if and only if $\mathbf{N} \cdot \overrightarrow{P_0P} = 0$.

Suppose M is a plane in space that passes through a point $P_0(x_0, y_0, z_0)$ and is perpendicular (normal) to the nonzero vector $\mathbf{N} = A\mathbf{i} + B\mathbf{j} + C\mathbf{k}$. Then M consists of all points $P(x, y, z)$ for which the vector $\overrightarrow{P_0P}$ is orthogonal to $\mathbf{N}$ (Fig. 11.39). That is, P lies on M if and only if

$$\mathbf{N} \cdot \overrightarrow{P_0P} = 0 \qquad (8)$$

or

$$A(x - x_0) + B(y - y_0) + C(z - z_0) = 0.$$

When rearranged, this becomes

$$Ax + By + Cz = Ax_0 + By_0 + Cz_0.$$

Equation for the Plane through $P_0(x_0, y_0, z_0)$ Perpendicular to $\mathbf{N} = A\mathbf{i} + B\mathbf{j} + C\mathbf{k}$.

$$Ax + By + Cz = D, \qquad D = Ax_0 + By_0 + Cz_0 \qquad (9)$$

Notice how the components of $\mathbf{N}$ become coefficients in the equation

$$Ax + By + Cz = D.$$

EXAMPLE 5 Find an equation for the plane through $P_0(-3, 0, 7)$ perpendicular to $\mathbf{N} = 5\mathbf{i} + 2\mathbf{j} - \mathbf{k}$.

Solution We use Eq. (9) to get

$$D = 5(-3) + 2(0) - 1(7) = -15 - 7 = -22$$

and

$$5x + 2y - z = -22.$$

■

EXAMPLE 6 Find an equation for the plane through the points $A(0, 0, 1)$, $B(2, 0, 0)$, and $C(0, 3, 0)$.

Solution We find a vector normal to the plane and use it with one of the points to write an equation for the plane.

The cross product

$$\overrightarrow{AB} \times \overrightarrow{AC} = \begin{vmatrix} \mathbf{i} & \mathbf{j} & \mathbf{k} \\ 2 & 0 & -1 \\ 0 & 3 & -1 \end{vmatrix} = 3\mathbf{i} + 2\mathbf{j} + 6\mathbf{k}$$

is normal to the plane. We substitute the components of this vector and the coordinates of the point $(0, 0, 1)$ into Eq. (9) to get $D = 3(0) + 2(0) + 6(1) = 6$ and

$$3x + 2y + 6z = 6$$

as an equation for the plane. ■

EXAMPLE 7 Find the point in which the line

$$x = \frac{8}{3} + 2t, \qquad y = -2t, \qquad z = 1 + t$$

meets the plane $3x + 2y + 6z = 6$.

Solution The point

$$\left(\frac{8}{3} + 2t,\ -2t,\ 1 + t\right)$$

will lie in the plane if its coordinates satisfy the equation of the plane; that is, if

$$3\left(\frac{8}{3} + 2t\right) + 2(-2t) + 6(1 + t) = 6$$

$$8 + 6t - 4t + 6 + 6t = 6$$

$$8t = -8$$

$$t = -1.$$

The point of intersection is

$$(x, y, z)_{t=-1} = \left(\frac{8}{3} - 2, 2, 1 - 1\right) = \left(\frac{2}{3}, 2, 0\right).$$

■

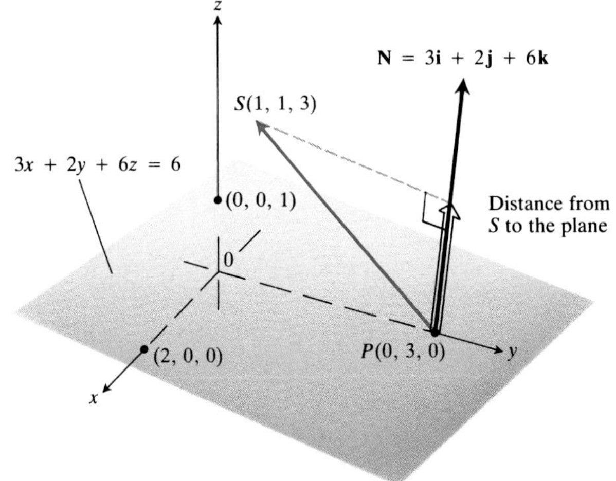

11.40 The distance from S to the plane is the length of the vector projection of $\overrightarrow{PS}$ onto $\mathbf{N}$.

EXAMPLE 8 Find the distance from the point $S(1, 1, 3)$ to the plane $3x + 2y + 6z = 6$.

To find the distance from a point S to a plane $Ax + By + Cz = D$, find

1. a point P on the plane,
2. $\overrightarrow{PS}$,
3. the direction of $\mathbf{N} = A\mathbf{i} + B\mathbf{j} + C\mathbf{k}$.

Then calculate the distance as

$$\left|\overrightarrow{PS} \cdot \frac{\mathbf{N}}{|\mathbf{N}|}\right|.$$

Solution We use the same approach we used in Section 11.3 to find the distance from a point to a line: We find a point P in the plane and calculate the length of the vector projection of $\overrightarrow{PS}$ onto a vector $\mathbf{N}$ normal to the plane (Fig. 11.40).

The coefficients in the equation $3x + 2y + 6z = 6$ give

$$\mathbf{N} = 3\mathbf{i} + 2\mathbf{j} + 6\mathbf{k}.$$

The points on the plane easiest to find from the plane's equation are the intercepts. If we take P to be the y-intercept $(0, 3, 0)$, then

$$\begin{aligned}\overrightarrow{PS} &= (1 - 0)\mathbf{i} + (1 - 3)\mathbf{j} + (3 - 0)\mathbf{k} \\ &= \mathbf{i} - 2\mathbf{j} + 3\mathbf{k}\end{aligned}$$

$$\begin{aligned}|\mathbf{N}| &= \sqrt{(3)^2 + (2)^2 + (6)^2} \\ &= \sqrt{49} \\ &= 7\end{aligned}$$

$$\begin{aligned}\text{Distance from } S \text{ to the plane} &= \left|\overrightarrow{PS} \cdot \frac{\mathbf{N}}{|\mathbf{N}|}\right| \\ &= \left|(\mathbf{i} - 2\mathbf{j} + 3\mathbf{k}) \cdot \left(\frac{3}{7}\mathbf{i} + \frac{2}{7}\mathbf{j} + \frac{6}{7}\mathbf{k}\right)\right| \\ &= \left|\frac{3}{7} - \frac{4}{7} + \frac{18}{7}\right| \\ &= \frac{17}{7}.\end{aligned}$$

■

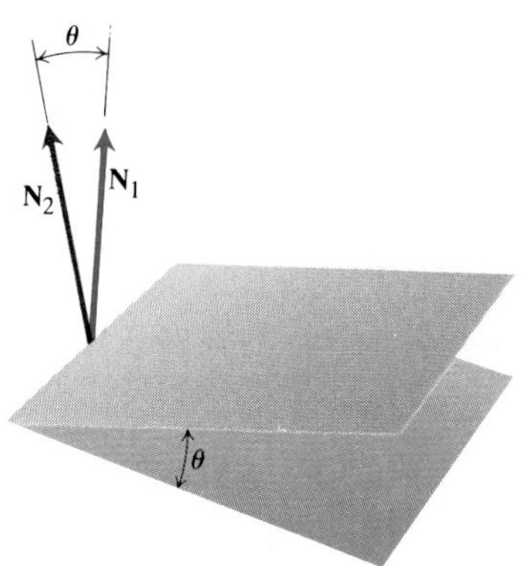

11.41 The angle between two planes is obtained from the angle between their normals.

Angles between Planes; Lines of Intersection

The angle between two intersecting planes is defined to be the (acute) angle made by their normal vectors (Fig. 11.41).

EXAMPLE 9 Find the angle between the planes $3x - 6y - 2z = 15$ and $2x + y - 2z = 5$.

Solution The vectors

$$\mathbf{N}_1 = 3\mathbf{i} - 6\mathbf{j} - 2\mathbf{k}, \qquad \mathbf{N}_2 = 2\mathbf{i} + \mathbf{j} - 2\mathbf{k}$$

are normals to the planes. The angle between them is

$$\begin{aligned} \theta &= \cos^{-1}\left(\frac{\mathbf{N}_1 \cdot \mathbf{N}_2}{|\mathbf{N}_1||\mathbf{N}_2|}\right) && \text{(Eq. 5, Section 11.3)} \\ &= \cos^{-1}\frac{4}{21} && \text{(Arithmetic omitted)} \\ &= 79^\circ. && \text{(Calculator, rounded)} \end{aligned}$$

EXAMPLE 10 Find a vector parallel to the line of intersection of the planes $3x - 6y - 2z = 15$ and $2x + y - 2z = 5$.

Solution Any vector parallel to the line of intersection will be parallel to both planes and therefore perpendicular to their normals. Conversely, any vector perpendicular to the planes' normals will be parallel to both planes and hence parallel to their line of intersection. These requirements are met by the vector

$$\begin{aligned} \mathbf{v} &= \mathbf{N}_1 \times \mathbf{N}_2 \\ &= \begin{vmatrix} \mathbf{i} & \mathbf{j} & \mathbf{k} \\ 3 & -6 & -2 \\ 2 & 1 & -2 \end{vmatrix} \\ &= 14\mathbf{i} + 2\mathbf{j} + 15\mathbf{k}. \end{aligned}$$

Any nonzero scalar multiple of $\mathbf{v}$ will do as well.

EXAMPLE 11 Find parametric equations for the line in which the planes $3x - 6y - 2z = 15$ and $2x + y - 2z = 5$ intersect.

Solution We find a vector parallel to the line and a point on the line and use Eqs. (4).

Example 10 gives a vector parallel to the line, namely, $\mathbf{v} = 14\mathbf{i} + 2\mathbf{j} + 15\mathbf{k}$. To find a point on the line, we find a point common to the two planes. Substituting $z = 0$ in the plane equations and solving for x and y simultaneously gives the point $(3, -1, 0)$. The line is

$$x = 3 + 14t, \qquad y = -1 + 2t, \qquad z = 15t.$$

Exercises 11.5

Find parametric equations for the lines in Exercises 1–12.

1. The line through $P(3, -4, -1)$ parallel to the vector $\mathbf{i} + \mathbf{j} + \mathbf{k}$
2. The line through $P(1, 2, -1)$ and $Q(-1, 0, 1)$
3. The line through $P(-2, 0, 3)$ and $Q(3, 5, -2)$
4. The line through $P(1, 2, 0)$ and $Q(1, 1, -1)$
5. The line through the origin parallel to the vector $2\mathbf{j} + \mathbf{k}$
6. The line through the point $(3, -2, 1)$ parallel to the line $x = 1 + 2t, y = 2 - t, z = 3t$
7. The line through $(1, 1, 1)$ parallel to the z-axis
8. The line through $(2, 4, 5)$ perpendicular to the plane $3x + 7y - 5z = 21$
9. The line through $(0, -7, 0)$ perpendicular to the plane $x + 2y + 2z = 13$
10. The line through $(2, 3, 0)$ perpendicular to the vectors $\mathbf{A} = \mathbf{i} + 2\mathbf{j} + 3\mathbf{k}$ and $\mathbf{B} = 3\mathbf{i} + 4\mathbf{j} + 5\mathbf{k}$
11. The x-axis
12. The z-axis

Find parametric equations for the line segments joining the points in Exercises 13–20. Draw coordinate axes and sketch each segment, indicating the direction of increasing t for your parametrization.

13. $(0, 0, 0)$, $(1, 1, 1)$
14. $(0, 0, 0)$, $(1, 0, 0)$
15. $(1, 0, 0)$, $(1, 1, 0)$
16. $(1, 1, 0)$, $(1, 1, 1)$
17. $(0, -1, 1)$, $(0, 1, 1)$
18. $(3, 0, 0)$, $(0, 2, 0)$
19. $(2, 2, 0)$, $(1, 2, -2)$
20. $(1, -1, -2)$, $(0, 2, 1)$

Find equations for the planes in Exercises 21–26.

21. The plane through $P_0(0, 2, -1)$ perpendicular to $\mathbf{N} = 3\mathbf{i} - 2\mathbf{j} - \mathbf{k}$
22. The plane through $(1, -1, 3)$ parallel to the plane $3x + y + z = 7$
23. The plane through $(1, 1, -1)$, $(2, 0, 2)$, and $(0, -2, 1)$
24. The plane through $(2, 4, 5)$, $(1, 5, 7)$, and $(-1, 6, 8)$
25. The plane through $P_0(2, 4, 5)$ perpendicular to the line
$$x = 5 + t, \quad y = 1 + 3t, \quad z = 4t$$
26. The plane through $A(1, -2, 1)$ perpendicular to the vector from the origin to A

In Exercises 27–30, find the distance from the point to the line.

27. $x = 4t, \quad y = -2t, \quad z = 2t; \quad P(0, 0, 12)$
28. $x = 5 + 3t, \quad y = 5 + 4t, \quad z = -3 - 5t; \quad P(0, 0, 0)$
29. $x = 2 + 2t, \quad y = 1 + 6t, \quad z = 3; \quad P(2, 1, 3)$
30. $x = 2t, \quad y = 1 + 2t, \quad z = 2t; \quad P(2, 1, -1)$

In Exercises 31–36, find the distance from the point to the plane.

31. $(2, -3, 4), \quad x + 2y + 2z = 13$
32. $(0, 0, 0), \quad 3x + 2y + 6z = 6$
33. $(0, 1, 1), \quad 4y + 3z = -12$
34. $(2, 2, 3), \quad 2x + y + 2z = 4$
35. $(0, -1, 0), \quad 2x + y + 2z = 4$
36. $(1, 0, -1), \quad -4x + y + z = 4$

In Exercises 37–40, find the point in which the line meets the plane.

37. $x = 1 - t, \quad y = 3t, \quad z = 1 + t; \quad 2x - y + 3z = 6$
38. $x = 2, \quad y = 3 + 2t, \quad z = -2 - 2t; \quad 6x + 3y - 4z = -12$
39. $x = 1 + 2t, \quad y = 1 + 5t, \quad z = 3t; \quad x + y + z = 2$
40. $x = -1 + 3t, \quad y = -2, \quad z = 5t; \quad 2x - 3z = 7$

Find the angles between the planes in Exercises 41–42 (calculator not needed).

41. $x + y = 1, \quad 2x + y - 2z = 2$
42. $5x + y - z = 10, \quad x - 2y + 3z = -1$

CALCULATOR Use a calculator to find the angles between the planes in Exercises 43–46 to the nearest tenth of a degree.

43. $2x + 2y + 2z = 3, \quad 2x - 2y - z = 5$
44. $x + y + z = 1, \quad z = 0$ (the xy-plane)
45. $2x + 2y - z = 3, \quad x + 2y + z = 2$
46. $4y + 3z = -12, \quad 3x + 2y + 6z = 6$

Find parametric equations for the lines in which the planes in Exercises 47–50 intersect.

47. $x + y + z = 1, \quad x + y = 2$
48. $3x - 6y - 2z = 3, \quad 2x + y - 2z = 2$
49. $x - 2y + 4z = 2, \quad x + y - 2z = 5$
50. $5x - 2y = 11, \quad 4y - 5z = -17$

11.6 Surfaces in Space

Just as we call the set of points in the plane whose coordinates satisfy an equation of the form $F(x, y) = 0$ a **curve,** we call the set of points in space whose coordinates satisfy an equation of the form $F(x, y, z) = 0$ a **surface.** We use such surfaces to describe the boundaries of solids. We also use them to model the membranes across which fluids flow, the plates over which electrical charges are distributed, and the walls of containers that are subjected to pressures of various kinds. We shall see all this and more in the chapters to come.

In the present section, our goal is to become acquainted with the surfaces most commonly used in the theory and applications of the calculus of functions of more than one variable. This means finding out what the surfaces look like, what their equations are, and how to draw them.

Cylinders

The simplest surfaces to draw and write equations for, besides planes, are cylinders.

DEFINITIONS

A **cylinder** is a surface composed of all the lines that pass through a plane curve parallel to a fixed line in space. The plane curve is the **generating curve** of the cylinder.

In solid geometry, *cylinder* usually means *circular cylinder,* but that is not the case here. The cross sections of the cylinder in our first example are parabolas, not circles. We allow our cylinders to have cross sections of any kind.

EXAMPLE 1 *A parabolic cylinder.* Find an equation for the cylinder made by the lines parallel to the z-axis that pass through the parabola $y = x^2, z = 0$ (Fig. 11.42).

Solution Suppose that the point $P_0(x_0, x_0^2, 0)$ lies on the parabola $y = x^2$ in the xy-plane. Then, for any value of z, the point $Q(x_0, x_0^2, z)$ will lie on the cylinder because it lies on the line $x = x_0, y = x_0^2$ through P_0 parallel to the z-axis. Conversely, any point $Q(x_0, x_0^2, z)$ whose y-coordinate is the square of its x-coordinate lies on the cylinder because it lies on the line $x = x_0, y = x_0^2$ through P_0 parallel to the z-axis (Fig. 11.43).

Regardless of the value of z, therefore, the points on the surface are the points whose coordinates satisfy the equation $y = x^2$. This makes $y = x^2$ an equation for the cylinder. ■

The cylinder $y = x^2$

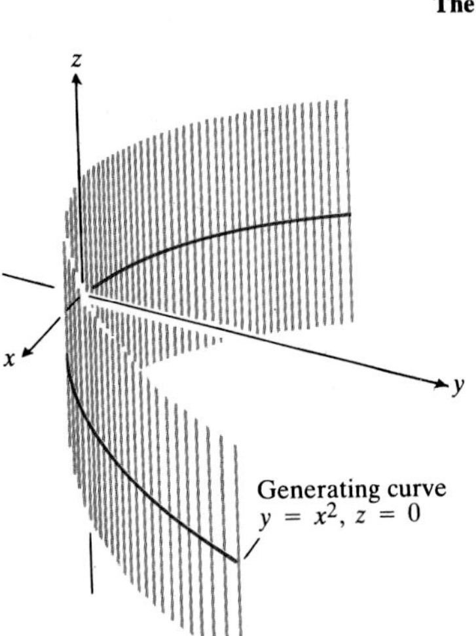

11.42 The cylinder generated by lines parallel to the z-axis and passing through the parabola $y = x^2$ in the xy-plane.

$Q_0(x_0, x_0^2, z)$

$P_0(x_0, x_0^2, 0)$

$y = x^2$

11.43 Every point of the cylinder in Fig. 11.43 has coordinates of the form (x_0, x_0^2, z).

As Example 1 suggests, any curve $f(x, y) = C$ in the xy-plane defines a cylinder parallel to the z-axis whose equation is also $f(x, y) = C$. The equation $x^2 + y^2 = 1$ defines the circular cylinder made by the lines parallel to the z-axis and passing through the circle $x^2 + y^2 = 1$ in the xy-plane. The equation $x^2 + 4y^2 = 9$ defines the elliptical cylinder made by the lines paral-

The elliptical cylinder $x^2 + 4z^2 = 4$

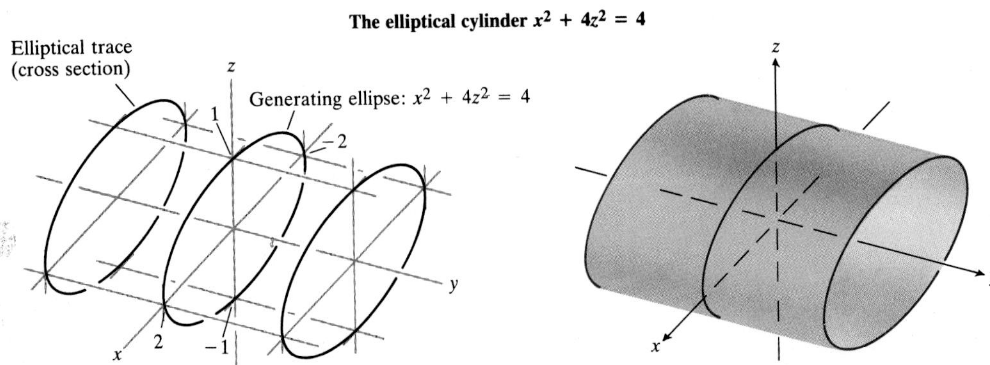

11.44 The elliptic cylinder $x^2 + 4z^2 = 4$ is made of lines parallel to the y-axis and passing through the ellipse $x^2 + 4z^2 = 4$ in the xz-plane. The cross sections or "traces" of the cylinder in planes perpendicular to the y-axis are ellipses congruent to the generating ellipse. The cylinder extends along the entire y-axis but we can draw only a finite portion of it.

lel to the z-axis and passing through the ellipse $x^2 + 4y^2 = 9$ in the xy-plane, and so on.

In a similar manner, any curve $g(x, z) = C$ in the xz-plane defines a cylinder parallel to the y-axis whose equation is also $g(x, z) = C$, and any curve $h(y, z) = C$ defines a cylinder parallel to the x-axis whose equation is also $h(y, z) = C$.

In short, an equation in any two of the three Cartesian coordinates defines a cylinder made of lines parallel to the axis of the third coordinate. See Figures 11.44 and 11.45.

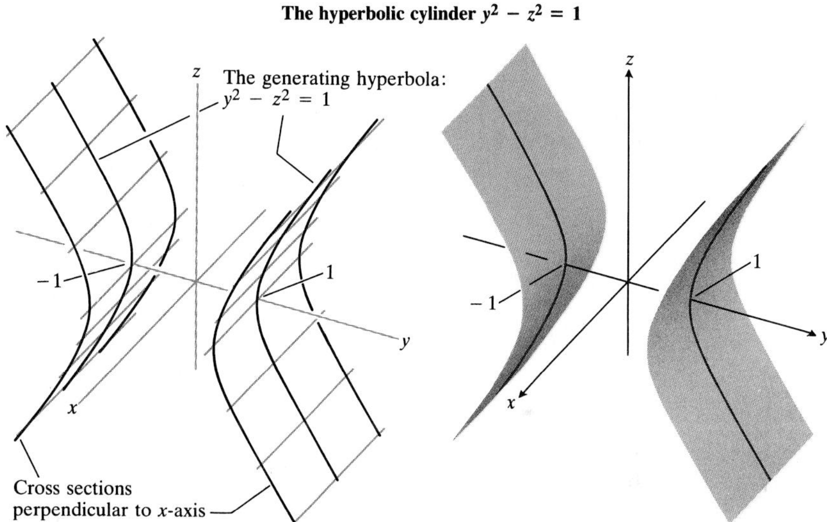

11.45 The hyperbolic cylinder $y^2 - z^2 = 1$ is made of lines parallel to the x-axis and passing through the hyperbola $y^2 - z^2 = 1$ in the yz-plane. The cross sections or "traces" of the cylinder in planes perpendicular to the x-axis are hyperbolas congruent to the generating hyperbola.

Drawing Cylinders Parallel to the Coordinate Axes

Here is some advice about drawing cylinders. As always, pencil is safer than pen because you can erase, but the advice applies no matter what medium you choose. Just determine which axis the cylinder is parallel to and carry out the steps shown.

How to Draw Cylinders Parallel to the Coordinate Axes

$x^2 + y^2 = 1$ $\qquad$ $z = y^2$

1.

Sketch all three coordinate axes very *lightly.*

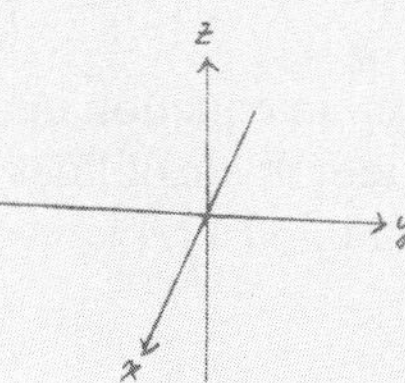

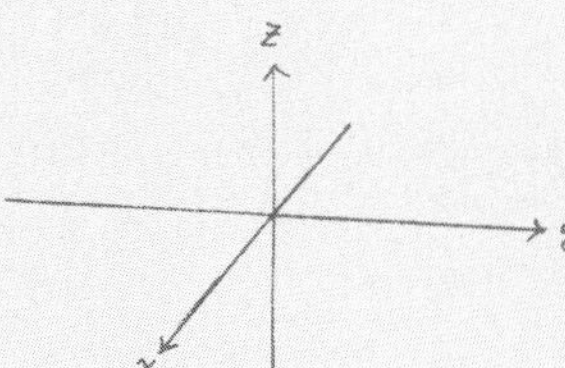

2.

Sketch the trace of the cylinder in the coordinate plane of the two variables that appear in the cylinder's equation. Sketch *very lightly.*

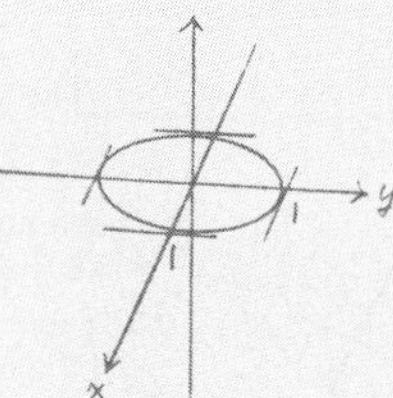

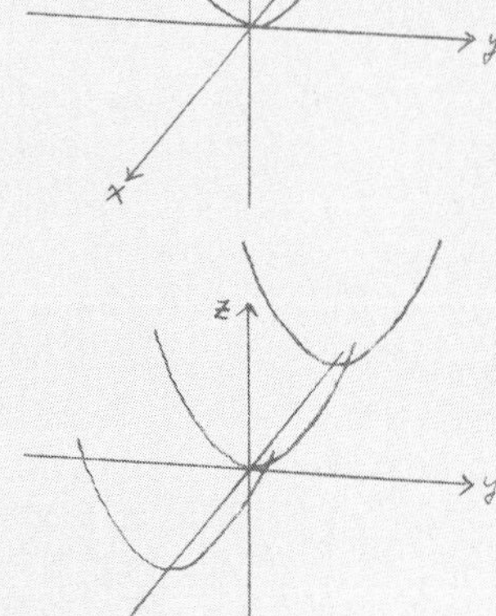

3.

Sketch traces in parallel planes on either side (again, lightly).

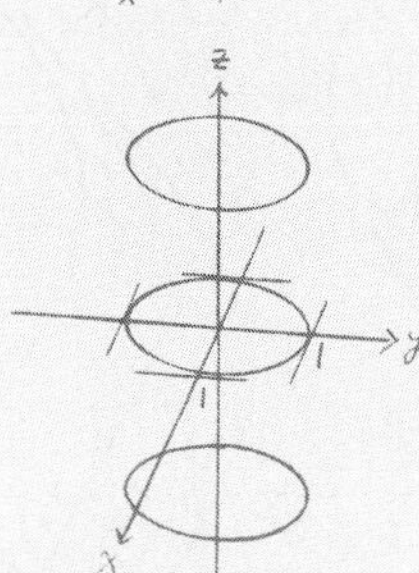

4.

Add parallel outer edges to give the shape definition.

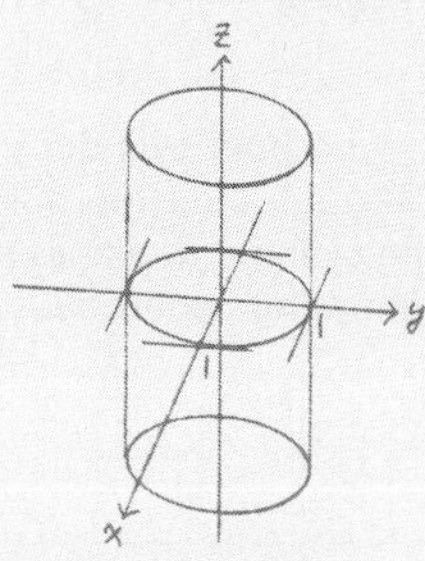

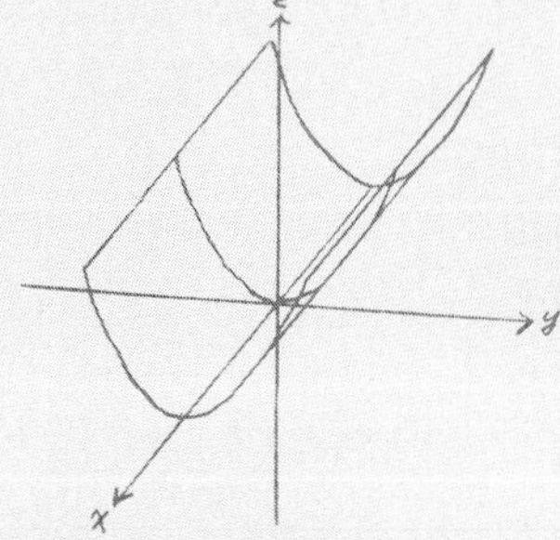

5.

If more definition is required, darken the parts of the lines that are exposed to view. Leave the hidden parts light. Use line breaks when you can.

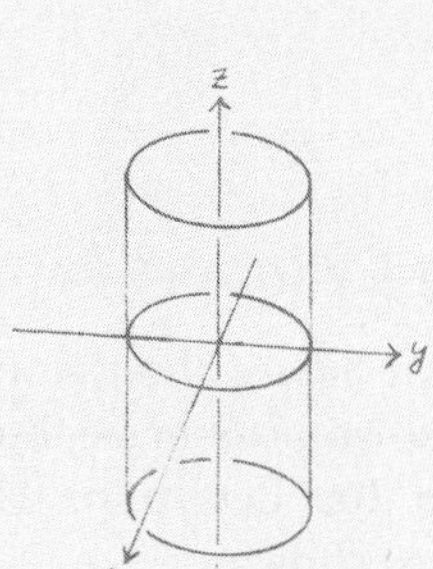

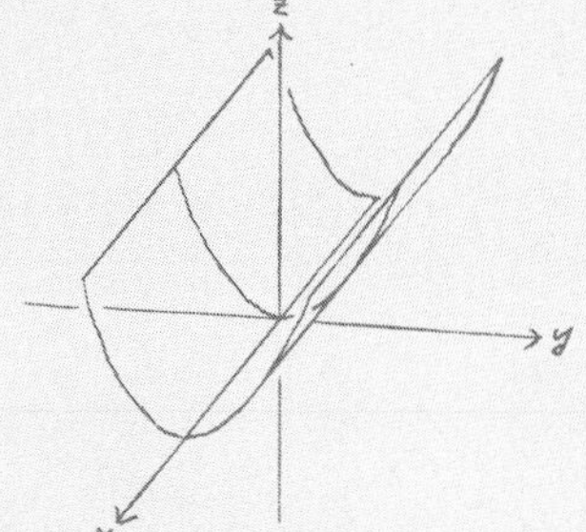

Quadric Surfaces

The cylinders in Figs. 11.43–11.45 are **quadric surfaces,** surfaces whose equations combine quadratic terms with linear terms and constants. A number of other important quadric surfaces are described in the examples that follow.

EXAMPLE 2 The **ellipsoid**

$$\frac{x^2}{a^2} + \frac{y^2}{b^2} + \frac{z^2}{c^2} = 1 \tag{1}$$

cuts the coordinate axes at $(\pm a, 0, 0,)$, $(0, \pm b, 0)$, and $(0, 0, \pm c)$ (Fig. 11.46). It lies inside the rectangular box

$$|x| \leq a, \qquad |y| \leq b, \qquad |z| \leq c.$$

Since only even powers of x, y, and z occur in the equation, this surface is symmetric with respect to each coordinate plane. The sections cut out by the coordinate planes are ellipses. For example,

$$\frac{x^2}{a^2} + \frac{y^2}{b^2} = 1 \qquad \text{when} \quad z = 0.$$

Each section cut out by a plane

$$z = z_0, \qquad |z_0| < c,$$

is an ellipse

$$\frac{x^2}{a^2[1 - (z_0^2/c^2)]} + \frac{y^2}{b^2[1 - (z_0^2/c^2)]} = 1. \tag{2}$$

When two of the three semiaxes a, b, and c are equal, the surface is an ellipsoid of revolution, and when all three are equal, it is a sphere. ■

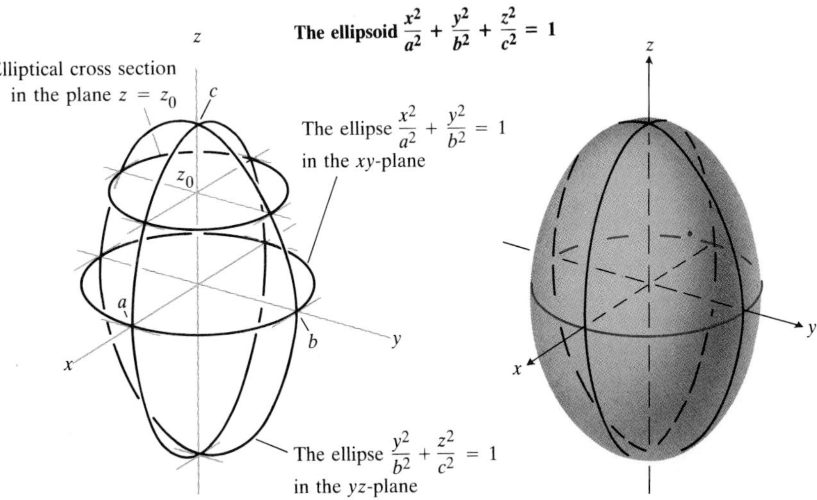

11.46 A typical ellipsoid.

The elliptical paraboloid $\frac{x^2}{a^2} + \frac{y^2}{b^2} = \frac{z}{c}$

The parabola $z = \frac{c}{a^2}x^2$ in the xz-plane

The ellipse $\frac{x^2}{a^2} + \frac{y^2}{b^2} = 1$ in the plane $z = c$

The parabola $z = \frac{c}{b^2}y^2$ in the yz-plane

11.47 The surface $(x^2/a^2) + (y^2/b^2) = z/c$ is an elliptic paraboloid. The cross sections perpendicular to the z-axis above the xy-plane are ellipses. The cross sections in the planes that contain the z-axis are parabolas.

EXAMPLE 3 The **elliptic paraboloid**

$$\frac{x^2}{a^2} + \frac{y^2}{b^2} = \frac{z}{c} \qquad (c > 0) \tag{3}$$

is symmetric with respect to the planes $x = 0$ and $y = 0$ (Fig. 11.47). The only intercept on the axes is at the origin. Except for this point, the surface lies above the xy-plane because z is positive whenever either x or y is different from zero. The sections cut by the coordinate planes are

$$\begin{aligned} x = 0: &\quad \text{the parabola } z = \frac{c}{b^2}y^2, \\ y = 0: &\quad \text{the parabola } z = \frac{c}{a^2}x^2, \\ z = 0: &\quad \text{the point } (0, 0, 0). \end{aligned} \tag{4}$$

Each plane $z = z_0$ above the xy-plane cuts the surface in the ellipse

$$\frac{x^2}{a^2} + \frac{y^2}{b^2} = \frac{z_0}{c}.$$

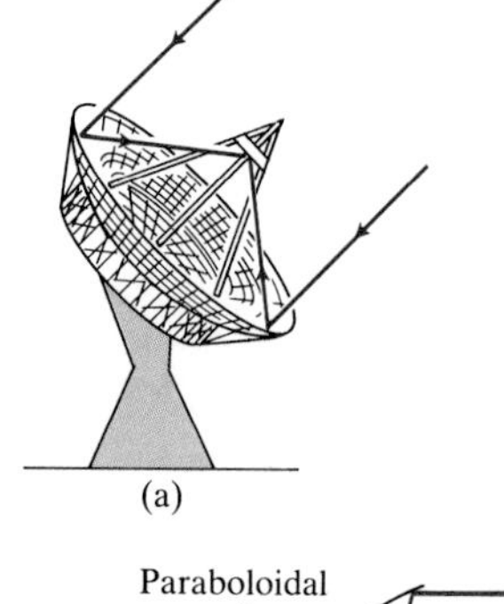

(a)

Paraboloidal reflector

Horn

Focus

(b) (c)

11.48 Many antennas are shaped like pieces of paraboloids of revolution. (a) Radio telescopes use the same principles as optical telescopes. (b) A "rectangular-cut" radar reflector. (c) The profile of a horn antenna in a microwave radio link.

EXAMPLE 4 The **circular paraboloid** or **paraboloid of revolution**

$$\frac{x^2}{a^2} + \frac{y^2}{a^2} = \frac{z}{c} \qquad (c > 0) \tag{5}$$

is obtained by taking $b = a$ in Eq. (3) for the elliptic paraboloid. The cross sections of the surface by planes perpendicular to the z-axis are circles centered on the z-axis. The cross sections by planes containing the z-axis are congruent parabolas with a common focus at the point $(0, 0, a^2/4c)$.

Shapes cut from circular paraboloids are used for antennas in radio telescopes, satellite trackers, and microwave radio links (Fig. 11.48).

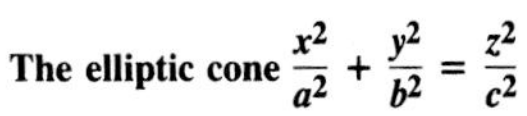

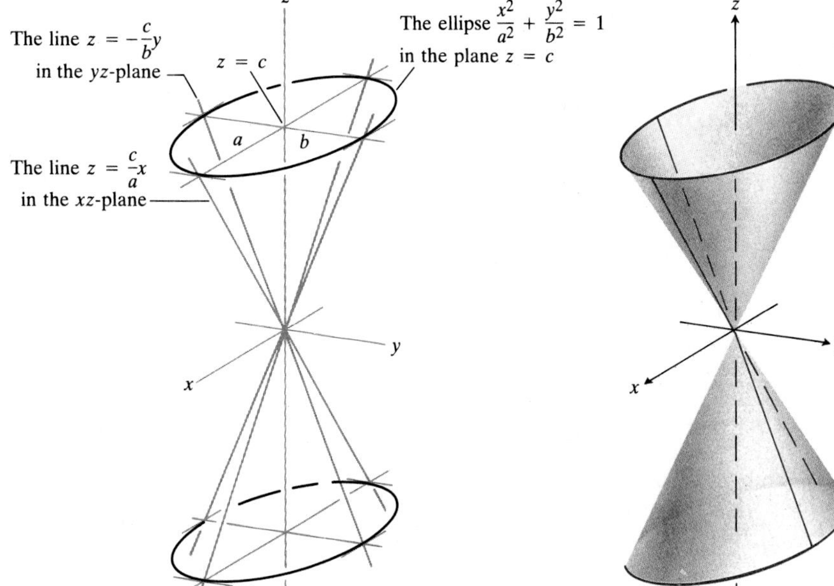

11.49 The surface $(x^2/a^2) + (y^2/b^2) = z^2/c^2$ is an elliptic cone. Planes perpendicular to the z-axis cut the cone in ellipses above and below the xy-plane. Vertical planes that contain the z-axis cut it in pairs of intersecting lines.

EXAMPLE 5 The **elliptic cone**

$$\frac{x^2}{a^2} + \frac{y^2}{b^2} = \frac{z^2}{c^2} \tag{6}$$

is symmetric with respect to the three coordinate planes (Fig. 11.49). The sections cut by the coordinate planes are

$$x = 0: \quad \text{the lines } z = \pm\frac{c}{b}y, \tag{7}$$

$$y = 0: \quad \text{the lines } z = \pm\frac{c}{a}x, \tag{8}$$

$$z = 0: \quad \text{the point } (0, 0, 0).$$

The sections cut by planes $z = z_0$ above and below the xy-plane are ellipses whose centers lie on the z-axis and whose vertices lie on the lines in Eqs. (7) and (8).

If $a = b$, the cone is a right circular cone. ■

EXAMPLE 6 The **hyperboloid of one sheet**

$$\frac{x^2}{a^2} + \frac{y^2}{b^2} - \frac{z^2}{c^2} = 1 \tag{9}$$

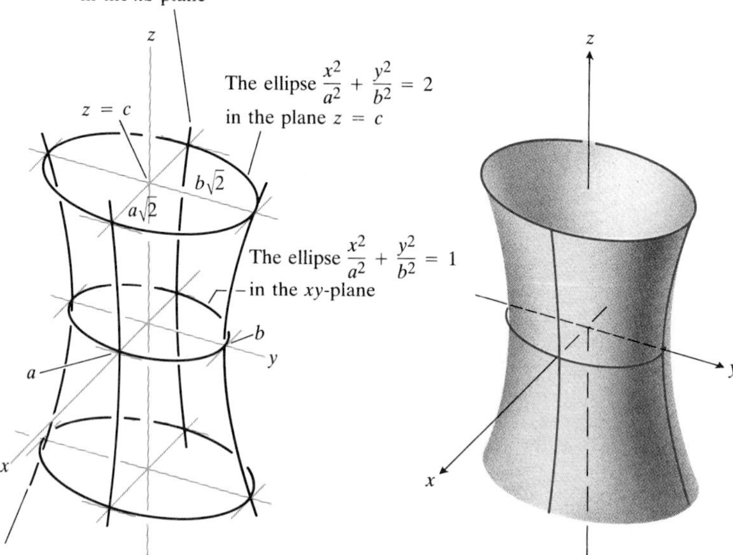

11.50 The surface $(x^2/a^2) + (y^2/b^2) - (z^2/c^2) = 1$ is an (elliptic) hyperboloid of one sheet. Planes perpendicular to the z-axis cut it in ellipses. Vertical planes containing the z-axis cut it in hyperbolas.

is symmetric with respect to each of the three coordinate planes (Fig. 11.50). The sections cut out by the coordinate planes are

$$\begin{aligned} x = 0&: \quad \text{the hyperbola } \frac{y^2}{b^2} - \frac{z^2}{c^2} = 1, \\ y = 0&: \quad \text{the hyperbola } \frac{x^2}{a^2} - \frac{z^2}{c^2} = 1, \qquad (10) \\ z = 0&: \quad \text{the ellipse } \frac{x^2}{a^2} + \frac{y^2}{b^2} = 1. \end{aligned}$$

The plane $z = z_0$ cuts the surface in an ellipse with center on the z-axis and vertices on one of the hyperbolas in (10).

The surface is connected, meaning that it is possible to travel from any point on it to any other point on it without leaving the surface. For this reason, it is said to have *one* sheet, in contrast to the hyperboloid in the next example, which has two sheets.

If $a = b$, the hyperboloid is a surface of revolution. ■

EXAMPLE 7 The **hyperboloid of two sheets**

$$\frac{z^2}{c^2} - \frac{x^2}{a^2} - \frac{y^2}{b^2} = 1 \qquad (11)$$

is symmetric with respect to the three coordinate planes (Fig. 11.51). The plane $z = 0$ does not intersect the surface; in fact, for a horizontal plane to

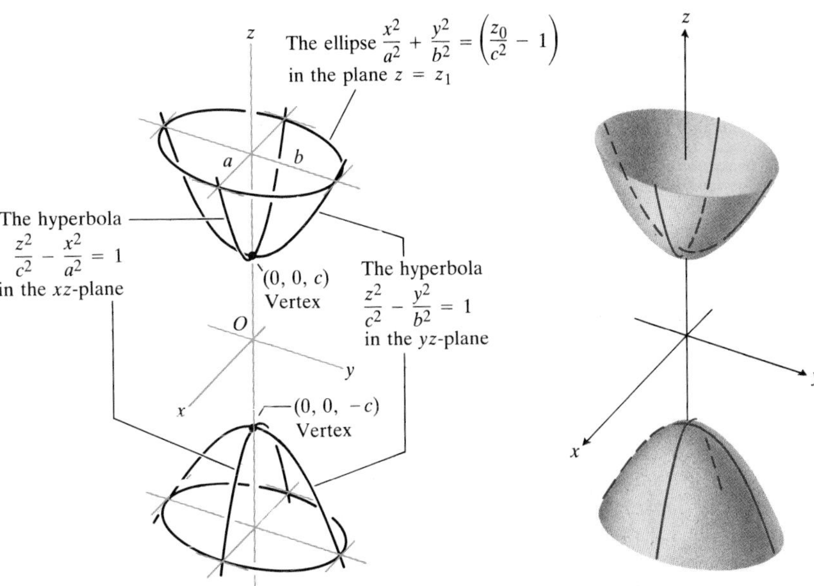

11.51 The surface $(z^2/c^2) - (x^2/a^2) - (y^2/b^2) = 1$ is an (elliptic) hyperboloid of two sheets. Planes perpendicular to the z-axis above and below the vertices cut it in ellipses. Vertical planes containing the z-axis cut it in hyperbolas.

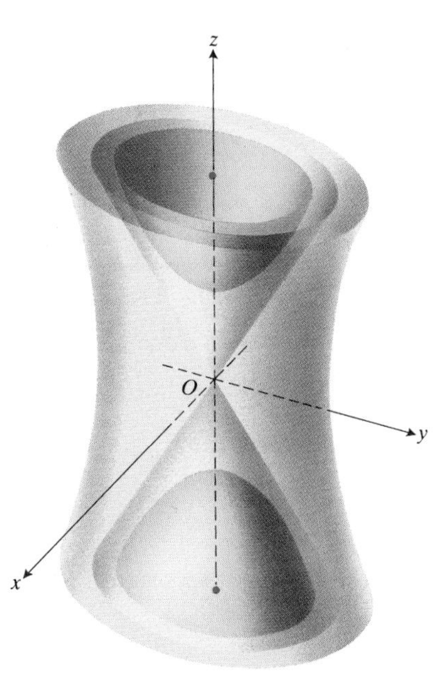

11.52 The cone is asymptotic to both hyperboloids.

intersect the surface, we must have $|z| \geq c$. The hyperbolic sections

$$x = 0: \quad \frac{z^2}{c^2} - \frac{y^2}{b^2} = 1,$$

$$y = 0: \quad \frac{z^2}{c^2} - \frac{x^2}{a^2} = 1,$$

have their vertices and foci on the z-axis. The surface is separated into two portions, one above the plane $z = c$ and the other below the plane $z = -c$. This accounts for its name.

Equations (9) and (11) have different numbers of negative terms. The number in each case is the same as the number of sheets of the hyperboloid. If we compare with Eq. (6), we see that replacing the one on the right side of either Eq. (9) or (11) by zero gives the equation of a cone. This cone (Fig. 11.52) is asymptotic to both of the hyperboloids (9) and (11) in the same way that the lines

$$\frac{x^2}{a^2} - \frac{y^2}{b^2} = 0$$

are asymptotic to the two hyperbolas

$$\frac{x^2}{a^2} - \frac{y^2}{b^2} = \pm 1$$

in the xy-plane. ■

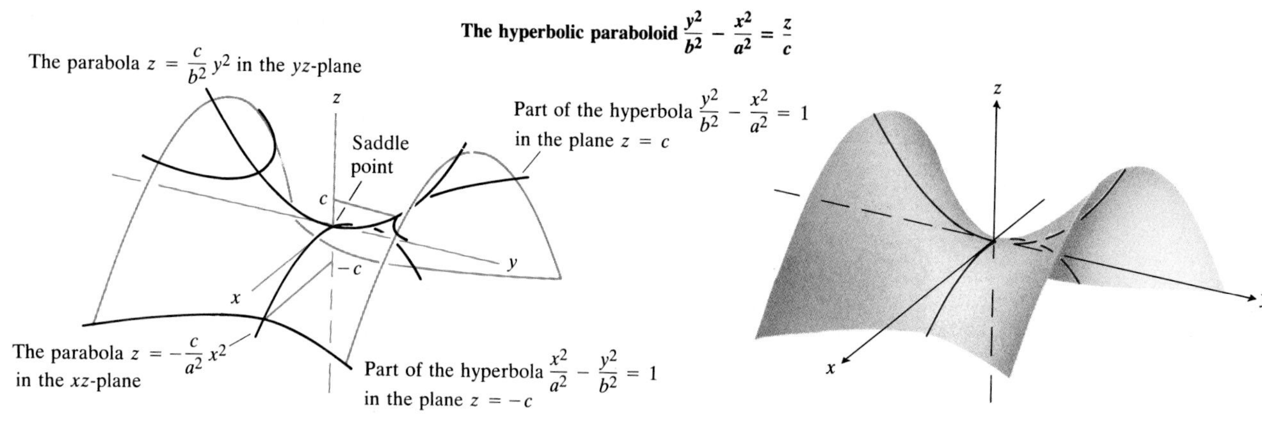

11.53 The surface $(y^2/b^2) - (x^2/a^2) = z/c$ is a hyperbolic paraboloid. The cross sections in planes perpendicular to the z-axis above and below the xy-plane are hyperbolas. The cross sections in planes perpendicular to the other axes are parabolas.

EXAMPLE 8 The **hyperbolic paraboloid**

$$\frac{y^2}{b^2} - \frac{x^2}{a^2} = \frac{z}{c} \tag{12}$$

has symmetry with respect to the planes $x = 0$ and $y = 0$ (Fig. 11.53). The sections in these planes are

$$x = 0: \quad \text{the parabola } z = \frac{c}{b^2}y^2, \tag{13}$$

$$y = 0: \quad \text{the parabola } z = -\frac{c}{a^2}x^2. \tag{14}$$

In the plane $x = 0$, the parabola opens upward from the origin. The parabola in the plane $y = 0$ opens downward.

If we cut the surface by a plane $z = z_0 > 0$, the section is a hyperbola,

$$\frac{y^2}{b^2} - \frac{x^2}{a^2} = \frac{z_0}{c}, \tag{15}$$

with its focal axis parallel to the y-axis and its vertices on the parabola in (13). If z_0 is negative, the focal axis is parallel to the x-axis and the vertices lie on the parabola in (14).

Near the origin, the surface is shaped like a saddle. To a person traveling along the surface in the yz-plane, the origin looks like a minimum. To a person traveling in the xz-plane, the origin looks like a maximum. Such a point is called a **minimax** or **saddle point** of a surface (Fig. 11.53). We shall discuss maximum and minimum points on surfaces in Chapter 13. ■

How to Draw Quadric Surfaces

$x^2 + \frac{y^2}{4} + z^2 = 1$

$z = 4 - x^2 - y^2$

1.

Lightly sketch the three coordinate axes.

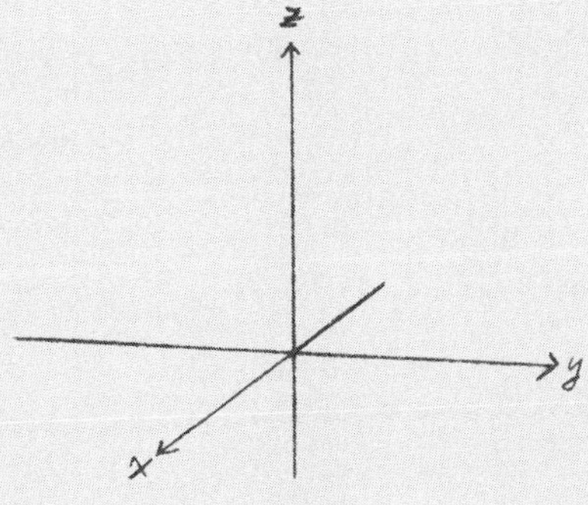

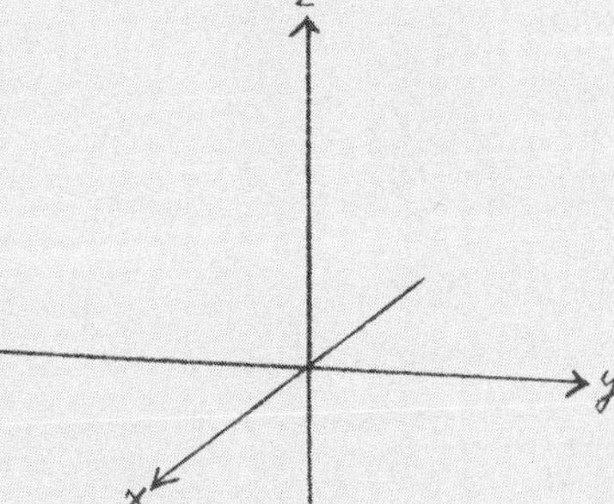

2.

Decide on a scale and mark the intercepts on the axes.

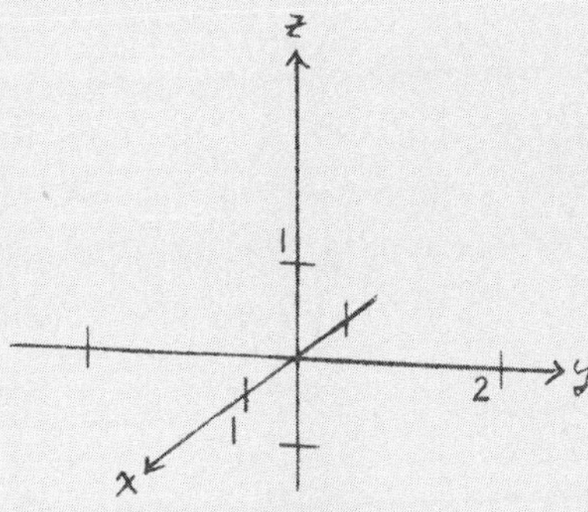

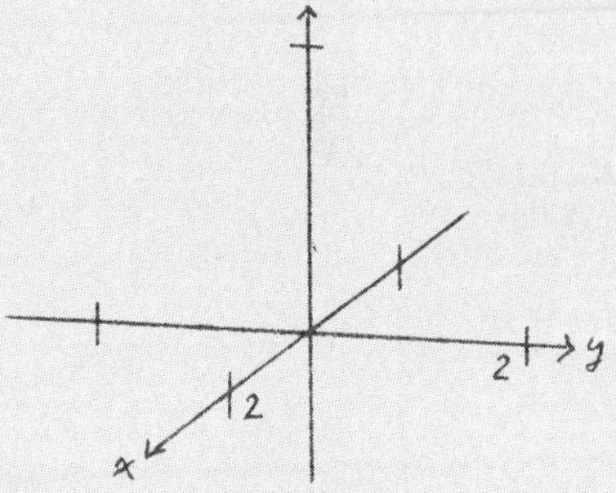

3.

Sketch cross sections in the coordinate planes and in a few parallel planes, but don't clutter the picture. Use tangent lines as guides.

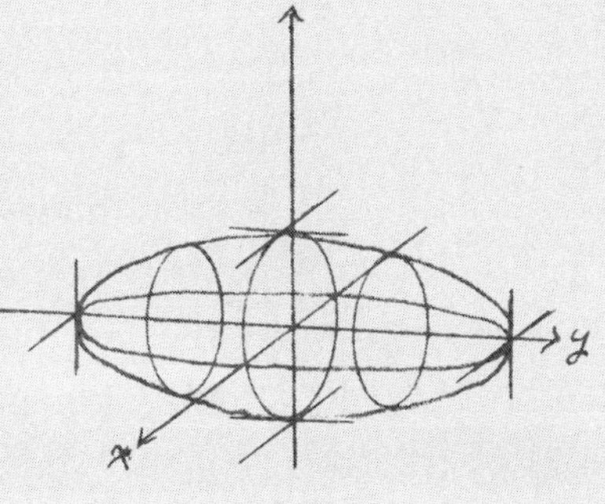

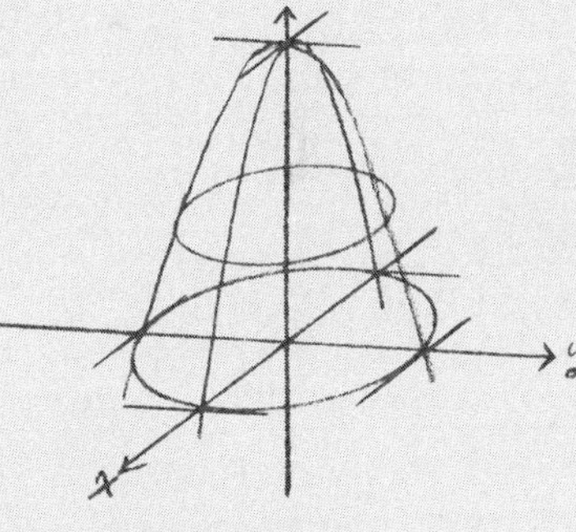

4.

If more is required, darken the parts exposed to view. Leave the rest light. Use line breaks when you can.

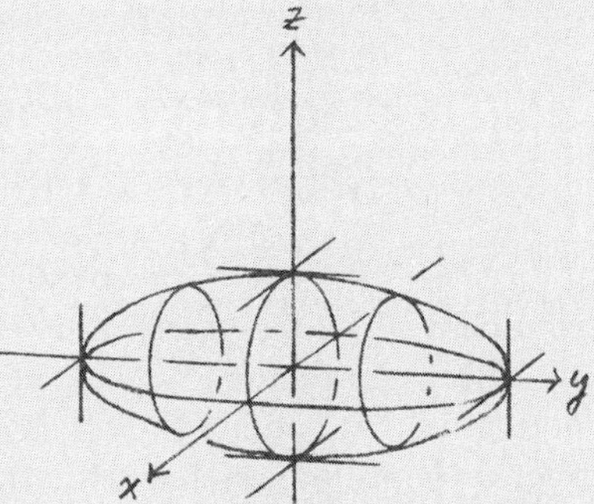

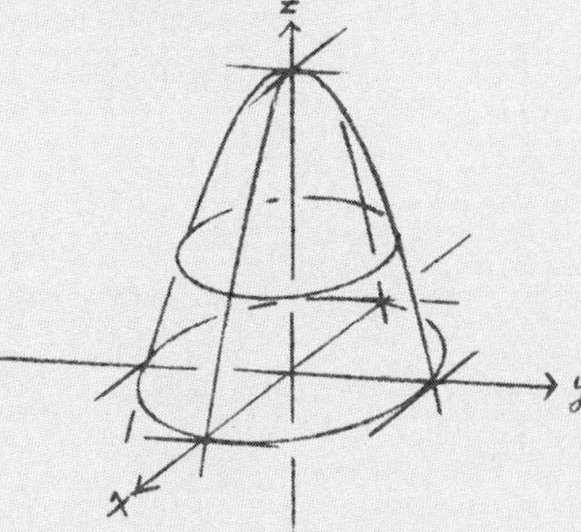

Exercises 11.6

Sketch the surfaces in Exercises 1–50.

Cylinders

1. $x^2 + y^2 = 4$
2. $x^2 + z^2 = 4$
3. $y^2 + z^2 = 1$
4. $z = y^2/4$
5. $z = y^2 - 1$
6. $x = y^2$
7. $z = 4 - x^2$
8. $x = 4 - y^2$
9. $y = x^2$
10. $y = x^2 - 2$
11. $y^2 + 4z^2 = 16$
12. $4x^2 + y^2 = 36$
13. $z^2 + 4y^2 = 9$
14. $y^2 - z^2 = 4$
15. $z^2 - y^2 = 1$
16. $yz = 1$

Ellipsoids

17. $9x^2 + y^2 + z^2 = 9$
18. $4x^2 + 4y^2 + z^2 = 16$
19. $x^2 + y^2 + z^2 = 4$
20. $9x^2 + 4y^2 + z^2 = 36$
21. $4x^2 + 9y^2 + 4z^2 = 36$
22. $9x^2 + 4y^2 + 36z^2 = 36$

Paraboloids

23. $x^2 + y^2 = z$
24. $x^2 + z^2 = y$
25. $x^2 + 4y^2 = z$
26. $z = x^2 + 9y^2$
27. $z = 8 - x^2 - y^2$
28. $z = 18 - x^2 - 9y^2$
29. $x = 4 - 4y^2 - z^2$
30. $y = 1 - x^2 - z^2$
31. $z = x^2 + y^2 + 1$
32. $z = 4x^2 + y^2 - 4$

Cones

33. $x^2 + y^2 = z^2$
34. $y^2 + z^2 = x^2$
35. $x^2 + z^2 = y^2$
36. $4x^2 + 9y^2 = z^2$
37. $9x^2 + 4y^2 = 36z^2$
38. $4x^2 + 9z^2 = 9y^2$

Hyperboloids

39. $x^2 + y^2 - z^2 = 1$
40. $y^2 + z^2 - x^2 = 1$
41. $(y^2/4) + (z^2/9) - (x^2/4) = 1$
42. $(x^2/4) + (y^2/4) - (z^2/9) = 1$
43. $(x^2/4) + y^2 - z^2 = 1$
44. $z^2 - x^2 - y^2 = 1$
45. $z^2 - (x^2/4) - y^2 = 1$
46. $(y^2/4) - (x^2/4) - z^2 = 1$
47. $x^2 - y^2 - (z^2/4) = 1$
48. $(x^2/4) - (z^2/4) - y^2 = 1$

Hyperbolic Paraboloids

49. $y^2 - x^2 = z$
50. $x^2 - y^2 = z$

Computer Grapher

If you have access to a computer grapher, try graphing the surfaces in Exercises 51–57.

51. $z = y^2$
52. $z = 1 - y^2$
53. $z = x^2 + y^2$
54. $z = x^2 + 2y^2$
55. $z = \sqrt{1 - x^2}$ (upper half of a circular cylinder)
56. $z = \sqrt{1 - (y^2/4)}$ (upper half of an elliptical cylinder)
57. $z = \sqrt{x^2 + 2y^2 + 4}$ (one sheet of an elliptic hyperboloid)

TOOLKIT PROGRAM

3D Grapher — Graphs equations of the form $z = f(x, y)$

11.7 Cylindrical and Spherical Coordinates

In this section we introduce two new systems of coordinates for space: the cylindrical coordinate system and the spherical coordinate system. In the cylindrical coordinate system, cylinders whose axes lie along the z-axis and planes that contain the z-axis have especially simple equations. In the spherical coordinate system, spheres centered at the origin and single cones at the origin whose axes lie along the z-axis have especially simple equations. When your work involves these shapes and you need a coordinate system, these may be the best systems to use, as we shall see in later chapters.

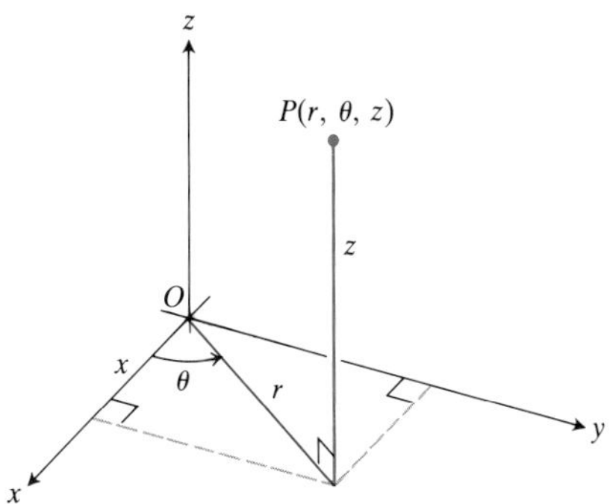

11.54 The cylindrical coordinates of a point in space are r, θ, and z.

Cylindrical Coordinates

We obtain cylindrical coordinates for space by combining polar coordinates in the xy-plane with the usual z-axis. This assigns to every point in space one or more coordinate triples of the form (r, θ, z), as shown in Fig. 11.54.

The values of x, y, r, and θ in cylindrical coordinates are related by the usual equations:

$$x = r\cos\theta, \quad r^2 = x^2 + y^2, \quad y = r\sin\theta, \quad \tan\theta = y/x. \tag{1}$$

Cylindrical coordinates are the coordinates we shall use to study planetary motion in Section 12.5.

In cylindrical coordinates, the equation $r = a$ describes not just a circle in the xy-plane but an entire cylinder about the z-axis (Fig. 11.55). The z-axis itself is given by the equation $r = 0$. The equation $\theta = \theta_0$ describes the plane that contains the z-axis and makes an angle of θ_0 radians with the positive x-axis.

EXAMPLE 1 Describe the points in space whose cylindrical coordinates satisfy the equations

$$r = 2, \qquad \theta = \frac{\pi}{4}.$$

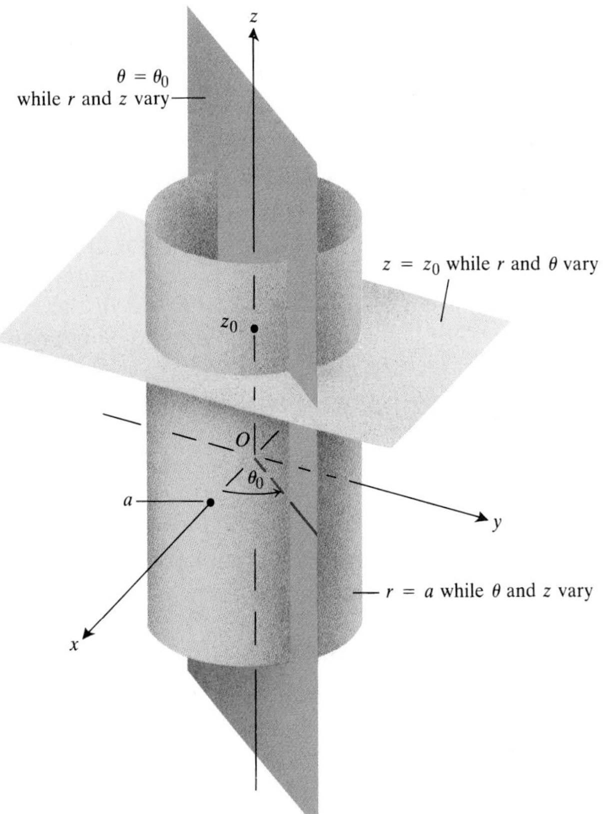

11.55 Planes and cylinders that have constant-coordinate equations in cylindrical coordinates.

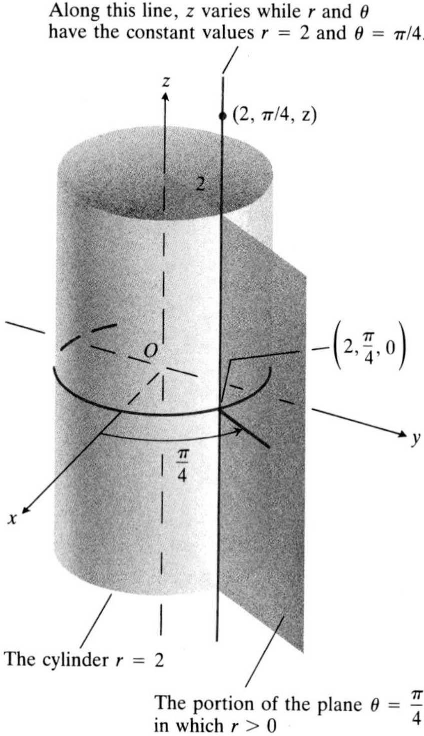

11.56 The points whose first two cylindrical coordinates are $r = 2$ and $\theta = \pi/4$ form a line parallel to the z-axis.

Generating curve $r = 1 + \cos\theta$ in the $r\theta$-plane (the plane $z = 0$)

11.57 The cylindrical-coordinate equation $r = 1 + \cos\theta$ defines a cylinder in space whose cross sections perpendicular to the z-axis are cardioids.

Solution These points make up the line in which the cylinder $r = 2$ cuts the portion of the plane $\theta = \pi/4$ in which r is positive (Fig. 11.56). This is the line through the point $(2, \pi/4, 0)$ parallel to the z-axis. ■

EXAMPLE 2 Sketch the surface $r = 1 + \cos\theta$.

Solution The equation involves only r and θ; the coordinate variable z is missing. Therefore, the surface is a cylinder of lines that pass through the cardioid $r = 1 + \cos\theta$ in the $r\theta$-plane and lie parallel to the z-axis. The rules for sketching the cylinder are the same as always: sketch the x-, y-, and z-axes, draw a few perpendicular cross sections, connect the cross sections with parallel lines, and darken the exposed parts (Fig. 11.57). ■

EXAMPLE 3 Find a Cartesian equation for the surface $z = r^2$ and identify the surface.

Solution From Eqs. (1) we have $z = r^2 = x^2 + y^2$. The surface is the circular paraboloid $x^2 + y^2 = z$. ■

EXAMPLE 4 Find an equation for the circular cylinder $4x^2 + 4y^2 = 9$ in cylindrical coordinates.

Solution The cylinder consists of the points whose distance from the z-axis is $\sqrt{x^2 + y^2} = 3/2$. The corresponding equation in cylindrical coordinates is $r = 3/2$. ■

11.58 The spherical coordinates ρ, ϕ, and θ and their relation to x, y, z, and r.

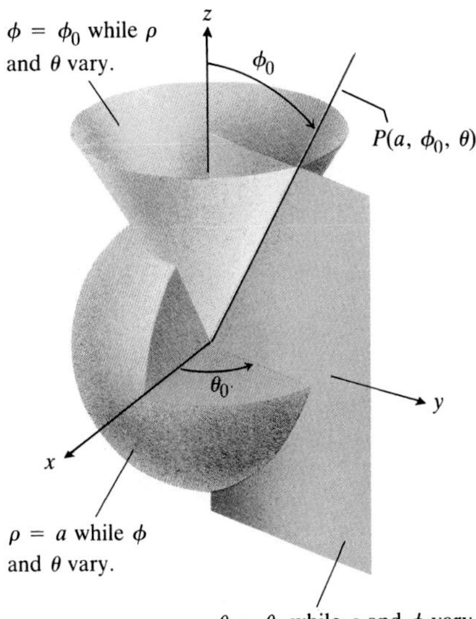

11.59 Spheres whose centers are at the origin, and single cones at the origin whose axes lie along the z-axis, and half-planes "hinged" along the z-axis have constant-coordinate equations in spherical coordinates.

Spherical Coordinates

Spherical coordinates locate points in space with two angles and a distance, as shown in Fig. 11.58.

The first coordinate, $\rho = |\overrightarrow{OP}|$, is the point's distance from the origin. Unlike r, the variable ρ is never negative. The second coordinate, ϕ, is the angle the vector $\overrightarrow{OP}$ makes with the positive z-axis. It is required to lie in the interval from 0 to π. The third coordinate is the angle θ from cylindrical coordinates.

The equation $\rho = a$ describes the sphere of radius a centered at the origin (Fig. 11.59). The equation $\phi = \phi_0$ describes a single cone whose vertex lies at the origin and whose axis lies along the z-axis. (We have to broaden our interpretation here to include the xy-plane as the cone $\phi = \pi/2$.) If ϕ_0 is greater than $\pi/2$, the cone $\phi = \phi_0$ opens downward.

A few books give spherical coordinates in the order (ρ, θ, ϕ), with the θ and ϕ reversed. Watch out for this when you read elsewhere.

Selected Equations Relating Cartesian (Rectangular), Cylindrical, and Spherical Coordinates

$$\begin{aligned} r &= \rho \sin \phi, \qquad x = r \cos \theta = \rho \sin \phi \cos \theta, \\ z &= \rho \cos \phi, \qquad y = r \sin \theta = \rho \sin \phi \sin \theta, \end{aligned} \tag{2}$$

$$\rho = \sqrt{x^2 + y^2 + z^2} = \sqrt{r^2 + z^2}$$

EXAMPLE 5 Find a spherical coordinate equation for the sphere

$$x^2 + y^2 + (z - 1)^2 = 1.$$

Solution From Eqs. (2) we find that the left side of the equation is

$$\rho^2\sin^2\phi(\cos^2\theta + \sin^2\theta) + \rho^2\cos^2\phi - 2\rho \cos \phi + 1 = \rho^2 - 2\rho \cos \phi + 1.$$

Hence the original equation transforms into

$$\begin{aligned} \rho^2 - 2\rho \cos \phi + 1 &= 1, \\ \rho^2 &= 2\rho \cos \phi, \\ \rho &= 2 \cos \phi. \end{aligned}$$

See Fig. 11.60. ■

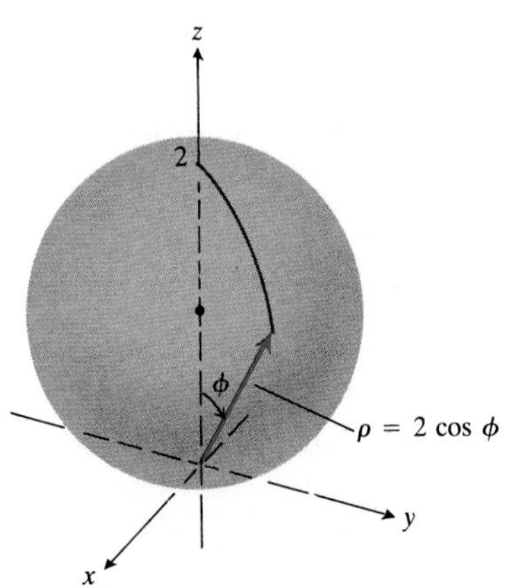

11.60 The sphere $\rho = 2 \cos \phi$. Notice that the equation restricts ϕ to lie in the interval $0 \le \phi \le \pi/2$ because the spherical coordinate ρ is not allowed to be negative.

Exercises 11.7

The following table gives the coordinates of specific points in space in one of three coordinate systems. In Exercises 1–10, find coordinates for each point in the other two systems. There may be more than one right answer because points in cylindrical and spherical coordinates may have more than one coordinate triple.

	Rectangular (x, y, z)	Cylindrical (r, θ, z)	Spherical (ρ, ϕ, θ)
1.	$(0, 0, 0)$		
2.	$(1, 0, 0)$		
3.	$(0, 1, 0)$		
4.	$(0, 0, 1)$		
5.		$(1, 0, 0)$	
6.		$(\sqrt{2}, 0, 1)$	
7.		$(1, \pi/2, 1)$	
8.			$(\sqrt{3}, \pi/3, -\pi/2)$
9.			$(2\sqrt{2}, \pi/2, 3\pi/2)$
10.			$(\sqrt{2}, \pi, 3\pi/2)$

In Exercises 11–20, translate the equations from the given coordinate system (rectangular, cylindrical, spherical) into equations in the other two systems. Also, identify the set of points defined by the equation.

11. $r = 0$

12. $x^2 + y^2 = 5$

13. $z = 0$

14. $z = -2$

15. $\rho \cos \phi = 3$

16. $\sqrt{x^2 + y^2} = z$

17. $\rho \sin \phi \cos \theta = 0$

18. $\tan^2\phi = 1$

19. $x^2 + y^2 + z^2 = 4$

20. $\rho = 6 \cos \phi$

In Exercises 21–28, describe the sets of points in space whose cylindrical coordinates satisfy the given equations or pairs of equations. Sketch.

21. $r = 4$

22. $r^2 + z^2 = 1$

23. $r = 1 - \cos \theta$

24. $r = 2 \cos \theta$

25. $r = 2, \quad z = 3$

26. $\theta = \pi/6, \quad z = r$

27. $r = 3, \quad z = \theta/2$

28. $r^2 = \cos 2\theta$

In Exercises 29–34, describe the sets of points in space whose spherical coordinates satisfy the given equations or pairs of equations. Sketch.

29. $\phi = \pi/6$

30. $\rho = 6, \quad \phi = \pi/6$

31. $\rho = 5, \quad \theta = \pi/4$

32. $\theta = \pi/4, \quad \phi = \pi/4$

33. $\theta = \pi/2, \quad \rho = 4 \sin \phi$

34. $\rho = 1 - \cos \phi$ (*Hint:* The absence of θ indicates symmetry with respect to the z-axis. What is the trace of the surface in the yz-plane?)

Review Questions

1. When are two vectors equal?
2. How are vectors added and subtracted?
3. How are the length and direction of a vector calculated?
4. If a vector is multiplied by a scalar, how is the result related to the original vector? What if the scalar is zero? Negative?
5. Define the *scalar* or *dot product* of two vectors. Which algebraic laws (commutative, associative, distributive) are satisfied by dot products and which, if any, are not? Give examples. When is the scalar product of two vectors equal to zero?
6. What is the vector projection of a vector **B** onto a vector **A**? How do you write **B** as the sum of a vector parallel to **A** and a vector orthogonal to **A**?
7. Define the *vector* or *cross product* of two vectors. Which algebraic laws (commutative, associative, distributive) are satisfied by cross products and which are not? Give examples. When is the vector product of two vectors equal to zero?
8. What is the determinant formula for evaluating the cross product of two vectors? Use it in an example.
9. How are vector and scalar products used to find equations for lines, line segments, planes? Give examples.
10. How can vectors be used to calculate the distance between a point and a line? A point and a plane? Give examples.
11. What is the geometric interpretation of $(\mathbf{A} \times \mathbf{B}) \cdot \mathbf{C}$ as a volume? How may the product be calculated from the components of **A**, **B**, and **C**?

12. What is a cylinder? Give examples of equations that define cylinders in Cartesian coordinates; in cylindrical coordinates. What advice can you give about drawing cylinders?

13. Give examples of ellipsoids, paraboloids, cones, and hyperboloids (equations and sketches). What advice can you give about sketching these surfaces?

14. How are cylindrical and spherical coordinates defined? Draw diagrams that show how cylindrical and spherical coordinates are related to rectangular coordinates. What sets have constant-coordinate equations (like $x = 1$, $r = 1$, or $\phi = \pi/3$) in the three coordinate systems?

Practice Exercises

1. Draw the unit vectors $\mathbf{u} = (\cos\theta)\mathbf{i} + (\sin\theta)\mathbf{j}$ for $\theta = 0, \pi/2, 2\pi/3, 5\pi/4$, and $5\pi/3$, together with the coordinate axes and unit circle.

2. Find the unit vector obtained by rotating
 a) $\mathbf{i}$ clockwise 45° b) $\mathbf{j}$ counterclockwise 120°

In Exercises 3–4, find the unit vectors that are tangent and normal to the curve at point P.

3. $y = \tan x$, $P(\pi/4, 1)$ 4. $x^2 + y^2 = 5$, $P(3, 4)$

Express the vectors in Exercises 5–8 in terms of their lengths and directions.

5. $\sqrt{2}\mathbf{i} + \sqrt{2}\mathbf{j}$ 6. $-\mathbf{i} - \mathbf{j}$

7. $2\mathbf{i} - 3\mathbf{j} + 6\mathbf{k}$ 8. $\mathbf{i} + 2\mathbf{j} - \mathbf{k}$

9. Find a vector 2 units long in the direction of $\mathbf{A} = 4\mathbf{i} - \mathbf{j} + 4\mathbf{k}$.

10. Find a vector 5 units long in the direction opposite to the direction of $\mathbf{A} = (3/5)\mathbf{i} + (4/5)\mathbf{k}$.

In Exercises 11–12, find $|\mathbf{A}|$, $|\mathbf{B}|$, $\mathbf{A}\cdot\mathbf{B}$, $\mathbf{B}\cdot\mathbf{A}$, $\mathbf{A}\times\mathbf{B}$, $\mathbf{B}\times\mathbf{A}$, $|\mathbf{A}\times\mathbf{B}|$, the acute angle between $\mathbf{A}$ and $\mathbf{B}$, the scalar component of $\mathbf{B}$ in the direction of $\mathbf{A}$, and the vector projection of $\mathbf{B}$ onto $\mathbf{A}$.

11. $\mathbf{A} = \mathbf{i} + \mathbf{j}$,
 $\mathbf{B} = 2\mathbf{i} + \mathbf{j} - 2\mathbf{k}$

12. $\mathbf{A} = 5\mathbf{i} + \mathbf{j} + \mathbf{k}$,
 $\mathbf{B} = \mathbf{i} - 2\mathbf{j} + 3\mathbf{k}$

In Exercises 13–14, write $\mathbf{B}$ as the sum of a vector parallel to $\mathbf{A}$ and a vector orthogonal to $\mathbf{A}$.

13. $\mathbf{A} = 2\mathbf{i} + \mathbf{j} - \mathbf{k}$,
 $\mathbf{B} = \mathbf{i} + \mathbf{j} - 5\mathbf{k}$

14. $\mathbf{A} = \mathbf{i} - 2\mathbf{j}$,
 $\mathbf{B} = \mathbf{i} + \mathbf{j} + \mathbf{k}$

In Exercises 15–16, draw coordinate axes and then sketch $\mathbf{A}$, $\mathbf{B}$, and $\mathbf{A}\times\mathbf{B}$ as vectors at the origin.

15. $\mathbf{A} = \mathbf{i}$, $\mathbf{B} = \mathbf{i} + \mathbf{j}$ 16. $\mathbf{A} = \mathbf{i} - \mathbf{j}$, $\mathbf{B} = \mathbf{i} + \mathbf{j}$

In Exercises 17–18, find the distance from the point to the line in the xy-plane.

17. $(3, 2)$, $3x + 4y = 2$ 18. $(-1, 1)$, $5x - 12y = 9$

In Exercises 19–20, find the distance from the point to the plane.

19. $(6, 0, -6)$, $x - y = 4$

20. $(3, 0, 10)$, $2x + 3y + z = 2$

21. Find an equation for the plane that passes through the point $(3, -2, 1)$ normal to the vector $\mathbf{N} = 2\mathbf{i} + \mathbf{j} - \mathbf{k}$.

22. Find an equation for the plane that passes through the point $(-1, 6, 0)$ perpendicular to the line $x = -1 + t$, $y = 6 - 2t$, $z = 3t$.

In Exercises 23–24, find an equation for the plane through points P, Q, and R.

23. $P(1, -1, 2)$, $Q(2, 1, 3)$, $R(-1, 2, -1)$

24. $P(1, 0, 0)$, $Q(0, 1, 0)$, $R(0, 0, 1)$

25. Find parametric equations for the line that passes through the point $(1, 2, 3)$ parallel to the vector $\mathbf{v} = -3\mathbf{i} + 7\mathbf{k}$.

26. Find the points in which the line $x = 1 + 2t$, $y = -1 - t$, $z = 3t$ meets the three coordinate planes.

27. Find the point in which the line through the origin perpendicular to the plane $2x - y - z = 4$ meets the plane $3x - 5y + 2z = 6$.

28. Find parametric equations for the line segment joining the points $P(1, 2, 0)$ and $Q(1, 3, -1)$.

29. Find parametric equations for the line in which the planes $x + 2y + z = 1$ and $x - y + 2z = -8$ intersect.

30. Find the acute angle of intersection of the planes $x + y = 1$ and $y + z = 1$.

In Exercises 31–32, find the distance from the point to the line.

31. $(2, -1, -10)$; $x = 4$, $y = 4 + 2t$, $z = 4t$

32. $(-1, 4, 3)$; $x = 10 + 4t$, $y = -3$, $z = 4t$

33. *Work.* Find the work done in pushing a car 800 ft with a force of magnitude 40 lbs directed 30° downward from the horizontal against the back of the car.

34. *Torque.* The operator's manual for the *Toro* 21-in. lawnmower says "tighten the spark plug to 15 ft-lb (20.4 N · m)." If you are installing the plug with a 10.5-in. socket wrench that places the center of your hand 9 in. from the axis of the spark plug, about how hard should you pull? Answer in pounds.

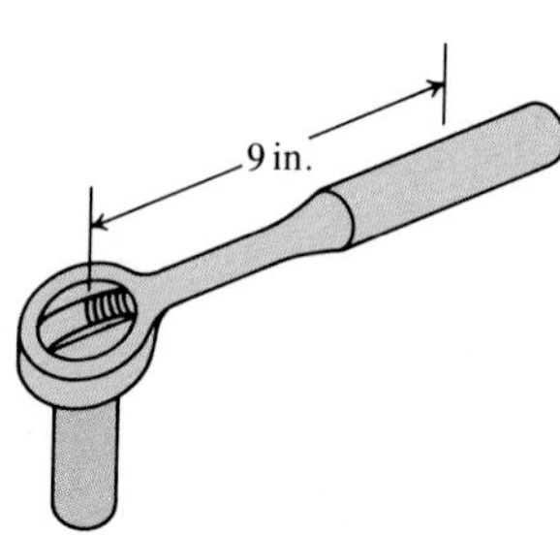

In Exercises 35–36, find (a) the area of the parallelogram determined by vectors **A** and **B**, (b) the volume of the parallelepiped determined by the vectors **A**, **B**, and **C**.

35. $\mathbf{A} = \mathbf{i} + \mathbf{j} - \mathbf{k}$, $\mathbf{B} = 2\mathbf{i} + \mathbf{j} + \mathbf{k}$, $\mathbf{C} = -\mathbf{i} - 2\mathbf{j} + 3\mathbf{k}$

36. $\mathbf{A} = \mathbf{i} + \mathbf{j}$, $\mathbf{B} = \mathbf{j}$, $\mathbf{C} = \mathbf{i} + \mathbf{j} + \mathbf{k}$

37. Which of the following are *always true* and which are *not always true*?

a) $|\mathbf{A}| = \sqrt{\mathbf{A} \cdot \mathbf{A}}$ b) $\mathbf{A} \cdot \mathbf{A} = |\mathbf{A}|$
c) $\mathbf{A} \times \mathbf{0} = \mathbf{0} \times \mathbf{A} = \mathbf{0}$ d) $\mathbf{A} \times (-\mathbf{A}) = \mathbf{0}$
e) $\mathbf{A} \times \mathbf{B} = \mathbf{B} \times \mathbf{A}$
f) $\mathbf{A} \times (\mathbf{B} + \mathbf{C}) = \mathbf{A} \times \mathbf{B} + \mathbf{A} \times \mathbf{C}$
g) $(\mathbf{A} \times \mathbf{B}) \cdot \mathbf{B} = 0$ h) $(\mathbf{A} \times \mathbf{B}) \cdot \mathbf{C} = \mathbf{A} \cdot (\mathbf{B} \times \mathbf{C})$

38. Which of the following are *always true* and which are *not always true*?

a) $\mathbf{A} \cdot \mathbf{B} = \mathbf{B} \cdot \mathbf{A}$ b) $\mathbf{A} \times \mathbf{B} = -(\mathbf{B} \times \mathbf{A})$
c) $(-\mathbf{A}) \times \mathbf{B} = -(\mathbf{A} \times \mathbf{B})$
d) $(c\mathbf{A}) \cdot \mathbf{B} = \mathbf{A} \cdot (c\mathbf{B}) = c(\mathbf{A} \cdot \mathbf{B})$ (any number c)
e) $c(\mathbf{A} \times \mathbf{B}) = (c\mathbf{A}) \times \mathbf{B} = \mathbf{A} \times (c\mathbf{B})$ (any number c)
f) $\mathbf{A} \cdot \mathbf{A} = |\mathbf{A}|^2$ g) $(\mathbf{A} \times \mathbf{A}) \cdot \mathbf{A} = 0$
h) $(\mathbf{A} \times \mathbf{B}) \cdot \mathbf{A} = \mathbf{B} \cdot (\mathbf{A} \times \mathbf{B})$

The equations in Exercises 39–48 define sets both in the plane and in three-dimensional space. Identify both sets for each equation.

Rectangular Coordinates

39. $x = 0$ **40.** $x + y = 1$ **41.** $x^2 + y^2 = 4$
42. $x^2 + 4y^2 = 16$ **43.** $x = y^2$ **44.** $y^2 - x^2 = 1$

Cylindrical Coordinates

45. $r = 1 - \cos\theta$ **46.** $r = \sin\theta$
47. $r^2 = 2\cos 2\theta$ **48.** $r = \cos 2\theta$

Describe the sets defined by the spherical-coordinate equations and inequalities in Exercises 49–54.

49. $\rho = 2$ **50.** $\theta = \pi/4$
51. $\phi = \pi/6$ **52.** $\rho = 1,\ \phi = \pi/2$
53. $\rho = 1,\ 0 \le \phi \le \pi/2$ **54.** $1 \le \rho \le 2$

The following table gives the coordinates of points in space in one of the three standard coordinate systems. In Exercises 55–60, find coordinates for each point in the other two systems. There may be more than one right answer because cylindrical and spherical coordinates are not unique.

	Rectangular (x, y, z)	Cylindrical (r, θ, z)	Spherical (ρ, ϕ, θ)
55.		$(1, 0, 0)$	
56.		$\left(1, \frac{\pi}{2}, 0\right)$	
57.			$\left(\sqrt{2}, \frac{\pi}{4}, \frac{\pi}{2}\right)$
58.			$\left(2, \frac{5\pi}{6}, 0\right)$
59.	$(1, 1, 1)$		
60.	$(0, -1, 1)$		

In Exercises 61–66, translate the equations from the given coordinate system (rectangular, cylindrical, spherical) into the other two systems. Identify the set of points defined by the equation.

Rectangular

61. $z = 2$ **62.** $z = \sqrt{x^2 + y^2}$

Cylindrical

63. $z = r^2$ **64.** $r = \cos\theta$

Spherical

65. $\rho = 4$ **66.** $\rho\cos\phi = 1$

Sketch the quadric surfaces in Exercises 67–78.

67. $x^2 + y^2 + z^2 = 4$ **68.** $x^2 + (y^2 - 1) + z^2 = 1$
69. $4x^2 + 4y^2 + z^2 = 4$ **70.** $36x^2 + 9y^2 + 4z^2 = 36$
71. $z = -(x^2 + y^2)$ **72.** $y = -(x^2 + z^2)$
73. $x^2 + y^2 = z^2$ **74.** $x^2 + z^2 = y^2$
75. $x^2 + y^2 - z^2 = 4$ **76.** $4y^2 + z^2 - 4x^2 = 4$
77. $y^2 - x^2 - z^2 = 1$ **78.** $z^2 - x^2 - y^2 = 1$

12 Vector-Valued Functions and Motion in Space

Comstock

We use vectors and their derivatives to describe motion in space.

OVERVIEW When a body travels through space, the three equations $x = f(t)$, $y = g(t)$, and $z = h(t)$ that give the body's coordinates as functions of time become parametric equations for the body's motion and path. With vector notation, we can condense these three equations into a single equation

$$\mathbf{r} = f(t)\mathbf{i} + g(t)\mathbf{j} + h(t)\mathbf{k}$$

that gives the body's position as a vector function of t.

In this chapter, we show how to use the calculus we know to differentiate and integrate vector functions and to study the paths, velocities, and accelerations of bodies in the plane and in space. As we go along, we shall also see how our work answers many of the standard questions about the paths and motions of projectiles, planets, and satellites.

12.1 Vector-Valued Functions and Curves in Space. Derivatives and Integrals

When a particle moves along a path in space, we can keep track of it by watching the vector $\mathbf{r}$ that runs from the origin out to the particle (Fig. 12.1). If the particle's position coordinates are twice-differentiable functions of time, then so is $\mathbf{r}$, and we can find the particle's velocity vector and acceleration vector as functions of time by differentiating $\mathbf{r}$. Conversely, if we know the particle's velocity vector or acceleration vector as a function of time, then under the right conditions we can find $\mathbf{r}$ by integration. This section gives the details.

12.1 The position vector $\mathbf{r} = \overrightarrow{OP}$ of a particle moving through space is a function of time.

Definitions

When a particle moves through space as a function of time, each of its rectangular coordinates is a function of time. We may describe the particle's path with a triple of equations

$$x = f(t), \qquad y = g(t), \qquad z = h(t), \qquad t \in I, \tag{1}$$

where I is the time interval in question. The set of points (x, y, z) defined by these equations is called a **curve** in space. The equations are **parametric equations** for the curve, t being the **parameter** and I the **parameter interval.** The point $P(f(t), g(t), h(t))$ is called the particle's **position** at time t.

The vector

$$\mathbf{r} = \overrightarrow{OP} = f(t)\mathbf{i} + g(t)\mathbf{j} + h(t)\mathbf{k} \tag{2}$$

is called the particle's **position vector,** and we think of the particle's path as the **curve traced by r.** We use Eq. (2) to name the curve, as "the curve $\mathbf{r} = f(t)\mathbf{i} + g(t)\mathbf{j} + h(t)\mathbf{k}$" or "the curve $\mathbf{r} = f\mathbf{i} + g\mathbf{j} + h\mathbf{k}$."

With Eq. (2), we can describe the particle's path with a single vector equation in place of three parametric equations. We can always get the parametric equations back if we need them by writing $x = f(t)$, $y = g(t)$, $z = h(t)$. This efficiency is one of the advantages of shifting to vector notation when we study curves.

Equation (2) defines a vector $\mathbf{r}$ as a function of a real variable t on an interval I. More generally, a **vector function** or **vector-valued function** on a domain D is a rule that assigns a unique vector in space to each element of D. In the present chapter, D will always be some interval of real numbers. Later on, in Chapter 15, we will study vector functions defined on regions in the plane and space and call such functions "vector fields."

EXAMPLE 1 *A helix.* The curve

$$\mathbf{r} = (\cos t)\mathbf{i} + (\sin t)\mathbf{j} + t\mathbf{k},$$

called a *helix* ("HEE-lix") from an old Greek word for spiral, winds around the cylinder $x^2 + y^2 = 1$ (Fig. 12.2). We know that the curve lies on the cylin-

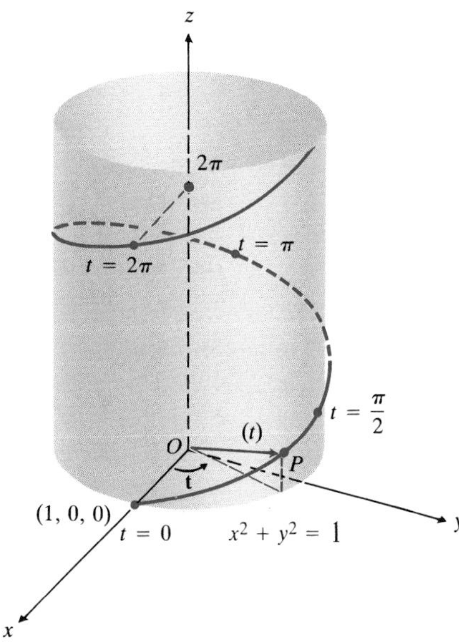

12.2 The helix $\mathbf{r} = (\cos t)\mathbf{i} + (\sin t)\mathbf{j} + t\mathbf{k}$ spirals up from the xy-plane as t increases from zero.

der because the x- and y-coordinates of the point $P(\cos t, \sin t, t)$ satisfy the cylinder's equation:

$$x^2 + y^2 = (\cos t)^2 + (\sin t)^2 = 1.$$

The curve rises steadily as its z-coordinate t increases, and completes one turn around the cylinder each time t increases by 2π. The parametric equations for the helix are

$$x = \cos t, \qquad y = \sin t, \qquad z = t.$$

You will find equations for other helices ("HEE-lih-sees") in Fig. 12.3. ■

Limits and Continuity

We define limits of vector functions in terms of their scalar components.

DEFINITION

If $\mathbf{L} = L_1\mathbf{i} + L_2\mathbf{j} + L_3\mathbf{k}$ is a vector and $\mathbf{f}$ is a vector-valued function of t defined by the rule $\mathbf{f}(t) = f(t)\mathbf{i} + g(t)\mathbf{j} + h(t)\mathbf{k}$, then **f has limit L as t approaches t_0,**

$$\lim_{t\to t_0} \mathbf{f}(t) = \mathbf{L},$$

if

$$\lim_{t\to t_0} f(t) = L_1, \qquad \lim_{t\to t_0} g(t) = L_2, \qquad \text{and} \qquad \lim_{t\to t_0} h(t) = L_3. \tag{3}$$

EXAMPLE 2 If $\mathbf{f}(t) = (\cos t)\mathbf{i} + (\sin t)\mathbf{j} + t\mathbf{k}$, then

$$\lim_{t\to \pi/4} \mathbf{f}(t) = \left(\lim_{t\to\pi/4} \cos t\right)\mathbf{i} + \left(\lim_{t\to\pi/4} \sin t\right)\mathbf{j} + \left(\lim_{t\to\pi/4} t\right)\mathbf{k}$$

$$= \frac{\sqrt{2}}{2}\mathbf{i} + \frac{\sqrt{2}}{2}\mathbf{j} + \frac{\pi}{4}\mathbf{k}.$$

■

12.3 Helices drawn by computer.

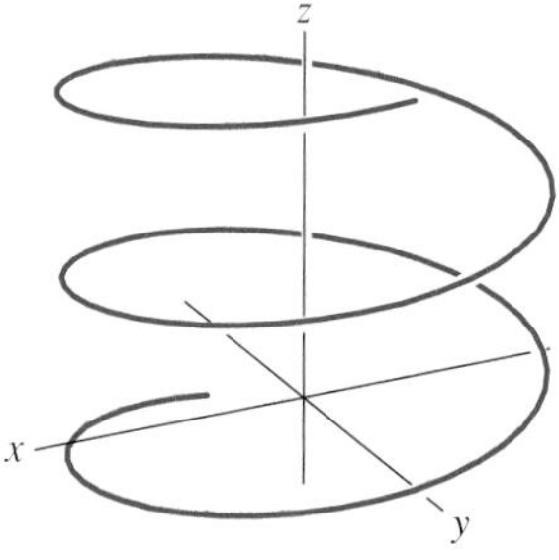

$\mathbf{r} = (\cos t)\mathbf{i} + (\sin t)\mathbf{j} + t\mathbf{k}$

$\mathbf{r} = (\cos t)\mathbf{i} + (\sin t)\mathbf{j} + (\sin 2t)\mathbf{k}$

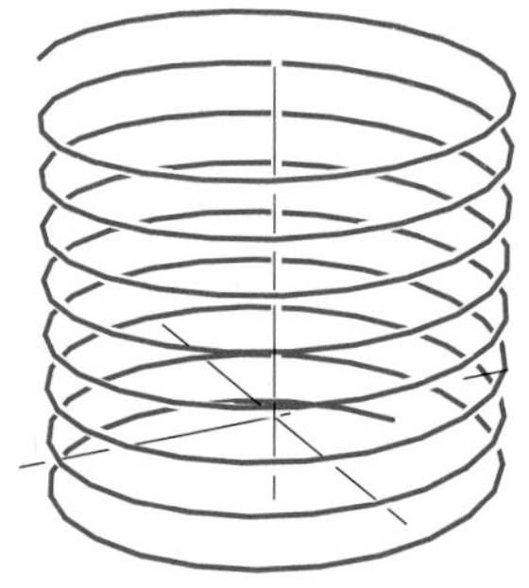

$\mathbf{r} = (\cos t)\mathbf{i} + (\sin t)\mathbf{j} + 0.3t\mathbf{k}$

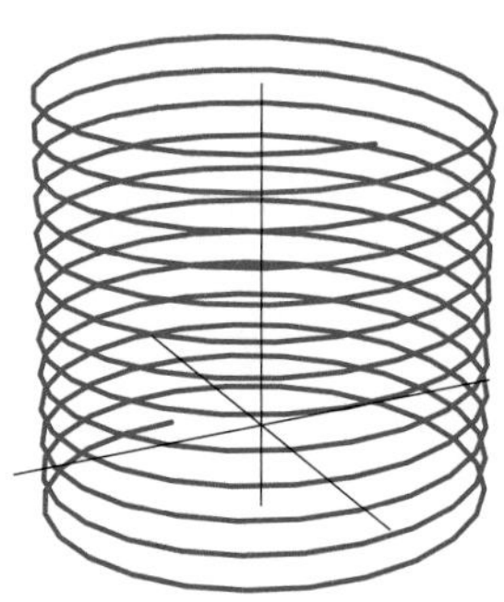

$\mathbf{r} = (\cos 5t)\mathbf{i} + (\sin 5t)\mathbf{j} + t\mathbf{k}$

(Generated by Mathematica)

Continuity is defined the same way for vector functions as it is for scalar functions.

DEFINITION

The vector function $\mathbf{f}$ is **continuous at a point** $t = t_0$ if

$$\lim_{t \to t_0} \mathbf{f}(t) = \mathbf{f}(t_0).$$

It is **continuous** if it is continuous at every point in its domain.

The fact that limits of vector functions are defined in terms of components leads to the following test for continuity.

Component Test for Continuity at a Point

A vector function $\mathbf{f}$ defined by the rule $\mathbf{f}(t) = f(t)\mathbf{i} + g(t)\mathbf{j} + h(t)\mathbf{k}$ is continuous at $t = t_0$ if and only if f, g, and h are continuous at t_0.

EXAMPLE 3 The function

$$\mathbf{r} = (\cos t)\mathbf{i} + (\sin t)\mathbf{j} + t\mathbf{k}$$

is continuous because the component functions $\cos t$, $\sin t$, and t are continuous. ■

EXAMPLE 4 The function

$$\mathbf{r} = (\cos t)\mathbf{i} + (\sin t)\mathbf{j} + \frac{1}{t}\mathbf{k}$$

is not continuous at $t = 0$ because the third component function, $h(t) = 1/t$, is not continuous at $t = 0$. ■

Derivatives and Motion

Suppose that

$$\mathbf{r} = f(t)\mathbf{i} + g(t)\mathbf{j} + h(t)\mathbf{k}$$

is the position vector of a particle moving along a curve in space and that f, g, and h are differentiable functions. Then the difference between the particle's positions at time t and a nearby time $t + \Delta t$ can be expressed as the vector difference

$$\Delta \mathbf{r} = \mathbf{r}(t + \Delta t) - \mathbf{r}(t) \tag{4}$$

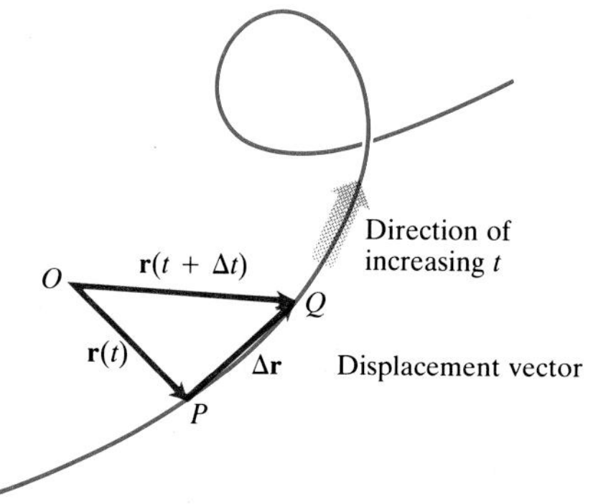

12.4 Between time t and time $t + \Delta t$, the particle moving along the path shown here undergoes the displacement $\overrightarrow{PQ} = \Delta\mathbf{r}$. The vector sum $\mathbf{r}(t) + \Delta\mathbf{r}$ gives the new position, $\mathbf{r}(t + \Delta t)$.

(Fig. 12.4). In terms of components,

$$\begin{aligned}\Delta\mathbf{r} &= \mathbf{r}(t + \Delta t) - \mathbf{r}(t)\\ &= [f(t + \Delta t)\mathbf{i} + g(t + \Delta t)\mathbf{j} + h(t + \Delta t)\mathbf{k}] - [f(t)\mathbf{i} + g(t)\mathbf{j} + h(t)\mathbf{k}]\\ &= [f(t + \Delta t) - f(t)]\mathbf{i} + [g(t + \Delta t) - g(t)]\mathbf{j} + [h(t + \Delta t) - h(t)]\mathbf{k}.\end{aligned} \tag{5}$$

As Δt approaches zero, three things happen simultaneously. First, Q approaches P along the curve. Second, the secant line PQ seems to approach a limiting position tangent to the curve at P. Third, the quotient $\Delta\mathbf{r}/\Delta t$ approaches the limit

$$\begin{aligned}\lim_{\Delta t\to 0}\frac{\Delta\mathbf{r}}{\Delta t} &= \left[\lim_{\Delta t\to 0}\frac{f(t + \Delta t) - f(t)}{\Delta t}\right]\mathbf{i} + \left[\lim_{\Delta t\to 0}\frac{g(t + \Delta t) - g(t)}{\Delta t}\right]\mathbf{j}\\ &\quad + \left[\lim_{\Delta t\to 0}\frac{h(t + \Delta t) - h(t)}{\Delta t}\right]\mathbf{k}\\ &= \left[\frac{df}{dt}\right]\mathbf{i} + \left[\frac{dg}{dt}\right]\mathbf{j} + \left[\frac{dh}{dt}\right]\mathbf{k}.\end{aligned} \tag{6}$$

We are therefore led by past experience to the following definition.

DEFINITION

Derivative of a Vector Function of a Real Variable

If f, g, and h are differentiable scalar functions of t and $\mathbf{r} = f\mathbf{i} + g\mathbf{j} + h\mathbf{k}$, then the vector

$$\mathbf{r}' = \frac{d\mathbf{r}}{dt} = \frac{df}{dt}\mathbf{i} + \frac{dg}{dt}\mathbf{j} + \frac{dh}{dt}\mathbf{k}$$

is the **derivative** of $\mathbf{r}$ with respect to t. The vector function $\mathbf{r} = f\mathbf{i} + g\mathbf{j} + h\mathbf{k}$ is said to be **differentiable at a point** $t = t_0$ if f, g, and h are differentiable at t_0, and is said to be **differentiable** if it is differentiable at every point in its domain.

Look once again at Fig. 12.4. We drew the figure for Δt positive, so that $\Delta\mathbf{r}$ points forward. The vector $\Delta\mathbf{r}/\Delta t$, not shown, also points forward, because it has the same direction as $\Delta\mathbf{r}$. If Δt had been negative, $\Delta\mathbf{r}$ would have pointed backward, against the direction of motion. The quotient $\Delta\mathbf{r}/\Delta t$, however, having a sign opposite to the sign of $\Delta\mathbf{r}$, would still have pointed forward. In either case, then, $\Delta\mathbf{r}/\Delta t$ points forward and we expect the vector $d\mathbf{r}/dt = \lim_{\Delta t\to 0}\Delta\mathbf{r}/\Delta t$, when different from $\mathbf{0}$, to do the same. This means that $d\mathbf{r}/dt$ has all the makings of a velocity vector—it gives the rate of change of position with respect to time and it points forward the way we want it to.

DEFINITIONS

If the position vector $\mathbf{r}$ of a particle moving in space is a differentiable function of time t, then the vector

$$\mathbf{v} = \frac{d\mathbf{r}}{dt}$$

is the particle's **velocity vector.** At any time t, the direction of $\mathbf{v}$ is the **direction of motion,** the magnitude of $\mathbf{v}$ is the particle's **speed,** and the derivative $\mathbf{a} = d\mathbf{v}/dt$, when it exists, is the particle's **acceleration vector.** In short,

1. Velocity is the derivative of position: $\mathbf{v} = \dfrac{d\mathbf{r}}{dt}$
2. Speed is the magnitude of velocity: Speed $= |\mathbf{v}|$
3. Acceleration is the derivative of velocity: $\mathbf{a} = \dfrac{d\mathbf{v}}{dt} = \dfrac{d^2\mathbf{r}}{dt^2}$
4. The vector $\mathbf{v}/|\mathbf{v}|$ is the direction of motion at time t.

We use the formula $\mathbf{A} = |\mathbf{A}| \cdot (\mathbf{A}/|\mathbf{A}|)$ from Section 11.2 to express the velocity of a moving particle as the product of its speed and direction.

$$\text{Velocity} = |\mathbf{v}| \cdot \frac{\mathbf{v}}{|\mathbf{v}|} = (\text{speed}) \cdot (\text{direction}) \tag{7}$$

EXAMPLE 5 The vector

$$\mathbf{r} = (3\cos t)\mathbf{i} + (3\sin t)\mathbf{j} + t^2\mathbf{k}$$

gives the position of a moving body at time t. Find the body's speed and direction when $t = 2$. At what times, if any, are the body's velocity and acceleration orthogonal?

Solution

$$\mathbf{r} = (3\cos t)\mathbf{i} + (3\sin t)\mathbf{j} + t^2\mathbf{k}$$

$$\mathbf{v} = \frac{d\mathbf{r}}{dt} = -(3\sin t)\mathbf{i} + (3\cos t)\mathbf{j} + 2t\mathbf{k}$$

$$\mathbf{a} = \frac{d^2\mathbf{r}}{dt^2} = -(3\cos t)\mathbf{i} - (3\sin t)\mathbf{j} + 2\mathbf{k}$$

At $t = 2$, the body's speed and direction are

Speed: $|\mathbf{v}(2)| = \sqrt{(-3\sin 2)^2 + (3\cos 2)^2 + (4)^2} = 5$

Direction: $\dfrac{\mathbf{v}(2)}{|\mathbf{v}(2)|} = -\left(\dfrac{3}{5}\sin 2\right)\mathbf{i} + \left(\dfrac{3}{5}\cos 2\right)\mathbf{j} + \dfrac{4}{5}\mathbf{k}.$

To find when $\mathbf{v}$ and $\mathbf{a}$ are orthogonal, we look for values of t for which

$$\mathbf{v} \cdot \mathbf{a} = 9\sin t\cos t - 9\cos t\sin t + 4t = 4t = 0.$$

The only value is $t = 0$. ■

Differentiation Rules

Because the derivatives of vector functions are defined component by component, the rules for differentiating vector functions have the same form as the rules for differentiating scalar functions.

Differentiation Rules for Vector Functions

Constant Function Rule: $\frac{d}{dt}\mathbf{C} = \mathbf{0}$ (any constant vector $\mathbf{C}$)

If $\mathbf{u}$ and $\mathbf{v}$ are differentiable functions of t, then

Scalar Multiple Rule: $\frac{d}{dt}(c\mathbf{u}) = c\frac{d\mathbf{u}}{dt}$ (any number c)

Sum Rule: $\frac{d}{dt}(\mathbf{u} + \mathbf{v}) = \frac{d\mathbf{u}}{dt} + \frac{d\mathbf{v}}{dt}$

Difference Rule: $\frac{d}{dt}(\mathbf{u} - \mathbf{v}) = \frac{d\mathbf{u}}{dt} - \frac{d\mathbf{v}}{dt}$

Dot-Product Rule: $\frac{d}{dt}(\mathbf{u} \cdot \mathbf{v}) = \frac{d\mathbf{u}}{dt} \cdot \mathbf{v} + \mathbf{u} \cdot \frac{d\mathbf{v}}{dt}$

Cross-Product Rule: $\frac{d}{dt}(\mathbf{u} \times \mathbf{v}) = \frac{d\mathbf{u}}{dt} \times \mathbf{v} + \mathbf{u} \times \frac{d\mathbf{v}}{dt}$

Chain Rule (Short Form): If $\mathbf{r}$ is a differentiable function of t and t is a differentiable function of s, then

$$\frac{d\mathbf{r}}{ds} = \frac{d\mathbf{r}}{dt}\frac{dt}{ds}.$$

When you use the Cross-Product Rule, remember to preserve the order of the factors. If $\mathbf{u}$ comes first on the left, it must also come first on the right or the signs will be wrong.

We shall prove the Dot-Product Rule and the Chain Rule but leave the rules for constants, scalar multiples, sums, and differences as exercises. The proof of the Cross-Product Rule is cumbersome and we shall omit it entirely.

PROOF of the Dot-Product Rule

Suppose that $\mathbf{u} = u_1(t)\mathbf{i} + u_2(t)\mathbf{j} + u_3(t)\mathbf{k}$

and $\mathbf{v} = v_1(t)\mathbf{i} + v_2(t)\mathbf{j} + v_3(t)\mathbf{k}.$

Then

$$\frac{d}{dt}(\mathbf{u} \cdot \mathbf{v}) = \frac{d}{dt}(u_1 v_1 + u_2 v_2 + u_3 v_3)$$

$$= \underbrace{u_1' v_1 + u_2' v_2 + u_3' v_3}_{\mathbf{u}' \cdot \mathbf{v}} + \underbrace{u_1 v_1' + u_2 v_2' + u_3 v_3'}_{\mathbf{u} \cdot \mathbf{v}'}.$$

■

PROOF of the Chain Rule Suppose that $\mathbf{r} = f(t)\mathbf{i} + g(t)\mathbf{j} + h(t)\mathbf{k}$ is a differentiable vector function of t and that t is a differentiable scalar function of some other variable s. Then f, g, and h are differentiable functions of s and the Chain Rule for differentiable real-valued functions gives

$$\begin{aligned}\frac{d\mathbf{r}}{ds} &= \frac{df}{ds}\mathbf{i} + \frac{dg}{ds}\mathbf{j} + \frac{dh}{ds}\mathbf{k}\\ &= \frac{df}{dt}\frac{dt}{ds}\mathbf{i} + \frac{dg}{dt}\frac{dt}{ds}\mathbf{j} + \frac{dh}{dt}\frac{dt}{ds}\mathbf{k}\\ &= \left(\frac{df}{dt}\mathbf{i} + \frac{dg}{dt}\mathbf{j} + \frac{dh}{dt}\mathbf{k}\right)\frac{dt}{ds}\\ &= \frac{d\mathbf{r}}{dt}\frac{dt}{ds}.\end{aligned}$$

■

Derivatives of Vectors of Constant Length

We might at first think that a vector whose length remained constant as time passed would have to have a zero derivative, but this is not the case. Think of a clock hand. Its length remains constant as time passes, but its direction still changes. What we *can* say about the derivative of a vector of constant length is that it is always orthogonal to the vector. Direction-changes, as it were, take place at right angles.

> If $\mathbf{u}$ is a differentiable vector function of t of constant length, then
>
> $$\mathbf{u}\cdot\frac{d\mathbf{u}}{dt} = 0. \tag{8}$$

To see why Eq. (8) holds, suppose $\mathbf{u}$ is a differentiable function of t and that $|\mathbf{u}|$ is constant. Then $\mathbf{u}\cdot\mathbf{u} = |\mathbf{u}|^2$ is constant and we may differentiate both sides of this equation to get

$$\frac{d}{dt}(\mathbf{u}\cdot\mathbf{u}) = \frac{d}{dt}|\mathbf{u}|^2 = 0$$

$$\frac{d\mathbf{u}}{dt}\cdot\mathbf{u} + \mathbf{u}\cdot\frac{d\mathbf{u}}{dt} = 0 \quad \text{(Dot-Product Rule with } \mathbf{v} = \mathbf{u}\text{)}$$

$$2\mathbf{u}\cdot\frac{d\mathbf{u}}{dt} = 0 \quad \text{(Dot multiplication is commutative.)}$$

$$\mathbf{u}\cdot\frac{d\mathbf{u}}{dt} = 0.$$

EXAMPLE 6 Show that

$$\mathbf{u}(t) = (\sin t)\mathbf{i} + (\cos t)\mathbf{j} + \sqrt{3}\mathbf{k}$$

has constant length and is orthogonal to its derivative.

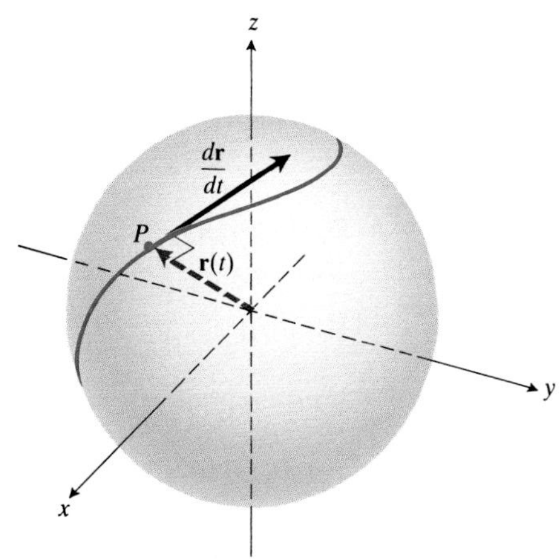

12.5 The velocity vector of a particle P that moves on the surface of a sphere is tangent to the sphere.

Solution

$$\mathbf{u}(t) = (\sin t)\mathbf{i} + (\cos t)\mathbf{j} + \sqrt{3}\mathbf{k}$$

$$|\mathbf{u}(t)| = \sqrt{(\sin t)^2 + (\cos t)^2 + (\sqrt{3})^2} = \sqrt{1+3} = 2$$

$$\frac{d\mathbf{u}}{dt} = (\cos t)\mathbf{i} - (\sin t)\mathbf{j}$$

$$\mathbf{u} \cdot \frac{d\mathbf{u}}{dt} = \sin t \cos t - \sin t \cos t = 0$$

■

EXAMPLE 7 Suppose a particle moves on a sphere centered at the origin in such a way that its position vector $\mathbf{r}(t)$ is differentiable (Fig. 12.5). Show that the particle's velocity is always orthogonal to $\mathbf{r}$.

Solution The length of $\mathbf{r}$ is constant (it equals the radius of the sphere). Hence, $\mathbf{r} \cdot (d\mathbf{r}/dt) = 0$. ■

Integrals of Vector Functions

If the scalar components of a vector function $\mathbf{f}$ are continuous functions of t throughout an interval $a \le t \le b$, we define the definite integral of $\mathbf{f}$ from a to b by the following formula.

DEFINITION

The **definite integral** of the function $\mathbf{f}$ defined by the rule $\mathbf{f}(t) = f(t)\mathbf{i} + g(t)\mathbf{j} + h(t)\mathbf{k}$ over the interval $a \le t \le b$ is

$$\int_a^b \mathbf{f}(t)\,dt = \left(\int_a^b f(t)\,dt\right)\mathbf{i} + \left(\int_a^b g(t)\,dt\right)\mathbf{j} + \left(\int_a^b h(t)\,dt\right)\mathbf{k}. \tag{9}$$

We call a differentiable function $\mathbf{F}$ an **antiderivative** of a function $\mathbf{f}$ if $d\mathbf{F}/dt = \mathbf{f}$. If $\mathbf{F}$ is an antiderivative of $\mathbf{f}$, it can be shown, working one component at a time, that every antiderivative of $\mathbf{f}$ has the form $\mathbf{F} + \mathbf{C}$ for some constant vector $\mathbf{C}$. We call the set of all antiderivatives of $\mathbf{f}$ the **indefinite integral** of $\mathbf{f}$ and denote it with an integral sign in the usual way.

DEFINITION

The **indefinite integral** of $\mathbf{f}$ with respect to t is the set of all antiderivatives of $\mathbf{f}$, denoted by $\int \mathbf{f}(t)\,dt$. If $\mathbf{F}$ is any antiderivative of $\mathbf{f}$, then

$$\int \mathbf{f}(t)\,dt = \mathbf{F}(t) + \mathbf{C}. \tag{10}$$

We can use the integral existence theorem for continuous scalar functions to show that every continuous vector function has an antiderivative, and all the usual arithmetic rules for definite and indefinite integrals apply.

EXAMPLE 8

$$\begin{aligned}\int_0^{\pi}((\cos t)\mathbf{i}+(\sin t)\mathbf{j}+t\mathbf{k})\,dt &= \left(\int_0^{\pi}\cos t\,dt\right)\mathbf{i}+\left(\int_0^{\pi}\sin t\,dt\right)\mathbf{j}+\left(\int_0^{\pi}t\,dt\right)\mathbf{k}\\ &= \Big[\sin t\Big]_0^{\pi}\mathbf{i}+\Big[-\cos t\Big]_0^{\pi}\mathbf{j}+\left[\frac{t^2}{2}\right]_0^{\pi}\mathbf{k}\\ &= [0-0]\mathbf{i}+[-(-1)+(1)]\mathbf{j}+\left[\frac{\pi^2}{2}-0\right]\mathbf{k}\\ &= 2\mathbf{j}+\frac{\pi^2}{2}\mathbf{k}.\end{aligned}$$

EXAMPLE 9 The velocity of a particle moving in space is

$$\frac{d\mathbf{r}}{dt}=(\cos t)\mathbf{i}-(\sin t)\mathbf{j}+\mathbf{k}.$$

Find the particle's position as a function of t if $\mathbf{r}=2\mathbf{i}+\mathbf{k}$ when $t=0$.

Solution Our goal is to solve the initial value problem that consists of

The differential equation: $\dfrac{d\mathbf{r}}{dt}=(\cos t)\mathbf{i}-(\sin t)\mathbf{j}+\mathbf{k}$

The initial condition: $\mathbf{r}=2\mathbf{i}+\mathbf{k}$ when $t=0$.

To solve it, we first use what we know about derivatives to find the general solution of the differential equation. Integrating both sides with respect to t gives

$$\begin{aligned}\mathbf{r} &= (\sin t+C_1)\mathbf{i}+(\cos t+C_2)\mathbf{j}+(t+C_3)\mathbf{k}\\ &= (\sin t)\mathbf{i}+(\cos t)\mathbf{j}+t\mathbf{k}+\underbrace{C_1\mathbf{i}+C_2\mathbf{j}+C_3\mathbf{k}}_{\mathbf{C}}\\ &= (\sin t)\mathbf{i}+(\cos t)\mathbf{j}+t\mathbf{k}+\mathbf{C}.\end{aligned}$$

We then use the initial condition to find the right value for $\mathbf{C}$:

$$\begin{aligned}(\sin 0)\mathbf{i}+(\cos 0)\mathbf{j}+(0)\mathbf{k}+\mathbf{C} &= 2\mathbf{i}+\mathbf{k} \qquad \left(\begin{array}{l}\text{The initial condition is}\\ \mathbf{r}(0)=2\mathbf{i}+\mathbf{k}.\end{array}\right)\\ \mathbf{j}+\mathbf{C} &= 2\mathbf{i}+\mathbf{k}\\ \mathbf{C} &= 2\mathbf{i}-\mathbf{j}+\mathbf{k}\end{aligned}$$

The particle's position as a function of t is

$$\mathbf{r}(t)=(\sin t+2)\mathbf{i}+(\cos t-1)\mathbf{j}+(t+1)\mathbf{k}.$$

As a check, we can see from this formula that

$$\frac{d\mathbf{r}}{dt}=(\cos t+0)\mathbf{i}+(-\sin t-0)\mathbf{j}+(1+0)\mathbf{k}=(\cos t)\mathbf{i}-(\sin t)\mathbf{j}+\mathbf{k}$$

and that

$$\mathbf{r}(0)=(\sin 0+2)\mathbf{i}+(\cos 0-1)\mathbf{j}+(0+1)\mathbf{k}=2\mathbf{i}+\mathbf{k}.$$

Motion in the Plane

When a particle moves in the xy-plane, its z-coordinate is zero and its position vector reduces to

$$\mathbf{r} = f(t)\mathbf{i} + g(t)\mathbf{j}. \tag{11}$$

EXAMPLE 10 The position vector of a particle moving along the parabola $y = x^2$ in the xy-plane is

$$\mathbf{r} = t\mathbf{i} + t^2\mathbf{j}.$$

Find the particle's velocity and acceleration vectors and sketch them on the parabola along with the position vector at time $t = 1$.

Solution

$$\mathbf{r} = t\mathbf{i} + t^2\mathbf{j}$$

$$\mathbf{v} = \frac{d\mathbf{r}}{dt} = \mathbf{i} + 2t\mathbf{j}$$

$$\mathbf{a} = \frac{d\mathbf{v}}{dt} = 0\mathbf{i} + 2\mathbf{j} = 2\mathbf{j}.$$

At time $t = 1$,

$$\mathbf{r} = \mathbf{i} + \mathbf{j}, \qquad \mathbf{v} = \mathbf{i} + 2\mathbf{j}, \qquad \mathbf{a} = 2\mathbf{j}.$$

We sketch the parabola in the xy-plane and draw in the vectors (Fig 12.6). Notice the convention of drawing the velocity and acceleration vectors at the tip of the position vector instead of at the origin. ■

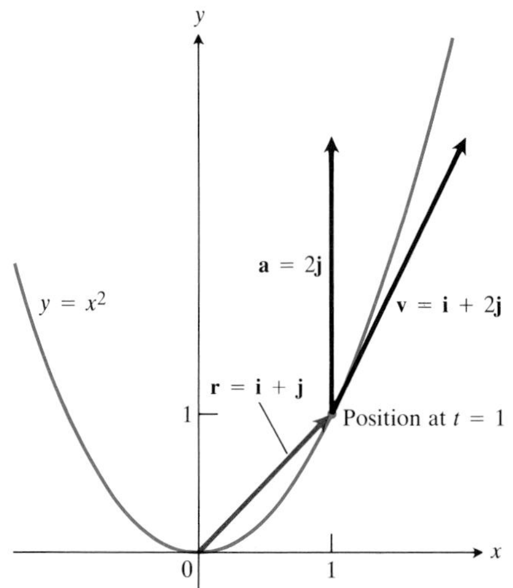

12.6 The position, velocity, and acceleration of the particle in Example 10 at $t = 1$. Notice the convention of drawing the velocity and acceleration as vectors at the point on the curve instead of at the origin.

Exercises 12.1

In Exercises 1–8, $\mathbf{r}$ is the position of a particle in space at time t. Find the particle's velocity and acceleration vectors. Then find the particle's speed and direction of motion at the given value of t and write the velocity at that time as the product of its speed and direction.

1. $\mathbf{r} = (2\cos t)\mathbf{i} + (3\sin t)\mathbf{j} + 4t\mathbf{k}, \quad t = \pi/2$

2. $\mathbf{r} = (t+1)\mathbf{i} + (t^2-1)\mathbf{j} + 2t\mathbf{k}, \quad t = 1$

3. $\mathbf{r} = (\cos 2t)\mathbf{j} + (2\sin t)\mathbf{k}, \quad t = 0$

4. $\mathbf{r} = e^t\mathbf{i} + \frac{2}{9}e^{2t}\mathbf{j}, \quad t = \ln 3$

5. $\mathbf{r} = (\sec t)\mathbf{i} + (\tan t)\mathbf{j} + \frac{4}{3}t\mathbf{k}, \quad t = \pi/6$

6. $\mathbf{r} = (2\ln(t+1))\mathbf{i} + t^2\mathbf{j} + \frac{t^2}{2}\mathbf{k}, \quad t = 1$

7. $\mathbf{r} = (e^{-t})\mathbf{i} + (2\cos 3t)\mathbf{j} + (2\sin 3t)\mathbf{k}, \quad t = 0$

8. $\mathbf{r} = (1+t)\mathbf{i} + \frac{t^2}{\sqrt{2}}\mathbf{j} + \frac{t^3}{3}\mathbf{k}, \quad t = 1$

In Exercises 9–12, $\mathbf{r}$ is the position of a particle in space at time t. Find the angle between the velocity and acceleration vectors at time $t = 0$.

9. $\mathbf{r} = (3t+1)\mathbf{i} + \sqrt{3}t\mathbf{j} + t^2\mathbf{k}$

10. $\mathbf{r} = \left(\frac{\sqrt{2}}{2}t\right)\mathbf{i} + \left(\frac{\sqrt{2}}{2}t - 16t^2\right)\mathbf{j}$

11. $\mathbf{r} = (\ln(t^2+1))\mathbf{i} + (\tan^{-1}t)\mathbf{j} + \sqrt{t^2+1}\,\mathbf{k}$

12. $\mathbf{r} = \frac{4}{9}(1+t)^{3/2}\mathbf{i} + \frac{4}{9}(1-t)^{3/2}\mathbf{j} + \frac{1}{3}t\mathbf{k}$

In Exercises 13–14, $\mathbf{r}$ is the position vector of a particle in space at time t. Find the time or times in the given time interval when the velocity and acceleration vectors are orthogonal.

13. $\mathbf{r} = (t - \sin t)\mathbf{i} + (1 - \cos t)\mathbf{j}, \quad 0 \leq t \leq 2\pi$

14. $\mathbf{r} = (\sin t)\mathbf{i} + t\mathbf{j} + (\cos t)\mathbf{k}, \quad t \geq 0$

Evaluate the integrals in Exercises 15–20.

15. $\displaystyle\int_0^1 [t^3\mathbf{i} + 7\mathbf{j} + (t + 1)\mathbf{k}]\,dt$

16. $\displaystyle\int_1^2 \left[(6 - 6t)\mathbf{i} + 3\sqrt{t}\mathbf{j} + \left(\frac{4}{t^2}\right)\mathbf{k}\right]dt$

17. $\displaystyle\int_{-\pi/4}^{\pi/4} [(\sin t)\mathbf{i} + (1 + \cos t)\mathbf{j} + (\sec^2 t)\mathbf{k}]\,dt$

18. $\displaystyle\int_0^{\pi/3} [(\sec t \tan t)\mathbf{i} + (\tan t)\mathbf{j} + (2 \sin t \cos t)\mathbf{k}]\,dt$

19. $\displaystyle\int_1^4 \left[\frac{1}{t}\mathbf{i} + \frac{1}{5 - t}\mathbf{j} + \frac{1}{2t}\mathbf{k}\right]dt$

20. $\displaystyle\int_0^1 \left[\frac{2}{\sqrt{1 - t^2}}\mathbf{i} + \frac{\sqrt{3}}{1 + t^2}\mathbf{k}\right]dt$

Exercises 21–24 give the position vectors of particles moving along various curves in the xy-plane. In each case, find the particle's velocity and acceleration vectors at the stated times and sketch them as vectors on the curve.

21. *Motion on the circle $x^2 + y^2 = 1$.*

$$\mathbf{r} = (\sin t)\mathbf{i} + (\cos t)\mathbf{j}; \quad t = \pi/4 \text{ and } \pi/2$$

22. *Motion on the circle $x^2 + y^2 = 4$.*

$$\mathbf{r} = \left(4 \cos \frac{t}{2}\right)\mathbf{i} + \left(4 \sin \frac{t}{2}\right)\mathbf{j}; \quad t = \pi \text{ and } 3\pi/2$$

23. *Motion on the cycloid $x = t - \sin t$, $y = 1 - \cos t$.*

$$\mathbf{r} = (t - \sin t)\mathbf{i} + (1 - \cos t)\mathbf{j}; \quad t = \pi \text{ and } 3\pi/2$$

24. *Motion on the parabola $y = x^2 + 1$.*

$$\mathbf{r} = t\mathbf{i} + (t^2 + 1)\mathbf{j}; \quad t = -1, 0, \text{ and } 1$$

Solve the initial value problems in Exercises 25–30 for $\mathbf{r}$ as a vector function of t.

25. Differential equation: $\dfrac{d\mathbf{r}}{dt} = -t\mathbf{i} - t\mathbf{j} - t\mathbf{k}$

Initial condition: $\mathbf{r} = \mathbf{i} + 2\mathbf{j} + 3\mathbf{k}$ when $t = 0$

26. Differential equation: $\dfrac{d\mathbf{r}}{dt} = (180t)\mathbf{i} + (180t - 16t^2)\mathbf{j}$

Initial condition: $\mathbf{r} = 100\mathbf{j}$ when $t = 0$

27. Differential equation: $\dfrac{d\mathbf{r}}{dt} = \dfrac{3}{2}(t + 1)^{1/2}\mathbf{i} + e^{-t}\mathbf{j} + \dfrac{1}{t + 1}\mathbf{k}$

Initial condition: $\mathbf{r} = \mathbf{k}$ when $t = 0$

28. Differential equation: $\dfrac{d\mathbf{r}}{dt} = (t^3 + 4t)\mathbf{i} + t\mathbf{j} + 2t^2\mathbf{k}$

Initial condition: $\mathbf{r} = \mathbf{i} + \mathbf{j}$ when $t = 0$

29. Differential equation: $\dfrac{d^2\mathbf{r}}{dt^2} = -32\mathbf{k}$

Initial conditions: $\mathbf{r} = 100\mathbf{k}$ and $\dfrac{d\mathbf{r}}{dt} = 8\mathbf{i} + 8\mathbf{j}$ when $t = 0$

30. Differential equation: $\dfrac{d^2\mathbf{r}}{dt^2} = -(\mathbf{i} + \mathbf{j} + \mathbf{k})$

Initial conditions: $\mathbf{r} = 10\mathbf{i} + 10\mathbf{j} + 10\mathbf{k}$ and $\dfrac{d\mathbf{r}}{dt} = \mathbf{0}$ when $t = 0$

31. A particle moves on a cycloid in the xy-plane in such a way that its position at time t is

$$\mathbf{r} = (t - \sin t)\mathbf{i} + (1 - \cos t)\mathbf{j}.$$

Find the maximum and minimum values of $|\mathbf{v}|$ and $|\mathbf{a}|$. (*Hint:* Find the extreme values of $|\mathbf{v}|^2$ and $|\mathbf{a}|^2$ first and take square roots later.)

32. A particle moves around the ellipse $(y/3)^2 + (z/2)^2 = 1$ in the yz-plane in such a way that its position at time t is

$$\mathbf{r} = (3 \cos t)\mathbf{j} + (2 \sin t)\mathbf{k}.$$

Find the maximum and minimum values of $|\mathbf{v}|$ and $|\mathbf{a}|$. (*Hint:* Find the extreme values of $|\mathbf{v}|^2$ and $|\mathbf{a}|^2$ first and take square roots later.)

33. *The Constant Function Rule.* Prove that if $\mathbf{f}$ is the vector function with the constant value $\mathbf{C}$, then $d\mathbf{f}/dt = \mathbf{0}$.

34. *The Scalar Multiple Rule.* Prove that if $\mathbf{f}$ is a differentiable function of t and c is any real number, then

$$\frac{d(c\mathbf{f})}{dt} = c\frac{d\mathbf{f}}{dt}.$$

35. *The Sum and Difference Rules.* Prove that if $\mathbf{u}$ and $\mathbf{v}$ are differentiable functions of t, then

$$\frac{d}{dt}(\mathbf{u} + \mathbf{v}) = \frac{d\mathbf{u}}{dt} + \frac{d\mathbf{v}}{dt}$$

and

$$\frac{d}{dt}(\mathbf{u} - \mathbf{v}) = \frac{d\mathbf{u}}{dt} - \frac{d\mathbf{v}}{dt}.$$

12.2 Modeling Projectile Motion

When we shoot a projectile into the air we usually want to know beforehand how far it will go (will it reach the target?), how high it will rise (will it clear the hill?), and when it will land (when do we get results?). We get this information from equations that calculate the answers we want from the direction and magnitude of the projectile's initial velocity vector, described in terms of the angle and speed at which the projectile is fired. The equations come from combining calculus and Newton's second law of motion in vector form. In this section, we derive these equations and show how to use them to get the information we want about projectile motion.

The Vector and Parametric Equations for Ideal Projectile Motion

To derive equations for projectile motion, we assume that the projectile behaves like a particle moving in a vertical coordinate plane and that the only force acting on the projectile during its flight is the constant force of gravity, which always points straight down. In practice, none of these assumptions really holds. The ground moves beneath the projectile as the earth turns, the air creates a frictional force that varies with the projectile's speed and altitude, and the force of gravity changes as the projectile moves along. All this must be taken into account by applying corrections to the predictions of the ideal equations we are about to derive. The corrections, however, are not the subject of this section.

We assume that our projectile is launched from the origin at time $t = 0$ into the first quadrant with an initial velocity $\mathbf{v}_0$ (Fig. 12.7). If α is the angle that $\mathbf{v}_0$ makes with the horizontal, then

$$\mathbf{v}_0 = (|\mathbf{v}_0| \cos \alpha)\mathbf{i} + (|\mathbf{v}_0| \sin \alpha)\mathbf{j}. \tag{1}$$

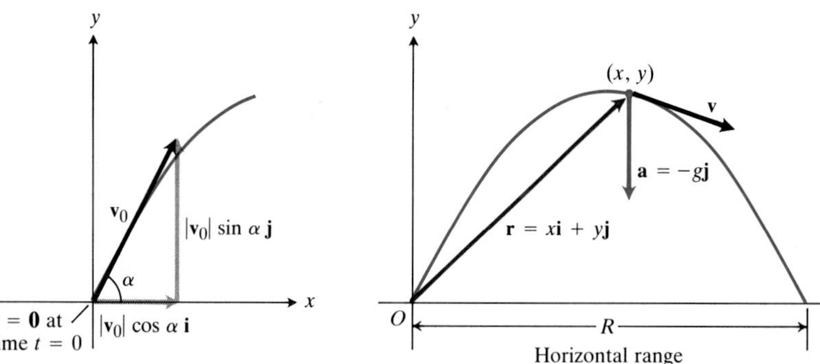

12.7 The flight of an ideal projectile.

If we use the simpler notation v_0 for the initial speed $|\mathbf{v}_0|$, then

$$\mathbf{v}_0 = (v_0 \cos \alpha)\mathbf{i} + (v_0 \sin \alpha)\mathbf{j}. \tag{2}$$

The projectile's initial position is

$$\mathbf{r}_0 = 0\mathbf{i} + 0\mathbf{j} = \mathbf{0}. \tag{3}$$

If the only force acting on the projectile during its flight is a constant downward acceleration of gravity of magnitude g, then

$$\frac{d^2\mathbf{r}}{dt^2} = -g\mathbf{j}. \tag{4}$$

We can find the projectile's position as a function of time t by solving the following initial value problem:

Differential equation: $\dfrac{d^2\mathbf{r}}{dt^2} = -g\mathbf{j}$

Initial conditions: $\mathbf{r} = \mathbf{0}$ and $\dfrac{d\mathbf{r}}{dt} = \mathbf{v}_0$ when $t = 0$.

The first integration gives

$$\frac{d\mathbf{r}}{dt} = -(gt)\mathbf{j} + \mathbf{v}_0. \tag{5}$$

A second integration gives

$$\mathbf{r} = -\frac{1}{2}gt^2\mathbf{j} + \mathbf{v}_0 t + \mathbf{r}_0. \tag{6}$$

Substituting the values of $\mathbf{v}_0$ and $\mathbf{r}_0$ from Eqs. (2) and (3) gives

$$\mathbf{r} = -\frac{1}{2}gt^2\mathbf{j} + \underbrace{(v_0 \cos \alpha)t\mathbf{i} + (v_0 \sin \alpha)t\mathbf{j}}_{\mathbf{v}_0 t} + \mathbf{0}$$

$$= (v_0 \cos \alpha)t\mathbf{i} + \left((v_0 \sin \alpha)t - \frac{1}{2}gt^2\right)\mathbf{j}. \tag{7}$$

Equations for the Motion of an Ideal Projectile Fired from the Origin at $t = 0$

Vector Form: $$\mathbf{r} = (v_0 \cos \alpha)t\mathbf{i} + \left((v_0 \sin \alpha)t - \frac{1}{2}gt^2\right)\mathbf{j} \tag{8}$$

Parametric Form: $$x = (v_0 \cos \alpha)t, \quad y = (v_0 \sin \alpha)t - \frac{1}{2}gt^2 \tag{9}$$

where α is the **firing angle (launch angle, angle of elevation)** and v_0 is the **initial speed.**

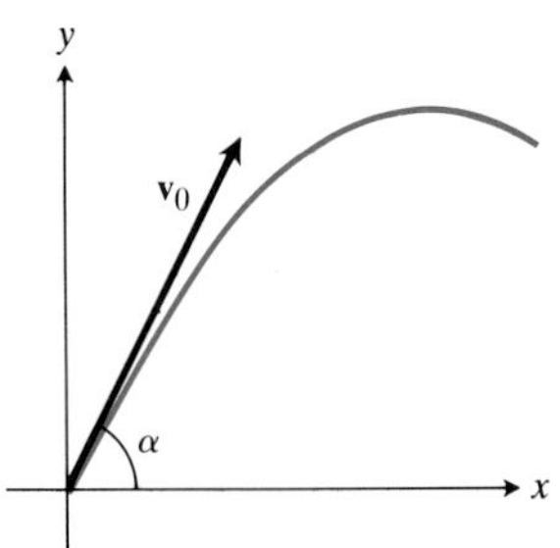

If we measure time in seconds and distance in meters, g is 9.8 m/sec^2 and the equations in (9) give x and y in meters. If we measure time in seconds and distance in feet, then g is 32 ft/sec^2 and the equations in (9) give x and y in feet.

EXAMPLE 1 A projectile is fired from the origin over horizontal ground at an initial speed of 500 m/sec at an angle of elevation of 60°. Where will the projectile be 10 sec later?

Solution We use Eqs. (9) with $v_0 = 500$, $\alpha = 60°$, $g = 9.8$, and $t = 10$ to find the projectile's coordinates to the nearest meter 10 sec after firing:

$$x = (v_0 \cos \alpha)t = 500 \cdot \frac{1}{2} \cdot 10 = 2500 \text{ m}$$

$$\begin{aligned} y &= (v_0 \sin \alpha)t - \frac{1}{2}gt^2 \\ &= 500 \cdot \frac{\sqrt{3}}{2} \cdot 10 - \frac{1}{2} \cdot 9.8 \cdot (10)^2 \\ &= 2500\sqrt{3} - 490 \\ &= 3840 \text{ m.} \qquad \text{(Calculator, rounded)} \end{aligned}$$

Ten seconds after firing, the projectile is 3840 m in the air and 2500 m downrange.

Height, Flight Time, and Range

Equations (9) enable us to answer most questions about an ideal projectile fired from the origin.

The projectile reaches its highest point when its vertical velocity component is zero, that is, when

$$\frac{dy}{dt} = v_0 \sin \alpha - gt = 0 \quad \text{or} \quad t = \frac{v_0 \sin \alpha}{g}.$$

For this value of t, the value of y is

$$y_{\max} = (v_0 \sin \alpha)\left(\frac{v_0 \sin \alpha}{g}\right) - \frac{1}{2}g\left(\frac{v_0 \sin \alpha}{g}\right)^2 = \frac{(v_0 \sin \alpha)^2}{2g}. \tag{10}$$

To find when the projectile lands, when fired over horizontal ground, we set y equal to zero in Eqs. (9) and solve for t:

$$(v_0 \sin \alpha)t - \frac{1}{2}gt^2 = 0$$

$$t\left(v_0 \sin \alpha - \frac{1}{2}gt\right) = 0$$

$$t = 0, \quad t = \frac{2v_0 \sin \alpha}{g}. \tag{11}$$

Since 0 is the time the projectile is fired, $(2v_0 \sin \alpha)/g$ must be the time when the projectile strikes the ground.

To find the projectile's range R, the distance from the origin to the point of impact on horizontal ground, we find the value of x when $t = (2v_0 \sin \alpha)/g$:

$$x = (v_0 \cos \alpha)t$$

$$R = (v_0 \cos \alpha)\left(\frac{2v_0 \sin \alpha}{g}\right) = \frac{v_0^2}{g}(2 \sin \alpha \cos \alpha) = \frac{v_0^2}{g} \sin 2\alpha. \tag{12}$$

Notice that the range is largest when $\sin 2\alpha = 1$ or $\alpha = 45°$.

Height, Flight Time, and Range

For an ideal projectile fired from the origin over horizontal ground:

Maximum height: $$y_{\max} = \frac{(v_0 \sin \alpha)^2}{2g} \tag{13}$$

Flight time (time to impact): $$t = \frac{2v_0 \sin \alpha}{g} \tag{14}$$

Range (distance to point of impact): $$R = \frac{v_0^2}{g} \sin 2\alpha \tag{15}$$

EXAMPLE 2 Find the maximum height, flight time, and range of a projectile fired from the origin over horizontal ground at an initial speed of 500 m/sec and a firing angle of 60° (same projectile as in Example 1).

Solution

Maximum height (Eq. 13): $$y_{\max} = \frac{(v_0 \sin \alpha)^2}{2g} = \frac{(500 \sin 60°)^2}{2(9.8)}$$

$$= 9566 \text{ m.} \qquad \text{(Calculator, rounded)}$$

Flight time (Eq. 14): $$t = \frac{2v_0 \sin \alpha}{g} = \frac{2(500) \sin 60°}{9.8}$$

$$= 88 \text{ sec.} \qquad \text{(Calculator, rounded)}$$

Range (Eq. 15): $$R = \frac{v_0^2}{g} \sin 2\alpha = \frac{(500)^2 \sin 120°}{9.8}$$

$$= 22{,}092 \text{ m.} \qquad \text{(Calculator, rounded)}$$

■

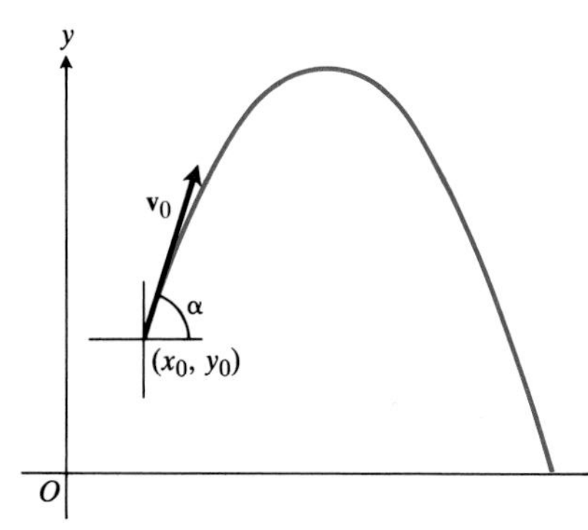

12.8 The path of a projectile fired from (x_0, y_0) with an initial velocity $\mathbf{v}_0$ at an angle of α degrees with the horizontal.

Firing from (x_0, y_0)

If we fire our ideal projectile from the point (x_0, y_0) instead of the origin (Fig. 12.8), the equations that replace Eqs. (9) are

$$x = x_0 + (v_0 \cos \alpha)t, \qquad y = y_0 + (v_0 \sin \alpha)t - \frac{1}{2}gt^2, \tag{16}$$

as you will be invited to show in Exercise 19.

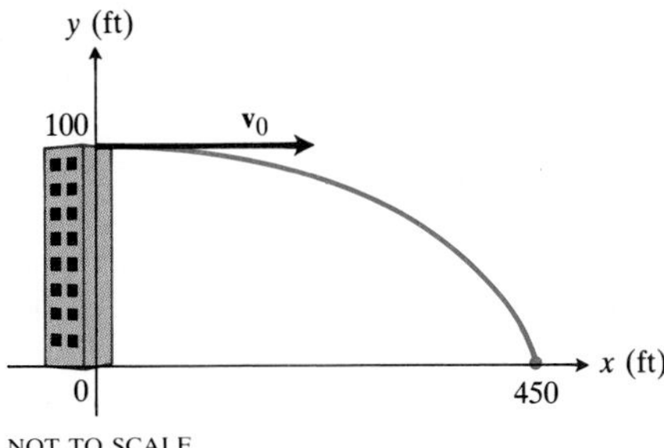

12.9 What initial horizontal speed gets the golf ball 450 ft downrange? To find out, see Example 3.

EXAMPLE 3 We are about to hit a golf ball horizontally off the top edge of a 100-ft building that stands on level ground. How fast will we have to launch the ball to land it 450 ft away?

Solution We use a coordinate system in which the x-axis lies along the ground and the ball's position coordinates at time $t = 0$ are

$$x_0 = 0, \qquad y_0 = 100$$

(Fig. 12.9). With these values for x_0 and y_0, and with $\alpha = 0°$ and $g = 32$ ft/sec^2, Eqs. (16) tell us that

$$x = 0 + (v_0 \cos 0)t = v_0 t,$$

$$y = 100 + (v_0 \sin 0)t - 16t^2 = 100 - 16t^2.$$

We find *when* the ball strikes the ground by setting y equal to zero and solving for t:

$$0 = 100 - 16t^2$$

$$t^2 = \frac{100}{16}$$

$$t = \sqrt{\frac{100}{16}} = \frac{10}{4} = 2.5. \quad \text{(No negative root because } t \geq 0\text{)}$$

We then set $x = 450$ and $t = 2.5$ in the equation $x = v_0 t$ to find the value of v_0 needed to make the ball hit the ground 450 ft downrange:

$$x = v_0 t$$

$$450 = v_0(2.5)$$

$$v_0 = \frac{450}{2.5} = 180 \text{ ft/sec}.$$

To land the ball 450 ft away, we must give it an initial horizontal speed of 180 ft/sec. ■

Ideal Trajectories Are Parabolic

It is often claimed that water from a hose traces a parabola in the air, but anyone who looks closely enough will see this isn't so. The air slows the water down and its forward progress is too slow at the end to match the rate at which it falls.

What is really being claimed is that ideal projectiles move along parabolas, and this we can see from Eqs. (9). If we substitute $t = x/(v_0 \cos \alpha)$ from the first equation into the second, we obtain the Cartesian-coordinate equation

$$y = -\left(\frac{g}{2v_0^2 \cos^2 \alpha}\right)x^2 + (\tan \alpha)x. \tag{17}$$

This equation is quadratic in x and linear in y, so its graph is a parabola.

Exercises 12.2

CALCULATOR The projectiles in the following exercises are to be treated as ideal projectiles whose behavior is faithfully portrayed by the equations derived in the text. Most of the arithmetic, however, is realistic and is best done with a **CALCULATOR.** All firing angles are assumed to be measured from the horizontal. All projectiles are assumed to be fired from the origin over horizontal ground, unless stated otherwise.

1. A projectile is fired at a speed of 840 m/sec at an angle of 60°. How long will it take to get 21 km downrange?

2. Find the muzzle speed of a gun whose maximum range is 24.5 km.

3. A projectile is fired over level ground with an initial speed of 500 m/sec at an angle of elevation of 45°.
a) When and how far away will the projectile strike?
b) How high overhead will the projectile be when it is 5 km downrange?
c) What is the highest the projectile will go?

4. A baseball is thrown from the stands 32 ft above the field at an angle of 30°. When and how far away will the ball strike the ground if its initial speed is 32 ft/sec?

5. Show that a projectile fired at an angle of α degrees, $0 < \alpha < 90$, has the same range as a projectile fired at the same speed at an angle of $(90 - \alpha)$ degrees.

6. What two angles of elevation will enable a projectile to reach a target 16 km downrange on the same level as the gun if the projectile's initial speed is 400 m/sec?

7. A spring gun at ground level fires a golf ball at an angle of 45°. The ball lands 10 m away. What was the ball's initial speed? For the same initial speed, find the two firing angles that make the range 6 m.

8. An electron in a TV tube is beamed horizontally at a speed of 5×10^6 m/sec toward the face of the tube 40 cm away. About how far will the electron drop before it hits?

9. Laboratory tests designed to find how far golf balls of different hardness go when hit with a driver showed that a 100-compression, Surlyn-covered, two-piece ball hit with a club-head speed of 100 mph at a launch angle of 9° carried 248.8 yd. What was the launch speed of the ball? (It was not 100 mph. The data are from "Does Compression Really Matter?" by Lew Fishman, *Golf Digest,* August 1986, pp. 35–37.)

10. Show that doubling the initial speed of a projectile multiplies its maximum height and range by 4. By about what percentage should you increase the initial speed to double the height and range?

11. Show that a projectile attains three-quarters of its maximum height in half the time it takes to reach the maximum height.

12. A human cannonball is to be fired with an initial speed of $v_0 = 80\sqrt{10}/3$ ft/sec. The circus performer (of the right caliber, naturally) hopes to land on a special cushion located 200 ft downrange. The circus is being held in a large room with a flat ceiling 75 ft high. Can the performer be fired to the cushion without striking the ceiling? If so, what should the cannon's angle of elevation be?

13. A golf ball leaves the ground at a 30° angle at a speed of 90 ft/sec. Will it clear the top of a 30-ft tree 135 ft away?

14. A golf ball is hit from the tee to an elevated green with an initial speed of 116 ft/sec at an angle of elevation of 45°, as shown in Fig. 12.10. Will the ball reach the pin?

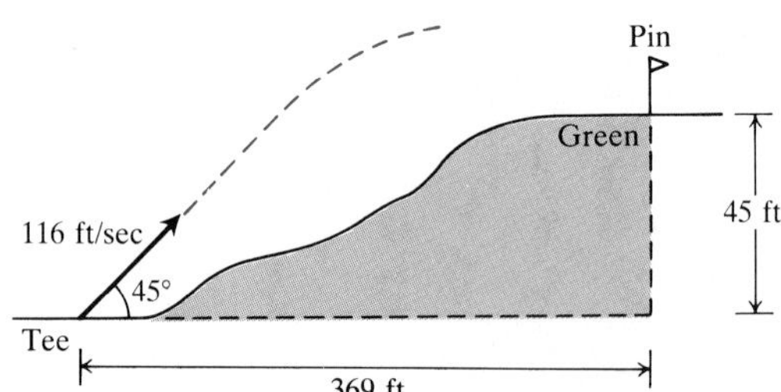

12.10 The tee and green in Exercise 14.

15. A baseball hit by a Boston Red Sox player at a 20° angle from 3 ft above the ground just cleared the left end of the "Green Monster," the left-field wall in Fenway Park (Fig. 12.11). The wall there is 37 ft high and 315 ft from home plate. About how fast was the ball going? How long did it take the ball to reach the wall?

16. The multiflash photograph in Figure 12.12 shows a model train engine moving at a constant speed on a straight track. As the engine moved along, a marble was fired into the air by a spring in the engine's smokestack. The marble, which continued to move with the same forward speed as the engine, rejoined the engine 1 second after it was fired. Measure the angle the marble's path makes with the horizontal and

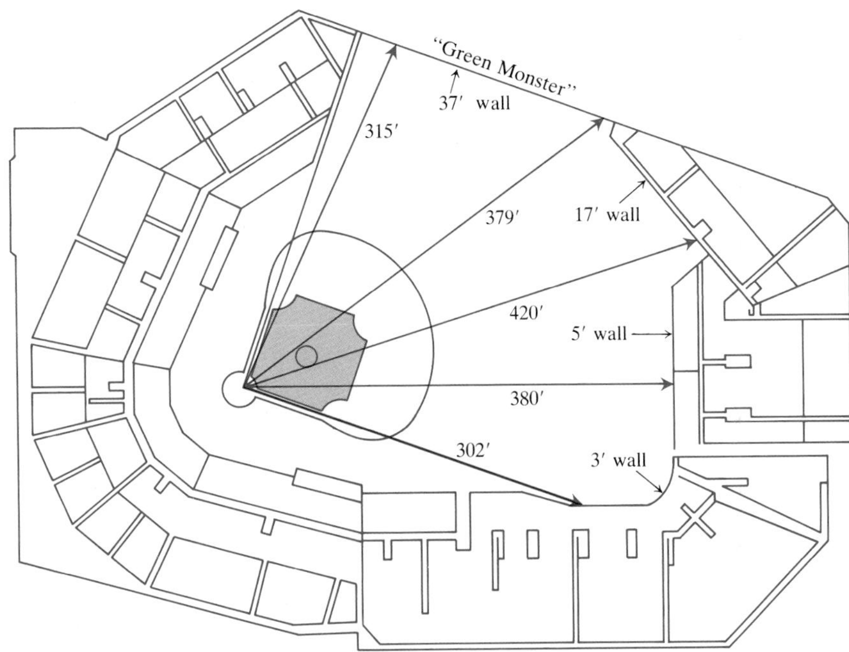

12.11 The Green Monster, the left-field wall at Fenway Park in Boston (Exercise 15).

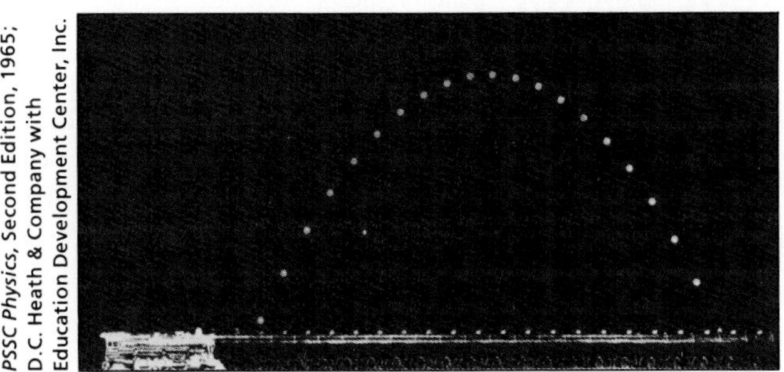

PSSC Physics, Second Edition, 1965; D.C. Heath & Company with Education Development Center, Inc.

12.12 The train in Exercise 16.

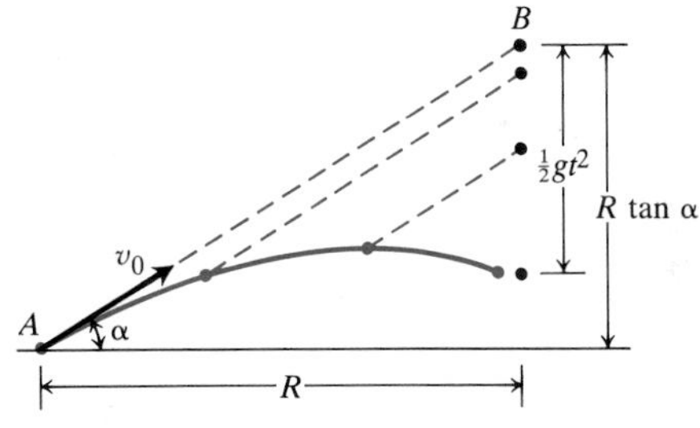

12.13 The marbles in Exercise 17.

use the information to find out how high the marble went and how fast the engine was moving.

17. Figure 12.13 shows an experiment with two marbles. Marble A is launched toward marble B at an initial speed v_0. The launch angle is α. At the same instant, marble B, suspended directly over a spot R units downrange from A, is released to fall from a height of $R \tan \alpha$. It is found that the marbles collide regardless of the value of v_0. Must this happen? Why?

18. An ideal projectile is fired down an inclined plane as shown in profile in Fig. 12.14. Show that the maximum downhill range is achieved when the initial

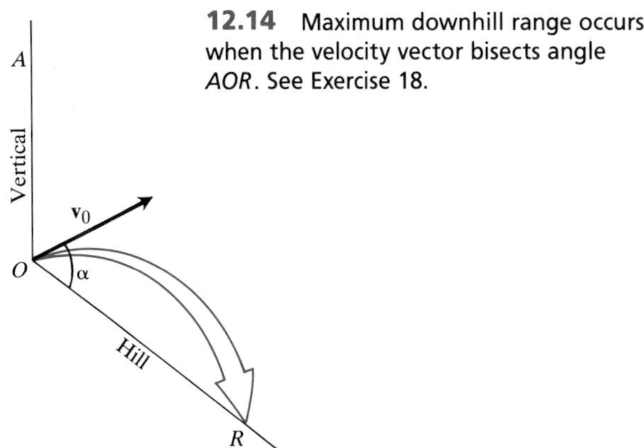

12.14 Maximum downhill range occurs when the velocity vector bisects angle *AOR*. See Exercise 18.

velocity vector bisects the angle AOR. If the projectile were fired uphill instead of down, what firing angle would maximize its range?

19. Derive the equations

$$x = x_0 + (v_0 \cos \alpha)t,$$
$$y = y_0 + (v_0 \sin \alpha)t - \frac{1}{2}gt^2$$

(Eqs. (16) in the text) by solving the following initial value problem for a vector $\mathbf{r}$ in the plane.

Differential equation: $\dfrac{d^2\mathbf{r}}{dt^2} = -g\mathbf{j}$

Initial conditions: $\mathbf{r} = x_0\mathbf{i} + y_0\mathbf{j}$ and $\dfrac{d\mathbf{r}}{dt} = (v_0 \cos \alpha)\mathbf{i} + (v_0 \sin \alpha)\mathbf{j}$ when $t = 0$.

20. COMPUTER GRAPHER If you have access to a parametric equation grapher, and have not yet done Exercise 39 in Section 10.3, do it now. It is about projectile motion.

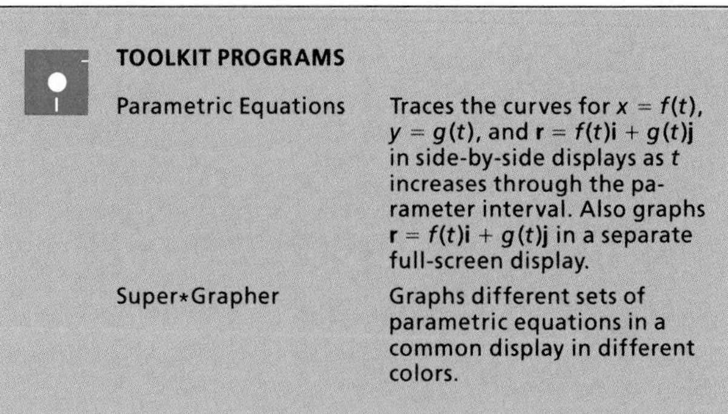

TOOLKIT PROGRAMS

Parametric Equations	Traces the curves for $x = f(t)$, $y = g(t)$, and $\mathbf{r} = f(t)\mathbf{i} + g(t)\mathbf{j}$ in side-by-side displays as t increases through the parameter interval. Also graphs $\mathbf{r} = f(t)\mathbf{i} + g(t)\mathbf{j}$ in a separate full-screen display.
Super*Grapher	Graphs different sets of parametric equations in a common display in different colors.

12.3 Directed Distance and the Unit Tangent Vector T

As you can imagine, differentiable curves, especially those with continuous first and second derivatives, have been subjects of intense study, for their mathematical interest as well as for their applications to motion in space. In this section and the next, we describe some of the features that account for the importance of these curves.

Distance along a Curve

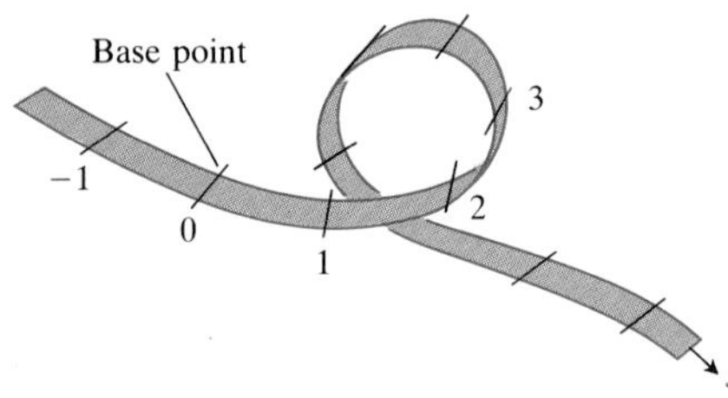

12.15 Continuously differentiable curves can be scaled like coordinate axes or like tape measures that include negative numbers as well as positive numbers. The coordinate given to each point is its directed distance from the base point.

One of the special features of space curves whose coordinate functions have continuous first derivatives is that, like plane curves, they are smooth enough to have a measurable length. This enables us to locate points along these curves by giving their directed distance s along the curve from some **base point,** the way we locate points on coordinate axes by giving their directed distance from the origin (Fig. 12.15). Although time is the natural parameter for describing a moving body's velocity and acceleration, the **directed distance coordinate** s is the natural parameter for studying a curve's geometry. The relationships between these parameters play an important role in calculations of space flight.

In the plane, we define the length of a parametrized curve $x = f(t)$, $y = g(t)$, $a \le t \le b$ by the formula

$$\text{Length} = \int_a^b \sqrt{\left(\frac{dx}{dt}\right)^2 + \left(\frac{dy}{dt}\right)^2}\, dt. \tag{1}$$

In space calculations, the analogous formula has an additional term for the third coordinate.

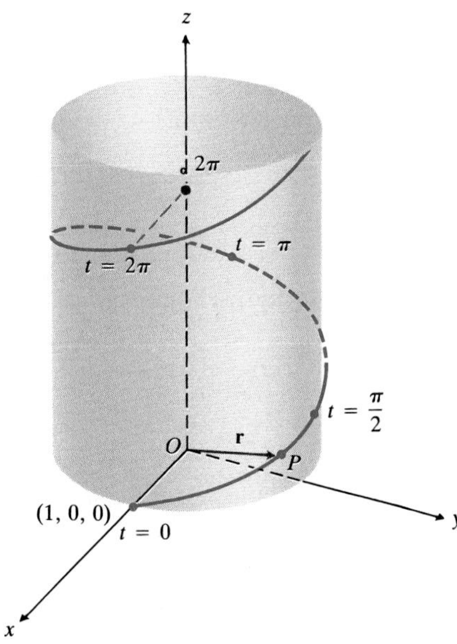

12.16 The helix $\mathbf{r} = (\cos t)\mathbf{i} + (\sin t)\mathbf{j} + t\mathbf{k}$ makes a complete turn during the time interval from $t = 0$ to $t = 2\pi$. We calculate the distance traveled by P by integrating $|\mathbf{v}|$ from $t = 0$ to $t = 2\pi$ (Example 1).

DEFINITION

If $x = f(t)$, $y = g(t)$, and $z = h(t)$ have continuous first derivatives with respect to t for $a \le t \le b$, and the position vector $\mathbf{r} = f(t)\mathbf{i} + g(t)\mathbf{j} + h(t)\mathbf{k}$ defined by these equations traces its curve exactly once as t moves from $t = a$ to $t = b$, then the length of the curve is given by the formula

$$\text{Length} = \int_a^b \sqrt{\left(\frac{dx}{dt}\right)^2 + \left(\frac{dy}{dt}\right)^2 + \left(\frac{dz}{dt}\right)^2}\, dt. \qquad (2)$$

Just as for plane curves, we can calculate the length of a curve in space from any convenient parametrization that meets the stated conditions. Again, we shall omit the proof.

Notice that the square root in Eq. (2) is $|\mathbf{v}|$, the length of the velocity vector $d\mathbf{r}/dt$. This lets us write the formula for length a shorter way.

Length Formula (Short Form)

$$\text{Length} = \int_a^b |\mathbf{v}|\, dt \qquad (3)$$

EXAMPLE 1 Find the length of one turn of the helix

$$\mathbf{r} = (\cos t)\mathbf{i} + (\sin t)\mathbf{j} + t\mathbf{k}.$$

Solution Since the sine and cosine both have periods of 2π, the helix makes one full turn as t runs from 0 to 2π (Fig. 12.16). The length of this portion of the curve is

$$\begin{aligned} \text{Length} &= \int_a^b |\mathbf{v}|\, dt && \text{(Eq. 3)} \\ &= \int_0^{2\pi} \sqrt{(-\sin t)^2 + (\cos t)^2 + (1)^2}\, dt \\ &= \int_0^{2\pi} \sqrt{2}\, dt = 2\pi\sqrt{2}. \end{aligned}$$

This is $\sqrt{2}$ times the length of the circle in the xy-plane over which the helix stands. ■

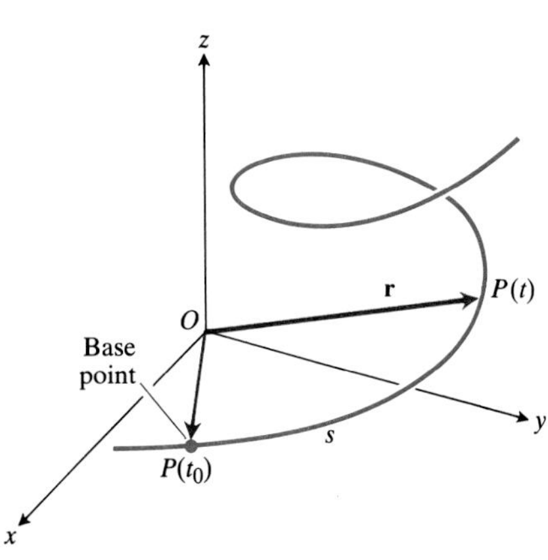

12.17 The directed distance along the curve from $P(t_0)$ to any point $P(t)$ is

$$s = \int_{t_0}^{t} |\mathbf{v}(\tau)|\, d\tau.$$

If $x = f(t)$, $y = g(t)$, and $z = h(t)$ have continuous first derivatives with respect to t and we choose a base point $P(t_0)$ on the curve $\mathbf{r} = f(t)\mathbf{i} + g(t)\mathbf{j} + h(t)\mathbf{k}$ (Fig. 12.17), the integral of $|\mathbf{v}|$ from t_0 to t gives the directed distance along the curve from $P(t_0)$ to $P(t)$. The distance is a function of t, and we denote it by $s(t)$.

Directed Distance along a Curve from t_0 to t

$$s(t) = \int_{t_0}^{t} \sqrt{x'(\tau)^2 + y'(\tau)^2 + z'(\tau)^2}\, d\tau = \int_{t_0}^{t} |\mathbf{v}(\tau)|\, d\tau \tag{4}$$

The value of s is positive if t is greater than t_0 and negative if t is less than t_0.

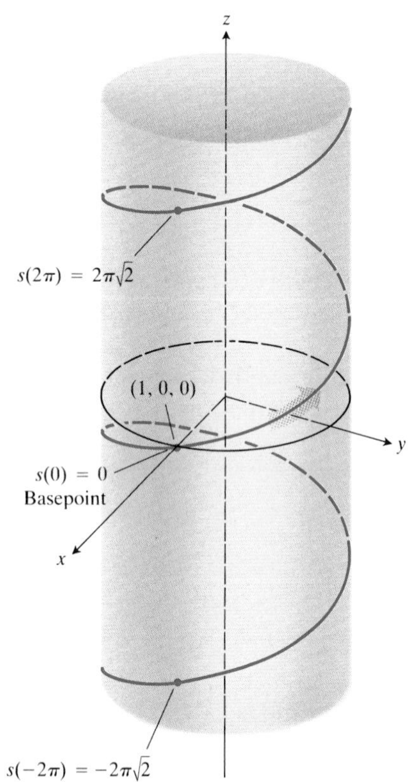

12.18 Directed distances along the helix $\mathbf{r} = (\cos t)\mathbf{i} + (\sin t)\mathbf{j} + t\mathbf{k}$ from the base point (1, 0, 0).

EXAMPLE 2 If $t_0 = 0$, the directed distance along the helix

$$\mathbf{r} = (\cos t)\mathbf{i} + (\sin t)\mathbf{j} + t\mathbf{k}$$

from t_0 to t is

$$\begin{aligned} s(t) &= \int_{t_0}^{t} |\mathbf{v}(\tau)|\, d\tau \quad \text{(Eq. 4)} \\ &= \int_{0}^{t} \sqrt{2}\, d\tau \qquad \text{(Value from Example 1)} \\ &= \sqrt{2}t. \end{aligned}$$

Thus, $s(2\pi) = 2\pi\sqrt{2}$, $s(-2\pi) = -2\pi\sqrt{2}$, and so on (Fig. 12.18). ■

EXAMPLE 3 *Distance along a line.* Show that if $\mathbf{u} = u_1\mathbf{i} + u_2\mathbf{j} + u_3\mathbf{k}$ is a unit vector then the directed distance along the line

$$\mathbf{r} = (x_0 + tu_1)\mathbf{i} + (y_0 + tu_2)\mathbf{j} + (z_0 + tu_3)\mathbf{k}$$

from the point $P_0(x_0, y_0, z_0)$ where $t = 0$ is t itself.

Solution

$$\mathbf{v} = \frac{d}{dt}(x_0 + tu_1)\mathbf{i} + \frac{d}{dt}(y_0 + tu_2)\mathbf{j} + \frac{d}{dt}(z_0 + tu_3)\mathbf{k} = u_1\mathbf{i} + u_2\mathbf{j} + u_3\mathbf{k} = \mathbf{u},$$

so

$$s(t) = \int_0^t |\mathbf{v}|\, d\tau = \int_0^t |\mathbf{u}|\, d\tau = \int_0^t 1\, d\tau = t.$$ ■

If the derivatives beneath the radical in Eq. (4) are continuous, the First Fundamental Theorem of Calculus (Section 5.4) tells us that s is a differentiable function of t whose derivative is

$$\frac{ds}{dt} = |\mathbf{v}(t)|. \tag{5}$$

As we expect, the speed with which a particle moves along a path is the magnitude of $\mathbf{v}$.

Notice that while the base point $P(t_0)$ plays a role in defining s in Eq. (4) it plays no role in Eq. (5). The rate at which a moving particle covers the distance along its path does not depend on how far away the base point is.

Notice also that as long as $|\mathbf{v}|$ is different from zero, *as we shall assume it to be in all examples from now on*, ds/dt is positive and s is an increasing function of t.

The Unit Tangent Vector T

Since ds/dt is positive for the curves we are considering from now on, s is one-to-one and has an inverse that gives t as a differentiable function of s (Section 7.1). The derivative of the inverse function is

$$\frac{dt}{ds} = \frac{1}{ds/dt} = \frac{1}{|\mathbf{v}|}. \tag{6}$$

This makes $\mathbf{r}$ a differentiable function of s whose derivative can be calculated with the Chain Rule for vector-valued functions to be

$$\frac{d\mathbf{r}}{ds} = \frac{d\mathbf{r}}{dt}\frac{dt}{ds} = \mathbf{v} \cdot \frac{1}{|\mathbf{v}|} = \frac{\mathbf{v}}{|\mathbf{v}|}. \tag{7}$$

This tells us that $d\mathbf{r}/ds$ has the constant length

$$\left|\frac{d\mathbf{r}}{ds}\right| = \frac{1}{|\mathbf{v}|}|\mathbf{v}| = 1. \tag{8}$$

Together, Eqs. (7) and (8) say that $d\mathbf{r}/ds$ is a unit vector that points in the direction of $\mathbf{v}$. We call $d\mathbf{r}/ds$ the unit tangent vector of the curve traced by $\mathbf{r}$ and denote it by $\mathbf{T}$ (Fig 12.19).

DEFINITION

The **unit tangent vector** of a differentiable curve $\mathbf{r} = \mathbf{f}(t)$ is

$$\mathbf{T} = \frac{d\mathbf{r}}{ds} = \frac{d\mathbf{r}/dt}{ds/dt} = \frac{\mathbf{v}}{|\mathbf{v}|}. \tag{9}$$

The unit tangent vector $\mathbf{T}$ is a differentiable function of t whenever $\mathbf{v}$ is a differentiable function of t. As we shall see in the next section, $\mathbf{T}$ is one of three unit vectors in a traveling reference frame that is used to describe the motion of space vehicles and other bodies moving in three dimensions.

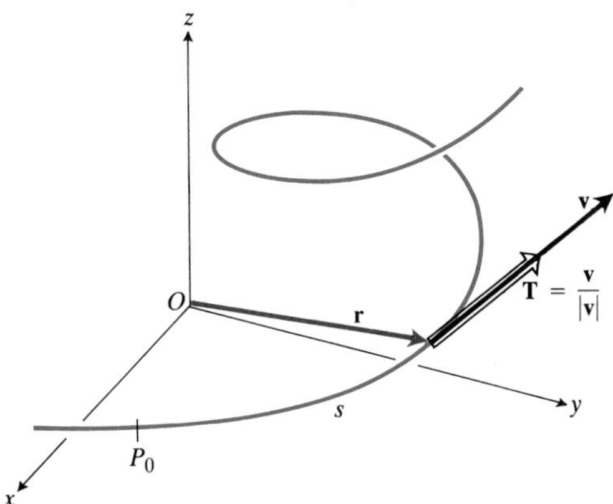

12.19 We get the unit tangent vector **T** by dividing **v** by $|\mathbf{v}|$.

EXAMPLE 4 Find the unit tangent vector of the helix

$$\mathbf{r} = (\cos t)\mathbf{i} + (\sin t)\mathbf{j} + t\mathbf{k}.$$

Solution

$$\mathbf{v} = (-\sin t)\mathbf{i} + (\cos t)\mathbf{j} + \mathbf{k}$$

$$|\mathbf{v}| = \sqrt{(-\sin t)^2 + (\cos t)^2 + (1)^2} = \sqrt{2}$$

$$\mathbf{T} = \frac{\mathbf{v}}{|\mathbf{v}|} = -\frac{\sin t}{\sqrt{2}}\mathbf{i} + \frac{\cos t}{\sqrt{2}}\mathbf{j} + \frac{1}{\sqrt{2}}\mathbf{k}.$$ ■

EXAMPLE 5 *The involute of a circle* (Fig. 12.20). Find the unit tangent vector of the curve

$$\mathbf{r} = (\cos t + t \sin t)\mathbf{i} + (\sin t - t \cos t)\mathbf{j}, \qquad t > 0.$$

Solution

$$\mathbf{v} = \frac{d\mathbf{r}}{dt} = (-\sin t + \sin t + t \cos t)\mathbf{i} + (\cos t - \cos t + t \sin t)\mathbf{j}$$

$$= (t \cos t)\mathbf{i} + (t \sin t)\mathbf{j}$$

$$|\mathbf{v}| = \sqrt{t^2\cos^2 t + t^2\sin^2 t} = \sqrt{t^2} = |t| = t \quad (|t| = t \text{ because } t > 0)$$

$$\mathbf{T} = \frac{\mathbf{v}}{|\mathbf{v}|} = \frac{\mathbf{v}}{t} = (\cos t)\mathbf{i} + (\sin t)\mathbf{j}$$ ■

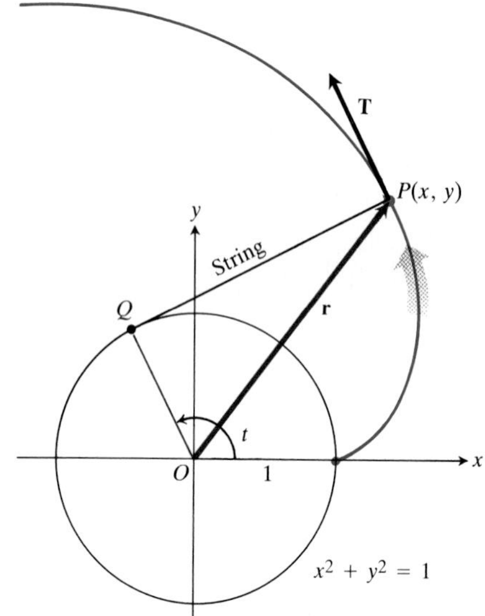

12.20 The *involute* of a circle is the path traced by the endpoint *P* of a string unwinding from a circle, here the unit circle in the *xy*-plane. The position vector of *P* can be shown to be

$$\mathbf{r} = (\cos t + t \sin t)\mathbf{i} + (\sin t - t \cos t)\mathbf{j},$$

where t is the angle from the positive x-axis to Q. Example 5 derives a formula for the curve's unit tangent vector **T**.

EXAMPLE 6 For the counterclockwise motion

$$\mathbf{r} = (\cos t)\mathbf{i} + (\sin t)\mathbf{j}$$

around the unit circle,

$$\mathbf{v} = (-\sin t)\mathbf{i} + (\cos t)\mathbf{j}$$

is already a unit vector, so $\mathbf{T} = \mathbf{v}$ (Fig. 12.21). ■

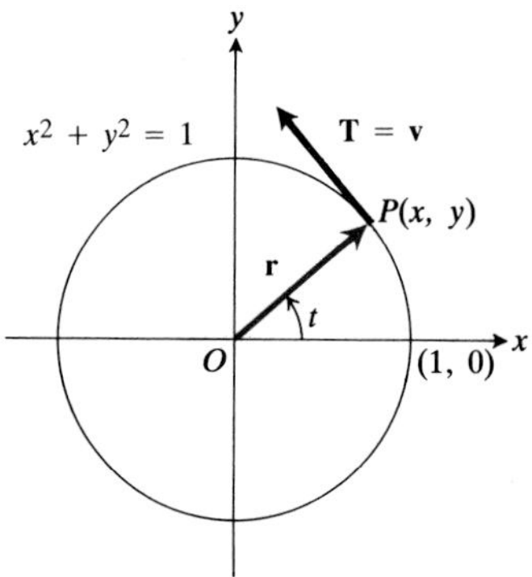

12.21 For the counterclockwise motion $\mathbf{r} = (\cos t)\mathbf{i} + (\sin t)\mathbf{j}$ about the unit circle, the unit tangent vector is **v** itself (Example 6).

Exercises 12.3

In Exercises 1–8, find the curve's unit tangent vector. Also, find the length of the indicated section of the curve.

1. $\mathbf{r} = (2\cos t)\mathbf{i} + (2\sin t)\mathbf{j} + \sqrt{5}t\mathbf{k}$, section from $t = 0$ to $t = \pi$
2. $\mathbf{r} = (6\sin 2t)\mathbf{i} + (6\cos 2t)\mathbf{j} + 5t\mathbf{k}$, section from $t = 0$ to $t = \pi$
3. $\mathbf{r} = t\mathbf{i} + (2/3)t^{3/2}\mathbf{k}$, section from $t = 0$ to $t = 8$
4. $\mathbf{r} = (\cos^3 t)\mathbf{j} + (\sin^3 t)\mathbf{k}$, section from $t = 0$ to $t = \pi/2$
5. $\mathbf{r} = (2 + t)\mathbf{i} - (t + 1)\mathbf{j} + t\mathbf{k}$, section from $t = 0$ to $t = 3$
6. $\mathbf{r} = 6t^3\mathbf{i} - 2t^3\mathbf{j} - 3t^3\mathbf{k}$, section from $t = -1$ to $t = 1$
7. $\mathbf{r} = (t\cos t)\mathbf{i} + (t\sin t)\mathbf{j} + (2\sqrt{2}/3)t^{3/2}\mathbf{k}$, section from $t = 0$ to $t = \pi$
8. $\mathbf{r} = (t\sin t + \cos t)\mathbf{i} + (t\cos t - \sin t)\mathbf{j}$, section from $t = 0$ to $t = \sqrt{2}$

In Exercises 9–12, find the directed distance along the curve from the point where $t = 0$ by evaluating the integral

$$s(t) = \int_0^t |\mathbf{v}(\tau)|\, d\tau.$$

Then find the length of the indicated section of the curve.

9. $\mathbf{r} = (4\cos t)\mathbf{i} + (4\sin t)\mathbf{j} + 3t\mathbf{k}$, section from $t = 0$ to $t = \pi/2$
10. $\mathbf{r} = (\cos t + t\sin t)\mathbf{i} + (\sin t - t\cos t)\mathbf{j}$, section from $t = \pi/2$ to $t = \pi$
11. $\mathbf{r} = (e^t\cos t)\mathbf{i} + (e^t\sin t)\mathbf{j} + e^t\mathbf{k}$, section from $t = 0$ to $t = \ln 4$
12. $\mathbf{r} = (1 + 2t)\mathbf{i} + (1 + 3t)\mathbf{j} + (6 - 6t)\mathbf{k}$, section from $t = 0$ to $t = 1$
13. Find the length of the curve
$$\mathbf{r} = (\sqrt{2}t)\mathbf{i} + (\sqrt{2}t)\mathbf{j} + (1 - t^2)\mathbf{k}$$
from $(0, 0, 1)$ to $(\sqrt{2}, \sqrt{2}, 0)$.
14. The length $2\pi\sqrt{2}$ of the turn of the helix in Example 1 is also the length of the diagonal of a square 2π units on a side. Show how to obtain this square by cutting away and flattening a portion of the cylinder around which the helix winds.
15. *Length is independent of parametrization.* To illustrate the fact that the length of a smooth space curve does not depend on the parametrization we use to compute it as long as the parametrization meets the conditions given with Eq. (2), calculate the length of one turn of the helix in Example 1 with the following parametrizations.
    a) $\mathbf{r} = (\cos 4t)\mathbf{i} + (\sin 4t)\mathbf{j} + 4t\mathbf{k}$, $0 \le t \le \pi/2$
    b) $\mathbf{r} = (\cos(t/2))\mathbf{i} + (\sin(t/2))\mathbf{j} + (t/2)\mathbf{k}$, $0 \le t \le 4\pi$
    c) $\mathbf{r} = (\cos t)\mathbf{i} - (\sin t)\mathbf{j} - t\mathbf{k}$, $-2\pi \le t \le 0$

12.4 Curvature, Torsion, and the TNB Frame

In this section we define a frame of mutually orthogonal unit vectors that always travels with a body moving along a curve in space (Fig. 12.22). The frame has three vectors. The first is $\mathbf{T}$, the unit tangent vector. The second is $\mathbf{N}$, the unit vector that gives the direction of $d\mathbf{T}/ds$. The third is $\mathbf{B} = \mathbf{T} \times \mathbf{N}$. These vectors and their derivatives, when available, give useful information about a vehicle's orientation in space and about how the vehicle's path turns and twists as the vehicle moves along.

For example, the magnitude of the derivative $d\mathbf{T}/ds$ tells how much a vehicle's path turns to the left or right as it moves along; it is called the *curvature* of the vehicle's path. The number $|d\mathbf{B}/ds|$ tells how much a vehicle's path rotates or twists as the vehicle moves along; it is called the *torsion*

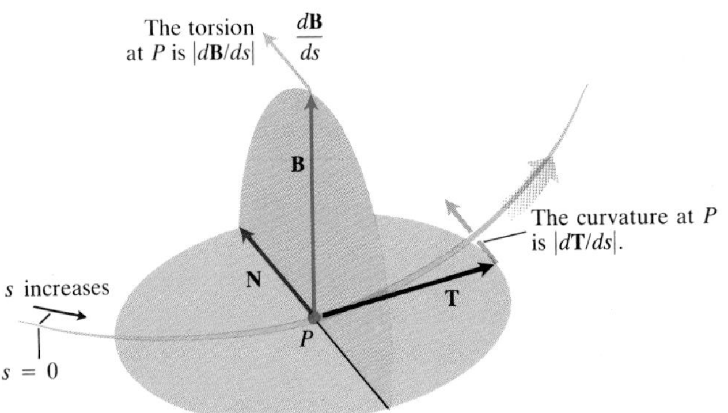

12.22 Every moving body travels with a **TNB** frame that describes how the body moves.

of the vehicle's path. Look at Fig. 12.22 again. If P were a train climbing up a curved track, the rate at which the headlight turned from side to side per unit distance would be the curvature of the track. The rate at which the engine rotated about its longitudinal axis (the line of $\mathbf{T}$) would be the torsion.

We begin with curves in the plane.

The Curvature of a Plane Curve

As we move along a differentiable curve in the plane, the unit tangent vector $\mathbf{T}$ turns as the curve bends. We measure the rate at which $\mathbf{T}$ turns by measuring the change in the angle ϕ that $\mathbf{T}$ makes with $\mathbf{i}$ (Fig. 12.23). At each point P, the absolute value of $d\phi/ds$, stated in radians per unit of length along the curve, is called the **curvature** at P. If $|d\phi/ds|$ is large, $\mathbf{T}$ turns sharply as we pass through P and the curvature at P is large. If $|d\phi/ds|$ is close to zero, $\mathbf{T}$ turns more slowly and the curvature at P is small. On circles and lines, the curvature is constant, as we shall see in Examples 1 and 2. On other curves, the curvature can vary from place to place. The traditional symbol for the curvature function is the Greek letter κ (kappa).

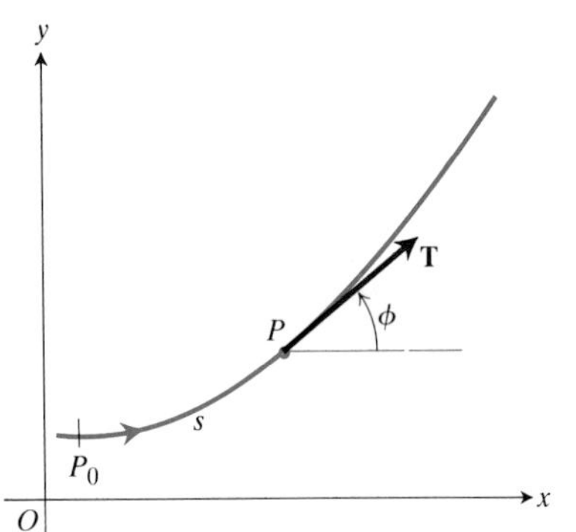

12.23 The value of $|d\phi/ds|$ at P is called the curvature of the curve at P.

If $x = f(t)$ and $y = g(t)$ are twice-differentiable functions of t, we can derive a formula for the curvature of the curve $\mathbf{r} = f(t)\mathbf{i} + g(t)\mathbf{j}$ in the following way. In Newton's dot notation, in which $\dot{y}$ ("y dot") means dy/dt, $\ddot{y}$ ("y double dot") means d^2y/dt^2, and so on, we have

$$\tan\phi = \frac{dy}{dx} = \frac{dy/dt}{dx/dt} = \frac{\dot{y}}{\dot{x}} \quad \text{and} \quad \phi = \tan^{-1}\left(\frac{\dot{y}}{\dot{x}}\right). \tag{1}$$

Hence,

$$\begin{aligned} \frac{d\phi}{ds} &= \frac{d\phi}{dt}\frac{dt}{ds} = \frac{1}{1 + (\dot{y}/\dot{x})^2}\frac{d}{dt}\left(\frac{\dot{y}}{\dot{x}}\right)\frac{1}{(\dot{x}^2 + \dot{y}^2)^{1/2}} \qquad \left(\frac{dt}{ds} = \frac{1}{|\mathbf{v}|}\right) \\ &= \frac{\dot{x}^2}{(\dot{x}^2 + \dot{y}^2)^{3/2}}\,\frac{\dot{x}\ddot{y} - \dot{y}\ddot{x}}{\dot{x}^2} \\ &= \frac{\dot{x}\ddot{y} - \dot{y}\ddot{x}}{|\mathbf{v}|^3}. \end{aligned} \tag{2}$$

The curvature, therefore, is

$$\left|\frac{d\phi}{ds}\right| = \frac{|\dot{x}\ddot{y} - \dot{y}\ddot{x}|}{|\mathbf{v}|^3}. \tag{3}$$

The observation that $|\dot{x}\ddot{y} - \dot{y}\ddot{x}|$ is the magnitude of the vector

$$\mathbf{v} \times \mathbf{a} = \begin{vmatrix} \mathbf{i} & \mathbf{j} & \mathbf{k} \\ \dot{x} & \dot{y} & 0 \\ \ddot{x} & \ddot{y} & 0 \end{vmatrix} \tag{4}$$

enables us to write Eq. (3) in the following compact vector form.

Curvature

$$\kappa = \frac{|\mathbf{v} \times \mathbf{a}|}{|\mathbf{v}|^3} \tag{5}$$

Equation (5) calculates the curvature, a geometric property of the curve, from the velocity and acceleration of any vector representation of the curve in which $|\mathbf{v}|$ is different from zero. Take a moment to think about how remarkable this really is: From any formula for motion along a curve, no matter how variable the motion may be (as long as $\mathbf{v}$ is never zero), we can calculate a physical property of the curve that seems to have nothing to do with the way the curve is traversed.

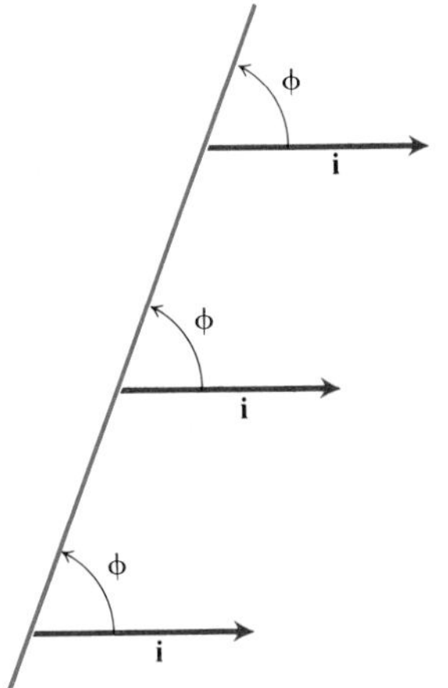

12.24 Along a line, the angle ϕ stays the same from point to point and the curvature, $d\phi/ds$, is zero.

EXAMPLE 1 *The curvature of a straight line is zero.* On a straight line, ϕ is constant (Fig. 12.24), so $d\phi/ds = 0$. ■

EXAMPLE 2 *The curvature of a circle of radius a is 1/a.* To see why, parametrize the circle as

$$\mathbf{r} = (a \cos t)\mathbf{i} + (a \sin t)\mathbf{j}.$$

Then

$$\mathbf{v} = -(a \sin t)\mathbf{i} + (a \cos t)\mathbf{j},$$

$$\mathbf{a} = -(a \cos t)\mathbf{i} - (a \sin t)\mathbf{j}.$$

Then

$$\mathbf{v} \times \mathbf{a} = \begin{vmatrix} \mathbf{i} & \mathbf{j} & \mathbf{k} \\ -a \sin t & a \cos t & 0 \\ -a \cos t & -a \sin t & 0 \end{vmatrix} = (a^2\sin^2 t + a^2\cos^2 t)\mathbf{k} = a^2\mathbf{k},$$

$$|\mathbf{v}|^3 = \left[\sqrt{(-a \sin t)^2 + (a \cos t)^2}\,\right]^3 = a^3,$$

and

$$\kappa = \frac{|\mathbf{v} \times \mathbf{a}|}{|\mathbf{v}|^3} = \frac{|a^2\mathbf{k}|}{a^3} = \frac{a^2|\mathbf{k}|}{a^3} = \frac{1}{a}. \quad ■$$

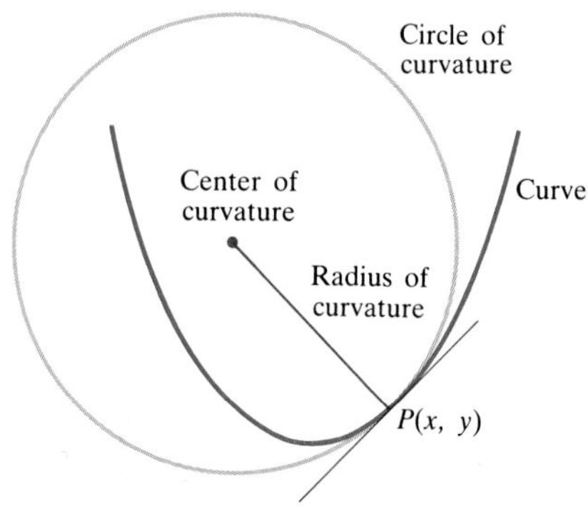

12.25 The osculating circle or circle of curvature at $P(x, y)$ lies toward the inner side of the curve.

Circle of Curvature and Radius of Curvature

The **circle of curvature** or **osculating circle** at a point P on a plane curve where $\kappa \neq 0$ is the circle in the plane of the curve that

1. is tangent to the curve at P (has the same tangent that the curve has);
2. has the same curvature the curve has at P; and
3. lies toward the concave or inner side of the curve (as in Fig. 12.25).

The **radius of curvature** of the curve at P is the radius of the circle of curvature, which, according to Example 2, is

$$\text{Radius of curvature} = \rho = \frac{1}{\kappa}. \tag{6}$$

To calculate ρ, we calculate κ and take its reciprocal.

The **center of curvature** of the curve at P is the center of the circle of curvature.

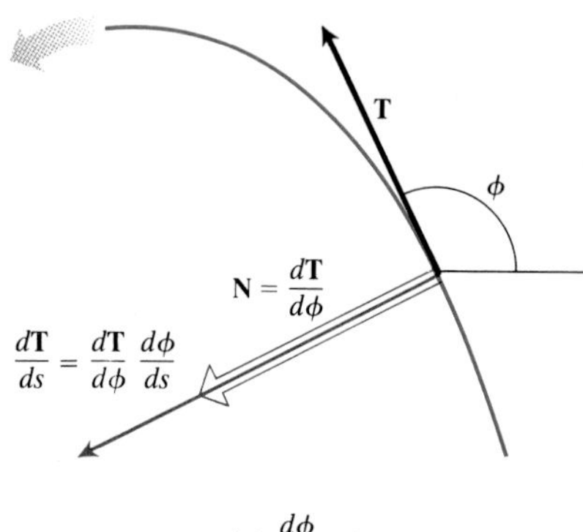

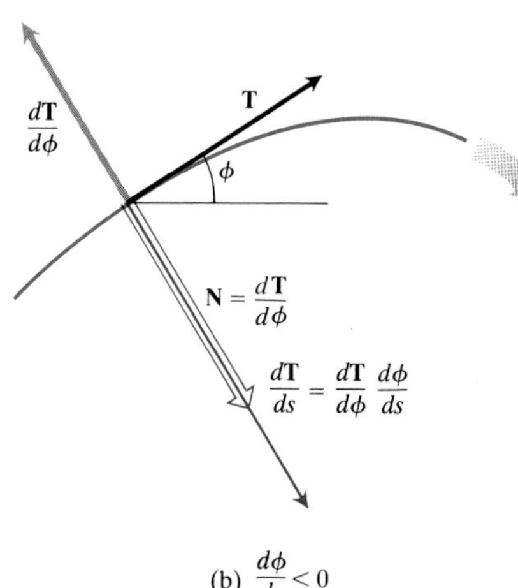

12.26 The vector $d\mathbf{T}/ds$, normal to the curve, always points inward. The principal unit normal vector $\mathbf{N}$ is the direction of $d\mathbf{T}/ds$.

The Principal Unit Normal Vector for Curves in the Plane

The vectors $d\mathbf{T}/ds$ and $d\mathbf{T}/d\phi$ are related by the Chain Rule equation

$$\frac{d\mathbf{T}}{ds} = \frac{d\mathbf{T}}{d\phi}\frac{d\phi}{ds}. \tag{7}$$

Furthermore, since $\mathbf{T}$ is a vector of constant length, $d\mathbf{T}/ds$ and $d\mathbf{T}/d\phi$ are both orthogonal to $\mathbf{T}$. They have the same direction if $d\phi/ds$ is positive and opposite directions if $d\phi/ds$ is negative. Now,

$$\mathbf{T} = (\cos\phi)\mathbf{i} + (\sin\phi)\mathbf{j}, \tag{8}$$

and

$$\frac{d\mathbf{T}}{d\phi} = -(\sin\phi)\mathbf{i} + (\cos\phi)\mathbf{j} = \cos\left(\phi + \frac{\pi}{2}\right)\mathbf{i} + \sin\left(\phi + \frac{\pi}{2}\right)\mathbf{j} \tag{9}$$

is the unit vector obtained by rotating $\mathbf{T}$ counterclockwise through $\pi/2$ radians. Therefore, if we stand at a point on the curve facing in the direction of $\mathbf{T}$, the vector $d\mathbf{T}/ds = (d\mathbf{T}/d\phi)(d\phi/ds)$ will point toward the left if $d\phi/ds$ is positive and toward the right if $d\phi/ds$ is negative. In other words, $d\mathbf{T}/ds$ will point toward the concave side of the curve (Fig. 12.26).

Since $d\mathbf{T}/d\phi$ is a unit vector, the magnitude of $d\mathbf{T}/ds$ at any point on the curve is the curvature at that point, as we can see from the equation

$$\left|\frac{d\mathbf{T}}{ds}\right| = \left|\frac{d\mathbf{T}}{d\phi}\right|\left|\frac{d\phi}{ds}\right| = (1)(\kappa) = \kappa. \tag{10}$$

When $d\mathbf{T}/ds \neq \mathbf{0}$, its direction is given by the unit vector

$$\mathbf{N} = \frac{d\mathbf{T}/ds}{|d\mathbf{T}/ds|} = \frac{1}{\kappa}\frac{d\mathbf{T}}{ds}. \tag{11}$$

Since $\mathbf{N}$ points the same way $d\mathbf{T}/ds$ does, $\mathbf{N}$ is always orthogonal to $\mathbf{T}$ and directed toward the concave side of the curve. The vector $\mathbf{N}$ is called the **principal unit normal vector** of the curve.

Because the directed distance on a curve $\mathbf{r} = f(t)\mathbf{i} + g(t)\mathbf{j}$ is defined with ds/dt positive, the Chain Rule gives

$$\begin{aligned}\mathbf{N} &= \frac{d\mathbf{T}/ds}{|d\mathbf{T}/ds|} = \frac{(d\mathbf{T}/dt)(dt/ds)}{|d\mathbf{T}/dt|\,|dt/ds|} \\ &= \frac{d\mathbf{T}/dt}{|d\mathbf{T}/dt|}.\end{aligned} \tag{12}$$

This formula enables us to find $\mathbf{N}$ without having to find ϕ, κ, or s first.

EXAMPLE 3 Find $\mathbf{T}$ and $\mathbf{N}$ for the circular motion

$$\mathbf{r} = (\cos 2t)\mathbf{i} + (\sin 2t)\mathbf{j}.$$

Solution We first find $\mathbf{T}$:

$$\mathbf{v} = -(2 \sin 2t)\mathbf{i} + (2 \cos 2t)\mathbf{j},$$

$$|\mathbf{v}| = \sqrt{4 \sin^2 2t + 4 \cos^2 2t} = 2,$$

$$\mathbf{T} = \frac{\mathbf{v}}{|\mathbf{v}|} = -(\sin 2t)\mathbf{i} + (\cos 2t)\mathbf{j}.$$

From this we find

$$\frac{d\mathbf{T}}{dt} = -(2 \cos 2t)\mathbf{i} - (2 \sin 2t)\mathbf{j},$$

$$\left|\frac{d\mathbf{T}}{dt}\right| = \sqrt{4 \cos^2 2t + 4 \sin^2 2t} = 2,$$

and

$$\mathbf{N} = \frac{d\mathbf{T}/dt}{|d\mathbf{T}/dt|} = -(\cos 2t)\mathbf{i} - (\sin 2t)\mathbf{j}. \qquad \text{(Eq. 12)}$$

■

Curvature for Curves in Space

In space there is no natural way to find an angle like ϕ with which to measure the change in $\mathbf{T}$ along a differentiable curve. But we still have s, the directed distance along the curve, and can define the curvature to be

$$\kappa = \left|\frac{d\mathbf{T}}{ds}\right|,$$

as it worked out to be for curves in the plane. The formula

$$\kappa = \frac{|\mathbf{v} \times \mathbf{a}|}{|\mathbf{v}|^3} \tag{13}$$

still holds, as you will see if you do the calculation in Exercise 23.

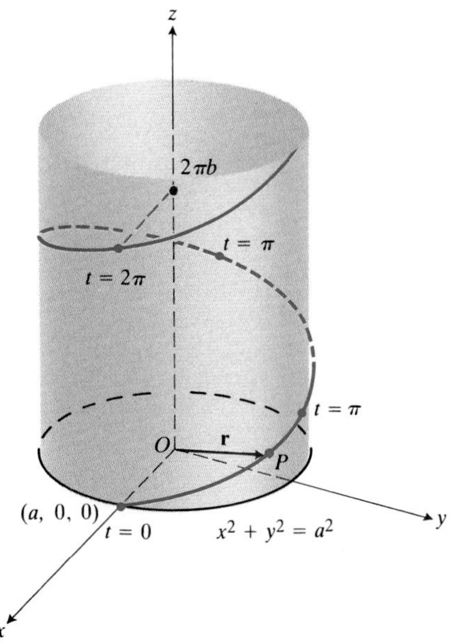

12.27 The helix $\mathbf{r} = (a \cos t)\mathbf{i} + (a \sin t)\mathbf{j} + bt\mathbf{k}$, drawn with a and b positive (Example 4).

EXAMPLE 4 How do the values of a and b control the curvature of the helix (Fig. 12.27)

$$\mathbf{r} = (a \cos t)\mathbf{i} + (a \sin t)\mathbf{j} + bt\mathbf{k}? \qquad (a, b \geq 0)$$

Solution We calculate the curvature with Eq. (13),

$$\mathbf{v} = -(a \sin t)\mathbf{i} + (a \cos t)\mathbf{j} + b\mathbf{k},$$

$$\mathbf{a} = -(a \cos t)\mathbf{i} - (a \sin t)\mathbf{j},$$

$$\mathbf{v} \times \mathbf{a} = \begin{vmatrix} \mathbf{i} & \mathbf{j} & \mathbf{k} \\ -a \sin t & a \cos t & b \\ -a \cos t & -a \sin t & 0 \end{vmatrix} = (ab \sin t)\mathbf{i} - (ab \cos t)\mathbf{j} + a^2\mathbf{k},$$

$$\kappa = \frac{|\mathbf{v} \times \mathbf{a}|}{|\mathbf{v}|^3} = \frac{\sqrt{a^2b^2 + a^4}}{(a^2 + b^2)^{3/2}} = \frac{a\sqrt{a^2 + b^2}}{(a^2 + b^2)^{3/2}} = \frac{a}{a^2 + b^2}. \tag{14}$$

From Eq. (14) we see that increasing b for a fixed a decreases the curvature. Decreasing a for a fixed b eventually decreases the curvature as well. In other words, stretching a spring tends to straighten it.

If $b = 0$, the helix reduces to a circle of radius a and its curvature reduces to $1/a$, as it should. If $a = 0$, the helix becomes the z-axis, and its curvature reduces to 0, again as it should. ■

N for Curves in Space

To define the principal unit normal vector of a curve in space, we use the same definition we use for a curve in the plane:

$$\mathbf{N} = \frac{d\mathbf{T}/ds}{|d\mathbf{T}/ds|} = \frac{1}{\kappa}\frac{d\mathbf{T}}{ds} = \frac{d\mathbf{T}/dt}{|d\mathbf{T}/dt|}. \tag{15}$$

As before, $d\mathbf{T}/ds$ and its unit-vector direction $\mathbf{N}$ are orthogonal to $\mathbf{T}$ because $\mathbf{T}$ has constant length, in this case 1.

EXAMPLE 5 Find $\mathbf{N}$ for the helix in Example 4.

Solution Using values from Example 4, we have

$$\mathbf{T} = \frac{\mathbf{v}}{|\mathbf{v}|} = \frac{-(a \sin t)\mathbf{i} + (a \cos t)\mathbf{j} + b\mathbf{k}}{\sqrt{a^2 + b^2}}$$

$$\frac{d\mathbf{T}}{dt} = -\frac{a}{\sqrt{a^2 + b^2}}((\cos t)\mathbf{i} + (\sin t)\mathbf{j})$$

$$\left|\frac{d\mathbf{T}}{dt}\right| = \frac{a}{\sqrt{a^2 + b^2}}\sqrt{\cos^2 t + \sin^2 t} = \frac{a}{\sqrt{a^2 + b^2}}$$

$$\mathbf{N} = \frac{d\mathbf{T}/dt}{|d\mathbf{T}/dt|} = -(\cos t)\mathbf{i} - (\sin t)\mathbf{j}.$$ ■

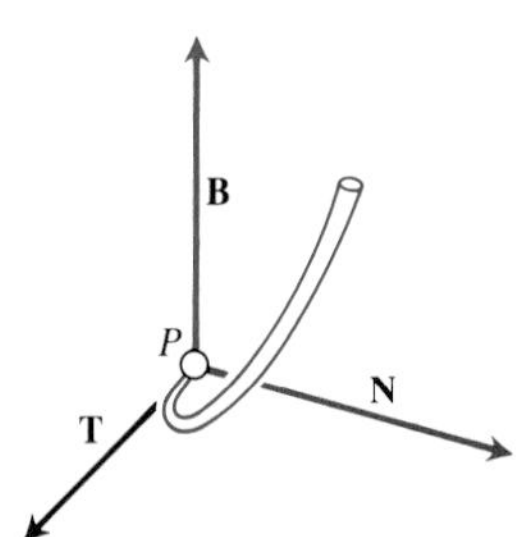

12.28 The vectors **T**, **N**, and **B** (in that order) make a right-handed frame of mutually orthogonal unit vectors in space. You can call it the *Frenet* ("fren *A*") *frame* (after Jean-Frédéric Frenet, 1816–1900) or you can call it the *TNB frame*.

Torsion and the Binormal Vector

The **binormal vector** of a curve in space is the vector $\mathbf{B} = \mathbf{T} \times \mathbf{N}$, a unit vector orthogonal to both $\mathbf{T}$ and $\mathbf{N}$ (Fig. 12.28). Together, $\mathbf{T}$, $\mathbf{N}$, and $\mathbf{B}$ define a moving right-handed vector frame that plays a significant role in calculating the flight paths of space vehicles.

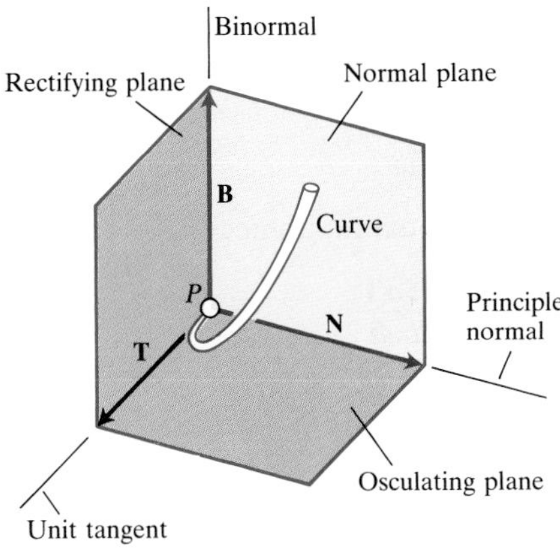

12.29 The three planes determined by **T**, **N**, and **B**.

The three planes determined by **T**, **N**, and **B** are shown in Fig. 12.29. The curvature $\kappa = |d\mathbf{T}/ds|$ can be thought of as the rate at which the normal plane turns as the point P moves along the curve. Similarly, the **torsion** $\tau = |d\mathbf{B}/ds|$ is the rate at which the osculating plane lifts as P moves along the curve. It is a measure of how much the curve twists.

The most widely used formula for torsion, derived in more advanced texts, is

$$\tau = \pm\frac{\begin{vmatrix} \dot{x} & \dot{y} & \dot{z} \\ \ddot{x} & \ddot{y} & \ddot{z} \\ \dddot{x} & \dddot{y} & \dddot{z} \end{vmatrix}}{|\mathbf{v}\times\mathbf{a}|^2} \quad (\text{if } \mathbf{v}\times\mathbf{a}\neq\mathbf{0}). \tag{16}$$

This formula calculates the torsion directly from the derivatives of the component functions $x = f(t)$, $y = g(t)$, $z = h(t)$, that make up $\mathbf{r}$. The determinant's first row comes from $\mathbf{v}$, the second row comes from $\mathbf{a}$, and the third row comes from $\dot{\mathbf{a}}$. The sign $(\pm)$ is chosen to keep τ from being negative. Exercise 24 gives a formula you can use to calculate τ if one or more of the third derivatives fails to exist (but that will not be a problem in calculations in this book).

EXAMPLE 6 Find the torsion of the helix

$$\mathbf{r} = (\cos t)\mathbf{i} + (\sin t)\mathbf{j} + t\mathbf{k}.$$

Solution We evaluate Eq. (16). We find the entries in the determinant by differentiating $\mathbf{r}$:

$$\mathbf{v} = -(\sin t)\mathbf{i} + (\cos t)\mathbf{j} + \mathbf{k}$$

$$\mathbf{a} = -(\cos t)\mathbf{i} - (\sin t)\mathbf{j}$$

$$\dot{\mathbf{a}} = (\sin t)\mathbf{i} - (\cos t)\mathbf{j}$$

Then,

$$\tau = \frac{\begin{vmatrix} \dot{x} & \dot{y} & \dot{z} \\ \ddot{x} & \ddot{y} & \ddot{z} \\ \dddot{x} & \dddot{y} & \dddot{z} \end{vmatrix}}{|\mathbf{v}\times\mathbf{a}|^2} = \frac{\begin{vmatrix} -\sin t & \cos t & 1 \\ -\cos t & -\sin t & 0 \\ \sin t & -\cos t & 0 \end{vmatrix}}{\left\|\begin{matrix} \mathbf{i} & \mathbf{j} & \mathbf{k} \\ -\sin t & \cos t & 1 \\ -\cos t & -\sin t & 0 \end{matrix}\right\|^2}$$

$$= \frac{\cos^2 t + \sin^2 t}{|(\sin t)\mathbf{i} - (\cos t)\mathbf{j} + \mathbf{k}|^2} = \frac{1}{2}.$$

The Tangential and Normal Components of Acceleration

When a moving body is accelerated by gravity, brakes, a combination of rocket motors, or whatever, we usually want to know how much of the acceleration acts to move the body straight ahead in the direction of motion, that is, in the tangential direction $\mathbf{T}$. We can find out if we use the Chain Rule to rewrite $\mathbf{v}$ as

$$\mathbf{v}=\frac{d\mathbf{r}}{dt}=\frac{d\mathbf{r}}{ds}\frac{ds}{dt}=\mathbf{T}\frac{ds}{dt} \tag{17}$$

and differentiate both ends of this string of equalities to get

$$\begin{aligned}\mathbf{a}&=\frac{d\mathbf{v}}{dt}=\frac{d}{dt}\left(\mathbf{T}\frac{ds}{dt}\right)=\frac{d^2s}{dt^2}\mathbf{T}+\frac{ds}{dt}\frac{d\mathbf{T}}{dt}\\&=\frac{d^2s}{dt^2}\mathbf{T}+\frac{ds}{dt}\left(\frac{d\mathbf{T}}{ds}\frac{ds}{dt}\right)=\frac{d^2s}{dt^2}\mathbf{T}+\frac{ds}{dt}\left(\kappa\mathbf{N}\frac{ds}{dt}\right)\\&=\frac{d^2s}{dt^2}\mathbf{T}+\kappa\left(\frac{ds}{dt}\right)^2\mathbf{N}.\end{aligned} \tag{18}$$

This is a remarkable equation. There is no $\mathbf{B}$ in it. No matter how the path of the moving body we are watching may appear to twist and turn, the acceleration $\mathbf{a}$ always lies in the plane of $\mathbf{T}$ and $\mathbf{N}$ orthogonal to $\mathbf{B}$. The equation also tells us exactly how much of the acceleration takes place tangent to the motion (d^2s/dt^2) and how much takes place normal to the motion ($\kappa(ds/dt)^2$).

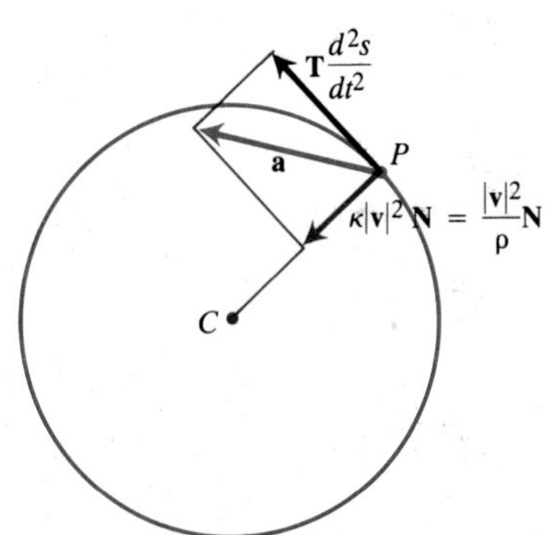

12.30 The tangential and normal components of the acceleration of a body speeding up as it moves counterclockwise around a circle of radius ρ.

> The **tangential** and **normal** scalar components of acceleration are
>
> $$a_{\mathrm{T}}=\frac{d^2s}{dt^2}=\frac{d}{dt}|\mathbf{v}| \quad \text{and} \quad a_{\mathrm{N}}=\kappa\left(\frac{ds}{dt}\right)^2=\kappa|\mathbf{v}|^2. \tag{19}$$
>
> That is,
>
> $$\mathbf{a}=\frac{d^2s}{dt^2}\mathbf{T}+\kappa\left(\frac{ds}{dt}\right)^2\mathbf{N}. \tag{20}$$

Notice that the normal scalar component of the acceleration is the curvature times the square of the speed. This explains why you have to hold on when your car makes a sharp (large κ), high-speed (large $|\mathbf{v}|$) turn.

If a body moves in a circle at a constant speed, d^2s/dt^2 is zero and all the acceleration points along $\mathbf{N}$ toward the circle's center. If the body is speeding up or slowing down, $\mathbf{a}$ has a nonzero tangential component (Fig. 12.30).

To calculate a_{N} we usually use the formula

$$a_{\mathrm{N}}=\sqrt{|\mathbf{a}|^2-a_{\mathrm{T}}^2}, \tag{21}$$

which comes from solving the equation $|\mathbf{a}|^2=\mathbf{a}\cdot\mathbf{a}=a_{\mathrm{T}}^2+a_{\mathrm{N}}^2$ for a_{N}. With this formula we can find a_{N} without having to calculate κ first.

EXAMPLE 7 Without finding $\mathbf{T}$ and $\mathbf{N}$, write the acceleration of the motion

$$\mathbf{r}=(\cos t+t\sin t)\mathbf{i}+(\sin t-t\cos t)\mathbf{j}, \quad t>0$$

in the form $\mathbf{a}=a_{\mathrm{T}}\mathbf{T}+a_{\mathrm{N}}\mathbf{N}$. (The path of the motion is the involute of the circle in Fig. 12.31.)

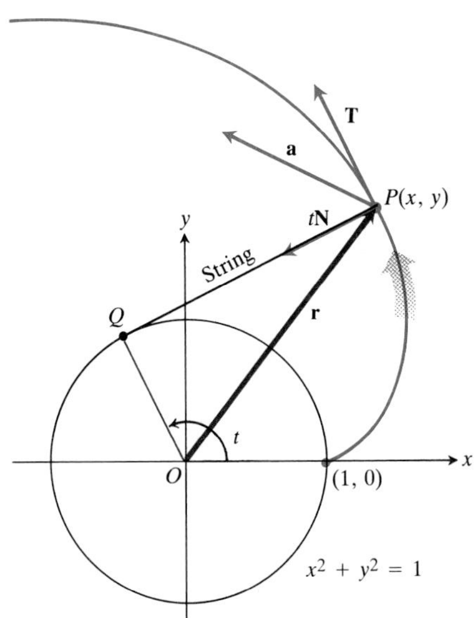

12.31 The tangential and normal components of the motion

$\mathbf{r} = (\cos t + t\sin t)\mathbf{i} + (\sin t - t\cos t)\mathbf{j}$

shown here are **T** and $t\mathbf{N}$ (Example 7).

Solution We use the first of Eqs. (19) to find a_{T}:

$$\mathbf{v} = (t\cos t)\mathbf{i} + (t\sin t)\mathbf{j} \qquad \begin{pmatrix}\text{Value from Section 12.3,}\\ \text{Example 5}\end{pmatrix}$$

$$|\mathbf{v}| = \sqrt{t^2\cos^2 t + t^2\sin^2 t} = \sqrt{t^2} = |t| = t \quad (t > 0)$$

$$a_{\mathrm{T}} = \frac{d}{dt}|\mathbf{v}| = \frac{d}{dt}(t) = 1 \qquad \text{(Eq. 19)}$$

Knowing a_{T}, we use Eq. (21) to find a_{N}:

$$\mathbf{a} = (\cos t - t\sin t)\mathbf{i} + (\sin t + t\cos t)\mathbf{j}$$

$$|\mathbf{a}|^2 = t^2 + 1 \quad \text{(after some algebra)}$$

$$a_{\mathrm{N}} = \sqrt{|\mathbf{a}|^2 - a_{\mathrm{T}}^2} = \sqrt{(t^2+1) - (1)} = \sqrt{t^2} = t.$$

We then use Eq. (20) to find **a**:

$$\mathbf{a} = a_{\mathrm{T}}\mathbf{T} + a_{\mathrm{N}}\mathbf{N} = (1)\mathbf{T} + (t)\mathbf{N} = \mathbf{T} + t\mathbf{N}.$$

See Fig. 12.31. ■

Motion Formulas ($|\mathbf{v}| \neq 0$)

Unit tangent vector	$\mathbf{T} = \dfrac{\mathbf{v}}{\lvert\mathbf{v}\rvert}$
Principal unit normal vector	$\mathbf{N} = \dfrac{d\mathbf{T}/dt}{\lvert d\mathbf{T}/dt\rvert}$
Binormal vector	$\mathbf{B} = \mathbf{T} \times \mathbf{N}$
Curvature	$\kappa = \left\lvert\dfrac{d\mathbf{T}}{ds}\right\rvert = \dfrac{\lvert\mathbf{v}\times\mathbf{a}\rvert}{\lvert\mathbf{v}\rvert^3}$
Torsion	$\tau = -\left\lvert\dfrac{d\mathbf{B}}{ds}\right\rvert = \pm\dfrac{\begin{vmatrix}\dot x & \dot y & \dot z\\ \ddot x & \ddot y & \ddot z\\ \dddot x & \dddot y & \dddot z\end{vmatrix}}{\lvert\mathbf{v}\times\mathbf{a}\rvert^2}$
Tangential and normal scalar components of acceleration	$\mathbf{a} = a_{\mathrm{T}}\mathbf{T} + a_{\mathrm{N}}\mathbf{N}$ $a_{\mathrm{T}} = \dfrac{d}{dt}\lvert\mathbf{v}\rvert$ $a_{\mathrm{N}} = \sqrt{\lvert\mathbf{a}\rvert^2 - a_{\mathrm{T}}^2}$

Exercises 12.4

Find **T**, **N**, and κ for the plane curves in Exercises 1–4.

1. $\mathbf{r} = t\mathbf{i} + (\ln\cos t)\mathbf{j}, \quad -\pi/2 < t < \pi/2$

2. $\mathbf{r} = (\ln\sec t)\mathbf{i} + t\mathbf{j}, \quad -\pi/2 < t < \pi/2$

3. $\mathbf{r} = (2t + 3)\mathbf{i} + (5 - t^2)\mathbf{j}$

4. $\mathbf{r} = (\cos t + t\sin t)\mathbf{i} + (\sin t - t\cos t)\mathbf{j}, \quad t > 0$

Find **T**, **N**, **B**, κ, and τ for the space curves in Exercises 5–8.

5. $\mathbf{r} = (3\sin t)\mathbf{i} + (3\cos t)\mathbf{j} + 4t\mathbf{k}$

6. $\mathbf{r} = (\cos t + t\sin t)\mathbf{i} + (\sin t - t\cos t)\mathbf{j} + 3\mathbf{k}$

7. $\mathbf{r} = (e^t\cos t)\mathbf{i} + (e^t\sin t)\mathbf{j} + 2\mathbf{k}$

8. $\mathbf{r} = (6\sin 2t)\mathbf{i} + (6\cos 2t)\mathbf{j} + 5t\mathbf{k}$

In Exercises 9–12, write **a** in the form $\mathbf{a} = a_T\mathbf{T} + a_N\mathbf{N}$ without finding **T** and **N**.

9. $\mathbf{r} = (2t + 3)\mathbf{i} + (t^2 - 1)\mathbf{j}$

10. $\mathbf{r} = \ln(t^2 + 1)\mathbf{i} + (t - 2\tan^{-1}t)\mathbf{j}$

11. $\mathbf{r} = (a\cos t)\mathbf{i} + (a\sin t)\mathbf{j} + bt\mathbf{k}$

12. $\mathbf{r} = (1 + 3t)\mathbf{i} + (t - 2)\mathbf{j} - 3t\mathbf{k}$

In Exercises 13–16, write **a** in the form $\mathbf{a} = a_T\mathbf{T} + a_N\mathbf{N}$ at the given value of t without finding **T** and **N**. You can save yourself some work by evaluating the vectors **v** and **a** at the given value of t *before* finding their lengths.

13. $\mathbf{r} = (t + 1)\mathbf{i} + 2t\mathbf{j} + t^2\mathbf{k}, \quad t = 1$

14. $\mathbf{r} = (t\cos t)\mathbf{i} + (t\sin t)\mathbf{j} + t^2\mathbf{k}, \quad t = 0$

15. $\mathbf{r} = t^2\mathbf{i} + (t + (1/3)t^3)\mathbf{j} + (t - (1/3)t^3)\mathbf{k}, \quad t = 0$

16. $\mathbf{r} = (e^t\cos t)\mathbf{i} + (e^t\sin t)\mathbf{j} + \sqrt{2}e^t\mathbf{k}, \quad t = 0$

In Exercises 17 and 18, find **r**, **T**, **N**, and **B** at the given value of t. Then find equations for the osculating, normal, and rectifying planes at that value of t.

17. $\mathbf{r} = (\cos t)\mathbf{i} + (\sin t)\mathbf{j} - \mathbf{k}, \quad t = \pi/4$

18. $\mathbf{r} = (\cos t)\mathbf{i} + (\sin t)\mathbf{j} + t\mathbf{k}, \quad t = 0$

19. Show that if a particle's speed is constant its acceleration is always normal to its path.

20. *A formula for the curvature of the graph of a function in the xy-plane.*

a) The graph of a function $y = f(x)$ in the xy-plane automatically has the parametrization $x = x$, $y = f(x)$ and the vector formula $\mathbf{r} = x\mathbf{i} + f(x)\mathbf{j}$. Use this formula to show that if f is a twice-differentiable function of x then

$$\kappa = \frac{|f''|}{(1 + (f')^2)^{3/2}}.$$

b) Use the curvature formula in (a) to find the curvature of the curve $y = \ln(\cos x)$, $-\pi/2 < x < \pi/2$. Compare your result with the result in Exercise 1.

21. Find an equation for the circle of curvature of the curve $\mathbf{r} = t\mathbf{i} + (\sin t)\mathbf{j}$ at the point $(\pi/2, 1)$. (The curve is the graph of $y = \sin x$ in the xy-plane.)

22. Find an equation for the circle of curvature of the curve $\mathbf{r} = (2\ln t)\mathbf{i} - (t + (1/t))\mathbf{j}$ in the xy-plane at the point $(0, -2)$ (the point where $t = 1$).

23. *How to derive the formula $\kappa = |\mathbf{v} \times \mathbf{a}|/|\mathbf{v}|^3$ for curves in space.* To derive the formula, carry out these two steps:

STEP 1: Use the equations $\mathbf{v} = \mathbf{T}(ds/dt)$ and $\mathbf{a} = (d^2s/dt^2)\mathbf{T} + \kappa(ds/dt)^2\mathbf{N}$ to find a formula for $|\mathbf{v} \times \mathbf{a}|$.

STEP 2: Solve the resulting equation for κ, assuming $|\mathbf{v} \times \mathbf{a}| \neq 0$.

24. *A formula that calculates torsion directly from **B** and **v**.* If we start with the definition $\tau = |d\mathbf{B}/ds|$ and apply the Chain Rule to rewrite $d\mathbf{B}/ds$ as

$$\frac{d\mathbf{B}}{ds} = \frac{d\mathbf{B}}{dt}\cdot\frac{dt}{ds} = \frac{d\mathbf{B}}{dt}\cdot\frac{1}{|\mathbf{v}|},$$

we arrive at the formula

$$\tau = \frac{|d\mathbf{B}/dt|}{|\mathbf{v}|}.$$

The advantage of this formula over the one in Eq. (16) is that it is easier to derive and state. The disadvantage is that it can take a lot of work to find $|d\mathbf{B}/dt|$. Use the new formula to find the torsion of the helix in Example 6.

25. *Maximizing the curvature of a helix.* In Example 4, we found the curvature of the helix $\mathbf{r} = (a\cos t)\mathbf{i} + (a\sin t)\mathbf{j} + bt\mathbf{k}$ $(a, b \geq 0)$ to be $\kappa = a/(a^2 + b^2)$. What is the largest value κ can have for a given value of b?

12.5 Planetary Motion and Satellites

This section introduces the standard vector equations for motion in polar and cylindrical coordinates and discusses Kepler's three laws of planetary motion.

Vector Equations for Motion in Polar and Cylindrical Coordinates

When a particle moves along a curve in the polar coordinate plane, we express its position, velocity, and acceleration in terms of the moving unit

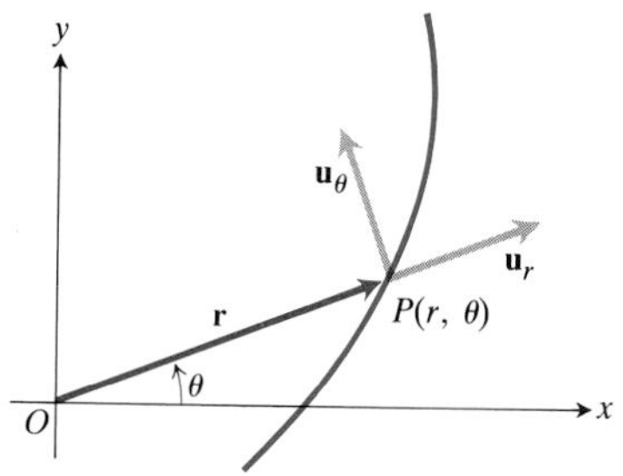

12.32 The length of the position vector **r** is the positive polar coordinate r of the point P. Thus, $\mathbf{u}_r$, which is $\mathbf{r}/|\mathbf{r}|$, is also $\mathbf{r}/r$. Equations (1) express $\mathbf{u}_r$ and $\mathbf{u}_\theta$ in terms of **i** and **j**.

vectors

$$\mathbf{u}_r = (\cos\theta)\mathbf{i} + (\sin\theta)\mathbf{j}, \qquad \mathbf{u}_\theta = -(\sin\theta)\mathbf{i} + (\cos\theta)\mathbf{j} \tag{1}$$

shown in Fig. 12.32. The vector $\mathbf{u}_r$ points along the position vector $\overrightarrow{OP}$, so $\mathbf{r} = r\mathbf{u}_r$. The vector $\mathbf{u}_\theta$, orthogonal to $\mathbf{u}_r$, points in the direction of increasing θ.

We find from (1) that

$$\begin{aligned} \frac{d\mathbf{u}_r}{d\theta} &= -(\sin\theta)\mathbf{i} + (\cos\theta)\mathbf{j} = \mathbf{u}_\theta, \\ \frac{d\mathbf{u}_\theta}{d\theta} &= -(\cos\theta)\mathbf{i} - (\sin\theta)\mathbf{j} = -\mathbf{u}_r. \end{aligned} \tag{2}$$

When we differentiate $\mathbf{u}_r$ and $\mathbf{u}_\theta$ with respect to t to find how they change with time, the Chain Rule gives

$$\frac{d\mathbf{u}_r}{dt} = \frac{d\mathbf{u}_r}{d\theta}\frac{d\theta}{dt} = \frac{d\theta}{dt}\mathbf{u}_\theta, \qquad \frac{d\mathbf{u}_\theta}{dt} = \frac{d\mathbf{u}_\theta}{d\theta}\frac{d\theta}{dt} = -\frac{d\theta}{dt}\mathbf{u}_r. \tag{3}$$

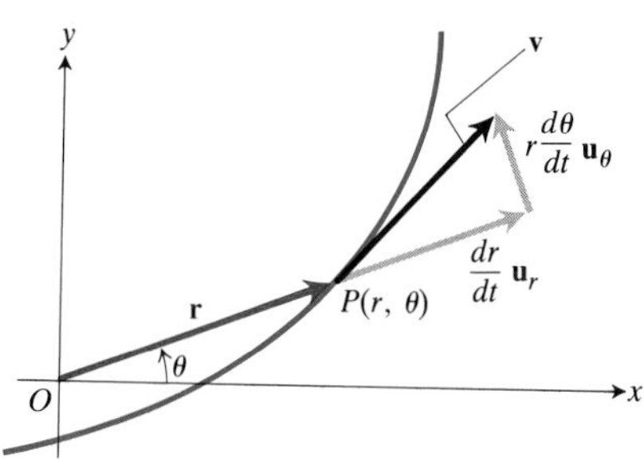

12.33 In polar coordinates, the velocity vector is

$\mathbf{v} = (dr/dt)\mathbf{u}_r + r(d\theta/dt)\mathbf{u}_\theta.$

Hence,

$$\mathbf{v} = \frac{d\mathbf{r}}{dt} = \frac{d}{dt}(r\mathbf{u}_r) = \frac{dr}{dt}\mathbf{u}_r + r\frac{d\mathbf{u}_r}{dt} = \frac{dr}{dt}\mathbf{u}_r + r\frac{d\theta}{dt}\mathbf{u}_\theta. \tag{4}$$

See Fig. 12.33.

The acceleration is

$$\mathbf{a} = \frac{d\mathbf{v}}{dt} = \left(\frac{d^2r}{dt^2}\mathbf{u}_r + \frac{dr}{dt}\frac{d\mathbf{u}_r}{dt}\right) + \left(r\frac{d^2\theta}{dt^2}\mathbf{u}_\theta + \frac{dr}{dt}\frac{d\theta}{dt}\mathbf{u}_\theta + r\frac{d\theta}{dt}\frac{d\mathbf{u}_\theta}{dt}\right). \tag{5}$$

When Eqs. (3) are used to evaluate $d\mathbf{u}_r/dt$ and $d\mathbf{u}_\theta/dt$ and the components are separated, the equation for acceleration becomes

$$\mathbf{a} = \left(\frac{d^2r}{dt^2} - r\left(\frac{d\theta}{dt}\right)^2\right)\mathbf{u}_r + \left(r\frac{d^2\theta}{dt^2} + 2\frac{dr}{dt}\frac{d\theta}{dt}\right)\mathbf{u}_\theta. \tag{6}$$

To extend the equations of motion to space, we add $z\mathbf{k}$ to the right-hand side of the equation $\mathbf{r} = r\mathbf{u}_r$. Then, in cylindrical coordinates,

$$\mathbf{r} = r\mathbf{u}_r + z\mathbf{k}, \tag{7}$$

$$\mathbf{v} = \frac{dr}{dt}\mathbf{u}_r + r\frac{d\theta}{dt}\mathbf{u}_\theta + \frac{dz}{dt}\mathbf{k}, \tag{8}$$

$$\mathbf{a} = \left(\frac{d^2r}{dt^2} - r\left(\frac{d\theta}{dt}\right)^2\right)\mathbf{u}_r + \left(r\frac{d^2\theta}{dt^2} + 2\frac{dr}{dt}\frac{d\theta}{dt}\right)\mathbf{u}_\theta + \frac{d^2z}{dt^2}\mathbf{k}. \tag{9}$$

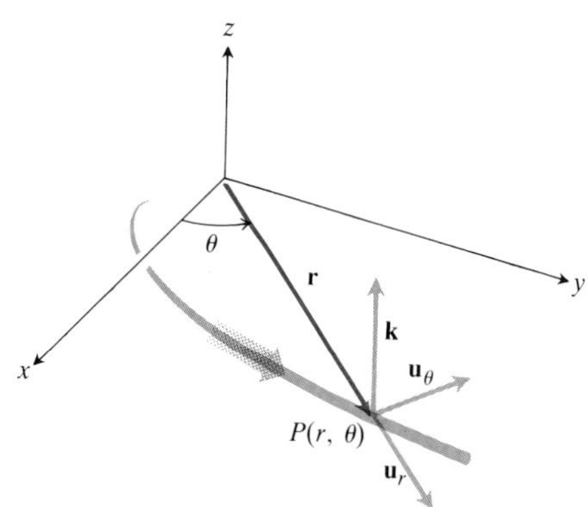

12.34 Motion in space with cylindrical coordinates.

The vectors $\mathbf{u}_r$, $\mathbf{u}_\theta$, and **k** make a right-handed frame (Fig. 12.34) in which

$$\mathbf{u}_r \times \mathbf{u}_\theta = \mathbf{k}, \qquad \mathbf{k} \times \mathbf{u}_r = \mathbf{u}_\theta, \qquad \mathbf{u}_\theta \times \mathbf{k} = \mathbf{u}_r. \tag{10}$$

Coordinates for Planetary Motion; Initial Conditions

Newton's law of gravitation says that if **r** is the radius vector from the center of a sun of mass M to the center of a planet of mass m, then the force **F** of

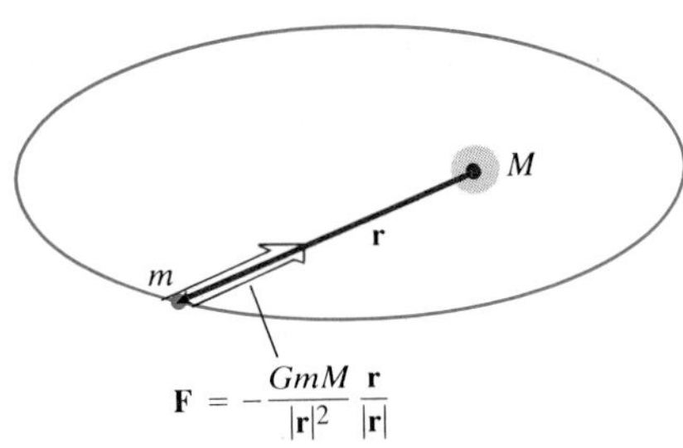

12.35 The force of gravity is directed along the line joining the centers of mass.

the gravitational attraction between the planet and sun is given by the equation

$$\mathbf{F} = -\frac{GmM}{|\mathbf{r}|^2}\frac{\mathbf{r}}{|\mathbf{r}|} \tag{11}$$

(Fig. 12.35). The constant G is called the (universal) **gravitational constant.** If we measure the masses m and M in kilograms, time in seconds, and distance in meters, then G is about $6.6720 \times 10^{-11}\mathrm{Nm^2kg^{-2}}$.

Combining Eq. (11) with Newton's second law of motion, $\mathbf{F} = m\, d^2\mathbf{r}/dt^2$, gives

$$m\frac{d^2\mathbf{r}}{dt^2} = -\frac{GmM}{|\mathbf{r}|^2}\frac{\mathbf{r}}{|\mathbf{r}|},$$

$$\frac{d^2\mathbf{r}}{dt^2} = -\frac{GM}{|\mathbf{r}|^3}\mathbf{r}. \tag{12}$$

This is typical of a **central force,** one that points toward a fixed center at all times—in this case the center of the sun.

Equation (12) says that $d^2\mathbf{r}/dt^2$ is a scalar multiple of $\mathbf{r}$ and hence tells us that

$$\mathbf{r} \times \frac{d^2\mathbf{r}}{dt^2} = \mathbf{0}. \tag{13}$$

The left-hand side of this equation is the derivative of $\mathbf{r} \times d\mathbf{r}/dt$:

$$\begin{aligned} \frac{d}{dt}\left(\mathbf{r} \times \frac{d\mathbf{r}}{dt}\right) &= \frac{d\mathbf{r}}{dt} \times \frac{d\mathbf{r}}{dt} + \mathbf{r} \times \frac{d^2\mathbf{r}}{dt^2} && \text{(Product Rule)} \\ &= \mathbf{0} + \mathbf{r} \times \frac{d^2\mathbf{r}}{dt^2} && (\mathbf{A} \times \mathbf{A} = \mathbf{0} \text{ for any vector}) \\ &= \mathbf{r} \times \frac{d^2\mathbf{r}}{dt^2}. \end{aligned} \tag{14}$$

Thus, Eq. (13) is equivalent to

$$\frac{d}{dt}\left(\mathbf{r} \times \frac{d\mathbf{r}}{dt}\right) = \mathbf{0},$$

which, in turn, says that

$$\mathbf{r} \times \frac{d\mathbf{r}}{dt} = \mathbf{C} \tag{15}$$

for some constant vector $\mathbf{C}$.

Equation (15) tells us that $\mathbf{r}$ and $d\mathbf{r}/dt$ always lie in a plane perpendicular to $\mathbf{C}$. Hence, the planet moves in a fixed plane through the center of its sun.

We now adjust the coordinate system for space, if necessary, to place the origin at the sun's center of mass and make the plane of the planet's motion the polar-coordinate plane. This makes $\mathbf{r}$ the planet's polar-coordinate position vector. We also position the z-axis in a way that makes $\mathbf{k}$ the direction of $\mathbf{C}$ and makes the planet's motion counterclockwise when viewed from the positive z-axis. This makes θ increase with t, so that $d\theta/dt$ is positive. Finally, we rotate the polar-coordinate plane about the z-axis, if

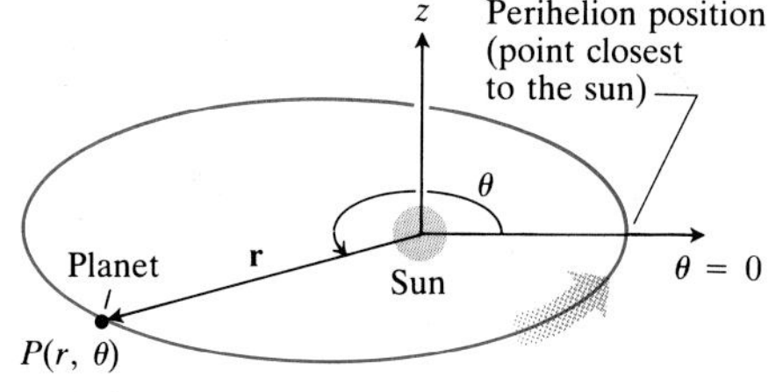

12.36 The coordinate system for planetary motion. The motion is counterclockwise when viewed from above, as it is here, and $d\theta/dt$ is positive.

necessary, to make the initial ray coincide with the direction **r** has when the planet is closest to the sun. This runs the initial ray through the **perihelion** position of the planet (Fig. 12.36).

If we now measure time so that $t = 0$ when the planet is at perihelion, we have the following initial conditions for the planet's motion:

1. $r = r_0$, the minimum radius, when $t = 0$
2. $\dfrac{dr}{dt} = 0$ when $t = 0$ (because r has a minimum value then)
3. $\theta = 0$ when $t = 0$
4. $|\mathbf{v}| = v_0$ when $t = 0$

Since

$$\begin{aligned} v_0 &= |\mathbf{v}|_{t=0} && \text{(Standard notation)} \\ &= \left|\frac{dr}{dt}\mathbf{u}_r + r\frac{d\theta}{dt}\mathbf{u}_\theta\right|_{t=0} && \text{(Eq. 4)} \\ &= \left|r\frac{d\theta}{dt}\mathbf{u}_\theta\right|_{t=0} && \left(\frac{dr}{dt} = 0 \text{ when } t = 0\right) \\ &= \left|r\frac{d\theta}{dt}\right|\left|\mathbf{u}_\theta\right|_{t=0} \\ &= \left|r\frac{d\theta}{dt}\right|_{t=0} && (|\mathbf{u}_\theta| = 1) \\ &= \left(r\frac{d\theta}{dt}\right)_{t=0}, && (r \text{ and } d\theta/dt \text{ are both positive}) \end{aligned}$$

we also know that

5. $r\dfrac{d\theta}{dt} = v_0$ when $t = 0$.

Statement of Kepler's First Law (The Conic-Section Law)

Kepler's first law says that a planet's path is a conic section with the sun at one focus. The eccentricity of the conic is

$$e = \frac{r_0 v_0^2}{GM} - 1 \tag{16}$$

and the polar equation is

$$r = \frac{(1 + e)r_0}{1 + e\cos\theta}. \tag{17}$$

The derivation is involved and we shall not give it here.

The German astronomer, mathematician, and physicist Johannes Kepler (1571–1630) was the first, and until Descartes the only, scientist to demand physical (as opposed to theological) explanations of celestial phenomena. His three laws of motion, the results of a lifetime of work, changed the course of astronomy forever and played a crucial role in the development of Newton's physics.

Kepler's Second Law (The Equal-Area Law)

Kepler's second law says that the radius vector from a sun to a planet (the position vector **r** in our model) sweeps out equal areas in equal times

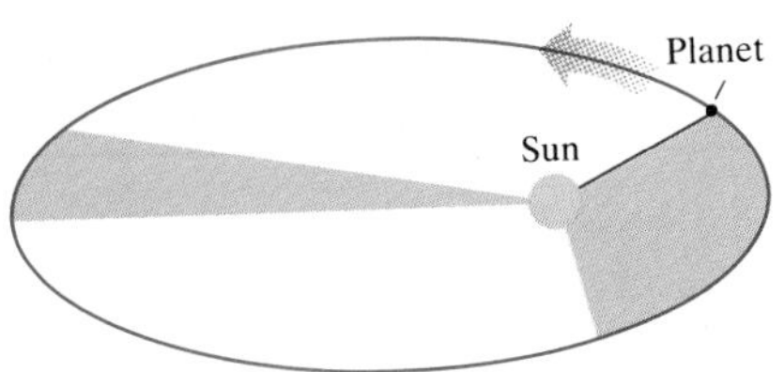

12.37 The line joining a planet to its sun sweeps over equal areas in equal times.

(Fig. 12.37). To derive this law, we use Eq. (4) to evaluate the cross product in Eq. (15):

$$
\begin{aligned}
\mathbf{C} &= \mathbf{r} \times \frac{d\mathbf{r}}{dt} && \text{(Eq. 15)} \\
&= \mathbf{r} \times \mathbf{v} && (\mathbf{v} = d\mathbf{r}/dt) \\
&= r\mathbf{u}_r \times \left(\frac{dr}{dt}\mathbf{u}_r + r\frac{d\theta}{dt}\mathbf{u}_\theta\right) && \text{(Eq. 4)} \\
&= r\frac{dr}{dt}\underbrace{\mathbf{u}_r \times \mathbf{u}_r}_{\mathbf{0}} + r\left(r\frac{d\theta}{dt}\right)\underbrace{\mathbf{u}_r \times \mathbf{u}_\theta}_{\mathbf{k}} \\
&= r\left(r\frac{d\theta}{dt}\right)\mathbf{k}.
\end{aligned}
\tag{18}
$$

Setting t equal to zero then shows that

$$\mathbf{C} = \left[r\left(r\frac{d\theta}{dt}\right)\right]_{t=0}\mathbf{k} = r_0 v_0 \mathbf{k}. \tag{19}$$

Substituting this value for $\mathbf{C}$ in Eq. (18) gives

$$r_0 v_0 \mathbf{k} = r^2 \frac{d\theta}{dt}\mathbf{k} \quad \text{or} \quad r^2\frac{d\theta}{dt} = r_0 v_0. \tag{20}$$

Here is where the area comes in. From Section 10.8, the area differential in polar coordinates is

$$dA = \frac{1}{2}r^2\,d\theta.$$

Accordingly, dA/dt has the constant value

$$\frac{dA}{dt} = \frac{1}{2}r^2\frac{d\theta}{dt} = \frac{1}{2}r_0 v_0, \tag{21}$$

which is Kepler's second law.

For Earth, r_0 is about 150,000,000 km, v_0 is about 30 km/sec, and dA/dt is about 2,250,000,000 km^2/sec. Every time your heart beats, Earth advances 30 km along its orbit and the line joining Earth to the sun sweeps out 2,250,000,000 square kilometers.

Statement of Kepler's Third Law (The Time–Distance Law)

The time T it takes a planet to go around its sun once is the planet's **orbital period.** The semimajor axis a of the planet's orbit is the planet's **mean distance** from the sun. Kepler's third law says that T and a are related by the equation

$$\frac{T^2}{a^3} = \frac{4\pi^2}{GM}. \tag{22}$$

Since the right-hand side of this equation is constant within a given solar system, the ratio of T^2 to a^3 is the same for every planet in the system. The derivation of Eq. (22) can be found, along with the derivation of Kepler's first law, in more advanced texts.

Kepler's third law is the starting point for working out the size of our solar system. It allows the semimajor axis of each planetary orbit to be expressed in astronomical units, Earth's semimajor axis being one unit. The distance between any two planets at any time can then be predicted in astronomical units and all that remains is to find one of these distances in kilometers. This can be done by bouncing radar waves off Venus, for example. The astronomical unit is now known, after a series of such measurements, to be 149,597,870 km.

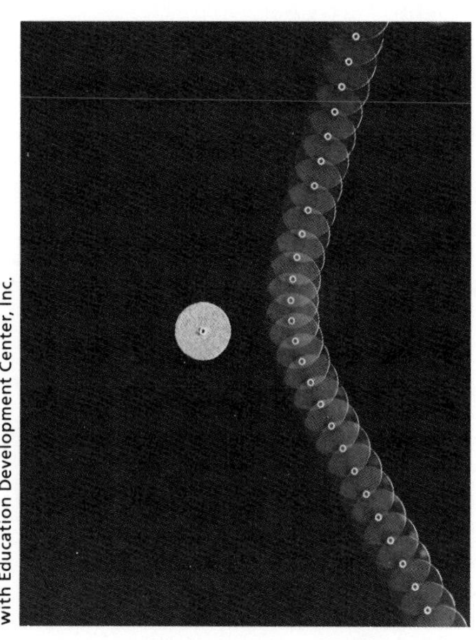

PSSC Physics, Second Edition, 1965; D.C. Heath & Company with Education Development Center, Inc.

12.38 This multiflash photograph shows a body being deflected by an inverse square law force. It moves along a hyperbola.

Astronomical Data

Although Kepler discovered his laws empirically and stated them only for the six planets known at the time, the modern derivations of Kepler's laws show that they apply to any body driven by a force that obeys an inverse square law. They apply to Halley's comet and the asteroid Icarus. They apply to the moon's orbit about Earth, and they applied to the orbit of the spacecraft Apollo 8 about the moon. They also applied to the air puck shown in Fig. 12.38 being deflected by an inverse square law force—its path is a hyperbola. Even charged particles fired at the nuclei of atoms scatter along hyperbolic paths.

Tables 12.1–12.3 give additional data for planetary orbits and for the orbits of seven of Earth's artificial satellites (Fig. 12.39). Vanguard 1 was nicknamed "Grapefruit." You will see why when you look at the

TABLE 12.1
Values of *a*, *e*, and *T* for the major planets

Planet	Semimajor axis $a^{\dagger}$	Eccentricity e	Period T
Mercury	57.95	0.2056	87.967 days
Venus	108.11	0.0068	224.701 days
Earth	149.57	0.0167	365.256 days
Mars	227.84	0.0934	1.8808 years
Jupiter	778.14	0.0484	11.8613 years
Saturn	1427.0	0.0543	29.4568 years
Uranus	2870.3	0.0460	84.0081 years
Neptune	4499.9	0.0082	164.784 years
Pluto	5909	0.2481	248.35 years

†Millions of kilometers

TABLE 12.2
Data on Earth's satellites

Name	Launch date	Time or expected time aloft	Weight at launch (kg)	Period (min)	Perigee height (km)	Apogee height (km)	Semimajor axis a (km)	Eccentricity
Sputnik 1	Oct. 1957	57.6 days	83.6	96.2	215	939	6,955	0.052
Vanguard 1	March 1958	300 years	1.47	138.5	649	4,340	8,872	0.208
Syncom 3	Aug. 1964	$> 10^6$ years	39	1436.2	35,718	35,903	42,189	0.002
Skylab 4	Nov. 1973	84.06 days	13,980	93.11	422	437	6,808	0.001
Tiros 11	Oct. 1978	500 years	734	102.12	850	866	7,236	0.001
GOES 4	Sept. 1980	$> 10^6$ years	627	1436.2	35,776	35,800	42,166	0.0003
Intelsat 5	Dec. 1980	$> 10^6$ years	1,928	1417.67	35,143	35,707	41,803	0.007

TABLE 12.3
Numerical data

Gravitational constant: $G = 6.6720 \times 10^{-11}\ \mathrm{Nm^2kg^{-2}}$
(When you use this value of G in a calculation, remember to express distance in meters, mass in kilograms, and time in seconds.)
Sun's mass: 1.99×10^{30} kg
Earth's mass: 5.975×10^{24} kg
Equatorial radius of Earth: 6378.533 km
Polar radius of Earth: 6356.912 km
Earth's rotational period: 1436.1 min
Earth's orbital period: 1 year = 365.256 days

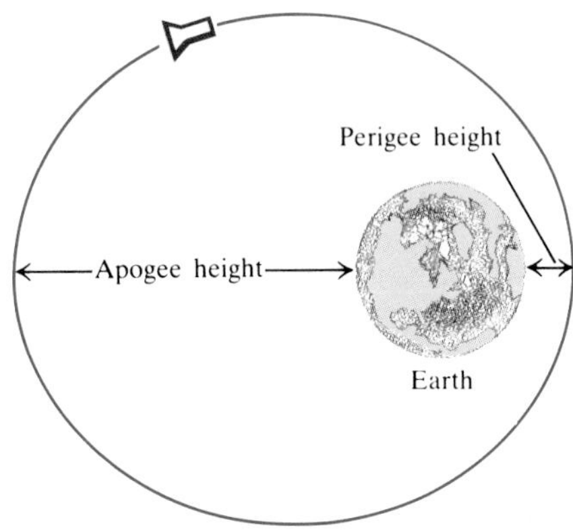

12.39 The orbit of an Earth satellite: $2a$ = Diameter of Earth + perigee height + apogee height.

12.40 Vanguard I, still in orbit, is so small it weighs only three and a quarter pounds.

photograph in Fig. 12.40. The data Vanguard sent back revealed differences between the levels of Earth's oceans and provided the first determination of the precise locations of some of the more isolated Pacific islands. The data also verified that the gravitation of the sun and moon would affect the orbits of Earth's satellites and that solar light could exert enough pressure to deform an orbit.

Syncom 3 is one of a series of U.S. Department of Defense telecommunications satellites. Tiros 11 (for "television infrared observation satellite") is one of a series of weather satellites. GOES 4 (for "geostationary operational environmental satellite") is one of a series of satellites designed to gather information about Earth's atmosphere. Its orbital period, 1436.2 minutes, is nearly the same as Earth's rotational period of 1436.1 minutes, and its orbit is nearly circular ($e = 0.0003$). Intelsat 5 is a heavy-capacity commercial telecommunications satellite.

Circular Orbits

For circular orbits, e is zero, $r = r_0$ is a constant, and Eq. (16) gives

$$r = r_0 = \frac{GM}{v_0^2} \tag{23}$$

which reduces to

$$r = \frac{GM}{v^2} \tag{24}$$

because v is constant as well (Exercise 12). Kepler's third law becomes

$$\frac{T^2}{r^3} = \frac{4\pi^2}{GM} \tag{25}$$

because $a = r$.

Exercises 12.5

Reminder: When a calculation involves the gravitational constant G, express distance in meters, mass in kilograms, and time in seconds.

CALCULATOR Exercises

1. Since the orbit of Skylab 4 had a semimajor axis of $a = 6808$ km, Kepler's third law with M equal to the earth's mass should give the period. Calculate it. Compare your result with the value in Table 12.2.
2. Earth's distance from the sun at perihelion is approximately 149,577,000 km, and the eccentricity of the earth's orbit about the sun is 0.0167. Compute the velocity v_0 of the earth in its orbit at perihelion. (Use Eq. 16).
3. In July 1965, the USSR launched Proton I, weighing 12,200 kg (at launch), with a perigee height of 183 km, an apogee height of 589 km, and a period of 92.25 minutes. Using the relevant data for the mass of the earth and the gravitational constant G, compute the semimajor axis a of the orbit from Eq. (22). Compare your answer with the number you get by adding the perigee and apogee heights to the diameter of the earth.
4. a) The Viking 1 orbiter, which surveyed Mars from August 1975 to June 1976, had a period of 1639 min. Use this and the fact that the mass of Mars is 6.418×10^{23} kg to find the semimajor axis of the Viking 1 orbit.

b) The Viking 1 orbiter was 1499 km from the surface of Mars at its closest point and 35,800 km from the surface at its farthest point. Use this information together with the value you obtained in part (a) to estimate the average diameter of Mars.

5. The Viking 2 orbiter, which surveyed Mars from September 1975 to August 1976, moved in an ellipse whose semimajor axis was 22,030 km. What was the orbital period? (Express your answer in minutes.)

6. If a satellite is to hold a geostationary orbit, what must the semimajor axis of its orbit be? Compare your result with the semimajor axes of the satellites in Table 12.2.

7. The mass of Mars is 6.418×10^{23} kg. If a satellite revolving about Mars is to hold a stationary orbit (have the same period as the period of Mars's rotation, which is 1477.4 min), what must the semimajor axis of its orbit be?

8. The period of the moon's rotation about the earth is 2.36055×10^6 sec. About how far away is the moon?

9. A satellite moves around the earth in a circular orbit. Express the satellite's speed as a function of the orbit's radius.

10. If T is measured in seconds and a in meters, what is the value of T^2/a^3 for planets in our solar system? For satellites orbiting the earth? For satellites orbiting the moon? (The moon's mass is 7.354×10^{22} kg.)

Non-calculator Exercises

11. For what values of v_0 in Eq. (16) is the orbit in Eq. (17) a circle? An ellipse? A hyperbola?

12. Show that a planet in a circular orbit moves with a constant speed. (*Hint:* This is a consequence of one of Kepler's laws.)

13. *Practice with* $\mathbf{u}_r$ *and* $\mathbf{u}_\theta$. A ball is placed in a frictionless tube that is pivoted at one end and rotates with constant angular velocity $d\theta/dt = 2$. In polar coordinates, the position of the ball at time t is $\theta = 2t$, $r = \cosh\theta$. Show that the $\mathbf{u}_r$ component of acceleration is always zero.

Review Questions

1. What are the rules for differentiating and integrating vector functions? Give examples.

2. How do you find the velocity, speed, direction of motion, and acceleration of a body moving along a smooth curve in space? Give examples.

3. What is special about the derivatives of vectors of constant length? Give an example.

4. What are the vector and parametric equations for ideal projectile motion? How do you find a projectile's maximum height, flight time, and range? Give examples.

5. How do you define and calculate the length of a segment of a space curve? Give an example. What mathematical assumptions are involved in the definition?

6. How do you measure distance along a curve in space from a preselected base point? Give an example of a directed-distance function.

7. What is a curve's unit tangent vector? Give an example.

8. How do you define the curvature of a curve in the plane? Give an example.

9. What is a plane curve's principal normal vector? Which way does it point? Give an example.

10. How do you define $\mathbf{N}$ and κ for curves in space? How are these quantities related? Give examples.

11. What is a curve's binormal vector? Give an example. What is the relation of this vector to the curve's torsion? Give an example.

12. What formulas are available for writing a moving body's acceleration as a sum of its tangential and normal components? Give an example. Why might one want to write the acceleration this way? What if the body moves at a constant speed? At a constant speed around a circle?

13. State Kepler's laws. To what do they apply?

Practice Exercises

Graph the curves in Exercises 1–2, and sketch their velocity and acceleration vectors at the given values of t.

1. $\mathbf{r} = (4\cos t)\mathbf{i} + (\sqrt{2}\sin t)\mathbf{j}, \quad t = 0$ and $\pi/4$

2. $\mathbf{r} = (\sqrt{3}\sec t)\mathbf{i} + (\sqrt{3}\tan t)\mathbf{j}, \quad t = 0$ and $\pi/6$

Evaluate the integrals in Exercises 3–4.

3. $\int_0^1 [(3 + 6t)\mathbf{i} + (4 + 8t)\mathbf{j} + (6\pi\cos \pi t)\mathbf{k}]\,dt$

4. $\int_e^{e^2} \left[\frac{2\ln t}{t}\mathbf{i} + \frac{1}{t\ln t}\mathbf{j} + \frac{1}{t}\mathbf{k}\right] dt$

Solve the initial value problems in Exercises. 5–8.

5. $\frac{d\mathbf{r}}{dt} = -(\sin t)\mathbf{i} + (\cos t)\mathbf{j} + \mathbf{k}, \quad \mathbf{r} = \mathbf{j}$ when $t = 0$

6. $\frac{d\mathbf{r}}{dt} = \frac{1}{t^2+1}\mathbf{i} - \frac{1}{\sqrt{1-t^2}}\mathbf{j} + \frac{1}{t\sqrt{t^2-1}}\mathbf{k}, \quad \mathbf{r} = \mathbf{j} + \mathbf{k}$ when $t = 0$

7. $\frac{d^2\mathbf{r}}{dt^2} = 2\mathbf{j}, \quad \frac{d\mathbf{r}}{dt} = \mathbf{k}$ and $\mathbf{r} = \mathbf{i}$ when $t = 0$

8. $\frac{d^2\mathbf{r}}{dt^2} = -2\mathbf{i} - 4\mathbf{j}, \quad \frac{d\mathbf{r}}{dt} = 4\mathbf{i}$ and $\mathbf{r} = 3\mathbf{i} + 3\mathbf{j}$ when $t = 1$

Find the lengths of the curves in Exercises 9–10.

9. $\mathbf{r} = (2\cos t)\mathbf{i} + (2\sin t)\mathbf{j} + t^2\mathbf{k}, \quad 0 \le t \le \pi/4$

10. $\mathbf{r} = (3\cos t)\mathbf{i} + (3\sin t)\mathbf{j} + 2t^{3/2}\mathbf{k}, \quad 0 \le t \le 3$

11. Find $\mathbf{T}$, $\mathbf{N}$, $\mathbf{B}$, κ, and τ at $t = 0$ if $\mathbf{r} = \frac{4}{9}(1+t)^{3/2}\mathbf{i} + \frac{4}{9}(1-t)^{3/2}\mathbf{j} + \frac{1}{3}t\mathbf{k}$.

12. Find $\mathbf{T}$, $\mathbf{N}$, $\mathbf{B}$, κ, and τ as functions of t if $\mathbf{r} = (\sin t)\mathbf{i} + (\sqrt{2}\cos t)\mathbf{j} + (\sin t)\mathbf{k}$.

In Exercises 13–14, write $\mathbf{a}$ in the form $\mathbf{a} = a_{\mathrm{T}}\mathbf{T} + a_{\mathrm{N}}\mathbf{N}$ at $t = 0$ without finding $\mathbf{T}$ and $\mathbf{N}$.

13. $\mathbf{r} = (2 + 3t + 3t^2)\mathbf{i} + (4t + 4t^2)\mathbf{j} - (6\cos t)\mathbf{k}$

14. $\mathbf{r} = (2 + t)\mathbf{i} + (t + 2t^2)\mathbf{j} + (1 + t^2)\mathbf{k}$

15. The position of a particle in the plane at time t is

$$\mathbf{r} = \frac{1}{\sqrt{1+t^2}}\mathbf{i} + \frac{t}{\sqrt{1+t^2}}\mathbf{j}.$$

Find the particle's highest speed.

16. Suppose $\mathbf{r} = (e^t\cos t)\mathbf{i} + (e^t\sin t)\mathbf{j}$. Show that the angle between $\mathbf{r}$ and $\mathbf{a}$ never changes. What *is* the angle?

17. At what times in the interval $0 \le t \le \pi$ are the velocity and acceleration vectors of the motion $\mathbf{r} = \mathbf{i} + (5\cos t)\mathbf{j} + (3\sin t)\mathbf{k}$ orthogonal?

18. The position of a particle moving in space at time $t \ge 0$ is

$$\mathbf{r} = 2\mathbf{i} + \left(4\sin\frac{t}{2}\right)\mathbf{j} + \left(3 - \frac{t}{\pi}\right)\mathbf{k}.$$

Find the first time $\mathbf{r}$ is perpendicular to the vector $\mathbf{i} - \mathbf{j}$.

19. *Shot-put.* A shot leaves the thrower's hand 8 ft above the ground at a 45° angle at 44 ft/sec. Where is it 3 sec later?

20. *Javelin.* A javelin leaves the thrower's hand 7 ft above the ground at a 45° angle at 80 ft/sec. How high does it go?

21. CALCULATOR *The Dictator.* The Civil War mortar *Dictator* weighed so much (17,120 lb) that it had to be mounted on a railroad car. It had a 13-in. bore and fired a 200-lb shell with a 20-lb powder charge. It was made by Mr. Charles Knapp in his ironworks in Pittsburgh, Pennsylvania, and was used by the Union army in 1864 in the siege of Petersburg, Virginia. How far did it shoot? Here we have a difference of opinion. The ordnance manual claimed 4,325 yards, while field officers claimed 4,752 yards. Assuming a 45° firing angle, what muzzle speeds are involved here?

Plate 75, Gardner's Photographic Sketch Book of the Civil War, Alexander Gardiner; © 1959: Dover Publications, Inc.

22. **CALCULATOR** *The world's record for popping a champagne cork.* The world's record for popping a champagne cork is 109 feet, 6 inches, held by Captain Michael Hill of the British Royal Artillery (of course). Assuming Capt. Hill held the bottle neck at ground level at a 45° angle, and the cork behaved like an ideal projectile, how fast was the cork going as it left the bottle?

23. Find parametric equations for the line that is tangent to the curve $\mathbf{r} = e^t\mathbf{i} + (\sin t)\mathbf{j} + \ln(1 - t)\mathbf{k}$ at $t = 0$.

24. Find parametric equations for the line tangent to the helix $\mathbf{r} = (\sqrt{2}\cos t)\mathbf{i} + (\sqrt{2}\sin t)\mathbf{j} + t\mathbf{k}$ at the point where $t = \pi/4$.

25. At point P, the velocity and acceleration of a particle moving in the plane are $\mathbf{v} = 3\mathbf{i} + 4\mathbf{j}$ and $\mathbf{a} = 5\mathbf{i} + 15\mathbf{j}$. Find the curvature of the particle's path at P.

26. Find the normal plane of the curve $\mathbf{r} = t\mathbf{i} + t\mathbf{j} + (2/3)t^{3/2}\mathbf{k}$ at the point $(1, 1, 2/3)$.

27. *The torsion of a helix.* Find the torsion of the helix $\mathbf{r} = (a\cos t)\mathbf{i} + (a\sin t)\mathbf{j} + bt\mathbf{k}$ of Example 4, Section 12.4. For a given value of a, what is the largest value the torsion can have?

28. *A useful formula for the curvature of a plane curve.* Equation (3) in Section 12.4 says that the curvature of a plane curve $\mathbf{r} = f(t)\mathbf{i} + g(t)\mathbf{j}$ in which the coordinate functions $x = f(t)$, $y = g(t)$ are twice differentiable is

$$\kappa = \frac{|\dot{x}\ddot{y} - \dot{y}\ddot{x}|}{(\dot{x}^2 + \dot{y}^2)^{3/2}}.$$

Use this formula to find the curvature of curve $\mathbf{r} = \ln(\sec t)\mathbf{i} + t\mathbf{j}, \quad -\pi/2 < t < \pi/2$.

29. *The view from Skylab 4.* What percentage of Earth's surface area could the astronauts see when Skylab 4 was at its apogee height, 437 km above the surface? To find out, model the visible surface as the surface generated by revolving the circular arc GT in Fig. 12.41 about the y-axis. Then carry out these steps:

1. Use similar triangles to show that $y_0/6380 = 6380/(6380 + 437)$. Solve for y_0.

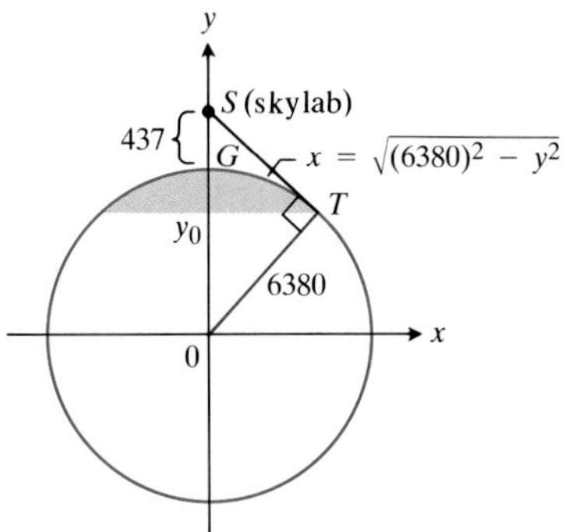

12.41 How much of the earth can you see from space at any one time? Exercise 29 shows how to find out.

2. Calculate the visible area as

$$VA = \int_{y_0}^{6380} 2\pi x\, ds.$$

3. Express the result as a percentage of Earth's surface area.

30. Use calculus to deduce from the orbit equation

$$r = \frac{(1 + e)r_0}{1 + e\cos\theta}$$

that a planet is closest to its sun when $\theta = 0$ and show that $r = r_0$ at that time.

31. **COMPUTER or PROGRAMMABLE CALCULATOR** *A Kepler equation.* The problem of locating a planet in its orbit at a given time and date eventually leads to solving "Kepler" equations of the form

$$f(x) = x - 1 - \frac{1}{2}\sin x = 0.$$

a) Show that this particular equation has a solution between $x = 0$ and $x = 2$.

b) With your machine in radian mode, use Newton's method to find the solution to as many places as you can.

13 Functions of Two or More Variables and Their Derivatives

The Concorde, flying at supersonic speed, creates a sonic boom heard by people on the ground. The width of the sonic boom carpet is a function of air temperatures and altitude.

OVERVIEW We now begin our study of functions that have more than one independent variable. Functions with two or more variables appear more often in science than functions of a single variable, and their calculus is even richer. Their derivatives are more varied and more interesting because of the different ways in which the variables can interact. Their integrals lead to a greater variety of applications. The studies of probability, statistics, fluid dynamics, and electricity, to mention only a few, all lead in natural ways to functions of more than one variable and the mathematical analysis associated with these functions is one of the finest achievements in science.

In the present chapter, we introduce (scalar) functions of two or more variables, define and calculate their derivatives, learn how to write the Chain Rule, and find out how to calculate the rates at which functions change as we move in different directions through their domains. We also learn how to estimate values of functions of two or more variables and how to calculate their extreme values in a given region.

The remaining chapters of the book build on what we do here. Chapter 14 deals with integrals of functions of two and three variables. These are the integrals with which we calculate the volumes and centers of mass of three-dimensional solids. In Chapter 15, we once more combine calculus with vectors, this time to derive the integral theorems that provide the mathematical foundation for the studies of fluid flow and electricity in physics and engineering. Finally, in Chapter 16, we examine various ways to solve equations of the form $dy/dx = f(x, y)$, equations in which the derivative of an unknown function y of x is given as a function of the two variables x and y.

13.1 Functions of Two or More Independent Variables

The values of many functions are determined by more than one independent variable. For example, the function $V = \pi r^2 h$ calculates the volume of a circular cylinder from its radius and height. The function $f(x, y) = x^2 + y^2$ calculates the height of the paraboloid $z = x^2 + y^2$ above the point $P(x, y)$ from the two coordinates of P. The function

$$w = \cos(1.7 \times 10^{-2}t - 0.2x)e^{-0.2x},$$

in Example 10, calculates the temperature x feet below ground on the tth day of the year as a fraction of the surface temperature on that day.

In this section, we define functions of more than one independent variable and discuss two ways to display their values graphically.

Functions and Variables

Real-valued functions of two or more real variables are defined much the way you would imagine from the single-variable case. The domains are sets of ordered pairs (triples, quadruples, whatever) of real numbers, and the ranges are sets of real numbers of the kind we have worked with all along.

DEFINITIONS

Suppose D is a set of n-tuples of real numbers $(x_1, x_2, \ldots, x_n)$. A **real-valued function** f on D is a rule that assigns a real number

$$w = f(x_1, x_2, \ldots, x_n)$$

to each element in D. The set D is the function's **domain.** The set of w-values taken on by f is the function's **range.** The symbol w is the **dependent variable** of f and f is said to be a function of the n **independent variables** x_1 to x_n. We also call the x's the function's **input variables** and call w the function's **output variable.**

If f is a function of two independent variables, we usually call the variables x and y and picture the domain of f as a region in the xy-plane. If f is a function of three independent variables, we call the variables x, y, and z and picture the domain of f as a region in space.

When functions of more than one variable arise in applications, we tend to use letters that remind us of what the variables stand for. To say that the volume of a right circular cylinder is a function of its radius and height, we might write $V = f(r, h)$. To be more specific, we might replace the notation $f(r, h)$ by the formula that calculates the value of V from the values of r and h, and write $V = \pi r^2 h$. In either case, r and h would be the independent variables and V the dependent variable of the function.

As usual, we evaluate functions defined by formulas by substituting the values of the independent variables in the formula and calculating the corresponding value of the dependent variable.

EXAMPLE 1 The value of the function $f(x, y, z) = \sqrt{x^2 + y^2 + z^2}$ at the point (3, 0, 4) is

$$f(3, 0, 4) = \sqrt{(3)^2 + (0)^2 + (4)^2} = \sqrt{25} = 5.$$

■

Domains

In defining functions of more than one variable, we follow the usual practice of excluding inputs that lead to complex numbers or division by zero. If $f(x, y) = \sqrt{y - x^2}$, we do not allow y to be less than x^2. If $f(x, y) = 1/xy$, we do not allow xy to be zero.

These are the only restrictions, however, and the domains of functions are otherwise assumed to be the largest sets for which the defining rules generate real numbers.

EXAMPLE 2 *Functions of two variables*

Function	Domain	Range
$w = \sqrt{y - x^2}$	$y \geq x^2$	$w \geq 0$
$w = \dfrac{1}{xy}$	$xy \neq 0$	$w \neq 0$
$w = \sin xy$	Entire plane	$-1 \leq w \leq 1$
$w = -\dfrac{1}{x^2 + y^2}$	$(x, y) \neq (0, 0)$	$-\infty < w < 0$

■

EXAMPLE 3 *Functions of three variables*

Function	Domain	Range
$w = \sqrt{x^2 + y^2 + z^2}$	Entire space	$w \geq 0$
$w = \dfrac{1}{x^2 + y^2 + z^2}$	$(x, y, z) \neq (0, 0, 0)$	$0 < w < \infty$
$w = xy \ln z$	Half-space $z > 0$	$-\infty < w < \infty$

■

R (x_0, y_0) R (x_0, y_0)

(a) Interior point (b) Boundary point

13.1 Interior points and boundary points of a plane region R. An interior point is necessarily a point of R. A boundary point of R need not belong to R.

The domains of functions defined on portions of the plane can have interior points and boundary points just the way the domains of functions defined on intervals of the real line can.

DEFINITIONS

A point (x_0, y_0) in a region (set) R in the xy-plane is an **interior point** of R if it is the center of a disk that lies entirely in R (Fig. 13.1). A point (x_0, y_0) is a **boundary point** of R if every disk centered at (x_0, y_0) contains points that lie outside of R as well as points that lie in R. (The boundary point itself need not belong to R.)

The interior points of a region, as a set, make up the **interior** of the region. The region's boundary points make up its **boundary.** A region is **open** if it consists entirely of interior points. A region is **closed** if it contains all of its boundary points.

EXAMPLE 4

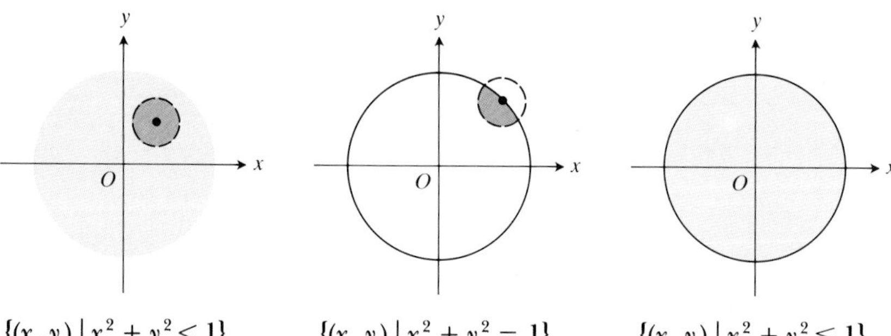

$\{(x, y) \mid x^2 + y^2 < 1\}$ Open unit disk. Every point an interior point.

$\{(x, y) \mid x^2 + y^2 = 1\}$ Boundary of unit disk. (The unit circle.)

$\{(x, y) \mid x^2 + y^2 \leq 1\}$ Closed unit disk. Contains all boundary points. ■

As with intervals of real numbers, some regions in the plane are neither open nor closed. If you start with the open disk in Example 4 and add to it some but not all of its boundary points, the resulting set is neither open nor closed. The boundary points that *are* there, keep the set from being open. The absence of the remaining boundary points keeps the set from being closed.

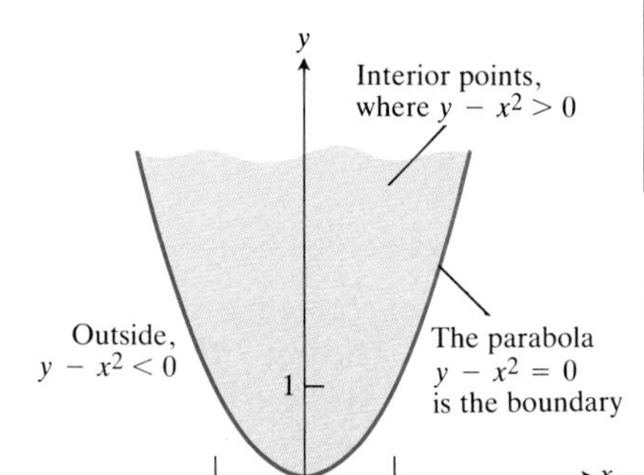

13.2 The domain of $f(x, y) = \sqrt{y - x^2}$ consists of the shaded region and its bounding parabola $y = x^2$.

DEFINITIONS

A region in the plane is **bounded** if all its points lie within some fixed radius of the origin; that is, if there is some limit to how far from the origin its points lie. A region is **unbounded** if it is not bounded.

EXAMPLE 5

Bounded sets in the plane:	Line segments, triangles, interiors of triangles, rectangles, disks
Unbounded sets in the plane:	Lines, coordinate axes, the graphs of functions defined on infinite intervals, quadrants, half-planes, the plane itself ■

EXAMPLE 6 The domain of the function $f(x, y) = \sqrt{y - x^2}$ is closed and unbounded (Fig. 13.2). The parabola $y = x^2$ is the boundary of the domain. The points inside the parabola make up the domain's interior. ■

The definitions of interior, boundary, open, closed, bounded, and unbounded for regions in space are similar to those for regions in the plane. To accommodate the extra dimension, we use solid spheres instead of disks.

DEFINITIONS

A point (x_0, y_0, z_0) in a region R in space is an **interior point** of R if it is the center of a solid sphere that lies entirely in R (Fig. 13.3). A point (x_0, y_0, z_0) is a **boundary point** of R if every sphere centered at (x_0, y_0, z_0) encloses points that lie outside of R as well as points that lie inside R. The **interior** of R is the set of interior points of R. The **boundary** of R is the set of boundary points of R.

A region R is **open** if it consists entirely of interior points. A region is **closed** if it contains its boundary.

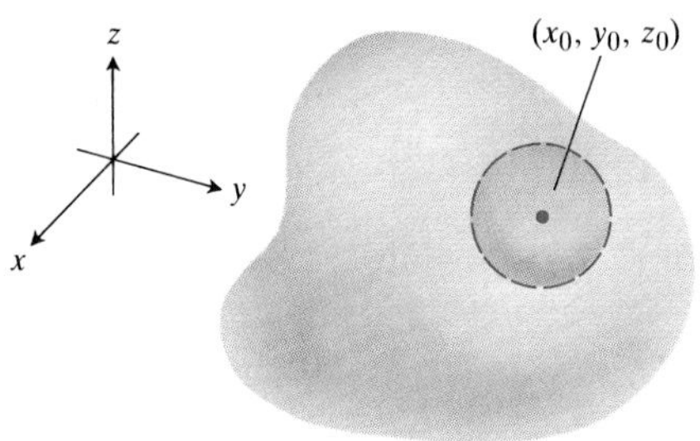

(a) Interior point

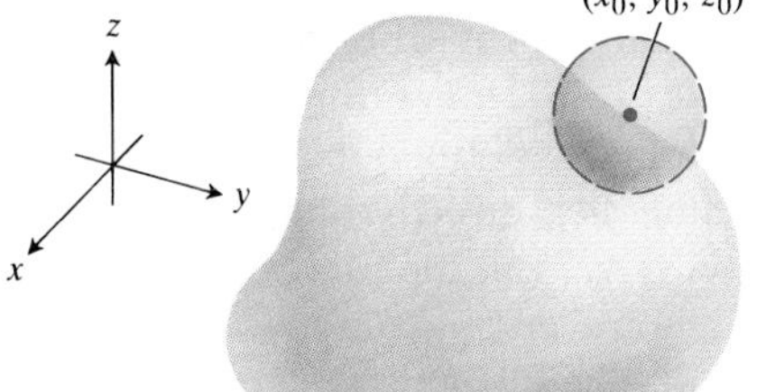

(b) Boundary point

13.3 Interior points and boundary points of a region in space.

EXAMPLE 7

Open sets in space:	Interior of a sphere; the open half-space $z > 0$; the first octant (bounding planes absent); space itself
Closed sets in space:	Lines; planes; the closed half-space $z \geq 0$; the first octant together with its bounding planes; space itself
Neither open nor closed:	A solid sphere with part of its boundary removed; solid cube with a missing face, edge, or corner point ■

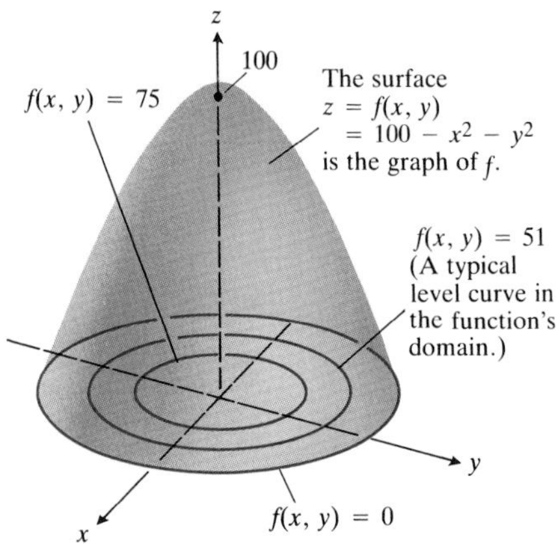

13.4 The graph and selected level curves of the function $f(x, y) = 100 - x^2 - y^2$.

Graphs and Level Curves of Functions of Two Variables

There are two standard ways to picture the values of a function $f(x, y)$. One is to draw and label curves in the domain on which f has a constant value. The other is to sketch the surface $z = f(x, y)$ in space.

DEFINITIONS

The set of points in the plane where a function $f(x, y)$ has a constant value $f(x, y) = c$ is called a **level curve** of f. The set of all points $(x, y, f(x, y))$ in space, for (x, y) in the domain of f, is called the **graph** of f. The graph of f is also called the **surface $z = f(x, y)$.**

EXAMPLE 8 Graph the function

$$f(x, y) = 100 - x^2 - y^2$$

in space and plot the level curves $f(x, y) = 0$, $f(x, y) = 51$, and $f(x, y) = 75$, in the domain of f in the plane.

Solution The domain of f is the entire xy-plane and the range of f is the set of real numbers less than or equal to 100. The graph is the paraboloid $z = 100 - x^2 - y^2$, a portion of which is shown in Fig. 13.4.

The level curve $f(x, y) = 0$ is the set of points in the xy-plane at which

$$f(x, y) = 100 - x^2 - y^2 = 0, \qquad \text{or} \qquad x^2 + y^2 = 100,$$

which is the circle of radius 10 centered at the origin (Fig. 13.4). Similarly, the level curves $f(x, y) = 51$ and $f(x, y) = 75$ are the circles

$$f(x, y) = 100 - x^2 - y^2 = 75, \qquad \text{or} \qquad x^2 + y^2 = 25,$$

$$f(x, y) = 100 - x^2 - y^2 = 51, \qquad \text{or} \qquad x^2 + y^2 = 49.$$

The level curve $f(x, y) = 100$ consists of the origin alone. (It is still called a level curve.) ■

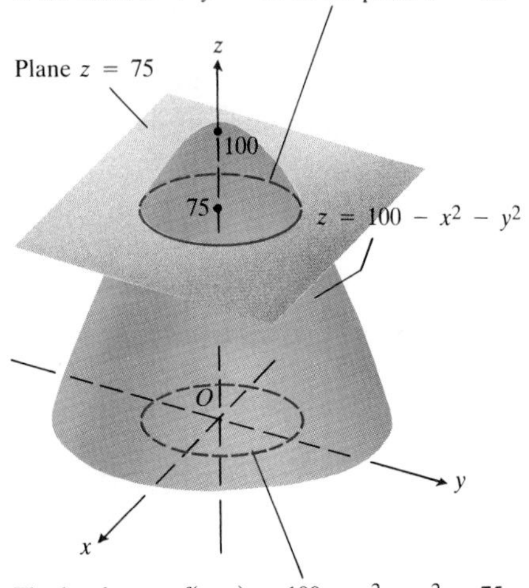

13.5 The graph of $f(x, y) = 100 - x^2 - y^2$ and its intersection with the plane $z = 75$.

Contour Lines

The curve in space in which the plane $z = c$ cuts a surface $z = f(x, y)$, is made up of the points that represent the function value $f(x, y) = c$. It is called the **contour line** $f(x, y) = c$ to distinguish it from the level curve $f(x, y) = c$ in the domain of f. Figure 13.5 shows the contour line $f(x, y) = 75$ on the surface $z = 100 - x^2 - y^2$ defined by the function $f(x, y) = 100 - x^2 - y^2$. The contour line lies directly above the circle $x^2 + y^2 = 25$, which is the level curve $f(x, y) = 75$ in the function's domain.

Not everyone makes this distinction, however, and you may wish to call both kinds of curves by a single name and rely on context to convey which one you have in mind. On most maps, for example, the curves that represent constant elevation (height above sea level) are called contours, not level curves (Fig. 13.6).

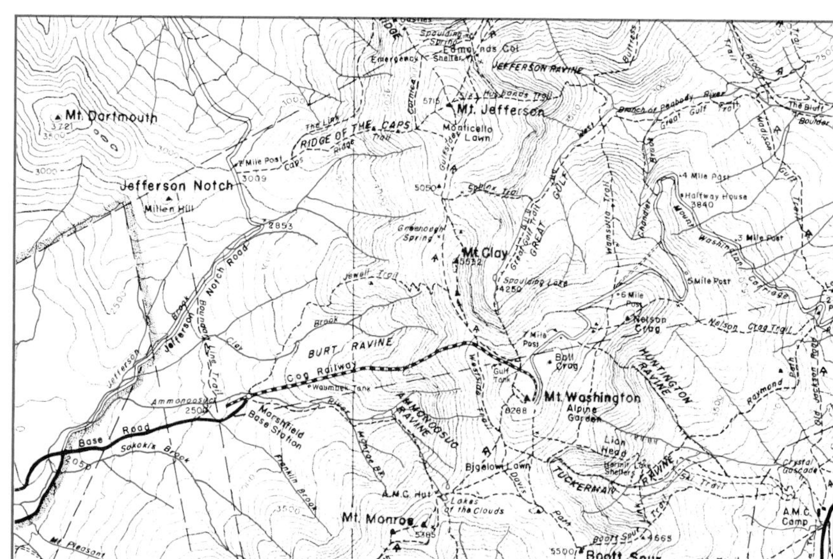

13.6 Contours on Mt. Washington in north-central New Hampshire. The streams, which follow paths of steepest descent, run perpendicular to the contours. So does the Cog Railway. (Based on a map by Louis F. Cutter, ©1987 Appalachian Mountain Club, Boston, Mass., by permission.)

Level Surfaces of Functions of Three Variables

In the plane, the points at which a function of two independent variables has a constant value $f(x, y) = c$ make a curve in the function's domain. In space, the points where a function of three independent variables has a constant value $f(x, y, z) = c$, make a surface in the function's domain.

DEFINITION

The set of points in space where a function of three independent variables has a constant value $f(x, y, z) = c$, is called a **level surface** of f.

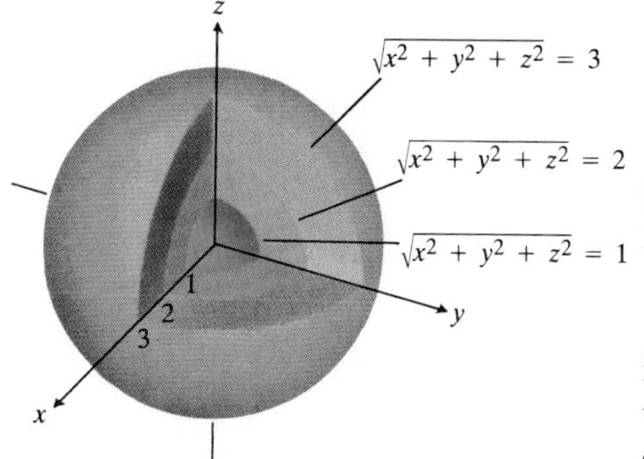

13.7 The level surfaces of the function $f(x, y, z) = \sqrt{x^2 + y^2 + z^2}$ are concentric spheres.

EXAMPLE 9 Describe the level surfaces of the function

$$f(x, y, z) = \sqrt{x^2 + y^2 + z^2}.$$

Solution The value of f is the distance from the origin to the point (x, y, z). Each level surface

$$\sqrt{x^2 + y^2 + z^2} = c, \qquad c > 0,$$

is a sphere of radius c centered at the origin. Figure 13.7 shows a cutaway view of three of these spheres. The level surface $\sqrt{x^2 + y^2 + z^2} = 0$ consists of the origin alone.

We are not graphing the function here. The graph of the function, made up of the points $(x, y, z, \sqrt{x^2 + y^2 + z^2})$, lies in a four-variable space. Instead, we are looking at surfaces in the function's domain.

Surfaces defined by functions of two variables

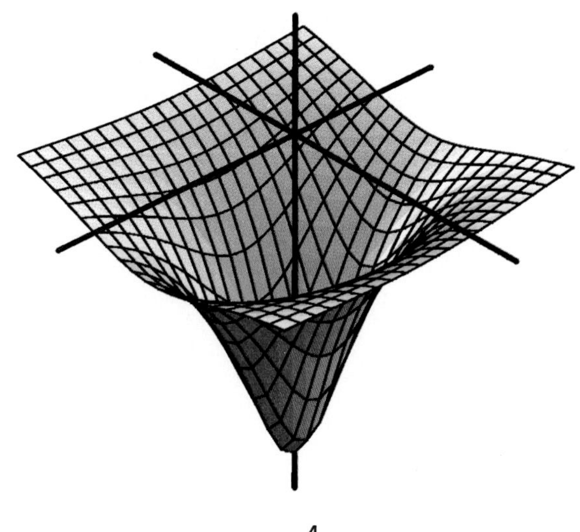

$$z = \frac{-4}{1 + x^2 + 2y^2}$$

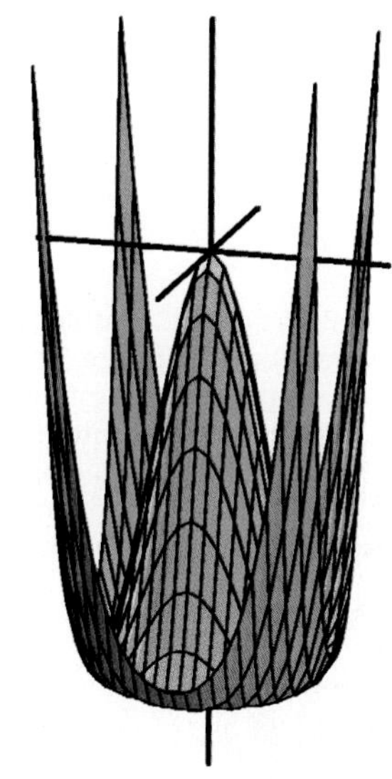

$$z = (x^2 + y^2)^{3/2} - 4(x^2 + y^2)$$

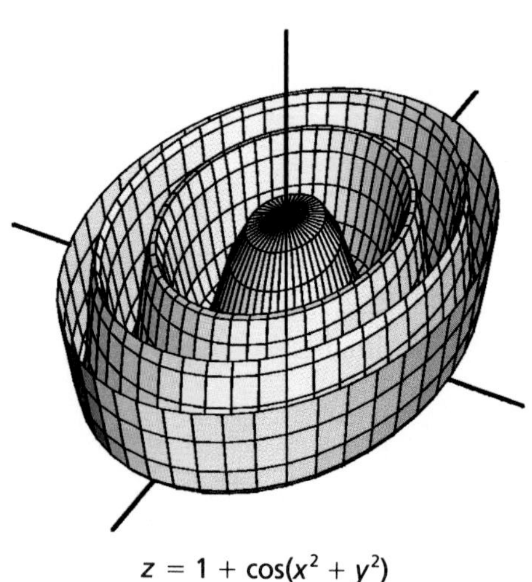

$$z = 1 + \cos(x^2 + y^2)$$

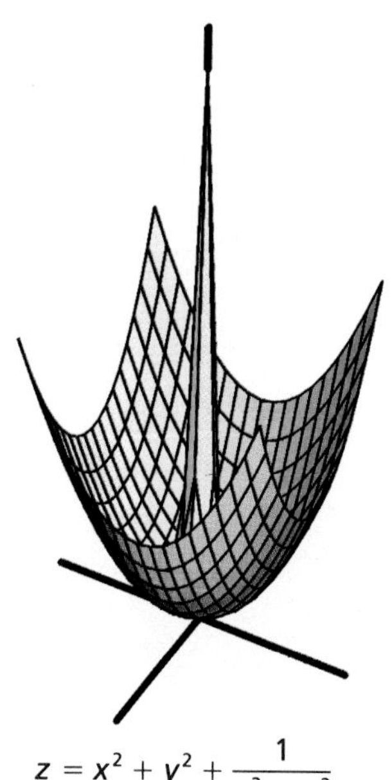

$$z = x^2 + y^2 + \frac{1}{x^2 + y^2}$$

(Generated by Mathematica)

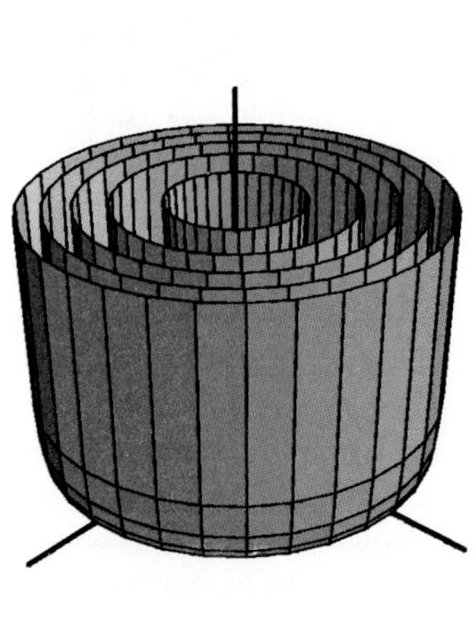

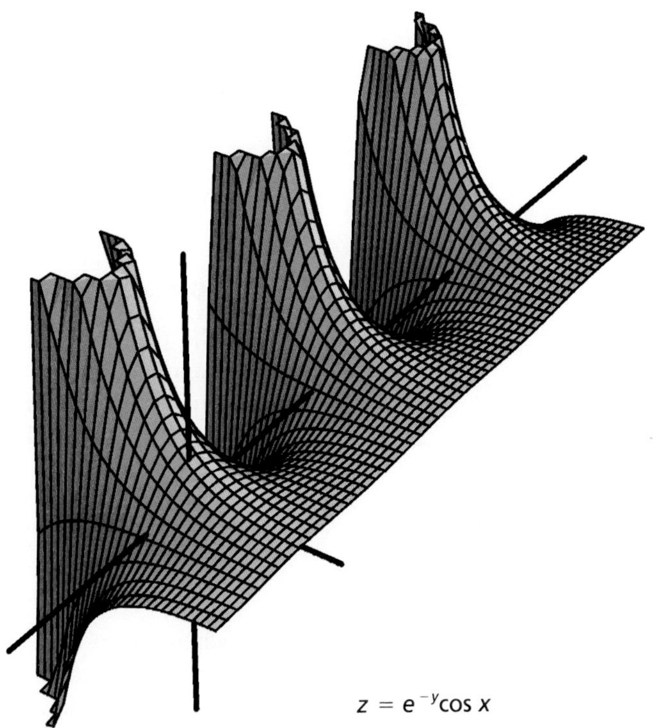

$z = e^{-y}\cos x$

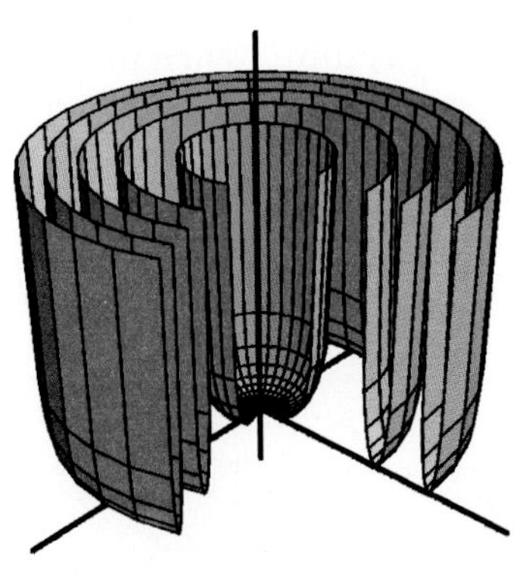

$z = \ln(\sec(x^2 + y^2))$

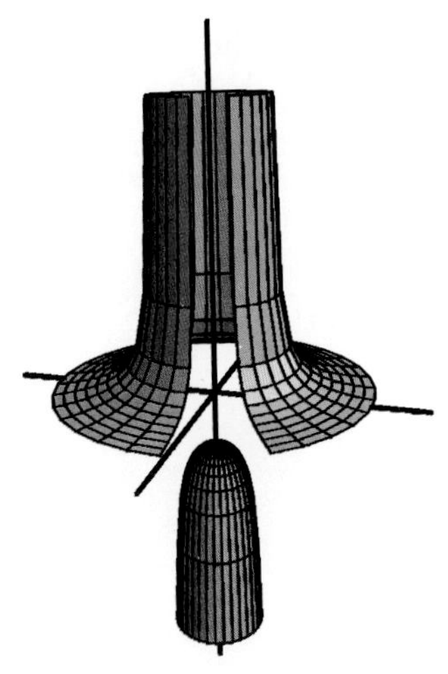

$$z = \frac{1}{x^2 + y^2 - 1}$$

The function's level surfaces show how the function's values change as we move around in its domain. If we remain on a sphere of radius c centered at the origin, the function maintains a constant value, namely c. If we move from one sphere to another, the function's value changes. It increases if we move away from the origin and decreases if we move toward the origin. The way the function's values change depends on the direction we take. The dependence of change on direction is important. We shall return to it in Section 13.5. ■

Computer Graphing

The three-dimensional graphing programs now available for computers make it possible to graph functions of two variables with only a few keystrokes. We can often get information about a function more quickly from one of these graphs than we can from a function's formula.

EXAMPLE 10 Figure 13.8 shows a computer-generated graph of the function

$$w = \cos(1.7 \times 10^{-2}t - 0.2x)e^{-0.2x},$$

where t is in days and x is in feet. The graph shows how the temperature beneath the earth's surface varies with time. The variation is given as a fraction of the variation at the surface. At a depth of 15 ft, the variation (change in vertical amplitude in the figure) is about 5 percent of the surface variation. At 30 ft, there is almost no variation in temperature during the year.

Another thing the graph shows that is not immediately apparent from the equation for f is that the temperature 15 ft below the surface is about half a year out of phase with the surface temperature. When the temperature is lowest on the surface (late January, say) it is at its highest 15 ft below. At 15 feet below, the seasons are reversed. ■

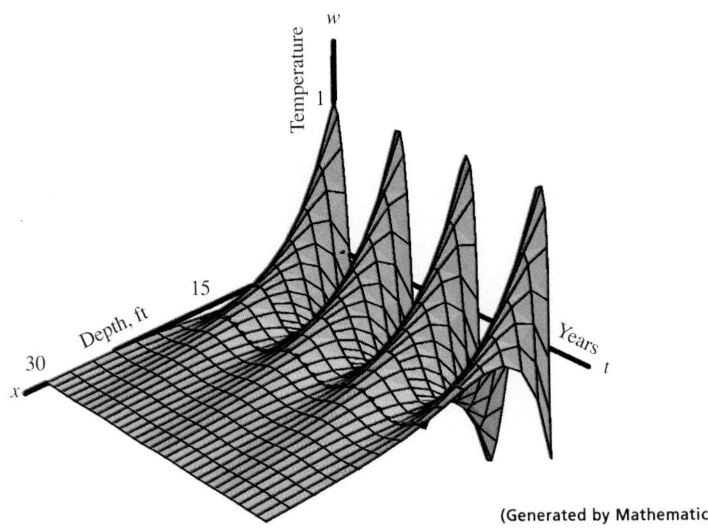

13.8 This computer-generated graph of

$$w = \cos(1.7 \times 10^{-2}t - 0.2x)e^{-0.2x}$$

shows the seasonal variation of the temperature below ground as a fraction of surface temperature. $\Delta x = 0.375$ ft, $\Delta t = 15.625$ days. At $x = 15$ ft the variation is only 5% of the variation at the surface. At $x = 30$ ft the variation is less than 0.25% of the surface variation. (Adapted from art provided by Norton Starr for G. C. Berresford's "Differential Equations and Root Cellars," *The UMAP Journal*, Volume 2, Number 3 (1981), pp. 53–75.)

Exercises 13.1

In Exercises 1–8, find the function's domain and range and describe its level curves.

1. $f(x, y) = y - x$ **2.** $f(x, y) = \sqrt{y - x}$

3. $f(x, y) = \ln(x^2 + y^2)$ **4.** $f(x, y) = \cos(x^2 + y^2)$

5. $f(x, y) = xy$ **6.** $f(x, y) = y/x^2$

7. $f(x, y) = 4x^2 + 9y^2$ **8.** $f(x, y) = x^2 - y^2$

Display the values of the functions in Exercises 9–16 in two ways: (a) by sketching the surface $z = f(x, y)$, and (b) by drawing an assortment of level curves in the function's domain. Label each level curve with its function value.

9. $f(x, y) = y^2$ **10.** $f(x, y) = 4 - y^2$

11. $f(x, y) = x^2 + y^2$ **12.** $f(x, y) = \sqrt{x^2 + y^2}$

13. $f(x, y) = -(x^2 + y^2)$ **14.** $f(x, y) = 4 - x^2 - y^2$

15. $f(x, y) = 4x^2 + y^2$ **16.** $f(x, y) = 4x^2 + y^2 + 1$

Exercises 17–22 show level curves for the functions graphed in (a)–(f). Match each set of curves with the appropriate function.

17.

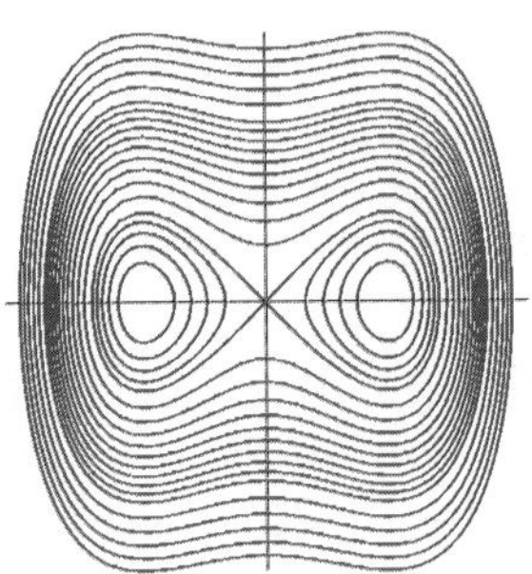

18.

19.

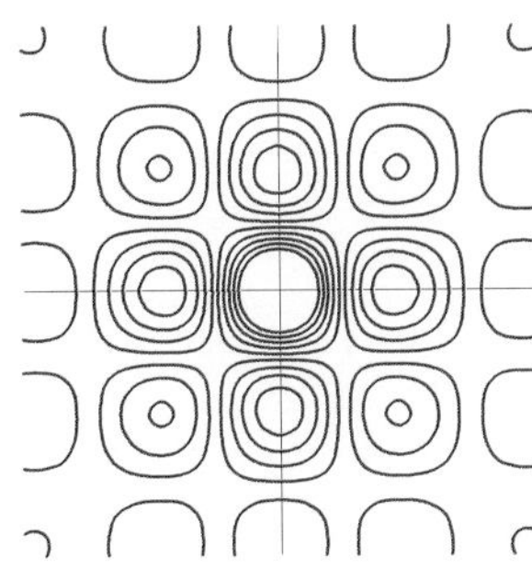

20.

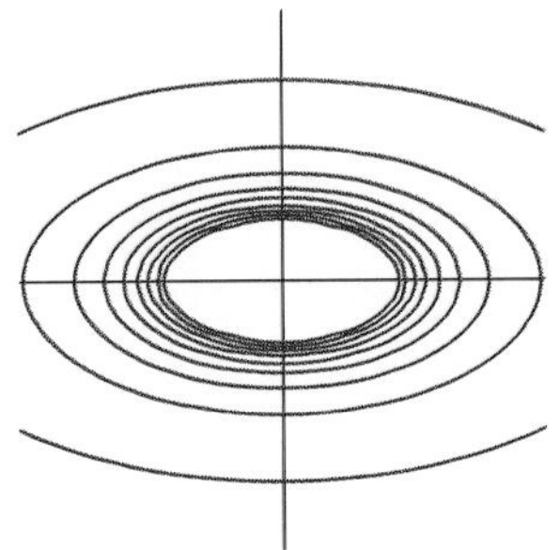

21.

22.

(a)

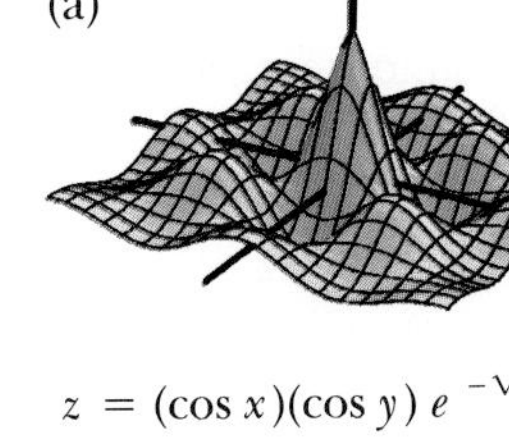

$z = (\cos x)(\cos y)\, e^{-\sqrt{x^2 + y^2}/4}$

(b)

$z = \dfrac{\left| |x| - |y| \right|}{2} - \dfrac{(|x| + |y|)}{2}$

(c)

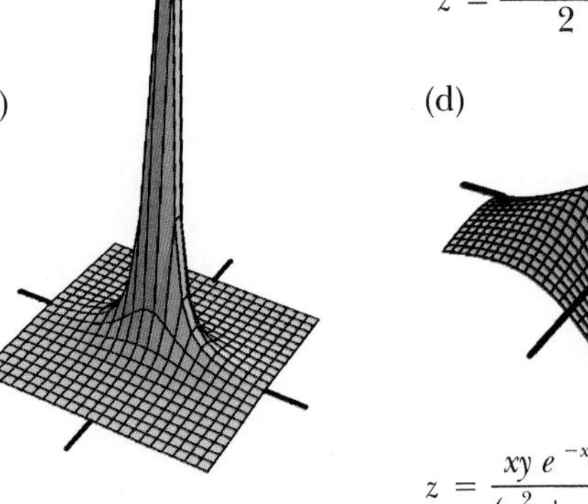

$z = \dfrac{1}{(x^2 + 4y^2)}$

(d)

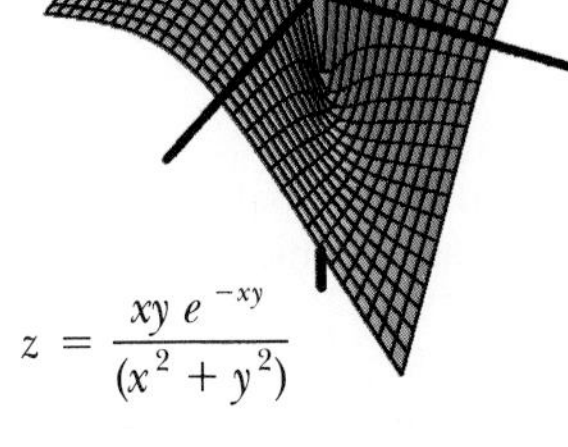

$z = \dfrac{xy\, e^{-xy}}{(x^2 + y^2)}$

(e)

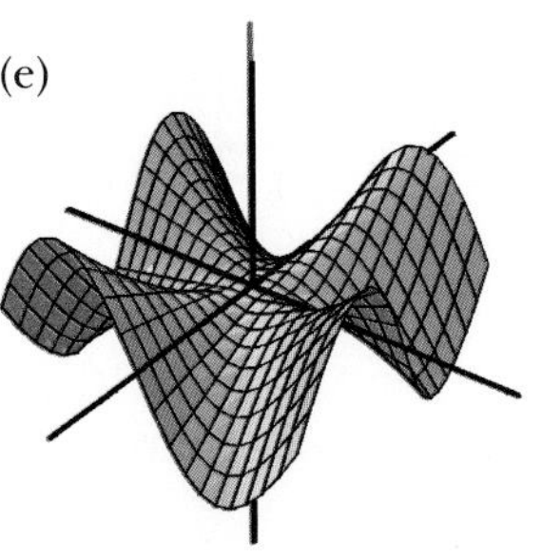

$z = \dfrac{xy\,(x^2 - y^2)}{(x^2 + y^2)}$

(Generated by Mathematica)

(f)

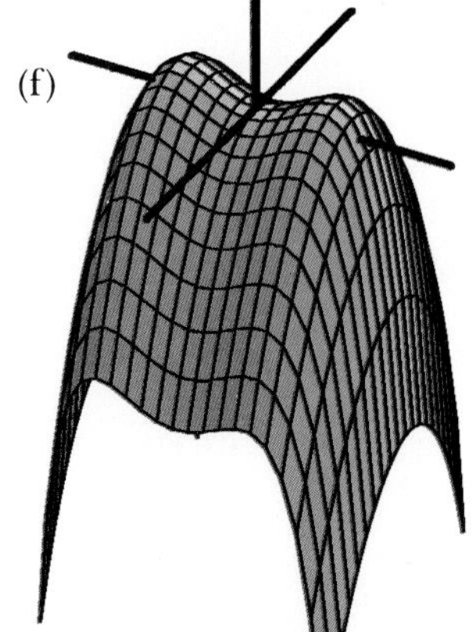

$z = x^2 - x^4 - y^2$

In Exercises 23–30, sketch a typical level surface for the function.

23. $f(x, y, z) = x^2 + y^2 + z^2$

24. $f(x, y, z) = \ln(x^2 + y^2 + z^2)$

25. $f(x, y, z) = x + z$

26. $f(x, y, z) = z$

27. $f(x, y, z) = x^2 + y^2$

28. $f(x, y, z) = y^2 + z^2$

29. $f(x, y, z) = z - x^2 - y^2$

30. $f(x, y, z) = (x^2/25) + (y^2/16) + (z^2/9)$

31. Find an equation for the level curve of $f(x, y) = 16 - x^2 - y^2$ that passes through the point $(2\sqrt{2}, \sqrt{2})$.

32. Find an equation for the level curve of the function $f(x, y) = \sqrt{x^2 - 1}$ that passes through the point $(1, 0)$.

33. *The Concorde's sonic booms.** The width w of the region in which people on the ground hear the Concorde's sonic boom directly, and not reflected from a layer in the atmosphere, is a function of

T = air temperature at ground level (in degrees Kelvin),

h = the Concorde's altitude (in km),

d = the vertical temperature gradient (temperature drop in degrees Kelvin per km).

The formula for w is

$$w = 4(Th/d)^{1/2}.$$

See Fig. 13.9.

The Washington-bound Concorde approaches the United States from Europe on a course that takes it south of Nantucket Island at an altitude of 16.8 km. If the surface temperature is 290K and the vertical temperature gradient is 5K/km, how far south of Nantucket must the plane be flown to keep its sonic boom carpet away from the island?

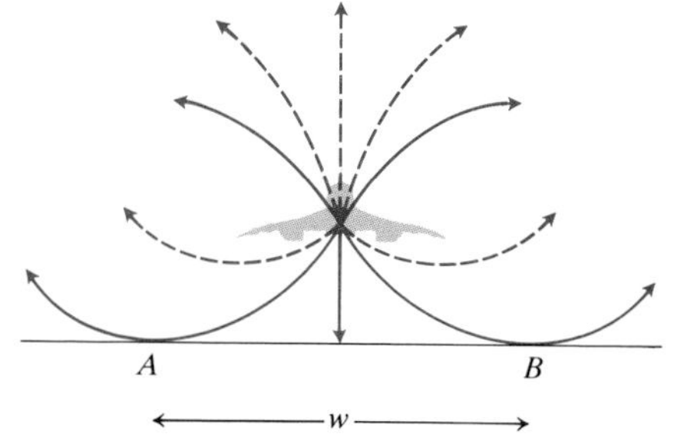

Sonic boom carpet

13.9 Sound waves from the Concorde bend as the temperature changes above and below the altitude at which the plane flies. The sonic boom carpet is the region on the ground that receives shock waves directly from the plane, not reflected from the atmosphere or diffracted along the ground. The carpet is determined by the grazing rays striking the ground from the point directly under the plane.

34. *The maximum value of a function on a line in space.* Does the function $f(x, y, z) = xyz$ have a maximum value on the line $x = 20 - t$, $y = t$, $z = 20$? If so, what is it? (*Hint:* Along the line, $w = f(x, y, z)$ is a differentiable function of t.)

*From N. K. Balachandra, W. L. Donn, and D. H. Rind, "Concorde Sonic Booms as an Atmospheric Probe," *Science*, 1 July 1977, Vol. 197, p. 47.

TOOLKIT PROGRAM

3D Grapher	Graphs surfaces defined by equations of the form $z = f(x, y)$ over rectangular regions in the xy-plane

13.2 Limits and Continuity

This section deals with limits and continuity of functions of more than one variable. The definitions are the same as for functions of a single variable, except that there are now more variables to watch.

Limits

If the values of a function $z = f(x, y)$ can be made as close as we like to a fixed number L by taking the point (x, y) close to the point (x_0, y_0), but not equal to (x_0, y_0), we say that L is the limit of f as (x, y) approaches (x_0, y_0). In symbols, we write

$$\lim_{(x,y)\to(x_0,y_0)} f(x, y) = L, \tag{1}$$

and we say "the limit of f as (x, y) approaches (x_0, y_0) equals L." This is like the limit of a function of one variable, except that two independent variables are involved instead of one. In addition, if (x_0, y_0) is an interior point of f's domain, (x, y) can approach (x_0, y_0) from any direction. The direction of approach may turn out to be an issue, as some of the following examples show.

For (x, y) to be "close" to (x_0, y_0) means that the Cartesian distance $\sqrt{(x - x_0)^2 + (y - y_0)^2}$ is small in some sense. Since

$$|x - x_0| = \sqrt{(x - x_0)^2} \le \sqrt{(x - x_0)^2 + (y - y_0)^2} \tag{2}$$

and

$$|y - y_0| = \sqrt{(y - y_0)^2} \le \sqrt{(x - x_0)^2 + (y - y_0)^2}, \tag{3}$$

the inequality

$$\sqrt{(x - x_0)^2 + (y - y_0)^2} < \delta$$

for any value of δ implies that

$$|x - x_0| < \delta \qquad \text{and} \qquad |y - y_0| < \delta.$$

Conversely, if for some $\delta > 0$ both $|x - x_0| < \delta$ and $|y - y_0| < \delta$, then

$$\sqrt{(x - x_0)^2 + (y - y_0)^2} < \sqrt{\delta^2 + \delta^2} = \sqrt{2}\delta,$$

which is small if δ is small (Fig. 13.10). Thus, in calculating limits we may think either in terms of the distance in the plane or in terms of differences in individual coordinates.

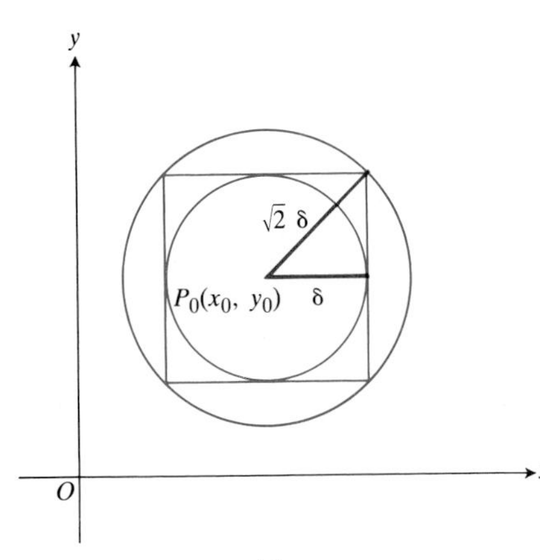

(a)

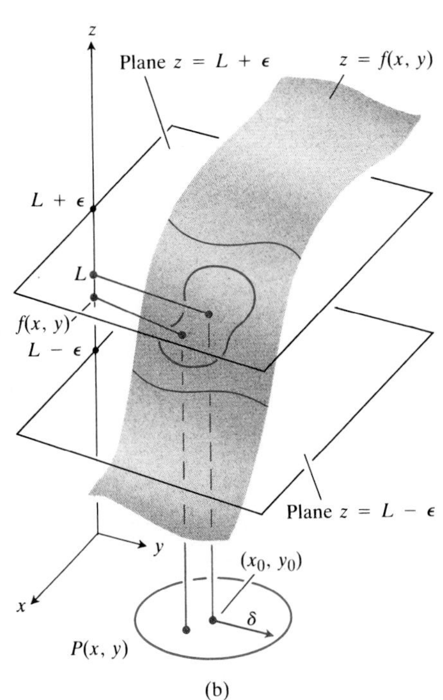

(b)

13.10 (a) The open square $|x - x_0| < \delta$, $|y - y_0| < \delta$ lies inside the open disk $\sqrt{(x - x_0)^2 + (y - y_0)^2} < \sqrt{2}\delta$ and contains the open disk $\sqrt{(x - x_0)^2 + (y - y_0)^2} < \delta$. Thus, we can make both $|x - x_0|$ and $|y - y_0|$ small by making $\sqrt{(x - x_0)^2 + (y - y_0)^2}$ small, and conversely. (b) For every point (x, y) within δ of (x_0, y_0), the value of $f(x,y)$ lies within ϵ of L.

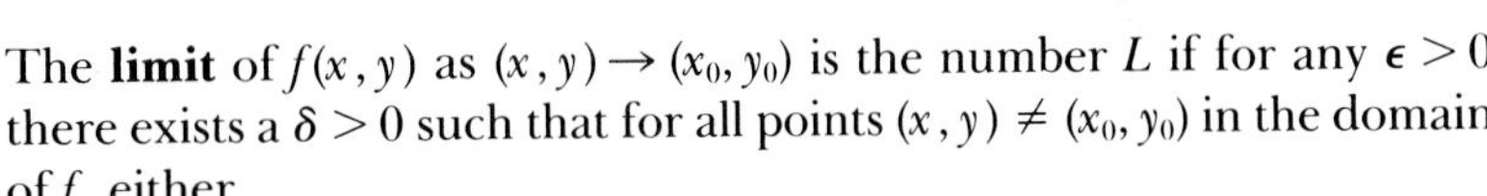

The **limit** of $f(x, y)$ as $(x, y) \to (x_0, y_0)$ is the number L if for any $\epsilon > 0$ there exists a $\delta > 0$ such that for all points $(x, y) \ne (x_0, y_0)$ in the domain of f, either

1. $\sqrt{(x - x_0)^2 + (y - y_0)^2} < \delta \quad \Rightarrow \quad |f(x, y) - L| < \epsilon$ or
2. $|x - x_0| < \delta$ and $|y - y_0| < \delta \quad \Rightarrow \quad |f(x, y) - L| < \epsilon$.

These definitions apply to boundary points as well as interior points of the domain of f. The only requirement is that the point (x, y) remain in the domain at all times.

It can be shown, as for functions of a single variable, that

$$\lim_{(x,y)\to(x_0,y_0)} x = x_0,$$

$$\lim_{(x,y)\to(x_0,y_0)} y = y_0, \tag{4}$$

$$\lim_{(x,y)\to(x_0,y_0)} k = k \qquad \text{(Any number } k\text{)}.$$

It can also be shown that the limit of the sum of two functions is the sum of their limits (when they both exist), with similar results for the limits of the differences, products, constant multiples, and quotients of functions.

THEOREM 1

Properties of Limits of Functions of Two Variables

If $\lim_{(x,y)\to(x_0,y_0)} f(x,y) = L_1$ and $\lim_{(x,y)\to(x_0,y_0)} g(x,y) = L_2$, then

1. *Sum Rule:* $\lim [f(x,y) + g(x,y)] = L_1 + L_2$
2. *Difference Rule:* $\lim [f(x,y) - g(x,y)] = L_1 - L_2$
3. *Product Rule:* $\lim f(x,y) \cdot g(x,y) = L_1 \cdot L_2$
4. *Constant Multiple Rule:* $\lim k\, g(x,y) = kL_2$ (Any number k)
5. *Quotient Rule:* $\lim \dfrac{f(x,y)}{g(x,y)} = \dfrac{L_1}{L_2}$ if $L_2 \neq 0$.

All limits are to be taken as $(x,y)\to(x_0,y_0)$, and L_1 and L_2 are to be real numbers.

When we combine Theorem 1 with the functions and limits in (4) we obtain the useful result that the limits of polynomials and rational functions as $(x,y)\to(x_0,y_0)$ may be calculated by evaluating the functions at the point (x_0,y_0). The only requirement is that the functions be defined there. In other words, polynomials and rational functions in two variables are *continuous* wherever they are defined, as we shall see in a moment.

EXAMPLE 1

a) $\lim_{(x,y)\to(3,-4)} (x^2+y^2) = (3)^2 + (-4)^2 = 9 + 16 = 25.$

b) $\lim_{(x,y)\to(0,1)} \dfrac{x - xy + 3}{x^2y + 5xy - y^3} = \dfrac{0 - 0(1) + 3}{(0)^2(1) + 5(0)(1) - (1)^3} = -3.$ ■

Continuity

The definition of continuity for functions of two variables is the same as for functions of a single variable.

DEFINITIONS

A function $f(x,y)$ is said to be **continuous at the point (x_0, y_0)** if

1. f is defined at (x_0,y_0),
2. $\lim_{(x,y)\to(x_0,y_0)} f(x,y)$ exists,
3. $\lim_{(x,y)\to(x_0,y_0)} f(x,y) = f(x_0,y_0)$.

A function is said to be **continuous** if it is continuous at every point of its domain.

As with the definition of limit, the definition of continuity applies at boundary points as well as interior points of the domain of f. The only requirement is that the point (x,y) remain in the domain at all times.

As you may have guessed, one of the consequences of Theorem 1 is that algebraic combinations of continuous functions are continuous at every point at which all the functions involved are defined. This means that sums, differences, products, and constant multiples of continuous functions are continuous at any shared domain point. Also, the quotient of two continuous functions is continuous at any point at which the quotient is defined. In particular, polynomials and rational functions of two variables are continuous at every point at which they are defined.

If $z = f(x, y)$ is a continuous function of x and y, and $w = g(z)$ is a continuous function of z, then the composite $w = g(f(x, y))$ is continuous. Thus, composites like

$$e^{x-y}, \qquad \cos\frac{xy}{x^2+1}, \qquad \ln(1 + x^2y^2)$$

are continuous at every point (x, y).

As with functions of a single variable, the general rule is that composites of continuous functions are continuous. The only requirement is that each function be continuous where it is applied.

EXAMPLE 2 Show that the function

$$f(x, y) = \begin{cases} \dfrac{2xy}{x^2 + y^2}, & (x, y) \neq (0, 0), \\ 0, & (x, y) = (0, 0), \end{cases}$$

is continuous at every point except the origin (Fig. 13.11).

13.11 The graph, in various stages of construction (a), and assorted level curves (b) of the function

$$f(x, y) = \begin{cases} \dfrac{2xy}{x^2 + y^2}, & (x, y) \neq (0, 0) \\ 0, & (x, y) = (0, 0). \end{cases}$$

The function is continuous at every point except the origin.

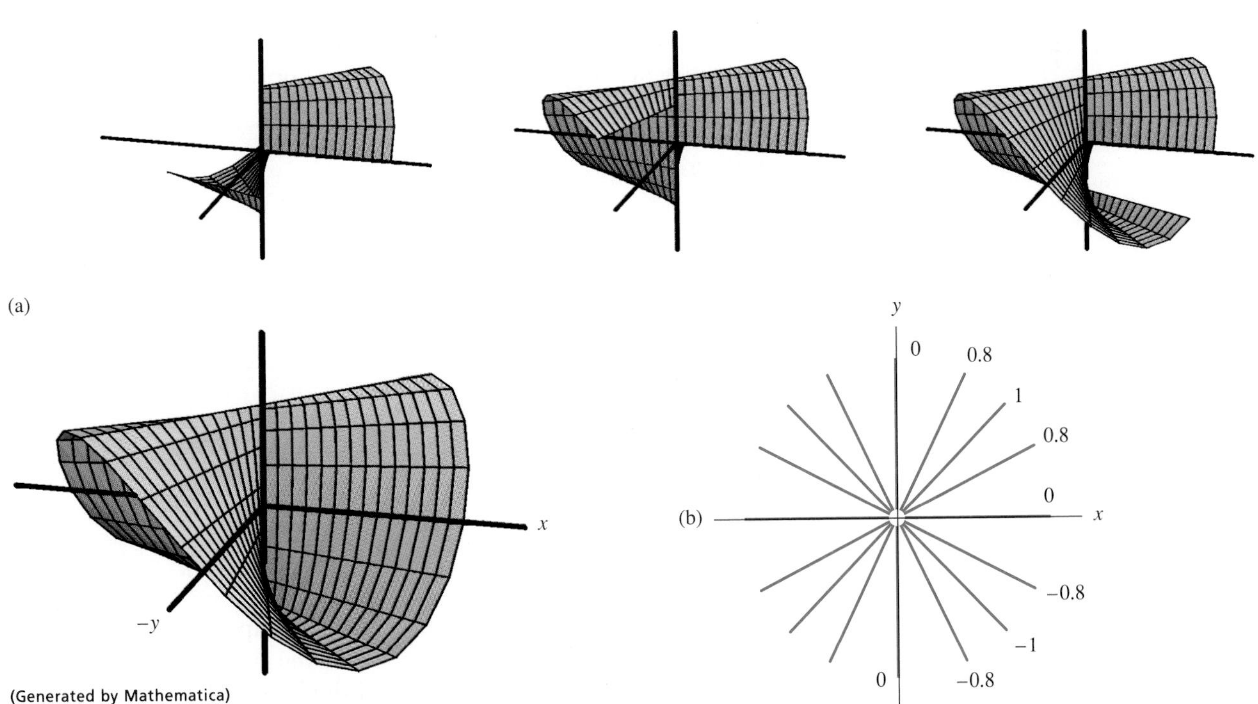

Solution The function f is continuous at any point $(x, y) \neq (0, 0)$ because its values are then given by a rational function of x and y.

At $(0, 0)$ the value of f is defined, but f, we claim, has no limit as $(x, y) \to (0, 0)$. The reason is that different paths of approach to the origin can lead to different results, as we shall now see.

For every value of m, the line $y = mx$ is a level curve of f because the value of f on the line has the constant value

$$f(x, y)\Big|_{y=mx} = \frac{2xy}{x^2 + y^2}\Big|_{y=mx} = \frac{2x(mx)}{x^2 + (mx)^2} = \frac{2mx^2}{x^2 + m^2x^2} = \frac{2m}{1 + m^2}.$$

Therefore, f has this number as its limit as (x, y) approaches the origin along the line:

$$\lim_{\substack{(x, y)\to(0, 0)\\ \text{along } y=mx}} f(x, y) = \lim_{(x, y)\to(0, 0)} \left[f(x, y)\Big|_{y=mx} \right] = \frac{2m}{1 + m^2}.$$

This limit changes with m. There is therefore no single number we may call the limit of f as (x, y) approaches the origin. The limit fails to exist, and the function is not continuous. ■

Example 2 illustrates an important point about limits of functions of two variables (or even more variables, for that matter). For a limit to exist at a point, the limit must be the same along every approach path. Therefore, if we ever find paths with different limits, we know the function has no limit at the point they approach.

The Two-Path Test for Discontinuity

If a function $f(x, y)$ has different limits along two different paths as (x, y) approaches (x_0, y_0) then $\lim_{(x, y)\to(x_0, y_0)} f(x, y)$ does not exist.

EXAMPLE 3 Show that the function

$$f(x, y) = \frac{2x^2y}{x^4 + y^2}$$

(Fig. 13.12) has no limit as (x, y) approaches $(0, 0)$.

Solution Along the curve $y = kx^2$, $x \neq 0$, the function has the constant value

$$f(x, y)\Big|_{y=kx^2} = \frac{2x^2y}{x^4 + y^2}\Big|_{y=kx^2} = \frac{2x^2(kx^2)}{x^4 + (kx^2)^2} = \frac{2kx^4}{x^4 + k^2x^4} = \frac{2k}{1 + k^2}.$$

Therefore,

$$\lim_{\substack{(x, y)\to(0, 0)\\ \text{along } y=kx^2}} f(x, y) = \lim_{(x, y)\to(0, 0)} \left[f(x, y)\Big|_{y=kx^2} \right] = \frac{2k}{1 + k^2}.$$

This limit varies with the path of approach. If (x, y) approaches $(0, 0)$ along the parabola $y = x^2$, for instance, $k = 1$ and the limit is 1. If (x, y) approaches $(0, 0)$ along the x-axis, $k = 0$ and the limit is 0. By the two-path test, f has no limit as (x, y) approaches $(0, 0)$.

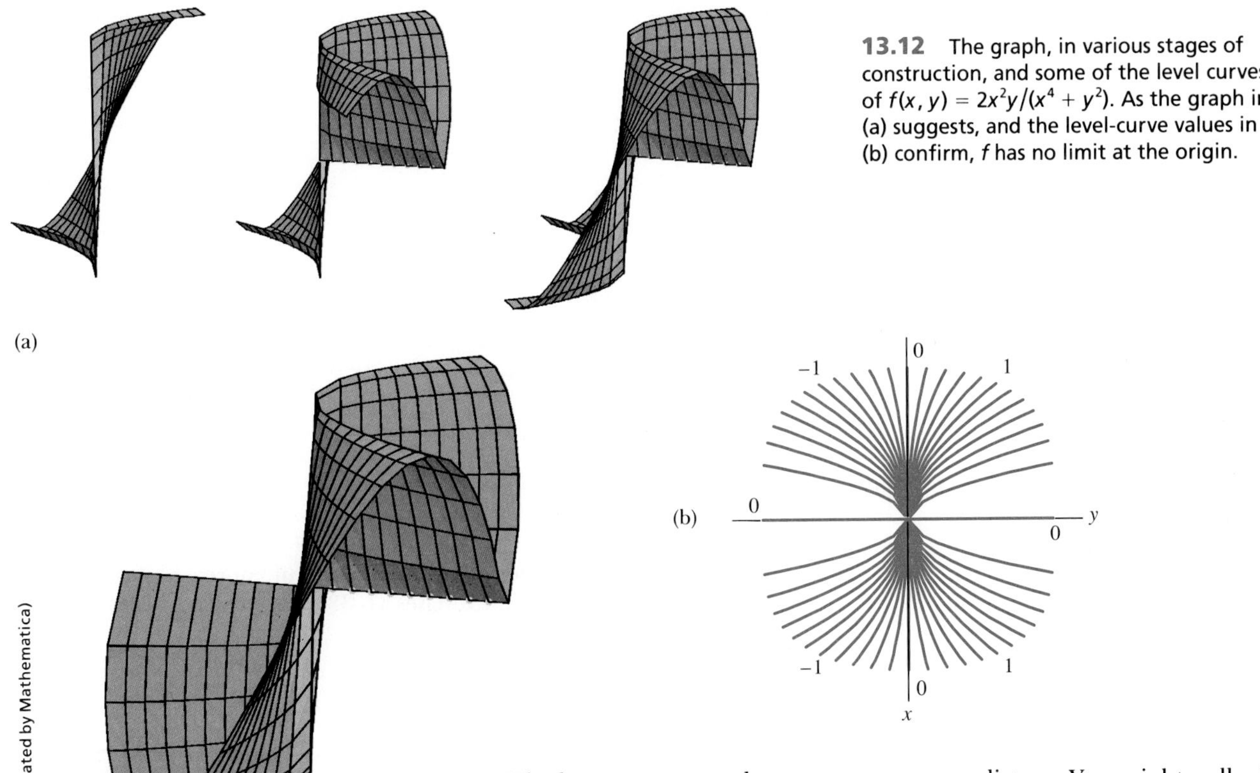

13.12 The graph, in various stages of construction, and some of the level curves of $f(x, y) = 2x^2y/(x^4 + y^2)$. As the graph in (a) suggests, and the level-curve values in (b) confirm, f has no limit at the origin.

The language we use here may seem contradictory. You might well ask, "What do you mean f has no limit as (x, y) approaches the origin—it has lots of limits." But that's the point. There is no path-independent limit, and therefore, by our definition, $\lim_{(x,y)\to(0,0)} f(x, y)$ does not exist. It is our translating this formal statement into the more colloquial "has no limit" that creates the apparent contradiction. The mathematics is fine. The problem arises in how we tend to talk about it. In the crunch, we need the formality to keep things straight. ■

Continuous Functions Defined on Closed, Bounded Regions

As we know, a function of a single variable that is continuous throughout a closed, bounded interval $[a, b]$ takes on an absolute maximum value and an absolute minimum value at least once in $[a, b]$. The same is true of a function $z = f(x, y)$ that is continuous on a closed, bounded set R in the plane (like a line segment, disc, or filled-in triangle). The function takes on an absolute maximum value at some point in R and an absolute minimum value at some point in R.

Similar theorems hold for functions of three or more variables. A continuous function $w = f(x, y, z)$, for example, must take on absolute maximum and minimum values on any closed, bounded set (solid ball or cube, spherical shell, rectangular plate) on which it is defined.

It is important to know how to find these extreme values and we shall show how to do it when we get to Section 13.8. But first, we need to know about derivatives, and that will be the topic of the next section.

Exercises 13.2

Find the limits in Exercises 1–12.

1. $\lim_{(x,y)\to(0,0)} \dfrac{3x^2 - y^2 + 5}{x^2 + y^2 + 2}$

2. $\lim_{(x,y)\to(1,1)} \ln|1 + x^2y^2|$

3. $\lim_{(x,y)\to(0,\ln 2)} e^{x-y}$

4. $\lim_{(x,y)\to(0,4)} \dfrac{x}{\sqrt{y}}$

5. $\lim_{P\to(1,3,4)} \sqrt{x^2 + y^2 + z^2 - 1}$

6. $\lim_{P\to(1,2,6)} \left(\dfrac{1}{x} + \dfrac{1}{y} + \dfrac{1}{z}\right)$

7. $\lim_{(x,y)\to(0,\pi/4)} \sec x \tan y$

8. $\lim_{(x,y)\to(0,0)} \cos \dfrac{x^2 + y^2}{x + y + 1}$

9. $\lim_{(x,y)\to(1,1)} \cos \sqrt[3]{|xy| - 1}$

10. $\lim_{(x,y)\to(1,0)} \dfrac{x \sin y}{x^2 + 1}$

11. $\lim_{(x,y)\to(0,0)} \dfrac{e^y \sin x}{x}$

12. $\lim_{(x,y)\to(0,0)} \tan^{-1}\left(1/\sqrt{x^2 + y^2}\right)$

Find the limits in Exercises 13–16 by rewriting the fractions first.

13. $\lim_{\substack{(x,y)\to(1,1)\\ x\neq y}} \dfrac{x^2 - 2xy + y^2}{x - y}$

14. $\lim_{\substack{(x,y)\to(1,1)\\ x\neq y}} \dfrac{x^2 - y^2}{x - y}$

15. $\lim_{\substack{(x,y)\to(1,1)\\ x\neq 1}} \dfrac{xy - y - 2x + 2}{x - 1}$

16. $\lim_{\substack{(x,y)\to(2,-4)\\ y\neq -4,\, x\neq x^2}} \dfrac{y + 4}{x^2y - xy + 4x^2 - 4x}$

Find the limits in Exercises 17–22. In each exercise, P stands for the point (x, y, z).

17. $\lim_{P\to(2,3,-6)} \sqrt{x^2 + y^2 + z^2}$

18. $\lim_{P\to(0,-2,0)} \ln\sqrt{x^2 + y^2 + z^2}$

19. $\lim_{P\to(3,3,0)} (\sin^2 x + \cos^2 y + \sec^2 z)$

20. $\lim_{P\to(\pi,0,3,)} ze^{-2y} \cos 2x$

21. $\lim_{P\to(-1/4,\pi/2,2)} \tan^{-1} xyz$

22. $\lim_{P\to(1,-1,-1)} \dfrac{2xy + yz}{x^2 + z^2}$

At what points (x, y) in the plane are the functions given by the formulas in Exercises 23–26 continuous?

23. a) $\sin(x + y)$ b) $\ln(x^2 + y^2)$

24. a) $\dfrac{x + y}{x - y}$ b) $\dfrac{y}{x^2 + 1}$

25. a) $\sin \dfrac{1}{xy}$ b) $\dfrac{x + y}{2 + \cos x}$

26. a) $\dfrac{x^2 + y^2}{x^2 - 3x + 2}$ b) $\dfrac{1}{x^2 - y}$

At what points (x, y, z) in space are the functions given by the formulas in Exercises 27–30 continuous?

27. a) $x^2 + y^2 - 2z^2$ b) $\sqrt{x^2 + y^2 - 1}$

28. a) $\ln xyz$ b) $e^{x+y} \cos z$

29. a) $xy \sin \dfrac{1}{z}$ b) $\dfrac{1}{x^2 + y^2 + z^2 - 1}$

30. a) $\dfrac{1}{|x| + |y| + |z|}$ b) $\dfrac{1}{|xy| + |z|}$

By considering different paths of approach, show that the functions in Exercises 31–38 have no limit as $(x, y) \to (0, 0)$.

31. $f(x, y) = \dfrac{x}{\sqrt{x^2 + y^2}}$

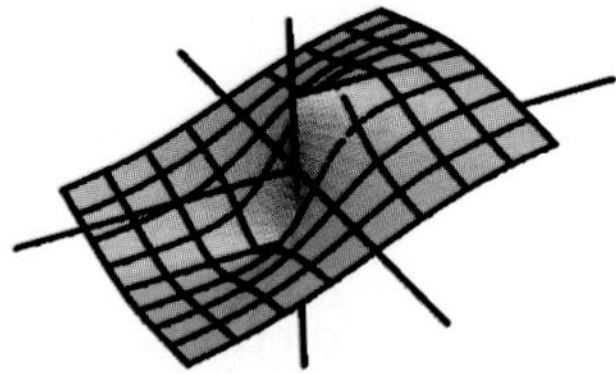

(Generated by Mathematica)

32. $f(x, y) = \dfrac{x^4}{x^4 + y^2}$

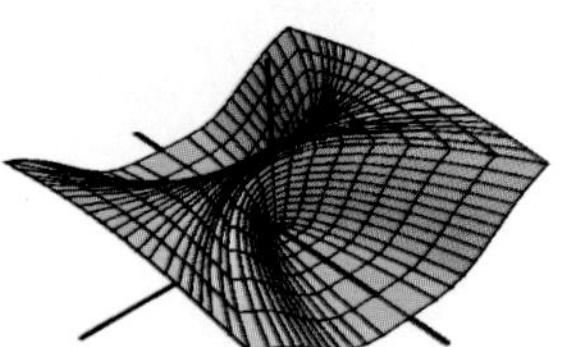

(Generated by Mathematica)

33. $f(x, y) = \dfrac{x^4 - y^2}{x^4 + y^2}$

34. $f(x, y) = \dfrac{xy}{|xy|}$

35. $f(x, y) = \dfrac{x - y}{x + y}$

36. $f(x, y) = \dfrac{x + y}{x - y}$

37. $f(x, y) = \dfrac{x^2 + y}{y}$

38. $f(x, y) = \dfrac{x^2}{x^2 - y}$

39. *Continuation of Example 2.*
   a) Reread Example 2. Then substitute $m = \tan\theta$ into the formula
   $$f(x, y)\Big|_{y = mx} = \frac{2m}{1 + m^2}$$
   and simplify the result to show how the value of f varies with the line's angle of inclination.
   b) Use the formula you obtained in (a) to show that the limit of f as $(x, y) \to (0, 0)$ along the line $y = mx$ varies from -1 to 1 depending on the angle of approach.

13.3 Partial Derivatives

Partial derivatives are what we get when we hold all but one of the independent variables of a function constant and differentiate with respect to that one variable. In this section, we show how partial derivatives arise geometrically and how we can calculate them by applying the rules we already know for differentiating functions of a single variable. We begin with functions of two independent variables.

Definitions and Notation

If (x_0, y_0) is a point in the domain of a function $f(x, y)$, the vertical plane $y = y_0$ will cut the surface $z = f(x, y)$ in the curve $z = f(x, y_0)$ (Fig. 13.13). This curve is the graph of the function $z = f(x, y_0)$ in the plane $y = y_0$. The horizontal coordinate in this plane is x; the vertical coordinate is z.

We define the partial derivative of f with respect to x at the point (x_0, y_0) as the ordinary derivative of $f(x, y_0)$ with respect to x at the point $x = x_0$.

DEFINITION

Partial Derivative with Respect to *x*

The partial derivative of $f(x, y)$ with respect to x at the point (x_0, y_0) is

$$\left.\frac{\partial f}{\partial x}\right|_{(x_0, y_0)} = \left.\frac{d}{dx} f(x, y_0)\right|_{x=x_0} = \lim_{h\to 0} \frac{f(x_0 + h, y_0) - f(x_0, y_0)}{h}, \tag{1}$$

provided the limit exists.

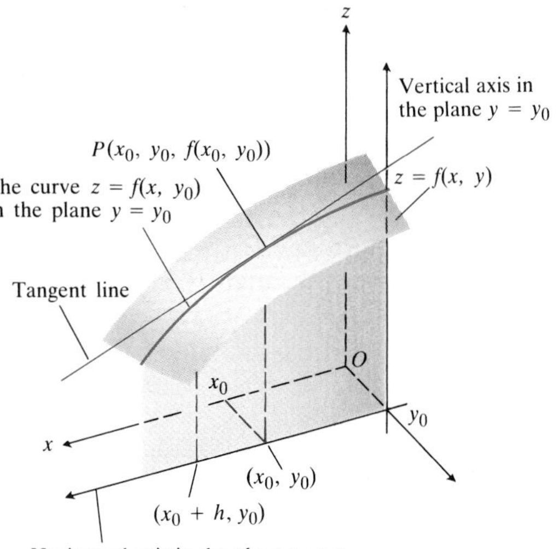

13.13 The intersection of the plane $y = y_0$ with the surface $z = f(x, y)$, viewed from a point above the first quadrant of the xy-plane.

The slope of the curve $z = f(x, y_0)$ at the point $P(x_0, y_0, f(x_0, y_0))$ in the plane $y = y_0$ is the value of the partial derivative of f with respect to x at (x_0, y_0). The tangent to the curve at P is the line in the plane $y = y_0$ that passes through P with this slope. The partial derivative $\partial f/\partial x$ at (x_0, y_0) gives the rate of change of f with respect to x when y is held fixed at the value y_0.

The notation we use for the partial derivative of f with respect to x depends on what aspect of the derivative we want to emphasize. The usual list looks like this:

$\frac{\partial f}{\partial x}(x_0, y_0)$ or $f_x(x_0, y_0)$	"Partial derivative of f with respect to x at (x_0, y_0)" or "f sub x at (x_0, y_0)." Convenient for stressing the point (x_0, y_0).
$\left.\frac{\partial z}{\partial x}\right\vert_{(x_0, y_0)}$	"Partial derivative of z with respect to x at (x_0, y_0)." Common in science and engineering when you are dealing with variables and do not mention the function explicitly.
f_x, $\frac{\partial f}{\partial x}$, z_x, or $\frac{\partial z}{\partial x}$	"Partial derivative of f (or z) with respect to x." Convenient when you regard the partial derivative as a function in its own right.

The letter ∂ is another Greek delta.

The definition of the partial derivative of $f(x, y)$ with respect to y at a point (x_0, y_0) is similar to the definition of the partial derivative of f with respect to x. We hold x fixed at the value x_0 and take the ordinary derivative of $f(x_0, y)$ with respect to y at y_0.

DEFINITION

Partial Derivative with Respect to *y*

The partial derivative of $f(x, y)$ with respect to y at the point (x_0, y_0) is

$$\left.\frac{\partial f}{\partial y}\right|_{(x_0, y_0)} = \left.\frac{d}{dy}f(x_0, y)\right|_{y=y_0} = \lim_{h\to 0}\frac{f(x_0, y_0 + h) - f(x_0, y_0)}{h}, \tag{2}$$

provided the limit exists.

The slope of the curve $z = f(x_0, y)$ at the point $P(x_0, y_0, f(x_0, y_0))$ in the vertical plane $x = x_0$ (Fig. 13.14) is the partial derivative of f with respect to y at (x_0, y_0). The tangent to the curve at P is the line in the plane $x = x_0$ that passes through P with this slope. The partial derivative gives the rate of change of f with respect to y at (x_0, y_0) when x is held fixed at the value x_0.

The partial derivative with respect to y is denoted the same way as the partial derivative with respect to x. We have

$$\frac{\partial f}{\partial y}(x_0, y_0), \quad f_y(x_0, y_0), \quad \frac{\partial f}{\partial y}, \quad f_y, \tag{3}$$

and so on.

Notice that we now have two tangent lines associated with the surface $z = f(x, y)$ at the point $P(x_0, y_0, f(x_0, y_0))$ (Fig. 13.15). Is the plane they

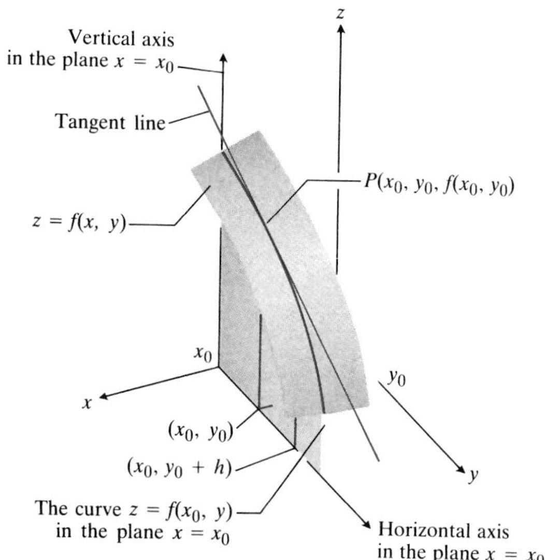

13.14 The intersection of the plane $x = x_0$ with the surface $z = f(x, y)$, viewed from above the first quadrant of the xy-plane.

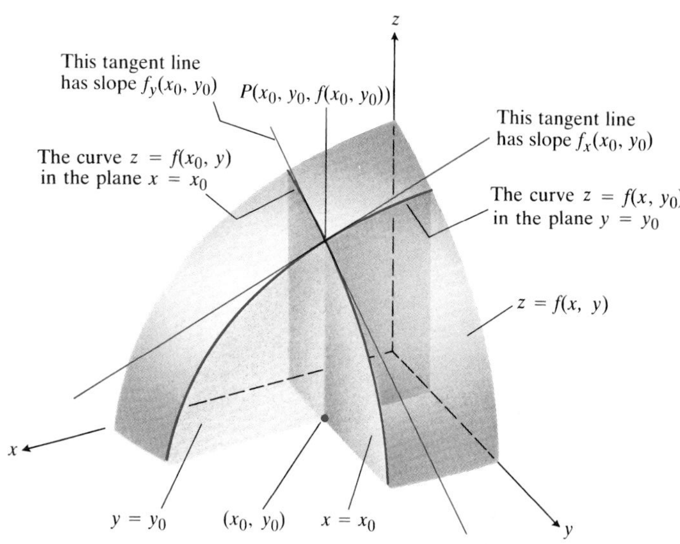

13.15 Figures 13.13 and 13.14 combined. The tangent lines at $(x_0, y_0, f(x_0, y_0))$ determine a plane that, in this picture at least, appears to be tangent to the surface.

determine tangent to the surface at P? It would be nice if it were, but we shall have to learn more about partial derivatives before we can find out.

Calculations

As Eq. (1) shows, we calculate $\partial f/\partial x$ by differentiating f with respect to x in the usual way while treating y as a constant. Similarly, from Eq. (2), we see that we can calculate $\partial f/\partial y$ by differentiating f with respect to y in the usual way while holding x constant.

EXAMPLE 1 Find the value of $\partial f/\partial x$ at the point (4, 5) if

$$f(x, y) = x^2 + 3xy + y - 1.$$

Solution We regard y as a constant and differentiate with respect to x:

$$\frac{\partial f}{\partial x} = \frac{\partial}{\partial x}(x^2 + 3xy + y - 1) = 2x + 3y + 0 - 0 = 2x + 3y.$$

The value of $\partial f/\partial x$ at (4, 5) is $2(4) + 3(5) = 23$. ■

EXAMPLE 2 Find the value of $\partial f/\partial y$ at the point (4, 5) if

$$f(x, y) = x^2 + 3xy + y - 1.$$

Solution We regard x as a constant and differentiate with respect to y:

$$\frac{\partial f}{\partial y} = \frac{\partial}{\partial y}(x^2 + 3xy + y - 1) = 0 + 3x + 1 - 0 = 3x + 1.$$

The value of $\partial f/\partial y$ at (4, 5) is $3(4) + 1 = 13$. ■

EXAMPLE 3 Find $\partial f/\partial y$ if

$$f(x, y) = y \sin xy.$$

Solution We treat x as a constant and f as a product of y and $\sin xy$:

$$\begin{aligned}\frac{\partial f}{\partial y} &= \frac{\partial}{\partial y}(y \sin xy) = y \frac{\partial}{\partial y} \sin xy + \sin xy \frac{\partial}{\partial y}(y)\\ &= y \cos xy \frac{\partial}{\partial y}(xy) + \sin xy = xy \cos xy + \sin xy.\end{aligned}$$

■

EXAMPLE 4 Find f_x and f_y if

$$f(x, y) = \frac{2y}{y + \cos x}.$$

Solution We treat f as the quotient of $2y$ divided by $(y + \cos x)$. With y held constant, we get

$$\begin{aligned}f_x &= \frac{\partial}{\partial x}\left(\frac{2y}{y + \cos x}\right) = \frac{(y + \cos x)\frac{\partial}{\partial x}(2y) - 2y\frac{\partial}{\partial x}(y + \cos x)}{(y + \cos x)^2}\\ &= \frac{(y + \cos x)(0) - 2y(-\sin x)}{(y + \cos x)^2} = \frac{2y \sin x}{(y + \cos x)^2}.\end{aligned}$$

With x held constant, we get

$$\begin{aligned}f_y &= \frac{\partial}{\partial y}\left(\frac{2y}{y + \cos x}\right) = \frac{(y + \cos x)\frac{\partial}{\partial y}(2y) - 2y\frac{\partial}{\partial y}(y + \cos x)}{(y + \cos x)^2}\\ &= \frac{(y + \cos x)(2) - 2y(1)}{(y + \cos x)^2} = \frac{2 \cos x}{(y + \cos x)^2}.\end{aligned}$$

■

Functions of More Than Two Variables

The definitions of the partial derivatives of functions of more than two independent variables are like the definitions for functions of two variables. They are ordinary derivatives with respect to one variable, taken while all the other independent variables are regarded as constants.

EXAMPLE 5 If x, y, and z are independent variables and

$$f(x, y, z) = x \sin(y + 3z),$$

then

$$\begin{aligned}\frac{\partial f}{\partial z} &= \frac{\partial}{\partial z}(x \sin(y + 3z)) = x \frac{\partial}{\partial z} \sin(y + 3z)\\ &= x \cos(y + 3z) \frac{\partial}{\partial z}(y + 3z) = 3x \cos(y + 3z).\end{aligned}$$

■

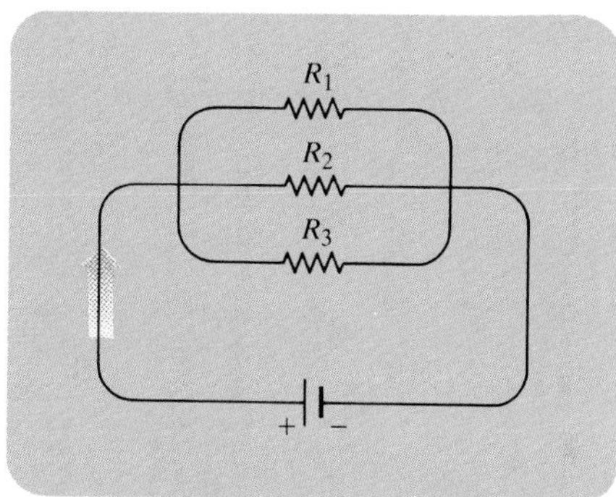
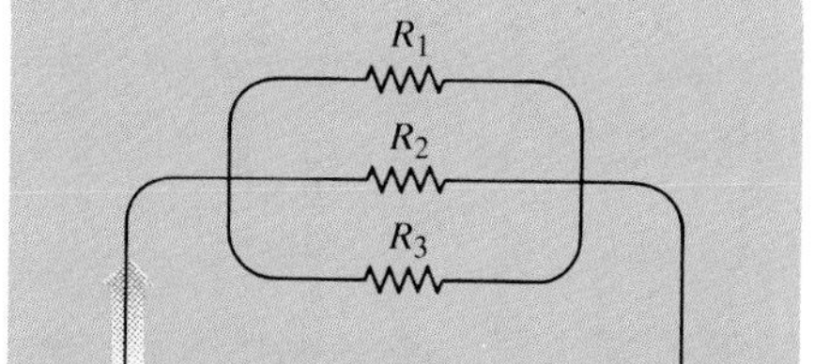

13.16 Resistors built into a circuit this way are said to be connected in parallel. Each resistor lets a portion of the current through. Their combined resistance R can be calculated from the formula

$$\frac{1}{R} = \frac{1}{R_1} + \frac{1}{R_2} + \frac{1}{R_3}.$$

EXAMPLE 6 *Electrical resistors in parallel.* If resistors of R_1, R_2, and R_3 ohms are connected in parallel to make an R-ohm resistor, the value of R can be found from the equation

$$\frac{1}{R} = \frac{1}{R_1} + \frac{1}{R_2} + \frac{1}{R_3} \tag{4}$$

(Fig. 13.16). Find the value of $\partial R/\partial R_2$ when $R_1 = 30$, $R_2 = 45$, and $R_3 = 90$ ohms.

Solution To find $\partial R/\partial R_2$, we treat R_1 and R_3 as constants and differentiate both sides of Eq. (4) with respect to R_2:

$$\frac{\partial}{\partial R_2}\left(\frac{1}{R}\right) = \frac{\partial}{\partial R_2}\left(\frac{1}{R_1} + \frac{1}{R_2} + \frac{1}{R_3}\right)$$

$$-\frac{1}{R^2}\frac{\partial R}{\partial R_2} = 0 - \frac{1}{R_2^2} + 0$$

$$\frac{\partial R}{\partial R_2} = \frac{R^2}{R_2^2} = \left(\frac{R}{R_2}\right)^2.$$

When $R_1 = 30$, $R_2 = 45$, and $R_3 = 90$,

$$\frac{1}{R} = \frac{1}{30} + \frac{1}{45} + \frac{1}{90} = \frac{3 + 2 + 1}{90} = \frac{6}{90} = \frac{1}{15},$$

so $R = 15$ and

$$\frac{\partial R}{\partial R_2} = \left(\frac{15}{45}\right)^2 = \left(\frac{1}{3}\right)^2 = \frac{1}{9}.$$

■

The Relationship Between the Existence of Partial Derivatives and Continuity?—There Is None

A function $f(x, y)$ can have partial derivatives with respect to both x and y at a point without being continuous at that point. This is very different from what we know about functions of a single variable, where the existence of a derivative implies continuity.

EXAMPLE 7 The function

$$f(x, y) = \begin{cases} 0, & xy \neq 0 \\ 1, & xy = 0 \end{cases}$$

(Fig. 13.17) is not continuous at the point $(0, 0)$. The limit of f as (x, y) approaches $(0, 0)$ along the line $y = x$ is 0, but $f(0, 0)$ itself is 1. The partial derivatives f_x and f_y, being the slopes of the horizontal lines L_1 and L_2 in Fig. 13.17, both exist at $(0, 0)$. ■

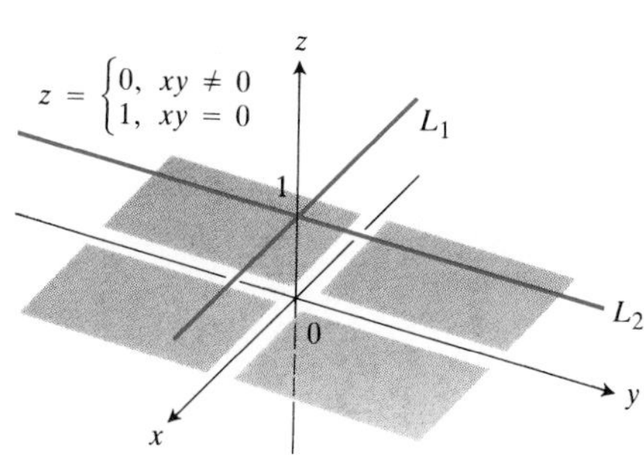

13.17 The graph of

$$z = \begin{cases} 0, & xy \neq 0 \\ 1, & xy = 0 \end{cases}$$

consists of the lines L_1 and L_2 and the four open quadrants of the xy-plane. The function has partial derivatives at the origin but is not continuous there.

However, if the partial derivatives of $f(x, y)$ are continuous throughout a disk centered at (x_0, y_0), then f is continuous at (x_0, y_0). We shall say more about this in the next section.

Second Order Partial Derivatives

When we differentiate a function $f(x, y)$ twice, we produce its second order derivatives. These derivatives are usually denoted by

$\dfrac{\partial^2 f}{\partial x^2}$	"d squared f d x squared"	or	f_{xx}	"f sub x x"
$\dfrac{\partial^2 f}{\partial y^2}$	"d squared f d y squared"		f_{yy}	"f sub y y"
$\dfrac{\partial^2 f}{\partial x\,\partial y}$	"d squared f d x d y"		f_{yx}	"f sub y x"
$\dfrac{\partial^2 f}{\partial y\,\partial x}$	"d squared f d y d x"		f_{xy}	"f sub x y."

The defining equations are

$$\frac{\partial^2 f}{\partial x^2} = \frac{\partial}{\partial x}\left(\frac{\partial f}{\partial x}\right), \qquad \frac{\partial^2 f}{\partial x\,\partial y} = \frac{\partial}{\partial x}\left(\frac{\partial f}{\partial y}\right),$$

and so on. Notice the order in which the derivatives are taken:

$\dfrac{\partial^2 f}{\partial x\,\partial y}$ Differentiate first with respect to y, then with respect to x.

$f_{yx} = (f_y)_x$ Means the same thing.

EXAMPLE 8 If $f(x, y) = x\cos y + ye^x$, then

$$\frac{\partial f}{\partial x} = \cos y + ye^x,$$

$$\frac{\partial^2 f}{\partial y\,\partial x} = \frac{\partial}{\partial y}\left(\frac{\partial f}{\partial x}\right) = -\sin y + e^x,$$

$$\frac{\partial^2 f}{\partial x^2} = \frac{\partial}{\partial x}\left(\frac{\partial f}{\partial x}\right) = ye^x.$$

Also,

$$\frac{\partial f}{\partial y} = -x\sin y + e^x$$

$$\frac{\partial^2 f}{\partial x\,\partial y} = \frac{\partial}{\partial x}\left(\frac{\partial f}{\partial y}\right) = -\sin y + e^x$$

$$\frac{\partial^2 f}{\partial y^2} = \frac{\partial}{\partial y}\left(\frac{\partial f}{\partial y}\right) = -x\cos y.$$

■

The Mixed Derivative Theorem

You may have noticed that the "mixed" second order derivatives

$$\frac{\partial^2 f}{\partial y\,\partial x} \quad \text{and} \quad \frac{\partial^2 f}{\partial x\,\partial y}$$

in Example 8 were equal. This was no mere coincidence. They have to be equal whenever f, f_x, f_y, f_{xy}, and f_{yx} are continuous.

THEOREM 2

The Mixed Derivative Theorem

If $f(x, y)$ and its partial derivatives f_x, f_y, f_{xy}, and f_{yx} are defined throughout an open region containing a point (a, b) and are all continuous at (a, b), then

$$f_{xy}(a, b) = f_{yx}(a, b). \tag{5}$$

You can find proofs of Theorem 2 in most advanced calculus texts.

Theorem 2 says that when we want to calculate a mixed second order derivative we may differentiate in either order. This can work to our advantage, as the next example shows.

EXAMPLE 9 Find $\partial^2 w/\partial x\,\partial y$ if

$$w = xy + \frac{e^y}{y^2+1}.$$

Solution The symbol $\partial^2 w/\partial x\,\partial y$ tells us to differentiate first with respect to y and then with respect to x. However, if we postpone the differentiation with respect to y and differentiate first with respect to x, we can get the answer more quickly. In two steps,

$$\frac{\partial w}{\partial x} = y \quad \text{and} \quad \frac{\partial^2 w}{\partial y\,\partial x} = 1.$$

We are in for more work if we differentiate first with respect to y. (Just try it.) ■

Partial Derivatives of Still Higher Order

Although we shall deal mostly with first and second order partial derivatives, because these appear most frequently in applications, there is no theoretical limit to how many times we can differentiate a function as long as the derivatives involved exist. Thus we get third and fourth order derivatives denoted by symbols like

$$\frac{\partial^3 f}{\partial x\,\partial y^2} = f_{yyx},$$

$$\frac{\partial^4 f}{\partial x^2\,\partial y^2} = f_{yyxx},$$

and so on. As with second order derivatives, the order of differentiation is immaterial as long as the function and its derivatives through the order in question are all defined and continuous where they should be.

Exercises 13.3

In Exercises 1–18, find $\partial f/\partial x$ and $\partial f/\partial y$.

1. $f(x,y) = 2x$
2. $f(x,y) = -3y$
3. $f(x,y) = -4$
4. $f(x,y) = 2x - 3y - 4$
5. $f(x,y) = x(y-1)$
6. $f(x,y) = x^2 + y^2$
7. $f(x,y) = x^2 - xy + y^2$
8. $f(x,y) = (x+2)(y+3)$
9. $f(x,y) = 5xy - 7x^2 - y^2 + 3x - 6y + 2$
10. $f(x,y) = 1/(x+y)$
11. $f(x,y) = \sqrt{x^2+y^2}$
12. $f(x,y) = x/(x^2+y^2)$
13. $f(x,y) = (x+y)/(xy-1)$
14. $f(x,y) = (y-x)e^x$
15. $f(x,y) = e^x \ln y$
16. $f(x,y) = \sin(x+y)$
17. $f(x,y) = e^x \sin(y+1)$
18. $f(x,y) = \tan^{-1}(y/x)$

In Exercises 19–24, find f_x, f_y, and f_z.

19. $f(x,y,z) = xy + yz + zx$
20. $f(x,y,z) = (x^2+y^2+z^2)^{-1/2}$
21. $f(x,y,z) = 1 + y^2 + 2z^2$
22. $f(x,y,z) = x - \sqrt{y^2+z^2}$
23. $f(x,y,z) = \sin(x+yz)$
24. $f(x,y,z) = \tan(x+y+z)$

In Exercises 25–30, find the partial derivative of the function with respect to each variable.

25. $f(t,\alpha) = \cos(2\pi t - \alpha)$
26. $g(u,v) = v^2 e^{(2u/v)}$
27. $h(\rho,\phi,\theta) = \rho \sin\phi \cos\theta$
28. $g(r,\theta,z) = r(1-\cos\theta) - z$
29. *Work done by the heart* (Section 3.7, Exercise 48).
$$W(P,V,\rho,v,g) = PV + \frac{V\rho v^2}{2g}$$
30. *Wilson lot size formula* (Section 4.5, Exercise 45).
$$A(c,h,k,m,q) = \frac{km}{q} + cm + \frac{hq}{2}$$

Find the second order partial derivatives of the functions in Exercises 31–36.

31. $f(x,y) = x + y + xy$
32. $f(x,y) = \sin xy$
33. $g(x,y) = x^2 y + \cos y + y \sin x$
34. $h(x,y) = xe^y + y + 1$
35. $r(x,y) = \ln(x+y)$
36. $s(x,y) = \tan^{-1}(y/x)$

In Exercises 37–40, verify that $w_{xy} = w_{yx}$.

37. $w = \ln(2x+3y)$
38. $w = e^x + x\ln y + y\ln x$
39. $w = xy^2 + x^2y^3 + x^3y^4$
40. $w = x\sin y + y\sin x + xy$
41. Which order of differentiation will calculate f_{xy} faster: x first or y first? Try to answer without writing anything down.
    a) $f(x,y) = x\sin y + e^y$
    b) $f(x,y) = 1/x$
    c) $f(x,y) = y + (x/y)$
    d) $f(x,y) = y + x^2y + 4y^3 - \ln(y^2+1)$
    e) $f(x,y) = x^2 + 5xy + \sin x + 7e^x$
    f) $f(x,y) = x\ln xy$
42. The fifth order partial derivative $\partial^5 f/\partial x^2 \partial y^3$ is zero for each of the following functions. To show this as quickly as possible, which variable would you differentiate with respect to first: x, or y? Try to answer without writing anything down.
    a) $f(x,y) = y^2x^4e^x + 2$
    b) $f(x,y) = y^2 + y(\sin x - x^4)$
    c) $f(x,y) = x^2 + 5xy + \sin x + 7e^x$
    d) $f(x,y) = xe^{y^2/2}$
43. Assume that the equation
$$xy + z^3x - 2yz = 0$$
defines z as a differentiable function of the two independent variables x and y. Find the value of $\partial z/\partial x$ at the point $(x,y,z) = (1,1,1)$.
44. Assume that the equation
$$xz + y\ln x + x^2 + 4 = 0$$
defines x as a differentiable function of the two independent variables y and z. Find the value of $\partial x/\partial z$ at the point $(x,y,z) = (1,-1,-3)$.

Laplace Equations

The *three-dimensional Laplace equation*
$$\frac{\partial^2 f}{\partial x^2} + \frac{\partial^2 f}{\partial y^2} + \frac{\partial^2 f}{\partial z^2} = 0 \tag{6}$$
is satisfied by steady-state temperature distributions $T = f(x,y,z)$ in space, by gravitational potentials, and by electrostatic potentials. The *two-dimensional Laplace equation*
$$\frac{\partial^2 f}{\partial x^2} + \frac{\partial^2 f}{\partial y^2} = 0, \tag{7}$$
obtained by dropping the $\partial^2 f/\partial z^2$ term from (6), describes potentials and steady-state temperature distributions in a plane (Fig. 13.18).

Show that each function in Exercises 45–50 satisfies a Laplace equation.

45. $f(x,y,z) = x^2 + y^2 - 2z^2$

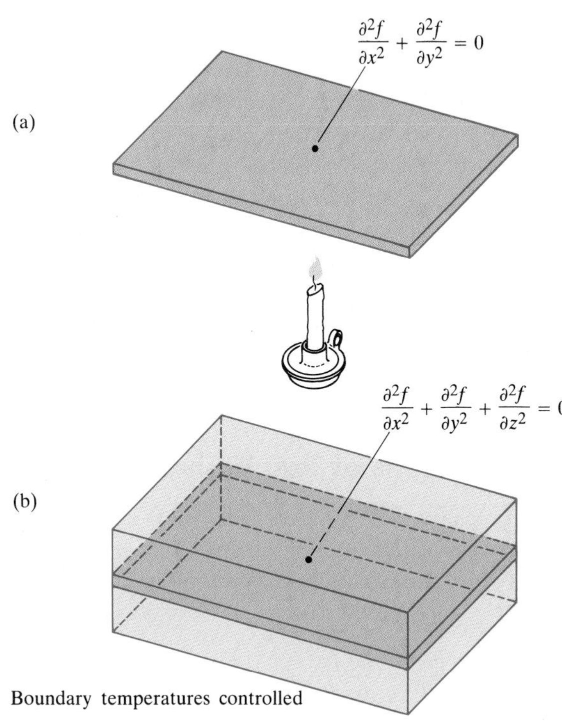

13.18 Steady-state temperature distributions in planes and solids satisfy Laplace equations. The plane (a) may be treated as a thin slice of the solid (b) perpendicular to the z-axis.

46. $f(x, y, z) = 2z^3 - 3(x^2 + y^2)z$

47. $f(x, y) = e^{-2y} \cos 2x$ **48.** $f(x, y) = \ln \sqrt{x^2 + y^2}$

49. $f(x, y, z) = (x^2 + y^2 + z^2)^{-1/2}$

50. $f(x, y, z) = e^{3x+4y} \cos 5z$

The Wave Equation

If we stand on an ocean shore and take a snapshot of the waves, the picture shows a regular pattern of peaks and valleys in an instant of time (Fig. 13.19). We see periodic vertical motion in space, with respect to distance. If we stand in the water, we can feel the rise and fall of the water as the waves go by. We see periodic vertical motion in time. In physics, this beautiful symmetry is expressed by the *one-dimensional wave equation*

$$\frac{\partial^2 w}{\partial t^2} = c^2 \frac{\partial^2 w}{\partial x^2}, \tag{8}$$

where w is the wave height, x is the distance variable, t is the time variable, and c is the velocity with which the waves are propagated.

In our example, x is the distance across the ocean's surface, but in other applications x might be the distance along a vibrating string, distance through air (sound waves), or distance through space (light waves). The number c varies with the medium and type of wave.

Show that the functions in Exercises 51–56 are all solutions of the wave equation.

51. $w = \sin(x + ct)$ **52.** $w = \cos(2x + 2ct)$

53. $w = \sin(x + ct) + \cos(2x + 2ct)$

54. $w = \ln(2x + 2ct)$

55. $w = \tan(2x - 2ct)$

56. $w = 5 \cos(3x + 3ct) + e^{x+ct}$

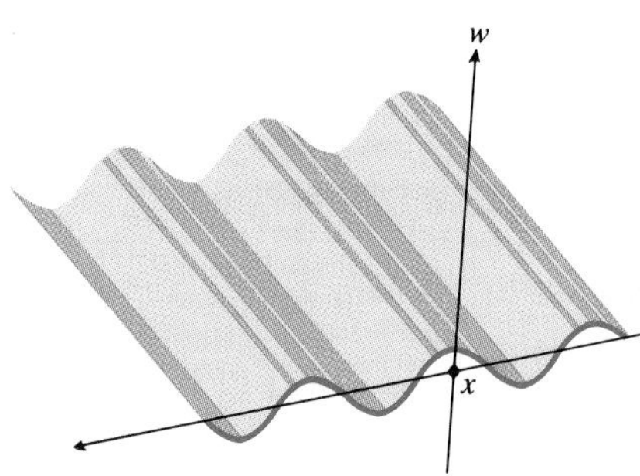

13.19 Waves in water at an instant in time. As time passes,

$$\frac{\partial^2 w}{\partial t^2} = c^2 \frac{\partial^2 w}{\partial x^2}.$$

13.4 The Chain Rule

When we are interested in the temperature $w = f(x, y, z)$ at points along a path

$$x = g(t), \quad y = h(t), \quad z = k(t)$$

in space, or in the pressure or density at points along a path through a gas or fluid, we may think of f as a function of the single variable t. For each value of t, the temperature at the point $(g(t), h(t), k(t))$ is the value of the composite $f(g(t), h(t), k(t))$. If we then wish to know the rate at which f changes with respect to t along the path, we have only to differentiate this composite with respect to t, provided, of course, the derivative exists.

Sometimes we can find the derivative by substituting the formulas for g, h, and k into the formula for f and differentiating directly with respect to t. But we often have to work with functions whose formulas are too complicated for convenient substitution or for which formulas are not readily available. To find a function's derivatives under circumstances like these, we use the Chain Rule. The form of the Chain Rule depends on how many variables are involved but, except for the presence of additional variables, it works just like the Chain Rule from Chapter 3.

The Chain Rule for Functions of Two Variables

In Chapter 3, we used the Chain Rule when $w = f(x)$ was a differentiable function of x and $x = g(t)$ was a differentiable function of t. This made w a differentiable function of t and the Chain Rule said that dw/dt could be calculated with the formula

$$\frac{dw}{dt} = \frac{dw}{dx}\frac{dx}{dt}. \tag{1}$$

The analogous formula for a function $w = f(x, y)$ of two independent variables, from advanced calculus, is this:

THEOREM 3

Chain Rule for $w = f(x, y)$

If f_x and f_y are continuous throughout an open region R in the xy-plane, and if x and y are differentiable functions of t, then w is a differentiable function of t and

$$\frac{dw}{dt} = \frac{\partial f}{\partial x}\frac{dx}{dt} + \frac{\partial f}{\partial y}\frac{dy}{dt} \tag{2}$$

throughout R.

The way to remember the Chain Rule is to picture the diagram below. To find dw/dt, start at w and read down each route to t, multiplying derivatives along the way. Then add the products.

Chain Rule

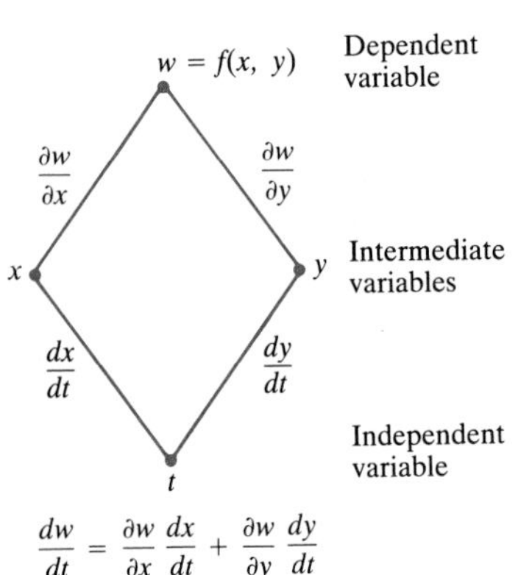

The derivation of Eq. (2) is much the same as in the single-variable case (Appendix 3). An equation similar to Eq. (12) in Section 3.7 says that if Δx, Δy, and Δw are the increments that result from changing t from t_0 to $t_0 + \Delta t$, then

$$\Delta w = \left(\frac{\partial w}{\partial x}\right)_{t_0} \Delta x + \left(\frac{\partial w}{\partial y}\right)_{t_0} \Delta y + \epsilon_1 \Delta x + \epsilon_2 \Delta y. \tag{3}$$

The ϵ_1 and ϵ_2 in this formula are like the ϵ in old Eq. (12) in that they both go to zero as Δt goes to zero. To Find the value of dw/dt at t_0, we divide Eq. (3) through by Δt and let Δt approach zero. The division gives

$$\frac{\Delta w}{\Delta t} = \left(\frac{\partial w}{\partial x}\right)_{t_0} \frac{\Delta x}{\Delta t} + \left(\frac{\partial w}{\partial y}\right)_{t_0} \frac{\Delta y}{\Delta t} + \epsilon_1 \frac{\Delta x}{\Delta t} + \epsilon_2 \frac{\Delta y}{\Delta t}, \tag{4}$$

and letting Δt approach zero gives the value of dw/dt at t_0:

$$\left(\frac{dw}{dt}\right)_{t_0} = \left(\frac{\partial w}{\partial x}\right)_{t_0}\left(\frac{dx}{dt}\right)_{t_0} + \left(\frac{\partial w}{\partial y}\right)_{t_0}\left(\frac{dy}{dt}\right)_{t_0} + 0 \cdot \left(\frac{dx}{dt}\right)_{t_0} + 0 \cdot \left(\frac{dy}{dt}\right)_{t_0}. \tag{5}$$

The last two expressions on the right are zero and the formula boils down to Eq. (2) when we leave off the subscript t_0 that tells where to evaluate the derivatives.

THEOREM 4

If the partial derivatives f_x and f_y of a function $f(x, y)$ are continuous throughout an open region R containing the point (x_0, y_0), then Eq. (3) holds for f at (x_0, y_0).

DEFINITIONS

We say that a function $f(x, y)$ is **differentiable at a point** (x_0, y_0) if Eq. (3) holds for f at (x_0, y_0) and we call f **differentiable** if it is differentiable at every point in its domain.

In light of this definition, we have the immediate corollary that a function with continuous partial derivatives is differentiable.

COROLLARY

If the partial derivatives f_x and f_y of a function $f(x, y)$ are continuous throughout an open region R, then f is differentiable at every point of R.

Since a function $f(x, y)$ must necessarily be continuous at any point (x_0, y_0) where Eq. (3) holds, we also have the following theorem.

THEOREM 5

If a function $f(x, y)$ is differentiable at (x_0, y_0) then f is continuous at (x_0, y_0).

EXAMPLE 1 Use the Chain Rule to find the derivative of

$$w = xy$$

with respect to t along the path $x = \cos t$, $y = \sin t$. What is the derivative's value when $t = \pi/2$?

Solution We evaluate the right-hand side of Eq. (2) with $w = xy$, $x = \cos t$, and $y = \sin t$:

$$\frac{\partial w}{\partial x} = y = \sin t, \quad \frac{\partial w}{\partial y} = x = \cos t, \quad \frac{dx}{dt} = -\sin t, \quad \frac{dy}{dt} = \cos t$$

$$\frac{dw}{dt} = \frac{\partial w}{\partial x}\frac{dx}{dt} + \frac{\partial w}{\partial y}\frac{dy}{dt} = (\sin t)(-\sin t) + (\cos t)(\cos t) \qquad \begin{pmatrix}\text{Eq. 2 with}\\ \text{values from}\\ \text{above}\end{pmatrix}$$

$$= -\sin^2 t + \cos^2 t = \cos 2t.$$

In this case we can check the result with a more direct calculation. As a function of t,

$$w = xy = \cos t \sin t = \frac{1}{2}\sin 2t,$$

so

$$\frac{dw}{dt} = \frac{d}{dt}\left(\frac{1}{2}\sin 2t\right) = \frac{1}{2}\cdot 2\cos 2t = \cos 2t.$$

In either case,

$$\left(\frac{dw}{dt}\right)_{t=\pi/2} = \cos\left(2\cdot\frac{\pi}{2}\right) = \cos \pi = -1.$$

The Chain Rule for Functions of Three Variables

To get the Chain Rule for functions of three variables, we add a term to Eq. (2).

Chain Rule for $w = f(x, y, z)$

If f_x, f_y, and f_z are continuous, and if x, y, and z are differentiable functions of t, then w is a differentiable function of t and

$$\frac{dw}{dt} = \frac{\partial f}{\partial x}\frac{dx}{dt} + \frac{\partial f}{\partial y}\frac{dy}{dt} + \frac{\partial f}{\partial z}\frac{dz}{dt}. \tag{6}$$

Here we have three routes from w to t instead of two. But the rule for evaluating dw/dt is still the same. Read down each route, multiplying derivatives along the way; then add.

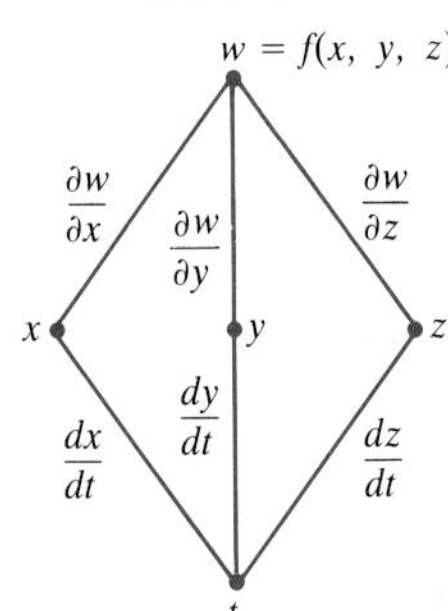

$$\frac{dw}{dt} = \frac{\partial w}{\partial x}\frac{dx}{dt} + \frac{\partial w}{\partial y}\frac{dy}{dt} + \frac{\partial w}{\partial z}\frac{dz}{dt}$$

The derivation is identical with the derivation of Eq. (2) except that there are now three intermediate variables instead of two. The diagram we use for remembering the new equation is similar as well.

EXAMPLE 2 *Changes in a function's values along a helix.* Find dw/dt if

$$w = xy + z, \quad x = \cos t, \quad y = \sin t, \quad z = t$$

(Fig. 13.20). What is the derivative's value at $t = 0$?

Solution

$$\frac{dw}{dt} = \frac{\partial w}{\partial x}\frac{dx}{dt} + \frac{\partial w}{\partial y}\frac{dy}{dt} + \frac{\partial w}{\partial z}\frac{dz}{dt} \qquad \text{(Eq. 6)}$$

$$= (y)(-\sin t) + (x)(\cos t) + (1)(1)$$

$$= (\sin t)(-\sin t) + (\cos t)(\cos t) + 1$$

$$= -\sin^2 t + \cos^2 t + 1 = 1 + \cos 2t,$$

$$\left(\frac{dw}{dt}\right)_{t=0} = 1 + \cos(0) = 2.$$

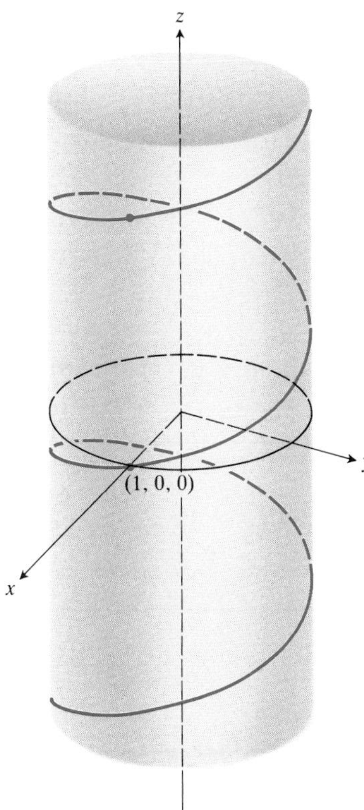

13.20 Example 2 shows how the values of the function $w = xy + z$ vary with t along this helix.

The Chain Rule for Functions Defined on Surfaces

If we are interested in the temperature $w = f(x, y, z)$ at points (x, y, z) on a globe in space, we might prefer to think of x, y, and z as functions of the variables r and s that give the points' longitudes and latitudes. If $x = g(r, s)$, $y = h(r, s)$, and $z = k(r, s)$, we could then express the temperature as a function of r and s with the composite function

$$w = f(g(r, s), h(r, s), k(r, s)).$$

Under the right conditions, w would have partial derivatives with respect to both r and s, that could be calculated in the following way.

Chain Rule for Two Independent Variables and Three Intermediate Variables

Suppose that $w = f(x, y, z)$, $x = g(r, s)$, $y = h(r, s)$, and $z = k(r, s)$. If all four functions are continuous and have continuous first partial derivatives with respect to all their independent variables, then w has partial derivatives with respect to r and s, given by the formulas

$$\frac{\partial w}{\partial r} = \frac{\partial w}{\partial x}\frac{\partial x}{\partial r} + \frac{\partial w}{\partial y}\frac{\partial y}{\partial r} + \frac{\partial w}{\partial z}\frac{\partial z}{\partial r}, \tag{7}$$

$$\frac{\partial w}{\partial s} = \frac{\partial w}{\partial x}\frac{\partial x}{\partial s} + \frac{\partial w}{\partial y}\frac{\partial y}{\partial s} + \frac{\partial w}{\partial z}\frac{\partial z}{\partial s}. \tag{8}$$

Equation (7) can be derived from Eq. (6) by holding s fixed and setting r equal to t. Similarly, Eq. (8) can be derived by holding r fixed and setting s equal to t. The tree diagrams for Eqs. (7) and (8) are shown in Fig. 13.21.

EXAMPLE 3 Express $\partial w/\partial r$ and $\partial w/\partial s$ in terms of r and s if

$$w = x + 2y + z^2, \quad x = \frac{r}{s}, \quad y = r^2 + \ln s, \quad z = 2r.$$

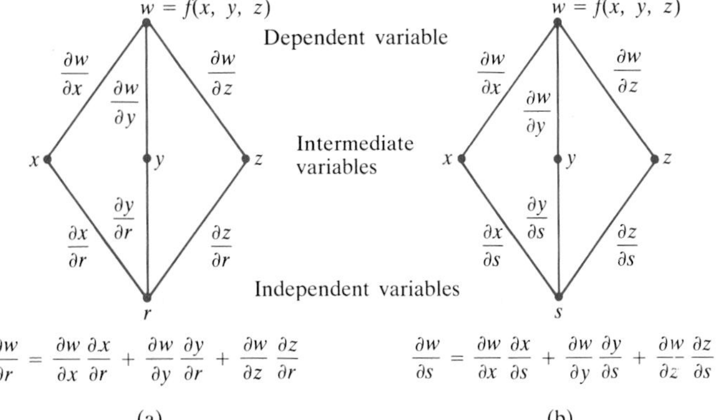

13.21 Tree diagrams for Eqs. (7) and (8).

Solution

$$\frac{\partial w}{\partial r} = \frac{\partial w}{\partial x}\frac{\partial x}{\partial r} + \frac{\partial w}{\partial y}\frac{\partial y}{\partial r} + \frac{\partial w}{\partial z}\frac{\partial z}{\partial r} \qquad \text{(Eq. 7)}$$

$$= (1)\left(\frac{1}{s}\right) + (2)(2r) + (2z)(2) = \frac{1}{s} + 4r + (4r)(2) = \frac{1}{s} + 12r,$$

$$\frac{\partial w}{\partial s} = \frac{\partial w}{\partial x}\frac{\partial x}{\partial s} + \frac{\partial w}{\partial y}\frac{\partial y}{\partial s} + \frac{\partial w}{\partial z}\frac{\partial z}{\partial s} \qquad \text{(Eq. 8)}$$

$$= (1)\left(-\frac{r}{s^2}\right) + 2\left(\frac{1}{s}\right) + (2z)(0) = \frac{2}{s} - \frac{r}{s^2}.$$

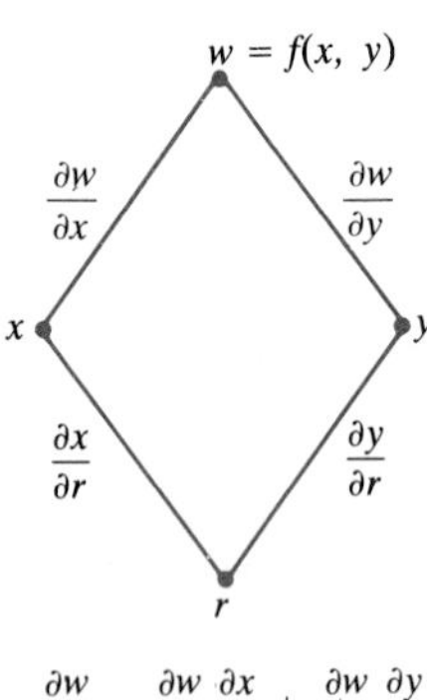

13.22 Tree diagram for Eq. (9).

If f is a function of two variables instead of three, Eqs. (7) and (8) become one term shorter.

If $w = f(x, y)$, $x = g(r, s)$, and $y = h(r, s)$ then

$$\frac{\partial w}{\partial r} = \frac{\partial w}{\partial x}\frac{\partial x}{\partial r} + \frac{\partial w}{\partial y}\frac{\partial y}{\partial r} \quad \text{and} \quad \frac{\partial w}{\partial s} = \frac{\partial w}{\partial x}\frac{\partial x}{\partial s} + \frac{\partial w}{\partial y}\frac{\partial y}{\partial s}. \qquad (9)$$

Figure 13.22 shows a tree diagram for Eq. (9). The diagram for the second equation in (9) is similar—just replace r with s.

EXAMPLE 4 Express $\partial w/\partial r$ and $\partial w/\partial s$ in terms of r and s if

$$w = x^2 + y^2, \quad x = r - s, \quad y = r + s.$$

Solution We use Eqs. (9):

$$\frac{\partial w}{\partial r} = \frac{\partial w}{\partial x}\frac{\partial x}{\partial r} + \frac{\partial w}{\partial y}\frac{\partial y}{\partial r} \qquad \frac{\partial w}{\partial s} = \frac{\partial w}{\partial x}\frac{\partial x}{\partial s} + \frac{\partial w}{\partial y}\frac{\partial y}{\partial s}$$

$$= (2x)(1) + (2y)(1) \qquad = (2x)(-1) + (2y)(1)$$

$$= 2(r - s) + 2(r + s) \qquad = -2(r - s) + 2(r + s)$$

$$= 4r, \qquad = 4s.$$

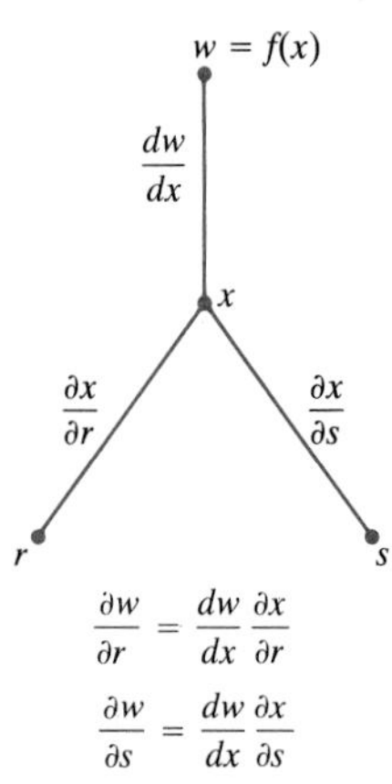

13.23 Tree diagram for Eqs. (10).

If f is a function of x alone, Eqs. (7) and (8) simplify still further.

If $w = f(x)$ and $x = g(r, s)$ then

$$\frac{\partial w}{\partial r} = \frac{dw}{dx}\frac{\partial x}{\partial r} \quad \text{and} \quad \frac{\partial w}{\partial s} = \frac{dw}{dx}\frac{\partial x}{\partial s}. \qquad (10)$$

Here dw/dx is the ordinary (single-variable) derivative (Fig. 13.23).

Implicit Differentiation (continued from Chapter 3)

Believe it or not, the two-variable Chain Rule in Eq. (2) leads to a formula that takes most of the work out of implicit differentiation. The formula comes up in the following way. Suppose:

1. The first partial derivatives with respect to x and y of a function $w = F(x, y)$ are continuous, and
2. The equation $F(x, y) = 0$ defines y implicitly as a differentiable function of x, say $y = h(x)$.

Now, x is also a differentiable function of x, so what we have, in effect, is a function $w = F(u, v)$ in which u and v are both differentiable functions of x. By Eq. (2), therefore,

$$\frac{dw}{dx} = \frac{\partial w}{\partial u}\frac{du}{dx} + \frac{\partial w}{\partial v}\frac{dv}{dx}. \qquad \text{(Eq. (2) with } x = u, y = v, t = x\text{)}$$

With $u = x$, and $v = y = h(x)$,

$$\frac{dw}{dx} = \frac{\partial w}{\partial x}\frac{dx}{dx} + \frac{\partial w}{\partial y}\frac{dy}{dx} = \frac{\partial w}{\partial x} \cdot 1 + \frac{\partial w}{\partial y}\frac{dy}{dx} \tag{11}$$

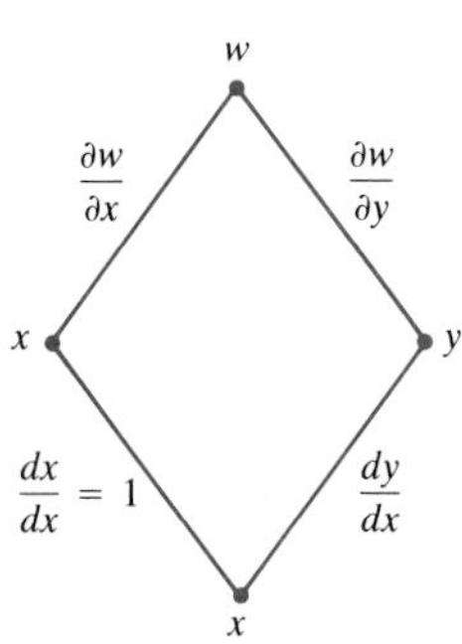

13.24 Tree diagram for Eq. (11).

(Fig. 13.24). Since $w = F(x, h(x)) = 0$ for all values of x (because h is defined by setting $F(x, y) = 0$), $dw/dx = 0$ and Eq. (11) reduces still further to

$$0 = \frac{\partial w}{\partial x} + \frac{\partial w}{\partial y}\frac{dy}{dx}. \tag{12}$$

If $\partial w/\partial y \neq 0$, we can solve Eq. (12) for dy/dx to get

$$\frac{dy}{dx} = -\frac{\partial w/\partial x}{\partial w/\partial y} = -\frac{F_x}{F_y}. \tag{13}$$

Suppose $F(x, y)$ and its partial derivatives F_x and F_y are continuous and the equation $F(x, y) = 0$ defines y as a differentiable function of x. Then, at any point where $F_y \neq 0$,

$$\frac{dy}{dx} = -\frac{F_x}{F_y}. \tag{14}$$

EXAMPLE 5 Find dy/dx if $x^2 + \sin y - 2y = 0$.

Solution Take $F(x, y) = x^2 + \sin y - 2y$. Then

$$\frac{dy}{dx} = -\frac{F_x}{F_y} = -\frac{2x}{\cos y - 2}. \qquad \text{(Eq. 14)}$$

This calculation is significantly shorter than the single-variable calculation with which we found dy/dx in Section 3.6, Example 3. ■

Remembering the Different Forms of the Chain Rule

How are we to remember all the different forms of the Chain Rule? The answer is that there is no need to remember them all. The best thing to do is to remember a few key equations, say (2), (6), and (7). Construct the others, when you need them, from tree diagrams. You can always draw the appropriate diagram by placing the dependent variable on top, the intermediate variables in the middle, and the selected independent variable at the bottom.

The Chain Rule for Functions of Many Variables

Suppose $w = f(x, y, \ldots, v)$ is a differentiable function of the variables $x, y, \ldots, v$ (a finite set) and the $x, y, \ldots, v$ are themselves differentiable functions of the variables $p, q, \ldots, t$ (another finite set). Suppose also that all the functions involved are continuous. Then w is a differentiable function of the variables p through t and the partial derivatives of w with respect to these variables are given by equations of the form

$$\frac{\partial w}{\partial p} = \frac{\partial w}{\partial x}\frac{\partial x}{\partial p} + \frac{\partial w}{\partial y}\frac{\partial y}{\partial p} + \cdots + \frac{\partial w}{\partial v}\frac{\partial v}{\partial p}. \tag{15}$$

The other equations are obtained by replacing p by $q, \ldots, t$, one at a time.

One way to remember Eq. (15) is to think of the right-hand side as the dot product of two vectors with components

$$\underbrace{\left(\frac{\partial w}{\partial x}, \frac{\partial w}{\partial y}, \ldots, \frac{\partial w}{\partial v}\right)}_{\substack{\text{Derivatives of } w \text{ with}\\ \text{respect to the}\\ \text{intermediate variables}}} \quad \text{and} \quad \underbrace{\left(\frac{\partial x}{\partial p}, \frac{\partial y}{\partial p}, \ldots, \frac{\partial v}{\partial p}\right)}_{\substack{\text{Derivatives of the intermediate}\\ \text{variables with respect to the}\\ \text{selected independent variable}}}. \tag{16}$$

Exercises 13.4

In Exercises 1–6, express dw/dt in terms of t. Then evaluate the derivative at the given value of t.

1. $w = x^2 + y^2, \quad x = \cos t, \quad y = \sin t; \quad t = \pi$

2. $w = x^2 + y^2, \quad x = \cos t + \sin t, \quad y = \cos t - \sin t; \quad t = 0$

3. $w = \dfrac{x}{z} + \dfrac{y}{z}, \quad x = \cos^2 t, \quad y = \sin^2 t, \quad z = 1/t; \quad t = 3$

4. $w = \ln(x^2 + y^2 + z^2), \quad x = \cos t, \quad y = \sin t, \quad z = 4\sqrt{t}; \quad t = 3$

5. $w = 2ye^x - \ln z, \quad x = \ln(t^2 + 1), \quad y = \tan^{-1} t, \quad z = e^t; \quad t = 1$

6. $w = z - \sin xy, \quad x = t, \quad y = \ln t, \quad z = e^{t-1}; \quad t = 1$

In Exercises 7–18, draw a tree diagram and write a Chain-Rule formula for each derivative.

7. $\dfrac{dz}{dt}$ for $z = f(x, y), \quad x = g(t), \quad y = h(t)$

8. $\dfrac{dz}{dt}$ for $z = f(u, v, w), \quad u = g(t), \quad v = h(t), \quad w = k(t)$

9. $\dfrac{\partial w}{\partial u}$ and $\dfrac{\partial w}{\partial v}$ for $w = h(x, y, z), \quad x = f(u, v), \quad y = g(u, v), \quad z = k(u, v)$

10. $\dfrac{\partial w}{\partial x}$ and $\dfrac{\partial w}{\partial y}$ for $w = f(r, s, t), \quad r = g(x, y), \quad s = h(x, y), \quad t = k(x, y)$

11. $\dfrac{\partial w}{\partial u}$ and $\dfrac{\partial w}{\partial v}$ for $w = g(x, y), \quad x = h(u, v), \quad y = k(u, v)$

12. $\dfrac{\partial w}{\partial x}$ and $\dfrac{\partial w}{\partial y}$ for $w = g(u, v), \quad u = h(x, y), \quad v = k(x, y)$

13. $\dfrac{\partial z}{\partial t}$ and $\dfrac{\partial z}{\partial s}$ for $z = f(x, y), \quad x = g(t, s), \quad y = h(t, s)$

14. $\dfrac{\partial y}{\partial r}$ for $y = f(u), \quad u = g(r, s)$

15. $\dfrac{\partial w}{\partial s}$ and $\dfrac{\partial w}{\partial t}$ for $w = g(u), \quad u = h(s, t)$

16. $\dfrac{\partial w}{\partial p}$ for $w = f(x, y, z, v), \quad x = g(p, q), \quad y = h(p, q), \quad z = j(p, q), \quad v = k(p, q)$

17. $\dfrac{\partial w}{\partial r}$ and $\dfrac{\partial w}{\partial s}$ for $w = f(x, y), \quad x = g(r), \quad y = h(s)$

18. $\dfrac{\partial w}{\partial s}$ for $w = g(x, y), \quad x = h(r, s, t), \quad y = k(r, s, t)$

Assuming that the equations in Exercises 19–22 define y as a differentiable function of x, use Eq. (14) to find the value of dy/dx at the given point.

19. $x^3 - 2y^2 + xy = 0, \quad (1, 1)$

20. $xy + y^2 - 3x - 3 = 0, \quad (-1, 1)$

21. $x^2 + xy + y^2 - 7 = 0, \quad (1, 2)$

22. $xe^y + \sin xy + y - \ln 2 = 0, \quad (0, \ln 2)$

Equation (14) can be generalized to functions of three variables and even more. The three-variable version goes like this:

If the equation $F(x, y, z) = 0$ determines z as a differentiable function of x and y, then, at points where $F_z \neq 0$,

$$\frac{\partial z}{\partial x} = -\frac{F_x}{F_z} \quad \text{and} \quad \frac{\partial z}{\partial y} = -\frac{F_y}{F_z}. \qquad (17)$$

Use these equations to find the values of $\partial z/\partial x$ and $\partial z/\partial y$ at the points in Exercises 23–26.

23. $z^3 - xy + yz + y^3 - 2 = 0, \quad (1, 1, 1)$

24. $\dfrac{1}{x} + \dfrac{1}{y} + \dfrac{1}{z} - 1 = 0, \quad (2, 3, 6)$

25. $\sin(x + y) + \sin(y + z) + \sin(x + z) = 0, \quad (\pi, \pi, \pi)$

26. $xe^y + ye^z + 2\ln x - 2 - 3\ln 2 = 0, \quad (1, \ln 2, \ln 3)$

27. Find $\partial w/\partial r$ when $r = 1, s = -1$ if $w = (x + y + z)^2$, $x = r - s, y = \cos(r + s), z = \sin(r + s)$.

28. Find $\partial w/\partial v$ when $u = -1, v = 2$ if $w = xy + \ln z$, $x = v^2/u, y = u + v, z = \sin u$.

29. Find $\partial w/\partial v$ when $u = 0, \ v = 0$ if $w = x^2 + \dfrac{y}{x}$, $x = u - 2v + 1, y = 2u + v - 2$.

30. Find $\partial z/\partial u$ when $u = 0, v = 1$ if $z = \sin xy + x \sin y$, $x = u^2 + v^2, y = uv$.

31. Find $\partial z/\partial u$ and $\partial z/\partial v$ when $u = \ln 2, v = 1$ if $z = 5\tan^{-1}x$ and $x = e^u + \ln v$.

32. Find $\partial w/\partial q$ when $p = 1, q = 4$ if $w = xy + yz + zu + uv, \quad x = pq, \quad y = q\tan^{-1}p, \quad z = \ln(q - 3p)$, $u = \sqrt{q + 5p}, v = p + q + 1$.

33. *Changing voltage in a circuit.* The voltage in a circuit that satisfies Ohm's law $V = IR$ is slowly dropping as the battery wears out. At the same time, the resistance is increasing as the resistor heats up. Use the equation

$$\frac{dV}{dt} = \frac{\partial V}{\partial I}\frac{dI}{dt} + \frac{\partial V}{\partial R}\frac{dR}{dt}$$

to find how the current is changing at the instant when $R = 600$ ohms, $I = 0.04$ amp, $dR/dt = 0.5$ ohms/sec, and $dV/dt = -0.01$ volts/sec.

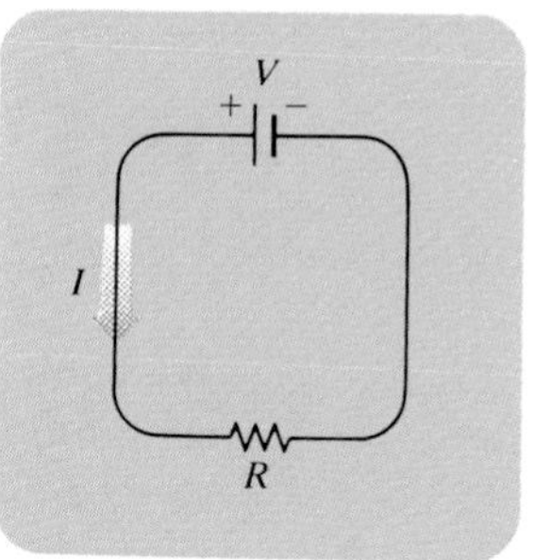

34. *Changing dimensions in a box.* The dimensions a, b, and c of a rectangular solid are changing with time. At the instant in question, $a = 13$ cm, $b = 9$ cm, $c = 5$ cm, $da/dt = dc/dt = 2$ cm/sec, and $db/dt = -5$ cm/sec.

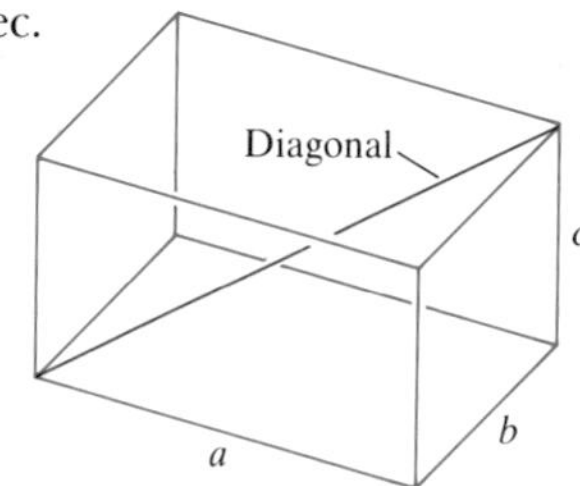

a) How fast are the volume $V = abc$ and the surface area $S = 4ac + 2bc$ changing at that instant? Are they increasing, or decreasing?

b) Is the solid's diagonal increasing, or decreasing?

Changes in Functions along Curves

35. Suppose that the partial derivatives of a function $f(x, y, z)$ at points on the helix $x = \cos t, y = \sin t, z = t$ are

$$f_x = \cos t, \quad f_y = \sin t, \quad f_z = t^2 + t - 2.$$

At what points on the curve, if any, can f take on extreme values?

36. Let $w = x^2e^{2y}\cos 3z$. Find the value of dw/dt at the point $(1, \ln 2, 0)$ on the curve $x = \cos t, y = \ln(t + 2), z = t$.

37. Let $T = f(x, y)$ be the temperature at the point (x, y) on the circle $x = \cos t, y = \sin t$ and suppose that

$$\frac{\partial T}{\partial x} = 8x - 4y, \quad \frac{\partial T}{\partial y} = 8y - 4x.$$

a) Find where the maximum and minimum temperatures on the circle occur by examining the derivatives dT/dt and d^2T/dt^2.

b) Suppose $T = 4x^2 - 4xy + 4y^2$. Find the maximum and minimum values of t on the circle.

13.5 Directional Derivatives and Gradient Vectors

We know from Section 13.4 that if $f(x, y, z)$ and its partial derivatives are continuous, then the rate at which the values of f change with respect to t along a differentiable curve

$$x = g(t), \qquad y = h(t), \qquad z = k(t)$$

may be calculated from the formula

$$\frac{df}{dt} = \frac{\partial f}{\partial x}\frac{dx}{dt} + \frac{\partial f}{\partial y}\frac{dy}{dt} + \frac{\partial f}{\partial z}\frac{dz}{dt}. \tag{1}$$

To use this formula at any particular point

$$P_0(x_0, y_0, z_0) = P_0(g(t_0), h(t_0), k(t_0)),$$

we evaluate the partial derivatives of f at P_0 and the derivatives of x, y, and z at t_0.

The derivative df/dt in Eq. (1) is the rate of change of f at P_0 with respect to increasing t and therefore depends, among other things, on the direction of motion along the curve. This observation is particularly important when the curve is a straight line and the parameter t is the distance along the line measured from P_0 in the direction of a given unit vector **u**. For then df/dt is the rate of change of f with respect to distance in its domain in the direction of **u**. By varying **u**, we can find the rates at which f changes with respect to distance as we move through P_0 in different directions. These "directional derivatives" are extremely useful in science and engineering (as well as in mathematics). The goal of this section is to develop a formula for calculating them.

Calculating Directional Derivatives

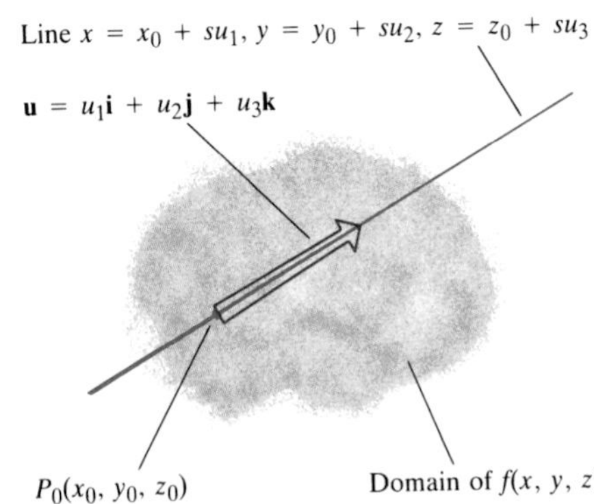

13.25 The derivative of f in the direction of the unit vector **u** at a point P_0 in its domain is the value of df/ds at P_0.

Suppose that the partial derivatives of a function $f(x, y, z)$ are continuous throughout some region R in space, that $P_0(x_0, y_0, z_0)$ is a point in R, and that $\mathbf{u} = u_1\mathbf{i} + u_2\mathbf{j} + u_3\mathbf{k}$ is a unit vector (Fig. 13.25). Then

$$x = x_0 + su_1, \qquad y = y_0 + su_2, \qquad z = z_0 + su_3 \tag{2}$$

are parametric equations for the line through P_0 parallel to **u**, and the parameter s measures the directed distance along this line from P_0 to the point $P_s(x_0 + su_1, y_0 + su_2, z_0 + su_3)$, as we saw in Example 3, Section 12.3. To calculate df/ds, the rate at which f changes with respect to distance in the direction of **u** at P_0, we substitute the derivatives

$$\frac{dx}{ds} = u_1, \qquad \frac{dy}{ds} = u_2, \qquad \frac{dz}{ds} = u_3$$

into the equation

$$\frac{df}{ds} = \frac{\partial f}{\partial x}\frac{dx}{ds} + \frac{\partial f}{\partial y}\frac{dy}{ds} + \frac{\partial f}{\partial z}\frac{dz}{ds}$$

and evaluate the partial derivatives of f at P_0. This gives

$$\frac{df}{ds} = \left(\frac{\partial f}{\partial x}\right)_{P_0} u_1 + \left(\frac{\partial f}{\partial y}\right)_{P_0} u_2 + \left(\frac{\partial f}{\partial z}\right)_{P_0} u_3. \tag{3}$$

The expression on the right-hand side of Eq. (3) is the dot product of $\mathbf{u}$ and the vector

$$\nabla f = \left(\frac{\partial f}{\partial x}\right)_{P_0} \mathbf{i} + \left(\frac{\partial f}{\partial y}\right)_{P_0} \mathbf{j} + \left(\frac{\partial f}{\partial z}\right)_{P_0} \mathbf{k}.$$

This vector is called the *gradient* of f at the point P_0.

It is customary to picture the gradient as a vector in the domain of f. Its components are calculated by evaluating the three partial derivatives of f at (x_0, y_0, z_0). The derivative on the left-hand side of Eq. (3) is called *the derivative of f at the point P_0 in the direction of* $\mathbf{u}$. It is often denoted by

$$(D_{\mathbf{u}} f)_{P_0} \quad \text{or} \quad \left(\frac{df}{ds}\right)_{\mathbf{u},\, P_0}$$

DEFINITIONS

If the partial derivatives of $f(x, y, z)$ are defined at $P_0(x_0, y_0, z_0)$, then the **gradient** of f at P_0 is the vector

$$\nabla f = \frac{\partial f}{\partial x}\mathbf{i} + \frac{\partial f}{\partial y}\mathbf{j} + \frac{\partial f}{\partial z}\mathbf{k} \tag{4}$$

in the domain of f obtained by evaluating the partial derivatives of f at P_0.

If $f(x, y, z)$ has continuous partial derivatives at $P_0(x_0, y_0, z_0)$ and $\mathbf{u}$ is a unit vector, then the **derivative of f at P_0 in the direction of $\mathbf{u}$** is the number

$$(D_{\mathbf{u}} f)_{P_0} = (\nabla f)_{P_0} \cdot \mathbf{u}, \tag{5}$$

which is the scalar product of the gradient f at P_0 and $\mathbf{u}$.

Another notation for the gradient of f is

$$\text{grad}\, f,$$

read the way it is written. The symbol ∇f is read "grad f" as well as "gradient of f" and "del f."

EXAMPLE 1 Find the derivative of

$$f(x, y, z) = x^3 - xy^2 - z$$

at $P_0(1, 1, 0)$ in the direction of the vector $\mathbf{A} = 2\mathbf{i} - 3\mathbf{j} + 6\mathbf{k}$.

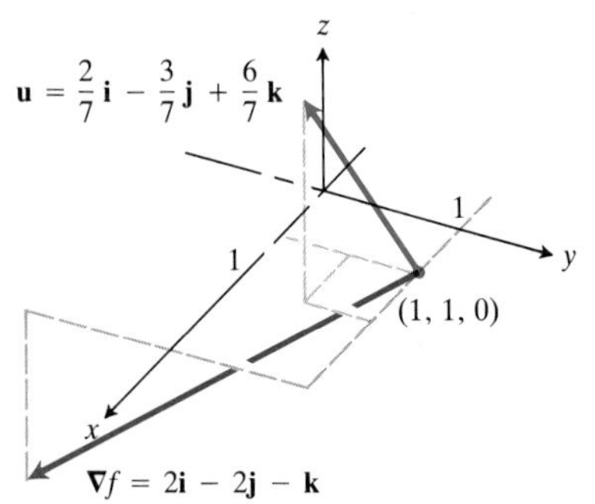

13.26 The change in the direction of **u** is $\nabla f \cdot \mathbf{u} = 4/7$ (Example 1).

Solution The direction of **A** is obtained by dividing **A** by its length:

$$|\mathbf{A}| = \sqrt{(2)^2 + (-3)^2 + (6)^2} = \sqrt{49} = 7,$$

$$\mathbf{u} = \frac{\mathbf{A}}{|\mathbf{A}|} = \frac{2}{7}\mathbf{i} - \frac{3}{7}\mathbf{j} + \frac{6}{7}\mathbf{k}.$$

The partial derivatives of f at P_0 are

$$f_x = 3x^2 - y^2|_{(1,1,0)} = 2, \qquad f_y = -2xy|_{(1,1,0)} = -2, \qquad f_z = -1|_{(1,1,0)} = -1.$$

The gradient of f at P_0 is

$$\nabla f|_{(1,1,0)} = 2\mathbf{i} - 2\mathbf{j} - \mathbf{k}$$

(Fig. 13.26). The derivative of f at P_0 in the direction of **A** is therefore

$$(D_{\mathbf{u}}f)|_{(1,1,0)} = \nabla f|_{(1,1,0)} \cdot \mathbf{u} = (2\mathbf{i} - 2\mathbf{j} - \mathbf{k}) \cdot \left(\frac{2}{7}\mathbf{i} - \frac{3}{7}\mathbf{j} + \frac{6}{7}\mathbf{k}\right)$$

$$= \frac{4}{7} + \frac{6}{7} - \frac{6}{7} = \frac{4}{7}.$$

■

Increments and Distance

The directional derivative plays the role of an ordinary derivative when we want to estimate how much a function f changes if we move a small distance ds from a point P_0 to a nearby point. If f were a function of a single variable, we would have

$$df = f'(P_0)\,ds. \qquad \text{(Ordinary derivative} \times \text{increment)}$$

For a function of two or more variables, we use the formula

$$df = (\nabla f|_{P_0} \cdot \mathbf{u})\,ds, \qquad \text{(Directional derivative} \times \text{increment)}$$

where **u** is the direction of the motion away from P_0.

Estimating the Change in *f* in a Direction u

To estimate the change in the value of a function f when we move a small distance ds from a point P_0 in a particular direction **u**, use the formula

$$df = \underbrace{(\nabla f|_{P_0} \cdot \mathbf{u})}_{\text{Directional derivative}} \cdot \underbrace{ds}_{\text{Distance increment}} \qquad (6)$$

EXAMPLE 2 Estimate how much the value of

$$f(x, y, z) = xe^y + yz$$

will change if the point $P(x, y, z)$ is moved from $P_0(2, 0, 0)$ straight toward $P_1(4, 1, -2)$ a distance of $ds = 0.1$ units (Fig. 13.27).

13.27 Example 2 shows how to estimate the change in the value of $f(x, y, z) = xe^y + yz$ as (x, y, z) moves a short distance $ds = 0.1$ units from P_0 along the line toward P_1.

Solution We first find the derivative of f at P_0 in the direction of the vector

$$\overrightarrow{P_0P_1} = 2\mathbf{i} + \mathbf{j} - 2\mathbf{k}.$$

The direction of this vector is

$$\mathbf{u} = \frac{\overrightarrow{P_0P_1}}{|\overrightarrow{P_0P_1}|} = \frac{\overrightarrow{P_0P_1}}{3} = \frac{2}{3}\mathbf{i} + \frac{1}{3}\mathbf{j} - \frac{2}{3}\mathbf{k}.$$

The gradient of f at P_0 is

$$\nabla f|_{(2,0,0)} = (e^y\mathbf{i} + (xe^y + z)\mathbf{j} + y\mathbf{k})|_{(2,0,0)} = \mathbf{i} + 2\mathbf{j}.$$

Therefore,

$$\nabla f|_{P_0} \cdot \mathbf{u} = (\mathbf{i} + 2\mathbf{j}) \cdot \left(\frac{2}{3}\mathbf{i} + \frac{1}{3}\mathbf{j} - \frac{2}{3}\mathbf{k}\right) = \frac{2}{3} + \frac{2}{3} = \frac{4}{3}.$$

The change df in f that results from moving $ds = 0.1$ units away from P_0 in the direction of $\mathbf{u}$ is approximately

$$df = \nabla f|_{P_0} \cdot \mathbf{u}\, ds = \left(\frac{4}{3}\right)(0.1) \approx 0.13.$$

■

We shall say more about estimating change in Section 13.7.

If we evaluate the dot product in the formula for the directional derivative, to obtain

$$D_{\mathbf{u}}f = \nabla f \cdot \mathbf{u} = |\nabla f|\,|\mathbf{u}| \cos\theta = |\nabla f| \cos\theta, \tag{7}$$

the following facts come to light:

Properties of the Directional Derivative $D_{\mathbf{u}}f = \nabla f \cdot \mathbf{u} = |\nabla f| \cos\theta$

1. The directional derivative has its largest positive value when $\cos\theta = 1$, or when $\mathbf{u}$ is the direction of the gradient. That is, f increases most rapidly at any point in its domain in the direction of ∇f. The derivative in this direction is $D_{\mathbf{u}}f = |\nabla f| \cos(0) = |\nabla f|$.
2. Similarly, f decreases most rapidly in the direction of $-\nabla f$. The derivative in this direction is $D_{\mathbf{u}}f = |\nabla f| \cos(\pi) = -|\nabla f|$.
3. $D_{-\mathbf{u}}f = \nabla f \cdot (-\mathbf{u}) = -\nabla f \cdot \mathbf{u} = -D_{\mathbf{u}}f$. That is, the derivative of f in the direction of $-\mathbf{u}$ is the negative of the derivative of f in the direction of $\mathbf{u}$.
4. The relationships of the partial derivatives of f to the directional derivative are

$$D_{\mathbf{i}}f = \nabla f \cdot \mathbf{i} = f_x, \qquad D_{\mathbf{j}}f = \nabla f \cdot \mathbf{j} = f_y, \qquad D_{\mathbf{k}}f = \nabla f \cdot \mathbf{k} = f_z.$$

Thus, f_x = derivative of f in the $\mathbf{i}$ direction,
f_y = derivative of f in the $\mathbf{j}$ direction,
f_z = derivative of f in the $\mathbf{k}$ direction.

Combining these results with Property 3 gives

$$D_{-\mathbf{i}}f = -f_x, \qquad D_{-\mathbf{j}}f = -f_y, \qquad D_{-\mathbf{k}}f = -f_z.$$

5. Any direction $\mathbf{u}$ normal (perpendicular) to the gradient is a direction of zero change in f because

$$D_{\mathbf{u}}f = |\nabla f| \cos(\pi/2) = |\nabla f| \cdot 0 = 0.$$

EXAMPLE 3 Find the directions in which $f(x, y, z) = x^2 + y^2 + z^2$ increases and decreases most rapidly at the point (1, 1, 1) in its domain. At what rates does f change in these directions?

Solution The most rapid increase is in the direction of

$$\nabla f = 2x\mathbf{i} + 2y\mathbf{j} + 2z\mathbf{k}.$$

At (1, 1, 1) we have

$$\nabla f = 2\mathbf{i} + 2\mathbf{j} + 2\mathbf{k}, \qquad |\nabla f| = \sqrt{4 + 4 + 4} = 2\sqrt{3}.$$

The direction of ∇f is

$$\mathbf{u} = \frac{\nabla f}{|\nabla f|} = \frac{1}{\sqrt{3}}(\mathbf{i} + \mathbf{j} + \mathbf{k}).$$

The derivative of f in this direction is $|\nabla f| = 2\sqrt{3}$.

The direction of most rapid decrease at (1, 1, 1) is

$$-\mathbf{u} = -\frac{1}{\sqrt{3}}(\mathbf{i} + \mathbf{j} + \mathbf{k})$$

and the derivative of f in this direction is $-|\nabla f| = -2\sqrt{3}$. ■

Results for Functions of Two Variables

For functions of two variables, we get results much like the ones for functions of three variables. We obtain the two-variable formulas by dropping the z-terms from the three-variable formulas. Thus, for a function $f(x, y)$ and a unit vector $\mathbf{u} = u_1\mathbf{i} + u_2\mathbf{j}$,

$$\nabla f = \frac{\partial f}{\partial x}\mathbf{i} + \frac{\partial f}{\partial y}\mathbf{j}, \tag{8}$$

and

$$D_{\mathbf{u}}f = \nabla f \cdot \mathbf{u} = \frac{\partial f}{\partial x}u_1 + \frac{\partial f}{\partial y}u_2. \tag{9}$$

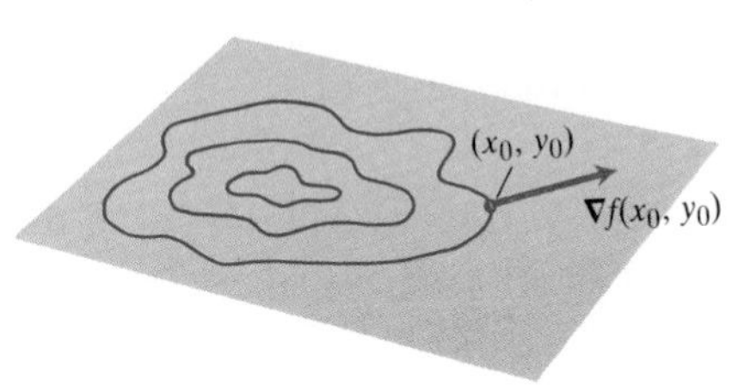

13.28 The gradient of a function of two variables is always normal to the function's level curves.

The five properties listed earlier for directional derivatives of functions of three variables hold equally well in the two-variable case. In addition, Property 5, which says that any direction normal to the gradient is a direction of zero change for the function, can now be turned around to say that the gradient of a function $f(x, y)$ is always normal to the function's level curves (Fig. 13.28). (See Exercise 33 for additional details.)

> At every point (x_0, y_0) in the domain of $f(x, y)$, the gradient of f is normal to the level curve through (x_0, y_0).

EXAMPLE 4

a) Find the derivative of

$$f(x, y) = x^2 + y^2$$

at the point $P_0(1, 1)$ in the direction of the unit vector $\mathbf{u} = u_1\mathbf{i} + u_2\mathbf{j}$.

b) In what direction in its domain (the xy-plane) does f increase most rapidly at (1, 1)? What is the derivative of f in this direction?

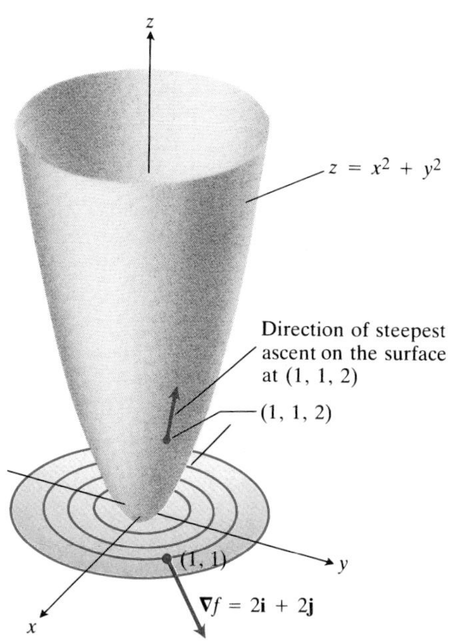

13.29 The gradient of $f(x, y) = x^2 + y^2$ at the point (1, 1) is the vector $\nabla f = 2\mathbf{i} + 2\mathbf{j}$ in the xy-plane. This vector points in the direction in which f increases most rapidly at the point (1, 1) and is normal to the level curve there. The corresponding direction on the surface $z = f(x, y)$ is the direction of steepest ascent at the point (1, 1, 2).

Solution

a) We have

$$f(x, y) = x^2 + y^2, \qquad f_x(1, 1) = 2x\big|_{(1,1)} = 2, \qquad f_y(1, 1) = 2y\big|_{(1,1)} = 2,$$
$$(\nabla f)_{(1,1)} = f_x(1, 1)\mathbf{i} + f_y(1, 1)\mathbf{j} = 2\mathbf{i} + 2\mathbf{j},$$

and

$$(D_{\mathbf{u}}f)_{(1,1)} = (\nabla f)_{(1,1)} \cdot \mathbf{u} = (2\mathbf{i} + 2\mathbf{j}) \cdot \mathbf{u} = 2u_1 + 2u_2.$$

The derivative of f at (1, 1) in the direction of $\mathbf{u}$ is $2u_1 + 2u_2$.

b) The function increases most rapidly at (1, 1) in the direction of ∇f at (1, 1), which is

$$\left(\frac{\nabla f}{|\nabla f|}\right)_{(1,1)} = \frac{2\mathbf{i} + 2\mathbf{j}}{\sqrt{2^2 + 2^2}} = \frac{2}{\sqrt{8}}\mathbf{i} + \frac{2}{\sqrt{8}}\mathbf{j} = \frac{1}{\sqrt{2}}(\mathbf{i} + \mathbf{j})$$

(Fig. 13.29). The derivative of f in this direction at (1, 1) is

$$|\nabla f|_{(1,1)} = 2\sqrt{2}.$$

■

Algebra Rules for Gradients

If we know the gradients of two functions f and g, we automatically know the gradients of their constant multiples and their sum, difference, and product.

Algebra Rules for Gradients

1. *Constant Multiple Rule:* $\nabla(kf) = k\nabla f$ (Any number k)
2. *Sum Rule:* $\nabla(f + g) = \nabla f + \nabla g$
3. *Difference Rule:* $\nabla(f - g) = \nabla f - \nabla g$
4. *Product Rule:* $\nabla(fg) = f\nabla g + g\nabla f$

Notice that these rules have the same form as the corresponding rules for derivatives, as they should. You will see where the rules come from if you do Exercise 34.

EXAMPLE 5 With

$$f(x, y, z) = e^x, \qquad g(x, y, z) = y - z,$$
$$\nabla f = e^x\mathbf{i}, \qquad \nabla g = \mathbf{j} - \mathbf{k},$$

we find

1. $\nabla(2f) = \nabla(2e^x) = 2e^x\mathbf{i} = 2\nabla f$,
2. $\nabla(f + g) = \nabla(e^x + y - z) = e^x\mathbf{i} + \mathbf{j} - \mathbf{k} = (e^x\mathbf{i}) + (\mathbf{j} - \mathbf{k}) = \nabla f + \nabla g$,
3. $\nabla(f - g) = \nabla(e^x - y + z) = e^x\mathbf{i} - \mathbf{j} + \mathbf{k} = e^x\mathbf{i} - (\mathbf{j} - \mathbf{k}) = \nabla f - \nabla g$,
4. $\nabla(fg) = \nabla(ye^x - ze^x) = (ye^x - ze^x)\mathbf{i} + e^x\mathbf{j} - e^x\mathbf{k}$
   $= (y - z)(e^x\mathbf{i}) + e^x(\mathbf{j} - \mathbf{k}) = g\nabla f + f\nabla g$.

■

Exercises 13.5

In Exercises 1–4, find ∇f at the given point.

1. $f(x, y, z) = x^2 + y^2 - 2z^2$, $(1, 1, 1)$

2. $f(x, y, z) = 2z^3 - 3(x^2 + y^2)z$, $(1, 1, 1)$

3. $f(x, y, z) = (x^2 + y^2 + z^2)^{-1/2}$, $(1, 2, -2)$

4. $f(x, y, z) = e^{x+y} \cos z$, $(0, 0, \pi/6)$

In Exercises 5–8, find ∇f at the given point. Then sketch ∇f together with the level curve that passes through the point.

5. $f(x, y) = y - x$, $(2, 1)$

6. $f(x, y) = \ln(x^2 + y^2)$, $(1, 1)$

7. $f(x, y) = y - x^2$, $(-1, 0)$

8. $f(x, y) = x^2 - y^2$, $(2, \sqrt{3})$

In Exercises 9–16, find the derivative of f at P_0 in the direction of $\mathbf{A}$.

9. $f(x, y, z) = xy + yz + zx$, $P_0(1, -1, 2)$, $\mathbf{A} = 3\mathbf{i} + 6\mathbf{j} - 2\mathbf{k}$

10. $f(x, y, z) = x^2 + 2y^2 - 3z^2$, $P_0(1, 1, 1)$, $\mathbf{A} = \mathbf{i} + \mathbf{j} + \mathbf{k}$

11. $f(x, y, z) = 3e^x \cos yz$, $P_0(0, 0, 0)$, $\mathbf{A} = 2\mathbf{i} + \mathbf{j} - 2\mathbf{k}$

12. $f(x, y, z) = \cos xy + e^{yz} + \ln zx$, $P_0(1, 0, 1/2)$, $\mathbf{A} = \mathbf{i} + 2\mathbf{j} + 2\mathbf{k}$

13. $f(x, y) = x - (y^2/x)$, $P_0(1, 1)$, $\mathbf{A} = 12\mathbf{i} + 5\mathbf{j}$

14. $f(x, y) = 2x^2 + y^2$, $P_0(-1, 1)$, $\mathbf{A} = 3\mathbf{i} - 4\mathbf{j}$

15. $f(x, y) = x^2 + 2xy - 3y^2$, $P_0(1, 1)$, $\mathbf{A} = 3\mathbf{i} + 4\mathbf{j}$

16. $f(x, y) = x \tan^{-1}(y/x)$, $P_0(1, 1)$, $\mathbf{A} = 2\mathbf{i} - \mathbf{j}$

In Exercises 17–22, find the directions in which f increases and decreases most rapidly at P_0. Then find the derivatives of f in these directions.

17. $f(x, y) = x^2 + \sin y$, $P_0(1, 0)$

18. $f(x, y) = x^2 + xy + y^2$, $P_0(-1, 1)$

19. $f(x, y, z) = xe^y + z^2$, $P_0(1, \ln 2, 1/2)$

20. $f(x, y, z) = (x/y) - yz$, $P_0(4, 1, 1)$

21. $f(x, y, z) = \sin x + \sin y + \sin z$, $P_0(0, 0, 0)$

22. $f(x, y, z) = \ln(x^2 + y^2 - 1) + y + 6z$, $P_0(1, 1, 0)$

23. Estimate how much the value of $f(x, y) = \cos \pi xy + xy^2$ will change if (x, y) is moved from $(-1, -1)$ a distance of $ds = 0.1$ units in the direction of $\mathbf{i} + \mathbf{j}$.

24. By about how much will

$$f(x, y, z) = \ln \sqrt{x^2 + y^2 + z^2}$$

change if the point (x, y, z) is moved from $(3, 4, 12)$ a distance of $ds = 0.1$ units in the direction of $3\mathbf{i} + 6\mathbf{j} - 2\mathbf{k}$?

25. By about how much will

$$f(x, y, z) = e^x \cos yz$$

change as the point $P(x, y, z)$ moves from the origin at a distance of $ds = 0.1$ units in the direction of $2\mathbf{i} + \mathbf{j} - 2\mathbf{k}$?

26. By about how much will

$$f(x, y, z) = x + x \cos z - y \sin z + y$$

change if the point $P(x, y, z)$ moves from $P_0(2, -1, 0)$ a distance of $ds = 0.2$ units toward the point $P_1(0, 1, 2)$?

27. In what two directions is the derivative of $f(x, y) = xy + y^2$ equal to zero at the point $(3, 2)$? (Fig. 13.30)

28. In what two directions is the derivative of $f(x, y) = (x^2 - y^2)/(x^2 + y^2)$ equal to zero at the point $(1, 1)$? (Fig. 13.31)

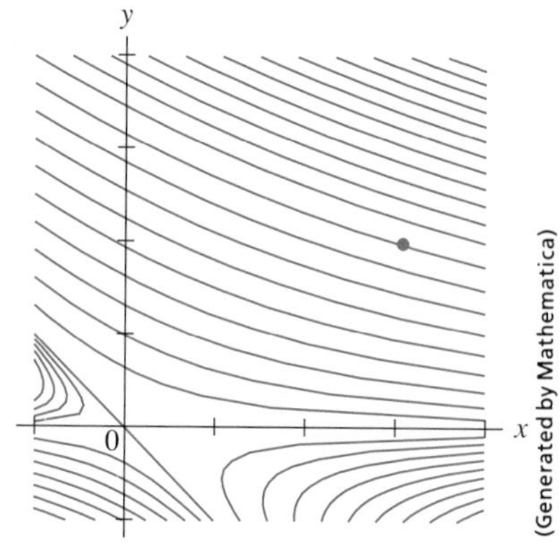

13.30 Level curves of $f(x, y) = xy + y^2$ (Exercise 27).

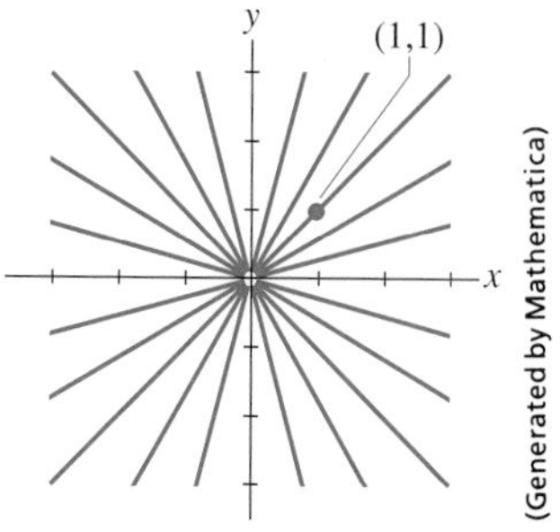

13.31 Level curves of $f(x, y) = (x^2 - y^2)/(x^2 + y^2)$ (Exercise 28).

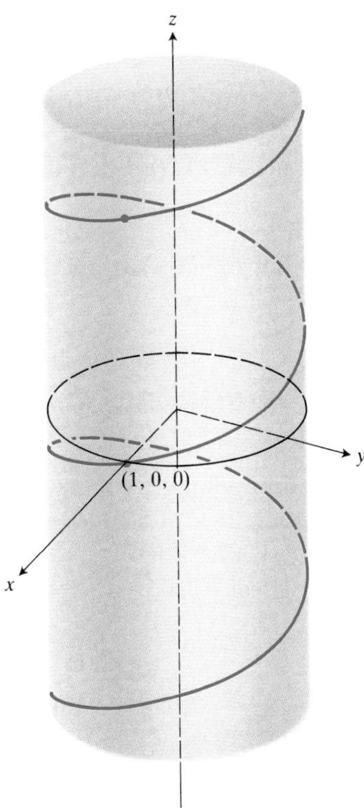

13.32 As you move up the helix, your distance from the origin decreases until you reach (1, 0, 0) and then increases as you rise above the *xy*-plane. Exercise 29 shows you how to calculate the rate at which the square of the distance to the origin changes with respect to distance traveled along the curve.

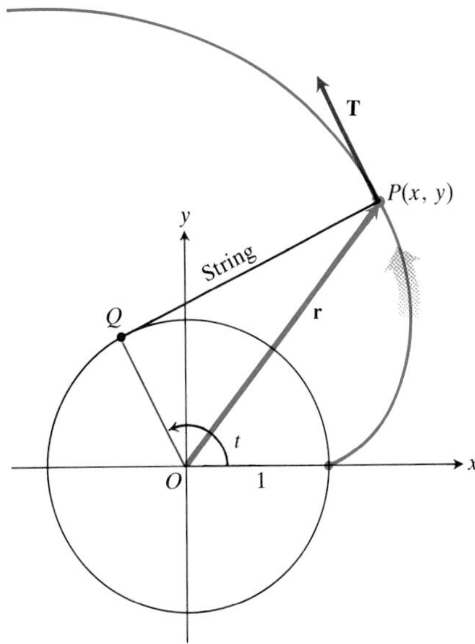

13.33 The involute of the unit circle from Section 12.3, Example 5. If you move out along the involute, covering distance along the curve at a steady rate, your distance from the origin will increase at a constant rate as well. (This is how to interpret the result of your calculation in Exercise 30.)

29. *Traveling on a helix.* Find the derivative of $f(x, y, z) = x^2 + y^2 + z^2$ in the direction of the unit tangent vector of the helix

$$\mathbf{r} = (\cos t)\mathbf{i} + (\sin t)\mathbf{j} + t\mathbf{k}$$

at the points where $t = -\pi/4$, 0, and $\pi/4$ (Fig. 13.32).

30. *Traveling on the involute of a circle.* Find the derivative of $f(x, y) = x^2 + y^2$ in the direction of the unit tangent vector of the curve

$$\mathbf{r} = (\cos t + t \sin t)\mathbf{i} + (\sin t - t \cos t)\mathbf{j}, \quad t > 0.$$

See Fig. 13.33.

31. The derivative of $f(x, y)$ at $P_0(1, 2)$ in the direction of the vector $\mathbf{i} + \mathbf{j}$ is $2\sqrt{2}$ and in the direction of the vector $-2\mathbf{j}$ is -3. What is the derivative of f in the direction of $-\mathbf{i} - 2\mathbf{j}$?

32. The derivative of $f(x, y, z)$ at a given point P is greatest in the direction of the vector $\mathbf{A} = \mathbf{i} + \mathbf{j} - \mathbf{k}$. In this direction the value of the derivative is $2\sqrt{3}$.

a) Find ∇f at P.

b) Find the derivative of f at P in the direction of $\mathbf{i} + \mathbf{j}$.

33. *Another way to see why gradients are always normal to level curves.* Suppose that $f(x, y)$ and its first partial derivatives are continuous and that f has the constant value c along the differentiable curve $x = g(t)$, $y = h(t)$ for all values of t. Differentiate both sides of the equation $f(g(t), h(t)) = c$ with respect to t to show that ∇f is normal to the curve's tangent vector at every point.

34. *The Algebra Rules for gradients.* Given a constant k and the gradients

$$\nabla f = \frac{\partial f}{\partial x}\mathbf{i} + \frac{\partial f}{\partial y}\mathbf{j} + \frac{\partial f}{\partial z}\mathbf{k} \quad \text{and} \quad \nabla g = \frac{\partial g}{\partial x}\mathbf{i} + \frac{\partial g}{\partial y}\mathbf{j} + \frac{\partial g}{\partial z}\mathbf{k},$$

use the facts that

$$\frac{\partial}{\partial x}(kf) = k\frac{\partial f}{\partial x}, \quad \frac{\partial}{\partial x}(f \pm g) = \frac{\partial f}{\partial x} \pm \frac{\partial g}{\partial x},$$

and so on, to establish the following rules:

a) $\nabla(kf) = k\nabla f$ b) $\nabla(f + g) = \nabla f + \nabla g$

c) $\nabla(f - g) = \nabla f - \nabla g$ d) $\nabla(fg) = f\nabla g + g\nabla f$

13.6 Tangent Planes and Normal Lines

Gradients provide the most natural and efficient way to define tangent planes and normal lines at points on surfaces that are level surfaces of functions of three variables. This section shows how it is done.

Equations for Tangent Planes and Normal Lines to Level Surfaces

To find equations for tangent planes and normal lines, we suppose that $P_0(x_0, y_0, z_0)$ is a point on the level surface $f(x, y, z) = c$ of a function f that has continuous first partial derivatives. If $x = g(t)$, $y = h(t)$, $z = k(t)$ is any differentiable curve on the surface then

$$f(g(t), h(t), k(t)) = c \tag{1}$$

for every value of t. Differentiating both sides of this equation with respect to t gives

$$\frac{\partial f}{\partial x}\frac{dx}{dt} + \frac{\partial f}{\partial y}\frac{dy}{dt} + \frac{\partial f}{\partial z}\frac{dz}{dt} = 0. \tag{2}$$

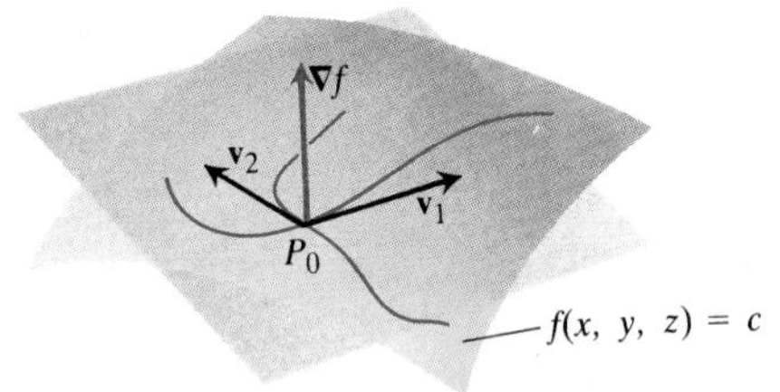

13.34 ∇f is perpendicular to the velocity vector of every differentiable curve in the surface through P_0. The velocity vectors at P_0 therefore lie in a common plane, which we call the tangent plane at P_0.

The left-hand side of this equation is the dot product of ∇f with the curve's velocity vector $\mathbf{v}$, so the equation can be written in the form

$$\nabla f \cdot \mathbf{v} = 0. \tag{3}$$

What Eq. (3) shows more clearly than Eq. (2) is that the gradient of f is orthogonal to the velocity vector at every point of the curve (Fig. 13.34).

Now let us restrict our attention to the differentiable curves that lie on the surface $f(x, y, z) = c$ and pass through P_0. All the velocity vectors at P_0 are orthogonal to ∇f at P_0, and hence all the tangent lines to these curves lie in the plane through P_0 normal to ∇f. We call this plane the tangent plane of the surface at P_0. We call the line through P_0 perpendicular to the plane the surface's normal line at P_0.

DEFINITIONS

The **tangent plane** at the point $P_0(x_0, y_0, z_0)$ on the level surface $f(x, y, z) = c$ is the plane

$$f_x(P_0)(x - x_0) + f_y(P_0)(y - y_0) + f_z(P_0)(z - z_0) = 0. \tag{4}$$

The **normal line** of the surface at P_0 is the line

$$x = x_0 + f_x(P_0)t, \qquad y = y_0 + f_y(P_0)t, \qquad z = z_0 + f_z(P_0)t. \tag{5}$$

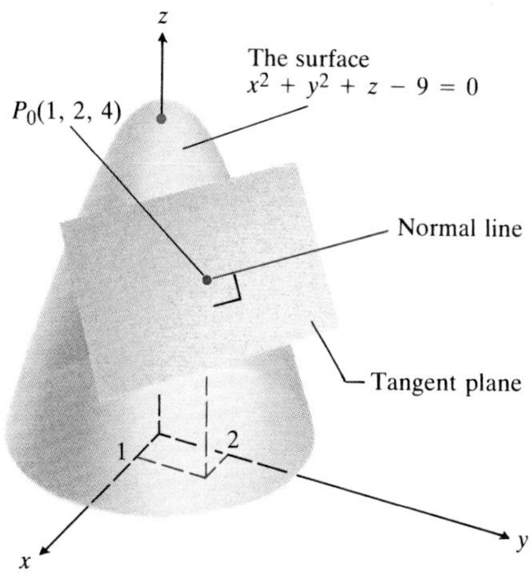

13.35 The tangent plane and normal line to the surface $x^2 + y^2 + z - 9 = 0$ at $P_0(1, 2, 4)$.

EXAMPLE 1 Find equations for the tangent plane and normal line of the surface

$$f(x, y, z) = x^2 + y^2 + z - 9 = 0 \qquad \text{(A circular paraboloid)}$$

at the point $P_0(1, 2, 4)$.

Solution The surface is shown in Fig. 13.35.

The tangent plane is the plane through P_0 perpendicular to the gradient of f at P_0. The gradient is

$$\nabla f|_{P_0} = (2x\mathbf{i} + 2y\mathbf{j} + \mathbf{k})_{(1,2,4)} = 2\mathbf{i} + 4\mathbf{j} + \mathbf{k}.$$

The plane is therefore the plane

$$2(x - 1) + 4(y - 2) + (z - 4) = 0 \qquad \text{or} \qquad 2x + 4y + z = 14.$$

The line normal to the surface at P_0 is

$$x = 1 + 2t, \qquad y = 2 + 4t, \qquad z = 4 + t.$$

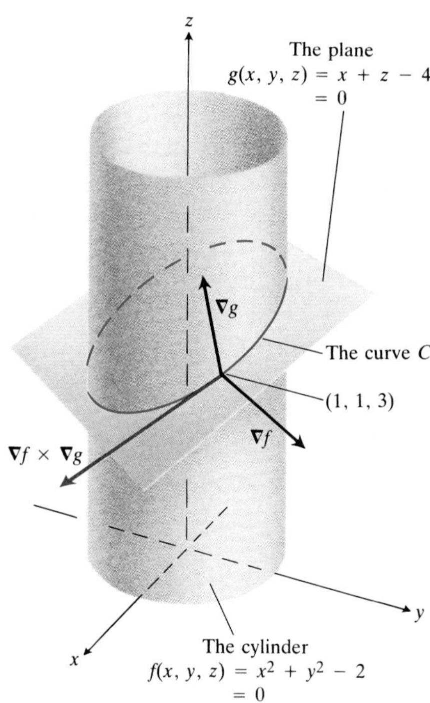

13.36 The cylinder $f(x, y, z) = x^2 + y^2 - 2 = 0$ and the plane $g(x, y, z) = x + z - 4 = 0$ intersect in a curve C. Any vector tangent to C will automatically be orthogonal to ∇f and ∇g and hence parallel to $\nabla f \times \nabla g$. This information enables us to write an equation for the tangent line in Example 2.

EXAMPLE 2 The surfaces

$$f(x, y, z) = x^2 + y^2 - 2 = 0 \qquad \text{(A cylinder)}$$

and

$$g(x, y, z) = x + z - 4 = 0 \qquad \text{(A plane)}$$

meet in a curve C (Fig. 13.36). Find parametric equations for the line tangent to C at the point $P_0(1, 1, 3)$.

Solution The tangent line is perpendicular to both ∇f and ∇g at P_0, so $\mathbf{v} = \nabla f \times \nabla g$ is parallel to the line. The components of $\mathbf{v}$ and the coordinates of P_0 give us equations for the line. We have

$$\nabla f_{(1,1,3)} = (2x\mathbf{i} + 2y\mathbf{j})_{(1,1,3)} = 2\mathbf{i} + 2\mathbf{j}$$

$$\nabla g_{(1,1,3)} = (\mathbf{i} + \mathbf{k})_{(1,1,3)} = \mathbf{i} + \mathbf{k}$$

$$\mathbf{v} = (2\mathbf{i} + 2\mathbf{j}) \times (\mathbf{i} + \mathbf{k}) = \begin{vmatrix} \mathbf{i} & \mathbf{j} & \mathbf{k} \\ 2 & 2 & 0 \\ 1 & 0 & 1 \end{vmatrix} = 2\mathbf{i} - 2\mathbf{j} - 2\mathbf{k}.$$

The line is

$$x = 1 + 2t, \qquad y = 1 - 2t, \qquad z = 3 - 2t.$$

Equations for Lines Tangent to Level Curves

The fact that the gradient of a function $f(x, y)$ is normal to the function's level curves in the plane enables us to use ∇f to write equations for tangent lines to level curves.

> The tangent to the level curve $f(x, y) = c$ at the point (x_0, y_0) is the line
>
> $$f_x(x_0, y_0)(x - x_0) + f_y(x_0, y_0)(y - y_0) = 0. \qquad (6)$$

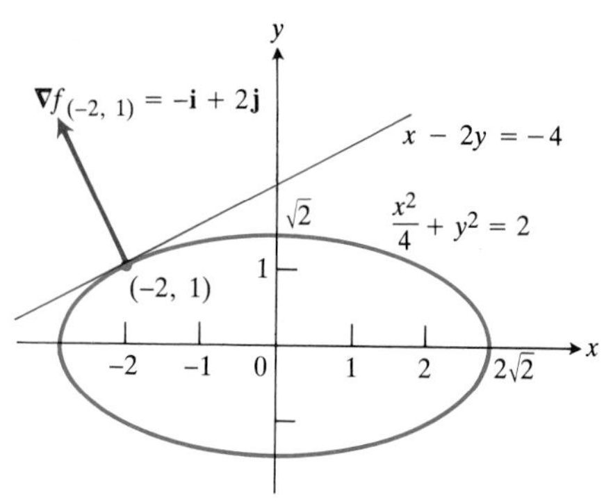

13.37 We can find the tangent to the ellipse $(x^2/4) + y^2 = 2$ by treating the ellipse as a level curve of the function $f(x, y) = (x^2/4) + y^2$ (Example 3).

EXAMPLE 3 Find an equation for the tangent to the ellipse

$$\frac{x^2}{4} + y^2 = 2$$

(Fig. 13.37) at the point $(-2, 1)$.

Solution The ellipse is a level curve of the function

$$f(x, y) = \frac{x^2}{4} + y^2.$$

The gradient of f at $(-2, 1)$ is

$$\nabla f|_{(-2,1)} = \left(\frac{x}{2}\mathbf{i} + 2y\mathbf{j}\right)_{(-2,1)} = -\mathbf{i} + 2\mathbf{j}.$$

The tangent is the line

$$(-1)(x + 2) + (2)(y - 1) = 0 \qquad \text{(Eq. 6)}$$

$$x - 2y = -4.$$

■

Exercises 13.6

In Exercises 1–12, find equations for the (a) tangent plane and (b) normal line at the point P_0 on the given surface.

1. $x^2 + y^2 + z^2 = 3, \quad P_0(1, 1, 1)$
2. $x^2 + y^2 - z^2 = 18, \quad P_0(3, 5, -4)$
3. $z^2 - x^2 - y^2 = 0, \quad P_0(3, 4, -5)$
4. $2z - x^2 = 0, \quad P_0(2, 0, 2)$
5. $z - \ln(x^2 + y^2) = 0, \quad P_0(1, 0, 0)$
6. $(x + y)^2 + z^2 = 25, \quad P_0(1, 2, 4)$
7. $x^2 + 2xy - y^2 + z^2 = 7, \quad P_0(1, -1, 3)$
8. $\cos \pi x - x^2y + e^{xz} + yz = 4, \quad P_0(0, 1, 2)$
9. $x^2 - xy - y^2 - z = 0, \quad P_0(1, 1, -1)$
10. $x + y + z = 1, \quad P_0(0, 1, 0)$
11. $x^2 + y^2 - 2xy - x + 3y - z = -4, \quad P_0(2, -3, 18)$
12. $x - y^3 + zy - 2z^2 = 0, \quad P_0(-4, -2, 1)$

In Exercises 13–18, sketch the surface $f(x, y, z) = c$ together with ∇f at the given point.

13. $x^2 + y^2 = 4, \quad (1, \sqrt{3}, 1)$
14. $x^2 + z^2 = 2, \quad (1, -1, 1)$
15. $x^2 + y^2 - z = 0, \quad (1, 1, 2)$
16. $x^2 + y^2 + z = 4, \quad (-1, -1, 2)$
17. $x^2 + y^2 - z^2/4 = 0, \quad (1, 2, -2\sqrt{5})$
18. $4y^2 + 9z^2 = 36, \quad (0, 3/2, \sqrt{3})$

In Exercises 19–24, find parametric equations for the line that is tangent to the curve of intersection of the surfaces at the given point.

19. Surfaces: $x + y^2 + 2z = 4, \quad x = 1$
    Point: $(1, 1, 1)$
20. Surfaces: $xyz = 1, \quad x^2 + 2y^2 + 3z^2 = 6$
    Point: $(1, 1, 1)$
21. Surfaces: $x^2 + 2y + 2z = 4, \quad y = 1$
    Point: $(1, 1, 1/2)$
22. Surfaces: $x + y^2 + z = 2, \quad y = 1$
    Point: $(1/2, 1, 1/2)$
23. Surfaces: $x^3 + 3x^2y^2 + y^3 + 4xy - z^2 = 0,$
    $x^2 + y^2 + z^2 = 11$
    Point: $(1, 1, 3)$
24. Surfaces: $x^2 + y^2 = 4, \quad x^2 + y^2 - z = 0$
    Point: $(\sqrt{2}, \sqrt{2}, 4)$

In Exercises 25–30 sketch the curve $f(x, y) = c$ together with ∇f and the tangent line at the given point. Then write an equation for the tangent line.

25. $x^2 + y^2 = 4, \quad (\sqrt{2}, \sqrt{2})$
26. $x^2 + y^2/2 = 3, \quad (1, 2)$
27. $4x^2 + 9y^2 = 36, \quad (2, 2\sqrt{5}/3)$
28. $x^2 - y = 1, \quad (\sqrt{2}, 1)$
29. $xy = -4, \quad (2, -2)$

30. $x^2 - xy + y^2 = 7, \quad (-1, 2)$ (This is the curve in Section 3.6, Example 4.)

Normal and Tangent Curves

A curve is said to be **normal to a surface** $f(x, y, z) = c$ at a point of intersection if the curve's velocity vector is a scalar multiple of ∇f at the point. The curve is said to be **tangent** to the surface at the point of intersection if its velocity vector is normal to ∇f there.

31. Show that the curve

$$\mathbf{r} = \sqrt{t}\,\mathbf{i} + \sqrt{t}\,\mathbf{j} - \frac{1}{4}(t + 3)\mathbf{k}$$

is normal to the surface $x^2 + y^2 - z = 3$ when $t = 1$.

32. Show that the curve

$$\mathbf{r} = \sqrt{t}\,\mathbf{i} + \sqrt{t}\,\mathbf{j} + (2t - 1)\mathbf{k}$$

is tangent to the surface $x^2 + y^2 - z = 1$ when $t = 1$.

13.7 Linearization and Differentials

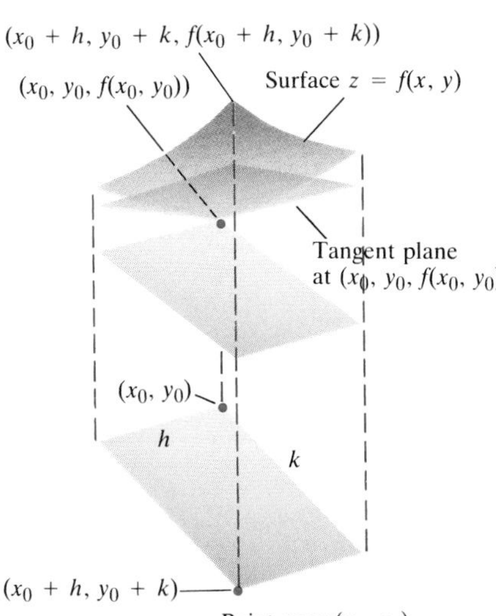

13.38 A portion of the surface $z = f(x, y)$ and its tangent plane at the point $(x_0, y_0, f(x_0, y_0))$. The linear approximation we develop for the values of f comes from noticing that the plane lies close to the surface near the point of tangency.

The functions of two variables that arise in science and mathematics can be quite complicated. Sometimes we can replace them with simpler functions that give the accuracy we want for specific applications and are not so hard to work with. We do this in a way that is similar to the way we found linear replacements for functions of a single variable in Section 3.7, except that now the linearization comes from a plane instead of a line. This section shows how to construct linearizations of functions of two variables and how to estimate the error involved in replacing a function by a linearization.

How to Linearize a Function $f(x, y)$

As you can see in Fig. 13.38, the plane tangent to the surface $z = f(x, y)$ at the point $(x_0, y_0, f(x_0, y_0))$ lies close to the surface near the point of tangency. For points $(x_0 + h, y_0 + k)$ near (x_0, y_0), the z-values on the tangent plane give good approximations of the corresponding z-values on the surface. We can therefore approximate the values of f near (x_0, y_0) by replacing the formula for f by the formula for the tangent plane.

The surface $z = f(x, y)$ is the level surface $F(x, y, z) = 0$ of the function $F(x, y, z) = f(x, y) - z$. The plane tangent to the surface at $P_0(x_0, y_0, f(x_0, y_0))$ is therefore the plane

$$F_x(P_0)(x - x_0) + F_y(P_0)(y - y_0) + F_z(P_0)(z - f(x_0, y_0)) = 0 \tag{1}$$

(Section 13.6, Eq. (4)). In terms of f, this becomes

$$f_x(x_0, y_0)(x - x_0) + f_y(x_0, y_0)(y - y_0) + (-1)(z - f(x_0, y_0)) = 0 \tag{2}$$

or

$$z = f(x_0, y_0) + f_x(x_0, y_0)(x - x_0) + f_y(x_0, y_0)(y - y_0). \tag{3}$$

The function on the right-hand side of Eq. (3) is called the linearization of f at (x_0, y_0). As long as the tangent plane lies close to the surface, the value of the linearization will give a good approximation to the value of f.

DEFINITIONS

The **linearization** of the function $f(x, y)$ at (x_0, y_0) is the function

$$L(x, y) = f(x_0, y_0) + f_x(x_0, y_0)(x - x_0) + f_y(x_0, y_0)(y - y_0). \tag{4}$$

The approximation

$$L(x, y) \approx f(x, y) \tag{5}$$

is the **standard linear approximation** of f near (x_0, y_0).

As usual, we assume that f has continuous first partial derivatives in the neighborhood of (x_0, y_0).

EXAMPLE 1 Find the linearization of

$$f(x, y) = x^2 - xy + \frac{1}{2}y^2 + 3$$

at the point (3, 2).

Solution We evaluate Eq. (4) with

$$f(x_0, y_0) = \left(x^2 - xy + \frac{1}{2}y^2 + 3\right)_{(3,\,2)} = 8, \qquad \text{(Arithmetic omitted)}$$

$$f_x(x_0, y_0) = \frac{\partial}{\partial x}\left(x^2 - xy + \frac{1}{2}y^2 + 3\right)_{(3,\,2)} = (2x - y)_{(3,\,2)} = 4,$$

$$f_y(x_0, y_0) = \frac{\partial}{\partial y}\left(x^2 - xy + \frac{1}{2}y^2 + 3\right)_{(3,\,2)} = (-x + y)_{(3,\,2)} = -1,$$

getting

$$\begin{aligned} L(x, y) &= f(x_0, y_0) + f_x(x_0, y_0)(x - x_0) + f_y(x_0, y_0)(y - y_0) \qquad \text{(Eq. 4)} \\ &= 8 + (4)(x - 3) + (-1)(y - 2) = 4x - y - 2. \end{aligned}$$

The linearization of f at (3, 2) is $L(x, y) = 4x - y - 2$. ■

How Accurate Is the Standard Linear Approximation?

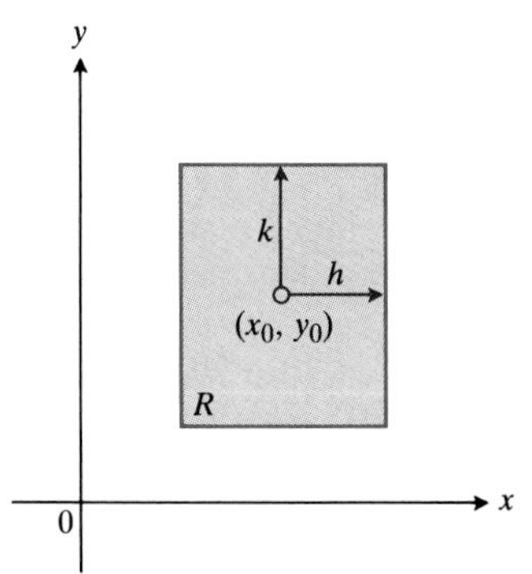

13.39 The rectangular region R: $|x - x_0| \le h$, $|y - y_0| \le k$ in the xy-plane. On this kind of region, we can find error bounds for our approximations.

To find the error in the approximation $f(x, y) \approx L(x, y)$, we need the second order partial derivatives of f. Suppose that the first and second order partial derivatives of f are continuous throughout a rectangular region R centered at (x_0, y_0) and given by the inequalities

$$|x - x_0| \le h, \qquad |y - y_0| \le k \tag{6}$$

(Fig. 13.39). Since R is closed and bounded, the second partial derivatives all have maximum values on R. A theorem from advanced calculus says that if B is the largest of these values then the error $E(x, y) = f(x, y) - L(x, y)$ in the standard linear approximation satisfies the inequality

$$|E(x, y)| \le \frac{1}{2}B\left(|x - x_0| + |y - y_0|\right)^2 \tag{7}$$

throughout R.

When we use this inequality to estimate E, we usually cannot find the values of f_{xx}, f_{yy}, and f_{xy} that determine B and we have to replace B itself with an upper-bound or "worst-case" value instead. If M is any common upper bound for $|f_{xx}|$, $|f_{yy}|$, and $|f_{xy}|$ on R, then B will be less than or equal to M and we will know that

$$|E(x, y)| \leq \frac{1}{2} M\left(|x - x_0| + |y - y_0|\right)^2. \tag{8}$$

This is the inequality we normally use in estimating E. When we need to make $|E(x, y)|$ small for a given M, we just make $|x - x_0|$ and $|y - y_0|$ small.

The Error in the Standard Linear Approximation of $f(x, y)$ near (x_0, y_0)

If f and its first and second partial derivatives are continuous throughout a rectangle R centered at (x_0, y_0), then the error $E(x, y)$ incurred in replacing $f(x, y)$ on R by its linearization

$$L(x, y) = f(x_0, y_0) + f_x(x_0, y_0)(x - x_0) + f_y(x_0, y_0)(y - y_0) \tag{9}$$

satisfies the inequality

$$|E(x, y)| \leq \frac{1}{2} M\left(|x - x_0| + |y - y_0|\right)^2, \tag{10}$$

where M is any upper bound for the values of $|f_{xx}|$, $|f_{yy}|$, and $|f_{xy}|$ on R.

EXAMPLE 2 In Example 1, we found the linearization of

$$f(x, y) = x^2 - xy + \frac{1}{2}y^2 + 3$$

at (3, 2) to be

$$L(x, y) = 4x - y - 2.$$

Find an upper bound for the error in the approximation $f(x, y) \approx L(x, y)$ over the rectangle

$$R: \quad |x - 3| \leq 0.1, \quad |y - 2| \leq 0.1.$$

Express the upper bound as a percentage of $f(3, 2)$, the value of f at the center of the rectangle.

Solution We use the inequality

$$|E(x, y)| \leq \frac{1}{2} M\left(|x - x_0| + |y - y_0|\right)^2. \qquad \text{(Eq. 10)}$$

To find a suitable value for M we calculate f_{xx}, f_{xy}, and f_{yy}, finding, after a routine differentiation (omitted), that all three derivatives are constant, with values

$$|f_{xx}| = |2| = 2, \quad |f_{xy}| = |-1| = 1, \quad |f_{yy}| = |1| = 1.$$

The largest of these is 2, so we may safely take M to be 2. With $(x_0, y_0) = (3, 2)$ we then know that, throughout R,

$$|E(x, y)| \leq \frac{1}{2}(2)\left(|x - 3| + |y - 2|\right)^2 = \left(|x - 3| + |y - 2|\right)^2.$$

Finally, since $|x - 3| \leq 0.1$ and $|y - 2| \leq 0.1$ on R, we have

$$|E(x, y)| \leq (0.1 + 0.1)^2 = (0.2)^2 = 0.04.$$

As a percentage of $f(3, 2) = 8$, the error is no greater than

$$\frac{0.04}{8} \times 100 = 0.5\%.$$

As long as (x, y) stays in R, the approximation $f(x, y) \approx L(x, y)$ will be in error by no more than 0.04, which is 1/2% of the value of f at the center of R. ■

Predicting Change with Differentials

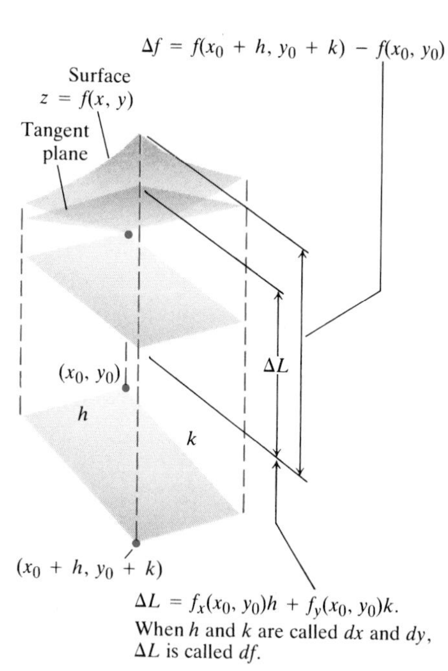

13.40 If h and k are small, ΔL will be nearly the same as Δf, giving us a practical way to estimate Δf.

Suppose we know the value of a differentiable function $f(x, y)$ at a particular point (x_0, y_0) and we want to predict how much this value will change if we move to a point $(x_0 + h, y_0 + k)$ nearby. If h and k are small, f and its linearization at (x_0, y_0) will change nearly the same amount. Since the values of L are always easy to calculate, calculating the change in L gives a practical way to estimate the change in f.

In the notation of Fig. 13.40, the change in f is

$$\Delta f = f(x_0 + h, y_0 + k) - f(x_0, y_0).$$

The corresponding change in L is

$$\begin{aligned} \Delta L &= L(x_0 + h, y_0 + k) - L(x_0, y_0) \\ &= f_x(x_0, y_0)h + f_y(x_0, y_0)k. \qquad \text{(After cancellations)} \end{aligned} \tag{11}$$

The formula for Δf is usually as hard to work with as the formula for f. The change in L, however, is just a constant times h plus a constant times k.

The change ΔL is usually described in the more suggestive notation

$$df = f_x(x_0, y_0)\,dx + f_y(x_0, y_0)\,dy, \tag{12}$$

in which df denotes the change in the linearization that results from the changes dx and dy in x and y. As usual, we call dx and dy differentials of x and y, and we call df the corresponding differential of f.

DEFINITIONS

> If we move (x_0, y_0) to a nearby point $(x_0 + dx, y_0 + dy)$, the resulting differential in f is
>
> $$df = f_x(x_0, y_0)\,dx + f_y(x_0, y_0)\,dy. \tag{13}$$
>
> This change in the linearization of f is called the **total differential of *f*.**

EXAMPLE 3 *Sensitivity to change.* Your company manufactures right circular cylindrical molasses storage tanks that are 25 ft high with a radius of 5 ft (Fig. 13.41). How sensitive are the tanks' volumes to small variations in height and radius?

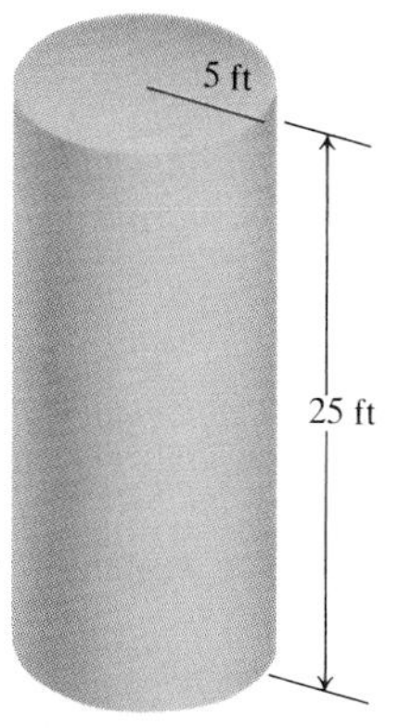

13.41 A small change in this tank's radius will change the volume 10 times as much as a change of the same size in the tank's height (Example 3).

Solution As a function of radius r and height h, the typical tank's volume is

$$V = \pi r^2 h.$$

The change in volume caused by small changes dr and dh in radius and height is approximately

$$\begin{aligned} dV &= V_r(5, 25)\, dr + V_h(5, 25)\, dh \qquad \begin{pmatrix}\text{Eq. 13 with } f = V \text{ and} \\ (x_0, y_0) = (5, 25)\end{pmatrix} \\ &= (2\pi rh)_{(5,25)}\, dr + (\pi r^2)_{(5,25)}\, dh \\ &= 250\pi\, dr + 25\pi\, dh. \end{aligned}$$

Thus, a one-unit change in r will change V by about 250π units. A one-unit change in h will change V by about 25π units. The tank's volume is 10 times more sensitive to a small change in r than it is to a small change of equal size in h. As a quality control engineer concerned with being sure the tanks have the right volume, you would want to pay special attention to their radii.

In contrast, if the values of r and h are reversed to make $r = 25$ and $h = 5$, then the total differential in V becomes

$$dV = (2\pi rh)_{(25,5)}\, dr + (\pi r^2)_{(25,5)}\, dh = 250\pi\, dr + 625\pi\, dh.$$

Now the volume is more sensitive to changes in h than it is to changes in r (Fig. 13.42).

The general rule to be learned from this example is that functions are most sensitive to small changes in the variables that give the largest partial derivatives. ■

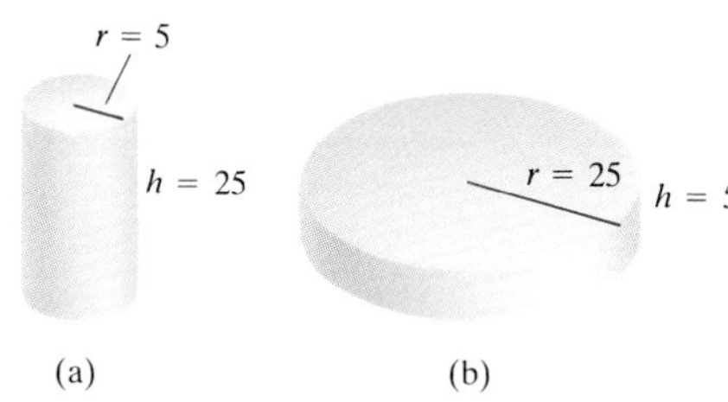

13.42 The volume of cylinder (a) is more sensitive to a small change in r than it is to an equally small change in h. The volume of cylinder (b) is more sensitive to small changes in h than it is to small changes in r.

Absolute, Relative, and Percentage Change

When we move from (x_0, y_0) to a nearby point, we can describe the corresponding change in the value of a function $f(x, y)$ in three different ways.

	True	Estimate	
Absolute change:	Δf	df	
Relative change:	$\dfrac{\Delta f}{f(x_0, y_0)}$	$\dfrac{df}{f(x_0, y_0)}$	(14)
Percentage change:	$\dfrac{\Delta f}{f(x_0, y_0)} \times 100$	$\dfrac{df}{f(x_0, y_0)} \times 100$	

EXAMPLE 4 Suppose that the variables r and h change from the initial values of $(r_0, h_0) = (1, 5)$ by the amounts $dr = 0.03$ and $dh = -0.1$. Estimate the resulting absolute, relative, and percentage changes in the values of the function $V = \pi r^2 h$.

Solution To estimate the absolute change in V we evaluate

$$dV = V_r(r_0, h_0)\, dr + V_h(r_0, h_0)\, dh \qquad (15)$$

to get

$$dV = 2\pi r_0 h_0 \, dr + \pi r_0^2 \, dh = 2\pi(1)(5)(0.03) + \pi(1)^2(-0.1)$$
$$= 0.3\pi - 0.1\pi = 0.2\pi.$$

We divide this by $V(r_0, h_0)$ to estimate the relative change:

$$\frac{dV}{V(r_0, h_0)} = \frac{0.2\pi}{\pi r_0^2 h_0} = \frac{0.2\pi}{\pi(1)^2(5)} = 0.04.$$

We multiply this by 100 to estimate the percentage change:

$$\frac{dV}{V(r_0, h_0)} \times 100 = 0.04 \times 100 = 4\%.$$

EXAMPLE 5 The volume $V = \pi r^2 h$ of a right circular cylinder is to be calculated from measured values of r and h. Suppose that r is measured with an error of no more than 2% and h with an error of no more than 0.5%. Estimate the resulting possible percentage error in the calculation of V.

Solution We are told that

$$\left|\frac{dr}{r} \times 100\right| \le 2 \qquad \text{and} \qquad \left|\frac{dh}{h} \times 100\right| \le 0.5.$$

Since

$$\frac{dV}{V} = \frac{2\pi rh \, dr + \pi r^2 \, dh}{\pi r^2 h} = \frac{2\,dr}{r} + \frac{dh}{h},$$

we have

$$\left|\frac{dV}{V} \times 100\right| = \left|2\frac{dr}{r} \times 100 + \frac{dh}{h} \times 100\right|$$
$$\le 2\left|\frac{dr}{r} \times 100\right| + \left|\frac{dh}{h} \times 100\right|$$
$$\le 2(2) + 0.5 = 4.5.$$

We estimate the error in the volume calculation to be at most 4.5%.

How accurately do we have to measure r and h to have a reasonable chance of calculating $V = \pi r^2 h$ with an error, say, of less than 2%? Questions like this are hard to answer for functions of more than one variable because there is usually no single right answer. Since

$$\frac{dV}{V} = 2\frac{dr}{r} + \frac{dh}{h},$$

we see that dV/V is controlled by a combination of dr/r and dh/h. If we can measure h with great accuracy, we might come out all right even if we are sloppy about measuring r. On the other hand, our measurement of h might have so large a dh that the resulting dV/V would be too crude an estimate of $\Delta V/V$ to be useful even if dr were zero.

What we do in such cases is look for a "reasonable" square about the measured values (r_0, h_0) in which V will not vary by more than the allowed amount from $V_0 = \pi r_0^2 h_0$. The next example shows how this is done.

13.43 A small square about the point (5, 12) in the rh-plane.

EXAMPLE 6 Find a reasonable square about the point $(r_0, h_0) = (5, 12)$ in which the value of $V = \pi r^2 h$ will not vary by more than ± 0.1.

Solution We approximate the variation ΔV by the differential

$$dV = 2\pi r_0 h_0\, dr + \pi r_0^2\, dh = 2\pi(5)(12)\, dr + \pi(5)^2\, dh = 120\pi\, dr + 25\pi\, dh.$$

Since the region to which we are restricting our attention is a square (Fig. 13.43), we may set $dh = dr$ to get

$$dV = 120\pi\, dr + 25\pi\, dr = 145\pi\, dr.$$

We then ask, How small must we take dr to be sure that $|dV|$ is no larger than 0.1? To answer, we start with the inequality

$$|dV| \le 0.1,$$

express dV in terms of dr,

$$|145\pi\, dr| \le 0.1,$$

and find a corresponding upper bound for dr:

$$|dr| \le \frac{0.1}{145\pi} \approx 2.1 \times 10^{-4}. \qquad \left(\begin{array}{l}\text{We round down to make sure}\\ dr \text{ won't accidentally be too big.}\end{array}\right)$$

With $dh = dr$, then, the square we want is described by the inequalities

$$|r - 5| \le 2.1 \times 10^{-4}, \qquad |h - 12| \le 2.1 \times 10^{-4}.$$

As long as (r, h) stays in this square, we may expect $|dV|$, and hence $|\Delta V|$, to be less than or equal to 0.1. ■

Exercises 13.7

In Exercises 1–6, find the linearization $L(x, y)$ of the given function at each point.

1. $f(x, y) = x^2 + y^2 + 1$ at (a) $(0, 0)$, (b) $(1, 1)$

2. $f(x, y) = x^3 y^4$ at (a) $(1, 1)$, (b) $(0, 0)$

3. $f(x, y) = e^x \cos y$ at (a) $(0, 0)$, (b) $(0, \pi/2)$

4. $f(x, y) = (x + y + 2)^2$ at (a) $(0, 0)$, (b) $(1, 2)$

5. $f(x, y) = 3x - 4y + 5$ at (a) $(0, 0)$, (b) $(1, 1)$

6. $f(x, y) = e^{2y-x}$ at (a) $(0, 0)$, (b) $(1, 2)$

In Exercises 7–12, find the linearization $L(x, y)$ of the function $f(x, y)$ at P_0. Then use Eq. (10) to find an upper bound for the magnitude $|E|$ of the error in the approximation $f(x, y) \approx L(x, y)$ over the rectangle R.

7. $f(x, y) = x^2 - 3xy + 5$ at $P_0(2, 1)$,
R: $|x - 2| \le 0.1$, $|y - 1| \le 0.1$

8. $f(x, y) = (1/2)x^2 + xy + (1/4)y^2 + 3x - 3y + 4$ at $P_0\,(2, 2)$,
R: $|x - 2| \le 0.1$, $|y - 2| \le 0.1$

9. $f(x, y) = 1 + y + x \cos y$ at $P_0(0, 0)$,
R: $|x| \le 0.2$, $|y| \le 0.2$
(Use $|\cos y| \le 1$ and $|\sin y| \le 1$ in estimating E.)

10. $f(x, y) = \ln x + \ln y$ at $P_0(1, 1)$,
R: $|x - 1| \le 0.2$, $|y - 1| \le 0.2$

11. $f(x, y) = e^x \cos y$ at $P_0(0, 0)$,
R: $|x| \le 0.1$, $|y| \le 0.1$
(Use $e^x \le 1.11$ and $|\cos y| \le 1$ in estimating E.)

12. $f(x, y) = xy^2 + y \cos(x - 1)$ at $P_0(1, 2)$,
R: $|x - 1| \le 0.1$, $|y - 2| \le 0.1$

13. You plan to calculate the area of a long, thin rectangle from measurements of its length and width. Which measurement should you be more careful with? Why?

14. a) Around the point $(1, 0)$, is $f(x, y) = x^2(y + 1)$ more sensitive to changes in x, or to changes in y?
b) What ratio of dx to dy will make df equal zero at $(1, 0)$?

15. Suppose T is to be found from the formula $T = x(e^y + e^{-y})$ where x and y are found to be 2 and $\ln 2$ with maximum possible errors of $|dx| = 0.1$ and $|dy| = 0.02$. Estimate the maximum possible error in the computed value of T.

16. About how accurately may $V = \pi r^2 h$ be calculated from measurements of r and h that are in error by 1%?

17. If $r = 5.0$ cm and $h = 12.0$ cm to the nearest millimeter, what should we expect the maximum percentage error in calculating $V = \pi r^2 h$ to be?

18. To estimate the volume of a cylinder of radius about 2 m and height about 3 m, about how accurately should the radius and height be measured so that the error in the volume estimate will not exceed 0.1 m^3? Assume that the possible error dr in measuring r is equal to the possible error dh in measuring h.

19. Give a reasonable square centered at (1, 1) over which the value of $f(x, y) = x^3y^4$ will not vary by more than ± 0.1.

20. *Variation in electrical resistance.* The resistance R produced by wiring resistors of R_1 and R_2 ohms in parallel (Fig. 13.44) can be calculated from the formula

$$\frac{1}{R} = \frac{1}{R_1} + \frac{1}{R_2}.$$

a) Show that

$$dR = \left(\frac{R}{R_1}\right)^2 dR_1 + \left(\frac{R}{R_2}\right)^2 dR_2.$$

b) You have designed a two-resistor circuit like the one in Fig. 13.44 to have resistances of $R_1 = 100$ ohms and $R_2 = 400$ ohms, but there is always some variation in manufacturing and the resistors received by your firm will probably not have these exact values. Will the value of R be more sensitive to variation in R_1, or to variation in R_2?

21. *Continuation of Exercise 20.* In another circuit like the one in Fig. 13.44, you plan to change R_1 from 20 to 20.1 ohms and R_2 from 25 to 24.9 ohms. By about what percentage will this change R?

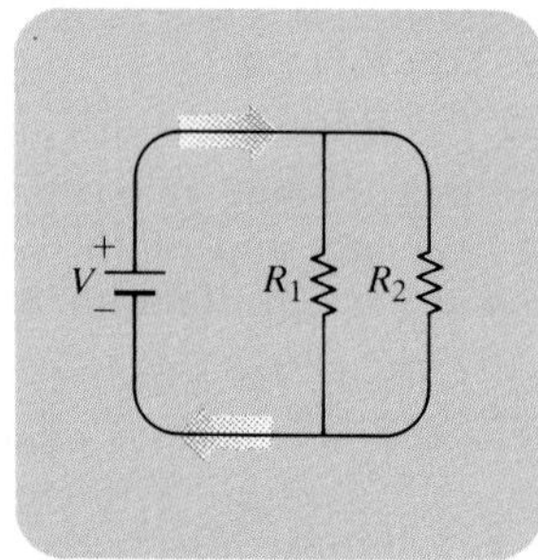

13.44 The circuit in Exercises 20–21.

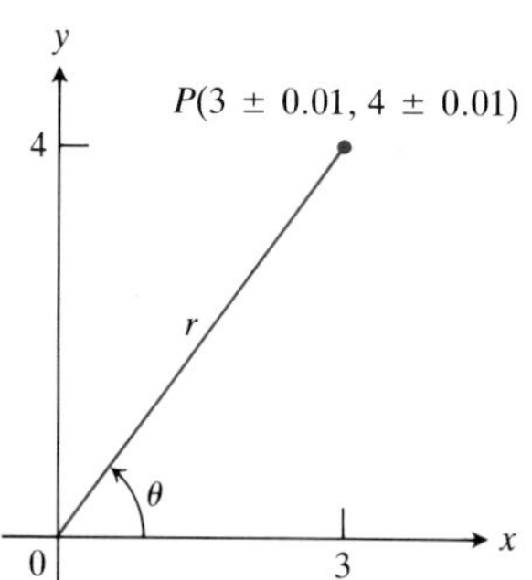

13.45 How much do the errors in the Cartesian coordinates of P affect the polar coordinates of P? See Exercise 22.

22. *Error carry-over in coordinate changes.*

a) If $x = 3 \pm 0.01$ and $y = 4 \pm 0.01$ (Fig. 13.45), with approximately what accuracy can you calculate the polar coordinates r and θ of the point $P(x, y)$ from the formulas $r^2 = x^2 + y^2$ and $\theta = \tan^{-1}(y/x)$? Express your estimates as percentage changes of the values that r and θ have at the point $(x_0, y_0) = (3, 4)$.

b) At the point $(x_0, y_0) = (3, 4)$, are the values of r and θ more sensitive to changes in x, or to changes in y?

Functions of Three Variables

To get the linearization and total-differential formulas for a differentiable function $f(x, y, z)$, we add z-terms to the two-variable formulas:

$$L(x, y, z) = f(x_0, y_0, z_0) + f_x(x_0, y_0, z_0)(x - x_0) + f_y(x_0, y_0, z_0)(y - y_0) + f_z(x_0, y_0, z_0)(z - z_0)$$

$$df = f_x(x_0, y_0, z_0)\,dx + f_y(x_0, y_0, z_0)\,dy + f_z(x_0, y_0, z_0)\,dz.$$

Similar results hold for functions of four or more variables.

Find the linearizations $L(x, y, z)$ of the functions in Exercises 23–28 at the given points.

23. $f(x, y, z) = xy + yz + xz$ at
a) (1, 1, 1) b) (1, 0, 0) c) (0, 0, 0)

24. $f(x, y, z) = x^2 + y^2 + z^2$ at
a) (1, 1, 1) b) (0, 1, 0) c) (1, 0, 0)

25. $f(x, y, z) = \sqrt{x^2 + y^2 + z^2}$ at
a) (1, 0, 0) b) (1, 1, 0) c) (1, 2, 2)

26. $f(x, y, z) = (\sin xy)/z$ at
a) $(\pi/2, 1, 1)$ b) (2, 0, 1)

27. $f(x, y, z) = e^x + \cos(y + z)$ at
a) (0, 0, 0) b) $\left(0, \frac{\pi}{2}, 0\right)$ c) $\left(0, \frac{\pi}{4}, \frac{\pi}{4}\right)$

28. $f(x, y, z) = \tan^{-1}(xyz)$ at
a) (1, 0, 0) b) (1, 1, 0) c) (1, 1, 1)

29. If $|a|$ is much greater than $|b|$, $|c|$, and $|d|$, to which of a, b, c, and d is the value of the determinant

$$f(a, b, c, d) = \begin{vmatrix} a & b \\ c & d \end{vmatrix}$$

most sensitive?

30. Estimate how strongly simultaneous errors of 2% in a, b, and c might affect the calculation of the product

$$p(a, b, c) = abc.$$

31. Estimate the amount of material in a hollow rectangular wooden box whose inside measurements are 5 ft long by 3 ft wide by 2 ft deep if the box is made of lumber 1/2-in. thick and the box has no top.

32. The area of a triangle is $(1/2)ab \sin C$, where a and b are the lengths of two sides of the triangle and C is the measure of the included angle. In surveying a triangular plot, you have measured a, b, and C to be 150 ft, 200 ft, and 60°, respectively. By about how much will your area calculation be in error if your values of a and b are off by half a foot each and your measurement of C is off by 2°? See Fig. 13.46. Remember to use radians.

33. Suppose that $u = xe^y + y \sin z$ and that x, y, and z can be measured with maximum possible errors of ± 0.2, ± 0.6, and $\pm \pi/180$, respectively. Estimate the maximum possible error in calculating u from the measured values $x = 2$, $y = \ln 3$, $z = \pi/2$.

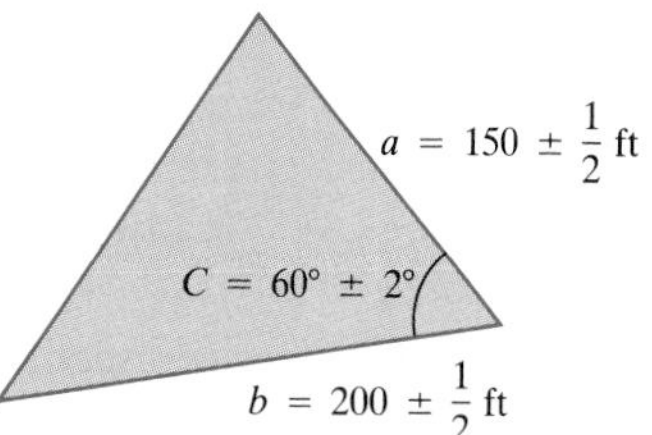

13.46 How accurately can you calculate the triangle's area from the measurement shown? See Exercise 32.

34. *The Wilson lot size formula.* The Wilson lot size formula in economics says that the most economical quantity Q of goods (radios, shoes, brooms, whatever) for a store to order is given by the formula $Q = \sqrt{2KM/h}$, where K is the cost of placing the order, M is the number of items sold per week, and h is the weekly holding cost for each item (cost of space, utilities, security, and so on). To which of the variables K, M, and h is Q most sensitive near the point $(K_0, M_0, h_0) = (2, 20, 0.05)$?

13.8 Maxima, Minima, and Saddle Points

As we mentioned at the end of Section 13.2, continuous functions defined on closed bounded regions in the xy-plane take on absolute maximum and minimum values on these domains (Figs. 13.47 and 13.48). It is important to be able to find these values and to know where they occur. For example, what is the highest temperature on a heated metal plate and where is it taken on? Where does a given surface attain its highest point above a given patch of the xy-plane? As we shall see in a moment, we can often answer questions like these by examining the partial derivatives of some appropriate function.

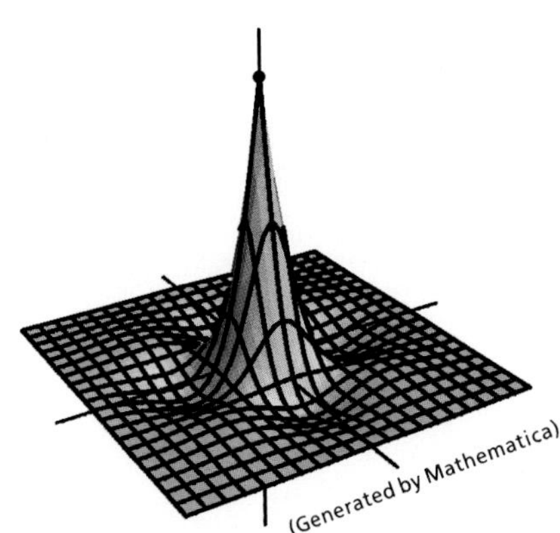

13.47 The function

$$z = (\cos x)(\cos y)e^{-\sqrt{x^2+y^2}}$$

has many local maxima and minima.

Our goal in this section is to show how to use partial derivatives to find the local maximum and minimum values of a continuous function $f(x, y)$ on a region R in the xy-plane. The first step is to use the function's first partial derivatives to make a (usually) short and comprehensive list of points where f can assume its local extreme values. What we do next then depends on whether R is closed and bounded. If it is, we appeal to a theorem from advanced calculus that a continuous function on a closed, bounded region R assumes an absolute maximum value and an absolute minimum value on R, and we look through the list to find what these values are. If R is not closed

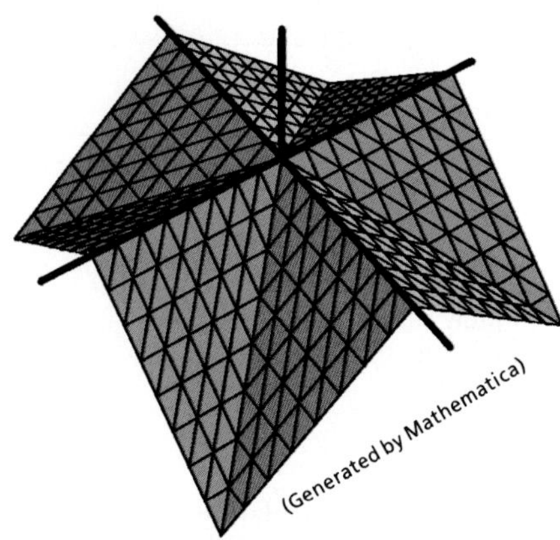

13.48 The "ridge" surface

$$z = \frac{1}{2}(||x| - |y|| - |x| - |y|)$$

viewed from the point (10, 15, 20).

and bounded, the function may not have absolute maximum and minimum values on R. However, we can still try to use the function's second partial derivatives to tell which points on the list, if any, give local maximum and minimum values. As you will see, the routine is much the same as the routine for functions of a single variable. The main difference is that the first and second derivative tests now involve more derivatives.

The Derivative Tests

To find the extreme values of a function of a single variable, we first look for points where the graph has a horizontal tangent line. At such a point we then look for a local maximum, a local minimum, or a point of inflection. We also examine the values of the function at the boundary points of its domain and at any points where the first derivative does not exist.

For a function $f(x, y)$ of two independent variables, we look for points where the surface $z = f(x, y)$ has a horizontal tangent plane. At such points we then look for a local maximum, a local minimum, or a **saddle point** (Fig. 13.49). We also examine the values of the function at the boundary points in its domain and at any points where the first partial derivatives f_x or f_y do not exist.

We organize the search for the extreme values assumed by a continuous function $f(x, y)$ on a region R into three steps:

STEP 1: *Make a list that includes the points where f has its local maxima and minima and evaluate f at all the points on the list.*

As we shall see later, the local maxima and minima of f can occur only at

i) boundary points of R,
ii) interior points of R where $f_x = f_y = 0$ **(the first derivative test)** and points where f_x or f_y fail to exist. (We call these the **critical points** of f.)

Therefore, to make the list, we find the maximum and minimum values of f on the boundary of R and find the values of f at the critical points.

In theory, such a list could be very long and might even include infinitely many points, but that is usually not the case in practice and definitely not the case in this book. The lists in the examples and problems will be short.

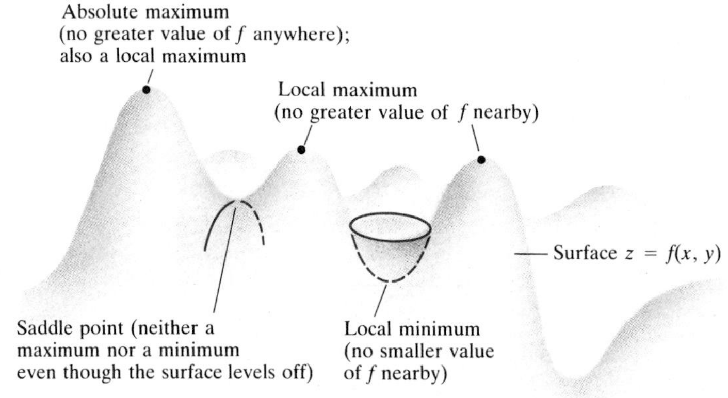

13.49 How values of $f(x, y)$ are classified. Saddle points are the three-dimensional analogues of points of inflection. Local values are often called *relative* values.

STEP 2: *If R is closed and bounded, look through the list for the maximum and minimum values of f. These will be the absolute maximum and minimum values of f on R.* As we mentioned in the introduction, a function that is continuous on a closed bounded region of the plane has an absolute maximum value on the region and an absolute minimum value on the region. Since absolute maxima and minima are also local maxima and minima, the absolute maximum and minimum values of f already appear somewhere in the list we made in Step 1. We have only to glance at the list to see what they are. If we then wish to learn which of the remaining values, if any, are local maxima and minima, we can go on to Step 3.

STEP 3: *If R is not closed and bounded, try the following second derivative test.*

The fact that $f_x = f_y = 0$ at an interior point (a, b) of R does not in itself guarantee that f will have an extreme value there. However, if f and its first and second partial derivatives are continuous on R, there is a second derivative test from advanced calculus that may identify the behavior of f at (a, b). The **second derivative test** goes like this:

If $f_x(a, b) = f_y(a, b) = 0$, then

i) f has a **local maximum** at (a, b) if $f_{xx} < 0$ and $f_{xx}f_{yy} - f_{xy}^2 > 0$ at (a, b);

ii) f has a **local minimum** at (a, b) if $f_{xx} > 0$ and $f_{xx}f_{yy} - f_{xy}^2 > 0$ at (a, b);

iii) f has a **saddle point** at (a, b) if $f_{xx}f_{yy} - f_{xy}^2 < 0$ at (a, b).

iv) The test is *inconclusive* at (a, b) if $f_{xx}f_{yy} - f_{xy}^2 = 0$ at (a, b). We must find some other way to determine the behavior of f at (a, b).

The expression $f_{xx}f_{yy} - f_{xy}^2$ is called the **discriminant** of f. It is sometimes easier to remember in the determinant form

$$f_{xx}f_{yy} - f_{xy}^2 = \begin{vmatrix} f_{xx} & f_{xy} \\ f_{xy} & f_{yy} \end{vmatrix}.$$

We shall now look at examples that show these tests at work. After that, we shall show why the condition $f_x = f_y = 0$ is a necessary condition for having an extreme value at an interior point of the domain of a differentiable function.

The Tests at Work

In the first example we look at the function $f(x, y) = x^2 + y^2$, whose behavior we already know from looking at the formula: Its value is zero at the origin and increases steadily as (x, y) moves away from the origin. The point of Example 1 is to show how the derivative tests reveal this behavior.

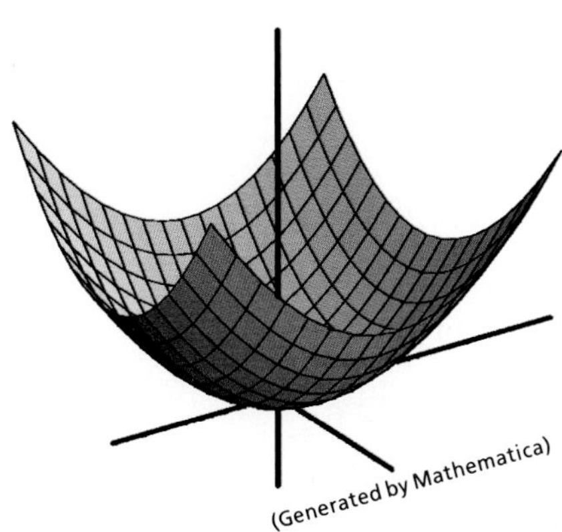

13.50 The graph of the function $f(x, y) = x^2 + y^2$ is the paraboloid $z = x^2 + y^2$. The function has only one extreme value, an absolute minimum value of 0 at the origin.

EXAMPLE 1 Find the extreme values of $f(x, y) = x^2 + y^2$.

Solution The domain of f has no boundary points, for it is the entire plane (Fig. 13.50). The derivatives $f_x = 2x$ and $f_y = 2y$ exist everywhere. Therefore, local maxima and minima can occur only where

$$f_x = 2x = 0 \quad \text{and} \quad f_y = 2y = 0.$$

The only possibility is the origin, where the value of f is zero. Since f is never negative, we see that zero is an absolute minimum.

We have not needed the second derivative test at all. Had we used it, we would have found

$$f_{xx} = 2, \qquad f_{yy} = 2, \qquad f_{xy} = 0$$

and

$$f_{xx}f_{yy} - f_{xy}^2 = (2)(2) - (0)^2 = 4 > 0,$$

identifying (0, 0) as a local minimum. This in itself does not identify (0, 0) as an absolute minimum. It takes more information to do that. ■

One virtue of the procedure we have described for finding extreme values is that it applies even to functions whose graphs are too complicated to draw. The next example illustrates this point.

EXAMPLE 2 Find the extreme values of the function

$$f(x, y) = xy - x^2 - y^2 - 2x - 2y + 4.$$

Solution The function is defined and differentiable for all x and y and its domain has no boundary points. The function therefore has extreme values only at the points where f_x and f_y are simultaneously zero. This leads to

$$f_x = y - 2x - 2 = 0, \qquad f_y = x - 2y - 2 = 0,$$

or

$$x = y = -2.$$

Therefore, the point $(-2, -2)$ is the only point where f may take on an extreme value. To see if it does so, we calculate

$$f_{xx} = -2, \qquad f_{yy} = -2, \qquad f_{xy} = 1.$$

The discriminant of f at $(a, b) = (-2, -2)$ is

$$f_{xx}f_{yy} - f_{xy}^2 = (-2)(-2) - (1)^2 = 4 - 1 = 3.$$

The combination

$$f_{xx} < 0 \qquad \text{and} \qquad f_{xx}f_{yy} - f_{xy}^2 > 0$$

tells us that f has a local maximum at $(-2, -2)$. The value of f at this point is $f(-2, -2) = 8$. ■

EXAMPLE 3 Find the extreme values of $f(x, y) = xy$.

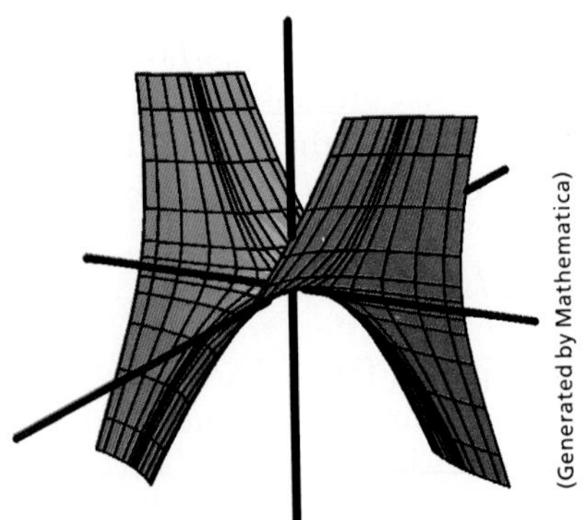

13.51 The surface $z = xy$ has a saddle point at the origin.

Solution Since the function is differentiable everywhere and its domain has no boundary points (Fig. 13.51), the function can assume extreme values only where

$$f_x = y = 0 \qquad \text{and} \qquad f_y = x = 0.$$

Thus, the origin is the only point where f might have an extreme value. To see what happens there, we calculate

$$f_{xx} = 0, \qquad f_{yy} = 0, \qquad f_{xy} = 1.$$

The discriminant,

$$f_{xx}f_{yy} - f_{xy}^2 = -1,$$

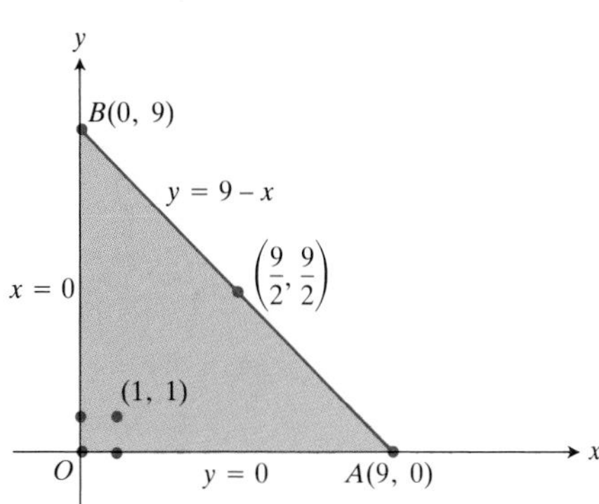

13.52 This triangular plate is the domain of the function in Example 4.

is negative. Therefore the function has a saddle point at (0, 0). We conclude that $f(x, y) = xy$ assumes no extreme values at all. ■

EXAMPLE 4 Find the absolute maximum and minimum values of

$$f(x, y) = 2 + 2x + 2y - x^2 - y^2$$

on the triangular plate in the first quadrant bounded by the lines $x = 0$, $y = 0$, $y = 9 - x$.

Solution The only places where f can assume these values are points on the boundary of the triangle (Fig. 13.52) and points inside the triangle where $f_x = f_y = 0$.

Boundary points. We take the triangle one side at a time:

1. On the segment OA, $y = 0$. The function

$$f(x, y) = f(x, 0) = 2 + 2x - x^2$$

may now be regarded as a function of x defined on the closed interval $0 \leq x \leq 9$. Its extreme values (we know from Chapter 4) may occur at the endpoints

$$x = 0 \qquad \text{where } f(0, 0) = 2,$$

$$x = 9 \qquad \text{where } f(9, 0) = 2 + 18 - 81 = -61,$$

and at the interior points where $f'(x, 0) = 2 - 2x = 0$. The only interior point where $f'(x, 0) = 0$ is $x = 1$, where

$$f(x, 0) = f(1, 0) = 3.$$

2. On the segment OB, $x = 0$ and

$$f(x, y) = f(0, y) = 2 + 2y - y^2.$$

We know from the symmetry of f in x and y and from the analysis we just carried out that the candidates on this segment are

$$f(0, 0) = 2, \qquad f(0, 9) = -61, \qquad f(0, 1) = 3.$$

3. We have already accounted for the values of f at the endpoints of AB, so we have only to look at the interior points of AB. With

$$y = 9 - x,$$

we have

$$f(x, y) = 2 + 2x + 2(9 - x) - x^2 - (9 - x)^2 = -61 + 18x - 2x^2.$$

Setting $f'(x, 9 - x) = 18 - 4x = 0$ gives

$$x = \frac{18}{4} = \frac{9}{2}.$$

At this value of x,

$$y = 9 - \frac{9}{2} = \frac{9}{2}, \qquad \text{and} \qquad f(x, y) = f\left(\frac{9}{2}, \frac{9}{2}\right) = -\frac{41}{2}.$$

Interior points. For these we have

$$f_x = 2 - 2x = 0, \qquad f_y = 2 - 2y = 0,$$

or

$$(x, y) = (1, 1),$$

where

$$f(1, 1) = 4.$$

Summary. We list all the candidates:

$$4, \quad 2, \quad -61, \quad 3, \quad -\frac{41}{2}.$$

The maximum is 4, which f assumes at (1, 1). The minimum is -61, which f assumes at (0, 9) and (9, 0). ■

The Condition $f_x(a, b) = f_y(a, b) = 0$

The assertion that a function $f(x, y)$ with defined first partial derivatives can have an extreme value at an interior point of its domain only if f_x and f_y are both zero at that point is called the first derivative test for local extreme values.

THEOREM 6

The First Derivative Test for Local Extreme Values

If $f(x, y)$ has a local maximum or local minimum value at an interior point (a, b) of its domain where f_x and f_y are both defined, then f_x and f_y are both zero at (a, b).

PROOF Suppose the value of f at (a, b) is a local maximum. Then

1. $x = a$ is an interior point of the domain of the curve $z = f(x, b)$ in which the plane $y = b$ cuts the surface $z = f(x, y)$ (Fig. 13.53).
2. The function $z = f(x, b)$ has a local maximum value at $x = a$.

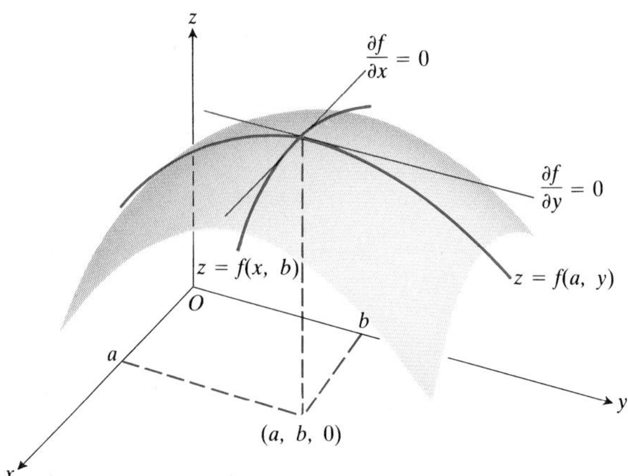

13.53 The maximum of f occurs at $x = a, y = b$.

3. The value of the derivative of $z = f(x, b)$ at $x = a$ is therefore zero (Theorem 1, Section 4.2).

Since this derivative is precisely $f_x(a, b)$, we conclude that $f_x(a, b) = 0$.

A similar argument with the function $z = f(a, y)$ shows that $f_y(a, b) = 0$.

This proves the theorem for local maximum values. To prove it for local minimum values, replace f by $-f$ and run through the argument again. ■

Despite the power of Theorem 6, we urge you to remember its limitations. It does not apply to boundary points of a function's domain, where it is possible for a function to have extreme values along with nonzero derivatives. And, it does not apply to points where either f_x or f_y fails to exist.

Summary of Max–Min Tests

The extreme values of $f(x, y)$ can occur only at

i) **boundary points** of the domain of f,

ii) **critical points** (interior points where $f_x = f_y = 0$ or points where f_x or f_y fail to exist).

If the first and second order partial derivatives of f are continuous throughout an open region containing a point (a, b) and $f_x(a, b) = f_y(a, b) = 0$, you may be able to classify (a, b) with the **second derivative test:**

i) $f_{xx} < 0$ and $f_{xx}f_{yy} - f_{xy}^2 > 0$ at (a, b) $\Rightarrow$ *local maximum,*

ii) $f_{xx} > 0$ and $f_{xx}f_{yy} - f_{xy}^2 > 0$ at (a, b) $\Rightarrow$ *local minimum,*

iii) $f_{xx}f_{yy} - f_{xy}^2 < 0$ at (a, b) $\Rightarrow$ *saddle point,*

iv) $f_{xx}f_{yy} - f_{xy}^2 = 0$ at (a, b) $\Rightarrow$ test is *inconclusive*.

Exercises 13.8

Test the functions in Exercises 1–24 for maxima, minima, and saddle points. Find the maximum and minimum values and say which maxima and minima are absolute.

1. $f(x, y) = x^2 + xy + y^2 + 3x - 3y + 4$
2. $f(x, y) = x^2 + 3xy + 3y^2 - 6x + 3y - 6$
3. $f(x, y) = 5xy - 7x^2 + 3x - 6y + 2$
4. $f(x, y) = 2xy - 5x^2 - 2y^2 + 4x + 4y - 4$
5. $f(x, y) = x^2 + xy + 3x + 2y + 5$
6. $f(x, y) = y^2 + xy - 2x - 2y + 2$
7. $f(x, y) = 2xy - 5x^2 - 2y^2 + 4x - 4$
8. $f(x, y) = 2xy - x^2 - 2y^2 + 3x + 4$
9. $f(x, y) = x^2 - 4xy + y^2 + 6y + 2$
10. $f(x, y) = 3x^2 + 6xy + 7y^2 - 2x + 4y$
11. $f(x, y) = 2x^2 + 3xy + 4y^2 - 5x + 2y$
12. $f(x, y) = 4x^2 - 6xy + 5y^2 - 20x + 26y$
13. $f(x, y) = x^2 - 4xy + y^2 + 5x - 2y$
14. $f(x, y) = x^2 - y^2 - 2x + 4y + 6$
15. $f(x, y) = x^2 - 2xy + 2y^2 - 2x + 2y + 1$
16. $f(x, y) = x^2 + 2xy$
17. $f(x, y) = 3 + 2x + 2y - 2x^2 - 2xy - y^2$
18. $f(x, y) = x^2 + xy + y^2 + x - 4y + 5$
19. $f(x, y) = x^3 - y^3 - 2xy + 6$
20. $f(x, y) = x^3 + y^3 + 3x^2 - 3y^2 - 8$
21. $f(x, y) = 6x^2 - 2x^3 + 3y^2 + 6xy$
22. $f(x, y) = 9x^3 + y^3/3 - 4xy$
23. $f(x, y) = x^3 + 3xy + y^3$
24. $f(x, y) = 4xy - x^4 - y^4$

In Exercises 25–32, find the absolute maxima and minima of the functions on the given domains.

25. $f(x, y) = 2x^2 - 4x + y^2 - 4y + 1$ on the closed triangular plate bounded by the lines $x = 0$, $y = 2$, $y = 2x$ in the first quadrant

26. $D(x, y) = x^2 - xy + y^2 + 1$ on the closed triangular plate in the first quadrant bounded by the lines $x = 0$, $y = 4$, $y = x$

27. $f(x, y) = x^2 + y^2$ on the closed triangular plate bounded by the lines $x = 0$, $y = 0$, $y + 2x = 2$ in the first quadrant

28. $T(x, y) = x^2 + xy + y^2 - 6x$ on the rectangular plate $0 \le x \le 5$, $-3 \le y \le 3$

29. $T(x, y) = x^2 + xy + y^2 - 6x + 2$ on the rectangular plate $0 \le x \le 5$, $-3 \le y \le 0$

30. $f(x, y) = 48xy - 32x^3 - 24y^2$ on the rectangular plate $0 \le x \le 1$, $0 \le y \le 1$

31. $f(x, y) = (x^2 - 4x) \cos y$ on the rectangular plate $1 \le x \le 3$, $-\pi/4 \le y \le \pi/4$.

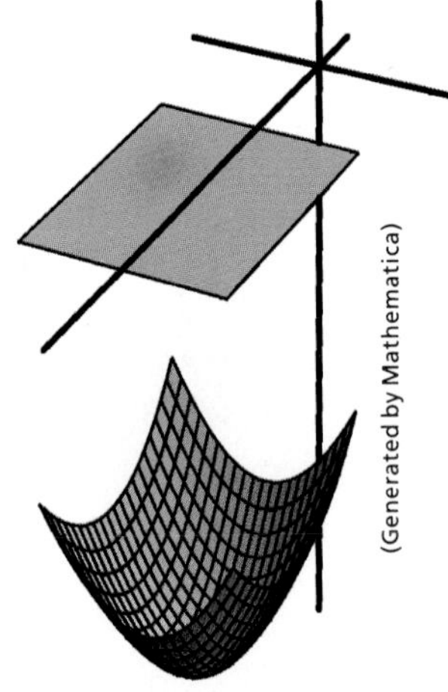

32. $f(x, y) = 4x - 8xy + 2y + 1$ on the triangular plate bounded by the lines $x = 0$, $y = 0$, $x + y = 1$ in the first quadrant.

33. *Temperatures.* A flat circular plate has the shape of the region $x^2 + y^2 \le 1$. The plate, including the boundary where $x^2 + y^2 = 1$, is heated so that the temperature at the point (x, y) is

$$T(x, y) = x^2 + 2y^2 - x.$$

Find the temperatures at the hottest and coldest points on the plate (Fig. 13.54).

34. Find the critical point of

$$f(x, y) = xy + 2x - \ln x^2y$$

in the open first quadrant ($x > 0, y > 0$) and show that f takes on a minimum there (Fig. 13.55).

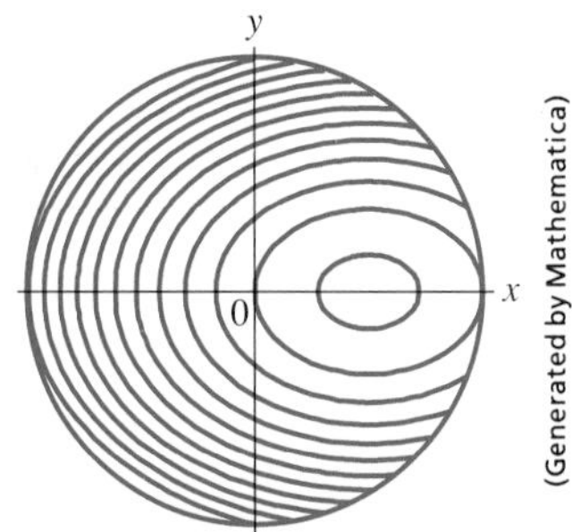

13.54 Curves of constant temperature are called isotherms. The figure shows isotherms of the temperature function $T(x, y) = x^2 + 2y^2 - x$ on the disk $x^2 + y^2 \le 1$ in the xy-plane. Exercise 33 asks you to locate the extreme temperatures.

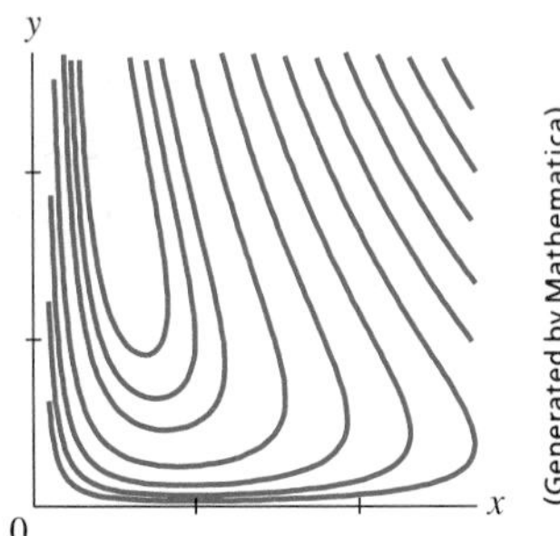

13.55 The function $f(x, y) = xy + 2x - \ln x^2y$ takes on a minimum value on the open first quadrant $x > 0$, $y > 0$. Exercise 34 asks you to find where.

35. Find the maxima, minima, and saddle points of $f(x, y)$, if any, given that

a) $f_x = 2x - 4y$ and $f_y = 2y - 4x$

b) $f_x = 2x - 2$ and $f_y = 2y - 4$

c) $f_x = 9x^2 - 9$ and $f_y = 2y + 4$

36. The discriminant $f_{xx}f_{yy} - f_{xy}^2$ is zero at the origin for each of the following functions, so the second derivative test fails there. Determine whether the function has a maximum, a minimum, or neither at the origin by imagining what the surface $z = f(x, y)$ looks like.

a) $f(x, y) = x^2y^2$ b) $f(x, y) = 1 - x^2y^2$

c) $f(x, y) = xy^2$ d) $f(x, y) = x^3y^2$

e) $f(x, y) = x^3y^3$ f) $f(x, y) = x^4y^4$

Extreme Values on Parametrized Curves

To find the extreme values of a function $f(x, y)$ on a curve $x = x(t)$, $y = y(t)$, we treat f as a function of the single variable t and use the Chain Rule to find where df/dt is zero. As in any other single-variable case, the extreme values of f are then found among the values at the

a) critical points (points where df/dt is zero or fails to exist), and

b) endpoints of the parameter domain.

37. Find the absolute maximum and minimum values of the following functions on
 i) the quarter-circle $x^2 + y^2 = 4$ in the first quadrant,
 ii) the half-circle $x^2 + y^2 = 4,\ \ y \geq 0$, and
 iii) the full circle $x^2 + y^2 = 4$.

 a) $f(x, y) = xy$ b) $f(x, y) = x + y$
 c) $f(x, y) = 2x^2 + y^2$

Use the parametrization $x = 2\cos t,\ y = 2\sin t$.

38. Find the absolute maximum and minimum values, if any, of the function $f(x, y) = xy$ on the following.
 a) The line $x = 2t,\ y = t + 1$
 b) The line segment $x = 2t,\ y = t + 1, \quad -1 \leq t \leq 0$
 c) The line segment $x = 2t,\ y = t + 1, \quad 0 \leq t \leq 1$

39. Find the absolute maximum and minimum values of the following functions on
 i) the quarter-ellipse $(x^2/9) + (y^2/4) = 1$ in the first quadrant,
 ii) the half-ellipse $(x^2/9) + (y^2/4) = 1,\ \ y \geq 0$, and
 iii) the full ellipse $(x^2/9) + (y^2/4) = 1$.

 a) $f(x, y) = x^2 + 3y^2$ b) $f(x, y) = 2x + 3y$

40. Find the absolute maximum and minimum values of $f(x, y) = xy$ on the ellipse $x^2 + 4y^2 = 8$.

Least Squares and Regression Lines

When we try to fit a line $y = mx + b$ to a set of numerical data points $(x_1, y_1), (x_2, y_2), \ldots, (x_n, y_n)$ (Fig. 13.56), we usually choose the line that minimizes the sum of the squares of the vertical distances from the points to the line. In theory, this means finding the values of m and b that minimize the value of the function

$$w = (mx_1 + b - y_1)^2 + \cdots + (mx_n + b - y_n)^2. \tag{1}$$

The values of m and b that do this are found with the first and second derivative tests to be

$$m = \frac{\left(\sum x_k\right)\left(\sum y_k\right) - n \sum x_k y_k}{\left(\sum x_k\right)^2 - n \sum x_k^{\,2}}, \tag{2}$$

$$b = \frac{1}{n}\left(\sum y_k - m \sum x_k\right), \tag{3}$$

with all sums running from $k = 1$ to $k = n$. Many scientific calculators have these formulas built in, enabling you to find m and b with only a few key presses after you have entered the data.

The line $y = mx + b$ determined by these values of m and b is called the **least squares line** or **regression line** for the data under study. Finding a least squares line lets you

1. summarize data with a simple expression,
2. predict values of y for other, experimentally untried values of x,
3. handle data analytically.

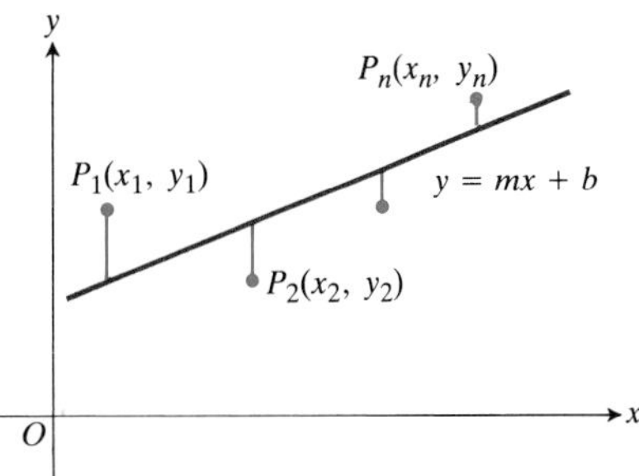

13.56 To fit a line to noncollinear points, we choose a line that minimizes the sum of the squares of the deviations.

EXAMPLE Find the least squares line for the points (0, 1), (1, 3), (2, 2), (3, 4), (4, 5).

Solution We organize the calculations in a table:

k	x_k	y_k	x_k^2	$x_k y_k$
1	0	1	0	0
2	1	3	1	3
3	2	2	4	4
4	3	4	9	12
5	4	5	16	20
Σ	10	15	30	39

Then we find

$$m = \frac{(10)(15) - 5(39)}{(10)^2 - 5(30)} = 0.9 \qquad \left(\begin{array}{l}\text{Eq. (2) with } n = 5 \text{ and} \\ \text{data from the table}\end{array}\right)$$

and use the value of m to find

$$b = \frac{1}{5}(15 - (0.9)(10)) = 1.2. \qquad \left(\begin{array}{l}\text{Eq (3) with } n = 5, \\ m = 0.9\end{array}\right)$$

The least squares line is $y = 0.9x + 1.2$ (Fig. 13.57). ■

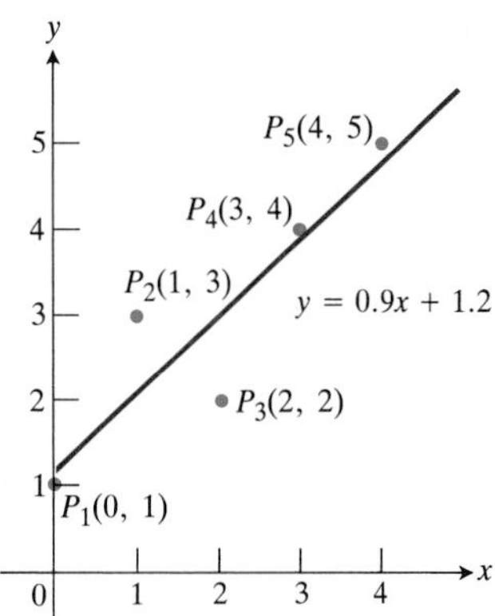

13.57 The least squares line for the data in the Example.

In Exercises 41–44, use Eqs. (2) and (3) to find the least squares line for each set of data points. Then use the linear equation you obtain to predict the value of y that would correspond to $x = 4$.

41. $(-1, 2)$, $(0, 1)$, $(3, -4)$

42. $(-2, 0)$, $(0, 2)$, $(2, 3)$

43. $(0, 0)$, $(1, 2)$, $(2, 3)$

44. $(0, 1)$, $(2, 2)$, $(3, 2)$

45. **CALCULATOR** Write a linear equation for the effect of irrigation on the yield of alfalfa by fitting a least squares line to the data in Table 13.1 (from the University of California Experimental Station. *Bulletin* No. 450, p. 8). Plot the data and draw the line.

46. **CALCULATOR** *Craters of Mars.* One theory of crater formation suggests that the frequency of large craters should fall off as the square of the diameter (Marcus, *Science,* June 21, 1968, p. 1334). Pictures from Mariner IV show the frequencies listed in Table 13.2. Fit a line of the form $F = m(1/D^2) + b$ to the data. Plot the data and draw the line.

TABLE 13.1
Growth of alfalfa

x (total seasonal depth of water applied (in.))	y (average alfalfa yield (tons/acre))
12	5.27
18	5.68
24	6.25
30	7.21
36	8.20
42	8.71

TABLE 13.2
Crater sizes on Mars

Diameter in km, D	$1/D^2$ (for left value of class interval)	Frequency, F
32–45	0.001	53
45–64	0.0005	22
64–90	0.00025	14
90–128	0.000125	3

13.9 Lagrange Multipliers

As we saw in Section 13.8, we sometimes need to find the maximum and minimum values of functions whose domains are confined to lie within some particular subset of the plane—a disk, for example, or a triangular region. But, as Fig. 13.58 suggests, functions may be subject to other kinds of constraints as well.

In this section, we explore a powerful method for finding the maxima and minima of constrained functions: the method of **Lagrange multipliers.** Lagrange developed the method in 1755 to solve sophisticated max-min problems in geometry. Today the method is important in economics, in engineering (where it is used in designing multistage rockets, for example), and in mathematics itself.

We begin with two examples, then describe the method in general terms and look at more examples.

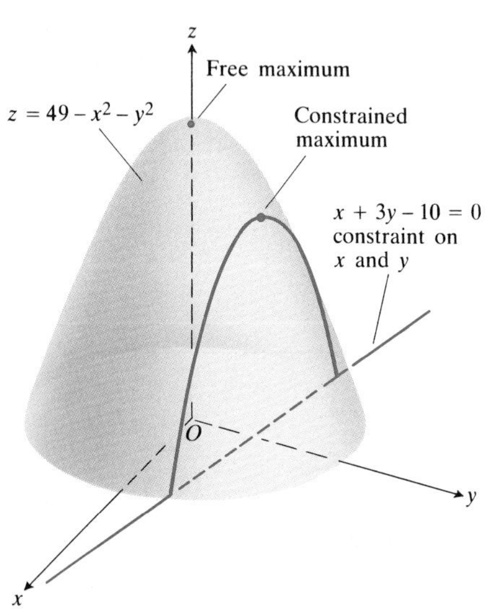

13.58 The function $f(x, y) = 49 - x^2 - y^2$, subject to the constraint $g(x, y) = x + 3y - 10 = 0$.

Constrained Maxima and Minima

EXAMPLE 1 Find the point $P(x, y, z)$ on the plane

$$2x + y - z - 5 = 0$$

that lies closest to the origin.

Lagrange Multipliers

The calculus of variations, invented by John Bernoulli (1667–1748), involves problems like that of finding a path $y = f(x)$ from a point $P_1(x_1, y_1)$ to a point $P_2(x_2, y_2)$ that minimizes the value of an integral

$$\int_{x_1}^{x_2} g(x, y)\, dx.$$

One such problem is that of finding a path from P_1 to P_2 that minimizes the time it takes for an object to slide from one point to the other under the force of gravity. Another is that of finding a path from P_1 to P_2 that sweeps out the minimum surface area when revolved about the x-axis. (It will be a catenary like the one in Section 7.9.)

Newton, Leibniz, and Euler all worked on variational problems as part of calculus, but the contributions of Joseph Louis Lagrange (1736–1813), beginning about 1760, forged the calculus of variations into a self-standing branch of mathematics. Euler considered Lagrange's work in this new field to be superior to his own and delayed publishing his own work so that Lagrange could receive full acclaim. While working on the calculus of variations Lagrange developed the elegant procedure that we know today as the method of Lagrange multipliers.

Solution The problem asks us to find the minimum value of the function

$$|\overrightarrow{OP}| = \sqrt{(x-0)^2 + (y-0)^2 + (z-0)^2} = \sqrt{x^2 + y^2 + z^2}$$

subject to the constraint that

$$2x + y - z - 5 = 0.$$

Since $|\overrightarrow{OP}|$ has a minimum value wherever the function

$$f(x, y, z) = x^2 + y^2 + z^2$$

has a minimum value, we may solve the problem by finding the minimum value of $f(x, y, z)$ subject to the constraint $2x + y - z - 5 = 0$. If we regard x and y as the independent variables in this equation and write z as

$$z = 2x + y - 5,$$

our problem reduces to one of finding the points (x, y) at which the function

$$h(x, y) = f(x, y, 2x + y - 5) = x^2 + y^2 + (2x + y - 5)^2$$

has its minimum value or values. Since the domain of h is the entire xy-plane, the first derivative test of Section 13.8 tells us that any minima that h might have must occur at points where

$$h_x = 2x + 2(2x + y - 5)(2) = 0, \qquad h_y = 2y + 2(2x + y - 5) = 0.$$

This leads to

$$10x + 4y = 20, \qquad 4x + 4y = 10,$$

and the solution

$$x = \frac{5}{3}, \qquad y = \frac{5}{6}.$$

We may apply either geometric arguments or the second derivative test to show that these values minimize h. The z-coordinate of the corresponding point on the plane $z = 2x + y - 5$ is

$$z = 2\left(\frac{5}{3}\right) + \frac{5}{6} - 5 = -\frac{5}{6}.$$

Therefore, the point we seek is

$$\text{Closest point:} \qquad P\left(\frac{5}{3}, \frac{5}{6}, -\frac{5}{6}\right).$$ ■

Attempts to solve a constrained maximum or minimum problem by substitution, as we might call the method of Example 1, do not always go smoothly, as the next example shows. This is one of the reasons for learning the new method of this section, which does not require us to decide in advance which of the constrained variables to regard as independent.

EXAMPLE 2 Find the points on the surface

$$x^2 - z^2 - 1 = 0$$

closest to the origin.

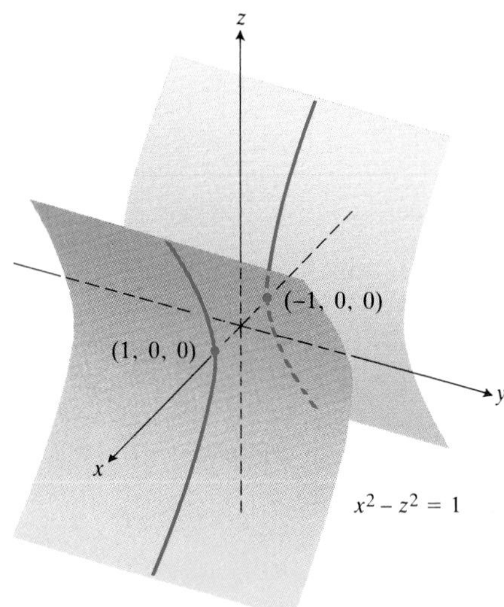

13.59 The hyperbolic cylinder $x^2 - z^2 - 1 = 0$ in Example 2.

Solution 1 The surface is shown in Fig. 13.59. We seek the points on the surface closest to the origin. These are the points whose coordinates minimize the value of the function

$$f(x, y, z) = x^2 + y^2 + z^2 \qquad \text{(Square of the distance)}$$

subject to the constraint that $x^2 - z^2 - 1 = 0$. If we regard x and y as independent variables in the constraint equation, then

$$z^2 = x^2 - 1$$

and the values of $f(x, y, z) = x^2 + y^2 + z^2$ on the surface are given by the function

$$h(x, y) = x^2 + y^2 + (x^2 - 1) = 2x^2 + y^2 - 1.$$

To find the points on the surface whose coordinates minimize f, we look for the points in the xy-plane whose coordinates minimize h. The only extreme value of h occurs where

$$h_x = 4x = 0 \qquad \text{and} \qquad h_y = 2y = 0,$$

i.e., at the point (0, 0). But now we're in trouble—there are no points on the surface where both x and y are zero. What went wrong?

What happened was that the first derivative test found (as it should have) the point *in the domain of* h where h has a minimum value. We, on the other hand, want the points *on the surface* where h has a minimum value. While the domain of h is the entire xy-plane, the domain from which we can select the first two coordinates of the points (x, y, z) on the surface is restricted to the "shadow" of the surface on the xy-plane; it does not include the band between the lines $x = -1$ and $x = 1$ (Fig. 13.60).

We can avoid this problem if we treat y and z as independent variables (instead of x and y) and express x in terms of y and z as

$$x^2 = z^2 + 1.$$

With this substitution, $f(x, y, z) = x^2 + y^2 + z^2$ becomes

$$k(y, z) = (z^2 + 1) + y^2 + z^2 = 1 + y^2 + 2z^2$$

and we look for the points where k takes on its smallest value. The domain of k in the yz-plane now matches the domain from which we select the

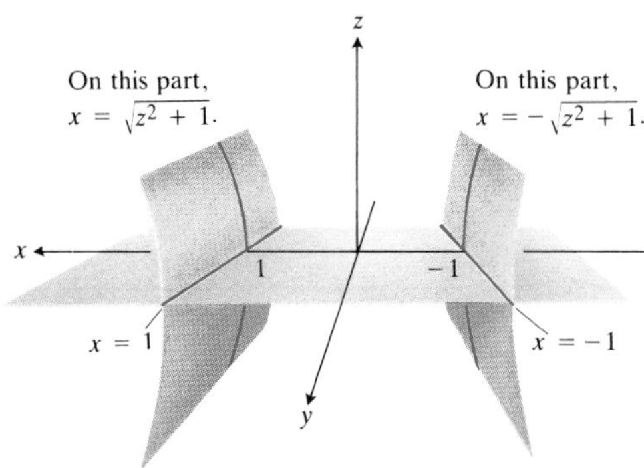

13.60 The region in the xy-plane from which the first two coordinates of the points (x, y, z) on the surface $x^2 - z^2 = 1$ are selected excludes the band $-1 < x < 1$ in the xy-plane.

y- and z-coordinates of the points (x, y, z) on the surface. Hence, the points that minimize k in the plane will have corresponding points on the hyperboloid. The smallest values of k occur where

$$k_y = 2y = 0 \qquad \text{and} \qquad k_z = 4z = 0,$$

or where $y = z = 0$. This leads to

$$x^2 = z^2 + 1 = 1, \qquad x = \pm 1.$$

The corresponding points on the surface are the points $(\pm 1, 0, 0)$. We can see from the inequality

$$k(y, z) = 1 + y^2 + 2z^2 \geq 1$$

that the points $(\pm 1, 0, 0)$ give a minimum value for k. We can also see that the minimum distance from the origin to a point on the surface is 1 unit.

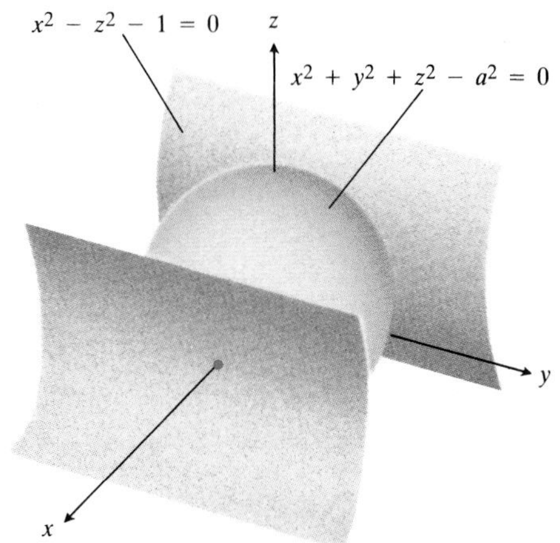

13.61 The sphere obtained by expanding a soap bubble centered at the origin until it just touches the hyperbolic cylinder $x^2 - z^2 - 1 = 0$. See Solution 2 of Example 2.

Solution 2 Another way to find the points on the surface closest to the origin is to imagine a small sphere centered at the origin expanding like a soap bubble until it just touches the surface (Fig. 13.61). At each point of contact, the surface and sphere have the same tangent plane and normal line. Therefore, if the sphere and surface are represented as level surfaces of the functions

$$f(x, y, z) = x^2 + y^2 + z^2 - a^2 = 0, \qquad g(x, y, z) = x^2 - z^2 - 1 = 0,$$

then the gradients ∇f and ∇g will be parallel where the surfaces touch. At any point of contact we will therefore be able to find a scalar λ such that

$$\nabla f = \lambda \nabla g,$$

or

$$2x\mathbf{i} + 2y\mathbf{j} + 2z\mathbf{k} = \lambda(2x\mathbf{i} - 2z\mathbf{k}).$$

Thus, the coordinates x, y, and z of any point of tangency will have to satisfy the three scalar equations

$$2x = 2\lambda x, \qquad 2y = 0, \qquad 2z = -2\lambda z. \tag{1}$$

For what values of λ will a point (x, y, z) whose coordinates satisfy the equations in (1) also lie on the surface $x^2 - z^2 - 1 = 0$? To answer this question, we use the fact that no point on the surface has a zero x-coordinate to conclude that $x \neq 0$ in the first equation in (1). This means that $2x = 2\lambda x$ only if

$$2 = 2\lambda \qquad \text{or} \qquad \lambda = 1.$$

For $\lambda = 1$, the equation $2z = -2\lambda z$ becomes $2z = -2z$. If this equation is to be satisfied as well, z must be zero. Since $y = 0$ also (from the equation $2y = 0$), we conclude that the points we seek all have coordinates of the form

$$(x, 0, 0).$$

What points on the surface $x^2 - z^2 = 1$ have coordinates of this form? The points $(x, 0, 0)$ for which

$$x^2 - (0)^2 = 1, \qquad x^2 = 1, \qquad \text{or} \qquad x = \pm 1.$$

The points on the surface closest to the origin are the points $(\pm 1, 0, 0)$. ■

The Method of Lagrange Multipliers

In Solution 2 above, we solved the problem in Example 2 by the **method of Lagrange multipliers.** In general terms, the method says that the extreme values of a function $f(x, y, z)$ whose variables are subject to a constraint $g(x, y, z) = 0$ are to be found on the surface $g = 0$ at the points where

$$\nabla f = \lambda \nabla g$$

for some scalar λ (called a **Lagrange multiplier**).

To explore the method further and see why it works, we first make the following observation, which we state as a theorem.

THEOREM 7

Suppose $f(x, y, z)$ has continuous first partial derivatives in a region whose interior contains the differentiable curve

$$C: \quad \mathbf{r} = g(t)\mathbf{i} + h(t)\mathbf{j} + k(t)\mathbf{k}.$$

If P_0 is a point on C where f has a local maximum or minimum relative to its values on C, then ∇f is perpendicular to C at P_0.

PROOF We show that ∇f is orthogonal to the curve's velocity vector at P_0. The values of f on C are given by the composite $f(g(t), h(t), k(t))$, whose derivative with respect to t is

$$\frac{df}{dt} = \frac{\partial f}{\partial x}\frac{dg}{dt} + \frac{\partial f}{\partial y}\frac{dh}{dt} + \frac{\partial f}{\partial z}\frac{dk}{dt} = \nabla f \cdot \mathbf{v}.$$

At any point P_0 where f has a local maximum or minimum relative to its values on the curve, $df/dt = 0$, so

$$\nabla f \cdot \mathbf{v} = 0.$$

■

By dropping the z-terms in Theorem 7, we obtain a similar result for functions of two variables.

COROLLARY

At the points on a differentiable curve $\mathbf{r} = g(t)\mathbf{i} + h(t)\mathbf{k}$ where a function $f(x, y)$ with continuous first partial derivatives takes on its local maxima and minima relative to its values on the curve, $\nabla f \cdot \mathbf{v} = 0$.

Theorem 7 is the key to why the method of Lagrange multipliers works, as we shall now see. Suppose that $f(x, y, z)$ and $g(x, y, z)$ have continuous first partial derivatives and that P_0 is a point on the surface $g(x, y, z) = 0$ where f has a local maximum or minimum value relative to its other values on the surface. Then f takes on a local maximum or minimum at P_0 relative to its values on every differentiable curve through P_0 on the surface $g(x, y, z) = 0$. Therefore, ∇f is perpendicular to the velocity vector of every such differentiable curve through P_0. But so is ∇g (because ∇g is perpendicular to the level surface $g = 0$, as we saw in Section 13.6). Therefore, at P_0, ∇f is some scalar multiple λ of ∇g.

The Method of Lagrange Multipliers

Suppose that $f(x, y, z)$ and $g(x, y, z)$ and their first partial derivatives are continuous. To find the local maximum and minimum values of f subject to the constraint $g(x, y, z) = 0$, find the values of x, y, z, and λ that simultaneously satisfy the equations

$$\nabla f = \lambda \nabla g \quad \text{and} \quad g(x, y, z) = 0. \tag{2}$$

For functions of two independent variables, the appropriate equations are

$$\nabla f = \lambda \nabla g \quad \text{and} \quad g(x, y) = 0. \tag{3}$$

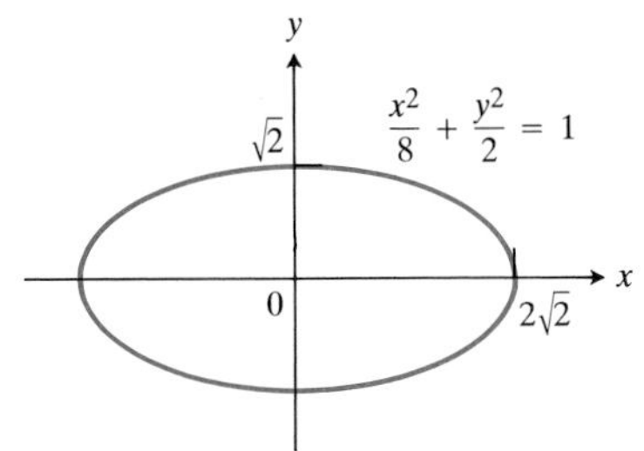

13.62 Example 3 shows how to find the largest and smallest values of the product *xy* on this ellipse.

EXAMPLE 3 Find the greatest and smallest values that the function

$$f(x, y) = xy$$

takes on the ellipse (Fig. 13.62)

$$\frac{x^2}{8} + \frac{y^2}{2} = 1.$$

Solution We are asked to find the extreme values of $f(x, y) = xy$ subject to the constraint

$$g(x, y) = \frac{x^2}{8} + \frac{y^2}{2} - 1 = 0.$$

To do so, we first find the values of x, y, and λ for which

$$\nabla f = \lambda \nabla g \quad \text{and} \quad g(x, y) = 0.$$

The gradient equation gives

$$y\mathbf{i} + x\mathbf{j} = \frac{\lambda}{4}x\mathbf{i} + \lambda y\mathbf{j},$$

from which we find

$$y = \frac{\lambda}{4}x, \qquad x = \lambda y, \qquad \text{and} \qquad y = \frac{\lambda}{4}(\lambda y) = \frac{\lambda^2}{4}y,$$

so that $y = 0$ or $\lambda = \pm 2$. We now consider these two cases.

CASE 1: If $y = 0$, then $x = y = 0$, but the point $(0, 0)$ is not on the ellipse. Hence, $y \neq 0$.

CASE 2: If $y \neq 0$, then $\lambda = \pm 2$ and $x = \pm 2y$. Substituting this in the equation $g(x, y) = 0$ gives

$$\frac{(\pm 2y)^2}{8} + \frac{y^2}{2} = 1, \qquad 4y^2 + 4y^2 = 8, \qquad \text{and} \qquad y = \pm 1.$$

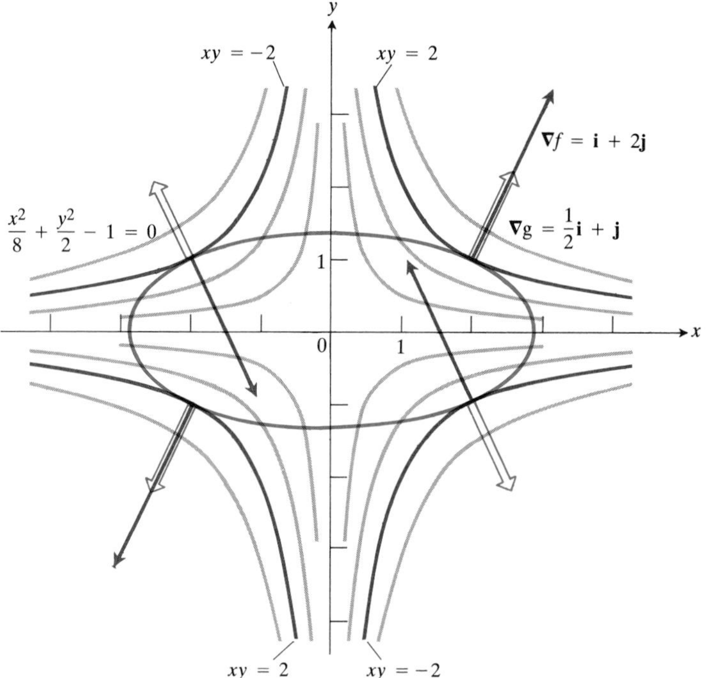

13.63 When subjected to the constraint $g(x, y) = x^2/8 + y^2/2 - 1 = 0$, the function $f(x, y) = xy$ takes on extreme values at the four points $(\pm 2, \pm 1)$. These are the points on the ellipse where ∇f (red) is a multiple of ∇g (blue).

The function $f(x, y) = xy$ therefore takes on its extreme values on the ellipse at the four points $(\pm 2, 1)$ $(\pm 2, -1)$. The extreme values are $xy = 2$ and $xy = -2$.

The Geometry of the Solution The level curves of the function $f(x, y) = xy$ are the hyperbolas $xy = c$ (Fig. 13.63). The farther the hyperbolas lie from the origin, the larger the absolute value of f. We want to find the extreme values of $f(x, y)$, given that the point (x, y) also lies on the ellipse $x^2 + 4y^2 = 8$. Which hyperbolas intersecting the ellipse lie farthest from the origin? The hyperbolas that just graze the ellipse, the ones that are tangent to it. At these points, any vector normal to the hyperbola is normal to the ellipse, so the gradient $\nabla f = y\mathbf{i} + x\mathbf{j}$ is a multiple $(\lambda = \pm 2)$ of the gradient $\nabla g = (x/4)\mathbf{i} + y\mathbf{j}$. At the point (2, 1), for example,

$$\nabla f = \mathbf{i} + 2\mathbf{j}, \qquad \nabla g = \frac{1}{2}\mathbf{i} + \mathbf{j}, \qquad \text{and} \qquad \nabla f = 2\nabla g.$$

At the point $(-2, 1)$,

$$\nabla f = \mathbf{i} - 2\mathbf{j}, \qquad \nabla g = -\frac{1}{2}\mathbf{i} + \mathbf{j}, \qquad \text{and} \qquad \nabla f = -2\nabla g.$$

■

EXAMPLE 4 Find the maximum and minimum values of the function $3x + 4y$ on the circle $x^2 + y^2 = 1$.

Solution We model this as a Lagrange multiplier problem with

$$f(x, y) = 3x + 4y, \qquad g(x, y) = x^2 + y^2 - 1$$

and look for the values of x, y, and λ that satisfy the equations

$$\begin{aligned} \nabla f = \lambda \nabla g: &\quad 3\mathbf{i} + 4\mathbf{j} = 2x\lambda\mathbf{i} + 2y\lambda\mathbf{j}, \\ g(x, y) = 0: &\quad x^2 + y^2 - 1 = 0. \end{aligned} \tag{4}$$

The gradient equation implies that $\lambda \neq 0$ and gives

$$x = \frac{3}{2\lambda}, \qquad y = \frac{2}{\lambda}.$$

These equations tell us, among other things, that x and y have the same sign. With these values for x and y, the equation $g(x, y) = 0$ gives

$$\left(\frac{3}{2\lambda}\right)^2 + \left(\frac{2}{\lambda}\right)^2 - 1 = 0,$$

so

$$\frac{9}{4\lambda^2} + \frac{4}{\lambda^2} = 1, \qquad 9 + 16 = 4\lambda^2, \qquad 4\lambda^2 = 25, \qquad \text{and} \qquad \lambda = \pm\frac{5}{2}.$$

Thus,

$$x = \frac{3}{2\lambda} = \pm\frac{3}{5}, \qquad y = \frac{2}{\lambda} = \pm\frac{4}{5},$$

and $f(x, y) = 3x + 4y$ has extreme values at the points $(x, y) = \pm\left(\frac{3}{5}, \frac{4}{5}\right)$. (There are two points instead of four because x and y have the same sign.)

By calculating the value of $3x + 4y$ at the points $\pm(3/5, 4/5)$, we see that its maximum and minimum values on the circle $x^2 + y^2 = 1$ are

$$3\left(\frac{3}{5}\right) + 4\left(\frac{4}{5}\right) = \frac{25}{5} = 5 \qquad \text{and} \qquad 3\left(-\frac{3}{5}\right) + 4\left(-\frac{4}{5}\right) = -\frac{25}{5} = -5.$$

The Geometry of the Solution (Fig. 13.64) The level curves of $f(x, y) = 3x + 4y$ are the lines $3x + 4y = c$. The farther the lines lie from the origin, the larger the absolute value of f. We want to find the extreme values of

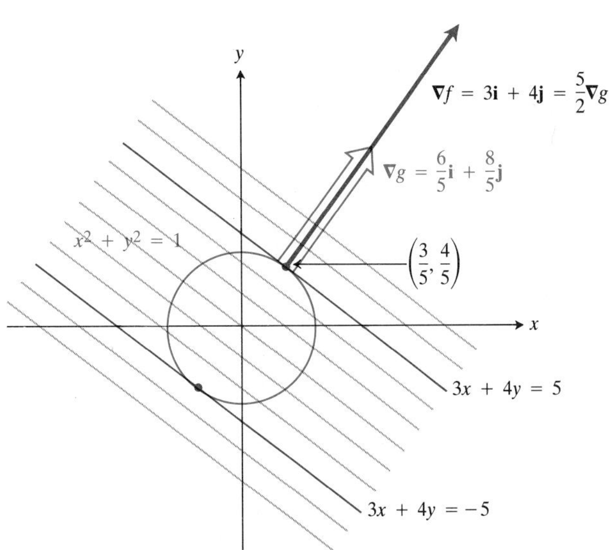

13.64 The function $f(x, y) = 3x + 4y$ takes on its largest value on the unit circle $g(x, y) = x^2 + y^2 - 1 = 0$ at the point $(3/5, 4/5)$ and its smallest value at the point $(-3/5, -4/5)$. At each of these points, ∇f is a scalar multiple of ∇g. The figure shows the gradients at the first point but not the second.

$f(x, y)$ given that the point (x, y) also lies on the circle $x^2 + y^2 = 1$. Which lines intersecting the circle lie farthest from the origin? The lines tangent to the circle. At the points of tangency, any vector normal to the line is normal to the circle, so the gradient $\nabla f = 3\mathbf{i} + 4\mathbf{j}$ is a multiple ($\lambda = \pm 5/2$) of the gradient $\nabla g = 2x\mathbf{i} + 2y\mathbf{j}$. At the point (3/5, 4/5), for example,

$$\nabla f = 3\mathbf{i} + 4\mathbf{j}, \qquad \nabla g = \frac{6}{5}\mathbf{i} + \frac{8}{5}\mathbf{j}, \qquad \text{and} \qquad \nabla f = \frac{5}{2}\nabla g.$$

Note: The method of Lagrange multipliers also applies to differentiable functions with more variables. See Appendix 8.

Exercises 13.9

1. Find the points on the ellipse $x^2 + 2y^2 = 1$ where $f(x, y) = xy$ has its extreme values.
2. Find the extreme values of $f(x, y) = xy$ subject to the constraint $g(x, y) = x^2 + y^2 - 10 = 0$.
3. Find the maximum value of $f(x, y) = 49 - x^2 - y^2$ on the line $x + 3y = 10$ (Fig. 13.58).
4. How close does the line $y = x + 1$ come to the parabola $y^2 = x$?
5. Find the extreme values of $f(x, y) = x^2y$ on the line $x + y = 3$.
6. Find the points on the curve $x^2y = 2$ nearest the origin.
7. Use the method of Lagrange multipliers to find
   a) the minimum value of $x + y$, subject to the constraints $xy = 16$, $x > 0$, $y > 0$;
   b) the maximum value of xy, subject to the constraint $x + y = 16$.

   Comment on the geometry of each solution.
8. Find the points on the curve $x^2 + xy + y^2 = 1$ in the xy-plane that are nearest to and farthest from the origin.
9. Find the dimensions of the closed right circular cylindrical can of smallest surface area whose volume is 16π cm^3.
10. Use the method of Lagrange multipliers to find the dimensions of the rectangle of greatest area that can be inscribed in the ellipse $x^2/16 + y^2/9 = 1$ with sides parallel to the coordinate axes.
11. The temperature at a point (x, y) on a metal plate is $T(x, y) = 4x^2 - 4xy + y^2$. An ant on the plate walks around the circle of radius 5 centered at the origin. What are the highest and lowest temperatures encountered by the ant?
12. Your firm has been asked to design a storage tank for liquid petroleum gas. The customer's specifications call for a cylindrical tank with hemispherical ends, and the tank is to hold 8000 m^3 of gas. The customer wants to use the smallest amount of material possible in building the tank. What radius and height do you recommend for the cylindrical portion of the tank?
13. Find the maximum and minimum values of $x^2 + y^2$ subject to the constraint $x^2 - 2x + y^2 - 4y = 0$.
14. Find the point on the plane $x + 2y + 3z = 13$ closest to the point (1, 1, 1).
15. Find the maximum and minimum values of
$$f(x, y, z) = x - 2y + 5z$$
on the sphere
$$x^2 + y^2 + z^2 = 30.$$
16. Find the minimum distance from the surface $x^2 + y^2 - z^2 = 1$ to the origin.
17. Find the point on the surface $z = xy + 1$ nearest the origin.
18. Find the points on the surface $z^2 = xy + 4$ closest to the origin.
19. Find the points on the sphere $x^2 + y^2 + z^2 = 25$ where $f(x, y, z) = x + 2y + 3z$ has its maximum and minimum values.
20. Find three real numbers whose sum is 9 and the sum of whose squares is as small as possible.
21. Find the largest product the positive numbers x, y, and z can have if $x + y + z^2 = 16$.
22. A space probe in the shape of the ellipsoid
$$4x^2 + y^2 + 4z^2 = 16$$
enters the earth's atmosphere and its surface begins to heat. After one hour, the temperature at the point (x, y, z) on the probe's surface is
$$T(x, y, z) = 8x^2 + 4yz - 16z + 600.$$
Find the hottest point on the probe's surface.

23. *An example from economics.* In economics, the usefulness or *utility* of amounts x and y of two capital goods G_1 and G_2 is sometimes measured by a function $U(x, y)$. For example, G_1 and G_2 might be two chemicals a pharmaceutical company needs to have on hand and $U(x, y)$ the gain from manufacturing a product whose synthesis requires different amounts of the chemicals depending on the process used. If G_1 costs a dollars per kilogram, G_2 costs b dollars per kilogram, and the total amount allocated for the purchase of G_1 and G_2 together is c dollars, then the company wants to maximize $U(x, y)$ given that $ax + by = c$. Thus, they need to solve a typical Lagrange multiplier problem.

Suppose

$$U(x, y) = xy + 2x$$

and that the equation $ax + by = c$ becomes

$$2x + y = 30$$

when reduced to lowest terms. Find the maximum value of U and the corresponding values of x and y subject to this latter constraint.

24. You are in charge of erecting a radio telescope on a newly discovered planet. To minimize interference, you want to place it where the magnetic field of the planet is weakest. The planet is spherical with a radius of 6 units. Based on a coordinate system whose origin is at the center of the planet, the strength of the magnetic field is given by $M(x, y, z) = 6x - y^2 + xz + 60$. Where should you locate the radio telescope?

25. *The condition $\nabla f = \lambda \nabla g$ is not sufficient.* While $\nabla f = \lambda \nabla g$ is a necessary condition for the occurrence of an extreme value of $f(x, y)$ subject to the condition $g(x, y) = 0$, it does not in itself guarantee that one exists. As a case in point, try using the method of Lagrange multipliers to find a maximum value of $f(x, y) = x + y$ subject to the constraint that $xy = 16$. The method will identify the two points $(4, 4)$ and $(-4, -4)$ as candidates for the location of extreme values. Yet the sum $(x + y)$ has no maximum value on the hyperbola $xy = 16$. The farther you go from the origin on this hyperbola in the first quadrant, the larger the sum $f(x, y) = x + y$ becomes.

Review Questions

1. Give examples of functions of two and three variables, and describe their domains and ranges.
2. Describe two ways to display the values of a function $f(x, y)$ graphically. Give examples.
3. What is a level surface of a function $f(x, y, z)$ of three independent variables? Give an example.
4. Give two equivalent definitions of

$$\lim_{(x, y) \to (x_0, y_0)} f(x, y) = L.$$

What is the basic theorem for calculating limits of sums, differences, products, constant multiples, and quotients of functions of two or more variables?

5. When is a function of two (three) variables continuous at a point in its domain? Give examples of functions that are continuous at some points but not continuous at others.
6. How are the partial derivatives $\partial f/\partial x$ and $\partial f/\partial y$ of a function $f(x, y)$ defined? What is their geometric meaning? What rates do they describe? How are they calculated? Give examples.
7. Give examples of the Chain Rule for functions of two and three variables. Draw tree diagrams for your examples.
8. Give examples of second order partial derivatives of functions of two variables. What is the basic theorem about mixed second order derivatives?
9. Define the gradient of a function $f(x, y, z)$ and describe the role the gradient plays in defining directional derivatives, tangent planes, and normal lines. What is the relation between ∇f and the directions in which f changes most rapidly? What are the analogous results for functions of two variables? Give examples.
10. What is the linearization $L(x, y)$ of a function $f(x, y)$ at a point (x_0, y_0)? How are linearizations used? Give an upper bound for the error incurred by approximating f by L in the neighborhood of (x_0, y_0). Give examples.
11. If (x, y) moves from (x_0, y_0) to a point $(x_0 + dx, y_0 + dy)$ nearby, how do we estimate the corresponding change, in a differentiable function $f(x, y)$? How do we estimate the relative change? The percentage change? Give an example.

12. Describe how to find the extreme values (if any) of a function $f(x, y)$ that has continuous first and second order partial derivatives. Give examples.

13. How do you find the extreme values of a differentiable function along a differentiable curve in its domain? Give an example.

14. Suppose $\mathbf{r} = g(t)\mathbf{i} + h(t)\mathbf{j} + k(t)\mathbf{k}$ is a differentiable curve in the domain of a differentiable function $f(x, y, z)$. Describe the relation between df/dt, ∇f, and $\mathbf{v} = d\mathbf{r}/dt$. What can be said about ∇f and $\mathbf{v}$ at points on the curve where f has extreme values relative to its other values on the curve?

15. Describe the method of Lagrange multipliers for finding the extreme values of a differentiable function $f(x, y)$ subject to a differentiable constraint $g(x, y) = 0$. What is going on geometrically? Give an example.

Practice Exercises

In Exercises 1–4, find the domain and range of f and identify the level curves. Sketch a typical level curve.

1. $f(x, y) = 9x^2 + y^2$
2. $f(x, y) = \sin(y - x)$
3. $f(x, y) = 1/xy$
4. $f(x, y) = \sqrt{x^2 - y}$

In Exercises 5–8, find the domain and range of f and identify the level surfaces. Sketch a typical level surface.

5. $f(x, y, z) = x^2 + y^2 - z$
6. $f(x, y, z) = x^2 + 4y^2 + 9z^2$
7. $f(x, y, z) = \dfrac{1}{x^2 + y^2 + z^2}$
8. $f(x, y, z) = \dfrac{1}{x^2 + y^2 + z^2 + 1}$

Find the limits in Exercises 9–12.

9. $\displaystyle\lim_{(x,y)\to(\pi, \ln 2)} e^y \cos x$
10. $\displaystyle\lim_{(x,y)\to(0,0)} \frac{2 + y}{x + \cos y}$
11. $\displaystyle\lim_{\substack{(x,y)\to(1,1)\\ x \neq y}} \frac{x^2 - y^2}{x - y}$
12. $\displaystyle\lim_{P\to(1,-1,e)} \ln|x + y + z|$

By considering different paths of approach, show that the limits in Exercises 13–14 do not exist.

13. $\displaystyle\lim_{\substack{(x,y)\to(0,0)\\ y \neq x^2}} \frac{y}{x^2 - y}$
14. $\displaystyle\lim_{\substack{(x,y)\to(0,0)\\ xy \neq 0}} \frac{x^2 + y^2}{xy}$

In Exercises 15–20, find the partial derivative of the function with respect to each variable.

15. $g(r, \theta) = r\cos\theta + r\sin\theta$
16. $f(x, y) = \dfrac{1}{2}\ln(x^2 + y^2) + \tan^{-1}\dfrac{y}{x}$
17. $f(R_1, R_2, R_3) = \dfrac{1}{R_1} + \dfrac{1}{R_2} + \dfrac{1}{R_3}$
18. $h(x, y, z) = \sin(2\pi x + y - 3z)$
19. $P(n, R, T, V) = \dfrac{nRT}{V}$ (the Ideal Gas Law)
20. $f(r, l, T, d) = \dfrac{1}{2rl}\sqrt{\dfrac{T}{\pi d}}$
(the frequency of a struck piano string, Chapter 3, Practice Exercise 64)

Find the second order partial derivatives of the functions in Exercises 21–24.

21. $f(x, y) = y + \dfrac{x}{y}$
22. $f(x, y) = e^x + y\sin x$
23. $f(x, y) = x + xy - 5x^3 + \ln(x^2 + 1)$
24. $f(x, y) = y^2 - 3xy + \cos y + 7e^y$
25. Find dw/dt at $t = 0$ if $w = \sin(xy + \pi)$, $x = e^t$, $y = \ln(t + 1)$.
26. Find dw/dt at $t = 1$ if $w = xe^y + y\sin z - \cos z$, $x = 2\sqrt{t}$, $y = t - 1 + \ln t$, $z = \pi t$.
27. Find $\partial w/\partial r$ and $\partial w/\partial s$ when $r = \pi$ and $s = 0$ if $w = \sin(2x - y)$, $x = r + \sin s$, $y = rs$.
28. Find $\partial w/\partial u$ and $\partial w/\partial v$ when $u = v = 0$ if $w = \ln\sqrt{1 + x^2} - \tan^{-1}x$ and $x = 2e^u\cos v$.

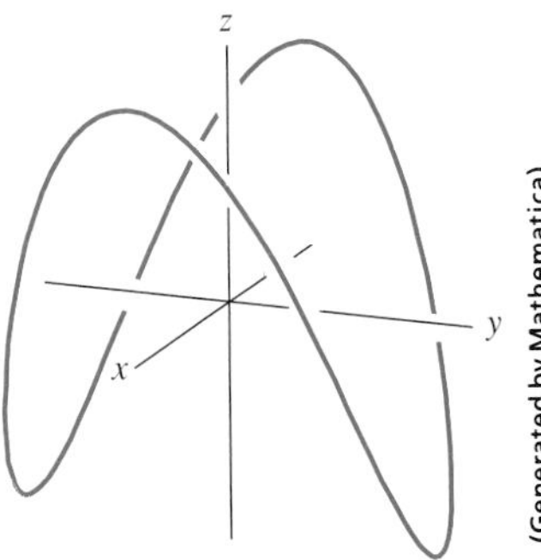

13.65 The curve in Practice Exercise 31.

Assuming that the equations in Exercises 29 and 30 define y as a differentiable function of x, find the value of dy/dx at the given point.

29. $1 - x - y^2 - \sin xy = 0, \quad (0, 1)$

30. $2xy + e^{x+y} - 2 = 0, \quad (0, \ln 2)$

31. Find the value of the derivative of $f(x, y, z) = xy + yz + xz$ with respect to t on the curve $x = \cos t$, $y = \sin t$, $z = \cos 2t$ at the point where $t = 1$ (Fig. 13.65).

32. Show that if $w = f(s)$ is any differentiable function of s whatever, and $s = y + 5x$, then

$$\frac{\partial w}{\partial x} - 5\frac{\partial w}{\partial y} = 0.$$

In Exercises 33–36, find the directions in which f increases and decreases most rapidly at P_0 and find the derivative of f in each direction. Also, find the derivative of f at P_0 in the direction of the vector $\mathbf{A}$.

33. $f(x, y) = \cos x \cos y, \quad P_0(\pi/4, \pi/4), \quad \mathbf{A} = 3\mathbf{i} + 4\mathbf{j}$

34. $f(x, y) = x^2 e^{-2y}, \quad P_0(1, 0), \quad \mathbf{A} = \mathbf{i} + \mathbf{j}$

35. $f(x, y, z) = \ln(2x + 3y + 6z), \quad P_0(-1, -1, 1),$ $\mathbf{A} = 2\mathbf{i} + 3\mathbf{j} + 6\mathbf{k}$

36. $f(x, y, z) = x^2 + 3xy - z^2 + 2y + z + 4, \quad P_0(0, 0, 0),$ $\mathbf{A} = \mathbf{i} + \mathbf{j} + \mathbf{k}$

In Exercises 37–38, sketch the surface $f(x, y, z) = c$ together with ∇f at the given points.

37. $x^2 + y + z^2 = 0; \quad (0, -1, \pm 1), (0, 0, 0)$

38. $y^2 + z^2 = 4; \quad (2, \pm 2, 0), (2, 0, \pm 2)$

In Exercises 39–40, find an equation for the plane tangent to the level surface $f(x, y, z) = c$ at the point P_0. Also, find parametric equations for the line that is normal to the surface at P_0.

39. $x^2 - y - 5z = 0, \quad P_0(2, -1, 1)$

40. $x^2 + y^2 + z = 4, \quad P_0(1, 1, 2)$

In Exercises 41–42, find equations for the lines that are tangent and normal to the level curve $f(x, y) = c$ at the point P_0. Then sketch the lines and level curve together with ∇f at P_0.

41. $y - \sin x = 1, \quad P_0(\pi, 1)$

42. $\dfrac{y^2}{2} - \dfrac{x^2}{2} = \dfrac{3}{2}, \quad P_0(1, 2)$

In Exercises 43–44, find the linearization $L(x, y)$ of the function $f(x, y)$ at the point P_0. Then find an upper bound for the magnitude of the error E in the approximation $f(x, y) \approx L(x, y)$ over the rectangle R.

43. $f(x, y) = \sin x \cos y, \quad P_0(\pi/4, \pi/4)$

$R: \quad \left|x - \dfrac{\pi}{4}\right| \le 0.1, \quad \left|y - \dfrac{\pi}{4}\right| \le 0.1$

44. $f(x, y) = xy - 3y^2 + 2, \quad P_0(1, 1)$

$R: \quad |x - 1| \le 0.1, \quad |y - 1| \le 0.2$

Find the linearizations of the functions in Exercises 45–46 at the given points.

45. $f(x, y, z) = xy + 2yz - 3xz$ at $(1, 0, 0)$ and $(1, 1, 0)$.

46. $f(x, y, z) = \sqrt{2} \cos x \sin(y + z)$ at $(0, 0, \pi/4)$ and $(\pi/4, \pi/4, 0)$.

47. You plan to calculate the volume inside a stretch of pipeline that is about 36 in. in diameter and 1 mile long. With which measurement should you be more careful—the length, or the diameter? Why?

48. Near the point $(1, 2)$, is $f(x, y) = x^2 - xy + y^2 - 3$ more sensitive to changes in x, or to changes in y? How do you know?

49. In a circuit that obeys Ohm's law, the current I (amperes) is related to the voltage V (volts) and the resistance R (ohms) by the equation $I = V/R$. If the voltage drops from 24 to 23 volts and the resistance drops from 100 to 80 ohms, will I increase, or decrease? By about how much? Express the changes in V and R and the estimated change in I as percentages of their original values.

50. If $a = 10$ cm and $b = 16$ cm to the nearest millimeter, what should you expect the maximum percentage error to be in the calculated area $A = \pi ab$ of the ellipse $x^2/a^2 + y^2/b^2 = 1$?

Test the functions in Exercises 51–56 for maxima, minima, and saddle points. Find the functions' values at these points.

51. $f(x, y) = x^2 - xy + y^2 + 2x + 2y - 4$

52. $f(x, y) = 5x^2 + 4xy - 2y^2 + 4x - 4y$

53. $f(x, y) = 6xy - x^3 - y^2$

54. $f(x, y) = 2x^3 + 3xy + y^3$

55. $f(x, y) = x^3 + y^3 - 3xy + 15$

56. $f(x, y) = x^3 + y^3 + 3x^2 - 3y^2$

In Exercises 57–62, find the absolute maximum and minimum values of f on the region R.

57. $f(x, y) = x^2 + xy + y^2 - 3x + 3y$
R: The triangular region cut from the first quadrant by the line $x + y = 4$

58. $f(x, y) = x^2 - y^2 - 2x + 4y + 1$
R: The rectangular region in the first quadrant bounded by the coordinate axes and the lines $x = 4$ and $y = 2$

59. $f(x, y) = y^2 - xy - 3y + 2x$
R: The square region enclosed by the lines $x = \pm 2$ and $y = \pm 2$

60. $f(x, y) = 2x + 2y - x^2 - y^2$
R: The square bounded by the coordinate axes and the lines $x = 2$, $y = 2$ in the first quadrant

61. $f(x, y) = x^2 - y^2 - 2x + 4y$
R: The triangular region bounded below by the x-axis, above by the line $y = x + 2$, and on the right by the line $x = 2$

62. $f(x, y) = 4xy - x^4 - y^4 + 16$
R: The triangular region bounded below by the line $y = -2$, above by the line $y = x$, and on the right by the line $x = 2$

63. Find the points nearest the origin on the curve $xy^2 = 54$.

64. Find the extreme values of $f(x, y) = x^3 + y^2$ on the cylinder $x^2 + y^2 = 1$.

65. Suppose that the temperature T (degrees) at the point (x, y, z) on the sphere $x^2 + y^2 + z^2 = 1$ is $T = 400\,xyz^2$. Locate the highest and lowest temperatures on the sphere.

66. Find the extreme values of $f(x, y) = x^2 + 3y^2 + 2y$ on the unit disk $x^2 + y^2 \leq 1$.

14 Multiple Integrals

Charles Fiel/F.P.G.

The Dulles International Airport in Reston, Virginia. We can calculate the volumes of many irregular shapes like these with multiple integrals.

OVERVIEW This chapter introduces the other half, so to speak, of the calculus of multivariable functions, complementing the previous chapter on differentiation. The problems we can solve by integrating functions of two and three variables are similar to the problems solved by single-variable integration, but more general. For example, in Chapter 6 we calculated volumes of solids of revolution, whereas now we will learn how to calculate volumes of other solids, in a variety of coordinate systems. Earlier, we calculated the moments and centers of mass of rods and thin plates; now we will be able to handle plates more easily, work with more general density functions, and treat solids as well. As in the previous chapter, we can perform the necessary calculations by drawing on our experience with functions of a single variable. In the next chapter, we shall combine multiple integrals with vector functions to derive the magnificent vector integral theorems with which we calculate surface area and fluid flow. Multiple integrals also play important roles in statistics, electrical engineering, and the physical theory of electromagnetism, but we shall not go into these applications here.

14.1 Double Integrals

We now show how to integrate a continuous function $f(x, y)$ over a bounded region in the xy-plane. We begin with rectangular regions and then proceed to bounded regions of a more general nature. There are many similarities between the "double" integrals we define here and the "single" integrals we defined in Chapter 5 for functions of a single variable. Indeed, the connection is very strong. The basic theorem for evaluating double integrals says that every double integral can be evaluated in stages, using the single-integration methods already at our command.

Double Integrals Over Rectangles

Suppose that $f(x, y)$ is defined on a rectangular region R given by

$$R: \quad a \leq x \leq b, \quad c \leq y \leq d.$$

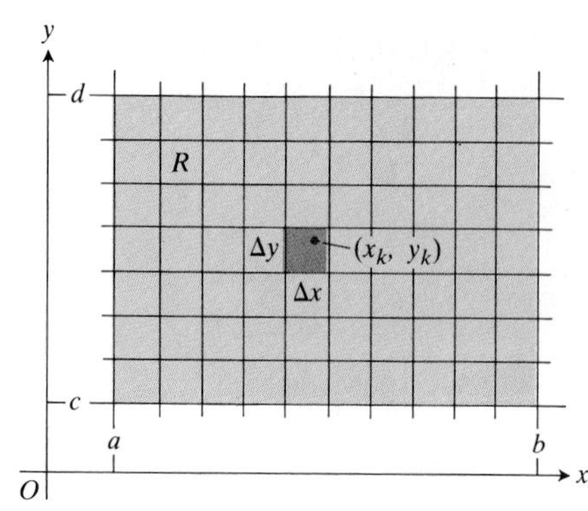

14.1 Rectangular grid subdividing the region R into small rectangles of area $\Delta A_k = \Delta x_k \Delta y_k$.

We imagine R to be covered by a network of lines parallel to the x- and y-axes (Fig. 14.1). These lines divide R into small pieces of area

$$\Delta A = \Delta x \, \Delta y.$$

We number these in some order $\Delta A_1, \Delta A_2, \ldots, \Delta A_n$, choose a point (x_k, y_k) in each piece ΔA_k, and form the sum

$$S_n = \sum_{k=1}^{n} f(x_k, y_k) \, \Delta A_k. \tag{1}$$

If f is continuous throughout R, then, as we refine the mesh width to make both Δx and Δy go to zero, the sums in (1) approach a limit called the **double integral** of f over R. The notation for it is

$$\iint_R f(x, y) \, dA \qquad \text{or} \qquad \iint_R f(x, y) \, dx \, dy.$$

Thus,

$$\iint_R f(x, y) \, dA = \lim_{\Delta A \to 0} \sum_{k=1}^{n} f(x_k, y_k) \, \Delta A_k. \tag{2}$$

As with functions of a single variable, the sums approach this limit no matter how the intervals $[a, b]$ and $[c, d]$ that determine R are subdivided, as long as the norms of the subdivisions both go to zero. The limit in (2) is also independent of the order in which the areas ΔA_k are numbered and independent of the choice of the point (x_k, y_k) within each ΔA_k. The values of the individual approximating sums S_n depend on these choices, but the sums approach the same limit in the end. The proof of the existence and uniqueness of this limit for a continuous function f is given in more advanced texts. The continuity of f is a sufficient condition for the existence of the double integral, but not a necessary one, and the limit in question exists for many discontinuous functions as well.

Properties of Double Integrals

Like single integrals, double integrals of continuous functions have algebraic properties that are useful in computations and applications. These properties hold because they hold for the sums with which the integrals are defined. Among these properties are the following.

1. $\displaystyle\iint_R kf(x,y)\,dA = k\iint_R f(x,y)\,dA \qquad \text{(any number } k\text{)}$

2. $\displaystyle\iint_R (f(x,y)+g(x,y))\,dA = \iint_R f(x,y)\,dA + \iint_R g(x,y)\,dA$

3. $\displaystyle\iint_R (f(x,y)-g(x,y))\,dA = \iint_R f(x,y)\,dA - \iint_R g(x,y)\,dA$

4. $\displaystyle\iint_R f(x,y)\,dA \geq 0 \text{ if } f(x,y) \geq 0 \text{ on } R$

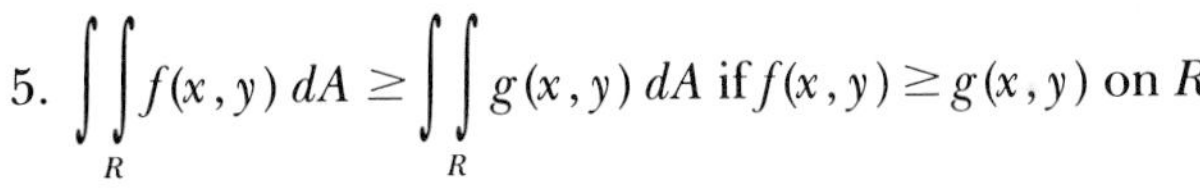

5. $\displaystyle\iint_R f(x,y)\,dA \geq \iint_R g(x,y)\,dA \text{ if } f(x,y) \geq g(x,y) \text{ on } R$

These are like the single-integral properties in Section 5.3. There is also a "domain additivity" property:

6. $\displaystyle\iint_R f(x,y)\,dA = \iint_{R_1} f(x,y)\,dA + \iint_{R_2} f(x,y)\,dA.$

It holds when R is the union of two nonoverlapping rectangles R_1 and R_2 (Fig. 14.2). Again, we shall omit the proof.

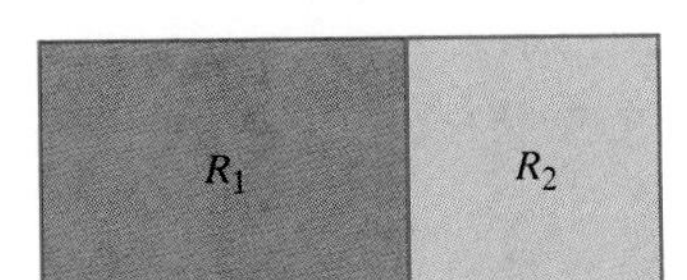

$$\iint_{R_1 \cup R_2} f(x, y)\,dA = \int_{R_1} f(x, y)\,dA + \int_{R_2} f(x, y)\,dA$$

14.2 Double integrals have the same kind of domain additivity property that single integrals have.

Interpreting the Double Integral as a Volume

When $f(x,y)$ is positive, we may interpret the double integral of f over a rectangular region R as the volume of the solid prism bounded below by R and above by the surface $z = f(x,y)$ (Fig. 14.3). Each term $f(x_k, y_k)\,\Delta A_k$ in the sum $S_n = \Sigma f(x_k, y_k)\,\Delta A_k$ is the volume of a vertical rectangular prism that approximates the volume of the portion of the solid that stands directly above the base ΔA_k. The sum S_n thus approximates what we want to call the total volume of the solid. We *define* this volume to be

$$\text{Volume} = \lim S_n = \iint_R f(x,y)\,dA. \tag{3}$$

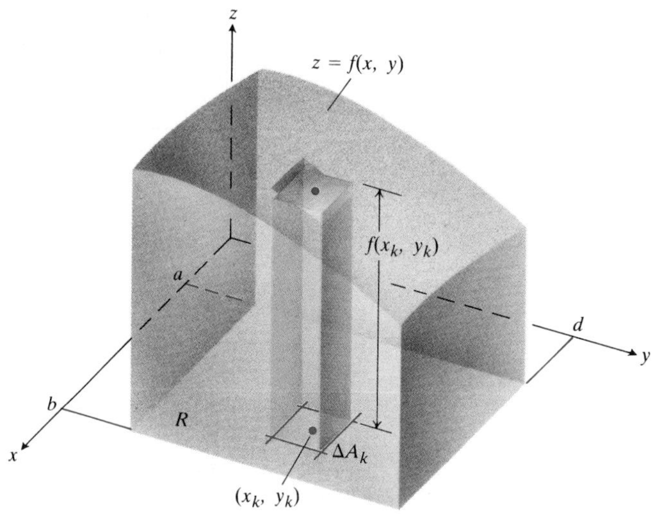

14.3 Approximating solids with rectangular prisms leads us to define the volumes of more general prisms as double integrals. The volume of the prism shown here is the double integral of $f(x, y)$ over the base region R.

As you might expect, this more general method of calculating volume agrees with the methods in Chapter 6 (it would be useless if it did not), but we shall not prove this here.

Fubini's Theorem for Calculating Double Integrals

We are now ready to calculate our first double integral.

Suppose we wish to calculate the volume under the plane $z = 4 - x - y$ over the rectangular region R: $0 \le x \le 2$, $0 \le y \le 1$ in the xy-plane. If we apply the method of slicing from Section 6.9, with slices perpendicular to the x-axis (Fig. 14.4), then the volume is

$$\int_{x=0}^{x=2} A(x)\,dx, \tag{4}$$

where $A(x)$ is the cross-section area at x. For each value of x we may calculate $A(x)$ as the integral

$$A(x) = \int_{y=0}^{y=1} (4 - x - y)\,dy, \tag{5}$$

which is the area under the curve $z = 4 - x - y$ in the plane of the cross section at x. In calculating $A(x)$, x is held fixed and the integration takes place with respect to y. Combining (4) and (5), we see that the volume of the entire solid is

$$\begin{aligned} \text{Volume} &= \int_{x=0}^{x=2} A(x)\,dx = \int_{x=0}^{x=2} \left(\int_{y=0}^{y=1} (4 - x - y)\,dy\right) dx \\ &= \int_{x=0}^{x=2} \left[4y - xy - \frac{y^2}{2}\right]_{y=0}^{y=1} dx = \int_{x=0}^{x=2} \left(\frac{7}{2} - x\right) dx = \left[\frac{7}{2}x - \frac{x^2}{2}\right]_0^2 = 5. \end{aligned} \tag{6}$$

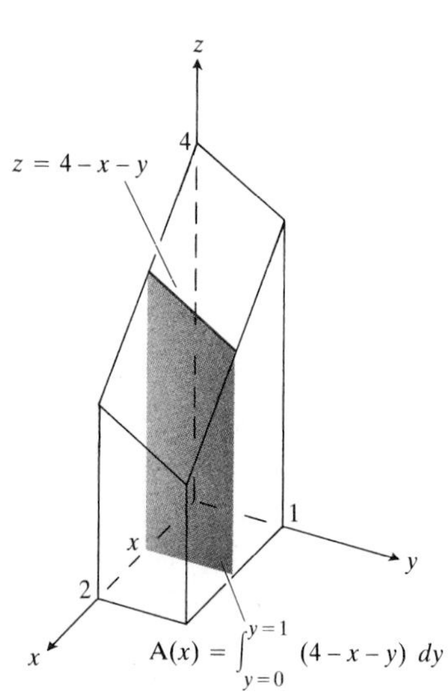

14.4 To obtain the cross-section area $A(x)$, we hold x fixed and integrate with respect to y.

If we had just wanted to write instructions for calculating the volume, without carrying out any of the integrations, we could have written

$$\text{Volume} = \int_0^2 \int_0^1 (4 - x - y)\,dy\,dx.$$

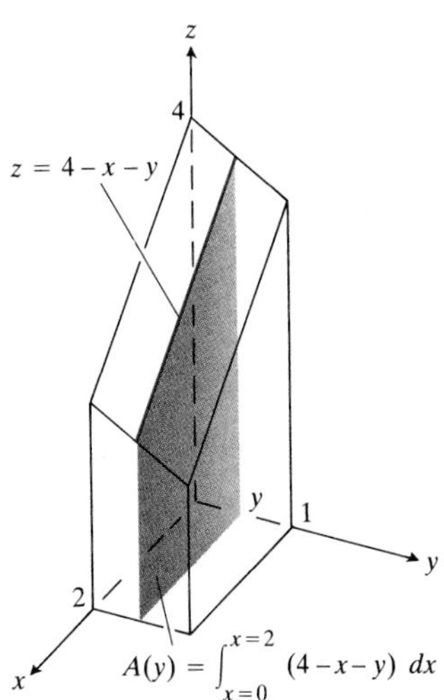

14.5 To obtain the cross-section area $A(y)$, we hold y fixed and integrate with respect to x.

The expression on the right, called an **iterated** or **repeated integral,** says that the volume is obtained by integrating $4 - x - y$ with respect to y from $y = 0$ to $y = 1$, holding x fixed, and then by integrating the resulting expression in x with respect to x from $x = 0$ to $x = 2$.

What would have happened if we had calculated the volume by slicing with planes perpendicular to the y-axis (Fig. 14.5)? As a function of y, the typical cross-section area is

$$A(y) = \int_{x=0}^{x=2} (4 - x - y)\, dx = \left[4x - \frac{x^2}{2} - xy\right]_{x=0}^{x=2} = 6 - 2y. \tag{7}$$

The volume of the entire solid is therefore

$$\text{Volume} = \int_{y=0}^{y=1} A(y)\, dy = \int_{y=0}^{y=1} (6 - 2y)\, dy = \left[6y - y^2\right]_0^1 = 5,$$

in agreement with our earlier calculation.

Again, we may give instructions for calculating the volume as an iterated integral by writing

$$\text{Volume} = \int_0^1 \int_0^2 (4 - x - y)\, dx\, dy.$$

The expression on the right says that the volume may be obtained by integrating $4 - x - y$ with respect to x from $x = 0$ to $x = 2$ (as we did in Eq. 7) and by integrating the result with respect to y from $y = 0$ to $y = 1$. In this iterated integral the order of integration is first x and then y, the reverse of the order we used in Eq. (6).

What do these two volume calculations with iterated integrals have to do with the double integral

$$\iint_R (4 - x - y)\, dA$$

over the rectangle $R\colon\ 0 \le x \le 2,\ 0 \le y \le 1$? The answer is that they both give the value of the double integral. A theorem proved by Guido Fubini (1879–1943) and published in 1907 says that the double integral of any continuous function over a rectangle can always be calculated as an iterated integral in either order of integration. (Fubini proved his theorem in greater generality, but this is how it translates into what we're doing now.)

THEOREM 1

Fubini's Theorem (First Form)

If $f(x, y)$ is continuous on the rectangular region $R: a \le x \le b, c \le y \le d$, then

$$\iint_R f(x, y)\, dA = \int_c^d \int_a^b f(x, y)\, dx\, dy = \int_a^b \int_c^d f(x, y)\, dy\, dx.$$

Fubini's theorem says that double integrals over rectangles can always be calculated as iterated integrals. This means that we can evaluate a double integral by integrating one variable at a time, using the integration techniques we already know for functions of a single variable.

Fubini's theorem also says that we may calculate the double integral by integrating in *either* order, a genuine convenience, as we shall see in Example 3. In particular, when we calculate a volume by slicing, we may use either planes perpendicular to the x-axis or planes perpendicular to the y-axis. We get the same answer both ways.

Even more important is the fact that Fubini's theorem holds for *any* continuous function $f(x, y)$. In particular, f may have negative values as well as positive values on R, and the integrals we calculate with Fubini's theorem may represent other things besides volumes (as we shall see later).

EXAMPLE 1 Calculate $\iint_R f(x, y)\, dA$ for

$$f(x, y) = 1 - 6x^2y \qquad \text{and} \qquad R:\ 0 \le x \le 2,\ -1 \le y \le 1.$$

Solution By Fubini's theorem,

$$\iint_R f(x, y)\, dA = \int_{-1}^{1}\int_0^2 (1 - 6x^2y)\, dx\, dy = \int_{-1}^{1}\Big[x - 2x^3y\Big]_{x=0}^{x=2} dy$$

$$= \int_{-1}^{1} (2 - 16y)\, dy = \Big[2y - 8y^2\Big]_{-1}^{1} = 4.$$

Reversing the order of integration gives the same answer:

$$\int_0^2\int_{-1}^{1} (1 - 6x^2y)\, dy\, dx = \int_0^2 \Big[y - 3x^2y^2\Big]_{y=-1}^{y=1} dx$$

$$= \int_0^2 \Big[(1 - 3x^2) - (-1 - 3x^2)\Big]\, dx = \int_0^2 2dx = 4. \qquad \blacksquare$$

Double Integrals over Bounded Nonrectangular Regions

To define the double integral of a function $f(x, y)$ over a bounded non-rectangular region, like the one shown in Fig. 14.6, we again imagine R to be covered by a rectangular grid, but we include in the partial sum only the small pieces of area $\Delta A = \Delta x\, \Delta y$ that lie entirely within the region (shaded in the figure). We number the pieces in some order, choose an arbitrary point (x_k, y_k) in each ΔA_k, and form the sum

$$S_n = \sum_{k=1}^{n} f(x_k, y_k)\, \Delta A_k.$$

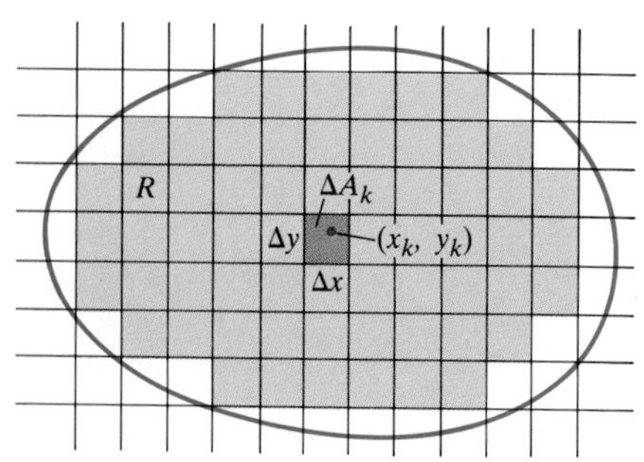

14.6 A rectangular grid subdividing a bounded nonrectangular region into cells.

The only difference between this sum and the one in Eq. (1) for rectangular regions is that now the areas ΔA_k may not cover all of R. But as the mesh becomes increasingly fine and the number of terms in S_n increases, more and more of R is included. If f is continuous and the boundary of R is made up of a finite number of line segments or smooth curves pieced together end to end, then the sums S_n will have a limit as Δx and Δy approach zero. We call the limit the *double integral* of f over R:

$$\iint_R f(x, y)\, dA = \lim_{\Delta A \to 0} \sum f(x_k, y_k)\, \Delta A_k.$$

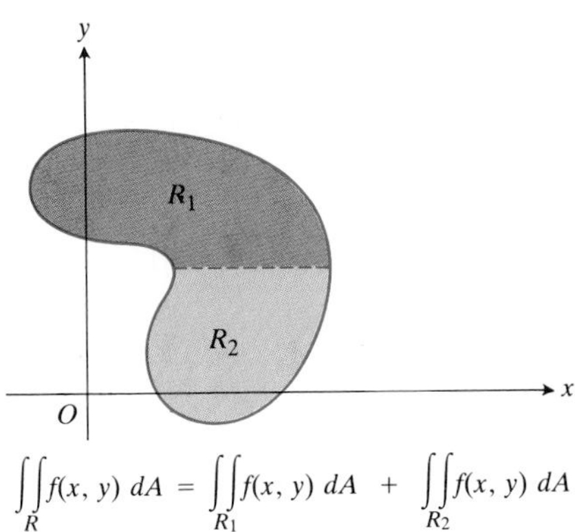

14.7 The domain additivity property stated earlier for rectangular regions holds for other regions as well.

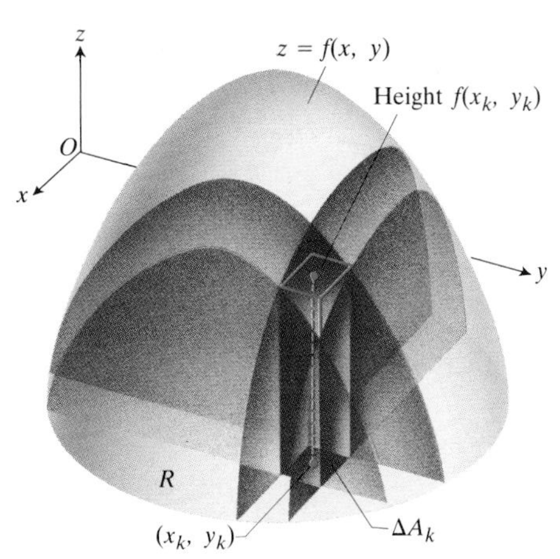

14.8 We define the volumes of solids with curved bases the same way we define the volumes of solids with rectangular bases.

This limit may also exist under less restrictive circumstances, but we shall not pursue the point here.

Double integrals of continuous functions over nonrectangular regions have all the algebraic properties listed earlier for integrals over rectangular regions. The domain additivity property corresponding to property 6 says that if R is decomposed into nonoverlapping regions R_1 and R_2 with boundaries that are again made of a finite number of line segments or smooth curves (see Fig. 14.7 for an example), then

$$6'.\quad \iint_R f(x,y)\,dA = \iint_{R_1} f(x,y)\,dA + \iint_{R_2} f(x,y)\,dA.$$

If $f(x,y)$ is positive and continuous over R (Fig. 14.8), we define the volume of the solid region between R and the surface $z = f(x,y)$ to be $\iint_R f(x,y)\,dA$, as before.

If R is a region like the one shown in the xy-plane in Fig. 14.9, bounded "above" and "below" by the curves $y = g_2(x)$ and $y = g_1(x)$ and on the sides by the lines $x = a$, $x = b$, we may again calculate the volume by the method of slicing. We first calculate the cross-section area

$$A(x) = \int_{y=g_1(x)}^{y=g_2(x)} f(x,y)\,dy$$

and then integrate $A(x)$ from $x = a$ to $x = b$ to get the volume as an iterated integral:

$$V = \int_a^b A(x)\,dx = \int_a^b \int_{g_1(x)}^{g_2(x)} f(x,y)\,dy\,dx. \tag{8}$$

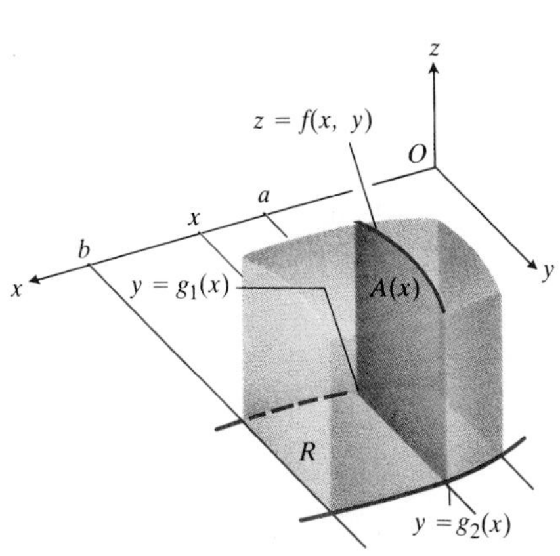

14.9 The area of the vertical slice shown here is

$$A(x) = \int_{g_1(x)}^{g_2(x)} f(x,y)\,dy.$$

To calculate the volume of the solid we integrate this area from $x = a$ to $x = b$.

Similarly, if R is a region like the one shown in Fig. 14.10, bounded by the curves $x = h_2(y)$ and $x = h_1(y)$ and the lines $y = c$ and $y = d$, then the volume calculated by slicing is given by the iterated integral

$$\text{Volume} = \int_c^d \int_{h_1(y)}^{h_2(y)} f(x,y)\,dx\,dy. \tag{9}$$

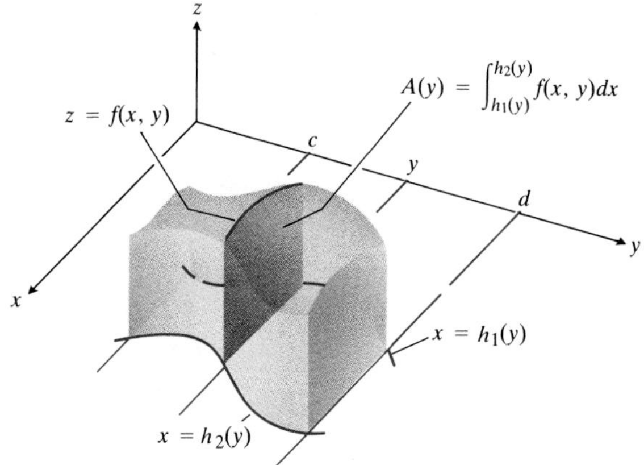

14.10 The volume of the solid shown here is

$$\int_c^d A(y)\,dy = \int_c^d \int_{h_1(y)}^{h_2(y)} f(x, y)\,dx\,dy.$$

The fact that the iterated integrals in Eqs. (8) and (9) both give the volume that we defined to be the double integral of f over R is a consequence of the following stronger form of Fubini's theorem.

THEOREM 2

Fubini's Theorem (Stronger Form)

Let $f(x, y)$ be continuous on a region R.

1. If R is defined by $a \le x \le b$, $g_1(x) \le y \le g_2(x)$, with g_1 and g_2 continuous on $[a, b]$, then
$$\iint_R f(x, y)\,dA = \int_a^b \int_{g_1(x)}^{g_2(x)} f(x, y)\,dy\,dx.$$
2. If R is defined by $c \le y \le d$, $h_1(y) \le x \le h_2(y)$, with h_1 and h_2 continuous on $[c, d]$, then
$$\iint_R f(x, y)\,dA = \int_c^d \int_{h_1(y)}^{h_2(y)} f(x, y)\,dx\,dy.$$

EXAMPLE 2 Find the volume of the prism whose base is the triangle in the xy-plane bounded by the x-axis and the lines $y = x$ and $x = 1$ and whose top lies in the plane

$$z = f(x, y) = 3 - x - y.$$

Solution See Fig. 14.11a. For any x between 0 and 1, y may vary from $y = 0$ to $y = x$ (Fig. 14.11b). Hence,

$$V = \int_0^1 \int_0^x (3 - x - y)\,dy\,dx = \int_0^1 \left[3y - xy - \frac{y^2}{2}\right]_{y=0}^{y=x} dx$$

$$= \int_0^1 \left(3x - \frac{3x^2}{2}\right) dx = \left[\frac{3x^2}{2} - \frac{x^3}{2}\right]_{x=0}^{x=1} = 1.$$

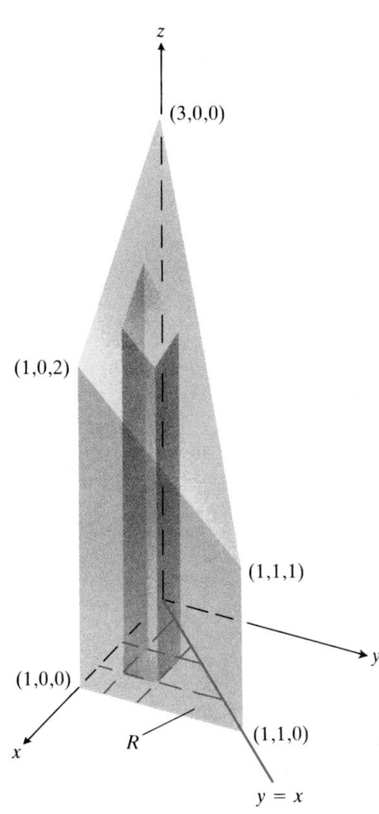

14.11 (a) Prism with a triangular base in the xy-plane. The volume of this prism is defined as a double integral over R. To evaluate it as an iterated integral, we may integrate first with respect to y and then with respect to x, or the other way around.

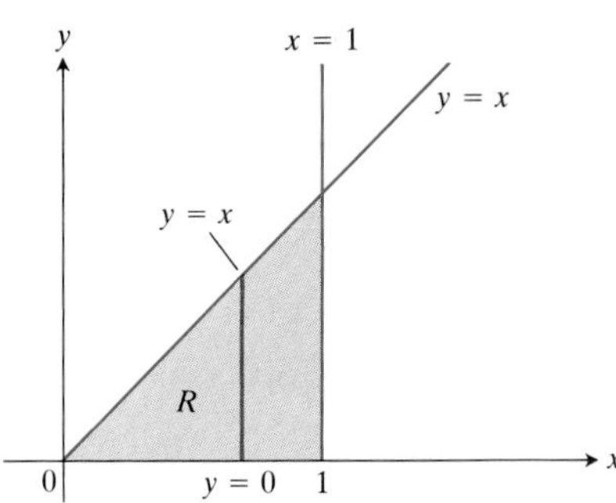

(b) Integration limits of

$$\int_{x=0}^{x=1}\int_{y=0}^{y=x} f(x,y)\,dy\,dx.$$

If we integrate first with respect to y, we integrate along a vertical line through R and then integrate to include all the vertical lines in R.

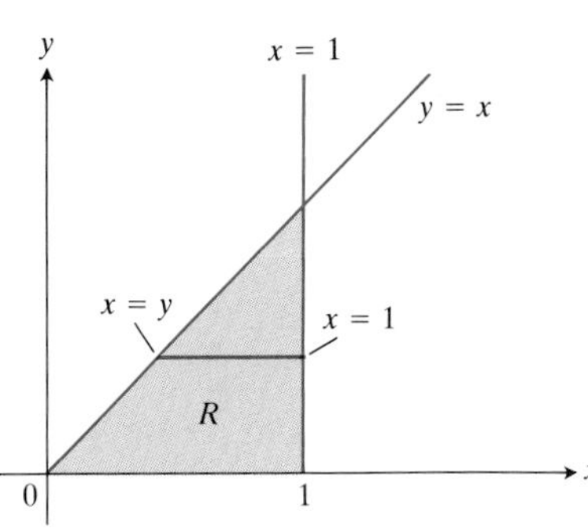

(c) Integration limits of

$$\int_{y=0}^{y=1}\int_{x=y}^{x=1} f(x,y)\,dx\,dy.$$

If we integrate first with respect to x, we integrate along a horizontal line through R and then integrate to include all the horizontal lines in R.

When the order of integration is reversed (Fig. 14.11c), the integral for the volume is

$$\begin{aligned}
V &= \int_0^1\int_y^1 (3 - x - y)\,dx\,dy = \int_0^1 \left[3x - \frac{x^2}{2} - xy\right]_{x=y}^{x=1} dy \\
&= \int_0^1 \left(3 - \frac{1}{2} - y - 3y + \frac{y^2}{2} + y^2\right) dy \\
&= \int_0^1 \left(\frac{5}{2} - 4y + \frac{3}{2}y^2\right) dy \\
&= \left[\frac{5}{2}y - 2y^2 + \frac{y^3}{2}\right]_{y=0}^{y=1} = 1.
\end{aligned}$$

The two integrals are equal, as they should be. ■

While Fubini's theorem assures us that a double integral may be calculated as an iterated integral in either order of integration, the value of one integral may be easier to find than the value of the other. The next example shows how this can happen.

EXAMPLE 3 Calculate

$$\iint_R \frac{\sin x}{x}\,dA,$$

where R is the triangle in the xy-plane bounded by the x-axis, the line $y = x$, and the line $x = 1$.

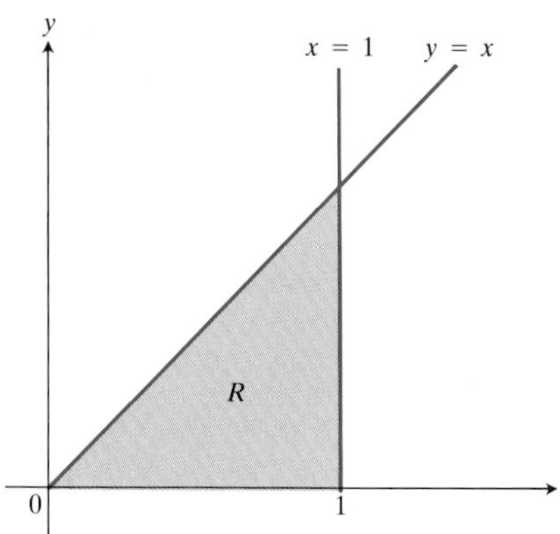

14.12 The region of integration in Example 3.

Solution The region of integration is shown in Fig. 14.12. If we integrate first with respect to y and then with respect to x, we find

$$\int_0^1 \left(\int_0^x \frac{\sin x}{x}\, dy \right) dx = \int_0^1 \left(y \frac{\sin x}{x} \bigg]_{y=0}^{y=x} \right) dx = \int_0^1 \sin x \, dx$$
$$= -\cos(1) + 1 \approx 0.46.$$

If we reverse the order of integration and attempt to calculate

$$\int_0^1 \int_y^1 \frac{\sin x}{x}\, dx\, dy,$$

we are stopped by the fact that $\int ((\sin x)/x)\, dx$ cannot be expressed in terms of elementary functions.

There is no general rule for predicting which order of integration will be the good one in circumstances like these, so don't worry about how to start your integrations. Just forge ahead and if the order you choose first doesn't work, try the other. ■

Finding the Limits of Integration

The hardest part of evaluating a double integral can be finding the limits of integration. Fortunately, there is a good procedure to follow.

Procedure for Finding Limits of Integration

A. To evaluate $\iint_R f(x, y)\, dA$ over a region R, integrating first with respect to y and then with respect to x, take the following steps:

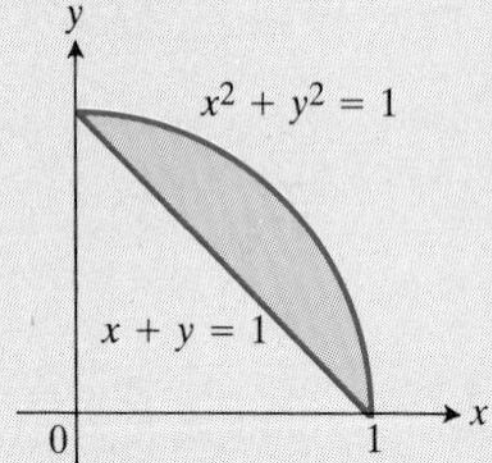

1. Sketch the region of integration and label the bounding curves.

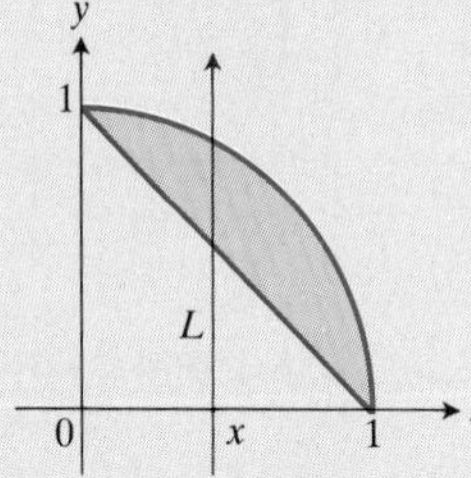

2. Imagine a vertical line L cutting through R in the direction of increasing y. Mark the y-values where L enters and leaves. These are the y-limits of integration.

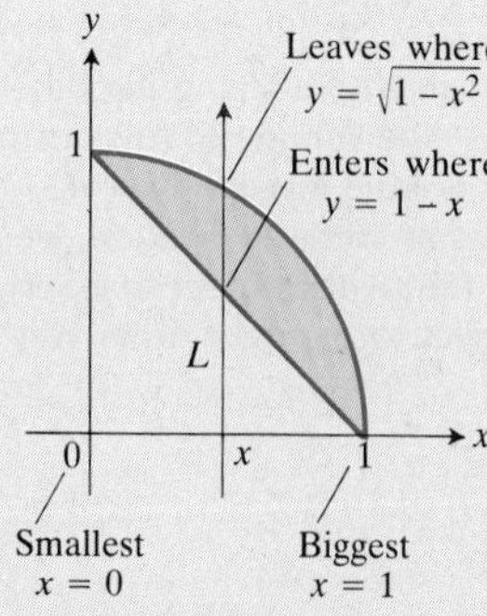

3. Choose x-limits that include all the vertical lines through R. The integral is

$$\iint_R f(x, y)\, dA = \int_{x=0}^{x=1} \int_{y=1-x}^{y=\sqrt{1-x^2}} f(x, y)\, dy\, dx.$$

B. To evaluate the same double integral as an iterated integral with the order of integration reversed, the procedure uses horizontal lines instead of vertical lines. The integral is

$$\iint_R f(x, y)\, dA = \int_0^1 \int_{1-y}^{\sqrt{1-y^2}} f(x, y)\, dx\, dy.$$

y
Largest→ 1
y = 1
Enters where
x = 1 − y
R
y
Leaves where
x = √(1 − y²)
Smallest→ 0
y = 0
1
x

14.13 To write

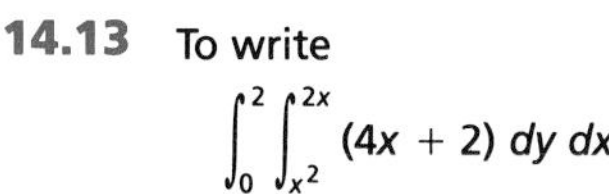

$$\int_0^2 \int_{x^2}^{2x} (4x + 2)\, dy\, dx$$

as a double integral with the order of integration reversed, we (a) sketch the region of integration and (b) work out what the new limits of integration have to be. Example 4 gives the details.

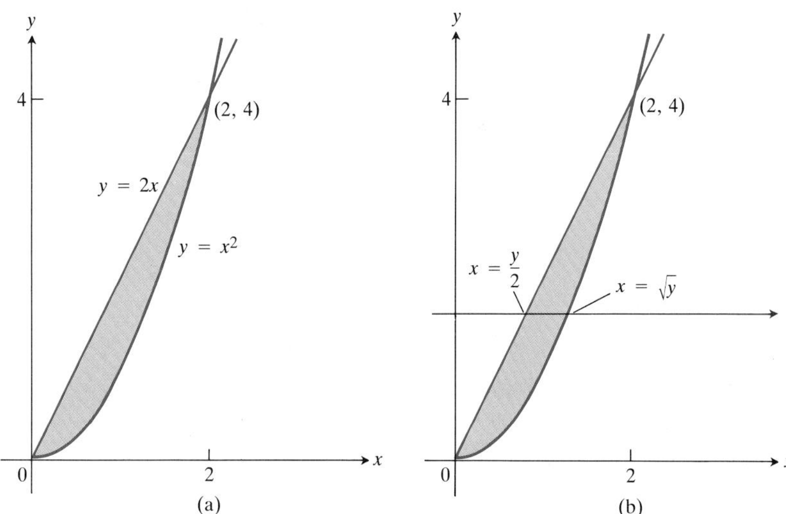

EXAMPLE 4 Sketch the region over which the integration

$$\int_0^2 \int_{x^2}^{2x} (4x + 2)\, dy\, dx$$

takes place and write an equivalent integral with the order of integration reversed.

Solution The region of integration is given by the inequalities $x^2 \le y \le 2x$ and $0 \le x \le 2$. It is therefore the region bounded by the curves $y = x^2$ and $y = 2x$ between $x = 0$ and $x = 2$ (Fig. 14.13a).

To find the limits for integrating in the reverse order, we imagine a horizontal line passing from left to right through the region. It enters at $x = y/2$ and leaves at $x = \sqrt{y}$. To include all such lines we let y run from $y = 0$ to $y = 4$ (Fig. 14.13b). The integral is

$$\int_0^4 \int_{y/2}^{\sqrt{y}} (4x + 2)\, dx\, dy.$$

The common value of these integrals, in case you are interested, is 8. ■

Exercises 14.1

Evaluate the integrals in Exercises 1–8 and sketch the regions of integration.

1. $\displaystyle\int_0^3 \int_0^2 (4 - y^2)\, dy\, dx$

2. $\displaystyle\int_{-1}^0 \int_{-1}^1 (x + y + 1)\, dx\, dy$

3. $\displaystyle\int_0^3 \int_{-2}^0 (x^2 y - 2xy)\, dy\, dx$

4. $\displaystyle\int_{\pi}^{2\pi} \int_0^{\pi} (\sin x + \cos y)\, dx\, dy$

5. $\displaystyle\int_0^{\pi} \int_0^{x} x \sin y\, dy\, dx$

6. $\displaystyle\int_1^{\ln 8}\int_0^{\ln y} e^{x+y}\,dx\,dy$

7. $\displaystyle\int_0^{\pi}\int_0^{\sin x} y\,dy\,dx$

8. $\displaystyle\int_1^{2}\int_y^{y^2} dx\,dy$

In Exercises 9–14, integrate the function $f(x, y)$ over the given region.

9. $f(x, y) = x/y$ over the region in the first quadrant bounded by the lines $y = x$, $y = 2x$, $x = 1$, $x = 2$

10. $f(x, y) = x^2 + y^2$ over the triangular region whose vertices are (0, 0), (1, 0) and (0, 1)

11. $f(x, y) = y - \sqrt{x}$ over the triangular region cut from the first quadrant by the line $x + y = 1$

12. $f(x, y) = x^2 + 3xy$ over the rectangle R: $0 \le x \le 1$, $0 \le y \le 1$

13. $f(x, y) = 1/xy$ over the rectangle R: $1 \le x \le 2$, $1 \le y \le 2$

14. $f(x, y) = y \cos xy$ over the rectangle R: $0 \le x \le \pi$, $0 \le y \le 1$

In Exercises 15–20, sketch the region of integration and write an equivalent integral with the order of integration reversed.

15. $\displaystyle\int_0^{2}\int_0^{4-2x} dy\,dx$

16. $\displaystyle\int_0^{1}\int_2^{4-2x} dy\,dx$

17. $\displaystyle\int_0^{1}\int_y^{\sqrt{y}} dx\,dy$

18. $\displaystyle\int_0^{4}\int_{\sqrt{y}}^{2} dx\,dy$

19. $\displaystyle\int_0^{2}\int_1^{e^x} dy\,dx$

20. $\displaystyle\int_0^{1}\int_{\sqrt{y}}^{1} dx\,dy$

In Exercises 21–24, write an equivalent iterated integral with the order of integration reversed. *Do not evaluate the integrals.* It will help to sketch the region of integration.

21. $\displaystyle\int_0^{1}\int_1^{e^x} dy\,dx$

22. $\displaystyle\int_0^{1}\int_{\sqrt{x}}^{1} \cos(x + y)\,dy\,dx$

23. $\displaystyle\int_0^{2}\int_0^{x^3} f(x, y)\,dy\,dx$

24. $\displaystyle\int_0^{1}\int_{-\sqrt{y}}^{\sqrt{y}} f(x, y)\,dx\,dy$

Evaluate the integrals in Exercises 25–28 by evaluating an equivalent integral obtained by reversing the order of integration.

25. $\displaystyle\int_0^{\pi}\int_x^{\pi} \frac{\sin y}{y}\,dy\,dx$

26. $\displaystyle\int_0^{1}\int_y^{1} x^2 e^{xy}\,dx\,dy$

27. $\displaystyle\int_0^{2}\int_x^{2} 2y^2 \sin xy\,dy\,dx$

28. $\displaystyle\int_0^{2}\int_0^{4-x^2} \frac{xe^{2y}}{4-y}\,dy\,dx$

29. Find the volume of the solid whose base is the region in the xy-plane that is bounded by the parabola $y = 4 - x^2$ and the line $y = 3x$, while the top of the solid is bounded by the plane $z = x + 4$.

30. Find the volume of the solid in the first octant bounded by the coordinate planes, the cylinder $x^2 + y^2 = 4$, and the plane $z + y = 3$.

31. Find the volume of the solid in the first octant bounded by the coordinate planes, the plane $x = 3$, and the parabolic cylinder $z = 4 - y^2$.

32. Find the volume of the solid cut from the first octant by the surface $z = 4 - x^2 - y$.

Computer Exercises

Just as there are methods for evaluating single integrals numerically, there are corresponding methods for the numerical evaluation of double integrals. If you have access to a double-integral evaluator, estimate the values of the following integrals.

33. $\displaystyle\int_1^{3}\int_1^{x} \frac{1}{xy}\,dy\,dx$

34. $\displaystyle\int_0^{1}\int_0^{1} e^{-(x^2+y^2)}\,dy\,dx$

35. $\displaystyle\int_0^{1}\int_0^{1} \tan^{-1} xy\,dy\,dx$

36. $\displaystyle\int_{-1}^{1}\int_0^{\sqrt{1-x^2}} 3\sqrt{1 - x^2 - y^2}\,dy\,dx$

TOOLKIT PROGRAM

Double Integral Evaluator

Evaluates integrals of the form

$$\int_a^b \int_{g(x)}^{h(x)} f(x, y)\,dy\,dx$$

by a variety of numerical methods.

14.2 Areas, Moments, and Centers of Mass

In this section we show how to use double integrals to define and calculate the areas of bounded regions in the plane and the masses, moments, centers of mass, and radii of gyration of thin plates covering these regions. The calculations are similar to the ones in Chapter 6, but we can now handle a greater variety of shapes.

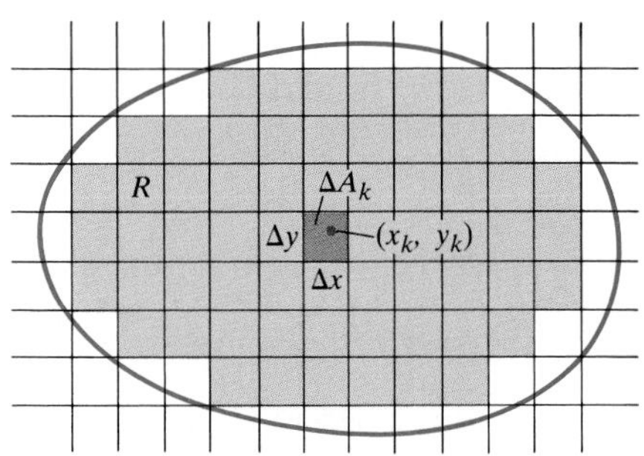

14.14 The first step in defining the area of a region is to divide the interior of the region into cells.

Areas of Bounded Regions in the Plane

If we take $f(x, y) = 1$ in the definition of the double integral over a region R in the preceding section, the partial sums reduce to

$$S_n = \sum_{k=1}^{n} f(x_k, y_k)\Delta A_k = \sum_{k=1}^{n} \Delta A_k. \tag{1}$$

This approximates what we would like to call the area of R. As Δx and Δy approach zero, the coverage of R by the ΔA_k's (Fig. 14.14) becomes increasingly complete, and we define the area of R to be the limit

$$\text{Area} = \lim \sum \Delta A_k = \iint_R dA. \tag{2}$$

DEFINITION

The **area** of a closed bounded plane region R is the value of the integral

$$\text{Area} = \iint_R dA. \tag{3}$$

As with the other definitions in this chapter, the definition here applies to a greater variety of regions than does the earlier single-variable definition of area, but it agrees with the earlier definition on regions to which they both apply.

To evaluate the area integral in (3), we integrate the constant function $f(x, y) = 1$ over R.

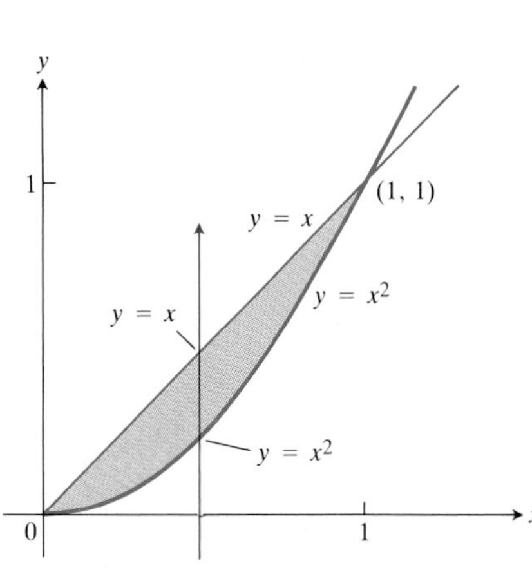

14.15 The area of the region between the parabola and the line is given by the double integral

$$\int_0^1 \int_{x^2}^{x} dy\, dx.$$

EXAMPLE 1 Find the area of the region R bounded by $y = x$ and $y = x^2$ in the first quadrant.

Solution We sketch the region (Fig. 14.15) and calculate the area as

$$A = \int_0^1 \int_{x^2}^{x} dy\, dx = \int_0^1 (x - x^2)\, dx = \left[\frac{x^2}{2} - \frac{x^3}{3}\right]_0^1 = \frac{1}{6}.$$ ■

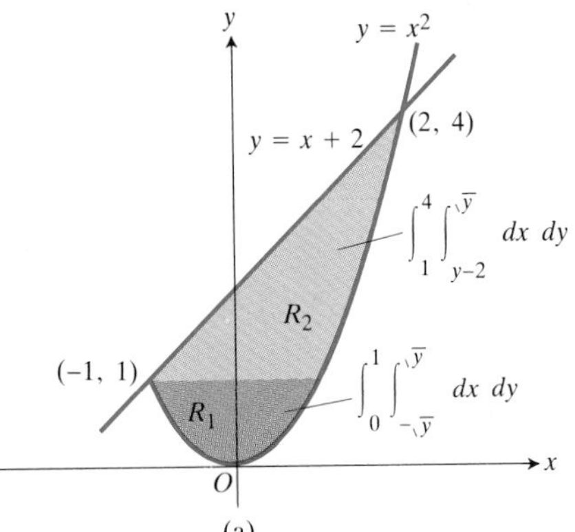

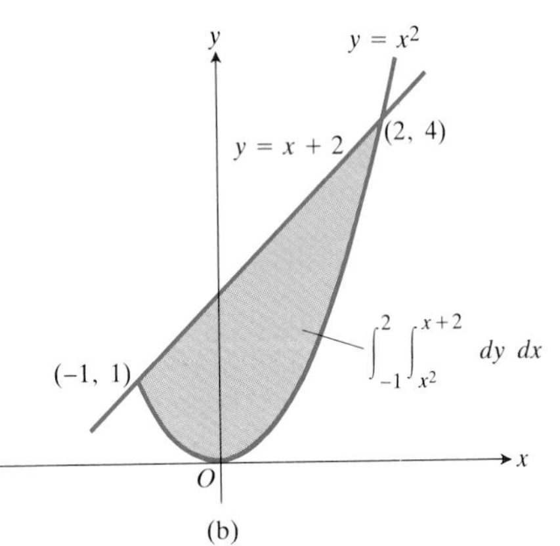

14.16 Calculating the area shown here takes (a) two integrals if the first integration is with respect to x, but (b) only one if the first integration is with respect to y.

EXAMPLE 2 Find the area of the region R enclosed by the parabola $y = x^2$ and the line $y = x + 2$.

Solution If we divide R into the regions R_1 and R_2 shown in Fig. 14.16a, we may calculate the area as

$$A = \iint_{R_1} dA + \iint_{R_2} dA = \int_0^1 \int_{-\sqrt{y}}^{\sqrt{y}} dx\,dy + \int_1^4 \int_{y-2}^{\sqrt{y}} dx\,dy.$$

On the other hand, reversing the order of integration (Fig. 14.16b) gives

$$A = \int_{-1}^{2} \int_{x^2}^{x+2} dy\,dx.$$

Clearly, this result is simpler and is the only one we would bother to write down in practice. Evaluation of this integral leads to the result

$$A = \int_{-1}^{2} \Big[y\Big]_{x^2}^{x+2} dx = \int_{-1}^{2} (x + 2 - x^2)\,dx = \left[\frac{x^2}{2} + 2x - \frac{x^3}{3}\right]_{-1}^{2} = \frac{9}{2}.$$ ■

First and Second Moments and Centers of Mass

To find the moments and centers of mass of thin sheets and plates, we use formulas similar to those in Chapter 6. The main difference is that now, with double integrals, we can accommodate a greater variety of shapes and density functions. The formulas are given in Table 14.1.

TABLE 14.1
Mass and moment formulas for thin plates covering regions in the *xy*-plane

Density: $\delta(x, y)$

Mass: $M = \iint \delta(x, y)\,dA$

First moments: $M_x = \iint y\delta(x, y)\,dA, \quad M_y = \iint x\delta(x, y)\,dA$

Center of mass: $\bar{x} = \dfrac{M_y}{M}, \quad \bar{y} = \dfrac{M_x}{M}$

Moments of inertia (second moments):

About the x-axis: $I_x = \iint y^2\delta(x, y)\,dA$

About the y-axis: $I_y = \iint x^2\delta(x, y)\,dA$

About the origin: $I_0 = \iint (x^2 + y^2)\delta(x, y)\,dA = I_x + I_y$

Radii of gyration:
About the x-axis: $R_x = \sqrt{I_x/M}$
About the y-axis: $R_y = \sqrt{I_y/M}$
About the origin: $R_0 = \sqrt{I_0/M}$

The examples that follow show how these formulas are used.

Moments of inertia are important in calculating the kinetic energies of rotating bodies. They also appear in the formulas that are used to calculate the stiffness of beams.

The difference between the **first moments** M_x and M_y and the **moments of inertia,** or **second moments,** I_x and I_y is that the second moments use the *squares* of the "lever-arm" distances x and y.

The moment I_0 is also called the **polar moment** of inertia about the origin. It is calculated by integrating the density $\delta(x, y)$ times $r^2 = x^2 + y^2$, the square of the distance from a representative point (x, y) to the origin. Notice that $I_0 = I_x + I_y$; once we find two, we get the third free. (The moment I_0 is sometimes called I_z, for moment of inertia about the z-axis. The identity $I_z = I_x + I_y$ is then called the **Perpendicular Axis Theorem.**)

The **radius of gyration** R_x is defined by the equation

$$I_x = MR_x^2.$$

It tells how far from the x-axis the entire mass of the plate might be concentrated to give the same I_x. The radius of gyration gives a convenient way to express the moment of inertia in terms of a mass and a length. The radii R_y and R_0 are defined in a similar way, by the equations

$$I_y = MR_y^2 \quad \text{and} \quad I_0 = MR_0^2.$$

We take square roots to get the formulas in Table 14.1.

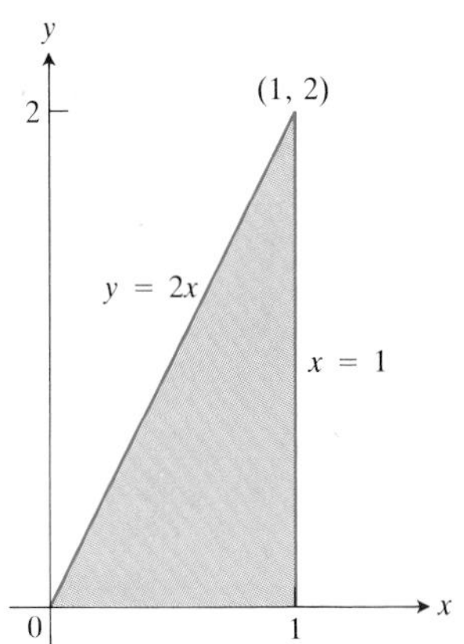

14.17 The triangular region covered by the plate in Example 3.

EXAMPLE 3 A thin plate covers the triangular region bounded by the x-axis and the lines $x = 1$ and $y = 2x$ in the first quadrant. The plate's density at the point (x, y) is $\delta(x, y) = 6x + 6y + 6$. Find the plate's mass, first moments, center of mass, moments of inertia, and radii of gyration about the coordinate axes.

Solution We sketch the plate and put in enough detail to determine the limits of integration for the integrals we have to evaluate (Fig. 14.17).

The plate's mass is

$$M = \int_0^1 \int_0^{2x} \delta(x, y)\, dy\, dx = \int_0^1 \int_0^{2x} (6x + 6y + 6)\, dy\, dx$$

$$= \int_0^1 \left[6xy + 3y^2 + 6y\right]_{y=0}^{y=2x} dx = \int_0^1 (24x^2 + 12x)\, dx = \left[8x^3 + 6x^2\right]_0^1 = 14.$$

The first moment about the x-axis is

$$M_x = \int_0^1 \int_0^{2x} y\delta(x, y)\, dy\, dx = \int_0^1 \int_0^{2x} (6xy + 6y^2 + 6y)\, dy\, dx$$

$$= \int_0^1 \left[3xy^2 + 2y^3 + 3y^2\right]_{y=0}^{y=2x} dx = \int_0^1 (28x^3 + 12x^2)\, dx$$

$$= \left[7x^4 + 4x^3\right]_0^1 = 11.$$

A similar calculation gives

$$M_y = \int_0^1 \int_0^{2x} x\delta(x, y)\, dy\, dx = 10.$$

The coordinates of the center of mass are therefore

$$\bar{x} = \frac{M_y}{M} = \frac{10}{14} = \frac{5}{7}, \qquad \bar{y} = \frac{M_x}{M} = \frac{11}{14}.$$

The moment of inertia about the x-axis is

$$I_x = \int_0^1 \int_0^{2x} y^2 \delta(x, y)\, dy\, dx = \int_0^1 \int_0^{2x} (6xy^2 + 6y^3 + 6y^2)\, dy\, dx$$

$$= \int_0^1 \left[2xy^3 + \frac{3}{2}y^4 + 2y^3\right]_{y=0}^{y=2x} dx = \int_0^1 (40x^4 + 16x^3)\, dx = \left[8x^5 + 4x^4\right]_0^1 = 12.$$

Similarly, the moment of inertia about the y-axis is

$$I_y = \int_0^1 \int_0^{2x} x^2 \delta(x, y)\, dy\, dx = \frac{39}{5}.$$

Since we know I_x and I_y, we do not need to evaluate an integral to find I_0; we can use the equation $I_0 = I_x + I_y$ instead:

$$I_0 = 12 + \frac{39}{5} = \frac{60 + 39}{5} = \frac{99}{5}.$$

The three radii of gyration are

$$R_x = \sqrt{I_x/M} = \sqrt{12/14} = \sqrt{6/7},$$

$$R_y = \sqrt{I_y/M} = \sqrt{\left(\frac{39}{5}\right)/14} = \sqrt{39/70},$$

$$R_0 = \sqrt{I_0/M} = \sqrt{\left(\frac{99}{5}\right)/14} = \sqrt{99/70}.$$ ■

Centroids of Geometric Figures

When the density of an object is constant, it cancels out of the numerator and denominator of the formulas for $\bar{x}$ and $\bar{y}$. As far as $\bar{x}$ and $\bar{y}$ are concerned, δ might as well be 1. Thus, when δ is constant, the location of the center of mass becomes a feature of the object's shape and not of the material of which it is made. In such cases, engineers may call the center of mass the **centroid** of the shape. To find a centroid, we just set δ equal to 1 and proceed to find $\bar{x}$ and $\bar{y}$ as before, by dividing first moments by masses.

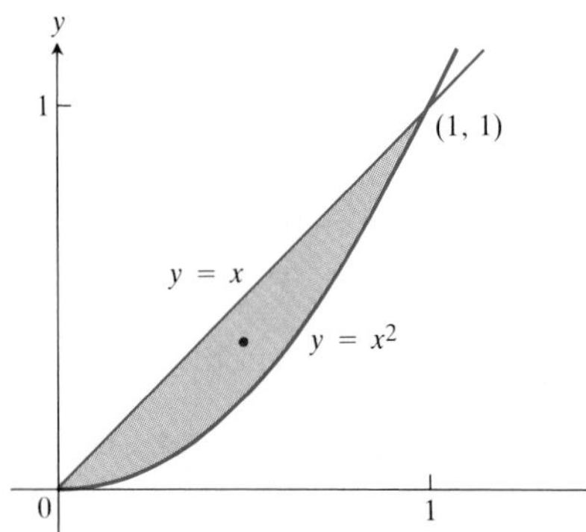

14.18 Example 4 calculates the coordinates of the centroid of the region shown here.

EXAMPLE 4 Find the centroid of the region in the first quadrant that is bounded above by the line $y = x$ and below by the parabola $y = x^2$.

Solution We sketch the region and include enough detail to determine the limits of integration (Fig. 14.18). We then set δ equal to 1 and evaluate the appropriate formulas from Table 14.1:

$$M = \int_0^1 \int_{x^2}^{x} 1\, dy\, dx = \int_0^1 \left[y\right]_{y=x^2}^{y=x} dx = \int_0^1 (x - x^2)\, dx = \left[\frac{x^2}{2} - \frac{x^3}{3}\right]_0^1 = \frac{1}{6},$$

$$M_x = \int_0^1 \int_{x^2}^{x} y\, dy\, dx = \int_0^1 \left[\frac{y^2}{2}\right]_{y=x^2}^{y=x} dx = \int_0^1 \left(\frac{x^2}{2} - \frac{x^4}{2}\right) dx = \left[\frac{x^3}{6} - \frac{x^5}{10}\right]_0^1 = \frac{1}{15},$$

$$M_y = \int_0^1 \int_{x^2}^{x} x\,dy\,dx = \int_0^1 \left[xy\right]_{y=x^2}^{y=x} dx = \int_0^1 (x^2 - x^3)\,dx = \left[\frac{x^3}{3} - \frac{x^4}{4}\right]_0^1 = \frac{1}{12}.$$

From these values of M, M_x, and M_y, we find

$$\bar{x} = \frac{M_y}{M} = \frac{1/12}{1/6} = \frac{1}{2} \quad \text{and} \quad \bar{y} = \frac{M_x}{M} = \frac{1/15}{1/6} = \frac{2}{5}.$$

The centroid is the point $\left(\frac{1}{2}, \frac{2}{5}\right)$. ■

Exercises 14.2

In Exercises 1–8, sketch the region bounded by the given lines and curves. Then find the region's area by double integration.

1. The coordinate axes and the line $x + y = 2$
2. The x-axis, the curve $y = e^x$, and the lines $x = 0$, $x = \ln 2$
3. The y-axis, the line $y = 2x$, and the line $y = 4$
4. The parabola $x = -y^2$ and the line $y = x + 2$
5. The parabolas $x = y^2$ and $x = 2y - y^2$
6. The parabola $x = y - y^2$ and the line $x + y = 0$
7. The semiellipse $y = 2\sqrt{1 - x^2}$ and the lines $x = \pm 1$, $y = -1$
8. Above by $y = x^2$, below by $y = -1$, on the left by $x = -2$, and on the right by $y = 2x - 1$

The integrals and sums of integrals in Exercises 9–14 give the areas of regions in the xy-plane. Sketch each region, label each bounding curve with its equation, and give the coordinates of the points where the curves intersect. Then find the area of the region.

9. $\displaystyle\int_0^6 \int_{y^2/3}^{2y} dx\,dy$
10. $\displaystyle\int_0^3 \int_{-x}^{x(2-x)} dy\,dx$
11. $\displaystyle\int_0^{\pi/4} \int_{\sin x}^{\cos x} dy\,dx$
12. $\displaystyle\int_{-1}^2 \int_{y^2}^{y+2} dx\,dy$
13. $\displaystyle\int_{-1}^0 \int_{-2x}^{1-x} dy\,dx + \int_0^2 \int_{-x/2}^{1-x} dy\,dx$
14. $\displaystyle\int_0^2 \int_{x^2-4}^{0} dy\,dx + \int_0^4 \int_0^{\sqrt{x}} dy\,dx$

Constant Density

15. Find the center of mass of a thin plate of density $\delta = 3$ bounded by the lines $x = 0$, $y = x$, and the parabola $y = 2 - x^2$ in the first quadrant.
16. Find the moments of inertia and radii of gyration about the coordinate axes of a thin rectangular plate of constant density δ bounded by the lines $x = 3$ and $y = 3$ in the first quadrant.
17. Find the centroid of the region in the first quadrant bounded by the x-axis, the parabola $y^2 = 2x$, and the line $x + y = 4$.
18. Find the centroid of the triangular region cut from the first quadrant by the line $x + y = 3$.
19. Find the centroid of the semicircular region bounded by the x-axis and the curve $y = \sqrt{1 - x^2}$.
20. The area of the region in the first quadrant bounded by the parabola $y = 6x - x^2$ and the line $y = x$ is $125/6$ square units. Find the centroid.
21. Find the centroid of the region cut from the first quadrant by the circle $x^2 + y^2 = a^2$.
22. Find the moment of inertia about the x-axis of a thin plate of density $\delta = 1$ bounded by the circle $x^2 + y^2 = 4$. Then use your result to find I_y and I_0 for the plate.
23. Find the centroid of the region between the x-axis and the arch $y = \sin x$, $0 \le x \le \pi$.
24. Find the moment of inertia with respect to the y-axis of a thin sheet of constant density $\delta = 1$ bounded by the curve $y = (\sin^2 x)/x^2$ and the interval $\pi \le x \le 2\pi$ of the x-axis.

Variable Density

25. Find the moment of inertia and radius of gyration about the x-axis of a thin plate bounded by the parabola $x = y - y^2$ and the line $x + y = 0$ if $\delta(x, y) = x + y$.

26. Find the mass of a thin plate occupying the smaller region cut from the ellipse $x^2 + 4y^2 = 12$ by the parabola $x = 4y^2$ if $\delta(x, y) = 5x$.

27. Find the center of mass of a thin triangular plate bounded by the y-axis and the lines $y = x$ and $y = 2 - x$ if $\delta(x, y) = 6x + 3y + 3$.

28. Find the center of mass and moment of inertia about the x-axis of a thin plate bounded by the curves $x = y^2$ and $x = 2y - y^2$ if the density at the point (x, y) is $\delta(x, y) = y + 1$.

29. Find the center of mass and the moment of inertia and radius of gyration about the y-axis of a thin rectangular plate cut from the first quadrant by the lines $x = 6$ and $y = 1$ if $\delta(x, y) = x + y + 1$.

30. Find the center of mass and the moment of inertia and radius of gyration about the y-axis of a thin plate bounded by the line $y = 1$ and the parabola $y = x^2$ if the density is $\delta(x, y) = y + 1$.

31. Find the center of mass and the moment of inertia and radius of gyration about the y-axis of a thin plate bounded by the x-axis, the lines $x = \pm 1$, and the parabola $y = x^2$ if $\delta(x, y) = 7y + 1$.

32. Find the center of mass and moment of inertia and radius of gyration about the x-axis of a thin rectangular plate bounded by the lines $x = 0$, $x = 20$, $y = -1$, and $y = 1$ if $\delta(x, y) = 1 + (x/20)$.

33. Find the center of mass, the moments of inertia and radii of gyration about the coordinate axes, and the polar moment of inertia and radius of gyration of a thin triangular plate bounded by the lines $y = x$, $y = -x$, and $y = 1$ if $\delta(x, y) = y + 1$.

34. Repeat Exercise 33 for $\delta(x, y) = 3x^2 + 1$.

Additional Problems

35. A rectangular plate of constant density $\delta(x, y) = 1$ occupies the region bounded by the lines $x = 4$ and $y = 2$ in the first quadrant. The moment of inertia I_a of the rectangle about the line $y = a$ is given by the integral

$$I_a = \int_0^4 \int_0^2 (y - a)^2 \, dy \, dx .$$

Find the value of a that minimizes I_a.

36. Find the centroid of the region in the xy-plane bounded by the curves $y = 1/\sqrt{1 - x^2}$, $y = -1/\sqrt{1 - x^2}$, and the lines $x = 0$, $x = 1$. (*Hint:* Notice that $\bar{y} = 0$, by symmetry.)

14.3 Double Integrals in Polar Form

Integrals are sometimes easier to evaluate if we change to polar coordinates. This section shows how to accomplish the change and how to evaluate integrals over regions whose boundaries are given by polar equations.

Integrals in Polar Coordinates

When we defined the double integral of a function over a region R in the xy-plane, we began by cutting R into rectangles whose sides were parallel to the coordinate axes. These were the natural shapes to use because their sides have either constant x-values or constant y-values. In polar coordinates, the natural shape is a "polar rectangle" of the kind we are about to describe.

Suppose that a function $f(r, \theta)$ is defined over a region R that is bounded by the rays $\theta = \alpha$ and $\theta = \beta$ and by the continuous curves $r = r_1(\theta)$ and $r = r_2(\theta)$. Suppose also that $0 \le r_1(\theta) \le r_2(\theta) \le a$ for every value of θ between α and β. Then R lies in a fan-shaped region Q defined by the inequalities $0 \le r \le a$ and $\alpha \le \theta \le \beta$. See Fig. 14.19.

We cover Q by a grid of circular arcs and rays. The arcs are cut from circles centered at the origin, with radii Δr, $2\Delta r$, $\ldots$, $m\Delta r$, where $\Delta r = a/m$. The

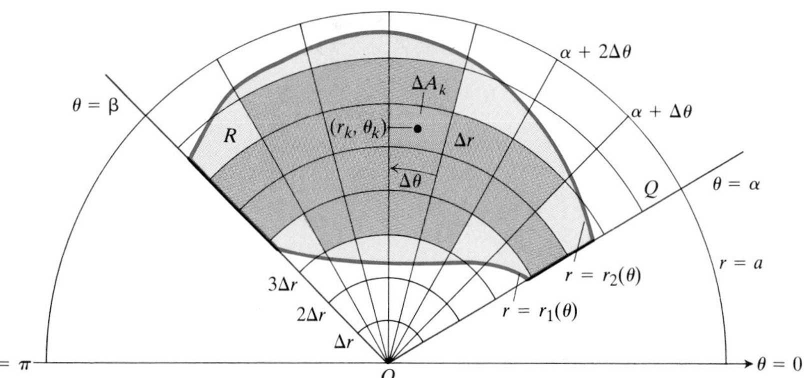

14.19 The region R: $r_1(\theta) \leq r \leq r_2(\theta)$, $\alpha \leq \theta \leq \beta$ is contained in the fan-shaped region Q: $0 \leq r \leq a$, $\alpha \leq \theta \leq \beta$. The subdivision of Q by circular arcs and rays induces a subdivision of R.

rays are given by

$$\theta = \alpha, \quad \theta = \alpha + \Delta\theta, \quad \theta = \alpha + 2\Delta\theta, \quad \ldots, \quad \theta = \alpha + m'\Delta\theta = \beta.$$

The arcs and rays divide Q into small patches called "polar rectangles."

We number the polar rectangles that lie inside R (the order doesn't matter), calling their areas

$$\Delta A_1, \quad \Delta A_2, \quad \ldots, \quad \Delta A_n.$$

We let (r_k, θ_k) be the center of the polar rectangle whose area is ΔA_k. By "center" we mean the point that lies halfway between the circular arcs on the ray that bisects the arcs. We then form the sum

$$S_n = \sum_{k=1}^{n} f(r_k, \theta_k)\, \Delta A_k. \tag{1}$$

If f is continuous throughout R, this sum will approach a limit as we refine the grid to make Δr and $\Delta\theta$ go to zero. The limit is called the double integral of f over R. In symbols,

$$\lim S_n = \iint_R f(r, \theta)\, dA.$$

To evaluate this limit, we first have to write the sum S_n in a way that expresses ΔA_k in terms of Δr and $\Delta\theta$. Here is how we do that. The radius of the inner arc bounding ΔA_k is $r_k - (\Delta r/2)$ (Fig. 14.20). The area of the circular sector subtended by this arc at the origin is therefore

$$\frac{1}{2}\left(r_k - \frac{\Delta r}{2}\right)^2 \Delta\theta. \tag{2}$$

Similarly, the radius of the outer boundary of ΔA_k is $r_k + (\Delta r/2)$. The area of the sector it subtends is

$$\frac{1}{2}\left(r_k + \frac{\Delta r}{2}\right)^2 \Delta\theta. \tag{3}$$

Therefore,

$$\begin{aligned}\Delta A_k &= \text{area of large sector} - \text{area of small sector}\\ &= \frac{\Delta\theta}{2}\left[\left(r_k + \frac{\Delta r}{2}\right)^2 - \left(r_k - \frac{\Delta r}{2}\right)^2\right] = \frac{\Delta\theta}{2}[2r_k\, \Delta r] = r_k\, \Delta r\, \Delta\theta.\end{aligned}$$

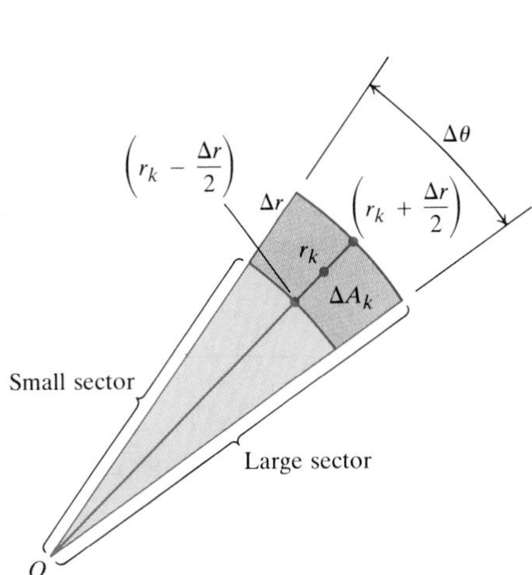

14.20 The observation that

$$\Delta A_k = \begin{pmatrix}\text{area of}\\ \text{large sector}\end{pmatrix} - \begin{pmatrix}\text{area of}\\ \text{small sector}\end{pmatrix}$$

leads to the formula $\Delta A_k = r_k \Delta r \Delta\theta$. The text explains why.

Combining this result with Eq. (1) gives

$$S_n = \sum_{k=1}^{n} f(r_k, \theta_k)\, r_k\, \Delta r\, \Delta\theta. \tag{4}$$

A version of Fubini's theorem now says that the limit approached by these sums can be evaluated by repeated single integrations with respect to r and θ as

$$\iint_R f(r, \theta)\, dA = \int_{\theta=\alpha}^{\theta=\beta} \int_{r=r_1(\theta)}^{r=r_2(\theta)} f(r, \theta)\, r\, dr\, d\theta. \tag{5}$$

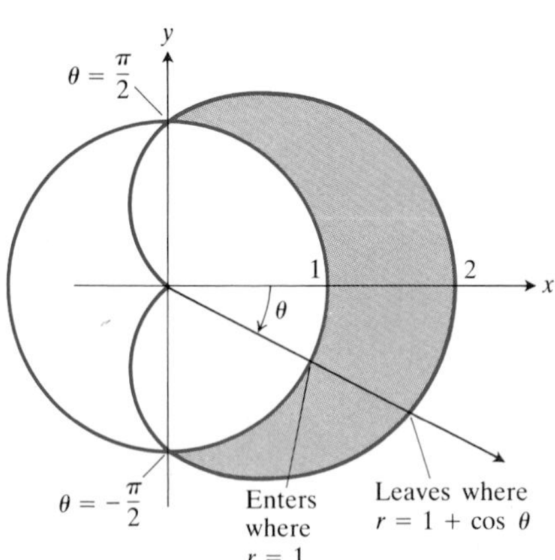

14.21 Example 1 describes the process by which we find limits of integration for calculating the area of the shaded region between the circle $r = 1$ and the cardioid $r = 1 + \cos\theta$.

Limits of Integration

The procedure we used for finding limits of integration for integrals in rectangular coordinates also works for polar coordinates.

EXAMPLE 1 *How to find limits of integration.* Find the limits of integration for integrating a function $f(r, \theta)$ over the region R that lies inside the cardioid $r = 1 + \cos\theta$ and outside the circle $r = 1$.

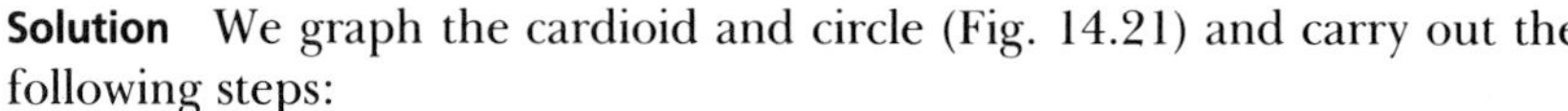

Solution We graph the cardioid and circle (Fig. 14.21) and carry out the following steps:

STEP 1: Holding θ fixed, let r increase to trace a ray out from the origin.

STEP 2: Integrate from the r-value where the ray enters R to the r-value where the ray leaves R.

STEP 3: Choose θ-limits to include all the rays from the origin that intersect R.

The result is the integral

$$\int_{-\pi/2}^{\pi/2} \int_{r=1}^{r=1+\cos\theta} f(r, \theta)\, r\, dr\, d\theta.$$

■

If $f(r, \theta)$ is the constant function whose value is 1, then the value of the integral of f over a region R is the area of R:

Area in Polar Coordinates

The area of a closed and bounded region R in the polar coordinate plane is given by the formula

$$\text{Area of } R = \iint_R r\, dr\, d\theta. \tag{6}$$

As you might expect, this formula for area is consistent with all earlier formulas, although we shall not prove the fact.

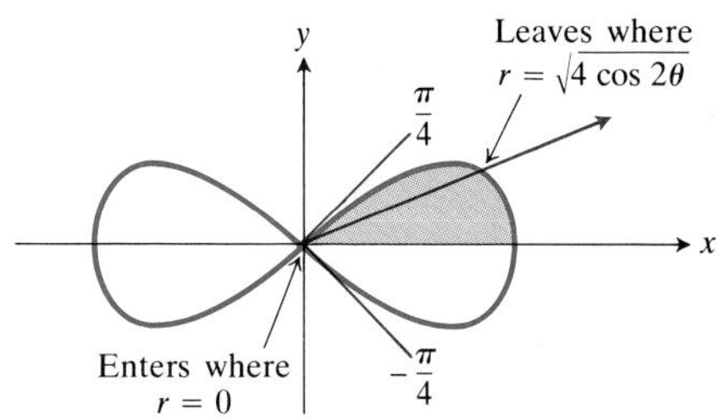

14.22 To integrate over the shaded region bounded by the lemniscate $r^2 = 4\cos 2\theta$, we run r from 0 to $\sqrt{4\cos 2\theta}$ and θ from 0 to $\pi/4$.

EXAMPLE 2 Find the area enclosed by the lemniscate $r^2 = 4\cos 2\theta$.

Solution We graph the lemniscate to determine the limits of integration (Fig. 14.22) and see that the total area is four times the first-quadrant portion.

$$\text{Area} = 4\int_0^{\pi/4}\int_0^{\sqrt{4\cos 2\theta}} r\,dr\,d\theta = 4\int_0^{\pi/4}\left[\frac{r^2}{2}\right]_{r=0}^{r=\sqrt{4\cos 2\theta}} d\theta$$

$$= 4\int_0^{\pi/4} 2\cos 2\theta\,d\theta = 4\sin 2\theta\Big]_0^{\pi/4} = 4.$$

Changing Cartesian Integrals into Polar Integrals

The procedure for changing a Cartesian integral

$$\iint_R f(x,y)\,dx\,dy$$

into a polar integral has two steps:

STEP 1: Substitute $x = r\cos\theta$ and $y = r\sin\theta$, and replace $dx\,dy$ by $r\,dr\,d\theta$ in the Cartesian integral.

STEP 2: Supply polar limits of integration for the boundary of R.

The Cartesian integral then becomes

$$\iint_R f(x,y)\,dx\,dy = \iint_G f(r\cos\theta, r\sin\theta)\,r\,dr\,d\theta, \tag{7}$$

where G denotes the region of integration in polar coordinates. This is like the substitution method in Chapter 5 except that there are now two variables to substitute for instead of one. Notice that $dx\,dy$ does not become $dr\,d\theta$; it becomes $r\,dr\,d\theta$. We shall go into the reasons for that briefly in Section 14.7.

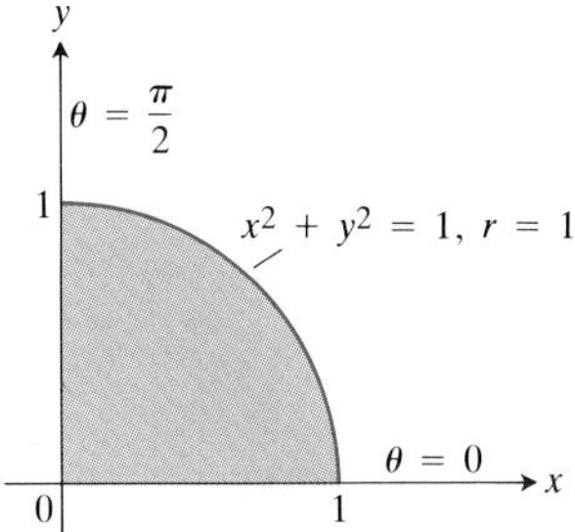

14.23 Example 3 evaluates the integral of a function over the region bounded by the quarter-circle $x^2 + y^2 = 1$ in the first quadrant. In polar coordinates, this region is described by simple inequalities:

$0 \le r \le 1$ and $0 \le \theta \le \pi/2$.

EXAMPLE 3 Find the polar moment of inertia about the origin of a thin plate of density $\delta(x,y) = 1$ bounded by the quarter-circle $x^2 + y^2 = 1$ in the first quadrant.

Solution We sketch the region of integration to determine the limits of integration (Fig. 14.23).

In Cartesian coordinates, the polar moment is the value of the integral

$$\int_0^1\int_0^{\sqrt{1-x^2}} (x^2 + y^2)\,dy\,dx.$$

Integration with respect to y gives

$$\int_0^1 \left(x^2\sqrt{1-x^2} + \frac{(1-x^2)^{3/2}}{3}\right) dx,$$

an integral difficult to evaluate without tables.

Things go better if we change the original integral to polar coordinates. Substituting

$$x = r\cos\theta, \qquad y = r\sin\theta, \qquad dx\,dy = r\,dr\,d\theta,$$

we get

$$\int_0^1\int_0^{\sqrt{1-x^2}} (x^2+y^2)\,dy\,dx = \int_0^{\pi/2}\int_0^1 (r^2)\,r\,dr\,d\theta$$

$$= \int_0^{\pi/2}\left[\frac{r^4}{4}\right]_{r=0}^{r=1} d\theta = \int_0^{\pi/2}\frac{1}{4}\,d\theta = \frac{\pi}{8}.$$

Why was the polar coordinate transformation so effective? One reason is that x^2+y^2 was simplified to r^2. Another is that the limits of integration became constants. ■

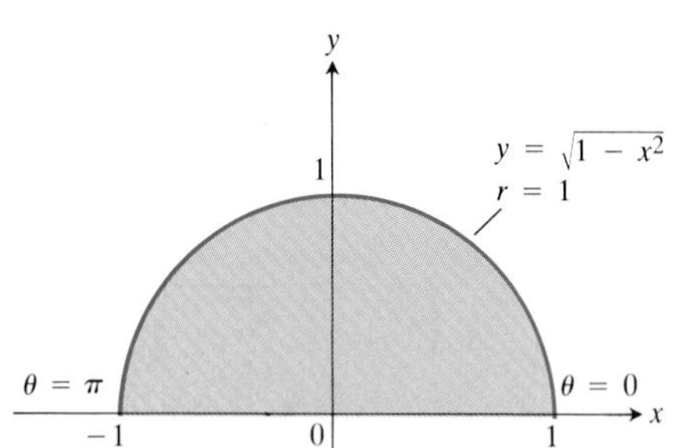

14.24 The semicircular region in Example 4 is described by the polar coordinate inequalities

$$0 \le r \le 1 \quad \text{and} \quad 0 \le \theta \le \pi.$$

EXAMPLE 4 Evaluate

$$\iint_R e^{x^2+y^2}\,dy\,dx,$$

where R is the semicircular region bounded by the x-axis and the curve $y = \sqrt{1-x^2}$ (Fig. 14.24).

Solution In Cartesian coordinates, the integral in question is a nonelementary integral and there is no direct way to integrate $e^{x^2+y^2}$ with respect to either x or y. Yet this integral and others like it are important in mathematics—in statistics, for example—and we must find a way to evaluate it. Polar coordinates save the day. The substitutions

$$x = r\cos\theta, \qquad y = r\sin\theta, \qquad dy\,dx = r\,dr\,d\theta$$

enable us to evaluate the integral as

$$\iint_R e^{x^2+y^2}\,dy\,dx = \int_0^{\pi}\int_0^1 e^{r^2}\,r\,dr\,d\theta = \int_0^{\pi}\left[\frac{1}{2}e^{r^2}\right]_0^1 d\theta$$

$$= \int_0^{\pi}\frac{1}{2}(e-1)\,d\theta = \frac{\pi}{2}(e-1).$$

The r in the $r\,dr\,d\theta$ was just what we needed to integrate e^{r^2}. Without it we would have been stuck, as we were at the beginning. ■

Exercises 14.3

In Exercises 1–12, change the Cartesian integral into an equivalent polar integral and evaluate the polar integral.

1. $\displaystyle\int_{-1}^{1}\int_0^{\sqrt{1-x^2}} dy\,dx$

2. $\displaystyle\int_{-1}^{1}\int_{-\sqrt{1-x^2}}^{\sqrt{1-x^2}} dy\,dx$

3. $\displaystyle\int_0^1\int_0^{\sqrt{1-y^2}} (x^2+y^2)\,dx\,dy$

4. $\displaystyle\int_{-1}^{1}\int_{-\sqrt{1-y^2}}^{\sqrt{1-y^2}} (x^2+y^2)\,dx\,dy$

5. $\int_{-a}^{a}\int_{-\sqrt{a^2-x^2}}^{\sqrt{a^2-x^2}} dy\,dx$

6. $\int_{0}^{a}\int_{0}^{\sqrt{4-y^2}} (x^2+y^2)\,dx\,dy$

7. $\int_{0}^{1}\int_{y}^{\sqrt{2-y^2}} x\,dx\,dy$

8. $\int_{0}^{2}\int_{0}^{x} y\,dy\,dx$

9. $\int_{0}^{3}\int_{0}^{\sqrt{3}x} \frac{dy\,dx}{\sqrt{x^2+y^2}}$

10. $\int_{0}^{2}\int_{0}^{\sqrt{4-x^2}} \frac{xy}{\sqrt{x^2+y^2}}\,dy\,dx$

11. $\int_{0}^{1}\int_{0}^{\sqrt{1-x^2}} 5\sqrt{x^2+y^2}\,dy\,dx$

12. $\int_{0}^{\infty}\int_{0}^{\infty} e^{-(x^2+y^2)}\,dx\,dy$

13. Find the area of the region cut from the first quadrant by the curve $r = 2(2 - \sin 2\theta)^{1/2}$.

14. Find the area of the region that lies inside the cardioid $r = 1 + \cos\theta$ and outside the circle $r = 1$.

15. Find the area enclosed by one leaf of the rose $r = 12\cos 3\theta$.

16. Find the area of the region enclosed by the positive x-axis and spiral $r = 4\theta/3,\ 0 \le \theta \le 2\pi$. The region looks like a snail shell.

17. Find the area of the region cut from the first quadrant by the cardioid $r = 1 + \sin\theta$.

18. Find the area of the region common to the interiors of the cardioids $r = 1 + \cos\theta$ and $r = 1 - \cos\theta$.

19. Integrate the function $f(x,y) = 1/(1 - x^2 - y^2)$ over the disk $x^2 + y^2 \le 3/4$.

20. Integrate the function $f(x,y) = (\ln(x^2+y^2))/(x^2+y^2)$ over the region between the circles $x^2 + y^2 = 1$ and $x^2 + y^2 = e^2$.

21. Find the moment about the x-axis of a thin plate of constant density $\delta(x,y) = 3$, bounded below by the x-axis and above by the cardioid $r = 1 - \cos\theta$.

22. Find the moment of inertia about the x-axis and the polar moment of inertia about the origin of a thin disk bounded by the circle $x^2 + y^2 = a^2$ if the density is $\delta(x,y) = k(x^2+y^2)$, k a constant.

23. Find the centroid of the region enclosed by the cardioid $r = 1 + \cos\theta$.

24. Find the moment of inertia about the origin of a thin plate bounded by the cardioid $r = 1 + \cos\theta$ if the density is $\delta(x,y) = 1$.

25. *An infinite region.* Find the centroid of the infinite region in the second quadrant enclosed by the coordinate axes and the curve $y = e^x$. (*Hint:* Use improper integrals in the mass–moment formulas.)

26. *Another infinite region.* Find the moment about the y-axis of the infinite region under the curve $y = e^{-x^2/2}$ in the first quadrant. (*Hint:* Use an improper integral.)

14.4 Triple Integrals in Rectangular Coordinates. Volumes and Average Values

Triple integrals are used to calculate the volumes of irregular three-dimensional shapes and the average values of functions over three-dimensional regions. They are also the integrals we use to calculate masses and moments of three-dimensional solids. When combined with vectors, as they will be in Chapter 15, they are the integrals we use to describe many of the phenomena associated with fluid flow and electromagnetism.

In the present section, we define triple integrals and use them to calculate volumes and average values.

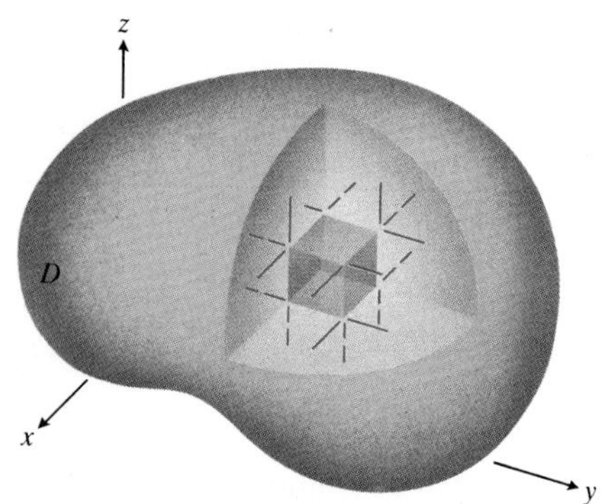

14.25 Partitioning a solid with rectangular cells of volume ΔV.

Triple Integrals

If $F(x, y, z)$ is a function defined on a closed bounded region D in space—the region occupied by a solid ball, for example, or a lump of clay—then the integral of F over D may be defined in the following way. We partition a rectangular region about D into rectangular cells by planes parallel to the coordinate planes (Fig. 14.25). The cells have dimensions Δx by Δy by Δz. We number the cells that lie inside D in some order

$$\Delta V_1, \quad \Delta V_2, \quad \ldots, \quad \Delta V_n,$$

choose a point (x_k, y_k, z_k) in each ΔV_k, and form the sum

$$S_n = \sum_{k=1}^{n} F(x_k, y_k, z_k)\,\Delta V_k. \tag{1}$$

If F is continuous and the bounding surface of D is made of smooth surfaces joined along continuous curves, then as Δx, Δy, and Δz all approach zero the sums S_n approach a limit

$$\lim_{n\to\infty} S_n = \iiint_D F(x, y, z)\,dV. \tag{2}$$

We call this limit the **triple integral of F over D.** The limit also exists for some discontinuous functions.

Properties of Triple Integrals

Triple integrals have the same algebraic properties as double integrals and single integrals. If $F = F(x, y, z)$ and $G = G(x, y, z)$ are both integrable, then

1. $\displaystyle\iiint_D kF\,dV = k\iiint_D F\,dV$ (any number k),

2. $\displaystyle\iiint_D (F + G)\,dV = \iiint_D F\,dV + \iiint_D G\,dV,$

3. $\displaystyle\iiint_D (F - G)\,dV = \iiint_D F\,dV - \iiint_D G\,dV,$

4. $\displaystyle\iiint_D F\,dV \ge 0$ if $F \ge 0$ on D,

5. $\displaystyle\iiint_D F\,dV \ge \iiint_D G\,dV$ if $F \ge G$ on D.

The integrals have these properties because the sums that approximate them have these properties.

Triple integrals also have a domain additivity property that proves useful in physics and engineering as well as in mathematics. If the domain D of a continuous function F is partitioned by smooth surfaces into a finite number of cells $D_1, D_2, \ldots, D_n$, then

$$6. \quad \iiint_D F\,dV = \iiint_{D_1} F\,dV + \iiint_{D_2} F\,dV + \cdots + \iiint_{D_n} F\,dV.$$

Volume of a Region in Space

If F is the constant function whose value is 1, then the sums in Eq. (1) reduce to

$$S_n = \sum F(x_k, y_k, z_k)\,\Delta V_k = \sum 1 \cdot \Delta V_k = \sum \Delta V_k. \tag{3}$$

As Δx, Δy, and Δz all approach zero, the cells ΔV_k become smaller and more numerous and fill up more and more of D. We therefore define the volume of D to be the triple integral

$$\lim \sum_{k=1}^{n} \Delta V_k = \iiint_D dV.$$

DEFINITION

The **volume** of a closed bounded region D in space is the value of the integral

$$\text{Volume of } D = \iiint_D dV. \tag{4}$$

As we shall see in a moment, this integral enables us to calculate the volumes of solids enclosed by curved surfaces.

Evaluation of Triple Integrals

The triple integral is seldom evaluated directly from its definition as a limit. Instead, we apply a three-dimensional version of Fubini's theorem to evaluate the integral by repeated single integrations.

For example, suppose we want to integrate a continuous function $F(x, y, z)$ over a region D that is bounded below by a surface $z = f_1(x, y)$, above by the surface $z = f_2(x, y)$, and on the side by a cylinder C parallel to the z-axis (Fig. 14.26). Let R denote the vertical projection of D onto the xy-plane, which is the region in the xy-plane enclosed by C. The integral of F over D is then evaluated as

$$\iiint_D F(x, y, z)\,dV = \iint_R \left(\int_{z=f_1(x,y)}^{z=f_2(x,y)} F(x, y, z)\,dz \right) dy\,dx,$$

or

$$\iiint_D F(x, y, z)\,dV = \iiint_R \int_{z=f_1(x,y)}^{z=f_2(x,y)} F(x, y, z)\,dz\,dy\,dx, \tag{5}$$

if we omit the large parentheses.

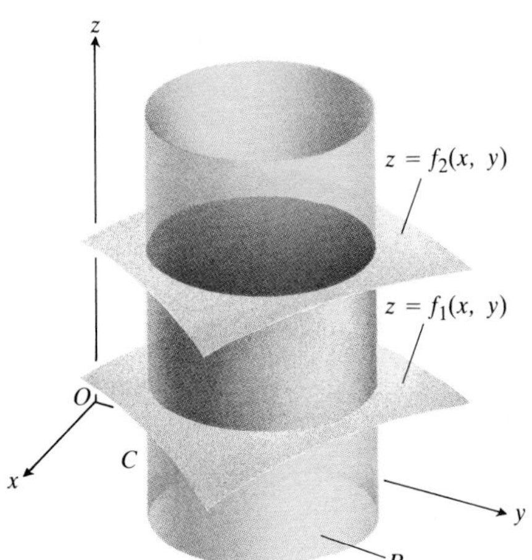

14.26 The enclosed volume can be found by evaluating

$$V = \iint_R \int_{z=f_1(x,y)}^{z=f_2(x,y)} dz\,dy\,dx.$$

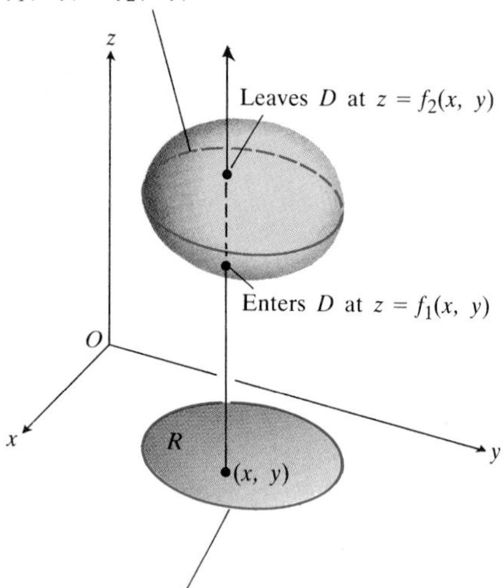

14.27 A schematic diagram for finding the limits of integration for a triple integral of a function F over a three-dimensional region D enclosed by two surfaces. For the region here,

$$\iiint_D F\,dV = \iint_R \int_{z=f_1(x,y)}^{z=f_2(x,y)} F\,dz\,dy\,dx.$$

The z-limits of integration indicate that for every (x, y) in the region R, z may extend from the lower surface $z = f_1(x, y)$ to the upper surface $z = f_2(x, y)$. The y- and x-limits of integration have not been given explicitly in Eq. (5) but are to be determined in the usual way from the boundaries of R.

In case the lateral surface of the cylinder reduces to zero, as in Fig. 14.27 and the example that follows, we may find the equation of the boundary of R by eliminating z between the two equations $z = f_1(x, y)$ and $z = f_2(x, y)$. This gives

$$f_1(x, y) = f_2(x, y),$$

an equation that contains no z and that defines the boundary of R in the xy-plane.

To supply the z-limits of integration in any particular instance we may use a procedure like the one for double integrals. We imagine a line L through a point (x, y) in R and parallel to the z-axis. As z increases, the line enters D at $z = f_1(x, y)$ and leaves D at $z = f_2(x, y)$. These give the lower and upper limits of the integration with respect to z. The result of this integration is now a function of x and y alone, which we integrate over R, supplying limits in the usual way.

EXAMPLE 1 Find the volume of the three-dimensional region enclosed by the surfaces

$$z = x^2 + 3y^2 \qquad \text{and} \qquad z = 8 - x^2 - y^2.$$

Solution The two surfaces (Fig. 14.28) intersect on the elliptic cylinder

$$x^2 + 3y^2 = 8 - x^2 - y^2, \qquad \text{or} \qquad x^2 + 2y^2 = 4.$$

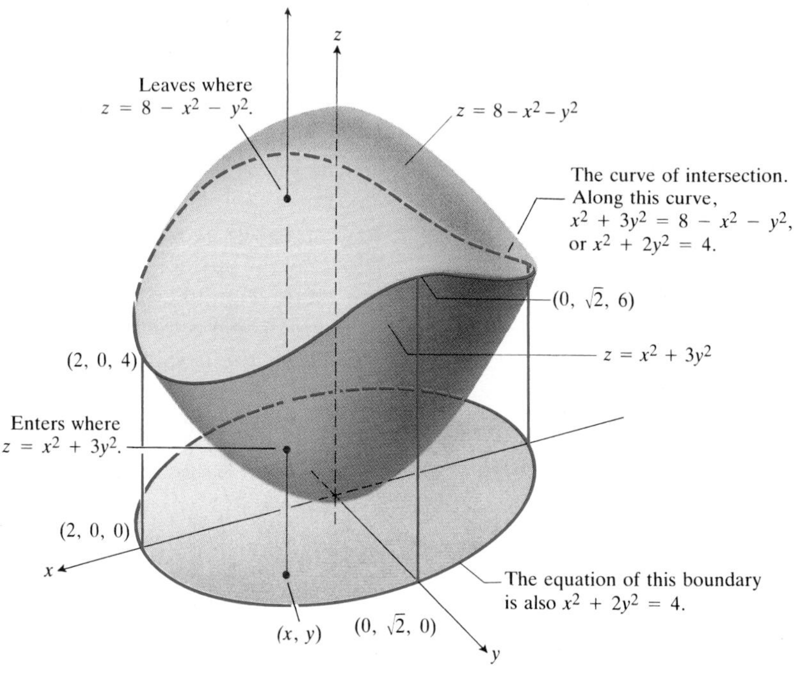

14.28 The volume of the region enclosed by these two paraboloids is calculated in Example 1.

The three-dimensional region projects into the two-dimensional region R in the xy-plane that is enclosed by the ellipse having this same equation. In the double integral with respect to y and x over R, if we integrate first with respect to y, holding x fixed, y varies from $-\sqrt{(4-x^2)/2}$ to $+\sqrt{(4-x^2)/2}$. Then x varies from -2 to $+2$. Thus we have

$$V = \int_{-2}^{2}\int_{-\sqrt{(4-x^2)/2}}^{\sqrt{(4-x^2)/2}}\int_{x^2+3y^2}^{8-x^2-y^2} dz\,dy\,dx = \int_{-2}^{2}\int_{-\sqrt{(4-x^2)/2}}^{\sqrt{(4-x^2)/2}} (8-2x^2-4y^2)\,dy\,dx$$

$$= \int_{-2}^{2}\left[(8-2x^2)y - \frac{4}{3}y^3\right]_{y=-\sqrt{(4-x^2)/2}}^{y=\sqrt{(4-x^2)/2}} dx$$

$$= \int_{-2}^{2}\left(2(8-2x^2)\sqrt{\frac{4-x^2}{2}} - \frac{8}{3}\left(\frac{4-x^2}{2}\right)^{3/2}\right) dx$$

$$= \int_{-2}^{2}\left(8\left(\frac{4-x^2}{2}\right)^{3/2} - \frac{8}{3}\left(\frac{4-x^2}{2}\right)^{3/2}\right) dx = \frac{4\sqrt{2}}{3}\int_{-2}^{2}(4-x^2)^{3/2}\,dx$$

$$= 8\pi\sqrt{2}. \qquad \left(\begin{array}{l}\text{After integration with the}\\ \text{substitution } x = 2\sin u\end{array}\right)$$

■

As we know, there are sometimes (but not always) two different orders in which the single integrations that evaluate a double integral may be worked. For triple integrals, there are sometimes (but not always) as many as *six* workable orders of integration. The next example shows an extreme case in which all six are workable.

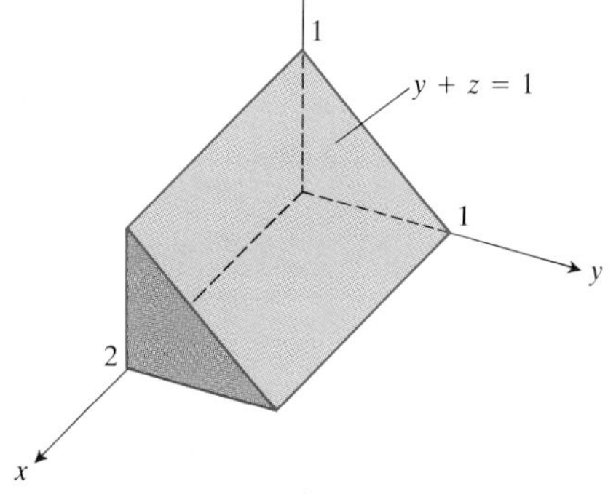

14.29 Example 2 shows how to calculate the volume of this prism with six different iterated triple integrals.

EXAMPLE 2 Each of the following integrals gives the volume of the solid shown in Fig. 14.29.

a) $\int_0^1\int_0^{1-z}\int_0^2 dx\,dy\,dz$ b) $\int_0^1\int_0^{1-y}\int_0^2 dx\,dz\,dy$

c) $\int_0^1\int_0^2\int_0^{1-z} dy\,dx\,dz$ d) $\int_0^2\int_0^1\int_0^{1-z} dy\,dz\,dx$

e) $\int_0^1\int_0^2\int_0^{1-y} dz\,dx\,dy$ f) $\int_0^2\int_0^1\int_0^{1-y} dz\,dy\,dx$ ■

Average Value of a Function in Space

The average value of a function F over a region R in space is defined by the formula

$$\text{Average value of } F \text{ over } R = \frac{1}{\text{volume of } R}\iiint_R F\,dV. \tag{6}$$

If $F(x, y, z) = \sqrt{x^2+y^2+z^2}$, then the average value of F over R is the average distance of points in R from the origin. If $F(x, y, z)$ is the density of a solid that occupies a region R in space, then the average value of F over R is the average density of the solid in units of mass per unit volume.

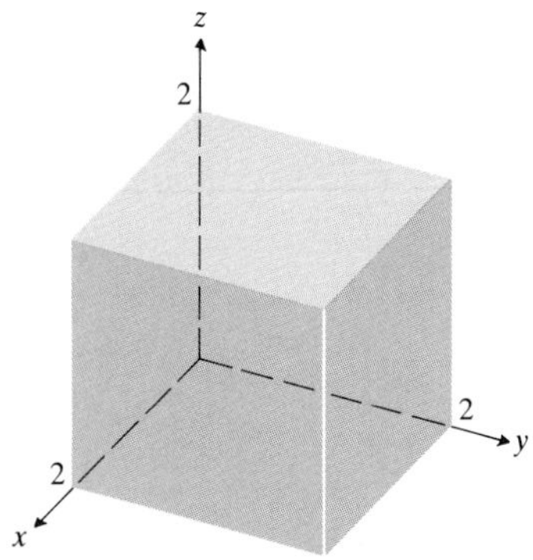

14.30 The solid cube bounded by the planes $x = 2$, $y = 2$, and $z = 2$ is the region of integration in Example 3.

EXAMPLE 3 Find the average value of $F(x, y, z) = xyz$ over the cube bounded by the coordinate planes and the planes $x = 2$, $y = 2$, and $z = 2$ in the first octant.

Solution We sketch the cube with enough detail to show the limits of integration (Fig. 14.30). We then use Eq. (6) to calculate the average value of F over the cube.

The volume of the cube is $(2)(2)(2) = 8$.

The value of the integral of F over the cube is

$$\int_0^2\int_0^2\int_0^2 xyz\,dx\,dy\,dz = \int_0^2\int_0^2 \left[\frac{x^2}{2}yz\right]_{x=0}^{x=2} dy\,dz = \int_0^2\int_0^2 2yz\,dy\,dz$$
$$= \int_0^2 \left[y^2 z\right]_{y=0}^{y=2} dz = \int_0^2 4z\,dz = \left[2z^2\right]_0^2 = 8.$$

With these values, Eq. (6) gives

$$\begin{matrix}\text{Average value of}\\ xyz \text{ over the cube}\end{matrix} = \frac{1}{\text{volume}}\iiint\limits_{\text{cube}} xyz\,dV = \left(\frac{1}{8}\right)(8) = 1.$$

In evaluating the integral, we chose the order dx, dy, dz, but any of the other five possible orders would have done as well. ■

Exercises 14.4

1. Find the common value of the integrals in Example 2.
2. Write six different iterated triple integrals for the volume of the rectangular solid in the first octant bounded by the coordinate planes and the planes $x = 1$, $y = 2$, and $z = 3$. Evaluate one of the integrals.
3. Write six different iterated triple integrals for the volume of the tetrahedron cut from the first octant by the plane $6x + 3y + 2z = 6$. Evaluate one of the integrals.
4. Write six different iterated triple integrals for the volume of the region in the first octant enclosed by the cylinder $x^2 + z^2 = 4$ and the plane $y = 3$. Evaluate one of the integrals.

Evaluate the integrals in Exercises 5–14.

5. $\displaystyle\int_0^1\int_0^1\int_0^1 (x^2 + y^2 + z^2)\,dz\,dy\,dx$

6. $\displaystyle\int_0^{\sqrt{2}}\int_0^{3y}\int_{x^2+3y^2}^{8-x^2-y^2} dz\,dx\,dy$

7. $\displaystyle\int_1^e\int_1^e\int_1^e \frac{1}{xyz}\,dx\,dy\,dz$

8. $\displaystyle\int_0^1\int_0^{3-3x}\int_0^{3-3x-y} dz\,dy\,dx$

9. $\displaystyle\int_0^1\int_0^{\pi}\int_0^{\pi} y\sin z\,dx\,dy\,dz$

10. $\displaystyle\int_{-1}^1\int_{-1}^1\int_{-1}^1 (x + y + z)\,dy\,dx\,dz$

11. $\displaystyle\int_0^3\int_0^{\sqrt{9-x^2}}\int_0^{\sqrt{9-x^2}} dz\,dy\,dx$

12. $\displaystyle\int_0^2\int_{-\sqrt{4-y^2}}^{\sqrt{4-y^2}}\int_0^{2x+y} dz\,dx\,dy$

13. $\displaystyle\int_0^1\int_0^{2-x}\int_0^{2-x-y} dz\,dy\,dx$

14. $\displaystyle\int_0^1\int_0^{1-x^2}\int_3^{4-x^2-y} x\,dz\,dy\,dx$

15. Figure 14.31 shows the region of integration of the integral
$$\int_{-1}^1\int_{x^2}^1\int_0^{1-y} dz\,dy\,dx.$$

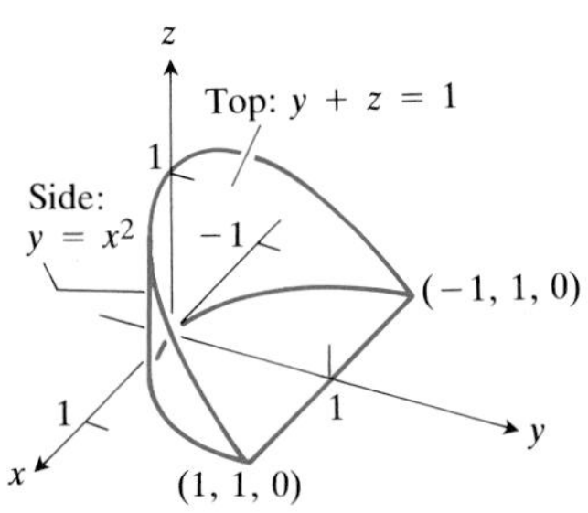

14.31 The region of integration in Exercise 15.

Rewrite the integral as an equivalent iterated integral in the order

a) $dy\,dz\,dx$, b) $dy\,dx\,dz$,
c) $dx\,dy\,dz$, d) $dx\,dz\,dy$,
e) $dz\,dx\,dy$.

16. Figure 14.32 shows the region of integration of the integral

$$\int_0^1 \int_{-1}^0 \int_0^{y^2} dz\, dy\, dx.$$

Rewrite the integral as an equivalent iterated integral in the order

a) $dy\,dz\,dx$, b) $dy\,dx\,dz$,
c) $dx\,dy\,dz$, d) $dx\,dz\,dy$,
e) $dz\,dx\,dy$.

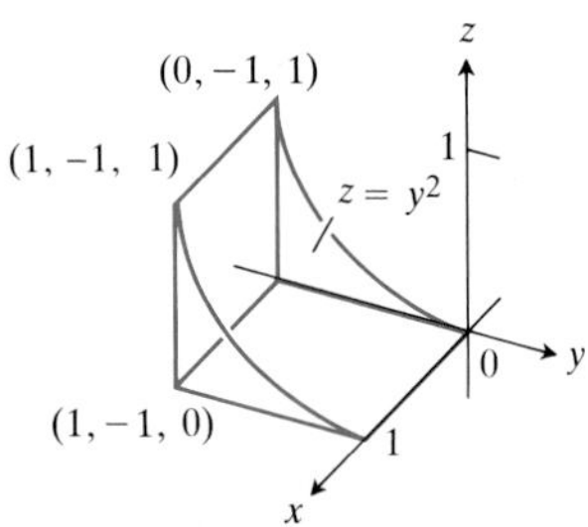

14.32 The region of integration in Exercise 16.

Find the volumes of the regions in Exercises 17–26.

17. The region between the cylinder $z = y^2$ and the xy-plane that is bounded by the four vertical planes $x = 0$, $x = 1$, $y = -1$, $y = 1$.

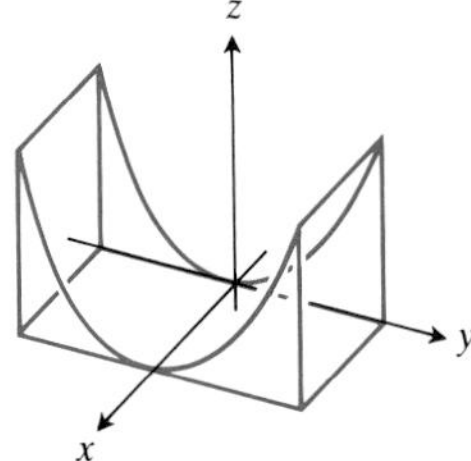

18. The region in the first octant bounded by the coordinate planes and the planes $x + z = 1$, $y + 2z = 2$.

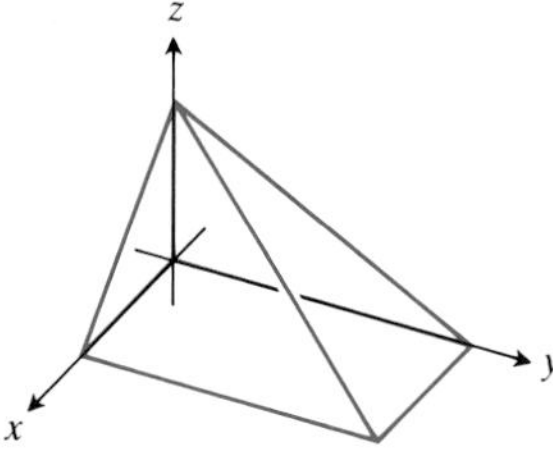

19. The region in the first octant bounded by the coordinate planes, the plane $y + z = 2$, and the cylinder $x = 4 - y^2$.

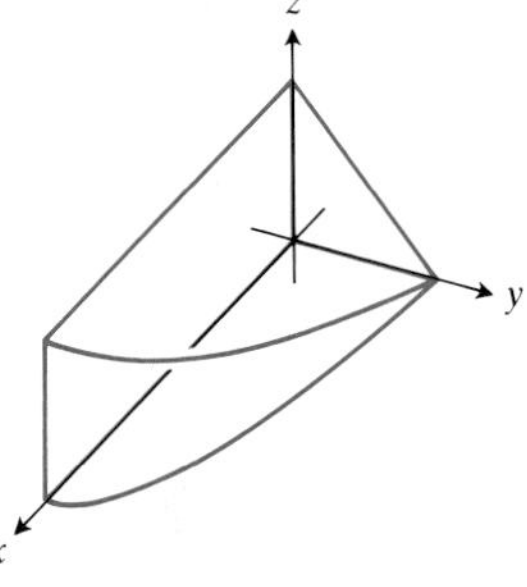

20. The wedge cut from the cylinder $x^2 + y^2 = 1$ by the planes $z = -y$ and $z = 0$.

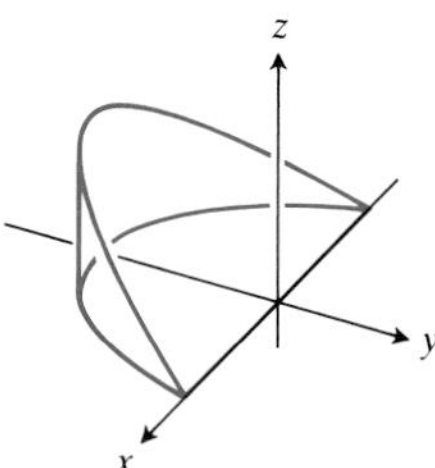

21. The tetrahedron in the first octant bounded by the coordinate planes and the plane $x + y/2 + z/3 = 1$.

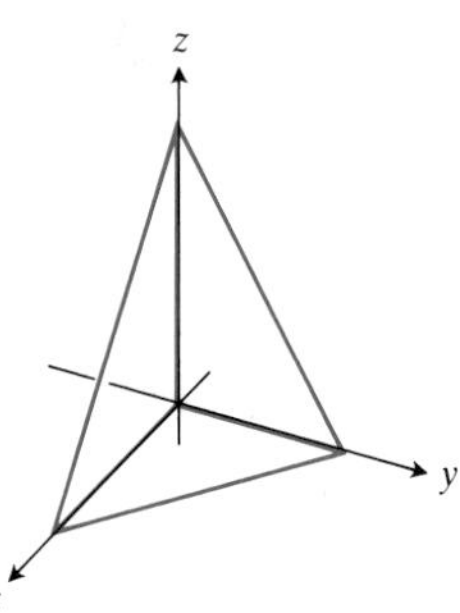

22. The region in the first octant bounded by the coordinate planes, the plane $y = 1 - x$, and the surface $z = \cos(\pi x/2)$, $0 \le x \le 1$.

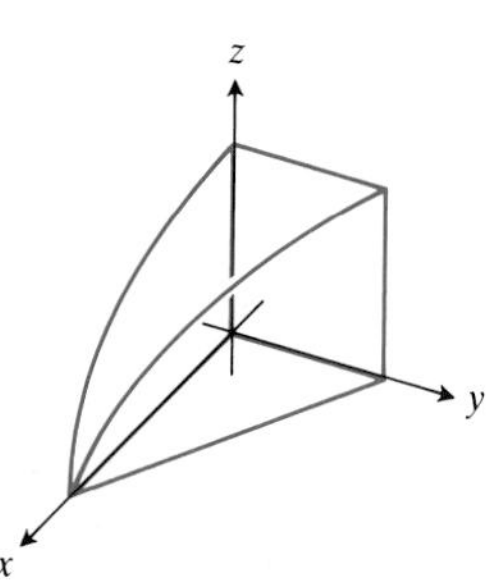

23. The region common to the interiors of the cylinders $x^2 + y^2 = 1$ and $x^2 + z^2 = 1$ (Fig. 14.33).

24. The region in the first octant bounded by the coordinate planes and the surface $z = 4 - x^2 - y$.

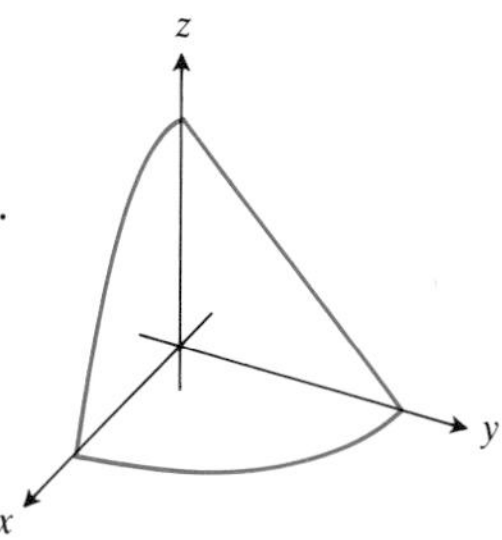

25. The region in the first octant bounded by the coordinate planes, the plane $x + y = 4$, and the cylinder $y^2 + 4z^2 = 16$.

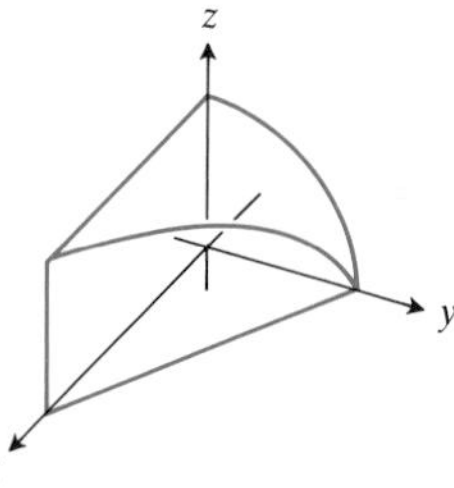

26. The region cut from the cylinder $x^2 + y^2 = 4$ by the plane $z = 0$ and the plane $x + z = 3$.

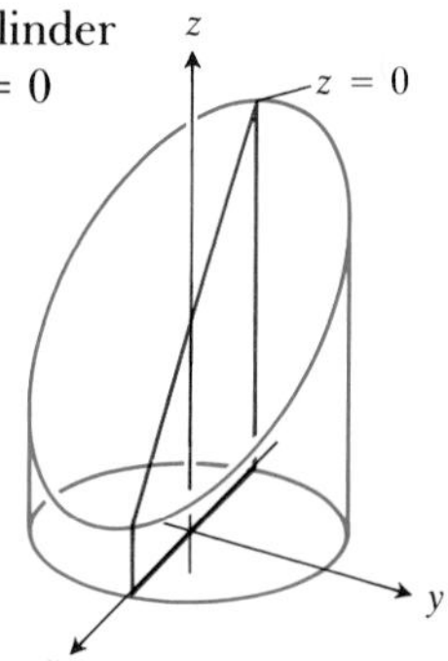

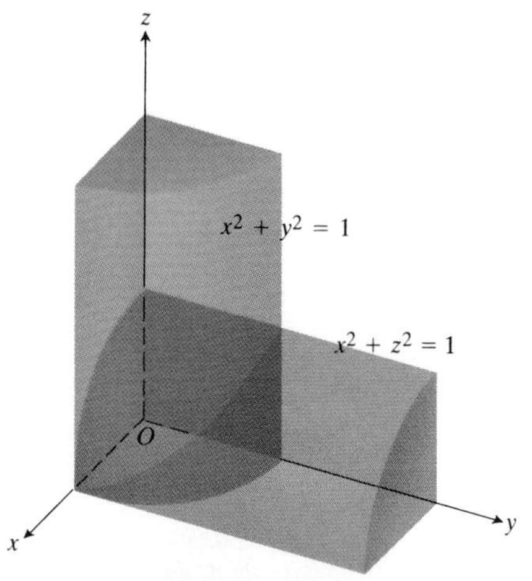

14.33 One eighth of the region common to the cylinders $x^2 + y^2 = 1$ and $x^2 + z^2 = 1$ in Exercise 23.

In Exercises 27–30, find the average value of $F(x, y, z)$ over the given region.

27. $F(x, y, z) = x^2 + 9$ over the cube in the first octant bounded by the coordinate planes and the planes $x = 2$, $y = 2$ and $z = 2$.

28. $F(x, y, z) = x + y - z$ over the rectangular solid in the first octant bounded by the coordinate planes and the planes $x = 1$, $y = 1$, and $z = 2$.

29. $F(x, y, z) = x^2 + y^2 + z^2$ over the cube in the first octant bounded by the coordinate planes and the planes $x = 1$, $y = 1$, and $z = 1$.

30. $F(x, y, z) = xyz$ over the cube in the first octant bounded by the coordinate planes and the planes $x = 2$, $y = 2$, and $z = 2$.

14.5 Masses and Moments in Three Dimensions

This section shows how we can calculate the masses and moments of three-dimensional objects in Cartesian coordinates. For calculations in spherical and cylindrical coordinates, see Section 14.6.

Masses and Moments

If $\delta(x, y, z)$ is the density of an object occupying a region R in space, then the integral of δ over R gives the mass of the object. To see why, imagine

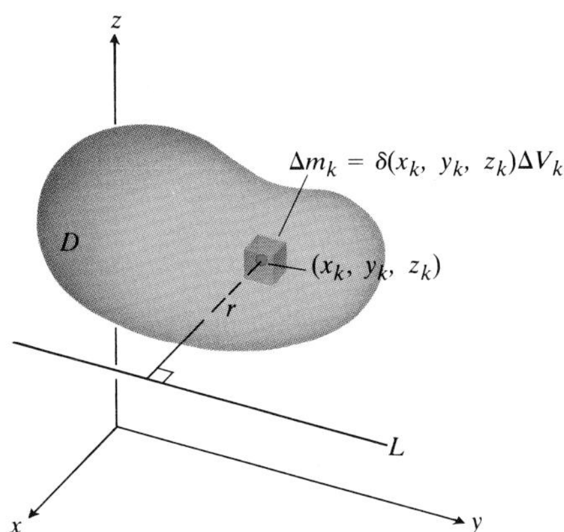

14.34 To define I_L we first imagine D to be subdivided into a finite number of mass elements Δm_k.

subdividing R as in Fig. 14.34. The object's mass is the limit

$$M = \lim \sum_k \Delta m_k = \lim \sum_k \delta(x_k, y_k, z_k)\Delta V_k = \iiint_R \delta(x, y, z)\, dV. \tag{1}$$

If $r(x, y, z)$ is the distance from the point (x, y, z) in R to a line L, then the moment of inertia of the mass

$$\Delta m_k = \delta(x_k, y_k, z_k)\Delta V_k$$

(shown in Fig. 14.34) about the line L is approximately

$$\Delta I_k = r^2(x_k, y_k, z_k)\Delta m_k.$$

The moment of inertia of the entire object about L is

$$I_L = \lim \sum_k \Delta I_k = \lim \sum_k r^2(x_k, y_k, z_k)\delta(x_k, y_k, z_k)\Delta V_k = \iiint_R r^2 \delta\, dV.$$

If L is the x-axis, then $r^2 = y^2 + z^2$ (Fig. 14.35) and

$$I_x = \iiint_R (y^2 + z^2)\delta\, dV.$$

Similarly,

$$I_y = \iiint_R (x^2 + z^2)\delta\, dV \quad \text{and} \quad I_z = \iiint_R (x^2 + y^2)\delta\, dV.$$

These and other useful formulas are summarized in Table 14.2.

TABLE 14.2
Mass and Moment Formulas for Objects in Space

Mass: $M = \iiint_R \delta\, dV \quad (\delta = \text{density})$

First moments about the coordinate planes:

$$M_{yz} = \iiint_R x\, \delta\, dV, \quad M_{xz} = \iiint_R y\, \delta\, dV, \quad M_{xy} = \iiint_R z\, \delta\, dV$$

Center of mass:

$$\bar{x} = \frac{\iiint x\, \delta\, dV}{M}, \quad \bar{y} = \frac{\iiint y\, \delta\, dV}{M}, \quad \bar{z} = \frac{\iiint z\, \delta\, dV}{M}$$

Moments of inertia (second moments):

$$I_x = \iiint (y^2 + z^2)\delta\, dV, \quad I_y = \iiint (x^2 + z^2)\delta\, dV,$$
$$I_z = \iiint (x^2 + y^2)\delta\, dV,$$

Moment of inertia about a line L:

$I_L = \iiint r^2\, \delta\, dV \quad$ ($r(x, y, z)$ = distance from point (x, y, z) to line L)

Radius of gyration about a line L:

$$R_L = \sqrt{I_L/M}$$

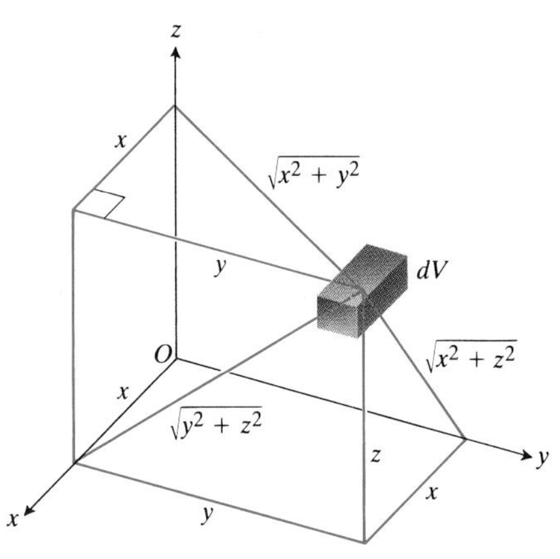

14.35 Distances from dV to the coordinate planes and axes.

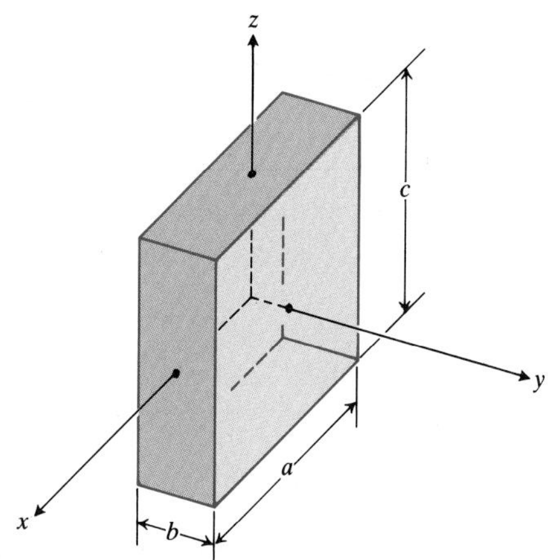

14.36 Example 1 calculates I_x, I_y, and I_z for the block shown here. The origin lies at the center of the block.

EXAMPLE 1 Find I_x, I_y, I_z for the rectangular solid of constant density δ shown in Fig. 14.36.

Solution The preceding formula for I_x gives

$$I_x = \int_{-c/2}^{c/2}\int_{-b/2}^{b/2}\int_{-a/2}^{a/2} (y^2 + z^2)\delta\, dx\, dy\, dz. \tag{2}$$

We can avoid some of the work of integration by observing that $(y^2 + z^2)\delta$ is an even function of x, y, and z and therefore

$$I_x = 8\int_0^{c/2}\int_0^{b/2}\int_0^{a/2} (y^2 + z^2)\delta\, dx\, dy\, dz = 4a\delta\int_0^{c/2}\int_0^{b/2} (y^2 + z^2)\, dy\, dz$$

$$= 4a\delta\int_0^{c/2}\left[\frac{y^3}{3} + z^2 y\right]_{y=0}^{y=b/2} dz$$

$$= 4a\delta\int_0^{c/2}\left(\frac{b^3}{24} + \frac{z^2 b}{2}\right) dz$$

$$= 4a\delta\left(\frac{b^3 c}{48} + \frac{c^3 b}{48}\right) = \frac{abc\delta}{12}(b^2 + c^2) = \frac{M}{12}(b^2 + c^2).$$

Similarly,

$$I_y = \frac{M}{12}(a^2 + c^2) \qquad \text{and} \qquad I_z = \frac{M}{12}(a^2 + b^2).$$

EXAMPLE 2 Find the center of mass of a solid of constant density δ bounded below by the disk $R: x^2 + y^2 \le 4$ in the plane $z = 0$ and above by the paraboloid $z = 4 - x^2 - y^2$ (Fig. 14.37).

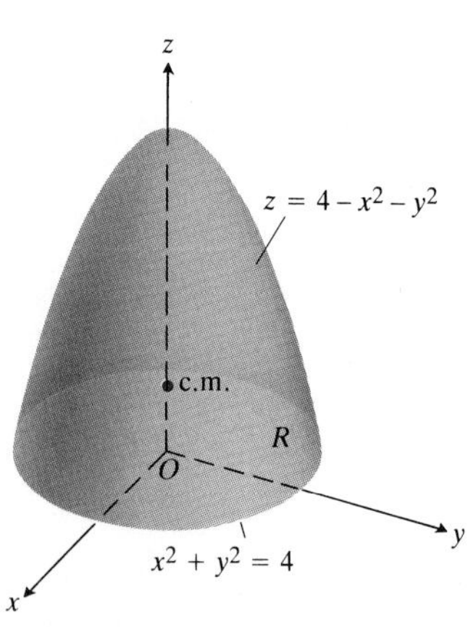

14.37 Example 2 calculates the coordinates of the center of mass of this solid.

Solution By symmetry, $\bar{x} = \bar{y} = 0$. To find $\bar{z}$ we first calculate

$$M_{xy} = \iiint_R \int_{z=0}^{z=4-x^2-y^2} z\,\delta\, dz\, dy\, dx = \iint_R \left[\frac{z^2}{2}\right]_{z=0}^{z=4-x^2-y^2} \delta\, dy\, dx$$

$$= \frac{\delta}{2}\iint_R (4 - x^2 - y^2)^2\, dy\, dx$$

$$= \frac{\delta}{2}\int_0^{2\pi}\int_0^2 (4 - r^2)^2\, r\, dr\, d\theta \qquad \left(\begin{matrix}\text{polar}\\ \text{coordinates}\end{matrix}\right)$$

$$= \frac{\delta}{2}\int_0^{2\pi}\left[-\frac{1}{6}(4 - r^2)^3\right]_{r=0}^{r=2} d\theta = \frac{16\delta}{3}\int_0^{2\pi} d\theta = \frac{32\pi\delta}{3}.$$

A similar calculation gives

$$M = \iiint_R \int_0^{4-x^2-y^2} \delta\, dz\, dy\, dx = 8\pi\delta.$$

Therefore $\bar{z} = (M_{xy}/M) = (4/3)$ and the center of mass is $(\bar{x}, \bar{y}, \bar{z}) = (0, 0, 4/3)$.

Exercises 14.5

Constant Density

The solids in Exercises 1–12 all have constant density $\delta = 1$.

1. Evaluate the integral for I_x in Eq. (2) directly to show that the shortcut in Example 1 gives the same answer. Use the results in Example 1 to find the radius of gyration of the rectangular solid about each coordinate axis.

2. The coordinate axes shown in Fig. 14.38 run through the centroid of a solid wedge parallel to its edges. Find I_x, I_y, and I_z if $a = b = 6$ and $c = 4$.

3. Find the moments of inertia of the rectangular solid shown in Fig. 14.39 with respect to its edges by calculating I_x, I_y, and I_z.

4. a) Find the centroid and the moments of inertia I_x, I_y, and I_z of the tetrahedron whose vertices are the points (0, 0, 0), (1, 0, 0), (0, 1, 0) and (0, 0, 1).
   b) Find the radius of gyration of the tetrahedron about the x-axis. Compare it with the distance from the centroid to the x-axis.

5. A solid "trough" of constant density is bounded below by the surface $z = 4y^2$, above by the plane $z = 4$, and on the ends by the planes $x = 1$ and $x = -1$. Find the center of mass and the moments of inertia with respect to the three axes.

6. A solid of constant density is bounded below by the plane $z = 0$, on the sides by the elliptic cylinder $x^2 + 4y^2 = 4$, and above by the plane $z = 2 - x$ (Fig. 14.40).
   a) Find $\bar{x}$ and $\bar{y}$.
   b) Evaluate the integral

$$M_{xy} = \int_{-2}^{2} \int_{-(1/2)\sqrt{4-x^2}}^{(1/2)\sqrt{4-x^2}} \int_{0}^{2-x} z\,dz\,dy\,dx,$$

   using integral tables to carry out the final integration with respect to x. Then divide M_{xy} by M to verify that $\bar{z} = 5/4$.

7. a) Find the center of mass of a solid of constant density bounded below by the paraboloid $z = x^2 + y^2$ and above by the plane $z = 4$.
   b) Find the plane $z = c$ that divides the solid into two parts of equal volume. This plane does not pass through the center of mass.

8. A solid cube, 2 units on a side, is bounded by the planes $x = \pm 1$, $z = \pm 1$, $y = 3$, and $y = 5$. Find the center of mass and the moments of inertia and radii of gyration about the coordinate axes. Suppose that the

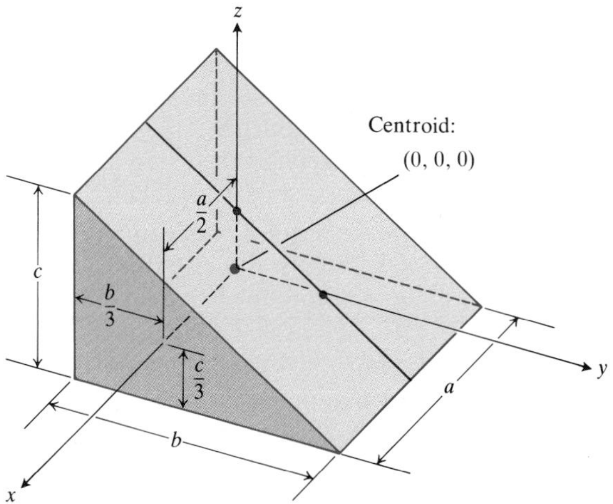

14.38 The figure for Exercise 2.

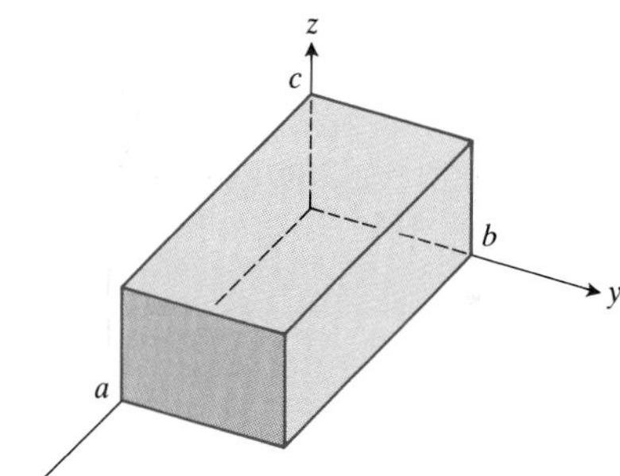

14.39 The rectangular solid in Exercise 3.

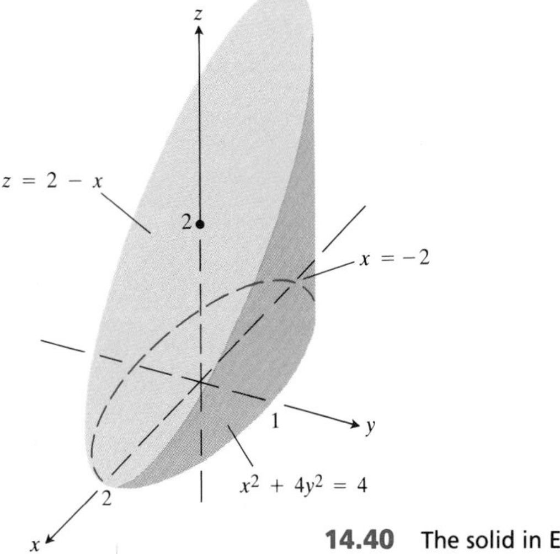

14.40 The solid in Exercise 6.

cube's mass is concentrated at the center of mass. How, if at all, does that change the moments of inertia and radii of gyration about the coordinate axes?

9. A wedge shaped like the one in Fig. 14.38 has $a = 4$, $b = 6$, and $c = 3$. Make a quick sketch to check for yourself that the square of the distance from a typical point (x, y, z) of the wedge to the line $L: z = 0, y = 6$ is $r^2 = (y - 6)^2 + z^2$. Then calculate the moment of inertia and radius of gyration of the wedge about L.

10. A wedge shaped like the one in Fig. 14.38 has $a = 4$, $b = 6$, and $c = 3$. Make a quick sketch to check for yourself that the square of the distance from a typical point (x, y, z) of the wedge to the line $L: x = 4, y = 0$ is $r^2 = (x - 4)^2 + y^2$. Then calculate the moment of inertia and radius of gyration of the wedge about L.

11. A rectangular solid like the one in Fig. 14.39 has $a = 4$, $b = 2$, and $c = 1$. Make a quick sketch to check for yourself that the square of the distance between a typical point (x, y, z) of the solid and the line $L: y = 2$, $z = 0$ is $r^2 = (y - 2)^2 + z^2$. Then find the moment of inertia and radius of gyration of the solid about L.

12. A rectangular solid like the one in Fig. 14.39 has $a = 4$, $b = 2$, and $c = 1$. Make a quick sketch to check for yourself that the square of the distance between a typical point (x, y, z) of the solid and the line $L: x = 4$, $y = 0$ is $r^2 = (x - 4)^2 + y^2$. Then find the moment of inertia and radius of gyration of the solid about L.

Variable Density

13. A solid region in the first octant is bounded by the coordinate planes and the plane $x + y + z = 2$. The density of the solid is $\delta(x, y, z) = 2x$. Find the center of mass.

14. A solid wedge shaped like the one in Fig. 14.38 has dimensions $a = 4$, $b = 6$, and $c = 3$. The density is $\delta(x, y, z) = x + y$. Find the center of mass and the moments of inertia and radii of gyration about the coordinate axes. Notice that if the density is constant, the center of mass will be $(2, 2, 1)$.

15. A solid cube in the first octant is bounded by the coordinate planes and by the planes $x = 1$, $y = 1$, and $z = 1$. Find its center of mass and the moments of inertia and radii of gyration about the coordinate axes if

$$\delta(x, y, z) = x + y + z + 1$$

16. A solid in the first octant is bounded by the planes $y = 0$ and $z = 0$ and by the surfaces $z = 4 - x^2$ and $x = y^2$ (Fig. 14.41). Its density function is $\delta(x, y, z) = kxy$.
a) Find the solid's mass. b) Find $\bar{x}$.
c) Find $\bar{y}$. d) Find $\bar{z}$.

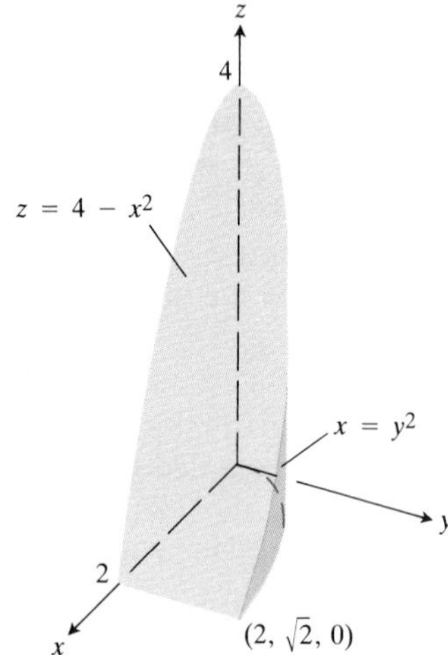

14.41 The solid in Exercise 16.

14.6 Triple Integrals in Cylindrical and Spherical Coordinates

If we are working with a solid like a cone or a cylindrical shell that has an axis of symmetry, we can often simplify our calculations by taking the axis to be the z-axis in a cylindrical coordinate system. In like manner, if we have to work with a shape like a ball or spherical shell that has a center of symmetry, we can usually save time by choosing the center to be the origin in a spherical coordinate system.

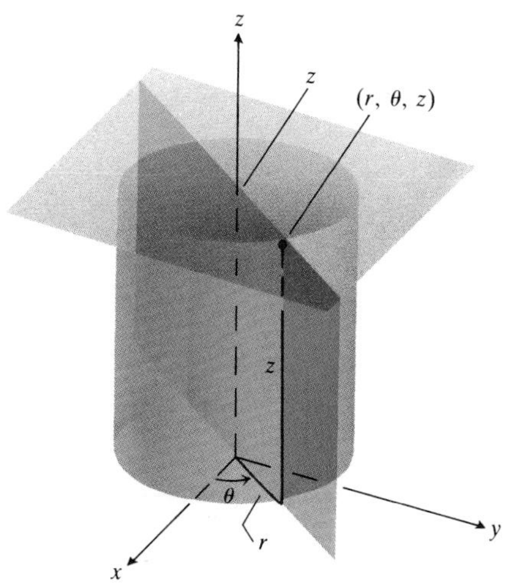

14.42 Cylindrical coordinates and typical surfaces of constant coordinate value.

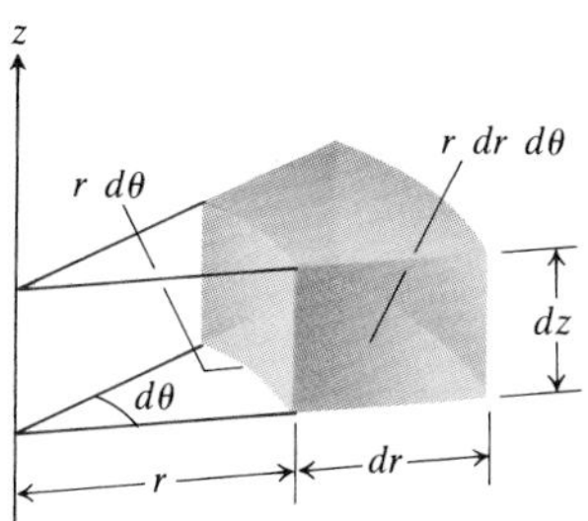

14.43 The volume element in cylindrical coordinates is $dV = dz\, r\, dr\, d\theta$.

This section shows how to calculate masses, moments, and volumes in cylindrical and spherical coordinates.

Cylindrical Coordinates

Cylindrical coordinates (Fig. 14.42) are good for describing cylinders whose axes run along the z-axis and planes that either contain the z-axis or lie perpendicular to the z-axis. As we saw in Section 11.7, surfaces like these have equations of constant coordinate value:

$$r = 4 \qquad \text{(cylinder, radius 4, axis the } z\text{-axis)},$$

$$\theta = \frac{\pi}{3} \qquad \text{(plane containing the } z\text{-axis)},$$

$$z = 2 \qquad \text{(plane perpendicular to the } z\text{-axis)}.$$

The volume element for subdividing a region in space with cylindrical coordinates is

$$dV = dz\, r\, dr\, d\theta \tag{1}$$

(Fig. 14.43). Triple integrals in cylindrical coordinates are then evaluated as iterated integrals, as in the following example.

How to Integrate in Cylindrical Coordinates

To integrate a continuous function $f(r, \theta, z)$ over a region given by inequalities

$$z_1(r, \theta) \le z \le z_2(r, \theta),$$
$$r_1(\theta) \le r \le r_2(\theta),$$
$$\theta_1 \le \theta \le \theta_2,$$

evaluate the iterated integral

$$\int_{\theta=\theta_1}^{\theta=\theta_2} \int_{r=r_1(\theta)}^{r=r_2(\theta)} \int_{z=z_1(r,\theta)}^{z=z_2(r,\theta)} f(r, \theta, z)\, dz\, r\, dr\, d\theta. \tag{2}$$

Integrate first with respect to z. Multiply by r and integrate with respect to r. Then integrate with respect to θ.

EXAMPLE 1 Find the centroid ($\delta = 1$) of the solid enclosed by the cylinder $x^2 + y^2 = 4$, bounded above by the paraboloid $z = x^2 + y^2$, and bounded below by the xy-plane.

Solution We sketch the figure (Fig. 14.44) and find the cylindrical-coordinate equations for the bounding surfaces. The coordinate inequali-

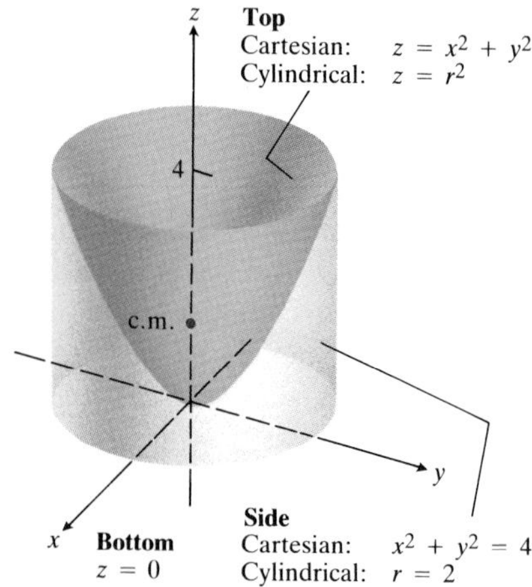

14.44 Example 1 shows how to locate the centroid of this solid.

ties for the region occupied by the solid are

$$z_1(r,\theta) \le z \le z_2(r,\theta):\quad 0 \le z \le r^2$$

$$r_1(\theta) \le r \le r_2(\theta):\quad 0 \le r \le 2$$

$$\theta_1 \le \theta \le \theta_2:\quad 0 \le \theta \le 2\pi.$$

The solid's center of mass lies on its axis of symmetry, in this case the z-axis, so $\bar{x} = \bar{y} = 0$. To find $\bar{z}$, we divide the moment M_{xy} by the mass M. The value of M_{xy} is

$$M_{xy} = \int_0^{2\pi}\int_0^2\int_0^{r^2} z\,dz\,r\,dr\,d\theta = \int_0^{2\pi}\int_0^2 \left[\frac{z^2}{2}\right]_0^{r^2} r\,dr\,d\theta$$

$$= \int_0^{2\pi}\int_0^2 \frac{r^5}{2}\,dr\,d\theta = \int_0^{2\pi}\left[\frac{r^6}{12}\right]_0^2 d\theta = \int_0^{2\pi}\frac{16}{3}\,d\theta = \frac{32\pi}{3}.$$

The value of M is

$$M = \int_0^{2\pi}\int_0^2\int_0^{r^2} dz\,r\,dr\,d\theta = \int_0^{2\pi}\int_0^2 \Big[z\Big]_0^{r^2} r\,dr\,d\theta$$

$$= \int_0^{2\pi}\int_0^2 r^3\,dr\,d\theta = \int_0^{2\pi}\left[\frac{r^4}{4}\right]_0^2 d\theta = \int_0^{2\pi} 4\,d\theta = 8\pi.$$

Therefore,

$$\bar{z} = \frac{M_{xy}}{M} = \frac{32\pi}{3}\frac{1}{8\pi} = \frac{4}{3},$$

and the solid's center of mass is the point $(0, 0, 4/3)$. ■

Spherical Coordinates

Spherical coordinates (Fig. 14.45) are good for describing spheres centered at the origin, half-planes hinged along the z-axis, and single-napped cones whose vertices lie at the origin and whose axes lie along the z-axis. Surfaces like these have equations of constant coordinate value:

$\rho = 4$ (sphere, radius 4, center at origin),

$\phi = \frac{\pi}{3}$ (cone opening up from the origin, making an angle of $\pi/3$ radians with the positive z-axis),

$\theta = \frac{\pi}{3}$ (half-plane, hinged along the z-axis, making an angle of $\pi/3$ radians with the positive x-axis).

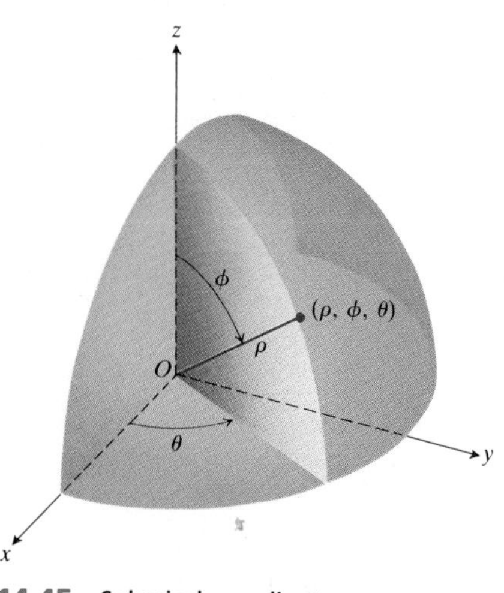

14.45 Spherical coordinates are measured with a distance and two angles.

The volume element in spherical coordinates is

$$dV = \rho^2 \sin\phi\,d\rho\,d\phi\,d\theta \tag{3}$$

(Fig. 14.46), and triple integrals take the form

$$\iiint F(\rho,\phi,\theta)\,dV = \iiint F(\rho,\phi,\theta)\,\rho^2\sin\phi\,d\rho\,d\phi\,d\theta. \tag{4}$$

To evaluate these integrals we usually integrate first with respect to ρ. The procedure for finding the limits of integration for a region R in space is shown in the following box.

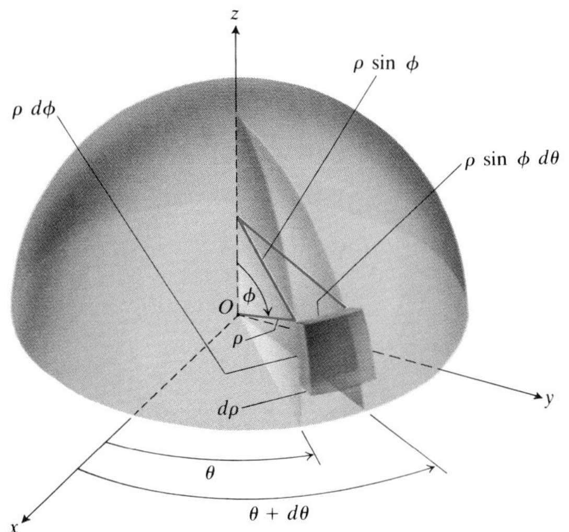

14.46 The volume element in spherical coordinates is

$$dV = d\rho \cdot \rho\, d\phi \cdot \rho \sin\phi\, d\theta$$
$$= \rho^2 \sin\phi\, d\rho\, d\phi\, d\theta.$$

How to Integrate in Spherical Coordinates

To integrate a continuous function $f(\rho, \phi, \theta)$ over a region given by the inequalities

$$\rho_1(\phi, \theta) \leq \rho \leq \rho_2(\phi, \theta),$$
$$\phi_1(\theta) \leq \phi \leq \phi_2(\theta),$$
$$\theta_1 \leq \theta \leq \theta_2,$$

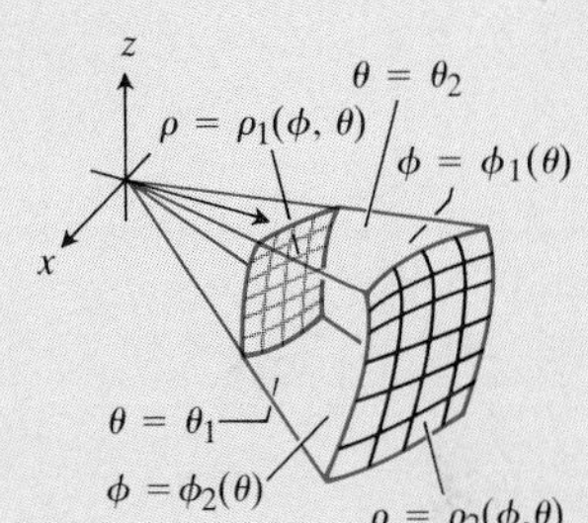

evaluate the iterated integral

$$\int_{\theta=\theta_1}^{\theta=\theta_2} \int_{\phi=\phi_1(\theta)}^{\phi=\phi_2(\theta)} \int_{\rho=\rho_1(\phi,\theta)}^{\rho=\rho_2(\phi,\theta)} f(\rho, \phi, \theta)\, \rho^2 \sin\phi\, d\rho\, d\phi\, d\theta. \qquad (5)$$

To do so, multiply f by ρ^2 and integrate with respect to ρ. Multiply the result by $\sin\phi$ and integrate with respect to ϕ. Then integrate with respect to θ.

EXAMPLE 2 Find the volume of the upper region cut from the solid sphere $\rho \leq 1$ by the cone $\phi = \pi/3$.

Solution We sketch the cone and sphere (Fig. 14.47). The coordinate inequalities for the region occupied by the solid are

$$\rho_1(\phi, \theta) \leq \rho \leq \rho_2(\phi, \theta): \quad 0 \leq \rho \leq 1$$

$$\phi_1(\theta) \leq \phi \leq \phi_2(\theta): \quad 0 \leq \phi \leq \frac{\pi}{3}$$

$$\theta_1 \leq \theta \leq \theta_2: \quad 0 \leq \theta \leq 2\pi.$$

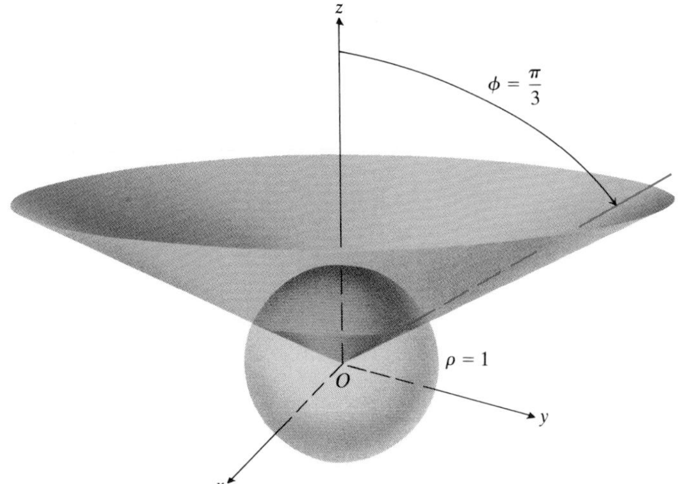

14.47 The region cut from the solid sphere $\rho \le 1$ by the cone $\phi = \pi/3$ (Example 2).

The volume is the integral of $f(\rho, \phi, \theta) = 1$ over this region:

$$V = \int_0^{2\pi}\int_0^{\pi/3}\int_0^1 \rho^2 \sin\phi \, d\rho \, d\phi \, d\theta = \int_0^{2\pi}\int_0^{\pi/3}\left[\frac{\rho^3}{3}\right]_0^1 \sin\phi \, d\phi \, d\theta$$

$$= \int_0^{2\pi}\int_0^{\pi/3} \frac{1}{3}\sin\phi \, d\phi \, d\theta = \int_0^{2\pi}\left[-\frac{1}{3}\cos\phi\right]_0^{\pi/3} d\theta$$

$$= \int_0^{2\pi}\left(-\frac{1}{6}+\frac{1}{3}\right) d\theta = \frac{1}{6}(2\pi) = \frac{\pi}{3}.$$

■

EXAMPLE 3 Find the moment of inertia about the z-axis of the region in Example 2.

Solution In rectangular coordinates, the moment is

$$I_z = \iiint (x^2 + y^2)\, dV.$$

In spherical coordinates, $x^2 + y^2 = (\rho \sin\phi \cos\theta)^2 + (\rho \sin\phi \sin\theta)^2 = \rho^2 \sin^2\phi$. Hence,

$$I_z = \iiint (\rho^2 \sin^2\phi)\, \rho^2 \sin\phi \, d\rho \, d\phi \, d\theta = \iiint \rho^4 \sin^3\phi \, d\rho \, d\phi \, d\theta.$$

For the region in Example 2, this becomes

$$I_z = \int_0^{2\pi}\int_0^{\pi/3}\int_0^1 \rho^4 \sin^3\phi \, d\rho \, d\phi \, d\theta = \int_0^{2\pi}\int_0^{\pi/3}\left[\frac{\rho^5}{5}\right]_0^1 \sin^3\phi \, d\phi \, d\theta$$

$$= \frac{1}{5}\int_0^{2\pi}\int_0^{\pi/3}(1-\cos^2\phi)\sin\phi \, d\phi \, d\theta = \frac{1}{5}\int_0^{2\pi}\left[-\cos\phi + \frac{\cos^3\phi}{3}\right]_0^{\pi/3} d\theta$$

$$= \frac{1}{5}\int_0^{2\pi}\left(-\frac{1}{2}+1+\frac{1}{24}-\frac{1}{3}\right) d\theta = \frac{1}{5}\int_0^{2\pi}\frac{5}{24}\, d\theta = \frac{1}{24}(2\pi) = \frac{\pi}{12}.$$

■

Coordinate Conversion Formulas (from Section 11.7)

Cylindrical to rectangular	Spherical to cylindrical	Spherical to rectangular
$x = r\cos\theta$	$r = \rho\sin\phi$	$x = \rho\sin\phi\cos\theta$
$y = r\sin\theta$	$z = \rho\cos\phi$	$y = \rho\sin\phi\sin\theta$
$z = z$	$\theta = \theta$	$z = \rho\cos\phi$

Volume: $dV = dx\,dy\,dz$ or $dz\,r\,dr\,d\theta$ or $\rho^2\sin\phi\,d\rho\,d\phi\,d\theta$

Exercises 14.6

Evaluate the cylindrical-coordinate integrals in Exercises 1–6.

1. $\displaystyle\int_0^{2\pi}\int_0^1\int_r^{\sqrt{2-r^2}} dz\,r\,dr\,d\theta$

2. $\displaystyle\int_0^{2\pi}\int_0^3\int_{r^2/3}^{\sqrt{18-r^2}} dz\,r\,dr\,d\theta$

3. $\displaystyle\int_0^{2\pi}\int_0^{\theta/2\pi}\int_0^{3+24r^2} dz\,r\,dr\,d\theta$

4. $\displaystyle\int_0^{\pi}\int_0^{\theta/\pi}\int_{-\sqrt{4-r^2}}^{3\sqrt{4-r^2}} z\,dz\,r\,dr\,d\theta$

5. $\displaystyle\int_0^{2\pi}\int_0^1\int_r^{1/\sqrt{2-r^2}} 3\,dz\,r\,dr\,d\theta$

6. $\displaystyle\int_0^{2\pi}\int_0^1\int_{-1/2}^{1/2} (r^2\sin^2\theta + z^2)\,dz\,r\,dr\,d\theta$

Evaluate the spherical-coordinate integrals in Exercises 7–12.

7. $\displaystyle\int_0^{\pi}\int_0^{\pi}\int_0^{2\sin\phi} \rho^2\sin\phi\,d\rho\,d\phi\,d\theta$

8. $\displaystyle\int_0^{2\pi}\int_0^{\pi/4}\int_0^{2} (\rho\cos\phi)\,\rho^2\sin\phi\,d\rho\,d\phi\,d\theta$

9. $\displaystyle\int_0^{2\pi}\int_0^{\pi}\int_0^{(1-\cos\phi)/2} \rho^2\sin\phi\,d\rho\,d\phi\,d\theta$

10. $\displaystyle\int_0^{3\pi/2}\int_0^{\pi}\int_0^{1} 5\rho^3\sin^3\phi\,d\rho\,d\phi\,d\theta$

11. $\displaystyle\int_0^{2\pi}\int_0^{\pi/3}\int_{\sec\phi}^{2} 3\rho^2\sin\phi\,d\rho\,d\phi\,d\theta$

12. $\displaystyle\int_0^{2\pi}\int_0^{\pi/2}\int_0^{\sec\phi} (\rho\cos\phi)\,\rho^2\sin\phi\,d\rho\,d\phi\,d\theta$

13. Set up an iterated triple integral for the volume of the sphere $x^2 + y^2 + z^2 = 4$ in (a) spherical, (b) cylindrical, and (c) rectangular coordinates.

14. Let R denote the region in the first octant that is bounded below by the cone $z = \sqrt{x^2 + y^2}$ and above by the sphere $x^2 + y^2 + z^2 = 9$. Express the volume of R as an iterated triple integral in (a) cylindrical and (b) spherical coordinates. Then (c) find V.

15. Give the limits of integration for evaluating the integral

$$\iiint f(r, \theta, z)\,dz\,r\,dr\,d\theta$$

as an iterated integral over the region that is bounded below by the plane $z = 0$, on the side by the cylinder $r = \cos\theta$, and on top by the paraboloid $z = 3r^2$.

16. Convert the integral

$$\int_{-1}^{1}\int_0^{\sqrt{1-y^2}}\int_0^{x} (x^2 + y^2)\,dz\,dx\,dy$$

to an equivalent integral in cylindrical coordinates and evaluate the result.

Cylindrical Coordinates

In exercises that ask for masses and moments, assume that the density is $\delta = 1$ unless stated otherwise.

17. Find the volume of the region bounded below by the plane $z = 0$, laterally by the cylinder $x^2 + y^2 = 1$, and above by the paraboloid $z = x^2 + y^2$.

18. Find the volume of the region bounded below by the paraboloid $z = x^2 + y^2$, laterally by the cylinder $x^2 + y^2 = 1$, and above by the paraboloid $z = x^2 + y^2 + 1$.

19. Find the volume of the region bounded below by the plane $z = 0$ and above by the paraboloid $z = 4 - x^2 - y^2$.

20. Find the volume of the region enclosed by the cylinder $x^2 + y^2 = 4$ and the planes $z = 0$ and $y + z = 4$. (*Hint:* In cylindrical coordinates, $z = 4 - y$ becomes $z = 4 - r \sin \theta$.)

21. Find the volume of the region bounded above by the paraboloid $z = 5 - x^2 - y^2$ and below by the paraboloid $z = 4x^2 + 4y^2$.

22. Find the volume of the region bounded above by the paraboloid $z = 9 - x^2 - y^2$, below by the xy-plane, and that lies *outside* the cylinder $x^2 + y^2 = 1$.

23. Find the volume of the region cut from the solid sphere $x^2 + y^2 + z^2 \leq 4$ by the cylinder $x^2 + y^2 = 1$.

24. Find the volume of the region bounded above by the sphere $x^2 + y^2 + z^2 = 2$ and below by the paraboloid $z = x^2 + y^2$.

25. Find the average value of the function $f(r, \theta, z) = r$ over the region bounded by the cylinder $r = 1$ between the planes $z = -1$ and $z = 1$.

26. Find the average value of the function $f(r, \theta, z) = r$ over the solid ball bounded by the sphere $r^2 + z^2 = 1$. (This is the sphere $x^2 + y^2 + z^2 = 1$.)

27. A solid is bounded below by the plane $z = 0$, above by the cone $z = r, r \geq 0$, and on the sides by the cylinder $r = 1$. Find the center of mass.

28. Find the centroid of the region that is bounded below by the surface $z = \sqrt{r}$ and above by the plane $z = 2$.

29. Find the moment of inertia and radius of gyration about the z-axis of a thick-walled right circular cylinder bounded on the inside by the cylinder $r = 1$, on the outside by the cylinder $r = 2$, and on the top and bottom by the planes $z = 4$ and $z = 0$.

30. A solid of constant density in the first octant is bounded above by the cone $z = \sqrt{x^2 + y^2}$, below by the plane $z = 0$, and on the sides by the cylinder $x^2 + y^2 = 4$ and the planes $x = 0$ and $y = 0$. Find the center of mass. (*Hint:* $\bar{x} = \bar{y}$.)

31. Find the moment of inertia of a solid circular cylinder of radius 1 and height 2 (a) about the axis of the cylinder, (b) about a line through the centroid perpendicular to the axis of the cylinder.

32. Find the moment of inertia of a right circular cone of base radius 1 and height 1 about an axis through the vertex parallel to the base.

33. Find the moment of inertia of a solid sphere of radius a about a diameter.

34. Find the centroid of the region cut from the solid ball $r^2 + z^2 \leq 1$ by the half-planes $\theta = -\pi/3, r \geq 0$ and $\theta = \pi/3, r \geq 0$.

Spherical Coordinates

In exercises that ask for masses and moments, assume that the density is $\delta = 1$ unless stated otherwise.

35. Find the volume of the region cut from the sphere $\rho = a$ by the cones $\phi = \pi/3$ and $\phi = 2\pi/3$.

36. Find the volume of the region cut from the sphere $\rho = a$ by the half-planes $\theta = 0$ and $\theta = \pi/6$ in the first octant.

37. Find the volume of the smaller region cut from the sphere $\rho = 2$ by the plane $z = 1$.

38. Find the volume of the region enclosed by the surface $\rho = 1 - \cos \phi$ (Fig. 14.48).

39. Find the average value of the function $f(\rho, \phi, \theta) = \rho$ over the solid ball $\rho \leq 1$.

40. Figure 14.49 shows a solid that was made by drilling a conical hole in a solid hemisphere. The equation of the cone is $\phi = \pi/3$. The radius of the hemisphere is $\rho = 2$. Find the center of mass.

41. Find the centroid of the solid bounded above by the sphere $\rho = a$ and below by the cone $\phi = \pi/4$.

42. A solid ball is bounded by the sphere $\rho = a$. Find the moment of inertia and radius of gyration about the z-axis if the density is

a) $\delta(\rho, \phi, \theta) = \rho^2$ b) $\delta(\rho, \phi, \theta) = r = \rho \sin \phi$

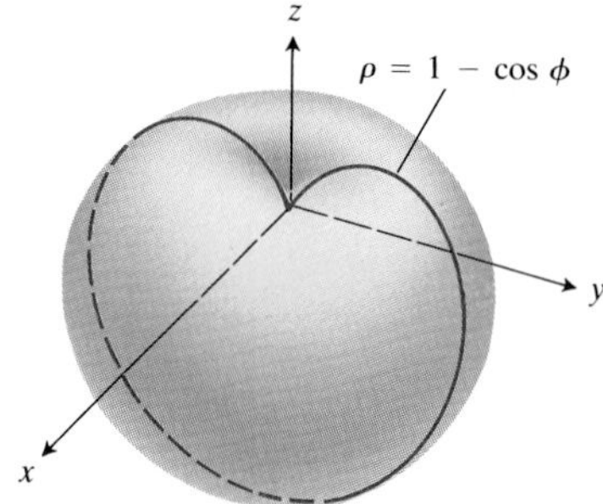

14.48 In spherical coordinates, the surface $\rho = 1 - \cos \phi$ is a cardioid of revolution (Exercise 38).

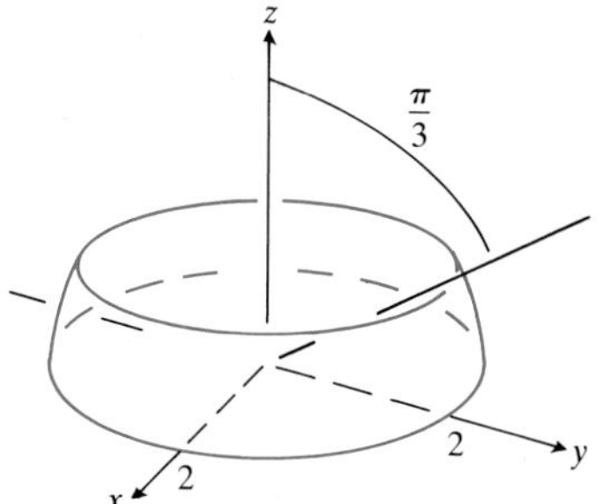

14.49 The solid in Exercise 40.

14.7 Substitutions in Multiple Integrals

This section shows how to evaluate multiple integrals by substitution. As in single integration, the goal of substitution is to replace complicated integrals by ones that are easier to evaluate. Substitutions can accomplish this by simplifying the integrand, by simplifying the limits of integration, or both.

Substitutions in Double Integrals

The polar coordinate substitution is a special case of a more general substitution method for double integrals, a method that pictures changes in variables as transformations of regions.

Suppose that a region G in the uv-plane is transformed one-to-one into the region R in the xy-plane by equations of the form

$$x = g(u, v), \qquad y = h(u, v),$$

as suggested in Fig. 14.50. Then any function $f(x, y)$ defined on R can be thought of as a function $f(g(u, v), h(u, v))$ defined on G as well. How is the integral of $f(x, y)$ over R related to the integral of $f(g(u, v), h(u, v))$ over G?

The answer is: If g, h, and f are continuous and have continuous partial derivatives, then the integrals are related by the equation

$$\iint_R f(x, y)\, dx\, dy = \iint_G f(g(u, v), h(u, v))\, |J(u, v)|\, du\, dv. \tag{1}$$

The factor $J(u, v)$, whose absolute value appears in this formula, is the determinant

$$J(u, v) = \begin{vmatrix} \dfrac{\partial x}{\partial u} & \dfrac{\partial x}{\partial v} \\ \dfrac{\partial y}{\partial u} & \dfrac{\partial y}{\partial v} \end{vmatrix} = \frac{\partial(x, y)}{\partial(u, v)}. \tag{2}$$

It is called the **Jacobian determinant** or **Jacobian** of the coordinate transformation $x = g(u, v)$, $y = h(u, v)$, named after the mathematician Carl Jacobi. The alternative notation $\partial(x, y)/\partial(u, v)$ may help you to remember how the determinant is constructed from the partial derivatives of x and y. The derivation of Eq. (1) is intricate and we shall not give it here.

For polar coordinates, we have r and θ in place of u and v. With $x = r\cos\theta$ and $y = r\sin\theta$, the Jacobian is

$$J(r, \theta) = \begin{vmatrix} \dfrac{\partial x}{\partial r} & \dfrac{\partial x}{\partial \theta} \\ \dfrac{\partial y}{\partial r} & \dfrac{\partial y}{\partial \theta} \end{vmatrix} = \begin{vmatrix} \cos\theta & -r\sin\theta \\ \sin\theta & r\cos\theta \end{vmatrix} = r(\cos^2\theta + \sin^2\theta) = r.$$

Notice the "reversed" order

The transforming equations $x = g(u, v)$ and $y = h(u, v)$ go from G to R, but we use them to change an integral over R into an integral over G.

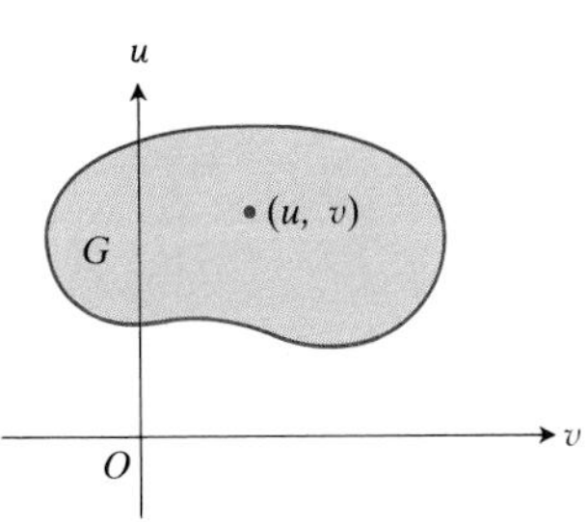

Cartesian uv-plane

$x = g(u, v)$
$y = h(u, v)$

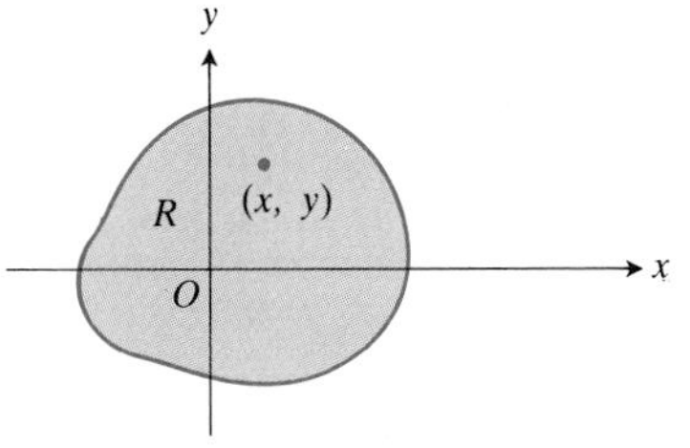

Cartesian xy-plane

14.50 The equations $x = g(u, v)$ and $y = h(u, v)$ allow us to write an integral over a region R in the xy-plane as an integral over a region G in the uv-plane.

Carl Gustav Jacob Jacobi

Jacobi (1804–1851), one of nineteenth-century Germany's most accomplished scientists, developed the theory of determinants and transformations into a powerful tool for evaluating multiple integrals and solving differential equations. He also applied transformation methods to study nonelementary integrals like the ones that arise in the calculation of arc length. Like Euler, Jacobi was a prolific writer and an even more prolific calculator, and worked in a variety of mathematical and applied fields.

Hence, Eq. (1) becomes

$$\begin{aligned}\iint_R f(x, y)\,dx\,dy &= \iint_G f(r\cos\theta, r\sin\theta)\,|r|\,dr\,d\theta \\ &= \iint_G f(r\cos\theta, r\sin\theta)\,r\,dr\,d\theta, \qquad (\text{If } r \geq 0)\end{aligned} \tag{3}$$

which is Eq. (7) in Section 14.3.

Figure 14.51 shows how the equations $x = r\cos\theta, y = r\sin\theta$ transform the rectangle $G\colon 0 \leq r \leq 1, 0 \leq \theta \leq \pi/2$ into the quarter-circle R bounded by $x^2 + y^2 = 1$ in the first quadrant of the xy-plane.

Notice that the integral on the right-hand side of Eq. (3) is not the integral of $f(r\cos\theta,\ r\sin\theta)$ over a region in the polar coordinate plane. It is the integral of the product of $f(r\cos\theta, r\sin\theta)$ and r over a region G in the *Cartesian* $r\theta$-plane.

Here is an example of another substitution.

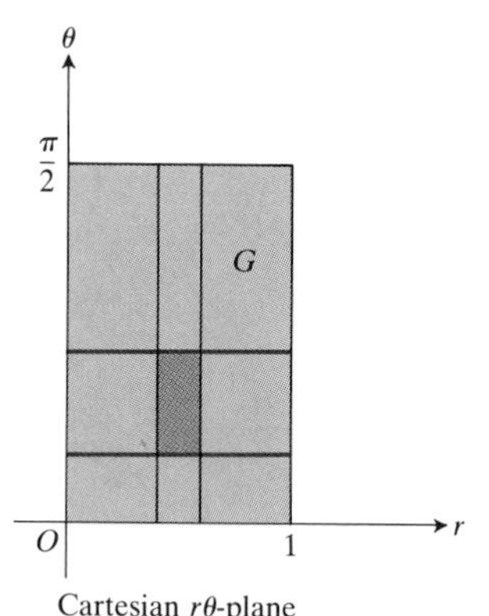

Cartesian $r\theta$-plane

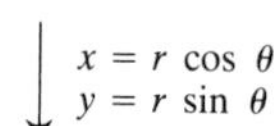

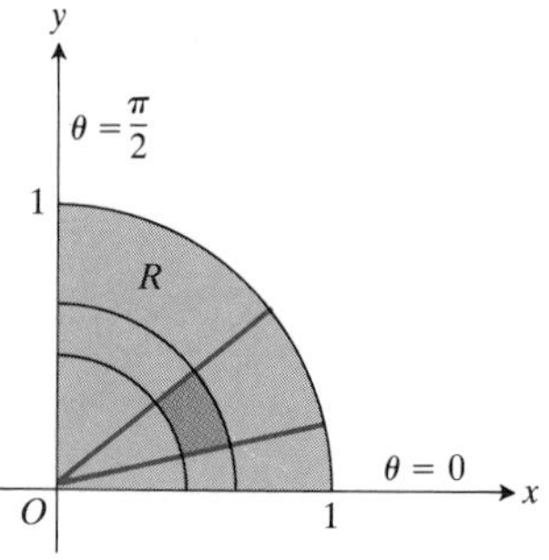

Cartesian xy-plane

14.51 The equations $x = r\cos\theta$, $y = r\sin\theta$ transform G into R.

EXAMPLE 1 Evaluate

$$\int_0^4 \int_{x=y/2}^{x=(y/2)+1} \frac{2x - y}{2}\,dx\,dy$$

by applying the transformation

$$u = \frac{2x - y}{2}, \qquad v = \frac{y}{2} \tag{4}$$

and integrating over an appropriate region in the uv-plane.

Solution We sketch the region R of integration in the xy-plane and identify its boundaries (Fig. 14.52).

To apply Eq. (1), we need to find the corresponding uv-region G and the Jacobian of the transformation. To find them, we first solve Eqs. (4) for x and y in terms of u and v. Routine algebra gives

$$x = u + v, \qquad y = 2v. \tag{5}$$

We then find the boundaries of G by substituting these expressions into the equations for the boundaries of R:

xy-equations for the boundary of R	Corresponding uv-equations for the boundary of G	Simplified uv-equations
$x = y/2$	$u + v = 2v/2 = v$	$u = 0$
$x = (y/2) + 1$	$u + v = (2v/2) + 1 = v + 1$	$u = 1$
$y = 0$	$2v = 0$	$v = 0$
$y = 4$	$2v = 4$	$v = 2$

See Fig. 14.52.

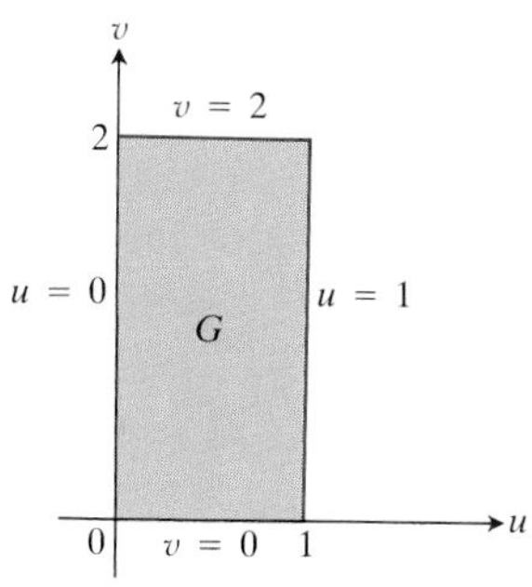

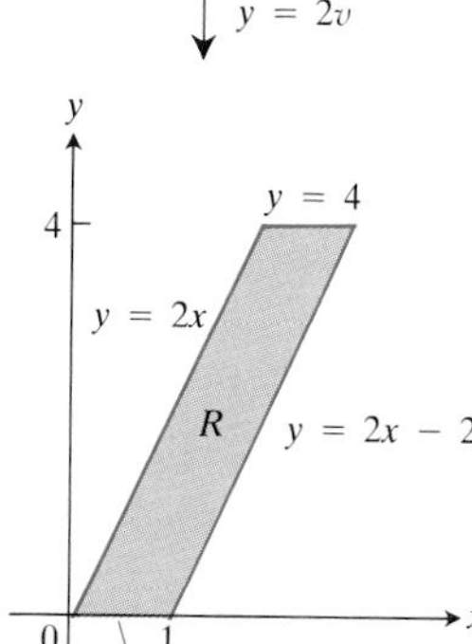

14.52 The equations $x = u + v$ and $y = 2v$ transform G into R. Reversing the transformation by the equations $u = (2x - y)/2$ and $v = y/2$ transforms R into G. See Example 1.

The Jacobian of the transformation (again from Eqs. 5) is

$$J(u, v) = \begin{vmatrix} \dfrac{\partial x}{\partial u} & \dfrac{\partial x}{\partial v} \\ \dfrac{\partial y}{\partial u} & \dfrac{\partial y}{\partial v} \end{vmatrix} = \begin{vmatrix} \dfrac{\partial}{\partial u}(u+v) & \dfrac{\partial}{\partial v}(u+v) \\ \dfrac{\partial}{\partial u}(2v) & \dfrac{\partial}{\partial v}(2v) \end{vmatrix} = \begin{vmatrix} 1 & 1 \\ 0 & 2 \end{vmatrix} = 2.$$

We now have everything we need to apply Eq. (1):

$$\int_0^4 \int_{x=y/2}^{x=(y/2)+1} \frac{2x - y}{2}\, dx\, dy = \int_{v=0}^{v=2} \int_{u=0}^{u=1} u\,|J(u, v)|\, du\, dv$$

$$= \int_0^2 \int_0^1 (u)(2)\, du\, dv = \int_0^2 \Big[u^2\Big]_0^1 dv = \int_0^2 dv = 2.$$

Exercise 1 asks you to evaluate the original xy-integral directly to show that its value is also 2. ■

Substitutions in Triple Integrals

The cylindrical and spherical coordinate substitutions are special cases of a substitution method that pictures changes of variables in triple integrals as transformations of three-dimensional regions. The method is just like the method for double integrals, except that now we work in three dimensions instead of two.

Suppose that a region G in uvw-space is transformed one-to-one into the region R in xyz-space by differentiable equations of the form

$$x = g(u, v, w), \qquad y = h(u, v, w), \qquad z = k(u, v, w),$$

as suggested in Fig. 14.53. Then any function $F(x, y, z)$ defined on R can be thought of as a function

$$F(g(u, v, w), h(u, v, w), k(u, v, w)) = H(u, v, w)$$

defined on G. If g, h, and k are continuous and have continuous first partial derivatives, then the integral of $F(x, y, z)$ over R is related to the integral of $H(u, v, w)$ over G by the equation

$$\iiint_R F(x, y, z)\, dx\, dy\, dz = \iiint_G H(u, v, w)\, |J(u, v, w)|\, du\, dv\, dw. \tag{6}$$

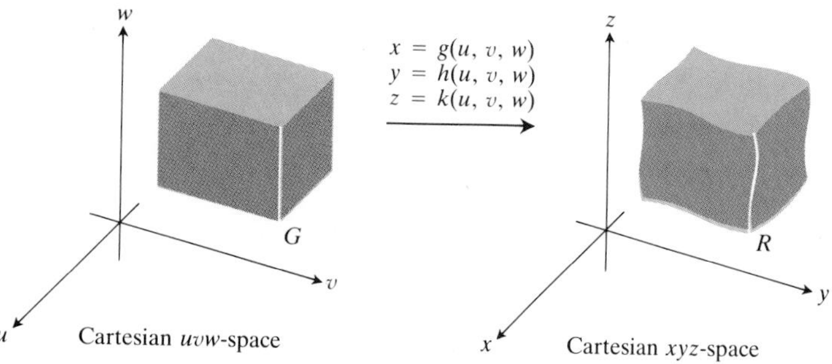

14.53 The equations $x = g(u, v, w)$, $y = h(u, v, w)$ and $z = k(u, v, w)$ allow us to write an integral over a region R in Cartesian xyz-space as an integral over a region G in Cartesian uvw-space.

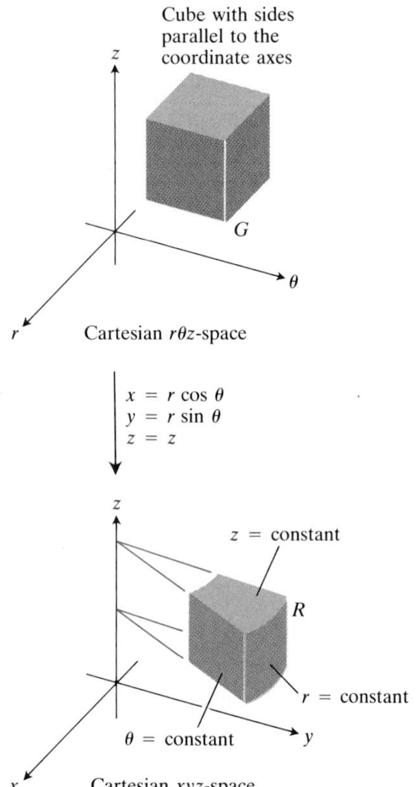

14.54 The equations $x = r\cos\theta$, $y = r\sin\theta$, and $z = z$ transform G into R.

The factor $J(u, v, w)$, whose absolute value appears in this equation, is the **Jacobian determinant**

$$J(u,v,w) = \begin{vmatrix} \dfrac{\partial x}{\partial u} & \dfrac{\partial x}{\partial v} & \dfrac{\partial x}{\partial w} \\ \dfrac{\partial y}{\partial u} & \dfrac{\partial y}{\partial v} & \dfrac{\partial y}{\partial w} \\ \dfrac{\partial z}{\partial u} & \dfrac{\partial z}{\partial v} & \dfrac{\partial z}{\partial w} \end{vmatrix} = \frac{\partial(x,y,z)}{\partial(u,v,w)}. \tag{7}$$

As in the two-dimensional case, the derivation of the change-of-variable formula in Eq. (6) is complicated and we shall not go into it here.

For cylindrical coordinates, we have r, θ, and z in place of u, v, and w. The transformation from Cartesian $r\theta z$-space to Cartesian xyz-space is given by the equations

$$x = r\cos\theta, \quad y = r\sin\theta, \quad z = z$$

(Fig. 14.54). The Jacobian of the transformation is

$$J(r,\theta,z) = \begin{vmatrix} \dfrac{\partial x}{\partial r} & \dfrac{\partial x}{\partial \theta} & \dfrac{\partial x}{\partial z} \\ \dfrac{\partial y}{\partial r} & \dfrac{\partial y}{\partial \theta} & \dfrac{\partial y}{\partial z} \\ \dfrac{\partial z}{\partial r} & \dfrac{\partial z}{\partial \theta} & \dfrac{\partial z}{\partial z} \end{vmatrix} = \begin{vmatrix} \cos\theta & -r\sin\theta & 0 \\ \sin\theta & r\cos\theta & 0 \\ 0 & 0 & 1 \end{vmatrix}$$

$$= r\cos^2\theta + r\sin^2\theta = r.$$

The corresponding version of Eq. (6) is

$$\iiint_R F(x,y,z)\,dx\,dy\,dz = \iiint_G H(r,\theta,z)\,|r|\,dr\,d\theta\,dz, \tag{8}$$

and we can drop the absolute value signs whenever $r \geq 0$.

For spherical coordinates, we have ρ, ϕ, and θ in place of u, v, and w. The transformation from Cartesian $\rho\phi\theta$-space to Cartesian xyz-space is given by the equations

$$x = \rho\sin\phi\cos\theta, \quad y = \rho\sin\phi\sin\theta, \quad z = \rho\cos\phi$$

(Fig. 14.55). The Jacobian of the transformation is

$$J(\rho,\phi,\theta) = \begin{vmatrix} \dfrac{\partial x}{\partial \rho} & \dfrac{\partial x}{\partial \phi} & \dfrac{\partial x}{\partial \theta} \\ \dfrac{\partial y}{\partial \rho} & \dfrac{\partial y}{\partial \phi} & \dfrac{\partial y}{\partial \theta} \\ \dfrac{\partial z}{\partial \rho} & \dfrac{\partial z}{\partial \phi} & \dfrac{\partial z}{\partial \theta} \end{vmatrix} = \rho^2\sin\phi, \tag{9}$$

as you will see if you do Exercise 9. The corresponding version of Eq. (6) is

$$\iiint_R F(x,y,z)\,dx\,dy\,dz = \iiint_G H(\rho,\phi,\theta)\,|\rho^2\sin\phi|\,d\rho\,d\phi\,d\theta, \tag{10}$$

and we can drop the absolute value signs because $\sin\phi$ is never negative.

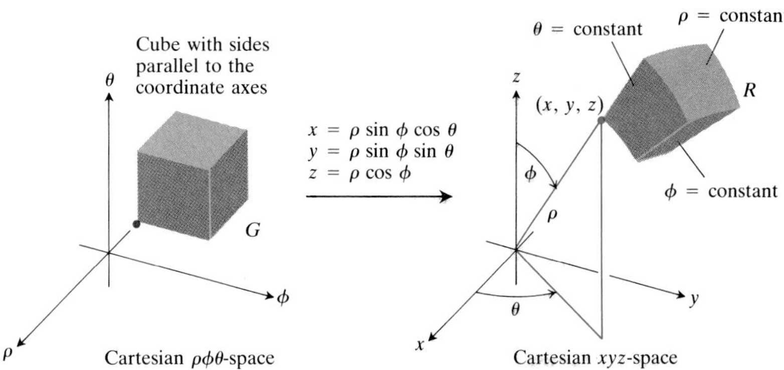

14.55 The equations $x = \rho \sin \phi \cos \theta$, $y = \rho \sin \phi \sin \theta$, and $z = \rho \cos \phi$ transform G into R.

Here is an example of another substitution.

EXAMPLE 2 Evaluate

$$\int_0^3 \int_0^4 \int_{x=y/2}^{x=(y/2)+1} \left(\frac{2x - y}{2} + \frac{z}{3} \right) dx\, dy\, dz$$

by applying the transformation

$$u = \frac{2x - y}{2}, \qquad v = \frac{y}{2}, \qquad w = \frac{z}{3} \tag{11}$$

and integrating over an appropriate region in uvw-space.

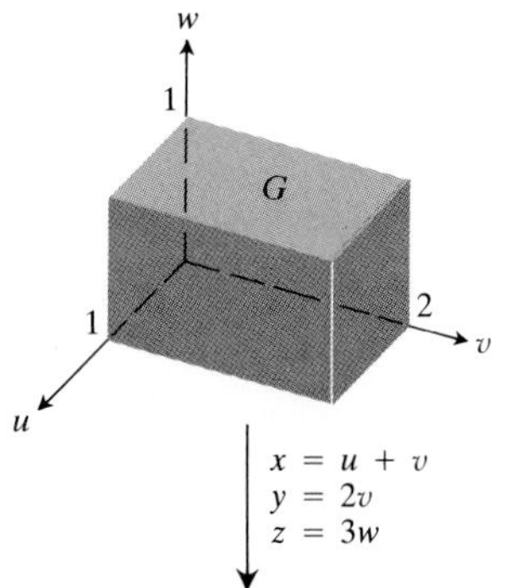

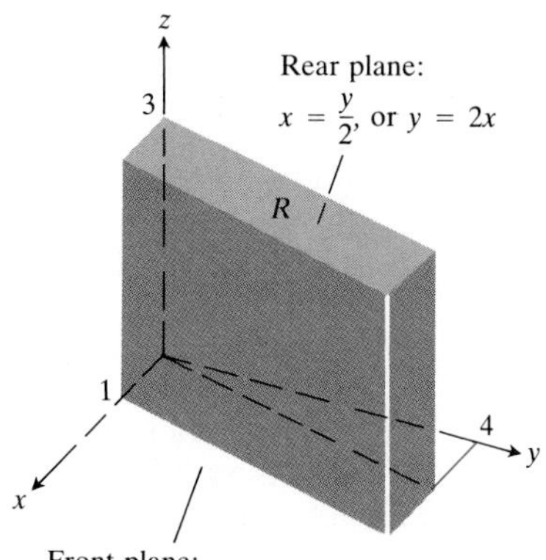

14.56 The equations $x = u + v$, $y = 2v$, and $z = 3w$ transform G into R. Reversing the transformation by the equations $u = (2x - y)/2$, $v = y/2$, and $w = z/3$ transforms R into G. See Example 2.

Solution We sketch the region R of integration in xyz-space and identify its boundaries (Fig. 14.56). In this case, the surfaces that bound R are all planes.

To apply Eq. (6), we need to find the corresponding uvw-region G and the Jacobian of the transformation. To find them, we first solve Eqs. (11) for x, y, and z in terms of u, v, and w. Routine algebra gives

$$\begin{aligned} x &= u + v, \\ y &= 2v, \\ z &= 3w. \end{aligned} \tag{12}$$

We then find the boundaries of G by substituting these expressions into the equations for the boundaries of R:

xyz-equations for the boundary of R	Corresponding uvw-equations for the boundary of G	Simplified uvw-equations
$x = y/2$	$u + v = 2v/2 = v$	$u = 0$
$x = (y/2) + 1$	$u + v = (2v/2) + 1 = v + 1$	$u = 1$
$y = 0$	$2v = 0$	$v = 0$
$y = 4$	$2v = 4$	$v = 2$
$z = 0$	$3w = 0$	$w = 0$
$z = 3$	$3w = 3$	$w = 1$

See Fig. 14.56.

The Jacobian of the transformation, again from Eqs. (12), is

$$J(u,v,w) = \begin{vmatrix} \frac{\partial x}{\partial u} & \frac{\partial x}{\partial v} & \frac{\partial x}{\partial w} \\ \frac{\partial y}{\partial u} & \frac{\partial y}{\partial v} & \frac{\partial y}{\partial w} \\ \frac{\partial z}{\partial u} & \frac{\partial z}{\partial v} & \frac{\partial z}{\partial w} \end{vmatrix} = \begin{vmatrix} 1 & 1 & 0 \\ 0 & 2 & 0 \\ 0 & 0 & 3 \end{vmatrix} = 6.$$

We now have everything we need to apply Eq. (6):

$$\int_0^3 \int_0^4 \int_{x=y/2}^{x=(y/2)+1} \left(\frac{2x-y}{2} + \frac{z}{3}\right) dx\, dy\, dz$$

$$= \int_0^1 \int_0^2 \int_0^1 (u+w)|J(u,v,w)|\, du\, dv\, dw$$

$$= \int_0^1 \int_0^2 \int_0^1 (u+w)(6)\, du\, dv\, dw = 6\int_0^1 \int_0^2 \left[\frac{u^2}{2} + uw\right]_0^1 dv\, dw$$

$$= 6\int_0^1 \int_0^2 \left(\frac{1}{2} + w\right) dv\, dw = 6\int_0^1 (1+2w)\, dw$$

$$= 6\Big[w + w^2\Big]_0^1 = 6(2) = 12.$$

Exercise 10 asks you to evaluate the original integral directly to show that its value is also 12. ■

Exercises 14.7

Double Integrals

1. Evaluate the integral

$$\int_0^4 \int_{x=y/2}^{x=(y/2)+1} \frac{2x-y}{2}\, dx\, dy$$

from Example 1 directly by integration with respect to x and y to confirm that its value is 2.

2. Find the Jacobian $\partial(x,y)/\partial(u,v)$ for the transformation
a) $x = u\cos v, \quad y = u\sin v$,
b) $x = u\sin v, \quad y = u\cos v$.

3. a) Solve the system

$$u = x - y, \quad v = 2x + y$$

for x and y in terms of u and v. Then find the value of the Jacobian $\partial(x,y)/\partial(u,v)$.

b) Let R be the region in the first quadrant bounded by the lines $y = -2x + 4$, $y = -2x + 7$, $y = x - 2$, and $y = x + 1$. Evaluate

$$\iint_R (2x^2 - xy - y^2)\, dx\, dy$$

by changing variables with the equations in (a) and integrating over a region G in the uv-plane.

4. a) Find the Jacobian of the transformation $x = u$, $y = uv$, and sketch the region G: $1 \le u \le 2$, $1 \le uv \le 2$ in the uv-plane.

b) Then use Eq. (1) to transform the integral

$$\int_1^2 \int_1^2 \frac{y}{x}\, dy\, dx$$

into an integral over G, and evaluate both integrals.

5. Let R be the region in the first quadrant of the xy-plane bounded by the hyperbolas $xy = 1$, $xy = 9$ and the lines $y = x$, $y = 4x$ (Fig. 14.57). Use the transformation $x = u/v$, $y = uv$ with $u > 0$ and $v > 0$ to rewrite

$$\iint_R \left(\sqrt{\frac{y}{x}} + \sqrt{xy}\right) dx\, dy$$

as an integral over an appropriate region G in the uv-plane. Then evaluate the uv-integral over G.

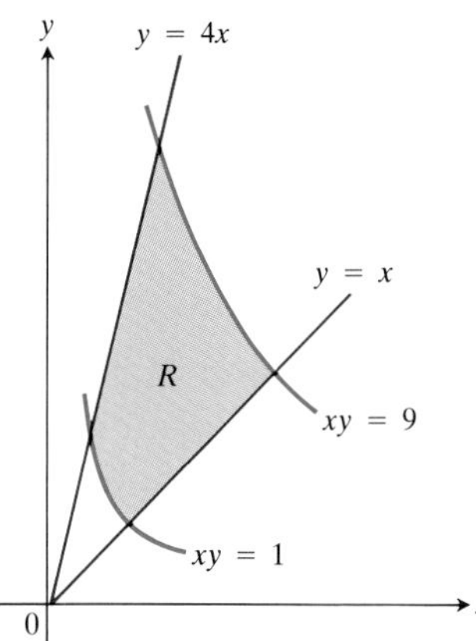

14.57 The *xy*-region in Exercise 5.

6. The area πab of the ellipse $x^2/a^2 + y^2/b^2 = 1$ can be found by integrating the function $f(x, y) = 1$ over the region bounded by the ellipse in the xy-plane. Evaluating the integral directly requires a trigonometric substitution. An easier way to evaluate the integral is to use the transformation $x = au, y = bv$ and evaluate the transformed integral over the disk $G: u^2 + v^2 \leq 1$ in the uv-plane. Find the area this way.

7. A thin plate of uniform thickness and density covers the region bounded by the ellipse $x^2/a^2 + y^2/b^2 = 1$ in the xy-plane. Find the moment of the plate about the origin. (*Hint:* Use the transformation $x = ar\cos\theta$, $y = br\sin\theta$.)

8. Use the transformation $x = u + 1/2v, y = v$ to evaluate the integral

$$\int_0^2 \int_{y/2}^{(y+4)/2} y^3(2x - y)e^{(2x-y)^2}\,dx\,dy$$

by first writing it as an integral over a region G in the uv-plane.

Triple Integrals

9. Evaluate the determinant in Eq. (9) to show that the Jacobian of the transformation from Cartesian $\rho\phi\theta$-space to Cartesian xyz-space is $\rho^2\sin\phi$.

10. Evaluate the integral in Example 2 by integrating with respect to x, y, and z.

11. Find the volume of the ellipsoid

$$\frac{x^2}{a^2} + \frac{y^2}{b^2} + \frac{z^2}{c^2} = 1.$$

(*Hint:* Let $x = au$, $y = bv$, and $z = cw$. Then find the volume of an appropriate region in uvw-space.)

12. Evaluate

$$\iiint |xyz|\,dx\,dy\,dz$$

over the solid ellipsoid

$$\frac{x^2}{a^2} + \frac{y^2}{b^2} + \frac{z^2}{c^2} \leq 1.$$

(*Hint:* Let $x = au$, $y = bv$, and $z = cw$. Then integrate over an appropriate region in uvw-space.)

13. Let R be the region in xyz-space defined by the inequalities

$$1 \leq x \leq 2, \quad 0 \leq xy \leq 2, \quad \text{and} \quad 0 \leq z \leq 1.$$

Evaluate

$$\iiint_R (x^2y + 3xyz)\,dx\,dy\,dz$$

by applying the transformation

$$u = x, \qquad v = xy, \qquad w = 3z$$

and integrating over an appropriate region G in uvw-space.

Review Questions

1. Define the double integral of a function of two variables over a bounded region in the coordinate plane.

2. How are double integrals evaluated? How are the limits of integration determined? Give an example.

3. How are double integrals used to calculate areas, masses, moments, centers of mass, and radii of gyration? Give examples.

4. How do you change a double integral from rectangular to polar coordinates? Give an example.

5. Define the triple integral of a function $f(x, y, z)$ over a bounded region in space.

6. How are triple integrals in Cartesian coordinates evaluated? How are the limits of integration determined? Give an example.

7. How are triple integrals in Cartesian coordinates used to calculate volumes, average values, masses, moments, centers of mass, and radii of gyration? Give examples.
8. How are triple integrals defined in cylindrical and spherical coordinates?
9. How are triple integrals in cylindrical coordinates evaluated? How are the limits of integration found? Give examples.
10. How are substitutions in double integrals pictured as transformations of two-dimensional regions? Give a sample calculation.
11. How are substitutions in triple integrals pictured as transformations of three-dimensional regions? Give a sample calculation.

Practice Exercises

Evaluate the integrals in Exercises 1–2 and sketch the regions of integration.

1. $\int_1^{10}\int_0^{1/y} y\,e^{xy}\,dx\,dy$

2. $\int_0^1\int_0^{x^3} e^{y/x}\,dy\,dx$

In Exercises 3–4, sketch the region of integration and write an equivalent iterated integral with the order of integration reversed.

3. $\int_0^1\int_{x^2}^{x} f(x,y)\,dy\,dx$

4. $\int_0^4\int_{y}^{2\sqrt{y}} f(x,y)\,dx\,dy$

In Exercises 5–6, sketch the region of integration and write an equivalent integral with the order of integration reversed. Then evaluate both integrals.

5. $\int_0^{3/2}\int_{-\sqrt{9-4y^2}}^{\sqrt{9-4y^2}} y\,dx\,dy$

6. $\int_0^2\int_0^{4-x^2} 2x\,dy\,dx$

Evaluate the integrals in Exercises 7–10.

7. $\int_0^1\int_{2y}^{2} 4\cos(x^2)\,dx\,dy$

8. $\int_0^2\int_{y/2}^{1} e^{x^2}\,dx\,dy$

9. $\int_0^8\int_{\sqrt[3]{x}}^{2} \frac{dy\,dx}{y^4+1}$

10. $\int_0^1\int_{\sqrt[3]{y}}^{1} \frac{2\pi\sin\pi x^2}{x^2}\,dx\,dy$

11. Find the volume under the paraboloid $z = x^2 + y^2$ above the triangle enclosed by the lines $y = x$, $x = 0$, and $x + y = 2$ in the xy-plane.
12. Find the volume under the parabolic cylinder $z = x^2$ above the region enclosed by the parabola $y = 6 - x^2$ and the line $y = x$ in the xy-plane.
13. Find the area of the region enclosed by the line $y = 2x + 4$ and the parabola $y = 4 - x^2$ in the xy-plane.
14. Find the area of the "triangular" region in the xy-plane that is bounded on the right by the parabola $y = x^2$, on the left by the line $x + y = 2$, and above by the line $y = 4$.

The average value of a function $f(x,y)$ over a region R in the xy-plane is defined to be

$$\text{Average value of } f \text{ over } R = \frac{1}{\text{Area of } R}\iint_R f(x,y)\,dA.$$

Find the average value of $f(x,y) = xy$ over the regions in Exercises 15–16.

15. The square bounded by the lines $x = 1$, $y = 1$ in the first quadrant.
16. The quarter circle $x^2 + y^2 \le 1$ in the first quadrant.
17. Find the centroid of the "triangular" region bounded by the lines $x = 2$, $y = 2$ and the hyperbola $xy = 2$ in the xy-plane.

18. Find the polar moment of inertia about the origin of a thin triangular plate of constant density $\delta = 3$, bounded by the y-axis and the lines $y = 2x$ and $y = 4$ in the xy-plane.

19. Find the polar moment of inertia about the center of a thin rectangular sheet of constant density $\delta = 1$ bounded by the lines

a) $x = \pm 2, \quad y = \pm 1$ in the xy-plane

b) $x = \pm a, \quad y = \pm b$ in the xy-plane

(*Hint:* Find I_x. Then use the formula for I_x to find I_y and add the two together to find I_0.)

20. Find the moment of inertia and radius of gyration about the x-axis of a thin plate of constant density δ covering the triangle with vertices (0, 0), (3, 0) and (3, 2) in the xy-plane.

21. Find the mass and first moments about the coordinate axes of a thin square plate bounded by the lines $x = \pm 1$, $y = \pm 1$ in the xy-plane if the density is $\delta(x, y) = x^2 + y^2 + 1/3$.

22. Find the center of mass and the moments of inertia and radii of gyration about the coordinate axes of a thin plate bounded by the line $y = x$ and the parabola $y = x^2$ in the xy-plane if the density is $\delta(x, y) = x + 1$.

Evaluate the integrals in Exercises 23–24 by changing to polar coordinates.

23. $\displaystyle\int_{-1}^{1}\int_{-\sqrt{1-x^2}}^{\sqrt{1-x^2}} \frac{2\, dy\, dx}{(1 + x^2 + y^2)^2}$

24. $\displaystyle\int_{-1}^{1}\int_{-\sqrt{1-y^2}}^{\sqrt{1-y^2}} \ln(x^2 + y^2 + 1)\, dx\, dy$

25. Find the centroid of the region of the plane region that lies inside the cardioid $r = 1 + \cos\theta$ and outside the circle $r = 1$.

26. Find the centroid of the region in the polar-coordinate plane defined by the inequalities $0 \le r \le 3$ and $-\pi/3 \le \theta \le \pi/3$.

27. Find the centroid of the region in the first quadrant bounded by the rays $\theta = 0$ and $\theta = \pi/2$ and the circles $r = 1$ and $r = 3$.

28. The usual way to evaluate the improper integral $I = \int_0^\infty e^{-x^2}\, dx$ is to calculate its square:

$$I^2 = \left(\int_0^\infty e^{-x^2}\, dx\right)\left(\int_0^\infty e^{-y^2}\, dy\right) = \int_0^\infty\int_0^\infty e^{-(x^2+y^2)}\, dx\, dy.$$

Introduce polar coordinates in the last expression and show that

$$I = \int_0^\infty e^{-x^2}\, dx = \frac{\sqrt{\pi}}{2}.$$

Evaluate the integrals in Exercises 29–30.

29. $\displaystyle\int_0^\pi\int_0^\pi\int_0^\pi \cos(x + y + z)\, dx\, dy\, dz$

30. $\displaystyle\int_0^1\int_0^{1-x^2}\int_{4x^2}^{5-x^2} (x - y + 1)\, dz\, dy\, dx$

31. Find the volume of the wedge-shaped region enclosed on the side by the cylinder $x = -\cos y$, $-\pi/2 \le y \le \pi/2$, on the top by the plane $z = -2x$, and below by the xy-plane.

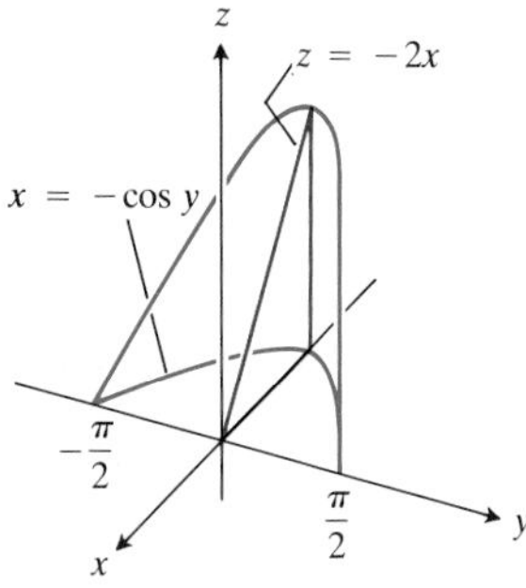

32. Find the volume of the solid that is bounded above by the cylinder $z = 4 - x^2$, on the sides by the cylinder $x^2 + y^2 = 4$, and below by the xy-plane.

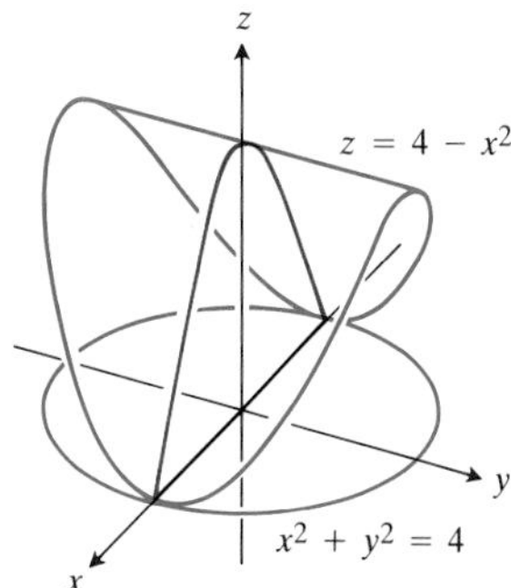

33. Let R be the smaller spherical cap cut from a solid ball of radius 2 by a plane 1 unit from the center of the sphere. Express the volume of R as an iterated triple integral in (a) rectangular, (b) cylindrical, and (c) spherical coordinates. *Do not evaluate the integrals.*

34. Express the moment of inertia I_z of the solid hemisphere bounded below by the plane $z = 0$ and above by the sphere $x^2 + y^2 + z^2 = 1$ as an iterated integral in (a) rectangular, (b) cylindrical, and (c) spherical coordinates. *Do not evaluate the integrals.*

35. Find the volume of the region in the first octant that lies between the cylinders $r = 1$ and $r = 2$ and that is bounded below by the xy-plane and above by the surface $z = xy$.

36. Find the volume of the region bounded above by the sphere $x^2 + y^2 + z^2 = 2$ and below by the paraboloid $z = x^2 + y^2$.

37. Find the volume of the region bounded above by the paraboloid $z = 3 - x^2 - y^2$ and below by the paraboloid $z = 2x^2 + 2y^2$.

38. A hole is bored through a solid sphere, the axis of the hole being a diameter of the sphere. The volume of the solid remaining is given by the integral

$$V = 2\int_0^{2\pi}\int_0^{\sqrt{3}}\int_1^{\sqrt{4-z^2}} r\,dr\,dz\,d\theta.$$

a) By inspecting the given integral, find the radius of the hole and the radius of the sphere.

b) Evaluate the integral. (See Example 4, Section 6.3.)

39. A solid is bounded on the top by the paraboloid $z = r^2$, on the bottom by the plane $z = 0$, and on the sides by the cylinder $r = 1$. Find the center of mass and the moment of inertia and radius of gyration about the z-axis if the density is (a) $\delta(r, \theta, z) = z$, (b) $\delta(r, \theta, z) = r$.

40. A solid is bounded below by the cone $z = \sqrt{x^2 + y^2}$ and above by the plane $z = 1$. Find the center of mass and the moment of inertia and radius of gyration about the z-axis if the density is (a) $\delta(r, \theta, z) = z$, (b) $\delta(r, \theta, z) = z^2$.

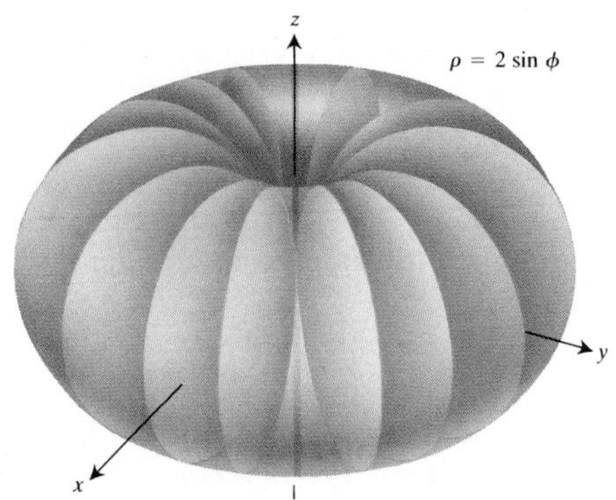

14.58 The spherical-coordinate surface $\rho = 2 \sin \phi$ is the surface generated by revolving a circle of radius 1 about one of its tangents (in this case, the z-axis).

41. Find the volume of the region enclosed by the spherical-coordinate surface $\rho = 2 \sin \phi$ (Fig. 14.58).

42. Find the moment of inertia about the z-axis of a solid of constant density $\delta = 1$, that is bounded above by the sphere $\rho = 2$ and below by the cone $\phi = \pi/3$ (spherical coordinates).

43. Use spherical coordinates to find the centroid of a solid hemisphere of radius a.

44. Find the moment of inertia and radius of gyration of a solid sphere of radius a and density $\delta = 1$ about a diameter of the sphere.

15 Vector Analysis

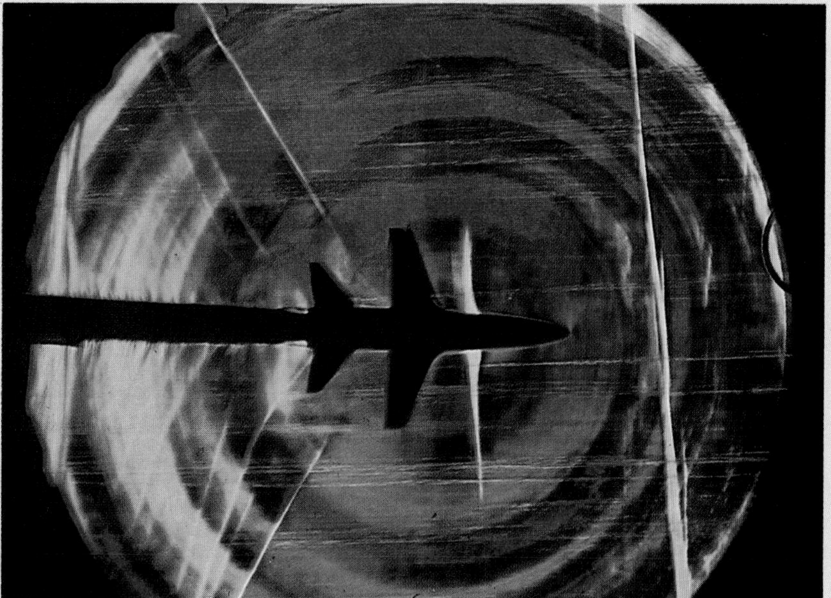

© Dr. Gary Settles/Photo Researchers

This schlieren photograph of a model plane in a wind tunnel reveals pressure variations created by the planes flight. We think of the gradient of the pressure function as a vector field in the surrounding air.

OVERVIEW This chapter brings together all our previous work with derivatives, integrals, and vector functions to study the calculus of vector fields. The field concept has proved to be one of the most useful ideas in all of physical science. The mathematics in this chapter is the mathematics we use today to describe the properties of electric charge; explain the motion of electromagnetic waves, including radio waves, X-rays, and visible light; predict the flow of air around airplane wings and rockets; explain the flow of heat in a room or in a star; calculate the amount of work it takes to put a satellite into orbit; and model storms in the atmosphere and currents in the oceans.

15.1 Line Integrals

When a curve $\mathbf{r} = g(t)\mathbf{i} + h(t)\mathbf{j} + k(t)\mathbf{k}$, $a \le t \le b$, passes through the domain of a function $f(x, y, z)$ in space, the values of f along the curve are given by the composite function $f(g(t), h(t), k(t))$. If we integrate this composite with respect to arc length from $t = a$ to $t = b$, we calculate the so-called line integral of f along the curve. Despite the three-dimensional geometry, the line integral is an ordinary integral of a real-valued function over an interval of real numbers. We have evaluated such integrals ever since Chapter 5. The usual techniques apply and nothing new is needed.

The importance of line integrals lies in their application. These are the integrals with which we calculate the moments and masses of springs and curved wires. We also combine them with vectors to calculate the work done by variable forces along paths in space and the rates at which fluids flow along curves and across boundaries. The present section shows how line integrals are defined, evaluated, and used to calculate moments and masses. The next section brings line integrals together with vectors to calculate work and measure fluid flow.

Definitions and Notation

Line integrals are integrals of functions over curves. We define them in the following way. Suppose that $f(x, y, z)$ is a function whose domain contains the curve

$$\mathbf{r} = g(t)\mathbf{i} + h(t)\mathbf{j} + k(t)\mathbf{k}, \quad a \le t \le b.$$

We partition the curve into a finite number of subarcs (Fig. 15.1). The typical subarc has length Δs_k. In each subarc we choose a point (x_k, y_k, z_k) and form the sum

$$S_n = \sum_{k=1}^{n} f(x_k, y_k, z_k)\, \Delta s_k. \tag{1}$$

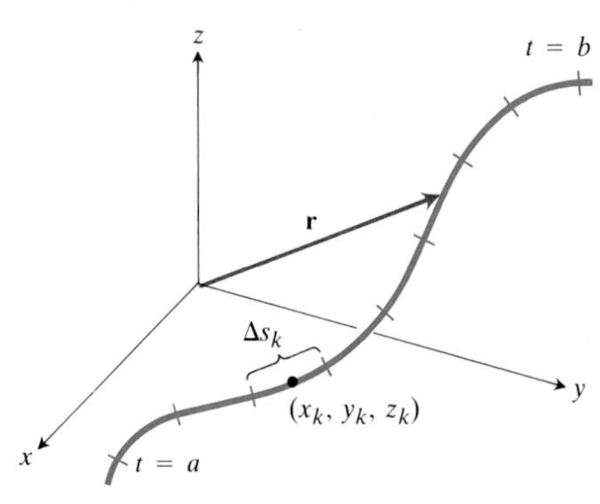

15.1 The curve $\mathbf{r} = g(t)\mathbf{i} + h(t)\mathbf{j} + k(t)\mathbf{k}$, partitioned into small arcs from $t = a$ to $t = b$. The length of a typical subarc is Δs_k.

If f is continuous and $\mathbf{r}$ is **continuously differentiable** (i.e., the functions g, h, and k have continuous first derivatives), then the sums in (1) approach a limit as n increases and the lengths Δs_k approach zero. We call this limit the **line integral of f over the curve from a to b.** If the curve is denoted by a single letter, C for example, the notation for the integral is

$$\int_C f(x, y, z)\, ds \qquad \text{("The integral of } f \text{ over } C\text{")} \tag{2}$$

Evaluation

To evaluate the line integral in Eq. (2), we express ds and the integrand in terms of the curve's parameter t. A remarkable theorem from advanced calculus tells us that we can find the value from any continuously differentiable parametrization $x = g(t)$, $y = h(t)$, $z = k(t)$ for which ds/dt is never zero by evaluating the integral

$$\int_{t=a}^{t=b} f(g(t), h(t), k(t))\frac{ds}{dt}\,dt = \int_a^b f(g(t), h(t), k(t))\sqrt{\left(\frac{dx}{dt}\right)^2 + \left(\frac{dy}{dt}\right)^2 + \left(\frac{dz}{dt}\right)^2}\,dt. \quad (3)$$

Since the square root in Eq. (3) is the length $|\mathbf{v}(t)|$ of the curve's velocity vector, the integral on the right-hand side can be simplified to give the following result.

The Evaluation Theorem for Line Integrals

The integral from $t = a$ to $t = b$ of a continuous function $f(x, y, z)$ over a continuously differentiable curve $\mathbf{r} = g(t)\mathbf{i} + h(t)\mathbf{j} + k(t)\mathbf{k}$ whose velocity vector never vanishes is given by the formula

$$\int_C f(x, y, z)\,ds = \int_{t=a}^{t=b} f(g(t), h(t), k(t))\,|\mathbf{v}(t)|\,dt. \quad (4)$$

Notice that if f has the constant value 1 then the integral of f over C from a to b gives the length of C from a to b. We calculated lengths in Section 12.3 and will not repeat the calculations here.

In our first example, we integrate a function over two different paths from (0, 0, 0) to (1, 1, 1).

EXAMPLE 1 Figure 15.2 shows two different paths from the origin to the point (1, 1, 1). One path is the union of the line segments C_1 and C_2. The other path is the line segment C_3. Integrate the function

$$f(x, y, z) = x - 3y^2 + z$$

along each path.

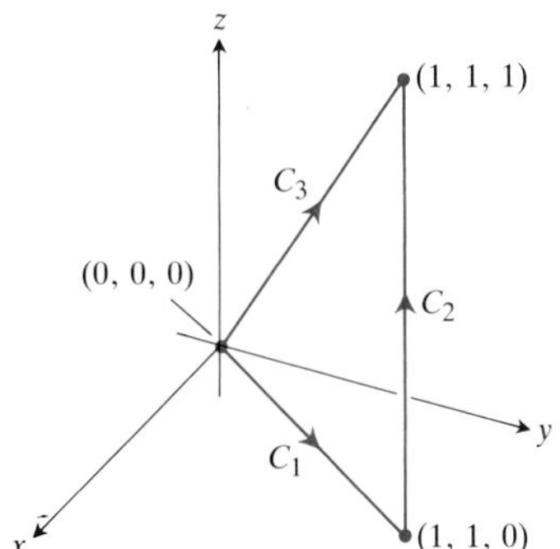

15.2 The paths of integration in Example 1. The arrows show the direction of increasing t in the parametrizations.

Solution We first parametrize the segments that make up the paths. According to the Evaluation Theorem, we can use any parametrization we want as long as $|\mathbf{v}(t)|$ is never zero and we integrate in the direction of increasing t so that $|\mathbf{v}(t)|\,dt$ is positive. We choose the simplest parametrizations we can think of, checking the lengths of the velocity vectors as we go along:

$$C_1: \quad \mathbf{r} = t\mathbf{i} + t\mathbf{j}, \quad 0 \le t \le 1; \qquad |\mathbf{v}| = \sqrt{1^2 + 1^2} = \sqrt{2}$$

$$C_2: \quad \mathbf{r} = \mathbf{i} + \mathbf{j} + t\mathbf{k}, \quad 0 \le t \le 1; \qquad |\mathbf{v}| = \sqrt{0^2 + 0^2 + 1^2} = 1$$

$$C_3: \quad \mathbf{r} = t\mathbf{i} + t\mathbf{j} + t\mathbf{k}, \quad 0 \le t \le 1; \qquad |\mathbf{v}| = \sqrt{1^2 + 1^2 + 1^2} = \sqrt{3}$$

Having chosen the parametrizations, we use Eq. (4) to integrate $f(x, y, z) = x - 3y^2 + z$ over each path in the direction of increasing t.

Path 1:
$$\int_{C_1 \cup C_2} f(x, y, z)\, ds = \int_{C_1} f(x, y, z)\, ds + \int_{C_2} f(x, y, z)\, ds$$
$$= \int_0^1 f(t, t, 0)\, \sqrt{2}\, dt + \int_0^1 f(1, 1, t)(1)\, dt$$
$$= \int_0^1 (t - 3t^2 + 0)\sqrt{2}\, dt + \int_0^1 (1 - 3 + t)(1)\, dt$$
$$= \sqrt{2}\int_0^1 (t - 3t^2)\, dt + \int_0^1 (t - 2)\, dt$$
$$= \sqrt{2}\left[\frac{t^2}{2} - t^3\right]_0^1 + \left[\frac{t^2}{2} - 2t\right]_0^1 = -\frac{\sqrt{2}}{2} - \frac{3}{2}.$$

Path 2:
$$\int_{C_3} f(x, y, z)\, ds = \int_0^1 f(t, t, t)(\sqrt{3})\, dt = \int_0^1 (t - 3t^2 + t)(\sqrt{3})\, dt$$
$$= \sqrt{3}\int_0^1 (2t - 3t^2)\, dt = \sqrt{3}\left[t^2 - t^3\right]_0^1 = 0 \qquad \blacksquare$$

There are three things to notice about the integrations in Example 1. First, as soon as the components of the appropriate curve are substituted in the formula for f and ds is replaced by the appropriate $|\mathbf{v}(t)|\, dt$ for each segment, the integration becomes a straightforward integration with respect to t. Second, the integral of f over Path 1 is obtained by integrating f over each section of the path and adding the results. Third, the integrals of f over Path 1 and Path 2 have different values. For most functions, the value of the line integral along a path joining two points changes when you change the path. For some functions, however, the value of the integral is the same for all paths joining the two points, as we shall see in Section 15.7.

Additivity

Line integrals have all the usual algebraic properties. The line integral of a sum of two functions is the sum of their line integrals, the line integral of a constant times a function is the constant times the line integral of the function, the line integral of a function from b to a is the negative of the line integral from a to b, and so on. Included in this list is the property that if a curve C is made by joining together a finite number of curves $C_1, C_2, \ldots, C_n$ end to end, then the integral of a function over C is the sum of the integrals over the curves that make it up:

$$\int_C f\, ds = \int_{C_1} f\, ds + \int_{C_2} f\, ds + \cdots + \int_{C_n} f\, ds. \tag{5}$$

This was illustrated in Example 1 when we evaluated the integral of f over the first path by adding the integrals over C_1 and C_2.

Mass and Moment Calculations

Suppose that a wire lies along a curve C in space and that the wire's density at the point (x, y, z) on the curve is $\delta(x, y, z)$. The wire's mass, center of mass, and moments can then be calculated with the formulas given in Table 15.1. The formulas apply to springs and thin rods as well as wires.

TABLE 15.1
Mass and moment formulas for wires, springs, and thin rods lying along a curve C in space

Mass: $$M = \int_C \delta(x, y, z)\, ds$$

First moments about the coordinate planes:

$$M_{yz} = \int_C x\,\delta\, ds, \qquad M_{xz} = \int_C y\,\delta\, ds, \qquad M_{xy} = \int_C z\,\delta\, ds$$

Coordinates of the center of mass:

$$\bar{x} = M_{yz}/M, \qquad \bar{y} = M_{xz}/M, \qquad \bar{z} = M_{xy}/M$$

Moments of inertia:

$$I_x = \int_C (y^2 + z^2)\,\delta\, ds, \qquad I_y = \int_C (x^2 + z^2)\,\delta\, ds,$$

$$I_z = \int_C (x^2 + y^2)\,\delta\, ds, \qquad I_L = \int_C r^2\,\delta\, ds,$$

$r(x, y, z) =$ distance from point (x, y, z) to line L

Radius of gyration about a line L: $R_L = \sqrt{I_L/M}$.

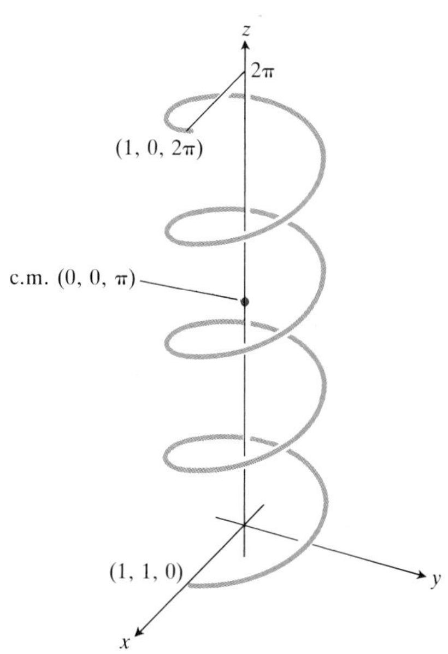

15.3 The helical spring in Example 2.

EXAMPLE 2 A coil spring lies along the helix

$$\mathbf{r} = (\cos 4t)\mathbf{i} + (\sin 4t)\mathbf{j} + t\mathbf{k}, \qquad 0 \le t \le 2\pi.$$

The spring's density is a constant, $\delta = 1$. Find the spring's mass, the coordinates of the spring's center of mass, and the spring's moment of inertia and radius of gyration about the z-axis.

Solution We sketch the spring (Fig. 15.3). Because of the symmetries involved, the center of mass lies at the point $(0, 0, \pi)$ on the z-axis.

For the remaining calculations, we first express ds in terms of t:

$$ds = \sqrt{\left(\frac{dx}{dt}\right)^2 + \left(\frac{dy}{dt}\right)^2 + \left(\frac{dz}{dt}\right)^2}\, dt$$

$$= \sqrt{(-4 \sin 4t)^2 + (4 \cos 4t)^2 + 1}\, dt = \sqrt{17}\, dt.$$

The formulas in Table 15.1 then give

$$M = \int_{\text{Helix}} \delta\, ds = \int_0^{2\pi} (1)\sqrt{17}\, dt = 2\pi\sqrt{17},$$

$$I_z = \int_{\text{Helix}} (x^2 + y^2)\delta\, ds = \int_0^{2\pi} (\cos^2 4t + \sin^2 4t)(1)\sqrt{17}\, dt = \int_0^{2\pi} \sqrt{17}\, dt = 2\pi\sqrt{17},$$

$$R_z = \sqrt{I_z/M} = \sqrt{2\pi\sqrt{17}/2\pi\sqrt{17}} = 1.$$

Notice that the radius of gyration about the z-axis is the radius of the cylinder around which the helix winds. ■

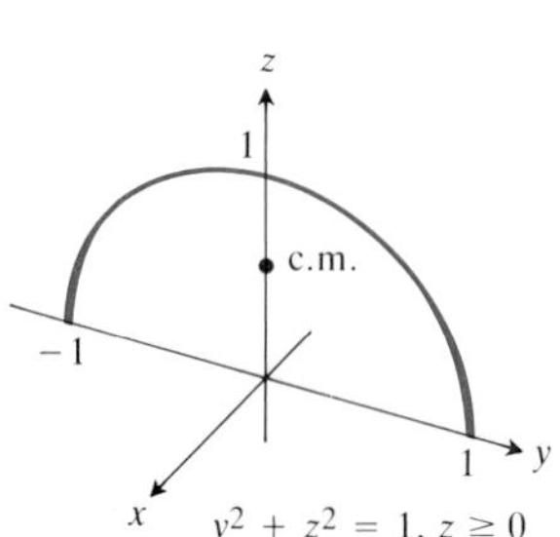

15.4 Example 3 shows how to find the center of mass of a circular arch of variable density.

EXAMPLE 3 A slender metal arch, thicker at the bottom than at the top, lies along the semicircle $y^2 + z^2 = 1$, $z \geq 0$, in the yz-plane (Fig. 15.4). Find the center of the arch's mass if the density at the point (x, y, z) on the arch is $\delta(x, y, z) = 2 - z$.

Solution We know that $\bar{x} = 0$ and $\bar{y} = 0$ because the arch lies in the yz-plane with its mass distributed symmetrically about the z-axis. To find $\bar{z}$, we parametrize the circle as

$$\mathbf{r} = (\cos t)\mathbf{j} + (\sin t)\mathbf{k}, \qquad 0 \leq t \leq \pi$$

and express ds in terms of dt:

$$ds = \sqrt{\left(\frac{dx}{dt}\right)^2 + \left(\frac{dy}{dt}\right)^2 + \left(\frac{dz}{dt}\right)^2}\, dt = \sqrt{(0)^2 + (-\sin t)^2 + (\cos t)^2}\, dt = dt.$$

The formulas in Table 15.1 then give

$$M = \int_C \delta\, ds = \int_C (2 - z)\, ds = \int_0^{\pi} (2 - \sin t)\, dt = 2\pi - 2,$$

$$M_{xy} = \int_C z\,\delta\, ds = \int_C z(2 - z)\, ds = \int_0^{\pi} \sin t(2 - \sin t)\, dt$$

$$= \int_0^{\pi} (2\sin t - \sin^2 t)\, dt = \frac{8 - \pi}{2},$$

$$\bar{z} = \frac{M_{xy}}{M} = \frac{8 - \pi}{2} \cdot \frac{1}{2\pi - 2} = \frac{8 - \pi}{4\pi - 4} = 0.57. \qquad \binom{\text{Calculator,}}{\text{rounded}}$$

To the nearest hundredth, the center of mass is $(0, 0, 0.57)$. ■

Exercises 15.1

Match the vector equations in Exercises 1–8 with the graphs in (a)–(h).

1. $\mathbf{r} = t\mathbf{i} + (1 - t)\mathbf{j}, \quad 0 \leq t \leq 1$
2. $\mathbf{r} = \mathbf{i} + \mathbf{j} + t\mathbf{k}, \quad -1 \leq t \leq 1$
3. $\mathbf{r} = (2\cos t)\mathbf{i} + (2\sin t)\mathbf{j}, \quad 0 \leq t \leq 2\pi$
4. $\mathbf{r} = t\mathbf{i}, \quad -1 \leq t \leq 1$
5. $\mathbf{r} = t\mathbf{i} + t\mathbf{j} + t\mathbf{k}, \quad 0 \leq t \leq 2$
6. $\mathbf{r} = t\mathbf{j} + (2 - 2t)\mathbf{k}, \quad 0 \leq t \leq 1$
7. $\mathbf{r} = (t^2 - 1)\mathbf{j} + 2t\mathbf{k}, \quad -1 \leq t \leq 1$
8. $\mathbf{r} = (2\cos t)\mathbf{i} + (2\sin t)\mathbf{k}, \quad 0 \leq t \leq \pi$

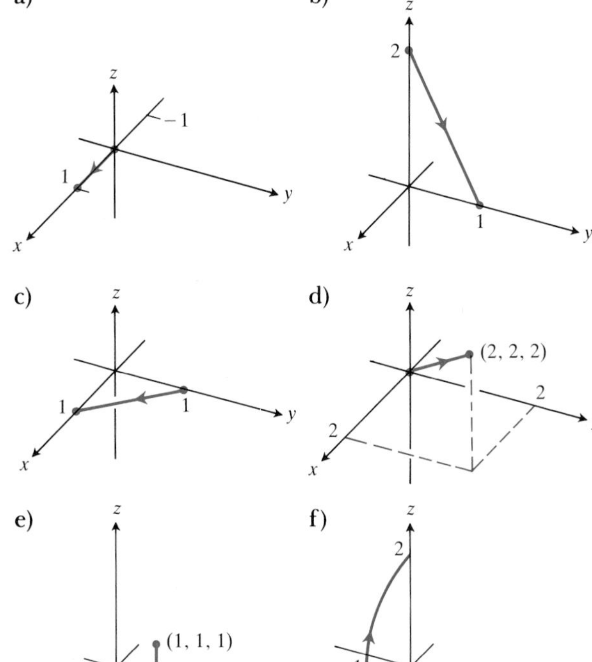

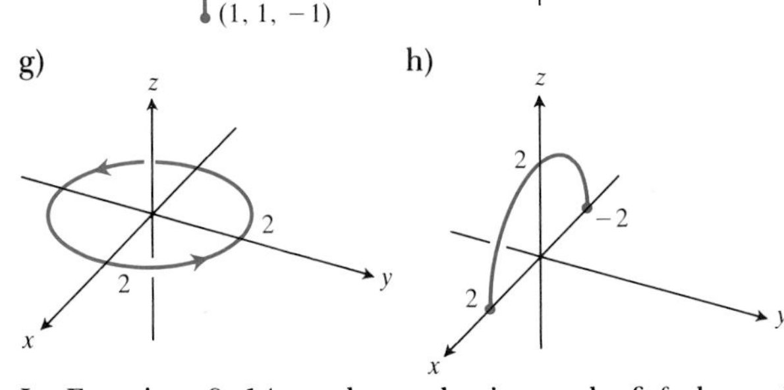

In Exercises 9–14, evaluate the integral of f along the given curve for the given parameter interval.

9. $f(x, y, z) = x + y$
$\mathbf{r} = t\mathbf{i} + (1 - t)\mathbf{j}, \quad 0 \leq t \leq 1$

10. $f(x, y, z) = x - y + z - 2$
$\mathbf{r} = t\mathbf{i} + (1 - t)\mathbf{j} + \mathbf{k}, \quad 0 \leq t \leq 1$

11. $f(x, y, z) = xy + y + z$
$\mathbf{r} = 2t\mathbf{i} + t\mathbf{j} + (2 - 2t)\mathbf{k}, \quad 0 \leq t \leq 1$

12. $f(x, y, z) = \sqrt{x^2 + y^2}$
$\mathbf{r} = (4 \cos t)\mathbf{i} + (4 \sin t)\mathbf{j} + 3t\mathbf{k}, \quad -2\pi \leq t \leq 2\pi$

13. $f(x, y, z) = 3z\sqrt{3x^2 + y^2 + z^2}$
$\mathbf{r} = \mathbf{i} + \mathbf{j} + t\mathbf{k}, \quad -1 \leq t \leq 1$

14. $f(x, y, z) = \sqrt{3}/(x^2 + y^2 + z^2)$
$\mathbf{r} = t\mathbf{i} + t\mathbf{j} + t\mathbf{k}, \quad 1 \leq t \leq 2$

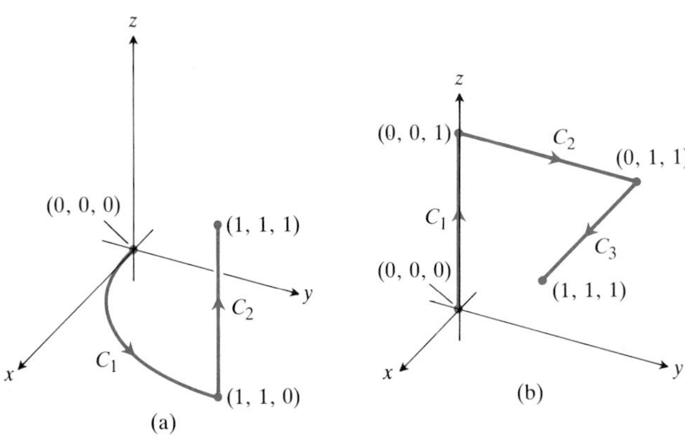

15.5 The paths of integration for Exercises 15 and 16.

15. Integrate $f(x, y, z) = x + \sqrt{y} - z^2$ over the path from (0, 0, 0) to (1, 1, 1) (Fig. 15.5a) given by
C_1: $\mathbf{r} = t\mathbf{i} + t^2\mathbf{j}, \quad 0 \leq t \leq 1$
C_2: $\mathbf{r} = \mathbf{i} + \mathbf{j} + t\mathbf{k}, \quad 0 \leq t \leq 1$

16. Integrate $f(x, y, z) = x + \sqrt{y} - z^2$ over the path from (0, 0, 0) to (1, 1, 1) (Fig. 15.5b) given by
C_1: $\mathbf{r} = t\mathbf{k}, \quad 0 \leq t \leq 1$
C_2: $\mathbf{r} = t\mathbf{j} + \mathbf{k}, \quad 0 \leq t \leq 1$
C_3: $\mathbf{r} = t\mathbf{i} + \mathbf{j} + \mathbf{k}, \quad 0 \leq t \leq 1$

17. Find I_x and R_x for the arch in Example 3.

18. Find the mass of a wire that lies along the curve $\mathbf{r} = (t^2 - 1)\mathbf{j} + 2t\mathbf{k}$, $0 \leq t \leq 1$, if the density is $\delta = (3/2)t$.

19. A circular wire hoop of constant density δ lies along the circle $x^2 + y^2 = a^2$ in the xy-plane. Find the hoop's moment of inertia and radius of gyration about the z-axis.

20. A slender rod of constant density lies along the line segment $\mathbf{r} = t\mathbf{j} + (2 - 2t)\mathbf{k}$, $0 \leq t \leq 1$ in the yz-plane. Find the moments of inertia and radii of gyration of the rod about the three coordinate axes.

21. A spring of constant density δ lies along the helix $\mathbf{r} = (\cos t)\mathbf{i} + (\sin t)\mathbf{j} + t\mathbf{k}$, $0 \leq t \leq 2\pi$.

a) Find I_z and R_z.

b) Suppose you have another spring of constant density δ that is twice as long as the spring in (a) and lies along the helix for $0 \leq t \leq 4\pi$. Do you expect I_z and R_z for the longer spring to be the same as those for the shorter one, or should they be different? Check your predictions by calculating I_z and R_z for the longer spring.

22. A wire of density $\delta(x, y, z) = 15\sqrt{y + 2}$ lies along the curve $\mathbf{r} = (t^2 - 1)\mathbf{j} + 2t\mathbf{k}$, $-1 \leq t \leq 1$. Find its center of mass and moments of inertia about the y- and z-axes.

15.2 Vector Fields, Work, Circulation, and Flux

In studies of physical phenomena represented by vectors, we introduce the idea of a vector field on a domain in space, a function that assigns a vector quantity to each point in the domain. For example, we might be studying the velocity field of a moving fluid or the electric field created by a distribution of electric charges. In situations like these, integrals of real valued functions over closed intervals are replaced by line integrals of functions over paths through the vector field, and the function being integrated is a scalar product of vectors. We can use these line integrals in vector fields to calculate the work done in moving an object along a path against a variable force. (The object might be a vehicle sent into space against Earth's gravitational field.) Or we might want to calculate the work done by a vector field in moving an object along a path through the field (for example, the work done by a particle accelerator in raising the energy of a particle). We can also use line integrals to calculate the rates at which fluids flow along and across curves.

This section shows how the calculations go.

Vector Fields

A **vector field** on a domain in the plane or in space is a function that assigns a vector to each point in the domain. A field of three-dimensional vectors might have a formula like

$$\mathbf{F} = M(x, y, z)\mathbf{i} + N(x, y, z)\mathbf{j} + P(x, y, z)\mathbf{k}. \tag{1}$$

The field is **continuous** if the **component functions** M, N, and P are continuous, **differentiable** if M, N, and P are differentiable, and so on. A field of two-dimensional vectors might have a formula like

$$\mathbf{F} = M(x, y)\mathbf{i} + N(x, y)\mathbf{j}. \tag{2}$$

If we attach a projectile's velocity vector to each point of the projectile's trajectory in the plane of motion, we have a two-dimensional field defined along the trajectory. If we attach the gradient vector of a scalar function to each point of a level surface of the function, we have a three-dimensional field on the surface. If we attach the velocity vector to each point of a flowing fluid, we have a three-dimensional field defined on a region in space. These and other fields are illustrated in Figures 15.6–15.14. Some of the illustrations give formulas for the fields as well.

To sketch the fields that had formulas, we picked a representative selection of domain points and sketched the vectors attached to them. Notice the convention that the arrows representing the vectors are drawn with their tails, not their heads, at the points where the vector functions are evaluated. This is different from the way we drew the position vectors of the planets and projectiles we studied in Chapter 12, with their tails at the origin and their heads at the planet's and projectile's locations.

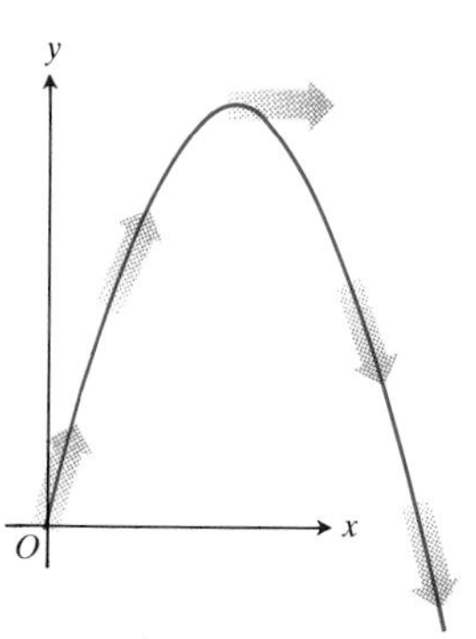

15.6 The velocity vectors $\mathbf{v}(t)$ of a projectile's motion make a vector field along the trajectory.

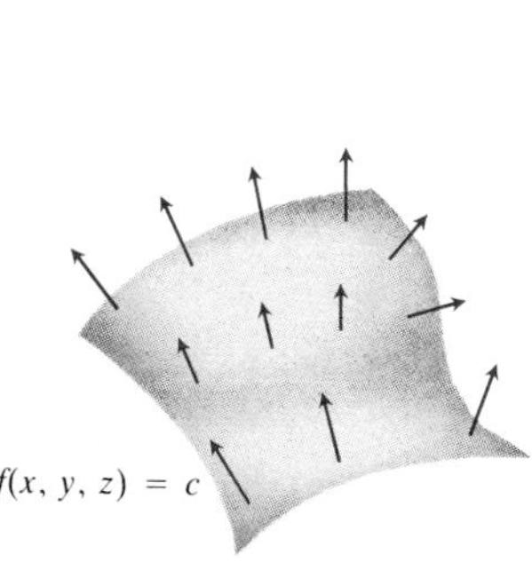

15.7 The field of gradient vectors ∇f on a surface $f(x, y, z) = c$.

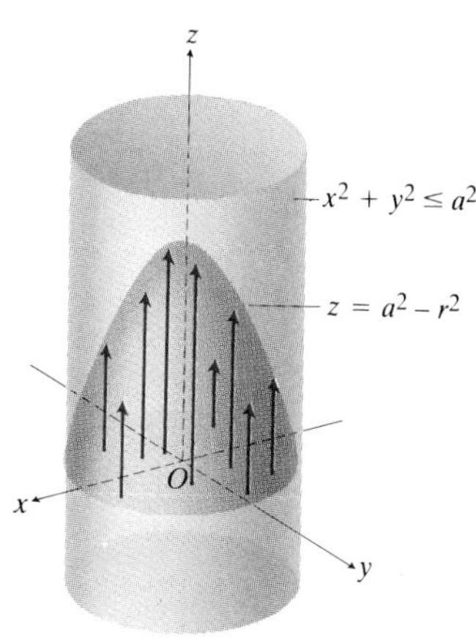

15.8 The flow of fluid in a long cylindrical pipe. The vectors $\mathbf{v} = (a^2 - r^2)\mathbf{k}$ inside the cylinder that have their bases in the xy-plane have their tips on the paraboloid $z = a^2 - r^2$.

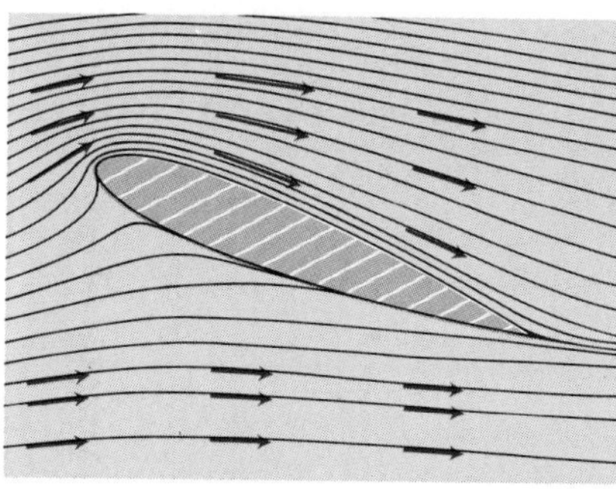

15.9 Velocity vectors of a flow around an airfoil in a wind tunnel. The streamlines were made visible by kerosene smoke. (Adapted from *NCFMF Book of Film Notes*, 1974; The MIT Press with Education Development Center, Inc., Newton, Massachusetts.)

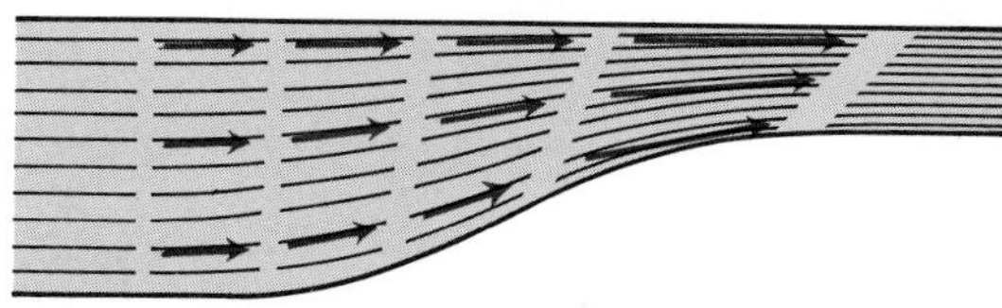

15.10 Streamlines in a contracting channel. The water speeds up as the channel narrows and the velocity vectors increase in length. (Adapted from *NCFMF Book of Film Notes*, 1974; The MIT Press with Education Development Center, Inc., Newton, Massachusetts.)

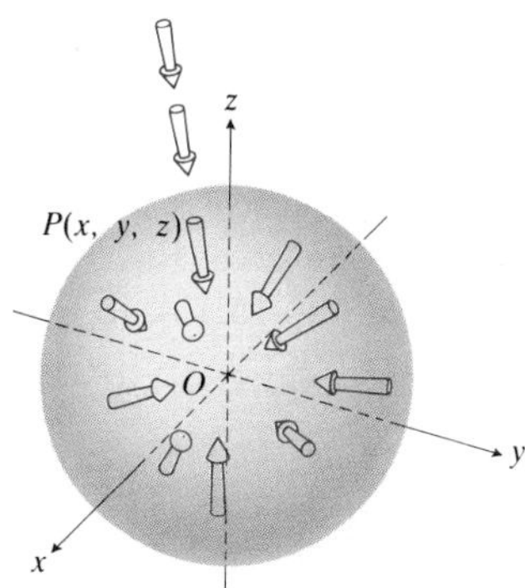

15.11 Vectors in the gravitational field

$$\mathbf{F} = -GM(x\mathbf{i} + y\mathbf{j} + z\mathbf{k})/(x^2 + y^2 + z^2)^{3/2}.$$

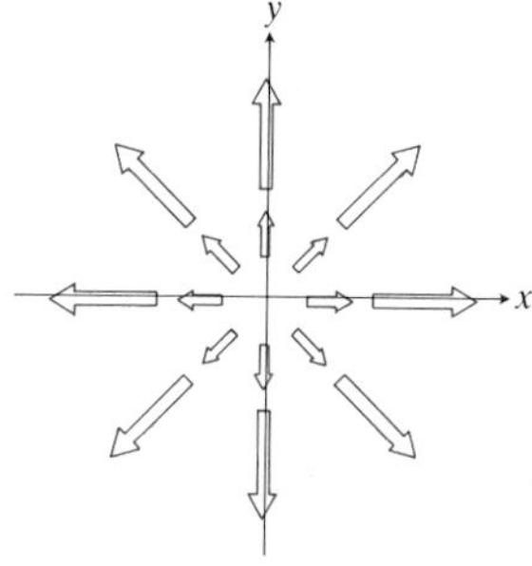

15.12 The field $\mathbf{F} = x\mathbf{i} + y\mathbf{j}$ of position vectors of points in the plane. Notice the convention that an arrow is drawn with its tail, not its head, at the point where $\mathbf{F}$ is evaluated.

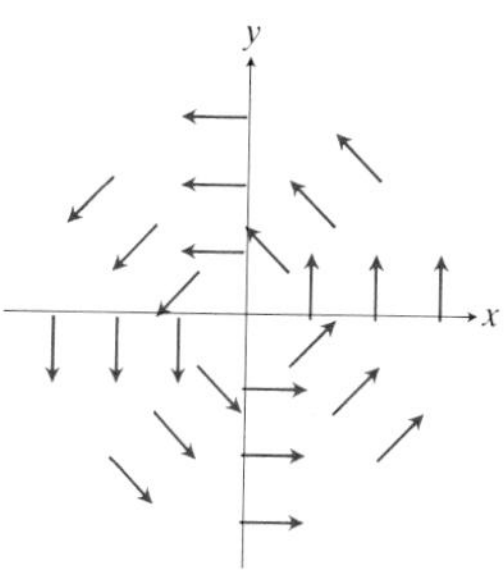

15.13 The two-dimensional field of unit vectors

$$\mathbf{F} = (-y\mathbf{i} + x\mathbf{j})/(x^2 + y^2)^{1/2}.$$

The field is not defined at the origin.

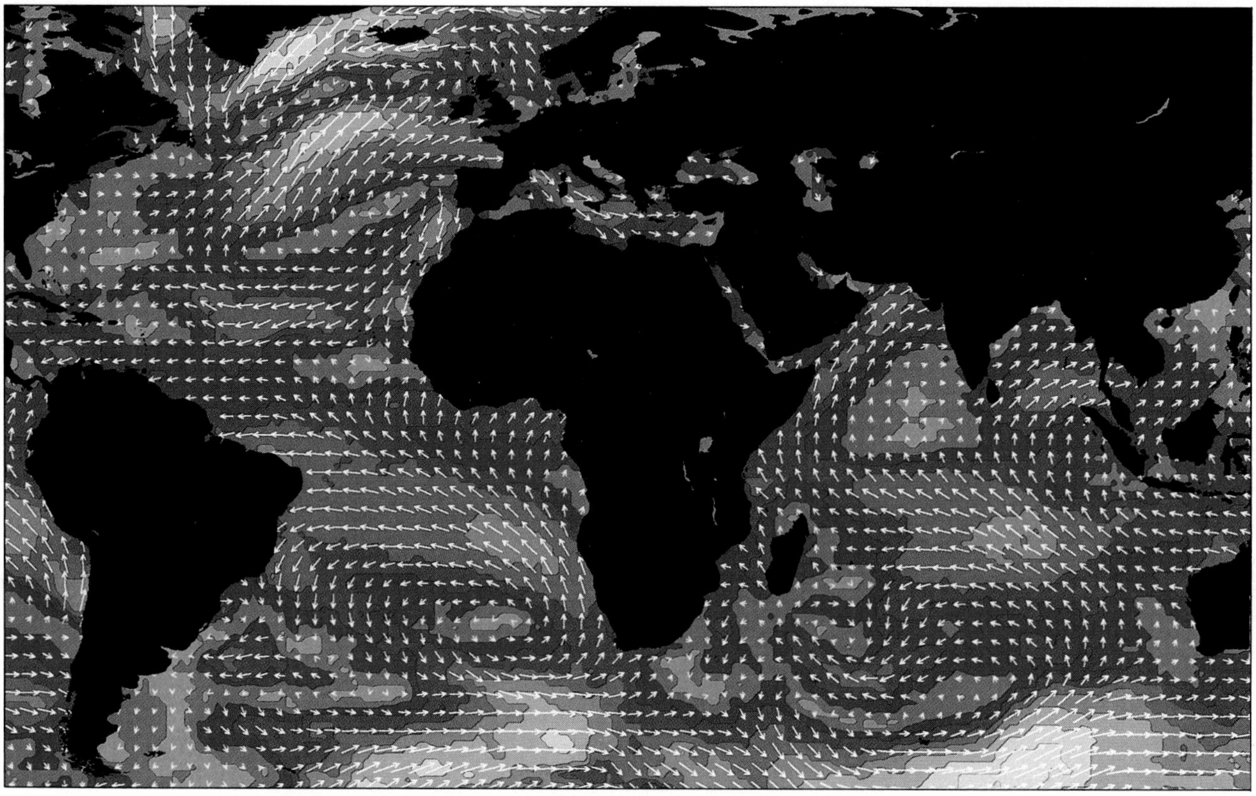

15.14 NASA's Seasat used radar during a 3-day period in September 1978 to take 350,000 wind measurements over the world's oceans. The arrows show wind direction; their length and the color contouring indicate speed. Notice the heavy storm south of Greenland.

The Work Done by a Force over a Curve in Space

Suppose that the vector field $\mathbf{F} = M(x, y, z)\mathbf{i} + N(x, y, z)\mathbf{j} + P(x, y, z)\mathbf{k}$ represents a force throughout a region in space (it might be the force of gravity or an electromagnetic force of some kind) and that

$$\mathbf{r} = g(t)\mathbf{i} + h(t)\mathbf{j} + k(t)\mathbf{k}, \quad a \le t \le b,$$

is a curve in the region. Then the line integral of $\mathbf{F} \cdot \mathbf{T}$, the scalar component of $\mathbf{F}$ in the direction of the curve's unit tangent vector, over the curve is called the work done by $\mathbf{F}$ over the curve from a to b (Fig. 15.15).

DEFINITION

The **work** done by a (continuous) force $\mathbf{F} = M(x, y, z)\mathbf{i} + N(x, y, z)\mathbf{j} + P(x, y, z)\mathbf{k}$ over a continuously differentiable curve $\mathbf{r} = g(t)\mathbf{i} + h(t)\mathbf{j} + k(t)\mathbf{k}$ from $t = a$ to $t = b$ is the value of the line integral

$$\text{Work} = \int_{t=a}^{t=b} \mathbf{F} \cdot \mathbf{T}\, ds. \tag{3}$$

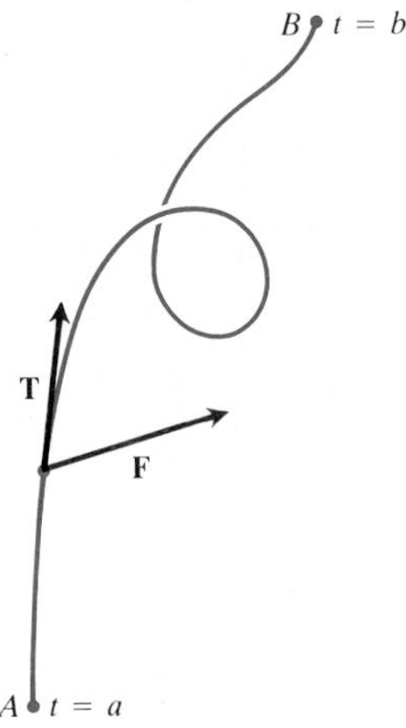

15.15 The work done by a continuous field **F** over a continuously differentiable path $\mathbf{r} = g(t)\mathbf{i} + h(t)\mathbf{j} + k(t)\mathbf{k}$ from A to B is the line integral of $\mathbf{F} \cdot \mathbf{T}$ from $t = a$ to $t = b$.

Notice how the numerical sign of the number we calculate with the integral in Eq. (3) depends on the direction in which the curve is traversed as t increases. If we reverse the direction, we reverse the direction of **T** and change the sign of $\mathbf{F} \cdot \mathbf{T}$.

Notation and Evaluation

There are six standard ways to write the work integral in Eq. (3).

Different Ways to Write the Work Integral

$$\text{Work} = \int_{t=a}^{t=b} \mathbf{F} \cdot \mathbf{T}\, ds \qquad \text{(The definition)}$$

$$= \int_{t=a}^{t=b} \mathbf{F} \cdot d\mathbf{r} \qquad \text{(Compact differential form)}$$

$$= \int_{a}^{b} \mathbf{F} \cdot \frac{d\mathbf{r}}{dt}\, dt \qquad \left(\begin{array}{l}\text{Expanded to include } dt.\\ \text{Emphasizes the}\\ \text{velocity vector } d\mathbf{r}/dt.\end{array}\right)$$

$$= \int_{a}^{b} \left(M \frac{dg}{dt} + N \frac{dh}{dt} + P \frac{dk}{dt}\right) dt \qquad \left(\begin{array}{l}\text{Emphasizes the}\\ \text{component functions}\end{array}\right)$$

$$= \int_{a}^{b} \left(M \frac{dx}{dt} + N \frac{dy}{dt} + P \frac{dz}{dt}\right) dt \qquad \left(\begin{array}{l}\text{Abbreviates the}\\ \text{components of } \mathbf{r}\end{array}\right)$$

$$= \int_{a}^{b} M\, dx + N\, dy + P\, dz \qquad \left(\begin{array}{l}dt\text{'s cancelled. The most}\\ \text{common differential form}\end{array}\right)$$

Despite their variety, these formulas are all evaluated the same way.

Evaluation

To evaluate the work integral, take these steps:

1. Evaluate **F** on the curve as a function of t.
2. Find $d\mathbf{r}/dt$.
3. Dot **F** with $d\mathbf{r}/dt$.
4. Integrate from $t = a$ to $t = b$.

EXAMPLE 1 Find the work done by the vector field

$$\mathbf{F} = (y - x^2)\mathbf{i} + (z - y^2)\mathbf{j} + (x - z^2)\mathbf{k}$$

over the curve

$$\mathbf{r} = t\mathbf{i} + t^2\mathbf{j} + t^3\mathbf{k}, \quad 0 \le t \le 1,$$

from (0, 0, 0) to (1, 1, 1) (Fig. 15.16).

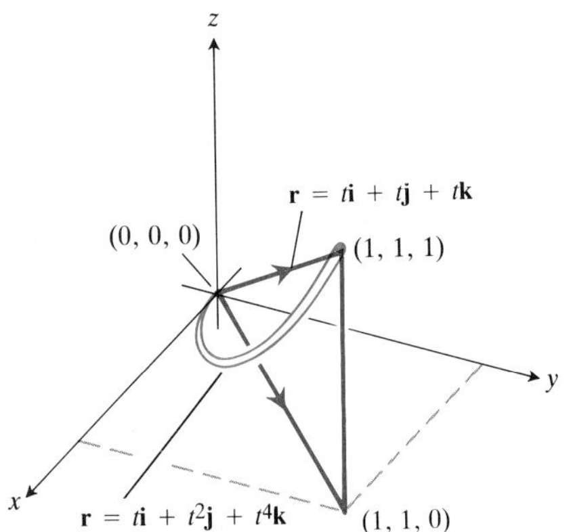

15.16 The curve in Example 1.

Solution

STEP 1: *Evaluate* $\mathbf{F}$ *on the curve.*

$$\mathbf{F} = (y - x^2)\mathbf{i} + (z - y^2)\mathbf{j} + (x - z^2)\mathbf{k} = \underbrace{(t^2 - t^2)}_{0}\mathbf{i} + (t^3 - t^4)\mathbf{j} + (t - t^6)\mathbf{k}$$

STEP 2: *Find* $d\mathbf{r}/dt$.

$$\frac{d\mathbf{r}}{dt} = \frac{d}{dt}(t\mathbf{i} + t^2\mathbf{j} + t^3\mathbf{k}) = \mathbf{i} + 2t\mathbf{j} + 3t^2\mathbf{k}$$

STEP 3: *Dot* $\mathbf{F}$ *with* $d\mathbf{r}/dt$.

$$\mathbf{F}\cdot\frac{d\mathbf{r}}{dt} = ((t^3 - t^4)\mathbf{j} + (t - t^6)\mathbf{k})\cdot(\mathbf{i} + 2t\mathbf{j} + 3t^2\mathbf{k})$$

$$= (t^3 - t^4)(2t) + (t - t^6)(3t^2) = 2t^4 - 2t^5 + 3t^3 - 3t^8$$

STEP 4: *Integrate from* $t = 0$ to $t = 1$.

$$\text{Work} = \int_0^1 (2t^4 - 2t^5 + 3t^3 - 3t^8)\,dt = \left[\frac{2}{5}t^5 - \frac{2}{6}t^6 + \frac{3}{4}t^4 - \frac{3}{9}t^9\right]_0^1 = \frac{29}{60}.$$ ■

Flow Integrals and Circulation

If instead of being a force field, the vector field $\mathbf{F} = M\mathbf{i} + N\mathbf{j} + P\mathbf{k}$ represents the velocity field of a fluid flowing through a region in space (a tidal basin, a river bed, or the turbine chamber of a hydroelectric generator, for example), then the line integral of $\mathbf{F}\cdot\mathbf{T}$ along a curve in the region is called the fluid's flow along the curve.

DEFINITIONS

If $\mathbf{r} = g(t)\mathbf{i} + h(t)\mathbf{j} + k(t)\mathbf{k}$, $a \le t \le b$, is a continuously differentiable curve in the domain of a continuous velocity field $\mathbf{F} = M(x, y, z)\mathbf{i} + N(x, y, z)\mathbf{j} + P(x, y, z)\mathbf{k}$, the **flow** along the curve from $t = a$ to $t = b$ is the integral of $\mathbf{F}\cdot\mathbf{T}$ over the curve from a to b,

$$\text{Flow} = \int_a^b \mathbf{F}\cdot\mathbf{T}\,ds. \tag{4}$$

The integral in this case is called a **flow integral.** If the curve is a closed loop, the flow is called the **circulation** around the curve.

We evaluate flow integrals the same way we evaluate work integrals.

EXAMPLE 2 A fluid's velocity field is

$$\mathbf{F} = x\mathbf{i} + z\mathbf{j} + y\mathbf{k}.$$

Find the flow along the helix

$$\mathbf{r} = (\cos t)\mathbf{i} + (\sin t)\mathbf{j} + t\mathbf{k}, \quad 0 \le t \le \pi/2.$$

Solution

STEP 1: *Evaluate* **F** *on the curve.*

$$\mathbf{F} = x\mathbf{i} + z\mathbf{j} + y\mathbf{k} = (\cos t)\mathbf{i} + t\mathbf{j} + (\sin t)\mathbf{k}$$

STEP 2: *Find* $d\mathbf{r}/dt$.

$$\frac{d\mathbf{r}}{dt} = (-\sin t)\mathbf{i} + (\cos t)\mathbf{j} + \mathbf{k}$$

STEP 3: *Find* $\mathbf{F} \cdot (d\mathbf{r}/dt)$.

$$\begin{aligned}\mathbf{F} \cdot \frac{d\mathbf{r}}{dt} &= (\cos t)(-\sin t) + (t)(\cos t) + (\sin t)(1)\\ &= -\sin t \cos t + t \cos t + \sin t\end{aligned}$$

STEP 4: *Integrate from* $t = a$ *to* $t = b$.

$$\begin{aligned}\text{Flow} &= \int_{t=a}^{t=b} \mathbf{F} \cdot \frac{d\mathbf{r}}{dt}\, dt = \int_0^{\pi/2} (-\sin t \cos t + t \cos t + \sin t)\, dt\\ &= \left[\frac{\cos^2 t}{2} + t \sin t\right]_0^{\pi/2} = \left(0 + \frac{\pi}{2}\right) - \left(\frac{1}{2} + 0\right) = \frac{\pi}{2} - \frac{1}{2}.\end{aligned}$$

■

EXAMPLE 3 Find the circulation of a fluid around the circle

$$\mathbf{r} = (\cos t)\mathbf{i} + (\sin t)\mathbf{j}, \quad 0 \le t \le 2\pi,$$

if the velocity field is $\mathbf{F} = (x - y)\mathbf{i} + x\mathbf{j}$.

Solution

1. On the circle, $\mathbf{F} = (x - y)\mathbf{i} + x\mathbf{j} = (\cos t - \sin t)\mathbf{i} + (\cos t)\mathbf{j}$.

2. $\dfrac{d\mathbf{r}}{dt} = (-\sin t)\mathbf{i} + (\cos t)\mathbf{j}$

3. $\mathbf{F} \cdot \dfrac{d\mathbf{r}}{dt} = -\sin t \cos t + \underbrace{\sin^2 t + \cos^2 t}_{1}$

4. $$\begin{aligned}\text{Circulation} &= \int_0^{2\pi} \mathbf{F} \cdot \frac{d\mathbf{r}}{dt}\, dt = \int_0^{2\pi} (1 - \sin t \cos t)\, dt\\ &= \left[t - \frac{\sin^2 t}{2}\right]_0^{2\pi} = 2\pi.\end{aligned}$$

■

Flux across a Plane Curve

To find the rate at which a fluid is entering or leaving a region enclosed by a curve C in the xy-plane, we calculate the line integral over C of $\mathbf{F} \cdot \mathbf{n}$, the scalar component of the fluid's velocity field in the direction of the curve's outward-pointing normal vector. The value of this integral is called the flux of **F** across C. **Flux** is Latin for *flow,* but many flux calculations involve no motion at all. If **F** were an electric field or a magnetic field, for instance, the integral of $\mathbf{F} \cdot \mathbf{n}$ would still be called the flux of the field across C.

DEFINITION

If C is a continuously differentiable closed curve in the domain of a continuous vector field $\mathbf{F} = M(x, y, z)\mathbf{i} + N(x, y, z)\mathbf{j}$ in the plane, and if $\mathbf{n}$ is the outward-pointing unit normal vector on C, the **flux** of $\mathbf{F}$ across C is given by the following line integral:

$$\text{Flux of } \mathbf{F} \text{ across } C = \int_C \mathbf{F} \cdot \mathbf{n}\, ds. \tag{5}$$

Notice the difference between flux and circulation. The flux of $\mathbf{F}$ across C is the line integral with respect to arc length of $\mathbf{F} \cdot \mathbf{n}$, the scalar component of $\mathbf{F}$ in the direction of the outward normal. The circulation of $\mathbf{F}$ around C is the line integral with respect to arc length of $\mathbf{F} \cdot \mathbf{T}$, the scalar component of $\mathbf{F}$ in the direction of the unit tangent vector. Flux is the integral of the normal component of $\mathbf{F}$; circulation is the integral of the tangential component of $\mathbf{F}$.

To evaluate the integral in (5), we begin with a parametrization

$$x = g(t), \qquad y = h(t), \qquad a \le t \le b,$$

that traces the curve C exactly once as t increases from a to b. We can find the outward unit normal vector $\mathbf{n}$ by crossing the curve's unit tangent vector $\mathbf{T}$ with the vector $\mathbf{k}$. But which order do we choose, $\mathbf{T} \times \mathbf{k}$ or $\mathbf{k} \times \mathbf{T}$? Which one points outward? It depends on which way C is traversed as the parameter t increases. If the motion is clockwise, then $\mathbf{k} \times \mathbf{T}$ points outward; if the motion is counterclockwise, then $\mathbf{T} \times \mathbf{k}$ points outward (Fig. 15.17). The usual choice is $\mathbf{n} = \mathbf{T} \times \mathbf{k}$, the choice that assumes counterclockwise motion. Thus, while the value of the arc-length integral in the definition of flux in Eq. (5) does not depend on which way C is traversed, the formulas we are about to derive for evaluating the integral in Eq. (5) will assume counterclockwise motion.

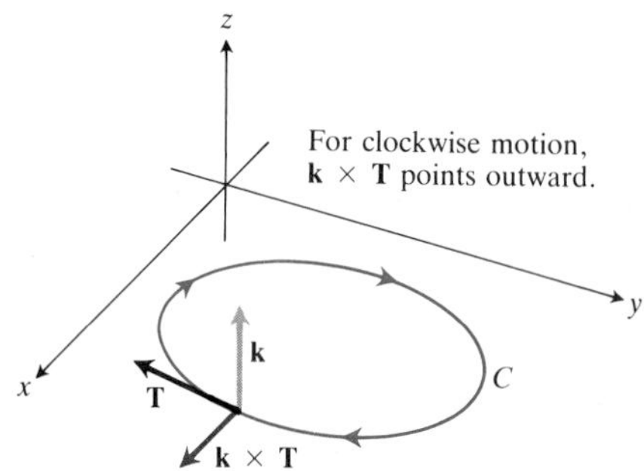

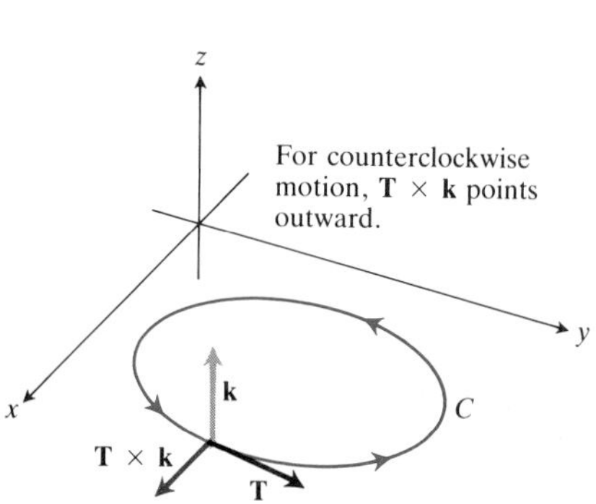

15.17 To find a unit outward normal vector for a curve C in the xy-plane that is traversed counterclockwise as t increases, we take $\mathbf{n} = \mathbf{T} \times \mathbf{k}$.

In terms of components,

$$\mathbf{n} = \mathbf{T} \times \mathbf{k} = \left(\frac{dx}{ds}\mathbf{i} + \frac{dy}{ds}\mathbf{j}\right) \times \mathbf{k} = \frac{dy}{ds}\mathbf{i} - \frac{dx}{ds}\mathbf{j}. \tag{6}$$

If $\mathbf{F} = M(x, y)\mathbf{i} + N(x, y)\mathbf{j}$, then

$$\mathbf{F} \cdot \mathbf{n} = M(x, y)\frac{dy}{ds} - N(x, y)\frac{dx}{ds}.$$

Hence,

$$\int_C \mathbf{F} \cdot \mathbf{n}\, ds = \int_C \left(M\frac{dy}{ds} - N\frac{dx}{ds}\right) ds = \oint_C M\, dy - N\, dx. \tag{7}$$

We put a directed circle ⟲ on the last integral as a reminder that the integration around the closed curve C is to be in the counterclockwise direction. To evaluate this integral, we express M, dy, N, and dx in terms of t and integrate from $t = a$ to $t = b$. Thus, strange as it may seem, we do not need to know either $\mathbf{n}$ or ds to find the flux.

The Formula for Calculating Flux across a Closed Plane Curve

$$\text{Flux of } \mathbf{F} = M\mathbf{i} + N\mathbf{j} \text{ across } C = \oint M\,dy - N\,dx \tag{8}$$

The integral can be evaluated from any parametrization $x = g(t)$, $y = h(t)$, $a \le t \le b$ that traces C counterclockwise exactly once. To evaluate the integral, express M, dy, N, and dx in terms of t and integrate from $t = a$ to $t = b$.

EXAMPLE 4 Find the flux of $\mathbf{F} = (x - y)\mathbf{i} + x\mathbf{j}$ across the circle $x^2 + y^2 = 1$ in the xy-plane.

Solution The parametrization

$$\mathbf{r} = (\cos t)\mathbf{i} + (\sin t)\mathbf{j}, \quad 0 \le t \le 2\pi,$$

traces the circle counterclockwise exactly once. We can therefore use this parametrization in Eq. (8). With

$$M = x - y = \cos t - \sin t \qquad dy = d(\sin t) = \cos t\,dt$$

$$N = x = \cos t \qquad dx = d(\cos t) = -\sin t\,dt,$$

we find

$$\text{Flux} = \int_C M\,dy - N\,dx = \int_0^{2\pi} (\cos^2 t - \sin t \cos t + \cos t \sin t)\,dt \qquad \text{(Eq. 8)}$$

$$= \int_0^{2\pi} \cos^2 t\,dt = \int_0^{2\pi} \frac{1 + \cos 2t}{2}\,dt = \left[\frac{t}{2} + \frac{\sin 2t}{4}\right]_0^{2\pi} = \pi.$$

The flux of $\mathbf{F}$ across the circle is π. Since the answer is positive, the net flow across the curve is outward. A net inward flow would give a negative flux. ■

Exercises 15.2

In Exercises 1–6, find the work done by the force $\mathbf{F}$ from $(0, 0, 0)$ to $(1, 1, 1)$ over each of the following paths:

a) the line segment $\mathbf{r} = t\mathbf{i} + t\mathbf{j} + t\mathbf{k}$, $0 \le t \le 1$

b) the curve $\mathbf{r} = t\mathbf{i} + t^2\mathbf{j} + t^4\mathbf{k}$, $0 \le t \le 1$

c) the path consisting of the line segment from $(0, 0, 0)$ to $(1, 1, 0)$ followed by the line segment from $(1, 1, 0)$ to $(1, 1, 1)$

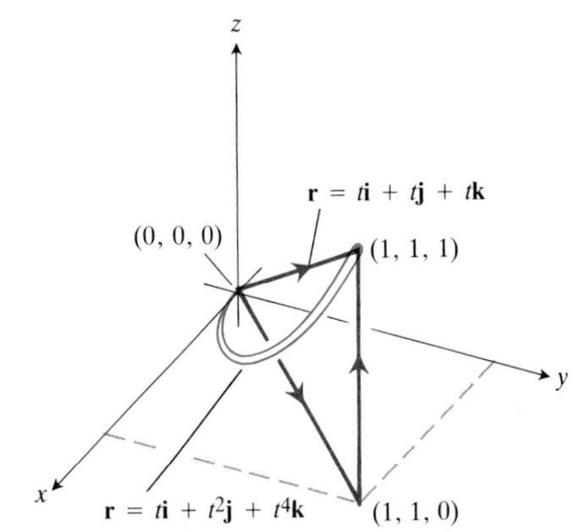

1. $\mathbf{F} = 3y\mathbf{i} + 2x\mathbf{j} + 4z\mathbf{k}$

2. $\mathbf{F} = (1/(x^2 + 1))\mathbf{j}$

3. $\mathbf{F} = \sqrt{z}\mathbf{i} - 2x\mathbf{j} + \sqrt{y}\mathbf{k}$

4. $\mathbf{F} = xy\mathbf{i} + yz\mathbf{j} + xz\mathbf{k}$

5. $\mathbf{F} = (3x^2 - 3x)\mathbf{i} + 3z\mathbf{j} + \mathbf{k}$

6. $\mathbf{F} = (y + z)\mathbf{i} + (z + x)\mathbf{j} + (x + y)\mathbf{k}$

In Exercises 7–10, find the work done by $\mathbf{F}$ over the curve in the direction of increasing t.

7. $\mathbf{F} = xy\mathbf{i} + y\mathbf{j} - yz\mathbf{k}$
$\mathbf{r} = t\mathbf{i} + t^2\mathbf{j} + t\mathbf{k}, \quad 0 \le t \le 1$

8. $\mathbf{F} = 2y\mathbf{i} + 3x\mathbf{j} + (x + y)\mathbf{k}$
$\mathbf{r} = (\cos t)\mathbf{i} + (\sin t)\mathbf{j} + (t/6)\mathbf{k}, \quad 0 \le t \le 2\pi$

9. $\mathbf{F} = z\mathbf{i} + x\mathbf{j} + y\mathbf{k}$
$\mathbf{r} = (\sin t)\mathbf{i} + (\cos t)\mathbf{j} + t\mathbf{k}, \quad 0 \le t \le 2\pi$

10. $\mathbf{F} = 6z\mathbf{i} + y^2\mathbf{j} + 12x\mathbf{k}$
$\mathbf{r} = (\sin t)\mathbf{i} + (\cos t)\mathbf{j} + (t/6)\mathbf{k}, \quad 0 \le t \le 2\pi$

In Exercises 11–14, $\mathbf{F}$ is the velocity field of a fluid flowing through a region in space. Find the flow along the given curve in the direction of increasing t.

11. $\mathbf{F} = -4xy\mathbf{i} + 8y\mathbf{j} + 2\mathbf{k}$
$\mathbf{r} = t\mathbf{i} + t^2\mathbf{j} + \mathbf{k}, \quad 0 \le t \le 2$

12. $\mathbf{F} = x^2\mathbf{i} + yz\mathbf{j} + y^2\mathbf{k}$
$\mathbf{r} = 3t\mathbf{j} + 4t\mathbf{k}, \quad 0 \le t \le 1$

13. $\mathbf{F} = (x - z)\mathbf{i} + x\mathbf{k}$
$\mathbf{r} = (\cos t)\mathbf{i} + (\sin t)\mathbf{k}, \quad 0 \le t \le \pi$

14. $\mathbf{F} = -y\mathbf{i} + x\mathbf{j} + 2\mathbf{k}$
$\mathbf{r} = (-2\cos t)\mathbf{i} + (2\sin t)\mathbf{j} + 2t\mathbf{k}, \quad 0 \le t \le 2\pi$

15. Find the circulation and flux of the fields

$$\mathbf{F}_1 = x\mathbf{i} + y\mathbf{j} \quad \text{and} \quad \mathbf{F}_2 = -y\mathbf{i} + x\mathbf{j}$$

around and across each of the following curves.
a) The circle $\mathbf{r} = (\cos t)\mathbf{i} + (\sin t)\mathbf{j}, \quad 0 \le t \le 2\pi$
b) The ellipse $\mathbf{r} = (\cos t)\mathbf{i} + (4\sin t)\mathbf{j}, \quad 0 \le t \le 2\pi$

16. Find the flux of the fields

$$\mathbf{F}_1 = 2x\mathbf{i} - 3y\mathbf{j} \quad \text{and} \quad \mathbf{F}_2 = 2x\mathbf{i} + (x - y)\mathbf{j}$$

across the circle

$$\mathbf{r} = (a\cos t)\mathbf{i} + (a\sin t)\mathbf{j}, \quad 0 \le t \le 2\pi.$$

In Exercises 17–20, find the circulation and flux of the field $\mathbf{F}$ around and across the closed semicircular path that consists of the semicircular arch $\mathbf{r}_1 = (a\cos t)\mathbf{i} + (a\sin t)\mathbf{j}$, $0 \le t \le \pi$, followed by the line segment $\mathbf{r}_2 = t\mathbf{i}$, $-a \le t \le a$.

17. $\mathbf{F} = x\mathbf{i} + y\mathbf{j}$

18. $\mathbf{F} = x^2\mathbf{i} + y^2\mathbf{j}$

19. $\mathbf{F} = -y\mathbf{i} + x\mathbf{j}$

20. $\mathbf{F} = -y^2\mathbf{i} + x^2\mathbf{j}$

21. Suppose that the function $f(t)$ is continuous and positive for $a \le t \le b$, that $\mathbf{r} = t\mathbf{i} + f(t)\mathbf{j}$, $a \le t \le b$, and that $\mathbf{F} = y\mathbf{i}$. Show that

$$\int_{t=a}^{t=b} \mathbf{F} \cdot d\mathbf{r}$$

is the area between the graph of f and the t-axis from $t = a$ to $t = b$. (*Hint:* Evaluate the integral.)

TOOLKIT PROGRAM

Vector Fields	Pictures fields of the form $\mathbf{F} = M(x, y)\mathbf{i} + N(x, y)\mathbf{j}$ and integrates $\mathbf{F} \cdot \mathbf{n}$ over polygonal paths.

15.3 Green's Theorem in the Plane

We now come to a theorem that describes the relationship between the way a fluid flows along or across the boundary of a plane region and the way the fluid moves around inside the region. We assume the fluid is incompressible—like water, for example, and not like a gas. The connection between the fluid's boundary behavior and its internal behavior is made possible by the notions of divergence and curl. The divergence of a fluid's velocity field is a measure of the rate at which fluid is being piped into or out of the region at any given point. The curl of the velocity field is a measure of the fluid's rate of rotation at each point. We'll be more precise about all this in a moment.

Green's theorem states that, under conditions usually met in practice, the outward flux of a vector field across the boundary of a plane region equals the double integral of the divergence of the field over the interior of the region. In another form, Green's theorem states that the counterclockwise circulation of a vector field around the boundary of a region equals the double integral of the curl of the vector field over the region.

The Green of Green's Theorem

The theorem is due to George Green (1793–1841), a self-taught mathematician who sought to use mathematics in his treatment of static electricity and magnetism. He published his work privately in 1828 as "An Essay on the Application of Analysis to the Theories of Electricity and Magnetism." This profound essay treated line and surface integrals as well, though Gauss's work in 1840 made the explicit connection of line integrals with work. Green's work went unnoticed until the 1850s, after Green's death, when the physicist William Thomson (later Lord Kelvin) published the work. Green's theorem was proved simultaneously by the Russian mathematician Mikhail Ostrogradsky in a paper delivered to the St. Petersburg Academy of Sciences in 1828. Of the two, Green is perhaps better known for this theorem because he inspired later mathematical physicists at Cambridge, such as William Thomson, Gabriel Stokes, Lord Rayleigh, and James Clerk Maxwell. Green was the first great English mathematician to join the mathematical work on the Continent in the eighteenth century, after the split between Newton and Leibniz, and to reintroduce the work into England.

Green's theorem is one of the great theorems of calculus. It is deep, surprising, and has far-reaching consequences. In pure mathematics, it ranks in importance with the Fundamental Theorems of calculus. In applied mathematics the generalizations of Green's theorem to three dimensions provide the foundation for important theorems about electricity, magnetism, and fluid flow. We shall explore the three-dimensional forms of Green's theorem in Sections 15.5 and 15.6.

Throughout our discussion of Green's theorem, we talk in terms of velocity fields of fluid flows. We do so because fluid flows are easy to picture and the notions of flux and circulation are easy to interpret. We would like you to be aware, however, that Green's theorem applies to any vector field satisfying certain mathematical conditions. It does not depend for its validity on the field's having a particular physical interpretation.

Flux Density at a Point. Divergence

We need two new ideas for Green's theorem. The first is the idea of the flux density of a vector field at a point, which in mathematics is called the *divergence* of the vector field. We obtain it in the following way.

Suppose that

$$\mathbf{F}(x, y) = M(x, y)\mathbf{i} + N(x, y)\mathbf{j}$$

is the velocity field of a fluid flow in the plane and that the first partial derivatives of M and N are continuous at each point of a region R. Let (x, y) be a point in R and let A be a small rectangle with one corner at (x, y) that, along with its interior, lies entirely in R (Fig. 15.18). The sides of the rectangle, parallel to the coordinate axes, have lengths Δx and Δy. The rate at which fluid leaves the rectangle across the bottom edge is approximately

$$\mathbf{F}(x, y) \cdot (-\mathbf{j})\,\Delta x = -N(x, y)\,\Delta x. \tag{1}$$

This is the scalar component of the velocity at (x, y) in the direction of the outward normal times the length of the segment. If the velocity is in meters per second, for example, the exit rate will be in meters per second times meters or square meters per second. The rates at which the fluid crosses the other three sides in the directions of their outward normals can be estimated in a similar way. All told, we have

$$\begin{aligned}
&\text{Top:} && \mathbf{F}(x, y + \Delta y) \cdot \mathbf{j}\,\Delta x = N(x, y + \Delta y)\,\Delta x,\\
&\text{Bottom:} && \mathbf{F}(x, y) \cdot (-\mathbf{j})\,\Delta x = -N(x, y)\,\Delta x,\\
&\text{Right:} && \mathbf{F}(x + \Delta x, y) \cdot \mathbf{i}\,\Delta y = M(x + \Delta x, y)\,\Delta y,\\
&\text{Left:} && \mathbf{F}(x, y) \cdot (-\mathbf{i})\,\Delta y = -M(x, y)\,\Delta y.
\end{aligned} \tag{2}$$

Combining opposite pairs gives

$$\text{Top and bottom:} \quad (N(x, y + \Delta y) - N(x, y))\,\Delta x \approx \left(\frac{\partial N}{\partial y}\,\Delta y\right)\Delta x, \tag{3}$$

$$\text{Right and left:} \quad (M(x + \Delta x, y) - M(x, y))\,\Delta y \approx \left(\frac{\partial M}{\partial x}\,\Delta x\right)\Delta y. \tag{4}$$

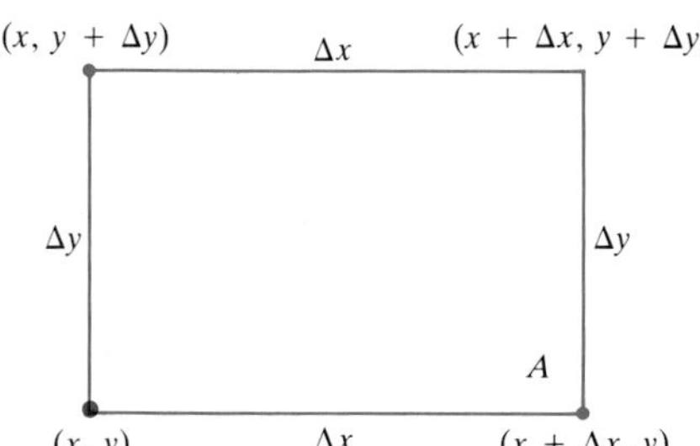

15.18 The rectangle for defining the flux density (divergence) of a vector field at a point (x, y).

Adding (3) and (4) gives

$$\text{Flux across rectangle boundary} \approx \left(\frac{\partial M}{\partial x} + \frac{\partial N}{\partial y}\right)\Delta x\,\Delta y. \tag{5}$$

We now divide by $\Delta x\,\Delta y$ to estimate the total flux per unit area or flux density for the rectangle:

$$\frac{\text{Flux across rectangle boundary}}{\text{Rectangle area}} \approx \left(\frac{\partial M}{\partial x} + \frac{\partial N}{\partial y}\right). \tag{6}$$

Finally, we let Δx and Δy approach zero to define what we call the *flux density* of **F** at the point (x, y).

In mathematics, we call the flux density the *divergence* of **F**. The symbol for it is div **F**, pronounced "divergence of **F**" or "div **F**."

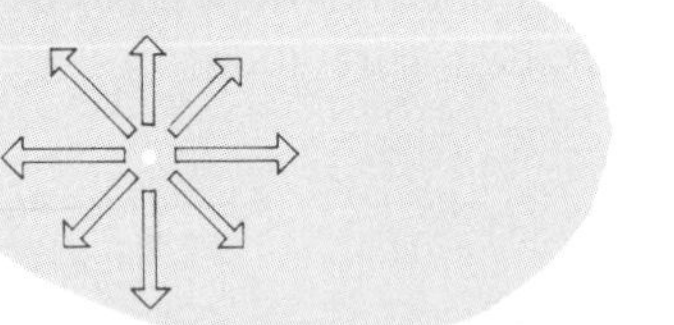

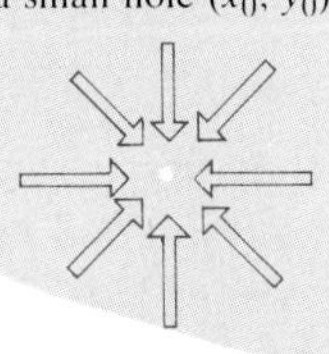

15.19 In the flow of an incompressible fluid across a plane region, the divergence is positive at a "source," a point where fluid enters the system, and negative at a "sink," a point where the fluid leaves the system.

DEFINITION

The **flux density** or **divergence** of a vector field $\mathbf{F} = M\mathbf{i} + N\mathbf{j}$ at the point (x, y) is

$$\text{div}\,\mathbf{F} = \frac{\partial M}{\partial x} + \frac{\partial N}{\partial y}. \tag{7}$$

If water were flowing into a region through a small hole at the point (x_0, y_0), the lines of flow would diverge there (hence the name) and div $\mathbf{F}(x_0, y_0)$ would be positive. If the water were draining out instead of flowing in, the divergence would be negative. See Fig. 15.19.

EXAMPLE 1 Find the divergence of the vector field

$$\mathbf{F}(x, y) = (x^2 - y)\mathbf{i} + (xy - y^2)\mathbf{j}.$$

Solution We use the formula in Eq. (7):

$$\begin{aligned}\text{div}\,\mathbf{F} &= \frac{\partial M}{\partial x} + \frac{\partial N}{\partial y} = \frac{\partial}{\partial x}(x^2 - y) + \frac{\partial}{\partial y}(xy - y^2)\\ &= 2x + x - 2y = 3x - 2y.\end{aligned}$$

■

Circulation Density at a Point. The Curl

The second of the two new ideas we need for Green's theorem is the idea of circulation density of a vector field **F** at a point, which in mathematics is called the *curl* of **F**. To obtain it, we return to the velocity field

$$\mathbf{F}(x, y) = M(x, y)\mathbf{i} + N(x, y)\mathbf{j} \tag{8}$$

and the rectangle A. The rectangle is redrawn here as Fig. 15.20.

The counterclockwise circulation of **F** around the boundary of A is the sum of flow rates along the sides. For the bottom edge, the flow rate is approximately

$$\mathbf{F}(x, y) \cdot \mathbf{i}\,\Delta x = M(x, y)\,\Delta x. \tag{9}$$

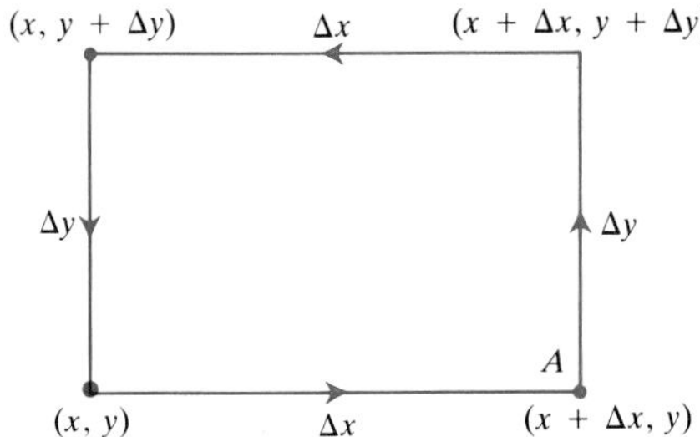

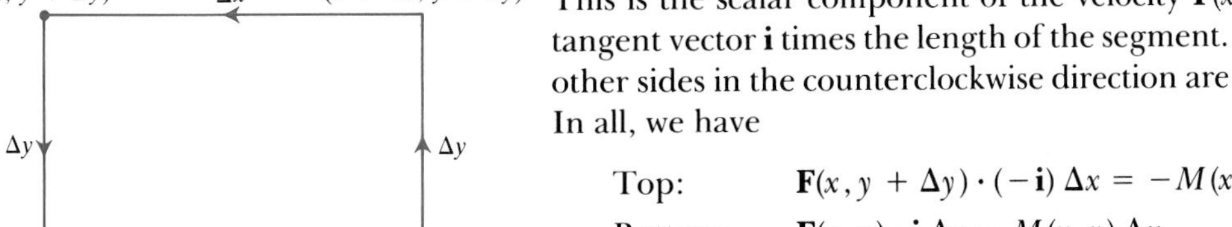

15.20 The rectangle for defining the circulation density (curl) of a vector field at a point (x, y).

This is the scalar component of the velocity $\mathbf{F}(x, y)$ in the direction of the tangent vector $\mathbf{i}$ times the length of the segment. The rates of flow along the other sides in the counterclockwise direction are expressed in a similar way. In all, we have

$$\begin{aligned} &\text{Top:} && \mathbf{F}(x, y+\Delta y)\cdot(-\mathbf{i})\,\Delta x = -M(x, y+\Delta y)\,\Delta x,\\ &\text{Bottom:} && \mathbf{F}(x, y)\cdot\mathbf{i}\,\Delta x = M(x, y)\,\Delta x,\\ &\text{Right:} && \mathbf{F}(x+\Delta x, y)\cdot\mathbf{j}\,\Delta y = N(x+\Delta x, y)\,\Delta y,\\ &\text{Left:} && \mathbf{F}(x, y)\cdot(-\mathbf{j})\,\Delta y = -N(x, y)\,\Delta y. \end{aligned} \tag{10}$$

We add opposite pairs to get

Top and bottom:

$$-(M(x, y+\Delta y) - M(x, y))\,\Delta x \approx -\left(\frac{\partial M}{\partial y}\,\Delta y\right)\Delta x, \tag{11}$$

Right and left:

$$(N(x+\Delta x, y) - N(x, y))\,\Delta y \approx \left(\frac{\partial N}{\partial x}\,\Delta x\right)\Delta y. \tag{12}$$

Adding (11) and (12) and dividing by $\Delta x\,\Delta y$ gives an estimate of the circulation density for the rectangle:

$$\frac{\text{Circulation around rectangle}}{\text{Rectangle area}} \approx \frac{\partial N}{\partial x} - \frac{\partial M}{\partial y}. \tag{13}$$

Finally, we let Δx and Δy approach zero to define what we call the *circulation density* of $\mathbf{F}$ at the point (x, y).

Vertical axis

(x_0, y_0)

curl $\mathbf{F}(x_0, y_0) > 0$
Counterclockwise circulation

DEFINITION

The **circulation density** or **curl** of a vector field $\mathbf{F} = M\mathbf{i} + N\mathbf{j}$ at the point (x, y) is

$$\text{curl}\,\mathbf{F} = \frac{\partial N}{\partial x} - \frac{\partial M}{\partial y}. \tag{14}$$

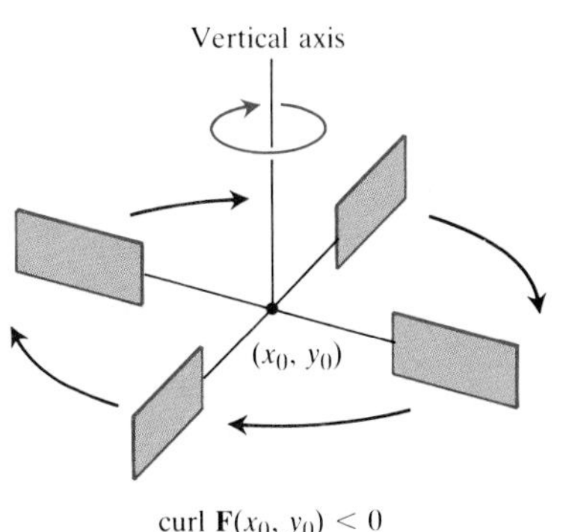

curl $\mathbf{F}(x_0, y_0) < 0$
Clockwise circulation

15.21 In the flow of an incompressible fluid over a plane region, the curl measures the rate of the fluid's rotation at a point. The curl is positive at points where the rotation is counterclockwise and negative where the rotation is clockwise.

If water is moving about a region in the xy-plane in a thin layer, then the circulation, or curl, at a point (x_0, y_0) gives a way to measure how fast and in what direction a small paddle wheel will spin if it is put into the water at (x_0, y_0) with its axis perpendicular to the plane (Fig. 15.21).

EXAMPLE 2 Find the curl of the vector field

$$\mathbf{F}(x, y) = (x^2 - y)\mathbf{i} + (xy - y^2)\mathbf{j}.$$

Solution We use the formula in Eq. (14):

$$\text{curl}\,\mathbf{F} = \frac{\partial N}{\partial x} - \frac{\partial M}{\partial y} = \frac{\partial}{\partial x}(xy - y^2) - \frac{\partial}{\partial y}(x^2 - y) = y + 1.$$ ■

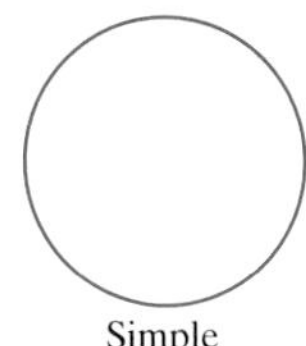
Simple

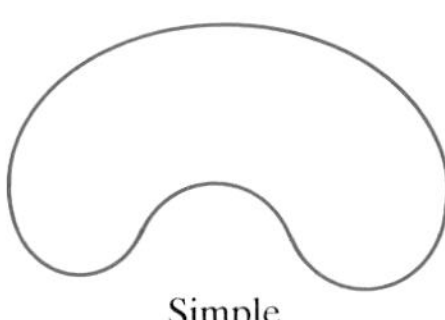
Simple

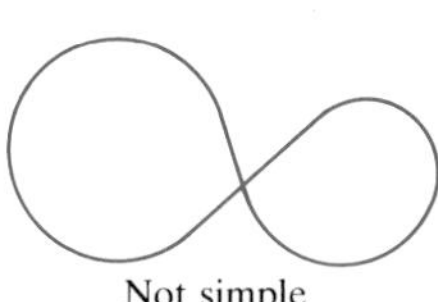
Not simple

15.22 In proving Green's theorem, we distinguish between two kinds of closed curves, simple and not simple. Simple curves do not cross themselves. A circle is simple but a figure eight is not.

Green's Theorem in the Plane

In one form, Green's theorem says that under suitable conditions the outward flux of a vector field across a simple (Fig. 15.22) closed curve in the plane equals the double integral of the divergence of the field over the region enclosed by the curve.

Green's Theorem (Flux–Divergence Form)

The outward flux of a vector field $\mathbf{F} = M\mathbf{i} + N\mathbf{j}$ across a simple closed curve C equals the double integral of div $\mathbf{F}$ over the region R enclosed by C.

$$\underbrace{\oint_C M\,dy - N\,dx}_{\text{Outward flux}} = \underbrace{\iint_R \left(\frac{\partial M}{\partial x} + \frac{\partial N}{\partial y}\right) dx\,dy}_{\text{Divergence integral}} \tag{15}$$

In another form, Green's theorem says that the counterclockwise circulation of a vector field around a simple closed curve is the double integral of the curl of the field over the region enclosed by the curve.

Green's Theorem (Circulation–Curl Form)

The counterclockwise circulation of a vector field $\mathbf{F} = M\mathbf{i} + N\mathbf{j}$ around a simple closed curve C in the plane equals the double integral of curl $\mathbf{F}$ over the region enclosed by C.

$$\underbrace{\oint_C M\,dx + N\,dy}_{\text{Counterclockwise circulation}} = \underbrace{\iint_R \left(\frac{\partial N}{\partial x} - \frac{\partial M}{\partial y}\right) dx\,dy}_{\text{Curl integral}} \tag{16}$$

The two forms of Green's theorem are equivalent. Applying Eq. (15) to the field $\mathbf{G}_1 = N\mathbf{i} - M\mathbf{j}$ gives Eq. (16), and applying Eq. (16) to $\mathbf{G}_2 = -N\mathbf{i} + M\mathbf{j}$ gives Eq. (15). We do not need to prove them both. We shall prove the curl form shortly.

We need two kinds of assumptions for Green's theorem to hold. First, we need conditions on M and N to ensure the existence of the integrals. The usual assumptions are that M, N, and their first partial derivatives are continuous at every point of some region containing C and R. Second, we need geometric conditions on the curve C. It must be simple, closed, and made up of pieces along which we can integrate M and N. The usual assumptions are that C consists of a finite number of continuously differentiable curves connected end to end. The proof we give for Green's theorem, however, assumes things about the shape of R as well. You can find less restrictive proofs in more advanced texts.

EXAMPLE 3 Verify both forms of Green's theorem for the field

$$\mathbf{F}(x, y) = (x - y)\mathbf{i} + x\mathbf{j}$$

and the region R bounded by the circle

$$C: \quad \mathbf{r} = (\cos t)\mathbf{i} + (\sin t)\mathbf{j}, \qquad 0 \le t \le 2\pi.$$

Solution We first express all functions, derivatives, and differentials in terms of t:

$$\begin{aligned} M &= \cos t - \sin t, & dx &= d(\cos t) = -\sin t\, dt, \\ N &= \cos t, & dy &= d(\sin t) = \cos t\, dt, \\ \frac{\partial M}{\partial x} &= 1, \quad \frac{\partial M}{\partial y} = -1, & \frac{\partial N}{\partial x} &= 1, \quad \frac{\partial N}{\partial y} = 0. \end{aligned} \tag{17}$$

The two sides of Eq. (15):

$$\begin{aligned} \oint_C M\, dy - N\, dx &= \int_{t=0}^{t=2\pi} (\cos t - \sin t)(\cos t\, dt) - (\cos t)(-\sin t\, dt) \\ &= \int_0^{2\pi} \cos^2 t\, dt = \pi, \end{aligned}$$

$$\iint_R \left(\frac{\partial M}{\partial x} + \frac{\partial N}{\partial y}\right) dx\, dy = \iint_R (1 + 0)\, dx\, dy = \iint_R dx\, dy = \pi.$$

The two sides of Eq. (16):

$$\begin{aligned} \oint_C M\, dx + N\, dy &= \int_{t=0}^{t=2\pi} (\cos t - \sin t)(-\sin t\, dt) + (\cos t)(\cos t\, dt) \\ &= \int_0^{2\pi} (-\sin t \cos t + 1)\, dt = 2\pi, \end{aligned}$$

$$\iint_R \left(\frac{\partial N}{\partial x} - \frac{\partial M}{\partial y}\right) dx\, dy = \iint_R (1 - (-1))\, dx\, dy = 2\iint_R dx\, dy = 2\pi. \quad \blacksquare$$

Using Green's Theorem to Evaluate Line Integrals

If we construct a closed curve C by piecing a number of different curves end to end, the process of evaluating a line integral over C can be a lengthy one because there are so many different integrals to evaluate. However, if C bounds a region R to which Green's theorem applies, we can use Green's theorem to change the line integral around C into a single double integral over R.

EXAMPLE 4 Evaluate the integral

$$\oint xy\, dy - y^2\, dx$$

around the square cut from the first quadrant by the lines $x = 1$ and $y = 1$.

Solution We can use either form of Green's theorem to change the line integral into a double integral over the square.

1. *With Eq. (15):* Taking $M = xy$, $N = y^2$, and C and R as the square's boundary and interior gives

$$\oint xy\,dy - y^2\,dx = \iint_R (y + 2y)\,dx\,dy = \int_0^1\int_0^1 3y\,dx\,dy$$

$$= \int_0^1 \left[3xy\right]_{x=0}^{x=1} dy = \int_0^1 3y\,dy = \frac{3}{2}y^2\bigg]_0^1 = \frac{3}{2}.$$

2. *With Eq. (16):* Taking $M = -y^2$ and $N = xy$ gives the same result:

$$\oint_C -y^2\,dx + xy\,dy = \iint_R (y - (-2y))\,dx\,dy = \frac{3}{2}.$$

EXAMPLE 5 Calculate the outward flux of the field $\mathbf{F}(x, y) = x\mathbf{i} + y^2\mathbf{j}$ across the square bounded by the lines $x = \pm 1$ and $y = \pm 1$.

Solution Calculating the flux with a line integral would take four integrations, one for each side of the square. With Green's theorem, we can change the line integral to a single double integral. With $M = x$, $N = y^2$, C the square, and R the square's interior, we have

$$\text{Flux} = \oint_C M\,dy - N\,dx = \iint_R \left(\frac{\partial M}{\partial x} + \frac{\partial N}{\partial y}\right) dx\,dy \qquad \text{(Green's theorem)}$$

$$= \int_{-1}^1\int_{-1}^1 (1 + 2y)\,dx\,dy = \int_{-1}^1 \left[x + 2xy\right]_{x=-1}^{x=1} dy$$

$$= \int_{-1}^1 (2 + 4y)\,dy = \left[2y + 2y^2\right]_{-1}^1 = 4.$$

A Proof of Green's Theorem (Special Regions)

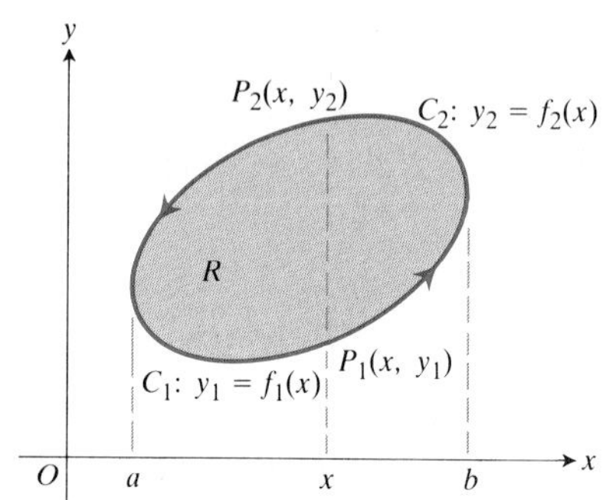

15.23 The boundary curve C is made up of C_1: $y = f_1(x)$ and C_2: $y = f_2(x)$.

Let C be a continuously differentiable simple closed curve in the xy-plane with the property that lines parallel to the axes cut it in no more than two points. Let M, N, and their first partial derivatives be continuous at every point of C and let R be the region enclosed by C. Then

$$\oint_C M\,dx + N\,dy = \iint_R \left(\frac{\partial N}{\partial x} - \frac{\partial M}{\partial y}\right) dx\,dy. \tag{18}$$

Figure 15.23 shows C made up of two parts:

$$C_1: \quad y = f_1(x), \quad a \le x \le b, \qquad C_2: \quad y = f_2(x), \quad b \ge x \ge a.$$

For any x between a and b, we can integrate $\partial M/\partial y$ with respect to y from $y = f_1(x)$ to $y = f_2(x)$ and obtain

$$\int_{f_1(x)}^{f_2(x)} \frac{\partial M}{\partial y}\,dy = M(x, y)\bigg]_{y=f_1(x)}^{y=f_2(x)} = M(x, f_2(x)) - M(x, f_1(x)). \tag{19}$$

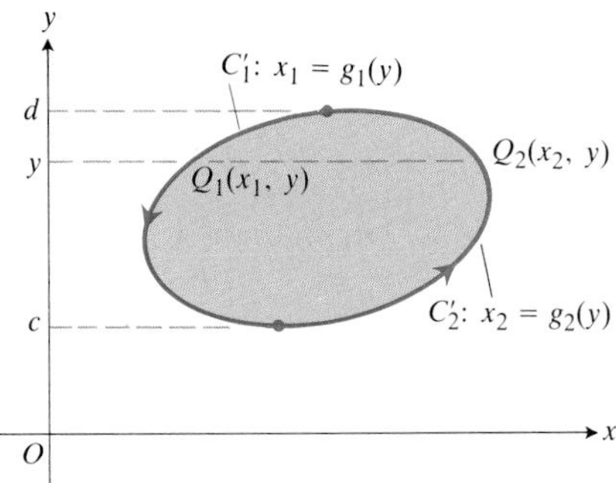

15.24 The boundary curve C is made up of C_1': $x = g_1(y)$ and C_2': $x = g_2(y)$.

We can then integrate this with respect to x from a to b:

$$\int_a^b \int_{f_1(x)}^{f_2(x)} \frac{\partial M}{\partial y}\, dy\, dx = \int_a^b \{M(x, f_2(x)) - M(x, f_1(x))\}\, dx$$

$$= -\int_b^a M(x, f_2(x))\, dx - \int_a^b M(x, f_1(x))\, dx$$

$$= -\int_{C_2} M\, dx - \int_{C_1} M\, dx = -\oint_C M\, dx.$$

Therefore

$$\oint_C M\, dx = \iint_R \left(-\frac{\partial M}{\partial y}\right) dx\, dy. \tag{20}$$

Equation (20) is half the result we need for Eq. (18). We derive the other half by integrating $\partial N/\partial x$ first with respect to x and then with respect to y, as suggested by Fig. 15.24. This shows the curve C of Fig. 15.23, decomposed into the two directed parts

$$C_1': \quad x = g_1(y), \quad d \geq y \geq c, \qquad C_2': \quad x = g_2(y), \quad c \leq y \leq d.$$

The result of this double integration is expressed by

$$\oint_C N\, dy = \iint_R \frac{\partial N}{\partial x}\, dx\, dy. \tag{21}$$

Combining Eqs. (20) and (21) gives Eq. (18). This concludes the proof. ■

Extending the Proof to Other Regions

The argument with which we just derived Green's equation does not apply directly to the rectangular region in Fig. 15.25 because the lines $x = a$, $x = b$, $y = c$, and $y = d$ meet the region's boundary in more than two points. However, if we divide the boundary C into four directed line segments,

$$C_1: \quad y = c, \quad a \leq x \leq b, \qquad C_2: \quad x = b, \quad c \leq y \leq d,$$
$$C_3: \quad y = d, \quad b \geq x \geq a, \qquad C_4: \quad x = a, \quad d \geq y \geq c,$$

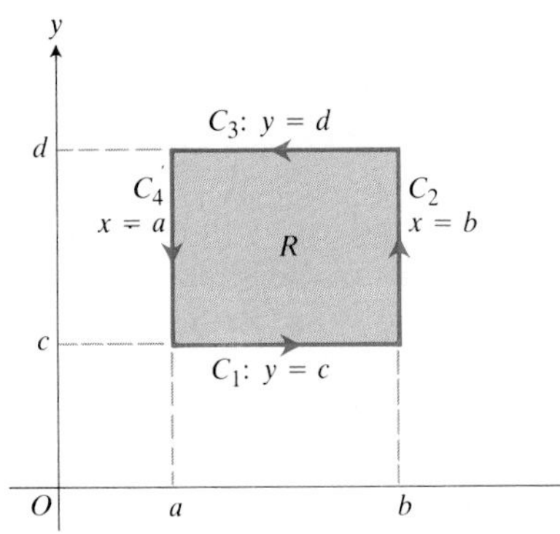

15.25 To prove Green's theorem for the rectangular region shown here, we divide the boundary into four directed line segments.

we can modify the argument in the following way.

Proceeding as in the proof of Eq. (21), we have

$$\int_c^d \int_a^b \frac{\partial N}{\partial x}\, dx\, dy = \int_c^d [N(b, y) - N(a, y)]\, dy = \int_c^d N(b, y)\, dy + \int_d^c N(a, y)\, dy$$

$$= \int_{C_2} N\, dy + \int_{C_4} N\, dy. \tag{22}$$

Because y is constant along C_1 and C_3,

$$\int_{C_1} N\, dy = \int_{C_3} N\, dy = 0,$$

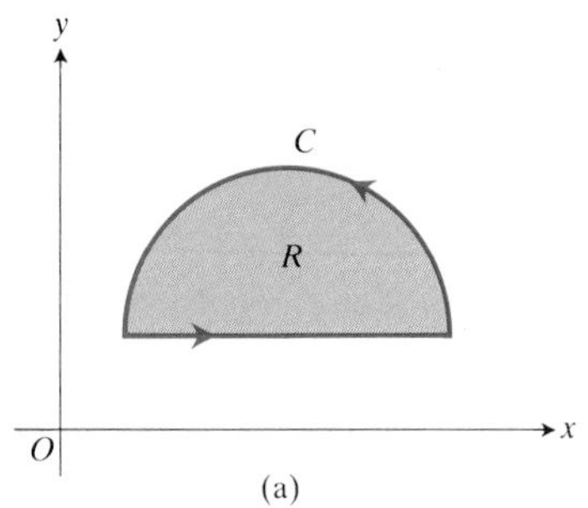

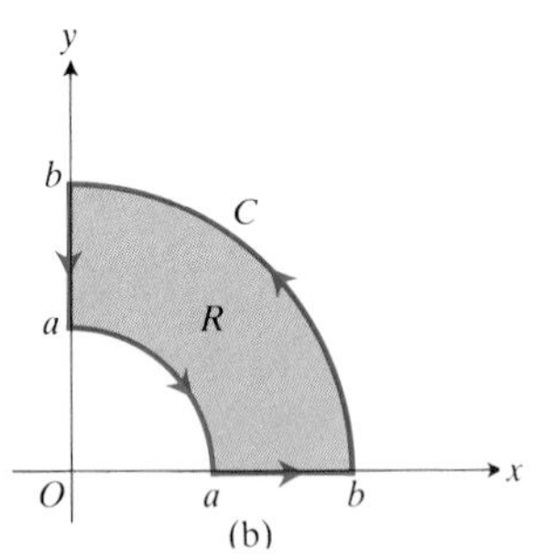

15.26 Other regions to which Green's theorem applies.

so we can add

$$\int_{C_1} N\,dy + \int_{C_3} N\,dy$$

to the right-hand side of Eq. (22) without changing the equality. Doing so, we have

$$\int_c^d \int_a^b \frac{\partial N}{\partial x}\,dx\,dy = \oint_C N\,dy. \tag{23}$$

Similarly, we could show that

$$\int_a^b \int_c^d \frac{\partial M}{\partial y}\,dy\,dx = -\oint_C M\,dx. \tag{24}$$

Subtracting (24) from (23), we again arrive at

$$\oint_C M\,dx + N\,dy = \iint_R \left(\frac{\partial N}{\partial x} - \frac{\partial M}{\partial y}\right) dx\,dy.$$

Regions like those in Fig. 15.26 can be handled with no greater difficulty. Equation (18) still applies. It also applies to the horseshoe-shaped region R shown in Fig. 15.27, as we see by putting together the regions R_1 and R_2 and their boundaries. Green's theorem applies to C_1, R_1, and to C_2, R_2, yielding

$$\int_{C_1} M\,dx + N\,dy = \iint_{R_1} \left(\frac{\partial N}{\partial x} - \frac{\partial M}{\partial y}\right) dx\,dy,$$

$$\int_{C_2} M\,dx + N\,dy = \iint_{R_2} \left(\frac{\partial N}{\partial x} - \frac{\partial M}{\partial y}\right) dx\,dy.$$

When we add, the line integral along the y-axis from b to a for C_1 cancels the integral over the same segment but in the opposite direction for C_2. Hence

$$\oint_C M\,dx + N\,dy = \iint_R \left(\frac{\partial N}{\partial x} - \frac{\partial M}{\partial y}\right) dx\,dy,$$

where C consists of the two segments of the x-axis from $-b$ to $-a$ and from a to b and of the two semicircles, and where R is the region inside C.

The device of adding line integrals over separate boundaries to build up an integral over a single boundary can be extended to any finite number of subregions. In Fig. 15.28(a), let C_1 be the boundary of the region R_1 in the first quadrant. Similarly for the other three quadrants: C_i is the boundary of the region R_i, $i = 1, 2, 3, 4$. By Green's theorem,

$$\oint_{C_i} M\,dx + N\,dy = \iint_{R_i} \left(\frac{\partial N}{\partial x} - \frac{\partial M}{\partial y}\right) dx\,dy. \tag{25}$$

We add Eqs. (25) for $i = 1, 2, 3, 4$, and get

$$\oint_{r=b} (M\,dx + N\,dy) + \oint_{r=a} (M\,dx + N\,dy) = \iint_{a \le r \le b} \left(\frac{\partial N}{\partial x} - \frac{\partial M}{\partial y}\right) dx\,dy. \tag{26}$$

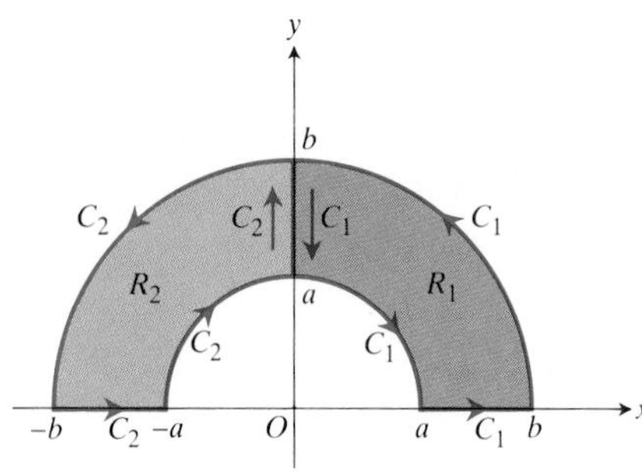

15.27 A region R that combines regions R_1 and R_2.

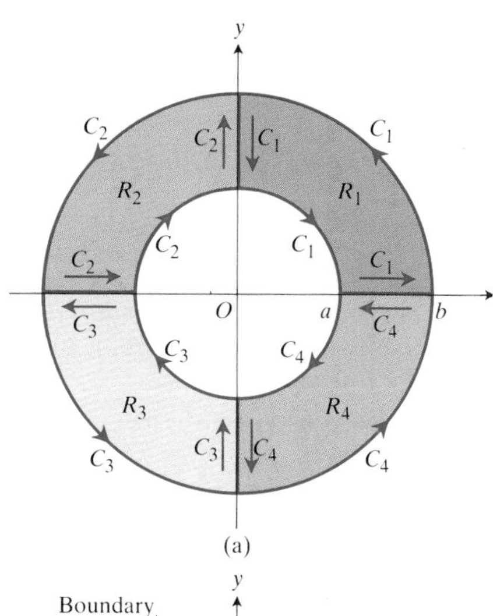

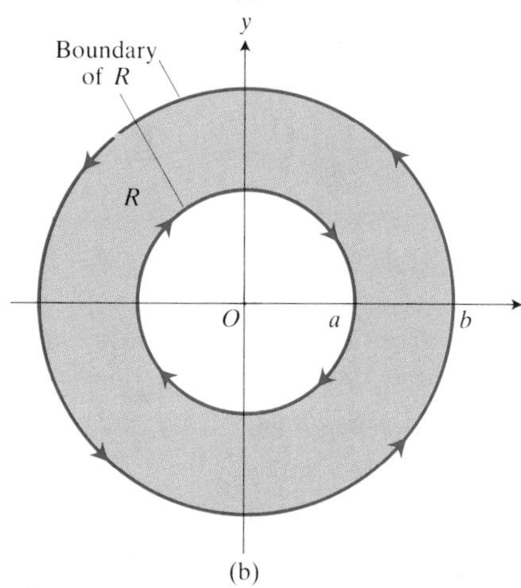

15.28 The annular region R shown here combines four smaller regions. In polar coordinates, we have $r = a$ for the inner circle, $r = b$ for the outer circle, and $a \le r \le b$ for the region itself.

Equation (26) says that the double integral of $(\partial N/\partial x - \partial M/\partial y)$ over the annular ring R is equal to the line integral of $M\,dx + N\,dy$ over the *entire* boundary of R, in the *direction* along the boundary that keeps the region R on our left as we progress (Fig. 15.28b).

The Normal and Tangential Forms of Green's Theorem

As we learned in Section 15.2, the flux of a two-dimensional vector field $\mathbf{F}$ across a closed curve C in the direction of its outer unit normal vector $\mathbf{n}$ is

$$\int_C \mathbf{F}\cdot\mathbf{n}\,ds. \qquad \text{(Flux)}$$

Similarly, the circulation around C in the direction of its unit tangent vector $\mathbf{T}$ is

$$\int_C \mathbf{F}\cdot\mathbf{T}\,ds. \qquad \text{(Circulation)}$$

Hence, the flux form of Green's theorem can be written as

$$\int_C \mathbf{F}\cdot\mathbf{n}\,ds = \iint_R \left(\frac{\partial M}{\partial x} + \frac{\partial N}{\partial y}\right) dx\,dy \qquad \text{(Normal form)} \tag{27}$$

and the circulation form of Green's theorem can be written as

$$\oint_C \mathbf{F}\cdot\mathbf{T}\,ds = \iint_R \left(\frac{\partial N}{\partial x} - \frac{\partial M}{\partial y}\right) dx\,dy. \qquad \text{(Tangential form)} \tag{28}$$

The left-hand sides of these equations are now in vector form. How about the right-hand sides? Can they be expressed in vector form too? The answer is yes, and we shall show how to do so with an ingenious vector operator ∇ ("del") in Sections 15.5 and 15.6.

Exercises 15.3

In Exercises 1–4, verify Green's theorem by evaluating both sides of Eqs. (15) and (16) for the field $\mathbf{F} = M\mathbf{i} + N\mathbf{j}$. Take the domains of integration in each case to be the disk $R: x^2 + y^2 \le a^2$ and its bounding circle $C: \mathbf{r} = (a\cos t)\mathbf{i} + (a\sin t)\mathbf{j},\ 0 \le t \le 2\pi$.

1. $\mathbf{F} = -y\mathbf{i} + x\mathbf{j}$

2. $\mathbf{F} = y\mathbf{i}$

3. $\mathbf{F} = 2x\mathbf{i} - 3y\mathbf{j}$

4. $\mathbf{F} = -x^2y\mathbf{i} + xy^2\mathbf{j}$

In Exercises 5–8, use Green's theorem to find the counterclockwise circulation and outward flux for the field $\mathbf{F}$ and curve C.

5. $\mathbf{F} = (x - y)\mathbf{i} + (y - x)\mathbf{j}$
C: The square bounded by $x = 0, x = 1, y = 0, y = 1$

6. $\mathbf{F} = (x^2 + 4y)\mathbf{i} + (x + y^2)\mathbf{j}$
C: The square bounded by $x = 0, x = 1, y = 0, y = 1$

7. $\mathbf{F} = (y^2 - x^2)\mathbf{i} + (x^2 + y^2)\mathbf{j}$
C: The triangle bounded by $y = 0, x = 3$, and $y = x$

8. $\mathbf{F} = (x + y)\mathbf{i} - (x^2 + y^2)\mathbf{j}$
C: The triangle bounded by $y = 0, x = 1$, and $y = x$

9. Find the counterclockwise circulation and outward flux of the field $\mathbf{F}(x, y) = xy\mathbf{i} + y^2\mathbf{j}$ over the boundary of the region enclosed by the parabola $y = x^2$ and the line $y = x$ in the first quadrant.

10. Find the counterclockwise circulation and the outward flux of the field $\mathbf{F}(x, y) = (-\sin y)\mathbf{i} + (x \cos y)\mathbf{j}$ over the square cut from the first quadrant by the lines $x = \pi/2$ and $y = \pi/2$.

Apply Green's theorem in one of its two forms to evaluate the line integrals in Exercises 11–16.

11. $\oint_C (y^2\,dx + x^2\,dy)$

C: The triangle bounded by $x = 0$, $x + y = 1$, $y = 0$.

12. $\oint_C (3y\,dx + 2x\,dy)$

C: The boundary of $0 \le x \le \pi$, $0 \le y \le \sin x$.

13. $\oint_C (6y + x)\,dx + (y + 2x)\,dy$

C: The circle $(x - 2)^2 + (y - 3)^2 = 4$.

14. $\oint_C (2x + y^2)\,dx + (2xy + 3y)\,dy$

C: Any simple closed curve in the plane for which Green's theorem holds.

15. $\oint_C 2xy^3\,dx + 4x^2y^2\,dy$

C: The boundary of the "triangular" region in the first quadrant enclosed by the x-axis, the line $x = 1$, and the curve $y = x^3$.

16. $\oint_C (4x - 2y)\,dx + (2x - 4y)\,dy$

C: The circle $(x - 2)^2 + (y - 2)^2 = 4$.

17. Let C be the boundary of a region on which Green's theorem holds. Use Green's theorem to calculate

a) $\oint_C f(x)\,dx + g(y)\,dy$,

b) $\oint_C k\,y\,dx + h\,x\,dy$ (k and h constants).

18. Show that

$$\oint_C 4x^3y\,dx + x^4\,dy = 0$$

for any closed curve C to which Green's theorem applies.

Calculating Area with Green's Theorem

If a simple closed curve C in the plane and the region R it encloses satisfy the hypotheses of Green's theorem, then the area of R is given by the following formula.

Green's Theorem Area Formula

$$\text{Area of } R = \frac{1}{2}\oint_C x\,dy - y\,dx. \tag{29}$$

The reason is that by Eq. (15), run backward,

$$\text{Area of } R = \iint_R dy\,dx = \iint_R \left(\frac{1}{2} + \frac{1}{2}\right) dy\,dx$$
$$= \oint_C \frac{1}{2}x\,dy - \frac{1}{2}y\,dx.$$

Use the Green's theorem area formula to find the areas of the regions enclosed by the curves in Exercises 19–22.

19. The circle $\mathbf{r} = (a \cos t)\mathbf{i} + (a \sin t)\mathbf{j}$, $0 \le t \le 2\pi$

20. The ellipse $\mathbf{r} = (a \cos t)\mathbf{i} + (b \sin t)\mathbf{j}$, $0 \le t \le 2\pi$

21. The astroid (Fig. 10.37) $\mathbf{r} = (\cos^3 t)\mathbf{i} + (\sin^3 t)\mathbf{j}$, $0 \le t \le 2\pi$

22. The curve (Fig. 10.75) $\mathbf{r} = t^2\mathbf{i} + ((t^3/3) - t)\mathbf{j}$, $-\sqrt{3} \le t \le \sqrt{3}$

TOOLKIT PROGRAM

Vector Fields	Displays two-dimensional vector fields and integrates the curl over rectangular regions. Also calculates flux across rectangular boundaries.

15.4 Surface Area and Surface Integrals

We know how to integrate a function over a flat region in the plane, but what if the function is defined over a curved surface instead? How do we calculate its integral then? The trick to evaluating one of these so-called surface integrals is to rewrite it as a double integral over a region in a coordinate plane beneath the surface (Fig. 15.29). This changes the surface integral into the kind of integral we already know how to evaluate.

Our first step is to find a double-integral formula for calculating the area of a curved surface. We will then see how the ideas that arise can be used again to define and evaluate surface integrals. With surface integrals under control, we shall be able to calculate the flux of a three-dimensional vector field through a surface and the masses and moments of thin shells of material. In Sections 15.5 and 15.6 we shall see how surface integrals provide just what we need to generalize the two forms of Green's theorem to three dimensions. One of these generalizations will enable us to express the flux of a vector field through a closed surface as a triple integral over the three-dimensional region enclosed by the surface. The other generalization will enable us to express the circulation of a vector field around a closed curve in space as an integral over a surface bounded by the curve. These results have far-reaching consequences in mathematics as well as in the theories of electromagnetism and fluid flow.

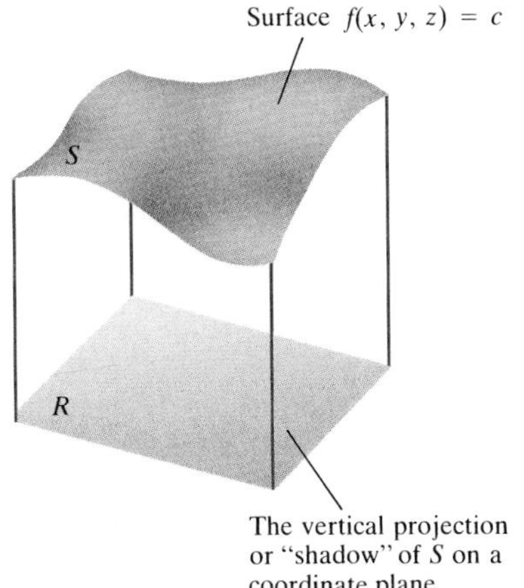

15.29 As we shall soon see, the integral of a function $g(x, y, z)$ over a surface S in space can be calculated by evaluating a closely related double integral over the vertical projection or "shadow" of S on a coordinate plane.

The Definition of Surface Area

Figure 15.30 shows a piece S of a surface $f(x, y, z) = c$ lying above its "shadow" region R in a plane directly beneath it. If f and its first partial derivatives are continuous, the area of S can be defined and calculated as a double integral over R. The derivation of the integral takes a while to describe, but the integral itself is easy to work with.

The first step in defining the area of S is to partition the region R into small rectangles ΔA_k of the kind we would use if we were defining an integral over R. Directly above each ΔA_k lies a patch of surface $\Delta \sigma_k$ that we may approximate with a portion ΔP_k of the tangent plane. To be specific, we suppose that ΔP_k is a portion of the plane that is tangent to the surface at the point (x_k, y_k, z_k) directly above the back corner C_k of ΔA_k. If the tangent plane is parallel to R, then ΔP_k will be congruent to ΔA_k. Otherwise, it will be a parallelogram whose area is somewhat larger than the area of ΔA_k.

Figure 15.31 gives a magnified view of $\Delta \sigma_k$ and ΔP_k, showing the gradient vector $\nabla f(x_k, y_k, z_k)$ and a unit vector $\mathbf{p}$ that is normal to R. The figure also shows the angle γ between ∇f and $\mathbf{p}$. The other vectors in the picture, $\mathbf{u}$ and $\mathbf{v}$, lie along the edges of the patch ΔP_k in the tangent plane. Thus, both $\mathbf{u} \times \mathbf{v}$ and ∇f are normal to the tangent plane.

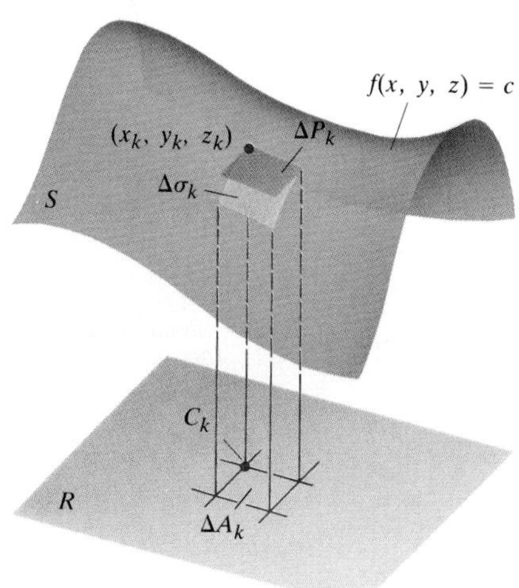

15.30 A surface S and its vertical projection onto a plane beneath it. You can think of R as the shadow of S on the plane. The tangent plate ΔP_k approximates the surface patch $\Delta \sigma_k$ above ΔA_k.

We now need the fact from advanced vector geometry that $|(\mathbf{u} \times \mathbf{v}) \cdot \mathbf{p}|$ is the area of the projection of the parallelogram determined by $\mathbf{u}$ and $\mathbf{v}$ onto any plane whose normal is $\mathbf{p}$. In our case, this translates into the statement

$$|(\mathbf{u} \times \mathbf{v}) \cdot \mathbf{p}| = \Delta A_k. \tag{1}$$

Now, $|\mathbf{u} \times \mathbf{v}|$ itself is the area ΔP_k (standard fact about cross products) so Eq. (1) becomes

$$\underbrace{|\mathbf{u} \times \mathbf{v}|}_{\Delta P_k} \; \underbrace{|\mathbf{p}|}_{1} \; \underbrace{|\cos(\text{angle between } \mathbf{u} \times \mathbf{v} \text{ and } \mathbf{p})|}_{\substack{\text{Same as } |\cos\gamma| \text{ because} \\ \nabla f \text{ and } \mathbf{u} \times \mathbf{v} \text{ are both} \\ \text{normal to the tangent plane}}} = \Delta A_k \tag{2}$$

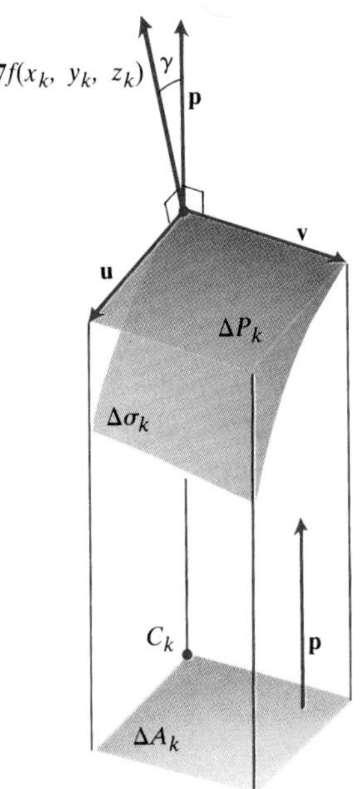

15.31 Magnified view from the preceding figure. The vector $\mathbf{u} \times \mathbf{v}$ (not shown) is parallel to the vector ∇f because both vectors are normal to the plane of ΔP_k.

or

$$\Delta P_k |\cos \gamma| = \Delta A_k$$

or

$$\Delta P_k = \frac{\Delta A_k}{|\cos \gamma|},$$

provided $\cos \gamma \neq 0$. We will have $\cos \gamma \neq 0$ as long as ∇f is not parallel to the ground plane and $\nabla f \cdot \mathbf{p} \neq 0$.

Since the patches ΔP_k approximate the surface patches $\Delta \sigma_k$ that fit together to make S, the sum

$$\sum \Delta P_k = \sum \frac{\Delta A_k}{|\cos \gamma|} \tag{3}$$

looks like an approximation of what we might like to call the surface area of S. It also looks as if the approximation would improve if we refined the partition of R. In fact, the sums on the right-hand side of Eq. (3) are approximating sums for the double integral

$$\iint_R \frac{1}{|\cos \gamma|}\, dA. \tag{4}$$

We therefore define the **area** of S to be the value of this integral whenever it exists.

A Practical Formula

For any particular surface $f(x, y, z) = c$,

$$|\nabla f \cdot \mathbf{p}| = |\nabla f|\,|\mathbf{p}|\,|\cos \gamma|,$$

so

$$\frac{1}{|\cos \gamma|} = \frac{|\nabla f|}{|\nabla f \cdot \mathbf{p}|}.$$

This combines with Eq. (4) to give a practical formula for area.

The Formula for Surface Area

The area of the surface $f(x, y, z) = c$ over a closed and bounded plane region R is

$$\text{Surface area} = \iint_R \frac{|\nabla f|}{|\nabla f \cdot \mathbf{p}|}\, dA, \tag{5}$$

where $\mathbf{p}$ is a unit vector normal to R and $\nabla f \cdot \mathbf{p} \neq 0$.

Thus, the area is the double integral over R of the magnitude of ∇f divided by the magnitude of the scalar component of ∇f normal to R.

We reached Eq. (5) under the assumption that $\nabla f \cdot \mathbf{p} \neq 0$ throughout R and that ∇f is continuous. Whenever the integral exists, however, we may define its value to be the area of the portion of the surface $f(x, y, z) = c$ that lies over R.

Equation (5) agrees with our other formulas for surface area, although we shall not prove this fact.

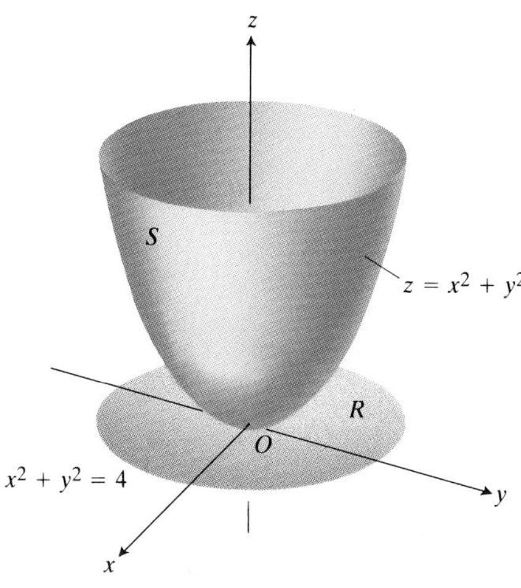

15.32 The area of the parabolic surface above is calculated in Example 1.

EXAMPLE 1 Find the area of the surface cut from the bottom of the paraboloid $x^2 + y^2 - z = 0$ by the plane $z = 4$.

Solution We sketch the surface S and the region R below it in the xy-plane (Fig. 15.32). The surface S is part of the level surface $f(x, y, z) = x^2 + y^2 - z = 0$, and R is the disk $x^2 + y^2 \leq 4$ in the xy-plane. To get a unit vector normal to the plane of R we can take $\mathbf{p} = \mathbf{k}$.

At any point (x, y, z) on the surface, we have

$$\begin{aligned} f(x, y, z) &= x^2 + y^2 - z, \\ \nabla f &= 2x\mathbf{i} + 2y\mathbf{j} - \mathbf{k}, \\ |\nabla f| &= \sqrt{(2x)^2 + (2y)^2 + (-1)^2} = \sqrt{4x^2 + 4y^2 + 1}, \\ |\nabla f \cdot \mathbf{p}| &= |\nabla f \cdot \mathbf{k}| = |-1| = 1. \end{aligned}$$

Therefore,

$$\begin{aligned} \text{Surface area} &= \iint_R \frac{|\nabla f|}{|\nabla f \cdot \mathbf{p}|}\, dA \qquad \text{(Eq. 5)} \\ &= \iint_{x^2+y^2\leq 4} \sqrt{4x^2 + 4y^2 + 1}\, dx\, dy \\ &= \int_0^{2\pi}\int_0^2 \sqrt{4r^2 + 1}\, r\, dr\, d\theta \qquad \text{(Polar coordinates)} \\ &= \int_0^{2\pi} \left[\frac{1}{12}(4r^2 + 1)^{3/2}\right]_0^2 d\theta \\ &= \int_0^{2\pi} \frac{1}{12}(17^{3/2} - 1)\, d\theta = \frac{\pi}{6}(17\sqrt{17} - 1). \end{aligned}$$

■

EXAMPLE 2 Find the area of the cap cut from the hemisphere $x^2 + y^2 + z^2 = 2$, $z \geq 0$, by the cylinder $x^2 + y^2 = 1$ (Fig. 15.33).

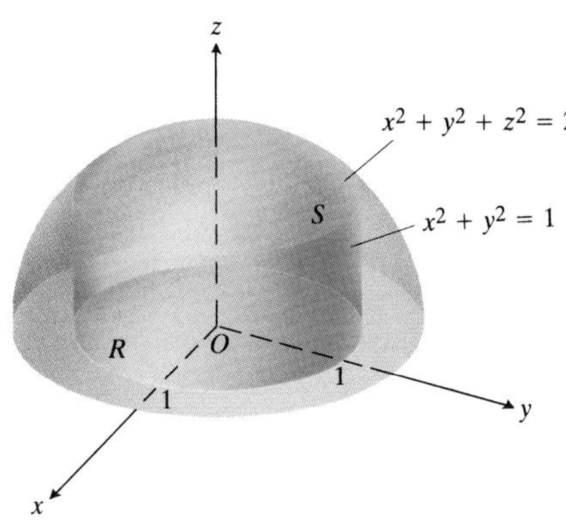

15.33 The cap cut from the hemisphere by the cylinder projects vertically onto the disk R: $x^2 + y^2 \leq 1$.

Solution The cap S is part of the level surface $f(x, y, z) = x^2 + y^2 + z^2 = 2$. It projects one-to-one onto the disk $R: x^2 + y^2 \leq 1$ in the xy-plane. The vector $\mathbf{p} = \mathbf{k}$ is normal to the plane of R.

At any point on the surface,

$$\begin{aligned} f(x, y, z) &= x^2 + y^2 + z^2, \\ \nabla f &= 2x\mathbf{i} + 2y\mathbf{j} + 2z\mathbf{k} \\ |\nabla f| &= 2\sqrt{x^2 + y^2 + z^2} = 2\sqrt{2}, \qquad \left(\begin{array}{l}\text{Because } x^2 + y^2 + z^2 = 2 \\ \text{at points of S}\end{array}\right) \\ |\nabla f \cdot \mathbf{p}| &= |\nabla f \cdot \mathbf{k}| = |2z| = 2z. \end{aligned}$$

Therefore,

$$\text{Surface area} = \iint_R \frac{|\nabla f|}{|\nabla f \cdot \mathbf{p}|}\, dA = \iint_R \frac{2\sqrt{2}}{2z}\, dA = \sqrt{2}\iint_R \frac{dA}{z}. \tag{6}$$

What do we do about the z?

Since z is the z-coordinate of a point on the sphere, we can express it in terms of x and y as

$$z = \sqrt{2 - x^2 - y^2}.$$

We continue the work of Eq. (6) with this substitution:

$$\begin{aligned}\text{Surface area} &= \sqrt{2}\iint_R \frac{dA}{z} = \sqrt{2}\iint_{x^2+y^2\le 1} \frac{dA}{\sqrt{2 - x^2 - y^2}} \\ &= \sqrt{2}\int_0^{2\pi}\int_0^1 \frac{r\,dr\,d\theta}{\sqrt{2 - r^2}} \qquad \text{(Polar coordinates)} \\ &= \sqrt{2}\int_0^{2\pi}\Big[-(2 - r^2)^{1/2}\Big]_{r=0}^{r=1} d\theta = \sqrt{2}\int_0^{2\pi}(\sqrt{2} - 1)\,d\theta \\ &= 2\pi(2 - \sqrt{2}).\end{aligned}$$

■

Surface Integrals

We now show how to integrate a function over a surface, using the ideas we just developed for calculating surface area.

Suppose, for example, that we have an electrical charge distributed over a surface $f(x, y, z) = c$ like the one shown in Fig. 15.34 and that the function $g(x, y, z)$ gives the charge per unit area (charge density) at each point on S. Then we may calculate the total charge on S as an integral in the following way.

We partition the shadow region R on the ground plane beneath the surface into small rectangles of the kind we would use if we were defining the surface area of S. Then directly above each ΔA_k lies a patch of surface $\Delta\sigma_k$ that we approximate with a parallelogram-shaped portion of tangent plane, ΔP_k.

Up to this point the construction proceeds as in the definition of surface area, but now we take one additional step: We evaluate g at (x_k, y_k, z_k) and approximate the total charge on the surface patch $\Delta\sigma_k$ by the product

$$g(x_k, y_k, z_k)\,\Delta P_k.$$

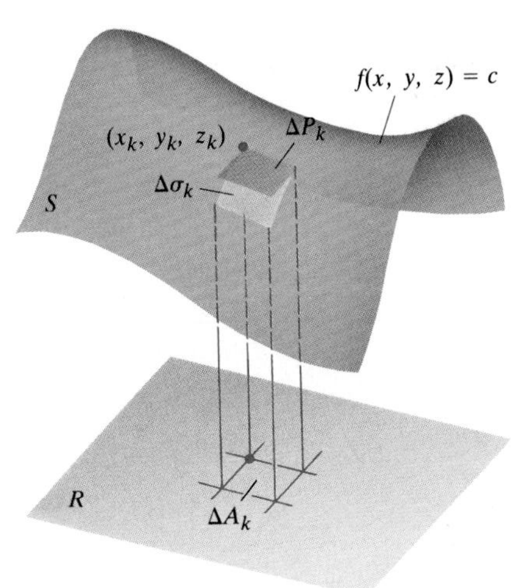

15.34 If we know how an electrical charge is distributed over a surface, we can find the total charge with a suitably modified surface integral.

The rationale is that when the partition of R is sufficiently fine, the value of g throughout $\Delta\sigma_k$ is nearly constant and ΔP_k is nearly the same as $\Delta\sigma_k$. The total charge over S is then approximated by the sum

$$\text{Total charge} \approx \sum g(x_k, y_k, z_k)\,\Delta P_k = \sum g(x_k, y_k, z_k)\frac{\Delta A_k}{|\cos\gamma|}. \tag{7}$$

If f, the function defining the surface S, and its first partial derivatives are continuous, and if g is continuous over S, then the sums on the right-hand side of Eq. (7) approach the limit

$$\iint_R g(x, y, z)\frac{dA}{|\cos\gamma|} = \iint_R g(x, y, z)\frac{|\nabla f|}{|\nabla f\cdot\mathbf{p}|}\,dA \tag{8}$$

as the rectangular subdivision of R is refined in the usual way. This limit is called the integral of g over the surface S and is calculated as a double integral over R. The value of the integral is the total charge on the surface S.

As you might expect, the formula in Eq. (8) defines the integral of *any* function g over the surface S as long as the integral exists.

DEFINITIONS

If R is the shadow region of a surface S defined by the equation $f(x, y, z) = c$ and g is a continuous function defined at the points of S, then the **integral of g over S** is the integral

$$\iint_R g(x, y, z)\frac{|\nabla f|}{|\nabla f \cdot \mathbf{p}|}\,dA, \tag{9}$$

where $\mathbf{p}$ is a unit vector normal to R and $\nabla f \cdot \mathbf{p} \neq 0$. The integral itself is called a **surface integral.**

The surface integral in (9) takes on different meanings in different applications. If g has the constant value 1, the integral gives the area of S. If g gives the mass density of a thin shell of material modeled by S, the integral gives the mass of the shell.

Algebraic Properties. The Surface Area Differential

We often abbreviate the integral in (9) by writing $d\sigma$ for $(|\nabla f| / |\nabla f \cdot \mathbf{p}|)\,dA$.

The Surface Area Differential and the Differential Form for Surface Integrals

$$d\sigma = \frac{|\nabla f|}{|\nabla f \cdot \mathbf{p}|}\,dA \qquad \iint_S g\,d\sigma \tag{10}$$

Surface area differential — Differential formula for surface integrals

Surface integrals have all the usual properties of double integrals, the surface integral of the sum of two functions being the sum of their surface integrals and so on. The domain additivity property takes the form

$$\iint_S g\,d\sigma = \iint_{S_1} g\,d\sigma + \iint_{S_2} g\,d\sigma + \cdots + \iint_{S_n} g\,d\sigma.$$

The idea is that if S is partitioned by smooth curves and line segments into a finite number of nonoverlapping smooth patches, then the integral of a function g over S is the sum of the integrals over the patches. Thus, the integral of a function over the surface of a cube is the sum of the integrals over the faces of the cube. We integrate over a turtle shell of welded plates by integrating one plate at a time and adding the results.

EXAMPLE 3 Integrate $g(x, y, z) = xyz$ over the surface of the cube cut from the first octant by the planes $x = 1$, $y = 1$, and $z = 1$ (Fig. 15.35).

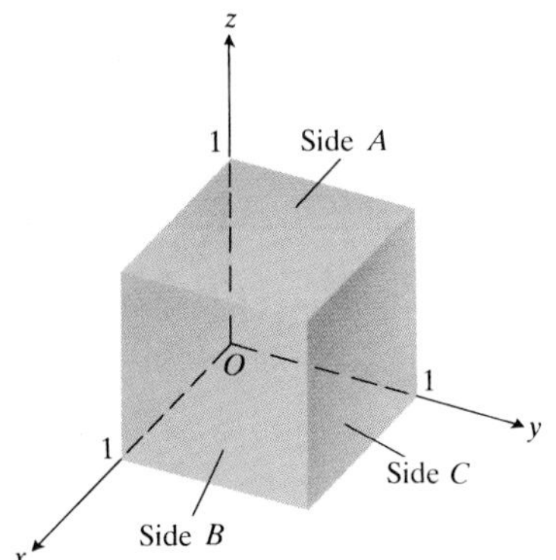

15.35 To integrate a function over the surface of a cube, we integrate over each face and add the results.

Solution We integrate xyz over each of the six sides and add the results. Since $xyz = 0$ on the sides that lie in the coordinate planes, the integral over the surface of the cube reduces to

$$\iint_{\substack{\text{Cube}\\\text{surface}}} xyz\, d\sigma = \iint_{\text{Side } A} xyz\, d\sigma + \iint_{\text{Side } B} xyz\, d\sigma + \iint_{\text{Side } C} xyz\, d\sigma.$$

Side A is the surface $f(x, y, z) = z = 1$ over the square region R_{xy}: $0 \le x \le 1,\ 0 \le y \le 1$, in the xy-plane. For this surface and region,

$$\mathbf{p} = \mathbf{k}, \qquad \nabla f = \mathbf{k}, \qquad |\nabla f| = 1, \qquad |\nabla f \cdot \mathbf{p}| = |\mathbf{k} \cdot \mathbf{k}| = 1,$$

$$d\sigma = \frac{|\nabla f|}{|\nabla f \cdot \mathbf{p}|}\, dA = \frac{1}{1}\, dx\, dy = dx\, dy,$$

$$xyz = xy(1) = xy,$$

and

$$\iint_{\text{Side } A} xyz\, d\sigma = \iint_{R_{xy}} xy\, dx\, dy = \int_0^1 \int_0^1 xy\, dx\, dy = \int_0^1 \frac{y}{2}\, dy = \frac{1}{4}.$$

Symmetry tells us that the integrals of xyz over sides B and C are also 1/4. Hence,

$$\iint_{\substack{\text{Cube}\\\text{surface}}} xyz\, d\sigma = \frac{1}{4} + \frac{1}{4} + \frac{1}{4} = \frac{3}{4}.$$

■

Orientation

We call a surface S **orientable** or **two-sided** if it is possible to find a unit normal vector $\mathbf{n}$ that varies continuously as its initial point moves about the surface. Spheres and other closed surfaces in space (surfaces that enclose solids) are orientable. By convention, we choose $\mathbf{n}$ on a closed surface to point outward. Once $\mathbf{n}$ is chosen, we say we have **oriented** the surface and we call the surface an **oriented surface.** The direction of $\mathbf{n}$ at any point is called the **positive direction** at that point (Fig. 15.36). Any patch or subportion of an orientable surface is also orientable. The Möbius band in Fig. 15.37 is not orientable. No matter where you start your candidate for $\mathbf{n}$ (shown as the shaft of a thumbtack in the figure), you will always be able to move it around the surface (without touching the boundary) in a way that will return it to its starting place pointing in the direction exactly opposite to its initial direction.

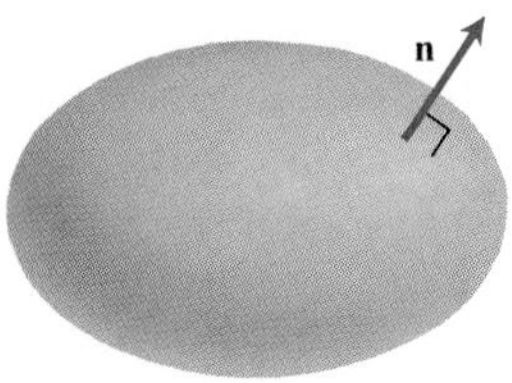

15.36 Closed surfaces in space are orientable.

The Surface Integral for Flux

Suppose $\mathbf{F}$ is a continuous vector field defined over an oriented surface S and that $\mathbf{n}$ is the chosen unit normal to the surface. We call the integral of $\mathbf{F} \cdot \mathbf{n}$ over S the flux across S in the positive direction. Thus, flux is the integral of the scalar component of $\mathbf{F}$ normal to the surface.

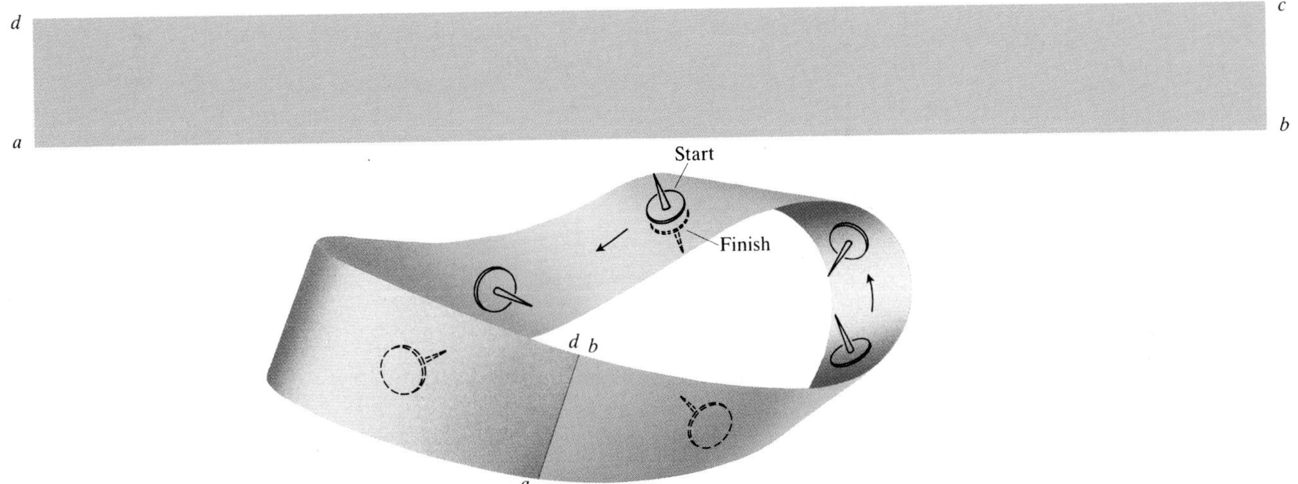

15.37 The Möbius band can be constructed by taking a rectangular strip of paper *abcd*, giving the end *bc* a single twist to interchange the positions of the vertices *b* and *c*, and then pasting the ends of the strip together to match *a* with *c* and *b* with *d*. The Möbius band is a nonorientable or one-sided surface.

DEFINITION

The **flux** of a three-dimensional vector field **F** across an oriented surface S in the direction of **n** is given by the formula

$$\text{Flux} = \iint_S \mathbf{F} \cdot \mathbf{n}\, d\sigma. \tag{11}$$

This definition is analogous to the flux of a two-dimensional field **F** across a plane curve C. In the plane (Section 15.2), the flux is

$$\int_C \mathbf{F} \cdot \mathbf{n}\, ds,$$

the integral of the scalar component of **F** normal to the curve.

If **F** is the velocity field of a three-dimensional fluid flow, the flux of **F** across S is the net rate at which fluid is crossing S in the chosen positive direction. We shall discuss such flows in more detail in Section 15.6.

If S is part of a level surface $g(x, y, z) = c$, then **n** may be taken to be one of the two vectors

$$\mathbf{n} = \pm \frac{\nabla g}{|\nabla g|}, \tag{12}$$

depending on which vector gives the preferred direction.

EXAMPLE 4 Find the flux of $\mathbf{F} = yz\mathbf{j} + z^2\mathbf{k}$ outward through the surface S cut from the semicircular cylinder $y^2 + z^2 = 1$, $z \geq 0$, by the planes $x = 0$ and $x = 1$.

Solution The outward normal to S (Fig. 15.38) may be calculated from the gradient of $g(x, y, z) = y^2 + z^2$ to be

$$\mathbf{n} = +\frac{\nabla g}{|\nabla g|} = \frac{2y\mathbf{j} + 2z\mathbf{k}}{\sqrt{4y^2 + 4z^2}} = \frac{2y\mathbf{j} + 2z\mathbf{k}}{2\sqrt{1}} = y\mathbf{j} + z\mathbf{k}.$$

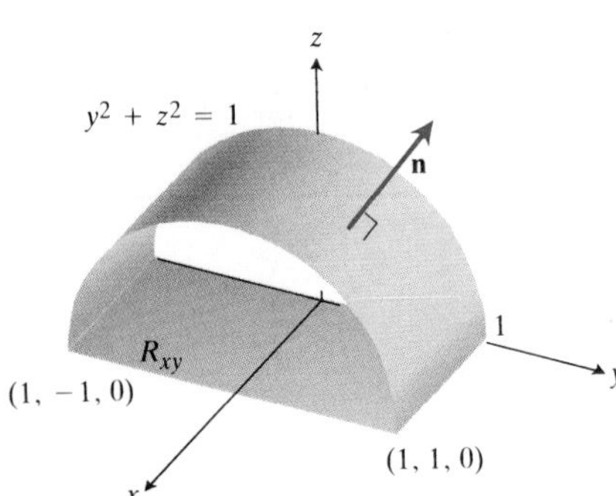

15.38 Example 4 calculates the flux of a vector field outward through this surface. The area of the shadow region R_{xy} is 2.

With $\mathbf{p} = \mathbf{k}$, we also have

$$d\sigma = \frac{|\nabla g|}{|\nabla g \cdot \mathbf{k}|}\,dA = \frac{2}{|2z|}\,dA = \frac{1}{z}\,dA.$$

We can drop the absolute value bars because $z \geq 0$ on S.

The value of $\mathbf{F} \cdot \mathbf{n}$ on the surface is given by the formula

$$\begin{aligned}\mathbf{F} \cdot \mathbf{n} &= (yz\mathbf{j} + z^2\mathbf{k}) \cdot (y\mathbf{j} + z\mathbf{k}) \\ &= y^2z + z^3 = z(y^2 + z^2) \\ &= z. \qquad\qquad (y^2 + z^2 = 1 \text{ on } S)\end{aligned}$$

Therefore, the flux of $\mathbf{F}$ outward through S is

$$\iint_S \mathbf{F} \cdot \mathbf{n}\,d\sigma = \iint_S (z)\left(\frac{1}{z}\,dA\right) = \iint_{R_{xy}} dA = \text{area}\,(R_{xy}) = 2.$$

■

Moments and Masses of Thin Shells

In engineering and physics, thin shells of material like bowls, metal drums, and domes are modeled with surfaces. Their moments and masses are calculated with surface integrals, using the formulas in Table 15.2.

EXAMPLE 5 Find the center of mass of a thin hemispherical shell of radius 3 and constant density δ.

Solution We model the shell with the hemisphere

$$f(x, y, z) = x^2 + y^2 + z^2 = 9, \quad z \geq 0$$

(Fig. 15.39). The symmetry of the surface about the z-axis tells us that $\bar{x} = \bar{y} = 0$. It remains only to find $\bar{z}$ from the formula $\bar{z} = M_{xy}/M$.

The mass of the shell is

$$M = \iint_S \delta\,d\sigma = \delta \iint_S d\sigma = (\delta)(\text{area of } S) = 18\pi\,\delta.$$

To evaluate the integral for M_{xy}, we take $\mathbf{p} = \mathbf{k}$ and calculate

$$\begin{aligned}|\nabla f| &= |2x\mathbf{i} + 2y\mathbf{j} + 2z\mathbf{k}| = 2\sqrt{x^2 + y^2 + z^2} = 2\sqrt{9} = 6, \\ |\nabla f \cdot \mathbf{p}| &= |\nabla f \cdot \mathbf{k}| = |2z| = 2z, \\ d\sigma &= \frac{|\nabla f|}{|\nabla f \cdot \mathbf{p}|}\,dA = \frac{6}{2z}\,dA = \frac{3}{z}\,dA.\end{aligned}$$

Then,

$$M_{xy} = \iint_S z\,\delta\,d\sigma = \delta \iint_R z\,\frac{3}{z}\,dA = 3\delta \iint_R dA = 3\delta(9\pi) = 27\pi\,\delta,$$

$$\bar{z} = \frac{M_{xy}}{M} = \frac{27\pi\,\delta}{18\pi\,\delta} = \frac{3}{2}.$$

The shell's center of mass is the point (0, 0, 3/2). ■

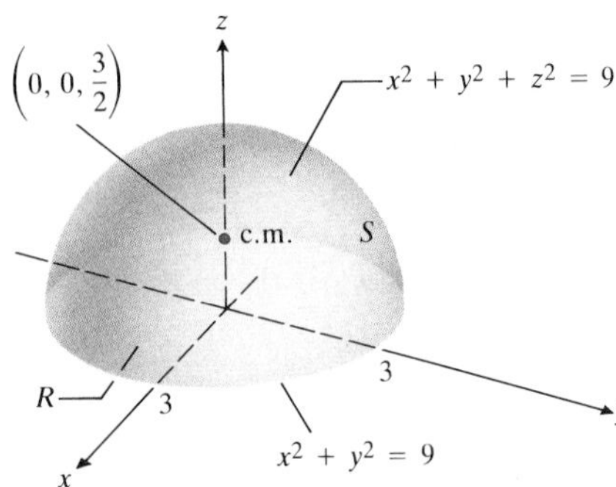

15.39 Example 5 calculates the coordinates of the center of mass of this thin hemispherical shell.

TABLE 15.2
Mass and moment formulas for very thin shells

Mass: $M = \iint_S \delta(x, y, z)\, d\sigma \qquad (\delta(x, y, z) = \text{density at } (x, y, z))$

First moments about the coordinate planes:

$$M_{yz} = \iint_S x\,\delta\, d\sigma, \qquad M_{xz} = \iint_S y\,\delta\, d\sigma, \qquad M_{xy} = \iint_S z\,\delta\, d\sigma$$

Coordinates of center of mass:

$$\bar{x} = M_{yz}/M, \qquad \bar{y} = M_{xz}/M, \qquad \bar{z} = M_{xy}/M$$

Moments of inertia:

$$I_x = \iint_S (y^2 + z^2)\,\delta\, d\sigma, \qquad I_y = \iint_S (x^2 + z^2)\,\delta\, d\sigma,$$

$$I_z = \iint_S (x^2 + y^2)\,\delta\, d\sigma, \qquad I_L = \iint_S r^2\,\delta\, d\sigma$$

$$r(x, y, z) = \text{distance from point } (x, y, z) \text{ to line } L$$

Radius of gyration about a line L: $R_L = \sqrt{I_L/M}$

Exercises 15.4

Surface Area

1. Find the area of the surface cut from the paraboloid $x^2 + y^2 - z = 0$ by the plane $z = 2$.
2. Find the area of the band cut from the paraboloid $x^2 + y^2 - z = 0$ by the planes $z = 2$ and $z = 6$.
3. Find the area of the region cut from the plane $x + 2y + 2z = 5$ by the cylinder whose walls are $x = y^2$ and $x = 2 - y^2$.
4. Find the area of the portion of the surface $x^2 - 2z = 0$ that lies above the triangle bounded by the lines $x = \sqrt{3}$, $y = 0$, and $y = x$ in the xy-plane.
5. Find the area of the surface $x^2 - 2y - 2z = 0$ that lies above the triangle bounded by the lines $x = 2$, $y = 0$, and $y = 3x$ in the xy-plane.
6. Find the area of the surface $x^2 - 2 \ln x + \sqrt{15}y - z = 0$ above the square $R: 0 \leq y \leq 1$, $1 \leq x \leq 2$ in the xy-plane.
7. Find the area of the cap cut from the sphere $x^2 + y^2 + z^2 = 2$ by the cone $z = \sqrt{x^2 + y^2}$.
8. Find the area of the ellipse cut from the plane $z = cx$ by the cylinder $x^2 + y^2 = 1$.
9. Find the area of the portion of the cylinder $x^2 + z^2 = 1$ that lies between the planes $x = \pm 1/2$ and $y = \pm 1/2$.
10. Find the area of the surface cut from the paraboloid $x^2 + y + z^2 = 1$ by the plane $y = 0$. (*Hint:* Project the surface onto the xz-plane.)

Surface Integrals

11. Integrate $g(x, y, z) = x + y + z$ over the surface of the cube cut from the first octant by the planes $x = 1$, $y = 1$, $z = 1$.
12. Integrate $g(x, y, z) = y + z$ over the surface of the wedge in Fig. 15.40.

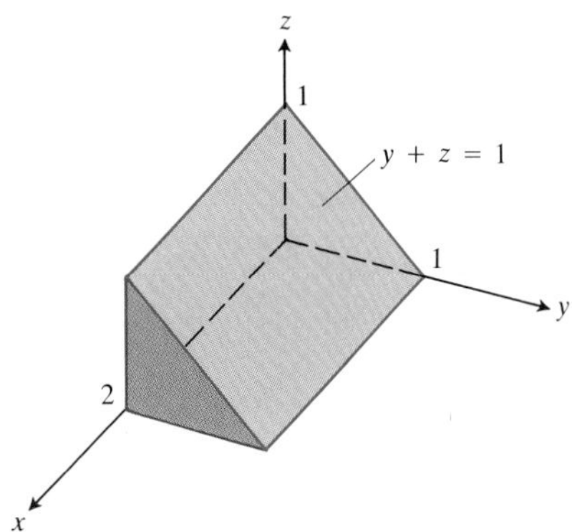

15.40 The wedge in Exercise 12.

13. Integrate $g(x, y, z) = xyz$ over the surface of the rectangular solid cut from the first octant by the planes $x = a$, $y = b$, and $z = c$.

14. Integrate $g(x, y, z) = xyz$ over the surface of the rectangular solid bounded by the planes $x = \pm a$, $y = \pm b$, and $z = \pm c$.

15. Integrate $g(x, y, z) = x + y + z$ over the portion of the plane $2x + 2y + z = 2$ that lies in the first octant.

16. Integrate $g(x, y, z) = x\sqrt{y^2 + 1}$ over the surface cut from the paraboloid $y^2 + 4z = 16$ by the planes $x = 0$, $x = 1$, and $z = 0$.

Flux across a Surface

In Exercises 17–22, find the flux of the field $\mathbf{F}$ across the portion of the sphere $x^2 + y^2 + z^2 = a^2$ in the first octant in the direction away from the origin.

17. $\mathbf{F}(x, y, z) = z\mathbf{k}$

18. $\mathbf{F}(x, y, z) = -y\mathbf{i} + x\mathbf{j}$

19. $\mathbf{F}(x, y, z) = y\mathbf{i} - x\mathbf{j} + \mathbf{k}$

20. $\mathbf{F}(x, y, z) = zx\mathbf{i} + zy\mathbf{j} + z^2\mathbf{k}$

21. $\mathbf{F}(x, y, z) = x\mathbf{i} + y\mathbf{j} + z\mathbf{k}$

22. $\mathbf{F}(x, y, z) = \dfrac{x\mathbf{i} + y\mathbf{j} + z\mathbf{k}}{\sqrt{x^2 + y^2 + z^2}}$

23. Find the flux of the field $\mathbf{F}(x, y, z) = z^2\mathbf{i} + x\mathbf{j} - 3z\mathbf{k}$ upward through the surface cut from the parabolic cylinder $z = 4 - y^2$ by the planes $x = 0$, $x = 1$, and $z = 0$.

24. Find the flux of the field $\mathbf{F}(x, y, z) = 4x\mathbf{i} + 4y\mathbf{j} + 2\mathbf{k}$ outward through the surface cut from the bottom of the paraboloid $z = x^2 + y^2$ by the plane $z = 1$.

25. Let S be the portion of the cylinder $y = e^x$ in the first octant that projects parallel to the x-axis onto the rectangle R_{yz}: $1 \leq y \leq 2$, $0 \leq z \leq 1$ in the yz-plane (Fig. 15.41). Let $\mathbf{n}$ be the unit vector normal to S that points away from the yz-plane. Find the flux of the field $\mathbf{F}(x, y, z) = -2\mathbf{i} + 2y\mathbf{j} + z\mathbf{k}$ across S in the direction of $\mathbf{n}$.

26. Let S be the portion of the cylinder $y = \ln x$ in the first octant whose projection parallel to the y-axis onto the xz-plane is the rectangle R_{xz}: $1 \leq x \leq e$, $0 \leq z \leq 1$. Let $\mathbf{n}$ be the unit vector normal to S that points away from the xz-plane. Find the flux of $\mathbf{F} = 2y\mathbf{j} + z\mathbf{k}$ through S in the direction of $\mathbf{n}$.

Moments and Masses

27. Find the centroid of the portion of the sphere $x^2 + y^2 + z^2 = 1$ in the first octant.

28. Find the centroid of the surface cut from the cylinder $y^2 + z^2 = 9$, $z \geq 0$, by the planes $x = 0$ and $x = 3$ (resembles the surface in Example 4).

29. Find the center of mass, and the moment of inertia and radius of gyration about the z-axis, of a thin shell of constant density δ cut from the cone $x^2 + y^2 - z^2 = 0$ by the planes $z = 1$ and $z = 2$.

30. Find the moment of inertia about the z-axis of a thin shell of constant density δ cut from the cone $4x^2 + 4y^2 - z^2 = 0$, $z \geq 0$, by the circular cylinder $x^2 + y^2 = 2x$ (Fig. 15.42).

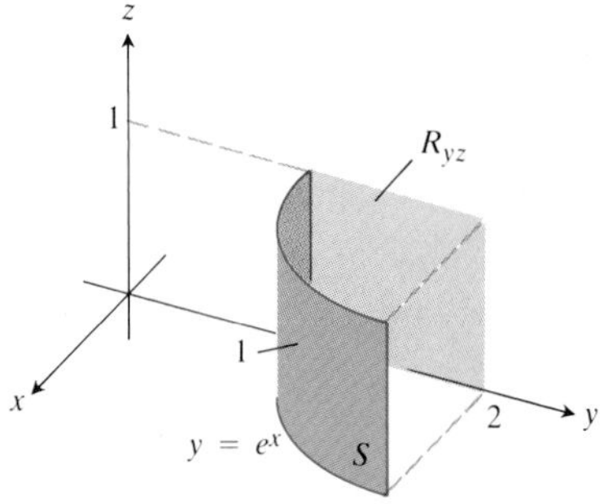

15.41 The surface and region in Exercise 25.

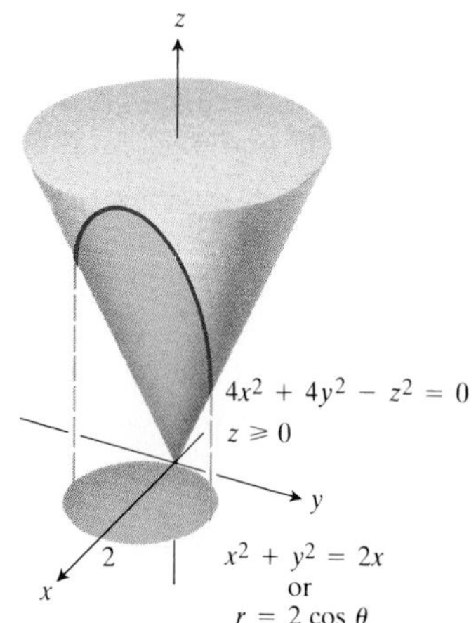

15.42 The surface in Exercise 30.

Special Formulas for Surface Area

If S is a surface defined by a function $z = f(x, y)$ that has continuous first partial derivatives throughout a region R_{xy} in the xy-plane, we may write

$$F(x, y, z) = f(x, y) - z$$

and regard S as the level surface $F(x, y, z) = 0$ of the function F.

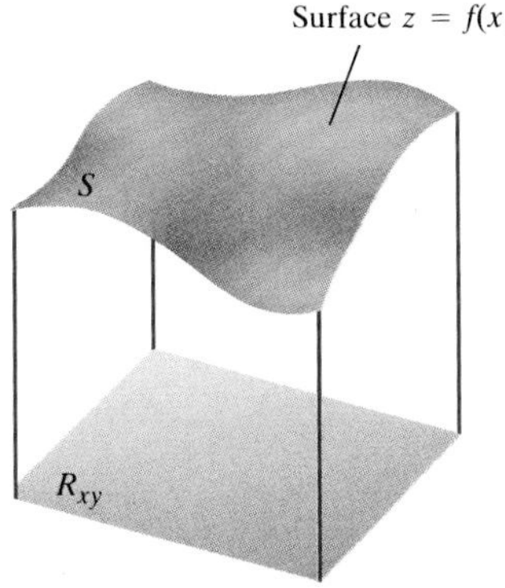

Taking the unit normal to R_{xy} to be $\mathbf{p} = \mathbf{k}$ then gives

$$|\nabla F| = |f_x\mathbf{i} + f_y\mathbf{j} - \mathbf{k}| = \sqrt{f_x^2 + f_y^2 + 1},$$

$$|\nabla F \cdot \mathbf{p}| = |(f_x\mathbf{i} + f_y\mathbf{j} - \mathbf{k}) \cdot \mathbf{k}| = |-1| = 1,$$

and

$$\text{Area of } S = \iint_{R_{xy}} \frac{|\nabla F|}{|\nabla F \cdot \mathbf{p}|}\, dA \qquad (13)$$

$$= \iint_{R_{xy}} \sqrt{f_x^2 + f_y^2 + 1}\, dx\, dy.$$

Similarly, the area of a smooth surface $x = f(y, z)$ over a region R_{yz} in the yz-plane is

$$\text{Area of } S = \iint_{R_{yz}} \sqrt{f_y^2 + f_z^2 + 1}\, dy\, dz, \qquad (14)$$

and the area of a smooth surface $y = f(x, z)$ over a region R_{xz} in the xz-plane is

$$\text{Area of } S = \iint_{R_{xz}} \sqrt{f_x^2 + f_z^2 + 1}\, dx\, dz. \qquad (15)$$

Use Eqs. (13)–(15) to find the areas of the surfaces in Exercises 31–36.

31. The surface cut from the bottom of the paraboloid $z = x^2 + y^2$ by the plane $z = \sqrt{3}$.

32. The portion of the cone $z = \sqrt{x^2 + y^2}$ that lies over the region between the circle $x^2 + y^2 = 1$ and the ellipse $9x^2 + 4y^2 = 36$ in the xy-plane. (*Hint:* Use formulas from geometry to find the area of the region.)

33. The surface cut from the bottom of the paraboloid $x = 1 - y^2 - z^2$ by the yz-plane.

34. The triangle cut from the plane $2x + 6y + 3z = 6$ by the bounding planes of the first octant. Calculate the area three ways, once with each area formula.

35. The surface in the first octant cut from the cylinder $2y = z^2$ by the planes $x = 1$ and $y = 2$.

36. The area of the plane $y + z = 4$ that lies above the region cut from the first quadrant of the xz-plane by the parabola $x = 4 - z^2$.

15.5 The Divergence Theorem

The divergence form of Green's theorem in the plane states that the net outward flux of a vector field across a simple closed curve in the plane can be calculated by integrating the divergence of the field over the region enclosed by the curve. The corresponding theorem in three dimensions, called the divergence theorem, states that the net outward flux of a vector field across a closed surface in space can be calculated by integrating the divergence of the field over the region enclosed by the surface. In this section, we prove the divergence theorem and show how it simplifies the calculation of flux. We also derive Gauss's law for flux in an electric field. The three-dimensional version of the circulation form of Green's theorem will be treated in the next section.

The Divergence Theorem

Mikhail Vassilievich Ostrogradsky (1801–1862) was the first mathematician to publish a proof of the divergence theorem. Having been denied his degree at Kharkhov University by the minister for religious affairs and national education (for being an atheist), Ostrogradsky left Russia for Paris in 1822. There he kept company with Laplace, Legendre, Fourier, Poisson, and Cauchy. While working on the theory of heat in the mid-1820s, he formulated the divergence theorem as a tool for turning volume integrals into surface integrals.

Carl Friedrich Gauss (1777–1855) had already proved the theorem while working on the theory of gravitation (although his notebooks were not published until many years later), and the theorem is sometimes known as Gauss's theorem. The list of Gauss's accomplishments in science and mathematics is truly astonishing, ranging from the invention of the electric telegraph (with Wilhelm Weber in 1833) to the development of a wonderfully accurate theory of planetary orbits and to work in non-Euclidean geometry that later became fundamental to Einstein's general theory of relativity.

Divergence in Three Dimensions

The **divergence** of a vector field $\mathbf{F} = M(x, y, z)\mathbf{i} + N(x, y, z)\mathbf{j} + P(x, y, z)\mathbf{k}$ is the scalar function

$$\operatorname{div} \mathbf{F} = \frac{\partial M}{\partial x} + \frac{\partial N}{\partial y} + \frac{\partial P}{\partial z}. \tag{1}$$

The symbol "div **F**" is read "divergence of **F**" or "div **F**." It has the same physical interpretation as the divergence of a two-dimensional vector field. When **F** is the velocity field of a fluid flow, the value of div **F** at a point (x, y, z) is the rate at which fluid is being piped in or drained away at (x, y, z). The divergence is the flux per unit volume or flux density at the point. We shall say more about this at the end of the section.

Del Notation

The divergence of a three-dimensional vector field **F** is usually expressed in terms of the symbolic operator ∇ ("del"), defined by the equation

$$\nabla = \mathbf{i}\frac{\partial}{\partial x} + \mathbf{j}\frac{\partial}{\partial y} + \mathbf{k}\frac{\partial}{\partial z}. \tag{2}$$

This operator can be applied to any differentiable vector field $\mathbf{F} = M\mathbf{i} + N\mathbf{j} + P\mathbf{k}$ to give the divergence of **F** as a symbolic dot product:

$$\nabla \cdot \mathbf{F} = \frac{\partial M}{\partial x} + \frac{\partial N}{\partial y} + \frac{\partial P}{\partial z} = \operatorname{div} \mathbf{F}. \tag{3}$$

The notation $\nabla \cdot \mathbf{F}$ is read "del dot **F**."

When applied to a differentiable scalar function $f(x, y, z)$, the del operator gives the gradient of f:

$$\nabla f = \frac{\partial f}{\partial x}\mathbf{i} + \frac{\partial f}{\partial y}\mathbf{j} + \frac{\partial f}{\partial z}\mathbf{k}. \tag{4}$$

This may now be read as "del f" as well as "grad f."

Since

$$\frac{\partial^2 f}{\partial x^2} + \frac{\partial^2 f}{\partial y^2} + \frac{\partial^2 f}{\partial z^2} = \nabla \cdot \nabla f = \nabla^2 f, \tag{5}$$

Laplace's equation,

$$\frac{\partial^2 f}{\partial x^2} + \frac{\partial^2 f}{\partial y^2} + \frac{\partial^2 f}{\partial z^2} = 0,$$

from Section 13.3, can now be written as

$$\operatorname{div} \operatorname{grad} f = 0 \qquad \text{or} \qquad \nabla^2 f = 0. \tag{6}$$

If we think of a two-dimensional vector field

$$\mathbf{F}(x, y) = M(x, y)\mathbf{i} + N(x, y)\mathbf{j}$$

as a three-dimensional field whose **k**-component is zero, then

$$\nabla \cdot \mathbf{F} = \frac{\partial M}{\partial x} + \frac{\partial N}{\partial y} = \operatorname{div} \mathbf{F}, \tag{7}$$

and the normal form of Green's theorem for two-dimensional fields (about flux and divergence; see Eq. 27 in Section 15.3) becomes

$$\int_C \mathbf{F} \cdot \mathbf{n}\, ds = \iint_R \nabla \cdot \mathbf{F}\, dA. \tag{8}$$

The tangential form of Green's theorem for two-dimensional fields can also be expressed in vector form in del notation, as we shall see in Section 15.6.

As these equations suggest, del notation provides an easy way to express and remember some of the most important relationships in mathematics, engineering, and physics. We shall see more of this as the chapter continues.

The Divergence Theorem

The divergence theorem says that under suitable conditions the outward flux of a vector field across a closed surface (oriented outward) equals the triple integral of the divergence of the field over the region enclosed by the surface.

The Divergence Theorem

The outward flux of a vector field $\mathbf{F} = M\mathbf{i} + N\mathbf{j} + P\mathbf{k}$ across an oriented surface S with outward unit normal $\mathbf{n}$ equals the integral of $\nabla \cdot \mathbf{F}$ over the region D enclosed by the surface.

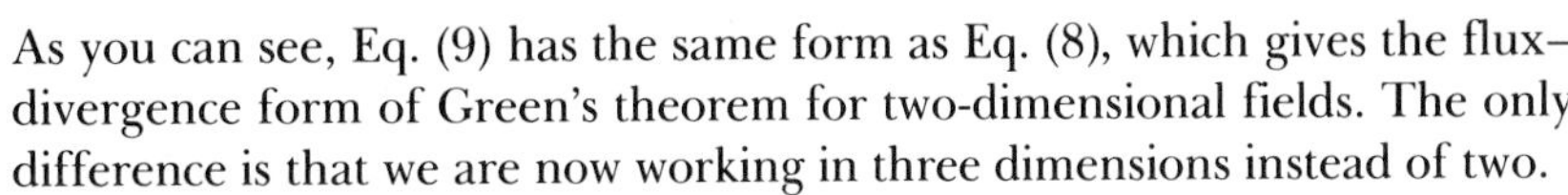

$$\iint_S \mathbf{F} \cdot \mathbf{n}\, d\sigma = \iiint_D \nabla \cdot \mathbf{F}\, dV \tag{9}$$

As you can see, Eq. (9) has the same form as Eq. (8), which gives the flux–divergence form of Green's theorem for two-dimensional fields. The only difference is that we are now working in three dimensions instead of two.

Proof for Special Regions

To prove the divergence theorem, we assume that the components of $\mathbf{F}$ have continuous partial derivatives. We also assume that D is a convex region with no holes or bubbles, such as a solid sphere, cube, or ellipsoid, and that S is a piecewise smooth surface. In addition, we assume that the projection of D into the xy-plane is a simply connected region R_{xy} and that any line perpendicular to the xy-plane at an interior point of R_{xy} intersects the surface S in exactly two points, producing surfaces S_1 and S_2:

$$S_1: \quad z = f_1(x, y), \quad (x, y) \text{ in } R_{xy},$$
$$S_2: \quad z = f_2(x, y), \quad (x, y) \text{ in } R_{xy},$$

with $f_1 \le f_2$. We make similar assumptions about the projection of D onto the other coordinate planes. See Fig. 15.43.

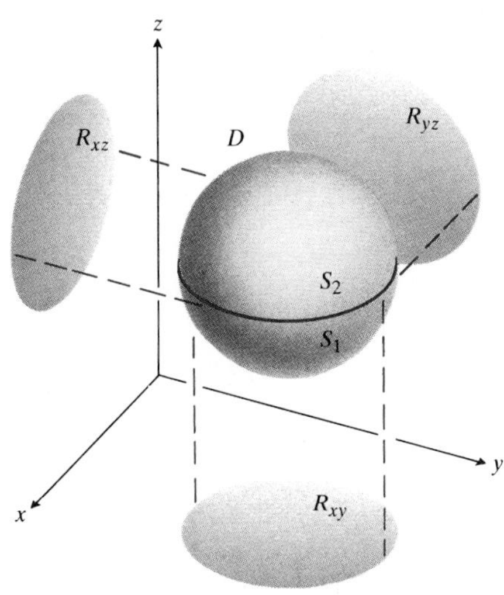

15.43 We first prove the divergence theorem for the kind of three-dimensional region shown here. We then extend the theorem to other regions.

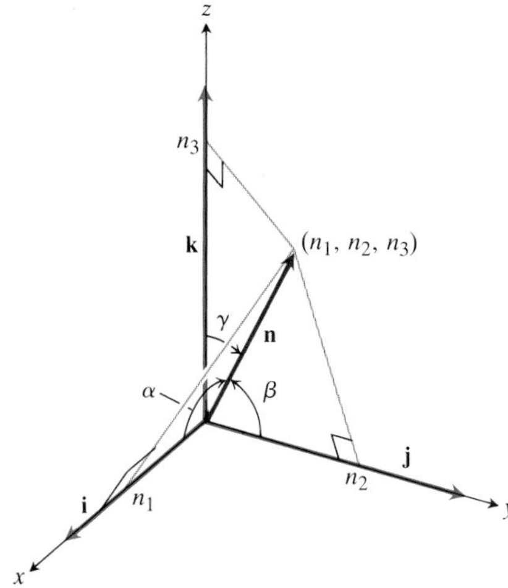

15.44 The scalar components of a unit normal vector **n** are the cosines of the angles α, β, and γ that it makes with **i**, **j**, and **k**.

The components of the unit normal vector $\mathbf{n} = n_1\mathbf{i} + n_2\mathbf{j} + n_3\mathbf{k}$ are the cosines of the angles α, β, and γ that $\mathbf{n}$ makes with $\mathbf{i}$, $\mathbf{j}$, and $\mathbf{k}$ (Fig. 15.44). This is true because all the vectors involved are unit vectors. We have

$$\begin{aligned} n_1 &= \mathbf{n}\cdot\mathbf{i} = |\mathbf{n}|\,|\mathbf{i}|\cos\alpha = \cos\alpha, \\ n_2 &= \mathbf{n}\cdot\mathbf{j} = |\mathbf{n}|\,|\mathbf{j}|\cos\beta = \cos\beta, \\ n_3 &= \mathbf{n}\cdot\mathbf{k} = |\mathbf{n}|\,|\mathbf{k}|\cos\gamma = \cos\gamma. \end{aligned} \tag{10}$$

Thus,

$$\mathbf{n} = (\cos\alpha)\,\mathbf{i} + (\cos\beta)\,\mathbf{j} + (\cos\gamma)\,\mathbf{k}$$

and

$$\mathbf{F}\cdot\mathbf{n} = M\cos\alpha + N\cos\beta + P\cos\gamma.$$

In component form, the divergence theorem states that

$$\iint_S (M\cos\alpha + N\cos\beta + P\cos\gamma)\,d\sigma = \iiint_D \left(\frac{\partial M}{\partial x} + \frac{\partial N}{\partial y} + \frac{\partial P}{\partial z}\right) dx\,dy\,dz. \tag{11}$$

We prove the theorem by proving three separate equalities that add up to Eq. (11):

$$\iint M\cos\alpha\,d\sigma = \iiint \frac{\partial M}{\partial x}\,dx\,dy\,dz, \tag{12}$$

$$\iint N\cos\beta\,d\sigma = \iiint \frac{\partial N}{\partial y}\,dx\,dy\,dz, \tag{13}$$

$$\iint P\cos\gamma\,d\sigma = \iiint \frac{\partial P}{\partial z}\,dx\,dy\,dz. \tag{14}$$

We prove Eq. (14) by converting the surface integral on the left to a double integral over the projection R_{xy} of D on the xy-plane (Fig. 15.45).

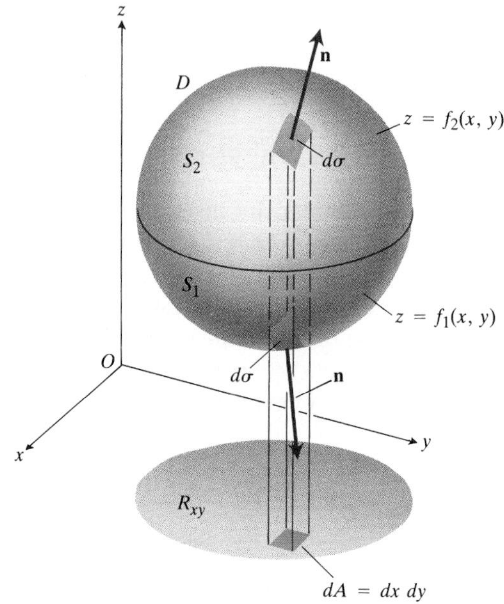

15.45 The three-dimensional region D enclosed by the surfaces S_1 and S_2 shown here projects vertically onto a two-dimensional region R_{xy} in the xy-plane.

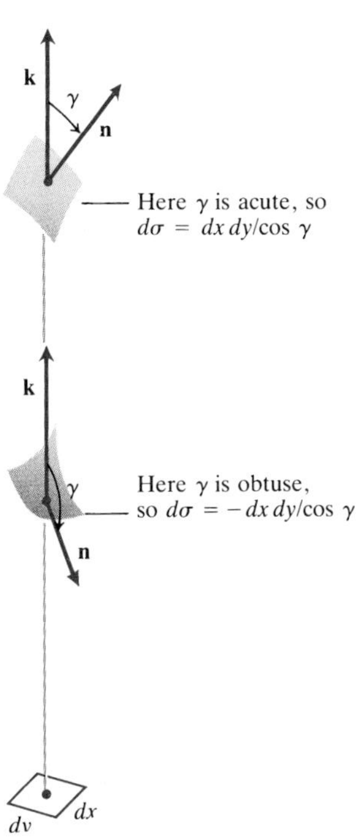

15.46 An enlarged view of the area patches in Fig. 15.45. The relations $d\sigma = \pm dx\, dy/\cos\gamma$ are derived in Section 15.4.

The surface S consists of an upper part S_2 whose equation is $z = f_2(x, y)$ and a lower part S_1 whose equation is $z = f_1(x, y)$. On S_2, the outer normal $\mathbf{n}$ has a positive $\mathbf{k}$-component and

$$\cos\gamma\, d\sigma = dx\, dy \qquad \text{because} \qquad d\sigma = \frac{dA}{|\cos\gamma|} = \frac{dx\, dy}{\cos\gamma}. \tag{15}$$

See Fig. 15.46. On S_1, the outer normal $\mathbf{n}$ has a negative $\mathbf{k}$-component and

$$\cos\gamma\, d\sigma = -dx\, dy. \tag{16}$$

Therefore,

$$\begin{aligned}
\iint_S P\cos\gamma\, d\sigma &= \iint_{S_2} P\cos\gamma\, d\sigma + \iint_{S_1} P\cos\gamma\, d\sigma \\
&= \iint_{R_{xy}} P(x, y, f_2(x, y))\, dx\, dy - \iint_{R_{xy}} P(x, y, f_1(x, y))\, dx\, dy \\
&= \iint_{R_{xy}} [P(x, y, f_2(x, y)) - P(x, y, f_1(x, y))]\, dx\, dy \\
&= \iint_{R_{xy}} \left[\int_{f_1(x,y)}^{f_2(x,y)} \frac{\partial P}{\partial z}\, dz\right] dx\, dy = \iiint_D \frac{\partial P}{\partial z}\, dz\, dx\, dy.
\end{aligned} \tag{17}$$

This proves Eq. (14).

The proofs for Eqs. (12) and (13) follow the same pattern; or just permute x, y, z; M, N, P; α, β, γ, in order, and get those results from Eq. (14).

EXAMPLE 1 Evalute both sides of Eq. (9) for the field $\mathbf{F} = x\mathbf{i} + y\mathbf{j} + z\mathbf{k}$ over the sphere $x^2 + y^2 + z^2 = a^2$.

Solution The outer unit normal to S, calculated from the gradient of $f(x, y, z) = x^2 + y^2 + z^2 - a^2$, is

$$\mathbf{n} = \frac{2(x\mathbf{i} + y\mathbf{j} + z\mathbf{k})}{\sqrt{4(x^2 + y^2 + z^2)}} = \frac{x\mathbf{i} + y\mathbf{j} + z\mathbf{k}}{a}.$$

Hence,

$$\mathbf{F}\cdot\mathbf{n}\, d\sigma = \frac{x^2 + y^2 + z^2}{a}\, d\sigma = \frac{a^2}{a}\, d\sigma = a\, d\sigma$$

because $x^2 + y^2 + z^2 = a^2$ on the surface. Therefore

$$\iint_S \mathbf{F}\cdot\mathbf{n}\, d\sigma = \iint_S a\, d\sigma = a\iint_S d\sigma = a(4\pi a^2) = 4\pi a^3.$$

The divergence of $\mathbf{F}$ is

$$\nabla\cdot\mathbf{F} = \frac{\partial}{\partial x}(x) + \frac{\partial}{\partial y}(y) + \frac{\partial}{\partial z}(z) = 3,$$

so

$$\iiint_D \nabla\cdot\mathbf{F}\, dV = \iiint_D 3\, dV = 3\left(\frac{4}{3}\pi a^3\right) = 4\pi a^3.$$

■

EXAMPLE 2 Calculate the flux of the field $\mathbf{F} = xy\mathbf{i} + yz\mathbf{j} + xz\mathbf{k}$ outward through the surface of the cube cut from the first octant by the planes $x = 1, y = 1$, and $z = 1$.

Solution The divergence theorem says that instead of calculating the flux as a sum of six separate integrals, one for each face of the cube, we can calculate the flux by integrating the divergence

$$\nabla \cdot \mathbf{F} = \frac{\partial}{\partial x}(xy) + \frac{\partial}{\partial y}(yz) + \frac{\partial}{\partial z}(xz) = y + z + x$$

over the cube's interior:

$$\begin{aligned} \text{Flux} &= \iint_{\substack{\text{Cube}\\ \text{surface}}} \mathbf{F} \cdot \mathbf{n}\, d\sigma \\ &= \iiint_{\substack{\text{Cube}\\ \text{interior}}} \nabla \cdot \mathbf{F}\, dV \qquad \text{(The divergence theorem)} \\ &= \int_0^1 \int_0^1 \int_0^1 (x + y + z)\, dx\, dy\, dz \\ &= \frac{3}{2}. \qquad \text{(After a routine integration)} \end{aligned}$$

■

The Divergence Theorem for Other Regions

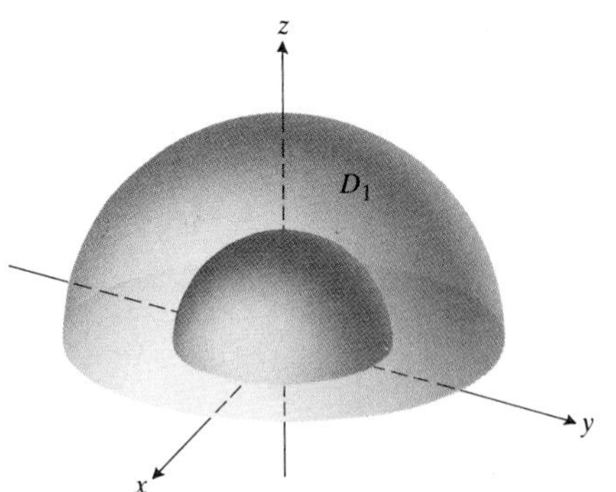

15.47 The upper half of the region between two spheres.

The divergence theorem can be extended to regions that can be split up into a finite number of simple regions of the type just discussed and to regions that can be defined as limits of simpler regions in certain ways. For example, suppose D is the region between two concentric spheres, and $\mathbf{F}$ has continuously differentiable components throughout D and on the bounding surfaces. Split D by an equatorial plane and apply the divergence theorem to each half separately. The top half, D_1, is shown in Fig. 15.47. The surface that bounds D_1 consists of an outer hemisphere, a plane washer-shaped base, and an inner hemisphere. The divergence theorem says that

$$\iint_{S_1} \mathbf{F} \cdot \mathbf{n}_1\, d\sigma_1 = \iiint_{D_1} \nabla \cdot \mathbf{F}\, dV_1. \tag{18}$$

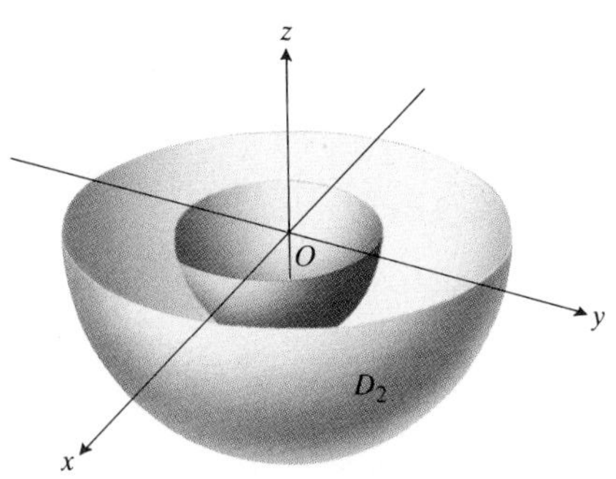

15.48 The lower half of the region between two spheres.

The unit normal $\mathbf{n}_1$ that points outward from D_1 points away from the origin along the outer surface, points down along the flat base, and points toward the origin along the inner surface. Next apply the divergence theorem to D_2, as shown in Fig. 15.48:

$$\iint_{S_2} \mathbf{F} \cdot \mathbf{n}_2\, d\sigma_2 = \iiint_{D_2} \nabla \cdot \mathbf{F}\, dV_2. \tag{19}$$

As we follow $\mathbf{n}_2$ over S_2, pointing outward from D_2, we see that $\mathbf{n}_2$ points upward along the flat surface in the xy-plane, points away from the origin on the outer sphere, and points toward the origin on the inner sphere. When we add Eqs. (18) and (19), the surface integrals over the flat base cancel because of the opposite signs of $\mathbf{n}_1$ and $\mathbf{n}_2$. We thus arrive at the result

$$\iint_S \mathbf{F} \cdot \mathbf{n}\, d\sigma = \iiint_D \nabla \cdot \mathbf{F}\, dV, \tag{20}$$

with D the region between the spheres, S the boundary of D consisting of two spheres, and $\mathbf{n}$ the unit normal to S directed outward from D.

EXAMPLE 3 Find the net outward flux of the field

$$\mathbf{F} = \frac{x\mathbf{i} + y\mathbf{j} + z\mathbf{k}}{\rho^3}, \qquad \rho = \sqrt{x^2 + y^2 + z^2}$$

across the boundary of the region $D: 0 < a^2 \le x^2 + y^2 + z^2 \le b^2$.

Solution The flux can be calculated by integrating $\nabla \cdot \mathbf{F}$ over D. We have

$$\frac{\partial \rho}{\partial x} = \frac{1}{2}(x^2 + y^2 + z^2)^{-1/2}(2x) = \frac{x}{\rho}$$

and

$$\frac{\partial M}{\partial x} = \frac{\partial}{\partial x}(x\rho^{-3}) = \rho^{-3} - 3x\rho^{-4}\frac{\partial \rho}{\partial x} = \frac{1}{\rho^3} - \frac{3x^2}{\rho^5}.$$

Similarly,

$$\frac{\partial N}{\partial y} = \frac{1}{\rho^3} - \frac{3y^2}{\rho^5} \quad \text{and} \quad \frac{\partial P}{\partial z} = \frac{1}{\rho^3} - \frac{3z^2}{\rho^5}.$$

Hence,

$$\text{div}\,\mathbf{F} = \frac{3}{\rho^3} - \frac{3}{\rho^5}(x^2 + y^2 + z^2) = \frac{3}{\rho^3} - \frac{3\rho^2}{\rho^5} = 0$$

and

$$\iiint_D \nabla \cdot \mathbf{F}\, dV = 0.$$

So the integral of $\nabla \cdot \mathbf{F}$ over D is zero and the net outward flux across the boundary of D is zero. But there is more to learn from this example. The flux leaving D across the inner sphere S_a is the negative of the flux leaving D across the outer sphere S_b (because the sum of these fluxes is zero). This means that the flux of $\mathbf{F}$ across S_a in the direction away from the origin equals the flux of $\mathbf{F}$ across S_b in the direction away from the origin. Thus, the flux of $\mathbf{F}$ across a sphere centered at the origin is independent of the radius of the sphere. What is this flux?

To find it, we evaluate the flux integral directly. The outward unit normal on the sphere of radius a is

$$\mathbf{n} = \frac{x\mathbf{i} + y\mathbf{j} + z\mathbf{k}}{\sqrt{x^2 + y^2 + z^2}} = \frac{x\mathbf{i} + y\mathbf{j} + z\mathbf{k}}{a}.$$

Hence, on the sphere,

$$\mathbf{F}\cdot\mathbf{n} = \frac{x\mathbf{i}+y\mathbf{j}+z\mathbf{k}}{a^3}\cdot\frac{x\mathbf{i}+y\mathbf{j}+z\mathbf{k}}{a} = \frac{x^2+y^2+z^2}{a^4} = \frac{a^2}{a^4} = \frac{1}{a^2}$$

and

$$\iint_{S_a} \mathbf{F}\cdot\mathbf{n}\,d\sigma = \frac{1}{a^2}\iint_{S_a} d\sigma = \frac{1}{a^2}(4\pi a^2) = 4\pi.$$

The outward flux of $\mathbf{F}$ across any sphere centered at the origin is 4π. ■

Gauss's Law—One of the Four Great Laws of Electromagnetic Theory

There is still more to be learned from Example 3. In electromagnetic theory, the electric field created by a point charge q located at the origin is the inverse square field

$$\mathbf{E}(x, y, z) = q\frac{\mathbf{r}}{|\mathbf{r}|^3} = q\frac{x\mathbf{i}+y\mathbf{j}+z\mathbf{k}}{\rho^3},$$

where $\mathbf{r}$ is the position vector of the point (x, y, z) and $\rho = |\mathbf{r}| = \sqrt{x^2+y^2+z^2}$. In the notation of Example 3,

$$\mathbf{E} = q\mathbf{F}.$$

The calculations in Example 3 show that the outward flux of $\mathbf{E}$ across any sphere centered at the origin is $4\pi q$. But this result is not confined to spheres. The outward flux of $\mathbf{E}$ across any surface S that encloses the origin (and to which the divergence theorem applies) is also $4\pi q$. To see why, we have only to imagine a large sphere S_a centered at the origin and enclosing the surface S. Since

$$\nabla\cdot\mathbf{E} = \nabla\cdot q\mathbf{F} = q\nabla\cdot\mathbf{F} = 0$$

when $\rho > 0$, the integral of $\nabla\cdot\mathbf{E}$ over the region D between S and S_a is zero. Hence, by the divergence theorem,

$$\iint_{\substack{\text{Boundary}\\ \text{of } D}} \mathbf{E}\cdot\mathbf{n}\,d\sigma = 0,$$

and the flux of $\mathbf{E}$ across S in the direction away from the origin must be the same as the flux of $\mathbf{E}$ across S_a in the direction away from the origin, which is $4\pi q$. This statement, called *Gauss's law,* applies to much more general charge distributions than the one we have assumed here, as you will see if you open nearly any college physics text.

Gauss's Law

$$\iint_S \mathbf{E}\cdot\mathbf{n}\,d\sigma = 4\pi q \qquad (21)$$

Exercises 15.5

In Exercises 1–12, use the divergence theorem to find the outward flux of $\mathbf{F}$ across the boundary of the region D.

1. $\mathbf{F} = (y - x)\mathbf{i} + (z - y)\mathbf{j} + (y - x)\mathbf{k}$
D: The cube bounded by the planes $x = \pm 1$, $y = \pm 1$, $z = \pm 1$

2. $\mathbf{F} = x^2\mathbf{i} + y^2\mathbf{j} + z^2\mathbf{k}$
D: The cube cut from the first octant by the planes $x = 1$, $y = 1$, and $z = 1$

3. $\mathbf{F} = x^2\mathbf{i} + y^2\mathbf{j} + z^2\mathbf{k}$
D: The cube bounded by the planes $x = \pm 1$, $y = \pm 1$, $z = \pm 1$

4. $\mathbf{F} = x^2\mathbf{i} + y^2\mathbf{j} + z^2\mathbf{k}$
D: The region cut from the solid cylinder $x^2 + y^2 \le 4$ by the planes $z = 0$ and $z = 1$.

5. $\mathbf{F} = y\mathbf{i} + xy\mathbf{j} - z\mathbf{k}$
D: The region inside the solid cylinder $x^2 + y^2 \le 4$ between the plane $z = 0$ and the paraboloid $z = x^2 + y^2$

6. $\mathbf{F} = x^2\mathbf{i} + xz\mathbf{j} - 3z\mathbf{k}$
D: The solid sphere $x^2 + y^2 + z^2 \le 4$

7. $\mathbf{F} = x^2\mathbf{i} - 2xy\mathbf{j} + 3xz\mathbf{k}$
D: The region cut from the first octant by the sphere $x^2 + y^2 + z^2 = 4$

8. $\mathbf{F} = (6x^2 + 2xy)\mathbf{i} + (2y + x^2z)\mathbf{j} + 4x^2y^3\mathbf{k}$
D: The region cut from the first octant by the cylinder $x^2 + y^2 = 4$ and the plane $z = 3$

9. $\mathbf{F} = 2xz\mathbf{i} - xy\mathbf{j} - z^2\mathbf{k}$
D: The wedge cut from the first octant by the plane $y + z = 4$ and the elliptical cylinder $4x^2 + y^2 = 16$

10. $\mathbf{F} = x^3\mathbf{i} + y^3\mathbf{j} + z^3\mathbf{k}$
D: The solid sphere $x^2 + y^2 + z^2 \le a^2$

11. $\mathbf{F} = \sqrt{x^2 + y^2 + z^2}\,(x\mathbf{i} + y\mathbf{j} + z\mathbf{k})$
D: The region $1 \le x^2 + y^2 + z^2 \le 2$

12. $\mathbf{F} = (x\mathbf{i} + y\mathbf{j} + z\mathbf{k})/\sqrt{x^2 + y^2 + z^2}$
D: The region $1 \le x^2 + y^2 + z^2 \le 4$

13. Let S be the closed cubelike surface in Fig. 15.49, with its base the unit square in the xy-plane, its four sides lying in the planes $x = 0$, $x = 1$, $y = 0$, $y = 1$, and its top an arbitrary smooth surface whose identity is unknown. Let $\mathbf{F} = x\mathbf{i} - 2y\mathbf{j} + (z + 3)\mathbf{k}$. Suppose the outward flux of $\mathbf{F}$ through Side A is 1 and through Side B is -3. Find the outward flux of $\mathbf{F}$ through the top.

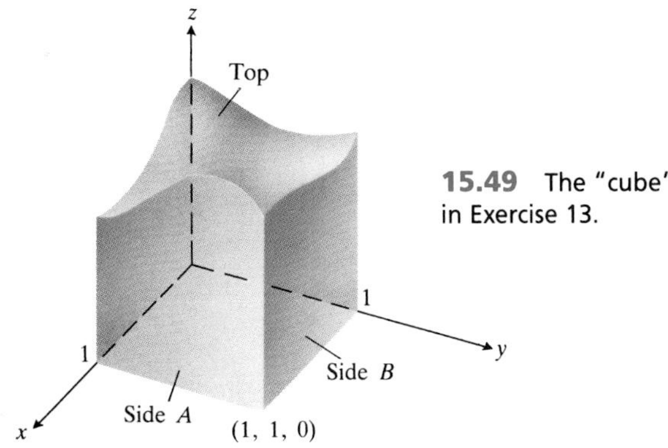

15.49 The "cube" in Exercise 13.

14. Let $\mathbf{F}$ be a field whose components have continuous first partial derivatives throughout a portion of space containing a closed and bounded region D and its smooth bounding surface S. Suppose that the length of the vector $\mathbf{F}$ never exceeds 1 on S. What bound can be placed on the numerical size of the integral

$$\iiint_D \nabla \cdot \mathbf{F}\, dV?$$

How do you know?

15. a) Show that the flux of the position vector field $\mathbf{F} = x\mathbf{i} + y\mathbf{j} + z\mathbf{k}$ outward through a smooth closed surface S is three times the volume of the region enclosed by the surface.

b) Let $\mathbf{n}$ be the outward unit normal vector on S. Show that it is not possible for $\mathbf{F}$ to be orthogonal to $\mathbf{n}$ at every point of S.

16. *Harmonic functions.* A function $f(x, y, z)$ is said to be **harmonic** in a region D in space if it satisfies the Laplace equation

$$\nabla^2 f = \nabla \cdot \nabla f = \frac{\partial^2 f}{\partial x^2} + \frac{\partial^2 f}{\partial y^2} + \frac{\partial^2 f}{\partial z^2} = 0$$

throughout D.

a) Suppose that f is harmonic throughout a bounded region D enclosed by a smooth surface S and that $\mathbf{n}$ is the chosen unit normal vector on S. Show that the integral over S of $\nabla f \cdot \mathbf{n}$, the derivative of f in the direction of $\mathbf{n}$, is zero.

b) Show that if f is harmonic on D, then

$$\iint_S f \nabla f \cdot \mathbf{n}\, d\sigma = \iiint_D |\nabla f|^2\, dV.$$

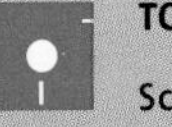

TOOLKIT PROGRAMS

Scalar Fields	Plots two-dimensional scalar fields and their associated gradient fields. Integrates div ∇f along line segments and rectangle boundaries.
Vector Fields	Displays two-dimensional vector fields and integrates the curl over rectangular regions. Also calculates flux across rectangular boundaries.

15.6 Stokes's Theorem

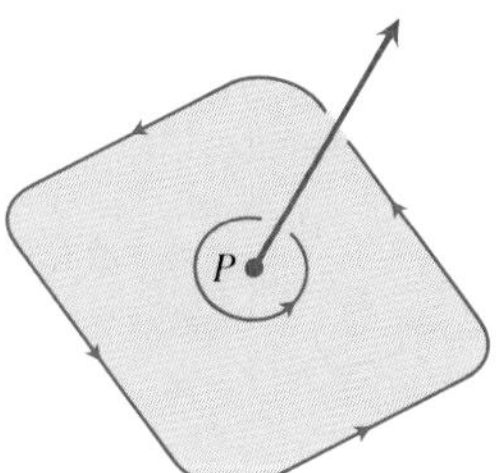

15.50 The circulation vector at a point P in a plane in a three-dimensional fluid flow. Notice its right-hand relation to the circulation line.

As we saw in Section 15.3, the circulation density or curl of a two-dimensional field $\mathbf{F} = M\mathbf{i} + N\mathbf{j}$ at a point (x, y) is described by the scalar quantity $(\partial N/\partial x - \partial M/\partial y)$. In three dimensions, the circulation around a point P in a plane is described not with a scalar but with a vector. This vector is normal to the plane of the circulation (Fig. 15.50) and points in the direction that gives it a right-hand relation to the circulation line. The length of the vector gives the rate of the fluid's rotation, which usually varies as the circulation plane is tilted about P from one position to another. It turns out that the vector of greatest circulation in a flow with velocity field $\mathbf{F} = M\mathbf{i} + N\mathbf{j} + P\mathbf{k}$ is the vector

$$\operatorname{curl} \mathbf{F} = \left(\frac{\partial P}{\partial y} - \frac{\partial N}{\partial z}\right)\mathbf{i} + \left(\frac{\partial M}{\partial z} - \frac{\partial P}{\partial x}\right)\mathbf{j} + \left(\frac{\partial N}{\partial x} - \frac{\partial M}{\partial y}\right)\mathbf{k}. \tag{1}$$

We get this information from Stokes's theorem, the generalization of the circulation–curl form of Green's theorem to space.

In this section we present Stokes's theorem and investigate what it says about circulation. We shall also see how Stokes's theorem sometimes enables us to evaluate a line integral around a loop in space by calculating an easier surface integral. This is similar to the way we used Green's theorem to evaluate a line integral over a complicated curve in the plane by calculating an easier double integral over the region enclosed by the curve. In addition, we shall state the circulation–curl or tangential form of Green's theorem in its final vector form. Stokes's theorem has important applications in electromagnetic theory, but we shall not go into them here.

Stokes's Theorem

In del notation, the curl of a three-dimensional vector field $\mathbf{F} = M\mathbf{i} + N\mathbf{j} + P\mathbf{k}$ is written as the symbolic cross product

$$\operatorname{curl} \mathbf{F} = \nabla \times \mathbf{F} = \begin{vmatrix} \mathbf{i} & \mathbf{j} & \mathbf{k} \\ \dfrac{\partial}{\partial x} & \dfrac{\partial}{\partial y} & \dfrac{\partial}{\partial z} \\ M & N & P \end{vmatrix}$$

$$= \left(\frac{\partial P}{\partial y} - \frac{\partial N}{\partial z}\right)\mathbf{i} + \left(\frac{\partial M}{\partial z} - \frac{\partial P}{\partial x}\right)\mathbf{j} + \left(\frac{\partial N}{\partial x} - \frac{\partial M}{\partial y}\right)\mathbf{k}. \tag{2}$$

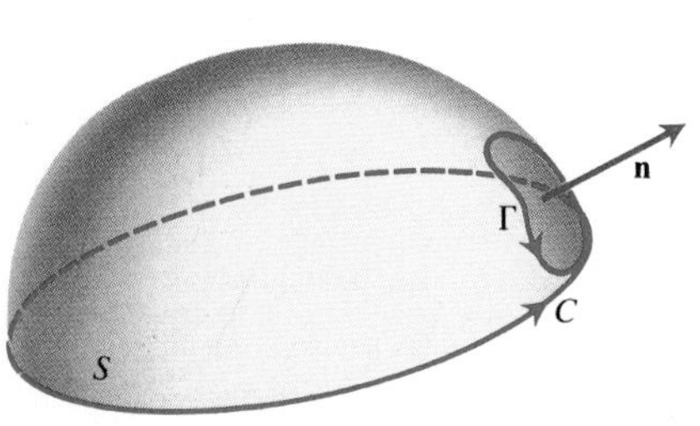

15.51 The orientation of the bounding curve C gives it a right-handed relation to $\mathbf{n}$.

Stokes's theorem says that, under conditions normally met in practice, the circulation of a vector field around the boundary of an oriented surface in space in the direction counterclockwise with respect to the surface's unit normal vector $\mathbf{n}$ (Fig. 15.51) is equal to the double integral of the curl of the field over the surface.

Stokes's Theorem

Sir George Gabriel Stokes (1819–1903) was Lucasian Professor of Mathematics at Cambridge University from 1849 until his death in 1903. One of the most influential scientific figures of his century, his theoretical and experimental investigations covered hydrodynamics, elasticity, light, gravity, sound, heat, meteorology, and solar physics. He left electricity and magnetism to his friend William Thomson, Baron Kelvin of Largs. It is another one of those delightful quirks of history that the theorem we call Stokes's theorem isn't his theorem at all. He learned of it from Thomson in 1850 and a few years later, thinking it would make a good examination question, put it on the Smith Prize examination. It has been known as Stokes's theorem ever since. As usual, things have balanced out. Stokes was the original discoverer of the principles of spectrum analysis that we now credit to Bunsen and Kirchhoff.

Stokes's Theorem

The circulation of $\mathbf{F} = M\mathbf{i} + N\mathbf{j} + P\mathbf{k}$ around the boundary C of an oriented surface S in the direction counterclockwise with respect to the surface's unit normal vector $\mathbf{n}$ equals the integral of $\nabla \times \mathbf{F} \cdot \mathbf{n}$ over S.

$$\underbrace{\oint_C \mathbf{F} \cdot d\mathbf{r}}_{\text{Counterclockwise circulation}} = \underbrace{\iint_S \nabla \times \mathbf{F} \cdot \mathbf{n}\, d\sigma}_{\text{Curl integral}} \tag{3}$$

Naturally, we need some mathematical restrictions on $\mathbf{F}$, C, and S to ensure the existence of the integrals in Stokes's equation. The usual restrictions are that all the functions and derivatives involved be continuous.

If C is a curve in the xy-plane, oriented counterclockwise, and R is the region bounded by C, then $d\sigma = dx\,dy$ and

$$(\nabla \times \mathbf{F}) \cdot \mathbf{n} = (\nabla \times \mathbf{F}) \cdot \mathbf{k} = \left(\frac{\partial N}{\partial x} - \frac{\partial M}{\partial y}\right). \tag{4}$$

Under these conditions, Stokes's equation becomes

$$\oint_C \mathbf{F} \cdot d\mathbf{r} = \iint_R \left(\frac{\partial N}{\partial x} - \frac{\partial M}{\partial y}\right) dx\,dy,$$

which is the circulation–curl form of the equation in Green's theorem. Conversely, by reversing these steps we can rewrite Green's theorem for two-dimensional fields in del notation as

$$\oint_C \mathbf{F} \cdot d\mathbf{r} = \iint_R \nabla \times \mathbf{F} \cdot \mathbf{k}\, dA. \tag{5}$$

See Fig. 15.52.

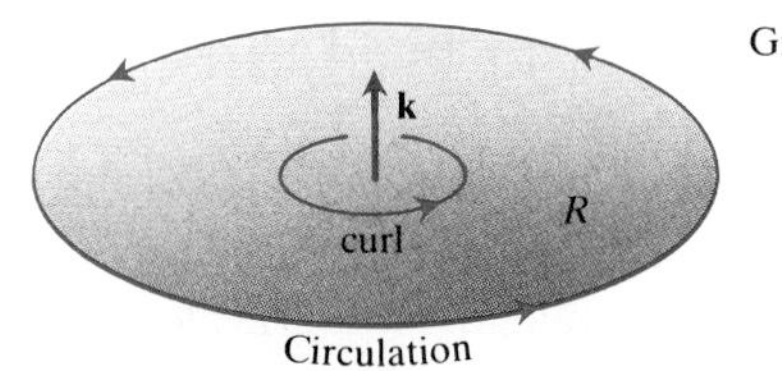

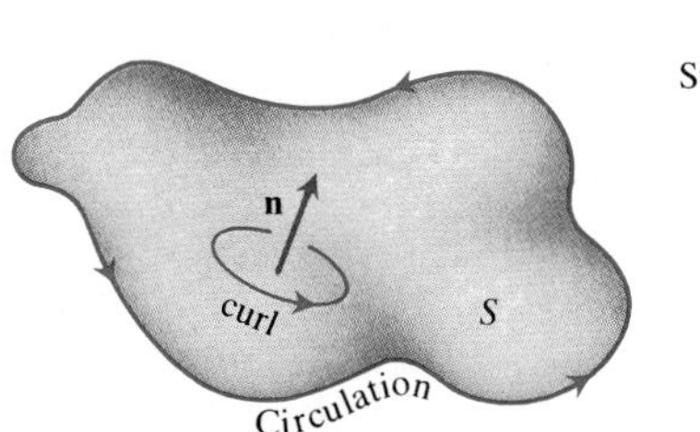

15.52 Green's theorem vs. Stokes's theorem.

EXAMPLE 1 Evaluate the two integrals in Stokes's theorem for the hemisphere

$$x^2 + y^2 + z^2 = 9, \qquad z \geq 0,$$

its bounding circle

$$x^2 + y^2 = 9, \qquad z = 0.$$

and the field

$$\mathbf{F} = y\mathbf{i} - x\mathbf{j}.$$

Solution The integrand of the circulation integral is

$$\mathbf{F} \cdot d\mathbf{r} = (y\mathbf{i} - x\mathbf{j}) \cdot (dx\mathbf{i} + dy\mathbf{j} + dz\mathbf{k}) = y\,dx - x\,dy.$$

By the circulation form of Green's theorem for the plane,

$$\oint_C \mathbf{F} \cdot d\mathbf{r} = \oint_C (y\,dx - x\,dy) = \iint_R \left(\frac{\partial N}{\partial x} - \frac{\partial M}{\partial y}\right) dx\,dy$$

$$= \iint_{x^2+y^2 \leq 9} -2\,dx\,dy = (-2)9\pi = -18\pi.$$

For the curl integral of $\mathbf{F} = y\mathbf{i} - x\mathbf{j}$, we have

$$\begin{aligned}\nabla \times \mathbf{F} &= \left(\frac{\partial P}{\partial y} - \frac{\partial N}{\partial z}\right)\mathbf{i} + \left(\frac{\partial M}{\partial z} - \frac{\partial P}{\partial x}\right)\mathbf{j} + \left(\frac{\partial N}{\partial x} - \frac{\partial M}{\partial y}\right)\mathbf{k} \\ &= (0-0)\mathbf{i} + (0-0)\mathbf{j} + (-1-1)\mathbf{k} = -2\mathbf{k},\end{aligned}$$

$$\mathbf{n} = \frac{x\mathbf{i} + y\mathbf{j} + z\mathbf{k}}{\sqrt{x^2+y^2+z^2}} = \frac{x\mathbf{i} + y\mathbf{j} + z\mathbf{k}}{3}, \qquad \text{(Outer unit normal)}$$

$$d\sigma = \frac{3}{z}\,dA \qquad \text{(As in Section 15.4, Example 5)}$$

$$\nabla \times \mathbf{F} \cdot \mathbf{n}\, d\sigma = -\frac{2z}{3}\frac{3}{z}\,dA = -2\,dA,$$

and

$$\iint_S \nabla \times \mathbf{F} \cdot \mathbf{n}\, d\sigma = \iint_{x^2+y^2 \le 9} -2\,dA = -18\pi.$$

The circulation around the circle equals the integral of the curl over the hemisphere, as it should. ■

Proof of Stokes's Theorem for Polyhedral Surfaces

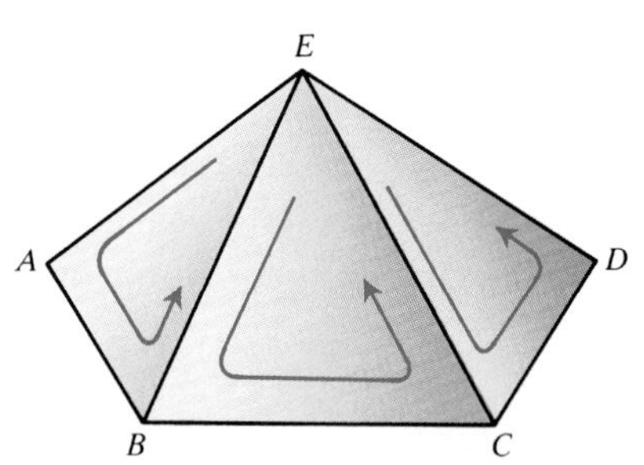

15.53 Part of a polyhedral surface.

Let the surface S be a polyhedral surface consisting of a finite number of plane regions. (Think of one of Buckminster Fuller's geodesic domes.) We apply Green's theorem to each separate panel of S. There are two types of panels:

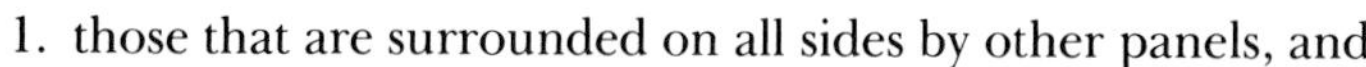

1. those that are surrounded on all sides by other panels, and
2. those that have one or more edges that are not adjacent to other panels.

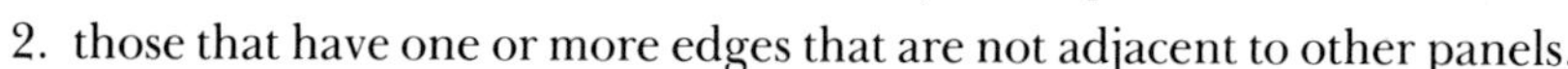

The boundary Δ of S consists of those edges of the type 2 panels that are not adjacent to other panels. In Fig. 15.53, the triangles EAB, BCE, and CDE represent a part of S, with $ABCD$ part of the boundary Δ. Applying Green's theorem to the three triangles in turn and adding the results, we get

$$\left(\oint_{EAB} + \oint_{BCE} + \oint_{CDE}\right) \mathbf{F} \cdot d\mathbf{r} = \left(\iint_{EAB} + \iint_{BCE} + \iint_{CDE}\right) \nabla \times \mathbf{F} \cdot \mathbf{n}\, d\sigma. \tag{6}$$

The three line integrals on the left-hand side of Eq. (6) combine into a single line integral taken around the periphery $ABCDE$ because the integrals along interior segments cancel in pairs. For example, the integral along the segment BE in triangle ABE is opposite in sign to the integral along the same segment in triangle EBC. Similarly for the segment CE. Hence (6) reduces to

$$\oint_{ABCDE} \mathbf{F} \cdot d\mathbf{r} = \iint_{ABCDE} \nabla \times \mathbf{F} \cdot \mathbf{n}\, d\sigma.$$

When we apply Green's theorem to all the panels and add the results, we get

$$\oint_{\Delta} \mathbf{F} \cdot d\mathbf{r} = \iint_S \nabla \times \mathbf{F} \cdot \mathbf{n}\, d\sigma. \tag{7}$$

This is Stokes's theorem for a polyhedral surface S. You can find proofs for more general surfaces in advanced calculus texts. ■

Stokes's Theorem for Surfaces with Holes

Stokes's theorem can also be extended to a surface S that has one or more holes in it (like a curved slice of Swiss cheese), in a way exactly analogous to Green's theorem: The surface integral over S of the normal component of $\nabla \times \mathbf{F}$ is equal to the sum of the line integrals around all the boundaries of S (including boundaries of the holes) of the tangential component of $\mathbf{F}$, where the boundary curves are to be traced in the positive direction induced by the positive orientation of S.

Circulation of a Fluid

Stokes's theorem provides the following vector interpretation for $\nabla \times \mathbf{F}$. As in the discussion of divergence, let $\mathbf{v}$ be the velocity field of a moving fluid, δ the density, and $\mathbf{F} = \delta \mathbf{v}$. Then

$$\oint_C \mathbf{F} \cdot d\mathbf{r}$$

is the circulation of the fluid around the closed curve C. By Stokes's theorem, the circulation is equal to the flux of $\nabla \times \mathbf{F}$ through a surface S spanning C:

$$\oint_C \mathbf{F} \cdot d\mathbf{r} = \iint_S \nabla \times \mathbf{F} \cdot \mathbf{n}\, d\sigma.$$

Suppose we fix a point Q in the domain of $\mathbf{F}$ and a direction $\mathbf{u}$ at Q. Let C be a circle of radius ρ, with center at Q, whose plane is normal to $\mathbf{u}$. If $\nabla \times \mathbf{F}$ is continuous at Q, then the average value of the $\mathbf{u}$-component of $\nabla \times \mathbf{F}$ over the circular disk bounded by C approaches the $\mathbf{u}$-component of $\nabla \times \mathbf{F}$ at Q as $\rho \to 0$:

$$(\nabla \times \mathbf{F} \cdot \mathbf{u})_Q = \lim_{\rho \to 0} \frac{1}{\pi \rho^2} \iint_S \nabla \times \mathbf{F} \cdot \mathbf{u}\, d\sigma. \tag{8}$$

If we replace the double integral on the right-hand side of Eq. (8) by the circulation, we get

$$(\nabla \times \mathbf{F} \cdot \mathbf{u})_Q = \lim_{\rho \to 0} \frac{1}{\pi \rho^2} \oint_C \mathbf{F} \cdot d\mathbf{r}. \tag{9}$$

The left-hand side of Eq. (9) is a maximum at Q when $\mathbf{u}$ has the same direction as $\nabla \times \mathbf{F}$. When ρ is small, the limit on the right-hand side of Eq. (9) is approximately equal to

$$\frac{1}{\pi \rho^2} \oint_C \mathbf{F} \cdot d\mathbf{r},$$

which is the circulation around C divided by the area of the disk (circulation density). Suppose that a small paddle wheel, of radius ρ, is introduced into

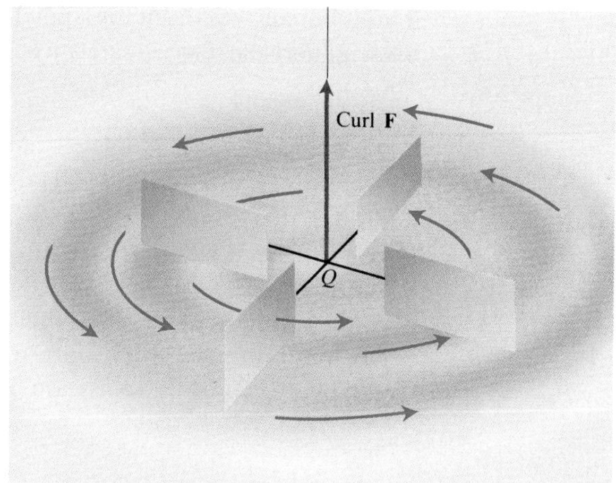

15.54 The paddle-wheel interpretation of curl **F**.

the fluid at Q, with its axle directed along $\mathbf{u}$. The circulation of the fluid around C will affect the rate of spin of the paddle wheel. The wheel will spin fastest when the circulation integral is maximized; therefore it will spin fastest when the axle of the paddle wheel points in the direction of $\nabla \times \mathbf{F}$ (Fig. 15.54).

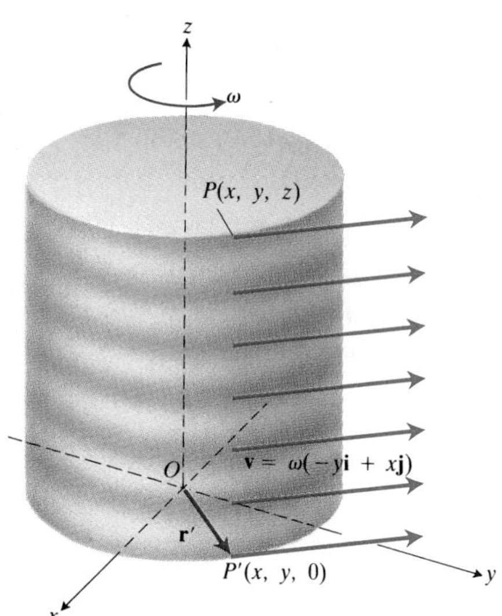

15.55 A steady rotational flow parallel to the xy-plane, with constant angular velocity ω in the positive (counterclockwise) direction.

EXAMPLE 2 A fluid of constant density δ rotates around the z-axis with velocity $\mathbf{v} = \omega(-y\mathbf{i} + x\mathbf{j})$, where ω is a positive constant called the *angular velocity* of the rotation (Fig. 15.55). If $\mathbf{F} = \delta\mathbf{v}$, find $\nabla \times \mathbf{F}$ and show its relation to the circulation density.

Solution With $\mathbf{F} = -\delta\omega y\mathbf{i} + \delta\omega x\mathbf{j}$,

$$\begin{aligned}\nabla \times \mathbf{F} &= \left(\frac{\partial P}{\partial y} - \frac{\partial N}{\partial z}\right)\mathbf{i} + \left(\frac{\partial M}{\partial z} - \frac{\partial P}{\partial x}\right)\mathbf{j} + \left(\frac{\partial N}{\partial x} - \frac{\partial M}{\partial y}\right)\mathbf{k} \\ &= (0 - 0)\mathbf{i} + (0 - 0)\mathbf{j} + (\delta\omega - (-\delta\omega))\mathbf{k} \\ &= 2\,\delta\omega\mathbf{k}.\end{aligned}$$

By Stokes's theorem, the circulation of $\mathbf{F}$ around a circle C of radius ρ is

$$\oint_C \mathbf{F}\cdot d\mathbf{r} = \iint_S \nabla \times \mathbf{F}\cdot\mathbf{n}\,d\sigma = \iint 2\,\delta\omega\mathbf{k}\cdot\mathbf{k}\,dx\,dy = (2\,\delta\omega)(\pi\rho^2).$$

Thus,

$$(\nabla \times \mathbf{F})\cdot\mathbf{k} = 2\,\delta\omega = \frac{1}{\pi\rho^2}\oint_C \mathbf{F}\cdot d\mathbf{r},$$

in agreement with Eq. (9) with $\mathbf{u} = \mathbf{k}$. ■

An Important Identity

The identity

$$\text{curl grad} f = \mathbf{0} \qquad \text{or} \qquad \nabla \times \nabla f = \mathbf{0} \tag{10}$$

arises frequently in mathematics and the physical sciences. (We shall use it ourselves in Section 15.7.) It holds for any continuous function $f(x, y, z)$ whose first and second partial derivatives are continuous. The proof goes like this:

$$\nabla \times \nabla f = \begin{vmatrix} \mathbf{i} & \mathbf{j} & \mathbf{k} \\ \dfrac{\partial}{\partial x} & \dfrac{\partial}{\partial y} & \dfrac{\partial}{\partial z} \\ \dfrac{\partial f}{\partial x} & \dfrac{\partial f}{\partial y} & \dfrac{\partial f}{\partial z} \end{vmatrix} = (f_{zy} - f_{yz})\mathbf{i} - (f_{zx} - f_{xz})\mathbf{j} + (f_{yx} - f_{xy})\mathbf{k}. \quad (11)$$

When f and its first and second partial derivatives are continuous, the mixed second derivatives in parentheses are equal and the vector is zero.

Summary of Integral Theorems in del Notation

We close with a table that lists the equations from Green's theorem and its generalizations to three dimensions.

TABLE 15.3
Green's theorem and its generalizations

Green's theorem:	$\oint_C \mathbf{F} \cdot \mathbf{n}\, ds = \iint_R \nabla \cdot \mathbf{F}\, dA$
	$\oint_C \mathbf{F} \cdot d\mathbf{r} = \iint_R \nabla \times \mathbf{F} \cdot \mathbf{k}\, dA$
The divergence theorem:	$\iint_S \mathbf{F} \cdot \mathbf{n}\, d\sigma = \iiint_D \nabla \cdot \mathbf{F}\, dV$
Stokes's theorem:	$\oint_C \mathbf{F} \cdot d\mathbf{r} = \iint_S \nabla \times \mathbf{F} \cdot \mathbf{n}\, d\sigma$

Exercises 15.6

In Exercises 1–6, use the surface integral in Stokes's theorem to calculate the circulation of the field $\mathbf{F}$ around the curve C in the indicated direction.

1. $\mathbf{F} = x^2\mathbf{i} + 2x\mathbf{j} + z^2\mathbf{k}$
C: The ellipse $4x^2 + y^2 = 4$, counterclockwise as viewed from above

2. $\mathbf{F} = 2y\mathbf{i} + 3x\mathbf{j} - z^2\mathbf{k}$
C: The circle $x^2 + y^2 = 9$ in the xy-plane, counterclockwise when viewed from above

3. $\mathbf{F} = y\mathbf{i} + xz\mathbf{j} + x^2\mathbf{k}$
C: The boundary of the triangle cut from the plane $x + y + z = 1$ by the first octant, counterclockwise when viewed from above

4. $\mathbf{F} = (y^2 + z^2)\mathbf{i} + (x^2 + z^2)\mathbf{j} + (x^2 + y^2)\mathbf{k}$
C: The boundary of the triangle cut from the plane $x + y + z = 1$ by the first octant, counterclockwise when viewed from above

5. $\mathbf{F} = (y^2 + z^2)\mathbf{i} + (x^2 + y^2)\mathbf{j} + (x^2 + y^2)\mathbf{k}$
   C: The square bounded by the lines $x = \pm 1$ and $y = \pm 1$ in the xy-plane, counterclockwise when viewed from above

6. $\mathbf{F} = x^2y^3\mathbf{i} + \mathbf{j} + z\mathbf{k}$
   C: The boundary of the circle $x^2 + y^2 = a^2$ in the xy-plane, counterclockwise when viewed from above

7. Let $\mathbf{n}$ be the outer unit normal of the elliptical shell

$$S:\quad 4x^2 + 9y^2 + 36z^2 = 36, \quad z \geq 0,$$

and let

$$\mathbf{F} = y\mathbf{i} + x^2\mathbf{j} + (x^2 + y^4)^{3/2} \sin e^{\sqrt{xyz}}\mathbf{k}.$$

Use Stokes's theorem to find the value of

$$\iint_S \nabla \times \mathbf{F} \cdot \mathbf{n}\, d\sigma.$$

(*Hint:* One parametrization of the ellipse at the base of the shell is $x = 3\cos t$, $y = 2\sin t$, $0 \leq t \leq 2\pi$.)

8. Let S be the cylinder $x^2 + y^2 = a^2$, $0 \leq z \leq h$, together with its top, $x^2 + y^2 \leq a^2$, $z = h$. Let $\mathbf{F} = -y\mathbf{i} + x\mathbf{j} + x^2\mathbf{k}$. Use Stokes's theorem to calculate the flux of $\nabla \times \mathbf{F}$ outward through S.

9. Show that

$$\iint_S \nabla \times \mathbf{F} \cdot \mathbf{n}\, d\sigma$$

has the same value for all oriented surfaces S that span C and that induce the same positive direction on C.

10. Show that if $\mathbf{F} = x\mathbf{i} + y\mathbf{j} + z\mathbf{k}$, then $\nabla \cdot \mathbf{F} = 3$ and $\nabla \times \mathbf{F} = \mathbf{0}$.

11. a) Show that if the necessary partial derivatives of the components of $\mathbf{F} = M\mathbf{i} + N\mathbf{j} + P\mathbf{k}$ are continuous then $\nabla \cdot \nabla \times \mathbf{F} = 0$ (div(curl $\mathbf{F}$) = 0).
    b) Use the result in (a) to show that

$$\iint_S \nabla \times \mathbf{F} \cdot \mathbf{n}\, d\sigma = 0$$

for any surface to which the divergence theorem applies.

12. Use the identity $\nabla \times \nabla f = 0$ and Stokes's theorem to show that the circulations of the following fields around the boundary of any smooth orientable surface in space are zero.
    a) $\mathbf{F} = 2x\mathbf{i} + 2y\mathbf{j} + 2z\mathbf{k}$
    b) $\mathbf{F} = \nabla(xy^2z^3)$
    c) $\mathbf{F} = \nabla \times (x\mathbf{i} + y\mathbf{j} + z\mathbf{k})$
    d) $\mathbf{F} = \nabla f$

13. Let $f(x, y, z) = (x^2 + y^2 + z^2)^{-1/2}$. Show that the clockwise circulation of the field $\mathbf{F} = \nabla f$ around the circle $x^2 + y^2 = a^2$ in the xy-plane is zero
    a) by taking $\mathbf{r} = (a\cos t)\mathbf{i} + (a\sin t)\mathbf{j}$, $0 \leq t \leq 2\pi$, and integrating $\mathbf{F} \cdot d\mathbf{r}$ over the circle, and
    b) by applying Stokes's theorem.

14. Let $g(x, y, z)$ be a differentiable scalar function and $\mathbf{F}$, $\mathbf{F}_1$, and $\mathbf{F}_2$ differentiable vector fields. Verify the following identities.
    a) $\operatorname{div}(g\mathbf{F}) = \nabla \cdot (g\mathbf{F}) = g\nabla \cdot \mathbf{F} + \mathbf{F} \cdot \nabla g$
    b) $\nabla \times (g\mathbf{F}) = g\nabla \times \mathbf{F} + \nabla g \times \mathbf{F}$
    c) $\nabla \cdot (\mathbf{F}_1 \times \mathbf{F}_2) = \mathbf{F}_2 \cdot \nabla \times \mathbf{F}_1 - \mathbf{F}_1 \cdot \nabla \times \mathbf{F}_2$

TOOLKIT PROGRAM

Vector Fields	Displays two-dimensional vector fields and integrates the curl over rectangular regions. Also calculates flux across rectangular boundaries.

15.7 Path Independence, Potential Functions, and Conservative Fields

This section introduces the notion of path independence of work integrals and discusses the remarkable properties of fields in which work integrals are path independent. For gravitational and electric fields, for example, this means that the amount of work it takes to move a mass or a charge from one point to another depends only on the initial and final positions and not on the path taken to get there. This leaves us free to choose the path we want. It also gives us a way to evaluate work integrals without bringing in parametrizations.

Path Independence

If A and B are two points in an open region D in space, the work $\int \mathbf{F} \cdot d\mathbf{r}$ done in moving a particle from A to B by a field $\mathbf{F}$ defined on D usually depends on the path taken. For some special fields, however, the integral's value depends only on the points A and B and is the same for all paths from A to B. If this is true for all points A and B in D, we say that the integral $\int \mathbf{F} \cdot d\mathbf{r}$ is path independent in D and that $\mathbf{F}$ is conservative on D.

DEFINITIONS

Let $\mathbf{F}$ be a field defined on an open region D in space, and suppose that for any two points A and B in D the work $\int_A^B \mathbf{F} \cdot d\mathbf{r}$ done in moving from A to B is the same over all paths from A to B. Then the integral $\int \mathbf{F} \cdot d\mathbf{r}$ is **path independent in D** and the field $\mathbf{F}$ is **conservative on D.**

The word *conservative* comes from physics, where it refers to fields in which the principle of conservation of energy holds (it does, in conservative fields).

Under conditions normally met in practice, a field $\mathbf{F}$ is conservative if and only if it is the gradient field of a scalar function f; that is, if and only if $\mathbf{F} = \nabla f$ for some f. The function f is then called a potential function for $\mathbf{F}$.

DEFINITIONS

If a scalar function f is differentiable throughout a domain D, the field of vectors ∇f is called the **gradient field** of f on D. If $\mathbf{F}$ is a field defined on D and $\mathbf{F} = \nabla f$ for some scalar function f on D, then f is called a **potential function** for $\mathbf{F}$.

An electric potential is a scalar function whose gradient field is an electric field. A gravitational potential is a scalar function whose gradient field is a gravitational field, and so on. As we shall see, once we have found a potential function f for a field $\mathbf{F}$, we can evaluate all work integrals in the domain of $\mathbf{F}$ by the rule

$$\int_A^B \mathbf{F} \cdot d\mathbf{r} = f(B) - f(A). \tag{1}$$

Conservative fields have other remarkable properties. Saying that $\mathbf{F}$ is conservative on D is equivalent to saying that the integral of $\mathbf{F}$ around every closed loop in D is zero. This, in turn, is equivalent in a great many domains to saying that $\nabla \times \mathbf{F} = \mathbf{0}$. We explain these relationships by discussing the implications in the following diagram.

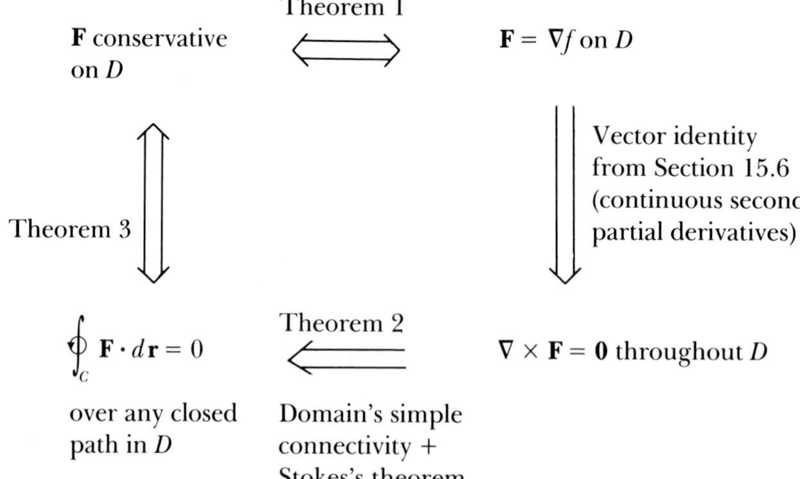

Naturally, we need to impose conditions on the curves, fields, and domains to make these implications hold.

We assume that all of the curves are **piecewise smooth,** i.e., made of continuously differentiable pieces connected end to end.

We also assume that the components of **F** are continuous and have continuous first and second partial derivatives. Without this assumption, $\mathbf{F} = \nabla f$ need not imply $\nabla \times \mathbf{F} = \mathbf{0}$. As we saw in the previous section, the implication requires the mixed second derivatives of f to be equal. We assume continuity to ensure that.

We assume D to be an *open* region in space. This means that every point in D is the center of a ball that lies entirely in D. We need an added restriction on D if $\nabla \times \mathbf{F} = \mathbf{0}$ is to imply that the integral of **F** around every closed path in D is zero. For this, we assume D to be **simply connected.** This means that every closed path in D can be contracted to a point in D without ever leaving D. If D consisted of space with one of the axes removed, for example, D would not be simply connected. There would be no way to contract a loop around the axis to a point without leaving D. (See Exercise 25.) On the other hand, space itself *is* simply connected.

Finally, we assume D to be **connected,** which in an open region means that every point can be connected to every other point by a smooth curve that lies in the region. See Fig. 15.56.

Connected and simply connected.

Connected but not simply connected.

Connected and simply connected.

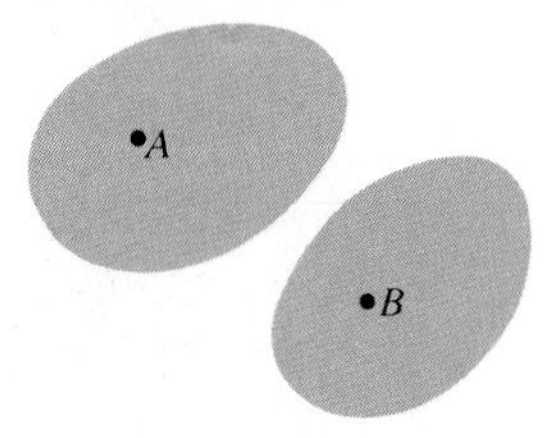

Simply connected but not connected.
No path from A to B lies entirely in the region.

15.56 Connectivity and simple connectivity are not the same thing. In fact, neither implies the other, as these pictures illustrate.

The Implications

We now examine the implications in the diagram above, starting with Theorem 1, then Theorem 2, and finally Theorem 3.

THEOREM 1

The Fundamental Theorem of Line Integrals

1. Let $\mathbf{F} = M\mathbf{i} + N\mathbf{j} + P\mathbf{k}$ be a vector field whose components are continuous throughout an open connected region D in space. Then there exists a differentiable function f such that

$$\mathbf{F} = \nabla f = \frac{\partial f}{\partial x}\mathbf{i} + \frac{\partial f}{\partial y}\mathbf{j} + \frac{\partial f}{\partial z}\mathbf{k}$$

if and only if for all points A and B in D the value of $\int_A^B \mathbf{F} \cdot d\mathbf{r}$ is independent of the path joining A to B in D.

2. If the integral is independent of the path from A to B, its value is

$$\int_A^B \mathbf{F} \cdot d\mathbf{r} = f(B) - f(A).$$

Proof that F = ∇*f* Implies Path Independence of the Integral Suppose that A and B are two points in D and that

$$C\colon\ \mathbf{r} = g(t)\mathbf{i} + h(t)\mathbf{j} + k(t)\mathbf{k}, \quad a \le t \le b,$$

is a smooth curve in D joining A and B. Along the curve, f is a differentiable function of t and

$$\begin{aligned}\frac{df}{dt} &= \frac{\partial f}{\partial x}\frac{dx}{dt} + \frac{\partial f}{\partial y}\frac{dy}{dt} + \frac{\partial f}{\partial z}\frac{dz}{dt} \qquad \text{(Chain Rule)} \\ &= \nabla f \cdot \left[\frac{dx}{dt}\mathbf{i} + \frac{dy}{dt}\mathbf{j} + \frac{dz}{dt}\mathbf{k}\right] = \nabla f \cdot \frac{d\mathbf{r}}{dt} = \mathbf{F} \cdot \frac{d\mathbf{r}}{dt}. \qquad \left(\begin{matrix}\text{Because} \\ \mathbf{F} = \nabla f\end{matrix}\right)\end{aligned} \tag{2}$$

Therefore,

$$\int_C \mathbf{F} \cdot d\mathbf{r} = \int_a^b \mathbf{F} \cdot \frac{d\mathbf{r}}{dt}\,dt = \int_a^b \frac{df}{dt}\,dt = f(g(t), h(t), k(t))\Big]_a^b = f(B) - f(A). \quad \text{(Eq. 2)}$$

Thus, the value of the work integral depends only on the values of f at A and B and not on the path in between. This proves Part 2 as well as the forward implication in Part 1. The proof of the reverse implication is somewhat more technical. You can find it in Appendix 9. ■

THEOREM 2

If curl $\mathbf{F} = \mathbf{0}$ at every point of a simply connected open region D in space, then

$$\oint_C \mathbf{F} \cdot d\mathbf{r} = 0$$

on any piecewise smooth closed path C in D.

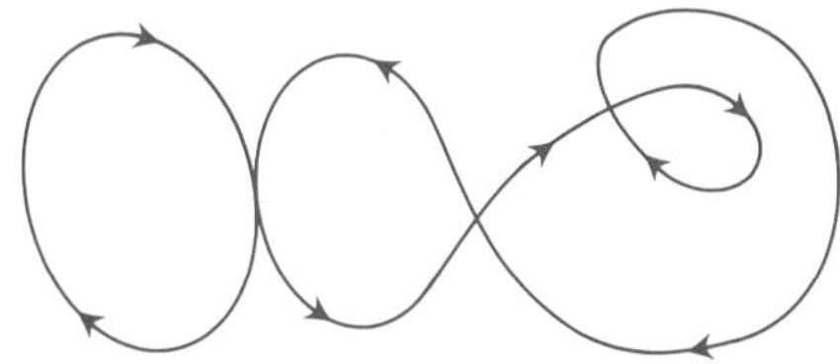

15.57 In a simply-connected open region, differentiable curves that cross themselves can be divided into loops to which Stokes's theorem applies.

Sketch of a Proof The theorem is usually proved in two steps. The first step is for simple closed curves. A theorem from topology, a branch of advanced mathematics, states that every differentiable simple closed curve C in a simply connected open region D is the boundary of a smooth two-sided surface S that also lies in D. Hence, by Stokes's theorem,

$$\oint_C \mathbf{F}\cdot d\mathbf{r} = \iint_S \nabla \times \mathbf{F}\cdot\mathbf{n}\, d\sigma = 0.$$

The second step is for curves that cross themselves, like the one in Fig. 15.57. The idea is to break these into simple loops spanned by orientable surfaces, apply Stokes's theorem one loop at a time, and add the results. ■

THEOREM 3

> The following statements are equivalent:
>
> 1. $\int \mathbf{F}\cdot d\mathbf{r} = 0$ around every closed loop in D.
> 2. The field $\mathbf{F}$ is conservative on D.

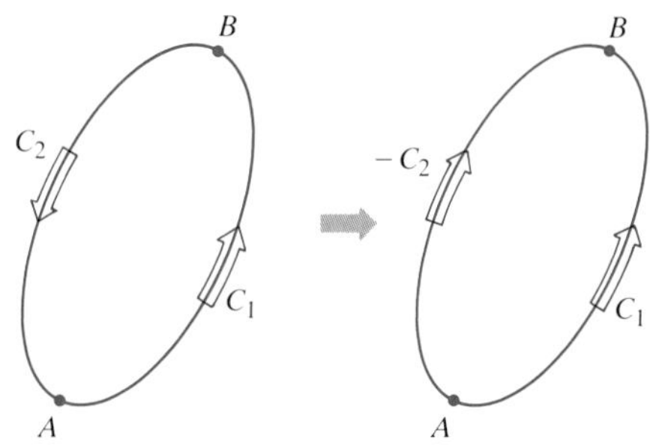

15.58 If we have two paths from A to B, one of them can be reversed to make a loop.

Proof that (1) ⇒ (2) We want to show that for any two points A and B in D the integral of $\mathbf{F}\cdot d\mathbf{r}$ has the same value over any two paths C_1 and C_2 from A to B. We reverse the direction on C_2 to make a path $-C_2$ from B to A (Fig. 15.58). Together, C_1 and $-C_2$ make a closed loop C, and

$$\int_{C_1} \mathbf{F}\cdot d\mathbf{r} - \int_{C_2} \mathbf{F}\cdot d\mathbf{r} = \int_{C_1} \mathbf{F}\cdot d\mathbf{r} + \int_{-C_2} \mathbf{F}\cdot d\mathbf{r} = \int_{C} \mathbf{F}\cdot d\mathbf{r} = 0.$$

Thus the integrals over C_1 and C_2 give the same value.

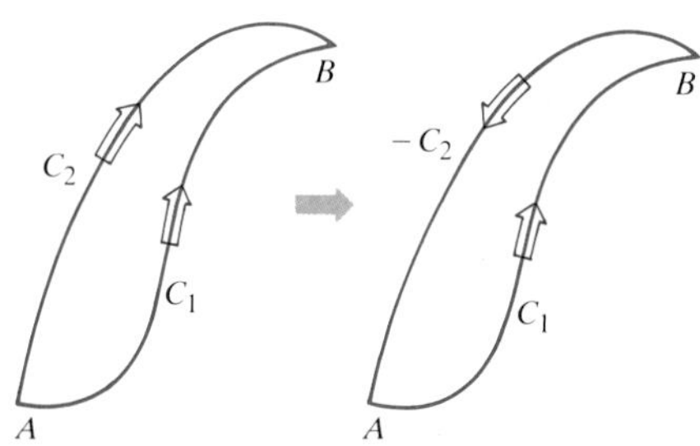

15.59 If A and B lie on a loop, we can reverse part of the loop to make two paths from A to B.

Proof that (2) ⇒ (1) We want to show that the integral of $\mathbf{F}\cdot d\mathbf{r}$ is zero over any closed loop C. We pick two points A and B on C and use them to break C into two pieces: C_1 from A to B followed by C_2 from B back to A (Fig. 15.59). Then

$$\oint_C \mathbf{F}\cdot d\mathbf{r} = \int_{C_1} \mathbf{F}\cdot d\mathbf{r} + \int_{C_2} \mathbf{F}\cdot d\mathbf{r} = \int_A^B \mathbf{F}\cdot d\mathbf{r} - \int_A^B \mathbf{F}\cdot d\mathbf{r} = 0. \quad ■$$

Finding Potentials for Conservative Fields

As we said earlier, conservative fields get their name from the fact that the principle of conservation of energy holds in such fields. The test for being conservative, as we have just seen, is that $\nabla \times \mathbf{F} = 0$. In terms of components, this gives us the following test.

> **Component Test for Conservative Fields**
>
> The field $\mathbf{F} = M(x, y, z)\mathbf{i} + N(x, y, z)\mathbf{j} + P(x, y, z)\mathbf{k}$ is conservative if and only if
>
> $$\frac{\partial P}{\partial y} = \frac{\partial N}{\partial z}, \quad \frac{\partial M}{\partial z} = \frac{\partial P}{\partial x}, \quad \text{and} \quad \frac{\partial N}{\partial x} = \frac{\partial M}{\partial y}. \tag{3}$$

When we know that $\mathbf{F}$ is conservative, we usually want to find a potential function for $\mathbf{F}$. This requires solving the equation

$$\nabla f = \frac{\partial f}{\partial x}\mathbf{i} + \frac{\partial f}{\partial y}\mathbf{j} + \frac{\partial f}{\partial z}\mathbf{k} = M\mathbf{i} + N\mathbf{j} + P\mathbf{k} \tag{4}$$

for f. We accomplish this by integrating the three equations

$$\frac{\partial f}{\partial x} = M, \qquad \frac{\partial f}{\partial y} = N, \qquad \frac{\partial f}{\partial z} = P \tag{5}$$

as in the following example.

EXAMPLE 1 Show that

$$\mathbf{F} = (e^x \cos y + yz)\mathbf{i} + (xz - e^x \sin y)\mathbf{j} + (xy + z)\mathbf{k}$$

is conservative and find a potential function for it.

Solution We apply the test in Eq. (3) to

$$M = e^x \cos y + yz, \qquad N = xz - e^x \sin y, \qquad P = xy + z$$

and calculate

$$\frac{\partial M}{\partial z} = y = \frac{\partial P}{\partial x}, \qquad \frac{\partial N}{\partial z} = x = \frac{\partial P}{\partial y}, \qquad \frac{\partial M}{\partial y} = -e^x \sin y + z = \frac{\partial N}{\partial x}.$$

Together, these equalities tell us that there is a function f with $\nabla f = \mathbf{F}$.

We find f by integrating the equations

$$\frac{\partial f}{\partial x} = e^x \cos y + yz, \qquad \frac{\partial f}{\partial y} = xz - e^x \sin y, \qquad \frac{\partial f}{\partial z} = xy + z. \tag{6}$$

We integrate the first equation with respect to x, holding y and z fixed, to get

$$f(x, y, z) = e^x \cos y + xyz + g(y, z).$$

We write the constant of integration as a function of y and z because its value may change if y and z change. We then calculate $\partial f/\partial y$ from this equation and match it with the expression for $\partial f/\partial y$ in Eq. (6). This gives

$$-e^x \sin y + xz + \frac{\partial g}{\partial y} = xz - e^x \sin y,$$

so

$$\frac{\partial g}{\partial y} = 0.$$

Therefore, g is a function of z alone, and

$$f(x, y, z) = e^x \cos y + xyz + h(z).$$

We now calculate $\partial f/\partial z$ from this equation and match it to the formula for $\partial f/\partial z$ in Eq. (6). This gives

$$xy + \frac{dh}{dz} = xy + z, \qquad \text{or} \qquad \frac{dh}{dz} = z,$$

so

$$h(z) = \frac{z^2}{2} + C.$$

Hence,

$$f(x, y, z) = e^x \cos y + xyz + \frac{z^2}{2} + C.$$

We have found infinitely many potential functions for **F**, one for each value of C. ■

EXAMPLE 2 Show that $\mathbf{F} = (2x - 3)\mathbf{i} - z\mathbf{j} + (\cos z)\mathbf{k}$ is not conservative.

Solution We apply the component test in Eq. (3) and find right away that

$$\frac{\partial P}{\partial y} = \frac{\partial}{\partial y}(\cos z) = 0, \qquad \frac{\partial N}{\partial z} = \frac{\partial}{\partial z}(-z) = -1.$$

The two are unequal, so **F** is not conservative. No further testing is required. ■

Exact Differential Forms

The forms $M\,dx + N\,dy + P\,dz$ that have been appearing in our line integrals are called differential forms. The most useful differential forms are differentials of scalar functions.

DEFINITIONS

The form

$$M(x, y, z)\,dx + N(x, y, z)\,dy + P(x, y, z)\,dz \tag{7}$$

is called a **differential form.** A differential form is **exact** on a domain D in space if

$$M\,dx + N\,dy + P\,dz = \frac{\partial f}{\partial x}dx + \frac{\partial f}{\partial y}dy + \frac{\partial f}{\partial z}dz = df \tag{8}$$

for some (scalar) function f throughout D.

Notice that if $M\,dx + N\,dy + P\,dz = df$ on D, then the field $\mathbf{F} = M\mathbf{i} + N\mathbf{j} + P\mathbf{k}$ is the gradient field of f on D. Conversely, if $\mathbf{F} = \nabla f$, then the form $M\,dx + N\,dy + P\,dz$ is exact. The test for the form's being exact is therefore the same as the test for **F**'s being conservative.

The Test for Exactness of *M dx + N dy + P dz*

The differential form $M\,dx + N\,dy + P\,dz$ is exact if and only if

$$\frac{\partial P}{\partial y} = \frac{\partial N}{\partial z}, \quad \frac{\partial M}{\partial z} = \frac{\partial P}{\partial x}, \quad \text{and} \quad \frac{\partial N}{\partial x} = \frac{\partial M}{\partial y}. \tag{9}$$

This is equivalent to saying that the field $\mathbf{F} = M\mathbf{i} + N\mathbf{j} + P\mathbf{k}$ is conservative or that

$$\nabla \times \mathbf{F} = \mathbf{0}. \tag{10}$$

EXAMPLE 3 Show that $y\,dx + x\,dy + 4\,dz$ is exact and evaluate the integral

$$\int_{(1,1,1)}^{(2,3,-1)} y\,dx + x\,dy + 4\,dz$$

over the line segment from $(1, 1, 1)$ to $(2, 3, -1)$.

Solution We let

$$M = y, \qquad N = x, \qquad P = 4$$

and apply the test of Eq. (9):

$$\frac{\partial M}{\partial z} = 0 = \frac{\partial P}{\partial x}, \qquad \frac{\partial N}{\partial z} = 0 = \frac{\partial P}{\partial y}, \qquad \frac{\partial M}{\partial y} = 1 = \frac{\partial N}{\partial x}.$$

These equalities tell us that $y\,dx + x\,dy + 4\,dz$ is exact, so

$$y\,dx + x\,dy + 4\,dz = df$$

for some function f, and the integral's value is $f(2, 3, -1) - f(1, 1, 1)$.

We find f up to a constant by integrating the equations

$$\frac{\partial f}{\partial x} = y, \qquad \frac{\partial f}{\partial y} = x, \qquad \frac{\partial f}{\partial z} = 4. \tag{11}$$

From the first equation we get

$$f(x, y, z) = xy + g(y, z).$$

The second equation tells us that

$$\frac{\partial f}{\partial y} = x + \frac{\partial g}{\partial y} = x, \qquad \text{or} \qquad \frac{\partial g}{\partial y} = 0.$$

Hence, g is a function of z alone, and

$$f(x, y, z) = xy + h(z).$$

The third equation in Eq. (9) tells us that

$$\frac{\partial f}{\partial z} = 0 + \frac{dh}{dz} = 4, \qquad \text{or} \qquad h(z) = 4z + C.$$

Therefore,

$$f(x, y, z) = xy + 4z + C,$$

and the value of the integral is

$$f(2, 3, -1) - f(1, 1, 1) = 2 + C - (5 + C) = -3.$$

Exercise 21 suggests another way to evaluate the integral. ■

In the next chapter, we shall see how exact differential forms arise in differential equations.

Exercises 15.7

Which fields in Exercises 1–6 are conservative and which are not?

1. $\mathbf{F} = yz\mathbf{i} + xz\mathbf{j} + xy\mathbf{k}$
2. $\mathbf{F} = (y \sin z)\mathbf{i} + (x \sin z)\mathbf{j} + (xy \cos z)\mathbf{k}$
3. $\mathbf{F} = y\mathbf{i} + (x + z)\mathbf{j} - y\mathbf{k}$
4. $\mathbf{F} = -y\mathbf{i} + x\mathbf{j}$
5. $\mathbf{F} = (z + y)\mathbf{i} + z\mathbf{j} + (y + x)\mathbf{k}$
6. $\mathbf{F} = (e^x \cos y)\mathbf{i} - (e^x \sin y)\mathbf{j} + z\mathbf{k}$

In Exercises 7–10, find a potential function f for the field $\mathbf{F}$.

7. $\mathbf{F} = 2x\mathbf{i} + 3y\mathbf{j} + 4z\mathbf{k}$
8. $\mathbf{F} = (y + z)\mathbf{i} + (x + z)\mathbf{j} + (x + y)\mathbf{k}$
9. $\mathbf{F} = e^{y+2z}(\mathbf{i} + x\mathbf{j} + 2x\mathbf{k})$
10. $\mathbf{F} = (y \sin z)\mathbf{i} + (x \sin z)\mathbf{j} + (xy \cos z)\mathbf{k}$

Evaluate the integrals in Exercises 11–20.

11. $\displaystyle\int_{(0,0,0)}^{(2,3,-6)} 2x\,dx + 2y\,dy + 2z\,dz$

12. $\displaystyle\int_{(1,1,2)}^{(3,5,0)} yz\,dx + xz\,dy + xy\,dz$

13. $\displaystyle\int_{(0,0,0)}^{(1,2,3)} 2xy\,dx + (x^2 - z^2)\,dy - 2yz\,dz$

14. $\displaystyle\int_{(1,1,1)}^{(1,2,3)} 3x^2\,dx + \frac{z^2}{y}\,dy + 2z \ln y\,dz$

15. $\displaystyle\int_{(1,0,0)}^{(0,1,1)} \sin y \cos x\,dx + \cos y \sin x\,dy + dz$

16. $\displaystyle\int_{(0,0,0)}^{(3,3,1)} 2x\,dx - y^2\,dy - \frac{4}{1 + z^2}\,dz$

17. $\displaystyle\int_{(0,2,1)}^{(1,\pi/2,2)} 2 \cos y\,dx + \left(\frac{1}{y} - 2x \sin y\right)dy + \frac{1}{z}\,dz$

18. $\displaystyle\int_{(1,2,1)}^{(2,1,1)} (2x \ln y - yz)\,dx + \left(\frac{x^2}{2} - xz\right)dy - xy\,dz$

19. $\displaystyle\int_{(1,1,1)}^{(2,2,2)} \frac{1}{y}\,dx + \left(\frac{1}{z} - \frac{x}{y^2}\right)dy - \frac{y}{z^2}\,dz$

20. $\displaystyle\int_{(-1,-1,-1)}^{(2,2,2)} \frac{2x\,dx + 2y\,dy + 2z\,dz}{x^2 + y^2 + z^2}$

21. Evaluate the integral

$$\int_{(1,1,1)}^{(2,3,-1)} y\,dx + x\,dy + 4\,dz$$

from Example 3 by finding parametric equations for the line segment from (1, 1, 1) to (2, 3, − 1) and evaluating the line integral of $\mathbf{F} = y\mathbf{i} + x\mathbf{j} + 4\mathbf{k}$ along the segment. This is permissible because $\mathbf{F}$ is conservative and the integral is independent of the path.

22. Evaluate

$$\int_C x^2\,dx + yz\,dy + y^2\,dz$$

along the line segment joining (0, 0, 0) to (0, 3, 4).

Show that the values of the integrals in Exercises 23–24 do not depend on the path taken from A to B.

23. $\displaystyle\int_A^B z^2\,dx + 2y\,dy + 2xz\,dz$

24. $\displaystyle\int_A^B \frac{x\,dx + y\,dy + z\,dz}{\sqrt{x^2 + y^2 + z^2}}$

25. Show that the curl of

$$\mathbf{F} = \frac{-y}{x^2 + y^2}\mathbf{i} + \frac{x}{x^2 + y^2}\mathbf{j} + z\mathbf{k}$$

is zero but that

$$\int_C \mathbf{F} \cdot d\mathbf{r}$$

is not zero if C is the circle $x^2 + y^2 = 1$ in the xy-plane. (Theorem 2 does not apply here because the domain of $\mathbf{F}$ is not simply connected. The field $\mathbf{F}$ is not defined along the z-axis so there is no way to contract C to a point without leaving the domain of $\mathbf{F}$.)

TOOLKIT PROGRAM

Scalar Fields — Plots two-dimensional scalar fields and their associated gradient fields. Integrates $\nabla f \cdot d\mathbf{r}$ along line segments parallel to the coordinate axes.

Review Questions

1. What is a line integral? What are line integrals used for? Give examples.
2. What is a vector field? Give examples.
3. How do you calculate the work done by a force in moving a particle along a curve?
4. What are the flux density and circulation density of a vector field? How are they related to the divergence and curl of the field?
5. What theorem relates flux to the divergence for a vector field defined in the plane? Give an example.
6. What theorem relates circulation to the curl for a vector field defined in the plane? Give an example.
7. How is surface area calculated? Give an example.
8. What is a surface integral? How are surface integrals used to calculate masses and moments? Give examples.
9. What theorem relates flux to the divergence of a vector field in space? Give an example.
10. What theorem relates circulation to the curl of a vector field in space? Give an example.
11. What does it mean for a vector field to be conservative on a region D in space?
12. What is a potential function for a conservative field? Give an example.
13. How can you evaluate $\int_A^B \mathbf{F} \cdot d\mathbf{r}$ over a path in the domain of a conservative field? Give an example.
14. What special properties do conservative fields have?
15. How do you find a potential function for a conservative field?
16. What is an exact differential form? Give an example.

Practice Exercises

1. Figure 15.60 shows two polygonal paths in space from the origin to the point (1, 1, 1). Integrate $f(x, y, z) = 2x - 3y^2 - 2z + 3$ over each path.
2. Figure 15.61 shows three polygonal paths from the origin to the point (1, 1, 1). Integrate $f(x, y, z) = x^2 + y - z$ over each path.

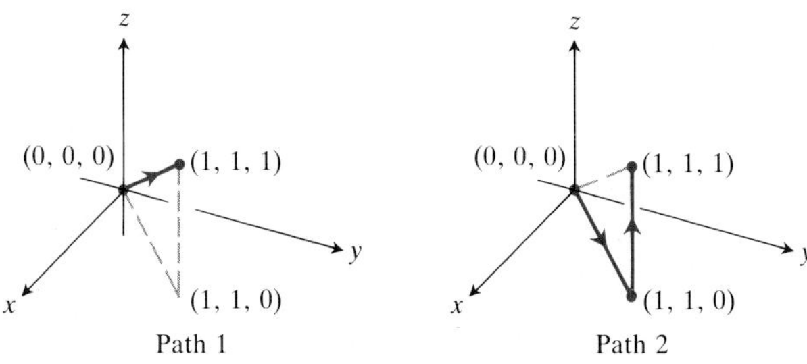

15.60 The paths in Exercise 1.

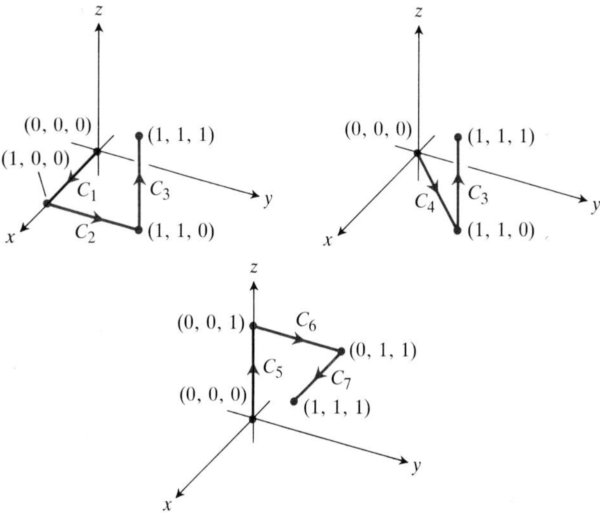

15.61 The paths in Exercise 2.

3. Integrate $f(x, y, z) = \sqrt{x^2 + z^2}$ over the circle $\mathbf{r} = (a \cos t)\mathbf{j} + (a \sin t)\mathbf{k}$, $0 \le t \le 2\pi$, in the direction of increasing t.

4. Integrate $f(x, y, z) = \sqrt{3}/(x^2 + y^2 + z^2)$ over the path $\mathbf{r} = t\mathbf{i} + t\mathbf{j} + t\mathbf{k}$, $0 \le t \le \infty$.

5. Find the mass of a thin wire lying along the curve $\mathbf{r} = \sqrt{2}\, t\mathbf{i} + \sqrt{2}\, t\mathbf{j} + (4 - t^2)\mathbf{k}$, $0 \le t \le 1$, if the density is (a) $\delta = 3t$, (b) $\delta = 1$.

6. Find the center of mass of a thin wire lying along the curve $\mathbf{r} = t\mathbf{i} + 2t\mathbf{j} + (2/3)t^{3/2}\mathbf{k}$, $0 \le t \le 2$, if the density is $\delta = 3\sqrt{5 + t}$.

7. Find the center of mass, and the moments of inertia and radii of gyration about the coordinate axes of a thin wire lying along the curve

$$\mathbf{r} = t\mathbf{i} + \frac{2\sqrt{2}}{3}t^{3/2}\mathbf{j} + \frac{t^2}{2}\mathbf{k}, \quad 0 \le t \le 2,$$

if the density is $\delta = 1/(t + 1)$.

8. A thin wire of constant density $\delta = 1$ lies along the curve $\mathbf{r} = (e^t \cos t)\mathbf{i} + (e^t \sin t)\mathbf{j} + e^t\mathbf{k}$, $0 \le t \le \ln 2$. Find $\bar{z}$, I_z, and R_z.

9. Suppose $\mathbf{F}(x, y) = (x + y)\mathbf{i} - (x^2 + y^2)\mathbf{j}$ is the velocity field of a fluid flowing across the xy-plane. Find the flow along each of the following paths from $(1, 0)$ to $(-1, 0)$.
   a) The upper half of the circle $x^2 + y^2 = 1$
   b) The line segment from $(1, 0)$ to $(-1, 0)$
   c) The line segment from $(1, 0)$ to $(0, -1)$ followed by the line segment from $(0, -1)$ to $(-1, 0)$

10. Find the circulation of $\mathbf{F} = 2x\mathbf{i} + 2z\mathbf{j} + 2y\mathbf{k}$ along the closed path consisting of the helix $\mathbf{r}_1 = (\cos t)\mathbf{i} + (\sin t)\mathbf{j} + t\mathbf{k}$, $0 \le t \le \pi/2$, followed by the line segments $\mathbf{r}_2 = \mathbf{j} + (\pi/2)(1 - t)\mathbf{k}$, $0 \le t \le 1$, and $\mathbf{r}_3 = t\mathbf{i} + (1 - t)\mathbf{j}$, $0 \le t \le 1$.

Use Green's theorem to find the counterclockwise circulation and outward flux for the fields and curves in Exercises 11–12.

11. $\mathbf{F} = (2xy + x)\mathbf{i} + (xy - y)\mathbf{j}$
    C: The square bounded by $x = 0$, $x = 1$, $y = 0$, $y = 1$

12. $\mathbf{F} = (y - 6x^2)\mathbf{i} + (x + y^2)\mathbf{j}$
    C: The triangle made by the lines $y = 0$, $y = x$, and $x = 1$

13. Show that

$$\oint_C 4x^3y\, dx + x^4\, dy = 0$$

for any closed curve C to which Green's theorem applies.

14. a) Show that the outward flux of the position vector field $\mathbf{F} = x\mathbf{i} + y\mathbf{j}$ across any closed curve to which Green's theorem applies is twice the area of the region enclosed by the curve.
    b) Let $\mathbf{n}$ be the outward unit normal vector to a closed curve to which Green's theorem applies. Show that it is not possible for $\mathbf{F} = x\mathbf{i} + y\mathbf{j}$ to be orthogonal to $\mathbf{n}$ at every point of C.

Apply Green's theorem in one of its two forms to evaluate the line integrals in Exercises 15–16.

15. $\displaystyle\int_C 8x \sin y\, dx - 8y \cos x\, dy$

    C is the square cut from the first quadrant by the lines $x = \pi/2$ and $y = \pi/2$.

16. $\displaystyle\int_C y^2\, dx + x^2\, dy$

    C is the circle $x^2 + y^2 = 4$.

17. Find the area of the elliptical region cut from the plane $x + y + z = 1$ by the cylinder $x^2 + y^2 = 1$.

18. Find the area of the surface cut from the end of the paraboloid $y^2 + z^2 = 3x$ by the plane $x = 1$.

19. Find the area of the cap cut from the sphere $x^2 + y^2 + z^2 = 1$ by the plane $z = \sqrt{2}/2$.

20. Find the area of the surface cut from the hemisphere $x^2 + y^2 + z^2 = 4$, $z \ge 0$, by the cylinder $x^2 + y^2 = 2x$ (Fig. 15.62).

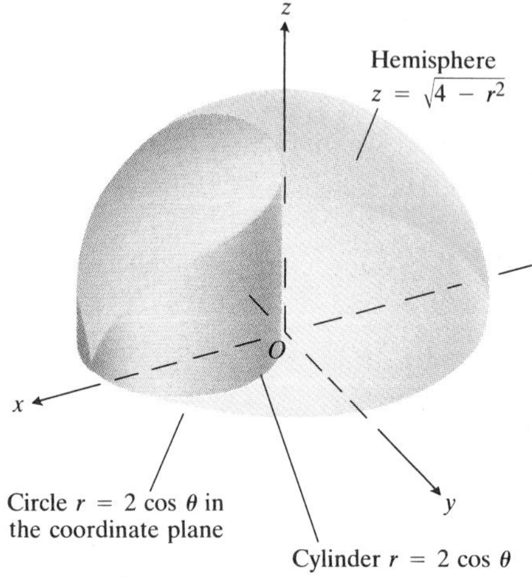

15.62 The surface cut from the hemisphere $x^2 + y^2 + z^2 = 4$, $z \ge 0$, by the cylinder $x^2 + y^2 = 2x$ in Exercise 20. In cylindrical coordinates, the equations of the hemisphere and cylinder are $z = \sqrt{4 - r^2}$ and $r = 2 \cos \theta$.

21. Integrate

a) $g(x, y, z) = \dfrac{yz}{\sqrt{4y^2 + 1}}$ b) $g(x, y, z) = \dfrac{z}{\sqrt{4y^2 + 1}}$

over the surface cut from the parabolic cylinder $y^2 - z = 1$ by the planes $x = 0$, $x = 3$, and $z = 0$.

22. Integrate $g(x, y, z) = x^4 y(y^2 + z^2)$ over the surface in the first quadrant cut from the cylinder $y^2 + z^2 = 25$ by the coordinate planes and the planes $x = 1$, $y = 4$, and $z = 3$.

23. Find the outward flux of the field $\mathbf{F} = 2xy\mathbf{i} + 2yz\mathbf{j} + 2xz\mathbf{k}$ across the surface of the cube cut from the first octant by the planes $x = 1$, $y = 1$, $z = 1$.

24. Find the outward flux of the field $\mathbf{F} = xz\mathbf{i} + yz\mathbf{j} + \mathbf{k}$ across the entire surface of the upper cap cut from the solid sphere $x^2 + y^2 + z^2 \leq 25$ by the plane $z = 3$.

25. Find I_z, R_z, and the center of mass of a thin shell of density $\delta(x, y, z) = z$ cut from the upper portion of the sphere $x^2 + y^2 + z^2 = 25$ by the plane $z = 3$.

26. Find the moment of inertia about the z-axis of the surface of the cube cut from the first octant by the planes $x = 1$, $y = 1$, and $z = 1$ if the density is $\delta = 1$.

In Exercises 27–30, find the outward flux of $\mathbf{F}$ across the boundary of D.

27. $\mathbf{F} = 2xy\mathbf{i} + 2yz\mathbf{j} + 2xz\mathbf{k}$
D: The cube cut from the first octant by the planes $x = 1$, $y = 1$, $z = 1$

28. $\mathbf{F} = xz\mathbf{i} + yz\mathbf{j} + \mathbf{k}$
D: The upper cap cut from the solid sphere $x^2 + y^2 + z^2 \leq 25$ by the plane $z = 3$

29. $\mathbf{F} = -2x\mathbf{i} - 3y\mathbf{j} + z\mathbf{k}$
D: The region cut from the solid sphere $x^2 + y^2 + z^2 \leq 2$ by the paraboloid $z = x^2 + y^2$

30. $\mathbf{F} = (6x + y)\mathbf{i} - (x + z)\mathbf{j} + 4yz\mathbf{k}$
D: The region in the first octant bounded by the cone $z = \sqrt{x^2 + y^2}$, the cylinder $x^2 + y^2 = 1$, and the coordinate planes

In Exercises 31–32, use the surface integral in Stokes's theorem to find the circulation of the field $\mathbf{F}$ around the curve C in the indicated direction.

31. $\mathbf{F} = y^2\mathbf{i} - y\mathbf{j} + 3z^2\mathbf{k}$
C: The ellipse in which the plane $2x + 6y - 3z = 6$ meets the cylinder $x^2 + y^2 = 1$, counterclockwise as viewed from above

32. $\mathbf{F} = (x^2 + y)\mathbf{i} + (x + y)\mathbf{j} + (4y^2 - z)\mathbf{k}$
C: The circle in which the plane $z = -y$ meets the sphere $x^2 + y^2 + z^2 = 4$, counterclockwise as viewed from above

Which of the fields in Exercises 33–36 are conservative and which are not?

33. $\mathbf{F} = x\mathbf{i} + y\mathbf{j} + z\mathbf{k}$

34. $\mathbf{F} = (x\mathbf{i} + y\mathbf{j} + z\mathbf{k})/(x^2 + y^2 + z^2)^{3/2}$

35. $\mathbf{F} = x e^y\mathbf{i} + y e^z\mathbf{j} + z e^x\mathbf{k}$

36. $\mathbf{F} = (\mathbf{i} + z\mathbf{j} + y\mathbf{k})/(x + yz)$

Find potential functions for the fields in Exercises 37–38.

37. $\mathbf{F} = 2\mathbf{i} + (2y + z)\mathbf{j} + (y + 1)\mathbf{k}$

38. $\mathbf{F} = (z \cos xz)\mathbf{i} + e^y\mathbf{j} + (x \cos xz)\mathbf{k}$

Evaluate the integrals in Exercises 39–40.

39. $\displaystyle\int_{(-1,1,1)}^{(4,-3,0)} \frac{dx + dy + dz}{\sqrt{x + y + z}}$

40. $\displaystyle\int_{(1,0,1)}^{(10,3,3)} dx - \sqrt{\frac{z}{y}}\,dy - \sqrt{\frac{y}{z}}\,dz$

In Exercises 41–42, find the work done by each field along the paths from (0, 0, 0) to (1, 1, 1) in Fig. 15.60.

41. $\mathbf{F} = 2xy\mathbf{i} + \mathbf{j} + x^2\mathbf{k}$

42. $\mathbf{F} = 2xy\mathbf{i} + x^2\mathbf{j} + \mathbf{k}$

Finding Areas with Green's Theorem

Use the Green's theorem area formula, Eq. (29) in the Section 15.3 exercises, to find the areas of the regions enclosed by the curves in Exercises 43–46.

43. The cardioid
$x = 2\cos t - \cos 2t$,
$y = 2\sin t - \sin 2t$,
$0 \leq t \leq 2\pi$

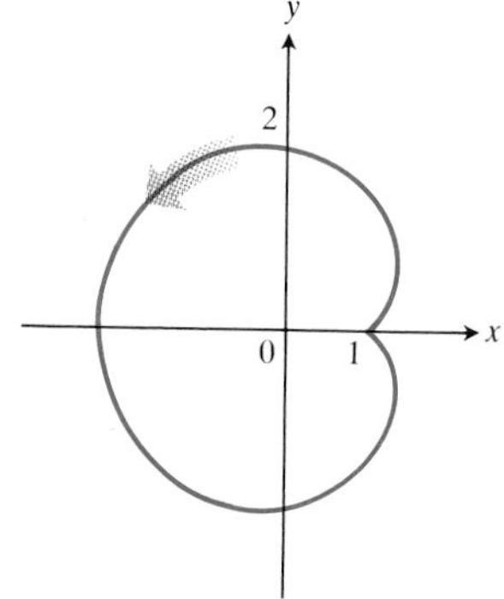

44. The deltoid
$x = 2\cos t + \cos 2t$,
$y = 2\sin t - \sin 2t$,
$0 \le t \le 2\pi$

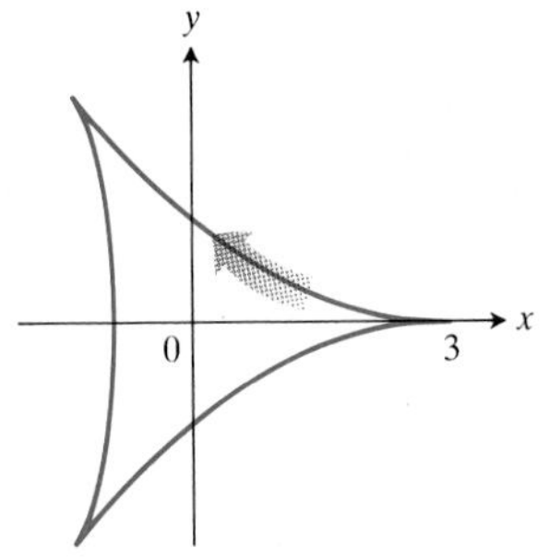

45. The eight curve
$x = (1/2)\sin 2t$,
$y = \sin t$,
$0 \le t \le \pi$ (one loop)

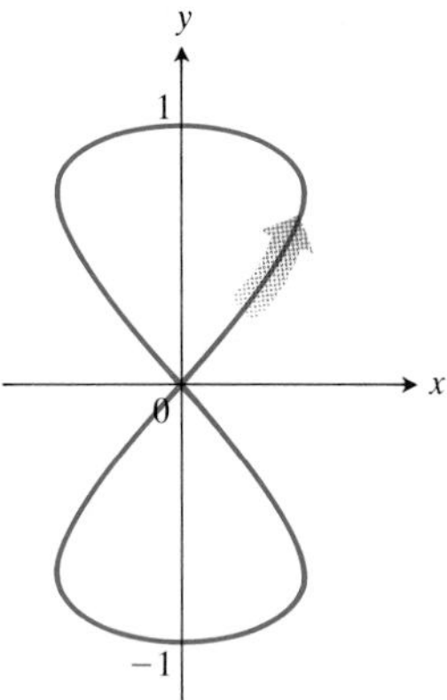

46. The teardrop
$x = 2a\cos t - a\sin 2t$,
$y = b\sin t$,
$0 \le t \le 2\pi$

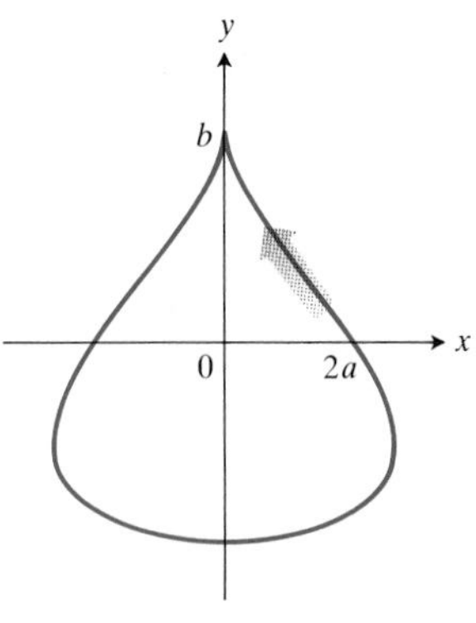

16 Preview of Differential Equations

This glass shattered after absorbing energy from sound vibrations in the surrounding air. Things tend to shake apart when their natural frequencies are in tune with the frequency of an external driving force. (The Shattering Glass is a registered trademark of Memorex Corporation. Licensed to Memtek Products. All rights reserved. Reprinted by permission.)

Memorex® Audio

OVERVIEW Differential equations can arise whenever we model the effects of change, motion, and growth. They can arise when we study moving particles, changing business conditions, oscillating voltages in neural networks, changing concentrations in chemical reactions, and flowing resources in a market economy. Our space program more or less runs on differential equations, and mathematical models of our environment involve huge systems of differential equations. The basic tool for solving differential equations is calculus. This chapter previews common types of differential equations and the methods we use to solve them.

16.1 Separable First Order Equations

We study differential equations for their mathematical importance and for what they tell us about reality. When we study unchecked bacterial growth, for example, and assume that the rate of increase in the population at any given time t is proportional to the number of bacteria then present, we are assuming that the population size y obeys the differential equation $dy/dt = ky$. If we also know that the size of the population is y_0 when $t = 0$, we can find a formula for the size of the population at any time $t > 0$, by solving the initial value problem

Differential equation: $\quad \dfrac{dy}{dt} = ky$

Initial condition: $\quad y = y_0 \quad$ when $t = 0$.

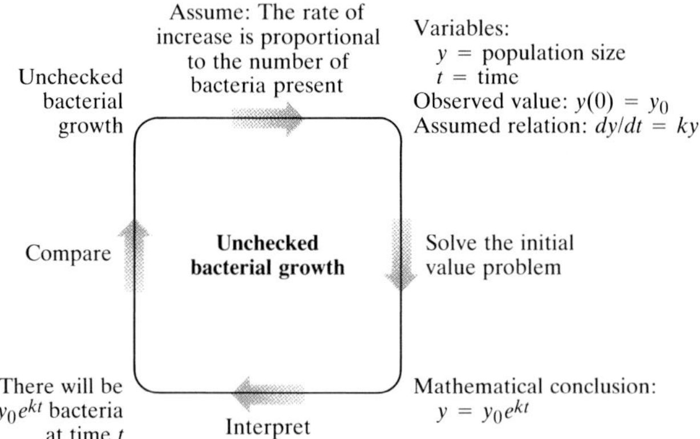

As we saw in Section 7.4, the general solution of the differential equation (the formula that gives all possible solutions) is

$y = Ae^{kt}$ (A an arbitrary constant)

and the particular solution that satisfies the initial condition $y(0) = y_0$ is

$$y = y_0e^{kt}.$$

We interpret this as saying that unchecked bacterial growth is exponential. There should be y_0e^{kt} bacteria present at time t. We know from experience that there are limits to all growth, and we do not expect this prediction to be accurate for large values of t. But for the early stages of growth the formula is a good predictor and gives valuable information about the population's size.

Definitions

A **differential equation** is an equation that contains one or more derivatives of a differentiable function. An equation with partial derivatives is called a

partial differential equation. The wave equation and the Laplace equations in Exercises 13.3 are partial differential equations. An equation with ordinary derivatives, that is, derivatives of a function of a single variable, is called an **ordinary differential equation.** The equations in this chapter are ordinary differential equations.

The **order** of a differential equation is the order of the equation's highest order derivative. A differential equation is **linear** if it can be put in the form

$$a_n(x)\frac{d^ny}{dx^n} + a_{n-1}(x)\frac{d^{n-1}y}{dx^{n-1}} + \cdots + a_1(x)\frac{dy}{dx} + a_0(x)y = F(x), \qquad (1)$$

where the a's are functions of x.

EXAMPLE 1

First order, linear: $\frac{dy}{dx} = 5y; \qquad 3\frac{dy}{dx} - \sin x = 0$

Third order, nonlinear: $\left(\frac{d^3y}{dx^3}\right)^2 + \left(\frac{d^2y}{dx^2}\right)^5 - \frac{dy}{dx} = e^x$ ■

Almost all the differential equations we have solved so far in this book have been linear equations of first or second order. This will continue to be true, but the equations will have more variety and model a broader range of applications.

We call a function $y = f(x)$ a **solution** of a differential equation if y and its derivatives satisfy the equation. To test whether a given function solves a particular equation, we substitute the function and its derivatives into the equation. If the equation then reduces to an identity, the function solves the equation; otherwise, it does not.

EXAMPLE 2 Show that for any values of the arbitrary constants C_1 and C_2 the function $y = C_1 \cos x + C_2 \sin x$ is a solution of the differential equation

$$\frac{d^2y}{dx^2} + y = 0.$$

Solution We differentiate the function twice to find d^2y/dx^2:

$$y = C_1 \cos x + C_2 \sin x$$

$$\frac{dy}{dx} = -C_1 \sin x + C_2 \cos x$$

$$\frac{d^2y}{dx^2} = -C_1 \cos x - C_2 \sin x.$$

We then substitute the expressions for y and d^2y/dx^2 into the differential equation to see whether the left-hand side reduces to zero, which is the right-hand side. It does, because

$$\frac{d^2y}{dx^2} + y = (-C_1 \cos x - C_2 \sin x) + (C_1 \cos x + C_2 \sin x) = 0.$$

So, the function is a solution of the differential equation. ■

It can be shown that the formula $y = C_1 \cos x + C_2 \sin x$ gives all possible solutions of the equation $d^2y/dx^2 + y = 0$. A formula that gives all the solutions of a differential equation is called the **general solution** of the equation. In this sense, $y = C_1 \cos x + C_2 \sin x$ is the general solution of $d^2y/dx^2 + y = 0$. To **solve** a differential equation means to find its general solution.

Notice that the equation $d^2y/dx^2 + y = 0$ has order two and that its general solution has two arbitrary constants. The general solution of an nth order ordinary differential equation can be expected to contain n arbitrary constants.

Separable Equations

A method that sometimes works for solving a first order differential equation involves treating the derivative dy/dx as a quotient of differentials and rearranging the equation to group the y terms alone with dy and the x terms alone with dx. We cannot always accomplish this, but when we can, the solution may then be found by separate integrations with respect to x and y.

DEFINITION

A first order differential equation is **separable** if it can be put in the form

$$M(x) + N(y)\frac{dy}{dx} = 0 \tag{2}$$

or in the equivalent differential form

$$M(x)\,dx + N(y)\,dy = 0. \tag{3}$$

When we write the equation this way we say that we have **separated the variables.**

We can solve Eq. (2) by integrating both sides with respect to x to get

$$\int M(x)\,dx + \int N(y)\frac{dy}{dx}\,dx = C. \tag{4}$$

However, the second integral in this equation is equivalent to

$$\int N(y)\,dy$$

and it usually saves time to integrate the equivalent equation

$$\int M(x)\,dx + \int N(y)\,dy = C \tag{5}$$

instead. The result of the integration will express y either explicitly or implicitly as a function of x that solves Eq. (2).

Steps for Solving a Separable First Order Differential Equation

1. Write the equation in the form $M(x)\,dx + N(y)\,dy = 0$.
2. Integrate M with respect to x and N with respect to y to obtain an equation that relates y and x.

EXAMPLE 3 Solve the differential equation

$$\frac{dy}{dx} = (1 + y^2)\, e^x.$$

Solution We use algebra to separate the variables and write the equation in the form $M(x)\,dx + N(y)\,dy = 0$, obtaining

$$e^x\,dx - \frac{1}{1 + y^2}\,dy = 0.$$

We then integrate to get

$$\int e^x\,dx - \int \frac{1}{1 + y^2}\,dy = C$$

$$e^x - \tan^{-1} y = C \qquad \text{(Constants of integration combined)}$$

$$\tan^{-1} y = e^x - C.$$

In this case, we can solve the resulting equation explicitly for y by taking the tangent of each side:

$$y = \tan(e^x - C).$$

■

Homogeneous Equations

We can sometimes use a change of variable to transform a differential equation whose variables cannot be separated into one whose variables can be separated. This is the case with homogeneous first order equations.

DEFINITION

A first order differential equation is **homogeneous** if it can be put into the form

$$\frac{dy}{dx} = F\left(\frac{y}{x}\right). \tag{6}$$

We can change Eq. (6) into a separable equation with the substitutions

$$y = vx, \qquad \frac{dy}{dx} = v + x\frac{dv}{dx}. \tag{7}$$

Then Eq. (6) becomes

$$v + x\frac{dv}{dx} = F(v), \tag{8}$$

which can be rearranged algebraically to give

$$\frac{dx}{x} + \frac{dv}{v - F(v)} = 0. \tag{9}$$

With the variables now separated, we can now solve the equation by integrating with respect to x and v. We can then return to x and y by substituting $v = y/x$.

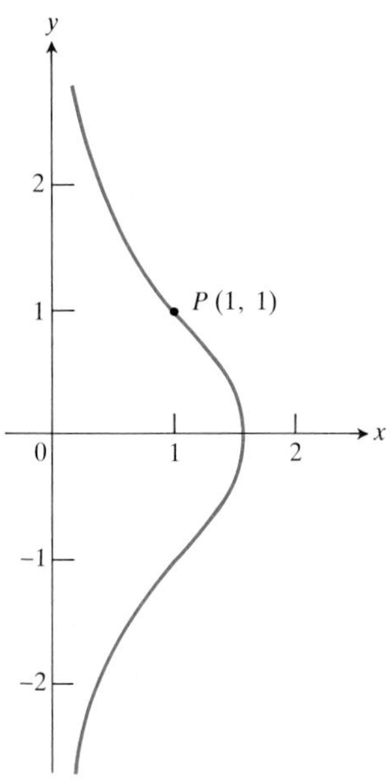

16.1 The solution of $(x^2 + y^2)\,dx + 2xy\,dy = 0$ with $y = 1$ when $x = 1$ is $x^3 + 3xy^2 = 4$.

EXAMPLE 4 Show that the equation

$$\frac{dy}{dx} = -\frac{x^2 + y^2}{2xy}$$

is homogeneous and find the solution that satisfies the condition $y(1) = 1$.

Solution Dividing the numerator and denominator of the right-hand side by x^2 gives

$$\frac{dy}{dx} = -\frac{1 + (y/x)^2}{2(y/x)}.$$

This has the form of Eq. (6) with

$$F(v) = -\frac{1 + v^2}{2v}, \quad \text{where } v = \frac{y}{x},$$

so the original equation is homogeneous. Equation (9) becomes

$$\frac{dx}{x} + \frac{dv}{v + \dfrac{1 + v^2}{2v}} = 0, \qquad \text{or} \qquad \frac{dx}{x} + \frac{2v\,dv}{1 + 3v^2} = 0.$$

The solution of this equation can be written as

$$\ln|x| + \frac{1}{3}\ln(1 + 3v^2) = C \qquad \text{or} \qquad x^3(1 + 3v^2) = \pm e^{3C} = C'.$$

We substitute $v = y/x$ to find the corresponding xy-equation, obtaining

$$x^3\left(1 + 3\frac{y^2}{x^2}\right) = C' \qquad \text{or} \qquad x^3 + 3xy^2 = C'.$$

The value of C' that gives $y = 1$ when $x = 1$ is

$$(1)^3 + 3(1)(1)^2 = C' \qquad \text{or} \qquad C' = 4.$$

The corresponding solution (Fig. 16.1) is $x^3 + 3xy^2 = 4$. ■

Exercises 16.1

Find the orders of the differential equations in Exercises 1–4. Which equations are linear and which are nonlinear?

1. $y' = x^2 + y^2$
2. $(y''')^2 - 3x^2y'' + y^5 = 10$
3. $y^{(4)} + y = 1$
4. $y'' - \frac{1}{x}y' + \frac{2}{x^2}y = e^x$

In Exercises 5–8, show that each function $y = f(x)$ is a solution of the accompanying differential equation.

5. $xy'' - y' = 0$
   a) $y = x^2$ b) $y = 1$ c) $y = C_1x^2 + C_2$
6. $y' + \frac{1}{x}y = 1$
   a) $y = \frac{x}{2}$ b) $y = \frac{1}{x} + \frac{x}{2}$ c) $y = \frac{C}{x} + \frac{x}{2}$
7. $2y' + 3y = e^{-x}$
   a) $y = e^{-x}$
   b) $y = e^{-x} + e^{-(3/2)x}$
   c) $y = e^{-x} + Ce^{-(3/2)x}$
8. $yy'' = 2(y')^2 - 2y'$
   a) $y = 1$
   b) $y = \tan x$

In Exercises 9–12, show that each function $y = f(x)$ is a solution of the accompanying initial value problem.

	Differential equation	Initial condition(s)	Solution candidate
9.	$y'' = -32$	$y(5) = 400$, $y'(5) = 0$	$y = 160x - 16x^2$
10.	$2\frac{dy}{dx} + 3y = 6$	$y(0) = 0$	$y = 2\left(1 - e^{-3t/2}\right)$
11.	$y'' + 4y = 0$	$y(0) = 3$, $y'(0) = -2$	$y = 3 \cos 2t - \sin 2t$
12.	$y'' - (y')^2 = 1$	$y(1) = 2$, $y'(1) = 0$	$y = 2 - \ln \cos(x - 1)$

13. *Continuous compounding.* You have \$1000 with which to open an account and plan to add \$1000 per year. All funds in the account will earn 10% interest per year compounded continuously. If the added deposits are also credited to your account continuously, the number of dollars x in your account at time t (years) will satisfy the initial value problem

Differential equation: $\frac{dx}{dt} = 1000 + 0.10x$

Initial condition: $x(0) = 1000.$

a) Solve the initial value problem for x as a function of t.

b) **CALCULATOR** About how many years will it take for the amount in your account to reach \$100,000?

14. **CALCULATOR** *Newton's law of cooling.* Newton's law of cooling assumes that the temperature T of a small hot object placed in a surrounding cooling medium of constant temperature T_s decreases at a rate proportional to $(T - T_s)$. An object cooled from 100°C to 40°C in 20 minutes when the surrounding temperature was 20°C. How long did it take the temperature to reach 60°C on the way down?

Homogeneous Equations

Show that the equations in Exercises 15–18 are homogeneous and find their general solutions.

15. $(x^2 + y^2)\,dx + xy\,dy = 0$

16. $(y^2 - xy)\,dx + x^2\,dy = 0$

17. $(xe^{y/x} + y)\,dx - x\,dy = 0$

18. $(x - y)\,dx + (x + y)\,dy = 0$

Solve the equations in Exercises 19 and 20 subject to the given initial conditions.

19. $\frac{dy}{dx} = \frac{y}{x} + \cos\frac{y - x}{x}, \quad y(2) = 2$

20. $\left(x \sin\frac{y}{x} - y \cos\frac{y}{x}\right) dx + x \cos\frac{y}{x}\,dy = 0, \quad y(2) = \pi$

Orthogonal Trajectories

If every member of one family of curves is a solution of the differential equation

$$M(x, y)\,dx + N(x, y)\,dy = 0$$

and every member of a second family of curves is a solution of the related equation

$$N(x, y)\,dx - M(x, y)\,dy = 0,$$

then each curve in each family is orthogonal to all the curves of the other family. Under these circumstances each family is said to be a family of **orthogonal trajectories** of the other family. If the curves in one family were electric field lines, the curves in the other family would be paths of constant electric potential.

Solve the differential equations in Exercises 21 and 22 to find equations for the families of orthogonal trajectories illustrated in Figures 16.2 and 16.3.

21. $x\,dx + 2y\,dy = 0$ and $2y\,dx - x\,dy = 0$ (Fig. 16.2)

22. $3x\,dx + 2y\,dy = 0$ and $2y\,dx - 3x\,dy = 0$ (Fig. 16.3)

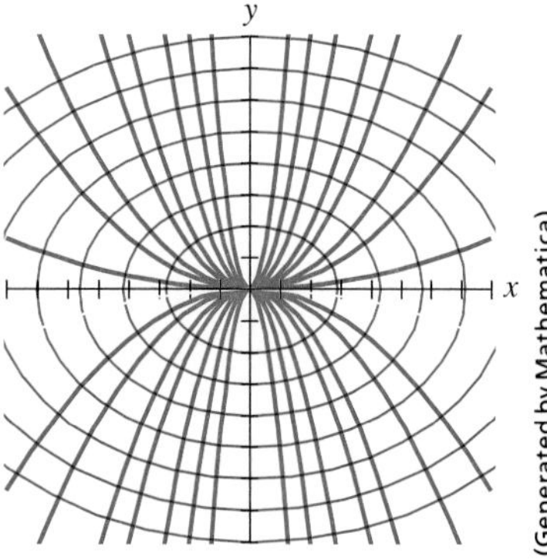

16.2 The curves in Exercise 21.

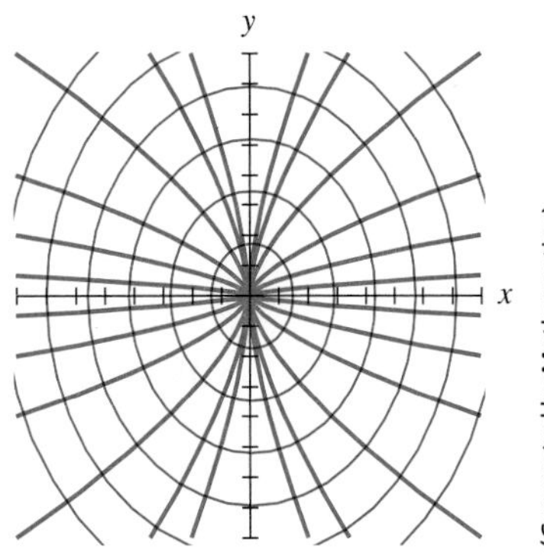

16.3 The curves in Exercise 22.

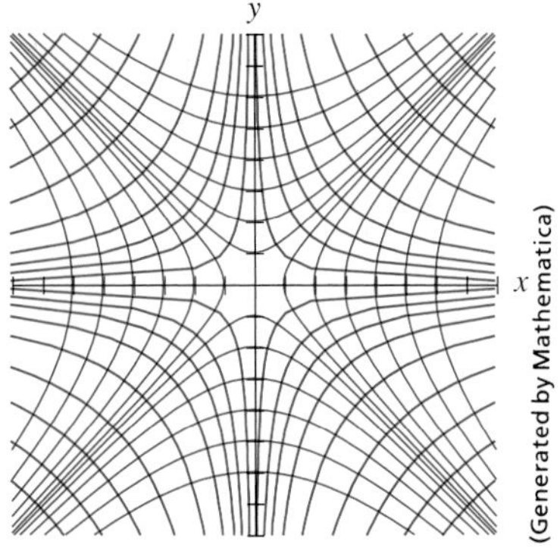

16.4 The curves in Exercise 23.

23. Identify the curves that satisfy the differential equation $y\,dx + x\,dy = 0$ and find equations for their orthogonal trajectories (Fig. 16.4).

TOOLKIT PROGRAM

Antiderivatives and Slope Fields — Graphs solutions of the initial value problem $y' = f(x, y)$, $y(x_0) = y_0$, for your choice of f, x_0, and y_0

16.2 Exact Differential Equations

If we write a (not necessarily separable) first order differential equation

$$M(x, y) + N(x, y)\frac{dy}{dx} = 0 \tag{1}$$

in the differential form

$$M(x, y)\,dx + N(x, y)\,dy = 0, \tag{2}$$

the left-hand side has the same form as the differential of a function of x and y. If the left-hand side *is* the differential df of a function $f(x, y)$, then Eq. (2) can be solved in terms of f:

$$\begin{aligned} M(x, y)\,dx + N(x, y)\,dy &= 0 \\ df(x, y) &= 0 \\ f(x, y) &= C. \end{aligned} \tag{3}$$

The equation $f(x, y) = C$ defines y implicitly as one or more differentiable functions of x. Each of these functions, it can be shown, is a solution of the original differential equation. In writing $f(x, y) = C$ we therefore say that we have solved the differential equation.

EXAMPLE 1 Since

$$d(\sin xy) = y\cos xy\,dx + x\cos xy\,dy = 0,$$

the equation

$$y\cos xy\,dx + x\cos xy\,dy = 0 \tag{4}$$

is equivalent to

$$d(\sin xy) = 0 \qquad \text{or} \qquad \sin xy = C.$$

The equation $\sin xy = C$ defines y implicitly as one or more functions of x that solve Eq. (4). ■

Not all differential forms are differentials of functions. Before we attempt to solve an equation this way we must know whether such a function f exists and then how to find it when it does. The test for the existence of f is the test for exactness in Section 15.7, repeated here for functions of two variables.

DEFINITION AND TEST

A differential form $M(x, y)\,dx + N(x, y)\,dy$ and the associated equation $M(x, y)\,dx + N(x, y)\,dy = 0$ are both said to be **exact** on a domain D in the xy-plane if for some function $f(x, y)$ defined on D,

$$M(x, y)\,dx + N(x, y)\,dy = \frac{\partial f}{\partial x}\,dx + \frac{\partial f}{\partial y}\,dy = df. \tag{5}$$

The form and equation are exact on D if and only if, throughout D,

$$\frac{\partial M}{\partial y} = \frac{\partial N}{\partial x}. \tag{6}$$

EXAMPLE 2

a) The equation $(x^2 + y^2)\,dx + (2xy + \cos y)\,dy = 0$ is exact because the partial derivatives

$$\frac{\partial M}{\partial y} = \frac{\partial}{\partial y}(x^2 + y^2) = 2y \qquad \text{and} \qquad \frac{\partial N}{\partial x} = \frac{\partial}{\partial x}(2xy + \cos y) = 2y$$

are equal.

b) The equation $(x + 3y)\,dx + (x^2 + \cos y)\,dy = 0$ is not exact because the partial derivatives

$$\frac{\partial M}{\partial y} = \frac{\partial}{\partial y}(x + 3y) = 3 \qquad \text{and} \qquad \frac{\partial N}{\partial x} = \frac{\partial}{\partial x}(x^2 + \cos y) = 2x$$

are not equal. ■

After we have tested an equation and found it to be exact, we can find the solution $f(x, y) = C$ by taking the steps described in the following example.

EXAMPLE 3 Solve the differential equation

$$(x^2 + y^2)\,dx + (2xy + \cos y)\,dy = 0.$$

Solution We already know that the equation is exact (Example 2), so it is safe to set about finding a function $f(x, y)$ whose differential is the equation's left-hand side. For the equality

$$\frac{\partial f}{\partial x}\,dx + \frac{\partial f}{\partial y}\,dy = (x^2 + y^2)\,dx + (2xy + \cos y)\,dy$$

to hold, we must have

$$\frac{\partial f}{\partial x} = x^2 + y^2 \quad \text{and} \quad \frac{\partial f}{\partial y} = 2xy + \cos y. \tag{7}$$

The partial derivative $\partial f/\partial x = x^2 + y^2$ was calculated by holding y fixed and differentiating f with respect to x. We may therefore find f by holding y at a constant value and integrating with respect to x. When we do so we get

$$f(x, y) = \int_{y \text{ const.}} (x^2 + y^2)\, dx = \frac{x^3}{3} + y^2x + k(y). \tag{8}$$

The constant of integration is written as a function of y because its value may change with each new y.

To find $k(y)$, we calculate $\partial f/\partial y$ from Eq. (8) and set the result equal to the known partial derivative $\partial f/\partial y = 2xy + \cos y$:

$$\frac{\partial}{\partial y}\left(\frac{x^3}{3} + y^2x + k(y)\right) = 2xy + \cos y$$

$$2xy + k'(y) = 2xy + \cos y$$

$$k'(y) = \cos y.$$

A single integration with respect to y then shows that $k(y) = \sin y$ plus a constant, so

$$f(x, y) = \frac{x^3}{3} + y^2x + \sin y + \text{a constant.}$$

If all we wanted was to find a function whose differential is $(x^2 + y^2)\, dx + (2xy + \cos y)\, dy$ we'd be done now. We have found infinitely many such functions, one for each possible constant. Our goal, however, is to solve the original differential equation, and for that we must take one more step. We must write down the equation $f(x, y) = C$, because it is this equation, and not the formula for f alone, that defines the differential equation's solution functions. The solution of the differential equation is

$$\frac{x^3}{3} + y^2x + \sin y = C. \qquad \text{(Constants combined)}$$

■

Steps for Solving an Equation You Know to Be Exact

1. Match the equation to the form $df = (\partial f/\partial x)\, dx + (\partial f/\partial y)\, dy$ to identify $\partial f/dx$ and $\partial f/dy$.
2. Integrate $\partial f/\partial x$ with respect to x, writing the constant of integration as $k(y)$.
3. Differentiate with respect to y and set the result equal to $\partial f/\partial y$ to find $k'(y)$.
4. Integrate to find $k(y)$ and determine f.
5. Write the solution of the exact equation as $f(x, y) = C$.

Integrating Factors

It can be shown that every nonexact differential equation $M(x, y)\, dx + N(x, y)\, dy = 0$ can be made exact by multiplying both sides by a suitable **integrating factor** $\rho(x, y)$. In other words, the equation

$$\rho(x, y)\, M(x, y)\, dx + \rho(x, y)\, N(x, y)\, dy = 0$$

is always exact for an appropriate choice of ρ.

EXAMPLE 4 The equation

$$2y\,dx + x\,dy = 0 \tag{9}$$

is not exact because $\partial(2y)/\partial y = 2$ while $\partial(x)/\partial x = 1$. The equation

$$2xy\,dx + x^2\,dy = 0,$$

obtained by multiplying both sides of Eq. (9) by x, *is* exact because $\partial(2xy)/\partial y = 2x$ and $\partial(x^2)/\partial x = 2x$. ■

Unfortunately, there is no general technique for finding an integrating factor when you need one and the search for one can be a frustrating experience. Practice does help, however, and Exercises 15–18 supply integrating factors for you to work with. As you will see there, integrating factors need not be unique.

Exercises 16.2

Test the equations in Exercises 1–8 for exactness.

1. $x\,dx - y\,dy = 0$

2. $y\,dx - x\,dy = 0$

3. $\dfrac{1}{y}\,dx - \dfrac{x}{y^2}\,dy = 0$

4. $(x + y^2)\,dx + (2xy + 1)\,dy = 0$

5. $(x + e^y)\,dx + (y + xe^y)\,dy = 0$

6. $y \cos xy\,dx + x \cos xy\,dy = 0$

7. $(x + 2y)\,dx - (2x - y)\,dy = 0$

8. $(x + 2y)\,dx - (2x + y)\,dy = 0$

Solve the equations in Exercises 9–14.

9. $(x + y)\,dx + (x + y^2)\,dy = 0$

10. $(2x\,e^y + e^x)\,dx + (x^2 + 1)e^y\,dy = 0$

11. $(2xy + y^2)\,dx + (x^2 + 2xy - y)\,dy = 0$

12. $(x^3 - y)\,dx - x\,dy = 0$

13. $\left(e^x + \ln y + \dfrac{y}{x}\right)dx + \left(\dfrac{x}{y} + \ln x + \sin y\right)dy = 0$

14. $(y \cos xy + 1)\,dx + (x \cos xy + 2y - 3)\,dy = 0$

Show that the integrating factors ρ in Exercises 15–18 make the differential equations exact. Then solve the equations using those integrating factors.

15. $(xy^2 + y)\,dx - x\,dy = 0; \quad \rho = 1/y^2$

16. $(x + 2y)\,dx - x\,dy = 0; \quad \rho = 1/x^3$

17. $y\,dx + x\,dy = 0,$

a) $\rho = \dfrac{1}{xy}$ b) $\rho = \dfrac{1}{(xy)^2}$

18. $y\,dx - x\,dy = 0,$

a) $\rho = \dfrac{1}{y^2}$ b) $\rho = \dfrac{1}{x^2}$

c) $\rho = \dfrac{1}{xy}$ d) $\rho = \dfrac{1}{x^2 + y^2}$

Solve the initial value problems in Exercises 19–22 for y as a function of x.

19. $(x + y^2)\,dx + (2xy + 1)\,dy = 0; \quad y(0) = 2$

20. $(x + e^y)\,dx + (y + xe^y)\,dy = 0; \quad y(2) = 0$

21. $\left(\dfrac{1}{x} - y\right)dx + \left(\dfrac{1}{y} - x\right)dy = 0; \quad y(1) = 1$

22. $(2x + y + 1)\,dx + (2y + x + 1)\,dy = 0; \quad y(1) = 5$

23. Find the value of a that makes the equation

$$(3x^2 + y^2)\,dx + a\,xy\,dy = 0$$

exact and solve the equation for this value of a.

24. The equation $(x^2 + by^2)\,dx + 6\,xy\,dy = 0$ is not exact but is known to have $\rho = 1/x^2$ as an integrating factor. Use this information to find the value of b and solve the equation.

TOOLKIT PROGRAM

Antiderivatives and Slope Fields — Graphs solutions of the initial value problem $y' = f(x, y)$, $y(x_0) = y_0$, for your choice of f, x_0, and y_0

16.3 Linear First Order Equations

The equation $dy/dx = ky$ we use to model bacterial growth, radioactive decay, and temperature change, is a special case of a more general equation called a linear first order equation. This section shows how to solve the more general equation and uses it to predict current flow in an electric circuit.

DEFINITIONS

A differential equation that can be written in the form

$$\frac{dy}{dx} + P(x)y = Q(x), \tag{1}$$

where P and Q are functions of x, is called a **linear first order equation.** Equation (1) is the equation's **standard form.**

EXAMPLE 1 The standard form of the equation

$$x\frac{dy}{dx} + 3y = x^2$$

is

$$\frac{dy}{dx} + \frac{3}{x}y = x.$$

This is Eq. (1) with $P(x) = 3/x$ and $Q(x) = x$. ■

The way we solve the equation

$$\frac{dy}{dx} + P(x)y = Q(x)$$

is to multiply both sides by an integrating factor $\rho(x)$ that turns the left-hand side of the multiplied equation,

$$\rho\frac{dy}{dx} + P\rho y = \rho Q, \tag{2}$$

into the derivative of the product ρy.

Once we have chosen ρ (we shall do so in a moment) and carried out the multiplication, we can solve Eq. (2) by integrating both sides with respect to x:

$$\rho\frac{dy}{dx} + P\rho y = \rho Q \qquad \text{(Eq. 2)}$$

$$\frac{d}{dx}(\rho y) = \rho Q \qquad \text{(Choice of } \rho\text{)}$$

$$\rho y = \int \rho Q\,dx \qquad \text{(Integration with respect to } x\text{)}$$

$$y = \frac{1}{\rho}\int \rho Q\,dx \qquad \text{(Solved for } y\text{)}$$

To find ρ, we find the function of x that satisfies the equations

$$\frac{d}{dx}(\rho y) = \rho\frac{dy}{dx} + P\rho y,$$

$$\rho\frac{dy}{dx} + y\frac{d\rho}{dx} = \rho\frac{dy}{dx} + P\rho y, \qquad \text{(Product Rule for derivatives)}$$

$$y\frac{d\rho}{dx} = P\rho y, \qquad \text{(The terms } \rho\frac{dy}{dx} \text{ cancel.)}$$

$$\frac{d\rho}{dx} = P\rho, \qquad \text{(The } y\text{'s cancel.)}$$

$$\frac{d\rho}{\rho} = P\,dx, \qquad \text{(Variables separated)}$$

$$\rho = C\,e^{\int P\,dx}. \qquad \text{(Equation integrated and solved for } \rho\text{)}$$

Since we do not need the most general function ρ, we may take C to be 1.

We summarize our conclusions with the following theorem.

THEOREM 1

The solution of the equation

$$\frac{dy}{dx} + P(x)y = Q(x) \tag{3}$$

is

$$y = \frac{1}{\rho(x)}\int \rho(x)\,Q(x)\,dx, \tag{4}$$

where

$$\rho(x) = e^{\int P(x)\,dx}. \tag{5}$$

In the formula for ρ we do not need the most general antiderivative of $P(x)$. Any antiderivative will do.

EXAMPLE 2 Solve the equation $x\dfrac{dy}{dx} - 3y = x^2, \quad x > 0.$

Solution We solve the equation in four steps.

Steps for Solving a Linear First Order Equation

1. Put it in standard form.
2. Find an antiderivative of $P(x)$.
3. Find $\rho = e^{\int P(x)\,dx}$.
4. Use Eq. (4) to find y.

STEP 1: *Put the equation in standard form and identify the functions P and Q.* To do so, we divide both sides of the equation by the coefficient of dy/dx, in this case x, obtaining

$$\frac{dy}{dx} - \frac{3}{x}y = x, \qquad P(x) = -\frac{3}{x}, \qquad Q(x) = x.$$

STEP 2: *Find an antiderivative of P(x)* (any one will do):

$$\int P(x)\,dx = \int -\frac{3}{x}\,dx = -3\int\frac{1}{x}\,dx = -3\ln|x| = -3\ln x \qquad (x > 0).$$

STEP 3: *Use Eq. (5) to find ρ:*

$$\rho(x) = e^{\int P(x)\,dx} = e^{-3\ln x} = e^{-\ln x^3} = \frac{1}{x^3}.$$

STEP 4: *Use Eq. (4) to find the solution:*

$$y = \frac{1}{\rho(x)} \int \rho(x)\, Q(x)\, dx = \frac{1}{(1/x^3)} \int \left(\frac{1}{x^3}\right)(x)\, dx \qquad \text{(Values from Steps 1–3)}$$

$$= x^3 \int \frac{1}{x^2}\, dx = x^3\left(-\frac{1}{x} + C\right) = Cx^3 - x^2.$$

The solution is the function $y = Cx^3 - x^2$. ■

R–L Circuits

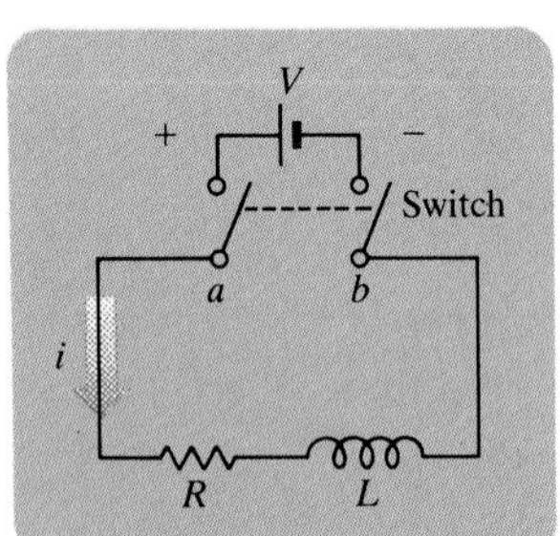

16.5 The *R–L* circuit in Example 3.

The diagram in Fig. 16.5 represents an electrical circuit whose total resistance is R ohms and whose self-inductance, shown schematically as a coil, is L henries (hence the name "R–L circuit"). There is also a switch whose terminals at a and b can be closed to connect a constant electrical source of V volts.

Ohm's law, $V = RI$, has to be modified for such a circuit. The modified form is

$$L\frac{di}{dt} + Ri = V, \tag{6}$$

where i is the intensity of the current in amperes and t is the time in seconds. By solving this equation, we can predict how the current will flow after the switch is closed.

EXAMPLE 3 The switch in the *R–L* circuit in Fig. 16.5 is closed at time $t = 0$. How will the current flow as a function of time?

Solution Equation (6) is a linear first order differential equation for i as a function of t. When we put it in the standard form

$$\frac{di}{dt} + \frac{R}{L}i = \frac{V}{L}$$

and carry out the solution steps with $P = R/L$ and $Q = V/L$ we find that

$$i = \frac{1}{e^{(R/L)t}} \int e^{(R/L)t}\left(\frac{V}{L}\right) dt = \frac{V}{R} + C\frac{V}{L}e^{-(R/L)t}.$$

Imposing the initial condition that $i(0)$ be 0 determines the value of C to be $-L/R$, so

$$i = \frac{V}{R} + \left(-\frac{L}{R}\right)\frac{V}{L}e^{-(R/L)t} = \frac{V}{R}\left(1 - e^{-(R/L)t}\right).$$

We see from this that the current is always less than V/R but that it approaches V/R as a **steady-state value:**

$$\lim_{t\to\infty} \frac{V}{R}\left(1 - e^{-Rt/L}\right) = \frac{V}{R}(1 - 0) = \frac{V}{R}. \tag{7}$$

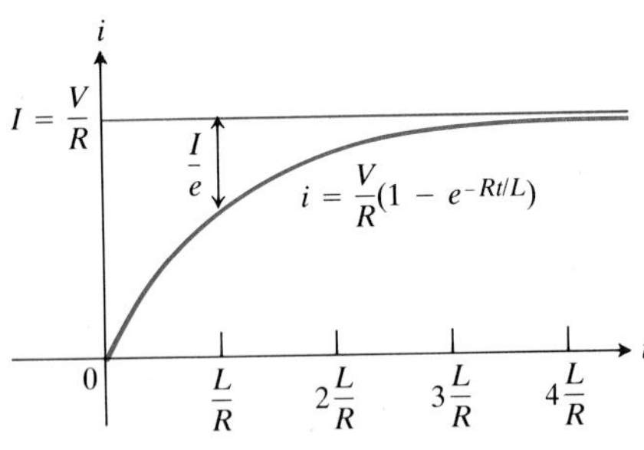

16.6 The growth of the current in the *R–L* circuit in Example 3. I is the current's steady-state value. The number $t = L/R$ is the time constant of the circuit. The current gets to within 5% of its steady-state value in 3 time constants (more about this in the exercises).

The current $I = V/R$ is the current that will flow in the circuit if either $L = 0$ (no inductance) or $di/dt = 0$ (steady current, $i =$ constant) (Fig. 16.6). ■

***R–L* Circuit Equations**

Current: $$i = \frac{V}{R}\left(1 - e^{-Rt/L}\right) \tag{8}$$

Steady-state value: $$I = \frac{V}{R} \tag{9}$$

Exercises 16.3

Solve the differential equations in Exercises 1–8.

1. $e^x \frac{dy}{dx} + 2e^x y = 1$
2. $2\frac{dy}{dx} - y = e^{x/2}$
3. $x\frac{dy}{dx} + 3y = \frac{\sin x}{x^2}$
4. $x\frac{dy}{dx} + y = x \cos x$
5. $(x-1)^3 \frac{dy}{dx} + 4(x-1)^2 y = x + 1$
6. $e^{2x}\frac{dy}{dx} + 2e^{2x} y = 2x$
7. $\sin x \frac{dy}{dx} + (\cos x)\, y = \tan x$
8. $\cosh x \frac{dy}{dx} + (\sinh x)\, y = e^{-x}$

Solve the initial value problems in Exercises 9–12 for y as a function of x.

	Differential equation	Initial condition
9.	$\frac{dy}{dx} + 2y = x$	$y(0) = 1$
10.	$x\frac{dy}{dx} + 2y = x^3$	$y(2) = 1$
11.	$x\frac{dy}{dx} + y = \sin x$	$y(\pi/2) = 1$
12.	$x\frac{dy}{dx} - 2y = x^3 \sec x \tan x$	$y(\pi/3) = 2$

13. Solve the initial value problem $dy/dx = ky$, $y(0) = y_0$, with Eq. (4) and see what you get.

14. *Blood sugar.* If glucose is fed intravenously at a constant rate, the change in the overall concentration $c(t)$ of glucose in the blood with respect to time may be described by the differential equation
$$\frac{dc}{dt} = \frac{G}{100\,V} - kc.$$
In this equation, G, V, and k are positive constants, G being the rate at which glucose is admitted, in milligrams per minute, and V the volume of blood in the body, in liters (around 5 liters for an adult). The concentration $c(t)$ is measured in milligrams per centiliter. The term $-kc$ is included because the glucose is assumed to be changing continually into other molecules at a rate proportional to its concentration.
a) Solve the equation for $c(t)$, using c_0 to denote $c(0)$.
b) Find the steady-state concentration, $\lim_{t\to\infty} c(t)$.

Second Order Equations Reducible to First Order

If the y-term is absent from a second order differential equation, as it is from the equation
$$x\frac{d^2y}{dx^2} + \frac{dy}{dx} = x^2,$$
we can treat the equation as a first order equation in the variable $p = dy/dx$ and its first derivative $dp/dx = d^2y/dx^2$. We can then solve the equation by solving
$$x\frac{dp}{dx} + p = x^2$$
to find p and solving
$$\frac{dy}{dx} = p$$
to find y.

Use the substitutions $dy/dx = p$, $d^2y/dx^2 = dp/dx$ to solve the differential equations in Exercises 15–18. Your solutions will each contain two arbitrary constants.

15. $\frac{d^2y}{dx^2} + \frac{dy}{dx} = 0$
16. $x\frac{d^2y}{dx^2} + 2\frac{dy}{dx} = 1$
17. $x\frac{d^2y}{dx^2} + \frac{dy}{dx} = 0$
18. $2\frac{d^2y}{dx^2} + \left(\frac{dy}{dx}\right)^2 = -1$

19. Solve the initial value problem
$$x\frac{d^2y}{dx^2} + \frac{dy}{dx} = x^2, \quad y = 0 \text{ and } y' = 1 \text{ when } x = 1.$$

20. Use the substitution $d^2y/dx^2 = q$ to solve the equation

$$x\frac{d^3y}{dx^3} - 2\frac{d^2y}{dx^2} = 0.$$

R–L Circuits

21. *Current in a closed R–L circuit.* How many seconds after the switch in an R–L circuit is closed will it take the current to reach half of its steady-state value? Notice that the time depends only on R and L and not on how much voltage is applied.

22. *Current in an open R–L circuit.* If the switch is thrown open after the current in an R–L circuit has built up to its steady-state value, the decaying current (Fig. 16.7) obeys the equation

$$L\frac{di}{dt} + Ri = 0, \tag{10}$$

which is Eq. (6) with $V = 0$.

a) Solve Eq. (10) to express i as a function of t.
b) How long after the switch is thrown will it take the current to fall to half its original value?
c) What is the value of the current when $t = L/R$? (The significance of this time is explained in the next exercise.)

23. **CALCULATOR** *Time constants.* Engineers call the number L/R the *time constant* of the R–L circuit in Fig. 16.5. The significance of the time constant is that the current will reach 95% of its final value within 3 time constants of the time the switch is closed. Thus, the time constant gives a built-in measure of how rapidly an individual circuit will reach equilibrium. Find the value of i in Eq. (8) that corresponds to $t = 3L/R$ and show that it is about 95% of the steady-state value $I = V/R$.

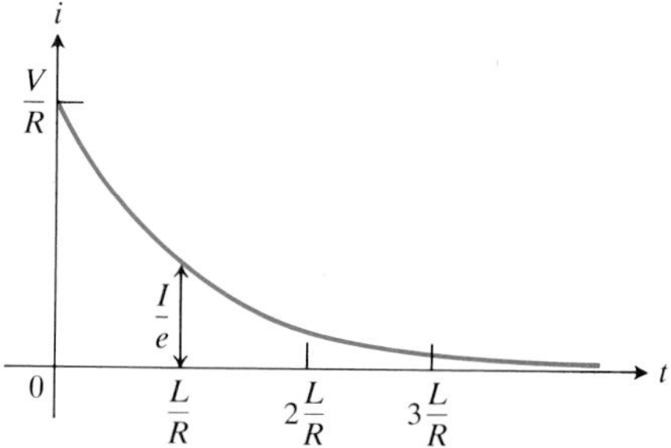

16.7 The current in an R–L circuit decays exponentially when the power is turned off. Exercise 22 asks you to find a formula for the current as a function of time.

TOOLKIT PROGRAM

Antiderivatives and Slope Fields	Graphs solutions of the initial value problem $y' = f(x, y), y(x_0) = y_0$, for your choice of f, x_0, and y_0

16.4 Second Order Linear Homogeneous Equations

If $F(x) = 0$, the linear equation

$$a_n(x)\frac{d^ny}{dx^n} + a_{n-1}(x)\frac{d^{n-1}y}{dx^{n-1}} + \cdots + a_1(x)\frac{dy}{dx} + a_0(x)y = F(x) \tag{1}$$

is called **homogeneous;** otherwise it is called **nonhomogeneous.** This section shows how to solve the homogeneous equation

$$\frac{d^2y}{dx^2} + 2a\frac{dy}{dx} + by = 0 \qquad (a \text{ and } b \text{ constant}).$$

Section 16.5 shows how to solve the nonhomogeneous equation

$$\frac{d^2y}{dx^2} + 2a\frac{dy}{dx} + by = F(x).$$

Section 16.6 shows how we can use these equations to model oscillation.

Linear Differential Operators

At this point, it is convenient to introduce the symbol D to represent the operation of differentiation with respect to x. That is, we write $Df(x)$ to mean $(d/dx)f(x)$. Furthermore, we define powers of D to mean taking successive derivatives:

$$D^2f(x) = D\{Df(x)\} = \frac{d^2f(x)}{dx^2}, \qquad D^3f(x) = D\{D^2f(x)\} = \frac{d^3f(x)}{dx^3},$$

and so on. A polynomial in D is to be interpreted as an operator that, when applied to $f(x)$, produces a linear combination of f and its successive derivatives. For example,

$$(D^2 + D - 2)f(x) = D^2f(x) + Df(x) - 2f(x) = \frac{d^2f(x)}{dx^2} + \frac{df(x)}{dx} - 2f(x).$$

DEFINITION

A polynomial in D is called a **linear differential operator.**

Linear differential operators are often denoted by single letters. If L_1 and L_2 are two such linear operators, their sum and product are defined by the equations

$$(L_1 + L_2)f(x) = L_1f(x) + L_2f(x), \qquad L_1L_2f(x) = L_1(L_2f(x)).$$

Linear differential operators that are polynomials in D with constant coefficients satisfy basic algebraic laws that make it possible for us to treat them like ordinary polynomials so far as addition, multiplication, and factoring are concerned. Thus,

$$(D^2 + D - 2)f(x) = (D + 2)(D - 1)f(x) = (D - 1)(D + 2)f(x).$$

Since this equation holds for any twice-differentiable function f, we can also write the equality between operators:

$$D^2 + D - 2 = (D + 2)(D - 1) = (D - 1)(D + 2).$$

The Characteristic Equation

In the remainder of this section, we consider only linear *second order* equations with constant real-number coefficients. Because the solutions of nonhomogeneous equations depend on the solutions of the corresponding homogeneous equations, we focus on equations of the form

$$\frac{d^2y}{dx^2} + 2a\frac{dy}{dx} + by = 0 \tag{2}$$

or, in operator notation,

$$(D^2 + 2aD + b)y = 0. \tag{3}$$

The usual method for solving Eq. (3) is to begin by factoring the operator:

$$D^2 + 2aD + b = (D - r_1)(D - r_2).$$

We do this by finding the two roots r_1 and r_2 of the equation $r^2 + 2ar + b = 0$.

DEFINITION

The equation

$$r^2 + 2ar + b = 0 \tag{4}$$

is the **characteristic equation** of the equation

$$\frac{d^2y}{dx^2} + 2a\frac{dy}{dx} + by = 0. \tag{5}$$

Equation (2) is equivalent to

$$(D - r_1)(D - r_2)y = 0. \tag{6}$$

If we now let

$$(D - r_2)y = u, \tag{7}$$

we get

$$(D - r_1)u = 0. \tag{8}$$

From here, we can solve Eq. (6) in two steps. From Eq. (8), which is first order separable in u and x, we find

$$u = C_1 e^{r_1 x}.$$

We substitute this into (7), which becomes

$$(D - r_2)y = C_1 e^{r_1 x} \qquad \text{or} \qquad \frac{dy}{dx} - r_2 y = C_1 e^{r_1 x}.$$

This equation is linear in y. Its integrating factor is $\rho = e^{-r_2 x}$, (Section 16.3), and its solution is

$$e^{-r_2 x}y = C_1 \int e^{(r_1 - r_2)x}\,dx + C_2. \tag{9}$$

How we proceed at this point depends on whether r_1 and r_2 are real and unequal, real and equal, or complex.

Real, Unequal Roots

If $r_1 \neq r_2$, the evaluation of the integral in Eq. (9) leads to

$$e^{-r_2 x}y = \frac{C_1}{r_1 - r_2}e^{(r_1 - r_2)x} + C_2 \qquad \text{or} \qquad y = \frac{C_1}{r_1 - r_2}e^{r_1 x} + C_2 e^{r_2 x}. \tag{10}$$

Since C_1 is an arbitrary constant, so is $C_1/(r_1 - r_2)$, and we can write the solution in Eq. (10) as

$$y = C_1 e^{r_1 x} + C_2 e^{r_2 x}. \tag{11}$$

EXAMPLE 1 Solve the equation $\dfrac{d^2y}{dx^2} + \dfrac{dy}{dx} - 2y = 0.$

Solution The characteristic equation $r^2 + r - 2 = 0$ has roots $r_1 = 1$, $r_2 = -2$. Hence, by Eq. (11), the solution of the differential equation is

$$y = C_1 e^x + C_2 e^{-2x}.$$

Real, Equal Roots

If $r_1 = r_2$, then $e^{(r_1 - r_2)x} = e^0 = 1$ and Eq. (9) reduces to $e^{-r_2 x}y = C_1 x + C_2$ or

$$y = (C_1 x + C_2)e^{r_2 x}. \tag{12}$$

EXAMPLE 2 Solve the equation $\dfrac{d^2y}{dx^2} + 4\dfrac{dy}{dx} + 4y = 0.$

Solution

$$r^2 + 4r + 4 = (r + 2)^2,$$

$$r_1 = r_2 = -2,$$

$$y = (C_1 x + C_2)e^{-2x}.$$

Complex Roots

From the theory of equations, we know that if a and b are real and the roots of the characteristic equation $r^2 + 2ar + b = 0$ are not real then the roots must be a pair of complex conjugate numbers

$$r_1 = \alpha + \beta i, \qquad r_2 = \alpha - \beta i. \tag{13}$$

If $\beta \neq 0$ then Eq. (11) applies once again to give

$$y = c_1 e^{(\alpha + i\beta)x} + c_2 e^{(\alpha - i\beta)x} = e^{\alpha x}\left(c_1 e^{i\beta x} + c_2 e^{-i\beta x}\right), \tag{14}$$

where c_1 and c_2 are complex constants. By Euler's formula (Section 9.7),

$$e^{i\beta x} = \cos \beta x + i \sin \beta x \qquad \text{and} \qquad e^{-i\beta x} = \cos \beta x - i \sin \beta x.$$

Hence, we may replace Eq. (14) by

$$y = e^{\alpha x}[(c_1 + c_2) \cos \beta x + i(c_1 - c_2) \sin \beta x]. \tag{15}$$

Finally, we introduce new arbitrary constants, $C_1 = c_1 + c_2$ and $C_2 = i(c_1 - c_2)$, to give the solution in Eq. (15) a shorter form:

$$y = e^{\alpha x}(C_1 \cos \beta x + C_2 \sin \beta x). \tag{16}$$

The constants C_1 and C_2 will be real as long as c_1 and c_2 are complex conjugates:

$$c_1 = \frac{1}{2}\left(C_1 - iC_2\right), \qquad c_2 = \frac{1}{2}\left(C_1 + iC_2\right).$$

EXAMPLE 3 Solve the equation

$$\frac{d^2y}{dx^2} + 4\frac{dy}{dx} + 6y = 0.$$

Solution We solve the characteristic equation $r^2 + 4r + 6 = 0$ with the quadratic formula to find its roots $r_1 = -2 + \sqrt{2}\,i$ and $r_2 = -2 - \sqrt{2}\,i$. With $\alpha = -2$ and $\beta = \sqrt{2}$, Eq. (16) gives

$$y = e^{-2x}(C_1 \cos \sqrt{2}\,x + C_2 \sin \sqrt{2}\,x).$$

■

EXAMPLE 4 *Imaginary roots.* Solve the equation

$$\frac{d^2y}{dx^2} + 4y = 0.$$

Solution The roots of the characteristic equation $r^2 + 4 = 0$ are $r = \pm 2i$. With $\alpha = 0$ and $\beta = 2$, Eq. (16) gives

$$y = C_1 \cos 2x + C_2 \sin 2x.$$

■

Solutions of $\dfrac{d^2y}{dx^2} + 2a\dfrac{dy}{dx} + by = 0$

Roots of $r^2 + 2ar + b = 0$	Solution
r_1, r_2 real and unequal	$y = C_1 e^{r_1x} + C_2 e^{r_2x}$
r_1, r_2 real and equal	$y = (C_1 x + C_2)e^{r_2x}$
r_1, r_2 complex conjugates, $\alpha \pm \beta i$	$y = e^{\alpha x}(C_1 \cos \beta x + C_2 \sin \beta x)$

Exercises 16.4

Solve the equations in Exercises 1–12.

1. $\dfrac{d^2y}{dx^2} + 2\dfrac{dy}{dx} = 0$
2. $\dfrac{d^2y}{dx^2} + 5\dfrac{dy}{dx} + 6y = 0$
3. $\dfrac{d^2y}{dx^2} + 6\dfrac{dy}{dx} + 5y = 0$
4. $\dfrac{d^2y}{dx^2} - 2\dfrac{dy}{dx} - 3y = 0$
5. $\dfrac{d^2y}{dx^2} - 4\dfrac{dy}{dx} + 4y = 0$
6. $\dfrac{d^2y}{dx^2} + 6\dfrac{dy}{dx} + 9y = 0$
7. $\dfrac{d^2y}{dx^2} - 10\dfrac{dy}{dx} + 25y = 0$
8. $\dfrac{d^2y}{dx^2} - 2\sqrt{2}\dfrac{dy}{dx} + 2y = 0$
9. $\dfrac{d^2y}{dx^2} + \dfrac{dy}{dx} + y = 0$
10. $\dfrac{d^2y}{dx^2} - 6\dfrac{dy}{dx} + 10y = 0$
11. $\dfrac{d^2y}{dx^2} - 2\dfrac{dy}{dx} + 4y = 0$
12. $\dfrac{d^2y}{dx^2} + 8\dfrac{dy}{dx} + 25y = 0$

Solve the initial value problems in Exercises 13–24 ($y'' = d^2y/dx^2$ and $y' = dy/dx$).

13. $y'' - y = 0,\ \ y(0) = 1,\ \ y'(0) = -2$
14. $2y'' - y' - y = 0,\ \ y(0) = -1,\ \ y'(0) = 0$
15. $y'' - 4y = 0,\ \ y(0) = 0,\ \ y'(0) = 3$
16. $y'' - 9y = 0,\ \ y(\ln 2) = 1,\ \ y'(\ln 2) = -3$
17. $y'' + 2y' + y = 0,\ \ y(0) = 0,\ \ y'(0) = 1$
18. $4y'' - 2y' + y = 0,\ \ y(0) = 4,\ \ y'(0) = 2$
19. $4y'' + 12y' + 9y = 0,\ \ y(0) = 0,\ \ y'(0) = -1$
20. $y'' = 0,\ \ y(0) = -3,\ \ y'(0) = 5$
21. $y'' + 4y = 0,\ \ y(0) = 0,\ \ y'(0) = 2$
22. $y'' + 9y = 0,\ \ y(0) = 0,\ \ y'(0) = \sqrt{3}$
23. $y'' - 2y' + 3y = 0,\ \ y(0) = 2,\ \ y'(0) = 1$
24. $y'' - 6y' + 10y = 0,\ \ y(0) = 7,\ \ y'(0) = 1$

16.5 Second Order Nonhomogeneous Linear Equations

This section shows how to solve nonhomogeneous equations of the form

$$\frac{d^2y}{dx^2} + 2a\frac{dy}{dx} + by = F(x) \qquad (a \text{ and } b \text{ real constants}). \tag{1}$$

The procedure has three basic steps. First, we use the techniques of Section 16.4 to find the general solution y_h (h stands for "homogeneous") of the **reduced equation**

$$\frac{d^2y}{dx^2} + 2a\frac{dy}{dx} + by = 0. \tag{2}$$

Then we find a particular solution y_p of the **complete** equation, Eq. (1). Then we add y_p to y_h to form the general solution of the complete equation. Why the procedure works and how we find y_p are the main topics of this section.

The Relation Between the Solutions of Eqs. (1) and (2)

The main idea is that we can obtain all solutions of the complete equation (1) by adding a particular solution of Eq. (1) to the general solution of the reduced equation (2). The crux of the matter is that if L is any linear differential operator (not necessarily one of second order or having constant coefficients), then

$$L(k_1 y_1 + k_2 y_2) = k_1 L(y_1) + k_2 L(y_2)$$

for any constants k_1 and k_2 and functions y_1 and y_2 to which the operator L can be applied. In the present context, we can take

$$L = D^2 + 2aD + b$$

and write Eqs. (1) and (2) as $L(y) = F(x)$ and $L(y) = 0$, respectively. The following theorem states the relation between solutions of homogeneous and nonhomogeneous linear equations.

THEOREM 2

Let $L(y) = F(x)$ be a linear nonhomogeneous differential equation, and $L(y) = 0$ be the reduced equation with the same operator L. If y_h is the general solution of the homogeneous equation and y_p is a particular solution of the nonhomogeneous equation, then the general solution of the nonhomogeneous equation is

$$y = y_h(x) + y_p(x).$$

PROOF Let $L(y_h(x)) = 0$ and $L(y_p(x)) = F(x)$. Then

$$L(y_h(x) + y_p(x)) = L(y_h(x)) + L(y_p(x)) = 0 + F(x) = F(x),$$

which establishes the fact that $y = y_h + y_p$ is a solution of the nonhomogeneous equation. To show that all solutions have this form, let y be any solution of the nonhomogeneous equation $L(y) = F(x)$. Then

$$L(y - y_p) = L(y) - L(y_p) = F(x) - F(x) = 0,$$

so that $y - y_p$ satisfies the homogeneous equation. Therefore, if y_h is the general solution of $(D^2 + 2aD + b)y = 0$, there is some choice of the constants C_1 and C_2 such that

$$y - y_p = C_1\, u_1(x) + C_2\, u_2(x),$$

which means that

$$y = y_p + C_1\, u_1(x) + C_2\, u_2(x).$$ ■

We now discuss three ways to find a particular solution of the complete equation (Eq. 1):

1. Inspired guessing (Try this first—it's a real time-saver when it works. Experience will help.)
2. The method of variation of parameters
3. The method of undetermined coefficients.

Inspired Guessing

In the following example, we guess that there exists a particular solution of the form $y_p(x) = C$. We find the right value of C by substituting y_p and its derivatives into the differential equation.

EXAMPLE 1 Find a particular solution of $\dfrac{d^2y}{dx^2} + 2\dfrac{dy}{dx} - 3y = 6.$

Solution We guess that there is a solution of the form $y_p = C$ and substitute y_p and its derivatives $y_p' = 0$, $y_p'' = 0$ into the equation. This leads to

$$\frac{d^2y_p}{dx^2} + 2\frac{dy_p}{dx} - 3y_p = 6, \qquad 0 + 2(0) - 3(C) = 6, \qquad C = -2.$$

The equation has $y_p = -2$ as a solution. ■

EXAMPLE 2 Find the general solution of $\dfrac{d^2y}{dx^2} + 2\dfrac{dy}{dx} - 3y = 6.$

Solution We find the solution in three steps.

STEP 1: *Find the general solution y_h of the reduced equation*

$$\frac{d^2y}{dx^2} + 2\frac{dy}{dx} - 3y = 0.$$

We always use these steps:

1. Find y_h.
2. Find y_p.
3. Add y_p to y_h.

The roots of the characteristic equation $r^2 + 2r - 3 = 0$ are $r_1 = 1$ and $r_2 = -3$, so

$$y_h = C_1 e^x + C_2 e^{-3x}.$$

STEP 2: *Find a particular solution y_p of the complete equation:*

$$y_p = -2. \qquad \text{(From Example 1)}$$

STEP 3: *Add y_p to y_h to form the general solution y of the complete equation:*

$$y = y_p + y_h = C_1 e^x + C_2 e^{-3x} - 2.$$

Variation of Parameters

When guessing doesn't work (it usually doesn't), we find a particular solution of the complete equation (Eq. 1) by a method called **variation of parameters.** The method assumes we already know the general solution

$$y_h = C_1 u_1(x) + C_2 u_2(x) \tag{3}$$

of the reduced equation (Eq. 2), which is why we make finding y_h our first solution step.

The method of variation of parameters consists of replacing the constants C_1 and C_2 on the right side of Eq. (3) by functions $v_1(x)$ and $v_2(x)$ and then requiring that the new expression

$$y = v_1u_1 + v_2u_2$$

be a solution of the complete equation (1). There are two functions v_1 and v_2 to be determined, and requiring that Eq. (1) be satisfied is only one condition. It turns out to simplify things if we also require that

$$v_1'u_1 + v_2'u_2 = 0. \tag{4}$$

Then we have

$$\begin{aligned} y &= v_1u_1 + v_2u_2, \\ \frac{dy}{dx} &= v_1u_1' + v_2u_2' + \underbrace{v_1'u_1 + v_2'u_2}_{0} = v_1u_1' + v_2u_2', \\ \frac{d^2y}{dx^2} &= v_1u_1'' + v_2u_2'' + v_1'u_1' + v_2'u_2'. \end{aligned}$$

If we substitute these expressions into the left-hand side of Eq. (1) and rearrange terms, we obtain

$$v_1\left[\frac{d^2u_1}{dx^2} + 2a\frac{du_1}{dx} + bu_1\right] + v_2\left[\frac{d^2u_2}{dx^2} + 2a\frac{du_2}{dx} + bu_2\right] + v_1'u_1' + v_2'u_2' = F(x).$$

The two bracketed terms are zero, since u_1 and u_2 are solutions of the reduced equation (2). Hence Eq. (1) is satisfied if, in addition to Eq. (4), we require that

$$v_1'u_1' + v_2'u_2' = F(x). \tag{5}$$

Equations (4) and (5) may be solved together as a pair,

$$v_1'u_1 + v_2'u_2 = 0, \qquad v_1'u_1' + v_2'u_2' = F(x),$$

for the unknown functions v_1' and v_2'. Cramer's rule (Appendix 7) gives

$$v_1' = \frac{\begin{vmatrix} 0 & u_2 \\ F(x) & u_2' \end{vmatrix}}{\begin{vmatrix} u_1 & u_2 \\ u_1' & u_2' \end{vmatrix}} = \frac{-u_2F(x)}{D}, \quad v_2' = \frac{\begin{vmatrix} u_1 & 0 \\ u_1' & F(x) \end{vmatrix}}{\begin{vmatrix} u_1 & u_2 \\ u_1' & u_2' \end{vmatrix}} = \frac{u_1F(x)}{D}, \tag{6}$$

where

$$D = \begin{vmatrix} u_1 & u_2 \\ u_1' & u_2' \end{vmatrix}$$

Now v_1 and v_2 can be found by integration.

How to Use the Method of Variation of Parameters

In applying the method of variation of parameters to solve the equation

$$\frac{d^2y}{dx^2} + 2a\frac{dy}{dx} + by = F(x), \tag{1}$$

we can work directly with the equations in Eq. (6). It is not necessary to rederive them. The steps to take are

STEP 1: Solve the reduced equation,

$$\frac{d^2y}{dx^2} + 2a\frac{dy}{dx} + by = 0,$$

to find the functions u_1 and u_2.

STEP 2: Calculate D and find v_1' and v_2' from Eqs. (6).

STEP 3: Integrate v_1' and v_2' to find v_1 and v_2.

STEP 4: Write down the general solution of Eq. (1) as $y = v_1u_1 + v_2u_2$.

EXAMPLE 3 Solve the equation $\dfrac{d^2y}{dx^2} + 2\dfrac{dy}{dx} - 3y = 6$ of Example 2 by variation of parameters.

Solution We first solve the reduced equation

$$\frac{d^2y}{dx^2} + 2\frac{dy}{dx} - 3y = 0$$

as in Example 2 to find

$$u_1(x) = e^{-3x}, \qquad u_2(x) = e^x.$$

Then, from the equations in (6), we have

$$D = \begin{vmatrix} e^{-3x} & e^x \\ -3e^{-3x} & e^x \end{vmatrix} = e^{-2x} + 3e^{-2x} = 4e^{-2x},$$

$$v_1' = \frac{-6e^x}{4e^{-2x}} = -\frac{3}{2}e^{3x}, \qquad v_2' = \frac{6e^{-3x}}{4e^{-2x}} = \frac{3}{2}e^{-x}. \tag{7}$$

Hence

$$v_1 = \int -\frac{3}{2}e^{3x}\,dx = -\frac{1}{2}e^{3x} + C_1, \qquad v_2 = \int \frac{3}{2}e^{-x}\,dx = -\frac{3}{2}e^{-x} + C_2,$$

and

$$y = v_1u_1 + v_2u_2 = \left(-\frac{1}{2}e^{3x} + C_1\right)e^{-3x} + \left(-\frac{3}{2}e^{-x} + C_2\right)e^{x}$$
$$= -2 + C_1e^{-3x} + C_2e^{x}.$$

■

Undetermined Coefficients

The method of variation of parameters is completely general, for it produces a particular solution of Eq. (1) for any continuous function $F(x)$. But the calculations involved can be complicated, and in special cases there may be easier methods to use. For instance, we do not need to use variation of parameters to find a particular solution of

$$\frac{d^2y}{dx^2} - \frac{dy}{dx} + 5y = 3, \tag{8}$$

if we can find the particular solution $y_p = 3/5$ by inspection. And even for an equation like

$$\frac{d^2y}{dx^2} + 3y = e^x, \tag{9}$$

we might guess that there is a solution of the form

$$y_p = Ae^x$$

and substitute $y_p = Ae^x$ and its second derivative into Eq. (9) to find A. If we do so, we find the solution $y_p = (1/4)e^x$.

Again, we might guess that the equation

$$\frac{d^2y}{dx^2} + y = 3x^2 + 4 \tag{10}$$

has a particular solution of the form

$$y_p = Cx^2 + Dx + E.$$

If we substitute this polynomial and its second derivative into Eq. (10) to look for appropriate values for the constants C, D, and E, we get

$$\begin{aligned} 2C + (Cx^2 + Dx + E) &= 3x^2 + 4, \\ Cx^2 + Dx + 2C + E &= 3x^2 + 4. \end{aligned} \tag{11}$$

This latter equation will hold for all values of x if its two sides are identical as polynomials in x; that is, if

$$C = 3, \qquad D = 0, \qquad \text{and} \qquad 2C + E = 4, \tag{12}$$

or

$$C = 3, \qquad D = 0, \qquad E = -2.$$

We conclude that

$$y_p = 3x^2 + 0\,x - 2 = 3x^2 - 2 \tag{13}$$

is a solution of Eq. (10).

In each of the foregoing examples, the particular solution we found resembled the function $F(x)$ on the right side of the given differential equation. This was no accident, for we guessed the form of the particular solution by looking at $F(x)$ first. The method of first guessing the form of the solution up to certain undetermined constants and then determining the values of these constants by using the differential equation is known as the **method of undetermined coefficients.** It depends for its success on our ability to recognize the form of a particular solution, and for this reason, among others, it lacks the generality of the method of variation of parameters. Nevertheless, its simplicity makes it the method of choice in a number of special cases.

We shall limit our discussion of the method of undetermined coefficients to selected equations in which the function $F(x)$ in Eq. (1) is the sum of one or more terms like

$$e^{rx}, \qquad \cos kx, \qquad \sin kx, \qquad ax^2 + bx + c.$$

Even so, we will find that the particular solutions of Eq. (1) do not always resemble $F(x)$ as closely as the ones we have seen.

EXAMPLE 4 Find a particular solution of

$$\frac{d^2y}{dx^2} - \frac{dy}{dx} = 2 \sin x. \tag{14}$$

Solution If we try to find a particular solution of the form

$$y_p = A \sin x$$

and substitute the derivatives of y_p in the given equation, we find that A must satisfy the equation

$$-A \sin x - A \cos x = 2 \sin x \tag{15}$$

for all values of x. Since this requires A to be equal to -2 and 0 at the same time, we conclude that Eq. (14) has no solution of the form $A \sin x$.

It turns out that the required form is the sum

$$y_p = A \sin x + B \cos x. \tag{16}$$

The result of substituting the derivatives of this new candidate into Eq. (14) is

$$\begin{aligned} -A \sin x - B \cos x - (A \cos x - B \sin x) &= 2 \sin x, \\ (B - A) \sin x - (A + B) \cos x &= 2 \sin x. \end{aligned} \tag{17}$$

Equation (17) will be an identity if

$$B - A = 2 \quad \text{and} \quad A + B = 0, \qquad \text{or} \qquad A = -1, \quad B = 1.$$

Our particular solution is $y_p = \cos x - \sin x$. ■

EXAMPLE 5 Find a particular solution of

$$\frac{d^2y}{dx^2} - 3\frac{dy}{dx} + 2y = 5e^x. \tag{18}$$

Solution If we substitute

$$y_p = Ae^x$$

and its derivatives in Eq. (18), we find that

$$Ae^x - 3Ae^x + 2Ae^x = 5e^x,$$
$$0 = 5e^x.$$

The trouble can be traced to the fact that $y = e^x$ is already a solution of the reduced equation

$$\frac{d^2y}{dx^2} - 3\frac{dy}{dx} + 2y = 0. \tag{19}$$

The characteristic equation of Eq. (19) is

$$r^2 - 3r + 2 = (r - 1)(r - 2) = 0,$$

which has $r = 1$ as a simple root. We may therefore expect Ae^x to "vanish" when substituted into the left-hand side of Eq. (18).

The appropriate way to modify the trial solution in this case is to replace Ae^x by Axe^x. When we substitute

$$y_p = Axe^x$$

and its derivatives into Eq. (18), we get

$$(Axe^x + 2Ae^x) - 3(Axe^x + Ae^x) + 2Axe^x = 5e^x$$
$$-Ae^x = 5e^x$$
$$A = -5.$$

Our solution is $y_p = -5xe^x$. ■

EXAMPLE 6 Find a particular solution of

$$\frac{d^2y}{dx^2} - 6\frac{dy}{dx} + 9y = e^{3x}. \tag{20}$$

Solution The characteristic equation,

$$r^2 - 6r + 9 = (r - 3)^2 = 0,$$

has $r = 3$ as a *double* root. The appropriate choice for y_p in this case is neither Ae^{3x} nor Axe^{3x}, but Ax^2e^{3x}. When we substitute

$$y_p = Ax^2e^{3x}$$

and its derivatives in the given differential equation, we get

$$(9Ax^2e^{3x} + 12Axe^{3x} + 2Ae^{3x}) - 6(3Ax^2e^{3x} + 2Axe^{3x}) + 9Ax^2e^{3x} = e^{3x}$$
$$2Ae^{3x} = e^{3x}$$
$$A = \frac{1}{2}.$$

Our solution is $y_p = (1/2)x^2e^{3x}$. ■

When we wish to find a particular solution of Eq. (1), and $F(x)$ has two or more terms, we include a trial function for each term in $F(x)$.

EXAMPLE 7 Solve the equation

$$\frac{d^2y}{dx^2} - \frac{dy}{dx} = 5e^x - \sin 2x. \tag{21}$$

Solution The roots of the characteristic equation $r^2 - r = 0$ are

$$r_1 = 1, \qquad r_2 = 0,$$

so the general solution of the reduced equation is

$$y_h = C_1\, e^x + C_2.$$

We now seek a particular solution y_p. That is, we seek a function that will produce $5e^x - \sin 2x$ when substituted into the left side of Eq. (21). One part of y_p is to produce $5e^x$, the other $-\sin 2x$.

Since any function of the form $C_1 e^x$ is a solution of the reduced equation, we choose our trial y_p to be the sum

$$y_p = Axe^x + B\cos 2x + C\sin 2x,$$

including xe^x where we might otherwise have included e^x. Substituting the derivatives of y_p in Eq. (21) gives

$$(Axe^x + 2Ae^x - 4B\cos 2x - 4C\sin 2x)$$
$$- (Axe^x + Ae^x - 2B\sin 2x + 2C\cos 2x) = 5e^x - \sin 2x,$$
$$Ae^x - (4B + 2C)\cos 2x + (2B - 4C)\sin 2x = 5e^x - \sin 2x.$$

These equations will hold if

$$A = 5, \qquad (4B + 2C) = 0, \qquad (2B - 4C) = -1,$$

or

$$A = 5, \qquad B = -\frac{1}{10}, \qquad C = \frac{1}{5}.$$

TABLE 16.1

The method of undetermined coefficients for selected equations of the form $\frac{d^2y}{dx^2} + 2a\frac{dy}{dx} + by = F(x)$

If $F(x)$ has a term that is a constant multiple of	And if	Then include this expression in the trial function for y_p
e^{rx}	r is not a root of the characteristic equation	Ae^{rx}
	r is a single root of the characteristic equation	Axe^{rx}
	r is a double root of the characteristic equation	Ax^2e^{rx}
$\sin kx$, $\cos kx$	ki is not a root of the characteristic equation	$B\cos kx + C\sin kx$
	ki is a root of the characteristic equation	$Bx\cos kx + Cx\sin kx$
$ax^2 + bx + c$	0 is not a root of the characteristic equation	$Dx^2 + Ex + F$ (chosen to match the degree of $ax^2 + bx + c$)
	0 is a single root of the characteristic equation	$Dx^3 + Ex^2 + Fx$ (degree one higher than the degree of $ax^2 + bx + c$)
	0 is a double root of the characteristic equation	$Dx^4 + Ex^3 + Fx^2$ (degree two higher than the degree of $ax^2 + bx + c$)

Our particular solution is

$$y_p = 5xe^x - \frac{1}{10}\cos 2x + \frac{1}{5}\sin 2x.$$

The complete solution of Eq. (21) is

$$y = y_h + y_p = C_1 e^x + C_2 + 5xe^x - \frac{1}{10}\cos 2x + \frac{1}{5}\sin 2x.$$

Exercises 16.5

Solve the equations in Exercises 1–12 by variation of parameters.

1. $\frac{d^2y}{dx^2} + \frac{dy}{dx} = x$
2. $\frac{d^2y}{dx^2} + y = \tan x, \quad -\frac{\pi}{2} < x < \frac{\pi}{2}$
3. $\frac{d^2y}{dx^2} + y = \sin x$
4. $\frac{d^2y}{dx^2} + 2\frac{dy}{dx} + y = e^x$
5. $\frac{d^2y}{dx^2} + 2\frac{dy}{dx} + y = e^{-x}$
6. $\frac{d^2y}{dx^2} - y = x$
7. $\frac{d^2y}{dx^2} - y = e^x$
8. $\frac{d^2y}{dx^2} - y = \sin x$
9. $\frac{d^2y}{dx^2} + 4\frac{dy}{dx} + 5y = 10$
10. $\frac{d^2y}{dx^2} - \frac{dy}{dx} = 2^x$
11. $\frac{d^2y}{dx^2} + y = \sec x, \quad -\frac{\pi}{2} < x < \frac{\pi}{2}$
12. $\frac{d^2y}{dx^2} - \frac{dy}{dx} = e^x \cos x, \quad x > 0$

Solve the equations in Exercises 13–28 by the method of undetermined coefficients.

13. $\frac{d^2y}{dx^2} - 3\frac{dy}{dx} - 10y = -3$
14. $\frac{d^2y}{dx^2} - 3\frac{dy}{dx} - 10y = 2x - 3$
15. $\frac{d^2y}{dx^2} - \frac{dy}{dx} = \sin x$
16. $\frac{d^2y}{dx^2} + 2\frac{dy}{dx} + y = x^2$
17. $\frac{d^2y}{dx^2} + y = \cos 3x$
18. $\frac{d^2y}{dx^2} + y = e^{2x}$
19. $\frac{d^2y}{dx^2} - \frac{dy}{dx} - 2y = 20\cos x$
20. $\frac{d^2y}{dx^2} + y = 2x + 3e^x$
21. $\frac{d^2y}{dx^2} - y = e^x + x^2$
22. $\frac{d^2y}{dx^2} + 2\frac{dy}{dx} + y = 6\sin 2x$
23. $\frac{d^2y}{dx^2} - \frac{dy}{dx} - 6y = e^{-x} - 7\cos x$
24. $\frac{d^2y}{dx^2} + 3\frac{dy}{dx} + 2y = e^{-x} + e^{-2x} - x$
25. $\frac{d^2y}{dx^2} + 5\frac{dy}{dx} = 15x^2$
26. $\frac{d^2y}{dx^2} - \frac{dy}{dx} = -8x + 3$
27. $\frac{d^2y}{dx^2} - 3\frac{dy}{dx} = e^{3x} - 12x$
28. $\frac{d^2y}{dx^2} + 7\frac{dy}{dx} = 42x^2 + 5x + 1$

In each of Exercises 29–31, the given differential equation has a particular solution y_p of the form given. Determine the coefficients in y_p. Then solve the differential equation.

29. $\frac{d^2y}{dx^2} - 5\frac{dy}{dx} = xe^{5x}, \quad y_p = A\,x^2e^{5x} + B\,xe^{5x}$
30. $\frac{d^2y}{dx^2} - \frac{dy}{dx} = \cos x + \sin x, \quad y_p = A\cos x + B\sin x$
31. $\frac{d^2y}{dx^2} + y = 2\cos x + \sin x, \quad y_p = A\,x\cos x + B\,x\sin x$

In Exercises 32–35, solve the given differential equations (a) by variation of parameters and (b) by the method of undetermined coefficients.

32. $\frac{d^2y}{dx^2} - 4\frac{dy}{dx} + 4y = 2e^{2x}$
33. $\frac{d^2y}{dx^2} - \frac{dy}{dx} = e^x + e^{-x}$
34. $\frac{d^2y}{dx^2} - 9\frac{dy}{dx} = 9e^{9x}$
35. $\frac{d^2y}{dx^2} - 4\frac{dy}{dx} - 5y = e^x + 4$

Solve the differential equations in Exercises 36–45. Some of the equations can be solved by the method of undetermined coefficients, but others cannot.

36. $\frac{d^2y}{dx^2} + y = \csc x, \quad 0 < x < \pi$
37. $\frac{d^2y}{dx^2} + y = \cot x, \quad 0 < x < \pi$
38. $\frac{d^2y}{dx^2} + 4y = \sin x$

39. $\dfrac{d^2y}{dx^2} - 8\dfrac{dy}{dx} = e^{8x}$

40. $\dfrac{d^2y}{dx^2} + 4\dfrac{dy}{dx} + 5y = x + 2$

41. $\dfrac{d^2y}{dx^2} - \dfrac{dy}{dx} = x^3$

42. $\dfrac{d^2y}{dx^2} + 9y = 9x - \cos x$

43. $\dfrac{d^2y}{dx^2} + 2\dfrac{dy}{dx} = x^2 - e^x$

44. $\dfrac{d^2y}{dx^2} - 3\dfrac{dy}{dx} + 2y = e^x - e^{2x}$

45. $\dfrac{d^2y}{dx^2} + y = \sec x \tan x, \quad -\dfrac{\pi}{2} < x < \dfrac{\pi}{2}$

The method of undetermined coefficients can sometimes be used to solve first order ordinary differential equations. Use the method to solve the equations in Exercises 46–49.

46. $\dfrac{dy}{dx} + 4y = x$

47. $\dfrac{dy}{dx} - 3y = e^x$

48. $\dfrac{dy}{dx} + y = \sin x$

49. $\dfrac{dy}{dx} - 3y = 5e^{3x}$

Solve the initial value problems in Exercises 50 and 51.

50. $\dfrac{d^2y}{dx^2} + y = e^{2x}; \quad y(0) = 0, \quad y'(0) = \dfrac{2}{5}$

51. $\dfrac{d^2y}{dx^2} + y = \sec^2 x, \quad -\dfrac{\pi}{2} < x < \dfrac{\pi}{2}; \quad y(0) = y'(0) = 1$

52. *Bernoulli's equation of order 2.* Solve the equation

$$\frac{dy}{dx} + y = (xy)^2$$

by carrying out the following steps: (1) divide both sides of the equation by y^2; (2) make the change of variable $u = y^{-1}$; (3) solve the resulting equation for u in terms of x; (4) let $y = u^{-1}$.

53. *An integral equation.* Solve the integral equation

$$y(x) + \int_0^x y(t)\,dt = x.$$

(*Hint:* Differentiate first.)

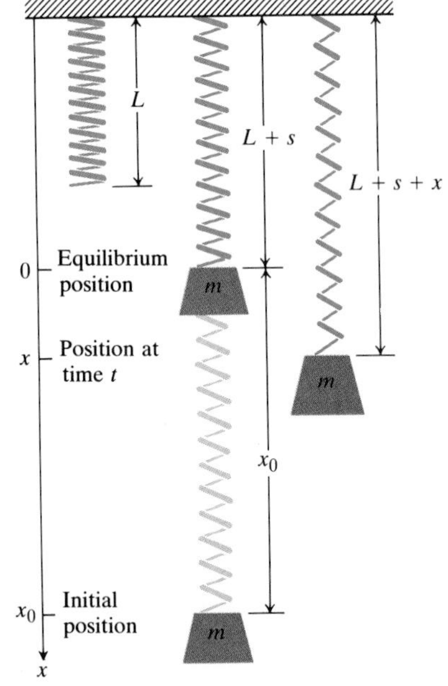

16.8 A spring of natural length L is stretched a distance s by the weight of a mass m. It is then stretched an additional distance x_0 and released. The position of the mass at any subsequent time is described by a differential equation.

16.6 Oscillation

Linear second order equations with constant coefficients are important because they model oscillation. Oscillation occurs, for example, in atoms and molecules, in machinery and electrical circuits, and in the chemical and electrical systems of our bodies. The classical example of mechanical oscillation is the motion of a weighted spring, and that is the example we shall study in this section. But all oscillations are modeled with the same basic equations, and the mathematics we present here comes up in all the other applications as well.

Undamped Oscillation (Simple Harmonic Motion)

Suppose we have a spring of natural length L and spring constant k, with its upper end fastened to a rigid support (Fig. 16.8). We hang a mass m from the spring. The weight of the mass stretches the spring to a length $L + s$ when allowed to come to rest in a new equilibrium position. By Hooke's law, the tension in the spring is ks. The force of gravity pulling down on the mass is mg. Equilibrium requires

$$ks = mg. \tag{1}$$

How will the mass behave if we pull it down an additional amount x_0 beyond the equilibrium position and release it? To find out, let x, positive direction downward, denote the displacement of the mass from equilibrium

t seconds after the motion has started. Then the forces acting on the mass are

$$+mg \qquad \text{(weight due to gravity)}$$

$$-k(s+x) \qquad \text{(spring tension).}$$

By Newton's second law, the sum of these forces is $m(d^2x/dt^2)$, so

$$m\frac{d^2x}{dt^2} = mg - ks - kx. \tag{2}$$

Since $mg = ks$ from Eq. (1), Eq. (2) simplifies to

$$m\frac{d^2x}{dt^2} + kx = 0. \tag{3}$$

In addition to satisfying this differential equation, the position of the mass satisfies the initial conditions

$$x = x_0 \quad \text{and} \quad \frac{dx}{dt} = 0 \quad \text{when } t = 0. \tag{4}$$

If we divide both sides of Eq. (3) by m and write ω for $\sqrt{k/m}$, the equation becomes

$$\frac{d^2x}{dt^2} + \omega^2 x = 0. \tag{5}$$

The roots of the characteristic equation $r^2 + \omega^2 = 0$ are $r = \pm\omega i$, so the general solution of Eq. (5) is

$$x = C_1 \cos \omega t + C_2 \sin \omega t. \tag{6}$$

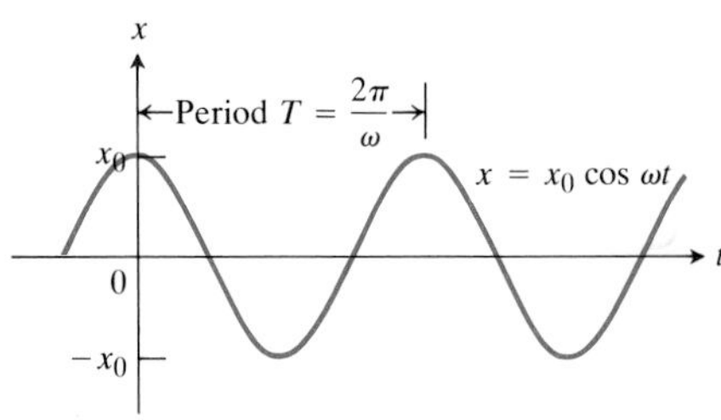

16.9 Simple harmonic motion.

Applying the initial conditions in Eq. (4) determines the constants to be

$$C_1 = x_0 \quad \text{and} \quad C_2 = 0.$$

The mass's displacement from equilibrium t seconds into the motion is

$$x = x_0 \cos \omega t. \tag{7}$$

This equation represents a **simple harmonic motion** of amplitude x_0 and period $T = 2\pi/\omega$ (Fig. 16.9).

We normally combine the two terms in the general solution in Eq. (6) into a single term, using the trigonometric identity

$$\sin(\omega t + \phi) = \cos \omega t \sin \phi + \sin \omega t \cos \phi.$$

To apply the identity, we take

$$C_1 = C \sin \phi, \qquad C_2 = C \cos \phi, \tag{8}$$

where

$$C = \sqrt{C_1^2 + C_2^2}, \qquad \phi = \tan^{-1}\frac{C_1}{C_2}, \tag{9}$$

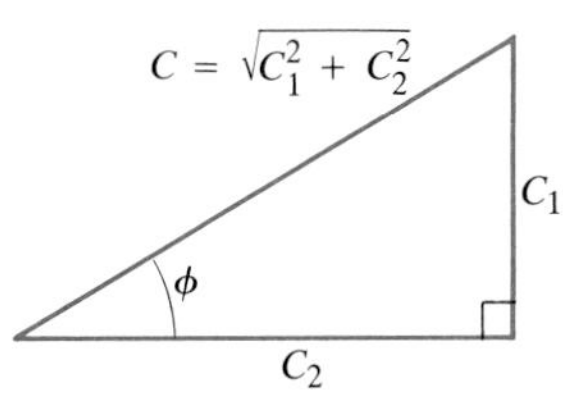

16.10 $C_1 = C \sin \phi$ and $C_2 = C \cos \phi$.

as in Fig. 16.10. With these substitutions, Eq. (6) becomes

$$x = C \sin(\omega t + \phi). \tag{10}$$

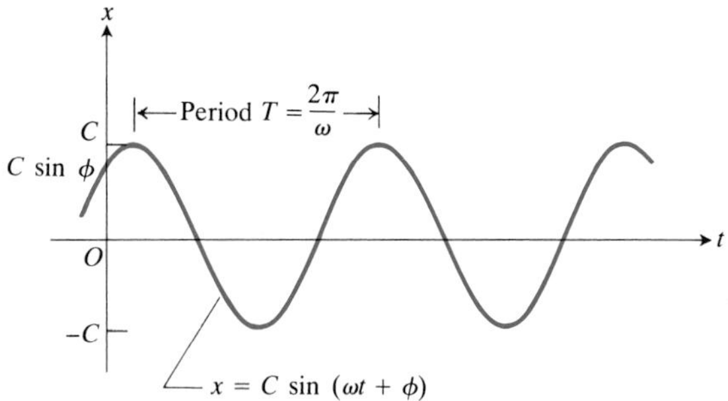

16.11 Graph of simple harmonic motion of a frictionless spring, referred to as undamped vibration.

We treat C and ϕ as two new arbitrary constants, replacing the arbitrary constants C_1 and C_2 of Eq. (6). Equation (10) represents a simple harmonic motion of amplitude C and period $T = 2\pi/\omega$ (Fig. 16.11). The angle ϕ is called the **phase angle** of the motion.

Damped Oscillation

Suppose that the mass is slowed by a frictional force $c(dx/dt)$ that is proportional to velocity, where c is a positive constant. Then the equation that replaces Eq. (3) is

$$m\frac{d^2x}{dt^2} + c\frac{dx}{dt} + kx = 0 \tag{11}$$

or

$$\frac{d^2x}{dt^2} + 2b\frac{dx}{dt} + \omega^2 x = 0, \tag{12}$$

where

$$2b = \frac{c}{m} \quad \text{and} \quad \omega = \sqrt{\frac{k}{m}}.$$

The roots of the characteristic equation $r^2 + 2br + \omega^2 = 0$ are

$$r_1 = -b + \sqrt{b^2 - \omega^2}, \qquad r_2 = -b - \sqrt{b^2 - \omega^2}. \tag{13}$$

As we shall now see, the mass behaves in three distinctly different ways, depending on the relative sizes of b and ω.

Critical Damping If $b = \omega$ the roots in (13) are equal and the solution of (12) is

$$x = (C_1 t + C_2)e^{-\omega t}. \tag{14}$$

As time passes, x approaches zero. The mass does not oscillate (Fig. 16.12a).

Overcritical Damping If $b > \omega$ the roots in (13) are real and unequal and the solution of (12) is

$$x = C_1 e^{r_1 t} + C_2 e^{r_2 t}. \tag{15}$$

Here again, the mass does not oscillate. Both r_1 and r_2 are negative and x approaches zero as time passes (Fig. 16.12b).

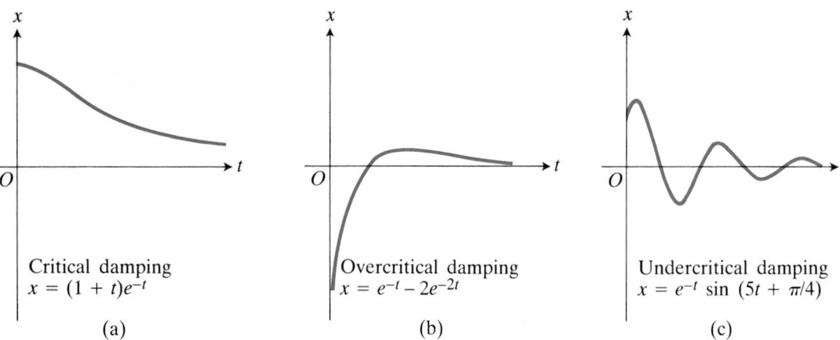

16.12 Three ways to damp vibration.

Undercritical Damping If $0 < b < \omega$, let $\omega^2 - b^2 = \alpha^2$. Then

$$r_1 = -b + \alpha i, \qquad r_2 = -b - \alpha i$$

and

$$x = e^{-bt}(C_1 \cos \alpha t + C_2 \sin \alpha t). \tag{16}$$

If we introduce the substitutions (8), we may also write Eq. (16) in the equivalent form

$$x = Ce^{-bt} \sin(\alpha t + \phi). \tag{17}$$

This equation represents damped oscillation. It is analogous to simple harmonic motion, of period $T = 2\pi/\alpha$, except that the amplitude is not constant but is given by Ce^{-bt}. Since this tends to zero as t increases, the oscillations die out as time goes on (Fig. 16.12c). Observe, however, that Eq. (17) reduces to Eq. (10) in the absence of friction. The effect of friction is twofold:

1. $b = c/(2m)$ appears in the exponential **damping factor** e^{-bt}. The larger b is, the more quickly the oscillations tend to become unnoticeable.
2. The period $T = 2\pi/\alpha = 2\pi/\sqrt{\omega^2 - b^2}$ is longer than the period $T_0 = 2\pi/\omega$ in the friction-free system.

Exercises 16.6

1. Suppose the motion of the mass in Fig. 16.8 is described by Eq. (3). Find x as a function of t if $x = x_0$ and $dx/dt = v_0$ when $t = 0$. Express your answer in two equivalent forms (Eqs. 6 and 10).

2. CALCULATOR A 5-lb mass is suspended from the lower end of a spring whose upper end is attached to a rigid support. The weight of the mass extends the spring by 6 in. If, after the mass has come to rest in its new equilibrium position, it is struck a sharp blow that starts it downward with a velocity of 4ft/sec, find its subsequent motion, assuming there is no friction.

3. *An R–L–C series circuit.* A simple electrical circuit shown in Fig. 16.13 contains a capacitor of capacitance C farads, a coil of inductance L henries, a resistance of R ohms, and a generator that produces an electromotive force v volts, in series. If the current

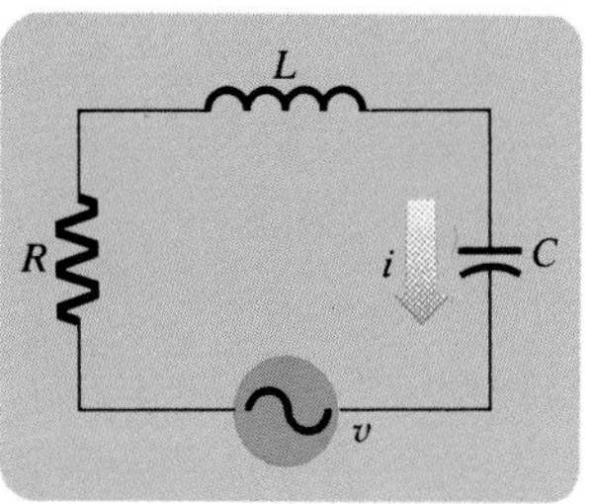

16.13 An *R–L–C* series circuit.

intensity in the circuit at time t is i amperes, the differential equation describing the current i is

$$L\frac{d^2i}{dt^2}+R\frac{di}{dt}+\frac{1}{C}i=\frac{dv}{dt}.$$

Find i as a function of t if

a) $R=0,\quad 1/(LC)=\omega^2,\quad v=\text{constant};$

b) $R=0,\quad 1/(LC)=\omega^2,\quad v=V\sin\alpha t,$
$V=\text{constant},\quad \alpha=\text{constant}\neq\omega;$

c) $R=0,\quad 1/(LC)=\omega^2,\quad v=V\sin\omega t,$
$V=\text{constant};$

d) $R=50,\quad L=5,\quad C=9\times10^{-6},\quad v=\text{constant}.$

4. A simple pendulum of length l makes an angle θ with the vertical. As it swings back and forth, its motion, neglecting friction, is described by the differential equation

$$\frac{d^2\theta}{dt^2}=-\frac{g}{l}\sin\theta,$$

where g is the (constant) acceleration of gravity (Fig. 16.14). Solve the differential equation of motion, under the assumption that θ is so small that $\sin\theta$ may be replaced by θ without appreciable error. Assume that $\theta=\theta_0$ and $d\theta/dt=0$ when $t=0$.

5. A circular disk of mass m and radius r is suspended by a thin wire attached to the center of one of its flat faces. If the disk is twisted through an angle θ, torsion in the wire tends to turn the disk back in the opposite direction. The differential equation for the motion is

$$\frac{1}{2}mr^2\frac{d^2\theta}{dt^2}=-k\theta,$$

where k is the coefficient of torsion of the wire. Find θ as a function of t if $\theta=\theta_0$ and $d\theta/dt=v_0$ at $t=0$.

6. **CALCULATOR** A cylindrical spar buoy, diameter 1 ft, weight 100 lb, floats partially submerged in an upright position. When it is depressed slightly from its equilibrium position and released, it bobs up and down according to the differential equation

$$\frac{100}{g}\frac{d^2x}{dt^2}=-16\pi x-c\frac{dx}{dt}.$$

Here x (ft) is the vertical displacement of the buoy from its equilibrium position, and $c(dx/dt)$ is the frictional resistance of the water. Find c if the period of oscillation is observed to be 1.6 sec. (Take $g=32$ ft/sec^2.)

7. Suppose the upper end of the spring in Fig. 16.8 is attached not to a rigid support but to a member that itself undergoes up and down motion given by a function of time t, say $y=f(t)$. If the positive direction of y is downward, the differential equation of motion is

$$m\frac{d^2x}{dt^2}+kx=kf(t).$$

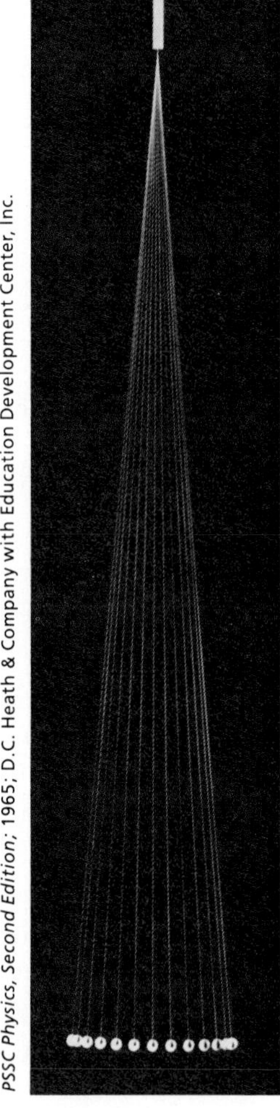

PSSC Physics, Second Edition; 1965; D.C. Heath & Company with Education Development Center, Inc.

16.14 Motion of a simple pendulum.

Let $x=x_0$ and $dx/dt=0$ when $t=0$, and solve for x

a) if $f(t)=A\sin\alpha t$ and $\alpha\neq\sqrt{k/m}$,

b) if $f(t)=A\sin\alpha t$ and $\alpha=\sqrt{k/m}$.

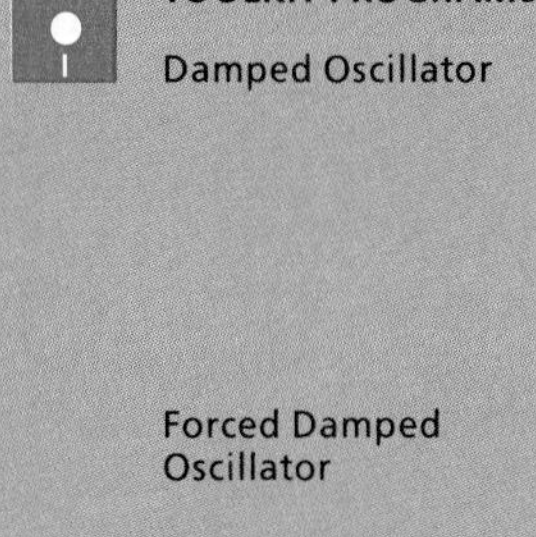

TOOLKIT PROGRAMS

Damped Oscillator	Enables you to explore solutions of the equation $x''+2bx'+\omega^2x=0$. The program can help you to learn more about amplitude, frequency, angular frequency (ω), period, and conservation of energy.
Forced Damped Oscillator	Enables you to explore solutions of the equation $x''+2bx'+\omega^2x=F(t)$ that models the behavior of an oscillator driven by an external force $F(t)$.

16.7 Numerical Methods

We complete our preview of differential equations by describing three numerical methods for solving the initial value problem $y' = f(x, y)$, $y(a) = y_0$ over an interval $a \le x \le b$. These methods do not produce a general solution of $y' = f(x, y)$; they produce tables of values of y for preselected values of x. Instead of being a drawback, however, this is a definite advantage, especially if we want to solve a differential equation like $y' = x^2 + y^2$ whose solution has no closed-form algebraic expression. Also, in solving equations of motion in real situations like launching a rocket or intercepting one in orbit, we find that numerical answers are much more useful than algebraic expressions would be. We shall keep our examples simple but the ideas can be extended to far more complicated equations and systems of equations. Naturally we would not attempt to solve such complicated systems with pencil and paper; instead, we would turn to well designed computer programs that had been carefully checked for accuracy.

We shall consider three numerical methods: first a method that dates back to Euler, then an improved Euler method, and finally an even more accurate method called a Runge–Kutta method. You will find BASIC programs for the three methods at the end of the section.

The Euler Method

The initial value problem

$$y' = f(x, y), \qquad y(x_0) = y_0, \tag{1}$$

provides us with a starting point, $P(x_0, y_0)$, and a slope $f(x_0, y_0)$. We know that the graph of the solution must be a curve through P with that slope. If we use the tangent through P to approximate the actual solution curve, the approximation may be fairly good from $x_0 - h$ to $x_0 + h$, for small values of h. Thus, we might choose $h = 0.1$, say, and move along the tangent line from P to $P'(x_1, y_1)$, where $x_1 = x_0 + h$ and $y_1 = y_0 + hf(x_0, y_0)$. If we think of P' as a new starting point, we can move from P' to $P''(x_2, y_2)$, where $x_2 = x_1 + h$ and $y_2 = y_1 + hf(x_1, y_1)$. If we replace h by $-h$, we move to the left from P instead of to the right. The process can be continued, but the errors are likely to accumulate as we take more steps (Fig. 16.15).

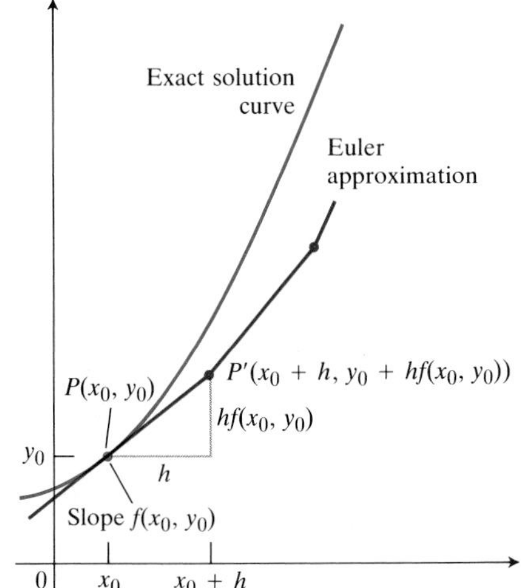

16.15 The Euler approximation to the solution of the initial value problem $y = f(x)$, $y = y_0$ when $x = x_0$. The errors involved may accumulate as we take more steps.

EXAMPLE 1 Take $h = 0.1$ and investigate the accuracy of the Euler approximation method for the initial value problem

$$y' = 1 + y, \qquad y(0) = 1, \tag{2}$$

over the interval $0 \le x \le 1$ by letting

$$x_{n+1} = x_n + h, \qquad y_{n+1} = y_n + h(1 + y_n). \tag{3}$$

TABLE 16.2
Euler solution of $y' = 1 + y$, $y(0) = 1$, step size 0.1

x	y (approx)	y (exact)	Error = y (exact) − y (approx)
0	1	1	0
0.1	1.2	1.2103	0.0103
0.2	1.42	1.4428	0.0228
0.3	1.662	1.6997	0.0377
0.4	1.9282	1.9836	0.0554
0.5	2.2210	2.2974	0.0764
0.6	2.5431	2.6442	0.1011
0.7	2.8974	3.0275	0.1301
0.8	3.2872	3.4511	0.1639
0.9	3.7159	3.9192	0.2033
1.0	4.1875	4.4366	0.2491

Solution The exact solution of Eqs. (2) is $y = 2e^x - 1$. Table 16.2 shows the results using Eqs. (3) and the exact results rounded to four decimals for comparison. By the time we get to $x = 1$, the error is about 5.6%. ■

The Improved Euler Method

This method first gets an estimate of y_{n+1}, as in the original Euler method, but calls the result z_{n+1}. We then take the average of $f(x_n, y_n)$ and $f(x_{n+1}, z_{n+1})$ in place of $f(x_n, y_n)$ in the next step. Thus

$$z_{n+1} = y_n + hf(x_n, y_n), \tag{4}$$

$$y_{n+1} = y_n + \frac{h}{2}[f(x_n, y_n) + f(x_{n+1}, z_{n+1})]. \tag{5}$$

If we apply this improved method to Example 1, again with $h = 0.1$, we get the following results at $x = 1$:

$$y \text{ (approx)} = 4.42816\ 1693,$$
$$y \text{ (exact)} = 4.43656\ 3656,$$
$$\text{Error} = y \text{ (exact)} - y \text{ (approx)} = 0.00840\ 1963,$$

and the error is less than 2/10 of 1%.

A Runge–Kutta Method*

The Runge–Kutta method we use requires four intermediate calculations, as given in the following equations:

$$k_1 = hf(x_n, y_n) \qquad k_2 = hf\left(x_n + \frac{h}{2}, y_n + \frac{k_1}{2}\right)$$
$$k_3 = hf\left(x_n + \frac{h}{2}, y_n + \frac{k_2}{2}\right) \qquad k_4 = hf(x_n + h, y_n + k_3). \tag{6}$$

*The method described here is one of many Runge–Kutta methods.

We then calculate y_{n+1} from the formula

$$y_{n+1} = y_n + \frac{1}{6}(k_1 + 2k_2 + 2k_3 + k_4). \tag{7}$$

When we apply this method to the problem of estimating $y(1)$ for the problem $y' = 1 + y, y(0) = 1$, still using $h = 0.1$, we get

$$y(1) = 4.43655\ 9490$$

with an error 0.00000 4166, which is less than 1/10,000 of 1%. This is clearly the most accurate of the three methods.

The next example shows that the error in the Runge–Kutta approximation did not continue to increase as the process was continued. In fact, with $h = 0.1$, the difference between the exact solutions and the approximations remained less than 10^{-6} for the two initial value problems:

a) $y' = x - y, \quad y(0) = 1,$ b) $y' = x - y, \quad y(0) = -2.$

The fact that the differential equation is linear in y is significant in discussing the accuracy of the Runge–Kutta approximation. Such accuracy is not attained for the initial value problem

$$y' = x^2 + y^2, \qquad y(0) = 0.$$

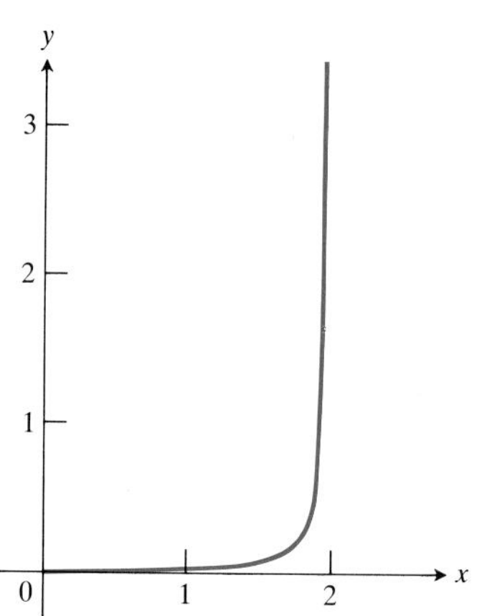

16.16 The graph of the Runge–Kutta solution of the initial value problem $y' = x^2 + y^2$, $y(0) = 0$, for $x > 0$. Data from Table 16.4.

The Runge–Kutta approximation to $y(2)$, using $h = 0.1$, is 1.47×10^{11}. For this problem, the value of y increases without bound as x increases beyond $x = 2$. The solution curve has a vertical asymptote at just beyond $x = 2$ (Fig. 16.16). No matter how small we take h, we cannot assert any accuracy for our approximations as the curve approaches this asymptote.

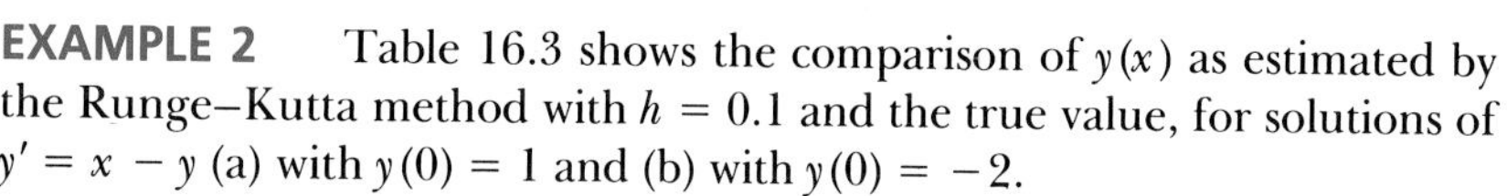

EXAMPLE 2 Table 16.3 shows the comparison of $y(x)$ as estimated by the Runge–Kutta method with $h = 0.1$ and the true value, for solutions of $y' = x - y$ (a) with $y(0) = 1$ and (b) with $y(0) = -2$.

TABLE 16.3

	x	**y (Runge–Kutta)**	**y (true value)**	**Difference**
a) $y' = x - y, \quad y(0) = 1$	0	1	1	0
	0.5	0.71306 1869	0.71306 1319	5.50×10^{-7}
	1.0	0.73575 9549	0.73575 8882	6.67×10^{-7}
	1.5	0.94626 0927	0.94626 0320	6.07×10^{-7}
	2.0	1.27067 1057	1.27067 0566	4.91×10^{-7}
	2.5	1.66417 0370	1.66416 9997	3.73×10^{-7}
	3.0	2.09957 4407	2.09957 4137	2.70×10^{-7}
b) $y' = x - y, \quad y(0) = -2$	0	−2	−2	0
	0.5	−1.10653 0935	−1.10653 0660	-2.75×10^{-7}
	1.0	−0.36787 9775	−0.36787 9441	-3.34×10^{-7}
	1.5	+0.27686 9537	+0.27686 9840	-3.03×10^{-7}
	2.0	0.86466 4472	0.86466 4717	-2.46×10^{-7}
	2.5	1.41791 4816	1.41791 5001	-1.85×10^{-7}
	3.0	1.95021 2796	1.95021 2932	-1.36×10^{-7}

TABLE 16.4
$y' = x^2 + y^2, \quad y(0) = 0$

x	y (Runge–Kutta)	y (actual)
0	0	0
0.5	0.04179 1288	0.04179 1146
1.0	0.35023 3742	0.35023 1844
1.5	1.51747 3414	1.51744 7544
2.0	71.57899 545	317.22440 0
2.1	1.47001 E + 11	
2.2	1.66667 E + **	(meaning, "you broke the bank!")

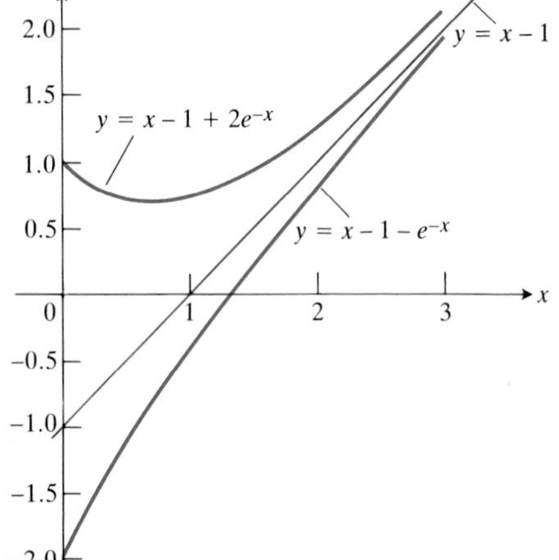

16.17 Two solutions of $y' = x - y$:
a) $y(0) = 1$, $y = x - 1 + 2e^{-x}$,
b) $y(0) = -2$, $y = x - 1 - e^{-x}$

More points were actually computed and plotted to give the graphs in Fig. 16.17. The upper curve, $y = x - 1 + 2e^{-x}$, is concave up and has a minimum when $x = y = \ln 2$. The lower curve is concave down, is always rising as x increases, and crosses the x-axis at a value of x near 1.3. Both curves approach the line $y = x - 1$ as $x \to \infty$. ■

EXAMPLE 3 Table 16.4 lists Runge–Kutta approximations for the initial value problem

$$y' = x^2 + y^2, \qquad y(0) = 0.$$

We obtained the approximations with a step size of $h = 0.1$. Figure 16.16 shows the graph of y as a function of x. ■

Computer Programs

Here are three computer programs for approximating the solutions of the initial value problem $y' = f(x, y)$, $y(x_0) = y_0$, over an interval starting at x_0.

Euler

PROGRAM	COMMENT
10 INPUT X	Initial value of x
20 INPUT Y	Initial value of y
30 INPUT H	Step size
40 INPUT N	The number of steps
50 DEF U = F(X, Y)	Key in the formula for $f(x, y)$. Use lines 50–59 as needed.
60 J = 0 TO N	Each value of J gives one Euler approximation.
70 PRINT X;Y	Prints the table
80 LET Y = Y + H*U	Next y
90 LET X = X + H	Next x
100 NEXT J	Returns to line 60 with new J.
110 END	

German scientist Carl Runge (1856–1927) was a mathematical physicist of Max Planck's caliber who developed numerical methods for solving the differential equations that arose in his studies of atomic spectra. He used so much mathematics in his research that physicists thought he was a mathematician, and did so much physics that mathematicians thought he was a physicist. Neither group claimed him as their own and it was years before anyone could find him a professorship. In 1904, Felix Klein finally persuaded his Göttingen colleagues to create for Runge Germany's only full professorship in applied mathematics. Runge was the professorship's first and only occupant, there being no one of his accomplishments to assume the post when he died.

Improved Euler

PROGRAM

```
10  INPUT X
20  INPUT Y
30  INPUT H
40  INPUT N
50  DEF U = F(X, Y)
60  DEF V = F(X, Z)
70  FOR J = 0 TO N
80  PRINT X;Y
90  LET D = H*U
100 LET Z = Y + D
110 LET X = X + H
120 LET Y = Y +
    (D + H*V)/2
130 NEXT J
140 END
```

Runge–Kutta

PROGRAM

```
10  DEF FX(Y) = F(X, Y)
20  INPUT X
30  INPUT Y
40  INPUT H
50  INPUT N
60  FOR J = 0 TO N
70  PRINT X;Y
80  K1 = H*FX(Y)
90  X = X + H/2
100 K2 = H*FX(Y + K1/2)
110 K3 = H*FX(Y + K2/2)
120 X = X + H/2
130 K4 = H*FX(Y + K3)
140 Y = Y + (K1 + 2*K2 + 2*K3 + K4)/6
150 NEXT J
160 END
```

Exercises 16.7

1. Use the Euler method with $h = 1/5$ to estimate $y(1)$ if $y' = y$ and $y(0) = 1$. What is the exact value of $y(1)$?
2. Show that the Euler method leads to the estimate $(1 + (1/n))^n$ for $y(1)$ if $h = 1/n$, $y' = y$, and $y(0) = 1$. What is the limit as $n \to \infty$?
3. Use the improved Euler method with $h = 1/5$ to estimate $y(1)$ if $y' = y$ and $y(0) = 1$.
4. Use the Runge–Kutta method with $h = 1/5$ to estimate $y(1)$ if $y' = y$ and $y(0) = 1$.
5. Show that the solution of the initial value problem $y' = x^2 + y^2$, $y(0) = 1$ increases faster on the interval $0 \leq x < 1$ than does the solution of the initial value problem $y' = y^2$, $y(0) = 1$. Solve the latter problem by separation of variables and thus show that the solution of the original problem becomes infinite at a value of x not greater than 1. (In case you are interested, the value is about 0.96981 0654.)
6. Solve the initial value problem $y' = 1 + y^2$, $y(0) = 0$, (a) by separation of variables, and (b) by using the substitution $y = -u'/u$ and solving the equivalent problem for u. (Notice the similarity with the initial value problem $y' = x^2 + y^2$, $y(0) = 0$.)

Computer Exercises (or Programmable Calculator)

Find computer solutions of the initial value problems in Exercises 7–10. Each exercise gives a differential equation in the form $y' = f(x, y)$, a solution interval $a \leq x \leq b$, the initial value $y(a)$, the step size, and the number of steps.

7. $y' = x/7$, $a = 0$, $b = 4$, $y(0) = 1$, $h = 0.1$, $n = 20$
8. $y' = -y^2/x$, $a = 2$, $b = 4$, $y(2) = 2$, $h = 0.1$, $n = 20$
9. $y' = (x^2 + y^2)/2y$, $a = 0$, $b = 2$, $y(0) = 0.1$, $h = 0.2$, $n = 10$
10. Same as Exercise 9, except with $h = 0.1$ and $n = 20$
11. *Evaluating a nonelementary integral.* The value of the nonelementary integral

$$\int_0^1 \sin(t^2)\,dt$$

is the value of the function

$$y(x) = \int_0^x \sin(t^2)\,dt$$

at $x = 1$. The function, in turn, is the solution of the initial value problem

$$y' = \sin(x^2), \qquad y(0) = 0.$$

Thus, by solving the initial value problem numerically on the interval $0 \leq x \leq 1$ we can find the value of the integral as the value of y that corresponds to $x = 1$.

Estimate the integral's value by solving the initial value problem with (a) $n = 20$ steps, and (b) $n = 40$ steps.

12. *Continuation of Exercise 11.* Estimate the value of

$$\int_0^1 x^2 e^{-x}\, dx$$

by solving an appropriate initial value problem on the interval [0, 1] with (a) $n = 20$ steps, and (b) $n = 40$ steps.

TOOLKIT PROGRAM

First Order Initial Value Problem — Provides numerical solutions of $y' = f(x, y)$, $y(a) = y_0$ over an interval $a \le x \le b$ with a Runge–Kutta method with up to 100 steps. Also prints tables and displays graphs of y and y'.

Review Questions

1. What is a differential equation?
2. What is a solution of a differential equation?
3. Describe the methods for solving first order equations that are (a) separable, (b) homogeneous, (c) exact, (d) linear. Give examples.
4. Describe the methods for solving second order linear equations with constant coefficients if the equations are (a) homogeneous, (b) nonhomogeneous. Give examples.
5. How are second order differential equations used to model and describe oscillation? Give examples.
6. Describe the Euler, improved Euler, and Runge–Kutta methods for solving the initial value problem $y' = f(x, y), y(a) = y_0$ over an interval $a \le x \le b$.

Practice Exercises

Solve the initial value problems in Exercises 1–20.

1. $e^{y-2}\, dx - e^{x+2y}\, dy = 0, \quad y(0) = -2$
2. $y \ln y\, dx + (1 + x^2)\, dy = 0, \quad y(0) = e$
3. $\dfrac{dy}{dx} = \dfrac{x^2 + y^2}{2xy}, \quad y(5) = 0$
4. $\dfrac{dy}{dx} = \dfrac{y(1 + \ln y - \ln x)}{x(\ln y - \ln x)}, \quad y(1) = 1$
5. $(x^2 + y)\, dx + (e^y + x)\, dy = 0, \quad y(3) = 0$
6. $(e^x + \ln y)\, dx + \left(\dfrac{x + y}{y}\right) dy = 0, \quad y(\ln 2) = 1$
7. $(x + 1)\dfrac{dy}{dx} + 2y = x, \quad y(0) = 1$
8. $x\dfrac{dy}{dx} + 2y = x^2 + 1, \quad y(1) = 1$
9. $\dfrac{d^2y}{dx^2} - \left(\dfrac{dy}{dx}\right)^2 = 1, \quad y(\pi/3) = 0, \quad y'(\pi/3) = \sqrt{3}$
10. $x^2\dfrac{d^2y}{dx^2} + x\dfrac{dy}{dx} = 1, \quad y(1) = 1, \quad y'(1) = 1$
11. $\dfrac{d^2y}{dx^2} - 4\dfrac{dy}{dx} + 3y = 0, \quad y(0) = 2, \quad y'(0) = -2$
12. $\dfrac{d^2y}{dx^2} + 5\dfrac{dy}{dx} + 6y = 0, \quad y(0) = 5/6, \quad y'(0) = -2$
13. $\dfrac{d^2y}{dx^2} + 4\dfrac{dy}{dx} + 4y = 0, \quad y(0) = 0, \quad y'(0) = 7$
14. $\dfrac{d^2y}{dx^2} - 8\dfrac{dy}{dx} + 16y = 0, \quad y(0) = 4, \quad y'(0) = -4$
15. $\dfrac{d^2y}{dx^2} + 2\dfrac{dy}{dx} + 2y = 0, \quad y(0) = 1, \quad y'(0) = 0$
16. $\dfrac{d^2y}{dx^2} - 2\dfrac{dy}{dx} - 4y = 0, \quad y(0) = 1, \quad y'(0) = -1$
17. $\dfrac{d^2y}{dx^2} + 2\dfrac{dy}{dx} = 4x, \quad y(0) = 1, \quad y'(0) = -3$
18. $\dfrac{d^2y}{dx^2} + y = \csc x, \quad y(\pi/2) = 1, \quad y'(\pi/2) = \pi/2$
19. $\dfrac{d^2y}{dx^2} - \dfrac{dy}{dx} - 2y = 3e^{2x}, \quad y(0) = -2, \quad y'(0) = 0$
20. $\dfrac{d^2y}{dx^2} - 2\dfrac{dy}{dx} + 5y = 4e^{-x}, \quad y(0) = 1, \quad y'(0) = -1/2$

3. Common Reference Triangles

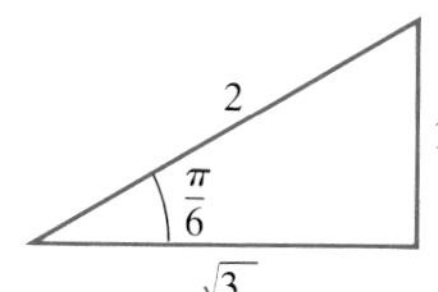

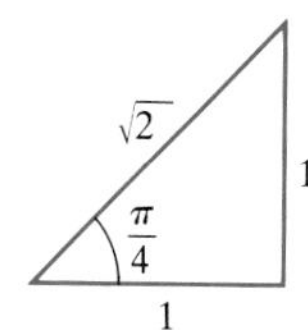

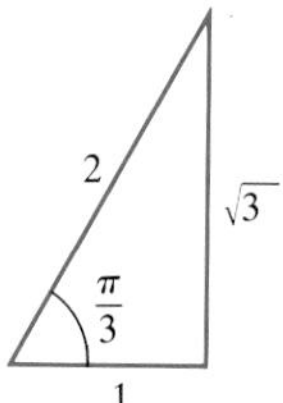

4. Angles and Sides of a Triangle

Law of cosines: $c^2 = a^2 + b^2 - 2ab \cos C$

Law of sines: $\dfrac{\sin A}{a} = \dfrac{\sin B}{b} = \dfrac{\sin C}{c}$

$$\text{Area} = \frac{1}{2}bc \sin A = \frac{1}{2}ac \sin B = \frac{1}{2}ab \sin C$$

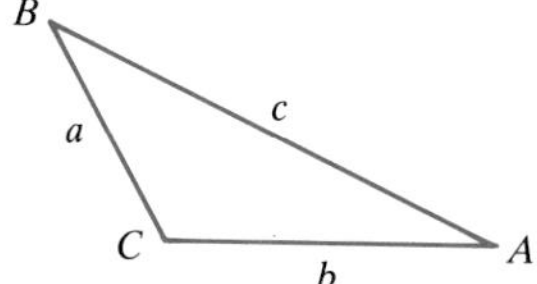

A.2 Proofs of the Limit Theorems in Chapter 2

This appendix furnishes the ϵ–δ proofs of the limit theorems in Chapter 2.

THEOREM 1

If $\lim_{x \to x_0} f_1(x) = L_1$ and $\lim_{x \to x_0} f_2(x) = L_2$, then

1. *Sum Rule:* $\lim (f_1(x) + f_2(x)) = L_1 + L_2$
2. *Difference Rule:* $\lim (f_1(x) - f_2(x)) = L_1 - L_2$
3. *Product Rule:* $\lim f_1(x) \cdot f_2(x) = L_1 \cdot L_2$
4. *Constant Multiple Rule:* $\lim k \cdot f_2(x) = k \cdot L_2$ (any number k)
5. *Quotient Rule:* $\lim \dfrac{f_1(x)}{f_2(x)} = \dfrac{L_1}{L_2}$ if $L_2 \neq 0$.

The limits are all to be taken as $x \to x_0$ in the appropriate domain (D), and L_1 and L_2 are to be real numbers.

We proved the Sum Rule in Section 2.5, and we obtain the Difference Rule by replacing $f_2(x)$ by $(-f_2(x))$ and L_2 by $(-L_2)$ in the Sum Rule. The Constant Multiple Rule is the special case $f_1(x) = k$ of the Product Rule. This leaves only the Product and Quotient Rules to prove.

PROOF of the Limit Product Rule We need to show that for any $\epsilon > 0$ there exists a $\delta > 0$ such that for all x in D

$$0 < |x - x_0| < \delta \quad \Rightarrow \quad |f_1(x)f_2(x) - L_1L_2| < \epsilon. \qquad (1)$$

Suppose then that ϵ is a positive number, and write $f_1(x)$ and $f_2(x)$ as

$$f_1(x) = L_1 + (f_1(x) - L_1) \qquad \text{and} \qquad f_2(x) = L_2 + (f_2(x) - L_2).$$

Multiply these expressions and subtract L_1L_2:

$$\begin{aligned} f_1(x) \cdot f_2(x) - L_1L_2 &= (L_1 + (f_1(x) - L_1))(L_2 + (f_2(x) - L_2)) - L_1L_2 \\ &= L_1L_2 + L_1(f_2(x) - L_2) + L_2(f_1(x) - L_1) \\ &\quad + (f_1(x) - L_1)(f_2(x) - L_2) - L_1L_2 \\ &= L_1(f_2(x) - L_2) + L_2(f_1(x) - L_1) + (f_1(x) - L_1)(f_2(x) - L_2). \end{aligned} \tag{2}$$

Since f_1 and f_2 have limits L_1 and L_2 as $x \to x_0$, there exist positive numbers δ_1, δ_2, δ_3, and δ_4 such that for all x

$$\begin{aligned} 0 < |x - x_0| < \delta_1 \quad &\Rightarrow \quad |f_1(x) - L_1| < \sqrt{\epsilon/3}, \\ 0 < |x - x_0| < \delta_2 \quad &\Rightarrow \quad |f_2(x) - L_2| < \sqrt{\epsilon/3}, \\ 0 < |x - x_0| < \delta_3 \quad &\Rightarrow \quad |f_1(x) - L_1| < \epsilon/(3(1 + |L_2|)), \\ 0 < |x - x_0| < \delta_4 \quad &\Rightarrow \quad |f_2(x) - L_2| < \epsilon/(3(1 + |L_1|)). \end{aligned} \tag{3}$$

All four of the inequalities on the right-hand side of (3) will hold for $0 < |x - x_0| < \delta$ if we take δ to be the smallest of the numbers δ_1 through δ_4. Therefore for all x, $0 < |x - x_0| < \delta$ implies

$$\begin{aligned} &|f_1(x) \cdot f_2(x) - L_1L_2| \\ &\qquad \le |L_1|\,|f_2(x) - L_2| + |L_2|\,|f_1(x) - L_1| + |f_1(x) - L_1|\,|f_2(x) - L_2| \\ &\qquad\qquad \text{(Triangle inequality applied to Eq. (2))} \\ &\qquad \le (1 + |L_1|)|f_2(x) - L_2| + (1 + |L_2|)|f_1(x) - L_1| + |f_1(x) - L_1|\,|f_2(x) - L_2| \\ &\qquad \le \frac{\epsilon}{3} + \frac{\epsilon}{3} + \sqrt{\frac{\epsilon}{3}}\sqrt{\frac{\epsilon}{3}} = \epsilon. \qquad \text{(Values from (3))} \end{aligned}$$

This completes the proof of the Limit Product Rule. ■

PROOF of the Limit Quotient Rule We show that

$$\lim_{x \to x_0} \frac{1}{f_2(x)} = \frac{1}{L_2}.$$

Then we can apply the Limit Product Rule to show that

$$\lim_{x \to x_0} \frac{f_1(x)}{f_2(x)} = \lim_{x \to x_0} f_1(x) \cdot \frac{1}{f_2(x)} = \lim_{x \to x_0} f_1(x) \cdot \lim_{x \to x_0} \frac{1}{f_2(x)} = L_1 \cdot \frac{1}{L_2} = \frac{L_1}{L_2}.$$

To show that $\lim_{x \to x_0}(1/f_2(x)) = 1/L_2$, we need to show that for any $\epsilon > 0$ there exists a $\delta > 0$ such that for all x

$$0 < |x - x_0| < \delta \quad \Rightarrow \quad \left|\frac{1}{f_2(x)} - \frac{1}{L_2}\right| < \epsilon.$$

Since $|L_2| > 0$, there exists a positive number δ_1 such that for all x

$$0 < |x - x_0| < \delta_1 \quad \Rightarrow \quad |f_2(x) - L_2| < \frac{|L_2|}{2}. \tag{4}$$

For any numbers A and B it can be shown that $|A| - |B| \leq |A - B|$ and $|B| - |A| \leq |A - B|$, from which it follows that

$$\big|\, |A| - |B| \,\big| \leq |A - B|. \tag{5}$$

With $A = f_2(x)$ and $B = L_2$, this gives

$$\big|\, |f_2(x)| - |L_2| \,\big| \leq |f_2(x) - L_2|,$$

which we can combine with the right-hand inequality in (4) to get, in turn,

$$\big|\, |f_2(x)| - |L_2| \,\big| < \frac{|L_2|}{2},$$

$$-\frac{|L_2|}{2} < |f_2(x)| - |L_2| < \frac{|L_2|}{2},$$

$$\frac{|L_2|}{2} < |f_2(x)| < \frac{3|L_2|}{2},$$

$$|L_2| < 2|f_2(x)| < 3|L_2|,$$

$$\frac{1}{|f_2(x)|} < \frac{2}{|L_2|} < \frac{3}{|f_2(x)|}. \tag{6}$$

Therefore $0 < |x - x_0| < \delta_1$ implies that

$$\left|\frac{1}{f_2(x)} - \frac{1}{L_2}\right| = \left|\frac{L_2 - f_2(x)}{L_2 f_2(x)}\right| \leq \frac{1}{|L_2|} \cdot \frac{1}{|f_2(x)|} \cdot |L_2 - f_2(x)|$$

$$< \frac{1}{|L_2|} \cdot \frac{2}{|L_2|} \cdot |L_2 - f_2(x)|. \qquad \text{(Eq. 6)}$$

Suppose now that ϵ is an arbitrary positive number. Then $\frac{1}{2}|L_2|^2 \epsilon > 0$, so there exists a number $\delta_2 > 0$ such that for all x

$$0 < |x - x_0| < \delta_2 \quad \Rightarrow \quad |L_2 - f_2(x)| < \frac{\epsilon}{2}|L_2|^2. \tag{7}$$

The conclusions in (6) and (7) both hold for all x such that $0 < |x - x_0| < \delta$ if we take δ to be the smaller of the positive values δ_1 and δ_2. Combining (6) and (7) then gives

$$0 < |x - x_0| < \delta \quad \Rightarrow \quad \left|\frac{1}{f_2(x)} - \frac{1}{L_2}\right| < \epsilon. \tag{8}$$

This concludes the proof of the Limit Quotient Rule. ■

THEOREM 2

The Sandwich Theorem
Suppose that $g(x) \leq f(x) \leq h(x)$ for all $x \neq x_0$ in some interval about x_0 and that $\lim_{x \to x_0} g(x) = \lim_{x \to x_0} h(x) = L$. Then $\lim_{x \to x_0} f(x) = L$.

PROOF for right-hand limits Suppose $\lim_{x \to x_0^+} g(x) = \lim_{x \to x_0^+} h(x) = L$. Then for any $\epsilon > 0$ there exists a $\delta > 0$ such that for all x the inequality $x_0 < x < x_0 + \delta$ implies

$$L - \epsilon < g(x) < L + \epsilon \quad \text{and} \quad L - \epsilon < h(x) < L + \epsilon. \tag{9}$$

These inequalities combine with the inequality $g(x) \le f(x) \le h(x)$ to give

$$L - \epsilon < g(x) \le f(x) \le h(x) < L + \epsilon, \tag{10}$$

$$L - \epsilon < f(x) < L + \epsilon,$$

$$-\epsilon < f(x) - L < \epsilon.$$

Therefore, for all x the inequality $x_0 < x < x_0 + \delta$ implies $|f(x) - L| < \epsilon$.

PROOF for left-hand limits Suppose $\lim_{x\to x_0^-} g(x) = \lim_{x\to x_0^-} h(x) = L$. Then for any $\epsilon > 0$ there exists a $\delta > 0$ such that for all x the inequality $x_0 - \delta < x < x_0$ implies

$$L - \epsilon < g(x) < L + \epsilon \quad \text{and} \quad L - \epsilon < h(x) < L + \epsilon.$$

We conclude as before that for all x the inequality $x_0 - \delta < x < x_0$ implies $|f(x) - L| < \epsilon$.

PROOF for two-sided limits If $\lim_{x\to x_0} g(x) = \lim_{x\to x_0} h(x) = L$, then $g(x)$ and $h(x)$ both approach L as $x \to x_0^+$ and as $x \to x_0^-$; so $\lim_{x\to x_0^+} f(x) = L$ and $\lim_{x\to x_0^-} f(x) = L$. Hence $\lim_{x\to x_0} f(x)$ exists and equals L. ■

Exercises A.2

1. Suppose that functions $f_1(x)$, $f_2(x)$, and $f_3(x)$ have limits L_1, L_2, and L_3, respectively, as $x \to x_0$. Show that their sum has limit $L_1 + L_2 + L_3$. Use mathematical induction (Appendix A.4) to generalize this result to the sum of any finite number of functions.
2. Use mathematical induction and the Limit Product Rule in Theorem 1 to show that if functions $f_1(x), f_2(x), \ldots, f_n(x)$ have limits $L_1, L_2, \ldots, L_n$ as $x \to x_0$, then $\lim_{x\to x_0} f_1(x) f_2(x) \cdot \cdots \cdot f_n(x) = L_1 \cdot L_2 \cdot \cdots \cdot L_n$.
3. Use the fact that $\lim_{x\to x_0} x = x_0$ and the result of Exercise 2 to show that $\lim_{x\to x_0} x^n = x_0^n$ for any integer $n > 1$.
4. Use the fact that $\lim_{x\to x_0} (k) = k$ for any number k together with the results of Exercises 1 and 3 to show that $\lim_{x\to x_0} f(x) = f(x_0)$ for any polynomial function
$$f(x) = a_0 x^n + a_1 x^{n-1} + \cdots + a_{n-1} x + a_n.$$
5. Use Theorem 1 and the result of Exercise 4 to show that if $f(x)$ and $g(x)$ are polynomial functions and $g(x_0) \neq 0$ then
$$\lim_{x\to x_0} \frac{f(x)}{g(x)} = \frac{f(x_0)}{g(x_0)}.$$
6. Figure A.1 gives the diagram for a proof that the composite of two continuous functions is continuous. Reconstruct the proof from the diagram. The statement to be proved is this: If g is continuous at $x = x_0$ and f is continuous at $g(x_0)$ then $f \circ g$ is continuous at x_0.

   Assume that x_0 is an interior point of the domain of g and that $g(x_0)$ is an interior point of the domain of f. This will make the limits involved two-sided. (The arguments for the cases that involve one-sided limits are similar.)

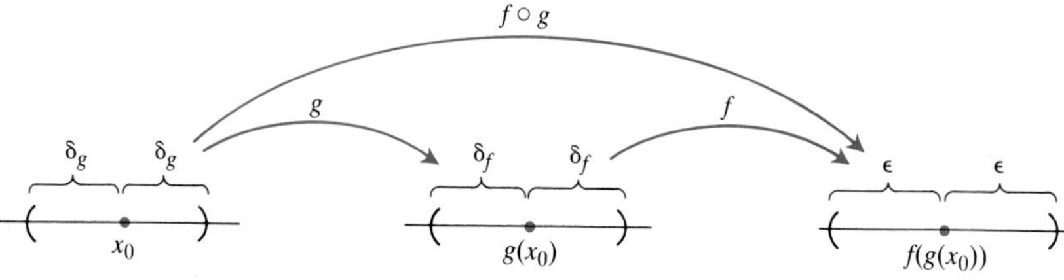

A.1 The diagram for a proof that the composite of two continuous functions is continuous. The continuity of composites holds for any finite number of functions. The only requirement is that each function be continuous where it is applied. In the figure, g is to be continuous at x_0, and f is to be continuous at $g(x_0)$.

A.3 A Proof of the Chain Rule for Functions of a Single Variable

Our goal is to show that if $f(u)$ is a differentiable function of u and $u = g(x)$ is a differentiable function of x then the composite $y = f(g(x))$ is a differentiable function of x. More precisely, if g is differentiable at x_0 and f is differentiable at $g(x_0)$, then the composite is differentiable at x_0 and

$$\left.\frac{dy}{dx}\right|_{x=x_0} = f'(g(x_0)) \cdot g'(x_0). \tag{1}$$

Suppose that Δx is an increment in x and that Δu and Δy are the corresponding increments in u and y. The following figure is drawn with the increments all positive, but Δx could be negative and Δu and Δy could be negative or even zero.

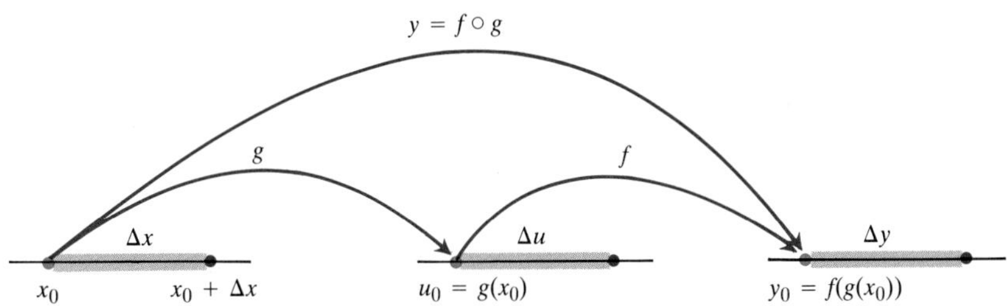

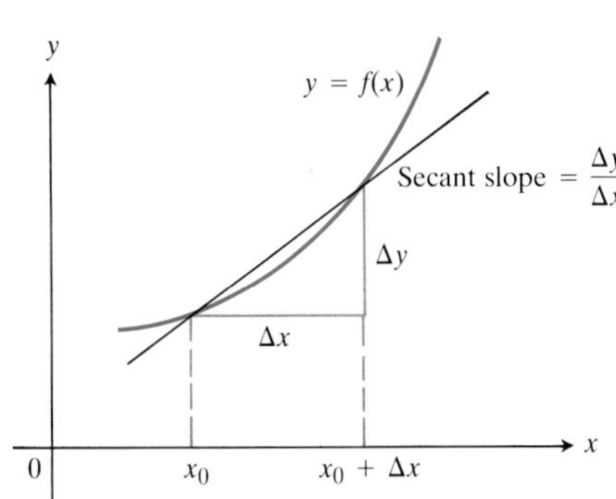

A.2 The graph of y as a function of x. The derivative of y with respect to x at $x = x_0$ is $\lim_{\Delta x\to 0} \Delta y/\Delta x$.

As you can see in Fig. A.2

$$\left.\frac{dy}{dx}\right|_{x=x_0} = \lim_{\Delta x\to 0} \frac{\Delta y}{\Delta x},$$

so our goal is to show that this limit is $f'(g(x_0)) \cdot g'(x_0)$.

By virtue of Eq. (12) in Section 3.7,

$$\Delta u = g'(x_0)\Delta x + \epsilon_1 \Delta x = [g'(x_0) + \epsilon_1]\, \Delta x, \tag{2}$$

where $\epsilon_1 \to 0$ as $\Delta x \to 0$. Similarly,

$$\Delta y = f'(u_0)\Delta u + \epsilon_2 \Delta u = [f'(u_0) + \epsilon_2]\, \Delta u, \tag{3}$$

where $\epsilon_2 \to 0$ as $\Delta u \to 0$. Combining the equation for Δu with the one for Δy gives

$$\Delta y = [f'(u_0) + \epsilon_2][g'(x_0) + \epsilon_1]\, \Delta x, \tag{4}$$

so

$$\frac{\Delta y}{\Delta x} = f'(u_0)g'(x_0) + \epsilon_2\, g'(x_0) + f'(u_0)\epsilon_1 + \epsilon_2\epsilon_1.$$

Since ϵ_1 and ϵ_2 go to zero as Δx goes to zero, the three terms on the right vanish in the limit and

$$\lim_{\Delta x\to 0} \frac{\Delta y}{\Delta x} = f'(u_0)g'(x_0) = f'(g(x_0)) \cdot g'(x_0).$$

This concludes the proof. ■

A.4 Mathematical Induction

Many formulas, like

$$1 + 2 + \cdots + n = \frac{n(n + 1)}{2},$$

can be shown to hold for every positive integer n by applying an axiom called the *mathematical induction principle.* A proof that uses this axiom is called a *proof by mathematical induction* or a *proof by induction.*

The steps in proving a formula by induction are

STEP 1: Check that it holds for $n = 1$.

STEP 2: Prove that if it holds for any positive integer $n = k$, then it also holds for $n = k + 1$.

Once these steps are completed (the axiom says), we know that the formula holds for all positive integers n. By Step 1 it holds for $n = 1$. By Step 2 it holds for $n = 2$, and therefore by Step 2 also for $n = 3$, and by Step 2 again for $n = 4$, and so on. If the first domino falls, and the kth domino always knocks over the $(k + 1)$st when it falls, all the dominoes fall.

From another point of view, suppose we have a sequence of statements

$$S_1, S_2, \ldots, S_n, \ldots,$$

one for each positive integer. Suppose we can show that assuming any one of the statements to be true implies that the next statement in line is true. Suppose that we can also show that S_1 is true. Then we may conclude that the statements are true from S_1 on.

EXAMPLE 1 Show that for every positive integer n

$$1 + 2 + \cdots + n = \frac{n(n + 1)}{2}.$$

Solution We accomplish the proof by carrying out the two steps of mathematical induction.

STEP 1: The formula holds for $n = 1$ because

$$1 = \frac{1(1 + 1)}{2}.$$

STEP 2: If the formula holds for $n = k$, does it also hold for $n = k + 1$? The answer is yes, and here's why: If

$$1 + 2 + \cdots + k = \frac{k(k + 1)}{2},$$

then

$$1 + 2 + \cdots + k + (k + 1) = \frac{k(k + 1)}{2} + (k + 1) = \frac{k^2 + k + 2k + 2}{2}$$

$$= \frac{(k + 1)(k + 2)}{2} = \frac{(k + 1)((k + 1) + 1)}{2}.$$

The last expression in this string of equalities is the expression $n(n+1)/2$ for $n = (k+1)$.

The mathematical induction principle now guarantees the original formula for all positive integers n.

Notice that all *we* have to do is carry out Steps 1 and 2. The mathematical induction principle does the rest. ■

EXAMPLE 2 Show that for all positive integers n

$$\frac{1}{2^1} + \frac{1}{2^2} + \cdots + \frac{1}{2^n} = 1 - \frac{1}{2^n}.$$

Solution We accomplish the proof by carrying out the two steps of mathematical induction.

STEP 1: The formula holds for $n = 1$ because

$$\frac{1}{2^1} = 1 - \frac{1}{2^1}.$$

STEP 2: If

$$\frac{1}{2^1} + \frac{1}{2^2} + \cdots + \frac{1}{2^k} = 1 - \frac{1}{2^k},$$

then

$$\frac{1}{2^1} + \frac{1}{2^2} + \cdots + \frac{1}{2^k} + \frac{1}{2^{k+1}} = 1 - \frac{1}{2^k} + \frac{1}{2^{k+1}} = 1 - \frac{1 \cdot 2}{2^k \cdot 2} + \frac{1}{2^{k+1}}$$

$$= 1 - \frac{2}{2^{k+1}} + \frac{1}{2^{k+1}} = 1 - \frac{1}{2^{k+1}}.$$

Thus, the original formula holds for $n = k + 1$ whenever it holds for $n = k$.

With these two steps verified, the mathematical induction principle now guarantees the formula for every positive integer n. ■

Other Starting Integers

Instead of starting at $n = 1$, some induction arguments start at another integer. The steps for such an argument are

STEP 1: Check that the formula holds for $n = n_1$ (whatever the appropriate first integer is).

STEP 2: Prove that if the formula holds for any integer $n = k \geq n_1$, then it also holds for $n = k + 1$.

Once these steps are completed, the mathematical induction principle will guarantee the formula for all $n \geq n_1$.

EXAMPLE 3 Show that $n! > 3^n$ if n is large enough.

Solution How large is large enough? We experiment:

n	1	2	3	4	5	6	7
$n!$	1	2	6	24	120	720	5040
3^n	3	9	27	81	243	729	2187

It looks as if $n! > 3^n$ for $n \geq 7$. To be sure, we apply mathematical induction. We take $n_1 = 7$ in Step 1 and try for Step 2.

Suppose $k! > 3^k$ for some $k \geq 7$. Then

$$(k+1)! = (k+1)(k!) > (k+1)3^k > 7 \cdot 3^k > 3^{k+1}.$$

Thus, for $k \geq 7$,

$$k! > 3^k \quad \Rightarrow \quad (k+1)! > 3^{k+1}.$$

The mathematical induction principle now guarantees $n! \geq 3^n$ for all $n \geq 7$. ■

Exercises A.4

1. Assuming that the triangle inequality $|a + b| \leq |a| + |b|$ holds for any two numbers a and b, show that

$$|x_1 + x_2 + \cdots + x_n| \leq |x_1| + |x_2| + \cdots + |x_n|$$

for any n numbers.

2. Show that if $r \neq 1$, then

$$1 + r + r^2 + \cdots + r^n = \frac{1 - r^{n+1}}{1 - r}$$

for all positive integers n.

3. Use the Product Rule

$$\frac{d}{dx}(uv) = u\frac{dv}{dx} + v\frac{du}{dx}$$

and the fact that

$$\frac{d}{dx}(x) = 1$$

to show that

$$\frac{d}{dx}(x^n) = nx^{n-1}$$

for all positive integers n.

4. Suppose that a function $f(x)$ has the property that $f(x_1x_2) = f(x_1) + f(x_2)$ for any two positive numbers x_1 and x_2. Show that

$$f(x_1x_2\cdots x_n) = f(x_1) + f(x_2) + \cdots + f(x_n)$$

for the product of any n positive numbers $x_1, x_2, \ldots, x_n$.

5. Show that

$$\frac{2}{3^1} + \frac{2}{3^2} + \cdots + \frac{2}{3^n} = 1 - \frac{1}{3^n}$$

for all positive integers n.

6. Show that $n! > n^3$ if n is large enough.

7. Show that $2^n > n^2$ if n is large enough.

8. Show that $2^n \geq 1/8$ for $n \geq -3$.

9. Show that the sum of the squares of the first n positive integers is $n(n+1)(2n+1)/6$.

10. Show that the sum of the cubes of the first n positive integers is $(n(n+1)/2)^2$.

11. Show that the following finite-sum rules hold for every positive integer n.

a) $\sum_{k=1}^{n} (a_k + b_k) = \sum_{k=1}^{n} a_k + \sum_{k=1}^{n} b_k$

b) $\sum_{k=1}^{n} (a_k - b_k) = \sum_{k=1}^{n} a_k - \sum_{k=1}^{n} b_k$

c) $\sum_{k=1}^{n} ca_k = c \cdot \sum_{k=1}^{n} a_k$ (Any number c)

d) $\sum_{k=1}^{n} a_k = n \cdot c$ if a_k has the constant value c.

A.5 Limits That Arise Frequently

This appendix verifies the limits in Table 9.1 of Section 9.1.

1. $\lim_{n\to\infty} \frac{\ln n}{n} = 0$	2. $\lim_{n\to\infty} \sqrt[n]{n} = 1$
3. $\lim_{n\to\infty} x^{1/n} = 1 \quad (x > 0)$	4. $\lim_{n\to\infty} x^n = 0 \quad (\lvert x\rvert < 1)$
5. $\lim_{n\to\infty} \left(1 + \frac{x}{n}\right)^n = e^x \quad (\text{Any } x)$	6. $\lim_{n\to\infty} \frac{x^n}{n!} = 0 \quad (\text{Any } x)$

In Eqs. (3)–(6), x remains fixed while n varies.

1. $\lim_{n\to\infty} \frac{\ln n}{n} = 0$ We proved this in Section 9.1, Example 9.

2. $\lim_{n\to\infty} \sqrt[n]{n} = 1$ Let $a_n = n^{1/n}$. Then

$$\ln a_n = \ln n^{1/n} = \frac{1}{n} \ln n \to 0. \tag{1}$$

Applying Theorem 3, Section 9.1, with $f(x) = e^x$ gives

$$n^{1/n} = a_n = e^{\ln a_n} = f(\ln a_n) \to f(0) = e^0 = 1. \tag{2}$$

3. If $x > 0$, $\lim_{n\to\infty} x^{1/n} = 1$ Let $a_n = x^{1/n}$. Then

$$\ln a_n = \ln x^{1/n} = \frac{1}{n} \ln x \to 0 \tag{3}$$

because x remains fixed as $n \to \infty$. Applying Theorem 3, Section 9.1, with $f(x) = e^x$ gives

$$x^{1/n} = a_n = e^{\ln a_n} \to e^0 = 1. \tag{4}$$

4. If $\lvert x\rvert < 1$, $\lim_{n\to\infty} x^n = 0$ We need to show that to each $\epsilon > 0$ there corresponds an integer N so large that $\lvert x^n\rvert < \epsilon$ for all n greater than N. Since $\epsilon^{1/n} \to 1$, while $\lvert x\rvert < 1$, there exists an integer N for which

$$\epsilon^{1/N} > \lvert x\rvert. \tag{5}$$

In other words,

$$\lvert x^N\rvert = \lvert x\rvert^N < \epsilon. \tag{6}$$

This is the integer we seek because, if $\lvert x\rvert < 1$, then

$$\lvert x^n\rvert < \lvert x^N\rvert \quad \text{for all } n > N. \tag{7}$$

Combining (6) and (7) produces

$$\lvert x^n\rvert < \epsilon \quad \text{for all } n > N, \tag{8}$$

and we're done.

5. For any number x, $\lim_{n\to\infty}\left(1+\frac{x}{n}\right)^n = e^x$ Let

$$a_n = \left(1+\frac{x}{n}\right)^n.$$

Then
$$\ln a_n = \ln\left(1+\frac{x}{n}\right)^n = n\ln\left(1+\frac{x}{n}\right)\to x,$$

as we can see by the following application of l'Hôpital's rule, in which we differentiate with respect to n:

$$\lim_{n\to\infty} n\ln\left(1+\frac{x}{n}\right) = \lim_{n\to\infty}\frac{\ln(1+x/n)}{1/n}$$

$$= \lim_{n\to\infty}\frac{\left(\dfrac{1}{1+x/n}\right)\cdot\left(-\dfrac{x}{n^2}\right)}{-1/n^2} = \lim_{n\to\infty}\frac{x}{1+x/n} = x.$$

Apply Theorem 3, Section 9.1, with $f(x) = e^x$ to conclude that

$$\left(1+\frac{x}{n}\right)^n = a_n = e^{\ln a_n}\to e^x.$$

6. For any number x, $\lim_{n\to\infty}\frac{x^n}{n!} = 0$ Since

$$-\frac{|x|^n}{n!}\le\frac{x^n}{n!}\le\frac{|x|^n}{n!},$$

all we need to show is that $|x|^n/n!\to 0$. We can then apply the Sandwich Theorem for Sequences (Section 9.1, Theorem 2) to conclude that $x^n/n!\to 0$.

The first step in showing that $|x|^n/n!\to 0$ is to choose an integer $M>|x|$, so that

$$\frac{|x|}{M}<1 \qquad\text{and}\qquad \left(\frac{|x|}{M}\right)^n\to 0.$$

We then restrict our attention to values of $n>M$. For these values of n, we can write

$$\frac{|x|^n}{n!} = \frac{|x|^n}{1\cdot 2\cdot\cdots\cdot M\cdot\underbrace{(M+1)(M+2)\cdot\cdots\cdot n}_{(n-M)\text{ factors}}}$$

$$\le\frac{|x|^n}{M!\,M^{n-M}} = \frac{|x|^n M^M}{M!\,M^n} = \frac{M^M}{M!}\left(\frac{|x|}{M}\right)^n.$$

Thus,
$$0\le\frac{|x|^n}{n!}\le\frac{M^M}{M!}\left(\frac{|x|}{M}\right)^n.$$

Now, the constant $M^M/M!$ does not change as n increases. Thus the Sandwich Theorem tells us that

$$\frac{|x|^n}{n!}\to 0 \qquad\text{because}\qquad \left(\frac{|x|}{M}\right)^n\to 0.$$

A.6 Complex Numbers

Complex numbers are numbers of the form $a + bi$ where a and b are real numbers and $i = \sqrt{-1}$. The number a is the **real part** of $a + bi$, and b is the **imaginary part.**

Two complex numbers $a + bi$ and $c + di$ are equal if and only if $a = c$ and $b = d$. In particular, $a + bi = 0$ if and only if $a = 0$ and $b = 0$, or, equivalently, if and only if $a^2 + b^2 = 0$.

We add and subtract complex numbers by adding and subtracting their real and imaginary parts:

$$(a + bi) + (c + di) = (a + c) + (b + d)i,$$

$$(a + bi) - (c + di) = (a - c) + (b - d)i.$$

We multiply complex numbers the way we multiply other binomials, using the fact that $i^2 = -1$ to simplify the final result:

$$\begin{aligned}(a + bi)(c + di) &= ac + adi + bci + bdi^2 \\ &= ac + adi + bci - bd && (i^2 = -1) \\ &= (ac - bd) + (ad + bc)i. && \left(\text{Real and imaginary parts combined}\right)\end{aligned}$$

To divide a complex number $c + di$ by a nonzero complex number $a + bi$, multiply the numerator and denominator of the quotient by $a - bi$ (the number $a - bi$ is the **complex conjugate** of $a + bi$):

$$\begin{aligned}\frac{c + di}{a + bi} &= \frac{c + di}{a + bi}\,\frac{a - bi}{a - bi} && \left(\text{Multiply numerator and denominator by the complex conjugate of } a + bi\right) \\ &= \frac{ac - bci + adi - bdi^2}{a^2 - abi + abi - b^2i^2} \\ &= \frac{(ac + bd) + (ad - bc)i}{a^2 + b^2} \\ &= \frac{ac + bd}{a^2 + b^2} + \frac{ad - bc}{a^2 + b^2}\,i.\end{aligned}$$

EXAMPLE

a) $(2 + 3i) + (6 - 2i) = (2 + 6) + (3 - 2)i = 8 + i$

b) $(2 + 3i) - (6 - 2i) = (2 - 6) + (3 - (-2))i = -4 + 5i$

c) $(2 + 3i)(6 - 2i) = (2)(6) + (2)(-2i) + (3i)(6) + (3i)(-2i)$
$= 12 - 4i + 18i - 6i^2 = 12 + 14i + 6 = 18 + 14i$

d) $$\begin{aligned}\frac{2 + 3i}{6 - 2i} &= \frac{2 + 3i}{6 - 2i}\,\frac{6 + 2i}{6 + 2i} \\ &= \frac{12 + 4i + 18i + 6i^2}{36 + 12i - 12i - 4i^2} \\ &= \frac{6 + 22i}{40} = \frac{3}{20} + \frac{11}{20}i\end{aligned}$$

■

A.7 Determinants and Cramer's Rule

A rectangular array of numbers like

$$A = \begin{bmatrix} 2 & 1 & 3 \\ 1 & 0 & -2 \end{bmatrix}$$

is called a **matrix.** We call A a 2 by 3 matrix because it has two rows and three columns. An m by n matrix has m rows and n columns, and the **entry** or **element** (number) in the ith row and jth column is often denoted by a_{ij}:

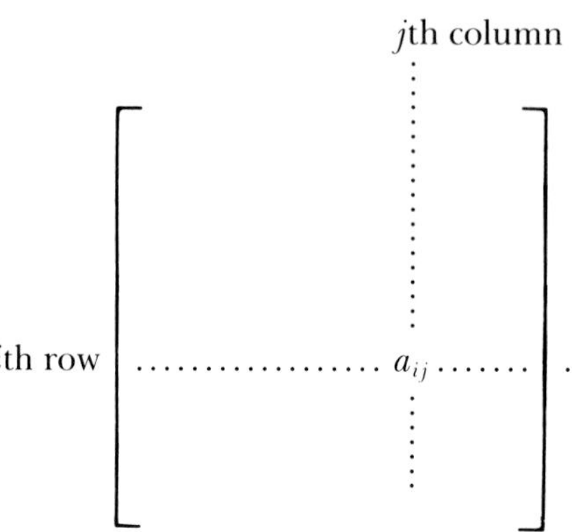

The matrix

$$A = \begin{bmatrix} 2 & 1 & 3 \\ 1 & 0 & -2 \end{bmatrix}$$

has

$$a_{11} = 2, \quad a_{12} = 1, \quad a_{13} = 3,$$
$$a_{21} = 1, \quad a_{22} = 0, \quad a_{23} = -2.$$

A matrix with the same number of rows as columns is a **square matrix.** It is a **matrix of order *n*** if the number of rows and columns is n.

With each square matrix A we associate a number $\det A$ or $|a_{ij}|$, called the **determinant** of A, calculated from the entries of A in the following way. (The vertical bars in the notation $|a_{ij}|$ do not mean absolute value.) For $n = 1$ and $n = 2$ we define

$$\det [a] = a, \tag{1}$$

$$\det \begin{bmatrix} a_{11} & a_{12} \\ a_{21} & a_{22} \end{bmatrix} = a_{11}a_{22} - a_{21}a_{12}. \tag{2}$$

For a matrix of order 3, we define

$$\det A = \det \begin{bmatrix} a_{11} & a_{12} & a_{13} \\ a_{21} & a_{22} & a_{23} \\ a_{31} & a_{32} & a_{33} \end{bmatrix} = \begin{array}{l}\text{Sum of all signed products} \\ \text{of the form } \pm a_{1i}a_{2j}a_{3k},\end{array} \tag{3}$$

where i, j, k is a permutation of 1, 2, 3 in some order. There are $3! = 6$ such permutations, so there are six terms in the sum. Half of these have plus signs and the other half have minus signs, according to the index of the permutation, where the index is the number we define next. The sign is positive when the index is even and negative when the index is odd.

DEFINITION

Index of a Permutation

Given any permutation of the numbers 1, 2, 3, . . . , n, denote the permutation by $i_1, i_2, i_3, \ldots, i_n$. In this arrangement, some of the numbers following i_1 may be less than i_1, and however many of these there are is called the **number of inversions** in the arrangement pertaining to i_1. Likewise, there is a number of inversions pertaining to each of the other i's; it is the number of indices that come after that particular one in the arrangement and are less than it. The **index** of the permutation is the sum of all of the numbers of inversions pertaining to the separate indices.

EXAMPLE 1 For $n = 5$, the permutation

$$5 \quad 3 \quad 1 \quad 2 \quad 4$$

has

4 inversions pertaining to the first element, 5,

2 inversions pertaining to the second element, 3,

and no further inversions, so the index is $4 + 2 = 6$. ■

The following table shows the permutations of 1, 2, 3, the index of each permutation, and the signed product in the determinant of Eq. (3).

Permutation	Index	Signed product
1 2 3	0	$+\,a_{11}a_{22}a_{33}$
1 3 2	1	$-\,a_{11}a_{23}a_{32}$
2 1 3	1	$-\,a_{12}a_{21}a_{33}$
2 3 1	2	$+\,a_{12}a_{23}a_{31}$
3 1 2	2	$+\,a_{13}a_{21}a_{32}$
3 2 1	3	$-\,a_{13}a_{22}a_{31}$

(4)

The sum of the six signed products is

$$a_{11}(a_{22}a_{33} - a_{23}a_{32}) - a_{12}(a_{21}a_{33} - a_{23}a_{31}) + a_{13}(a_{21}a_{32} - a_{22}a_{31})$$

$$= a_{11}\begin{vmatrix} a_{22} & a_{23} \\ a_{32} & a_{33} \end{vmatrix} - a_{12}\begin{vmatrix} a_{21} & a_{23} \\ a_{31} & a_{33} \end{vmatrix} + a_{13}\begin{vmatrix} a_{21} & a_{22} \\ a_{31} & a_{32} \end{vmatrix} = \begin{vmatrix} a_{11} & a_{12} & a_{13} \\ a_{21} & a_{22} & a_{23} \\ a_{31} & a_{32} & a_{33} \end{vmatrix}. \qquad (5)$$

The formula

$$\begin{vmatrix} a_{11} & a_{12} & a_{13} \\ a_{21} & a_{22} & a_{23} \\ a_{31} & a_{32} & a_{33} \end{vmatrix} = a_{11}\begin{vmatrix} a_{22} & a_{23} \\ a_{32} & a_{33} \end{vmatrix} - a_{12}\begin{vmatrix} a_{21} & a_{23} \\ a_{31} & a_{33} \end{vmatrix} + a_{13}\begin{vmatrix} a_{21} & a_{22} \\ a_{31} & a_{32} \end{vmatrix} \qquad (6)$$

reduces the calculation of a 3 by 3 determinant to the calculation of three 2 by 2 determinants.

Many people prefer to remember the following scheme for calculating the six signed products in the determinant of a 3 by 3 matrix:

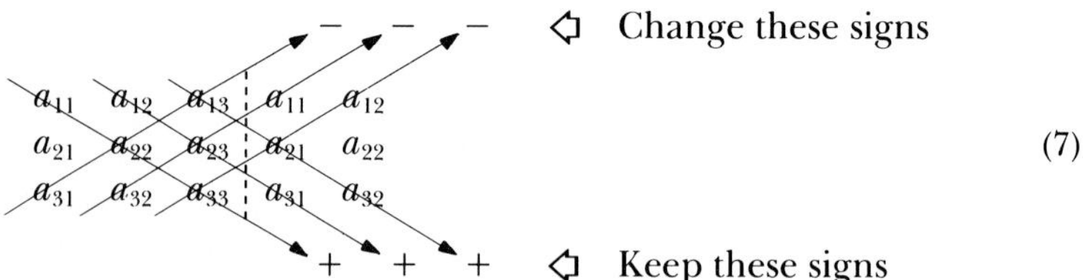

(7)

Minors and Cofactors

The second order determinants on the right-hand side of Eq. (6) are called the **minors** (short for minor determinant) of the entries they multiply. Thus,

$$\begin{vmatrix} a_{22} & a_{23} \\ a_{32} & a_{33} \end{vmatrix} \text{ is the minor of } a_{11},$$

$$\begin{vmatrix} a_{21} & a_{23} \\ a_{31} & a_{33} \end{vmatrix} \text{ is the minor of } a_{12},$$

and so on. The minor of the element a_{ij} in a matrix A is the determinant of the matrix that remains after we delete the row and column containing a_{ij}:

$$\begin{vmatrix} a_{11} & a_{12} & a_{13} \\ a_{21} & a_{22} & a_{23} \\ a_{31} & a_{32} & a_{33} \end{vmatrix}. \quad \text{The minor of } a_{22} \text{ is } \begin{vmatrix} a_{11} & a_{13} \\ a_{31} & a_{33} \end{vmatrix}.$$

$$\begin{vmatrix} a_{11} & a_{12} & a_{13} \\ a_{21} & a_{22} & a_{23} \\ a_{31} & a_{32} & a_{33} \end{vmatrix}. \quad \text{The minor of } a_{23} \text{ is } \begin{vmatrix} a_{11} & a_{12} \\ a_{31} & a_{32} \end{vmatrix}.$$

The **cofactor** A_{ij} of a_{ij} is $(-1)^{i+j}$ times the minor of a_{ij}. Thus,

$$A_{22} = (-1)^{2+2} \begin{vmatrix} a_{11} & a_{13} \\ a_{31} & a_{33} \end{vmatrix} = \begin{vmatrix} a_{11} & a_{13} \\ a_{31} & a_{33} \end{vmatrix},$$

$$A_{23} = (-1)^{2+3} \begin{vmatrix} a_{11} & a_{12} \\ a_{31} & a_{32} \end{vmatrix} = - \begin{vmatrix} a_{11} & a_{12} \\ a_{31} & a_{32} \end{vmatrix}.$$

The effect of the factor $(-1)^{i+j}$ is to change the sign of the minor when the sum $i+j$ is odd. There is a checkerboard pattern for remembering these sign changes.

$$\begin{matrix} + & - & + \\ - & + & - \\ + & - & + \end{matrix}$$

In the upper left corner, $i = 1, j = 1$ and $(-1)^{1+1} = +1$. In going from any cell to an adjacent cell in the same row or column, we change i by 1 or j by 1, but not both, so we change the exponent from even to odd or from odd to even, which changes the sign from + to − or from − to + .

When we rewrite Eq. (6) in terms of cofactors we get

$$\det A = a_{11}A_{11} + a_{12}A_{12} + a_{13}A_{13}. \tag{8}$$

EXAMPLE 2 Find the determinant of the matrix

$$A = \begin{bmatrix} 2 & 1 & 3 \\ 3 & -1 & -2 \\ 2 & 3 & 1 \end{bmatrix}.$$

Solution 1 The cofactors are

$$A_{11} = (-1)^{1+1}\begin{vmatrix} -1 & -2 \\ 3 & 1 \end{vmatrix}, \quad A_{12} = (-1)^{1+2}\begin{vmatrix} 3 & -2 \\ 2 & 1 \end{vmatrix},$$

$$A_{13} = (-1)^{1+3}\begin{vmatrix} 3 & -1 \\ 2 & 3 \end{vmatrix}.$$

To find det A, we multiply each element of the first row of A by its cofactor and add:

$$\det A = 2\begin{vmatrix} -1 & -2 \\ 3 & 1 \end{vmatrix} + (-1)\begin{vmatrix} 3 & -2 \\ 2 & 1 \end{vmatrix} + 3\begin{vmatrix} 3 & -1 \\ 2 & 3 \end{vmatrix}$$

$$= 2(-1 + 6) - 1(3 + 4) + 3(9 + 2) = 10 - 7 + 33 = 36.$$

Solution 2 From (7) we find

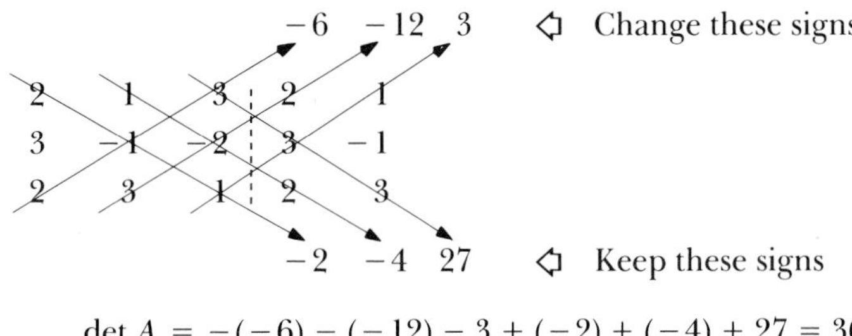

$$\det A = -(-6) - (-12) - 3 + (-2) + (-4) + 27 = 36.$$ ■

Expanding by Columns or by Other Rows

The determinant of a square matrix can be calculated from the cofactors of any row or any column.

If we were to expand the determinant in Example 2 by cofactors according to elements of its third column, say, we would get

$$+3\begin{vmatrix} 3 & -1 \\ 2 & 3 \end{vmatrix} - (-2)\begin{vmatrix} 2 & 1 \\ 2 & 3 \end{vmatrix} + 1\begin{vmatrix} 2 & 1 \\ 3 & -1 \end{vmatrix}$$

$$= 3(9 + 2) + 2(6 - 2) + 1(-2 - 3) = 33 + 8 - 5 = 36.$$

Useful Facts About Determinants

FACT 1: If two rows (or columns) of a matrix are identical, the determinant is zero.

FACT 2: Interchanging two rows (or columns) of a matrix changes the sign of its determinant.

FACT 3: The determinant of a matrix is the sum of the products of the elements of the ith row (or column) by their cofactors, for any i.

FACT 4: The determinant of the **transpose** of a matrix is the same as the determinant of the original matrix. ("Transpose" means to write the rows as columns.)

FACT 5: Multiplying each element of some row (or column) of a matrix by a constant c multiplies the determinant by c.

FACT 6: If all elements of a matrix above the main diagonal (or all below it) are zero, the determinant of the matrix is the product of the elements on the main diagonal. (The **main diagonal** is the diagonal from upper left to lower right.)

EXAMPLE 3

$$\begin{vmatrix} 3 & 4 & 7 \\ 0 & -2 & 5 \\ 0 & 0 & 5 \end{vmatrix} = (3)(-2)(5) = -30.$$

■

FACT 7: If the elements of any row of a matrix are multiplied by the cofactors of the corresponding elements of a different row and these products are summed, the sum is zero.

EXAMPLE 4 If A_{11}, A_{12}, A_{13} are the cofactors of the elements of the first row of $A = (a_{ij})$, then the sums

$$a_{21}A_{11} + a_{22}A_{12} + a_{23}A_{13}$$

(elements of second row times cofactors of elements of first row) and

$$a_{31}A_{11} + a_{32}A_{12} + a_{33}A_{13}$$

are both zero. ■

FACT 8: If the elements of any column of a matrix are multiplied by the cofactors of the corresponding elements of a different column and these products are summed, the sum is zero.

FACT 9: If each element of a row of a matrix is multiplied by a constant c and the results added to a different row, the determinant is not changed. A similar result holds for columns.

EXAMPLE 5 If we start with

$$A = \begin{bmatrix} 2 & 1 & 3 \\ 3 & -1 & -2 \\ 2 & 3 & 1 \end{bmatrix}$$

and add -2 times row 1 to row 2 (subtract 2 times row 1 from row 2) we get

$$B = \begin{bmatrix} 2 & 1 & 3 \\ -1 & -3 & -8 \\ 2 & 3 & 1 \end{bmatrix}.$$

Since det $A = 36$ (Example 2), we should find that det $B = 36$ as well. Indeed we do, as the following calculation shows:

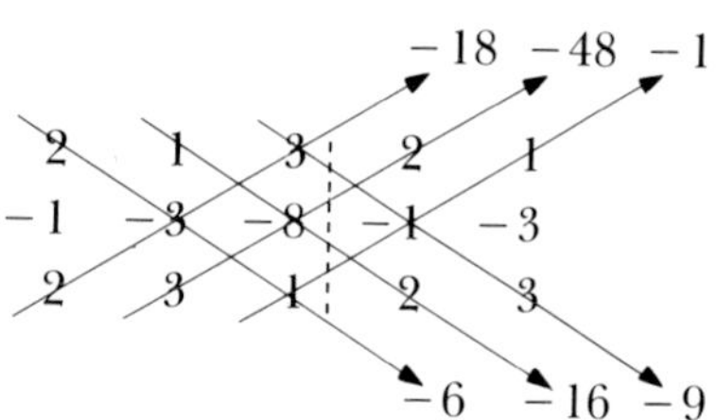

$$\det B = -(-18) - (-48) - (-1) + (-6) + (-16) + (-9)$$
$$= 18 + 48 + 1 - 6 - 16 - 9 = 67 - 31 = 36.$$

EXAMPLE 6 Evaluate the fourth order determinant

$$D = \begin{vmatrix} 1 & -2 & 3 & 1 \\ 2 & 1 & 0 & 2 \\ -1 & 2 & 1 & -2 \\ 0 & 1 & 2 & 1 \end{vmatrix}.$$

Solution We subtract 2 times row 1 from row 2 and add row 1 to row 3 to get

$$D = \begin{vmatrix} 1 & -2 & 3 & 1 \\ 0 & 5 & -6 & 0 \\ 0 & 0 & 4 & -1 \\ 0 & 1 & 2 & 1 \end{vmatrix}.$$

We then multiply the elements of the first column by their cofactors to get

$$D = \begin{vmatrix} 5 & -6 & 0 \\ 0 & 4 & -1 \\ 1 & 2 & 1 \end{vmatrix} = 5(4 + 2) - (-6)(0 + 1) + 0 = 36.$$

Cramer's Rule

If the determinant

$$D = \det A = \begin{vmatrix} a_{11} & a_{12} \\ a_{21} & a_{22} \end{vmatrix}$$

of the coefficient matrix of the system

$$\begin{aligned} a_{11}x + a_{12}y &= b_1, \\ a_{21}x + a_{22}y &= b_2 \end{aligned} \tag{9}$$

of linear equations is 0, the system has either infinitely many solutions or no solution at all. The system

$$\begin{aligned} x + y &= 0, \\ 2x + 2y &= 0 \end{aligned}$$

whose determinant is

$$D = \begin{vmatrix} 1 & 1 \\ 2 & 2 \end{vmatrix} = 2 - 2 = 0$$

has infinitely many solutions. We can find an x to match any given y. The system

$$x + y = 0,$$
$$2x + 2y = 2$$

has no solution. If $x + y = 0$, then $2x + 2y = 2(x + y)$ cannot be 2.

If $D \neq 0$, the system (9) has a unique solution, and Cramer's rule states that it may be found from the formulas

$$x = \frac{\begin{vmatrix} b_1 & a_{12} \\ b_2 & a_{22} \end{vmatrix}}{D}, \qquad y = \frac{\begin{vmatrix} a_{11} & b_1 \\ a_{21} & b_2 \end{vmatrix}}{D}. \tag{10}$$

The numerator in the formula for x comes from replacing the first column in A (the x-column) by the column of constants b_1 and b_2 (the b-column). Replacing the y-column by the b-column gives the numerator of the y-solution.

EXAMPLE 7 Solve the system

$$3x - y = 9,$$
$$x + 2y = -4.$$

Solution We use Eqs. (10). The determinant of the coefficient matrix is

$$D = \begin{vmatrix} 3 & -1 \\ 1 & 2 \end{vmatrix} = 6 + 1 = 7.$$

Hence,

$$x = \frac{\begin{vmatrix} 9 & -1 \\ -4 & 2 \end{vmatrix}}{D} = \frac{18 - 4}{7} = \frac{14}{7} = 2,$$

$$y = \frac{\begin{vmatrix} 3 & 9 \\ 1 & -4 \end{vmatrix}}{D} = \frac{-12 - 9}{7} = \frac{-21}{7} = -3.$$ ■

Systems of three equations in three unknowns work the same way. If the determinant

$$D = \det A = \begin{vmatrix} a_{11} & a_{12} & a_{13} \\ a_{21} & a_{22} & a_{23} \\ a_{31} & a_{32} & a_{33} \end{vmatrix}$$

of the system

$$\begin{aligned} a_{11}x + a_{12}y + a_{13}z &= b_1, \\ a_{21}x + a_{22}y + a_{23}z &= b_2, \\ a_{31}x + a_{32}y + a_{33}z &= b_3 \end{aligned} \tag{11}$$

is zero, the system has either infinitely many solutions or no solution at all. If $D \neq 0$, the system has a unique solution, given by Cramer's rule:

$$x = \frac{1}{D}\begin{vmatrix} b_1 & a_{12} & a_{13} \\ b_2 & a_{22} & a_{23} \\ b_3 & a_{32} & a_{33} \end{vmatrix}, \quad y = \frac{1}{D}\begin{vmatrix} a_{11} & b_1 & a_{13} \\ a_{21} & b_2 & a_{23} \\ a_{31} & b_3 & a_{33} \end{vmatrix}, \quad z = \frac{1}{D}\begin{vmatrix} a_{11} & a_{12} & b_1 \\ a_{21} & a_{22} & b_2 \\ a_{31} & a_{32} & b_3 \end{vmatrix}.$$

The pattern continues in higher dimensions.

Exercises A.7

Evaluate the following determinants.

1. $\begin{vmatrix} 2 & 3 & 1 \\ 4 & 5 & 2 \\ 1 & 2 & 3 \end{vmatrix}$

2. $\begin{vmatrix} 2 & -1 & -2 \\ -1 & 2 & 1 \\ 3 & 0 & -3 \end{vmatrix}$

3. $\begin{vmatrix} 1 & 2 & 3 & 4 \\ 0 & 1 & 2 & 3 \\ 0 & 0 & 2 & 1 \\ 0 & 0 & 3 & 2 \end{vmatrix}$

4. $\begin{vmatrix} 1 & -1 & 2 & 3 \\ 2 & 1 & 2 & 6 \\ 1 & 0 & 2 & 3 \\ -2 & 2 & 0 & -5 \end{vmatrix}$

Evaluate the following determinants by expanding according to the cofactors of (a) the third row and (b) the second column.

5. $\begin{vmatrix} 2 & -1 & 2 \\ 1 & 0 & 3 \\ 0 & 2 & 1 \end{vmatrix}$

6. $\begin{vmatrix} 1 & 0 & -1 \\ 0 & 2 & -2 \\ 2 & 0 & 1 \end{vmatrix}$

7. $\begin{vmatrix} 1 & 1 & 0 & 0 \\ 0 & 0 & -2 & 1 \\ 0 & -1 & 0 & 7 \\ 3 & 0 & 2 & 1 \end{vmatrix}$

8. $\begin{vmatrix} 0 & 1 & 0 & 0 \\ 0 & 1 & 1 & 0 \\ 1 & 1 & 1 & 1 \\ 1 & 1 & 0 & 0 \end{vmatrix}$

Solve the following systems of equations by Cramer's rule.

9. $\begin{aligned} x + 8y &= 4 \\ 3x - y &= -13 \end{aligned}$

10. $\begin{aligned} 2x + 3y &= 5 \\ 3x - y &= 2 \end{aligned}$

11. $\begin{aligned} 4x - 3y &= 6 \\ 3x - 2y &= 5 \end{aligned}$

12. $\begin{aligned} x + y + z &= 2 \\ 2x - y + z &= 0 \\ x + 2y - z &= 4 \end{aligned}$

13. $\begin{aligned} 2x + y - z &= 2 \\ x - y + z &= 7 \\ 2x + 2y + z &= 4 \end{aligned}$

14. $\begin{aligned} 2x - 4y \phantom{{}+7z} &= 6 \\ x + y + z &= 1 \\ 5y + 7z &= 10 \end{aligned}$

15. $\begin{aligned} x \phantom{{}+2y} - z &= 3 \\ 2y - 2z &= 2 \\ 2x \phantom{{}+2y} + z &= 3 \end{aligned}$

16. $\begin{aligned} x_1 + x_2 - x_3 + x_4 &= 2 \\ x_1 - x_2 + x_3 + x_4 &= -1 \\ x_1 + x_2 + x_3 - x_4 &= 2 \\ x_1 \phantom{{}+x_2} + x_3 + x_4 &= -1 \end{aligned}$

17. Find values of h and k for which the system

$$2x + hy = 8, \qquad x + 3y = k$$

has (a) infinitely many solutions, (b) no solution at all.

18. For what value of x will

$$\begin{vmatrix} x & x & 1 \\ 2 & 0 & 5 \\ 6 & 7 & 1 \end{vmatrix} = 0?$$

19. Suppose u, v, and w are twice-differentiable functions of x that satisfy the relation $au + bv + cw = 0$, where a, b, and c are constants, not all zero. Show that

$$\begin{vmatrix} u & v & w \\ u' & v' & w' \\ u'' & v'' & w'' \end{vmatrix} = 0.$$

20. *Partial fractions.* Expanding the quotient

$$\frac{ax + b}{(x - r_1)(x - r_2)}$$

by partial fractions calls for finding the values of C and D that make the equation

$$\frac{ax + b}{(x - r_1)(x - r_2)} = \frac{C}{x - r_1} + \frac{D}{x - r_2}$$

hold for all x.

a) Find a system of linear equations that determines C and D.

b) Under what circumstances does the system of equations in part (a) have a unique solution? That is, when is the determinant of the coefficient matrix of the system different from zero?

A.8 Lagrange Multipliers with Two Constraints

In many applied problems we need to find the extreme values of a function $f(x, y, z)$ whose variables are subject to two constraints. This appendix continues the discussion in Section 13.9 to show how to find the extreme values with Lagrange multipliers if the functions involved have continuous first partial derivatives.

If the constraints on x, y, and z are

$$g_1(x, y, z) = 0 \quad \text{and} \quad g_2(x, y, z) = 0,$$

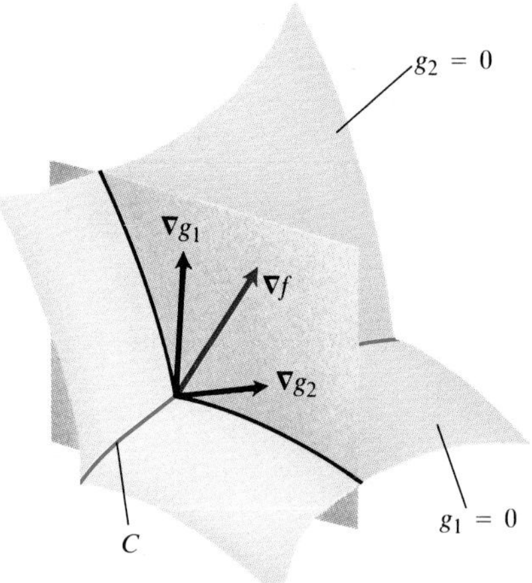

A.3 The vectors ∇g_1 and ∇g_2 lie in a plane perpendicular to the curve C because ∇g_1 is normal to the surface $g_1 = 0$ and ∇g_2 is normal to the surface $g_2 = 0$.

we find the constrained local maxima and minima of f by introducing two Lagrange multipliers λ and μ (mu, pronounced "mew"). That is, we locate the points $P(x, y, z)$ where f takes on its constrained extreme values by finding the values of x, y, z, λ, and μ that simultaneously satisfy the equations

$$\nabla f = \lambda \nabla g_1 + \mu \nabla g_2, \qquad g_1(x, y, z) = 0, \qquad g_2(x, y, z) = 0. \tag{1}$$

The equations in (1) have a nice geometric interpretation. The surfaces $g_1 = 0$ and $g_2 = 0$ (usually) intersect in a differentiable curve, say C (Fig. A.3), and along this curve we seek the points where f has local maximum and minimum values relative to its other values on the curve. These are the points where ∇f is normal to C, as we saw in Section 13.9, Theorem 7. But ∇g_1 and ∇g_2 are also normal to C at these points because C lies in the surfaces $g_1 = 0$ and $g_2 = 0$. Therefore ∇f lies in the plane determined by ∇g_1 and ∇g_2, which means that

$$\nabla f = \lambda \nabla g_1 + \mu \nabla g_2$$

for some λ and μ. Since the points we seek also lie in both surfaces, their coordinates must satisfy the equations $g_1(x, y, z) = 0$ and $g_2(x, y, z) = 0$, which are the remaining requirements in Eq. (1).

EXAMPLE 1 The plane $x + y + z = 1$ cuts the cylinder $x^2 + y^2 = 1$ in an ellipse (Fig. A.4). Find the points on the ellipse that lie closest to and farthest from the origin.

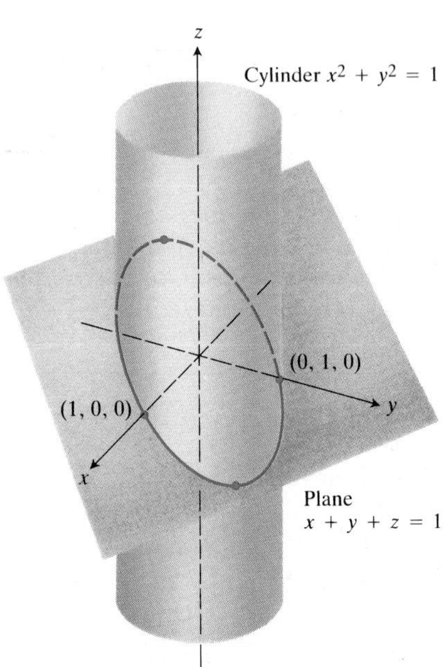

A.4 On the ellipse where the plane and cylinder meet, what are the points closest to and farthest from the origin (Example 1)?

Solution We model this as a Lagrange multiplier problem in which we find the extreme values of

$$f(x, y, z) = x^2 + y^2 + z^2$$

(the square of the distance from (x, y, z) to the origin) subject to the two constraints

$$g_1(x, y, z) = x^2 + y^2 - 1 = 0, \tag{2}$$

$$g_2(x, y, z) = x + y + z - 1 = 0. \tag{3}$$

The gradient equation in (1) then gives

$$\nabla f = \lambda \nabla g_1 + \mu \nabla g_2 \qquad \text{(Eq. 1)}$$

$$2x\mathbf{i} + 2y\mathbf{j} + 2z\mathbf{k} = \lambda(2x\mathbf{i} + 2y\mathbf{j}) + \mu(\mathbf{i} + \mathbf{j} + \mathbf{k}) \tag{4}$$

$$2x\mathbf{i} + 2y\mathbf{j} + 2z\mathbf{k} = (2\lambda x + \mu)\mathbf{i} + (2\lambda y + \mu)\mathbf{j} + \mu\mathbf{k} \tag{5}$$

or

$$2x = 2\lambda x + \mu, \qquad 2y = 2\lambda y + \mu, \qquad 2z = \mu. \tag{6}$$

The scalar equations in (6) yield

$$\begin{aligned} 2x &= 2\lambda x + 2z \quad \Rightarrow \quad (1 - \lambda)x = z, \\ 2y &= 2\lambda y + 2z \quad \Rightarrow \quad (1 - \lambda)y = z. \end{aligned} \tag{7}$$

Equations (7) are satisfied simultaneously if either $\lambda = 1$ and $z = 0$ or if $\lambda \neq 1$ and $x = y = z/(1 - \lambda)$.

If $z = 0$, then solving Eqs. (2) and (3) simultaneously to find the corresponding points on the ellipse gives (algebra omitted) the two points $(1, 0, 0)$ and $(0, 1, 0)$. This makes sense when you look at Fig. A.4.

If $x = y$, then Eqs. (2) and (3) give

$$\begin{aligned} x^2 + x^2 - 1 &= 0 & x + x + z - 1 &= 0 \\ 2x^2 &= 1 & z &= 1 - 2x \\ x &= \pm\frac{\sqrt{2}}{2} & z &= 1 \mp \sqrt{2}. \end{aligned} \tag{8}$$

The corresponding points on the ellipse are

$$\left(\frac{\sqrt{2}}{2}, \frac{\sqrt{2}}{2}, 1 - \sqrt{2}\right) \quad \text{and} \quad \left(-\frac{\sqrt{2}}{2}, -\frac{\sqrt{2}}{2}, 1 + \sqrt{2}\right). \tag{9}$$

Again this makes sense when you look at Fig. A.4.

The points on the ellipse closest to the origin are $(1, 0, 0)$ and $(0, 1, 0)$. The points on the ellipse farthest from the origin are the two points displayed in (9). ■

Exercises A.8

1. Find the point closest to the origin on the line of intersection of the planes $y + 2z = 12$ and $x + y = 6$.
2. Find the maximum value that $f(x, y, z) = x^2 + 2y - z^2$ can have on the line of intersection of the planes $2x - y = 0$ and $y + z = 0$.
3. Find extreme values of $f(x, y, z) = x^2yz + 1$ on the intersection of the plane $z = 1$ with the sphere $x^2 + y^2 + z^2 = 10$.
4. a) Find the maximum value of $w = xyz$ on the line of intersection of the two planes $x + y + z = 40$ and $x + y - z = 0$.
   b) Give a geometric argument to support your claim that you have found a maximum, and not a minimum, value of w.
5. Find the extreme values of the function $f(x, y, z) = xy + z^2$ on the circle in which the plane $y - x = 0$ intersects the sphere $x^2 + y^2 + z^2 = 4$.

A.9 Path Independence of $\int \mathbf{F} \cdot d\mathbf{r}$ Implies $\mathbf{F} = \nabla f$

This appendix completes the proof of Theorem 1 in Section 15.7 by showing that, under the mathematical hypotheses discussed there, path independence of the integral $\int \mathbf{F} \cdot d\mathbf{r}$ throughout a domain D implies that $\mathbf{F}$ is the gradient of a scalar function f defined on D.

We need to show that if $\mathbf{F} = M\mathbf{i} + N\mathbf{j} + P\mathbf{k}$ there exists a scalar function f defined on D for which

$$\frac{\partial f}{\partial x} = M, \quad \frac{\partial f}{\partial y} = N, \quad \text{and} \quad \frac{\partial f}{\partial z} = P. \tag{1}$$

To define f, we choose an arbitrary basepoint A in D. We define the value of f at A to be zero and the value of f at every other point B of D to be the value of the integral of $\mathbf{F} \cdot d\mathbf{r}$ along some smooth path in D from A to B. Such paths exist because D is a connected open region, and all such paths assign

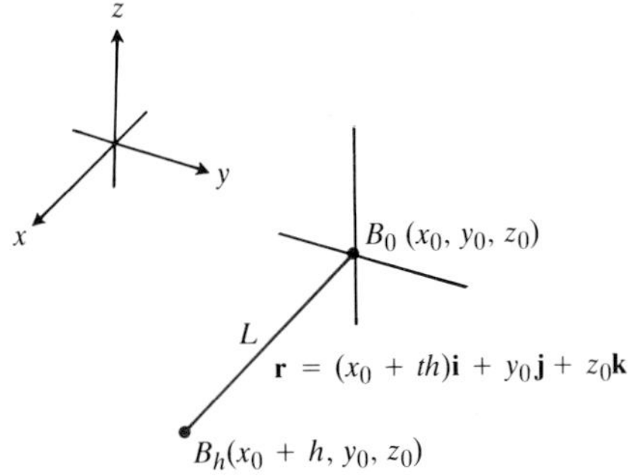

A.5 The line segment L from B_0 to B_h lies parallel to the x-axis. The partial derivative with respect to x of the function

$$f(x, y, z) = \int_A^{(x,y,z)} \mathbf{F}\cdot d\mathbf{r}$$

at B_0 is the limit of $(1/h)\int_L \mathbf{F}\cdot d\mathbf{r}$ as $h \to 0$.

the same value to $f(B)$ because the integral of $\mathbf{F}\cdot d\mathbf{r}$ is path independent in D. Since the integral of $\mathbf{F}\cdot d\mathbf{r}$ from A to A is zero, the values of f are all defined by the single equation

$$f(B) = \int_A^B \mathbf{F}\cdot d\mathbf{r}. \tag{2}$$

One consequence of this definition is that for any points B and B_0 in D,

$$f(B) - f(B_0) = \int_A^B \mathbf{F}\cdot d\mathbf{r} - \int_A^{B_0} \mathbf{F}\cdot d\mathbf{r} = \int_{B_0}^A \mathbf{F}\cdot d\mathbf{r} + \int_A^B \mathbf{F}\cdot d\mathbf{r} = \int_{B_0}^B \mathbf{F}\cdot d\mathbf{r}. \tag{3}$$

Let us now show that f has partial derivatives that give the components of $\mathbf{F}$ at every point $B_0(x_0, y_0, z_0)$ in D.

Since D is open, B_0 is the center of a solid sphere that lies entirely in D. We can therefore find a positive number h (it might be small) for which the point $B_h(x_0 + h, y_0, z_0)$ and the line segment L joining B_0 to B_h both lie in D (Fig. A.5). When we calculate the value of the integral of $\mathbf{F}\cdot d\mathbf{r}$ along L with the parametrization

$$\mathbf{r} = (x_0 + th)\mathbf{i} + y_0\mathbf{j} + z_0\mathbf{k}, \quad 0 \le t \le 1, \tag{4}$$

we find that

$$\int_{B_0}^{B_h} \mathbf{F}\cdot d\mathbf{r} = \int_0^1 (M\mathbf{i} + N\mathbf{j} + P\mathbf{k}) \cdot (h\,dt\,\mathbf{i}) = h\int_0^1 M(x_0 + th, y_0, z_0)\,dt. \tag{5}$$

The difference quotient for defining $\partial f/\partial x$ at B_0 is therefore

$$\begin{aligned} \frac{f(B_h) - f(B_0)}{h} &= \frac{f(x_0 + h, y_0, z_0) - f(x_0, y_0, z_0)}{h} \\ &= \frac{1}{h}\int_{B_0}^{B_h} \mathbf{F}\cdot d\mathbf{r} && \text{(Eq. 3)} \\ &= \int_0^1 M(x_0 + th, y_0, z_0)\,dt. && \text{(Eq. 5)} \end{aligned}$$

Since M is continuous, given any $\epsilon > 0$ there exists a $\delta > 0$ such that

$$|th| < \delta \quad \Rightarrow \quad |M(x_0 + th, y_0, z_0) - M(x_0, y_0, z_0)| < \epsilon.$$

This implies that whenever $|h| < \delta$

$$\begin{aligned} \left|\int_0^1 M(x_0 + h, y_0, z_0)\,dt - M(x_0, y_0, z_0)\right| &= \left|\int_0^1 (M(x_0 + h, y_0, z_0) - M(x_0, y_0, z_0))\,dt\right| \\ &\le \int_0^1 \left|M(x_0 + h, y_0, z_0) - M(x_0, y_0, z_0)\right| dt < \int_0^1 \epsilon\,dt = \epsilon. \end{aligned}$$

We can find such a positive δ for every positive ϵ, so

$$\left.\frac{\partial f}{\partial x}\right|_{(x_0, y_0, z_0)} = \lim_{h\to 0}\int_0^1 M(x_0 + th, y_0, z_0)\,dt = M(x_0, y_0, z_0). \tag{6}$$

In other words, the partial derivative of f with respect to x exists at B_0 and equals M at B_0. Since B_0 was chosen arbitrarily, we can conclude that $\partial f/\partial x$ exists and equals M at every point of D.

The equations $\partial f/\partial y = N$ and $\partial f/\partial z = P$ are derived in a similar way. This concludes the proof. ■

A.10 Tables for sin x, cos x, tan x, e^x, e^{-x}, and ln x

TABLE 1
Natural trigonometric functions

Angle					Angle				
Degree	Radian	Sine	Cosine	Tangent	Degree	Radian	Sine	Cosine	Tangent
0°	0.000	0.000	1.000	0.000					
1°	0.017	0.017	1.000	0.017	46°	0.803	0.719	0.695	1.036
2°	0.035	0.035	0.999	0.035	47°	0.820	0.731	0.682	1.072
3°	0.052	0.052	0.999	0.052	48°	0.838	0.743	0.669	1.111
4°	0.070	0.070	0.998	0.070	49°	0.855	0.755	0.656	1.150
5°	0.087	0.087	0.996	0.087	50°	0.873	0.766	0.643	1.192
6°	0.105	0.105	0.995	0.105	51°	0.890	0.777	0.629	1.235
7°	0.122	0.122	0.993	0.123	52°	0.908	0.788	0.616	1.280
8°	0.140	0.139	0.990	0.141	53°	0.925	0.799	0.602	1.327
9°	0.157	0.156	0.988	0.158	54°	0.942	0.809	0.588	1.376
10°	0.175	0.174	0.985	0.176	55°	0.960	0.819	0.574	1.428
11°	0.192	0.191	0.982	0.194	56°	0.977	0.829	0.559	1.483
12°	0.209	0.208	0.978	0.213	57°	0.995	0.839	0.545	1.540
13°	0.227	0.225	0.974	0.231	58°	1.012	0.848	0.530	1.600
14°	0.244	0.242	0.970	0.249	59°	1.030	0.857	0.515	1.664
15°	0.262	0.259	0.966	0.268	60°	1.047	0.866	0.500	1.732
16°	0.279	0.276	0.961	0.287	61°	1.065	0.875	0.485	1.804
17°	0.297	0.292	0.956	0.306	62°	1.082	0.883	0.469	1.881
18°	0.314	0.309	0.951	0.325	63°	1.100	0.891	0.454	1.963
19°	0.332	0.326	0.946	0.344	64°	1.117	0.899	0.438	2.050
20°	0.349	0.342	0.940	0.364	65°	1.134	0.906	0.423	2.145
21°	0.367	0.358	0.934	0.384	66°	1.152	0.914	0.407	2.246
22°	0.384	0.375	0.927	0.404	67°	1.169	0.921	0.391	2.356
23°	0.401	0.391	0.921	0.424	68°	1.187	0.927	0.375	2.475
24°	0.419	0.407	0.914	0.445	69°	1.204	0.934	0.358	2.605
25°	0.436	0.423	0.906	0.466	70°	1.222	0.940	0.342	2.748
26°	0.454	0.438	0.899	0.488	71°	1.239	0.946	0.326	2.904
27°	0.471	0.454	0.891	0.510	72°	1.257	0.951	0.309	3.078
28°	0.489	0.469	0.883	0.532	73°	1.274	0.956	0.292	3.271
29°	0.506	0.485	0.875	0.554	74°	1.292	0.961	0.276	3.487
30°	0.524	0.500	0.866	0.577	75°	1.309	0.966	0.259	3.732
31°	0.541	0.515	0.857	0.601	76°	1.326	0.970	0.242	4.011
32°	0.559	0.530	0.848	0.625	77°	1.344	0.974	0.225	4.332
33°	0.576	0.545	0.839	0.649	78°	1.361	0.978	0.208	4.705
34°	0.593	0.559	0.829	0.675	79°	1.379	0.982	0.191	5.145
35°	0.611	0.574	0.819	0.700	80°	1.396	0.985	0.174	5.671
36°	0.628	0.588	0.809	0.727	81°	1.414	0.988	0.156	6.314
37°	0.646	0.602	0.799	0.754	82°	1.431	0.990	0.139	7.115
38°	0.663	0.616	0.788	0.781	83°	1.449	0.993	0.122	8.144
39°	0.681	0.629	0.777	0.810	84°	1.466	0.995	0.105	9.514
40°	0.698	0.643	0.766	0.839	85°	1.484	0.996	0.087	11.43
41°	0.716	0.656	0.755	0.869	86°	1.501	0.998	0.070	14.30
42°	0.733	0.669	0.743	0.900	87°	1.518	0.999	0.052	19.08
43°	0.750	0.682	0.731	0.933	88°	1.536	0.999	0.035	28.64
44°	0.768	0.695	0.719	0.966	89°	1.553	1.000	0.017	57.29
45°	0.785	0.707	0.707	1.000	90°	1.571	1.000	0.000	

TABLE 2
Exponential functions

x	e^x	e^{-x}	x	e^x	e^{-x}
0.00	1.0000	1.0000	2.5	12.182	0.0821
0.05	1.0513	0.9512	2.6	13.464	0.0743
0.10	1.1052	0.9048	2.7	14.880	0.0672
0.15	1.1618	0.8607	2.8	16.445	0.0608
0.20	1.2214	0.8187	2.9	18.174	0.0550
0.25	1.2840	0.7788	3.0	20.086	0.0498
0.30	1.3499	0.7408	3.1	22.198	0.0450
0.35	1.4191	0.7047	3.2	24.533	0.0408
0.40	1.4918	0.6703	3.3	27.113	0.0369
0.45	1.5683	0.6376	3.4	29.964	0.0334
0.50	1.6487	0.6065	3.5	33.115	0.0302
0.55	1.7333	0.5769	3.6	36.598	0.0273
0.60	1.8221	0.5488	3.7	40.447	0.0247
0.65	1.9155	0.5220	3.8	44.701	0.0224
0.70	2.0138	0.4966	3.9	49.402	0.0202
0.75	2.1170	0.4724	4.0	54.598	0.0183
0.80	2.2255	0.4493	4.1	60.340	0.0166
0.85	2.3396	0.4274	4.2	66.686	0.0150
0.90	2.4596	0.4066	4.3	73.700	0.0136
0.95	2.5857	0.3867	4.4	81.451	0.0123
1.0	2.7183	0.3679	4.5	90.017	0.0111
1.1	3.0042	0.3329	4.6	99.484	0.0101
1.2	3.3201	0.3012	4.7	109.95	0.0091
1.3	3.6693	0.2725	4.8	121.51	0.0082
1.4	4.0552	0.2466	4.9	134.29	0.0074
1.5	4.4817	0.2231	5	148.41	0.0067
1.6	4.9530	0.2019	6	403.43	0.0025
1.7	5.4739	0.1827	7	1096.6	0.0009
1.8	6.0496	0.1653	8	2981.0	0.0003
1.9	6.6859	0.1496	9	8103.1	0.0001
2.0	7.3891	0.1353	10	22026	0.00005
2.1	8.1662	0.1225			
2.2	9.0250	0.1108			
2.3	9.9742	0.1003			
2.4	11.023	0.0907			

TABLE 3
Natural logarithms

x	$\log_e x$	x	$\log_e x$	x	$\log_e x$
0.0	*	4.5	1.5041	9.0	2.1972
0.1	7.6974	4.6	1.5261	9.1	2.2083
0.2	8.3906	4.7	1.5476	9.2	2.2192
0.3	8.7960	4.8	1.5686	9.3	2.2300
0.4	9.0837	4.9	1.5892	9.4	2.2407
0.5	9.3069	5.0	1.6094	9.5	2.2513
0.6	9.4892	5.1	1.6292	9.6	2.2618
0.7	9.6433	5.2	1.6487	9.7	2.2721
0.8	9.7769	5.3	1.6677	9.8	2.2824
0.9	9.8946	5.4	1.6864	9.9	2.2925
1.0	0.0000	5.5	1.7047	10	2.3026
1.1	0.0953	5.6	1.7228	11	2.3979
1.2	0.1823	5.7	1.7405	12	2.4849
1.3	0.2624	5.8	1.7579	13	2.5649
1.4	0.3365	5.9	1.7750	14	2.6391
1.5	0.4055	6.0	1.7918	15	2.7081
1.6	0.4700	6.1	1.8083	16	2.7726
1.7	0.5306	6.2	1.8245	17	2.8332
1.8	0.5878	6.3	1.8405	18	2.8904
1.9	0.6419	6.4	1.8563	19	2.9444
2.0	0.6931	6.5	1.8718	20	2.9957
2.1	0.7419	6.6	1.8871	25	3.2189
2.2	0.7885	6.7	1.9021	30	3.4012
2.3	0.8329	6.8	1.9169	35	3.5553
2.4	0.8755	6.9	1.9315	40	3.6889
2.5	0.9163	7.0	1.9459	45	3.8067
2.6	0.9555	7.1	1.9601	50	3.9120
2.7	0.9933	7.2	1.9741	55	4.0073
2.8	1.0296	7.3	1.9879	60	4.0943
2.9	1.0647	7.4	2.0015	65	4.1744
3.0	1.0986	7.5	2.0149	70	4.2485
3.1	1.1314	7.6	2.0281	75	4.3175
3.2	1.1632	7.7	2.0412	80	4.3820
3.3	1.1939	7.8	2.0541	85	4.4427
3.4	1.2238	7.9	2.0669	90	4.4998
3.5	1.2528	8.0	2.0794	95	4.5539
3.6	1.2809	8.1	2.0919	100	4.6052
3.7	1.3083	8.2	2.1041		
3.8	1.3350	8.3	2.1163		
3.9	1.3610	8.4	2.1282		
4.0	1.3863	8.5	2.1401		
4.1	1.4110	8.6	2.1518		
4.2	1.4351	8.7	2.1633		
4.3	1.4586	8.8	2.1748		
4.4	1.4816	8.9	2.1861		

*Subtract 10 from $\log_e x$ entries for $x < 1.0$.

Answers

Chapter 1

Section 1.1, pp. 8–9

1. $Q = (3, -1), R = (-3, 1), S = (-3, -1)$
3. $Q = (-2, -2), R = (2, 2), S = (2, -2)$
5. $Q = (-1, 1), R = (1, -1), S = (1, 1)$
7. $Q = (1, 0), R = (-1, 0), S = (-1, 0)$
9. $Q = (\sqrt{3}, -\sqrt{3}), R = (-\sqrt{3}, \sqrt{3}), S = (-\sqrt{3}, -\sqrt{3})$
11. $Q = (0, -\pi), R = (0, \pi), S = (0, -\pi)$ **13.** $d = \sqrt{2}$
15. $d = 6$ **17.** $d = \sqrt{a^2 + b^2}$ **19.** Symmetric to y-axis only
21. Symmetric to y-axis only **23.** Symmetric to origin only
25. Symmetric to x-axis, y-axis, and origin

27. x-intercept $= -1$
y-intercept $= 1$

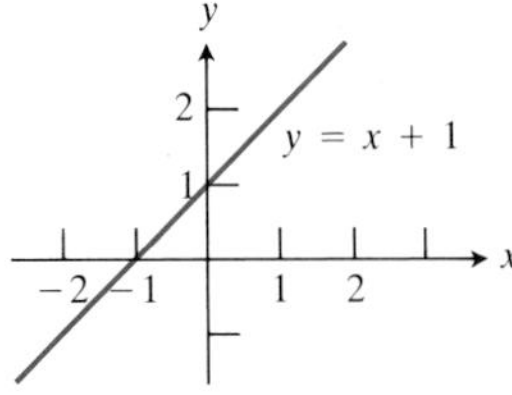

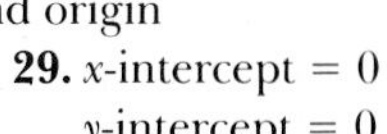
29. x-intercept $= 0$
y-intercept $= 0$

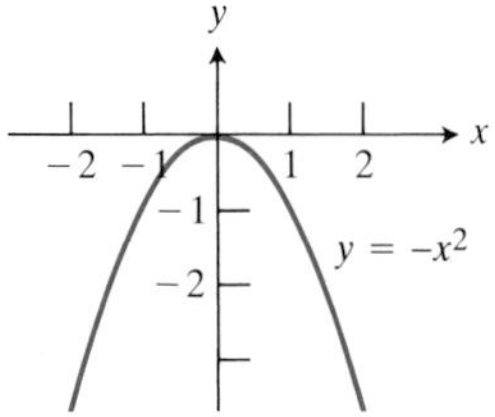

31. x-intercept $= 0$
y-intercept $= 0$

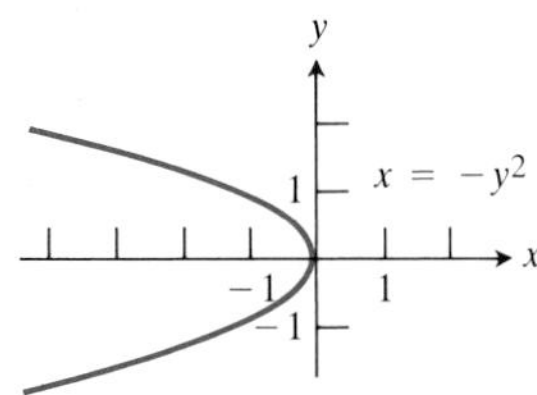

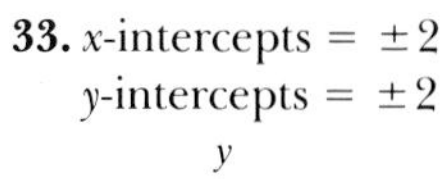
33. x-intercepts $= \pm 2$
y-intercepts $= \pm 2$

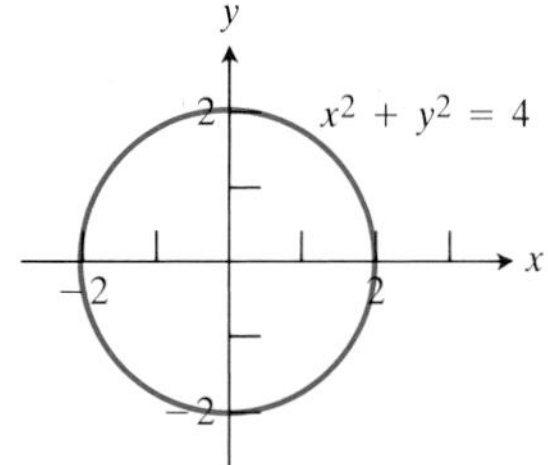

35. x-intercept $= 0$
y-intercept $= 0$

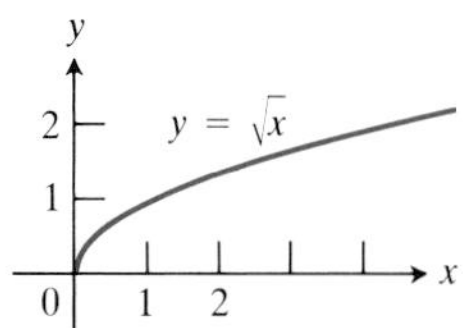

37. A 45° angle
39. $A = (-12, 2)$,
$B = (-12, -5)$,
$C = (9, -5)$
41. $b = 2$ **43.** Yes

Section 1.2, pp. 18–19

1. $\Delta x = 2$, $\Delta y = 1$ **3.** $\Delta x = 2$, $\Delta y = -4$ **5.** $\Delta x = -5$, $\Delta y = 0$ **7.** $m = -1$, $m_{\perp} = 1$
9. $m = -1/3, m_{\perp} = 3$ **11.** $m = 0$, $m_{\perp}$ does not exist
13. a) $x = 2$ b) $y = 3$ **15.** a) $x = -7$ b) $y = -2$
17. a) $x = 0$ b) $y = -\sqrt{2}$ **19.** $y = x$ **21.** $y = x + 2$
23. $y = 2x + b$ **25.** $y = \dfrac{3}{2}x$ **27.** $x = 1$ **29.** $x = -2$
31. $y = 3x - 2$ **33.** $y = x + \sqrt{2}$ **35.** $y = -5x + 2.5$
37. x-intercept $= 4$, y-intercept $= 3$ **39.** x-intercept $= 3$, y-intercept $= -4$ **41.** x-intercept $= -2$, y-intercept $= 4$
43. $y = x, d = \sqrt{2}$ **45.** $y = 2x, d = 2\sqrt{5}/5$ **47.** $y = x + 3$, $d = 3\sqrt{2}$ **49.** $y = x - 1$ **51.** $y = -2x + 2$
53. $(3, -3)$ **55.** $(-2, -9)$ **57.** Answers may vary slightly.
a) $-2.5°$/in. b) $-16.1°$/in. c) $-8.3°$/in. **59.** 5.97 atm at 50 m
61. $-40°$ **63.** $(-1, 4), (5, 2), (-1, -2)$ **65.** $-2, 2$

Section 1.3, pp. 30–31

1. D: $x \geq 0$, R: $y \geq 0$ **3.** D: $x \geq 0$, R: $y \leq 0$ **5.** D: $x \geq -4$, R: $y \geq 0$
7. D: $\{x \mid x \neq 2\}$
R: $\{y \mid y \neq 0\}$
9. Odd **11.** Odd **13.** Neither **15.** Even
17. Even **19.** Odd

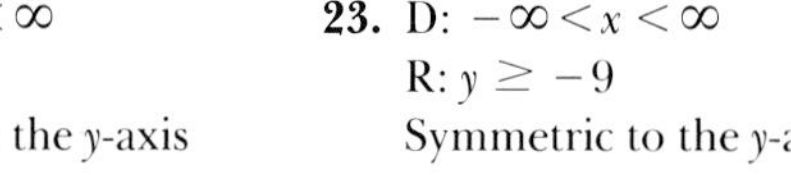
21. D: $-\infty < x < \infty$
R: $y \geq 0$
Symmetric to the y-axis

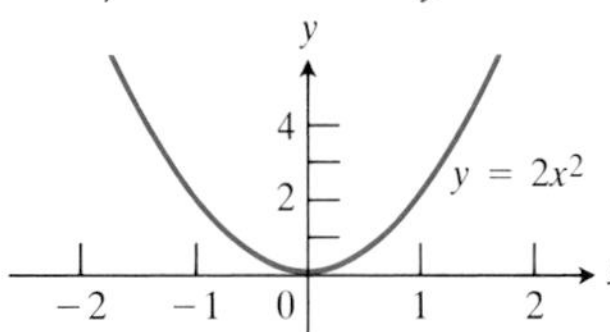

23. D: $-\infty < x < \infty$
R: $y \geq -9$
Symmetric to the y-axis

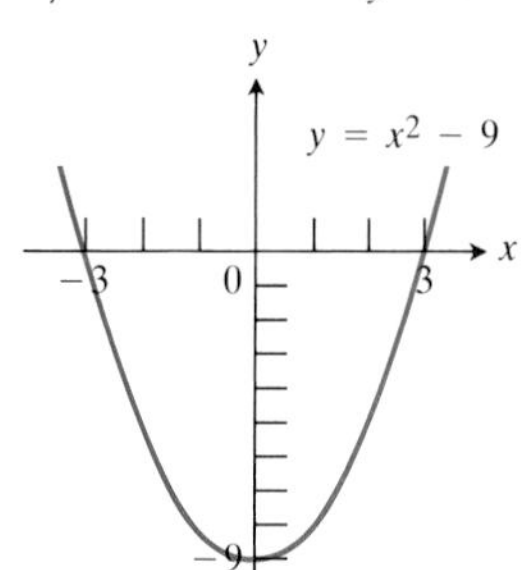

25. D: $-\infty < x < \infty$
R: $-\infty < y < \infty$
Symmetric to origin

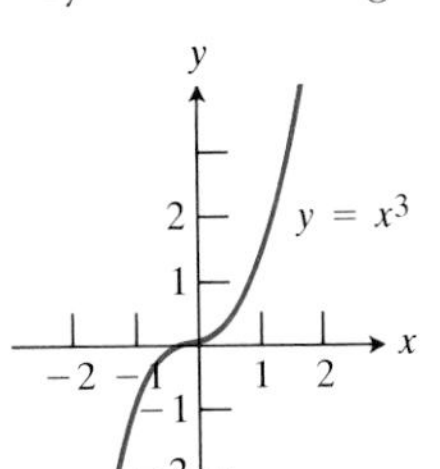

27. D: $x \geq -1$
R: $y \geq 0$
No symmetry

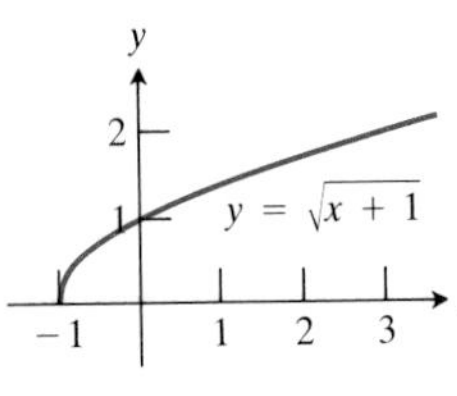

29. D: $\{x \mid x \neq 0\}$
R: $\{y \mid y \neq 0\}$
Symmetric to origin

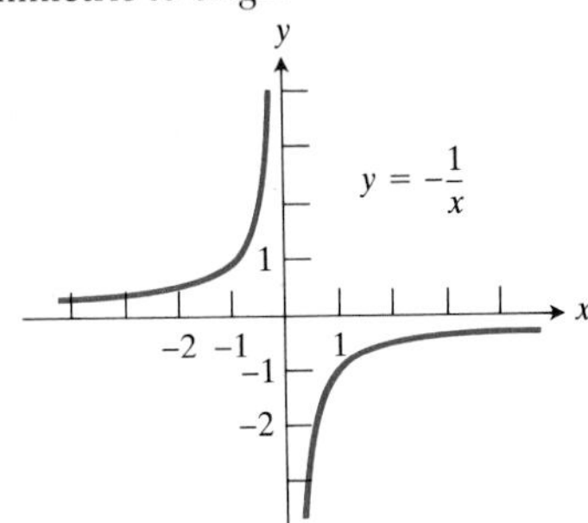

31. D: $\{x \mid x \neq 0\}$
R: $\{y \mid y \neq 1\}$
No symmetry

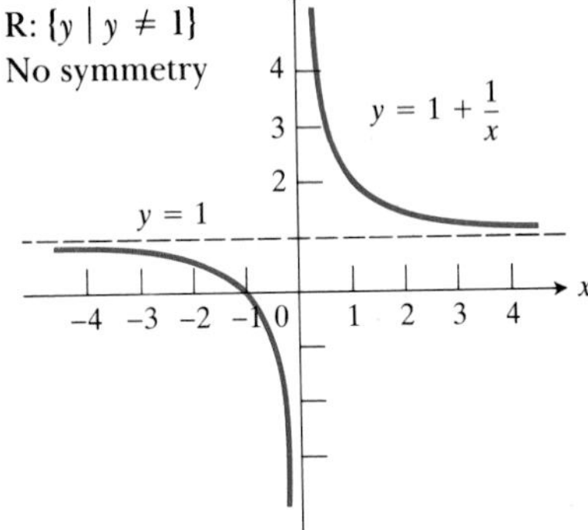

33. a) No b) No c) D: $x > 0$
35. a) i b) iv
37. a) $0 \leq x < 1$ b) $-1 < x \leq 0$
39. a)

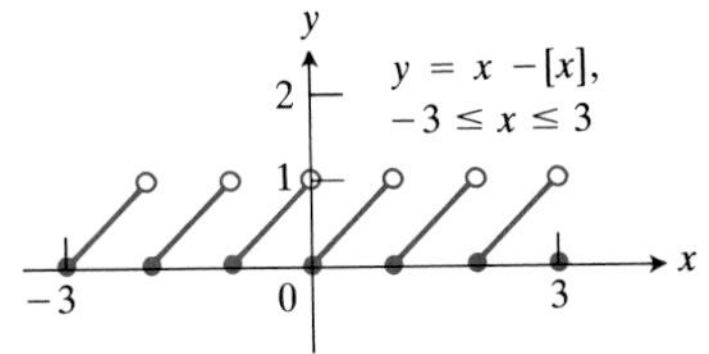

b) $y = \lfloor x \rfloor - \lceil x \rceil$

41.

x	0	1	2
y	0	1	0

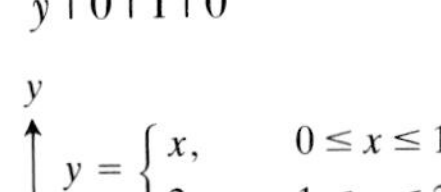

43.

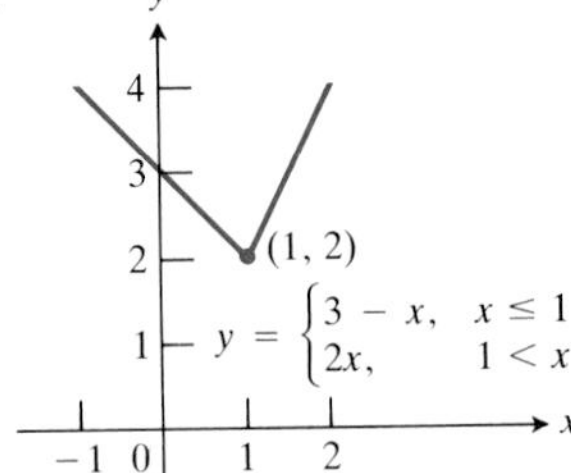

45.

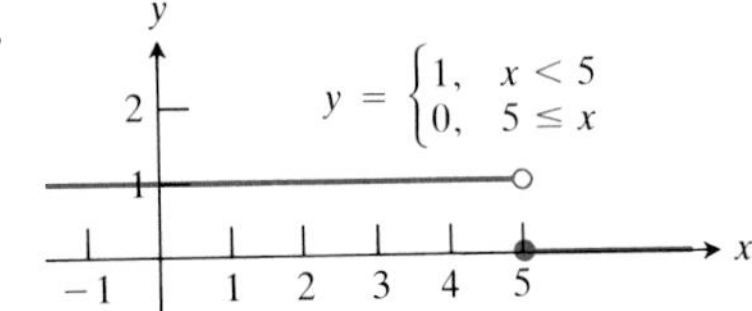

47. a) $y = \begin{cases} x, & 0 \leq x \leq 1 \\ 2 - x, & 1 \leq x \leq 2 \end{cases}$

b) $y = \begin{cases} 2, & 0 \leq x < 1 \text{ or } 2 \leq x < 3 \\ 0, & 1 \leq x < 2 \text{ or } 3 \leq x \leq 4 \end{cases}$

49. D_f: $-\infty < x < \infty$ D_g: $x \geq 1$
$D_{f+g} = D_{f-g} = D_{fg} = D_{g/f}$: $x \geq 1$ $D_{f/g}$: $x > 1$

51. a) 2 b) 22 c) $x^2 + 2$ d) $x^2 + 10x + 22$ e) 5 f) -2
g) $x + 10$ h) $x^4 - 6x^2 + 6$

53.

	$g(x)$	$f(x)$	$f \circ g(x)$
a)	$x - 7$	$\sqrt{x}$	$\sqrt{x - 7}$
b)	$x + 2$	$3x$	$3x + 6$
c)	x^2	$\sqrt{x - 5}$	$\sqrt{x^2 - 5}$
d)	$\dfrac{x}{x - 1}$	$\dfrac{x}{x - 1}$	x
e)	$\dfrac{1}{x - 1}$	$1 + \dfrac{1}{x}$	x
f)	$\dfrac{1}{x}$	$\dfrac{1}{x}$	x

Section 1.4, pp. 40–42

1. a) $y = (x + 4)^2$ b) $y = (x - 7)^2$
3. a) Position 4 b) Position 1 c) Position 2 d) Position 3
5.

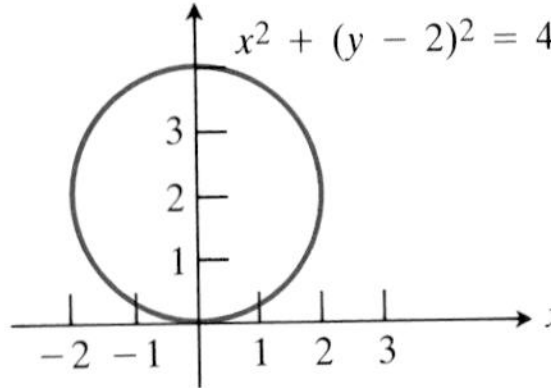

7.

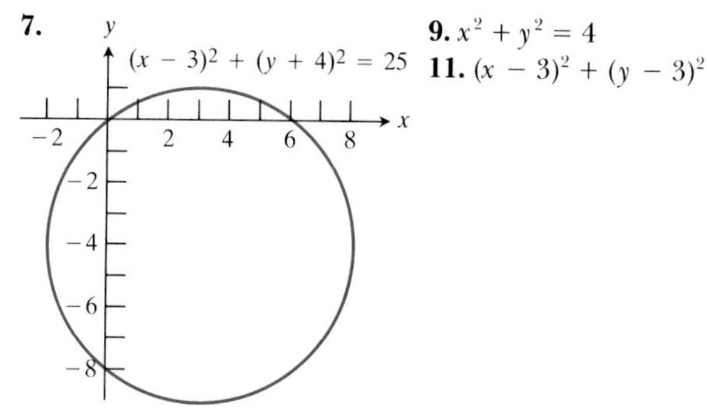

9. $x^2 + y^2 = 4$ **11.** $(x - 3)^2 + (y - 3)^2 = 9$

13. a) Exterior of circle with center (0, 0) and radius 1
b) Interior of circle with center (0, 0) and radius 2
c) Interior of concentric ring centered at (0, 0) between circles in a) and b)

15. $(x + 2)^2 + (y + 1)^2 < 6$ **17.** $y = (1/16)x^2$

19. $y = (-1/12)x^2$

21. Focus: $(0, 1/16)$
Directrix: $y = -1/16$

23. Focus: $(0, -1/12)$
Directrix: $y = 1/12$

25. $y + 3 = (x + 2)^2$ **27.** $y = x^2$ **29.** $y = \sqrt{x + 4}$

31. $y = \frac{1}{2}x$ **33.** $y = \sqrt{-(x - 9)}$ **35.** $y - y_0 = m(x - x_0)$

37. $(-1, 1), (1, 1)$ **39.** $(4, 2), (1, -1)$

Section 1.5, pp. 54–55

1. $2\pi/3$ **3.** $3\pi/2$ **5.** $-\pi/4$ **7.** $-\pi$ **9.** 30° **11.** 225°
13. −270° **15.** −630°

	Angle	sin	cos	tan	cot	sec	csc
17. a)	$\frac{\pi}{6}$	$\frac{1}{2}$	$\frac{\sqrt{3}}{2}$	$\frac{1}{\sqrt{3}}$	$\sqrt{3}$	$\frac{2}{\sqrt{3}}$	2
b)	$-\frac{\pi}{6}$	$-\frac{1}{2}$	$\frac{\sqrt{3}}{2}$	$-\frac{1}{\sqrt{3}}$	$-\sqrt{3}$	$\frac{2}{\sqrt{3}}$	−2
19. a)	$\frac{\pi}{3}$	$\frac{\sqrt{3}}{2}$	$\frac{1}{2}$	$\sqrt{3}$	$\frac{1}{\sqrt{3}}$	2	$\frac{2}{\sqrt{3}}$
b)	$-\frac{\pi}{3}$	$-\frac{\sqrt{3}}{2}$	$\frac{1}{2}$	$-\sqrt{3}$	$-\frac{1}{\sqrt{3}}$	2	$-\frac{2}{\sqrt{3}}$
21. a)	$\frac{3\pi}{4}$	$\frac{\sqrt{2}}{2}$	$-\frac{\sqrt{2}}{2}$	−1	−1	$-\sqrt{2}$	$\sqrt{2}$
b)	$-\frac{3\pi}{4}$	$-\frac{\sqrt{2}}{2}$	$-\frac{\sqrt{2}}{2}$	1	1	$-\sqrt{2}$	$-\sqrt{2}$
23. a)	0	0	1	0	und.	1	und.
b)	π	0	−1	0	und.	−1	und.

25.

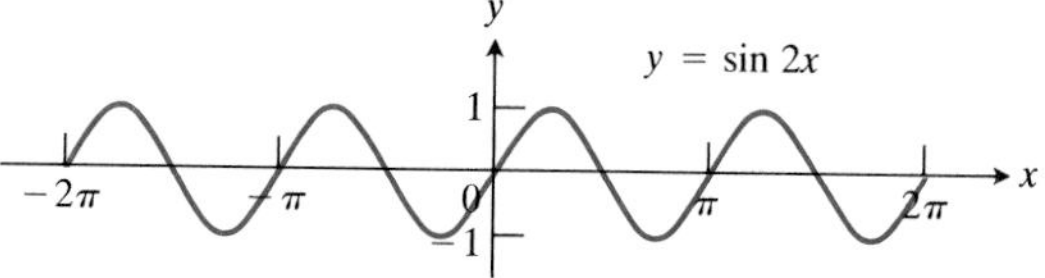

27.

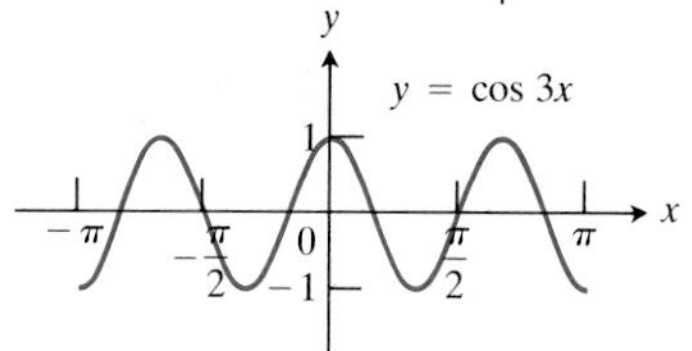

29.

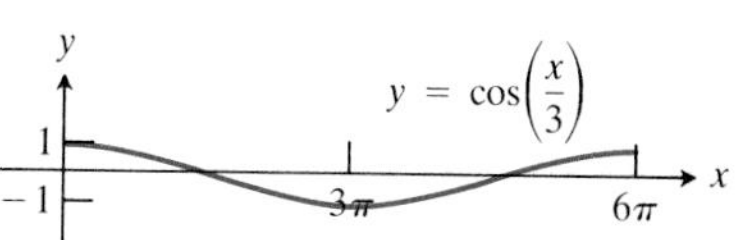

31.

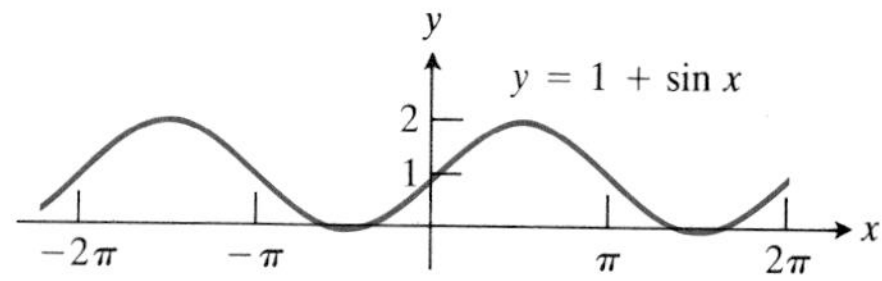

33.

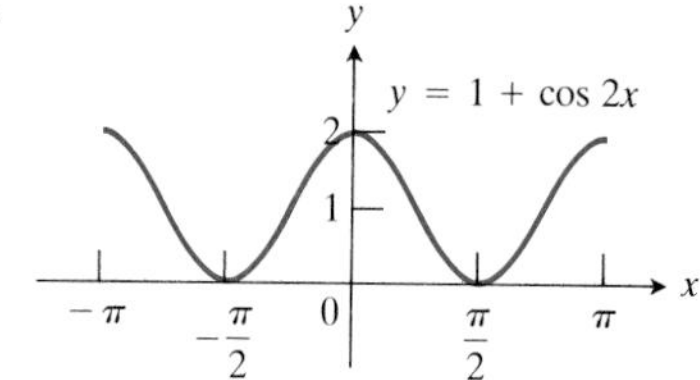

35.

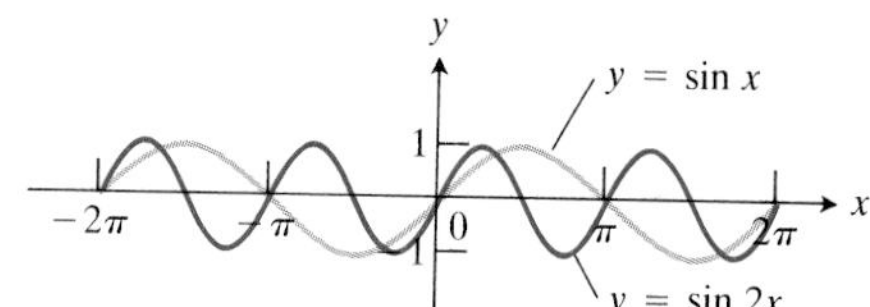

37.

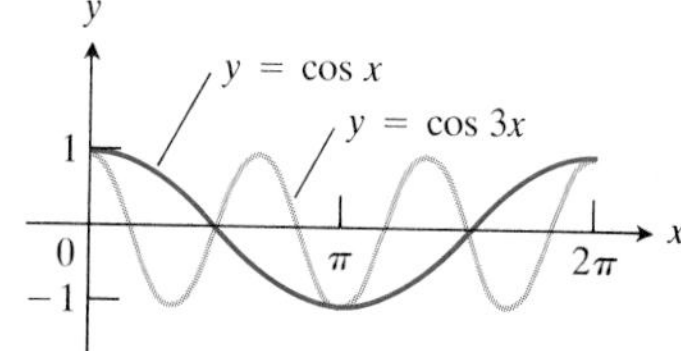

39.

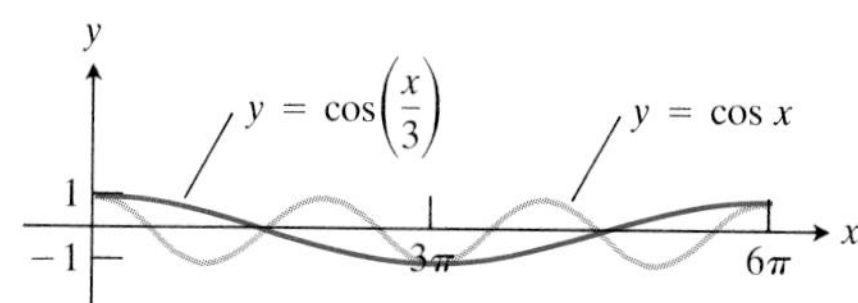

41.

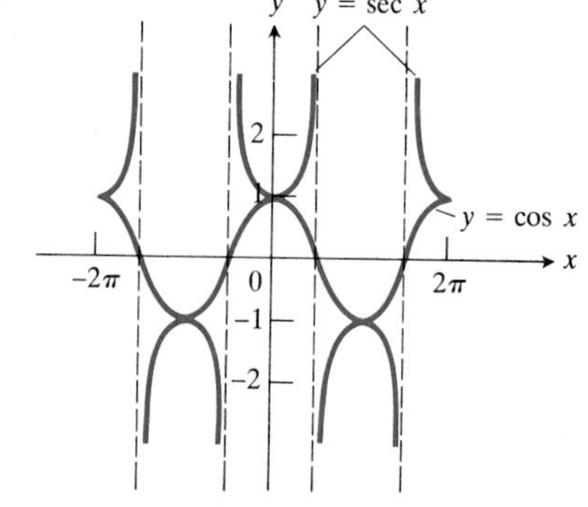

43. a, j; b, d, g; e, c, i, m, n; f, k, l, h **45.** Cos: to y-axis only, sin: to origin only, tan: to origin only
47. a) Yes b) 1 and -1 c) 1 and 0 d) Domain: $-\infty < x < \infty$, Range: $0 \le y \le 1$ **49.** a) 37 b) 365 c) Right 100 d) Up 25
51. If $A = B$, then $\cos(0) = 1$. **53.** If $B = \pi/2$, then $\cos(A + \pi/2) = -\sin A$ and $\sin(A + \pi/2) = \cos A$.
55. $(\sqrt{6} + \sqrt{2})/4$ **57.** $(\sqrt{2} + \sqrt{6})/4$ **59.** $(2 + \sqrt{2})/4$
61. $(2 - \sqrt{3})/4$

Section 1.6, pp. 63–65

1. a) False b) True c) True d) True e) True f) True g) True h) True **3.** 3 **5.** 5 **7.** ± 2 **9.** $-1/2, -9/2$
11. $-1/3, 17/3$ **13.** g **15.** e **17.** h **19.** b **21.** $-2 < y < 2$
23. $-1 \le y \le 3$ **25.** $5/3 < y < 3$ **27.** $-30 \le y \le 30$
29. $0 \le y \le 4$ **31.** $11/10 > y > 9/10$ **33.** $|x - 6| < 3$
35. $|x + 1| < 4$ **37.** $\sqrt{99.9} < x < \sqrt{100.1}$ **39.** $22.21 < x < 23.81$ **41.** $30 > x > 20$ **43.** $|x - 3| < .5$ **45.** $|x - 6| < 1$
47. $|x - 3.385| \le .002$ **49.** a) iii b) iv c) i d) ii
51. **53.**

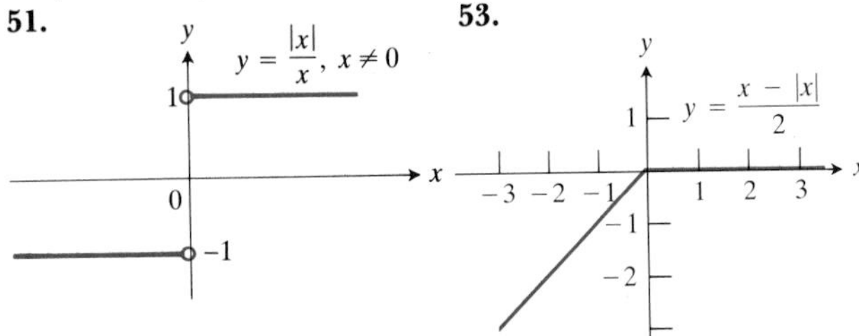

55. a) a any negative b) $a \ge 0$ **57.** Domains: $-\infty < x < \infty$ versus $x \ge 0$, Ranges: Both are $y \ge 0$ **59.** $g(x) = \sqrt{x}$
61.

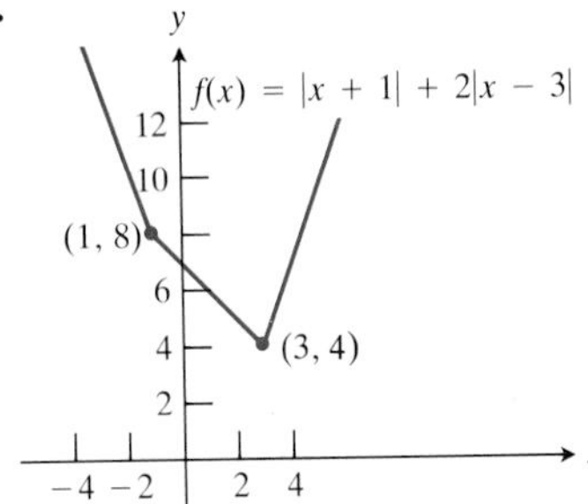

63.

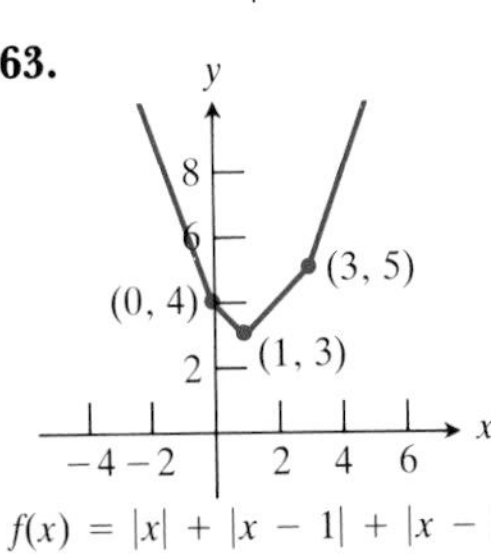

65. $d = 7$ at $x = 2$

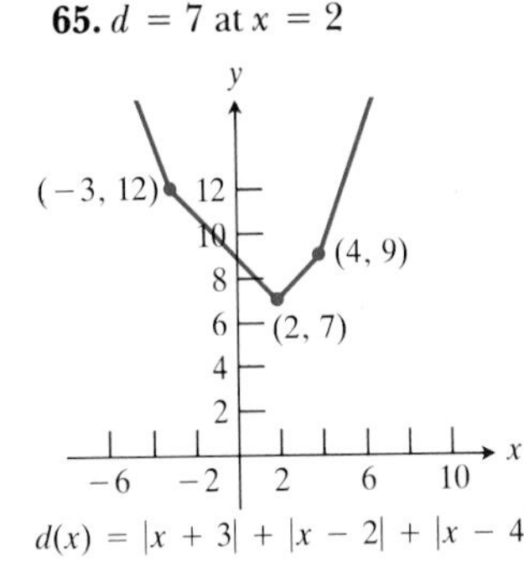

67. $d = 17$ at $x = 2$

Section 1.7, pp. 70–72

1. a) Answers vary b) $e^{-1} = 0.367879441$, $e^{-10} = 0.000045399$, $e^{-100} = 3.7200759 \times 10^{-44}$, $e^{-1000} = 0$ (answers vary) **3.** Answers vary, $x^* = \pi$ **5.** $y = x$ **7.** $x \ne 1$
9. $x \ne n\pi/2$, n an odd integer **11.** $x > 0$
13. a) $\sqrt{3} = 1.732050808$
$\sqrt{1.732050808} = 1.316074013$
$\sqrt{1.316074013} = 1.14720269$, etc.
b) $\sqrt{5} = 2.236067978$
$\sqrt{2.236067978} = 1.495348781$
$\sqrt{1.495348781} = 1.222844545$, etc.
15. $\sqrt{2} = 1.071773463$
$\sqrt{1.070773463} = 1.00695555$
$\sqrt{1.00695555} = 1.000693387$
$\sqrt{1.000693387} = 1.000069317$, etc.
17. $x_0 = 1$
$x_1 = 1.540302306$
$x_2 = 1.570791601$
$x_3 = 1.570796327$
$x_4 = 1.570796327$, etc.

19.

Δx	$\dfrac{\sin \Delta x}{\Delta x}$
0.1	0.017453283
0.01	0.017453292
0.001	0.017453292, etc.

$m_{\text{sec}} = 1.000000845$

21. 0.8391 **23.** Min: 17.35°, Max: 48.37°

Chapter 1 Practice Exercises, pp. 73–76

1. a) $(1, -4)$ b) $(-1, 4)$ c) $(-1, -4)$ **3.** a) $(-4, -2)$ b) $(4, 2)$ c) $(4, -2)$ **5.** a) Origin b) y-axis **7.** a) Symmetric to all b) None **9.** $x = 1$; $y = 3$ **11.** $x = 0$; $y = -3$ **13.** $y = 2x - 1$
15. $y = -x + 1$ **17.** $y = 3x - 9$ **19.** $y = (-x/2) + (3/2)$
21. $y = (5x/3) + (4/3)$ **23.** $y = (5x/2) - 6$ **25.** $y = (\frac{1}{2})x + 2$
27. $y = -2x - 1$ **29.** a) $2x - y = 12$ b) $x + 2y = 6$, $d = 14\sqrt{5}/5$ **31.** a) $4x + 3y = -20$ b) $3x - 4y = 60$, $d = 32/5$

33. D: $-\infty < x < \infty$
R: $y \ge 0$

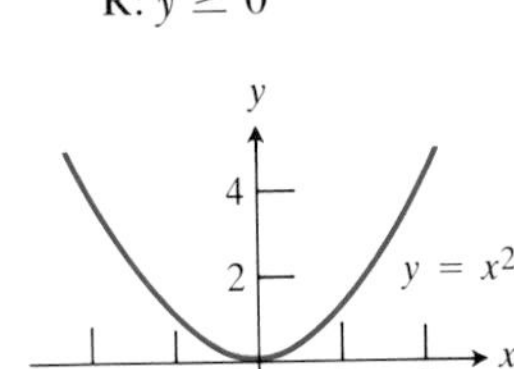

35. D: $-\infty < x < \infty$
R: $y \ge 0$

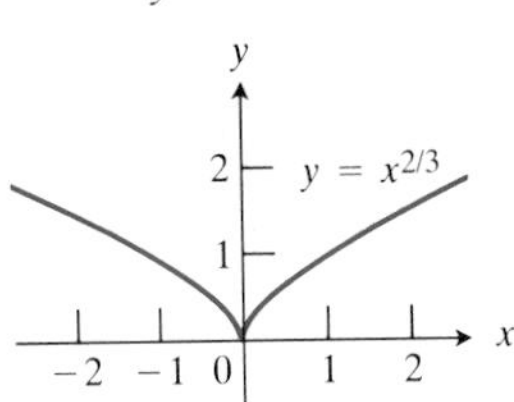

37. D: $-\infty < x < \infty$
R: $-\infty < y < \infty$

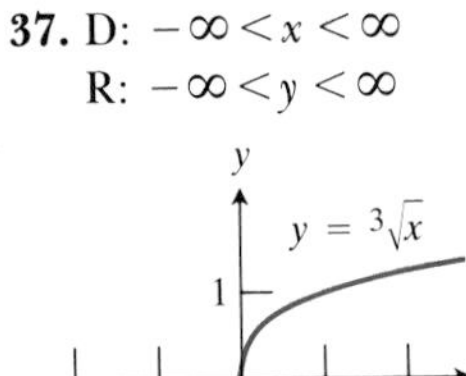
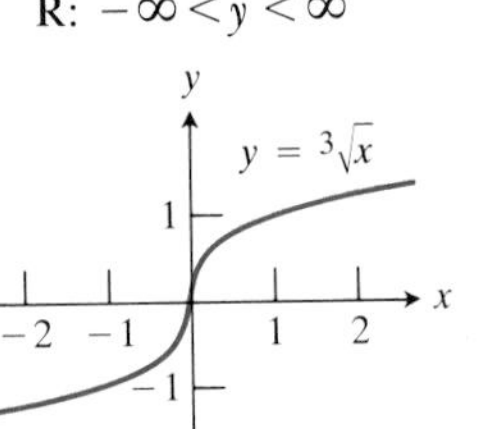

39. D: $\{x \mid x \ne 0\}$
R: $y > 0$

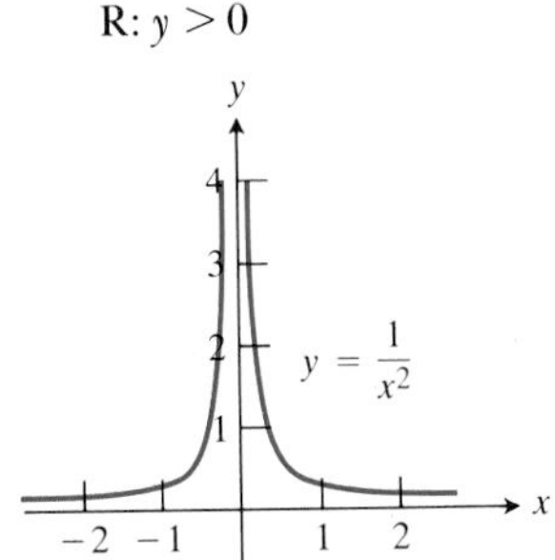

41. D: $-\infty < x < \infty$
R: $\{y \mid y \text{ is an integer}\}$

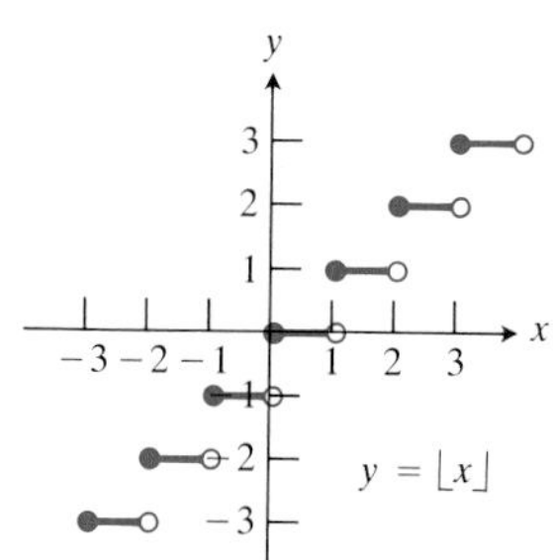

43. D: $-\infty < x < \infty$
R: $-1 \le y \le 1$

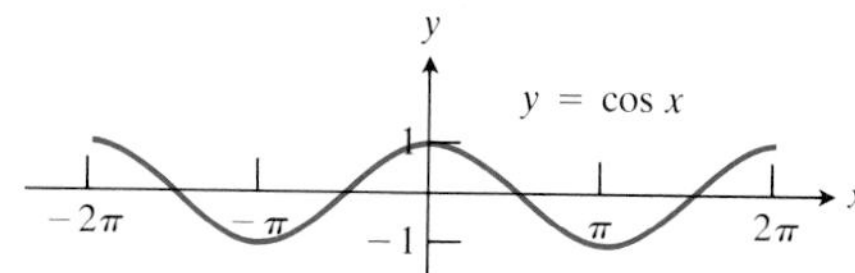

45. D: $\{x \mid x \ne n\pi/2, n \text{ an odd integer}\}$
R: $-\infty < y < \infty$

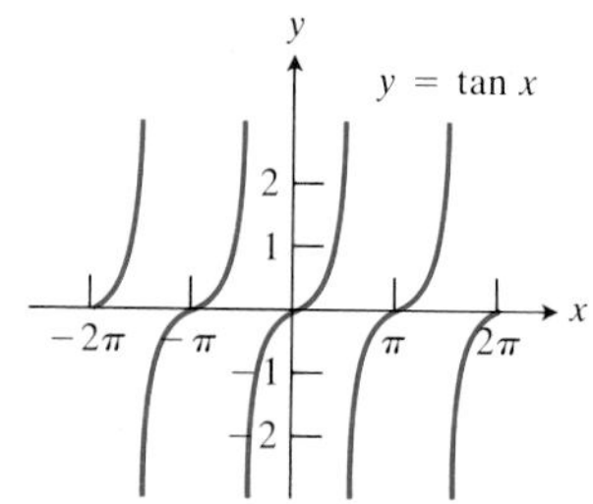

47. D: $\{x \mid x \ne n\pi/2, n \text{ an odd integer}\}$
R: $(-\infty, -1] \cup [1, \infty)$

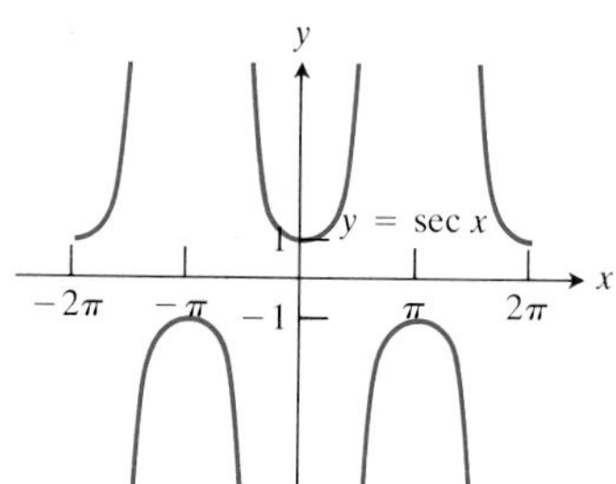

49. a) Even b) Even c) Even **51.** a) Even b) Odd c) Odd
53. a) Even b) Odd c) Odd **55.** Yes, 1

57.

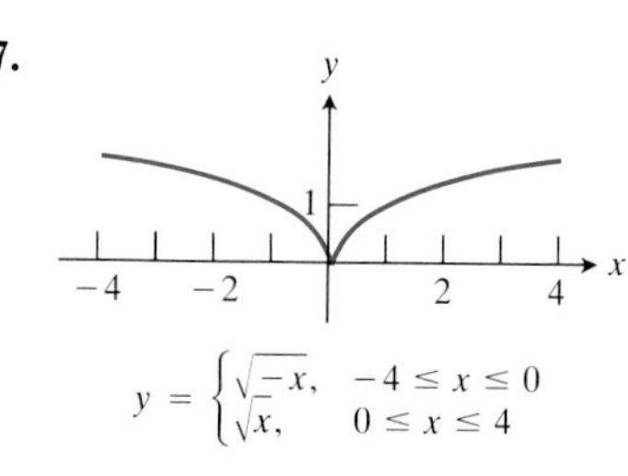

59.

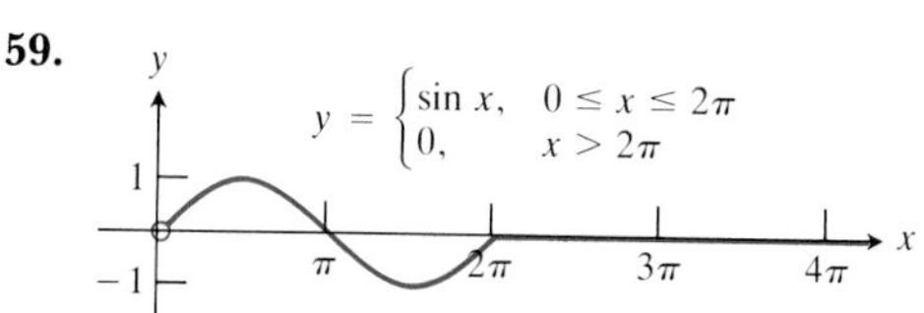

61. $y = \begin{cases} -x + 1, & 0 \le x < 1 \\ -x + 2, & 1 \le x \le 2 \end{cases}$

63. $D_f: x \ne 0 \quad R_f: y \ne 0$
$D_g: x > 0 \quad R_g: y > 0$
$D_{f+g} = D_{fg} = D_{f/g} = D_{g/f}: x > 0$
$R_{f+g} = R_{fg} = R_{f/g} = R_{g/f}: y > 0$
$D_{f\circ g} = D_{g\circ f}: x > 0 \quad R_{f\circ g} = R_{g\circ f}: y > 0$

65. a) $y + 1 = (x - 2)^2$ b) $y = (x + 2)^2$
67. a) $y = \sin(x - \pi/2)$ b) $y - 1 = \sin(x - \pi)$
69. a) $y = \sqrt{x - 2}$ b) $y = \sqrt{x + 2}$
71. $(x - 1)^2 + (y - 1)^2 = 1$ **73.** $(x - 2)^2 + (y + 3)^2 = \frac{1}{4}$
75. $(3, -5)$; 4 **77.** $(-1, 7)$; 11 **79.** a) $x^2 + y^2 < 1$
b) $x^2 + y^2 \le 1$ **81.** $y = x^2/8$ **83.** $y = -x^2$
85. $(0, \frac{1}{2})$; $y = -\frac{1}{2}$ **87.** $(0, -1)$; $y = 1$ **89.** a) $\pi/6$ b) $\pi/4$
c) $-2\pi/3$ d) $-5\pi/6$

91.

	Angle	sin	cos	tan	cot	sec	csc
a)	$\pi/3$	$\sqrt{3}/2$	$1/2$	$\sqrt{3}$	$1/\sqrt{3}$	2	$2/\sqrt{3}$
b)	$-\pi/3$	$-\sqrt{3}/2$	$1/2$	$-\sqrt{3}$	$-1/\sqrt{3}$	2	$-2/\sqrt{3}$
c)	$2\pi/3$	$\sqrt{3}/2$	$-1/2$	$-\sqrt{3}$	$-1/\sqrt{3}$	-2	$2/\sqrt{3}$
d)	$-2\pi/3$	$-\sqrt{3}/2$	$-1/2$	$\sqrt{3}$	$1/\sqrt{3}$	-2	$-2/\sqrt{3}$

93. a)

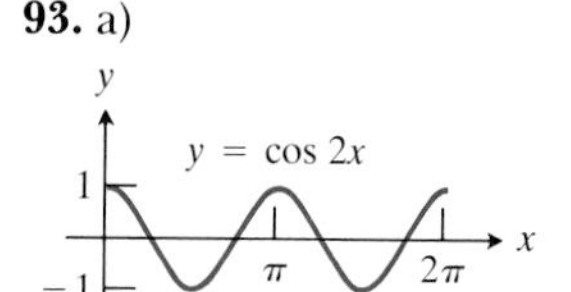

b)

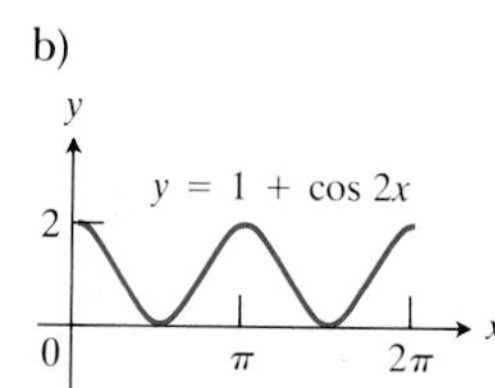

c)

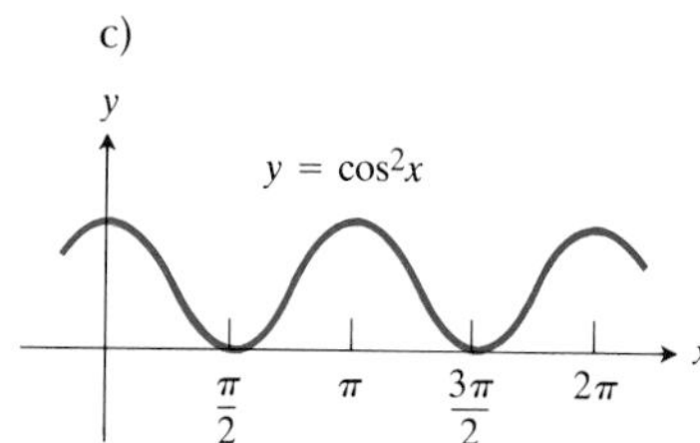

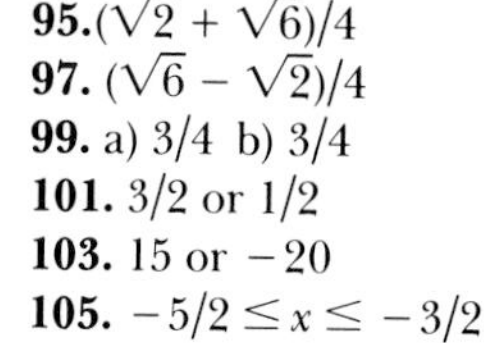

95. $(\sqrt{2} + \sqrt{6})/4$
97. $(\sqrt{6} - \sqrt{2})/4$
99. a) 3/4 b) 3/4
101. 3/2 or 1/2
103. 15 or -20
105. $-5/2 \le x \le -3/2$

107. $-1/5 < y < 1$ **109.** $|x - 7| < 4$ **111.** $|y - 3| < 4$
113. $|x - 2| < 1$ **115.** $|x - 15| < 8$

117.

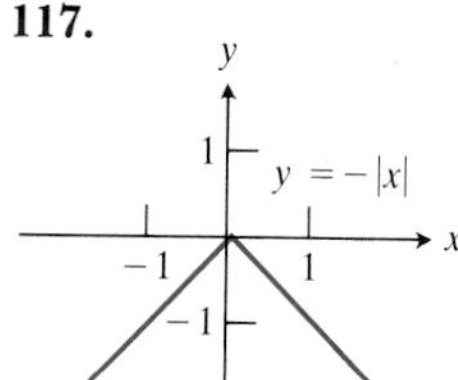

119.

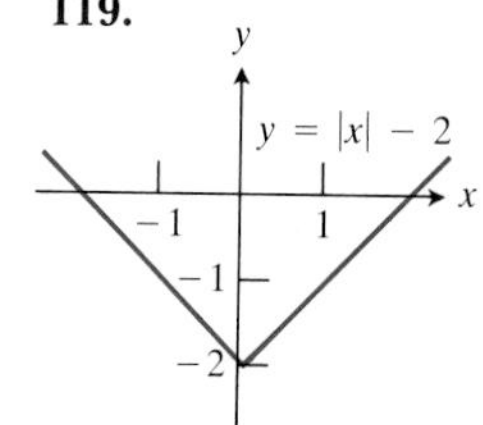

121.

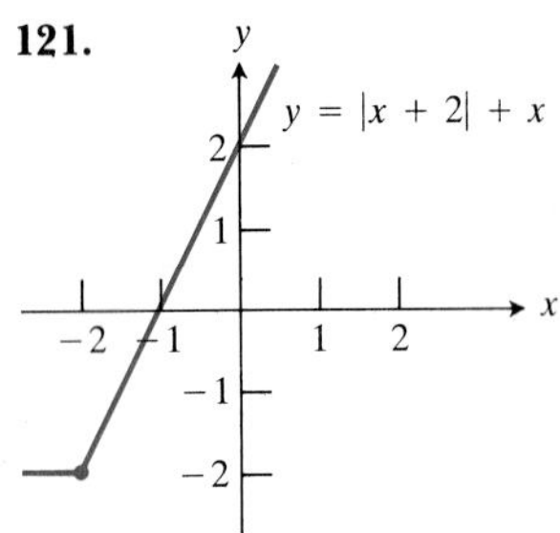

123. a) iv b) i c) ii d) iii **125.** For $E = 0.2$, $1.28 < x < 2.88$
For $E = 0.1$, $1.62 < x < 2.42$

Chapter 2

Section 2.1, pp. 86–88

1. 4 **3.** 4 **5.** 2 **7.** 25 **9.** −5 **11.** 9 **13.** 1 **15.** 45
17. 4 **19.** −15 **21.** 5/8 **23.** −2 **25.** 5 **27.** 0 **29.** 11/4
31. 1/2 **33.** −7 **35.** 3/4 **37.** −1/2 **39.** 1/10 **41.** 1/2
43. 3 **45.** a) $\lim_{x\to 2^+} f(x) = 2$, $\lim_{x\to 2^-} f(x) = 1$
b) No, $\lim_{x\to 2^+} f(x) \neq \lim_{x\to 2^-} f(x)$. **47.** a) False b) True
c) False d) True e) True f) True g) False h) False
i) False j) False
49. a)

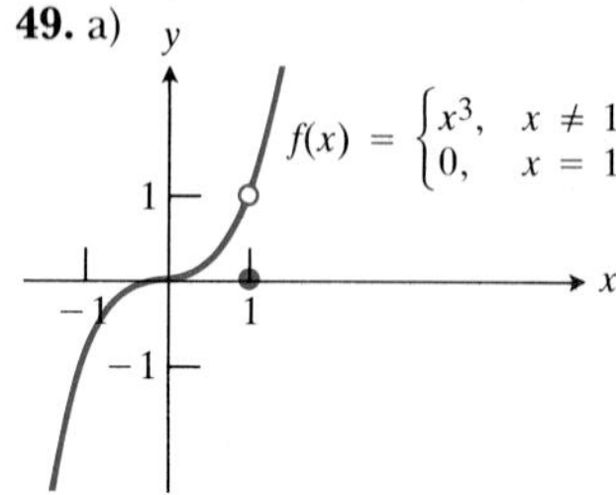

b) $\lim_{x\to 1^-} f(x) = 1$, $\lim_{x\to 1^+} f(x) = 1$
c) Yes, $\lim_{x\to 1^-} f(x) = \lim_{x\to 1^+} f(x) = 1$.
51. a) $(0, 1) \cup (1, 2)$ b) $x = 2$ c) $x = 0$

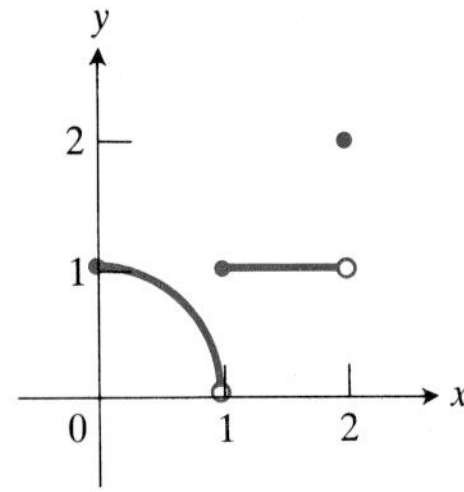

53. a) No b) Yes, 0. c) No, $\lim_{x\to 0^+} f(x)$ does not exist.
55. 0 **57.** 0 **59.** 1 **61.** a) −10 b) −20 **63.** a) 4 b) −21
c) −12 d) −7/3

65. a)

x	$\dfrac{(x^2-1)}{(x-1)}$
1.1	2.1
1.01	2.01
1.001	2.001 etc.

b) 2

67.

x	$\dfrac{\ln(x^2)}{\ln x}$
1.1	2
1.01	2
1.001	2

Section 2.2, pp. 92–93

1. 1 **3.** 1/2 **5.** 1 **7.** 2 **9.** 1 **11.** −1
13. $\lim_{x\to 0}(1 - x^2/6) = 1$, $\lim_{x\to 0} 1 = 1$. Therefore,
$\lim_{x\to 0}(\sin x/x) = 1$.

15. a)

x	$\dfrac{(1-\cos x)}{x^2}$
0.1	0.4995583474
0.01	0.4999958
0.001	0.5

b) $\lim_{x\to 0}(1/2 - x^2/24) = 1/2$, $\lim_{x\to 0}(1/2) = 1/2 \Rightarrow$
$\lim_{x\to 0}(1 - \cos x)/x^2 = 1/2$

Section 2.3, pp. 100–101

1. a) 2/5 b) 2/5 **3.** a) 0 b) 0 **5.** a) −3 b) −3 **7.** a) $+\infty$
b) $-\infty$ **9.** a) 0 b) 0 **11.** a) 7 b) 7 **13.** a) −2/3 b) −2/3
15. a) −1 b) −1 **17.** a) −1 b) −1 **19.** $+\infty$ **21.** $+\infty$
23. $+\infty$ **25.** $+\infty$ **27.** $-\infty$ **29.** 1/2 **31.** 1 **33.** $+\infty$
35. −1/2 **37.** $+\infty, -\infty, -\infty, +\infty$ **39.** $+\infty, -\infty$
41. 0 **43.** 1 **45.** $+\infty$ **47.** $-\infty$ **49.** $\lim_{x\to -\infty} f(x) = 0$,
$\lim_{x\to 0^-} f(x) = -\infty$, $\lim_{x\to 0^+} f(x) = -1$, $\lim_{x\to\infty} f(x) = -1$
51. 2 **53.** 2 **55.** 0 **57.** $\lim_{x\to\infty} f(x) = 2$, $\lim_{x\to -\infty} f(x) = 2$

59. a)

x	$\dfrac{\ln(x+1)}{\ln x}$
10	1.041392685
100	1.002160687
1000	1.000144693 etc.

b)

x	$\dfrac{\ln(x+999)}{\ln x}$
10	3.003891166
100	1.520498846
1000	1.100270931
10000	1.010338301
100000	1.000863415 etc.

c) 2

d)

x	$\dfrac{\ln x}{\log x}$
10	2.302585093
100	2.302585093
1000	2.302585093 etc.

Section 2.4, pp. 111–112

1. a) Yes b) 1, Yes c) Yes d) Yes **3.** a) No b) No **5.** a) 0
b) 0 **7.** All except $x = 2$ **9.** $[-1, 0) \cup (0, 1) \cup (1, 2]$
11. All except $x = 1$ **13.** At all x except 0,1 **15.** $x = 2$
17. $x = -1$ **19.** Continuous everywhere **21.** $x = \pm 1$
23. $x = 0$ **25.** Yes, $\lim_{x\to 1} f(x) = f(1)$. **27.** $h(2) = 7$
29. $g(4) = 8/5$ **31.** 4/3 **33.** 1 **35.** 0 **37.** $x = 2, 3$. No
minimum because $y \to 0$ as $x \to 1^-$, but $y \neq 0$ for any x.

39. No maximum because $y \to 1$ as $x \to 1^-$ or $x \to 1^+$, but $y \neq 1$ for any x in $(-1, 1)$; minimum is 0 at $x = 0$.
41. Intermediate Value Theorem

Section 2.5, pp. 121–123

1. $\delta = 2$ **3.** $\delta = 1/2$ **5.** $\delta = 2$ **7.** $\delta = 1/18$ **9.** $\delta = 0.1$ **11.** $\delta = \sqrt{5} - 2$ **13.** $\delta = 7/16$ **15.** $2.99 < x < 3.01$ **17.** $\sqrt{8.95} < x < \sqrt{9.05}$ **19.** $3.2391 > x > 2.7591$ **21.** $1/3.9 > x > 1/4.1$ **23.** $L = 5$; $\delta = 0.005$ **25.** $L = 0$; $\delta = 0.005$ **27.** $L = 4$; $\delta = 0.05$ **29.** $L = 2$; $\delta = 0.0399$ **31.** $L = 2$; $\delta = 1/3$
33. $\epsilon = 0.01 \Rightarrow \delta = 0.01$
$\epsilon = 0.001 \Rightarrow \delta = 0.001$
$\epsilon = 0.0001 \Rightarrow \delta = 0.0001$
$\delta = \epsilon$
35. $\delta = \epsilon^2$, $\lim_{x \to 5^+} \sqrt{x - 5} = 0$
37. $\delta = \epsilon/6$

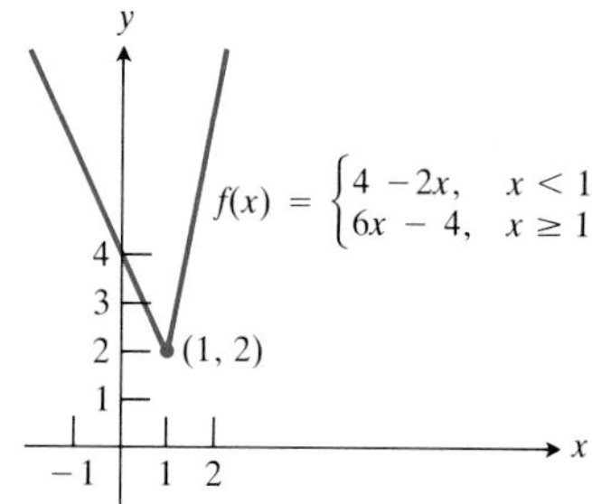

41. $\delta = \sqrt{4 + \epsilon} - 2$. As $\epsilon \to 0$, $\delta \to 0$.

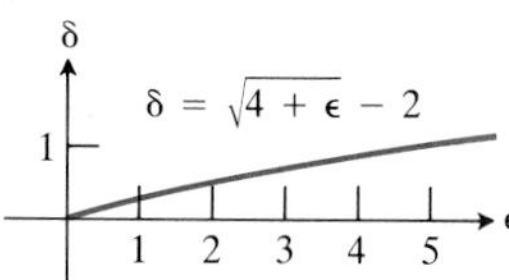

Chapter 2 Practice Exercises, pp. 123–126

1. 1 **3.** -4 **5.** $2/3$ **7.** 0 **9.** -1 **11.** -3 **13.** 2 **15.** $1/5$ **17.** $2/5$ **19.** 0 **21.** $-\infty$ **23.** $+\infty$ **25.** $+\infty$ **27.** $1/2$ **29.** 1 **31.** a) $+\infty$ b) $-\infty$ **33.** a) 1, 1, 0, 0, 1, -1 b) 1, 0, $\lim_{x \to 1} f(x)$ does not exist because $\lim_{x \to 1^-} f(x) \neq \lim_{x \to 1^+} f(x)$.
35. a)

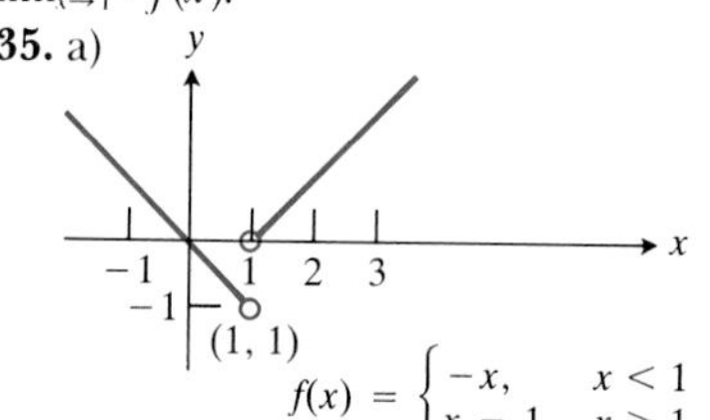

b) 0, -1 c) None

37. a) -21 b) 49 c) 0 d) 1 e) 1 f) 7
39. 0
41. 0
43. 1
45. $k = 8$

47. No, interval is not closed. **49.** Intermediate Value Theorem says since f is continuous on $[1, 2]$ and since $f(1) = 0, f(2) = 3$, f takes on all values between 0 and 3, including 2.5, for some x in $(1, 2)$. **53.** $\lim_{x \to 0} x^2 = 0$. But as $x \to 0$, x^2 gets closer to -0.1. But $L \neq -0.1$. **55.** $\delta \leq \epsilon/2$ **57.** $\delta = 0.01$ **59.** $\delta = 0.01$ **61.** $\delta = 3$ **63.** $\delta = 2/11$

Chapter 3

Section 3.1 pp. 140–141

1. $6x$, $m = 18$, $y = 18x - 27$ **3.** $4x$, $m = 12$, $y = 12x - 23$ **5.** $2x - 6$, $m = 0$, $y = -9$ **7.** $-2/x^2$, $m = -2/9$, $y = (-2/9)x + 4/3$ **9.** $1/(x + 1)^2$, $m = 1/16$, $y = (1/16)x + (9/16)$ **11.** $1 - \frac{9}{x^2}$, $m = 0$, $y = 6$ **13.** $\frac{1}{2\sqrt{x}}$, $m = \frac{1}{2\sqrt{3}}$, $y = \frac{1}{2\sqrt{3}}x + 1 + \frac{\sqrt{3}}{2}$ **15.** $\frac{1}{\sqrt{2x}}$, $m = \frac{1}{\sqrt{6}}$, $y = \frac{1}{\sqrt{6}}x + \frac{3}{\sqrt{6}}$ **17.** $\frac{-1}{2\sqrt{x^3}}$, $m = \frac{-1}{6\sqrt{3}}$, $y = \frac{-1}{6\sqrt{3}}x + \frac{\sqrt{3}}{2}$ **19.** $3x^2$, $m = 27$, $y = 27x - 54$
21.

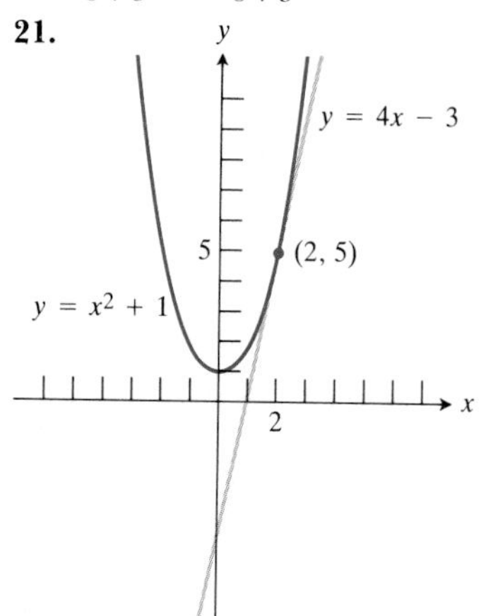

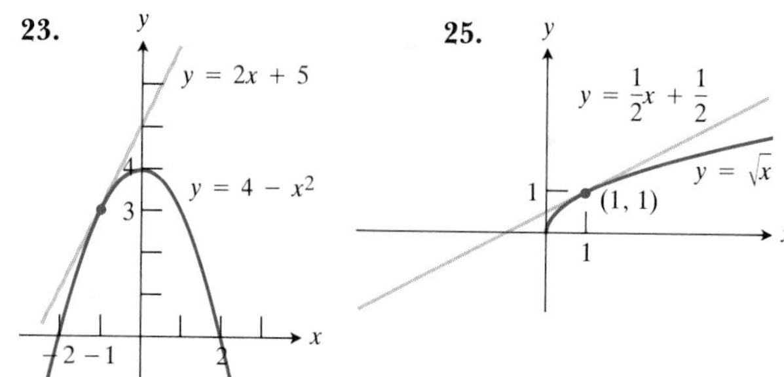

27. a) Yes b) No c) No d) No **29.** a) Yes b) Yes c) Yes d) No **31.** Since $\lim_{x \to 0^+} f'(x) = 1$ while $\lim_{x \to 0^-} f'(x) = 0$, $f(x)$ is not differentiable at $x = 0$. **33.** Since $\lim_{x \to 1^+} f'(x) = 2$ while $\lim_{x \to 1^-} f'(x) = 1/2$, $f(x)$ is not differentiable at $x = 1$.

Section 3.2, pp. 150–151

1. $y' = 1$, $y'' = 0$ **3.** $y' = 2x$, $y'' = 2$
5. $y' = -2x$, $y'' = -2$ **7.** $y' = 2$, $y'' = 0$
9. $y' = x^2 + x + 1$, $y'' = 2x + 1$
11. $y' = 4x^3 - 21x^2 + 4x$, $y'' = 12x^2 - 42x + 4$
13. $y' = 8x - 8$, $y'' = 8$ **15.** $y' = 8x^3 - 8x$, $y'' = 24x^2 - 8$
17. $y' = 2x - 1$, $y'' = 2$
19. $y' = 2x^3 - 3x - 1$, $y'' = 6x^2 - 3$, $y''' = 12x$, $y^{(4)} = 12$
21. a) $3(2x + 5) + 2(3x - 1)$ b) $12x + 13$

23. a) $1(x^2+1)+2x(x+1)$ b) $3x^2+2x+1$
25. a) $2x(x^3-1)+x^2(3x^2)$ b) $5x^4-2x$
27. a) $1(x^2+x+1)+(2x+1)(x-1)$ b) $3x^2$ **29.** $8/(x+7)^2$
31. $\frac{-2(x^2+x+1)}{(x^2-1)^2}$ **33.** $\frac{x^2-2x-1}{(1+x^2)^2}$ **35.** $\frac{x^4+2x}{(1-x^3)^2}$
37. $y'=-6/x^3, y''=18/x^4$ **39.** $y'=-1/x^3, y''=3/x^4$
41. $y'=-20/x^5, y''=100/x^6$ **43.** $y'=1-1/x^2, y''=2/x^3$
45. $y'=2x-7/x^2$ **47.** $y'=3/x^4$ **49.** a) 13 b) −7 c) 7/25
d) 20 **51.** c **53.** $y=\frac{-x}{9}+\frac{29}{9}$ **55.** (2, 0) and (−1, 27)
57. $y=4x$ at (0, 0), $y=2$ at (1, 2)
59. $\frac{dP}{dV}=\frac{-nRT}{(V-nb)^2}+\frac{2an^2}{V^3}$ **61.** $\frac{dR}{dM}=CM-M^2$

Section 3.3, pp. 162–164

1. a) 3.2 m, 1.6 m/sec b) 3.2 m/sec, 1.6 m/sec² **3.** a) 18 m, 9 m/sec b) 13 m/sec, 4 m/sec² **5.** a) −8 m, −4 m/sec b) −6 m/sec, −2 m/sec² **7.** a) 8 m, 4 m/sec b) 4 m/sec, 0 m/sec² **9.** On Mars 4.46 sec; on Jupiter 0.726 sec **11.** $\frac{1440}{49}$ m **13.** a) 10^4 bacteria/hr b) 0 bacteria/hr c) -10^4 bacteria/hr **15.** a) \$110/machine b) \$80 c) \$79.90 **17.** −6 m/sec², 6 m/sec² **19.** a) 190 ft/sec b) 2 sec c) 8 sec, 0 ft/sec d) 10.8 sec, 90 ft/sec e) 8.8 sec f) Greatest acceleration at 2 sec; constant acceleration between 2 and 10.8 sec **21.** a) 0,0 b) 1700 rabbits, 1400 rabbits **23.** b **25.** d
27. a)

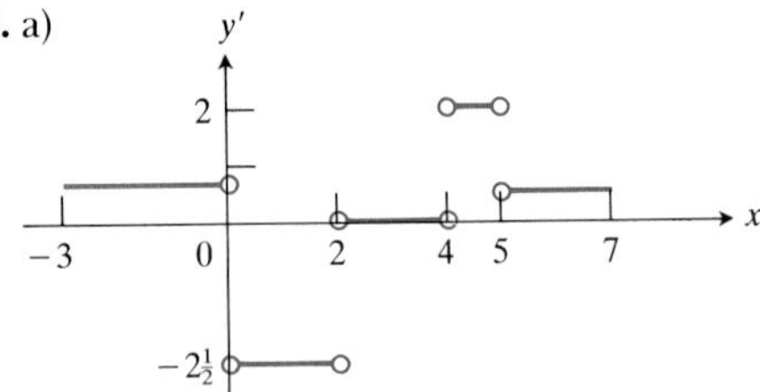

b) 0, 2, 4, and 5
29. a)

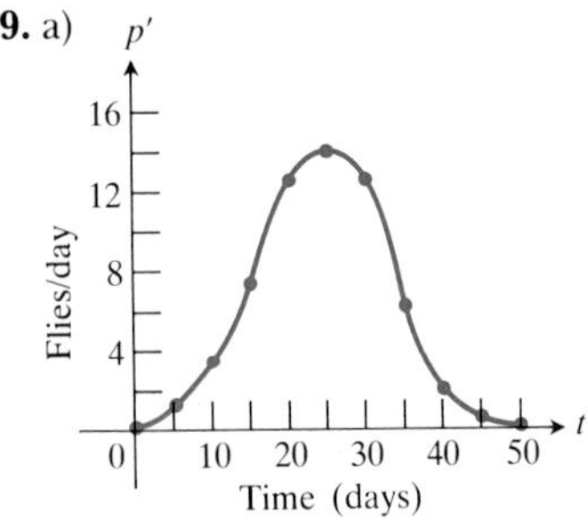

b) Fastest on day 25; slowest on first and last days

Section 3.4, pp. 170–171

1. $\frac{dy}{dx}=1+\sin x$ **3.** $\frac{dy}{dx}=\frac{-1}{x^2}+5\cos x$
5. $\frac{dy}{dx}=-\csc x\cot x-5$ **7.** $\frac{dy}{dx}=\sec x+x\sec x\tan x$
9. $\frac{dy}{dx}=2x\cot x-x^2\csc^2 x$ **11.** $\frac{dy}{dx}=3+\tan x+x\sec^2 x$
13. $\frac{dy}{dx}=\sec^2 x$ **15.** $\frac{dy}{dx}=0$ **17.** $\frac{dy}{dx}=4\sec x\tan x$
19. $\frac{-(x\sin x+\cos x)}{x^2}$ **21.** $\frac{1+\cos x+x\sin x}{(1+\cos x)^2}$
23. $\frac{-\csc^2 x}{(1+\cot x)^2}$ **25.** $2\csc^3 x-\csc x$ **27.** Tangent, $y=x$; normal, $y=-x$ **29.** Tangent, $y=-1$; normal $x=\pi$
31. $y'(0)=\sec 0\tan 0=0, y'(0)=-\sin 0=0$
33. Yes; at $x=\pi$ **35.** Yes; at $x=\frac{\pi}{2}$ **37.** a) $-\sqrt{2}, \sqrt{2}, \sqrt{2}$
b) $0, 0, -\sqrt{2}$ **39.** Tangent, $y=-x+\frac{\pi}{4}+1$; normal, $y=x+1-\frac{\pi}{4}$ **41.** $y=-1$

43.

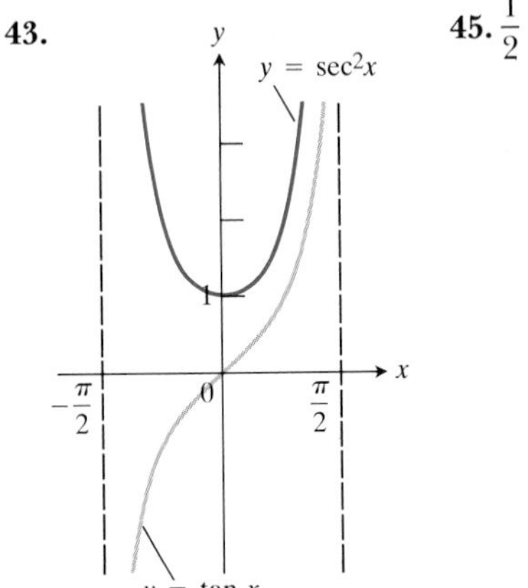

45. $\frac{1}{2}$

Section 3.5, pp. 179–181

1. $\frac{dy}{dx}=\cos(x+1)$ **3.** $\frac{dy}{dx}=-5\sin 5x$
5. $\frac{dy}{dx}=\frac{2\pi}{5}\cos\frac{2\pi x}{5}$ **7.** $\frac{dy}{dx}=-\sec^2(2-x)$
9. $\frac{dy}{dx}=2\sec(2x-1)\tan(2x-1)$
11. $\frac{dy}{dx}=-\csc(x^2+7x)[\cot(x^2+7x)](2x+7)$
13. $\frac{dy}{dx}=-3\csc^2(3x+\pi)$ **15.** $\frac{dy}{dx}=0$
17. $\frac{dy}{dx}=3(\sin^2 x)\cos x$ **19.** $\frac{dy}{dx}=10(2x+1)^4$
21. $\frac{dy}{dx}=-3(x+1)^{-4}$ **23.** $\frac{dy}{dx}=\left(1-\frac{x}{7}\right)^{-8}$
25. $\frac{dy}{dx}=3\left(1+x-\frac{1}{x}\right)^2\left(1+\frac{1}{x^2}\right)$
27. $\frac{dy}{dx}=6(x^2+2x+3)^2(x+1)$
29. $\frac{dy}{dx}=-[\sin(\sin x)](\cos x)$ **31.** $\frac{dy}{dx}=1+(2\sec^2 x)\tan x$

33. $\frac{dy}{dx} = \csc x(\cot x + \csc x)^{-1}$

35. $\frac{dy}{dx} = \cos\left(\frac{x-2}{x+3}\right)\left(\frac{5}{(x+3)^2}\right)$

37. $\frac{dy}{dx} = 6\,[\sin(3x-2)]\cos(3x-2)$

39. $\frac{dy}{dx} = (-4\sin 2x)(1+\cos 2x)$

41. $\frac{dy}{dx} = -2[\cos(\cos(2x-5))]\sin(2x-5)$

43. $y'' = 2(\sec^2 x)\tan x$ **45.** $y'' = 2(\csc^2 x)\cot x$

47. $\frac{5}{2}$ **49.** $\frac{-\pi}{4}$ **51.** 0 **53.** a) $[-\sin(6x+2)]6$ b) $[-2\sin(6x+2)]3$ **55.** a) 1 b) 1 **57.** 5 **59.** $\frac{1}{2}$

61. Tangent, $y = \pi x + 2 - \pi$; normal, $y = -\frac{1}{\pi}x + 2 + \frac{1}{\pi}$

63. a) $\frac{2}{3}$ b) $2\pi + 5$ c) $15 - 8\pi$ d) $\frac{37}{6}$ e) -1 f) $\frac{\sqrt{2}}{24}$ g) $\frac{5}{32}$ h) $\frac{-5}{3\sqrt{17}}$ **65.** 200π cm^3/sec

Section 3.6, pp. 188–189

1. $\frac{dy}{dx} = \frac{9}{4}x^{5/4}$ **3.** $\frac{dy}{dx} = \frac{1}{3\sqrt[3]{x^2}}$ **5.** $\frac{dy}{dx} = -(2x+5)^{-3/2}$

7. $\frac{dy}{dx} = \frac{2x^2+1}{\sqrt{x^2+1}}$ **9.** $\frac{dy}{dx} = \frac{-2xy - y^2}{x^2 + 2xy}$ **11.** $\frac{dy}{dx} = \frac{1-2y}{2x+2y-1}$

13. $\frac{dy}{dx} = \frac{x(1-y^2)}{y(x^2-1)}$ **15.** $\frac{dy}{dx} = \frac{1}{y(x+1)^2}$

17. $\frac{dy}{dx} = \frac{-1}{4\sqrt{x}\sqrt{1-\sqrt{x}}}$ **19.** $\frac{dy}{dx} = \frac{-\sin 2x}{\sqrt{1+\cos 2x}}$

21. $\frac{dy}{dx} = \frac{-9}{2}(\csc x)^{3/2}\cot x$ **23.** $\frac{dy}{dx} = \cos^2 y$

25. $\frac{dy}{dx} = \frac{-\cos^2(xy) - y}{x}$ **27.** $\frac{dy}{dx} = \frac{-x}{y}, \frac{d^2y}{dx^2} = \frac{-y^2 - x^2}{y^3}$

29. $\frac{dy}{dx} = \frac{x+1}{y}, \frac{d^2y}{dx^2} = \frac{y^2 - (x+1)^2}{y^3}$

31. $\frac{dy}{dx} = \frac{\sqrt{y}}{\sqrt{y}+1}, \frac{d^2y}{dx^2} = \frac{1}{2(\sqrt{y}+1)^3}$ **33.** a) $y = \frac{7}{4}x - \frac{1}{2}$ b) $y = \frac{-4}{7}x + \frac{29}{7}$ **35.** a) $y = 3x + 6$ b) $y = \frac{-1}{3}x + \frac{8}{3}$

37. $(\sqrt{7}, 0), (-\sqrt{7}, 0)$; the common slope is -2.

39. $\frac{-\pi}{2}$ **41.** $m = -1$ at $\left(\frac{\sqrt{3}}{4}, \frac{\sqrt{3}}{2}\right)$; $m = \sqrt{3}$ at $\left(\frac{\sqrt{3}}{4}, \frac{1}{2}\right)$.

43. b, c, and d **45.** $\frac{2}{5}$ m/sec, $\frac{-4}{125}$ m/sec^2

Section 3.7, pp. 199–201

1. $4x - 3$ **3.** $2x - 2$
5. $(1/4)x + 1$ **7.** $2x$
9. -5 **11.** $(1/12)x + (4/3)$

13. $L(x) = x$,

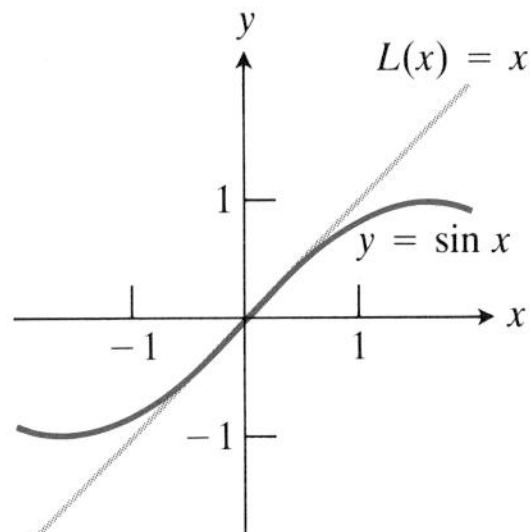

15. $L(x) = -x + \pi$

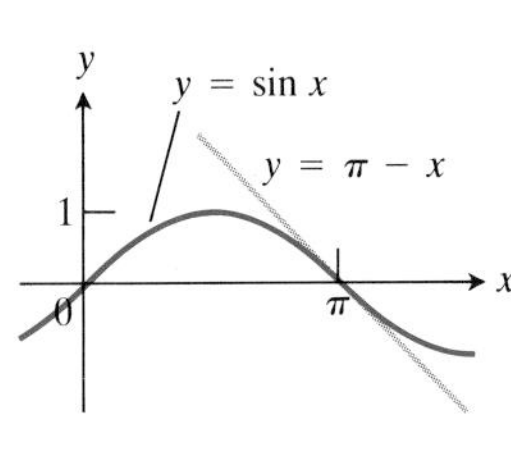

17. $L(x) = 2x + 1 - \frac{\pi}{2}$

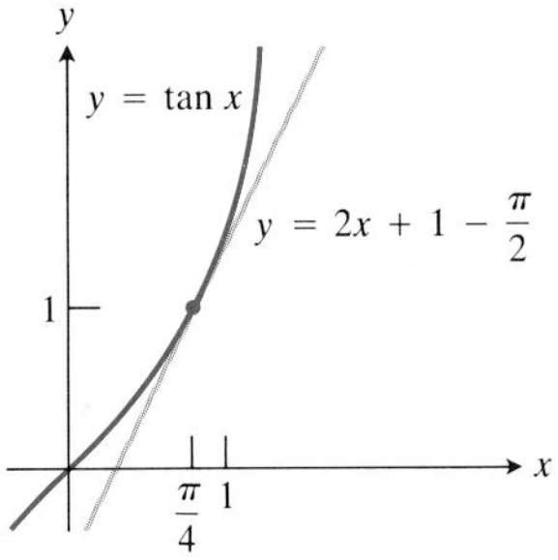

y = tan x
$y = 2x + 1 - \frac{\pi}{2}$
1
$\frac{\pi}{4}$ 1
x

19. a) $1 + 2x$ b) $1 - 5x$ c) $2(1 + x)$ d) $1 - 6x$ e) $3 + x$ f) $1 - \frac{1}{2}x$

21. $\frac{3}{2}x + 1$; it is equal to their sum.

25. a) .21 b) .20 c) .01

27. a) .231 b) .2 c) .031

29. a) $\frac{-1}{3}$ b) $\frac{-2}{5}$ c) $\frac{1}{15}$

31. $dV = 4\pi r^2 dr$ **33.** $dV = 3x^2 dx$ **35.** $dV = 2\pi rh\,dr$

37. a) $.08\pi$m^2 b) 2% **39.** 3% **41.** 3% **43.** $\frac{1}{3}$% **45.** .05%

49. b) $28\sqrt{3} - 45$ **51.** $dy = (3x^2 - 3)\,dx$

53. $dy = \frac{2-2x^2}{(1+x^2)^2}dx$ **55.** $dy = \frac{1-y}{1+x}dx$ **57.** $dy = 5\cos(5x)\,dx$

59. $dy = 2\sec^2\left(\frac{x}{2}\right)dx$ **61.** $dy = \csc\left(1 - \frac{x}{3}\right)\cot\left(1 - \frac{x}{3}\right)dx$

Section 3.8, pp. 208–210

1. $-1.618034, 0.618034$ **3.** $x_2 = -1.645161$, -1.452626; $x_2 = 1.165419, 1.164035$ **5.** $x_2 = 1.1935$, 1.189207 **7.** If $f'(x_0) \neq 0$ then $x_0 = x_1 = x_2 = \dots$.
11. Because $f(1) < 0$ and $f(2) > 0$, $f(x)$ has a root between 1 and 2; 1.17951 **13.** 3.14159 **15.** 0.682328 **17.** 1.165561
19. $-1.306563, -0.541196, 0.541196, 1.306563$

Chapter 3 Practice Exercises, pp. 211–216

1. $5x^4 - \frac{1}{4}x + \frac{1}{4}$ **3.** $2(x+1)(2x^2 + 4x + 1)$

5. $\frac{dy}{dx} = 2\cos 2x$ **7.** $\frac{dy}{dx} = \frac{1}{(x+1)^2}$

9. $\frac{dy}{dx} = -4x^2(x^3+1)^{-7/3}$ **11.** $\frac{dy}{dx} = 2\sin(1-2x)$

13. $\frac{dy}{dx} = 3(x^2 + x + 1)^2(2x + 1)$

15. $\frac{dy}{dx} = \frac{4(x+2)}{\sqrt{4x^2 + 16x + 15}}$

17. $\frac{dy}{dx} = \frac{-y}{x+2y}$ **19.** $\frac{dy}{dx} = \frac{5-2x-y}{x+2y}$ **21.** $\frac{dy}{dx} = \frac{-1}{3(xy)^{1/5}}$
23. $\frac{dy}{dx} = \frac{1}{2y(x+1)^2}$ **25.** $\frac{(5x^2+2x)^{1/2}(5x+1)}{2y}$ **27.** $\frac{dy}{dx} = \frac{x-1}{2\sqrt{x^3}}$
29. $\frac{dy}{dx} = 3[\sec(1+3x)]\tan(1+3x)$ **31.** $\frac{dy}{dx} = -2x\csc^2 x^2$
33. $\frac{dy}{dx} = \frac{(x^2-2x-1)}{2\sqrt{(1+x^2)^3(1-x)}}$
35. a) $y = \begin{cases} x, & 0 \le x \le 1 \\ 2-x, & 1 < x \le 2 \end{cases}$ b) Yes c) No, since $\lim_{x\to 1^-} f'(x) = 1 \neq -1 = \lim_{x\to 1^+} f'(x)$.

37. $(2, 0), (-1, 27)$ **39.** a) $\frac{10}{\sqrt{2}}$ b) Left, -10; right, 10 c) Left: $v = 0, a = 10$; right: $v = 0, a = -10$ d) When $t = \frac{\pi}{4}$, $v = -10, |v| = 10, a = 0$. **41.** a) $\frac{4}{7}$ sec, 280 cm/sec b) 560 cm/sec, 980 cm/sec^2 c) .0336 sec/flash
43. a) iii b) i c) ii **45.**

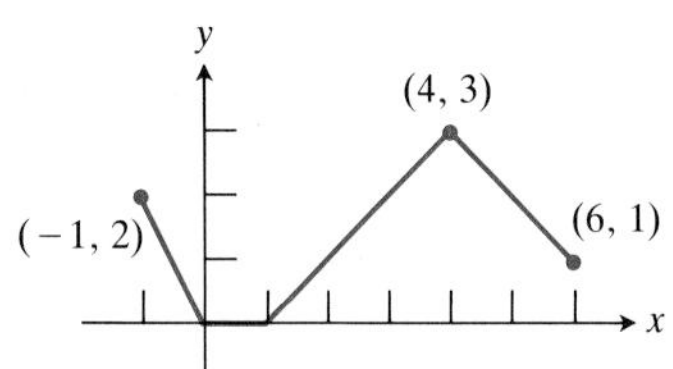

47. $\pi(20x - x^2)$ in^3 **49.** a) $y = 4 - \sqrt{3}$ b) $y = -x + \frac{\pi+4}{2}$
51. Yes, $x = 0 + n(2\pi)$ when n is an integer.
53. a) 1 b) 6 c) 1 d) $\frac{-1}{9}$ e) $\frac{-40}{3}$ f) 2 g) $\frac{-4}{9}$
55. a) Yes b) No, $x^2 - 2x - 8 = 0$ is not an identity.
57. $\frac{9}{2}$ **59.** $(0, 1), (-4, 0)$
61. a) Tangent, $y = \frac{-1}{4}x + \frac{9}{4}$; normal $y = 4x - 2$
b) Tangent, $y = \frac{-3}{2}x + \frac{5}{2}$; normal $y = \frac{2}{3}x + \frac{1}{3}$
c) Tangent, $y = 2x - 4$; normal $y = \frac{-1}{2}x + \frac{7}{2}$
63. 3 ft **65.** a) $\frac{-2xy^3 - 2x^4}{y^5}$ b) $\frac{-2xy^2 - 1}{x^4y^3}$
67. a) $\frac{-1}{\sqrt{(2x+7)^3}}$ b) $\frac{-y^2 - x^2}{y^3}$ **69.** a) $L(x) = 2x + \frac{\pi - 2}{2}$ b) $L(x) = -\sqrt{2}x + \frac{\sqrt{2}(4-\pi)}{4}$ **71.** a) Since $f\left(\frac{-\pi}{4}\right) \approx -0.74$, $f(0) = 0.5$, and by the Intermediate Value Property $f(x) = 0$ has a solution between $\frac{-\pi}{4}$ and 0. b) $-\frac{1}{3}$ c) -0.0107

73. $\frac{dV}{dt} = \frac{2}{3}\pi r_0 h \frac{dr}{dt}$ **75.** a) Have less than 1% error b) 3%
77. $\pm\frac{20\pi}{27}$ ft **81.** 1.587401 **83.** 1.318116 **85.** 0.828360

Chapter 4

Section 4.1, pp. 221–225

1. $\frac{dA}{dt} = 2\pi r\frac{dr}{dt}$ **3.** $\frac{dV}{dt} = 3x^2\frac{dx}{dt}$ **5.** $\frac{dV}{dt} = \frac{1}{3}\pi r^2\frac{dh}{dt}$
7. πcm^2/min **9.** a) 14 cm^2/sec, increasing b) 0, constant c) $\frac{14}{13}$ cm/sec, decreasing **11.** -12 ft/sec, -59.5 ft^2/sec
13. 0.0076π in^3/min **17.** 2.5 ft/sec **19.** $\frac{10}{9\pi}$in/min, $-\frac{8}{5\pi}$ in/min **21.** Increasing at $\frac{446}{1681}$ L/min **23.** 1 rad/sec, 0 rad/sec **25.** a) 8 ft/sec b) -3 ft/sec
27. 20 ft/sec **29.** 80 mph **31.** 29.5 knots

Section 4.2, pp. 234–235

5. a) The zeros of $y' = 0$ are between the zeros of $y = 0$

i) −2, 0, 2
ii) −5, −4, −3
iii) −1, 0, 2
iv) 0, 4, 9, 18, 24

7. $\frac{1}{2}$ **9.** 1 **19.** 1.1

Section 4.3, pp. 242–243

1. Maximum of $\frac{3}{2}$ at $x = -1$; minimum of -3 at $x = 2$; point of inflection at $\left(\frac{1}{2}, \frac{-3}{4}\right)$; rising on $(-\infty, -1)$ and $(2, \infty)$; falling on $(-1, 2)$; concave upward on $\left(\frac{1}{2}, \infty\right)$; concave downward on $\left(-\infty, \frac{1}{2}\right)$

3. Maximum of 1 at $x = \frac{-\pi}{2}$ and $x = \frac{\pi}{2}$; minimum of -1 at $x = \frac{-3\pi}{2}$ and $x = \frac{3\pi}{2}$, local minimum of 0 at $x = 0$; points of inflection at $(-\pi, 0)$ and $(\pi, 0)$; rising on $\left(-\frac{3\pi}{2}, -\frac{\pi}{2}\right)$, $\left(0, \frac{\pi}{2}\right)$ and $\left(\frac{3\pi}{2}, 2\pi\right)$; falling on $\left(-2\pi, -\frac{3\pi}{2}\right)$, $\left(-\frac{\pi}{2}, 0\right)$ and $\left(\frac{\pi}{2}, \frac{3\pi}{2}\right)$; concave upward on $(-2\pi, -\pi)$ and $(\pi, 2\pi)$; concave downward on $(-\pi, \pi)$

5.

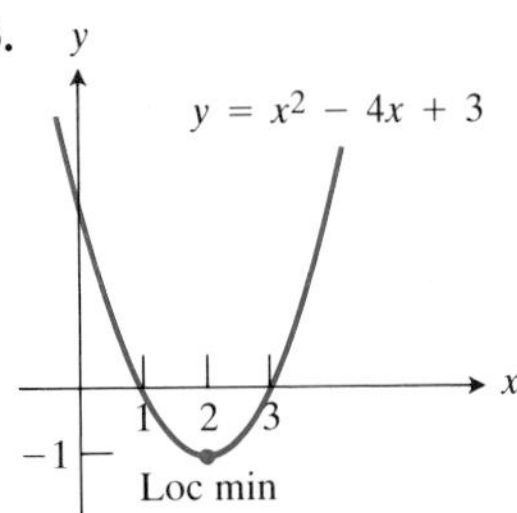

7.

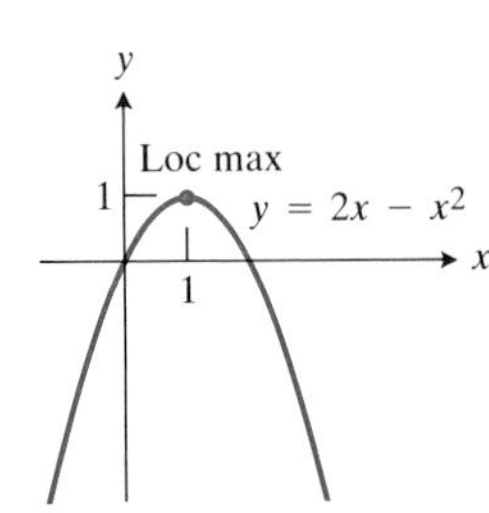

9.

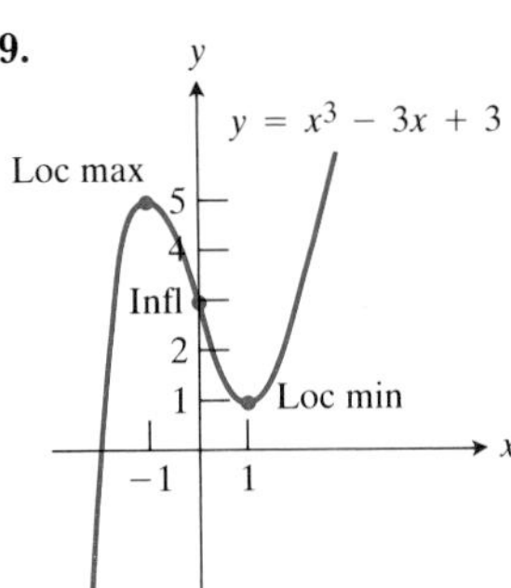

11.

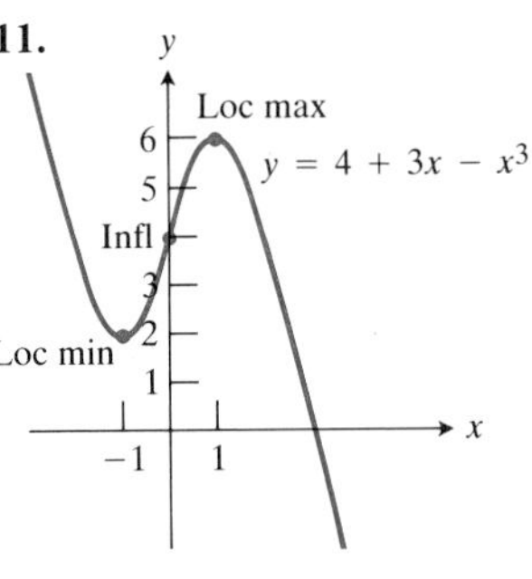

13.

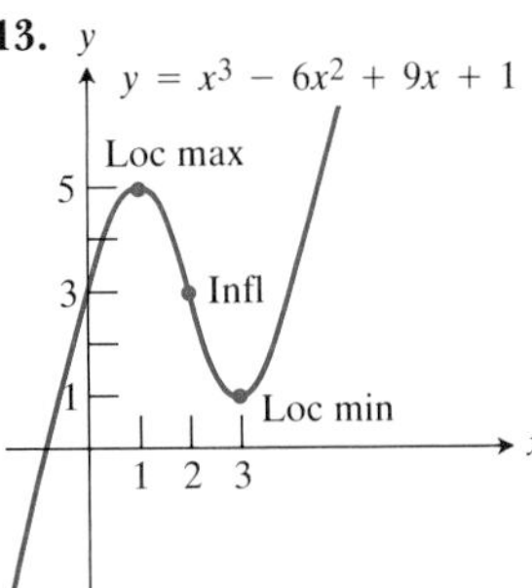

15.

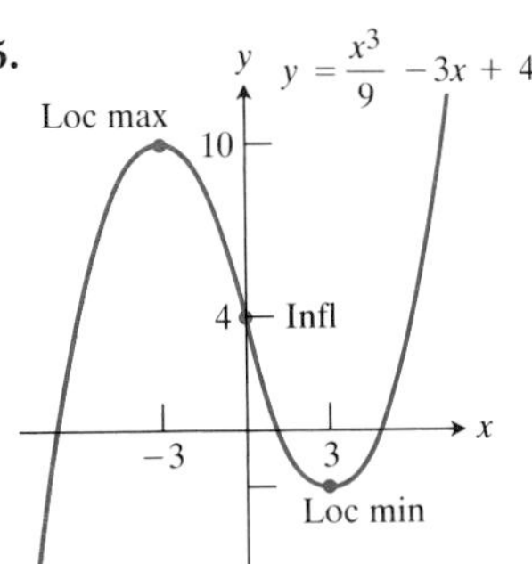

17.

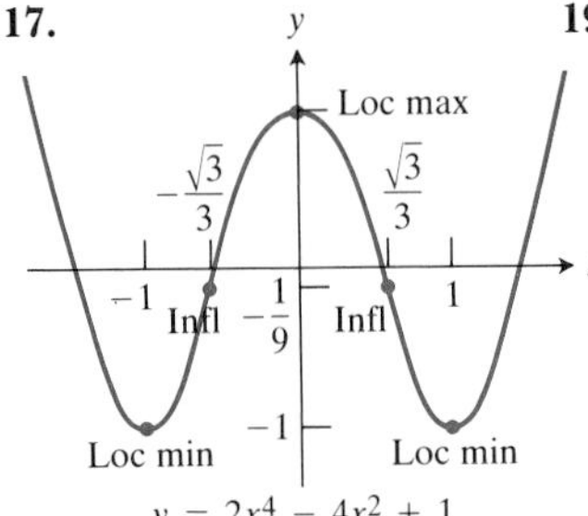

19.

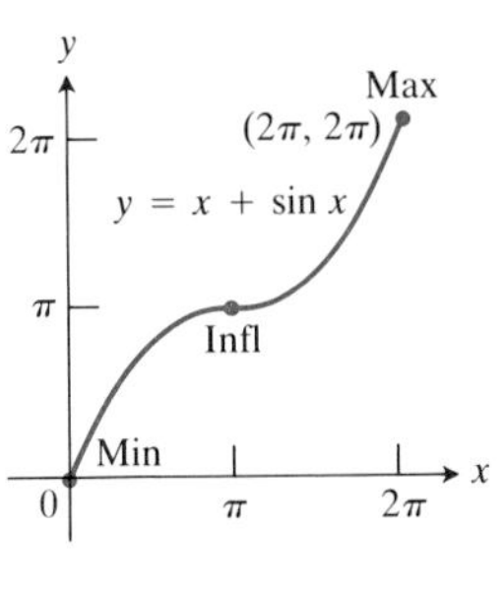

21. a) 2 sec, 6 sec, 9.5 sec b) 4 sec, 8 sec, 12.5 sec

23. $\left(-1, \frac{3}{4}\right)$ and $\left(1, \frac{3}{4}\right)$

25.

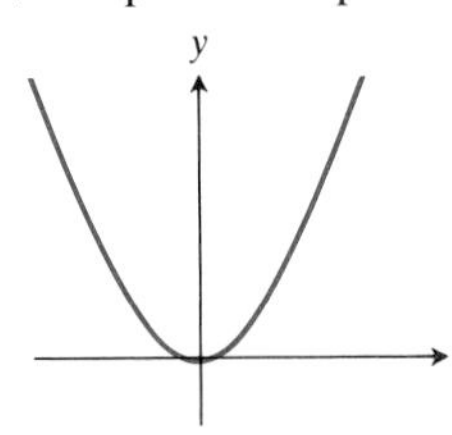

27.

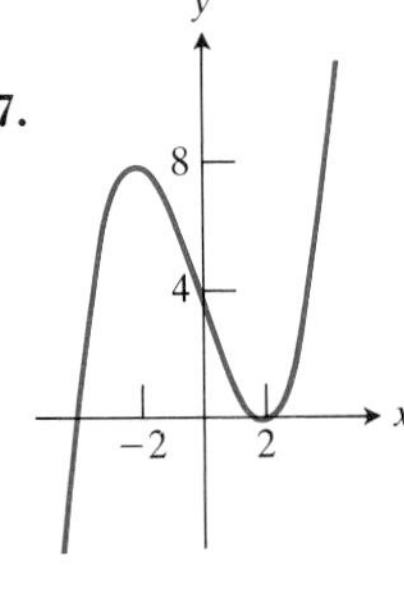

29. Local minimum at $x = 2$; points of inflection at $x = 1$ and $x = \frac{5}{3}$ **31.** Local maximum of -2 at $x = -1$; local minimum of 2 at $x = 1$ **35.** True

37. a) Local maximum of -1 at $x = 0$; local minimum of -4 at $x = -\sqrt{3}$ and $x = \sqrt{3}$; points of inflection at $\left(-1, -\frac{8}{3}\right)$ and $\left(1, -\frac{8}{3}\right)$ b) -2.54245976 and 2.54245976

Section 4.4, p. 249

1. **3.**

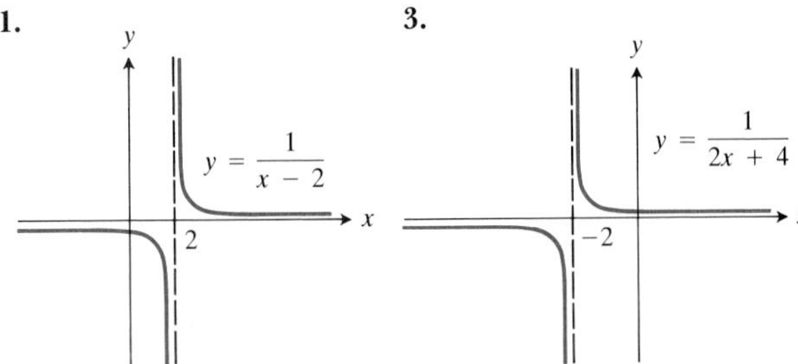

5.

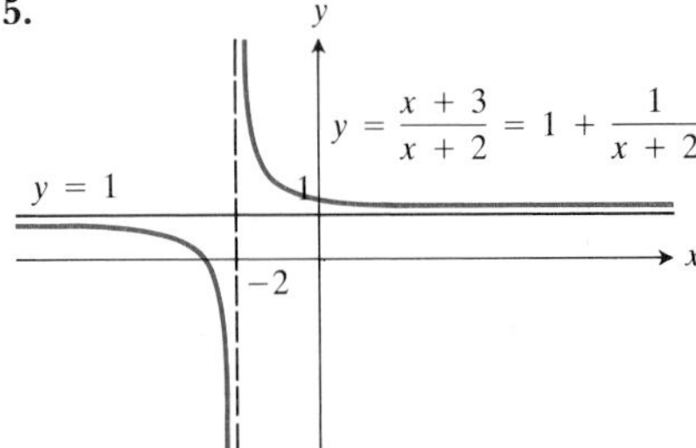

7.

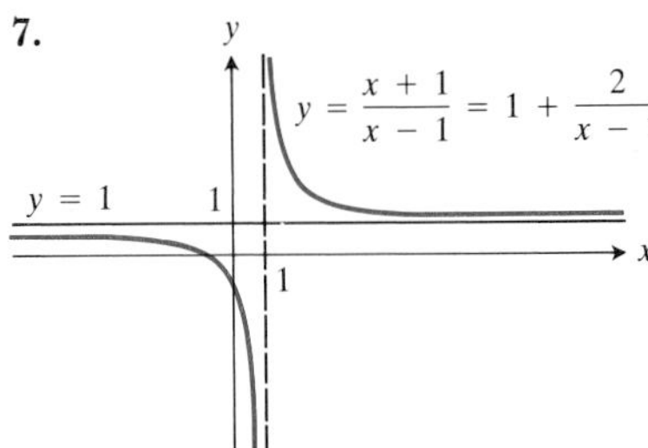

9.

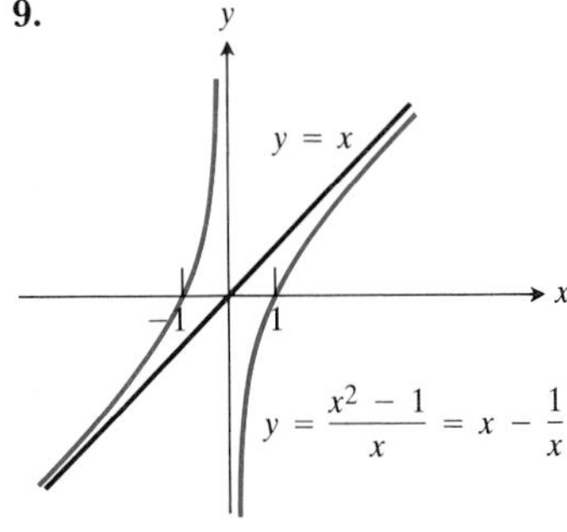

11.

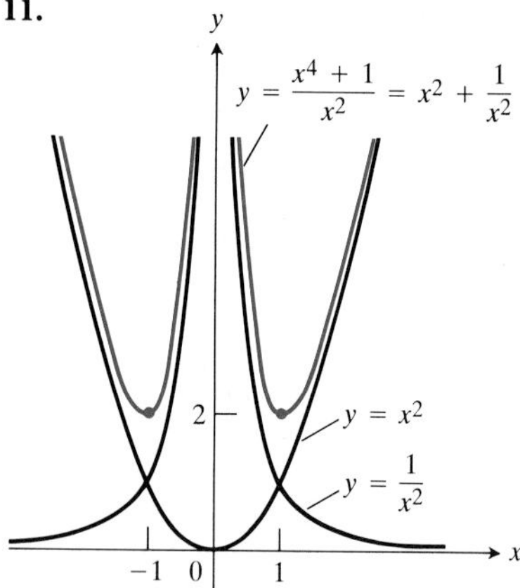

13.

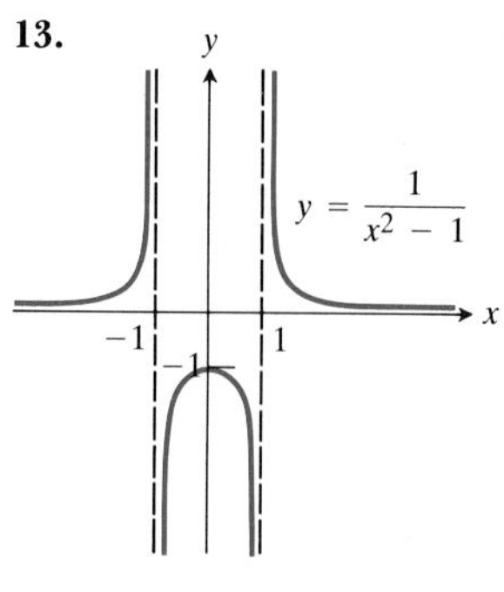

15.

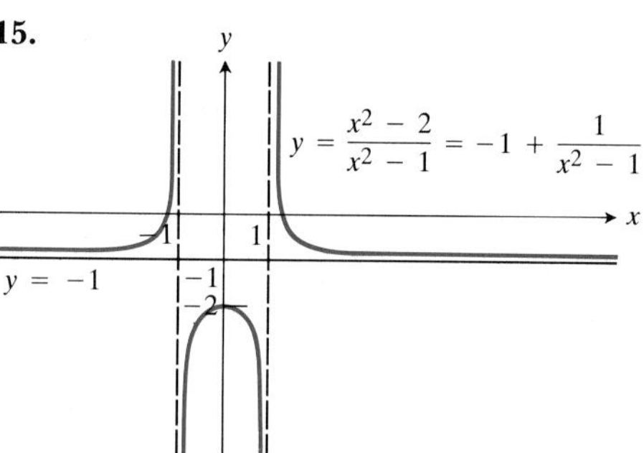

17.

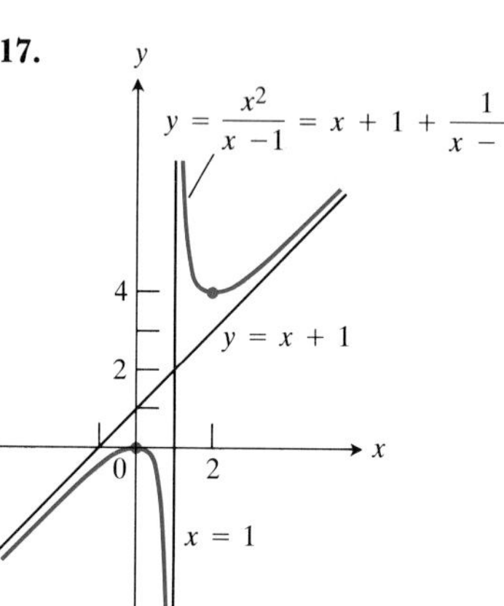

19.

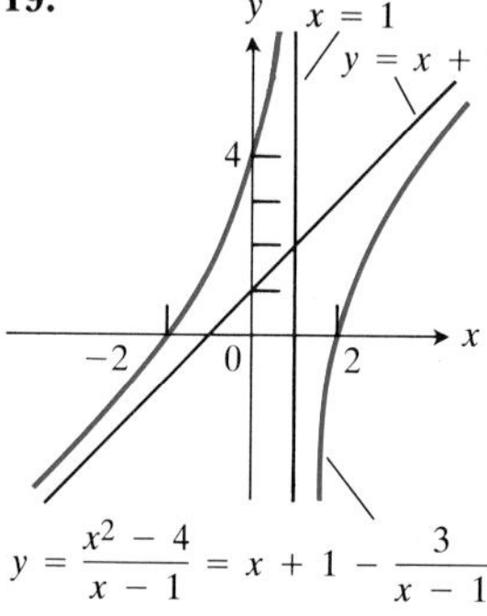

21.

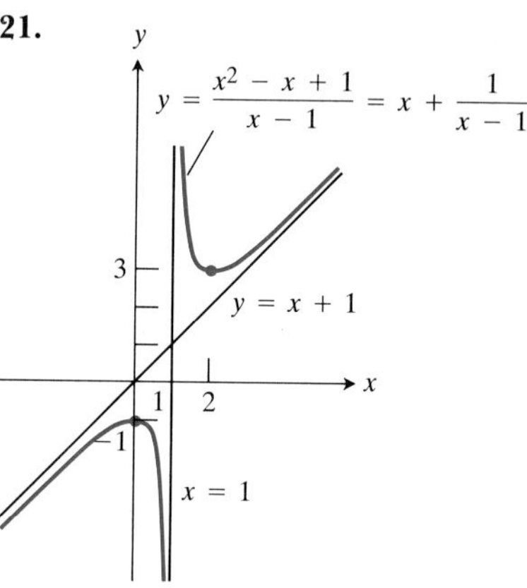

23.

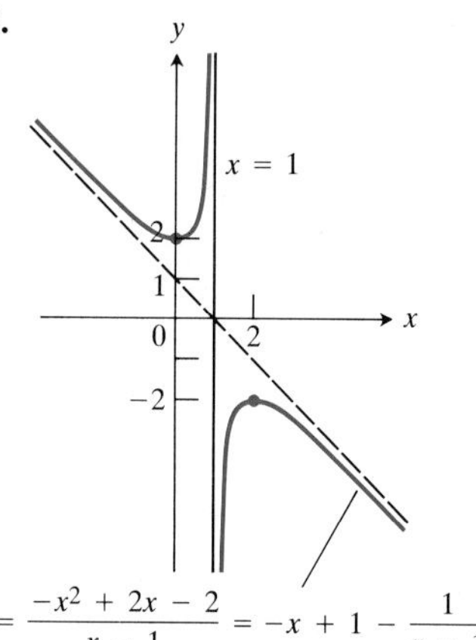

25.

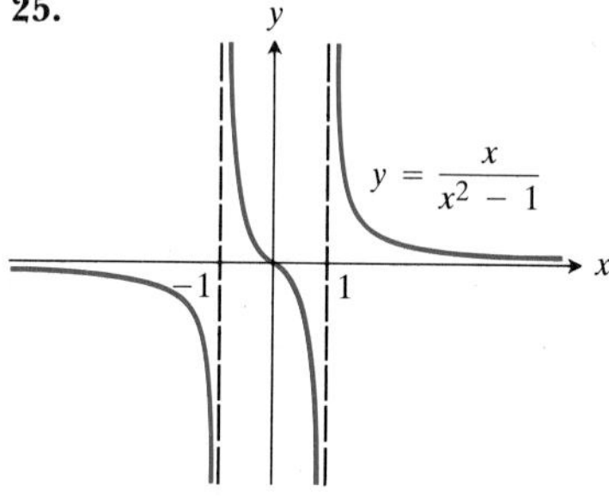

27.

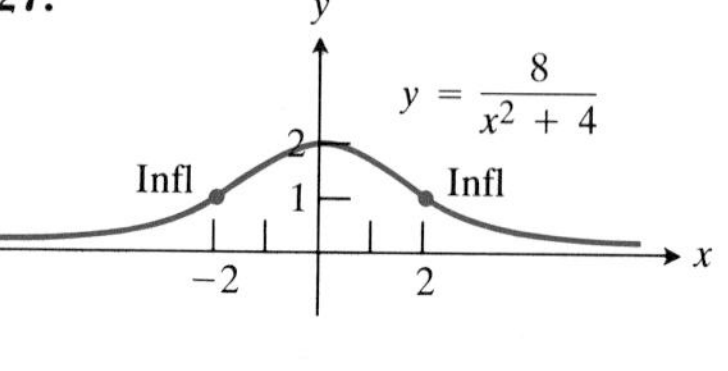

Section 4.5, pp. 259–263

1. a) 0 and 20 b) $\frac{1}{4}$ and $19\frac{3}{4}$ **3.** 16 in.

5. a) $(x, 1 - x)$ b) $2x(1 - x)$ c) $\frac{1}{2}$ square units

7. $\frac{14}{3}$ in., $\frac{35}{3}$ in., $\frac{5}{3}$ in. **9.** 80,000 m^2

11. Base 10 ft, height 5 ft **13.** 18 in., 9 in.

15. $\frac{\pi}{2}$ **17.** $h = r = \frac{10}{\sqrt[3]{\pi}}$ cm **19.** $18 \times 18 \times 36$ inches
21. a) 12 cm, 6 cm b) 12 cm, 6 cm **23.** a) 16 b) -1
25. a) $a = -3$ and $b = -9$ b) $a = -3$ and $b = -24$
27. Width $4\sqrt{3}$ in., depth $4\sqrt{6}$ in. **29.** -1
31. $\frac{\sqrt{5}}{2}$ **33.** No, minimum of $\frac{3}{4}$ and concave upward
35. $\frac{\pi}{6}$ **39.** $\frac{a}{2}, \frac{ka^2}{4}$ **43.** $50 + \frac{c}{2}$ **45.** $\sqrt{\frac{2km}{h}}$

Section 4.6, pp. 273–275

1. a) $x^2 + C$ b) $3x + C$ c) $x^2 + 3x + C$ **3.** a) $x^3 + C$
b) $\frac{x^3}{3} + C$ c) $\frac{x^3}{3} + x^2 + x + C$ **5.** a) $x^{-3} + C$ b) $-\frac{x^{-3}}{3} + C$
c) $-\frac{x^{-3}}{3} + x^2 + 3x + C$ **7.** a) $\sqrt{x^3} + C$ b) $\frac{8\sqrt{x^3}}{3} + C$
c) $\frac{x^3}{3} + \frac{8\sqrt{x^3}}{3} + C$ **9.** a) $x^{2/3} + C$ b) $x^{1/3} + C$ c) $x^{-1/3} + C$
11. a) $\frac{\cos 3x}{3} + C$ b) $-3\cos x + C$ c) $-3\cos x + \frac{\cos 3x}{3} + C$
13. a) $\tan x + C$ b) $\tan 5x + C$ c) $\frac{\tan 5x}{5} + C$
15. a) $\sec x + C$ b) $\sec 2x + C$ c) $2\sec 2x + C$
17. $x + \frac{\cos 2x}{2} + C$ **19.** a) $-\sqrt{x} + C$ b) $x + C$ c) $\sqrt{x} + C$
d) $-x + C$ e) $-\sqrt{x} + x + C$ f) $-3\sqrt{x} - 2x + C$
g) $\frac{x^2}{2} - \sqrt{x} + C$ h) $-3x + C$ **21.** $y = x^2 - 7x + 10$
23. $y = \frac{x^3}{3} + x + 1$ **25.** $y = \frac{5}{x} + 2$ **27.** $y = x^3 + x^2 + x - 3$
29. $y = x + \sin x + 4$ **31.** $y = x^2 - x^3 + 4x + 1$
33. $y = 4.9t^2 + 10$ **35.** $s = 16t^2 + 20t$
37. $y = 2x^{3/2} - 50$ **39.** $r(x) = x^3 - 3x^2 + 12x$
41. 48 m/sec **43.** 14m/sec
45. a) $2y^{1/2} + C_1 = -kt + C_2$ b) $y = \frac{(C - kt)^2}{4}$ **47.** d

Chapter 4 Practice Exercises, pp. 276–280

1. -40 m²/sec **3.** 1 cm/min **5.** a) $r = \frac{2h}{5}$ b) $\frac{-125}{144\pi}$ ft/min
7. a) -2 rad/sec b) $\sqrt{2}$ rad/sec

13.

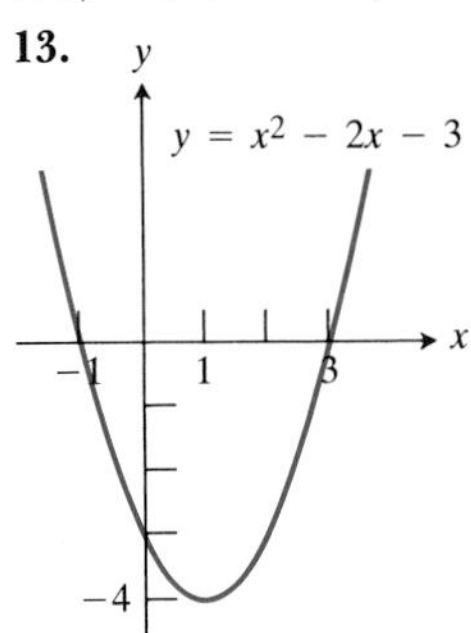

15.

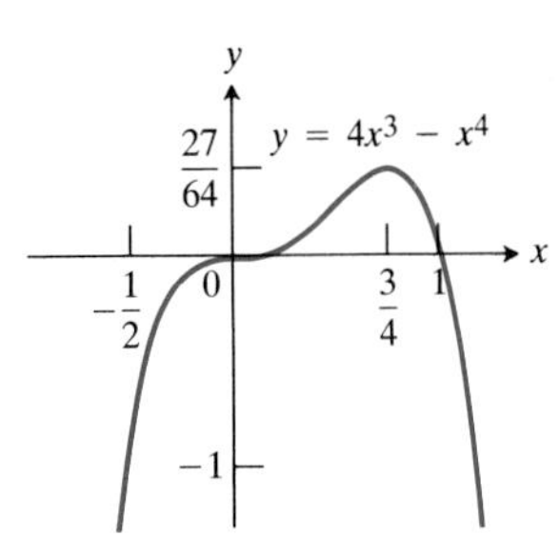

17.

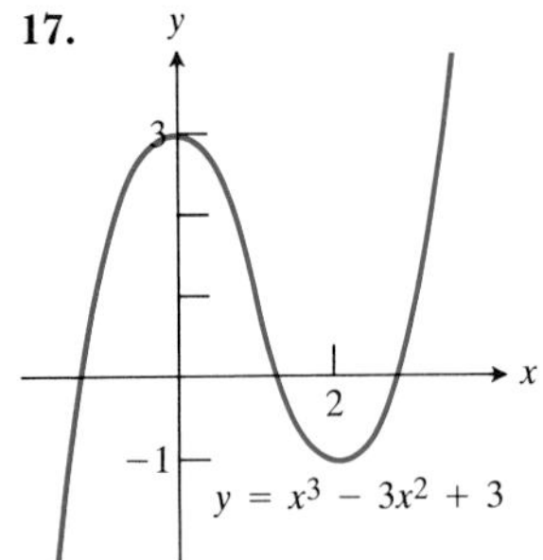

19.

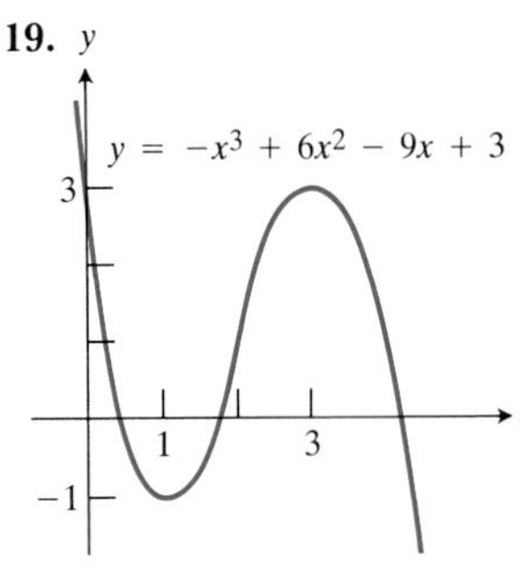

21. Local minimum at $(-1, f(-1))$; points of inflection at $(0, f(0))$ and $(2, f(2))$ **23.** $b = -3$ **25.** a) T b) P
27. Minimum of 0 at $x = 0$ and $x = \frac{3}{2}$; maximum of 2 at $x = \frac{1}{2}$ and $x = 2$ **29.** $r = 25$ ft, $s = 50$ ft **31.** Side base is 6 ft, height is 3 ft **33.** Height = 2, radius = $\sqrt{2}$ **35.** $x = 15$ mi, $y = 9$ mi **37.** $x = 5 - \sqrt{5}$ hundred ≈ 276 tires; $y = 2(5 - \sqrt{5})$ hundred ≈ 553 tires **41.** a) C b) $x + C$
c) $\frac{x^2}{2} + C$ d) $\frac{x^3}{3} + C$ e) $\frac{x^{11}}{11} + C$ f) $-x^{-1} + C$
g) $\frac{-x^{-4}}{4} + C$ h) $\frac{2x^{7/2}}{7} + C$ i) $\frac{3x^{7/3}}{7} + C$ j) $\frac{4x^{7/4}}{7} + C$
k) $\frac{2x^{3/2}}{3} + C$ l) $2x^{1/2} + C$ m) $\frac{7x^{4/7}}{4} + C$ n) $\frac{-3x^{-4/3}}{4} + C$
43. $x^3 + \frac{5x^2}{2} - 7x + C$ **45.** $\frac{2\sqrt[4]{x^3}}{3} + 2\sqrt{x} + C$
47. $\frac{3\sin(5x)}{5} + C$ **49.** $\tan(3x) + C$ **51.** $\frac{x}{2} - \sin(x) + C$
53. $3\sec\left(\frac{x}{3}\right) + 5x + C$ **55.** $\tan(x) - x + C$
57. $x - \frac{(\sin 2x)}{2} + C$ **59.** $y = x + \frac{x^2}{2} + \frac{x^3}{6} + 1$
61. $y = x - \frac{1}{x} - 1$ **63.** $y = \sin(x)$ **65.** $y = x$
67. You better duck. **69.** 0.682

Chapter 5

Section 5.1, pp. 288–289

1. $\frac{8}{3}$ **3.** $\frac{16}{3}$ **5.** $\frac{4}{3}$ **7.** $\frac{27}{4}$ **9.** 1 **11.** 1 **13.** $\frac{2}{\pi}$
15. $\sqrt{3} + 1$ **17.** $\frac{b^{n+1}}{n+1}$

Section 5.2, pp. 293–294

1. $1 + \frac{1}{2} + \frac{1}{3} + \frac{1}{4} = \frac{25}{12}$ **3.** $(1 + 2) + (2 + 2) + (3 + 2) = 12$
5. $\frac{0}{4} + \frac{1}{4} + \frac{2}{4} + \frac{3}{4} + \frac{4}{4} = \frac{5}{2}$ **7.** $(2(1) - 2) + (2(2) - 2) + (2(3) - 2) + (2(4) - 2) + (2(5) - 2) = 20$

9. $\frac{6(1)}{1+1} + \frac{6(2)}{2+1} = 7$ **11.** $1(1-1)(1-2) + 2(2-1)(2-2) + 3(3-1)(3-2) + 4(4-1)(4-2) + 5(5-1)(5-2) = 90$
13. $\cos(1)\pi + \cos(2)\pi + \cos(3)\pi + \cos(4)\pi = 0$
15. $(-1)^1 + (-1)^2 + (-1)^3 + (-1)^4 = 0$ **17.** $\sum_{k=1}^{6} k$
19. $\sum_{k=1}^{4} \frac{1}{2^k}$ **21.** $\sum_{k=1}^{5} (-1)^{k+1}\frac{1}{k}$ **23.** $\sum_{k=1}^{5} (-1)^{k+1}\frac{k}{5}$ **25.** 55
27. -91 **29.** -20 **31.** 1 000 000 **33.** All of them
35. a) -15 b) 1 c) 1 d) -11 e) 16 **37.** $(12 \cdot 13)/2 = 78$
41. Sum in rows of triangle = Sum in rows of square
43. Sum of rows in square + Sum in rows of triangle = Sum in rows of trapezoid

Section 5.3, pp. 306–308

1. $\int_0^2 x^2\,dx$ **3.** $\int_{-7}^{5} (x^2 - 3x)\,dx$ **5.** $\int_2^3 \frac{1}{1-x}\,dx$ **7.** $\int_0^4 \cos x\,dx$
9. $\int_{-\pi/4}^{0} \sec x\,dx$ **11.** $\int_{-\pi}^{0} \sin 3x\,dx$ **13.** $\int_{1/\pi}^{2/\pi} \sin\frac{1}{x}\,dx$
15. $-\int_{\pi/2}^{3\pi/2} \cos x\,dx$ **17.** $\int_0^5 \sqrt{25 - x^2}\,dx$

19. $\frac{\pi}{2}$

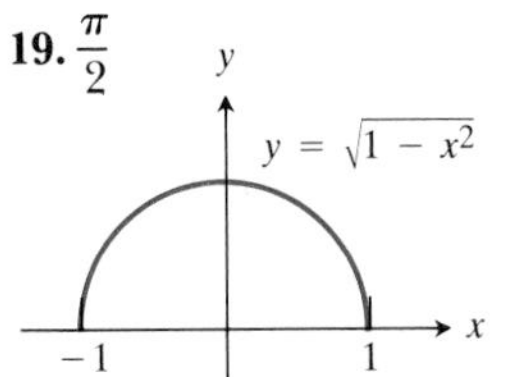

21. 1
23. a) 0 b) -8 c) -12 d) 10 e) -2 f) 16
25. a) 5 b) 5 c) -5

27. Upper bound = 1; lower bound = $\frac{1}{2}$

31.

n	Upper endpoint sum	Lower endpoint sum
4	0.625	0.375
10	0.55	0.45
20	0.525	0.475
50	0.51	0.49

33.

n	Upper endpoint sum	Lower endpoint sum
4	3.14159551	−3.14157979
10	1.25664474	−1.25662538
20	0.628327454	−0.628307606
50	0.251337007	−0.251317017

35.

n	Upper endpoint sum	Lower endpoint sum
4	1.5	0.5
10	1.2	0.8
20	1.1	0.9
50	1.04	0.96

Section 5.4, pp. 316–317

1. 8 **3.** 3 **5.** 1 **7.** $\frac{5}{2}$ **9.** 2 **11.** $2\sqrt{3}$ **13.** 0 **15.** $\frac{8}{3}$ **17.** $\frac{5}{2}$
19. $\frac{1}{2}$ **21.** $\frac{5}{6}$ **23.** π **25.** $\cos x$ **27.** $\frac{dy}{dx} = \sqrt{1+x^2}$
29. $\frac{1}{2}x^{-1/2}\sin x$ **31.** d **33.** b **37.** a) \$9.00 b) \$10.00
39. $I_{av} = 300$; \$6.00 **41.** $x = a$ **43.** 1

Section 5.5, pp. 325–326

1. $\frac{x^4}{4} + C$ **3.** $\frac{x^2}{2} + x + C$ **5.** $2x^{3/2} + C$ **7.** $\frac{3}{2}x^{2/3} + C$
9. $\frac{5x^3}{3} + x^2 + C$ **11.** $\frac{x^4}{2} - \frac{5x^2}{2} + 7x + C$ **13.** $2\sin x + C$
15. $-3\cos\frac{x}{3} + C$ **17.** $-3\cot x + C$ **19.** $-\frac{1}{2}\csc x + C$
21. $4\sec x - 2\tan x + C$ **23.** $-\frac{1}{2}\cos 2x + \cot x + C$
25. $2y - \sin 2y + C$ **27.** $-\frac{1}{4}\cos 2x + C$ **29.** $\tan\theta + C$
35. a) Wrong b) Wrong c) Right **37.** $y = 2x^{3/2} - 50$
39. $y = \cos\pi x - 1$ **41.** $y = 2x$ **43.** $y = \frac{x^3}{16}$ **45.** Step 1: $s = \frac{-kt^2}{2} + 88t$; Step 2: $t = \frac{88}{k}$ since $t > 0$; Step 3: $k = 16$ ft/sec^2 **47.** $t = \sqrt{\frac{4}{2.6}} \approx 1.24$ sec since $t > 0$

Section 5.6, pp. 333–335

1. $-\frac{1}{3}\cos 3x + C$ **3.** $\frac{1}{2}\sec 2x + C$ **5.** $(7x-2)^4 + C$
7. $-6(1 - r^3)^{1/2} + C$ **9.** a) $\frac{\cot^2 2\theta}{4} + C$ b) $-\frac{\csc^2 2\theta}{4} + C$
11. $\frac{3}{16}$ **13.** $\frac{1}{2}$ **15.** 0 **17.** $\frac{2}{3}$ **19.** $\frac{1}{1-x} + C$
21. $\tan(x+2) + C$ **23.** $3(r^2 - 1)^{4/3} + C$
25. $\sec\left(\theta + \frac{\pi}{2}\right) + C$
27. $2(1 + x^4)^{3/4} + C$ **29.** a) $\frac{14}{3}$ b) $\frac{2}{3}$ **31.** a) $\frac{1}{2}$ b) $-\frac{1}{2}$
33. a) $\frac{\sqrt{10} - 3}{2}$ b) $\frac{3 - \sqrt{10}}{2}$ **35.** a) $\frac{45}{8}$ b) $-\frac{45}{8}$ **37.** a) $\frac{1}{6}$ b) $\frac{1}{6}$ **39.** a) 0 b) 0 **41.** $2\sqrt{3}$ **43.** 0 **45.** 8 **47.** $\frac{38}{3}$ **49.** $\frac{16}{3}$
51. $s = (3t^2 - 1)^4 - 1$ **53.** $s = -6\cos(t + \pi) - 6$ **55.** a) 1 b) 1 c) 1 **57.** $C_1 = C_2 - 1$, $C_2 = C_3 + \frac{1}{2}$, $C_1 = C_3 - \frac{1}{2}$

Section 5.7, pp. 344–347

1. a) 2 b) 2 c) 2 **3.** a) $4\frac{1}{4}$ b) 4 c) 4 **5.** a) $\frac{5}{2} + \sqrt{2} + \sqrt{3} \approx 5.646$ b) $2 + \frac{2}{3}\sqrt{2} + \frac{4}{3}\sqrt{3} \approx 5.252$ c) $\frac{16}{3}$

7. $|E_T| \le .00167$ **9.** a) $n = 1$ b) $n = 1$ **11.** a) 283 b) 1 **13.** a) 75 b) 11 **15.** 1013 **17.** $1400/3 \approx 467$ in^2 **19.** 56.86
21. a) $|E_s| \le 0.0002075$ b) 1.370792988 c) 0.01513% **23.** 1.08942941 **25.** 0.828116331

Section 5.8, pp. 362–365

1. $\ln 3 - \ln 2$ **3.** $2\ln 2 - 2\ln 3$ **5.** $-\ln 2$ **7.** $2\ln 3 - \ln 2$ **9.** $\ln 3 + \frac{1}{2}\ln 2$ **11.** 7 **13.** 2 **15.** $3e^2$ **17.** $\frac{1}{x}$ **19.** $\frac{2}{x}$ **21.** $-\frac{1}{x}$ **23.** $\frac{1}{x+2}$ **25.** $\frac{\sin x}{2-\cos x}$ **27.** $\frac{1}{x\ln x}$ **29.** $2e^x$ **31.** $-e^{-x}$ **33.** $\frac{2}{3}e^{2x/3}$ **35.** xe^x **37.** $\frac{e^{\sqrt{x}}}{2\sqrt{x}}$ **39.** $\ln 4$ **41.** 8 **43.** $\frac{-3+2e+e^2}{2}$ **45.** $e^2 - e$ **47.** 2 **49.** 1 **51.** $\ln\frac{3}{2}$ **53.** 0 **55.** $\frac{\ln 4}{2}$ **57.** $\frac{\ln 2}{10}$ **59.** 0 **61.** $-10\ln 3$ **63.** $-10\ln 5$ **65.** $y = e^{2t+4}$ **67.** $y = e^{5t} + 40$ **69.** 2.8147×10^{14} **71.** a) 8.00 b) $t > 32.02$ years **73.** 4.50% **75.** 262 090 086 **77.** 2013 **79.** 1990: 0.786; 1992: 0.720; 1994: 0.659 **81.** 32 days **83.** 16.09 years
85. a) Food: 13.52 years b) Food: 23.3 years
Rent: 8.11 years Rent: 14 years
Medical: 6.34 years Medical: 10.9 years
87. a) $p(t) = 100\, e^{\int_0^t (1+1.3\tau)\,d\tau}$ b) $p(1) = 520.7\%$; $p(2) = 9948.4\%$

Chapter 5 Practice Exercises, pp. 366–370

1. a) 5050 b) 338 350 c) 25 502 500 **3.** a) 90 b) 112 **5.** a) 7 b) 5 c) 8 d) $\frac{11}{30}$ **7.** 91 **9.** $\int_1^2 \frac{1}{x}\,dx$ **11.** a) True b) True c) False **13.** $\pi - 2$ **15.** 1 **17.** 10 **19.** 16 **21.** 2 **23.** 1 **25.** 8 **27.** $\frac{2}{5}$ **29.** $\sqrt{3}$ **31.** $6\sqrt{3} - 2\pi$ **33.** -1 **35.** 2 **37.** $\ln 2$ **39.** $\frac{1}{2}\ln\frac{9}{5}$ **41.** 1 **43.** $e - 1$ **45.** $\frac{17}{3}$ **47.** 8 **49.** -2 **51.** 1 **53.** $\frac{1}{2x}$ **55.** $\frac{2x}{x^2+2}$ **57.** $\frac{-1}{e^x}$ **59.** e **65.** a) $2x$ b) $(\ln 2) + 1$ c) -1 **67.** $y = e^x - 1$ **69.** $\ln 2 + \ln x$ **73.** a) Yes b) No c) Yes **75.** $f(x) = \cos x$ **77.** $F(1) - F(0)$ **79.** 10 **81.** Step 1: $s(t) = \frac{-kt^2}{2} + 44t$; Step 2: $t^* = \frac{44}{k}$; Step 3: $k = \frac{44^2}{90} \approx 21.5$ ft/sec^2 **83.** b) $240\sqrt{2} \approx 339$ volts **87.** $n > 15$; $h \le \frac{2}{15}$ **89.** $T = \pi$; $S = \pi$ **91.** Cost $\approx$ \$12,518.10; No **93.** $x \approx 59.8$ ft **95.** a) 3.11% b) 255.3

Chapter 6

Section 6.1, pp. 378–379

1. 2 **3.** 2 **5.** $\frac{\pi}{2}$ **7.** $\frac{32}{3}$ **9.** $\frac{32}{3}$ **11.** $\frac{1}{6}$ **13.** $\frac{9}{2}$ **15.** $\frac{9}{2}$ **17.** 4 **19.** $\frac{1}{6}$ **21.** $\frac{4}{3} - \frac{4}{\pi}$ **23.** $\frac{1}{12}$ **25.** 1 **27.** $\sqrt{2} - 1$ **29.** a) $c = \sqrt[3]{16}$ b) $c = \sqrt[3]{16}$ **31.** 4

Section 6.2, pp. 389–390

1. $\frac{8\pi}{3}$ **3.** 36π **5.** $\frac{128\pi}{7}$ **7.** π **9.** $\frac{32\pi}{3}$ **11.** 2π **13.** 4π **15.** $4\pi\ln 4$ **17.** $\frac{2\pi}{3}$ **19.** $\frac{128\pi}{5}$ **21.** $\frac{117\pi}{5}$ **23.** $\pi(\pi - 2)$ **25.** $\frac{4\pi}{3}$ **27.** 8π **29.** $\frac{500\pi}{3}$ **31.** $\pi^2 - 2\pi$ **33.** $\frac{\pi}{3}$ **35.** a) 8π b) $\frac{32\pi}{5}$ c) $\frac{8\pi}{3}$ d) $\frac{224\pi}{15}$ **37.** a) $\frac{16\pi}{15}$ b) $\frac{56\pi}{15}$ c) $\frac{64\pi}{15}$ **39.** $\frac{2}{\pi}$ **41.** 1053π cm^3

Section 6.3, pp. 397–398

1. 8π **3.** $\frac{3\pi}{2}$ **5.** 3π **7.** $\frac{4\pi}{3}$ **9.** $\frac{16\pi}{3}$ **11.** $\frac{8\pi}{3}$ **13.** $\frac{14\pi}{3}$ **15.** 2π **17.** a) $\frac{5\pi}{3}$ b) $\frac{4\pi}{3}$ **19.** a) $\frac{11\pi}{15}$ b) $\frac{97\pi}{105}$ c) $\frac{121\pi}{210}$ d) $\frac{23\pi}{30}$ **21.** a) $\frac{512\pi}{21}$ b) $\frac{832\pi}{21}$ **23.** a) $\frac{\pi}{6}$ b) $\frac{\pi}{6}$

Section 6.4, pp. 404–405

1. 12 **3.** $\frac{14}{3}$ **5.** $\frac{53}{6}$ **7.** $\frac{123}{32}$ **9.** $\frac{3}{2}$ **11.** 6 **13.** $y = \sqrt{x}$ **15.** $y = \ln(x)$ **17.** $\int_0^{1/2} \frac{1+x^2}{1-x^2}\,dx$ **19.** 21.07

Section 6.5, pp. 410–411

1. $4\sqrt{5}\pi$ **3.** $3\sqrt{5}\pi$ **5.** $\frac{98\pi}{81}$ **7.** 4π **9.** $\frac{\pi(2\sqrt{2}-1)}{9}$ **11.** $\frac{2\pi(2\sqrt{2}-1)}{3}$ **13.** $\frac{253\pi}{20}$ **15.** 4π **17.** $\frac{12\pi}{5}$ **19.** 21.07 in.

Section 6.6, pp. 416–419

1. 400 ft-lb **3.** 925 N · m **5.** 64800 ft-lb **7.** 1.2 N · m **9.** a) 104.17 ft-lb b) 312.5 ft-lb **11.** a) 1500000 ft-lb b) 1 hr 40 min **13.** 245436.93 ft-lb **15.** 7238229.48 ft-lb **17.** a) 34582.65 ft-lb b) 53482.5 ft-lb **19.** 21446605.85 N · m **21.** 967610.54 ft-lb; \$4,838.05 **23.** 5.144×10^{10} N · m

Section 6.7, pp. 423–425

1. 2812.5 lb **3.** 375 lb; no, the length is of no concern. **5.** 1166.67 lb **7.** $\frac{125}{3}$ lb **9.** 1309 lb **11.** 1161 lb **13.** 1034 ft^3

Section 6.8, pp. 435–436

1. 4 ft **3.** $M_0 = 8, \bar{x} = 1$ **5.** $M_0 = \frac{68}{3}, \bar{x} = \frac{17}{7}$
7. $\bar{x} = 0, \bar{y} = \frac{2}{3}$ **9.** $\bar{x} = \frac{16}{105}, \bar{y} = \frac{8}{15}$ **11.** $\bar{x} = \frac{3}{5}, \bar{y} = 1$
13. $\bar{x} = 0, \bar{y} = \frac{\pi}{8}$ **15.** $\bar{x} = 1, \bar{y} = -\frac{4}{5}$ **17.** $\bar{x} = \bar{y} = \frac{2}{4 - \pi}$
19. $\bar{x} = 0, \bar{y} = 1$ **21.** $\bar{x} = \bar{y} = \frac{a}{3}$ **23.** $\bar{x} = \frac{14}{9}, \bar{y} = \frac{\ln 4}{3}$
25. $V = 32\pi, S = 32\sqrt{2}\pi$ **27.** $4\pi^2$ **29.** $\bar{x} = 0, \bar{y} = \frac{2a}{\pi}$
31. $\bar{x} = 0, \bar{y} = \frac{4a}{3\pi}$ **33.** $\frac{\sqrt{2}\pi a^3(4 + 3\pi)}{6}$

Section 6.9, pp. 445–447

1. 16 **3.** $\frac{16}{3}$ **5.** $\pi \ln 2$ **7.** $\frac{8}{3}$ **9.** s^2h, s^2h

11. a) Positive for $0 < t < \pi/2$ and $3\pi/2 < t < 2\pi$; negative for $\pi/2 < t < 3\pi/2$
b) 20 m c) 0 m

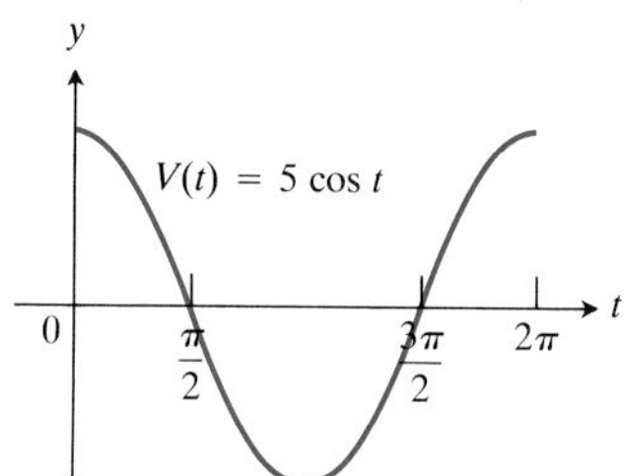

13. a) Positive for $0 < t < \pi/3$; negative for $\pi/3 < x < 2\pi/3$
b) 6 m c) 2 m

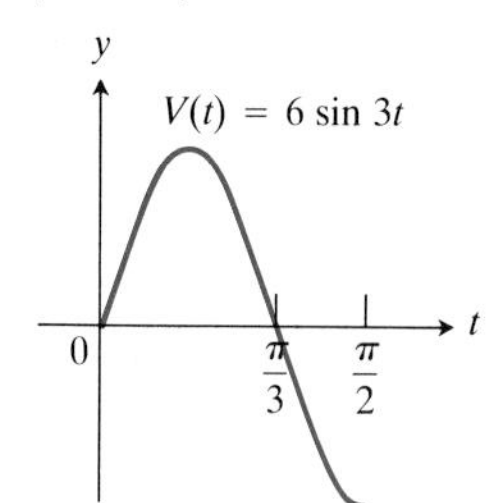

15. a) Positive for $0 < t < 5$; negative for $5 < t < 10$
b) 245 m c) 0 m

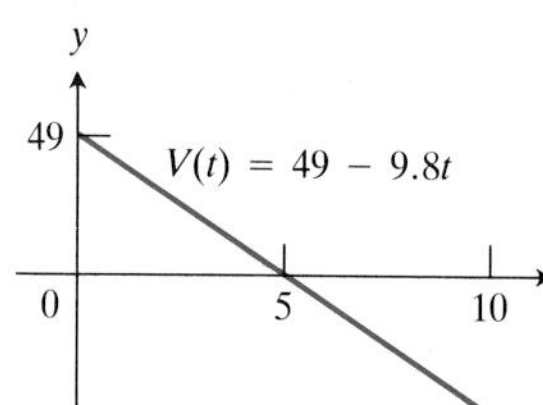

17. a) Positive for $0 < t < 1$; negative for $1 < t < 2$
b) 6 m c) 4 m

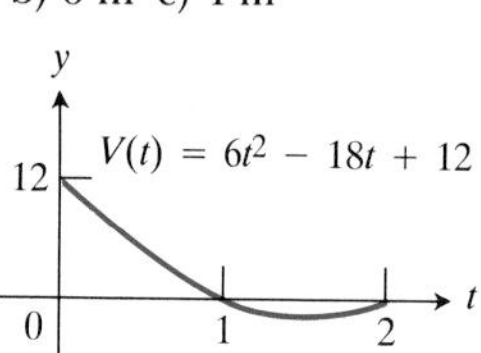

19. a) Distance = 2, shift = 2 b) Distance = 4, shift 0
c) Distance = 4, shift 4 d) Distance = 2, shift 2
21. 0000 **23.** $\sqrt{3}\,\pi$

Chapter 6 Practice Exercises, pp. 448–450

1. 1 **3.** $\frac{9}{2}$ **5.** 18 **7.** $\frac{9}{8}$ **9.** $\frac{-32 + 16\sqrt{2} + \pi^2}{32}$
11. 4 **13.** $\frac{8\sqrt{2} - 7}{6}$ **15.** 3 **17.** a) 2π b) π

19. a) 8π b) $\frac{1088\pi}{15}$ c) $\frac{512\pi}{15}$ **21.** $\frac{\pi(3\sqrt{3} - \pi)}{3}$ **23.** $\pi \ln 16$
25. $\frac{28\pi}{3}$ **27.** $2\sqrt{3}$ **29.** $6 + \ln 2$ **31.** $\frac{4\sqrt{2}\pi[\sqrt{2197} - 1]}{3}$
33. 3π **35.** $\frac{3\pi}{2}$ **37.** 4560 N · m **39.** 10 ft-lb, 30 ft-lb
41. 418879.02 ft-lb **43.** 333.33 lb **45.** 2200 lb
47. $\bar{x} = 0, \bar{y} = \frac{8}{5}$ **49.** $\bar{x} = \frac{3}{2}, \bar{y} = \frac{12}{5}$ **51.** $\bar{x} = 0, \bar{y} = \frac{8}{15}$
53. $\bar{x} = 7, \bar{y} = \frac{\ln 2}{3}$ **55.** $\frac{9\pi}{280}$ **57.** π^2 **59.** 18

61. a) Positive for $2 < t < 6$; negative for $0 < t < 2$
b) 5 ft c) 3 ft

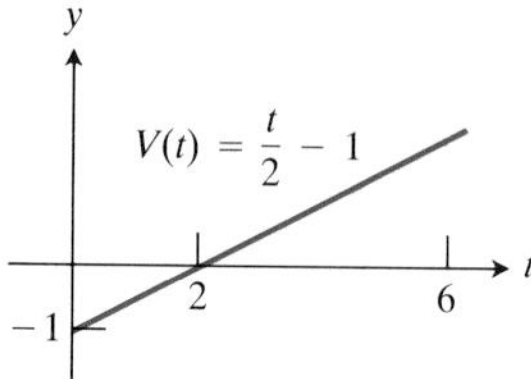

63. a) Positive for $0 < t < \pi/2$; negative for $\pi/2 < t < 3\pi/2$
b) 15 ft c) −5 ft

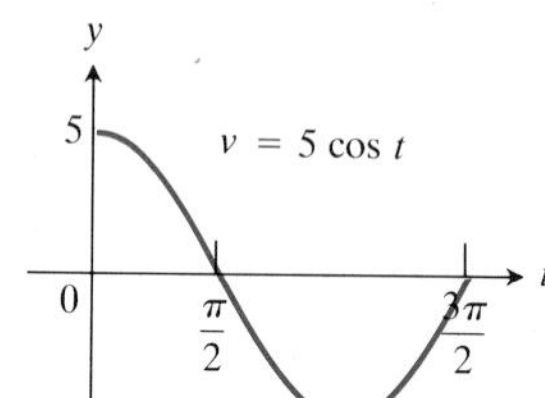

Chapter 7

Section 7.1, p. 456

1. a) $f^{-1}(x) = \frac{x - 3}{2}$ b)

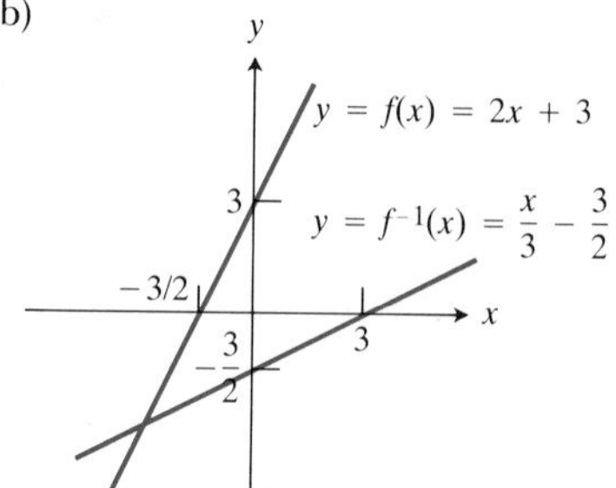

c) $\left.\frac{df}{dx}\right|_{x=-1} = 2; \left.\frac{df^{-1}}{dx}\right|_{x=1} = \frac{1}{2}$

3. a) $f^{-1}(x) = 5x - 35$ b)

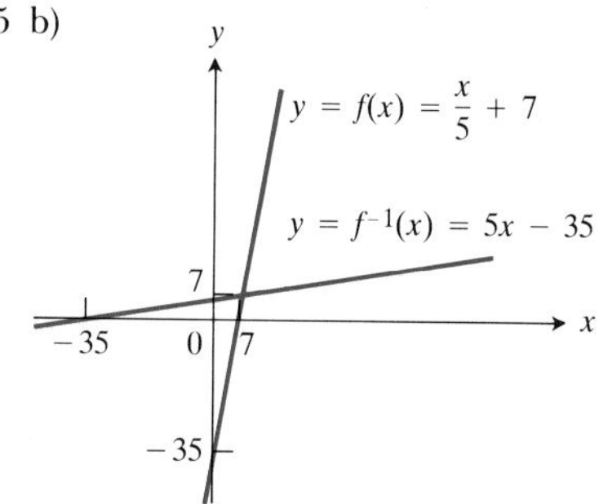

c) $\left.\frac{df}{dx}\right|_{x=-1} = \frac{1}{5}; \left.\frac{df^{-1}}{dx}\right|_{x=\frac{34}{5}} = 5$

5. $f^{-1}(x) = \sqrt{x-1}$ **7.** $f^{-1}(x) = \sqrt[3]{x+1}$ **9.** $f^{-1}(x) = \sqrt[5]{x}$
11. $f^{-1}(x) = \sqrt[3]{x-1}$ **13.** $f^{-1}(x) = \dfrac{1}{\sqrt{x}}$ **15.** $f^{-1}(x) = \sqrt{x} - 1$
17. $y = 0, x = 0$ **19.** $1/4$

Section 7.2, pp. 464–465

1. $\frac{1}{2}\sqrt{x(x+1)}\left(\frac{1}{x} + \frac{1}{x+1}\right)$
3. $\sqrt{x+3}(\sin x)\left(\frac{1}{2(x+3)} + \cot x\right)$
5. $x(x+1)(x+2)\left(\frac{1}{x} + \frac{1}{x+1} + \frac{1}{x+2}\right)$
7. $\frac{x+5}{x\cos x}\left(\frac{1}{x+5} - \frac{1}{x} + \tan x\right)$
9. $\frac{x\sqrt{x^2+1}}{(x+1)^{2/3}}\left(\frac{1}{x} + \frac{x}{x^2+1} - \frac{2}{3(x+1)}\right)$
11. $\frac{1}{3}\sqrt[3]{\frac{x(x-2)}{x^2+1}}\left(\frac{1}{x} + \frac{1}{x-2} - \frac{2x}{x^2+1}\right)$
13. $\ln\frac{2}{3}$ **15.** $\ln\frac{2}{5}$ **17.** $-\ln 2$ **19.** $\ln\frac{16}{25}$ **21.** $\ln 3$
23. $(\ln 2)^2$ **25.** 1 **27.** $\ln 2$ **29.** $\ln 27$ **31.** $-\infty$
33. 0 **35.** $\ln 2$ **37.** $\ln 2$ **39.** $L(x) = x + 1$
43. $\frac{1}{2e}$ **45.** $e^{\sin x} - 1$ **47.** $\ln\left(\frac{81}{16}\right) + 1$ **49.** 2.71828183

Section 7.3, pp. 472–474

1. $\pi x^{\pi-1}$ **3.** $-\sqrt{2}x^{(-\sqrt{2}-1)}$ **5.** $2^x \ln 2$ **7.** $-2^{-x}\ln 2$
9. $2^{\sec x}(\ln 2)\sec x \tan x$ **11.** $x^{\ln x}\left(\frac{\ln x^2}{x}\right)$
13. $(\sin x)^{(\tan x)}((\sec^2 x)(\ln \sin x) + 1)$ **15.** $\frac{3}{\sqrt{3}+1}$ **17.** $3^{(\sqrt{2}+1)}$
19. $\frac{4}{\ln 5}$ **21.** $\frac{1}{\ln 4}$ **23.** $\frac{3}{2}$ **25.** $\frac{1}{\ln 2}$ **27.** 2 **29.** -2 **31.** 2
33. $\frac{\ln 3}{\ln 2}$ **35.** $\frac{1}{2}$ **37.** 12 **39.** $\frac{1}{x\ln 4}$ **41.** $\frac{1}{\ln 10}$ **43.** $\frac{1}{x}$
45. $\frac{1}{(\ln 100)(x+1)}$ **47.** $-\frac{\ln 2}{x(\ln x)^2}$ **49.** $\frac{2}{(\ln 5)(x+1)}$
51. $\frac{\ln 10}{2}$ **53.** $2(\ln 2)^2$ **55.** $\frac{3\ln 2}{2}$ **57.** $\ln 10$ **59.** a) ∞
b) $-\infty$ **61.** a) ∞ b) 0 **63.** a) $2^{\ln x}\left(\frac{\ln 2}{x}\right)$ b) $\ln 2$ c) $\frac{2}{x}$
d) $\frac{2\ln x}{x}$ **65.** $(10^{-7.44}, 10^{-7.37})$ **67.** $k = 10$

Section 7.4, pp. 479–481

1. a) -0.00001 b) 10,536 years c) 82% **3.** 54.88 g
5. 0.585 days **7.** a) 550 ft b) 77.28 sec **9.** a) 17.5 min
b) 13.26 min **11.** -3°C **13.** 4.2 sec **15.** 6,658.3 yr
17. 41.22 yr

Section 7.5, pp. 486–487

1. $\frac{1}{4}$ **3.** $\frac{3}{11}$ **5.** 0 **7.** $\ln 3$ **9.** $\frac{5}{7}$ **11.** $\frac{1}{4}$ **13.** 0 **15.** $-\frac{5}{3}$
17. 0 **19.** e **21.** e **23.** b **25.** a) 1 b) $\frac{\pi}{2}$ c) π

Section 7.6, pp. 492–493

1. a, b, c, e, f, g, h **3.** a, b, c, h **7.** a) False b) False
c) True d) True e) False f) True g) False h) True
11. b) $e^{17000000}$ **13.** n

Section 7.7, pp. 497–498

1. a) $\frac{\pi}{4}$ b) $\frac{\pi}{3}$ c) $\frac{\pi}{6}$ **3.** a) $\frac{\pi}{6}$ b) $\frac{\pi}{4}$ c) $\frac{\pi}{3}$ **5.** a) $\frac{\pi}{3}$ b) $\frac{\pi}{4}$ c) $\frac{\pi}{6}$
7. a) $\frac{\pi}{4}$ b) $\frac{\pi}{6}$ c) $\frac{\pi}{3}$ **9.** a) $\frac{\pi}{4}$ b) $\frac{\pi}{3}$ c) $\frac{\pi}{6}$ **11.** a) $\frac{\pi}{4}$ b) $\frac{\pi}{6}$ c) $\frac{\pi}{3}$
13. $\cos\alpha = \frac{\sqrt{3}}{2}$, $\tan\alpha = \frac{1}{\sqrt{3}}$, $\sec\alpha = \frac{2}{\sqrt{3}}$, $\csc\alpha = 2$
15. $\frac{1}{\sqrt{2}}$ **17.** 2 **19.** $\frac{2}{\sqrt{3}}$ **21.** $\frac{1}{\sqrt{2}}$ **23.** 0 **25.** 0 **27.** π
29. $-\frac{\pi}{3}$ **31.** 0.6 **33.** $\frac{2\pi}{3}$ **35.** $\frac{\pi}{2}$ **37.** $\frac{\pi}{2}$ **39.** $\frac{\pi}{2}$
41. 0 **45.** $\arccos\frac{1}{\sqrt{3}} \approx 54.7°$ **47.** $\cot^{-1}2 = 26.57°$,
$\sec^{-1}(1.5) = 48.19°$, $\csc^{-1}(1.5) = 41.81°$

Section 7.8, pp. 503–504

1. $\frac{-2x}{\sqrt{1-x^4}}$ **3.** $\frac{15}{1+9x^2}$ **5.** $\frac{1}{\sqrt{4-x^2}}$ **7.** $\frac{1}{|x|\sqrt{25x^2-1}}$
9. $\frac{-2x}{(x^2+1)\sqrt{x^4+2x^2}}$ **11.** 0 **13.** $\frac{-1}{2x\sqrt{x-1}}$ **15.** $\frac{x|x|-1}{|x|\sqrt{x^2-1}}$
17. $2\arctan x$ **19.** $\frac{\pi}{6}$ **21.** $\frac{\pi}{12}$ **23.** π **25.** $\frac{\pi}{12}$ **27.** $\frac{\pi}{6}$
29. $\frac{\pi}{3}$ **31.** $\frac{\pi}{12}$ **33.** $2\left(\ln\frac{\pi}{3} - \ln\frac{\pi}{4}\right) = 2\ln\frac{4}{3}$ **35.** $\frac{\pi}{12}$
37. 1 **39.** 1 **41.** $\frac{\pi^2}{2}$ **43.** $3\sqrt{5}$ ft **45.** The maximum of θ is $\frac{\pi}{2}$ when $x = 1$. **47.** $\sec^{-1}|x| + \frac{2\pi}{3}$ **49.** $\cos^{-1}x - \pi/4$
51. 0.643517104
53. Yes, $\sin^{-1}x$ and $-\cos^{-1}x$ differ by the constant $\frac{\pi}{2}$.

Section 7.9, pp. 512–515

1. $\cosh x = \frac{5}{4}$, $\tanh x = -\frac{3}{5}$, $\coth x = -\frac{5}{3}$, $\operatorname{sech} x = \frac{4}{5}$, $\operatorname{csch} x = -\frac{4}{3}$ **3.** $\sinh x = \frac{8}{15}$, $\tanh x = \frac{8}{17}$, $\coth x = \frac{17}{8}$, $\operatorname{sech} x = \frac{15}{17}$, $\operatorname{csch} x = \frac{15}{8}$ **5.** $x + \frac{1}{x}$ **7.** e^{5x} **9.** e^{-3x}

13. $3\cosh 3x$ **15.** $\operatorname{sech}^2\frac{x}{2}$ **17.** $-\tanh x$ **19.** $-\operatorname{csch} x$

21. $\operatorname{csch} 2x$ **23.** a) $\sinh 2x$ b) $\sinh 2x$ c) $\sinh 2x$

25. $\frac{2}{\sqrt{1+4x^2}}$ **27.** $\frac{1}{1+x} - \tanh^{-1}x$ **29.** $\operatorname{sech}^{-1}x - \frac{1}{\sqrt{1-x^2}}$

31. $\sec x$ **33.** $\sec x$ **35.** $-\csc x$ **37.** $\frac{e^5}{5} - \frac{1}{5e^5}$ **39.** 0

41. e **43.** $\frac{3}{4}$ **45.** $\frac{4}{5}$ **47.** $2\sinh 2$ **49.** 0 **51.** $\ln 3$

53. $\frac{\ln 3}{2}$ **55.** $\ln 3$ **57.** $\ln 3$ **59.** $\ln 5$

61. a) $\sinh^{-1}(1)$ b) $\ln(1+\sqrt{2})$

63. a) $\cosh^{-1}\left(\frac{5}{3}\right) - \cosh^{-1}\left(\frac{5}{4}\right)$ b) $\ln\left(\frac{3}{2}\right)$

65. a) $\coth^{-1}(2) - \coth^{-1}\left(\frac{5}{4}\right)$ b) $\frac{\ln(1/3)}{2}$

67. a) $\frac{1}{2}\left(\operatorname{csch}^{-1}\left(\frac{1}{2}\right) - \operatorname{csch}^{-1}(1)\right)$ b) $\frac{1}{2}\ln\left(\frac{2+\sqrt{5}}{1+\sqrt{2}}\right)$

69. 2π **71.** $\bar{x} = 0, \bar{y} = \frac{3}{2\pi}$ **75.** b) $\sqrt{\frac{mg}{k}}$ c) $80\sqrt{5}$ ft/sec

77. 99 **79.** $\pi\left(\ln(64) + \frac{63}{4}\right)$

Chapter 7 Practice Exercises, pp. 516–520

1. $-\tan x$ **3.** $\frac{-1}{\sqrt{1-x^2}\cos^{-1}x}$ **5.** $\frac{2}{(\ln 2)x}$ **7.** $e^{\tan^{-1}x}\left(\frac{1}{1+x^2}\right)$

9. $8^{-x}\ln\frac{1}{8}$ **11.** $\frac{-1}{2\sqrt{x-x^2}}$ **13.** 2 **15.** $\tan^{-1}x + \frac{x}{1+x^2} - \frac{1}{2x}$

17. $\frac{\sec^{-1}\sqrt{x}}{\sqrt{x-1}} + \frac{1}{x}$ **19.** $\frac{2(x^2+1)}{\sqrt{\cos 2x}}\left(\frac{2x}{x^2+1} + \tan 2x\right)$

21. $5\left(\frac{(x+5)(x-1)}{(x-2)(x+3)}\right)^5\left(\frac{1}{x+5} + \frac{1}{x-1} - \frac{1}{x-2} - \frac{1}{x-3}\right)$

23. $(2x+1)e^{\tan^{-1}x}$ **25.** $1 + \operatorname{csch}^2 x$ **27.** $-x\operatorname{csch}^2 x$

29. $\operatorname{sech} x$ **31.** $\frac{x\sinh^{-1}x}{\sqrt{1+x^2}} + 1$ **33.** $\frac{-1}{1-x^2}$ **35.** $2\sec 2x$

37. $-\frac{1}{3}\ln 7$ **39.** $\ln 8$ **41.** $\ln\left(\frac{9}{25}\right)$ **43.** $\ln(\sqrt{2}+1)$

45. $\frac{9}{4}\ln 2$ **47.** $\frac{1}{\ln 3}$ **49.** π **51.** $\frac{\pi}{2}$ **53.** $\frac{\pi}{6}$ **55.** $3 + \ln 4$

57. $8\sqrt{2}$ **59.** $5\operatorname{csch}^2(2) - 5\operatorname{csch}^2(4)$ **61.** a) $\sinh^{-1}(1)$ b) 0

63. a) $2\left(\tanh^{-1}\left(\frac{1}{2}\right)\right)^2 - \left(\tanh^{-1}\left(\frac{1}{5}\right)^2\right)$ b) $\ln\left(\frac{9}{2}\right)\ln\sqrt{2}$

65. a) $\left(\operatorname{sech}^{-1}\left(\frac{3}{5}\right)\right)^2 - \left(\operatorname{sech}^{-1}\left(\frac{4}{5}\right)\right)^2$ b) $(\ln 6)\left(\ln\left(\frac{3}{2}\right)\right)$

67. $\frac{1}{3}$ **69.** $\ln\left(\frac{5}{3}\right)$ **71.** $\frac{1}{\ln 2}, \frac{1}{\ln 4}$, 1:2 **73.** 18,935 yr

75. 92 min **77.** ∞ **79.** 2 **81.** $\ln 2$ **83.** 1 **85.** e^3

87. a) Same b) Same c) Same

89. a) True b) False c) True

91. $\pi/2$ **93.** 17.54 m **95.** $y = \operatorname{sech}^{-1}x - \sqrt{1-x^2}$

Chapter 8

Section 8.1, pp. 527–528

1. 4 **3.** $\ln 5$ **5.** $2\ln|\sec x^2| + C$ **7.** $\ln(2+\sqrt{3})^6$

9. $\ln(\sqrt{2}+1)$ **11.** 1 **13.** $\frac{2}{\ln 3}$ **15.** π **17.** $\frac{\pi}{18}$ **19.** π

21. $\sin^{-1}(x-2) + C$ **23.** 2π

25. $\sec^{-1}|x+1| + C$, when $|x+1| > 1$

27. $\frac{8-\pi}{2}$ **29.** 0 **31.** $1 - \ln 2$ **33.** $7 + \ln 8$ **35.** $\frac{2\pi - 3}{6}$

37. $\sqrt{2}$ **39.** $\ln(2+\sqrt{3})$ **41.** $\bar{x} = 0, \bar{y} = \frac{1}{\ln(2\sqrt{2}+3)}$

43. $25\ln\left(\frac{\sqrt{2}+1}{\sqrt{3}}\right)$ cm **45.** 2 **47.** $-\ln 3$ **49.** $\frac{\ln(\sqrt{2}+1)}{\pi}$

51. $e^{\sqrt{3}} - 1$ **53.** $\frac{2}{\ln 2}$ **55.** π **57.** $\frac{\pi}{6}$ **59.** $\frac{\pi}{12}$

Section 8.2, p. 534

1. $-x\cos x + \sin x + C$

3. $-x^2\cos x + 2x\sin x + 2\cos x + C$

5. $\ln(4) - \frac{3}{4}$

7. $x\tan^{-1}(x) - \ln\sqrt{1+x^2} + C$

9. $x\tan x + \ln|\cos x| + C$ **11.** $(x^3 - 3x^2 + 6x - 6)e^x + C$

13. $(x^2 - 7x + 7)e^x + C$

15. $(x^5 - 5x^4 + 20x^3 - 60x^2 + 120x - 120)e^x + C$

17. $\frac{\pi^2 - 4}{8}$ **19.** $\frac{2\pi}{3} - \frac{\sqrt{3}}{2}$

21. $\frac{e^x}{2}(\sin x - \cos x) + C$

23. $\frac{e^{2x}}{13}(3\sin(3x) + 2\cos(3x)) + C$

25. a) π b) 3π **27.** $2\pi - \frac{4\pi}{e}$ **29.** $\pi^2 + \pi - 4$

Section 8.3, pp. 541–542

1. $\frac{8}{15}$ **3.** $\frac{4}{3}$ **5.** $\frac{16}{35}$ **7.** 3π **9.** π **11.** 2 **13.** 1 **15.** 4

17. 2 **19.** $\ln(3+2\sqrt{2})$ **21.** $\sqrt{2}$ **23.** $2\sqrt{3} + \ln(2-\sqrt{3})$

25. $\frac{4}{3}$ **27.** $\frac{4}{3}$ **29.** $2 - \ln 4$ **31.** $\frac{4}{3} - \ln\sqrt{3}$ **33.** $-\frac{6}{5}$

35. π **37.** 0 **39.** d, f, and h are not zero.

41. d, f, and h; $\int_{-\pi/2}^{\pi/2} x \sin x\, dx = 2$, $\int_{-\pi/2}^{\pi/2} \cos^3 x\, dx = \frac{4}{3}$, $\int_{-\pi/2}^{\pi/2} \sin x \sin 2x\, dx = \frac{4}{3}$

Section 8.4, pp. 547–548

1. $\frac{\pi}{4}$ **3.** $\frac{\pi}{6}$ **5.** $\ln|x + \sqrt{x^2 - 4}| + C$ **7.** $\frac{\pi}{3}$ **9.** $\frac{\pi}{8}$

11. $\frac{\pi}{3}$ **13.** $\ln(\sqrt{2} + 1)$ **15.** π **17.** $\frac{\pi}{8}$ **19.** $\frac{2 - \sqrt{3}}{2}$

21. $\ln|x - 1 + \sqrt{x^2 - 2x}| + C$

23. 2 **25.** $\frac{4x}{\sqrt{1 - x^2}} - 4\sin^{-1}(x) + C$ **27.** $\frac{3\pi}{4}$ **29.** 0.6435

Section 8.5, pp. 553–554

1. $\frac{2}{x - 3} + \frac{3}{x - 2}$ **3.** $\frac{1}{x + 1} + \frac{3}{(x + 1)^2}$

5. $-\frac{2}{x} - \frac{1}{x^2} + \frac{2}{x - 1}$ **7.** $1 - \frac{12}{x - 2} + \frac{17}{x - 3}$

9. $\ln\sqrt{3}$ **11.** $\frac{1}{7}\ln|(x + 6)^2(x - 1)^5| + C$ **13.** $\ln\sqrt{15}$

15. $-\frac{1}{2}\ln|t| + \frac{1}{6}\ln|t + 2| + \frac{1}{3}\ln|t - 1| + C$

17. $\frac{x^2}{2} - 2x + 3\ln|x + 1| + \frac{1}{x + 1} + C$

19. $\frac{1}{4}\ln\left|\frac{x + 1}{x - 1}\right| - \frac{x}{2(x^2 - 1)} + C$

21. $4 - \ln 3$ **23.** $\ln\frac{2\sqrt{2}}{\sqrt{5}}$

25. $5\sqrt{5} - \frac{5\pi}{3}$

27. $\ln\left(\frac{x^4}{(x^2 + 1)^2}\right) - \frac{4}{x} - 4\tan^{-1}(x) + C$

29. $\frac{2 - \pi}{4}$ **31.** $\frac{2 + \pi - \ln 4}{4}$

33. $\frac{\pi + 2}{4}$ **35.** $2 + \ln 3$ **37.** $3\pi\ln 25$

39. a) $x = \frac{1000\, e^{4t}}{499 + e^{4t}}$ b) $\frac{\ln 499}{4} = 1.55$ days **41.** 1

43. $\frac{2}{1 - \tan(x/2)} + C$ **45.** $-\cot(x/2) - x + C$

47. $\frac{1}{\sqrt{2}}\ln\left|\frac{\tan(x/2) + 1 - \sqrt{2}}{\tan(x/2) + 1 + \sqrt{2}}\right| + C$

Section 8.6, pp. 559–560

1. $\frac{\sqrt{\pi}}{2}$ **3.** $\frac{2}{\sqrt{3}}[\tan^{-1}(\sqrt{2}) - \frac{\pi}{4}]$

5. $\frac{x}{18(9 - x^2)} + \frac{1}{108}\ln\left|\frac{x + 3}{x - 3}\right| + C$

7. $\frac{44 - 24\sqrt{2}}{231}$ **9.** $\sqrt{2}(\frac{\pi}{4} - 1)$

11. $-\frac{1}{6}\ln\left|\frac{5 + 4\sin 2x + 3\cos 2x}{4 + 5\sin 2x}\right| + C$

13. $\frac{(2x - 3)^{3/2}(x + 1)}{5} + C$ **15.** 10!

17. $\frac{\pi}{4}$ **19.** $\frac{\pi - 2}{4}$ **21.** $\bar{x} = \frac{4}{3}, \bar{y} = \ln\sqrt{2}$ **27.** $\frac{3\pi}{4}$

29. $\frac{3\pi}{8}$ **31.** $\frac{1 - \ln 2}{4}$ **33.** $\frac{3\pi - 8}{6}$ **35.** $4\sqrt{3}$ **37.** $\frac{4}{3}$

39. $324(\ln 3)^2 - 162(\ln 3) + 40$

41. $\frac{1}{\sqrt{2}}$ **43.** $\frac{735 + 384\ln 2}{1024}$

Section 8.7, pp. 569–570

1. $\frac{\pi}{2}$ **3.** 6 **5.** 4 **7.** 1000 **9.** ln 4 **11.** Diverges

13. Converges **15.** Diverges **17.** Diverges

19. Converges **21.** Diverges **23.** Diverges

25. Diverges **27.** Converges **29.** Diverges

31. Diverges **33.** Converges **35.** Diverges

37. Converges **39.** Converges **41.** 0.8862

45. 1 **47.** 2π **49.** ln 2

Chapter 8 Practice Exercises, pp. 571–572

1. $2\sqrt{2} - 2$ **3.** 0

5. $\frac{\pi^2}{32}$ **7.** $\frac{\ln\sqrt{3}}{2}$

9. $\ln\sqrt{x^2 + 1} + 4\tan^{-1}x + C$ **11.** $\frac{x^3\ln x}{3} - \frac{x^3}{9} + C$

13. $-x^5\cos x + 5x^4\sin x + 20x^3\cos x - 60x^2\sin x - 120x\cos x + 120\sin x + C$

15. $\frac{2e^x\sin 2x}{5} + \frac{e^x\cos 2x}{5} + C$

17. $-\cos y + \frac{\cos^3 y}{3} + C$

19. $-\frac{\sin^3 x\cos x}{24} + \frac{x}{16} - \frac{\sin 2x}{32} + \frac{\sin^5 x\cos x}{6} + C$

21. 0 **23.** $\frac{3 - \ln 4}{2}$ **25.** $\frac{3}{80}$ **27.** $-\frac{\sqrt{1 - x^2}}{x} + C$

29. 5 **31.** $\frac{\pi}{8}$ **33.** $\frac{\pi}{3}$ **35.** $\frac{48 + \ln 400}{3}$

37. $\ln|x - 1| - \frac{1}{x - 1} + C$ **39.** $\ln\left|\frac{x}{\sqrt{x^2 + 4}}\right| + C$

41. ln 3 **43.** 1

45. $\frac{32\pi}{35}$ **47.** $x = \frac{a x_0\, e^{(akt)}}{a - x_0 + x_0\, e^{(akt)}}$ **49.** 2π

Chapter 9

Section 9.1, pp. 582–584

1. $a_1 = 0, a_2 = -\frac{1}{4}, a_3 = -\frac{2}{9}, a_4 - \frac{3}{16}$

3. $a_1 = 1, a_2 = -\frac{1}{3}, a_3 = \frac{1}{5}, a_4 - \frac{1}{7}$

5. $1, \frac{3}{2}, \frac{7}{4}, \frac{15}{8}, \frac{31}{16}, \frac{63}{32}$ **7.** $2, 1, \frac{1}{2}, \frac{1}{4}, \frac{1}{8}, \frac{1}{16}$

9. 1, 1, 2, 3, 5, 8 **11.** Converges to 2

13. Converges to -1 **15.** Diverges

17. Converges to -5 **19.** Diverges

21. Converges to 5 **23.** Diverges

25. Converges to $\frac{1}{2}$

27. 0 **29.** 0 **31.** $\sqrt{2}$ **33.** $\frac{\pi}{2}$ **35.** 0 **37.** 0

39. Converges to 1 **41.** Converges to e^7 **43.** Diverges **45.** 1 **47.** 1 **49.** Diverges **51.** 0 **53.** 0 **55.** e^{-1} **57.** Diverges **59.** 693 **61.** 66

Section 9.2, pp. 591–593

1. 3 **3.** $\frac{2}{3}$ **5.** $\frac{1}{2}$ **7.** $\frac{4}{3}$ **9.** $\frac{7}{3}$ **11.** $\frac{23}{2}$ **13.** $\frac{13}{4}$ **15.** 1

17. $\frac{1}{9}$ **19.** Converges to $2 + \sqrt{2}$ **21.** Converges to 1

23. Diverges **25.** Converges to $\frac{e^2}{e^2 - 1}$ **27.** Diverges

29. $\frac{3}{2}$ **31.** Diverges **33.** $a = 1, r = -x$ **35.** 28 m

37. $\frac{26}{111}$

39. a) $\sum_{n=-2}^{\infty} \frac{1}{(n+4)(n+5)}$ b) $\sum_{n=0}^{\infty} \frac{1}{(n+2)(n+3)}$

c) $\sum_{n=5}^{\infty} \frac{1}{(n-3)(n-2)}$

43. Take $\sum_{n=1}^{\infty} n$ and $\sum_{n=1}^{\infty} (-n)$

45. $\sum_{n=0}^{\infty} \left[\frac{1}{5}\right] n$ converges to $\frac{5}{4}$ and $\sum_{n=0}^{\infty} \left[\frac{1}{2}\right] n$ converges to 2

but, $\sum_{n=0}^{\infty} \left[\frac{(1/5)^n}{(1/2)^n}\right]$ converges to $\frac{5}{3} \neq \frac{5/4}{2} = \frac{5}{8}$

Section 9.3, pp. 601–602

1. Converges; geometric series, $r = \frac{1}{10} < 1$ **3.** Converges; compare with $\sum \frac{1}{2^n}$ **5.** Converges; compare with $\sum \frac{2}{n^2}$

7. Diverges; compare with $\sum \frac{1}{n}$ **9.** Converges; geometric series, $r = \frac{2}{3} < 1$ **11.** Diverges; compare with $\sum \frac{1}{n+1}$

13. Diverges; $\lim_{x\to\infty} \frac{2^n}{n+1} \neq 0$ **15.** Converges; compare with $\sum \frac{1}{n^{3/2}}$ **17.** Diverges; compare with $\sum \frac{1}{n}$ **19.** Diverges; $\lim_{x\to\infty} \left[1 + \frac{1}{n}\right]^n \neq 0$ **21.** Converges; sum of two convergent series $\sum \frac{1}{n\,2^n}$ and $\sum \frac{-1}{2^n}$ **23.** Converges; compare with $\sum \frac{1}{3^{(n-1)}}$ **25.** 41.55

Section 9.4, p. 607

1. Converges; ratio test **3.** Converges; ratio test

5. Diverges; $\lim_{n\to\infty} \left[\frac{n-2}{n}\right]^n = e^{-2} \neq 0$ **7.** Diverges; ratio test

9. Diverges; $\lim_{n\to\infty} \left[1 - \frac{3}{n}\right]^n = e^{-3} \neq 0$

11. Converges; compare with $\sum \frac{1}{n^2}$

13. Diverges; compare with $\sum \frac{1}{n}$

15. Converges; ratio test

17. Converges; ratio test **19.** Converges; ratio test

21. Converges; root test

23. Converges; compare with $\sum \frac{1}{n^2}$

25. Converges; ratio test **27.** Converges; ratio test

29. Diverges; compare with $\sum \frac{1}{n}$

31. Converges; ratio test **33.** Converges; ratio test

35. Converges; compare with $\sum \frac{1}{n^4}$

Section 9.5, pp. 613–614

1. Converges **3.** Diverges

5. Converges **7.** Converges

9. Diverges **11.** Converges absolutely

13. Converges absolutely **15.** Converges conditionally

17. Diverges **19.** Converges conditionally

21. Converges absolutely **23.** Converges absolutely

25. Diverges **27.** Diverges

29. Converges absolutely **31.** Converges absolutely

33. Converges absolutely **35.** Converges conditionally

37. 0.2 **39.** 2×10^{-11}

41. 0.54030

43. a) $a_n \geq a_{n+1}$ fails b) $-\frac{1}{2}$ **45.** 0.6687714

Section 9.6, pp. 622–623

1. a) $-1 < x < 1$ b) $-1 < x < 1$
3. a) $-2 < x < 0$ b) $-2 < x < 0$
5. a) $-8 < x < 12$ b) $-8 < x < 12$
7. a) $-1 < x < 1$ b) $-1 < x < 1$
9. a) $-1 \le x \le 1$ b) $-1 \le x \le 1$
11. a) For all x b) For all x
13. a) For all x b) For all x
15. a) $-1 \le x < 1$ b) $-1 < x < 1$
17. a) $-1 \le x < 1$ b) $-1 < x < 1$
19. a) $-3 < x < 3$ b) $-3 < x < 3$
21. a) $-1 < x < 1$ b) $-1 < x < 1$
23. a) $x = 0$ b) $x = 0$
25. $1 < x < 5$; $\dfrac{2}{x-1}$
27. a) $\dfrac{x^2}{2} + \dfrac{x^4}{12} + \dfrac{x^6}{45} + \dfrac{17x^8}{2520}$; $-\dfrac{\pi}{2} < x < \dfrac{\pi}{2}$
b) $1 + x^2 + \dfrac{2x^4}{3} + \dfrac{17x^6}{45}$; $-\dfrac{\pi}{2} < x < \dfrac{\pi}{2}$

Section 9.7, pp. 634–636

1. $P_0(x) = 0$
$P_1(x) = x - 1$
$P_2(x) = (x-1) - \frac{1}{2}(x-1)^2$
$P_3(x) = (x-1) - \frac{1}{2}(x-1)^2 + \frac{1}{3}(x-1)^3$

3. $P_0(x) = \frac{1}{2}$
$P_1(x) = \frac{1}{2} - \frac{1}{4}(x-2)$
$P_2(x) = \frac{1}{2} - \frac{1}{4}(x-2) + \frac{1}{8}(x-2)^2$
$P_3(x) = \frac{1}{2} - \frac{1}{4}(x-2) + \frac{1}{8}(x-2)^2 - \frac{1}{16}(x-2)^3$

5. $P_0(x) = \frac{\sqrt{2}}{2}$
$P_1(x) = \frac{\sqrt{2}}{2} + \frac{\sqrt{2}}{2}\left(x - \frac{\pi}{4}\right)$
$P_2(x) = \frac{\sqrt{2}}{2} + \frac{\sqrt{2}}{2}\left(x - \frac{\pi}{4}\right) - \frac{\sqrt{2}}{4}\left(x - \frac{\pi}{4}\right)^2$
$P_3(x) = \frac{\sqrt{2}}{2} + \frac{\sqrt{2}}{2}\left(x - \frac{\pi}{4}\right) - \frac{\sqrt{2}}{4}\left(x - \frac{\pi}{4}\right)^2 - \frac{\sqrt{2}}{12}\left(x - \frac{\pi}{4}\right)^3$

7. $P_0(x) = 2$
$P_1(x) = 2 + \frac{1}{4}(x-4)$
$P_2(x) = 2 + \frac{1}{4}(x-4) - \frac{1}{64}(x-4)^2$
$P_3(x) = 2 + \frac{1}{4}(x-4) - \frac{1}{64}(x-4)^2 + \frac{1}{512}(x-4)^3$

9. $1 - x + \dfrac{x^2}{2!} - \dfrac{x^3}{3!} + \dfrac{x^4}{4!} - \cdots$ **11.** $3x - \dfrac{(3x)^3}{3!} + \dfrac{(3x)^5}{5!} - \cdots$

13. $1 - \dfrac{x^2}{2!} + \dfrac{x^4}{4!} - \dfrac{x^6}{6!} + \cdots$ **15.** $1 + \dfrac{x^2}{2!} + \dfrac{x^4}{4!} + \dfrac{x^6}{6!} + \cdots$
17. $\dfrac{x^4}{4!} - \dfrac{x^6}{6!} + \dfrac{x^8}{8!} - \dfrac{x^{10}}{10!} + \cdots$ **19.** $1 - x + x^2 + R_2(x)$
21. $x - \dfrac{x^2}{2} + R_2(x)$ **23.** $x + R_2(x)$
27. $-0.5697 < x < 0.5697$ **29.** 1.67×10^{-10}; $-10^{-3} < x < 0$ **31.** 0.000184 **33.** 0.00014
35. $\sin x$, when $x = 0.01$; the sum is 0.0998
39. $x + x^2 + \dfrac{x^3}{3} - \dfrac{x^5}{30} - \dfrac{7x^6}{360}$ **43.** a) -1 b) $\dfrac{1}{\sqrt{2}} + \dfrac{i}{\sqrt{2}}$ c) $-i$
47. $\displaystyle\int e^{ax}\cos(bx)\,dx = \frac{e^{ax}(a\cos(bx) + b\sin(bx))}{a^2+b^2} + C_1$
$\displaystyle\int e^{ax}\sin(bx)\,dx = \frac{e^{ax}(a\sin(bx) - b\cos(bx))}{a^2+b^2} + C_2$

Section 9.8, p. 642

1. $\dfrac{1}{2} - \dfrac{\sqrt{3}}{2}\left(x - \dfrac{\pi}{3}\right) - \dfrac{1}{4}\left(x - \dfrac{\pi}{3}\right)^2 + \dfrac{\sqrt{3}}{12}\left(x - \dfrac{\pi}{3}\right)^3$
3. $1 + x + \dfrac{x^2}{2!} + \dfrac{x^3}{3!}$
5. $1 - \dfrac{(x-22\pi)^2}{2!} + \dfrac{(x-22\pi)^4}{4!} - \dfrac{(x-22\pi)^6}{6!}$
7. 0.099944461 **9.** 0.100001
13. 501 terms **15.** 3 terms

Chapter 9 Practice Exercises, pp. 643–644

1. Converges to 1 **3.** Diverges **5.** Converges to 0
7. Converges to 3 **9.** Converges to 0 **11.** Diverges
13. Diverges **15.** $\dfrac{e}{e-1}$
17. Diverges; a p-series where $p = \dfrac{1}{2}$
19. Converges conditionally; Alternating Series Theorem
21. Converges conditionally; Alternating Series Theorem
23. Converges absolutely; compare with $\sum \dfrac{1}{n^2}$
25. Converges absolutely; ratio test **27.** Converges absolutely; ratio test **29.** a) $-5 \le x < 1$ b) $-5 < x < 1$
31. a) For all x b) For all x **33.** a) $0 \le x < 2$ b) $0 \le x \le 2$
35. $\dfrac{1}{1+x}$ when $x = \dfrac{1}{4}$; the sum is $\dfrac{4}{5}$
37. $\sin x$ when $x = \pi$; the sum is 0
39. e^x when $x = \ln 2$; the sum is 2
41. $2 - \dfrac{(x+1)}{(2)\,1!} + \dfrac{3(x+1)^2}{(2)^3\,2!} + \dfrac{9(x+1)^3}{(2)^5\,3!}$
43. $2x - \dfrac{(2x)^3}{3!} + \dfrac{(2x)^5}{5!} - \dfrac{(2x)^7}{7!}$ **45.** 0.48491676
47. a) 0; for all x, $x = 0$ b) $R_n(x) = e^{(-1/x^2)}$

Chapter 10

Section 10.1, pp. 658–660

1. $x = \frac{y^2}{8}$ **3.** $y = -\frac{x^2}{6}$ **5.** $\frac{x^2}{4} - \frac{y^2}{9} = 1$ **7.** $\frac{x^2}{2} + y^2 = 1$

9. $\frac{x^2}{25} + \frac{y^2}{16} = 1$

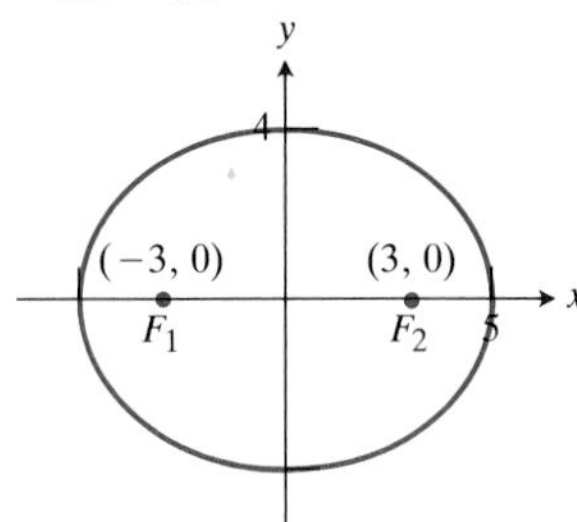

11. $x^2 + \frac{y^2}{2} = 1$; $e = \frac{1}{\sqrt{2}}$

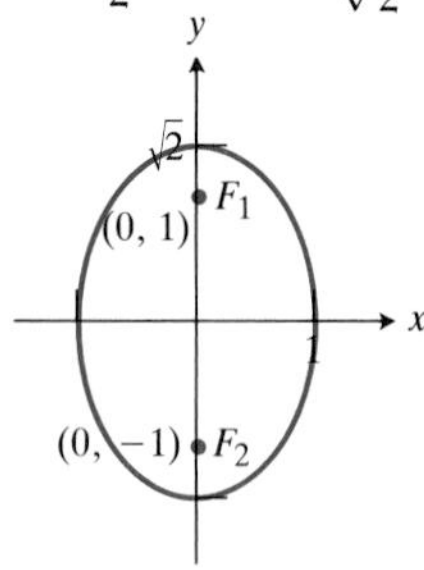

13. $\frac{x^2}{2} + \frac{y^2}{3} = 1$; $e = \frac{1}{\sqrt{3}}$

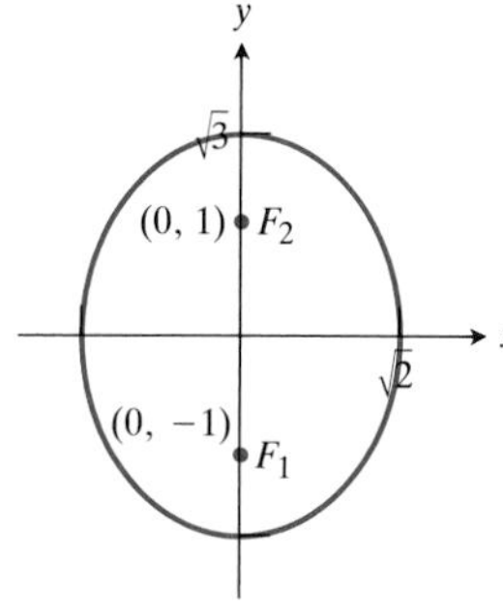

15. $\frac{x^2}{9} + \frac{y^2}{6} = 1$; $e = \frac{1}{\sqrt{3}}$

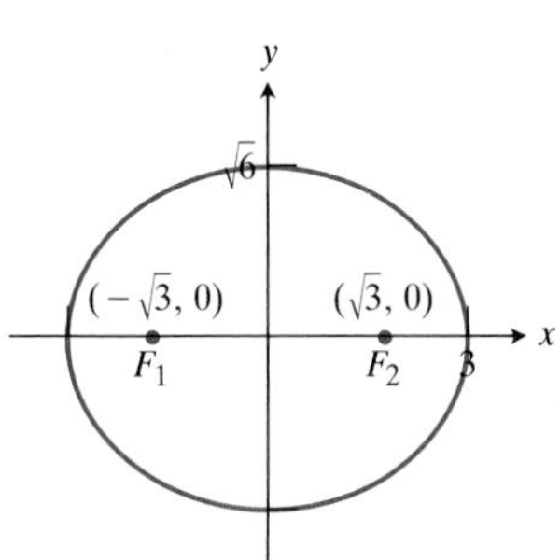

17. $x^2 - y^2 = 1$; $e = \sqrt{2}$; $y = \pm x$

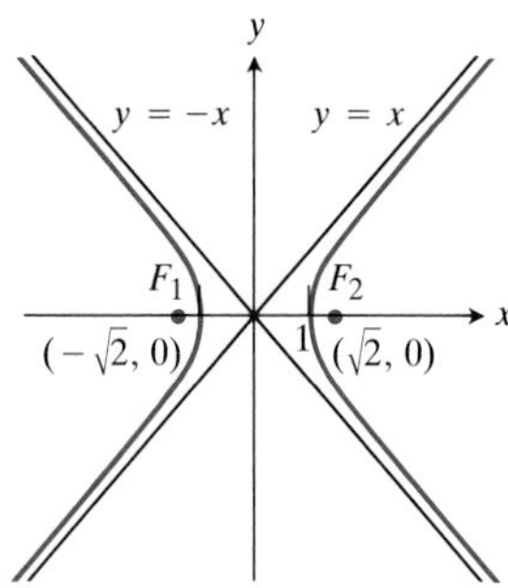

19. $\frac{y^2}{8} - \frac{x^2}{8} = 1$; $e = \sqrt{2}$; $y = \pm x$

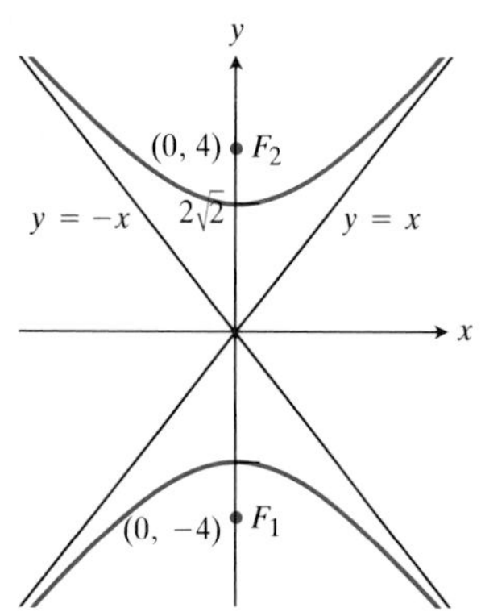

21. $\frac{x^2}{2} - \frac{y^2}{8} = 1$; $e = \sqrt{5}$; $y = \pm 2x$

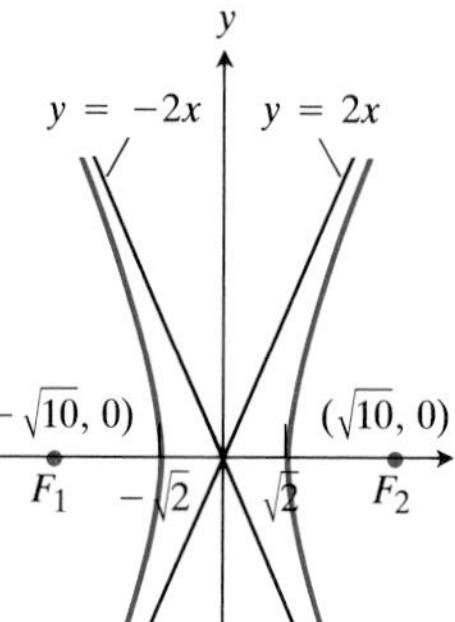

23. $\frac{y^2}{2} - \frac{x^2}{8} = 1$; $e = \sqrt{5}$; $y = \pm\frac{x}{2}$

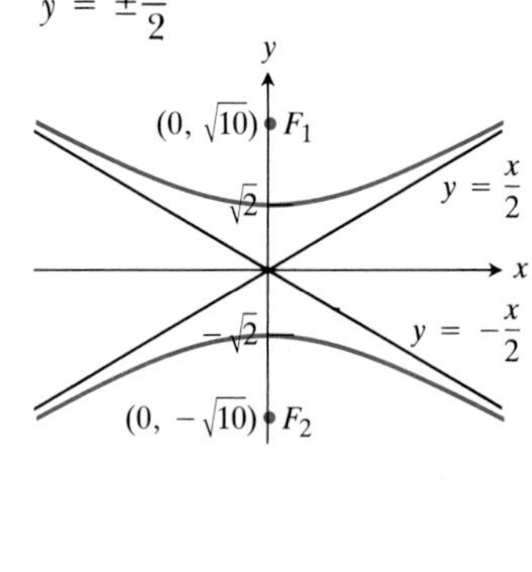

27. $\frac{x^2}{25} + \frac{y^2}{9} = 1$ **29.** $L = 2\sqrt{2}$, $W = \sqrt{2}$, area = 4 **31.** 24π

Section 10.2, pp. 665–666

1. Hyperbola **3.** Parabola **5.** Ellipse **7.** Parabola **9.** Hyperbola **11.** Hyperbola **13.** Ellipse **15.** Ellipse **17.** $x'^2 - y'^2 = 4$; hyperbola **19.** $4x'^2 + \frac{y'^2}{4} + 16y' = 0$; ellipse **21.** $3y'^2 = 2$; parallel lines **23.** $2\sqrt{2}x'^2 + 8\sqrt{2}y' = 0$; parabola **25.** $4x'^2 + 2y'^2 = 19$; ellipse **27.** a) $x'^2 - y'^2 = 2$ b) $x'^2 - y'^2 = 2a$

Section 10.3, pp. 670–672

1. $x^2 + y^2 = 1$

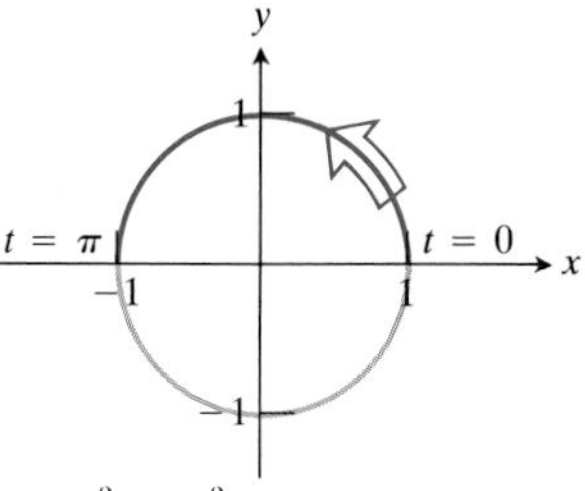

3. $x^2 + y^2 = 1$

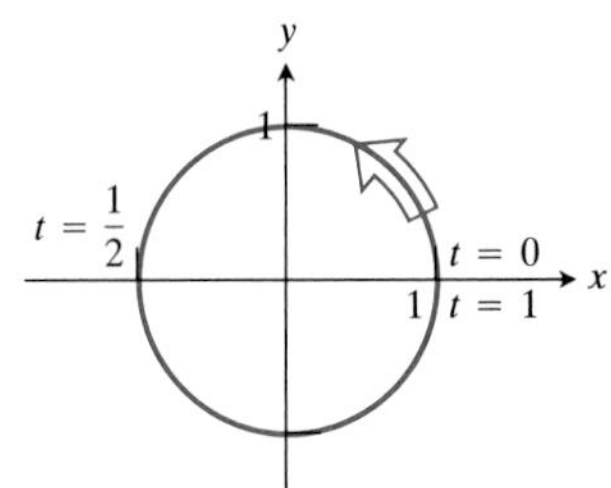

5. $\frac{x^2}{16} + \frac{y^2}{4} = 1$

7. $\frac{x^2}{16} + \frac{y^2}{25} = 1$

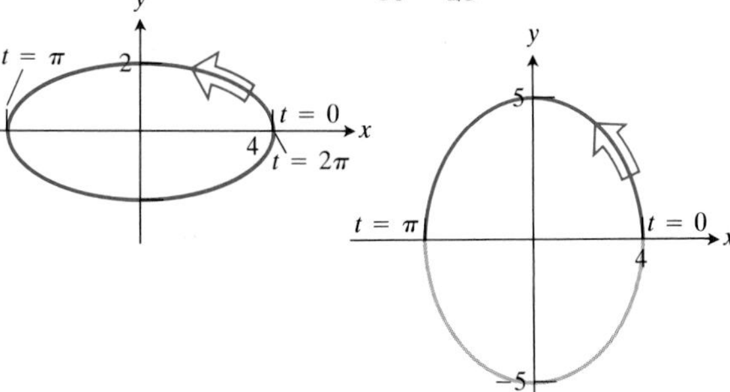

9. $y = x^2$ or $y = \sqrt{x}$

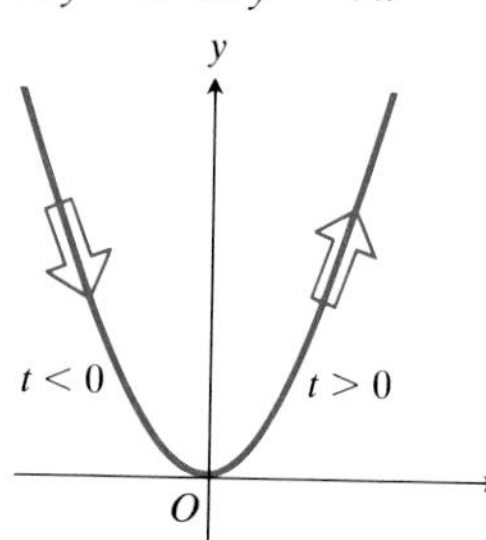

11. $x = y^2$

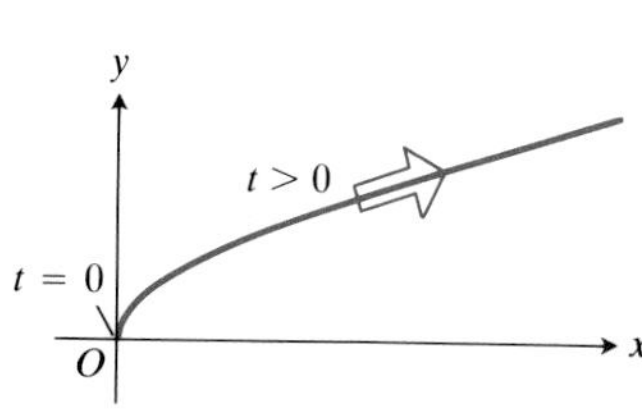

13. $x^2 - y^2 = 1$

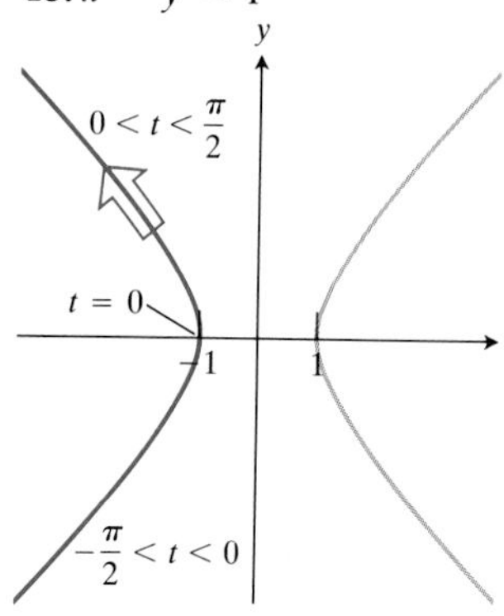

15. $y = 2x + 3$

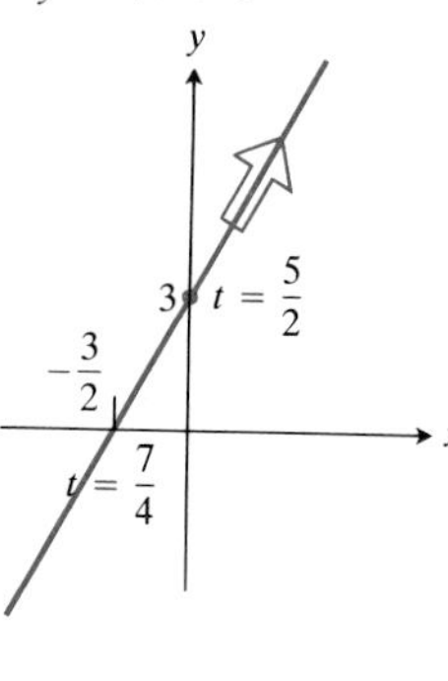

17. $y = 1 - x$

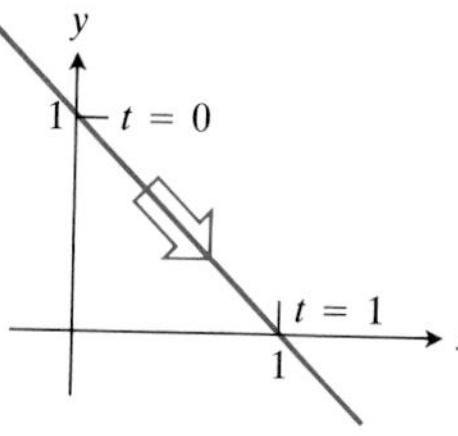

19. $y = \sqrt{4 - x^2}$

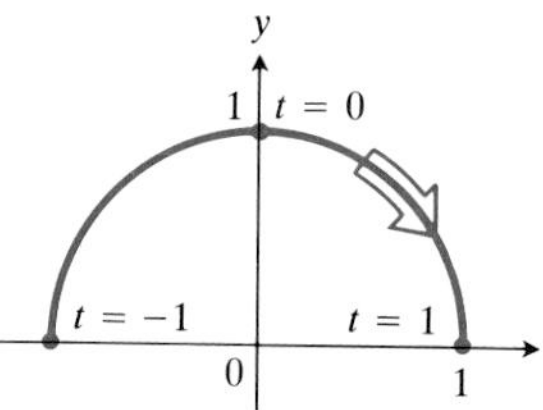

21. $y^2 - x^2 = 1$ **23.** $x^2 - y^2 = 1$

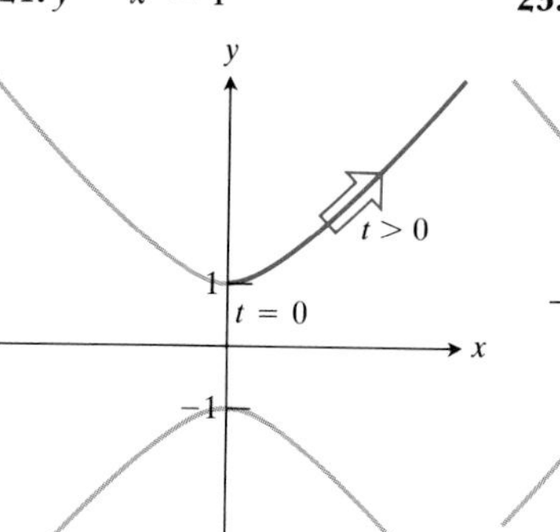

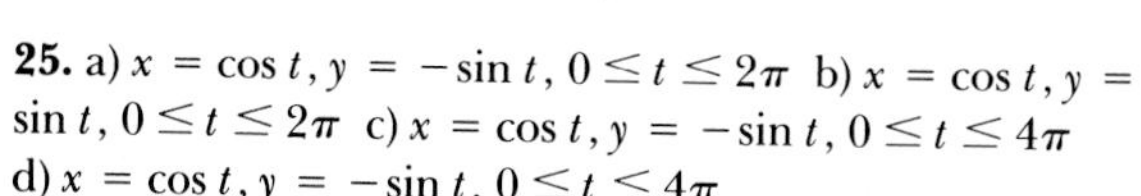

25. a) $x = \cos t, y = -\sin t, 0 \le t \le 2\pi$ b) $x = \cos t, y = \sin t, 0 \le t \le 2\pi$ c) $x = \cos t, y = -\sin t, 0 \le t \le 4\pi$ d) $x = \cos t, y = -\sin t, 0 \le t \le 4\pi$

27. $x = \cos t + t \sin t, y = \sin t - t \cos t$ **29.** $\sqrt[3]{\frac{1}{2}}, \sqrt[3]{\frac{1}{4}}$

31. b) $x = x_1 t, y = y_1 t$ (answer not unique) c) With $(x_0, y_0) = (-1, 0)$ and $(x_1, y_1) = (0, 1)$, we get $x = -1 + t, y = t$

Section 10.4, pp. 678–679

1. $y = -x + 2\sqrt{2}, \frac{d^2y}{dx^2} = 1$ **3.** $y = -\frac{1}{2}x + 2\sqrt{2}, \frac{d^2y}{dx^2} = -\frac{1}{2}$ **5.** $y = x + \frac{1}{4}, \frac{d^2y}{dx^2} = -\frac{1}{4}t^{-3/2}$ **7.** $y = 2x - \sqrt{3}, \frac{d^2y}{dx^2} = 2 \sec t$ **9.** $y = x - 4, \frac{d^2y}{dx^2} = 3t$ **11.** $y = \sqrt{3}x - \frac{\pi\sqrt{3}}{3} + 2$, $\frac{d^2y}{dx^2} = \frac{\cos t}{1 - \cos t}$ **13.** 4 **15.** 12 **17.** π^2 **19.** $8\pi^2$ **21.** $\frac{52}{3}\pi$ **23.** $3\sqrt{5}\pi$ **25.** a) π b) π **27.** $\left(\frac{\sqrt{2}}{2}, 1\right)$; $y = 2x$ at $t = 0$, $y = -2x$ at $t = \pi$

Section 10.5, p. 685

1. a, c; b, d; e, k; g, j; h, f; i, l; m, o; n, p

3. a) $\left(2, \frac{\pi}{2} + 2n\pi\right), \left(-2, \frac{\pi}{2} + (2n+1)\pi\right)$, n an integer

b) $(2, 2n\pi), (-2, (2n+1)\pi)$, n an integer

c) $\left(2, \frac{3\pi}{2} + 2n\pi\right), \left(-2, \frac{3\pi}{2} + (2n+1)\pi\right)$, n an integer

d) $(2, (2n+1)\pi), (-2, 2n\pi)$, n an integer

5. a) $(1, 1)$ b) $(1, 0)$ c) $(0, 0)$ d) $(-1, -1)$ e) $\left(\frac{3}{2}, -\sqrt{3}\right)$ f) $(3, 4)$ g) $(1, 0)$ h) $(-\sqrt{3}, 3)$

7.

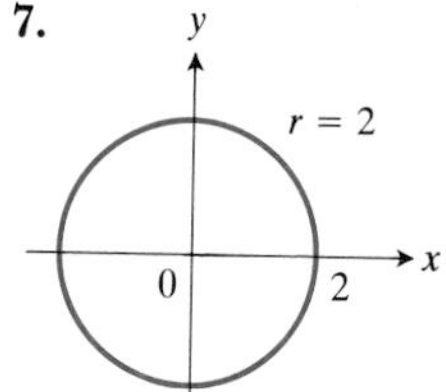

9.

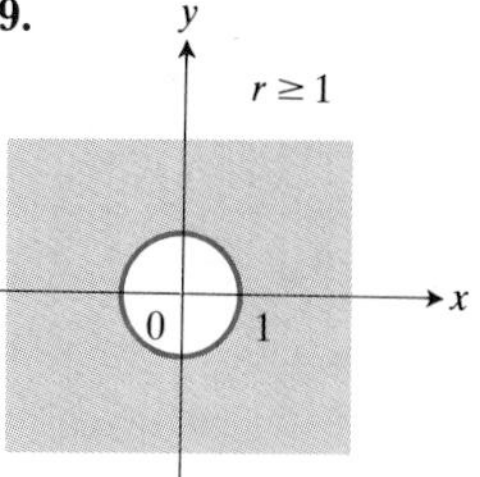

11.

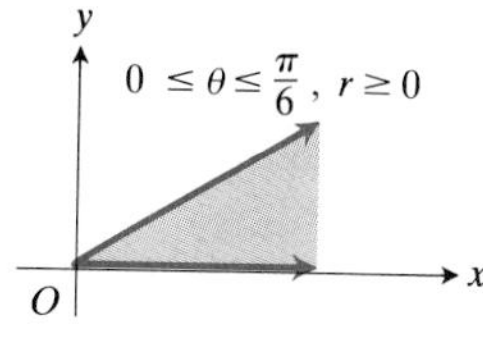

13.

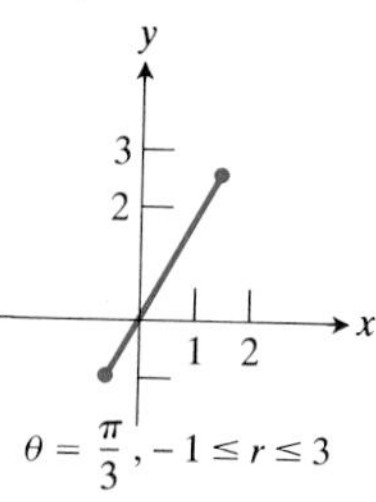

15.

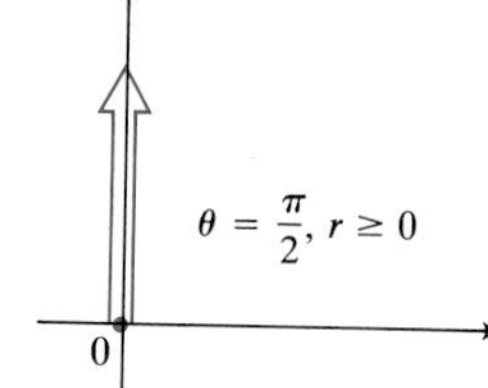

17.

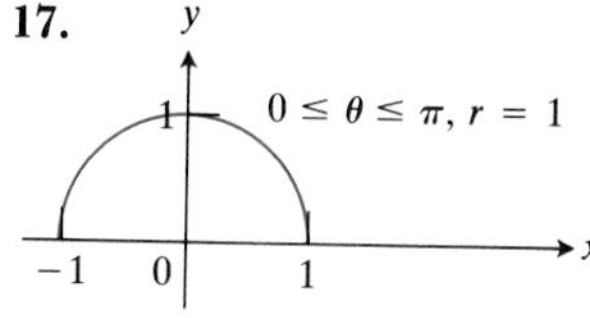

19.

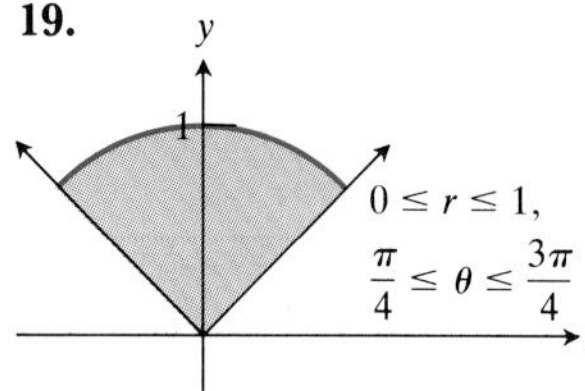

21.

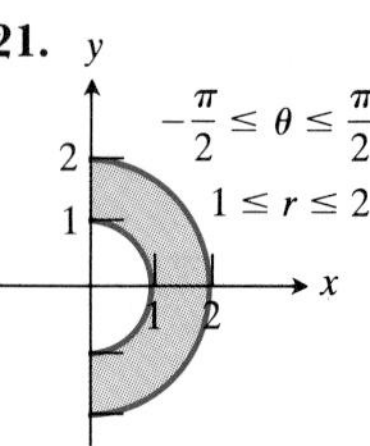

23. $x = 2$, vertical line through $(2, 0)$

25. $y = 4$, horizontal line through $(0, 4)$ **27.** $y = 0$, x-axis

29. $x + y = 1$, straight line with $m = 1, b = 1$

31. $x^2 + y^2 = 1$, circle with center $(0, 0)$, radius $= 1$

33. $y - 2x = 5$, straight line with $m = 2, b = 5$

35. $r\cos\theta = 7$ **37.** $r\cos\theta = r\sin\theta$ **39.** $r = 2$

41. $4r^2\cos^2\theta + 9r^2\sin^2\theta = 36$ **43.** $r^2\sin^2\theta = 4r\cos\theta$

Section 10.6, pp. 692–693

1.

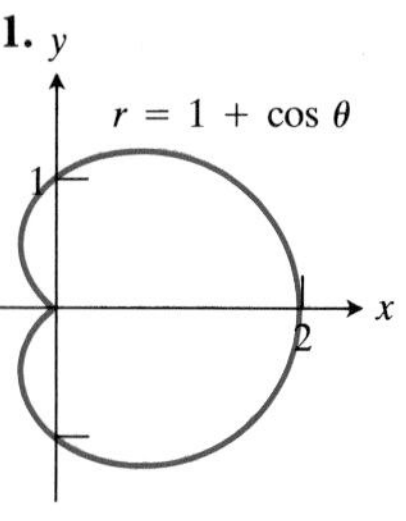

3.

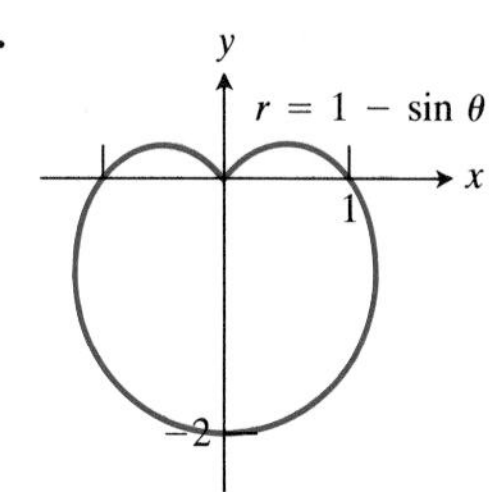

5.

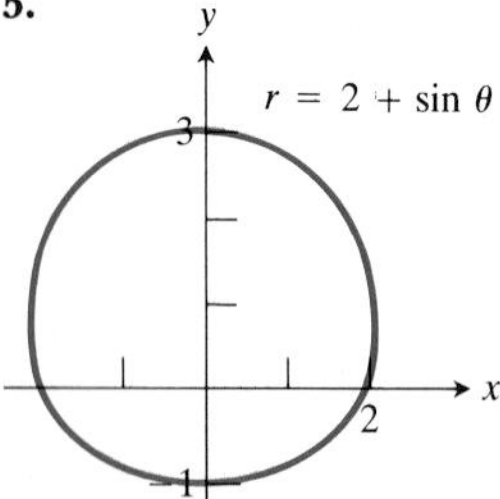

7.

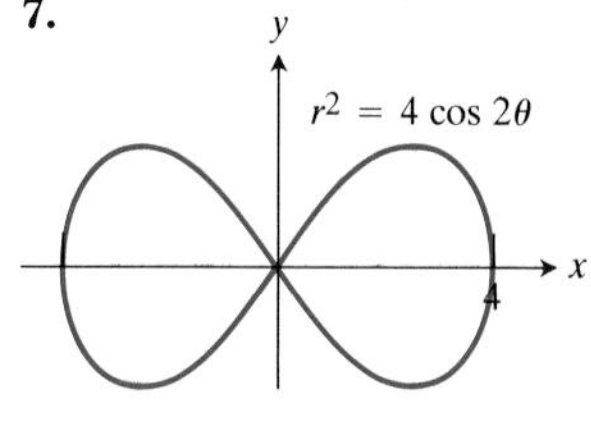

9.

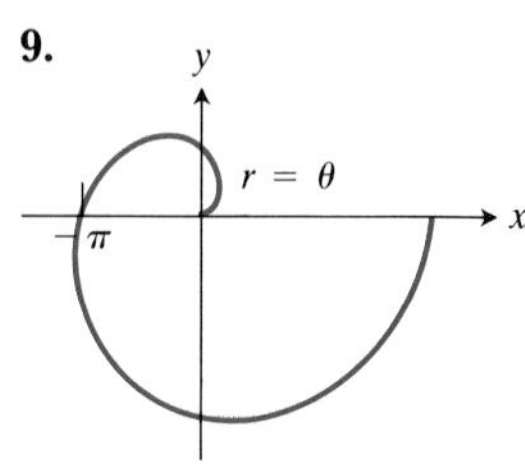

11. $\text{Slope}\big|_{\left(-1, \frac{\pi}{2}\right)} = -1$, $\text{Slope}\big|_{\left(-1, -\frac{\pi}{2}\right)} = 1$

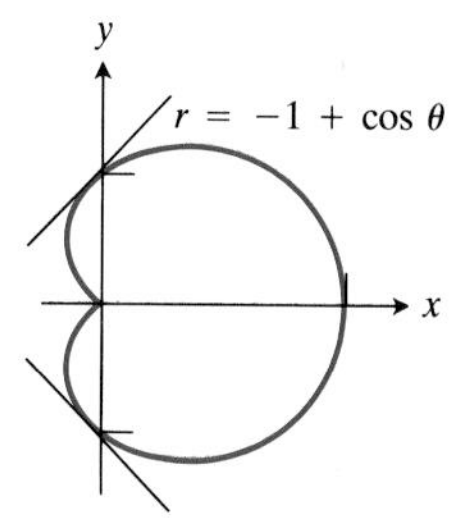

13. $\text{Slope}\big|_{\left(1, \frac{\pi}{4}\right)} = -1$, $\text{Slope}\big|_{\left(-1, -\frac{\pi}{4}\right)} = 1$, $\text{Slope}\big|_{\left(-1, \frac{3\pi}{4}\right)} = 1$, $\text{Slope}\big|_{\left(1, -\frac{3\pi}{4}\right)} = -1$, $\text{Slope}\big|_{\theta = 0;\, \pi} = 0$, $\text{Slope}\big|_{\theta = \frac{\pi}{2}, \frac{3\pi}{2}}$ does not exist

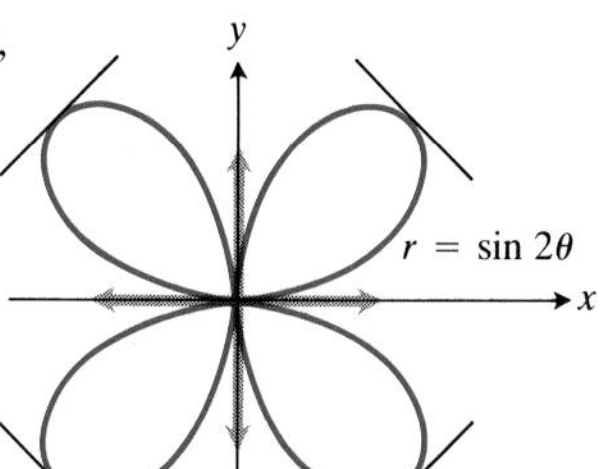

15.

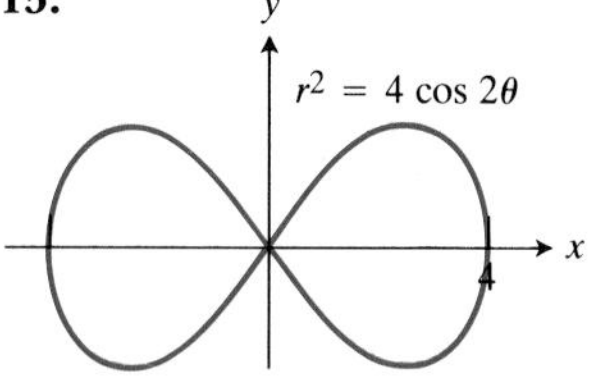

17. a)

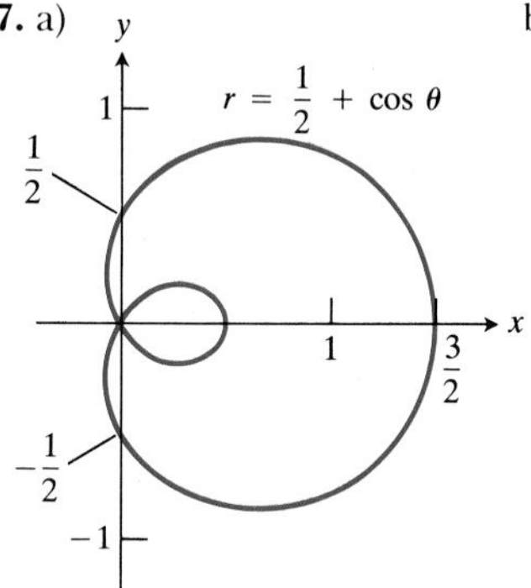

b)

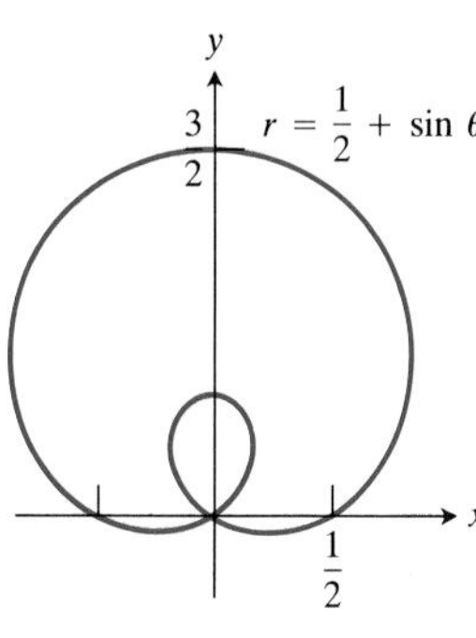

19. a)

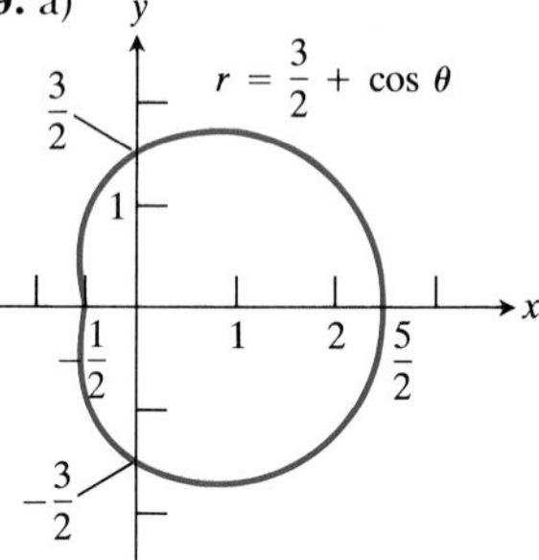

b)

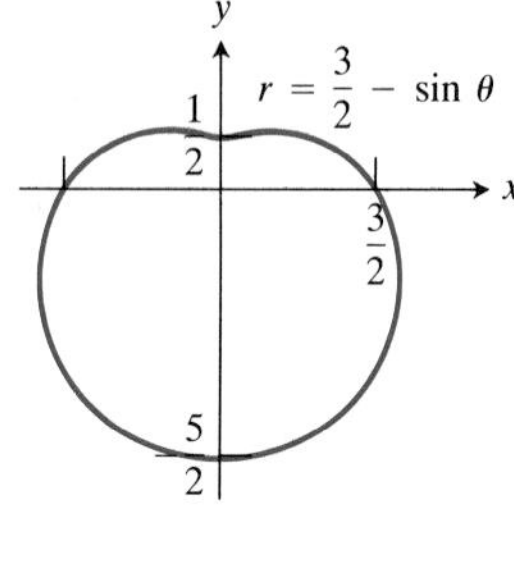

21. $0 \le r \le 2 - 2\cos\theta$

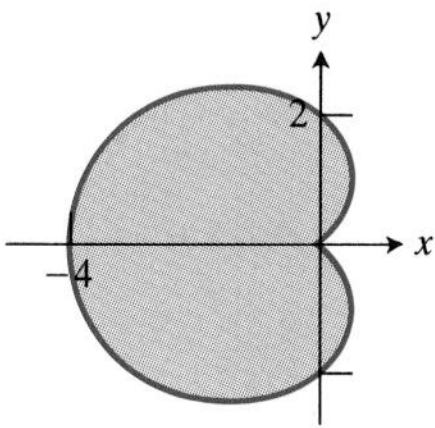

25. $(0, 0), \left(1, \frac{\pi}{2}\right), \left(1, \frac{3\pi}{2}\right)$

27. $(-2 + 2\sqrt{2}, \sin^{-1}(3 - 2\sqrt{2})), (0, 0), \left(2, \frac{3\pi}{2}\right)$

31. $\left(\pm 1/\sqrt[4]{2}, \frac{\pi}{8}\right), (0, 0)$

33. $\theta = \frac{\pi}{12}, \frac{5\pi}{12}, \frac{7\pi}{12}, \frac{11\pi}{12}, \frac{13\pi}{12}, \frac{17\pi}{12}, \frac{19\pi}{12}, \frac{23\pi}{12}$ and $r = 1$

Section 10.7, pp. 699–700

1. $r\cos\left(\theta - \frac{\pi}{6}\right) = 5$; $\sqrt{3}x + y = 10$

3. $r\cos\left(\theta - \frac{4\pi}{3}\right) = 3$; $-x - \sqrt{3}y = 6$ **5.**

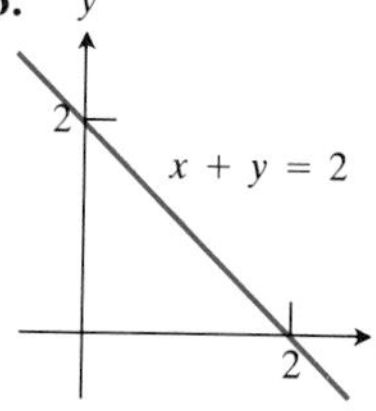

7.

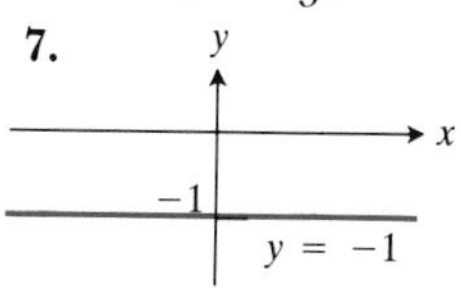

9. $r = 8\cos\theta$ **11.** $r = 2\sqrt{2}\sin\theta$

13. $r = 4\cos\theta$ **15.** $r = -2\cos\theta$ **17.** $r = \frac{2}{1 + \cos\theta}$

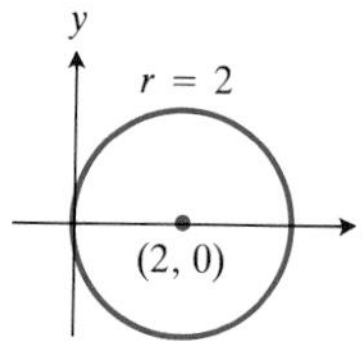

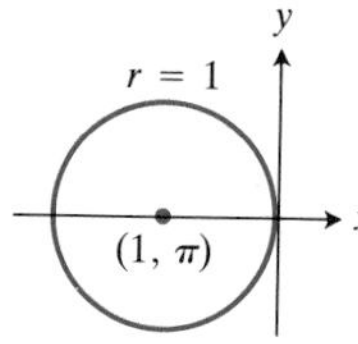

19. $r = \frac{8}{1 + 2\cos\theta}$ **21.** $r = \frac{1}{2 + 2\cos\theta}$ **23.** $r = \frac{-10}{5 - \sin\theta}$

25. $r = \frac{1}{1 + \cos\theta}$ **27.** $r = \frac{25}{10 - 5\cos\theta}$

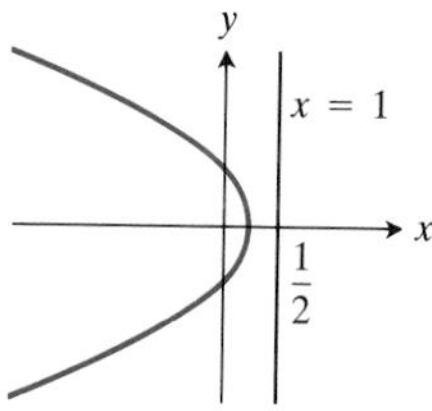

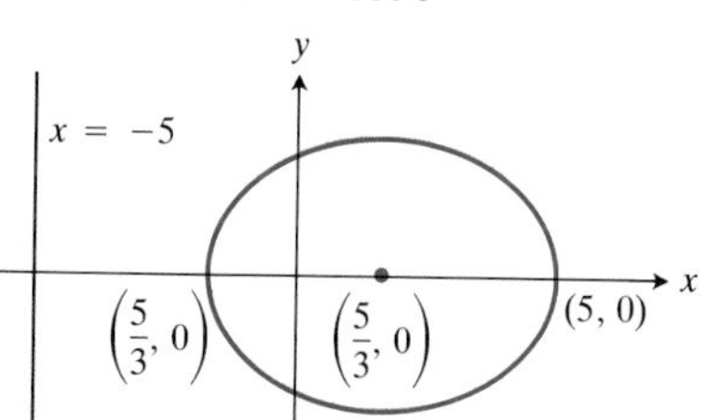

29. $r = \frac{400}{16 + 8\sin\theta}$ **31.** $r = \frac{8}{2 - 2\sin\theta}$

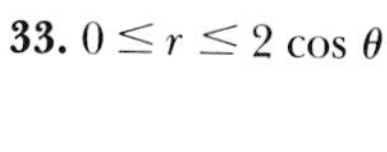

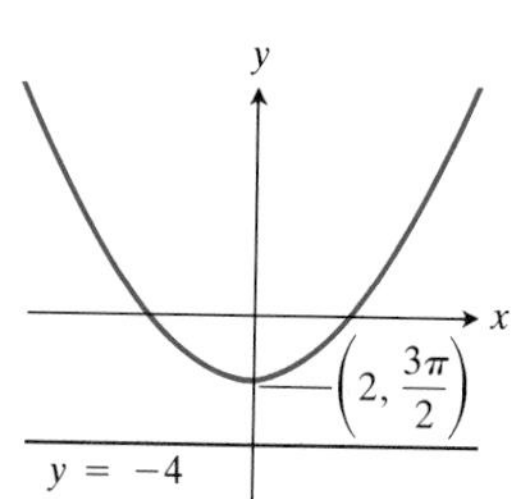

33. $0 \le r \le 2\cos\theta$

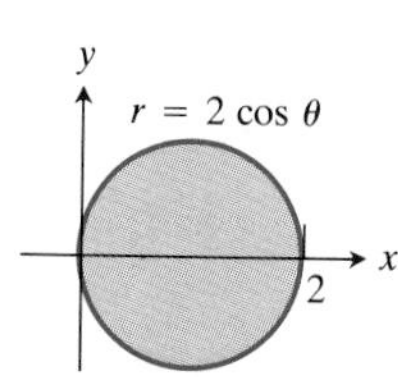

35. b)

Planet	Perihelion	Aphelion
Mercury	0.3075 AU	0.4667 AU
Venus	0.7184 AU	0.7282 AU
Earth	0.9833 AU	1.0167 AU
Mars	1.3817 AU	1.6663 AU
Jupiter	4.9512 AU	5.4548 AU
Saturn	9.0210 AU	10.0570 AU
Uranus	18.2977 AU	20.0623 AU
Neptune	29.8135 AU	30.3065 AU
Pluto	29.6549 AU	49.2251 AU

37. a) $x^2 + (y - 1)^2 = 1$; $y = 0$ b)

39. $r = \frac{4}{1 + \cos\theta}$

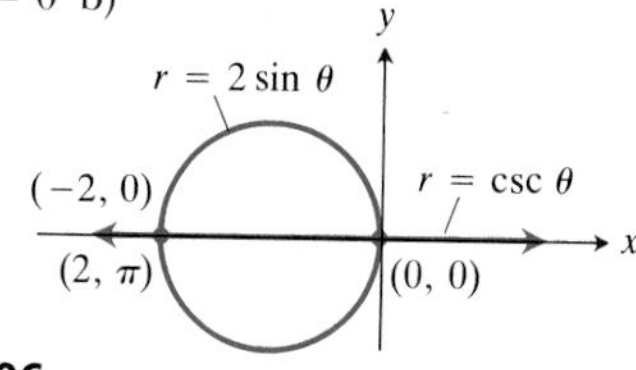

Section 10.8, pp. 705–706

1. $\frac{\pi}{16} + \frac{1}{8}$ **3.** 18π **5.** $\frac{\pi}{8}$ **7.** $\frac{1}{2}a^2$ **9.** 4 **11.** $\frac{4\pi - 3\sqrt{3}}{6}$

13. $6\pi - 16$ **15.** πa **17.** a) $2\pi + \frac{3\sqrt{3}}{2}$ b) $\pi + 3\sqrt{3}$

19. $\frac{19}{3}$ **21.** 1 **23.** 8 **25.** $\pi\sqrt{2}$ **27.** $2\pi(2 - \sqrt{2})$

29. $\bar{x} = \frac{5}{6}a$, $\bar{y} = 0$

Chapter 10 Practice Exercises, pp. 707–710

1. $x = \frac{y^2}{8}$ **3.** $\frac{x^2}{7} + \frac{y^2}{10} = 1$

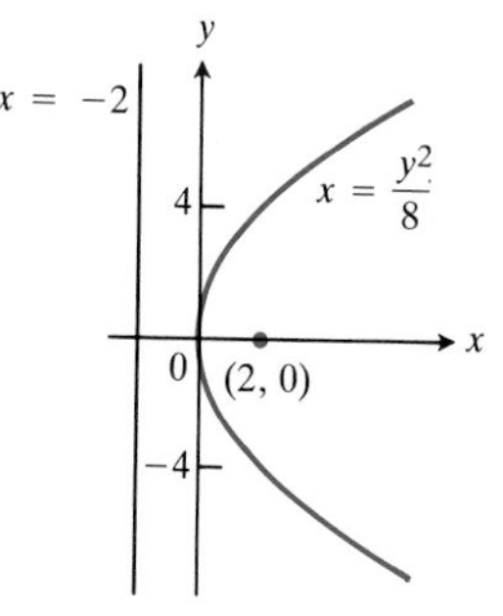

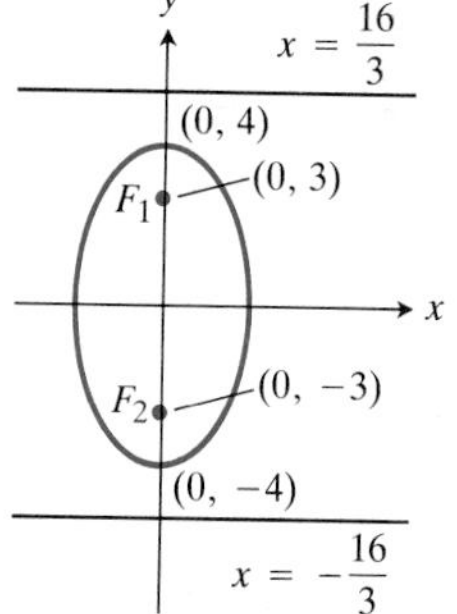

5. $x^2 - \frac{y^2}{3} = 1$

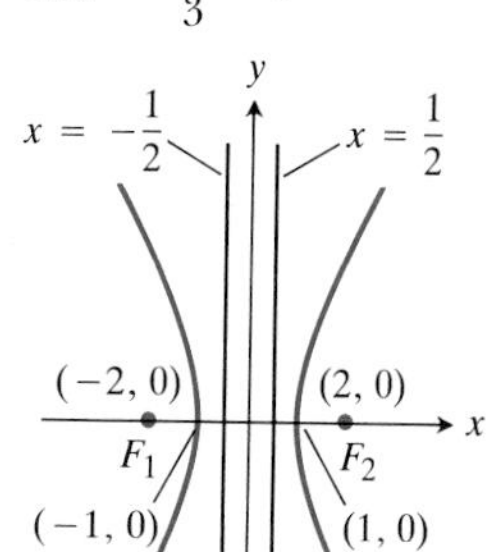

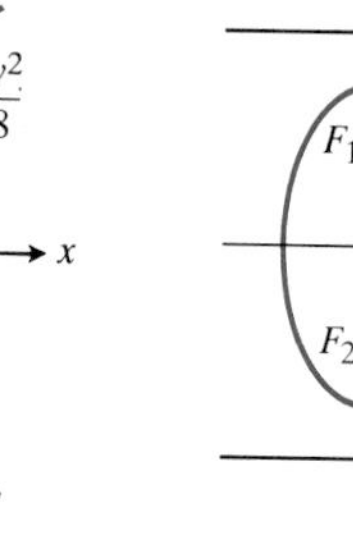

7. Ellipse
9. Parabola
11. Ellipse; $5x'^2 + 3y'^2 - 30 = 0$
13. $\sqrt{2}$
15. a) 24π b) 16π
17. $(2, 1)$, $(-2, 1)$

19.

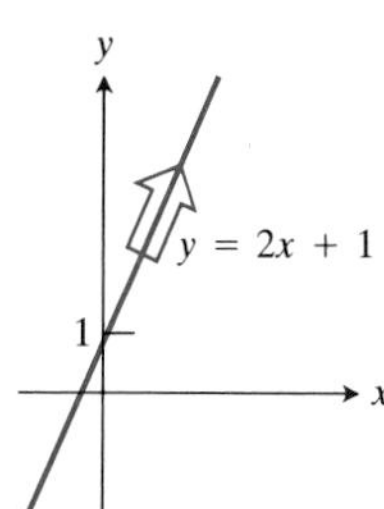

21.

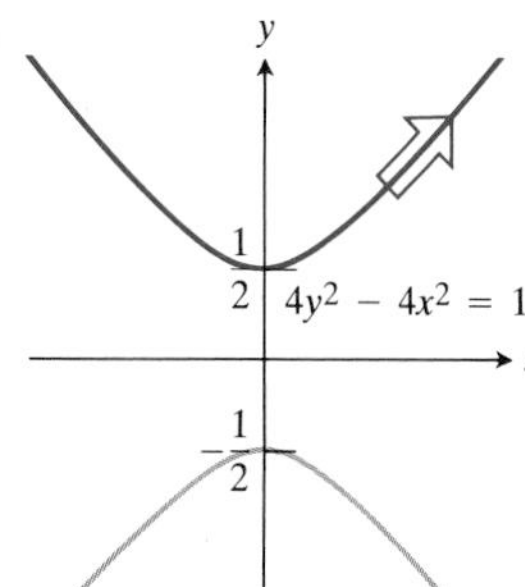

23.

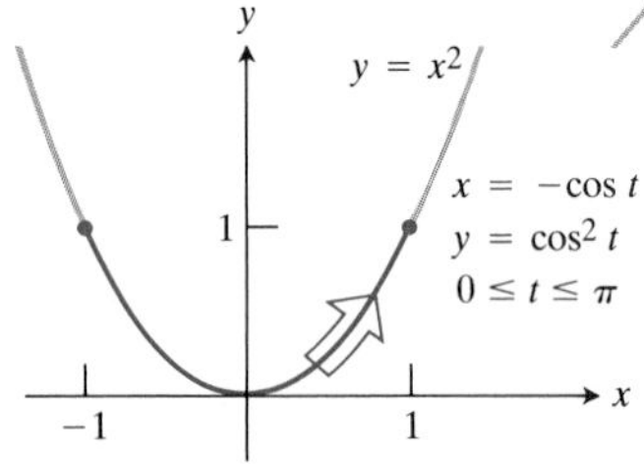

25. $x = 3\cos t,\ y = 4\sin t,\ 0 \le t \le 2\pi$

27. $y = \dfrac{2}{\sqrt{3}}x,\ \dfrac{d^2y}{dx^2} = 2\sec t$ **29.** $3 + \dfrac{\ln 2}{8}$ **31.** $\dfrac{76\pi}{3}$

33. d **35.** l **37.** k **39.** i

41.

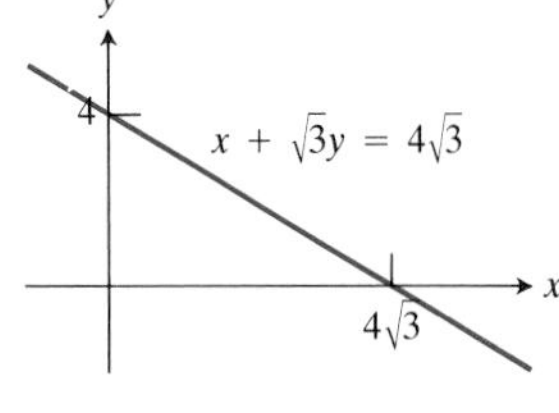

43. Center (0, 1), radius = 1

45. $0 \le r \le 6\cos\theta$

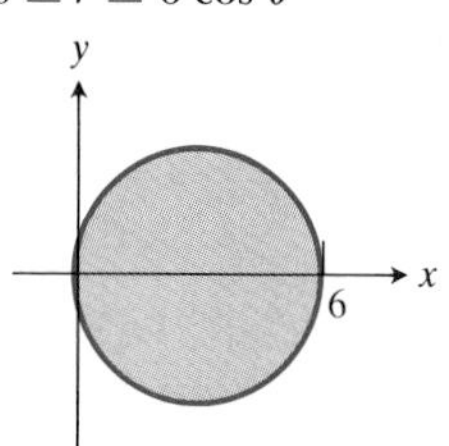

47. (0, 0)

49. Intersect at all points on $r = 1 + \sin\theta$

51. $r = \dfrac{2}{1 + \cos\theta}$

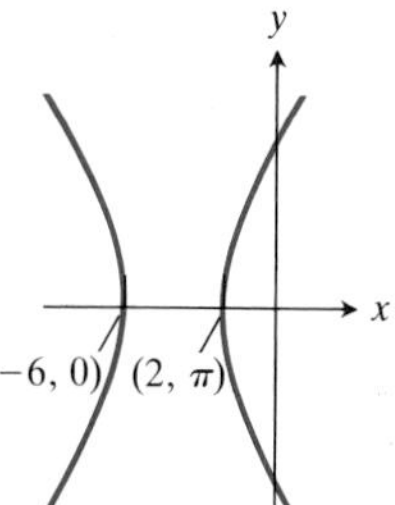

53. $r = \dfrac{6}{1 - 2\cos\theta}$

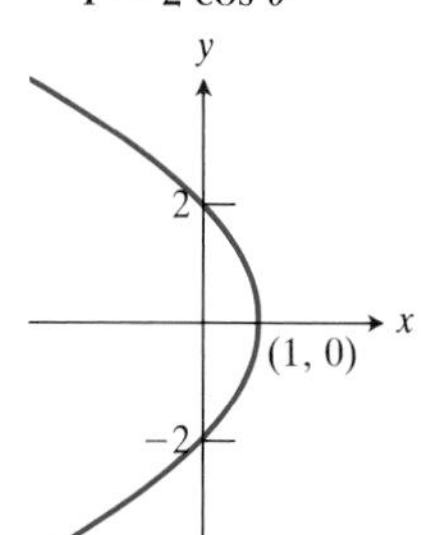

55. $r = \dfrac{4}{1 + 2\cos\theta}$ **57.** $r = \dfrac{2}{2 + \sin\theta}$ **59.** $y = x,\ y = -x$

61. $r\cos\left(\theta - \dfrac{\pi}{4}\right) = 1;\ r\cos\left(\theta - \dfrac{3\pi}{4}\right) = 1;\ r\cos\left(\theta - \dfrac{5\pi}{4}\right) = 1;$ $r\cos\left(\theta - \dfrac{7\pi}{4}\right) = 1$ **63.** $\dfrac{9}{2}\pi$ **65.** $2 + \dfrac{\pi}{4}$ **67.** $2\pi\left(1 - \dfrac{\sqrt{2}}{2}\right)$

69. 8 **71.** $\dfrac{\pi + 3}{8}$ **73.** a) a b) a c) $\dfrac{2a}{\pi}$

Chapter 11

Section 11.1, p. 719

1. (a)

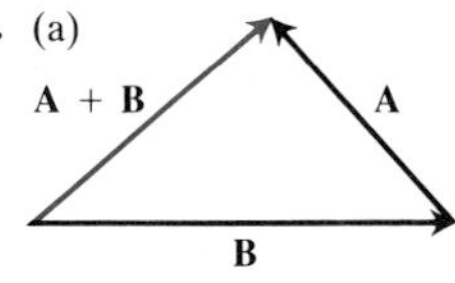

(b)

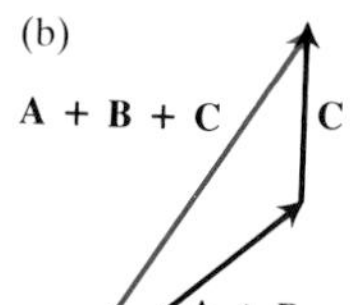

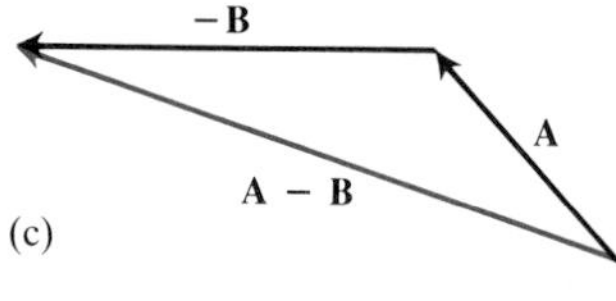

(c)

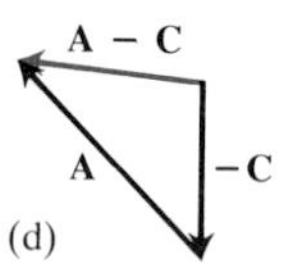

(d)

3. $1\mathbf{i} - 4\mathbf{j}$

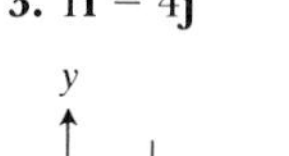

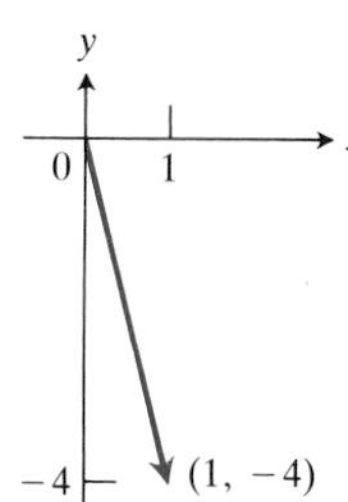

5. $-2\mathbf{i} - 3\mathbf{j}$

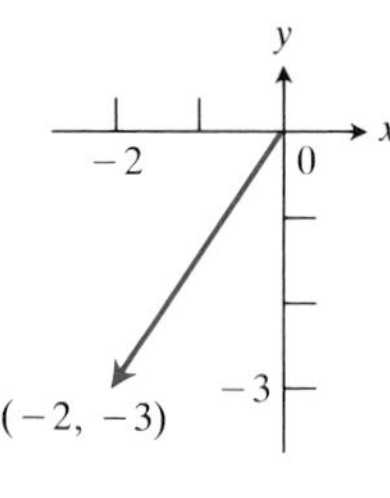

7. $\theta = \dfrac{\pi}{6},\ \mathbf{u} = \dfrac{\sqrt{3}}{2}\mathbf{i} + \dfrac{1}{2}\mathbf{j};$ $\theta = \dfrac{2\pi}{3},\ \mathbf{u} = -\dfrac{1}{2}\mathbf{i} + \dfrac{\sqrt{3}}{2}\mathbf{j}$

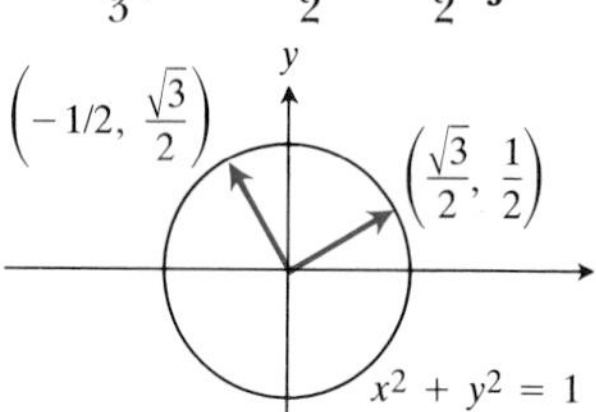

9. $\dfrac{\sqrt{3}}{2}\mathbf{i} - \dfrac{1}{2}\mathbf{j}$

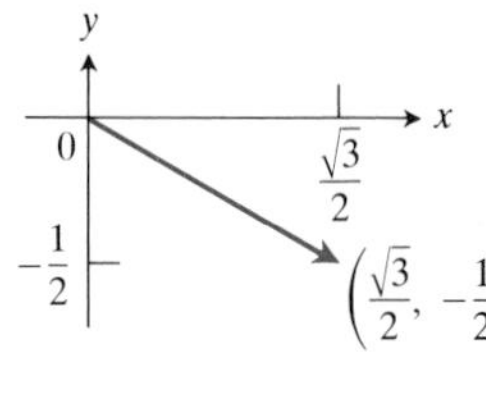

11. Tangents, $\mathbf{u} = \dfrac{1}{\sqrt{17}}\mathbf{i} + \dfrac{4}{\sqrt{17}}\mathbf{j}$ or $-\mathbf{u} = \dfrac{-1}{\sqrt{17}}\mathbf{i} - \dfrac{-4}{\sqrt{17}}\mathbf{j};$ normals, $\mathbf{n} = \dfrac{-4}{\sqrt{17}}\mathbf{i} + \dfrac{1}{\sqrt{17}}\mathbf{j}$ or $-\mathbf{n} = \dfrac{4}{\sqrt{17}}\mathbf{i} - \dfrac{1}{\sqrt{17}}\mathbf{j}$

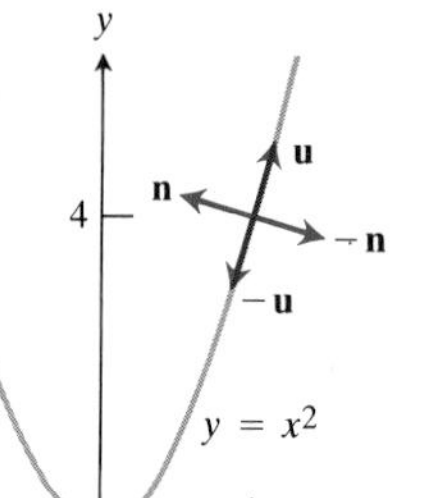

13. $\sqrt{2}\left(\dfrac{1}{\sqrt{2}}\mathbf{i} + \dfrac{1}{\sqrt{2}}\mathbf{j}\right)$

15. $2\left(\dfrac{\sqrt{3}}{2}\mathbf{i} + \dfrac{1}{2}\mathbf{j}\right)$

17. $13\left(\dfrac{5}{13}\mathbf{i} + \dfrac{12}{13}\mathbf{j}\right)$

19.

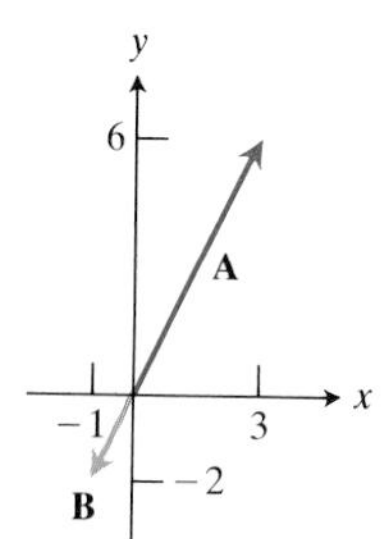

Section 11.2, pp. 729–730

1. A line through the point (2, 3, 0) parallel to the z-axis **3.** The x-axis **5.** The circle $x^2 + y^2 = 4$ in the xy-plane **7.** The circle $x^2 + z^2 = 4$ in the xz-plane **9.** The circle $y^2 + z^2 = 1$ in the yz-plane **11.** The circle $x^2 + y^2 = 16$ in the xy-plane **13.** a) The first quadrant of the xy-plane b) The fourth quadrant of the xy-plane **15.** a) All points that have a distance of 1 or less from the origin b) All points that have a distance greater than 1 from the origin **17.** a) The upper hemisphere of a sphere with a radius of 1 centered at the origin b) A solid upper hemisphere of a sphere with a radius of 1 centered at the origin **19.** a) $x = 3$ b) $y = -1$ c) $z = -2$ **21.** a) $z = 1$ b) $x = 3$ c) $y = -1$ **23.** a) $x^2 + (y-2)^2 = 4$ b) $(y-2)^2 + z^2 = 4$ c) $x^2 + z^2 = 4$ **25.** a) $y = 3, z = -1$ b) $x = 1, z = -1$ c) $x = 1, y = 3$ **27.** $z = 3, x^2 + y^2 + z^2 = 25$ **29.** $0 \le z \le 1$ **31.** $z \le 0$ **33.** a) $(x-1)^2 + (y-1)^2 + (z-1)^2 < 1$ b) $(x-1)^2 + (y-1)^2 + (z-1)^2 > 1$ **35.** Length, 3; direction, $\frac{2}{3}\mathbf{i} + \frac{1}{3}\mathbf{j} - \frac{2}{3}\mathbf{k}$ **37.** Length, 9; direction, $\frac{1}{9}\mathbf{i} + \frac{4}{9}\mathbf{j} - \frac{8}{9}\mathbf{k}$ **39.** Length, 5; direction $\mathbf{k}$ **41.** Length, 4; direction $-\mathbf{j}$ **43.** Length, $\frac{5}{12}$; direction, $-\frac{4}{5}\mathbf{i} + \frac{3}{5}\mathbf{k}$ **45.** Length, $\frac{1}{\sqrt{2}}$; direction, $\frac{1}{\sqrt{3}}\mathbf{i} - \frac{1}{\sqrt{3}}\mathbf{j} - \frac{1}{\sqrt{3}}\mathbf{k}$ **47.** Length, 3; direction, $\frac{2}{3}\mathbf{i} + \frac{2}{3}\mathbf{j} - \frac{1}{3}\mathbf{k}$ **49.** Length, 7; direction, $\frac{3}{7}\mathbf{i} - \frac{6}{7}\mathbf{j} + \frac{2}{7}\mathbf{k}$ **51.** Length, $2\sqrt{3}$; direction, $\frac{1}{\sqrt{3}}\mathbf{i} - \frac{1}{\sqrt{3}}\mathbf{j} - \frac{1}{\sqrt{3}}\mathbf{k}$ **53.** a) $2\mathbf{i}$ b) $-4\mathbf{j}$ c) $\sqrt{3}\mathbf{k}$ d) $\frac{3}{10}\mathbf{j} + \frac{2}{5}\mathbf{k}$ e) $6\mathbf{i} + 2\mathbf{j} + 3\mathbf{k}$ f) $au_1\mathbf{i} + au_2\mathbf{j} + au_3\mathbf{k}$ **55.** $5\left(\frac{1}{\sqrt{3}}\mathbf{i} + \frac{1}{\sqrt{3}}\mathbf{j} + \frac{1}{\sqrt{3}}\mathbf{k}\right)$

57. a) Center, $(-2, 0, 2)$; radius, $2\sqrt{2}$

b) Center, $\left(-\frac{1}{2}, -\frac{1}{2}, -\frac{1}{2}\right)$; radius, $\frac{\sqrt{21}}{2}$

c) Center, $(\sqrt{2}, \sqrt{2}, -\sqrt{2})$; radius, $\sqrt{2}$

d) Center, $\left(0, -\frac{1}{3}, \frac{1}{3}\right)$; radius $\frac{\sqrt{29}}{3}$

59. a) $\sqrt{x^2 + z^2}$ b) $\sqrt{x^2 + z^2}$ c) $\sqrt{x^2 + y^2}$

Section 11.3, pp. 737–738

	$\mathbf{A}\cdot\mathbf{B}$	$\lvert\mathbf{A}\rvert$	$\lvert\mathbf{B}\rvert$	$\cos\theta$	$\lvert\mathbf{B}\rvert\cos\theta$	$\text{Proj}_{\mathbf{A}}\mathbf{B}$
1.	10	$\sqrt{13}$	$\sqrt{26}$	$5\sqrt{2}/13$	$10/\sqrt{13}$	$(30/13)\mathbf{i} + (20/13)\mathbf{j}$
3.	4	$\sqrt{14}$	2	$\sqrt{14}/7$	$2\sqrt{14}/7$	$(6/7)\mathbf{i} - (4/7)\mathbf{j} - (2/7)\mathbf{k}$
5.	2	$\sqrt{34}$	$\sqrt{3}$	$2/\sqrt{102}$	$2/\sqrt{34}$	$(5/17)\mathbf{j} - (3/17)\mathbf{k}$
7.	$-\sqrt{2}+\sqrt{3}$	$\sqrt{2}$	3	$(\sqrt{6}-2)/6$	$(\sqrt{6}-2)/2$	$((\sqrt{3}-\sqrt{2})/2)(-\mathbf{i}+\mathbf{j})$
9.	−25	5	5	−1	−5	$-2\mathbf{i} + 4\mathbf{j} - \sqrt{5}\mathbf{k}$
11.	25	15	5	1/3	5/3	$(10/9)\mathbf{i} + (11/9)\mathbf{j} - (2/9)\mathbf{k}$

13. $\mathbf{B} = \left(\frac{3}{2}\mathbf{i} + \frac{3}{2}\mathbf{j}\right) + \left(-\frac{3}{2}\mathbf{i} + \frac{3}{2}\mathbf{j} + 4\mathbf{k}\right)$

15. $\mathbf{B} = \left(\frac{14}{3}\mathbf{i} + \frac{28}{3}\mathbf{j} - \frac{14}{3}\mathbf{k}\right) + \left(\frac{10}{3}\mathbf{i} - \frac{16}{3}\mathbf{j} - \frac{22}{3}\mathbf{k}\right)$

17.

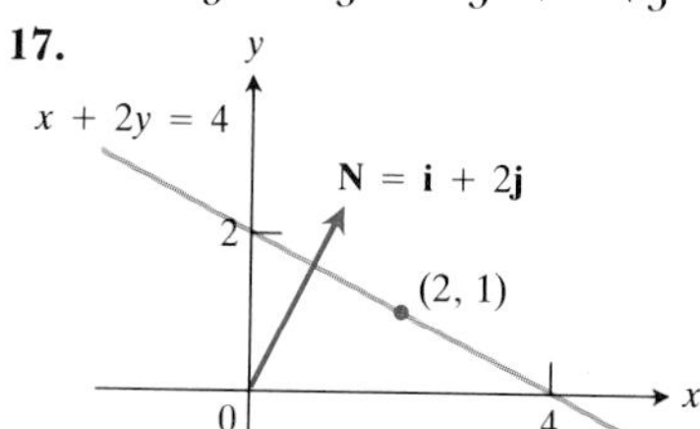

19.

y
y = −2x
(−1, 2)
2
−2
1
x
−1
N = −2i − j

21. $2\sqrt{10}$ **23.** $\sqrt{2}$ **25.** Because $\mathbf{A}\cdot\mathbf{B} = \mathbf{A}\cdot\mathbf{C} = \mathbf{B}\cdot\mathbf{C} = 0$ the vectors are mutually orthogonal. **27.** When $\mathbf{A} = \mathbf{i} + \mathbf{j} + \mathbf{k}$, $\mathbf{B}_1 = \mathbf{i}$, and $\mathbf{B}_2 = \mathbf{j}$ then $\mathbf{A}\cdot\mathbf{B} = \mathbf{B}\cdot\mathbf{A}$ but $\mathbf{B}_1 \ne \mathbf{B}_2$. **29.** $A = 71.068°$, $B = 37.86°$, $C = 71.068°$ **31.** 35.3° **33.** −5nm **35.** $2000\sqrt{3}$nm **37.** $\frac{\pi}{4}$ **39.** $\frac{\pi}{3}$ **41.** $\frac{\pi}{2}$ at (0, 0); $\frac{\pi}{4}$ at (1, 1)

Section 11.4, pp. 744–745

1. $\lvert\mathbf{A}\times\mathbf{B}\rvert = 3$, direction $\frac{2}{3}\mathbf{i} + \frac{1}{3}\mathbf{j} + \frac{2}{3}\mathbf{k}$; $\lvert\mathbf{B}\times\mathbf{A}\rvert = 3$, direction $-\frac{2}{3}\mathbf{i} - \frac{1}{3}\mathbf{j} - \frac{2}{3}\mathbf{k}$
3. $\lvert\mathbf{A}\times\mathbf{B}\rvert = \lvert\mathbf{B}\times\mathbf{A}\rvert = 0$, $\mathbf{0}$ has no direction
5. $\lvert\mathbf{A}\times\mathbf{B}\rvert = 6$, direction $-\mathbf{k}$; $\lvert\mathbf{B}\times\mathbf{A}\rvert = 6$, direction $\mathbf{k}$
7. $\lvert\mathbf{A}\times\mathbf{B}\rvert = 6\sqrt{5}$, direction $\frac{1}{\sqrt{5}}\mathbf{i} - \frac{2}{\sqrt{5}}\mathbf{k}$; $\lvert\mathbf{B}\times\mathbf{A}\rvert = 6\sqrt{5}$, direction $-\frac{1}{\sqrt{5}}\mathbf{i} + \frac{2}{\sqrt{5}}\mathbf{k}$

9. $\mathbf{i} \times \mathbf{j} = \mathbf{k}$

11. $(\mathbf{i} - \mathbf{k}) \times (\mathbf{j} + \mathbf{k}) = \mathbf{i} - \mathbf{j} + \mathbf{k}$

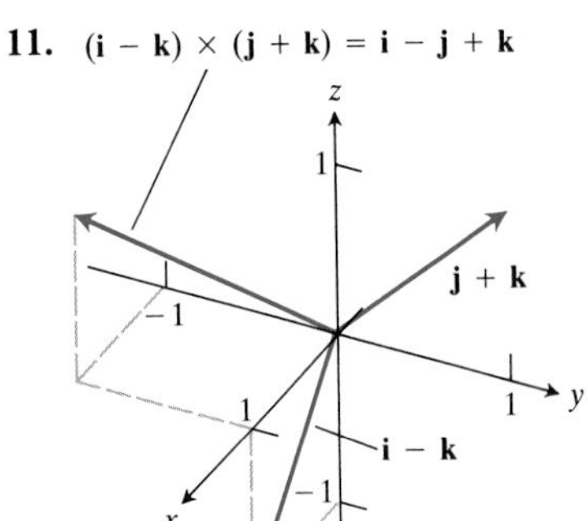

13.

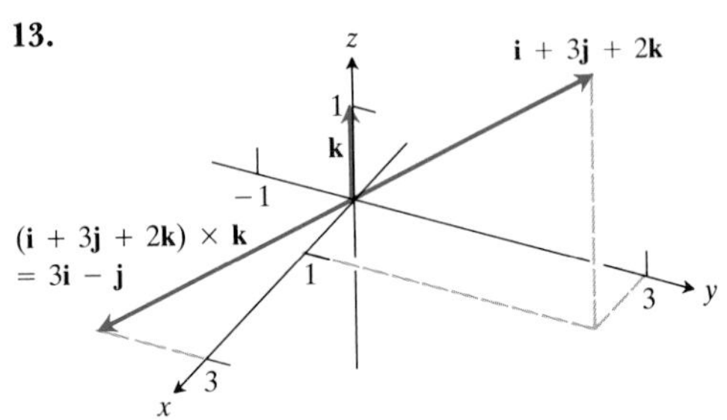

15. a) $(\pm 1)(8\mathbf{i} + 4\mathbf{j} + 4\mathbf{k})$ b) $2\sqrt{6}$
c) $\pm 4\sqrt{6}\left(\frac{2}{\sqrt{6}}\mathbf{i} + \frac{1}{\sqrt{6}}\mathbf{j} + \frac{1}{\sqrt{6}}\mathbf{k}\right)$

17. a) $(\pm 1)(-\mathbf{i} + \mathbf{j})$ b) $\frac{\sqrt{2}}{2}$ c) $\pm\sqrt{2}\left(-\frac{1}{\sqrt{2}}\mathbf{i} + \frac{1}{\sqrt{2}}\mathbf{j}\right)$

19. a) None b) **A** and **C**

21. $(\mathbf{A} \times \mathbf{B}) \cdot \mathbf{A} = 0$, $(\mathbf{A} \times \mathbf{B}) \cdot \mathbf{B} = 0$

23. a) $\text{Proj}_{\mathbf{B}}\mathbf{A} = \frac{\mathbf{A} \cdot \mathbf{B}}{\mathbf{B} \cdot \mathbf{B}}\mathbf{B}$ b) $(\pm 1)(\mathbf{A} \times \mathbf{B})$ c) $\sqrt{\mathbf{A} \cdot \mathbf{A}}\frac{\mathbf{B}}{\sqrt{\mathbf{B} \cdot \mathbf{B}}}$
d) $(\pm 1)((\mathbf{A} \times \mathbf{B}) \times \mathbf{C})$ e) $(\pm 1)((\mathbf{B} \times \mathbf{C}) \times \mathbf{A})$

25. $10\sqrt{3}$ ft-lb **27.** 8 **29.** 7

Section 11.5, p. 752

1. $x = 3 + t, y = -4 + t, z = -1 + t$
3. $x = -2 + 5t, y = 5t, z = 3 - 5t$
5. $x = 0, y = 2t, z = t$ **7.** $x = 1, y = 1, z = 1 + t$
9. $x = t, y = -7 + 2t, z = 2t$
11. $x = 0, y = 0, z = t$
13. $x = t, y = t, z = t$ where $0 \le t \le 1$

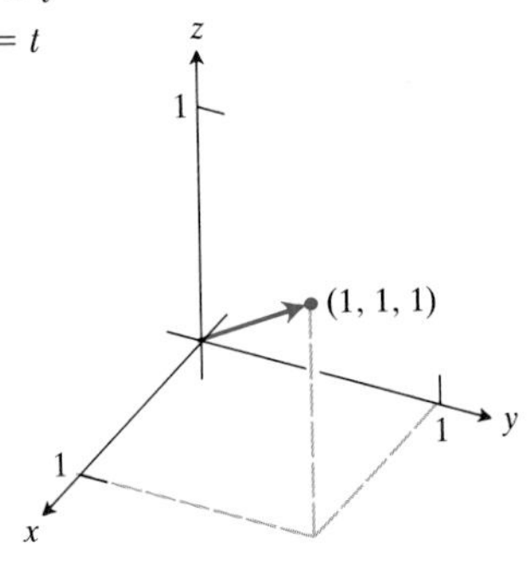

15. $x = 1, y = 1 + t, z = 0$ where $-1 \le t \le 0$

(1, 0, 0) (0, 1, 0) (1, 1, 0)

17. $x = 0, y = -1 + 2t$, $z = 1$ where $0 \le t \le 1$

19. $x = 2 - t, y = 2, z = -2t$ where $0 \le t < 1$

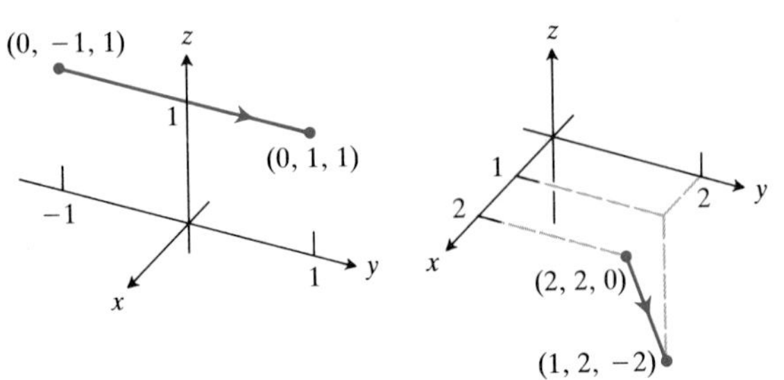

21. $3x - 2y - z = -3$ **23.** $7x - 5y - 4z = 6$
25. $x + 3y + 4z = 34$ **27.** $2\sqrt{30}$ **29.** 0 **31.** 3
33. $\frac{19}{5}$ **35.** $\frac{5}{3}$ **37.** $\left(\frac{3}{2}, -\frac{3}{2}, \frac{1}{2}\right)$ **39.** $(1, 1, 0)$ **41.** $\frac{\pi}{4}$
43. $101.1°$ **45.** $47.1°$ **47.** $x = 1 - t, y = 1 + t, z = -1$
49. $x = 4, y = 3 + 6t, z = 1 + 3t$

Section 11.6, p. 764

1. $x^2 + y^2 = 4$

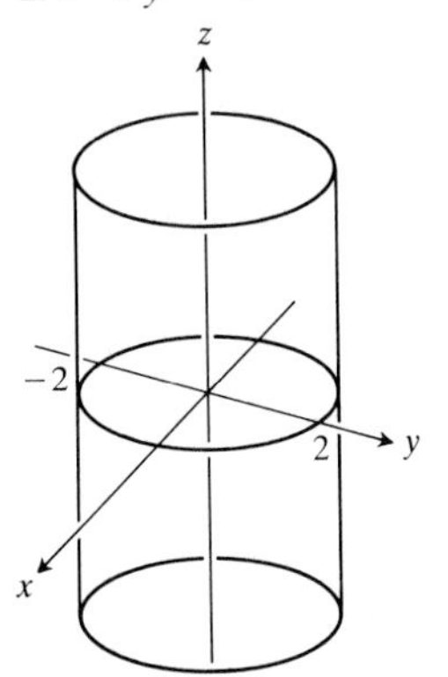

3. $y^2 + z^2 = 1$

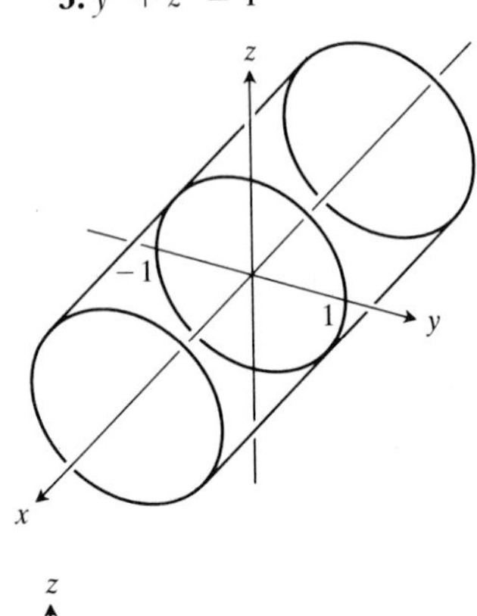

5. $z = y^2 - 1$

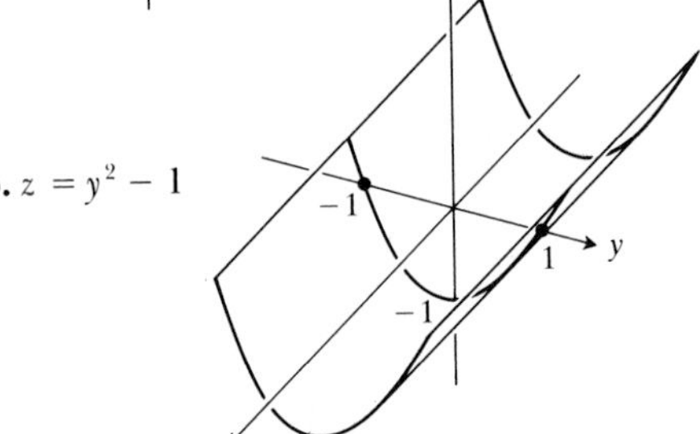

7. $z = 4 - x^2$

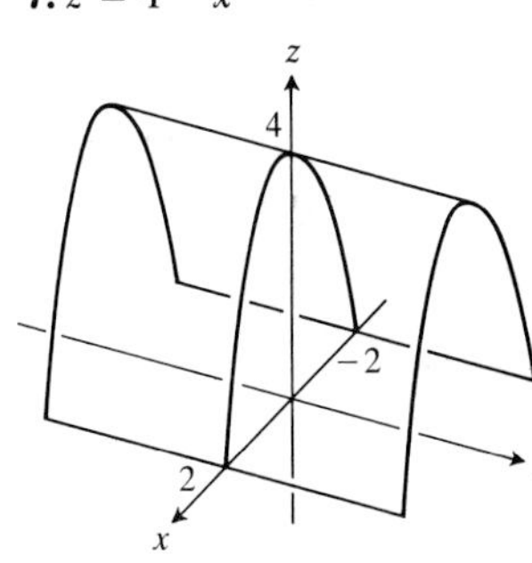

9. $y = x^2$

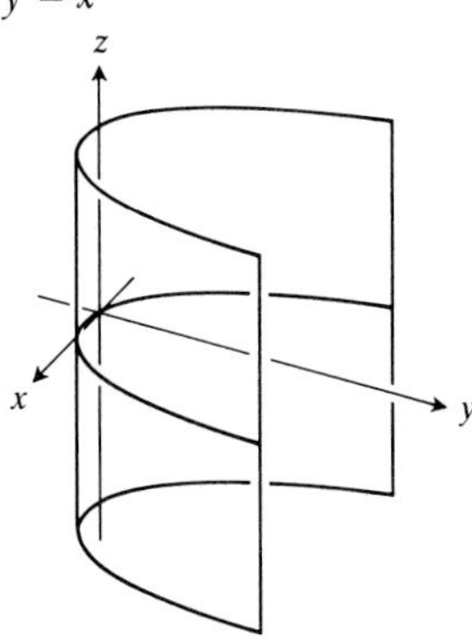

11. $y^2 + 4z^2 = 16$

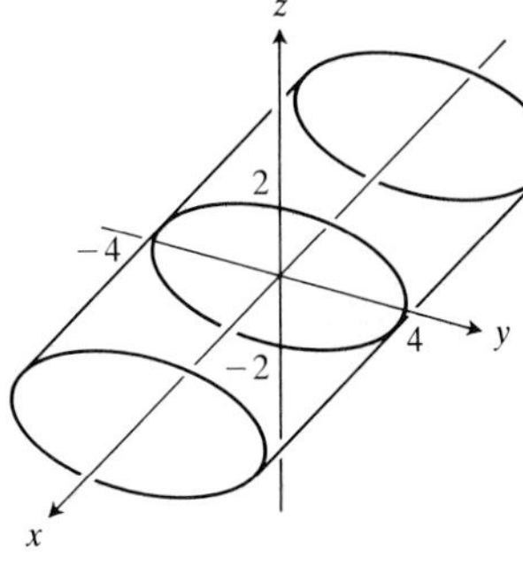

13. $z^2 + 4y^2 = 9$

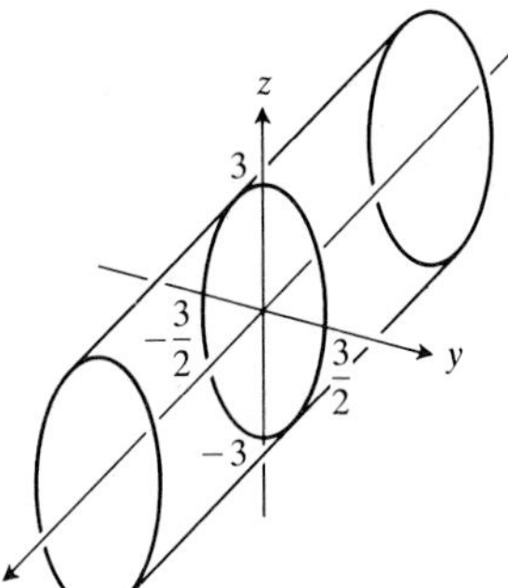

15. $z^2 - y^2 = 1$

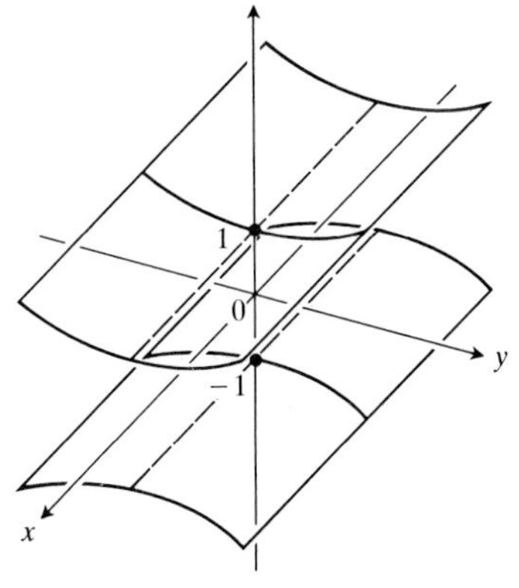

17. $9x^2 + y^2 + z^2 = 9$

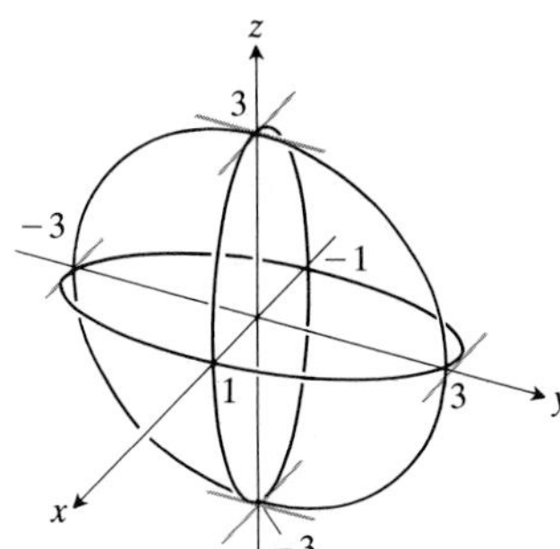

19. $x^2 + y^2 + z^2 = 4$

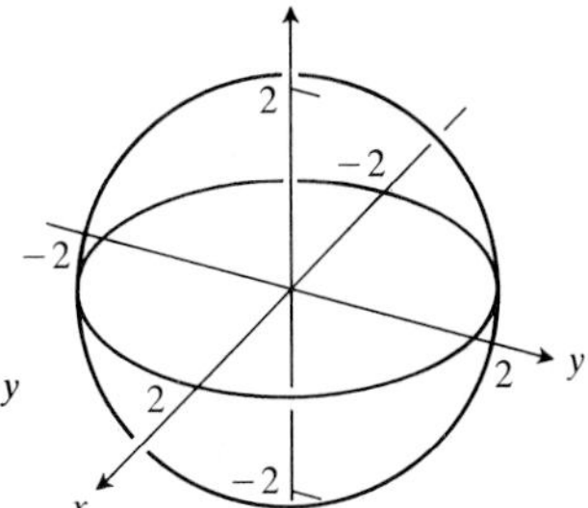

21. $4x^2 + 9y^2 + 4z^2 = 36$

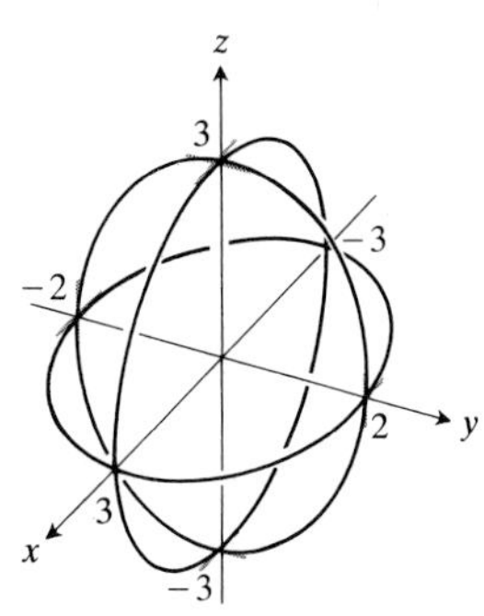

23. $x^2 + y^2 = z$

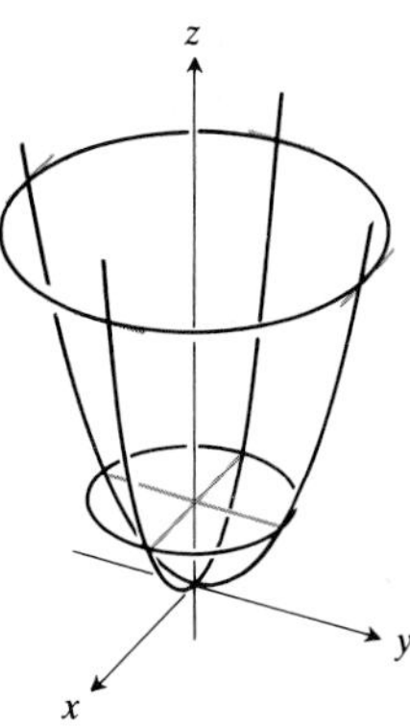

25. $x^2 + 4y^2 = z$

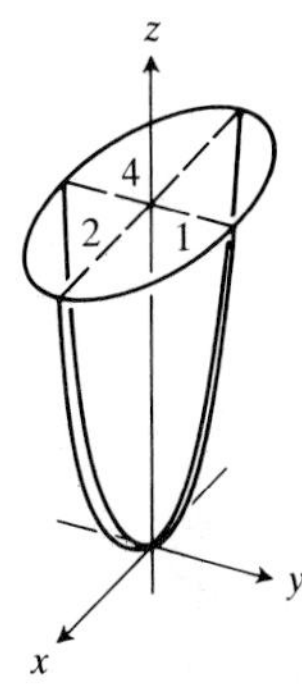

27. $z = 8 - x^2 - y^2$

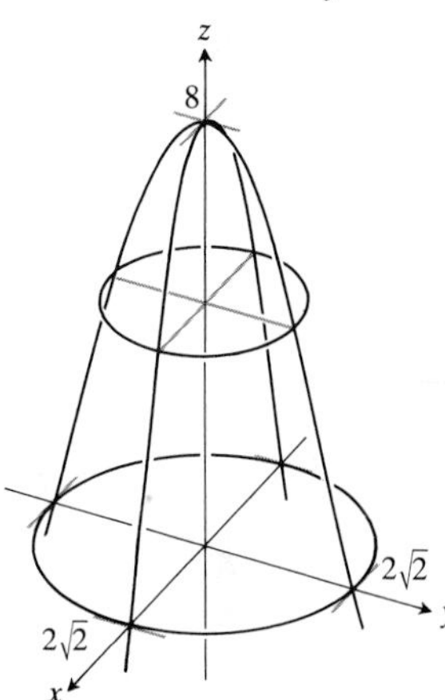

29. $x = 4 - 4y^2 - z^2$

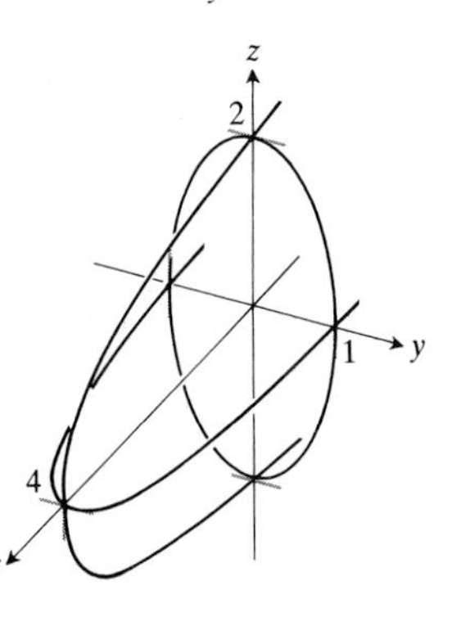

31. $z = x^2 + y^2 + 1$

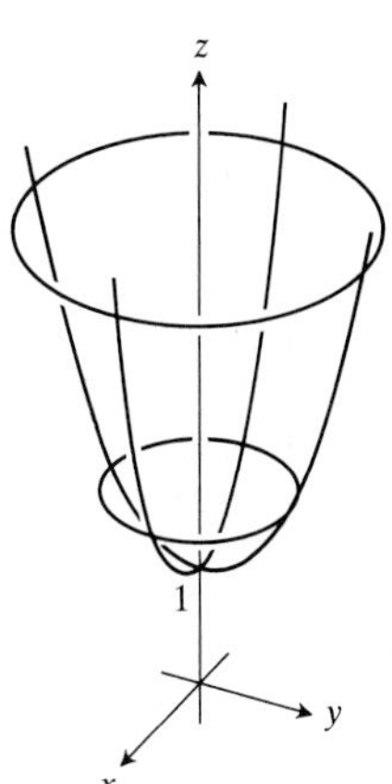

33. $x^2 + y^2 = z^2$

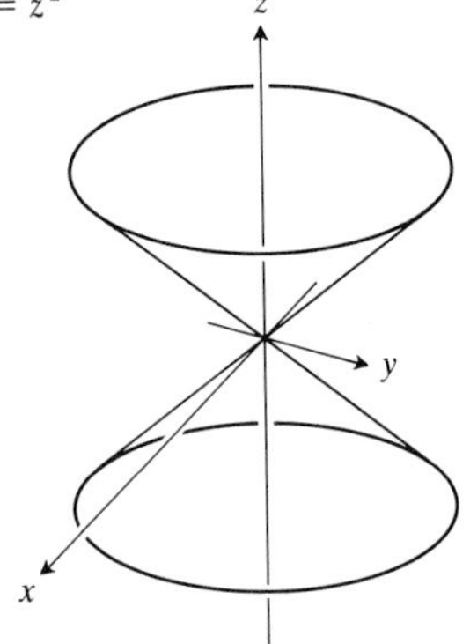

35. $x^2 + z^2 = y^2$

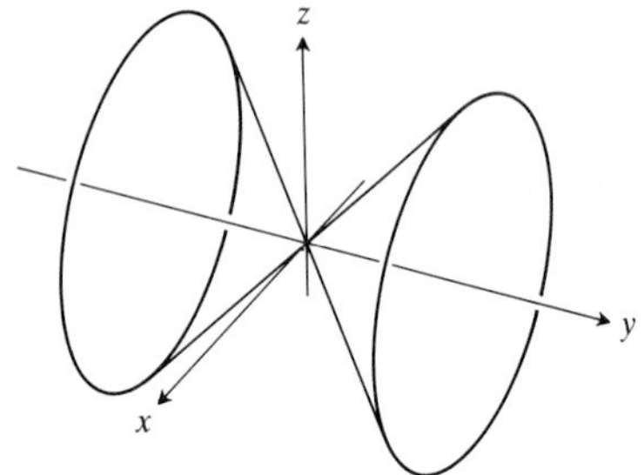

37. $9x^2 + 4y^2 = 36z^2$

39. $x^2 + y^2 - z^2 = 1$

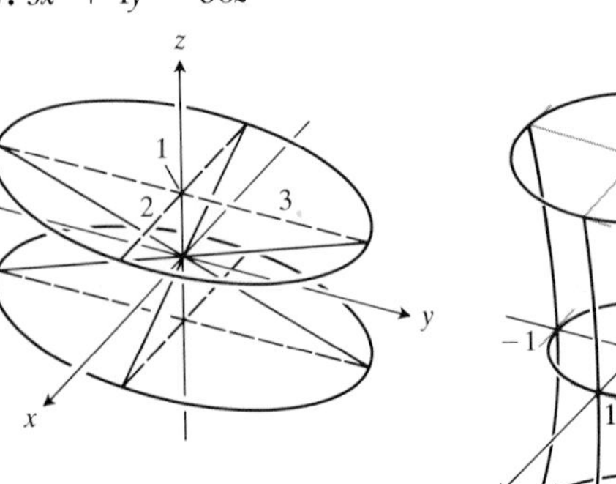

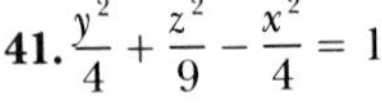
41. $\frac{y^2}{4} + \frac{z^2}{9} - \frac{x^2}{4} = 1$

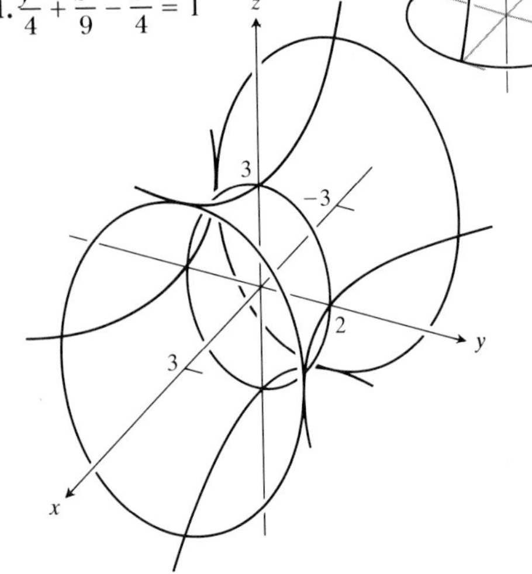

43. $\frac{x^2}{4} + y^2 - z^2 = 1$

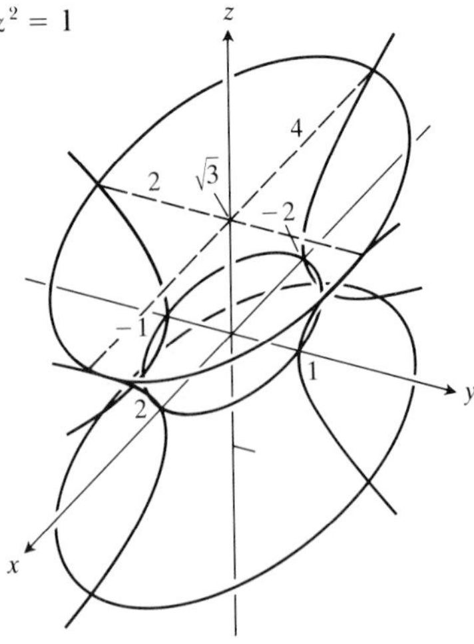

45. $z^2 - \frac{x^2}{4} - y^2 = 1$

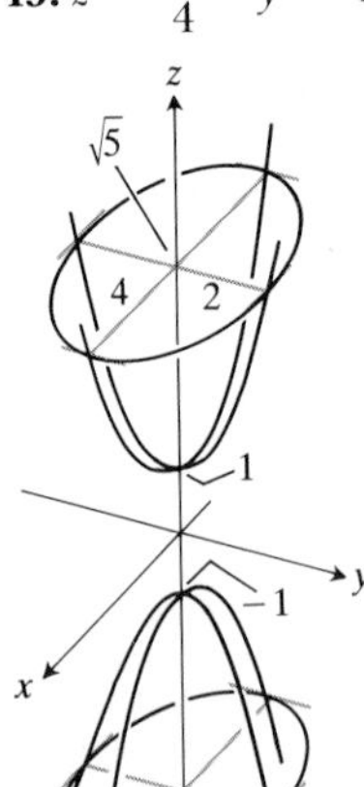

47. $x^2 - y^2 - \frac{z^2}{4} = 1$

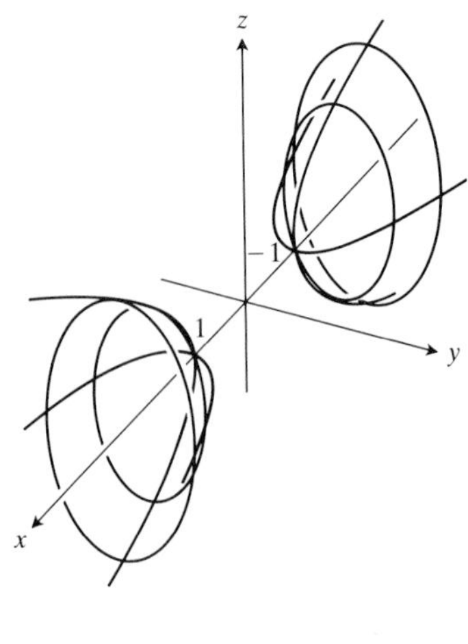

49. $y^2 - x^2 = z^2$

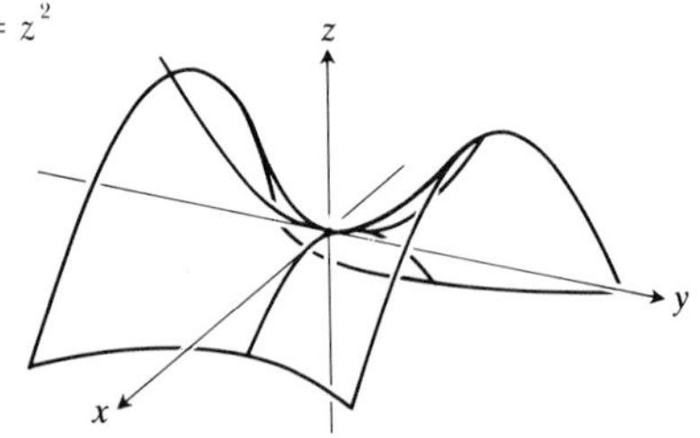

Section 11.7, p. 768

	(x, y, z)	(r, θ, z)	(ρ, ϕ, θ)
1.	$(0, 0, 0)$	$(0, 0, 0)$	$(0, 0, 0)$
3.	$(0, 1, 0)$	$(1, \pi/2, 0)$	$(1, \pi/2, \pi/2)$
5.	$(1, 0, 0)$	$(1, 0, 0)$	$(1, \pi/2, 0)$
7.	$(0, 1, 1)$	$(1, \pi/2, 1)$	$(\sqrt{2}, \pi/4, \pi/2)$
9.	$(0, -2\sqrt{2}, 0)$	$(2\sqrt{2}, 3\pi/2, 0)$	$(2\sqrt{2}, \pi/2, 3\pi/2)$

11. Rectangular, $x^2 + y^2 = 0$; spherical, $\phi = 0$ and $\phi = \pi$; the z-axis

13. Cylindrical, $z = 0$; spherical, $\phi = \frac{\pi}{2}$; the xy-plane

15. Rectangular, $z = 3$; cylindrical, $z = 3$; the plane $z = 3$

17. Rectangular, $x = 0$; cylindrical $\phi = \frac{\pi}{2}$; the yz-plane

19. Cylindrical, $r^2 + z^2 = 4$; spherical, $\rho = 2$; a sphere centered at the origin with a radius of 2

21. A right circular cylinder whose generating curve is a circle centered at the origin with a radius of 2 in the $r\theta$-plane

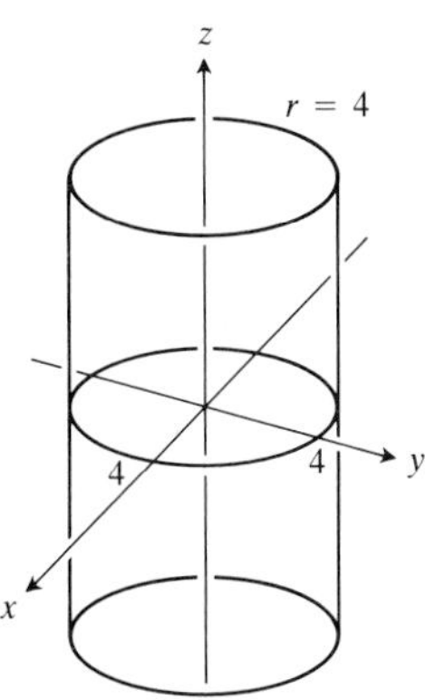

23. A cylinder whose generating curve is a cardioid in the $r\theta$-plane

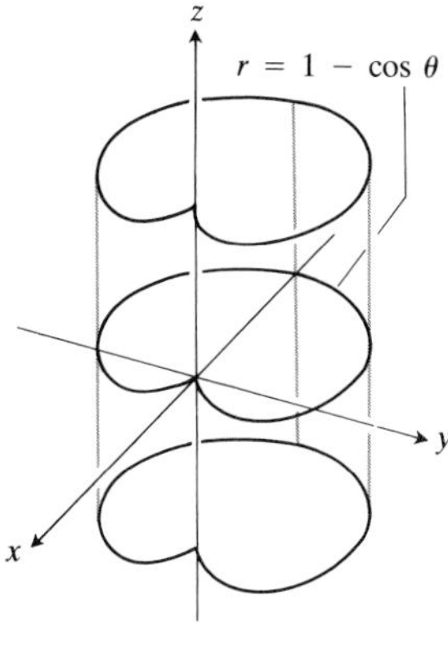

25. A circle contained in the plane $z = 3$, having a radius of $\sqrt{2}$ and center at (0, 0, 3)

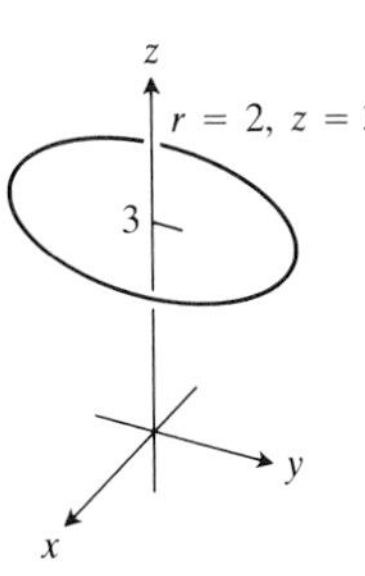

27. Helix

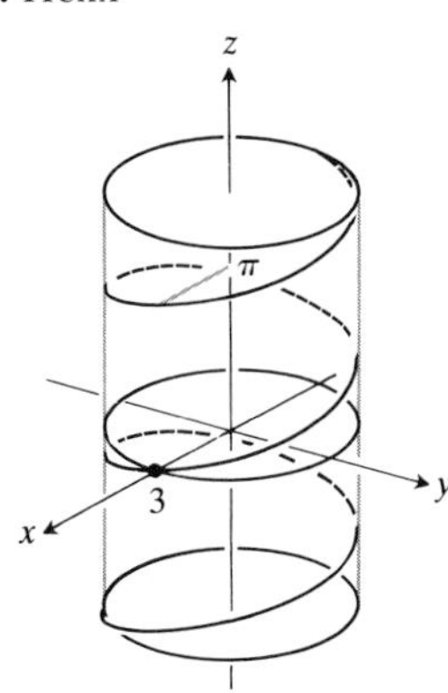

The helix $r = 3$, $z = \theta/2$

29. The upper nappe of a cone

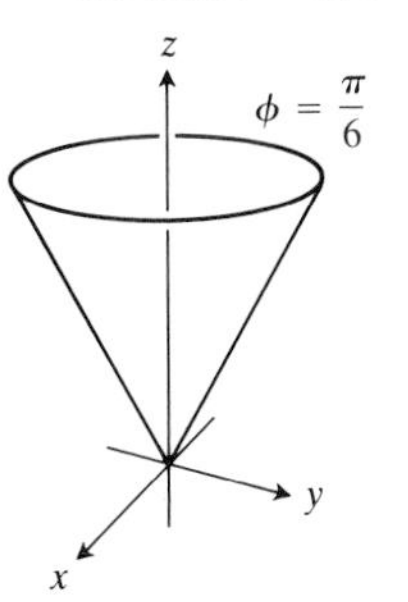

31. A vertical semicircle in the plane $y = x$

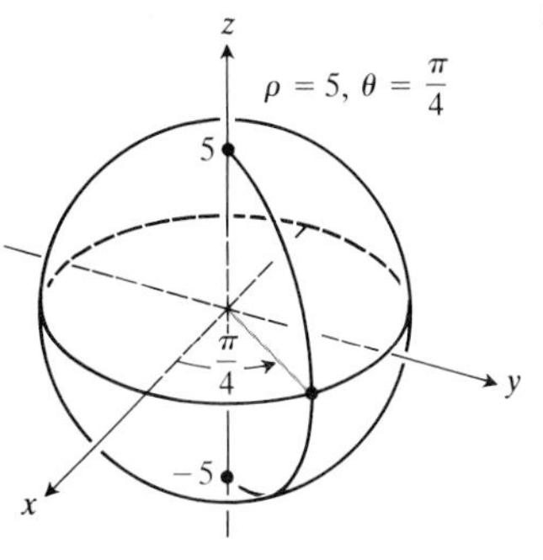

33. The intersection of the sphere $x^2 + y^2 + (z - 2)^2 = 4$ and the yz-plane when $y \geq 0$

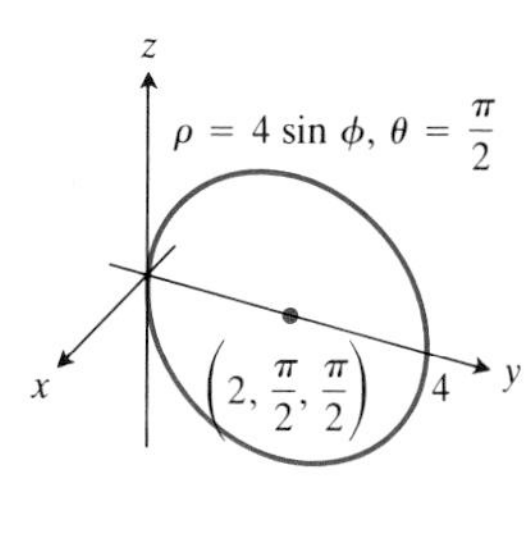

Chapter 11 Practice Exercises, pp. 769–770

1. $\theta = 0, \mathbf{u} = \mathbf{i}$; $\theta = \frac{\pi}{2}, \mathbf{u} = \mathbf{j}$;
$\theta = \frac{2\pi}{3}, \mathbf{u} = -\frac{1}{2}\mathbf{i} + \frac{\sqrt{3}}{2}\mathbf{j}$;
$\theta = \frac{5\pi}{4}, \mathbf{u} = -\frac{1}{\sqrt{2}}\mathbf{i} - \frac{1}{\sqrt{2}}\mathbf{j}$;
$\theta = \frac{5\pi}{3}, \mathbf{u} = \frac{1}{2}\mathbf{i} - \frac{\sqrt{3}}{2}\mathbf{j}$

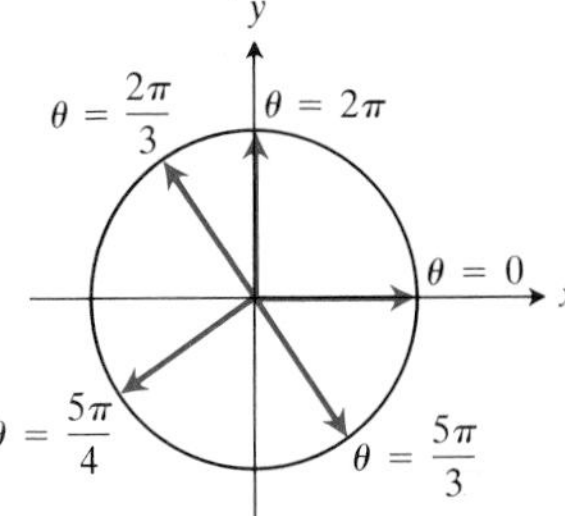

3. Unit tangent, $(\pm 1)\left(\frac{1}{\sqrt{5}}\mathbf{i} + \frac{2}{\sqrt{5}}\mathbf{j}\right)$; unit normal, $(\pm 1)\left(-\frac{2}{\sqrt{5}}\mathbf{i} + \frac{1}{\sqrt{5}}\mathbf{j}\right)$ **5.** $2\left(\frac{1}{\sqrt{2}}\mathbf{i} + \frac{1}{\sqrt{2}}\mathbf{j}\right)$

7. $\sqrt{49}\left(\frac{2}{\sqrt{49}}\mathbf{i} - \frac{3}{\sqrt{49}}\mathbf{j} + \frac{6}{\sqrt{49}}\mathbf{k}\right)$ **9.** $\frac{2}{\sqrt{33}}(4\mathbf{i} - \mathbf{j} + 4\mathbf{k})$

11. $|\mathbf{A}| = \sqrt{2}$; $|\mathbf{B}| = 3$; $\mathbf{A} \cdot \mathbf{B} = 3$; $\mathbf{B} \cdot \mathbf{A} = 3$; $\mathbf{A} \times \mathbf{B} = -2\mathbf{i} + 2\mathbf{j} - \mathbf{k}$; $\mathbf{B} \times \mathbf{A} = 2\mathbf{i} - 2\mathbf{j} + \mathbf{k}$; $|\mathbf{A} \times \mathbf{B}| = 3$; $\theta = \frac{\pi}{4}$; $|\mathbf{B}| \cos\theta = \frac{3}{\sqrt{2}}$; $\text{proj}_{\mathbf{A}}\mathbf{B} = \frac{3}{2}(\mathbf{i} + \mathbf{j})$

13. $\mathbf{B} = \frac{4}{3}(2\mathbf{i} + \mathbf{j} - \mathbf{k}) - \frac{1}{3}(5\mathbf{i} + \mathbf{j} + 11\mathbf{k})$

15.

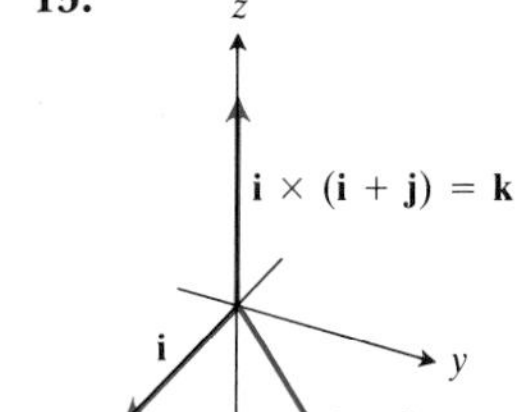

17. 3
19. $\sqrt{2}$
21. $2x + y - z = 3$
23. $-9x + y + 7z = 4$
25. $x = 1 - 3t, y = 2, z = 3 + 7t$
27. $\left(\frac{4}{3}, -\frac{2}{3}, -\frac{2}{3}\right)$

29. $x = -5 + 15t, y = 3 - 3t, z = -9t$ **31.** 2
33. $16{,}000\sqrt{3}$ ft-lb **35.** a) $\sqrt{14}$ b) 1 **37.** a) True b) False c) True d) True e) False f) True g) True h) True

39. The y-axis in the xy-plane; the yz-plane in three-dimensional space **41.** A circle centered at (0, 0) with a radius of 2 in the xy-plane; a cylinder parallel with the z-axis in three-dimensional space with the circle as its generating curve **43.** A horizontal parabola opening to the right with its vertex at (0, 0) in the xy-plane; a cylinder parallel with the z-axis in three-dimensional space with the parabola as its generating curve **45.** A horizontal cardioid in the $r\theta$-plane; a cylinder parallel with the z-axis in three-dimensional space with the cardioid as its generating curve **47.** A horizontal lemniscate of length $2\sqrt{2}$ in the $r\theta$-plane; a cylinder parallel with the z-axis in three-dimensional space with the lemniscate as its generating curve **49.** A sphere with radius 2 centered at the origin **51.** The upper nappe of a cone whose surface makes a $\frac{\pi}{6}$ angle with the z-axis **53.** The upper hemisphere of a sphere with a radius of 1 centered at the origin **55.** $(1, 0, 0), \left(1, \frac{\pi}{2}, 0\right)$ **57.** $(0, 1, 1), \left(1, \frac{\pi}{2}, 1\right)$ **59.** $\left(1, \frac{\pi}{4}, 1\right), \left(\sqrt{3}, \tan^{-1}\frac{\sqrt{2}}{2}, \frac{\pi}{4}\right)$ **61.** A plane parallel to the xy-plane; cylindrical, $z = 2$; spherical, $\rho \cos \phi = 2$ **63.** A paraboloid symmetric to the z-axis which opens upward; rectangular, $z = x^2 + y^2$; spherical, $\rho = 0$ or $\rho = \frac{\cos\phi}{\sin^2\phi}$ when $0 < \phi < \frac{\pi}{2}$ **65.** A sphere with a radius 4 centered at the origin; rectangular, $x^2 + y^2 + z^2 = 16$; cylindrical, $r^2 + z^2 = 16$

67. $x^2 + y^2 + z^2 = 4$

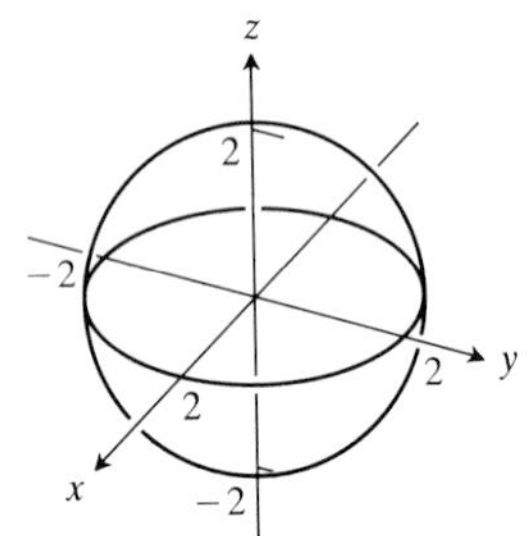

69. $4x^2 + 4y^2 + z^2 = 4$

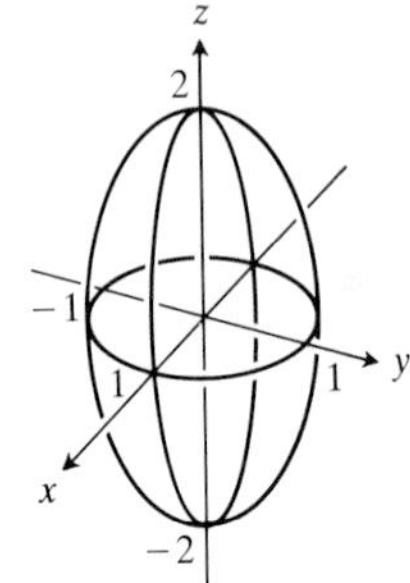

71. $z = -(x^2 + y^2)$

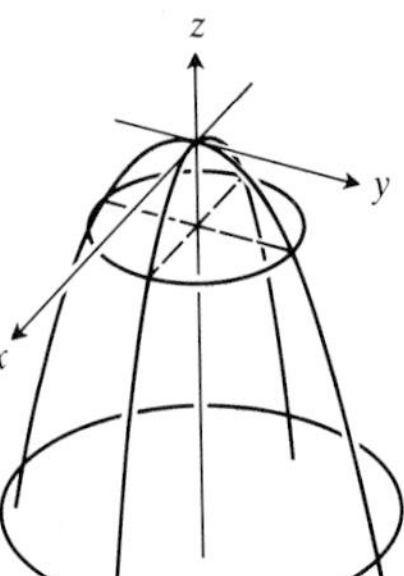

73. $x^2 + y^2 = z^2$

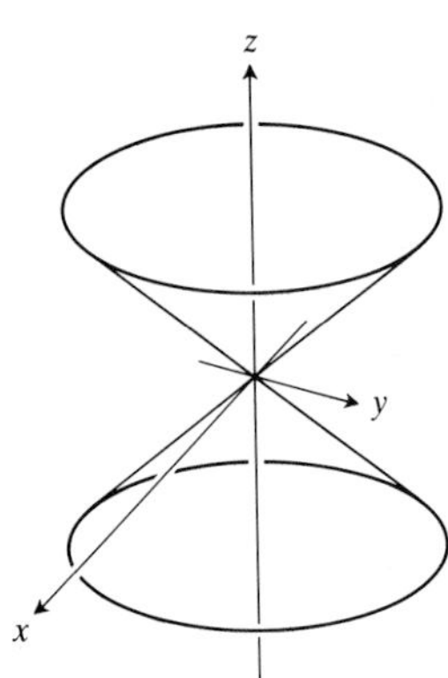

75. $x^2 + y^2 - z^2 = 4$

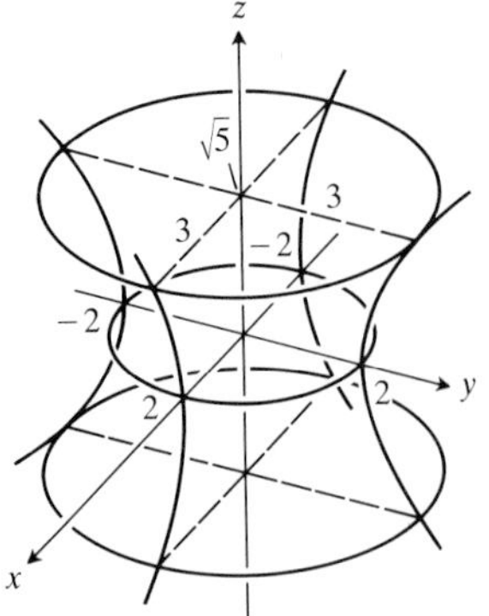

77. $y^2 - x^2 - z^2 = 1$

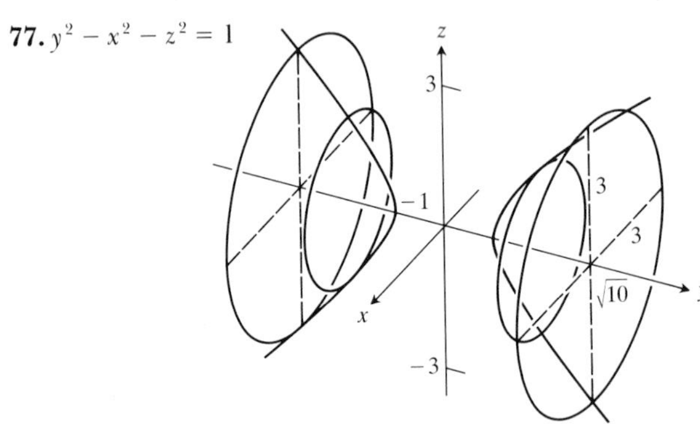

Chapter 12

Section 1, pp. 781–782

1. $\mathbf{v} = \frac{d\mathbf{r}}{dt} = (-2 \sin t)\mathbf{i} + (3 \cos t)\mathbf{j} + 4\mathbf{k}$; $\mathbf{a} = \frac{d^2\mathbf{r}}{ds^2} = (-2 \cos t)\mathbf{i} - (3 \sin t)\mathbf{j}$; speed: $2\sqrt{5}$; direction: $(-1/\sqrt{5})\mathbf{i} + (2/\sqrt{5})\mathbf{k}$; $\mathbf{v}(\pi/2) = 2\sqrt{5}[(-1/\sqrt{5})\mathbf{i} + (2/\sqrt{5})\,\mathbf{k}]$ **3.** $\mathbf{v} = (-2 \sin 2t)\mathbf{j} + (2 \cos t)\mathbf{k}$; $\mathbf{a} = (-4 \cos 2t)\mathbf{j} + (-2 \sin t)\mathbf{k}$; speed: 4; direction: $\frac{1}{2}\mathbf{k}$; $\mathbf{v}(0) = 4\left(\frac{1}{2}\mathbf{k}\right)$ **5.** $\mathbf{v} = (\sec t \tan t)\mathbf{i} + (\sec^2 t)\mathbf{j} + \frac{4}{3}\mathbf{k}$; $\mathbf{a} = (\sec t \tan^2 + \sec^3 t)\mathbf{i} + (2 \sec^2 t \tan t)\mathbf{j}$; speed: $\frac{2\sqrt{123}}{9}$; direction: $\frac{3}{\sqrt{123}}\mathbf{i} + \frac{6}{\sqrt{123}}\mathbf{j} + \frac{6}{\sqrt{123}}\mathbf{k}$; $\mathbf{v}\left(\frac{\pi}{6}\right) = \frac{2\sqrt{123}}{9}\left(\frac{3}{\sqrt{123}}\mathbf{i} + \frac{6}{\sqrt{123}}\mathbf{j} + \frac{6}{\sqrt{123}}\mathbf{k}\right)$ **7.** $\mathbf{v} = (-e^{-t})\mathbf{i} - (6 \sin 3t)\mathbf{j} + (6 \cos 3t)\mathbf{k}$; $\mathbf{a} = (e^{-t})\mathbf{i} - (18 \cos 3t)\mathbf{j} - (18 \sin 3t)\mathbf{k}$; speed: $5\sqrt{13}$; direction: $\frac{-1}{5\sqrt{13}}\mathbf{i} + \frac{6}{5\sqrt{13}}\mathbf{k}$; $\mathbf{v}(0) = 5\sqrt{13}\left(\frac{-1}{5\sqrt{13}}\mathbf{i} + \frac{6}{5\sqrt{13}}\mathbf{k}\right)$ **9.** $\frac{\pi}{2}$ **11.** $\frac{\pi}{2}$ **13.** $0, 2\pi, \frac{\pi}{3}, \frac{5\pi}{3}$ **15.** $\frac{1}{4}\mathbf{i} + 7\mathbf{j} + \frac{3}{2}\mathbf{k}$ **17.** $\left(\frac{\pi + 2\sqrt{2}}{2}\right)\mathbf{j} + 2\mathbf{k}$ **19.** $(\ln 4)\mathbf{i} + (\ln 4)\mathbf{j} + (\ln 2)\mathbf{k}$

21. $t = \frac{\pi}{4}$: $\mathbf{v} = \frac{\sqrt{2}}{2}\mathbf{i} - \frac{\sqrt{2}}{2}\mathbf{j}$, $\mathbf{a} = \frac{-\sqrt{2}}{2}\mathbf{i} - \frac{\sqrt{2}}{2}\mathbf{j}$;

$t = \frac{\pi}{2}$: $\mathbf{v} = -\mathbf{j}$, $\mathbf{a} = -\mathbf{i}$

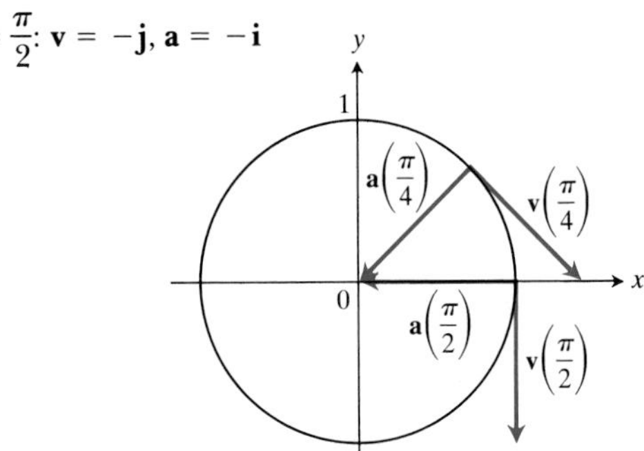

23. $t = \pi$: $\mathbf{v} = 2\mathbf{i}$, $\mathbf{a} = -\mathbf{j}$; $t = \frac{3\pi}{2}$: $\mathbf{v} = \mathbf{i} - \mathbf{j}$, $\mathbf{a} = -\mathbf{i}$

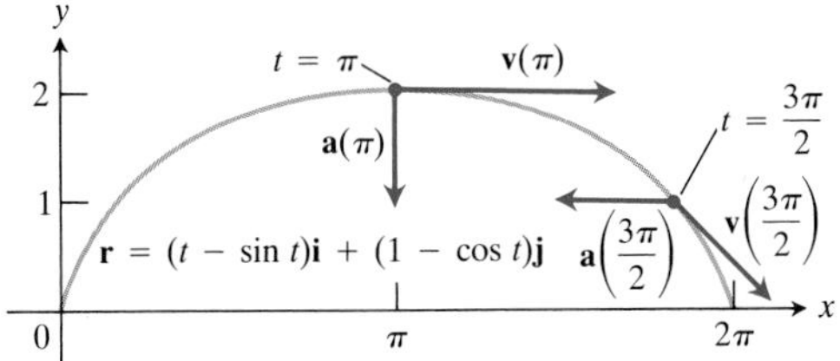

25. $\mathbf{r} = \left(\frac{-t^2}{2} + 1\right)\mathbf{i} + \left(\frac{-t^2}{2} + 2\right)\mathbf{j} + \left(\frac{-t^2}{2} + 3\right)\mathbf{k}$

27. $\mathbf{r} = ((t + 1)^{3/2} - 1)\mathbf{i} + (-e^{-t} + 1)\mathbf{j} + (\ln(t + 1) + 1)\mathbf{k}$

29. $\mathbf{r} = 8t\mathbf{i} + 8t\mathbf{j} + (-16t^2 + 100)\mathbf{k}$

31. max $|\mathbf{v}| = 2$; max $|\mathbf{a}| = 1$

Section 12.2, pp. 788–790

1. $t = 50$ **3.** a) $t = 72.2$ sec, $R = 25510.2$ m b) 4020.3 m c) 6377.6 m **7.** $v_0 = 9.9$ m/sec; $\alpha = 18.44°$ or $71.56°$ **9.** $v_0 = 278.01$ ft/sec $= 189.6$ mph **13.** It will clip the leaves at the top. **15.** $t \approx 2.25$ sec; $v_0 \approx 148.98$ ft/sec

Section 12.3, p. 795

1. $\mathbf{T} = \left(-\frac{2}{3}\sin t\right)\mathbf{i} + \left(\frac{2}{3}\cos t\right)\mathbf{j} + \frac{\sqrt{5}}{3}\mathbf{k}$, length $= 3\pi$

3. $\mathbf{T} = \frac{1}{\sqrt{1+t}}\mathbf{i} + \frac{\sqrt{t}}{\sqrt{1+t}}\mathbf{k}$, length $= \frac{52}{3}$ **5.** $\mathbf{T} = \frac{1}{\sqrt{3}}\mathbf{i} - \frac{1}{\sqrt{3}}\mathbf{j} + \frac{1}{\sqrt{3}}\mathbf{k}$, length $= 3\sqrt{3}$ **7.** $\mathbf{T} = \left(\frac{\cos t - t\sin t}{t+1}\right)\mathbf{i} + \left(\frac{\sin t + t\cos t}{t+1}\right)\mathbf{j} + \left(\frac{\sqrt{2}\,t^{1/2}}{t+1}\right)\mathbf{k}$, length $= \frac{\pi^2}{2} + \pi$

9. $s(t) = 5t$, length $= \frac{5\pi}{2}$ **11.** $s(t) = \sqrt{3}e^t - \sqrt{3}$, length $= 3\sqrt{3}$ **13.** $\sqrt{2} + \ln(1 + \sqrt{2})$ **15.** a) $2\pi\sqrt{2}$ b) $2\pi\sqrt{2}$ c) $2\pi\sqrt{2}$

Section 12.4, pp. 803–804

1. $\mathbf{T} = \cos t\,\mathbf{i} - \sin t\,\mathbf{j}$, $\mathbf{N} = -\sin t\,\mathbf{i} - \cos t\,\mathbf{j}$, $\kappa = \cos t$

3. $\mathbf{T} = \frac{1}{\sqrt{1+t^2}}\mathbf{i} - \frac{t}{\sqrt{1+t^2}}\mathbf{j}$, $\mathbf{N} = \frac{-t}{\sqrt{1+t^2}}\mathbf{i} - \frac{1}{\sqrt{1+t^2}}\mathbf{j}$, $\kappa = \frac{1}{2(\sqrt{1+t^2})^3}$ **5.** $\kappa = \frac{3}{25}$, $\mathbf{T} = \frac{3\cos t}{5}\mathbf{i} - \frac{3\sin t}{5}\mathbf{j} + \frac{4}{5}\mathbf{k}$, $\mathbf{N} = (-\sin t)\mathbf{i} - (\cos t)\mathbf{j}$, $\mathbf{B} = \left(\frac{4}{5}\cos t\right)\mathbf{i} - \left(\frac{4}{5}\sin t\right)\mathbf{j} - \frac{3}{5}\mathbf{k}$, $\tau = \frac{4}{25}$ **7.** $\kappa = \frac{1}{\sqrt{2}e^t}$, $\mathbf{T} = \left(\frac{\cos t - \sin t}{\sqrt{2}}\right)\mathbf{i} + \left(\frac{\sin t + \cos t}{\sqrt{2}}\right)\mathbf{j}$, $\mathbf{N} = \left(\frac{-\cos t - \sin t}{\sqrt{2}}\right)\mathbf{i} + \left(\frac{-\sin t + \cos t}{\sqrt{2}}\right)\mathbf{j}$, $\mathbf{B} = \mathbf{k}$, $\tau = 0$

9. $\mathbf{a} = \frac{2t}{\sqrt{1+t^2}}\mathbf{T} + \frac{2}{\sqrt{1+t^2}}\mathbf{N}$ **11.** $\mathbf{a} = |a|\mathbf{N}$

13. $\mathbf{a} = \frac{4}{3}\mathbf{T} + \frac{2\sqrt{5}}{3}\mathbf{N}$ **15.** $\mathbf{a} = 2\mathbf{N}$ **17.** $\mathbf{T}\left(\frac{\pi}{4}\right) = -\frac{\sqrt{2}}{2}\mathbf{i} + \frac{\sqrt{2}}{2}\mathbf{j}$, $\mathbf{N}\left(\frac{\pi}{4}\right) = -\frac{\sqrt{2}}{2}\mathbf{i} - \frac{\sqrt{2}}{2}\mathbf{j}$, $\mathbf{B}\left(\frac{\pi}{4}\right) = \mathbf{k}$, $z = -1$, $-x + y = 0$, $x + y = \sqrt{2}$

21. $\left(x - \frac{\pi}{2}\right) + y^2 = 1$ **25.** $\frac{1}{2b}$ is maximum value of κ

Section 12.5, pp. 811–812

1. $T = 93.17$ min **3.** $a = 6765$ km **5.** $T = 1655$ min **7.** $a = 2.043 \times 10^7$m **9.** $|v| = 1.9966 \times 10^7 r^{-1/2}$ **11.** Circle:

$v_0 = \sqrt{\frac{GM}{r_0}}$; ellipse: $\sqrt{\frac{GM}{r_0}} < v_0 < \sqrt{\frac{GM}{r_0}}$;

parabola: $v_0 = \sqrt{\frac{2GM}{r_0}}$; hyperbola: $v_0 > \sqrt{\frac{GM}{r_0}}$

Chapter 12 Practice Exercises, pp. 813–814

1.

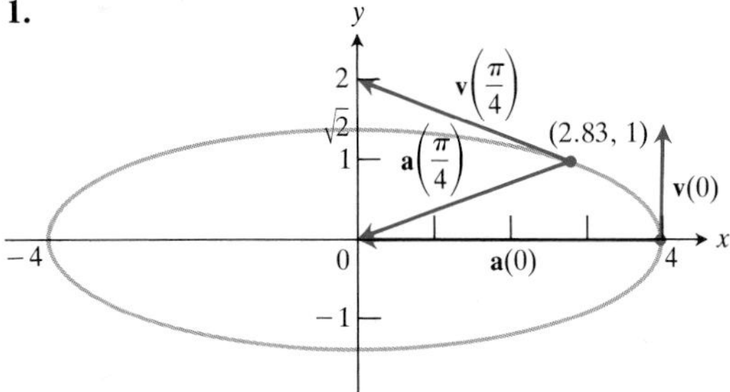

3. $6\mathbf{i} + 8\mathbf{j}$ **5.** $\mathbf{r} = ((\cos t) - 1)\mathbf{i} + ((\sin t) + 1)\mathbf{j} + t\mathbf{k}$

7. $\mathbf{r} = \mathbf{i} + t^2\mathbf{j} + t\mathbf{k}$ **9.** Length $= \frac{\pi}{4}\sqrt{1 + \frac{\pi^2}{16}} + \ln\left|\frac{\pi}{4} + \sqrt{1 + \frac{\pi^2}{16}}\right|$ **11.** $\mathbf{T}(0) = \frac{2}{3}\mathbf{i} + \frac{2}{3}\mathbf{j} + \frac{1}{3}\mathbf{k}$, $\mathbf{N}(0) = \frac{1}{\sqrt{2}}\mathbf{i} + \frac{1}{\sqrt{2}}\mathbf{j}$, $\mathbf{B}(0) = -\frac{1}{3\sqrt{2}}\mathbf{i} + \frac{1}{3\sqrt{2}}\mathbf{j}$, $\kappa = \frac{\sqrt{2}}{9}$, $\tau = 0$ **13.** $\mathbf{a} = 10\mathbf{T} + 6\mathbf{N}$

15. $|\mathbf{v}|_{\max} = 1$ **17.** $0, \pi, \frac{\pi}{2}$ **19.** Shot put is on the ground.
21. 644.36 ft/sec; 675.42 ft/sec **23.** $x = 1 + t, y = t, z = -t$
25. $\kappa = 5$ **27.** $\tau_{\max} = 1/2a$ **29.** 16 395 469, % visible = 3.21%
31. b) ≈ 1.49870113

Chapter 13

Section 13.1, pp. 825–826

1. Domain: All points in the xy-plane
Range: All reals
Level curves are straight lines parallel to the xy-plane.

3. Domain: $(x, y) \neq (0, 0)$
Range: All reals
Level curves are circles with centers on the z-axis and radii from 0 to $+\infty$.

5. Domain: All points in the xy-plane
Range: All reals
Level curves are hyperbolas.

7. Domain: All points in the xy-plane
Range: $z \geq 0$
Level curves are ellipses with center on the z-axis.

9. a)

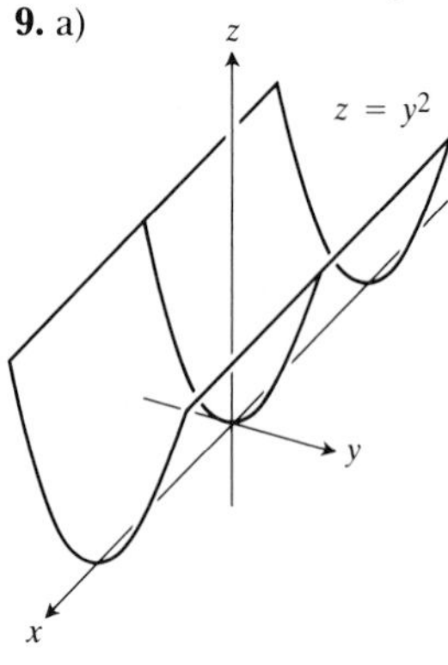

b)

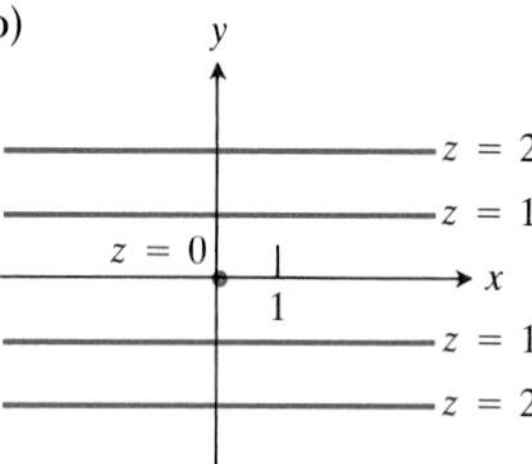

11. a)

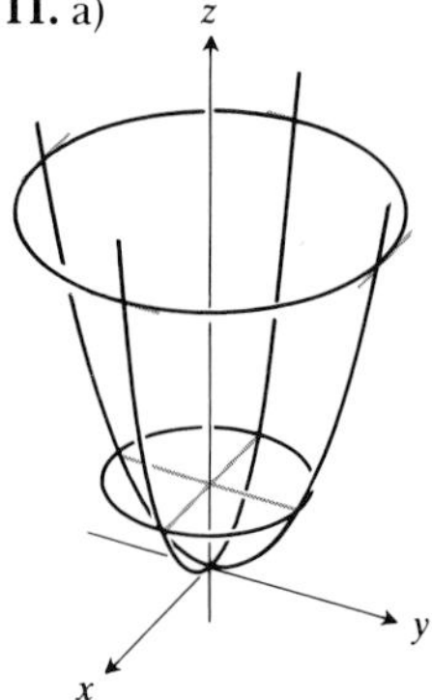

b)

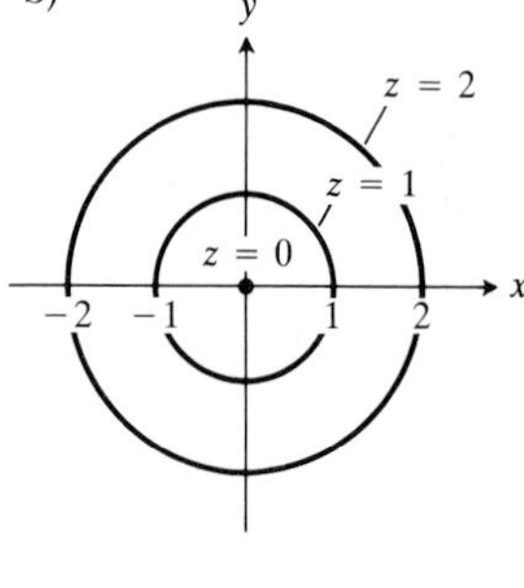

13. a)

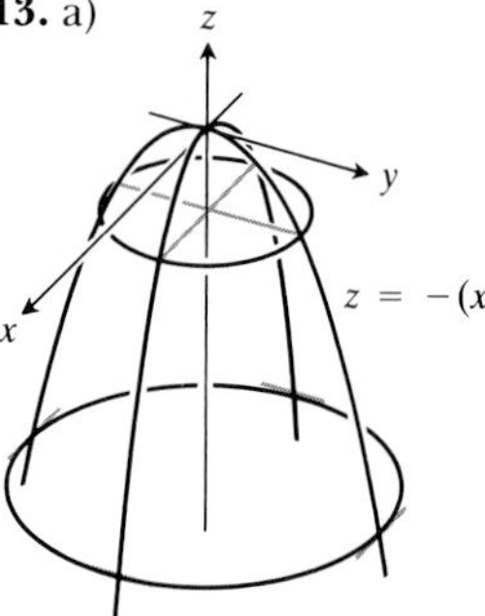

b)

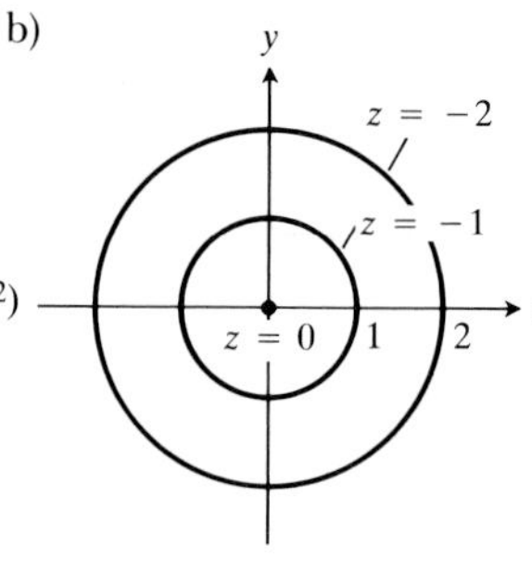

15. a)

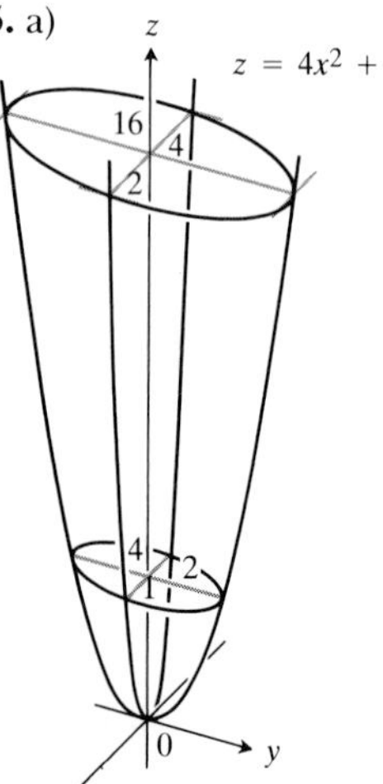

b)

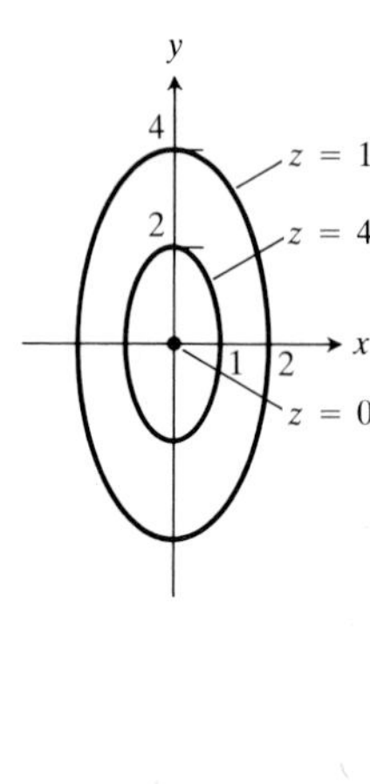

17. f **19.** a **21.** d

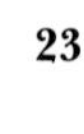

23.

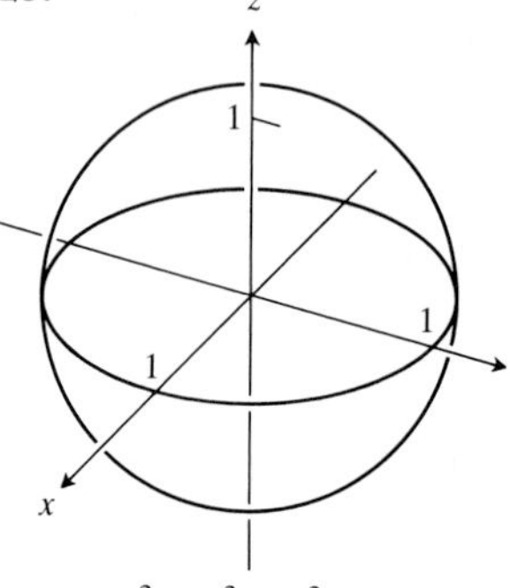

25.

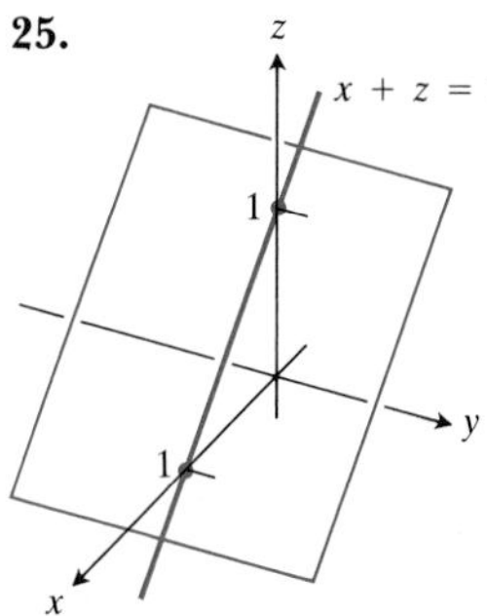

27.

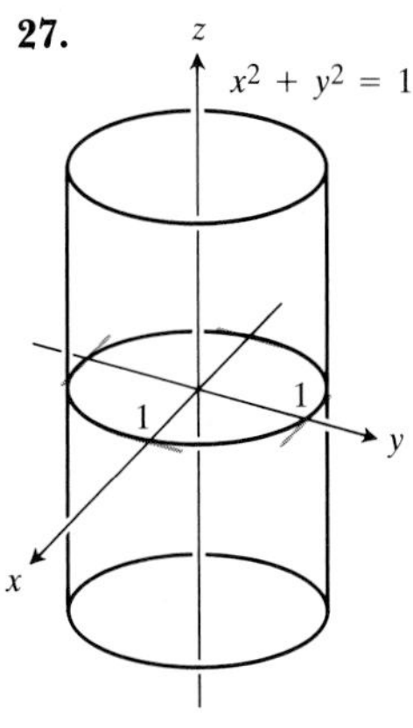

29.

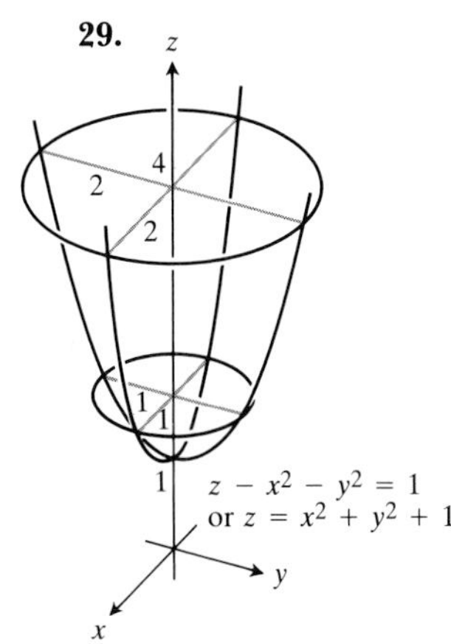

31. $x^2 + y^2 = 10$ **33.** 62.43 km

Section 13.2, p. 832

1. 5/2 **3.** −2 **5.** 5 **7.** 1 **9.** 1 **11.** 1 **13.** 0 **15.** −1 **17.** 7 **19.** 2 **21.** $\tan^{-1}\left(-\frac{\pi}{4}\right)$ **23.** a) All (x, y) b) All (x, y) except $(0, 0)$ **25.** a) All (x, y) except where $x = 0$ or $y = 0$ b) All (x, y) **27.** a) All (x, y, z) b) All (x, y, z) except the interior of the cylinder $x^2 + y^2 = 1$ **29.** a) All $(x, y, z) \neq (x, y, 0)$ b) All (x, y, z) except those on the sphere $x^2 + y^2 + z^2 = 1$ **31.** Consider paths along $y = x, x > 0$, and along $y = x, x < 0$ **33.** Consider the paths $y = kx^2$, k a constant **35.** Consider the paths $y = mx$, m a constant **37.** Consider the paths $y = kx^2$, k a constant **39.** a) $f(x, y)|_{y=mx} = \tan 2\theta$ where $\theta = m$

Section 13.3, pp. 840–841

1. $\frac{\partial f}{\partial x} = 2, \frac{\partial f}{\partial y} = 0$ **3.** $\frac{\partial f}{\partial x} = 0, \frac{\partial f}{\partial y} = 0$ **5.** $\frac{\partial f}{\partial x} = y - 1, \frac{\partial f}{\partial y} = x$ **7.** $\frac{\partial f}{\partial x} = 2x - y, \frac{\partial f}{\partial y} = -x + 2y$ **9.** $\frac{\partial f}{\partial x} = 5y - 14x + 3, \frac{\partial f}{\partial y} = 5x - 2y - 6$ **11.** $\frac{\partial f}{\partial x} = x(x^2 + y^2)^{-1/2}, \frac{\partial f}{\partial y} = y(x^2 + y^2)^{-1/2}$ **13.** $\frac{\partial f}{\partial x} = \frac{-y^2 - 1}{(xy - 1)^2}, \frac{\partial f}{\partial y} = \frac{-x^2 - 1}{(xy - 1)^2}$ **15.** $\frac{\partial f}{\partial x} = e^x \ln y, \frac{\partial f}{\partial y} = \frac{e^x}{y}$ **17.** $\frac{\partial f}{\partial x} = e^x \sin(y + 1), \frac{\partial f}{\partial y} = e^x \cos(y + 1)$ **19.** $\frac{\partial f}{\partial x} = y + z, \frac{\partial f}{\partial y} = x + z, \frac{\partial f}{\partial z} = y + x$ **21.** $\frac{\partial f}{\partial x} = 0, \frac{\partial f}{\partial y} = 2y, \frac{\partial f}{\partial z} = 4z$ **23.** $\frac{\partial f}{\partial x} = \cos(x + yz), \frac{\partial f}{\partial y} = z\cos(x + yz), \frac{\partial f}{\partial z} = y\cos(x + yz)$ **25.** $\frac{\partial f}{\partial t} = -2\pi\sin(2\pi t - \alpha), \frac{\partial f}{\partial \alpha} = \sin(2\pi t - \alpha)$ **27.** $\frac{\partial h}{\partial \rho} = \sin\phi\cos\theta, \frac{\partial h}{\partial \phi} = \rho\cos\phi\cos\theta, \frac{\partial h}{\partial \theta} = -\rho\sin\phi\sin\theta$ **29.** $\frac{\partial W}{\partial P} = V, \frac{\partial W}{\partial V} = P + \frac{\rho v^2}{2g}, \frac{\partial W}{\partial \rho} = \frac{Vv^2}{2g}, \frac{\partial W}{\partial v} = \frac{V\rho v}{g}, \frac{\partial W}{\partial g} = \frac{-V\rho v^2}{2g^2}$ **31.** $\frac{\partial^2 f}{\partial x^2} = 0, \frac{\partial^2 f}{\partial y^2} = 0, \frac{\partial^2 f}{\partial y \partial x} = \frac{\partial^2 f}{\partial x \partial y} = 1$ **33.** $\frac{\partial^2 g}{\partial x^2} = 2y - y\sin x, \frac{\partial^2 g}{\partial y^2} = -\cos y, \frac{\partial^2 g}{\partial y \partial x} = \frac{\partial^2 g}{\partial x \partial y} = 2x + \cos x$ **35.** $\frac{\partial^2 r}{\partial x^2} = \frac{-1}{(x + y)^2}, \frac{\partial^2 r}{\partial y^2} = \frac{-1}{(x + y)^2}, \frac{\partial^2 r}{\partial y \partial x} = \frac{\partial^2 r}{\partial x \partial y} = \frac{-1}{(x + y)^2}$ **37.** $\frac{\partial^2 W}{\partial y \partial x} = \frac{\partial^2 W}{\partial x \partial y} = \frac{-6}{(2x + 3y)^2}$ **39.** $\frac{\partial^2 W}{\partial y \partial x} = \frac{\partial^2 W}{\partial x \partial y} = 2y + 6xy^2 + 12x^2y^3$ **41.** a) x first b) y first c) x first d) x first e) y first f) y first **43.** $\left.\frac{\partial z}{\partial x}\right|_{(1,1,1)} = -2$

Section 13.4, pp. 848–849

1. $\frac{dw}{dt} = 0, \left.\frac{dw}{dt}\right|_{t=\pi} = 0$ **3.** $\frac{dw}{dt} = 1, \left.\frac{dw}{dt}\right|_{t=3} = 1$ **5.** $\frac{dw}{dt} = \frac{4t\tan^{-1}t}{t^2 + 1}, \left.\frac{dw}{dt}\right|_{t=1} = \frac{\pi}{2}$

7. $\frac{dw}{dt} = \frac{\partial w}{\partial x}g'(t) + \frac{\partial w}{\partial y}h'(t)$

9. $\frac{\partial w}{\partial u} = \frac{\partial w}{\partial x}\frac{\partial x}{\partial u} + \frac{\partial w}{\partial y}\frac{\partial y}{\partial u} + \frac{\partial w}{\partial z}\frac{\partial z}{\partial u}$

$\frac{\partial w}{\partial v} = \frac{\partial w}{\partial x}\frac{\partial x}{\partial v} + \frac{\partial w}{\partial y}\frac{\partial y}{\partial v} + \frac{\partial w}{\partial z}\frac{\partial z}{\partial v}$

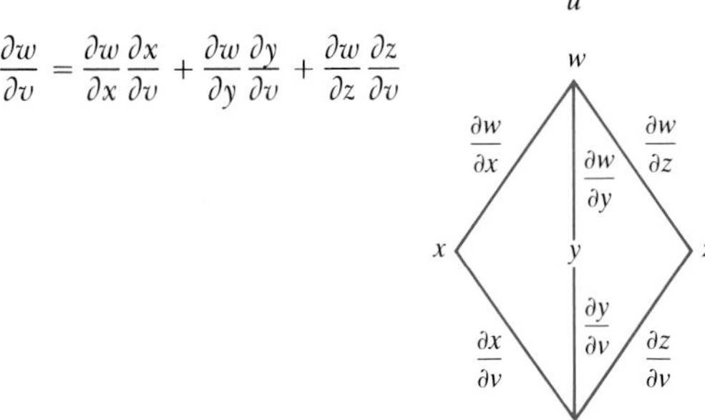

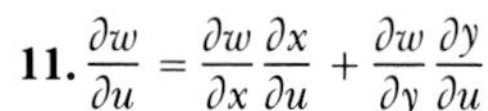

11. $\dfrac{\partial w}{\partial u} = \dfrac{\partial w}{\partial x}\dfrac{\partial x}{\partial u} + \dfrac{\partial w}{\partial y}\dfrac{\partial y}{\partial u}$ $\qquad \dfrac{\partial w}{\partial v} = \dfrac{\partial w}{\partial x}\dfrac{\partial x}{\partial v} + \dfrac{\partial w}{\partial y}\dfrac{\partial y}{\partial v}$

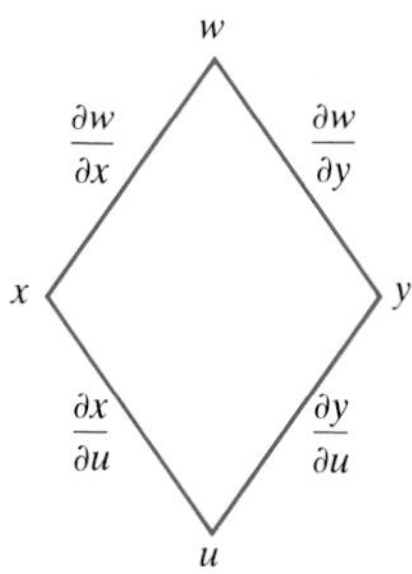

13. $\dfrac{\partial z}{\partial t} = \dfrac{\partial z}{\partial x}\dfrac{\partial x}{\partial t} + \dfrac{\partial z}{\partial y}\dfrac{\partial y}{\partial t}$ $\qquad \dfrac{\partial z}{\partial s} = \dfrac{\partial z}{\partial x}\dfrac{\partial x}{\partial s} + \dfrac{\partial z}{\partial y}\dfrac{\partial y}{\partial s}$

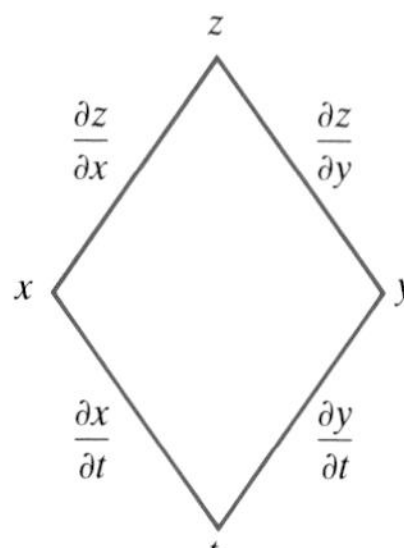

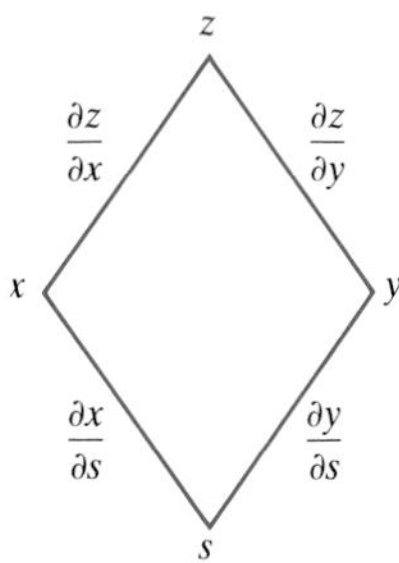

15. $\dfrac{\partial w}{\partial s} = \dfrac{dw}{du}\dfrac{\partial u}{\partial s}$

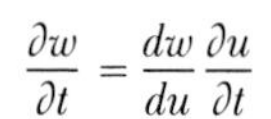

$\dfrac{\partial w}{\partial t} = \dfrac{dw}{du}\dfrac{\partial u}{\partial t}$

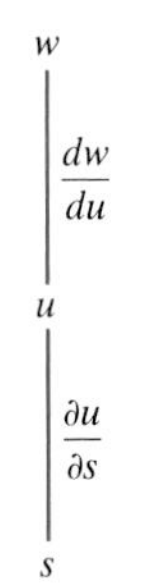

17. $\dfrac{\partial w}{\partial r} = \dfrac{\partial w}{\partial x}\dfrac{dx}{dr}$ $\qquad \dfrac{\partial w}{\partial s} = \dfrac{\partial w}{\partial y}\dfrac{dy}{ds}$

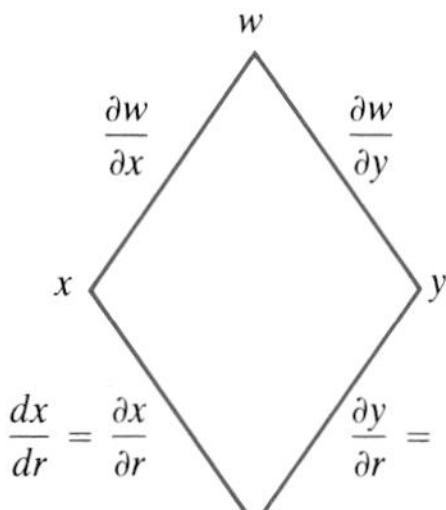

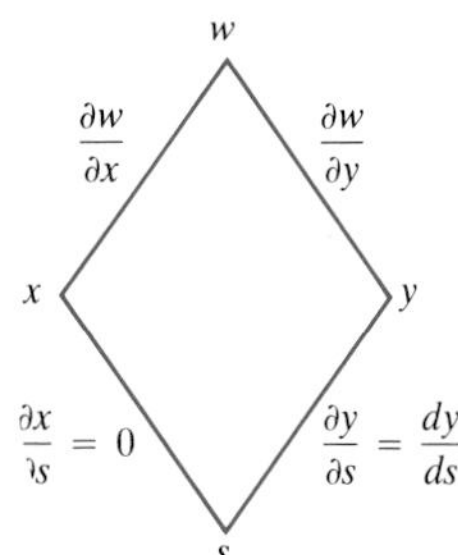

23. $\left.\dfrac{\partial z}{\partial x}\right|_{(1,1,1)} = \dfrac{1}{4}$, $\left.\dfrac{\partial z}{\partial y}\right|_{(1,1,1)} = -\dfrac{3}{4}$ **25.** $\left.\dfrac{\partial z}{\partial x}\right|_{(\pi,\pi,\pi)} = -1$, $\left.\dfrac{\partial z}{\partial y}\right|_{(\pi,\pi,\pi)} = -1$

27. $\left.\dfrac{\partial w}{\partial r}\right|_{(r,s)=(1,1)} = 12$ **29.** $\left.\dfrac{\partial w}{\partial v}\right|_{(u,v)=(0,0)} = -7$

31. $\left.\dfrac{\partial z}{\partial u}\right|_{(u,v)=(\ln 2,1)} = 0$, $\left.\dfrac{\partial z}{\partial v}\right|_{(u,v)=(\ln 2,1)} = 10$ **33.** $\dfrac{dv}{dt} = -0.01$ volts/sec

35. $x = \cos(-2)$, $y = \sin(-2)$, $z = -2$
$x = \cos(1)$, $y = \sin(1)$, $z = 1$

37. a) T has minimum at $t = \dfrac{\pi}{4}, \dfrac{5\pi}{4}$; T has maximum at $t = \dfrac{3\pi}{4}, \dfrac{7\pi}{4}$ b) $T_{\max} = 6$, $T_{\min} = 2$

Section 13.5, pp. 856–857

1. $\nabla f = 2\mathbf{i} + 2\mathbf{j} - 4\mathbf{k}$ **3.** $\nabla f = \dfrac{-1}{27}\mathbf{i} - \dfrac{-2}{27}\mathbf{j} + \dfrac{2}{27}\mathbf{k}$

5. $\nabla f = -\mathbf{i} + \mathbf{j}$ **7.** $\nabla f = 2\mathbf{i} + \mathbf{j}$

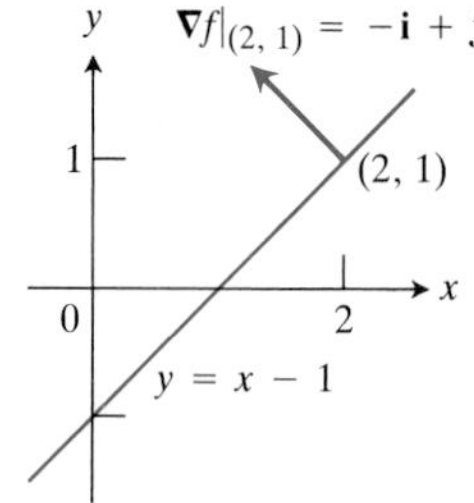

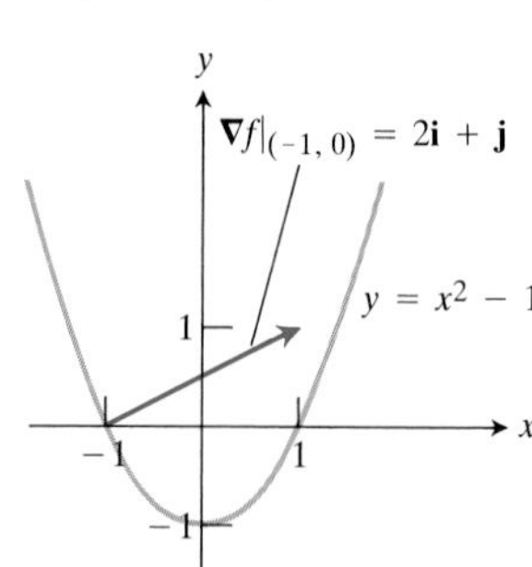

9. $(D_{\mathbf{u}}f)_{P_0} = 3$

11. $(D_{\mathbf{u}}f)_{P_0} = 2$ **13.** $(D_{\mathbf{u}}f)_{P_0} = \dfrac{14}{13}$ **15.** $(D_{\mathbf{u}}f)_{P_0} = \dfrac{-4}{5}$

17. f increases most rapidly in the direction $\mathbf{u} = \dfrac{2}{\sqrt{5}}\mathbf{i} + \dfrac{1}{\sqrt{5}}\mathbf{j}$; decreases most rapidly in the direction $-\mathbf{u} = -\dfrac{2}{\sqrt{5}}\mathbf{i} - \dfrac{1}{\sqrt{5}}\mathbf{j}$. $(D_{\mathbf{u}}f) = \sqrt{5}$, $(D_{-\mathbf{u}}f) = -\sqrt{5}$ **19.** f increases most rapidly in the direction $\mathbf{u} = \dfrac{2}{\sqrt{6}}\mathbf{i} + \dfrac{1}{\sqrt{6}}\mathbf{j} + \dfrac{1}{\sqrt{6}}\mathbf{k}$; decreases most rapidly in the direction $-\mathbf{u} = -\dfrac{2}{\sqrt{6}}\mathbf{i} - \dfrac{1}{\sqrt{6}}\mathbf{j} - \dfrac{1}{\sqrt{6}}\mathbf{k}$. $(D_{\mathbf{u}}f) = \sqrt{6}$, $(D_{-\mathbf{u}}f) = -\sqrt{6}$

21. f increases most rapidly in the direction $\mathbf{u} = \dfrac{1}{\sqrt{3}}\mathbf{i} + \dfrac{1}{\sqrt{3}}\mathbf{j} + \dfrac{1}{\sqrt{3}}\mathbf{k}$; decreases most rapidly in the direction $-\mathbf{u} = -\dfrac{1}{\sqrt{3}}\mathbf{i} - \dfrac{1}{\sqrt{3}}\mathbf{j} - \dfrac{1}{\sqrt{3}}\mathbf{k}$. $(D_{\mathbf{u}}f) = \sqrt{3}$, $(D_{-\mathbf{u}}f) = -\sqrt{3}$.

23. $df = 0.15\sqrt{2}$ **25.** $df = \dfrac{0.1\sqrt{6}}{3}$ **27.** $\mathbf{u} = \dfrac{7}{\sqrt{58}}\mathbf{i} - \dfrac{3}{\sqrt{58}}\mathbf{j}$; $-\mathbf{u} = -\dfrac{7}{\sqrt{58}}\mathbf{i} + \dfrac{3}{\sqrt{58}}\mathbf{j}$ **29.** $(D_{\mathbf{T}}f)_{t=-\pi/4} = \dfrac{-\pi}{2\sqrt{2}}$, $(D_{\mathbf{T}}f)_{t=0} = 0$, $(D_{\mathbf{T}}f)_{t=\pi/4} = \dfrac{\pi}{2\sqrt{2}}$ **31.** $D_{\mathbf{u}}f = -\dfrac{7}{\sqrt{5}}$

Section 13.6, pp. 860–861

1. Tangent: $x + y + z = 3$
Normal line: $x = 1 + 2t, y = 1 + 2t, z = 1 + 2t$
3. Tangent: $3x + 4y + 5z = 0$
Normal line: $x = 3 - 6t, y = 4 - 8t, z = -5 - 10t$
5. Tangent: $-2x + z = -2$
Normal line: $x = 1 - 2t, y = 0, z = t$
7. Tangent: $2y + 3z = 7$
Normal line: $x = 1, y = -1 + 4t, z = 3 + 6t$
9. Tangent: $x - 3y - z = -1$
Normal line: $x = 1 + t, y = 1 - 3t, z = -1 - t$
11. Tangent: $9x - 7y - z = 21$
Normal line: $x = 2 + 9t, y = -3 - 7t, z = 18 - t$

13.

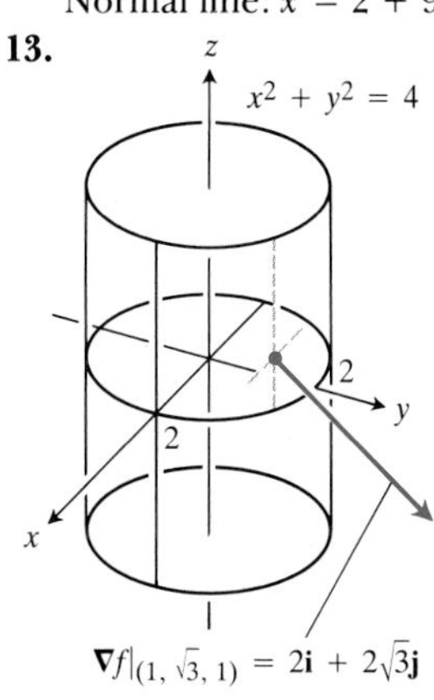

15.

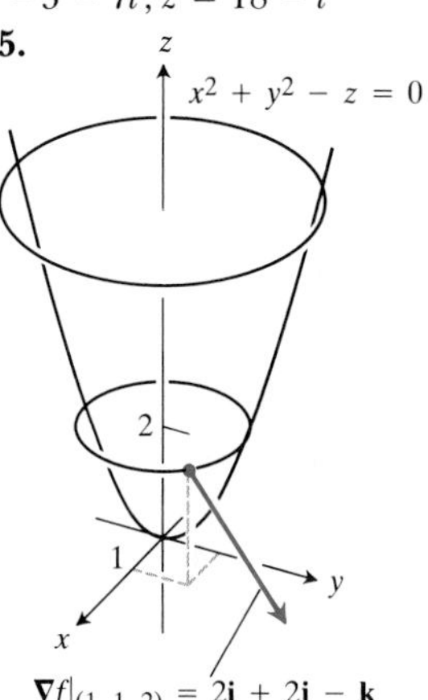

17.

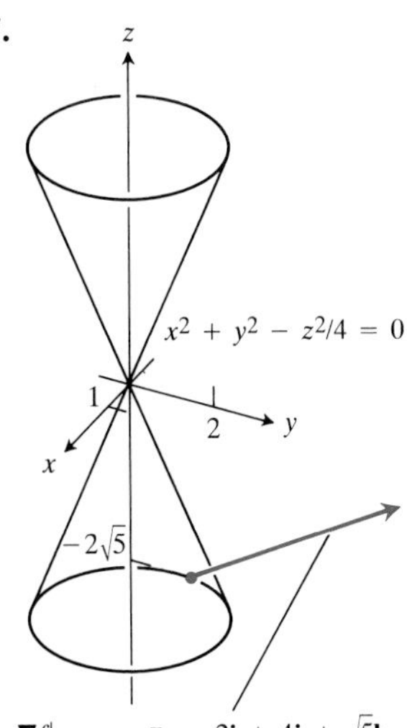

19. $x = 1, y = 1 + 2t, z = 1 - 2t$ **21.** $x = 1 - 2t, y = 1, z = (1/2) + 2t$ **23.** $x = 1 + 126t, y = 1 - 90t, z = 3 - 12t$

25.

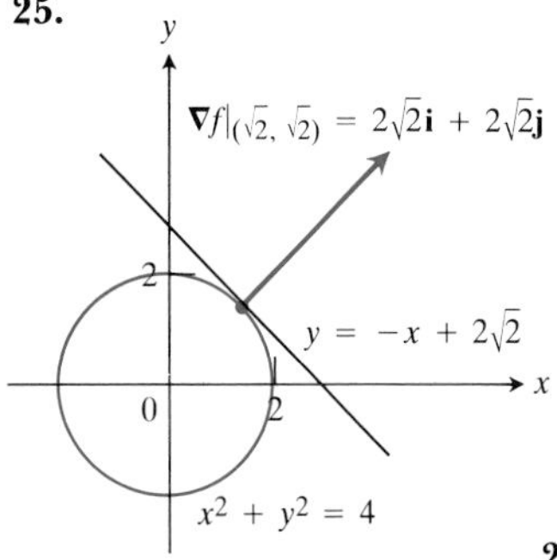

27. **29.**

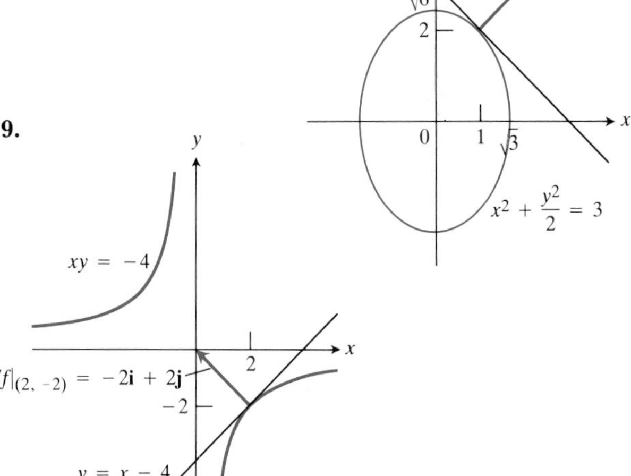

Section 13.7, pp. 867–869

1. a) $L(0, 0) = 1$ b) $L(1, 1) = x + y + 1$ **3.** $L(0, 0) = 1 + x$
b) $L\left(0, \dfrac{\pi}{2}\right) = -y + \dfrac{\pi}{2}$ **5.** a) $L(0, 0) = 5 + 3x - 4y$
b) $L(1, 1) = 3x - 4y + 5$
7. $L(2, 1) = 7 + x - 6y$
$|E(x, y)| \leq 0.06$
9. $L(0, 0) = x + y$
$|E(x, y)| \leq 0.08$
11. $L(0, 0) = 1 + x$
$|E(x, y)| \leq 0.022$
13. Because $w > 1$, dA is more sensitive to a change in w than in l. Therefore pay more attention to the width.
15. Maximum error (estimate) ≤ 0.32 in magnitude
17. Maximum percentage error $= \pm 4.83\%$
19. Let $|x - 1| \leq 0.014$, $|y - 1| \leq 0.014$ **21.** Percentage change $= 0.12\%$ **23.** a) $L(x, y, z) = 2x + 2y + 2z - 3$
b) $L(x, y, z) = y + z$ c) $L(x, y, z) = 0$ **25.** a) $L(x, y, z) = x$
b) $L(x, y, z) = \dfrac{1}{\sqrt{e2}}x + \dfrac{1}{\sqrt{2}}y$ c) $L(x, y, z) = \dfrac{1}{3}x + \dfrac{2}{3}y + \dfrac{2}{3}z$

27. a) $L(x, y, z) = 2 + x$ b) $L(x, y, z) = x - y - z + \frac{\pi}{2} + 1$ c) $L(x, y, z) = x - y - z + \frac{\pi}{2} + 1$ **29.** f is most sensitive to a change in d. **31.** $\frac{47}{24}$ ft^3 of wood **33.** Magnitude of maximum possible error ≤ 4.8

Section 13.8, pp. 875–878

1. $f(-3, 3) = -5$, local and absolute minimum **3.** $f\left(\frac{6}{5}, \frac{69}{25}\right)$, saddle point **5.** $f(-2, 1)$, saddle point **7.** $f\left(\frac{4}{9}, \frac{2}{9}\right) = -\frac{252}{81}$, local and absolute maximum **9.** $f(2, 1)$ saddle point
11. $f(2, -1) = -6$, local and absolute minimum
13. $f\left(\frac{1}{6}, \frac{4}{3}\right)$, saddle point **15.** $f(1, 0) = 0$, local and absolute minimum **17.** $f(0, 1) = 4$, local absolute maximum
19. $f(0, 0)$, saddle point
$f\left(-\frac{2}{3}, \frac{2}{3}\right) = \frac{170}{27}$, local maximum
21. $f(0, 0) = 0$, local minimum
$f(-1, 1)$, saddle point
23. $f(0, 0)$, saddle point
$f(-1, -1) = 1$, local maximum
25. Absolute maximum is 1 at (0, 0), absolute minimum is -5 at (1, 2). **27.** Absolute maximum is 4 at (0, 2), absolute minimum is 0 at (0, 0). **29.** Absolute maximum is 19 at (5, 3), absolute minimum is -12 at $(4, -2)$. **31.** Absolute maximum is 0 at (1, 0), absolute minimum is -4 at (2, 0).
33. Hottest is $2\frac{1}{4}^\circ$ at $\left(-\frac{1}{2}, \frac{\sqrt{3}}{2}\right)$ and $\left(-\frac{1}{2}, -\frac{\sqrt{3}}{2}\right)$; coldest is $-\frac{1}{4}^\circ$ at $\left(\frac{1}{2}, 0\right)$. **35.** a) Saddle point at (0, 0) b) Local minimum at (1, 2) c) Local minimum at (1, 2); saddle point at $(-1, -2)$
37. a) i) On the quarter circle: Absolute minimum of 0 at $t = 0, \frac{\pi}{2}$; absolute maximum of 2 at $t = \frac{\pi}{4}$
ii) On the half circle: absolute minimum of -2 at $t = \frac{3\pi}{4}$; absolute maximum of 2 at $t = \frac{\pi}{4}$
iii) On the full circle: absolute minimum of -2 at $t = \frac{3\pi}{4}, \frac{7\pi}{4}$; absolute maximum of 2 at $t = \frac{\pi}{4}, \frac{5\pi}{4}$
b) i) On $\left[0, \frac{\pi}{2}\right]$: absolute minimum of 2 at $t = 0, \frac{\pi}{2}$; absolute maximum of $2\sqrt{2}$ at $t = \frac{\pi}{4}$
ii) On $[0, \pi]$: absolute minimum of -2 at $t = \pi$; absolute maximum of $2\sqrt{2}$ at $t = \frac{\pi}{4}$
iii) On $[0, 2\pi]$: absolute minimum of $-2\sqrt{2}$ at $t = \frac{5\pi}{4}$; absolute maximum of $2\sqrt{2}$ at $t = \frac{\pi}{4}$
c) i) On $\left[0, \frac{\pi}{2}\right]$: absolute minimum of 1 at $t = \frac{\pi}{2}$; absolute maximum of 8 at $t = 0$
ii) On $[0, \pi]$: absolute minimum of 1 at $t = \frac{\pi}{2}$; absolute maximum of 8 at $t = 0, \pi$
iii) On $[0, 2\pi]$: absolute minimum of 1 at $t = \frac{\pi}{2}, \frac{3\pi}{2}$; absolute maximum of 8 at $t = 0, \pi, 2\pi$
39. a) i) On the quarter ellipse: absolute minimum of 9 at $t = 0$; absolute maximum of 12 at $t = \frac{\pi}{2}$
ii) On the half ellipse: absolute minimum of 9 at $t = 0, \pi$; absolute maximum of 12 at $t = \frac{\pi}{2}$
iii) On the full ellipse: absolute minimum of 9 at $t = 0, \pi, 2\pi$; absolute maximum of 12 at $t = \frac{\pi}{2}, \frac{3\pi}{2}$
b) i) On $\left[0, \frac{\pi}{2}\right]$: absolute minimum of 6 at $t = 0, \frac{\pi}{2}$; absolute maximum of $6\sqrt{2}$ at $t = \frac{\pi}{4}$
ii) On $[0, \pi]$: absolute minimum of -6 at $t = \pi$; absolute maximum of $6\sqrt{2}$ at $t = \frac{\pi}{4}$
iii) On $[0, 2\pi]$: absolute minimum of $-6\sqrt{2}$ at $t = \frac{5\pi}{4}$; absolute maximum of $6\sqrt{2}$ at $t = \frac{\pi}{4}$
41. $y = -1.5x + 0.7, y|_{x=4} = -5.3$
43. $y = 1.5x + 0.2, y|_{x=4} = 6.2$
45. $y = 0.122x + 3.59$

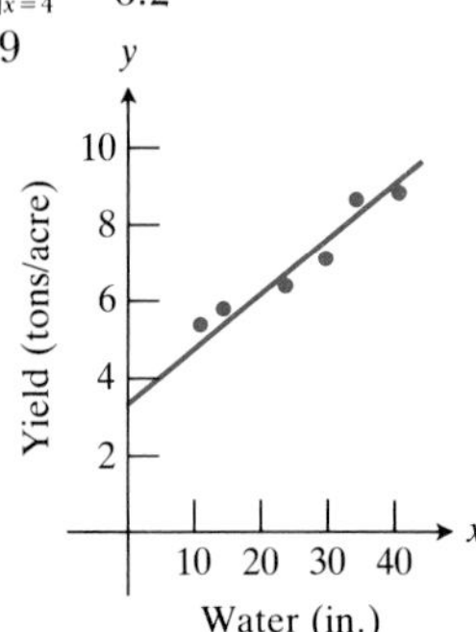

Section 13.9, pp. 886–887

1. f takes on extreme values at $\left(\pm\sqrt{2}, \pm\frac{1}{2}\right)$ **3.** $f(1, 3) = 39$, maximum **5.** $f(0, 3) = 0$, minimum; $f(2, 1) = 4$, maximum
7. a) Minimum is 8 at $x = 4, y = 4$ b) $f(8, 8) = 64$, maximum
9. $r = 2$ cm, $h = 4$ cm **11.** 0° is minimum, 125° is maximum

13. $f(0, 0) = 0$, minimum; $f(2, 4) = 20$, maximum
15. Maximum $= 30$, minimum $= -30$ **17.** $(0, 0, 1)$
19. $5\sqrt{14}$ maximum, $-5\sqrt{14}$ minimum **21.** $\dfrac{4096}{25\sqrt{5}}$
23. $x = 8, y = 14$, \$128

Chapter 13 Practice Exercises, pp. 888–890

1. Domain: All points in the xy plane
Range: $f(x, y) \geq 0$
Level curves are ellipses with major axis along the y-axis, minor axis along the x-axis.

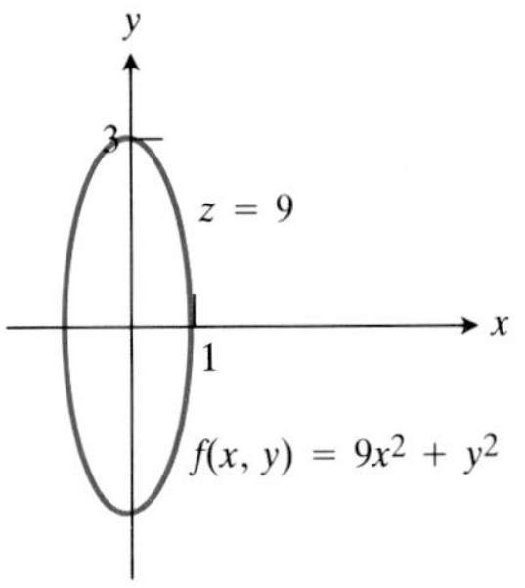

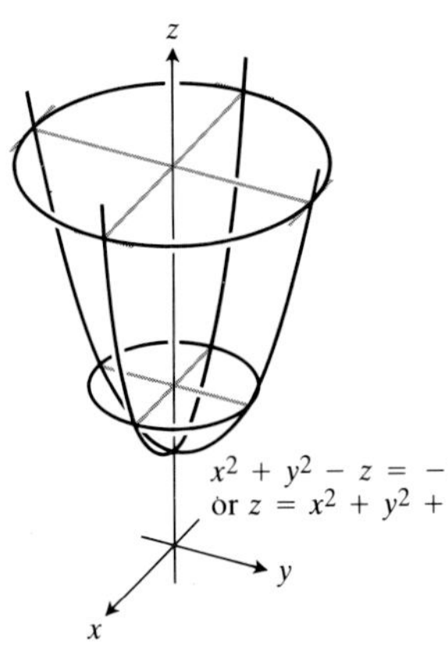

3. Domain: All $(x, y) \neq (0, y)$ and $\neq (x, 0)$
Range: $f(x, y) \neq 0$
Level curves are hyperbolas rotated $\frac{\pi}{4}$ or $\frac{3\pi}{4}$ radians from the positive x-axis.

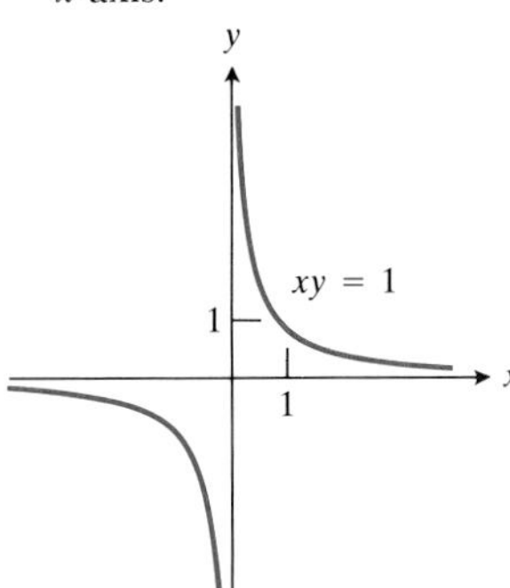

5. Domain: All points (x, y, z) in space
Range: All real numbers
Level surfaces are paraboloids of revolution with z-axis as axis.

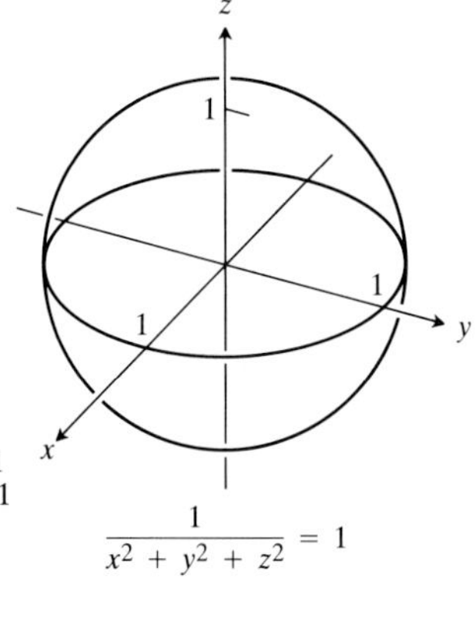

7. Domain: $(x, y, z) \neq (0, 0, 0)$
Range: $w > 0$
Level surfaces are spheres with center $(0, 0, 0)$ and radius > 0.

9. -2 **11.** 2 **13.** Let $y = kx^2, k \neq 1$

15. $\dfrac{\partial g}{\partial r} = \cos\theta + \sin\theta, \dfrac{\partial g}{\partial \theta} = -r\sin\theta + r\cos\theta$

17. $\dfrac{\partial f}{\partial R_1} = -\dfrac{1}{R_1^2}, \dfrac{\partial f}{\partial R_2} = -\dfrac{1}{R_2^2}, \dfrac{\partial f}{\partial R_3} = -\dfrac{1}{R_3^2}$

17. $\dfrac{\partial f}{\partial R_1} = -\dfrac{1}{R_1^2}, \dfrac{\partial f}{\partial R_2} = -\dfrac{1}{R_2^2}, \dfrac{\partial f}{\partial R_3} = -\dfrac{1}{R_3^2}$

19. $\dfrac{\partial P}{\partial n} = \dfrac{RT}{V}, \dfrac{\partial P}{\partial R} = \dfrac{nT}{V}, \dfrac{\partial P}{\partial T} = \dfrac{nR}{V}, \dfrac{\partial P}{\partial V} = -\dfrac{nRT}{V^2}$

21. $\dfrac{\partial^2 f}{\partial x^2} = 0, \dfrac{\partial^2 f}{\partial y^2} = -\dfrac{2x}{y^3}$
$\dfrac{\partial^2 f}{\partial y \partial x} = \dfrac{\partial^2 f}{\partial x \partial y} = -\dfrac{1}{y^2}$

23. $\dfrac{\partial^2 f}{\partial x^2} = -30x + \dfrac{2 - 3x^2}{(x^2 + 1)^2}$
$\dfrac{\partial^2 f}{\partial y^2} = 0, \dfrac{\partial^2 f}{\partial y \partial x} = \dfrac{\partial^2 f}{\partial x \partial y} = 1$

25. $\left.\dfrac{dw}{dt}\right|_{t=0} = -1$

27. $\left.\dfrac{\partial w}{\partial r}\right|_{(r,s)=(\pi,0)} = 2$
$\left.\dfrac{\partial w}{\partial s}\right|_{(r,s)=(\pi,0)} = 2 - \pi$

29. $\left.\dfrac{dy}{dx}\right|_{(x,y)=(0,1)} = -1$

31. $\left.\dfrac{df}{dt}\right|_{t=1} = -(\sin 1 + \cos 2)\sin 1 + (\cos 1 + \cos 2)\cos 1 - 2(\sin 1 + \cos 1)\sin 2$ **33.** f increases most rapidly in the direction $\mathbf{u} = -\dfrac{\sqrt{2}}{2}\mathbf{i} - \dfrac{\sqrt{2}}{2}\mathbf{j}$; decreases most rapidly in the direction $-\mathbf{u} = \dfrac{\sqrt{2}}{2}\mathbf{i} + \dfrac{\sqrt{2}}{2}\mathbf{j}$; $D_{\mathbf{u}_1}f = -\dfrac{7}{10}$ **35.** f increases most rapidly in the direction $\mathbf{u} = \dfrac{2}{7}\mathbf{i} + \dfrac{3}{7}\mathbf{j} + \dfrac{6}{7}\mathbf{k}$; decreases most rapidly in the direction $-\mathbf{u} = -\dfrac{2}{7}\mathbf{i} - \dfrac{3}{7}\mathbf{j} - \dfrac{6}{7}\mathbf{k}$; $D_{\mathbf{u}}f = 7$

37.

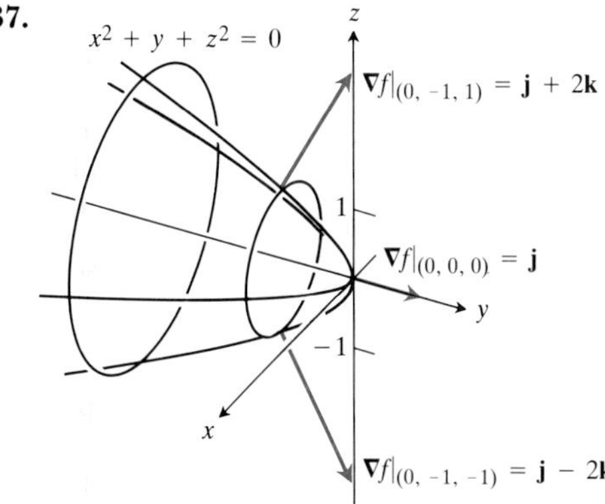

39. Tangent: $4x - y - 5z = 34$
Normal line: $x = 2 + 4t, y = -1 - t, z = 1 - 5t$
41. Tangent: $x + y = \pi + 1$
Normal: $y = x - \pi + 1$

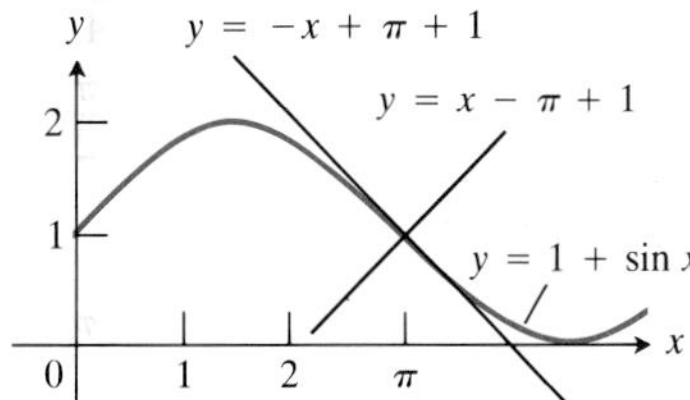

43. $L(x, y) = \frac{1}{2} + \frac{1}{2}x - \frac{1}{2}y$
$|E(x, y)| \leq 0.01$
45. $L(x, y, z) = y - 3z$ **47.** Diameter **49.** Increases by 0.038 amps; percentage change in $V = -4.17\%$; percentage change in $R = -20\%$; Estimated percentage change in $I = 15.83\%$ **51.** Local minimum of -8 at $(-2, -2)$ **53.** Saddle point at (0, 0); absolute maximum of 108 at (6, 18) **55.** Saddle point at (0, 0); absolute minimum of 14 at (1, 1) **57.** Absolute maximum of 28 at (0, 4); absolute minimum of $-\frac{9}{4}$ at $\left(\frac{3}{2}, 0\right)$ **59.** Absolute maximum of 18 at (2, −2); absolute minimum of $-\frac{17}{4}$ at $\left(-2, \frac{1}{2}\right)$
61. Absolute maximum of 4 at (0, 2) and (2, 2); absolute minimum of −1 at (1, 0) **63.** $(3, 3\sqrt{2}), (3, -3\sqrt{2})$ **65.** 50 is maximum, −50 is minimum **67.** 4 is maximum, no minimum

Chapter 14

Section 14.1, pp. 901–902

1. 16 **3.** 0 **5.** $\frac{\pi^2}{2} + 2$ **7.** $\frac{\pi}{4}$ **9.** $\frac{\ln 8}{2}$ **11.** $\frac{-1}{10}$ **13.** $(\ln 2)^2$

15. $\int_0^4 \int_0^{(4-y)/2} dx\, dy$

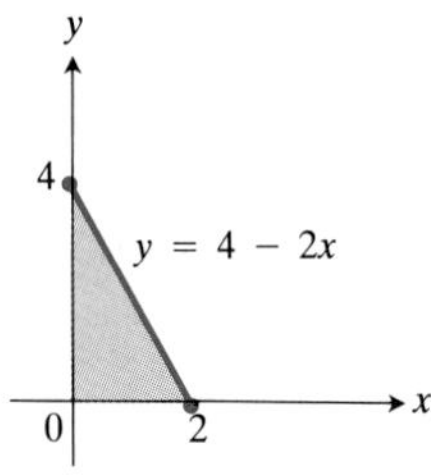

17. $\int_0^1 \int_{x^2}^{x} dy\, dx$

y
1
(1, 1)
y = x
y = x²
0
1
x

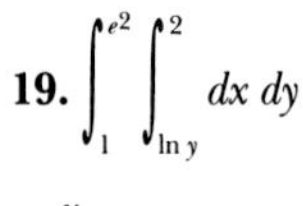

19. $\int_1^{e^2} \int_{\ln y}^{2} dx\, dy$

y
(2, e²)
y = eˣ
1
(2, 1)
0
2
x

21. $\int_1^e \int_{\ln y}^{1} dx\, dy$

23. $\int_0^8 \int_{\sqrt[3]{y}}^{2} f(x, y)\, dx\, dy$

25. 2
27. 4 − sin 4
29. 625/12
31. 16
33. 0.603
35. 0.233

Section 14.2, pp. 907–908

1. 2

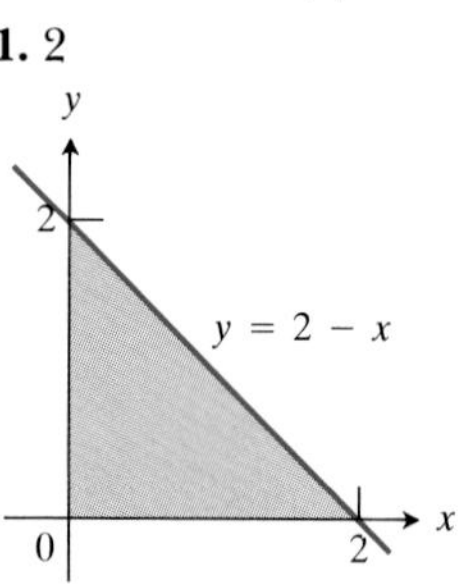

3. 4

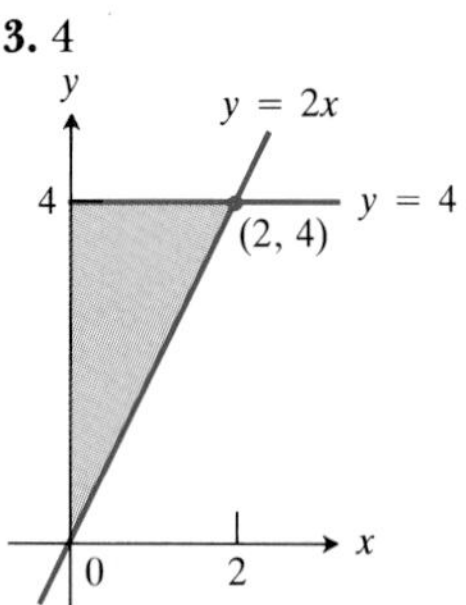

5. $\frac{1}{3}$

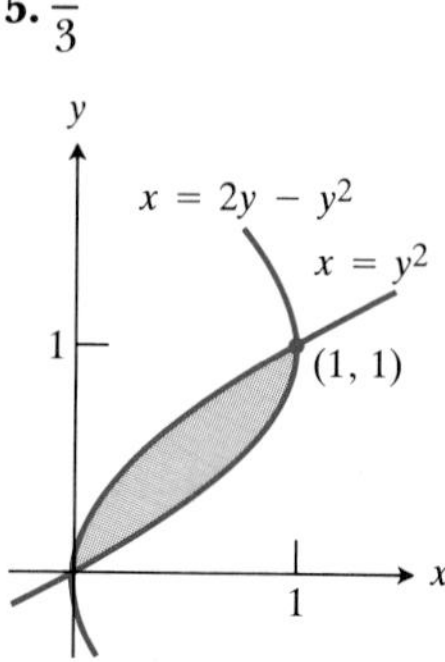

7. $\pi + 2$

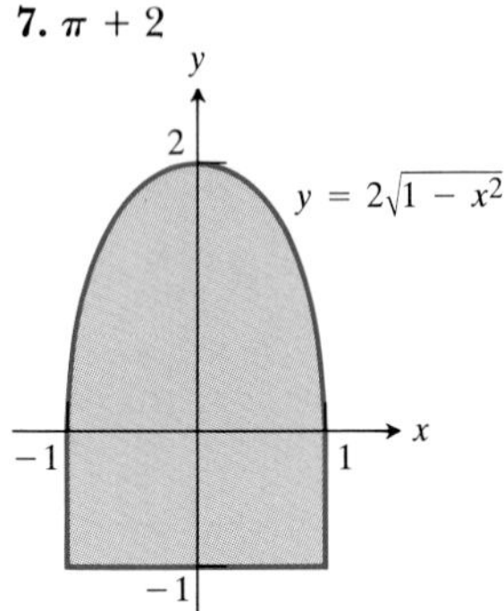

9. 12

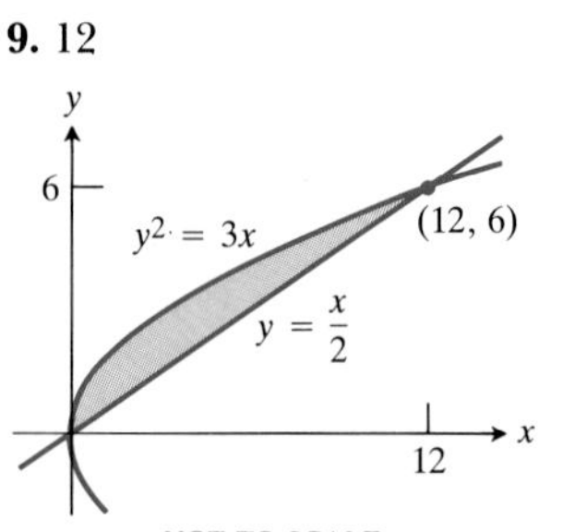

11. $\sqrt{2} - 1$

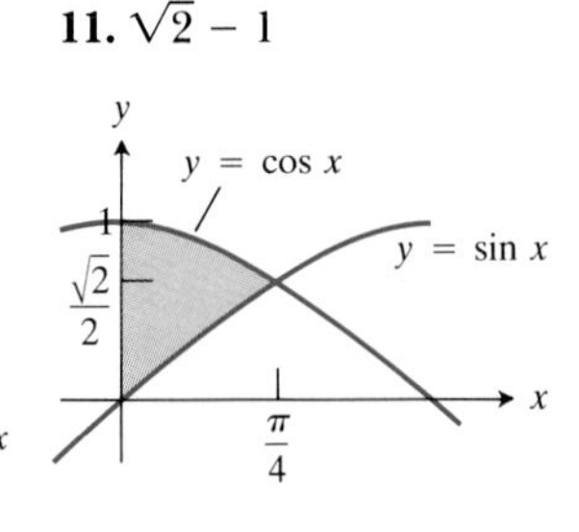

13. $\frac{3}{2}$

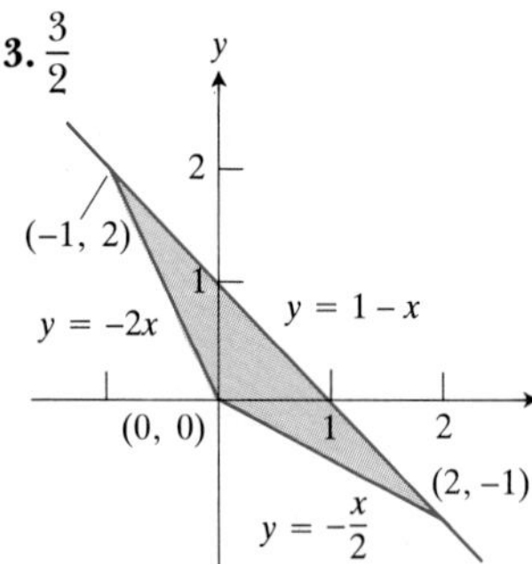

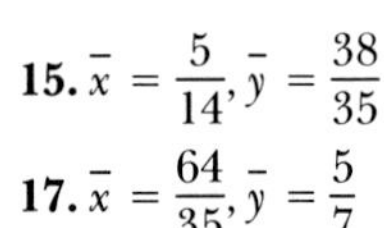

15. $\bar{x} = \frac{5}{14}, \bar{y} = \frac{38}{35}$
17. $\bar{x} = \frac{64}{35}, \bar{y} = \frac{5}{7}$
19. $\bar{x} = 0, \bar{y} = \frac{4}{3\pi}$

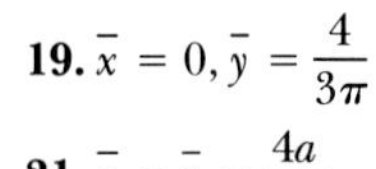

21. $\bar{x} = \bar{y} = \frac{4a}{3\pi}$
23. $\bar{x} = \frac{\pi}{2}, \bar{y} = \frac{\pi}{8}$

25. $I_x = \frac{64}{105}, R_x = 8\sqrt{\frac{2}{7}}$ **27.** $\bar{x} = \frac{3}{8}, \bar{y} = \frac{17}{16}$
29. $\bar{x} = \frac{11}{3}, \bar{y} = \frac{14}{27}; I_y = 432; R_y = 4$
31. $\bar{x} = \bar{y} = 0; I_y = \frac{7}{5}; R_y = \sqrt{\frac{21}{31}}$

33. $\bar{x} = 0, \bar{y} = \frac{7}{10}; I_x = \frac{9}{10}; I_y = \frac{3}{10}; I_0 = \frac{6}{5}; R_x = \frac{3\sqrt{6}}{10};$
$R_y = \frac{3\sqrt{2}}{10}; R_0 = \frac{3\sqrt{2}}{5}$ **35.** $a = 1$

Section 14.3, pp. 912–913

1. $\frac{\pi}{2}$ **3.** $\frac{\pi}{8}$ **5.** πa^2 **7.** $\frac{2}{3}$ **9.** $3\ln(2+\sqrt{3})$ **11.** $\frac{5\pi}{6}$
13. $2(\pi - 1)$ **15.** 12π **17.** $\frac{3\pi}{8} + 1$ **19.** $\pi \ln 4$ **21.** 4
23. $\bar{x} = \frac{5}{6}, \bar{y} = 0$ **25.** $\bar{x} = \frac{-1}{e}, \bar{y} = \frac{1}{4e}$

Section 14.4, pp. 918–920

1. 1

3. $\int_0^1 \int_0^{2-2x} \int_0^{\frac{6-6x-3y}{2}} dz\, dy\, dx$

$\int_0^2 \int_0^{x-\frac{y}{2}} \int_0^{\frac{6-6x-3y}{2}} dz\, dx\, dy$

$\int_0^1 \int_0^{3-3x} \int_0^{\frac{6-6x-2z}{3}} dy\, dz\, dx$

$\int_0^3 \int_0^{1-\frac{z}{3}} \int_0^{\frac{6-6x-2z}{3}} dy\, dx\, dz$

$\int_0^2 \int_0^{3-\frac{3y}{2}} \int_0^{\frac{6-3y-2z}{6}} dx\, dz\, dy$

$\int_0^3 \int_0^{2-\frac{2z}{3}} \int_0^{\frac{6-3y-2z}{6}} dx\, dy\, dz$

The value of all six integrals is 1.

5. 1 **7.** 1 **9.** $\frac{\pi^3}{2}(1 - \cos 1)$ **11.** 18 **13.** $\frac{7}{6}$

15. a) $\int_{-1}^1 \int_0^{1-x^2} \int_{x^2}^{1-z} dy\, dz\, dx$ b) $\int_0^1 \int_{-\sqrt{1-z}}^{\sqrt{1-z}} \int_{x^2}^{1-z} dy\, dx\, dz$

c) $\int_0^1 \int_0^{1-z} \int_{-\sqrt{y}}^{\sqrt{y}} dx\, dy\, dz$ d) $\int_0^1 \int_0^{1-y} \int_{-\sqrt{y}}^{\sqrt{y}} dx\, dz\, dy$

e) $\int_0^1 \int_{-\sqrt{y}}^{\sqrt{y}} \int_0^{1-y} dz\, dx\, dy$

17. $\frac{2}{3}$ **19.** $\frac{20}{3}$ **21.** 1 **23.** $\frac{16}{3}$ **25.** 32π **27.** $\frac{31}{3}$ **29.** 1

Section 14.5, pp. 923–924

1. $R_x = \sqrt{\frac{b^2+c^2}{12}}, R_y = \sqrt{\frac{a^2+c^2}{12}}, R_z = \sqrt{\frac{a^2+b^2}{12}}$
3. $I_x = \frac{M}{3}(b^2+c^2), I_y = \frac{M}{3}(a^2+c^2), I_z = \frac{M}{3}(a^2+b^2)$
5. $\bar{x} = \bar{y} = 0, \bar{z} = \frac{12}{5}$ **7.** a) $\bar{x} = \bar{y} = 0, \bar{z} = \frac{8}{3}$ b) $c = 2\sqrt{2}$
9. $I_L = 1386, R_L = \sqrt{\frac{77}{2}}$ **11.** $I_L = \frac{40}{3}, R_L = \sqrt{\frac{5}{3}}$
13. $\bar{x} = \frac{4}{5}, \bar{y} = \bar{z} = \frac{2}{5}$
15. $\bar{x} = \bar{y} = \bar{z} = \frac{8}{15}; I_x = I_y = I_z = \frac{11}{6}; R_x = R_y = R_z = \sqrt{\frac{11}{15}}$

Section 14.6, pp. 929–930

1. $\frac{4\pi(\sqrt{2}-1)}{3}$ **3.** $\frac{17\pi}{5}$ **5.** $\pi(6\sqrt{2}-8)$ **7.** π^2
9. $\frac{\pi}{3}$ **11.** 5π

13. a) $4\int_0^{\pi/2} \int_0^{\pi/2} \int_0^2 \rho^2 \sin\phi\, d\rho\, d\phi\, d\theta$

b) $4\int_0^{\pi/2} \int_0^2 \int_0^{\sqrt{4-r^2}} dz\, r\, dr\, d\theta$

c) $4\int_0^2 \int_0^{\sqrt{4-x^2}} \int_0^{\sqrt{4-x^2-y^2}} dz\, dy\, dx$

15. $\int_{-\pi/2}^{\pi/2} \int_0^{\cos\theta} \int_0^{3r^2} f(r, \theta, z)\, dz\, r\, dr\, d\theta$

17. $\frac{\pi}{2}$ **19.** 8π **21.** $\frac{5\pi}{2}$ **23.** $\frac{4\pi(8-3\sqrt{3})}{3}$ **25.** $\frac{2}{3}$
27. $\bar{x} = \bar{y} = 0, \bar{z} = \frac{3}{8}$ **29.** $I_z = 30\pi, R_z = \sqrt{\frac{5}{2}}$
31. a) π b) $\frac{7\pi}{6}$ **33.** $\frac{8\pi a^2}{15}$ **35.** $\frac{2\pi a^3}{3}$ **37.** $\frac{5\pi}{3}$ **39.** $\frac{3}{4}$
41. $\bar{x} = \bar{z} = 0, \bar{z} = \frac{3(2+\sqrt{2})a}{16}$

Section 14.7, pp. 936–937

3. a) $x = \frac{u+v}{3}, y = \frac{v-2u}{3}; J(u,v) = \frac{1}{3}$ b) $\frac{33}{4}$
5. $8 + \frac{16\ln 4}{3}$ **7.** $\frac{\pi ab(a^2+b^2)}{4}$
11. $\frac{4\pi abc}{3}$ **13.** $2 + \ln 8$

Chapter 14 Practice Exercises, pp. 938–940.

1. $9e - 9$

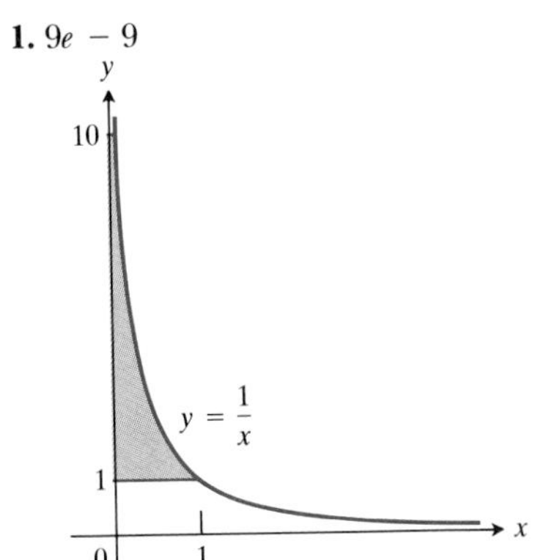

NOT TO SCALE

3. 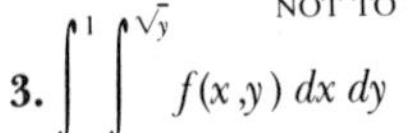 $\int_0^1 \int_y^{\sqrt{y}} f(x,y)\,dx\,dy$

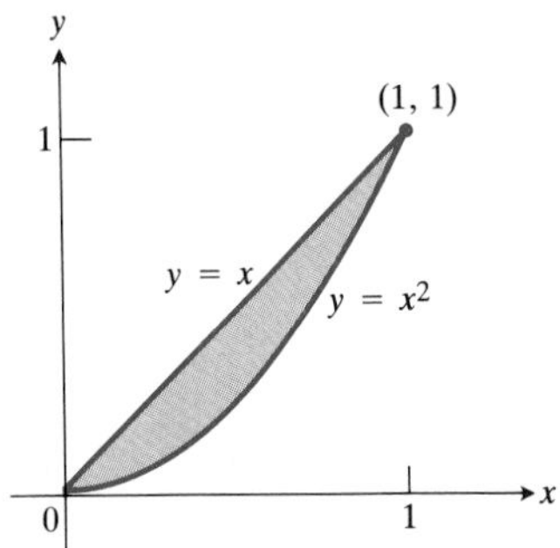

5. $\int_{-3}^{3} \int_0^{\sqrt{9-x^2}/2} y\,dy\,dx,\ \frac{9}{2}$

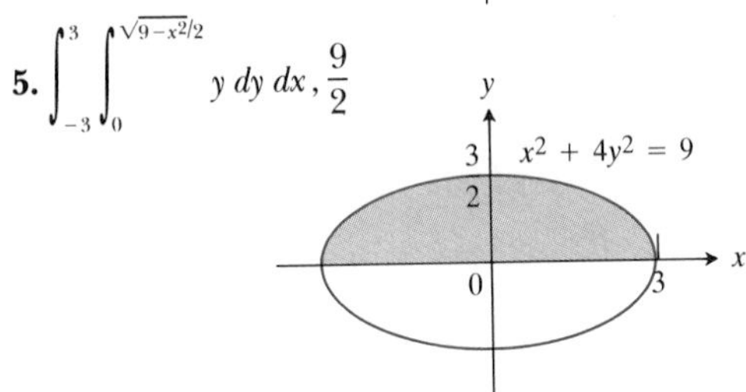

7. $\sin 4$ **9.** $\frac{\ln 17}{4}$ **11.** $\frac{4}{3}$ **13.** $\frac{4}{3}$ **15.** $\frac{1}{4}$

17. $\bar{x} = \bar{y} = \frac{1}{2 - \ln 4}$ **19.** a) $I_0 = \frac{40}{3}$ b) $I_0 = \frac{4ab(b^2 + a^2)}{3}$

21. $M = 4, M_x = 0, M_y = 0$ **23.** π

25. $\bar{x} = \frac{15\pi + 32}{6\pi + 48}, \bar{y} = 0$ **27.** $\bar{x} = \bar{y} = \frac{13}{3\pi}$ **29.** 0 **31.** $\frac{\pi}{2}$

33. a) $4\int_0^{\sqrt{3}} \int_0^{\sqrt{3-x^2}} \int_0^{\sqrt{4-x^2-y^2}} dz\,dy\,dx$

b) $4\int_0^{\pi/2} \int_0^{\sqrt{3}} \int_0^{\sqrt{4-r^2}} r\,dz\,dr\,d\theta$

c) $4\int_0^{\pi/2} \int_0^{\pi/3} \int_{\sec\phi}^{2} \rho^2 \sin\phi\,d\rho\,d\phi\,d\theta$

35. $\frac{15}{8}$ **37.** $\frac{3\pi}{2}$

39. a) $\bar{x} = \bar{y} = 0, \bar{z} = \frac{1}{2}; I_z = \frac{\pi}{8}; R_z = \frac{\sqrt{3}}{2}$

b) $\bar{x} = \bar{y} = 0, \bar{z} = \frac{5}{14}; I_z = \frac{2\pi}{7}; R_z = \sqrt{\frac{5}{7}}$

41. $2\pi^2$ **43.** $\bar{x} = \bar{y} = 0, \bar{z} = \frac{3a}{8}$

Chapter 15

Section 15.1, pp. 946–947

1. c **3.** g **5.** d **7.** f **9.** $\sqrt{2}$ **11.** $\frac{13}{2}$ **13.** 0

15. $\frac{5}{6}\sqrt{5} + \frac{3}{2}$ **17.** $I_x = 2\pi - 2, R_x = 1$

19. $I_z = 2\pi a^3\delta, R_z = a$ **21.** a) $I_z = 2\pi\delta\sqrt{2}, R_z = 1$
b) $I_z = 4\pi\delta\sqrt{2}, R_z = 1$

Section 15.2, pp. 955–956

1. a) $\frac{9}{2}$ b) $\frac{13}{3}$ c) $\frac{11}{3}$ **3.** a) $\frac{1}{3}$ b) $-\frac{1}{5}$ c) 0 **5.** a) 2 b) $\frac{3}{2}$ c) $\frac{1}{2}$

7. $\frac{1}{2}$ **9.** $-\pi$ **11.** 48 **13.** π

15. a) $\text{Circ}_1 = 0$; $\text{Circ}_2 = \pi$; $\text{Flux}_1 = 2\pi$; $\text{Flux}_2 = 3\pi$ b) $\text{Circ}_1 = 0$; $\text{Circ}_2 = 4\pi$; $\text{Flux}_1 = 8\pi$; $\text{Flux}_2 = 4\pi$

17. Circ = 0; Flux = $a^2\pi$ **19.** Circ = $a^2\pi$; Flux = 0

Section 15.3, pp. 965–966

1. 0; $2\pi a^2$ **3.** $-\pi a^2$; 0 **5.** Flux = 2; Circ = 0

7. Flux = -9; Circ = 9 **9.** Flux = $\frac{1}{5}$; Circ = $-\frac{1}{12}$

11. 0 **13.** $-16\pi - 96$ **15.** 2/33

17. a) 0 b) $(h - k)$(area of region) **19.** πa^2 **21.** $3\pi/8$

Section 15.4, pp. 975–977

1. $13\pi/3$ **3.** 4 **5.** $6\sqrt{6} - 2\sqrt{2}$ **7.** $2\pi(2 - \sqrt{2})$ **9.** $2\pi/3$

11. 9 **13.** $\frac{abc(ab + ac + bc)}{4}$ **15.** 2 **17.** $\frac{a^3\pi}{6}$ **19.** $\frac{\pi a^2}{4}$

21. $\frac{a^3\pi}{2}$ **23.** -32 **25.** -4 **27.** $\left(\frac{1}{2}, \frac{1}{2}, \frac{1}{2}\right)$ **29.** $(\bar{x}, \bar{y}, \bar{z}) = \left(0, 0, \frac{14}{9}\right), I_z = \frac{15\pi\sqrt{2}}{2}\delta, R_z = \frac{\sqrt{10}}{2}$ **31.** $\frac{\pi}{6}(13\sqrt{13} - 1)$

33. $\frac{\pi}{6}(5\sqrt{5} - 1)$ **35.** $\sqrt{5} + \frac{1}{2}\ln(2 + \sqrt{5})$

Section 15.5, p. 985

1. -16 **3.** 0 **5.** -8π **7.** 3π **9.** $-40/3$ **11.** 0 **13.** 5

Section 15.6, pp. 991–992

1. 4π **3.** $-5/6$ **5.** 0 **7.** -6π

Section 15.7, p. 1000

1. Conservative **3.** Not conservative **5.** Not conservative **7.** $f(x, y, z) = x^2 + \frac{3y^2}{2} + 2z^2 + C$ **9.** $f(x, y, z) = xe^{y+2z} + C$ **11.** 49 **13.** -16 **15.** 1 **17.** $\ln(\pi/2)$ **19.** 0 **21.** -3

Chapter 15 Practice Exercises, pp 1001–1004

1. Path 1: $2\sqrt{3}$; Path 2: $3\sqrt{2} + 1$ **3.** $2\pi a^2$ **5.** a) $4\sqrt{2} - 2$ b) $\sqrt{2} + \ln(1 + \sqrt{2})$ **7.** 2 **9.** a) $-\pi/2$ b) 1/2 c) 1 **11.** Flux $= \frac{3}{2}$; Circ $= -\frac{1}{2}$ **15.** 0 **17.** $\pi\sqrt{3}$ **19.** $2\pi - \pi\sqrt{3}$ **21.** a) 0 b) -4 **23.** 3 **25.** $(\bar{x}, \bar{y}, \bar{z}) = \left(0, 0, \frac{49}{12}\right)$, $I_z = 640\pi$, $R_z = 2\sqrt{2}$ **27.** 3 **29.** $-\frac{7\pi}{6} + \frac{4\pi}{3}\sqrt{2}$ **31.** 0 **33.** Conservative **35.** Not conservative **37.** $f(x, y, z) = 2x + y^2 + zy + z + C$ **39.** 0 **41.** Path 1: 7/3; Path 2: 8/3 **43.** 3π **45.** 4/3

Chapter 16

Section 16.1, pp. 1010–1012

1. First order, linear **3.** Fourth order, linear **13.** a) $x = 11\,000e^{0.1t} - 10\,000$ b) $t \approx 23.03$ **15.** $x^4 + 2x^2y^2 = C$ **17.** $\ln|x| + e^{-y/x} = C$ **19.** $x = 2\left(\sec\left(\frac{y - x}{x}\right) + \tan\left(\frac{y - x}{x}\right)\right)$ **21.** $x^2 + 2y^2 = C_1$, $x^2 = C_2y$ **23.** Both are families of hyperbolas.

Section 16.2, p. 1015

1. Exact **3.** Exact **5.** Exact **7.** Not Exact **9.** $3x^2 + 6xy + 2y^3 = C_3$ **11.** $2x^2y + 2xy^2 - y^2 = C_3$ **13.** $e^x + x\ln y + y\ln x - \cos y = C_2$ **15.** $\frac{x^2}{2} + \frac{x}{y} = C_2$ **17.** a) $xy = C_3$ b) $xy = C_3$ **19.** $x^2 + 2xy^2 + 2y = 4$ **21.** $xy = e^{xy-1}$ **23.** $x^3 + xy^2 = C_2$

Section 16.3, pp. 1019–1020

1. $y = e^{-x} + Ce^{-2x}$ **3.** $y = -\frac{\cos x}{x^3} + \frac{C}{x^3}$ **5.** $y = \frac{x^3}{3(x-1)^4} - \frac{x}{(x-1)^4} + \frac{C}{(x-1)^4}$ **7.** $y = \csc x(\ln|\sec x| + C)$ **9.** $y = x - \frac{1}{2} + \frac{3}{2}e^{-2x}$ **11.** $y = -\frac{1}{x}\cos x + \frac{\pi}{2x}$ **13.** $y = y_0 e^{kx}$ **15.** $y = C_3e^{-x} + C_2$ **17.** $y = C_2 \ln x + C_3$ if $x > 0$ **19.** $y = x - 1$ **21.** $t = \frac{L}{R}(\ln 2)$

Section 16.4, p. 1024

1. $y = C_1 + C_2e^{-2x}$ **3.** $y = C_1 e^{-5x} + C_2e^{-x}$ **5.** $y = (C_1x + C_2)e^{2x}$ **7.** $y = (C_1x + C_2)e^{5x}$ **9.** $y = e^{-x/2}\left(C_1 \cos\frac{\sqrt{3}}{2}x + iC_2 \sin\frac{\sqrt{3}}{2}x\right)$ **11.** $y = e^x(C_1 \cos \sqrt{3}\,x + C_2 \sin \sqrt{3}\,x)$ **13.** $y = -\frac{1}{2}e^x + \frac{3}{2}e^{-x}$ **15.** $y = \frac{3}{4}e^{2x} - \frac{3}{4}e^{-2x}$ **17.** $y = xe^{-x}$ **19.** $y = -xe^{-3x/2}$ **21.** $y = \frac{\sqrt{3}}{2}\sin 2x$ **23.** $y = e^x\left(2\cos\sqrt{2}x - \frac{1}{\sqrt{2}}\sin\sqrt{2}x\right)$

Section 16.5, pp. 1033–1034

1. $y = \frac{x^2}{2} - x + C_3 + C_2e^{-x}$ **3.** $y = -\frac{1}{2}x\cos x + C_1\cos x + C_3 \sin x$ **5.** $y = C_1 xe^{-x} + C_2 e^{-x} + \frac{1}{2}x^2e^{-x}$ **7.** $y = \frac{1}{2}xe^x + C_3 e^x + C_2 e^{-x}$ **9.** $y = 2 + C_1e^{-2x}\cos x + C_2 e^{-2x}\sin x$ **11.** $y = \cos x \ln(\cos x) + x\sin x + C_1\cos x + C_2 \sin x$ **13.** $y = \frac{3}{10} + C_1e^{5x} + C_2 e^{-2x}$ **15.** $y = \frac{1}{2}\cos x - \frac{1}{2}\sin x + C_1 + C_2 e^x$ **17.** $y = -\frac{1}{8}\cos 3x + C_1 \cos x + C_2 \sin x$ **19.** $y = -6\cos x - 2\sin x + C_1e^{2x} + C_2 e^{-x}$ **21.** $y = \frac{1}{2}xe^x - x^2 - 2 + C_1e^x + C_2 e^{-x}$ **23.** $y = -\frac{1}{4}e^{-x} + \frac{49}{50}\cos x + \frac{7}{50}\sin x + C_1e^{3x} + C_2 e^{-2x}$ **25.** $y = x^3 - \frac{3}{5}x^2 + \frac{6}{25}x + C_1 + C_2 e^{-5x}$ **27.** $y = \frac{1}{3}xe^{3x} + 2x^2 + \frac{4}{3}x + C_1 + C_2 e^{3x}$ **29.** $y = C_1 + C_2 e^{5x} + \frac{1}{10}x^2e^{5x} - \frac{1}{25}xe^{5x}$ **31.** $y = C_1 \cos x + C_2 \sin x - \frac{1}{2}x\cos x + x\sin x$ **33.** a) $y = C_3 e^x + \frac{1}{2}e^{-x} + xe^x + C_1$ b) $y_p = xe^x + \frac{1}{2}e^{-x} + C_1 + C_2 e^x$ **35.** a) $y = -\frac{1}{8}e^x - \frac{4}{5} + C_1e^{5x} + C_2 e^{-x}$ b) $y = -\frac{1}{8}e^x - \frac{4}{5} + C_1e^{5x} + C_2 e^{-x}$ **37.** $y = C_1 \cos x + C_2 \sin x - \sin x \ln|\csc x + \cot x|$ **39.** $y = \frac{1}{8}xe^{8x} + C_1 + C_2 e^{8x}$ **41.** $y = -\frac{1}{4}x^4 - x^3 - 3x^2 - 6x + C_1 + C_2 e^x$ **43.** $y = \frac{1}{6}x^3 - \frac{1}{4}x^2 + \frac{1}{4}x - \frac{1}{3}e^x + C_1 + C_2 e^{-2x}$ **45.** $y = x\cos x + \sin x \ln(\sec x) + C_1 \cos x + C_2 \sin x$ **47.** $y = C_1 e^{3x} - \frac{1}{2}e^x$ **49.** $y = C_1 e^{3x} + 5x\,e^{3x}$ **51.** $y = 2\cos x + \sin x - 1 + \sin x \ln|\sec x + \tan x|$ **53.** $y = -e^{-x} + 1$

Section 16.6, pp. 1037–1038

1. $x = x_0 \cos \omega t + \frac{v_0}{\omega}\sin \omega t$

$$x = \frac{\sqrt{(\omega x_0)^2 + v_0^2}}{\omega}\sin(\omega t + \phi) \text{ where } f = \tan^{-1}\left(\frac{\omega x_0}{v_0}\right)$$

3. a) $i = C_1 \cos \omega t + C_2 \sin \omega t$ b) $i = C_1 \cos \omega t + C_2 \sin \omega t + \dfrac{V\alpha}{L(\omega^2 - \alpha^2)} \cos \alpha t$ c) $i = C_1 \cos \omega t + C_2 \sin \omega t + \dfrac{V}{2L} t \sin \omega t$ d) $i = e^{-5t}(C_1 \cos(148.99\, t) + C_2 \sin(148.99\, t))$

5. $\theta = \theta_0 \cos\sqrt{\dfrac{2k}{mr^2}}\, t + v_0\sqrt{\dfrac{mr^2}{2k}} \sin\sqrt{\dfrac{2k}{mr^2}}\, t$ **7.** a) $x = x_0 \cos\left(\sqrt{\dfrac{k}{m}}\, t\right) - \dfrac{A\sqrt{km}\,\alpha}{k - m\alpha^2} \sin\left(\sqrt{\dfrac{k}{m}}\, t\right) + \dfrac{Ak}{k - m\alpha^2} \sin \alpha t$
b) $x = \left(x_0 - \dfrac{A\alpha}{2}\right)\cos \alpha t + \dfrac{A}{2} \sin \alpha t$

Section 16.7, pp. 1043–1044

1. $y_{\text{estimate}} = 2.48832$, $y_{\text{exact}} = 2.718281828$ **3.** 2.690681891 **7.** 2.14285714 **9.** 0.390271217 **11.** a) 0.31026827 b) 0.3102683

Chapter 16 Practice Exercises, p. 1044

1. $e^{-x} + e^{y+2} = 2$ **3.** $\dfrac{y^2}{x} - x = -5$ **5.** $\dfrac{x^3}{3} + xy + e^y = 10$ **7.** $y = \dfrac{1}{2}x - \dfrac{1}{4} + \dfrac{5}{4}e^{-2x}$ **9.** $y = \ln|\sec x| - \ln 2$ **11.** $y = -2e^{3x} + 4e^x$ **13.** $y = 7xe^{-2x}$ **15.** $y = e^{-x}(\cos x - \sin x)$ **17.** $y = 2e^{-x} + x^2 - x - 1$ **19.** $y = \dfrac{1}{3}e^{2x} + \dfrac{5}{3}e^{-x} + xe^{2x}$

Appendixes

Appendix A.4, pp.

7. Statement true for $n \geq 5$

Appendix A.7, pp.

1. -5 **3.** 1 **5.** a) -7 b) -7 **7.** a) 38 b) 38 **9.** $x = -4$, $y = 1$ **11.** $x = 3, y = 2$ **13.** $x = 3, y = -2, z = 2$
15. $x = 2, y = 0, z = -1$
17. a) $h = 6, k = 4$ b) $h = 6, k \neq 4$

Appendix A.8, pp.

1. (2, 4, 4)
3. $f(\sqrt{6}, \sqrt{3}, 1) = f(-\sqrt{6}, \sqrt{3}, 1) = 1 + \sqrt{6}$, max
$f(\sqrt{6}, -\sqrt{3}, 1) = f(-\sqrt{6}, -\sqrt{3}, 1) = 1 - \sqrt{6}$, min
5. $f(0, 0, 2) = f(0, 0, -2) = 4$, max
$f(2, 2, 0) = f(-2, -2, 0) = 2$, min

Index

Note: Numbers in parentheses refer to problems on the pages indicated.

A Brief Table of Integrals

1. $\int u\,dv = uv - \int v\,du$

2. $\int a^u\,du = \dfrac{a^u}{\ln a} + C, \quad a \neq 1, \quad a > 0$

3. $\int \cos u\,du = \sin u + C$

4. $\int \sin u\,du = -\cos u + C$

5. $\int (ax+b)^n\,dx = \dfrac{(ax+b)^{n+1}}{a(n+1)} + C, \quad n \neq -1$

6. $\int (ax+b)^{-1}\,dx = \dfrac{1}{a}\ln|ax+b| + C$

7. $\int x(ax+b)^n\,dx = \dfrac{(ax+b)^{n+1}}{a^2}\left[\dfrac{ax+b}{n+2} - \dfrac{b}{n+1}\right] + C, \quad n \neq -1, -2$

8. $\int x(ax+b)^{-1}\,dx = \dfrac{x}{a} - \dfrac{b}{a^2}\ln|ax+b| + C$

9. $\int x(ax+b)^{-2}\,dx = \dfrac{1}{a^2}\left[\ln|ax+b| + \dfrac{b}{ax+b}\right] + C$

10. $\int \dfrac{dx}{x(ax+b)} = \dfrac{1}{b}\ln\left|\dfrac{x}{ax+b}\right| + C$

11. $\int (\sqrt{ax+b})^n\,dx = \dfrac{2}{a}\dfrac{(\sqrt{ax+b})^{n+2}}{n+2} + C, \quad n \neq -2$

12. $\int \dfrac{\sqrt{ax+b}}{x}\,dx = 2\sqrt{ax+b} + b\int \dfrac{dx}{x\sqrt{ax+b}}$

13. (a) $\int \dfrac{dx}{x\sqrt{ax+b}} = \dfrac{2}{\sqrt{-b}}\tan^{-1}\sqrt{\dfrac{ax+b}{-b}} + C, \quad \text{if } b < 0$

(b) $\int \dfrac{dx}{x\sqrt{ax+b}} = \dfrac{1}{\sqrt{b}}\ln\left|\dfrac{\sqrt{ax+b}-\sqrt{b}}{\sqrt{ax+b}+\sqrt{b}}\right| + C, \quad \text{if } b > 0$

14. $\int \dfrac{\sqrt{ax+b}}{x^2}\,dx = -\dfrac{\sqrt{ax+b}}{x} + \dfrac{a}{2}\int \dfrac{dx}{x\sqrt{ax+b}} + C$

15. $\int \dfrac{dx}{x^2\sqrt{ax+b}} = -\dfrac{\sqrt{ax+b}}{bx} - \dfrac{a}{2b}\int \dfrac{dx}{x\sqrt{ax+b}} + C$

16. $\int \dfrac{dx}{a^2+x^2} = \dfrac{1}{a}\tan^{-1}\dfrac{x}{a} + C$

17. $\int \dfrac{dx}{(a^2+x^2)^2} = \dfrac{x}{2a^2(a^2+x^2)} + \dfrac{1}{2a^3}\tan^{-1}\dfrac{x}{a} + C$

18. $\int \dfrac{dx}{a^2-x^2} = \dfrac{1}{2a}\ln\left|\dfrac{x+a}{x-a}\right| + C$

19. $\int \dfrac{dx}{(a^2-x^2)^2} = \dfrac{x}{2a^2(a^2-x^2)} + \dfrac{1}{2a^2}\int \dfrac{dx}{a^2-x^2}$

20. $\int \dfrac{dx}{\sqrt{a^2+x^2}} = \sinh^{-1}\dfrac{x}{a} + C = \ln|x+\sqrt{a^2+x^2}| + C$

Continued

21. $\int \sqrt{a^2+x^2}\,dx = \frac{x}{2}\sqrt{a^2+x^2} + \frac{a^2}{2}\sinh^{-1}\frac{x}{a} + C$

22. $\int x^2\sqrt{a^2+x^2}\,dx = \frac{x(a^2+2x^2)\sqrt{a^2+x^2}}{8} - \frac{a^4}{8}\sinh^{-1}\frac{x}{a} + C$

23. $\int \frac{\sqrt{a^2+x^2}}{x}\,dx = \sqrt{a^2+x^2} - a\sinh^{-1}\left|\frac{a}{x}\right| + C$

24. $\int \frac{\sqrt{a^2+x^2}}{x^2}\,dx = \sinh^{-1}\frac{x}{a} - \frac{\sqrt{a^2+x^2}}{x} + C$

25. $\int \frac{x^2}{\sqrt{a^2+x^2}}\,dx = -\frac{a^2}{2}\sinh^{-1}\frac{x}{a} + \frac{x\sqrt{a^2+x^2}}{2} + C$

26. $\int \frac{dx}{x\sqrt{a^2+x^2}} = -\frac{1}{a}\ln\left|\frac{a+\sqrt{a^2+x^2}}{x}\right| + C$

27. $\int \frac{dx}{x^2\sqrt{a^2+x^2}} = -\frac{\sqrt{a^2+x^2}}{a^2x} + C$

28. $\int \frac{dx}{\sqrt{a^2-x^2}} = \sin^{-1}\frac{x}{a} + C$

29. $\int \sqrt{a^2-x^2}\,dx = \frac{x}{2}\sqrt{a^2-x^2} + \frac{a^2}{2}\sin^{-1}\frac{x}{a} + C$

30. $\int x^2\sqrt{a^2-x^2}\,dx = \frac{a^4}{8}\sin^{-1}\frac{x}{a} - \frac{1}{8}x\sqrt{a^2-x^2}\,(a^2-2x^2) + C$

31. $\int \frac{\sqrt{a^2-x^2}}{x}\,dx = \sqrt{a^2-x^2} - a\ln\left|\frac{a+\sqrt{a^2-x^2}}{x}\right| + C$

32. $\int \frac{\sqrt{a^2-x^2}}{x^2}\,dx = -\sin^{-1}\frac{x}{a} - \frac{\sqrt{a^2-x^2}}{x} + C$

33. $\int \frac{x^2}{\sqrt{a^2-x^2}}\,dx = \frac{a^2}{2}\sin^{-1}\frac{x}{a} - \frac{1}{2}x\sqrt{a^2-x^2} + C$

34. $\int \frac{dx}{x\sqrt{a^2-x^2}} = -\frac{1}{a}\ln\left|\frac{a+\sqrt{a^2-x^2}}{x}\right| + C$

35. $\int \frac{dx}{x^2\sqrt{a^2-x^2}} = -\frac{\sqrt{a^2-x^2}}{a^2x} + C$

36. $\int \frac{dx}{\sqrt{x^2-a^2}} = \cosh^{-1}\frac{x}{a} + C = \ln|x+\sqrt{x^2-a^2}| + C$

37. $\int \sqrt{x^2-a^2}\,dx = \frac{x}{2}\sqrt{x^2-a^2} - \frac{a^2}{2}\cosh^{-1}\frac{x}{a} + C$

38. $\int (\sqrt{x^2-a^2})^n\,dx = \frac{x(\sqrt{x^2-a^2})^n}{n+1} - \frac{na^2}{n+1}\int (\sqrt{x^2-a^2})^{n-2}\,dx, \quad n \neq -1$

39. $\int \frac{dx}{(\sqrt{x^2-a^2})^n} = \frac{x(\sqrt{x^2-a^2})^{2-n}}{(2-n)a^2} - \frac{n-3}{(n-2)a^2}\int \frac{dx}{(\sqrt{x^2-a^2})^{n-2}}, \quad n \neq 2$

40. $\int x(\sqrt{x^2-a^2})^n\,dx = \frac{(\sqrt{x^2-a^2})^{n+2}}{n+2} + C, \quad n \neq -2$

41. $\int x^2\sqrt{x^2-a^2}\,dx = \frac{x}{8}(2x^2-a^2)\sqrt{x^2-a^2} - \frac{a^4}{8}\cosh^{-1}\frac{x}{a} + C$

42. $\int \frac{\sqrt{x^2-a^2}}{x}\,dx = \sqrt{x^2-a^2} - a\sec^{-1}\left|\frac{x}{a}\right| + C$

43. $\displaystyle\int \frac{\sqrt{x^2-a^2}}{x^2}\,dx = \cosh^{-1}\frac{x}{a} - \frac{\sqrt{x^2-a^2}}{x} + C$

44. $\displaystyle\int \frac{x^2}{\sqrt{x^2-a^2}}\,dx = \frac{a^2}{2}\cosh^{-1}\frac{x}{a} + \frac{x}{2}\sqrt{x^2-a^2} + C$

45. $\displaystyle\int \frac{dx}{x\sqrt{x^2-a^2}} = \frac{1}{a}\sec^{-1}\left|\frac{x}{a}\right| + C = \frac{1}{a}\cos^{-1}\left|\frac{a}{x}\right| + C$

46. $\displaystyle\int \frac{dx}{x^2\sqrt{x^2-a^2}} = \frac{\sqrt{x^2-a^2}}{a^2x} + C$

47. $\displaystyle\int \frac{dx}{\sqrt{2ax-x^2}} = \sin^{-1}\left(\frac{x-a}{a}\right) + C$

48. $\displaystyle\int \sqrt{2ax-x^2}\,dx = \frac{x-a}{2}\sqrt{2ax-x^2} + \frac{a^2}{2}\sin^{-1}\left(\frac{x-a}{a}\right) + C$

49. $\displaystyle\int (\sqrt{2ax-x^2})^n\,dx = \frac{(x-a)(\sqrt{2ax-x^2})^n}{n+1} + \frac{na^2}{n+1}\int (\sqrt{2ax-x^2})^{n-2}\,dx,$

50. $\displaystyle\int \frac{dx}{(\sqrt{2ax-x^2})^n} = \frac{(x-a)(\sqrt{2ax-x^2})^{2-n}}{(n-2)a^2} + \frac{(n-3)}{(n-2)a^2}\int \frac{dx}{(\sqrt{2ax-x^2})^{n-2}}$

51. $\displaystyle\int x\sqrt{2ax-x^2}\,dx = \frac{(x+a)(2x-3a)\sqrt{2ax-x^2}}{6} + \frac{a^3}{2}\sin^{-1}\frac{x-a}{a} + C$

52. $\displaystyle\int \frac{\sqrt{2ax-x^2}}{x}\,dx = \sqrt{2ax-x^2} + a\sin^{-1}\frac{x-a}{a} + C$

53. $\displaystyle\int \frac{\sqrt{2ax-x^2}}{x^2}\,dx = -2\sqrt{\frac{2a-x}{x}} - \sin^{-1}\left(\frac{x-a}{a}\right) + C$

54. $\displaystyle\int \frac{x\,dx}{\sqrt{2ax-x^2}} = a\sin^{-1}\frac{x-a}{a} - \sqrt{2ax-x^2} + C$

55. $\displaystyle\int \frac{dx}{x\sqrt{2ax-x^2}} = -\frac{1}{a}\sqrt{\frac{2a-x}{x}} + C$

56. $\displaystyle\int \sin ax\,dx = -\frac{1}{a}\cos ax + C$

57. $\displaystyle\int \cos ax\,dx = \frac{1}{a}\sin ax + C$

58. $\displaystyle\int \sin^2 ax\,dx = \frac{x}{2} - \frac{\sin 2ax}{4a} + C$

59. $\displaystyle\int \cos^2 ax\,dx = \frac{x}{2} + \frac{\sin 2ax}{4a} + C$

60. $\displaystyle\int \sin^n ax\,dx = \frac{-\sin^{n-1} ax\cos ax}{na} + \frac{n-1}{n}\int \sin^{n-2} ax\,dx$

61. $\displaystyle\int \cos^n ax\,dx = \frac{\cos^{n-1} ax\sin ax}{na} + \frac{n-1}{n}\int \cos^{n-2} ax\,dx$

62. (a) $\displaystyle\int \sin ax\cos bx\,dx = -\frac{\cos(a+b)x}{2(a+b)} - \frac{\cos(a-b)x}{2(a-b)} + C, \quad a^2 \neq b^2$

(b) $\displaystyle\int \sin ax\sin bx\,dx = \frac{\sin(a-b)x}{2(a-b)} - \frac{\sin(a+b)x}{2(a+b)}, \quad a^2 \neq b^2$

(c) $\displaystyle\int \cos ax\cos bx\,dx = \frac{\sin(a-b)x}{2(a-b)} + \frac{\sin(a+b)x}{2(a+b)}, \quad a^2 \neq b^2$

Continued

63. $\int \sin ax \cos ax\,dx = -\frac{\cos 2ax}{4a} + C$

64. $\int \sin^n ax \cos ax\,dx = \frac{\sin^{n+1} ax}{(n+1)a} + C, \quad n \neq -1$

65. $\int \frac{\cos ax}{\sin ax}\,dx = \frac{1}{a}\ln|\sin ax| + C$

66. $\int \cos^n ax \sin ax\,dx = -\frac{\cos^{n+1} ax}{(n+1)a} + C, \quad n \neq -1$

67. $\int \frac{\sin ax}{\cos ax}\,dx = -\frac{1}{a}\ln|\cos ax| + C$

68. $\int \sin^n ax \cos^m ax\,dx = -\frac{\sin^{n-1} ax \cos^{m+1} ax}{a(m+n)} + \frac{n-1}{m+n}\int \sin^{n-2} ax \cos^m ax\,dx,$

$n \neq -m$ (If $n = -m$, use No. 86.)

69. $\int \sin^n ax \cos^m ax\,dx = \frac{\sin^{n+1} ax \cos^{m-1} ax}{a(m+n)} + \frac{m-1}{m+n}\int \sin^n ax \cos^{m-2} ax\,dx,$

$m \neq -n$ (If $m = -n$, use No. 87.)

70. $\int \frac{dx}{b + c\sin ax} = \frac{-2}{a\sqrt{b^2 - c^2}}\tan^{-1}\left[\sqrt{\frac{b-c}{b+c}}\tan\left(\frac{\pi}{4} - \frac{ax}{2}\right)\right] + C, \quad b^2 > c^2$

71. $\int \frac{dx}{b + c\sin ax} = \frac{-1}{a\sqrt{c^2 - b^2}}\ln\left|\frac{c + b\sin ax + \sqrt{c^2 - b^2}\cos ax}{b + c\sin ax}\right| + C, \quad b^2 < c^2$

72. $\int \frac{dx}{1 + \sin ax} = -\frac{1}{a}\tan\left(\frac{\pi}{4} - \frac{ax}{2}\right) + C$

73. $\int \frac{dx}{1 - \sin ax} = \frac{1}{a}\tan\left(\frac{\pi}{4} + \frac{ax}{2}\right) + C$

74. $\int \frac{dx}{b + c\cos ax} = \frac{2}{a\sqrt{b^2 - c^2}}\tan^{-1}\left[\sqrt{\frac{b-c}{b+c}}\tan\frac{ax}{2}\right] + C, \quad b^2 > c^2$

75. $\int \frac{dx}{b + c\cos ax} = \frac{1}{a\sqrt{c^2 - b^2}}\ln\left|\frac{c + b\cos ax + \sqrt{c^2 - b^2}\sin ax}{b + c\cos ax}\right| + C, \quad b^2 < c^2$

76. $\int \frac{dx}{1 + \cos ax} = \frac{1}{a}\tan\frac{ax}{2} + C$

77. $\int \frac{dx}{1 - \cos ax} = -\frac{1}{a}\cot\frac{ax}{2} + C$

78. $\int x\sin ax\,dx = \frac{1}{a^2}\sin ax - \frac{x}{a}\cos ax + C$

79. $\int x\cos ax\,dx = \frac{1}{a^2}\cos ax + \frac{x}{a}\sin ax + C$

80. $\int x^n \sin ax\,dx = -\frac{x^n}{a}\cos ax + \frac{n}{a}\int x^{n-1}\cos ax\,dx$

81. $\int x^n \cos ax\,dx = \frac{x^n}{a}\sin ax - \frac{n}{a}\int x^{n-1}\sin ax\,dx$

82. $\int \tan ax\,dx = \frac{1}{a}\ln|\sec ax| + C$

83. $\int \cot ax\,dx = \frac{1}{a}\ln|\sin ax| + C$

84. $\int \tan^2 ax\,dx = \frac{1}{a}\tan ax - x + C$

85. $\int \cot^2 ax\,dx = -\frac{1}{a}\cot ax - x + C$

86. $\int \tan^n ax\,dx = \frac{\tan^{n-1} ax}{a(n-1)} - \int \tan^{n-2} ax\,dx, \quad n \neq 1$

87. $\int \cot^n ax\,dx = -\frac{\cot^{n-1} ax}{a(n-1)} - \int \cot^{n-2} ax\,dx, \quad n \neq 1$

88. $\int \sec ax\,dx = \frac{1}{a}\ln|\sec ax + \tan ax| + C$

89. $\int \csc ax\,dx = -\frac{1}{a}\ln|\csc ax + \cot ax| + C$

90. $\int \sec^2 ax\, dx = \frac{1}{a} \tan ax + C$

91. $\int \csc^2 ax\, dx = -\frac{1}{a} \cot ax + C$

92. $\int \sec^n ax\, dx = \frac{\sec^{n-2} ax \tan ax}{a(n-1)} + \frac{n-2}{n-1} \int \sec^{n-2} ax\, dx, \quad n \neq 1$

93. $\int \csc^n ax\, dx = -\frac{\csc^{n-2} ax \cot ax}{a(n-1)} + \frac{n-2}{n-1} \int \csc^{n-2} ax\, dx, \quad n \neq 1$

94. $\int \sec^n ax \tan ax\, dx = \frac{\sec^n ax}{na} + C, \quad n \neq 0$

95. $\int \csc^n ax \cot ax\, dx = -\frac{\csc^n ax}{na} + C, \quad n \neq 0$

96. $\int \sin^{-1} ax\, dx = x \sin^{-1} ax + \frac{1}{a} \sqrt{1 - a^2x^2} + C$

97. $\int \cos^{-1} ax\, dx = x \cos^{-1} ax - \frac{1}{a} \sqrt{1 - a^2x^2} + C$

98. $\int \tan^{-1} ax\, dx = x \tan^{-1} ax - \frac{1}{2a} \ln (1 + a^2x^2) + C$

99. $\int x^n \sin^{-1} ax\, dx = \frac{x^{n+1}}{n+1} \sin^{-1} ax - \frac{a}{n+1} \int \frac{x^{n+1}\, dx}{\sqrt{1 - a^2x^2}}, \quad n \neq -1$

100. $\int x^n \cos^{-1} ax\, dx = \frac{x^{n+1}}{n+1} \cos^{-1} ax + \frac{a}{n+1} \int \frac{x^{n+1}\, dx}{\sqrt{1 - a^2x^2}}, \quad n \neq -1$

101. $\int x^n \tan^{-1} ax\, dx = \frac{x^{n+1}}{n+1} \tan^{-1} ax - \frac{a}{n+1} \int \frac{x^{n+1}\, dx}{1 + a^2x^2}, \quad n \neq -1$

102. $\int e^{ax}\, dx = \frac{1}{a} e^{ax} + C$

103. $\int b^{ax}\, dx = \frac{1}{a} \frac{b^{ax}}{\ln b} + C, \quad b > 0,\ b \neq 1$

104. $\int xe^{ax}\, dx = \frac{e^{ax}}{a^2} (ax - 1) + C$

105. $\int x^n e^{ax}\, dx = \frac{1}{a} x^n e^{ax} - \frac{n}{a} \int x^{n-1} e^{ax}\, dx$

106. $\int x^n b^{ax}\, dx = \frac{x^n b^{ax}}{a \ln b} - \frac{n}{a \ln b} \int x^{n-1} b^{ax}\, dx, \quad b > 0,\ b \neq 1$

107. $\int e^{ax} \sin bx\, dx = \frac{e^{ax}}{a^2 + b^2} (a \sin bx - b \cos bx) + C$

108. $\int e^{ax} \cos bx\, dx = \frac{e^{ax}}{a^2 + b^2} (a \cos bx + b \sin bx) + C$

109. $\int \ln ax\, dx = x \ln ax - x + C$

110. $\int x^n \ln ax\, dx = \frac{x^{n+1}}{n+1} \ln ax - \frac{x^{n+1}}{(n+1)^2} + C, \quad n \neq -1$

111. $\int x^{-1} \ln ax\, dx = \frac{1}{2} (\ln ax)^2 + C$

112. $\int \frac{dx}{x \ln ax} = \ln |\ln ax| + C$

113. $\int \sinh ax\, dx = \frac{1}{a} \cosh ax + C$

114. $\int \cosh ax\, dx = \frac{1}{a} \sinh ax + C$

115. $\int \sinh^2 ax\, dx = \frac{\sinh 2ax}{4a} - \frac{x}{2} + C$

116. $\int \cosh^2 ax\, dx = \frac{\sinh 2ax}{4a} + \frac{x}{2} + C$

117. $\int \sinh^n ax\, dx = \frac{\sinh^{n-1} ax \cosh ax}{na} - \frac{n-1}{n} \int \sinh^{n-2} ax\, dx, \quad n \neq 0$

Continued

118. $\int \cosh^n ax\,dx = \frac{\cosh^{n-1} ax \sinh ax}{na} + \frac{n-1}{n}\int \cosh^{n-2} ax\,dx, \qquad n \neq 0$

119. $\int x \sinh ax\,dx = \frac{x}{a}\cosh ax - \frac{1}{a^2}\sinh ax + C$

120. $\int x \cosh ax\,dx = \frac{x}{a}\sinh ax - \frac{1}{a^2}\cosh ax + C$

121. $\int x^n \sinh ax\,dx = \frac{x^n}{a}\cosh ax - \frac{n}{a}\int x^{n-1}\cosh ax\,dx$

122. $\int x^n \cosh ax\,dx = \frac{x^n}{a}\sinh ax - \frac{n}{a}\int x^{n-1}\sinh ax\,dx$

123. $\int \tanh ax\,dx = \frac{1}{a}\ln(\cosh ax) + C$

124. $\int \coth ax\,dx = \frac{1}{a}\ln|\sinh ax| + C$

125. $\int \tanh^2 ax\,dx = x - \frac{1}{a}\tanh ax + C$

126. $\int \coth^2 ax\,dx = x - \frac{1}{a}\coth ax + C$

127. $\int \tanh^n ax\,dx = -\frac{\tanh^{n-1} ax}{(n-1)a} + \int \tanh^{n-2} ax\,dx, \qquad n \neq 1$

128. $\int \coth^n ax\,dx = -\frac{\coth^{n-1} ax}{(n-1)a} + \int \coth^{n-2} ax\,dx, \qquad n \neq 1$

129. $\int \operatorname{sech} ax\,dx = \frac{1}{a}\sin^{-1}(\tanh ax) + C$

130. $\int \operatorname{csch} ax\,dx = \frac{1}{a}\ln\left|\tanh\frac{ax}{2}\right| + C$

131. $\int \operatorname{sech}^2 ax\,dx = \frac{1}{a}\tanh ax + C$

132. $\int \operatorname{csch}^2 ax\,dx = -\frac{1}{a}\coth ax + C$

133. $\int \operatorname{sech}^n ax\,dx = \frac{\operatorname{sech}^{n-2} ax \tanh ax}{(n-1)a} + \frac{n-2}{n-1}\int \operatorname{sech}^{n-2} ax\,dx, \qquad n \neq 1$

134. $\int \operatorname{csch}^n ax\,dx = -\frac{\operatorname{csch}^{n-2} ax \coth ax}{(n-1)a} - \frac{n-2}{n-1}\int \operatorname{csch}^{n-2} ax\,dx, \qquad n \neq 1$

135. $\int \operatorname{sech}^n ax \tanh ax\,dx = -\frac{\operatorname{sech}^n ax}{na} + C, \qquad n \neq 0$

136. $\int \operatorname{csch}^n ax \coth ax\,dx = -\frac{\operatorname{csch}^n ax}{na} + C, \qquad n \neq 0$

137. $\int e^{ax}\sinh bx\,dx = \frac{e^{ax}}{2}\left[\frac{e^{bx}}{a+b} - \frac{e^{-bx}}{a-b}\right] + C, \qquad a^2 \neq b^2$

138. $\int e^{ax}\cosh bx\,dx = \frac{e^{ax}}{2}\left[\frac{e^{bx}}{a+b} + \frac{e^{-bx}}{a-b}\right] + C, \qquad a^2 \neq b^2$

139. $\int_0^\infty x^{n-1}e^{-x}\,dx = \Gamma(n) = (n-1)!, \qquad n > 0.$

140. $\int_0^\infty e^{-ax^2}\,dx = \frac{1}{2}\sqrt{\frac{\pi}{a}}, \qquad a > 0$

141. $$\int_0^{\pi/2} \sin^n x\,dx = \int_0^{\pi/2} \cos^n x\,dx = \begin{cases} \dfrac{1\cdot 3\cdot 5\cdots(n-1)}{2\cdot 4\cdot 6\cdots n}\cdot\dfrac{\pi}{2}, & \text{if } n \text{ is an even integer} \geq 2, \\ \dfrac{2\cdot 4\cdot 6\cdots(n-1)}{3\cdot 5\cdot 7\cdots n}, & \text{if } n \text{ is an odd integer} \geq 3 \end{cases}$$